Student Study and Solutions Manual

Algebra and Trigonometry

NINTH EDITION

Ron Larson
The Pennsylvania State University,
The Behrend College

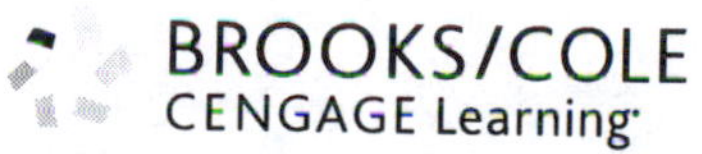

Australia • Brazil • Japan • Korea • Mexico • Singapore • Spain • United Kingdom • United States

For product information and technology assistance, contact us at **Cengage Learning Customer & Sales Support, 1-800-354-9706**

For permission to use material from this text or product, submit all requests online at **www.cengage.com/permissions**
Further permissions questions can be emailed to **permissionrequest@cengage.com**

ISBN-13: 978-1-133-95441-5
ISBN-10: 1-133-95441-3

Brooks/Cole
20 Channel Center Street
Boston, MA 02210
USA

Cengage Learning is a leading provider of customized learning solutions with office locations around the globe, including Singapore, the United Kingdom, Australia, Mexico, Brazil, and Japan. Locate your local office at: **www.cengage.com/global**

Cengage Learning products are represented in Canada by Nelson Education, Ltd.

To learn more about Brooks/Cole, visit **www.cengage.com/brookscole**

Purchase any of our products at your local college store or at our preferred online store **www.cengagebrain.com**

Printed in the United States of America
1 2 3 4 5 20 19 18 17 16

CONTENTS

CHAPTER P
Prerequisites

CHAPTER P
Prerequisites

Section P.1 Review of Real Numbers and Their Properties

1. irrational

3. absolute value

5. terms

7. $-9, -\frac{7}{2}, 5, \frac{2}{3}, \sqrt{2}, 0, 1, -4, 2, -11$

(a) Natural numbers: 5, 1, 2

(b) Whole numbers: 0, 5, 1, 2

(c) Integers: $-9, 5, 0, 1, -4, 2, -11$

(d) Rational numbers: $-9, -\frac{7}{2}, 5, \frac{2}{3}, 0, 1, -4, 2, -11$

(e) Irrational numbers: $\sqrt{2}$

9. $2.01, 0.666\ldots, -13, 0.010110111\ldots, 1, -6$

(a) Natural numbers: 1

(b) Whole numbers: 1

(c) Integers: $-13, 1, -6$

(d) Rational numbers: $2.01, 0.666\ldots, -13, 1, -6$

(e) Irrational numbers: $0.010110111\ldots$

11. (a)

(b)

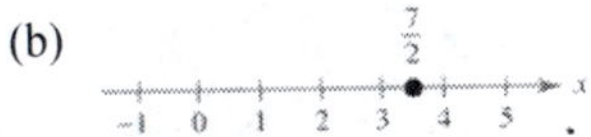

(c)

(d)

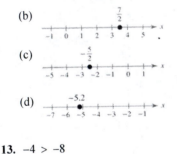

13. $-4 > -8$

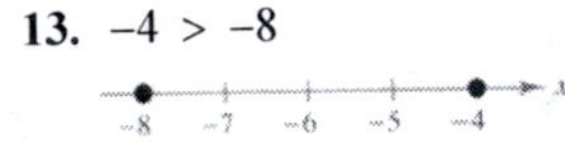

15. $\frac{5}{6} > \frac{2}{3}$

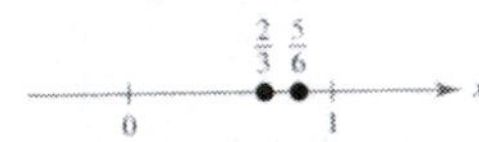

17. (a) The inequality $x \le 5$ denotes the set of all real numbers less than or equal to 5.

(b)

(c) The interval is unbounded.

19. (a) The interval $[4, \infty)$ denotes the set of all real numbers greater than or equal to 4.

(b)

(c) The interval is unbounded.

21. (a) The inequality $-2 < x < 2$ denotes the set of all real numbers greater than -2 and less than 2.

(b)

(c) The interval is bounded.

23. (a) The interval $[-5, 2)$ denotes the set of all real numbers greater than or equal to -5 and less than 2.

(b)

(c) The interval is bounded.

25. $y \ge 0; [0, \infty)$

27. $10 \le t \le 22; [10, 22]$

29. $W > 65; (65, \infty)$

31. $|-10| = -(-10) = 10$

33. $|3 - 8| = |-5| = -(-5) = 5$

35. $|-1| - |-2| = 1 - 2 = -1$

37. $\dfrac{-5}{|-5|} = \dfrac{-5}{-(-5)} = \dfrac{-5}{5} = -1$

39. If $x < -2$, then $x + 2$ is negative.

So, $\dfrac{|x + 2|}{x + 2} = \dfrac{-(x + 2)}{x + 2} = -1.$

41. $|-4| = |4|$ because $|-4| = 4$ and $|4| = 4$.

43. $-|-6| < |-6|$ because $|-6| = 6$ and $-|-6| = -(6) = -6$.

45. $d(126, 75) = |75 - 126| = 51$

47. $d\left(-\frac{5}{2}, 0\right) = \left|0 - \left(-\frac{5}{2}\right)\right| = \frac{5}{2}$

49. $d\left(\frac{16}{5}, \frac{112}{75}\right) = \left|\frac{112}{75} - \frac{16}{5}\right| = \frac{128}{75}$

51. $d(x, 5) = |x - 5|$ and $d(x, 5) \le 3$, so $|x - 5| \le 3$.

53. $d(y, a) = |y - a|$ and $d(y, a) \le 2$, so $|y - a| \le 2$.

	Receipts, R	Expenditures, E	$\lvert R - E \rvert$
55.	\$1880.1	\$2292.8	$\lvert 1880.1 - 2292.8 \rvert$ = \$412.7 billion
57.	\$2524.0	\$2982.5	$\lvert 2524.0 - 2982.5 \rvert$ = \$458.5 billion

59. $7x + 4$

Terms: $7x, 4$

Coefficient: 7

61. $4x^3 + \frac{x}{2} - 5$

Terms: $4x^3, \frac{x}{2}, -5$

Coefficients: $4, \frac{1}{2}$

63. $4x - 6$

(a) $4(-1) - 6 = -4 - 6 = -10$

(b) $4(0) - 6 = 0 - 6 = -6$

65. $-x^2 + 5x - 4$

(a) $-(-1)^2 + 5(-1) - 4 = -1 - 5 - 4 = -10$

(b) $-(1)^2 + 5(1) - 4 = -1 + 5 - 4 = 0$

67. $\frac{1}{(h + 6)}(h + 6) = 1, h \ne -6$

Multiplicative Inverse Property

69. $2(x + 3) = 2 \cdot x + 2 \cdot 3$

Distributive Property

71. $x(3y) = (x \cdot 3)y$ Associative Property of Multiplication

$= (3x)y$ Commutative Property of Multiplication

73. $\frac{5}{8} - \frac{5}{12} + \frac{1}{6} = \frac{15}{24} - \frac{10}{24} + \frac{4}{24} = \frac{9}{24} = \frac{3}{8}$

75. $\frac{2x}{3} - \frac{x}{4} = \frac{8x}{12} - \frac{3x}{12} = \frac{5x}{12}$

77. (a) Because $A > 0, -A < 0$.
The expression is negative.

(b) Because $B < A, B - A < 0$.
The expression is negative.

(c) Because $C < 0, -C > 0$.
The expression is positive.

(d) Because $A > C, A - C > 0$.
The expression is positive.

79. False. Because 0 is nonnegative but not positive, not every nonnegative number is positive.

81. (a)

n	0.0001	0.01	1	100	10,000
$5/n$	50,000	500	5	0.05	0.0005

(b) (i) As n approaches 0, the value of $5/n$ increases without bound (approaches infinity).

(ii) As n increases without bound (approaches infinity), the value of $5/n$ approaches 0.

Section P.2 Exponents and Radicals

1. exponent; base

3. square root

5. like radicals

7. rationalizing

9. (a) $3 \cdot 3^3 = 3^4 = 81$

(b) $\frac{3^2}{3^4} = 3^{-2} = \frac{1}{3^2} = \frac{1}{9}$

11. (a) $(2^3 \cdot 3^2)^2 = 2^{3\cdot 2} \cdot 3^{2\cdot 2}$

$= 2^6 \cdot 3^4 = 64 \cdot 81 = 5184$

(b) $\left(-\frac{3}{5}\right)^3\left(\frac{5}{3}\right)^2 = (-1)^3\frac{3^3}{5^3} \cdot \frac{5^2}{3^2} = -1 \cdot 3^{3-2} \cdot 5^{2-3}$

$= -3 \cdot 5^{-1} = -\frac{3}{5}$

13. (a) $\frac{4 \cdot 3^{-2}}{2^{-2} \cdot 3^{-1}} = 4 \cdot 2^2 \cdot 3^{-2-(-1)} = 4 \cdot 4 \cdot 3^{-1} = \frac{16}{3}$

(b) $(-2)^0 = 1$

15. When $x = 2$,

$-3x^3 = -3(2)^3 = -24.$

17. When $x = 10$,

$6x^0 = 6(10)^0 = 6(1) = 6.$

19. When $x = -2$,

$-3x^4 = -3(-2)^4 = -3(16) = -48.$

21. (a) $(-5z)^3 = (-5)^3 z^3 = -125z^3$

(b) $5x^4(x^2) = 5x^{4+2} = 5x^6$

23. (a) $6y^2(2y^0)^2 = 6y^2(2 \cdot 1)^2 = 6y^2(4) = 24y^2$

(b) $(-z)^3(3z^4) = (-1)^3(z^3)3z^4$

$= -1 \cdot 3 \cdot z^{3+4} = -3z^7$

25. (a) $\left(\frac{4}{y}\right)^3\left(\frac{3}{y}\right)^4 = \frac{4^3}{y^3} \cdot \frac{3^4}{y^4} = \frac{64 \cdot 81}{y^{3+4}} = \frac{5184}{y^7}$

(b) $\left(\frac{b^{-2}}{a^{-2}}\right)\left(\frac{b}{a}\right)^2 = \left(\frac{a^2}{b^2}\right)\left(\frac{b^2}{a^2}\right) = 1,\ a \neq 0,\ b \neq 0$

27. (a) $(x + 5)^0 = 1,\ x \neq -5$

(b) $(2x^2)^{-2} = \frac{1}{(2x^2)^2} = \frac{1}{4x^4}$

29. (a) $\left(\frac{x^{-3}y^4}{5}\right)^{-3} = \left(\frac{5x^3}{y^4}\right)^3 = \frac{125x^4}{y^{12}}$

(b) $\left(\frac{a^{-2}}{b^{-2}}\right)\left(\frac{b}{a}\right)^3 = \left(\frac{b^2}{a^2}\right)\left(\frac{b^3}{a^3}\right) = \frac{b^5}{a^5}$

31. $10{,}250.4 = 1.02504 \times 10^4$

33. $0.00003937 = 3.937 \times 10^{-5}$ inch

35. $-1.801 \times 10^5 = -180{,}100$

37. $9.46 \times 10^{12} = 9{,}460{,}000{,}000{,}000$ kilometers

39. (a) $(2.0 \times 10^9)(3.4 \times 10^{-4}) = 6.8 \times 10^5$

(b) $(1.2 \times 10^7)(5.0 \times 10^{-3}) = 6.0 \times 10^4$

41. (a) $\sqrt{9} = 3$

(b) $\sqrt[3]{\frac{27}{8}} = \frac{\sqrt[3]{27}}{\sqrt[3]{8}} = \frac{3}{2}$

43. (a) $\left(\sqrt[5]{2}\right)^5 = 2^{5/5} = 2^1 = 2$

(b) $\sqrt[5]{32x^5} = \sqrt[5]{(2x)^5} = 2x$

45. (a) $\sqrt{20} = \sqrt{4 \cdot 5}$

$= \sqrt{4}\sqrt{5} = 2\sqrt{5}$

(b) $\sqrt[3]{128} = \sqrt[3]{64 \cdot 2}$

$= \sqrt[3]{64}\sqrt[3]{2} = 4\sqrt[3]{2}$

47. (a) $\sqrt{72x^3} = \sqrt{36x^2 \cdot 2x}$

$= 6x\sqrt{2x}$

(b) $\sqrt{\frac{18^2}{z^3}} = \frac{\sqrt{18^2}}{\sqrt{z^2 \cdot z}} = \frac{18}{z\sqrt{z}} = \frac{18\sqrt{z}}{z^2}$

49. (a) $\sqrt[3]{16x^5} = \sqrt[3]{8x^3 \cdot 2x^2}$

$= 2x\sqrt[3]{2x^2}$

(b) $\sqrt{75x^2y^{-4}} = \sqrt{\frac{75x^2}{y^4}}$

$= \frac{\sqrt{25x^2 \cdot 3}}{\sqrt{y^4}}$

$= \frac{5|x|\sqrt{3}}{y^2}$

51. (a) $10\sqrt{32} - 6\sqrt{18} = 10\sqrt{16 \cdot 2} - 6\sqrt{9 \cdot 2}$

$= 10(4\sqrt{2}) - 6(3\sqrt{2})$

$= 40\sqrt{2} - 18\sqrt{2}$

$= 22\sqrt{2}$

(b) $\sqrt[3]{16} + 3\sqrt[3]{54} = \sqrt[3]{2 \cdot 2^3} + 3\sqrt[3]{2 \cdot 3^3}$

$= 2\sqrt[3]{2} + 3 \cdot (3\sqrt[3]{2})$

$= 2\sqrt[3]{2} + 9\sqrt[3]{2}$

$= 11\sqrt[3]{2}$

53. (a) $-3\sqrt{48x^2} + 7\sqrt{75x^2} = -3\sqrt{3 \cdot 4^2 \cdot x^2} + 7\sqrt{3 \cdot 5^2 \cdot x^2}$

$= -3 \cdot \left(4|x|\sqrt{3}\right) + 7 \cdot \left(5|x|\sqrt{3}\right)$

$= -12|x|\sqrt{3} + 35|x|\sqrt{3} = 23|x|\sqrt{3}$

(b) $7\sqrt{80x} - 2\sqrt{125x} = 7\sqrt{16 \cdot 5x} - 2\sqrt{25 \cdot 5x} = 7\left(4\sqrt{5x}\right) - 2\left(5\sqrt{5x}\right) = 28\sqrt{5x} - 10\sqrt{5x} = 18\sqrt{5x}$

55. $\dfrac{1}{\sqrt{3}} = \dfrac{1}{\sqrt{3}} \cdot \dfrac{\sqrt{3}}{\sqrt{3}} = \dfrac{\sqrt{3}}{3}$

57. $\dfrac{5}{\sqrt{14} - 2} = \dfrac{5}{\sqrt{14} - 2} \cdot \dfrac{\sqrt{14} + 2}{\sqrt{14} + 2} = \dfrac{5\left(\sqrt{14} + 2\right)}{\left(\sqrt{14}\right)^2 - (2)^2} = \dfrac{5\left(\sqrt{14} + 2\right)}{14 - 4} = \dfrac{5\left(\sqrt{14} + 2\right)}{10} = \dfrac{\sqrt{14} + 2}{2}$

59. $\dfrac{\sqrt{8}}{2} = \dfrac{\sqrt{4 \cdot 2}}{2} = \dfrac{2\sqrt{2}}{2} = \dfrac{\sqrt{2}}{1} \cdot \dfrac{\sqrt{2}}{\sqrt{2}} = \dfrac{2}{\sqrt{2}}$

61. $\dfrac{\sqrt{5} + \sqrt{3}}{3} = \dfrac{\sqrt{5} + \sqrt{3}}{3} \cdot \dfrac{\sqrt{5} - \sqrt{3}}{\sqrt{5} - \sqrt{3}} = \dfrac{5 - 3}{3\left(\sqrt{5} - \sqrt{3}\right)} = \dfrac{2}{3\left(\sqrt{5} - \sqrt{3}\right)}$

	Radical Form	**Rational Exponent Form**
63.	$\sqrt[3]{64} = 4$, Given	$64^{1/3} = 4$, Answer
65.	$\dfrac{3}{\sqrt[3]{x^2}}, x \neq 0$	$3x^{-2/3} = \dfrac{3}{x^{2/3}}$
67.	$x\sqrt{3xy}$, Given	$x \cdot 3^{1/2}x^{1/2}y^{1/2} = 3^{1/2}x^{3/2}y^{1/2}$, Answer

69. (a) $32^{-3/5} = \dfrac{1}{32^{3/5}} = \dfrac{1}{\left(\sqrt[5]{32}\right)^3} = \dfrac{1}{(2)^3} = \dfrac{1}{8}$

(b) $\left(\dfrac{16}{81}\right)^{-3/4} = \left(\dfrac{81}{16}\right)^{3/4} = \left(\sqrt[4]{\dfrac{81}{16}}\right)^3 = \left(\dfrac{3}{2}\right)^3 = \dfrac{27}{8}$

71. (a) $\dfrac{\left(2x^2\right)^{3/2}}{2^{1/2}x^4} = \dfrac{2^{3/2}\left(x^2\right)^{3/2}}{2^{1/2}x^4} = \dfrac{2^{3/2}|x|^3}{2^{1/2}x^4} = 2^{3/2-1/2}|x|^{3-4} = 2^1|x|^{-1} = \dfrac{2}{|x|}$

(b) $\dfrac{x^{4/3}y^{2/3}}{(xy)^{1/3}} = \dfrac{x^{4/3}y^{2/3}}{x^{1/3}y^{1/3}} = x^{3/3}y^{1/3} = xy^{1/3}$

73. (a) $\sqrt[4]{3^2} = 3^{2/4} = 3^{1/2} = \sqrt{3}$

(b) $\sqrt[6]{(x + 1)^4} = (x + 1)^{4/6} = (x + 1)^{2/3} = \sqrt[3]{(x + 1)^2}$

75. (a) $\sqrt{\sqrt{32}} = \left(32^{1/2}\right)^{1/2}$

$= 32^{1/4} = \sqrt[4]{32} = \sqrt[4]{16 \cdot 2} = 2\sqrt[4]{2}$

(b) $\sqrt{\sqrt[4]{2x}} = \left((2x)^{1/4}\right)^{1/2} = (2x)^{1/8} = \sqrt[8]{2x}$

77. (a) $(x - 1)^{1/3}(x - 1)^{2/3} = (x - 1)^{3/3} = (x - 1)$

(b) $(x - 1)^{1/3}(x - 1)^{-4/3} = (x - 1)^{-3/3} = (x - 1)^{-1} = \dfrac{1}{x - 1}$

79. $t = 0.03\left[12^{5/2} - (12 - h)^{5/2}\right], 0 \le h \le 12$

(a)

h (in centimeters)	t (in seconds)
0	0
1	2.93
2	5.48
3	7.67
4	9.53
5	11.08
6	12.32
7	13.29
8	14.00
9	14.50
10	14.80
11	14.93
12	14.96

(b) As h approaches 12, t approaches
$0.03\left(12^{5/2}\right) = 8.64\sqrt{3} = 14.96$ seconds.

81. True. When dividing variables, you subtract exponents.

83. False. When a sum is raised to a power, you multiply the sum by itself using the Distributive Property.

$(a + b)^2 = a^2 + 2ab + b^2 \ne a^2 + b^2$

Section P.3 Polynomials and Special Products

1. n; a_n; a_0

3. like terms

5. (a) Standard form: $-\frac{1}{2}x^5 + 14x$

(b) Degree: 5
Leading coefficient: $-\frac{1}{2}$

(c) Binomial

7. (a) Standard form: $-x^6 + 3$

(b) Degree: 6
Leading coefficient: -1

(c) Binomial

9. (a) Standard form: 3

(b) Degree: 0
Leading coefficient: 3

(c) Monomial

11. (a) Standard form: $-4x^5 + 6x^4 + 1$

(b) Degree: 5
Leading coefficient: -4

(c) Trinomial

13. (a) Standard form: $4x^3y$

(b) Degree: 4 (add the exponents on x and y)
Leading coefficient: 4

(c) Monomial

15. $2x - 3x^3 + 8$ *is* a polynomial.

Standard form: $-3x^3 + 2x + 8$

17. $\dfrac{3x + 4}{x} = 3 + \dfrac{4}{x} = 3 + 4x^{-1}$ is *not* a polynomial because it includes a term with a negative exponent.

19. $y^2 - y^4 + y^3$ *is* a polynomial.

Standard form: $-y^4 + y^3 + y^2$

21.
$$\begin{aligned}(6x + 5) - (8x + 15) &= 6x + 5 - 8x - 15\\ &= (6x - 8x) + (5 - 15)\\ &= -2x - 10\end{aligned}$$

23.
$$\begin{aligned}-\left(t^3 - 1\right) + \left(6t^3 - 5t\right) &= -t^3 + 1 + 6t^3 - 5t\\ &= \left(-t^3 + 6t^3\right) - 5t + 1\\ &= 5t^3 - 5t + 1\end{aligned}$$

25. $(15x^2 - 6) - (-8.3x^3 - 14.7x^2 - 17) = 15x^2 - 6 + 8.3x^3 + 14.7x^2 + 17$
$= 8.3x^3 + (15x^2 + 14.7x^2) + (-6 + 17)$
$= 8.3x^3 + 29.7x^2 + 11$

27. $5z - [3z - (10z + 8)] = 5z - (3z - 10z - 8)$
$= 5z - 3z + 10z + 8$
$= (5z - 3z + 10z) + 8$
$= 12z + 8$

29. $3x(x^2 - 2x + 1) = 3x(x^2) + 3x(-2x) + 3x(1)$
$= 3x^3 - 6x^2 + 3x$

31. $-5z(3z - 1) = -5z(3z) + (-5z)(-1)$
$= -15z^2 + 5z$

33. $(1 - x^3)(4x) = 1(4x) - x^3(4x)$
$= 4x - 4x^4$
$= -4x^4 + 4x$

35. $(1.5t^2 + 5)(-3t) = (1.5t^2)(-3t) + (5)(-3t)$
$= -4.5t^3 - 15t$

37. $-2x(0.1x + 17) = (-2x)(0.1x) + (-2x)(17)$
$= -0.2x^2 - 34x$

39. $(7x^3 - 2x^2 + 8) + (-3x^2 - 4) = (7x^3 - 3x^3) + (-2x^2) + (8 - 4)$
$= 4x^3 - 2x^2 + 4$

41. $(5x^2 - 3x + 8) - (x - 3) = 5x^2 - 3x + 8 - x + 3$
$= 5x^2 + (-3x - x) + (8 + 3)$
$= 5x^2 - 4x + 11$

43. $(x + 7)(2x + 3) = 2x^2 + 3x + 14x + 21$ FOIL
$= 2x^2 + 17x + 21$

45. $(x + 3)(x + 4) = x^2 + 4x + 3x + 12$ FOIL
$= x^2 + 7x + 12$

47. $(3x - 5)(2x + 1) = 6x^2 + 3x - 10x - 5$ FOIL
$= 6x^2 - 7x - 5$

49. $(x^2 - x + 1)(x^2 + x + 1)$

Multiply:
$$\begin{array}{r} x^2 - x + 1 \\ \underline{x^2 + x + 1} \\ x^4 - x^3 + x^2 \qquad\qquad \\ x^3 - x^2 + x \qquad \\ \underline{x^2 - x + 1} \\ x^4 - 0x^3 + x^2 + 0x + 1 = x^4 + x^2 + 1 \end{array}$$

51. $(x + 10)(x - 10) = x^2 - 10^2 = x^2 - 100$

53. $(x + 2y)(x - 2y) = x^2 - (2y)^2 = x^2 - 4y^2$

55. $(2x + 3)^2 = (2x)^2 + 2(2x)(3) + 3^2$
$= 4x^2 + 12x + 9$

57. $(4x^3 - 3)^2 = (4x^3)^2 - 2(4x^3)(3) + (3)^2$
$= 16x^6 - 24x^3 + 9$

59. $(x + 1)^3 = x^3 + 3x^2(1) + 3x(1^2) + 1^3$
$= x^3 + 3x^2 + 3x + 1$

61. $(2x - y)^3 = (2x)^3 - 3(2x)^2y + 3(2x)y^2 - y^3$
$= 8x^3 - 12x^2y + 6xy^2 - y^3$

63. $[(m - 3) + n][(m - 3) - n] = (m - 3)^2 - (n)^2$
$= m^2 - 6m + 9 - n^2$
$= m^2 - n^2 - 6m + 9$

65. $[(x - 3) + y]^2 = (x - 3)^2 + 2y(x - 3) + y^2$
$= x^2 - 6x + 9 + 2xy - 6y + y^2$
$= x^2 + 2xy + y^2 - 6x - 6y + 9$

67. $(\frac{1}{5}x - 3)(\frac{1}{5}x + 3) = (\frac{1}{5}x)^2 - (3)^2$
$= \frac{1}{25}x^2 - 9$

69. $(1.5x - 4)(1.5x + 4) = (1.5x)^2 - (4)^2$
$= 2.25x^2 - 16$

71. $\left(\frac{1}{4}x - 5\right)^2 = \left(\frac{1}{4}x\right)^2 - 2\left(\frac{1}{4}x\right)(5) + (-5)^2$
$= \frac{1}{16}x^2 - \frac{5}{2}x + 25$

73. $5x(x + 1) - 3x(x + 1) = 2x(x + 1)$
$= 2x^2 + 2x$

75. $(u + 2)(u - 2)(u^2 + 4) = (u^2 - 4)(u^2 + 4)$
$= u^4 - 16$

77. $\left(\sqrt{x} + \sqrt{y}\right)\left(\sqrt{x} - \sqrt{y}\right) = \left(\sqrt{x}\right)^2 - \left(\sqrt{y}\right)^2$
$= x - y$

79. $\left(x - \sqrt{5}\right)^2 = x^2 - 2(x)\left(\sqrt{5}\right) + \left(\sqrt{5}\right)^2$
$= x^2 - 2\sqrt{5}x + 5$

81. (a) Profit = Revenue − Cost
Profit $= 95x - (73x + 25{,}000)$
$= 95x - 73x - 25{,}000$
$= 22x - 25{,}000$

(b) For $x = 5000$:
Profit $= 22(5000) - 25{,}000$
$= 110{,}000 - 25{,}000$
$= \$85{,}000$

83. $A = (18 + 2x)(14 + x) = 252 + 18x + 28x + 2x^2 = 2x^2 + 46x + 252$

85. Area of shaded region = Area of outer rectangle − Area of inner rectangle
$A = 2x(2x + 6) - x(x + 4)$
$= 4x^2 + 12x - x^2 - 4x$
$= 3x^2 + 8x$

87. The area of the shaded region is the difference between the area of the larger triangle and the area of the smaller triangle.
$A = \frac{1}{2}(10x)(10x) - \frac{1}{2}(4x)(4x) = 50x^2 - 8x^2 = 42x^2$

89. The area of the shaded region is the difference between the area of the larger square and the area of the smaller square.
$A = (4x + 2)^2 - (x - 1)^2 = 16x^2 + 16x + 4 - (x^2 - 2x + 1) = 15x^2 + 18x + 3$

91. (a) $V = l \cdot w \cdot h = (26 - 2x)(18 - 2x)(x)$
$= 2(13 - x)(2)(9 - x)(x)$
$= 4x(-1)(x - 13)(-1)(x - 9)$
$= 4x(x - 13)(x - 9)$
$= 4x^3 - 88x^2 + 468x$

(b)

x (cm)	1	2	3
V (cm^3)	384	616	720

93. (a) Estimates will vary. Actual safe loads for $x = 12$:

$S_6 = \left(0.06(12)^2 - 2.42(12) + 38.71\right)^2 = 335.2561$ (using a calculator)

$S_8 = \left(0.08(12)^2 - 3.30(12) + 51.93\right)^2 = 568.8225$ (using a calculator)

Difference in safe loads $= 568.8225 - 335.2561 = 233.5664$ pounds

(b) The difference in safe loads decreases in magnitude as the span increases.

95. False. $(4x^2 + 1)(3x + 1) = 12x^3 + 4x^2 + 3x + 1$

97. Because $x^m x^n = x^{m+n}$, the degree of the product is $m + n$.

99. $(x - 3)^2 \neq x^2 + 9$

The student did not remember the middle term when squaring the binomial. The correct method for squaring the binomial is:

$(x - 3)^2 = (x)^2 - 2(x)(3) + (3)^2 = x^2 - 6x + 9$

101. No; $(x^2 + 1) + (-x^2 + 3) = 4$, which is not a second-degree polynomial.

103. $(x + y)^2 \neq x^2 + y^2$

Let $x = 3$ and $y = 4$.

$(3 + 4)^2 = (7)^2 = 49$
$3^2 + 4^2 = 9 + 16 = 25$ > Not Equal

If either x or y is zero, then $(x + y)^2$ would equal $x^2 + y^2$.

Section P.4 Factoring Polynomials

1. factoring

3. perfect square binomial

5. $2x^3 - 6x = 2x(x^2 - 3)$

7. $3x(x - 5) + 8(x - 5) = (x - 5)(3x + 8)$

9. $\frac{1}{2}x^3 + 2x^2 - 5x = \frac{1}{2}x^3 + \frac{4}{2}x^2 - \frac{10}{2}x$
$= \frac{1}{2}x(x^2 + 4x - 10)$

11. $\frac{2}{3}x(x - 3) - 4(x - 3) = \frac{2}{3}x(x - 3) - \frac{12}{3}(x - 3)$
$= \frac{2}{3}(x - 3)(x - 6)$

13. $x^2 - 81 = x^2 - 9^2 = (x + 9)(x - 9)$

15. $48y^2 - 27 = 3(16y^2 - 9)$
$= 3((4y)^2 - 3^2)$
$= 3(4y + 3)(4y - 3)$

17. $16x^2 - \frac{1}{9} = (4x)^2 - \left(\frac{1}{3}\right)^2 = \left(4x + \frac{1}{3}\right)\left(4x - \frac{1}{3}\right)$

19. $(x - 1)^2 - 4 = (x - 1)^2 - (2)^2$
$= [(x - 1) + 2][(x - 1) - 2]$
$= (x + 1)(x - 3)$

21. $81u^4 - 1 = (9u^2 + 1)(9u^2 - 1)$
$= (9u^2 + 1)(3u + 1)(3u - 1)$

23. $x^2 - 4x + 4 = x^2 - 2(2)x + 2^2 = (x - 2)^2$

25. $9u^2 + 24uv + 16v^2 = (3u)^2 + 2(3u)(4v) + (4v)^2$
$= (3u + 4v)^2$

27. $z^2 + z + \frac{1}{4} = z^2 + 2(z)\left(\frac{1}{2}\right) + \left(\frac{1}{2}\right)^2 = \left(z + \frac{1}{2}\right)^2$

29. $x^3 - 8 = x^3 - 2^3 = (x - 2)(x^2 + 2x + 4)$

31. $y^3 + 64 = y^3 + 4^3 = (y + 4)(y^2 - 4y + 16)$

33. $8t^3 - 1 = (2t)^3 - 1^3 = (2t - 1)(4t^2 + 2t + 1)$

35. $u^3 + 27v^3 = u^3 + (3v)^3 = (u + 3v)(u^2 - 3uv + 9v^2)$

37. $x^2 + x - 2 = (x + 2)(x - 1)$

39. $s^2 - 5s + 6 = (s - 3)(s - 2)$

41. $20 - y - y^2 = -(y^2 + y - 20) = -(y + 5)(y - 4)$

43. $3x^2 - 5x + 2 = (3x - 2)(x - 1)$

45. $5x^2 + 26x + 5 = (5x + 1)(x + 5)$

47. $x^3 - x^2 + 2x - 2 = x^2(x - 1) + 2(x - 1)$
$= (x - 1)(x^2 + 2)$

49. $2x^3 - x^2 - 6x + 3 = x^2(2x - 1) - 3(2x - 1)$
$= (2x - 1)(x^2 - 3)$

51. $x^5 + 2x^3 + x^2 + 2 = x^3(x^2 + 2) + (x^2 + 2)$
$= (x^2 + 2)(x^3 + 1)$
$= (x^2 + 2)(x + 1)(x^2 - x + 1)$

53. $a \cdot c = (2)(9) = 18$. Rewrite the middle term,

$9x = 6x + 3x$, because $(6)(3) = 18$ and $6 + 3 = 9$.

$2x^2 + 9x + 9 = 2x^2 + 6x + 3x + 9$
$= 2x(x + 3) + 3(x + 3)$
$= (x + 3)(2x + 3)$

55. $a \cdot c = (6)(-15) = -90$. Rewrite the middle term, $-x = -10x + 9x$, because $(-10)(9) = -90$ and $-10 + 9 = -1$.

$$\begin{aligned} 6x^2 - x - 15 &= 6x^2 - 10x + 9x - 15 \\ &= 2x(3x - 5) + 3(3x - 5) \\ &= (2x + 3)(3x - 5) \end{aligned}$$

57. $6x^2 - 54 = 6(x^2 - 9) = 6(x + 3)(x - 3)$

59. $x^3 - x^2 = x^2(x - 1)$

61. $x^2 - 2x + 1 = (x - 1)^2$

63. $1 - 4x + 4x^2 = (1 - 2x)^2$

65. $$\begin{aligned} 2x^2 + 4x - 2x^3 &= -2x(-x - 2 + x^2) \\ &= -2x(x^2 - x - 2) \\ &= -2x(x + 1)(x - 2) \end{aligned}$$

67. $$\begin{aligned} \tfrac{1}{8}x^2 - \tfrac{1}{96}x - \tfrac{1}{16} &= \tfrac{1}{96}(12x^2 - x - 6) \\ &= \tfrac{1}{96}(4x - 3)(3x + 2) \end{aligned}$$

69. $$\begin{aligned} (x^2 + 1)^2 - 4x^2 &= \left[(x^2 + 1) + 2x\right]\left[(x^2 + 1) - 2x\right] \\ &= (x^2 + 2x + 1)(x^2 - 2x + 1) \\ &= (x + 1)^2(x - 1)^2 \end{aligned}$$

71. $$\begin{aligned} 5 - x + 5x^2 - x^3 &= 1(5 - x) + x^2(5 - x) \\ &= (5 - x)(1 + x^2) \end{aligned}$$

73. $$\begin{aligned} 2x^3 + x^2 - 8x - 4 &= x^2(2x + 1) - 4(2x + 1) \\ &= (2x + 1)(x^2 - 4) \\ &= (2x + 1)(x + 2)(x - 2) \end{aligned}$$

75. $2t^3 - 16 = 2(t^3 - 8) = 2(t - 2)(t^2 + 2t + 4)$

77. $$\begin{aligned} 5(3 - 4x)^2 - 8(3 - 4x)(5x - 1) &= (3 - 4x)\left[5(3 - 4x) - 8(5x - 1)\right] \\ &= (3 - 4x)[15 - 20x - 40x + 8] \\ &= (3 - 4x)(23 - 60x) \end{aligned}$$

79. $$\begin{aligned} 7(3x + 2)^2(1 - x)^2 + (3x + 2)(1 - x)^3 &= (3x + 2)(1 - x)^2\left[7(3x + 2) + (1 - x)\right] \\ &= (3x + 2)(1 - x)^2(21x + 14 + 1 - x) \\ &= (3x + 2)(1 - x)^2(20x + 15) \\ &= 5(3x + 2)(1 - x)^2(4x + 3) \end{aligned}$$

81. $$\begin{aligned} x^4(4)(2x + 1)^3(2x) + (2x + 1)^4(4x^3) &= 2x^3(2x + 1)^3\left[4x^2 + 2(2x + 1)\right] \\ &= 2x^3(2x + 1)^3(4x^2 + 4x + 2) \\ &= 4x^3(2x + 1)^3(2x^2 + 2x + 1) \end{aligned}$$

83. $$\begin{aligned} &(2x - 5)^4(3)(5x - 4)^2(5) + (5x - 4)^3(4)(2x - 5)^3(2) \\ &= (2x - 5)^3(5x - 4)^2\left[15(2x - 5) + 8(5x - 4)\right] \\ &= (2x - 5)^3(5x - 4)^2(30x - 75 + 40x - 32) \\ &= (2x - 5)^3(5x - 4)^2(70x - 107) \end{aligned}$$

85. $x^2 + 3x + 2 = (x + 2)(x + 1)$

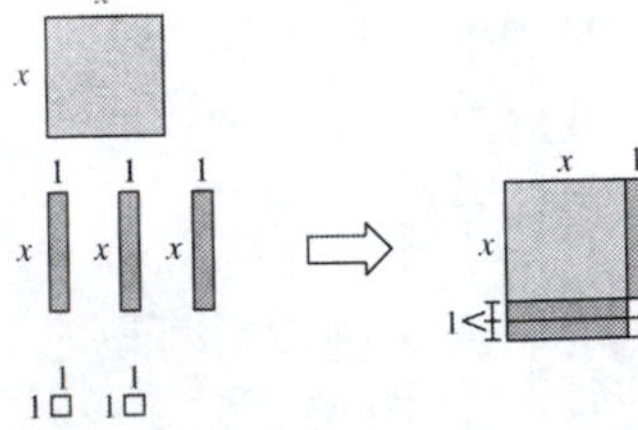

87. $2x^2 + 7x + 3 = (2x + 1)(x + 3)$

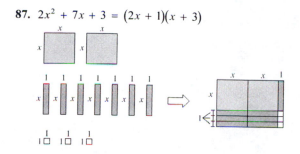

89. $A = \pi(r + 2)^2 - \pi r^2$

$= \pi\left[(r + 2)^2 - r^2\right]$

$= \pi\left[r^2 + 4r + 4 - r^2\right]$

$= \pi(4r + 4)$

$= 4\pi(r + 1)$

91. (a) $V = \pi R^2 h - \pi r^2 h$

$= \pi h(R^2 - r^2)$

$= \pi h(R + r)(R - r)$

(b) Let w = thickness of the shell and let p = average radius of the shell.

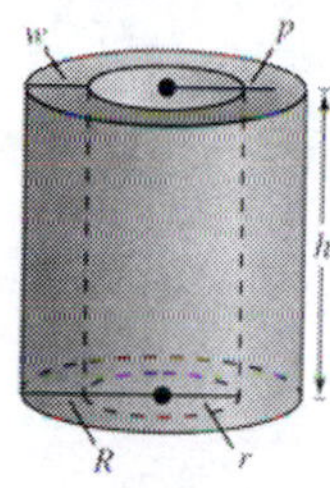

So, $R = p + \frac{1}{2}w$ and $r = p - \frac{1}{2}w$

$V = \pi h(R + r)(R - r)$

$= \pi h\left[\left(p + \frac{1}{2}w\right) + \left(p - \frac{1}{2}w\right)\right]\left[\left(p + \frac{1}{2}w\right) - \left(p - \frac{1}{2}w\right)\right]$

$= \pi h(2p)(w)$

$= 2\pi pwh$

$= 2\pi(\text{average radius})(\text{thickness of shell})\,h$

93. For $x^2 + bx - 15$ to be factorable, b must equal $m + n$ where $mn = -15$.

Factors of -15	Sum of factors
$(15)(-1)$	$15 + (-1) = 14$
$(-15)(1)$	$-15 + 1 = -14$
$(3)(-5)$	$3 + (-5) = -2$
$(-3)(5)$	$-3 + 5 = 2$

The possible b-values are $14, -14, -2$, or 2.

95. For $2x^2 + 5x + c$ to be factorable, the factors of $2c$ must add up to 5.

Possible c-values	$2c$	Factors of $2c$ that add up to 5
2	4	$(1)(4) = 4$ and $1 + 4 = 5$
3	6	$(2)(3) = 6$ and $2 + 3 = 5$
−3	−6	$(6)(-1) = -6$ and $6 + (-1) = 5$
−7	−14	$(7)(-2) = -14$ and $7 + (-2) = 5$
−12	−24	$(8)(-3) = -24$ and $8 + (-3) = 5$

These are a few possible c-values. There are *many* correct answers.

If $c = 2$: $2x^2 + 5x + 2 = (2x + 1)(x + 2)$

If $c = 3$: $2x^2 + 5x + 3 = (2x + 3)(x + 1)$

If $c = -3$: $2x^2 + 5x - 3 = (2x - 1)(x + 3)$

If $c = -7$: $2x^2 + 5x - 7 = (2x + 7)(x - 1)$

If $c = -12$: $2x^2 + 5x - 12 = (2x - 3)(x + 4)$

97. True. $a^2 - b^2 = (a + b)(a - b)$

99. $9x^2 - 9x - 54 = 9(x^2 - x - 6) = 9(x + 2)(x - 3)$

The error on the problem in the book was that 3 factored out of the first binomial but not out of the second binomial.

$(3x + 6)(3x - 9) = 3(x + 2)(3)(x - 3) = 9(x + 2)(x - 2)$

101. $x^{2n} - y^{2n} = (x^n)^2 - (y^n)^2 = (x^n + y^n)(x^n - y^n)$

This is not completely factored unless $n = 1$.

For $n = 2$: $(x^2 + y^2)(x^2 - y^2) = (x^2 + y^2)(x + y)(x - y)$

For $n = 3$: $(x^3 + y^3)(x^3 - y^3) = (x + y)(x^2 - xy + y^2)(x - y)(x^2 + xy + y^2)$

For $n = 4$: $(x^4 + y^4)(x^4 - y^4) = (x^4 + y^4)(x^2 + y^2)(x + y)(x - y)$

103. Answers will vary. Sample answer: $x^2 - 3$

105.
$$\begin{aligned} u^6 - v^6 &= (u^3)^2 - (v^3)^2 \\ &= (u^3 + v^3)(u^3 - v^3) \\ &= \left[(u + v)(u^2 - uv + v^2)\right]\left[(u - v)(u^2 + uv + v^2)\right] \\ &= (u + v)(u - v)(u^2 + uv + v^2)(u^2 - uv + v^2) \end{aligned}$$

$x^6 - 1 = (x + 1)(x - 1)(x^2 + x + 1)(x^2 - x + 1)$

$x^6 - 64 = x^6 - 2^6 = (x + 2)(x - 2)(x^2 + 2x + 4)(x^2 - 2x + 4)$

Section P.5 Rational Expressions

1. domain

3. complex

5. The domain of the polynomial $3x^2 - 4x + 7$ is the set of all real numbers.

7. The domain of $\frac{1}{3 - x}$ is the set of all real numbers x such that $x \neq 3$.

9. The domain of $\frac{x^2 - 1}{x^2 - 2x + 1} = \frac{(x + 1)(x - 1)}{(x - 1)(x - 1)}$ is the set of all real numbers x such that $x \neq 1$.

11. The domain of $\frac{x^2 - 2x - 3}{x^2 - 6x + 9} = \frac{(x - 3)(x + 1)}{(x - 3)(x - 3)}$ is the set of all real numbers x such that $x \neq 3$.

13. The domain of $\sqrt{4 - x}$ is the set of all real numbers x such that $x \leq 4$.

15. The domain of $\frac{1}{\sqrt{x - 3}}$ is the set of all real numbers x such that $x > 3$.

17. $\frac{15x^2}{10x} = \frac{5x(3x)}{5x(2)} = \frac{3x}{2}, x \neq 0$

19. $\frac{3xy}{xy + x} = \frac{x(3y)}{x(y + 1)} = \frac{3y}{y + 1}, x \neq 0$

21. $\frac{x - 5}{10 - 2x} = \frac{x - 5}{-2(x - 5)}$

$= -\frac{1}{2}, x \neq 5$

23. $\frac{y^2 - 16}{y + 4} = \frac{(y + 4)(y - 4)}{y + 4}$

$= y - 4, y \neq -4$

25. $\frac{x^3 + 5x^2 + 6x}{x^2 - 4} = \frac{x(x + 2)(x + 3)}{(x + 2)(x - 2)} = \frac{x(x + 3)}{x - 2}, x \neq -2$

27. $\frac{2 - x + 2x^2 - x^3}{x^2 - 4} = \frac{(2 - x) + x^2(2 - x)}{(x + 2)(x - 2)}$

$= \frac{(2 - x)(1 + x^2)}{(x + 2)(x - 2)}$

$= \frac{-(x - 2)(x^2 + 1)}{(x + 2)(x - 2)}$

$= -\frac{x^2 + 1}{x + 2}, x \neq 2$

29. $\frac{z^3 - 8}{z^2 + 2z + 4} = \frac{(z - 2)(z^2 + 2z + 4)}{z^2 + 2z + 4} = z - 2$

31. $\frac{5x^3}{2x^3 + 4} = \frac{5x^3}{2(x^3 + 2)}$

There are no common factors so this expression cannot be simplified. In this case, factors of terms were incorrectly cancelled.

33. $\frac{5}{x - 1} \cdot \frac{x - 1}{25(x - 2)} = \frac{1}{5(x - 2)}, x \neq 1$

35. $\frac{4y - 16}{5y + 15} \div \frac{4 - y}{2y + 6} = \frac{4y - 16}{5y + 15} \cdot \frac{2y + 6}{4 - y} = \frac{4(y - 4)}{5(y + 3)} \cdot \frac{2(y + 3)}{(-1)(y - 4)}$

$= \frac{8}{-5} = -\frac{8}{5}, y \neq -3, 4$

37. $\frac{x^2 + xy - 2y^2}{x^3 + x^2y} \cdot \frac{x}{x^2 + 3xy + 2y^2} = \frac{(x + 2y)(x - y)}{x^2(x + y)} \cdot \frac{x}{(x + 2y)(x + y)} = \frac{x - y}{x(x + y)^2}, x \neq -2y$

39. $\frac{x^2 - 14x + 49}{x^2 - 49} \div \frac{3x - 21}{x + 7} = \frac{(x - 7)(x - 7)}{(x + 7)(x - 7)} \cdot \frac{x + 7}{3(x - 7)}$

$= \frac{1}{3}, x \neq \pm 7$

41. $\frac{3}{x - 2} + \frac{5}{2 - x} = \frac{3}{x - 2} - \frac{5}{x - 2} = -\frac{2}{x - 2}$

43. $\frac{4}{2x + 1} - \frac{x}{x + 2} = \frac{4(x + 2)}{(2x + 1)(x + 2)} - \frac{x(2x + 1)}{(x + 2)(2x + 1)}$

$= \frac{4x + 8 - 2x^2 - x}{(x + 2)(2x + 1)}$

$= \frac{-2x^2 + 3x + 8}{(x + 2)(2x + 1)}$

45. $-\dfrac{1}{x} + \dfrac{2}{x^2 + 1} + \dfrac{1}{x^3 + x} = \dfrac{-(x^2 + 1)}{x(x^2 + 1)} + \dfrac{2x}{x(x^2 + 1)} + \dfrac{1}{x(x^2 + 1)}$

$$= \frac{-x^2 - 1 + 2x + 1}{x(x^2 + 1)}$$

$$= \frac{-x^2 + 2x}{x(x^2 + 1)}$$

$$= \frac{-x(x - 2)}{x(x^2 + 1)}$$

$$= -\frac{x - 2}{x^2 + 1}$$

$$= \frac{2 - x}{x^2 + 1}, x \neq 0$$

47. $\dfrac{x + 4}{x + 2} - \dfrac{3x - 8}{x + 2} = \dfrac{(x + 4) - (3x - 8)}{x + 2} = \dfrac{x + 4 - 3x + 8}{x + 2} = \dfrac{-2x + 12}{x + 2} = \dfrac{-2(x - 6)}{x + 2}$

The error was incorrect subtraction in the numerator.

49. $\dfrac{\left(\dfrac{x}{2} - 1\right)}{(x - 2)} = \dfrac{\left(\dfrac{x}{2} - \dfrac{2}{2}\right)}{\left(\dfrac{x - 2}{1}\right)}$

$$= \frac{x - 2}{2} \cdot \frac{1}{x - 2}$$

$$= \frac{1}{2}, x \neq 2$$

51. $\dfrac{\left[\dfrac{x^2}{(x + 1)^2}\right]}{\left[\dfrac{x}{(x + 1)^3}\right]} = \dfrac{x^2}{(x + 1)^2} \cdot \dfrac{(x + 1)^3}{x}$

$$= x(x + 1), x \neq -1, 0$$

53. $\dfrac{\left(\sqrt{x} - \dfrac{1}{2\sqrt{x}}\right)}{\sqrt{x}} = \dfrac{\left(\sqrt{x} - \dfrac{1}{2\sqrt{x}}\right)}{\sqrt{x}} \cdot \dfrac{2\sqrt{x}}{2\sqrt{x}}$

$$= \frac{2x - 1}{2x}, x > 0$$

55. $x^5 - 2x^{-2} = x^{-2}(x^7 - 2) = \dfrac{x^7 - 2}{x^2}$

57. $x^2(x^2 + 1)^{-5} - (x^2 + 1)^{-4} = (x^2 + 1)^{-5}\left[x^2 - (x^2 + 1)\right]$

$$= -\frac{1}{(x^2 + 1)^5}$$

59. $2x^2(x - 1)^{1/2} - 5(x - 1)^{-1/2} = (x - 1)^{-1/2}\left[2x^2(x - 1)^1 - 5\right] = \dfrac{2x^3 - 2x^2 - 5}{(x - 1)^{1/2}}$

61. $\dfrac{3x^{1/3} - x^{-2/3}}{3x^{-2/3}} = \dfrac{3x^{1/3} - x^{-2/3}}{3x^{-2/3}} \cdot \dfrac{x^{2/3}}{x^{2/3}} = \dfrac{3x^1 - x^0}{3x^0} = \dfrac{3x - 1}{3}, x \neq 0$

63. $\dfrac{\left(\dfrac{1}{x + h} - \dfrac{1}{x}\right)}{h} = \dfrac{\left(\dfrac{1}{x + h} - \dfrac{1}{x}\right)}{h} \cdot \dfrac{x(x + h)}{x(x + h)} = \dfrac{x - (x + h)}{hx(x + h)} = \dfrac{-h}{hx(x + h)} = -\dfrac{1}{x(x + h)}, h \neq 0$

65. $\dfrac{\left(\dfrac{1}{x+h-4} - \dfrac{1}{x-4}\right)}{h} = \dfrac{\left(\dfrac{1}{x+h-4} - \dfrac{1}{x-4}\right)}{h} \cdot \dfrac{(x-4)(x+h-4)}{(x-4)(x+h-4)}$

$= \dfrac{(x-4)-(x+h-4)}{h(x-4)(x+h-4)}$

$= \dfrac{-h}{h(x-4)(x+h-4)}$

$= -\dfrac{1}{(x-4)(x+h-4)},\ h \neq 0$

67. $\dfrac{\sqrt{x+2}-\sqrt{x}}{2} = \dfrac{\sqrt{x+2}-\sqrt{x}}{2} \cdot \dfrac{\sqrt{x+2}+\sqrt{x}}{\sqrt{x+2}+\sqrt{x}}$

$= \dfrac{(x+2)-x}{2\left(\sqrt{x+2}+\sqrt{x}\right)}$

$= \dfrac{2}{2\left(\sqrt{x+2}+\sqrt{x}\right)}$

$= \dfrac{1}{\sqrt{x+2}+\sqrt{x}}$

69. $\dfrac{\sqrt{t+3}-\sqrt{3}}{t} = \dfrac{\sqrt{t+3}-\sqrt{3}}{t} \cdot \dfrac{\sqrt{t+3}+\sqrt{3}}{\sqrt{t+3}+\sqrt{3}}$

$= \dfrac{(t+3)-3}{t\left(\sqrt{t+3}+\sqrt{3}\right)}$

$= \dfrac{t}{t\left(\sqrt{t+3}+\sqrt{3}\right)}$

$= \dfrac{1}{\sqrt{t+3}+\sqrt{3}}$

71. $\dfrac{\sqrt{x+h+1}-\sqrt{x+1}}{h} = \dfrac{\sqrt{x+h+1}-\sqrt{x+1}}{h} \cdot \dfrac{\sqrt{x+h+1}+\sqrt{x+1}}{\sqrt{x+h+1}+\sqrt{x+1}}$

$= \dfrac{(x+h+1)-(x+1)}{h\left(\sqrt{x+h+1}+\sqrt{x+1}\right)}$

$= \dfrac{h}{h\left(\sqrt{x+h+1}+\sqrt{x+1}\right)}$

$= \dfrac{1}{\sqrt{x+h+1}+\sqrt{x+1}},\ h \neq 0$

73. $T = 10\left(\dfrac{4t^2+16t+75}{t^2+4t+10}\right)$

(a)

t	0	2	4	6	8	10	12	14	16	18	20	22
T	75°	55.9°	48.3°	45°	43.3°	42.3°	41.7°	41.3°	41.1°	40.9°	40.7°	40.6°

(b) T is approaching 40°.

75. Probability $= \dfrac{\text{Shaded area}}{\text{Total area}} = \dfrac{x(x/2)}{x(2x+1)} = \dfrac{x/2}{2x+1} \cdot \dfrac{2}{2} = \dfrac{x}{2(2x+1)}$

77. (a)

Year	Banking using model (in millions)	Paying Bills using model (in millions)
2005	46.9	17
2006	57.6	25.6
2007	63.5	27.3
2008	67.3	28.8
2009	69.9	30.8
2010	71.9	33.7

(b) The values given by the models are close to the actual data.

(c) $$\frac{\text{Number of households paying bills online}}{\text{Number of households banking online}}$$

$$= \frac{\dfrac{0.307t^2 - 6.54t + 24.6}{0.015t^2 - 0.28t + 1.0}}{\dfrac{-33.74t + 121.8}{-0.40t + 1.0}}$$

$$= \frac{0.307t^2 - 6.54t + 24.6}{0.015t^2 - 0.28t + 1.0} \cdot \frac{-0.40t + 1.0}{-33.74t + 121.8}$$

$$= \frac{(0.307t^2 - 6.54t + 24.6)(-0.40t + 1.0)}{(0.015t^2 - 0.28t + 1.0)(-33.74t + 121.8)}$$

(d) When $t = 5$,

$$\frac{\left[0.307(5)^2 - 6.54(5) + 24.6\right]\left[-0.40(5) + 1.0\right]}{\left[0.015(5)^2 - 0.28(5) + 1.0\right]\left[-33.74(5) + 121.8\right]} \approx 0.362$$

When $t = 6$,

$$\frac{\left[0.307(6)^2 - 6.54(6) + 24.6\right]\left[-0.40(6) + 1.0\right]}{\left[0.015(6)^2 - 0.28(6) + 1.0\right]\left[-33.74(6) + 121.8\right]} \approx 0.445$$

When $t = 7$,

$$\frac{\left[0.307(7)^2 - 6.54(7) + 24.6\right]\left[-0.40(7) + 1.0\right]}{\left[0.015(7)^2 - 0.28(7) + 1.0\right]\left[-33.74(7) + 121.8\right]} \approx 0.429$$

When $t = 8$,

$$\frac{\left[0.307(8)^2 - 6.54(8) + 24.6\right]\left[-0.40(8) + 1.0\right]}{\left[0.015(8)^2 - 0.28(8) + 1.0\right]\left[-33.74(8) + 121.8\right]} \approx 0.428$$

When $t = 9$,

$$\frac{\left[0.307(9)^2 - 6.54(9) + 24.6\right]\left[-0.40(9) + 1.0\right]}{\left[0.015(9)^2 - 0.28(9) + 1.0\right]\left[-33.74(9) + 121.8\right]} \approx 0.440$$

When $t = 10$,

$$\frac{\left[0.307(10)^2 - 6.54(10) + 24.6\right]\left[-0.40(10) + 1.0\right]}{\left[0.015(10)^2 - 0.28(10) + 1.0\right]\left[-33.74(10) + 121.8\right]} \approx 0.468$$

For each year, the ratio is about the same.

79. $R_T = \dfrac{1}{\dfrac{1}{R_1} + \dfrac{1}{R_2}} = \dfrac{1}{\dfrac{R_2 + R_1}{R_1 R_2}} = \dfrac{R_1 R_2}{R_1 + R_2}$

81. False. In order for the simplified expression to be equivalent to the original expression, the domain of the simplified expression needs to be restricted. If n is even, $x \neq \pm 1$. If n is odd, $x \neq 1$.

Section P.6 The Rectangular Coordinate System and Graphs

1. Cartesian

3. Distance Formula

5.

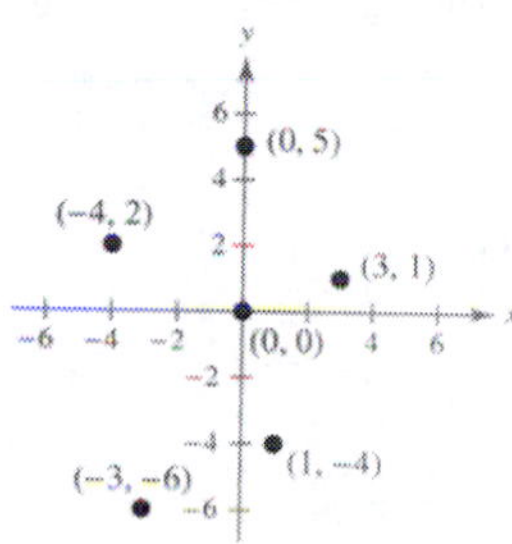

7. $(-3, 4)$

9. $x > 0$ and $y < 0$ in Quadrant IV.

11. $x = -4$ and $y > 0$ in Quadrant II.

13. $(x, -y)$ is in the second Quadrant means that (x, y) is in Quadrant III.

15.

Year, x	Number of Stores, y
2003	4906
2004	5289
2005	6141
2006	6779
2007	7262
2008	7720
2009	8416
2010	8970

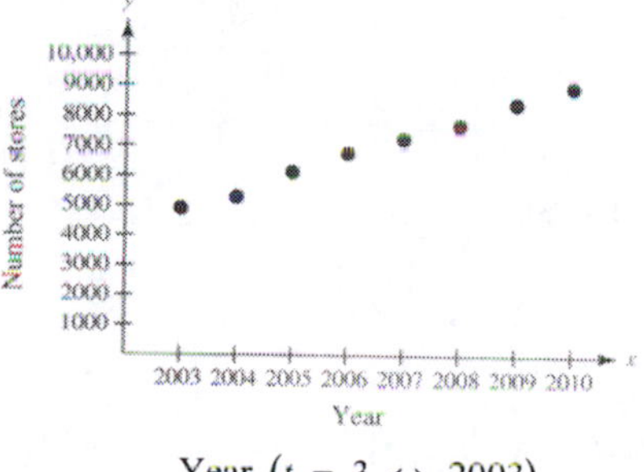

Year ($t = 3 \leftrightarrow 2003$)

17.
$$\begin{aligned} d &= \sqrt{(x_2 - x_1)^2 + (y_2 - y_1)^2} \\ &= \sqrt{(3 - (-2))^2 + (-6 - 6)^2} \\ &= \sqrt{(5)^2 + (-12)^2} \\ &= \sqrt{25 + 144} \\ &= 13 \text{ units} \end{aligned}$$

19.
$$\begin{aligned} d &= \sqrt{(x_2 - x_1)^2 + (y_2 - y_1)^2} \\ &= \sqrt{(-5 - 1)^2 + (-1 - 4)^2} \\ &= \sqrt{(-6)^2 + (-5)^2} \\ &= \sqrt{36 + 25} \\ &= \sqrt{61} \text{ units} \end{aligned}$$

21.
$$\begin{aligned} d &= \sqrt{(x_2 - x_1)^2 + (y_2 - y_1)^2} \\ &= \sqrt{\left(2 - \frac{1}{2}\right)^2 + \left(-1 - \frac{4}{3}\right)^2} \\ &= \sqrt{\left(\frac{3}{2}\right)^2 + \left(-\frac{7}{3}\right)^2} \\ &= \sqrt{\frac{9}{4} + \frac{49}{9}} \\ &= \sqrt{\frac{277}{36}} \\ &= \frac{\sqrt{277}}{6} \text{ units} \end{aligned}$$

23. (a) $(1, 0), (13, 5)$

$\text{Distance} = \sqrt{(13 - 1)^2 + (5 - 0)^2}$
$= \sqrt{12^2 + 5^2} = \sqrt{169} = 13$

$(13, 5), (13, 0)$

$\text{Distance} = |5 - 0| = |5| = 5$

$(1, 0), (13, 0)$

$\text{Distance} = |1 - 13| = |-12| = 12$

(b) $5^2 + 12^2 = 25 + 144 = 169 = 13^2$

25. $d_1 = \sqrt{(4-2)^2 + (0-1)^2} = \sqrt{4+1} = \sqrt{5}$

$d_2 = \sqrt{(4+1)^2 + (0+5)^2} = \sqrt{25+25} = \sqrt{50}$

$d_3 = \sqrt{(2+1)^2 + (1+5)^2} = \sqrt{9+36} = \sqrt{45}$

$(\sqrt{5})^2 + (\sqrt{45})^2 = (\sqrt{50})^2$

27. $d_1 = \sqrt{(1-3)^2 + (-3-2)^2} = \sqrt{4+25} = \sqrt{29}$

$d_2 = \sqrt{(3+2)^2 + (2-4)^2} = \sqrt{25+4} = \sqrt{29}$

$d_3 = \sqrt{(1+2)^2 + (-3-4)^2} = \sqrt{9+49} = \sqrt{58}$

$d_1 = d_2$

29. (a)

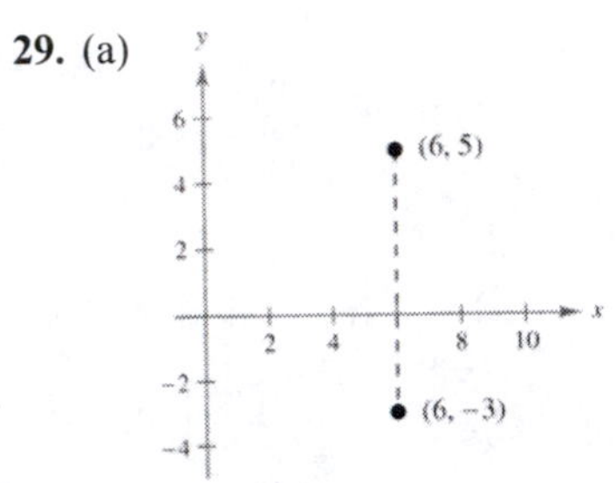

(b) $d = \sqrt{(5-(-3))^2 + (6-6)^2} = \sqrt{64} = 8$

(c) $\left(\frac{6+6}{2}, \frac{5+(-3)}{2}\right) = (6, 1)$

31. (a)

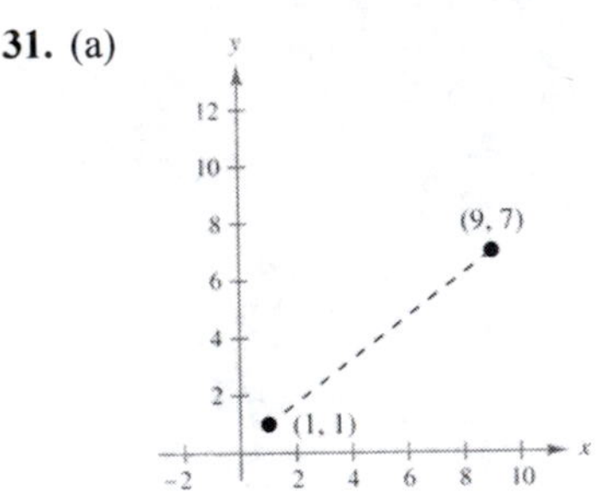

(b) $d = \sqrt{(9-1)^2 + (7-1)^2} = \sqrt{64+36} = 10$

(c) $\left(\frac{9+1}{2}, \frac{7+1}{2}\right) = (5, 4)$

33. (a)

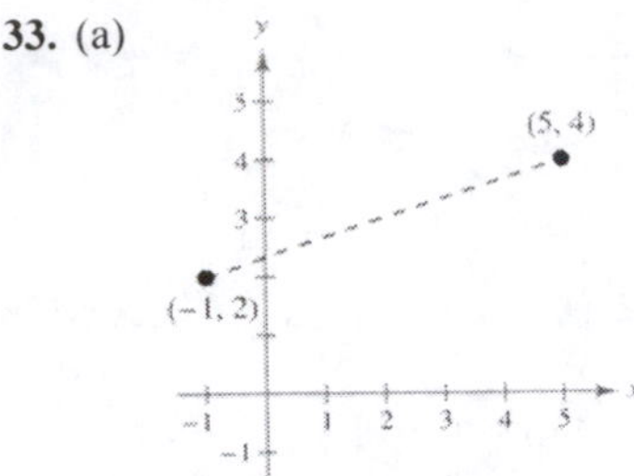

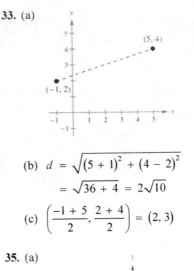

(b) $d = \sqrt{(5+1)^2 + (4-2)^2}$

$= \sqrt{36+4} = 2\sqrt{10}$

(c) $\left(\frac{-1+5}{2}, \frac{2+4}{2}\right) = (2, 3)$

35. (a)

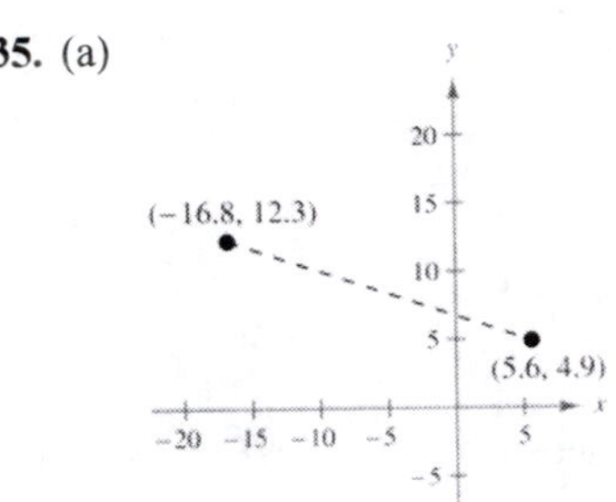

(b) $d = \sqrt{(-16.8-5.6)^2 + (12.3-4.9)^2}$

$= \sqrt{501.76 + 54.76} = \sqrt{556.52}$

(c) $\left(\frac{-16.8+5.6}{2}, \frac{12.3+4.9}{2}\right) = (-5.6, 8.6)$

37. $d = \sqrt{120^2 + 150^2}$

$= \sqrt{36{,}900}$

$= 30\sqrt{41}$

≈ 192.09

The plane flies about 192 kilometers.

39. midpoint $= \left(\frac{2002+2010}{2}, \frac{19{,}564+35{,}123}{2}\right)$

$= (2006, 27{,}343.5)$

In 2006, the sales for the Coca-Cola Company were about \$27,343.5 million.

41. $(-2+2, -4+5) = (0, 1)$

$(2+2, -3+5) = (4, 2)$

$(-1+2, -1+5) = (1, 4)$

43. $(-7+4, -2+8) = (-3, 6)$

$(-2+4, 2+8) = (2, 10)$

$(-2+4, -4+8) = (2, 4)$

$(-7+4, -4+8) = (-3, 4)$

45. (a) The minimum wage had the greatest increase from 2000 to 2010.

(b) Minimum wage in 1990: \$3.80

Minimum wage in 1995: \$4.25

Percent increase: $\left(\frac{4.25 - 3.80}{3.80}\right)(100) \approx 11.8\%$

Minimum wage in 1995: \$4.25

Minimum wage in 2011: \$7.25

Percent increase: $\left(\frac{7.25 - 4.25}{4.25}\right)(100) \approx 70.6\%$

So, the minimum wage increased 11.8% from 1990 to 1995 and 70.6% from 1995 to 2011.

(c) $\begin{pmatrix}\text{Minimum wage}\\ \text{in 2016}\end{pmatrix} = \begin{pmatrix}\text{Minimum wage}\\ \text{in 2011}\end{pmatrix} + \begin{pmatrix}\text{Percent}\\ \text{increase}\end{pmatrix}\begin{pmatrix}\text{Minimum wage}\\ \text{in 2011}\end{pmatrix} \approx \$7.25 + 0.706(\$7.25) \approx \$12.37$

So, the minimum wage will be about \$12.37 in the year 2016.

(d) Answer will vary. Sample answer: No, the prediction is too high because it is likely that the percent increase over a 4-year period (2011–2016) will be less than the percent increase over a 16-year period (1995–2011).

47. Because $x_m = \frac{x_1 + x_2}{2}$ and $y_m = \frac{y_1 + y_2}{2}$ we have:

$$2x_m = x_1 + x_2 \qquad 2y_m = y_1 + y_2$$

$$2x_m - x_1 = x_2 \qquad 2y_m - y_1 = y_2$$

So, $(x_2, y_2) = (2x_m - x_1, 2y_m - y_1)$.

49. The midpoint of the given line segment is $\left(\frac{x_1 + x_2}{2}, \frac{y_1 + y_2}{2}\right)$.

The midpoint between (x_1, y_1) and $\left(\frac{x_1 + x_2}{2}, \frac{y_1 + y_2}{2}\right)$ is $\left(\frac{x_1 + \frac{x_1 + x_2}{2}}{2}, \frac{y_1 + \frac{y_1 + y_2}{2}}{2}\right) = \left(\frac{3x_1 + x_2}{4}, \frac{3y_1 + y_2}{4}\right)$.

The midpoint between $\left(\frac{x_1 + x_2}{2}, \frac{y_1 + y_2}{2}\right)$ and (x_2, y_2) is $\left(\frac{\frac{x_1 + x_2}{2} + x_2}{2}, \frac{\frac{y_1 + y_2}{2} + y_2}{2}\right) = \left(\frac{x_1 + 3x_2}{4}, \frac{y_1 + 3y_2}{4}\right)$.

So, the three points are $\left(\frac{3x_1 + x_2}{4}, \frac{3y_1 + y_2}{4}\right)$, $\left(\frac{x_1 + x_2}{2}, \frac{y_1 + y_2}{2}\right)$, and $\left(\frac{x_1 + 3x_2}{4}, \frac{y_1 + 3y_2}{4}\right)$.

51.

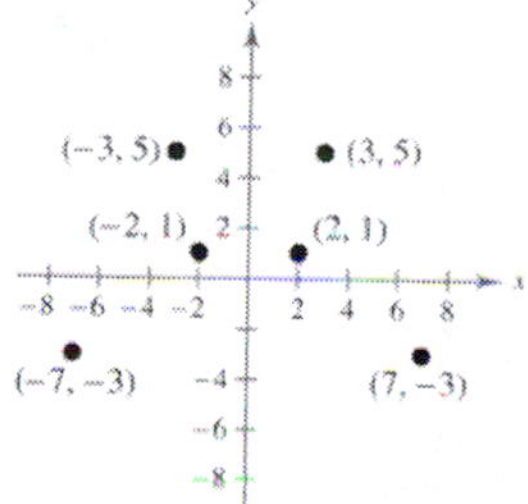

(a) The point is reflected through the y-axis.

(b) The point is reflected through the x-axis.

(c) The point is reflected through the origin.

53. No. It depends on the magnitude of the quantities measured.

55. False, you would have to use the Midpoint Formula 15 times.

57. False. The polygon could be a rhombus. For example, consider the points $(4, 0)$, $(0, 6)$, $(-4, 0)$, and $(0, -6)$.

59. Use the Midpoint Formula to prove the diagonals of the parallelogram bisect each other.

$$\left(\frac{b + a}{2}, \frac{c + 0}{2}\right) = \left(\frac{a + b}{2}, \frac{c}{2}\right)$$

$$\left(\frac{a + b + 0}{2}, \frac{c + 0}{2}\right) = \left(\frac{a + b}{2}, \frac{c}{2}\right)$$

Review Exercises for Chapter P

1. $\left\{11, -14, -\frac{8}{9}, \frac{5}{2}, \sqrt{6}, 0.4\right\}$

(a) Natural numbers: 11

(b) Whole numbers: 0, 11

(c) Integers: 11, −14

(d) Rational numbers: $11, -14, -\frac{8}{9}, \frac{5}{2}, 0.4$

(e) Irrational numbers: $\sqrt{6}$

3. 

$\frac{5}{4} > \frac{7}{8}$

5. $x \le 7$

The set consists of all real numbers less than or equal to 7.

7. $d(-74, 48) = |48 - (-74)| = 122$

9. $d(x, 7) = |x - 7|$ and $d(x, 7) \ge 4$, thus $|x - 7| \ge 4$.

11. $d(y, -30) = |y - (-30)| = |y + 30|$ and $d(y, -30) < 5$, thus $|y + 30| < 5$.

13. $12x - 7$

(a) $12(0) - 7 = -7$

(b) $12(-1) - 7 = -19$

15. $-x^2 + x - 1$

(a) $-(1)^2 + 1 - 1 = -1$

(b) $-(-1)^2 + (-1) - 1 = -3$

17. $2x + (3x - 10) = (2x + 3x) - 10$

Illustrates the Associative Property of Addition

19. $0 + (a - 5) = a - 5$

Illustrates the Additive Identity Property

21. $(t^2 + 1) + 3 = 3 + (t^2 + 1)$

Illustrates the Commutative Property of Addition

23. $|-3| + 4(-2) - 6 = 3 - 8 - 6 = -11$

25. $\frac{5}{18} \div \frac{10}{3} = \frac{5}{18} \cdot \frac{3}{10} = \frac{1}{12}$

27. $6[4 - 2(6 + 8)] = 6[4 - 2(14)]$
$= 6[4 - 28]$
$= 6(-24)$
$= -144$

29. $\frac{x}{5} + \frac{7x}{12} = \frac{12x}{60} + \frac{35x}{60} = \frac{47x}{60}$

31. (a) $3x^2(4x^3)^3 = 3x^2(64x^9) = 192x^{11}$

(b) $\frac{5y^6}{10y} = \frac{y^{6-1}}{2} = \frac{y^5}{2}, \quad y \ne 0$

33. (a) $(-2z)^3 = -8z^3$

(b) $\frac{(8y)^0}{y^2} = \frac{1}{y^2}$

35. (a) $\frac{a^2}{b^{-2}} = a^2b^2$

(b) $(a^2b^2)(3ab^{-2}) = 3a^{2+1}b^{4-2} = 3a^3b^2$

37. (a) $\frac{(5a)^{-2}}{(5a)^2} = (5a)^{-2-2} = (5a)^{-4} = \frac{1}{(5a)^4} = \frac{1}{625a^4}$

(b) $\frac{4(x^{-1})^{-3}}{4^{-2}(x^{-1})^{-1}} = \frac{4^{1-(-2)}x^3}{x} = 4^3x^{3-1} = 64x^2$

39. $168,500,000 = \$1.685 \times 10^8$

41. $4.84 \times 10^8 = 484,000,000$

43. (a) $\sqrt[3]{27^2} = \left(\sqrt[3]{27}\right)^2 = (3)^2 = 9$

(b) $\sqrt{49^3} = \left(\sqrt{49}\right)^3 = (7)^3 = 343$

45. (a) $\left(\sqrt[3]{216}\right)^3 = \left(\sqrt[3]{6^3}\right)^3 = (6)^3 = 216$

(b) $\sqrt[4]{32^4} = \left(\sqrt[4]{32}\right)^4 = 32$

47. (a) $\sqrt{50} - \sqrt{18} = \sqrt{25 \cdot 2} - \sqrt{9 \cdot 2}$
$= 5\sqrt{2} - 3\sqrt{2}$
$= 2\sqrt{2}$

(b) $2\sqrt{32} + 3\sqrt{72} = 2\sqrt{16 \cdot 2} + 3\sqrt{36 \cdot 2}$
$= 2\left(4\sqrt{2}\right) + 3\left(6\sqrt{2}\right)$
$= 8\sqrt{2} + 18\sqrt{2}$
$= 26\sqrt{2}$

49. These are not like terms. Radicals cannot be combined by addition or subtraction unless the index and the radicand are the same.

51. $\dfrac{3}{4\sqrt{3}} = \dfrac{3}{4\sqrt{3}} \cdot \dfrac{\sqrt{3}}{\sqrt{3}} = \dfrac{3\sqrt{3}}{4(3)} = \dfrac{\sqrt{3}}{4}$

53. $\dfrac{1}{2 - \sqrt{3}} = \dfrac{1}{2 - \sqrt{3}} \cdot \dfrac{2 + \sqrt{3}}{2 + \sqrt{3}} = \dfrac{2 + \sqrt{3}}{2^2 - \left(\sqrt{3}\right)^2} = \dfrac{2 + \sqrt{3}}{4 - 3} = \dfrac{2 + \sqrt{3}}{1} = 2 + \sqrt{3}$

55. $\dfrac{\sqrt{7} + 1}{2} = \dfrac{\sqrt{7} + 1}{2} \cdot \dfrac{\sqrt{7} - 1}{\sqrt{7} - 1} = \dfrac{\left(\sqrt{7}\right)^2 - 1^2}{2\left(\sqrt{7} - 1\right)} = \dfrac{7 - 1}{2\left(\sqrt{7} - 1\right)} = \dfrac{6}{2\left(\sqrt{7} - 1\right)} = \dfrac{3}{\sqrt{7} - 1}$

57. $16^{3/2} = \sqrt{16^3} = \left(\sqrt{16}\right)^3 = (4)^3 = 64$

59. $\left(3x^{2/5}\right)\left(2x^{1/2}\right) = 6x^{2/5+1/2} = 6x^{9/10}$

61. Standard form: $-11x^2 + 3$

Degree: 2

Leading coefficient: -11

63. Standard form: $-12x^2 - 4$

Degree: 2

Leading coefficient: -12

65. $-\left(3x^2 + 2x\right) + (1 - 5x) = -3x^2 - 2x + 1 - 5x$
$= -3x^2 - 7x + 1$

67. $2x\left(x^2 - 5x + 6\right) = (2x)\left(x^2\right) + (2x)(-5x) + (2x)(6)$
$= 2x^3 - 10x^2 + 12x$

69. $\left(2x^3 - 5x^2 + 10x - 7\right) + \left(4x^2 - 7x - 2\right) = 2x^3 + \left(-5x^2 + 4x^2\right) + (10x - 7x) + (-7 - 2) = 2x^3 - x^2 + 3x - 9$

71. $(3x - 6)(5x + 1) = 15x^2 + 3x - 30x - 6$
$= 15x^2 - 27x - 6$

73. $(2x - 3)^2 = (2x)^2 - 2(2x)(3) + 3^3$
$= 4x^2 - 12x + 9$

75. $(3\sqrt{5} + 2x)(3\sqrt{5} - 2x) = (3\sqrt{5})^2 - (2x)^2$
$= 45 - 4x^2$

77. $2500(1 + r)^2 = 2500(r + 1)^2$
$= 2500\left(r^2 + 2r + 1\right)$
$= 2500r^2 + 5000r + 2500$

79. Area $= (x + 12)(x + 16)$
$= x^2 + 16x + 12x + 192$
$= x^2 + 28x + 192$ square feet

81. $x^3 - x = x\left(x^2 - 1\right) = x(x + 1)(x - 1)$

83. $25x^2 - 49 = (5x)^2 - 7^2 = (5x + 7)(5x - 7)$

85. $x^3 - 64 = x^3 - 4^3 = (x - 4)\left(x^2 + 4x + 16\right)$

87. $2x^2 + 21x + 10 = (2x + 1)(x + 10)$

89. $x^3 - x^2 + 2x - 2 = x^2(x - 1) + 2(x - 1)$
$= (x - 1)\left(x^2 + 2\right)$

91. The domain of $\dfrac{1}{x + 6}$ is the set of all real numbers except $x = -6$.

93. $\dfrac{x^2 - 64}{5(3x + 24)} = \dfrac{(x + 8)(x - 8)}{5 \cdot 3(x + 8)} = \dfrac{x - 8}{15}, \quad x \neq -8$

95. $\dfrac{x^2 - 4}{x^4 - 2x^2 - 8} \cdot \dfrac{x^2 + 2}{x^2} = \dfrac{\left(x^2 - 4\right)\left(x^2 + 2\right)}{\left(x^2 - 4\right)\left(x^2 + 2\right)x^2}$
$= \dfrac{1}{x^2}, \quad x \neq \pm 2$

97. $\dfrac{1}{x-1} + \dfrac{1-x}{x^2+x+1} = \dfrac{x^2+x+1+(1-x)(x-1)}{(x-1)(x^2+x+1)} = \dfrac{x^2+x+1+x-1-x^2+x}{(x-1)(x^2+x+1)} = \dfrac{3x}{(x-1)(x^2+x+1)}$

99. $\dfrac{\left(\dfrac{\dfrac{3a}{a^2}-1}{x}\right)}{\left(\dfrac{a}{x}-1\right)} = \dfrac{\left(\dfrac{\dfrac{3a}{a^2-x}}{x}\right)}{\left(\dfrac{a-x}{x}\right)} = \dfrac{3a}{1}\cdot\dfrac{x}{a^2-x}\cdot\dfrac{x}{a-x} = \dfrac{3ax^2}{(a^2-x)(a-x)}$

101. $\dfrac{\dfrac{1}{2(x+h)}-\dfrac{1}{2x}}{h} = \dfrac{\dfrac{x-(x+6)}{2x(x+h)}}{h} = \dfrac{-h}{2x(x+h)}\cdot\dfrac{1}{h} = \dfrac{-1}{2x(x+h)},\quad h \neq 0$

103.

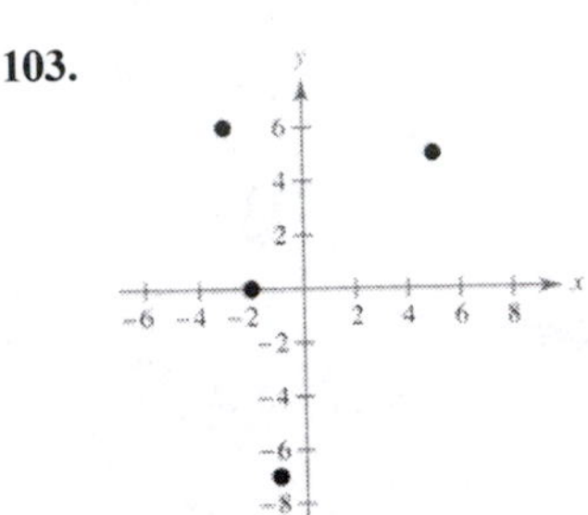

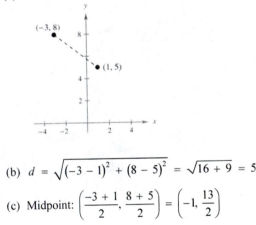

105. $x > 0$ and $y = -2$ in Quadrant IV.

107. (a)

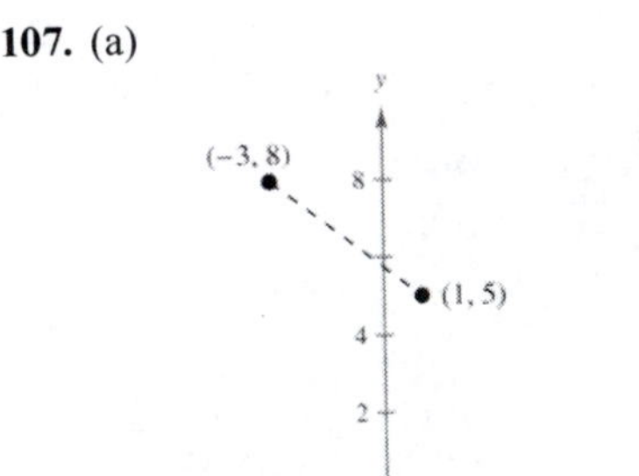

(b) $d = \sqrt{(-3-1)^2 + (8-5)^2} = \sqrt{16+9} = 5$

(c) Midpoint: $\left(\dfrac{-3+1}{2}, \dfrac{8+5}{2}\right) = \left(-1, \dfrac{13}{2}\right)$

109. (a)

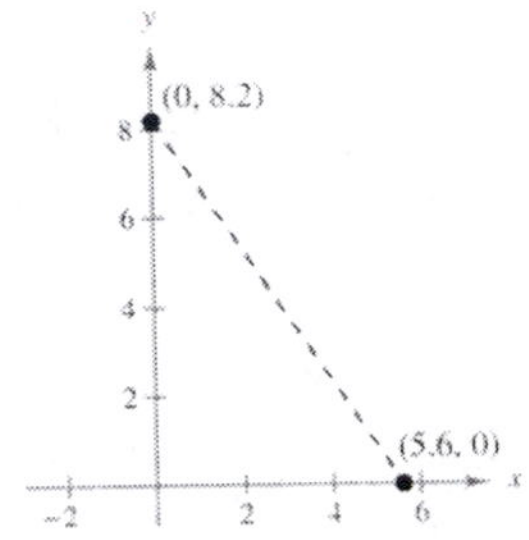

(b) $d = \sqrt{(5.6-0)^2 + (0-8.2)^2}$
$= \sqrt{31.36 + 67.24} = \sqrt{98.6}$

(c) Midpoint: $\left(\dfrac{0+5.6}{2}, \dfrac{8.2+0}{2}\right) = (2.8, 4.1)$

111. $(4-4, 8-8) = (0, 0)$
$(6-4, 8-8) = (2, 0)$
$(4-4, 3-8) = (0, -5)$
$(6-4, 3-8) = (2, -5)$

113. midpoint $= \left(\dfrac{x_1+x_2}{2}, \dfrac{y_1+y_2}{2}\right)$
$= \left(\dfrac{2008+2010}{2}, \dfrac{5.1+7.0}{2}\right)$
$= (2009, 6.05)$

In 2009, Barnes & Noble had annual sales of \$6.05 billion.

115. False, $(a+b)^2 = a^2 + 2ab + b^2 \neq a^2 + b^2$

There is also a cross-product term when a binomial sum is squared.

117. $\dfrac{ax-b}{b-ax} = -1$ for all real nonzero numbers a and b except for $x = \dfrac{b}{a}$. This makes the expression undefined.

Problem Solving for Chapter P

1. (a) Men's

Minimum Volume: $V = \frac{4}{3}\pi(55)^3 \approx 696{,}910 \text{ mm}^3$

Maximum Volume: $V = \frac{4}{3}\pi(65)^3 \approx 1{,}150{,}347 \text{ mm}^3$

Women's

Minimum Volume: $V = \frac{4}{3}\pi\left(\frac{95}{2}\right)^3 \approx 448{,}921 \text{ mm}^3$

Maximum Volume: $V = \frac{4}{3}\pi(55)^3 \approx 696{,}910 \text{ mm}^3$

(b) Men's

Minimum density: $\dfrac{7.26}{1{,}150{,}347} \approx 6.31 \times 10^{-6} \text{ kg/mm}^3$

Maximum density: $\dfrac{7.26}{696{,}910} \approx 1.04 \times 10^{-5} \text{ kg/mm}^3$

Women's

Minimum density: $\dfrac{4.00}{696{,}910} \approx 5.74 \times 10^{-6} \text{ kg/mm}^3$

Maximum density: $\dfrac{4.00}{448{,}921} \approx 8.91 \times 10^{-6} \text{ kg/mm}^3$

(c) No. The weight would be different. Cork is much lighter than iron so it would have a much smaller density.

3. One golf ball: $\dfrac{1.6 \times 10^7}{1.58 \times 10^8} \approx 0.101 \text{ pound}$

$0.101(16) = 1.62 \text{ ounces}$

5. To say that a number has n significant digits means that the number has n digits with the leftmost non-zero digit and ending with the rightmost non-zero digit. For example; 28,000, 1.400, 0.00079 each have two significant digits.

7. $r = 1 - \left(\dfrac{3225}{12{,}000}\right)^{1/4} \approx 0.280 \text{ or } 28\%$

9.

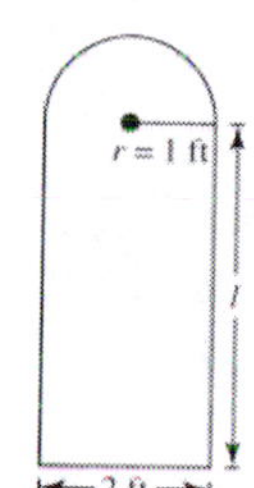

Perimeter: $P = 2l + w + \pi r$

$13.14 = 2l + 2 + \pi(1)$

$l = \dfrac{13.14 - 2 - \pi}{2} \approx 4 \text{ feet}$

Amount of glass = Area of window

$A = lw + \frac{1}{2}\pi r^2$

$= (4)(2) + \frac{1}{2}\pi(1)^2$

$\approx 9.57 \text{ square feet}$

11. $y_1 = 2x\sqrt{1-x^2} - \dfrac{x^2}{\sqrt{1-x^2}}$ $\quad y_2 = \dfrac{2-3x^2}{\sqrt{1-x^2}}$

When $x = 0$, $y_1 = 0$. When $x = 0$, $y_2 = 2$.

Thus, $y_1 \neq y_2$.

$$\begin{aligned}
y_1 &= \frac{2x\sqrt{1-x^2}}{1} - \frac{x^3}{\sqrt{1-x^2}} \\
&= \frac{2x\sqrt{1-x^2}}{1} \cdot \frac{\sqrt{1-x^2}}{\sqrt{1-x^2}} - \frac{x^3}{\sqrt{1-x^2}} \\
&= \frac{2x(1-x^2) - x^3}{\sqrt{1-x^2}} \\
&= \frac{2x - 2x^3 - x^3}{\sqrt{1-x^2}} \\
&= \frac{2x - 3x^3}{\sqrt{1-x^2}} \\
&= \frac{x(2-3x^2)}{\sqrt{1-x^2}}
\end{aligned}$$

Let $y_2 = \dfrac{x(2-3x^2)}{\sqrt{1-x^2}}$. Then $y_1 = y_2$.

13. (a) $(1, -2)$ and $(4, 1)$

The points of trisection are:

$$\left(\frac{2(1)+4}{3}, \frac{2(-2)+1}{3}\right) = (2, -1)$$

$$\left(\frac{1+2(4)}{3}, \frac{-2+2(1)}{3}\right) = (3, 0)$$

(b) $(-2, -3)$ and $(0, 0)$

The points of trisection are:

$$\left(\frac{2(-2)+0}{3}, \frac{2(-3)+0}{3}\right) = \left(-\frac{4}{3}, -2\right)$$

$$\left(\frac{-2+2(0)}{3}, \frac{-3+2(0)}{3}\right) = \left(-\frac{2}{3}, -1\right)$$

Practice Test for Chapter P

1. Evaluate $\dfrac{|-42| - 20}{15 - |-4|}$.

2. Simplify $\dfrac{x}{z} - \dfrac{z}{y}$.

3. The distance between x and 7 is no more than 4. Use absolute value notation to describe this expression.

4. Evaluate $10(-x)^3$ for $x = 5$.

5. Simplify $(-4x^3)(-2x^{-5})(\frac{1}{16}x)$.

6. Change 0.0000412 to scientific notation.

7. Evaluate $125^{2/3}$.

8. Simplify $\sqrt[4]{64x^7y^9}$.

9. Rationalize the denominator and simplify $\dfrac{6}{\sqrt{12}}$.

10. Simplify $3\sqrt{80} - 7\sqrt{500}$.

11. Simplify $(8x^4 - 9x^2 + 2x - 1) - (3x^3 + 5x + 4)$.

12. Multiply $(x - 3)(x^2 + x - 7)$.

13. Multiply $[(x - 2) - y]^2$.

14. Factor $16x^4 - 1$.

15. Factor $6x^2 + 5x - 4$.

16. Factor $x^3 - 64$.

17. Combine and simplify $-\dfrac{3}{x} + \dfrac{x}{x^2 + 2}$.

18. Combine and simplify $\dfrac{x - 3}{4x} \div \dfrac{x^2 - 9}{x^2}$.

19. Simplify $\dfrac{1 - \left(\dfrac{1}{x}\right)}{1 - \dfrac{1}{1 - \left(\dfrac{1}{x}\right)}}$.

20. (a) Plot the points $(-3, 6)$ and $(5, -1)$,
    (b) find the distance between the points, and
    (c) find the midpoint of the line segment joining the points.

CHAPTER 1
Equations, Inequalities, and Mathematical Modeling

CHAPTER 1
Equations, Inequalities, and Mathematical Modeling

Section 1.1 Graphs of Equations

1. solution or solution point

3. intercepts

5. circle; (h, k); r

7. (a) $(0, 2)$: $2 \stackrel{?}{=} \sqrt{0 + 4}$

$2 = 2$

Yes, the point *is* on the graph.

(b) $(5, 3)$: $3 \stackrel{?}{=} \sqrt{5 + 4}$

$3 \stackrel{?}{=} \sqrt{9}$

$3 = 3$

Yes, the point *is* on the graph.

9. (a) $(2, 0)$: $(2)^2 - 3(2) + 2 \stackrel{?}{=} 0$

$4 - 6 + 2 \stackrel{?}{=} 0$

$0 = 0$

Yes, the point *is* on the graph.

(b) $(-2, 8)$: $(-2)^2 - 3(-2) + 2 \stackrel{?}{=} 8$

$4 + 6 + 2 \stackrel{?}{=} 8$

$12 \neq 8$

No, the point *is not* on the graph.

11. (a) $(2, 3)$: $3 \stackrel{?}{=} |2 - 1| + 2$

$3 \stackrel{?}{=} 1 + 2$

$3 = 3$

Yes, the point *is* on the graph.

(b) $(-1, 0)$: $0 \stackrel{?}{=} |-1 - 1| + 2$

$0 \stackrel{?}{=} 2 + 2$

$0 \neq 4$

No, the point *is not* on the graph.

13. (a) $(3, -2)$: $(3)^2 + (-2)^2 \stackrel{?}{=} 20$

$9 + 4 \stackrel{?}{=} 20$

$13 \neq 20$

No, the point *is not* on the graph.

(b) $(-4, 2)$: $(-4)^2 + (2)^2 \stackrel{?}{=} 20$

$16 + 4 \stackrel{?}{=} 20$

$20 = 20$

Yes, the point *is* on the graph.

15. $y = -2x + 5$

x	-1	0	1	2	$\frac{5}{2}$
y	7	5	3	1	0
(x, y)	$(-1, 7)$	$(0, 5)$	$(1, 3)$	$(2, 1)$	$\left(\frac{5}{2}, 0\right)$

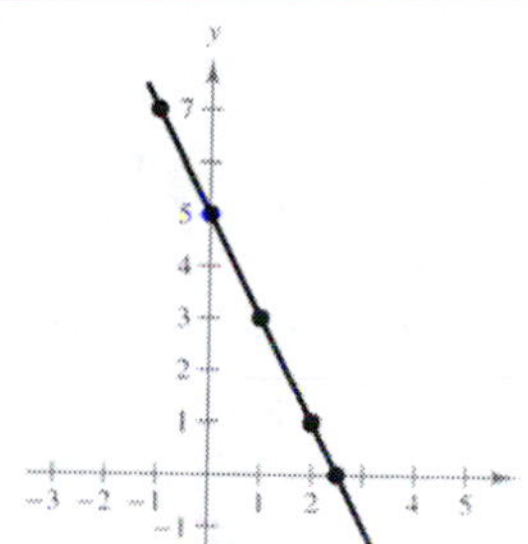

17. $y = x^2 - 3x$

x	-1	0	1	2	3
y	4	0	-2	-2	0
(x, y)	$(-1, 4)$	$(0, 0)$	$(1, -2)$	$(2, -2)$	$(3, 0)$

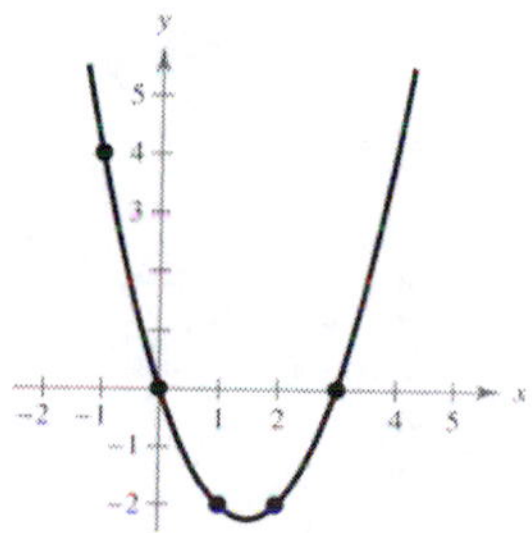

19. x-intercept: $(3, 0)$

y-intercept: $(0, 9)$

21. x-intercept: $(-2, 0)$

y-intercept: $(0, 2)$

23. x-intercept: $(1, 0)$

y-intercept: $(0, 2)$

25. $x^2 - y = 0$

$(-x)^2 - y = 0 \Rightarrow x^2 - y = 0 \Rightarrow$ y-axis symmetry

$x^2 - (-y) = 0 \Rightarrow x^2 + y = 0 \Rightarrow$ No x-axis symmetry

$(-x)^2 - (-y) = 0 \Rightarrow x^2 + y = 0 \Rightarrow$ No origin symmetry

27. $y = x^3$

$y = (-x)^3 \Rightarrow y = -x^3 \Rightarrow$ No y-axis symmetry

$-y = x^3 \Rightarrow y = -x^3 \Rightarrow$ No x-axis symmetry

$-y = (-x)^3 \Rightarrow -y = -x^3 \Rightarrow y = x^3 \Rightarrow$ Origin symmetry

29. $y = \dfrac{x}{x^2 + 1}$

$y = \dfrac{-x}{(-x)^2 + 1} \Rightarrow y = \dfrac{-x}{x^2 + 1} \Rightarrow$ No y-axis symmetry

$-y = \dfrac{x}{x^2 + 1} \Rightarrow y = \dfrac{-x}{x^2 + 1} \Rightarrow$ No x-axis symmetry

$-y = \dfrac{-x}{(-x)^2 + 1} \Rightarrow -y = \dfrac{-x}{x^2 + 1} \Rightarrow y = \dfrac{x}{x^2 + 1} \Rightarrow$ Origin symmetry

31. $xy^2 + 10 = 0$

$(-x)y^2 + 10 = 0 \Rightarrow -xy^2 + 10 = 0 \Rightarrow$ No y-axis symmetry

$x(-y)^2 + 10 = 0 \Rightarrow xy^2 + 10 = 0 \Rightarrow$ x-axis symmetry

$(-x)(-y)^2 + 10 = 0 \Rightarrow -xy^2 + 10 = 0 \Rightarrow$ No origin symmetry

33.

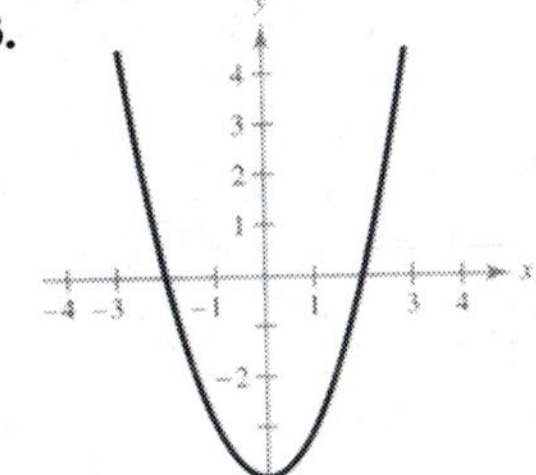

35.

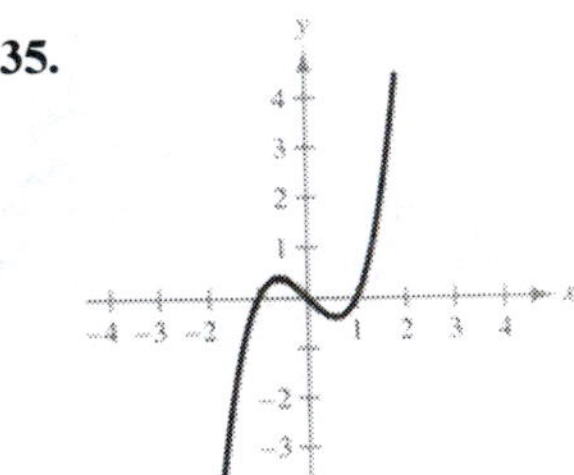

37. $y = -3x + 1$

x-intercept: $\left(\frac{1}{3}, 0\right)$

y-intercept: $(0, 1)$

No symmetry

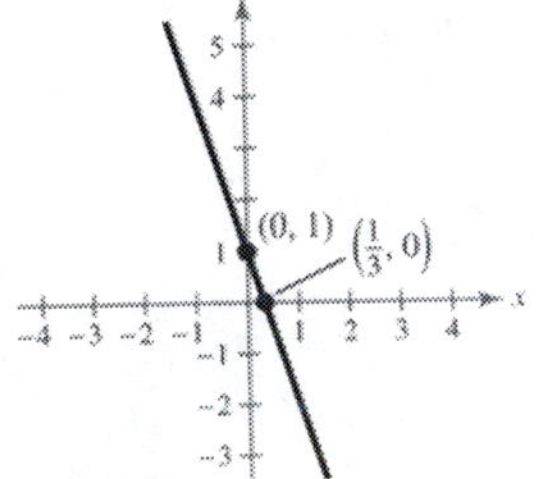

39. $y = x^2 - 2x$

x-intercepts: $(0, 0), (2, 0)$

y-intercept: $(0, 0)$

No symmetry

x	−1	0	1	2	3
y	3	0	−1	0	3

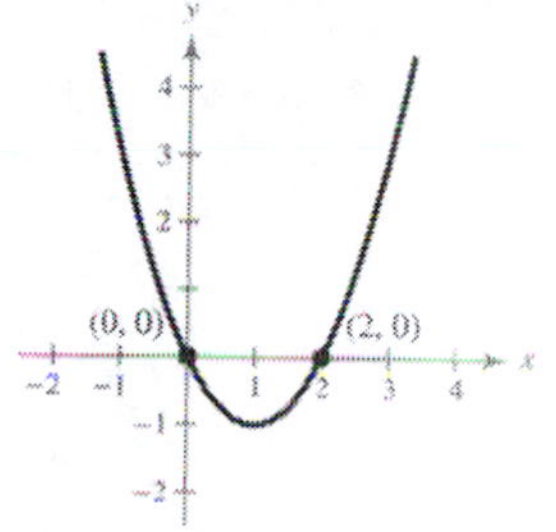

41. $y = x^3 + 3$

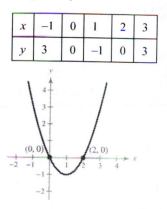

x-intercept: $\left(\sqrt[3]{-3}, 0\right)$

y-intercept: $(0, 3)$

No symmetry

x	−2	−1	0	1	2
y	−5	2	3	4	11

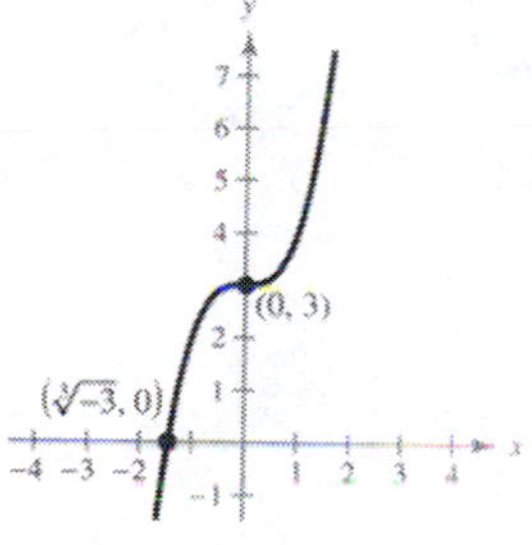

43. $y = \sqrt{x - 3}$

x-intercept: $(3, 0)$

y-intercept: none

No symmetry

x	3	4	7	12
y	0	1	2	3

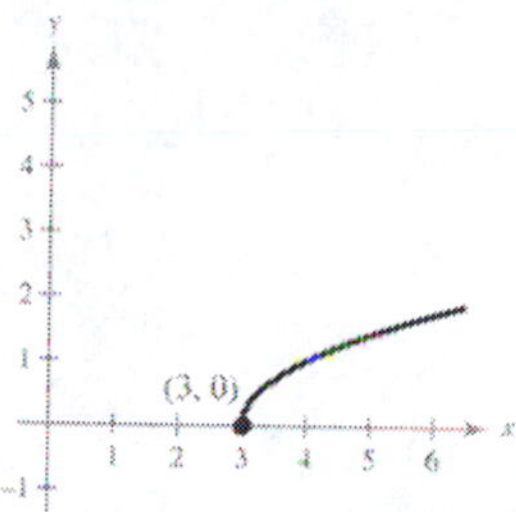

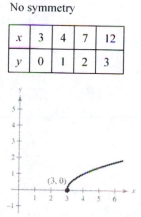

45. $y = |x - 6|$

x-intercept: $(6, 0)$

y-intercept: $(0, 6)$

No symmetry

x	−2	0	2	4	6	8	10
y	8	6	4	2	0	2	4

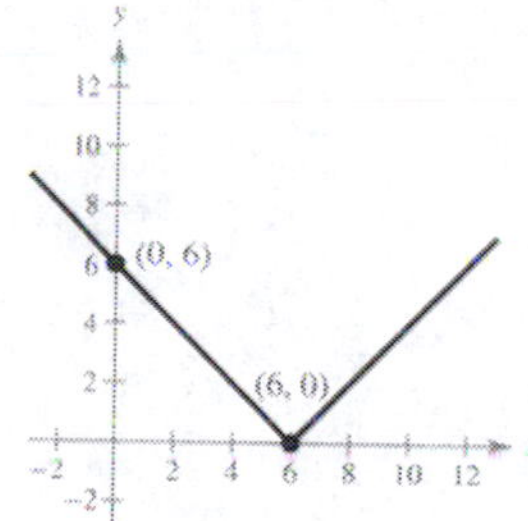

47. $x = y^2 - 1$

x-intercept: $(-1, 0)$

y-intercepts: $(0, -1), (0, 1)$

x-axis symmetry

x	-1	0	3
y	0	±1	±2

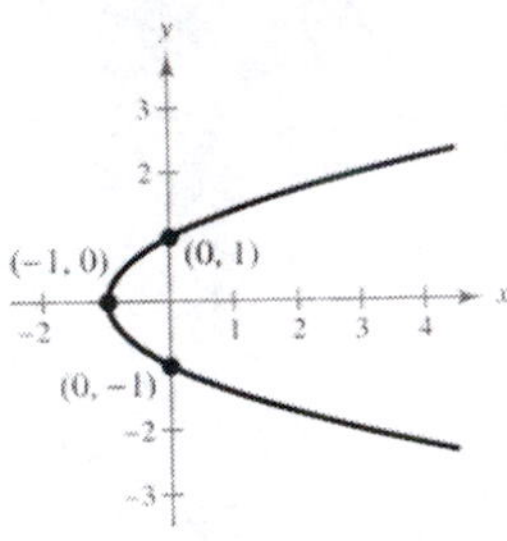

49. $y = 5 - \frac{1}{2}x$

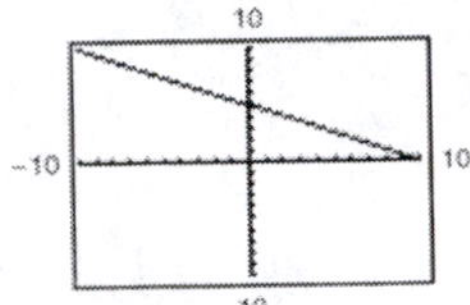

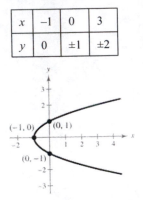

Intercepts: $(10, 0), (0, 5)$

51. $y = x^2 - 4x + 3$

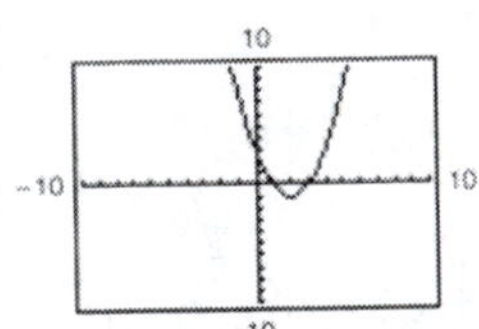

Intercepts: $(3, 0), (1, 0), (0, 3)$

53. $y = \dfrac{2x}{x - 1}$

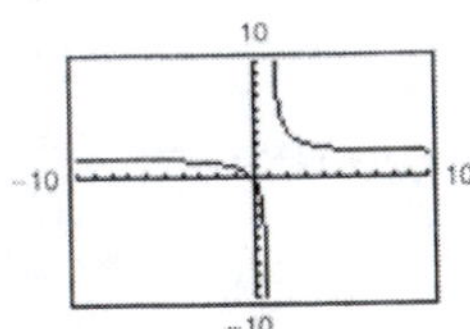

Intercept: $(0, 0)$

55. $y = \sqrt[3]{x} + 2$

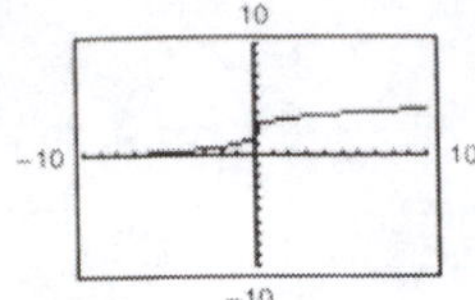

Intercepts: $(-8, 0), (0, 2)$

57. $y = x\sqrt{x + 6}$

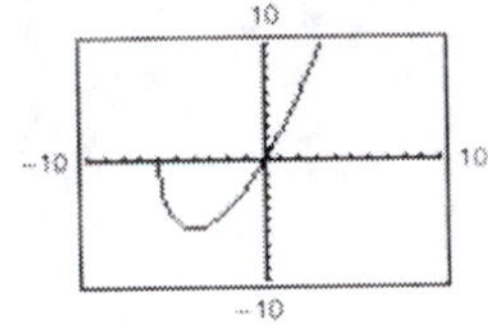

Intercepts: $(0, 0), (-6, 0)$

59. $y = |x + 3|$

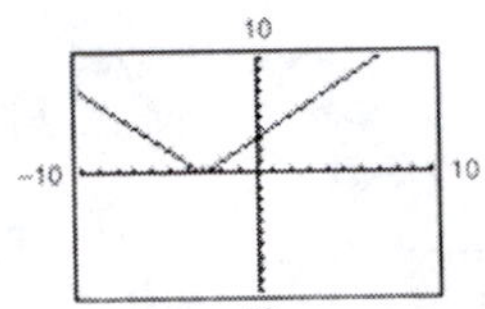

Intercepts: $(-3, 0), (0, 3)$

61. Center: $(0, 0)$; Radius: 4

$$(x - 0)^2 + (y - 0)^2 = 4^2$$
$$x^2 + y^2 = 16$$

63. Center: $(2, -1)$; Radius: 4

$$(x - 2)^2 + (y - (-1))^2 = 4^2$$
$$(x - 2)^2 + (y + 1)^2 = 16$$

65. Center: $(-1, 2)$; Solution point: $(0, 0)$

$$(x - (-1))^2 + (y - 2)^2 = r^2$$
$$(0 + 1)^2 + (0 - 2)^2 = r^2 \Rightarrow 5 = r^2$$
$$(x + 1)^2 + (y - 2)^2 = 5$$

67. Endpoints of a diameter: $(0, 0), (6, 8)$

Center: $\left(\dfrac{0+6}{2}, \dfrac{0+8}{2}\right) = (3, 4)$

$$(x-3)^2 + (y-4)^2 = r^2$$

$$(0-3)^2 + (0-4)^2 = r^2 \Rightarrow 25 = r^2$$

$$(x-3)^2 + (y-4)^2 = 25$$

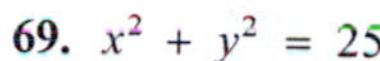

69. $x^2 + y^2 = 25$

Center: $(0, 0)$, Radius: 5

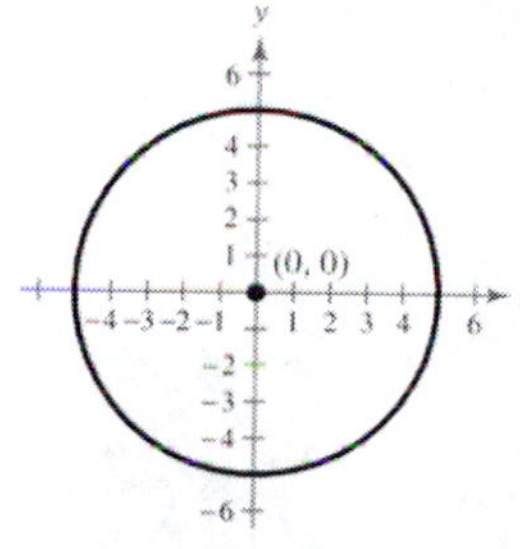

71. $(x-1)^2 + (y+3)^2 = 9$

Center: $(1, -3)$, Radius: 3

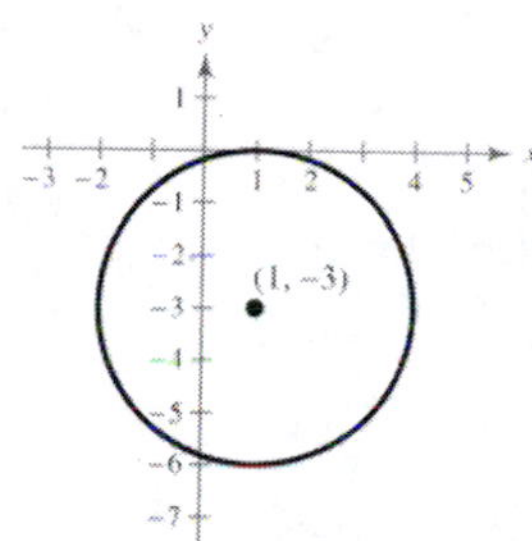

73. $\left(x - \frac{1}{2}\right)^2 + \left(y - \frac{1}{2}\right)^2 = \frac{9}{4}$

Center: $\left(\frac{1}{2}, \frac{1}{2}\right)$, Radius: $\frac{3}{2}$

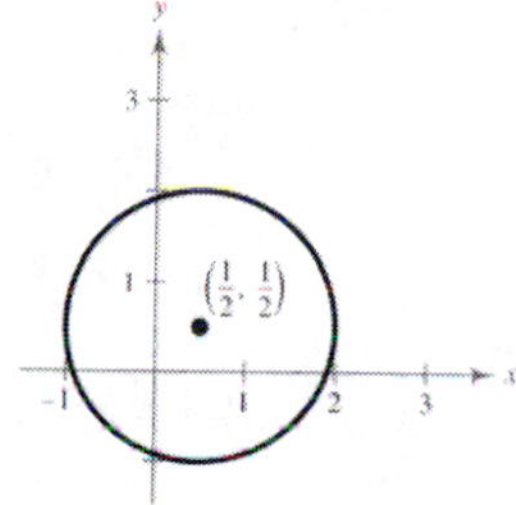

75. $y = 500{,}000 - 40{,}000t,\ 0 \le t \le 8$

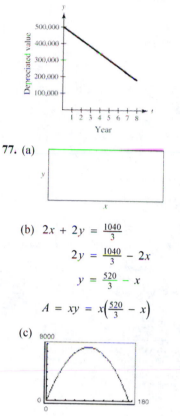

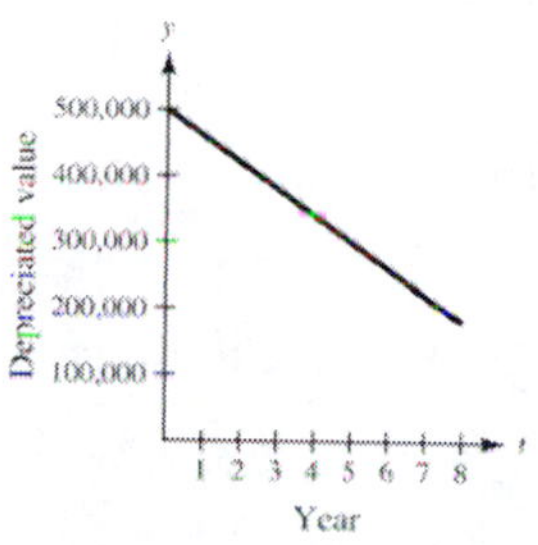

77. (a)

(b) $2x + 2y = \frac{1040}{3}$

$$2y = \frac{1040}{3} - 2x$$

$$y = \frac{520}{3} - x$$

$$A = xy = x\left(\frac{520}{3} - x\right)$$

(c)

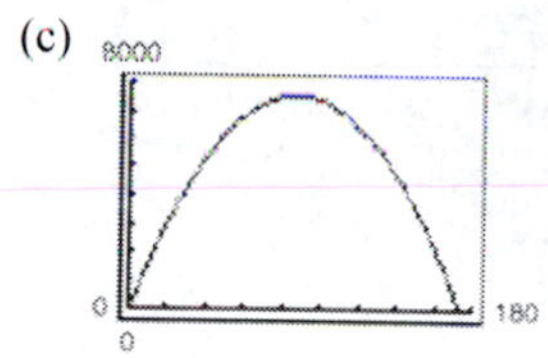

(d) When $x = y = 86\frac{2}{3}$ yards, the area is a maximum of $7511\frac{1}{9}$ square yards.

(e) A regulation NFL playing field is 120 yards long and $53\frac{1}{3}$ yards wide. The actual area is 6400 square yards.

79. (a) 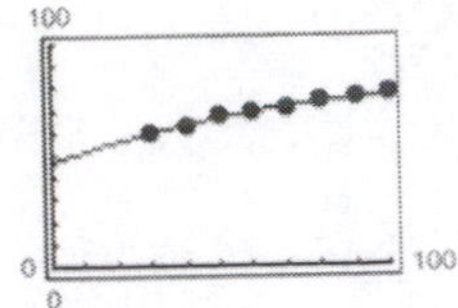

Because the line is close to the points, the model fits the data well.

(b) Graphically: The point $(90, 75.4)$ represents a life expectancy of 75.4 years in 1990.

Algebraically:
$$\begin{aligned} y &= -0.002t^2 + 0.5t + 46.6 \\ &= -0.002(90)^2 + 0.5(90) + 46.6 \\ &= 75.4 \end{aligned}$$

So, the life expectancy in 1990 was about 75.4 years.

(c) Graphically: The point $(94.6, 76.0)$ represents a life expectancy of 76 years during the year 1994.

Algebraically:
$$\begin{aligned} y &= -0.002t^2 + 0.5t + 46.6 \\ 76.0 &= -0.002t^2 + 0.5t + 46.6 \\ 0 &= -0.002t^2 + 0.5t - 29.4 \end{aligned}$$

Use the quadratic formula to solve.

$$\begin{aligned} t &= \frac{-b \pm \sqrt{b^2 - 4ac}}{2a} \\ &= \frac{-(0.5) \pm \sqrt{(0.5)^2 - 4(-0.002)(-29.4)}}{2(-0.002)} \\ &= \frac{-0.5 \pm \sqrt{0.0148}}{-0.004} \\ &= 125 \pm 30.4 \end{aligned}$$

So, $t = 94.6$ or $t = 155.4$. Since 155.4 is not in the domain, the solution is $t = 94.6$, which is the year 1994.

(d) When $t = 115$:

$$\begin{aligned} y &= -0.002t^2 + 0.5t + 46.6 \\ &= -0.002(115)^2 + (0.5)(115) + 46.6 \\ &= 77.65 \end{aligned}$$

The life expectancy using the model is 77.65 years, which is slightly less than the given projection of 78.9 years.

(e) Answers will vary. *Sample answer:* No. Because the model is quadratic, the life expectancies begin to decrease after a certain point.

81. $y = ax^2 + bx^3$

(a)
$$\begin{aligned} y &= a(-x)^2 + b(-x)^3 \\ &= ax^2 - bx^3 \end{aligned}$$

To be symmetric with respect to the y-axis; a can be any non-zero real number, b must be zero.

(b)
$$\begin{aligned} -y &= a(-x)^2 + b(-x)^3 \\ -y &= ax^2 - bx^3 \\ y &= -ax^2 + bx^3 \end{aligned}$$

To be symmetric with respect to the origin; a must be zero, b can be any non-zero real number.

Section 1.2 Linear Equations in One Variable

1. equation

3. $ax + b = 0$

5. rational

7. $2(x - 1) = 2x - 2$ is an *identity* by the Distributive Property. It is true for all real values of x.

9. $-6(x - 3) + 5 = -2x + 10$ is *conditional*. There are real values of x for which the equation is not true.

11. $4(x + 1) - 2x = 4x + 4 - 2x = 2x + 4 = 2(x + 2)$

This is an *identity* by simplification. It is true for all real values of x.

13. $2(x - 1) = 2x - 1$ is *conditional*. There are real values of x for which the equation is not true.

15.
$$\begin{aligned} x + 11 &= 15 \\ x + 11 - 11 &= 15 - 11 \\ x &= 4 \end{aligned}$$

17.
$$\begin{aligned} 7 - 2x &= 25 \\ 7 - 7 - 2x &= 25 - 7 \\ -2x &= 18 \\ \frac{-2x}{-2} &= \frac{18}{-2} \\ x &= -9 \end{aligned}$$

19.
$$\begin{aligned} 3x - 5 &= 2x + 7 \\ 3x - 2x - 5 &= 2x - 2x + 7 \\ x - 5 &= 7 \\ x - 5 + 5 &= 7 + 5 \\ x &= 12 \end{aligned}$$

21.
$$\begin{aligned} 4y + 2 - 5y &= 7 - 6y \\ 4y - 5y + 2 &= 7 - 6y \\ -y + 2 &= 7 - 6y \\ -y + 6y + 2 &= 7 - 6y + 6y \\ 5y + 2 &= 7 \\ 5y + 2 - 2 &= 7 - 2 \\ 5y &= 5 \\ \frac{5y}{5} &= \frac{5}{5} \\ y &= 1 \end{aligned}$$

23.
$$\begin{aligned} x - 3(2x + 3) &= 8 - 5x \\ x - 6x - 9 &= 8 - 5x \\ -5x - 9 &= 8 - 5x \\ -5x + 5x - 9 &= 8 - 5x + 5x \\ -9 &\neq 8 \end{aligned}$$

No solution

25.
$$\begin{aligned} 0.25x + 0.75(10 - x) &= 3 \\ 0.25x + 7.5 - 0.75x &= 3 \\ -0.50x + 7.5 &= 3 \\ -0.50x &= -4.5 \\ x &= 9 \end{aligned}$$

27.
$$\begin{aligned} \frac{3x}{8} - \frac{4x}{3} &= 4 \\ \frac{9x}{24} - \frac{32x}{24} &= 4 \\ -\frac{23x}{24} &= 4 \\ -\frac{23x}{24}\left(-\frac{24}{23}\right) &= 4\left(-\frac{24}{23}\right) \\ x &= -\frac{96}{23} \end{aligned}$$

or

$$\begin{aligned} \frac{3x}{8} - \frac{4x}{3} &= 4 \\ 24\left(\frac{3x}{8} - \frac{4x}{3}\right) &= 24(4) \\ 9x - 32x &= 96 \\ -23x &= 96 \\ x &= -\frac{96}{23} \end{aligned}$$

The second method is easier. The fractions are eliminated in the first step.

29.
$$\begin{aligned} \frac{5x}{4} + \frac{1}{2} &= x - \frac{1}{2} \\ 4\left(\frac{5x}{4} + \frac{1}{2}\right) &= 4\left(x - \frac{1}{2}\right) \\ 4\left(\frac{5x}{4}\right) + 4\left(\frac{1}{2}\right) &= 4(x) - 4\left(\frac{1}{2}\right) \\ 5x + 2 &= 4x - 2 \\ x &= -4 \end{aligned}$$

31.
$$\begin{aligned} \frac{5x - 4}{5x + 4} &= \frac{2}{3} \\ 3(5x - 4) &= 2(5x + 4) \\ 15x - 12 &= 10x + 8 \\ 5x &= 20 \\ x &= 4 \end{aligned}$$

33.
$$\begin{aligned} 10 - \frac{13}{x} &= 4 + \frac{5}{x} \\ \frac{10x - 13}{x} &= \frac{4x + 5}{x} \\ 10x - 13 &= 4x + 5 \\ 6x &= 18 \\ x &= 3 \end{aligned}$$

35.
$$\begin{aligned} 3 &= 2 + \frac{2}{z+2} \\ 3(z+2) &= \left(2 + \frac{2}{z+2}\right)(z+2) \\ 3z + 6 &= 2z + 4 + 2 \\ z &= 0 \end{aligned}$$

37.
$$\begin{aligned} \frac{x}{x+4} + \frac{4}{x+4} + 2 &= 0 \\ \frac{x+4}{x+4} + 2 &= 0 \\ 1 + 2 &= 0 \\ 3 &\neq 0 \end{aligned}$$

Contradiction; no solution

39.
$$\begin{aligned} \frac{2}{(x-4)(x-2)} &= \frac{1}{x-4} + \frac{2}{x-2} \quad \text{Multiply both sides by } (x-4)(x-2). \\ 2 &= 1(x-2) + 2(x-4) \\ 2 &= x - 2 + 2x - 8 \\ 2 &= 3x - 10 \\ 12 &= 3x \\ 4 &= x \end{aligned}$$

A check reveals that $x = 4$ is an extraneous solution—it makes the denominator zero. There is no real solution.

41.
$$\begin{aligned} \frac{1}{x-3} + \frac{1}{x+3} &= \frac{10}{x^2 - 9} \\ \frac{1}{x-3} + \frac{1}{x+3} &= \frac{10}{(x+3)(x-3)} \quad \text{Multiply both sides by } (x+3)(x-3). \\ 1(x+3) + 1(x-3) &= 10 \\ 2x &= 10 \\ x &= 5 \end{aligned}$$

43.
$$\begin{aligned} \frac{3}{x^2 - 3x} + \frac{4}{x} &= \frac{1}{x-3} \\ \frac{3}{x(x-3)} + \frac{4}{x} &= \frac{1}{x-3} \quad \text{Multiply both sides by } x(x-3). \\ 3 + 4(x-3) &= x \\ 3 + 4x - 12 &= x \\ 3x &= 9 \\ x &= 3 \end{aligned}$$

A check reveals that $x = 3$ is an extraneous solution since it makes the denominator zero, so there is no solution.

45.
$$\begin{aligned} y &= 12 - 5x & \qquad y &= 12 - 5x \\ 0 &= 12 - 5x & y &= 12 - 5(0) \\ 5x &= 12 & y &= 12 \\ x &= \tfrac{12}{5} \end{aligned}$$

The x-intercept is $\left(\frac{12}{5}, 0\right)$ and the y-intercept is $(0, 12)$.

47.
$$\begin{aligned} y &= -3(2x + 1) & \qquad y &= -3(2x + 1) \\ 0 &= -3(2x + 1) & y &= -3(2(0) + 1) \\ 0 &= 2x + 1 & y &= -3 \\ x &= -\tfrac{1}{2} \end{aligned}$$

The x-intercept is $\left(-\frac{1}{2}, 0\right)$ and the y-intercept is $(0, -3)$.

49.

$$\begin{aligned} 2x + 3y &= 10 \\ 2x + 3(0) &= 10 \\ 2x &= 10 \\ x &= 5 \end{aligned} \qquad \begin{aligned} 2x + 3y &= 10 \\ 2(0) + 3y &= 10 \\ 3y &= 10 \\ y &= \tfrac{10}{3} \end{aligned}$$

The x-intercept is $(5, 0)$ and the y-intercept is $\left(0, \frac{10}{3}\right)$.

51.

$$\begin{aligned} 4y - 0.75x + 1.2 &= 0 \\ 4(0) - 0.75x + 1.2 &= 0 \\ -0.75 + 1.2 &= 0 \\ x = \frac{1.2}{0.75} &= 1.6 \end{aligned} \qquad \begin{aligned} 4y - 0.75x + 1.2 &= 0 \\ 4y - 0.75(0) + 1.2 &= 0 \\ 4y + 1.2 &= 0 \\ y = \frac{-1.2}{4} &= -0.3 \end{aligned}$$

The x-intercept is $(1.6, 0)$ and the y-intercept is $(0, -0.3)$.

53. $\dfrac{2x}{5} + 8 - 3y = 0 \Rightarrow 2x + 40 - 15y = 0$ Multiply both sides by 5.

$$\begin{aligned} 2x + 40 - 15(0) &= 0 \\ 2x + 40 &= 0 \\ x &= -20 \end{aligned} \qquad \begin{aligned} 2x + 40 - 15y &= 0 \\ 2(0) + 40 - 15y &= 0 \\ 40 - 15y &= 0 \\ y = \frac{40}{15} &= \frac{8}{3} \end{aligned}$$

The x-intercept is $(-20, 0)$ and the y-intercept is $\left(0, \frac{8}{3}\right)$.

55. $y = 2(x - 1) - 4$

$$\begin{aligned} 0 &= 2(x - 1 - 4) \\ 0 &= 2x - 2 - 4 \\ 0 &= 2x - 6 \\ 6 &= 2x \\ 3 &= x \\ x &= 3 \end{aligned}$$

The x-intercept is at 3. The solution of $0 = 2(x - 1) - 4$ and the x-intercept of $y = 2(x - 1) - 4$ are the same. They are both $x = 3$. The x-intercept is $(3, 0)$.

57. $y = 20 - (3x - 10)$

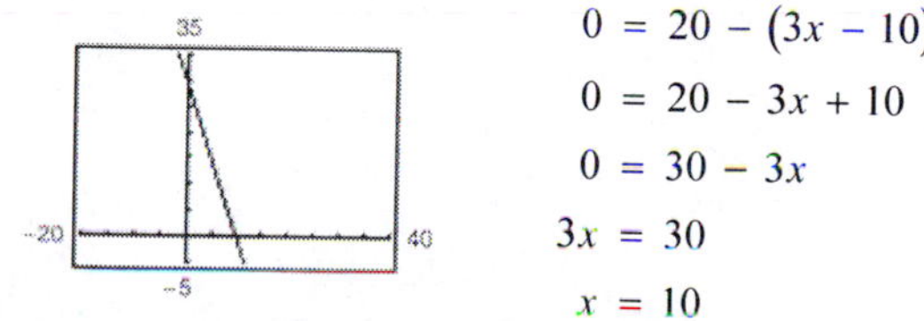

$$\begin{aligned} 0 &= 20 - (3x - 10) \\ 0 &= 20 - 3x + 10 \\ 0 &= 30 - 3x \\ 3x &= 30 \\ x &= 10 \end{aligned}$$

The x-intercept is at 10. The solution of $0 = 2 - (3x - 10)$ and the x-intercept of $y = 20 - (3x - 10)$ are the same. They are both $x = 10$. The x-intercept is $(10, 0)$.

59. $y = -38 + 5(9 - x)$

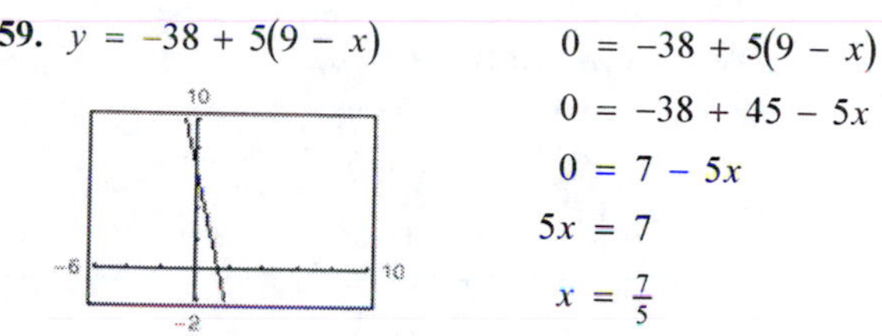

$$\begin{aligned} 0 &= -38 + 5(9 - x) \\ 0 &= -38 + 45 - 5x \\ 0 &= 7 - 5x \\ 5x &= 7 \\ x &= \tfrac{7}{5} \end{aligned}$$

The x-intercept is at $\frac{7}{5}$. The solution of $0 = -38 + 5(9 - x)$ and the x-intercept of $y = -38 + 5(9 - x)$ are the same. They are both $x = \frac{7}{5}$. The x-intercept is $\left(\frac{7}{5}, 0\right)$.

61.

$$\begin{aligned} 0.275x + 0.725(500 - x) &= 300 \\ 0.275x + 362.5 - 0.725x &= 300 \\ -0.45x &= -62.5 \\ x &= \frac{62.5}{0.45} \approx 138.889 \end{aligned}$$

63. $\dfrac{2}{7.398} - \dfrac{4.405}{x} = \dfrac{1}{x}$ Multiply both sides by $7.398x$.

$$\begin{aligned} 2x - (4.405)(7.398) &= 7.398 \\ 2x &= (4.405)(7.398) + 7.398 \\ 2x &= (5.405)(7.398) \\ x &= \frac{(5.405)(7.398)}{2} \approx 19.993 \end{aligned}$$

65.
$$471 = 2\pi(25) + 2\pi(5h)$$
$$471 = 50\pi + 10\pi h$$
$$471 - 50\pi = 10\pi h$$
$$h = \frac{471 - 50\pi}{10\pi} = \frac{471 - 50(3.14)}{10(3.14)} = 10$$
$$h = 10 \text{ feet}$$

67. Let $y = 18$:
$$y = 0.432x - 10.44$$
$$18 = 0.432x - 10.44$$
$$28.44 = 0.432x$$
$$\frac{28.44}{0.432} = x$$
$$65.8 \approx x$$

So, the height of the female is about 65.8 inches or 5 feet 6 inches.

69. (a) The y-intercept is about $(0, 1480)$.

(b) Let $t = 0$:
$$y = -8.37t + 1480.6$$
$$= -8.37(0) + 1480.6$$
$$= 1480.6$$

The y-intercept is (0, 1480.6).

So, there were about 1481 newspapers in the United States in 2000.

(c) Let $y = 1355$:
$$y = -8.37t + 1480.6$$
$$1355 = -8.37t + 1480.6$$
$$-125.6 = -8.37t$$
$$\frac{125.6}{8.37} = t$$
$$15.0 \approx t$$

In 2015, there will be about 1355 daily newspapers.; Answer will vary. *Sample answer:* Yes. This answer does seem reasonable because the amount of daily newspapers will most likely continue to decline.

71. Let $c = 10{,}000$:
$$c = 0.37m + 2600$$
$$10{,}000 = 0.37m + 2600$$
$$7400 = 0.37m$$
$$\frac{7400}{0.37} = m$$
$$m = 20{,}000$$

So, the number of miles is 20,000.

73. False. $x(3 - x) = 10 \Rightarrow 3x - x^2 = 10$

This is a quadratic equation. The equation cannot be written in the form $ax + b = 0$.

75.
$$2(x - 3) + 1 = 2x - 5$$
$$2x - 6 + 1 = 2x - 5$$
$$2x - 5 = 2x - 5$$

False. The equation is an identity, so every real number is a solution.

77.
$$2 - \frac{1}{x - 2} = \frac{3}{x - 2}$$
$$(x - 2)\left(2 - \frac{1}{x - 2}\right) = (x - 2)\left(\frac{3}{x - 2}\right)$$
$$2(x - 2) - 1 = 3$$
$$2x - 4 - 1 = 3$$
$$2x - 5 = 3$$
$$2x = 8$$
$$x = 4$$

False. $x = 4$ is a solution.

79. (a)

x	-1	0	1	2	3	4
$3.2x - 5.8$	-9	-5.8	-2.6	0.6	3.8	7

(b) Since the sign changes from negative at 1 to positive at 2, the root is somewhere between 1 and 2.

$1 < x < 2$

(c)

x	1.5	1.6	1.7	1.8	1.9	2
$3.2x - 5.8$	-1	-0.68	-0.36	-0.04	0.28	0.6

(d) Since the sign changes from negative at 1.8 to positive at 1.9, the root is somewhere between 1.8 and 1.9.

$1.8 < x < 1.9$

To improve accuracy, evaluate the expression at subintervals within this interval and determine where the sign changes.

81. (a)

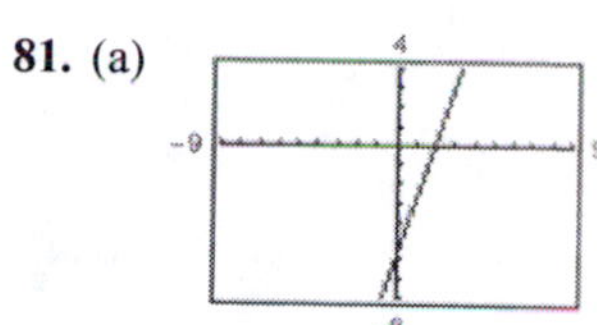

(b) x-intercept: $(2, 0)$

(c) The x-intercept is the solution of the equation $3x - 6 = 0$.

83. (a) To find the x-intercept, let $y = 0$, and solve for x.

$$ax + by = c$$
$$ax + b(0) = c$$
$$ax = c$$
$$x = \frac{c}{a}$$

The x-intercept is $\left(\frac{c}{a}, 0\right)$.

(b) To find the y-intercept, let $x = 0$, and solve for y.

$$a(0) + by = c$$
$$by = c$$
$$y = \frac{c}{b}$$

The y-intercept is $\left(0, \frac{c}{b}\right)$.

(c) x-intercept:

$$x = \frac{c}{a}$$
$$= \frac{11}{2}$$

The x-intercept is $\left(\frac{11}{2}, 0\right)$.

y-intercept:

$$y = \frac{c}{b}$$
$$= \frac{11}{7}$$

The y-intercept is $\left(0, \frac{11}{7}\right)$.

Section 1.3 Modeling with Linear Equations

1. mathematical modeling

3. $x + 4$

The sum of a number and 4
A number increased by 4

5. $\frac{u}{5}$

The ratio of a number and 5
The quotient of a number and 5
A number divided by 5

7. $\frac{y - 4}{5}$

The difference of a number and 4 is divided by 5.
A number decreased by 4 is divided by 5.

9. $-3(b + 2)$

The product of −3 and the sum of a number and 2.
Negative 3 is multiplied by a number increased by 2.

11. $\frac{4(p - 1)}{p}$

The product of 4 and the difference of a number and 1 is divided by the number.
A number decreased by 1 is multiplied by 4 and divided by the number.

13. *Verbal Model:*

(Sum) = (first number) + (second number)

Labels:

Sum = S, first number = n, second number = $n + 1$

Expression: $S = n + (n + 1) = 2n + 1$

15. *Verbal Model:* Product = (first odd integer) · (second odd integer)

Labels: Product = P, first odd integer = $2n - 1$, second odd integer = $2n - 1 + 2 = 2n + 1$

Expression: $P = (2n - 1)(2n + 1) = 4n^2 - 1$

17. *Verbal Model:* (distance) = (rate) · (time)

Labels: Distance = d, rate = 55 mph, time = t

Expression: $d = 55t$

19. *Verbal Model:* (Amount of acid) = 20% · (amount of solution)

Labels: Amount of acid (in gallons) = A, amount of solution (in gallons) = x

Expression: $A = 0.20x$

21. *Verbal Model:* Perimeter = 2(width) + 2(length)

Labels: Perimeter = P, width = x, length = 2(width) = $2x$

Expression: $P = 2x + 2(2x) = 6x$

23. *Verbal Model:* (Total cost) = (unit cost)(number of units) + (fixed cost)

Labels: Total cost = C, unit cost = \$40, number of units = x, fixed cost = \$2500

Expression: $C = 40x + 2500$

25. *Verbal Model:* Thirty percent of the list price L.

Expression: $0.30L$

27. *Verbal Model:* Percent of 672 that is represented by the number N

Expression: $N = p(672)$, p is in decimal form

29.

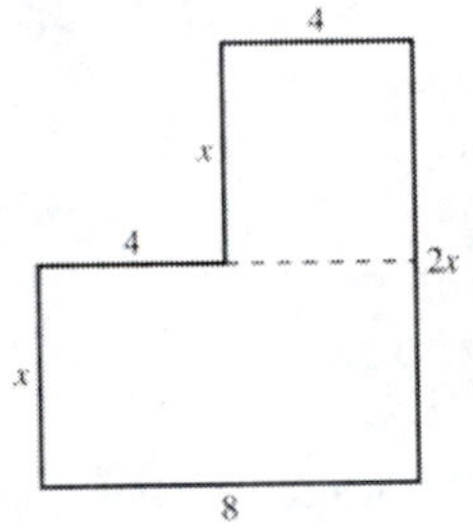

Area = Area of top rectangle + Area of bottom rectangle

$A = 4x + 8x = 12x$

31. *Verbal Model:* Sum = (first number) + (second number)

Labels: Sum = 525, first number = n, second number = $n + 1$

Equation:

$$525 = n + (n + 1)$$
$$525 = 2n + 1$$
$$524 = 2n$$
$$n = 262$$

Answer: First number = n = 262, second number = $n + 1$ = 263

33. *Verbal Model:* Difference = (one number) − (another number)

Labels: Difference = 148, one number = $5x$, another number = x

Equation:

$$148 = 5x - x$$
$$148 = 4x$$
$$x = 37$$
$$5x = 185$$

Answer: The two numbers are 37 and 185.

35. *Verbal Model:* Product = (smaller number) · (larger number) = (smaller number)^2 − 5

Labels: Smaller number = n, larger number = $n + 1$

Equation:

$$n(n + 1) = n^2 - 5$$
$$n^2 + n = n^2 - 5$$
$$n = -5$$

Answer: Smaller number = $n = -5$, larger number = $n + 1 = -4$

37. *Verbal Model:* (first paycheck) + (second paycheck) = total

Labels: second paycheck = x, first paycheck = $0.85x$, total = \$1125

Equation:

$$0.85x + x = 1125$$
$$1.85x = 1125$$
$$x \approx 608.11$$
$$0.85x \approx 516.89$$

Answer: The first salesperson's weekly paycheck is \$516.89 and the second salesperson's weekly paycheck is \$608.11.

39. *Verbal Model:* (Loan payments) = (Percent) · (Annual Income)

Labels: Loan payments = 15,680 (dollars)

Percent = 0.32

Annual income = I (dollars)

Equation:

$$15{,}680 = 0.32I$$
$$\frac{15{,}680}{0.32} = \frac{0.32I}{0.32}$$
$$49{,}000 = I$$

Answer: The family's annual income is \$49,000.

41. (a)

w

l

(b) $l = 1.5w$

$$P = 2l + 2w$$
$$= 2(1.5w) + 2w$$
$$= 5w$$

(c) $25 = 5w$

$$5 = w$$

Width: $w = 5$ meters

Length: $l = 1.5w = 7.5$ meters

Dimensions: 7.5 meters × 5 meters

43. *Verbal Model:*

$$\text{Average} = \frac{(\text{test \#1}) + (\text{test \#2}) + (\text{test \#3}) + (\text{test \#4})}{4}$$

Labels: Average = 90, test #1 = 87, test #2 = 92, test #3 = 84, test #4 = x

Equation:

$$90 = \frac{87 + 92 + 84 + x}{4}$$

Answer: You must score 97 or better on test #4 to earn an A for the course.

45. $\text{Rate} = \dfrac{\text{distance}}{\text{time}} = \dfrac{50 \text{ kilometers}}{\frac{1}{2} \text{ hour}} = 100 \text{ kilometers/hour}$

$\text{Total time} = \dfrac{\text{total distance}}{\text{rate}} = \dfrac{500 \text{ kilometers}}{100 \text{ kilometers/hour}} = 5 \text{ hours}$

47. *Verbal Model:* (Distance) = (rate)(time)

Labels: Distance $= 1.5 \times 10^{11}$ (meters)

Rate $= 3.0 \times 10^{8}$ (meters per second)

Time $= t$

Equation: $1.5 \times 10^{11} = (3.0 \times 10^{8})t$

$500 = t$

Light from the sun travels to the Earth in 500 seconds or approximately 8.33 minutes.

49. *Verbal Model:* $\dfrac{\text{(Height of building)}}{\text{(Length of building's shadow)}} = \dfrac{\text{(Height of post)}}{\text{(Length of post's shadow)}}$

Labels: Height of building $= x$ (feet)

Length of building's shadow $= 105$ (feet)

Height of post $= 3 \cdot 12 = 36$ (inches)

Length of post's shadow $= 4$ (inches)

Equation: $\dfrac{x}{105} = \dfrac{36}{4}$

$x = 945$

One Liberty Place is 945 feet tall.

51. (a)

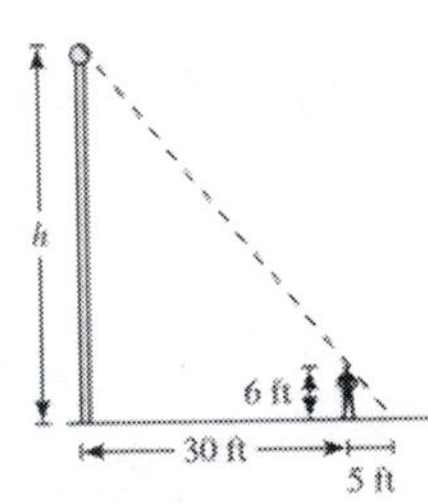

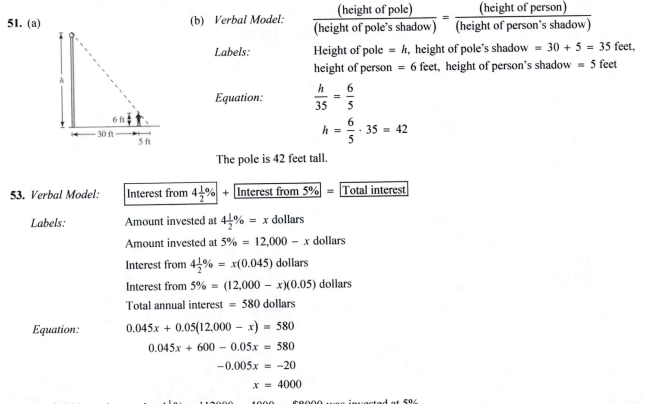

(b) *Verbal Model:* $\dfrac{\text{(height of pole)}}{\text{(height of pole's shadow)}} = \dfrac{\text{(height of person)}}{\text{(height of person's shadow)}}$

Labels: Height of pole $= h$, height of pole's shadow $= 30 + 5 = 35$ feet, height of person $= 6$ feet, height of person's shadow $= 5$ feet

Equation: $\dfrac{h}{35} = \dfrac{6}{5}$

$h = \dfrac{6}{5} \cdot 35 = 42$

The pole is 42 feet tall.

53. *Verbal Model:* [Interest from $4\frac{1}{2}\%$] + [Interest from 5%] = [Total interest]

Labels: Amount invested at $4\frac{1}{2}\% = x$ dollars

Amount invested at 5% $= 12{,}000 - x$ dollars

Interest from $4\frac{1}{2}\% = x(0.045)$ dollars

Interest from 5% $= (12{,}000 - x)(0.05)$ dollars

Total annual interest $= 580$ dollars

Equation: $0.045x + 0.05(12{,}000 - x) = 580$

$0.045x + 600 - 0.05x = 580$

$-0.005x = -20$

$x = 4000$

So, \$4000 was invested at $4\frac{1}{2}\%$ and $12000 - 4000 = \$8000$ was invested at 5%.

55. *Verbal Model:* (Profit from dogwood trees) + (profit from red maple trees) = (total profit)

Labels: Inventory of dogwood trees $= x$, inventory of red maple trees $= 40{,}000 - x$, profit from dogwood trees $= 0.25x$, profit from red maple trees $= 0.17(40{,}000 - x)$, total profit $= 0.20(40{,}000) = 8000$

Equation:

$$
\begin{aligned}
0.25x + 0.17(40{,}000) &= 8000 \\
0.25x + 6800 - 0.17x &= 8000 \\
0.08x &= 1200 \\
x &= 15{,}000
\end{aligned}
$$

The amount invested in dogwood trees was \$15,000 and the amount invested in red maple trees was $40{,}000 - 15{,}000 = \$25{,}000$.

57. *Verbal Model:* [Amount of gasoline in mixture] + [Amount of gasoline to add] = [Amount of gasoline in final mixture]

Labels: Amount of gasoline in mixture $= \frac{32}{33}(2)$ (gallons)

Amount of gasoline to add $= x$ (gallons)

Amount of gasoline in final mixture $= \frac{50}{51}(2 + x)$ (gallons)

Equation:

$$
\begin{aligned}
\frac{64}{33} + x &= \frac{50}{51}(2 + x) \\
\frac{64}{33} + x &= \frac{100}{51} + \frac{50}{51}x \\
3264 + 1683x &= 3300 + 1650x \\
33x &= 36 \\
x &\approx 1.09
\end{aligned}
$$

The forester should add about 1.09 gallons of gasoline to the mixture.

59.
$$
\begin{aligned}
A &= \frac{1}{2}bh \\
2A &= bh \\
\frac{2A}{b} &= h
\end{aligned}
$$

61.
$$
\begin{aligned}
S &= C + RC \\
S &= C(1 + R) \\
\frac{S}{1 + R} &= C
\end{aligned}
$$

63.
$$
\begin{aligned}
A &= P + Prt \\
A - P &= Prt \\
\frac{A - P}{Pt} &= r
\end{aligned}
$$

65.
$$
\begin{aligned}
W_1x &= W_2(L - x) \\
50x &= 75(10 - x) \\
50x &= 750 - 75x \\
125x &= 750 \\
x &= 6 \text{ feet from 50-pound child}
\end{aligned}
$$

67.
$$
\begin{aligned}
V &= \frac{4}{3}\pi r^3 \\
5.96 &= \frac{4}{3}\pi r^3 \\
17.88 &= 4\pi r^3 \\
\frac{17.88}{4\pi} &= r^3 \\
r &= \sqrt[3]{\frac{4.47}{\pi}} \approx 1.12 \text{ inches}
\end{aligned}
$$

69.
$$
\begin{aligned}
C &= \tfrac{5}{9}(F - 32) \\
&= \tfrac{5}{9}(64.4 - 32) \\
&= \tfrac{5}{9}(32.4) \\
&= 18
\end{aligned}
$$

The average daily temperature is 18° C.

71. $F = \frac{9}{5}C + 32$

$= \frac{9}{5}(50) + 32$

$= 90 + 32$

$= 122$

The highest recorded temperature was 122°F.

73. False, it should be written as $\frac{z^3 - 8}{z^2 - 9}$.

75. False

Cube: $V = s^3 = 9.5^3 = 857.375 \text{ in.}^3$

Sphere: $V = \frac{4}{3}\pi r^3 = \frac{4}{3}\pi(5.9)^3 \approx 860.290 \text{ in.}^3$

Section 1.4 Quadratic Equations and Applications

1. quadratic equation

3. factoring; square roots; completing; square; Quadratic Formula

5. position equation

7. $2x^2 = 3 - 5x$

General form: $2x^2 + 5x - 3 = 0$

9. $(x - 3)^2 = 3$

$x^2 - 6x + 9 = 3$

General form: $x^2 - 6x + 6 = 0$

11. $\frac{1}{5}(3x^2 - 10) = 12x$

$3x^2 - 10 = 60x$

General form: $3x^2 - 60x - 10 = 0$

13. $6x^2 + 3x = 0$

$3x(2x + 1) = 0$

$3x = 0$ or $2x + 1 = 0$

$x = 0$ or $x = -\frac{1}{2}$

15. $x^2 - 2x - 8 = 0$

$(x - 4)(x + 2) = 0$

$x - 4 = 0$ or $x + 2 = 0$

$x = 4$ or $x = -2$

17. $x^2 + 10x + 25 = 0$

$(x + 5)(x + 5) = 0$

$x + 5 = 0$

$x = -5$

19. $3 + 5x - 2x^2 = 0$

$(3 - x)(1 + 2x) = 0$

$3 - x = 0$ or $1 + 2x = 0$

$x = 3$ or $x = -\frac{1}{2}$

21. $x^2 + 4x = 12$

$x^2 + 4x - 12 = 0$

$(x + 6)(x - 2) = 0$

$x + 6 = 0$ or $x - 2 = 0$

$x = -6$ or $x = 2$

23. $\frac{3}{4}x^2 + 8x + 20 = 0$

$4\left(\frac{3}{4}x^2 + 8x + 20\right) = 4(0)$

$3x^2 + 32x + 80 = 0$

$(3x + 20)(x + 4) = 0$

$3x + 20 = 0$ or $x + 4 = 0$

$x = -\frac{20}{3}$ or $x = -4$

25. $x^2 = 49$

$x = \pm 7$

27. $x^2 = 11$

$x = \pm\sqrt{11}$

29. $3x^2 = 81$

$x^2 = 27$

$x = \pm 3\sqrt{3}$

31. $(x - 12)^2 = 16$

$x - 12 = \pm 4$

$x = 12 \pm 4$

$x = 16$ or $x = 8$

33. $(x + 2)^2 = 14$

$x + 2 = \pm\sqrt{14}$

$x = -2 \pm \sqrt{14}$

35. $(2x - 1)^2 = 18$

$2x - 1 = \pm\sqrt{18}$

$2x = 1 \pm 3\sqrt{2}$

$x = \frac{1 \pm 3\sqrt{2}}{2}$

37. $(x - 7)^2 = (x + 3)^2$

$$x - 7 = \pm(x + 3)$$

$$x - 7 = x + 3 \quad \text{or} \quad x - 7 = -x - 3$$

$$-7 \neq 3 \quad \text{or} \quad 2x = 4$$

$$x = 2$$

The only solution of the equation is $x = 2$.

39. $x^2 + 4x - 32 = 0$

$$x^2 + 4x = 32$$

$$x^2 + 4x + 2^2 = 32 + 2^2$$

$$(x + 2)^2 = 36$$

$$x + 2 = \pm 6$$

$$x = -2 \pm 6$$

$$x = 4 \quad \text{or} \quad x = -8$$

41. $x^2 + 6x + 2 = 0$

$$x^2 + 6x = -2$$

$$x^2 + 6x + 3^2 = -2 + 3^2$$

$$(x + 3)^2 = 7$$

$$x + 3 = \pm\sqrt{7}$$

$$x = -3 \pm \sqrt{7}$$

43. $9x^2 - 18x = -3$

$$x^2 - 2x = -\frac{1}{3}$$

$$x^2 - 2x + 1^2 = -\frac{1}{3} + 1^2$$

$$(x - 1)^2 = \frac{2}{3}$$

$$x - 1 = \pm\sqrt{\frac{2}{3}}$$

$$x = 1 \pm \sqrt{\frac{2}{3}}$$

$$x = 1 \pm \frac{\sqrt{6}}{3}$$

45. $7 + 2x - x^2 = 0$

$$-x^2 + 2x + 7 = 0$$

$$x^2 - 2x - 7 = 0$$

$$x^2 - 2x = 7$$

$$x^2 - 2x + (-1)^2 = 7 + (-1)^2$$

$$(x - 1)^2 = 8$$

$$x - 1 = \pm 2\sqrt{2}$$

$$x = 1 \pm 2\sqrt{2}$$

47. $2x^2 + 5x - 8 = 0$

$$2x^2 + 5x = 8$$

$$x^2 + \frac{5}{2}x = 4$$

$$x^2 + \frac{5}{2}x + \left(\frac{5}{4}\right)^2 = 4 + \left(\frac{5}{4}\right)^2$$

$$\left(x + \frac{5}{4}\right)^2 = \frac{89}{16}$$

$$x + \frac{5}{4} = \pm\frac{\sqrt{89}}{4}$$

$$x = -\frac{5}{4} \pm \frac{\sqrt{89}}{4}$$

$$x = \frac{-5 \pm \sqrt{89}}{4}$$

49. $\dfrac{1}{x^2 + 2x + 5} = \dfrac{1}{x^2 + 2x + 1^2 - 1^2 + 5}$

$$= \frac{1}{(x + 1)^2 + 4}$$

51. $\dfrac{4}{x^2 + 10x + 74} = \dfrac{4}{x^2 + 10x + (5)^2 - (5)^2 + 74}$

$$= \frac{4}{x^2 + 10x + 25 + 49}$$

$$= \frac{4}{(x + 5)^2 + 49}$$

53. $\dfrac{1}{\sqrt{3 + 2x - x^2}} = \dfrac{1}{\sqrt{-1(x^2 - 2x - 3)}}$

$$= \frac{1}{\sqrt{-1\left[x^2 - 2x + (1)^2 - (1)^2 - 3\right]}}$$

$$= \frac{1}{\sqrt{-1(x^2 - 2x + 1) + 4}}$$

$$= \frac{1}{\sqrt{4 - (x - 1)^2}}$$

55. $\dfrac{1}{\sqrt{12 + 4x - x^2}} = \dfrac{1}{\sqrt{-1(x^2 - 4x - 12)}}$

$= \dfrac{1}{\sqrt{-1\left[x^2 - 4x + (2)^2 - (2)^2 - 12\right]}}$

$= \dfrac{1}{\sqrt{-1\left[(x^2 - 4x + 4) - 16\right]}}$

$= \dfrac{1}{\sqrt{16 - (x - 2)^2}}$

57. (a) $y = (x + 3)^2 - 4$

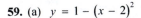

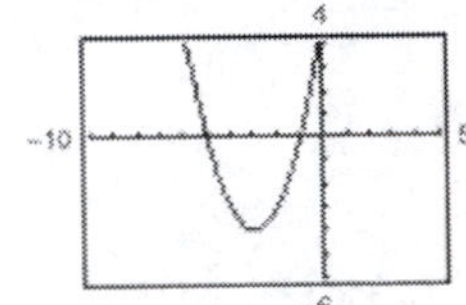

(b) The x-intercepts are $(-1, 0)$ and $(-5, 0)$.

(c)
$$0 = (x + 3)^2 - 4$$
$$4 = (x + 3)^2$$
$$\pm\sqrt{4} = x + 3$$
$$-3 \pm 2 = x$$
$$x = -1 \text{ or } x = -5$$

(d) The x-intercepts of the graphs are solutions of the equation $0 = (x + 3)^2 - 4$.

59. (a) $y = 1 - (x - 2)^2$

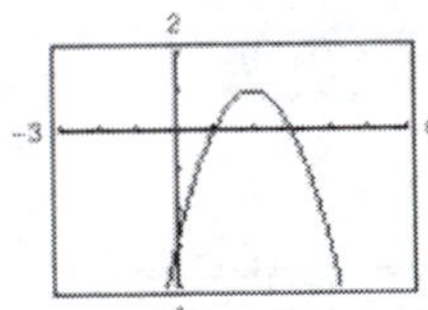

(b) The x-intercepts are $(1, 0)$ and $(3, 0)$.

(c)
$$0 = 1 - (x - 2)^2$$
$$(x - 2)^2 = 1$$
$$x - 2 = \pm 1$$
$$x = 2 \pm 1$$
$$x = 3 \text{ or } x = 1$$

(d) The x-intercepts of the graphs are solutions of the equation $0 = 1 - (x - 2)^2$.

61. (a) $y = -4x^2 + 4x + 3$

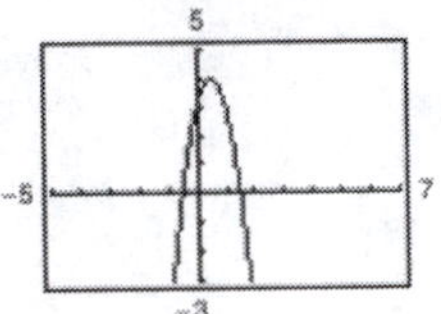

(b) The x-intercepts are $\left(-\frac{1}{2}, 0\right)$ and $\left(\frac{3}{2}, 0\right)$.

(c)
$$0 = -4x^2 + 4x + 3$$
$$4x^2 - 4x = 3$$
$$4\left(x^2 - x\right) = 3$$
$$x^2 - x = \frac{3}{4}$$
$$x^2 - x + \left(\frac{1}{2}\right)^2 = \frac{3}{4} + \left(\frac{1}{2}\right)^2$$
$$\left(x - \frac{1}{2}\right)^2 = 1$$
$$x - \frac{1}{2} = \pm\sqrt{1}$$
$$x = \frac{1}{2} \pm 1$$
$$x = \frac{3}{2} \text{ or } x = -\frac{1}{2}$$

(d) The x-intercepts of the graphs are solutions of the equation $0 = -4x^2 + 4x + 3$.

63. (a) $y = x^2 + 3x - 4$

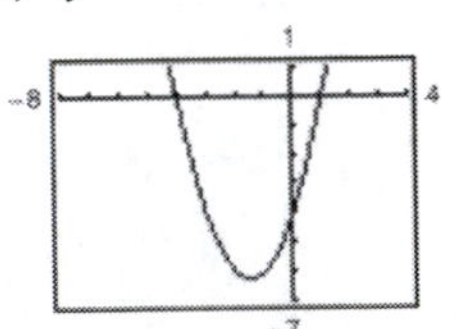

(b) The x-intercepts are $(-4, 0)$ and $(1, 0)$.

(c)
$$0 = x^2 + 3x - 4$$
$$0 = (x + 4)(x - 1)$$
$$x + 4 = 0 \text{ or } x - 1 = 0$$
$$x = -4 \text{ or } x = 1$$

(d) The x-intercepts of the graphs are solutions of the equation $0 = x^2 + 3x - 4$.

65. $2x^2 - 5x + 5 = 0$

$b^2 - 4ac = (-5)^2 - 4(2)(5) = -15 < 0$

No real solution

67. $2x^2 - z - 1 = 0$

$b^2 - 4ac = (-1)^2 - 4(2)(-1) = 9 > 0$

Two real solutions

69. $\frac{1}{3}x^2 - 5x + 25 = 0$

$b^2 - 4ac = (-5)^2 - 4\left(\frac{1}{3}\right)(25) = -\frac{25}{3} < 0$

No real solution

71. $0.2x^2 + 1.2x - 8 = 0$

$b^2 - 4ac = (1.2)^2 - 4(0.2)(-8) = 7.84 > 0$

Two real solutions

73. $2x^2 + x - 1 = 0$

$$x = \frac{-b \pm \sqrt{b^2 - 4ac}}{2a}$$
$$= \frac{-1 \pm \sqrt{1^2 - 4(2)(-1)}}{2(2)}$$
$$= \frac{-1 \pm 3}{4} = \frac{1}{2}, -1$$

75. $16x^2 + 8x - 3 = 0$

$$x = \frac{-b \pm \sqrt{b^2 - 4ac}}{2a}$$
$$= \frac{-8 \pm \sqrt{8^2 - 4(16)(-3)}}{2(16)}$$
$$= \frac{-8 \pm 16}{32} = \frac{1}{4}, -\frac{3}{4}$$

77. $2 + 2x - x^2 = 0$

$-x^2 + 2x + 2 = 0$

$$x = \frac{-b \pm \sqrt{b^2 - 4ac}}{2a}$$
$$= \frac{-2 \pm \sqrt{2^2 - 4(-1)(2)}}{2(-1)}$$
$$= \frac{-2 \pm 2\sqrt{3}}{-2} = 1 \pm \sqrt{3}$$

79. $x^2 + 12x + 16 = 0$

$$x = \frac{-b \pm \sqrt{b^2 - 4ac}}{2a}$$
$$= \frac{-12 \pm \sqrt{12^2 - 4(1)(16)}}{2(1)}$$
$$= \frac{-12 \pm 4\sqrt{5}}{2}$$
$$= -6 \pm 2\sqrt{5}$$

81. $x^2 + 8x - 4 = 0$

$$x = \frac{-b \pm \sqrt{b^2 - 4ac}}{2a}$$
$$= \frac{-8 \pm \sqrt{8^2 - 4(1)(-4)}}{2(1)}$$
$$= \frac{-8 \pm 4\sqrt{5}}{2} = -4 \pm 2\sqrt{5}$$

83. $12x - 9x^2 = -3$

$-9x^2 + 12x + 3 = 0$

$$x = \frac{-b \pm \sqrt{b^2 - 4ac}}{2a}$$
$$= \frac{-12 \pm \sqrt{12^2 - 4(-9)(3)}}{2(-9)}$$
$$= \frac{-12 \pm 6\sqrt{7}}{-18} = \frac{2}{3} \pm \frac{\sqrt{7}}{3}$$

85. $9x^2 + 30x + 25 = 0$

$$x = \frac{-b \pm \sqrt{b^2 - 4ac}}{2a}$$
$$= \frac{-30 \pm \sqrt{30^2 - 4(9)(25)}}{2(9)}$$
$$= \frac{-30 \pm 0}{18} = -\frac{5}{3}$$

87. $4x^2 + 4x = 7$

$4x^2 + 4x - 7 = 0$

$$x = \frac{-b \pm \sqrt{b^2 - 4ac}}{2a}$$
$$= \frac{-4 \pm \sqrt{4^2 - 4(4)(-7)}}{2(4)}$$
$$= \frac{-4 \pm 8\sqrt{2}}{8} = -\frac{1}{2} \pm \sqrt{2}$$

89.
$$28x - 49x^2 = 4$$
$$-49x^2 + 28x - 4 = 0$$
$$x = \frac{-b \pm \sqrt{b^2 - 4ac}}{2a}$$
$$= \frac{-28 \pm \sqrt{28^2 - 4(-49)(-4)}}{2(-49)}$$
$$= \frac{-28 \pm 0}{-98} = \frac{2}{7}$$

91.
$$8t = 5 + 2t^2$$
$$-2t^2 + 8t - 5 = 0$$
$$t = \frac{-b \pm \sqrt{b^2 - 4ac}}{2a}$$
$$= \frac{-8 \pm \sqrt{8^2 - 4(-2)(-5)}}{2(-2)}$$
$$= \frac{-8 \pm 2\sqrt{6}}{-4} = 2 \pm \frac{\sqrt{6}}{2}$$

93.
$$(y - 5)^2 = 2y$$
$$y^2 - 12y + 25 = 0$$
$$y = \frac{-b \pm \sqrt{b^2 - 4ac}}{2a}$$
$$= \frac{-(-12) \pm \sqrt{(-12)^2 - 4(1)(25)}}{2(1)}$$
$$= \frac{12 \pm 2\sqrt{11}}{2} = 6 \pm \sqrt{11}$$

95.
$$\frac{1}{2}x^2 + \frac{3}{8}x = 2$$
$$4x^2 + 3x = 16$$
$$4x^2 + 3x - 16 = 0$$
$$x = \frac{-b \pm \sqrt{b^2 - 4ac}}{2a}$$
$$= \frac{-3 \pm \sqrt{3^2 - 4(4)(-16)}}{2(4)}$$
$$= \frac{-3 \pm \sqrt{265}}{8} = -\frac{3}{8} \pm \frac{\sqrt{265}}{8}$$

97. $5.1x^2 - 1.7x - 3.2 = 0$
$$x = \frac{1.7 \pm \sqrt{(-1.7)^2 - 4(5.1)(-3.2)}}{2(5.1)}$$
$$\approx 0.976, -0.643$$

99. $-0.067x^2 - 0.852x + 1.277 = 0$
$$x = \frac{-(-0.852) \pm \sqrt{(-0.852)^2 - 4(-0.067)(1.277)}}{2(-0.067)}$$
$$\approx -14.071, 1.355$$

101. $422x^2 - 506x - 347 = 0$
$$x = \frac{506 \pm \sqrt{(-506)^2 - 4(422)(-347)}}{2(422)}$$
$$\approx 1.687, -0.488$$

103. $12.67x^2 + 31.55x + 8.09 = 0$
$$x = \frac{-31.55 \pm \sqrt{(31.55)^2 - 4(12.67)(8.09)}}{2(12.67)}$$
$$\approx 2.200, -0.290$$

105. $x^2 - 2x - 1 = 0$ Complete the square.
$$x^2 - 2x = 1$$
$$x^2 - 2x + 1^2 = 1 + 1^2$$
$$(x - 1)^2 = 2$$
$$x - 1 = \pm\sqrt{2}$$
$$x = 1 \pm \sqrt{2}$$

107. $(x + 3)^2 = 81$ Extract square roots.
$$x + 3 = \pm 9$$
$$x + 3 = 9 \quad \text{or} \quad x + 3 = -9$$
$$x = 6 \quad \text{or} \quad x = -12$$

109. $x^2 - x - \frac{11}{4} = 0$ Complete the square.
$$x^2 - x = \frac{11}{4}$$
$$x^2 - x + \left(\frac{1}{2}\right)^2 = \frac{11}{4} + \left(\frac{1}{2}\right)^2$$
$$\left(x - \frac{1}{2}\right)^2 = \frac{12}{4}$$
$$x - \frac{1}{2} = \pm\sqrt{\frac{12}{4}}$$
$$x = \frac{1}{2} \pm \sqrt{3}$$

111. $(x + 1)^2 = x^2$ Extract square roots.
$$x^2 = (x + 1)^2$$
$$x = \pm(x + 1)$$
For $x = +(x + 1)$:
$$0 \neq 1 \quad \text{No solution}$$
For $x = -(x + 1)$:
$$2x = -1$$
$$x = -\frac{1}{2}$$

113. (a) $w(w + 14) = 1632$

(b) $w^2 + 14w - 1632 = 0$

$(w + 48)(w - 34) = 0$

$w = -48$ or $w = 34$

Because w must be greater than zero, $w = 34$ feet and the length is $w + 14 = 48$ feet.

115. $S = x^2 + 4xh$

$108 = x^2 + 4x(3)$

$0 = x^2 + 12x - 108$

$0 = (x + 18)(x - 6)$

$x = -18$ or $x = 6$

Because x must be positive, $x = 6$ inches. The dimensions of the box are 6 inches × 6 inches × 3 inches.

117. $(\text{Area mowed by first landscaper}) = \frac{1}{2}(\text{Total area of the lawn})$

$$(125 - 2x)(150 - 2x) = \frac{1}{2}(125)(150)$$

$$18750 - 550x + 4x^2 = 9375$$

$$4x^2 - 550x + 9375 = 0$$

150, 125, x, $125 - 2x$, x, x, $150 - 2x$, x

Let $a = 4$, $b = -550$, and $c = 9375$. Use the Quadratic Formula:

$$x = \frac{-(-550) \pm \sqrt{(-550)^2 - 4(4)(93750)}}{2(4)} = \frac{550 \pm \sqrt{302500 - 150{,}000}}{8} = \frac{550 \pm \sqrt{152{,}500}}{8}$$

$x \approx 19.936$ and $x \approx 117.564$

Since the lot is only 125 feet wide, the only possible solution is $x \approx 19.936$ feet.

The first landscaper must go around the lot $\frac{19.936 \text{ feet}}{24 \text{ inches}} = \frac{19.936 \text{ feet}}{2 \text{ feet}} \approx 10$ times.

119. (a) $s = -16t^2 + v_0t + s_0$

$s = -16t^2 + 550$

Let $s = 0$ and solve for t.

$0 = -16t^2 + 550$

$16t^2 = 550$

$t^2 = \frac{550}{16}$

$t = \sqrt{\frac{550}{16}}$

$t \approx 5.86$

The supply package will take about 5.86 seconds to reach the ground.

(b) *Verbal Model:* $(\text{Distance}) = (\text{Rate}) \cdot (\text{Time})$

Labels: Distance $= d$

Rate = 138 miles per hour

Time = $\frac{5.86 \text{ seconds}}{3600 \text{ seconds per hour}} \approx 0.0016$ hour

Equation: $d = (138)(0.0016)$

≈ 0.22 mile

The supply package will travel about 0.22 mile.

121. (a) $s = -16t^2 + v_0t + s_0$

$$v_0 = 100 \text{ mph} = \frac{(100)(5280)}{3600} = 146\frac{2}{3} \text{ ft/sec}$$

$$s_0 = 6\frac{1}{4} \text{ feet}$$

$$s = -16t^2 + 146\frac{2}{3}t + 6\frac{1}{4}$$

(b) When $t = 3$: $s(3) = 302.25$ feet

When $t = 4$: $s(4) \approx 336.92$ feet

When $t = 5$: $s(5) \approx 339.58$ feet

During the interval $3 \le t \le 5$, the baseball's speed decreased due to gravity.

(c) The ball hits the ground when $s = 0$.

$$-16t^2 + 146\tfrac{2}{3}t + 6\tfrac{1}{4} = 0$$

By the Quadratic Formula, $t \approx -0.042$ or $t \approx 9.209$. Assuming that the ball is not caught and drops to the ground, it will be in the air for approximately 9.209 seconds.

123. (a)

t	5	6	7	8	9	10	11
D	7.7	8.0	8.6	9.5	10.7	12.2	14.0

Sometime during the year 2008, the total public dept reached $10 trillion.

(b) *Algebraically:* $D = 0.157t^2 - 1.46t + 11.1$

$$10 = 0.157t^2 - 1.46t + 11.1$$

$$0 = 0.157t^2 - 1.46t + 1.1$$

Let $a = 0.157$, $b = -1.46$, and $c = 1.1$.

Use the Quadratic Formula:

$$t = \frac{-(-1.46) \pm \sqrt{(-1.46)^2 - 4(0.157)(1.1)}}{2(0.157)}$$

$$= \frac{1.46 \pm \sqrt{1.4408}}{0.314}$$

$t \approx 8.47$ and $t \approx 0.827$

Because the domain of the model is $5 \le t \le 11$, $t \approx 8.47$ is the only solution. So, the total public debt reached $10 trillion during 2008.

Graphically: Use a graphing utility to graph $y_1 = 0.157t^2 - 1.46t + 11.1$ and $y_2 = 10$ in the same viewing window. Then use the intersect feature to find that the graphs intersect when $t \approx 0.827$ and $t \approx 8.47$. Choose $t \approx 8.47$ because it is in the domain.

So, the total public debt reached $10 trillion during 2008.

(c) For 2020, let $t = 20$

$$D = 0.157t^2 - 1.46t + 11.0$$

$$= 0.157(20)^2 - 1.46(20) + 11.1$$

$$= 44.7$$

Using the model, in 2020, the total public debt will be $44.7 trillion.

Answers will vary. *Sample Answer*: No, it is unlikely the total public debt will increase nearly $31 trillion over 9 years.

125. $L = -0.270t^2 + 3.59t + 83.1$

$$93 = -0.270t^2 + 3.59t + 83.1$$

$$0 = -0.270t^2 + 3.59t - 9.9$$

$$0 = 0.270t^2 - 3.59t + 9.9$$

Using the Quadratic Formula,

$$t = \frac{-(-3.59) \pm \sqrt{(-3.59)^2 - 4(0.270)(9.9)}}{2(0.270)} = \frac{3.59 \pm \sqrt{2.1961}}{0.54}$$

$t \approx 3.9$ and $t \approx 9.4$

Because the domain of the model is $2 \le t \le 7$, $t \approx 3.9$ is the only solution. The patient's blood oxygen level was 93% at approximately 4:00 P.M.

127. (a) *Model:* $(\text{winch})^2 + (\text{distance to dock})^2 = (\text{length of rope})^2$

Labels: winch = 15, distance to dock = x, length of rope = l

Equation: $15^2 + x^2 = l^2$

(b) When $l = 75$: $15^2 + x^2 = 75^2$

$$x^2 = 5625 - 225 = 5400$$

$$x = \sqrt{5400} = 30\sqrt{5} \approx 73.5$$

The boat is approximately 73.5 feet from the dock when there is 75 feet of rope out.

129. False.

$b^2 - 4ac = (-1)^2 - 4(-3)(-10) < 0,$

So, the quadratic equation has no real solutions.

131. *Sample answer:* $(x - 3)(x - (-5)) = 0$

$$(x - 3)(x + 5) = 0$$

$$x^2 + 2x - 15 = 0$$

133. One possible equation is:

$(x - 8)(x - 14) = 0$

$x^2 - 22x + 112 = 0$

Any non-zero multiple of this equation would also have these solutions.

135. One possible equation is:

$$\left[x - \left(1 + \sqrt{2}\right)\right]\left[x - \left(1 - \sqrt{2}\right)\right] = 0$$

$$\left[(x - 1) - \sqrt{2}\right]\left[(x - 1) + \sqrt{2}\right] = 0$$

$$(x - 1)^2 - \left(\sqrt{2}\right)^2 = 0$$

$$x^2 - 2x + 1 - 2 = 0$$

$$x^2 - 2x - 1 = 0$$

Any non-zero multiple of this equation would also have these solutions.

137. Yes, the vertex of the parabola would be on the x-axis.

Section 1.5 Complex Numbers

1. real

3. pure imaginary

5. principal square

7. $a + bi = -12 + 7i$

$a = -12$

$b = 7$

9. $(a - 1) + (b + 3)i = 5 + 8i$

$a - 1 = 5 \Rightarrow a = 6$

$b + 3 = 8 \Rightarrow b = 5$

11. $8 + \sqrt{-25} = 8 + 5i$

13. $2 - \sqrt{-27} = 2 - \sqrt{27}i$

$= 2 - 3\sqrt{3}i$

15. $\sqrt{-80} = 4\sqrt{5}i$

17. $14 = 14 + 0i = 14$

19. $-10i + i^2 = -10i - 1 = -1 - 10i$

21. $\sqrt{-0.09} = \sqrt{0.09}i$

$= 0.3i$

23. $(7 + i) + (3 - 4i) = 10 - 3i$

25. $(9 - i) - (8 - i) = 1$

27. $\left(-2 + \sqrt{-8}\right) + \left(5 - \sqrt{-50}\right) = -2 + 2\sqrt{2}i + 5 - 5\sqrt{2}i$

$= 3 - 3\sqrt{2}i$

29. $13i - (14 - 7i) = 13i - 14 + 7i$

$= -14 + 20i$

31. $-\left(\frac{3}{2} + \frac{5}{2}i\right) + \left(\frac{5}{3} + \frac{11}{3}i\right) = -\frac{3}{2} - \frac{5}{2}i + \frac{5}{3} + \frac{11}{3}i$

$= -\frac{9}{6} - \frac{15}{6}i + \frac{10}{6} + \frac{22}{6}i$

$= \frac{1}{6} + \frac{7}{6}i$

33. $(1 + i)(3 - 2i) = 3 - 2i + 3i - 2i^2$
$= 3 + i + 2 = 5 + i$

35. $12i(1 - 9i) = 12i - 108i^2$
$= 12i + 108$
$= 108 + 12i$

37. $\left(\sqrt{14} + \sqrt{10}i\right)\left(\sqrt{14} - \sqrt{10}i\right) = 14 - 10i^2$
$= 14 + 10 = 24$

39. $(6 + 7i)^2 = 36 + 84i + 49i^2$
$= 36 + 84i - 49$
$= -13 + 84i$

41. $(2 + 3i)^2 + (2 - 3i)^2 = 4 + 12i + 9i^2 + 4 - 12i + 9i^2$
$= 4 + 12i - 9 + 4 - 12i - 9$
$= -10$

43. The complex conjugate of $9 + 2i$ is $9 - 2i$.
$(9 + 2i)(9 - 2i) = 81 - 4i^2$
$= 81 + 4$
$= 85$

45. The complex conjugate of $-1 - \sqrt{5}i$ is $-1 + \sqrt{5}i$.
$\left(-1 - \sqrt{5}i\right)\left(-1 + \sqrt{5}i\right) = 1 - 5i^2$
$= 1 + 5 = 6$

47. The complex conjugate of $\sqrt{-20} = 2\sqrt{5}i$ is $-2\sqrt{5}i$.
$\left(2\sqrt{5}i\right)\left(-2\sqrt{5}i\right) = -20i^2 = 20$

49. The complex conjugate of $\sqrt{6}$ is $\sqrt{6}$.
$\left(\sqrt{6}\right)\left(\sqrt{6}\right) = 6$

51. $\dfrac{3}{i} \cdot \dfrac{-i}{-i} = \dfrac{-3i}{-i^2} = -3i$

53. $\dfrac{2}{4 - 5i} = \dfrac{2}{4 - 5i} \cdot \dfrac{4 + 5i}{4 + 5i}$
$= \dfrac{2(4 + 5i)}{16 + 25} = \dfrac{8 + 10i}{41} = \dfrac{8}{41} + \dfrac{10}{41}i$

55. $\dfrac{5 + i}{5 - i} \cdot \dfrac{(5 + i)}{(5 + i)} = \dfrac{25 + 10i + i^2}{25 - i^2}$
$= \dfrac{24 + 10i}{26} = \dfrac{12}{13} + \dfrac{5}{13}i$

57. $\dfrac{9 - 4i}{i} \cdot \dfrac{-i}{-i} = \dfrac{-9i + 4i^2}{-i^2} = -4 - 9i$

59. $\dfrac{3i}{(4 - 5i)^2} = \dfrac{3i}{16 - 40i + 25i^2} = \dfrac{3i}{-9 - 40i} \cdot \dfrac{-9 + 40i}{-9 + 40i}$
$= \dfrac{-27i + 120i^2}{81 + 1600} = \dfrac{-120 - 27i}{1681}$
$= -\dfrac{120}{1681} - \dfrac{27}{1681}i$

61. $\dfrac{2}{1 + i} - \dfrac{3}{1 - i} = \dfrac{2(1 - i) - 3(1 + i)}{(1 + i)(1 - i)}$
$= \dfrac{2 - 2i - 3 - 3i}{1 + 1}$
$= \dfrac{-1 - 5i}{2}$
$= -\dfrac{1}{2} - \dfrac{5}{2}i$

63. $\dfrac{i}{3 - 2i} + \dfrac{2i}{3 + 8i} = \dfrac{i(3 + 8i) + 2i(3 - 2i)}{(3 - 2i)(3 + 8i)}$
$= \dfrac{3i + 8i^2 + 6i - 4i^2}{9 + 24i - 6i - 16i^2}$
$= \dfrac{4i^2 + 9i}{9 + 18i + 16}$
$= \dfrac{-4 + 9i}{25 + 18i} \cdot \dfrac{25 - 18i}{25 - 18i}$
$= \dfrac{-100 + 72i + 225i - 162i^2}{625 + 324}$
$= \dfrac{62 + 297i}{949} = \dfrac{62}{949} + \dfrac{297}{949}i$

65. $\sqrt{-6} \cdot \sqrt{-2} = \left(\sqrt{6}i\right)\left(\sqrt{2}i\right) = \sqrt{12}i^2 = \left(2\sqrt{3}\right)(-1)$
$= -2\sqrt{3}$

67. $\left(\sqrt{-15}\right)^2 = \left(\sqrt{15}i\right)^2 = 15i^2 = -15$

69. $\left(3 + \sqrt{-5}\right)\left(7 - \sqrt{-10}\right) = \left(3 + \sqrt{5}i\right)\left(7 - \sqrt{10}i\right)$
$= 21 - 3\sqrt{10}i + 7\sqrt{5}i - \sqrt{50}i^2$
$= \left(21 + \sqrt{50}\right) + \left(7\sqrt{5} - 3\sqrt{10}\right)i$
$= \left(21 + 5\sqrt{2}\right) + \left(7\sqrt{5} - 3\sqrt{10}\right)i$

71. $x^2 - 2x + 2 = 0;\ a = 1,\ b = -2,\ c = 2$

$$x = \frac{-(-2) \pm \sqrt{(-2)^2 - 4(1)(2)}}{2(1)} = \frac{2 \pm \sqrt{-4}}{2} = \frac{2 \pm 2i}{2} = 1 \pm i$$

73. $4x^2 + 16x + 17 = 0;\ a = 4,\ b = 16,\ c = 17$

$$x = \frac{-16 \pm \sqrt{(16)^2 - 4(4)(17)}}{2(4)} = \frac{-16 \pm \sqrt{-16}}{8} = \frac{-16 \pm 4i}{8} = -2 \pm \frac{1}{2}i$$

75. $4x^2 + 16x + 15 = 0;\ a = 4,\ b = 16,\ c = 15$

$$x = \frac{-16 \pm \sqrt{(16)^2 - 4(4)(15)}}{2(4)} = \frac{-16 \pm \sqrt{16}}{8} = \frac{-16 \pm 4}{8}$$

$$x = -\frac{12}{8} = -\frac{3}{2} \text{ or } x = -\frac{20}{8} = -\frac{5}{2}$$

77. $\frac{3}{2}x^2 - 6x + 9 = 0$ Multiply both sides by 2.

$3x^2 - 12x + 18 = 0;\ a = 3,\ b = -12,\ c = 18$

$$x = \frac{-(-12) \pm \sqrt{(-12)^2 - 4(3)(18)}}{2(3)} = \frac{12 \pm \sqrt{-72}}{6} = \frac{12 \pm 6\sqrt{2}i}{6} = 2 \pm \sqrt{2}i$$

79. $1.4x^2 - 2x - 10 = 0$ Multiply both sides by 5.

$7x^2 - 10x - 50 = 0;\ a = 7,\ b = -10,\ c = -50$

$$x = \frac{-(-10) \pm \sqrt{(-10)^2 - 4(7)(-50)}}{2(7)} = \frac{10 \pm \sqrt{1500}}{14} = \frac{10 \pm 10\sqrt{15}}{14} = \frac{5}{7} \pm \frac{5\sqrt{15}}{7}$$

81. $-6i^3 + i^2 = -6i^2i + i^2 = -6(-1)i + (-1) = 6i - 1 = -1 + 6i$

83. $-14i^5 = -14i^2i^2i = -14(-1)(-1)(i) = -14i$

85. $\left(\sqrt{-72}\right)^3 = \left(6\sqrt{2}i\right)^3 = 6^3\left(\sqrt{2}\right)^3 i^3 = 216\left(2\sqrt{2}\right)i^2i = 432\sqrt{2}(-1)i = -432\sqrt{2}i$

87. $\frac{1}{i^3} = \frac{1}{i^2i} = \frac{1}{-i} = \frac{1}{-i} \cdot \frac{i}{i} = \frac{i}{-i^2} = i$

89. $(3i)^4 = 81i^4 = 81i^2i^2 = 81(-1)(-1) = 81$

91. (a) $z_1 = 9 + 16i,\ z_2 = 20 - 10i$

(b) $\frac{1}{z} = \frac{1}{z_1} + \frac{1}{z_2} = \frac{1}{9 + 16i} + \frac{1}{20 - 10i} = \frac{20 - 10i + 9 + 16i}{(9 + 16i)(20 - 10i)} = \frac{29 + 6i}{340 + 230i}$

$$z = \left(\frac{340 + 230i}{29 + 6i}\right)\left(\frac{29 - 6i}{29 - 6i}\right) = \frac{11{,}240 + 4630i}{877} = \frac{11{,}240}{877} + \frac{4630}{877}i$$

93. False.

If $b = 0$ then $a + bi = a - bi = a$.

That is, if the complex number is real, the number equals its conjugate.

95. False.

$$\begin{aligned} i^{44} + i^{150} - i^{74} - i^{109} + i^{61} &= \left(i^2\right)^{22} + \left(i^2\right)^{75} - \left(i^2\right)^{37} - \left(i^2\right)^{54} i + \left(i^2\right)^{30} i \\ &= (-1)^{22} + (-1)^{75} - (-1)^{37} - (-1)^{54} i + (-1)^{30} i \\ &= 1 - 1 + 1 - i + i = 1 \end{aligned}$$

97.

$$\begin{aligned} i &= i \\ i^2 &= -1 \\ i^3 &= -i \\ i^4 &= 1 \\ i^5 &= i^4 i = i \\ i^6 &= i^4 i^2 = -1 \\ i^7 &= i^4 i^3 = -i \\ i^8 &= i^4 i^4 = 1 \\ i^9 &= i^4 i^4 i = i \\ i^{10} &= i^4 i^4 i^2 = -1 \\ i^{11} &= i^4 i^4 i^3 = -i \\ i^{12} &= i^4 i^4 i^4 = 1 \end{aligned}$$

The pattern $i, -1, -i, 1$ repeats. Divide the exponent by 4.

If the remainder is 1, the result is i.

If the remainder is 2, the result is -1.

If the remainder is 3, the result is $-i$.

If the remainder is 0, the result is 1.

99. $\sqrt{-6}\sqrt{-6} = \sqrt{6}i\sqrt{6}i = 6i^2 = -6$

101. $(a_1 + b_1 i) + (a_2 + b_2 i) = (a_1 + a_2) + (b_1 + b_2)i$

The complex conjugate of this sum is $(a_1 + a_2) - (b_1 + b_2)i.$

The sum of the complex conjugates is $(a_1 - b_1 i) + (a_2 - b_2 i) = (a_1 + a_2) - (b_1 + b_2)i.$

So, the complex conjugate of the sum of two complex numbers is the sum of their complex conjugates.

Section 1.6 Other Types of Equations

1. polynomial

3. rational

5.

$$\begin{aligned} 6x^4 - 14x^2 &= 0 \\ 2x^2\left(3x^2 - 7\right) &= 0 \\ 2x^2 = 0 &\Rightarrow x = 0 \\ 3x^2 - 7 = 0 &\Rightarrow x = \pm\frac{\sqrt{21}}{3} \end{aligned}$$

7.

$$\begin{aligned} x^4 - 81 &= 0 \\ \left(x^2 + 9\right)(x + 3)(x - 3) &= 0 \\ x^2 + 9 = 0 &\Rightarrow x = \pm 3i \\ x + 3 = 0 &\Rightarrow x = -3 \\ x - 3 = 0 &\Rightarrow x = 3 \end{aligned}$$

9.

$$\begin{aligned} x^3 + 512 &= 0 \\ x^3 + 8^3 &= 0 \\ (x + 8)\left(x^2 - 8x + 64\right) &= 0 \\ x + 8 = 0 &\Rightarrow x = -8 \\ x^2 - 8x + 64 = 0 &\Rightarrow x = 4 \pm 4\sqrt{3}i \end{aligned}$$

11.

$$\begin{aligned} 5x^3 + 3 - x^2 + 45x &= 0 \\ 5x\left(x^2 + 6x + 9\right) &= 0 \\ 5x(x + 3)^2 &= 0 \\ 5x = 0 &\Rightarrow x = 0 \\ x + 3 = 0 &\Rightarrow x = -3 \end{aligned}$$

13.
$$\begin{aligned} x^3 - 3x^2 - x + 3 &= 0 \\ x^2(x-3) - (x-3) &= 0 \\ (x-3)(x^2-1) &= 0 \\ (x-3)(x+3)(x-1) &= 0 \\ x - 3 = 0 &\Rightarrow x = 3 \\ x + 1 = 0 &\Rightarrow x = -1 \\ x - 1 = 0 &\Rightarrow x = 1 \end{aligned}$$

15.
$$\begin{aligned} x^4 - x^3 + x - 1 &= 0 \\ x^3(x-1) + (x-1) &= 0 \\ (x-1)(x^3+1) &= 0 \\ (x-1)(x+1)(x^2-x+1) &= 0 \\ x - 1 = 0 &\Rightarrow x = 1 \\ x + 1 = 0 &\Rightarrow x = -1 \\ x^2 - x + 1 = 0 &\Rightarrow x = \frac{1}{2} \pm \frac{\sqrt{3}}{2}i \end{aligned}$$

17.
$$\begin{aligned} x^4 - 4x^2 + 3 &= 0 \\ (x^2-3)(x^2-1) &= 0 \\ (x+\sqrt{3})(x-\sqrt{3})(x+1)(x-1) &= 0 \\ x + \sqrt{3} = 0 &\Rightarrow x = -\sqrt{3} \\ x - \sqrt{3} = 0 &\Rightarrow x = \sqrt{3} \\ x + 1 = 0 &\Rightarrow x = -1 \\ x - 1 = 0 &\Rightarrow x = 1 \end{aligned}$$

19.
$$\begin{aligned} 4x^4 - 65x^2 + 16 &= 0 \\ (4x^2-1)(x^2-16) &= 0 \\ (2x+1)(2x-1)(x+4)(x-4) &= 0 \\ 2x + 1 = 0 &\Rightarrow x = -\tfrac{1}{2} \\ 2x - 1 = 0 &\Rightarrow x = \tfrac{1}{2} \\ x + 4x = 0 &\Rightarrow x = -4 \\ x - 4 = 0 &\Rightarrow x = 4 \end{aligned}$$

21.
$$\begin{aligned} x^6 + 7x^3 - 8 &= 0 \\ (x^3+8)(x^3-1) &= 0 \\ (x+2)(x^2-2x+4)(x-1)(x^2+x+1) &= 0 \\ x + 2 = 0 &\Rightarrow x = -2 \\ x^2 - 2x + 4 = 0 &\Rightarrow x = 1 \pm \sqrt{3}i \\ x - 1 = 0 &\Rightarrow x = 1 \\ x^2 + x + 1 = 0 &\Rightarrow x = -\frac{1}{2} \pm \frac{\sqrt{3}}{2}i \end{aligned}$$

23.
$$\begin{aligned} \frac{1}{x^2} + \frac{8}{x} + 15 &= 0 \\ 1 + 8x + 15x^2 &= 0 \\ (1+3x)(1+5x) &= 0 \\ 1 + 3x = 0 &\Rightarrow x = -\frac{1}{3} \\ 1 + 5x = 0 &\Rightarrow x = -\frac{1}{5} \end{aligned}$$

25.
$$\begin{aligned} 2\left(\frac{x}{x+2}\right)^2 - 3\left(\frac{x}{x+2}\right) - 2 &= 0 \\ 2x^2 - 3x(x+2) - 2(x+2)^2 &= 0 \\ 2x^2 - 3x^2 - 6x - 2x^2 - 8x - 8 &= 0 \\ -3x^2 - 14x - 8 &= 0 \\ 3x^2 + 14x + 8 &= 0 \\ (3x+2)(x+4) &= 0 \\ 3x + 2 = 0 &\Rightarrow x -\frac{2}{3} \\ x + 4 = 0 &\Rightarrow x = -4 \end{aligned}$$

27.
$$\begin{aligned} 2x + 9\sqrt{x} &= 5 \\ 2x + 9\sqrt{x} - 5 &= 0 \\ 2(\sqrt{x})^2 + 9\sqrt{x} - 5 &= 0 \\ (2\sqrt{x} - 1)(\sqrt{x} + 5) &= 0 \\ \sqrt{x} = \tfrac{1}{2} &\Rightarrow x = \tfrac{1}{4} \end{aligned}$$

$(\sqrt{x} = -5$ is not a solution.$)$

29.
$$\begin{aligned} 3x^{1/3} + 2x^{2/3} &= 5 \\ 2x^{2/3} + 3x^{1/3} - 5 &= 0 \\ 2\left(x^{1/3}\right)^2 + 3x^{1/3} - 5 &= 0 \\ \left(2x^{1/3} + 5\right)\left(x^{1/3} - 1\right) &= 0 \end{aligned}$$
$$2x^{1/3} + 5 = 0 \Rightarrow x^{1/3} = -\tfrac{5}{2} \Rightarrow x = \left(-\tfrac{5}{2}\right)^3 = -\tfrac{125}{8}$$
$$x^{1/3} - 1 = 0 \Rightarrow x^{1/3} = 1 \Rightarrow x = (1)^3 = 1$$

31.
$$\begin{aligned} \sqrt{3x} - 12 &= 0 \\ \sqrt{3x} &= 12 \\ 3x &= 144 \\ x &= 48 \end{aligned}$$

33.
$$\begin{aligned} \sqrt{x - 10} - 4 &= 0 \\ \sqrt{x - 10} &= 0 \\ x - 10 &= 16 \\ x &= 26 \end{aligned}$$

35.
$$\begin{aligned} \sqrt[3]{2x + 5} + 3 &= 0 \\ \sqrt[3]{2x + 5} &= -3 \\ 2x + 5 &= -27 \\ 2x &= -32 \\ x &= -16 \end{aligned}$$

37.
$$\begin{aligned} -\sqrt{26 - 11x} + 4 &= x \\ 4 - x &= \sqrt{26 - 11x} \\ 16 - 8x + x^2 &= 26 - 11x \\ x^2 + 3x - 10 &= 0 \\ (x + 5)(x - 2) &= 0 \end{aligned}$$
$$x + 5 = 0 \Rightarrow x = -5$$
$$x - 2 = 0 \Rightarrow x = 2$$

39.
$$\begin{aligned} \sqrt{x} - \sqrt{x - 5} &= 1 \\ \sqrt{x} &= 1 + \sqrt{x - 5} \\ \left(\sqrt{x}\right)^2 &= \left(1 + \sqrt{x - 5}\right)^2 \\ x &= 1 + 2\sqrt{x - 5} + x - 5 \\ 4 &= 2\sqrt{x - 5} \\ 2 &= \sqrt{x - 5} \\ 4 &= x - 5 \\ 9 &= x \end{aligned}$$

41.
$$\begin{aligned} \sqrt{x + 5} + \sqrt{x - 5} &= 10 \\ \sqrt{x + 5} &= 10 - \sqrt{x - 5} \\ \left(\sqrt{x + 5}\right)^2 &= \left(10 - \sqrt{x - 5}\right)^2 \\ x + 5 &= 100 - 20\sqrt{x - 5} + x - 5 \\ -90 &= -20\sqrt{x - 5} \\ 9 &= 2\sqrt{x - 5} \\ 81 &= 4(x - 5) \\ 81 &= 4x - 20 \\ 101 &= 4x \\ \tfrac{101}{4} &= x \end{aligned}$$

43.
$$\begin{aligned} \sqrt{4\sqrt{4x + 9}} &= \sqrt{8x + 2} \\ \left(\sqrt{4\sqrt{4x + 9}}\right)^2 &= \left(\sqrt{8x + 2}\right)^2 \\ 4\sqrt{4x + 9} &= 8x + 2 \\ 2\sqrt{4x + 9} &= 4x + 1 \\ \left(2\sqrt{4x + 9}\right)^2 &= (4x + 1)^2 \\ 4(4x + 9) &= 16x^2 + 8x + 1 \\ 16x + 36 &= 16x^2 + 8x + 1 \\ 0 &= 16x^2 - 8x - 35 \\ 0 &= (4x + 5)(4x - 7) \end{aligned}$$
$$4x + 5 = 0 \Rightarrow x = -\frac{5}{4}, \text{ extraneous}$$
$$4x - 7 = 0 \Rightarrow x = \frac{7}{4}$$

45.
$$\begin{aligned} (x - 5)^{3/2} &= 8 \\ (x - 5)^3 &= 8^2 \\ x - 5 &= \sqrt[3]{64} \\ x &= 5 + 4 = 9 \end{aligned}$$

47. $(x^2 - 5)^{3/2} = 27$

$$(x^2 - 5)^3 = 27^2$$
$$x^2 - 5 = \sqrt[3]{27^2}$$
$$x^2 = 5 + 9$$
$$x^2 = 14$$
$$x = \pm\sqrt{14}$$

49. $3x(x - 1)^{1/2} + 2(x - 1)^{3/2} = 0$

$$(x - 1)^{1/2}[3x + 2(x - 1)] = 0$$
$$(x - 1)^{1/2}(5x - 2) = 0$$
$$(x - 1)^{1/2} = 0 \Rightarrow x - 1 = 0 \Rightarrow x = 1$$
$$5x - 2 = 0 \Rightarrow x = \tfrac{2}{5}, \text{ extraneous}$$

51. $x = \dfrac{3}{x} + \dfrac{1}{2}$

$$(2x)(x) = (2x)\left(\frac{3}{x}\right) + (2x)\left(\frac{1}{2}\right)$$
$$2x^2 = 6 + x$$
$$2x^2 - x - 6 = 0$$
$$(2x + 3)(x - 2) = 0$$
$$2x + 3 = 0 \Rightarrow x = -\frac{3}{2}$$
$$x - 2 = 0 \Rightarrow x = 2$$

53. $\dfrac{1}{x} - \dfrac{1}{x + 1} = 3$

$$x(x + 1)\frac{1}{x} - x(x + 1)\frac{1}{x + 1} = x(x + 1)(3)$$
$$x + 1 - x = 3x(x + 1)$$
$$1 = 3x^2 + 3x$$
$$0 = 3x^2 + 3x - 1$$

$a = 3, b = 3, c = -1$

$$x = \frac{-3 \pm \sqrt{(3)^2 - 4(3)(-1)}}{2(3)} = \frac{-3 \pm \sqrt{21}}{6}$$

55. $\dfrac{30 - x}{x} = x$

$$30 - x = x^2$$
$$0 = x^2 + x - 30$$
$$0 = (x + 6)(x - 5)$$
$$x + 6 = 0 \Rightarrow x = -6$$
$$x - 5 = 0 \Rightarrow x = 5$$

57. $\dfrac{x}{x^2 - 4} + \dfrac{1}{x + 2} = 3$

$$(x + 2)(x - 2)\frac{x}{x^2 - 4} + (x + 2)(x - 2)\frac{1}{x + 2} = 3(x + 2)(x - 2)$$
$$x + x - 2 = 3x^2 - 12$$
$$3x^2 - 2x - 10 = 0$$

$a = 3, b = -2, c = -10$

$$x = \frac{-(-2) \pm \sqrt{(-2)^2 - 4(3)(-10)}}{2(3)}$$
$$= \frac{2 \pm \sqrt{124}}{6} = \frac{2 \pm 2\sqrt{31}}{6} = \frac{1 \pm \sqrt{31}}{3}$$

59. $|2x - 5| = 11$

$2x - 5 = 11 \Rightarrow x = 8$

$-(2x - 5) = 11 \Rightarrow x = -3$

61. $|x| = x^2 + x - 24$

First equation:

$$x = x^2 + x - 24$$
$$x^2 - 24 = 0$$
$$x^2 = 24$$
$$x = \pm 2\sqrt{6}$$

Second equation:

$$-x = x^2 + x - 24$$
$$x^2 + 2x - 24 = 0$$
$$(x + 6)(x - 4) = 0$$
$$x + 6 = 0 \Rightarrow x = -6$$
$$x - 4 = 0 \Rightarrow x = 4$$

Only $x = 2\sqrt{6}$ and $x = -6$ are solutions of the original equation. $x = -2\sqrt{6}$ and $x = 4$ are extraneous.

63. $|x + 1| = x^2 - 5$

First equation:

$$x + 1 = x^2 - 5$$
$$x^2 - x - 6 = 0$$
$$(x - 3)(x + 2) = 0$$
$$x - 3 = 0 \Rightarrow x = 3$$
$$x + 2 = 0 \Rightarrow x = -2$$

Second equation:

$$-(x + 1) = x^2 - 5$$
$$-x - 1 = x^2 - 5$$
$$x^2 + x - 4 = 0$$
$$x = \frac{-1 + \sqrt{17}}{2}$$

Only $x = 3$ and $x = \dfrac{-1 - \sqrt{17}}{2}$ are solutions of the original equation. $x = -2$ and $x = \dfrac{-1 + \sqrt{17}}{2}$ are extraneous.

65. (a)

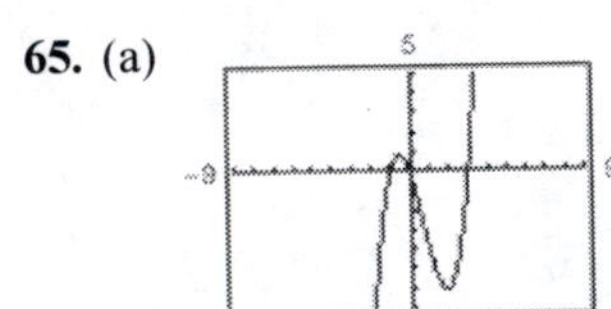

(b) x-intercepts: $(-1, 0), (0, 0), (3, 0)$

(c)
$$0 = x^3 - 2x^2 - 3x$$
$$0 = x(x + 1)(x - 3)$$
$$x = 0$$
$$x + 1 = 0 \Rightarrow x = -1$$
$$x - 3 = 0 \Rightarrow x = 3$$

(d) The x-intercepts of the graph are the same as the solutions of the equation.

67. (a)

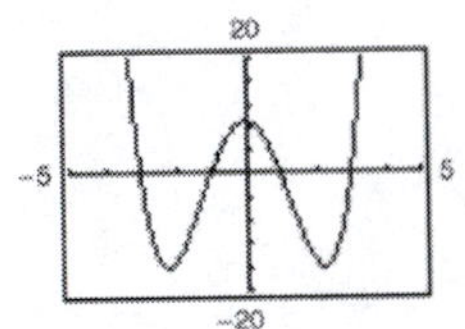

(b) x-intercepts: $(\pm 1, 0), (\pm 3, 0)$

(c)
$$0 = x^4 - 10x^2 + 9$$
$$0 = (x^2 - 1)(x^2 - 9)$$
$$0 = (x + 1)(x - 1)(x + 3)(x - 3)$$
$$x + 1 = 0 \Rightarrow x = -1$$
$$x - 1 = 0 \Rightarrow x = 1$$
$$x + 3 = 0 \Rightarrow x = -3$$
$$x - 3 = 0 \Rightarrow x = 3$$

(d) The x-intercepts of the graph are the same as the solutions of the equation.

69. (a)

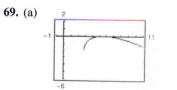

(b) x-intercepts: $(5, 0), (6, 0)$

(c)
$$\begin{aligned} 0 &= \sqrt{11x - 30} - x \\ x &= \sqrt{11x - 30} \\ x^2 &= 11x - 30 \\ x^2 - 11x + 30 &= 0 \\ (x - 5)(x - 6) &= 0 \\ x - 5 = 0 &\Rightarrow x = 5 \\ x - 6 = 0 &\Rightarrow x = 6 \end{aligned}$$

(d) The x-intercepts of the graph are the same as the solutions of the equation.

71. (a)

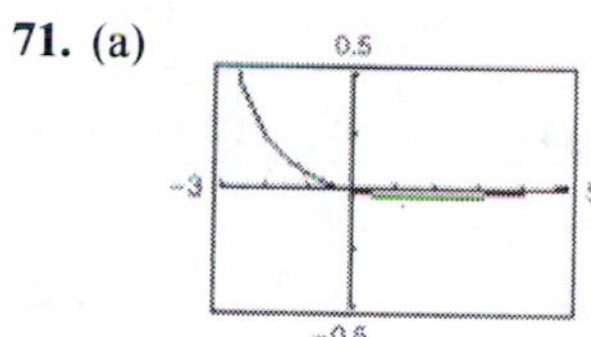

(b) x-intercepts: $(0, 0), (4, 0)$

(c)
$$\begin{aligned} 0 &= \sqrt{7x + 26} - \sqrt{5x + 16} - 2 \\ -\sqrt{7x + 36} &= -\sqrt{5x + 16} - 2 \\ \sqrt{7x + 36} &= 2 + \sqrt{5x + 16} \\ \left(\sqrt{7x + 36}\right)^2 &= \left(2 + \sqrt{5x + 16}\right)^2 \\ 7x + 36 &= 4 + 4\sqrt{5x + 16} + 5x + 16 \\ 7x + 36 &= 5x + 20 + 4\sqrt{5x + 16} \\ 2x + 16 &= 4\sqrt{5x + 16} \\ x + 8 &= 2\sqrt{5x + 16} \\ (x + 8)^2 &= \left(2\sqrt{5x + 16}\right)^2 \\ x^2 + 16x + 64 &= 4(5x + 16) \\ x^2 + 16x + 64 &= 20x + 64 \\ x^2 - 4x &= 0 \\ x(x - 4) &= 0 \\ x &= 0 \\ x - 4 = 0 &\Rightarrow x = 4 \end{aligned}$$

(d) The x-intercepts of the graph are the same as the solutions of the equation.

73. (a)

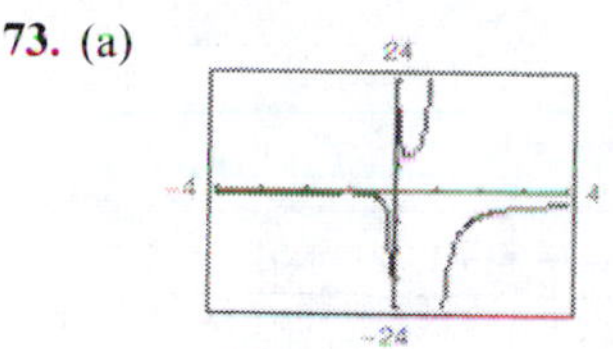

(b) x-intercept: $(-1, 0)$

(c)
$$\begin{aligned} 0 &= \frac{1}{x} - \frac{4}{x - 1} - 1 \\ 0 &= (x - 1) - 4x - x(x - 1) \\ 0 &= x - 1 - 4x - x^2 + x \\ 0 &= -x^2 - 2x - 1 \\ 0 &= x^2 + 2x + 1 \\ 0 &= (x + 1)^2 \\ x + 1 = 0 &\Rightarrow x = -1 \end{aligned}$$

(d) The x-intercept of the graph is the same as the solution of the equation.

75. (a)

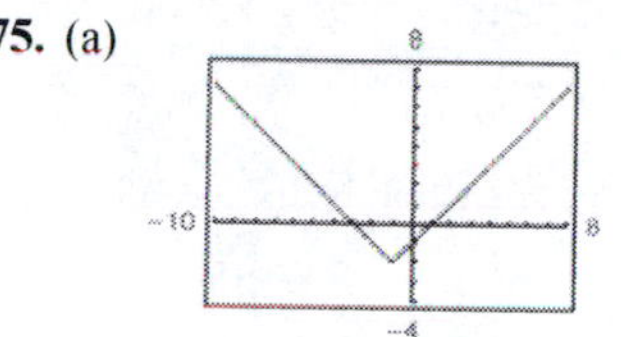

(b) x-intercepts: $(1, 0), (-3, 0)$

(c)
$$\begin{aligned} 0 &= |x + 1| - 2 \\ 2 &= |x + 1| \end{aligned}$$

$x + 1 = 2$ or $-(x + 1) = 2$

$x = 1$ or $-x - 1 = 2$

$-x = 3$

$x = -3$

(d) The x-intercepts of the graph are the same as the solutions of the equation.

77. $3.2x^4 - 1.5x^2 - 2.1 = 0$

$$x^2 = \frac{1.5 \pm \sqrt{1.5^2 - 4(3.2)(-2.1)}}{2(3.2)}$$

Using the positive value for x^2, we have

$$x = \pm\sqrt{\frac{1.5 + \sqrt{29.13}}{6.4}} \approx \pm 1.038.$$

79. $7.08x^6 + 4.15x^3 - 9.6 = 0$

$a = 7.8, b = 4.15, c = -9.6$

$$x^3 = \frac{-4.15 \pm \sqrt{(4.15)^2 - 4(7.08)(-9.6)}}{2(7.08)}$$

$$= \frac{-4.15 \pm \sqrt{2.89.0945}}{14.16}$$

$$x = \sqrt[3]{\frac{-4.15 + \sqrt{289.0945}}{14.16}} \approx 0.968$$

$$x = \sqrt[3]{\frac{-4.15 - \sqrt{289.0945}}{14.16}} \approx -1.143$$

81. $1.8x - 6\sqrt{x} - 5.6 = 0$ Given equation

$1.8\left(\sqrt{x}\right)^2 - 6\sqrt{x} - 5.6 = 0$

Use the Quadratic Formula with $a = 1.8, b = -6$, and $c = -5.6$.

$$\sqrt{x} = \frac{6 \pm \sqrt{36 - 4(1.8)(-5.6)}}{2(1.8)} \approx \frac{6 \pm 8.7361}{3.6}$$

Considering only the positive value for $\sqrt{x}$, we have:

$\sqrt{x} \approx 4.0934$

$x \approx 16.756.$

83. $4x^{2/3} + 8x^{1/3} + 3.6 = 0$

$a = 4, b = 8, c = 3.6$

$$x^{1/3} = \frac{-8 \pm \sqrt{8^2 - 4(4)(3.6)}}{2(4)}$$

$$x = \left[\frac{-8 + \sqrt{6.4}}{8}\right]^3 \approx -0.320$$

$$x = \left[\frac{-8 - \sqrt{6.4}}{8}\right]^3 \approx -2.280$$

85. $-4, 7$

Sample answer: $(x - (-4))(x - 7) = 0$

$(x + 4)(x - 7) = 0$

$x^2 - 3x - 28 = 0$

87. $-\frac{7}{3}, \frac{6}{7}$

One possible equation is:

$x = -\frac{7}{3} \Rightarrow 3x = -7 \Rightarrow 3x + 7$ is a factor.

$x = \frac{6}{7} \Rightarrow 7x = 6 \Rightarrow 7x - 6$ is a factor.

$(3x + 7)(7x - 6) = 0$

$21x^2 + 31x - 42 = 0$

Any non-zero multiple of this equation would also have these solutions.

89. $\sqrt{3}, -\sqrt{3}$, and 4

One possible equation is:

$\left(x - \sqrt{3}\right)\left(x - \left(-\sqrt{3}\right)\right)(x - 4) = 0$

$\left(x - \sqrt{3}\right)\left(x + \sqrt{3}\right)(x - 4) = 0$

$\left(x^2 - 3\right)(x - 4) = 0$

$x^3 - 4x^2 - 3x + 12 = 0$

Any non-zero multiple of this equation would also have these solutions.

91. $i, -i$

Sample answer: $(x - i)(x - (-i)) = 0$

$(x - i)(x + i) = 0$

$x^2 - i^2 = 0$

$x^2 + 1 = 0$

93. $-1, 1, i$, and $-i$

One possible equation is:

$(x - (-1))(x - 1)(x - i)(x - (-i)) = 0$

$(x + 1)(x - 1)(x - i)(x + i) = 0$

$\left(x^2 - 1\right)\left(x^2 + 1\right) = 0$

$x^4 - 1 = 0$

Any non-zero multiple of this equation would also have these solutions.

95. Let x = the number of students in the original group. Then, $\dfrac{1700}{x}$ = the original cost per student.

When six more students join the group, the cost per student becomes $\dfrac{1700}{x} - 7.50$.

Model: (Cost per student) · (Number of students) = (Total cost)

$$\left(\frac{1700}{x} - 7.5\right)(x + 6) = 1700$$

$$(3400 - 15x)(x + 6) = 3400x \quad \text{Multiply both sides by } 2x \text{ to clear fraction.}$$

$$-15x^2 - 90x + 20{,}400 = 0$$

$$x = \frac{90 \pm \sqrt{(-90)^2 - 4(-15)(20{,}400)}}{2(-15)} = \frac{90 \pm 1110}{-30}$$

Using the positive value for x we conclude that the original number was $x = 34$ students.

97. *Model:* $\text{Time} = \dfrac{\text{Distance}}{\text{Rate}}$

Labels: Let x = average speed of the plane. Then we have a travel time of $t = 145/x$. If the average speed is increased by 40 mph, then

$$t - \frac{12}{60} = \frac{145}{x + 40}$$

$$t = \frac{145}{x + 40} + \frac{1}{5}.$$

Now, we equate these two equations and solve for x.

Equation:

$$\frac{145}{x} = \frac{145}{x + 40} + \frac{1}{5}$$

$$145(5)(x + 40) = 145(5)x + x(x + 40)$$

$$725x + 29{,}000 = 725x + x^2 + 40x$$

$$0 = x^2 + 40x - 29{,}000$$

Using the positive value for x found by the Quadratic Formula, we have $x \approx 151.5$ mph and $x + 40 = 191.5$ mph. The airspeed required to obtain the decrease in travel time is 191.5 miles per hour.

99.

$$A = P\left(1 + \frac{r}{n}\right)^{nt}$$

$$3052.49 = 2500\left(1 + \frac{r}{12}\right)^{(12)(5)}$$

$$1.220996 = \left(1 + \frac{r}{12}\right)^{60}$$

$$(1.220996)^{1/60} = 1 + \frac{r}{12}$$

$$\left[(1.220996)^{1/60} - 1\right](12) = r$$

$$r \approx 0.04 = 4\%$$

101. When $C = 2.5$ we have:

$$2.5 = \sqrt{0.2x + 1}$$

$$6.25 = 0.2x + 1$$

$$5.25 = 0.2x$$

$$x = 26.25 = 26{,}250 \text{ passengers}$$

103. $T = 75.82 - 2.11x + 43.51\sqrt{x},\ 5 \le x \le 40$

(a) $212 = 75.82 - 2.11x + 43.51\sqrt{x}$

$$0 = -2.11x + 43.51\sqrt{x} - 136.18$$

By the Quadratic Formula, we have

$$\sqrt{x} \approx 16.77928 \Rightarrow x \approx 281.333$$

$$\sqrt{x} \approx 3.84787 \Rightarrow x \approx 14.806.$$

Since x is restricted to $5 \le x \le 40$, let $x = 14.806$ pounds per square inch.

(b)

300

(14.806119, 212)

5 0 40

105.

$$37.55 = 40 - \sqrt{0.01x + 1}$$

$$\sqrt{0.01x + 1} = 2.45$$

$$0.01x + 1 = 6.0025$$

$$0.01x = 5.0025$$

$$x = 500.25$$

Rounding x to the nearest whole unit yields $x \approx 500$ units.

107.
$$\frac{1}{t} + \frac{1}{t+3} = \frac{1}{y}$$
$$\frac{1}{t} + \frac{1}{t+3} = \frac{1}{2}$$
$$2t(t+3)\frac{1}{t} + 2t(t+3)\frac{1}{t+3} = 2t(t+3)\frac{1}{2}$$
$$2(t+3) + 2t = t(t+3)$$
$$2t + 6 + 2t = t^2 + 3t$$
$$0 = t^2 - t - 6$$
$$0 = (t-3)(t+2)$$
$$t - 3 = 0 \Rightarrow t = 3$$
$$t + 2 = 0 \Rightarrow t = -2$$

Since t represents time, $t = 3$ is the only solution. It takes 3 hours for you working alone to tile the floor.

109.
$$v = \sqrt{\frac{gR}{\mu s}}$$
$$v^2 = \frac{gR}{\mu s}$$
$$v^2\mu s = gR$$
$$\frac{v^2\mu s}{R} = g$$

111. False—See Example 7 on page 125.

113. The distance between $(1, 2)$ and $(x, -10)$ is 13.
$$\sqrt{(x-1)^2 + (-10-2)^2} = 13$$
$$(x-1)^2 + (-12)^2 = 13^2$$
$$x^2 - 2x + 1 + 144 = 169$$
$$x^2 - 2x - 24 = 0$$
$$(x+4)(x-6) = 0$$
$$x + 4 = 0 \Rightarrow x = -4$$
$$x - 6 = 0 \Rightarrow x = 6$$

Both $(-4, -10)$ and $(6, -10)$ are a distance of 13 from $(1, 2)$.

115. The distance between $(0, 0)$ and $(8, y)$ is 17.
$$(8-0)^2 + (y-0)^2 = 17$$
$$(8)^2 + (y)^2 = 17^2$$
$$64 + y^2 = 289$$
$$y^2 = 225$$
$$y = \pm\sqrt{225}$$
$$= \pm 15$$

Both $(8, 15)$ and $(8, -15)$ are a distance of 17 from $(0, 0)$.

121. The quadratic equation was not written in general form before the values of a, b, and c were substituted in the Quadratic Formula.

117.
$$9 + |9 - a| = b$$
$$|9 - a| = b - 9$$

$$9 - a = b - 9 \quad \text{or} \quad 9 - a = -(b - 9)$$
$$-a = b - 18 \qquad 9 - a = -b + 9$$
$$a = 18 - b \qquad -a = -b$$
$$a = b$$

Thus, $a = 18 - b$ *or* $a = b$. From the original equation we know that $b \geq 9$.

Some possibilities are: $b = 9, a = 9$
$b = 10, a = 8$ *or* $a = 10$
$b = 11, a = 7$ *or* $a = 11$
$b = 12, a = 6$ *or* $a = 12$
$b = 13, a = 5$ *or* $a = 13$
$b = 14, a = 4$ *or* $a = 14$

119.
$$20 + \sqrt{20 - 1} = b$$
$$\sqrt{20 - a} = b - 20$$
$$20 - a = b^2 - 40b + 400$$
$$-a = b^2 - 40b + 380$$
$$a = -b^2 + 40b - 380$$

This formula gives the relationship between a and b. From the original equation we know that $a \leq 20$ and $b \geq 20$. Choose a b value, where $b \geq 20$ and then solve for a, keeping in mind that $a \leq 20$.

Some possibilities are: $b = 20, a = 20$
$b = 21, a = 19$
$b = 22, a = 16$
$b = 23, a = 11$
$b = 24, a = 4$
$b = 25, a = -5$

Section 1.7 Linear Inequalities in One Variable

1. solution set

3. double

5. Interval: $[0, 9)$

(a) Inequality: $0 \le x \le 9$

(b) The interval is bounded.

7. Interval: $[-1, 5]$

(a) Inequality: $-1 \le x \le 5$

(b) The interval is bounded.

9. Interval: $(11, \infty)$

(a) Inequality: $x > 11$

(b) The interval is unbounded.

11. Interval: $(-\infty, -2)$

(a) Inequality: $x < -2$

(b) The interval is unbounded.

13. $4x < 12$

$\frac{1}{4}(4x) < \frac{1}{4}(12)$

$x < 3$

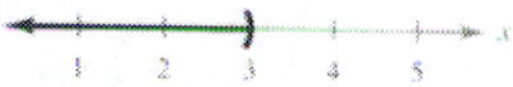

15. $-2x > -3$

$-\frac{1}{2}(-2x) < \left(-\frac{1}{2}\right)(-3)$

$x < \frac{3}{2}$

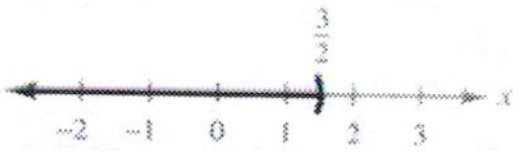

17. $x - 5 \ge 7$

$x \ge 12$

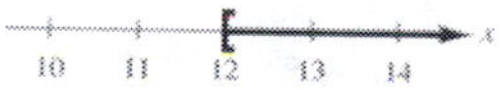

19. $2x + 7 < 3 + 4x$

$-2x < -4$

$x > 2$

21. $2x - 1 \ge 1 - 5x$

$7x \ge 2$

$x \ge \frac{2}{7}$

23. $4 - 2x < 3(3 - x)$

$4 - 2x < 9 - 3x$

$x < 5$

25. $\frac{3}{4}x - 6 \le x - 7$

$-\frac{1}{4}x \le -1$

$x \ge 4$

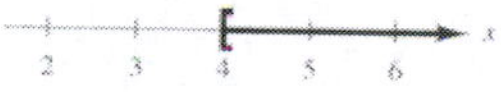

27. $\frac{1}{2}(8x + 1) \ge 3x + \frac{5}{2}$

$4x + \frac{1}{2} \ge 3x + \frac{5}{2}$

$x \ge 2$

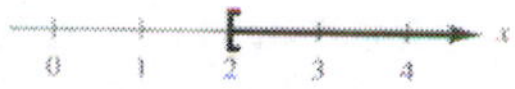

29. $3.6x + 11 \ge -3.4$

$3.6x \ge 14.4$

$x \ge -4$

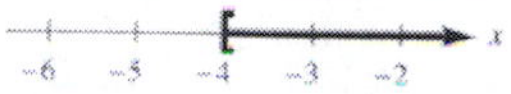

31. $1 < 2x + 3 < 9$

$-2 < 2x < 6$

$-1 < x < 3$

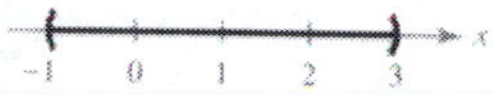

33. $0 < 3(x + 7) \le 20$

$0 < x + 7 \le \frac{20}{3}$

$-7 < x \le -\frac{1}{3}$

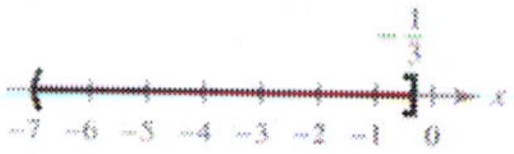

35. $-4 < \frac{2x - 3}{3} < 4$

$-12 < 2x - 3 < 12$

$-9 < 2x < 15$

$-\frac{9}{2} < x < \frac{15}{2}$

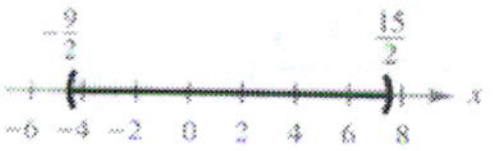

37. $-1 < \frac{-x - 2}{3} \le 1$

$-3 < -x - 2 \le 3$

$-1 < -x \le 5$

$1 > x \ge -5$

$-5 \le x < 1$

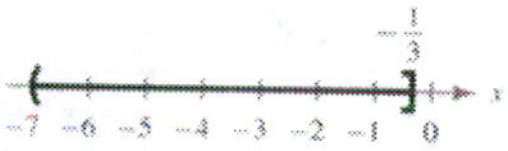

39. $\frac{3}{4} > x + 1 > \frac{1}{4}$

$-\frac{1}{4} > x > -\frac{3}{4}$

$-\frac{3}{4} < x < -\frac{1}{4}$

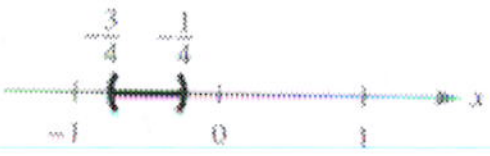

41. $3.2 \le 0.4x - 1 \le 4.4$

$4.2 \le 0.4x \le 5.4$

$10.5 \le x \le 13.5$

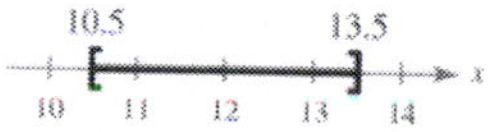

43. $|x| < 5$

$-5 < x < 5$

45. $\left|\frac{x}{2}\right| > 1$

$\frac{x}{2} < -1$ or $\frac{x}{2} > 1$

$x < -2$ $\quad$ $x > 2$

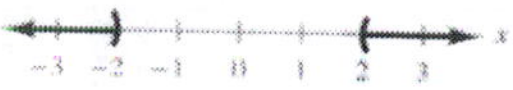

47. $|x - 5| < -1$

No solution. The absolute value of a number cannot be less than a negative number.

49. $|x - 20| \le 6$

$-6 \le x - 20 \le 6$

$14 \le x \le 26$

51. $|3 - 4x| \ge 9$

$3 - 4x \le -9$ or $3 - 4x \ge 9$

$-4x \le -12$ $\quad -4x \ge 6$

$x \ge 3$ $\quad x \le -\frac{3}{2}$

53. $\left|\dfrac{x - 3}{2}\right| \ge 4$

$\dfrac{x-3}{2} \le -4$ or $\dfrac{x-3}{2} \ge 4$

$x - 3 \le -8$ $\quad x - 3 \ge 8$

$x \le -5$ $\quad x \ge 11$

55. $|9 - 2x| - 2 < -1$

$|9 - 2x| < 1$

$-1 < 9 - 2x < 1$

$-10 < -2x < -8$

$5 > x > 4$

$4 < x < 5$

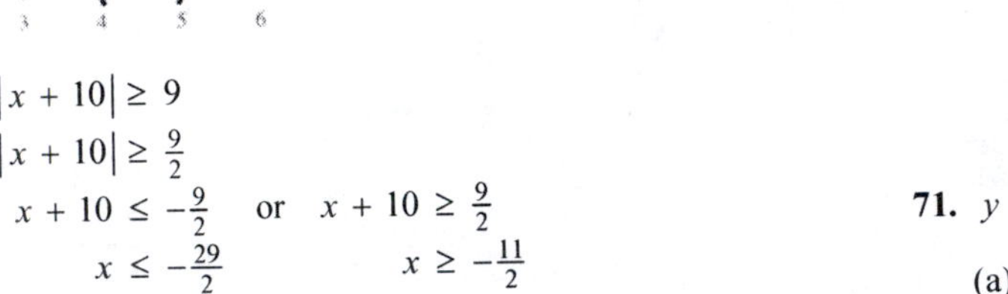

57. $2|x + 10| \ge 9$

$|x + 10| \ge \frac{9}{2}$

$x + 10 \le -\frac{9}{2}$ or $x + 10 \ge \frac{9}{2}$

$x \le -\frac{29}{2}$ $\quad x \ge -\frac{11}{2}$

59. $6x > 12$

$x > 2$

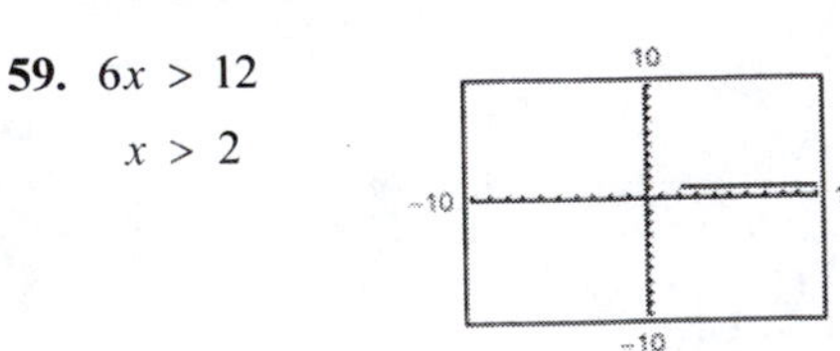

61. $5 - 2x \ge 1$

$-2x \ge -4$

$x \le 2$

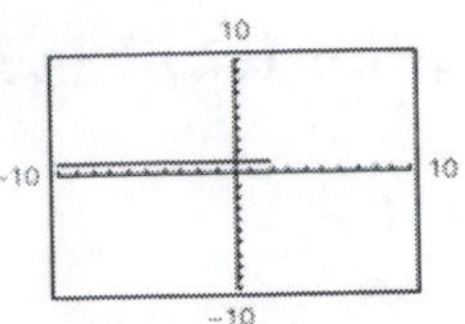

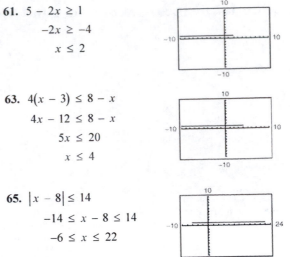

63. $4(x - 3) \le 8 - x$

$4x - 12 \le 8 - x$

$5x \le 20$

$x \le 4$

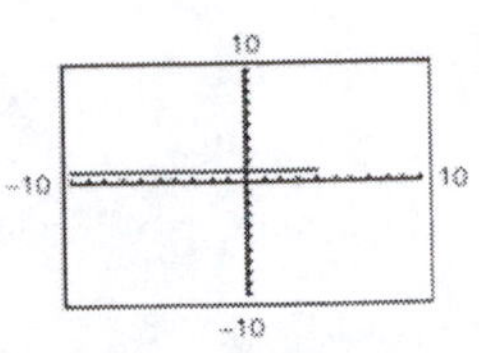

65. $|x - 8| \le 14$

$-14 \le x - 8 \le 14$

$-6 \le x \le 22$

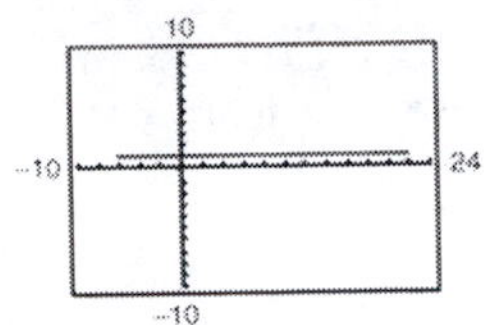

67. $2|x + 7| \ge 13$

$|x + 7| \ge \frac{13}{2}$

$x + 7 \le -\frac{13}{2}$ or $x + 7 \ge \frac{13}{2}$

$x \le -\frac{27}{2}$ $\quad x \ge -\frac{1}{2}$

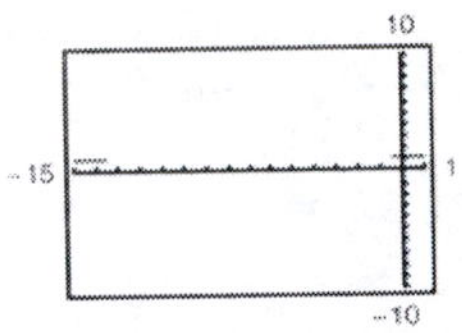

69. $y = 2x - 3$

(a) $y \ge 1$

$2x - 3 \ge 1$

$2x \ge 4$

$x \ge 2$

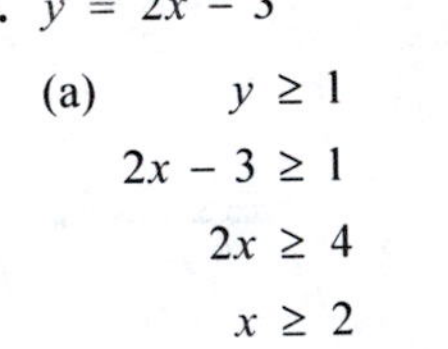

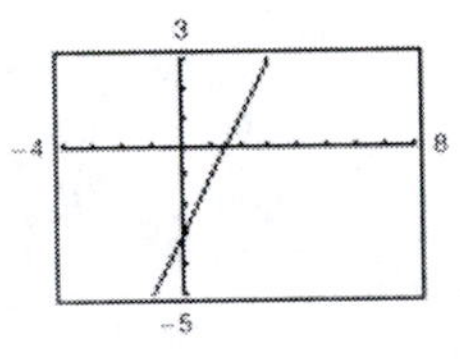

(b) $y \le 0$

$2x - 3 \le 0$

$2x \le 3$

$x \le \frac{3}{2}$

71. $y = -\frac{1}{2}x + 2$

(a) $0 \le y \le 3$

$0 \le -\frac{1}{2}x + 2 \le 3$

$-2 \le -\frac{1}{2}x \le 1$

$4 \ge x \ge -2$

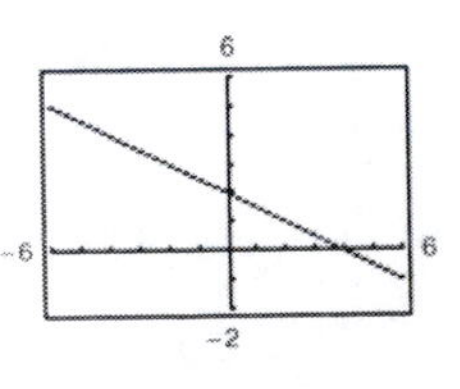

(b) $y \ge 0$

$-\frac{1}{2}x + 2 \ge 0$

$-\frac{1}{2}x \ge -2$

$x \le 4$

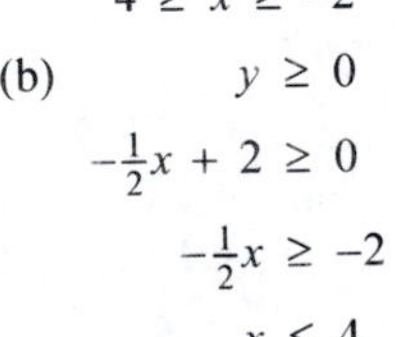

73. $y = |x - 3|$

(a) $$\begin{aligned} y &\le 2 \\ |x - 3| &\le 2 \\ -2 \le x - 3 &\le 2 \\ 1 \le x &\le 5 \end{aligned}$$

(b) $$\begin{aligned} y &\ge 4 \\ |x - 3| &\ge 4 \\ x - 3 \le -4 \quad &\text{or} \quad x - 3 \ge 4 \\ x \le -1 \quad &\text{or} \quad x \ge 7 \end{aligned}$$

75. $x - 5 \ge 0$

$x \ge 5$

$[5, \infty)$

77. $x + 3 \ge 0$

$x \ge -3$

$[-3, \infty)$

79. $7 - 2x \ge 0$

$-2x \ge -7$

$x \le \frac{7}{2}$

$\left(-\infty, \frac{7}{2}\right]$

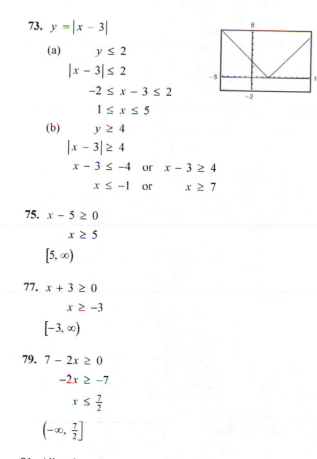

81. All real numbers less than 8 units from 10.

83. The midpoint of the interval $[-3, 3]$ is 0. The interval represents all real numbers x no more than 3 units from 0.

$|x - 0| \le 3$

$|x| \le 3$

85. The graph shows all real numbers at least 3 units from 7.

$|x - 7| \ge 3$

87. All real numbers at least 10 units from 12

$|x - 12| \ge 10$

89. All real numbers more than 4 units from -3

$|x - (-3)| > 4$

$|x + 3| > 4$

91. $\$4.10 \le E \le \$4.25$

93. $r \le 0.08$

95. $r = 220 - A = 220 - 20 = 200$ beats per minute

$0.50(200) \le r \le 0.85(200)$

$100 \le r \le 170$

The target heart rate is at least 100 beats per minute and at most 170 beats per minute.

97. $9.00 + 0.75x > 13.50$

$0.75x > 4.50$

$x > 6$

You must produce at least 6 units each hour in order to yield a greater hourly wage at the second job.

99. $1000(1 + r(2)) > 1062.50$

$1 + 2r > 1.0625$

$2r > 0.0625$

$r > 0.03125$

$r > 3.125\%$

101. $R > C$

$115.95x > 95x + 750$

$20.95x > 750$

$x \ge 35.7995$

$x \ge 36$ units

103. Let x = number of dozen doughnuts sold per day.

Revenue: $R = 7.95x$

Cost: $C = 1.45x + 165$

$$\begin{aligned} P &= R - C \\ &= 7.95x - (1.45x + 165) \\ &= 6.50x - 165 \end{aligned}$$

$400 \le P \le 1200$

$400 \le 6.50x - 165 \le 1200$

$565 \le 6.50x \le 1365$

$86.9 \le x \le 210$

The daily sales vary between 87 and 210 dozen doughnuts per day.

105. (a)

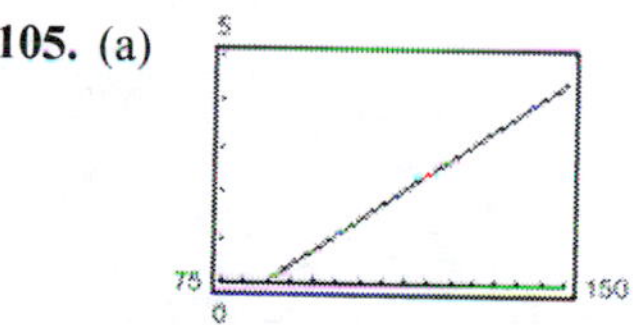

(b) From the graph you see that $y \ge 3$ when $x \ge 129$.

(c) Algebraically:

$3 \le 0.067x - 5.638$

$8.638 \le 0.067x$

$x \ge 129$

(d) IQ scores are not a good predictor of GPAs. Other factors include study habits, class attendance, and attitude.

107. (a) $S = 1.36t + 41.1$

$45 \le 1.36t + 41.1 \le 50$

$3.9 \le 1.36t \le 8.9$

$2.9 \le t \le 6.5$

Between the years 2002 and 2006 the average salary was between \$45,000 and \$50,000.

(b) $1.36t + 41.1 \ge 62$

$1.36t \ge 20.9$

$t \ge 15.4$

The average salary will exceed \$62,000 sometime during the year 2015.

109. $|s - 10.4| \le \frac{1}{16}$

$-\frac{1}{16} \le s - 10.4 \le \frac{1}{16}$

$-0.0625 \le s - 10.4 \le 0.0625$

$10.3375 \le s \le 10.4625$

Because $A = s^2$,

$(10.3375)^2 \le \text{area} \le (10.4625)^2$

$106.864 \text{ in.}^2 \le \text{area} \le 109.464 \text{ in.}^2$.

111. $\frac{1}{10}(3.61) \approx 0.361$

You might have been undercharged or overcharged by \$0.36.

113. $\left|\frac{t - 15.6}{1.9}\right| < 1$

$-1 < \frac{t - 15.6}{1.9} < 1$

$-1.9 < t - 15.6 < 1.9$

$13.7 < t < 17.5$

Two-thirds of the workers could perform the task in the time interval between 13.7 minutes and 17.5 minutes.

115. True. This is the Addition of a Constant Property of Inequality.

117. False. If $-10 \le x \le 8$, then $10 \ge -x$ and $-x \ge -8$.

Section 1.8 Other Types of Inequalities

1. positive; negative

3. zeros; undefined values

5. $x^2 - 3 < 0$

(a) $x = 3$

$(3)^2 - 3 \stackrel{?}{<} 0$

$6 \not< 0$

No, $x = 3$ *is not* a solution.

(b) $x = 0$

$(0)^2 - 3 \stackrel{?}{<} 0$

$-3 < 0$

Yes, $x = 0$ *is* a solution.

(c) $x = \frac{3}{2}$

$\left(\frac{3}{2}\right)^2 - 3 \stackrel{?}{<} 0$

$-\frac{3}{4} < 0$

Yes, $x = \frac{3}{2}$ *is* a solution.

(d) $x = -5$

$(-5)^2 - 3 \stackrel{?}{<} 0$

$22 \not< 0$

No, $x = -5$ *is not* a solution.

7. $\frac{x + 2}{x - 4} \ge 3$

(a) $x = 5$

$\frac{5 + 2}{5 - 4} \stackrel{?}{\ge} 3$

$7 \ge 3$

Yes, $x = 5$ *is* a solution.

(b) $x = 4$

$\frac{4 + 2}{4 - 4} \stackrel{?}{\ge} 3$

$\frac{6}{0}$ is undefined.

No, $x = 4$ *is not* a solution.

(c) $x = -\frac{9}{2}$

$\frac{-\frac{9}{2} + 2}{-\frac{9}{2} - 4} \stackrel{?}{\ge} 3$

$\frac{5}{17} \not\ge 3$

No, $x = -\frac{9}{2}$ *is not* a solution.

(d) $x = \frac{9}{2}$

$\frac{\frac{9}{2} + 2}{\frac{9}{2} - 4} \stackrel{?}{\ge} 3$

$13 \ge 3$

Yes, $x = \frac{9}{2}$ *is* a solution.

9. $3x^2 - x - 2 = (3x + 2)(x - 1)$

$3x + 2 = 0 \Rightarrow x = -\frac{2}{3}$

$x - 1 = 0 \Rightarrow x = 1$

The key numbers are $-\frac{2}{3}$ and 1.

11. $\dfrac{1}{x - 5} + 1 = \dfrac{1 + 1(x - 5)}{x - 5}$

$= \dfrac{x - 4}{x - 5}$

$x - 4 = 0 \Rightarrow x = 4$

$x - 5 = 0 \Rightarrow x = 5$

The key numbers are 4 and 5.

13. $x^2 < 9$

$x^2 - 9 < 0$

$(x + 3)(x - 3) < 0$

Key numbers: $x = \pm 3$

Test intervals: $(-\infty, -3), (-3, 3), (3, \infty)$

Test: Is $(x + 3)(x - 3) < 0$?

Interval	x-Value	Value of $x^2 - 9$	Conclusion
$(-\infty, -3)$	-4	7	Positive
$(-3, 3)$	0	-9	Negative
$(3, \infty)$	4	7	Positive

Solution set: $(-3, 3)$

15. $(x + 2)^2 \le 25$

$x^2 + 4x + 4 \le 25$

$x^2 + 4x - 21 \le 0$

$(x + 7)(x - 3) \le 0$

Key numbers: $x = -7, x = 3$

Test intervals: $(-\infty, -7), (-7, 3), (3, \infty)$

Test: Is $(x + 7)(x - 3) \le 0$?

Interval	x-Value	Value of $(x + 7)(x - 3)$	Conclusion
$(-\infty, -7)$	-8	$(-1)(-11) = 11$	Positive
$(-7, 3)$	0	$(7)(-3) = -21$	Negative
$(3, \infty)$	4	$(11)(1) = 11$	Positive

Solution set: $[-7, 3]$

17. $x^2 + 4x + 4 \ge 9$

$x^2 + 4x - 5 \ge 0$

$(x + 5)(x - 1) \ge 0$

Key numbers: $x = -5, x = 1$

Test intervals: $(-\infty, -5), (-5, 1), (1, \infty)$

Test: Is $(x + 5)(x - 1) \ge 0$?

Interval	x-Value	Value of $(x + 5)(x - 1)$	Conclusion
$(-\infty, -5)$	-6	$(-1)(-7) = 7$	Positive
$(-5, 1)$	0	$(5)(-1) = -5$	Negative
$(1, \infty)$	2	$(7)(1) = 7$	Positive

Solution set: $(-\infty, -5] \cup [1, \infty)$

19. $x^2 + x < 6$

$x^2 + x - 6 < 0$

$(x + 3)(x - 2) < 0$

Key numbers: $x = -3, x = 2$

Test intervals: $(-\infty, -3), (-3, 2), (2, \infty)$

Test: Is $(x + 3)(x - 2) < 0$?

Interval	x-Value	Value of $(x + 3)(x - 2)$	Conclusion
$(-\infty, -3)$	-4	$(-1)(-6) = 6$	Positive
$(-3, 2)$	0	$(3)(-2) = -6$	Negative
$(2, \infty)$	3	$(6)(1) = 6$	Positive

Solution set: $(-3, 2)$

21. $x^2 + 2x - 3 < 0$

$(x + 3)(x - 1) < 0$

Key numbers: $x = -3, x = 1$

Test intervals: $(-\infty, -3), (-3, 1), (1, \infty)$

Test: Is $(x + 3)(x - 1) < 0$?

Interval	x-Value	Value of $(x + 3)(x - 1)$	Conclusion
$(-\infty, -3)$	-4	$(-1)(-5) = 5$	Positive
$(-3, 1)$	0	$(3)(-1) = -3$	Negative
$(1, \infty)$	2	$(5)(1) = 5$	Positive

Solution set: $(-3, 1)$

23. $3x^2 - 11x > 20$

$3x^2 - 11x - 20 > 0$

$(3x + 4)(x - 5) > 0$

Key numbers: $x = 5, x = -\frac{4}{3}$

Test intervals: $\left(-\infty, -\frac{4}{5}\right), \left(-\frac{4}{3}, 5\right), (5, \infty)$

Test: Is $(3x + 4)(x - 5) > 0$?

Interval	x-Value	Value of $(3x + 4)(x - 5)$	Conclusion
$\left(-\infty, -\frac{4}{3}\right)$	-3	$(-5)(-8) = 40$	Positive
$\left(-\frac{4}{3}, 5\right)$	0	$(4)(-5) = -20$	Negative
$(5, \infty)$	6	$(22)(1) = 22$	Positive

Solution set: $\left(-\infty, -\frac{4}{3}\right) \cup (5, \infty)$

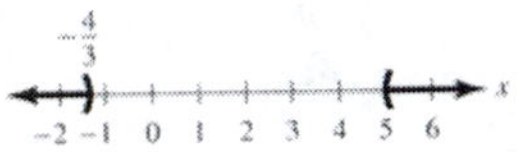

25. $x^2 - 3x - 18 > 0$

$(x + 3)(x - 6) > 0$

Key numbers: $x = -3, x = 6$

Test intervals: $(-\infty, -3), (-3, 6), (6, \infty)$

Test: Is $(x + 3)(x - 6) > 0$?

Interval	x-Value	Value of $(x + 3)(x - 6)$	Conclusion
$(-\infty, -3)$	-4	$(-1)(-10) = 10$	Positive
$(-3, 6)$	0	$(3)(-6) = -18$	Negative
$(6, \infty)$	7	$(10)(1) = 10$	Positive

Solution set: $(-\infty, -3) \cup (6, \infty)$

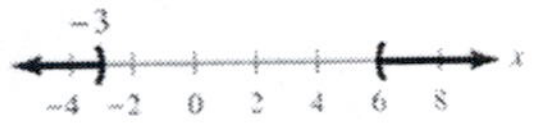

27. $x^3 - 3x^2 - x > -3$

$x^3 - 3x^2 - x + 3 > 0$

$x^2(x - 3) - (x - 3) > 0$

$(x - 3)(x^2 - 1) > 0$

$(x - 3)(x + 1)(x - 1) > 0$

Key numbers: $x = -1, x = 1, x = 3$

Test intervals: $(-\infty, -1), (-1, 1), (1, 3), (3, \infty)$

Test: Is $(x - 3)(x + 1)(x - 1) > 0$?

Interval	x-Value	Value of $(x - 3)(x + 1)(x - 1)$	Conclusion
$(-\infty, -1)$	-2	$(-5)(-1)(-3) = -15$	Negative
$(-1, 1)$	0	$(-3)(1)(-1) = 3$	Positive
$(1, 3)$	2	$(-1)(3)(1) = -3$	Negative
$(3, \infty)$	4	$(1)(5)(3) = 15$	Positive

Solution set: $(-1, 1) \cup (3, \infty)$

29. $4x^3 - 6x^2 < 0$

$2x^2(2x - 3) < 0$

Key numbers: $x = 0, x = \frac{3}{2}$

Test intervals: $(-\infty, 0) \Rightarrow 2x^2(2x - 3) < 0$

$\left(0, \frac{3}{2}\right) \Rightarrow 2 \Rightarrow 2x^2(2x - 3) < 0$

$\left(\frac{3}{2}, \infty\right) \Rightarrow 2x^2(2x - 3) > 0$

Solution set: $(-\infty, 0) \cup \left(0, \frac{3}{2}\right)$

31. $x^3 - 4x \geq 0$

$x(x + 2)(x - 2) \geq 0$

Key numbers: $x = 0, x = \pm 2$

Test intervals: $(-\infty, -2) \Rightarrow x(x + 2)(x - 2) < 0$

$(-2, 0) \Rightarrow x(x + 2)(x - 2) > 0$

$(0, 2) \Rightarrow x(x + 2)(x - 2) < 0$

$(2, \infty) \Rightarrow x(x + 2)(x - 2) > 0$

Solution set: $[-2, 0] \cup [2, \infty)$

33. $(x - 1)^2(x + 2)^3 \geq 0$

Key numbers: $x = 1, x = -2$

Test intervals: $(-\infty, -2) \Rightarrow (x - 1)^2(x + 2)^3 < 0$

$(-2, 1) \Rightarrow (x - 1)^2(x + 2)^3 > 0$

$(1, \infty) \Rightarrow (x - 1)^2(x + 2)^3 > 0$

Solution set: $[-2, \infty)$

35. $4x^2 - 4x + 1 \leq 0$

$(2x - 1)^2 \leq 0$

Key number: $x = \dfrac{1}{2}$

Test Interval	x-Value	Polynomial Value	Conclusion
$\left(-\infty, \frac{1}{2}\right]$	$x = 0$	$[2(0) - 1]^2 = 1$	Positive
$\left(\frac{1}{2}, \infty\right)$	$x = 1$	$[2(1) - 1]^2 = 1$	Positive

The solution set consists of the single real number $\dfrac{1}{2}$.

37. $x^2 - 6x + 12 \leq 0$

Using the Quadratic Formula, you can determine that the key numbers are $x = 3 \pm \sqrt{3}i$.

Test Interval	x-Value	Polynomial Value	Conclusion
$(-\infty, \infty)$	$x = 0$	$(0)^2 - 6(0) + 12 = 12$	Positive

The solution set is empty, that is there are no real solutions.

39. $\dfrac{4x-1}{x} > 0$

Key numbers: $x = 0, x = \dfrac{1}{4}$

Test intervals: $(-\infty, 0), \left(0, \frac{1}{4}\right), \left(\frac{1}{4}, \infty\right)$

Test: Is $\dfrac{4x-1}{x} > 0$?

Interval	x-Value	Value of $\dfrac{4x-1}{x}$	Conclusion
$(-\infty, 0)$	-1	$\dfrac{-5}{-1} = 5$	Positive
$\left(0, \dfrac{1}{4}\right)$	$\dfrac{1}{8}$	$\dfrac{-\frac{1}{2}}{\frac{1}{8}} = -4$	Negative
$\left(\dfrac{1}{4}, \infty\right)$	1	$\dfrac{3}{1} = 3$	Positive

Solution set: $(-\infty, 0) \cup \left(\dfrac{1}{4}, \infty\right)$

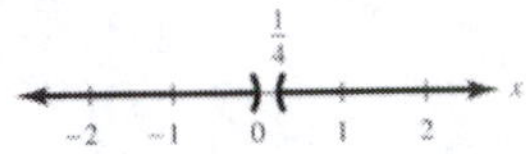

41. $\dfrac{3x-5}{x-5} \geq 0$

Key numbers: $x = \dfrac{5}{3}, x = 5$

Test intervals: $\left(-\infty, \dfrac{5}{3}\right), \left(\dfrac{5}{3}, 5\right), (5, \infty)$

Test: Is $\dfrac{3x-5}{x-5} \geq 0$?

Interval	x-Value	Value of $\dfrac{3x-5}{x-5}$	Conclusion
$\left(-\infty, \dfrac{5}{3}\right)$	0	$\dfrac{-5}{-5} = 1$	Positive
$\left(\dfrac{5}{3}, 5\right)$	2	$\dfrac{6-5}{2-5} = -\dfrac{1}{3}$	Negative
$(5, \infty)$	6	$\dfrac{18-5}{6-5} = 13$	Positive

Solution set: $\left(-\infty, \dfrac{5}{3}\right] \cup (5, \infty)$

43.
$$\frac{x+6}{x+1} - 2 < 0$$
$$\frac{x+6-2(x+1)}{x+1} < 0$$
$$\frac{4-x}{x+1} < 0$$

Key numbers: $x = -1, x = 4$

Test intervals: $(-\infty, -1) \Rightarrow \dfrac{4-x}{x+1} < 0$

$(-1, 4) \Rightarrow \dfrac{4-x}{x+1} > 0$

$(4, \infty) \Rightarrow \dfrac{4-x}{x+1} < 0$

Solution set: $(-\infty, -1) \cup (4, \infty)$

45.
$$\frac{2}{x+5} > \frac{1}{x-3}$$
$$\frac{2}{x+5} - \frac{1}{x-3} > 0$$
$$\frac{2(x-3) - 1(x+5)}{(x+5)(x-3)} > 0$$
$$\frac{x-11}{(x+5)(x-3)} > 0$$

Key numbers: $x = -5, x = 3, x = 11$

Test intervals: $(-\infty, -5) \Rightarrow \dfrac{x-11}{(x+5)(x-3)} < 0$

$(-5, 3) \Rightarrow \dfrac{x-11}{(x+5)(x-3)} > 0$

$(3, 11) \Rightarrow \dfrac{x-11}{(x+5)(x-3)} < 0$

$(11, \infty) \Rightarrow \dfrac{x-11}{(x+5)(x-3)} > 0$

Solution set: $(-5, 3) \cup (11, \infty)$

47.
$$\frac{1}{x-3} \le \frac{9}{4x+3}$$
$$\frac{1}{x-3} - \frac{9}{4x+3} \le 0$$
$$\frac{4x+3-9(x-3)}{(x-3)(4x+3)} \le 0$$
$$\frac{30-5x}{(x-3)(4x+3)} \le 0$$

Key numbers: $x = 3, x = -\frac{3}{4}, x = 6$

Test intervals: $\left(-\infty, -\frac{3}{4}\right) \Rightarrow \frac{30-5x}{(x-3)(4x+3)} > 0$

$\left(-\frac{3}{4}, 3\right) \Rightarrow \frac{30-5x}{(x-3)(4x+3)} < 0$

$(3, 6) \Rightarrow \frac{30-5x}{(x-3)(4x+3)} > 0$

$(6, \infty) \Rightarrow \frac{30-5x}{(x-3)(4x+3)} < 0$

Solution set: $\left(-\frac{3}{4}, 3\right) \cup [6, \infty)$

49.
$$\frac{x^2+2x}{x^2-9} \le 0$$
$$\frac{x(x+2)}{(x+3)(x-3)} \le 0$$

Key numbers: $x = 0, x = -2, x = \pm 3$

Test intervals: $(-\infty, -3) \Rightarrow \frac{x(x+2)}{(x+3)(x-3)} > 0$

$(-3, -2) \Rightarrow \frac{x(x+2)}{(x+3)(x-3)} < 0$

$(-2, 0) \Rightarrow \frac{x(x+2)}{(x+3)(x-3)} > 0$

$(0, 3) \Rightarrow \frac{x(x+2)}{(x+3)(x-3)} < 0$

$(3, \infty) \Rightarrow \frac{x(x+2)}{(x+3)(x-3)} > 0$

Solution set: $(-3, -2] \cup [0, 3)$

51.
$$\frac{3}{x-1} + \frac{2x}{x+1} > -1$$
$$\frac{3(x+1) + 2x(x-1) + 1(x+1)(x-1)}{(x-1)(x+1)} > 0$$
$$\frac{3x^2+x+2}{(x-1)(x+1)} > 0$$

Key numbers: $x = -1, x = 1$

Test intervals: $(-\infty, -1) \Rightarrow \frac{3x^2+x+2}{(x-1)(x+1)} > 0$

$(-1, 1) \Rightarrow \frac{3x^2+x+2}{(x-1)(x+1)} < 0$

$(1, \infty) \Rightarrow \frac{3x^2+x+2}{(x-1)(x+1)} > 0$

Solution set: $(-\infty, -1) \cup (1, \infty)$

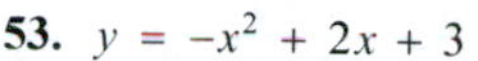

53. $y = -x^2 + 2x + 3$

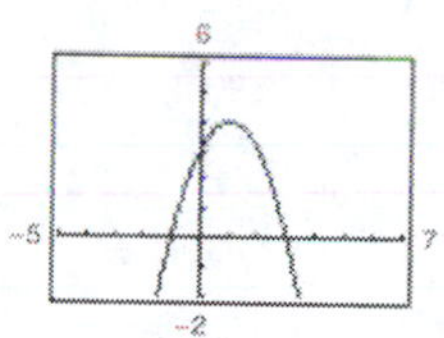

(a) $y \le 0$ when $x \le -1$ or $x \ge 3$.

(b) $y \ge 3$ when $0 \le x \le 2$.

55. $y = \frac{1}{8}x^3 - \frac{1}{2}x$

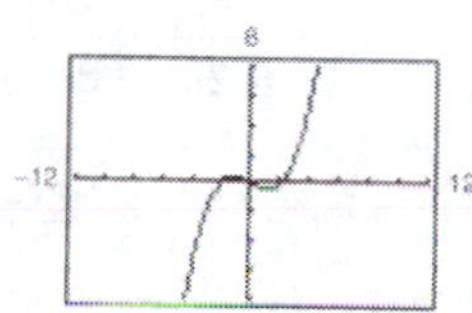

(a) $y \ge 0$ when $-2 \le x \le 0$ or $2 \le x < \infty$.

(b) $y \le 6$ when $x \le 4$.

57. $y = \frac{3x}{x-2}$

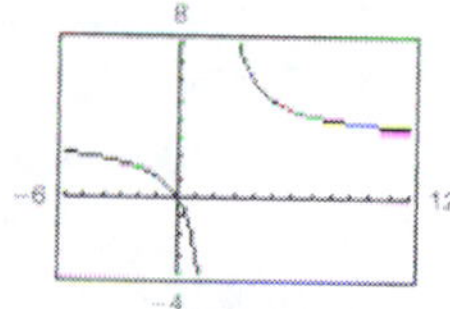

(a) $y \le 0$ when $0 \le x < 2$.

(b) $y \ge 6$ when $2 < x \le 4$.

59. $y = \dfrac{2x^2}{x^2 + 4}$

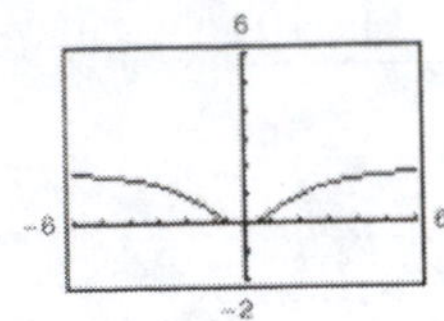

(a) $y \ge 1$ when $x \le -2$ or $x \ge 2$.

This can also be expressed as $|x| \ge 2$.

(b) $y \le 2$ for all real numbers x.

This can also be expressed as $-\infty < x < \infty$.

61. $4 - x^2 \ge 0$

$(2 + x)(2 - x) \ge 0$

Key numbers: $x = \pm 2$

Test intervals: $(-\infty, -2) \Rightarrow 4 - x^2 < 0$

$(-2, 2) \Rightarrow 4 - x^2 > 0$

$(2, \infty) \Rightarrow 4 - x^2 < 0$

Domain: $[-2, 2]$

63. $x^2 - 9x + 20 \ge 0$

$(x - 4)(x - 5) \ge 0$

Key numbers: $x = 4, x = 5$

Test intervals: $(-\infty, 4), (4, 5), (5, \infty)$

Interval	x-Value	Value of $(x - 4)(x - 5)$	Conclusion
$(-\infty, 4)$	0	$(-4)(-5) = 20$	Positive
$(4, 5)$	$\frac{9}{2}$	$\left(\frac{1}{2}\right)\left(-\frac{1}{2}\right) = -\frac{1}{4}$	Negative
$(5, \infty)$	6	$(2)(1) = 2$	Positive

Domain: $(-\infty, 4] \cup [5, \infty)$

65. $\dfrac{x}{x^2 - 2x - 35} \ge 0$

$\dfrac{x}{(x + 5)(x - 7)} \ge 0$

Key numbers: $x = 0, x = -5, x = 7$

Test intervals: $(-\infty, -5) \Rightarrow \dfrac{x}{(x + 5)(x - 7)} < 0$

$(-5, 0) \Rightarrow \dfrac{x}{(x + 5)(x - 7)} > 0$

$(0, 7) \Rightarrow \dfrac{x}{(x + 5)(x - 7)} < 0$

$(7, \infty) \Rightarrow \dfrac{x}{(x + 5)(x - 7)} > 0$

Domain: $(-5, 0] \cup (7, \infty)$

67. $0.4x^2 + 5.26 < 10.2$

$0.4x^2 - 4.94 < 0$

$0.4(x^2 - 12.35) < 0$

Key numbers: $x \approx \pm 3.51$

Test intervals: $(-\infty, -3.51), (-3.51, 3.51), (3.51, \infty)$

Solution set: $(-3.51, 3.51)$

69. $-0.5x^2 + 12.5x + 1.6 > 0$

Key numbers: $x \approx -0.13, x \approx 25.13$

Test intervals: $(-\infty, -0.13), (-0.13, 25.13), (25.13, \infty)$

Solution set: $(-0.13, 25.13)$

71. $\dfrac{1}{2.3x - 5.2} > 3.4$

$\dfrac{1}{2.3x - 5.2} - 3.4 > 0$

$\dfrac{1 - 3.4(2.3x - 5.2)}{2.3x - 5.2} > 0$

$\dfrac{-7.82x + 18.68}{2.3x - 5.2} > 0$

Key numbers: $x \approx 2.39, x \approx 2.26$

Test intervals: $(-\infty, 2.26), (2.26, 2.39), (2.39, \infty)$

Solution set: $(2.26, 2.39)$

73. $s = -16t^2 + v_0t + s_0 = -16t^2 + 160t$

(a) $-16t^2 + 160t = 0$

$-16t(t - 10) = 0$

$t = 0, t = 10$

It will be back on the ground in 10 seconds.

(b) $-16t^2 + 160t > 384$

$-16t^2 + 160t - 384 > 0$

$-16(t^2 - 10t + 24) > 0$

$t^2 - 10t + 24 < 0$

$(t - 4)(t - 6) < 0$

Key numbers: $t = 4, t = 6$

Test intervals: $(-\infty, 4), (4, 6), (6, \infty)$

Solution set: 4 seconds $< t <$ 6 seconds

75. $2L + 2W = 100 \Rightarrow W = 50 - L$

$LW \ge 500$

$L(50 - L) \ge 500$

$-L^2 + 50L - 500 \ge 0$

By the Quadratic Formula you have:

Key numbers: $L = 25 \pm 5\sqrt{5}$

Test: Is $-L^2 + 50L - 500 \ge 0$?

Solution set: $25 - 5\sqrt{5} \le L \le 25 + 5\sqrt{5}$

13.8 meters $\le L \le$ 36.2 meters

77. $R = x(75 - 0.0005x)$ and $C = 30x + 250{,}000$

$P = R - C$

$= (75x - 0.0005x^2) - (30x + 250{,}000)$

$= -0.0005x^2 + 45x - 250{,}000$

$P \ge 750{,}000$

$-0.0005x^2 + 45x - 250{,}000 \ge 750{,}000$

$-0.0005x^2 + 45x - 1{,}000{,}000 \ge 0$

Key numbers: $x = 40{,}000$, $x = 50{,}000$

(These were obtained by using the Quadratic Formula.)

Test intervals:

$(0, 40{,}000), (40{,}000, 50{,}000), (50{,}000, \infty)$

The solution set is $[40{,}000, 50{,}000]$ or $40{,}000 \le x \le 50{,}000$. The price per unit is

$p = \dfrac{R}{x} = 75 - 0.0005x.$

For $x = 40{,}000$, $p = \$55$. For $x = 50{,}000$, $p = \$50$. So, for $40{,}000 \le x \le 50{,}000$, $\$50.00 \le p \le \$55.00$.

79. (a)

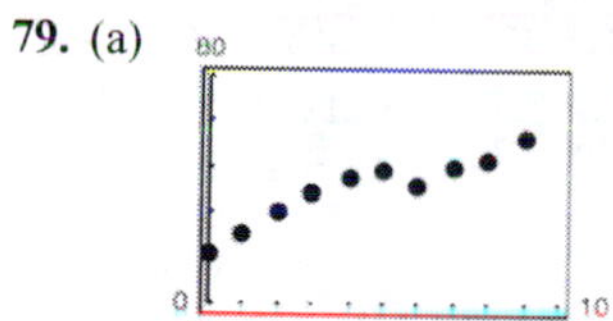

(b) $N = 0.00406t^4 - 0.0564t^3 + 0.147t^2 + 0.86t + 72.2$

(c)

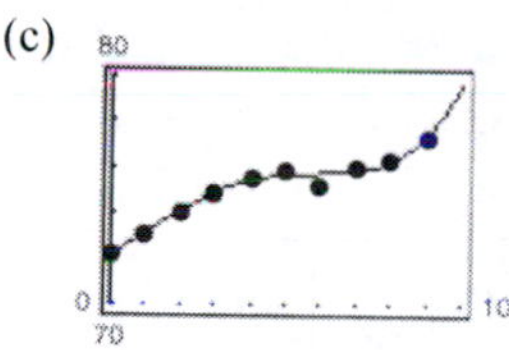

The model fits the data well.

(d) Using the zoom and trace features, the number of students enrolled in schools exceeded 74 million in the year 2001.

(e) No. The model can be used to predict enrollments for years close to those in its domain but when you project too far into the future, the numbers predicted by the model increase too rapidly to be considered reasonable.

81. $$\frac{1}{R} = \frac{1}{R_1} + \frac{1}{2}$$
$$2R_1 = 2R + RR_1$$
$$2R_1 = R(2 + R_1)$$
$$\frac{2R_1}{2 + R_1} = R$$

Because $R \geq 1$,

$$\frac{2R_1}{2 + R_1} \geq 1$$
$$\frac{2R_1}{2 + R_1} - 1 \geq 0$$
$$\frac{R_1 - 2}{2 + R_1} \geq 0.$$

Because $R_1 > 0$, the only key number is $R_1 = 2$. The inequality is satisfied when $R_1 \geq 2$ ohms.

83. True.

$x^3 - 2x^2 - 11x + 12 = (x + 3)(x - 1)(x - 4)$

The test intervals are $(-\infty, -3)$, $(-3, 1)$, $(1, 4)$, and $(4, \infty)$.

85. $x^2 + bx + 4 = 0$

(a) To have at least one real solution, $b^2 - 4ac \geq 0$.

$$b^2 - 4(1)(4) \geq 0$$
$$b^2 - 16 \geq 0$$

Key numbers: $b = -4$, $b = 4$

Test intervals: $(-\infty, -4) \Rightarrow b^2 - 16 > 0$

$(-4, 4) \Rightarrow b^2 - 16 < 0$

$(4, \infty) \Rightarrow b^2 - 16 > 0$

Solution set: $(-\infty, -4] \cup [4, \infty]$

(b) $b^2 - 4ac \geq 0$

Key numbers: $b = -2\sqrt{ac}$, $b = 2\sqrt{ac}$

Similar to part (a), if $a > 0$ and $c > 0$, $b \leq -2\sqrt{ac}$ or $b \geq 2ac$.

87. $3x^2 + bx + 10 = 0$

(a) To have at least one real solution, $b^2 - 4ac \geq 0$.

$$b^2 - 4(3)(10) \geq 0$$
$$b^2 - 120 \geq 0$$

Key numbers: $b = -2\sqrt{30}$, $b = 2\sqrt{30}$

Test intervals: $\left(-\infty, -2\sqrt{30}\right) \Rightarrow b^2 - 120 > 0$

$\left(-2\sqrt{30}, 2\sqrt{30}\right) \Rightarrow b^2 - 120 < 0$

$\left(2\sqrt{30}, \infty\right) \Rightarrow b^2 - 120 > 0$

Solution set: $\left(-\infty, -2\sqrt{30}\right] \cup \left[2\sqrt{30}, \infty\right]$

(b) $b^2 - 4ac \geq 0$

Similar to part (a), if $a > 0$ and $c > 0$, $b \leq -2\sqrt{ac}$ or $b \geq 2ac$.

89.

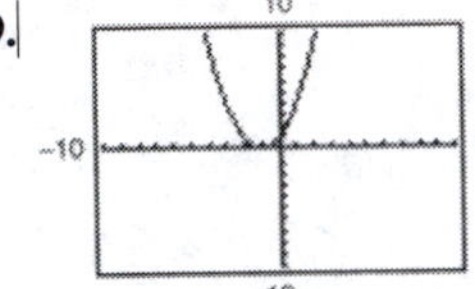

For part (b), the y-values that are less than or equal to 0 occur only at $x = -1$.

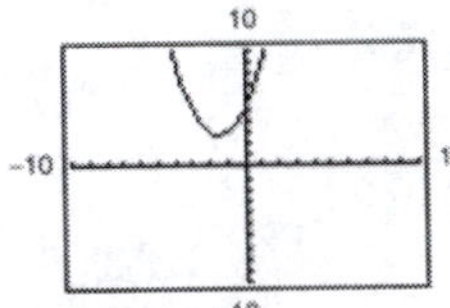

For part (c), there are no y-values that are less than 0.

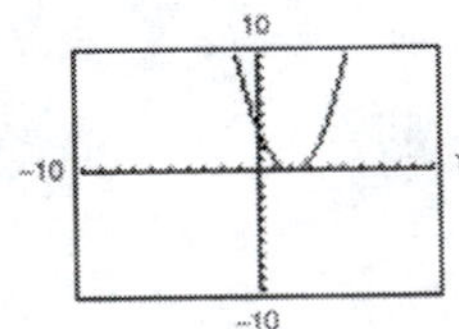

For part (d), the y-values that are greater than 0 occur for all values of x except 2

Review Exercises for Chapter 1

1. $y = -4x + 1$

x	-2	-1	0	1	2
y	9	5	1	-3	-7

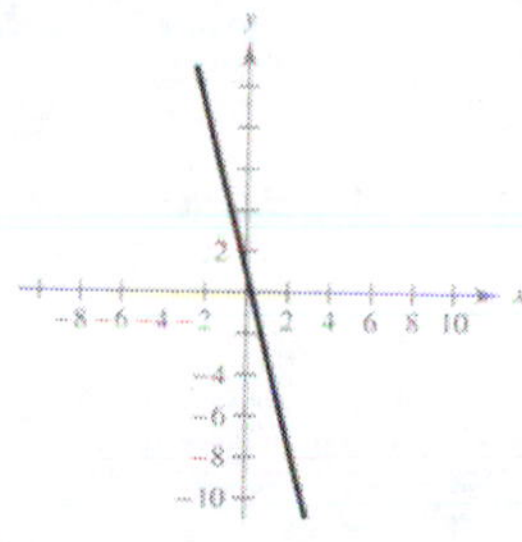

3. x-intercepts: $(1, 0), (5, 0)$

y-intercept: $(0, 5)$

5. $y = -4x + 1$

Intercepts: $\left(\frac{1}{4}, 0\right), (0, 1)$

$y = -4(-x) + 1 \Rightarrow y = 4x + 1 \Rightarrow$ No y-axis symmetry

$-y = -4x + 1 \Rightarrow y = 4x - 1 \Rightarrow$ No x-axis symmetry

$-y = -4(-x) + 1 \Rightarrow y = -4x - 1 \Rightarrow$ No origin symmetry

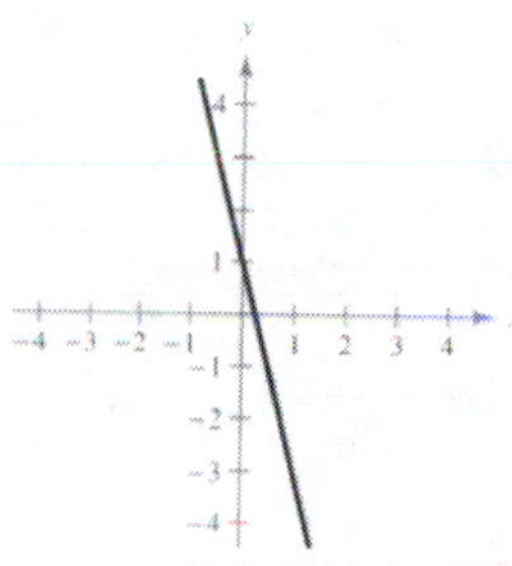

7. $y = 7 - x^2$

Intercepts: $\left(\pm\sqrt{7}, 0\right), (0, 7)$

$y = 7 - (-x)^2 \Rightarrow y = 7 - x^2 \Rightarrow$ y-axis symmetry

$y = 7 - x^2 \Rightarrow y = -7 + x^2 \Rightarrow$ No x-axis symmetry

$-y = 7 - (-x)^2 \Rightarrow y = -7 + x^2 \Rightarrow$ No origin symmetry

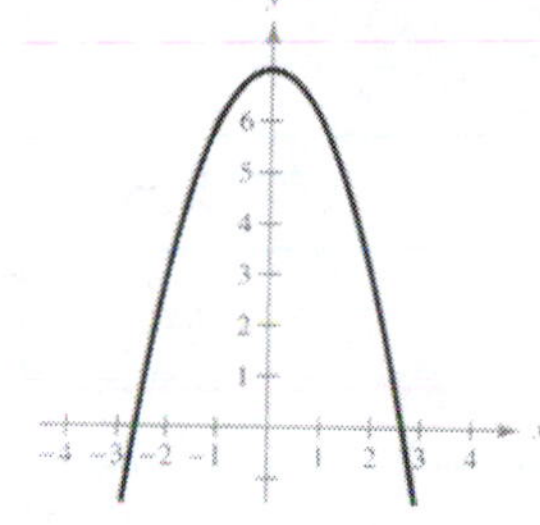

9. $y = x^3 + 3$

Intercepts: $\left(-\sqrt[3]{3}, 0\right), (0, 3)$

$y = (-x)^3 + 3 \Rightarrow y = -x^3 + 3 \Rightarrow$ No y-axis symmetry

$-y = x^3 + 3 \Rightarrow y = -x^3 - 3 \Rightarrow$ No x-axis symmetry

$-y = (-x)^3 + 3 \Rightarrow y = x^3 - 3 \Rightarrow$ No origin symmetry

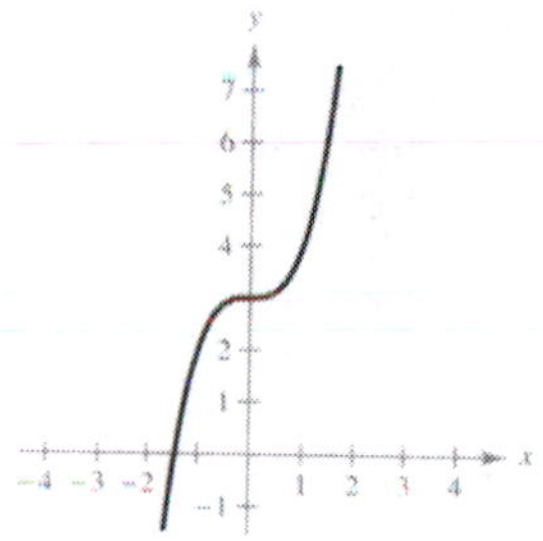

11. $y = -|x| - 2$

Intercept: $(0, -2)$

$y = -|-x| - 2 \Rightarrow y = -|x| - 2 \Rightarrow$ y-axis symmetry

$-y = -|x| - 2 \Rightarrow y = |x| + 2 \Rightarrow$ No x-axis symmetry

$-y = -|-x| - 2 \Rightarrow y = |x| + 2 \Rightarrow$ No origin symmetry

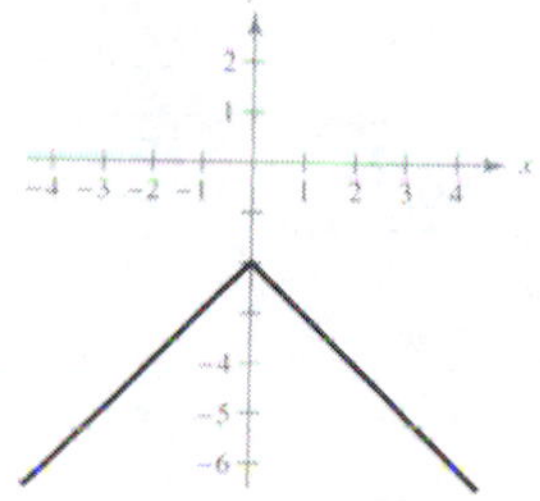

13. $x^2 + y^2 = 9$

Center: $(0, 0)$

Radius: 3

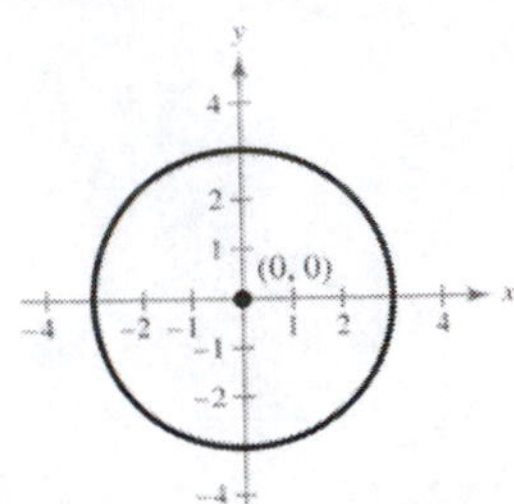

15. $(x + 2)^2 + y^2 = 16$

$(x - (-2))^2 + (y - 0)^2 = 4^2$

Center: $(-2, 0)$

Radius: 4

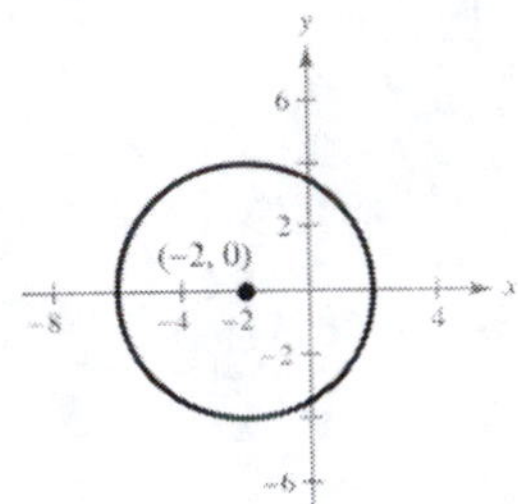

17. Endpoints of a diameter: $(0, 0)$ and $(4, -6)$

Center: $\left(\dfrac{0 + 4}{2}, \dfrac{0 + (-6)}{2}\right) = (2, -3)$

Radius: $r = \sqrt{(2 - 0)^2 + (-3 - 0)^2} = \sqrt{4 + 9} = \sqrt{13}$

Standard form: $(x - 2)^2 + (y - (-3))^2 = \left(\sqrt{13}\right)^2$

$$(x - 2)^2 + (y + 3)^2 = 13$$

19. (a)

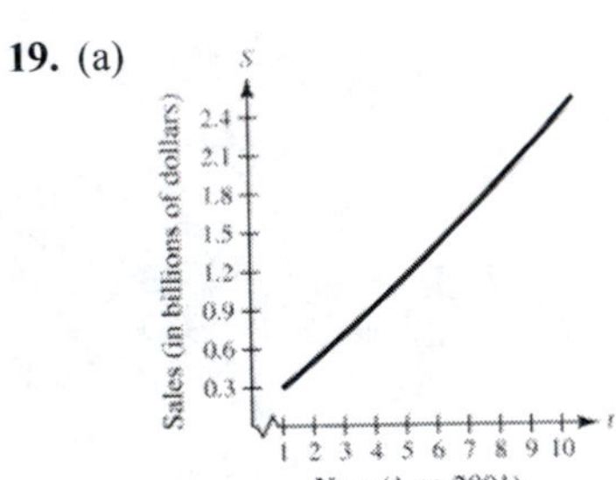

(b) Using the zoom and trace features, sales were $1.2 million during the year 2005.

21. $2(x - 2) = 2x - 4$

$2x - 4 = 2x - 4$

$0 = 0$ Identity

All real numbers are solutions.

23. $2(x + 3) = 2x - 2$

$2x + 6 = 2x - 2$

$6 = -2$ Contradiction

No solution

25. $8x - 5 = 3x + 20$

$5x = 25$

$x = 5$

27. $2(x + 5) - 7 = 3(x - 2)$

$2x + 10 - 7 = 3x - 6$

$2x + 3 = 3x - 6$

$-x = -9$

$x = 9$

29. $\dfrac{x}{5} - 3 = \dfrac{2x}{2} + 1$

$5\left(\dfrac{x}{5} - 3\right) = (x + 1)5$

$x - 15 = 5x + 5$

$-4x = 20$

$x = -5$

31. $y = 3x - 1$

x-intercept: $0 = 3x - 1 \Rightarrow x = \frac{1}{3}$

y-intercept: $y = 3(0) - 1 \Rightarrow y = -1$

The x-intercept is $\left(\frac{1}{3}, 0\right)$ and the y-intercept is $(0, -1)$.

33. $y = 2(x - 4)$

x-intercept: $0 = 2(x - 4) \Rightarrow x = 4$

y-intercept: $y = 2(0 - 4) \Rightarrow y = -8$

The x-intercept is $(4, 0)$ and the y-intercept is $(0, -8)$.

35. $y = -\frac{1}{2}x + \frac{2}{3}$

x-intercept: $0 = -\frac{1}{2}x + \frac{2}{3} \Rightarrow x = \frac{2/3}{1/2} = \frac{4}{3}$

y-intercept: $y = -\frac{1}{2}(0) + \frac{2}{3} \Rightarrow y = \frac{2}{3}$

The x-intercept is $\left(\frac{4}{3}, 0\right)$ and the y-intercept is $\left(0, \frac{2}{3}\right)$.

37. $244.92 = 2(3.14)(3)^2 + 2(3.14)(3)h$

$$244.92 = 56.52 + 18.84h$$
$$188.40 = 18.84h$$
$$10 = h$$

The height is 10 inches.

39. *Verbal Model:* September's profit + October's profit = 689,000

Labels: Let x = September's profit. Then $x + 0.12x$ = October's profit.

Equation:
$$x + (x + 0.12x) = 689{,}000$$
$$2.12x = 689{,}000$$
$$x = 325{,}000$$
$$x + 0.12x = 364{,}000$$

So, September profit was \$325,000 and October profit was $325{,}000 + 0.12(325{,}000) = \$364{,}000$.

41. Let x = the number of original investors.

Each person's share is $\frac{90{,}000}{x}$. If three more people invest, each person's share is $\frac{90{,}000}{x+3}$.

Since this is \$2500 less than the original cost, we have:

$$\frac{90{,}000}{x} - 2500 = \frac{90{,}000}{x+3}$$
$$90{,}000(x + 3) - 2500x(x + 3) = 90{,}000x$$
$$90{,}000x + 270{,}000 - 2500x^2 - 7500x = 90{,}000x$$
$$-2500x^2 - 7500x + 270{,}000 = 0$$
$$-2500(x^2 + 3x - 108) = 0$$
$$-2500(x + 12)(x - 9) = 0$$

$x = -12$, extraneous or $x = 9$

There are currently nine investors.

43. Let x = the number of liters of pure antifreeze.

$$30\% \text{ of } (10 - x) + 100\% \text{ of } x = 50\% \text{ of } 10$$
$$0.30(10 - x) + 1.00x = 0.50(10)$$
$$3 - 0.30x + 1.00x = 5$$
$$0.70x = 2$$
$$x = \frac{2}{0.70} = \frac{20}{7} = 2\frac{6}{7} \text{ liters}$$

45.
$$V = \frac{1}{3}\pi r^2 h$$
$$3V = \pi r^2 h$$
$$\frac{3V}{\pi r^2} = h$$

47. $15 + x - 2x^2 = 0$

$$0 = 2x^2 - x - 15$$
$$0 = (2x + 5)(x - 3)$$
$$2x + 5 = 0 \Rightarrow x = -\frac{5}{2}$$
$$x - 3 = 0 \Rightarrow x = 3$$

49.
$$\begin{aligned} 6 &= 3x^2 \\ 2 &= x^2 \\ \pm\sqrt{2} &= x \end{aligned}$$

51.
$$\begin{aligned} (x + 13)^2 &= 25 \\ x + 13 &= \pm 5 \\ x &= -13 \pm 5 \\ x &= -18 \text{ or } x = -8 \end{aligned}$$

53. $x^2 + 12x + 250 = 0$
$$\begin{aligned} x &= \frac{-12 \pm \sqrt{12^2 - 4(1)(25)}}{2(1)} \\ &= \frac{-12 \pm 2\sqrt{11}}{2} \\ &= -6 \pm \sqrt{11} \end{aligned}$$

55. $-2x^2 - 5x + 27 = 0$
$$\begin{aligned} 2x^2 + 5x - 27 &= 0 \\ x &= \frac{-5 \pm \sqrt{5^2 - 4(2)(-27)}}{2(2)} \\ &= \frac{-5 \pm \sqrt{241}}{4} \end{aligned}$$

57. $M = 500x(20 - x)$

(a) $500x(20 - x) = 0$ when $x = 0$ feet and $x = 20$ feet.

(b)

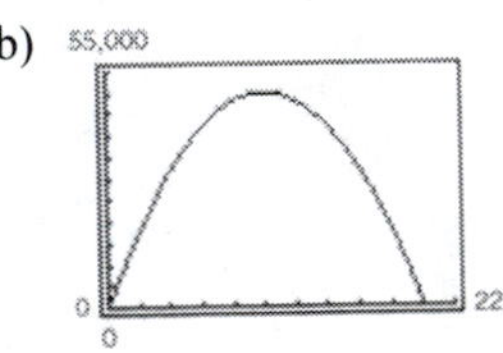

(c) The bending moment is greatest when $x = 10$ feet.

59. $4 + \sqrt{-9} = 4 + 3i$

61. $i^2 + 3i = -1 + 3i$

63. $(7 + 5i) + (-4 + 2i) = (7 - 4) + (5i + 2i) = 3 + 7i$

65. $6i(5 - 2i) = 30i - 12i^2 = 12 + 30i$

67.
$$\begin{aligned} \frac{6 - 5i}{i} &= \frac{6 - 5i}{i} \cdot \frac{-i}{-i} \\ &= \frac{-6i + 5i^2}{-i^2} \\ &= -5 - 6i \end{aligned}$$

69.
$$\begin{aligned} \frac{4}{2 - 3i} + \frac{2}{1 + i} &= \frac{4}{2 - 3i} \cdot \frac{2 + 3i}{2 + 3i} + \frac{2}{1 + i} \cdot \frac{1 - i}{1 - i} \\ &= \frac{8 + 12i}{4 + 9} + \frac{2 - 2i}{1 + 1} \\ &= \frac{8}{13} + \frac{12}{13}i + 1 - i \\ &= \left(\frac{8}{13} + 1\right) + \left(\frac{12}{13}i - i\right) \\ &= \frac{21}{13} - \frac{1}{13}i \end{aligned}$$

71. $x^2 - 2x + 10 = 0$
$$\begin{aligned} x &= \frac{-b \pm \sqrt{b^2 - 4ac}}{2a} \\ &= \frac{-(-2) \pm \sqrt{(-2)^2 - 4(1)(10)}}{2(1)} \\ &= \frac{2 \pm \sqrt{-36}}{2} \\ &= \frac{2 \pm 6i}{2} \\ &= 1 \pm 3i \end{aligned}$$

73. $4x^2 + 4x + 7 = 0$
$$\begin{aligned} x &= \frac{-b \pm \sqrt{b^2 - 4ac}}{2a} \\ &= \frac{-4 \pm \sqrt{(4)^2 - 4(4)(7)}}{2(4)} \\ &= \frac{-4 \pm \sqrt{-96}}{8} \\ &= \frac{-4 \pm 4\sqrt{6}i}{8} \\ &= -\frac{1}{2} \pm \frac{\sqrt{6}}{2}i \end{aligned}$$

75.
$$\begin{aligned} 5x^4 - 12x^3 &= 0 \\ x^3(5x - 12) &= 0 \\ x^3 = 0 \text{ or } 5x - 12 &= 0 \\ x = 0 \text{ or } \quad x &= \tfrac{12}{5} \end{aligned}$$

77.
$$\begin{aligned} x^4 - 5x^2 + 6 &= 0 \\ (x^2 - 2)(x^2 - 3) &= 0 \end{aligned}$$
$$\begin{aligned} x^2 - 2 &= 0 \quad \text{or} \quad & x^2 - 3 &= 0 \\ x^2 &= 2 & x^2 &= 3 \\ x &= \pm\sqrt{2} & x &= \pm\sqrt{3} \end{aligned}$$

79. $\sqrt{2x+3} + \sqrt{x-2} = 2$

$$\left(\sqrt{2x+3}\right)^2 = \left(2 - \sqrt{x-2}\right)^2$$
$$2x + 3 = 4 - 4\sqrt{x-2} + x - 2$$
$$x + 1 = -4\sqrt{x-2}$$
$$(x+1)^2 = \left(-4\sqrt{x-2}\right)^2$$
$$x^2 + 2x + 1 = 16(x-2)$$
$$x^2 - 14x + 33 = 0$$
$$(x-3)(x-11) = 0$$

$x = 3$, extraneous or $x = 11$, extraneous

No solution

81. $(x-1)^{2/3} - 25 = 0$

$$(x-1)^{2/3} = 25$$
$$(x-1)^2 = 25^3$$
$$x - 1 = \pm\sqrt{25^3}$$
$$x = 1 \pm 125$$
$$x = 126 \text{ or } x = -124$$

83. $\dfrac{5}{x} = 1 + \dfrac{3}{x+2}$

$$5(x+2) = 1(x)(x+2) + 3x$$
$$5x + 10 = x^2 + 2x + 3x$$
$$10 = x^2$$
$$\pm\sqrt{10} = x$$

85. $|x - 5| = 10$

$x - 5 = -10$ or $x - 5 = 10$

$x = -5 \qquad x = 15$

87. $|x^2 - 3| = 2x$ or $x^2 - 3 = 2x$

$x^2 - 2x - 3 = 0$	$x2 + 2x - 3 = 0$
$(x-3)(x+1) = 0$	$(x+3)(x-1) = 0$
$x = 3$ or $x = -1$	$x = -3$ or $x = 1$

The only solutions of the original equation are $x = 3$ or $x = 1$. ($x = 3$ and $x = -1$ are extraneous.)

89.

$$29.95 = 42 - \sqrt{0.001x + 2}$$
$$-12.05 = -\sqrt{0.001x + 2}$$
$$\sqrt{0.001x + 2} = 12.05$$
$$0.001x + 2 = 145.2025$$
$$0.001x = 143.2025$$
$$x = 143{,}202.5$$
$$\approx 143{,}203 \text{ units}$$

91. Interval: $(-7, 2]$

Inequality: $-7 < x \le 2$

The interval is bounded.

93. Interval: $(-\infty, -10]$

Inequality: $x \le -10$

The interval is unbounded.

95. $3(x+2) + 7 < 2x - 5$

$$3x + 6 + 7 < 2x - 5$$
$$3x + 13 < 2x - 5$$
$$x < -18$$

$(-\infty\ -18)$

97. $4(5 - 2x) \le \frac{1}{2}(8 - x)$

$$20 - 8x \le 4 - \tfrac{1}{2}x$$
$$-\tfrac{15}{2}x \le -16$$
$$x \ge \tfrac{32}{15}$$

$\left[\frac{32}{15}, \infty\right)$

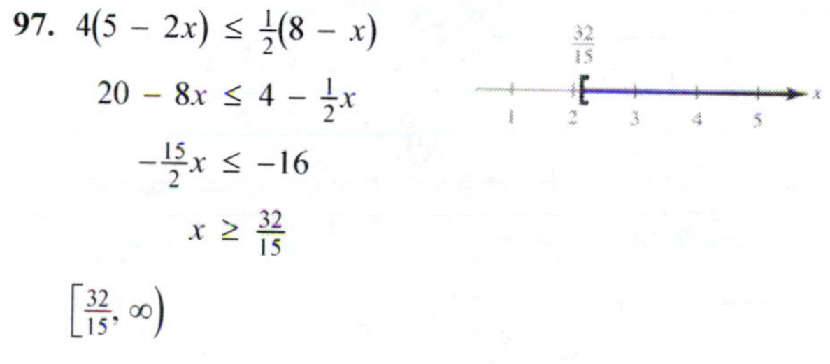

99. $|x - 3| > 4$

$x - 3 < -4$ or $x - 3 > 4$

$x < -1$ or $x > 7$

$(-\infty, -1) \cup (7, \infty)$

101. If the side is 19.3 cm, then with the possible error of 0.5 cm we have:

$$18.8 \le \text{side} \le 19.8$$
$$353.44 \text{ cm}^2 \le \text{area} \le 392.04 \text{ cm}^2$$

103. $x^2 - 6x - 27 < 0$

$(x + 3)(x - 9) < 0$

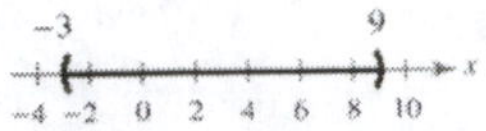

Key numbers: $x = -3, x = 9$

Test intervals: $(-\infty, -3), (-3, 9), (9, \infty)$

Test: Is $(x + 3)(x - 9) < 0$?

By testing an x-value in each test interval in the inequality, we see that the solution set is $(-3, 9)$.

105. $6x^2 + 5x < 4$

$6x^2 + 5x - 4 < 0$

$(3x + 4)(2x - 1) < 0$

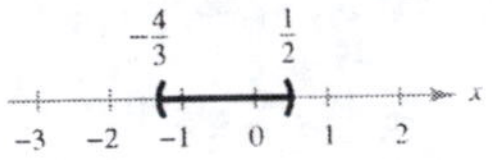

Key numbers: $x = -\frac{4}{3}, x = \frac{1}{2}$

Test intervals: $\left(-\infty, -\frac{4}{3}\right), \left(-\frac{4}{3}, \frac{1}{2}\right), \left(\frac{1}{2}, \infty\right)$

Test: Is $(3x + 4)(2x - 1) < 0$?

By testing an x-value in each test interval in the inequality, we see that the solution set is $\left(-\frac{4}{3}, \frac{1}{2}\right)$.

107. $\dfrac{2}{x + 1} \le \dfrac{3}{x - 1}$

$\dfrac{2(x - 1) - 3(x + 1)}{(x + 1)(x - 1)} \le 0$

$\dfrac{2x - 2 - 3x - 3}{(x + 1)(x - 1)} \le 0$

$\dfrac{-(x + 5)}{(x + 1)(x - 1)} \le 0$

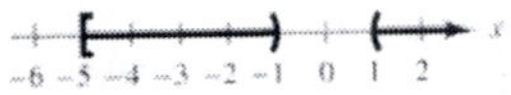

Key numbers: $x = -5, x = -1, x = 1$

Test intervals: $(-5, -1), (-1, 1), (1, \infty)$

Test: Is $\dfrac{-(x + 5)}{(x + 1)(x - 1)} \le 0$?

By testing an x-value in each test interval in the inequality, we see that the solution set is $[-5, -1) \cup (1, \infty)$.

109. $5000(1 + r)^2 > 5500$

$(1 + r)^2 > 1.1$

$1 + r > 1.0488$

$r > 0.0488$

$r > 4.9\%$

111. False.

$\sqrt{-18}\sqrt{-2} = \left(\sqrt{18}i\right)\left(\sqrt{2}i\right) = \sqrt{36}i^2 = -6$

$\sqrt{(-8)(-2)} = \sqrt{36} = 6$

113. Rational equations, equations involving radicals, and absolute value equations, may have "solutions" that are extraneous. So checking solutions, in the original equations, is crucial to eliminate these extraneous values.

Problem Solving for Chapter 1

1.

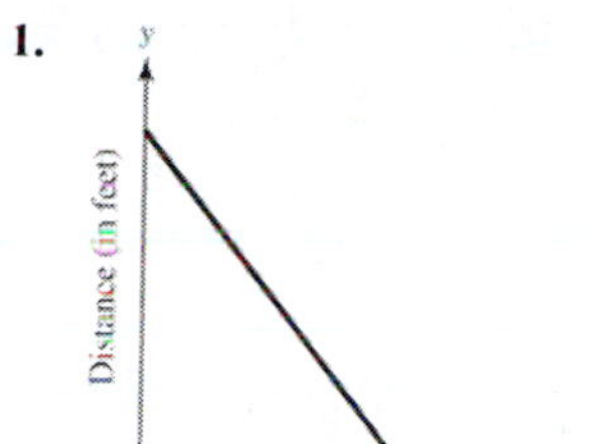

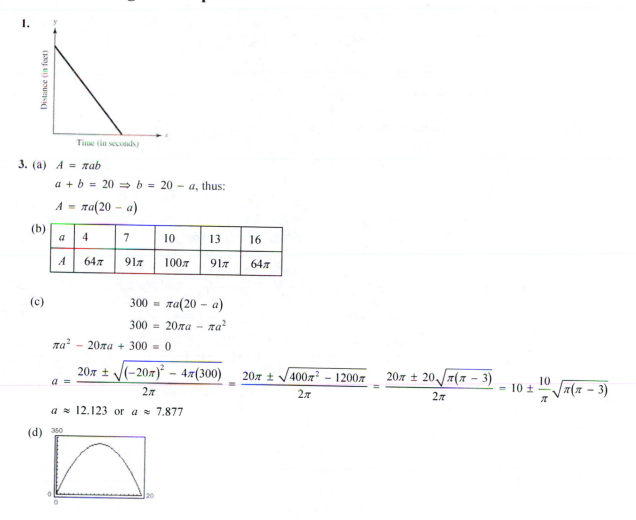

3. (a) $A = \pi ab$

$a + b = 20 \Rightarrow b = 20 - a$, thus:

$A = \pi a(20 - a)$

(b)

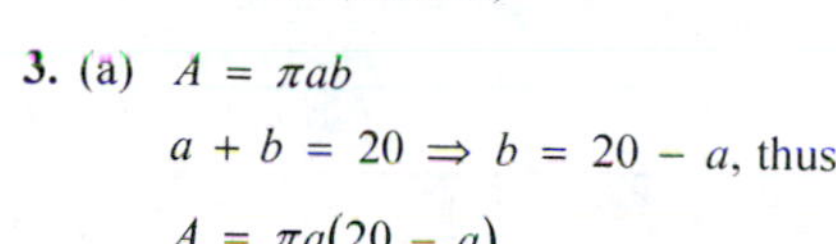

a	4	7	10	13	16
A	64π	91π	100π	91π	64π

(c)

$$300 = \pi a(20 - a)$$

$$300 = 20\pi a - \pi a^2$$

$$\pi a^2 - 20\pi a + 300 = 0$$

$$a = \frac{20\pi \pm \sqrt{(-20\pi)^2 - 4\pi(300)}}{2\pi} = \frac{20\pi \pm \sqrt{400\pi^2 - 1200\pi}}{2\pi} = \frac{20\pi \pm 20\sqrt{\pi(\pi - 3)}}{2\pi} = 10 \pm \frac{10}{\pi}\sqrt{\pi(\pi - 3)}$$

$a \approx 12.123$ or $a \approx 7.877$

(d)

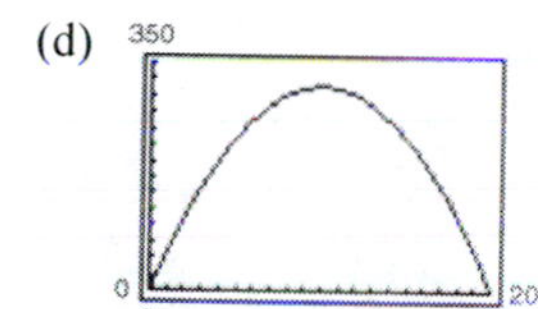

(e) The a-intercepts occur at $a = 0$ and $a = 20$. Both yield an area of 0. When $a = 0, b = 20$ and you have a vertical line of length 40. Likewise when $a = 20, b = 0$ and you have a horizontal line of length 40. They represent the minimum and maximum values of a.

(f) The maximum value of A is $100\pi \approx 314.159$. This occurs when $a = b = 10$ and the ellipse is actually a circle.

5. $h = \left(\sqrt{h_0} - \frac{2\pi d^2\sqrt{3}}{lw}t\right)^2$

$l = 60'', w = 30'', h_0 = 25'', d = 2''$

$$h = \left(5 - \frac{8\pi\sqrt{3}}{1800}t\right)^2 = \left(5 - \frac{\pi\sqrt{3}}{225}t\right)^2$$

(a) $12.5 = \left(5 - \frac{\pi\sqrt{3}}{225}t\right)^2$

$\sqrt{12.5} = 5 - \frac{\pi\sqrt{3}}{225}t$

$t = \frac{225}{\pi\sqrt{3}}\left(5 - \sqrt{12.5}\right) \approx 60.6$ seconds

(b) $0 = \left(\sqrt{12.5} - \frac{\pi\sqrt{3}}{225}t\right)^2$

$t = \frac{225\sqrt{12.5}}{\pi\sqrt{3}} \approx 146.2$ seconds

(c) The speed at which the water drains decreases as the amount of the water in the bathtub decreases.

7. (a) 5, 12, and 13; 8, 15, and 17

7, 24, and 25

(b) $5 \cdot 12 \cdot 13 = 780$ which is divisible by 3, 4, and 5.

$8 \cdot 15 \cdot 17 = 2040$ which is divisible by 3, 4, and 5.

$7 \cdot 24 \cdot 25 = 4200$ which is also divisible by 3, 4, and 5.

(c) Conjecture: If $a^2 + b^2 = c^2$ where a, b, and c are positive integers, then abc is divisible by 60.

9. (a)
$$S = \frac{-b + \sqrt{b^2 - 4ac}}{2a} + \frac{-b - \sqrt{b^2 - 4ac}}{2a}$$
$$= \frac{-2a}{2a}$$
$$= -\frac{b}{a}$$

(b)
$$P = \left(\frac{-b + \sqrt{b^2 - 4ac}}{2a}\right)\left(\frac{-b - \sqrt{b^2 - 4ac}}{2a}\right)$$
$$= \frac{b^2 - (b^2 - 4ac)}{4a^2}$$
$$= \frac{4ac}{4a^2}$$
$$= \frac{c}{a}$$

11. (a)
$$z_m = \frac{1}{z}$$
$$= \frac{1}{1 + i} = \frac{1}{1 + i} \cdot \frac{1 - i}{1 - i}$$
$$= \frac{1 - i}{2} = \frac{1}{2} - \frac{1}{2}i$$

(b)
$$z_m = \frac{1}{z}$$
$$= \frac{1}{3 - i} = \frac{1}{3 - i} \cdot \frac{3 + i}{3 + i}$$
$$= \frac{3 + i}{10} = \frac{3}{10} + \frac{1}{10}i$$

(c)
$$z_m = \frac{1}{z}$$
$$= \frac{1}{-2 + 8i}$$
$$= \frac{1}{-2 + 8i} \cdot \frac{-2 - 8i}{-2 - 8i}$$
$$= \frac{-2 - 8i}{68} = -\frac{1}{34} - \frac{2}{17}i$$

13. (a) $c = 1$

The terms are:
$i, -1 + i, -i, -1 + i, -i, -1 + i, -i, -1 + i, -i, \ldots$

The sequence is bounded so $c = i$ *is* in the Mandelbrot Set.

(b) $c = -2$

The terms are:
$1 + i, 1 + 3i, -7 + 7i, 1 - 97i, -9407 - 1931i, \ldots$

The sequence is unbounded so $c = 1 + i$ *is not* in the Mandelbrot Set.

(c) $c = -2$

The terms are: $-2, 2, 2, 2, 2, \ldots$

The sequence is bounded so $c = -2$ *is* in the Mandelbrot Set.

15.
$$y = x^4 - x^3 - 6x^2 + 4x + 8$$
$$= (x - 2)^2(x + 1)(x + 2)$$

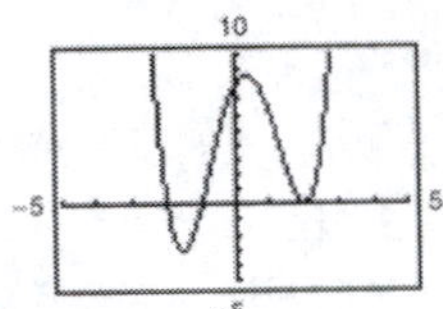

From the graph you see that $x^4 - x^3 - 6x^2 + 4x + 8 > 0$ on the intervals $(-\infty, -2) \cup (-1, 2) \cup (2, \infty)$.

Practice Test for Chapter 1

1. Graph $3x - 5y = 15$.

2. Graph $y = \sqrt{9 - x}$.

3. Solve $5x + 4 = 7x - 8$.

4. Solve $\frac{x}{3} - 5 = \frac{x}{5} + 1$.

5. Solve $\frac{3x + 1}{6x - 7} = \frac{2}{5}$.

6. Solve $(x - 3)^2 + 4 = (x + 1)^2$.

7. Solve $A = \frac{1}{2}(a + b)h$ for a.

8. 301 is what percent of 4300?

9. Cindy has \$6.05 in quarter and nickels. How many of each coin does she have if there are 53 coins in all?

10. Ed has \$15,000 invested in two fund paying $9\frac{1}{2}\%$ and 11% simple interest, respectively. How much is invested in each if the yearly interest is \$1582.50?

11. Solve $28 + 5x - 3x^2 = 0$ by factoring.

12. Solve $(x - 2)^2 = 24$ by taking the square root of both sides.

13. Solve $x^2 - 4x - 9 = 0$ by completing the square.

14. Solve $x^2 + 5x - 1 = 0$ by the Quadratic Formula.

15. Solve $3x^2 - 2x + 4 = 0$ by the Quadratic Formula.

16. The perimeter of a rectangle is 1100 feet. Find the dimensions so that the enclosed area will be 60,000 square feet.

17. Find two consecutive even positive integers whose product is 624.

18. Solve $x^3 - 10x^2 + 24x = 0$ by factoring.

19. Solve $\sqrt[3]{6 - x} = 4$.

20. Solve $(x^2 - 8)^{2/5} = 4$.

21. Solve $x^4 - x^2 - 12 = 0$.

22. Solve $4 - 3x > 16$.

23. Solve $\left|\frac{x - 3}{2}\right| < 5$.

24. Solve $\frac{x + 1}{x - 3} < 2$.

25. Solve $|3x - 4| \geq 9$.

CHAPTER 2
Functions and Their Graphs

CHAPTER 2

Functions and Their Graphs

Section 2.1 Linear Equations in Two Variables

1. linear 

3. point-slope

5. perpendicular

7. linear extrapolation

9. (a) $m = \frac{2}{3}$. Because the slope is positive, the line rises. Matches L_2.

   (b) m is undefined. The line is vertical. Matches L_3.

   (c) $m = -2$. The line falls. Matches L_1.

11.

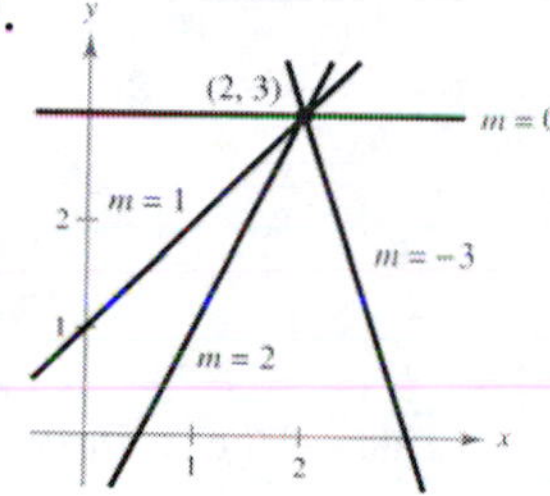

13. Two points on the line: $(0, 0)$ and $(4, 6)$

$$\text{Slope} = \frac{y_2 - y_1}{x_2 - x_1} = \frac{6}{4} = \frac{3}{2}$$

15. $y = 5x + 3$

    Slope: $m = 5$

    y-intercept: $(0, 3)$

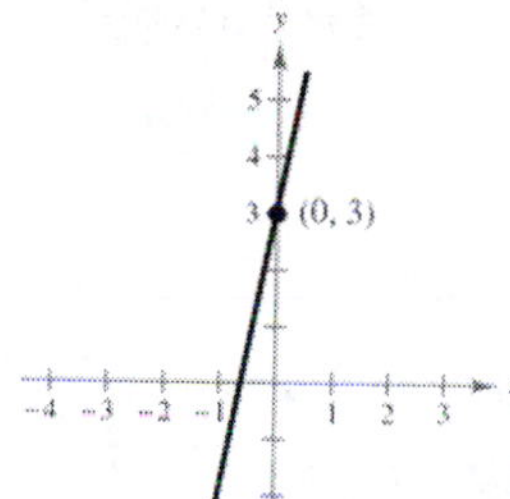

17. $y = -\frac{1}{2}x + 4$

    Slope: $m = -\frac{1}{2}$

    y-intercept: $(0, 4)$

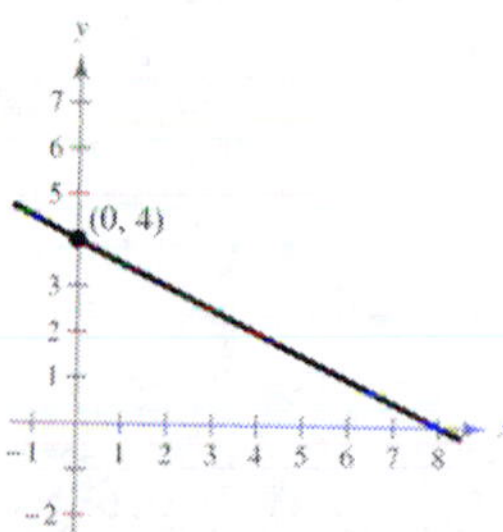

19. $y - 3 = 0$

    $y = 3$, horizontal line

    Slope: $m = 0$

    y-intercept: $(0, 3)$

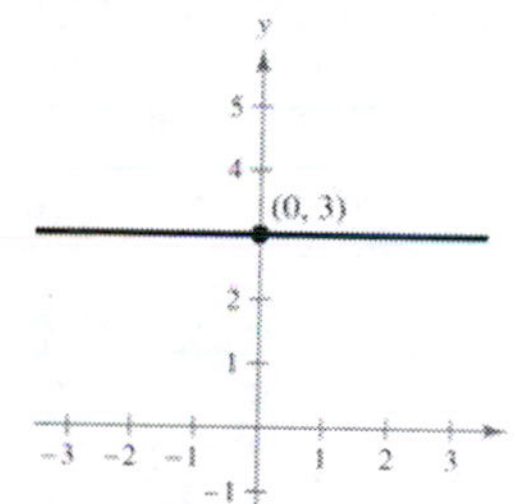

21. $5x - 2 = 0$

    $x = \frac{2}{5}$, vertical line

    Slope: undefined

    No y-intercept

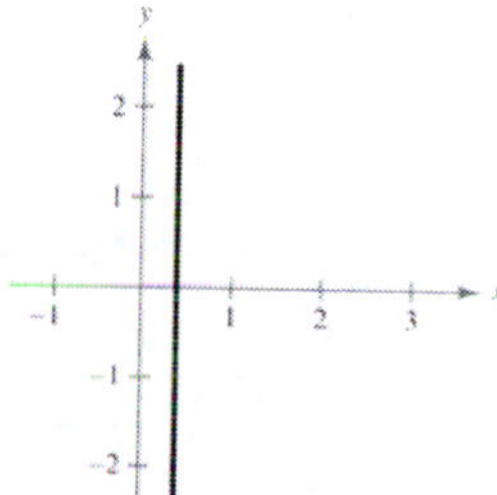

23. $7x - 6y = 30$

$-6y = -7x + 30$

$y = \frac{7}{6}x - 5$

Slope: $m = \frac{7}{6}$

y-intercept: $(0, -5)$

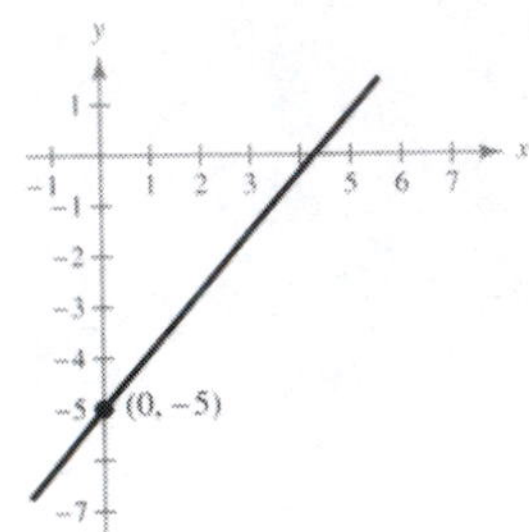

25. $m = \dfrac{0 - 9}{6 - 0} = \dfrac{-9}{6} = -\dfrac{3}{2}$

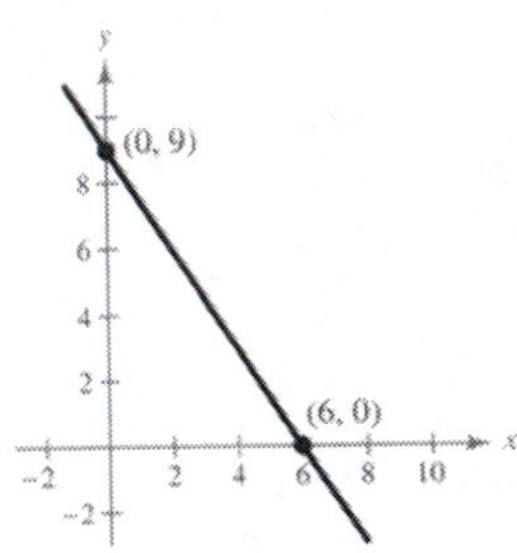

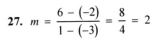

27. $m = \dfrac{6 - (-2)}{1 - (-3)} = \dfrac{8}{4} = 2$

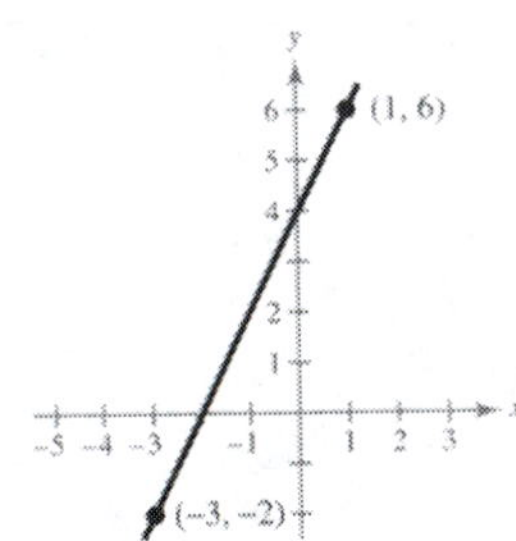

29. $m = \dfrac{-7 - (-7)}{8 - 5} = \dfrac{0}{3} = 0$

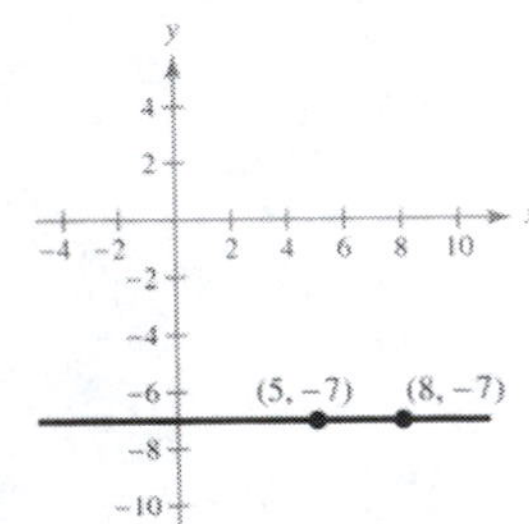

31. $m = \dfrac{4 - (-1)}{-6 - (-6)} = \dfrac{5}{0}$

m is undefined.

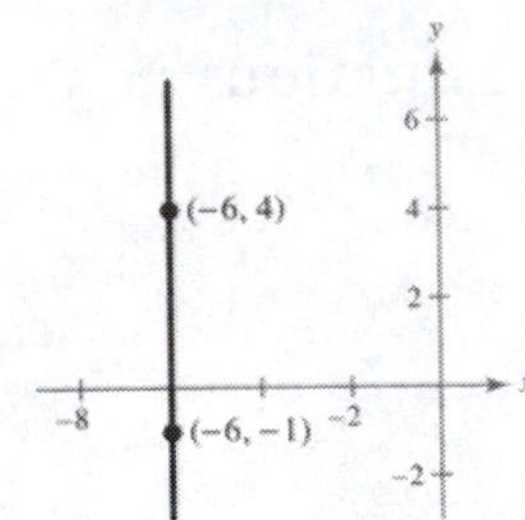

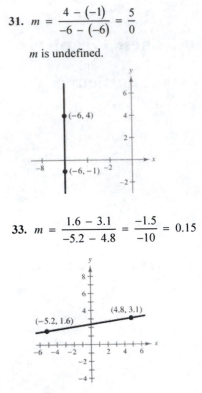

33. $m = \dfrac{1.6 - 3.1}{-5.2 - 4.8} = \dfrac{-1.5}{-10} = 0.15$

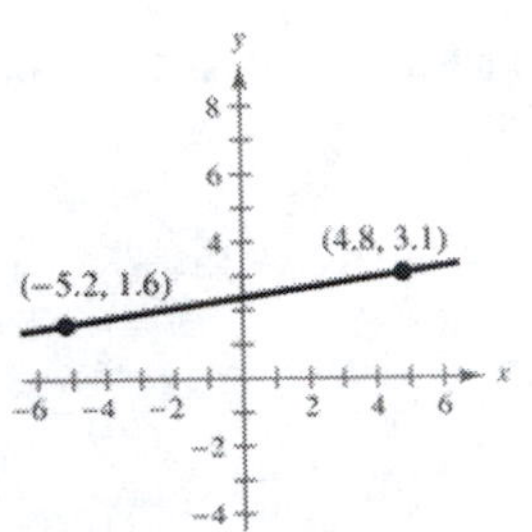

35. Point: $(2, 1)$, Slope: $m = 0$

Because $m = 0$, y does not change. Three points are $(0, 1)$, $(3, 1)$, and $(-1, 1)$.

37. Point: $(-8, 1)$, Slope is undefined.

Because m is undefined, x does not change. Three points are $(-8, 0)$, $(-8, 2)$, and $(-8, 3)$.

39. Point: $(-5, 4)$, Slope: $m = 2$

Because $m = 2 = \frac{2}{1}$, y increases by 2 for every one unit increase in x. Three additional points are $(-4, 6)$, $(-3, 8)$, and $(-2, 10)$.

41. Point: $(-1, -6)$, Slope: $m = -\frac{1}{2}$

Because $m = -\frac{1}{2}$, y decreases by 1 unit for every two unit increase in x. Three additional points are $(1, -7)$, $(3, -8)$, and $(-13, 0)$.

43. Point: $(0, -2)$; $m = 3$

$$y + 2 = 3(x - 0)$$
$$y = 3x - 2$$

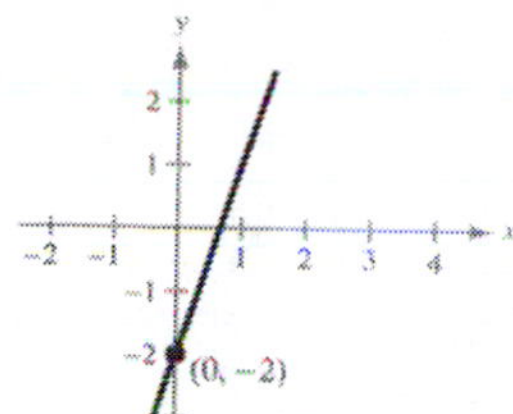

45. Point: $(-3, 6)$; $m = -2$

$$y - 6 = -2(x + 3)$$
$$y = -2x$$

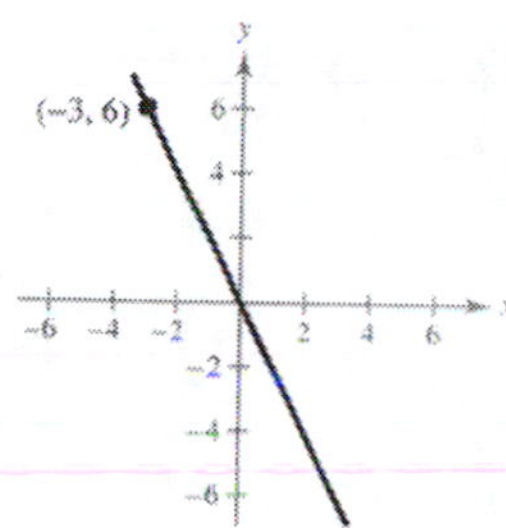

47. Point: $(4, 0)$; $m = -\frac{1}{3}$

$$y - 0 = -\tfrac{1}{3}(x - 4)$$
$$y = -\tfrac{1}{3}x + \tfrac{4}{3}$$

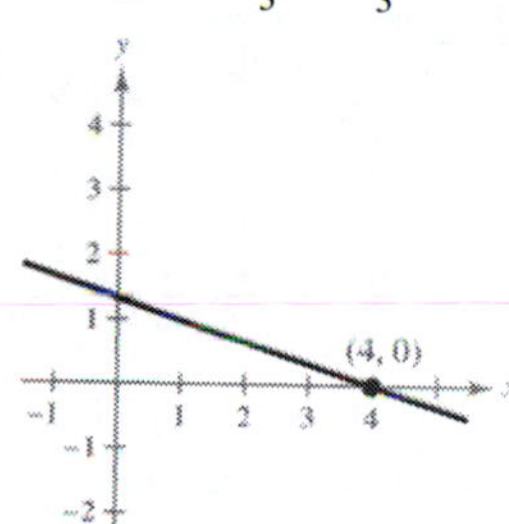

49. Point: $(2, -3)$; $m = -\dfrac{1}{2}$

$$y - (-3) = -\frac{1}{2}(x - 2)$$
$$y + 3 = -\frac{1}{2}x + 1$$
$$y = -\frac{1}{2}x - 2$$

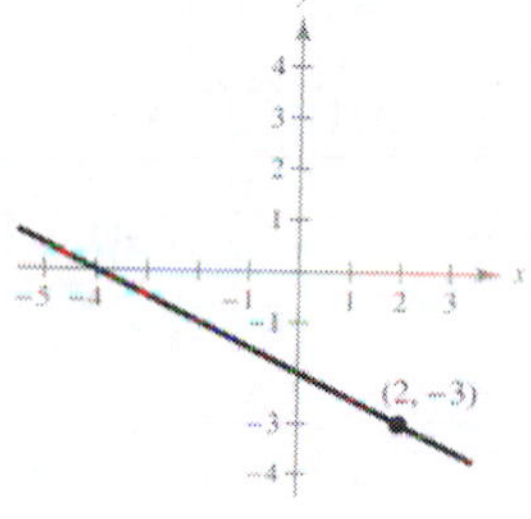

51. Point: $(6, -1)$; m is undefined.

Because the slope is undefined, the line is a vertical line.

$x = 6$

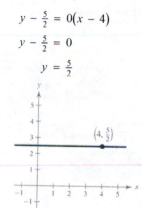

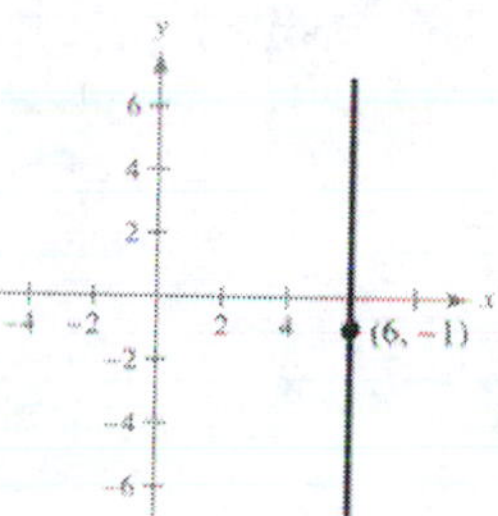

53. Point: $\left(4, \frac{5}{2}\right)$; $m = 0$

$$y - \tfrac{5}{2} = 0(x - 4)$$
$$y - \tfrac{5}{2} = 0$$
$$y = \tfrac{5}{2}$$

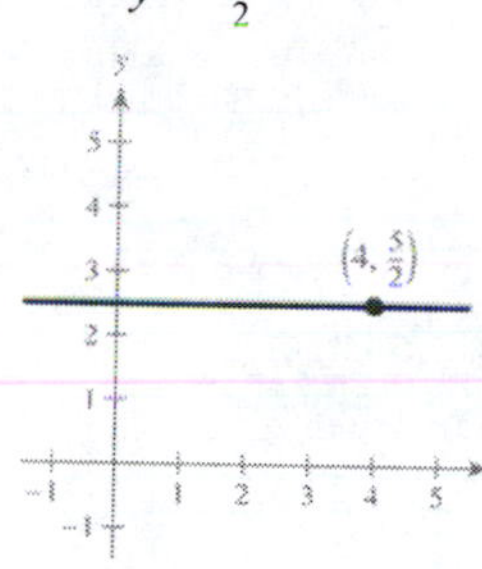

55. $(5, -1), (-5, 5)$

$$y + 1 = \frac{5 + 1}{-5 - 5}(x - 5)$$
$$y = -\frac{3}{5}(x - 5) - 1$$
$$y = -\frac{3}{5}x + 2$$

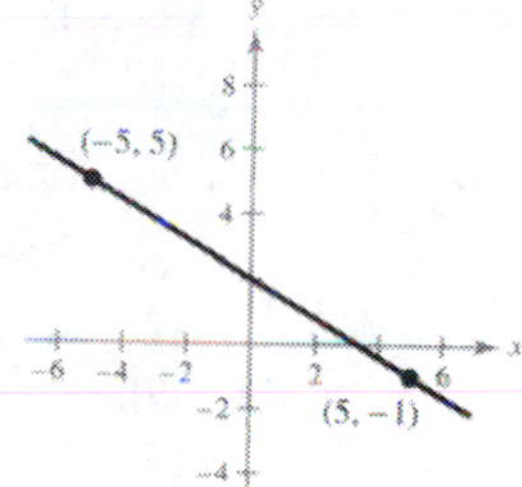

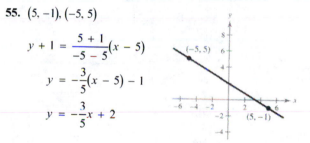

57. $(-8, 1), (-8, 7)$

Because both points have $x = -8$, the slope is undefined, and the line is vertical.

$x = -8$

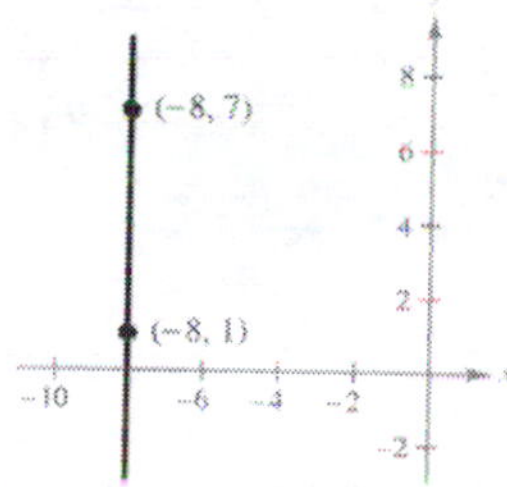

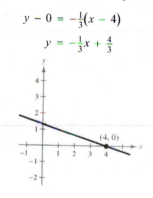

59. $\left(2, \frac{1}{2}\right), \left(\frac{1}{2}, \frac{5}{4}\right)$

$$y - \frac{1}{2} = \frac{\frac{5}{4} - \frac{1}{2}}{\frac{1}{2} - 2}(x - 2)$$

$$y = -\frac{1}{2}(x - 2) + \frac{1}{2}$$

$$y = -\frac{1}{2}x + \frac{3}{2}$$

61. $(1, 0.6), (-2, -0.6)$

$$y - 0.6 = \frac{-0.6 - 0.6}{-2 - 1}(x - 1)$$

$$y = 0.4(x - 1) + 0.6$$

$$y = 0.4x + 0.2$$

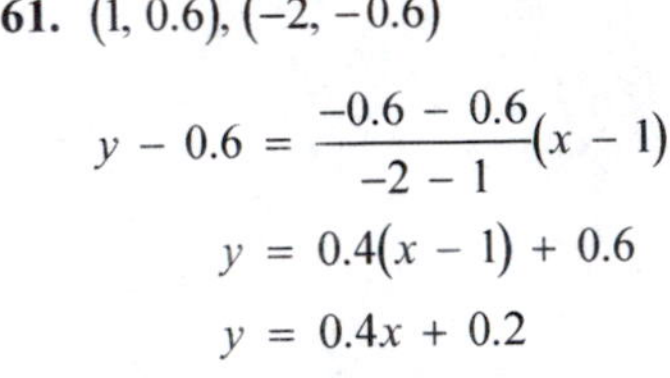

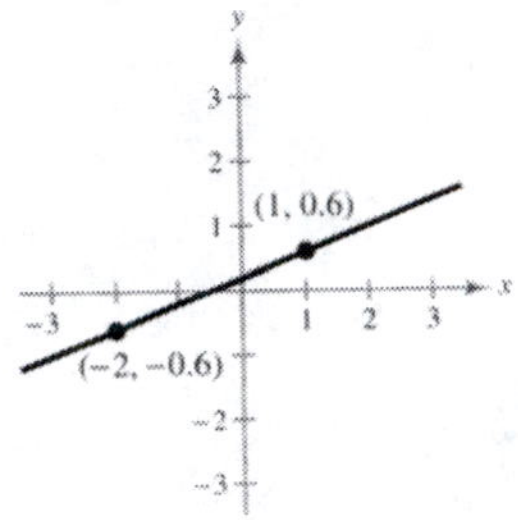

63. $(2, -1), \left(\frac{1}{3}, -1\right)$

$$y + 1 = \frac{-1 - (-1)}{\frac{1}{3} - 2}(x - 2)$$

$$y + 1 = 0$$

$$y = -1$$

The line is horizontal.

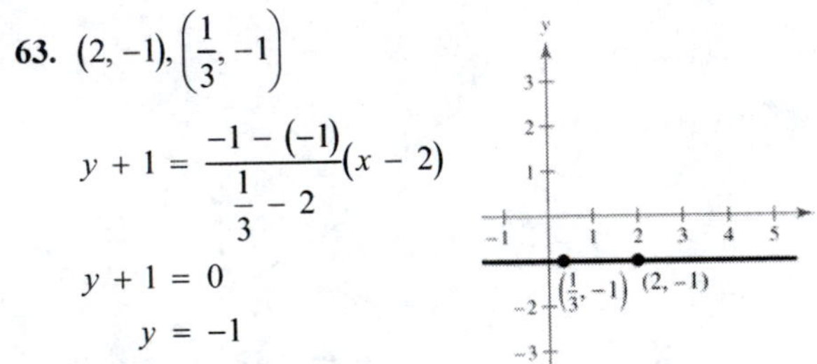

65. $L_1 : y = \frac{1}{3}x - 2$

$$m_1 = \frac{1}{3}$$

$L_2 : y = \frac{1}{3}x + 3$

$$m_2 = \frac{1}{3}$$

The lines are parallel.

67. $L_1 : y = \frac{1}{2}x - 3$

$$m_1 = \frac{1}{2}$$

$L_2 : y = -\frac{1}{2}x + 1$

$$m_2 = -\frac{1}{2}$$

The lines are neither parallel nor perpendicular.

69. $L_1 : (0, -1), (5, 9)$

$$m_1 = \frac{9 + 1}{5 - 0} = 2$$

$L_2 : (0, 3), (4, 1)$

$$m_2 = \frac{1 - 3}{4 - 0} = -\frac{1}{2}$$

The lines are perpendicular.

71. $L_1 : (3, 6), (-6, 0)$

$$m_1 = \frac{0 - 6}{-6 - 3} = \frac{2}{3}$$

$L_2 : (0, -1), \left(5, \frac{7}{3}\right)$

$$m_2 = \frac{\frac{7}{3} + 1}{5 - 0} = \frac{2}{3}$$

The lines are parallel.

73. $4x - 2y = 3$

$$y = 2x - \frac{3}{2}$$

Slope: $m = 2$

(a) $(2, 1), m = 2$

$$y - 1 = 2(x - 2)$$

$$y = 2x - 3$$

(b) $(2, 1), m = -\frac{1}{2}$

$$y - 1 = -\frac{1}{2}(x - 2)$$

$$y = -\frac{1}{2}x + 2$$

75. $3x + 4y = 7$

$$y = -\frac{3}{4}x + \frac{7}{4}$$

Slope: $m = -\frac{3}{4}$

(a) $\left(-\frac{2}{3}, \frac{7}{8}\right), m = -\frac{3}{4}$

$$y - \frac{7}{8} = -\frac{3}{4}\left(x - \left(-\frac{2}{3}\right)\right)$$

$$y = -\frac{3}{4}x + \frac{3}{8}$$

(b) $\left(-\frac{2}{3}, \frac{7}{8}\right), m = \frac{4}{3}$

$$y - \frac{7}{8} = \frac{4}{3}\left(x - \left(-\frac{2}{3}\right)\right)$$

$$y = \frac{4}{3}x + \frac{127}{72}$$

77. $y + 3 = 0$

$y = -3$

Slope: $m = 0$

(a) $(-1, 0), m = 0$

$y = 0$

(b) $(-1, 0)$, m is undefined.

$x = -1$

79. $x - y = 4$

$y = x - 4$

Slope: $m = 1$

(a) $(2.5, 6.8), m = 1$

$y - 6.8 = 1(x - 2.5)$

$y = x + 4.3$

(b) $(2.5, 6.8), m = -1$

$y - 6.8 = (-1)(x - 2.5)$

$y = -x + 9.3$

81. $\frac{x}{2} + \frac{y}{3} = 1$

$3x + 2y - 6 = 0$

83. $\frac{x}{-1/6} + \frac{y}{-2/3} = 1$

$6x + \frac{3}{2}y = -1$

$12x + 3y + 2 = 0$

85. $\frac{x}{c} + \frac{y}{c} = 1, c \neq 0$

$x + y = c$

$1 + 2 = c$

$3 = c$

$x + y = 3$

$x + y - 3 = 0$

87. (a) $m = 135$. The sales are increasing 135 units per year.

(b) $m = 0$. There is no change in sales during the year.

(c) $m = -40$. The sales are decreasing 40 units per year.

89. $y = \frac{6}{100}x$

$y = \frac{6}{100}(200) = 12$ feet

91. $(10, 2540), m = -125$

$V - 2540 = -125(t - 10)$

$V - 2540 = -125t + 1250$

$V = -125t + 3790, 5 \le t \le 10$

93. The C-intercept measures the fixed costs of manufacturing when zero bags are produced.

The slope measures the cost to produce one laptop bag.

95. Using the points $(0, 875)$ and $(5, 0)$, where the first coordinate represents the year t and the second coordinate represents the value V, you have

$m = \frac{0 - 875}{5 - 0} = -175$

$V = -175t + 875, 0 \le t \le 5.$

97. Using the points $(0, 32)$ and $(100, 212)$, where the first coordinate represents a temperature in degrees Celsius and the second coordinate represents a temperature in degrees Fahrenheit, you have

$m = \frac{212 - 32}{100 - 0} = \frac{180}{100} = \frac{9}{5}.$

Since the point $(0, 32)$ is the F- intercept, $b = 32$, the equation is $F = \frac{9}{5}C + 32.$

99. (a) Total Cost = cost for fuel and maintainance + cost for operator + purchase cost

$C = 9.5t + 11.5t + 42{,}000$

$C = 21.0t + 42{,}000$

(b) Revenue = Rate per hour · Hours

$R = 45t$

(c) $P = R - C$

$P = 45t - (21t + 42{,}000)$

$P = 24t - 42{,}000$

(d) Let $P = 0$, and solve for t.

$0 = 24t - 42{,}000$

$42{,}000 = 24t$

$1750 = t$

The equipment must be used 1750 hours to yield a profit of 0 dollars.

101. False. The slope with the greatest magnitude corresponds to the steepest line.

103. Find the slope of the line segments between the points A and B, and B and C.

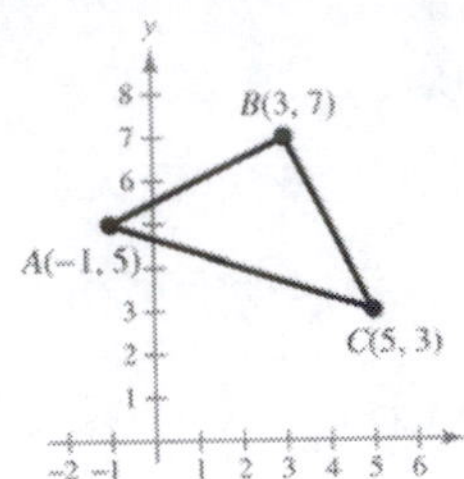

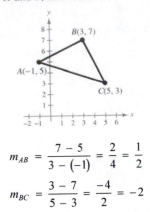

$$m_{AB} = \frac{7-5}{3-(-1)} = \frac{2}{4} = \frac{1}{2}$$

$$m_{BC} = \frac{3-7}{5-3} = \frac{-4}{2} = -2$$

Since the slopes are negative reciprocals, the line segments are perpendicular and therefore intersect to form a right angle. So, the triangle is a right triangle.

105. No. The slope cannot be determined without knowing the scale on the y-axis. The slopes will be the same if the scale on the y-axis of (a) is $2\frac{1}{2}$ and the scale on the y-axis of (b) is 1. Then the slope of both is $\frac{5}{4}$.

107. No, the slopes of two perpendicular lines have opposite signs. (Assume that neither line is vertical or horizontal.)

109. The line $y = 4x$ rises most quickly.

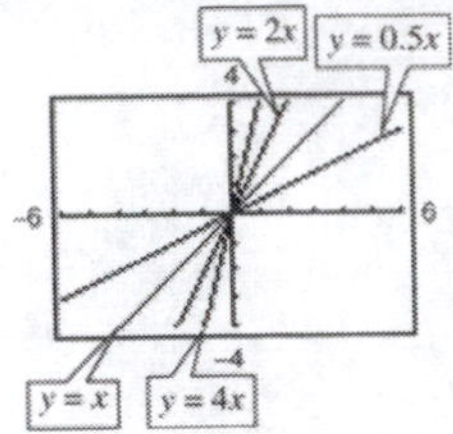

The line $y = -4x$ falls most quickly.

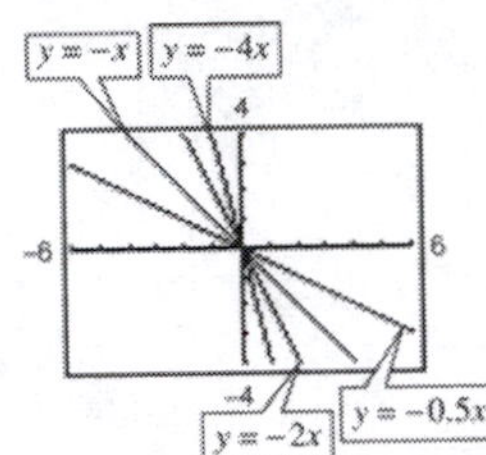

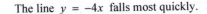

The greater the magnitude of the slope (the absolute value of the slope), the faster the line rises or falls.

111. Set the distance between $(4, -1)$ and (x, y) equal to the distance between $(-2, 3)$ and (x, y).

$$\sqrt{(x-4)^2 + [y-(-1)]^2} = \sqrt{[x-(-2)]^2 + (y-3)^2}$$
$$(x-4)^2 + (y+1)^2 = (x+2)^2 + (y-3)^2$$
$$x^2 - 8x + 16 + y^2 + 2y + 1 = x^2 + 4x + 4 + y^2 - 6y + 9$$
$$-8x + 2y + 17 = 4x - 6y + 13$$
$$0 = 12x - 8y - 4$$
$$0 = 4(3x - 2y - 1)$$
$$0 = 3x - 2y - 1$$

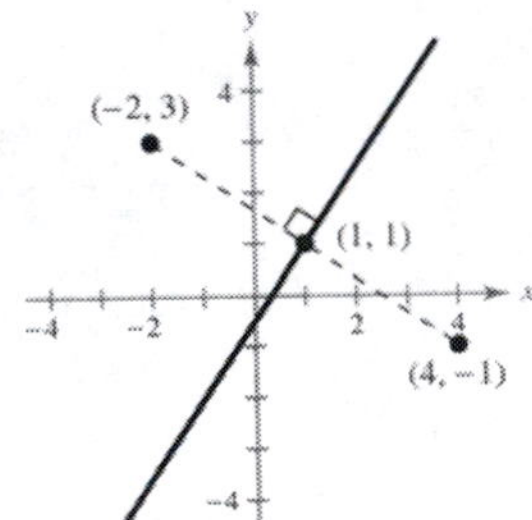

This line is the perpendicular bisector of the line segment connecting $(4, -1)$ and $(-2, 3)$.

113. Set the distance between $\left(3, \frac{5}{2}\right)$ and (x, y) equal to the distance between $(-7, 1)$ and (x, y).

$$\sqrt{(x-3)^2 + \left(y-\frac{5}{2}\right)^2} = \sqrt{[x-(-7)]^2 + (y-1)^2}$$
$$(x-3)^2 + \left(y-\frac{5}{2}\right)^2 = (x+7)^2 + (y-1)^2$$
$$x^2 - 6x + 9 + y^2 - 5y + \frac{25}{4} = x^2 + 14x + 49 + y^2 - 2y + 1$$
$$-6x - 5y + \frac{61}{4} = 14x - 2y + 50$$
$$-24x - 20y + 61 = 56x - 8y + 200$$
$$80x + 12y + 139 = 0$$

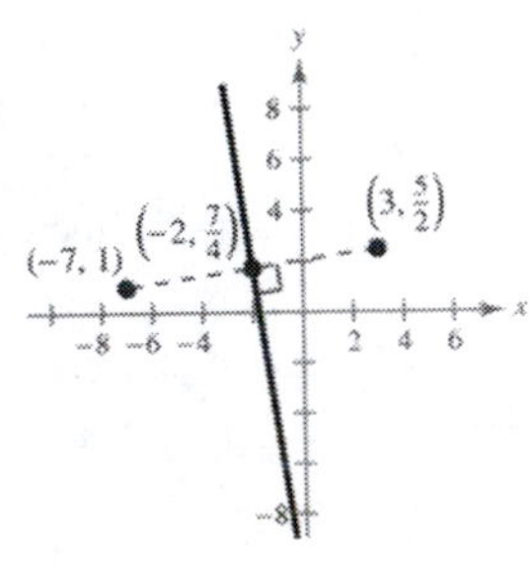

This line is the perpendicular bisector of the line segment connecting $\left(3, \frac{5}{2}\right)$ and $(-7, 1)$.

Section 2.2 Functions

1. domain; range; function

3. implied domain

5. Yes, the relationship is a function. Each domain value is matched with exactly one range value.

7. No, it does not represent a function. The input values of 10 and 7 are each matched with two output values.

9. (a) Each element of A is matched with exactly one element of B, so it does represent a function.
   (b) The element 1 in A is matched with two elements, -2 and 1 of B, so it does not represent a function.
   (c) Each element of A is matched with exactly one element of B, so it does represent a function.
   (d) The element 2 in A is not matched with an element of B, so the relation does not represent a function.

11. $x^2 + y^2 = 4 \Rightarrow y = \pm\sqrt{4 - x^2}$

    No, y *is not* a function of x.

13. $2x + 3y = 4 \Rightarrow y = \frac{1}{3}(4 - 2x)$

    Yes, y *is* a function of x.

15. $y = \sqrt{16 - x^2}$

    Yes, y *is* a function of x.

17. $y = |4 - x|$

    Yes, y *is* a function of x.

19. $y = -75$ or $y = -75 + 0x$

    Yes, y *is* a function of x.

21. $f(x) = 2x - 3$
   (a) $f(1) = 2(1) - 3 = -1$
   (b) $f(-3) = 2(-3) - 3 = -9$
   (c) $f(x - 1) = 2(x - 1) - 3 = 2x - 5$

23. $g(t) = 4t^2 - 3t + 5$
   (a) $g(2) = 4(2)^2 - 3(2) + 5$
   $= 15$
   (b) $g(t - 2) = 4(t - 2)^2 - 3(t - 2) + 5$
   $= 4t^2 - 19t + 27$
   (c) $g(t) - g(2) = 4t^2 - 3t + 5 - 15$
   $= 4t^2 - 3t - 10$

25. $f(y) = 3 - \sqrt{y}$
   (a) $f(4) = 3 - \sqrt{4} = 1$
   (b) $f(0.25) = 3 - \sqrt{0.25} = 2.5$
   (c) $f(4x^2) = 3 - \sqrt{4x^2} = 3 - 2|x|$

27. $q(x) = \dfrac{1}{x^2 - 9}$
   (a) $q(0) = \dfrac{1}{0^2 - 9} = -\dfrac{1}{9}$
   (b) $q(3) = \dfrac{1}{3^2 - 9}$ is undefined.
   (c) $q(y + 3) = \dfrac{1}{(y + 3)^2 - 9} = \dfrac{1}{y^2 + 6y}$

29. $f(x) = \dfrac{|x|}{x}$
   (a) $f(2) = \dfrac{|2|}{2} = 1$
   (b) $f(-2) = \dfrac{|-2|}{-2} = -1$
   (c) $f(x - 1) = \dfrac{|x - 1|}{x - 1} = \begin{cases} -1, & \text{if } x < 1 \\ 1, & \text{if } x > 1 \end{cases}$

31. $f(x) = \begin{cases} 2x + 1, & x < 0 \\ 2x + 2, & x \ge 0 \end{cases}$
   (a) $f(-1) = 2(-1) + 1 = -1$
   (b) $f(0) = 2(0) + 2 = 2$
   (c) $f(2) = 2(2) + 2 = 6$

33. $f(x) = x^2 - 3$

    $f(-2) = (-2)^2 - 3 = 1$

    $f(-1) = (-1)^2 - 3 = -2$

    $f(0) = (0)^2 - 3 = -3$

    $f(1) = (1)^2 - 3 = -2$

    $f(2) = (2)^2 - 3 = 1$

x	-2	-1	0	1	2
$f(x)$	1	-2	-3	-2	1

35. $f(x) = \begin{cases} -\frac{1}{2}x + 4, & x \le 0 \\ (x-2)^2, & x > 0 \end{cases}$

$f(-2) = -\frac{1}{2}(-2) + 4 = 5$

$f(-1) = -\frac{1}{2}(-1) + 4 = 4\frac{1}{2} = \frac{9}{2}$

$f(0) = -\frac{1}{2}(0) + 4 = 4$

$f(1) = (1-2)^2 = 1$

$f(2) = (2-2)^2 = 0$

x	-2	-1	0	1	2
$f(x)$	5	$\frac{9}{2}$	4	1	0

37. $15 - 3x = 0$

$3x = 15$

$x = 5$

39. $\dfrac{3x - 4}{5} = 0$

$3x - 4 = 0$

$x = \dfrac{4}{3}$

41. $x^2 - 9 = 0$

$x^2 = 9$

$x = \pm 3$

43. $x^3 - x = 0$

$x(x^2 - 1) = 0$

$x(x+1)(x-1) = 0$

$x = 0, x = -1$, or $x = 1$

45. $f(x) = g(x)$

$x^2 = x + 2$

$x^2 - x - 2 = 0$

$(x-2)(x+1) = 0$

$x - 2 = 0 \quad x + 1 = 0$

$x = 2 \quad x = -1$

47. $f(x) = g(x)$

$x^4 - 2x^2 = 2x^2$

$x^4 - 4x^2 = 0$

$x^2(x^2 - 4) = 0$

$x^2(x+2)(x-2) = 0$

$x^2 = 0 \Rightarrow x = 0$

$x + 2 = 0 \Rightarrow x = -2$

$x - 2 = 0 \Rightarrow x = 2$

49. $f(x) = 5x^2 + 2x - 1$

Because $f(x)$ is a polynomial, the domain is all real numbers x.

51. $h(t) = \dfrac{4}{t}$

The domain is all real numbers t except $t = 0$.

53. $g(y) = \sqrt{y - 10}$

Domain: $y - 10 \ge 0$

$y \ge 10$

The domain is all real numbers y such that $y \ge 10$.

55. $g(x) = \dfrac{1}{x} - \dfrac{3}{x+2}$

The domain is all real numbers x except $x = 0, x = -2$.

57. $f(s) = \dfrac{\sqrt{s-1}}{s-4}$

Domain: $s - 1 \ge 0 \Rightarrow s \ge 1$ and $s \ne 4$

The domain consists of all real numbers s, such that $s \ge 1$ and $s \ne 4$.

59. $f(x) = \dfrac{x - 4}{\sqrt{x}}$

The domain is all real numbers x such that $x > 0$ or $(0, \infty)$.

61. (a)

Height, x	Volume, V
1	484
2	800
3	972
4	1024
5	980
6	864

The volume is maximum when $x = 4$ and $V = 1024$ cubic centimeters.

(b)

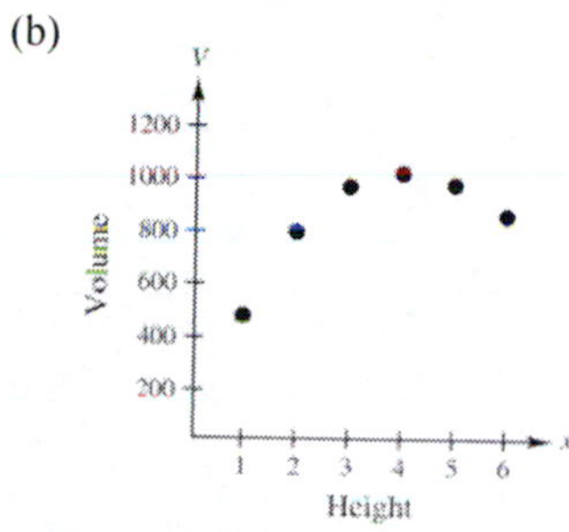

V is a function of x.

(c) $V = x(24 - 2x)^2$

Domain: $0 < x < 12$

63. $A = s^2$ and $P = 4s \Rightarrow \frac{P}{4} = s$

$$A = \left(\frac{P}{4}\right)^2 = \frac{P^2}{16}$$

65. $y = -\frac{1}{10}x^2 + 3x + 6$

$$y(30) = -\tfrac{1}{10}(30)^2 + 3(30) + 6 = 6 \text{ feet}$$

If the child holds a glove at a height of 5 feet, then the ball *will* be over the child's head because it will be at a height of 6 feet.

67. $A = \frac{1}{2}bh = \frac{1}{2}xy$

Because $(0, y)$, $(2, 1)$, and $(x, 0)$ all lie on the same line, the slopes between any pair are equal.

$$\frac{1 - y}{2 - 0} = \frac{0 - 1}{x - 2}$$

$$\frac{1 - y}{2} = \frac{-1}{x - 2}$$

$$y = \frac{2}{x - 2} + 1$$

$$y = \frac{x}{x - 2}$$

So, $A = \frac{1}{2}x\left(\frac{x}{x - 2}\right) = \frac{x^2}{2(x - 2)}$.

The domain of A includes x-values such that $x^2/[2(x - 2)] > 0$. By solving this inequality, the domain is $x > 2$.

69. For 2004 through 2007, use

$$p(t) = 4.57t + 27.3.$$

2004: $p(4) = 4.57(4) + 27.3 = 45.58\%$

2005: $p(5) = 4.57(5) + 27.3 = 50.15\%$

2006: $p(6) = 4.57(6) + 27.3 = 54.72\%$

2007: $p(7) = 4.57(7) + 27.3 = 59.29\%$

For 2008 through 2010, use

$$p(t) = 3.35t + 37.6.$$

2008: $p(8) = 3.35(8) + 37.6 = 64.4\%$

2009: $p(9) = 3.35(9) + 37.6 = 67.75\%$

2010: $p(10) = 3.35(10) + 37.6 = 71.1\%$

71. (a) Cost = variable costs + fixed costs

$$C = 12.30x + 98{,}000$$

(b) Revenue = price per unit × number of units

$$R = 17.98x$$

(c) Profit = Revenue − Cost

$$P = 17.98x - (12.30x + 98{,}000)$$

$$P = 5.68x - 98{,}000$$

73. (a)

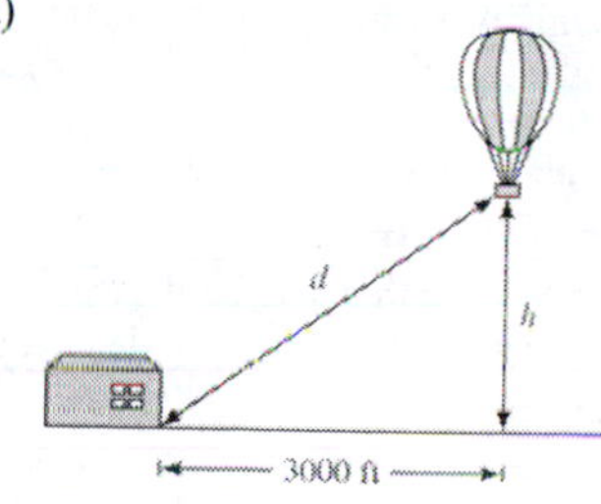

(b) $(3000)^2 + h^2 = d^2$

$$h = \sqrt{d^2 - (3000)^2}$$

Domain: $d \geq 3000$ (because both $d \geq 0$ and $d^2 - (3000)^2 \geq 0$)

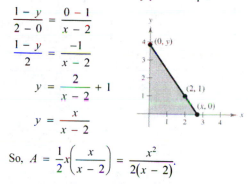

75. (a) $R = n(\text{rate}) = n[8.00 - 0.05(n - 80)],\ n \geq 80$

$$R = 12.00n - 0.05n^2 = 12n - \frac{n^2}{20} = \frac{240n - n^2}{20},\ n \geq 80$$

(b)

n	90	100	110	120	130	140	150
$R(n)$	\$675	\$700	\$715	\$720	\$715	\$700	\$675

The revenue is maximum when 120 people take the trip.

77.
$$f(x) = x^2 - x + 1$$
$$f(2 + h) = (2 + h)^2 - (2 + h) + 1 = 4 + 4h + h^2 - 2 - h + 1 = h^2 + 3h + 3$$
$$f(2) = (2)^2 - 2 + 1 = 3$$
$$f(2 + h) - f(2) = h^2 + 3h$$
$$\frac{f(2 + h) - f(2)}{h} = \frac{h^2 + 3h}{h} = h + 3,\ h \neq 0$$

79.
$$f(x) = x^3 + 3x$$
$$f(x + h) = (x + h)^3 + 3(x + h)$$
$$= x^3 + 3x^2h + 3xh^2 + h^3 + 3x + 3h$$
$$\frac{f(x + h) - f(x)}{h} = \frac{(x^3 + 3x^2h + 3xh^2 + h^3 + 3x + 3h) - (x^3 + 3x)}{h}$$
$$= \frac{h(3x^2 + 3xh + h^2 + 3)}{h}$$
$$= 3x^2 + 3xh + h^2 + 3,\ h \neq 0$$

81.
$$g(x) = \frac{1}{x^2}$$
$$\frac{g(x) - g(3)}{x - 3} = \frac{\frac{1}{x^2} - \frac{1}{9}}{x - 3}$$
$$= \frac{9 - x^2}{9x^2(x - 3)}$$
$$= \frac{-(x + 3)(x - 3)}{9x^2(x - 3)}$$
$$= -\frac{x + 3}{9x^2},\ x \neq 3$$

83. $f(x) = \sqrt{5x}$

$$\frac{f(x) - f(5)}{x - 5} = \frac{\sqrt{5x} - 5}{x - 5},\ x \neq 5$$

85. By plotting the points, we have a parabola, so $g(x) = cx^2$. Because $(-4, -32)$ is on the graph, you have $-32 = c(-4)^2 \Rightarrow c = -2$. So, $g(x) = -2x^2$.

87. Because the function is undefined at 0, we have $r(x) = c/x$. Because $(-4, -8)$ is on the graph, you have $-8 = c/-4 \Rightarrow c = 32$. So, $r(x) = 32/x$.

89. False. The equation $y^2 = x^2 + 4$ is a relation between x and y. However, $y = \pm\sqrt{x^2 + 4}$ does not represent a function.

91. False. The range is $[-1, \infty)$.

93. $f(x) = \sqrt{x - 1}$ Domain: $x \geq 1$

$g(x) = \dfrac{1}{\sqrt{x - 1}}$ Domain: $x > 1$

The value 1 may be included in the domain of $f(x)$ as it is possible to find the square root of 0. However, 1 cannot be included in the domain of $g(x)$ as it causes a zero to occur in the denominator which results in the function being undefined.

95. No; x is the independent variable, f is the name of the function.

97. (a) Yes. The amount that you pay in sales tax will increase as the price of the item purchased increases.

(b) No. The length of time that you study the night before an exam does not necessarily determine your score on the exam.

Section 2.3 Analyzing Graphs of Functions

1. Vertical Line Test

3. decreasing

5. average rate of change; secant

7. Domain: $(-\infty, \infty)$; Range: $[-4, \infty)$

(a) $f(-2) = 0$

(b) $f(-1) = -1$

(c) $f\left(\frac{1}{2}\right) = 0$

(d) $f(1) = -2$

9. Domain: $(-\infty, \infty)$; Range: $(-2, \infty)$

(a) $f(2) = 0$

(b) $f(1) = 1$

(c) $f(3) = 2$

(d) $f(-1) = 3$

11. $y = \frac{1}{4}x^3$

A vertical line intersects the graph at most once, so y *is* a function of x.

13. $x^2 + y^2 = 25$

A vertical line intersects the graph more than once, so y *is not* a function of x.

15.

$$f(x) = 2x^2 - 7x - 30$$
$$2x^2 - 7x - 30 = 0$$
$$(2x + 5)(x - 6) = 0$$
$$2x + 5 = 0 \quad \text{or} \quad x - 6 = 0$$
$$x = -\tfrac{5}{2} \qquad\qquad x = 6$$

17.

$$f(x) = \frac{x}{9x^2 - 4}$$
$$\frac{x}{9x^2 - 4} = 0$$
$$x = 0$$

19.

$$f(x) = \tfrac{1}{2}x^3 - x$$
$$\tfrac{1}{2}x^3 - x = 0$$
$$x^3 - 2x = 2(0)$$
$$x(x^2 - 2) = 0$$
$$x = 0 \quad \text{or} \quad x^2 - 2 = 0$$
$$x^2 = 2$$
$$x = \pm\sqrt{2}$$

21.

$$f(x) = 4x^3 - 24x^2 - x + 6$$
$$4x^3 - 24x^2 - x + 6 = 0$$
$$4x^2(x - 6) - 1(x - 6) = 0$$
$$(x - 6)(4x^2 - 1) = 0$$
$$(x - 6)(2x + 1)(2x - 1) = 0$$
$$x - 6 = 0 \quad \text{or} \quad 2x + 1 = 0 \quad \text{or} \quad 2x - 1 = 0$$
$$x = 6 \qquad x = -\tfrac{1}{2} \qquad x = \tfrac{1}{2}$$

23.

$$f(x) = \sqrt{2x} - 1$$
$$\sqrt{2x} - 1 = 0$$
$$\sqrt{2x} = 1$$
$$2x = 1$$
$$x = \tfrac{1}{2}$$

25. (a)

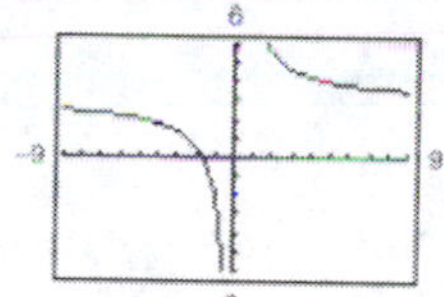

Zero: $x = -\dfrac{5}{3}$

(b)

$$f(x) = 3 + \tfrac{5}{x}$$
$$3 + \tfrac{5}{x} = 0$$
$$3x + 5 = 0$$
$$x = -\tfrac{5}{3}$$

27. (a)

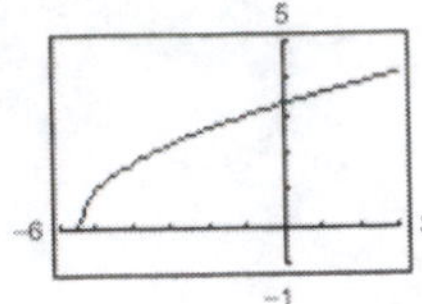

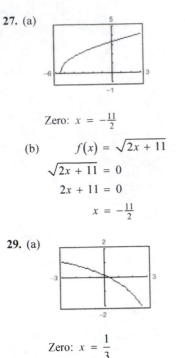

Zero: $x = -\frac{11}{2}$

(b)
$$f(x) = \sqrt{2x + 11}$$
$$\sqrt{2x + 11} = 0$$
$$2x + 11 = 0$$
$$x = -\frac{11}{2}$$

29. (a)

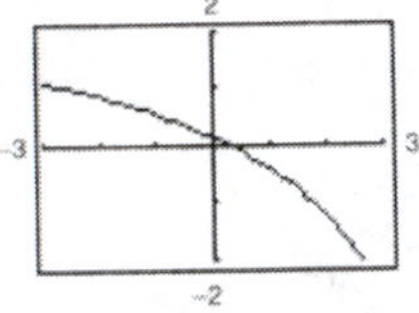

Zero: $x = \frac{1}{3}$

(b)
$$f(x) = \frac{3x - 1}{x - 6}$$
$$\frac{3x - 1}{x - 6} = 0$$
$$3x - 1 = 0$$
$$x = \frac{1}{3}$$

31. $f(x) = \frac{3}{2}x$

The function is increasing on $(-\infty, \infty)$.

33. $f(x) = x^3 - 3x^2 + 2$

The function is increasing on $(-\infty, 0)$ and $(2, \infty)$ and decreasing on $(0, 2)$.

35. $f(x) = |x + 1| + |x - 1|$

The function is increasing on $(1, \infty)$.

The function is constant on $(-1, 1)$.

The function is decreasing on $(-\infty, -1)$.

37. $f(x) = \begin{cases} x + 3, & x \le 0 \\ 3, & 0 < x \le 2 \\ 2x + 1, & x > 2 \end{cases}$

The function is increasing on $(-\infty, 0)$ and $(2, \infty)$.

The function is constant on $(0, 2)$.

39. $f(x) = 3$

(a)

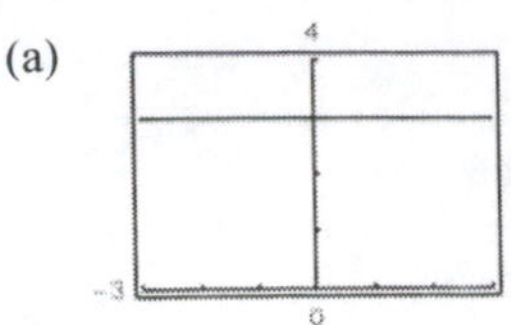

Constant on $(-\infty, \infty)$

(b)

x	-2	-1	0	1	2
$f(x)$	3	3	3	3	3

41. $g(s) = \frac{s^2}{4}$

(a)

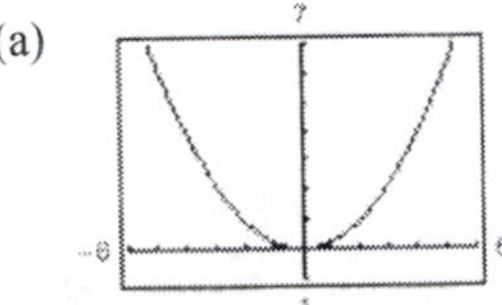

Decreasing on $(-\infty, 0)$; Increasing on $(0, \infty)$

(b)

s	-4	-2	0	2	4
$g(s)$	4	1	0	1	4

43. $f(x) = \sqrt{1 - x}$

(a)

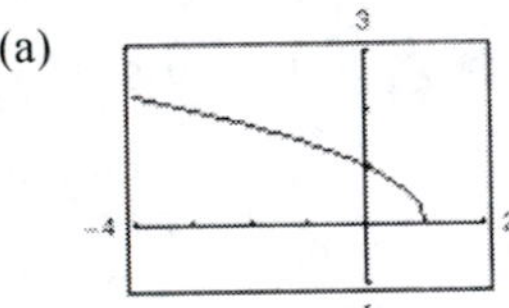

Decreasing on $(-\infty, 1)$

(b)

x	-3	-2	-1	0	1
$f(x)$	2	$\sqrt{3}$	$\sqrt{2}$	1	0

45. $f(x) = x^{3/2}$

(a)

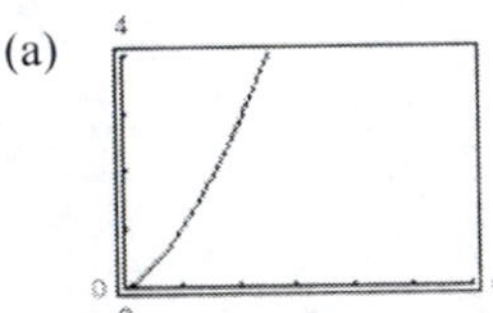

Increasing on $(0, \infty)$

(b)

x	0	1	2	3	4
$f(x)$	0	1	2.8	5.2	8

47. $f(x) = 3x^2 - 2x - 5$

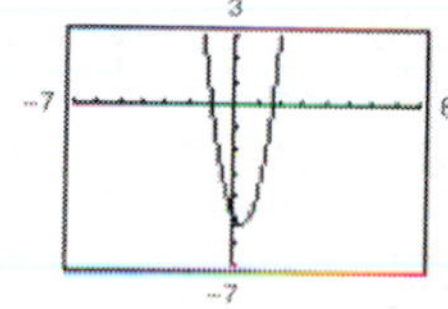

Relative minimum: $\left(\frac{1}{3}, -\frac{16}{3}\right)$ or $(0.33, -5.33)$

49. $f(x) = -2x^2 + 9x$

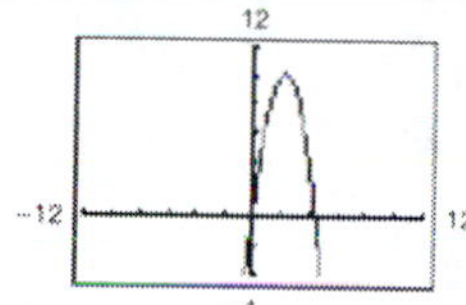

Relative maximum: $(2.25, 10.125)$

51. $f(x) = x^3 - 3x^2 - x + 1$

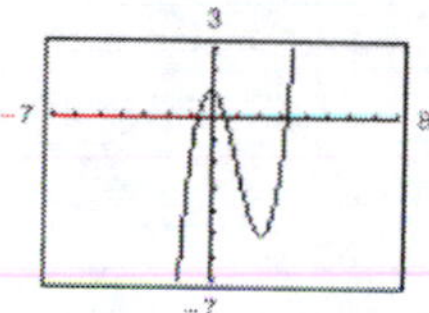

Relative maximum: $(-0.15, 1.08)$

Relative minimum: $(2.15, -5.08)$

53. $h(x) = (x - 1)\sqrt{x}$

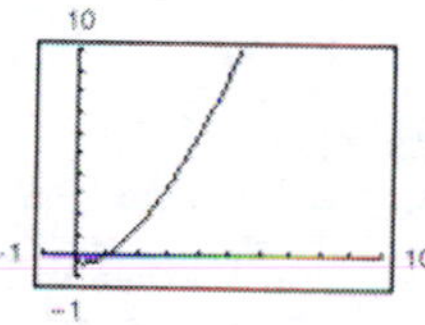

Relative minimum: $(0.33, -0.38)$

55. $f(x) = 4 - x$

$f(x) \geq 0$ on $(-\infty, 4]$

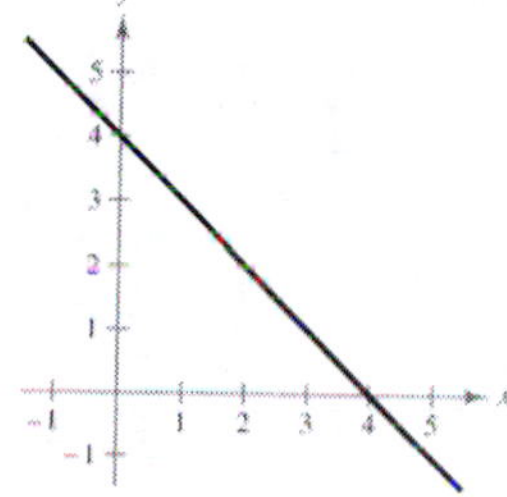

57. $f(x) = 9 - x^2$

$f(x) \geq 0$ on $[-3, 3]$

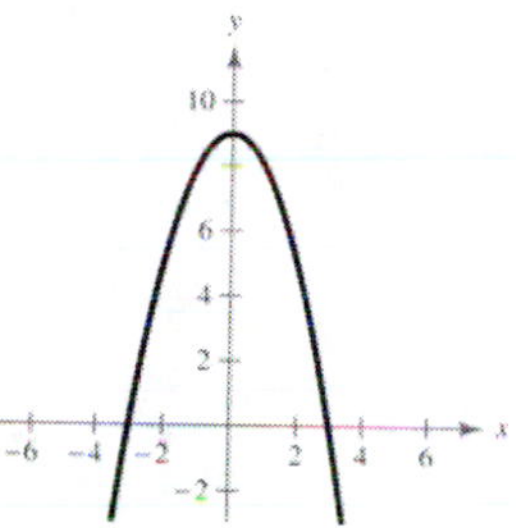

59. $f(x) = \sqrt{x - 1}$

$f(x) \geq 0$ on $[1, \infty)$

$\sqrt{x - 1} \geq 0$

$x - 1 \geq 0$

$x \geq 1$

$[1, \infty)$

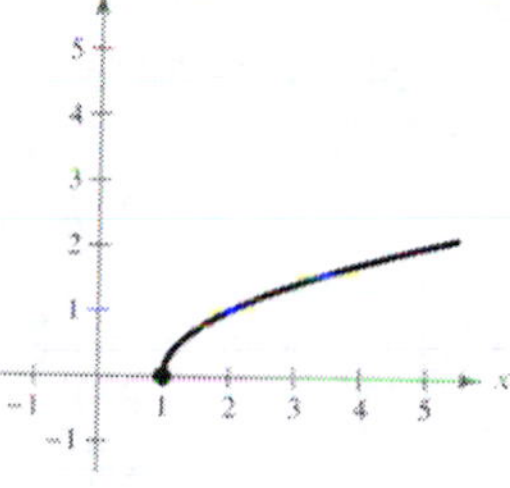

61. $f(x) = -2x + 15$

$$\frac{f(3) - f(0)}{3 - 0} = \frac{9 - 15}{3} = -2$$

The average rate of change from $x_1 = 0$ to $x_2 = 3$ is -2.

63. $f(x) = x^3 - 3x^2 - x$

$$\frac{f(3) - f(1)}{3 - 1} = \frac{-3 - (-3)}{2} = 0$$

The average rate of change from $x_1 = 1$ to $x_2 = 3$ is 0.

65. (a)

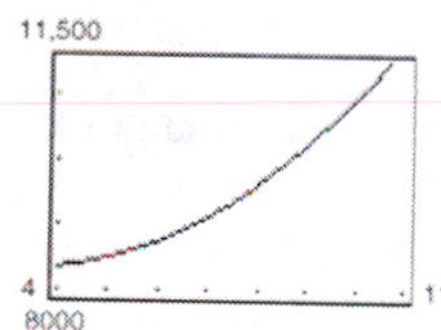

(b) To find the average rate of change of the amount the U.S. Department of Energy spent for research and development from 2005 to 2010, find the average rate of change from $(5, f(5))$ to $(10, f(10))$.

$$\frac{f(10) - f(5)}{10 - 5} = \frac{10{,}925 - 8501.25}{5} = 484.75$$

The amount the U.S. Department of Energy spent for research and development increased by about \$484.75 million each year from 2005 to 2010.

67. $s_0 = 6, v_0 = 64$

(a) $s = -16t^2 + 64t + 6$

(b) 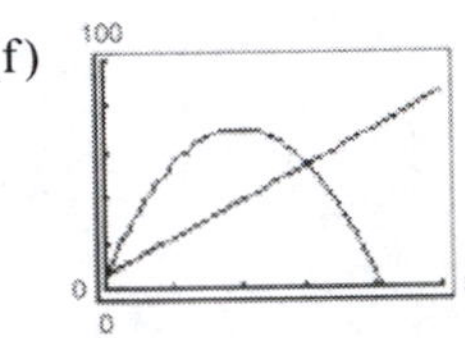

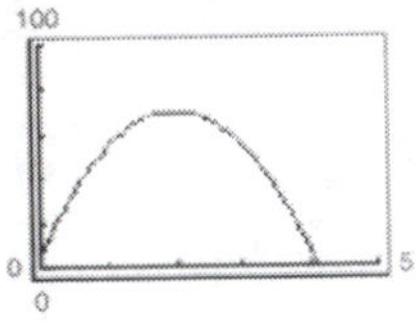

(c) $\dfrac{s(3) - s(0)}{3 - 0} = \dfrac{54 - 6}{3} = 16$

(d) The slope of the secant line is positive.

(e) $s(0) = 6, m = 16$

Secant line: $y - 6 = 16(t - 0)$

$y = 16t + 6$

(f) 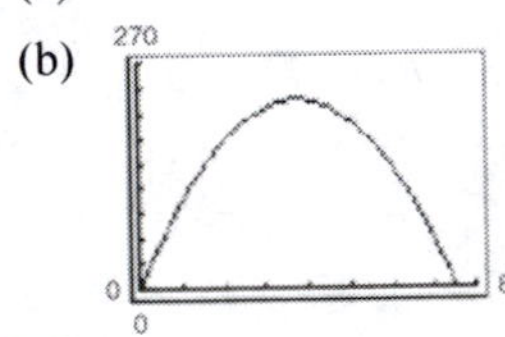

69. $v_0 = 120, s_0 = 0$

(a) $s = -16t^2 + 120t$

(b) 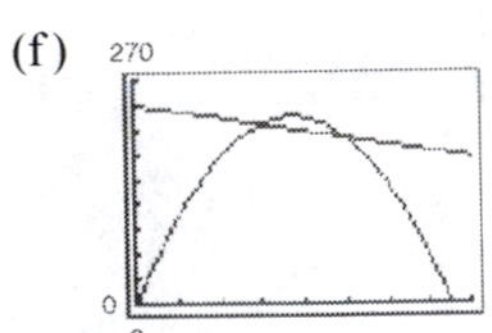

(c) The average rate of change from $t = 3$ to $t = 5$:

$\dfrac{s(5) - s(3)}{5 - 3} = \dfrac{200 - 216}{2} = -\dfrac{16}{2} = -8$ feet per second

(d) The slope of the secant line through $(3, s(3))$ and $(5, s(5))$ is negative.

(e) The equation of the secant line: $m = -8$

Using $(5, s(5)) = (5, 200)$ we have

$y - 200 = -8(t - 5)$

$y = -8t + 240.$

(f) 270 0 0 8

71. $f(x) = x^6 - 2x^2 + 3$

$f(-x) = (-x)^6 - 2(-x)^2 + 3$

$= x^6 - 2x^2 + 3$

$= f(x)$

The function is even. y-axis symmetry.

73. $h(x) = x\sqrt{x + 5}$

$h(-x) = (-x)\sqrt{-x + 5}$

$= -x\sqrt{5 - x}$

$\neq h(x)$

$\neq -h(x)$

The function is neither odd nor even. No symmetry.

75. $f(s) = 4s^{3/2} = 4(-s)^{3/2} \neq f(s) \neq -f(s)$

The function is neither odd nor even. No symmetry.

77.

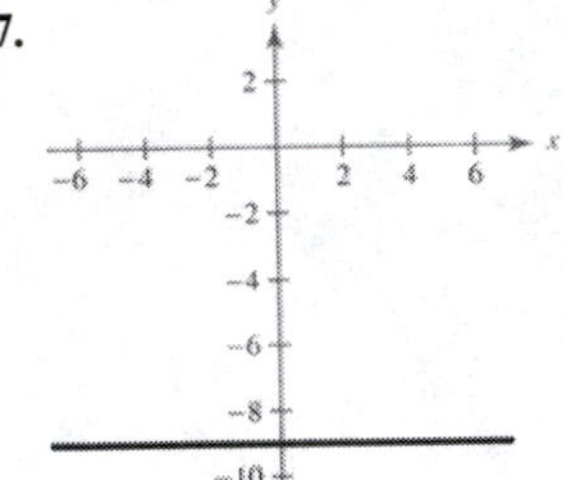

The graph of $f(x) = -9$ is symmetric to the y-axis, which implies $f(x)$ is even.

$f(-x) = -9 = f(x)$

The function is even.

79. $f(x) = -|x - 5|$

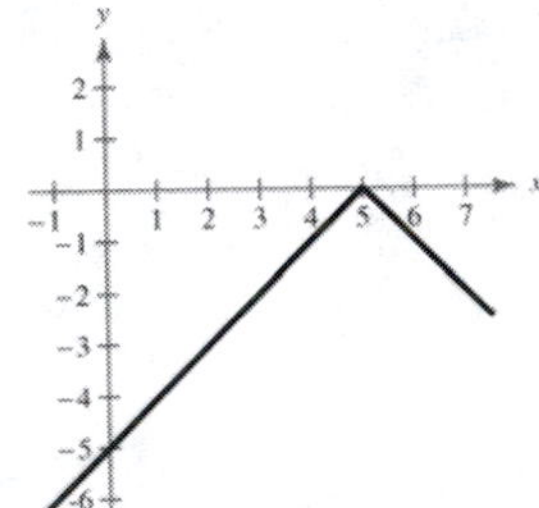

The graph displays no symmetry, which implies $f(x)$ is neither odd nor even.

$f(x) = -|(-x) - 5|$

$= -|-x - 5|$

$\neq f(x)$

$\neq -f(x)$

The function is neither even nor odd.

81. $f(x) = \sqrt{1 - x}$

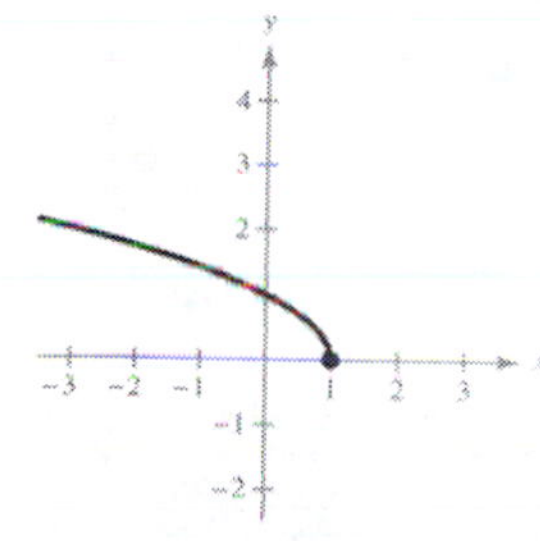

The graph displays no symmetry, which implies $f(x)$ is neither odd nor even.

$$\begin{aligned} f(-x) &= \sqrt{1 - (-x)} \\ &= \sqrt{1 + x} \\ &\neq f(x) \\ &\neq -f(x) \end{aligned}$$

The function is neither even nor odd.

83. $h = \text{top} - \text{bottom}$

$$\begin{aligned} &= 3 - (4x - x^2) \\ &= 3 - 4x + x^2 \end{aligned}$$

85. $L = \text{right} - \text{left}$

$$= 2 - \sqrt[3]{2y}$$

87. $L = -0.294x^2 + 97.744x - 664.875,\ 20 \leq x \leq 90$

(a)

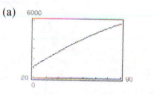

(b) $L = 2000$ when $x \approx 29.9645 \approx 30$ watts.

89. (a) For the average salaries of college professors, a scale of $10,000 would be appropriate.

(b) For the population of the United States, use a scale of 10,000,000.

(c) For the percent of the civilian workforce that is unemployed, use a scale of 1%.

91. (a) $y = x$

(b) $y = x^2$

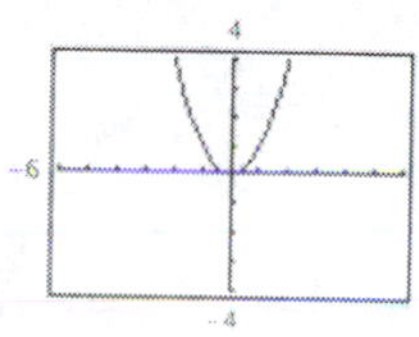

(c) $y = x^3$

(d) $y = x^4$

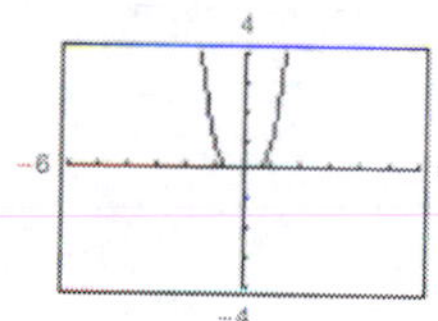

(e) $y = x^5$

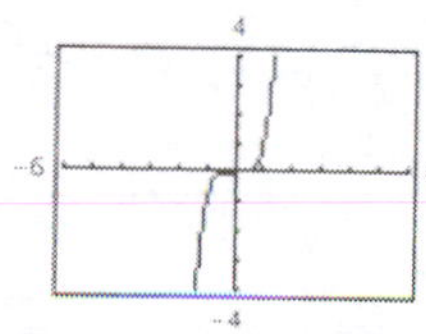

(f) $y = x^6$

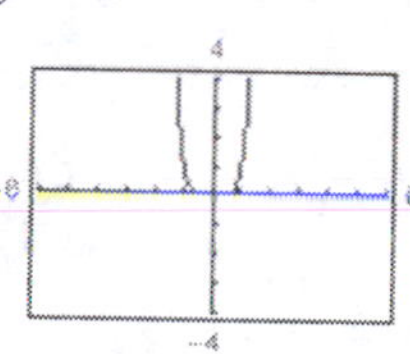

All the graphs pass through the origin. The graphs of the odd powers of x are symmetric with respect to the origin and the graphs of the even powers are symmetric with respect to the y-axis. As the powers increase, the graphs become flatter in the interval $-1 < x < 1$.

93. False. The function $f(x) = \sqrt{x^2 + 1}$ has a domain of all real numbers.

95. $\left(-\frac{5}{3}, -7\right)$

(a) If f is even, another point is $\left(\frac{5}{3}, -7\right)$.

(b) If f is odd, another point is $\left(\frac{5}{3}, 7\right)$.

97.

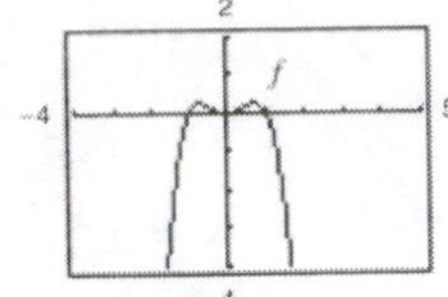

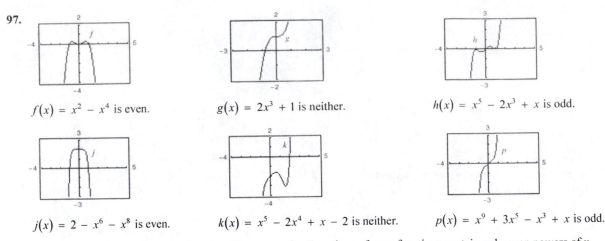

$f(x) = x^2 - x^4$ is even.

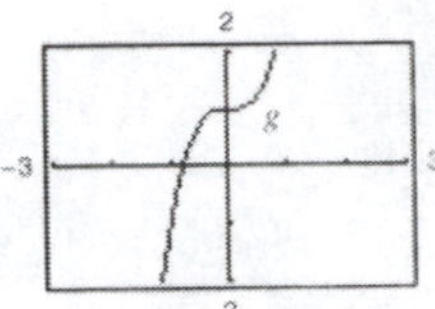

$g(x) = 2x^3 + 1$ is neither.

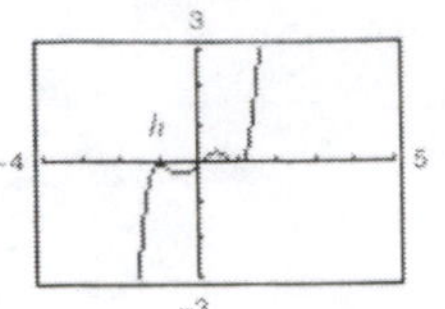

$h(x) = x^5 - 2x^3 + x$ is odd.

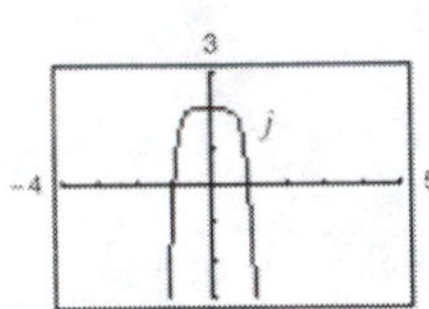

$j(x) = 2 - x^6 - x^8$ is even.

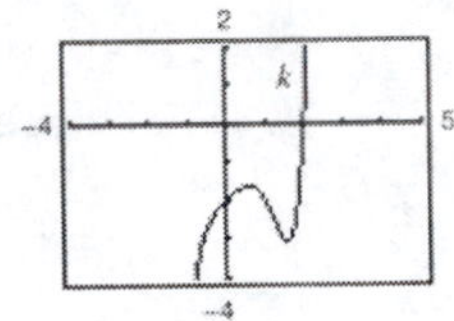

$k(x) = x^5 - 2x^4 + x - 2$ is neither.

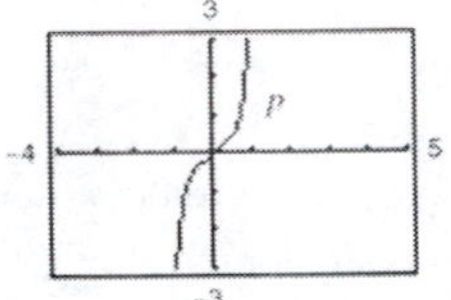

$p(x) = x^9 + 3x^5 - x^3 + x$ is odd.

Equations of odd functions contain only odd powers of x. Equations of even functions contain only even powers of x. A function that has variables raised to even and odd powers is neither odd nor even.

Section 2.4 A Library of Parent Functions

1. $f(x) = [\![x]\!]$

(g) greatest integer function

3. $f(x) = \dfrac{1}{x}$

(h) reciprocal function

5. $f(x) = \sqrt{x}$

(b) square root function

7. $f(x) = |x|$

(f) absolute value function

9. $f(x) = ax + b$

(d) linear function

11. (a) $f(1) = 4,\ f(0) = 6$

$(1, 4), (0, 6)$

$$m = \frac{6 - 4}{0 - 1} = -2$$

$$y - 6 = -2(x - 0)$$

$$y = -2x + 6$$

$$f(x) = -2x + 6$$

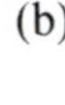

(b)

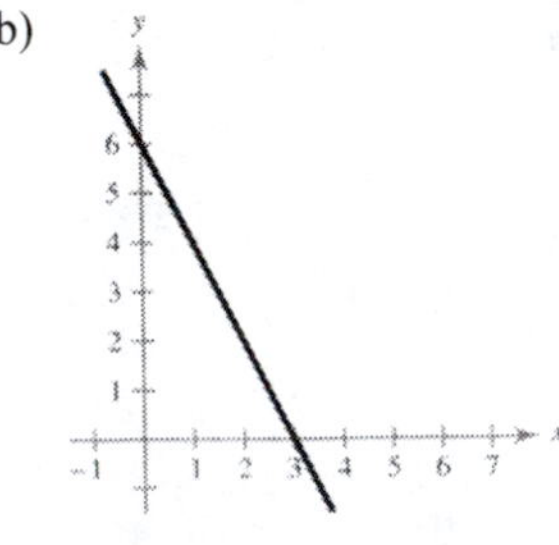

13. (a) $f(-5) = -1,\ f(5) = -1$

$(-5, -1), (5, -1)$

$$m = \frac{-1 - (-1)}{5 - (-5)} = \frac{0}{10} = 0$$

$$y - (-1) = 0(x - (-5))$$

$$y = -1$$

$$f(x) = -1$$

(b)

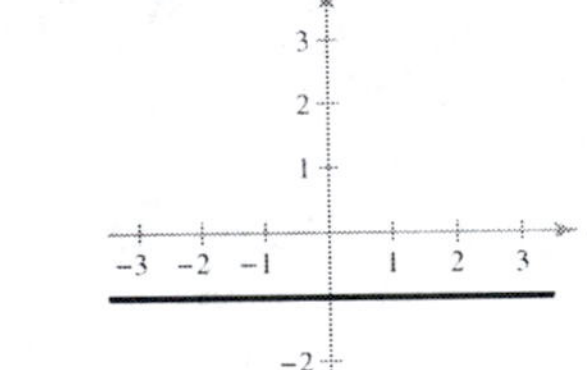

15. $f(x) = 2.5x - 4.25$

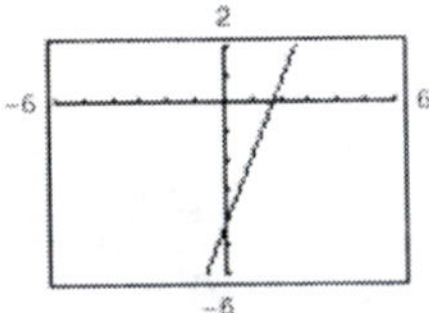

17. $g(x) = -2x^2$

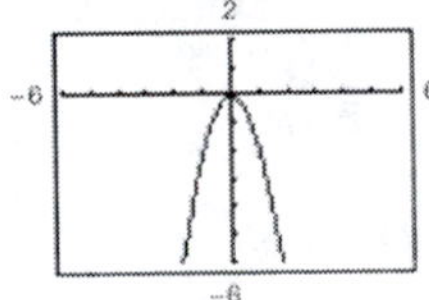

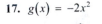

81. $f(x) = \sqrt{1 - x}$

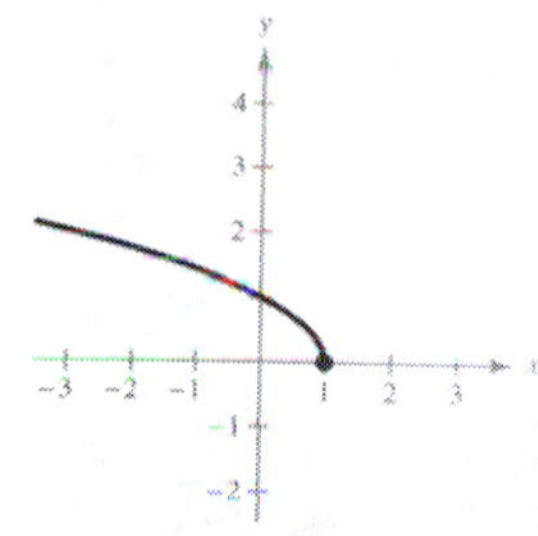

The graph displays no symmetry, which implies $f(x)$ is neither odd nor even.

$$\begin{aligned} f(-x) &= \sqrt{1 - (-x)} \\ &= \sqrt{1 + x} \\ &\neq f(x) \\ &\neq -f(x) \end{aligned}$$

The function is neither even nor odd.

83.
$$\begin{aligned} h &= \text{top} - \text{bottom} \\ &= 3 - \left(4x - x^2\right) \\ &= 3 - 4x + x^2 \end{aligned}$$

85.
$$\begin{aligned} L &= \text{right} - \text{left} \\ &= 2 - \sqrt[3]{2y} \end{aligned}$$

87. $L = -0.294x^2 + 97.744x - 664.875,\ 20 \le x \le 90$

(a)

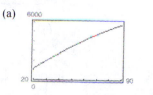

(b) $L = 2000$ when $x \approx 29.9645 \approx 30$ watts.

89. (a) For the average salaries of college professors, a scale of $10,000 would be appropriate.

(b) For the population of the United States, use a scale of 10,000,000.

(c) For the percent of the civilian workforce that is unemployed, use a scale of 1%.

91. (a) $y = x$

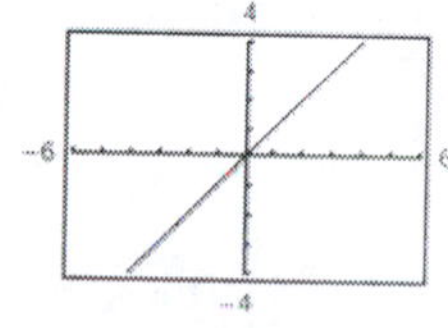

(b) $y = x^2$

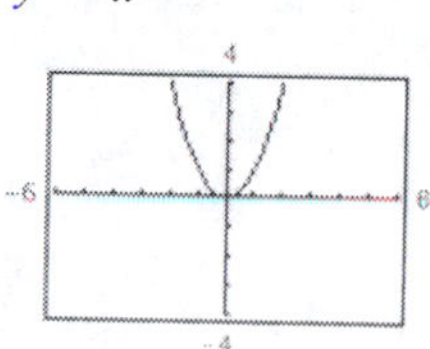

(c) $y = x^3$

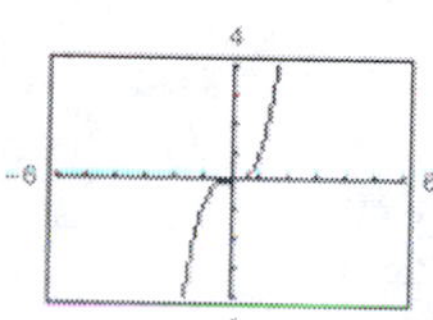

(d) $y = x^4$

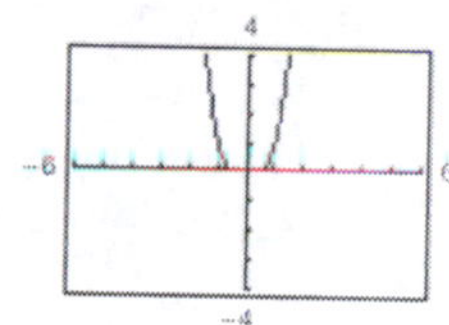

(e) $y = x^5$

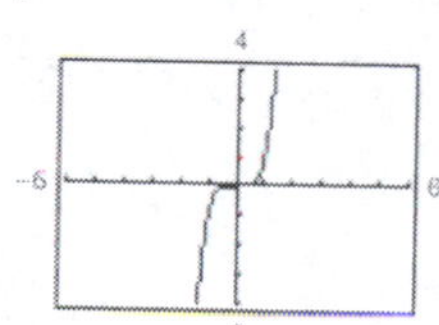

(f) $y = x^6$

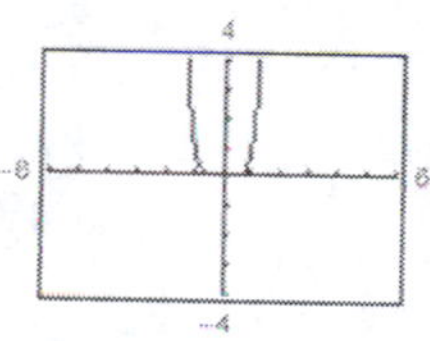

All the graphs pass through the origin. The graphs of the odd powers of x are symmetric with respect to the origin and the graphs of the even powers are symmetric with respect to the y-axis. As the powers increase, the graphs become flatter in the interval $-1 < x < 1$.

93. False. The function $f(x) = \sqrt{x^2 + 1}$ has a domain of all real numbers.

95. $\left(-\frac{5}{3}, -7\right)$

(a) If f is even, another point is $\left(\frac{5}{3}, -7\right)$.

(b) If f is odd, another point is $\left(\frac{5}{3}, 7\right)$.

97.

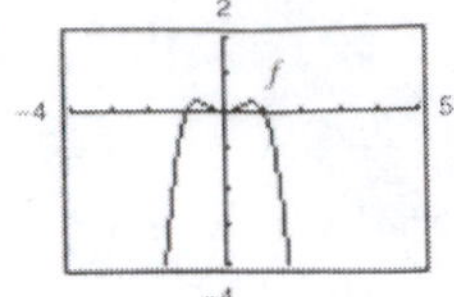

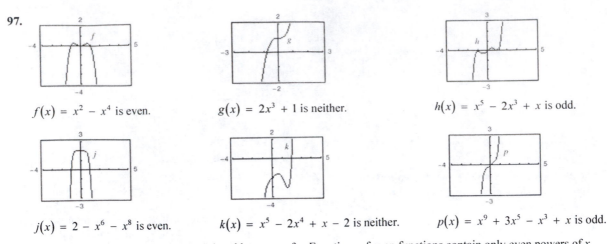

$f(x) = x^2 - x^4$ is even.

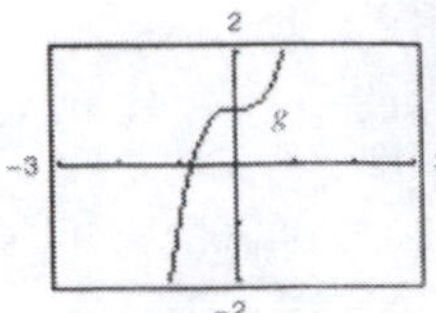

$g(x) = 2x^3 + 1$ is neither.

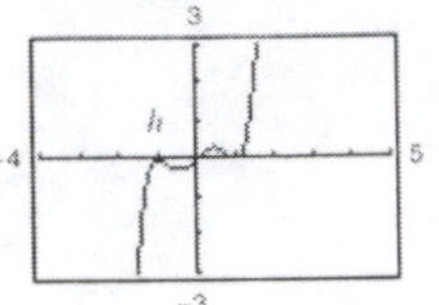

$h(x) = x^5 - 2x^3 + x$ is odd.

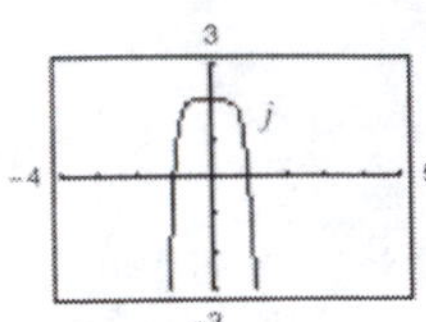

$j(x) = 2 - x^6 - x^8$ is even.

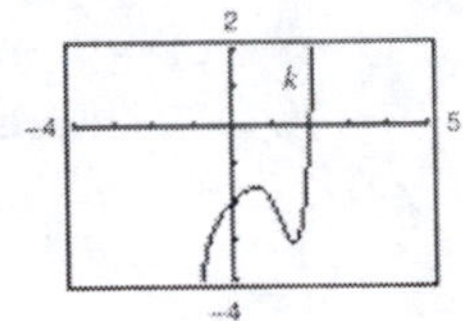

$k(x) = x^5 - 2x^4 + x - 2$ is neither.

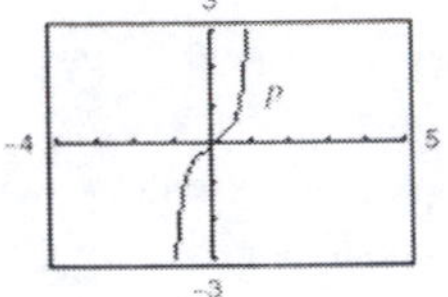

$p(x) = x^9 + 3x^5 - x^3 + x$ is odd.

Equations of odd functions contain only odd powers of x. Equations of even functions contain only even powers of x. A function that has variables raised to even and odd powers is neither odd nor even.

Section 2.4 A Library of Parent Functions

1. $f(x) = [\![x]\!]$

(g) greatest integer function

3. $f(x) = \dfrac{1}{x}$

(h) reciprocal function

5. $f(x) = \sqrt{x}$

(b) square root function

7. $f(x) = |x|$

(f) absolute value function

9. $f(x) = ax + b$

(d) linear function

11. (a) $f(1) = 4,\ f(0) = 6$

$(1, 4), (0, 6)$

$$m = \frac{6 - 4}{0 - 1} = -2$$

$$y - 6 = -2(x - 0)$$

$$y = -2x + 6$$

$$f(x) = -2x + 6$$

(b) 

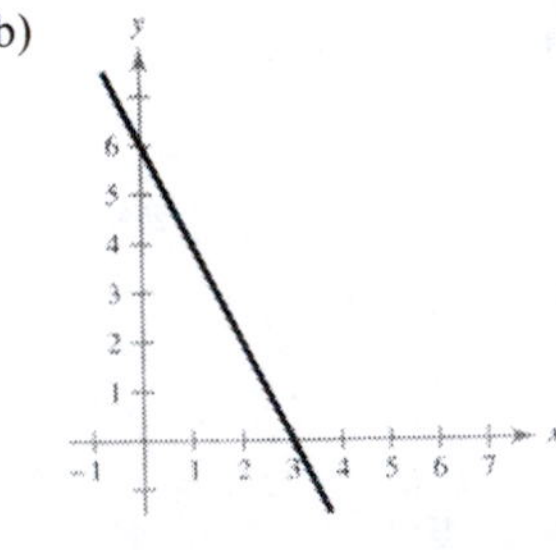

13. (a) $f(-5) = -1,\ f(5) = -1$

$(-5, -1), (5, -1)$

$$m = \frac{-1 - (-1)}{5 - (-5)} = \frac{0}{10} = 0$$

$$y - (-1) = 0(x - (-5))$$

$$y = -1$$

$$f(x) = -1$$

(b)

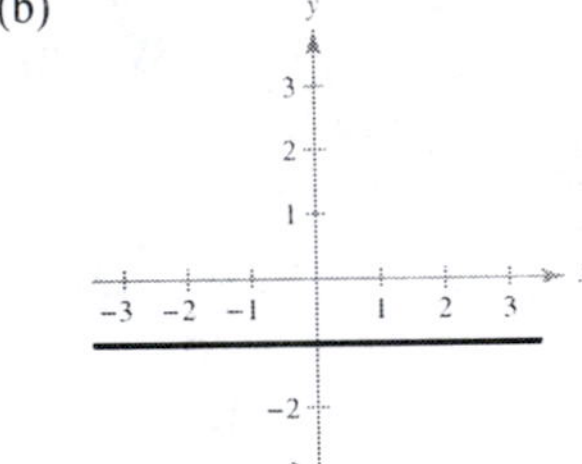

15. $f(x) = 2.5x - 4.25$

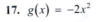

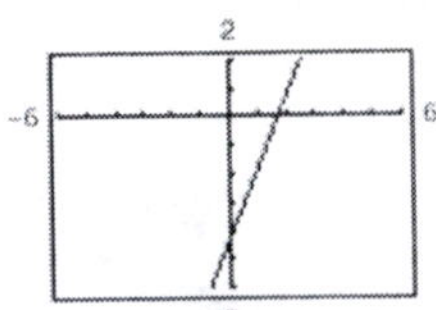

17. $g(x) = -2x^2$

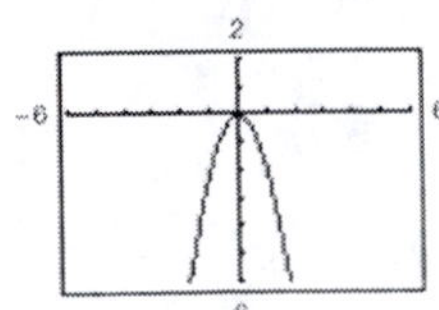

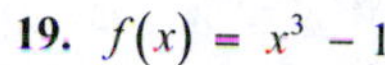

19. $f(x) = x^3 - 1$

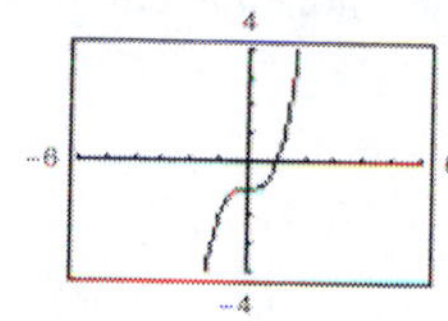

21. $f(x) = 4 - 2\sqrt{x}$

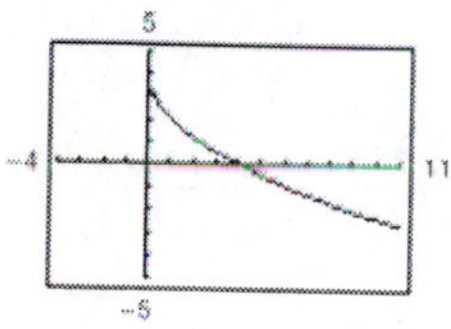

23. $f(x) = 4 + \dfrac{1}{x}$

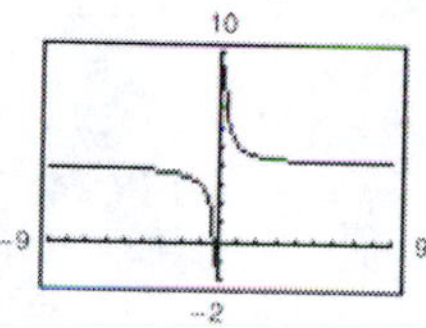

25. $g(x) = |x| - 5$

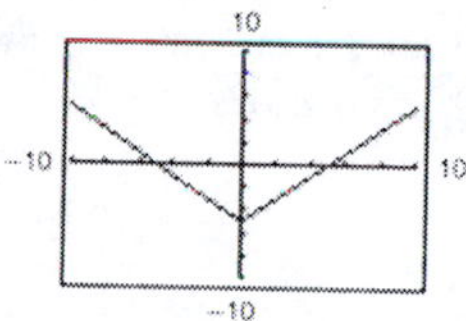

27. $f(x) = [\![x]\!]$

(a) $f(2.1) = 2$

(b) $f(2.9) = 2$

(c) $f(-3.1) = -4$

(d) $f\left(\frac{7}{2}\right) = 3$

29. $k(x) = \left[\!\left[\frac{1}{2}x + 6\right]\!\right]$

(a) $k(5) = \left[\!\left[\frac{1}{2}(5) + 6\right]\!\right] = [\![8.5]\!] = 8$

(b) $k(-6.1) = \left[\!\left[\frac{1}{2}(-6.1) + 6\right]\!\right] = [\![2.95]\!] = 2$

(c) $k(0.1) = \left[\!\left[\frac{1}{2}(0.1) + 6\right]\!\right] = [\![6.05]\!] = 6$

(d) $k(15) = \left[\!\left[\frac{1}{2}(15) + 6\right]\!\right] = [\![13.5]\!] = 13$

31. $g(x) = -[\![x]\!]$

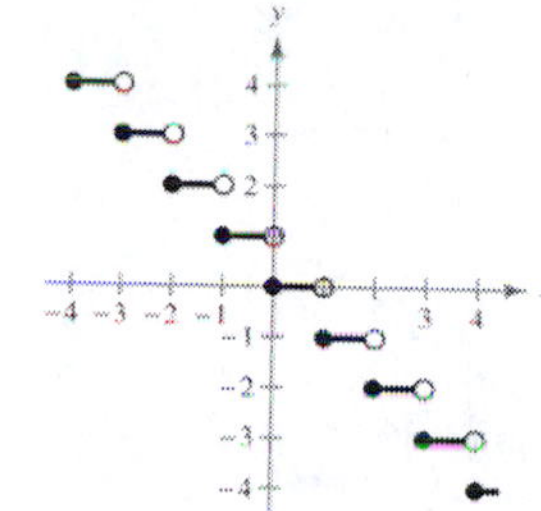

33. $g(x) = [\![x]\!] - 1$

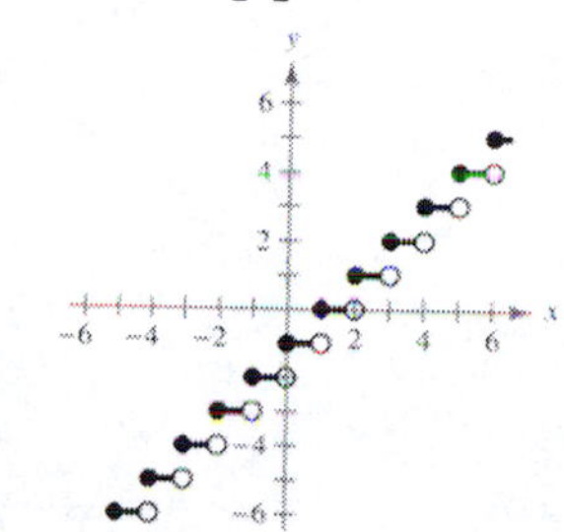

35. $g(x) = \begin{cases} x + 6, & x \le -4 \\ \frac{1}{2}x - 4, & x > -4 \end{cases}$

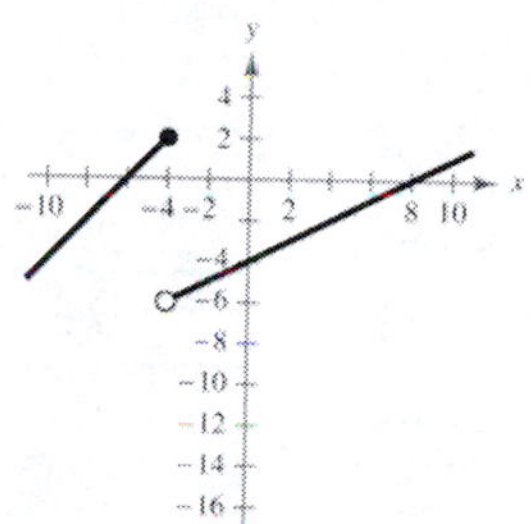

37. $f(x) = \begin{cases} 1 - (x - 1)^2, & x \le 2 \\ \sqrt{x - 2}, & x > 2 \end{cases}$

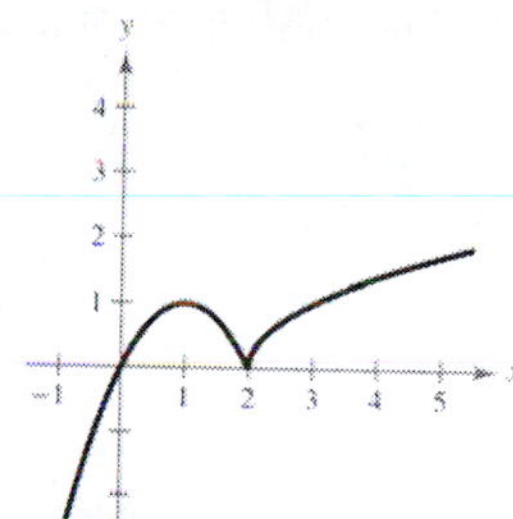

39. $h(x) = \begin{cases} 4 - x^2, & x < -2 \\ 3 + x, & -2 \le x < 0 \\ x^2 + 1, & x \ge 0 \end{cases}$

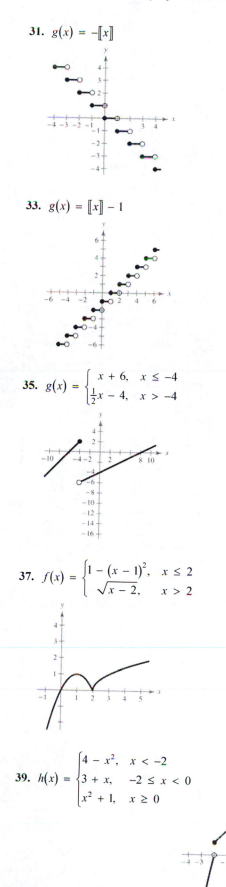

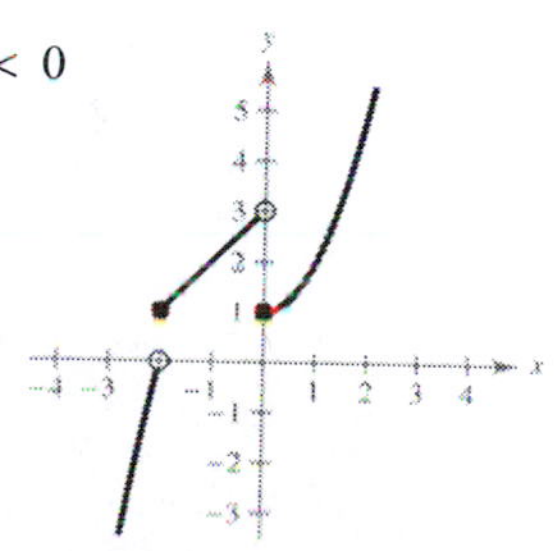

41. $s(x) = 2\left(\frac{1}{4}x - \left[\!\left[\frac{1}{4}x\right]\!\right]\right)$

(a)

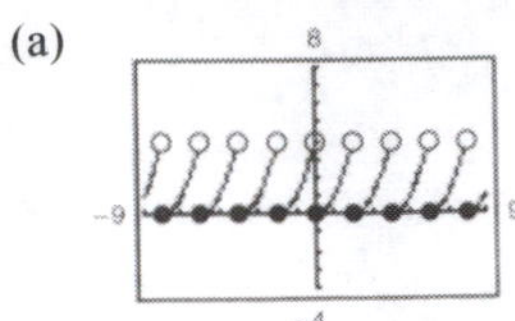

(b) Domain: $(-\infty, \infty)$; Range: $[0, 2)$

43. (a) $W(30) = 14(30) = 420$

$W(40) = 14(40) = 560$

$W(45) = 21(45 - 40) + 560 = 665$

$W(50) = 21(50 - 40) + 560 = 770$

(b) $W(h) = \begin{cases} 14h, & 0 < h \le 45 \\ 21(h - 45) + 630, & h > 45 \end{cases}$

45. Answers will vary. *Sample answer:*

Interval	*Input Pipe*	*Drain Pipe 1*	*Drain Pipe 2*
[0, 5]	Open	Closed	Closed
[5, 10]	Open	Open	Closed
[10, 20]	Closed	Closed	Closed
[20, 30]	Closed	Closed	Open
[30, 40]	Open	Open	Open
[40, 45]	Open	Closed	Open
[45, 50]	Open	Open	Open
[50, 60]	Open	Open	Closed

47. For the first two hours the slope is 1. For the next six hours, the slope is 2. For the final hour, the slope is $\frac{1}{2}$.

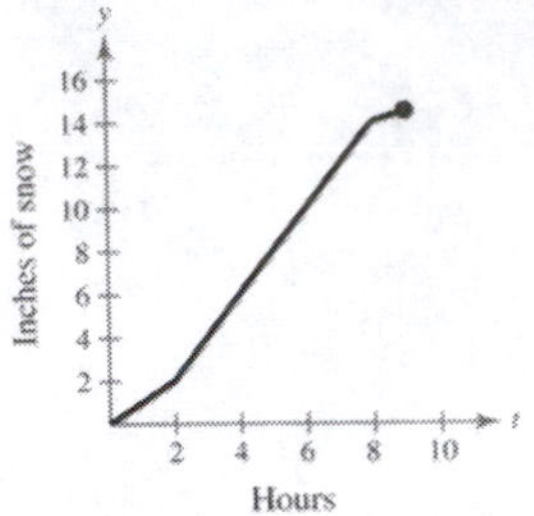

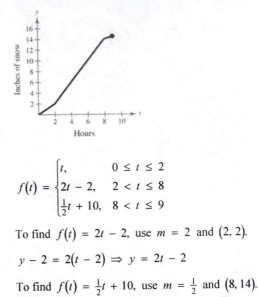

$$f(t) = \begin{cases} t, & 0 \le t \le 2 \\ 2t - 2, & 2 < t \le 8 \\ \frac{1}{2}t + 10, & 8 < t \le 9 \end{cases}$$

To find $f(t) = 2t - 2$, use $m = 2$ and $(2, 2)$.

$y - 2 = 2(t - 2) \Rightarrow y = 2t - 2$

To find $f(t) = \frac{1}{2}t + 10$, use $m = \frac{1}{2}$ and $(8, 14)$.

$y - 14 = \frac{1}{2}(t - 8) \Rightarrow y = \frac{1}{2}t + 10$

Total accumulation = 14.5 inches

49. False. A piecewise-defined function is a function that is defined by two or more equations over a specified domain. That domain may or may not include x- and y-intercepts.

Section 2.5 Transformations of Functions

1. rigid

3. vertical stretch; vertical shrink

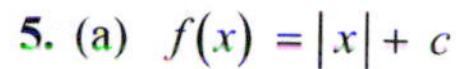

5. (a) $f(x) = |x| + c$ Vertical shifts

$c = -1$: $f(x) = |x| - 1$ 1 unit down

$c = 1$: $f(x) = |x| + 1$ 1 unit up

$c = 3$: $f(x) = |x| + 3$ 3 units up

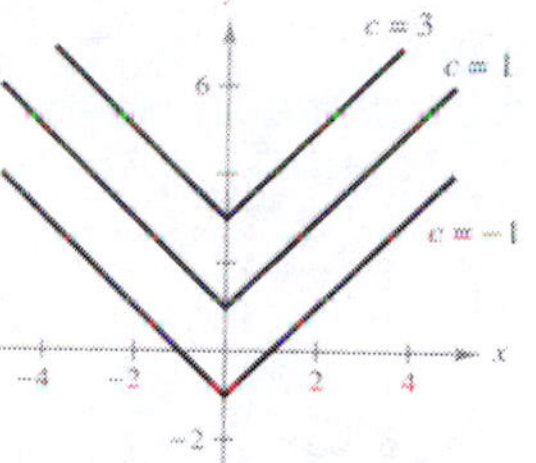

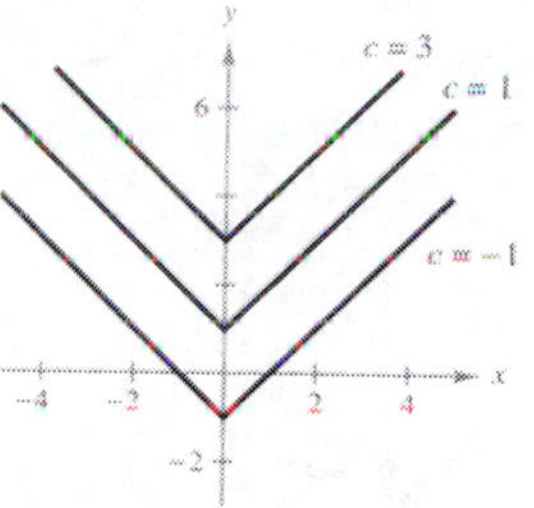

(b) $f(x) = |x - c|$ Horizontal shifts

$c = -1$: $f(x) = |x + 1|$ 1 unit left

$c = 1$: $f(x) = |x - 1|$ 1 unit right

$c = 3$: $f(x) = |x - 3|$ 3 units right

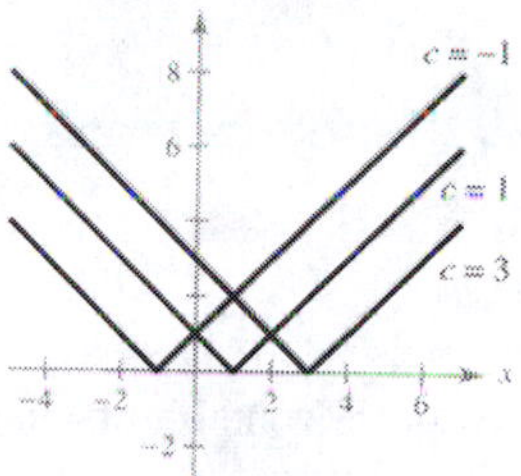

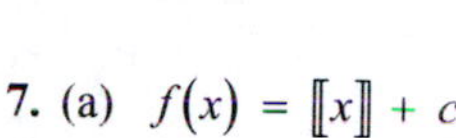

7. (a) $f(x) = [\![x]\!] + c$ Vertical shifts

$c = -2$: $f(x) = [\![x]\!] - 2$ 2 units down

$c = 0$: $f(x) = [\![x]\!]$ Parent function

$c = 2$: $f(x) = [\![x]\!] + 2$ 2 units up

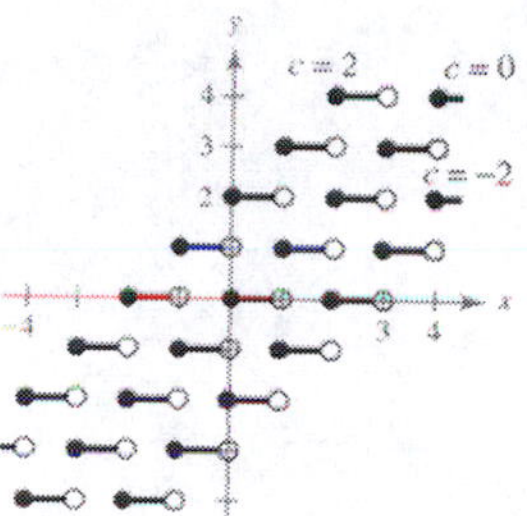

(b) $f(x) = [\![x + c]\!]$ Horizontal shifts

$c = -2$: $f(x) = [\![x - 2]\!]$ 2 units right

$c = 0$: $f(x) = [\![x]\!]$ Parent function

$c = 2$: $f(x) = [\![x + 2]\!]$ 2 units left

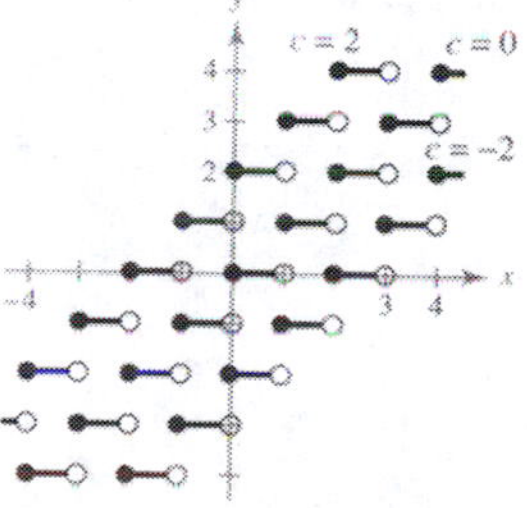

9. (a) $y = f(-x)$

Reflection in the y-axis

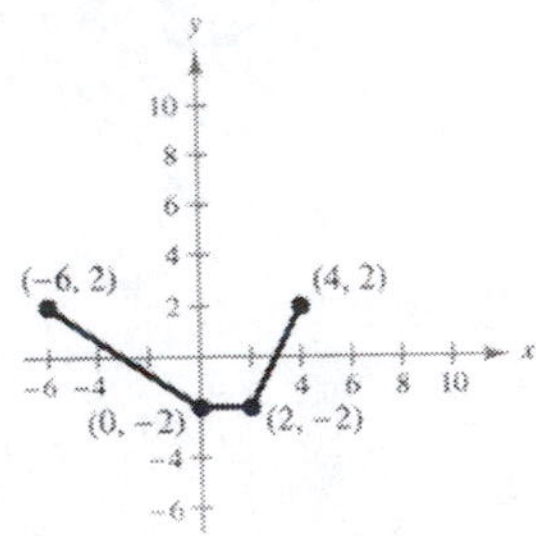

(b) $y = f(x) + 4$

Vertical shift 4 units upward

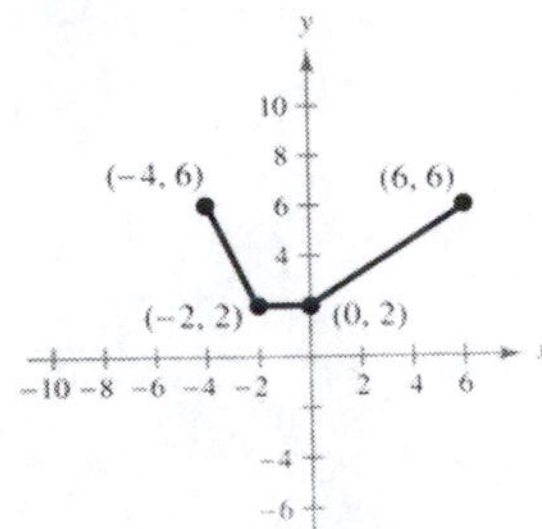

(c) $y = 2f(x)$

Vertical stretch (each y-value is multiplied by 2)

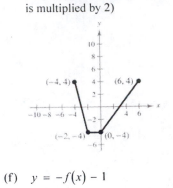

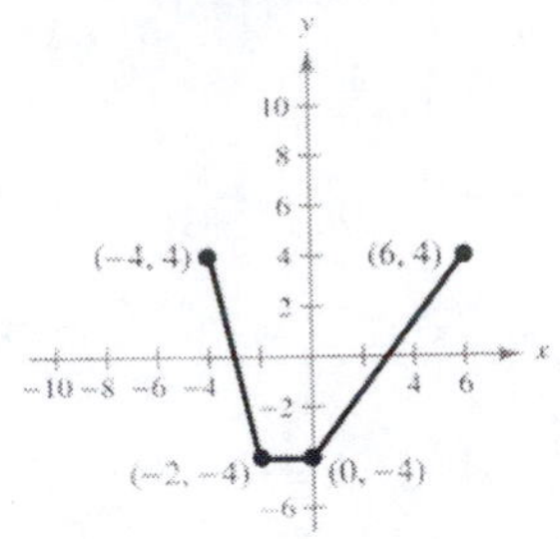

(d) $y = -f(x - 4)$

Reflection in the x-axis and a horizontal shift 4 units to the right

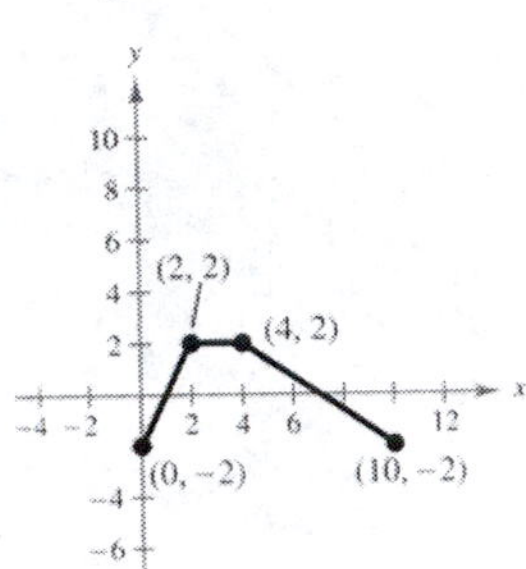

(e) $y = f(x) - 3$

Vertical shift 3 units downward

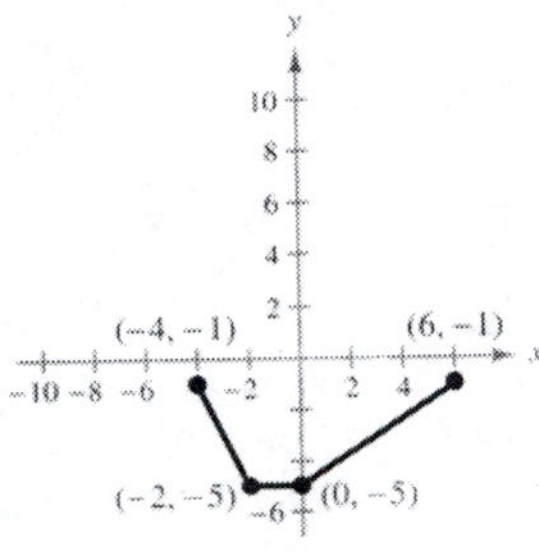

(f) $y = -f(x) - 1$

Reflection in the x-axis and a vertical shift 1 unit downward

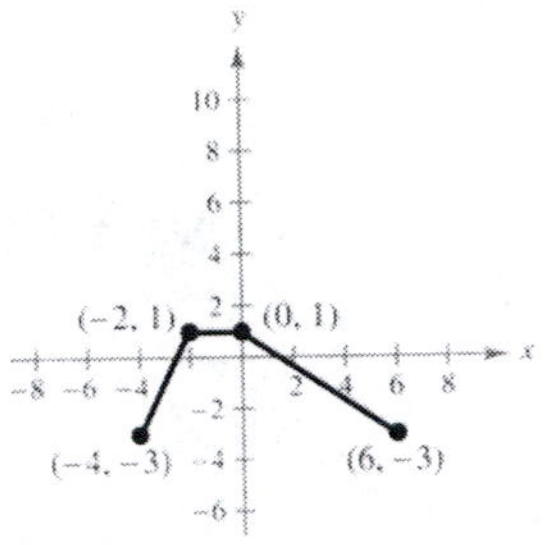

(g) $y = f(2x)$

Horizontal shrink
(each x-value is divided by 2)

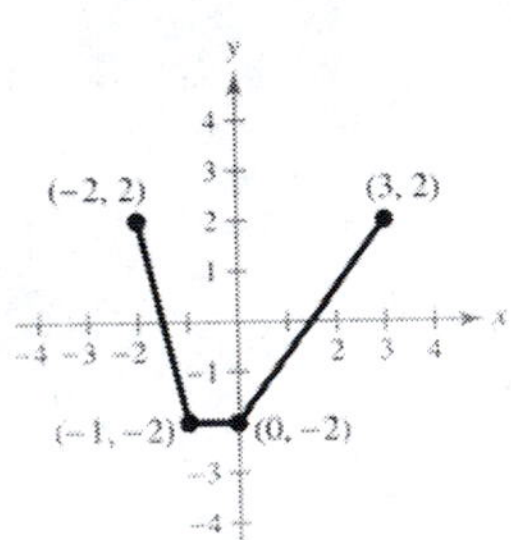

11. Parent function: $f(x) = x^2$

(a) Vertical shift 1 unit downward

$g(x) = x^2 - 1$

(b) Reflection in the x-axis, horizontal shift 1 unit to the left, and a vertical shift 1 unit upward

$g(x) = -(x + 1)^2 + 1$

13. Parent function: $f(x) = |x|$

(a) Reflection in the x-axis and a horizontal shift 3 units to the left

$g(x) = -|x + 3|$

(b) Horizontal shift 2 units to the right and a vertical shift 4 units downward

$g(x) = |x - 2| - 4$

15. Parent function: $f(x) = x^3$

Horizontal shift 2 units to the right

$y = (x - 2)^3$

17. Parent function: $f(x) = x^2$

Reflection in the x-axis

$y = -x^2$

19. Parent function: $f(x) = \sqrt{x}$

Reflection in the x-axis and a vertical shift 1 unit upward

$y = -\sqrt{x} + 1$

21. $g(x) = 12 - x^2$

(a) Parent function: $f(x) = x^2$

(b) Reflection in the x-axis and a vertical shift 12 units upward

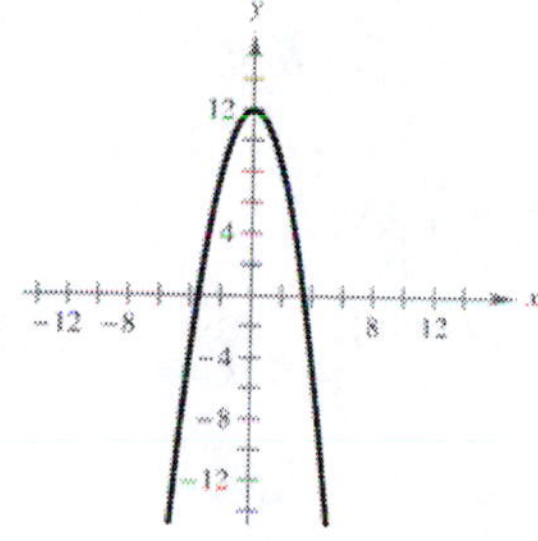

(d) $g(x) = 12 - f(x)$

23. $g(x) = x^3 + 7$

(a) Parent function: $f(x) = x^3$

(b) Vertical shift 7 units upward

(c)

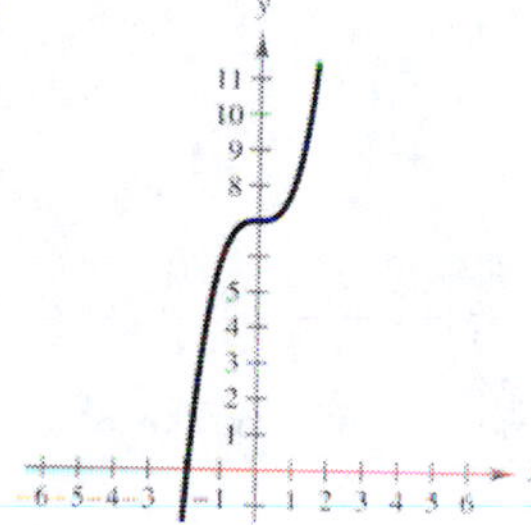

(d) $g(x) = f(x) + 7$

25. $g(x) = \frac{2}{3}x^2 + 4$

(a) Parent function: $f(x) = x^2$

(b) Vertical shrink of two-thirds, and a vertical shift 4 units upward

(c)

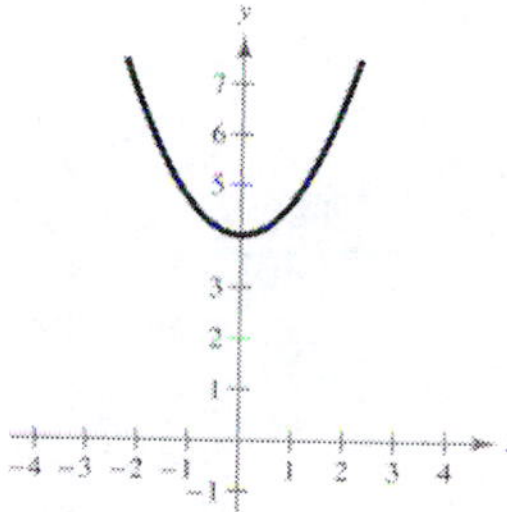

(d) $g(x) = \frac{2}{3}f(x) + 4$

27. $g(x) = 2 - (x + 5)^2$

(a) Parent function: $f(x) = x^2$

(b) Reflection in the x-axis, horizontal shift 5 units to the left, and a vertical shift 2 units upward

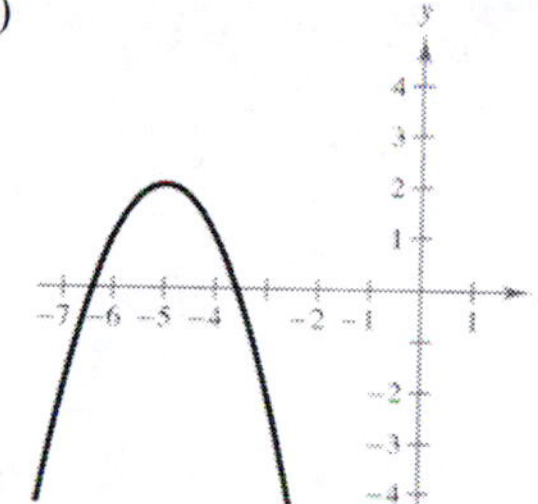

(d) $g(x) = 2 - f(x + 5)$

29. $g(x) = \sqrt{3x}$

(a) Parent function: $f(x) = \sqrt{x}$

(b) Horizontal shrink by $\frac{1}{3}$

(c)

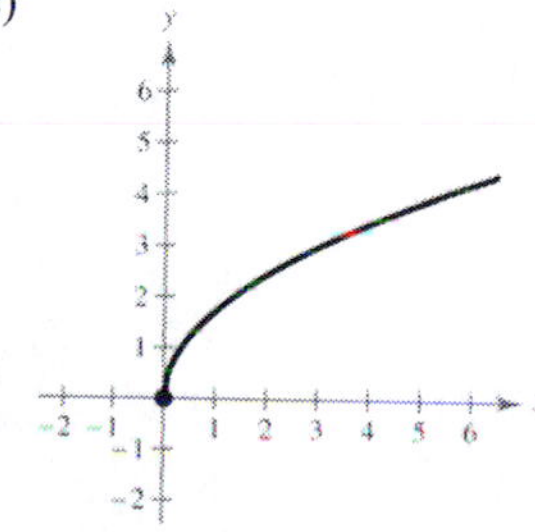

(d) $g(x) = f(3x)$

31. $g(x) = (x - 1)^3 + 2$

(a) Parent function: $f(x) = x^3$

(b) Horizontal shift 1 unit to the right and a vertical shift 2 units upward

(c)

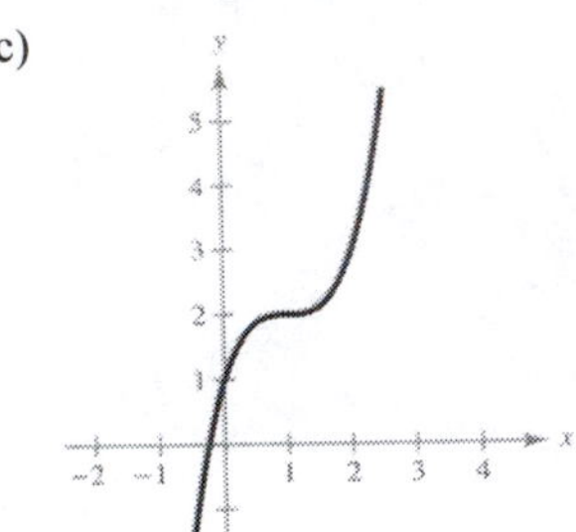

(d) $g(x) = f(x - 1) + 2$

33. $g(x) = 3(x - 2)^3$

(a) Parent function: $f(x) = x^3$

(b) Horizontal shift 2 units to the right, vertical stretch (each y-value is multiplied by 3)

(c)

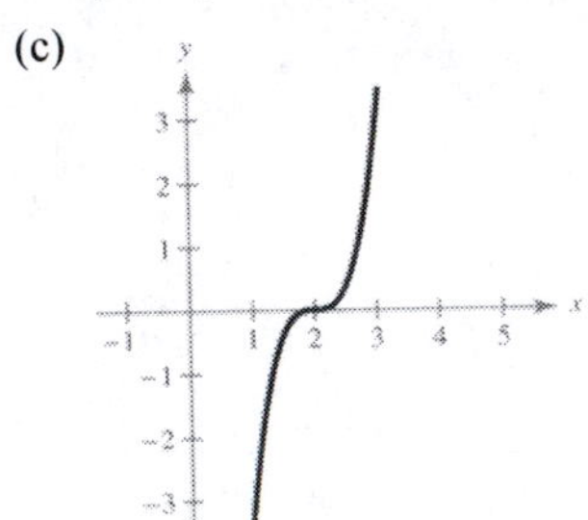

(d) $g(x) = 3f(x - 2)$

35. $g(x) = -|x| - 2$

(a) Parent function: $f(x) = |x|$

(b) Reflection in the x-axis, vertical shift 2 units downward

(c)

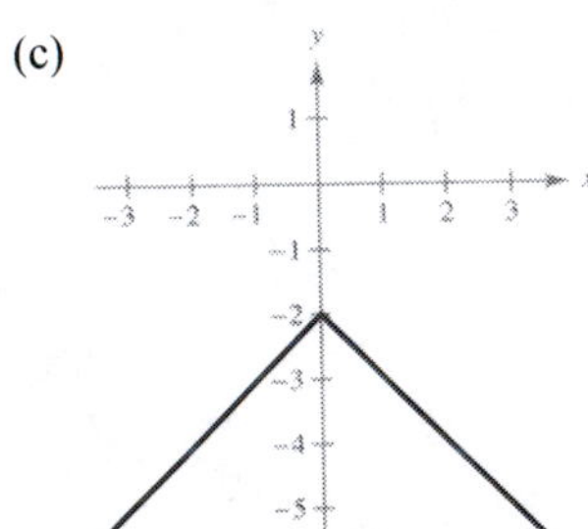

(d) $g(x) = -f(x) - 2$

37. $g(x) = -|x + 4| + 8$

(a) Parent function: $f(x) = |x|$

(b) Reflection in the x-axis, horizontal shift 4 units to the left, and a vertical shift 8 units upward

(c)

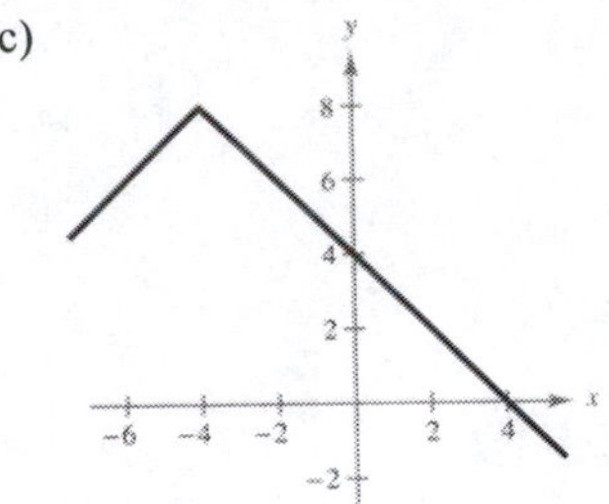

(d) $g(x) = -f(x + 4) + 8$

39. $g(x) = -2|x - 1| - 4$

(a) Parent function: $f(x) = |x|$

(b) Horizontal shift one unit to the right, vertical stretch, reflection in the x-axis, vertical shift four units downward

(c)

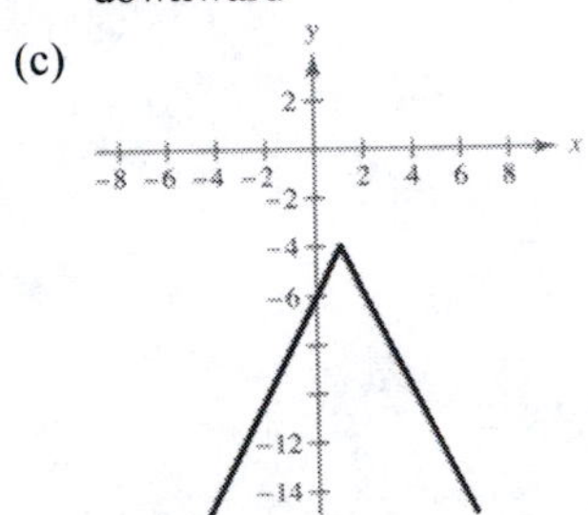

(d) $g(x) = -2f(x - 1) - 4$

41. $g(x) = 3 - [\![x]\!]$

(a) Parent function: $f(x) = [\![x]\!]$

(b) Reflection in the x-axis and a vertical shift 3 units upward

(c)

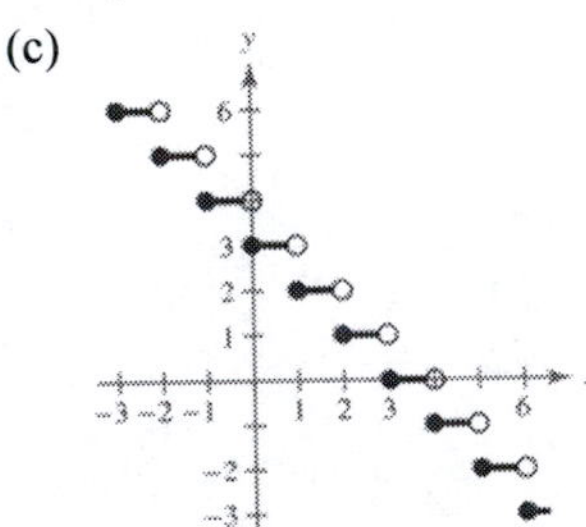

(d) $g(x) = 3 - f(x)$

43. $g(x) = \sqrt{x - 9}$

(a) Parent function: $f(x) = \sqrt{x}$

(b) Horizontal shift 9 units to the right

(c)

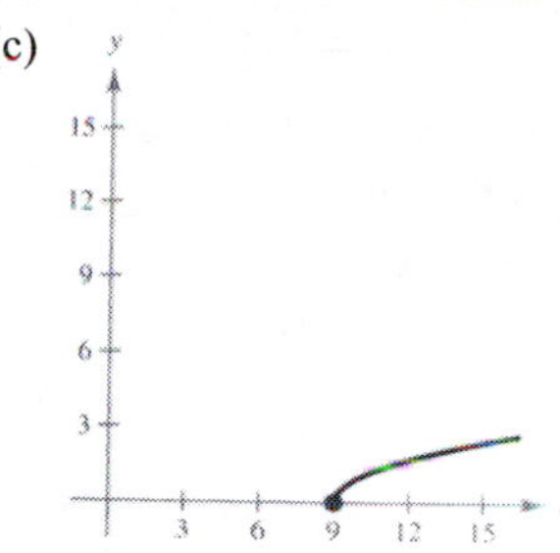

(d) $g(x) = f(x - 9)$

45. $g(x) = \sqrt{7 - x} - 2$ or $g(x) = \sqrt{-(x - 7)} - 2$

(a) Parent function: $f(x) = \sqrt{x}$

(b) Reflection in the y-axis, horizontal shift 7 units to the right, and a vertical shift 2 units downward

(c)

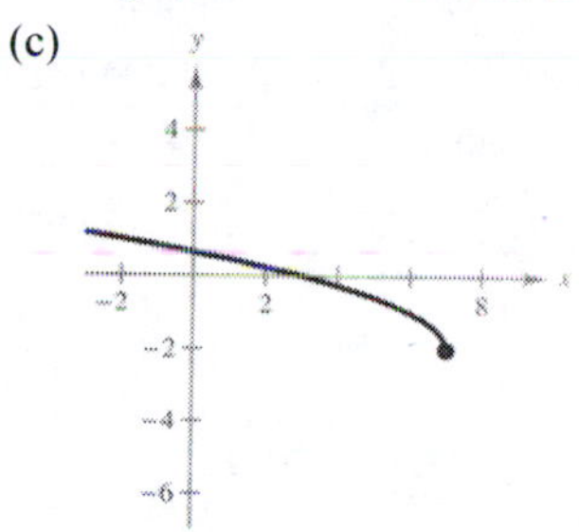

(d) $g(x) = f(7 - x) - 2$

47. $g(x) = (x - 3)^2 - 7$

49. $f(x) = x^3$ moved 13 units to the right

$g(x) = (x - 13)^3$

51. $g(x) = -|x| + 12$

53. $f(x) = \sqrt{x}$ moved 6 units to the left and reflected in both the x- and y-axes

$g(x) = -\sqrt{-x + 6}$

55. $f(x) = x^2$

(a) Reflection in the x-axis and a vertical stretch (each y-value is multiplied by 3)

$g(x) = -3x^2$

(b) Vertical shift 3 units upward and a vertical stretch (each y-value is multiplied by 4)

$g(x) = 4x^2 + 3$

57. $f(x) = |x|$

(a) Reflection in the x-axis and a vertical shrink $\left(\text{each } y\text{-value is multiplied by } \frac{1}{2}\right)$

$g(x) = -\frac{1}{2}|x|$

(b) Vertical stretch (each y-value is multiplied by 3) and a vertical shift 3 units downward

$g(x) = 3|x| - 3$

59. Parent function: $f(x) = x^3$

Vertical stretch (each y-value is multiplied by 2)

$g(x) = 2x^3$

61. Parent function: $f(x) = x^2$

Reflection in the x-axis, vertical shrink $\left(\text{each } y\text{-value is multiplied by } \frac{1}{2}\right)$

$g(x) = -\frac{1}{2}x^2$

63. Parent function: $f(x) = \sqrt{x}$

Reflection in the y-axis, vertical shrink $\left(\text{each } y\text{-value is multiplied by } \frac{1}{2}\right)$

$g(x) = \frac{1}{2}\sqrt{-x}$

65. Parent function: $f(x) = x^3$

Reflection in the x-axis, horizontal shift 2 units to the right and a vertical shift 2 units upward

$g(x) = -(x - 2)^3 + 2$

67. Parent function: $f(x) = \sqrt{x}$

Reflection in the x-axis and a vertical shift 3 units downward

$g(x) = -\sqrt{x} - 3$

69. (a)

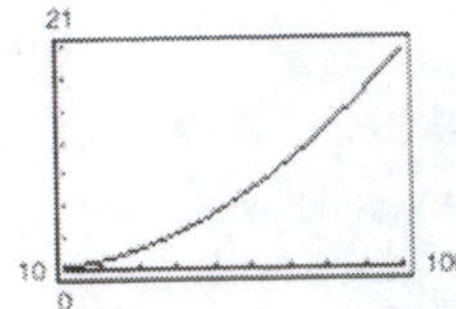

(b) $H(x) = 0.002x^2 + 0.005x - 0.029$

$$H\left(\frac{x}{1.6}\right) = 0.002\left(\frac{x}{1.6}\right)^2 + 0.005\left(\frac{x}{1.6}\right) - 0.029$$

$$= 0.002\left(\frac{x^2}{2.56}\right) + 0.005\left(\frac{x}{1.6}\right) - 0.029$$

$$= 0.00078125x^2 + 0.003125x - 0.029$$

The graph of $H\left(\frac{x}{1.6}\right)$ is a horizontal stretch of the graph of $H(x)$.

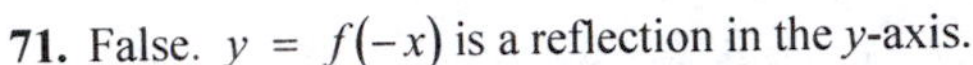

71. False. $y = f(-x)$ is a reflection in the y-axis.

73. True. Because $|x| = |-x|$, the graphs of $f(x) = |x| + 6$ and $f(x) = |-x| + 6$ are identical.

75. $y = f(x + 2) - 1$

Horizontal shift 2 units to the left and a vertical shift 1 unit downward

$(0, 1) \to (0 - 2, 1 - 1) = (-2, 0)$

$(1, 2) \to (1 - 2, 2 - 1) = (-1, 1)$

$(2, 3) \to (2 - 2, 3 - 1) = (0, 2)$

77. (a)

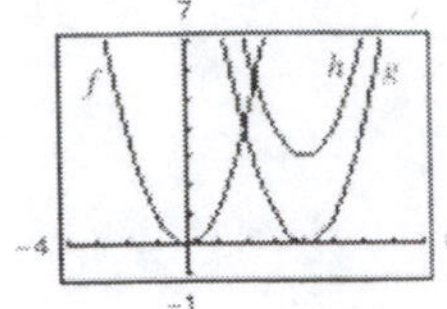

(b)

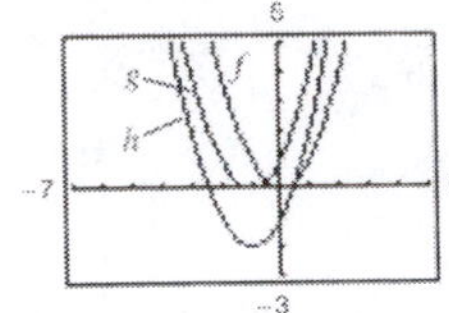

(c)

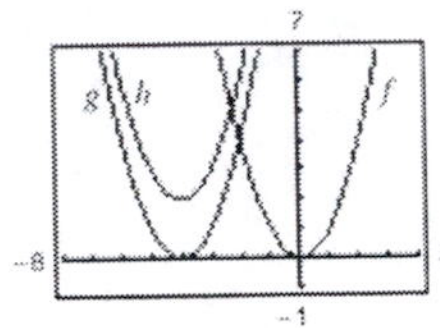

79. (a) The profits were only $\frac{3}{4}$ as large as expected:

$g(t) = \frac{3}{4}f(t)$

(b) The profits were \$10,000 greater than predicted:

$g(t) = f(t) + 10{,}000$

(c) There was a two-year delay: $g(t) = f(t - 2)$

Section 2.6 Combinations of Functions: Composite Functions

1. addition; subtraction; multiplication; division

3.

x	0	1	2	3
f	2	3	1	2
g	−1	0	$\frac{1}{2}$	0
$f + g$	1	3	$\frac{3}{2}$	2

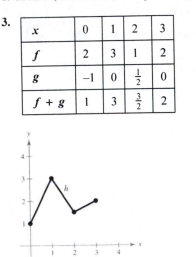

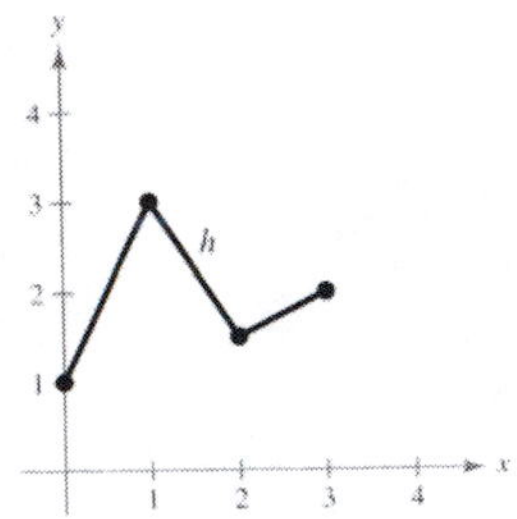

5. $f(x) = x + 2,\ g(x) = x - 2$

(a) $(f + g)(x) = f(x) + g(x)$
$= (x + 2) + (x - 2)$
$= 2x$

(b) $(f - g)(x) = f(x) - g(x)$
$= (x + 2) - (x - 2)$
$= 4$

(c) $(fg)(x) = f(x) \cdot g(x)$
$= (x + 2)(x - 2)$
$= x^2 - 4$

(d) $\left(\frac{f}{x}\right)(x) = \frac{f(x)}{g(x)} = \frac{x + 2}{x - 2}$

Domain: all real numbers x except $x = 2$

7. $f(x) = x^2,\ g(x) = 4x - 5$

(a) $(f + g)(x) = f(x) + g(x) = x^2 + (4x - 5) = x^2 + 4x - 5$

(b) $(f - g)(x) = f(x) - g(x) = x^2 - (4x - 5) = x^2 - 4x + 5$

(c) $(fg)(x) = f(x) \cdot g(x) = x^2(4x - 5) = 4x^3 - 5x^2$

(d) $\left(\frac{f}{g}\right)(x) = \frac{f(x)}{g(x)} = \frac{x^2}{4x - 5}$

Domain: all real numbers x except $x = \frac{5}{4}$

9. $f(x) = x^2 + 6,\ g(x) = \sqrt{1 - x}$

(a) $(f + g)(x) = f(x) + g(x) = x^2 + 6 + \sqrt{1 - x}$

(b) $(f - g)(x) = f(x) - g(x) = x^2 + 6 - \sqrt{1 - x}$

(c) $(fg)(x) = f(x) \cdot g(x) = (x^2 + 6)\sqrt{1 - x}$

(d) $\left(\frac{f}{g}\right)(x) = \frac{f(x)}{g(x)} = \frac{x^2 + 6}{\sqrt{1 - x}} = \frac{(x^2 + 6)\sqrt{1 - x}}{1 - x}$

Domain: $x < 1$

11. $f(x) = \frac{1}{x},\ g(x) = \frac{1}{x^2}$

(a) $(f + g)(x) = f(x) + g(x) = \frac{1}{x} + \frac{1}{x^2} = \frac{x + 1}{x^2}$

(b) $(f - g)(x) = f(x) - g(x) = \frac{1}{x} - \frac{1}{x^2} = \frac{x - 1}{x^2}$

(c) $(fg)(x) = f(x) \cdot g(x) = \frac{1}{x}\left(\frac{1}{x^2}\right) = \frac{1}{x^3}$

(d) $\left(\frac{f}{g}\right)(x) = \frac{f(x)}{g(x)} = \frac{1/x}{1/x^2} = \frac{x^2}{x} = x$

Domain: all real numbers x except $x = 0$

For Exercises 13–23, $f(x) = x^2 + 1$ and $g(x) = x - 4$.

13. $(f + g)(2) = f(2) + g(2) = (2^2 + 1) + (2 - 4) = 3$

15. $(f - g)(0) = f(0) - g(0) = (0^2 + 1) - (0 - 4) = 5$

17. $(f - g)(3t) = f(3t) - g(3t) = \left[(3t)^2 + 1\right] - (3t - 4) = 9t^2 - 3t + 5$

19. $(fg)(6) = f(6)g(6) = (6^2 + 1)(6 - 4) = 74$

21. $\left(\frac{f}{g}\right)(5) = \frac{f(5)}{g(5)} = \frac{5^2 + 1}{5 - 4} = 26$

23. $\left(\frac{f}{g}\right)(-1) - g(3) = \frac{f(-1)}{g(-1)} - g(3) = \frac{(-1)^2 + 1}{-1 - 4} - (3 - 4) = -\frac{2}{5} + 1 = \frac{3}{5}$

25. $f(x) = \frac{1}{2}x,\ g(x) = x - 1$

$(f + g)(x) = \frac{3}{2}x - 1$

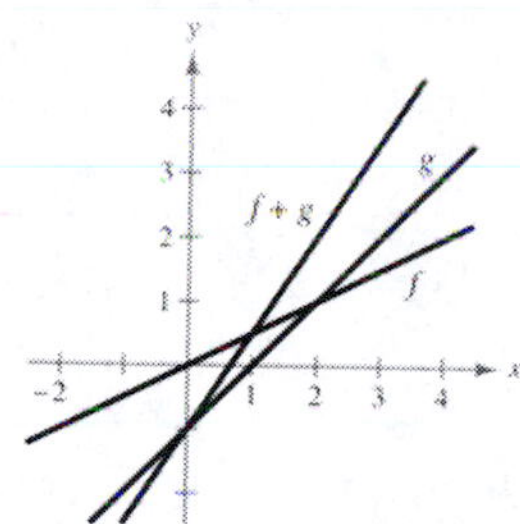

27. $f(x) = 3x,\ g(x) = -\frac{x^3}{10}$

$(f + g)(x) = 3x - \frac{x^3}{10}$

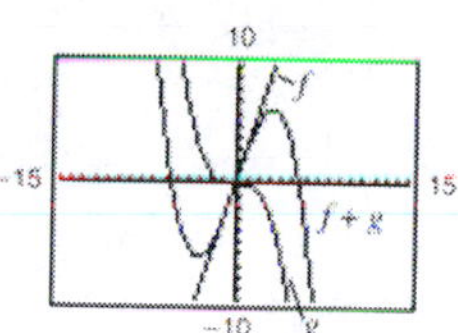

For $0 \le x \le 2$, $f(x)$ contributes most to the magnitude.

For $x > 6$, $g(x)$ contributes most to the magnitude.

29. $f(x) = 3x + 2,\ g(x) = -\sqrt{x + 5}$

$(f + g)x = 3x - \sqrt{x + 5} + 2$

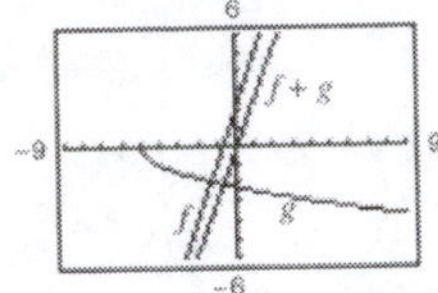

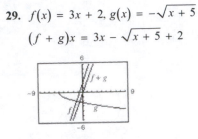

For $0 \le x \le 2$, $f(x)$ contributes most to the magnitude.

For $x > 6$, $f(x)$ contributes most to the magnitude.

31. $f(x) = x^2,\ g(x) = x - 1$

(a) $(f \circ g)(x) = f(g(x)) = f(x - 1) = (x - 1)^2$

(b) $(g \circ f)(x) = g(f(x)) = g(x^2) = x^2 - 1$

(c) $(g \circ g)(x) = g(g(x)) = g(x - 1) = x - 2$

33. $f(x) = \sqrt[3]{x - 1},\ g(x) = x^3 + 1$

(a) $(f \circ g)(x) = f(g(x))$

$= f(x^3 + 1)$

$= \sqrt[3]{(x^3 + 1) - 1}$

$= \sqrt[3]{x^3} = x$

(b) $(g \circ f)(x) = g(f(x))$

$= g(\sqrt[3]{x - 1})$

$= (\sqrt[3]{x - 1})^3 + 1$

$= (x - 1) + 1 = x$

(c) $(g \circ g)(x) = g(g(x))$

$= g(x^3 + 1)$

$= (x^3 + 1)^3 + 1$

$= x^9 + 3x^6 + 3x^3 + 2$

35. $f(x) = \sqrt{x + 4}$ Domain: $x \ge -4$

$g(x) = x^2$ Domain: all real numbers x

(a) $(f \circ g)(x) = f(g(x)) = f(x^2) = \sqrt{x^2 + 4}$

Domain: all real numbers x

(b) $(g \circ f)(x) = g(f(x))$

$= g(\sqrt{x + 4}) = (\sqrt{x + 4})^2 = x + 4$

Domain: $x \ge -4$

37. $f(x) = x^2 + 1$ Domain: all real numbers x

$g(x) = \sqrt{x}$ Domain: $x \ge 0$

(a) $(f \circ g)(x) = f(g(x))$

$= f(\sqrt{x})$

$= (\sqrt{x})^2 + 1$

$= x + 1$

Domain: $x \ge 0$

(b) $(g \circ f)(x) = g(f(x)) = g(x^2 + 1) = \sqrt{x^2 + 1}$

Domain: all real numbers x

39. $f(x) = |x|$ Domain: all real numbers x

$g(x) = x + 6$ Domain: all real numbers x

(a) $(f \circ g)(x) = f(g(x)) = f(x + 6) = |x + 6|$

Domain: all real numbers x

(b) $(g \circ f)(x) = g(f(x)) = g(|x|) = |x| + 6$

Domain: all real numbers x

41. $f(x) = \dfrac{1}{x}$ Domain: all real numbers x except $x = 0$

$g(x) = x + 3$ Domain: all real numbers x

(a) $(f \circ g)(x) = f(g(x)) = f(x + 3) = \dfrac{1}{x + 3}$

Domain: all real numbers x except $x = -3$

(b) $(g \circ f)(x) = g(f(x)) = g\left(\dfrac{1}{x}\right) = \dfrac{1}{x} + 3$

Domain: all real numbers x except $x = 0$

43. (a) $(f + g)(3) = f(3) + g(3) = 2 + 1 = 3$

(b) $\left(\frac{f}{g}\right)(2) = \frac{f(2)}{g(2)} = \frac{0}{2} = 0$

45. (a) $(f \circ g)(2) = f(g(2)) = f(2) = 0$

(b) $(g \circ f)(2) = g(f(2)) = g(0) = 4$

47. $h(x) = (2x^2 + 1)^2$

One possibility: Let $f(x) = x^2$ and $g(x) = 2x + 1$, then $(f \circ g)(x) = h(x)$.

49. $h(x) = \sqrt[3]{x^2 - 4}$

One possibility: Let $f(x) = \sqrt[3]{x}$ and $g(x) = x^2 - 4$, then $(f \circ g)(x) = h(x)$.

51. $h(x) = \frac{1}{x + 2}$

One possibility: Let $f(x) = 1/x$ and $g(x) = x + 2$, then $(f \circ g)(x) = h(x)$.

53. $h(x) = \frac{-x^2 + 3}{4 - x^2}$

One possibility: Let $f(x) = \frac{x + 3}{4 + x}$ and $g(x) = -x^2$, then $(f \circ g)(x) = h(x)$.

55. (a) $T(x) = R(x) + B(x) = \frac{3}{4}x + \frac{1}{15}x^2$

(b)

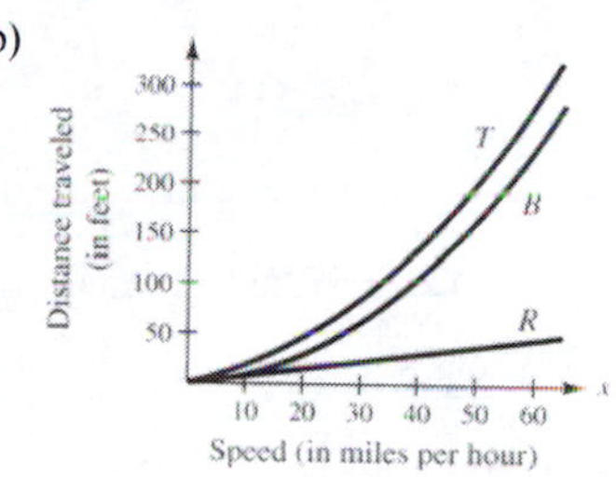

(c) $B(x)$; As x increases, $B(x)$ increases at a faster rate.

57. (a) $p(t) = d(t) + c(t)$

(b) $p(5)$ represents the number of dogs and cats in 2005.

(c) $h(t) = \frac{p(t)}{n(t)} = \frac{d(t) + c(t)}{n(t)}$

$h(t)$ represents the number of dogs and cats at time t compared to the population at time t or the number of dogs and cats per capita.

59. (a) $r(x) = \frac{x}{2}$

(b) $A(r) = \pi r^2$

(c) $(A \circ r)(x) = A(r(x)) = A\left(\frac{x}{2}\right) = \pi\left(\frac{x}{2}\right)^2$

$(A \circ r)(x)$ represents the area of the circular base of the tank on the square foundation with side length x.

61. (a) $f(g(x)) = f(0.03x) = 0.03x - 500{,}000$

(b) $g(f(x)) = g(x - 500{,}000) = 0.03(x - 500{,}000)$

$g(f(x))$ represents your bonus of 3% of an amount over \$500,000.

63. Let

O = oldest sibling, M = middle sibling,

Y = youngest sibling.

Then the ages of each sibling can be found using the equations:

$O = 2M$

$M = \frac{1}{2}Y + 6$

(a) $O(M(Y)) = 2\left(\frac{1}{2}(Y) + 6\right) = 12 + Y$; Answers will vary.

(b) Oldest sibling is 16: $O = 16$

Middle sibling: $O = 2M$

$16 = 2M$

$M = 8$ years old

Youngest sibling: $M = \frac{1}{2}Y + 6$

$8 = \frac{1}{2}Y + 6$

$2 = \frac{1}{2}Y$

$Y = 4$ years old

65. False. $(f \circ g)(x) = 6x + 1$ and $(g \circ f)(x) = 6x + 6$

67. Let $f(x)$ and $g(x)$ be two odd functions and define $h(x) = f(x)g(x)$. Then

$$\begin{aligned} h(-x) &= f(-x)g(-x) \\ &= [-f(x)][-g(x)] \quad \text{because } f \text{ and } g \text{ are odd} \\ &= f(x)g(x) \\ &= h(x). \end{aligned}$$

So, $h(x)$ is even.

Let $f(x)$ and $g(x)$ be two even functions and define $h(x) = f(x)g(x)$. Then

$$\begin{aligned} h(-x) &= f(-x)g(-x) \\ &= f(x)g(x) \quad \text{because } f \text{ and } g \text{ are even} \\ &= h(x). \end{aligned}$$

So, $h(x)$ is even.

69. Let $f(x)$ be an odd function, $g(x)$ be an even function, and define $h(x) = f(x)g(x)$. Then

$$\begin{aligned} h(-x) &= f(-x)g(-x) \\ &= [-f(x)]g(x) \quad \text{because } f \text{ is odd and } g \text{ is even} \\ &= -f(x)g(x) \\ &= -h(x). \end{aligned}$$

So, h is odd and the product of an odd function and an even function is odd.

Section 2.7 Inverse Functions

1. inverse

3. range; domain

5. one-to-one

7. $f(x) = 6x$

$$f^{-1}(x) = \frac{x}{6} = \frac{1}{6}x$$

$$f(f^{-1}(x)) = f\left(\frac{x}{6}\right) = 6\left(\frac{x}{6}\right) = x$$

$$f^{-1}(f(x)) = f^{-1}(6x) = \frac{6x}{6} = x$$

9. $f(x) = 3x + 1$

$$f^{-1}(x) = \frac{x - 1}{3}$$

$$f(f^{-1}(x)) = f\left(\frac{x - 1}{3}\right) = 3\left(\frac{x - 1}{3}\right) + 1 = x$$

$$f^{-1}(f(x)) = f^{-1}(3x + 1) = \frac{(3x + 1) - 1}{3} = x$$

11. $f(x) = \sqrt[3]{x}$

$$f^{-1}(x) = x^3$$

$$f(f^{-1}(x)) = f(x^3) = \sqrt[3]{x^3} = x$$

$$f^{-1}(f(x)) = f^{-1}(\sqrt[3]{x}) = (\sqrt[3]{x})^3 = x$$

13. $(f \circ g)(x) = f(g(x)) = f\left(-\dfrac{2x + 6}{7}\right) = -\dfrac{7}{2}\left(-\dfrac{2x + 6}{7}\right) - 3 = x + 3 - 3 = x$

$$(g \circ f)(x) = g(f(x)) = g\left(-\frac{7}{2}x - 3\right) = -\frac{2\left(-\frac{7}{2}x - 3\right) + 6}{7} = \frac{-(-7x)}{7} = x$$

15. $(f \circ g)(x) = f(g(x)) = f\left(\sqrt[3]{x - 5}\right) = \left(\sqrt[3]{x - 5}\right)^3 + 5 = x - 5 + 5 = x$

$$(g \circ f)(x) = g(f(x)) = g(x^3 + 5) = \sqrt[3]{x^3 + 5 - 5} = \sqrt[3]{x^3} = x$$

17.

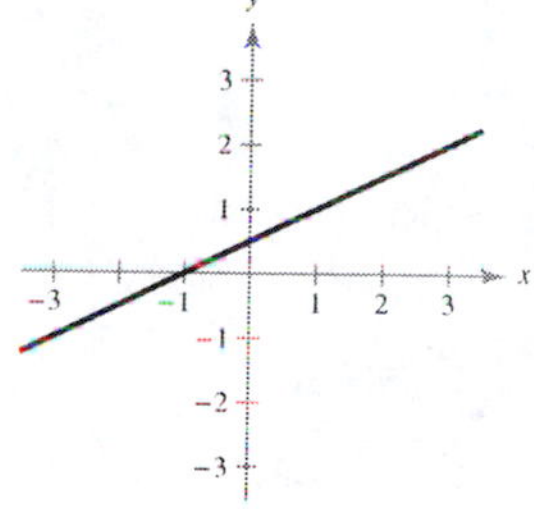

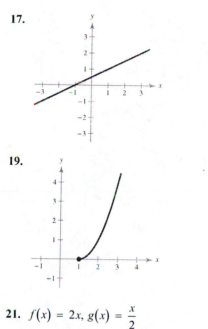

19.

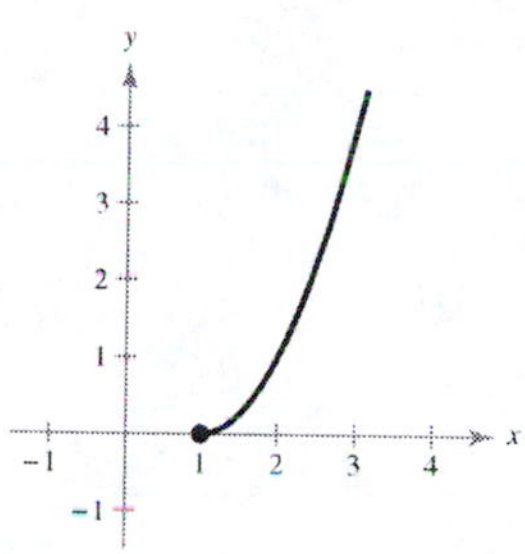

21. $f(x) = 2x,\ g(x) = \dfrac{x}{2}$

(a) $f(g(x)) = f\left(\dfrac{x}{2}\right) = 2\left(\dfrac{x}{2}\right) = x$

$g(f(x)) = g(2x) = \dfrac{2x}{2} = x$

(b)

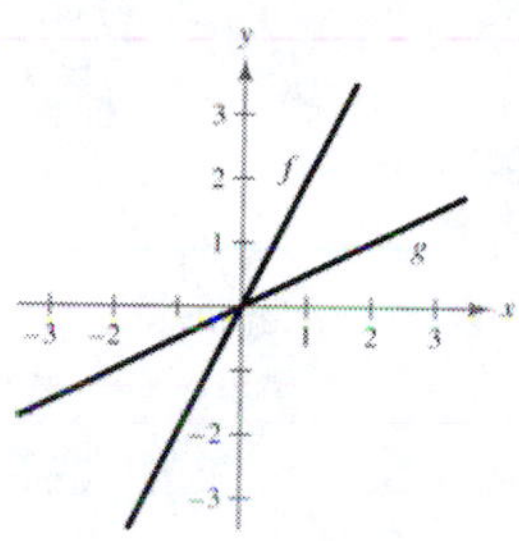

23. $f(x) = 7x + 1,\ g(x) = \dfrac{x-1}{7}$

(a) $f(g(x)) = f\left(\dfrac{x-1}{7}\right) = 7\left(\dfrac{x-1}{7}\right) + 1 = x$

$g(f(x)) = g(7x+1) = \dfrac{(7x+1)-1}{7} = x$

(b)

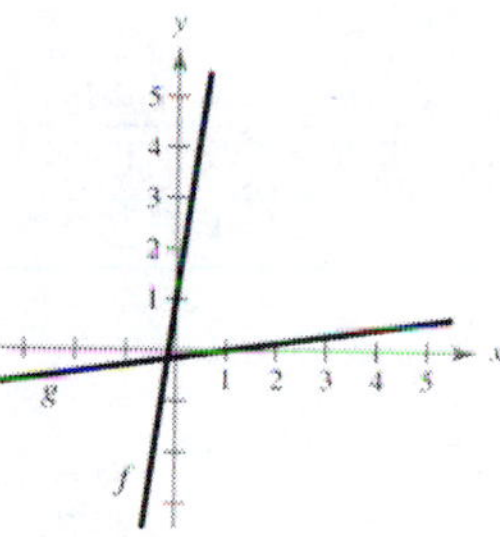

25. $f(x) = \dfrac{x^3}{8},\ g(x) = \sqrt[3]{8x}$

(a) $f(g(x)) = f\left(\sqrt[3]{8x}\right) = \dfrac{\left(\sqrt[3]{8x}\right)^3}{8} = \dfrac{8x}{8} = x$

$g(f(x)) = g\left(\dfrac{x^3}{8}\right) = \sqrt[3]{8\left(\dfrac{x^3}{8}\right)} = \sqrt[3]{x^3} = x$

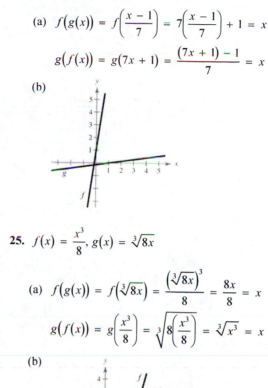

(b)

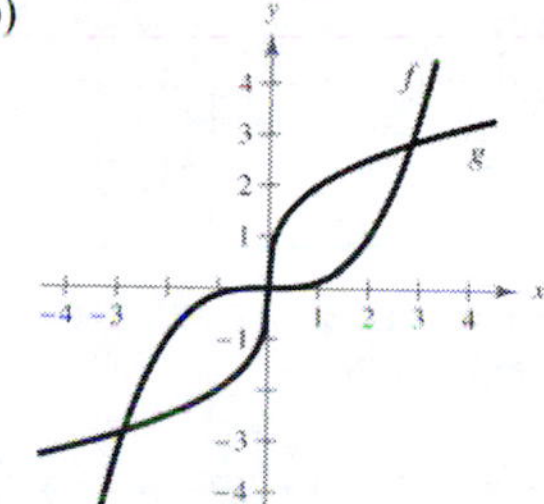

27. $f(x) = \sqrt{x-4},\ g(x) = x^2 + 4,\ x \ge 0$

(a) $f(g(x)) = f(x^2+4),\ x \ge 0 = \sqrt{(x^2+4)-4} = x$

$g(f(x)) = g\left(\sqrt{x-4}\right) = \left(\sqrt{x-4}\right)^2 + 4 = x$

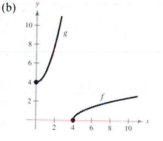

(b)

29. $f(x) = 9 - x^2,\ x \ge 0;\ g(x) = \sqrt{9-x},\ x \le 9$

(a) $f(g(x)) = f\left(\sqrt{9-x}\right),\ x \le 9 = 9 - \left(\sqrt{9-x}\right)^2 = x$

$g(f(x)) = g(9-x^2),\ x \ge 0 = \sqrt{9-(9-x^2)} = x$

(b)

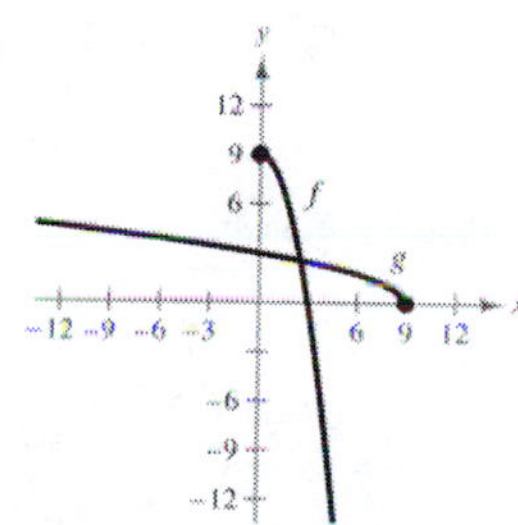

31. $f(x) = \dfrac{x-1}{x+5}$, $g(x) = -\dfrac{5x+1}{x-1}$

(a) $f(g(x)) = f\left(-\dfrac{5x+1}{x-1}\right) = \dfrac{\left(-\dfrac{5x+1}{x-1} - 1\right)}{\left(-\dfrac{5x+1}{x-1} + 5\right)} \cdot \dfrac{x-1}{x-1} = \dfrac{-(5x+1)-(x-1)}{-(5x+1)+5(x-1)} = \dfrac{-6x}{-6} = x$

$g(f(x)) = g\left(\dfrac{x-1}{x+5}\right) = -\dfrac{\left[5\left(\dfrac{x-1}{x+5}\right)+1\right]}{\left[\dfrac{x-1}{x+5} - 1\right]} \cdot \dfrac{x+5}{x+5} = -\dfrac{5(x-1)+(x+5)}{(x-1)-(x+5)} = -\dfrac{6x}{-6} = x$

(b)

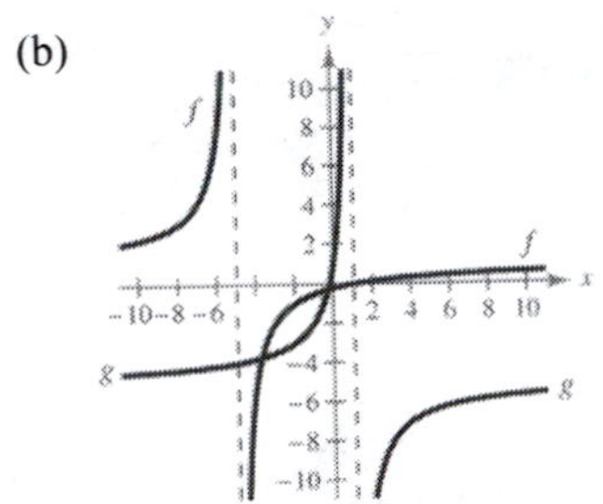

33. No, $\{(-2,-1), (1,0), (2,1), (1,2), (-2,3), (-6,4)\}$ does not represent a function. -2 and 1 are paired with two different values.

35.

x	-2	0	2	4	6	8
$f^{-1}(x)$	-2	-1	0	1	2	3

37. Yes, because no horizontal line crosses the graph of f at more than one point, f *has* an inverse.

39. No, because some horizontal lines cross the graph of f twice, f *does not* have an inverse.

41. $g(x) = (x+5)^3$

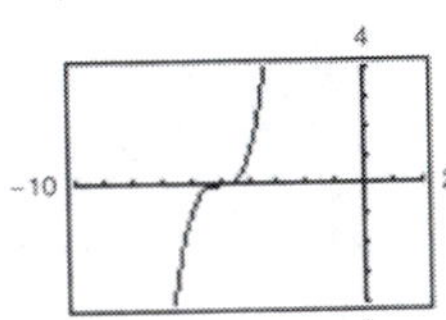

g passes the Horizontal Line Test, so g *has* an inverse.

43. $f(x) = -2x\sqrt{16 - x^2}$

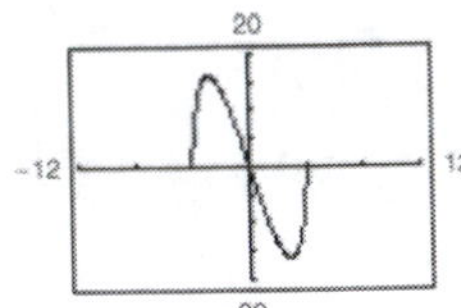

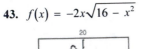

f does not pass the Horizontal Line Test, so f *does not* have an inverse.

45. (a) $f(x) = 2x - 3$

$y = 2x - 3$

$x = 2y - 3$

$y = \dfrac{x+3}{2}$

$f^{-1}(x) = \dfrac{x+3}{2}$

(b)

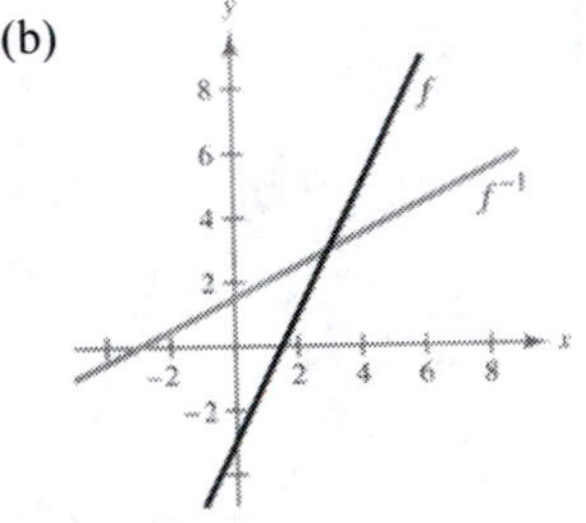

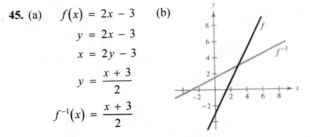

(c) The graph of f^{-1} is the reflection of the graph of f in the line $y = x$.

(d) The domains and ranges of f and f^{-1} are all real numbers.

47. (a) $f(x) = x^5 - 2$

$y = x^5 - 2$

$x = y^5 - 2$

$y = \sqrt[5]{x+2}$

$f^{-1}(x) = \sqrt[5]{x+2}$

(b)

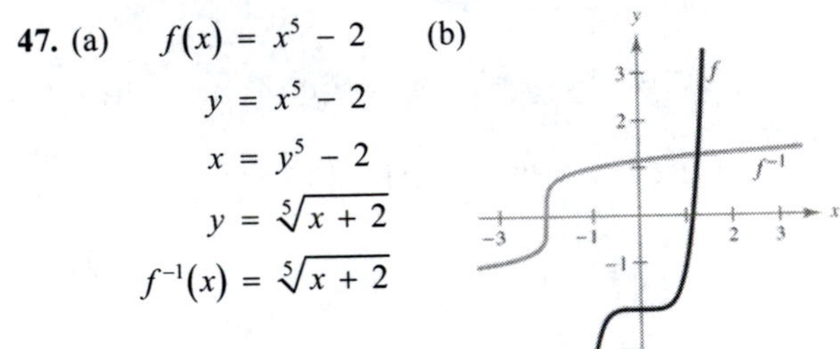

(c) The graph of f^{-1} is the reflection of the graph of f in the line $y = x$.

(d) The domains and ranges of f and f^{-1} are all real numbers.

49. (a) $f(x) = \sqrt{4 - x^2}, 0 \le x \le 2$

$$y = \sqrt{4 - x^2}$$
$$x = \sqrt{4 - y^2}$$
$$x^2 = 4 - y^2$$
$$y^2 = 4 - x^2$$
$$y = \sqrt{4 - x^2}$$
$$f^{-1}(x) = \sqrt{4 - x^2}, 0 \le x \le 2$$

(b)

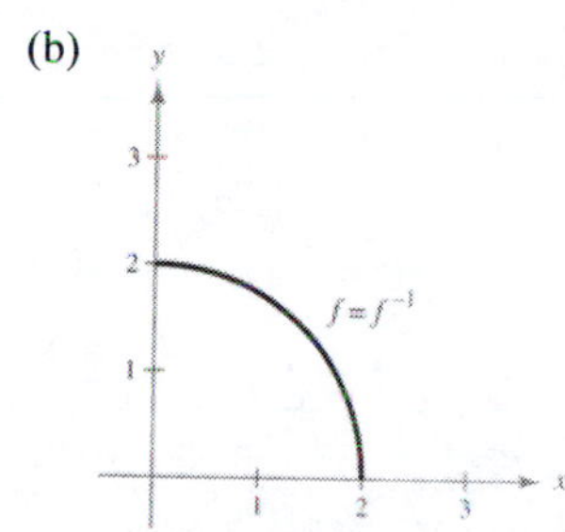

(c) The graph of f^{-1} is the same as the graph of *f*.

(d) The domains and ranges of *f* and f^{-1} are all real numbers *x* such that $0 \le x \le 2$.

51. (a) $f(x) = \dfrac{4}{x}$

$$y = \frac{4}{x}$$
$$x = \frac{4}{y}$$
$$xy = 4$$
$$y = \frac{4}{x}$$
$$f^{-1}(x) = \frac{4}{x}$$

(b)

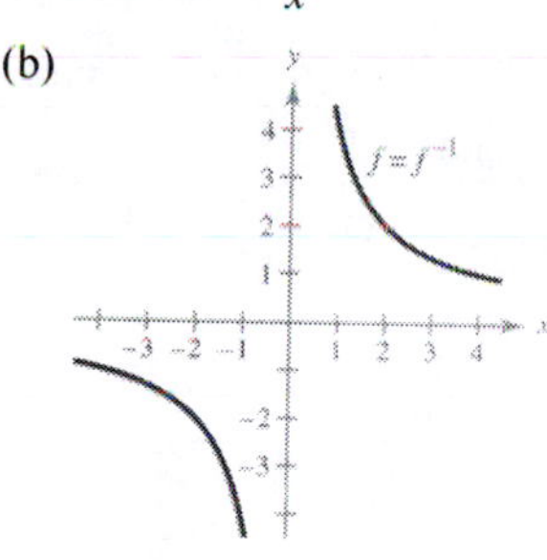

(c) The graph of f^{-1} is the same as the graph of *f*.

(d) The domains and ranges of *f* and f^{-1} are all real numbers except for 0.

53. (a) $f(x) = \dfrac{x + 1}{x - 2}$

$$y = \frac{x + 1}{x - 2}$$
$$x = \frac{y + 1}{y - 2}$$
$$x(y - 2) = y + 1$$
$$xy - 2x = y + 1$$
$$xy - y = 2x + 1$$
$$y(x - 1) = 2x + 1$$
$$y = \frac{2x + 1}{x - 1}$$
$$f^{-1}(x) = \frac{2x + 1}{x - 1}$$

(b)

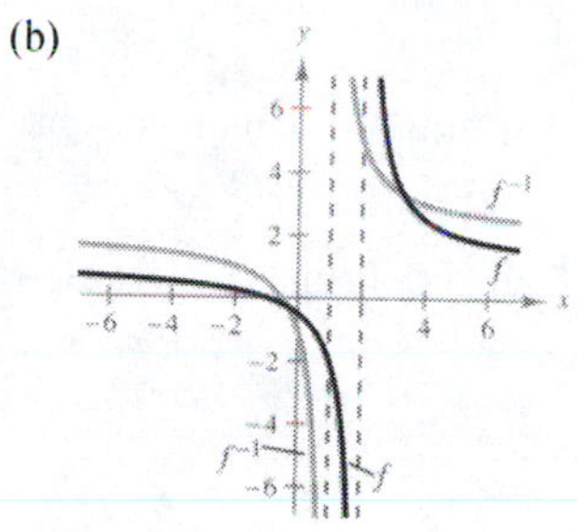

(c) The graph of f^{-1} is the reflection of graph of *f* in the line $y = x$.

(d) The domain of *f* and the range of f^{-1} is all real numbers except 2.

The range of *f* and the domain of f^{-1} is all real numbers except 1.

55. (a) $f(x) = \sqrt[3]{x-1}$

$y = \sqrt[3]{x-1}$

$x = \sqrt[3]{y-1}$

$x^3 = y - 1$

$y = x^3 + 1$

$f^{-1}(x) = x^3 + 1$

(b)

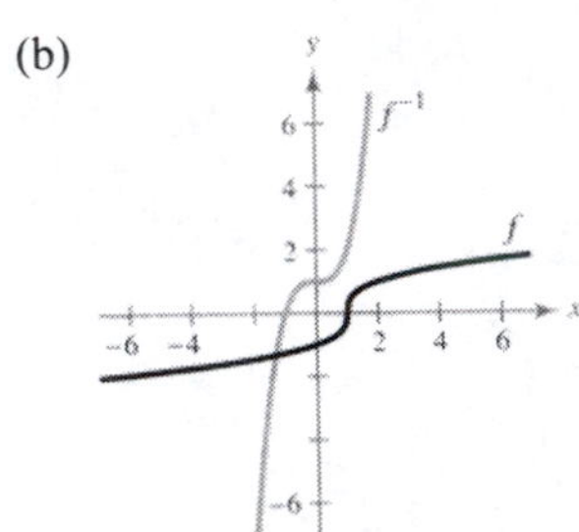

(c) The graph of f^{-1} is the reflection of the graph of f in the line $y = x$.

(d) The domains and ranges of f and f^{-1} are all real numbers.

57. $f(x) = x^4$

$y = x^4$

$x = y^4$

$y = \pm\sqrt[4]{x}$

This does not represent y as a function of x. f does not have an inverse.

59. $g(x) = \dfrac{x}{8}$

$y = \dfrac{x}{8}$

$x = \dfrac{y}{8}$

$y = 8x$

This is a function of x, so g has an inverse.

$g^{-1}(x) = 8x$

61. $p(x) = -4$

$y = -4$

Because $y = -4$ for all x, the graph is a horizontal line and fails the Horizontal Line Test. p does not have an inverse.

63. $f(x) = (x+3)^2,\ x \ge -3 \Rightarrow y \ge 0$

$y = (x+3)^2,\ x \ge -3,\ y \ge 0$

$x = (y+3)^2,\ y \ge -3,\ x \ge 0$

$\sqrt{x} = y + 3,\ y \ge -3,\ x \ge 0$

$y = \sqrt{x} - 3,\ x \ge 0,\ y \ge -3$

This is a function of x, so f has an inverse.

$f^{-1}(x) = \sqrt{x} - 3,\ x \ge 0$

65. $f(x) = \begin{cases} x + 3, & x < 0 \\ 6 - x, & x \ge 0 \end{cases}$

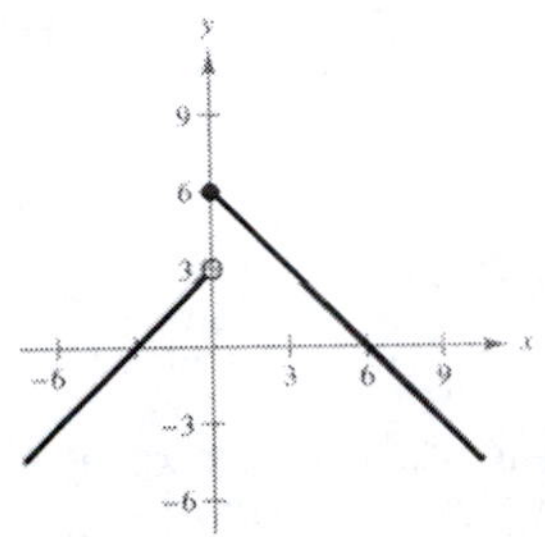

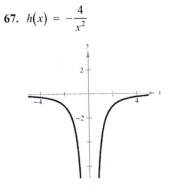

This graph fails the Horizontal Line Test, so f does not have an inverse.

67. $h(x) = -\dfrac{4}{x^2}$

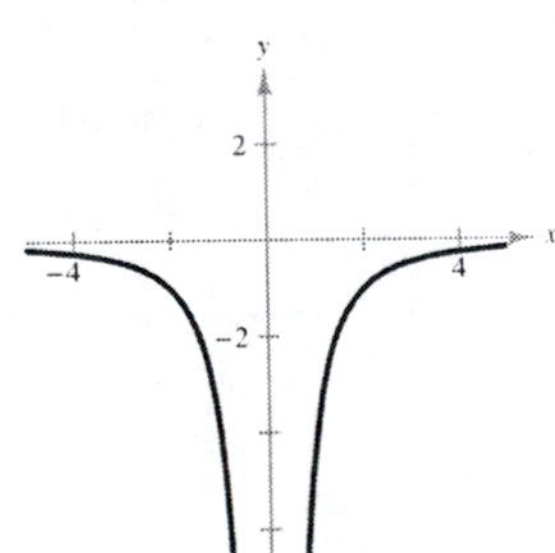

The graph fails the Horizontal Line Test so h does not have an inverse.

69. $f(x) = \sqrt{2x+3} \Rightarrow x \ge -\dfrac{3}{2},\ y \ge 0$

$y = \sqrt{2x+3},\ x \ge -\dfrac{3}{2},\ y \ge 0$

$x = \sqrt{2y+3},\ y \ge -\dfrac{3}{2},\ x \ge 0$

$x^2 = 2y + 3,\ x \ge 0,\ y \ge -\dfrac{3}{2}$

$y = \dfrac{x^2 - 3}{2},\ x \ge 0,\ y \ge -\dfrac{3}{2}$

This is a function of x, so f has an inverse.

$f^{-1}(x) = \dfrac{x^2 - 3}{2},\ x \ge 0$

71. $f(x) = \dfrac{6x + 4}{4x + 5}$

$$y = \frac{6x + 4}{4x + 5}$$
$$x = \frac{6y + 4}{4y + 5}$$
$$x(4y + 5) = 6y + 4$$
$$4xy + 5x = 6y + 4$$
$$4xy - 6y = -5x + 4$$
$$y(4x - 6) = -5x + 4$$
$$y = \frac{-5x + 4}{4x - 6}$$
$$= \frac{5x - 4}{6 - 4x}$$

This is a function of x, so f has an inverse.

$$f^{-1}(x) = \frac{5x - 4}{6 - 4x}$$

73. $f(x) = (x - 2)^2$

domain of $f: x \geq 2$, range of $f: y \geq 0$

$$f(x) = (x - 2)^2$$
$$y = (x - 2)^2$$
$$x = (y - 2)^2$$
$$\sqrt{x} = y - 2$$
$$\sqrt{x} + 2 = y$$

So, $f^{-1}(x) = \sqrt{x} + 2$.

domain of $f^{-1}: x \geq 0$, range of $f^{-1}: x \geq 2$

75. $f(x) = |x + 2|$

domain of $f: x \geq -2$, range of $f: y \geq 0$

$$f(x) = |x + 2|$$
$$y = |x + 2|$$
$$x = y + 2$$
$$x - 2 = y$$

So, $f^{-1}(x) = x - 2$.

domain of $f^{-1}: x \geq 0$, range of $f^{-1}: y \geq -2$

77. $f(x) = (x + 6)^2$

domain of $f: x \geq -6$, range of $f: y \geq 0$

$$f(x) = (x + 6)^2$$
$$y = (x + 6)^2$$
$$x = (y + 6)^2$$
$$\sqrt{x} = y + 6$$
$$\sqrt{x} - 6 = y$$

So, $f^{-1}(x) = \sqrt{x} - 6$.

domain of $f^{-1}: x \geq 0$, range of $f^{-1}: y \geq -6$

79. $f(x) = -2x^2 + 5$

domain of $f: x \geq 0$, range of $f: y \leq 5$

$$f(x) = -2x^2 + 5$$
$$y = -2x^2 + 5$$
$$x = -2y^2 + 5$$
$$x - 5 = -2y^2$$
$$5 - x = 2y^2$$
$$\sqrt{\frac{5 - x}{2}} = y$$
$$\frac{\sqrt{5 - x}}{\sqrt{2}} \cdot \frac{\sqrt{2}}{\sqrt{2}} = y$$
$$\frac{\sqrt{2(5 - x)}}{2} = y$$

So, $f^{-1}(x) = \dfrac{\sqrt{-2(x - 5)}}{2}$.

domain of $f^{-1}(x): x \leq 5$, range of $f^{-1}(x): y \geq 0$

81. $f(x) = |x - 4| + 1$

domain of $f: x \geq 4$, range of $f: y \geq 1$

$$f(x) = |x - 4| + 1$$
$$y = x - 3$$
$$x = y - 3$$
$$x + 3 = y$$

So, $f^{-1}(x) = x + 3$.

domain of $f^{-1}: x \geq 1$, range of $f^{-1}: y \geq 4$

In Exercises 83–87, $f(x) = \frac{1}{8}x - 3$, $f^{-1}(x) = 8(x + 3)$, $g(x) = x^3$, $g^{-1}(x) = \sqrt[3]{x}$.

83. $(f^{-1} \circ g^{-1})(1) = f^{-1}(g^{-1}(1))$

$= f^{-1}(\sqrt[3]{1})$

$= 8(\sqrt[3]{1} + 3) = 32$

85. $(f^{-1} \circ f^{-1})(6) = f^{-1}(f^{-1}(6))$

$= f^{-1}(8[6 + 3])$

$= 8[8(6 + 3) + 3] = 600$

87. $(f \circ g)(x) = f(g(x)) = f(x^3) = \frac{1}{8}x^3 - 3$

$y = \frac{1}{8}x^3 - 3$

$x = \frac{1}{8}y^3 - 3$

$x + 3 = \frac{1}{8}y^3$

$8(x + 3) = y^3$

$\sqrt[3]{8(x + 3)} = y$

$(f \circ g)^{-1}(x) = 2\sqrt[3]{x + 3}$

In Exercises 89–91, $f(x) = x + 4$, $f^{-1}(x) = x - 4$, $g(x) = 2x - 5$, $g^{-1}(x) = \dfrac{x + 5}{2}$.

89. $(g^{-1} \circ f^{-1})(x) = g^{-1}(f^{-1}(x))$

$= g^{-1}(x - 4)$

$= \dfrac{(x - 4) + 5}{2}$

$= \dfrac{x + 1}{2}$

91. $(f \circ g)(x) = f(g(x))$

$= f(2x - 5)$

$= (2x - 5) + 4$

$= 2x - 1$

$(f \circ g)^{-1}(x) = \dfrac{x + 1}{2}$

Note: Comparing Exercises 89 and 91, $(f \circ g)^{-1}(x) = (g^{-1} \circ f^{-1})(x)$.

93. (a) $y = 10 + 0.75x$

$x = 10 + 0.75y$

$x - 10 = 0.75y$

$\dfrac{x - 10}{0.75} = y$

So, $f^{-1}(x) = \dfrac{x - 10}{0.75}$.

x = hourly wage, y = number of units produced

(b) $y = \dfrac{24.25 - 10}{0.75} = 19$

So, 19 units are produced.

95. False. $f(x) = x^2$ is even and does not have an inverse.

97.

x	1	3	4	6
f	1	2	6	7

x	1	2	6	7
$f^{-1}(x)$	1	3	4	6

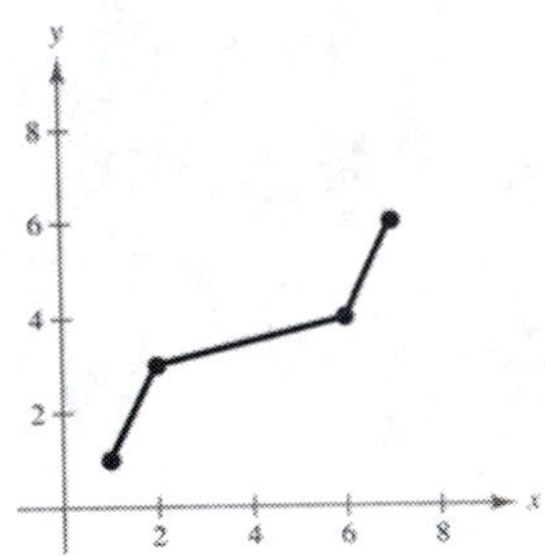

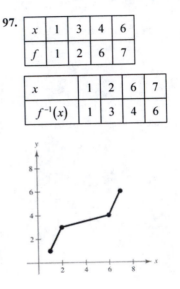

99. Let $(f \circ g)(x) = y$. Then $x = (f \circ g)^{-1}(y)$. Also,

$(f \circ g)(x) = y \Rightarrow f(g(x)) = y$

$g(x) = f^{-1}(y)$

$x = g^{-1}(f^{-1}(y))$

$x = (g^{-1} \circ f^{-1})(y)$.

Because f and g are both one-to-one functions, $(f \circ g)^{-1} = g^{-1} \circ f^{-1}$.

101. If $f(x) = k(2 - x - x^3)$ has an inverse and $f^{-1}(3) = -2$, then $f(-2) = 3$. So,

$f(-2) = k(2 - (-2) - (-2)^3) = 3$

$k(2 + 2 + 8) = 3$

$12k = 3$

$k = \frac{3}{12} = \frac{1}{4}$.

So, $k = \frac{1}{4}$.

103.

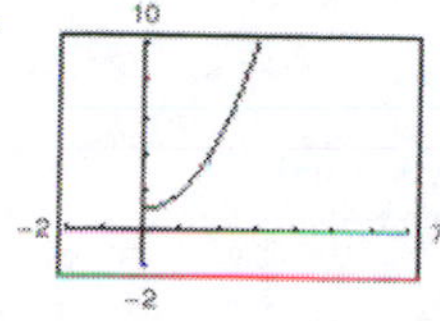

There is an inverse function $f^{-1}(x) = \sqrt{x - 1}$ because the domain of f is equal to the range of f^{-1} and the range of f is equal to the domain of f^{-1}.

105. This situation could be represented by a one-to-one function if the runner does not stop to rest. The inverse function would represent the time in hours for a given number of miles completed.

Review Exercises for Chapter 2

1. $y = -2x - 7$

Slope: $m = -2 = -\frac{2}{1}$

y-intercept: $(0, -7)$

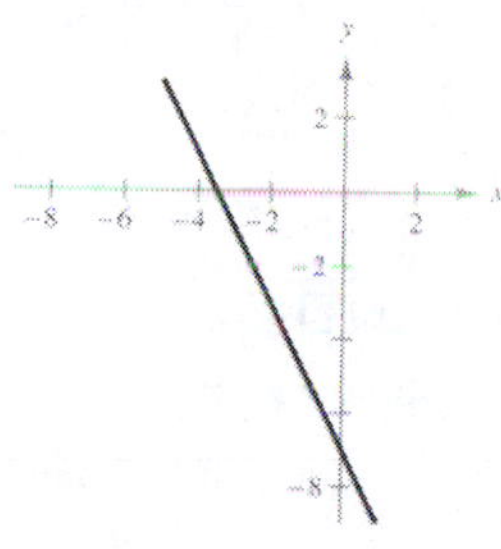

3. $y = 6$

Slope: $m = 0$

y-intercept: $(0, 6)$

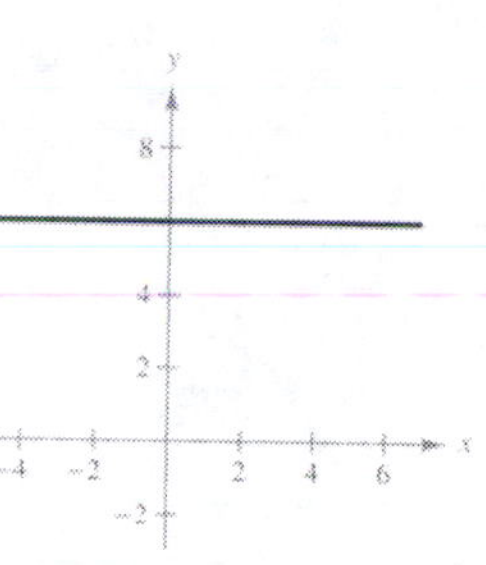

5. $(6, 4), (-3, -4)$

$$m = \frac{4 - (-4)}{6 - (-3)} = \frac{4 + 4}{6 + 3} = \frac{8}{9}$$

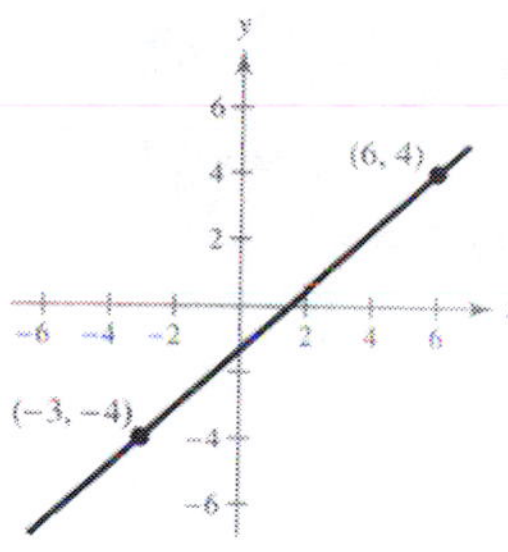

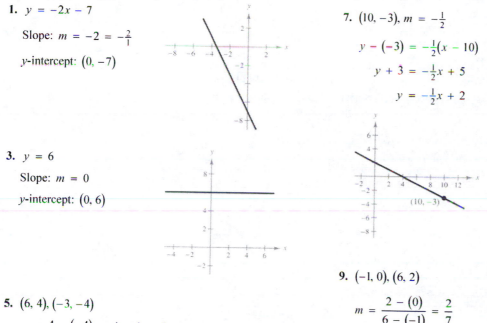

7. $(10, -3), m = -\frac{1}{2}$

$$y - (-3) = -\frac{1}{2}(x - 10)$$
$$y + 3 = -\frac{1}{2}x + 5$$
$$y = -\frac{1}{2}x + 2$$

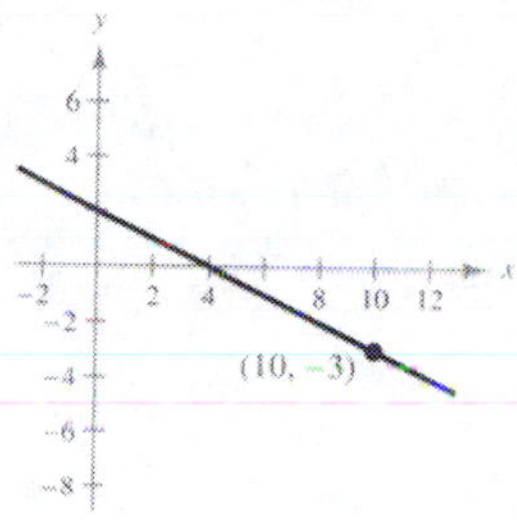

9. $(-1, 0), (6, 2)$

$$m = \frac{2 - (0)}{6 - (-1)} = \frac{2}{7}$$

$$y - 0 = \frac{2}{7}(x - (-1))$$
$$y = \frac{2}{7}(x + 1)$$
$$y = \frac{2}{7}x + \frac{2}{7}$$

11. Point: $(3, -2)$

$$5x - 4y = 8$$
$$y = \frac{5}{4}x - 2$$

(a) Parallel slope: $m = \frac{5}{4}$

$$y - (-2) = \frac{5}{4}(x - 3)$$
$$y + 2 = \frac{5}{4}x - \frac{15}{4}$$
$$y = \frac{5}{4}x - \frac{23}{4}$$

(b) Perpendicular slope: $m = -\frac{4}{5}$

$$y - (-2) = -\frac{4}{5}(x - 3)$$
$$y + 2 = -\frac{4}{5}x + \frac{12}{5}$$
$$y = -\frac{4}{5}x + \frac{2}{5}$$

13. *Verbal Model*: Sale price = (List price) − (Discount)

Labels: Sale price = S

List price = L

Discount = 20% of $L = 0.2L$

Equation: $S = L - 0.2L$

$S = 0.8L$

15. $16x - y^4 = 0$

$$y^4 = 16x$$
$$y = \pm 2\sqrt[4]{x}$$

No, y is not a function of x. Some x-values correspond to two y-values.

17. $y = \sqrt{1 - x}$

Yes, the equation represents y as a function of x. Each x-value, $x \leq 1$, corresponds to only one y-value.

19. $g(x) = x^{4/3}$

(a) $g(8) = 8^{4/3} = 2^4 = 16$

(b) $g(t + 1) = (t + 1)^{4/3}$

(c) $(-27)^{4/3} = (-3)^4 = 81$

(d) $g(-x) = (-x)^{4/3} = x^{4/3}$

21. $f(x) = \sqrt{25 - x^2}$

Domain: $25 - x^2 \geq 0$

$$(5 + x)(5 - x) \geq 0$$

Critical numbers: $x = \pm 5$

Test intervals: $(-\infty, -5), (-5, 5), (5, \infty)$

Test: Is $25 - x^2 \geq 0$?

Solution set: $-5 \leq x \leq 5$

Domain: all real numbers x such that $-5 \leq x \leq 5$, or $[-5, 5]$

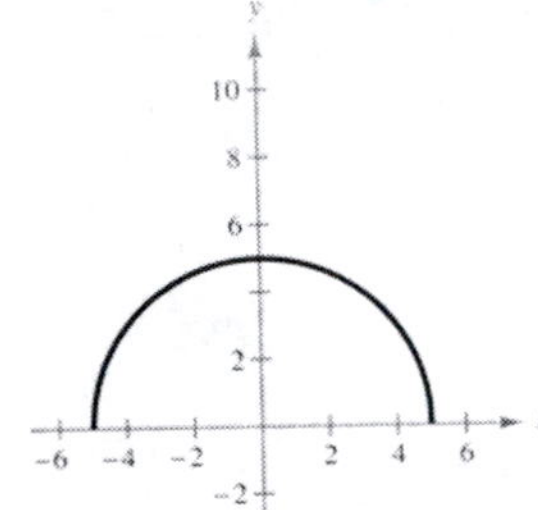

23. $v(t) = -32t + 48$

$v(1) = 16$ feet per second

25. $f(x) = 2x^2 + 3x - 1$

$$\frac{f(x + h) - f(x)}{h} = \frac{\left[2(x + h)^2 + 3(x + h) - 1\right] - (2x^2 + 3x - 1)}{h}$$
$$= \frac{2x^2 + 4xh + 2h^2 + 3x + 3h - 1 - 2x^2 - 3x + 1}{h}$$
$$= \frac{h(4x + 2h + 3)}{h}$$
$$= 4x + 2h + 3, \quad h \neq 0$$

27. $y = (x - 3)^2$

A vertical line intersects the graph no more than once, so y is a function of x.

29. $f(x) = 5x^2 + 4x - 1$

$$5x^2 + 4x - 1 = 0$$
$$(5x - 1)(x + 1) = 0$$
$$5x - 1 = 0 \Rightarrow x = \frac{1}{5}$$
$$x + 1 = 0 \Rightarrow x = -1$$

31. $f(x) = \sqrt{2x + 1}$

$$\sqrt{2x+1} = 0$$
$$2x + 1 = 0$$
$$2x = -1$$
$$x = \frac{1}{2}$$

33. $f(x) = |x| + |x + 1|$

f is increasing on $(0, \infty)$.

f is decreasing on $(-\infty, -1)$.

f is constant on $(-1, 0)$.

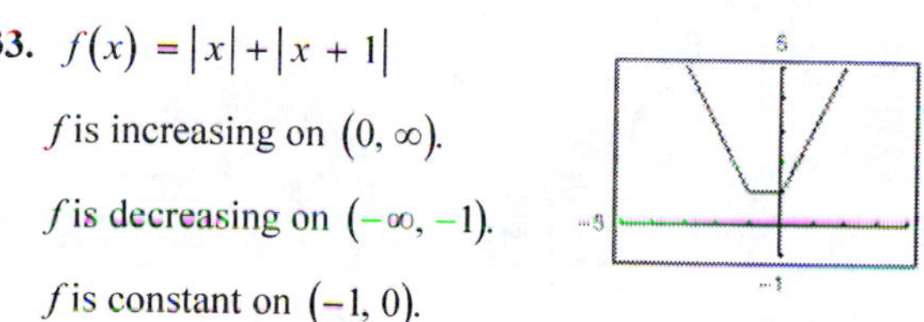

35. $f(x) = -x^2 + 2x + 1$

Relative maximum: $(1, 2)$

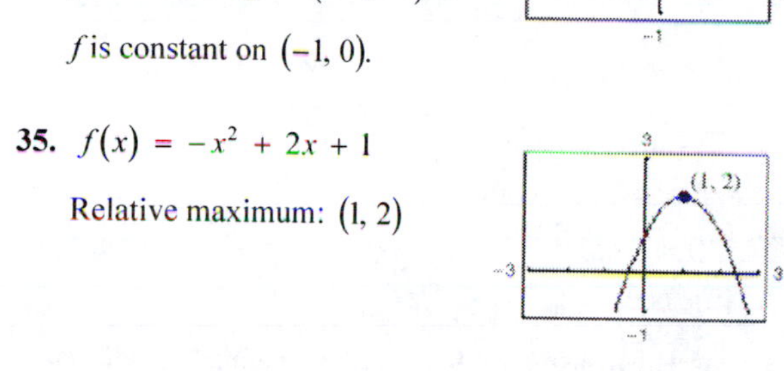

37. $f(x) = -x^2 + 8x - 4$

$$\frac{f(4) - f(0)}{4 - 0} = \frac{12 - (-4)}{4} = 4$$

The average rate of change of f from $x_1 = 0$ to $x_2 = 4$ is 4.

39. $f(x) = x^5 + 4x - 7$

$$f(-x) = (-x)^5 + 4(-x) - 7$$
$$= -x^5 - 4x - 7$$
$$\neq f(x)$$
$$\neq -f(x)$$

Neither even nor odd

41. $f(x) = 2x\sqrt{x^2 + 3}$

$$f(-x) = 2(-x)\sqrt{(-x)^2 + 3}$$
$$= -2x\sqrt{x^2 + 3}$$
$$= -f(x)$$

The function is odd.

43. (a) $f(2) = -6$, $f(-1) = 3$

Points: $(2, -6), (-1, 3)$

$$m = \frac{3 - (-6)}{-1 - 2} = \frac{9}{-3} = -3$$

$$y - (-6) = -3(x - 2)$$
$$y + 6 = -3x + 6$$
$$y = -3x$$
$$f(x) = -3x$$

(b)

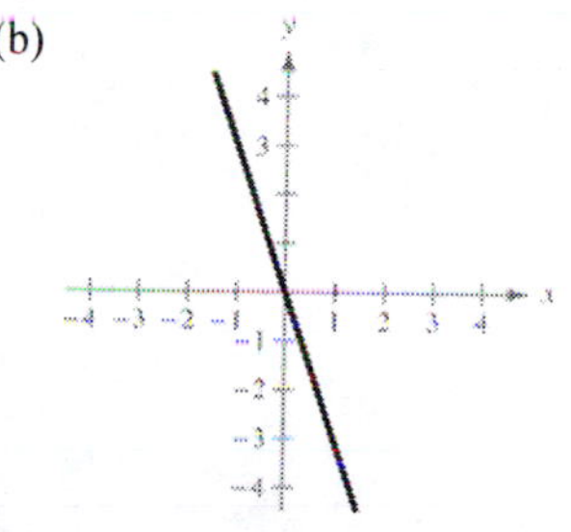

45. $f(x) = x^2 + 5$

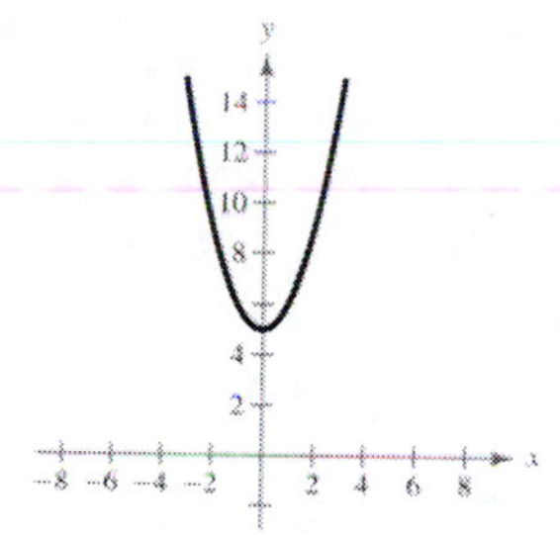

47. $f(x) = \sqrt{x + 1}$

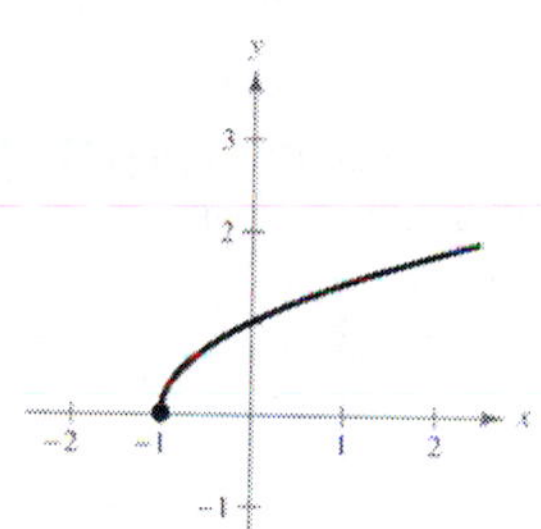

49. $g(x) = [\![x + 4]\!]$

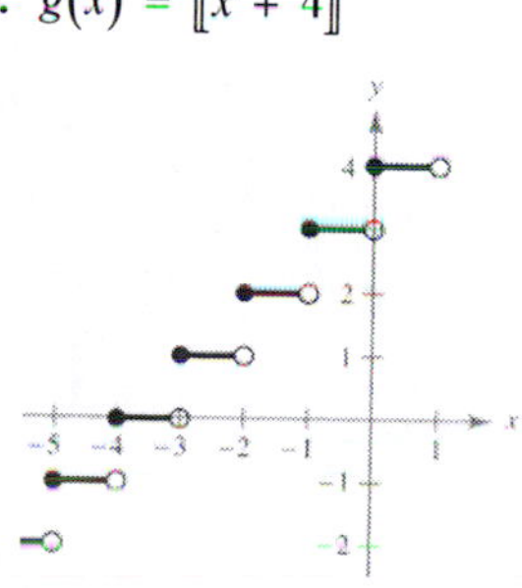

51. (a) $f(x) = x^2$

(b) $h(x) = x^2 - 9$

Vertical shift 9 units downward

(c)

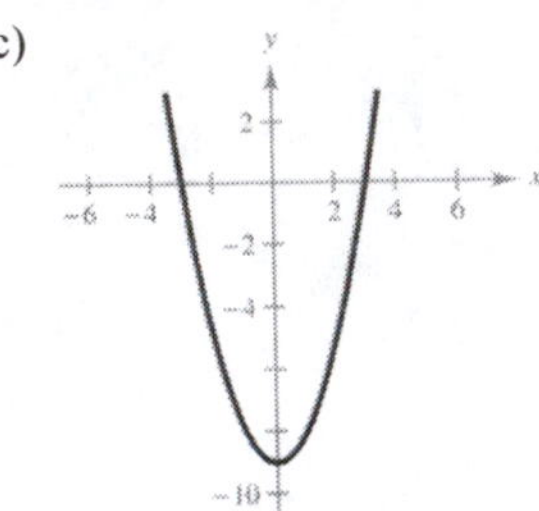

(d) $h(x) = f(x) - 9$

53. (a) $f(x) = \sqrt{x}$

(b) $h(x) = -\sqrt{x} + 4$

Vertical shift 4 units upward, reflection in the x-axis

(c)

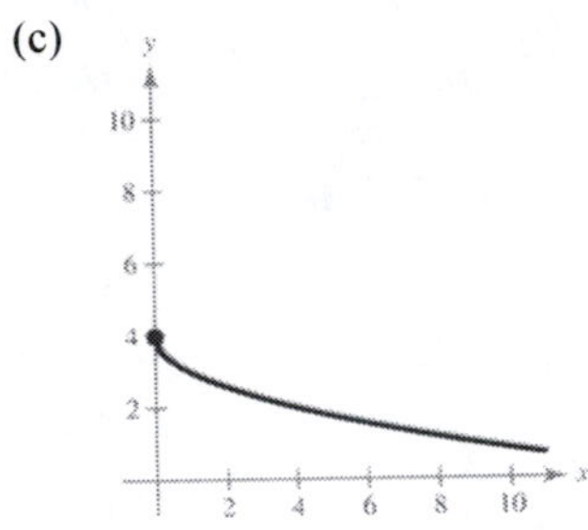

(d) $h(x) = -f(x) + 4$

55. (a) $f(x) = x^2$

(b) $h(x) = -(x + 2)^2 + 3$

Horizontal shift two units to the left, vertical shift 3 units upward, reflection in the x-axis.

(c)

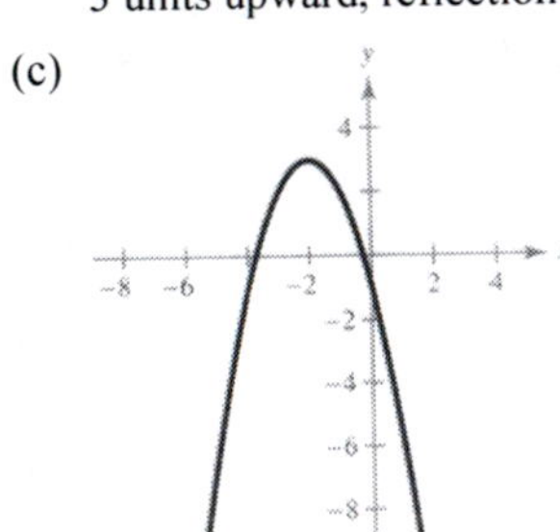

(d) $h(x) = -f(x + 2) + 3$

57. (a) $f(x) = [\![x]\!]$

(b) $h(x) = -[\![x]\!] + 6$

Reflection in the x-axis and a vertical shift 6 units upward

(c)

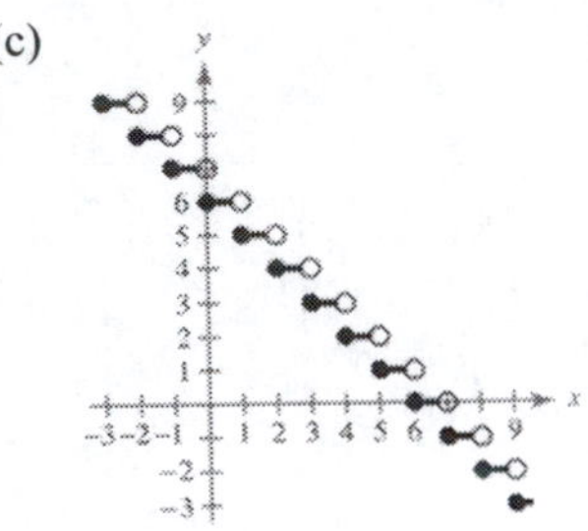

(d) $h(x) = -f(x) + 6$

59. (a) $f(x) = [\![x]\!]$

(b) $h(x) = 5[\![x - 9]\!]$

Horizontal shift 9 units to the right and a vertical stretch (each y-value is multiplied by 5)

(c)

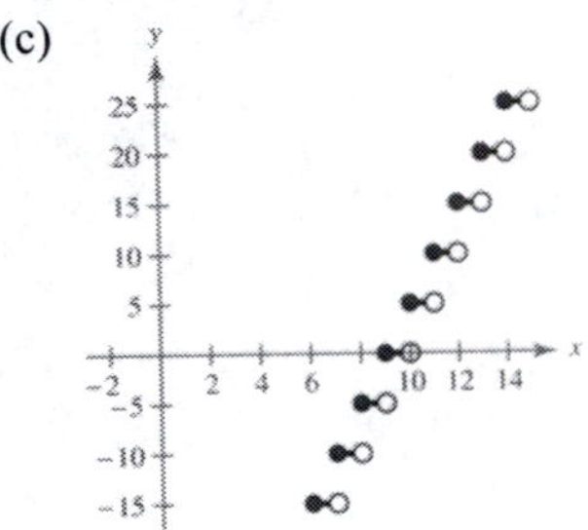

(d) $h(x) = 5f(x - 9)$

61. $f(x) = x^2 + 3,\ g(x) = 2x - 1$

(a) $(f + g)(x) = (x^2 + 3) + (2x - 1) = x^2 + 2x + 2$

(b) $(f - g)(x) = (x^2 + 3) - (2x - 1) = x^2 - 2x + 4$

(c) $(fg)(x) = (x^2 + 3)(2x - 1) = 2x^3 - x^2 + 6x - 3$

(d) $\left(\dfrac{f}{g}\right)(x) = \dfrac{x^2 + 3}{2x - 1}$, Domain: $x \neq \dfrac{1}{2}$

63. $f(x) = \frac{1}{3}x - 3,\ g(x) = 3x + 1$

The domains of f and g are all real numbers.

(a) $(f \circ g)(x) = f(g(x)) = f(3x + 1) = \frac{1}{3}(3x + 1) - 3 = x + \frac{1}{3} - 3 = x - \frac{8}{3}$

Domain: all real numbers

(b) $(g \circ f)(x) = g(f(x)) = g\left(\frac{1}{3}x - 3\right) = 3\left(\frac{1}{3}x - 3\right) + 1 = x - 9 + 1 = x - 8$

Domain: all real numbers

65.
$$\begin{aligned} N(T(t)) &= 25(2t + 1)^2 - 50(2t + 1) + 300,\ 2 \le t \le 20 \\ &= 25\left(4t^2 + 4t + 1\right) - 100t - 50 + 300 \\ &= 100t^2 + 100t + 25 - 100t + 250 \\ &= 100t^2 + 275 \end{aligned}$$

The composition $N(T(t))$ represents the number of bacteria in the food as a function of time.

67.
$$\begin{aligned} f(x) &= \frac{x - 4}{5} \\ y &= \frac{x - 4}{5} \\ x &= \frac{y - 4}{5} \\ 5x &= y - 4 \\ y &= 5x + 4 \end{aligned}$$

So, $f^{-1}(x) = 5x + 4$.

$$f\left(f^{-1}(x)\right) = f(5x + 4) = \frac{5x + 4 - 4}{5} = \frac{5x}{5} = x$$

$$f^{-1}\left(f(x)\right) = f^{-1}\left(\frac{x - 4}{5}\right) = 5\left(\frac{x - 4}{5}\right) + 4 = x - 4 + 4 = x$$

69. $f(x) = (x - 1)^2$

No, the function does not have an inverse because some horizontal lines intersect the graph twice.

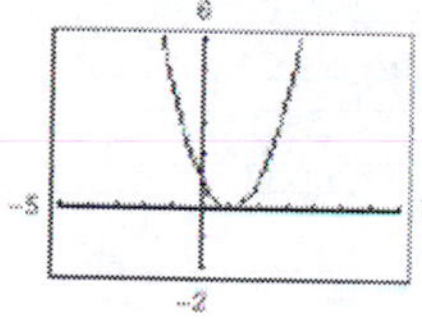

71. (a)
$$\begin{aligned} f(x) &= \frac{1}{2}x - 3 \\ y &= \frac{1}{2}x - 3 \\ x &= \frac{1}{2}y - 3 \\ x + 3 &= \frac{1}{2}y \\ 2(x + 3) &= y \\ f^{-1}(x) &= 2x + 6 \end{aligned}$$

(b)

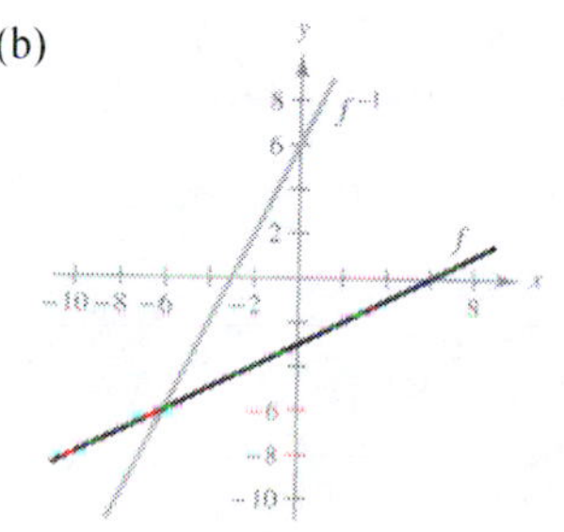

(c) The graph of f^{-1} is the reflection of the graph of f in the line $y = x$.

(d) The domains and ranges of f and f^{-1} are the set of all real numbers.

73. $f(x) = 2(x - 4)^2$ is increasing on $(4, \infty)$.

Let $f(x) = 2(x - 4)^2,\ x > 4$ and $y > 0$.

$$\begin{aligned} y &= 2(x - 4)^2 \\ x &= 2(y - 4)^2,\ x > 0,\ y > 4 \\ \frac{x}{2} &= (y - 4)^2 \\ \sqrt{\frac{x}{2}} &= y - 4 \\ \sqrt{\frac{x}{2}} + 4 &= y \\ f^{-1}(x) &= \sqrt{\frac{x}{2}} + 4,\ x > 0 \end{aligned}$$

75. False. The graph is reflected in the x-axis, shifted 9 units to the left, then shifted 13 units downward.

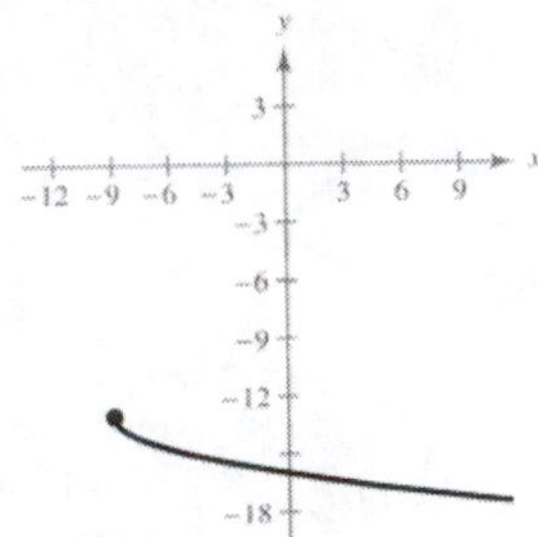

Problem Solving for Chapter 2

1. (a) $W_1 = 0.07S + 2000$

(b) $W_2 = 0.05S + 2300$

(c)

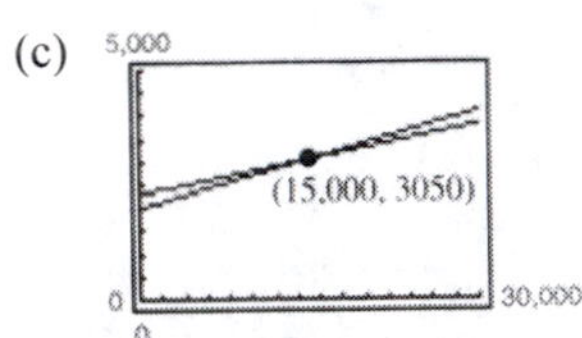

Point of intersection: $(15{,}000, 3050)$

Both jobs pay the same, \$3050, if you sell \$15,000 per month.

(d) No. If you think you can sell \$20,000 per month, keep your current job with the higher commission rate. For sales over \$15,000 it pays more than the other job.

3. (a) Let $f(x)$ and $g(x)$ be two even functions.

Then define $h(x) = f(x) \pm g(x)$.

$$\begin{aligned} h(-x) &= f(-x) \pm g(-x) \\ &= f(x) \pm g(x) \text{ because } f \text{ and } g \text{ are even} \\ &= h(x) \end{aligned}$$

So, $h(x)$ is also even.

(b) Let $f(x)$ and $g(x)$ be two odd functions.

Then define $h(x) = f(x) \pm g(x)$.

$$\begin{aligned} h(-x) &= f(-x) \pm g(-x) \\ &= -f(x) \pm g(x) \text{ because } f \text{ and } g \text{ are odd} \\ &= -h(x) \end{aligned}$$

So, $h(x)$ is also odd. $\left(\text{If } f(x) \neq g(x)\right)$

(c) Let $f(x)$ be odd and $g(x)$ be even. Then define $h(x) = f(x) \pm g(x)$.

$$\begin{aligned} h(-x) &= f(-x) \pm g(-x) \\ &= -f(x) \pm g(x) \text{ because } f \text{ is odd and } g \text{ is even} \\ &\neq h(x) \\ &\neq -h(x) \end{aligned}$$

So, $h(x)$ is neither odd nor even.

5. $f(x) = a_{2n}x^{2n} + a_{2n-2}x^{2n-2} + \cdots + a_2x^2 + a_0$

$f(-x) = a_{2n}(-x)^{2n} + a_{2n-2}(-x)^{2n-2} + \cdots + a_2(-x)^2 + a_0 = a_{2n}x^{2n} + a_{2n-2}x^{2n-2} + \cdots + a_2x^2 + a_0 = f(x)$

So, $f(x)$ is even.

7. (a) April 11: 10 hours

April 12: 24 hours

April 13: 24 hours

April 14: $23\frac{2}{3}$ hours

Total: $81\frac{2}{3}$ hours

(b) $\text{Speed} = \dfrac{\text{distance}}{\text{time}} = \dfrac{2100}{81\frac{2}{3}} = \dfrac{180}{7} = 25\frac{5}{7}$ mph

(c) $D = -\dfrac{180}{7}t + 3400$

Domain: $0 \le t \le \dfrac{1190}{9}$

Range: $0 \le D \le 3400$

(d)

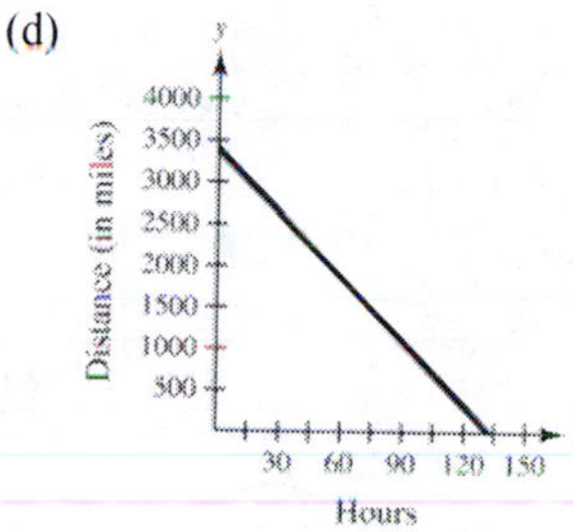

9. (a)–(d) Use $f(x) = 4x$ and $g(x) = x + 6$.

(a) $(f \circ g)(x) = f(x + 6) = 4(x + 6) = 4x + 24$

(b) $(f \circ g)^{-1}(x) = \dfrac{x - 24}{4} = \dfrac{1}{4}x - 6$

(c) $f^{-1}(x) = \dfrac{1}{4}x$

$g^{-1}(x) = x - 6$

(d) $(g^{-1} \circ f^{-1})(x) = g^{-1}\left(\dfrac{1}{4}x\right) = \dfrac{1}{4}x - 6$

(e) $f(x) = x^3 + 1$ and $g(x) = 2x$

$(f \circ g)(x) = f(2x) = (2x)^3 + 1 = 8x^3 + 1$

$(f \circ g)^{-1}(x) = \sqrt[3]{\dfrac{x - 1}{8}} = \dfrac{1}{2}\sqrt[3]{x - 1}$

$f^{-1}(x) = \sqrt[3]{x - 1}$

$g^{-1}(x) = \dfrac{1}{2}x$

$(g^{-1} \circ f^{-1})(x) = g^{-1}\left(\sqrt[3]{x - 1}\right) = \dfrac{1}{2}\sqrt[3]{x - 1}$

(f) Answers will vary.

(g) Conjecture: $(f \circ g)^{-1}(x) = (g^{-1} \circ f^{-1})(x)$

11. $H(x) = \begin{cases} 1, & x \geq 0 \\ 0, & x < 0 \end{cases}$

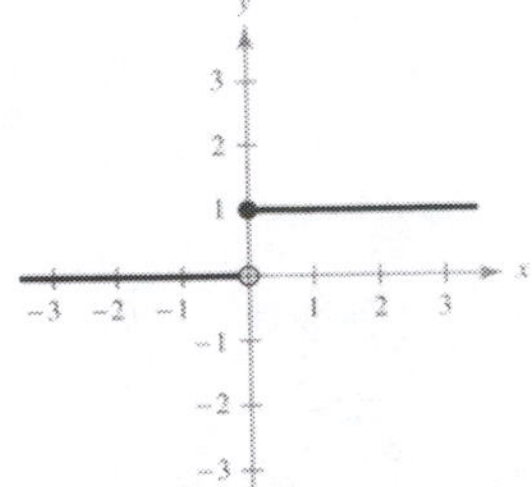

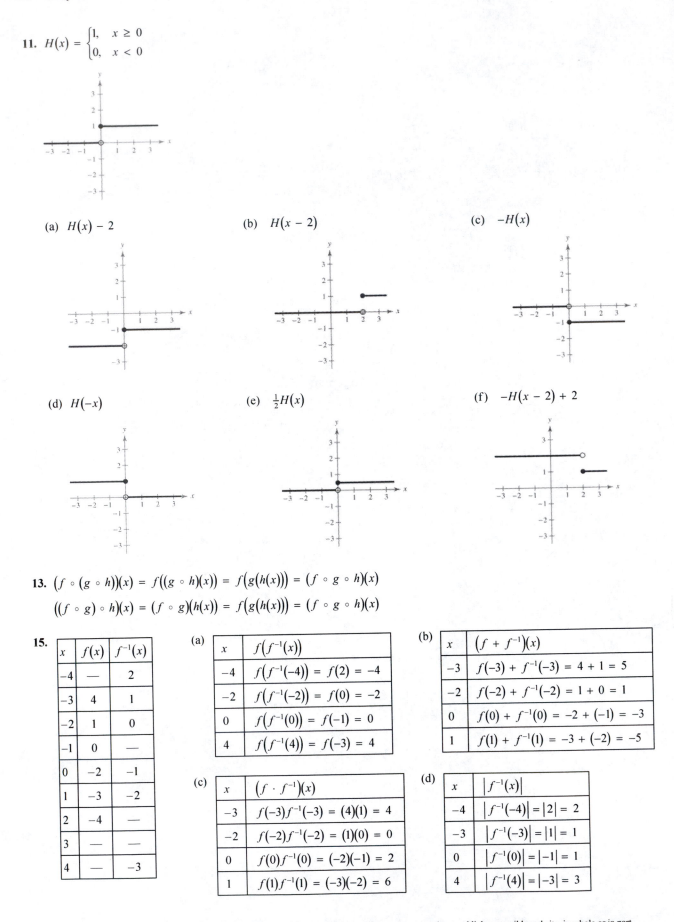

(a) $H(x) - 2$

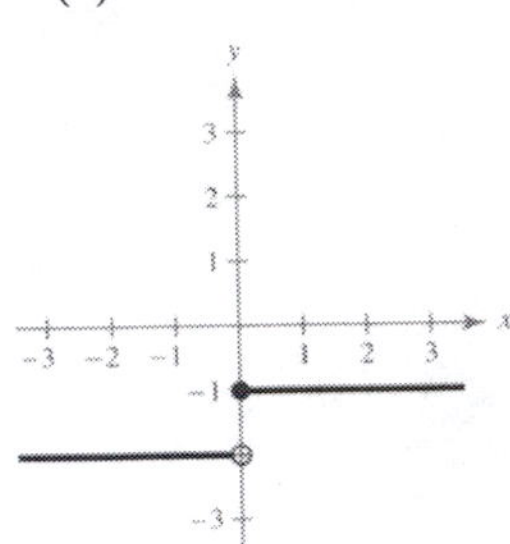

(b) $H(x - 2)$

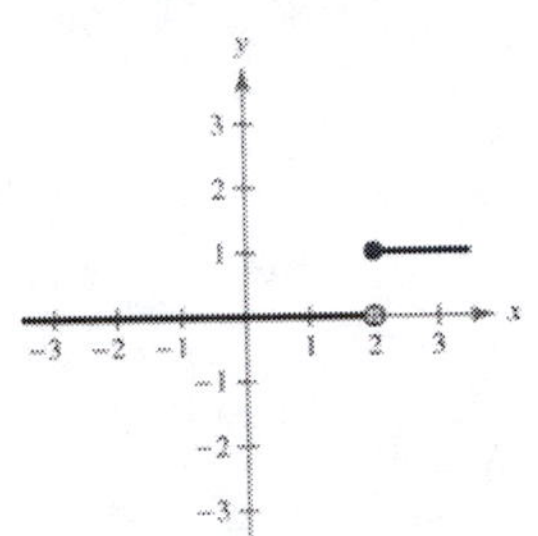

(c) $-H(x)$

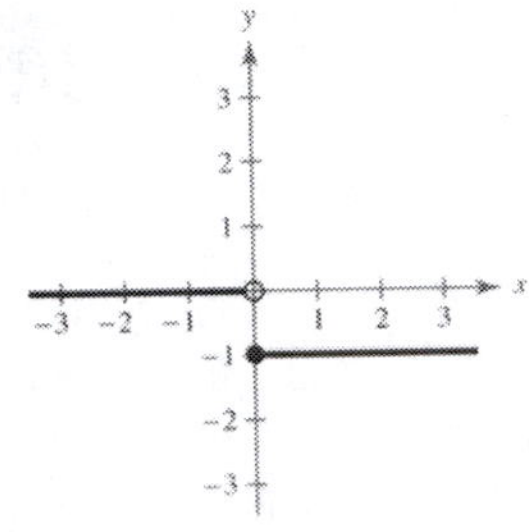

(d) $H(-x)$

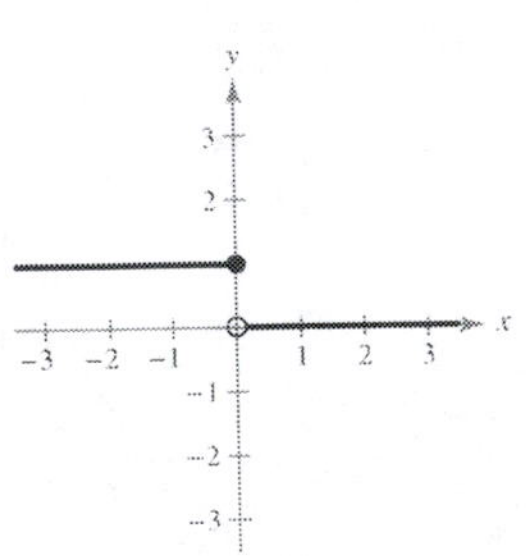

(e) $\frac{1}{2}H(x)$

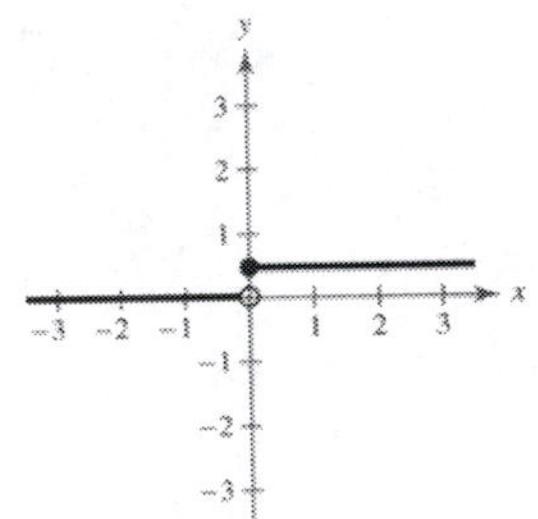

(f) $-H(x - 2) + 2$

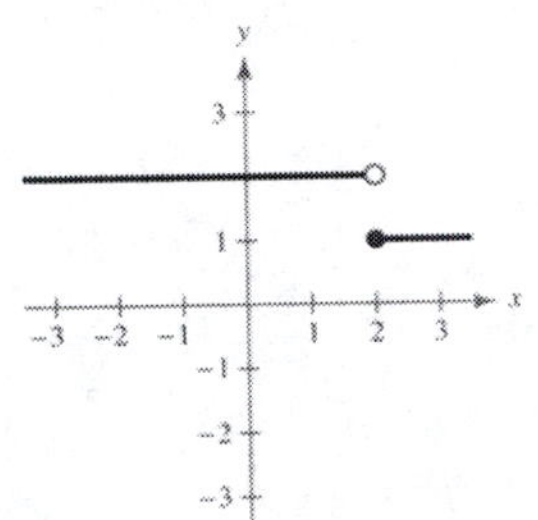

13. $(f \circ (g \circ h))(x) = f((g \circ h)(x)) = f(g(h(x))) = (f \circ g \circ h)(x)$

$((f \circ g) \circ h)(x) = (f \circ g)(h(x)) = f(g(h(x))) = (f \circ g \circ h)(x)$

15.

x	$f(x)$	$f^{-1}(x)$
−4	—	2
−3	4	1
−2	1	0
−1	0	—
0	−2	−1
1	−3	−2
2	−4	—
3	—	—
4	—	−3

(a)

x	$f(f^{-1}(x))$
−4	$f(f^{-1}(-4)) = f(2) = -4$
−2	$f(f^{-1}(-2)) = f(0) = -2$
0	$f(f^{-1}(0)) = f(-1) = 0$
4	$f(f^{-1}(4)) = f(-3) = 4$

(b)

x	$(f + f^{-1})(x)$
−3	$f(-3) + f^{-1}(-3) = 4 + 1 = 5$
−2	$f(-2) + f^{-1}(-2) = 1 + 0 = 1$
0	$f(0) + f^{-1}(0) = -2 + (-1) = -3$
1	$f(1) + f^{-1}(1) = -3 + (-2) = -5$

(c)

x	$(f \cdot f^{-1})(x)$
−3	$f(-3)f^{-1}(-3) = (4)(1) = 4$
−2	$f(-2)f^{-1}(-2) = (1)(0) = 0$
0	$f(0)f^{-1}(0) = (-2)(-1) = 2$
1	$f(1)f^{-1}(1) = (-3)(-2) = 6$

(d)

x	$\lvert f^{-1}(x)\rvert$
−4	$\lvert f^{-1}(-4)\rvert = \lvert 2\rvert = 2$
−3	$\lvert f^{-1}(-3)\rvert = \lvert 1\rvert = 1$
0	$\lvert f^{-1}(0)\rvert = \lvert -1\rvert = 1$
4	$\lvert f^{-1}(4)\rvert = \lvert -3\rvert = 3$

Practice Test for Chapter 2

1. Find the equation of the line through $(2, 4)$ and $(3, -1)$.

2. Find the equation of the line with slope $m = 4/3$ and y-intercept $b = -3$.

3. Find the equation of the line through $(4, 1)$ perpendicular to the line $2x + 3y = 0$.

4. If it costs a company \$32 to produce 5 units of a product and \$44 to produce 9 units, how much does it cost to produce 20 units? (Assume that the cost function is linear.)

5. Given $f(x) = x^2 - 2x + 1$, find $f(x - 3)$.

6. Given $f(x) = 4x - 11$, find $\dfrac{f(x) - f(3)}{x - 3}$

7. Find the domain and range of $f(x) = \sqrt{36 - x^2}$.

8. Which equations determine y as a function of x?

   (a) $6x - 5y + 4 = 0$

   (b) $x^2 + y^2 = 9$

   (c) $y^3 = x^2 + 6$

9. Sketch the graph of $f(x) = x^2 - 5$.

10. Sketch the graph of $f(x) = |x + 3|$.

11. Sketch the graph of $f(x) = \begin{cases} 2x + 1, & \text{if } x \ge 0, \\ x^2 - x, & \text{if } x < 0. \end{cases}$

12. Use the graph of $f(x) = |x|$ to graph the following:

    (a) $f(x + 2)$

    (b) $-f(x) + 2$

13. Given $f(x) = 3x + 7$ and $g(x) = 2x^2 - 5$, find the following:

    (a) $(g - f)(x)$

    (b) $(fg)(x)$

14. Given $f(x) = x^2 - 2x + 16$ and $g(x) = 2x + 3$, find $f(g(x))$.

15. Given $f(x) = x^3 + 7$, find $f^{-1}(x)$.

16. Which of the following functions have inverses?

    (a) $f(x) = |x - 6|$

    (b) $f(x) = ax + b, a \neq 0$

    (c) $f(x) = x^3 - 19$

17. Given $f(x) = \sqrt{\dfrac{3 - x}{x}}, 0 < x \le 3,$ find $f^{-1}(x)$.

Exercises 18–20, true or false?

18. $y = 3x + 7$ and $y = \frac{1}{3}x - 4$ are perpendicular.

19. $(f \circ g)^{-1} = g^{-1} \circ f^{-1}$

20. If a function has an inverse, then it must pass both the Vertical Line Test and the Horizontal Line Test.

CHAPTER 3
Polynomial Functions

CHAPTER 3
Polynomial Functions

Section 3.1 Quadratic Functions and Models

1. polynomial

3. quadratic; parabola

5. positive; minimum

7. $f(x) = (x - 2)^2$ opens upward and has vertex $(2, 0)$. Matches graph (e).

8. $f(x) = (x + 4)^2$ opens upward and has vertex $(-4, 0)$. Matches graph (c).

9. $f(x) = x^2 - 2$ opens upward and has vertex $(0, -2)$. Matches graph (b).

10. $f(x) = (x + 1)^2 - 2$ opens upward and has vertex $(-1, -2)$. Matches graph (a).

11. $f(x) = 4 - (x - 2)^2 = -(x - 2)^2 + 4$ opens downward and has vertex $(2, 4)$. Matches graph (f).

12. $f(x) = -(x - 4)^2$ opens downward and has vertex $(4, 0)$. Matches graph (d).

13. (a) $y = \frac{1}{2}x^2$

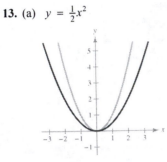

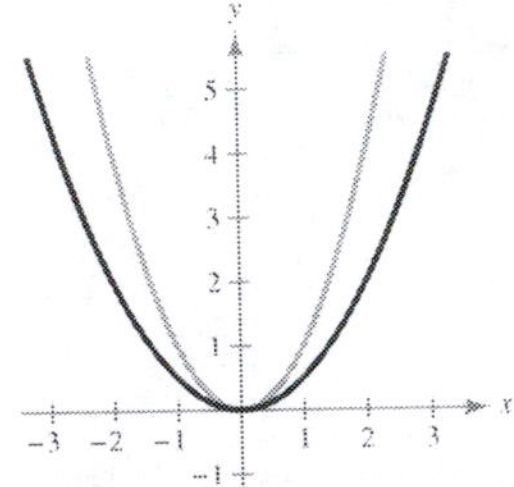

Vertical shrink

(b) $y = -\frac{1}{8}x^2$

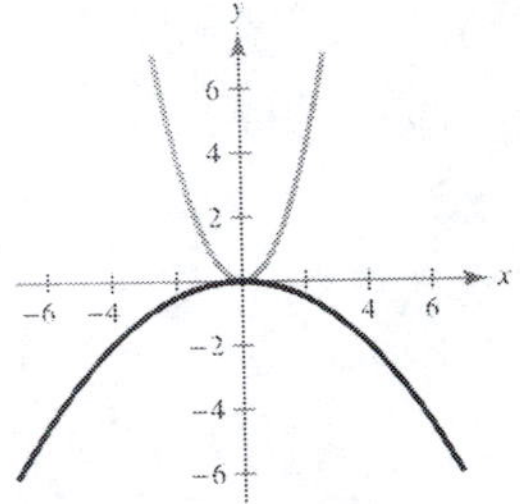

Vertical shrink and reflection in the x-axis

(c) $y = \frac{3}{2}x^2$

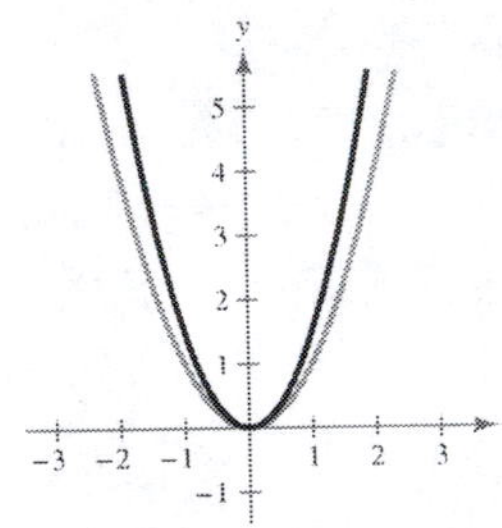

Vertical stretch

(d) $y = -3x^2$

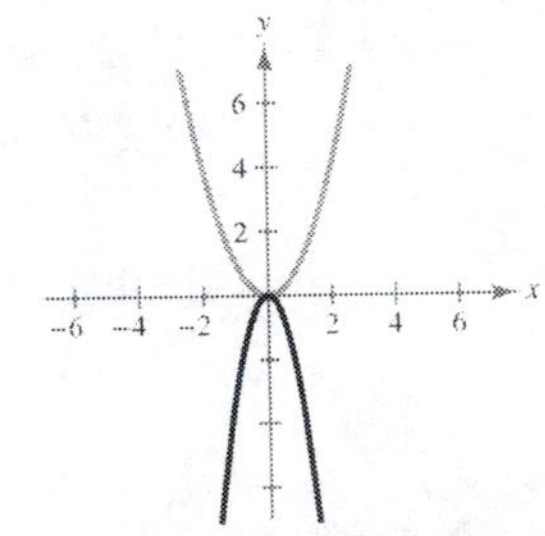

Vertical stretch and reflection in the x-axis

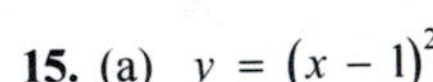

15. (a) $y = (x - 1)^2$

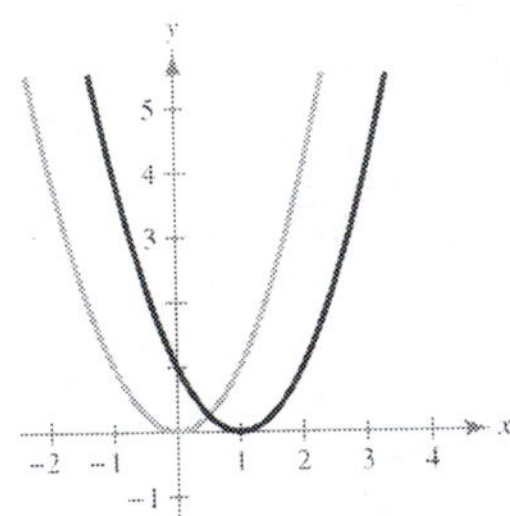

Horizontal shift one unit to the right

(b) $y = (3x)^2 + 1$

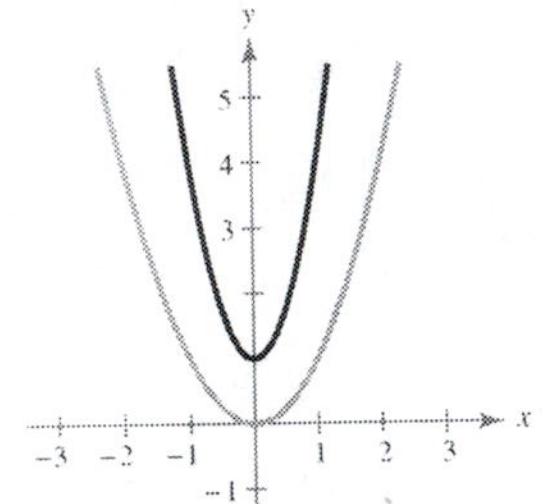

Horizontal shrink and a vertical shift one unit upward

(c) $y = \left(\frac{1}{3}x\right)^2 - 3$

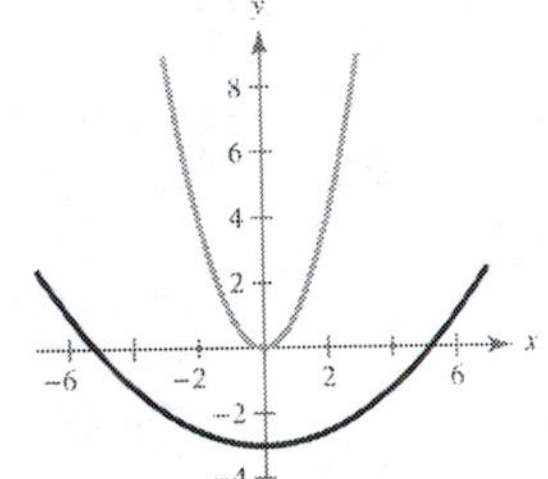

Horizontal stretch and a vertical shift three units downward

(d) $y = (x + 3)^2$

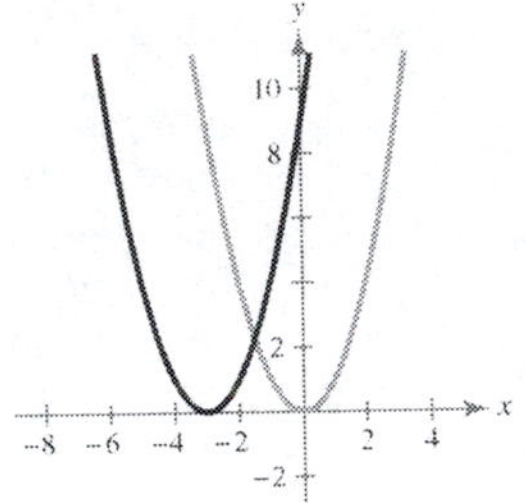

Horizontal shift three units to the left

17. $f(x) = x^2 - 6x$

$= (x^2 - 6x + 9) - 9$

$= (x - 3)^2 - 9$

Vertex: $(3, -9)$

Axis of symmetry: $x = 3$

Find x-intercepts:

$x^2 - 6x = 0$

$x(x - 6) = 0$

$x = 0$

$x - 6 = 0 \Rightarrow x = 6$

x-intercepts: $(0, 0), (6, 0)$

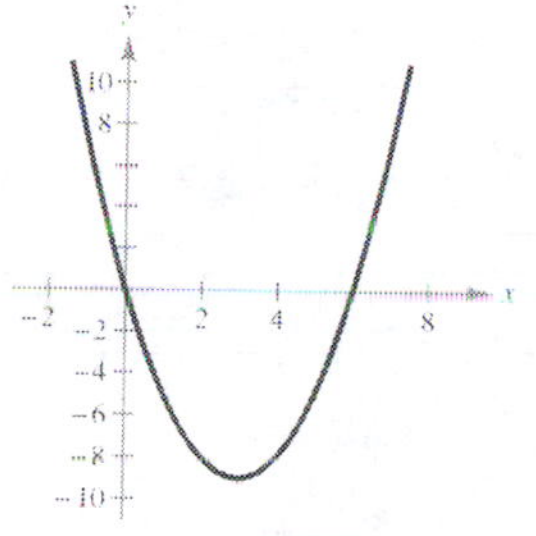

19. $h(x) = x^2 - 8x + 16 = (x - 4)^2$

Vertex: $(4, 0)$

Axis of symmetry: $x = 4$

Find x-intercepts:

$(x - 4)^2 = 0$

$x - 4 = 0$

$x = 4$

x-intercept: $(4, 0)$

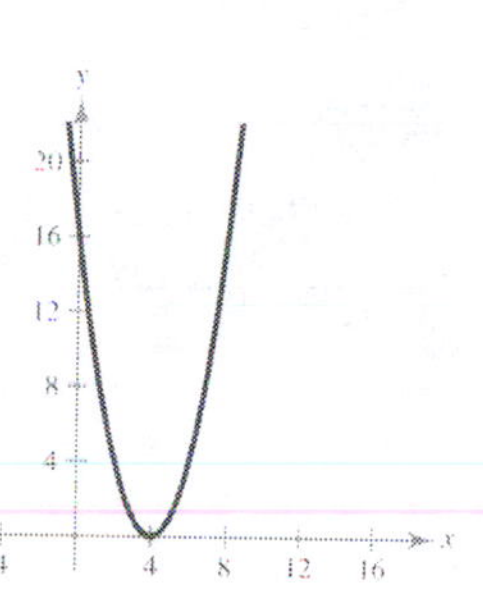

21. $f(x) = x^2 + 8x + 13$

$= (x^2 + 8x + 16) - 16 + 13$

$= (x + 4)^2 - 3$

Vertex: $(-4, -3)$

Axis of symmetry: $x = -4$

Find x-intercepts:

$x^2 + 8x + 13 = 0$

$x^2 + 8x = -13$

$x^2 + 8x + 16 = 16 - 13$

$(x + 4)^2 = 3$

$x + 4 = \pm\sqrt{3}$

$x = -4 \pm\sqrt{3}$

x-intercepts: $\left(-4 \pm\sqrt{3}, 0\right)$

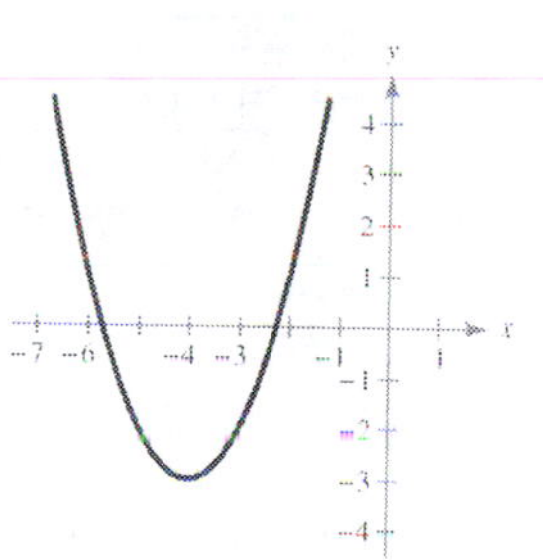

23. $f(x) = x^2 - 14x + 54$

$= (x^2 - 14x + 49) - 49 + 54$

$= (x - 7)^2 + 5$

Vertex: $(7, 5)$

Axis of symmetry: $x = 7$

Find x-intercepts:

$x^2 - 14x + 54 = 0$

$x^2 - 14x = -54$

$x^2 - 14x + 49 = -54 + 49$

$(x - 7)^2 = -5$

$x - 7 = \pm\sqrt{-5}$

$x = 7 \pm\sqrt{5}i$

Not a real number

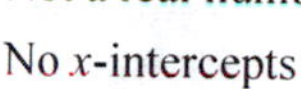

No x-intercepts

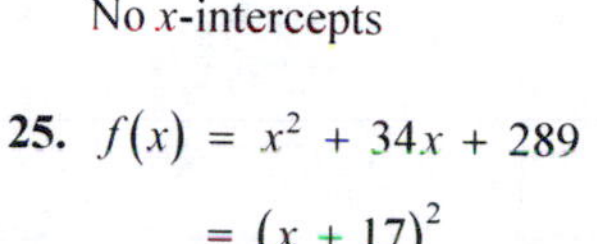

25. $f(x) = x^2 + 34x + 289$

$= (x + 17)^2$

Vertex: $(-17, 0)$

Axis of symmetry: $x = -17$

Find x-intercepts:

$x^2 + 34x + 289 = 0$

$(x + 17)^2 = 0$

$x + 17 = 0$

$x = -17$

x-intercept: $(-17, 0)$

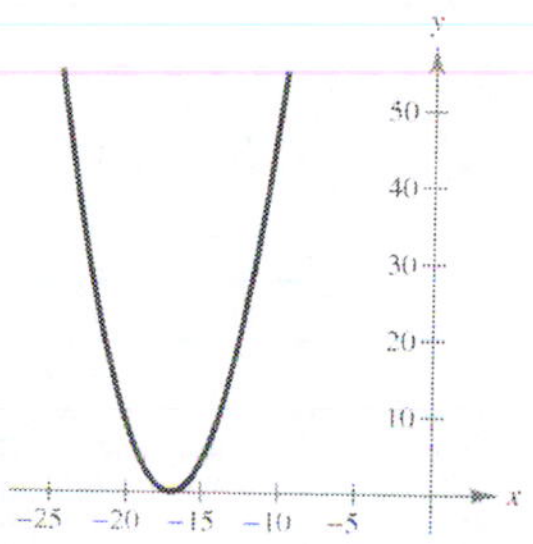

27. $f(x) = x^2 - x + \dfrac{5}{4}$

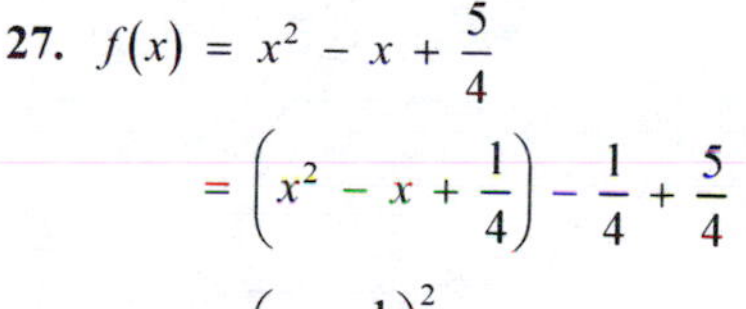

$= \left(x^2 - x + \dfrac{1}{4}\right) - \dfrac{1}{4} + \dfrac{5}{4}$

$= \left(x - \dfrac{1}{2}\right)^2 + 1$

Vertex: $\left(\dfrac{1}{2}, 1\right)$

Axis of symmetry: $x = \dfrac{1}{2}$

Find x-intercepts:

$x^2 - x + \dfrac{5}{4} = 0$

$x = \dfrac{1 \pm \sqrt{1 - 5}}{2}$

Not a real number

No x-intercepts

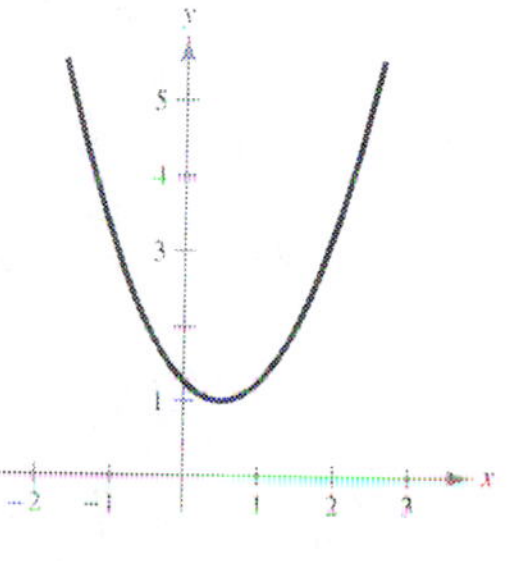

29. $f(x) = -x^2 + 2x + 5$

$= -(x^2 - 2x + 1) - (-1) + 5$

$= -(x - 1)^2 + 6$

Vertex: $(1, 6)$

Axis of symmetry: $x = 1$

Find x-intercepts:

$-x^2 + 2x + 5 = 0$

$x^2 - 2x - 5 = 0$

$$x = \frac{2 \pm \sqrt{4 + 20}}{2}$$

$$= 1 \pm \sqrt{6}$$

x-intercepts: $\left(1 - \sqrt{6}, 0\right), \left(1 + \sqrt{6}, 0\right)$

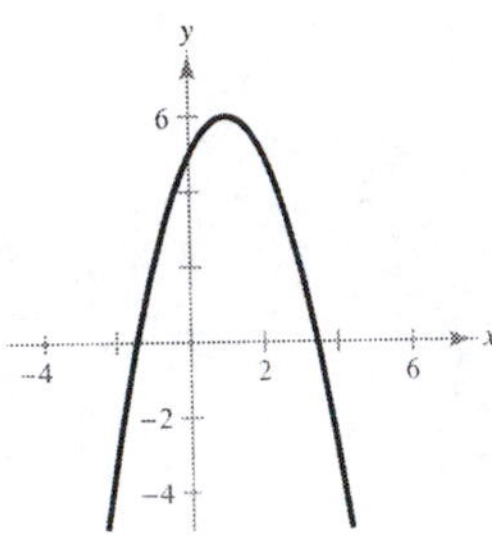

31. $h(x) = 4x^2 - 4x + 21$

$$= 4\left(x^2 - x + \frac{1}{4}\right) - 4\left(\frac{1}{4}\right) + 21$$

$$= 4\left(x - \frac{1}{2}\right)^2 + 20$$

Vertex: $\left(\frac{1}{2}, 20\right)$

Axis of symmetry: $x = \frac{1}{2}$

Find x-intercepts:

$4x^2 - 4x + 21 = 0$

$$x = \frac{4 \pm \sqrt{16 - 336}}{2(4)}$$

Not a real number

No x-intercepts

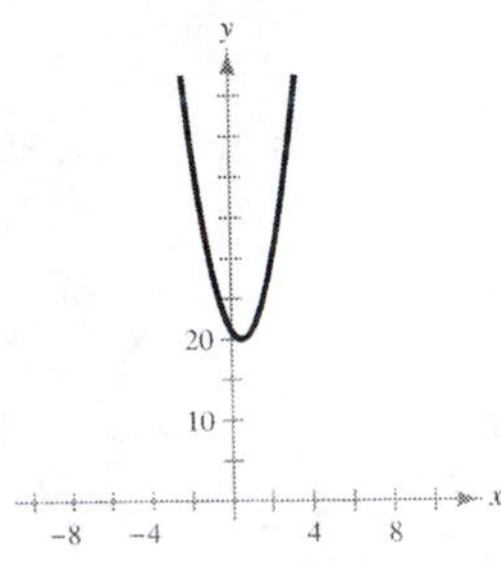

33. $f(x) = \frac{1}{4}x^2 - 2x - 12$

$= \frac{1}{4}(x^2 - 8x + 16) - \frac{1}{4}(16) - 12$

$= \frac{1}{4}(x - 4)^2 - 16$

Vertex: $(4, -16)$

Axis of symmetry: $x = 4$

Find x-intercepts:

$\frac{1}{4}x^2 - 2x - 12 = 0$

$x^2 - 8x - 48 = 0$

$(x + 4)(x - 12) = 0$

$x = -4$ or $x = 12$

x-intercepts: $(-4, 0), (12, 0)$

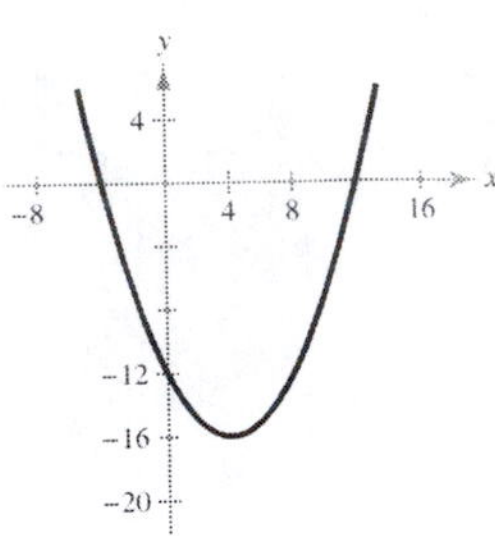

35. $f(x) = -(x^2 + 2x - 3) = -(x + 1)^2 + 4$

Vertex: $(-1, 4)$

Axis of symmetry: $x = -1$

x-intercepts: $(-3, 0), (1, 0)$

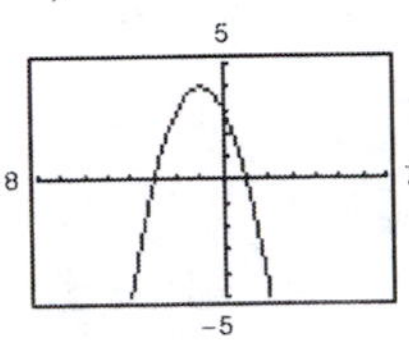

37. $g(x) = x^2 + 8x + 11 = (x + 4)^2 - 5$

Vertex: $(-4, -5)$

Axis of symmetry: $x = -4$

x-intercepts: $\left(-4 \pm \sqrt{5}, 0\right)$

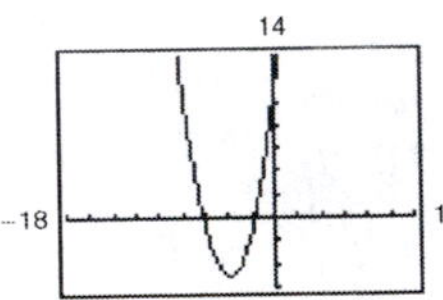

39. $f(x) = 2x^2 - 16x + 32$

$= 2(x^2 - 8x + 16)$

$= 2(x - 4)^2$

Vertex: $(4, 0)$

Axis of symmetry: $x = 4$

x-intercepts: $(4, 0)$

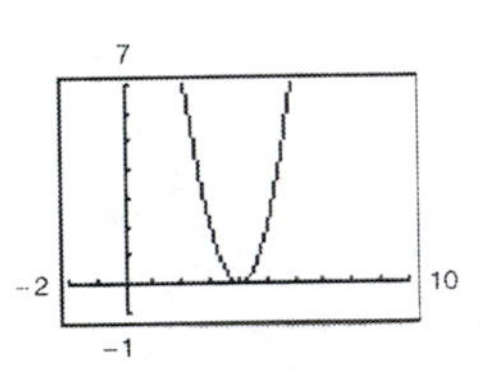

41. $g(x) = \frac{1}{2}(x^2 + 4x - 2) = \frac{1}{2}(x + 2)^2 - 3$

Vertex: $(-2, -3)$

Axis of symmetry: $x = -2$

x-intercepts: $\left(-2 \pm \sqrt{6}, 0\right)$

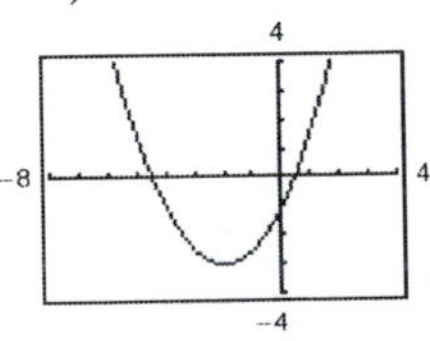

43. $(-1, 4)$ is the vertex.

$y = a(x + 1)^2 + 4$

Because the graph passes through $(1, 0)$,

$0 = a(1 + 1)^2 + 4$

$-4 = 4a$

$-1 = a.$

So, $y = -1(x + 1)^2 + 4 = -(x + 1)^2 + 4.$

45. $(-2, 2)$ is the vertex.

$y = a(x + 2)^2 + 2$

Because the graph passes through $(-1, 0)$,

$0 = a(-1 + 2)^2 + 2$

$-2 = a.$

So, $y = -2(x + 2)^2 + 2.$

47. $(-2, 5)$ is the vertex.

$f(x) = a(x + 2)^2 + 5$

Because the graph passes through $(0, 9)$,

$9 = a(0 + 2)^2 + 5$

$4 = 4a$

$1 = a.$

So, $f(x) = 1(x + 2)^2 + 5 = (x + 2)^2 + 5.$

49. $(1, -2)$ is the vertex.

$f(x) = a(x - 1)^2 - 2$

Because the graph passes through $(-1, 14)$,

$14 = a(-1 - 1)^2 - 2$

$14 = 4a - 2$

$16 = 4a$

$4 = a.$

So, $f(x) = 4(x - 1)^2 - 2.$

51. $(5, 12)$ is the vertex.

$f(x) = a(x - 5)^2 + 12$

Because the graph passes through $(7, 15)$,

$15 = a(7 - 5)^2 + 12$

$3 = 4a \Rightarrow a = \frac{3}{4}.$

So, $f(x) = \frac{3}{4}(x - 5)^2 + 12.$

53. $\left(-\frac{1}{4}, \frac{3}{2}\right)$ is the vertex.

$f(x) = a\left(x + \frac{1}{4}\right)^2 + \frac{3}{2}$

Because the graph passes through $(-2, 0)$,

$0 = a\left(-2 + \frac{1}{4}\right)^2 + \frac{3}{2}$

$-\frac{3}{2} = \frac{49}{16}a \Rightarrow a = -\frac{24}{49}.$

So, $f(x) = -\frac{24}{49}\left(x + \frac{1}{4}\right)^2 + \frac{3}{2}.$

55. $\left(-\frac{5}{2}, 0\right)$ is the vertex.

$f(x) = a\left(x + \frac{5}{2}\right)^2$

Because the graph passes through $\left(-\frac{7}{2}, -\frac{16}{3}\right)$,

$-\frac{16}{3} = a\left(-\frac{7}{2} + \frac{5}{2}\right)^2$

$-\frac{16}{3} = a.$

So, $f(x) = -\frac{16}{3}\left(x + \frac{5}{2}\right)^2.$

57. $y = x^2 - 4x - 5$

x-intercepts: $(5, 0), (-1, 0)$

$0 = x^2 - 4x - 5$

$0 = (x - 5)(x + 1)$

$x = 5$ or $x = -1$

59. $f(x) = x^2 - 4x$

x-intercepts: $(0, 0), (4, 0)$

$0 = x^2 - 4x$

$0 = x(x - 4)$

$x = 0$ or $x = 4$

The x-intercepts and the solutions of $f(x) = 0$ are the same.

61. $f(x) = x^2 - 9x + 18$

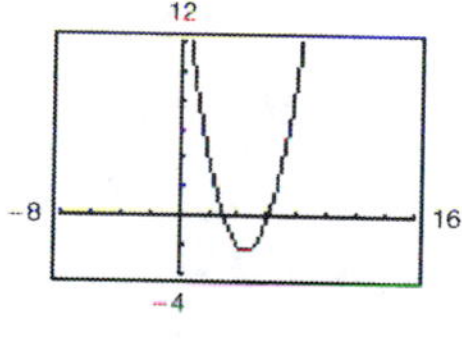

x-intercepts: $(3, 0), (6, 0)$

$0 = x^2 - 9x + 18$

$0 = (x - 3)(x - 6)$

$x = 3$ or $x = 6$

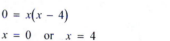

The x-intercepts and the solutions of $f(x) = 0$ are the same.

63. $f(x) = 2x^2 - 7x - 30$

x-intercepts: $\left(-\frac{5}{2}, 0\right), (6, 0)$

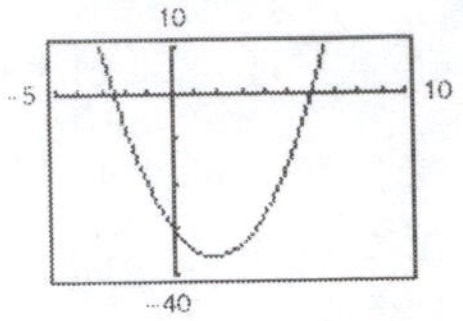

$0 = 2x^2 - 7x - 30$

$0 = (2x + 5)(x - 6)$

$x = -\frac{5}{2}$ or $x = 6$

The x-intercepts and the solutions of $f(x) = 0$ are the same.

65. $f(x) = [x - (-1)](x - 3)$ opens upward

$= (x + 1)(x - 3)$

$= x^2 - 2x - 3$

$g(x) = -[x - (-1)](x - 3)$ opens downward

$= -(x + 1)(x - 3)$

$= -(x^2 - 2x - 3)$

$= -x^2 + 2x + 3$

Note: $f(x) = a(x + 1)(x - 3)$ has x-intercepts $(-1, 0)$ and $(3, 0)$ for all real numbers $a \neq 0$.

67. $f(x) = (x - 0)(x - 10)$ opens upward

$= x^2 - 10x$

$g(x) = -(x - 0)(x - 10)$ opens downward

$= -x^2 + 10x$

Note: $f(x) = a(x - 0)(x - 10) = ax(x - 10)$ has x-intercepts $(0, 0)$ and $(10, 0)$ for all real numbers $a \neq 0$.

69. $f(x) = [x - (-3)]\left[x - \left(-\frac{1}{2}\right)\right](2)$ opens upward

$= (x + 3)\left(x + \frac{1}{2}\right)(2)$

$= (x + 3)(2x + 1)$

$= 2x^2 + 7x + 3$

$g(x) = -(2x^2 + 7x + 3)$ opens downward

$= -2x^2 - 7x - 3$

Note: $f(x) = a(x + 3)(2x + 1)$ has x-intercepts $(-3, 0)$ and $\left(-\frac{1}{2}, 0\right)$ for all real numbers $a \neq 0$.

71. Let x = the first number and y = the second number. Then the sum is

$x + y = 110 \Rightarrow y = 110 - x.$

The product is $P(x) = xy = x(110 - x) = 110x - x^2$.

$$P(x) = -x^2 + 110x$$
$$= -(x^2 - 110x + 3025 - 3025)$$
$$= -\left[(x - 55)^2 - 3025\right]$$
$$= -(x - 55)^2 + 3025$$

The maximum value of the product occurs at the vertex of $P(x)$ and is 3025. This happens when $x = y = 55$.

73. Let x = the first number and y = the second number. Then the sum is

$x + 2y = 24 \Rightarrow y = \dfrac{24 - x}{2}.$

The product is $P(x) = xy = x\left(\dfrac{24 - x}{2}\right)$.

$$P(x) = \frac{1}{2}(-x^2 + 24x)$$
$$= -\frac{1}{2}(x^2 - 24x + 144 - 144)$$
$$= -\frac{1}{2}\left[(x - 12)^2 - 144\right] = -\frac{1}{2}(x - 12)^2 + 72$$

The maximum value of the product occurs at the vertex of $P(x)$ and is 72. This happens when $x = 12$ and $y = (24 - 12)/2 = 6$. So, the numbers are 12 and 6.

75. $y = -\frac{4}{9}x^2 + \frac{24}{9}x + 12$

The vertex occurs at $-\dfrac{b}{2a} = \dfrac{-24/9}{2(-4/9)} = 3$. The maximum height is

$y(3) = -\frac{4}{9}(3)^2 + \frac{24}{9}(3) + 12 = 16$ feet.

77. $C = 800 - 10x + 0.25x^2 = 0.25x^2 - 10x + 800$

The vertex occurs at $x = -\dfrac{b}{2a} = -\dfrac{-10}{2(0.25)} = 20.$

The cost is minimum when $x = 20$ fixtures.

79. $R(p) = -25p^2 + 1200p$

(a) $R(20) = \$14,000$ thousand $= \$14,000,000$

$R(25) = \$14,375$ thousand $= \$14,375,000$

$R(30) = \$13,500$ thousand $= \$13,500,000$

(b) The revenue is a maximum at the vertex.

$$-\frac{b}{2a} = \frac{-1200}{2(-25)} = 24$$

$$R(24) = 14,400$$

The unit price that will yield a maximum revenue of \$14,400 thousand is \$24.

81. 

(a) 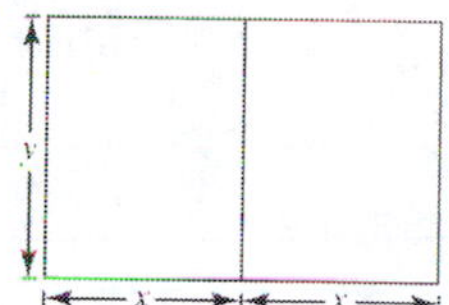

$$4x + 3y = 200 \Rightarrow y = \frac{1}{3}(200 - 4x) = \frac{4}{3}(50 - x)$$

$$A = 2xy = 2x\left[\frac{4}{3}(50 - x)\right] = \frac{8}{3}x(50 - x) = \frac{8x(50 - x)}{3}$$

(b)

x	A
5	600
10	$1066\frac{2}{3}$
15	1400
20	1600
25	$1666\frac{2}{3}$
30	1600

This area is maximum when $x = 25$ feet and $y = \frac{100}{3} = 33\frac{1}{3}$ feet.

(c) 

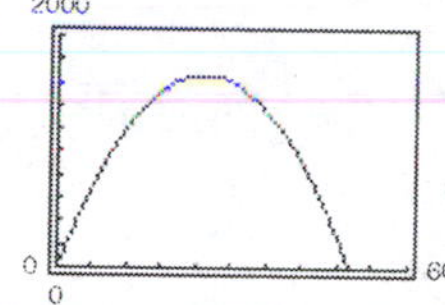

This area is maximum when $x = 25$ feet and $y = \frac{100}{3} = 33\frac{1}{3}$ feet.

(d)
$$\begin{aligned} A &= \frac{8}{3}x(50 - x) \\ &= -\frac{8}{3}(x^2 - 50x) \\ &= -\frac{8}{3}(x^2 - 50x + 625 - 625) \\ &= -\frac{8}{3}\left[(x - 25)^2 - 625\right] \\ &= -\frac{8}{3}(x - 25)^2 + \frac{5000}{3} \end{aligned}$$

The maximum area occurs at the vertex and is $5000/3$ square feet. This happens when $x = 25$ feet and $y = (200 - 4(25))/3 = 100/3$ feet. The dimensions are $2x = 50$ feet by $33\frac{1}{3}$ feet.

(e) They are all identical.

$x = 25$ feet and $y = 33\frac{1}{3}$ feet

83. (a) Revenue = (number of tickets sold)(price per ticket)

Let y = attendance, or the number of tickets sold.

$m = -100, (20, 1500)$

$$y - 1500 = -100(x - 20)$$
$$y - 1500 = -100x + 2000$$
$$y = -100x + 3500$$
$$R(x) = (y)(x)$$
$$R(x) = (-100x + 3500)(x)$$
$$R(x) = -100x^2 + 3500x$$

(b) The revenue is at a maximum at the vertex.

$$-\frac{b}{2a} = \frac{-3500}{2(-100)} = 17.5$$

$$R(17.5) = -100(17.5)^2 + 3500(17.5) = \$30,625$$

A ticket price of \$17.50 will yield a maximum revenue of \$30,625.

85. (a)

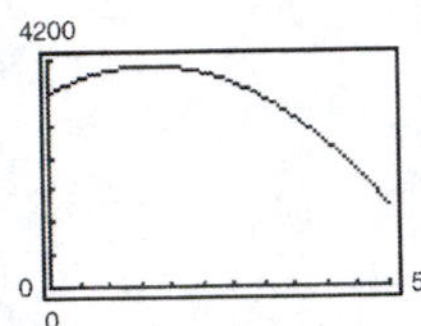

(b) The maximum annual consumption occurs at the point $(16.9, 4074.813)$.

4075 cigarettes

$1966 \rightarrow t = 16$

The maximum consumption occurred in 1966. After that year, the consumption decreases. It is likely that the warning was responsible for the decrease in consumption.

(c) $$\text{Annual consumption per smoker} = \frac{\text{Annual consumption in 2005} \cdot \text{total population}}{\text{total number of smokers in 2005}} = \frac{1487.9(296{,}329{,}000)}{59{,}858{,}458} = 7365.8$$

About 7366 cigarettes per smoker annually

$$\text{Daily consumption per smoker} = \frac{\text{Number of cigarettes per year}}{\text{Number of days per year}} = \frac{7366}{365} \approx 20.2$$

About 20 cigarettes per day

87. True. The equation $-12x^2 - 1 = 0$ has no real solution, so the graph has no x-intercepts.

89. $f(x) = -x^2 + bx - 75$, maximum value: 25

The maximum value, 25, is the y-coordinate of the vertex. Find the x-coordinate of the vertex:

$$x = -\frac{b}{2a} = -\frac{b}{2(-1)} = \frac{b}{2}$$

$$f(x) = -x^2 + bx - 75$$
$$f\left(\frac{b}{2}\right) = -\left(\frac{b}{2}\right)^2 + b\left(\frac{b}{2}\right) - 75$$
$$25 = -\frac{b^2}{4} + \frac{b^2}{2} - 75$$
$$100 = \frac{b^2}{4}$$
$$400 = b^2$$
$$\pm 20 = b$$

91. $f(x) = x^2 + bx + 26$, minimum value: 10

The minimum value, 10, is the y-coordinate of the vertex.

Find the x-coordinate of the vertex:

$$x = -\frac{b}{2a} = -\frac{b}{2(1)} = -\frac{b}{2}$$

$$f(x) = x^2 + bx + 26$$

$$f\left(-\frac{b}{2}\right) = \left(-\frac{b}{2}\right)^2 + b\left(-\frac{b}{2}\right) + 26$$

$$10 = \frac{b^2}{4} - \frac{b^2}{2} + 26$$

$$-16 = -\frac{b^2}{4}$$

$$64 = b^2$$

$$\pm 8 = b$$

93. $f(x) = ax^2 + bx + c$

$$= a\left(x^2 + \frac{b}{a}x\right) + c$$

$$= a\left(x^2 + \frac{b}{a}x + \frac{b^2}{4a^2} - \frac{b^2}{4a^2}\right) + c$$

$$= a\left(x + \frac{b}{2a}\right)^2 - \frac{b^2}{4a} + c$$

$$= a\left(x + \frac{b}{2a}\right)^2 + \frac{4ac - b^2}{4a}$$

$$f\left(-\frac{b}{2a}\right) = a\left(\frac{b^2}{4a^2}\right) + b\left(-\frac{b}{2a}\right) + c$$

$$= \frac{b^2}{4a} - \frac{b^2}{2a} + c$$

$$= \frac{b^2 - 2b^2 + 4ac}{4a} = \frac{4ac - b^2}{4a}$$

So, the vertex occurs at

$$\left(-\frac{b}{2a}, \frac{4ac - b^2}{4a}\right) = \left(-\frac{b}{2a}, f\left(-\frac{b}{2a}\right)\right).$$

95. If $f(x) = ax^2 + bx + c$ has two real zeros, then by the Quadratic Formula they are

$$x = \frac{-b \pm \sqrt{b^2 - 4ac}}{2a}.$$

The average of the zeros of f is

$$\frac{\dfrac{-b - \sqrt{b^2 - 4ac}}{2a} + \dfrac{-b + \sqrt{b^2 - 4ac}}{2a}}{2} = \frac{\dfrac{-2b}{2a}}{2} = -\frac{b}{2a}.$$

This is the x-coordinate of the vertex of the graph.

Section 3.2 Polynomial Functions of Higher Degree

1. continuous

3. n; $n - 1$

5. touches; crosses

7. standard

9. $f(x) = -2x^2 - 5x$ is a parabola with x-intercepts $(0, 0)$ and $\left(-\frac{5}{2}, 0\right)$ and opens downward. Matches graph (h).

10. $f(x) = 2x^3 - 3x + 1$ has intercepts $(0, 1), (1, 0), \left(-\frac{1}{2} - \frac{1}{2}\sqrt{3}, 0\right)$ and $\left(-\frac{1}{2} + \frac{1}{2}\sqrt{3}, 0\right)$. Matches graph (f).

11. $f(x) = -\frac{1}{4}x^4 + 3x^2$ has intercepts $(0, 0)$ and $\left(\pm 2\sqrt{3}, 0\right)$. Matches graph (a).

12. $f(x) = -\frac{1}{3}x^3 + x^2 - \frac{4}{3}$ has y-intercept $\left(0, -\frac{4}{3}\right)$. Matches graph (e).

13. $f(x) = x^4 + 2x^3$ has intercepts $(0, 0)$ and $(-2, 0)$. Matches graph (d).

14. $f(x) = \frac{1}{5}x^5 - 2x^3 + \frac{9}{5}x$ has intercepts $(0, 0), (1, 0), (-1, 0), (3, 0), (-3, 0)$. Matches graph (b).

15. $y = x^3$

(a) $f(x) = (x - 4)^3$

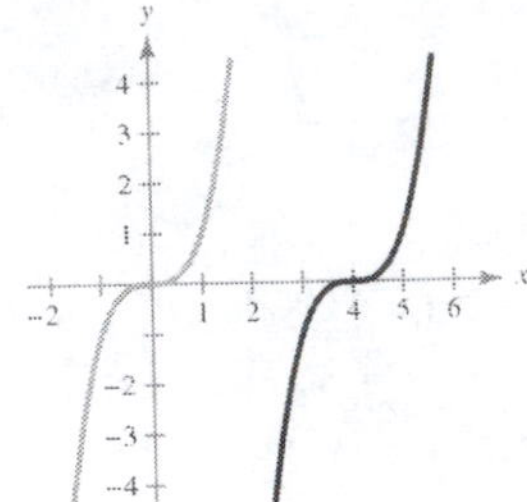

Horizontal shift four units to the right

(b) $f(x) = x^3 - 4$

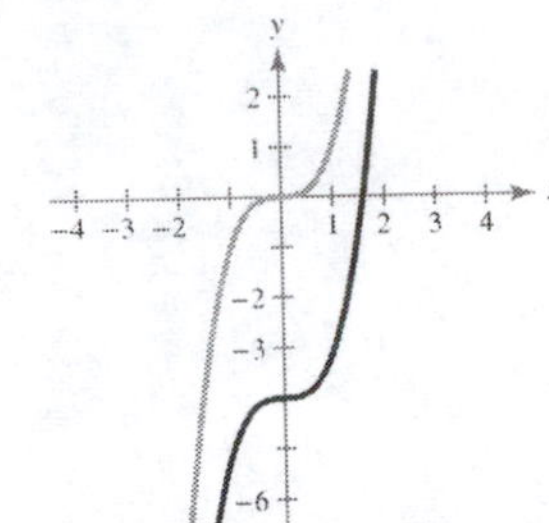

Vertical shift four units downward

(c) $f(x) = -\frac{1}{4}x^3$

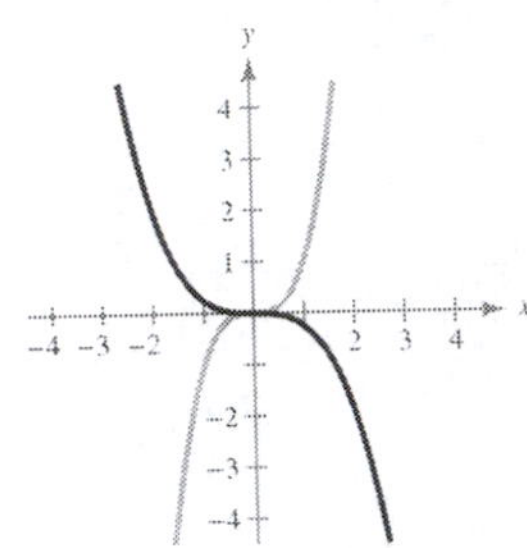

Reflection in the x-axis and a vertical shrink (each y-value is multiplied by $\frac{1}{4}$)

(d) $f(x) = (x - 4)^3 - 4$

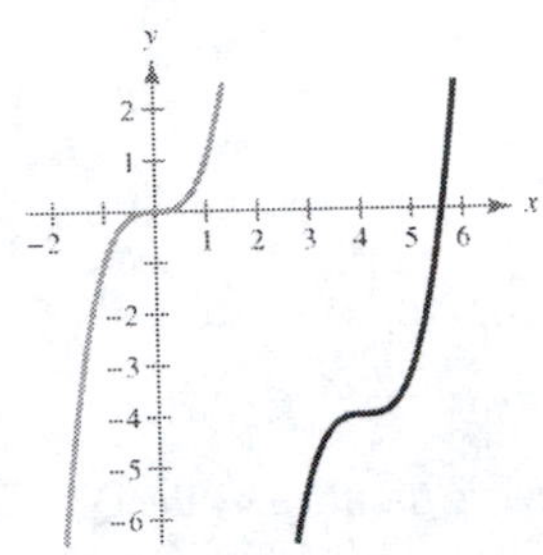

Horizontal shift four units to the right and vertical shift four units downward

17. $y = x^4$

(a) $f(x) = (x + 3)^4$

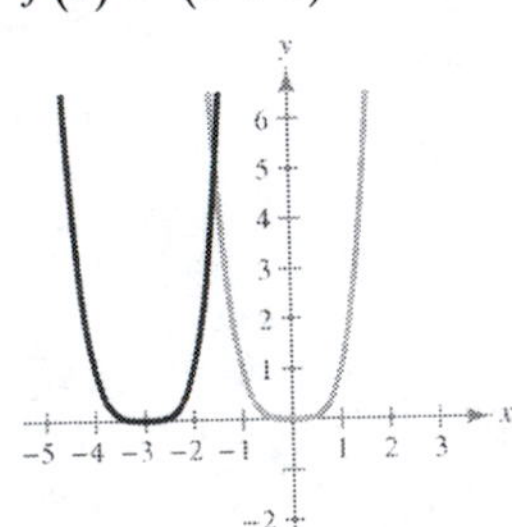

Horizontal shift three units to the left

(b) $f(x) = x^4 - 3$

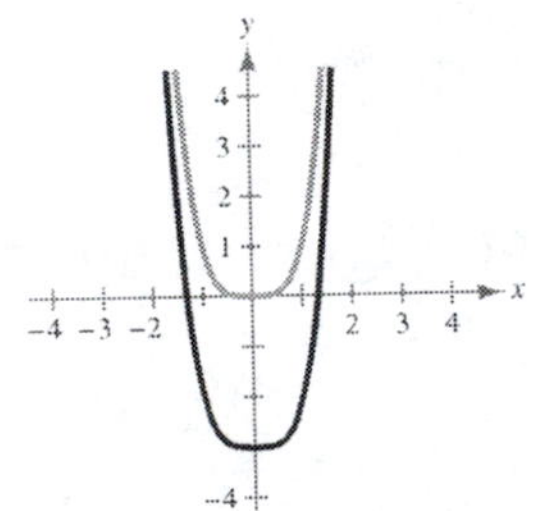

Vertical shift three units downward

(c) $f(x) = 4 - x^4$

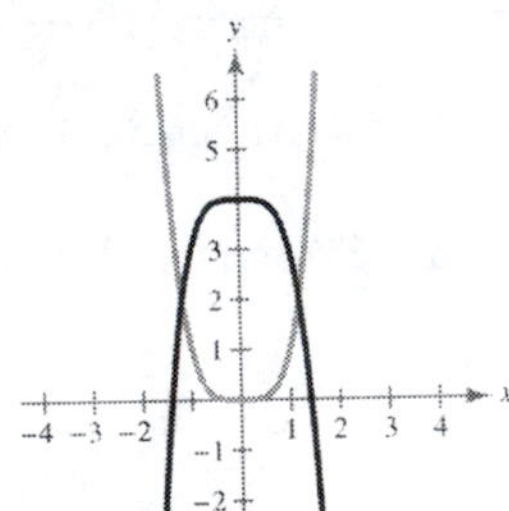

Reflection in the x-axis and then a vertical shift four units upward

(d) $f(x) = \frac{1}{2}(x - 1)^4$

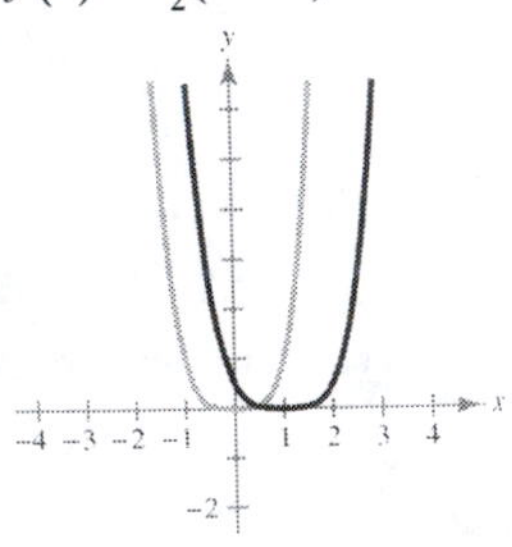

Horizontal shift one unit to the right and a vertical shrink (each y-value is multiplied by $\frac{1}{2}$)

(e) $f(x) = (2x)^4 + 1$

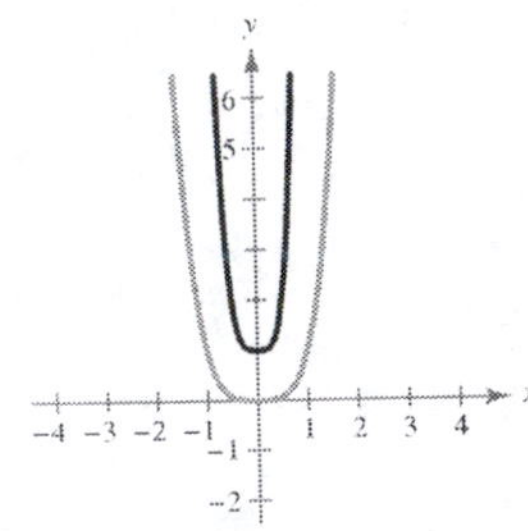

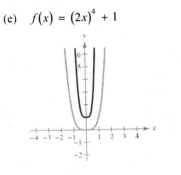

Vertical shift one unit upward and a horizontal shrink (each y-value is multiplied by 16)

(f) $f(x) = \left(\frac{1}{2}x\right)^4 - 2$

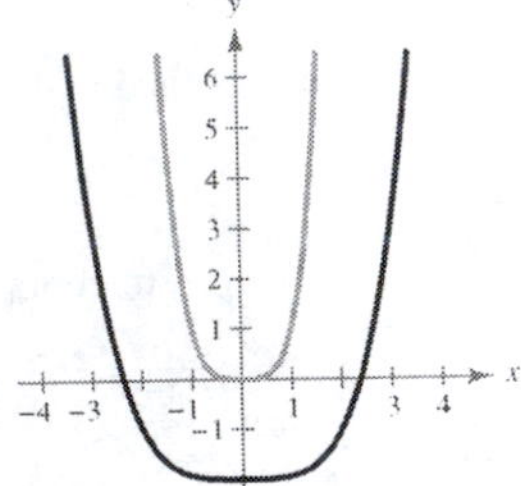

Vertical shift two units downward and a horizontal stretch (each y-value is multipied by $\frac{1}{16}$)

19. $f(x) = \frac{1}{5}x^3 + 4x$

Degree: 3

Leading coefficient: $\frac{1}{5}$

The degree is odd and the leading coefficient is positive. The graph falls to the left and rises to the right.

21. $g(x) = 5 - \frac{7}{2}x - 3x^2$

Degree: 2

Leading coefficient: -3

The degree is even and the leading coefficient is negative. The graph falls to the left and falls to the right.

23. $g(x) = -x^3 + 3x^2$

Degree: 3

Leading coefficient: -1

The degree is odd and the leading coefficient is negative. The graph rises to the left and falls to the right.

25. $f(x) = -2.1x^5 + 4x^3 - 2$

Degree: 5

Leading coefficient: -2.1

The degree is odd and the leading coefficient is negative. The graph rises to the left and falls to the right.

27. $f(x) = 6 - 2x + 4x^2 - 5x^3$

Degree: 3

Leading coefficient: -5

The degree is odd and the leading coefficient is negative. The graph rises to the left and falls to the right.

29. $h(x) = -\dfrac{3}{4}(t^2 - 3t + 6)$

Degree: 2

Leading coefficient: $-\dfrac{3}{4}$

The degree is even and the leading coefficient is negative. The graph falls to the left and falls to the right.

31. $f(x) = 3x^3 - 9x + 1;\ g(x) = 3x^3$

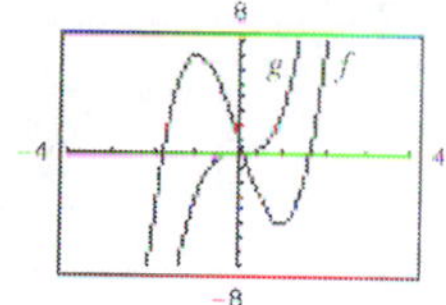

33. $f(x) = -(x^4 - 4x^3 + 16x);\ g(x) = -x^4$

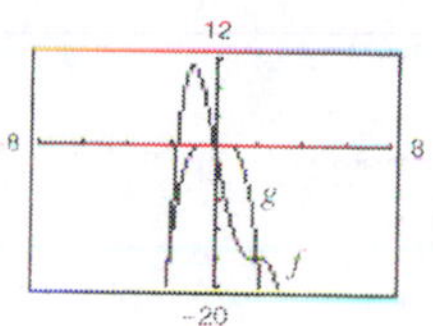

35. $f(x) = x^2 - 36$

(a) $0 = x^2 - 36$

$0 = (x + 6)(x - 6)$

$x + 6 = 0 \qquad x - 6 = 0$

$x = -6 \qquad x = 6$

Zeros: ± 6

(b) Each zero has a multiplicity of one (odd multiplicity).

(c) Turning points: 1 (the vertex of the parabola)

(d)

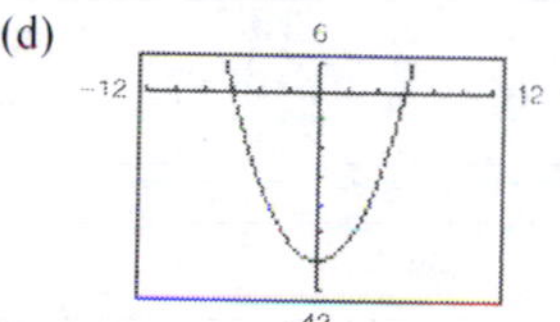

37. $h(t) = t^2 - 6t + 9$

(a) $0 = t^2 - 6t + 9 = (t - 3)^2$

Zero: $t = 3$

(b) $t = 3$ has a multiplicity of 2 (even multiplicity).

(c) Turning points: 1 (the vertex of the parabola)

(d)

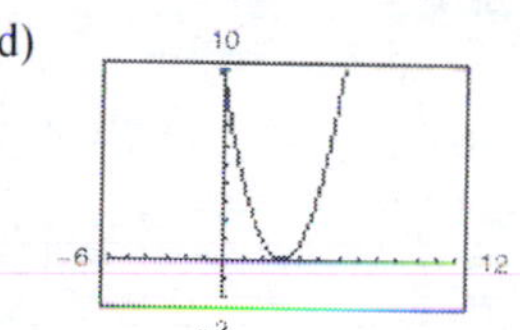

39. $f(x) = \frac{1}{3}x^2 + \frac{1}{3}x - \frac{2}{3}$

(a) $0 = \frac{1}{3}x^2 + \frac{1}{3}x - \frac{2}{3}$

$= \frac{1}{3}(x^2 + x - 2)$

$= \frac{1}{3}(x + 2)(x - 1)$

Zeros: $x = -2,\ x = 1$

(b) Each zero has a multiplicity of 1 (odd multiplicity).

(c) Turning points: 1 (the vertex of the parabola)

(d)

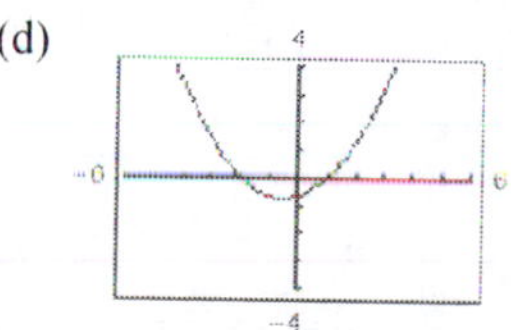

41. $f(x) = 3x^3 - 12x^2 + 3x$

(a) $0 = 3x^3 - 12x^2 + 3x = 3x(x^2 - 4x + 1)$

Zeros: $x = 0, x = 2 \pm \sqrt{3}$ (by the Quadratic Formula)

(b) Each zero has a multiplicity of 1 (odd multiplicity).

(c) Turning points: 2

(d)

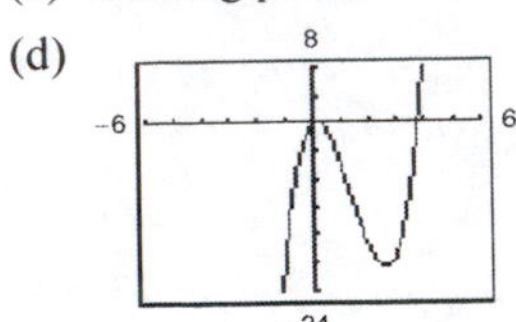

43. $f(t) = t^3 - 8t^2 + 16t$

(a) $0 = t^3 - 8t^2 + 16t$

$0 = t(t^2 - 8t + 16)$

$0 = t(t - 4)(t - 4)$

$t = 0 \quad t - 4 = 0 \quad t - 4 = 0$

$t = 0 \quad t = 4 \quad t = 4$

Zeros: $t = 0, t = 4$

(b) The multiplicity of $t = 0$ is 1 (odd multiplicity).

The multiplicity of $t = 4$ is 2 (even multiplicity).

(c) Turning points: 2

(d)

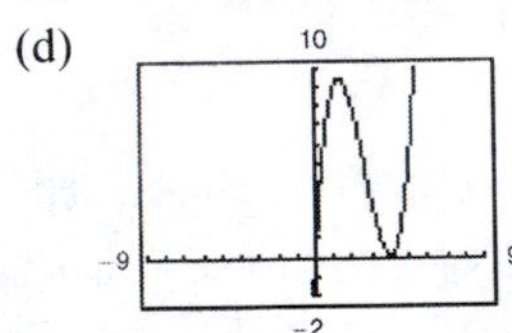

45. $g(t) = t^5 - 6t^3 + 9t$

(a) $0 = t^5 - 6t^3 + 9t = t(t^4 - 6t^2 + 9) = t(t^2 - 3)^2$

$= t(t + \sqrt{3})^2(t - \sqrt{3})^2$

Zeros: $t = 0, t = \pm\sqrt{3}$

(b) $t = 0$ has a multiplicity of 1 (odd multiplicity).

$t = \pm\sqrt{3}$ each have a multiplicity of 2 (even multiplicity).

(c) Turning points: 4

(d)

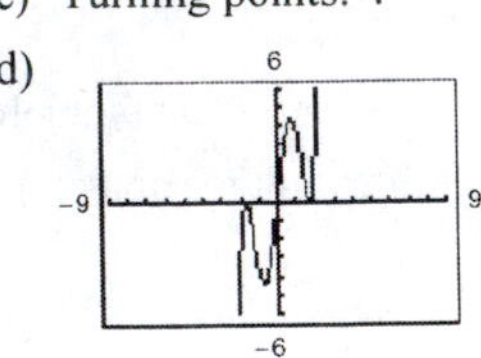

47. $f(x) = 3x^4 + 9x^2 + 6$

(a) $0 = 3x^4 + 9x^2 + 6$

$0 = 3(x^4 + 3x^2 + 2)$

$0 = 3(x^2 + 1)(x^2 + 2)$

(b) No real zeros

(c) Turning points: 1

(d)

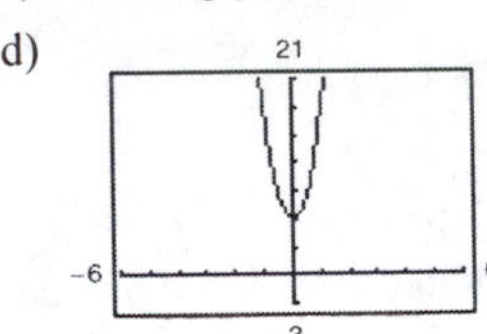

49. $g(x) = x^3 + 3x^2 - 4x - 12$

(a) $0 = x^3 + 3x^2 - 4x - 12 = x^2(x + 3) - 4(x + 3)$

$= (x^2 - 4)(x + 3) = (x - 2)(x + 2)(x + 3)$

Zeros: $x = \pm 2, x = -3$

(b) Each zero has a multiplicity of 1 (odd multiplicity).

(c) Turning points: 2

(d)

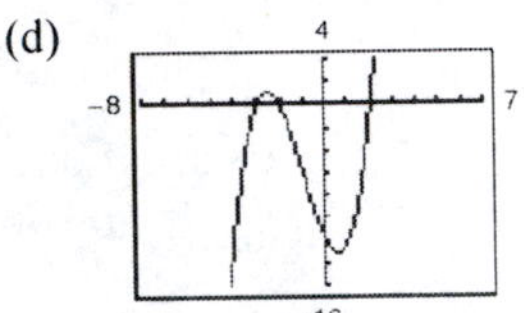

51. $y = 4x^3 - 20x^2 + 25x$

(a)

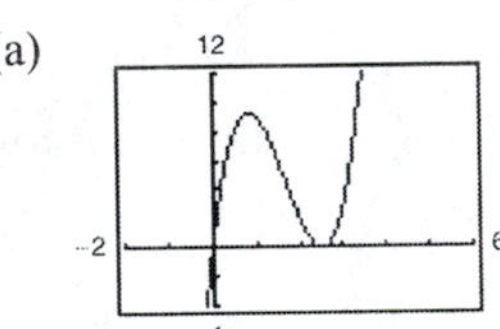

(b) x-intercepts: $(0, 0), \left(\frac{5}{2}, 0\right)$

(c) $0 = 4x^3 - 20x^2 + 25x$

$0 = x(4x^2 - 20x + 25)$

$0 = x(2x - 5)^2$

$x = 0, \frac{5}{2}$

(d) The solutions are the same as the x-coordinates of the x-intercepts.

53. $y = x^5 - 5x^3 + 4x$

(a)

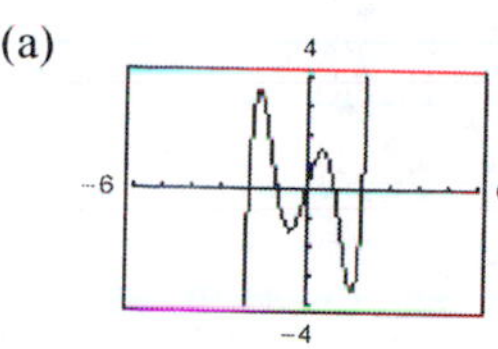

(b) x-intercepts: $(0, 0), (\pm 1, 0), (\pm 2, 0)$

(c) $0 = x^5 - 5x^3 + 4x$

$0 = x(x^2 - 1)(x^2 - 4)$

$0 = x(x + 1)(x - 1)(x + 2)(x - 2)$

$x = 0, \pm 1, \pm 2$

(d) The solutions are the same as the x-coordinates of the x-intercepts.

55. $f(x) = (x - 0)(x - 8)$

$= x^2 - 8x$

Note: $f(x) = ax(x - 8)$ has zeros 0 and 8 for all real numbers $a \neq 0$.

57. $f(x) = (x - 2)(x + 6)$

$= x^2 + 4x - 12$

Note: $f(x) = a(x - 2)(x + 6)$ has zeros 2 and -6 for all real numbers $a \neq 0$.

59. $f(x) = (x - 0)(x + 4)(x + 5)$

$= x(x^2 + 9x + 20)$

$= x^3 + 9x^2 + 20x$

Note: $f(x) = ax(x + 4)(x + 5)$ has zeros $0, -4,$ and -5 for all real numbers $a \neq 0$.

61. $f(x) = (x - 4)(x + 3)(x - 3)(x - 0)$

$= (x - 4)(x^2 - 9)x$

$= x^4 - 4x^3 - 9x^2 + 36x$

Note: $f(x) = a(x^4 - 4x^3 - 9x^2 + 36x)$ has zeros $4, -3, 3,$ and 0 for all real numbers $a \neq 0$.

63. $f(x) = \left[x - (1 + \sqrt{3})\right]\left[x - (1 - \sqrt{3})\right]$

$= \left[(x - 1) - \sqrt{3}\right]\left[(x - 1) + \sqrt{3}\right]$

$= (x - 1)^2 - (\sqrt{3})^2$

$= x^2 - 2x + 1 - 3$

$= x^2 - 2x - 2$

Note: $f(x) = a(x^2 - 2x - 2)$ has zeros $1 + \sqrt{3}$ and $1 - \sqrt{3}$ for all real numbers $a \neq 0$.

65. $f(x) = (x + 3)(x + 3) = x^2 + 6x + 9$

Note: $f(x) = a(x^2 + 6x + 9), a \neq 0,$ has degree 2 and zero $x = -3$.

67. $f(x) = (x - 0)(x + 5)(x - 1)$

$= x(x^2 + 4x - 5)$

$= x^3 + 4x^2 - 5x$

Note: $f(x) = ax(x^2 + 4x - 5), a \neq 0,$ has degree 3 and zeros $x = 0, -5,$ and 1.

69. $f(x) = (x - 0)(x - \sqrt{3})(x - (-\sqrt{3}))$

$= x(x - \sqrt{3})(x + \sqrt{3}) = x^3 - 3x$

Note: $f(x) = a(x^3 - 3x), a \neq 0,$ has degree 3 and zeros $x = 0, \sqrt{3},$ and $-\sqrt{3}$.

71. $f(x) = (x - (-5))^2(x - 1)(x - 2) = x^4 + 7x^3 - 3x^2 - 55x + 50$

or $f(x) = (x - (-5))(x - 1)^2(x - 2) = x^4 + x^3 - 15x^2 + 23x - 10$

or $f(x) = (x - (-5))(x - 1)(x - 2)^2 = x^4 - 17x^2 + 36x - 20$

Note: Any nonzero scalar multiple of these functions would also have degree 4 and zeros $x = -5, 1,$ and 2.

73. $f(x) = x^4(x + 4) = x^5 + 4x^4$

or $f(x) = x^3(x + 4)^2 = x^5 + 8x^4 + 16x^3$

or $f(x) = x^2(x + 4)^3 = x^5 + 12x^4 + 48x^3 + 64x^2$

or $f(x) = x(x + 4)^4 = x^5 + 16x^4 + 96x^3 + 256x^2 + 256x$

Note: Any nonzero scalar multiple of these functions would also have degree 5 and zeros $x = 0$ and -4.

75. $f(x) = x^3 - 25x = x(x + 5)(x - 5)$

(a) Falls to the left; rises to the right

(b) Zeros: 0, −5, 5

(c)

x	−2	−1	0	1	2
$f(x)$	42	24	0	−24	−42

(d)

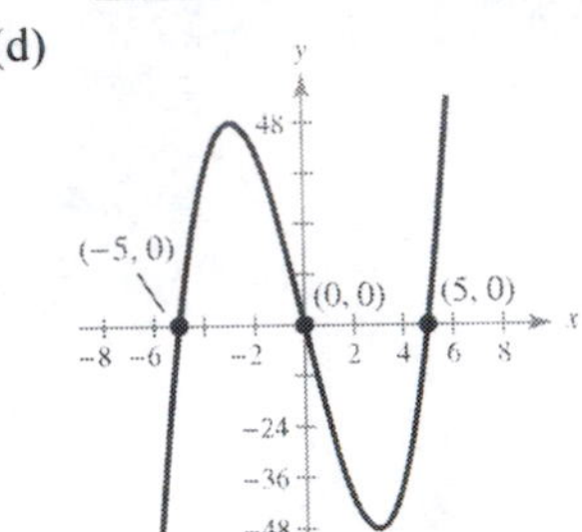

77. $f(t) = \frac{1}{4}(t^2 - 2t + 15) = \frac{1}{4}(t - 1)^2 + \frac{7}{2}$

(a) Rises to the left; rises to the right

(b) No real zeros (no x-intercepts)

(c)

t	−1	0	1	2	3
$f(t)$	4.5	3.75	3.5	3.75	4.5

(d) The graph is a parabola with vertex $\left(1, \frac{7}{2}\right)$.

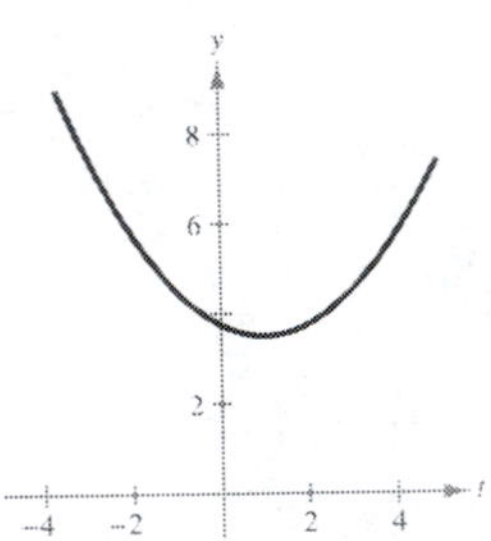

79. $f(x) = x^3 - 2x^2 = x^2(x - 2)$

(a) Falls to the left; rises to the right

(b) Zeros: 0, 2

(c)

x	−1	0	$\frac{1}{2}$	1	2	3
$f(x)$	−3	0	$-\frac{3}{8}$	−1	0	9

(d)

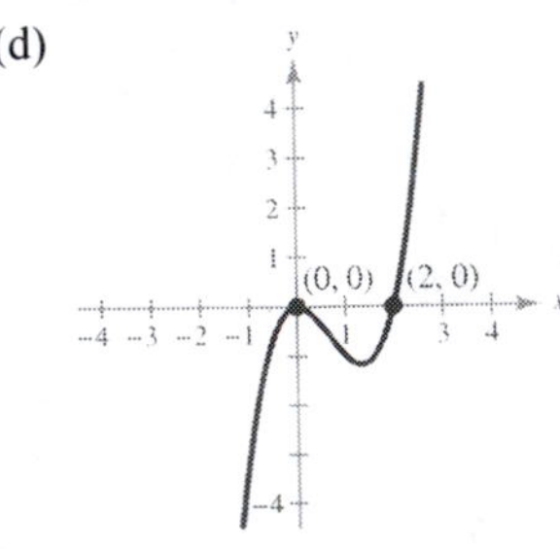

81. $f(x) = 3x^3 - 15x^2 + 18x = 3x(x - 2)(x - 3)$

(a) Falls to the left; rises to the right

(b) Zeros: 0, 2, 3

(c)

x	0	1	2	2.5	3	3.5
$f(x)$	0	6	0	−1.875	0	7.875

(d)

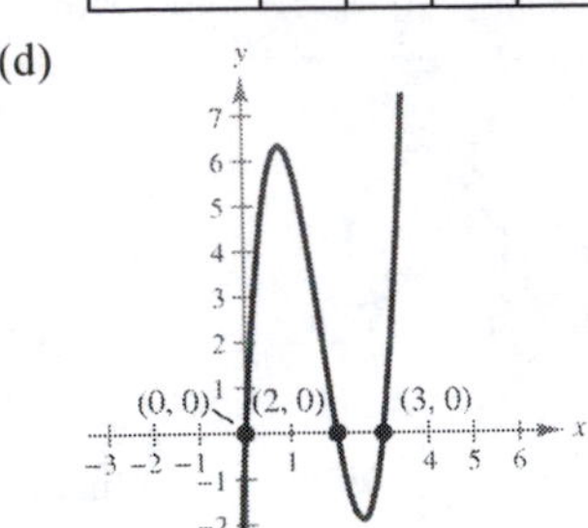

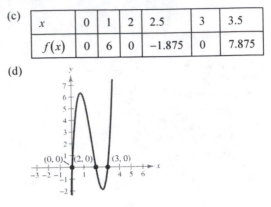

83. $f(x) = -5x^2 - x^3 = -x^2(5 + x)$

(a) Rises to the left; falls to the right

(b) Zeros: 0, −5

(c)

x	−5	−4	−3	−2	−1	0	1
$f(x)$	0	−16	−18	−12	−4	0	−6

(d)

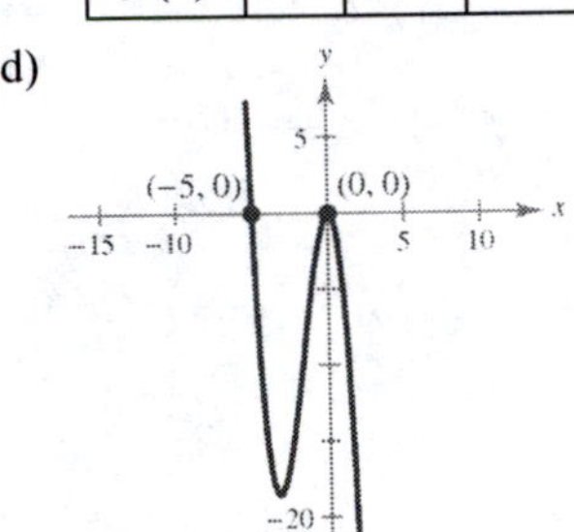

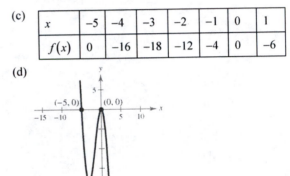

85. $f(x) = x^2(x - 4)$

(a) Falls to the left; rises to the right

(b) Zeros: 0, 4

(c)

x	−1	0	1	2	3	4	5
$f(x)$	−5	0	−3	−8	−9	0	25

(d)

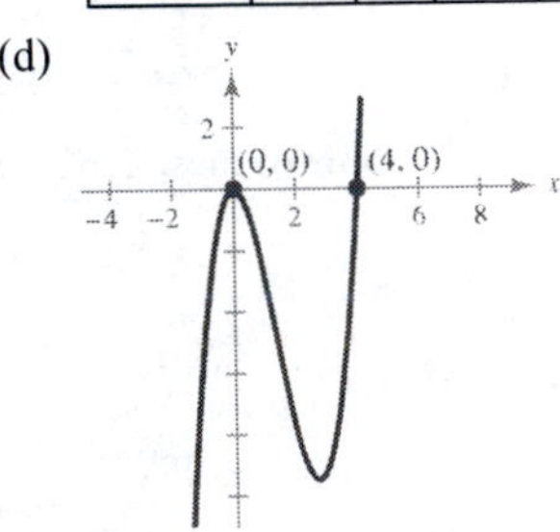

87. $g(t) = -\frac{1}{4}(t-2)^2(t+2)^2$

(a) Falls to the left; falls to the right

(b) Zeros: 2, – 2

(c)

t	–3	–2	–1	0	1	2	3
$g(t)$	$-\frac{25}{4}$	0	$-\frac{9}{4}$	–4	$-\frac{9}{4}$	0	$-\frac{25}{4}$

(d)

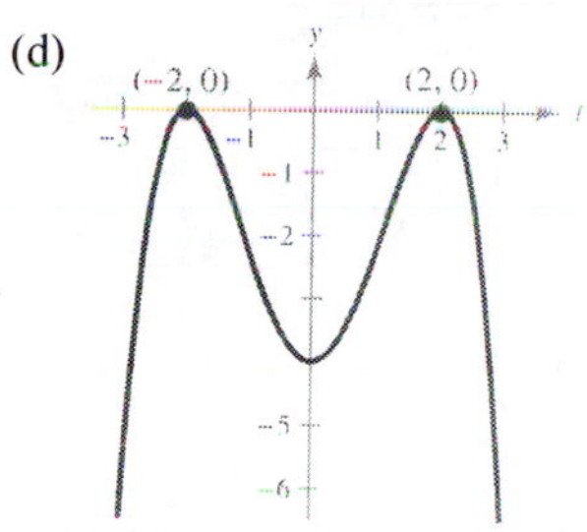

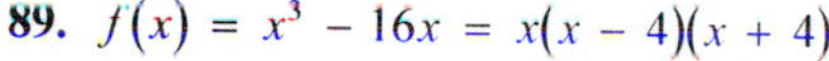

89. $f(x) = x^3 - 16x = x(x-4)(x+4)$

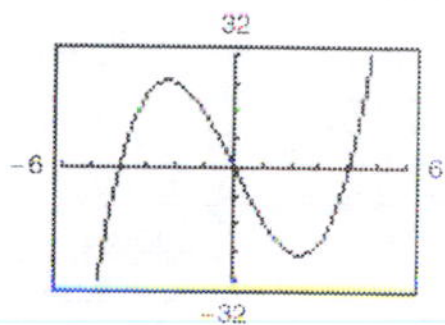

Zeros: 0 of multiplicity 1; 4 of multiplicity 1; and –4 of multiplicity 1

91. $g(x) = \frac{1}{5}(x+1)^2(x-3)(2x-9)$

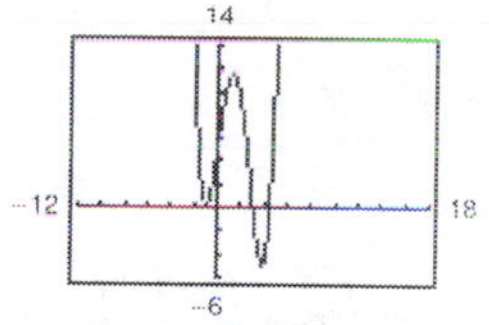

Zeros: –1 of multiplicity 2; 3 of multiplicity 1; $\frac{9}{2}$ of multiplicity 1

93. $f(x) = x^3 - 3x^2 + 3$

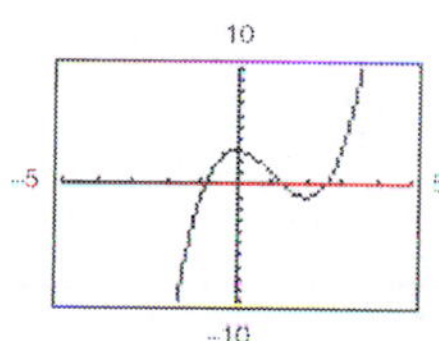

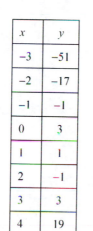

x	y
–3	–51
–2	–17
–1	–1
0	3
1	1
2	–1
3	3
4	19

The function has three zeros. They are in the intervals $[-1, 0]$, $[1, 2]$, and $[2, 3]$. They are $x \approx -0.879, 1.347, 2.532$.

95. $g(x) = 3x^4 + 4x^3 - 3$

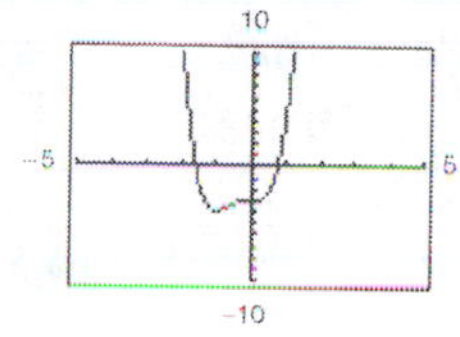

x	y
–4	509
–3	132
–2	13
–1	–4
0	–3
1	4
2	77
3	348

The function has two zeros. They are in the intervals $[-2, -1]$ and $[0, 1]$. They are $x \approx -1.585, 0.779$.

97. (a) Volume $= l \cdot w \cdot h$

height $= x$

length $=$ width $= 36 - 2x$

So, $V(x) = (36-2x)(36-2x)(x) = x(36-2x)^2$.

(b) Domain: $0 < x < 18$

The length and width must be positive.

(c)

Box Height	Box Width	Box Volume, V
1	$36 - 2(1)$	$1[36-2(1)]^2 = 1156$
2	$36 - 2(2)$	$2[36-2(2)]^2 = 2048$
3	$36 - 2(3)$	$3[36-2(3)]^2 = 2700$
4	$36 - 2(4)$	$4[36-2(4)]^2 = 3136$
5	$36 - 2(5)$	$5[36-2(5)]^2 = 3380$
6	$36 - 2(6)$	$6[36-2(6)]^2 = 3456$
7	$36 - 2(7)$	$7[36-2(7)]^2 = 3388$

The volume is a maximum of 3456 cubic inches when the height is 6 inches and the length and width are each 24 inches. So the dimensions are $6 \times 24 \times 24$ inches.

(d)

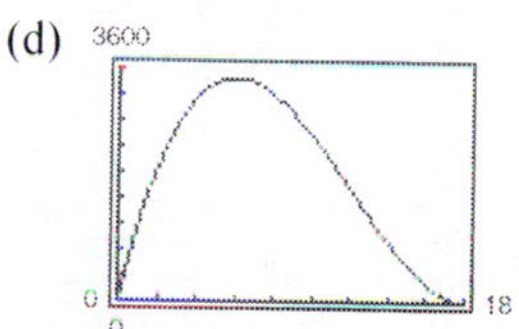

The maximum point on the graph occurs at $x = 6$. This agrees with the maximum found in part (c).

99. (a) $A = l \cdot w = (12 - 2x)(x) = -2x^2 + 12x$

(b) 16 feet = 192 inches

$$V = l \cdot w \cdot h$$
$$= (12 - 2x)(x)(192)$$
$$= -384x^2 + 2304x$$

(c) Because x and $12 - 2x$ cannot be negative, we have $0 < x < 6$ inches for the domain.

(d)

x	V
0	0
1	1920
2	3072
3	3456
4	3072
5	1920
6	0

When $x = 3$, the volume is a maximum with $V = 3456$ in.^{3}. The dimensions of the gutter cross-section are 3 inches × 6 inches × 3 inches.

(e)

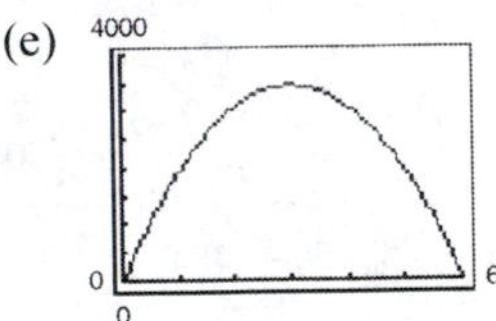

Maximum: (3, 3456)

The maximum value is the same.

(f) No. The volume is a product of the constant length and the cross-sectional area. The value of x would remain the same; only the value of V would change if the length was changed.

101. (a)

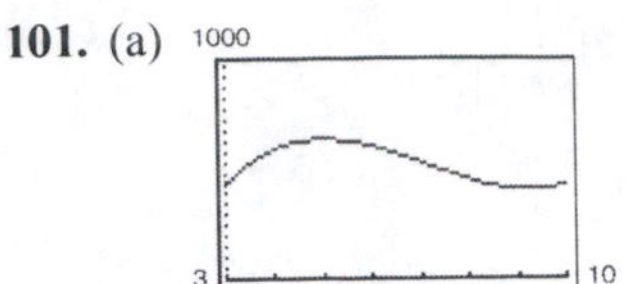

Relative maximum: (5.01, 655.75)

Relative minimum: (9.25, 417.42)

(b) The revenue is increasing over (3, 5.01) and decreasing over (5.01, 9.25), and then increasing over (9.25, 10).

(c) The revenue for this company is increasing from 2003 to 2005, when it reached a (relative) maximum of \$655.75 million. From 2005 to 2009, revenue was decreasing when it dropped to \$417.42 million. From 2009 to 2010, revenue began to increase again.

103. $R = \dfrac{1}{100{,}000}\left(-x^3 + 600x^2\right)$

The point of diminishing returns (where the graph changes from curving upward to curving downward) occurs when $x = 200$. The point is (200, 160) which corresponds to spending \$2,000,000 on advertising to obtain a revenue of \$160 million.

105. False. A fifth-degree polynomial can have at most four turning points.

107. False. The function $f(x) = (x - 2)^2$ has one turning point and two real (repeated) zeros.

109. False. $f(x) = -x^3$ rises to the left.

111. True. A polynomial of degree 7 with a negative leading coefficient rises to the left and falls to the right.

113. Answers will vary. *Sample answers:*

$a_4 < 0$

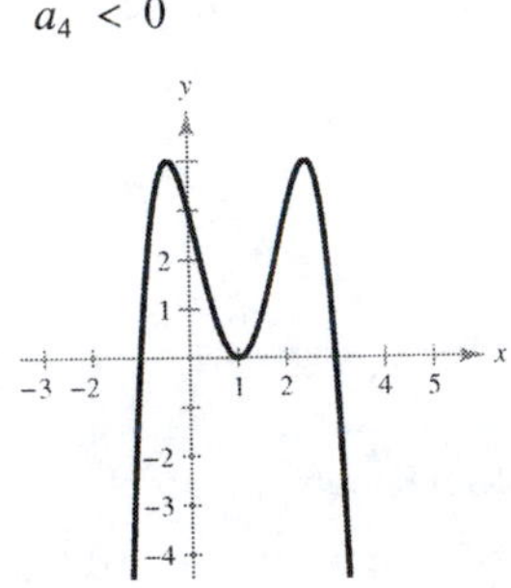

$a_4 > 0$

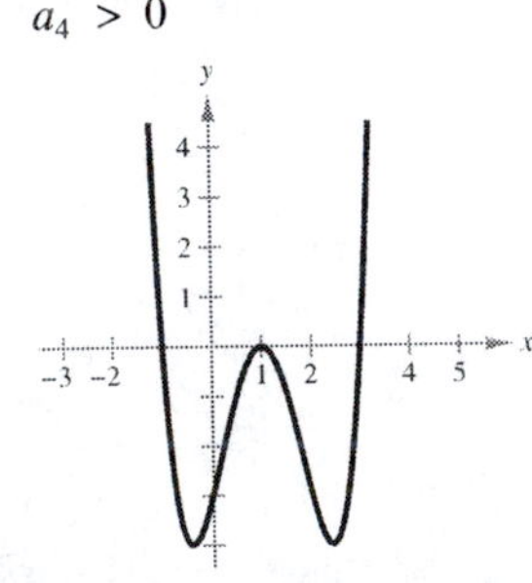

115. $f(x) = x^4$; $f(x)$ is even.

(a) $g(x) = f(x) + 2$

Vertical shift two units upward

$$g(-x) = f(-x) + 2$$
$$= f(x) + 2$$
$$= g(x)$$

Even

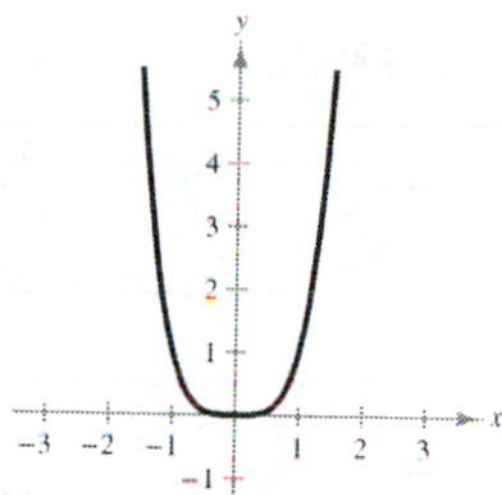

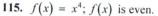

(b) $g(x) = f(x + 2)$

Horizontal shift two units to the left

Neither odd nor even

(c) $g(x) = f(-x) = (-x)^4 = x^4$

Reflection in the y-axis. The graph looks the same.

Even

(d) $g(x) = -f(x) = -x^4$

Reflection in the x-axis

Even

(e) $g(x) = f\left(\frac{1}{2}x\right) = \frac{1}{16}x^4$

Horizontal stretch

Even

(f) $g(x) = \frac{1}{2}f(x) = \frac{1}{2}x^4$

Vertical shrink

Even

(g) $g(x) = f\left(x^{3/4}\right) = \left(x^{3/4}\right)^4 = x^3, x \ge 0$

Neither odd nor even

(h) $g(x) = (f \circ f)(x) = f(f(x)) = f\left(x^4\right) = \left(x^4\right)^4 = x^{16}$

Even

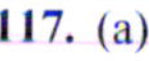

117. (a)

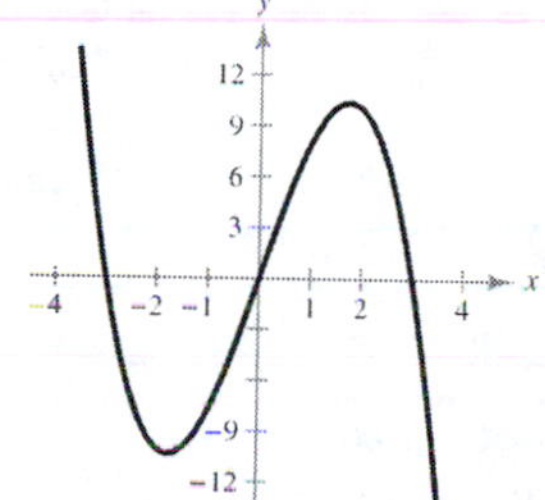

Zeros: 3

Relative minimum: 1

Relative maximum: 1

The number of zeros is the same as the degree and the number of extrema is one less than the degree.

(b)

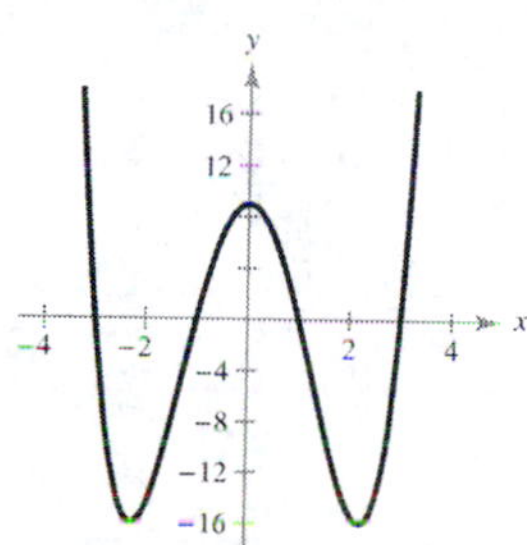

Zeros: 4

Relative minima: 2

Relative maximum: 1

The number of zeros is the same as the degree and the number of extrema is one less than the degree.

(c)

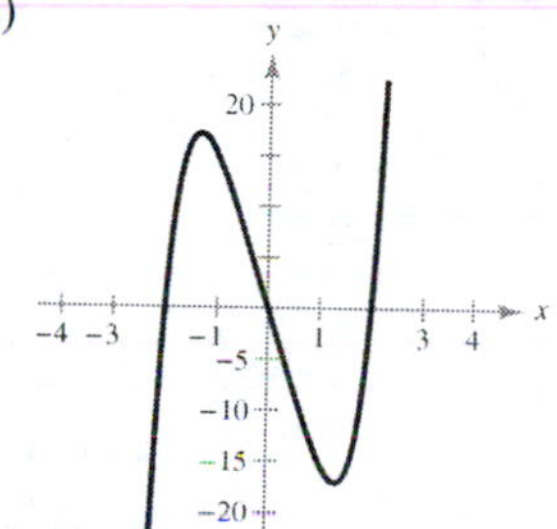

Zeros: 3

Relative minimum: 1

Relative maximum: 1

The number of zeros and the number of extrema are both less than the degree.

Section 3.3 Polynomial and Synthetic Division

1. $f(x)$ is the dividend; $d(x)$ is the divisor: $q(x)$ is the quotient: $r(x)$ is the remainder

3. improper

5. Factor

7. $y_1 = \dfrac{x^2}{x+2}$ and $y_2 = x - 2 + \dfrac{4}{x+2}$

$$\begin{array}{r}
x - 2 \\
x + 2\overline{)x^2 + 0x + 0} \\
\underline{x^2 + 2x} \\
-2x + 0 \\
\underline{-2x - 4} \\
4
\end{array}$$

So, $\dfrac{x^2}{x+2} = x - 2 + \dfrac{4}{x+2}$ and $y_1 = y_2$.

9. $y_1 = \dfrac{x^2 + 2x - 1}{x+3}$, $y_2 = x - 1 + \dfrac{2}{x+3}$

(a) and (b)

(c)
$$\begin{array}{r}
x - 1 \\
x + 3\overline{)x^2 + 2x - 1} \\
\underline{x^2 + 3x} \\
-x - 1 \\
\underline{-x - 3} \\
2
\end{array}$$

So, $\dfrac{x^2 + 2x - 1}{x+3} = x - 1 + \dfrac{2}{x+3}$ and $y_1 = y_2$.

11.
$$\begin{array}{r}
2x + 4 \\
x + 3\overline{)2x^2 + 10x + 12} \\
\underline{2x^2 + 6x} \\
4x + 12 \\
\underline{4x + 12} \\
0
\end{array}$$

$$\frac{2x^2 + 10x + 12}{x+3} = 2x + 4,\ x \neq 3$$

13.
$$\begin{array}{r}
x^2 - 3x + 1 \\
4x + 5\overline{)4x^3 - 7x^2 - 11x + 5} \\
\underline{4x^3 + 5x^2} \\
-12x^2 - 11x \\
\underline{-12x^2 - 15x} \\
4x + 5 \\
\underline{4x + 5} \\
0
\end{array}$$

$$\frac{4x^3 - 7x^2 - 11x + 5}{4x+5} = x^2 - 3x + 1,\ x \neq -\frac{5}{4}$$

15.
$$\begin{array}{r}
x^3 + 3x^2 \qquad -1 \\
x + 2\overline{)x^4 + 5x^3 + 6x^2 - x - 2} \\
\underline{x^4 + 2x^3} \\
3x^3 + 6x^2 \\
\underline{3x^3 + 6x^2} \\
-x - 2 \\
\underline{-x - 2} \\
0
\end{array}$$

$$\frac{x^4 + 5x^3 + 6x^2 - x - 2}{x+2} = x^3 + 3x^2 - 1,\ x \neq -2$$

17.
$$\begin{array}{r}
x^2 + 3x + 9 \\
x - 3\overline{)x^3 + 0x^2 + 0x - 27} \\
\underline{x^3 - 3x^2} \\
3x^2 + 0x \\
\underline{3x^2 - 9x} \\
9x - 27 \\
\underline{9x - 27} \\
0
\end{array}$$

$$\frac{x^3 - 27}{x-3} = x^2 + 3x + 9,\ x \neq 3$$

19.
$$\begin{array}{r}
7 \\
x + 2\overline{)7x + 3} \\
\underline{7x + 14} \\
-11
\end{array}$$

$$\frac{7x+3}{x+2} = 7 - \frac{11}{x+2}$$

21.
$$\begin{array}{r}
x \\
x^2 + 0x + 1\overline{)x^3 + 0x^2 + 0x - 9} \\
\underline{x^3 + 0x^2 + x} \\
-x - 9
\end{array}$$

$$\frac{x^3 - 9}{x^2 + 1} = x - \frac{x+9}{x^2+1}$$

23.
$$\begin{array}{r}
2x - 8 \\
x^2 + 0x + 1\overline{)2x^3 - 8x^2 + 3x - 9} \\
\underline{2x^3 + 0x^2 + 2x} \\
-8x^2 + x - 9 \\
\underline{-8x^2 - 0x - 8} \\
x - 1
\end{array}$$

$$\frac{2x^3 - 8x^2 + 3x - 9}{x^2 + 1} = 2x - 8 + \frac{x-1}{x^2+1}$$

25.
$$\begin{array}{r}
x + 3 \\
x^3 - 3x^2 + 3x - 1\overline{)x^4 + 0x^3 + 0x^2 + 0x + 0} \\
\underline{x^4 - 3x^3 + 3x^2 - x} \\
3x^3 - 3x^2 + x + 0 \\
\underline{3x^3 - 9x^2 + 9x - 3} \\
6x^2 - 8x + 3
\end{array}$$

$$\frac{x^4}{(x-1)^3} = x + 3 + \frac{6x^2 - 8x + 3}{(x-1)^3}$$

27.

5	3	−17	15	−25
		15	−10	25
	3	−2	5	0

$$\frac{3x^3 - 17x^2 + 15x - 25}{x - 5} = 3x^2 - 2x + 5, x \neq 5$$

29.

3	6	7	−1	26
		18	75	222
	6	25	74	248

$$\frac{6x^3 + 7x^2 - x + 26}{x - 3} = 6x^2 + 25x + 74 + \frac{248}{x - 3}$$

31.

−2	4	8	−9	−18
		−8	0	18
	4	0	−9	0

$$\frac{4x^3 + 8x^2 - 9x - 18}{x + 2} = 4x^2 - 9, x \neq -2$$

37.

6	10	−50	0	0	−800
		60	60	360	2160
	10	10	60	360	1360

$$\frac{10x^4 - 50x^3 - 800}{x - 6} = 10x^3 + 10x^2 + 60x + 360 + \frac{1360}{x - 6}$$

39.

−8	1	0	0	512
		−8	64	−512
	1	−8	64	0

$$\frac{x^3 + 512}{x + 8} = x^2 - 8x + 64, x \neq -8$$

41.

2	−3	0	0	0	0
		−6	−12	−24	−48
	−3	−6	−12	−24	−48

$$\frac{-3x^4}{x - 2} = -3x^3 - 6x^2 - 12x - 24 - \frac{48}{x - 2}$$

43.

6	−1	0	0	180	0
		−6	−36	−216	−216
	−1	−6	−36	−36	−216

$$\frac{180x - x^4}{x - 6} = -x^3 - 6x^2 - 36x - 36 - \frac{216}{x - 6}$$

45.

$-\frac{1}{2}$	4	16	−23	−15
		−2	−7	15
	4	14	−30	0

$$\frac{4x^3 + 16x^2 - 23x - 15}{x + \frac{1}{2}} = 4x^2 + 14x - 30, x \neq -\frac{1}{2}$$

33.

−10	−1	0	75	−250
		10	−100	250
	−1	10	−25	0

$$\frac{-x^3 + 75x - 250}{x + 10} = -x^2 + 10x - 25, x \neq -10$$

35.

4	5	−6	0	8
		20	56	224
	5	14	56	232

$$\frac{5x^3 - 6x^2 + 8}{x - 4} = 5x^2 + 14x + 56 + \frac{232}{x - 4}$$

47. $f(x) = x^3 - x^2 - 14x + 11, k = 4$

4	1	−1	−14	11
		4	12	−8
	1	3	−2	3

$$f(x) = (x - 4)(x^2 + 3x - 2) + 3$$

$$f(4) = 4^3 - 4^2 - 14(4) + 11 = 3$$

49. $f(x) = 15x^4 + 10x^3 - 6x^2 + 14, k = -\frac{2}{3}$

$-\frac{2}{3}$	15	10	−6	0	14
		−10	0	4	$-\frac{8}{3}$
	15	0	−6	4	$\frac{34}{3}$

$$f(x) = \left(x + \tfrac{2}{3}\right)(15x^3 - 6x + 4) + \tfrac{34}{3}$$

$$f\left(-\tfrac{2}{3}\right) = 15\left(-\tfrac{2}{3}\right)^4 + 10\left(-\tfrac{2}{3}\right)^3 - 6\left(-\tfrac{2}{3}\right)^2 + 14 = \tfrac{34}{3}$$

51. $f(x) = x^3 + 3x^2 - 2x - 14, k = \sqrt{2}$

$\sqrt{2}$	1	3	−2	−14
		$\sqrt{2}$	$2 + 3\sqrt{2}$	6
	1	$3 + \sqrt{2}$	$3\sqrt{2}$	−8

$$f(x) = \left(x - \sqrt{2}\right)\left[x^2 + \left(3 + \sqrt{2}\right)x + 3\sqrt{2}\right] - 8$$

$$f\left(\sqrt{2}\right) = \left(\sqrt{2}\right)^3 + 3\left(\sqrt{2}\right)^2 - 2\sqrt{2} - 14 = -8$$

53. $f(x) = -4x^3 + 6x^2 + 12x + 4,\ k = 1 - \sqrt{3}$

$$\begin{array}{r|rrrr} 1-\sqrt{3} & -4 & 6 & 12 & 4 \\ & & -4+4\sqrt{3} & -10+2\sqrt{3} & -4 \\ \hline & -4 & 2+4\sqrt{3} & 2+2\sqrt{3} & 0 \end{array}$$

$$f(x) = \left(x - 1 + \sqrt{3}\right)\left[-4x^2 + \left(2 + 4\sqrt{3}\right)x + \left(2 + 2\sqrt{3}\right)\right]$$

$$f\left(1 - \sqrt{3}\right) = -4\left(1 - \sqrt{3}\right)^3 + 6\left(1 - \sqrt{3}\right)^2 + 12\left(1 - \sqrt{3}\right) + 4 = 0$$

55. $f(x) = 2x^3 - 7x + 3$

(a) Using the Remainder Theorem:

$$f(1) = 2(1)^3 - 7(1) + 3 = -2$$

Using synthetic division:

$$\begin{array}{r|rrrr} 1 & 2 & 0 & -7 & 3 \\ & & 2 & 2 & -5 \\ \hline & 2 & 2 & -5 & -2 \end{array}$$

Verify using long division:

$$\begin{array}{r} 2x^2 + 2x - 5 \\ x - 1\overline{)2x^3 + 0x^2 - 7x + 3} \\ \underline{2x^3 - 2x^2} \\ 2x^2 - 7x \\ \underline{2x^2 - 2x} \\ -5x + 3 \\ \underline{-5x + 5} \\ -2 \end{array}$$

(b) Using the Remainder Theorem:

$$f(-2) = 2(-2)^3 - 7(-2) + 3 = 1$$

Using synthetic division:

$$\begin{array}{r|rrrr} -2 & 2 & 0 & -7 & 3 \\ & & -4 & 8 & -2 \\ \hline & 2 & -4 & 1 & 1 \end{array}$$

Verify using long division:

$$\begin{array}{r} 2x^2 - 4x + 1 \\ x + 2\overline{)2x^3 + 0x^2 - 7x + 3} \\ \underline{2x^3 + 4x^2} \\ -4x^2 - 7x \\ \underline{-4x^2 - 8x} \\ x + 3 \\ \underline{x + 2} \\ 1 \end{array}$$

(c) Using the Remainder Theorem:

$$f\left(\frac{1}{2}\right) = 2\left(\frac{1}{2}\right)^3 - 7\left(\frac{1}{2}\right) + 3 = -\frac{1}{4}$$

Using synthetic division:

$$\begin{array}{r|rrrr} \frac{1}{2} & 2 & 0 & -7 & 3 \\ & & 1 & \frac{1}{2} & -\frac{13}{4} \\ \hline & 2 & 1 & -\frac{13}{2} & -\frac{1}{4} \end{array}$$

Verify using long division:

$$\begin{array}{r} 2x^2 + x - \frac{13}{2} \\ x - \frac{1}{2}\overline{)2x^3 + 0x^2 - 7x + 3} \\ \underline{2x^3 - x^2} \\ x^2 - 7x \\ \underline{x^2 - \frac{1}{2}x} \\ -\frac{13}{2}x + 3 \\ \underline{-\frac{13}{2}x + \frac{13}{4}} \\ -\frac{1}{4} \end{array}$$

(d) Using the Remainder Theorem:

$$f(2) = 2(2)^3 - 7(2) + 3 = 5$$

Using synthetic division:

$$\begin{array}{r|rrrr} 2 & 2 & 0 & -7 & 3 \\ & & 4 & 8 & 2 \\ \hline & 2 & 4 & 1 & 5 \end{array}$$

Verify using long division:

$$\begin{array}{r} 2x^2 + 4x + 1 \\ x - 2\overline{)2x^3 + 0x^2 - 7x + 3} \\ \underline{2x^3 - 4x^2} \\ 4x^2 - 7x \\ \underline{4x^2 - 8x} \\ x + 3 \\ \underline{x - 2} \\ 5 \end{array}$$

57. $h(x) = x^3 - 5x^2 - 7x + 4$

(a) Using the Remainder Theorem:

$h(3) = (3)^3 - 5(3)^2 - 7(3) + 4 = -35$

Using synthetic division:

$$\begin{array}{r|rrrr} 3 & 1 & -5 & -7 & 4 \\ & & 3 & -6 & -39 \\ \hline & 1 & -2 & -13 & -35 \end{array}$$

Verify using long division:

$$\begin{array}{r} x^2 - 2x - 13 \\ x - 3\overline{)x^3 - 5x^2 - 7x + 4} \\ \underline{x^3 - 3x^2} \qquad\qquad \\ -2x^2 - 7x \qquad \\ \underline{-2x^2 + 6x} \qquad \\ -13x + 4 \\ \underline{-13x + 39} \\ -35 \end{array}$$

(b) Using the Remainder Theorem:

$h(2) = (2)^3 - 5(2)^2 - 7(2) + 4 = -22$

Using synthetic division:

$$\begin{array}{r|rrrr} 2 & 1 & -5 & -7 & 4 \\ & & 2 & -6 & -26 \\ \hline & 1 & -3 & -13 & -22 \end{array}$$

Verify using long division:

$$\begin{array}{r} x^2 - 3x - 13 \\ x - 2\overline{)x^3 - 5x^2 - 7x + 4} \\ \underline{x^3 - 2x^2} \qquad\qquad \\ -3x^2 - 7x \qquad \\ \underline{-3x^2 + 6x} \qquad \\ -13x + 4 \\ \underline{-13x + 26} \\ -22 \end{array}$$

(c) Using the Remainder Theorem:

$h(-2) = (-2)^3 - 5(-2)^2 - 7(-2) + 4 = -10$

Using synthetic division:

$$\begin{array}{r|rrrr} -2 & 1 & -5 & -7 & 4 \\ & & -2 & 14 & -14 \\ \hline & 1 & -7 & 7 & -10 \end{array}$$

Verify using long division:

$$\begin{array}{r} x^2 - 7x + 7 \\ x + 2\overline{)x^3 - 5x^2 - 7x + 4} \\ \underline{x^3 + 2x^2} \qquad\qquad \\ -7x^2 - 7x \qquad \\ \underline{-7x^2 - 14x} \qquad \\ 7x + 4 \\ \underline{7x + 14} \\ -10 \end{array}$$

(d) Using the Remainder Theorem:

$h(-5) = (-5)^3 - 5(-5)^2 - 7(-5) + 4 = -211$

Using synthetic division:

$$\begin{array}{r|rrrr} -5 & 1 & -5 & -7 & 4 \\ & & -5 & 50 & -215 \\ \hline & 1 & -10 & 43 & -211 \end{array}$$

Verify using long division:

$$\begin{array}{r} x^2 - 10x + 43 \\ x + 5\overline{)x^3 - 5x^2 - 7x + 4} \\ \underline{x^3 + 5x^2} \qquad\qquad \\ -10x^2 - 7x \qquad \\ \underline{-10x^2 - 50x} \qquad \\ 43x + 4 \\ \underline{43x + 215} \\ -211 \end{array}$$

59.

$$\begin{array}{r|rrrr} 2 & 1 & 0 & -7 & 6 \\ & & 2 & 4 & -6 \\ \hline & 1 & 2 & -3 & 0 \end{array}$$

$$\begin{aligned} x^3 - 7x + 6 &= (x - 2)(x^2 + 2x - 3) \\ &= (x - 2)(x + 3)(x - 1) \end{aligned}$$

Zeros: $2, -3, 1$

61.

$$\begin{array}{r|rrrr} \frac{1}{2} & 2 & -15 & 27 & -10 \\ & & 1 & -7 & 10 \\ \hline & 2 & -14 & 20 & 0 \end{array}$$

$$\begin{aligned} 2x^3 - 15x^2 + 27x - 10 &= \left(x - \tfrac{1}{2}\right)(2x^2 - 14x + 20) \\ &= (2x - 1)(x - 2)(x - 5) \end{aligned}$$

Zeros: $\frac{1}{2}, 2, 5$

63.

$$\begin{array}{r|rrrr} \sqrt{3} & 1 & 2 & -3 & -6 \\ & & \sqrt{3} & 3 + 2\sqrt{3} & 6 \\ \hline & 1 & 2 + \sqrt{3} & 2\sqrt{3} & 0 \end{array}$$

$$\begin{array}{r|rrr} -\sqrt{3} & 1 & 2 + \sqrt{3} & 2\sqrt{3} \\ & & -\sqrt{3} & -2\sqrt{3} \\ \hline & 1 & 2 & 0 \end{array}$$

$x^3 + 2x^2 - 3x - 6 = (x - \sqrt{3})(x + \sqrt{3})(x + 2)$

Zeros: $-\sqrt{3}, \sqrt{3}, -2$

65.
$$\begin{array}{r|rrrr} 1+\sqrt{3} & 1 & -3 & 0 & 2 \\ & & 1+\sqrt{3} & 1-\sqrt{3} & -2 \\ \hline & 1 & -2+\sqrt{3} & 1-\sqrt{3} & 0 \end{array}$$

$$\begin{array}{r|rrr} 1-\sqrt{3} & 1 & -2+\sqrt{3} & 1-\sqrt{3} \\ & & 1-\sqrt{3} & -1+\sqrt{3} \\ \hline & 1 & -1 & 0 \end{array}$$

$$x^3 - 3x^2 + 2 = \left[x - \left(1 + \sqrt{3}\right)\right]\left[x - \left(1 - \sqrt{3}\right)\right](x - 1)$$
$$= (x - 1)\left(x - 1 - \sqrt{3}\right)\left(x - 1 + \sqrt{3}\right)$$

Zeros: $1, 1 - \sqrt{3}, 1 + \sqrt{3}$

67. $f(x) = 2x^3 + x^2 - 5x + 2$; Factors: $(x + 2), (x - 1)$

(a)
$$\begin{array}{r|rrrr} -2 & 2 & 1 & -5 & 2 \\ & & -4 & 6 & -2 \\ \hline & 2 & -3 & 1 & 0 \end{array}$$

$$\begin{array}{r|rrr} 1 & 2 & -3 & 1 \\ & & 2 & -1 \\ \hline & 2 & -1 & 0 \end{array}$$

Both are factors of $f(x)$ because the remainders are zero.

(b) The remaining factor of $f(x)$ is $(2x - 1)$.

(c) $f(x) = (2x - 1)(x + 2)(x - 1)$

(d) Zeros: $\frac{1}{2}, -2, 1$

(e)

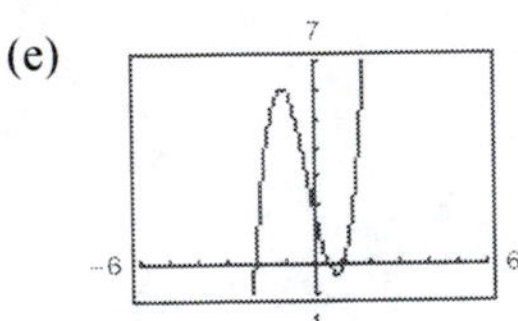

69. $f(x) = x^4 - 4x^3 - 15x^2 + 58x - 40$;
Factors: $(x - 5), (x + 4)$

(a)
$$\begin{array}{r|rrrrr} 5 & 1 & -4 & -15 & 58 & -40 \\ & & 5 & 5 & -50 & 40 \\ \hline & 1 & 1 & -10 & 8 & 0 \end{array}$$

$$\begin{array}{r|rrrr} -4 & 1 & 1 & -10 & 8 \\ & & -4 & 12 & -8 \\ \hline & 1 & -3 & 2 & 0 \end{array}$$

Both are factors of $f(x)$ because the remainders are zero.

(b) $x^2 - 3x + 2 = (x - 1)(x - 2)$

The remaining factors are $(x - 1)$ and $(x - 2)$.

(c) $f(x) = (x - 1)(x - 2)(x - 5)(x + 4)$

(d) Zeros: $1, 2, 5, -4$

(e)

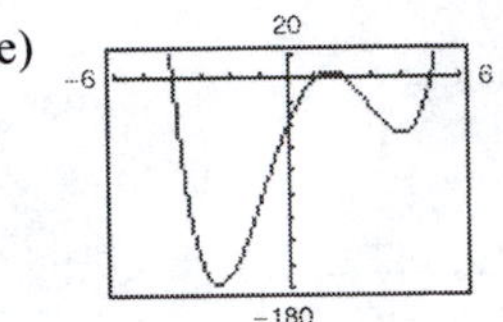

71. $f(x) = 6x^3 + 41x^2 - 9x - 14$;
Factors: $(2x + 1), (3x - 2)$

(a)
$$\begin{array}{r|rrrr} -\frac{1}{2} & 6 & 41 & -9 & -14 \\ & & -3 & -19 & 14 \\ \hline & 6 & 38 & -28 & 0 \end{array}$$

$$\begin{array}{r|rrr} \frac{2}{3} & 6 & 38 & -28 \\ & & 4 & 28 \\ \hline & 6 & 42 & 0 \end{array}$$

Both are factors of $f(x)$ because the remainders are zero.

(b) $6x + 42 = 6(x + 7)$

This shows that $\dfrac{f(x)}{\left(x + \frac{1}{2}\right)\left(x - \frac{2}{3}\right)} = 6(x + 7)$,

so $\dfrac{f(x)}{(2x + 1)(3x - 2)} = x + 7$.

The remaining factor is $(x + 7)$.

(c) $f(x) = (x + 7)(2x + 1)(3x - 2)$

(d) Zeros: $-7, -\frac{1}{2}, \frac{2}{3}$

(e)

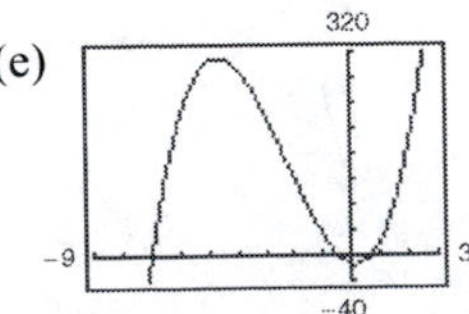

73. $f(x) = 2x^3 - x^2 - 10x + 5$;

Factors: $(2x - 1), \left(x + \sqrt{5}\right)$

(a)
$$\begin{array}{r|rrrr} \frac{1}{2} & 2 & -1 & -10 & 5 \\ & & 1 & 0 & -5 \\ \hline & 2 & 0 & -10 & 0 \end{array}$$

$$\begin{array}{r|rrr} -\sqrt{5} & 2 & 0 & -10 \\ & & -2\sqrt{5} & 10 \\ \hline & 2 & -2\sqrt{5} & 0 \end{array}$$

Both are factors of $f(x)$ because the remainders are zero.

(b) $2x - 2\sqrt{5} = 2\left(x - \sqrt{5}\right)$

This shows that $\dfrac{f(x)}{\left(x - \frac{1}{2}\right)\left(x + \sqrt{5}\right)} = 2\left(x - \sqrt{5}\right)$,

so $\dfrac{f(x)}{(2x - 1)\left(x + \sqrt{5}\right)} = x - \sqrt{5}$.

The remaining factor is $\left(x - \sqrt{5}\right)$.

(c) $f(x) = \left(x + \sqrt{5}\right)\left(x - \sqrt{5}\right)(2x - 1)$

(d) Zeros: $-\sqrt{5}, \sqrt{5}, \frac{1}{2}$

(e)

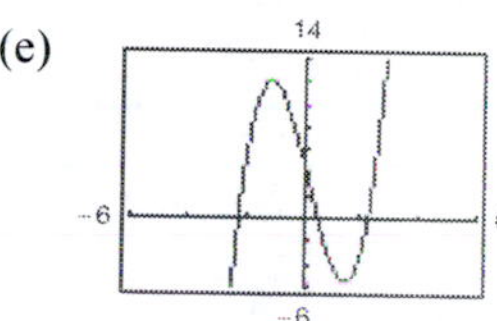

75. $f(x) = x^3 - 2x^2 - 5x + 10$

(a) The zeros of f are $x = 2$ and $x \approx \pm 2.236$.

(b) An exact zero is $x = 2$.

(c)
$$\begin{array}{r|rrrr} 2 & 1 & -2 & -5 & 10 \\ & & 2 & 0 & -10 \\ \hline & 1 & 0 & -5 & 0 \end{array}$$

$$\begin{aligned} f(x) &= (x - 2)\left(x^2 - 5\right) \\ &= (x - 2)\left(x - \sqrt{5}\right)\left(x + \sqrt{5}\right) \end{aligned}$$

77. $h(t) = t^3 - 2t^2 - 7t + 2$

(a) The zeros of h are $t = -2, t \approx 3.732, t \approx 0.268$.

(b) An exact zero is $t = -2$.

(c)
$$\begin{array}{r|rrrr} -2 & 1 & -2 & -7 & 2 \\ & & -2 & 8 & -2 \\ \hline & 1 & -4 & 1 & 0 \end{array}$$

$$h(t) = (t + 2)\left(t^2 - 4t + 1\right)$$

By the Quadratic Formula, the zeros of $t^2 - 4t + 1$ are $2 \pm \sqrt{3}$. Thus,

$$h(t) = (t + 2)\left[t - \left(2 + \sqrt{3}\right)\right]\left[t - \left(2 - \sqrt{3}\right)\right].$$

79. $h(x) = x^5 - 7x^4 + 10x^3 + 14x^2 - 24x$

(a) The zeros of h are $x = 0, x = 3, x = 4, x \approx 1.414, x \approx -1.414$.

(b) An exact zero is $x = 4$.

(c)
$$\begin{array}{r|rrrrr} 4 & 1 & -7 & 10 & 14 & -24 \\ & & 4 & -12 & -8 & 24 \\ \hline & 1 & -3 & -2 & 6 & 0 \end{array}$$

$$\begin{aligned} h(x) &= (x - 4)\left(x^4 - 3x^3 - 2x^2 + 6x\right) \\ &= x(x - 4)(x - 3)\left(x + \sqrt{2}\right)\left(x - \sqrt{2}\right) \end{aligned}$$

81. $\dfrac{4x^3 - 8x^2 + x + 3}{2x - 3}$

$$\begin{array}{r|rrrr} \frac{3}{2} & 4 & -8 & 1 & 3 \\ & & 6 & -3 & -3 \\ \hline & 4 & -2 & -2 & 0 \end{array}$$

$$\frac{4x^3 - 8x^2 + x + 3}{x - \frac{3}{2}} = 4x^2 - 2x - 2 = 2\left(2x^2 - x - 1\right)$$

So, $\dfrac{4x^3 - 8x^2 + x + 3}{2x - 3} = 2x^2 - x - 1, x \neq \dfrac{3}{2}$.

83. $\dfrac{x^4 + 6x^3 + 11x^2 + 6x}{x^2 + 3x + 2} = \dfrac{x^4 + 6x^3 + 11x^2 + 6x}{(x + 1)(x + 2)}$

$$\begin{array}{r|rrrrr} -1 & 1 & 6 & 11 & 6 & 0 \\ & & -1 & -5 & -6 & 0 \\ \hline & 1 & 5 & 6 & 0 & 0 \end{array}$$

$$\begin{array}{r|rrrr} -2 & 1 & 5 & 6 & 0 \\ & & -2 & -6 & 0 \\ \hline & 1 & 3 & 0 & 0 \end{array}$$

$$\frac{x^4 + 6x^3 + 11x^2 + 6x}{(x + 1)(x + 2)} = x^2 + 3x, x \neq -2, -1$$

85. (a)

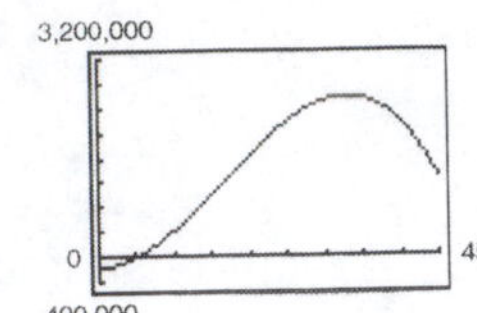

(b) Using the trace and zoom features, when $x = 25$, an advertising expense of about \$250,000 would produce the same profit of \$2,174,375.

(c) $x = 25$

$$\begin{array}{r|rrrr} 25 & -152 & 7545 & 0 & -169{,}625 \\ & & -3800 & 93{,}625 & 2{,}340{,}625 \\ \hline & -152 & 3745 & 93{,}625 & 2{,}171{,}000 \end{array}$$

So, an advertising expense of \$250,000 yields a profit of \$2,171,000, which is close to \$2,174,375.

87. False. If $(7x + 4)$ is a factor of f, then $-\frac{4}{7}$ is a zero of f.

89. True. The degree of the numerator is greater than the degree of the denominator.

91.

$$\begin{array}{r} x^{2n} + 6x^n + 9 \\ x^n + 3 \overline{\big) x^{3n} + 9x^{2n} + 27x^n + 27} \\ \underline{x^{3n} + 3x^{2n} \qquad\qquad\quad} \\ 6x^{2n} + 27x^n \qquad \\ \underline{6x^{2n} + 18x^n \qquad} \\ 9x^n + 27 \\ \underline{9x^n + 27} \\ 0 \end{array}$$

$$\frac{x^{3n} + 9x^{2n} + 27x^n + 27}{x^n + 3} = x^{2n} + 6x^n + 9,\ x^n \neq -3$$

93. A divisor divides evenly into a dividend if the remainder is zero.

95.

$$\begin{array}{r|rrrr} 5 & 1 & 4 & -3 & c \\ & & 5 & 45 & 210 \\ \hline & 1 & 9 & 42 & c + 210 \end{array}$$

To divide evenly, $c + 210$ must equal zero. So, c must equal -210.

97. If $x - 4$ is a factor of $f(x) = x^3 - kx^2 + 2kx - 8$, then $f(4) = 0$.

$$f(4) = (4)^3 - k(4)^2 + 2k(4) - 8$$
$$0 = 64 - 16k + 8k - 8$$
$$-56 = -8k$$
$$7 = k$$

Section 3.4 Zeros of Polynomial Functions

1. Fundamental Theorem of Algebra

3. Rational Zero

5. linear; quadratic; quadratic

7. Descartes's Rule of Signs

9. Since f is a 1st degree polynomial function, there is one zero.

11. Since f is a 3rd degree polynomial function, there are three zeros.

13. Since f is a 2nd degree polynomial function, there are two zeros.

15. $f(x) = x^3 + 2x^2 - x - 2$

Possible rational zeros: $\pm 1, \pm 2$

Zeros shown on graph: $-2, -1, 1, 2$

17. $f(x) = 2x^4 - 17x^3 + 35x^2 + 9x - 45$

Possible rational zeros: $\pm 1, \pm 3, \pm 5, \pm 9, \pm 15, \pm 45, \pm\frac{1}{2}, \pm\frac{3}{2}, \pm\frac{5}{2}, \pm\frac{9}{2}, \pm\frac{15}{2}, \pm\frac{45}{2}$

Zeros shown on graph: $-1, \frac{3}{2}, 3, 5$

19. $f(x) = x^3 - 7x - 6$

Possible rational zeros: $\pm 1, \pm 2, \pm 3, \pm 6$

$$\begin{array}{r|rrrr} 3 & 1 & 0 & -7 & -6 \\ & & 3 & 9 & 6 \\ \hline & 1 & 3 & 2 & 0 \end{array}$$

$$f(x) = (x - 3)(x^2 + 3x + 2)$$
$$= (x - 3)(x + 2)(x + 1)$$

So, the rational zeros are $-2, -1$, and 3.

21. $g(x) = x^3 - 4x^2 - x + 4$

$= x^2(x - 4) - 1(x - 4)$

$= (x - 4)(x^2 - 1)$

$= (x - 4)(x - 1)(x + 1)$

So, the rational zeros are 4, 1, and -1.

23. $h(t) = t^3 + 8t^2 + 13t + 6$

Possible rational zeros: $\pm 1, \pm 2, \pm 3, \pm 6$

$$\begin{array}{r|rrrr} -6 & 1 & 8 & 13 & 6 \\ & & -6 & -12 & -6 \\ \hline & 1 & 2 & 1 & 0 \end{array}$$

$t^3 + 8t^2 + 13t + 6 = (t + 6)(t^2 + 2t + 1)$

$= (t + 6)(t + 1)(t + 1)$

So, the rational zeros are -1 and -6.

25. $C(x) = 2x^3 + 3x^2 - 1$

Possible rational zeros: $\pm 1, \pm\frac{1}{2}$

$$\begin{array}{r|rrrr} -1 & 2 & 3 & 0 & -1 \\ & & -2 & -1 & 1 \\ \hline & 2 & 1 & -1 & 0 \end{array}$$

$2x^3 + 3x^2 - 1 = (x + 1)(2x^2 + x - 1)$

$= (x + 1)(x + 1)(2x - 1)$

$= (x + 1)^2(2x - 1)$

So, the rational zeros are -1 and $\frac{1}{2}$.

27. $f(x) = 9x^4 - 9x^3 - 58x^2 + 4x + 24$

Possible rational zeros:
$\pm 1, \pm 2, \pm 3, \pm 4, \pm 6, \pm 8, \pm 12, \pm 24,$
$\pm\frac{1}{3}, \pm\frac{2}{3}, \pm\frac{4}{3}, \pm\frac{8}{3}, \pm\frac{1}{9}, \pm\frac{2}{9}, \pm\frac{4}{9}, \pm\frac{8}{9}$

$$\begin{array}{r|rrrrr} -2 & 9 & -9 & -58 & 4 & 24 \\ & & -18 & 54 & 8 & -24 \\ \hline & 9 & -27 & -4 & 12 & 0 \end{array}$$

$$\begin{array}{r|rrrr} 3 & 9 & -27 & -4 & 12 \\ & & 27 & 0 & -12 \\ \hline & 9 & 0 & -4 & 0 \end{array}$$

$f(x) = (x + 2)(x - 3)(9x^2 - 4)$

$= (x + 2)(x - 3)(3x - 2)(3x + 2)$

So, the rational zeros are $-2, 3, \frac{2}{3}$, and $-\frac{2}{3}$.

29. $z^4 + z^3 + z^2 + 3z - 6 = 0$

Possible rational zeros: $\pm 1, \pm 2, \pm 3, \pm 6$

$$\begin{array}{r|rrrrr} 1 & 1 & 1 & 1 & 3 & -6 \\ & & 1 & 2 & 3 & 6 \\ \hline & 1 & 2 & 3 & 6 & 0 \end{array}$$

$(z - 1)(z^3 + 2z^2 + 3z + 6) = 0$

$(z - 1)(z^2 + 3)(z + 2) = 0$

So, the real zeros are -2 and 1.

31. $2y^4 + 3y^3 - 16y^2 + 15y - 4 = 0$

Possible rational zeros: $\pm\frac{1}{2}, \pm 1, \pm 2, \pm 4$

$$\begin{array}{r|rrrrr} 1 & 2 & 3 & -16 & 15 & -4 \\ & & 2 & 5 & -11 & 4 \\ \hline & 2 & 5 & -11 & 4 & 0 \end{array}$$

$$\begin{array}{r|rrrr} 1 & 2 & 5 & -11 & 4 \\ & & 2 & 7 & -4 \\ \hline & 2 & 7 & -4 & 0 \end{array}$$

$(y - 1)(y - 1)(2y^2 + 7y - 4) = 0$

$(y - 1)(y - 1)(2y - 1)(y + 4) = 0$

So, the real zeros are $-4, \frac{1}{2}$ and 1.

33. $f(x) = x^3 + x^2 - 4x - 4$

(a) Possible rational zeros: $\pm 1, \pm 2, \pm 4$

(b)

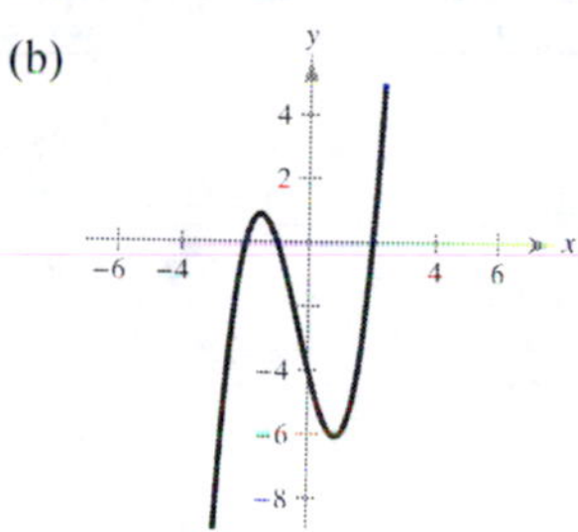

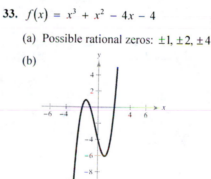

(c) Real zeros: $-2, -1, 2$

35. $f(x) = -4x^3 + 15x^2 - 8x - 3$

(a) Possible rational zeros: $\pm 1, \pm 3, \pm\frac{1}{2}, \pm\frac{3}{2}, \pm\frac{1}{4}, \pm\frac{3}{4}$

(b)

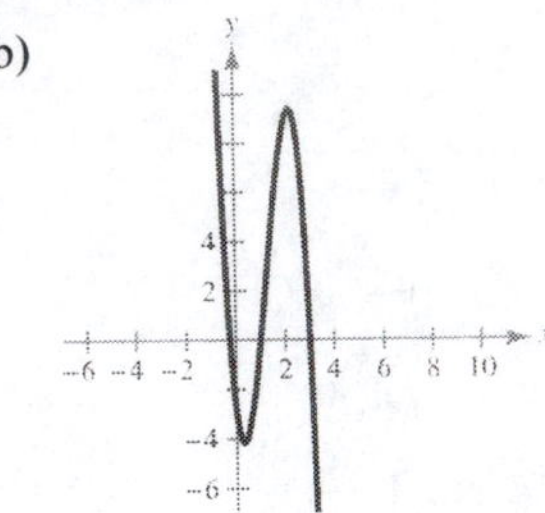

(c) Real zeros: $-\frac{1}{4}, 1, 3$

37. $f(x) = -2x^4 + 13x^3 - 21x^2 + 2x + 8$

(a) Possible rational zeros: $\pm 1, \pm 2, \pm 4, \pm 8, \pm\frac{1}{2}$

(b)

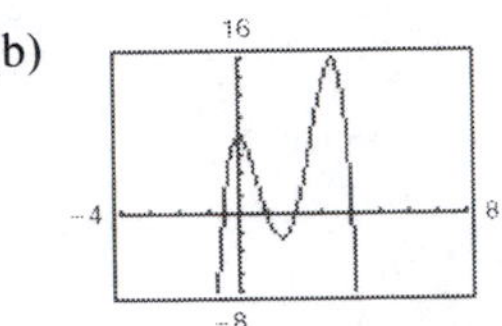

(c) Real zeros: $-\frac{1}{2}, 1, 2, 4$

39. $f(x) = 32x^3 - 52x^2 + 17x + 3$

(a) Possible rational zeros: $\pm 1, \pm 3, \pm\frac{1}{2}, \pm\frac{3}{2}, \pm\frac{1}{4}, \pm\frac{3}{4}, \pm\frac{1}{8}, \pm\frac{3}{8}, \pm\frac{1}{16}, \pm\frac{3}{16}, \pm\frac{1}{32}, \pm\frac{3}{32}$

(b)

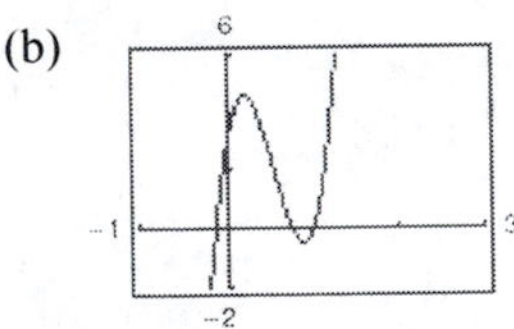

(c) Real zeros: $-\frac{1}{8}, \frac{3}{4}, 1$

41. $f(x) = x^4 - 3x^2 + 2$

(a) $x = \pm 1$, about ± 1.414

(b) An exact zero is $x = 1$.

$$\begin{array}{r|rrrrr} 1 & 1 & 0 & -3 & 0 & 2 \\ & & 1 & 1 & -2 & -2 \\ \hline & 1 & 1 & -2 & -2 & 0 \end{array}$$

(c)

$$\begin{array}{r|rrrr} -1 & 1 & 1 & -2 & -2 \\ & & -1 & 0 & 2 \\ \hline & 1 & 0 & -2 & 0 \end{array}$$

$$\begin{aligned} f(x) &= (x - 1)(x + 1)(x^2 - 2) \\ &= (x - 1)(x + 1)(x - \sqrt{2})(x + \sqrt{2}) \end{aligned}$$

43. $h(x) = x^5 - 7x^4 + 10x^3 + 14x^2 - 24x$

(a) $h(x) = x(x^4 - 7x^3 + 10x^2 + 14x - 24)$

$x = 0, 3, 4$, about ± 1.414

(b) An exact zero is $x = 3$.

$$\begin{array}{r|rrrrr} 3 & 1 & -7 & 10 & 14 & -24 \\ & & 3 & -12 & -6 & 24 \\ \hline & 1 & -4 & -2 & 8 & 0 \end{array}$$

(c)

$$\begin{array}{r|rrrr} 4 & 1 & -4 & -2 & 8 \\ & & 4 & 0 & -8 \\ \hline & 1 & 0 & -2 & 0 \end{array}$$

$$\begin{aligned} h(x) &= x(x - 3)(x - 4)(x^2 - 2) \\ &= x(x - 3)(x - 4)(x - \sqrt{2})(x + \sqrt{2}) \end{aligned}$$

45.
$$\begin{aligned} f(x) &= (x - 1)(x - 5i)(x + 5i) \\ &= (x - 1)(x^2 + 25) \\ &= x^3 - x^2 + 25x - 25 \end{aligned}$$

Note: $f(x) = a(x^3 - x^2 + 25x - 25)$, where a is any nonzero real number, has the zeros 1 and $\pm 5i$.

47. If $5 + i$ is a zero, so is its conjugate, $5 - i$.

$$\begin{aligned} f(x) &= (x - 2)(x - (5 + i))(x - (5 - i)) \\ &= (x - 2)(x^2 - 10x + 26) \\ &= x^3 - 12x^2 + 46x - 52 \end{aligned}$$

Note: $f(x) = a(x^3 - 12x^2 + 46x - 52)$, where a is any nonzero real number, has the zeros 2 and $5 \pm i$.

49. If $3 + \sqrt{2}i$ is a zero, so is its conjugate, $3 - \sqrt{2}i$.

$$\begin{aligned} f(x) &= (3x - 2)(x + 1)\left[x - \left(3 + \sqrt{2}i\right)\right]\left[x - \left(3 - \sqrt{2}i\right)\right] \\ &= (3x - 2)(x + 1)\left[(x - 3) - \sqrt{2}i\right]\left[(x - 3) + \sqrt{2}i\right] \\ &= \left(3x^2 + x - 2\right)\left[(x - 3)^2 - \left(\sqrt{2}i\right)^2\right] \\ &= \left(3x^2 + x - 2\right)\left(x^2 - 6x + 9 + 2\right) \\ &= \left(3x^2 + x - 2\right)\left(x^2 - 6x + 11\right) \\ &= 3x^4 - 17x^3 + 25x^2 + 23x - 22 \end{aligned}$$

Note: $f(x) = a\left(3x^4 - 17x^3 + 25x^2 + 23x - 22\right)$, where a is any nonzero real number, has the zeros $\frac{2}{3}, -1,$ and $3 \pm \sqrt{2}i$.

51. $f(x) = x^4 + 6x^2 - 27$

(a) $f(x) = \left(x^2 + 9\right)\left(x^2 - 3\right)$

(b) $f(x) = \left(x^2 + 9\right)\left(x + \sqrt{3}\right)\left(x - \sqrt{3}\right)$

(c) $f(x) = (x + 3i)(x - 3i)\left(x + \sqrt{3}\right)\left(x - \sqrt{3}\right)$

53. $f(x) = x^4 - 4x^3 + 5x^2 - 2x - 6$

$$\begin{array}{r} x^2 - 2x + 3 \\ x^2 - 2x - 2 \overline{\big) x^4 - 4x^3 + 5x^2 - 2x - 6} \\ \underline{x^4 - 2x^3 - 2x^2} \qquad\qquad \\ -2x^3 + 7x^2 - 2x \qquad \\ \underline{-2x^3 + 4x^2 + 4x} \qquad \\ 3x^2 - 6x - 6 \\ \underline{3x^2 - 6x - 6} \\ 0 \end{array}$$

(a) $f(x) = \left(x^2 - 2x - 2\right)\left(x^2 - 2x + 3\right)$

(b) $f(x) = \left(x - 1 + \sqrt{3}\right)\left(x - 1 - \sqrt{3}\right)\left(x^2 - 2x + 3\right)$

(c) $f(x) = \left(x - 1 + \sqrt{3}\right)\left(x - 1 - \sqrt{3}\right)\left(x - 1 + \sqrt{2}i\right)\left(x - 1 - \sqrt{2}i\right)$

Note: Use the Quadratic Formula for (b) and (c).

55. $f(x) = x^3 - x^2 + 4x - 4$

Because $2i$ is a zero, so is $-2i$.

$$\begin{array}{r|rrrr} 2i & 1 & -1 & 4 & -4 \\ & & 2i & -4 - 2i & 4 \\ \hline & 1 & 2i - 1 & -2i & 0 \end{array}$$

$$\begin{array}{r|rrr} -2i & 1 & 2i - 1 & -2i \\ & & -2i & 2i \\ \hline & 1 & -1 & 0 \end{array}$$

$f(x) = (x - 2i)(x + 2i)(x - 1)$

The zeros of $f(x)$ are $x = 1, \pm 2i$.

Alternate Solution:

Because $x = \pm 2i$ are zeros of $f(x)$,

$(x + 2i)(x - 2i) = x^2 + 4$ is a factor of $f(x)$.

By long division, you have:

$$\begin{array}{r} x - 1 \\ x^2 + 0x + 4 \overline{\big) x^3 - x^2 + 4x - 4} \\ \underline{x^3 + 0x^2 + 4x} \qquad \\ -x^2 + 0x - 4 \\ \underline{-x^2 + 0x - 4} \\ 0 \end{array}$$

$f(x) = \left(x^2 + 4\right)(x - 1)$

The zeros of $f(x)$ are $x = 1, \pm 2i$.

57. $f(x) = 2x^4 - x^3 + 49x^2 - 25x - 25$

Because $5i$ is a zero, so is $-5i$.

$$\begin{array}{r|rrrrr} 5i & 2 & -1 & 49 & -25 & -25 \\ & & 10i & -5i - 50 & -5i + 25 & 25 \\ \hline & 2 & -1 + 10i & -1 - 5i & -5i & 0 \end{array}$$

$$\begin{array}{r|rrrr} -5i & 2 & -1 + 10i & -1 - 5i & -5i \\ & & -10i & 5i & 5i \\ \hline & 2 & -1 & -1 & 0 \end{array}$$

$$\begin{aligned} f(x) &= (x - 5i)(x + 5i)(2x^2 - x - 1) \\ &= (x - 5i)(x + 5i)(2x + 1)(x - 1) \end{aligned}$$

The zeros of $f(x)$ are $x = \pm 5i, -\frac{1}{2}, 1$.

Alternate Solution:

Because $x = \pm 5i$ are zeros of $f(x)$, $(x - 5i)(x + 5i) = x^2 + 25$ is a factor of $f(x)$.

By long division, you have:

$$\begin{array}{r@{}l} & \quad\; 2x^2 - \quad x - \; 1 \\ x^2 + 0x + 25 & \overline{\big) 2x^4 - x^3 + 49x^2 - 25x - 25} \\ & \underline{2x^4 + 0x^3 + 50x^2} \\ & \qquad -x^3 - x^2 - 25x \\ & \qquad \underline{-x^3 + 0x^2 - 25x} \\ & \qquad\qquad -x^2 + 0x - 25 \\ & \qquad\qquad \underline{-x^2 + 0x - 25} \\ & \qquad\qquad\qquad\qquad 0 \end{array}$$

$$f(x) = (x^2 + 25)(2x^2 - x - 1)$$

The zeros of $f(x)$ are $x = \pm 5i, -\frac{1}{2}, 1$.

59. $g(x) = 4x^3 + 23x^2 + 34x - 10$

Because $-3 + i$ is a zero, so is $-3 - i$.

$$\begin{array}{r|rrrr} -3 + i & 4 & 23 & 34 & -10 \\ & & -12 + 4i & -37 - i & 10 \\ \hline & 4 & 11 + 4i & -3 - i & 0 \end{array}$$

$$\begin{array}{r|rrr} -3 - i & 4 & 11 + 4i & -3 - i \\ & & -12 - 4i & 3 + i \\ \hline & 4 & -1 & 0 \end{array}$$

The zero of $4x - 1$ is $x = \frac{1}{4}$. The zeros of $g(x)$ are $x = -3 \pm i, \frac{1}{4}$.

Alternate Solution

Because $-3 \pm i$ are zeros of $g(x)$,

$$\begin{aligned} \left[x - (-3 + i)\right]\left[x - (-3 - i)\right] &= \left[(x + 3) - i\right]\left[(x + 3) + i\right] \\ &= (x + 3)^3 - i^2 \\ &= x^2 + 6x + 10 \end{aligned}$$

is a factor of $g(x)$. By long division, you have:

$$\begin{array}{r@{}l} & \qquad\qquad 4x - \; 1 \\ x^2 + 6x + 10 & \overline{\big) 4x^3 + 23x^2 + 34x - 10} \\ & \underline{4x^3 + 24x^2 + 40x} \\ & \qquad -x^2 - 6x - 10 \\ & \qquad \underline{-x^2 - 6x - 10} \\ & \qquad\qquad\qquad 0 \end{array}$$

$$g(x) = (x^2 + 6x + 10)(4x - 1)$$

The zeros of $g(x)$ are $x = -3 \pm i, \frac{1}{4}$.

61. $f(x) = x^4 + 3x^3 - 5x^2 - 21x + 22$

Because $-3 + \sqrt{2}i$ is a zero, so is $-3 - \sqrt{2}i$, and

$$\begin{aligned}\left[x - \left(-3 + \sqrt{2}i\right)\right]\left[x - \left(-3 - \sqrt{2}i\right)\right] &= \left[(x + 3) - \sqrt{2}i\right]\left[(x + 3) + \sqrt{2}i\right]\\ &= (x + 3)^2 - \left(\sqrt{2}i\right)^2\\ &= x^2 + 6x + 11\end{aligned}$$

is a factor of $f(x)$. By long division, you have:

$$\begin{array}{r} x^2 - 3x + 2 \\ x^2 + 6x + 11\overline{\big) x^4 + 3x^3 - 5x^2 - 21x + 22} \\ \underline{x^4 + 6x^3 + 11x^2} \\ -3x^3 - 16x^2 - 21x \\ \underline{-3x^3 - 18x^2 - 33x} \\ 2x^2 + 12x + 22 \\ \underline{2x^2 + 12x + 22} \\ 0 \end{array}$$

$$\begin{aligned}f(x) &= \left(x^2 + 6x + 11\right)\left(x^2 - 3x + 2\right)\\ &= \left(x^2 + 6x + 11\right)(x - 1)(x - 2)\end{aligned}$$

The zeros of $f(x)$ are $x = -3 \pm \sqrt{2}i, 1, 2$.

63. $f(x) = x^2 + 36$

$= (x + 6i)(x - 6i)$

The zeros of $f(x)$ are $x = \pm 6i$.

65. $h(x) = x^2 - 2x + 17$

By the Quadratic Formula, the zeros of $f(x)$ are

$$x = \frac{2 \pm \sqrt{4 - 68}}{2} = \frac{2 \pm \sqrt{-64}}{2} = 1 \pm 4i.$$

$$\begin{aligned}f(x) &= (x - (1 + 4i))(x - (1 - 4i))\\ &= (x - 1 - 4i)(x - 1 + 4i)\end{aligned}$$

67. $f(x) = x^4 - 16$

$$\begin{aligned}&= \left(x^2 - 4\right)\left(x^2 + 4\right)\\ &= (x - 2)(x + 2)(x - 2i)(x + 2i)\end{aligned}$$

Zeros: $\pm 2, \pm 2i$

69. $f(z) = z^2 - 2z + 2$

By the Quadratic Formula, the zeros of $f(z)$ are

$$z = \frac{2 \pm \sqrt{4 - 8}}{2} = 1 \pm i.$$

$$\begin{aligned}f(z) &= \left[z - (1 + i)\right]\left[z - (1 - i)\right]\\ &= (z - 1 - i)(z - 1 + i)\end{aligned}$$

71. $g(x) = x^3 - 3x^2 + x + 5$

Possible rational zeros: $\pm 1, \pm 5$

$$\begin{array}{r|rrrr} -1 & 1 & -3 & 1 & 5 \\ & & -1 & 4 & -5 \\ \hline & 1 & -4 & 5 & 0 \end{array}$$

By the Quadratic Formula, the zeros of $x^2 - 4x + 5$ are: $x = \dfrac{4 \pm \sqrt{16 - 20}}{2} = 2 \pm i$

Zeros: $-1, 2 \pm i$

$g(x) = (x + 1)(x - 2 - i)(x - 2 + i)$

73. $h(x) = x^3 - x + 6$

Possible rational zeros: $\pm 1, \pm 2, \pm 3, \pm 6$

$$\begin{array}{r|rrrr} -2 & 1 & 0 & -1 & 6 \\ & & -2 & 4 & -6 \\ \hline & 1 & -2 & 3 & 0 \end{array}$$

By the Quadratic Formula, the zeros of $x^2 - 2x + 3$ are

$$x = \frac{2 \pm \sqrt{4 - 12}}{2} = 1 \pm \sqrt{2}i.$$

Zeros: $-2, 1 \pm \sqrt{2}i$

$$\begin{aligned}h(x) &= (x + 2)\left[x - \left(1 + \sqrt{2}i\right)\right]\left[x - \left(1 - \sqrt{2}i\right)\right]\\ &= (x + 2)\left(x - 1 - \sqrt{2}i\right)\left(x - 1 + \sqrt{2}i\right)\end{aligned}$$

75. $f(x) = 5x^3 - 9x^2 + 28x + 6$

Possible rational zeros:

$\pm1, \pm2, \pm3, \pm6, \pm\frac{1}{5}, \pm\frac{2}{5}, \pm\frac{3}{5}, \pm\frac{6}{5}$

$$\begin{array}{r|rrrr} -\frac{1}{5} & 5 & -9 & 28 & 6 \\ & & -1 & 2 & -6 \\ \hline & 5 & -10 & 30 & 0 \end{array}$$

By the Quadratic Formula, the zeros of $5x^2 - 10x + 30 = 5(x^2 - 2x + 6)$ are

$$x = \frac{2 \pm \sqrt{4 - 24}}{2} = 1 \pm \sqrt{5}i.$$

Zeros: $-\frac{1}{5}, 1 \pm \sqrt{5}i$

$$f(x) = \left[x - \left(-\frac{1}{5}\right)\right](5)\left[x - \left(1 + \sqrt{5}i\right)\right]\left[x - \left(1 - \sqrt{5}i\right)\right]$$
$$= (5x + 1)\left(x - 1 - \sqrt{5}i\right)\left(x - 1 + \sqrt{5}i\right)$$

77. $g(x) = x^4 - 4x^3 + 8x^2 - 16x + 16$

Possible rational zeros: $\pm1, \pm2, \pm4, \pm8, \pm16$

$$\begin{array}{r|rrrrr} 2 & 1 & -4 & 8 & -16 & 16 \\ & & 2 & -4 & 8 & -16 \\ \hline & 1 & -2 & 4 & -8 & 0 \end{array}$$

$$\begin{array}{r|rrrr} 2 & 1 & -2 & 4 & -8 \\ & & 2 & 0 & 8 \\ \hline & 1 & 0 & 4 & 0 \end{array}$$

$$g(x) = (x - 2)(x - 2)(x^2 + 4)$$
$$= (x - 2)^2(x + 2i)(x - 2i)$$

Zeros: $2, \pm 2i$

79. $f(x) = x^4 + 10x^2 + 9$
$$= (x^2 + 1)(x^2 + 9)$$
$$= (x + i)(x - i)(x + 3i)(x - 3i)$$

Zeros: $\pm i, \pm 3i$

81. $f(x) = x^3 + 24x^2 + 214x + 740$

Possible rational zeros: $\pm1, \pm2, \pm4, \pm5, \pm10, \pm20, \pm37, \pm74, \pm148, \pm185, \pm370, \pm740$

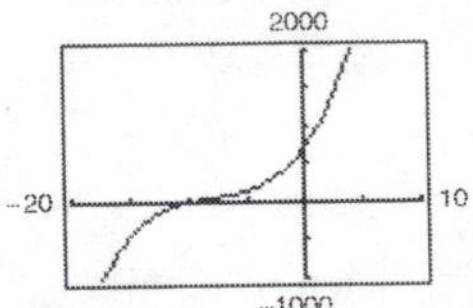

Based on the graph, try $x = -10$.

$$\begin{array}{r|rrrr} -10 & 1 & 24 & 214 & 740 \\ & & -10 & -140 & -740 \\ \hline & 1 & 14 & 74 & 0 \end{array}$$

By the Quadratic Formula, the zeros of $x^2 + 14x + 74$ are $x = \frac{-14 \pm \sqrt{196 - 296}}{2} = -7 \pm 5i.$

The zeros of $f(x)$ are $x = -10$ and $x = -7 \pm 5i$.

83. $f(x) = 16x^3 - 20x^2 - 4x + 15$

Possible rational zeros:

$\pm1, \pm3, \pm5, \pm15, \pm\frac{1}{2}, \pm\frac{3}{2}, \pm\frac{5}{2}, \pm\frac{15}{2}, \pm\frac{1}{4}, \pm\frac{3}{4}, \pm\frac{5}{4}, \pm\frac{15}{4}, \pm\frac{1}{8}, \pm\frac{3}{8}, \pm\frac{5}{8}, \pm\frac{15}{8}, \pm\frac{1}{16}, \pm\frac{3}{16}, \pm\frac{5}{16}, \pm\frac{15}{16}$

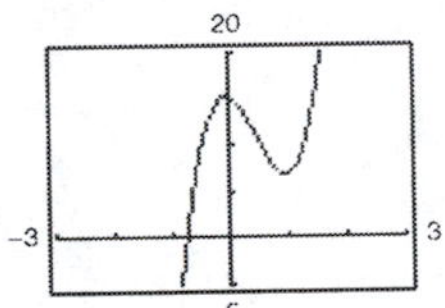

Based on the graph, try $x = -\frac{3}{4}$.

$$\begin{array}{r|rrrr} -\frac{3}{4} & 16 & -20 & -4 & 15 \\ & & -12 & 24 & -15 \\ \hline & 16 & -32 & 20 & 0 \end{array}$$

By the Quadratic Formula, the zeros of $16x^2 - 32x + 20 = 4(4x^2 - 8x + 5)$ are

$$x = \frac{8 \pm \sqrt{64 - 80}}{8} = 1 \pm \frac{1}{2}i.$$

The zeros of $f(x)$ are $x = -\frac{3}{4}$ and $x = 1 \pm \frac{1}{2}i$.

85. $f(x) = 2x^4 + 5x^3 + 4x^2 + 5x + 2$

Possible rational zeros: $\pm 1, \pm 2, \pm\frac{1}{2}$

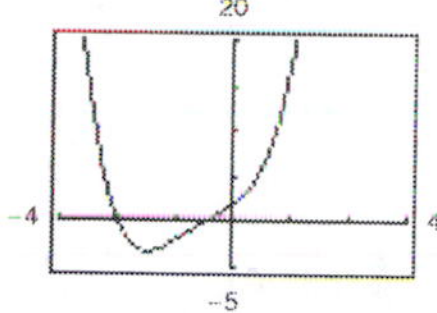

Based on the graph, try $x = -2$ and $x = -\frac{1}{2}$.

$$\begin{array}{r|rrrrr} -2 & 2 & 5 & 4 & 5 & 2 \\ & & -4 & -2 & -4 & -2 \\ \hline & 2 & 1 & 2 & 1 & 0 \end{array}$$

$$\begin{array}{r|rrrr} -\frac{1}{2} & 2 & 1 & 2 & 1 \\ & & -1 & 0 & -1 \\ \hline & 2 & 0 & 2 & 0 \end{array}$$

The zeros of $2x^2 + 2 = 2(x^2 + 1)$ are $x = \pm i$.

The zeros of $f(x)$ are $x = -2, x = -\frac{1}{2}$, and $x = \pm i$.

87. $g(x) = 2x^3 - 3x^2 - 3$

Sign variations: 1, positive zeros: 1

$g(-x) = -2x^3 - 3x^2 - 3$

Sign variations: 0, negative zeros: 0

89. $h(x) = 2x^3 + 3x^2 + 1$

Sign variations: 0, positive zeros: 0

$h(-x) = -2x^3 + 3x^2 + 1$

Sign variations: 1, negative zeros: 1

91. $g(x) = 5x^5 - 10x = 5x(x^4 - 2)$

Let $g(x) = x^4 - 2$.

Sign variations: 1, positive zeros: 1

$g(-x) = x^4 - 2$

Sign variations: 1, negative zeros: 1

93. $f(x) = -5x^3 + x^2 - x + 5$

Sign variations: 3, positive zeros: 3 or 1

$f(-x) = 5x^3 + x^2 + x + 5$

Sign variations: 0, negative zeros: 0

95. $f(x) = x^3 + 3x^2 - 2x + 1$

(a)
$$\begin{array}{r|rrrr} 1 & 1 & 3 & -2 & 1 \\ & & 1 & 4 & 2 \\ \hline & 1 & 4 & 2 & 3 \end{array}$$

1 is an upper bound.

(b)
$$\begin{array}{r|rrrr} -4 & 1 & 3 & -2 & 1 \\ & & -4 & 4 & -8 \\ \hline & 1 & -1 & 2 & -7 \end{array}$$

−4 is a lower bound.

97. $f(x) = x^4 - 4x^3 + 16x - 16$

(a)
$$\begin{array}{r|rrrrr} 5 & 1 & -4 & 0 & 16 & -16 \\ & & 5 & 5 & 25 & 205 \\ \hline & 1 & 1 & 5 & 41 & 189 \end{array}$$

5 is an upper bound.

(b)
$$\begin{array}{r|rrrrr} -3 & 1 & -4 & 0 & 16 & -16 \\ & & -3 & 21 & -63 & 141 \\ \hline & 1 & -7 & 21 & -47 & 125 \end{array}$$

−3 is a lower bound.

99. $f(x) = 4x^3 - 3x - 1$

Possible rational zeros: $\pm 1, \pm\frac{1}{2}, \pm\frac{1}{4}$

$$\begin{array}{r|rrrr} 1 & 4 & 0 & -3 & -1 \\ & & 4 & 4 & 1 \\ \hline & 4 & 4 & 1 & 0 \end{array}$$

$$4x^3 - 3x - 1 = (x - 1)(4x^2 + 4x + 1)$$
$$= (x - 1)(2x + 1)^2$$

So, the zeros are 1 and $-\frac{1}{2}$.

101. $f(y) = 4y^3 + 3y^2 + 8y + 6$

Possible rational zeros: $\pm 1, \pm 2, \pm 3, \pm 6, \pm\frac{1}{2}, \pm\frac{3}{2}, \pm\frac{1}{4}, \pm\frac{3}{4}$

$$\begin{array}{r|rrrr} -\frac{3}{4} & 4 & 3 & 8 & 6 \\ & & -3 & 0 & -6 \\ \hline & 4 & 0 & 8 & 0 \end{array}$$

$$4y^3 + 3y^2 + 8y + 6 = \left(y + \tfrac{3}{4}\right)(4y^2 + 8)$$
$$= \left(y + \tfrac{3}{4}\right)4(y^2 + 2)$$
$$= (4y + 3)(y^2 + 2)$$

So, the only real zero is $-\frac{3}{4}$.

103. $P(x) = x^4 - \frac{25}{4}x^2 + 9$
$= \frac{1}{4}(4x^4 - 25x^2 + 36)$
$= \frac{1}{4}(4x^2 - 9)(x^2 - 4)$
$= \frac{1}{4}(2x + 3)(2x - 3)(x + 2)(x - 2)$

The rational zeros are $\pm\frac{3}{2}$ and ± 2.

105. $f(x) = x^3 - \frac{1}{4}x^2 - x + \frac{1}{4}$
$= \frac{1}{4}(4x^3 - x^2 - 4x + 1)$
$= \frac{1}{4}[x^2(4x - 1) - 1(4x - 1)]$
$= \frac{1}{4}(4x - 1)(x^2 - 1)$
$= \frac{1}{4}(4x - 1)(x + 1)(x - 1)$

The rational zeros are $\frac{1}{4}$ and ± 1.

107. $f(x) = x^3 - 1 = (x - 1)(x^2 + x + 1)$

Rational zeros: $1\ (x = 1)$

Irrational zeros: 0

Matches (d).

108. $f(x) = x^3 - 2$
$= (x - \sqrt[3]{2})(x^2 + \sqrt[3]{2}x + \sqrt[3]{4})$

Rational zeros: 0

Irrational zeros: $1\ (x = \sqrt[3]{2})$

Matches (a).

109. $f(x) = x^3 - x = x(x + 1)(x - 1)$

Rational zeros: $3\ (x = 0, \pm 1)$

Irrational zeros: 0

Matches (b).

110. $f(x) = x^3 - 2x$
$= x(x^2 - 2)$
$= x(x + \sqrt{2})(x - \sqrt{2})$

Rational zeros: $1\ (x = 0)$

Irrational zeros: $2\ (x = \pm\sqrt{2})$

Matches (c).

111. (a)

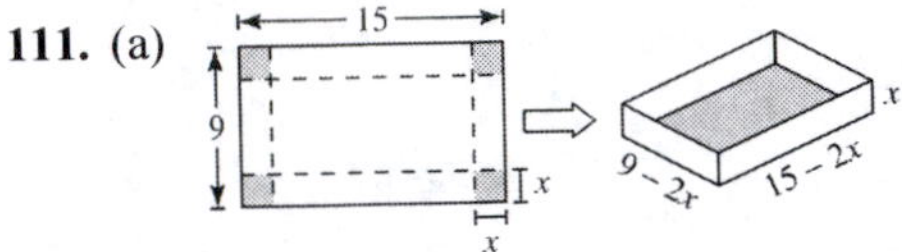

(b) $V = l \cdot w \cdot h = (15 - 2x)(9 - 2x)x$
$= x(9 - 2x)(15 - 2x)$

Because length, width, and height must be positive, you have $0 < x < \frac{9}{2}$ for the domain.

(c)

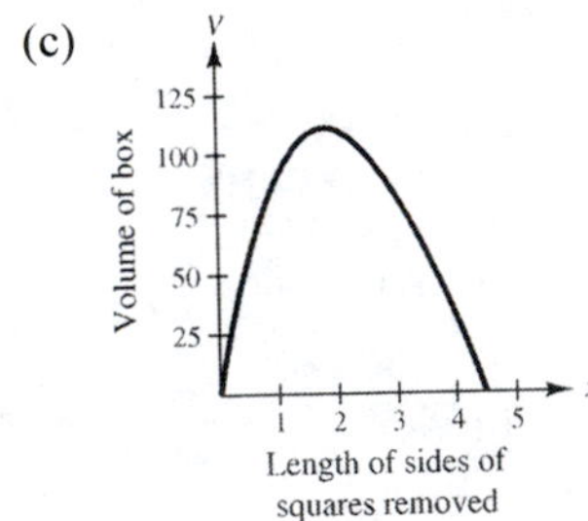

The volume is maximum when $x \approx 1.82$.

The dimensions are: length $\approx 15 - 2(1.82) = 11.36$
width $\approx 9 - 2(1.82) = 5.36$
height $= x \approx 1.82$

1.82 cm × 5.36 cm × 11.36 cm

(d) $56 = x(9 - 2x)(15 - 2x)$
$56 = 135x - 48x^2 + 4x^3$
$0 = 4x^3 - 48x^2 + 135x - 56$

The zeros of this polynomial are $\frac{1}{2}, \frac{7}{2}$, and 8.

x cannot equal 8 because it is not in the domain of V. [The length cannot equal -1 and the width cannot equal -7. The product of $(8)(-1)(-7) = 56$ so it showed up as an extraneous solution.]

So, the volume is 56 cubic centimeters when $x = \frac{1}{2}$ centimeter or $x = \frac{7}{2}$ centimeters.

113. (a) Current bin: $V = 2 \times 3 \times 4 = 24$ cubic feet
New bin: $V = 5(24) = 120$ cubic feet
$V(x) = (2 + x)(3 + x)(4 + x) = 120$

(b) $x^3 + 9x^2 + 26x + 24 = 120$
$x^3 + 9x^2 + 26x - 96 = 0$

The only real zero of this polynomial is $x = 2$. All the dimensions should be increased by 2 feet, so the new bin will have dimensions of 4 feet by 5 feet by 6 feet.

115. False. The most complex zeros it can have is two, and the Linear Factorization Theorem guarantees that there are three linear factors, so one zero must be real.

117. $g(x) = -f(x)$. This function would have the same zeros as $f(x)$, so $r_1, r_2,$ and r_3 are also zeros of $g(x)$.

119. $g(x) = f(x - 5)$. The graph of $g(x)$ is a horizontal shift of the graph of $f(x)$ five units of the right, so the zeros of $g(x)$ are $5 + r_1, 5 + r_2,$ and $5 + r_3$.

121. $g(x) = 3 + f(x)$. Because $g(x)$ is a vertical shift of the graph of $f(x)$, the zeros of $g(x)$ cannot be determined.

123. Zeros: $-2, \frac{1}{2}, 3$

$$f(x) = -(x + 2)(2x - 1)(x - 3)$$
$$= -2x^3 + 3x^2 + 11x - 6$$

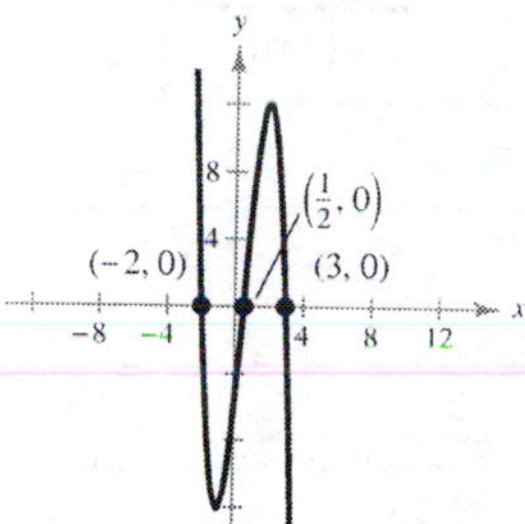

Any nonzero scalar multiple of f would have the same three zeros. Let $g(x) = af(x), a > 0$. There are infinitely many possible functions for f.

125. Because $1 + i$ is a zero of f, so is $1 - i$. From the graph, 1 is also a zero.

$$f(x) = (x - (1 + i))(x - (1 - i))(x - 1)$$
$$= (x^2 - 2x + 2)(x - 1)$$
$$= x^3 - 3x^2 + 4x - 2$$

127. Because $f(i) = f(2i) = 0$, then i and $2i$ are zeros of f. Because i and $2i$ are zeros of f, so are $-i$ and $-2i$.

$$f(x) = (x - i)(x + i)(x - 2i)(x + 2i)$$
$$= (x^2 + 1)(x^2 + 4)$$
$$= x^4 + 5x^2 + 4$$

129. (a) $f(x) = (x - \sqrt{bi})(x + \sqrt{bi}) = x^2 + b$

(b) $f(x) = [x - (a + bi)][x - (a - bi)]$
$$= [(x - a) - bi][(x - a) + bi]$$
$$= (x - a)^2 - (bi)^2$$
$$= x^2 - 2ax + a^2 + b^2$$

131. (a) $f(x)$ cannot have this graph because it also has a zero at $x = 0$.

(b) $g(x)$ cannot have this graph because it is a quadratic function. Its graph is a parabola.

(c) $h(x)$ is the correct function. It has two real zeros, $x = 2$ and $x = 3.5$, and it has a degree of four, needed to yield three turning points.

(d) $k(x)$ cannot have this graph because it also has a zero at $x = -1$. In addition, because it is only of degree three, it would have at most two turning points.

Section 3.5 Mathematical Modeling and Variation

1. variation; regression

3. least squares regression

5. directly proportional

7. directly proportional

9. combined

11.

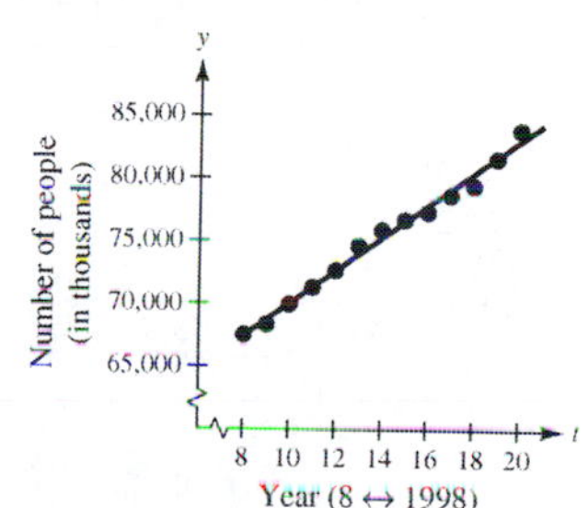

The model fits the data well.

13.

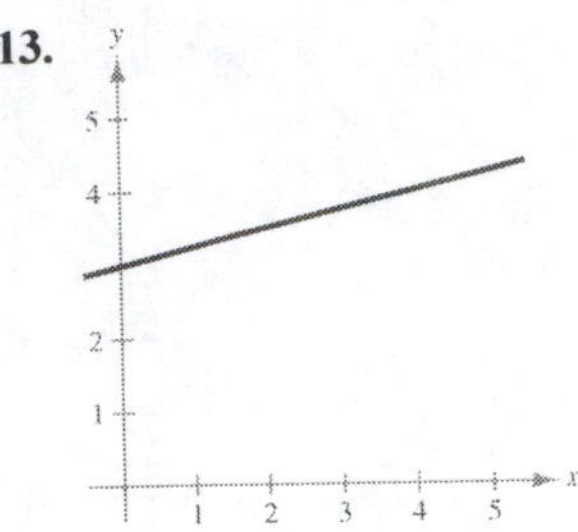

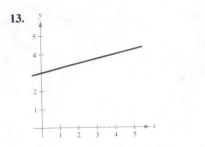

Using the point $(0, 3)$ and $(4, 4)$, $y = \frac{1}{4}x + 3$.

15.

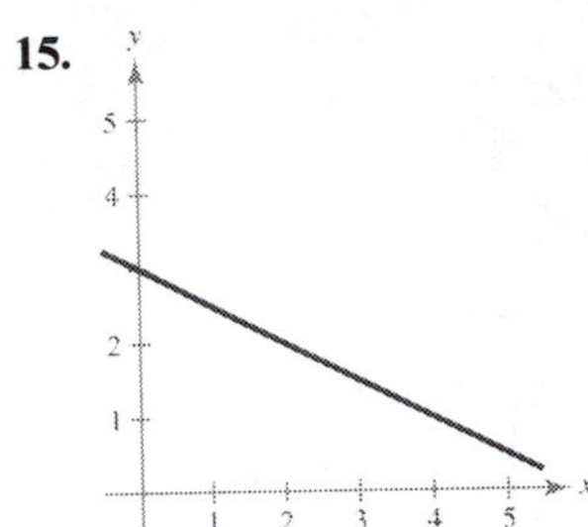

Using the points $(2, 2)$ and $(4, 1)$, $y = -\frac{1}{2}x + 3$.

17. (a)

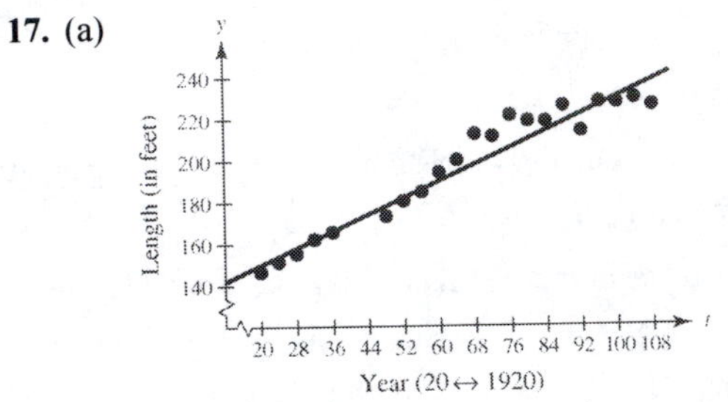

(b) Using the points $(32, 162.3)$ and $(96, 227.7)$:

$$m = \frac{227.7 - 162.3}{96 - 32}$$
$$\approx 1.02$$
$$y - 162.3 = 1.02(t - 32)$$
$$y = 1.02t + 129.66$$

(c) $y \approx 1.01t + 130.82$

(d) The models are similar.

$2012 \rightarrow$ use $t = 112$

Model from part (b):

$y = 1.02(112) + 129.66 = 243.9$ feet

Model from part (c):

$y = 1.01(112) + 130.82 = 243.94$ feet

19.
$$y = kx$$
$$14 = k(2)$$
$$7 = k$$
$$y = 7x$$

21.
$$y = kx$$
$$2050 = k(10)$$
$$205 = k$$
$$y = 205x$$

23.
$$y = kx$$
$$1 = k(5)$$
$$\frac{1}{5} = k$$
$$y = \frac{1}{5}x$$

25.
$$y = kx$$
$$8\pi = k(4)$$
$$\pi = k$$
$$y = \frac{\pi}{2}x$$

27. $k = 1$

x	2	4	6	8	10
$y = kx^2$	4	16	36	64	100

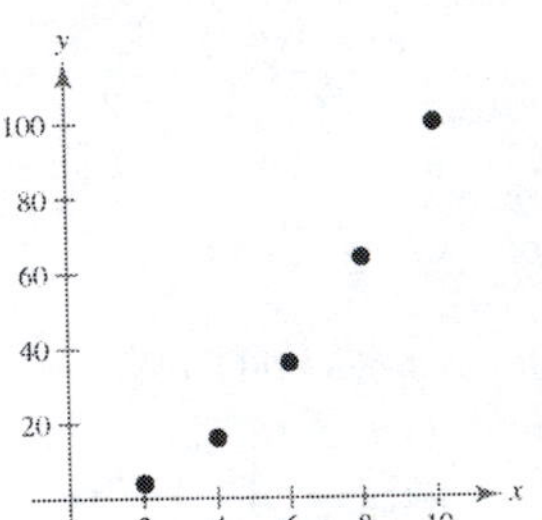

29. $k = \frac{1}{2}$

x	2	4	6	8	10
$y = \frac{1}{2}x^3$	4	32	108	256	500

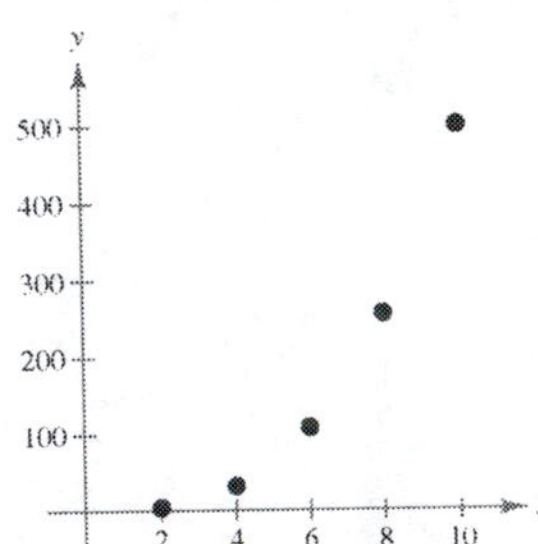

31. $k = 2, n = 1$

x	2	4	6	8	10
$y = \frac{2}{x}$	1	$\frac{1}{2}$	$\frac{1}{3}$	$\frac{1}{4}$	$\frac{1}{5}$

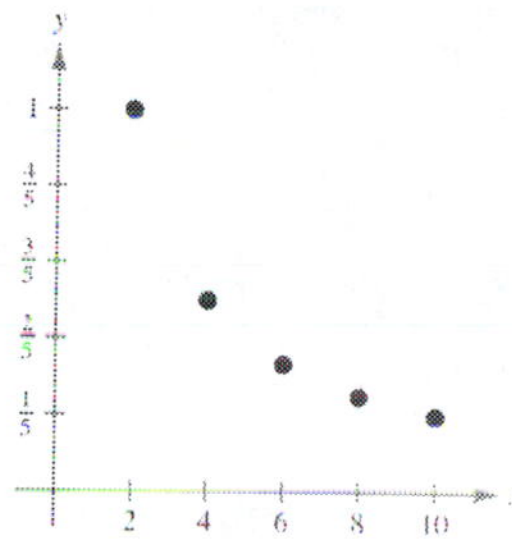

33. $k = 10$

x	2	4	6	8	10
$y = \frac{k}{x^2}$	$\frac{5}{2}$	$\frac{5}{8}$	$\frac{5}{18}$	$\frac{5}{32}$	$\frac{1}{10}$

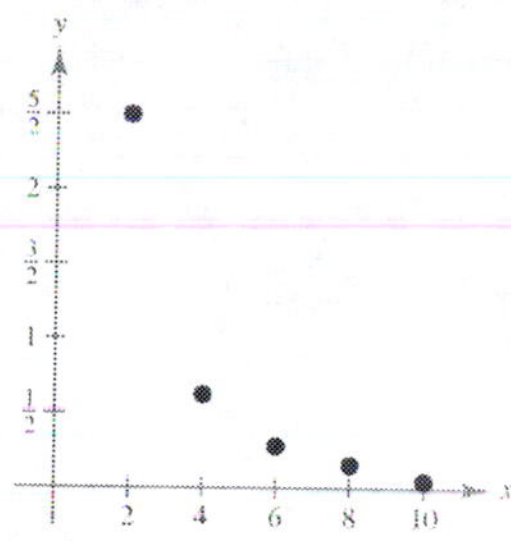

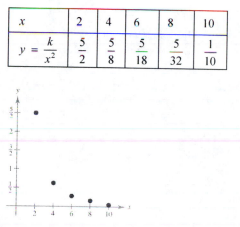

35. The graph appears to represent $y = 4/x$, so y varies inversely as x.

37. $y = \frac{k}{x}$

$1 = \frac{k}{5}$

$5 = k$

$y = \frac{5}{x}$

This equation checks with the other points given in the table.

39. $y = kx$

$-7 = k(10)$

$-\frac{7}{10} = k$

$y = -\frac{7}{10}x$

This equation checks with the other points given in the table.

41. $A = kr^2$

43. $y = \frac{k}{x^2}$

45. $F = \frac{kg}{r^2}$

47. $R = k(T - T_e)$

49. $R = kS(S - L)$

51. $S = 4\pi r^2$

The surface area of a sphere varies directly as the square of the radius r.

53. $A = \frac{1}{2}bh$

The area of a triangle is jointly proportional to its base and height.

55. $A = kr^2$

$9\pi = k(3)^2$

$\pi = k$

$A = \pi r^2$

57. $y = \frac{k}{x}$

$7 = \frac{k}{4}$

$28 = k$

$y = \frac{28}{x}$

59. $F = krs^3$

$4158 = k(11)(3)^3$

$k = 14$

$F = 14rs^3$

61. $z = \frac{kx^2}{y}$

$6 = \frac{k(6)^2}{4}$

$\frac{24}{36} = k$

$\frac{2}{3} = k$

$z = \frac{2/3x^2}{y} = \frac{2x^2}{3y}$

63. $I = kP$

$113.75 = k(3250)$

$0.035 = k$

$I = 0.035P$

65. $y = kx$

$33 = k(13)$

$\frac{33}{13} = k$

$y = \frac{33}{13}x$

When $x = 10$ inches, $y \approx 25.4$ centimeters.

When $x = 20$ inches, $y \approx 50.8$ centimeters.

67. $d = kF$

$0.12 = k(220)$

$\frac{3}{5500} = k$

$d = \frac{3}{5500}F$

$0.16 = \frac{3}{5500}F$

$\frac{880}{3} = F$

The required force is $293\frac{1}{3}$ newtons.

69. $d = kF$

$1.9 = k(25) \Rightarrow k = 0.076$

$d = 0.076F$

When the distance compressed is 3 inches, we have

$3 = 0.076F$

$F \approx 39.47.$

No child over 39.47 pounds should use the toy.

71. $d = kv^2$

$0.02 = k\left(\frac{1}{4}\right)^2$

$k = 0.32$

$d = 0.32v^2$

$0.12 = 0.32v^2$

$v^2 = \frac{0.12}{0.32} = \frac{3}{8}$

$v = \frac{\sqrt{3}}{2\sqrt{2}} = \frac{\sqrt{6}}{4} \approx 0.61$ mi/hr

73. $W = kmh$

$2116.8 = k(120)(1.8)$

$k = \frac{2116.8}{(120)(1.8)} = 9.8$

$W = 9.8mh$

When $m = 100$ kilograms and $h = 1.5$ meters, we have $W = 9.8(100)(1.5) = 1470$ joules.

75. (a)

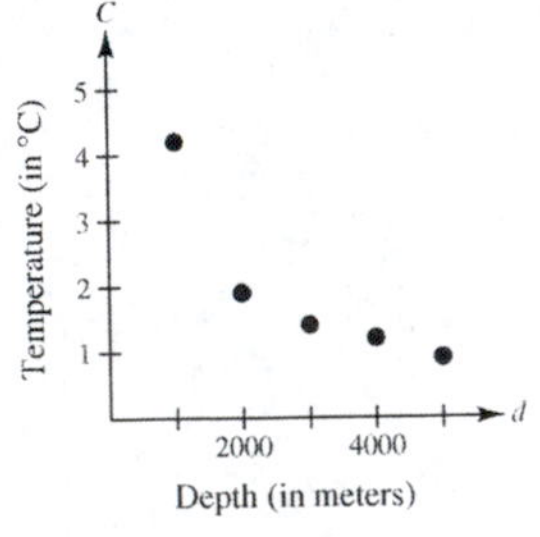

(b) Yes, the data appears to be modeled (approximately) by the inverse proportion model.

$4.2 = \frac{k_1}{1000}$ $\quad 1.9 = \frac{k_2}{2000}$ $\quad 1.4 = \frac{k_3}{3000}$ $\quad 1.2 = \frac{k_4}{4000}$ $\quad 0.9 = \frac{k_5}{5000}$

$4200 = k_1$ $\quad 3800 = k_2$ $\quad 4200 = k_3$ $\quad 4800 = k_4$ $\quad 4500 = k_5$

(c) Mean: $k = \frac{4200 + 3800 + 4200 + 4800 + 4500}{5} = 4300$, Model: $C = \frac{4300}{d}$

(d)

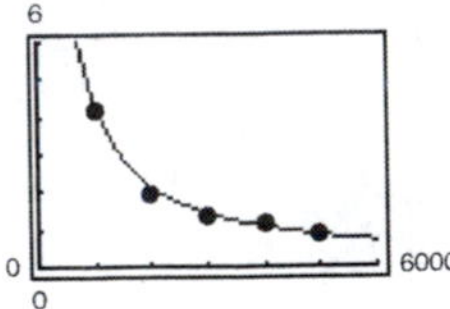

(e) $3 = \frac{4300}{d}$

$d = \frac{4300}{3} = 1433\frac{1}{3}$ meters

77. False. π is a constant, not a variable. So, the area A varies directly as the square of the radius, r.

79. (a) y will change by a factor of one-fourth.

(b) y will change by a factor of four.

Review Exercises for Chapter 3

1. (a) $y = 2x^2$

Vertical stretch

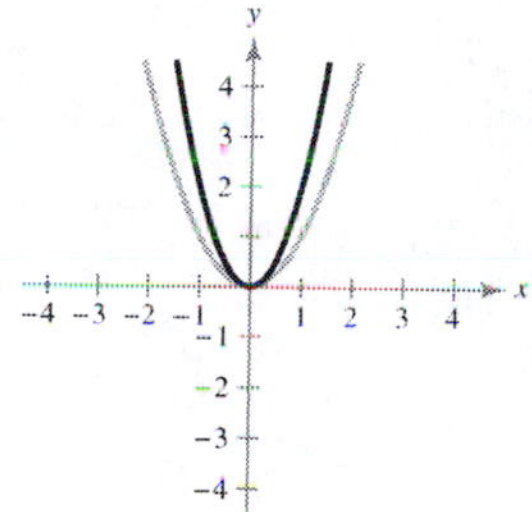

(b) $y = -2x^2$

Vertical stretch and a reflection in the x-axis

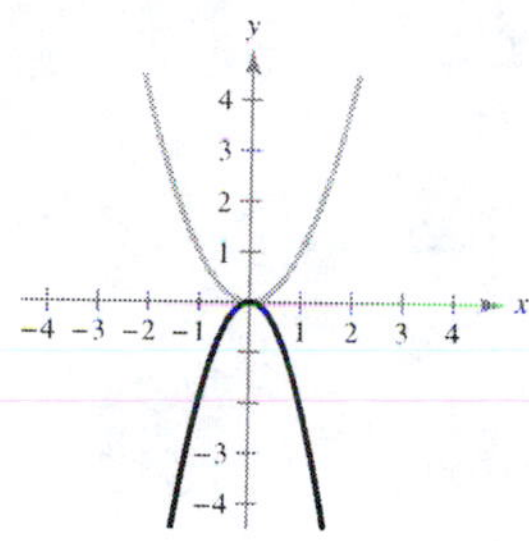

(c) $y = x^2 + 2$

Vertical shift two units upward

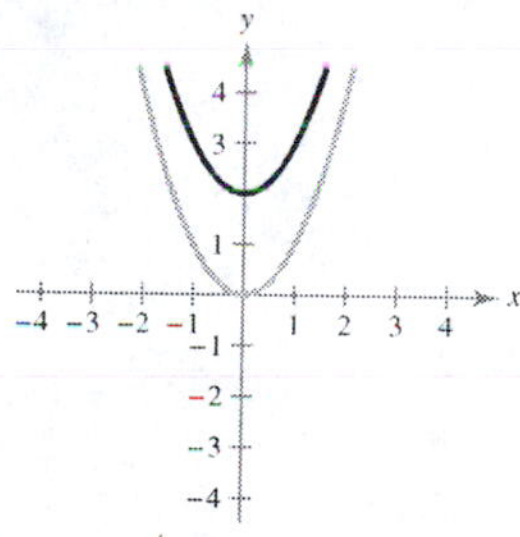

(d) $y = (x + 2)^2$

Horizontal shift two units to the left

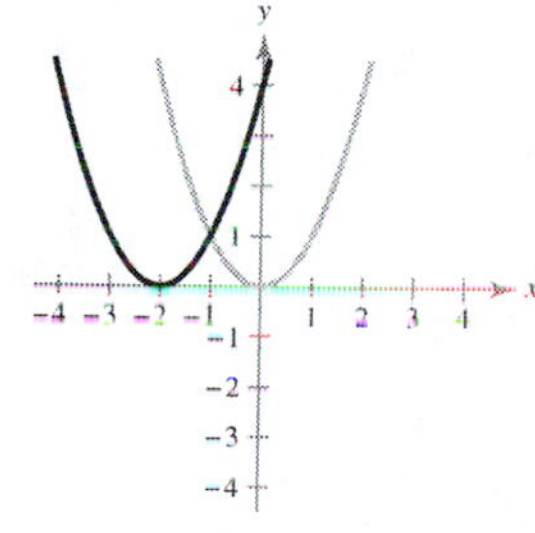

3. $g(x) = x^2 - 2x$

$= x^2 - 2x + 1 - 1$

$= (x - 1)^2 - 1$

Vertex: $(1, -1)$

Axis of symmetry: $x = 1$

$0 = x^2 - 2x = x(x - 2)$

x-intercepts: $(0, 0), (2, 0)$

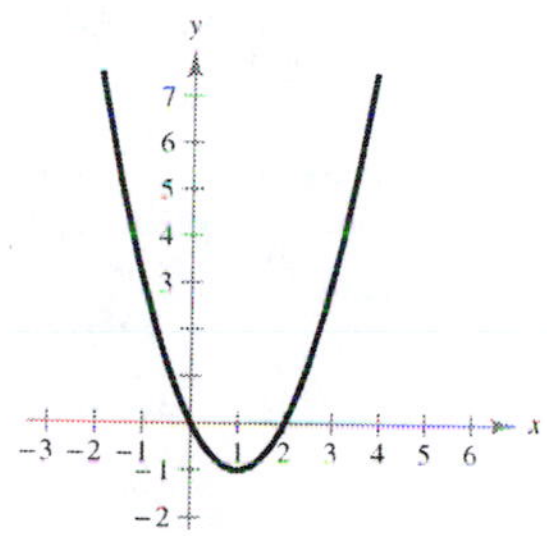

5. $f(x) = x^2 + 8x + 10$

$= x^2 + 8x + 16 - 16 + 10$

$= (x + 4)^2 - 6$

Vertex: $(-4, -6)$

Axis of symmetry: $x = -4$

$0 = (x + 4)^2 - 6$

$(x + 4)^2 = 6$

$x + 4 = \pm\sqrt{6}$

$x = -4 \pm \sqrt{6}$

x-intercepts: $\left(-4 \pm \sqrt{6}, 0\right)$

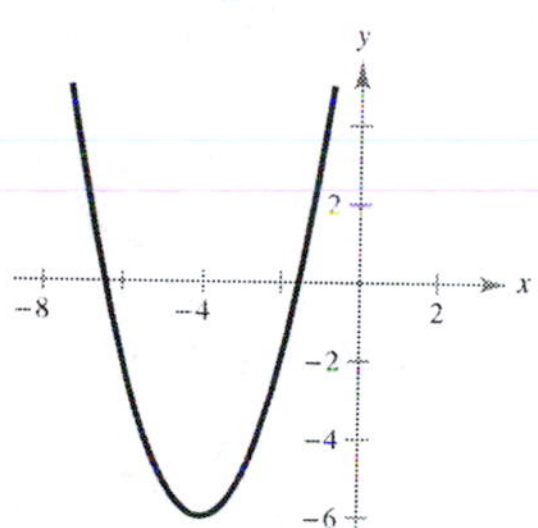

7. $f(t) = -2t^2 + 4t + 1$

$= -2(t^2 - 2t + 1 - 1) + 1$

$= -2\left[(t - 1)^2 - 1\right] + 1$

$= -2(t - 1)^2 + 3$

Vertex: $(1, 3)$

Axis of symmetry: $t = 1$

$0 = -2(t - 1)^2 + 3$

$2(t - 1)^2 = 3$

$t - 1 = \pm\sqrt{\frac{3}{2}}$

$t = 1 \pm \frac{\sqrt{6}}{2}$

t-intercepts: $\left(1 \pm \frac{\sqrt{6}}{2}, 0\right)$

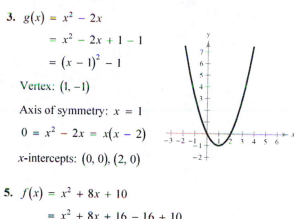

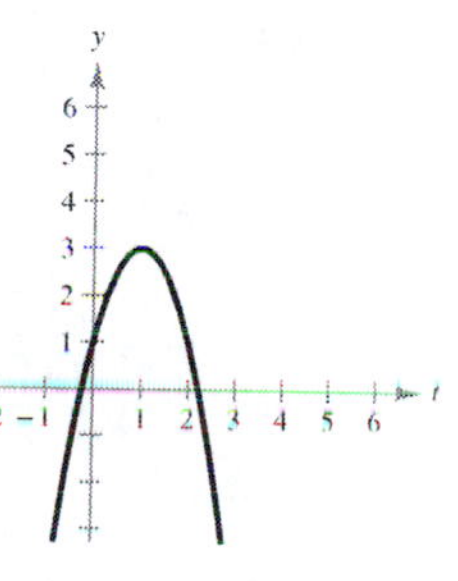

9. $h(x) = 4x^2 + 4x + 13$

$= 4(x^2 + x) + 13$

$= 4\left(x^2 + x + \frac{1}{4} - \frac{1}{4}\right) + 13$

$= 4\left(x^2 + x + \frac{1}{4}\right) - 1 + 13$

$= 4\left(x + \frac{1}{2}\right)^2 + 12$

Vertex: $\left(-\frac{1}{2}, 12\right)$

Axis of symmetry: $x = -\frac{1}{2}$

$0 = 4\left(x + \frac{1}{2}\right)^2 + 12$

$\left(x + \frac{1}{2}\right)^2 = -3$

No real zeros

x-intercepts: none

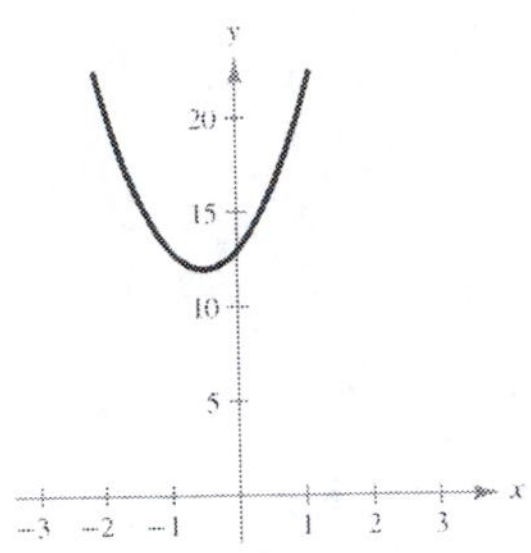

11. $h(x) = x^2 + 5x - 4$

$= x^2 + 5x + \frac{25}{4} - \frac{25}{4} - 4$

$= \left(x + \frac{5}{2}\right)^2 - \frac{25}{4} - \frac{16}{4}$

$= \left(x + \frac{5}{2}\right)^2 - \frac{41}{4}$

Vertex: $\left(-\frac{5}{2}, -\frac{41}{4}\right)$

Axis of symmetry:

$x = -\frac{5}{2}$

$0 = x^2 + 5x - 4$

$x = \frac{-5 \pm \sqrt{5^2 - 4(1)(-4)}}{2(1)}$

$= \frac{-5 \pm \sqrt{41}}{2}$

x-intercepts: $\left(\frac{-5 \pm \sqrt{41}}{2}, 0\right)$

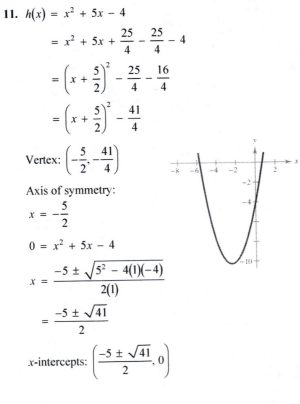

13. $f(x) = \frac{1}{3}(x^2 + 5x - 4)$

$= \frac{1}{3}\left(x^2 + 5x + \frac{25}{4} - \frac{25}{4} - 4\right)$

$= \frac{1}{3}\left[\left(x + \frac{5}{2}\right)^2 - \frac{41}{4}\right]$

$= \frac{1}{3}\left(x + \frac{5}{2}\right)^2 - \frac{41}{12}$

Vertex: $\left(-\frac{5}{2}, -\frac{41}{12}\right)$

Axis of symmetry: $x = -\frac{5}{2}$

$0 = x^2 + 5x - 4$

$x = \frac{-5 \pm \sqrt{5^2 - 4(1)(-4)}}{2(1)} = \frac{-5 \pm \sqrt{41}}{2}$

x-intercepts: $\left(\frac{-5 \pm \sqrt{41}}{2}, 0\right)$

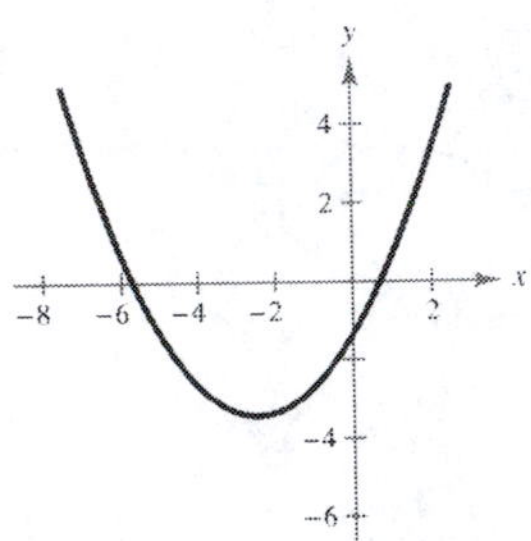

15. Vertex: $(4, 1) \Rightarrow f(x) = a(x - 4)^2 + 1$

Point: $(2, -1) \Rightarrow -1 = a(2 - 4)^2 + 1$

$-2 = 4a$

$-\frac{1}{2} = a$

$f(x) = -\frac{1}{2}(x - 4)^2 + 1$

17. Vertex: $(1, -4) \Rightarrow f(x) = a(x - 1)^2 - 4$

Point: $(2, -3) \Rightarrow -3 = a(2 - 1)^2 - 4$

$1 = a$

$f(x) = (x - 1)^2 - 4$

19. Vertex: $\left(-\frac{3}{2}, 0\right) \Rightarrow f(x) = a\left(x + \frac{3}{2}\right)^2$

Point: $\left(-\frac{9}{2}, -\frac{11}{4}\right) \Rightarrow -\frac{11}{4} = a\left(-\frac{9}{2} + \frac{3}{2}\right)^2$

$-\frac{11}{4} = 9a$

$-\frac{11}{36} = a$

$f(x) = -\frac{11}{36}\left(x + \frac{3}{2}\right)^2$

21. (a) $A = xy = x\left(\frac{8 - x}{2}\right)$ since

$x + 2y - 8 = 0 \Rightarrow y = \frac{8 - x}{2}.$

(b) Since the figure is in the first quadrant and x and y must be positive, the domain of

$A = x\left(\frac{8 - x}{2}\right)$ is $0 < x < 8$.

(c)

x	y	Area
1	$4 - \frac{1}{2}(1)$	$(1)\left[4 - \frac{1}{2}(1)\right] = \frac{7}{2}$
2	$4 - \frac{1}{2}(2)$	$(2)\left[4 - \frac{1}{2}(2)\right] = 6$
3	$4 - \frac{1}{2}(3)$	$(3)\left[4 - \frac{1}{2}(3)\right] = \frac{15}{2}$
4	$4 - \frac{1}{2}(4)$	$(4)\left[4 - \frac{1}{2}(4)\right] = 8$
5	$4 - \frac{1}{2}(5)$	$(5)\left[4 - \frac{1}{2}(5)\right] = \frac{15}{2}$
6	$4 - \frac{1}{2}(6)$	$(6)\left[4 - \frac{1}{2}(6)\right] = 6$

Using the table, when $x = 4$ and $y = 2$, the area A is 8 square units.

(d)

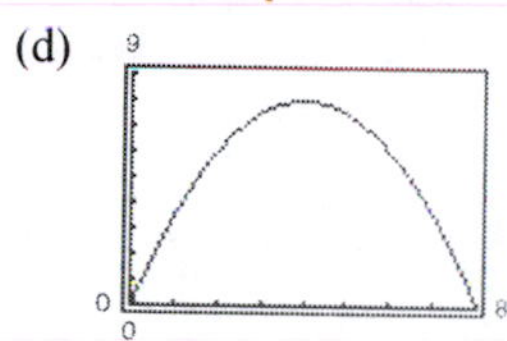

The maximum area of 8 occurs at the vertex when

$x = 4$ and $y = \frac{8 - 4}{2} = 2.$

(e) $$\begin{aligned} A &= x\left(\frac{8 - x}{2}\right) \\ &= \frac{1}{2}(8x - x^2) \\ &= -\frac{1}{2}(x^2 - 8x) \\ &= -\frac{1}{2}(x^2 - 8x + 16 - 16) \\ &= -\frac{1}{2}\left[(x - 4)^2 - 16\right] \\ &= -\frac{1}{2}(x - 4)^2 + 8 \end{aligned}$$

The maximum area of 8 occurs when $x = 4$ and

$y = \frac{8 - 4}{2} = 2.$

23. $R = -10p^2 + 800p$

(a) $R(20) = \$12{,}000$

$R(25) = \$13{,}750$

$R(30) = \$15{,}000$

(b) The maximum revenue occurs at the vertex of the parabola.

$$-\frac{b}{2a} = \frac{-800}{2(-10)} = \$40$$

$R(40) = \$16{,}000$

The revenue is maximum when the price is \$40 per unit.

The maximum revenue is \$16,000.

Any price greater or less than \$40 per unit will not yield as much revenue.

25. $C = 70{,}000 - 120x + 0.055x^2$

The minimum cost occurs at the vertex of the parabola.

Vertex: $-\frac{b}{2a} = -\frac{-120}{2(0.055)} \approx 1091$ units

About 1091 units should be produced each day to yield a minimum cost.

27. $y = x^3$, $f(x) = -(x - 2)^3$

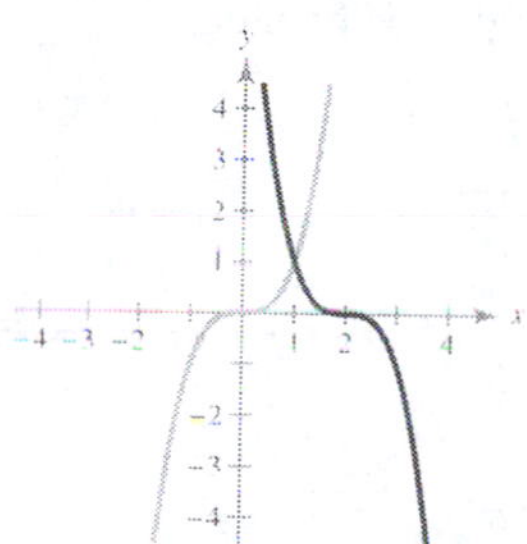

Transformation: Reflection in the x-axis and a horizontal shift two units to the right

29. $y = x^4$, $f(x) = 6 - x^4$

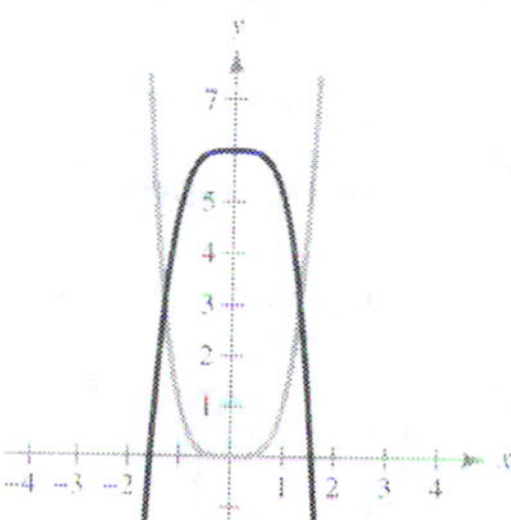

Transformation: Reflection in the x-axis and a vertical shift six units upward

31. $y = x^5$, $f(x) = (x - 5)^5$

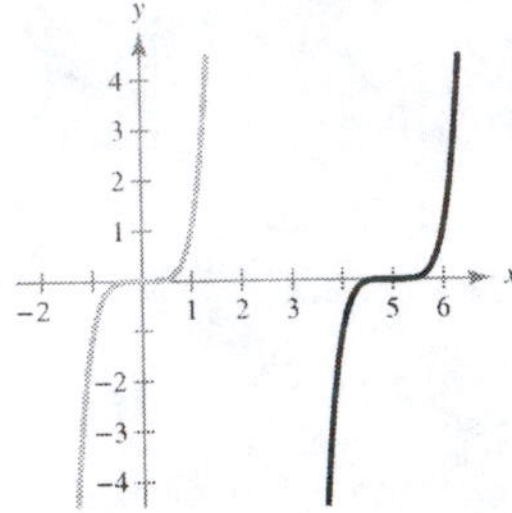

Transformation: Horizontal shift five units to the right

33. $f(x) = -2x^2 - 5x + 12$

The degree is even and the leading coefficient is negative. The graph falls to the left and falls to the right.

35. $g(x) = \frac{3}{4}(x^4 + 3x^2 + 2)$

The degree is even and the leading coefficient is positive. The graph rises to the left and rises to the right.

37. $f(x) = 3x^2 + 20x - 32$

$0 = 3x^2 + 20x - 32$

$0 = (3x - 4)(x + 8)$

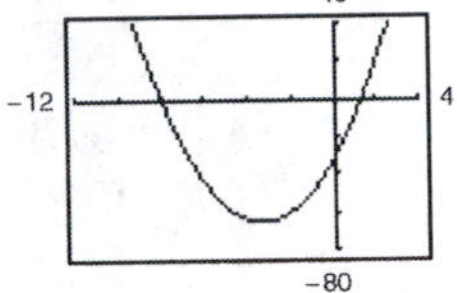

Zeros: $x = \frac{4}{3}$ and $x = -8$, both of multiplicity 1 (odd multiplicity)

Turning points: 1

39. $f(t) = t^3 - 3t$

$0 = t^3 - 3t$

$0 = t(t^2 - 3)$

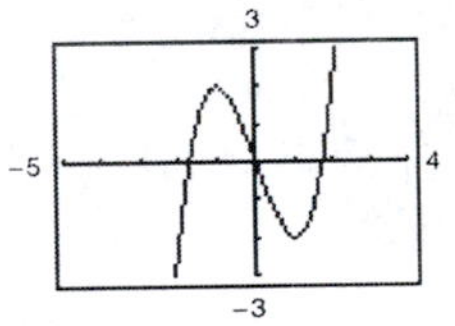

Zeros: $t = 0, \pm\sqrt{3}$, all of multiplicity 1 (odd multiplicity)

Turning points: 2

41. $f(x) = -18x^3 + 12x^2$

$0 = -18x^3 + 12x^2$

$0 = -6x^2(3x - 2)$

Zeros: $x = \frac{2}{3}$ of multiplicity 1 (odd multiplicity)

$x = 0$ of multiplicity 2 (even multiplicity)

Turning points: 2

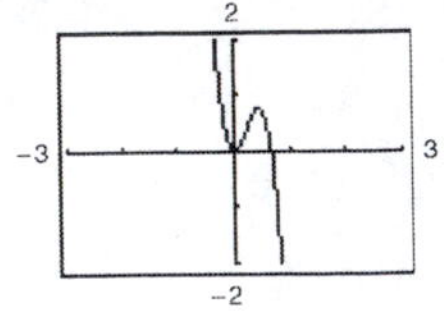

43. $f(x) = -x^3 + x^2 - 2$

(a) The degree is odd and the leading coefficient is negative. The graph rises to the left and falls to the right.

(b) Zero: $x = -1$

(c)

x	-3	-2	-1	0	1	2
$f(x)$	34	10	0	-2	-2	-6

(d)

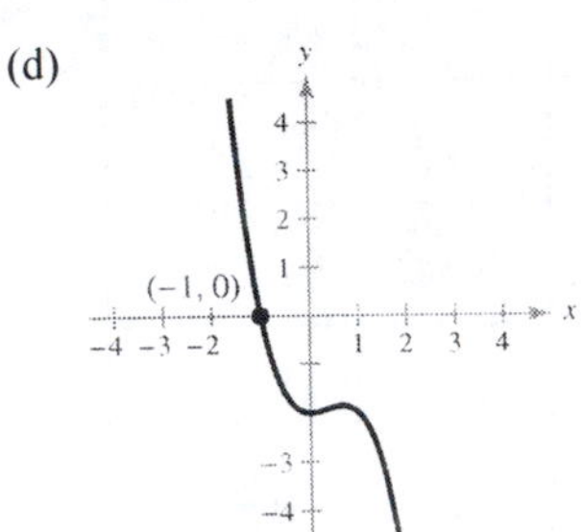

45. $f(x) = x(x^3 + x^2 - 5x + 3)$

(a) The degree is even and the leading coefficient is positive. The graph rises to the left and rises to the right.

(b) Zeros: $x = 0, 1, -3$

(c)

x	-4	-3	-2	-1	0	1	2	3
$f(x)$	100	0	-18	-8	0	0	10	72

(d)

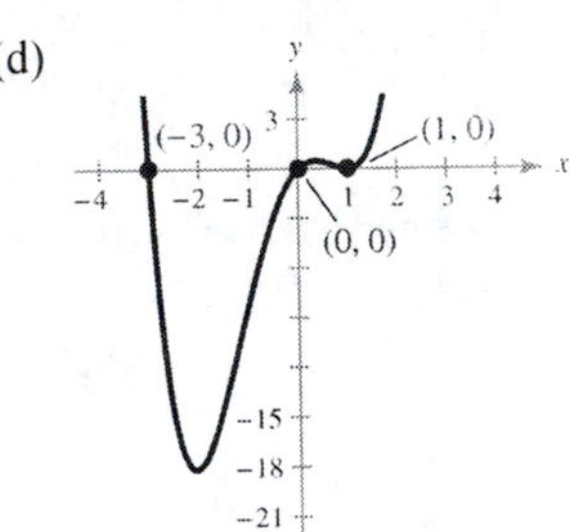

47. (a) $f(x) = 3x^3 - x^2 + 3$

x	-3	-2	-1	0	1	2	3
$f(x)$	-87	-25	-1	3	5	23	75

The zero is in the interval $[-1, 0]$.

(b) Zero: $x \approx -0.900$

49. (a) $f(x) = x^4 - 5x - 1$

x	-3	-2	-1	0	1	2	3
$f(x)$	95	25	5	-1	-5	5	65

There are zeros in the intervals $[-1, 0]$ and $[1, 2]$.

(b) Zeros: $x \approx -0.200$, $x \approx 1.772$

51. $$\begin{array}{r} 6x + 3 \\ 5x - 3\overline{)30x^2 - 3x + 8} \\ \underline{30x^2 - 18x} \\ 15x + 8 \\ \underline{15x - 9} \\ 17 \end{array}$$

$$\frac{30x^2 - 3x + 8}{5x - 3} = 6x + 3 + \frac{17}{5x - 3}$$

53. $$\begin{array}{r} 5x + 4 \\ x^2 - 5x - 1\overline{)5x^3 - 21x^2 - 25x - 4} \\ \underline{5x^3 - 25x^2 - 5x} \\ 4x^2 - 20x - 4 \\ \underline{4x^2 - 20x - 4} \\ 0 \end{array}$$

$$\frac{5x^3 - 21x^2 - 25x - 4}{x^2 - 5x - 1} = 5x + 4,\ x \neq \frac{5}{2} \pm \frac{\sqrt{29}}{2}$$

55. $$\begin{array}{r} x^2 - 3x + 2 \\ x^2 + 0x + 2\overline{)x^4 - 3x^3 + 4x^2 - 6x + 3} \\ \underline{x^4 + 0x^3 + 2x^2} \\ -3x^3 + 2x^2 - 6x \\ \underline{-3x^3 + 0x^2 - 6x} \\ 2x^2 + 0x + 3 \\ \underline{2x^2 + 0x + 4} \\ -1 \end{array}$$

$$\frac{x^4 - 3x^3 + 4x^2 - 6x + 3}{x^2 + 2} = x^2 - 3x + 2 - \frac{1}{x^2 + 2}$$

57. $$\begin{array}{r|rrrrr} 2 & 6 & -4 & -27 & 18 & 0 \\ & & 12 & 16 & -22 & -8 \\ \hline & 6 & 8 & -11 & -4 & -8 \end{array}$$

$$\frac{6x^4 - 4x^3 - 27x^2 + 18x}{x - 2} = 6x^3 + 8x^2 - 11x - 4 - \frac{8}{x - 2}$$

59. $$\begin{array}{r|rrrr} 8 & 2 & -25 & 66 & 48 \\ & & 16 & -72 & -48 \\ \hline & 2 & -9 & -6 & 0 \end{array}$$

$$\frac{2x^3 - 25x^2 + 66x + 48}{x - 8} = 2x^2 - 9x - 6, \quad x \neq 8$$

61. $f(x) = x^4 + 10x^3 - 24x^2 + 20x + 44$

(a) Remainder Theorem:

$$f(-3) = (-3)^4 + 10(-3)^3 - 24(-3)^2 + 20(-3) + 44 = -421$$

Synthetic Division:

$$\begin{array}{r|rrrrr} -3 & 1 & 10 & -24 & 20 & 44 \\ & & -3 & -21 & 135 & -465 \\ \hline & 1 & 7 & -45 & 155 & -421 \end{array}$$

So, $f(-3) = -421$.

(b) Remainder Theorem:

$$f(-1) = (-1)^4 + 10(-1)^3 - 24(-1)^2 + 20(-1) + 44 = -9$$

Synthetic Division:

$$\begin{array}{r|rrrrr} -1 & 1 & 10 & -24 & 20 & 44 \\ & & -1 & -9 & 33 & -53 \\ \hline & 1 & 9 & -33 & 53 & -9 \end{array}$$

So, $f(-1) = -9$.

63. $f(x) = 20x^4 + 9x^3 - 14x^2 - 3x$

(a)
$$\begin{array}{r|rrrrr} -1 & 20 & 9 & -14 & -3 & 0 \\ & & -20 & 11 & 3 & 0 \\ \hline & 20 & -11 & -3 & 0 & 0 \end{array}$$

Yes, $x = -1$ is a zero of f.

(b)
$$\begin{array}{r|rrrrr} \frac{3}{4} & 20 & 9 & -14 & -3 & 0 \\ & & 15 & 18 & 3 & 0 \\ \hline & 20 & 24 & 4 & 0 & 0 \end{array}$$

Yes, $x = \frac{3}{4}$ is a zero of f.

(c)
$$\begin{array}{r|rrrrr} 0 & 20 & 9 & -14 & -3 & 0 \\ & & 0 & 0 & 0 & 0 \\ \hline & 20 & 9 & -14 & -3 & 0 \end{array}$$

Yes, $x = 0$ is a zero of f.

(d)
$$\begin{array}{r|rrrrr} 1 & 20 & 9 & -14 & -3 & 0 \\ & & 20 & 29 & 15 & 12 \\ \hline & 20 & 29 & 15 & 12 & 12 \end{array}$$

No, $x = 1$ is not a zero of f.

65. $f(x) = x^3 + 4x^2 - 25x - 28$; Factor: $(x - 4)$

(a)
$$\begin{array}{r|rrrr} 4 & 1 & 4 & -25 & -28 \\ & & 4 & 32 & 28 \\ \hline & 1 & 8 & 7 & 0 \end{array}$$

Yes, $(x - 4)$ is a factor of $f(x)$.

(b) $x^2 + 8x + 7 = (x + 7)(x + 1)$

The remaining factors of f are $(x + 7)$ and $(x + 1)$.

(c) $f(x) = x^3 + 4x^2 - 25x - 28$
$= (x + 7)(x + 1)(x - 4)$

(d) Zeros: $-7, -1, 4$

(e)

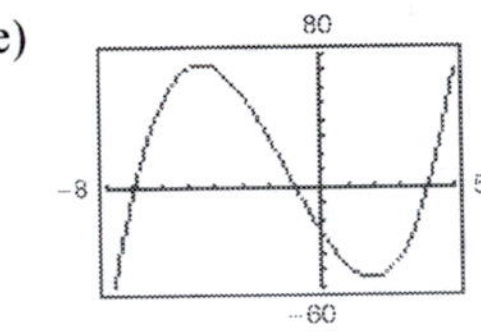

67. $f(x) = x^4 - 4x^3 - 7x^2 + 22x + 24$

Factors: $(x + 2), (x - 3)$

(a)
$$\begin{array}{r|rrrrr} -2 & 1 & -4 & -7 & 22 & 24 \\ & & -2 & 12 & -10 & -24 \\ \hline & 1 & -6 & 5 & 12 & 0 \end{array}$$

$$\begin{array}{r|rrrr} 3 & 1 & -6 & 5 & 12 \\ & & 3 & -9 & -12 \\ \hline & 1 & -3 & -4 & 0 \end{array}$$

Yes, $(x + 2)$ and $(x - 3)$ are both factors of $f(x)$.

(b) $x^2 - 3x - 4 = (x + 1)(x - 4)$

The remaining factors are $(x + 1)$ and $(x - 4)$.

(c) $f(x) = (x + 1)(x - 4)(x + 2)(x - 3)$

(d) Zeros: $-2, -1, 3, 4$

(e)

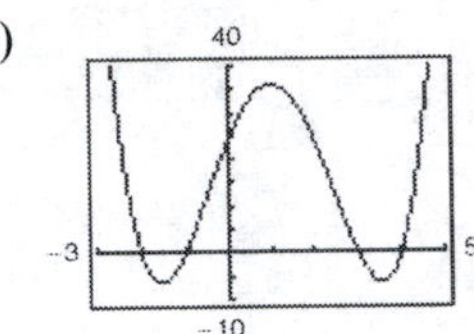

69. Since $f(x) = x - 6$ is a 1st degree polynomial function, it has one zero.

71. Since $h(t) = t^2 - t^5$ is a 5th degree polynomial function, it has five zeros.

73. Since $f(x) = (x - 8)^3 = x^3 - 24x^2 + 19x^2 - 512$ is a 3rd degree polynomial function, it has three zeros.

75. $f(x) = -4x^3 + 8x^2 - 3x + 15$

Possible rational zeros:
$\pm1, \pm3, \pm5, \pm15, \pm\frac{1}{2}, \pm\frac{3}{2}, \pm\frac{5}{2}, \pm\frac{15}{2}, \pm\frac{1}{4}, \pm\frac{3}{4}, \pm\frac{5}{4}, \pm\frac{15}{4}$

77. $f(x) = x^3 + 3x^2 - 28x - 60$

Possible rational zeros:
$\pm1, \pm2, \pm3, \pm4, \pm5, \pm6, \pm10, \pm12, \pm15, \pm20, \pm30, \pm60$

$$\begin{array}{r|rrrr} -2 & 1 & 3 & -28 & -60 \\ & & -2 & -2 & 60 \\ \hline & 1 & 1 & -30 & 0 \end{array}$$

$x^3 + 3x^2 - 28x - 60 = (x + 2)(x^2 + x - 30)$
$= (x + 2)(x + 6)(x - 5)$

The zeros of $f(x)$ are $x = -2$, $x = -6$, and $x = 5$.

79. $f(x) = x^3 - 10x^2 + 17x - 8$

Possible rational zeros: $\pm1, \pm2, \pm4, \pm8$

$$\begin{array}{r|rrrr} 1 & 1 & -10 & 17 & -8 \\ & & 1 & -9 & 8 \\ \hline & 1 & -9 & 8 & 0 \end{array}$$

$$\begin{aligned} x^3 - 10x^2 + 17x - 8 &= (x - 1)(x^2 - 9x + 8) \\ &= (x - 1)(x - 1)(x - 8) \\ &= (x - 1)^2(x - 8) \end{aligned}$$

The zeros of $f(x)$ are $x = 1$ and $x = 8$.

81. $f(x) = x^4 + x^3 - 11x^2 + x - 12$

Possible rational zeros: $\pm1, \pm2, \pm3, \pm4, \pm6, \pm12$

$$\begin{array}{r|rrrrr} 3 & 1 & 1 & -11 & 1 & -12 \\ & & 3 & 12 & 3 & 12 \\ \hline & 1 & 4 & 1 & 4 & 0 \end{array}$$

$$\begin{array}{r|rrrr} -4 & 1 & 4 & 1 & 4 \\ & & -4 & 0 & -4 \\ \hline & 1 & 0 & 1 & 0 \end{array}$$

$$x^4 + x^3 - 11x^2 + x - 12 = (x - 3)(x + 4)(x^2 + 1)$$

The zeros of $f(x)$ are $x = 3$ and $x = -4$.

83. Because $\sqrt{3}i$ is a zero, so is $-\sqrt{3}i$.

Multiply by 3 to clear the fraction.

$$\begin{aligned} f(x) &= 3\left(x - \tfrac{2}{3}\right)(x - 4)\left(x - \sqrt{3}i\right)\left(x + \sqrt{3}i\right) \\ &= (3x - 2)(x - 4)(x^2 + 3) \\ &= (3x^2 - 14x + 8)(x^2 + 3) \\ &= 3x^4 - 14x^3 + 17x^2 - 42x + 24 \end{aligned}$$

Note: $f(x) = a(3x^4 - 14x^3 + 17x^2 - 42x + 24)$,

where a is any real nonzero number, has zeros $\frac{2}{3}$, 4, and $\pm\sqrt{3}i$.

85. $f(x) = x^3 - 4x^2 + x - 4$, Zero: i

Because i is a zero, so is $-i$.

$$\begin{array}{r|rrrr} i & 1 & -4 & 1 & -4 \\ & & i & -1 - 4i & 4 \\ \hline & 1 & -4 + i & -4i & 0 \end{array}$$

$$\begin{array}{r|rrr} -i & 1 & -4 + i & -4i \\ & & -i & 4i \\ \hline & 1 & -4 & 0 \end{array}$$

$$f(x) = (x - i)(x + i)(x - 4)$$

Zeros: $x = \pm i, 4$

87. $g(x) = 2x^4 - 3x^3 - 13x^2 + 37x - 15$, Zero: $2 + i$

Because $2 + i$ is a zero, so is $2 - i$.

$$\begin{array}{r|rrrrr} 2 + i & 2 & -3 & -13 & 37 & -15 \\ & & 4 + 2i & 5i & -31 - 3i & 15 \\ \hline & 2 & 1 + 2i & -13 + 5i & 6 - 3i & 0 \end{array}$$

$$\begin{array}{r|rrrr} 2 - i & 2 & 1 + 2i & -13 + 5i & 6 - 3i \\ & & 4 - 2i & 10 - 5i & -6 + 3i \\ \hline & 2 & 5 & -3 & 0 \end{array}$$

$$\begin{aligned} g(x) &= [x - (2 + i)][x - (2 - i)](2x^2 + 5x - 3) \\ &= (x - 2 - i)(x - 2 + i)(2x - 1)(x + 3) \end{aligned}$$

Zeros: $x = 2 \pm i, \frac{1}{2}, -3$

89. $$\begin{aligned} f(x) &= x^3 + 4x^2 - 5x \\ &= x(x^2 + 4x - 5) \\ &= x(x + 5)(x - 1) \end{aligned}$$

Zeros: $x = 0, -5, 1$

91. $g(x) = x^4 + 4x^3 - 3x^2 + 40x + 208$, Zero: $x = -4$

$$\begin{array}{r|rrrrr} -4 & 1 & 4 & -3 & 40 & 208 \\ & & -4 & 0 & 12 & -208 \\ \hline & 1 & 0 & -3 & 52 & 0 \end{array}$$

$$\begin{array}{r|rrrr} -4 & 1 & 0 & -3 & 52 \\ & & -4 & 16 & -52 \\ \hline & 1 & -4 & 13 & 0 \end{array}$$

$g(x) = (x + 4)^2(x^2 - 4x + 13)$

By the Quadratic Formula the zeros of $x^2 - 4x + 13$ are $x = 2 \pm 3i$. The zeros of $g(x)$ are $x = -4$ and $x = 2 \pm 3i$.

$$\begin{aligned} g(x) &= (x + 4)^2[x - (2 + 3i)][x - (2 - 3i)] \\ &= (x + 4)^2(x - 2 - 3i)(x - 2 + 3i) \end{aligned}$$

93. $f(x) = x^3 - 16x^2 + x - 16$

Possible rational zeros: $\pm 1, \pm 2, \pm 4, \pm 8, \pm 16$

$$\begin{array}{r|rrrr} 16 & 1 & -16 & 1 & -16 \\ & & 16 & 0 & 16 \\ \hline & 1 & 0 & 1 & 0 \end{array}$$

$$x^3 - 16x^2 + x - 16 = (x - 16)(x^2 + 1)$$

$$\begin{aligned} x^2 + 1 &= 0 \\ x^2 &= -1 \\ x &= \pm\sqrt{-1} \\ x &= \pm i \end{aligned}$$

The zero of $f(x)$ are $x = 16$ and $x = \pm i$.

95. $g(x) = x^4 - 3x^3 - 14x^2 - 12x - 72$

Possible rational zeros: $\pm 1, \pm 2, \pm 3, \pm 4, \pm 6, \pm 8, \pm 9, \pm 12, \pm 18, \pm 24, \pm 36, \pm 72$

$$\begin{array}{r|rrrrr} -3 & 1 & -3 & -14 & -12 & -72 \\ & & -3 & 18 & -12 & 72 \\ \hline & 1 & -6 & 4 & -24 & 0 \end{array}$$

$$\begin{array}{r|rrrr} 6 & 1 & -6 & 4 & -24 \\ & & 6 & 0 & 24 \\ \hline & 1 & 0 & 4 & 0 \end{array}$$

$$x^4 - 3x^3 - 14x^2 - 12x - 72 = (x + 3)(x - 6)(x^2 + 4)$$

$$\begin{aligned} x^2 + 4 &= 0 \\ x^2 &= -4 \\ x &= \pm\sqrt{-4} \\ x &= \pm 2i \end{aligned}$$

The zeros of $g(x)$ are $x = -3$, $x = 6$, and $x = \pm 2i$.

97. $g(x) = 5x^3 + 3x^2 - 6x + 9$

$g(x)$ has two variations in sign, so g has either two or no positive real zeros.

$g(-x) = -5x^3 + 3x^2 + 6x + 9$

$g(-x)$ has one variation in sign, so g has one negative real zero.

99. $f(x) = 4x^3 - 3x^2 + 4x - 3$

(a)
$$\begin{array}{r|rrrr} 1 & 4 & -3 & 4 & -3 \\ & & 4 & 1 & 5 \\ \hline & 4 & 1 & 5 & 2 \end{array}$$

Because the last row has all positive entries, $x = 1$ is an upper bound.

(b)
$$\begin{array}{r|rrrr} -\frac{1}{4} & 4 & -3 & 4 & -3 \\ & & -1 & 1 & -\frac{5}{4} \\ \hline & 4 & -4 & 5 & -\frac{17}{4} \end{array}$$

Because the last row entries alternate in sign, $x = -\frac{1}{4}$ is a lower bound.

101.

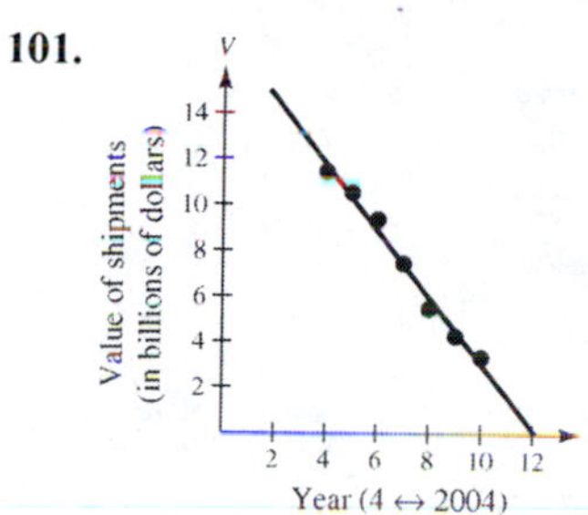

The model fits the data well.

103. $D = km$

$4 = 2.5k$

$1.6 = k$

$D = \frac{8}{5}m$ or $D = 1.6m$

In 2 miles:

$D = 1.6(2) = 3.2$ kilometers

In 10 miles:

$D = 1.6(10) = 16$ kilometers

105. $F = ks^2$

If speed is doubled,

$F = k(2s)^2$

$F = 4ks^2.$

So, the force will be changed by a factor of 4.

106. $x = \dfrac{k}{p}$

$800 = \dfrac{k}{5}$

$k = 4000$

$x = \dfrac{4000}{6} \approx 667$ boxes

107. $T = \dfrac{k}{r}$

$3 = \dfrac{k}{65}$

$k = 3(65) = 195$

$T = \dfrac{195}{r}$

When $r = 80$ mph,

$T = \dfrac{195}{80} = 2.4375$ hours

≈ 2 hours, 26 minutes.

109. False. A fourth-degree polynomial can have at most four zeros, and complex zeros occur in conjugate pairs.

111. The maximum (or minimum) value of a quadratic function is located at its graph's vertex. To find the vertex, either write the equation in standard form or use the formula

$$\left(-\frac{b}{2a}, f\left(-\frac{b}{2a}\right)\right).$$

If the leading coefficient is positive, the vertex is a minimum. If the leading coefficient is negative, the vertex is a maximum.

Problem Solving for Chapter 3

1. (a) (i) $g(x) = x^2 - 4x - 12$

$0 = (x - 6)(x + 2)$

Zeros: 6, −2

(ii) $g(x) = x^2 + 5x$

$0 = x(x + 5)$

Zeros: 0, −5

(iii) $g(x) = x^2 + 3x - 10$

$0 = (x + 5)(x - 2)$

Zeros: −5, 2

(iv) $g(x) = x^2 - 4x + 4$

$0 = (x - 2)(x - 2)$

Zeros: 2, 2

(v) $g(x) = x^2 - 2x - 6$

$0 = x^2 - 2x - 6$

By the Quadratic Formula, $x = 1 \pm \sqrt{7}$.

Zeros: $1 \pm \sqrt{7}$

(vi) $g(x) = x^2 + 3x + 4$

$0 = x^2 + 3x + 4$

By the Quadratic Formula, $x = \dfrac{-3 \pm \sqrt{7}i}{2}$.

Zeros: $\dfrac{-3 \pm \sqrt{7}i}{2}$

(b) (i) $f(x) = (x - 2)(x^2 - 4x - 12)$

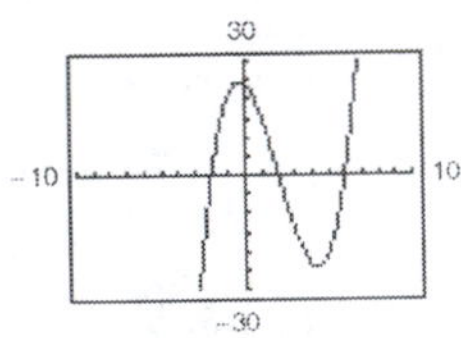

(ii) $f(x) = (x - 2)(x^2 + 5x)$

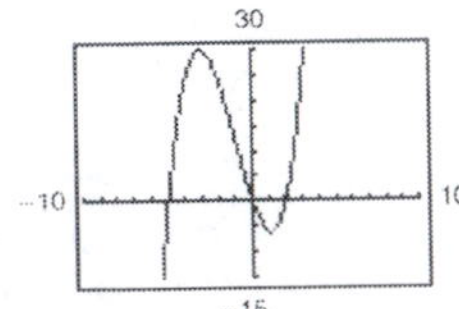

(iii) $f(x) = (x - 2)(x^2 + 3x - 10)$

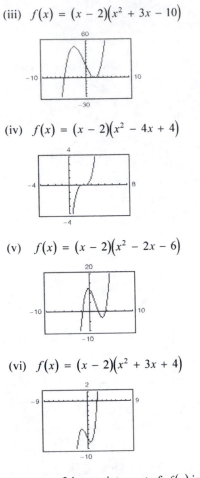

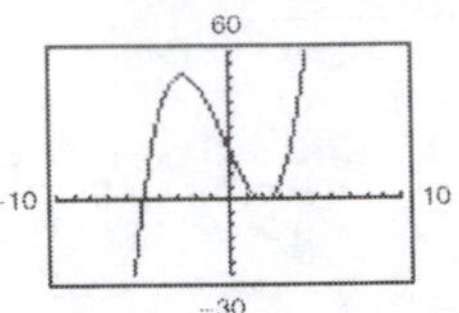

(iv) $f(x) = (x - 2)(x^2 - 4x + 4)$

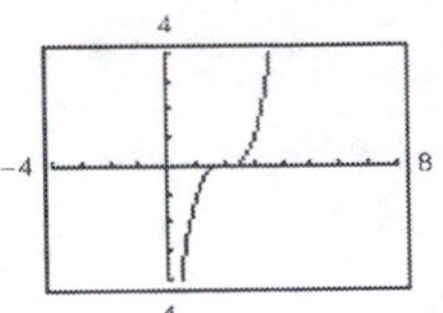

(v) $f(x) = (x - 2)(x^2 - 2x - 6)$

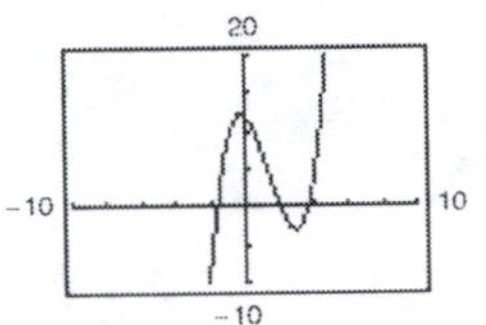

(vi) $f(x) = (x - 2)(x^2 + 3x + 4)$

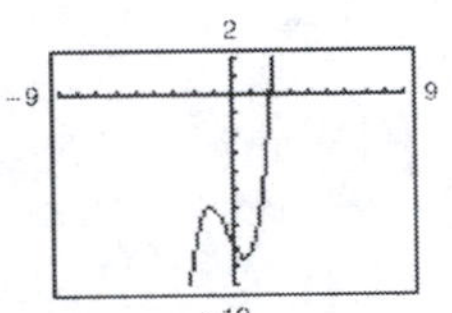

$x = 2$ is an x-intercept of $f(x)$ in all six graphs. All the graphs, except (iii), cross the x-axis at $x = 2$.

(c) (i) other x-intercepts: $(-2, 0), (6, 0)$

(ii) other x-intercepts: $(-5, 0), (0, 0)$

(iii) other x-intercepts: $(-5, 0)$

(iv) other x-intercepts: No additional x-intercepts

(v) other x-intercepts: $(-1.6, 0), (3.6, 0)$

(vi) other x-intercepts: No additional x-intercepts

(d) The x-intercepts of $f(x) = (x - 2)g(x)$ are the same as the zeros of $g(x)$ plus $x = 2$.

3. $f(x) = ax^3 + bx^2 + cx + d$

$$\begin{array}{r} ax^2 + (ak + b)x + (ak^2 + bk + c) \\ x - k \overline{)ax^3 + bx^2 \quad\quad + cx \quad\quad + d} \\ \underline{ax^3 - akx^2} \quad\quad\quad\quad\quad\quad \\ (ak + b)x^2 + cx \quad\quad\quad \\ \underline{(ak + b)x^2 - (ak^2 + bk)x} \\ (ak^2 + bk + c)x + d \\ \underline{(ak^2 + bk + c)x - (ak^3 + bk^2 + ck)} \\ (ak^3 + bk^2 + ck + d) \end{array}$$

So, $f(x) = ax^3 + bx^2 + cx + d = (x - k)\left[ax^2 + (ak + b)x + (ak^2 + bx + c)\right] + ak^3 + bk^2 + ck + d$ and $f(x) = ak^3 + bk^2 + ck + d$. Because the remainder is $r = ak^3 + bk^2 + ck + d$, $f(k) = r$.

5. $V = l \cdot w \cdot h = x^2(x + 3)$

$x^2(x + 3) = 20$

$x^3 + 3x^2 - 20 = 0$

Possible rational zeros: $\pm 1, \pm 2, \pm 4, \pm 5, \pm 10, \pm 20$

$$\begin{array}{r|rrrr} 2 & 1 & 3 & 0 & -20 \\ & & 2 & 10 & 20 \\ \hline & 1 & 5 & 10 & 0 \end{array}$$

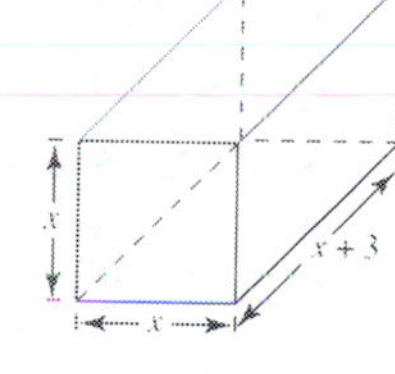

$(x - 2)(x^2 + 5x + 10) = 0$

$x = 2$ or $x = \dfrac{-5 \pm \sqrt{15}i}{2}$

Choosing the real positive value for x we have: $x = 2$ and $x + 3 = 5$.

The dimensions of the mold are 2 inches × 2 inches × 5 inches.

7. False. Since $f(x) = d(x)q(x) + r(x)$, we have

$$\frac{f(x)}{d(x)} = q(x) + \frac{r(x)}{d(x)}.$$

The statement should be corrected to read $f(-1) = 2$ since $\dfrac{f(x)}{x + 1} = q(x) + \dfrac{f(-1)}{x + 1}$.

9. (a) Slope $= \dfrac{9 - 4}{3 - 2} = 5$

Slope of tangent line is less than 5.

(b) Slope $= \dfrac{4 - 1}{2 - 1} = 3$

Slope of tangent line is greater than 3.

(c) Slope $= \dfrac{4.41 - 4}{2.1 - 2} = 4.1$

Slope of tangent line is less than 4.1.

(d) $$\begin{aligned} \text{Slope} &= \frac{f(2 + h) - f(2)}{(2 + h) - 2} \\ &= \frac{(2 + h)^2 - 4}{h} \\ &= \frac{4h + h^2}{h} \\ &= 4 + h, h \neq 0 \end{aligned}$$

(e) Slope $= 4 + h, \quad h \neq 0$

$4 + (-1) = 3$

$4 + 1 = 5$

$4 + 0.1 = 4.1$

The results are the same as in (a)–(c).

(f) Letting h get closer and closer to 0, the slope approaches 4. So, the slope at $(2, 4)$ is 4.

11. Let x = length of the wire used to form the square. Then $100 - x$ = length of wire used to form the circle.

(a) Let s = the side of the square. Then $4s = x \Rightarrow s = x/4$ and the area of the square is $s^2 = (x/4)^2$.

Let r = the radius of the circle. Then

$2\pi r = 100 - x \Rightarrow r = \dfrac{100 - x}{2\pi}$ and the area of the circle is $\pi r^2 = \pi\left(\dfrac{100 - x}{2\pi}\right)^2$.

The combined area is:

$$\begin{aligned}
A(x) &= \left(\frac{x}{4}\right)^2 + \pi\left(\frac{100 - x}{2\pi}\right)^2 \\
&= \frac{x^2}{16} + \pi\left(\frac{10{,}000 - 200x + x^2}{4\pi^2}\right) \\
&= \frac{x^2}{16} + \frac{2500}{\pi} = \frac{50x}{\pi} + \frac{x^2}{4\pi} \\
&= \left(\frac{1}{16} + \frac{1}{4\pi}\right)x^2 - \frac{50x}{\pi} + \frac{2500}{\pi} \\
&= \left(\frac{x + 4}{16\pi}\right)x^2 - \frac{50}{\pi}x + \frac{2500}{\pi}
\end{aligned}$$

(b) Domain: Since the wire is 100 cm, $0 \le x \le 100$.

(c)
$$\begin{aligned}
A(x) &= \left(\frac{x + 4}{16\pi}\right)x^2 - \frac{50}{\pi}x + \frac{2500}{\pi} \\
&= \left(\frac{x + 4}{16\pi}\right)\left(x^2 - \frac{800}{\pi + 4}x\right) + \frac{2500}{\pi} \\
&= \left(\frac{x + 4}{16\pi}\right)\left[x^2 - \frac{800}{\pi + 4}x + \left(\frac{400}{\pi + 4}\right)^2 - \left(\frac{400}{\pi + 4}\right)^2\right] + \frac{2500}{\pi} \\
&= \left(\frac{x + 4}{16\pi}\right)\left[x - \left(\frac{400}{\pi + 4}\right)\right]^2 - \left(\frac{\pi + 4}{16\pi}\right)\left(\frac{400}{\pi + 4}\right)^2 + \frac{2500}{\pi} \\
&= \left(\frac{x + 4}{16\pi}\right)\left[x - \left(\frac{400}{\pi + 4}\right)\right]^2 - \frac{10{,}000}{\pi(\pi + 4)} + \frac{2500}{\pi} \\
&= \left(\frac{x + 4}{16\pi}\right)\left[x - \left(\frac{400}{\pi + 4}\right)\right]^2 + \frac{2500}{\pi + 4}
\end{aligned}$$

The minimum occurs at the vertex when $x = 400/(\pi + 4) \approx 56$ cm and $A(x) \approx 350$ cm^2. The maximum occurs at one of the endpoints of the domain. When $x = 0$, $A(x) \approx 796$ cm^2. When $x = 100$, $A(x) \approx 625$ cm^2. Thus, the area is maximum when $x = 0$ cm.

Answers will vary. Graph $A(x)$ to see where the minimum and maximum values occur.

Practice Test for Chapter 3

1. Sketch the graph of $f(x) = x^2 - 6x + 5$ and identify the vertex and the intercepts.

2. Find the number of units x that produce a minimum cost C if
$C = 0.01x^2 - 90x + 15{,}000$.

3. Find the quadratic function that has a maximum at $(1, 7)$ and passes through the point $(2, 5)$.

4. Find two quadratic functions that have x-intercepts $(2, 0)$ and $\left(\frac{4}{3}, 0\right)$.

5. Use the leading coefficient test to determine the right and left end behavior of the graph of the polynomial function $f(x) = -3x^5 + 2x^3 - 17$.

6. Find all the real zeros of $f(x) = x^5 - 5x^3 + 4x$.

7. Find a polynomial function with 0, 3, and -2 as zeros.

8. Sketch $f(x) = x^3 - 12x$.

9. Divide $3x^4 - 7x^2 + 2x - 10$ by $x - 3$ using long division.

10. Divide $x^3 - 11$ by $x^2 + 2x - 1$.

11. Use synthetic division to divide $3x^5 + 13x^4 + 12x - 1$ by $x + 5$.

12. Use synthetic division to find $f(-6)$ given $f(x) = 7x^3 + 40x^2 - 12x + 15$.

13. Find the real zeros of $f(x) = x^3 - 19x - 30$.

14. Find the real zeros of $f(x) = x^4 + x^3 - 8x^2 - 9x - 9$.

15. List all possible rational zeros of the function $f(x) = 6x^3 - 5x^2 + 4x - 15$.

16. Find the rational zeros of the polynomial $f(x) = x^3 - \frac{20}{3}x^2 + 9x - \frac{10}{3}$.

17. Write $f(x) = x^4 + x^3 + 5x - 10$ as a product of linear factors.

18. Find a polynomial with real coefficients that has $2, 3 + i$, and $3 - 2i$ as zeros.

19. Use synthetic division to show that $3i$ is a zero of $f(x) = x^3 + 4x^2 + 9x + 36$.

20. Find a mathematical model for the statement, "z varies directly as the square of x and inversely as the square of x and inversely as the square root of y."

CHAPTER 4
Rational Functions and Conics

CHAPTER 4
Rational Functions and Conics

Section 4.1 Rational Functions and Asymptotes

1. rational functions

3. horizontal asymptote

5. Because the denominator is zero when $x - 1 = 0$, the domain of f is all real numbers except $x = 1$.

x	0	0.5	0.9	0.99	$\to 1$
$f(x)$	-1	-2	-10	-100	$\to -\infty$

x	$1 \leftarrow$	1.01	1.1	1.5	2
$f(x)$	$\infty \leftarrow$	100	10	2	1

As x approaches 1 from the left, $f(x)$ decreases without bound. As x approaches 1 from the right, $f(x)$ increases without bound.

7. Because the denominator is zero when $x + 2 = 0$, the domain of f is all real numbers except $x = -2$.

x	-3	-2.5	-2.1	-2.01	$\to -2$
$f(x)$	15	25	105	1005	$\to \infty$

x	$-2 \leftarrow$	-1.99	-1.9	-1.5	-1
$f(x)$	$-\infty \leftarrow$	-955	-95	-15	-5

As x approaches -2 from the left, $f(x)$ increases without bound. As x approaches -2 from the right, $f(x)$ increases without bound.

9. Because the denominator is zero when $x^2 - 1 = 0$, the domain of f is all real numbers except $x = -1$ and $x = 1$.

x	-2	-1.5	-1.1	-1.01	$\to -1$
$f(x)$	4	5.4	17.3	152.3	$-\infty$

x	$-1 \leftarrow$	-0.99	-0.9	-0.5	0
$f(x)$	$-\infty \leftarrow$	-147.8	-12.8	-1	0

As x approaches -1 from the left, $f(x)$ increases without bound. As x approaches -1 from the right, $f(x)$ decreases without bound.

x	0	0.5	0.9	0.99	$\to 1$
$f(x)$	0	-1	-12.8	-147.8	$\to -\infty$

x	$1 \leftarrow$	1.01	1.1	1.5	2
$f(x)$	$\infty \leftarrow$	152.3	17.3	5.4	4

As x approaches 1 from the left, $f(x)$ increases without bound. As x approaches 1 from the right, $f(x)$ decreases without bound.

11. Because the denominator is zero when $x^2 - 2x + 1 = (x - 1)^2 = 0$, the domain of f is all real numbers except $x = 1$.

x	0	0.5	0.9	0.99	1
$f(x)$	2	15	551	59,501	Undef.

x	1.01	1.1	1.5	2
$f(x)$	60,501	651	35	12

As x approaches 1 from the left, $f(x)$ increases without bound. As x approaches 1 from the right, $f(x)$ increases without bound.

13. $f(x) = \dfrac{4}{x^2}$

Domain: all real numbers except $x = 0$

Vertical asymptote: $x = 0$

Horizontal asymptote: $y = 0$

$\left[\text{Degree of } N(x) < \text{degree of } D(x)\right]$

15. $f(x) = \dfrac{5 + x}{5 - x} = \dfrac{x + 5}{-x + 5}$

Domain: all real numbers except $x = 5$

Vertical asymptote: $x = 5$

Horizontal asymptote: $y = -1$

$\left[\text{Degree of } N(x) = \text{degree of } D(x)\right]$

17. $f(x) = \dfrac{x^3}{x^2 - 1}$

Domain: all real numbers except $x = \pm 1$

Vertical asymptotes: $x = \pm 1$

Horizontal asymptote: None

$\left[\text{Degree of } N(x) > \text{degree of } D(x)\right]$

19. $f(x) = \dfrac{3x^2 + 1}{x^2 + x + 9}$

Domain: All real numbers. The denominator has no real zeros. [Try the Quadratic Formula on the denominator.]

Vertical asymptote: None

Horizontal asymptote: $y = 3$

$\left[\text{Degree of } N(x) = \text{degree of } D(x)\right]$

21. $f(x) = \dfrac{x - 4}{x^2 - 16} = \dfrac{1}{x + 4}, x \neq 4$

Horizontal asymptote: $y = 0$

$\left(\text{Degree of } N(x) < \text{degree of } D(x)\right)$

Vertical asymptote: $x = -4$

(Because $x - 4$ is a common factor of $N(x)$ and $D(x)$, $x = 4$ is not a vertical asymptote of $f(x)$.)

23. $f(x) = \dfrac{x^2 - 1}{x^2 - 2x - 3}$

$= \dfrac{(x + 1)(x - 1)}{(x + 1)(x - 3)}$

$= \dfrac{x - 1}{x - 3}, x \neq -1$

Horizontal asymptote:

$y = 1 \left(\text{Degree of } N(x) = \text{degree of } D(x)\right)$

Vertical asymptote: $x = 3$

(Because $x + 1$ is a common factor of $N(x)$ and $D(x)$, $x = -1$ is not a vertical asymptote of $f(x)$.)

25. $f(x) = \dfrac{x^2 - 3x - 4}{2x^2 + x - 1}$

$= \dfrac{(x + 1)(x - 4)}{(2x - 1)(x + 1)}$

$= \dfrac{x - 4}{2x - 1}, x \neq -1$

Horizontal asymptote: $y = \dfrac{1}{2}$

$\left(\text{Degree of } N(x) = \text{degree of } D(x)\right)$

Vertical asymptote: $x = \dfrac{1}{2}$

(Because $x + 1$ is a common factor of $N(x)$ and $D(x)$, $x = -1$ is not a vertical asymptote of $f(x)$.)

27. $f(x) = \dfrac{6x^2 + 5x - 6}{3x^2 - 8x + 4}$

$= \dfrac{(3x - 2)(2x + 3)}{(3x - 2)(x - 2)}$

$= \dfrac{2x + 3}{x - 2}, x \neq \dfrac{2}{3}$

Horizontal asymptote:

$y = 2 \left(\text{Degree of } N(x) = \text{degree of } D(x)\right)$

Vertical asymptote: $x = 2$

(Because $3x - 2$ is a common factor of $N(x)$ and $D(x)$, $x = \dfrac{2}{3}$ is not a vertical asymptote of $f(x)$.)

29. g

30. e

31. c

32. b

33. a

34. d

35. f

36. h

37. $f(x) = \frac{x^2 - 4}{x + 2}$, $g(x) = x - 2$

$f(x) = \frac{(x + 2)(x - 2)}{x + 2}$

(a) Domain of f: all real numbers except $x = -2$

Domain of g: all real numbers

(b) $f(x) = x - 2$

Because $x + 2$ is a common factor of both the numerator and the denominator of $f(x)$, $x = -2$ is not a vertical asymptote of f. f has no vertical asymptotes.

(c)

x	-4	-3	-2.5	-2	-1.5	-1	0
$f(x)$	-6	-5	-4.5	Undef.	-3.5	-3	-2
$g(x)$	-6	-5	-4.5	-4	-3.5	-3	-2

(d) f and g differ only at $x = -2$, where f is undefined and g is defined.

39. $f(x) = \frac{2x - 1}{2x^2 - x}$, $g(x) = \frac{1}{x}$

$f(x) = \frac{2x - 1}{x(2x - 1)}$

(a) Domain of f: all real numbers except $x = 0$ and $x = \frac{1}{2}$

Domain of g: all real numbers except $x = 0$

(b) $f(x) = \frac{1}{x}$

Because $2x - 1$ is a common factor of both the numerator and the denominator of f, $x = 0.5$ is not a vertical asymptote of f. The only vertical asymptote is $x = 0$.

(c)

x	-1	-0.5	0	0.5	2	3	4
$f(x)$	-1	-2	Undef.	Undef.	$\frac{1}{2}$	$\frac{1}{3}$	$\frac{1}{4}$
$g(x)$	-1	-2	Undef.	2	$\frac{1}{2}$	$\frac{1}{3}$	$\frac{1}{4}$

(d) f and g differ only at $x = 0.5$, where f is undefined and g is defined.

41. $C = \frac{25{,}000p}{100 - p}$, $0 \le p < 100$

(a)

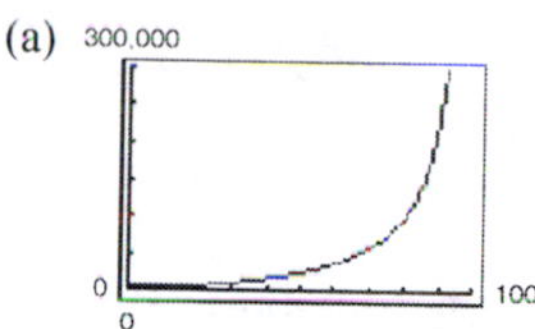

(b) $C = \frac{25{,}000(15)}{100 - 15} \approx \$4411.76$

$C = \frac{25{,}000(50)}{100 - 50} = \$25{,}000$

$C = \frac{25{,}000(90)}{100 - 90} = \$225{,}000$

(c) $C \to \infty$ as $x \to 100$. No, it would not be possible to supply bins to 100% of the residents because the model is undefined for $p = 100$.

43. $N = \frac{20(5 + 3t)}{1 + 0.04t}$, $t \ge 0$

(a)

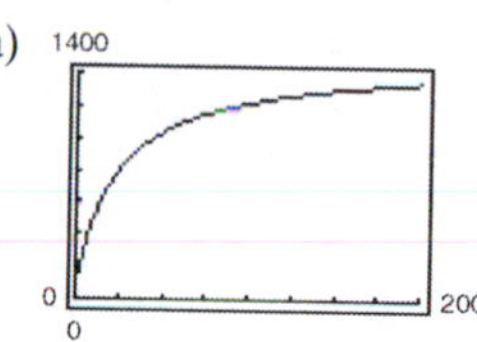

(b) $N(5) \approx 333$ deer

$N(10) = 500$ deer

$N(25) = 800$ deer

(c) The herd is limited by the horizontal asymptote:

$N = \frac{60}{0.04} = 1500$ deer

45. $P = \frac{0.5 + 0.9(n - 1)}{1 + 0.9(n - 1)}$, $n > 0$

(a)

n	1	2	3	4	5
P	0.50	0.74	0.82	0.86	0.89

n	6	7	8	9	10
P	0.91	0.92	0.93	0.94	0.95

P approaches 1 as n increases.

(b) $P = \frac{0.9n - 0.4}{0.9n + 0.1}$

The percentage of correct responses is limited by the horizontal asymptote:

$P = \frac{0.9}{0.9} = 1 = 100\%$

47. False. Polynomial functions do not have vertical asymptotes.

49. $f(x) = 4 - \frac{1}{x}$

(a) As $x \to \pm\infty$, $f(x) \to 4$.

(b) As $x \to \infty$, $f(x) \to 4$ but is less than 4.

(c) As $x \to -\infty$, $f(x) \to 4$ but is greater than 4.

51. $f(x) = \frac{2x - 1}{x - 3}$

(a) As $x \to \pm\infty$, $f(x) \to 2$.

(b) As $x \to \infty$, $f(x) \to 2$ but is greater than 2.

(c) As $x \to -\infty$, $f(x) \to 2$ but is less than 2.

53. Vertical asymptote: None $\Rightarrow$ The denominator is not zero for any value of x (unless the numerator is also zero there).

Horizontal asymptote: $y = 2 \Rightarrow$ The degree of the numerator equals the degree of the denominator.

$f(x) = \frac{2x^2}{x^2 + 1}$ is one possible function. There are many correct answers.

55. Domain: All real numbers

Example: $f(x) = \frac{1}{x^2 + 2}$

Domain: All real numbers except $x = 15$

Example: $f(x) = \frac{1}{x - 15}$

There are many correct answers.

Section 4.2 Graphs of Rational Functions

1. slant asymptote

3. $g(x) = \frac{2}{x} + 4$

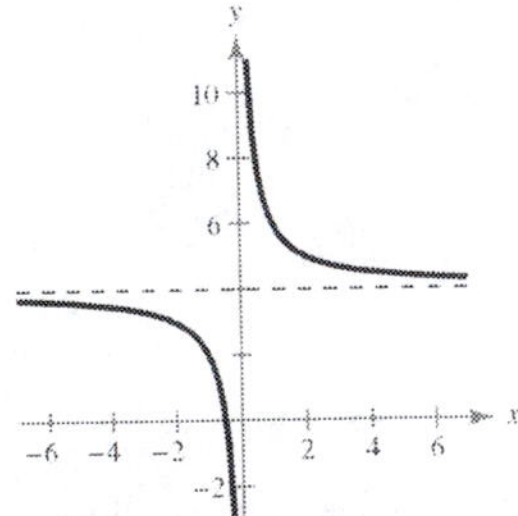

Vertical shift four units upward

5. $g(x) = -\frac{2}{x}$

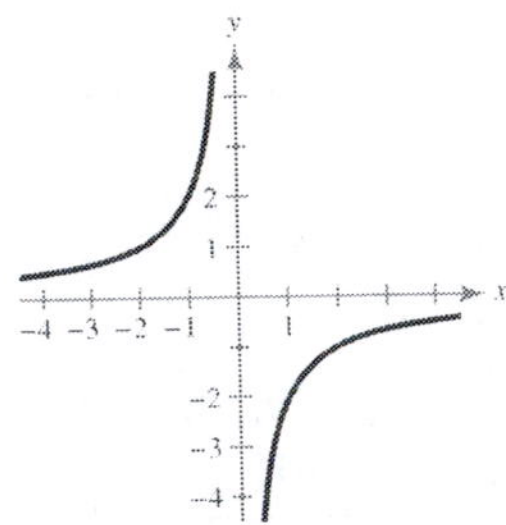

Reflection in the x-axis

7. $g(x) = \frac{3}{x^2} - 1$

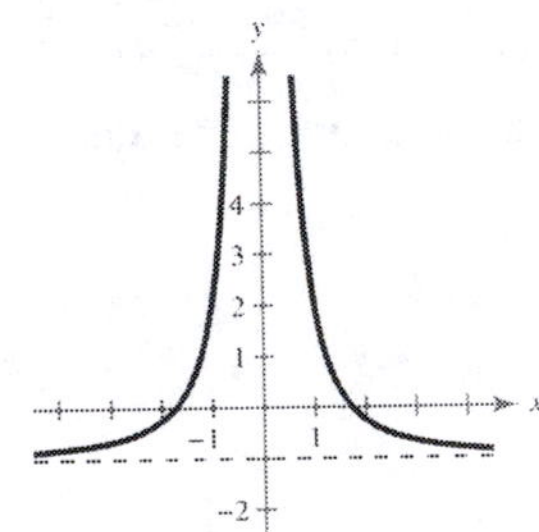

Vertical shift one unit downward

9. $g(x) = \frac{3}{(x - 1)^2}$

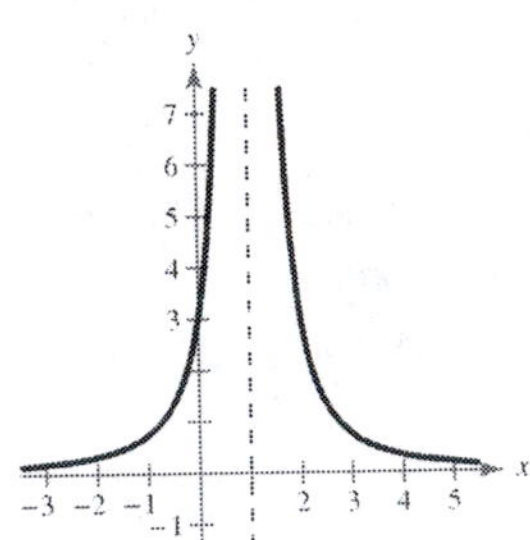

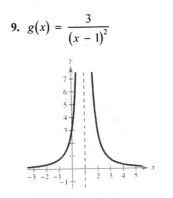

Horizontal shift one unit to the right

11. $g(x) = \dfrac{4}{(x+2)^3}$

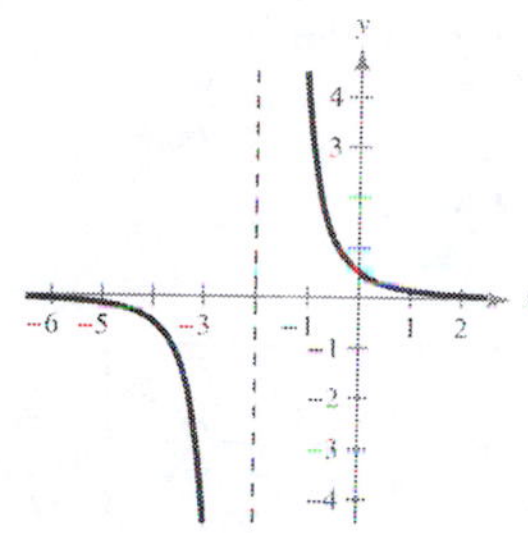

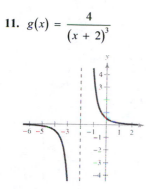

Horizontal shift two units to the left

13. $g(x) = -\dfrac{4}{x^3}$

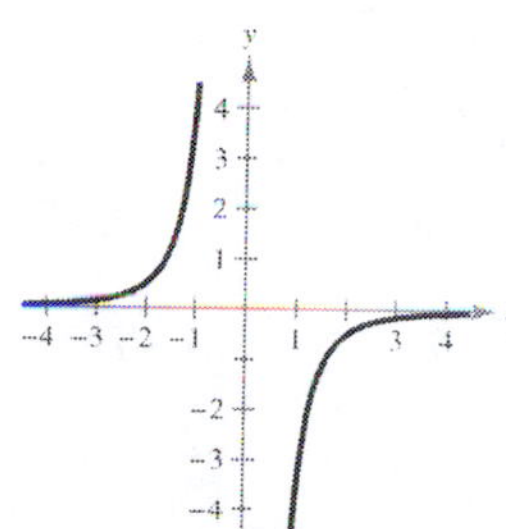

Reflection in the x-axis

15. $f(x) = \dfrac{1}{x+2}$

(a) Domain: all real numbers x except $x = -2$

(b) y-intercept: $\left(0, \dfrac{1}{2}\right)$

(c) Vertical asymptote: $x = -2$

Horizontal asymptote: $y = 0$

(d)

x	-4	-3	-1	0	1
$f(x)$	$-\frac{1}{2}$	-1	1	$\frac{1}{2}$	$\frac{1}{3}$

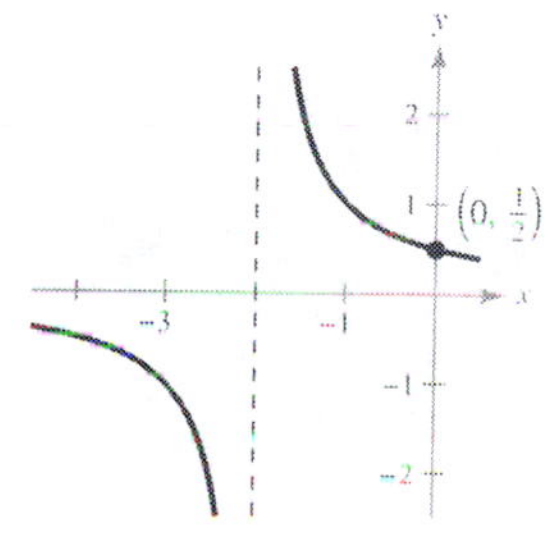

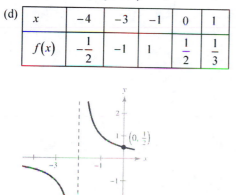

17. $h(x) = \dfrac{-1}{x+4}$

(a) Domain: all real numbers x except $x = -4$

(b) y-intercept: $\left(0, -\dfrac{1}{4}\right)$

(c) Vertical asymptote: $x = -4$

Horizontal asymptote: $y = 0$

(d)

x	-6	-5	-3	-2	-1	0
$h(x)$	$\frac{1}{2}$	1	-1	$-\frac{1}{2}$	$-\frac{1}{3}$	$-\frac{1}{4}$

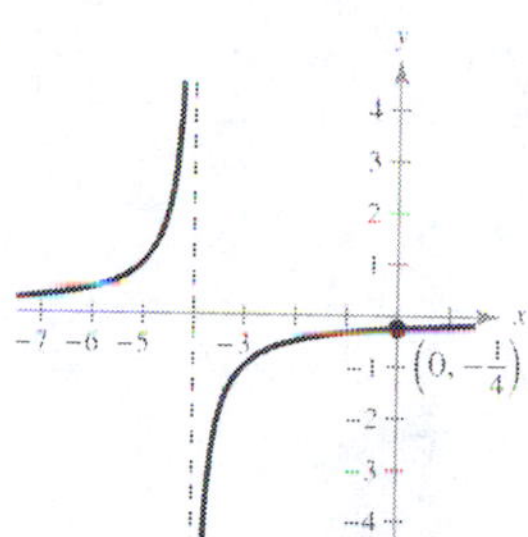

19. $C(x) = \dfrac{7+2x}{2+x}$

(a) Domain: all real numbers x except $x = -2$

(b) x-intercept: $\left(-\dfrac{7}{2}, 0\right)$

y-intercept: $\left(0, \dfrac{7}{2}\right)$

(c) Vertical asymptote: $x = -2$

Horizontal asymptote: $y = 2$

(d)

x	-4	-3	-1	0	1
$C(x)$	$\frac{1}{2}$	-1	5	$\frac{7}{2}$	3

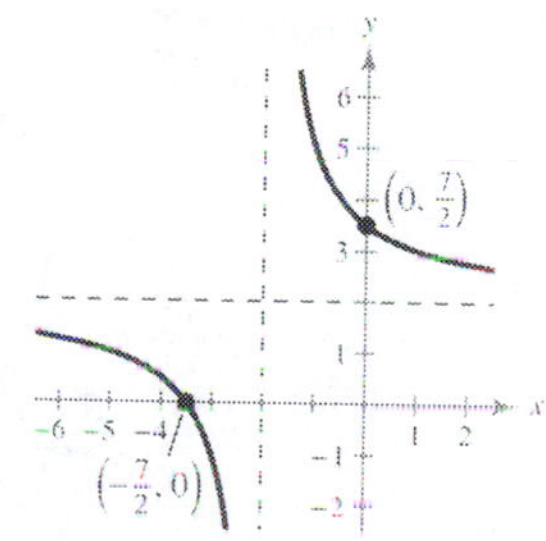

21. $g(x) = \dfrac{1}{x+2} + 2 = \dfrac{2x+5}{x+2}$

(a) Domain: all real numbers x except $x = -2$

(b) x-intercept: $\left(-\dfrac{5}{2}, 0\right)$

y-intercept: $\left(0, \dfrac{5}{2}\right)$

(c) Vertical asymptote: $x = -2$

Horizontal asymptote: $y = 2$

(d)

x	-4	-3	-1	0	1
$g(x)$	$\frac{3}{2}$	1	3	$\frac{5}{2}$	$\frac{7}{3}$

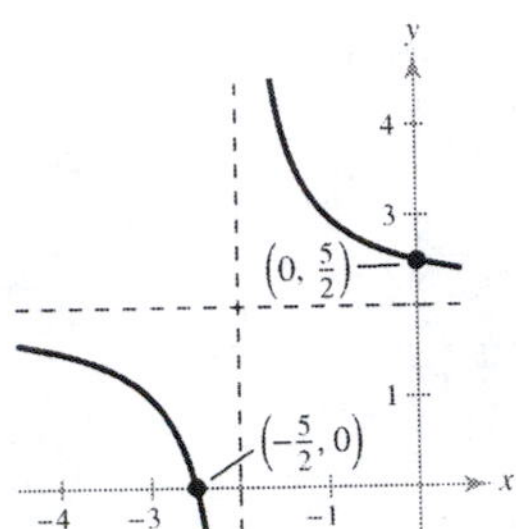

23. $f(x) = \dfrac{x^2}{x^2+9}$

(a) Domain: all real numbers x

(b) Intercept: $(0, 0)$

(c) Horizontal asymptote: $y = 1$

(d)

x	± 1	± 2	± 3
$f(x)$	$\frac{1}{10}$	$\frac{4}{13}$	$\frac{1}{2}$

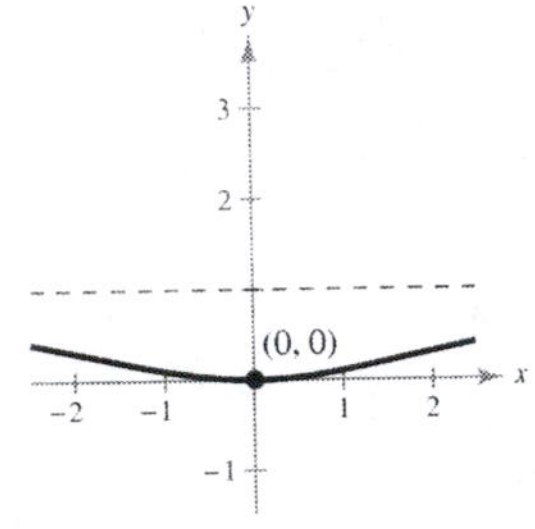

25. $h(x) = \dfrac{x^2}{x^2-9}$

(a) Domain: all real numbers x except $x = \pm 3$

(b) Intercept: $(0, 0)$

(c) Vertical asymptotes: $x = \pm 3$

Horizontal asymptote: $y = 1$

(d)

x	± 5	± 4	± 2	± 1	0
$h(x)$	$\frac{25}{16}$	$\frac{16}{7}$	$-\frac{4}{5}$	$-\frac{1}{8}$	0

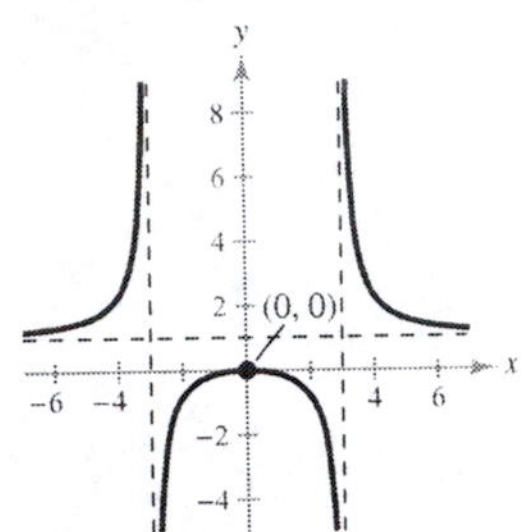

27. $g(s) = \dfrac{4s}{s^2+4}$

(a) Domain: all real numbers s

(b) Intercept: $(0, 0)$

(c) Horizontal asymptote: $y = 0$

(d)

s	-2	-1	0	1	2
$g(s)$	-1	$-\frac{4}{5}$	0	$\frac{4}{5}$	1

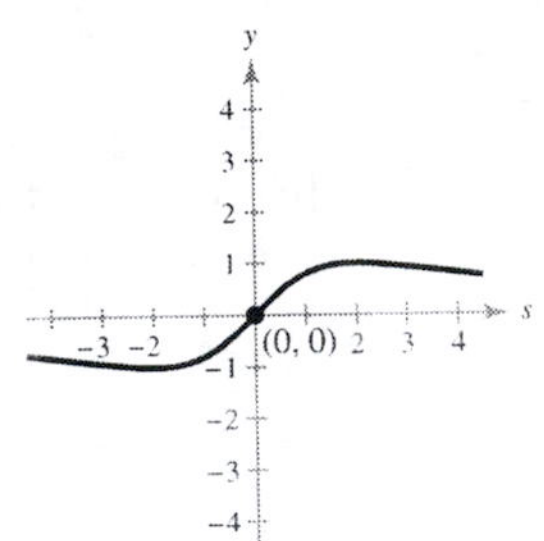

29. $g(x) = \dfrac{4(x + 1)}{x(x - 4)}$

(a) Domain: all real numbers x except $x = 0$ and $x = 4$

(b) x-intercept: $(-1, 0)$

(c) Vertical asymptotes: $x = 0, x = 4$

Horizontal asymptote: $y = 0$

(d)

x	-2	-1	1	2	3	5	6
$g(x)$	$-\frac{1}{3}$	0	$-\frac{8}{3}$	-3	$-\frac{16}{3}$	$\frac{24}{5}$	$\frac{7}{3}$

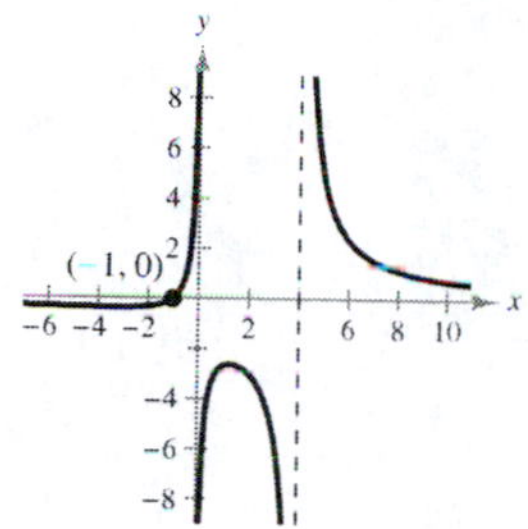

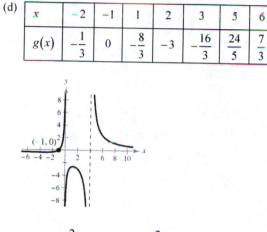

31. $f(x) = \dfrac{2x}{x^2 - 3x - 4} = \dfrac{2x}{(x - 4)(x + 1)}$

(a) Domain: all real numbers x except $x = 4$ and $x = -1$

(b) Intercept: $(0, 0)$

(c) Vertical asymptotes: $x = 4, x = -1$

Horizontal asymptote: $y = 0$

(d)

x	-3	-2	0	1	2	3	5
$f(x)$	$-\frac{3}{7}$	$-\frac{2}{3}$	0	$-\frac{1}{3}$	$-\frac{2}{3}$	$-\frac{3}{2}$	$\frac{5}{3}$

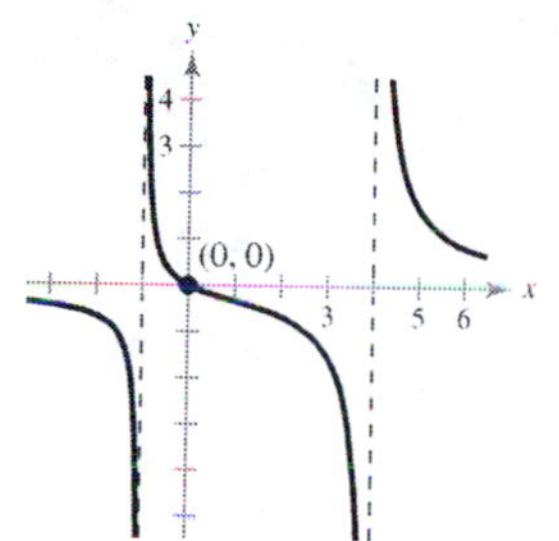

33. $h(x) = \dfrac{x^2 - 5x + 4}{x^2 - 4} = \dfrac{(x - 1)(x - 4)}{(x + 2)(x - 2)}$

(a) Domain: all real numbers x except $x = \pm 2$

(b) x-intercepts: $(1, 0), (4, 0)$

y-intercept: $(0, -1)$

(c) Vertical asymptotes: $x = -2, x = 2$

Horizontal asymptote: $y = 1$

(d)

x	-4	-3	-1	0	1	3	4
$h(x)$	$\frac{10}{3}$	$\frac{28}{5}$	$-\frac{10}{3}$	-1	0	$-\frac{2}{5}$	0

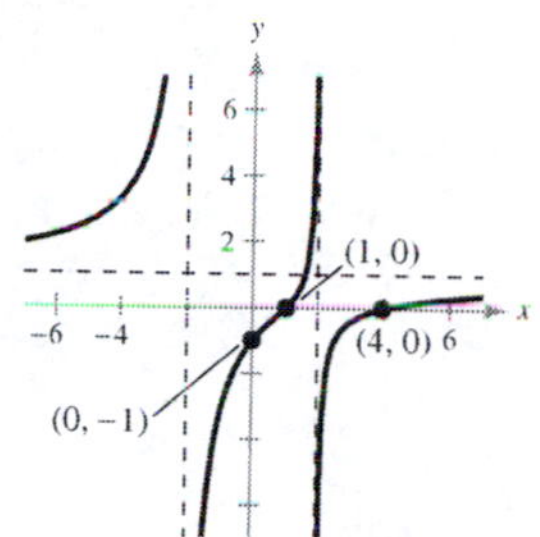

35. $f(x) = \dfrac{6x}{x^2 - 5x - 14} = \dfrac{6x}{(x + 2)(x - 7)}$

(a) Domain: all real numbers x except $x = -2$ and $x = 7$

(b) Intercept: $(0, 0)$

(c) Vertical asymptotes: $x = -2, x = 7$

Horizontal asymptote: $y = 0$

(d)

x	-6	-4	0	2	4	6	8	10
$f(x)$	$-\frac{9}{13}$	$-\frac{12}{11}$	0	$-\frac{3}{5}$	$-\frac{4}{3}$	$-\frac{9}{2}$	$\frac{24}{5}$	$\frac{5}{3}$

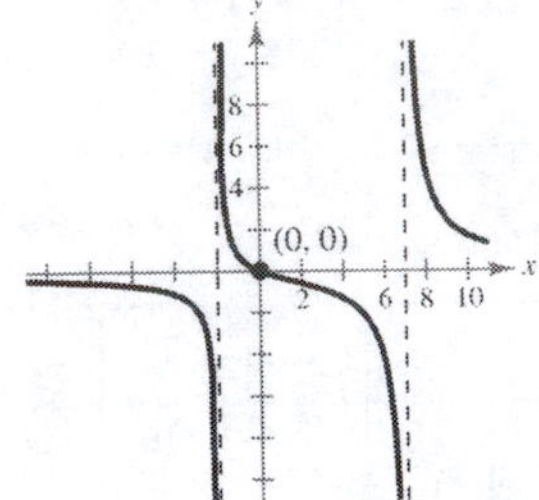

37. $f(x) = \dfrac{2x^2 - 5x - 3}{x^3 - 2x^2 - x + 2} = \dfrac{(2x + 1)(x - 3)}{(x - 2)(x + 1)(x - 1)}$

(a) Domain: all real numbers x except $x = 2, x = \pm 1$

(b) x-intercepts: $\left(-\frac{1}{2}, 0\right), (3, 0)$

y-intercept: $\left(0, -\frac{3}{2}\right)$

(c) Vertical asymptotes: $x = 2, x = -1$, and $x = 1$

Horizontal asymptote: $y = 0$

(d)

x	-3	-2	0	$\frac{3}{2}$	3	4
$f(x)$	$-\frac{3}{4}$	$-\frac{5}{4}$	$-\frac{3}{2}$	$\frac{48}{5}$	0	$\frac{3}{10}$

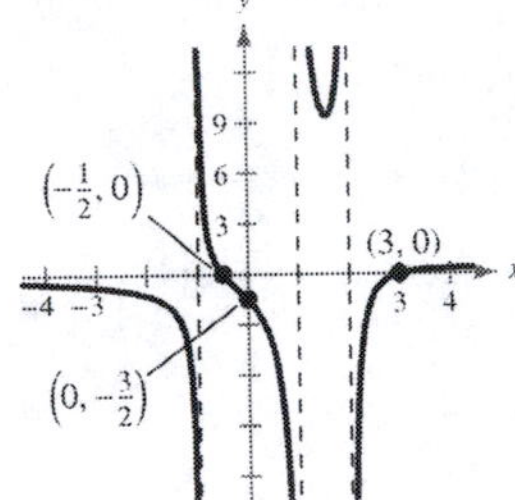

39. $f(x) = \dfrac{x^2 + 3x}{x^2 + x - 6} = \dfrac{x(x + 3)}{(x + 3)(x - 2)} = \dfrac{x}{x - 2}, \quad x \neq -3$

(a) Domain: all real numbers x except $x = -3$ and $x = 2$

(b) Intercept: $(0, 0)$

(c) Vertical asymptote: $x = 2$

Horizontal asymptote: $y = 1$

(d)

x	-1	0	1	3	4
$f(x)$	$\frac{1}{3}$	0	-1	3	2

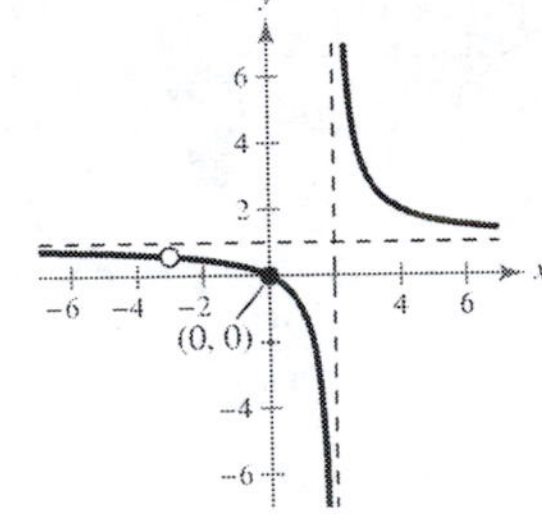

41. $f(x) = \dfrac{2x^2 - 5x + 2}{2x^2 - x - 6} = \dfrac{(2x - 1)(x - 2)}{(2x + 3)(x - 2)} = \dfrac{2x - 1}{2x + 3}, \quad x \neq 2$

(a) Domain: all real numbers x except $x = 2$ and $x = -\dfrac{3}{2}$

(b) x-intercept: $\left(\dfrac{1}{2}, 0\right)$

y-intercept: $\left(0, -\dfrac{1}{3}\right)$

(c) Vertical asymptote: $x = -\dfrac{3}{2}$

Horizontal asymptote: $y = 1$

(d)

x	-3	-2	-1	0	1
$f(x)$	$\frac{7}{3}$	5	-3	$-\frac{1}{3}$	$\frac{1}{5}$

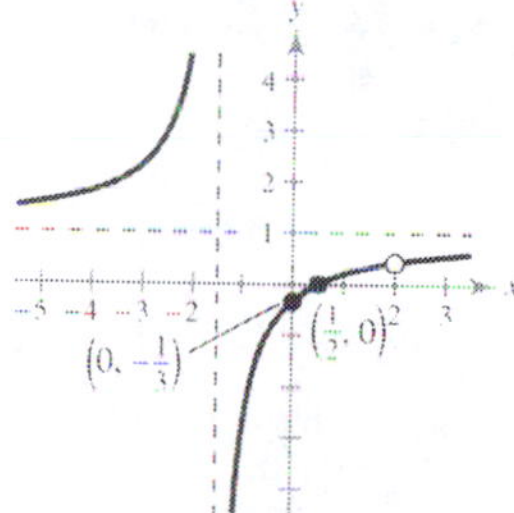

43. $f(t) = \dfrac{t^2 - 1}{t - 1} = \dfrac{(t + 1)(t - 1)}{t - 1} = t + 1, \quad t \neq 1$

(a) Domain: all real numbers t except $t = 1$

(b) t-intercept: $(-1, 0)$

y-intercept: $(0, 1)$

(c) No asymptotes

(d)

t	-3	-2	-1	0	1	2
$f(t)$	-2	-1	0	1	Undef.	3

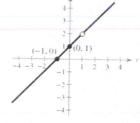

45. (a) Domain of f: all real numbers x except $x = -1$

Domain of g: all real numbers x

(b)

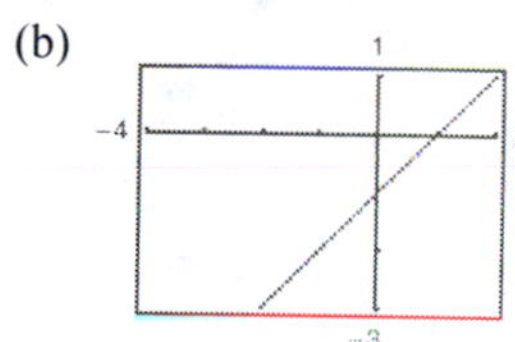

(c) Because there are only finitely many pixels, the graphing utility may not attempt to evaluate the function where it does not exist.

47. $h(x) = \dfrac{x^2 - 9}{x} = x - \dfrac{9}{x}$

(a) Domain: all real numbers x except $x = 0$

(b) x-intercepts: $(-3, 0), (3, 0)$

(c) Vertical asymptote: $x = 0$

Slant asymptote: $y = x$

(d)

x	-6	-4	-3	-2	2	3	4	6
$h(x)$	$-\frac{9}{2}$	$-\frac{7}{4}$	0	$\frac{5}{2}$	$-\frac{5}{2}$	0	$\frac{7}{4}$	$\frac{9}{2}$

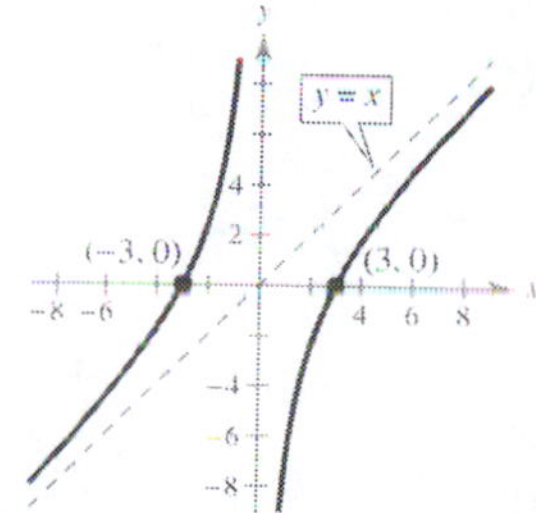

49. $f(x) = \dfrac{2x^2 + 1}{x} = 2x + \dfrac{1}{x}$

(a) Domain: all real numbers x except $x = 0$

(b) No intercepts

(c) Vertical asymptote: $x = 0$

Slant asymptote: $y = 2x$

(d)

x	-4	-2	2	4	6
$f(x)$	$-\frac{33}{4}$	$-\frac{9}{2}$	$\frac{9}{2}$	$\frac{33}{4}$	$\frac{73}{6}$

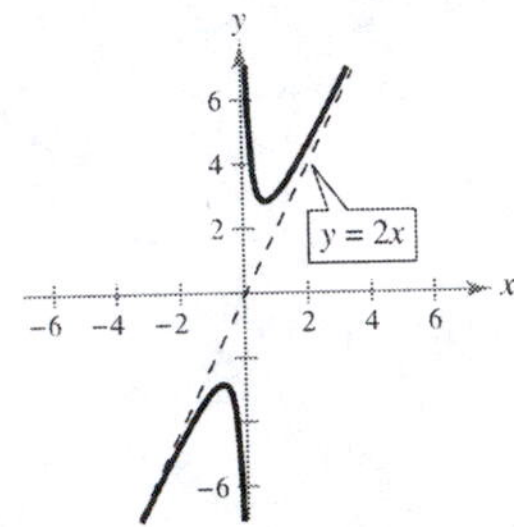

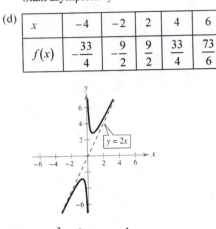

51. $g(x) = \dfrac{x^2 + 1}{x} = x + \dfrac{1}{x}$

(a) Domain: all real numbers x except $x = 0$

(b) No intercepts

(c) Vertical asymptote: $x = 0$

Slant asymptote: $y = x$

(d)

x	-4	-2	2	4	6
$g(x)$	$-\frac{17}{4}$	$-\frac{5}{2}$	$\frac{5}{2}$	$\frac{17}{4}$	$\frac{37}{6}$

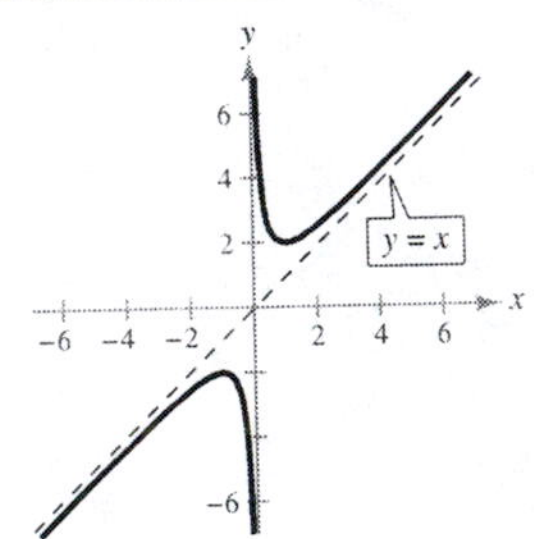

53. $f(t) = \dfrac{t^2 + 1}{t + 5} = -t + 5 - \dfrac{26}{t + 5}$

(a) Domain: all real numbers t except $t = -5$

(b) Intercept: $\left(0, -\frac{1}{5}\right)$

(c) Vertical asymptote: $t = -5$

Slant asymptote: $y = -t + 5$

(d)

t	-7	-6	-4	-3	0
$f(t)$	25	37	-17	-5	$-\frac{1}{5}$

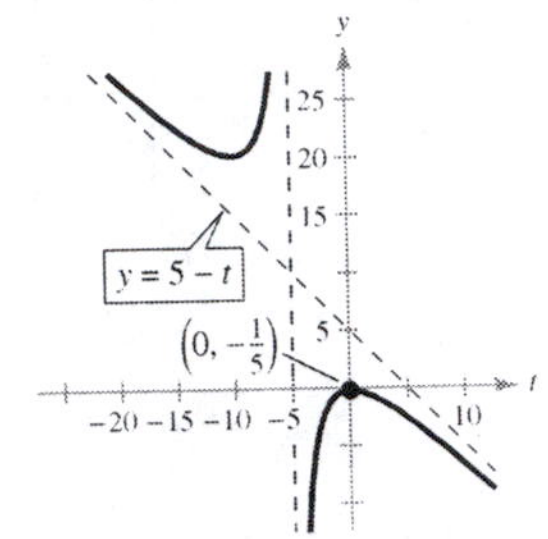

55. $f(x) = \dfrac{x^3}{x^2 - 4} = x + \dfrac{4x}{x^2 - 4}$

(a) Domain: all real numbers x except $x = \pm 2$

(b) Intercept: $(0, 0)$

(c) Vertical asymptotes: $x = \pm 2$

Slant asymptote: $y = x$

(d)

x	-6	-4	-1	0	1	4	6
$f(x)$	$-\frac{27}{4}$	$-\frac{16}{3}$	$\frac{1}{3}$	0	$-\frac{1}{3}$	$\frac{16}{3}$	$\frac{27}{4}$

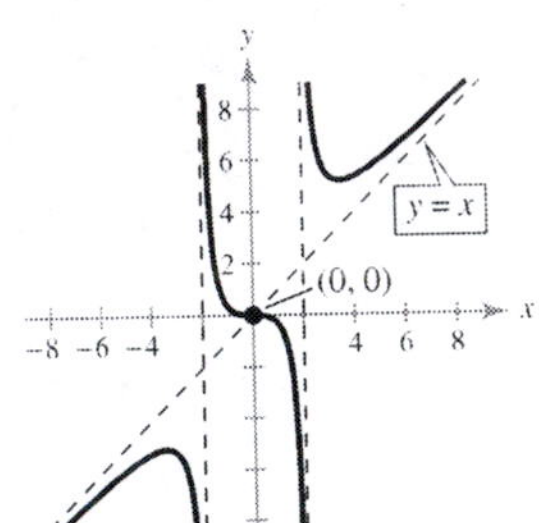

57. $f(x) = \dfrac{x^3 - 1}{x^2 - x} = \dfrac{(x-1)(x^2+x+1)}{x(x-1)} = \dfrac{x^2+x+1}{x}$

$= x + 1 + \dfrac{1}{x}, \quad x \neq 1$

(a) Domain: all real numbers x except $x = 0$ and $x = 1$

(b) Intercepts: none

(c) Vertical asymptote: $x = 0$

Note: $x = 1$ is not a vertical asymptote since it also makes the numerator zero.

Slant asymptote: $y = x + 1$

(d)

x	-2	-1	$\frac{1}{2}$	2
$f(x)$	$-\frac{3}{2}$	-1	$\frac{7}{2}$	$\frac{7}{2}$

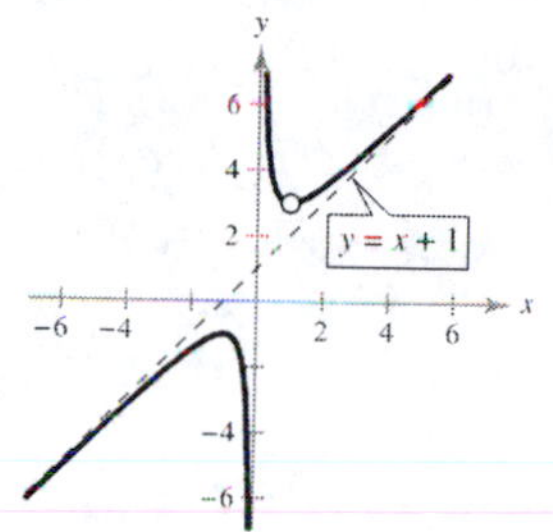

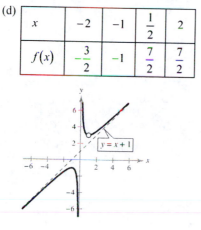

59. $f(x) = \dfrac{x^2 - x + 1}{x - 1} = x + \dfrac{1}{x-1}$

(a) Domain: all real numbers x except $x = 1$

(b) y-intercept: $(0, -1)$

(c) Vertical asymptote: $x = 1$

Slant asymptote: $y = x$

(d)

x	-4	-2	0	2	4
$f(x)$	$-\frac{21}{5}$	$-\frac{7}{3}$	-1	3	$\frac{13}{3}$

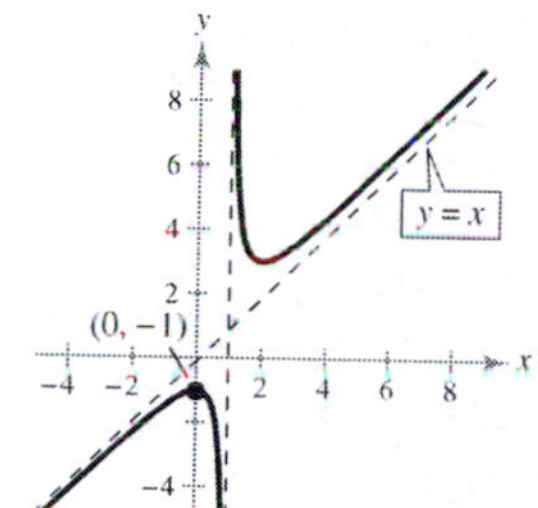

61. $f(x) = \dfrac{2x^3 - x^2 - 2x + 1}{x^2 + 3x + 2} = \dfrac{(2x-1)(x+1)(x-1)}{(x+1)(x+2)} = \dfrac{(2x-1)(x-1)}{x+2}, \quad x \neq -1$

$= \dfrac{2x^2 - 3x + 1}{x + 2} = 2x - 7 + \dfrac{15}{x+2}, \quad x \neq -1$

(a) Domain: all real numbers x except $x = -1$ and $x = -2$

(b) y-intercept: $\left(0, \dfrac{1}{2}\right)$

x-intercepts: $\left(\dfrac{1}{2}, 0\right)$, $(1, 0)$

(c) Vertical asymptote: $x = -2$

Slant asymptote: $y = 2x - 7$

(d)

x	-4	-3	$-\frac{3}{2}$	0	1
$f(x)$	$-\frac{45}{2}$	-28	20	$\frac{1}{2}$	0

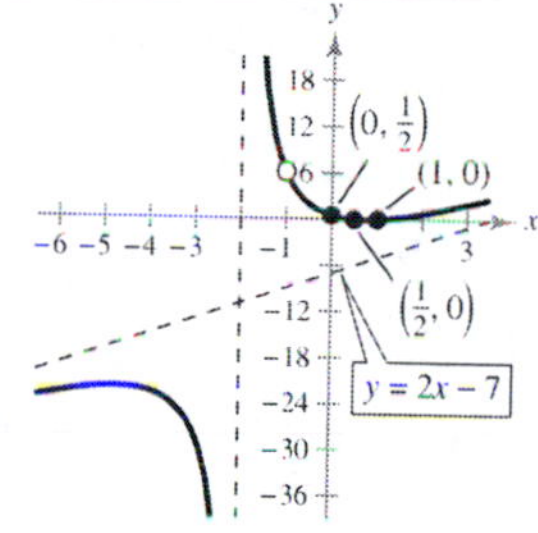

63. $f(x) = \dfrac{x^2 + 5x + 8}{x + 3} = x + 2 + \dfrac{2}{x + 3}$

Domain: all real numbers x except $x = -3$

y-intercept: $\left(0, \dfrac{8}{3}\right)$

Vertical asymptote: $x = -3$

Slant asymptote: $y = x + 2$

Line: $y = x + 2$

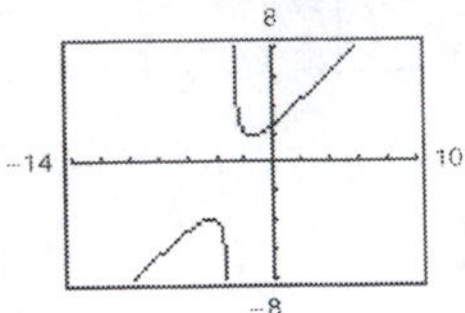

65. $g(x) = \dfrac{1 + 3x^2 - x^3}{x^2} = \dfrac{1}{x^2} + 3 - x = -x + 3 + \dfrac{1}{x^2}$

Domain: all real numbers x except $x = 0$

Vertical asymptote: $x = 0$

Slant asymptote: $y = -x + 3$

Line: $y = -x + 3$

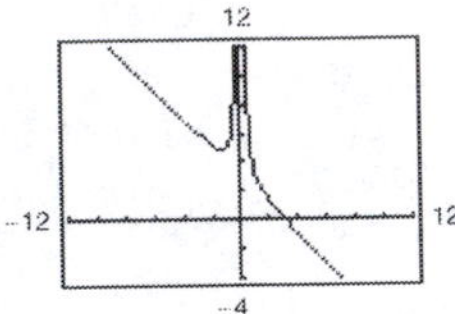

67. $y = \dfrac{x + 1}{x - 3}$

(a) x-intercept: $(-1, 0)$

(b) $0 = \dfrac{x + 1}{x - 3}$

$0 = x + 1$

$-1 = x$

69. $y = \dfrac{1}{x} - x$

(a) x-intercepts: $(-1, 0), (1, 0)$

(b) $0 = \dfrac{1}{x} - x$

$x = \dfrac{1}{x}$

$x^2 = 1$

$x = \pm 1$

71. $y = \dfrac{1}{x + 5} + \dfrac{4}{x}$

(a)

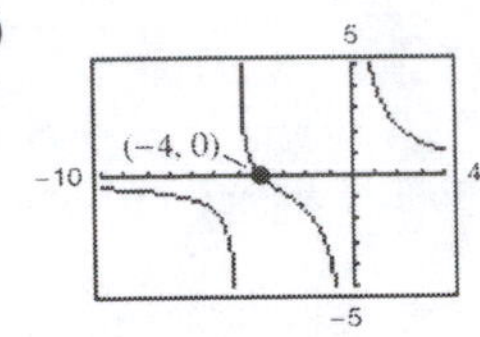

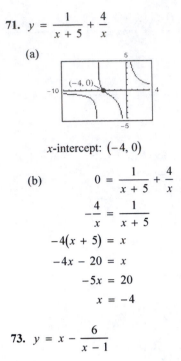

x-intercept: $(-4, 0)$

(b) $0 = \dfrac{1}{x + 5} + \dfrac{4}{x}$

$-\dfrac{4}{x} = \dfrac{1}{x + 5}$

$-4(x + 5) = x$

$-4x - 20 = x$

$-5x = 20$

$x = -4$

73. $y = x - \dfrac{6}{x - 1}$

(a)

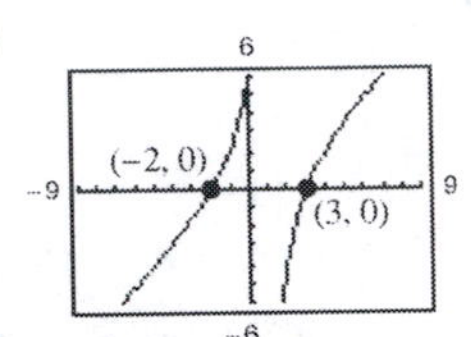

x-intercepts: $(-2, 0), (3, 0)$

(b) $0 = x - \dfrac{6}{x - 1}$

$\dfrac{6}{x - 1} = x$

$6 = x(x - 1)$

$0 = x^2 - x - 6$

$0 = (x + 2)(x - 3)$

$x = -2, \ x = 3$

75. (a) $0.25(50) + 0.75(x) = C(50 + x)$

$$C = \frac{12.50 + 0.75x}{50 + x} \cdot \frac{4}{4} = \frac{50 + 3x}{4(50 + x)} = \frac{3x + 50}{4(x + 50)}$$

(b) Domain: $x \geq 0$ and $x \leq 1000 - 50$

Thus, $0 \leq x \leq 950$. Using interval notation, the domain is $[0, 950]$.

(c)

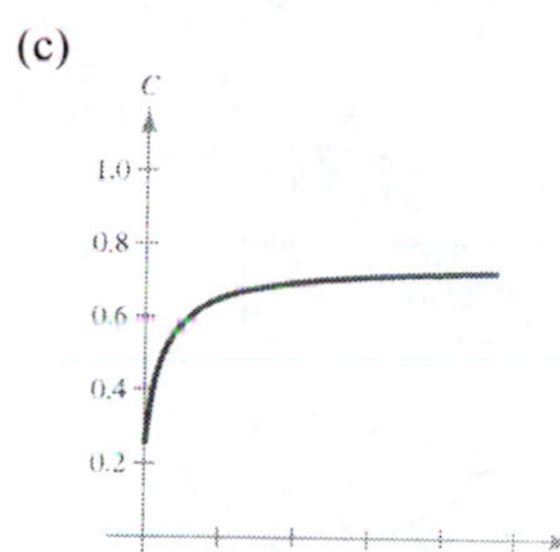

(d) As the tank is filled, the concentration increases more slowly. It approaches the horizontal asymptote of $C = \frac{3}{4} = 0.75$.

77. (a) $A = xy$ and

$$(x - 4)(y - 2) = 30$$

$$y - 2 = \frac{30}{x - 4}$$

$$y = 2 + \frac{30}{x - 4} = \frac{2x + 22}{x - 4}$$

Thus, $A = xy = x\left(\frac{2x + 22}{x - 4}\right) = \frac{2x(x + 11)}{x - 4}$.

(b) Domain: Since the margins on the left and right are each 2 inches, $x > 4$. In interval notation, the domain is $(4, \infty)$.

(c)

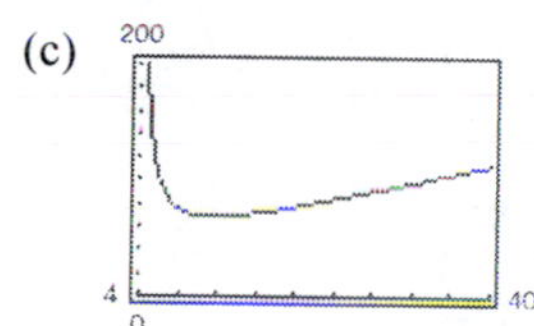

The area is minimum when $x \approx 11.75$ inches and $y \approx 5.87$ inches.

x	5	6	7	8	9	10	11	12	13	14	15
y_1 (Area)	160	102	84	76	72	70	69.143	69	69.333	70	70.909

The area is minimum when x is approximately 12.

79. $f(x) = \frac{3(x + 1)}{x^2 + x + 1}$

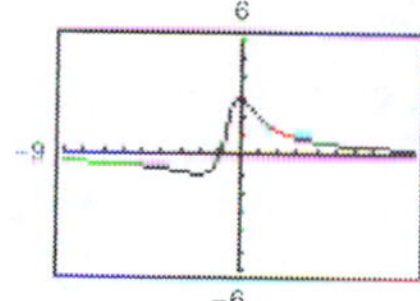

Relative minimum: $(-2, -1)$

Relative maximum: $(0, 3)$

81. $C = 100\left(\frac{200}{x^2} + \frac{x}{x + 30}\right), x \geq 1$

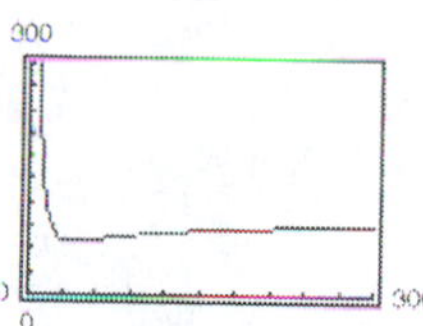

The minimum occurs when $x \approx 40.45$, or 4045 components.

83. (a) Let t_1 = time from Akron to Columbus and t_2 = time from Columbus back to Akron.

$$xt_1 = 100 \Rightarrow t_1 = \frac{100}{x}$$

$$yt_2 = 100 \Rightarrow t_2 = \frac{100}{y}$$

$$50(t_1 + t_2) = 200$$
$$t_1 + t_2 = 4$$
$$\frac{100}{x} + \frac{100}{y} = 4$$
$$100y + 100x = 4xy$$
$$25y + 25x = xy$$
$$25x = xy - 25y$$
$$25x = y(x - 25)$$

Thus, $y = \dfrac{25x}{x - 25}$.

(b) Vertical asymptote: $x = 25$

Horizontal asymptote: $y = 25$

(c)

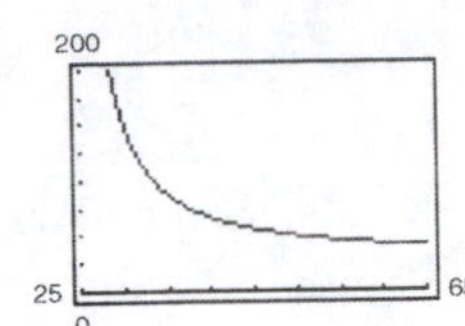

(d)

x	30	35	40	45	50	55	60
y	150	87.5	66.7	56.3	50	45.8	42.9

(e) Sample answer: No. You might expect the average speed for the round trip to be the average of the average speeds for the two parts of the trip.

(f) No. At 20 miles per hour you would use more time in one direction than is required for the round trip at an average speed of 50 miles per hour.

85. False. There are two distinct branches of the graph.

87. False. The degree of the numerator is two more than the degree of the denominator. To have a slant asymptote, it has to be exactly one more.

89. $h(x) = \dfrac{6 - 2x}{3 - x} = \dfrac{2(3 - x)}{3 - x}$

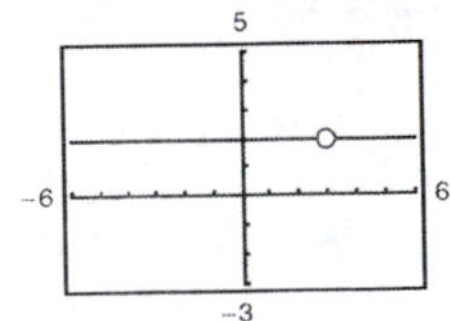

Since $h(x)$ is not reduced and $3 - x$ is a factor of both the numerator and the denominator, $x = 3$ is not a vertical asymptote.

91. (a) If the degree of the numerator is exactly one more than the degree of the denominator, then the graph of the function has a slant asymptote.

(b) To find the equation of a slant asymptote, use long division to expand the function.

93. No, given $f(x) = \dfrac{a_n x^n + \cdots + a_0}{b_m x^m + \cdots b_0}$, if $n > m$, there is no horizontal asymptote and n must be greater than m for a slant asymptote to occur.

Section 4.3 Conics

1. conic or conic section

3. parabola; directrix; focus

5. axis

7. major axis; center

9. hyperbola; foci

11. $x^2 = -2y$

Parabola opening downward

Matches (b).

12. $y^2 = 2x$

Parabola opening to the right

Matches (c).

13. $x^2 + 9y^2 = 9$

$$\frac{x^2}{9} + \frac{y^2}{1} = 1$$

Ellipse with horizontal major axis

Matches (f).

14. $9x^2 - y^2 = 9$

$$\frac{x^2}{1} - \frac{y^2}{9} = 1$$

Hyperbola with horizontal transverse axis

Matches (d).

15. $y^2 - 9x^2 = 9$

$$\frac{y^2}{9} - \frac{x^2}{1} = 1$$

Hyperbola with vertical transverse axis

Matches (a).

16. $x^2 + y^2 = 16$

Circle with radius 4

Matches (e).

17. $y = \frac{1}{2}x^2$

$x^2 = 2y = 4\left(\frac{1}{2}\right)y;\ p = \frac{1}{2}$

Focus: $\left(0, \frac{1}{2}\right)$

Directrix: $y = -\frac{1}{2}$

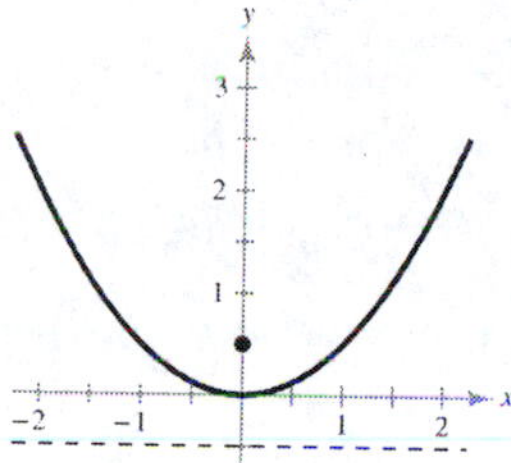

19. $y^2 = -6x$

$y^2 = 4\left(-\frac{3}{2}\right)x;\ p = -\frac{3}{2}$

Focus: $\left(-\frac{3}{2}, 0\right)$

Directrix: $x = \frac{3}{2}$

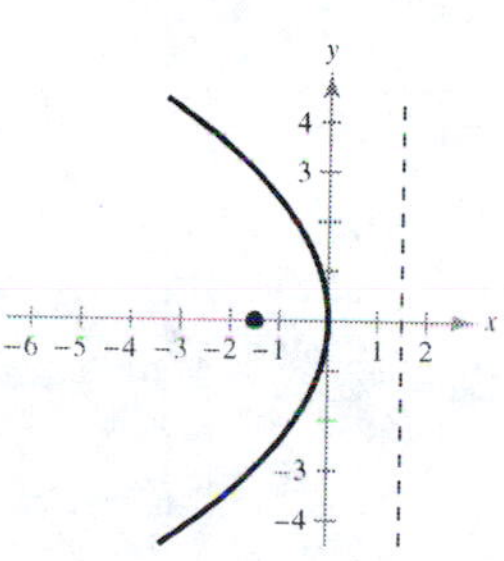

21. $x^2 + 12y = 0$

$x^2 = 4(-3)y;\ p = -3$

Focus: $(0, -3)$

Directrix: $y = 3$

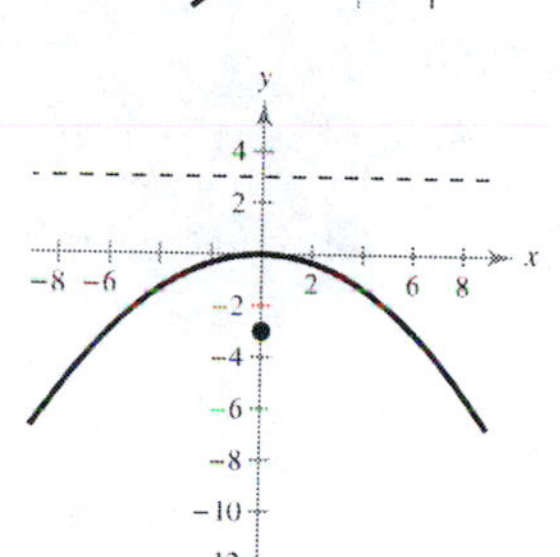

23. Focus: $(-2, 0)$

$$y^2 = 4(-2)x$$
$$y^2 = -8x$$

25. Directrix: $y = -2$

$$x^2 = 4(2)y$$
$$x^2 = 8y$$

27. $y^2 = 4px$

$$6^2 = 4p(4)$$
$$36 = 16p$$
$$p = \frac{9}{4}$$
$$y^2 = 4\left(\frac{9}{4}\right)x$$
$$y^2 = 9x$$

29. $x^2 = 4py$

$$3^2 = 4p(6)$$
$$9 = 24p$$
$$\frac{3}{8} = p$$
$$x^2 = 4\left(\frac{3}{8}\right)y$$
$$x^2 = \frac{3}{2}y$$

Focus: $\left(0, \frac{3}{8}\right)$

31. $y^2 = 4px,\ p = 1.5$

$$y^2 = 4(1.5)x$$
$$y^2 = 6x$$

33. (a)

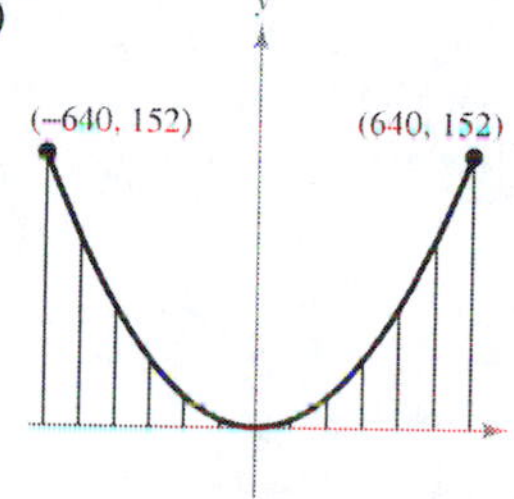

(b) $x^2 = 4py$

$$(640)^2 = 4p(152) \Rightarrow p = \frac{409{,}600}{608} = \frac{12{,}800}{19}$$
$$x^2 = 4\left(\frac{12{,}800}{19}\right)y$$
$$y = \frac{19x^2}{51{,}200}$$

(c)

Distance, x	Height, y
0	0
200	$14\frac{27}{32}$
400	$59\frac{3}{8}$
500	$92\frac{99}{128}$
600	$133\frac{19}{32}$

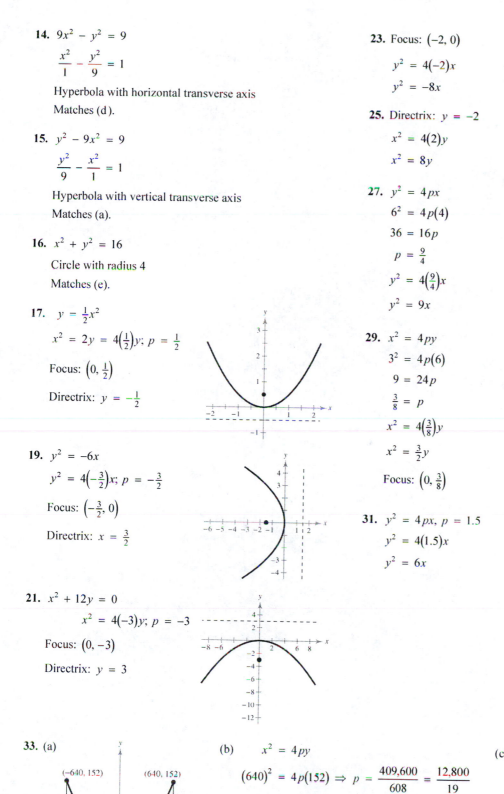

35. Vertices: $(0, \pm 2) \Rightarrow a = 2$

Minor axis of length $2 \Rightarrow b = 1$

Vertical major axis

$$\frac{x^2}{b^2} + \frac{y^2}{a^2} = 1$$

$$\frac{x^2}{1} + \frac{y^2}{4} = 1$$

37. Vertices: $(\pm 5, 0) \Rightarrow a = 5$

Foci: $(\pm 2, 0) \Rightarrow c = 2$

$b = \sqrt{5^2 - 2^2} = \sqrt{21}$

Horizontal major axis

$$\frac{x^2}{a^2} + \frac{y^2}{b^2} = 1$$

$$\frac{x^2}{25} + \frac{y^2}{21} = 1$$

39. Foci: $(\pm 5, 0) \Rightarrow c = 5$

Major axis of length $14 \Rightarrow a = 7$

$b = \sqrt{7^2 - 5^2} = \sqrt{24}$

Horizontal major axis

$$\frac{x^2}{a^2} + \frac{y^2}{b^2} = 1$$

$$\frac{x^2}{49} + \frac{y^2}{24} = 1$$

41. Vertices: $(\pm 9, 0) \Rightarrow a = 9$

Minor axis of length $6 \Rightarrow b = 3$

Horizontal major axis

$$\frac{x^2}{a^2} + \frac{y^2}{b^2} = 1$$

$$\frac{x^2}{9^2} + \frac{y^2}{3^2} = 1$$

$$\frac{x^2}{81} + \frac{y^2}{9} = 1$$

43. Vertices: $(0, \pm 5) \Rightarrow a = 5$

Vertical major axis

$$\frac{x^2}{b^2} + \frac{y^2}{25} = 1$$

Passes through $(4, 2)$

$$\frac{(4)^2}{b^2} + \frac{(2)^2}{25} = 1$$

$$\frac{16}{b^2} = \frac{21}{25}$$

$$\frac{400}{21} = b^2$$

$$\frac{21x^2}{400} + \frac{y^2}{25} = 1$$

45. $\dfrac{x^2}{25} + \dfrac{y^2}{16} = 1$

Horizontal major axis

$a = 5, b = 4$

Center: $(0, 0)$

Vertices: $(\pm 5, 0)$

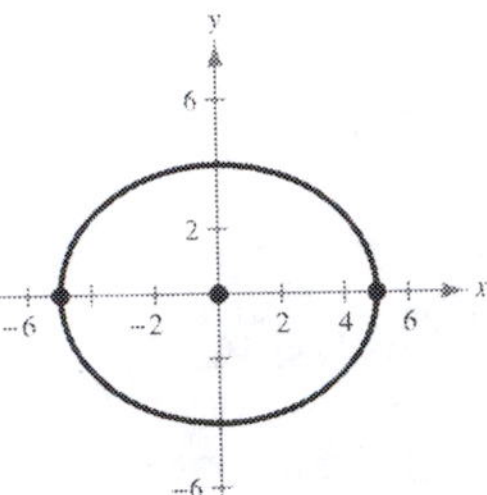

47. $\dfrac{x^2}{25/9} + \dfrac{y^2}{16/9} = 1$

Horizontal major axis

$a = \dfrac{5}{3}, b = \dfrac{4}{3}$

Center: $(0, 0)$

Vertices: $\left(\pm\dfrac{5}{3}, 0\right)$

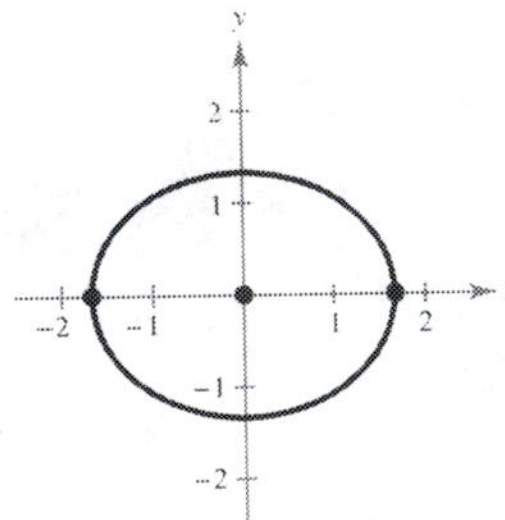

49. $\dfrac{x^2}{36} + \dfrac{y^2}{7} = 1$

Horizontal major axis

$a = 6, b = \sqrt{7}$

Center: $(0, 0)$

Vertices: $(\pm 6, 0)$

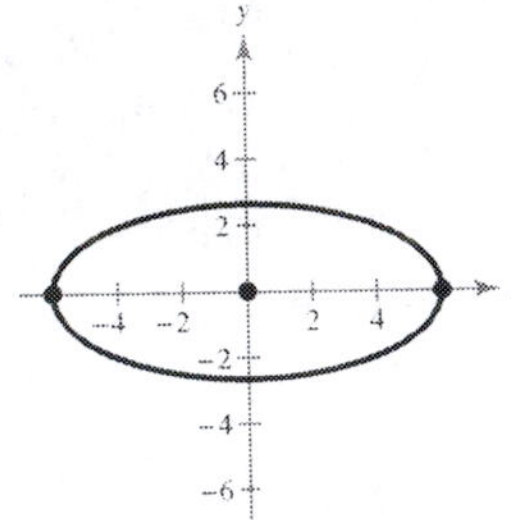

51. $4x^2 + y^2 = 1$

$$\frac{x^2}{1/4} + y^2 = 1$$

Vertical major axis

$a = 1, b = \dfrac{1}{2}$

Center: $(0, 0)$

Vertices: $(0, \pm 1)$

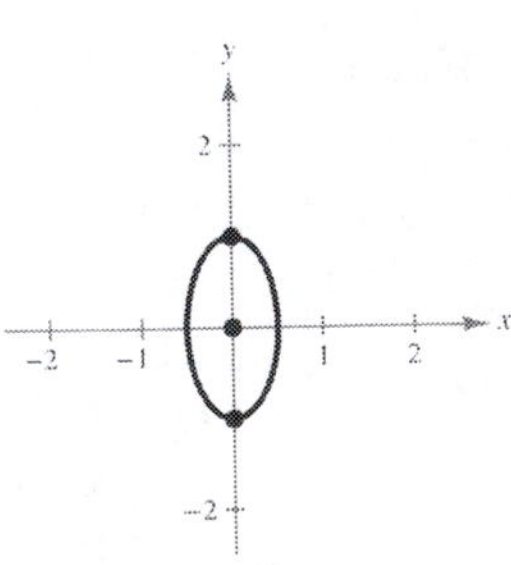

53. $\frac{1}{16}x^2 + \frac{1}{81}y^2 = 1$

$\frac{x^2}{16} + \frac{y^2}{81} = 1$

Vertical major axis

$a = 9, b = 4$

Center: $(0, 0)$

Vertices: $(0, \pm 9)$

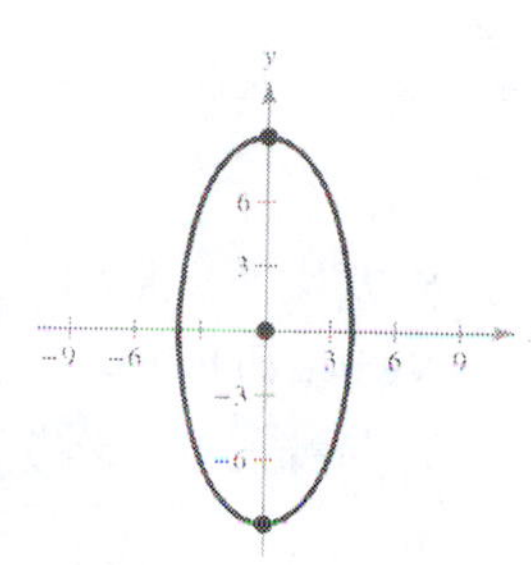

55. The length of string needed is $2a$ or $2(3) = 6$ feet. The positions of the tacks are $(\pm c, 0)$ where c is given by

$c^2 = a^2 - b^2$

$= 3^2 - 2^2 = 9 - 4 = 5 \Rightarrow c = \pm\sqrt{5}.$

Positions: $(\pm\sqrt{5}, 0)$

57. (a)

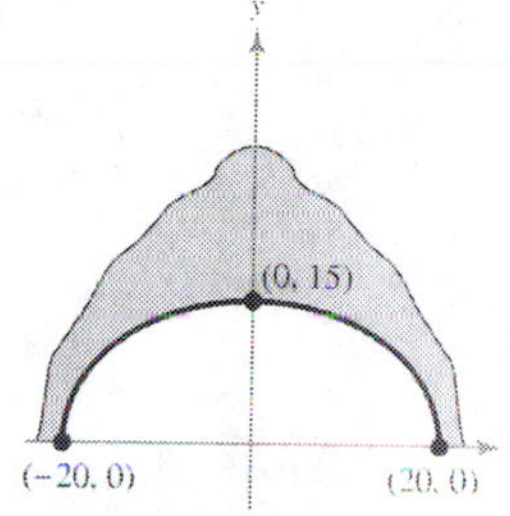

(b) Horizontal major axis

$a = 20, b = 15$

$\frac{x^2}{400} + \frac{y^2}{225} = 1 \Rightarrow y^2 = 225\left(1 - \frac{x^2}{400}\right)$

$y = \pm\sqrt{\frac{225}{400}(400 - x^2)}$

$y = \frac{3}{4}\sqrt{400 - x^2}$ Top half of ellipse

(c) When $x = 5$:

$y = \frac{3}{4}\sqrt{400 - 5^2}$

$y \approx 14.52$ feet

Yes, the truck can clear the tunnel with 0.52 foot clearance.

59. $\frac{x^2}{4} + \frac{y^2}{1} = 1$

$a = 2, b = 1, c = \sqrt{3}$

Points on the ellipse: $(\pm 2, 0), (0, \pm 1)$

Length of each latus rectum: $\frac{2b^2}{a} = \frac{2(1)}{2} = 1$

Additional points: $\left(\sqrt{3}, \pm\frac{1}{2}\right), \left(-\sqrt{3}, \pm\frac{1}{2}\right)$

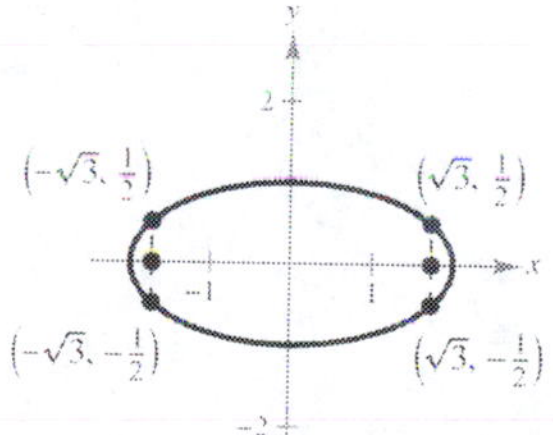

61. $9x^2 + 4y^2 = 36$

$\frac{x^2}{4} + \frac{y^2}{9} = 1$

$a = 3, b = 2, c = \sqrt{5}$

Points on the ellipse: $(\pm 2, 0), (0, \pm 3)$

Length of each latus rectum: $\frac{2b^2}{a} = \frac{2(2)^2}{3} = \frac{8}{3}$

Additional points: $\left(\pm\frac{4}{3}, -\sqrt{5}\right), \left(\pm\frac{4}{3}, \sqrt{5}\right)$

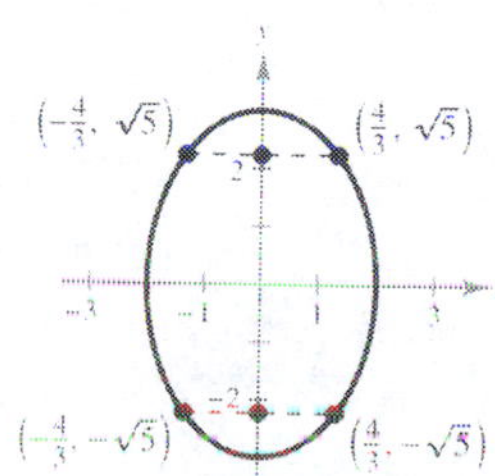

63. Vertices: $(0, \pm 2) \Rightarrow a = 2$

Foci: $(0, \pm 6) \Rightarrow c = 6$

Vertical transverse axis

$$b^2 = c^2 - a^2 = 32$$

$$\frac{y^2}{a^2} - \frac{x^2}{b^2} = 1$$

$$\frac{y^2}{4} - \frac{x^2}{32} = 1$$

65. Vertices: $(\pm 1, 0) \Rightarrow a = 1$

Asymptotes: $y = \pm 3x$

Horizontal transverse axis

$$3 = \frac{b}{a} = \frac{b}{1} \Rightarrow b = 3$$

$$\frac{x^2}{a^2} - \frac{y^2}{b^2} = 1$$

$$\frac{x^2}{1} - \frac{y^2}{9} = 1$$

67. Foci: $(0, \pm 8) \Rightarrow c = 8$

Asymptotes: $y = \pm 4x$

Vertical transverse axis

$$4 = \frac{a}{b} \Rightarrow a = 4b$$

$$a^2 + b^2 = c^2 \Rightarrow 16b^2 + b^2 = (8)^2$$

$$b^2 = \frac{64}{17} \Rightarrow a^2 = \frac{1024}{17}$$

$$\frac{y^2}{a^2} - \frac{x^2}{b^2} = 1$$

$$\frac{y^2}{1024/17} - \frac{x^2}{64/17} = 1$$

$$\frac{17y^2}{1024} - \frac{17x^2}{64} = 1$$

69. Vertices: $(0, \pm 3) \Rightarrow a = 3$

Vertical transverse axis

$$\frac{y^2}{9} - \frac{x^2}{b^2} = 1$$

Point on the graph: $(-2, 5)$

$$\frac{5^2}{9} - \frac{(-2)^2}{b^2} = 1$$

$$b^2 = \frac{9}{4}$$

$$\frac{y^2}{9} - \frac{x^2}{9/4} = 1$$

71. $x^2 - y^2 = 1$

$a = 1, b = 1$

Center: $(0, 0)$

Vertices: $(\pm 1, 0)$

Asymptotes: $y = \pm x$

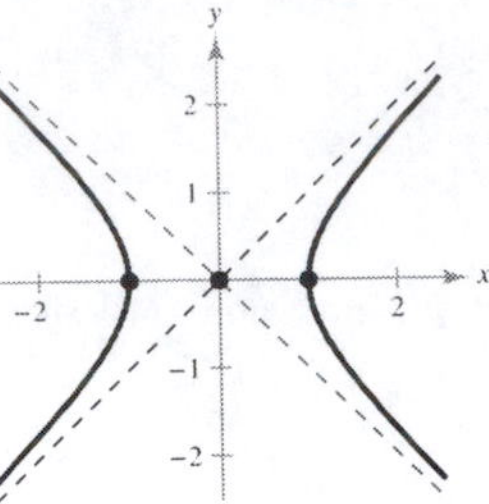

73. $\dfrac{y^2}{1} - \dfrac{x^2}{4} = 1$

$a = 1, b = 2$

Center: $(0, 0)$

Vertices: $(0, \pm 1)$

Asymptotes: $y = \pm \frac{1}{2}x$

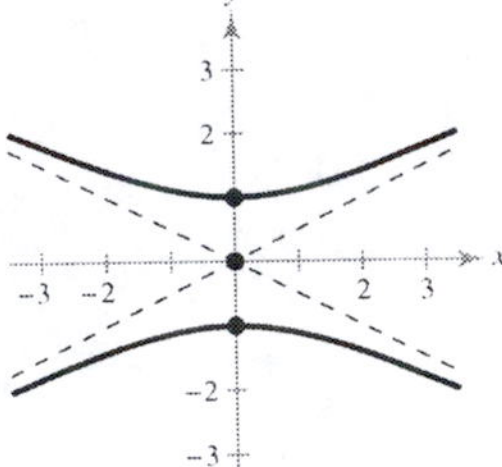

75. $\dfrac{y^2}{49} - \dfrac{x^2}{196} = 1$

$a = 7, b = 14$

Center: $(0, 0)$

Vertices: $(0, \pm 7)$

Asymptotes: $y = \pm \frac{1}{2}x$

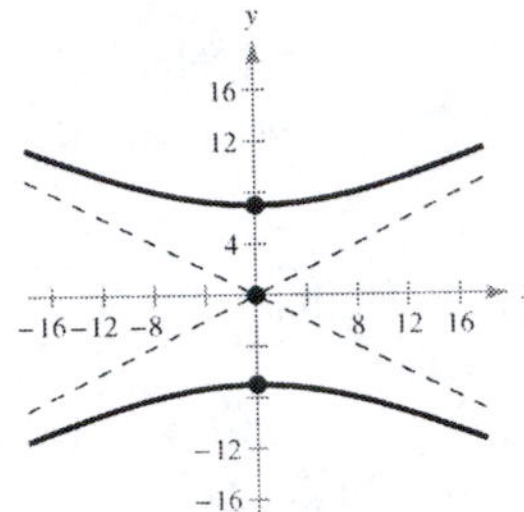

77. $4y^2 - x^2 = 1$

$$\frac{y^2}{1/4} - \frac{x^2}{1} = 0$$

$a = \frac{1}{2}, b = 1$

Center: $(0, 0)$

Vertices: $\left(0, \pm\frac{1}{2}\right)$

Asymptotes: $y = \pm \frac{1}{2}x$

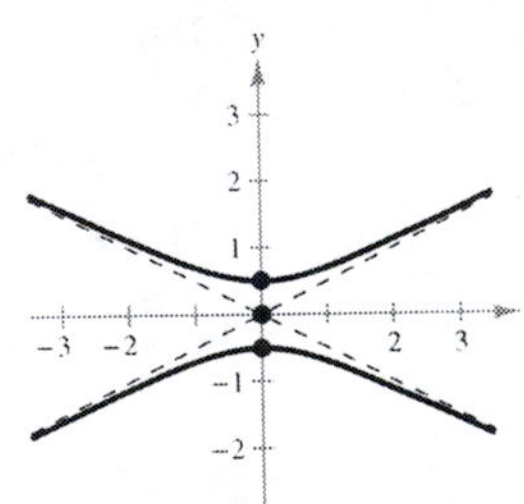

79. $\dfrac{1}{36}y^2 - \dfrac{1}{100}x^2 = 1$

$$\frac{y^2}{36} - \frac{x^2}{100} = 1$$

$a = 6, b = 10$

Center: $(0, 0)$

Vertices: $(0, \pm 6)$

Asymptotes: $y = \pm \frac{3}{5}x$

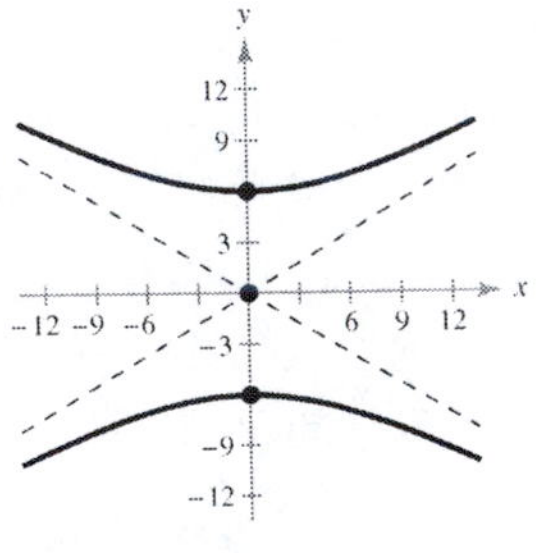

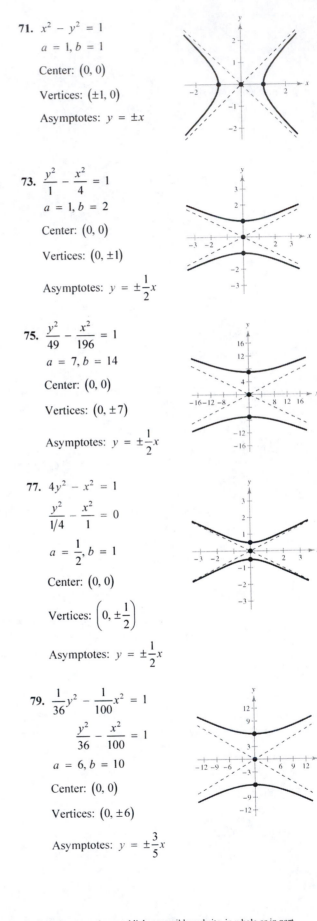

81. (a) Vertices: $(\pm 1, 0) \Rightarrow a = 1$

Horizontal transverse axis

$$\frac{x^2}{a^2} - \frac{y^2}{b^2} = 1$$

Point on the graph: $(2, 13)$

$$\frac{2^2}{1^2} - \frac{13^2}{b^2} = 1$$

$$4 - \frac{169}{b^2} = 1$$

$$3b^2 = 169$$

$$b^2 = \frac{169}{3}$$

Thus, we have $\frac{x^2}{1} - \frac{3y^2}{169} = 1.$

(b) When $y = 5$:

$$x^2 = 1 + \frac{3(5^2)}{169}$$

$$x \approx \sqrt{1 + \frac{75}{169}} \approx 1.2016$$

Width: $2x \approx 2.403$ feet

83. $\frac{x^2}{100} - \frac{y^2}{4} = 1$

The shortest horizontal distance would be the distance between the center and a vertex of the hyperbola, that is, 10 miles.

85. False. The equation represents a hyperbola.

$$\frac{x^2}{144} - \frac{y^2}{144} = 1$$

87. False. If the graph crossed the directrix, there would exist points nearer the directrix than the focus.

89. (a) $a + b = 20 \Rightarrow b = 20 - a$

$A = \pi ab = \pi a(20 - a)$

(b) $264 = \pi a(20 - a)$

$\pi a^2 - 20\pi a + 264 = 0$

$a \approx 14$ or $a \approx 6$ by the Quadratic Formula

$b = 6$ $\quad b = 14$

Since $a > b$ we choose $a = 14$ and $b = 6$.

$$\frac{x^2}{14^2} + \frac{y^2}{6^2} = 1, \frac{x^2}{196} + \frac{y^2}{36} = 1$$

(c)

a	8	9	10	11	12	13
A	301.6	311.0	314.2	311.0	301.6	285.9

Conjecture: Area is maximum when $a = b = 10$ and the shape is a circle.

(d)

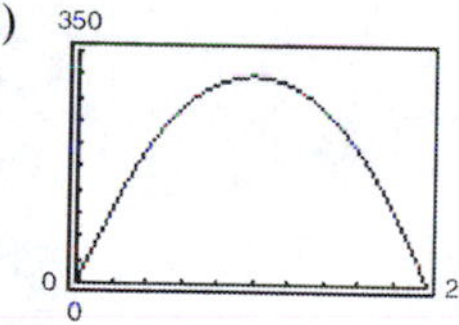

The area is maximum when $a = b = 10$ and the shape is a circle.

91. An ellipse is a circle if $a = b$.

93. $x^2 - y^2 = 0$

$y^2 = x^2$

$y = \pm x$

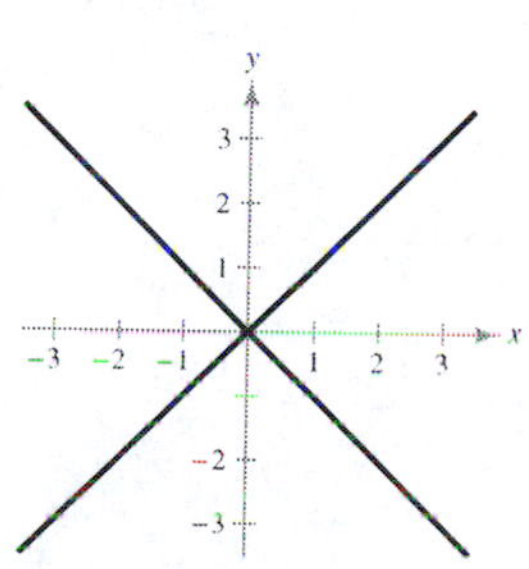

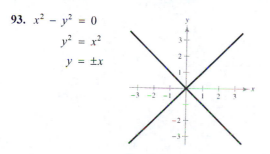

Two intersecting lines

95. Answers will vary. From the exercise set, we have examples of hyperbolas in optics, art, navigation, and aeronautics.

97. Let (x, y) be such that the sum of the distances from $(-c, 0)$ and $(c, 0)$ is $2a$.

(We are only deriving the form where the major axis is horizontal.)

$$\sqrt{(x + c)^2 + y^2} + \sqrt{(x - c)^2 + y^2} = 2a$$
$$\sqrt{(x + c)^2 + y^2} = 2a - \sqrt{(x - c)^2 + y^2}$$
$$(x + c)^2 + y^2 = 4a^2 - 4a\sqrt{(x - c)^2 + y^2} + (x - c)^2 + y^2$$
$$4a\sqrt{(x - c)^2 + y^2} = 4a^2 + (x - c)^2 - (x + c)^2$$
$$4a\sqrt{(x - c)^2 + y^2} = 4a^2 - 4cx$$
$$a\sqrt{(x - c)^2 + y^2} = a^2 - cx$$
$$a^2\left[(x - c)^2 + y^2\right] = a^4 - 2a^2cx + c^2x^2$$
$$a^2x^2 - 2a^2cx + a^2c^2 + a^2y^2 = a^4 - 2a^2cx + c^2x^2$$
$$a^2x^2 + a^2c^2 + a^2y^2 = a^4 + c^2x^2$$
$$a^2x^2 - c^2x^2 + a^2y^2 = a^4 - a^2c^2$$
$$x^2\left(a^2 - c^2\right) + a^2y^2 = a^2\left(a^2 - c^2\right)$$

Let $b^2 = a^2 - c^2$. Then $x^2b^2 + a^2y^2 = a^2b^2 \Rightarrow \dfrac{x^2}{a^2} + \dfrac{y^2}{b^2} = 1.$

99. Sample answer: The smaller the distance between the thumbtacks, the more circular the ellipse becomes. The larger the distance between the thumbtacks, the more long and narrow the ellipse becomes.

Section 4.4 Translations of Conics

1. Hyperbola with horizontal transverse axis $\Rightarrow \dfrac{(x - h)^2}{a^2} - \dfrac{(y - k)^2}{b^2} = 1$

Matches (b).

2. Ellipse with vertical major axis $\Rightarrow \dfrac{(x - h)^2}{b^2} + \dfrac{(y - k)^2}{a^2} = 1$

Matches (d).

3. Parabola with vertical axis $\Rightarrow (x - h)^2 = 4p(y - k)$

Matches (e).

4. Hyperbola with vertical transverse axis $\Rightarrow \dfrac{(y - k)^2}{a^2} - \dfrac{(x - h)^2}{b^2} = 1$

Matches (c).

5. Ellipse with horizontal major axis $\Rightarrow \dfrac{(x - h)^2}{a^2} + \dfrac{(y - k)^2}{b^2} = 1$

Matches (a).

6. Parabola with horizontal axis $\Rightarrow (y - k)^2 = 4p(x - h)$

Matches (f).

7. The graph of $(x + 2)^2 + (y - 1)^2 = 4$ is a circle.

The graph has been shifted two units to the left and one unit upward from standard position.

9. The graph of $y - 2 = 4(-1)(x + 1)^2$, rewritten as $(x + 1)^2 = 4\left(-\frac{1}{16}\right)(y - 2)$, is a parabola.

The graph has been reflected in the x-axis, shifted one unit to the left and two units up from standard position.

11. The graph of $\dfrac{(y + 3)^2}{4} - (x - 1)^2 = 1$ is a hyperbola.

The graph has been shifted one unit to the right and three units downward from standard position.

13. The graph of $\dfrac{(x + 4)^2}{9} + \dfrac{(y + 2)^2}{16} = 1$ is an ellipse.

The graph has been shifted four units to the left and two units downward from standard position.

15. $x^2 + y^2 = 49$

Center: $(0, 0)$

Radius: 7

17. $(x - 4)^2 + (y - 5)^2 = 36$

Center: $(4, 5)$

Radius: 6

19. $(x - 1)^2 + y^2 = 10$

Center: $(1, 0)$

Radius: $\sqrt{10}$

21.

$$\begin{aligned} x^2 + y^2 - 2x + 6y + 9 &= 0 \\ \left(x^2 - 2x\right) + \left(y^2 + 6y\right) &= -9 \\ \left(x^2 - 2x + 1\right) + \left(y^2 + 6y + 9\right) &= -9 + 1 + 9 \\ (x - 1)^2 + (y + 3)^2 &= 1 \end{aligned}$$

Center: $(1, -3)$

Radius: 1

23.

$$\begin{aligned} x^2 + y^2 - 8x &= 0 \\ \left(x^2 - 8x\right) + y^2 &= 0 \\ \left(x^2 - 8x + 16\right) + y^2 &= 16 \\ (x - 4)^2 + y^2 &= 16 \end{aligned}$$

Center: $(4, 0)$

Radius: 4

25.

$$\begin{aligned} 4x^2 + 4y^2 + 12x - 24y + 41 &= 0 \\ x^2 + y^2 + 3x - 6y + \tfrac{41}{4} &= 0 \\ \left(x^2 + 3x\right) + \left(y^2 - 6y\right) &= -\tfrac{41}{4} \\ \left(x^2 + 3x + \tfrac{9}{4}\right) + \left(y^2 - 6y + 9\right) &= -\tfrac{41}{4} + \tfrac{9}{4} + 9 \\ \left(x + \tfrac{3}{2}\right)^2 + (y - 3)^2 &= 1 \end{aligned}$$

Center: $\left(-\frac{3}{2}, 3\right)$

Radius: 1

27. $(x - 1)^2 + 8(y + 2) = 0$

$(x - 1)^2 = 4(-2)(y + 2);\ p = -2$

Vertex: $(1, -2)$

Focus: $(1, -4)$

Directrix: $y = 0$

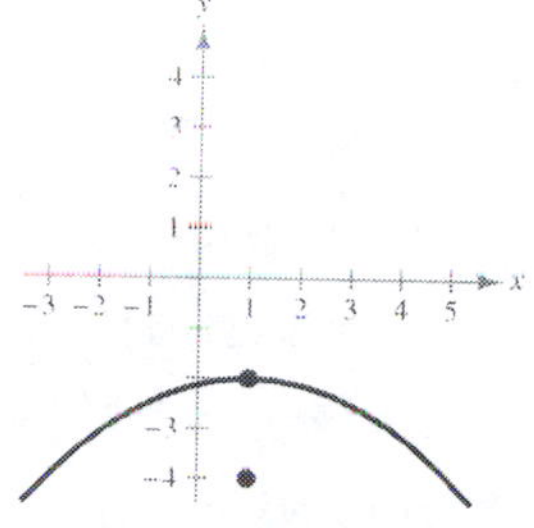

29. $\left(y + \frac{1}{2}\right)^2 = 2(x - 5)$

$\left(y + \frac{1}{2}\right)^2 = 4\left(\frac{1}{2}\right)(x - 5);\ p = \frac{1}{2}$

Vertex: $\left(5, -\frac{1}{2}\right)$

Focus: $\left(\frac{11}{2}, -\frac{1}{2}\right)$

Directrix: $x = \frac{9}{2}$

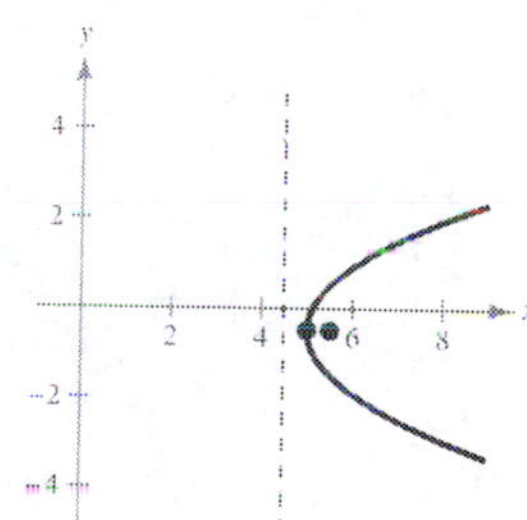

31.

$$\begin{aligned} y &= \tfrac{1}{4}\left(x^2 - 2x + 5\right) \\ x^2 - 2x &= 4y - 5 \\ x^2 - 2x + 1 &= 4y - 5 + 1 \\ (x - 1)^2 &= 4y - 4 \\ (x - 1)^2 &= 4(1)(y - 1);\ p = 1 \end{aligned}$$

Vertex: $(1, 1)$

Focus: $(1, 2)$

Directrix: $y = 0$

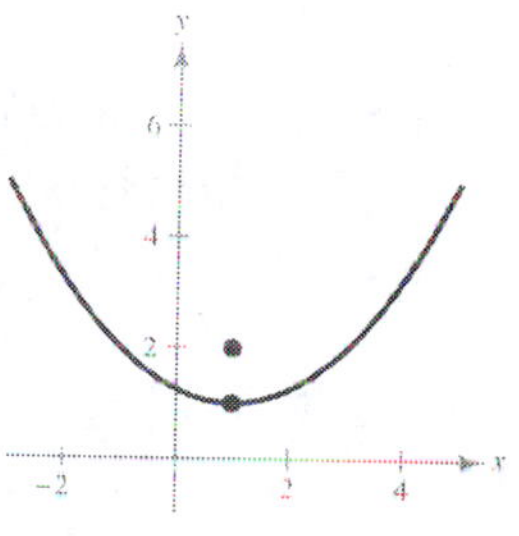

33. $y^2 + 6y + 8x + 25 = 0$

$$y^2 + 6y = -8x - 25$$
$$y^2 + 6y + 9 = -8x - 25 + 9$$
$$(y + 3)^2 = -8x - 16$$
$$(y + 3)^2 = 4(-2)(x + 2);\ p = -2$$

Vertex: $(-2, -3)$

Focus: $(-4, -3)$

Directrix: $x = 0$

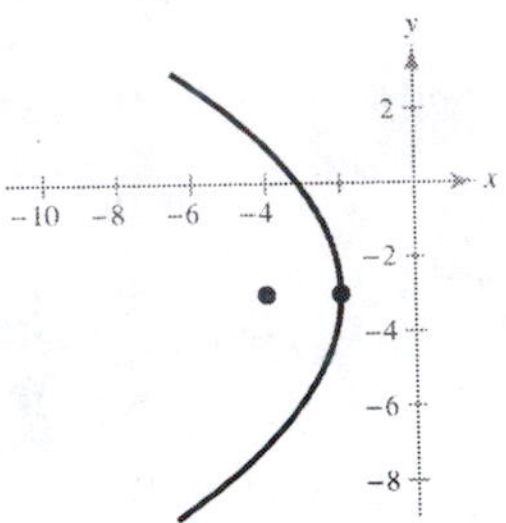

35. Vertex: $(3, 2)$

Focus: $(1, 2)$

Horizontal axis

$p = 1 - 3 = -2$

$(y - 2)^2 = 4(-2)(x - 3)$

$(y - 2)^2 = -8(x - 3)$

37. Vertex: $(0, 4)$

Directrix: $y = 2$

Vertical axis

$p = 4 - 2 = 2$

$(x - 0)^2 = 4(2)(y - 4)$

$x^2 = 8(y - 4)$

39. Focus: $(4, 4)$

Directrix: $x = -4$

Horizontal axis

Vertex: $(0, 4)$

$p = 4 - 0 = 4$

$(y - 4)^2 = 4(4)(x - 0)$

$(y - 4)^2 = 16x$

41. (a) $V = 17{,}500\sqrt{2}$ mi/h

$\approx 24{,}750$ mi/h

(b) $p = -4100, (h, k) = (0, 4100)$

$(x - 0)^2 = 4(-4100)(y - 4100)$

$x^2 = -16{,}400(y - 4100)$

43. (a) To write the equation, let $s = 100$ and $v_0 = 28$.

$$x^2 = -\frac{v^2}{16}(y - s)$$
$$x^2 = -\frac{28^2}{16}(y - 100)$$
$$x^2 = -49(y - 100)$$

(b) To find the distance traveled horizontally, before the ball strikes the ground, let $y = 0$ and solve for x.

$$x^2 = -49(0 - 100)$$
$$x^2 = 4900$$
$$x = \sqrt{4900}$$
$$= 70$$

The ball travels 70 feet horizontally before striking the ground.

45. $\dfrac{(x - 1)^2}{9} + \dfrac{(y - 5)^2}{25} = 1$

$a = 5, b = 3, c = \sqrt{a^2 - b^2} = 4$

Center: $(1, 5)$

Foci: $(1, 1), (1, 9)$

Vertices: $(1, 0), (1, 10)$

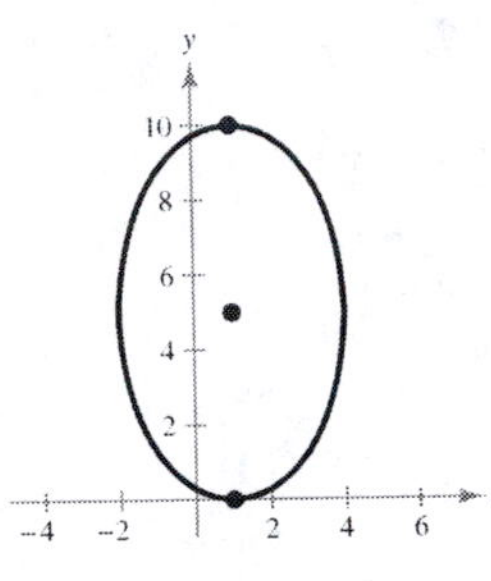

47. $\dfrac{(x + 2)^2}{1} + \dfrac{(y + 4)^2}{1/4} = 1$

$a = 1, b = \dfrac{1}{2}, c = \sqrt{a^2 - b^2} = \sqrt{\dfrac{3}{4}} = \dfrac{\sqrt{3}}{2}$

Center: $(-2, -4)$

Foci: $\left(-2 \pm \dfrac{\sqrt{3}}{2}, -4\right)$

Vertices: $(-3, -4), (-1, -4)$

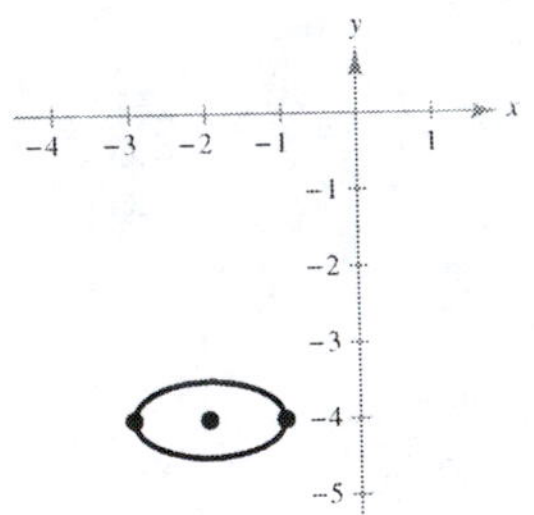

49. $9x^2 + 25y^2 - 36x - 50y + 52 = 0$

$$9(x^2 - 4x) + 25(y^2 - 2y) = -52$$

$$9(x^2 - 4x + 4) + 25(y^2 - 2y + 1) = -52 + 9(4) + 25(1)$$

$$9(x - 2)^2 + 25(y - 1)^2 = 9$$

$$(x - 2)^2 + \frac{(y - 1)^2}{9/25} = 1$$

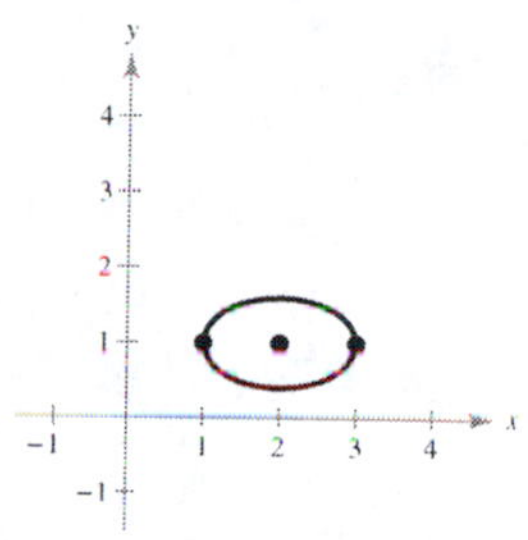

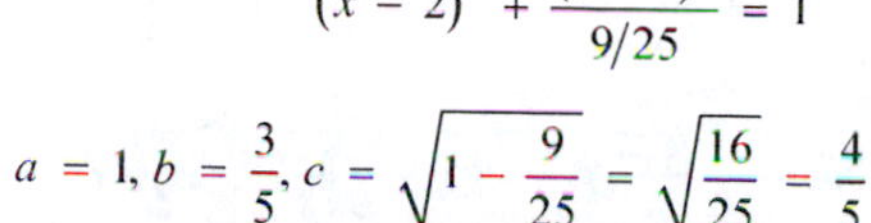

$$a = 1, b = \frac{3}{5}, c = \sqrt{1 - \frac{9}{25}} = \sqrt{\frac{16}{25}} = \frac{4}{5}$$

Horizontal major axis

Center: $(2, 1)$

Foci: $\left(\frac{14}{5}, 1\right), \left(\frac{6}{5}, 1\right)$

Vertices: $(1, 1), (3, 1)$

51. $9x^2 + 4y^2 + 36x - 16y + 16 = 0$

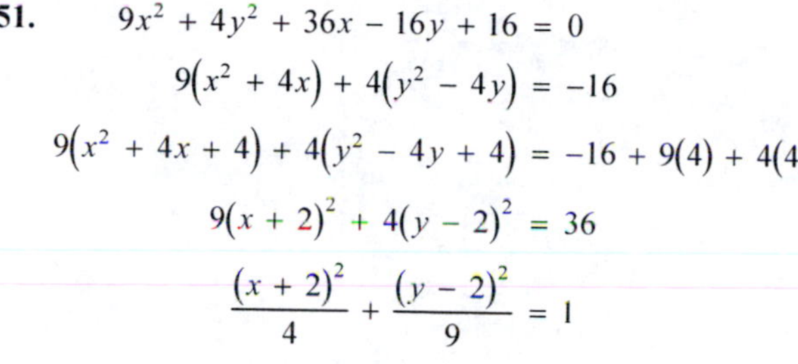

$$9(x^2 + 4x) + 4(y^2 - 4y) = -16$$

$$9(x^2 + 4x + 4) + 4(y^2 - 4y + 4) = -16 + 9(4) + 4(4)$$

$$9(x + 2)^2 + 4(y - 2)^2 = 36$$

$$\frac{(x + 2)^2}{4} + \frac{(y - 2)^2}{9} = 1$$

$$a = 2, b = 3, c = \sqrt{a^2 - 6^2} = \sqrt{5}$$

Vertical major axis

Center: $(-2, 2)$

Foci: $\left(-2, 2 \pm \sqrt{5}\right)$

Vertices: $(-2, -1), (-2, 5)$

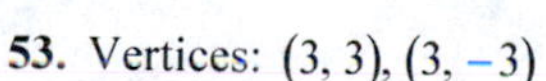

53. Vertices: $(3, 3), (3, -3)$

Minor axis of length $2 \Rightarrow b = 1$

Center: $(3, 0) \Rightarrow a = 3$

Vertical major axis

$$\frac{(x - 3)^2}{1} + \frac{y^2}{9} = 1$$

55. Foci: $(0, 0), (4, 0) \Rightarrow c = 2$

Major axis of length $8 \Rightarrow a = 4$

$b^2 = a^2 - c^2 = 12$

Center: $(2, 0)$

Horizontal major axis

$$\frac{(x - 2)^2}{16} + \frac{y^2}{12} = 1$$

57. Center: $(0, 4)$

$a = 2c$

Vertices: $(-4, 4), (4, 4) \Rightarrow a = 4$

$c = 2, b^2 = a^2 - c^2 = 12$

Horizontal major axis

$$\frac{x^2}{16} + \frac{(y - 4)^2}{12} = 1$$

59. Vertices: $(-3, 0), (7, 0) \Rightarrow a = 5$

Foci: $(0, 0), (4, 0) \Rightarrow c = 2$

Center: $(2, 0)$

Horizontal major axis

$a^2 = b^2 + c^2 \Rightarrow b^2 = 5^2 - 2^2 = 21$

$$\frac{(x - 2)^2}{25} + \frac{y^2}{21} = 1$$

61. Vertices: $(\pm 5, 0)$

$$e = \frac{c}{a} = \frac{3}{5}$$

$$a = 5, c = 3, b = 4$$

$$\frac{x^2}{25} + \frac{y^2}{16} = 1$$

63. $a = 3.67 \times 10^9$

$$e = \frac{c}{a} = 0.249$$

$$c = 913{,}830{,}000$$

Smallest distance: $a - c = 2{,}756{,}170{,}000$ miles

Greatest distance: $a + c = 4{,}583{,}830{,}000$ miles

65. $\dfrac{(x-2)^2}{16} - \dfrac{(y+1)^2}{9} = 1$

$a = 4, b = 3, c = \sqrt{a^2 + b^2} = 5$

Center: $(2, -1)$

Horizontal transverse axis

Vertices: $(6, -1), (-2, -1)$

Foci: $(7, -1), (-3, -1)$

Asymptotes: $y = \pm\frac{3}{4}(x - 2) - 1$

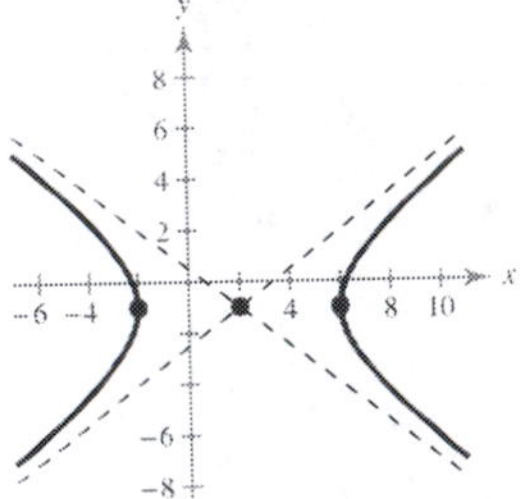

67. $(y + 6)^2 - (x - 2)^2 = 1$

$a = 1, b = 1, c = \sqrt{a^2 + b^2} = \sqrt{2}$

Center: $(2, -6)$

Vertical transverse axis

Vertices: $(2, -7), (2, -5)$

Foci: $\left(2, -6 \pm \sqrt{2}\right)$

Asymptotes:
$y = \pm(x - 2) - 6$

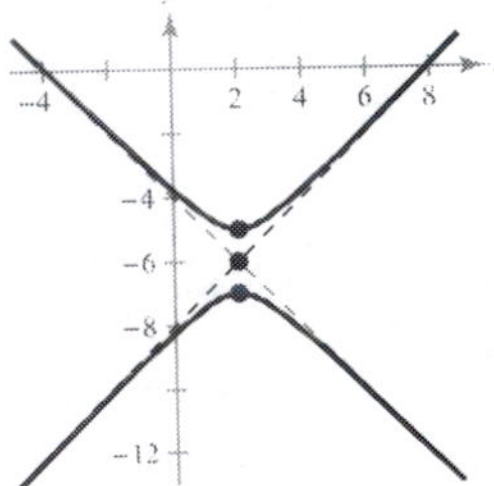

69.

$$\begin{aligned} x^2 - 9y^2 + 2x - 54y - 85 &= 0 \\ \left(x^2 + 2x\right) - 9\left(y^2 + 6y\right) &= 85 \\ \left(x^2 + 2x + 1\right) - 9\left(y^2 + 6y + 9\right) &= 85 + 1 - 9(9) \\ (x+1)^2 - 9(y+3)^2 &= 5 \\ \frac{(x+1)^2}{5} - \frac{(y+3)^2}{5/9} &= 1 \end{aligned}$$

$a = \sqrt{5}, b = \dfrac{\sqrt{5}}{3}, c = \sqrt{a^2 + b^2} = \dfrac{5\sqrt{2}}{3}$

Center: $(-1, -3)$

Horizontal transverse axis

Vertices: $\left(-1 \pm \sqrt{5}, -3\right)$

Foci: $\left(-1 \pm \dfrac{5\sqrt{2}}{3}, -3\right)$

Asymptotes: $y = \pm\frac{1}{3}(x + 1) - 3$

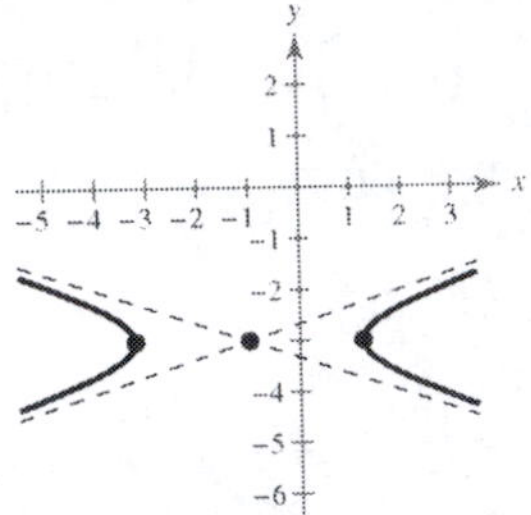

71.

$$\begin{aligned} 9x^2 - y^2 - 36x - 6y + 18 &= 0 \\ 9\left(x^2 - 4x\right) - \left(y^2 + 6y\right) &= -18 \\ 9\left(x^2 - 4x + 4\right) - \left(y^2 + 6y + 9\right) &= -18 + 9(4) - 9 \\ 9(x-2)^2 - (y+3)^2 &= 9 \\ (x-2)^2 - \frac{(y+3)^2}{9} &= 1 \end{aligned}$$

$a = 1, b = 3, c = \sqrt{a^2 + b^2} = \sqrt{10}$

Center: $(2, -3)$

Horizontal transverse axis

Vertices: $(1, -3), (3, -3)$

Foci: $\left(2 \pm \sqrt{10}, -3\right)$

Asymptotes:
$y = \pm 3(x - 2) - 3$

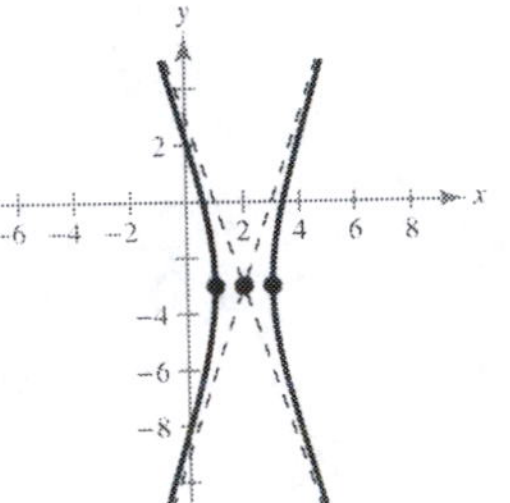

73. Vertices: $(0, 0), (0, 2)$

Foci: $(0, -1), (0, 3)$

Center: $(0, 1)$

Vertical transverse axis

$a = 1, c = 2, b^2 = c^2 - a^2 = 3$

$$\frac{(y - 1)^2}{1} - \frac{x^2}{3} = 1$$

75. Vertices: $(2, 0), (6, 0)$

Foci: $(0, 0), (8, 0)$

Center: $(4, 0)$

Horizontal transverse axis

$a = 2, c = 4, b^2 = c^2 - a^2 = 12$

$$\frac{(x - 4)^2}{4} - \frac{y^2}{12} = 1$$

77. Vertices: $(2, 3), (2, -3)$

Passes through the point $(0, 5)$

Center: $(2, 0)$

Vertical transverse axis

$a = 3$

$$\frac{y^2}{9} - \frac{(x - 2)^2}{b^2} = 1$$

$$\frac{5^2}{9} - \frac{(0 - 2)^2}{b^2} = 1$$

$$b^2 = \frac{9}{4}$$

$$\frac{y^2}{9} - \frac{4(x - 2)^2}{9} = 1$$

79. Vertices: $(0, 2), (6, 2)$

Asymptotes: $y = \frac{2}{3}x, y = 4 - \frac{2}{3}x$

Center: $(3, 2)$

Horizontal transverse axis

$a = 3, b = 2$

$$\frac{(x - 3)^2}{9} - \frac{(y - 2)^2}{4} = 1$$

81.
$$x^2 + y^2 - 6x + 4y + 9 = 0$$
$$(x^2 - 6x) + (y^2 + 4y) = -9$$
$$(x^2 - 6x + 9) + (y^2 + 4y + 4) = -9 + 9 + 4$$
$$(x - 3)^2 + (y + 2)^2 = 4$$

Circle

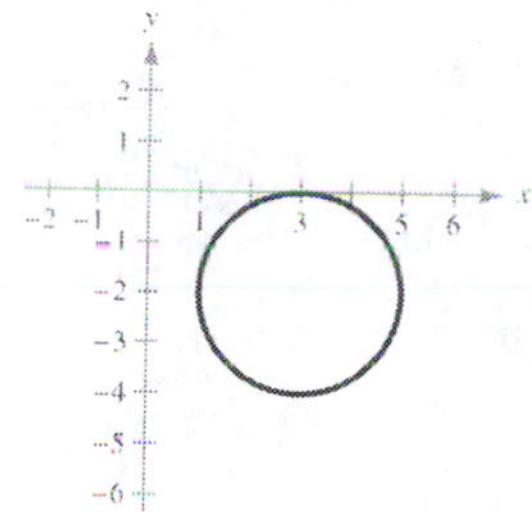

83.
$$y^2 - x^2 + 4y = 0$$
$$(y^2 + 4y + 4) - x^2 = 4$$
$$(y + 2)^2 - x^2 = 4$$
$$\frac{(y + 2)^2}{4} - \frac{x^2}{4} = 1$$

Hyperbola

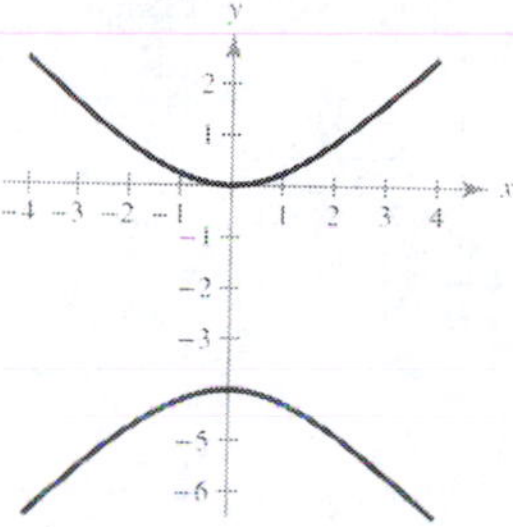

85. $16y^2 + 128x + 8y - 7 = 0$
$$16\left(y^2 + \tfrac{1}{2}y\right) = -128x + 7$$
$$16\left(y^2 + \tfrac{1}{2}y + \tfrac{1}{16}\right) = -128x + 7 + 16\left(\tfrac{1}{16}\right)$$
$$16\left(y + \tfrac{1}{4}\right)^2 = -128x + 8$$
$$\left(y + \tfrac{1}{4}\right)^2 = -8x + \tfrac{1}{2}$$
$$\left(y + \tfrac{1}{4}\right)^2 = -8\left(x - \tfrac{1}{16}\right)$$

Parabola

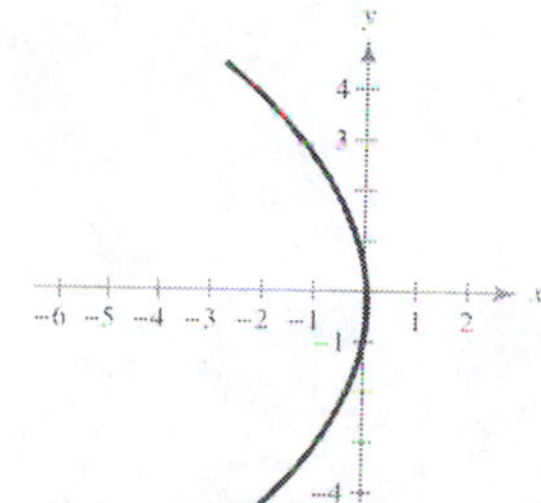

87.
$$9x^2 + 16y^2 + 36x + 128y + 148 = 0$$
$$9(x^2 + 4x) + 16(y^2 + 8y) = -148$$
$$9(x^2 + 4x + 4) + 16(y^2 + 8y + 16) = -148 + 9(4) + 16(16)$$
$$9(x + 2)^2 + 16(y + 4)^2 = 144$$
$$\frac{(x + 2)^2}{16} + \frac{(y + 4)^2}{9} = 1$$

Ellipse

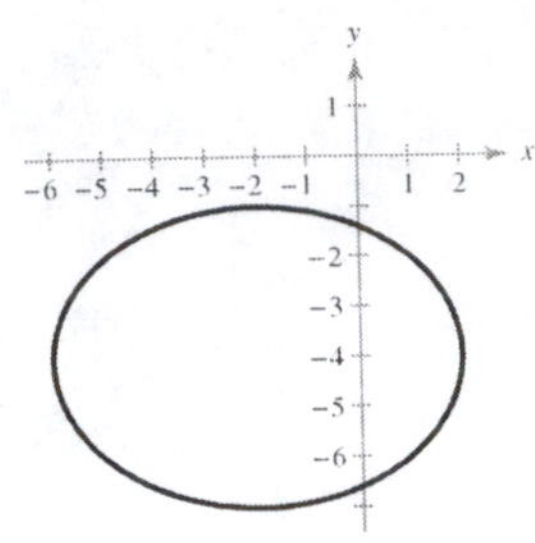

89.
$$16x^2 + 16y^2 - 16x + 24y - 3 = 0$$
$$16(x^2 - x) + 16\left(y^2 + \tfrac{3}{2}y\right) = 3$$
$$16\left(x^2 - x + \tfrac{1}{4}\right) + 16\left(y^2 + \tfrac{3}{2}y + \tfrac{9}{16}\right) = 3 + 16\left(\tfrac{1}{4}\right) + 16\left(\tfrac{9}{16}\right)$$
$$16\left(x - \tfrac{1}{2}\right)^2 + 16\left(y + \tfrac{3}{4}\right)^2 = 16$$
$$\left(x - \tfrac{1}{2}\right)^2 + \left(y + \tfrac{3}{4}\right)^2 = 1$$

Circle

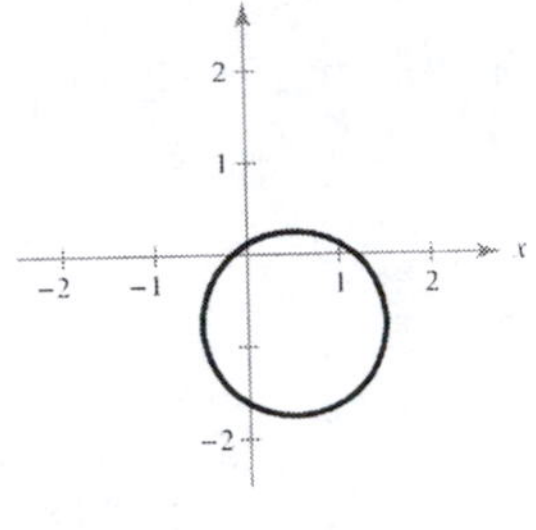

91.
$$3x^2 + 2y^2 - 18x - 16y + 58 = 0$$
$$3(x^2 - 6x) + 2(y^2 - 8y) = -58$$
$$3(x^2 - 6x + 9) + 2(y^2 - 8y + 16) = -58 + 27 + 32$$
$$3(x - 3)^2 + 2(y - 4)^2 = 1$$
$$\frac{(x - 3)^2}{1/3} + \frac{(y - 4)^2}{1/2} = 1$$

True. The equation in standard form is
$$\frac{(x - 3)^2}{1/3} + \frac{(y - 4)^2}{1/2} = 1.$$

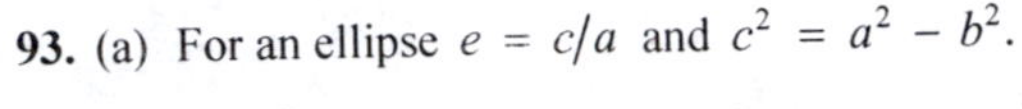

93. (a) For an ellipse $e = c/a$ and $c^2 = a^2 - b^2$.

$$e = \frac{c}{a} \Rightarrow ea = c$$
$$e^2a^2 = c^2$$
$$e^2a^2 = a^2 - b^2$$
$$b^2 = a^2 - e^2a^2$$
$$b^2 = a^2(1 - e^2)$$

Thus, by substitution, $\dfrac{(x - h)^2}{a^2} + \dfrac{(y - k)^2}{b^2} = 1$ is equivalent to $\dfrac{(x - h)^2}{a^2} + \dfrac{(y - k)^2}{a^2(1 - e^2)} = 1.$

(b)

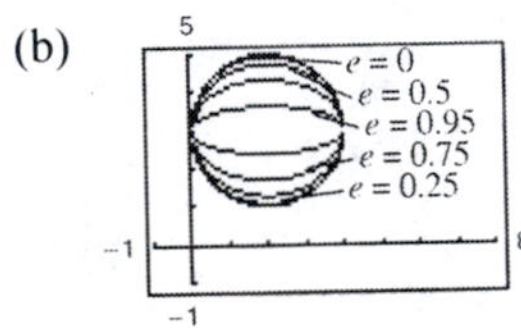

As e approaches 0, the ellipse becomes a circle.

Review Exercises for Chapter 4

1. Because the denominator is zero when $x + 10 = 0$, the domain of f is all real numbers except $x = -10$.

x	-11	-10.5	-10.1	-10.01	-10.001	$\to -10$
$f(x)$	33	63	303	3003	30,003	$\to \infty$

x	$-10 \leftarrow$	-9.999	-9.99	-9.9	-9.5	-9
$f(x)$	$-\infty$	$-29,997$	-2997	-297	-57	-27

As x approaches -10 from the left, $f(x)$ increases without bound.

As x approaches -10 from the right, $f(x)$ decreases without bound.

3. Because the denominator is zero when $x^2 - 10x + 24 = (x - 4)(x - 6) = 0$, the domain of f is all real numbers except $x = 4$ and $x = 6$.

x	3	3.5	3.9	3.99	3.999	$\to 4$
$f(x)$	2.667	6.4	38.095	398.010	3998.001	$\to \infty$

x	$4 \leftarrow$	4.001	4.01	4.1	4.5	5
$f(x)$	$-\infty \leftarrow$	-4002.001	-402.010	-42.105	-10.67	-8

As x approaches 4 from the left, $f(x)$ increases without bound.

As x approaches 4 from the right, $f(x)$ decreases without bound.

x	5	5.5	5.9	5.99	5.999	$\to 6$
$f(x)$	-8	-10.67	-42.015	-402.010	-4002.001	$\to -\infty$

x	$6 \leftarrow$	6.001	6.01	6.1	6.5	7
$f(x)$	$\infty \leftarrow$	3998.001	398.010	38.095	6.4	2.667

As x approaches 6 from the left, $f(x)$ decreases without bound.

As x approaches 6 from the right, $f(x)$ increases without bound.

5. $f(x) = \dfrac{4}{x + 3}$

Vertical asymptote: $x = -3$

Horizontal asymptote: $y = 0$

7. $g(x) = \dfrac{x^2}{x^2 - 4}$

Vertical asymptotes: $x = -2, x = 2$

Horizontal asymptote: $y = 1$

9. $h(x) = \dfrac{5x + 20}{x^2 - 2x - 24}$

$$= \frac{5(x + 4)}{(x - 6)(x + 4)}$$

$$= \frac{5}{x - 6}, \quad x \neq -4$$

Vertical asymptote: $x = 6$

Horizontal asymptote: $y = 0$

11. $\overline{C} = \frac{C}{x} = \frac{0.5x + 500}{x}, \ 0 < x$

Horizontal asymptote: $\overline{C} = \frac{0.5}{1} = 0.5$

As x increases, the average cost per unit approaches the horizontal asymptote, $\overline{C} = 0.5 = \$0.50$.

13. $f(x) = \frac{-3}{2x^2}$

(a) Domain: all real numbers x except $x = 0$

(b) No intercepts

(c) Vertical asymptote: $x = 0$

Horizontal asymptote: $y = 0$

(d)

x	-3	-2	-1	1	2	3
$f(x)$	$-\frac{1}{6}$	$-\frac{3}{8}$	$-\frac{3}{2}$	$-\frac{3}{2}$	$-\frac{3}{8}$	$-\frac{1}{6}$

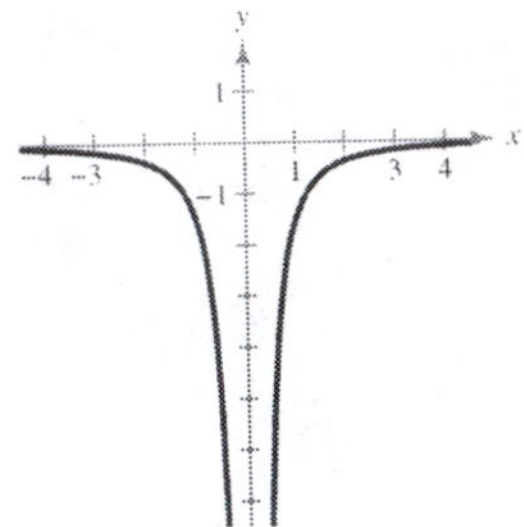

15. $g(x) = \frac{2 + x}{1 - x} = -\frac{x + 2}{x - 1}$

(a) Domain: all real numbers x except $x = 1$

(b) x-intercept: $(-2, 0)$

y-intercept: $(0, 2)$

(c) Vertical asymptote: $x = 1$

Horizontal asymptote: $y = -1$

(d)

x	-1	0	2	3
$g(x)$	$\frac{1}{2}$	2	-4	$-\frac{5}{2}$

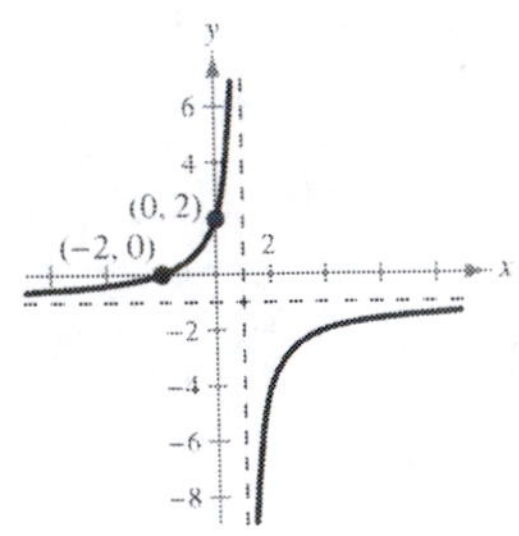

17. $p(x) = \frac{5x^2}{4x^2 + 1}$

(a) Domain: all real numbers x

(b) Intercept: $(0, 0)$

(c) Horizontal asymptote: $y = \frac{5}{4}$

(d)

x	-3	-2	-1	0	1	2	3
$p(x)$	$\frac{45}{37}$	$\frac{20}{17}$	1	0	1	$\frac{20}{17}$	$\frac{45}{37}$

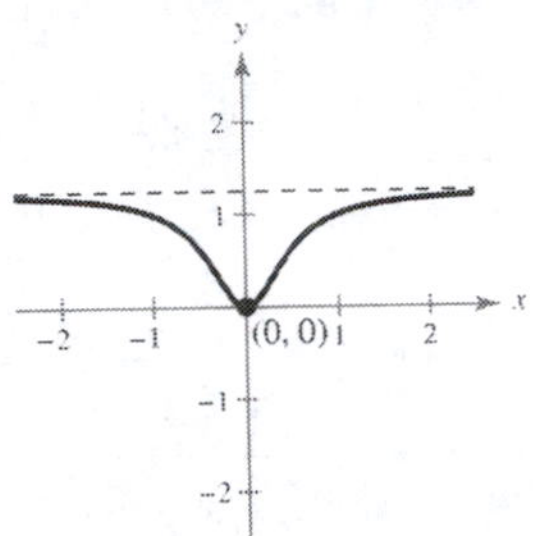

19. $f(x) = \frac{x}{x^2 + 1}$

(a) Domain: all real numbers x

(b) Intercept: $(0, 0)$

(c) Horizontal asymptote: $y = 0$

(d)

x	-2	-1	0	1	2
$f(x)$	$-\frac{2}{5}$	$-\frac{1}{2}$	0	$\frac{1}{2}$	$\frac{2}{5}$

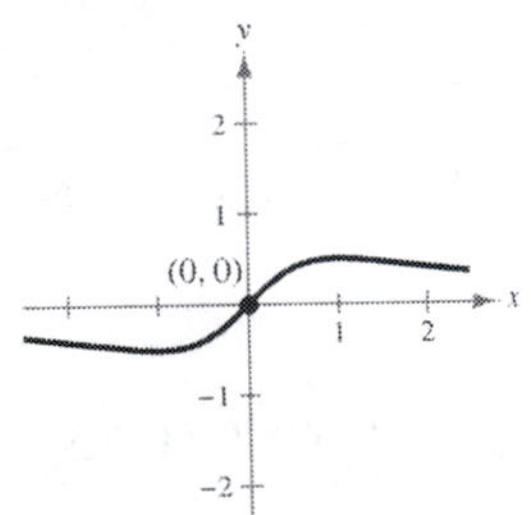

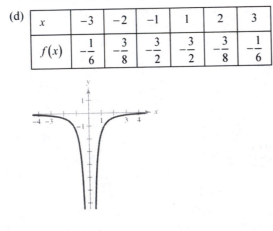

21. $f(x) = \dfrac{-6x^2}{x^2 + 1}$

(a) Domain: all real numbers x

(b) Intercept: $(0, 0)$

(c) Horizontal asymptote: $y = -6$

(d)

x	± 3	± 2	± 1	0
$f(x)$	$-\frac{27}{5}$	$-\frac{24}{5}$	-3	0

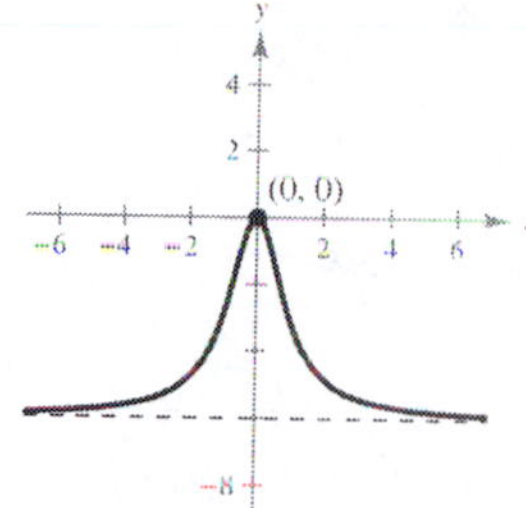

23. $f(x) = \dfrac{6x^2 - 11x + 3}{3x^2 - x}$

$= \dfrac{(3x - 1)(2x - 3)}{x(3x - 1)} = \dfrac{2x - 3}{x}, \; x \neq \dfrac{1}{3}$

(a) Domain: all real numbers x except $x = 0$ and $x = \dfrac{1}{3}$

(b) x-intercept: $\left(\dfrac{3}{2}, 0\right)$

(c) Vertical asymptote: $x = 0$

Horizontal asymptote: $y = 2$

(d)

x	-2	-1	1	2	3	4
$f(x)$	$\frac{7}{2}$	5	-1	$\frac{1}{2}$	1	$\frac{5}{4}$

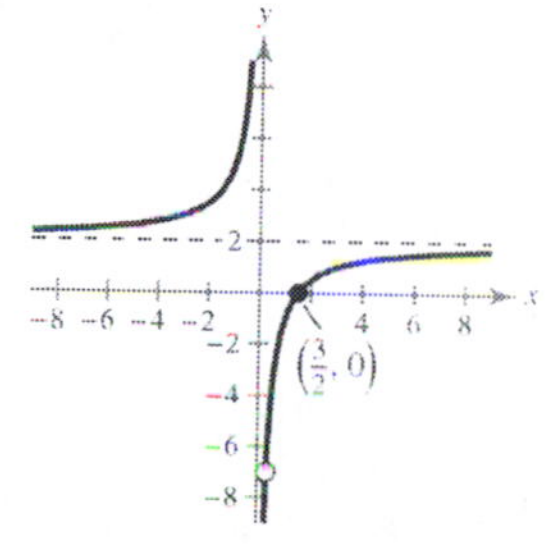

25. $f(x) = \dfrac{2x^3}{x^2 + 1} = 2x - \dfrac{2x}{x^2 + 1}$

(a) Domain: all real numbers x

(b) Intercept: $(0, 0)$

(c) Slant asymptote: $y = 2x$

(d)

x	-2	-1	0	1	2
$f(x)$	$-\frac{16}{5}$	-1	0	1	$\frac{16}{5}$

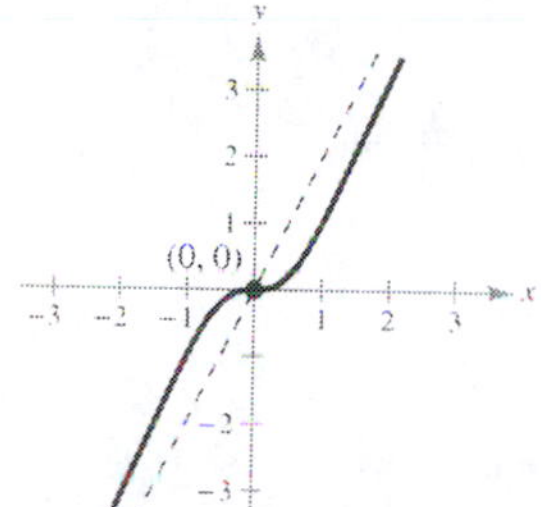

27. $f(x) = \dfrac{x^2 + 3x - 10}{x + 2} = x + 1 - \dfrac{12}{x + 2}$

(a) Domain: all real numbers x except $x = -2$

(b) y-intercept: $(0, -5)$

x-intercepts: $(2, 0), (-5, 0)$

(c) Vertical asymptote: $x = -2$

Slant asymptote: $y = x + 1$

(d)

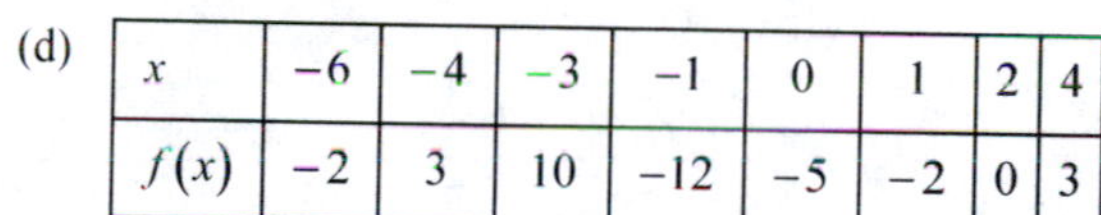

x	-6	-4	-3	-1	0	1	2	4
$f(x)$	-2	3	10	-12	-5	-2	0	3

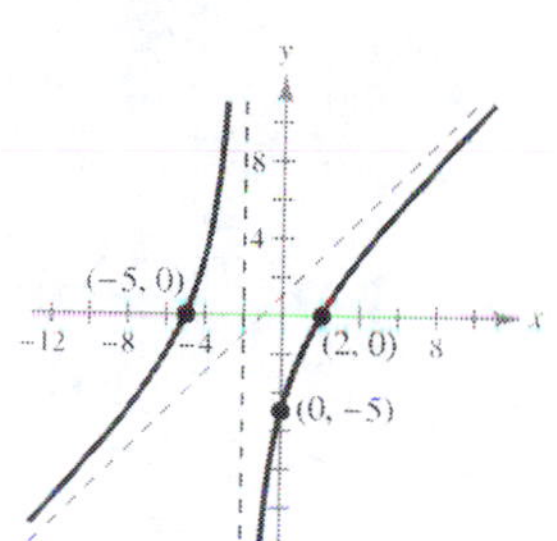

29. $f(x) = \dfrac{3x^3 - 2x^2 - 3x + 2}{3x^2 - x - 4}$

$= \dfrac{(3x - 2)(x + 1)(x - 1)}{(3x - 4)(x + 1)}$

$= \dfrac{(3x - 2)(x - 1)}{3x - 4}$

$= x - \dfrac{1}{3} + \dfrac{2/3}{3x - 4}, \quad x \neq -1$

(a) Domain: all real numbers x except $x = -1$ and $x = \dfrac{4}{3}$

(b) x-intercepts: $(1, 0), \left(\dfrac{2}{3}, 0\right)$

y-intercept: $\left(0, -\dfrac{1}{2}\right)$

(c) Vertical asymptote: $x = \dfrac{4}{3}$

Slant asymptote: $y = x - \dfrac{1}{3}$

(d)

x	-3	-2	0	1	2	3
$f(x)$	$-\dfrac{44}{13}$	$-\dfrac{12}{5}$	$-\dfrac{1}{2}$	0	2	$\dfrac{14}{5}$

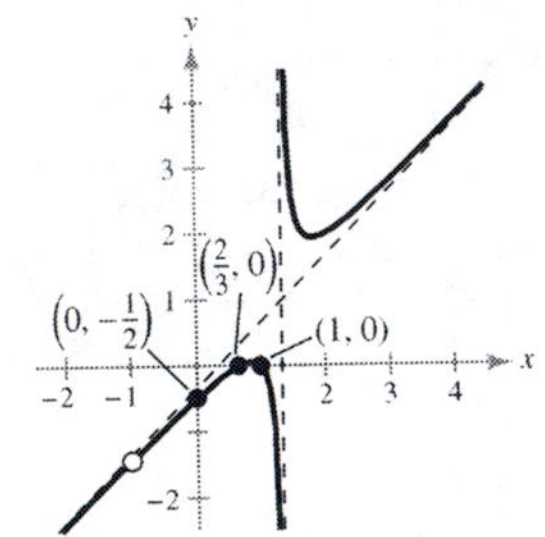

31. $\overline{C} = \dfrac{100{,}000 + 0.9x}{x}, \quad x > 0$

(a)

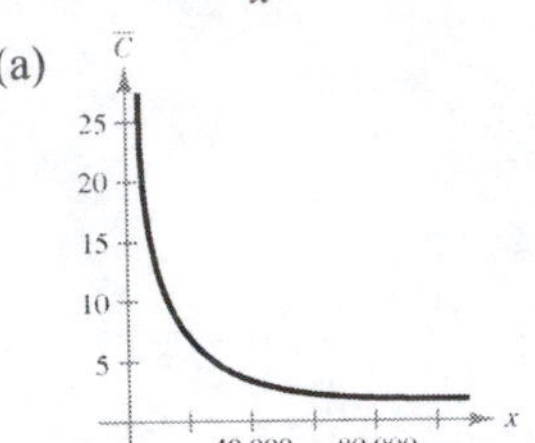

(b) When $x = 1000$,

$C = \dfrac{100{,}000 + 900}{1000} = \$100.90.$

When $x = 10{,}000$,

$C = \dfrac{100{,}000 + 9000}{10{,}000} = \$10.90.$

When $x = 100{,}000$,

$C = \dfrac{100{,}000 + 90{,}000}{100{,}000} = \$1.90.$

(c) The average cost per unit is always greater than \$0.90 because \$0.90 is the horizontal asymptote of the function.

33. $y = \dfrac{18.47x - 2.96}{0.23x + 1}, \quad 0 < x$

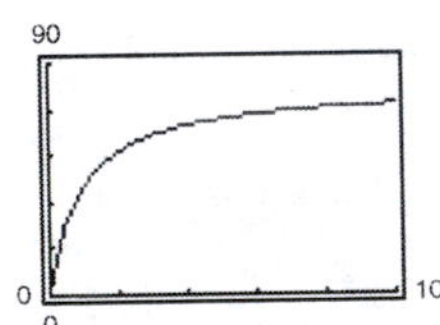

The limiting amount of CO_2 uptake is determined by the horizontal asymptote, $y = \dfrac{18.47}{0.23} \approx 80.3$ mg/dm^2/hr.

35. $y^2 = -16x$

$y^2 = 4(-4)x$

Parabola

37. $\dfrac{x^2}{64} - \dfrac{y^2}{4} = 1$

Hyperbola

39. $x^2 + 20y = 0$

$x^2 = -20y$

$x^2 = 4(-5)y$

Parabola

41. $\dfrac{y^2}{49} - \dfrac{x^2}{144} = 1$

Hyperbola

43. Vertex: $(0, 0)$

Focus: $(-6, 0) \Rightarrow p = -6$

Horizontal axis

$y^2 = 4px$

$y^2 = 4(-6)x$

$y^2 = -24x$

45. Vertex: $(0, 0)$

Directrix: $y = -3 \Rightarrow p = 3$

Vertical axis

$x^2 = 4py$

$x^2 = 4(3)y$

$x^2 = 12y$

47. Vertex: $(0, 0)$

Point: $(3, 6)$

Horizontal axis

$y^2 = 4px$

$6^2 = 4p(3) \Rightarrow p = 3$

$y^2 = 4(3)x$

$y^2 = 12x$

49. $y = \dfrac{x^2}{200}, \ -100 \le x \le 100$

Vertex: $(0, 0)$

$x^2 = 200y$

$x^2 = 4(50)y$

$p = 50$

Focus: $(0, 50)$

51. Vertices: $(0, 6), (0, -6) \Rightarrow a = 6$

Ends of minor axis: $(5, 0), (-5, 0) \Rightarrow b = 5$

Center: $(0, 0)$

Vertical major axis

$$\frac{(x - 0)^2}{5^2} + \frac{(y - 0)^2}{6^2} = 1$$

$$\frac{x^2}{25} + \frac{y^2}{36} = 1$$

53. Vertices: $(\pm 7, 0) \Rightarrow a = 7$

Foci: $(\pm 6, 0) \Rightarrow c = 6$

$$\Rightarrow b = \sqrt{a^2 - c^2} = \sqrt{49 - 36} = \sqrt{13}$$

Center: $(0, 0)$

Horizontal major axis

$$\frac{x^2}{49} + \frac{y^2}{13} = 1$$

55. Foci: $(\pm 14, 0) \Rightarrow c = 14$

Center: $(0, 0)$

Minor axis of length $10 \Rightarrow b = 5$

$c^2 = a^2 - b^2$

$196 = a^2 - 25$

$a^2 = 221$

$$\frac{x^2}{221} + \frac{y^2}{25} = 1$$

57. $a = 5, b = 4, c = \sqrt{a^2 - b^2} = 3$

The foci should be placed 3 feet on either side of the center and have the same height as the pillars.

59. Vertices: $(0, \pm 1)$

Foci: $(0, \pm 5)$

Vertical transverse axis

Center: $(0, 0)$

$a = 1, c = 5$

$b = \sqrt{25 - 1} = \sqrt{24}$

$$\frac{y^2}{a^2} - \frac{x^2}{b^2} = 1$$

$$\frac{y^2}{1} - \frac{x^2}{24} = 1$$

61. Center: $(0, 0)$

Horizontal transverse axis

Vertices: $(\pm 1, 0) \Rightarrow a = 1$

Asymptotes: $y = \pm 2x \Rightarrow \dfrac{b}{a} = 2 \Rightarrow b = 2$

$$\frac{x^2}{1} - \frac{y^2}{4} = 1$$

63. Vertex: $(4, 2)$

Focus: $(4, 0)$

Vertical axis, $p = -2$

$(x - 4)^2 = 4(-2)(y - 2)$

$(x - 4)^2 = -8(y - 2)$

65. Vertex: $(-8, 8)$

Directrix: $y = 1$

Vertical axis

$p = 8 - 1 = 7$

$(x + 8)^2 = 4(7)(y - 8)$

$(x + 8)^2 = 28(y - 8)$

67. Vertices: $(0, 2), (4, 2) \Rightarrow a = 2$

Minor axis of length $2 \Rightarrow b = 1$

Center: $\left(\dfrac{0 + 4}{2}, \dfrac{2 + 2}{2}\right) = (2, 2)$

Horizontal major axis

$$\frac{(x - 2)^2}{2^2} + \frac{(y - 2)^2}{1} = 1$$

$$\frac{(x - 2)^2}{4} + (y - 2)^2 = 1$$

69. Vertices: $(2, -2), (2, 8) \Rightarrow a = 5$

Foci: $(2, 0), (2, 6) \Rightarrow c = 3$

Center: $\left(\dfrac{2 + 2}{2}, \dfrac{-2 + 8}{2}\right) = (2, 3)$

$$c^2 = a^2 - b^2 \Rightarrow b^2 = a^2 - c^2$$
$$b^2 = 25 - 9$$
$$b^2 = 16$$

Vertical major axis

$$\frac{(x - 2)^2}{4^2} + \frac{(y - 3)^2}{5^2} = 1$$

$$\frac{(x - 2)^2}{16} + \frac{(y - 3)^2}{25} = 1$$

71. Vertices: $(-10, 3), (6, 3) \Rightarrow a = 8$

Foci: $(-12, 3), (8, 3) \Rightarrow c = 10 \Rightarrow b^2 = \sqrt{c^2 - a^2} = \sqrt{100 - 64} = 6$

Horizontal transverse axis

Center: $(-2, 3)$

$$\frac{(x + 2)^2}{64} - \frac{(y - 3)^2}{36} = 1$$

73. Vertices: $(0, 0), (0, -4) \Rightarrow a = 2$

Passes through: $\left(2, 2\left(\sqrt{5} - 1\right)\right)$

Vertical transverse axis

Center: $(0, -2)$

$$\frac{(y + 2)^2}{4} - \frac{x^2}{b^2} = 1$$

$$\frac{\left(2\left(\sqrt{5} - 1\right) + 2\right)^2}{4} - \frac{2^2}{b^2} = 1$$

$$5 - \frac{4}{b^2} = 1$$

$$\frac{4}{b^2} = 4$$

$$b^2 = 1$$

$$\frac{(y + 2)^2}{4} - x^2 = 1$$

75. Foci: $(0, 0), (8, 0) \Rightarrow c = 4$

Asymptotes: $y = \pm 2(x - 4)$

Horizontal transverse axis

Center: $(4, 0)$

$$\frac{b}{a} = 2 \Rightarrow b = 2a$$

$$a^2 + b^2 = c^2$$

$$a^2 + (2a)^2 = 4^2$$

$$a^2 = \frac{16}{5}$$

$$b^2 = \frac{64}{5}$$

$$\frac{(x - h)^2}{a^2} - \frac{(y - k)^2}{b^2} = 1$$

$$\frac{5(x - 4)^2}{16} - \frac{5y^2}{64} = 1$$

77. $x^2 - 6x + 2y + 9 = 0$

$$(x - 3)^2 = -2y$$

$$(x - 3)^2 = 4\left(-\frac{1}{2}\right)y \Rightarrow p = -\frac{1}{2}$$

Parabola

Vertex: $(3, 0)$

Focus: $\left(3, -\frac{1}{2}\right)$

The graph of $x^2 = -2y$ has been shifted to the right three units.

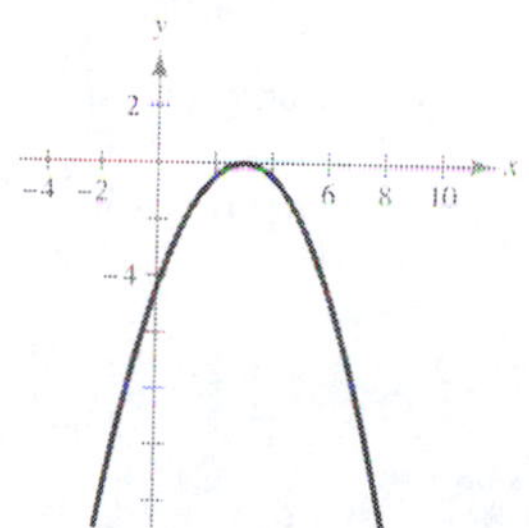

79.

$$x^2 + y^2 - 2x - 4y + 5 = 0$$

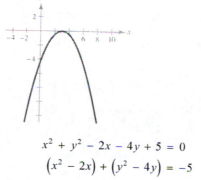

$$\left(x^2 - 2x\right) + \left(y^2 - 4y\right) = -5$$

$$\left(x^2 - 2x + 1\right) + \left(y^2 - 4y + 4\right) = -5 + 1 + 4$$

$$(x - 1)^2 + (y - 2)^2 = 0$$

Point: $(1, 2)$

Note: This is a degenerate conic, a circle of radius zero. The graph has been shifted to the right one unit and upward two units from the origin.

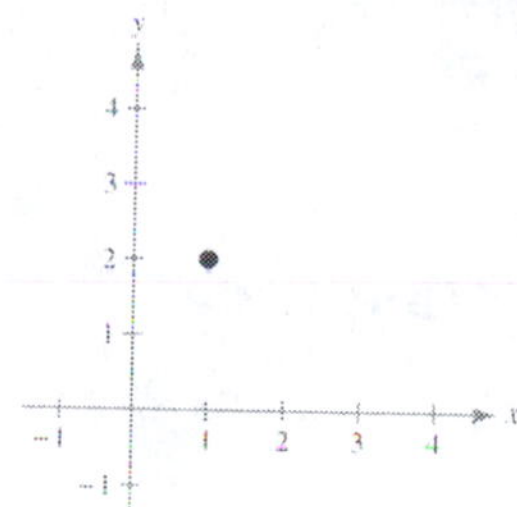

81.

$$x^2 + 9y^2 + 10x - 18y + 25 = 0$$

$$\left(x^2 + 10x\right) + 9\left(y^2 - 2y\right) = -25$$

$$\left(x^2 + 10x + 25\right) + 9\left(y^2 - 2y + 1\right) = -25 + 25 + 9$$

$$(x + 5)^2 + 9(y - 1)^2 = 9$$

$$\frac{(x + 5)^2}{9} + (y - 1)^2 = 1$$

Ellipse

Center: $(-5, 1)$

Vertices: $(-8, 1), (-2, 1)$

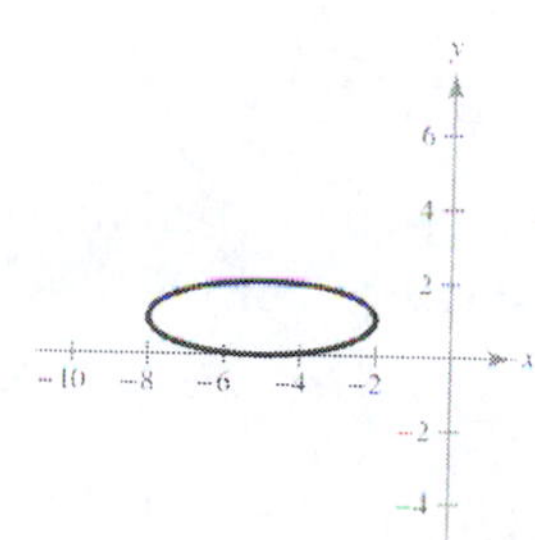

The graph of $\frac{x^2}{9} + y^2 = 1$ has been shifted left five units and upward one unit.

83.
$$9x^2 - y^2 - 72x + 8y + 119 = 0$$
$$9(x^2 - 8x) - (y^2 - 8y) = -119$$
$$9(x^2 - 8x + 16) - (y^2 - 8y + 16) = -119 + 144 - 16$$
$$9(x - 4)^2 - (y - 4)^2 = 9$$
$$\frac{(x - 4)^2}{1} - \frac{(y - 4)^2}{9} = 1$$

Hyperbola

Center: $(4, 4)$

Vertices: $(3, 4), (5, 4)$

The graph of $x^2 - \frac{y^2}{9} = 1$ has been shifted right four units and upward four units from standard position.

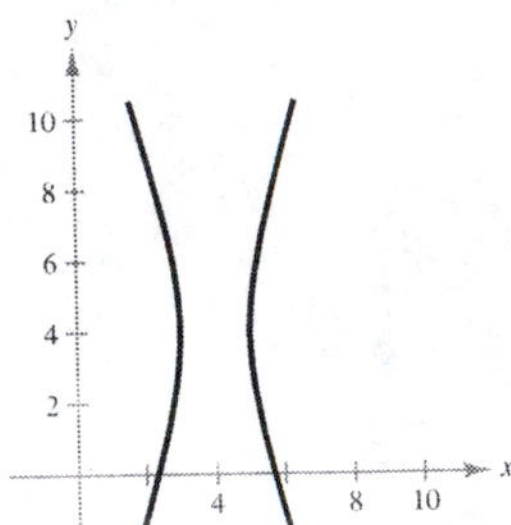

85.
$$x^2 = 4p(y - 12)$$
$$(\pm 4)^2 = 4p(10 - 12)$$
$$16 = -8p$$
$$-2 = p$$
$$x^2 = 4(-2)(y - 12)$$
$$x^2 = -8(y - 12)$$

When $y = 0$, we have:

$$x^2 = 96$$
$$x = \pm\sqrt{96} = \pm 4\sqrt{6}$$

At ground level, the width is $2x = 8\sqrt{6} \approx 19.6$ meters.

87. $x^2 + y^2 - 200x - 52{,}500 = 0$

(a) The conic is a circle because x^2 and y^2 have identical coefficients.

(b)
$$x^2 - 200x + 10{,}000 + y^2 = 52{,}500 + 10{,}000$$
$$(x - 100)^2 + y^2 = 62{,}500$$

Center: $(100, 0)$

Radius: 250

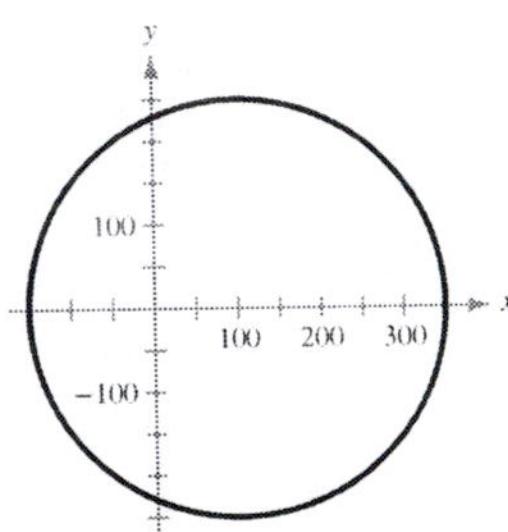

89. True. (See Exercise 79.)

Problem Solving for Chapter 4

1. $f(x) = \dfrac{ax + b}{cx + d}$

Vertical asymptote: $x = -\dfrac{d}{c}$

Horizontal asymptote: $y = \dfrac{a}{c}$

(i) $a > 0, b < 0, c > 0, d < 0$

Both the vertical asymptote and the horizontal asymptote are positive. Matches graph (d).

(ii) $a > 0, b > 0, c < 0, d < 0$

Both the vertical asymptote and the horizontal asymptote are negative. Matches graph (b).

(iii) $a < 0, b > 0, c > 0, d < 0$

The vertical asymptote is positive and the horizontal asymptote is negative. Matches graph (a).

(iv) $a > 0, b < 0, c > 0, d > 0$

The vertical asymptote is negative and the horizontal asymptote is positive. Matches graph (c).

3. (a)

Age, x	Near point, y
16	3.0
32	4.7
44	9.8
50	19.7
60	39.4

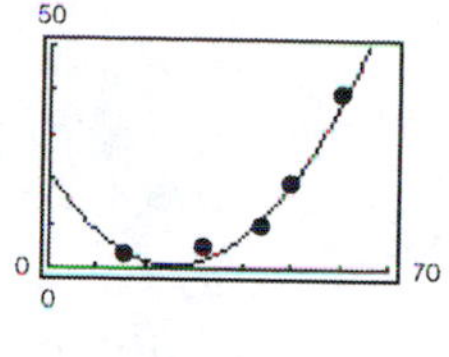

$y \approx 0.031x^2 - 1.59x + 21.0$

(b) $\frac{1}{y} \approx -0.007x + 0.44$

$y \approx \frac{1}{-0.007x + 0.44}$

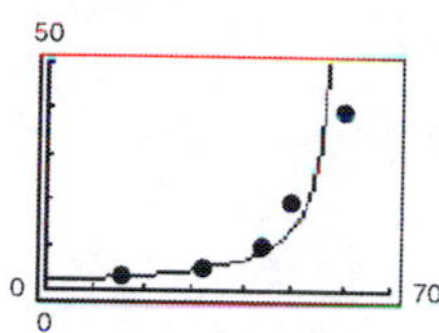

(c)

Age, x	Near point, y	Quadratic Model	Rational Model
16	3.0	3.66	3.05
32	4.7	2.32	4.63
44	9.8	11.83	7.58
50	19.7	19.97	11.11
60	39.4	38.54	50.00

The models are fairly good fits to the data. The quadratic model seems to be a better fit for older ages and the rational model a better fit for younger ages.

(d) For $x = 25$, the quadratic model yields $y \approx 0.625$ inch and the rational model yields $y \approx 3.774$ inches.

(e) The reciprocal model cannot be used to predict the near point for a person who is 70 years old because it results in a negative value $(y \approx -20)$. The quadratic model yields $y \approx 63.37$ inches.

5. A hyperbola is the set of all points (x, y) in a plane, the difference of whose distances from two distinct fixed points (foci) is a positive constant.

Using the vertex $(a, 0)$, the distance to $(-c, 0)$ is $a + c$, and the distance to $(c, 0)$ is $(c - a)$.

$d_2 - d_1 = (a + c) - (c - a) = 2a$

By definition, $|d_2 - d_1|$ is constant for any point (x, y) on the hyperbola. Thus, $|d_2 - d_1| = 2a$.

7.

(2.04, 3.5)

7 in.

(−2.04, 3.5)

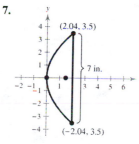

Place the vertex at $(0, 0)$. Then, $y^2 = 4px$ and $p = 1.5 \Rightarrow y^2 = 6x$.

When $y = 3.5$, we have $x = \frac{(3.5)^2}{6} \approx 2.04$.

The reflector is approximately 2.04 inches deep.

9. Tangent line: $y - y_1 = \dfrac{x_1}{2p}(x - x_1)$

(a) The slope is $m = \dfrac{x_1}{2p}$.

(b) For $x^2 = 4y$ or $y = x^2/4$, the endpoints of the chord are $(\pm 2, 1)$.

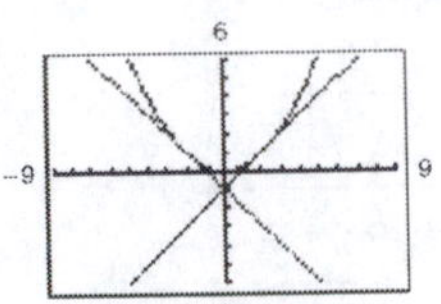

The tangent lines are:

$$y - 1 = \frac{2}{2(1)}(x - 2) \Rightarrow y = x - 1$$

$$y - 1 = \frac{-2}{2(1)}(x + 2) \Rightarrow y = -x - 1$$

For $x^2 = 8y$ or $y = x^2/8$, the endpoints of the chord are $(\pm 4, 2)$.

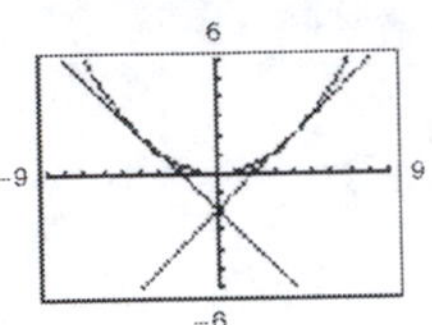

The tangent lines are:

$$y - 2 = \frac{4}{2(2)}(x - 4) \Rightarrow y = x - 2$$

$$y - 2 = \frac{-4}{2(2)}(x + 4) \Rightarrow y = -x - 2$$

For $x^2 = 12y$ or $y = x^2/12$, the endpoints of the chord are $(\pm 6, 3)$.

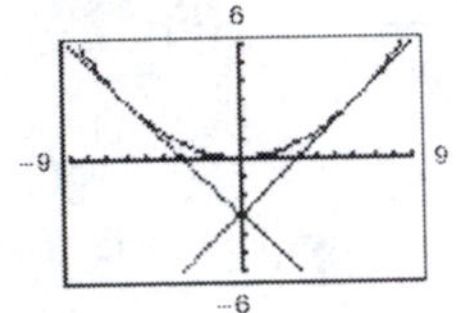

The tangent lines are:

$$y - 3 = \frac{6}{2(3)}(x - 6) \Rightarrow y = x - 3$$

$$y - 3 = \frac{-6}{2(3)}(x + 6) \Rightarrow y = -x - 3$$

For $x^2 = 16y$ or $y = x^2/16$, the endpoints of the chord are $(\pm 8, 4)$.

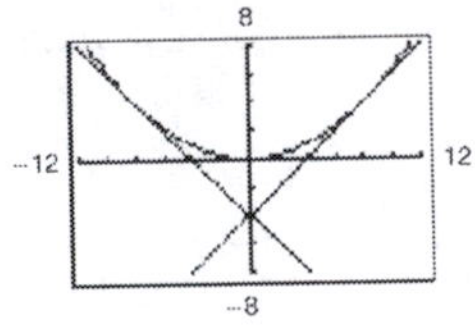

The tangent lines are:

$$y - 4 = \frac{8}{2(4)}(x - 8) \Rightarrow y = x - 4$$

$$y - 4 = \frac{-8}{2(4)}(x + 8) \Rightarrow y = -x - 4$$

11. $Ax^2 + Cy^2 + Dx + Ey + F = 0$

Assume that the conic is *not* degenerate.

(a) $A = C, A \neq 0$

$$Ax^2 + Ay^2 + Dx + Ey + F = 0$$

$$x^2 + y^2 + \frac{D}{A}x + \frac{E}{A}y + \frac{F}{A} = 0$$

$$\left(x^2 + \frac{D}{A}x + \frac{D^2}{4A^2}\right) + \left(y^2 + \frac{E}{A}y + \frac{E^2}{4A^2}\right) = -\frac{F}{A} + \frac{D^2}{4A^2} + \frac{E^2}{4A^2}$$

$$\left(x + \frac{D}{2A}\right)^2 + \left(y + \frac{E}{2A}\right)^2 = \frac{D^2 + E^2 - 4AF}{4A^2}$$

This is a circle with center $\left(-\frac{D}{2A}, -\frac{E}{2A}\right)$ and radius $\frac{\sqrt{D^2 + E^2 - 4AF}}{2|A|}$.

(b) $A = 0$ or $C = 0$ (but not both). Let $C = 0$.

$Ax^2 + Dx + Ey + F = 0$

$$x^2 + \frac{D}{A}x = -\frac{E}{A}y - \frac{F}{A}$$

$$x^2 + \frac{D}{A}x + \frac{D^2}{4A^2} = -\frac{E}{A}y - \frac{F}{A} + \frac{D^2}{4A^2}$$

$$\left(x + \frac{D}{2A}\right)^2 = -\frac{E}{A}\left(y + \frac{F}{E} - \frac{D^2}{4AE}\right)$$

This is a parabola with vertex $\left(-\frac{D}{2A}, \frac{D^2 - 4AF}{4AE}\right)$. $A = 0$ yields a similar result.

(c) $AC > 0 \Rightarrow A$ and C are either both positive or are both negative (if that is the case, move the terms to the other side of the equation so that they are both positive).

$$Ax^2 + Cy^2 + Dx + Ey + F = 0$$

$$A\left(x^2 + \frac{D}{A}x + \frac{D^2}{4A^2}\right) + C\left(y^2 + \frac{E}{C}y + \frac{E^2}{4C^2}\right) = -F + \frac{D^2}{4A} + \frac{E^2}{4C}$$

$$A\left(x + \frac{D}{2A}\right)^2 + C\left(y + \frac{E}{2C}\right)^2 = \frac{CD^2 + AE^2 - 4ACF}{4AC}$$

$$\frac{\left(x + \frac{D}{2A}\right)^2}{\frac{CD^2 + AE^2 - 4ACF}{4A^2C}} + \frac{\left(y + \frac{E}{2C}\right)^2}{\frac{CD^2 + AE^2 - 4ACF}{4AC^2}} = 1$$

Since A and C are both positive, $4A^2C$ and $4AC^2$ are both positive. $CD^2 + AE^2 - 4ACF$ must be positive or the conic is degenerate. Thus, we have an ellipse with center $(-D/2A, -E/2C)$.

(d) $AC < 0 \Rightarrow A$ and C have opposite signs. Let's assume that A is positive and C is negative. (If A is negative and C is positive, move the terms to the other side of the equation.) From part (c) we have

$$\frac{\left(x + \frac{D}{2A}\right)^2}{\frac{CD^2 + AE^2 - 4ACF}{4A^2C}} + \frac{\left(y + \frac{E}{2C}\right)^2}{\frac{CD^2 + AE^2 - 4ACF}{4AC^2}} = 1.$$

Since $A > 0$ and $C < 0$, the first denominator is positive if $CD^2 + AE^2 - 4ACF < 0$ and is negative if $CD^2 + AE^2 - 4ACF > 0$, since $4A^2C$ is negative. The second denominator would have the *opposite* sign since $4AC^2 > 0$. Thus, we have a hyperbola with center $(-D/2A, -E/2C)$.

Practice Test for Chapter 4

1. Sketch the graph of $f(x) = \dfrac{x - 1}{2x}$ and label all intercepts and asymptotes.

2. Sketch the graph of $f(x) = \dfrac{3x^2 - 4}{x}$ and label all intercepts and asymptotes.

3. Find all the asymptotes of $f(x) = \dfrac{8x^2 - 9}{x^2 + 1}$.

4. Find all the asymptotes of $f(x) = \dfrac{4x^2 - 2x + 7}{x - 1}$.

5. Sketch the graph of $f(x) = \dfrac{x - 5}{(x - 5)^2}$.

6. Find the vertex, focus, and directrix of the parabola $x^2 = 20y$.

7. Find the equation of the parabola with vertex $(0, 0)$ and focus $(7, 0)$.

8. Find the center, foci, and vertices of the ellipse $\dfrac{x^2}{144} + \dfrac{y^2}{25} = 1$.

9. Find the equation of the ellipse with foci $(\pm 4, 0)$ and minor axis of length 6.

10. Find the center, vertices, foci, and asymptotes of the hyperbola $\dfrac{y^2}{144} - \dfrac{x^2}{169} = 1$.

11. Find the equation of the hyperbola with vertices $(\pm 4, 0)$ and asymptotes of $y = \pm\frac{1}{2}x$.

12. Find the equation of the parabola with vertex $(6, -1)$ and focus $(6, 3)$.

13. Find the center, foci, and vertices of the ellipse $16x^2 + 9y^2 - 96x + 36y + 36 = 0$.

14. Find the equation of the ellipse with vertices $(-1, 1)$ and $(7, 1)$ and minor axis of length 2.

15. Find the center, vertices, foci, and asymptotes of the hyperbola $4(x + 3)^2 - 9(y - 1)^2 = 1$.

16. Find the equation of the hyperbola with vertices $(3, 4)$ and $(3, -4)$ and foci $(3, 7)$ and $(3, -7)$.

CHAPTER 5
Exponential and Logarithmic Functions

CHAPTER 5
Exponential and Logarithmic Functions

Section 5.1 Exponential Functions and Their Graphs

1. algebraic

3. One-to-One

5. $A = P\left(1 + \frac{r}{n}\right)^{nt}$

7. $f(1.4) = (0.9)^{1.4} \approx 0.863$

9. $f(-\pi) = 5^{-\pi} \approx 0.006$

11. $g(x) = 5000(2^x) = 5000(2^{-1.5})$
≈ 1767.767

13. $f(x) = 2^x$
Increasing
Asymptote: $y = 0$
Intercept: $(0, 1)$
Matches graph (d).

14. $f(x) = 2^x + 1$
Increasing
Asymptote: $y = 1$
Intercept: $(0, 2)$
Matches graph (c).

15. $f(x) = 2^{-x}$
Decreasing
Asymptote: $y = 0$
Intercept: $(0, 1)$
Matches graph (a).

16. $f(x) = 2^{x-2}$
Increasing
Asymptote: $y = 0$
Intercept: $\left(0, \frac{1}{4}\right)$
Matches graph (b).

17. $f(x) = \left(\frac{1}{2}\right)^x$

x	−2	−1	0	1	2
$f(x)$	4	2	1	0.5	0.25

Asymptote: $y = 0$

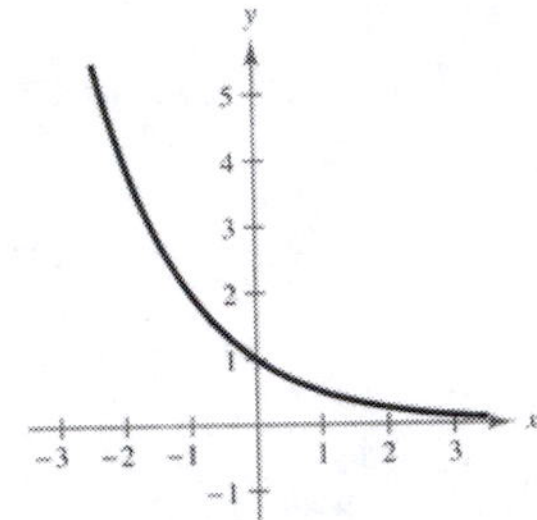

19. $f(x) = 6^{-x}$

x	−2	−1	0	1	2
$f(x)$	36	6	1	0.167	0.028

Asymptote: $y = 0$

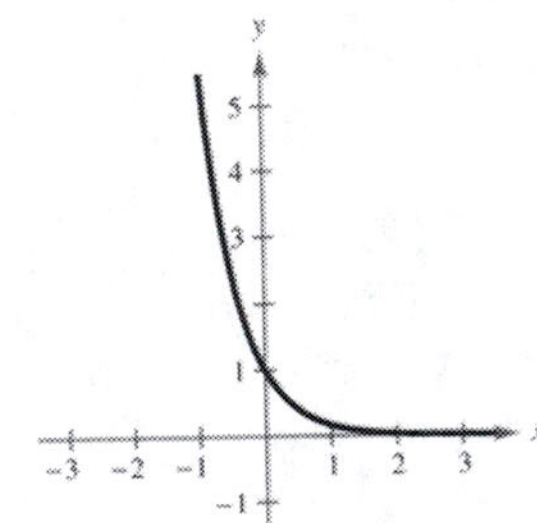

21. $f(x) = 2^{x-1}$

x	−2	−1	0	1	2
$f(x)$	0.125	0.25	0.5	1	2

Asymptote: $y = 0$

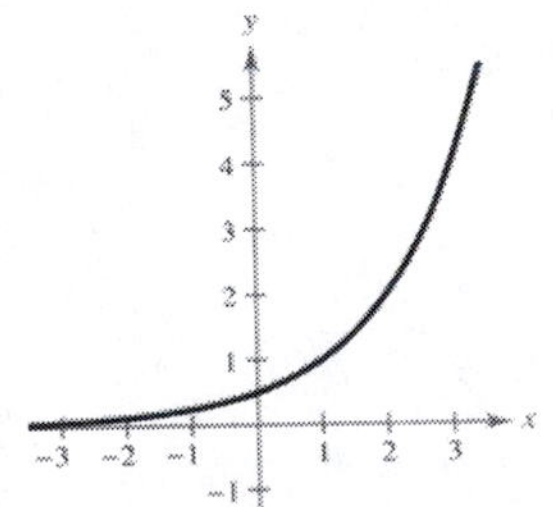

23. $3^{x+1} = 27$

$3^{x+1} = 3^3$

$x + 1 = 3$

$x = 2$

25. $\left(\frac{1}{2}\right)^x = 32$

$\left(\frac{1}{2}\right)^x = \left(\frac{1}{2}\right)^{-5}$

$x = -5$

27. $f(x) = 3^x,\ g(x) = 3^x + 1$

Because $g(x) = f(x) + 1$, the graph of g can be obtained by shifting the graph of f one unit upward.

29. $f(x) = \left(\frac{7}{2}\right)^x,\ g(x) = -\left(\frac{7}{2}\right)^{-x}$

Because $g(x) = -f(-x)$, the graph of g can be obtained by reflecting the graph of f in the x-axis and y-axis.

31. $y = 2^{-x^2}$

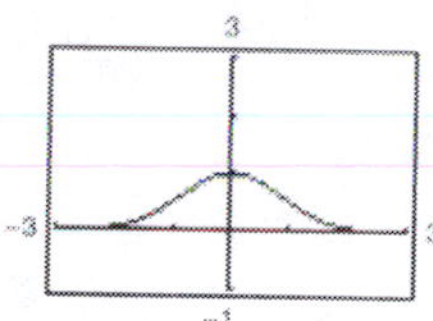

33. $f(x) = 3^{x-2} + 1$

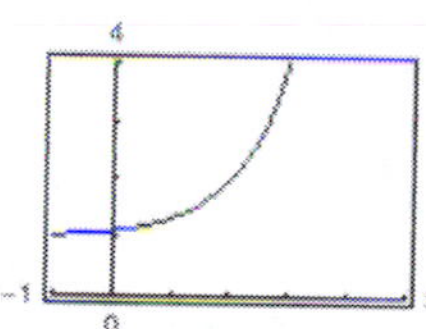

35. $f(x) = e^x = e^{3.2} \approx 24.533$

37. $f(6) = 5000e^{0.06(6)} \approx 7166.647$

39. $f(x) = e^x$

x	−2	−1	0	1	2
$f(x)$	0.135	0.368	1	2.718	7.389

Asymptote: $y = 0$

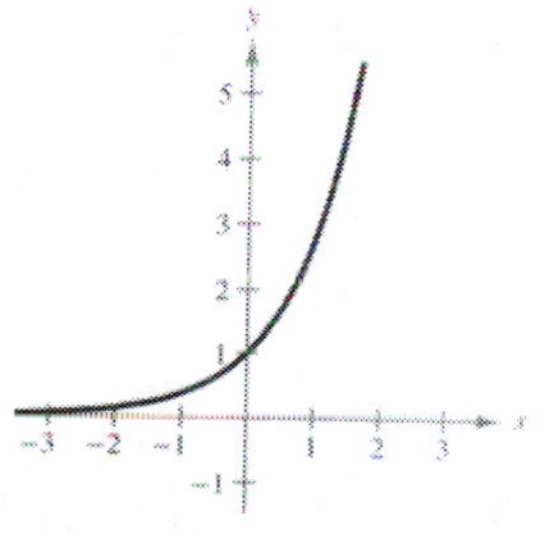

41. $f(x) = 3e^{x+4}$

x	−8	−7	−6	−5	−4
$f(x)$	0.055	0.149	0.406	1.104	3

Asymptote: $y = 0$

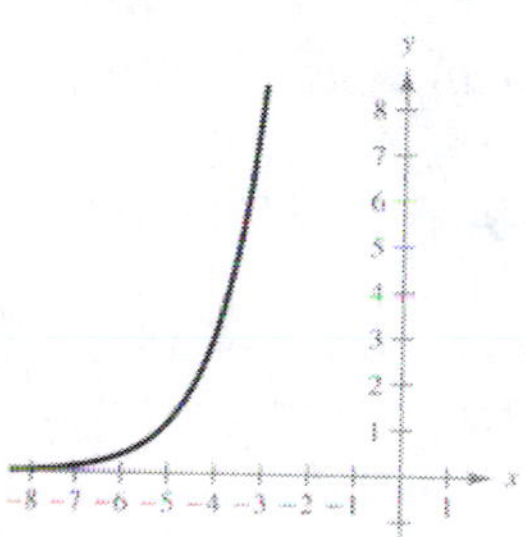

43. $f(x) = 2e^{x-2} + 4$

x	−2	−1	0	1	2
$f(x)$	4.037	4.100	4.271	4.736	6

Asymptote: $y = 4$

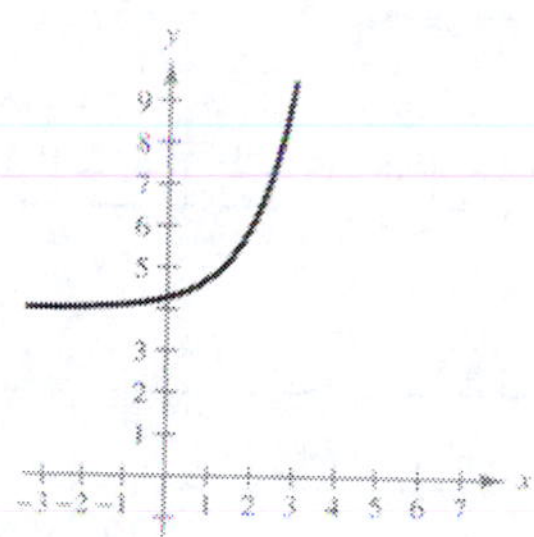

45. $y = 1.08e^{-5x}$

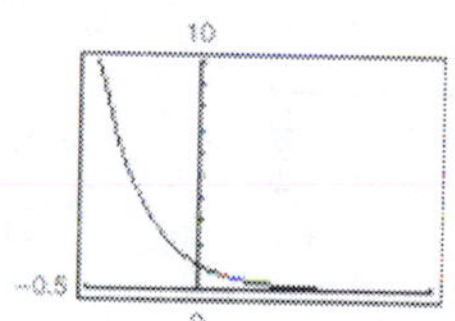

47. $s(t) = 2e^{0.12t}$

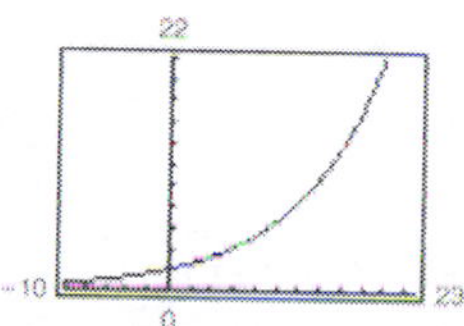

49. $g(x) = 1 + e^{-x}$

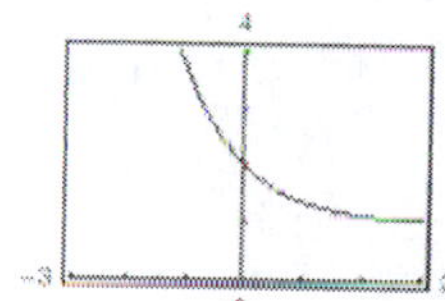

51. $e^{3x+2} = e^3$

$3x + 2 = 3$

$3x = 1$

$x = \frac{1}{3}$

53. $e^{x^2-3} = e^{2x}$

$x^2 - 3 = 2x$

$x^2 - 2x - 3 = 0$

$(x - 3)(x + 1) = 0$

$x = 3$ or $x = -1$

55. $P = \$1500, r = 2\%, t = 10$ years

Compounded n times per year: $A = P\left(1 + \frac{r}{n}\right)^{nt} = 1500\left(1 + \frac{0.02}{n}\right)^{10n}$

Compounded continuously: $A = Pe^{rt} = 1500e^{0.02(10)}$

n	1	2	4	12	365	Continuous
A	\$1828.49	\$1830.29	\$1831.19	\$1831.80	\$1832.09	\$1832.10

57. $P = \$2500, r = 4\%, t = 20$ years

Compounded n times per year: $A = P\left(1 + \frac{r}{n}\right)^{nt} = 2500\left(1 + \frac{0.04}{n}\right)^{20n}$

Compounded continuously: $A = Pe^{rt} = 2500e^{0.04(20)}$

n	1	2	4	12	365	Continuous
A	\$5477.81	\$5520.10	\$5541.79	\$5556.46	\$5563.61	\$5563.85

59. $A = Pe^{rt} = 12{,}000e^{0.04t}$

t	10	20	30	40	50
A	\$17,901.90	\$26,706.49	\$39,841.40	\$59,436.39	\$88,668.67

61. $A = Pe^{rt} = 12{,}000e^{0.065t}$

t	10	20	30	40	50
A	\$22,986.49	\$44,031.56	\$84,344.25	\$161,564.86	\$309,484.08

63. $A = 30{,}000e^{(0.05)(25)} \approx \$104{,}710.29$

65. $C(10) = 23.95(1.04)^{10} \approx \$35.45$

67. (a)

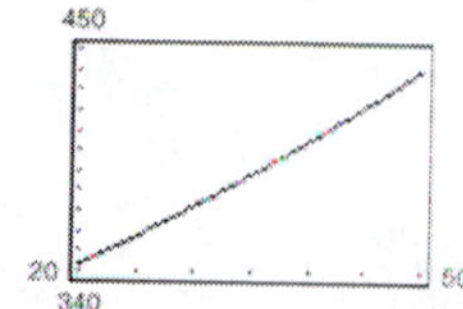

(b)

t	20	21	22	23
P (in millions)	342.748	345.604	348.485	351.389

t	24	25	26	27
P (in millions)	354.318	357.271	360.249	363.251

t	28	29	30	31
P (in millions)	366.279	369.331	372.410	375.513

t	32	33	34	35
P (in millions)	378.643	381.799	384.981	388.190

t	36	37	38	39
P (in millions)	391.425	394.687	397.977	401.294

t	40	41	42	43
P (in millions)	404.639	408.011	411.412	414.840

t	44	45	46	47
P (in millions)	418.298	421.784	425.300	428.844

t	48	49	50
P (in millions)	432.419	436.023	439.657

(c) Using the table of values created by the model, the population will exceed 400 million in 2038.

69. $Q = 16\left(\frac{1}{2}\right)^{t/24{,}100}$

(a) $Q(0) = 16$ grams

(b) $Q(75{,}000) \approx 1.85$ grams

(c)

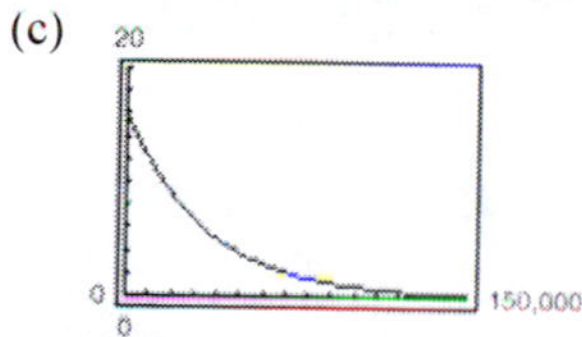

71. (a) $V(t) = 49{,}810\left(\frac{7}{8}\right)^t$ where t is the number of years since it was purchased.

(b) $V(4) = 49{,}810\left(\frac{7}{8}\right)^4 \approx 29{,}197.71$

After 4 years, the value of the van is about $29,198.

73. True. The line $y = -2$ is a horizontal asymptote for the graph of $f(x) = 10^x - 2$.

75.
$$\begin{aligned} f(x) &= 3^{x-2} \\ &= 3^x 3^{-2} \\ &= 3^x\left(\frac{1}{3^2}\right) \\ &= \frac{1}{9}\left(3^x\right) \\ &= h(x) \end{aligned}$$

So, $f(x) \neq g(x)$, but $f(x) = h(x)$.

77.
$$\begin{aligned} f(x) &= 16\left(4^{-x}\right) \\ &= 4^2\left(4^{-x}\right) \\ &= 4^{2-x} \\ &= \left(\frac{1}{4}\right)^{-(2-x)} \\ &= \left(\frac{1}{4}\right)^{x\;2} \\ &= g(x) \end{aligned}$$

and

$$\begin{aligned} f(x) &= 16\left(4^{-x}\right) \\ &= 16\left(2^2\right)^{-x} \\ &= 16\left(2^{-2x}\right) \\ &= h(x) \end{aligned}$$

So, $f(x) = g(x) = h(x)$.

79. $y = 3^x$ and $y = 4^x$

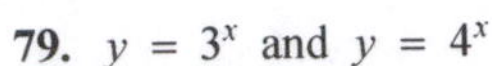

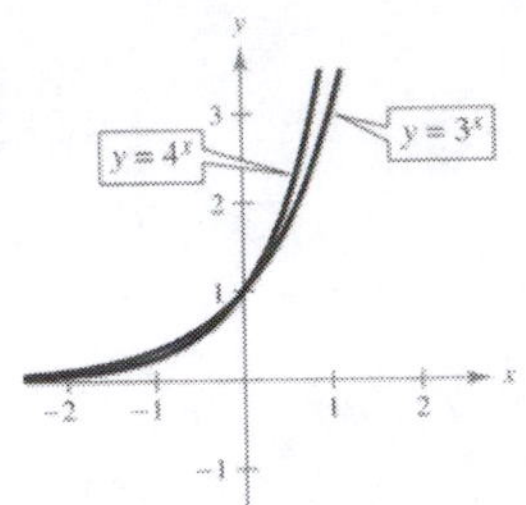

x	-2	-1	0	1	2
3^x	$\frac{1}{9}$	$\frac{1}{3}$	1	3	9
4^x	$\frac{1}{16}$	$\frac{1}{4}$	1	4	16

(a) $4^x < 3^x$ when $x < 0$.

(b) $4^x > 3^x$ when $x > 0$.

81.

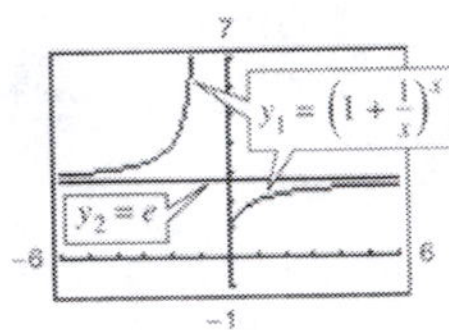

As x increases, the graph of y_1 approaches e, which is y_2.

83. (a)

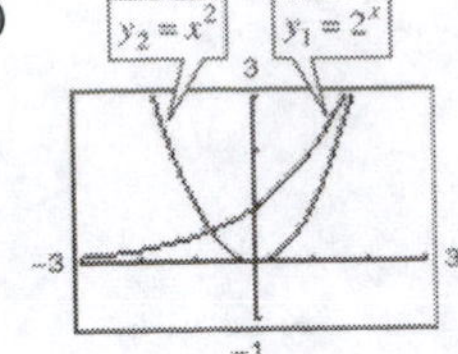

At $x = 2$, both functions have a value of 4. The function y_1 increases for all values of x. The function y_2 is symmetric with respect to the y-axis.

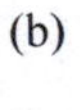

(b)

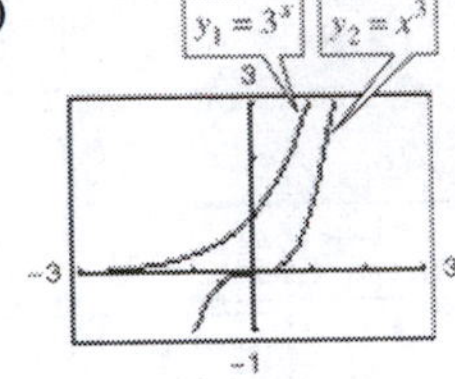

Both functions are increasing for all values of x. For $x > 0$, both functions have a similar shape. The function y_2 is symmetric with respect to the origin.

85. The functions (c) 3^x and (d) 2^{-x} are exponential.

Section 5.2 Logarithmic Functions and Their Graphs

1. logarithmic

3. natural; e

5. $x = y$

7. $\log_4 16 = 2 \Rightarrow 4^2 = 16$

9. $\log_{32} 4 = \frac{2}{5} \Rightarrow 32^{2/5} = 4$

11. $5^3 = 125 \Rightarrow \log_5 125 = 3$

13. $4^{-3} = \frac{1}{64} \Rightarrow \log_4 \frac{1}{64} = -3$

15. $f(x) = \log_2 x$

$f(64) = \log_2 64 = 6$ because $2^6 = 64$

17. $f(x) = \log_8 x$

$f(1) = \log_8 1 = 0$ because $8^0 = 1$

19. $g(x) = \log_a x$

$g(a^2) = \log_a a^2$

$= 2$ by the Inverse Property

21. $f(x) = \log x$

$f\left(\frac{7}{8}\right) = \log\left(\frac{7}{8}\right) \approx -0.058$

23. $f(x) = \log x$

$f(12.5) = \log 12.5 \approx 1.097$

25. $\log_{11} 11^7 = 7$ because $11^7 = 11^7$

27. $\log_\pi \pi = 1$ because $\pi^1 = \pi$.

29. $\log_5(x + 1) = \log_5 6$

$x + 1 = 6$

$x = 5$

31. $\log(2x + 1) = \log 15$

$2x + 1 = 15$

$x = 7$

33.

x	-2	-1	0	1	2
$f(x) = 7^x$	$\frac{1}{49}$	$\frac{1}{7}$	1	7	49

x	$\frac{1}{49}$	$\frac{1}{7}$	1	7	49
$g(x) = \log_7 x$	-2	-1	0	1	2

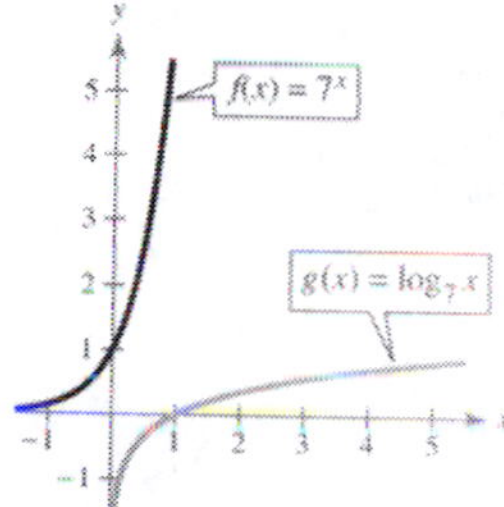

35.

x	-2	-1	0	1	2
$f(x) = 5^x$	$\frac{1}{25}$	$\frac{1}{5}$	1	5	25

x	$\frac{1}{36}$	$\frac{1}{6}$	1	6	36
$g(x) = \log_6 x$	-2	-1	0	1	2

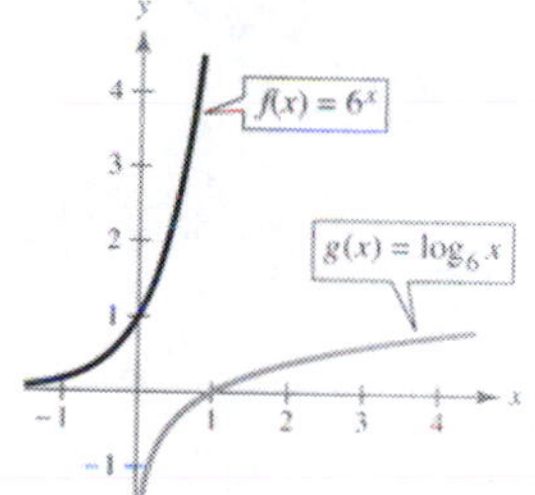

37. $f(x) = -\log_3(x + 2)$

Asymptote: $x = -2$

Point on graph: $(-1, 0)$

Matches graph (c).

The graph of $f(x)$ is obtained by reflecting the graph of $g(x)$ in the x-axis and shifting the graph two units to the left.

38. $f(x) = \log_3(x - 1)$

Asymptote: $x = 1$

Point on graph: $(2, 0)$

Matches graph (d).

$f(x)$ shifts $g(x)$ one unit to the right.

39. $f(x) = \log_3(1 - x) = \log_3[-(x - 1)]$

Asymptote: $x = 1$

Point on graph: $(0, 0)$

Matches graph (b).

The graph of $f(x)$ is obtained by reflecting the graph of $g(x)$ in the y-axis and shifting the graph one unit to the right.

40. $f(x) = -\log_3(-x)$

Asymptote: $x = 0$

Point on graph: $(-1, 0)$

Matches graph (a).

$f(x)$ reflects $g(x)$ in the x-axis the reflects that graph in the y-axis.

41. $f(x) = \log_4 x$

Domain: $(0, \infty)$

x-intercept: $(1, 0)$

Vertical asymptote: $x = 0$

$y = \log_4 x \Rightarrow 4^y = x$

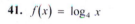

x	$\frac{1}{4}$	1	4	2
$f(x)$	-1	0	1	$\frac{1}{2}$

43. $y = -\log_3 x + 2$

Domain: $(0, \infty)$

x-intercept:

$-\log_3 x + 2 = 0$

$2 = \log_3 x$

$3^2 = x$

$9 = x$

The x-intercept is $(9, 0)$.

Vertical asymptote: $x = 0$

$y = -\log_3 x + 2$

$\log_3 x = 2 - y \Rightarrow 3^{2-y} = x$

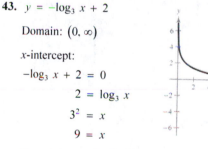

x	27	9	3	1	$\frac{1}{3}$
y	-1	0	1	2	3

45. $f(x) = -\log_6(x + 2)$

Domain: $x + 2 > 0 \Rightarrow x > -2$

The domain is $(-2, \infty)$.

x-intercept:

$0 = -\log_6(x + 2)$

$0 = \log_6(x + 2)$

$6^0 = x + 2$

$1 = x + 2$

$-1 = x$

The x-intercept is $(-1, 0)$.

Vertical asymptote: $x + 2 = 0 \Rightarrow x = -2$

$y = -\log_6(x + 2)$

$-y = \log_6(x + 2)$

$6^{-y} - 2 = x$

x	4	-1	$-1\frac{5}{6}$	$-1\frac{35}{36}$
$f(x)$	-1	0	1	2

47. $y = \log\left(\dfrac{x}{7}\right)$

Domain: $\dfrac{x}{7} > 0 \Rightarrow x > 0$

The domain is $(0, \infty)$.

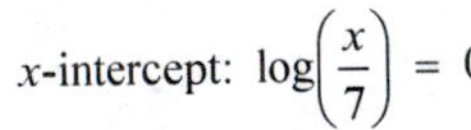

x-intercept: $\log\left(\dfrac{x}{7}\right) = 0$

$\dfrac{x}{7} = 10^0$

$\dfrac{x}{7} = 1$

$x = 7$

The x-intercept is $(7, 0)$.

Vertical asymptote: $\dfrac{x}{7} = 0 \Rightarrow x = 0$

The vertical asymptote is the y-axis.

x	1	2	3	4	5
y	-0.85	-0.54	-0.37	-0.24	-0.15

x	6	7	8
y	-0.069	0	0.06

49. $\ln \frac{1}{2} = -0.693\ldots \Rightarrow e^{-0.693\ldots} = \frac{1}{2}$

51. $\ln 250 = 5.521\ldots \Rightarrow e^{5.521\ldots} = 250$

53. $e^2 = 7.3890\ldots \Rightarrow \ln 7.3890\ldots = 2$

55. $e^{-0.9} = 0.406\ldots \Rightarrow \ln 0.406\ldots = -0.9$

57. $f(x) = \ln x$

$f(18.42) = \ln 18.42 \approx 2.913$

59. $g(x) = 8 \ln x$

$g(0.05) = 8 \ln 0.05 \approx -23.966$

61. $g(x) = \ln x$

$g(e^5) = \ln e^5 = 5$ by the Inverse Property

63. $g(x) = \ln x$

$g(e^{-5/6}) = \ln e^{-5/6} = -\frac{5}{6}$ by the Inverse Property

65. $f(x) = \ln(x - 4)$

Domain: $x - 4 > 0 \Rightarrow x > 4$

The domain is $(4, \infty)$.

x-intercept: $0 = \ln(x - 4)$

$e^0 = x - 4$

$5 = x$

The x-intercept is $(5, 0)$.

Vertical asymptote: $x - 4 = 0 \Rightarrow x = 4$

x	4.5	5	6	7
$f(x)$	-0.69	0	0.69	1.10

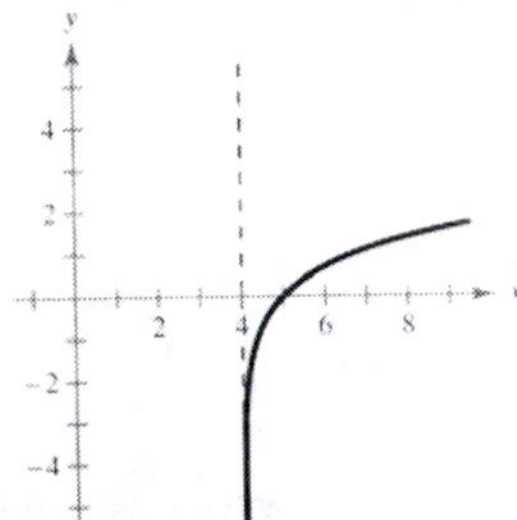

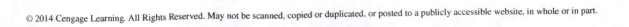

67. $g(x) = \ln(-x)$

Domain: $-x > 0 \Rightarrow x < 0$

The domain is $(-\infty, 0)$.

x-intercept:

$0 = \ln(-x)$

$e^0 = -x$

$-1 = x$

The x-intercept is $(-1, 0)$.

Vertical asymptote: $-x = 0 \Rightarrow x = 0$

x	-0.5	-1	-2	-3
$g(x)$	-0.69	0	0.69	1.10

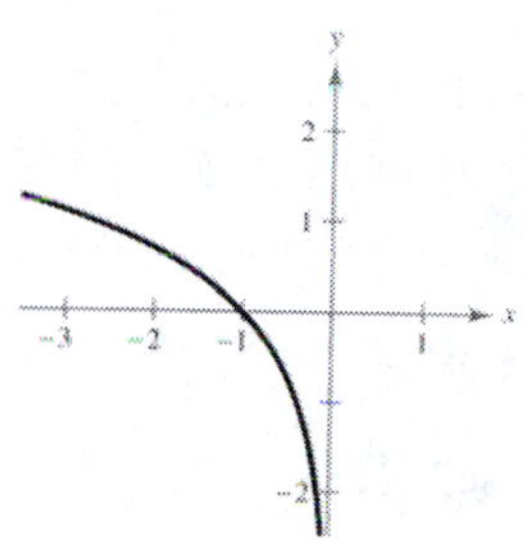

69. $f(x) = \ln(x - 1)$

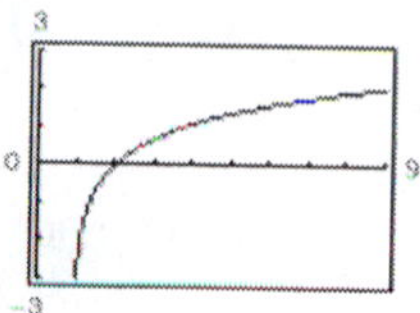

71. $f(x) = \ln x + 8$

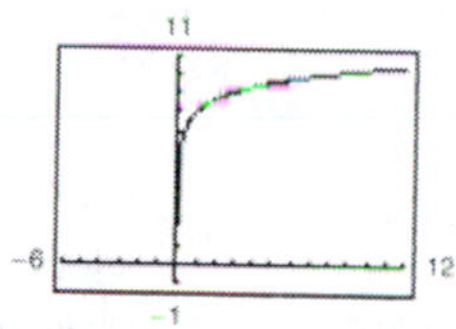

73. $\ln(x + 4) = \ln 12$

$x + 4 = 12$

$x = 8$

75. $\ln(x^2 - 2) = \ln 23$

$x^2 - 2 = 23$

$x^2 = 25$

$x = \pm 5$

77. $t = 16.625 \ln\left(\dfrac{x}{x - 750}\right), x > 750$

(a) When $x = \$897.72$: $t = 16.625 \ln\left(\dfrac{897.72}{897.72 - 750}\right) \approx 30$ years

When $x = \$1659.24$: $t = 16.625 \ln\left(\dfrac{1659.24}{1659.24 - 750}\right) \approx 10$ years

(b) Total amounts: $(897.72)(12)(30) = \$323{,}179.20 \approx \$323{,}179$

$(1659.24)(12)(10) = \$199{,}108.80 \approx \$199{,}109$

(c) Interest charges: $323{,}179.20 - 150{,}000 = \$173{,}179.20 \approx \$173{,}179$

$199{,}108.80 - 150{,}000 = \$49{,}108.80 \approx \$49{,}109$

(d) The vertical asymptote is $x = 750$. The closer the payment is to \$750 per month, the longer the length of the mortgage will be. Also, the monthly payment must be greater than \$750.

79. $t = \dfrac{\ln 2}{r}$

(a)

r	0.005	0.010	0.015	0.020	0.025	0.030
t	138.6	69.3	46.2	34.7	27.7	23.1

(b)

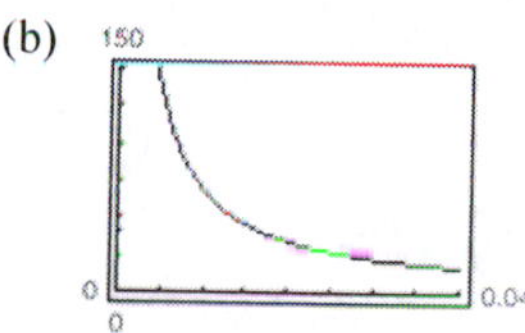

81. $f(t) = 80 - 17 \log(t + 1), 0 \le t \le 12$

(a)

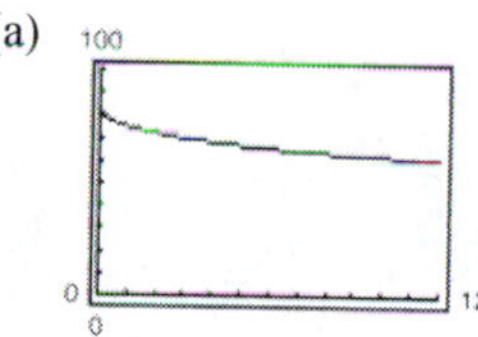

(b) $f(0) = 80 - 17 \log 1 = 80.0$

(c) $f(4) = 80 - 17 \log 5 \approx 68.1$

(d) $f(10) = 80 - 17 \log 11 \approx 62.3$

83. False. Reflecting $g(x)$ about the line $y = x$ will determine the graph of $f(x)$.

85. (a) $f(x) = \ln x, g(x) = \sqrt{x}$

The natural log function grows at a slower rate than the square root function.

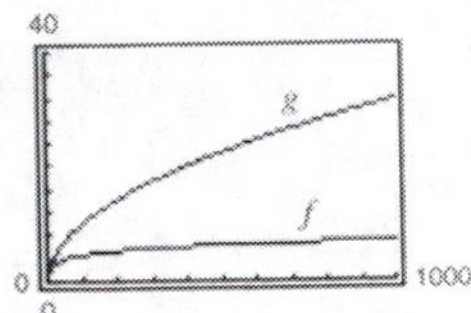

(b) $f(x) = \ln x, g(x) = \sqrt[4]{x}$

The natural log function grows at a slower rate than the fourth root function.

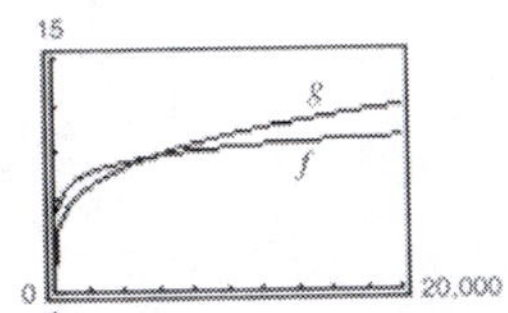

87. (a) False. If y were an exponential function of x, then $y = a^x$, but $a^1 = a$, not 0. Because one point is $(1, 0)$, y is not an exponential function of x.

(b) True. $y = \log_a x$

For $a = 2, y = \log_2 x$.

$x = 1, \log_2 1 = 0$

$x = 2, \log_2 2 = 1$

$x = 8, \log_2 8 = 3$

(c) True. $x = a^y$

For $a = 2, x = 2^y$.

$y = 0, 2^0 = 1$

$y = 1, 2^1 = 2$

$y = 3, 2^3 = 8$

(d) False. If y were a linear function of x, the slope between $(1, 0)$ and $(2, 1)$ and the slope between $(2, 1)$ and $(8, 3)$ would be the same. However,

$$m_1 = \frac{1 - 0}{2 - 1} = 1 \text{ and } m_2 = \frac{3 - 1}{8 - 2} = \frac{2}{6} = \frac{1}{3}.$$

So, y is not a linear function of x.

89. $y = \log_a x \Rightarrow a^y = x$, so, for example, if $a = -2$, there is no value of y for which $(-2)^y = -4$. If $a = 1$, then every power of a is equal to 1, so x could only be 1. So, $\log_a x$ is defined only for $0 < a < 1$ and $a > 1$.

Section 5.3 Properties of Logarithms

1. change-of-base

3. $\dfrac{1}{\log_b a}$

4. $\log_a(uv) = \log_a u + \log_a v$

This is the Product Property. Matches (c).

5. $\ln u^n = n \ln u$

This is the Power Property. Matches (a).

6. $\log_a \dfrac{u}{v} = \log_a u - \log_a v$

This is the Quotient Property. Matches (b).

7. (a) $\log_5 16 = \dfrac{\log 16}{\log 5}$

(b) $\log_5 16 = \dfrac{\ln 16}{\ln 5}$

9. (a) $\log_x \dfrac{3}{10} = \dfrac{\log(3/10)}{\log x}$

(b) $\log_x \dfrac{3}{10} = \dfrac{\ln(3/10)}{\ln x}$

11. $\log_3 7 = \dfrac{\log 7}{\log 3} = \dfrac{\ln 7}{\ln 3} \approx 1.771$

13. $\log_9 0.1 = \dfrac{\log 0.1}{\log 9} = \dfrac{\ln 0.1}{\ln 9} \approx -1.048$

15. $\log_4 8 = \dfrac{\log_2 8}{\log_2 4} = \dfrac{\log_2 2^3}{\log_2 2^2} = \dfrac{3}{2}$

17.
$$\begin{aligned}\log_5 \tfrac{1}{250} &= \log_5\left(\tfrac{1}{125} \cdot \tfrac{1}{2}\right)\\ &= \log_5 \tfrac{1}{125} + \log_5 \tfrac{1}{2}\\ &= \log_5 5^{-3} + \log_5 2^{-1}\\ &= -3 - \log_5 2\end{aligned}$$

19.
$$\begin{aligned}\ln(5e^6) &= \ln 5 + \ln e^6\\ &= \ln 5 + 6\\ &= 6 + \ln 5\end{aligned}$$

21. $\log_3 9 = 2 \log_3 3 = 2$

23. $\log_2 \sqrt[4]{8} = \frac{1}{4} \log_2 2^3 = \frac{3}{4} \log_2 2 = \frac{3}{4}(1) = \frac{3}{4}$

25. $\log_4 16^2 = 2 \log_4 16 = 2 \log_4 4^2 = 2(2) = 4$

27. $\log_2(-2)$ is undefined. -2 is not in the domain of $\log_2 x$.

29. $\ln e^{4.5} = 4.5$

31. $$\ln \frac{1}{\sqrt{e}} = \ln 1 - \ln \sqrt{e} = 0 - \frac{1}{2}\ln e = 0 - \frac{1}{2}(1) = -\frac{1}{2}$$

33. $\ln e^2 + \ln e^5 = 2 + 5 = 7$

35. $$\log_5 75 - \log_5 3 = \log_5 \tfrac{75}{3} = \log_5 25 = \log_5 5^2 = 2\log_5 5 = 2$$

37. $\ln 4x = \ln 4 + \ln x$

39. $\log_8 x^4 = 4\log_8 x$

41. $$\log_5 \frac{5}{x} = \log_5 5 - \log_5 x = 1 - \log_5 x$$

43. $\ln\sqrt{z} = \ln z^{1/2} = \frac{1}{2}\ln z$

45. $$\ln xyz^2 = \ln x + \ln y + \ln z^2 = \ln x + \ln y + 2\ln z$$

47. $$\ln z(z-1)^2 = \ln z + \ln(z-1)^2 = \ln z + 2\ln(z-1),\ z > 1$$

49. $$\log_2 \frac{\sqrt{a-1}}{9} = \log_2\sqrt{a-1} - \log_2 9 = \frac{1}{2}\log_2(a-1) - \log_2 3^2 = \frac{1}{2}\log_2(a-1) - 2\log_2 3,\ a > 1$$

51. $$\ln\sqrt[3]{\frac{x}{y}} = \frac{1}{3}\ln\frac{x}{y} = \frac{1}{3}[\ln x - \ln y] = \frac{1}{3}\ln x - \frac{1}{3}\ln y$$

53. $$\ln x^2\sqrt{\frac{y}{z}} = \ln x^2 + \ln\sqrt{\frac{y}{z}} = \ln x^2 + \frac{1}{2}\ln\frac{y}{z} = \ln x^2 + \frac{1}{2}[\ln y - \ln z] = 2\ln x + \frac{1}{2}\ln y - \frac{1}{2}\ln z$$

55. $$\log_5\left(\frac{x^2}{y^2z^3}\right) = \log_5 x^2 - \log_5 y^2z^3 = \log_5 x^2 - \left(\log_5 y^2 + \log_5 z^3\right) = 2\log_5 x - 2\log_5 y - 3\log_5 z$$

57. $$\ln\sqrt[4]{x^3\left(x^2+3\right)} = \tfrac{1}{4}\ln x^3\left(x^2+3\right) = \tfrac{1}{4}\left[\ln x^3 + \ln\left(x^2+3\right)\right] = \tfrac{1}{4}\left[3\ln x + \ln\left(x^2+3\right)\right] = \tfrac{3}{4}\ln x + \tfrac{1}{4}\ln\left(x^2+3\right)$$

59. $$\log_b 10 = \log_b 2\cdot5 = \log_b 2 + \log_b 5 \approx 0.3562 + 0.8271 = 1.1833$$

61. $$\log_b 8 = \log_b 2^3 = 3\log_b 2 \approx 3(0.3562) = 1.0686$$

63. $$\log_b 45 = \log_b 9\cdot5 = \log_b 9 + \log_b 5 = \log_b 3^2 + \log_b 5 = 2\log_b 3 + \log_b 5 \approx 2(0.5646) + 0.8271 = 1.9563$$

65. $$\log_b 3b^2 = \log_b 3 + \log_b b^2 = \log_b 3 + 2\log_b b = \log_b 3 + 2(1) \approx 0.5646 + 2 = 2.5646$$

67. $\ln 2 + \ln x = \ln 2x$

69. $2\log_2 x + 4\log_2 y = \log_2 x^2 + \log_2 y^4 = \log_2 x^2y^4$

71. $\frac{1}{4}\log_3 5x = \log_3(5x)^{1/4} = \log_3\sqrt[4]{5x}$

73. $\log x - 2\log(x+1) = \log x - \log(x+1)^2 = \log\dfrac{x}{(x+1)^2}$

75. $\log x - 2\log y + 3\log z = \log x - \log y^2 + \log z^3 = \log\dfrac{x}{y^2} + \log z^3 = \log\dfrac{xz^3}{y^2}$

77. $\ln x - \left[\ln(x+1) + \ln(x-1)\right] = \ln x - \ln(x+1)(x-1) = \ln\dfrac{x}{(x+1)(x-1)}$

79.
$$\begin{aligned}\frac{1}{3}\left[2\ln(x+3) + \ln x - \ln(x^2-1)\right] &= \frac{1}{3}\left[\ln(x+3)^2 + \ln x - \ln(x^2-1)\right]\\ &= \frac{1}{3}\left[\ln x(x+3)^2 - \ln(x^2-1)\right]\\ &= \frac{1}{3}\ln\frac{x(x+3)^2}{x^2-1}\\ &= \ln\sqrt[3]{\frac{x(x+3)^2}{x^2-1}}\end{aligned}$$

81.
$$\begin{aligned}\frac{1}{3}\left[\log_8 y + 2\log_8(y+4)\right] - \log_8(y-1) &= \frac{1}{3}\left[\log_8 y + \log_8(y+4)^2\right] - \log_8(y-1)\\ &= \frac{1}{3}\log_8 y(y+4)^2 - \log_8(y-1)\\ &= \log_8\sqrt[3]{y(y+4)^2} - \log_8(y-1)\\ &= \log_8\left(\frac{\sqrt[3]{y(y+4)^2}}{y-1}\right)\end{aligned}$$

83. $\log_2\dfrac{32}{4} = \log_2 32 - \log_2 4 \neq \dfrac{\log_2 32}{\log_2 4}$

The second and third expressions are equal by Property 2.

85. $\beta = 10\log\left(\dfrac{I}{10^{-12}}\right) = 10\left[\log I - \log 10^{-12}\right]$

$= 10\left[\log I + 12\right] = 120 + 10\log I$

When $I = 10^{-6}$:

$\beta = 120 + 10\log 10^{-6} = 120 + 10(-6) = 60$ decibels

87. $\beta = 10\log\left(\dfrac{I}{10^{-12}}\right)$

$$\begin{aligned}\text{Difference} &= 10\log\left(\frac{10^{-4}}{10^{-12}}\right) - 10\log\left(\frac{10^{-11}}{10^{-12}}\right)\\ &= 10\left[\log 10^8 - \log 10\right]\\ &= 10(8-1)\\ &= 10(7)\\ &= 70\text{ dB}\end{aligned}$$

89.

x	1	2	3	4	5	6
y	1.000	1.189	1.316	1.414	1.495	1.565
$\ln x$	0	0.693	1.099	1.386	1.609	1.792
$\ln y$	0	0.173	0.275	0.346	0.402	0.448

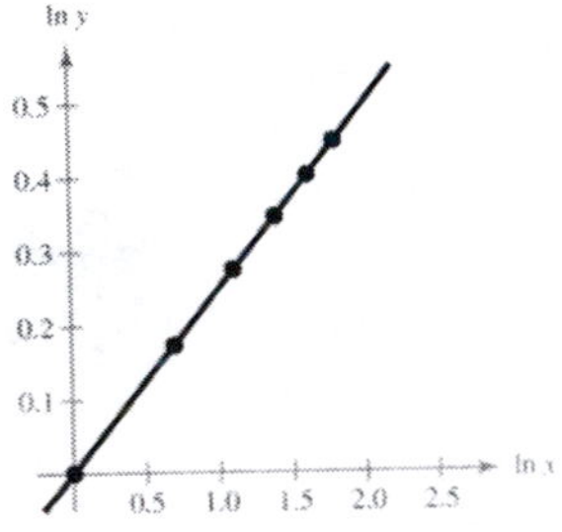

The slope of the line is $\frac{1}{4}$. So, $\ln y = \frac{1}{4}\ln x$

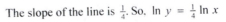

91.

x	1	2	3	4	5	6
y	2.500	2.102	1.900	1.768	1.672	1.597
$\ln x$	0	0.693	1.099	1.386	1.609	1.792
$\ln y$	0.916	0.743	0.642	0.570	0.514	0.468

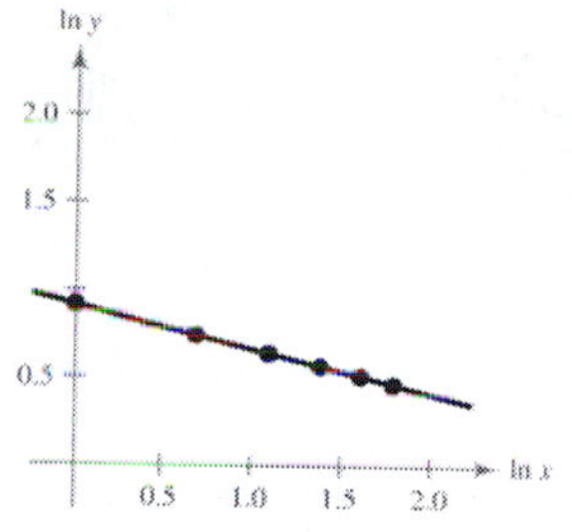

The slope of the line is $-\frac{1}{4}$. So, $\ln y = -\frac{1}{4}\ln x + \ln\frac{5}{2}$.

93.

Weight, x	25	35	50	75	500	1000
Galloping Speed, y	191.5	182.7	173.8	164.2	125.9	114.2
$\ln x$	3.219	3.555	3.912	4.317	6.215	6.908
$\ln y$	5.255	5.208	5.158	5.101	4.835	4.738

$y = 256.24 - 20.8 \ln x$

95. (a)

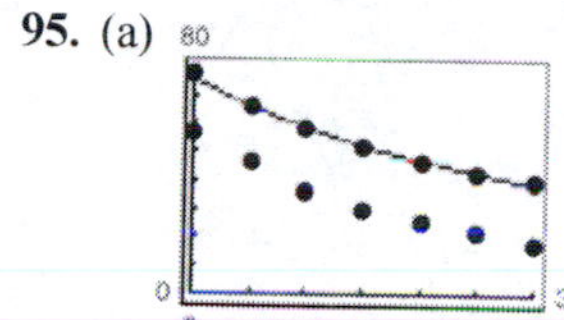

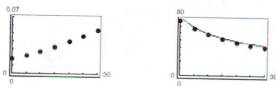

(b) $T - 21 = 54.4(0.964)^t$

$T = 54.4(0.964)^t + 21$

See graph in (a).

(c)

t (in minutes)	T (°C)	$T - 21$ (°C)	$\ln(T - 21)$	$1/(T - 21)$
0	78	57	4.043	0.0175
5	66	45	3.807	0.0222
10	57.5	36.5	3.597	0.0274
15	51.2	30.2	3.408	0.0331
20	46.3	25.3	3.231	0.0395
25	42.5	21.5	3.068	0.0465
30	39.6	18.6	2.923	0.0538

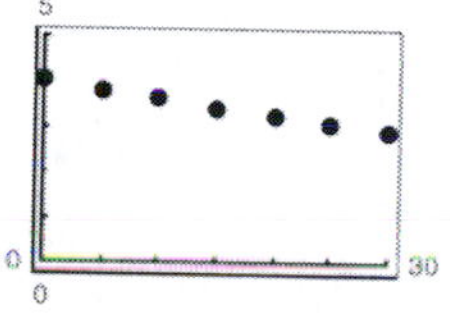

$\ln(T - 21) = -0.037t + 4$

$T = e^{-0.037t + 3.997} + 21$

This graph is identical to T in (b).

(d) $\dfrac{1}{T - 21} = 0.0012t + 0.016$

$T = \dfrac{1}{0.001t + 0.016} + 21$

(e) Taking logs of temperatures led to a linear scatter plot because the log function increases very slowly as the x-values increase. Taking the reciprocals of the temperatures led to a linear scatter plot because of the asymptotic nature of the reciprocal function.

97. $f(x) = \ln x$

False, $f(0) \neq 0$ because 0 is not in the domain of $f(x)$.

$f(1) = \ln 1 = 0$

99. False.

$f(x) - f(2) = \ln x - \ln 2 = \ln \frac{x}{2} \neq \ln(x - 2)$

101. False.

$f(u) = 2f(v) \Rightarrow \ln u = 2 \ln v \Rightarrow \ln u = \ln v^2 \Rightarrow u = v^2$

103. $f(x) = \log_2 x = \dfrac{\log x}{\log 2} = \dfrac{\ln x}{\ln 2}$

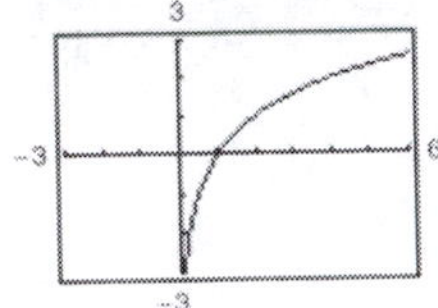

105. $f(x) = \log_{1/4} x$

$= \dfrac{\log x}{\log(1/4)} = \dfrac{\ln x}{\ln(1/4)}$

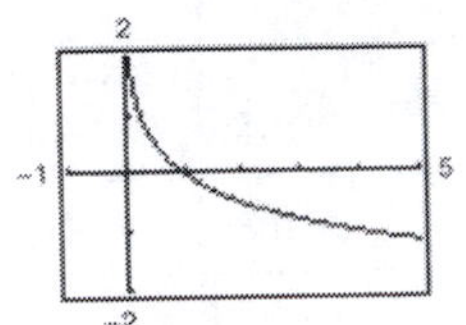

107. *Sample answers:*

(a) $\ln(1 + 3) \stackrel{?}{=} \ln 1 + \ln 3$

$1.39 \neq 0 + 1.10$

$\ln(u + v) = \ln(uv)$, but $\ln(u + v) \neq \ln u + \ln v$.

(b) $\ln(3 - 1) \stackrel{?}{=} \ln 3 - \ln 1$

$0.69 \neq 1.10 - 0$

$\ln u - \ln v = \ln \frac{u}{v}$, but $\ln(u - v) \neq \ln u - \ln v$.

(c) $(\ln 2)^3 \stackrel{?}{=} 3(\ln 2)$

$0.33 \neq 2.08$

$n(\ln u) = \ln u^n$, but $(\ln u)^n \neq n(\ln u)$.

109. $\ln 2 \approx 0.6931$, $\ln 3 \approx 1.0986$, $\ln 5 \approx 1.6094$

$\ln 1 = 0$

$\ln 2 \approx 0.6931$

$\ln 3 \approx 1.0986$

$\ln 4 = \ln(2 \cdot 2) = \ln 2 + \ln 2 \approx 0.6931 + 0.6931 = 1.3862$

$\ln 5 \approx 1.6094$

$\ln 6 = \ln(2 \cdot 3) = \ln 2 + \ln 3 \approx 0.6931 + 1.0986 = 1.7917$

$\ln 8 = \ln 2^3 = 3 \ln 2 \approx 3(0.6931) = 2.0793$

$\ln 9 = \ln 3^2 = 2 \ln 3 \approx 2(1.0986) = 2.1972$

$\ln 10 = \ln(5 \cdot 2) = \ln 5 + \ln 2 \approx 1.6094 + 0.6931 = 2.3025$

$\ln 12 = \ln(2^2 \cdot 3) = \ln 2^2 + \ln 3 = 2 \ln 2 + \ln 3 \approx 2(0.6931) + 1.0986 = 2.4848$

$\ln 15 = \ln(5 \cdot 3) = \ln 5 + \ln 3 \approx 1.6094 + 1.0986 = 2.7080$

$\ln 16 = \ln 2^4 = 4 \ln 2 \approx 4(0.6931) = 2.7724$

$\ln 18 = \ln(3^2 \cdot 2) = \ln 3^2 + \ln 2 = 2 \ln 3 + \ln 2 \approx 2(1.0986) + 0.6931 = 2.8903$

$\ln 20 = \ln(5 \cdot 2^2) = \ln 5 + \ln 2^2 = \ln 5 + 2 \ln 2 \approx 1.6094 + 2(0.6931) = 2.9956$

Section 5.4 Exponential and Logarithmic Equations

1. (a) $x = y$

(b) $x = y$

(c) x

(d) x

3. $4^{2x-7} = 64$

(a) $x = 5$

$4^{2(5)-7} = 4^3 = 64$

Yes, $x = 5$ *is* a solution.

(b) $x = 2$

$4^{2(2)-7} = 4^{-3} = \frac{1}{64} \neq 64$

No, $x = 2$ *is not* a solution.

5. $\log_2(x + 3) = 10$

(a) $x = 1021$

$\log_2(1021 + 3) = \log_2(1024)$

Because $2^{10} = 1024$, $x = 1021$ *is* a solution.

(b) $x = 17$

$\log_2(17 + 3) = \log_2(20)$

Because $2^{10} \neq 20$, $x = 17$ *is not* a solution.

(c) $x = 10^2 - 3 = 97$

$\log_2(97 + 3) = \log_2(100)$

Because $2^{10} \neq 100$, $10^2 - 3$ *is not* a solution.

7. $4^x = 16$

$4^x = 4^2$

$x = 2$

9. $\ln x - \ln 2 = 0$

$\ln x = \ln 2$

$x = 2$

11. $\ln x = -1$

$e^{\ln x} = e^{-1}$

$x = e^{-1}$

$x \approx 0.368$

13. $\log_4 x = 3$

$4^{\log_4 x} = 4^3$

$x = 4^3$

$x = 64$

15. $f(x) = g(x)$

$2^x = 8$

$2^x = 2^3$

$x = 3$

Point of intersection:

$(3, 8)$

17. $e^x = e^{x^2-2}$

$x = x^2 - 2$

$0 = x^2 - x - 2$

$0 = (x + 1)(x - 2)$

$x = -1, x = 2$

19. $4(3^x) = 20$

$3^x = 5$

$\log_3 3^x = \log_3 5$

$x = \log_3 5 = \dfrac{\log 5}{\log 3}$ or $\dfrac{\ln 5}{\ln 3}$

$x \approx 1.465$

21. $e^x - 9 = 19$

$e^x = 28$

$\ln e^x = \ln 28$

$x = \ln 28 \approx 3.332$

23. $3^{2x} = 80$

$\ln 3^{2x} = \ln 80$

$2x \ln 3 = \ln 80$

$x = \dfrac{\ln 80}{2 \ln 3} \approx 1.994$

25. $2^{3-x} = 565$

$\ln 2^{3-x} = \ln 565$

$(3 - x) \ln 2 = \ln 565$

$3 \ln 2 - x \ln 2 = \ln 565$

$-x \ln 2 = \ln 565 - 3 \ln 2$

$x \ln 2 = 3 \ln 2 - \ln 565$

$x = \dfrac{3 \ln 2 - \ln 565}{\ln 2}$

$x = 3 - \dfrac{\ln 565}{\ln 2} \approx -6.142$

27. $8(10^{3x}) = 12$

$$10^{3x} = \tfrac{12}{8}$$
$$\log 10^{3x} = \log\left(\tfrac{3}{2}\right)$$
$$3x = \log\left(\tfrac{3}{2}\right)$$
$$x = \tfrac{1}{3}\log\left(\tfrac{3}{2}\right)$$
$$x \approx 0.059$$

29. $e^{3x} = 12$

$$3x = \ln 12$$
$$x = \frac{\ln 12}{3} \approx 0.828$$

31. $7 - 2e^{x} = 5$

$$-2e^{x} = -2$$
$$e^{x} = 1$$
$$x = \ln 1 = 0$$

33. $6(2^{3x-1}) - 7 = 9$

$$6(2^{3x-1}) = 16$$
$$2^{3x-1} = \frac{8}{3}$$
$$\log_2 2^{3x-1} = \log_2\left(\frac{8}{3}\right)$$
$$3x - 1 = \log_2\left(\frac{8}{3}\right) = \frac{\log(8/3)}{\log 2} \text{ or } \frac{\ln(8/3)}{\ln 2}$$
$$x = \frac{1}{3}\left[\frac{\log(8/3)}{\log 2} + 1\right] \approx 0.805$$

35. $2^{x} = 3^{x+1}$

$$\ln 2^{x} = \ln 3^{x+1}$$
$$x \ln 2 = (x + 1)\ln 3$$
$$x \ln 2 = x \ln 3 + \ln 3$$
$$x \ln 2 - x \ln 3 = \ln 3$$
$$x(\ln 2 - \ln 3) = \ln 3$$
$$x = \frac{\ln 3}{\ln 2 - \ln 3} \approx -2.710$$

37. $4^{x} = 5^{x^2}$

$$\ln 4^{x} = \ln 5^{x^2}$$
$$x \ln 4 = x^2 \ln 5$$
$$x^2 \ln 5 - x \ln 4 = 0$$
$$x(x \ln 5 - \ln 4) = 0$$
$$x = 0$$
$$x \ln 5 - \ln 4 = 0 \Rightarrow x = \frac{\ln 4}{\ln 5} \approx 0.861$$

39. $e^{2x} - 4e^{x} - 5 = 0$

$$(e^{x} + 1)(e^{x} - 5) = 0$$
$$e^{x} = -1 \quad \text{or} \quad e^{x} = 5$$
$$\text{(No solution)} \quad x = \ln 5 \approx 1.609$$

41. $\dfrac{500}{100 - e^{x/2}} = 20$

$$500 = 20(100 - e^{x/2})$$
$$25 = 100 - e^{x/2}$$
$$e^{x/2} = 75$$
$$\frac{x}{2} = \ln 75$$
$$x = 2\ln 75 \approx 8.635$$

43. $\left(1 + \dfrac{0.065}{365}\right)^{365t} = 4$

$$\ln\left(1 + \frac{0.065}{365}\right)^{365t} = \ln 4$$
$$365t \ln\left(1 + \frac{0.065}{365}\right) = \ln 4$$
$$t = \frac{\ln 4}{365 \ln\left(1 + \frac{0.065}{365}\right)} \approx 21.330$$

45. $\ln x = -3$

$$x = e^{-3} \approx 0.050$$

47. $2.1 = \ln 6x$

$$e^{2.1} = 6x$$
$$\frac{e^{2.1}}{6} = x$$
$$1.361 \approx x$$

49. $3 \ln 5x = 10$

$$\ln 5x = \frac{10}{3}$$
$$5x = e^{10/3}$$
$$x = \frac{e^{10/3}}{5} \approx 5.606$$

51. $2 - 6\ln x = 10$

$$-6\ln x = 8$$
$$\ln x = -\tfrac{4}{3}$$
$$e^{\ln x} = e^{-4/3}$$
$$x = e^{-4/3}$$
$$x \approx 0.264$$

53. $6\log_3(0.5x) = 11$

$$\log_3(0.5x) = \tfrac{11}{6}$$
$$3^{\log_3(0.5x)} = 3^{11/6}$$
$$0.5x = 3^{11/6}$$
$$x = 2\left(3^{11/6}\right) \approx 14.988$$

55. $\ln x - \ln(x+1) = 2$

$$\ln\left(\frac{x}{x+1}\right) = 2$$
$$\frac{x}{x+1} = e^2$$
$$x = e^2(x+1)$$
$$x = e^2x + e^2$$
$$x - e^2x = e^2$$
$$x\left(1 - e^2\right) = e^2$$
$$x = \frac{e^2}{1-e^2} \approx -1.157$$

This negative value is extraneous. The equation has no solution.

57. $\ln(x+5) = \ln(x-1) - \ln(x+1)$

$$\ln(x+5) = \ln\left(\frac{x-1}{x+1}\right)$$
$$x + 5 = \frac{x-1}{x+1}$$
$$(x+5)(x+1) = x - 1$$
$$x^2 + 6x + 5 = x - 1$$
$$x^2 + 5x + 6 = 0$$
$$(x+2)(x+3) = 0$$
$$x = -2 \quad \text{or} \quad x = -3$$

Both of these solutions are extraneous, so the equation has no solution.

59. $\log(3x+4) = \log(x-10)$

$$3x + 4 = x - 10$$
$$2x = -14$$
$$x = -7$$

The negative value is extraneous.

The equation has no solution.

61. $\log_4 x - \log_4(x-1) = \frac{1}{2}$

$$\log_4\left(\frac{x}{x-1}\right) = \frac{1}{2}$$
$$4^{\log_4[x/(x-1)]} = 4^{1/2}$$
$$\frac{x}{x-1} = 4^{1/2}$$
$$x = 2(x-1)$$
$$x = 2x - 2$$
$$-x = -2$$
$$x = 2$$

63. $f(x) = 5^x - 212$

Algebraically:

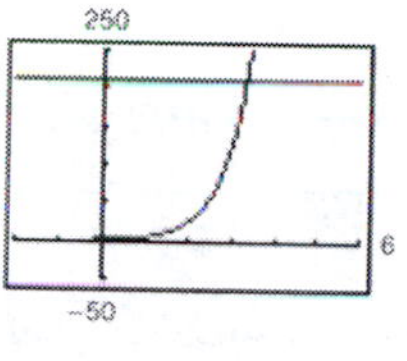

$$5^x = 212$$
$$\ln 5^x = \ln 212$$
$$x\ln 5 = \ln 212$$
$$x = \frac{\ln 212}{\ln 5}$$
$$x \approx 3.328$$

The zero is $x \approx 3.328$.

65. $g(x) = 8e^{-2x/3} - 11$

Algebraically:

$$8e^{-2x/3} = 11$$
$$e^{-2x/3} = 1.375$$
$$-\frac{2x}{3} = \ln 1.375$$
$$x = -1.5\ln 1.375$$
$$x \approx -0.478$$

The zero is $x \approx -0.478$.

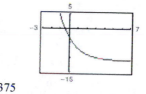

67. $y_1 = 3$

$y_2 = \ln x$

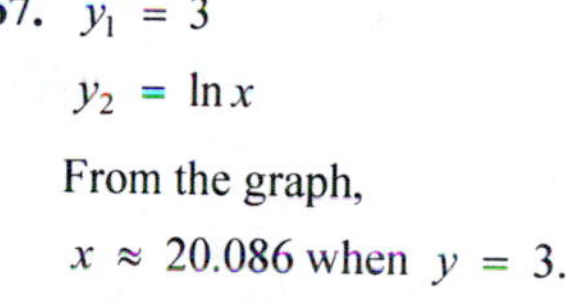

From the graph,

$x \approx 20.086$ when $y = 3$.

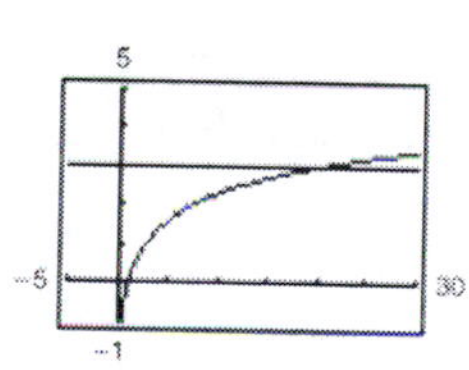

Algebraically:

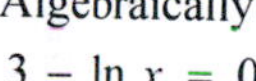

$$3 - \ln x = 0$$
$$\ln x = 3$$
$$x = e^3 \approx 20.086$$

69. $y_1 = 2\ln(x + 3)$

$y_2 = 3$

From the graph, $x \approx 1.482$ when $y = 3$.

Algebraically:

$$2\ln(x + 3) = 3$$
$$\ln(x + 3) = \frac{3}{2}$$
$$x + 3 = e^{3/2}$$
$$x = e^{3/2} - 3 \approx 1.482$$

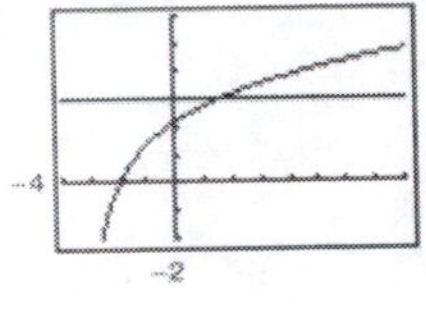

71. (a)

$$r = 0.025$$
$$A = Pe^{rt}$$
$$5000 = 2500e^{0.025t}$$
$$2 = e^{0.025t}$$
$$\ln 2 = 0.025t$$
$$\frac{\ln 2}{0.025} = t$$
$$t \approx 27.73 \text{ years}$$

(b)

$$r = 0.025$$
$$A = Pe^{rt}$$
$$7500 = 2500e^{0.025t}$$
$$3 = e^{0.025t}$$
$$\ln 3 = 0.025t$$
$$\frac{\ln 3}{0.025} = t$$
$$t \approx 43.94 \text{ years}$$

73. $2x^2e^{2x} + 2xe^{2x} = 0$

$$(2x^2 + 2x)e^{2x} = 0$$
$$2x^2 + 2x = 0 \quad (\text{because } e^{2x} \neq 0)$$
$$2x(x + 1) = 0$$
$$x = 0, -1$$

75. $-xe^{-x} + e^{-x} = 0$

$$(-x + 1)e^{-x} = 0$$
$$-x + 1 = 0 \quad (\text{because } e^{-x} \neq 0)$$
$$x = 1$$

77. $2x\ln x + x = 0$

$$x(2\ln x + 1) = 0$$
$$2\ln x + 1 = 0 \quad (\text{because } x > 0)$$
$$\ln x = -\frac{1}{2}$$
$$x = e^{-1/2} \approx 0.607$$

79. $\dfrac{1 + \ln x}{2} = 0$

$$1 + \ln x = 0$$
$$\ln x = -1$$
$$x = e^{-1} = \frac{1}{e} \approx 0.368$$

81. (a)

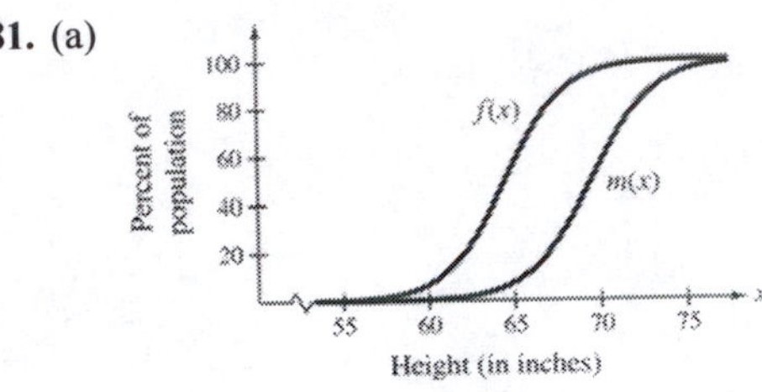

From the graph you see horizontal asymptotes at $y = 0$ and $y = 100$.

These represent the lower and upper percent bounds; the range falls between 0% and 100%.

(b) Males:

$$50 = \frac{100}{1 + e^{-0.5536(x - 69.51)}}$$
$$1 + e^{-0.5536(x - 69.51)} = 2$$
$$e^{-0.5536(x - 69.51)} = 1$$
$$-0.5536(x - 69.51) = \ln 1$$
$$-0.5536(x - 69.51) = 0$$
$$x = 69.51$$

The average height of an American male is 69.51 inches.

Females:

$$50 = \frac{100}{1 + e^{-0.5834(x - 64.49)}}$$
$$1 + e^{-0.5834(x - 64.49)} = 2$$
$$e^{-0.5834(x - 64.49)} = 1$$
$$-0.5834(x - 64.49) = \ln 1$$
$$-0.5834(x - 64.49) = 0$$
$$x = 64.49$$

The average height of an American female is 64.49 inches.

83. $N = 68\left(10^{-0.04x}\right)$

When $N = 21$:

$$21 = 68\left(10^{-0.04x}\right)$$
$$\frac{21}{68} = 10^{-0.04x}$$
$$\log_{10}\frac{21}{68} = -0.04x$$
$$x = -\frac{\log_{10}(21/68)}{0.04} \approx 12.76 \text{ inches}$$

85. $y = -3.00 + 11.88 \ln x + \dfrac{36.94}{x}$

(a)

x	0.2	0.4	0.6	0.8	1.0
y	162.6	78.5	52.5	40.5	33.9

(b)

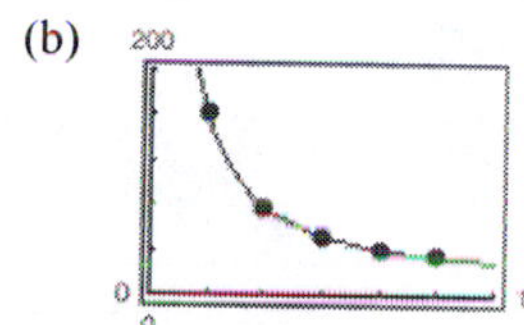

The model seems to fit the data well.

(c) When $y = 30$:

$$30 = -3.00 + 11.88 \ln x + \frac{36.94}{x}$$

Add the graph of $y = 30$ to the graph in part (a) and estimate the point of intersection of the two graphs.

You find that $x \approx 1.20$ meters.

(d) No, it is probably not practical to lower the number of gs experienced during impact to less than 23 because the required distance traveled at $y = 23$ is $x \approx 2.27$ meters. It is probably not practical to design a car allowing a passenger to move forward 2.27 meters (or 7.45 feet) during an impact.

87. $\log_a(uv) = \log_a u + \log_a v$

True by Property 1 in Section 3.3.

89. $\log_a(u - v) = \log_a u - \log_a v$

False.

$$1.95 = \log(100 - 10)$$
$$\neq \log 100 - \log 10 = 1$$

91. Yes, a logarithmic equation can have more than one extraneous solution. See Exercise 103.

93. $A = Pe^{rt}$

(a) $A = (2P)e^{rt} = 2(Pe^{rt})$ This doubles your money.

(b) $A = Pe^{(2r)t} = Pe^{rt}e^{rt} = e^{rt}(Pe^{rt})$

(c) $A = Pe^{r(2t)} = Pe^{rt}e^{rt} = e^{rt}(Pe^{rt})$

Doubling the interest rate yields the same result as doubling the number of years.

If $2 > e^{rt}$ (i.e., $rt < \ln 2$), then doubling your investment would yield the most money. If $rt > \ln 2$, then doubling either the interest rate or the number of years would yield more money.

95. (a) $P = 1000, r = 0.07$, compounded annually, $n = 1$

Effective yield: $A = P\left(1 + \dfrac{r}{n}\right)^{nt} = 1000\left(1 + \dfrac{0.07}{1}\right)^{1} = \$1070$

$$\frac{1070 - 1000}{1000} = 7\%$$

The effective yield is 7%.

Balance after 5 years: $A = P\left(1 + \dfrac{r}{n}\right)^{nt} = 1000\left(1 + \dfrac{0.07}{1}\right)^{1(5)} \approx \$1402.55$

(b) $P = 1000, r = 0.07$, compounded continuously

Effective yield: $A = Pe^{rt} = 1000e^{0.07(1)} \approx \$1072.51$

$$\frac{1072.51 - 1000}{1000} = 7.25\%$$

The effective yield is about 7.25%.

Balance after 5 years: $A = Pe^{rt} = 1000e^{0.07(5)} \approx \$1419.07$

(c) $P = 1000, r = 0.07$, compounded quarterly, $n = 4$

Effective yield: $A = P\left(1 + \dfrac{r}{n}\right)^{nt} = 1000\left(1 + \dfrac{0.07}{4}\right)^{4(1)} \approx \$1071.86$

$$\frac{1071.86 - 1000}{1000} = 7.19\%$$

The effective yield is about 7.19%.

Balance after 5 years: $A = P\left(1 + \dfrac{r}{n}\right)^{nt} = 1000\left(1 + \dfrac{0.07}{4}\right)^{4(5)} \approx \$1414.78$

(d) $P = 1000, r = 0.0725$, compounded quarterly, $n = 4$

Effective yield: $A = P\left(1 + \frac{r}{n}\right)^{nt} = 1000\left(1 + \frac{0.0725}{4}\right)^{4(1)} \approx \$1074.50$

$\frac{1074.50 - 1000}{1000} \approx 7.45\%$

The effective yield is about 7.45%.

Balance after 5 years: $A = P\left(1 + \frac{r}{n}\right)^{nt} = 1000\left(1 + \frac{0.0725}{4}\right)^{4(5)} \approx \$1432.26$

Savings plan (d) has the greatest effective yield and the highest balance after 5 years.

Section 5.5 Exponential and Logarithmic Models

1. $y = ae^{bx}$; $y = ae^{-bx}$

3. normally distributed

5. (a) $A = Pe^{rt}$

$\frac{A}{e^{rt}} = P$

(b) $A = Pe^{rt}$

$\frac{A}{P} = e^{rt}$

$\ln \frac{A}{P} = \ln e^{rt}$

$\ln \frac{A}{P} = rt$

$\frac{\ln(A/P)}{r} = t$

7. Because $A = 1000e^{0.035t}$, the time to double is given by $2000 = 1000e^{0.035t}$ and you have

$2 = e^{0.035t}$

$\ln 2 = \ln e^{0.035t}$

$\ln 2 = 0.035t$

$t = \frac{\ln 2}{0.035} \approx 19.8$ years.

Amount after 10 years: $A = 1000e^{0.35} \approx \$1419.07$

9. Because $A = 750e^{rt}$ and $A = 1500$ when $t = 7.75$, you have

$1500 = 750e^{7.75r}$

$2 = e^{7.75r}$

$\ln 2 = \ln e^{7.75r}$

$\ln 2 = 7.75r$

$r = \frac{\ln 2}{7.75} \approx 0.089438 = 8.9438\%.$

Amount after 10 years: $A = 750e^{0.089438(10)} \approx \$1834.37$

11. Because $A = Pe^{0.045t}$ and $A = 10{,}000.00$ when $t = 10$, you have

$10{,}000.00 = Pe^{0.045(10)}$

$\frac{10{,}000.00}{e^{0.045(10)}} = P = \$6376.28.$

The time to double is given by

$t = \frac{\ln 2}{0.045} \approx 15.40$ years.

13. $A = 500{,}000, r = 0.05, n = 12, t = 10$

$A = P\left(1 + \frac{r}{n}\right)^{nt}$

$500{,}000 = P\left(1 + \frac{0.05}{12}\right)^{12(10)}$

$P = \frac{500{,}000}{\left(1 + \frac{0.05}{12}\right)^{12(10)}}$

$\approx \$303{,}580.52$

15. $P = 1000, r = 0.1, A = 2000$

$$A = P\left(1 + \frac{r}{n}\right)^{nt}$$

$$2000 = 1000\left(1 + \frac{0.1}{n}\right)^{nt}$$

$$2 = \left(1 + \frac{0.1}{n}\right)^{nt}$$

(a) $n = 1$

$$(1 + 0.1)^t = 2$$
$$(1.1)^t = 2$$
$$\ln(1.1)^t = \ln 2$$
$$t \ln 1.1 = \ln 2$$
$$t = \frac{\ln 2}{\ln 1.1} \approx 7.27 \text{ years}$$

(b) $n = 12$

$$\left(1 + \frac{0.1}{12}\right)^{12t} = 2$$
$$\ln\left(\frac{12.1}{12}\right)^{12t} = \ln 2$$
$$12t \ln\left(\frac{12.1}{12}\right) = \ln 2$$
$$12t = \frac{\ln 2}{\ln(12.1/12)}$$
$$t = \frac{\ln 2}{12 \ln(12.1/12)} \approx 6.96 \text{ years}$$

(c) $n = 365$

$$\left(1 + \frac{0.1}{365}\right)^{365t} = 2$$
$$\ln\left(\frac{365.1}{365}\right)^{365t} = \ln 2$$
$$365t \ln\left(\frac{365.1}{365}\right) = \ln 2$$
$$365t = \frac{\ln 2}{\ln(365.1/365)}$$
$$t = \frac{\ln 2}{365 \ln(365.1/365)} \approx 6.93 \text{ years}$$

(d) Compounded continuously

$$A = Pe^{rt}$$
$$2000 = 1000e^{0.1t}$$
$$2 = e^{0.1t}$$
$$\ln 2 = \ln e^{0.1t}$$
$$0.1t = \ln 2$$
$$t = \frac{\ln 2}{0.1} \approx 6.93 \text{ years}$$

17. (a)

$$3P = Pe^{rt}$$
$$3 = e^{rt}$$
$$\ln 3 = rt$$
$$\frac{\ln 3}{r} = t$$

r	2%	4%	6%	8%	10%	12%
$t = \frac{\ln 3}{r}$ (years)	54.93	27.47	18.31	13.73	10.99	9.16

(b)

$$3P = P(1 + r)^t$$
$$3 = (1 + r)^t$$
$$\ln 3 = \ln (1 + r)^t$$
$$\frac{\ln 3}{\ln (1 + r)} = t$$

r	2%	4%	6%	8%	10%	12%
$t = \frac{\ln 3}{\ln (1 + r)}$ (years)	55.48	28.01	18.85	14.27	11.53	9.69

19. Continuous compounding results in faster growth.

$A = 1 + 0.075[\![t]\!]$ and $A = e^{0.07t}$

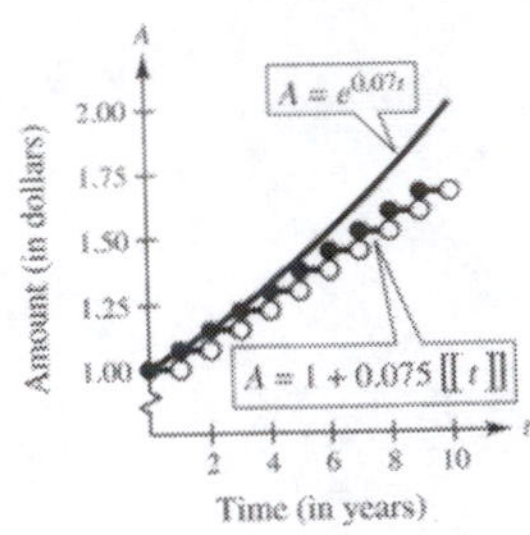

21. $a = 10,\ y = \frac{1}{2}(10) = 5,\ t = 1599$

$$\begin{aligned} y &= ae^{-bt} \\ 5 &= 10e^{-b(1599)} \\ 0.5 &= e^{-1599b} \\ \ln 0.5 &= \ln e^{-1599b} \\ \ln 0.5 &= -1599b \\ b &= -\frac{\ln 0.5}{1599} \end{aligned}$$

Given an initial quantity of 10 grams, after 1000 years, you have

$y = 10e^{-[-(\ln 0.5)/1599](1000)} \approx 6.48$ grams.

23. $y = 2,\ a = 2(2) = 4,\ t = 5715$

$$\begin{aligned} y &= ae^{-bt} \\ 2 &= 4e^{-b(5715)} \\ 0.5 &= e^{-5715b} \\ \ln 0.5 &= \ln e^{-5715b} \\ \ln 0.5 &= -5715b \\ b &= -\frac{\ln 0.5}{5715} \end{aligned}$$

Given 2 grams after 1000 years, the initial amount is

$2 = ae^{-[-(\ln 0.5)/5715](1000)}$

$a \approx 2.26$ grams.

25.

$$\begin{aligned} y &= ae^{bx} \\ 1 &= ae^{b(0)} \Rightarrow 1 = a \\ 10 &= e^{b(3)} \\ \ln 10 &= 3b \\ \frac{\ln 10}{3} &= b \Rightarrow b \approx 0.7675 \end{aligned}$$

So, $y = e^{0.7675x}$.

27.

$$\begin{aligned} y &= ae^{bx} \\ 5 &= ae^{b(0)} \Rightarrow 5 = a \\ 1 &= 5e^{b(4)} \\ \frac{1}{5} &= e^{4b} \\ \ln\left(\frac{1}{5}\right) &= 4b \\ \frac{\ln(1/5)}{4} &= b \Rightarrow b \approx -0.4024 \end{aligned}$$

So, $y = 5e^{-0.4024x}$.

29. (a)

Year	1980	1990	2000	2010
Population	106.1	143.15	196.25	272.37

(b) Let $P = 350$, and solve for t.

$$\begin{aligned} 350 &= 20.6 + 85.5e^{0.0360t} \\ 329.4 &= 85.5e^{0.0360t} \\ \frac{329.4}{85.5} &= e^{0.0360t} \\ \ln\left(\frac{329.4}{85.5}\right) &= 0.0360t \\ \frac{1}{0.0360}\ln\left(\frac{329.4}{85.5}\right) &= t \\ 37.4 &\approx t \end{aligned}$$

(c) No; The population will not continue to grow at such a quick rate.

The population will reach 350,000 people in the year 2017.

31. $y = 4080e^{kt}$

When $t = 3$, $y = 10{,}000$:

$$\begin{aligned} 10{,}000 &= 4080e^{k(3)} \\ \frac{10{,}000}{4080} &= e^{3k} \\ \ln\left(\frac{10{,}000}{4080}\right) &= 3k \\ k &= \frac{\ln(10{,}000/4080)}{3} \approx 0.2988 \end{aligned}$$

When $t = 24$: $y = 4080e^{0.2988(24)} \approx 5{,}309{,}734$ hits

33. $y = ae^{bt}$

When $t = 3$, $y = 100$:

$$100 = ae^{3b}$$

$$\frac{100}{e^{3b}} = a$$

When $t = 5$, $y = 400$:

$$400 = ae^{5b}$$

Substitute $\frac{100}{e^{3b}}$ for a in the equation on the right.

$$400 = \frac{100}{e^{3b}}e^{5b}$$

$$400 = 100e^{2b}$$

$$4 = e^{2b}$$

$$\ln 4 = 2b$$

$$\ln 2^2 = 2b$$

$$2 \ln 2 = 2b$$

$$\ln 2 = b$$

$$a = \frac{100}{e^{3b}} = \frac{100}{e^{3 \ln 2}} = \frac{100}{e^{\ln 2^3}} = \frac{100}{2^3} = \frac{100}{8} = 12.5$$

$$y = 12.5e^{(\ln 2)t}$$

After 6 hours, there are $y = 12.5e^{(\ln 2)(6)} = 800$ bacteria.

35. $(0, 1150)$, $(2, 550)$

(a) $m = \dfrac{550 - 1150}{2 - 0} = -300$

$V = -300t + 1150$

(b) $550 = 1150e^{k(2)}$

$\ln\left(\dfrac{550}{1150}\right) = 2k \Rightarrow k \approx -0.369$

$V = 1150e^{-0.368799t}$

(c) 

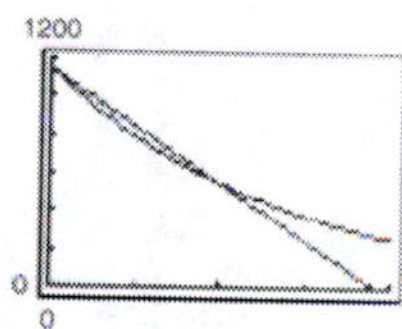

The exponential model depreciates faster in the first two years.

(d)

t	1	3
$V = -300t + 1100$	\$850	\$250
$V = 1150e^{-0.369t}$	\$795	\$380

(e) The slope of the linear model means that the computer depreciates \$300 per year, then loses all value in the third year. The exponential model depreciates faster in the first two years but maintains value longer.

37. $R = \frac{1}{10^{12}}e^{-t/8223}$

(a)
$$R = \frac{1}{8^{14}}$$
$$\frac{1}{10^{12}}e^{-t/8223} = \frac{1}{8^{14}}$$
$$e^{-t/8223} = \frac{10^{12}}{8^{14}}$$
$$-\frac{t}{8223} = \ln\left(\frac{10^{12}}{8^{14}}\right)$$
$$t = -8223\ln\left(\frac{10^{12}}{8^{14}}\right) \approx 12{,}180 \text{ years old}$$

(b)
$$\frac{1}{10^{12}}e^{-t/8223} = \frac{1}{13^{11}}$$
$$e^{-t/8223} = \frac{10^{12}}{13^{11}}$$
$$-\frac{t}{8223} = \ln\left(\frac{10^{12}}{13^{11}}\right)$$
$$t = -8223\ln\left(\frac{10^{12}}{13^{11}}\right) \approx 4797 \text{ years old}$$

39. $y = 0.0266e^{-(x-100)^2/450},\ 70 \le x \le 116$

(a)

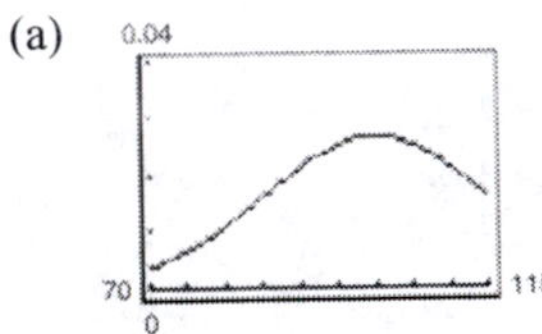

(b) The average IQ score of an adult student is 100.

41. (a) 1998: $t = 18,\ y = \frac{269{,}573}{1 + 985e^{-0.308(18)}}$
$\approx 55{,}557$ sites

2003: $t = 23,\ y = \frac{269{,}573}{1 + 985e^{-0.308(23)}}$
$\approx 147{,}644$ sites

2006: $t = 26,\ y = \frac{269{,}573}{1 + 985e^{-0.308(26)}}$
$\approx 203{,}023$ sites

(b) and (c)

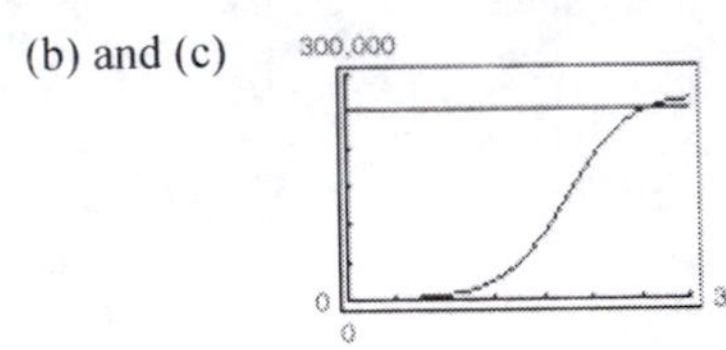

Because the lines intersect at $(30.6, 250, 000)$, the number of cell sites will reach 250,000 in the year 2010.

(d) Let $y = 250, 000$ and solve for t.
$$250{,}000 = \frac{269,573}{1 + 985e^{-0.308t}}$$
$$1 + 985e^{-0.308t} = \frac{269,573}{250{,}000}$$
$$985e^{-0.308t} = 0.078292$$
$$e^{-0.308t} \approx 0.000079484$$
$$-0.308t \approx \ln(0.00007948)$$
$$t \approx 30.6$$

The number of cell sites will reach 250,000 during the year 2010.

43. $p(t) = \frac{1000}{1 + 9e^{-0.1656t}}$

(a) $p(5) = \frac{1000}{1 + 9e^{-0.1656(5)}} \approx 203$ animals

(b)
$$500 = \frac{1000}{1 + 9e^{-0.1656t}}$$
$$1 + 9e^{-0.1656t} = 2$$
$$9e^{-0.1656t} = 1$$
$$e^{-0.1656t} = \frac{1}{9}$$
$$t = -\frac{\ln(1/9)}{0.1656} \approx 13 \text{ months}$$

(c)

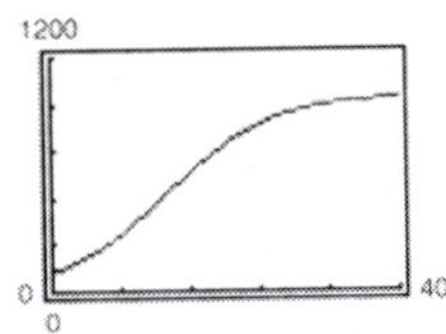

The horizontal asymptotes are $p = 0$ and $p = 1000$. The asymptote with the larger p-value, $p = 1000$, indicates that the population size will approach 1000 as time increases.

45. $R = \log\frac{I}{I_0} = \log I$ because $I_0 = 1$.

(a)
$$R = 6.6$$
$$6.6 = \log I$$
$$10^{6.6} = 10^{\log I}$$
$$3{,}981{,}072 \approx I$$

(b)
$$R = 5.6$$
$$5.6 = \log I$$
$$10^{5.6} = 10^{\log I}$$
$$10^{5.6} = I$$
$$398{,}107 \approx I$$

(c)
$$R = 7.1$$
$$7.1 = \log I$$
$$10^{7.1} = 10^{\log I}$$
$$10^{7.1} = I$$
$$12{,}589{,}254 \approx I$$

47. $\beta = 10 \log \dfrac{I}{I_0}$ where $I_0 = 10^{-12}$ watt/m^2.

(a) $\beta = 10 \log \dfrac{10^{-10}}{10^{-12}} = 10 \log 10^2 = 20$ decibels

(b) $\beta = 10 \log \dfrac{10^{-5}}{10^{-12}} = 10 \log 10^7 = 70$ decibels

(c) $\beta = 10 \log \dfrac{10^{-8}}{10^{-12}} = 10 \log 10^4 = 40$ decibels

(d) $\beta = 10 \log \dfrac{1}{10^{-12}} = 10 \log 10^{12} = 120$ decibels

49.
$$\beta = 10 \log \frac{I}{I_0}$$
$$\frac{\beta}{10} = \log \frac{I}{I_0}$$
$$10^{\beta/10} = 10^{\log I/I_0}$$
$$10^{\beta/10} = \frac{I}{I_0}$$
$$I = I_0 10^{\beta/10}$$
$$\% \text{ decrease} = \frac{I_0 10^{9.3} - I_0 10^{8.0}}{I_0 10^{9.3}} \times 100 \approx 95\%$$

51. $\text{pH} = -\log\left[H^+\right]$

$-\log\left(2.3 \times 10^{-5}\right) \approx 4.64$

53.
$$5.8 = -\log\left[H^+\right]$$
$$-5.8 = \log\left[H^+\right]$$
$$10^{-5.8} = 10^{\log\left[H^+\right]}$$
$$10^{-5.8} = \left[H^+\right]$$
$$\left[H^+\right] \approx 1.58 \times 10^{-6} \text{ moles per liter}$$

55.
$$2.9 = -\log\left[H^+\right]$$
$$-2.9 = \log\left[H^+\right]$$
$\left[H^+\right] = 10^{-2.9}$ for the apple juice
$$8.0 = -\log\left[H^+\right]$$
$$-8.0 = \log\left[H^+\right]$$
$\left[H^+\right] = 10^{-8}$ for the drinking water

$\dfrac{10^{-2.9}}{10^{-8}} = 10^{5.1}$ times the hydrogen ion concentration of drinking water

57. $t = -10 \ln \dfrac{T - 70}{98.6 - 70}$

At 9:00 A.M. you have:

$t = -10 \ln \dfrac{85.7 - 70}{98.6 - 70} \approx 6$ hours

From this you can conclude that the person died at 3:00 A.M.

59. $u = 120{,}000\left[\dfrac{0.075t}{1 - \left(\dfrac{1}{1 + 0.075/12}\right)^{12t}} - 1\right]$

(a)

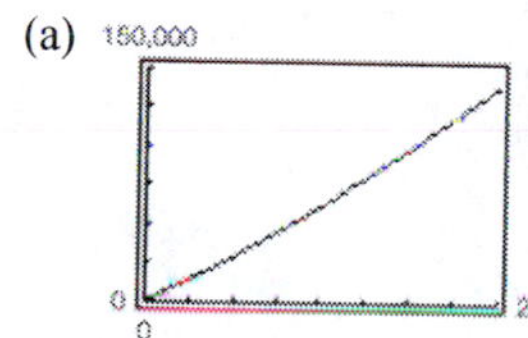

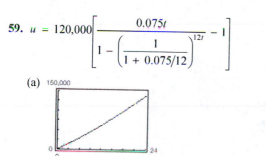

(b) From the graph, $u = \$120{,}000$ when $t \approx 21$ years. It would take approximately 37.6 years to pay \$240,000 in interest. Yes, it is possible to pay twice as much in interest charges as the size of the mortgage. It is especially likely when the interest rates are higher.

61. False. The domain can be the set of real numbers for a logistic growth function.

63. False. The graph of $f(x)$ is the graph of $g(x)$ shifted upward five units.

65. Answers will vary.

Review Exercises for Chapter 5

1. $f(x) = 0.3^x$

$f(1.5) = 0.3^{1.5} \approx 0.164$

3. $f(x) = 2^{-0.5x}$

$f(\pi) = 2^{-0.5(\pi)} \approx 0.337$

5. $f(x) = 7(0.2^x)$

$f(-\sqrt{11}) = 7(0.2^{-\sqrt{11}})$

≈ 1456.529

7. $f(x) = 5^x,\ g(x) = 5^x + 1$

Because $g(x) = f(x) + 1$, the graph of g can be obtained by shifting the graph of f one unit upward.

9. $f(x) = 3^x,\ g(x) = 1 - 3^x$

Because $g(x) = 1 - f(x)$, the graph of g can be obtained by reflecting the graph of f in the x-axis and shifting the graph one unit upward. (**Note:** This is equivalent to shifting the graph of f one unit upward and then reflecting the graph in the x-axis.)

11. $f(x) = 4^{-x} + 4$

Horizontal asymptote: $y = 4$

x	-1	0	1	2	3
$f(x)$	8	5	4.25	4.063	4.016

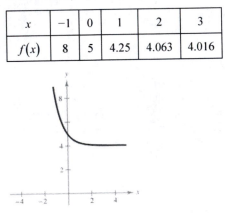

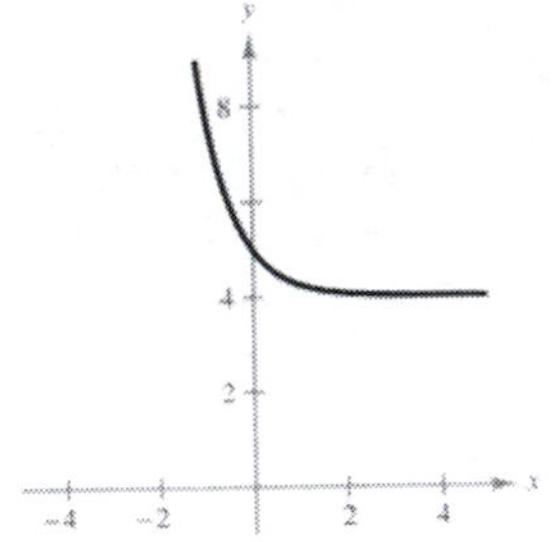

13. $f(x) = 5^{x-2} + 4$

Horizontal asymptote: $y = 4$

x	-1	0	1	2	3
$f(x)$	4.008	4.04	4.2	5	9

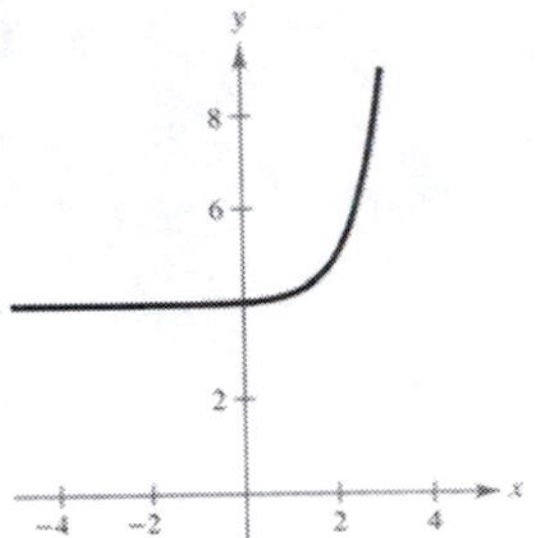

15. $f(x) = \left(\frac{1}{2}\right)^{-x} + 3 = 2^x + 3$

Horizontal asymptote: $y = 3$

x	-2	-1	0	1	2
$f(x)$	3.25	3.5	4	5	7

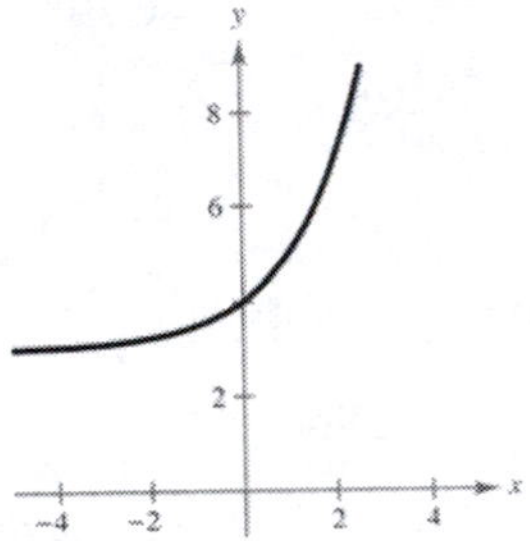

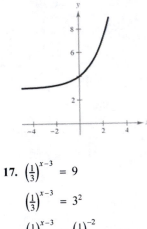

17. $\left(\frac{1}{3}\right)^{x-3} = 9$

$\left(\frac{1}{3}\right)^{x-3} = 3^2$

$\left(\frac{1}{3}\right)^{x-3} = \left(\frac{1}{3}\right)^{-2}$

$x - 3 = -2$

$x = 1$

19. $e^{3x-5} = e^7$

$3x - 5 = 7$

$3x = 12$

$x = 4$

21. $e^8 \approx 2980.958$

23. $e^{-1.7} \approx 0.183$

25. $h(x) = e^{-x/2}$

x	-2	-1	0	1	2
$h(x)$	2.72	1.65	1	0.61	0.37

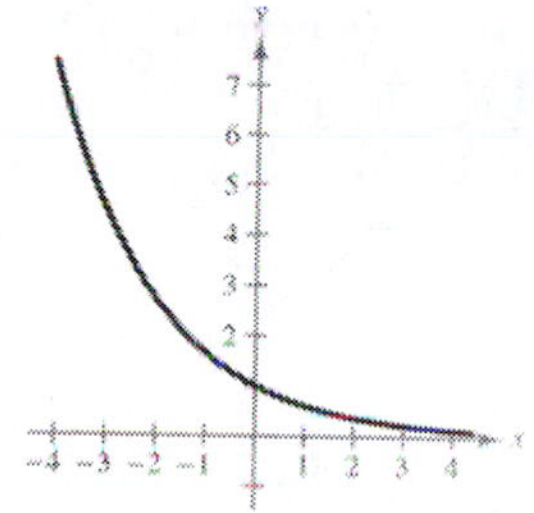

27. $f(x) = e^{x+2}$

x	-3	-2	-1	0	1
$f(x)$	0.37	1	2.72	7.39	20.09

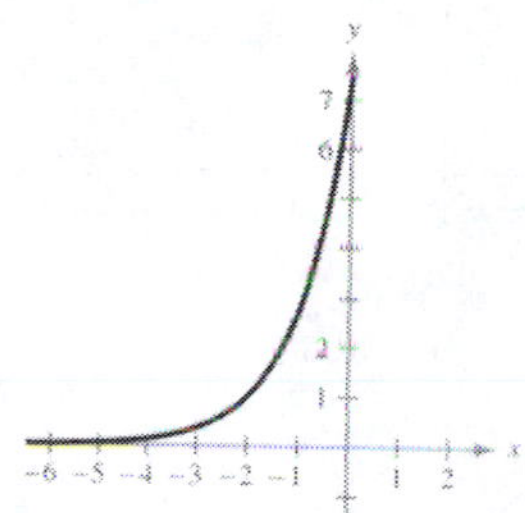

29. $F(t) = 1 - e^{-t/3}$

(a) $F\left(\frac{1}{2}\right) \approx 0.154$

(b) $F(2) \approx 0.487$

(c) $F(5) \approx 0.811$

31. $P = \$5000, r = 3\%, t = 10$ years

Compounded n times per year: $A = P\left(1 + \frac{r}{n}\right)^{nt} = 5000\left(1 + \frac{0.03}{n}\right)^{10n}$

Compounded continuously: $A = Pe^{rt} = 5000e^{0.03(10)}$

n	1	2	4	12	365	Continuous
A	\$6719.58	\$6734.28	\$6741.74	\$6746.77	\$6749.21	\$6749.29

33. $3^3 = 27$

$\log_3 27 = 3$

35. $e^{0.8} = 2.2255\ldots$

$\ln 2.2255\ldots = 0.8$

37. $f(x) = \log x$

$f(1000) = \log 1000$

$= \log 10^3 = 3$

39. $g(x) = \log_2 x$

$g\left(\frac{1}{4}\right) = \log_2 \frac{1}{4}$

$= \log_2 2^{-2} = -2$

41. $\log_4(x + 7) = \log_4 14$

$x + 7 = 14$

$x = 7$

43. $\ln(x + 9) = \ln 4$

$x + 9 = 4$

$x = -5$

45. $g(x) = \log_7 x \Rightarrow x = 7^y$

Domain: $(0, \infty)$

x-intercept: $(1, 0)$

Vertical asymptote: $x = 0$

x	$\frac{1}{7}$	1	7	49
$g(x)$	-1	0	1	2

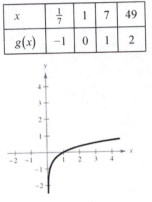

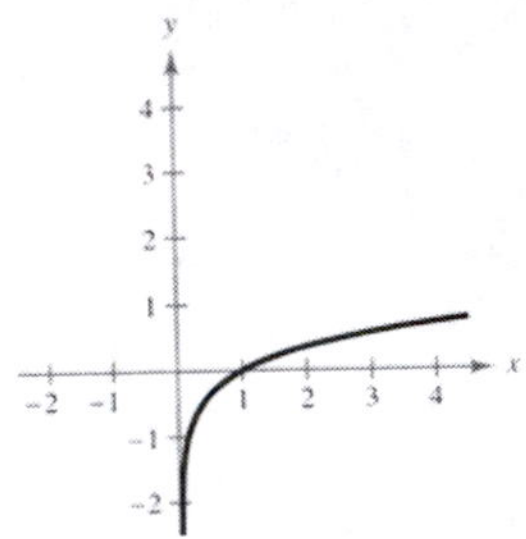

47. $f(x) = 4 - \log(x + 5)$

Domain: $(-5, \infty)$

Because

$$\begin{aligned} 4 - \log(x + 5) = 0 \Rightarrow \log(x + 5) &= 4 \\ x + 5 &= 10^4 \\ x &= 10^4 - 5 \\ &= 9995. \end{aligned}$$

x-intercept: $(9995, 0)$

Vertical asymptote: $x = -5$

x	-4	-3	-2	-1	0	1
$f(x)$	4	3.70	3.52	3.40	3.30	3.22

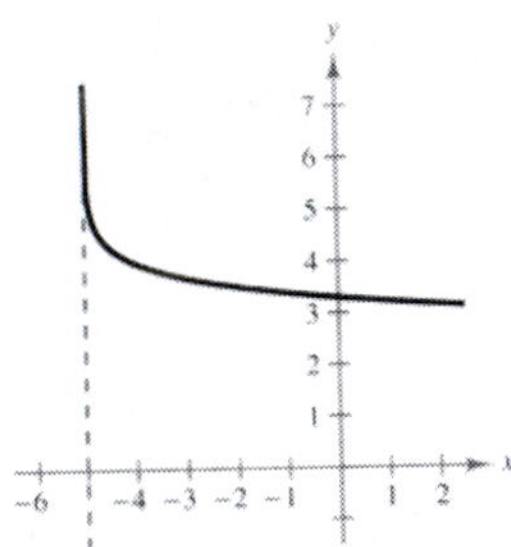

49. $f(22.6) = \ln 22.6 \approx 3.118$

51. $f(\sqrt{e}) = \frac{1}{2}\ln\sqrt{e} = 0.25$

53. $f(x) = \ln x + 3$

Domain: $(0, \infty)$

$$\begin{aligned} \ln x + 3 &= 0 \\ \ln x &= -3 \\ x &= e^{-3} \end{aligned}$$

x-intercept: $(e^{-3}, 0)$

Vertical asymptote: $x = 0$

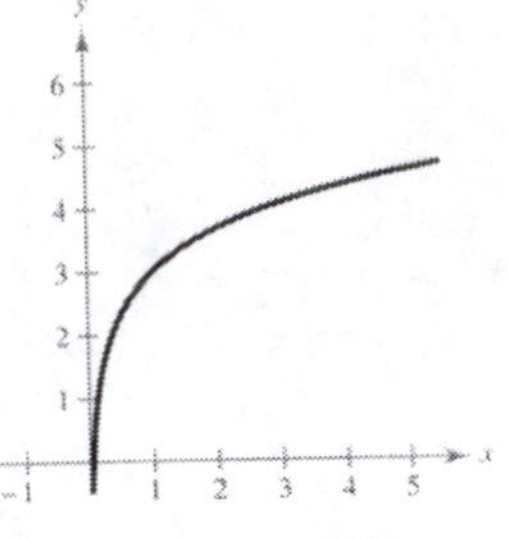

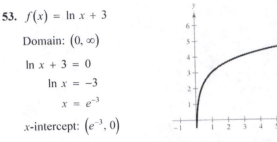

x	1	2	3	$\frac{1}{2}$	$\frac{1}{4}$
$f(x)$	3	3.69	4.10	2.31	1.61

55. $h(x) = \ln(x^2) = 2\ln|x|$

Domain: $(-\infty, 0) \cup (0, \infty)$

x-intercepts: $(\pm 1, 0)$

Vertical asymptote: $x = 0$

x	± 0.5	± 1	± 2	± 3	± 4
$h(x)$	-1.39	0	1.39	2.20	2.77

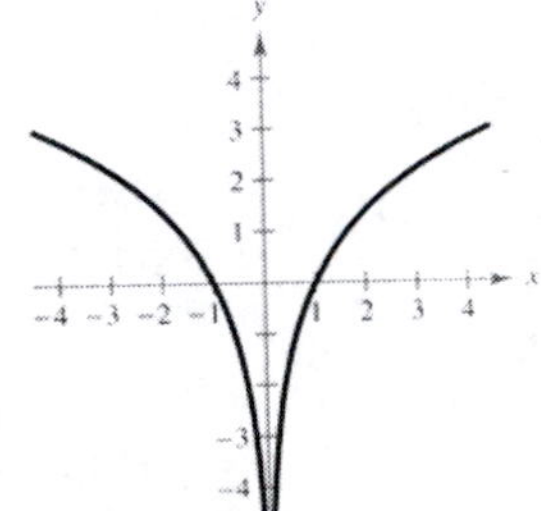

57. $h = 116\log(a + 40) - 176$

$h(55) = 116\log(55 + 40) - 176 \approx 53.4$ inches

59. (a) $\log_2 6 = \dfrac{\log 6}{\log 2} \approx 2.585$

(b) $\log_2 6 = \dfrac{\ln 6}{\ln 2} \approx 2.585$

61. (a) $\log_{1/2} 5 = \dfrac{\log 5}{\log(1/2)} \approx -2.322$

(b) $\log_{1/2} 5 = \dfrac{\ln 5}{\ln(1/2)} \approx -2.322$

63. $\log 18 = \log(2 \cdot 3^2) = \log 2 + 2\log 3 \approx 1.255$

65. $\ln 20 = \ln(2^2 \cdot 5) = 2\ln 2 + \ln 5 \approx 2.996$

67. $\log_5 5x^2 = \log_5 5 + \log_5 x^2 = 1 + 2\log_5 x$

69. $\log_3 \frac{9}{\sqrt{x}} = \log_3 9 - \log_3 \sqrt{x}$

$= \log_3 3^2 - \log_3 x^{1/2}$

$= 2 - \frac{1}{2}\log_3 x$

71. $\ln x^2y^2z = \ln x^2 + \ln y^2 + \ln z$

$= 2\ln x + 2\ln y + \ln z$

73. $\log_2 5 + \log_2 x = \log_2 5x$

75. $\ln x - \frac{1}{4}\ln y = \ln x - \ln \sqrt[4]{y} = \ln \frac{x}{\sqrt[4]{y}}$

77. $\frac{1}{2}\log_3 x - 2\log_3(y + 8) = \log_3 x^{1/2} - \log_3(y + 8)^2$

$= \log_3 \sqrt{x} - \log_3(y + 8)^2$

$= \log_3 \frac{\sqrt{x}}{(y + 8)^2}$

79. $t = 50 \log \frac{18{,}000}{18{,}000 - h}$

(a) Domain: $0 \le h < 18{,}000$

(b)

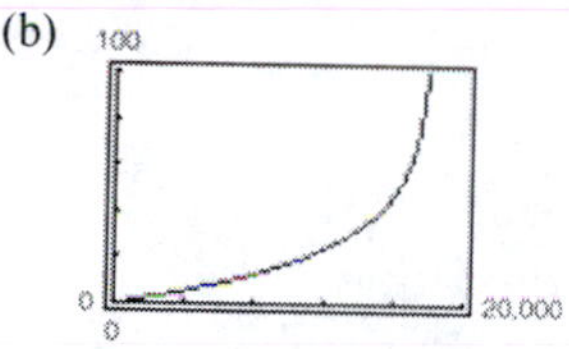

Vertical asymptote: $h = 18{,}000$

(c) As the plane approaches its absolute ceiling, it climbs at a slower rate, so the time required increases.

(d) $50 \log \frac{18{,}000}{18{,}000 - 4000} \approx 5.46$ minutes

81. $5^x = 125$

$5^x = 5^3$

$x = 3$

83. $e^x = 3$

$x = \ln 3 \approx 1.099$

85. $\ln x = 4$

$x = e^4 \approx 54.598$

87. $e^{4x} = e^{x^2+3}$

$4x = x^2 + 3$

$0 = x^2 - 4x + 3$

$0 = (x - 1)(x - 3)$

$x = 1, x = 3$

89. $2^x - 3 = 29$

$2^x = 32$

$2^x = 2^5$

$x = 5$

91. $25e^{-0.3x} = 12$

Graph $y_1 = 25e^{-0.3x}$ and $y_2 = 12$.

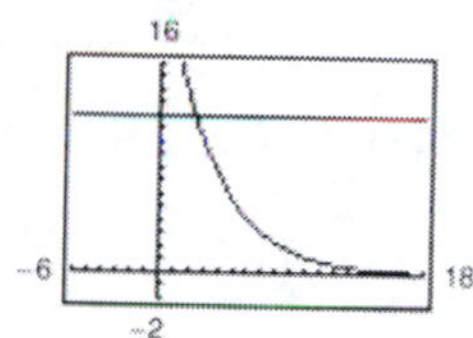

The graphs intersect at $x \approx 2.447$.

93. $\ln 3x = 8.2$

$e^{\ln 3x} = e^{8.2}$

$3x = e^{8.2}$

$x = \frac{e^{8.2}}{3} \approx 1213.650$

95. $\ln x - \ln 3 = 2$

$\ln \frac{x}{3} = 2$

$e^{\ln(x/3)} = e^2$

$\frac{x}{3} = e^2$

$x = 3e^2 \approx 22.167$

97. $\log_8(x - 1) = \log_8(x - 2) - \log_8(x + 2)$

$\log_8(x - 1) = \log_8\left(\frac{x - 2}{x + 2}\right)$

$x - 1 = \frac{x - 2}{x + 2}$

$(x - 1)(x + 2) = x - 2$

$x^2 + x - 2 = x - 2$

$x^2 = 0$

$x = 0$

Because $x = 0$ is not in the domain of $\log_8(x - 1)$ or of $\log_8(x - 2)$, it is an extraneous solution. The equation has no solution.

99. $\log(1 - x) = -1$

$$1 - x = 10^{-1}$$
$$1 - \frac{1}{10} = x$$
$$x = 0.900$$

101. $2\ln(x + 3) - 3 = 0$

Graph $y_1 = 2\ln(x + 3) - 3.$

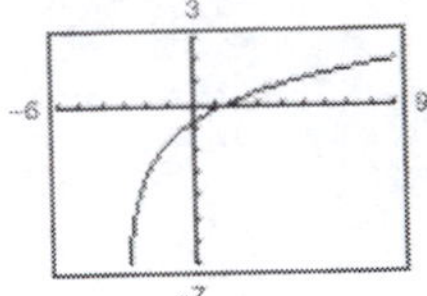

The x-intercept is at $x \approx 1.482$.

103. $6\log(x^2 + 1) - x = 0$

Graph $y_1 = 6\log(x^2 + 1) - x.$

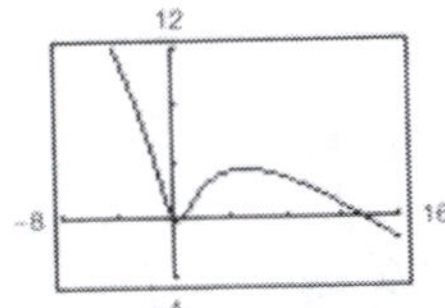

The x-intercepts are at $x = 0, x \approx 0.416,$ and $x \approx 13.627.$

105. $P = 8500, A = 3(8500) = 25{,}500, r = 3.5\%$

$$A = Pe^{rt}$$
$$25{,}500 = 8500e^{0.035t}$$
$$3 = e^{0.035t}$$
$$\ln 3 = 0.035t$$
$$t = \frac{\ln 3}{0.035} \approx 31.4 \text{ years}$$

107. $y = 3e^{-2x/3}$

Exponential decay model

Matches graph (e).

108. $y = 4e^{2x/3}$

Exponential growth model

Matches graph (b).

109. $y = \ln(x + 3)$

Logarithmic model

Vertical asymptote: $x = -3$

Graph includes $(-2, 0)$

Matches graph (f).

110. $y = 7 - \log(x + 3)$

Logarithmic model

Vertical asymptote: $x = -3$

Matches graph (d).

111. $y = 2e^{-(x+4)^2/3}$

Gaussian model

Matches graph (a).

112. $y = \dfrac{6}{1 + 2e^{-2x}}$

Logistics growth model

Matches graph (c).

113. $y = ae^{bx}$

Using the point $(0, 2)$, you have

$$2 = ae^{b(0)}$$
$$2 = ae^0$$
$$2 = a(1)$$
$$2 = a$$

Then, using the point $(4, 3)$, you have

$$3 = 2e^{b(4)}$$
$$3 = 2e^{4b}$$
$$\tfrac{3}{2} = e^{4b}$$
$$\ln \tfrac{3}{2} = 4b$$
$$\tfrac{1}{4}\ln\left(\tfrac{3}{2}\right) = b$$

So, $y = 2e^{\frac{1}{4}\ln\left(\frac{3}{2}\right)x}$

or

$y = 2e^{0.1014x}$

115. $y = 0.0499e^{-(x-71)^2/128}, \; 40 \le x \le 100$

Graph $y_1 = 0.0499e^{-(x-71)^2/128}.$

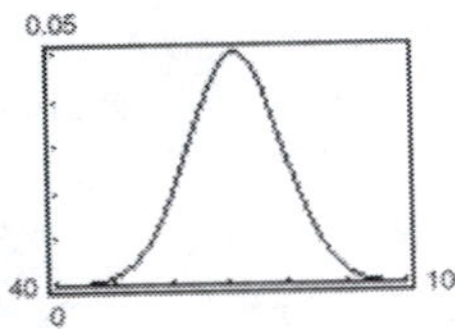

The average test score is 71.

117. $\beta = 10\log\left(\dfrac{I}{10^{-12}}\right)$

$\dfrac{\beta}{10} = \log\left(\dfrac{I}{10^{-12}}\right)$

$10^{\beta/10} = \dfrac{I}{10^{-12}}$

$I = 10^{\beta/10 - 12}$

(a) $\beta = 60$

$I = 10^{60/10-12}$

$= 10^{-6}$ watt/m^2

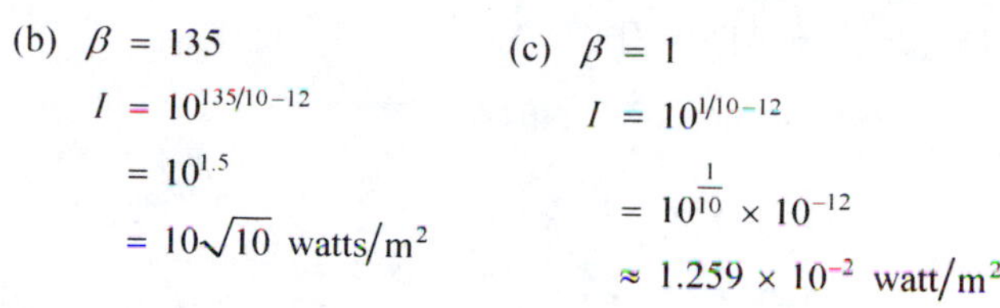

(b) $\beta = 135$

$I = 10^{135/10-12}$

$= 10^{1.5}$

$= 10\sqrt{10}$ watts/m^2

(c) $\beta = 1$

$I = 10^{1/10-12}$

$= 10^{\frac{1}{10}} \times 10^{-12}$

$\approx 1.259 \times 10^{-2}$ watt/m^2

119. True. By the inverse properties, $\log_b b^{2x} = 2x$.

Problem Solving for Chapter 5

1. $y = a^x$

$y_1 = 0.5^x$

$y_2 = 1.2^x$

$y_3 = 2.0^x$

$y_4 = x$

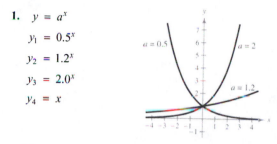

The curves $y = 0.5^x$ and $y = 1.2^x$ cross the line $y = x$. From checking the graphs it appears that $y = x$ will cross $y = a^x$ for $0 \le a \le 1.44$.

3. The exponential function, $y = e^x$, increases at a faster rate than the polynomial $y = x^n$.

5. (a) $f(u + v) = a^{u+v} = a^u \cdot a^v = f(u)\cdot f(v)$

(b) $f(2x) = a^{2x} = (a^x)^2 = [f(x)]^2$

7. (a)

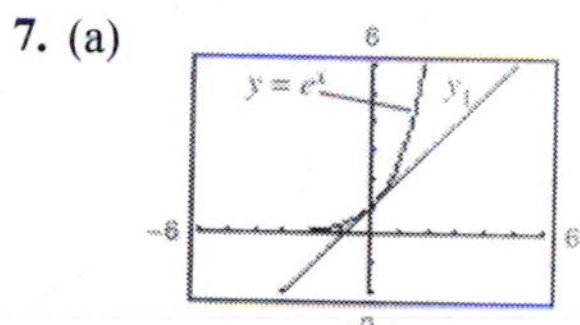

(b)

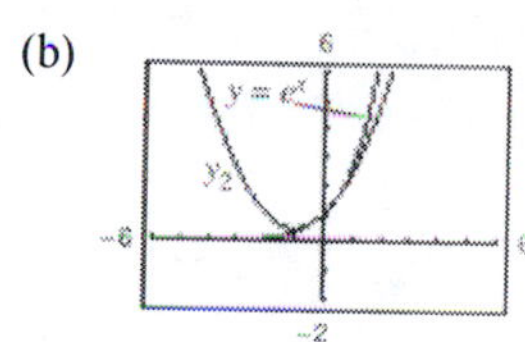

(c)

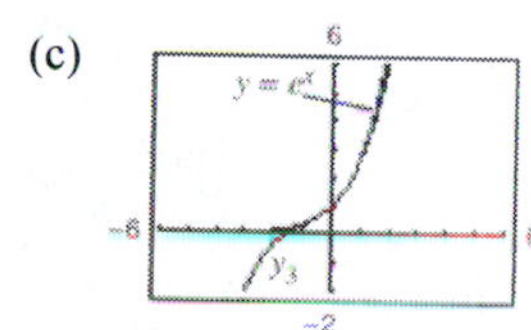

9.

$f(x) = e^x - e^{-x}$

$y = e^x - e^{-x}$

$x = e^y - e^{-y}$

$x = \dfrac{e^{2y} - 1}{e^y}$

$xe^y = e^{2y} - 1$

$e^{2y} - xe^y - 1 = 0$

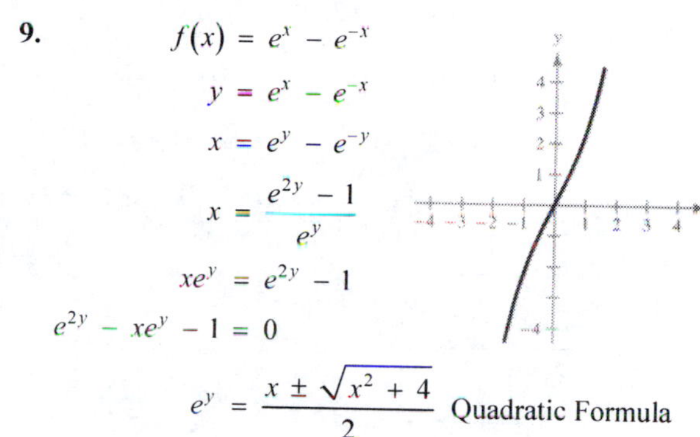

$e^y = \dfrac{x \pm \sqrt{x^2 + 4}}{2}$ Quadratic Formula

Choosing the positive quantity for e^y you have

$y = \ln\left(\dfrac{x + \sqrt{x^2 + 4}}{2}\right)$. So,

$f^{-1}(x) = \ln\left(\dfrac{x + \sqrt{x^2 + 4}}{2}\right)$.

11. Answer (c). $y = 6\left(1 - e^{-x^2/2}\right)$

The graph passes through (0, 0) and neither (a) nor (b) pass through the origin. Also, the graph has y-axis symmetry and a horizontal asymptote at $y = 6$.

13. $y_1 = c_1\left(\dfrac{1}{2}\right)^{t/k_1}$ and $y_2 = c_2\left(\dfrac{1}{2}\right)^{t/k_2}$

$c_1\left(\dfrac{1}{2}\right)^{t/k_1} = c_2\left(\dfrac{1}{2}\right)^{t/k_2}$

$\dfrac{c_1}{c_2} = \left(\dfrac{1}{2}\right)^{(t/k_2 - t/k_1)}$

$\ln\left(\dfrac{c_1}{c_2}\right) = \left(\dfrac{t}{k_2} - \dfrac{t}{k_1}\right)\ln\left(\dfrac{1}{2}\right)$

$\ln c_1 - \ln c_2 = t\left(\dfrac{1}{k_2} - \dfrac{1}{k_1}\right)\ln\left(\dfrac{1}{2}\right)$

$t = \dfrac{\ln c_1 - \ln c_2}{[(1/k_2) - (1/k_1)]\ln(1/2)}$

15. (a) $y_1 \approx 252.606(1.0310)^t$

(b) $y_2 \approx 400.88t^2 - 1464.6t + 291{,}782$

(c)

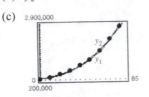

(d) The exponential model is a better fit for the data, but neither would be reliable to predict the population of the United States in 2015. The exponential model approaches infinity rapidly.

19. $y_4 = (x-1) - \frac{1}{2}(x-1)^2 + \frac{1}{3}(x-1)^3 - \frac{1}{4}(x-1)^4$

The pattern implies that

$\ln x = (x-1) - \frac{1}{2}(x-1)^2 + \frac{1}{3}(x-1)^3 - \frac{1}{4}(x-1)^4 + \dots$

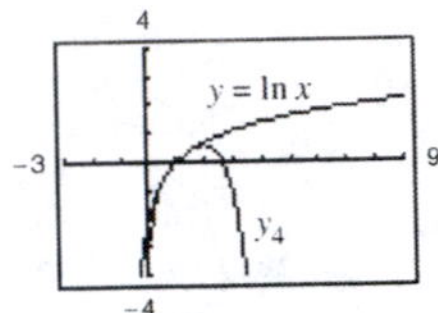

21. $y = 80.4 - 11 \ln x$

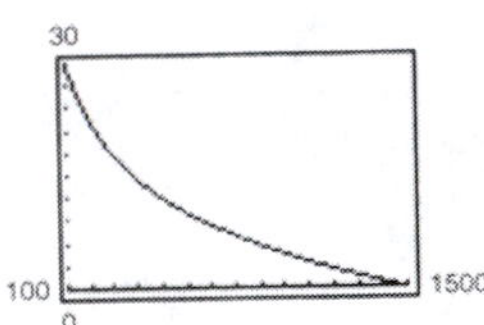

$y(300) = 80.4 - 11 \ln 300 \approx 17.7 \text{ ft}^3/\text{min}$

23. (a)

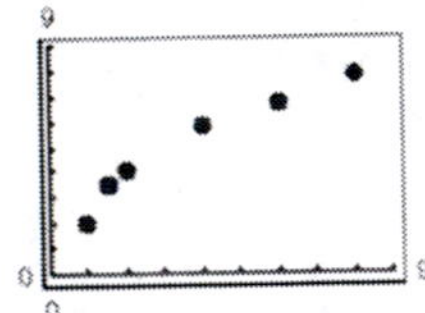

(b) The data could best be modeled by a logarithmic model.

(c) The shape of the curve looks much more logarithmic than linear or exponential.

(d) $y \approx 2.1518 + 2.7044 \ln x$

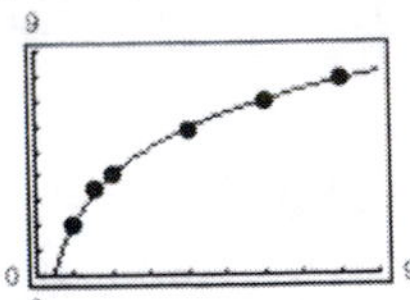

(e) The model is a good fit to the actual data.

17.

$$(\ln x)^2 = \ln x^2$$
$$(\ln x)^2 - 2 \ln x = 0$$
$$\ln x(\ln x - 2) = 0$$
$$\ln x = 0 \text{ or } \ln x = 2$$
$$x = 1 \text{ or } x = e^2$$

25. (a)

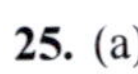

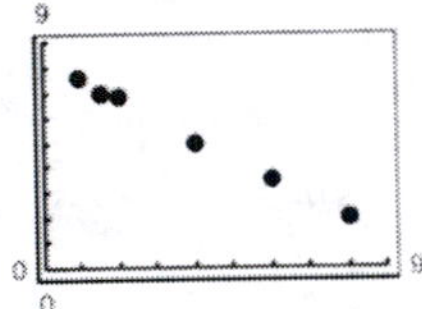

(b) The data could best be modeled by a linear model.

(c) The shape of the curve looks much more linear than exponential or logarithmic.

(d) $y \approx -0.7884x + 8.2566$

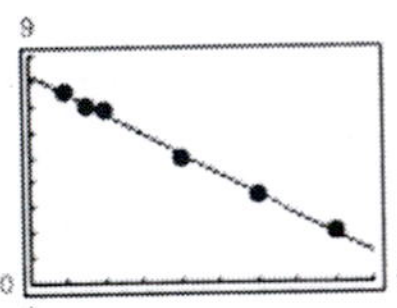

(e) The model is a good fit to the actual data.

Practice Test for Chapter 5

1. Solve for x: $x^{3/5} = 8$.

2. Solve for x: $3^{x-1} = \frac{1}{81}$.

3. Graph $f(x) = 2^{-x}$.

4. Graph $g(x) = e^x + 1$.

5. If \$5000 is invested at 9% interest, find the amount after three years if the interest is compounded

(a) monthly.

(b) quarterly.

(c) continuously.

6. Write the equation in logarithmic form: $7^{-2} = \frac{1}{49}$.

7. Solve for x: $x - 4 = \log_2 \frac{1}{64}$.

8. Given $\log_b 2 = 0.3562$ and $\log_b 5 = 0.8271$, evaluate $\log_b \sqrt[4]{8/25}$.

9. Write $5 \ln x - \frac{1}{2} \ln y + 6 \ln z$ as a single logarithm.

10. Using your calculator and the change of base formula, evaluate $\log_9 28$.

11. Use your calculator to solve for N: $\log_{10} N = 0.6646$

12. Graph $y = \log_4 x$.

13. Determine the domain of $f(x) = \log_3(x^2 - 9)$.

14. Graph $y = \ln(x - 2)$.

15. True or false: $\dfrac{\ln x}{\ln y} = \ln(x - y)$

16. Solve for x: $5^x = 41$

17. Solve for x: $x - x^2 = \log_5 \frac{1}{25}$

18. Solve for x: $\log_2 x + \log_2(x - 3) = 2$

19. Solve for x: $\dfrac{e^x + e^{-x}}{3} = 4$

20. Six thousand dollars is deposited into a fund at an annual interest rate of 13%. Find the time required for the investment to double if the interest is compounded continuously.

CHAPTER 6
Trigonometry

CHAPTER 6
Trigonometry

Section 6.1 Angles and Their Measures

1. Trigonometry
3. coterminal
5. acute; obtuse
7. radian
9. angular

11. 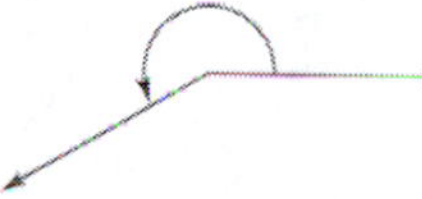

The angle shown is approximately 210°.

13. 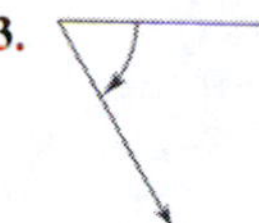

The angle shown is approximately −60°.

15. (a) Since $90° < 130° < 180°$; 130° lies in Quadrant II.

(b) Since $270° < 285° < 360°$; 285° lies in Quadrant IV.

17. (a) Since $-180° < -132°50' < -90°$; $-132°50'$ lies in Quadrant III.

(b) Since $-360° < -336° < -270°$; −336° lies in Quadrant I.

19. (a) 30°

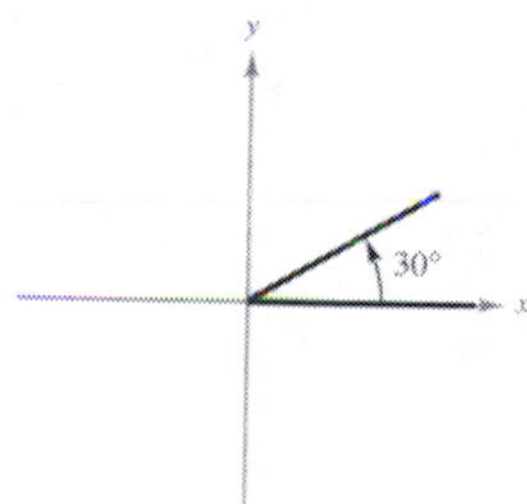

(b) 150°

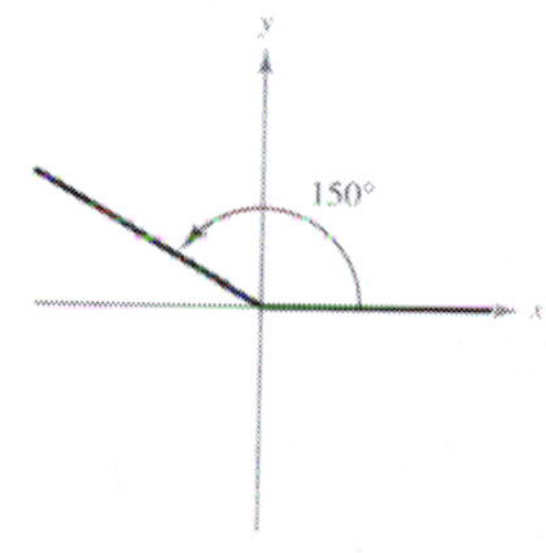

21. (a) 405°

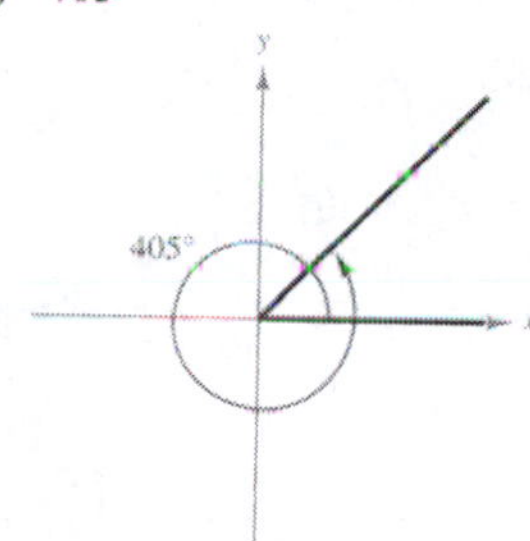

(b) 480°

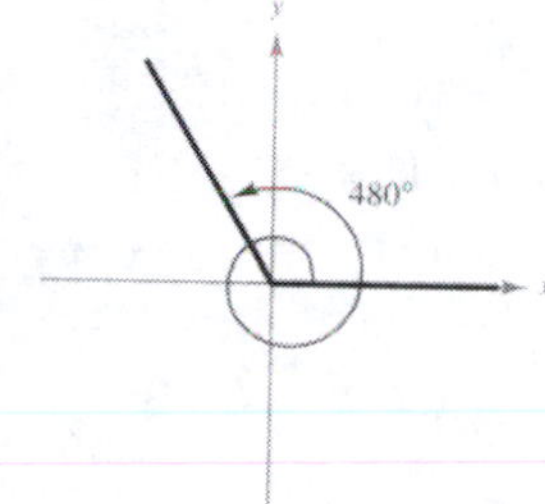

23. (a) Coterminal angles for 45°

$45° + 360° = 405°$

$45° - 360° = -315°$

(b) Coterminal angles for −36°

$-36° + 360° = 324°$

$-36° - 360° = -396°$

25. (a) Coterminal angles for 300°

$300° + 360° = 660°$

$300° - 360° = -60°$

(b) Coterminal angles for 740°

$740° - 2(360°) = 20°$

$20° - 360° = -340°$

27. (a) $54°45' = 54° + \left(\frac{45}{60}\right)^\circ = 54.75°$

(b) $-128°30' = -128° - \left(\frac{30}{60}\right)^\circ = -128.5°$

29. (a) $85°18'30'' = \left(85 + \frac{18}{60} + \frac{30}{3600}\right)^\circ \approx 85.308°$

(b) $330°25'' = \left(330 + \frac{25}{3600}\right)^\circ \approx 330.007°$

31. (a) $240.6° = 240° + 0.6(60)' = 240°36'$

(b) $-145.8° = -[145° + 0.8(60')] = -145°48'$

33. (a) $2.5° = 2° + 0.5(60)' = 2°30'$

(b) $-3.58° = -\left[3° + 0.58(60)'\right]$

$= -[3°34.8']$

$= -\left[3°34' + 0.8(60)''\right]$

$= -3°34'48''$

35. (a) Complement: $90° - 18° = 72°$

Supplement: $180° - 18° = 162°$

(b) Complement: $90° - 85° = 5°$

Supplement: $180° - 85° = 95°$

37. (a) Complement: $90° - 24° = 66°$

Supplement: $180° - 24° = 156°$

(b) Complement: Not possible. $126°$ is greater than $90°$.

Supplement: $180° - 126° = 54°$

39.

The angle shown is approximately 2 radians.

41.

The angle shown is approximately -3 radians.

43.

The angle shown is approximately 1 radian.

45. (a) Because $0 < \frac{\pi}{4} < \frac{\pi}{2}$; $\frac{\pi}{4}$ lies in Quadrant I.

(b) Because $\pi < \frac{5\pi}{4} < \frac{3\pi}{2}$; $\frac{5\pi}{4}$ lies in Quadrant III.

47. (a) Because $0 < \frac{\pi}{5} < \frac{\pi}{2}$; $\frac{\pi}{5}$ lies in Quadrant I.

(b) Because $\pi < \frac{7\pi}{5} < \frac{3\pi}{2}$; $\frac{7\pi}{5}$ lies in Quadrant III.

49. (a) Because $-\frac{\pi}{2} < -1 < 0$; -1 lies in Quadrant IV.

(b) Because $-\pi < -2 < -\frac{\pi}{2}$; -2 lies in Quadrant III.

51. (a) $\frac{\pi}{3}$

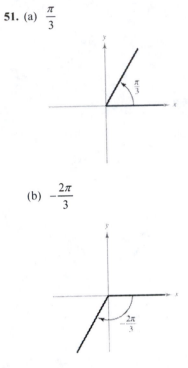

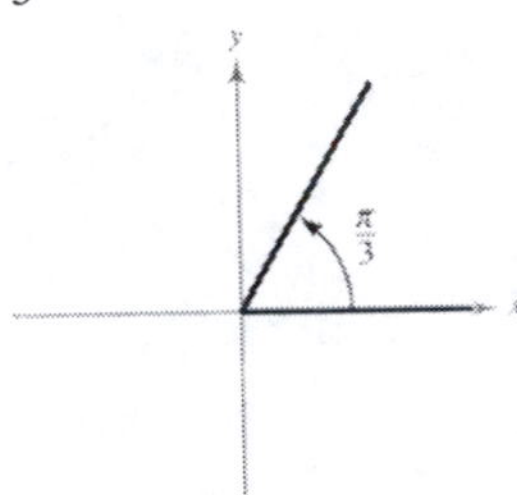

(b) $-\frac{2\pi}{3}$

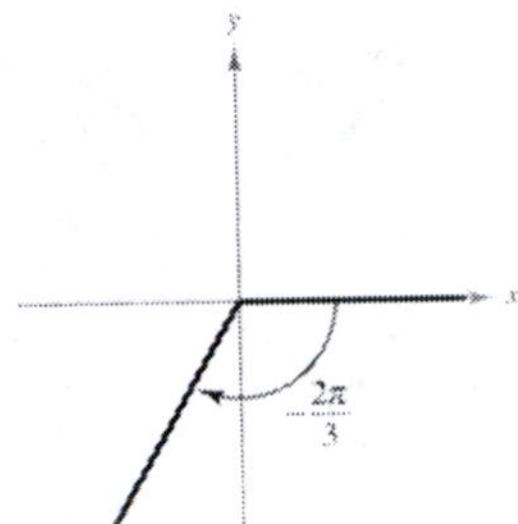

53. (a) $\frac{11\pi}{6}$

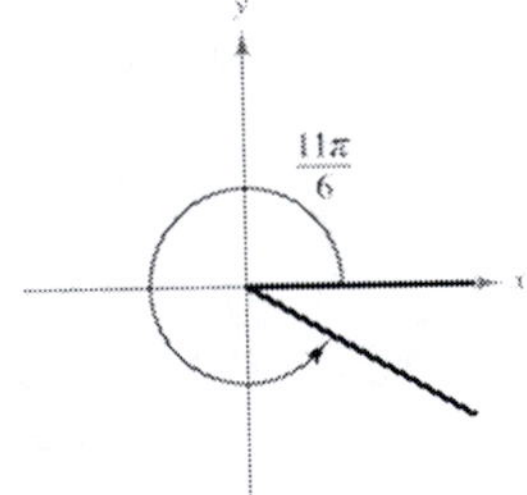

(b) -3

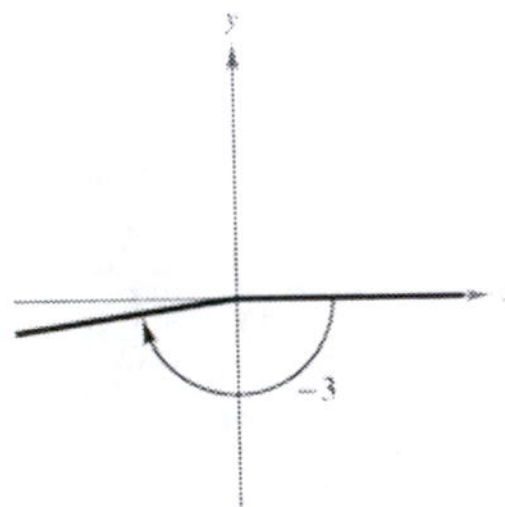

55. (a) Coterminal angles for $\frac{\pi}{6}$

$$\frac{\pi}{6} + 2\pi = \frac{13\pi}{6}$$
$$\frac{\pi}{6} - 2\pi = -\frac{11\pi}{6}$$

(b) Coterminal angles for $\frac{5\pi}{6}$

$$\frac{5\pi}{6} + 2\pi = \frac{17\pi}{6}$$
$$\frac{5\pi}{6} - 2\pi = -\frac{7\pi}{6}$$

57. (a) Coterminal angles for $\frac{2\pi}{3}$

$$\frac{2\pi}{3} + 2\pi = \frac{8\pi}{3}$$
$$\frac{2\pi}{3} - 2\pi = -\frac{4\pi}{3}$$

(b) Coterminal angles for $-\frac{\pi}{12}$

$$-\frac{\pi}{12} + 2\pi = \frac{23\pi}{12}$$
$$-\frac{\pi}{12} - 2\pi = -\frac{25\pi}{12}$$

59. (a) Coterminal angles for $-\frac{9\pi}{4}$

$$-\frac{9\pi}{4} + 4\pi = \frac{7\pi}{4}$$
$$-\frac{9\pi}{4} + 2\pi = -\frac{\pi}{4}$$

(b) Coterminal angles for $-\frac{2\pi}{15}$

$$-\frac{2\pi}{15} + 2\pi = \frac{28\pi}{15}$$
$$-\frac{2\pi}{15} - 2\pi = -\frac{32\pi}{15}$$

61. (a) Complement: $\frac{\pi}{2} - \frac{\pi}{12} = \frac{5\pi}{12}$

Supplement: $\pi - \frac{\pi}{12} = \frac{11\pi}{12}$

(b) Complement: Not possible; $\frac{11\pi}{12}$ is greater than $\frac{\pi}{2}$

Supplement: $\pi - \frac{11\pi}{12} = \frac{\pi}{12}$

63. (a) $30° = 30\left(\frac{\pi}{180°}\right) = \frac{\pi}{6}$

(b) $45° = 45\left(\frac{\pi}{180°}\right) = \frac{\pi}{4}$

65. (a) $-20° = -20\left(\frac{\pi}{180°}\right) = -\frac{\pi}{9}$

(b) $-60° = -60\left(\frac{\pi}{180°}\right) = -\frac{\pi}{3}$

67. (a) $\frac{3\pi}{2} = \frac{3\pi}{2}\left(\frac{180°}{\pi}\right) = 270°$

(b) $\frac{7\pi}{6} = \frac{7\pi}{6}\left(\frac{180°}{\pi}\right) = 210°$

69. (a) $\frac{5\pi}{4} = \frac{5\pi}{4}\left(\frac{180°}{\pi}\right) = 225°$

(b) $-\frac{7\pi}{3} = -\frac{7\pi}{3}\left(\frac{180°}{\pi}\right) = -420°$

71. $45° = 45\left(\frac{\pi}{180°}\right) \approx 0.785$ radian

73. $-216.35° = -216.35\left(\frac{\pi}{180°}\right) \approx -3.776$ radians

75. $532° = 532\left(\frac{\pi}{180°}\right) \approx 9.285$ radians

77. $-0.83° = -0.83\left(\frac{\pi}{180°}\right) \approx -0.014$ radian

79. $\frac{\pi}{7} = \frac{\pi}{7}\left(\frac{180°}{\pi}\right) \approx 25.714°$

81. $\frac{15\pi}{8} = \frac{15\pi}{8}\left(\frac{180°}{\pi}\right) = 337.500°$

83. $-2 = -2\left(\frac{180°}{\pi}\right) \approx -114.592°$

85. $s = r\theta$

$6 = 5\theta$

$\theta = \frac{6}{5}$ radians

87. $s = r\theta$

$32 = 7\theta$

$\theta = \frac{32}{7}$ radians

89. $r = 4$ inches, $s = 18$ inches

$18 = 4\theta$

$\theta = \frac{9}{2}$ radians

91. $s = r\theta$

$25 = 14.5\theta$

$\theta = \dfrac{25}{14.5} = \dfrac{50}{29}$ radians

93. $r = 15$ inches, $\theta = 120° = \dfrac{2\pi}{3}$

$s = r\theta = 15\left(\dfrac{2\pi}{3}\right) = 10\pi$ inches ≈ 31.42 inches

95. $s = r\theta$, θ in radians

$s = 3(1) = 3$ meters

97. $r = 6$ inches, $\theta = \dfrac{\pi}{3}$

$A = \dfrac{1}{2}r^2\theta = \dfrac{1}{2}(6)^2\left(\dfrac{\pi}{3}\right) = 6\pi \text{ in.}^2 \approx 18.85 \text{ in.}^2$

99. $A = \dfrac{1}{2}r^2\theta$

$A = \dfrac{1}{2}(2.5)^2(225)\left(\dfrac{\pi}{180}\right) \approx 12.27$ square feet

101. $\theta = 41°15'50'' - 32°47'9''$

$\approx 8.47806°$

≈ 0.14797 radian

$s = r\theta \approx 4000(0.14782) \approx 592$ miles

103. $\theta = \dfrac{s}{r} = \dfrac{450}{6378} \approx 0.071 \text{ radian} \approx 4.04°$

105. $\theta = \dfrac{s}{r} = \dfrac{2.5}{6} = \dfrac{25}{60} = \dfrac{5}{12}$ radian

107. Linear speed $= \dfrac{s}{t} = \dfrac{r\theta}{t}$

$= \dfrac{(6378 \text{ km} + 1250 \text{ km})(2\pi)}{110 \text{ minutes}}$

$= \dfrac{7628\,(2\pi) \text{ kilometers}}{110 \text{ minutes}}$

$\approx 435.7 \dfrac{\text{kilometers}}{\text{minutes}}$

109. (a) Angular speed $= \dfrac{(5200)(2\pi) \text{ radians}}{1 \text{ minute}} = 10{,}400\pi$ radians per minute

(b) Linear speed $= \dfrac{\left(\dfrac{7.25}{2} \text{ in.}\right)\left(\dfrac{1 \text{ ft}}{12 \text{ in.}}\right)(5200)(2\pi) \text{ feet}}{1 \text{ minute}} = 3141\frac{2}{3}\pi$ feet per minute ≈ 164.5 feet per second

111. (a) $(200)(2\pi) \le$ Angular speed $\le (500)(2\pi)$ radians per minute

Interval: $[400\pi, 1000\pi]$ radians per minute

(b) $(6)(200)(2\pi) \le$ Linear speed $\le (6)(500)(2\pi)$ centimeters per minute

Interval: $[2400\pi, 6000\pi]$ centimeters per minute

113. (a) Arc length of larger sprocket in feet:

$$s = r\theta$$

$$s = \frac{1}{3}(2\pi) = \frac{2\pi}{3} \text{ feet}$$

Therefore, the chain moves $\frac{2\pi}{3}$ feet as does the smaller rear sprocket.

Thus, the angle θ of the smaller sprocket is

$$\theta = \frac{s}{r} = \frac{\frac{2\pi}{3}\text{ ft}}{\frac{2}{12}\text{ ft}} = 4\pi \quad \left(r = 2 \text{ inches} = \frac{2}{12} \text{ feet}\right)$$

and the arc length of the tire in feet is:

$$s = \theta r$$

$$s = (4\pi)\left(\frac{14}{12}\right) = \frac{14\pi}{3} \text{ feet}$$

$$\text{Speed} = \frac{s}{t} = \frac{\frac{14\pi}{3}}{1 \text{ sec}} = \frac{14\pi}{3} \text{ feet per second}$$

$$\frac{14\pi \text{ feet}}{3 \text{ seconds}} \times \frac{3600 \text{ seconds}}{1 \text{ hour}} \times \frac{1 \text{ mile}}{5280 \text{ feet}} \approx 10 \text{ miles per hour}$$

(b) Since the arc length of the tire is $\frac{14\pi}{3}$ feet and the cyclist is pedaling at a rate of one revolution per second, we have:

$$\text{Distance} = \left(\frac{14\pi}{3} \frac{\text{feet}}{\text{revolutions}}\right)\left(\frac{1 \text{ mile}}{5280 \text{ feet}}\right)(n \text{ revolutions})$$

$$= \frac{7\pi}{7920} n \text{ miles}$$

(c) Distance = Rate · Time

$$= \left(\frac{14\pi}{3} \text{ feet/second}\right)\left(\frac{1 \text{ mile}}{5280 \text{ feet}}\right)(t \text{ seconds})$$

$$= \frac{7\pi}{7920} t \text{ miles}$$

(d) The functions are both linear.

115. $A = \frac{1}{2}\theta\left(R^2 - r^2\right)$

$R = 25$

$r = 25 - 14 = 11$

$A = \frac{1}{2}\left(\frac{125}{180}\right)^{\circ} \pi \cdot \left(25^2 - 11^2\right) \approx 175\pi$

$\approx 549.8 \text{ in.}^2$

25
125°
r
14

117. False. A measurement of 4π radians corresponds to two complete revolutions from the initial to the terminal side of an angle.

119. False. The terminal side of $-1260°$ lies on the negative x-axis.

121. The speed increases, since the linear speed is proportional to the radius.

123. Since the arc length s is given by $s = r\theta$, if the central angle θ is fixed while the radius r increases, then s increases in proportion to r.

Section 6.2 Right Triangle Trigonometry

1. (i) $\dfrac{\text{hypotenuse}}{\text{adjacent}} = \sec\theta$ (e) (ii) $\dfrac{\text{adjacent}}{\text{opposite}} = \cot\theta$ (f) (iii) $\dfrac{\text{hypotenuse}}{\text{opposite}} = \csc\theta$ (d)

(iv) $\dfrac{\text{adjacent}}{\text{hypotenuse}} = \cos\theta$ (b) (v) $\dfrac{\text{opposite}}{\text{hypotenuse}} = \sin\theta$ (a) (vi) $\dfrac{\text{opposite}}{\text{adjacent}} = \tan\theta$ (c)

3. Complementary

5. $\text{hyp} = \sqrt{6^2 + 8^2} = \sqrt{36 + 64} = \sqrt{100} = 10$

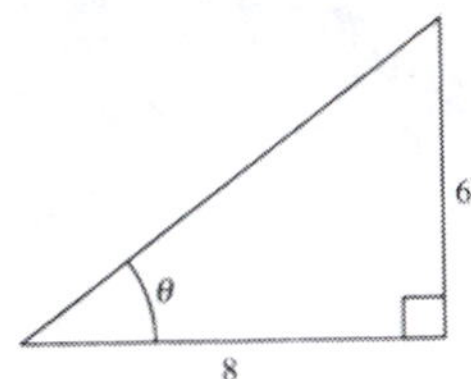

$\sin\theta = \dfrac{\text{opp}}{\text{hyp}} = \dfrac{6}{10} = \dfrac{3}{5}$ $\csc\theta = \dfrac{\text{hyp}}{\text{opp}} = \dfrac{10}{6} = \dfrac{5}{3}$

$\cos\theta = \dfrac{\text{adj}}{\text{hyp}} = \dfrac{8}{10} = \dfrac{4}{5}$ $\sec\theta = \dfrac{\text{hyp}}{\text{adj}} = \dfrac{10}{8} = \dfrac{5}{4}$

$\tan\theta = \dfrac{\text{opp}}{\text{adj}} = \dfrac{6}{8} = \dfrac{3}{4}$ $\cot\theta = \dfrac{\text{adj}}{\text{opp}} = \dfrac{8}{6} = \dfrac{4}{3}$

7. $\text{adj} = \sqrt{41^2 - 9^2} = \sqrt{1681 - 81} = \sqrt{1600} = 40$

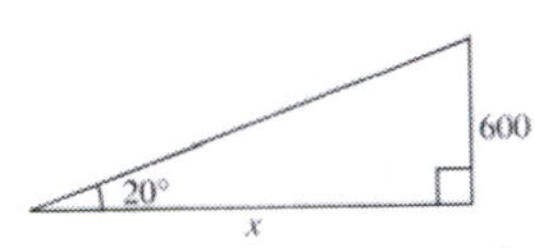

$\sin\theta = \dfrac{\text{opp}}{\text{hyp}} = \dfrac{9}{41}$ $\csc\theta = \dfrac{\text{hyp}}{\text{opp}} = \dfrac{41}{9}$

$\cos\theta = \dfrac{\text{adj}}{\text{hyp}} = \dfrac{40}{41}$ $\sec\theta = \dfrac{\text{hyp}}{\text{adj}} = \dfrac{41}{40}$

$\tan\theta = \dfrac{\text{opp}}{\text{adj}} = \dfrac{9}{40}$ $\cot\theta = \dfrac{\text{adj}}{\text{opp}} = \dfrac{40}{9}$

9. $\text{hyp} = \sqrt{15^2 + 8^2} = \sqrt{289} = 17$

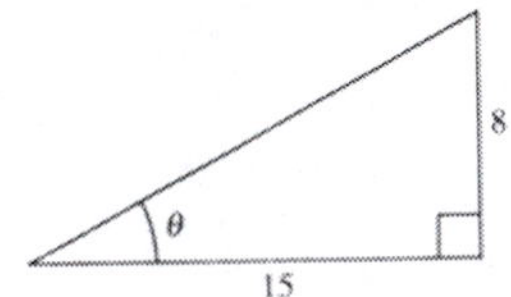

$\sin\theta = \dfrac{\text{opp}}{\text{hyp}} = \dfrac{8}{17}$ $\csc\theta = \dfrac{\text{hyp}}{\text{opp}} = \dfrac{17}{8}$

$\cos\theta = \dfrac{\text{adj}}{\text{hyp}} = \dfrac{15}{17}$ $\sec\theta = \dfrac{\text{hyp}}{\text{adj}} = \dfrac{17}{15}$

$\tan\theta = \dfrac{\text{opp}}{\text{adj}} = \dfrac{8}{15}$ $\cot\theta = \dfrac{\text{adj}}{\text{opp}} = \dfrac{15}{8}$

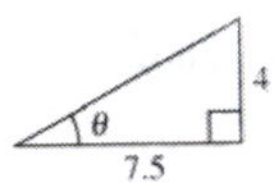

$\text{hyp} = \sqrt{7.5^2 + 4^2} = \dfrac{17}{2}$

$\sin\theta = \dfrac{\text{opp}}{\text{hyp}} = \dfrac{4}{(17/2)} = \dfrac{8}{17}$ $\csc\theta = \dfrac{\text{hyp}}{\text{opp}} = \dfrac{(17/2)}{4} = \dfrac{17}{8}$

$\cos\theta = \dfrac{\text{adj}}{\text{hyp}} = \dfrac{7.5}{(17/2)} = \dfrac{15}{17}$ $\sec\theta = \dfrac{\text{hyp}}{\text{adj}} = \dfrac{(17/2)}{7.5} = \dfrac{17}{15}$

$\tan\theta = \dfrac{\text{opp}}{\text{adj}} = \dfrac{4}{7.5} = \dfrac{8}{15}$ $\cot\theta = \dfrac{\text{adj}}{\text{opp}} = \dfrac{7.5}{4} = \dfrac{15}{8}$

The function values are the same because the triangles are similar, and corresponding sides are proportional.

11. $\text{adj} = \sqrt{3^2 - 1^2} = \sqrt{8} = 2\sqrt{2}$

$\sin\theta = \dfrac{\text{opp}}{\text{hyp}} = \dfrac{1}{3}$ $\qquad \csc\theta = \dfrac{\text{hyp}}{\text{opp}} = 3$

$\cos\theta = \dfrac{\text{adj}}{\text{hyp}} = \dfrac{2\sqrt{2}}{3}$ $\qquad \sec\theta = \dfrac{\text{hyp}}{\text{adj}} = \dfrac{3}{2\sqrt{2}} = \dfrac{3\sqrt{2}}{4}$

$\tan\theta = \dfrac{\text{opp}}{\text{adj}} = \dfrac{1}{2\sqrt{2}} = \dfrac{\sqrt{2}}{4}$ $\qquad \cot\theta = \dfrac{\text{adj}}{\text{opp}} = 2\sqrt{2}$

$\text{adj} = \sqrt{6^2 - 2^2} = \sqrt{32} = 4\sqrt{2}$

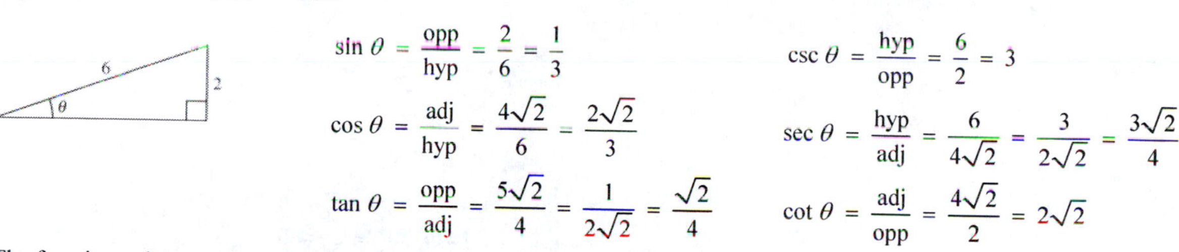

$\sin\theta = \dfrac{\text{opp}}{\text{hyp}} = \dfrac{2}{6} = \dfrac{1}{3}$ $\qquad \csc\theta = \dfrac{\text{hyp}}{\text{opp}} = \dfrac{6}{2} = 3$

$\cos\theta = \dfrac{\text{adj}}{\text{hyp}} = \dfrac{4\sqrt{2}}{6} = \dfrac{2\sqrt{2}}{3}$ $\qquad \sec\theta = \dfrac{\text{hyp}}{\text{adj}} = \dfrac{6}{4\sqrt{2}} = \dfrac{3}{2\sqrt{2}} = \dfrac{3\sqrt{2}}{4}$

$\tan\theta = \dfrac{\text{opp}}{\text{adj}} = \dfrac{5\sqrt{2}}{4} = \dfrac{1}{2\sqrt{2}} = \dfrac{\sqrt{2}}{4}$ $\qquad \cot\theta = \dfrac{\text{adj}}{\text{opp}} = \dfrac{4\sqrt{2}}{2} = 2\sqrt{2}$

The function values are the same since the triangles are similar and the corresponding sides are proportional.

13. Given: $\tan\theta = \dfrac{3}{4} = \dfrac{\text{opp}}{\text{adj}}$

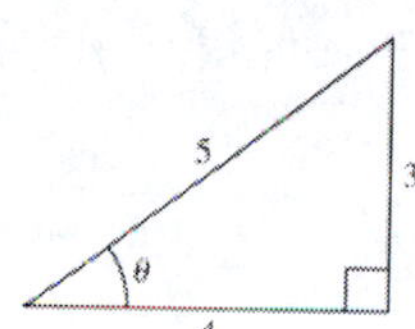

$3^2 + 4^2 = (\text{hyp})^2$

$\text{hyp} = 5$

$\sin\theta = \dfrac{\text{opp}}{\text{hyp}} = \dfrac{3}{5}$

$\cos\theta = \dfrac{\text{adj}}{\text{hyp}} = \dfrac{4}{5}$

$\csc\theta = \dfrac{\text{hyp}}{\text{opp}} = \dfrac{5}{3}$

$\sec\theta = \dfrac{\text{hyp}}{\text{adj}} = \dfrac{5}{4}$

$\cot\theta = \dfrac{\text{adj}}{\text{opp}} = \dfrac{4}{3}$

15. Given: $\sec\theta = \dfrac{3}{2} = \dfrac{\text{hyp}}{\text{adj}}$

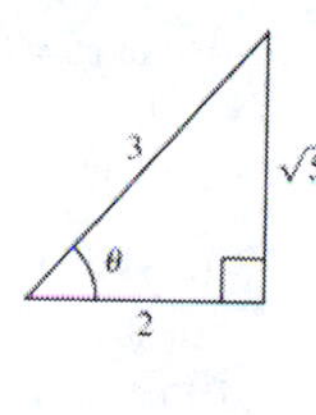

$(\text{opp})^2 + 2^2 = 3^2$

$\text{opp} = \sqrt{5}$

$\sin\theta = \dfrac{\text{opp}}{\text{hyp}} = \dfrac{\sqrt{5}}{3}$

$\cos\theta = \dfrac{\text{adj}}{\text{hyp}} = \dfrac{2}{3}$

$\tan\theta = \dfrac{\text{opp}}{\text{adj}} = \dfrac{\sqrt{5}}{2}$

$\csc\theta = \dfrac{\text{hyp}}{\text{opp}} = \dfrac{3\sqrt{5}}{5}$

$\cot\theta = \dfrac{\text{adj}}{\text{opp}} = \dfrac{2\sqrt{5}}{5}$

17. Given: $\sin\theta = \dfrac{1}{5} = \dfrac{\text{opp}}{\text{hyp}}$

$1^2 + (\text{adj})^2 = 5^2$

$\text{adj} = \sqrt{24} = 2\sqrt{6}$

$\cos\theta = \dfrac{\text{adj}}{\text{hyp}} = \dfrac{2\sqrt{6}}{5}$

$\tan\theta = \dfrac{\text{opp}}{\text{adj}} = \dfrac{\sqrt{6}}{12}$

$\csc\theta = \dfrac{\text{hyp}}{\text{opp}} = 5$

$\sec\theta = \dfrac{\text{hyp}}{\text{adj}} = \dfrac{5\sqrt{6}}{12}$

$\cot\theta = \dfrac{\text{adj}}{\text{opp}} = 2\sqrt{6}$

19. Given: $\cot\theta = 3 = \dfrac{3}{1} = \dfrac{\text{adj}}{\text{opp}}$

$1^2 + 3^2 = (\text{hyp})^2$

$\text{hyp} = \sqrt{10}$

$\sin\theta = \dfrac{\text{opp}}{\text{hyp}} = \dfrac{\sqrt{10}}{10}$

$\cos\theta = \dfrac{\text{adj}}{\text{hyp}} = \dfrac{3\sqrt{10}}{10}$

$\tan\theta = \dfrac{\text{opp}}{\text{adj}} = \dfrac{1}{3}$

$\csc\theta = \dfrac{\text{hyp}}{\text{opp}} = \sqrt{10}$

$\sec\theta = \dfrac{\text{hyp}}{\text{adj}} = \dfrac{\sqrt{10}}{3}$

21.

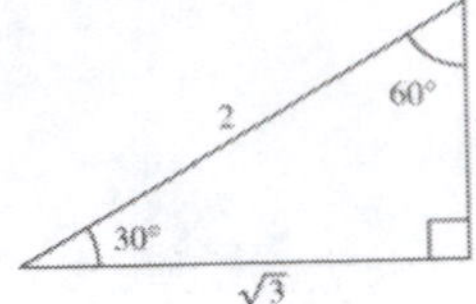

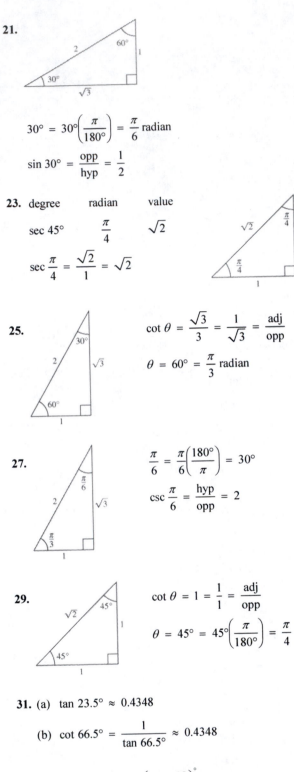

$$30° = 30°\left(\frac{\pi}{180°}\right) = \frac{\pi}{6} \text{ radian}$$

$$\sin 30° = \frac{\text{opp}}{\text{hyp}} = \frac{1}{2}$$

23.

degree	radian	value
sec 45°	$\frac{\pi}{4}$	$\sqrt{2}$

$$\sec\frac{\pi}{4} = \frac{\sqrt{2}}{1} = \sqrt{2}$$

25.

$$\cot\theta = \frac{\sqrt{3}}{3} = \frac{1}{\sqrt{3}} = \frac{\text{adj}}{\text{opp}}$$

$$\theta = 60° = \frac{\pi}{3} \text{ radian}$$

27.

$$\frac{\pi}{6} = \frac{\pi}{6}\left(\frac{180°}{\pi}\right) = 30°$$

$$\csc\frac{\pi}{6} = \frac{\text{hyp}}{\text{opp}} = 2$$

29.

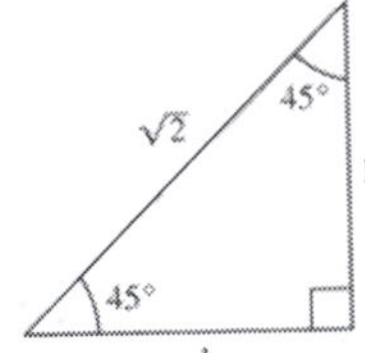

$$\cot\theta = 1 = \frac{1}{1} = \frac{\text{adj}}{\text{opp}}$$

$$\theta = 45° = 45°\left(\frac{\pi}{180°}\right) = \frac{\pi}{4}$$

31. (a) $\tan 23.5° \approx 0.4348$

(b) $\cot 66.5° = \dfrac{1}{\tan 66.5°} \approx 0.4348$

33. (a) $\cos 16°18' = \cos\left(16 + \dfrac{18}{60}\right)^{\circ} \approx 0.9598$

(b) $\sin 73°56' = \sin\left(73 + \dfrac{56}{60}\right)^{\circ} \approx 0.9609$

35. Make sure that your calculator is in radian mode.

(a) $\cot\dfrac{\pi}{16} = \dfrac{1}{\tan\dfrac{\pi}{16}} \approx 5.0273$

(b) $\tan\dfrac{\pi}{16} \approx 0.1989$

37. Make sure that your calculator is in radian mode.

(a) $\csc 1 = \dfrac{1}{\sin 1} \approx 1.1884$

(b) $\tan\dfrac{1}{2} \approx 0.5463$

39. $\sin 60° = \dfrac{\sqrt{3}}{2}, \cos 60° = \dfrac{1}{2}$

(a) $\sin 30° = \cos 60° = \dfrac{1}{2}$

(b) $\cos 30° = \sin 60° = \dfrac{\sqrt{3}}{2}$

(c) $\tan 60° = \dfrac{\sin 60°}{\cos 60°} = \sqrt{3}$

(d) $\cot 60° = \dfrac{\cos 60°}{\sin 60°} = \dfrac{1}{\sqrt{3}} = \dfrac{\sqrt{3}}{3}$

41. $\cos\theta = \dfrac{1}{3}$

(a)
$$\sin^2\theta + \cos^2\theta = 1$$
$$\sin^2\theta + \left(\frac{1}{3}\right)^2 = 1$$
$$\sin^2\theta = \frac{8}{9}$$
$$\sin\theta = \frac{2\sqrt{2}}{3}$$

(b) $\tan\theta = \dfrac{\sin\theta}{\cos\theta} = \dfrac{\frac{2\sqrt{2}}{3}}{\frac{1}{3}} = 2\sqrt{2}$

(c) $\sec\theta = \dfrac{1}{\cos\theta} = 3$

(d) $\csc(90° - \theta) = \sec\theta = 3$

43. $\cot\alpha = 5$

(a) $\tan\alpha = \dfrac{1}{\cot\alpha} = \dfrac{1}{5}$

(b) $\csc^2\alpha = 1 + \cot^2\alpha$

$\csc^2\alpha = 1 + 5^2$

$\csc^2\alpha = 26$

$\csc\alpha = \sqrt{26}$

(c) $\cot(90° - \alpha) = \tan\alpha = \dfrac{1}{5}$

(d) $\sec^2\alpha = 1 + \tan^2\alpha$

$\sec^2\alpha = 1 + \left(\dfrac{1}{5}\right)^2$

$\sec^2\alpha = \dfrac{26}{25}$

$\sec\alpha = \dfrac{\sqrt{26}}{5}$

$\cos\alpha = \dfrac{1}{\sec\alpha} = \dfrac{5\sqrt{26}}{26}$

45. $\tan\theta\cot\theta = \tan\theta\left(\dfrac{1}{\tan\theta}\right) = 1$

47. $\tan\alpha\cos\alpha = \left(\dfrac{\sin\alpha}{\cos\alpha}\right)\cos\alpha = \sin\alpha$

49. $(1 + \sin\theta)(1 - \sin\theta) = 1 - \sin^2\theta = \cos^2\theta$

51. $(\sec\theta + \tan\theta)(\sec\theta - \tan\theta) = \sec^2\theta - \tan^2\theta$

$= (1 + \tan^2\theta) - \tan^2\theta$

$= 1$

53. $\dfrac{\sin\theta}{\cos\theta} + \dfrac{\cos\theta}{\sin\theta} = \dfrac{\sin^2\theta + \cos^2\theta}{\sin\theta\cos\theta}$

$= \dfrac{1}{\sin\theta\cos\theta}$

$= \dfrac{1}{\sin\theta} \cdot \dfrac{1}{\cos\theta}$

$= \csc\theta\sec\theta$

55. (a) $\sin\theta = \dfrac{1}{2} \Rightarrow \theta = 30° = \dfrac{\pi}{6}$

(b) $\csc\theta = 2 \Rightarrow \theta = 30° = \dfrac{\pi}{6}$

57. (a) $\sec\theta = 2 \Rightarrow \theta = 60° = \dfrac{\pi}{3}$

(b) $\cot\theta = 1 \Rightarrow \theta = 45° = \dfrac{\pi}{4}$

59. (a) $\csc\theta = \dfrac{2\sqrt{3}}{3} \Rightarrow \theta = 60° = \dfrac{\pi}{3}$

(b) $\sin\theta = \dfrac{\sqrt{2}}{2} \Rightarrow \theta = 45° = \dfrac{\pi}{4}$

61. $\cos 60° = \dfrac{x}{18}$

$x = 18\cos 60° = 18\left(\dfrac{1}{2}\right) = 9$

$\sin 60° = \dfrac{y}{18}$

$y = 18\sin 60° = 18\dfrac{\sqrt{3}}{2} = 9\sqrt{3}$

63. $\tan 60° = \dfrac{32}{x}$

$\sqrt{3} = \dfrac{32}{x}$

$\sqrt{3}x = 32$

$x = \dfrac{32}{\sqrt{3}} = \dfrac{32\sqrt{3}}{3}$

$\sin 60° = \dfrac{32}{r}$

$r = \dfrac{32}{\sin 60°}$

$r = \dfrac{32}{\frac{\sqrt{3}}{2}} = \dfrac{64\sqrt{3}}{3}$

65.

y
x
82°
45 m

$\tan 82° = \dfrac{x}{45}$

$x = 45\tan 82°$

Height of the building:

$123 + 45\tan 82° \approx 443.2$ meters

Distance between friends:

$\cos 82° = \dfrac{45}{y} \Rightarrow y = \dfrac{45}{\cos 82°}$

≈ 323.34 meters

67. $\sin\theta = \dfrac{1250}{2500} = \dfrac{1}{2}$

$\theta = 30° = \dfrac{\pi}{6}$

2500 ft
1250 ft
θ

69. (a) $\sin 43° = \dfrac{150}{x}$

$x = \dfrac{150}{\sin 43°} \approx 219.9$ ft

(b) $\tan 43° = \dfrac{150}{y}$

$y = \dfrac{150}{\tan 43°} \approx 160.9$ ft

x
150
43°
y

71.

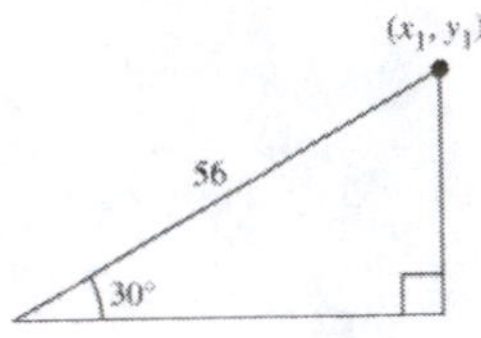

$\sin 30° = \dfrac{y_1}{56}$

$y_1 = (\sin 30°)(56) = \left(\dfrac{1}{2}\right)(56) = 28$

$\cos 30° = \dfrac{x_1}{56}$

$x_1 = \cos 30°(56) = \dfrac{\sqrt{3}}{2}(56) = 28\sqrt{3}$

$(x_1, y_1) = \left(28\sqrt{3}, 28\right)$

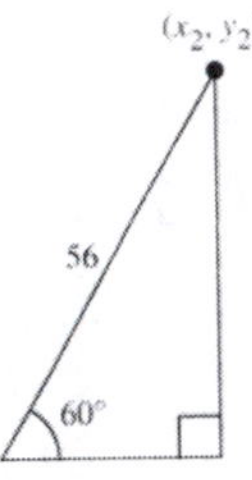

$\sin 60° = \dfrac{y_2}{56}$

$y_2 = \sin 60°(56) = \left(\dfrac{\sqrt{3}}{2}\right)(56) = 28\sqrt{3}$

$\cos 60° = \dfrac{x_2}{56}$

$x_2 = (\cos 60°)(56) = \left(\dfrac{1}{2}\right)(56) = 28$

$(x_2, y_2) = \left(28, 28\sqrt{3}\right)$

73. $x \approx 9.397,\ y \approx 3.420$

$\sin 20° = \dfrac{y}{10} \approx 0.34$

$\cos 20° = \dfrac{x}{10} \approx 0.94$

$\tan 20° = \dfrac{y}{x} \approx 0.36$

$\cot 20° = \dfrac{x}{y} \approx 2.75$

$\sec 20° = \dfrac{10}{x} \approx 1.06$

$\csc 20° = \dfrac{10}{y} \approx 2.92$

75. (a)

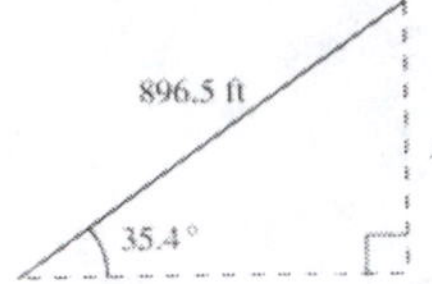

$\sin 35.4° = \dfrac{x}{896.5}$

$x = 896.5 \sin 35.4° \approx 519.33$ feet

(b) Because the top of the incline is 1693.5 feet above sea level and the vertical rise of the inclined plane is 519.33 feet, the elevation of the lower end of the inclined plan is about
$1693.5 - 519.33 = 1174.17$ feet.

(c) Ascent time: $d = rt$

$896.5 = 300t$

$3 \approx t$

It takes about 3 minutes for the cars to get from the bottom to the top.

Vertical rate: $d = rt$

$519.33 = r(3)$

$r = 173.11$ ft/min

77. $\sin 60° \csc 60° = 1$

True,

$\csc x = \dfrac{1}{\sin x} \Rightarrow \sin 60° \csc 60° = \sin 60°\left(\dfrac{1}{\sin 60°}\right)$

$= 1$

79. $\sin 45° + \sin 45° = 1$

False, $\dfrac{\sqrt{2}}{2} + \dfrac{\sqrt{2}}{2} = \sqrt{2} \neq 1$

81. $\dfrac{\sin 60°}{\sin 30°} = \sin 2°$

False, $\dfrac{\sin 60°}{\sin 30°} = \dfrac{\cos 30°}{\sin 30°} = \cot 30° \approx 1.7321$

$\sin 2° \approx 0.0349$

83. Yes. Given $\tan\theta$, $\sec\theta$ can be found from the identity $1 + \tan^2\theta = \sec^2\theta$.

85. (a)

θ	0	0.3	0.6	0.9	1.2	1.5
$\sin\theta$	0	0.2955	0.5646	0.7833	0.9320	0.9975
$\cos\theta$	1	0.9553	0.8253	0.6216	0.3624	0.0707

(b) On $[0, 1.5]$, $\sin\theta$ is an increasing function.

(c) On $[0, 1.5]$, $\cos\theta$ is a decreasing function.

(d) As the angle increases the length of the side opposite the angle increases relative to the length of the hypotenuse and the length of the side adjacent to the angle decreases relative to the length of the hypotenuse. Thus the sine increases and the cosine decreases.

Section 6.3 Trigonometric Functions of Any Angle

1. reference

3. period

5. (a) $(x, y) = (4, 3)$

$r = \sqrt{16 + 9} = 5$

$\sin\theta = \dfrac{y}{r} = \dfrac{3}{5}$ $\quad\csc\theta = \dfrac{r}{y} = \dfrac{5}{3}$

$\cos\theta = \dfrac{x}{r} = \dfrac{4}{5}$ $\quad\sec\theta = \dfrac{r}{x} = \dfrac{5}{4}$

$\tan\theta = \dfrac{y}{x} = \dfrac{3}{4}$ $\quad\cot\theta = \dfrac{x}{y} = \dfrac{4}{3}$

(b) $(x, y) = (-8, 15)$

$r = \sqrt{64 + 225} = 17$

$\sin\theta = \dfrac{y}{r} = \dfrac{15}{17}$ $\quad\csc\theta = \dfrac{r}{y} = \dfrac{17}{15}$

$\cos\theta = \dfrac{x}{r} = -\dfrac{8}{17}$ $\quad\sec\theta = \dfrac{r}{x} = -\dfrac{17}{8}$

$\tan\theta = \dfrac{y}{x} = -\dfrac{15}{8}$ $\quad\cot\theta = \dfrac{x}{y} = -\dfrac{8}{15}$

7. (a) $(x, y) = \left(-\sqrt{3}, -1\right)$

$r = \sqrt{3 + 1} = 2$

$\sin\theta = \dfrac{y}{r} = -\dfrac{1}{2}$ $\quad\csc\theta = \dfrac{r}{y} = -2$

$\cos\theta = \dfrac{x}{r} = -\dfrac{\sqrt{3}}{2}$ $\quad\sec\theta = \dfrac{r}{x} = -\dfrac{2\sqrt{3}}{3}$

$\tan\theta = \dfrac{y}{x} = \dfrac{\sqrt{3}}{3}$ $\quad\cot\theta = \dfrac{x}{y} = \sqrt{3}$

(b) $(x, y) = (4, -1)$

$r = \sqrt{16 + 1} = \sqrt{17}$

$\sin\theta = \dfrac{y}{r} = -\dfrac{1}{\sqrt{17}} = -\dfrac{\sqrt{17}}{17}$ $\quad\csc\theta = \dfrac{r}{y} = -\sqrt{17}$

$\cos\theta = \dfrac{x}{r} = \dfrac{4}{\sqrt{17}} = \dfrac{4\sqrt{17}}{17}$ $\quad\sec\theta = \dfrac{r}{x} = \dfrac{\sqrt{17}}{4}$

$\tan\theta = \dfrac{y}{x} = -\dfrac{1}{4}$ $\quad\cot\theta = \dfrac{x}{y} = -4$

9. $(x, y) = (5, 12)$

$r = \sqrt{25 + 144} = 13$

$\sin\theta = \dfrac{y}{r} = \dfrac{12}{13}$ $\quad\csc\theta = \dfrac{r}{y} = \dfrac{13}{12}$

$\cos\theta = \dfrac{x}{r} = \dfrac{5}{13}$ $\quad\sec\theta = \dfrac{r}{x} = \dfrac{13}{5}$

$\tan\theta = \dfrac{y}{x} = \dfrac{12}{5}$ $\quad\cot\theta = \dfrac{x}{y} = \dfrac{5}{12}$

11. $x = -5$, $y = -2$

$r = \sqrt{(-5)^2 + (-2)^2} = \sqrt{29}$

$\sin\theta = \dfrac{y}{r} = \dfrac{-2}{\sqrt{29}} = -\dfrac{2\sqrt{29}}{29}$

$\cos\theta = \dfrac{x}{r} = \dfrac{-5}{\sqrt{29}} = -\dfrac{5\sqrt{29}}{29}$

$\tan\theta = \dfrac{y}{x} = \dfrac{-2}{-5} = \dfrac{2}{5}$

$\csc\theta = \dfrac{r}{y} = \dfrac{\sqrt{29}}{-2} = -\dfrac{\sqrt{29}}{2}$

$\sec\theta = \dfrac{r}{x} = \dfrac{\sqrt{29}}{-5} = -\dfrac{\sqrt{29}}{5}$

$\cot\theta = \dfrac{x}{y} = \dfrac{-5}{-2} = \dfrac{5}{2}$

13. $(x, y) = (-5.4, 7.2)$

$r = \sqrt{29.16 + 51.84} = 9$

$\sin\theta = \frac{y}{r} = \frac{7.2}{9} = \frac{4}{5}$

$\cos\theta = \frac{x}{r} = -\frac{5.4}{9} = -\frac{3}{5}$

$\tan\theta = \frac{y}{x} = -\frac{7.2}{5.4} = -\frac{4}{3}$

$\csc\theta = \frac{r}{y} = \frac{9}{7.2} = \frac{5}{4}$

$\sec\theta = \frac{r}{x} = -\frac{9}{5.4} = -\frac{5}{3}$

$\tan\theta = \frac{x}{y} = -\frac{5.4}{7.2} = -\frac{3}{4}$

15. $\sin\theta > 0 \Rightarrow \theta$ lies in Quadrant I or in Quadrant II.

$\cos\theta > 0 \Rightarrow \theta$ lies in Quadrant I or in Quadrant IV.

$\sin\theta > 0$ and $\cos\theta > 0 \Rightarrow \theta$ lies in Quadrant I.

17. $\sin\theta > 0 \Rightarrow \theta$ lies in Quadrant I or in Quadrant II.

$\cos\theta < 0 \Rightarrow \theta$ lies in Quadrant II or in Quadrant III.

$\sin\theta > 0$ and $\cos\theta < 0 \Rightarrow \theta$ lies in Quadrant II.

19. $\tan\theta < 0$ and $\sin\theta > 0 \Rightarrow \theta$ is in Quadrant II

$\Rightarrow x < 0$ and $y > 0$.

$\tan\theta = \frac{y}{x} = \frac{15}{-8} \Rightarrow r = 17$

$\sin\theta = \frac{y}{r} = \frac{15}{17}$ $\quad\csc\theta = \frac{r}{y} = \frac{17}{15}$

$\cos\theta = \frac{x}{r} = -\frac{8}{17}$ $\quad\sec\theta = \frac{r}{x} = -\frac{17}{8}$

$\tan\theta = \frac{y}{x} = -\frac{15}{8}$ $\quad\cot\theta = \frac{x}{y} = -\frac{8}{15}$

21. $\sin\theta = \frac{y}{r} = \frac{3}{5} \Rightarrow x^2 = 25 - 9 = 16$

θ in Quadrant II $\Rightarrow x = -4$

$\sin\theta = \frac{y}{r} = \frac{3}{5}$ $\quad\csc\theta = \frac{r}{y} = \frac{5}{3}$

$\cos\theta = \frac{x}{r} = -\frac{4}{5}$ $\quad\sec\theta = \frac{r}{x} = -\frac{5}{4}$

$\tan\theta = \frac{y}{x} = -\frac{3}{4}$ $\quad\cot\theta = \frac{x}{y} = -\frac{4}{3}$

23. $\cot\theta = \frac{x}{y} = -\frac{3}{1} = \frac{3}{-1}$

$\cos\theta > 0 \Rightarrow \theta$ is in Quadrant IV $\Rightarrow x$ is positive;

$x = 3, y = -1, r = \sqrt{10}$

$\sin\theta = \frac{y}{r} = -\frac{\sqrt{10}}{10}$ $\quad\csc\theta = \frac{r}{y} = -\sqrt{10}$

$\cos\theta = \frac{x}{r} = \frac{3\sqrt{10}}{10}$ $\quad\sec\theta = \frac{r}{x} = \frac{\sqrt{10}}{3}$

$\tan\theta = \frac{y}{x} = -\frac{1}{3}$ $\quad\cot\theta = \frac{x}{y} = -3$

25. $\sec\theta = \frac{r}{x} = \frac{2}{-1} \Rightarrow y^2 = 4 - 1 = 3$

$\sin\theta < 0 \Rightarrow \theta$ is in Quadrant III $\Rightarrow y = -\sqrt{3}$

$\sin\theta = \frac{y}{r} = -\frac{\sqrt{3}}{2}$ $\quad\csc\theta = \frac{r}{y} = -\frac{2}{\sqrt{3}} = -\frac{2\sqrt{3}}{3}$

$\cos\theta = \frac{x}{r} = -\frac{1}{2}$ $\quad\sec\theta = \frac{r}{x} = -2$

$\tan\theta = \frac{y}{x} = \sqrt{3}$ $\quad\cot\theta = \frac{x}{y} = \frac{1}{\sqrt{3}} = \frac{\sqrt{3}}{3}$

27. $\cot\theta$ is undefined,

$\frac{\pi}{2} \le \theta \le \frac{3\pi}{2} \Rightarrow y = 0 \Rightarrow \theta = \pi$

$\sin\theta = 0$ $\quad\csc\theta$ is undefined.

$\cos\theta = -1$ $\quad\sec\theta = -1$

$\tan\theta = 0$ $\quad\cot\theta$ is undefined.

29. To find a point on the terminal side of θ, use any point on the line $y = -x$ that lies in Quadrant II. $(-1, 1)$ is one such point.

$x = -1, y = 1, r = \sqrt{2}$

$\sin\theta = \frac{1}{\sqrt{2}} = \frac{\sqrt{2}}{2}$ $\quad\csc\theta = \sqrt{2}$

$\cos\theta = -\frac{1}{\sqrt{2}} = -\frac{\sqrt{2}}{2}$ $\quad\sec\theta = -\sqrt{2}$

$\tan\theta = -1$ $\quad\cot\theta = -1$

31. To find a point on the terminal side of θ, use any point on the line $y = 2x$ that lies in Quadrant III. $(-1, -2)$ is one such point.

$x = -1, y = -2, y = \sqrt{5}$

$\sin\theta = -\frac{2}{\sqrt{5}} = -\frac{2\sqrt{5}}{5}$ $\quad\csc\theta = \frac{\sqrt{5}}{-2} = -\frac{\sqrt{5}}{2}$

$\cos\theta = -\frac{1}{\sqrt{5}} = -\frac{\sqrt{5}}{5}$ $\quad\sec\theta = \frac{\sqrt{5}}{-1} = -\sqrt{5}$

$\tan\theta = \frac{-2}{-1} = 2$ $\quad\cot\theta = \frac{-1}{-2} = \frac{1}{2}$

33. $(x, y) = (-1, 0), r = 1$

$$\sin \pi = \frac{y}{r} = \frac{0}{1} = 0$$

35. $(x, y) = (0, -1), r = 1$

$$\sec \frac{3\pi}{2} = \frac{r}{x} = \frac{1}{0} \Rightarrow \text{undefined}$$

37. $(x, y) = (0, 1), r = 1$

$$\sin \frac{\pi}{2} = \frac{y}{r} = \frac{1}{1} = 1$$

39. $(x, y) = (-1, 0), r = 1$

$$\csc \pi = \frac{r}{y} = \frac{1}{0} \Rightarrow \text{undefined}$$

41. $\theta = 160°$

$\theta' = 180° - 160° = 20°$

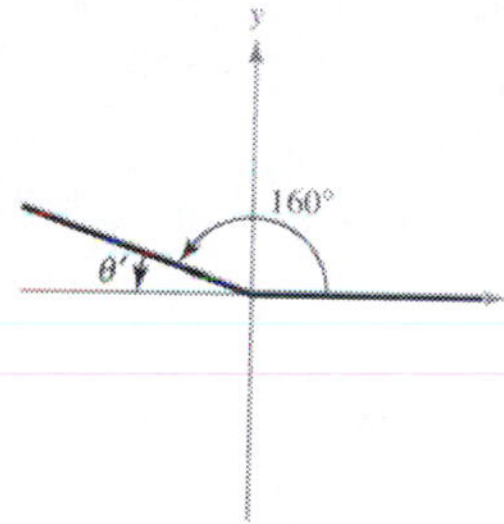

43. $\theta = -125°$

$360° - 125° = 235°$ (coterminal angle)

$\theta' = 235° - 180° = 55°$

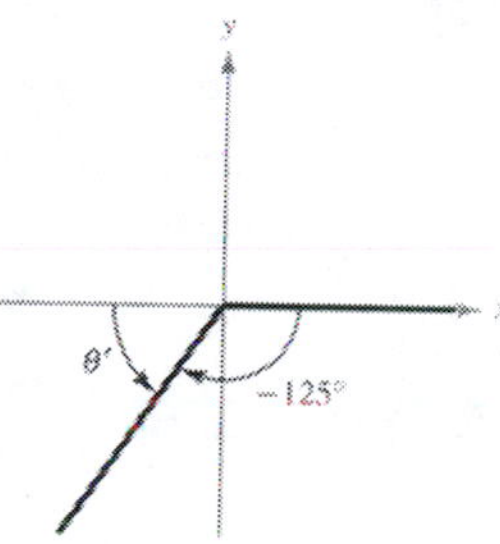

45. $\theta = \frac{2\pi}{3}$

$$\theta' = \pi - \frac{2\pi}{3} = \frac{\pi}{3}$$

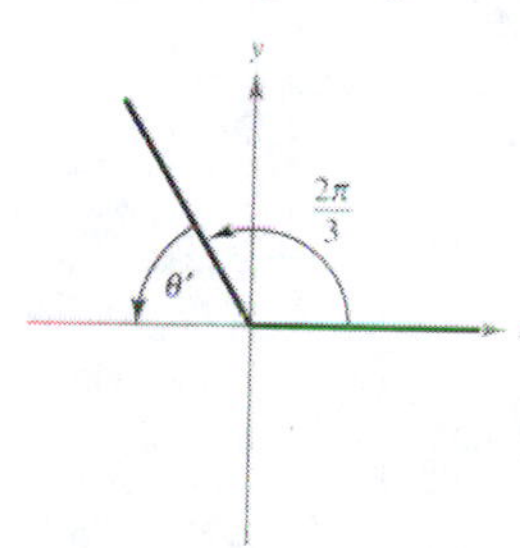

47. $\theta = 4.8$

$\theta' = 2\pi - 4.8\pi \approx 1.4832$

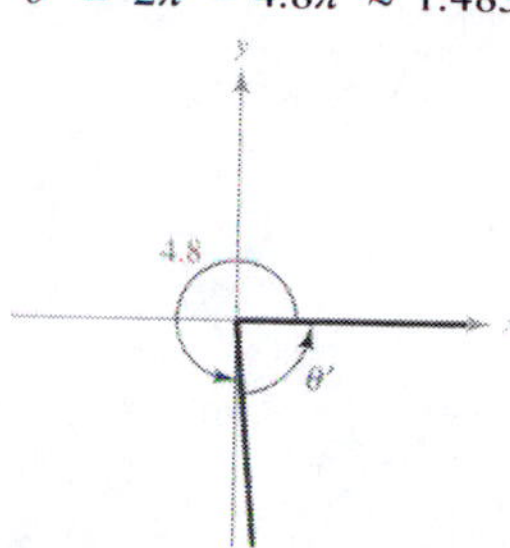

49. $\theta = 225°, \theta' = 45°$, Quadrant III

$$\sin 225° = -\sin 45° = -\frac{\sqrt{2}}{2}$$

$$\cos 225° = -\cos 45° = -\frac{\sqrt{2}}{2}$$

$$\tan 225° = \tan 45° = 1$$

51. $\theta = 750°, \theta' = 30°$, Quadrant I

$$\sin 750° = \sin 30° = \frac{1}{2}$$

$$\cos 750° = \cos 30° = \frac{\sqrt{3}}{2}$$

$$\tan 750° = \tan 30° = \frac{\sqrt{3}}{3}$$

53. $\theta = -150°, \theta' = 30°$, Quadrant III

$$\sin(-150°) = -\sin 30° = -\frac{1}{2}$$

$$\cos(-150°) = -\cos 30° = -\frac{\sqrt{3}}{2}$$

$$\tan(-150°) = \tan 30° = \frac{\sqrt{3}}{3}$$

55. $\theta = \frac{2\pi}{3}, \theta' = \frac{\pi}{3}$ in Quadrant II

$$\sin \frac{2\pi}{3} = \sin \frac{\pi}{3} = \frac{\sqrt{3}}{2}$$

$$\cos \frac{2\pi}{3} = -\cos \frac{\pi}{3} = -\frac{1}{2}$$

$$\tan \frac{2\pi}{3} = -\tan \frac{\pi}{3} = -\sqrt{3}$$

57. $\theta = \frac{5\pi}{4}, \theta' = \frac{\pi}{4}$ in Quadrant III

$$\sin \frac{5\pi}{4} = -\sin \frac{\pi}{4} = -\frac{\sqrt{2}}{2}$$

$$\cos \frac{5\pi}{4} = -\cos \frac{\pi}{4} = -\frac{\sqrt{2}}{2}$$

$$\tan \frac{5\pi}{4} = \tan \frac{\pi}{4} = 1$$

59. $\theta = -\frac{\pi}{6}, \theta' = \frac{\pi}{6}$, Quadrant IV

$$\sin\left(-\frac{\pi}{6}\right) = -\sin\frac{\pi}{6} = -\frac{1}{2}$$

$$\cos\left(-\frac{\pi}{6}\right) = \cos\frac{\pi}{6} = \frac{\sqrt{3}}{2}$$

$$\tan\left(-\frac{\pi}{6}\right) = -\tan\frac{\pi}{6} = -\frac{\sqrt{3}}{3}$$

61. $\theta = \frac{11\pi}{4}, \theta' = \frac{\pi}{4}$, Quadrant II

$$\sin\frac{11\pi}{4} = \sin\frac{\pi}{4} = \frac{\sqrt{2}}{2}$$

$$\cos\frac{11\pi}{4} = -\cos\frac{\pi}{4} = -\frac{\sqrt{2}}{2}$$

$$\tan\frac{11\pi}{4} = -\tan\frac{\pi}{4} = -1$$

63. $\theta = \frac{9\pi}{4}, \theta' = \frac{\pi}{4}$ in Quadrant I

$$\sin\frac{9\pi}{4} = \sin\frac{\pi}{4} = \frac{\sqrt{2}}{2}$$

$$\cos\frac{9\pi}{4} = \cos\frac{\pi}{4} = \frac{\sqrt{2}}{2}$$

$$\tan\frac{9\pi}{4} = \tan\frac{\pi}{4} = 1$$

65. $\sin\theta = -\frac{3}{5}$

$$\sin^2\theta + \cos^2\theta = 1$$

$$\cos^2\theta = 1 - \sin^2\theta$$

$$\cos^2\theta = 1 - \left(-\frac{3}{5}\right)^2$$

$$\cos^2\theta = 1 - \frac{9}{25}$$

$$\cos^2\theta = \frac{16}{25}$$

$\cos\theta > 0$ in Quadrant IV.

$$\cos\theta = \frac{4}{5}$$

67. $\tan\theta = \frac{3}{2}$

$$\sec^2\theta = 1 + \tan^2\theta$$

$$\sec^2\theta = 1 + \left(\frac{3}{2}\right)^2$$

$$\sec^2\theta = 1 + \frac{9}{4}$$

$$\sec^2\theta = \frac{13}{4}$$

$\sec\theta < 0$ in Quadrant III.

$$\sec\theta = -\frac{\sqrt{13}}{2}$$

69. $\cos\theta = \frac{5}{8}$

$$\cos\theta = \frac{1}{\sec\theta} \Rightarrow \sec\theta = \frac{1}{\cos\theta}$$

$$\sec\theta = \frac{1}{\frac{5}{8}} = \frac{8}{5}$$

71. $\sin 10° \approx 0.1736$

73. $\cos(-110°) \approx -0.3420$

75. $\tan 304° \approx -1.4826$

77. $\sec 72° = \frac{1}{\cos 72°} \approx 3.2361$

79. $\tan 4.5 \approx 4.6373$

81. $\tan\left(\frac{\pi}{9}\right) \approx 0.3640$

83. $\sin(-0.65) \approx -0.6052$

85. $\cot\left(-\frac{11\pi}{8}\right) = \frac{1}{\tan(-11\pi/8)} \approx -0.4142$

87. (a) $\sin\theta = \frac{1}{2} \Rightarrow$ reference angle is 30° or $\frac{\pi}{6}$ and θ is in Quadrant I or Quadrant II.

Values in degrees: 30°, 150°

Values in radian: $\frac{\pi}{6}, \frac{5\pi}{6}$

(b) $\sin\theta = \frac{1}{2} \Rightarrow$ reference angle is 30° or $\frac{\pi}{6}$ and θ is in Quadrant III or Quadrant IV.

Values in degrees: 210°, 330°

Values in radians: $\frac{7\pi}{6}, \frac{11\pi}{6}$

89. (a) $\csc\theta = \frac{2\sqrt{3}}{3} \Rightarrow$ reference angle is $60°$ or $\frac{\pi}{3}$ and θ is in Quadrant I or Quadrant II.

Values in degrees: $60°, 120°$

Values in radians: $\frac{\pi}{3}, \frac{2\pi}{3}$

(b) $\cot\theta = -1 \Rightarrow$ reference angle is $45°$ or $\frac{\pi}{4}$ and θ is in Quadrant II or Quadrant IV.

Values in degrees: $135°, 315°$

Values in radians: $\frac{3\pi}{4}, \frac{7\pi}{4}$

91. (a) $\tan\theta = 1 \Rightarrow$ reference angle is $45°$ or $\frac{\pi}{4}$ and θ is in Quadrant I or Quadrant III.

Values in degrees: $45°, 225°$

Values in radians: $\frac{\pi}{4}, \frac{5\pi}{4}$

(b) $\cot\theta = -\sqrt{3} \Rightarrow$ reference angle is $30°$ or $\frac{\pi}{6}$ and θ is in Quadrant II or Quadrant IV.

Values in degrees: $150°, 330°$

Values in radians: $\frac{5\pi}{6}, \frac{11\pi}{6}$

93. $\left(\frac{\sqrt{2}}{2}, \frac{\sqrt{2}}{2}\right)$ corresponds to $t = \frac{\pi}{4}$ on the unit circle.

$\sin\frac{\pi}{4} = \frac{\sqrt{2}}{2}$ since $\sin t = y$.

$\cos\frac{\pi}{4} = \frac{\sqrt{2}}{2}$ since $\cos t = x$.

$\tan\frac{\pi}{4} = 1$ since $\tan t = \frac{y}{x}$.

95. $\left(-\frac{\sqrt{3}}{2}, \frac{1}{2}\right)$ corresponds to $t = \frac{5\pi}{6}$ on the unit circle.

$\sin\frac{5\pi}{6} = \frac{1}{2}$ since $\sin t = y$.

$\cos\frac{5\pi}{6} = -\frac{\sqrt{3}}{2}$ since $\cos t = x$.

$\tan\frac{5\pi}{6} = -\frac{\sqrt{3}}{3}$ since $\tan t = \frac{y}{x}$.

97. $\left(-\frac{1}{2}, -\frac{\sqrt{3}}{2}\right)$ corresponds to $t = \frac{4\pi}{3}$ on the unit circle.

$\sin\frac{4\pi}{3} = -\frac{\sqrt{3}}{2}$ since $\sin t = y$.

$\cos\frac{4\pi}{3} = -\frac{1}{2}$ since $\cos t = x$.

$\tan\frac{4\pi}{3} = \sqrt{3}$ since $\tan t = \frac{y}{x}$.

99. $t = \frac{\pi}{2}, (x, y) = (0, 1)$

$\sin\frac{\pi}{2} = 1$ because $\sin t = y$.

$\cos\frac{\pi}{2} = 0$ because $\cos t = x$.

$\tan\frac{\pi}{2}$ is undefined because $\tan t = \frac{y}{x} = \frac{1}{0}$.

101. (a) $\sin 5 \approx -1$

(b) $\cos 2 \approx -0.4$

103. (a) $\sin t = 0.25$

$t \approx 0.25$ or 2.89

(b) $\cos t = -0.25$

$t \approx 1.82$ or 4.46

105. (a) New York City:

$N \approx 22.1\sin(0.52t - 2.22) + 55.01$

Fairbanks: $F \approx 36.6\sin(0.50t - 1.83) + 25.61$

(b)

Month	New York City	Fairbanks
February	35°	−1°
March	41°	14°
May	63°	48°
June	72°	59°
August	76°	56°
September	69°	42°
November	47°	7°

(c) The periods are about the same for both models, approximately 12 months.

107. $y(t) = 2\cos 6t$

(a) $y(0) = 2\cos 0 = 2$ centimeters

(b) $y\left(\frac{1}{4}\right) = 2\cos\left(\frac{3}{2}\right) \approx 0.14$ centimeter

(c) $y\left(\frac{1}{2}\right) = 2\cos 3 \approx -1.98$ centimeters

109. $I = 5e^{-2t}\sin t$

$I(0.7) = 5e^{-1.4}\sin 0.7 \approx 0.79$

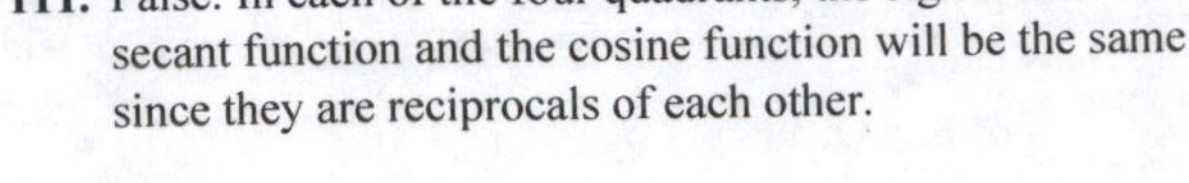

111. False. In each of the four quadrants, the sign of the secant function and the cosine function will be the same since they are reciprocals of each other.

113. $h(t) = f(t)g(t)$

$h(-t) = f(-t)g(-t) = -f(t)g(t) = -h(t)$

Therefore, $h(t)$ is odd.

115. (a)

θ	$0°$	$20°$	$40°$	$60°$	$80°$
$\sin\theta$	0	0.342	0.643	0.866	0.985
$\sin(180° - \theta)$	0	0.342	0.643	0.866	0.985

(b) Conjecture: $\sin\theta = \sin(180° - \theta)$

117. $y = \sin x$

Domain: All real numbers x

Range: $[-1, 1]$

Period: 2π

Zeros: $n\pi$

The function is odd.

$y = \cos x$

Domain: All real numbers x

Range: $[-1, 1]$

Period: 2π

Zeros: $n\pi + \dfrac{\pi}{2}$

The function is even.

$y = \tan x$

Domain: All real numbers x except $x = n\pi + \dfrac{\pi}{2}$

Range: $(-\infty, \infty)$

Period: π

Zeros: $n\pi$

The function is odd.

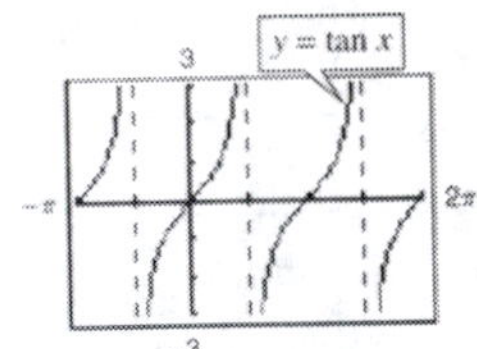

$y = \csc x$

Domain: All real numbers x except $x = n\pi$

Range: $(-\infty, -1] \cup [1, \infty)$

Period: 2π

Zeros: none

The function is odd.

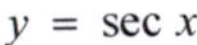

$y = \sec x$

Domain: All real numbers x except $x = n\pi + \dfrac{\pi}{2}$

Range: $(-\infty, -1] \cup [1, \infty)$

Period: 2π

Zeros: none

The function is even.

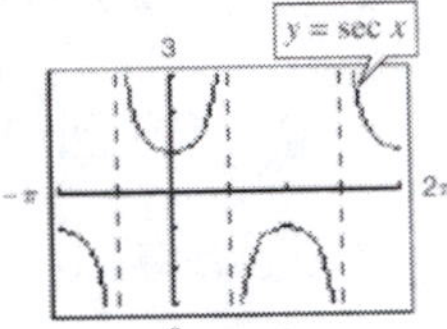

$y = \cot x$

Domain: All real numbers x except $x = n\pi$

Range: $(-\infty, \infty)$

Period: π

Zeros: $n\pi + \dfrac{\pi}{2}$

The function is odd.

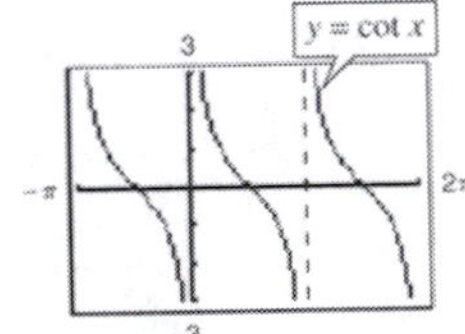

The secant function is similar to the tangent function because they both have vertical asymptotes at

$x = n\pi + \dfrac{\pi}{2}.$

The cotangent function and the cosecant function both have vertical asymptotes at $x = n\pi$. A maximum point on the sine curve corresponds to a relative minimum on the cosecant curve. The maximum points of sine and cosine are interchanged with the minimum points of cosecant and secant. The x-intercepts of the sine and cosine functions become vertical asymptotes of the cosecant and secant functions.

The graphs of sine and cosine may be translated left or right (respectively) to $\pi/2$ to coincide with each other.

119. (a) The points have y-axis symmetry.

(b) $\sin t_1 = \sin(\pi - t_1)$ because they have the same y-value.

(c) $\cos(\pi - t_1) = -\cos t_1$ because the x-values have opposite signs.

Section 6.4 Graphs of Sine and Cosine Functions

1. cycle

3. phase shift

5. $y = 2 \sin 5x$

Period: $\frac{2\pi}{5}$

Amplitude: $|2| = 2$

7. $y = \frac{3}{4} \cos \frac{x}{2}$

Period: $\frac{2\pi}{1/2} = 4\pi$

Amplitude: $\left|\frac{3}{4}\right| = \frac{3}{4}$

9. $y = \frac{1}{2} \sin \frac{\pi x}{3}$

Period: $\frac{2\pi}{\pi/3} = 6$

Amplitude: $\left|\frac{1}{2}\right| = \frac{1}{2}$

11. $y = -4 \sin x$

Period: $\frac{2\pi}{1} = 2\pi$

Amplitude: $|-4| = 4$

13. $y = 3 \sin 10x$

Period: $\frac{2\pi}{10} = \frac{\pi}{5}$

Amplitude: $|3| = 3$

15. $y = \frac{5}{3} \cos \frac{4x}{5}$

Period: $\frac{2\pi}{4/5} = \frac{10\pi}{4} = \frac{5\pi}{2}$

Amplitude: $\left|\frac{5}{3}\right| = \frac{5}{3}$

17. $y = \frac{1}{4} \sin 2\pi x$

Period: $\frac{2\pi}{2\pi} = 1$

Amplitude: $\left|\frac{1}{4}\right| = \frac{1}{4}$

19. $f(x) = \sin x$

$g(x) = \sin(x - \pi)$

g is a horizontal shift to the right π units of the graph of f (a phase shift).

21. $f(x) = \cos 2x$

$g(x) = -\cos 2x$

g is a reflection in the x-axis of the graph of f.

23. $f(x) = \cos x$

$g(x) = \cos 2x$

The period of f is twice that of g.

25. $f(x) = \sin 2x$

$f(x) = 3 + \sin 2x$

g is a vertical shift three units upward of the graph of f.

27. The graph of g has twice the amplitude as the graph of f. The period is the same.

29. The graph of g is a horizontal shift π units to the right of the graph of f.

31. $f(x) = -2 \sin x$

Period: $\frac{2\pi}{b} = \frac{2\pi}{1} = 2\pi$

Amplitude: 2

Symmetry: origin

Key points: Intercept	Minimum	Intercept	Maximum	Intercept
$(0, 0)$	$\left(\frac{\pi}{2}, -2\right)$	$(\pi, 0)$	$\left(\frac{3\pi}{2}, 0\right)$	$(2\pi, 0)$

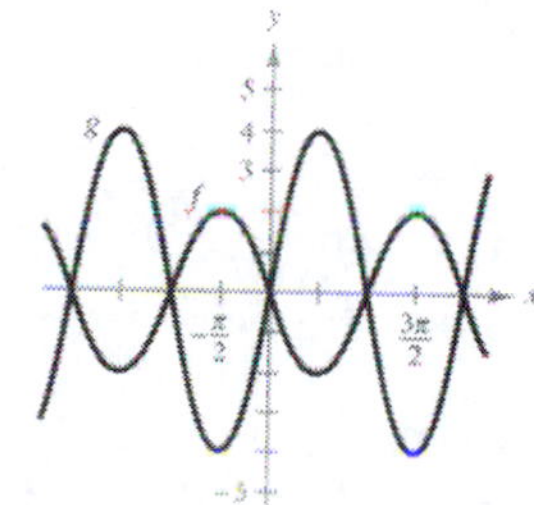

Because $g(x) = 4 \sin x = (-2)f(x)$, generate key points for the graph of $g(x)$ by multiplying the y-coordinate of each key point of $f(x)$ by -2.

33. $f(x) = \cos x$

Period: $\dfrac{2\pi}{b} = \dfrac{2\pi}{1} = 2\pi$

Amplitude: $|1| = 1$

Symmetry: y-axis

Key points: Maximum	Intercept	Minimum	Intercept	Maximum
$(0, 1)$	$\left(\dfrac{\pi}{2}, 0\right)$	$(\pi, -1)$	$\left(\dfrac{3\pi}{2}, 0\right)$	$(2\pi, 1)$

Because $g(x) = 2 + \cos x = f(x) + 2$, the graph of $g(x)$ is the graph of $f(x)$, but translated upward by two units. Generate key points of $g(x)$ by adding 2 to the y-coordinate of each key point of $f(x)$.

35. 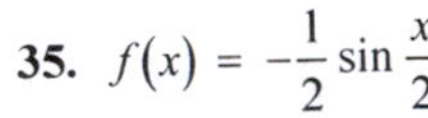 $f(x) = -\dfrac{1}{2}\sin\dfrac{x}{2}$

Period: $\dfrac{2\pi}{b} = \dfrac{2\pi}{1/2} = 4\pi$

Amplitude: $\dfrac{1}{2}$

Symmetry: origin

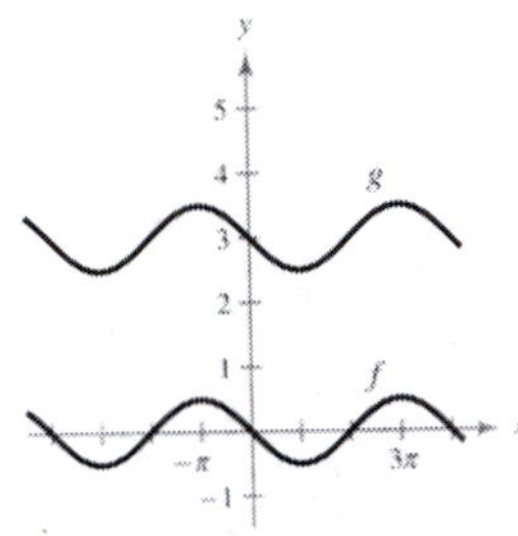

Key points: Intercept	Minimum	Intercept	Maximum	Intercept
$(0, 0)$	$\left(\pi, -\dfrac{1}{2}\right)$	$(2\pi, 0)$	$\left(3\pi, \dfrac{1}{2}\right)$	$(4\pi, 0)$

Because $g(x) = 3 - \dfrac{1}{2}\sin\dfrac{x}{2} = 3 - f(x)$, the graph of $g(x)$ is the graph of $f(x)$, but translated upward by three units. Generate key points for the graph of $g(x)$ by adding 3 to the y-coordinate of each key point of $f(x)$.

37. $f(x) = 2\cos x$

Period: $\dfrac{2\pi}{b} = \dfrac{2\pi}{1} = 2\pi$

Amplitude: 2

Symmetry: y-axis

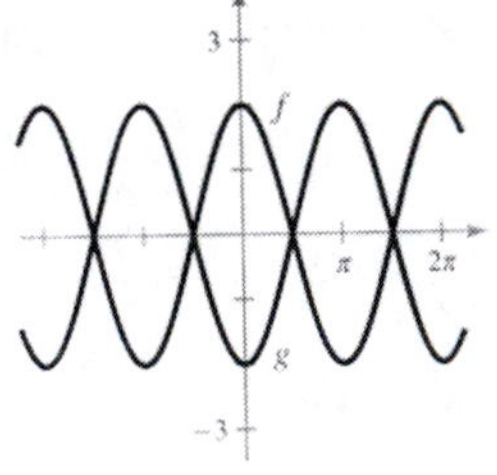

Key points: Maximum	Intercept	Minimum	Intercept	Maximum
$(0, 2)$	$\left(\dfrac{\pi}{2}, 0\right)$	$(\pi, -2)$	$\left(\dfrac{3\pi}{2}, 0\right)$	$(2\pi, 2)$

Because $g(x) = 2\cos(x + \pi) = f(x + \pi)$, the graph of $g(x)$ is the graph of $f(x)$, but with a phase shift (horizontal translation) of $-\pi$. Generate key points for the graph of $g(x)$ by shifting each key point of $f(x)$ π units to the left.

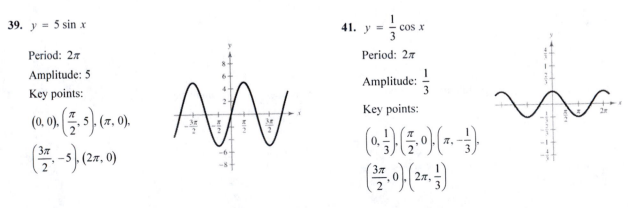

39. $y = 5\sin x$

Period: 2π

Amplitude: 5

Key points:

$(0, 0), \left(\dfrac{\pi}{2}, 5\right), (\pi, 0),$

$\left(\dfrac{3\pi}{2}, -5\right), (2\pi, 0)$

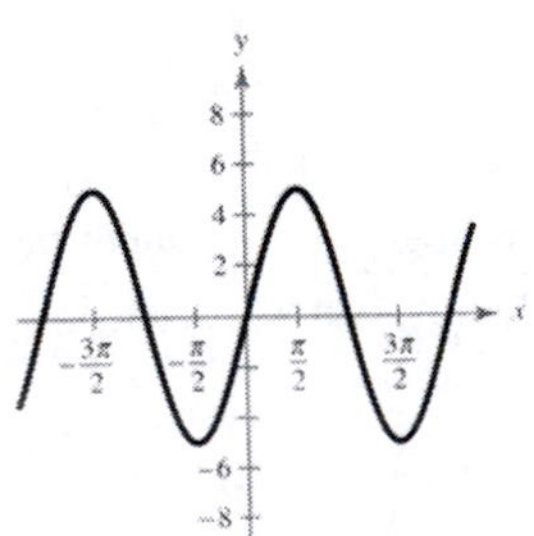

41. $y = \dfrac{1}{3}\cos x$

Period: 2π

Amplitude: $\dfrac{1}{3}$

Key points:

$\left(0, \dfrac{1}{3}\right), \left(\dfrac{\pi}{2}, 0\right), \left(\pi, -\dfrac{1}{3}\right),$

$\left(\dfrac{3\pi}{2}, 0\right), \left(2\pi, \dfrac{1}{3}\right)$

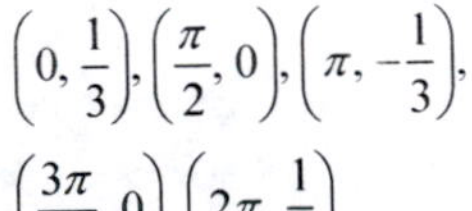

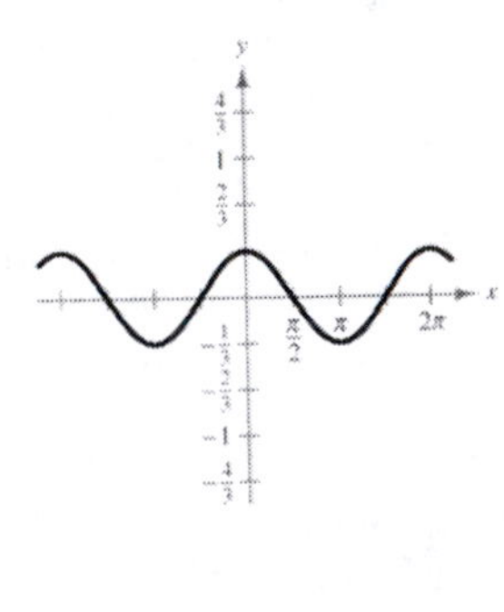

43. $y = \cos\dfrac{y}{2}$

Period $\dfrac{2\pi}{1/2} = 4\pi$

Amplitude: 1

Key points:

$(0, 1), (\pi, 0), (2\pi, -1),$

$(3\pi, 0), (4\pi, 1)$

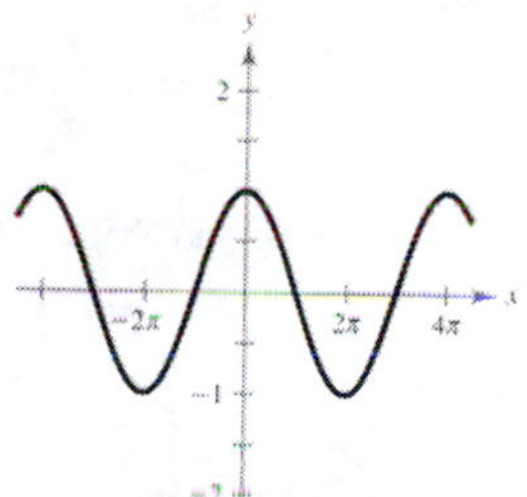

45. $y = \cos 2\pi x$

Period: $\dfrac{2\pi}{2\pi} = 1$

Amplitude: 1

Key points:

$(0, 1), \left(\dfrac{1}{4}, 0\right), \left(\dfrac{1}{2}, -1\right), \left(\dfrac{3}{4}, 0\right)$

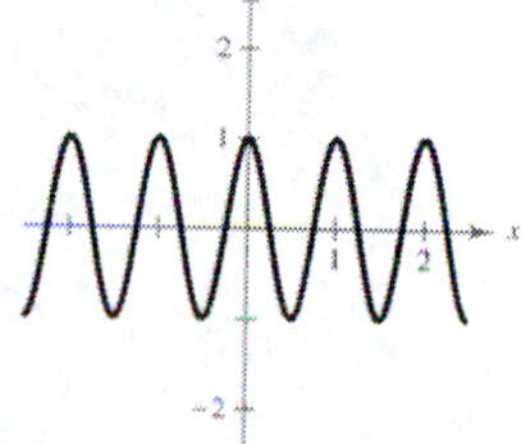

47. $y = -\sin\dfrac{2\pi x}{3}$

Period: $\dfrac{2\pi}{2\pi/3} = 3$

Amplitude: 1

Key points:

$(0, 0), \left(\dfrac{3}{4}, -1\right), \left(\dfrac{3}{2}, 0\right),$

$\left(\dfrac{9}{4}, 1\right), (3, 0)$

49. $y = 3\cos(x + \pi)$

Period: 2π

Amplitude: 3

Shift: Set $x + \pi = 0$ and $x + \pi = 2\pi$

$x = -\pi$ $\qquad x = \pi$

Key points: $(-\pi, 3), \left(-\dfrac{\pi}{2}, 0\right), (0, -3), \left(\dfrac{\pi}{2}, 0\right), (\pi, 3)$

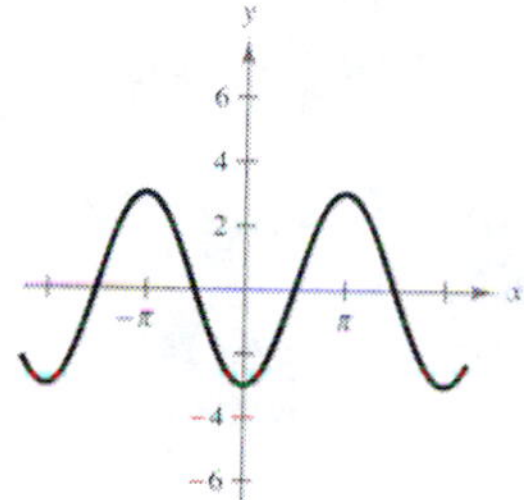

51. $y = \sin\left(x - \dfrac{\pi}{2}\right)$

Period: 2π

Amplitude: 1

Shift: Set $x - \dfrac{\pi}{2} = 0$ and $x - \dfrac{\pi}{2} = 2\pi$

$x = \dfrac{\pi}{2}$ $\qquad x = \dfrac{5\pi}{2}$

Key points: $\left(\dfrac{\pi}{2}, 0\right), (\pi, 1), \left(\dfrac{3\pi}{2}, 0\right), (2\pi, -1), \left(\dfrac{5\pi}{2}, 0\right)$

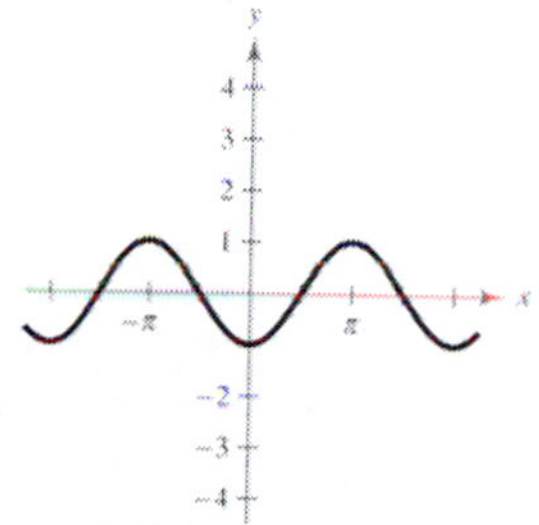

53. $y = 2 - \sin\dfrac{2\pi x}{3}$

Period: $\dfrac{2\pi}{2\pi/3} = 3$

Amplitude: 1

Key points:

$(0, 2), \left(\dfrac{3}{4}, 1\right), \left(\dfrac{3}{2}, 2\right),$

$\left(\dfrac{9}{4}, 3\right), (3, 2)$

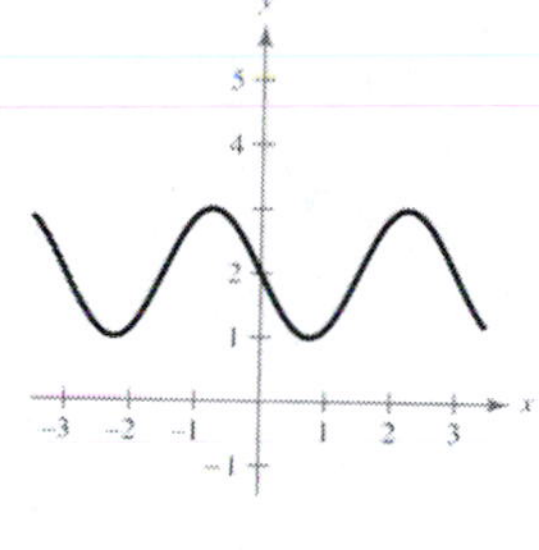

55. $y = 2 + \dfrac{1}{10}\cos 60\pi x$

Period: $\dfrac{2\pi}{60\pi} = \dfrac{1}{30}$

Amplitude: $\dfrac{1}{10}$

Vertical shift two units upward

Key points:

$(0, 2.1), \left(\dfrac{1}{120}, 2\right)\left(\dfrac{1}{60}, 1.9\right),$

$\left(\dfrac{1}{40}, 2\right), \left(\dfrac{1}{30}, 2.1\right)$

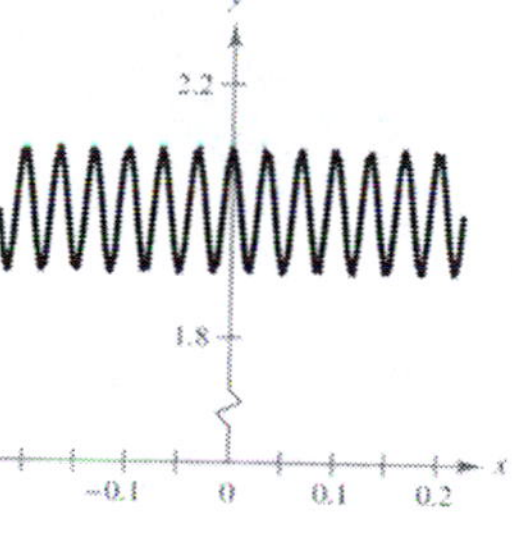

57. $y = 3\cos(x + \pi) - 3$

Period: 2π

Amplitude: 3

Shift: Set $x + \pi = 0$ and $x + \pi = 2\pi$

$x = -\pi$ $\quad$ $x = \pi$

Key points: $(-\pi, 0), \left(-\frac{\pi}{2}, -3\right), (0, -6), \left(\frac{\pi}{2}, -3\right), (\pi, 0)$

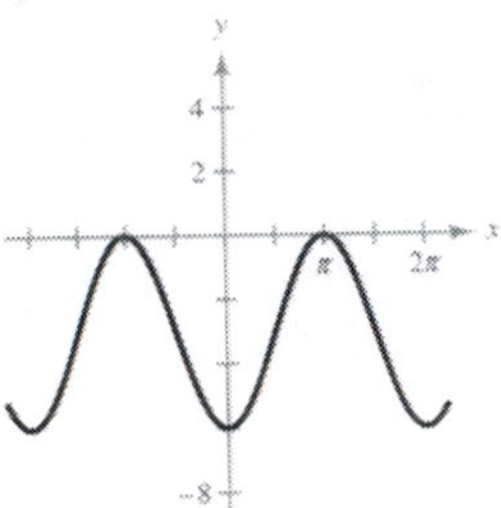

59. $y = \frac{2}{3}\cos\left(\frac{x}{2} - \frac{\pi}{4}\right)$

Period: $\frac{2\pi}{1/2} = 4\pi$

Amplitude: $\frac{2}{3}$

Shift: $\frac{x}{2} - \frac{\pi}{4} = 0$ and $\frac{\pi}{2} - \frac{\pi}{4} = 2\pi$

$x = \frac{\pi}{2}$ $\quad$ $x = \frac{9\pi}{2}$

Key points:

$\left(\frac{\pi}{2}, \frac{2}{3}\right), \left(\frac{3\pi}{2}, 0\right), \left(\frac{5\pi}{2}, \frac{-2}{3}\right), \left(\frac{7\pi}{2}, 0\right), \left(\frac{9\pi}{2}, \frac{2}{3}\right)$

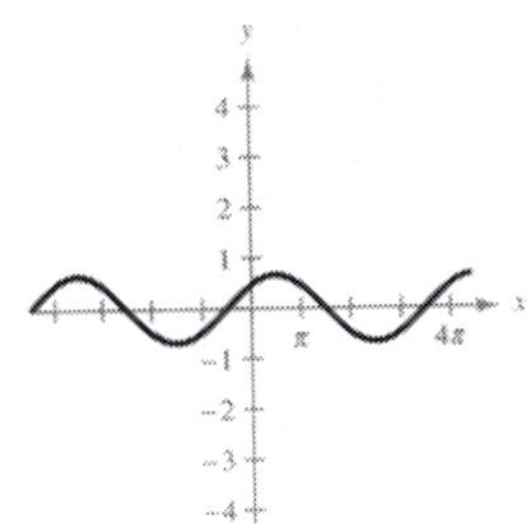

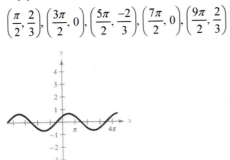

61. $g(x) = \sin(4x - \pi)$

(a) $g(x)$ is obtained by a horizontal shrink of four and a phase shift of $\frac{\pi}{4}$; and one cycle of $g(x)$ corresponds to the interval $\left[\frac{\pi}{4}, \frac{3\pi}{4}\right]$.

(b)

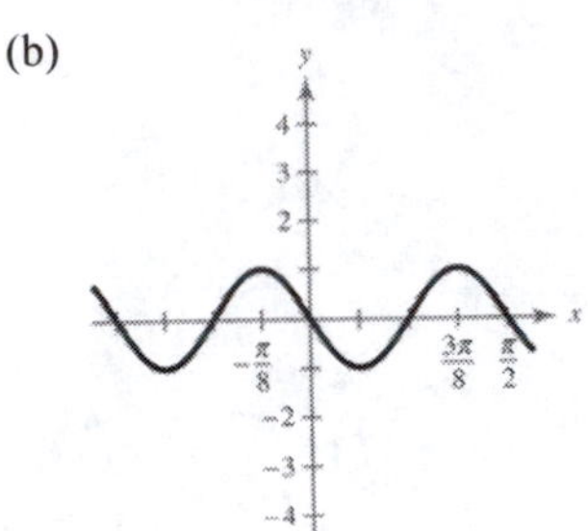

(c) $g(x) = f(4x - \pi)$ where $f(x) = \sin x$.

63. $g(x) = \cos(x - \pi) + 2$

(a) $g(x)$ is obtained by shifting $f(x)$ two units upward and a phase shift of π; and one cycle of $g(x)$ corresponds to the interval $[\pi, 3\pi]$.

(b)

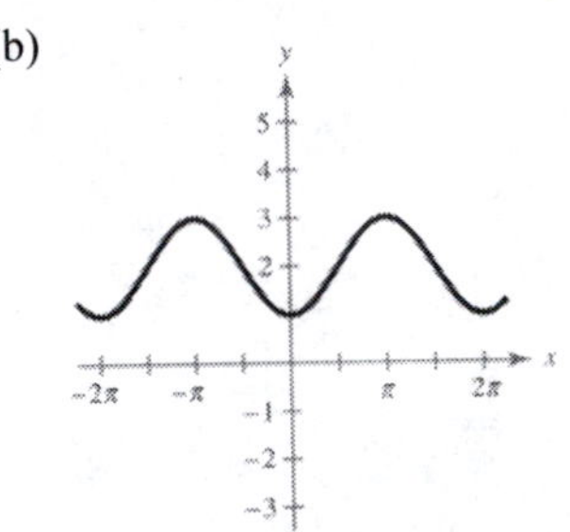

(c) $g(x) = f(x - \pi) + 2$ where $f(x) = \cos x$

65. $g(x) = 2\sin(4x - \pi) - 3$

(a) $g(x)$ is obtained by a horizontal shrink of four, a phase shift of $\frac{\pi}{4}$, shifting $f(x)$ three units downward, and has an amplitude of two. One cycle of $g(x)$ corresponds to the interval $\left[\frac{\pi}{4}, \frac{3\pi}{4}\right]$.

(b)

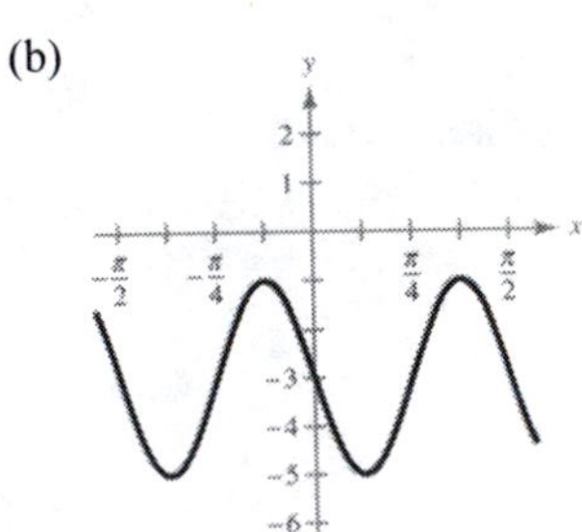

(c) $g(x) = 2f(4x - \pi) - 3$ where $f(x) = \sin x$

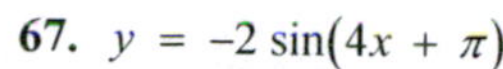

67. $y = -2\sin(4x + \pi)$

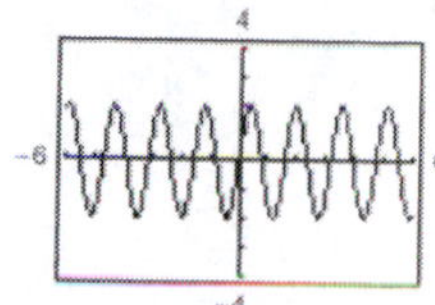

69. $y = \cos\left(2\pi x - \dfrac{\pi}{2}\right) + 1$

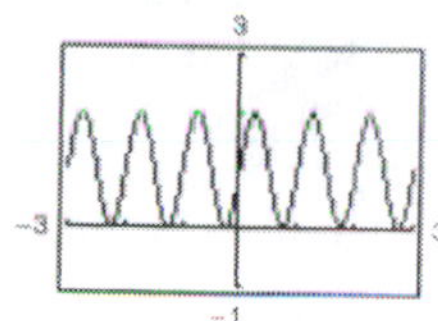

71. $y = -0.1\sin\left(\dfrac{\pi x}{10} + \pi\right)$

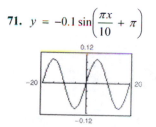

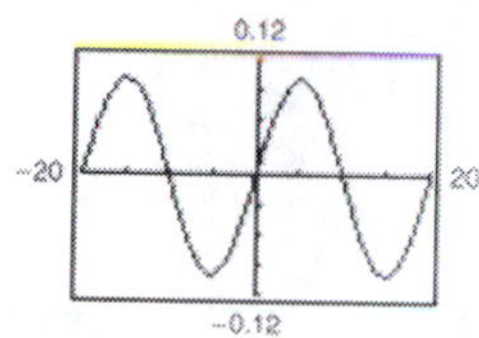

73. $f(x) = a\cos x + d$

Amplitude: $\frac{1}{2}[3 - (-1)] = 2 \Rightarrow a = 2$

$3 = 2\cos 0 + d$

$d = 3 - 2 = 1$

$a = 2, d = 1$

75. $f(x) = a\cos x + d$

Amplitude: $\frac{1}{2}[8 - 0] = 4$

Reflected in the x-axis: $a = -4$

$0 = -4\cos 0 + d$

$d = 4$

$a = -4, d = 4$

77. $y = a\sin(bx - c)$

Amplitude: $|a| = |3|$

Since the graph is reflected in the x-axis, we have $a = -3$.

Period: $\dfrac{2\pi}{b} = \pi \Rightarrow b = 2$

Phase shift: $c = 0$

$a = -3, b = 2, c = 0$

79. $y = a\sin(bx - c)$

Amplitude: $a = 2$

Period: $2\pi \Rightarrow b = 1$

Phase shift: $bx - c = 0$ when $x = -\dfrac{\pi}{4}$

$$(1)\left(-\frac{\pi}{4}\right) - c = 0 \Rightarrow c = -\frac{\pi}{4}$$

$a = 2, b = 1, c = -\dfrac{\pi}{4}$

81. $y_1 = \sin x$

$y_2 = -\dfrac{1}{2}$

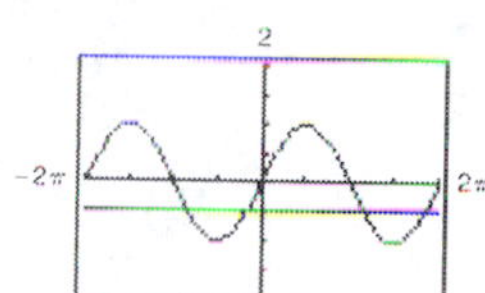

In the interval $[-2\pi, 2\pi]$,

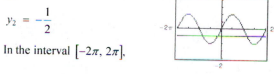

$y_1 = y_2$ when $x = -\dfrac{5\pi}{6}, -\dfrac{\pi}{6}, \dfrac{7\pi}{6}, \dfrac{11\pi}{6}$.

Answers for 83–85 are sample answers.

83. $f(x) = 2\sin(2x - \pi) + 1$

85. $f(x) = \cos(2x + 2\pi) - \frac{3}{2}$

87. $v = 1.75\sin\dfrac{\pi t}{2}$

(a) Period $= \dfrac{2\pi}{\pi/2} = 4$ seconds

(b) $\dfrac{1 \text{ cycle}}{4 \text{ seconds}} \cdot \dfrac{60 \text{ seconds}}{1 \text{ minute}} = 15$ cycles per minute

(c)

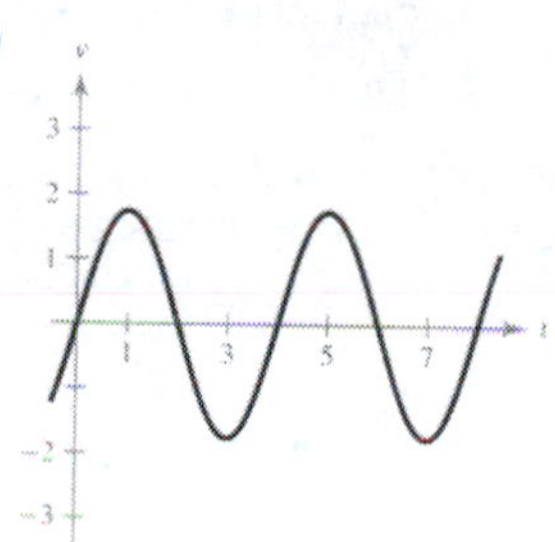

89. (a) $y = a\cos(bt - c) + d$

Amplitude: $a = \frac{1}{2}[\text{max temp} - \text{min temp}] = \frac{1}{2}[78.6 - 13.8] = 32.4$

Period: $p = 2[\text{month of max temp} - \text{month of min temp}] = 2[7 - 1] = 12$

$$b = \frac{2\pi}{p} = \frac{2\pi}{12} = \frac{\pi}{6}$$

Because the maximum temperature occurs in the seventh month,

$\frac{c}{b} = 7$ so $c \approx 3.67$.

The average temperature is $\frac{1}{2}(78.6 + 13.8) = 46.2°$, so $d = 46.2$.

So, $I(t) = 32.4\cos\left(\frac{\pi}{6}t - 3.67\right) + 46.2$.

(b) 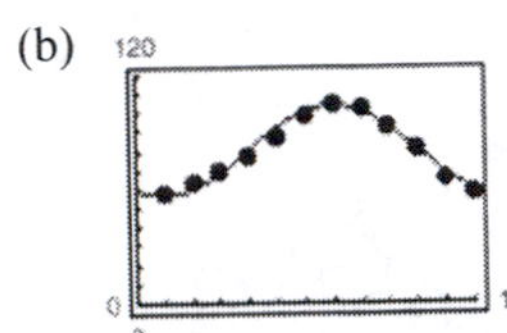

The model fits the data well.

(c) 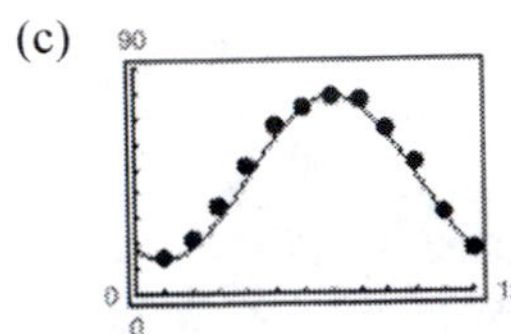

The model fits the data well.

(d) The d value in each model represents the average temperature.

Las Vegas: 80.6°; International Falls: 46.2°

(e) The period of each model is 12. This is what you would expect because the time period is one year (twelve months).

(f) International Falls has the greater temperature variability. The amplitude determines the variability. The greater the amplitude, the greater the temperature varies.

91. $y = 0.001\sin 880\pi t$

(a) Period: $\frac{2\pi}{880\pi} = \frac{1}{440}$ seconds

(b) $f = \frac{1}{p} = 440$ cycles per second

93. (a) Period $= \frac{2\pi}{\left(\frac{\pi}{10}\right)} = 20$ seconds

The wheel takes 20 seconds to revolve once.

(b) Amplitude: 50 feet

The radius of the wheel is 50 feet.

(c) 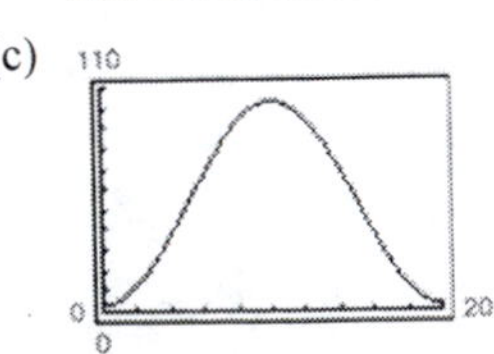

95. False. The graph of $\sin(x + 2\pi)$ is the graph of $\sin(x)$ translated to the *left* by one period, and the graphs are indeed identical.

97.

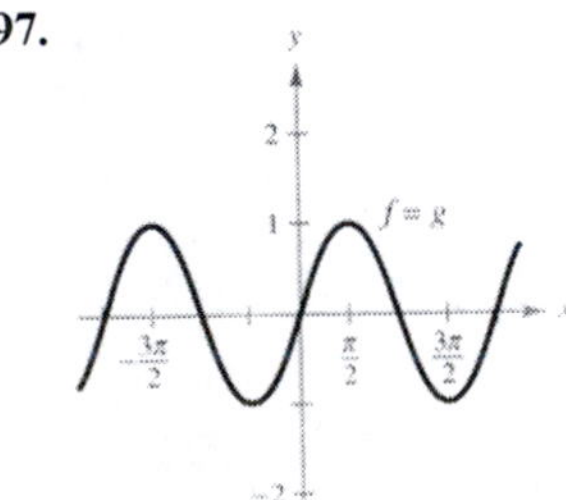

Because the graphs are the same, the conjecture is that $\sin(x) = \cos\left(x - \frac{\pi}{2}\right)$.

99.

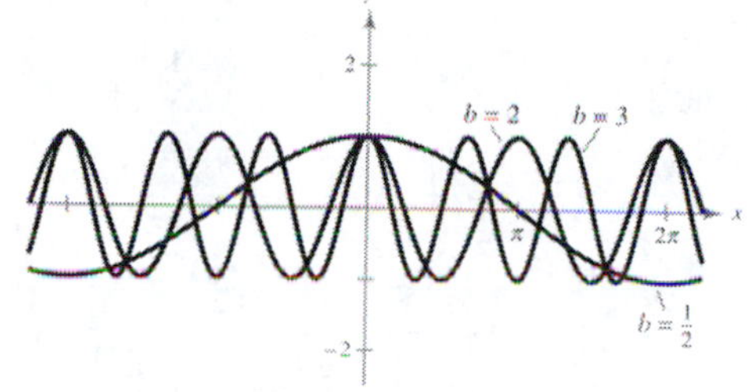

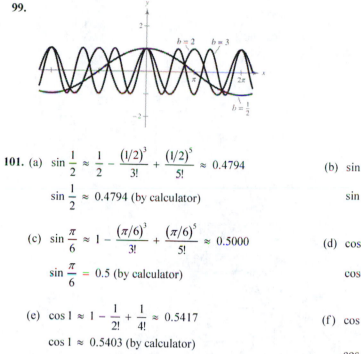

As the value of b increases, the period decreases.

$b = \frac{1}{2} \rightarrow \frac{1}{2}$ cycle

$b = 2 \rightarrow 2$ cycles

$b = 3 \rightarrow 3$ cycles

101. (a) $\sin\frac{1}{2} \approx \frac{1}{2} - \frac{(1/2)^3}{3!} + \frac{(1/2)^5}{5!} \approx 0.4794$

$\sin\frac{1}{2} \approx 0.4794$ (by calculator)

(b) $\sin 1 \approx 1 - \frac{1}{3!} + \frac{1}{5!} \approx 0.8417$

$\sin 1 \approx 0.8415$ (by calculator)

(c) $\sin\frac{\pi}{6} \approx 1 - \frac{(\pi/6)^3}{3!} + \frac{(\pi/6)^5}{5!} \approx 0.5000$

$\sin\frac{\pi}{6} = 0.5$ (by calculator)

(d) $\cos(-0.5) \approx 1 - \frac{(-0.5)^2}{2!} + \frac{(-0.5)^4}{4!} \approx 0.8776$

$\cos(-0.5) \approx 0.8776$ (by calculator)

(e) $\cos 1 \approx 1 - \frac{1}{2!} + \frac{1}{4!} \approx 0.5417$

$\cos 1 \approx 0.5403$ (by calculator)

(f) $\cos\frac{\pi}{4} \approx 1 - \frac{(\pi/4)^2}{2!} + \frac{(\pi/4)^2}{4!} = 0.7074$

$\cos\frac{\pi}{4} \approx 0.7071$ (by calculator)

The error in the approximation is not the same in each case. The error appears to increase as x moves farther away from 0.

Section 6.5 Graphs of Other Trigonometric Functions

1. odd; origin

3. reciprocal

5. π

7. $(-\infty, -1] \cup [1, \infty)$

9. $y = \sec 2x$

Period: $\frac{2\pi}{2} = \pi$

Matches graph (e).

10. $y = \tan\frac{x}{2}$

Period: $\frac{\pi}{b} = \frac{\pi}{1/2} = 2\pi$

Asymptotes: $x = -\pi, x = \pi$

Matches graph (c).

11. $y = \frac{1}{2}\cot \pi x$

Period: $\frac{\pi}{\pi} = 1$

Matches graph (a).

12. $y = -\csc x$

Period: 2π

Matches graph (d).

13. $y = \frac{1}{2}\sec\frac{\pi x}{2}$

Period: $\frac{2\pi}{b} = \frac{2\pi}{\pi/2} = 4$

Asymptotes: $x = -1, x = 1$

Matches graph (f).

14. $y = -2\sec\frac{\pi x}{2}$

Period: $\frac{2\pi}{b} = \frac{2\pi}{\pi/2} = 4$

Asymptotes: $x = -1, x = 1$

Reflected in x-axis

Matches graph (b).

15. $y = \frac{1}{3}\tan x$

Period: π

Two consecutive asymptotes:

$x = -\frac{\pi}{2}$ and $x = \frac{\pi}{2}$

x	$-\frac{\pi}{4}$	0	$\frac{\pi}{4}$
y	$-\frac{1}{3}$	0	$\frac{1}{3}$

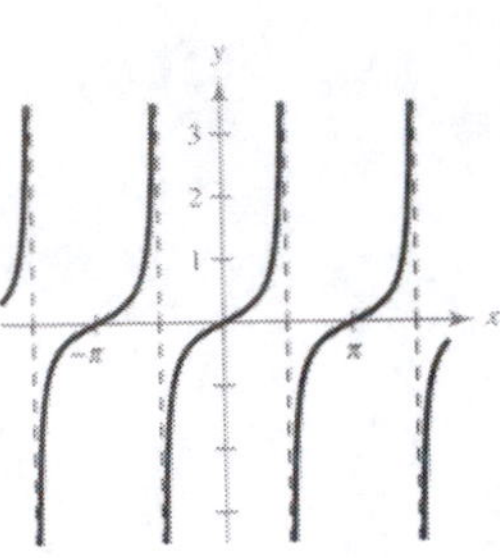

17. $y = -2\tan 3x$

Period: $\frac{\pi}{3}$

Two consecutive asymptotes:

$x = -\frac{\pi}{6}, x = \frac{\pi}{6}$

x	$-\frac{\pi}{3}$	0	$\frac{\pi}{3}$
y	0	0	0

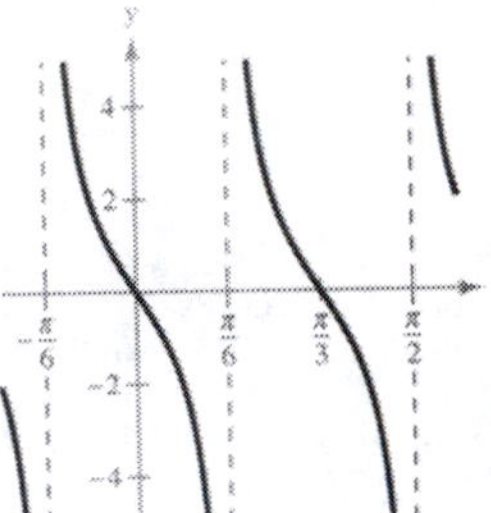

19. $y = -\frac{1}{2}\sec x$

Period: 2π

Two consecutive asymptotes:

$x = -\frac{\pi}{2}, x = \frac{\pi}{2}$

x	$-\frac{\pi}{3}$	0	$\frac{\pi}{3}$
y	-1	$-\frac{1}{2}$	-1

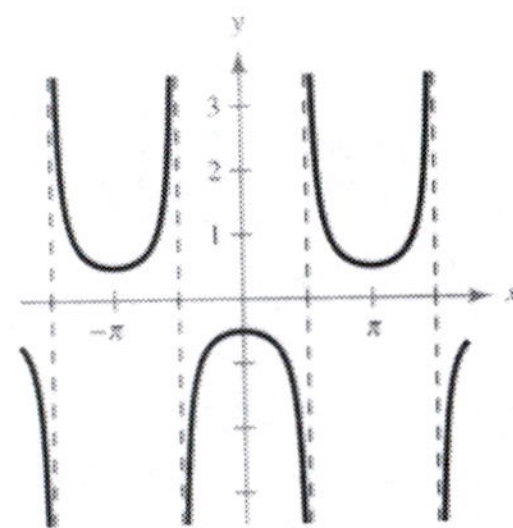

21. $y = \csc \pi x$

Period: $\frac{2\pi}{\pi} = 2$

Two consecutive asymptotes:

$x = 0, x = 1$

x	$\frac{1}{6}$	$\frac{1}{2}$	$\frac{5}{6}$
y	2	1	2

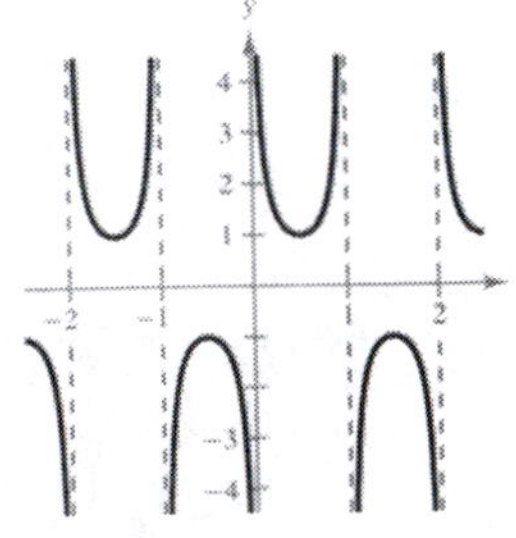

23. $y = \frac{1}{2}\sec \pi x$

Period: 2

Two consecutive asymptotes:

$x = -\frac{1}{2}, x = \frac{1}{2}$

x	-1	0	1
y	$-\frac{1}{2}$	$\frac{1}{2}$	$-\frac{1}{2}$

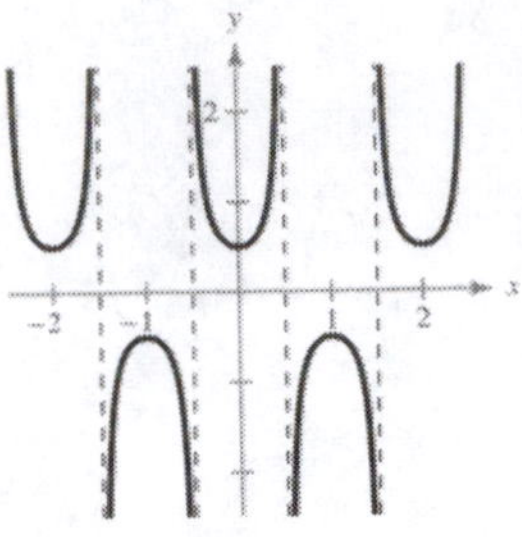

25. $y = \csc \frac{x}{2}$

Period: $\frac{2\pi}{1/2} = 4\pi$

Two consecutive asymptotes:

$x = 0, x = 2\pi$

x	$\frac{\pi}{3}$	π	$\frac{5\pi}{3}$
y	2	1	2

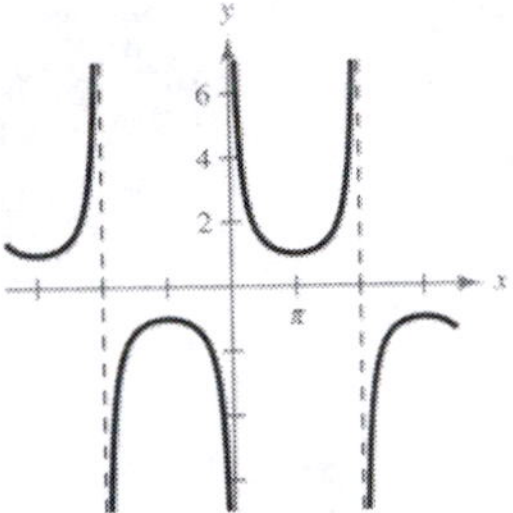

27. $y = 3\cot 2x$

Period: $\frac{\pi}{2}$

Two consecutive asymptotes:

$x = -\frac{\pi}{2}, x = \frac{\pi}{2}$

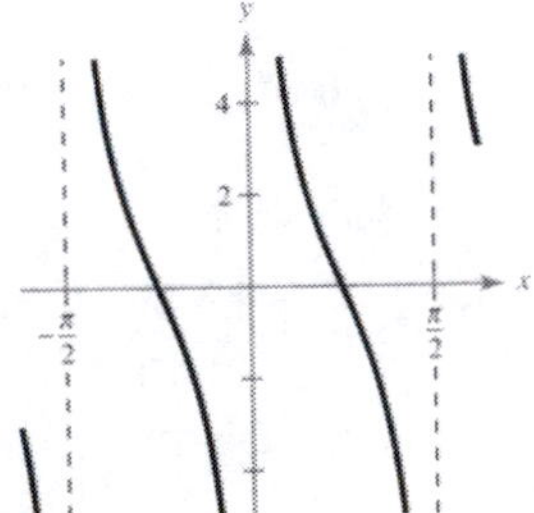

x	$-\frac{\pi}{6}$	$-\frac{\pi}{8}$	$\frac{\pi}{8}$	$\frac{\pi}{6}$
y	$-3\sqrt{3}$	-3	3	$3\sqrt{3}$

29. $y = 2\sec 3x$

Period: $\frac{2\pi}{3}$

Two consecutive asymptotes:

$x = -\frac{\pi}{6}, x = \frac{\pi}{6}$

x	$-\frac{\pi}{3}$	0	$\frac{\pi}{3}$
y	-2	2	-2

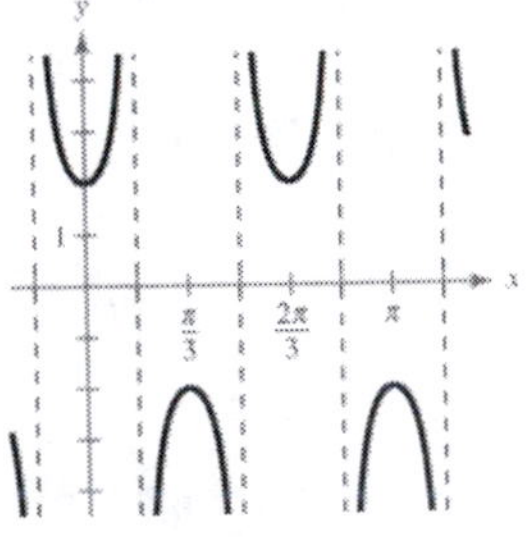

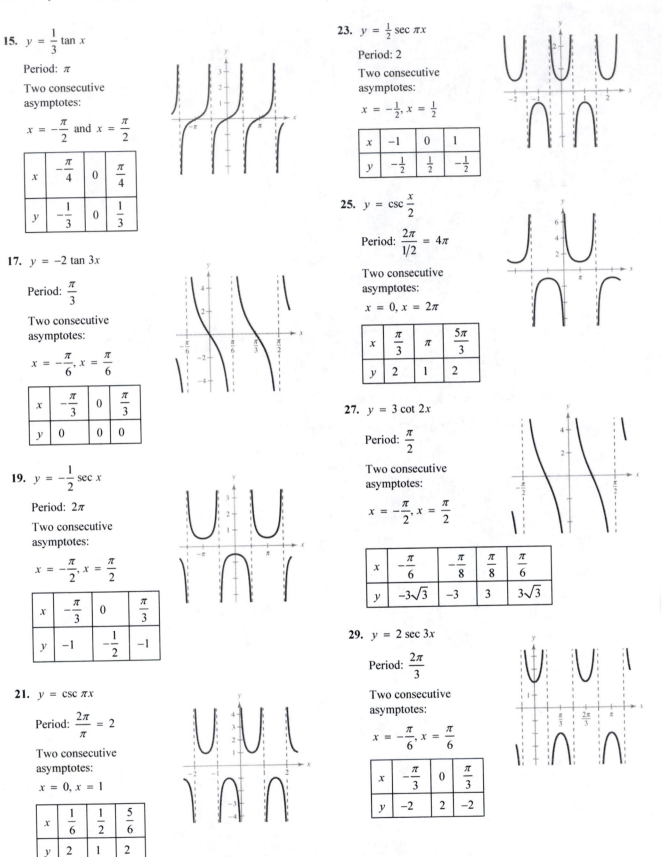

31. $y = \tan \frac{\pi x}{4}$

Period: $\frac{\pi}{\pi/4} = 4$

Two consecutive asymptotes:

$\frac{\pi x}{4} = -\frac{\pi}{2} \Rightarrow x = -2$

$\frac{\pi x}{4} = \frac{\pi}{2} \Rightarrow x = 2$

x	-1	0	1
y	-1	0	1

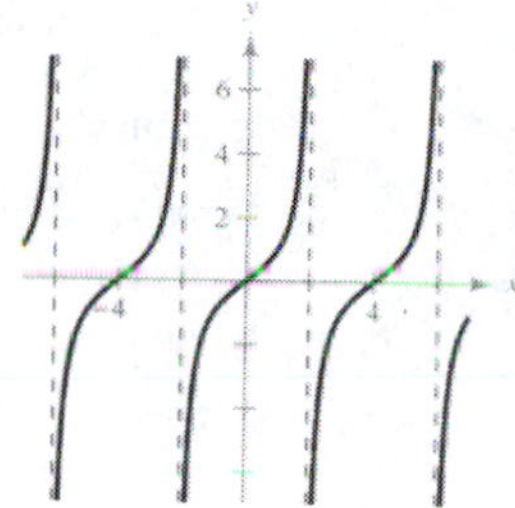

33. $y = 2 \csc(x - \pi)$

Period: 2π

Two consecutive asymptotes:

$x = -\pi, x = \pi$

x	$-\frac{\pi}{2}$	$\frac{\pi}{2}$	$\frac{3\pi}{2}$
y	2	-2	-2

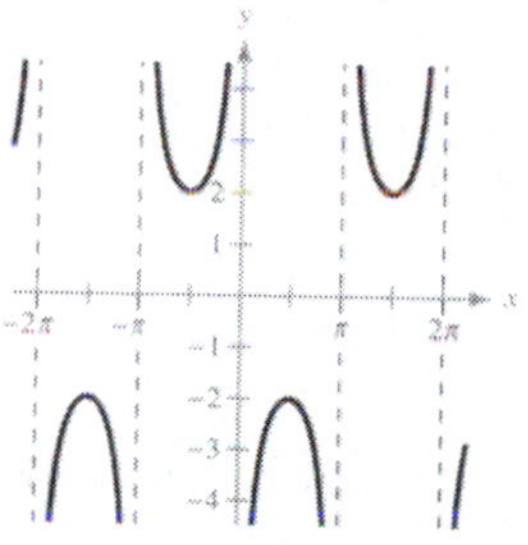

35. $y = 2 \sec(x + \pi)$

Period: 2π

Two consecutive asymptotes:

$x = -\frac{\pi}{2}, x = \frac{\pi}{2}$

x	$-\frac{\pi}{3}$	0	$\frac{\pi}{3}$
y	-4	-2	-4

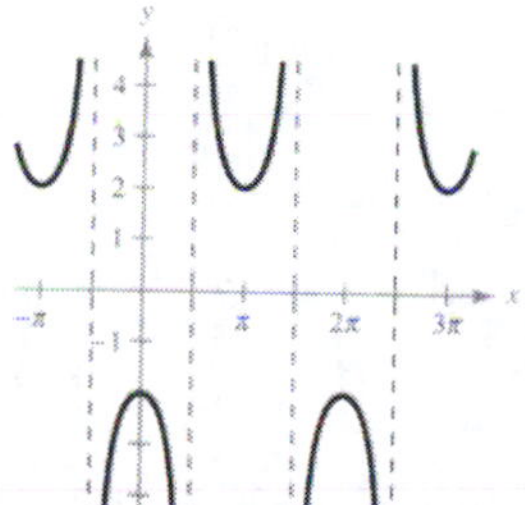

37. $y = \frac{1}{4} \csc\left(x + \frac{\pi}{4}\right)$

Period: 2π

Two consecutive asymptotes:

$x = -\frac{\pi}{4}, x = \frac{3\pi}{4}$

x	$-\frac{\pi}{12}$	$\frac{\pi}{4}$	$\frac{7\pi}{12}$
y	$\frac{1}{2}$	$\frac{1}{4}$	$\frac{1}{2}$

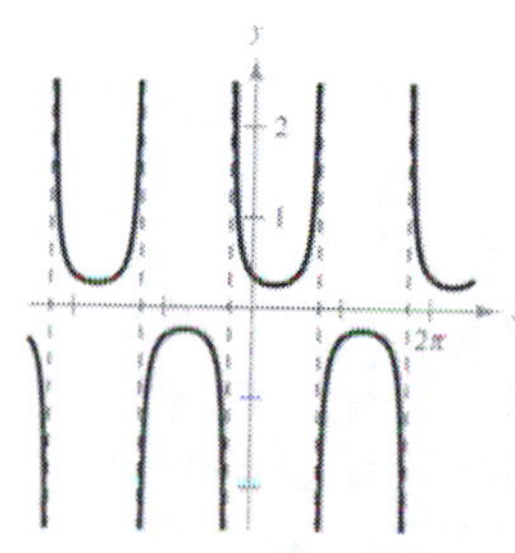

39. $y = \tan \frac{x}{3}$

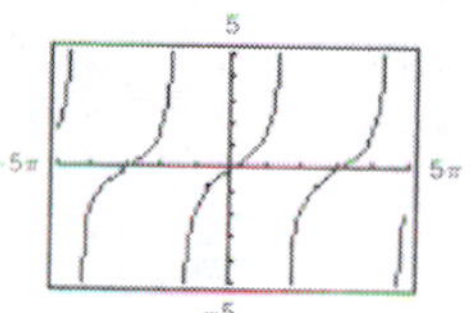

41. $y = -2 \sec 4x = \frac{-2}{\cos 4x}$

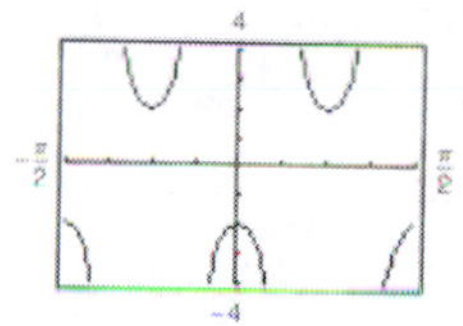

43. $y = \tan\left(x - \frac{\pi}{4}\right)$

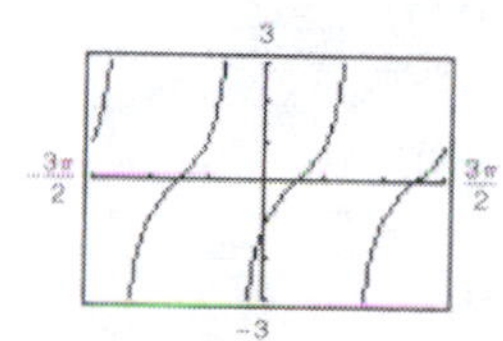

45. $y = -\csc(4x - \pi)$

$y = \frac{-1}{\sin(4x - \pi)}$

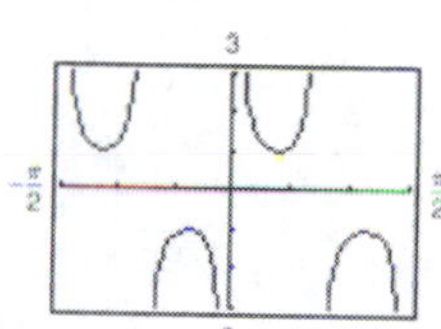

47. $y = 0.1 \tan\left(\frac{\pi x}{4} + \frac{\pi}{4}\right)$

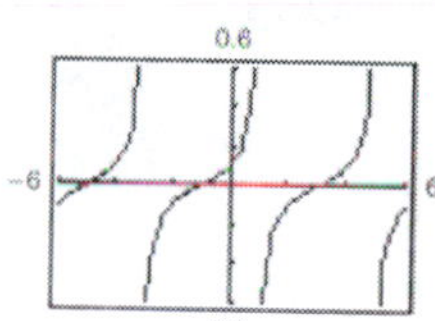

49. $\tan x = 1$

$x = -\frac{7\pi}{4}, -\frac{3\pi}{4}, \frac{\pi}{4}, \frac{5\pi}{4}$

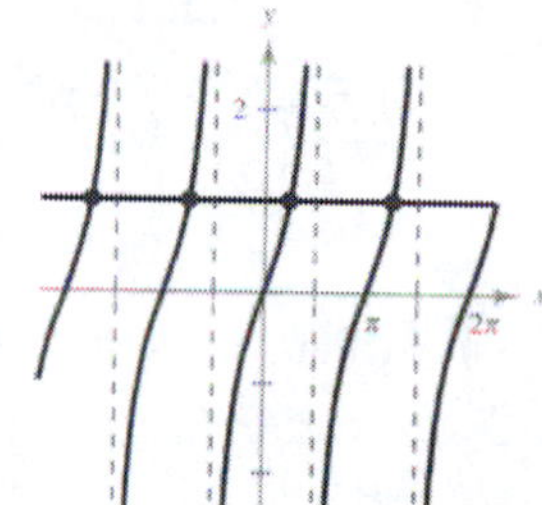

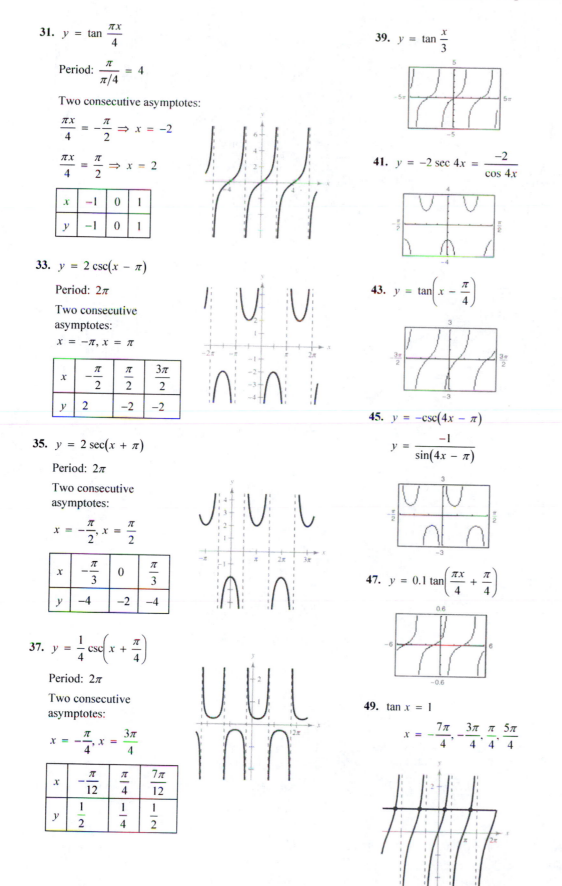

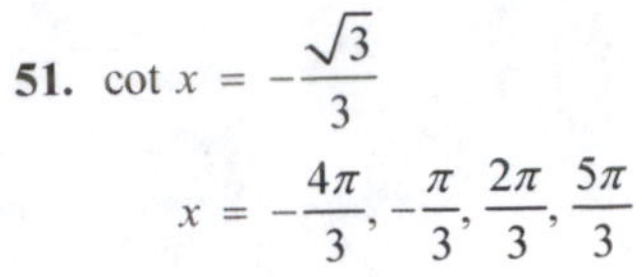

51. $\cot x = -\dfrac{\sqrt{3}}{3}$

$x = -\dfrac{4\pi}{3}, -\dfrac{\pi}{3}, \dfrac{2\pi}{3}, \dfrac{5\pi}{3}$

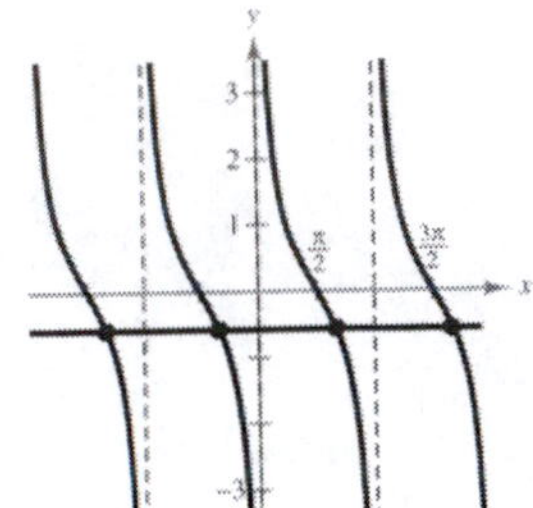

53. $\sec x = -2$

$x = \dfrac{2\pi}{3}, \dfrac{4\pi}{3}, -\dfrac{2\pi}{3}, -\dfrac{4\pi}{3}$

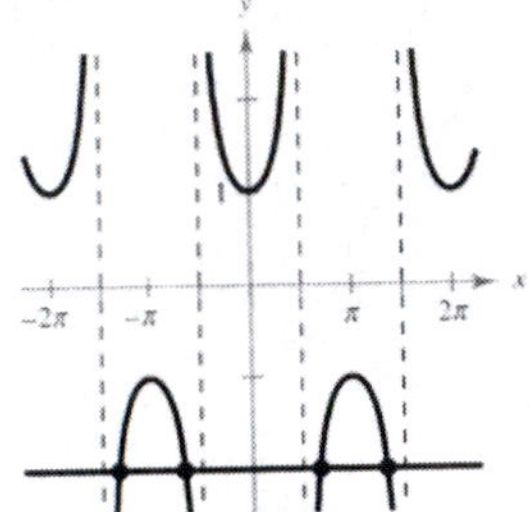

55. $\csc x = \sqrt{2}$

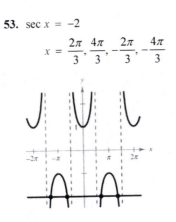

$x = -\dfrac{7\pi}{4}, -\dfrac{5\pi}{4}, \dfrac{\pi}{4}, \dfrac{3\pi}{4}$

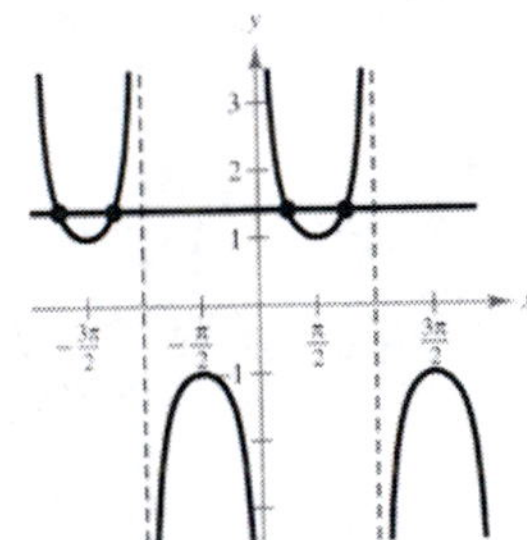

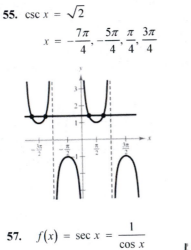

57. $f(x) = \sec x = \dfrac{1}{\cos x}$

$f(-x) = \sec(-x)$

$= \dfrac{1}{\cos(-x)}$

$= \dfrac{1}{\cos x}$

$= f(x)$

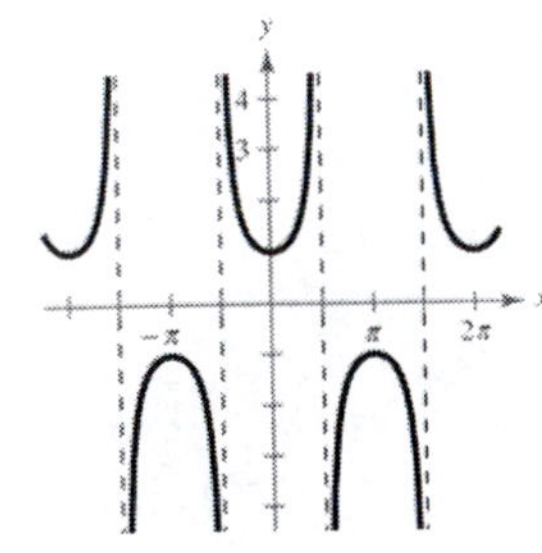

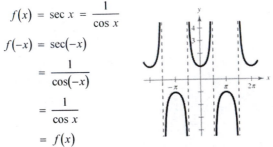

So, $f(x) = \sec x$ is an even function and the graph has y-axis symmetry.

59. $g(x) = \cot x = \dfrac{1}{\tan x}$

$g(-x) = \cot(-x)$

$= \dfrac{1}{\tan(-x)}$

$= -\dfrac{1}{\tan x}$

$= -g(x)$

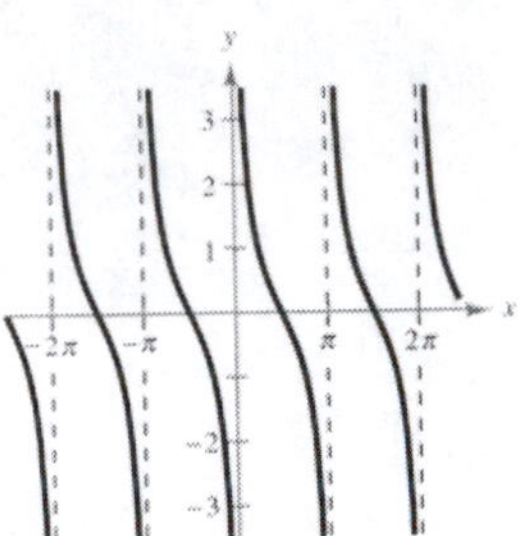

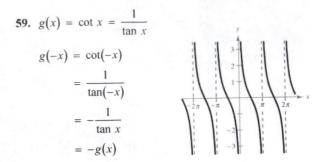

So, $g(x) = \cot x$ is an odd function and the graph has origin symmetry.

61. $f(x) = x + \tan x$

$f(-x) = (-x) + \tan(-x)$

$= -x - \tan x$

$= -(x + \tan x)$

$= -f(x)$

So, $f(x) = x + \tan x$ is an odd function and the graph has origin symmetry.

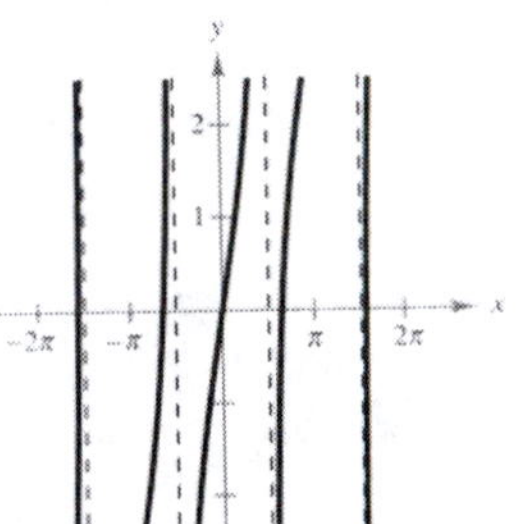

63. $g(x) = x \csc x = \dfrac{x}{\sin x}$

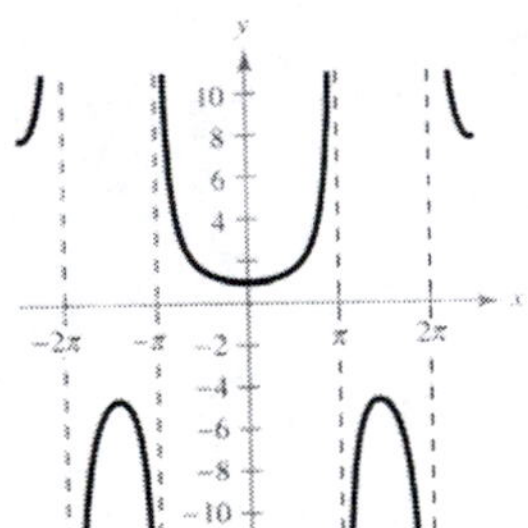

$g(-x) = (-x)\csc(-x)$

$= \dfrac{-x}{\sin(-x)}$

$= \dfrac{-x}{-\sin x}$

$= \dfrac{x}{\sin x}$

$= x \csc x$

$= g(x)$

So, $g(x) = x \csc x$ is an even function and the graph has y-axis symmetry.

65. $f(x) = |x \cos x|$

Matches graph (d).

As $x \to 0$, $f(x) \to 0$.

66. $f(x) = x \sin x$

Matches graph (a)

As $x \to 0$, $f(x) \to 0$.

67. $g(x) = |x| \sin x$

Matches graph (b).

As $x \to 0$, $g(x) \to 0$.

68. $g(x) = |x| \cos x$

Matches graph (c).

As $x \to 0$, $g(x) \to 0$.

69. $f(x) = \sin x + \cos\left(x + \dfrac{\pi}{2}\right)$

$g(x) = 0$

$f(x) = g(x)$

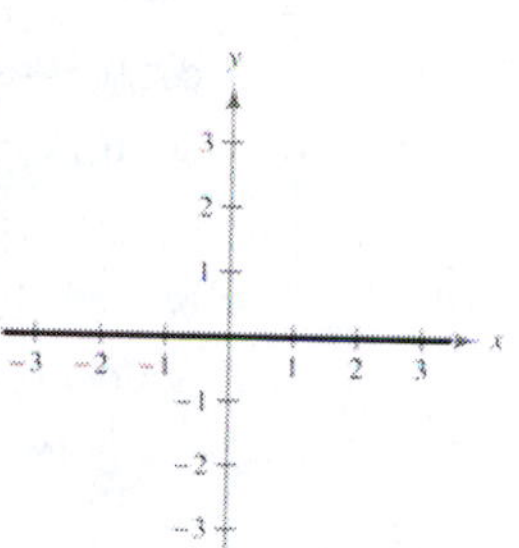

71. $f(x) = \sin^2 x$

$g(x) = \frac{1}{2}(1 - \cos 2x)$

$f(x) = g(x)$

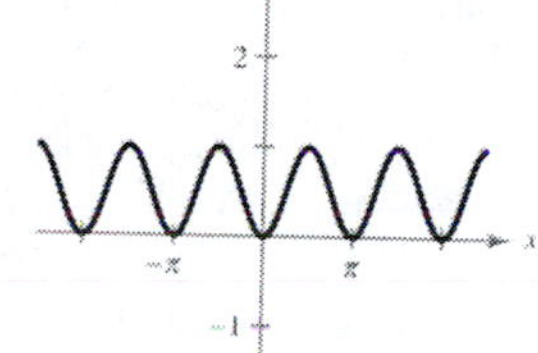

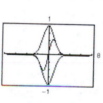

73. $g(x) = e^{-x^2/2} \sin x$

Damping factor: $e^{-x^2/2}$

As $x \to \infty$, $g(x) \to 0$.

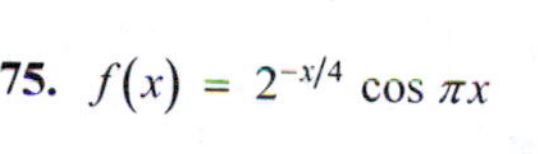

75. $f(x) = 2^{-x/4} \cos \pi x$

Damping factor: $y = 2^{-x/4}$

As $x \to \infty$, $f(x) \to 0$.

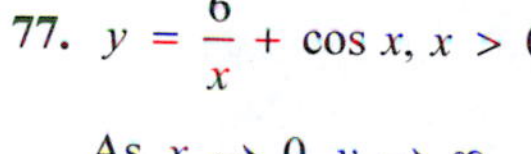

77. $y = \dfrac{6}{x} + \cos x,\ x > 0$

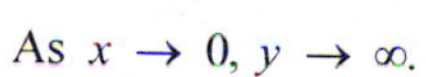

As $x \to 0$, $y \to \infty$.

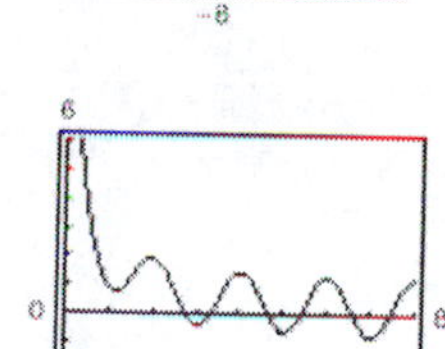

79. $g(x) = \dfrac{\sin x}{x}$

As $x \to 0$, $g(x) \to 1$.

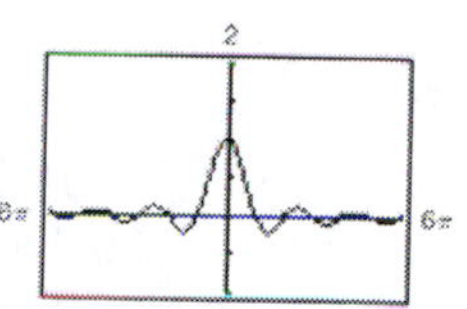

81. $f(x) = \sin \dfrac{1}{x}$

As $x \to 0$, $f(x)$ oscillates between -1 and 1.

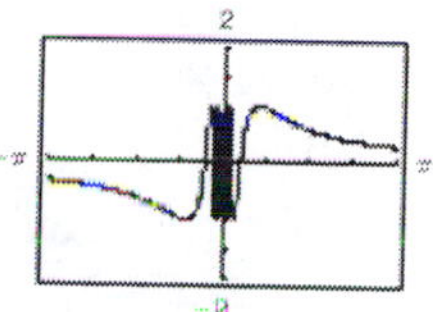

83. (a) Period of $\cos \dfrac{\pi t}{6} = \dfrac{2\pi}{\pi/6} = 12$

Period of $\sin \dfrac{\pi t}{6} = \dfrac{2\pi}{\pi/6} = 12$

The period of $H(t)$ is 12 months.

The period of $L(t)$ is 12 months.

(b) From the graph, it appears that the greatest difference between high and low temperatures occurs in the summer. The smallest difference occurs in the winter.

(c) The highest high and low temperatures appear to occur about half of a month after the time when the sun is northernmost in the sky.

85. $\tan x = \dfrac{7}{d}$

$d = \dfrac{7}{\tan x} = 7 \cot x$

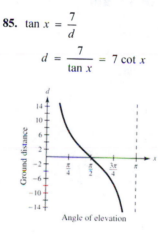

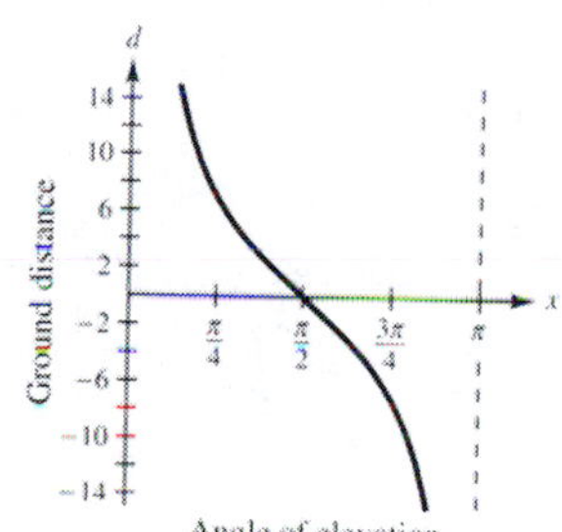

87. True.

$y = \sec x = \dfrac{1}{\cos x}$

If the reciprocal of $y = \sin x$ is translated $\pi/2$ units to the left, then

$$y = \frac{1}{\sin\left(x + \dfrac{\pi}{2}\right)} = \frac{1}{\cos x} = \sec x.$$

89. $f(x) = \csc x$

(a) $x \to 0^+, f(x) \to \infty$

(b) $x \to 0^-, f(x) \to -\infty$

(c) $x \to \pi^+, f(x) \to -\infty$

(d) $x \to \pi^-, f(x) \to \infty$

91. $f(x) = \sec x$

(a) $x \to \frac{\pi}{2}^+, f(x) \to -\infty$

(b) $x \to \frac{\pi}{2}^-, f(x) \to \infty$

(c) $x \to -\frac{\pi}{2}^+, f(x) \to \infty$

(d) $x \to -\frac{\pi}{2}^-, f(x) \to -\infty$

93. $f(x) = x - \cos x$

(a)

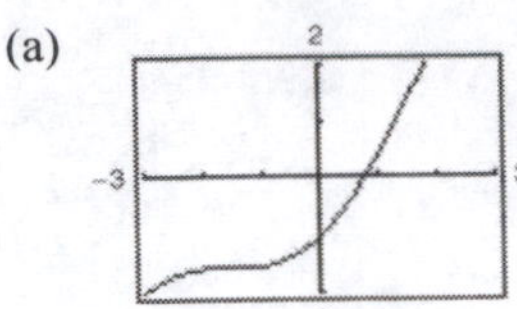

The zero between 0 and 1 occurs at $x \approx 0.7391$.

(b) $x_n = \cos(x_{n-1})$

$x_0 = 1$

$x_1 = \cos 1 \approx 0.5403$

$x_2 = \cos 0.5403 \approx 0.8576$

$x_3 = \cos 0.8576 \approx 0.6543$

$x_4 = \cos 0.6543 \approx 0.7935$

$x_5 = \cos 0.7935 \approx 0.7014$

$x_6 = \cos 0.7014 \approx 0.7640$

$x_7 = \cos 0.7640 \approx 0.7221$

$x_8 = \cos 0.7221 \approx 0.7504$

$x_9 = \cos 0.7504 \approx 0.7314$

$\vdots$

This sequence appears to be approaching the zero of $f: x \approx 0.7391$.

Section 6.6 Inverse Trigonometric Functions

Function	*Alternative Notation*	*Domain*	*Range*
1. $y = \arcsin x$	$y = \sin^{-1} x$	$-1 \le x \le 1$	$-\frac{\pi}{2} \le y \le \frac{\pi}{2}$
3. $y = \arctan x$	$y = \tan^{-1} x$	$-\infty < x < \infty$	$-\frac{\pi}{2} < y < \frac{\pi}{2}$

5. $y = \arcsin \frac{1}{2} \Rightarrow \sin y = \frac{1}{2}$ for $-\frac{\pi}{2} \le y \le \frac{\pi}{2} \Rightarrow y = \frac{\pi}{6}$

7. $y = \arccos \frac{1}{2} \Rightarrow \cos y = \frac{1}{2}$ for $0 \le y \le \pi \Rightarrow y = \frac{\pi}{3}$

9. $y = \arctan \frac{\sqrt{3}}{3} \Rightarrow \tan y = \frac{\sqrt{3}}{3}$ for $-\frac{\pi}{2} < y < \frac{\pi}{2} \Rightarrow y = \frac{\pi}{6}$

11. $y = \cos^{-1}\left(-\frac{\sqrt{3}}{2}\right) \Rightarrow \cos y = -\frac{\sqrt{3}}{2}$ for $0 \le y \le \pi \Rightarrow y = \frac{5\pi}{6}$

13. $y = \arctan\left(-\sqrt{3}\right) \Rightarrow \tan y = -\sqrt{3}$ for $-\frac{\pi}{2} < y < \frac{\pi}{2} \Rightarrow y = -\frac{\pi}{3}$

15. $y = \arccos\left(-\frac{1}{2}\right) \Rightarrow \cos y = -\frac{1}{2}$ for $0 \le y \le \pi \Rightarrow y = \frac{2\pi}{3}$

17. $y = \sin^{-1} -\frac{\sqrt{3}}{2} \Rightarrow \sin y = -\frac{\sqrt{3}}{2}$ for $-\frac{\pi}{2} \le y \le \frac{\pi}{2} \Rightarrow y = -\frac{\pi}{3}$

19. $f(x) = \cos x$

$g(x) = \arccos x$

$y = x$

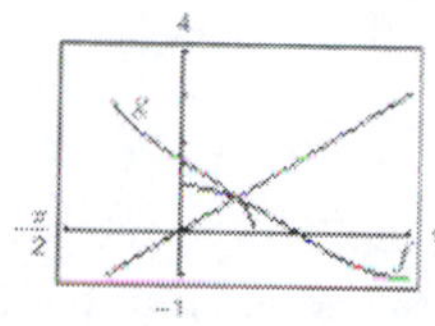

21. $\arccos 0.37 = \cos^{-1}(0.37) \approx 1.19$

23. $\arcsin(-0.75) = \sin^{-1}(-0.75) \approx -0.85$

25. $\arctan(-3) = \tan^{-1}(-3) \approx -1.25$

27. $\sin^{-1} 0.31 = \sin^{-1} 0.31 \approx 0.32$

29. $\arccos(-0.41) = \cos^{-1}(-0.41) \approx 1.99$

31. $\arctan 0.92 = \tan^{-1} 0.92 \approx 0.74$

33. $\arcsin \frac{7}{8} = \sin^{-1}\left(\frac{7}{8}\right) \approx 1.07$

35. $\tan^{-1}\left(\frac{19}{4}\right) \approx 1.36$

37. $\tan^{-1}\left(-\sqrt{372}\right) \approx -1.52$

39. $\arctan\left(-\sqrt{3}\right) = -\dfrac{\pi}{3}$

$\tan\left(-\dfrac{\pi}{6}\right) = -\dfrac{\sqrt{3}}{3}$

$\tan\left(\dfrac{\pi}{4}\right) = 1$

41. $\tan \theta = \dfrac{x}{4}$

$\theta = \arctan \dfrac{x}{4}$

43. $\sin \theta = \dfrac{x + 2}{5}$

$\theta = \arcsin\left(\dfrac{x + 2}{5}\right)$

45. $\cos \theta = \dfrac{x + 3}{2x}$

$\theta = \arccos \dfrac{x + 3}{2x}$

47. $\sin(\arcsin 0.3) = 0.3$

49. $\cos\left[\arccos(-0.1)\right] = -0.1$

51. $\arcsin(\sin 3\pi) = \arcsin(0) = 0$

Note: 3π is not in the range of the arcsine function.

53. Let $y = \arctan \dfrac{3}{4}$.

$\tan y = \dfrac{3}{4}, 0 < y < \dfrac{\pi}{2},$

$\sin\left(\arctan \dfrac{3}{4}\right) = \sin y = \dfrac{3}{5}$

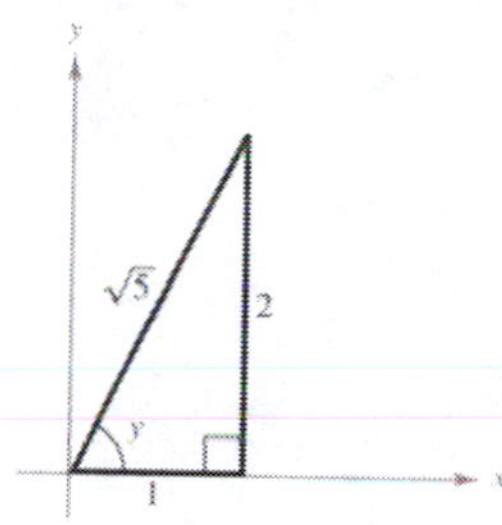

55. Let $y = \tan^{-1} 2,$

$\tan y = 2 = \dfrac{2}{1}, 0 < y < \dfrac{\pi}{2},$

$\cos\left(\tan^{-1} 2\right) = \cos y = \dfrac{1}{\sqrt{5}} = \dfrac{\sqrt{5}}{5}.$

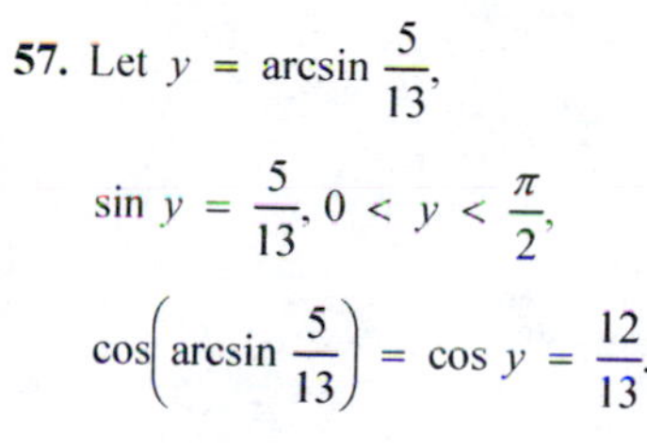

57. Let $y = \arcsin \dfrac{5}{13},$

$\sin y = \dfrac{5}{13}, 0 < y < \dfrac{\pi}{2},$

$\cos\left(\arcsin \dfrac{5}{13}\right) = \cos y = \dfrac{12}{13}.$

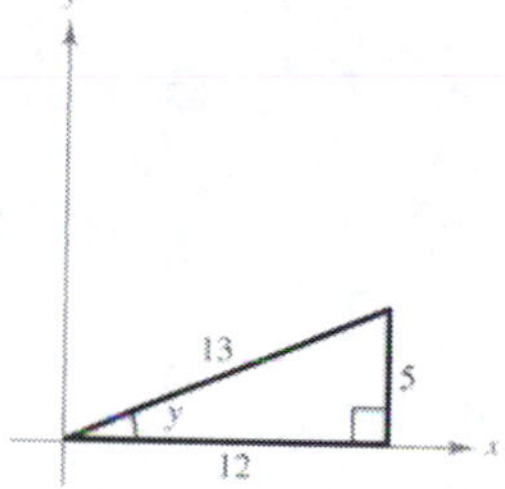

59. Let $y = \arctan\left(-\frac{3}{5}\right)$,

$\tan y = -\frac{3}{5}, -\frac{\pi}{2} < y < 0,$

$\sec\left[\arctan\left(-\frac{3}{5}\right)\right] = \sec y = \frac{\sqrt{34}}{5}.$

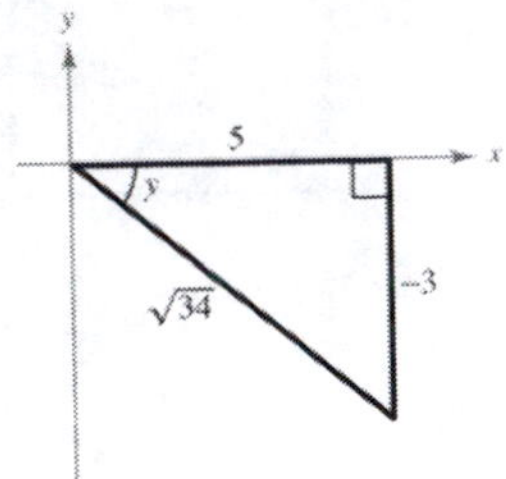

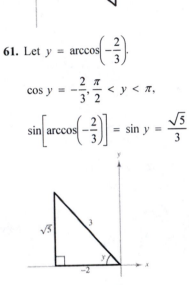

61. Let $y = \arccos\left(-\frac{2}{3}\right)$.

$\cos y = -\frac{2}{3}, \frac{\pi}{2} < y < \pi,$

$\sin\left[\arccos\left(-\frac{2}{3}\right)\right] = \sin y = \frac{\sqrt{5}}{3}$

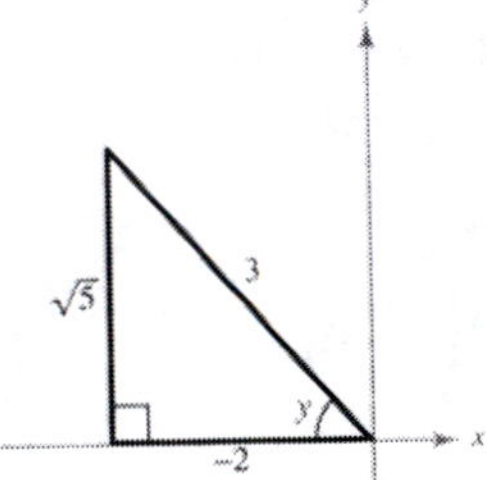

63. Let $u = \cos^{-1}\frac{\sqrt{3}}{2}$.

$\cos u = \frac{\sqrt{3}}{2}, 0 < u < \frac{\pi}{2},$

$\csc\left[\cos^{-1}\frac{\sqrt{3}}{2}\right] = \csc u = 2.$

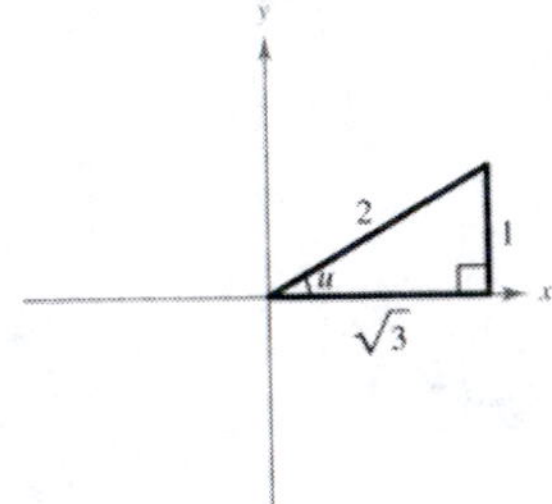

65. Let $y = \arctan x$.

$\tan y = x = \frac{x}{1},$

$\cot(\arctan x) = \cot y = \frac{1}{x}$

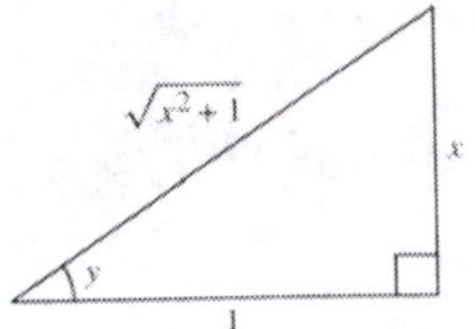

67. Let $y = \arcsin(2x)$.

$\sin y = 2x = \frac{2x}{1},$

$\cos(\arcsin 2x) = \cos y = \sqrt{1 - 4x^2}$

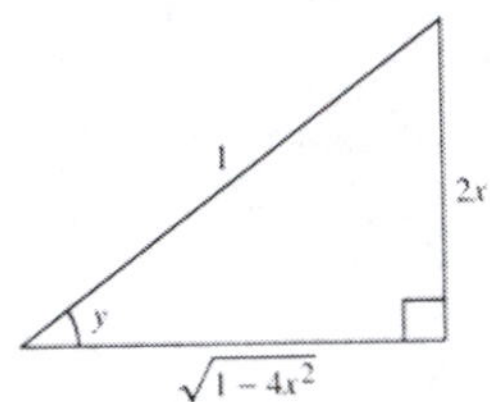

69. Let $y = \arccos x$.

$\cos y = x = \frac{x}{1},$

$\sin(\arccos x) = \sin y = \sqrt{1 - x^2}$

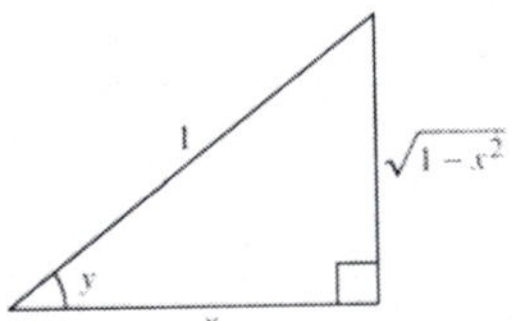

71. Let $y = \arccos\left(\frac{x}{3}\right)$.

$\cos y = \frac{x}{3},$

$\tan\left(\arccos\frac{x}{3}\right) = \tan y = \frac{\sqrt{9 - x^2}}{x}$

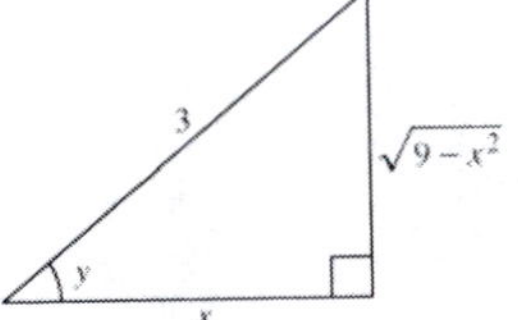

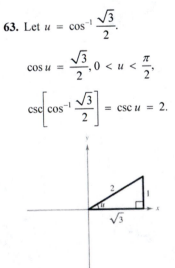

73. Let $y = \arctan \dfrac{x}{\sqrt{2}}$.

$\tan y = \dfrac{x}{\sqrt{2}}$,

$\csc\left(\arctan \dfrac{x}{\sqrt{2}}\right) = \csc y = \dfrac{\sqrt{x^2 + 2}}{x}$

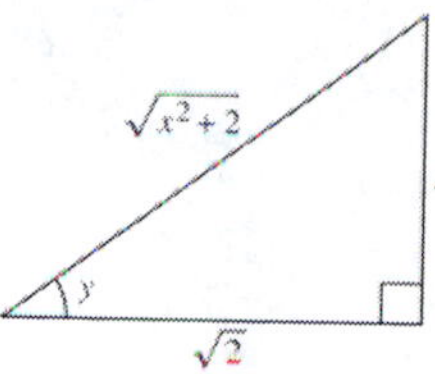

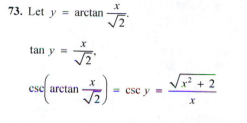

75. $f(x) = \sin(\arctan 2x)$, $g(x) = \dfrac{2x}{\sqrt{1 + 4x^2}}$

They are equal. Let $y = \arctan 2x$,

$\tan y = 2x = \dfrac{2x}{1}$,

and $\sin y = \dfrac{2x}{\sqrt{1 + 4x^2}}$.

$g(x) = \dfrac{2x}{\sqrt{1 + 4x^2}} = f(x)$

The graph has horizontal asymptotes at $y = \pm 1$.

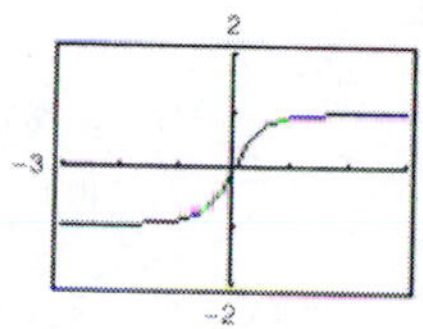

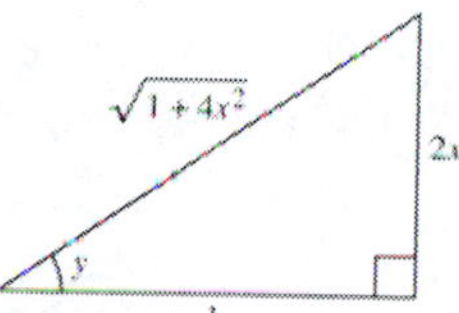

77. Let $y = \arctan \dfrac{9}{x}$.

$\tan y = \dfrac{9}{x}$ and $\sin y = \dfrac{9}{\sqrt{x^2 + 81}}$, $x > 0$

So,

$\arctan \dfrac{9}{x} = \arcsin \dfrac{9}{\sqrt{x^2 + 81}}$, $x > 0$.

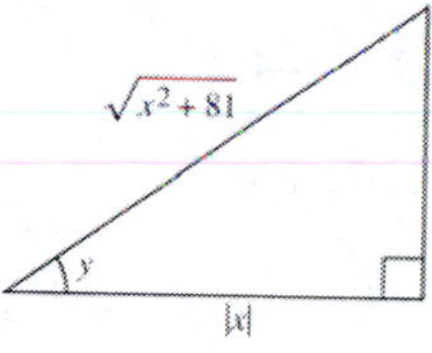

79. Let $y = \arccos \dfrac{3}{\sqrt{x^2 - 2x + 10}}$. Then,

$\cos y = \dfrac{3}{\sqrt{x^2 - 2x + 10}} = \dfrac{3}{\sqrt{(x - 1)^2 + 9}}$

and $\sin y = \dfrac{|x - 1|}{\sqrt{(x - 1)^2 + 9}}$.

So, $y = \arcsin \dfrac{|x - 1|}{\sqrt{x^2 - 2x + 10}}$.

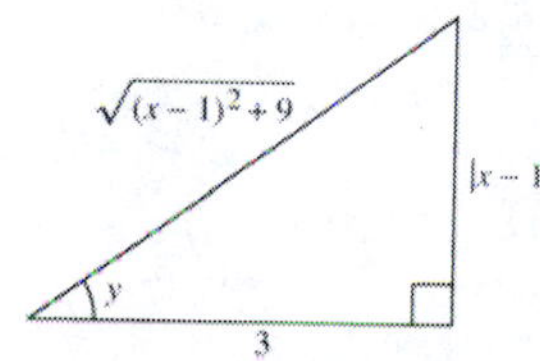

81. $g(x) = \arcsin(x - 1)$

Domain: $0 \le x \le 2$

Range: $-\dfrac{\pi}{2} \le y \le \dfrac{\pi}{2}$

This is the graph of $f(x) = \arcsin(x)$ shifted one unit to the right.

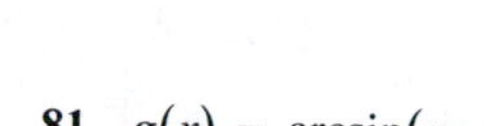

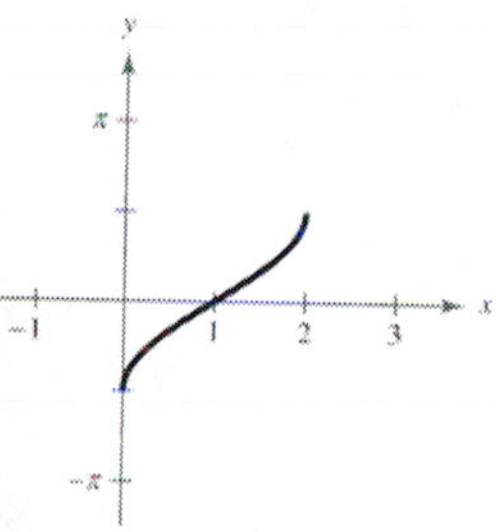

83. $y = 2 \arccos x$

Domain: $-1 \le x \le 1$

Range: $0 \le y \le 2\pi$

This is the graph of $f(x) = \arccos x$ with a factor of 2.

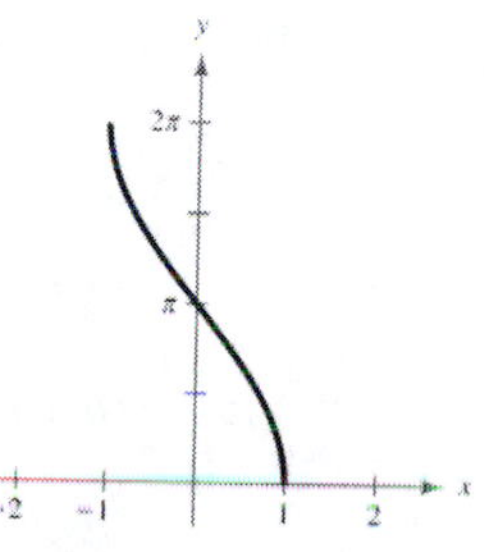

85. $f(x) = \arctan 2x$

Domain: all real numbers

Range: $-\frac{\pi}{2} < y < \frac{\pi}{2}$

This is the graph of $g(x) = \arctan(x)$ with a horizontal shrink of a factor of 2.

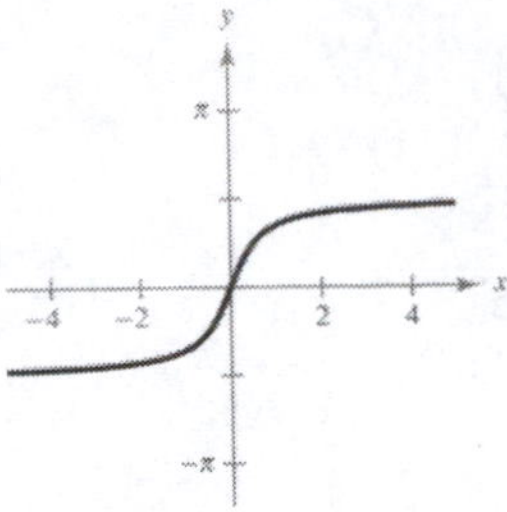

87. $h(v) = \arccos \frac{v}{2}$

Domain: $-2 \le v \le 2$

Range: $0 \le y \le \pi$

This is the graph of $h(v) = \arccos v$ with a horizontal stretch of a factor of 2.

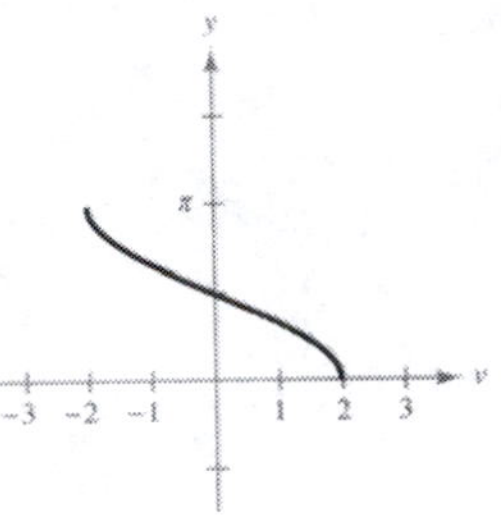

89. $f(x) = 2 \arccos(2x)$

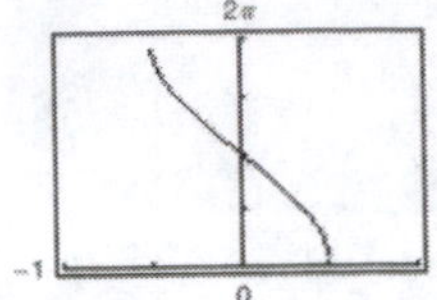

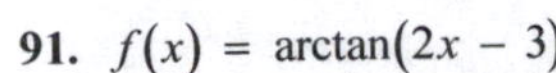

91. $f(x) = \arctan(2x - 3)$

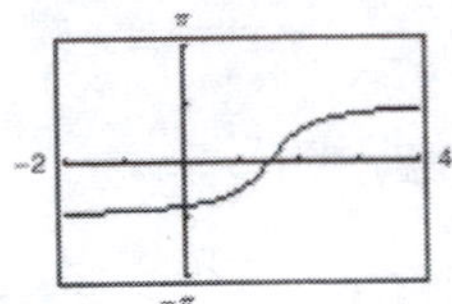

93. $f(x) = \pi - \sin^{-1}\left(\frac{2}{3}\right) \approx 2.412$

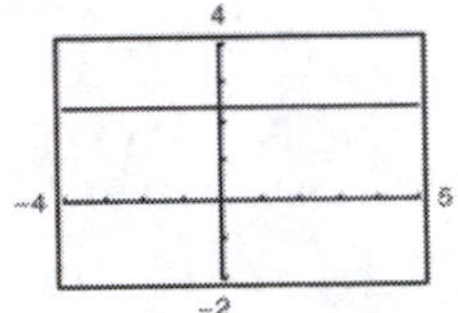

95. $f(t) = 3 \cos 2t + 3 \sin 2t = \sqrt{3^2 + 3^2} \sin\left(2t + \arctan \frac{3}{3}\right)$

$= 3\sqrt{2} \sin(2t + \arctan 1)$

$= 3\sqrt{2} \sin\left(2t + \frac{\pi}{4}\right)$

The graph implies that the identity is true.

97. $\frac{\pi}{2}$

99. $\frac{\pi}{2}$

101. π

103. (a) $\sin \theta = \frac{5}{s}$

$\theta = \arcsin \frac{5}{s}$

(b) $s = 40$: $\theta = \arcsin \frac{5}{40} \approx 0.13$

$s = 20$: $\theta = \arcsin \frac{5}{20} \approx 0.25$

105. $\beta = \arctan \frac{3x}{x^2 + 4}$

(a)

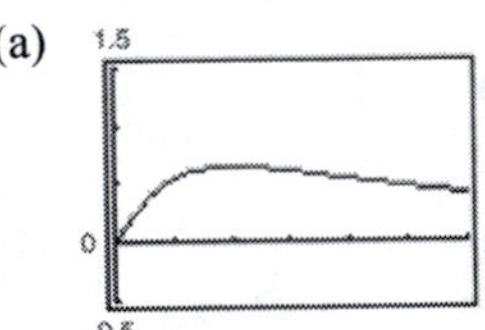

(b) β is maximum when $x = 2$ feet.

(c) The graph has a horizontal asymptote at $\beta = 0$. As x increases, β decreases.

107.

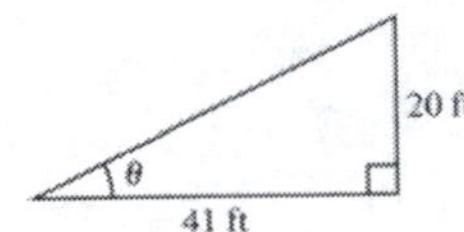

(a) $\tan \theta = \frac{20}{41}$

$\theta = \arctan\left(\frac{20}{41}\right) \approx 26.0°$

(b) $\tan 26° = \frac{h}{50}$

$h = 50 \tan 26° \approx 24.4$ feet

109. (a) $\tan\theta = \dfrac{x}{20}$

$\theta = \arctan\dfrac{x}{20}$

(b) $x = 5$: $\theta = \arctan\dfrac{5}{20} \approx 14.0°$

$x = 12$: $\theta = \arctan\dfrac{12}{20} \approx 31.0°$

111. False.

$\dfrac{5\pi}{4}$ is not in the range of the arctangent function.

$\arctan 1 = \dfrac{\pi}{4}$

113. False. $\sin^{-1} x \neq \dfrac{1}{\sin x}$

The function $\sin^{-1} x$ is equivalent to $\arcsin x$, which is the inverse sine function. The expression, $\dfrac{1}{\sin x}$ is the reciprocal of the sine function and is equivalent to $\csc x$.

115. $y = \operatorname{arccot} x$ if and only if $\cot y = x$.

Domain: $(-\infty, \infty)$

Range: $(0, \pi)$

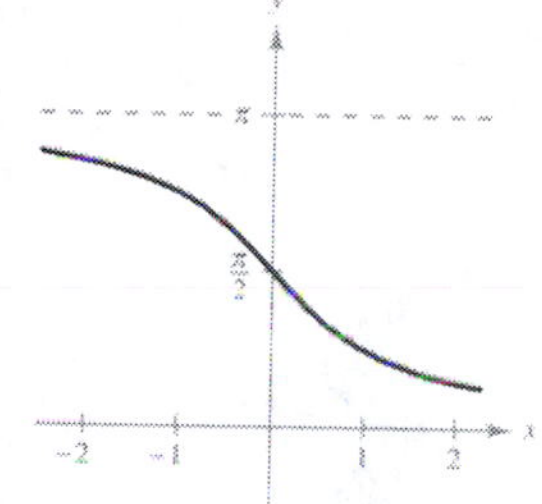

117. $y = \operatorname{arccsc} x$ if and only if $\csc y = x$.

Domain: $(-\infty, -1] \cup [1, \infty)$

Range: $\left[-\dfrac{\pi}{2}, 0\right) \cup \left(0, \dfrac{\pi}{2}\right]$

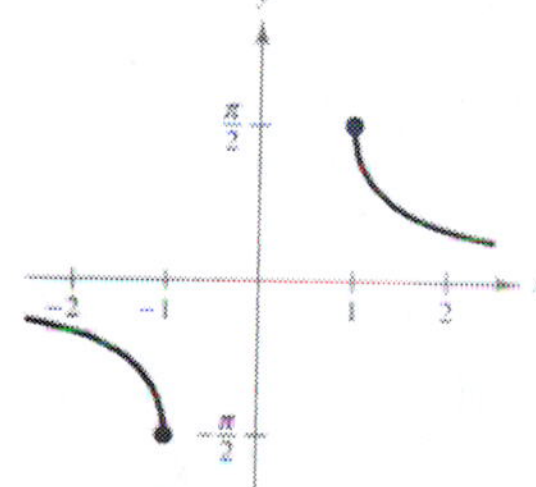

119. $y = \operatorname{arcsec}\sqrt{2} \Rightarrow \sec y = \sqrt{2}$ and

$0 \le y < \dfrac{\pi}{2} \cup \dfrac{\pi}{2} < y \le \pi \Rightarrow y = \dfrac{\pi}{4}$

121. $y = \operatorname{arccot}(-1) \Rightarrow \cot y = -1$ and

$0 < y < \pi \Rightarrow y = \dfrac{3\pi}{4}$

123. $y = \operatorname{arccsc} 2 \Rightarrow \csc y = 2$ and

$-\dfrac{\pi}{2} \le y < 0 \cup 0 < y \le \dfrac{\pi}{2} \Rightarrow y = \dfrac{\pi}{6}$

125. $y = \operatorname{arccsc}\left(\dfrac{2\sqrt{3}}{3}\right) \Rightarrow \csc y = \dfrac{2\sqrt{3}}{3}$ and

$-\dfrac{\pi}{2} \le y < 0 \cup 0 < y \le \dfrac{\pi}{2} \Rightarrow y = \dfrac{\pi}{3}$

127. $\operatorname{arcsec} 2.54 = \arccos\left(\dfrac{1}{2.54}\right) \approx 1.17$

129. $\operatorname{arccot} 5.25 = \arctan\left(\dfrac{1}{5.25}\right) \approx 0.19$

131. $\operatorname{arccot}\left(\frac{5}{3}\right) = \arctan\left(\frac{3}{5}\right) \approx 0.54$

133. $\operatorname{arccsc}\left(-\frac{25}{3}\right) = \arcsin\left(-\frac{3}{25}\right) \approx -0.12$

135. Area $= \arctan b - \arctan a$

(a) $a = 0, b = 1$

Area $= \arctan 1 - \arctan 0 = \dfrac{\pi}{4} - 0 = \dfrac{\pi}{4}$

(b) $a = -1, b = 1$

Area $= \arctan 1 - \arctan(-1)$

$= \dfrac{\pi}{4} - \left(-\dfrac{\pi}{4}\right) = \dfrac{\pi}{2}$

(c) $a = 0, b = 3$

Area $= \arctan 3 - \arctan 0$

$\approx 1.25 - 0 = 1.25$

(d) $a = -1, b = 3$

Area $= \arctan 3 - \arctan(-1)$

$\approx 1.25 - \left(-\dfrac{\pi}{4}\right) \approx 2.03$

137. $f(x) = \sin(x),\ f^{-1}(x) = \arcsin(x)$

(a) $f \circ f^{-1} = \sin(\arcsin x)$ $\qquad f^{-1} \circ f = \arcsin(\sin x)$

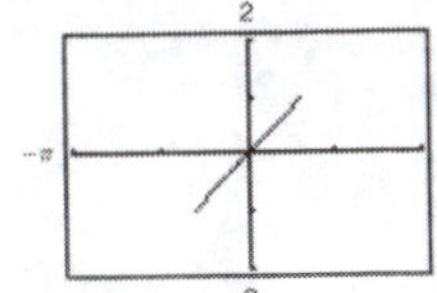

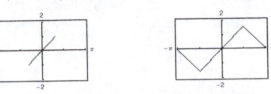

(b) The graphs coincide with the graph of $y = x$ only for certain values of x.

$f \circ f^{-1} = x$ over its entire domain, $-1 \le x \le 1$.

$f^{-1} \circ f = x$ over the region $-\dfrac{\pi}{2} \le x \le \dfrac{\pi}{2}$, corresponding to the region where $\sin x$ is one-to-one and has an inverse.

Section 6.7 Applications and Models

1. bearing

3. period

5. Given: $A = 30°, b = 3$

$$\tan A = \frac{a}{b} \Rightarrow a = b \tan A = 3 \tan 30° \approx 1.73$$

$$\cos A = \frac{b}{c} \Rightarrow c = \frac{b}{\cos A} = \frac{3}{\cos 30°} \approx 3.46$$

$$B = 90° - 30° = 60°$$

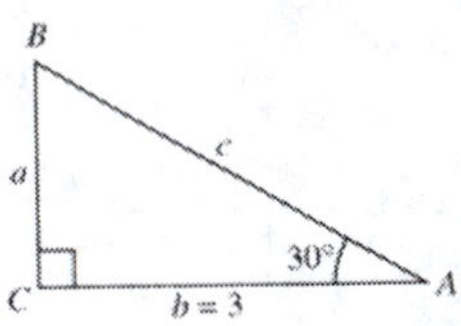

7. Given: $B = 71°, b = 24$

$$\tan B = \frac{b}{a} \Rightarrow a = \frac{b}{\tan B} = \frac{24}{\tan 71°} \approx 8.26$$

$$\sin B = \frac{b}{c} \Rightarrow c = \frac{b}{\sin B} = \frac{24}{\sin 71°} \approx 25.38$$

$$A = 90° - 71° = 19°$$

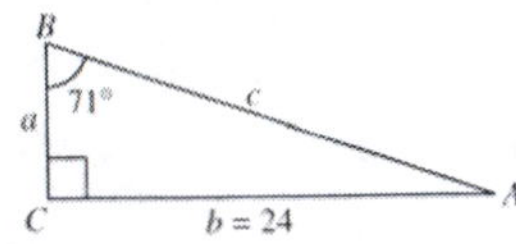

9. Given: $a = 3, b = 4$

$$a^2 + b^2 = c^2 \Rightarrow c^2 = (3)^2 + (4)^2 \Rightarrow c = 5$$

$$\tan A = \frac{a}{b} \Rightarrow A = \tan^{-1}\left(\frac{a}{b}\right) = \tan^{-1}\left(\frac{3}{4}\right) \approx 36.87°$$

$$B = 90° - 36.87° = 53.13°$$

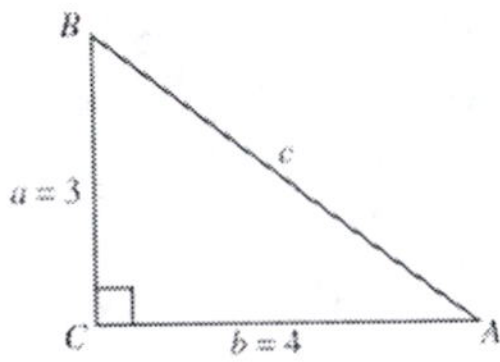

11. Given: $b = 16, c = 52$

$$a = \sqrt{52^2 - 16^2}$$
$$= \sqrt{2448} = 12\sqrt{17} \approx 49.48$$

$$\cos A = \tfrac{16}{52}$$

$$A = \arccos \tfrac{16}{52} \approx 72.80°$$

$$B = 90° - 72.08° \approx 17.92°$$

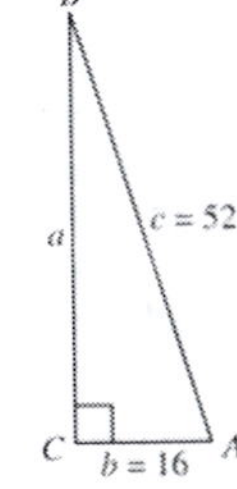

13. Given: $A = 12°15', c = 430.5$

$$B = 90° - 12°15' = 77°45'$$

$$\sin 12°15' = \frac{a}{430.5}$$

$$a = 430.5 \sin 12°15' \approx 91.34$$

$$\cos 12°15' = \frac{b}{430.5}$$

$$b = 430.5 \cos 12°15' \approx 420.70$$

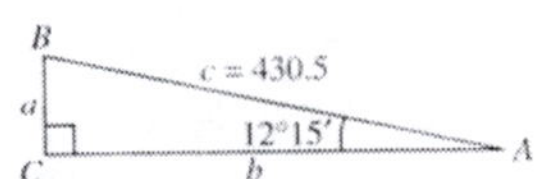

15. $\theta = 45°, b = 6$

$$\tan\theta = \frac{h}{(1/2)b} \Rightarrow h = \frac{1}{2}b\tan\theta$$

$$h = \frac{1}{2}(6)\tan 45° = 3.00 \text{ units}$$

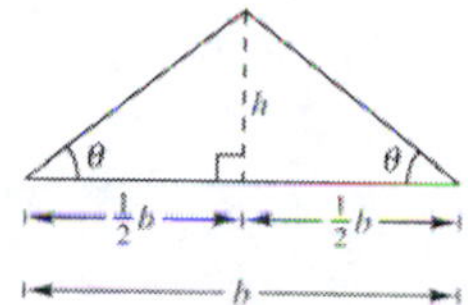

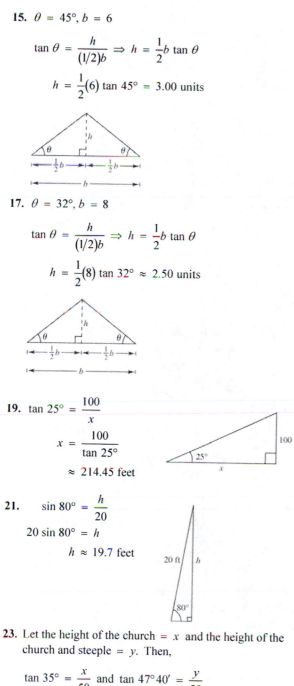

17. $\theta = 32°, b = 8$

$$\tan\theta = \frac{h}{(1/2)b} \Rightarrow h = \frac{1}{2}b\tan\theta$$

$$h = \frac{1}{2}(8)\tan 32° \approx 2.50 \text{ units}$$

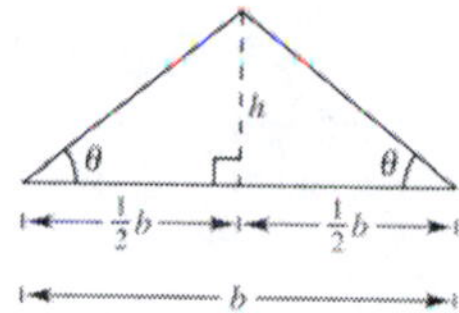

19. $\tan 25° = \dfrac{100}{x}$

$$x = \frac{100}{\tan 25°}$$

$$\approx 214.45 \text{ feet}$$

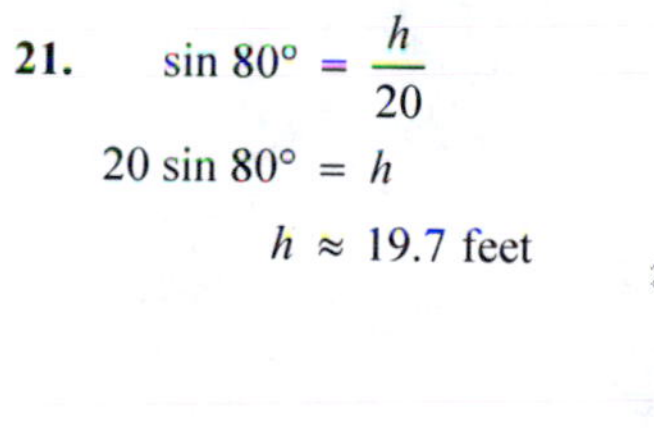

21. $\sin 80° = \dfrac{h}{20}$

$$20\sin 80° = h$$

$$h \approx 19.7 \text{ feet}$$

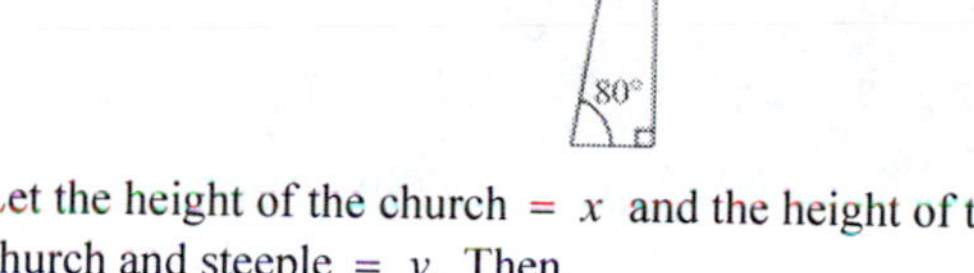

23. Let the height of the church $= x$ and the height of the church and steeple $= y$. Then,

$$\tan 35° = \frac{x}{50} \text{ and } \tan 47°40' = \frac{y}{50}$$

$$x = 50\tan 35° \text{ and } y = 50\tan 47°40'$$

$$h = y - x = 50(\tan 47°40' - \tan 35°).$$

$$h \approx 19.9 \text{ feet}$$

25. $\cot 55 = \dfrac{d}{10} \Rightarrow d \approx 7$ kilometers

$\cot 28° = \dfrac{D}{10} \Rightarrow D \approx 18.8$ kilometers

Distance between towns:

$D - d = 18.8 - 7 = 11.8$ kilometers

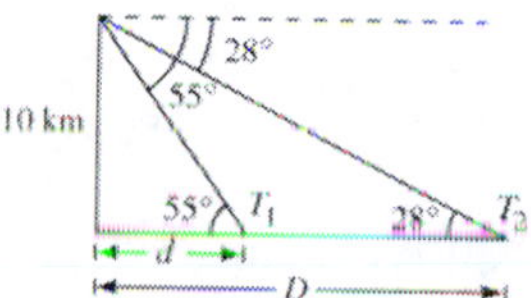

27. $\tan\theta = \frac{75}{50}$

$$\theta = \arctan\tfrac{3}{2} \approx 56.3°$$

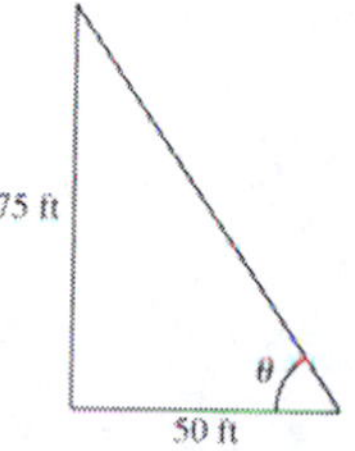

29. 1200 feet + 150 feet − 400 feet = 950 feet

$$5 \text{ miles} = 5 \text{ miles}\left(\frac{5280 \text{ feet}}{1 \text{ mile}}\right) = 26{,}400 \text{ feet}$$

$$\tan\theta = \frac{950}{26{,}400}$$

$$\theta = \arctan\left(\frac{950}{26{,}400}\right) \approx 2.06°$$

Not drawn to scale

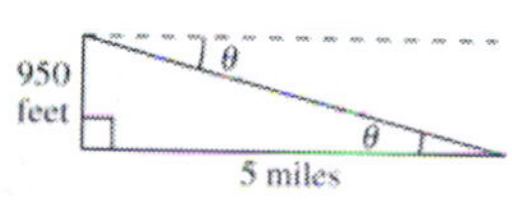

31. (a) $l^2 = (h + 17)^2 + 100^2$

$$l = \sqrt{(h + 17)^2 + 10{,}000}$$

$$= \sqrt{h^2 + 34h + 10{,}289}$$

(b) $\cos\theta = \dfrac{100}{l}$

$$\theta = \arccos\left(\frac{100}{l}\right)$$

(c) $\cos\theta = \dfrac{100}{l}$

$$\cos 35° = \frac{100}{l}$$

$$l \approx 122.077$$

$$l^2 = 100^2 + (h + 17)^2$$

$$l^2 = h^2 + 34h + 10.289$$

$$0 = h^2 + 34h - 4613.794$$

$$h \approx 53.02 \text{ feet}$$

33. (a) $l^2 = (200)^2 + (150)^2$

$l = 250$ feet

$\tan A = \frac{150}{200} \Rightarrow A = \arctan\left(\frac{150}{200}\right) \approx 36.87°$

$\tan B = \frac{200}{150} \Rightarrow B = \arctan\left(\frac{200}{150}\right) \approx 53.13°$

(b) $250 \text{ ft} \times \frac{\text{mile}}{5280 \text{ ft}} \times \frac{\text{hour}}{35 \text{ miles}} \times \frac{3600 \text{ sec}}{\text{hour}} \approx 4.87$ seconds

35. The plane has traveled $1.5(600) = 900$ miles.

$\sin 38° = \frac{a}{900} \Rightarrow a \approx 554$ miles north

$\cos 38° = \frac{b}{900} \Rightarrow b \approx 709$ miles east

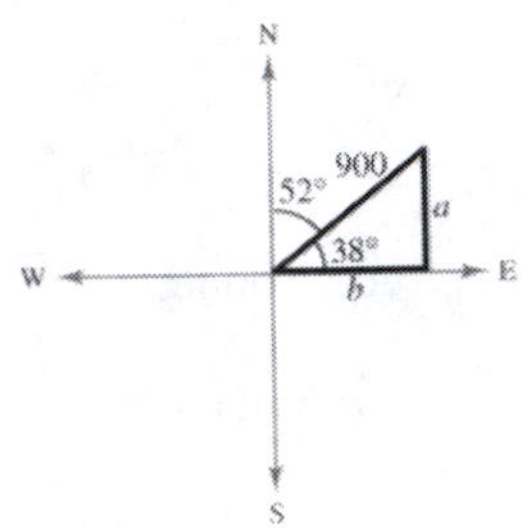

37.

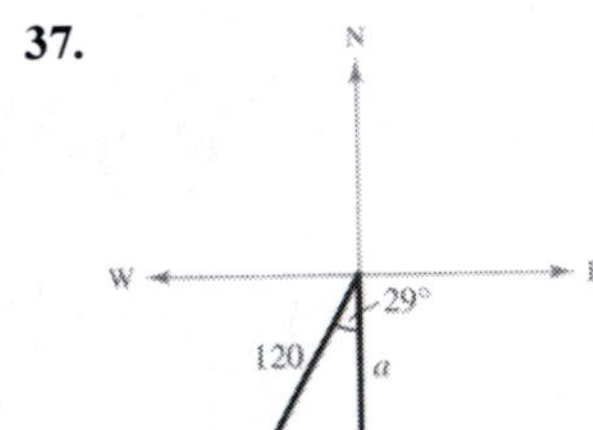

(a) $\cos 29° = \frac{a}{120} \Rightarrow a \approx 104.95$ nautical miles south

$\sin 29° = \frac{b}{120} \Rightarrow b \approx 58.18$ nautical miles west

(b) $\tan \theta = \frac{20 + b}{a} \approx \frac{78.18}{104.95} \Rightarrow \theta \approx 36.7°$

Bearing: S 36.7° W

Distance: $d \approx \sqrt{104.95^2 + 78.18^2}$

≈ 130.9 nautical miles from port

39. $\tan \theta = \frac{45}{30} \Rightarrow \theta \approx 56.3°$

Bearing: N 56.31°

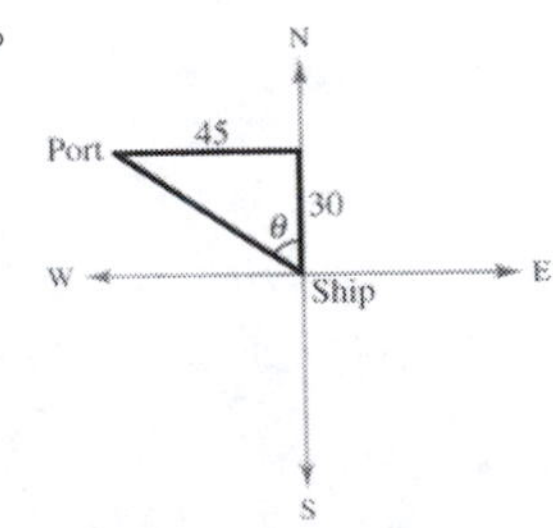

41. $\theta = 32°, \phi = 68°$

(a) $\alpha = 90° - 32° = 58°$

Bearing from A to C: N 58° E

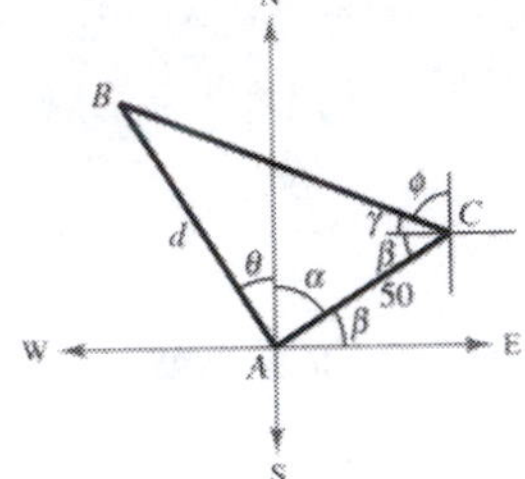

(b) $\beta = \theta = 32°$

$\gamma = 90° - \phi = 22°$

$C = \beta + \gamma = 54°$

$\tan C = \frac{d}{50} \Rightarrow \tan 54°$

$= \frac{d}{50} \Rightarrow d \approx 68.82$ meters

43. The diagonal of the base has a length of $\sqrt{a^2 + a^2} = \sqrt{2}a$. Now, you have

$\tan \theta = \frac{a}{\sqrt{2}a} = \frac{1}{\sqrt{2}}$

$\theta = \arctan \frac{1}{\sqrt{2}}$

$\theta \approx 35.3°.$

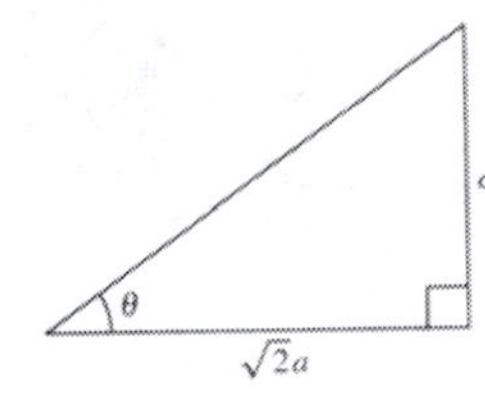

45. $\sin 36° = \frac{d}{25} \Rightarrow d \approx 14.69$

Length of side: $2d \approx 29.4$ inches

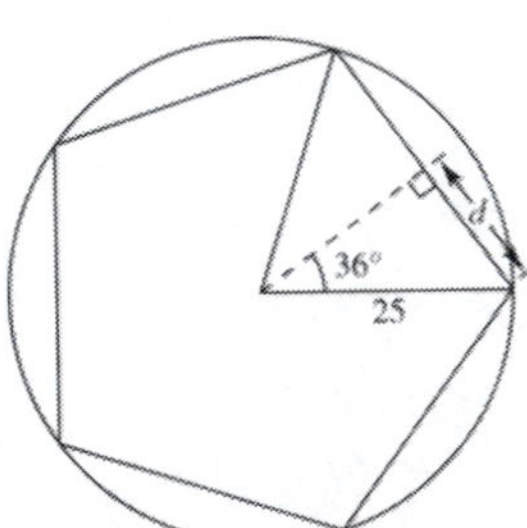

47. Use $d = a \sin \omega t$ because $d = 0$ when $t = 0$.

Period: $\frac{2\pi}{\omega} = 2 \Rightarrow \omega = \pi$

So, $d = 4 \sin(\pi t)$.

49. Use $d = a\cos\omega t$ because $d = 3$ when $t = 0$.

Period: $\dfrac{2\pi}{\omega} = 1.5 \Rightarrow \omega = \dfrac{4\pi}{3}$

So, $d = 3\cos\left(\dfrac{4\pi}{3}t\right) = 3\cos\left(\dfrac{4\pi t}{3}\right)$.

51.

$$d = a\sin\omega t$$

$$\text{Frequency} = \frac{\omega}{2\pi}$$

$$264 = \frac{\omega}{2\pi}$$

$$\omega = 2\pi(264) = 528\pi$$

53. $d = 9\cos\dfrac{6\pi}{5}t$

(a) Maximum displacement = amplitude = 9

(b) $\text{Frequency} = \dfrac{\omega}{2\pi} = \dfrac{\frac{6\pi}{5}}{2\pi}$

$= \dfrac{3}{5}$ cycle per unit of time

(c) $d = 9\cos\dfrac{6\pi}{5}(5) = 9$

(d) $9\cos\dfrac{6\pi}{5}t = 0$

$$\cos\frac{6\pi}{5}t = 0$$

$$\frac{6\pi}{5}t = \arccos 0$$

$$\frac{6\pi}{5}t = \frac{\pi}{2}$$

$$t = \frac{5}{12}$$

55. $d = \dfrac{1}{4}\sin 6\pi t$

(a) Maximum displacement = amplitude = $\dfrac{1}{4}$

(b) $\text{Frequency} = \dfrac{\omega}{2\pi} = \dfrac{6\pi}{2\pi}$

$= 3$ cycles per unit of time

(c) $d = \dfrac{1}{4}\sin 30\pi \approx 0$

(d) $\dfrac{1}{4}\sin 6\pi t = 0$

$$\sin 6\pi t = 0$$

$$6\pi t = \arcsin 0$$

$$6\pi t = \pi$$

$$t = \frac{1}{6}$$

57. $y = \dfrac{1}{4}\cos 16t,\ t > 0$

(a)

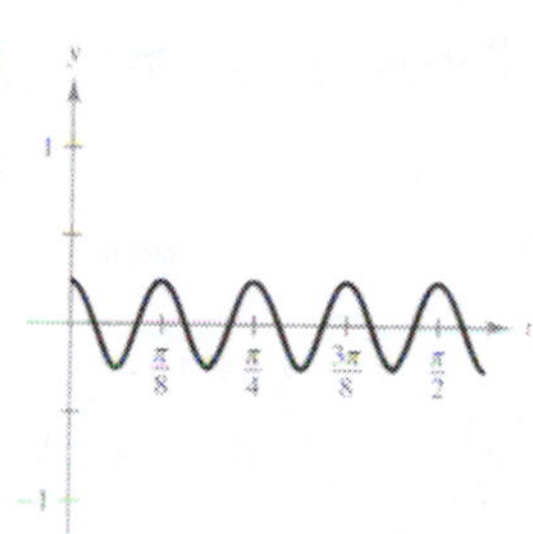

(b) Period: $\dfrac{2\pi}{16} = \dfrac{\pi}{8}$

(c) $\dfrac{1}{4}\cos 16t = 0$ when $16t = \dfrac{\pi}{2} \Rightarrow t = \dfrac{\pi}{32}$

59. (a)

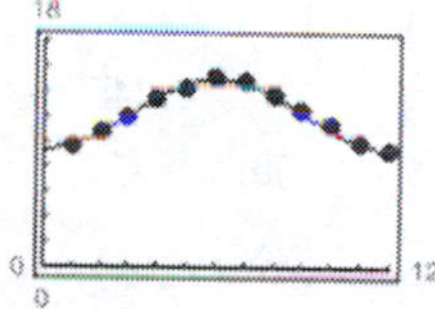

(b) $\text{Period} = \dfrac{2\pi}{n} = \dfrac{2\pi}{\pi/6} = 12$

The period is what you expect as the model examines the number of hours of daylight over one year (12 months).

(c) Amplitude $= |2.77| = 2.77$

The amplitude represents the maximum displacement from the average number of hours of daylight.

61. False. The tower isn't vertical and so the triangle formed is not a right triangle.

Review Exercises for Chapter 6

1.  The angle shown is 60°.

3. 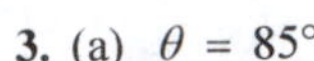 (a) $\theta = 85°$

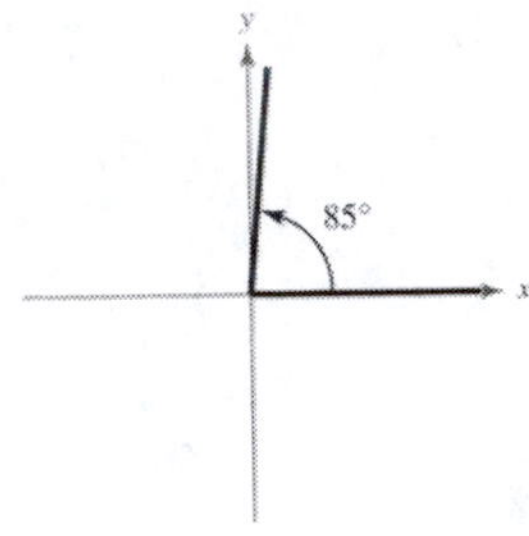

(b) The angle lies in Quadrant I.

(c) Coterminal angles: $85° + 360° = 445°$

$85° - 360° = -275°$

5. $\theta = -110°$

(a)

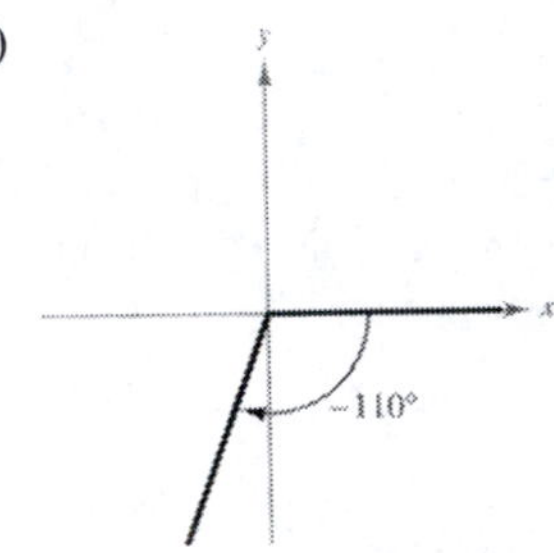

(b) Quadrant III

(c) Coterminal angles:

$-110° + 360° = 250°$

$-110° - 360° = -470°$

7. $\theta = \dfrac{15\pi}{4}$

(a)

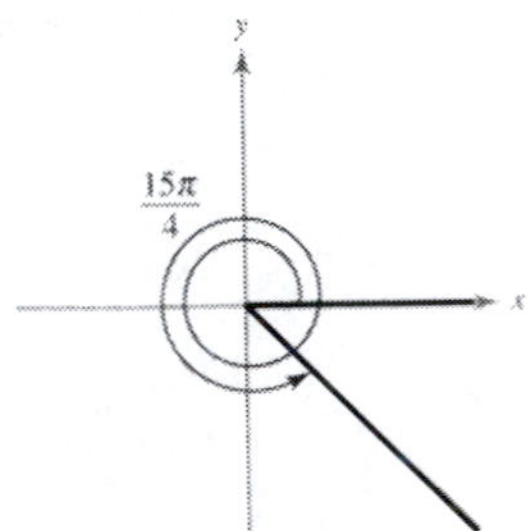

(b) Quadrant IV

(c) $\dfrac{15\pi}{4} - 2\pi = \dfrac{7\pi}{4}$

$\dfrac{7\pi}{4} - 2\pi = -\dfrac{\pi}{4}$

9. $\theta = -\dfrac{4\pi}{3}$

(a)

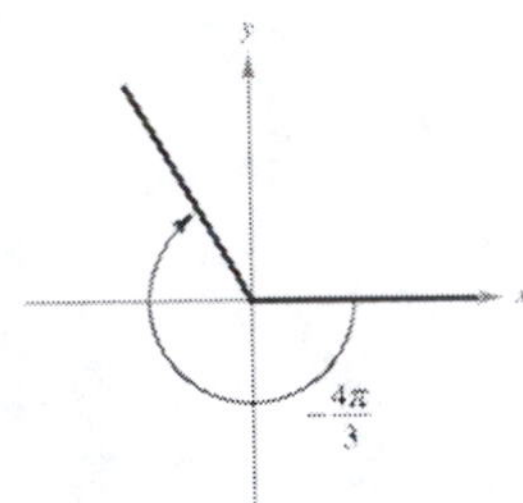

(b) Quadrant II

(c) Coterminal angles:

$-\dfrac{4\pi}{3} + 2\pi = \dfrac{2\pi}{3}$

$-\dfrac{4\pi}{3} - 2\pi = -\dfrac{10\pi}{3}$

11. $450° = 450° \cdot \dfrac{\pi \text{ rad}}{180°} = \dfrac{5\pi}{2} \approx 7.854 \text{ rad}$

13. $-16.5° = -16.5° \cdot \dfrac{\pi \text{ rad}}{180°} \approx -0.288 \text{ rad}$

15. $-33°45' = -33.75° = -33.75° \cdot \dfrac{\pi \text{ rad}}{180°}$

$= -\dfrac{3\pi}{16} \text{ rad} \approx -0.589 \text{ rad}$

17. $-84°15' = 84.25° = 84.25° \cdot \dfrac{\pi \text{ rad}}{180°} \approx 1.470 \text{ rad}$

19. $\dfrac{3\pi}{10} = \dfrac{3\pi}{10} \cdot \dfrac{180°}{\pi \text{ rad}} = 54.000°$

21. $-\dfrac{3\pi}{5} \text{ rad} = -\dfrac{3\pi \text{ rad}}{5} \cdot \dfrac{180°}{\pi \text{ rad}} = -108°$

23. $-3.5 \text{ rad} = -3.5 \text{ rad} \cdot \dfrac{180°}{\pi \text{ rad}} \approx -200.535°$

25. $4.75 \text{ rad} = \dfrac{4.75}{1} \text{ rad} \cdot \dfrac{180°}{\pi \text{ rad}} \approx 272.155°$

27. $138° = \dfrac{138\pi}{180} = \dfrac{23\pi}{30}$ radians

$s = r\theta = 20\left(\dfrac{23\pi}{30}\right) \approx 48.17$ inches

29. $120° = \dfrac{120\pi}{180} = \dfrac{2\pi}{3}$ radians

$A = \dfrac{1}{2}r^2\theta = \dfrac{1}{2}(18)^2\left(\dfrac{2\pi}{3}\right) \approx 339.29$ square inches

31. opp = 4, adj = 5, hyp = $\sqrt{4^2 + 5^2} = \sqrt{41}$

$\sin\theta = \dfrac{\text{opp}}{\text{hyp}} = \dfrac{4}{\sqrt{41}} = \dfrac{4\sqrt{41}}{41}$ $\qquad \csc\theta = \dfrac{\text{hyp}}{\text{opp}} = \dfrac{\sqrt{41}}{4}$

$\cos\theta = \dfrac{\text{adj}}{\text{hyp}} = \dfrac{5}{\sqrt{41}} = \dfrac{5\sqrt{41}}{41}$ $\qquad \sec\theta = \dfrac{\text{hyp}}{\text{adj}} = \dfrac{\sqrt{41}}{5}$

$\tan\theta = \dfrac{\text{opp}}{\text{adj}} = \dfrac{4}{5}$ $\qquad \cot\theta = \dfrac{\text{adj}}{\text{opp}} = \dfrac{5}{4}$

33. $\tan 41^\circ \approx 0.8693$

35. $\cos 38.9^\circ \approx 0.7782$

37. $\cot 25^\circ 13' \approx \cot 25.2167^\circ = \dfrac{1}{\tan 25.2167^\circ} \approx 2.1235$

39. $\cos\dfrac{\pi}{18} \approx 0.9848$

41. $\sin\theta = \dfrac{1}{3}$

(a) $\csc\theta = \dfrac{1}{\sin\theta} = 3$

(b) $\sin^2\theta + \cos^2\theta = 1$

$$\left(\frac{1}{3}\right)^2 + \cos^2\theta = 1$$

$$\cos^2\theta = 1 - \frac{1}{9}$$

$$\cos^2\theta = \frac{8}{9}$$

$$\cos\theta = \sqrt{\frac{8}{9}}$$

$$\cos\theta = \frac{2\sqrt{2}}{3}$$

(c) $\sec\theta = \dfrac{1}{\cos\theta} = \dfrac{3}{2\sqrt{2}} = \dfrac{3\sqrt{2}}{4}$

(d) $\tan\theta = \dfrac{\sin\theta}{\cos\theta} = \dfrac{1/3}{\left(2\sqrt{2}\right)/3} = \dfrac{1}{2\sqrt{2}} = \dfrac{\sqrt{2}}{4}$

43. $\csc\theta = 4$

(a) $\sin\theta = \dfrac{1}{\csc\theta} = \dfrac{1}{4}$

(b) $\sin^2\theta + \cos^2\theta = 1$

$$\left(\frac{1}{4}\right)^2 + \cos^2\theta = 1$$

$$\cos^2\theta = 1 - \frac{1}{16}$$

$$\cos^2\theta = \frac{15}{16}$$

$$\cos\theta = \sqrt{\frac{15}{16}}$$

$$\cos\theta = \frac{\sqrt{15}}{4}$$

(c) $\sec\theta = \dfrac{1}{\cos\theta} = \dfrac{4}{\sqrt{15}} = \dfrac{4\sqrt{15}}{15}$

(d) $\tan\theta = \dfrac{\sin\theta}{\cos\theta} = \dfrac{1/4}{\sqrt{15}/4} = \dfrac{1}{\sqrt{15}} = \dfrac{\sqrt{15}}{15}$

45. $\sin 1^\circ 10' = \dfrac{x}{3.5}$

$x = 3.5 \sin 1^\circ 10' \approx 0.07$ kilometer or 71.3 meters

3.5 km
1° 10′
x

Not drawn to scale

47. $x = 12,\ y = 16,\ r = \sqrt{144 + 256} = \sqrt{400} = 20$

$\sin\theta = \dfrac{y}{r} = \dfrac{4}{5}$ $\qquad \csc\theta = \dfrac{r}{y} = \dfrac{5}{4}$

$\cos\theta = \dfrac{x}{r} = \dfrac{3}{5}$ $\qquad \sec\theta = \dfrac{r}{x} = \dfrac{5}{3}$

$\tan\theta = \dfrac{y}{x} = \dfrac{4}{3}$ $\qquad \cot\theta = \dfrac{x}{y} = \dfrac{3}{4}$

49. $x = \frac{2}{3}, y = \frac{5}{2}$

$$r = \sqrt{\left(\frac{2}{3}\right)^2 + \left(\frac{5}{2}\right)^2} = \frac{\sqrt{241}}{6}$$

$$\sin\theta = \frac{y}{r} = \frac{5/2}{\sqrt{241}/6} = \frac{15}{\sqrt{241}} = \frac{15\sqrt{241}}{241} \qquad \csc\theta = \frac{r}{y} = \frac{\sqrt{241}/6}{5/2} = \frac{2\sqrt{241}}{30} = \frac{\sqrt{241}}{15}$$

$$\cos\theta = \frac{x}{r} = \frac{2/3}{\sqrt{241}/6} = \frac{4}{\sqrt{241}} = \frac{4\sqrt{241}}{241} \qquad \sec\theta = \frac{r}{x} = \frac{\sqrt{241}/6}{2/3} = \frac{\sqrt{241}}{4}$$

$$\tan\theta = \frac{y}{x} = \frac{5/2}{2/3} = \frac{15}{4} \qquad \cot\theta = \frac{x}{y} = \frac{2/3}{5/2} = \frac{4}{15}$$

51. $x = -0.5, y = 4.5$

$$r = \sqrt{(-0.5)^2 + (4.5)^2} = \sqrt{20.5} = \frac{\sqrt{82}}{2}$$

$$\sin\theta = \frac{y}{r} = \frac{4.5}{\sqrt{82}/2} = \frac{9\sqrt{82}}{82} \qquad \csc\theta = \frac{r}{y} = \frac{\sqrt{82}/2}{4.5} = \frac{\sqrt{82}}{9}$$

$$\cos\theta = \frac{x}{r} = \frac{-0.5}{\sqrt{82}/2} = \frac{-\sqrt{82}}{82} \qquad \sec\theta = \frac{r}{x} = \frac{\sqrt{82}/2}{-0.5} = -\sqrt{82}$$

$$\tan\theta = \frac{y}{x} = \frac{4.5}{-0.5} = -9 \qquad \cot\theta = \frac{x}{y} = \frac{-0.5}{4.5} = -\frac{1}{9}$$

53. $(x, 4x), x > 0$

$x' = x, y' = 4x$

$$r = \sqrt{x^2 + (4x)^2} = \sqrt{17}x$$

$$\sin\theta = \frac{y'}{r} = \frac{4x}{\sqrt{17}x} = \frac{4\sqrt{17}}{17} \qquad \csc\theta = \frac{r}{y'} = \frac{\sqrt{17}x}{4x} = \frac{\sqrt{17}}{4}$$

$$\cos\theta = \frac{x'}{r} = \frac{x}{\sqrt{17}x} = \frac{\sqrt{17}}{17} \qquad \sec\theta = \frac{r}{x'} = \frac{\sqrt{17}x}{x} = \sqrt{17}$$

$$\tan\theta = \frac{y'}{x'} = \frac{4x}{x} = 4 \qquad \cot\theta = \frac{x'}{y'} = \frac{x}{4x} = \frac{1}{4}$$

55. $\sec\theta = \frac{6}{5}, \tan\theta < 0 \Rightarrow \theta$ is in Quadrant IV.

$r = 6, x = 5, y = -\sqrt{36 - 25} = -\sqrt{11}$

$$\sin\theta = \frac{y}{r} = -\frac{\sqrt{11}}{6}$$

$$\cos\theta = \frac{x}{r} = \frac{5}{6}$$

$$\tan\theta = \frac{y}{x} = -\frac{\sqrt{11}}{5}$$

$$\csc\theta = \frac{r}{y} = -\frac{6\sqrt{11}}{11}$$

$$\sec\theta = \frac{6}{5}$$

$$\cot\theta = -\frac{5\sqrt{11}}{11}$$

57. $\tan\theta = \frac{7}{3}, \cos\theta < 0 \Rightarrow \theta$ is in Quadrant III.

$y = -7, x = -3, r = \sqrt{58}$

$$\sin\theta = \frac{y}{r} = -\frac{7\sqrt{58}}{58}$$

$$\cos\theta = \frac{x}{r} = -\frac{3\sqrt{58}}{58}$$

$$\tan\theta = \frac{y}{x} = \frac{7}{3}$$

$$\csc\theta = \frac{r}{y} = -\frac{\sqrt{58}}{7}$$

$$\sec\theta = \frac{r}{x} = -\frac{\sqrt{58}}{3}$$

$$\cot\theta = \frac{x}{y} = \frac{3}{7}$$

59. $\tan\theta = \frac{y}{x} = -\frac{40}{9} \Rightarrow r = 41$

$\sin\theta > 0 \Rightarrow \theta$ is in Quadrant II $\Rightarrow x = -9, y = 40$

$\sin\theta = \frac{y}{r} = \frac{40}{41}$

$\cos\theta = \frac{x}{r} = -\frac{9}{41}$

$\tan\theta = \frac{y}{x} = -\frac{40}{9}$

$\csc\theta = \frac{r}{y} = \frac{41}{40}$

$\sec\theta = \frac{r}{x} = -\frac{41}{9}$

$\cot\theta = \frac{x}{y} = -\frac{9}{40}$

61. $\theta = 264°$

$\theta' = 264° - 180° = 84°$

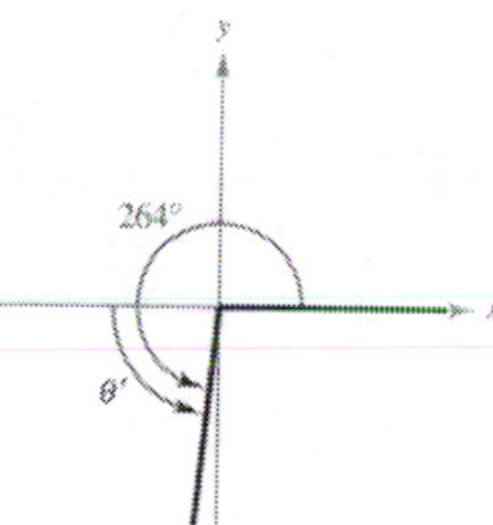

63. $\theta = -\frac{6\pi}{5}$

$-\frac{6\pi}{5} + 2\pi = \frac{4\pi}{5}$

$\theta' = \pi - \frac{4\pi}{5} = \frac{\pi}{5}$

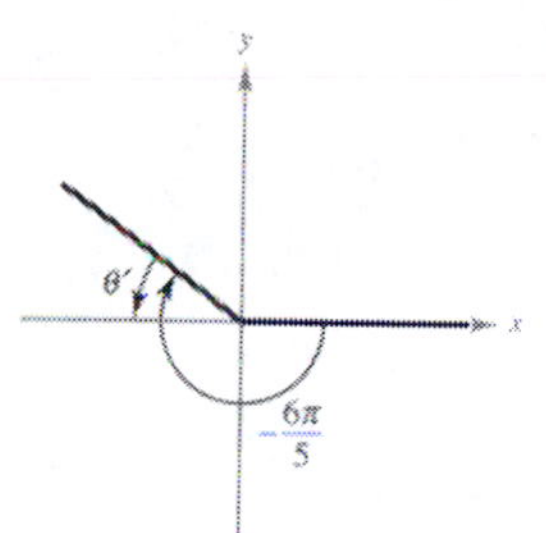

65. $\sin\frac{\pi}{3} = \frac{\sqrt{3}}{2}$

$\cos\frac{\pi}{3} = \frac{1}{2}$

$\tan\frac{\pi}{3} = \sqrt{3}$

67. $\sin\frac{5\pi}{6} = \sin\left(\pi - \frac{5\pi}{6}\right) = \sin\frac{\pi}{6} = \frac{1}{2}$

$\cos\frac{5\pi}{6} = -\cos\left(\pi - \frac{5\pi}{6}\right) = -\cos\frac{\pi}{6} = -\frac{\sqrt{3}}{2}$

$\tan\frac{5\pi}{6} = -\tan\left(\pi - \frac{5\pi}{6}\right) = -\tan\frac{\pi}{6} = -\frac{\sqrt{3}}{3}$

69. $\sin\left(-\frac{7\pi}{3}\right) = -\sin\frac{\pi}{3} = -\frac{\sqrt{3}}{2}$

$\cos\left(-\frac{7\pi}{3}\right) = \cos\frac{\pi}{3} = \frac{1}{2}$

$\tan\left(-\frac{7\pi}{3}\right) = -\tan\frac{\pi}{3} = -\sqrt{3}$

71. $\sin 495° = \sin 45° = \frac{\sqrt{2}}{2}$

$\cos 495° = -\cos 45° = -\frac{\sqrt{2}}{2}$

$\tan 495° = -\tan 45° = -1$

73. $\sin(-150°) = -\frac{1}{2}$

$\cos(-150°) = -\frac{\sqrt{3}}{2}$

$\tan(-150°) = \frac{-1/2}{-\sqrt{3}/2} = \frac{\sqrt{3}}{3}$

75. $\sin 10 \approx -0.5440$

77. $\sec 2.8 = \frac{1}{\cos 2.8} \approx -1.0613$

79. $\sin\left(-\frac{17\pi}{15}\right) \approx 0.4067$

81. $t = \frac{2\pi}{3}, (x, y) = \left(-\frac{1}{2}, \frac{\sqrt{3}}{2}\right)$

$\sin\frac{2\pi}{3} = y = \frac{\sqrt{3}}{2}$

$\cos\frac{2\pi}{3} = x = -\frac{1}{2}$

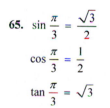

$\tan\frac{2\pi}{3} = \frac{y}{x} = \frac{\sqrt{3}/2}{-1/2} = -\sqrt{3}$

83. $t = \frac{7\pi}{6}, (x, y) = \left(-\frac{\sqrt{3}}{2}, -\frac{1}{2}\right)$

$\sin\left(\frac{7\pi}{6}\right) = y = -\frac{1}{2}$

$\cos\left(\frac{7\pi}{6}\right) = x = -\frac{\sqrt{3}}{2}$

$\tan\left(\frac{7\pi}{6}\right) = \frac{x}{y} = \frac{-1/2}{-\sqrt{3}/2} = \frac{1}{\sqrt{3}} = \frac{\sqrt{3}}{3}$

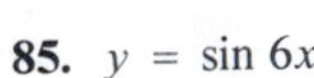

85. $y = \sin 6x$

Amplitude: 1

Period: $\frac{2\pi}{6} = \frac{\pi}{3}$

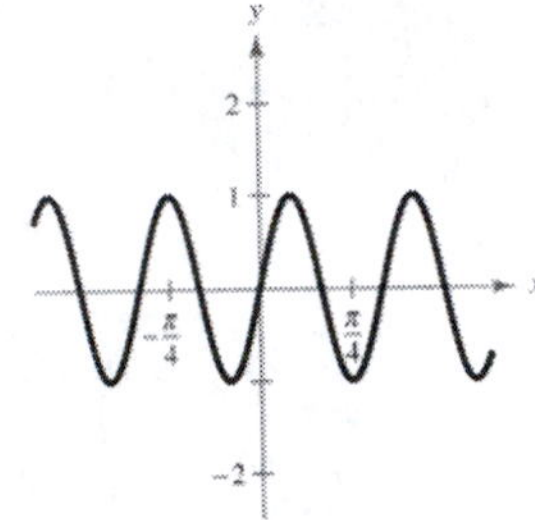

87. $y = 3\cos 2\pi x$

Amplitude: 3

Period: $\frac{2\pi}{2\pi} = 1$

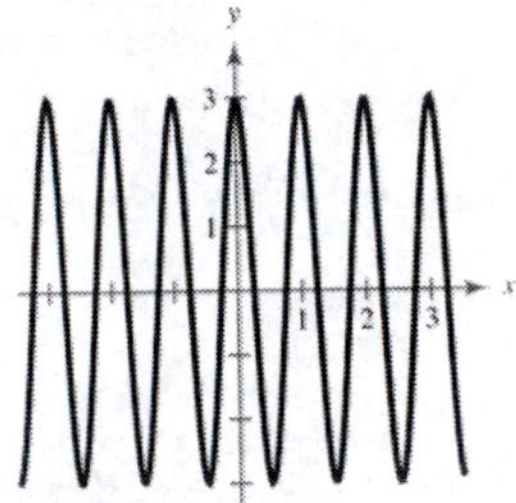

89. $f(x) = 5 \sin \frac{2x}{5}$

Amplitude: 5

Period: $\frac{2\pi}{2/5} = 5\pi$

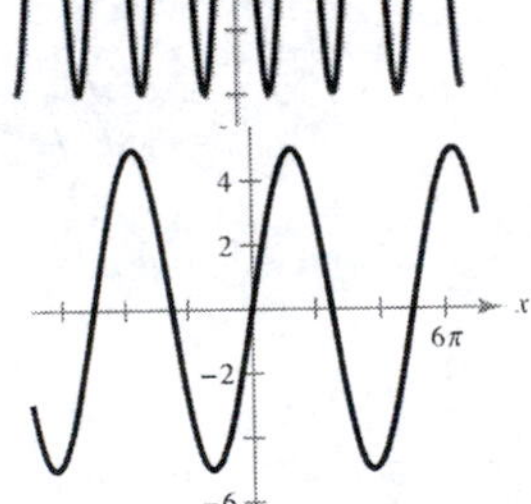

91. $y = 5 + \sin x$

Amplitude: 1

Period: 2π

Shift the graph of $y = \sin x$ 5 units upward

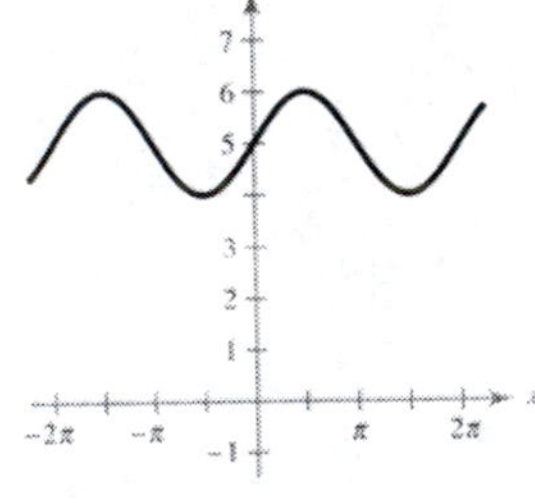

93. $g(t) = \frac{5}{2}\sin(t - \pi)$

Amplitude: $\frac{5}{2}$

Period: 2π

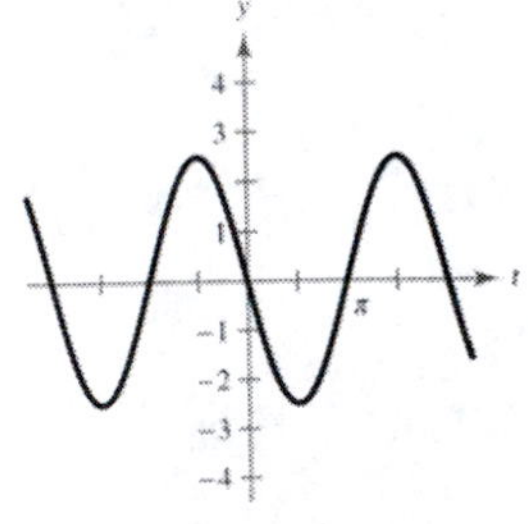

95. $y = a \sin bx$

(a) $a = 2$,

$\frac{2\pi}{b} = \frac{1}{264} \Rightarrow b = 528\pi$

$y = 2 \sin 528\pi x$

(b) $f = \frac{1}{1/264} = 264$ cycles per second

97. $f(x) = 3 \tan 2x$

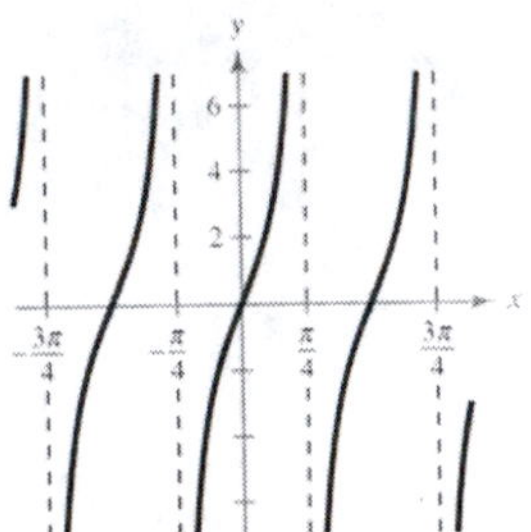

99. $f(x) = \frac{1}{2}\cot x$

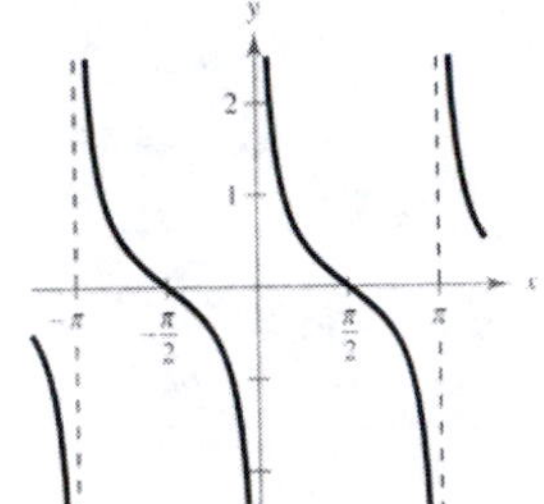

101. $f(x) = 3 \sec x$

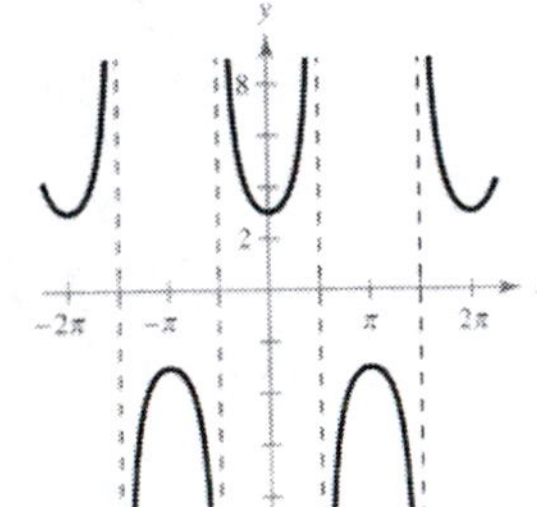

103. $f(x) = \frac{1}{2}\csc\frac{x}{2}$

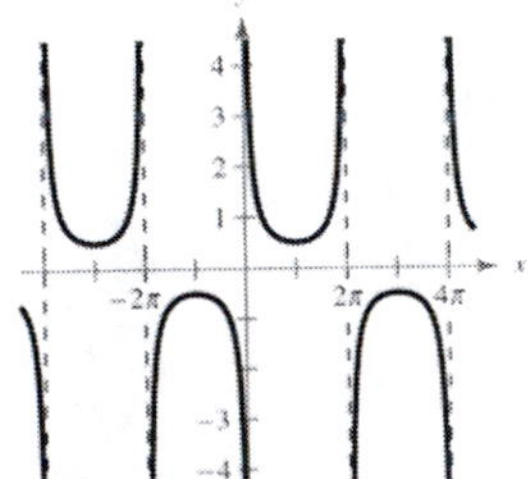

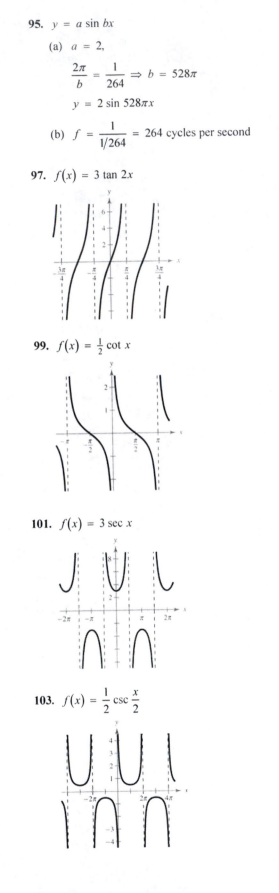

105. $f(x) = x \cos x$

Damping factor: x

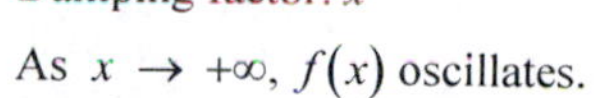

As $x \to +\infty$, $f(x)$ oscillates.

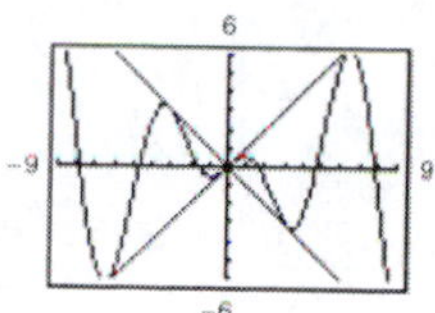

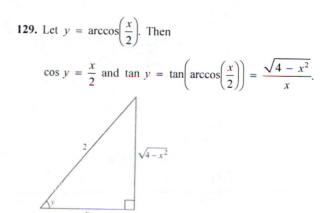

107. $\arcsin\left(-\frac{1}{2}\right) = -\arcsin\frac{1}{2} = -\frac{\pi}{6}$

109. $\arccos\left(-\frac{\sqrt{2}}{2}\right) = \frac{3\pi}{4}$

111. $\cos^{-1}(-1) = \pi$

113. $\arcsin 0.4 \approx 0.41$ radian

115. $\sin^{-1}(-0.44) \approx -0.46$ radian

117. $\arccos 0.425 \approx 1.13$ radians

119. $\tan^{-1}(-1.5) \approx -0.98$ radian

121. $f(x) = 2\arcsin\frac{x}{2} = 2\sin^{-1}\left(\frac{x}{2}\right)$

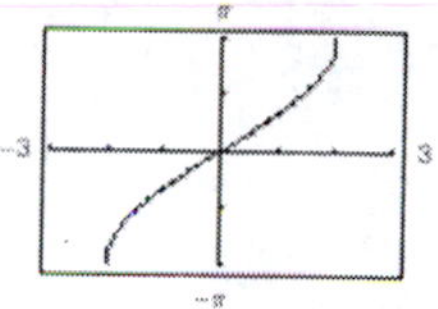

123. $f(x) = \arctan\left(\frac{x}{2}\right) = \tan^{-1}\left(\frac{x}{2}\right)$

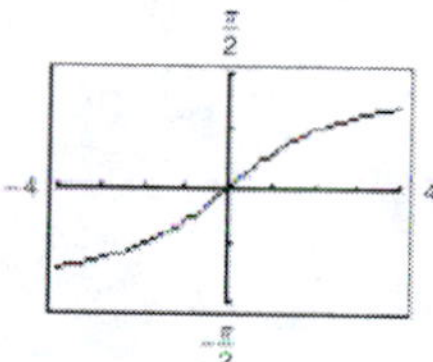

125. Let $u = \arctan\frac{3}{4}$ then $\tan u = \frac{3}{4}$.

$\cos\left(\arctan\frac{3}{4}\right) = \frac{4}{5}$

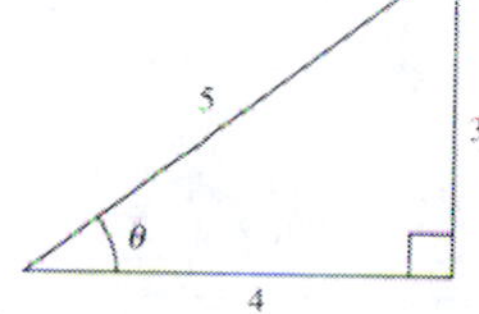

127. Let $u = \arctan\frac{12}{5}$

then $\tan u = \frac{12}{5}$.

$\sec\left(\arctan\frac{12}{5}\right) = \frac{13}{5}$

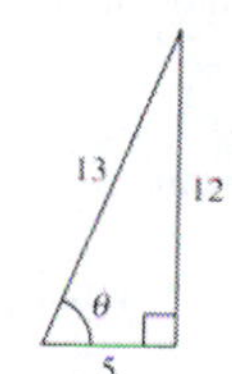

129. Let $y = \arccos\left(\frac{x}{2}\right)$. Then

$\cos y = \frac{x}{2}$ and $\tan y = \tan\left(\arccos\left(\frac{x}{2}\right)\right) = \frac{\sqrt{4 - x^2}}{x}$.

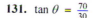

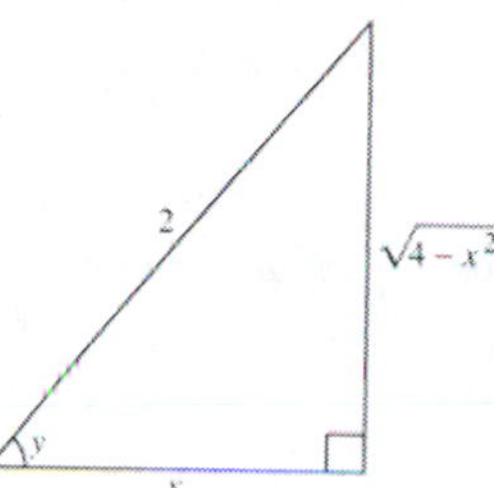

131. $\tan\theta = \frac{70}{30}$

$\theta = \arctan\left(\frac{70}{30}\right) \approx 66.8°$

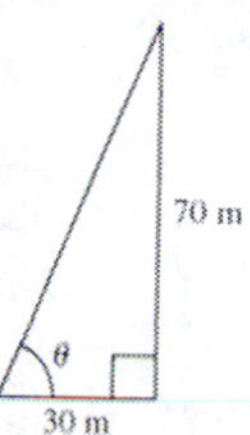

133. $\cos 45° = \frac{a}{60} \Rightarrow$

$a \approx 42.43$ nautical miles north

$\sin 45° = \frac{b}{60} \Rightarrow$

$b \approx 42.43$ nautical miles east

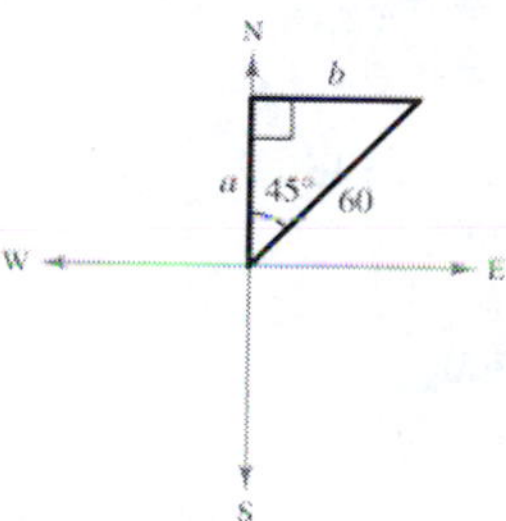

135. False. For each θ there corresponds exactly one value of y.

137. $f(\theta) = \sec\theta$ is undefined at the zeros of $g(\theta) = \cos\theta$ because $\sec\theta = \frac{1}{\cos\theta}$.

139. The ranges for the other four trigonometric functions are not bounded. For $y = \tan x$ and $y = \cot x$, the range is $(-\infty, \infty)$. For $y = \sec x$ and $y = \csc x$, the range is $(-\infty, -1] \cup [1, \infty)$.

141. (a)

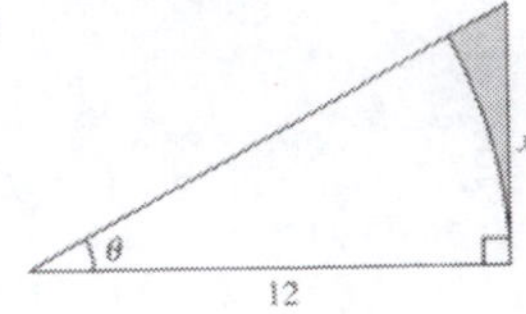

$$\tan\theta = \frac{x}{12}$$

$$x = 12\tan\theta$$

$$\text{Area} = \text{Area of triangle} - \text{Area of sector}$$

$$= \left(\frac{1}{2}bh\right) - \left(\frac{1}{2}r^2\theta\right)$$

$$= \frac{1}{2}(12)(12\tan\theta) - \frac{1}{2}(12^2)(\theta)$$

$$= 72\tan\theta - 72\theta$$

$$= 72(\tan\theta - \theta)$$

(b)

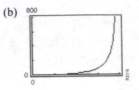

800
0
0
$\frac{\pi}{2}$

As θ approaches $\frac{\pi}{2}$, the area increases without bound.

Problem Solving for Chapter 6

1. (a) 8:57 − 6:45 = 2 hours 12 minutes = 132 minutes

$$\frac{132}{48} = \frac{11}{4} \text{ revolutions}$$

$$\theta = \left(\frac{11}{4}\right)(2\pi) = \frac{11\pi}{2} \text{ radians or } 990^\circ$$

(b) $s = r\theta = 47.25(5.5\pi) \approx 816.42$ feet

3. (a) $\sin 39^\circ = \dfrac{3000}{d}$

$$d = \frac{3000}{\sin 39^\circ} \approx 4767 \text{ feet}$$

(b) $\tan 39^\circ = \dfrac{3000}{x}$

$$x = \frac{3000}{\tan 39^\circ} \approx 3705 \text{ feet}$$

(c) $\tan 63^\circ = \dfrac{w + 3705}{3000}$

$$3000\tan 63^\circ = w + 3705$$

$$w = 3000\tan 63^\circ - 3705 \approx 2183 \text{ feet}$$

5. (a) 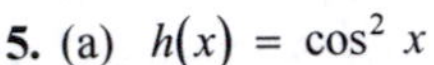 $h(x) = \cos^2 x$

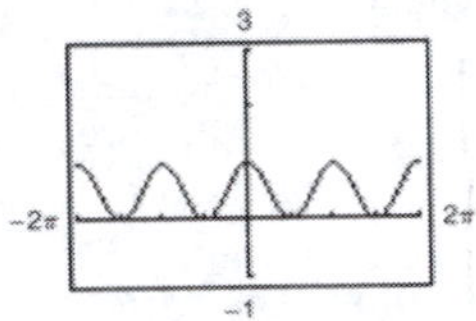

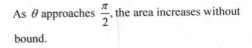

h is even.

(b) $h(x) = \sin^2 x$

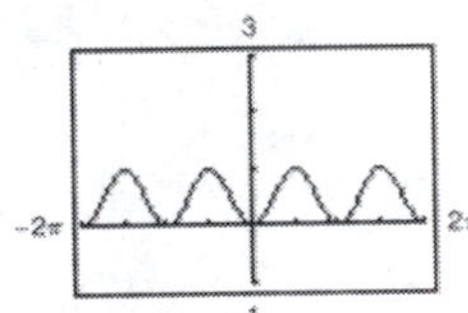

h is even.

7. If you alter the model so that $h = 1$ when $t = 0$, you can use either a sine or a cosine model.

$$a = \frac{1}{2}[\max - \min] = \frac{1}{2}[101 - 1] = 50$$

$$d = \frac{1}{2}[\max + \min] = \frac{1}{2}[101 + 1] = 51$$

$$b = 8\pi$$

Cosine model: $h = 51 - 50\cos(8\pi t)$

Sine model: $h = 51 - 50\sin\left(8\pi t + \dfrac{\pi}{2}\right)$

Notice that you needed the horizontal shift so that the sine value was one when $t = 0$.

Another model would be: $h = 51 + 50\sin\left(8\pi t + \dfrac{3\pi}{2}\right)$

Here you wanted the sine value to be 1 when $t = 0$.

9. $P = 100 - 20\cos\left(\frac{8\pi}{3}t\right)$

(a)

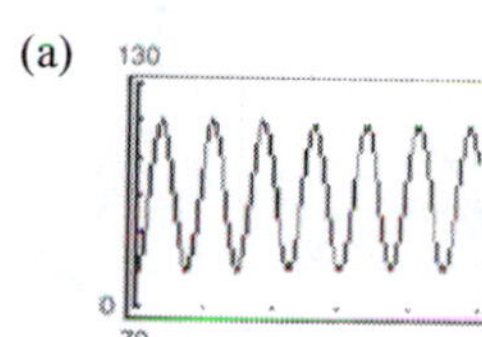

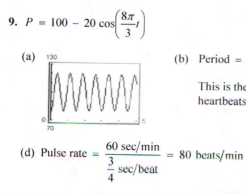

(b) Period $= \dfrac{2\pi}{8\pi/3} = \dfrac{6}{8} = \dfrac{3}{4}$ sec

This is the time between heartbeats.

(c) Amplitude: 20

The blood pressure ranges between $100 - 20 = 80$ and $100 + 20 = 120$.

(d) Pulse rate $= \dfrac{60 \text{ sec/min}}{\frac{3}{4} \text{ sec/beat}} = 80$ beats/min

(e) Period $= \dfrac{60}{64} = \dfrac{15}{16}$ sec

$$64 = \frac{60}{2\pi/b} \Rightarrow b = \frac{64}{60} \cdot 2\pi = \frac{32}{15}\pi$$

11. $f(x) = 2\cos 2x + 3\sin 3x$

$g(x) = 2\cos 2x + 3\sin 4x$

(a)

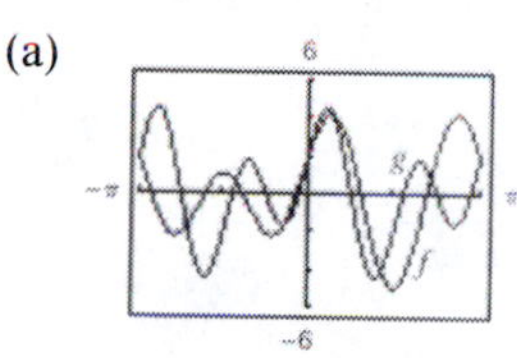

(b) The period of $f(x)$ is 2π.

The period of $g(x)$ is π.

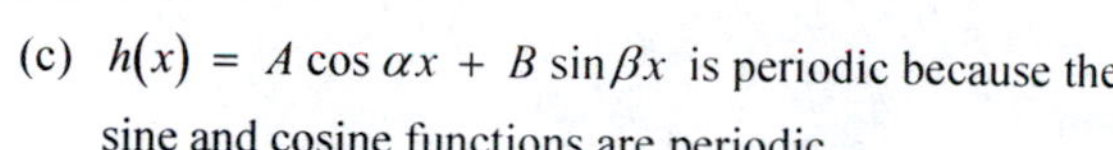

(c) $h(x) = A\cos\alpha x + B\sin\beta x$ is periodic because the sine and cosine functions are periodic.

13.

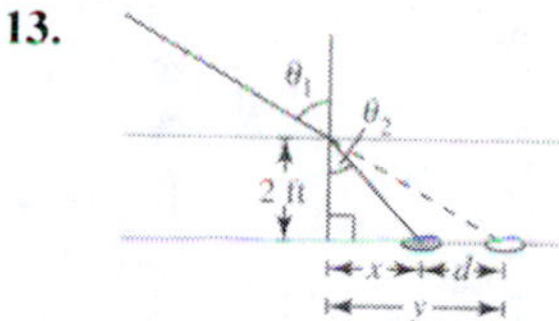

(a) $\dfrac{\sin\theta_1}{\sin\theta_2} = 1.333$

$$\sin\theta_2 = \frac{\sin\theta_1}{1.333} = \frac{\sin 60^\circ}{1.333} \approx 0.6497$$

$$\theta_2 = 40.5^\circ$$

(b) $\tan\theta_2 = \dfrac{x}{2} \Rightarrow x = 2\tan 40.52^\circ \approx 1.71$ feet

$\tan\theta_1 = \dfrac{y}{2} \Rightarrow y = 2\tan 60^\circ \approx 3.46$ feet

(c) $d = y - x = 3.46 - 1.71 = 1.75$ feet

(d) As you move closer to the rock, θ_1 decreases, which causes y to decrease, which in turn causes d to decrease.

Practice Test for Chapter 6

1. Express $350°$ in radian measure.

2. Express $(5\pi)/9$ in degree measure.

3. Convert $135° 14' 12''$ to decimal form.

4. Convert $-22.569°$ to $D° M' S''$ form.

5. If $\cos\theta = \frac{2}{3}$, use the trigonometric identities to find $\tan\theta$.

6. Find θ given $\sin\theta = 0.9063$.

7. Solve for x in the figure below.

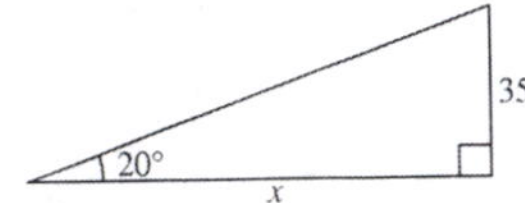

8. Find the magnitude of the reference angle for $\theta = (6\pi)/5$.

9. Evaluate $\csc 3.92$.

10. Find $\sec\theta$ given that θ lies in Quadrant III and $\tan\theta = 6$.

11. Graph $y = 3\sin\frac{x}{2}$.

12. Graph $y = -2\cos(x - \pi)$.

13. Graph $y = \tan 2x$.

14. Graph $y = -\csc\left(x + \frac{\pi}{4}\right)$.

15. Graph $y = 2x + \sin x$, using a graphing calculator.

16. Graph $y = 3x\cos x$, using a graphing calculator.

17. Evaluate $\arcsin 1$.

18. Evaluate $\arctan(-3)$.

19. Evaluate $\sin\left(\arccos\frac{4}{\sqrt{35}}\right)$.

20. Write an algebraic expression for $\cos\left(\arcsin\frac{x}{4}\right)$.

For Exercises 21–23, solve the right triangle.

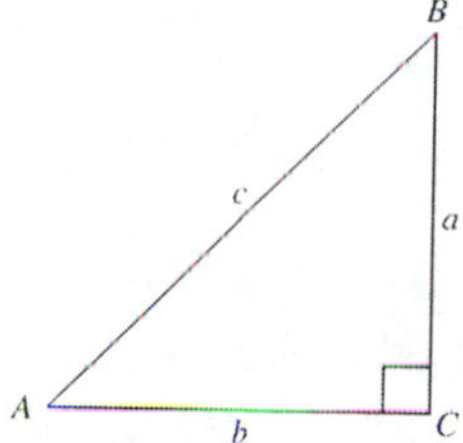

21. $A = 40°, c = 12$

22. $B = 6.84°, a = 21.3$

23. $a = 5, b = 9$

24. A 20-foot ladder leans against the side of a barn. Find the height of the top of the ladder if the angle of elevation of the ladder is 67°.

25. An observer in a lighthouse 250 feet above sea level spots a ship off the shore. If the angle of depression to the ship is 5°, how far out is the ship?

CHAPTER 7
Analytic Trigonometry

CHAPTER 7
Analytic Trigonometry

Section 7.1 Using Fundamental Identities

1. $\tan u$

3. $\cot u$

5. $\cot^2 u$

7. $\sin x = \frac{1}{2}, \cos x = \frac{\sqrt{3}}{2} \Rightarrow x$ is in Quadrant I.

$$\tan x = \frac{\sin x}{\cos x} = \frac{1/2}{\sqrt{3}/2} = \frac{1}{\sqrt{3}} = \frac{\sqrt{3}}{3}$$

$$\cot x = \frac{1}{\tan x} = \frac{1}{1/\sqrt{3}} = \sqrt{3}$$

$$\sec x = \frac{1}{\cos x} = \frac{1}{\sqrt{3}/2} = \frac{2}{\sqrt{3}} = \frac{2\sqrt{3}}{3}$$

$$\csc x = \frac{1}{\sin x} = \frac{1}{1/2} = 2$$

9. $\cos\left(\frac{\pi}{2} - x\right) = \frac{3}{5}, \cos x = \frac{4}{5} \Rightarrow x$ is in Quadrant I.

$$\sin x = \sqrt{1 - \left(\frac{4}{5}\right)^2} = \frac{3}{5}$$

$$\tan x = \frac{\sin x}{\cos x} = \frac{3}{5} \cdot \frac{5}{4} = \frac{3}{4}$$

$$\csc x = \frac{1}{\sin x} = \frac{5}{3}$$

$$\sec x = \frac{1}{\cos x} = \frac{5}{4}$$

$$\cot x = \frac{1}{\tan x} = \frac{4}{3}$$

11. $\sec x = 4, \sin x > 0 \Rightarrow x$ is in Quadrant I.

$$\cos x = \frac{1}{\sec x} = \frac{1}{4}$$

$$\sin x = \sqrt{1 - \left(\frac{1}{4}\right)^2} = \frac{\sqrt{15}}{4}$$

$$\tan x = \frac{\sin x}{\cos x} = \frac{\sqrt{15}}{4} \cdot \frac{4}{1} = \sqrt{15}$$

$$\csc x = \frac{1}{\sin x} = \frac{4}{\sqrt{15}} = \frac{4\sqrt{15}}{15}$$

$$\cot x = \frac{1}{\tan x} = \frac{1}{\sqrt{15}} = \frac{\sqrt{15}}{15}$$

13. $\sin\theta = -1, \cot\theta = 0 \Rightarrow \theta = \frac{3\pi}{2}$

$$\cos\theta = \sqrt{1 - \sin^2\theta} = 0$$

$\sec\theta$ is undefined.

$\tan\theta$ is undefined.

$\csc\theta = -1$

15. $\sec x \cos x = \left(\frac{1}{\cancel{\cos x}}\right)\cancel{\cos x}$

$= 1$

Matches (c).

16. $\cot^2 x - \csc^2 x = \left(\csc^2 x - 1\right) - \csc^2 x$

$= -1$

Matches (b).

17. $\sec^4 x - \tan^4 x = \left(\sec^2 x + \tan^2 x\right)\left(\sec^2 x - \tan^2 x\right)$

$= \left(\sec^2 x + \tan^2 x\right)(1)$

$= \sec^2 x + \tan^2 x$

Matches (f).

18. $\cot x \sec x = \frac{\cos x}{\sin x} \cdot \frac{1}{\cos x} = \frac{1}{\sin x} = \csc x$

Matches (a).

19. $\frac{\sec^2 x - 1}{\sin^2 x} = \frac{\tan^2 x}{\sin^2 x} = \frac{\sin^2 x}{\cos^2 x} \cdot \frac{1}{\sin^2 x} = \sec^2 x$

Matches (e).

20. $\frac{\cos^2\left[(\pi/2) - x\right]}{\cos x} = \frac{\sin^2 x}{\cos x} = \frac{\sin x}{\cos x}\sin x = \tan x \sin x$

Matches (d).

19. $\frac{\sec^2 x - 1}{\sin^2 x} = \frac{\tan^2 x}{\sin^2 x} = \frac{\sin^2 x}{\cos^2 x} \cdot \frac{1}{\sin^2 x} = \sec^2 x$

Matches (e).

21. $\tan^2 x - \tan^2 x \sin^2 x = \tan^2 x\left(1 - \sin^2 x\right)$

$= \tan^2 x \cos^2 x$

$= \frac{\sin^2 x}{\cos^2 x} \cdot \cos^2 x$

$= \sin^2 x$

23. $\dfrac{\sec^2 x - 1}{\sec x - 1} = \dfrac{(\sec x + 1)(\sec x - 1)}{\sec x - 1}$

$= \sec x + 1$

25. $1 - 2\cos^2 x + \cos^4 x = (1 - \cos^2 x)^2$

$= (\sin^2 x)^2$

$= \sin^4 x$

27. $\cot^3 x + \cot^2 x + \cot x + 1 = \cot^2 x(\cot x + 1) + (\cot x + 1)$

$= (\cot x + 1)(\cot^2 x + 1)$

$= (\cot x + 1)\csc^2 x$

29. $3\sin^2 x - 5\sin x - 2 = (3\sin x + 1)(\sin x - 2)$

31. $\cot^2 x + \csc x - 1 = (\csc^2 x - 1) + \csc x - 1$

$= \csc^2 x + \csc x - 2$

$= (\csc x - 1)(\csc x + 2)$

33. $(\sin x + \cos x)^2 = \sin^2 x + 2\sin x\cos x + \cos^2 x$

$= (\sin^2 x + \cos^2 x) + 2\sin x\cos x$

$= 1 + 2\sin x\cos x$

35. $\cot\theta\sec\theta = \dfrac{\cos\theta}{\sin\theta}\cdot\dfrac{1}{\cos\theta} = \dfrac{1}{\sin\theta} = \csc\theta$

37. $\sin\phi(\csc\phi - \sin\phi) = (\sin\phi)\dfrac{1}{\sin\phi} - \sin^2\phi$

$= 1 - \sin^2\phi = \cos^2\phi$

39. $\dfrac{1 - \sin^2 x}{\csc^2 x - 1} = \dfrac{\cos^2 x}{\cot^2 x} = \cos^2 x\tan^2 x = (\cos^2 x)\dfrac{\sin^2 x}{\cos^2 x}$

$= \sin^2 x$

41. $\cos\left(\dfrac{\pi}{2} - x\right)\sec x = (\sin x)(\sec x)$

$= (\sin x)\left(\dfrac{1}{\cos x}\right)$

$= \dfrac{\sin x}{\cos x}$

$= \tan x$

43. $\sin\beta\tan\beta + \cos\beta = (\sin\beta)\dfrac{\sin\beta}{\cos\beta} + \cos\beta$

$= \dfrac{\sin^2\beta}{\cos\beta} + \dfrac{\cos^2\beta}{\cos\beta}$

$= \dfrac{\sin^2\beta + \cos^2\beta}{\cos\beta}$

$= \dfrac{1}{\cos\beta}$

$= \sec\beta$

45. $\dfrac{1}{1 + \cos x} + \dfrac{1}{1 - \cos x} = \dfrac{1 - \cos x + 1 + \cos x}{(1 + \cos x)(1 - \cos x)}$

$= \dfrac{2}{1 - \cos^2 x}$

$= \dfrac{2}{\sin^2 x}$

$= 2\csc^2 x$

47. $\tan x - \dfrac{\sec^2 x}{\tan x} = \dfrac{\tan^2 x - \sec^2 x}{\tan x}$

$= \dfrac{-1}{\tan x} = -\cot x$

49. $\dfrac{\sin^2 y}{1 - \cos y} = \dfrac{1 - \cos^2 y}{1 - \cos y}$

$= \dfrac{(1 + \cos y)(1 - \cos y)}{1 - \cos y} = 1 + \cos y$

51. $y_1 = \cos x\cot x + \sin x = \csc x$

$\cos x\cot x + \sin x = \cos x\left(\dfrac{\cos x}{\sin x}\right) + \sin x$

$= \dfrac{\cos^2 x}{\sin x} + \dfrac{\sin^2 x}{\sin x}$

$= \dfrac{\cos^2 x + \sin^2 x}{\sin x} = \dfrac{1}{\sin x} = \csc x$

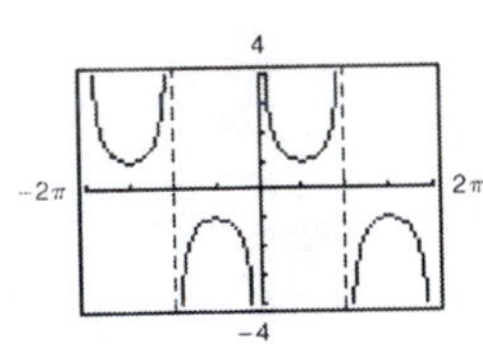

53. Let $x = 3\cos\theta$.

$$\begin{aligned}\sqrt{9 - x^2} &= \sqrt{9 - (3\cos\theta)^2}\\ &= \sqrt{9 - 9\cos^2\theta}\\ &= \sqrt{9(1 - \cos^2\theta)}\\ &= \sqrt{9\sin^2\theta} = 3\sin\theta\end{aligned}$$

55. Let $x = 2\sec\theta$.

$$\begin{aligned}\sqrt{x^2 - 4} &= \sqrt{(2\sec\theta)^2 - 4}\\ &= \sqrt{4(\sec^2\theta - 1)}\\ &= \sqrt{4\tan^2\theta}\\ &= 2\tan\theta\end{aligned}$$

57. Let $x = 3\sin\theta$.

$$\begin{aligned}\sqrt{9 - x^2} &= 3\\ \sqrt{9 - (3\sin\theta)^2} &= 3\\ \sqrt{9 - 9\sin^2\theta} &= 3\\ \sqrt{9(1 - \sin^2\theta)} &= 3\\ \sqrt{9\cos^2\theta} &= 3\\ 3\cos\theta &= 3\\ \cos\theta &= 1\\ \sin\theta = \sqrt{1 - \cos^2\theta} &= \sqrt{1 - (1)^2} = 0\end{aligned}$$

59. $\sin\theta = \sqrt{1 - \cos^2\theta}$

Let $y_1 = \sin x$ and $y_2 = \sqrt{1 - \cos^2 x}, 0 \le x \le 2\pi$.

$y_1 = y_2$ for $0 \le x \le \pi$.

So, $\sin\theta = \sqrt{1 - \cos^2\theta}$ for $0 \le \theta \le \pi$.

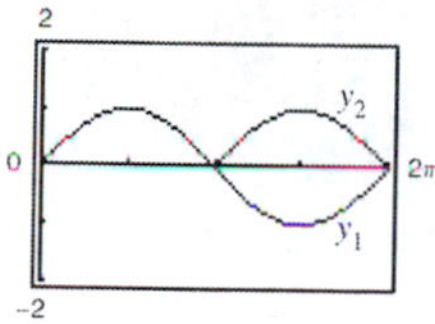

61.
$$\begin{aligned}\ln|\sin x| + \ln|\cot x| &= \ln|\sin x \cot x|\\ &= \ln\left|\sin x \cdot \frac{\cos x}{\sin x}\right|\\ &= \ln|\cos x|\end{aligned}$$

63.
$$\begin{aligned}\ln|\cot t| + \ln(1 + \tan^2 t) &= \ln\left[|\cot t|(1 + \tan^2 t)\right]\\ &= \ln|\cot t \sec^2 t|\\ &= \ln\left|\frac{\cot t}{\sin t} \cdot \frac{1}{\cos^2 t}\right|\\ &= \ln\left|\frac{1}{\sin t \cos t}\right|\\ &= \ln|\csc t \sec t|\end{aligned}$$

65.
$$\begin{aligned}\mu W\cos\theta &= W\sin\theta\\ \mu &= \frac{W\sin\theta}{W\cos\theta} = \tan\theta\end{aligned}$$

67. True. For example, $\sin(-x) = -\sin x$ means that the graph of $\sin x$ is symmetric about the origin.

69. As $x \to \frac{\pi}{2}^-$, $\tan x \to \infty$ and $\cot x \to 0$.

71. $\dfrac{\sin k\theta}{\cos k\theta} = \tan\theta$ *is not* an identity.

$$\frac{\sin k\theta}{\cos k\theta} = \tan k\theta$$

73. Let $u = a\tan\theta$, then

$$\begin{aligned}\sqrt{a^2 + u^2} &= \sqrt{a^2 + (a\tan\theta)^2}\\ &= \sqrt{a^2 + a^2\tan^2\theta}\\ &= \sqrt{a^2(1 + \tan^2\theta)}\\ &= \sqrt{a^2\sec^2\theta}\\ &= a\sec\theta.\end{aligned}$$

75. Because $\sin^2\theta + \cos^2\theta = 1$, then $\cos^2\theta = 1 - \sin^2\theta$.

$$\cos\theta = \pm\sqrt{1 - \sin\theta}$$

$$\tan\theta = \frac{\sin\theta}{\cos\theta} = \frac{\sin\theta}{\pm\sqrt{1 - \sin^2\theta}}$$

$$\cot\theta = \frac{\cos\theta}{\sin\theta} = \frac{\pm\sqrt{1 - \sin^2\theta}}{\sin\theta}$$

$$\sec\theta = \frac{1}{\cos\theta} = \frac{1}{\pm\sqrt{1 - \sin^2\theta}}$$

$$\csc\theta = \frac{1}{\sin\theta}$$

Section 7.2 Verifying Trigonometric Identities

1. identity

3. $\tan u$

5. $\cos^2 u$

7. $-\csc u$

9. $\tan t \cot t = \dfrac{\sin t}{\cos t} \cdot \dfrac{\cos t}{\sin t} = 1$

11. $\cot^2 y(\sec^2 y - 1) = \cot^2 y \tan^2 y = 1$

13. $(1 + \sin\alpha)(1 - \sin\alpha) = 1 - \sin^2\alpha = \cos^2\alpha$

15. $\cos^2\beta - \sin^2\beta = (1 - \sin^2\beta) - \sin^2\beta$
$= 1 - 2\sin^2\beta$

17. $\dfrac{\tan^2\theta}{\sec\theta} = \dfrac{(\sin\theta/\cos\theta)\tan\theta}{1/\cos\theta} = \sin\theta\tan\theta$

19. $\dfrac{\cot^2 t}{\csc t} = \dfrac{\cos^2 t/\sin^2 t}{1/\sin t} = \dfrac{\cos^2 t}{\sin t} = \dfrac{1 - \sin^2 t}{\sin t}$

21. $\sin^{1/2} x\cos x - \sin^{5/2} x\cos x = \sin^{1/2} x\cos x(1 - \sin^2 x) = \sin^{1/2} x\cos x \cdot \cos^2 x = \cos^3 x\sqrt{\sin x}$

23. $\dfrac{\cot x}{\sec x} = \dfrac{\cos x/\sin x}{1/\cos x} = \dfrac{\cos^2 x}{\sin x} = \dfrac{1 - \sin^2 x}{\sin x} = \dfrac{1}{\sin x} - \dfrac{\sin^2 x}{\sin x} = \csc x - \sin x$

25. $\sec x - \cos x = \dfrac{1}{\cos x} - \cos x$
$= \dfrac{1 - \cos^2 x}{\cos x}$
$= \dfrac{\sin^2 x}{\cos x}$
$= \sin x \cdot \dfrac{\sin x}{\cos x}$
$= \sin x \tan x$

27. $\dfrac{1}{\tan x} + \dfrac{1}{\cot x} = \dfrac{\cot x + \tan x}{\tan x \cot x}$
$= \dfrac{\cot x + \tan x}{1}$
$= \tan x + \cot x$

29. $\dfrac{1 + \sin\theta}{\cos\theta} + \dfrac{\cos\theta}{1 + \sin\theta} = \dfrac{(1 + \sin\theta)^2 + \cos^2\theta}{\cos\theta(1 + \sin\theta)}$
$= \dfrac{1 + 2\sin\theta + \sin^2\theta + \cos^2\theta}{\cos\theta(1 + \sin\theta)}$
$= \dfrac{2 + 2\sin\theta}{\cos\theta(1 + \sin\theta)}$
$= \dfrac{2(1 + \sin\theta)}{\cos\theta(1 + \sin\theta)}$
$= \dfrac{2}{\cos\theta}$
$= 2\sec\theta$

31. $\dfrac{1}{\cos x + 1} + \dfrac{1}{\cos x - 1} = \dfrac{\cos x - 1 + \cos x + 1}{(\cos x + 1)(\cos x - 1)}$
$= \dfrac{2\cos x}{\cos^2 x - 1}$
$= \dfrac{2\cos x}{-\sin^2 x}$
$= -2 \cdot \dfrac{1}{\sin x} \cdot \dfrac{\cos x}{\sin x}$
$= -2\csc x \cot x$

33. $\tan\left(\dfrac{\pi}{2} - \theta\right)\tan\theta = \cot\theta\tan\theta$
$= \left(\dfrac{1}{\tan\theta}\right)\tan\theta$
$= 1$

35. $\dfrac{\tan x \cot x}{\cos x} = \dfrac{1}{\cos x} = \sec x$

37. $(1 + \sin y)[1 + \sin(-y)] = (1 + \sin y)(1 - \sin y)$
$= 1 - \sin^2 y$
$= \cos^2 y$

39. $\dfrac{\tan x + \cot y}{\tan x \cot y} = \dfrac{\dfrac{1}{\cot x} + \dfrac{1}{\tan y}}{\dfrac{1}{\cot x} \cdot \dfrac{1}{\tan y}} \cdot \dfrac{\cot x \tan y}{\cot x \tan y}$
$= \tan y + \cot x$

41. $\sqrt{\dfrac{1 + \sin\theta}{1 - \sin\theta}} = \sqrt{\dfrac{1 + \sin\theta}{1 - \sin\theta} \cdot \dfrac{1 + \sin\theta}{1 + \sin\theta}}$

$$= \sqrt{\frac{(1 + \sin\theta)^2}{1 - \sin^2\theta}}$$

$$= \sqrt{\frac{(1 + \sin\theta)^2}{\cos^2\theta}}$$

$$= \frac{1 + \sin\theta}{|\cos\theta|}$$

43. $\cos^2\beta + \cos^2\left(\dfrac{\pi}{2} - \beta\right) = \cos^2\beta + \sin^2\beta = 1$

45. $\sin t \csc\left(\dfrac{\pi}{2} - t\right) = \sin t \sec t = \sin t\left(\dfrac{1}{\cos t}\right)$

$$= \frac{\sin t}{\cos t} = \tan t$$

47. Let $\theta = \sin^{-1} x \Rightarrow \sin\theta = x = \dfrac{x}{1}$.

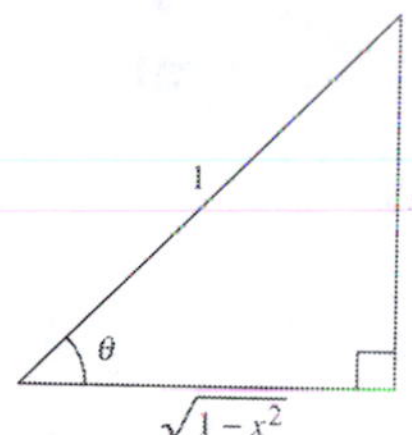

From the diagram,

$$\tan(\sin^{-1} x) = \tan\theta = \frac{x}{\sqrt{1 - x^2}}.$$

49. Let $\theta = \sin^{-1}\dfrac{x - 1}{4} \Rightarrow \sin\theta = \dfrac{x - 1}{4}$.

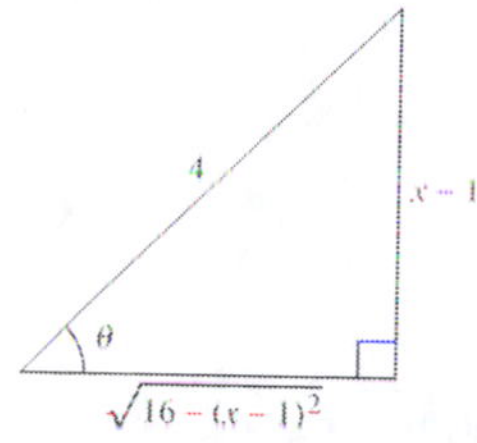

From the diagram,

$$\tan\left(\sin^{-1}\frac{x - 1}{4}\right) = \tan\theta = \frac{x - 1}{\sqrt{16 - (x - 1)^2}}.$$

51. The first line claims that $\cot(-x) = \cot x$, which is not true. The correct substitution is $\cot(-x) = -\cot x$.

53. (a)

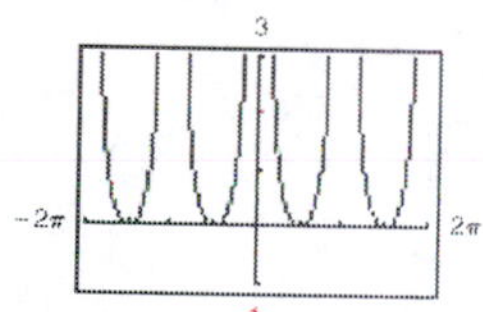

Identity

(b)

X	Y₁	Y₂
-3	49.214	49.214
-2	.20945	.20945
-1	.41228	.41228
0	ERROR	ERROR
1	.41228	.41228
2	.20945	.20945
3	49.214	49.214

X=-3

Identity

(c) $(1 + \cot^2 x)(\cos^2 x) = \csc^2 x \cos^2 x = \dfrac{1}{\sin^2 x} \cdot \cos^2 x = \cot^2 x$

55. (a)

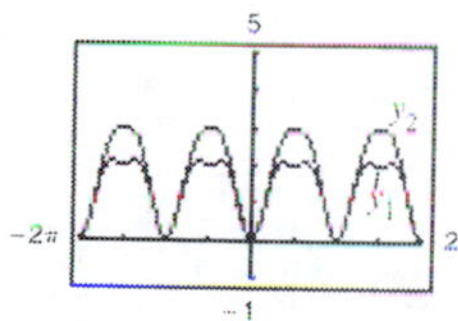

Not an identity

(b)

X	Y₁	Y₂
-4.712	2	3
-3.142	0	0
-1.571	2	3
0	0	0
1.5708	2	3
3.1416	0	0
4.7124	2	3

X=-4.71238898038

Not an identity

(c) $2 + \cos^2 x - 3\cos^4 x = (1 - \cos^2 x)(2 + 3\cos^2 x) = \sin^2 x(2 + 3\cos^2 x) \neq \sin^2 x(3 + 2\cos^2 x)$

57. (a) 

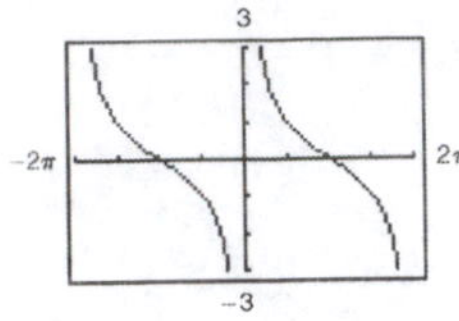

Identity

(b)

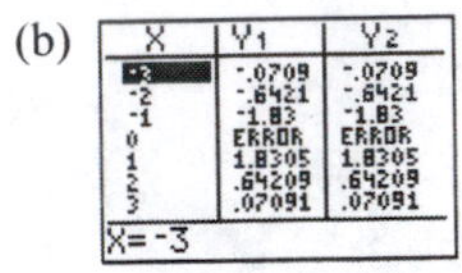

X	Y_1	Y_2
-3	-.0709	-.0709
-2	-.6421	-.6421
-1	-1.83	-1.83
0	ERROR	ERROR
1	1.8305	1.8305
2	.64209	.64209
3	.07091	.07091

X=-3

Identity

(c) $$\frac{1+\cos x}{\sin x} = \frac{(1+\cos x)(1-\cos x)}{\sin x(1-\cos x)}$$
$$= \frac{1-\cos^2 x}{\sin x(1-\cos x)}$$
$$= \frac{\sin^2 x}{\sin x(1-\cos x)}$$
$$= \frac{\sin x}{1-\cos x}$$

59. $$\tan^3 x \sec^2 x - \tan^3 x = \tan^3 x(\sec^2 x - 1)$$
$$= \tan^3 x \tan^2 x$$
$$= \tan^5 x$$

61. $$(\sin^2 x - \sin^4 x)\cos x = \sin^2 x(1 - \sin^2 x)\cos x$$
$$= \sin^2 x \cos^2 x \cos x$$
$$= \sin^2 x \cos^3 x$$

63. $$\sin^2 25° + \sin^2 65° = \sin^2 25° + \cos^2(90° - 65°)$$
$$= \sin^2 25° + \cos^2 25°$$
$$= 1$$

65. $$\cos x - \csc x \cot x = \cos x - \frac{1}{\sin x}\frac{\cos x}{\sin x}$$
$$= \cos x\left(1 - \frac{1}{\sin^2 x}\right)$$
$$= \cos x(1 - \csc^2 x)$$
$$= -\cos x(\csc^2 x - 1)$$
$$= -\cos x \cot^2 x$$

67. True. You can use many different techniques to verify a trigonometric identity.

69. False. Because $\sin x^2 = \sin(x \cdot x)$ and $\sin^2 x = (\sin x)(\sin x)$, $\sin x^2 \neq \sin^2 x$.

71. Because $\sin^2\theta = 1 - \cos^2\theta$, then $\sin\theta = \pm\sqrt{1-\cos^2\theta}$; $\sin\theta \neq \sqrt{1-\cos^2\theta}$ if θ lies in Quadrant III or IV.

One such angle is $\theta = \dfrac{7\pi}{4}$.

73. $$1 - \cos\theta = \sin\theta$$
$$(1-\cos\theta)^2 = (\sin\theta)^2$$
$$1 - 2\cos\theta + \cos^2\theta = \sin^2\theta$$
$$1 - 2\cos\theta + \cos^2\theta = 1 - \cos^2\theta$$
$$2\cos^2\theta - 2\cos\theta = 0$$
$$2\cos\theta(\cos\theta - 1) = 0$$

The equation is not an identity because it is only true when $\cos\theta = 0$ or $\cos\theta = 1$. So, one angle for which the equation is not true is $-\dfrac{\pi}{2}$.

Section 7.3 Solving Trigonometric Equations

1. isolate

3. quadratic

5. $\tan x - \sqrt{3} = 0$

(a) $x = \dfrac{\pi}{3}$

$$\tan\frac{\pi}{3} - \sqrt{3} = \sqrt{3} - \sqrt{3} = 0$$

(b) $x = \dfrac{4\pi}{3}$

$$\tan\frac{4\pi}{3} - \sqrt{3} = \sqrt{3} - \sqrt{3} = 0$$

7. $3\tan^2 2x - 1 = 0$

(a) $x = \frac{\pi}{12}$

$$3\left[\tan 2\left(\frac{\pi}{12}\right)\right]^2 - 1 = 3\tan^2\frac{\pi}{6} - 1$$
$$= 3\left(\frac{1}{\sqrt{3}}\right)^2 - 1$$
$$= 0$$

(b) $x = \frac{5\pi}{12}$

$$3\left[\tan 2\left(\frac{5\pi}{12}\right)\right]^2 - 1 = 3\tan^2\frac{5\pi}{6} - 1$$
$$= 3\left(-\frac{1}{\sqrt{3}}\right)^2 - 1$$
$$= 0$$

9. $2\sin^2 x - \sin x - 1 = 0$

(a) $x = \frac{\pi}{2}$

$$2\sin^2\frac{\pi}{2} - \sin\frac{\pi}{2} - 1 = 2(1)^2 - 1 - 1$$
$$= 0$$

(b) $x = \frac{7\pi}{6}$

$$2\sin^2\frac{7\pi}{6} - \sin\frac{7\pi}{6} - 1 = 2\left(-\frac{1}{2}\right)^2 - \left(-\frac{1}{2}\right) - 1$$
$$= \frac{1}{2} + \frac{1}{2} - 1$$
$$= 0$$

11. $\sqrt{3}\csc x - 2 = 0$

$$\sqrt{3}\csc x = 2$$
$$\csc x = \frac{2}{\sqrt{3}}$$
$$x = \frac{\pi}{3} + 2n\pi$$
$$\text{or } x = \frac{2\pi}{3} + 2n\pi$$

13. $\cos x + 1 = -\cos x$

$$2\cos x + 1 = 0$$
$$\cos x = -\frac{1}{2}$$
$$x = \frac{2\pi}{3} + 2n\pi \text{ or } x = \frac{4\pi}{3} + 2n\pi$$

15. $3\sec^2 x - 4 = 0$

$$\sec^2 x = \frac{4}{3}$$
$$\sec x = \pm\frac{2}{\sqrt{3}}$$
$$x = \frac{\pi}{6} + n\pi$$
$$\text{or } x = \frac{5\pi}{6} + n\pi$$

17. $4\cos^2 x - 1 = 0$

$$\cos^2 x = \frac{1}{4}$$
$$\cos^2 x = \pm\frac{1}{2}$$
$$x = \frac{\pi}{3} + n\pi \quad \text{or} \quad x = \frac{2\pi}{3} + n\pi$$

19. $2\sin^2 2x = 1$

$$\sin 2x = \pm\frac{1}{\sqrt{2}} = \pm\frac{\sqrt{2}}{2}$$
$$2x = \frac{\pi}{4} + 2n\pi,\ 2x = \frac{3\pi}{4} + 2n\pi,$$
$$2x = \frac{5\pi}{4} + 2n\pi,\ 2x = \frac{7\pi}{4} + 2n\pi$$

So, $x = \frac{\pi}{8} + n\pi, \frac{3\pi}{8} + n\pi, \frac{5\pi}{8} + n\pi, \frac{7\pi}{8} + n\pi.$

You can combine these as follows:

$$x = \frac{\pi}{8} + \frac{n\pi}{2},\ x = \frac{3\pi}{8} + \frac{n\pi}{2}$$

21. $\tan 3x(\tan x - 1) = 0$

$\tan 3x = 0$ or $\tan x - 1 = 0$

$3x = n\pi \qquad \tan x = 1$

$x = \frac{n\pi}{3} \qquad x = \frac{\pi}{4} + n\pi$

23. $\sin x(\sin x + 1) = 0$

$\sin x = 0$ or $\sin x = -1$

$x = n\pi \qquad x = \frac{3\pi}{2} + 2n\pi$

25.

$$\cos^3 x = \cos x$$
$$\cos^3 x - \cos x = 0$$
$$\cos x(\cos^2 x - 1) = 0$$

$\cos x = 0$ or $\cos^2 x - 1 = 0$

$x = \frac{\pi}{2}, \frac{3\pi}{2} \qquad \cos x = \pm 1$

$x = 0, \pi$

27. $3\tan^3 x - \tan x = 0$

$\tan x(3\tan^2 x - 1) = 0$

$\tan x = 0$ or $3\tan^2 x - 1 = 0$

$x = 0, \pi$ $\qquad \tan x = \pm\dfrac{\sqrt{3}}{3}$

$x = \dfrac{\pi}{6}, \dfrac{5\pi}{6}, \dfrac{7\pi}{6}, \dfrac{11\pi}{6}$

29. $\sec^2 x - \sec x - 2 = 0$

$(\sec x - 2)(\sec x + 1) = 0$

$\sec x - 2 = 0$ or $\sec x + 1 = 0$

$\sec x = 2 \qquad \sec x = -1$

$x = \dfrac{\pi}{3}, \dfrac{5\pi}{3} \qquad x = \pi$

31. $2\sin x + \csc x = 0$

$2\sin x + \dfrac{1}{\sin x} = 0$

$2\sin^2 x + 1 = 0$

$\sin^2 x = -\dfrac{1}{2} \Rightarrow$ No solution

33. $2\cos^2 x + \cos x - 1 = 0$

$(2\cos x - 1)(\cos x + 1) = 0$

$2\cos x - 1 = 0$ or $\cos x + 1 = 0$

$\cos x = \dfrac{1}{2} \qquad \cos x = -1$

$x = \dfrac{\pi}{3}, \dfrac{5\pi}{3} \qquad x = \pi$

35. $2\sec^2 x + \tan^2 x - 3 = 0$

$2(\tan^2 x + 1) + \tan^2 x - 3 = 0$

$3\tan^2 x - 1 = 0$

$\tan x = \pm\dfrac{\sqrt{3}}{3}$

$x = \dfrac{\pi}{6}, \dfrac{5\pi}{6}, \dfrac{7\pi}{6}, \dfrac{11\pi}{6}$

37. $\csc x + \cot x = 1$

$(\csc x + \cot x)^2 = 1^2$

$\csc^2 x + 2\csc x\cot x + \cot^2 x = 1$

$\cot^2 x + 1 + 2\csc x\cot x + \cot^2 x = 1$

$2\cot^2 x + 2\csc x\cot x = 0$

$2\cot x(\cot x + \csc x) = 0$

$2\cot x = 0$ or $\cot x + \csc x = 0$

$x = \dfrac{\pi}{2}, \dfrac{3\pi}{2} \qquad \dfrac{\cos x}{\sin x} = -\dfrac{1}{\sin x}$

$\left(\dfrac{3\pi}{2}\text{ is extraneous.}\right) \qquad \cos x = -1$

$x = \pi$

(π is extraneous.)

$x = \pi/2$ is the only solution.

39. $2\cos 2x - 1 = 0$

$\cos 2x = \dfrac{1}{2}$

$2x = \dfrac{\pi}{3} + 2n\pi$ or $2x = \dfrac{5\pi}{3} + 2n\pi$

$x = \dfrac{\pi}{6} + n\pi \qquad x = \dfrac{5\pi}{6} + n\pi$

41. $\tan 3x - 1 = 0$

$\tan 3x = 1$

$3x = \dfrac{\pi}{4} + n\pi$

$x = \dfrac{\pi}{12} + \dfrac{n\pi}{3}$

43. $2\cos\dfrac{x}{2} = \sqrt{2} = 0$

$\cos\dfrac{x}{2} = \dfrac{\sqrt{2}}{2}$

$\dfrac{x}{2} = \dfrac{\pi}{4} + 2n\pi$ or $\dfrac{x}{2} = \dfrac{7\pi}{4} + 2n\pi$

$x = \dfrac{\pi}{2} + 4n\pi \qquad x = \dfrac{7\pi}{2} + 4n\pi$

45. $y = \sin\dfrac{\pi x}{2} + 1$

$\sin\left(\dfrac{\pi x}{2}\right) + 1 = 0$

$\sin\left(\dfrac{\pi x}{2}\right) = -1$

$\dfrac{\pi x}{2} = \dfrac{3\pi}{2} + 2n\pi$

$x = 3 + 4n$

For $-2 < x < 4$, the intercepts are -1 and 3.

47. $y = \tan^2\left(\dfrac{\pi x}{6}\right) - 3$

$$\tan^2\left(\frac{\pi x}{6}\right) - 3 = 0$$

$$\tan^2\left(\frac{\pi x}{6}\right) = 3$$

$$\tan\left(\frac{\pi x}{6}\right) = \pm\sqrt{3}$$

$$\frac{\pi x}{6} = \pm\frac{\pi}{3} + n\pi$$

$$x = \pm 2 + 6n$$

For $-3 < x < 3$, the intercepts are -2 and 2.

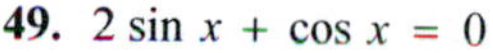

49. $2 \sin x + \cos x = 0$

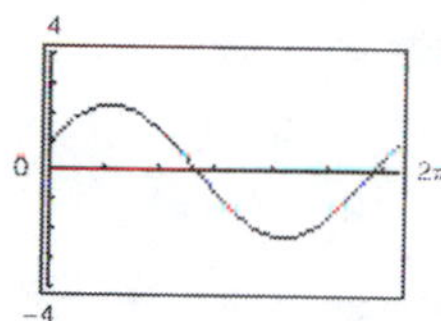

$x \approx 2.678$ and $x \approx 5.820$

51. $\dfrac{1 + \sin x}{\cos x} + \dfrac{\cos x}{1 + \sin x} - 4 = 0$

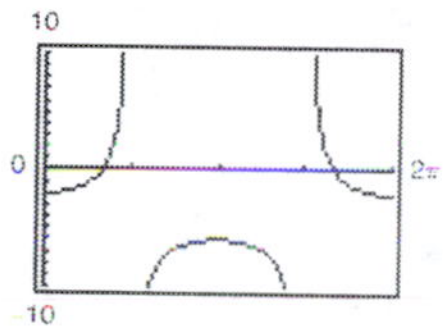

$x = \dfrac{\pi}{3} \approx 1.047$ and $x = \dfrac{5\pi}{3} \approx 5.236$

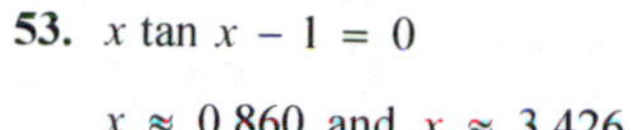

53. $x \tan x - 1 = 0$

$x \approx 0.860$ and $x \approx 3.426$

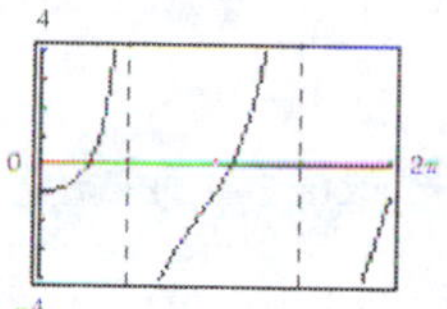

55. $\sec^2 x + 0.5 \tan x - 1 = 0$

$x = 0,\ x \approx 2.678,$

$x = \pi \approx 3.142$

$x \approx 5.820$

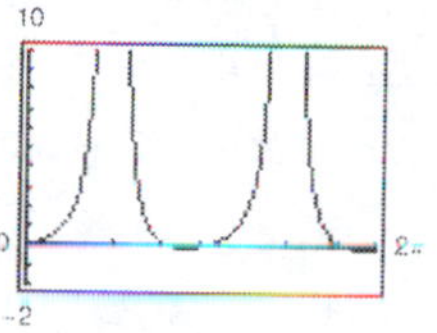

57. $2 \tan^2 x + 7 \tan x - 15 = 0$

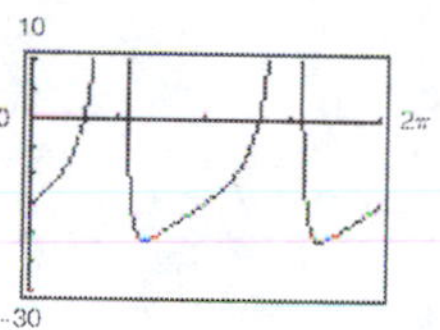

$x \approx 0.983,\ x \approx 1.768,\ x \approx 4.124$ and $x \approx 4.910$

59. $12 \sin^2 x - 13 \sin x + 3 = 0$

$$\sin x = \frac{-(-13) \pm \sqrt{(-13)^2 - 4(12)(3)}}{2(12)} = \frac{13 \pm 5}{24}$$

$\sin x = \dfrac{1}{3}$ or $\sin x = \dfrac{3}{4}$

$x \approx 0.3398, 2.8018$ $\quad x \approx 0.8481, 2.2935$

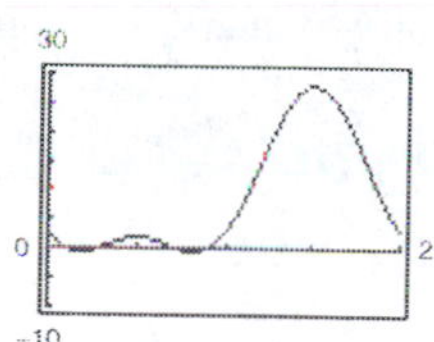

The x-intercepts occur at $x \approx 0.3398$, $x \approx 0.8481$, $x \approx 2.2935$, and $x \approx 2.8018$.

61. $\tan^2 x + 3 \tan x + 1 = 0$

$$\tan x = \frac{-3 \pm \sqrt{3^2 - 4(1)(1)}}{2(1)} = \frac{-3 \pm \sqrt{5}}{2}$$

$\tan x = \dfrac{-3 - \sqrt{5}}{2}$ or $\tan x = \dfrac{-3 + \sqrt{5}}{2}$

$x \approx 1.9357, 5.0773$ $\quad x \approx 2.7767, 5.9183$

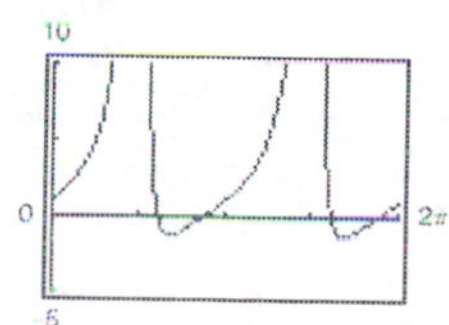

The x-intercepts occur at $x \approx 1.9357$, $x \approx 2.7767$, $x \approx 5.0773$, and $x \approx 5.9183$.

63. $\tan^2 x + \tan x - 12 = 0$

$(\tan x + 4)(\tan x - 3) = 0$

$\tan x + 4 = 0$ or $\tan x - 3 = 0$

$\tan x = -4$ $\quad$ $\tan x = 3$

$x = \arctan(-4) + \pi, \arctan(-4) + 2\pi$ $\quad$ $x = \arctan 3, \arctan 3 + \pi$

65. $\sec^2 x - 6 \tan x = -4$

$1 + \tan^2 x - 6 \tan x + 4 = 0$

$\tan^2 x - 6 \tan x + 5 = 0$

$(\tan x - 1)(\tan x - 5) = 0$

$\tan x - 1 = 0$ $\quad$ $\tan x - 5 = 0$

$\tan x = 1$ $\quad$ $\tan x = 5$

$x = \dfrac{\pi}{4}, \dfrac{5\pi}{4}$ $\quad$ $x = \arctan 5, \arctan 5 + \pi$

67. $2 \sin^2 x + 5 \cos x = 4$

$2(1 - \cos^2 x) + 5 \cos x - 4 = 0$

$-2 \cos^2 x + 5 \cos x - 2 = 0$

$-(2 \cos x - 1)(\cos x - 2) = 0$

$2 \cos x - 1 = 0$ or $\cos x - 2 = 0$

$\cos x = \dfrac{1}{2}$ $\quad$ $\cos x = 2$

$x = \dfrac{\pi}{3}, \dfrac{5\pi}{3}$ $\quad$ No solution

69. $\cot^2 x - 9 = 0$

$\cot^2 x = 9$

$\frac{1}{9} = \tan^2 x$

$\pm\frac{1}{3} = \tan x$

$x = \arctan \frac{1}{3}, \arctan \frac{1}{3} + \pi, \arctan\left(-\frac{1}{3}\right) + \pi, \arctan\left(-\frac{1}{3}\right) + 2\pi$

71. $\sec^2 x - 4 \sec x = 0$

$\sec x(\sec x - 4) = 0$

$\sec x = 0$ $\quad$ $\sec x - 4 = 0$

No solution $\quad$ $\sec x = 4$

$\dfrac{1}{4} = \cos x$

$x = \arccos \dfrac{1}{4}, -\arccos \dfrac{1}{4} + 2\pi$

73. $\csc^2 x + 3\csc x - 4 = 0$

$(\csc x + 4)(\csc x - 1) = 0$

$\csc x + 4 = 0$ or $\csc x - 1 = 0$

$\csc x = -4$

$-\dfrac{1}{4} = \sin x$

$x = \arcsin\left(\dfrac{1}{4}\right) + \pi, \arcsin\left(-\dfrac{1}{4}\right) + 2\pi$

$\csc x = 1$

$1 = \sin x$

$x = \dfrac{\pi}{2}$

75. $3\tan^2 x + 5\tan x - 4 = 0, \left[-\dfrac{\pi}{2}, \dfrac{\pi}{2}\right]$

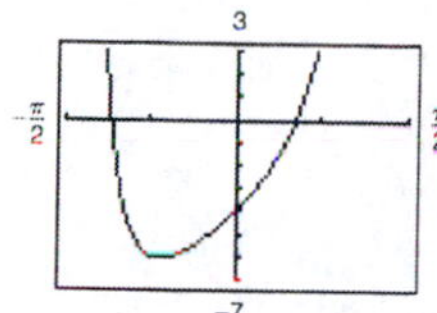

$x \approx -1.154, 0.534$

77. $4\cos^2 x - 2\sin x + 1 = 0, \left[-\dfrac{\pi}{2}, \dfrac{\pi}{2}\right]$

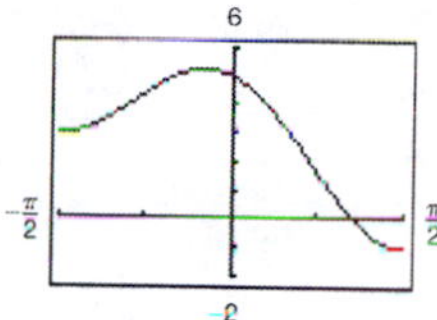

$x \approx 1.110$

79. (a) $f(x) = \sin^2 x + \cos x$

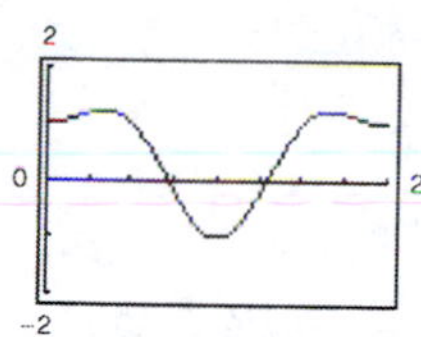

Maximum: $(1.0472, 1.25)$

Maximum: $(5.2360, 1.25)$

Minimum: $(0, 1)$

Minimum: $(3.1416, -1)$

(b) $2\sin x\cos x - \sin x = 0$

$\sin x(2\cos x - 1) = 0$

$\sin x = 0$ or $2\cos x - 1 = 0$

$x = 0, \pi$

$\approx 0, 3.1416$

$\cos x = \dfrac{1}{2}$

$x = \dfrac{\pi}{3}, \dfrac{5\pi}{3}$

$\approx 1.0472, 5.2360$

81. (a) $f(x) = \sin x + \cos x$

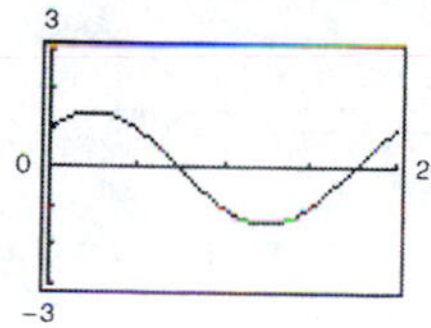

Maximum: $(0.7854, 1.4142)$

Minimum: $(3.9270, -1.4142)$

(b) $\cos x - \sin x = 0$

$\cos x = \sin x$

$1 = \dfrac{\sin x}{\cos x}$

$\tan x = 1$

$x = \dfrac{\pi}{4}, \dfrac{5\pi}{4}$

$\approx 0.7854, 3.9270$

83. (a) $f(x) = \sin x \cos x$

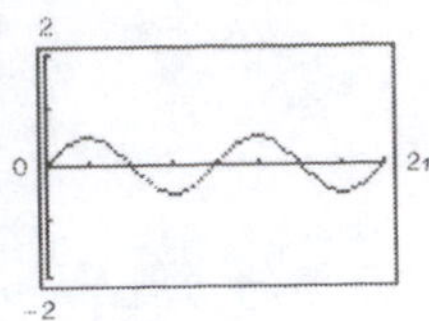

Maximum: $(0.7854, 0.5)$

Maximum: $(3.9270, 0.5)$

Minimum: $(2.3562, -0.5)$

Minimum: $(5.4978, -0.5)$

(b)
$$\begin{aligned} -\sin^2 x + \cos^2 x &= 0 \\ -\sin^2 x + 1 - \sin^2 x &= 0 \\ -2\sin^2 x + 1 &= 0 \\ \sin^2 x &= \frac{1}{2} \end{aligned}$$

$$\begin{aligned} \sin x &= \pm\sqrt{\frac{1}{2}} = \pm\frac{\sqrt{2}}{2} \\ x &= \frac{\pi}{4}, \frac{3\pi}{4}, \frac{5\pi}{4}, \frac{7\pi}{4} \\ &\approx 0.7854, 2.3562, 3.9270, 5.4978 \end{aligned}$$

85. The graphs of $y_1 = 2\sin x$ and $y_2 = 3x + 1$ appear to have one point of intersection. This implies there is one solution to the equation $2\sin x = 3x + 1$.

87. $f(x) = \dfrac{\sin x}{x}$

(a) Domain: all real numbers except $x = 0$.

(b) The graph has y-axis symmetry.

(c) As $x \to 0$, $f(x) \to 1$.

(d) $\dfrac{\sin x}{x} = 0$ has four solutions in the interval $[-8, 8]$.

$$\begin{aligned} \sin x\left(\frac{1}{x}\right) &= 0 \\ \sin x &= 0 \\ x &= -2\pi, -\pi, \pi, 2\pi \end{aligned}$$

89.
$$\begin{aligned} y &= \frac{1}{12}(\cos 8t - 3\sin 8t) \\ \frac{1}{12}(\cos 8t - 3\sin 8t) &= 0 \\ \cos 8t &= 3\sin 8t \\ \frac{1}{3} &= \tan 8t \\ 8t &\approx 0.32175 + n\pi \\ t &\approx 0.04 + \frac{n\pi}{8} \end{aligned}$$

In the interval $0 \le t \le 1$, $t \approx 0.04, 0.43,$ and 0.83.

91. Graph $y_1 = 58.3 + 32\cos\left(\dfrac{\pi t}{6}\right)$

$y_2 = 75.$

Left point of intersection: $(1.95, 75)$

Right point of intersection: $(10.05, 75)$

So, sales exceed 7500 in January, November, and December.

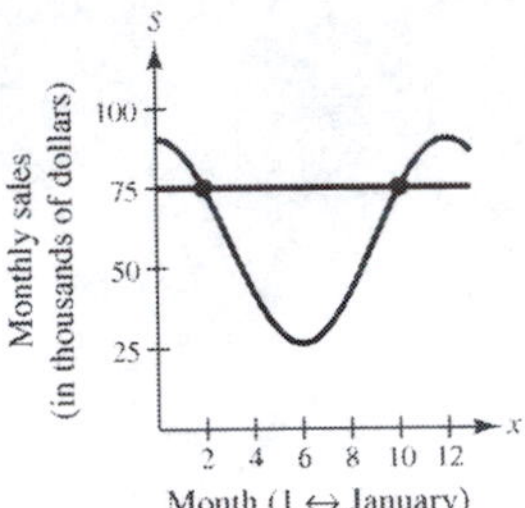

93. (a) and (c)

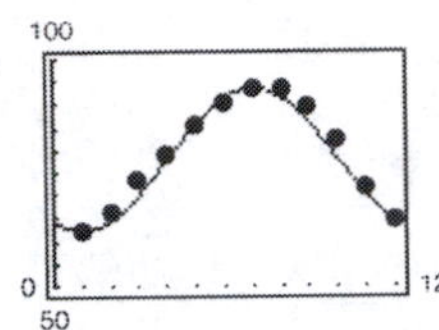

The model is a good fit.

(b) $H = a\cos(bt - c) + d$

$$\begin{aligned} a &= \frac{1}{2}[\text{high} - \text{low}] = \frac{1}{2}[93.6 - 62.3] = 15.65 \\ p &= 2[\text{high time} - \text{low time}] = 2[7 - 1] = 12 \\ b &= \frac{2\pi}{p} = \frac{2\pi}{12} = \frac{\pi}{6} \\ \frac{c}{b} &= 7 \Rightarrow c = 7\left(\frac{\pi}{6}\right) = \frac{7\pi}{6} \\ d &= \frac{1}{2}[\text{high} + \text{low}] = \frac{1}{2}[93.6 + 62.3] = 77.95 \\ H &= 15.65\cos\left(\frac{\pi}{6}t - \frac{7\pi}{6}\right) + 77.95 \end{aligned}$$

(d) The constant term, d, gives the average maximum temperature.

The average maximum temperature in Houston is 77.95°F.

(e) The average maximum temperature is above 86°F from June to September. The average maximum temperature is below 86°F from October to May.

95. $A = 2x\cos x, 0 < x < \dfrac{\pi}{2}$

(a)

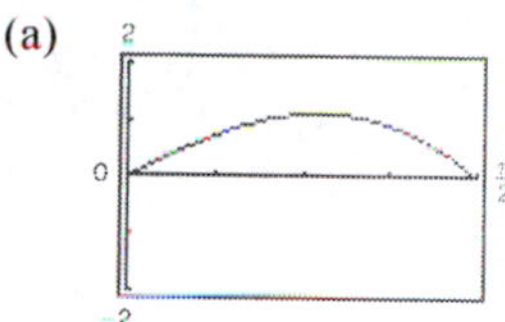

The maximum area of $A \approx 1.12$ occurs when $x \approx 0.86$.

(b) $A \geq 1$ for $0.6 < x < 1.1$

97. $f(x) = \tan \dfrac{\pi x}{4}$

Because $\tan \pi/4 = 1$, $x = 1$ is the smallest nonnegative fixed point.

99. True. The period of $2\sin 4t - 1$ is $\dfrac{\pi}{2}$ and the period of $2\sin t - 1$ is 2π.

In the interval $[0, 2\pi)$ the first equation has four cycles whereas the second equation has only one cycle, so the first equation has four times the x-intercepts (solutions) as the second equation.

101. $\cot x \cos^2 x = 2\cot x$

$\cos^2 x = 2$

$\cos x = \pm\sqrt{2}$

No solution

Because you solved this problem by first dividing by $\cot x$, you do not get the same solution as Example 3.

When solving equations, you do not want to divide each side by a variable expression that will cancel out because you may accidentally remove one of the solutions.

103. (a)

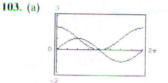

The graphs intersect when $x = \dfrac{\pi}{2}$ and $x = \pi$.

(b)

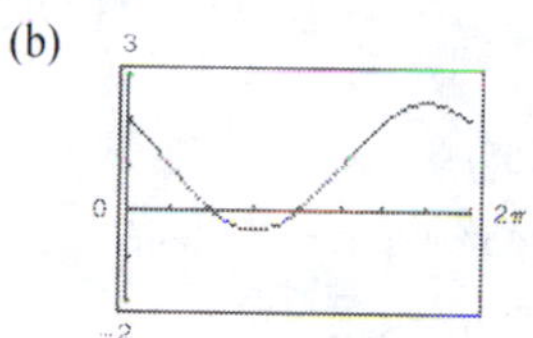

The x-intercepts are $\left(\dfrac{\pi}{2}, 0\right)$ and $(\pi, 0)$.

Both methods produce the same x-values.

Answers will vary on which method is preferred.

Section 7.4 Sum and Difference Formulas

1. $\sin u \cos v - \cos u \sin v$

3. $\dfrac{\tan u + \tan v}{1 - \tan u \tan v}$

5. $\cos u \cos v + \sin u \sin v$

7. (a) $\cos\left(\dfrac{\pi}{4} + \dfrac{\pi}{3}\right) = \cos\dfrac{\pi}{4}\cos\dfrac{\pi}{3} - \sin\dfrac{\pi}{4}\sin\dfrac{\pi}{3}$

$$= \frac{\sqrt{2}}{2}\cdot\frac{1}{2} - \frac{\sqrt{2}}{2}\cdot\frac{\sqrt{3}}{2}$$

$$= \frac{\sqrt{2} - \sqrt{6}}{4}$$

(b) $\cos\dfrac{\pi}{4} + \cos\dfrac{\pi}{3} = \dfrac{\sqrt{2}}{2} + \dfrac{1}{2} = \dfrac{\sqrt{2} + 1}{2}$

9. (a) $\sin(135° - 30°) = \sin 135° \cos 30° - \cos 135° \sin 30°$

$$= \left(\frac{\sqrt{2}}{2}\right)\left(\frac{\sqrt{3}}{2}\right) - \left(-\frac{\sqrt{2}}{2}\right)\left(\frac{1}{2}\right)$$

$$= \frac{\sqrt{6} + \sqrt{2}}{4}$$

(b) $\sin 135° - \cos 30° = \dfrac{\sqrt{2}}{2} - \dfrac{\sqrt{3}}{2} = \dfrac{\sqrt{2} - \sqrt{3}}{2}$

11. $\sin\dfrac{11\pi}{12} = \sin\left(\dfrac{3\pi}{4} + \dfrac{\pi}{6}\right) = \sin\dfrac{3\pi}{4}\cos\dfrac{\pi}{6} + \cos\dfrac{3\pi}{4}\sin\dfrac{\pi}{6} = \dfrac{\sqrt{2}}{2}\cdot\dfrac{\sqrt{3}}{2} + \left(-\dfrac{\sqrt{2}}{2}\right)\dfrac{1}{2} = \dfrac{\sqrt{2}}{4}\left(\sqrt{3} - 1\right)$

$\cos\dfrac{11\pi}{12} = \cos\left(\dfrac{3\pi}{4} + \dfrac{\pi}{6}\right) = \cos\dfrac{3\pi}{4}\cos\dfrac{\pi}{6} - \sin\dfrac{3\pi}{4}\sin\dfrac{\pi}{6} = -\dfrac{\sqrt{2}}{2}\cdot\dfrac{\sqrt{3}}{2} - \dfrac{\sqrt{2}}{2}\cdot\dfrac{1}{2} = -\dfrac{\sqrt{2}}{4}\left(\sqrt{3} + 1\right)$

$\tan\dfrac{11\pi}{4} = \tan\left(\dfrac{3\pi}{4} + \dfrac{\pi}{6}\right) = \dfrac{\tan\dfrac{3\pi}{4} + \tan\dfrac{\pi}{6}}{1 - \tan\dfrac{3\pi}{4}\tan\dfrac{\pi}{6}} = \dfrac{-1 + \dfrac{\sqrt{3}}{3}}{1 - (-1)\dfrac{\sqrt{3}}{3}} = \dfrac{-3 + \sqrt{3}}{3 + \sqrt{3}}\cdot\dfrac{3 - \sqrt{3}}{3 - \sqrt{3}} = \dfrac{-12 + 6\sqrt{3}}{6} = -2 + \sqrt{3}$

13. $\sin\dfrac{17\pi}{12} = \sin\left(\dfrac{9\pi}{4} - \dfrac{5\pi}{6}\right) = \sin\dfrac{9\pi}{4}\cos\dfrac{5\pi}{6} - \cos\dfrac{9\pi}{4}\sin\dfrac{5\pi}{6} = \dfrac{\sqrt{2}}{2}\left(-\dfrac{\sqrt{3}}{2}\right) - \left(\dfrac{\sqrt{2}}{2}\right)\left(\dfrac{1}{2}\right) = -\dfrac{\sqrt{2}}{4}\left(\sqrt{3} + 1\right)$

$\cos\dfrac{17\pi}{12} = \cos\left(\dfrac{9\pi}{4} - \dfrac{5\pi}{6}\right) = \cos\dfrac{9\pi}{4}\cos\dfrac{5\pi}{6} + \sin\dfrac{9\pi}{4}\sin\dfrac{5\pi}{6} = \dfrac{\sqrt{2}}{2}\left(-\dfrac{\sqrt{3}}{2}\right) + \dfrac{\sqrt{2}}{2}\left(\dfrac{1}{2}\right) = \dfrac{\sqrt{2}}{4}\left(1 - \sqrt{3}\right)$

$\tan\dfrac{17\pi}{12} = \tan\left(\dfrac{9\pi}{4} - \dfrac{5\pi}{6}\right) = \dfrac{\tan(9\pi/4) - \tan(5\pi/6)}{1 + \tan(9\pi/4)\tan(5\pi/6)} = \dfrac{1 - \left(-\sqrt{3}/3\right)}{1 + \left(-\sqrt{3}/3\right)} = \dfrac{3 + \sqrt{3}}{3 - \sqrt{3}}\cdot\dfrac{3 + \sqrt{3}}{3 + \sqrt{3}} = \dfrac{12 + 6\sqrt{3}}{6} = 2 + \sqrt{3}$

15. $\sin 105^\circ = \sin(60^\circ + 45^\circ)$

$= \sin 60^\circ\cos 45^\circ + \cos 60^\circ\sin 45^\circ$

$= \dfrac{\sqrt{3}}{2}\cdot\dfrac{\sqrt{2}}{2} + \dfrac{1}{2}\cdot\dfrac{\sqrt{2}}{2}$

$= \dfrac{\sqrt{2}}{4}\left(\sqrt{3} + 1\right)$

$\cos 105^\circ = \cos(60^\circ + 45^\circ)$

$= \cos 60^\circ\cos 45^\circ - \sin 60^\circ\sin 45^\circ$

$= \dfrac{1}{2}\cdot\dfrac{\sqrt{2}}{2} - \dfrac{\sqrt{3}}{2}\cdot\dfrac{\sqrt{2}}{2}$

$= \dfrac{\sqrt{2}}{4}\left(1 - \sqrt{3}\right)$

$\tan 105^\circ = \tan(60^\circ + 45^\circ)$

$= \dfrac{\tan 60^\circ + \tan 45^\circ}{1 - \tan 60^\circ\tan 45^\circ}$

$= \dfrac{\sqrt{3} + 1}{1 - \sqrt{3}} = \dfrac{\sqrt{3} + 1}{1 - \sqrt{3}}\cdot\dfrac{1 + \sqrt{3}}{1 + \sqrt{3}}$

$= \dfrac{4 + 2\sqrt{3}}{-2} = -2 - \sqrt{3}$

17. $\sin 195^\circ = \sin(225^\circ - 30^\circ)$

$= \sin 225^\circ\cos 30^\circ - \cos 225^\circ\sin 30^\circ$

$= -\sin 45^\circ\cos 30^\circ + \cos 45^\circ\sin 30^\circ$

$= -\dfrac{\sqrt{2}}{2}\cdot\dfrac{\sqrt{3}}{2} + \dfrac{\sqrt{2}}{2}\cdot\dfrac{1}{2}$

$= \dfrac{\sqrt{2}}{4}\left(1 - \sqrt{3}\right)$

$\cos 195^\circ = \cos(225^\circ - 30^\circ)$

$= \cos 225^\circ\cos 30^\circ + \sin 225^\circ\sin 30^\circ$

$= -\cos 45^\circ\cos 30^\circ - \sin 45^\circ\sin 30^\circ$

$= -\dfrac{\sqrt{2}}{2}\cdot\dfrac{\sqrt{3}}{2} - \dfrac{\sqrt{2}}{2}\cdot\dfrac{1}{2}$

$= -\dfrac{\sqrt{2}}{4}\left(\sqrt{3} + 1\right)$

$\tan 195^\circ = \tan(225^\circ - 30^\circ)$

$= \dfrac{\tan 225^\circ - \tan 30^\circ}{1 + \tan 225^\circ\tan 30^\circ}$

$= \dfrac{\tan 45^\circ - \tan 30^\circ}{1 + \tan 45^\circ\tan 30^\circ}$

$= \dfrac{1 - \left(\dfrac{\sqrt{3}}{3}\right)}{1 + \left(\dfrac{\sqrt{3}}{3}\right)} = \dfrac{3 - \sqrt{3}}{3 + \sqrt{3}}\cdot\dfrac{3 - \sqrt{3}}{3 - \sqrt{3}}$

$= \dfrac{12 - 6\sqrt{3}}{6} = 2 - \sqrt{3}$

19. $\dfrac{13\pi}{12} = \dfrac{3\pi}{4} + \dfrac{\pi}{3}$

$$\sin\frac{13\pi}{12} = \sin\left(\frac{3\pi}{4} + \frac{\pi}{3}\right)$$
$$= \sin\frac{3\pi}{4}\cos\frac{\pi}{3} + \cos\frac{3\pi}{4}\sin\frac{\pi}{3}$$
$$= \frac{\sqrt{2}}{2}\cdot\frac{1}{2} + \left(-\frac{\sqrt{2}}{2}\right)\left(\frac{\sqrt{3}}{2}\right)$$
$$= \frac{\sqrt{2}}{4}\left(1 - \sqrt{3}\right)$$

$$\cos\frac{13\pi}{12} = \cos\left(\frac{3\pi}{4} + \frac{\pi}{3}\right)$$
$$= \cos\frac{3\pi}{4}\cos\frac{\pi}{3} - \sin\frac{3\pi}{4}\sin\frac{\pi}{3}$$
$$= -\frac{\sqrt{2}}{2}\cdot\frac{1}{2} - \frac{\sqrt{2}}{2}\cdot\frac{\sqrt{3}}{2}$$
$$= -\frac{\sqrt{2}}{4}\left(1 + \sqrt{3}\right)$$

$$\tan\frac{13\pi}{12} = \tan\left(\frac{3\pi}{4} + \frac{\pi}{3}\right)$$
$$= \frac{\tan\left(\frac{3\pi}{4}\right) + \tan\left(\frac{\pi}{3}\right)}{1 - \tan\left(\frac{3\pi}{4}\right)\tan\left(\frac{\pi}{3}\right)}$$
$$= \frac{-1 + \sqrt{3}}{1 - (-1)\left(\sqrt{3}\right)}$$
$$= -\frac{1 - \sqrt{3}}{1 + \sqrt{3}}\cdot\frac{1 - \sqrt{3}}{1 - \sqrt{3}}$$
$$= -\frac{4 - 2\sqrt{3}}{-2}$$
$$= 2 - \sqrt{3}$$

21. $-\dfrac{13\pi}{12} = -\left(\dfrac{3\pi}{4} + \dfrac{\pi}{3}\right)$

$$\sin\left[-\left(\frac{3\pi}{4} + \frac{\pi}{3}\right)\right] = -\sin\left(\frac{3\pi}{4} + \frac{\pi}{3}\right)$$
$$= -\left[\sin\frac{3\pi}{4}\cos\frac{\pi}{3} + \cos\frac{3\pi}{4}\sin\frac{\pi}{3}\right]$$
$$= -\left[\frac{\sqrt{2}}{2}\left(\frac{1}{2}\right) + \left(-\frac{\sqrt{2}}{2}\right)\left(\frac{\sqrt{3}}{2}\right)\right]$$
$$= -\frac{\sqrt{2}}{4}\left(1 - \sqrt{3}\right)$$
$$= \frac{\sqrt{2}}{4}\left(\sqrt{3} - 1\right)$$

$$\cos\left[-\left(\frac{3\pi}{4} + \frac{\pi}{3}\right)\right] = \cos\left(\frac{3\pi}{4} + \frac{\pi}{3}\right)$$
$$= \cos\frac{3\pi}{4}\cos\frac{\pi}{3} - \sin\frac{3\pi}{4}\sin\frac{\pi}{3}$$
$$= -\frac{\sqrt{2}}{2}\left(\frac{1}{2}\right) - \frac{\sqrt{2}}{2}\left(\frac{\sqrt{3}}{2}\right)$$
$$= -\frac{\sqrt{2}}{4}\left(\sqrt{3} + 1\right)$$

$$\tan\left[-\left(\frac{3\pi}{4} + \frac{\pi}{3}\right)\right] = -\tan\left(\frac{3\pi}{4} + \frac{\pi}{3}\right)$$
$$= -\frac{\tan\frac{3\pi}{4} + \tan\frac{\pi}{3}}{1 - \tan\frac{3\pi}{4}\tan\frac{\pi}{3}}$$
$$= -\frac{-1 + \sqrt{3}}{1 - \left(-\sqrt{3}\right)}$$
$$= \frac{1 - \sqrt{3}}{1 + \sqrt{3}}\cdot\frac{1 - \sqrt{3}}{1 - \sqrt{3}}$$
$$= \frac{4 - 2\sqrt{3}}{-2}$$
$$= -2 + \sqrt{3}$$

23. $285° = 225° + 60°$

$$\sin 285° = \sin(225° + 60°) = \sin 225°\cos 60° + \cos 225°\sin 60°$$
$$= -\frac{\sqrt{2}}{2}\left(\frac{1}{2}\right) - \frac{\sqrt{2}}{2}\left(\frac{\sqrt{3}}{2}\right) = -\frac{\sqrt{2}}{4}\left(\sqrt{3} + 1\right)$$

$$\cos 285° = \cos(225° + 60°) = \cos 225°\cos 60° - \sin 225°\sin 60°$$
$$= -\frac{\sqrt{2}}{2}\left(\frac{1}{2}\right) - \left(-\frac{\sqrt{2}}{2}\right)\left(\frac{\sqrt{3}}{2}\right) = \frac{\sqrt{2}}{4}\left(\sqrt{3} - 1\right)$$

$$\tan 285° = \tan(225° + 60°) = \frac{\tan 225° + \tan 60°}{1 - \tan 225°\tan 60°}$$
$$= \frac{1 + \sqrt{3}}{1 - \sqrt{3}}\cdot\frac{1 + \sqrt{3}}{1 + \sqrt{3}} = \frac{4 + 2\sqrt{3}}{-2} = -2 - \sqrt{3} = -\left(2 + \sqrt{3}\right)$$

25. $-165° = -(120° + 45°)$

$$\sin(-165°) = \sin[-(120° + 45°)] = -\sin(120° + 45°) = -[\sin 120° \cos 45° + \cos 120° \sin 45°]$$

$$= -\left[\frac{\sqrt{3}}{2} \cdot \frac{\sqrt{2}}{2} - \frac{1}{2} \cdot \frac{\sqrt{2}}{2}\right] = -\frac{\sqrt{2}}{4}(\sqrt{3} - 1)$$

$$\cos(-165°) = \cos[-(120° + 45°)] = \cos(120° + 45°) = \cos 120° \cos 45° - \sin 120° \sin 45°$$

$$= -\frac{1}{2} \cdot \frac{\sqrt{2}}{2} - \frac{\sqrt{3}}{2} \cdot \frac{\sqrt{2}}{2} = -\frac{\sqrt{2}}{4}(1 + \sqrt{3})$$

$$\tan(-165°) = \tan[-(120° + 45°)] = -\tan(120° + \tan 45°) = -\frac{\tan 120° + \tan 45°}{1 - \tan 120° \tan 45°}$$

$$= -\frac{-\sqrt{3} + 1}{1 - (-\sqrt{3})(1)} = -\frac{1 - \sqrt{3}}{1 + \sqrt{3}} \cdot \frac{1 - \sqrt{3}}{1 - \sqrt{3}} = -\frac{4 - 2\sqrt{3}}{-2} = 2 - \sqrt{3}$$

27. $\sin 3 \cos 1.2 - \cos 3 \sin 1.2 = \sin(3 - 1.2) = \sin 1.8$

29. $\sin 60° \cos 15° + \cos 60° \sin 15° = \sin(60° + 15°)$
$= \sin 75°$

31. $\dfrac{\tan 45° - \tan 30°}{1 + \tan 45° \tan 30°} = \tan(45° - 30°) = \tan 15°$

33. $\cos 3x \cos 2y + \sin 3x \sin 2y = \cos(3x - 2y)$

35. $\sin \dfrac{\pi}{12} \cos \dfrac{\pi}{4} + \cos \dfrac{\pi}{12} \sin \dfrac{\pi}{4} = \sin\left(\dfrac{\pi}{12} + \dfrac{\pi}{4}\right)$
$= \sin \dfrac{\pi}{3}$
$= \dfrac{\sqrt{3}}{2}$

37. $\sin 120° \cos 60° - \cos 120° \sin 60° = \sin(120° - 60°)$
$= \sin 60°$
$= \dfrac{\sqrt{3}}{2}$

39. $\dfrac{\tan(5\pi/6) - \tan(\pi/6)}{1 + \tan(5\pi/6)\tan(\pi/6)} = \tan\left(\dfrac{5\pi}{6} - \dfrac{\pi}{6}\right)$
$= \tan \dfrac{2\pi}{3}$
$= -\sqrt{3}$

For Exercises 41–45, you have:

$\sin u = \frac{5}{13}$, u in Quadrant II $\Rightarrow$ $\cos u = -\frac{12}{13}$, $\tan u = -\frac{5}{12}$

$\cos v = -\frac{3}{5}$, v in Quadrant II $\Rightarrow$ $\sin v = \frac{4}{5}$, $\tan v = -\frac{4}{3}$

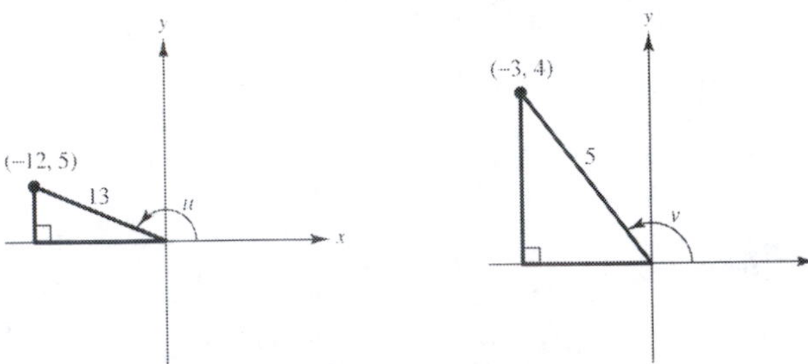

Figures for Exercises 41–45

41. $\sin(u + v) = \sin u \cos v + \cos u \sin v = \left(\frac{5}{13}\right)\left(-\frac{3}{5}\right) + \left(-\frac{12}{13}\right)\left(\frac{4}{5}\right) = -\frac{63}{65}$

43. $\tan(u + v) = \dfrac{\tan u + \tan v}{1 - \tan u \tan v} = \dfrac{-\dfrac{5}{12} + \left(-\dfrac{4}{3}\right)}{1 - \left(-\dfrac{5}{12}\right)\left(-\dfrac{4}{3}\right)} = \dfrac{-\dfrac{21}{12}}{1 - \dfrac{5}{9}} = \left(-\dfrac{7}{4}\right)\left(\dfrac{9}{4}\right) = -\dfrac{63}{16}$

45. $\sec(v-u) = \dfrac{1}{\cos(v-u)} = \dfrac{1}{\cos v \cos u + \sin v \sin u} = \dfrac{1}{\left(-\dfrac{3}{5}\right)\left(-\dfrac{12}{13}\right) + \left(\dfrac{4}{5}\right)\left(\dfrac{5}{13}\right)} = \dfrac{1}{\left(\dfrac{36}{65}\right) + \left(\dfrac{20}{65}\right)} = \dfrac{1}{\dfrac{56}{65}} = \dfrac{65}{56}$

For Exercises 47–51, you have:

$\sin u = -\frac{7}{25}$, u in Quadrant III $\Rightarrow \cos u = -\frac{24}{25}$, $\tan u = \frac{7}{24}$

$\cos v = -\frac{4}{5}$, v in Quadrant III $\Rightarrow \sin v = -\frac{3}{5}$, $\tan v = \frac{3}{4}$

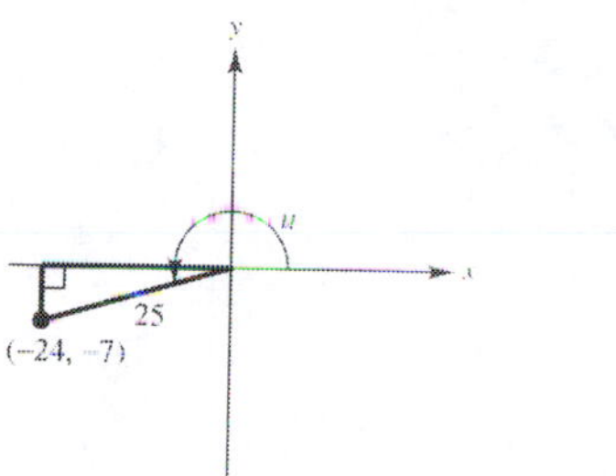

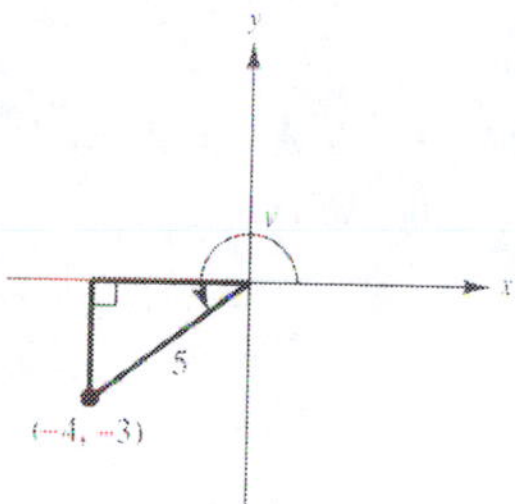

Figures for Exercises 47–51

47. $\cos(u+v) = \cos u \cos v - \sin u \sin v$

$= \left(-\frac{24}{25}\right)\left(-\frac{4}{5}\right) - \left(-\frac{7}{25}\right)\left(-\frac{3}{5}\right)$

$= \frac{3}{5}$

49. $\tan(u-v) = \dfrac{\tan u - \tan v}{1 + \tan u \tan v}$

$= \dfrac{\dfrac{7}{24} - \dfrac{3}{4}}{1 + \left(\dfrac{7}{24}\right)\left(\dfrac{3}{4}\right)} = \dfrac{-\dfrac{11}{24}}{\dfrac{39}{32}} = -\dfrac{44}{117}$

51. $\csc(u-v) = \dfrac{1}{\sin(u-v)} = \dfrac{1}{\sin u \cos v - \cos u \sin v}$

$= \dfrac{1}{\left(-\dfrac{7}{25}\right)\left(-\dfrac{4}{5}\right) - \left(-\dfrac{24}{25}\right)\left(-\dfrac{3}{5}\right)}$

$= \dfrac{1}{-\dfrac{44}{125}}$

$= -\dfrac{125}{44}$

53. $\sin(\arcsin x + \arccos x) = \sin(\arcsin x)\cos(\arccos x) + \sin(\arccos x)\cos(\arcsin x)$

$= x \cdot x + \sqrt{1-x^2} \cdot \sqrt{1-x^2}$

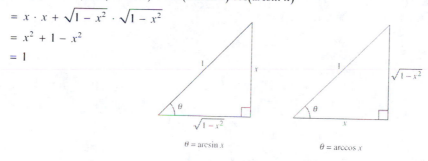

$= x^2 + 1 - x^2$

$= 1$

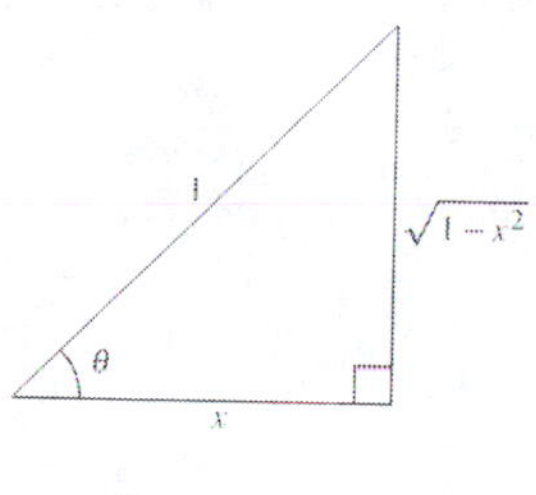

55. $\cos(\arccos x + \arcsin x) = \cos(\arccos x)\cos(\arcsin x) - \sin(\arccos x)\sin(\arcsin x)$

$= x \cdot \sqrt{1-x^2} - \sqrt{1-x^2} \cdot x$

$= 0$

(Use the triangles in Exercise 53.)

57. $\sin\left(\dfrac{\pi}{2} - x\right) = \sin\dfrac{\pi}{2}\cos x - \cos\dfrac{\pi}{2}\sin x$

$= (1)(\cos x) - (0)(\sin x)$

$= \cos x$

59. $\sin\left(\dfrac{\pi}{6} + x\right) = \sin\dfrac{\pi}{6}\cos x + \cos\dfrac{\pi}{6}\sin x$

$= \dfrac{1}{2}\left(\cos x + \sqrt{3}\sin x\right)$

61. $\cos(\pi - \theta) + \sin\left(\dfrac{\pi}{2} + \theta\right) = \cos \pi \cos \theta + \sin \pi \sin \theta + \sin \dfrac{\pi}{2} \cos \theta + \cos \dfrac{\pi}{2} \sin \theta$

$= (-1)(\cos \theta) + (0)(\sin \theta) + (1)(\cos \theta) + (\sin \theta)(0)$

$= -\cos \theta + \cos \theta$

$= 0$

63. $\cos(x + y)\cos(x - y) = (\cos x \cos y - \sin x \sin y)(\cos x \cos y + \sin x \sin y)$

$= \cos^2 x \cos^2 y - \sin^2 x \sin^2 y$

$= \cos^2 x(1 - \sin^2 y) - \sin^2 x \sin^2 y$

$= \cos^2 x - \cos^2 x \sin^2 y - \sin^2 x \sin^2 y$

$= \cos^2 x - \sin^2 y(\cos^2 x + \sin^2 x)$

$= \cos^2 x - \sin^2 y$

65. $\cos\left(\dfrac{3\pi}{2} - x\right) = \cos \dfrac{3\pi}{2} \cos x + \sin \dfrac{3\pi}{2} \sin x$

$= (0)(\cos x) + (-1)(\sin x)$

$= -\sin x$

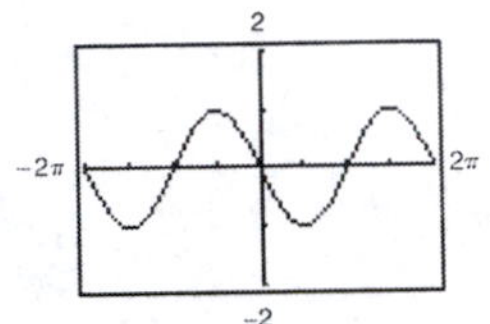

67. $\sin\left(\dfrac{3\pi}{2} + \theta\right) = \sin \dfrac{3\pi}{2} \cos \theta + \cos \dfrac{3\pi}{2} \sin \theta$

$= (-1)(\cos \theta) + (0)(\sin \theta)$

$= -\cos \theta$

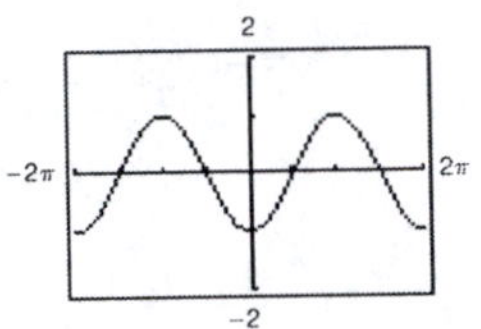

69.

$$\begin{aligned} \sin(x + \pi) - \sin x + 1 &= 0 \\ \sin x \cos \pi + \cos x \sin \pi - \sin x + 1 &= 0 \\ (\sin x)(-1) + (\cos x)(0) - \sin x + 1 &= 0 \\ -2 \sin x + 1 &= 0 \\ \sin x &= \frac{1}{2} \\ x &= \frac{\pi}{6}, \frac{5\pi}{6} \end{aligned}$$

71.

$$\begin{aligned} \cos\left(x + \frac{\pi}{4}\right) - \cos\left(x - \frac{\pi}{4}\right) &= 1 \\ \cos x \cos \frac{\pi}{4} - \sin x \sin \frac{\pi}{4} - \left(\cos x \cos \frac{\pi}{4} + \sin x \sin \frac{\pi}{4}\right) &= 1 \\ -2 \sin x\left(\frac{\sqrt{2}}{2}\right) &= 1 \\ -\sqrt{2} \sin x &= 1 \\ \sin x &= -\frac{1}{\sqrt{2}} \\ \sin x &= -\frac{\sqrt{2}}{2} \\ x &= \frac{5\pi}{4}, \frac{7\pi}{4} \end{aligned}$$

73.
$$\tan(x + \pi) + 2\sin(x + \pi) = 0$$
$$\frac{\tan x + \tan \pi}{1 - \tan x \tan \pi} + 2(\sin x \cos \pi + \cos x \sin \pi) = 0$$
$$\frac{\tan x + 0}{1 - \tan x(0)} + 2[\sin x(-1) + \cos x(0)] = 0$$
$$\frac{\tan x}{1} - 2\sin x = 0$$
$$\frac{\sin x}{\cos x} = 2\sin x$$
$$\sin x = 2\sin x \cos x$$
$$\sin x(1 - 2\cos x) = 0$$
$$\sin x = 0 \quad \text{or} \quad \cos x = \frac{1}{2}$$
$$x = 0, \pi \qquad x = \frac{\pi}{3}, \frac{5\pi}{3}$$

75. $\cos\left(x + \frac{\pi}{4}\right) + \cos\left(x - \frac{\pi}{4}\right) = 1$

Graph $y_1 = \cos\left(x + \frac{\pi}{4}\right) + \cos\left(x - \frac{\pi}{4}\right)$ and $y_2 = 1$.

$x = \frac{\pi}{4}, \frac{7\pi}{4}$

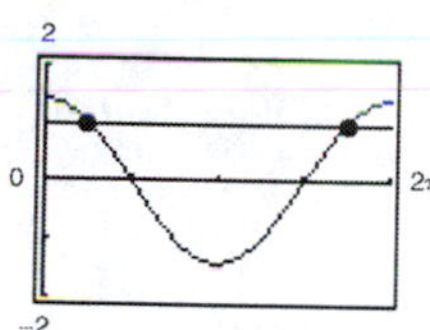

77. $\sin\left(x + \frac{\pi}{2}\right) + \cos^2 x = 0$

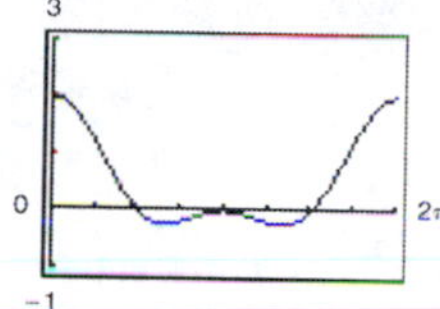

$x = \frac{\pi}{2}, \pi, \frac{3\pi}{2}$

79. $y = \frac{1}{3}\sin 2t + \frac{1}{4}\cos 2t$

(a) $a = \frac{1}{3}, b = \frac{1}{4}, B = 2$

$C = \arctan\frac{b}{a} = \arctan\frac{3}{4} \approx 0.6435$

$y \approx \sqrt{\left(\frac{1}{3}\right)^2 + \left(\frac{1}{4}\right)^2}\sin(2t + 0.6435) = \frac{5}{12}\sin(2t + 0.6435)$

(b) Amplitude: $\frac{5}{12}$ feet

(c) Frequency: $\frac{1}{\text{period}} = \frac{B}{2\pi} = \frac{2}{2\pi} = \frac{1}{\pi}$ cycle per second

81. True.

$\sin(u + v) = \sin u \cos v + \cos u \sin v$

$\sin(u - v) = \sin u \cos v - \cos u \sin v$

So, $\sin(u \pm v) = \sin u \cos v \pm \cos u \sin v$.

83. False.

$$\tan\left(x - \frac{\pi}{4}\right) = \frac{\tan x - \tan(\pi/4)}{1 + \tan x \tan(\pi/4)} = \frac{\tan x - 1}{1 + \tan x}$$

85. (a) The domains of f and g are the same, all real numbers h, except $h = 0$.

(b)

h	0.5	0.2	0.1	0.05	0.02	0.01
$f(h)$	0.267	0.410	0.456	0.478	0.491	0.496
$g(h)$	0.267	0.410	0.456	0.478	0.491	0.496

(c)

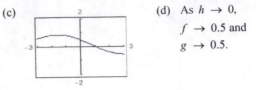

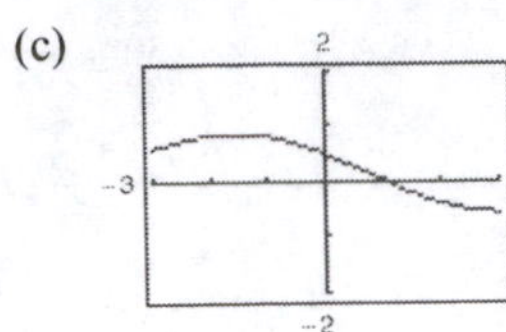

(d) As $h \to 0$,
$f \to 0.5$ and
$g \to 0.5$.

87.
$$\begin{aligned}\cos(n\pi + \theta) &= \cos n\pi \cos \theta - \sin n\pi \sin \theta \\ &= (-1)^n(\cos \theta) - (0)(\sin \theta) \\ &= (-1)^n(\cos \theta), \text{ where } n \text{ is an integer.}\end{aligned}$$

89. $C = \arctan \dfrac{b}{a} \Rightarrow \sin C = \dfrac{b}{\sqrt{a^2 + b^2}}, \cos C = \dfrac{a}{\sqrt{a^2 + b^2}}$

$$\sqrt{a^2 + b^2} \sin(B\theta + C) = \sqrt{a^2 + b^2}\left(\sin B\theta \cdot \frac{a}{\sqrt{a^2 + b^2}} + \frac{b}{\sqrt{a^2 + b^2}} \cdot \cos B\theta\right) = a \sin B\theta + b \cos B\theta$$

91. $\sin \theta + \cos \theta$

$a = 1, b = 1, B = 1$

(a) $C = \arctan \dfrac{b}{a} = \arctan 1 = \dfrac{\pi}{4}$

$$\begin{aligned}\sin \theta + \cos \theta &= \sqrt{a^2 + b^2} \sin(B\theta + C) \\ &= \sqrt{2} \sin\left(\theta + \frac{\pi}{4}\right)\end{aligned}$$

(b) $C = \arctan \dfrac{a}{b} = \arctan 1 = \dfrac{\pi}{4}$

$$\begin{aligned}\sin \theta + \cos \theta &= \sqrt{a^2 + b^2} \cos(B\theta - C) \\ &= \sqrt{2} \cos\left(\theta - \frac{\pi}{4}\right)\end{aligned}$$

93. $12 \sin 3\theta + 5 \cos 3\theta$

$a = 12, b = 5, B = 3$

(a) $C = \arctan \dfrac{b}{a} = \arctan \dfrac{5}{12} \approx 0.3948$

$$\begin{aligned}12 \sin 3\theta + 5 \cos 3\theta &= \sqrt{a^2 + b^2} \sin(B\theta + C) \\ &\approx 13 \sin(3\theta + 0.3948)\end{aligned}$$

(b) $C = \arctan \dfrac{a}{b} = \arctan \dfrac{12}{5} \approx 1.1760$

$$\begin{aligned}12 \sin 3\theta + 5 \cos 3\theta &= \sqrt{a^2 + b^2} \cos(B\theta - C) \\ &\approx 13 \cos(3\theta - 1.1760)\end{aligned}$$

95. $C = \arctan \dfrac{b}{a} = \dfrac{\pi}{4} \Rightarrow a = b, a > 0, b > 0$

$\sqrt{a^2 + b^2} = 2 \Rightarrow a = b = \sqrt{2}$

$B = 1$

$2 \sin\left(\theta + \dfrac{\pi}{4}\right) = \sqrt{2} \sin \theta + \sqrt{2} \cos \theta$

97.

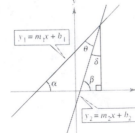

$m_1 = \tan \alpha$ and $m_2 = \tan \beta$

$\beta + \delta = 90° \Rightarrow \delta = 90° - \beta$

$\alpha + \theta + \delta = 90° \Rightarrow \alpha + \theta + (90° - \beta) = 90° \Rightarrow \theta = \beta - \alpha$

So, $\theta = \arctan m_2 - \arctan m_1$.

For $y = x$ and $y = \sqrt{3}x$ you have $m_1 = 1$ and $m_2 = \sqrt{3}$.

$\theta = \arctan\sqrt{3} - \arctan 1 = 60° - 45° = 15°$

99. $y_1 = \cos(x + 2),\ y_2 = \cos x + \cos 2$

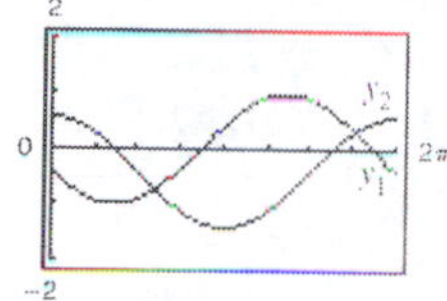

No, $y_1 \neq y_2$ because their graphs are different.

101. (a) To prove the identity for $\sin(u + v)$ you first need to prove the identity for $\cos(u - v)$.
Assume $0 < v < u < 2\pi$ and locate u, v, and $u - v$ on the unit circle.

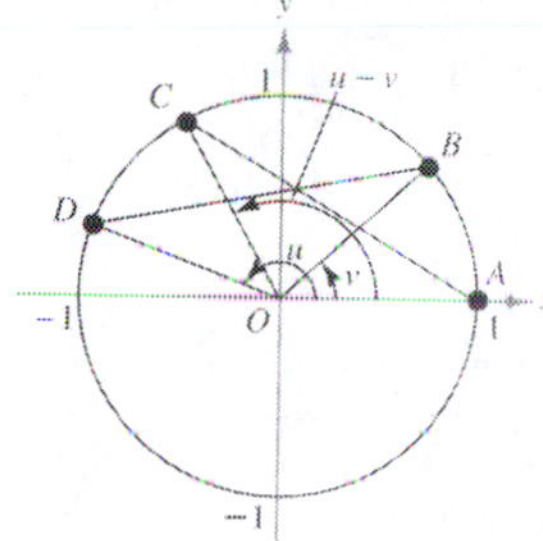

The coordinates of the points on the circle are:

$A = (1, 0)$, $B = (\cos v, \sin v)$, $C = (\cos(u - v), \sin(u - v))$, and $D = (\cos u, \sin u)$.

Because $\angle DOB = \angle COA$, chords AC and BD are equal. By the Distance Formula:

$$\sqrt{[\cos(u - v) - 1]^2 + [\sin(u - v) - 0]^2} = \sqrt{(\cos u - \cos v)^2 + (\sin u - \sin v)^2}$$

$$\cos^2(u - v) - 2\cos(u - v) + 1 + \sin^2(u - v) = \cos^2 u - 2\cos u \cos v + \cos^2 v + \sin^2 u - 2\sin u \sin v + \sin^2 v$$

$$[\cos^2(u - v) + \sin^2(u - v)] + 1 - 2\cos(u - v) = (\cos^2 u + \sin^2 u) + (\cos^2 v + \sin^2 v) - 2\cos u \cos v - 2\sin u \sin v$$

$$2 - 2\cos(u - v) = 2 - 2\cos u \cos v - 2\sin u \sin v$$

$$-2\cos(u - v) = -2(\cos u \cos v + \sin u \sin v)$$

$$\cos(u - v) = \cos u \cos v + \sin u \sin v$$

Now, to prove the identity for $\sin(u + v)$, use cofunction identities.

$$\begin{aligned}\sin(u + v) &= \cos\left[\frac{\pi}{2} - (u + v)\right] = \cos\left[\left(\frac{\pi}{2} - u\right) - v\right] \\ &= \cos\left(\frac{\pi}{2} - u\right)\cos v + \sin\left(\frac{\pi}{2} - u\right)\sin v \\ &= \sin u \cos v + \cos u \sin v\end{aligned}$$

(b) First, prove $\cos(u - v) = \cos u \cos v + \sin u \sin v$ using the figure containing points

$A(1, 0)$

$B(\cos(u - v), \sin(u - v))$

$C(\cos v, \sin v)$

$D(\cos u, \sin u)$

on the unit circle.

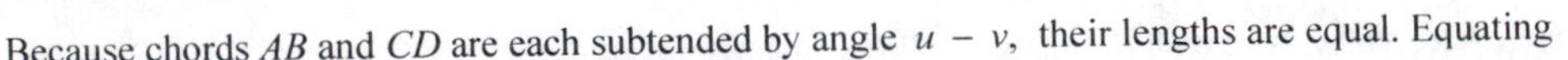

Because chords AB and CD are each subtended by angle $u - v$, their lengths are equal. Equating $[d(A, B)]^2 = [d(C, D)]^2$ you have $(\cos(u - v) - 1)^2 + \sin^2(u - v) = (\cos u - \cos v)^2 + (\sin u - \sin v)^2$.

Simplifying and solving for $\cos(u - v)$, you have $\cos(u - v) = \cos u \cos v + \sin u \sin v$.

Using $\sin\theta = \cos\left(\frac{\pi}{2} - \theta\right)$,

$$\sin(u - v) = \cos\left[\frac{\pi}{2} - (u - v)\right] = \cos\left[\left(\frac{\pi}{2} - u\right) - (-v)\right] = \cos\left(\frac{\pi}{2} - u\right)\cos(-v) + \sin\left(\frac{\pi}{2} - u\right)\sin(-v)$$

$$= \sin u \cos v - \cos u \sin v$$

Section 7.5 Multiple-Angle and Product-to-Sum Formulas

1. $2 \sin u \cos u$

3. $\tan^2 u$

5. $\frac{1}{2}[\sin(u + v) + \sin(u - v)]$

7.

$$\sin 2x - \sin x = 0$$
$$2 \sin x \cos x - \sin x = 0$$
$$\sin x(2 \cos x - 1) = 0$$

$\sin x = 0$ or $2 \cos x - 1 = 0$

$x = 0, \pi$ $\qquad \cos x = \frac{1}{2}$

$x = \frac{\pi}{3}, \frac{5\pi}{3}$

$$x = 0, \frac{\pi}{3}, \pi, \frac{5\pi}{3}$$

9.

$$\cos 2x - \cos x = 0$$
$$\cos 2x = \cos x$$
$$\cos^2 x - \sin^2 x = \cos x$$
$$\cos^2 x - (1 - \cos^2 x) - \cos x = 0$$
$$2 \cos^2 x - \cos x - 1 = 0$$
$$(2 \cos x + 1)(\cos x - 1) = 0$$

$2 \cos x + 1 = 0$ or $\cos x - 1 = 0$

$\cos x = -\frac{1}{2}$ $\qquad \cos x = 1$

$x = \frac{2\pi}{3}, \frac{4\pi}{3}$ $\qquad x = 0$

11.

$$\sin 4x = -2 \sin 2x$$
$$\sin 4x + 2 \sin 2x = 0$$
$$2 \sin 2x \cos 2x + 2 \sin 2x = 0$$
$$2 \sin 2x(\cos 2x + 1) = 0$$

$2 \sin 2x = 0$ or $\cos 2x + 1 = 0$

$\sin 2x = 0$ $\qquad \cos 2x = -1$

$2x = n\pi$ $\qquad 2x = \pi + 2n\pi$

$x = \frac{n}{2}\pi$ $\qquad x = \frac{\pi}{2} + n\pi$

$x = 0, \frac{\pi}{2}, \pi, \frac{3\pi}{2}$ $\qquad x = \frac{\pi}{2}, \frac{3\pi}{2}$

13.

$$\tan 2x - \cot x = 0$$
$$\frac{2\tan x}{1 - \tan^2 x} = \cot x$$
$$2\tan x = \cot x(1 - \tan^2 x)$$
$$2\tan x = \cot x - \cot x \tan^2 x$$
$$2\tan x = \cot x - \tan x$$
$$3\tan x = \cot x$$
$$3\tan x - \cot x = 0$$
$$3\tan x - \frac{1}{\tan x} = 0$$
$$\frac{3\tan^2 x - 1}{\tan x} = 0$$
$$\frac{1}{\tan x}(3\tan^2 x - 1) = 0$$
$$\cot x(3\tan^2 x - 1) = 0$$

$\cot x = 0$ or $3\tan^2 x - 1 = 0$

$x = \frac{\pi}{2}, \frac{3\pi}{2}$ $\qquad \tan^2 x = \frac{1}{3}$

$\tan x = \pm\frac{\sqrt{3}}{3}$

$x = \frac{\pi}{6}, \frac{5\pi}{6}, \frac{7\pi}{6}, \frac{11\pi}{6}$

$$x = \frac{\pi}{6}, \frac{\pi}{2}, \frac{5\pi}{6}, \frac{7\pi}{6}, \frac{3\pi}{2}, \frac{11\pi}{6}$$

15. $6\sin x\cos x = 3(2\sin x\cos x)$

$= 3\sin 2x$

17. $6\cos^2 x - 3 = 3(2\cos^2 x - 1)$

$= 3\cos 2x$

19. $4 - 8\sin^2 x = 4(1 - 2\sin^2 x)$

$= 4\cos 2x$

21. $\sin u = -\frac{3}{5}, \frac{3\pi}{2} < u < 2\pi$

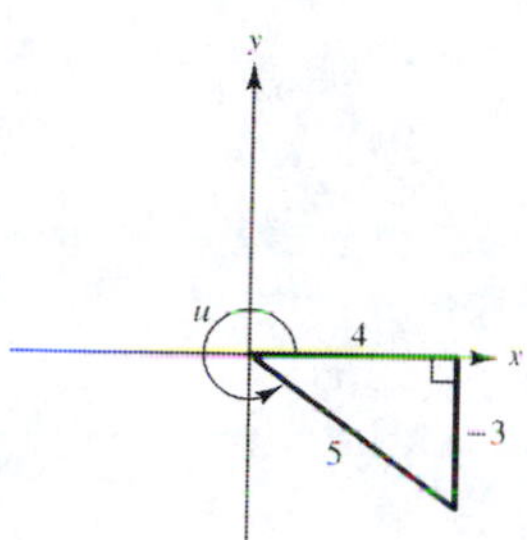

$$\sin 2u = 2\sin u\cos u = 2\left(-\frac{3}{5}\right)\left(\frac{4}{5}\right) = -\frac{24}{25}$$
$$\cos 2u = \cos^2 u - \sin^2 u = \frac{16}{25} - \frac{9}{25} = \frac{7}{25}$$
$$\tan 2u = \frac{2\tan u}{1 - \tan^2 u} = \frac{2\left(-\frac{3}{4}\right)}{1 - \frac{9}{16}} = -\frac{3}{2}\left(\frac{16}{7}\right) = -\frac{24}{7}$$

23. $\tan u = \frac{3}{5}, 0 < u < \frac{\pi}{2}$

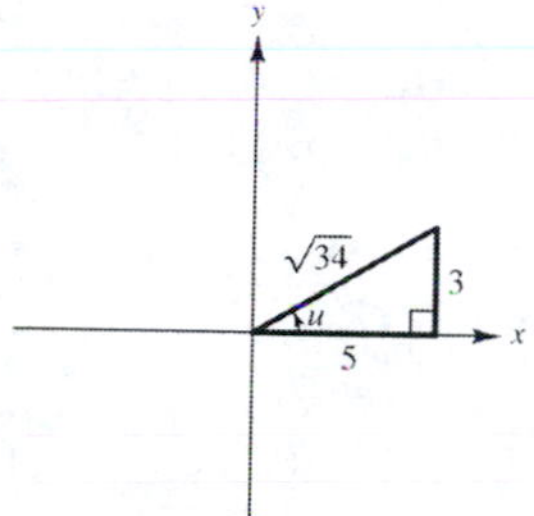

$$\sin 2u = 2\sin u\cos u = 2\left(\frac{3}{\sqrt{34}}\right)\left(\frac{5}{\sqrt{34}}\right) = \frac{15}{17}$$
$$\cos 2u = \cos^2 u - \sin^2 u = \frac{25}{34} - \frac{9}{34} = \frac{8}{17}$$
$$\tan 2u = \frac{2\tan u}{1 - \tan^2 u} = \frac{2\left(\frac{3}{5}\right)}{1 - \frac{9}{25}} = \frac{6}{5}\left(\frac{25}{16}\right) = \frac{15}{8}$$

25.
$$\cos 4x = \cos(2x + 2x)$$
$$= \cos 2x\cos 2x - \sin 2x\sin 2x$$
$$= \cos^2 2x - \sin^2 2x$$
$$= \cos^2 2x - (1 - \cos^2 2x)$$
$$= 2\cos^2 2x - 1$$
$$= 2(\cos 2x)^2 - 1$$
$$= 2(2\cos^2 x - 1)^2 - 1$$
$$= 2(4\cos^4 x - 4\cos x + 1) - 1$$
$$= 8\cos^4 x - 8\cos x + 1$$

27. $\cos^4 x = (\cos^2 x)(\cos^2 x) = \left(\dfrac{1 + \cos 2x}{2}\right)\left(\dfrac{1 + \cos 2x}{2}\right) = \dfrac{1 + 2\cos 2x + \cos^2 2x}{4}$

$$= \frac{1 + 2\cos 2x + \dfrac{1 + \cos 4x}{2}}{4}$$

$$= \frac{2 + 4\cos 2x + 1 + \cos 4x}{8}$$

$$= \frac{3 + 4\cos 2x + \cos 4x}{8}$$

$$= \frac{1}{8}(3 + 4\cos 2x + \cos 4x)$$

29. $\tan^4 2x = (\tan^2 2x)^2$

$$= \left(\frac{1 - \cos 4x}{1 + \cos 4x}\right)^2$$

$$= \frac{1 - 2\cos 4x + \cos^2 4x}{1 + 2\cos 4x + \cos^2 4x}$$

$$= \frac{1 - 2\cos 4x + \dfrac{1 + \cos 8x}{2}}{1 + 2\cos 4x + \dfrac{1 + \cos 8x}{2}}$$

$$= \frac{\frac{1}{2}(2 - 4\cos 4x + 1 + \cos 8x)}{\frac{1}{2}(2 + 4\cos 4x + 1 + \cos 8x)}$$

$$= \frac{3 - 4\cos 4x + \cos 8x}{3 + 4\cos 4x + \cos 8x}$$

31. $\sin^2 2x \cos^2 2x = \left(\dfrac{1 - \cos 4x}{2}\right)\left(\dfrac{1 + \cos 4x}{2}\right)$

$$= \frac{1}{4}(1 - \cos^2 4x)$$

$$= \frac{1}{4}\left(1 - \frac{1 + \cos 8x}{2}\right)$$

$$= \frac{1}{4} - \frac{1}{8} - \frac{1}{8}\cos 8x$$

$$= \frac{1}{8} - \frac{1}{8}\cos 8x$$

$$= \frac{1}{8}(1 - \cos 8x)$$

33. $\sin 75° = \sin\left(\dfrac{1}{2} \cdot 150°\right) = \sqrt{\dfrac{1 - \cos 150°}{2}} = \sqrt{\dfrac{1 + (\sqrt{3}/2)}{2}} = \dfrac{1}{2}\sqrt{2 + \sqrt{3}}$

$\cos 75° = \cos\left(\dfrac{1}{2} \cdot 150°\right) = \sqrt{\dfrac{1 + \cos 150°}{2}} = \sqrt{\dfrac{1 - (\sqrt{3}/2)}{2}} = \dfrac{1}{2}\sqrt{2 - \sqrt{3}}$

$\tan 75° = \tan\left(\dfrac{1}{2} \cdot 150°\right) = \dfrac{\sin 150°}{1 + \cos 150°} = \dfrac{1/2}{1 - (\sqrt{3}/2)} = \dfrac{1}{2 - \sqrt{3}} \cdot \dfrac{2 + \sqrt{3}}{2 + \sqrt{3}} = \dfrac{2 + \sqrt{3}}{4 - 3} = 2 + \sqrt{3}$

35. $\sin \dfrac{\pi}{8} = \sin\left[\dfrac{1}{2}\left(\dfrac{\pi}{4}\right)\right] = \sqrt{\dfrac{1 - \cos \frac{\pi}{4}}{2}} = \dfrac{1}{2}\sqrt{2 - \sqrt{2}}$

$\cos \dfrac{\pi}{8} = \cos\left[\dfrac{1}{2}\left(\dfrac{\pi}{4}\right)\right] = \sqrt{\dfrac{1 + \cos \frac{\pi}{4}}{2}} = \dfrac{1}{2}\sqrt{2 + \sqrt{2}}$

$\tan \dfrac{\pi}{8} = \tan\left[\dfrac{1}{2}\left(\dfrac{\pi}{4}\right)\right] = \dfrac{\sin \frac{\pi}{4}}{1 + \cos \frac{\pi}{4}} = \dfrac{\frac{\sqrt{2}}{2}}{1 + \frac{\sqrt{2}}{2}} = \sqrt{2} - 1$

37. $\cos u = \frac{7}{25}, 0 < u < \frac{\pi}{2}$

(a) Because u is in Quadrant I, $\frac{u}{2}$ is also in Quadrant I.

(b) $\sin\frac{u}{2} = \sqrt{\frac{1 - \cos u}{2}} = \sqrt{\frac{1 - \frac{7}{25}}{2}} = \sqrt{\frac{9}{25}} = \frac{3}{5}$

$\cos\frac{u}{2} = \sqrt{\frac{1 + \cos u}{2}} = \sqrt{\frac{1 + \frac{7}{25}}{2}} = \sqrt{\frac{16}{25}} = \frac{4}{5}$

$\tan\frac{u}{2} = \frac{1 - \cos u}{\sin u} = \frac{1 - \frac{7}{25}}{\frac{24}{25}} = \frac{3}{4}$

39. $\tan u = -\frac{5}{12}, \frac{3\pi}{2} < u < 2\pi$

(a) Because u is in Quadrant IV, $\frac{u}{2}$ is in Quadrant II.

(b) $\sin\frac{u}{2} = \sqrt{\frac{1 - \cos u}{2}} = \sqrt{\frac{1 - \frac{12}{13}}{2}} = \sqrt{\frac{1}{26}} = \frac{\sqrt{26}}{26}$

$\cos\frac{u}{2} = -\sqrt{\frac{1 + \cos u}{2}} = -\sqrt{\frac{1 + \frac{12}{13}}{2}} = -\sqrt{\frac{25}{26}} = -\frac{5\sqrt{26}}{26}$

$\tan\frac{u}{2} = \frac{1 - \cos u}{\sin u} = \frac{1 - \frac{12}{13}}{\left(-\frac{5}{13}\right)} = -\frac{1}{5}$

41. $\sqrt{\frac{1 - \cos 6x}{2}} = |\sin 3x|$

43. $-\sqrt{\frac{1 - \cos 8x}{1 + \cos 8x}} = -\frac{\sqrt{\frac{1 - \cos 8x}{2}}}{\sqrt{\frac{1 + \cos 8x}{2}}}$

$= -\left|\frac{\sin 4x}{\cos 4x}\right|$

$= -|\tan 4x|$

45. $\sin\frac{x}{2} + \cos x = 0$

$\pm\sqrt{\frac{1 - \cos x}{2}} = -\cos x$

$\frac{1 - \cos x}{2} = \cos^2 x$

$0 = 2\cos^2 x + \cos x - 1$

$= (2\cos x - 1)(\cos x + 1)$

$\cos x = \frac{1}{2}$ or $\cos x = -1$

$x = \frac{\pi}{3}, \frac{5\pi}{3}$ $\quad x = \pi$

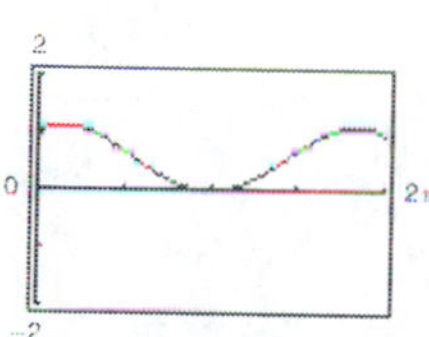

By checking these values in the original equation, $x = \pi/3$ and $x = 5\pi/3$ are extraneous, and $x = \pi$ is the only solution.

47.

$$\cos\frac{x}{2} - \sin x = 0$$
$$\pm\sqrt{\frac{1+\cos x}{2}} = \sin x$$
$$\frac{1+\cos x}{2} = \sin^2 x$$
$$1 + \cos x = 2\sin^2 x$$
$$1 + \cos x = 2 - 2\cos^2 x$$
$$2\cos^2 x + \cos x - 1 = 0$$
$$(2\cos x - 1)(\cos x + 1) = 0$$

$2\cos x - 1 = 0$ or $\cos x + 1 = 0$

$\cos x = \frac{1}{2}$ $\qquad$ $\cos x = -1$

$x = \frac{\pi}{3}, \frac{5\pi}{3}$ $\qquad$ $x = \pi$

$$x = \frac{\pi}{3}, \pi, \frac{5\pi}{3}$$

$\pi/3$, π, and $5\pi/3$ are all solutions to the equation.

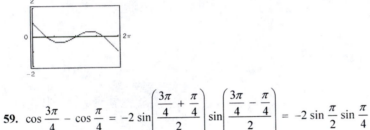

49. $\sin 5\theta \sin 3\theta = \frac{1}{2}[\cos(5\theta - 3\theta) - \cos(5\theta + 3\theta)]$

$= \frac{1}{2}(\cos 2\theta - \cos 8\theta)$

51. $\cos 2\theta \cos 4\theta = \frac{1}{2}[\cos(2\theta - 4\theta) + \cos(2\theta + 4\theta)]$

$= \frac{1}{2}[\cos(-2\theta) + \cos 6\theta]$

53. $\sin 5\theta - \sin 3\theta = 2\cos\left(\frac{5\theta + 3\theta}{2}\right)\sin\left(\frac{5\theta - 3\theta}{2}\right)$

$= 2\cos 4\theta \sin\theta$

55. $\cos 6x + \cos 2x = 2\cos\left(\frac{6x + 2x}{2}\right)\cos\left(\frac{6x - 2x}{2}\right)$

$= 2\cos 4x \cos 2x$

57. $\sin 75° + \sin 15° = 2\sin\left(\frac{75° + 15°}{2}\right)\cos\left(\frac{75° - 15°}{2}\right)$

$= 2\sin 45° \cos 30°$

$= 2\left(\frac{\sqrt{2}}{2}\right)\left(\frac{\sqrt{3}}{2}\right)$

$= \frac{\sqrt{6}}{2}$

59. $\cos\frac{3\pi}{4} - \cos\frac{\pi}{4} = -2\sin\left(\frac{\frac{3\pi}{4} + \frac{\pi}{4}}{2}\right)\sin\left(\frac{\frac{3\pi}{4} - \frac{\pi}{4}}{2}\right) = -2\sin\frac{\pi}{2}\sin\frac{\pi}{4}$

$\cos\frac{3\pi}{4} - \cos\frac{\pi}{4} = -\frac{\sqrt{2}}{2} - \frac{\sqrt{2}}{2} = -\sqrt{2}$

61.

$$\sin 6x + \sin 2x = 0$$
$$2\sin\left(\frac{6x + 2x}{2}\right)\cos\left(\frac{6x - 2x}{2}\right) = 0$$
$$2(\sin 4x)\cos 2x = 0$$

$\sin 4x = 0$ or $\cos 2x = 0$

$4x = n\pi$ $\qquad$ $2x = \frac{\pi}{2} + n\pi$

$x = \frac{n\pi}{4}$ $\qquad$ $x = \frac{\pi}{4} + \frac{n\pi}{2}$

In the interval $[0, 2\pi)$

$x = 0, \frac{\pi}{4}, \frac{\pi}{2}, \frac{3\pi}{4}, \pi, \frac{5\pi}{4}, \frac{3\pi}{2}, \frac{7\pi}{4}.$

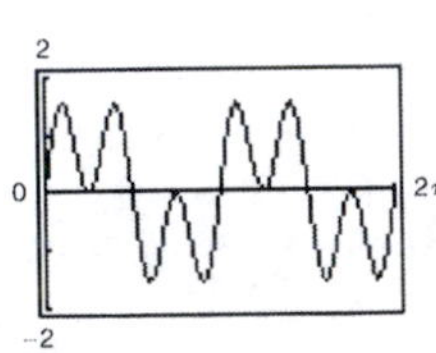

63.
$$\frac{\cos 2x}{\sin 3x - \sin x} - 1 = 0$$
$$\frac{\cos 2x}{\sin 3x - \sin x} = 1$$
$$\frac{\cos 2x}{2\cos 2x \sin x} = 1$$
$$2\sin x = 1$$
$$\sin x = \frac{1}{2}$$
$$x = \frac{\pi}{6}, \frac{5\pi}{6}$$

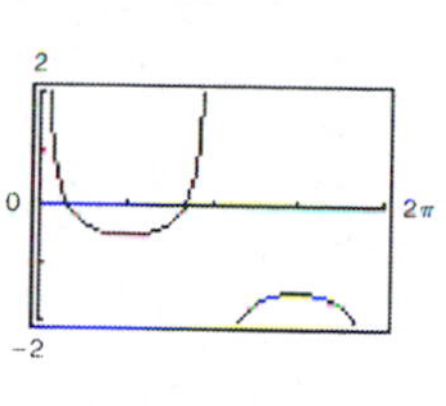

65.
$$\csc 2\theta = \frac{1}{\sin 2\theta}$$
$$= \frac{1}{2\sin\theta\cos\theta}$$
$$= \frac{1}{\sin\theta} \cdot \frac{1}{2\cos\theta}$$
$$= \frac{\csc\theta}{2\cos\theta}$$

67.
$$1 + \cos 10y = 1 + \cos^2 5y - \sin^2 5y$$
$$= 1 + \cos^2 5y - \left(1 - \cos^2 5y\right)$$
$$= 2\cos^2 5y$$

69.
$$(\sin x + \cos x)^2 = \sin^2 x + 2\sin x\cos x + \cos^2 x$$
$$= \left(\sin^2 x + \cos^2 x\right) + 2\sin x\cos x$$
$$= 1 + \sin 2x$$

71.
$$\frac{\sin x \pm \sin y}{\cos x + \cos y} = \frac{2\sin\left(\frac{x \pm y}{2}\right)\cos\left(\frac{x \mp y}{2}\right)}{2\cos\left(\frac{x + y}{2}\right)\cos\left(\frac{x - y}{2}\right)}$$
$$= \tan\left(\frac{x \pm y}{2}\right)$$

73. (a)
$$\sin\left(\frac{\theta}{2}\right) = \pm\sqrt{\frac{1 - \cos\theta}{2}} = \frac{1}{M}$$
$$\left(\pm\sqrt{\frac{1 - \cos\theta}{2}}\right)^2 = \left(\frac{1}{M}\right)^2$$
$$\frac{1 - \cos\theta}{2} = \frac{1}{M^2}$$
$$M^2(1 - \cos\theta) = 2$$
$$1 - \cos\theta = \frac{2}{M^2}$$
$$-\cos\theta = \frac{2}{M^2} - 1$$
$$\cos\theta = 1 - \frac{2}{M^2}$$
$$\cos\theta = \frac{M^2 - 2}{M^2}$$

(b) When $M = 1$, $\cos\theta = \frac{1 - 2}{1} = -1$. So, $\theta = \pi$.

(c) When $M = 4.5$, $\cos\theta = \frac{(4.5)^2 - 2}{(4.5)^2}$

$$\cos\theta \approx 0.901235.$$

So, $\theta \approx 0.4482$ radian.

(d) When $M = 1$,
$$\frac{\text{speed of object}}{\text{speed of sound}} = M$$
$$\frac{\text{speed of object}}{760 \text{ mph}} = 1$$
$$\text{speed of object} = 760 \text{ mph}.$$

When $M = 4.5$,
$$\frac{\text{speed of object}}{\text{speed of sound}} = M$$
$$\frac{\text{speed of object}}{760 \text{ mph}} = 4.5$$
$$\text{speed of object} = 3420 \text{ mph}.$$

75.
$$\frac{x}{2} = 2r\sin^2\frac{\theta}{2} = 2r\left(\frac{1 - \cos\theta}{2}\right)$$
$$= r(1 - \cos\theta)$$

So, $x = 2r(1 - \cos\theta)$.

77. False. For $u < 0$,
$$\sin 2u = -\sin(-2u)$$
$$= -2\sin(-u)\cos(-u)$$
$$= -2(-\sin u)\cos u$$
$$= 2\sin u\cos u.$$

Review Exercises for Chapter 7

1. $\dfrac{\sin x}{\cos x} = \tan x$

3. $\dfrac{1}{\tan x} = \cot x$

5. $\tan\theta = \dfrac{2}{3}$, $\sec\theta = \dfrac{\sqrt{13}}{3}$

θ is in Quadrant I.

$$\cos\theta = \frac{1}{\sec\theta} = \frac{3}{\sqrt{13}} = \frac{3\sqrt{13}}{13}$$

$$\sin\theta = \sqrt{1-\cos^2\theta} = \sqrt{1-\frac{9}{13}} = \sqrt{\frac{4}{13}} = \frac{2\sqrt{13}}{13}$$

$$\csc\theta = \frac{1}{\sin\theta} = \frac{\sqrt{13}}{2}$$

$$\cot\theta = \frac{1}{\tan\theta} = \frac{3}{2}$$

7. $\dfrac{1}{\cot^2 x + 1} = \dfrac{1}{\csc^2 x} = \sin^2 x$

9.
$$\begin{aligned}\tan^2 x(\csc^2 x - 1) &= \tan^2 x(\cot^2 x)\\ &= \tan^2 x\left(\frac{1}{\tan^2 x}\right)\\ &= 1\end{aligned}$$

11. $\dfrac{\cot\left(\dfrac{\pi}{2} - u\right)}{\cos u} = \dfrac{\tan u}{\cos u} = \tan u \sec u$

13.
$$\begin{aligned}\cos^2 x + \cos^2 x\cot^2 x &= \cos^2 x(1 + \cot^2 x)\\ &= \cos^2 x(\csc^2 x)\\ &= \cos^2 x\left(\frac{1}{\sin^2 x}\right)\\ &= \frac{\cos^2 x}{\sin^2 x}\\ &= \cot^2 x\end{aligned}$$

15.
$$\begin{aligned}\frac{1}{\csc\theta + 1} - \frac{1}{\csc\theta - 1} &= \frac{(\csc\theta - 1) - (\csc\theta + 1)}{(\csc\theta + 1)(\csc\theta - 1)}\\ &= \frac{-2}{\csc^2\theta - 1}\\ &= \frac{-2}{\cot^2\theta}\\ &= -2\tan^2\theta\end{aligned}$$

17. Let $x = 5\sin\theta$, then

$$\begin{aligned}\sqrt{25 - x^2} &= \sqrt{25 - (5\sin\theta)^2}\\ &= \sqrt{25 - 25\sin^2\theta}\\ &= \sqrt{25(1 - \sin^2\theta)}\\ &= \sqrt{25\cos^2\theta}\\ &= 5\cos\theta.\end{aligned}$$

19.
$$\begin{aligned}\cos x(\tan^2 x + 1) &= \cos x\sec^2 x\\ &= \frac{1}{\sec x}\sec^2 x\\ &= \sec x\end{aligned}$$

21. $\sec\left(\dfrac{\pi}{2} - \theta\right) = \csc\theta$

23. $\dfrac{1}{\tan\theta\csc\theta} = \dfrac{1}{\dfrac{\sin\theta}{\cos\theta}\cdot\dfrac{1}{\sin\theta}} = \cos\theta$

25.
$$\begin{aligned}\sin^5 x\cos^2 x &= \sin^4 x\cos^2 x\sin x\\ &= (1 - \cos^2 x)^2\cos^2 x\sin x\\ &= (1 - 2\cos^2 x + \cos^4 x)\cos^2 x\sin x\\ &= (\cos^2 x - 2\cos^4 x + \cos^6 x)\sin x\end{aligned}$$

27.
$$\begin{aligned}\sin x &= \sqrt{3} - \sin x\\ \sin x &= \frac{\sqrt{3}}{2}\\ x &= \frac{\pi}{3} + 2\pi n, \frac{2\pi}{3} + 2\pi n\end{aligned}$$

29.
$$\begin{aligned}3\sqrt{3}\tan u &= 3\\ \tan u &= \frac{1}{\sqrt{3}}\\ u &= \frac{\pi}{6} + n\pi\end{aligned}$$

31. $3\csc^2 x = 4$

$$\csc^2 x = \frac{4}{3}$$

$$\sin x = \pm\frac{\sqrt{3}}{2}$$

$$x = \frac{\pi}{3} + 2\pi n, \frac{2\pi}{3} + 2\pi n, \frac{4\pi}{3} + 2\pi n, \frac{5\pi}{3} + 2\pi n$$

These can be combined as:

$$x = \frac{\pi}{3} + n\pi \quad \text{or} \quad x = \frac{2\pi}{3} + n\pi$$

33. $2\cos^2 x - \cos x = 1$

$$2\cos^2 x - \cos x - 1 = 0$$

$$(2\cos x + 1)(\cos x - 1) = 0$$

$2\cos x + 1 = 0 \qquad \cos x - 1 = 0$

$\cos x = -\frac{1}{2} \qquad \cos x = 1$

$x = \frac{2\pi}{3}, \frac{4\pi}{3} \qquad x = 0$

35. $\cos^2 x + \sin x = 1$

$$1 - \sin^2 x + \sin x - 1 = 0$$

$$-\sin x(\sin x - 1) = 0$$

$\sin x = 0 \qquad \sin x - 1 = 0$

$x = 0, \pi \qquad \sin x = 1$

$x = \frac{\pi}{2}$

37. $2\sin 2x - \sqrt{2} = 0$

$$\sin 2x = \frac{\sqrt{2}}{2}$$

$$2x = \frac{\pi}{4} + 2\pi n, \frac{3\pi}{4} + 2\pi n$$

$$x = \frac{\pi}{8} + \pi n, \frac{3\pi}{8} + \pi n$$

$$x = \frac{\pi}{8}, \frac{3\pi}{8}, \frac{9\pi}{8}, \frac{11\pi}{8}$$

39. $3\tan^2\left(\frac{x}{3}\right) - 1 = 0$

$$\tan^2\left(\frac{x}{3}\right) = \frac{1}{3}$$

$$\tan\frac{x}{3} = \pm\sqrt{\frac{1}{3}}$$

$$\tan\frac{x}{3} = \pm\frac{\sqrt{3}}{3}$$

$$\frac{x}{3} = \frac{\pi}{6}, \frac{5\pi}{6}, \frac{7\pi}{6}$$

$$x = \frac{\pi}{2}, \frac{5\pi}{2}, \frac{7\pi}{2}$$

$\frac{5\pi}{2}$ and $\frac{7\pi}{2}$ are greater than 2π, so they are not solutions. The solution is $x = \frac{\pi}{2}$.

41. $\cos 4x(\cos x - 1) = 0$

$\cos 4x = 0 \qquad \cos x - 1 = 0$

$4x = \frac{\pi}{2} + 2\pi n, \frac{3\pi}{2} + 2\pi n \qquad \cos x = 1$

$x = \frac{\pi}{8} + \frac{\pi}{2}n, \frac{3\pi}{8} + \frac{\pi}{2}n \qquad x = 0$

$$x = 0, \frac{\pi}{8}, \frac{3\pi}{8}, \frac{5\pi}{8}, \frac{7\pi}{8}, \frac{9\pi}{8}, \frac{11\pi}{8}, \frac{13\pi}{8}, \frac{15\pi}{8}$$

43. $\tan^2 x - 2\tan x = 0$

$$\tan x(\tan x - 2) = 0$$

$\tan x = 0$ or $\tan x - 2 = 0$

$x = 0, \pi \qquad \tan x = 2$

$x = \arctan 2, \arctan 2 + \pi$

$$x = 0, \pi, \arctan 2, \arctan 2 + \pi$$

45. $\tan^2\theta + \tan\theta - 6 = 0$

$$(\tan\theta + 3)(\tan\theta - 2) = 0$$

$\tan\theta + 3 = 0$ or $\tan\theta - 2 = 0$

$\tan\theta = -3 \qquad \tan\theta = 2$

$\theta = \arctan(-3) + \pi, \arctan(-3) + 2\pi \qquad \theta = \arctan 2, \arctan 2 + \pi$

47. $\sin 285° = \sin(315° - 30°)$

$= \sin 315° \cos 30° - \cos 315° \sin 30°$

$= \left(-\frac{\sqrt{2}}{2}\right)\left(\frac{\sqrt{3}}{2}\right) - \left(\frac{\sqrt{2}}{2}\right)\left(\frac{1}{2}\right)$

$= -\frac{\sqrt{2}}{4}\left(\sqrt{3} + 1\right)$

$\cos 285° = \cos(315° - 30°)$

$= \cos 315° \cos 30° + \sin 315° \sin 30°$

$= \left(\frac{\sqrt{2}}{2}\right)\left(\frac{\sqrt{3}}{2}\right) + \left(-\frac{\sqrt{2}}{2}\right)\left(\frac{1}{2}\right)$

$= \frac{\sqrt{2}}{4}\left(\sqrt{3} - 1\right)$

$\tan 285° = \tan(315° - 30°) = \frac{\tan 315° - \tan 30°}{1 + \tan 315° \tan 30°}$

$= \frac{(-1) - \left(\frac{\sqrt{3}}{3}\right)}{1 + (-1)\left(\frac{\sqrt{3}}{3}\right)} = -2 - \sqrt{3}$

49. $\sin \frac{25\pi}{12} = \sin\left(\frac{11\pi}{6} + \frac{\pi}{4}\right)$

$= \sin \frac{11\pi}{6} \cos \frac{\pi}{4} + \cos \frac{11\pi}{6} \sin \frac{\pi}{4}$

$= \left(-\frac{1}{2}\right)\left(\frac{\sqrt{2}}{2}\right) + \left(\frac{\sqrt{3}}{2}\right)\left(\frac{\sqrt{2}}{2}\right)$

$= \frac{\sqrt{2}}{4}\left(\sqrt{3} - 1\right)$

$\cos \frac{25\pi}{12} = \cos\left(\frac{11\pi}{6} + \frac{\pi}{4}\right)$

$= \cos \frac{11\pi}{6} \cos \frac{\pi}{4} - \sin \frac{11\pi}{6} \sin \frac{\pi}{4}$

$= \left(\frac{\sqrt{3}}{2}\right)\left(\frac{\sqrt{2}}{2}\right) - \left(-\frac{1}{2}\right)\left(\frac{\sqrt{2}}{2}\right)$

$= \frac{\sqrt{2}}{4}\left(\sqrt{3} + 1\right)$

$\tan \frac{25\pi}{12} = \tan\left(\frac{11\pi}{6} + \frac{\pi}{4}\right)$

$= \frac{\tan \frac{11\pi}{6} + \tan \frac{\pi}{4}}{1 - \tan \frac{11\pi}{6} \tan \frac{\pi}{4}}$

$= \frac{\left(-\frac{\sqrt{3}}{3}\right) + 1}{1 - \left(-\frac{\sqrt{3}}{3}\right)(1)}$

$= 2 - \sqrt{3}$

51. $\sin 60° \cos 45° - \cos 60° \sin 45° = \sin(60° - 45°)$

$= \sin 15°$

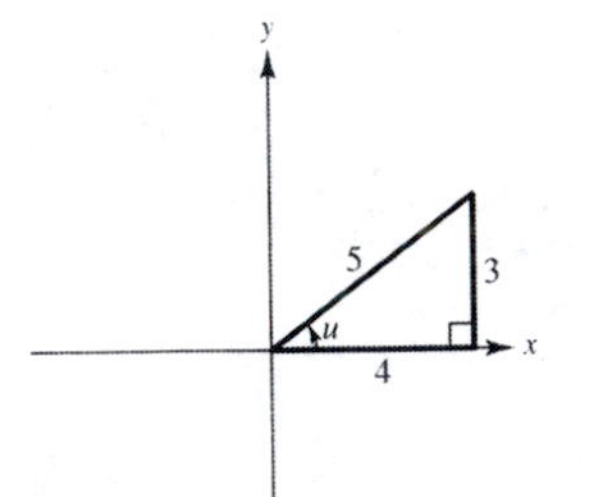

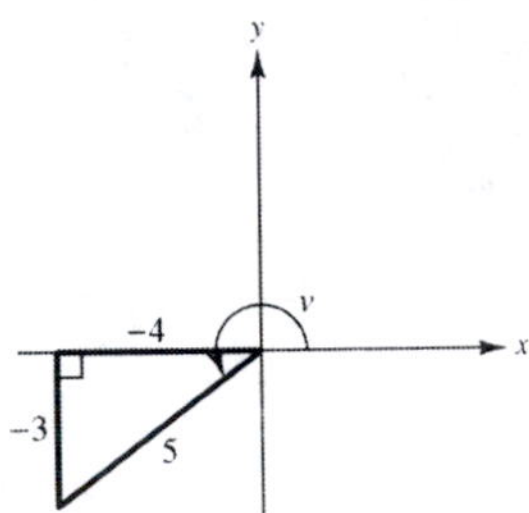

Figures for Exercises 53–57

53. $\sin(u + v) = \sin u \cos v + \cos u \sin v = \frac{3}{5}\left(-\frac{4}{5}\right) + \frac{4}{5}\left(-\frac{3}{5}\right) = -\frac{24}{25}$

55. $\cos(u - v) = \cos u \cos v + \sin u \sin v = \frac{4}{5}\left(-\frac{4}{5}\right) + \frac{3}{5}\left(-\frac{3}{5}\right) = -1$

57. $\cos\left(x + \frac{\pi}{2}\right) = \cos x \cos \frac{\pi}{2} - \sin x \sin \frac{\pi}{2} = \cos x(0) - \sin x(1) = -\sin x$

59. $\tan(\pi - x) = \dfrac{\tan \pi - \tan x}{1 - \tan \pi \tan x} = -\tan x$

61. $\sin\left(x + \dfrac{\pi}{4}\right) - \sin\left(x - \dfrac{\pi}{4}\right) = 1$

$$2 \cos x \sin \frac{\pi}{4} = 1$$

$$\cos x = \frac{\sqrt{2}}{2}$$

$$x = \frac{\pi}{4}, \frac{7\pi}{4}$$

63. $\sin u = -\dfrac{4}{5}, \pi < u < \dfrac{3\pi}{2}$

$$\cos u = -\sqrt{1 - \sin^2 u} = \frac{-3}{5}$$

$$\tan u = \frac{\sin u}{\cos u} = \frac{4}{3}$$

$$\sin 2u = 2 \sin u \cos u = 2\left(-\frac{4}{5}\right)\left(-\frac{3}{5}\right) = \frac{24}{25}$$

$$\cos 2u = \cos^2 u - \sin^2 u = \left(-\frac{3}{5}\right)^2 - \left(-\frac{4}{5}\right)^2 = -\frac{7}{25}$$

$$\tan 2u = \frac{2 \tan u}{1 - \tan^2 u} = \frac{2\left(\frac{4}{3}\right)}{1 - \left(\frac{4}{3}\right)^2} = -\frac{24}{7}$$

65. $\sin 4x = 2 \sin 2x \cos 2x$

$$= 2\left[2 \sin x \cos x\left(\cos^2 x - \sin^2 x\right)\right]$$

$$= 4 \sin x \cos x\left(2 \cos^2 x - 1\right)$$

$$= 8 \cos^3 x \sin x - 4 \cos x \sin x$$

67. $\tan^2 2x = \dfrac{\sin^2 2x}{\cos^2 2x} = \dfrac{\dfrac{1 - \cos 4x}{2}}{\dfrac{1 + \cos 4x}{2}} = \dfrac{1 - \cos 4x}{1 + \cos 4x}$

69. $\sin(-75^\circ) = -\sqrt{\dfrac{1 - \cos 150^\circ}{2}} = -\sqrt{\dfrac{1 - \left(-\frac{\sqrt{3}}{2}\right)}{2}} = -\dfrac{\sqrt{2 + \sqrt{3}}}{2} = -\dfrac{1}{2}\sqrt{2 + \sqrt{3}}$

$$\cos(-75^\circ) = -\sqrt{\frac{1 + \cos 150^\circ}{2}} = \sqrt{\frac{1 + \left(-\frac{\sqrt{3}}{2}\right)}{2}} = \frac{\sqrt{2 - \sqrt{3}}}{2} = \frac{1}{2}\sqrt{2 - \sqrt{3}}$$

$$\tan(-75^\circ) = -\left(\frac{1 - \cos 150^\circ}{\sin 150^\circ}\right) = -\left(\frac{1 - \left(-\frac{\sqrt{3}}{2}\right)}{\frac{1}{2}}\right) = -\left(2 + \sqrt{3}\right) = -2 - \sqrt{3}$$

71. $\tan u = \frac{4}{3}, \pi < u < \frac{3\pi}{2}$

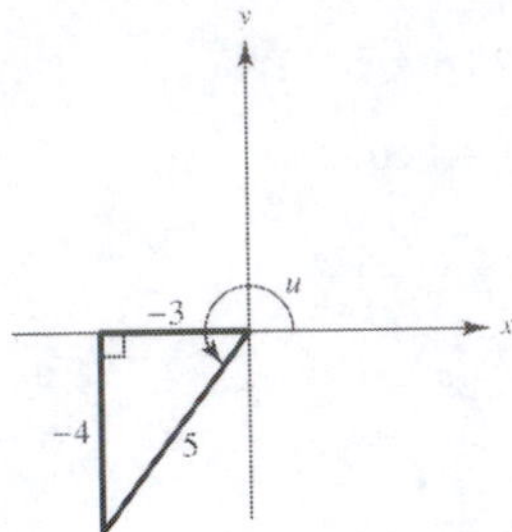

(a) Because u is in Quadrant III, $\frac{u}{2}$ is in Quadrant II.

(b) $\sin\frac{u}{2} = \sqrt{\frac{1-\cos u}{2}} = \sqrt{\frac{1-\left(-\frac{3}{5}\right)}{2}} = \sqrt{\frac{4}{5}}$

$= \frac{2\sqrt{5}}{5}$

$\cos\frac{u}{2} = -\sqrt{\frac{1+\cos u}{2}} = -\sqrt{\frac{1+\left(-\frac{3}{5}\right)}{2}} = -\sqrt{\frac{1}{5}}$

$= -\frac{\sqrt{5}}{5}$

$\tan\frac{u}{2} = \frac{1-\cos u}{\sin u} = \frac{1-\left(-\frac{3}{5}\right)}{\left(-\frac{4}{5}\right)} = -2$

73. $-\sqrt{\frac{1+\cos 10x}{2}} = -\left|\cos\frac{10x}{2}\right| = -\left|\cos 5x\right|$

75. $\cos 4\theta \sin 6\theta = \frac{1}{2}\left[\sin(4\theta + 6\theta) - \sin(4\theta - 6\theta)\right]$

$= \frac{1}{2}\left(\sin 10\theta - \sin(-2\theta)\right)$

77. $\cos 6\theta + \cos 5\theta = 2\cos\left(\frac{6\theta + 5\theta}{2}\right)\cos\left(\frac{6\theta - 5\theta}{2}\right)$

$= 2\cos\frac{11\theta}{2}\cos\frac{\theta}{2}$

79. $r = \frac{1}{32}v_0^2 \sin 2\theta$

range $= 100$ feet

$v_0 = 80$ feet per second

$r = \frac{1}{32}(80)^2 \sin 2\theta = 100$

$\sin 2\theta = 0.5$

$2\theta = 30°$

$\theta = 15°$ or $\frac{\pi}{12}$

81. False. If $\frac{\pi}{2} < \theta < \pi$, then $\frac{\pi}{4} < \frac{\theta}{2} < \frac{\pi}{2}$ and $\frac{\theta}{2}$ is in Quadrant I.

$\cos\frac{\theta}{2} > 0$

83. True. $4\sin(-x)\cos(-x) = 4(-\sin x)\cos x$

$= -4\sin x \cos x$

$= -2(2\sin x \cos x)$

$= -2\sin 2x$

85. No. For an equation to be an identity, the equation must be true for all real numbers. $\sin\theta = \frac{1}{2}$ has an infinite number of solutions but is not an identity.

Problem Solving for Chapter 7

1. $\sin\theta = \pm\sqrt{1-\cos^2\theta}$

$\tan\theta = \frac{\sin\theta}{\cos\theta} = \pm\frac{\sqrt{1-\cos^2\theta}}{\cos\theta}$

$\csc\theta = \frac{1}{\sin\theta} = \pm\frac{1}{\sqrt{1-\cos^2\theta}}$

$\sec\theta = \frac{1}{\cos\theta}$

$\cot\theta = \frac{1}{\tan\theta} = \pm\frac{\cos\theta}{\sqrt{1-\cos^2\theta}}$

You also have the following relationships:

$\sin\theta = \cos\left(\frac{\pi}{2} - \theta\right)$

$\tan\theta = \frac{\cos\left[(\pi/2) - \theta\right]}{\cos\theta}$

$\csc\theta = \frac{1}{\cos\left[(\pi/2) - \theta\right]}$

$\sec\theta = \frac{1}{\cos\theta}$

$\cot\theta = \frac{\cos\theta}{\cos\left[(\pi/2) - \theta\right]}$

3. $\sin\left[\dfrac{(12n+1)\pi}{6}\right] = \sin\left[\dfrac{1}{6}(12n\pi + \pi)\right]$

$$= \sin\left(2n\pi + \frac{\pi}{6}\right)$$

$$= \sin\frac{\pi}{6}$$

$$= \frac{1}{2}$$

So, $\sin\left[\dfrac{(12n+1)\pi}{6}\right] = \dfrac{1}{2}$ for all integers n

5. From the figure, it appears that $u + v = w$. Assume that u, v, and w are all in Quadrant I.

From the figure:

$$\tan u = \frac{s}{3s} = \frac{1}{3}$$

$$\tan v = \frac{s}{2s} = \frac{1}{2}$$

$$\tan w = \frac{s}{s} = 1$$

$$\tan(u+v) = \frac{\tan u + \tan v}{1 - \tan u \tan v}$$

$$= \frac{1/3 + 1/2}{1 - (1/3)(1/2)}$$

$$= \frac{5/6}{1 - (1/6)}$$

$$= 1$$

$$= \tan w.$$

So, $\tan(u+v) = \tan w$. Because u, v, and w are all in Quadrant I, you have

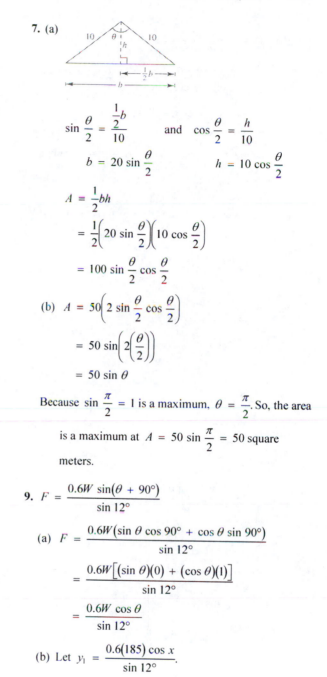

$\arctan[\tan(u+v)] = \arctan[\tan w] u + v = w.$

7. (a)

$$\sin\frac{\theta}{2} = \frac{\frac{1}{2}b}{10} \quad \text{and} \quad \cos\frac{\theta}{2} = \frac{h}{10}$$

$$b = 20\sin\frac{\theta}{2} \qquad h = 10\cos\frac{\theta}{2}$$

$$A = \frac{1}{2}bh$$

$$= \frac{1}{2}\left(20\sin\frac{\theta}{2}\right)\left(10\cos\frac{\theta}{2}\right)$$

$$= 100\sin\frac{\theta}{2}\cos\frac{\theta}{2}$$

(b) $A = 50\left(2\sin\dfrac{\theta}{2}\cos\dfrac{\theta}{2}\right)$

$$= 50\sin\left(2\left(\frac{\theta}{2}\right)\right)$$

$$= 50\sin\theta$$

Because $\sin\dfrac{\pi}{2} = 1$ is a maximum, $\theta = \dfrac{\pi}{2}$. So, the area is a maximum at $A = 50\sin\dfrac{\pi}{2} = 50$ square meters.

9. $F = \dfrac{0.6W\sin(\theta + 90°)}{\sin 12°}$

(a) $F = \dfrac{0.6W(\sin\theta\cos 90° + \cos\theta\sin 90°)}{\sin 12°}$

$$= \frac{0.6W[(\sin\theta)(0) + (\cos\theta)(1)]}{\sin 12°}$$

$$= \frac{0.6W\cos\theta}{\sin 12°}$$

(b) Let $y_1 = \dfrac{0.6(185)\cos x}{\sin 12°}$.

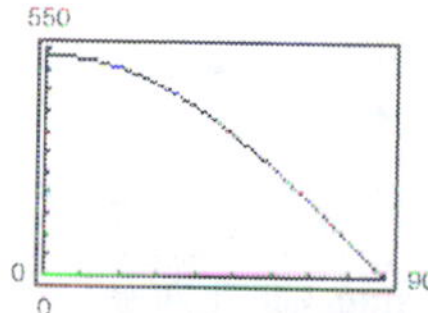

(c) The force is maximum (533.88 pounds) when $\theta = 0°$.

The force is minimum (0 pounds) when $\theta = 90°$.

11. $d = 35 - 28\cos\frac{\pi}{6.2}t$ when $t = 0$ corresponds to 12:00 A.M.

(a) The high tides occur when $\cos\frac{\pi}{6.2}t = -1$. Solving yields $t = 6.2$ or $t = 18.6$.

These t-values correspond to 6:12 A.M. and 6:36 P.M.

The low tide occurs when $\cos\frac{\pi}{6.2}t = 1$. Solving yields $t = 0$ and $t = 12.4$ which corresponds to 12:00 A.M. and 12:24 P.M.

(b) The water depth is never 3.5 feet. At low tide, the depth is $d = 35 - 28 = 7$ feet.

(c)

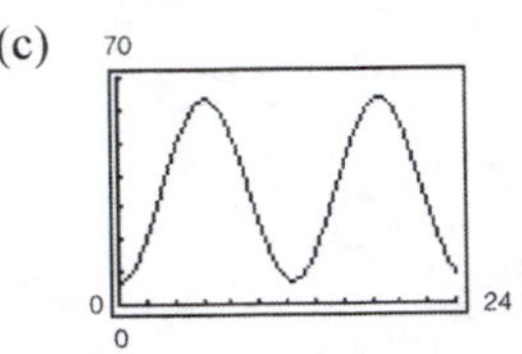

13. (a) $n = \dfrac{\sin\left(\frac{\theta}{2} + \frac{\alpha}{2}\right)}{\sin\frac{\theta}{2}}$

$$= \frac{\sin\left(\frac{\theta}{2}\right)\cos\left(\frac{\alpha}{2}\right) + \cos\left(\frac{\theta}{2}\right)\sin\left(\frac{\alpha}{2}\right)}{\sin\left(\frac{\theta}{2}\right)}$$

$$= \cos\left(\frac{\alpha}{2}\right) + \cot\left(\frac{\theta}{2}\right)\sin\left(\frac{\alpha}{2}\right)$$

For $\alpha = 60^\circ$, $n = \cos 30^\circ + \cot\left(\frac{\theta}{2}\right)\sin 30^\circ$

$$n = \frac{\sqrt{3}}{2} + \frac{1}{2}\cot\left(\frac{\theta}{2}\right).$$

(b) For glass, $n = 1.50$.

$$1.50 = \frac{\sqrt{3}}{2} + \frac{1}{2}\cot\left(\frac{\theta}{2}\right)$$

$$2\left(1.50 - \frac{\sqrt{3}}{2}\right) = \cot\left(\frac{\theta}{2}\right)$$

$$\frac{1}{3 - \sqrt{3}} = \tan\left(\frac{\theta}{2}\right)$$

$$\theta = 2\tan^{-1}\left(\frac{1}{3 - \sqrt{3}}\right)$$

$$\theta \approx 76.5^\circ$$

15. (a) Let $y_1 = \sin x$ and $y_2 = 0.5$.

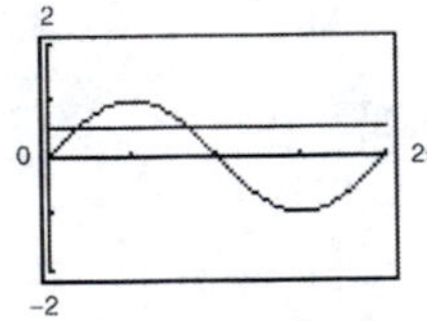

$\sin x \ge 0.5$ on the interval $\left[\frac{\pi}{6}, \frac{5\pi}{6}\right]$.

(b) Let $y_1 = \cos x$ and $y_2 = -0.5$.

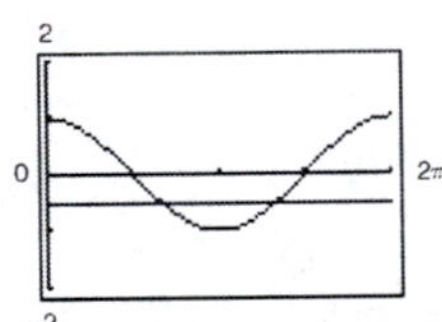

$\cos x \le -0.5$ on the interval $\left[\frac{2\pi}{3}, \frac{4\pi}{3}\right]$.

(c) Let $y_1 = \tan x$ and $y_2 = \sin x$.

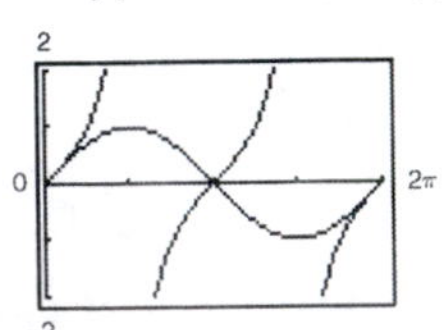

$\tan x < \sin x$ on the intervals $\left(\frac{\pi}{2}, \pi\right)$ and $\left(\frac{3\pi}{2}, 2\pi\right)$.

(d) Let $y_1 = \cos x$ and $y_2 = \sin x$.

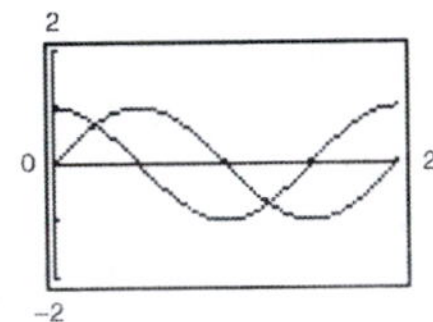

$\cos x \ge \sin x$ on the intervals $\left[0, \frac{\pi}{4}\right]$ and $\left[\frac{5\pi}{4}, 2\pi\right)$.

Practice Test for Chapter 7

1. Find the value of the other five trigonometric functions, given $\tan x = \frac{4}{11}$, $\sec x < 0$.

2. Simplify $\dfrac{\sec^2 x + \csc^2 x}{\csc^2 x(1 + \tan^2 x)}$.

3. Rewrite as a single logarithm and simplify $\ln|\tan\theta| - \ln|\cot\theta|$.

4. True or false:

$$\cos\left(\frac{\pi}{2} - x\right) = \frac{1}{\csc x}$$

5. Factor and simplify: $\sin^4 x + (\sin^2 x)\cos^2 x$

6. Multiply and simplify: $(\csc x + 1)(\csc x - 1)$

7. Rationalize the denominator and simplify:

$$\frac{\cos^2 x}{1 - \sin x}$$

8. Verify:

$$\frac{1 + \cos\theta}{\sin\theta} + \frac{\sin\theta}{1 + \cos\theta} = 2\csc\theta$$

9. Verify:

$$\tan^4 x + 2\tan^2 x + 1 = \sec^4 x$$

10. Use the sum or difference formulas to determine:

(a) $\sin 105^\circ$

(b) $\tan 15^\circ$

11. Simplify: $(\sin 42^\circ)\cos 38^\circ - (\cos 42^\circ)\sin 38^\circ$

12. Verify $\tan\left(\theta + \dfrac{\pi}{4}\right) = \dfrac{1 + \tan\theta}{1 - \tan\theta}$.

13. Write $\sin(\arcsin x - \arccos x)$ as an algebraic expression in x.

14. Use the double-angle formulas to determine:

(a) $\cos 120^\circ$

(b) $\tan 300^\circ$

15. Use the half-angle formulas to determine:

(a) $\sin 22.5^\circ$

(b) $\tan \dfrac{\pi}{12}$

16. Given $\sin\theta = 4/5$, θ lies in Quadrant II, find $\cos(\theta/2)$.

17. Use the power-reducing identities to write $(\sin^2 x)\cos^2 x$ in terms of the first power of cosine.

18. Rewrite as a sum: $6(\sin 5\theta)\cos 2\theta$.

19. Rewrite as a product: $\sin(x + \pi) + \sin(x - \pi)$.

20. Verify $\dfrac{\sin 9x + \sin 5x}{\cos 9x - \cos 5x} = -\cot 2x$.

21. Verify:

$(\cos u)\sin v = \frac{1}{2}[\sin(u + v) - \sin(u - v)]$.

22. Find all solutions in the interval $[0, 2\pi)$:

$4\sin^2 x = 1$

23. Find all solutions in the interval $[0, 2\pi)$:

$\tan^2\theta + (\sqrt{3} - 1)\tan\theta - \sqrt{3} = 0$

24. Find all solutions in the interval $[0, 2\pi)$:

$\sin 2x = \cos x$

25. Use the quadratic formula to find all solutions in the interval $[0, 2\pi)$:

$\tan^2 x - 6\tan x + 4 = 0$

CHAPTER 8
Additional Topics in Trigonometry

CHAPTER 8
Additional Topics in Trigonometry

Section 8.1 Law of Sines

1. oblique

3. angles; side

5.

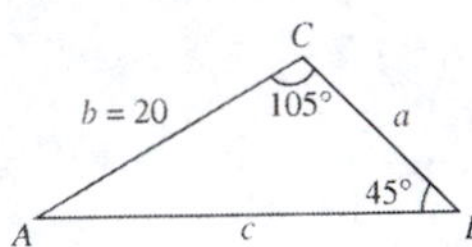

Given: $B = 45°, C = 105°, b = 20$

$A = 180° - B - C = 30°$

$$a = \frac{b}{\sin B}(\sin A) = \frac{20 \sin 30°}{\sin 45°} = 10\sqrt{2} \approx 14.14$$

$$C = \frac{b}{\sin B}(\sin C) = \frac{20 \sin 105°}{\sin 45°} \approx 27.32$$

7.

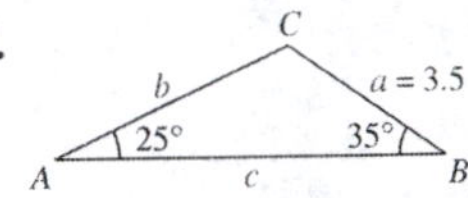

Given: $A = 25°, B = 35°, a = 3.5$

$C = 180° - A - B = 120°$

$$b = \frac{a}{\sin A}(\sin B) = \frac{3.5}{\sin 25°}(\sin 35°) \approx 4.75$$

$$c = \frac{a}{\sin A}(\sin C) = \frac{3.5}{\sin 25°}(\sin 120°) \approx 7.17$$

9. Given: $A = 102.4°, C = 16.7°, a = 21.6$

$B = 180° - A - C = 60.9°$

$$b = \frac{a}{\sin A}(\sin B) = \frac{21.6}{\sin 102.4°}(\sin 60.9°) \approx 19.32$$

$$c = \frac{a}{\sin A}(\sin C) = \frac{21.6}{\sin 102.4°}(\sin 16.7°) \approx 6.36$$

11. Given: $A = 83°20', C = 54.6°, c = 18.1$

$B = 180° - A - C = 180° - 83°20' - 54°36' = 42°4'$

$$a = \frac{c}{\sin C}(\sin A) = \frac{18.1}{\sin 54.6°}(\sin 83°20') \approx 22.05$$

$$b = \frac{c}{\sin C}(\sin B) = \frac{18.1}{\sin 54.6°}(\sin 42°4') \approx 14.88$$

13. Given: $A = 35°, B = 65°, c = 10$

$C = 180° - A - B = 80°$

$$a = \frac{c}{\sin C}(\sin A) = \frac{10 \sin 35°}{\sin 80°} \approx 5.82$$

$$b = \frac{c}{\sin C}(\sin B) = \frac{10 \sin 65°}{\sin 80°} \approx 9.20$$

15. Given: $A = 55°, B = 42°, c = \frac{3}{4}$

$C = 180° - A - B = 83°$

$$a = \frac{c}{\sin C}(\sin A) = \frac{0.75}{\sin 83°}(\sin 55°) \approx 0.62$$

$$b = \frac{c}{\sin C}(\sin B) = \frac{0.75}{\sin 83°}(\sin 42°) \approx 0.51$$

17. Given: $A = 36°, a = 8, b = 5$

$$\sin B = \frac{b \sin A}{a} = \frac{5 \sin 36°}{8} \approx 0.36737 \Rightarrow B \approx 21.55°$$

$C = 180° - A - B \approx 180° - 36° - 21.55 = 122.45°$

$$c = \frac{a}{\sin A}(\sin C) = \frac{8}{\sin 36°}(\sin 122.45°) \approx 11.49$$

19. Given: $B = 15°30', a = 4.5, b = 6.8$

$$\sin A = \frac{a \sin B}{b} = \frac{4.5 \sin 15°30'}{6.8} \approx 0.17685 \Rightarrow A \approx 10°11'$$

$C = 180° - A - B \approx 180° - 10°11' - 15°30' = 154°19'$

$$c = \frac{b}{\sin B}(\sin C) = \frac{6.8}{\sin 15°30'}(\sin 154°19') \approx 11.03$$

21. Given: $A = 145°$, $a = 14$, $b = 4$

$$\sin B = \frac{b \sin A}{a} = \frac{4 \sin 145°}{14} \approx 0.1639 \Rightarrow B \approx 9.43°$$

$$C = 180° - A - B \approx 25.57°$$

$$c = \frac{a}{\sin A}(\sin C) \approx \frac{14 \sin 25.57°}{\sin 145°} \approx 10.53$$

23. Given: $A = 110°15'$, $a = 48$, $b = 16$

$$\sin B = \frac{b \sin A}{a} = \frac{16 \sin 110°15'}{48} \approx 0.31273 \Rightarrow B \approx 18°13'$$

$$C = 180° - A - B \approx 180° - 110°15' - 18°13' = 51°32'$$

$$c = \frac{a}{\sin A}(\sin C) = \frac{48}{\sin 110°15'}(\sin 15°32') \approx 40.06$$

25. Given: $A = 110°$, $a = 125$, $b = 100$

$$\sin B = \frac{b \sin A}{a} = \frac{100 \sin 110°}{125} \approx 0.75175 \Rightarrow B \approx 48.74°$$

$$C = 180° - A - B \approx 21.26°$$

$$c = \frac{a \sin C}{\sin A} \approx \frac{125 \sin 21.26°}{\sin 110°} \approx 48.23$$

27. Given: $a = 18$, $b = 20$, $A = 76°$

$$h = 20 \sin 76° \approx 19.41$$

Because $a < h$, no triangle is formed.

29. Given: $A = 58°$, $a = 11.4$, $c = 12.8$

$$\sin B = \frac{b \sin A}{a} = \frac{12.8 \sin 58°}{11.4} \approx 0.9522 \Rightarrow B \approx 72.21° \text{ or } B \approx 107.79°$$

Case 1

$B \approx 72.21°$

$C = 180° - A - B \approx 49.79°$

$$c = \frac{a}{\sin A}(\sin C) \approx \frac{11.4 \sin 49.79°}{\sin 58°} \approx 10.27$$

Case 2

$B \approx 107.79°$

$C = 180° - A - B \approx 14.21°$

$$c = \frac{a}{\sin A}(\sin C) \approx \frac{11.4 \sin 14.21°}{\sin 58°} \approx 3.30$$

31. Given: $A = 120°$, $a = b = 25$

No triangle is formed because A is obtuse and $a = b$.

33. Given: $A = 45°$, $a = b = 1$

Because $a = b = 1$, $B = 45°$.

$C = 180° - A - B = 90°$

$$c = \frac{a}{\sin A}(\sin C) = \frac{1 \sin 90°}{\sin 45°} = \sqrt{2} \approx 1.41$$

35. Given: $A = 36°$, $a = 5$

(a) One solution if $b \le 5$ or $b = \dfrac{5}{\sin 36°}$.

(b) Two solutions if $5 < b < \dfrac{5}{\sin 36°}$.

(c) No solution if $b > \dfrac{5}{\sin 36°}$.

37. Given: $A = 10°, a = 10.8$

(a) One solution if $b \le 10.8$ or $b = \dfrac{10.8}{\sin 10°}$.

(b) Two solutions if $10.8 < b < \dfrac{10.8}{\sin 10°}$.

(c) No solution if $b > \dfrac{10.8}{\sin 10°}$.

39. Area $= \frac{1}{2}ab \sin C = \frac{1}{2}(4)(6)\sin 120° \approx 10.4$

41. Area $= \frac{1}{2}bc \sin A = \frac{1}{2}(8)(10)\sin 150° = 20$

43. Area $= \frac{1}{2}bc \sin A = \frac{1}{2}(57)(85)\sin 43°45' \approx 1675.2$

45. Area $= \frac{1}{2}ac \sin B = \frac{1}{2}(105)(64)\sin(72°30') \approx 3204.5$

47. $C = 180° - 94° - 30° = 56°$

$$h = \frac{40}{\sin 56°}(\sin 30°) \approx 24.1 \text{ meters}$$

49. $\dfrac{\sin(42° - \theta)}{10} = \dfrac{\sin 48°}{17}$

$$\sin(42° - \theta) \approx 0.43714$$

$$42° - \theta \approx 25.9°$$

$$\theta \approx 16.1°$$

51.

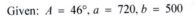

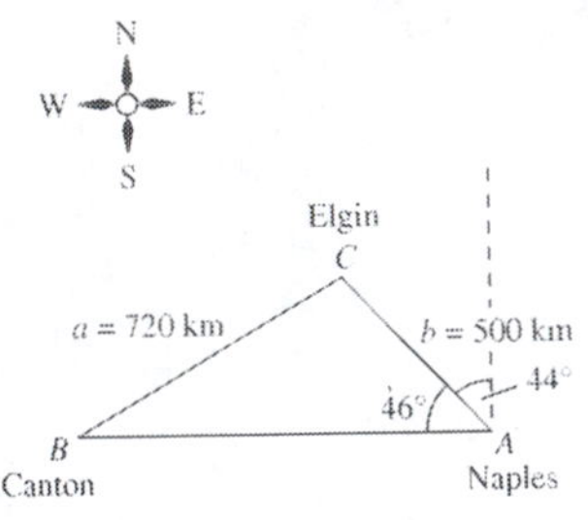

Given: $A = 46°, a = 720, b = 500$

$$\sin B = \frac{b \sin A}{a} = \frac{500 \sin 46°}{720} \approx 0.50 \Rightarrow B \approx 30°$$

The bearing from C to B is $240°$.

53.

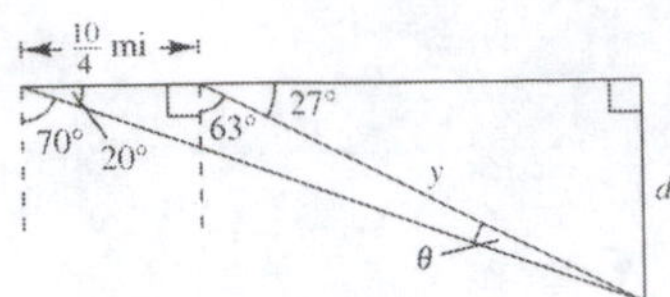

In 15 minutes the boat has traveled

$$(10 \text{ mph})\left(\frac{1}{4} \text{ hr}\right) = \frac{10}{4} \text{ miles.}$$

$$\theta = 180° - 20° - (90° + 63°)$$

$$\theta = 7°$$

$$\frac{10/4}{\sin 7°} = \frac{y}{\sin 20°}$$

$$y \approx 7.0161$$

$$\sin 27° = \frac{d}{7.0161}$$

$$d \approx 3.2 \text{ miles}$$

55. $\alpha = 180 - \theta - (180 - \phi) = \phi - \theta$

$$\frac{d}{\sin \theta} = \frac{2}{\sin \alpha}$$

$$d = \frac{2 \sin \theta}{\sin(\phi - \theta)}$$

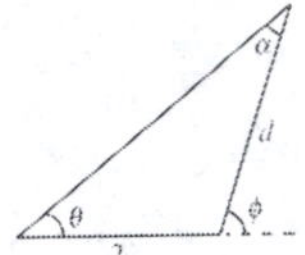

57. True. If one angle of a triangle is obtuse, then there is less than $90°$ left for the other two angles, so it cannot contain a right angle. It must be oblique.

59. False. To solve an oblique triangle using the Law of Sines, you need to know two angles and any side, or two sides and an angle opposite one of them.

61. (a) $A = \frac{1}{2}(30)(20)\sin\left(\theta + \frac{\theta}{2}\right) - \frac{1}{2}(8)(20)\sin\frac{\theta}{2} - \frac{1}{2}(8)(30)\sin\theta$

$$= 300 \sin\frac{3\theta}{2} - 80 \sin\frac{\theta}{2} - 120 \sin\theta$$

$$= 20\left[15 \sin\frac{3\theta}{2} - 4 \sin\frac{\theta}{2} - 6 \sin\theta\right]$$

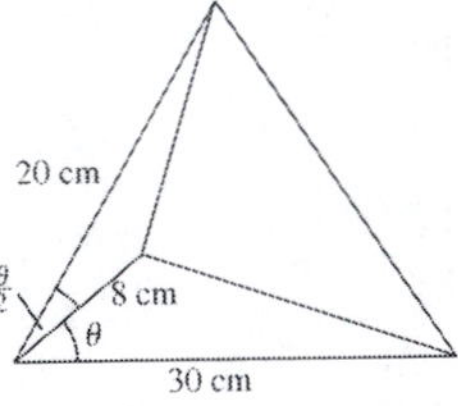

(b)

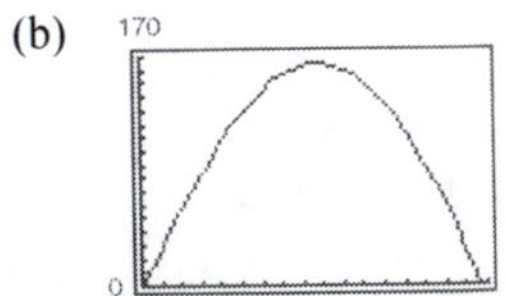

(c) Domain: $0 \le \theta \le 1.6690$

The domain would increase in length and the area would have a greater maximum value if the 8-centimeter line segment were decreased.

Section 8.2 Law of Cosines

1. Cosines

3. $b^2 = a^2 + c^2 - 2ac\cos B$

5. Given: $a = 10, b = 12, c = 16$

$$\cos C = \frac{a^2 + b^2 - c^2}{2ab} = \frac{100 + 144 - 256}{2(10)(12)} = -0.05 \Rightarrow C \approx 92.87°$$

$$\sin B = \frac{b\sin C}{c} \approx \frac{12\sin 92.87°}{16} \approx 0.7491 \Rightarrow B \approx 48.51°$$

$$A \approx 180° - 48.51° - 92.87° = 38.62°$$

7. Given: $a = 6, b = 8, c = 12$

$$\cos C = \frac{a^2 + b^2 + c^2}{2ab} = \frac{6^2 + 8^2 - 12^2}{2(6)(8)} \approx -0.4583 \Rightarrow C \approx 117.28°$$

$$\sin B = b\left(\frac{\sin C}{c}\right) = 8\left(\frac{\sin 117.28°}{12}\right) \approx 0.5925 \Rightarrow B \approx 36.33°$$

$$A = 180° - B - C \approx 180° - 36.33° - 117.28° = 26.39°$$

9. Given: $A = 30°, b = 15, c = 30$

$$a^2 = b^2 + c^2 - 2bc\cos A$$
$$= 225 + 900 - 2(15)(30)\cos 30° \approx 345.5771$$
$$a \approx 18.59$$

$$\cos C = \frac{a^2 + b^2 - c^2}{2ab} \approx \frac{(18.59)^2 + 15^2 - 30^2}{2(18.59)(15)} \approx -0.5907 \Rightarrow C \approx 126.21°$$

$$B \approx 180° - 30° - 126.21° = 23.79°$$

11. Given: $A = 50°, b = 15, c = 30$

$$a^2 = b^2 + c^2 - 2bc\cos A = 15^2 + 30^2 - 2(15)(30)\cos 50° \approx 546.4912^2 \Rightarrow a \approx 23.38$$

$$\sin C = c\left(\frac{\sin A}{a}\right) \approx 30\left(\frac{\sin 50°}{23.38}\right) \approx 0.9829$$

There are two angles between 0° and 180° whose sine is 0.9829, $C_1 \approx 79.39°$ and $C_2 \approx 180° - 79.39° \approx 100.61°$.

Because side c is the longest side of the triangle, C must be the largest angle of the triangle. So, $C \approx 100.61°$ and $B = 180° - A - C \approx 180° - 50° - 100.56° = 29.39°$.

13. Given: $a = 11, b = 15, c = 21$

$$\cos C = \frac{a^2 + b^2 - c^2}{2ab} = \frac{121 + 225 - 441}{2(11)(15)} \approx -0.2879 \Rightarrow C \approx 106.73°$$

$$\sin B = \frac{b\sin C}{c} = \frac{15\sin 106.73°}{21} \approx 0.6841 \Rightarrow B \approx 43.16°$$

$$A \approx 180° - 43.16° - 106.73° = 30.11°$$

15. Given: $a = 75.4, b = 52, c = 52$

$$\cos A = \frac{b^2 + c^2 - a^2}{2bc} = \frac{52^2 + 52^2 - 75.4^2}{2(52)(52)} = -0.05125 \Rightarrow A \approx 92.94°$$

$$\sin B = \frac{b\sin A}{a} \approx \frac{52(0.9987)}{75.4} \approx 0.68876 \Rightarrow B \approx 43.53°$$

$$C = B \approx 43.53°$$

17. Given: $A = 120°, b = 6, c = 7$

$$a^2 = b^2 + c^2 - 2bc \cos A = 36 + 49 - 2(6)(7) \cos 120° = 127 \Rightarrow a \approx 11.27$$

$$\sin B = \frac{b \sin A}{a} \approx \frac{6 \sin 120°}{11.27} \approx 0.4611 \Rightarrow B \approx 27.46°$$

$$C \approx 180° - 120° - 27.46° = 32.54°$$

19. Given: $B = 10° 35', a = 40, c = 30$

$$b^2 = a^2 + c^2 - 2ac \cos B = 1600 + 900 - 2(40)(30) \cos 10° 35' \approx 140.8268 \Rightarrow b \approx 11.87$$

$$\sin C = \frac{c \sin B}{b} = \frac{30 \sin 10° 35'}{11.87} \approx 0.4642 \Rightarrow C \approx 27.66° \approx 27° 40'$$

$$A \approx 180° - 10° 35' - 27° 40' = 141° 45'$$

21. Given: $B = 125° 40', a = 37, c = 37$

$$b^2 = a^2 + c^2 - 2ac \cos B = 1369 + 1369 - 2(37)(37) \cos 125° 40' \approx 4334.4420 \Rightarrow b \approx 65.84$$

$$A = C \Rightarrow 2A = 180 - 125° 40' = 54° 20' \Rightarrow A = C = 27° 10'$$

23. $C = 43°, a = \frac{4}{9}, b = \frac{7}{9}$

$$c^2 = a^2 + b^2 - 2ab \cos C = \left(\frac{4}{9}\right)^2 + \left(\frac{7}{9}\right)^2 - 2\left(\frac{4}{9}\right)\left(\frac{7}{9}\right) \cos 43° \approx 0.2968 \Rightarrow c \approx 0.54$$

$$\sin A = \frac{a \sin C}{c} = \frac{(4/9) \sin 43°}{0.5448} \approx 0.5564 \Rightarrow A \approx 33.80°$$

$$B \approx 180° - 43° - 33.8° = 103.20°$$

25. $d^2 = 5^2 + 8^2 - 2(5)(8) \cos 45° \approx 32.4315 \Rightarrow d \approx 5.69$

$2\phi = 360° - 2(45°) = 270° \Rightarrow \phi = 135°$

$c^2 = 5^2 + 8^2 - 2(5)(8) \cos 135° \approx 145.5685 \Rightarrow c \approx 12.07$

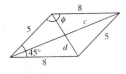

27.

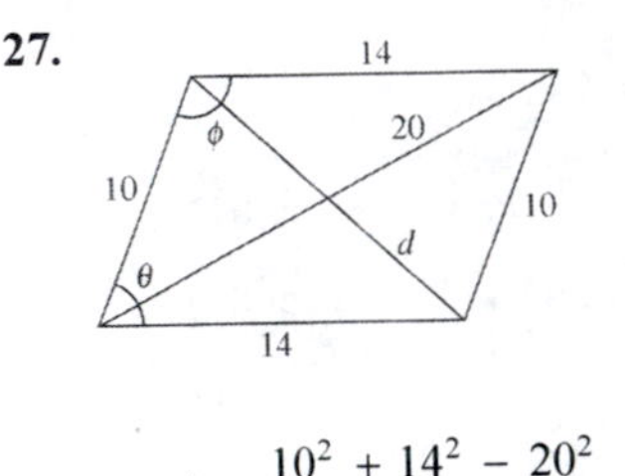

$$\cos \phi = \frac{10^2 + 14^2 - 20^2}{2(10)(14)}$$

$$\phi \approx 111.8°$$

$$2\theta \approx 360° - 2(111.8°)$$

$$\theta = 68.2°$$

$$d^2 = 10^2 + 14^2 - 2(10)(14) \cos 68.2°$$

$$d \approx 13.86$$

29. $\cos\alpha = \dfrac{(12.5)^2 + (15)^2 - 10^2}{2(12.5)(15)} = 0.75 \Rightarrow \alpha \approx 41.41°$

$\cos\beta = \dfrac{10^2 + 15^2 - (12.5)^2}{2(10)(15)} = 0.5625 \Rightarrow \beta \approx 55.77°$

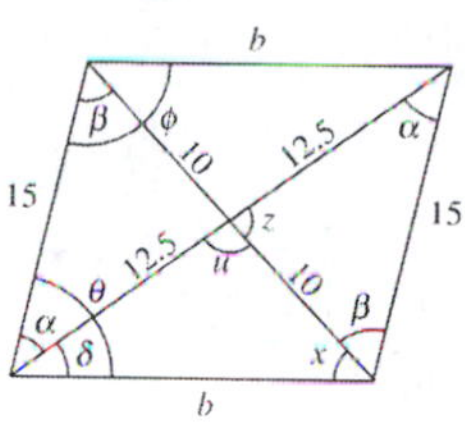

$z = 180° - \alpha - \beta = 82.82°$

$u = 180° - z = 97.18°$

$b^2 = 12.5^2 + 10^2 - 2(12.5)(10)\cos 97.18° \approx 287.4967 \Rightarrow b \approx 16.96$

$\cos\delta = \dfrac{12.5^2 + 16.96^2 - 10^2}{2(12.5)(16.96)} \approx 0.8111 \Rightarrow \delta \approx 35.80°$

$\theta = \alpha + \delta = 41.41° + 35.80° = 77.1°$

$2\phi = 360° - 2\theta \Rightarrow \phi = \dfrac{360° - 2(77.21°)}{2} = 102.79°$

31. Given: $a = 8, c = 5, B = 40°$

Given two sides and included angle, use the Law of Cosines.

$b^2 = a^2 + c^2 - 2ac\cos B = 64 + 25 - 2(8)(5)\cos 40° \approx 27.7164 \Rightarrow b \approx 5.26$

$\cos A = \dfrac{b^2 + c^2 - a^2}{2bc} \approx \dfrac{(5.26)^2 + 25 - 64}{2(5.26)(5)} \approx -0.2154 \Rightarrow A \approx 102.44°$

$C \approx 180° - 102.44° - 40° = 37.56°$

33. Given: $A = 24°, a = 4, b = 18$

Given two sides and an angle opposite one of them, use the Law of Sines.

$h = b\sin A = 18\sin 24° \approx 7.32$

Because $a < h$, no triangle is formed.

35. Given: $A = 42°, B = 35°, c = 1.2$

Given two angles and a side, use the Law of Sines.

$C = 180° - 42° - 35° = 103°$

$a = \dfrac{c\sin A}{\sin C} = \dfrac{1.2\sin 42°}{\sin 103°} \approx 0.82$

$b = \dfrac{c\sin B}{\sin C} = \dfrac{1.2\sin 35°}{\sin 103°} \approx 0.71$

37. $a = 8, b = 12, c = 17$

$s = \dfrac{a + b + c}{2} = \dfrac{8 + 12 + 17}{2} = 18.5$

$\text{Area} = \sqrt{s(s - a)(s - b)(s - c)}$

$= \sqrt{18.5(10.5)(6.5)(1.5)}$

≈ 43.52

39. $a = 2.5, b = 10.2, c = 9$

$s = \dfrac{a + b + c}{2} = 10.85$

$\text{Area} = \sqrt{s(s - a)(s - b)(s - c)}$

$= \sqrt{10.85(8.35)(0.65)(1.85)}$

≈ 10.4

41. Given: $a = 12.32, b = 8.46, c = 15.05$

$s = \dfrac{a + b + c}{2} = 17.915$

$\text{Area} = \sqrt{s(s - a)(s - b)(s - c)}$

$= \sqrt{17.915(5.595)(9.455)(2.865)}$

≈ 52.11

43. Given: $a = 1, b = \dfrac{1}{2}, c = \dfrac{3}{4}$

$s = \dfrac{a + b + c}{2} = \dfrac{1 + \frac{1}{2} + \frac{3}{4}}{2} = \dfrac{9}{8}$

$\text{Area} = \sqrt{s(s - a)(s - b)(s - c)}$

$= \sqrt{\dfrac{9}{8}\left(\dfrac{1}{8}\right)\left(\dfrac{5}{8}\right)\left(\dfrac{3}{8}\right)}$

≈ 0.18

45. $\cos B = \dfrac{1700^2 + 3700^2 - 3000^2}{2(1700)(3700)} \Rightarrow B \approx 52.9°$

Bearing: $90° - 52.9° = $ N 37.1° E

$\cos C = \dfrac{1700^2 + 3000^2 - 3700^2}{2(1700)(3000)} \Rightarrow C \approx 100.2°$

Bearing: $90° - 26.9° = $ S 63.1° E

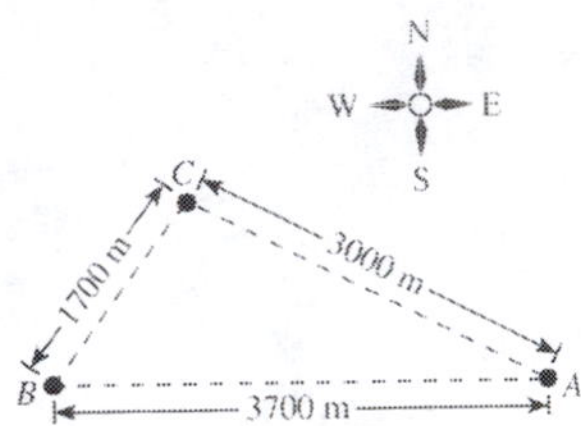

47. $b^2 = 220^2 + 250^2 - 2(220)(250)\cos 105° \Rightarrow b \approx 373.3$ meters

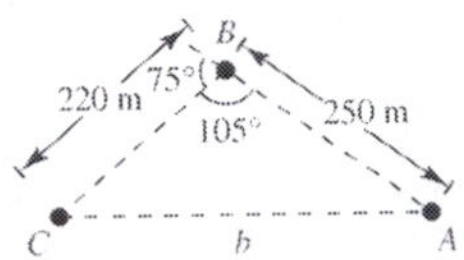

49. $C = 180° - 53° - 67° = 60°$

$c^2 = a^2 + b^2 - 2ab\cos C$

$= 36^2 + 48^2 - 2(36)(48)(0.5)$

$= 1872$

$c \approx 43.3$ mi

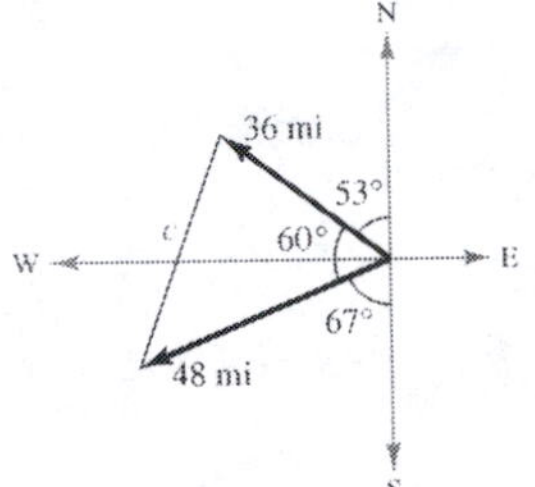

51.

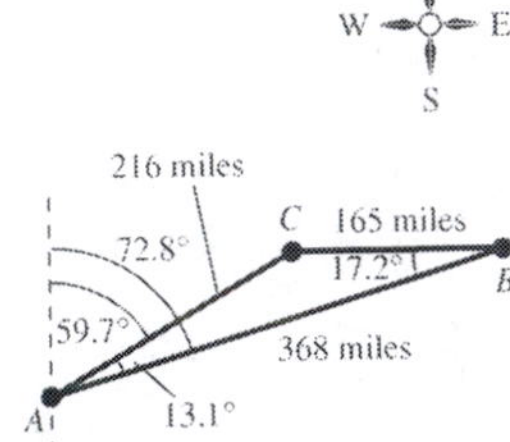

$a = 165, b = 216, c = 368$

$\cos B = \dfrac{165^2 + 368^2 - 216^2}{2(165)(368)} \approx 0.9551$

$B \approx 17.2°$

$\cos A = \dfrac{216^2 + 368^2 - 165^2}{2(216)(368)} \approx 0.9741$

$A \approx 13.1°$

(a) Bearing of Minneapolis (C) from Phoenix (A)

N $(90° - 17.2° - 13.1°)$ E

N 59.7° E

(b) Bearing of Albany (B) from Phoenix (A)

N $(90° - 17.2°)$ E

N 72.8° E

53. $d^2 = 60.5^2 + 90^2 - 2(60.5)(90)\cos 45° \approx 4059.8572 \Rightarrow d \approx 63.7$ ft

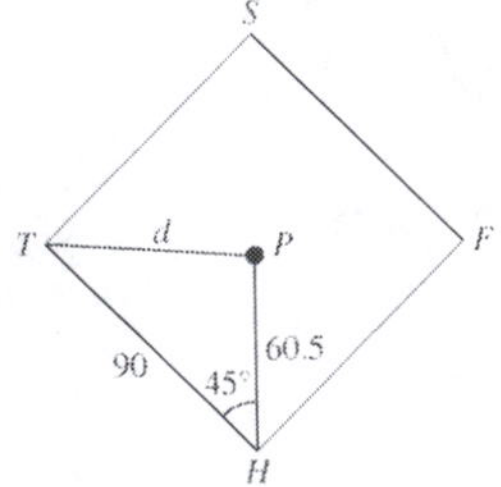

55.

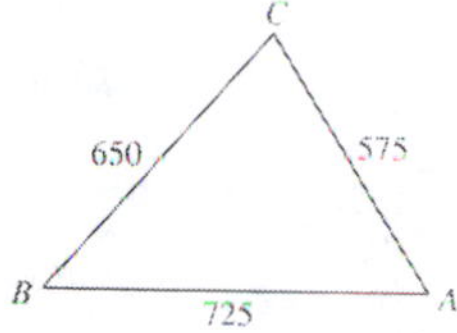

The largest angle is across from the largest side.

$$\cos C = \frac{650^2 + 575^2 - 725^2}{2(650)(575)}$$

$$C \approx 72.3°$$

57. $a = 200$

$b = 500$

$$c = 600 \Rightarrow s = \frac{200 + 500 + 600}{2} = 650$$

$$\text{Area} = \sqrt{650(450)(150)(50)} \approx 46{,}837.5 \text{ square feet}$$

59. $$s = \frac{510 + 840 + 1120}{2} = 1235$$

$$\text{Area} = \sqrt{1235(1235 - 510)(1235 - 840)(1235 - 1120)}$$

$$\approx 201{,}674 \text{ square yards}$$

$$\text{Cost} \approx \left(\frac{201{,}674}{4840}\right)(2000) \approx \$83{,}336.36$$

61. False. The average of the three sides of a triangle is $\frac{a + b + c}{3}$, not $\frac{a + b + c}{2} = s$.

63.

$$\begin{aligned} c^2 &= a^2 + b^2 - 2ab \cos C \\ &= a^2 + b^2 - 2ab \cos 90° \\ &= a^2 + b^2 - 2ab(0) \\ &= a^2 + b^2 \end{aligned}$$

When $C = 90°$, you obtain the Pythagorean Theorem. The Pythagorean Theorem is a special case of the Law of Cosines.

65. There is no method that can be used to solve the no-solution case of SSA.

The Law of Cosines can be used to solve the single-solution case of SSA. You can substitute values into $a^2 = b^2 + c^2 - 2bc \cos A$. The simplified quadratic equation in terms of c can be solved, with one positive solution and one negative solution. The negative solution can be discarded because length is positive. You can use the positive solution to solve the triangle.

67.

$$\begin{aligned} \frac{1}{2}bc(1 + \cos A) &= \frac{1}{2}bc\left[1 + \frac{b^2 + c^2 - a^2}{2bc}\right] \\ &= \frac{1}{2}bc\left[\frac{2bc + b^2 + c^2 - a^2}{2bc}\right] \\ &= \frac{1}{4}\left[(b + c)^2 - a^2\right] \\ &= \frac{1}{4}\left[(b + c) + a\right]\left[(b + c) - a\right] \\ &= \frac{b + c + a}{2} \cdot \frac{b + c - a}{2} \\ &= \frac{a + b + c}{2} \cdot \frac{-a + b + c}{2} \end{aligned}$$

Section 8.3 Vectors in the Plane

1. directed line segment

3. magnitude

5. magnitude; direction

7. unit vector

9. resultant

11. $\|\mathbf{u}\| = \sqrt{(6 - 2)^2 + (5 - 4)^2} = \sqrt{17}$

$\|\mathbf{v}\| = \sqrt{(4 - 0)^2 + (1 - 0)^2} = \sqrt{17}$

$$\text{slope}_{\mathbf{u}} = \frac{5 - 4}{6 - 2} = \frac{1}{4}$$

$$\text{slope}_{\mathbf{v}} = \frac{1 - 0}{4 - 0} = \frac{1}{4}$$

$\mathbf{u}$ and $\mathbf{v}$ have the same magnitude and direction so they are equal.

13. Initial point: $(0, 0)$

Terminal point: $(1, 3)$

$\mathbf{v} = \langle 1 - 0, 3 - 0 \rangle = \langle 1, 3 \rangle$

$\|\mathbf{v}\| = \sqrt{1^2 + 3^2} = \sqrt{10}$

15. Initial point: $(-1, -1)$

Terminal point: $(3, 5)$

$\mathbf{v} = \langle 3 - (-1), 5 - (-1) \rangle = \langle 4, 6 \rangle$

$\|\mathbf{v}\| = \sqrt{4^2 + 6^2} = \sqrt{52} = 2\sqrt{13}$

17. Initial point: $(3, -2)$

Terminal point: $(3, 3)$

$\mathbf{v} = \langle 3 - 3, 3 - (-2) \rangle = \langle 0, 5 \rangle$

$\|\mathbf{v}\| = \sqrt{0^2 + 5^2} = \sqrt{25} = 5$

19. Initial point: $(-3, -5)$

Terminal point: $(5, 1)$

$\mathbf{v} = \langle 5 - (-3), 1 - (-5) \rangle = \langle 8, 6 \rangle$

$\|\mathbf{v}\| = \sqrt{8^2 + 6^2} = \sqrt{100} = 10$

21. Initial point: $(1, 3)$

Terminal point: $(-8, -9)$

$\mathbf{v} = \langle -8 - 1, -9 - 3 \rangle = \langle -9, -12 \rangle$

$\|\mathbf{v}\| = \sqrt{(-9)^2 + (-12)^2} = \sqrt{225} = 15$

23. Initial point: $(-1, 5)$

Terminal point: $(15, 12)$

$\mathbf{v} = \langle 15 - (-1), 12 - 5 \rangle = \langle 16, 7 \rangle$

$\|\mathbf{v}\| = \sqrt{16^2 + 7^2} = \sqrt{305}$

25. $-\mathbf{v}$

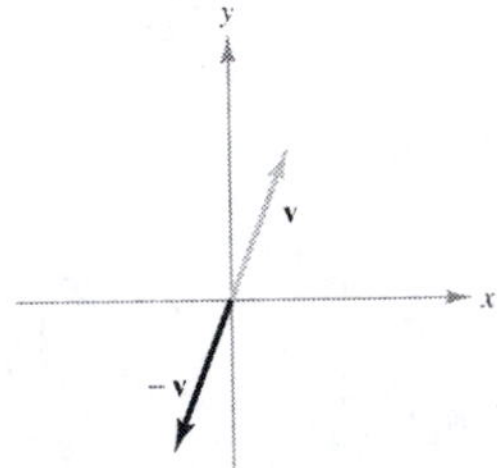

27. $\mathbf{u} + \mathbf{v}$

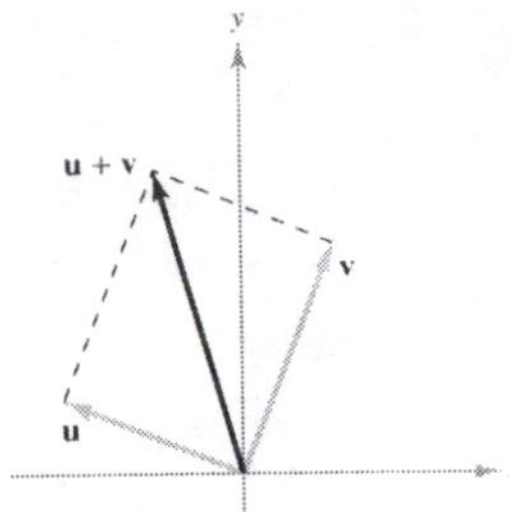

29. $\mathbf{u} - \mathbf{v}$

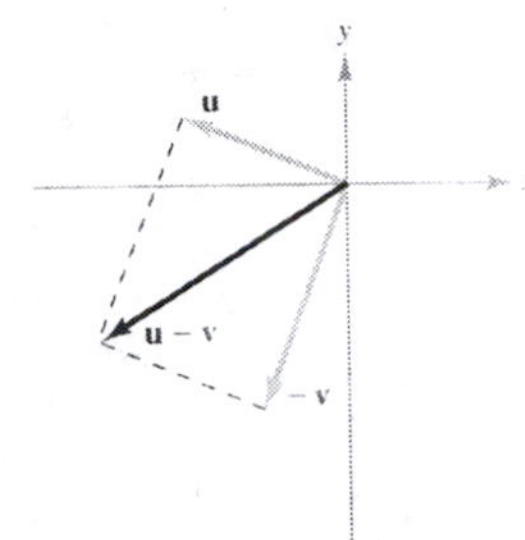

31. $\mathbf{u} = \langle 2, 1 \rangle$, $\mathbf{v} = \langle 1, 3 \rangle$

(a) $\mathbf{u} + \mathbf{v} = \langle 3, 4 \rangle$

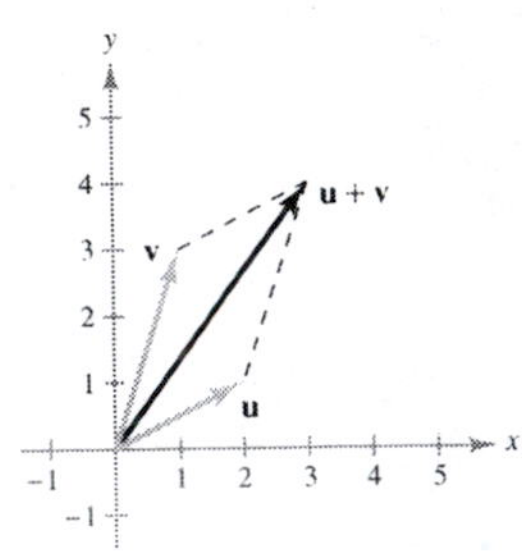

(b) $\mathbf{u} - \mathbf{v} = \langle 1, -2 \rangle$

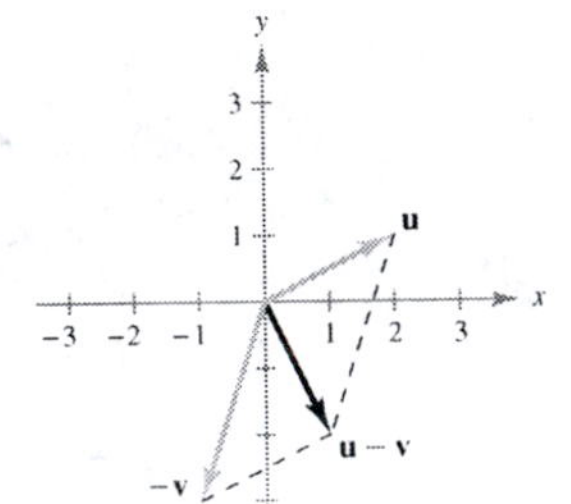

(c) $2\mathbf{u} - 3\mathbf{v} = \langle 4, 2 \rangle - \langle 3, 9 \rangle = \langle 1, -7 \rangle$

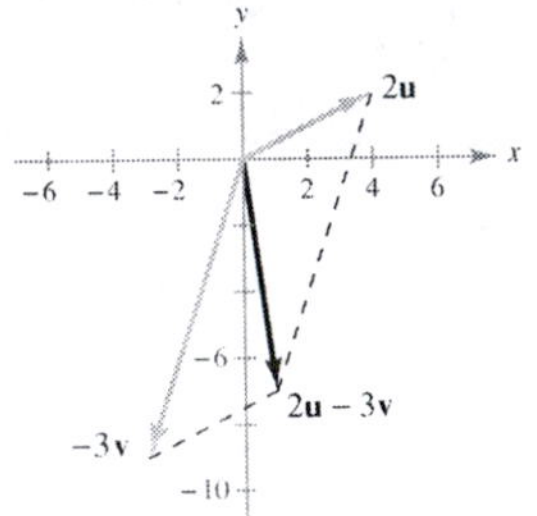

33. $\mathbf{u} = \langle -5, 3 \rangle$, $\mathbf{v} = \langle 0, 0 \rangle$

(a) $\mathbf{u} + \mathbf{v} = \langle -5, 3 \rangle = \mathbf{u}$

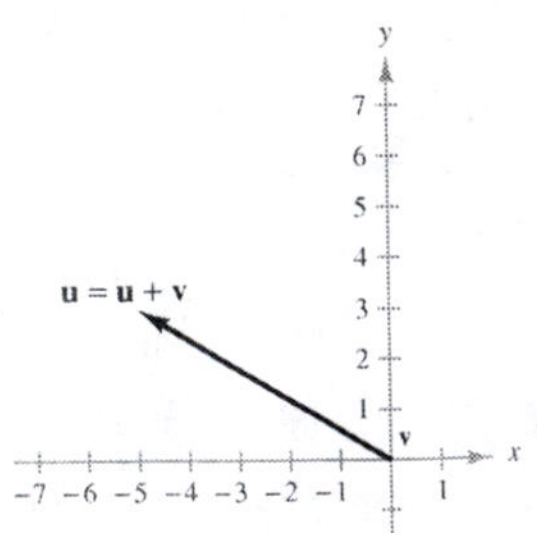

(b) $\mathbf{u} - \mathbf{v} = \langle -5, 3 \rangle = \mathbf{u}$

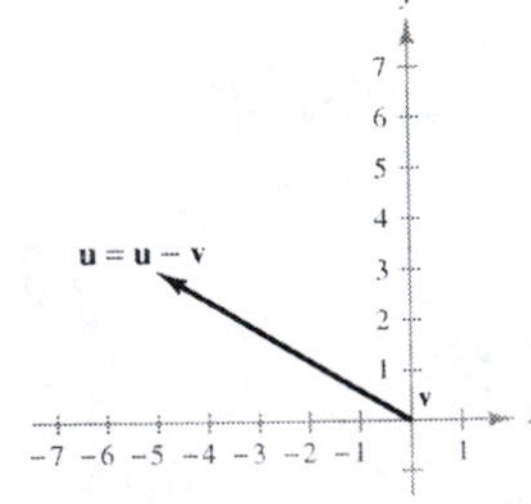

(c) $2\mathbf{u} - 3\mathbf{v} = \langle -10, 6 \rangle = 2\mathbf{u}$

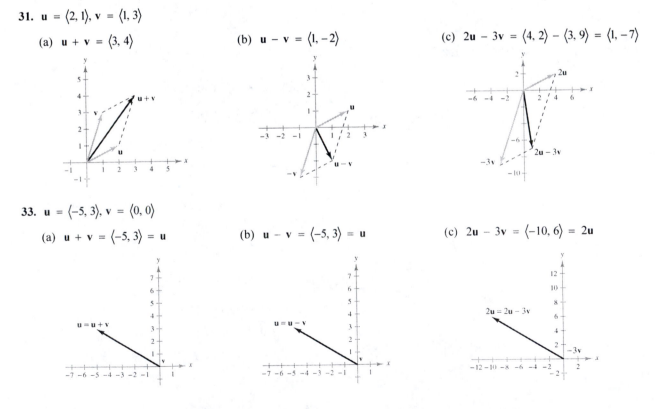

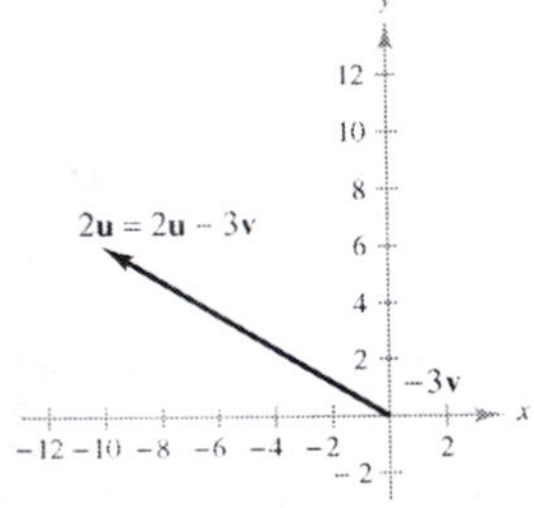

35. $\mathbf{u} = \mathbf{i} + \mathbf{j}, \mathbf{v} = 2\mathbf{i} - 3\mathbf{j}$

(a) $\mathbf{u} + \mathbf{v} = 3\mathbf{i} - 2\mathbf{j}$

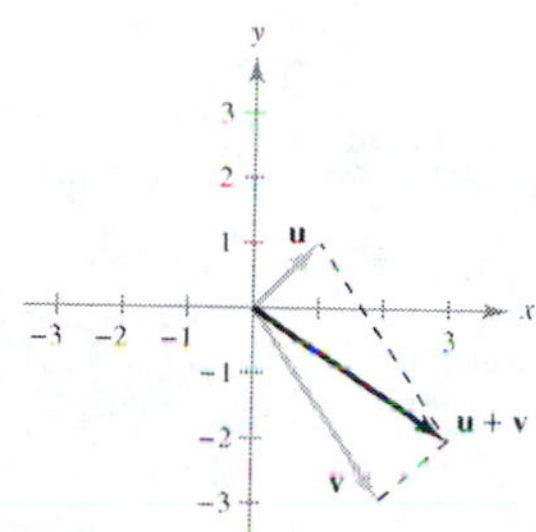

(b) $\mathbf{u} - \mathbf{v} = -\mathbf{i} + 4\mathbf{j}$

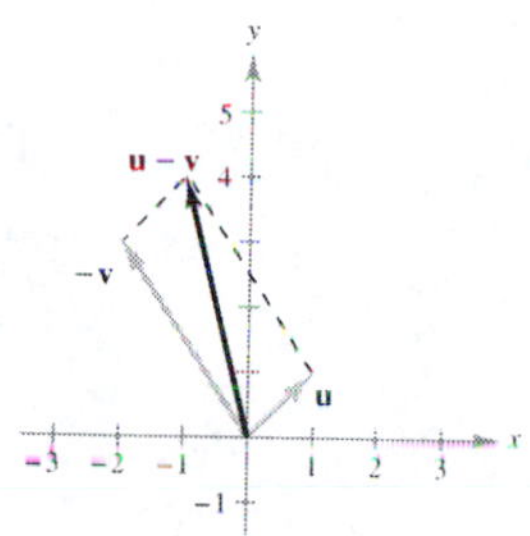

(c) $2\mathbf{u} - 3\mathbf{v} = (2\mathbf{i} + 2\mathbf{j}) - (6\mathbf{i} - 9\mathbf{j})$
$= -4\mathbf{i} + 11\mathbf{j}$

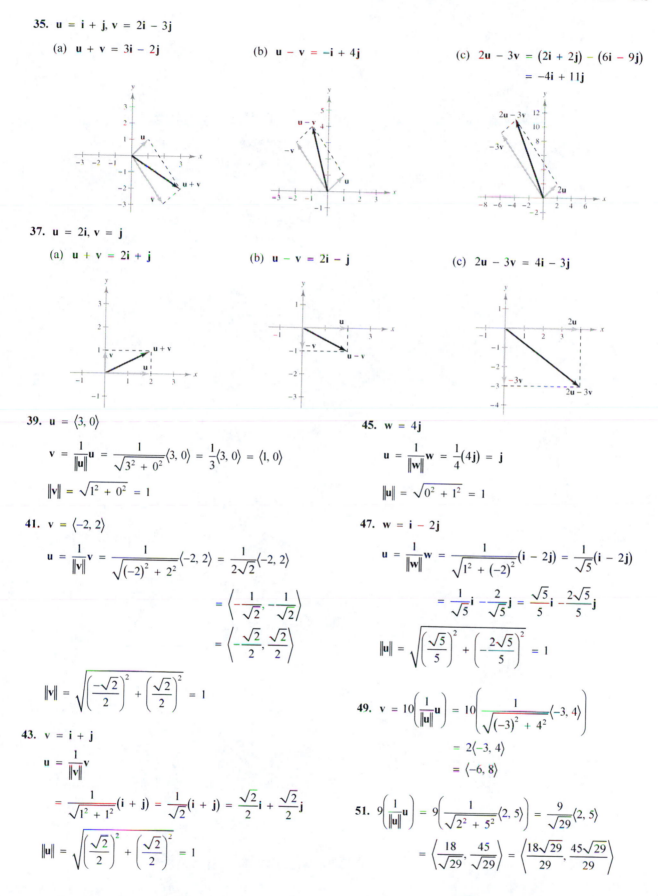

37. $\mathbf{u} = 2\mathbf{i}, \mathbf{v} = \mathbf{j}$

(a) $\mathbf{u} + \mathbf{v} = 2\mathbf{i} + \mathbf{j}$

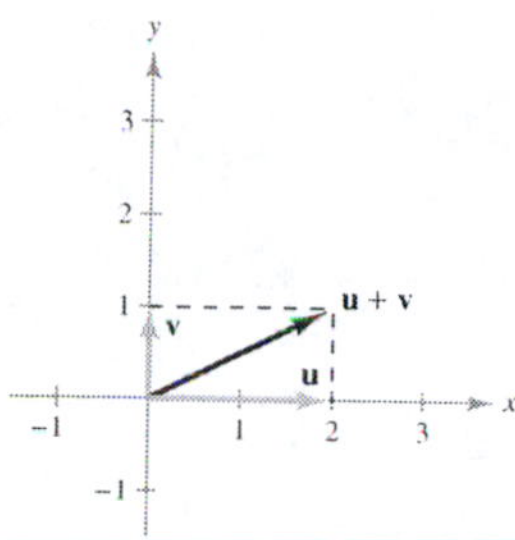

(b) $\mathbf{u} - \mathbf{v} = 2\mathbf{i} - \mathbf{j}$

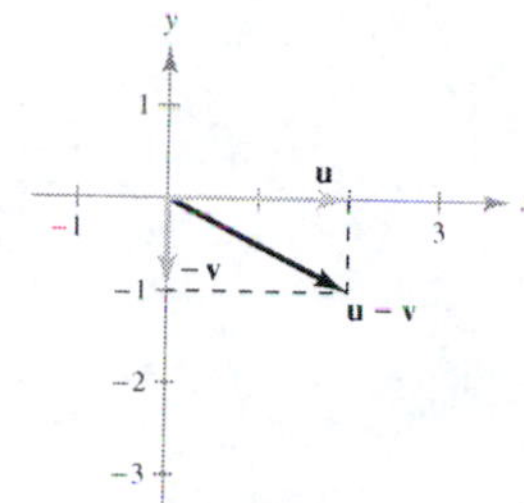

(c) $2\mathbf{u} - 3\mathbf{v} = 4\mathbf{i} - 3\mathbf{j}$

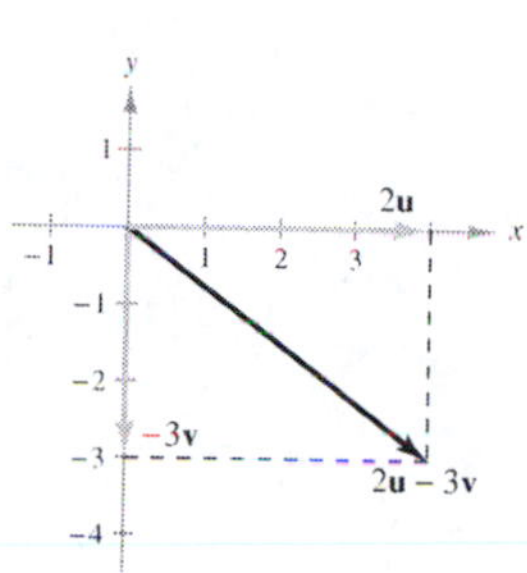

39. $\mathbf{u} = \langle 3, 0\rangle$

$$\mathbf{v} = \frac{1}{\|\mathbf{u}\|}\mathbf{u} = \frac{1}{\sqrt{3^2 + 0^2}}\langle 3, 0\rangle = \frac{1}{3}\langle 3, 0\rangle = \langle 1, 0\rangle$$

$$\|\mathbf{v}\| = \sqrt{1^2 + 0^2} = 1$$

41. $\mathbf{v} = \langle -2, 2\rangle$

$$\mathbf{u} = \frac{1}{\|\mathbf{v}\|}\mathbf{v} = \frac{1}{\sqrt{(-2)^2 + 2^2}}\langle -2, 2\rangle = \frac{1}{2\sqrt{2}}\langle -2, 2\rangle$$

$$= \left\langle -\frac{1}{\sqrt{2}}, -\frac{1}{\sqrt{2}}\right\rangle$$

$$= \left\langle -\frac{\sqrt{2}}{2}, \frac{\sqrt{2}}{2}\right\rangle$$

$$\|\mathbf{v}\| = \sqrt{\left(\frac{-\sqrt{2}}{2}\right)^2 + \left(\frac{\sqrt{2}}{2}\right)^2} = 1$$

43. $\mathbf{v} = \mathbf{i} + \mathbf{j}$

$$\mathbf{u} = \frac{1}{\|\mathbf{v}\|}\mathbf{v}$$

$$= \frac{1}{\sqrt{1^2 + 1^2}}(\mathbf{i} + \mathbf{j}) = \frac{1}{\sqrt{2}}(\mathbf{i} + \mathbf{j}) = \frac{\sqrt{2}}{2}\mathbf{i} + \frac{\sqrt{2}}{2}\mathbf{j}$$

$$\|\mathbf{u}\| = \sqrt{\left(\frac{\sqrt{2}}{2}\right)^2 + \left(\frac{\sqrt{2}}{2}\right)^2} = 1$$

45. $\mathbf{w} = 4\mathbf{j}$

$$\mathbf{u} = \frac{1}{\|\mathbf{w}\|}\mathbf{w} = \frac{1}{4}(4\mathbf{j}) = \mathbf{j}$$

$$\|\mathbf{u}\| = \sqrt{0^2 + 1^2} = 1$$

47. $\mathbf{w} = \mathbf{i} - 2\mathbf{j}$

$$\mathbf{u} = \frac{1}{\|\mathbf{w}\|}\mathbf{w} = \frac{1}{\sqrt{1^2 + (-2)^2}}(\mathbf{i} - 2\mathbf{j}) = \frac{1}{\sqrt{5}}(\mathbf{i} - 2\mathbf{j})$$

$$= \frac{1}{\sqrt{5}}\mathbf{i} - \frac{2}{\sqrt{5}}\mathbf{j} = \frac{\sqrt{5}}{5}\mathbf{i} - \frac{2\sqrt{5}}{5}\mathbf{j}$$

$$\|\mathbf{u}\| = \sqrt{\left(\frac{\sqrt{5}}{5}\right)^2 + \left(-\frac{2\sqrt{5}}{5}\right)^2} = 1$$

49. $\mathbf{v} = 10\left(\frac{1}{\|\mathbf{u}\|}\mathbf{u}\right) = 10\left(\frac{1}{\sqrt{(-3)^2 + 4^2}}\langle -3, 4\rangle\right)$

$$= 2\langle -3, 4\rangle$$

$$= \langle -6, 8\rangle$$

51. $9\left(\frac{1}{\|\mathbf{u}\|}\mathbf{u}\right) = 9\left(\frac{1}{\sqrt{2^2 + 5^2}}\langle 2, 5\rangle\right) = \frac{9}{\sqrt{29}}\langle 2, 5\rangle$

$$= \left\langle \frac{18}{\sqrt{29}}, \frac{45}{\sqrt{29}}\right\rangle = \left\langle \frac{18\sqrt{29}}{29}, \frac{45\sqrt{29}}{29}\right\rangle$$

53. $\mathbf{u} = \langle 3 - (-2), -2 - 1 \rangle$
$= \langle 5, -3 \rangle$
$= 5\mathbf{i} - 3\mathbf{j}$

55. $\mathbf{u} = \langle 0 - (-6), 1 - 4 \rangle$
$\mathbf{u} = \langle 6, -3 \rangle$
$\mathbf{u} = 6\mathbf{i} - 3\mathbf{j}$

57. $\mathbf{v} = \frac{3}{2}\mathbf{u}$
$= \frac{3}{2}(2\mathbf{i} - \mathbf{j})$
$= 3\mathbf{i} - \frac{3}{2}\mathbf{j} = \langle 3, -\frac{3}{2} \rangle$

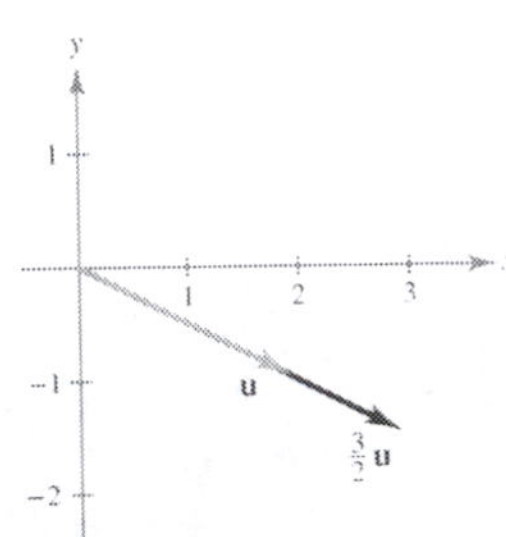

59. $\mathbf{v} = \mathbf{u} + 2\mathbf{w}$
$= (2\mathbf{i} - \mathbf{j}) + 2(\mathbf{i} + 2\mathbf{j})$
$= 4\mathbf{i} + 3\mathbf{j} = \langle 4, 3 \rangle$

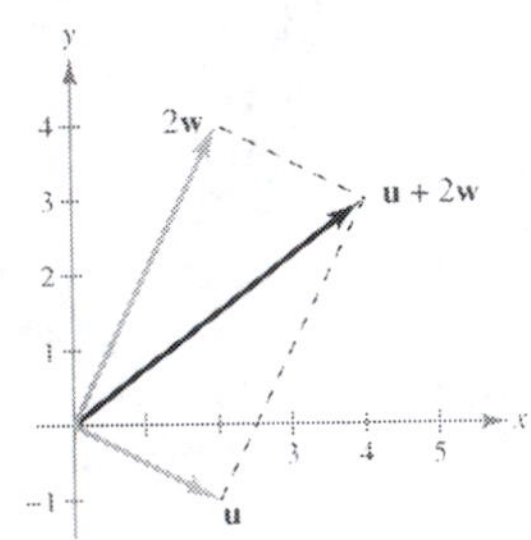

61. $\mathbf{v} = \frac{1}{2}(3\mathbf{u} + \mathbf{w})$
$= \frac{1}{2}(6\mathbf{i} - 3\mathbf{j} + \mathbf{i} + 2\mathbf{j})$
$= \frac{7}{2}\mathbf{i} - \frac{1}{2}\mathbf{j} = \langle \frac{7}{2}, -\frac{1}{2} \rangle$

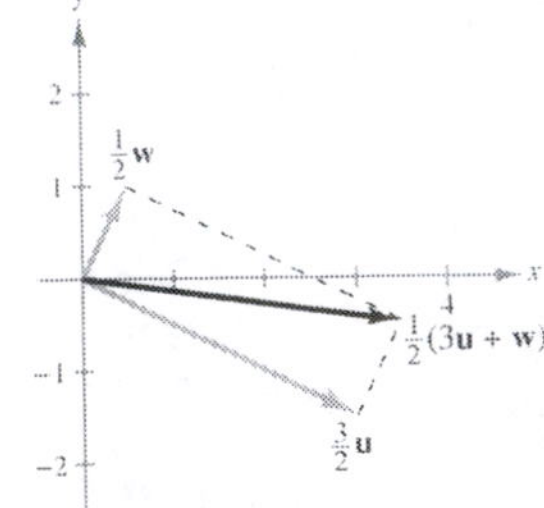

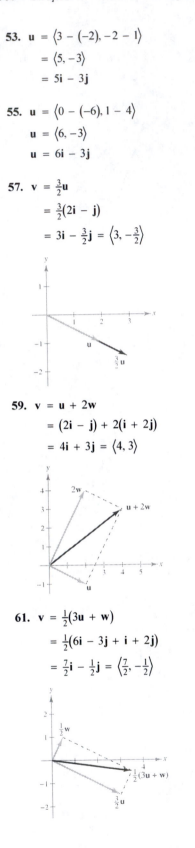

63. $\mathbf{v} = 6\mathbf{i} - 6\mathbf{j}$
$\|\mathbf{v}\| = \sqrt{6^2 + (-6)^2} = \sqrt{72} = 6\sqrt{2}$
$\tan\theta = \dfrac{-6}{6} = -1$
Since $\mathbf{v}$ lies in Quadrant IV, $\theta = 315°$.

65. $\mathbf{v} = 3(\cos 60°\mathbf{i} + \sin 60°\mathbf{j})$
$\|\mathbf{v}\| = 3, \theta = 60°$

67. $\mathbf{v} = \langle 3\cos 0°, 3\sin 0° \rangle$
$= \langle 3, 0 \rangle$

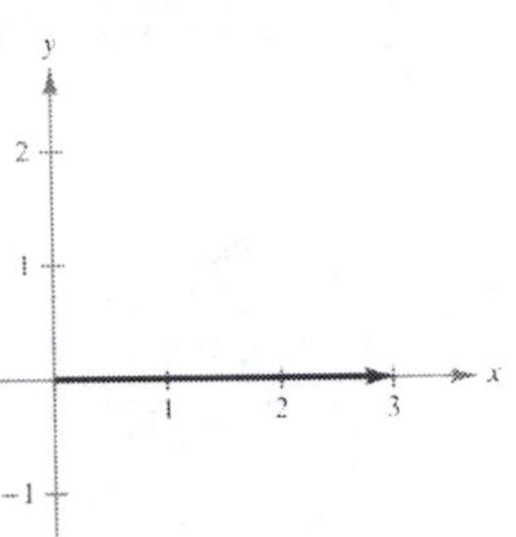

69. $\mathbf{v} = \left\langle \dfrac{7}{2}\cos 150°, \dfrac{7}{2}\sin 150° \right\rangle$
$= \left\langle -\dfrac{7\sqrt{3}}{4}, \dfrac{7}{4} \right\rangle$

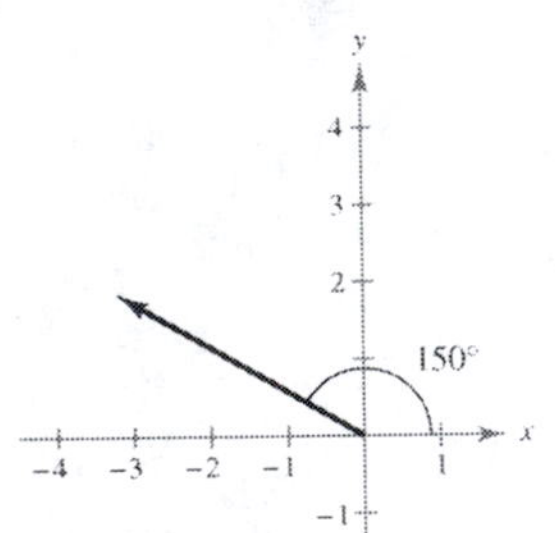

71. $\mathbf{v} = \langle 2\sqrt{3}\cos 45°, 2\sqrt{3}\sin 45° \rangle$
$= \langle \sqrt{6}, \sqrt{6} \rangle$

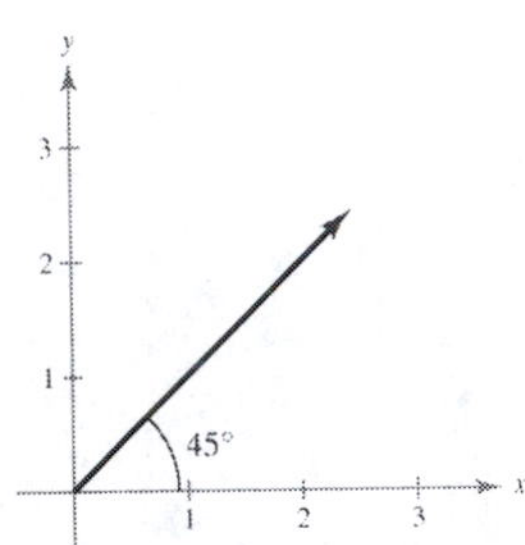

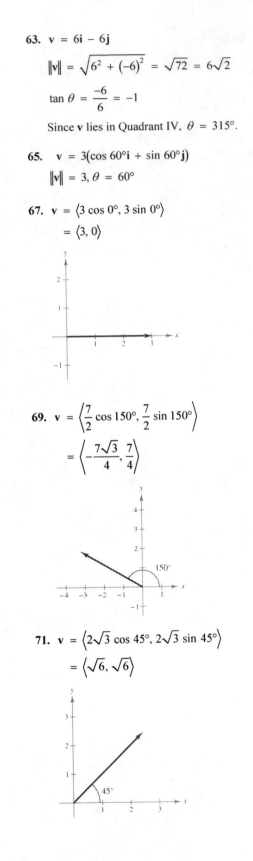

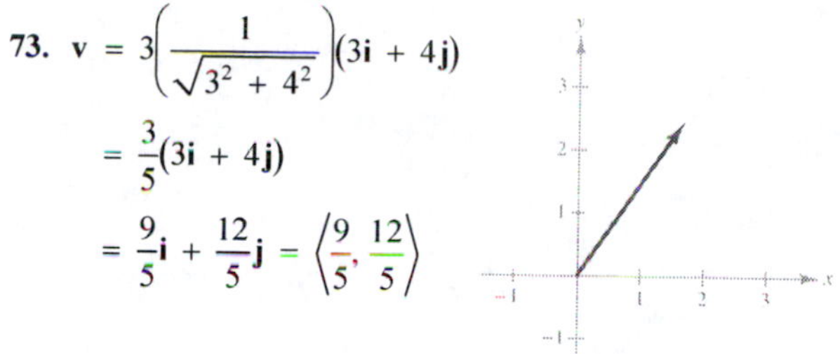

73. $\mathbf{v} = 3\left(\dfrac{1}{\sqrt{3^2 + 4^2}}\right)(3\mathbf{i} + 4\mathbf{j})$

$= \dfrac{3}{5}(3\mathbf{i} + 4\mathbf{j})$

$= \dfrac{9}{5}\mathbf{i} + \dfrac{12}{5}\mathbf{j} = \left\langle \dfrac{9}{5}, \dfrac{12}{5} \right\rangle$

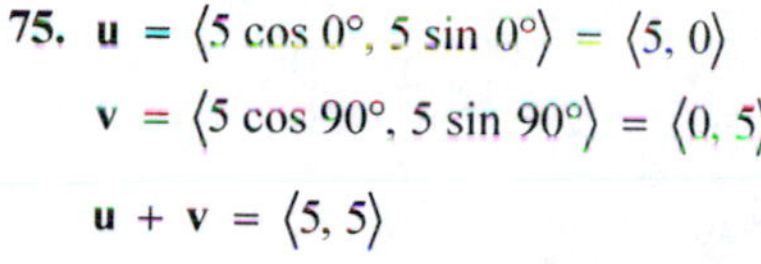

75. $\mathbf{u} = \langle 5\cos 0°, 5\sin 0° \rangle = \langle 5, 0 \rangle$

$\mathbf{v} = \langle 5\cos 90°, 5\sin 90° \rangle = \langle 0, 5 \rangle$

$\mathbf{u} + \mathbf{v} = \langle 5, 5 \rangle$

77. $\mathbf{u} = \langle 20\cos 45°, 20\sin 45° \rangle = \langle 10\sqrt{2}, 10\sqrt{2} \rangle$

$\mathbf{v} = \langle 50\cos 180°, 50\sin 180° \rangle = \langle -50, 0 \rangle$

$\mathbf{u} + \mathbf{v} = \langle 10\sqrt{2} - 50, 10\sqrt{2} \rangle$

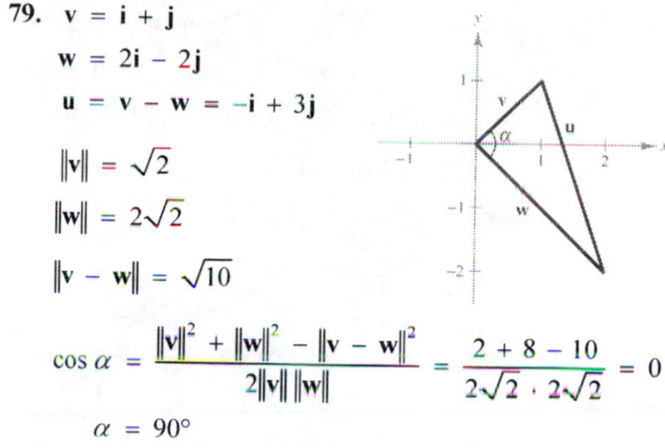

79. $\mathbf{v} = \mathbf{i} + \mathbf{j}$

$\mathbf{w} = 2\mathbf{i} - 2\mathbf{j}$

$\mathbf{u} = \mathbf{v} - \mathbf{w} = -\mathbf{i} + 3\mathbf{j}$

$\|\mathbf{v}\| = \sqrt{2}$

$\|\mathbf{w}\| = 2\sqrt{2}$

$\|\mathbf{v} - \mathbf{w}\| = \sqrt{10}$

$$\cos\alpha = \frac{\|\mathbf{v}\|^2 + \|\mathbf{w}\|^2 - \|\mathbf{v} - \mathbf{w}\|^2}{2\|\mathbf{v}\|\,\|\mathbf{w}\|} = \frac{2 + 8 - 10}{2\sqrt{2} \cdot 2\sqrt{2}} = 0$$

$\alpha = 90°$

81. Force One: $\mathbf{u} = 45\mathbf{i}$

Force Two: $\mathbf{v} = 60\cos\theta\mathbf{i} + 60\sin\theta\mathbf{j}$

Resultant Force: $\mathbf{u} + \mathbf{v} = (45 + 60\cos\theta)\mathbf{i} + 60\sin\theta\mathbf{j}$

$$\|\mathbf{u} + \mathbf{v}\| = \sqrt{(45 + 60\cos\theta)^2 + (60\sin\theta)^2} = 90$$

$$2025 + 5400\cos\theta + 3600 = 8100$$

$$5400\cos\theta = 2475$$

$$\cos\theta = \tfrac{2475}{5400} \approx 0.4583$$

$$\theta \approx 62.7°$$

83. Horizontal component of velocity: $1200\cos 6° \approx 1193.4$ ft/sec

Vertical component of velocity: $1200\sin 6° \approx 125.4$ ft/sec

85. $\mathbf{u} = 300\mathbf{i}$

$$\mathbf{v} = (125\cos 45°)\mathbf{i} + (125\sin 45°)\mathbf{j} = \frac{125}{\sqrt{2}}\mathbf{i} + \frac{125}{\sqrt{2}}\mathbf{j}$$

$$\mathbf{u} + \mathbf{v} = \left(300 + \frac{125}{\sqrt{2}}\right)\mathbf{i} + \frac{125}{\sqrt{2}}\mathbf{j}$$

$$\|\mathbf{u} + \mathbf{v}\| = \sqrt{\left(300 + \frac{125}{\sqrt{2}}\right)^2 + \left(\frac{125}{\sqrt{2}}\right)^2} \approx 398.32 \text{ newtons}$$

$$\tan\theta = \frac{\dfrac{125}{\sqrt{2}}}{300 + \left(\dfrac{125}{\sqrt{2}}\right)} \Rightarrow \theta \approx 12.8°$$

87.

$$\mathbf{u} = (75\cos 30°)\mathbf{i} + (75\sin 30°)\mathbf{j} \approx 64.95\mathbf{i} + 37.5\mathbf{j}$$
$$\mathbf{v} = (100\cos 45°)\mathbf{i} + (100\sin 45°)\mathbf{j} \approx 70.71\mathbf{i} + 70.71\mathbf{j}$$
$$\mathbf{w} = (125\cos 120°)\mathbf{i} + (125\sin 120°)\mathbf{j} \approx -62.5\mathbf{i} + 108.3\mathbf{j}$$
$$\mathbf{u} + \mathbf{v} + \mathbf{w} \approx 73.16\mathbf{i} + 216.5\mathbf{j}$$
$$\|\mathbf{u} + \mathbf{v} + \mathbf{w}\| \approx 228.5 \text{ pounds}$$
$$\tan\theta \approx \frac{216.5}{73.16} \approx 2.9593$$
$$\theta \approx 71.3°$$

89. Left crane: $\mathbf{u} = \|\mathbf{u}\|(\cos 155.7°\mathbf{i} + \sin 155.7°\mathbf{j})$

Right crane: $\mathbf{v} = \|\mathbf{v}\|(\cos 44.5°\mathbf{i} + \sin 44.5°\mathbf{j})$

Resultant: $\mathbf{u} + \mathbf{v} = -20{,}240\mathbf{j}$

System of equations:

$\|\mathbf{u}\|\cos 155.7° + \|\mathbf{v}\|\cos 44.5° = 0$

$\|\mathbf{u}\|\sin 155.7° + \|\mathbf{v}\|\sin 44.5° = 20{,}240$

Solving this system of equations yields the following:

Left crane $= \|\mathbf{u}\| \approx 15{,}484$ pounds

Right crane $= \|\mathbf{v}\| \approx 19{,}786$ pounds

91. Cable $\overrightarrow{AC}$: $\mathbf{u} = \|\mathbf{u}\|(\cos 50°\mathbf{i} - \sin 50°\mathbf{j})$

Cable $\overrightarrow{BC}$: $\mathbf{v} = \|\mathbf{v}\|(-\cos 30°\mathbf{i} - \sin 30°\mathbf{j})$

Resultant: $\mathbf{u} + \mathbf{v} = -2000\mathbf{j}$

$\|\mathbf{u}\|\cos 50° - \|\mathbf{v}\|\cos 30° = 0$

$-\|\mathbf{u}\|\sin 50° - \|\mathbf{v}\|\sin 30° = -2000$

Solving this system of equations yields:

$T_{AC} = \|\mathbf{u}\| \approx 1758.8$ pounds

$T_{BC} = \|\mathbf{v}\| \approx 1305.4$ pounds

93. Towline 1: $\mathbf{u} = \|\mathbf{u}\|(\cos 18°\mathbf{i} + \sin 18°\mathbf{j})$

Towline 2: $\mathbf{v} = \|\mathbf{u}\|(\cos 18°\mathbf{i} - \sin 18°\mathbf{j})$

Resultant: $\mathbf{u} + \mathbf{v} = 6000\mathbf{i}$

$\|\mathbf{u}\|\cos 18° + \|\mathbf{u}\|\cos 18° = 6000$

$\|\mathbf{u}\| \approx 3154.4$

So, the tension on each towline is $\|\mathbf{u}\| \approx 3154.4$ pounds.

95. $W = 100, \theta = 12°$

$$\sin\theta = \frac{F}{W}$$
$$F = W\sin\theta = 100\sin 12° \approx 20.8 \text{ pounds}$$

97. $F = 5000, W = 15{,}000$

$$\sin\theta = \frac{F}{W}$$
$$\sin\theta = \frac{5000}{15{,}000}$$
$$\theta = \sin^{-1}\frac{1}{3} \approx 19.5°$$

99. $W = FD = (100\cos 50°)(30) \approx 1928.4$ foot-pounds

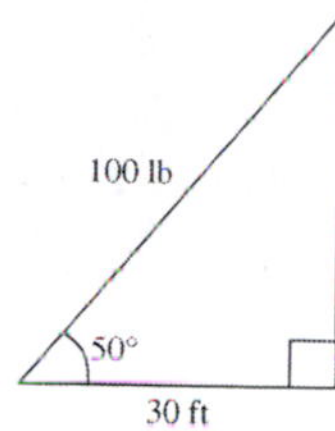

101. Airspeed: $\mathbf{u} = (875\cos 58°)\mathbf{i} - (875\sin 58°)\mathbf{j}$

Groundspeed: $\mathbf{v} = (800\cos 50°)\mathbf{i} - (800\sin 50°)\mathbf{j}$

Wind: $\mathbf{w} = \mathbf{v} - \mathbf{u} = (800\cos 50° - 875\cos 58°)\mathbf{i} + (-800\sin 50° + 875\sin 58°)\mathbf{j}$

$\approx 50.5507\mathbf{i} + 129.2065\mathbf{j}$

Wind speed: $\|\mathbf{w}\| \approx \sqrt{(50.5507)^2 + (129.2065)^2} \approx 138.7$ kilometers per hour

Wind direction: $\tan\theta \approx \dfrac{129.2065}{50.5507}$

$\theta \approx 68.6°;\ 90° - \theta = 21.4°$

Bearing: N 21.4° E

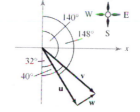

103. True. Two directed line segments that have the same magnitude and direction are equivalent.

105. True. If $\mathbf{v} = a\mathbf{i} + b\mathbf{j} = 0$ is the zero vector, then $a + b = 0$. So, $a = -b$.

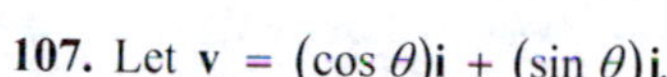

107. Let $\mathbf{v} = (\cos\theta)\mathbf{i} + (\sin\theta)\mathbf{j}$.

$\|\mathbf{v}\| = \sqrt{\cos^2\theta + \sin^2\theta} = \sqrt{1} = 1$

So, $\mathbf{v}$ is a unit vector for any value of θ.

109. $\mathbf{u} = \langle 5 - 1, 2 - 6\rangle = \langle 4, -4\rangle$

$\mathbf{v} = \langle 9 - 4, 4 - 5\rangle = \langle 5, -1\rangle$

$\mathbf{u} - \mathbf{v} = \langle -1, -3\rangle$ or $\mathbf{v} - \mathbf{u} = \langle 1, 3\rangle$

111. $\mathbf{F}_1 = \langle 10, 0\rangle,\ \mathbf{F}_2 = 5\langle \cos\theta, \sin\theta\rangle$

(a) $\mathbf{F}_1 + \mathbf{F}_2 = \langle 10 + 5\cos\theta, 5\sin\theta\rangle$

$$\begin{aligned}\|\mathbf{F}_1 + \mathbf{F}_2\| &= \sqrt{(10 + 5\cos\theta)^2 + (5\sin\theta)^2}\\ &= \sqrt{100 + 100\cos\theta + 25\cos^2\theta + 25\sin^2\theta}\\ &= 5\sqrt{4 + 4\cos\theta + \cos^2\theta + \sin^2\theta}\\ &= 5\sqrt{4 + 4\cos\theta + 1}\\ &= 5\sqrt{5 + 4\cos\theta}\end{aligned}$$

(b)

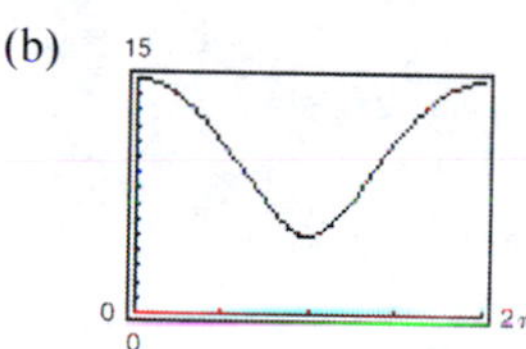

(c) Range: $[5, 15]$

Maximum is 15 when $\theta = 0$.

Minimum is 5 when $\theta = \pi$.

(d) The magnitude of the resultant is never 0 because the magnitudes of $\mathbf{F}_1$ and $\mathbf{F}_2$ are not the same.

113. *Sample answer*:

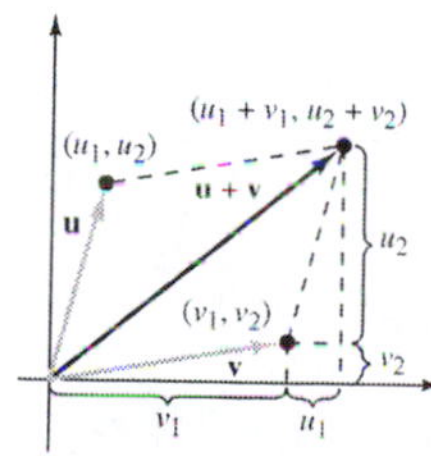

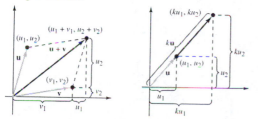

Section 8.4 Vectors and Dot Products

1. dot product

3. $\dfrac{\mathbf{u}\cdot\mathbf{v}}{\|\mathbf{u}\|\,\|\mathbf{v}\|}$

5. $\left(\dfrac{\mathbf{u}\cdot\mathbf{v}}{\|\mathbf{v}\|^2}\right)\mathbf{v}$

7. $\mathbf{u} = \langle 7, 1\rangle, \mathbf{v} = \langle -3, 2\rangle$

$\mathbf{u}\cdot\mathbf{v} = 7(-3) + 1(2) = -19$

9. $\mathbf{u} = \langle -4, 1\rangle, \mathbf{v} = \langle 2, -3\rangle$

$\mathbf{u}\cdot\mathbf{v} = -4(2) + 1(-3) = -11$

11. $\mathbf{u} = 4\mathbf{i} - 2\mathbf{j}, \mathbf{v} = \mathbf{i} - \mathbf{j}$

$\mathbf{u}\cdot\mathbf{v} = 4(1) + (-2)(-1) = 6$

13. $\mathbf{u} = 3\mathbf{i} + 2\mathbf{j}, \mathbf{v} = -2\mathbf{i} - 3\mathbf{j}$

$\mathbf{u}\cdot\mathbf{v} = 3(-2) + 2(-3) = -12$

15. $\mathbf{u} = \langle 3, 3\rangle$

$\mathbf{u}\cdot\mathbf{u} = 3(3) + 3(3) = 18$

The result is a scalar.

17. $\mathbf{u} = \langle 3, 3\rangle, \mathbf{v} = \langle -4, 2\rangle$

$$\begin{aligned}(\mathbf{u}\cdot\mathbf{v})\mathbf{v} &= [3(-4) + 3(2)]\langle -4, 2\rangle\\ &= -6\langle -4, 2\rangle\\ &= \langle 24, -12\rangle\end{aligned}$$

The result is a vector.

19. $\mathbf{u} = \langle 3, 3\rangle, \mathbf{v} = \langle -4, 2\rangle, \mathbf{w} = \langle 3, -1\rangle$

$$\begin{aligned}(3\mathbf{w}\cdot\mathbf{v})\mathbf{u} &= 3[3(-4) + (-1)(2)]\langle 3, 3\rangle\\ &= 3(-14)\langle 3, 3\rangle\\ &= -42\langle 3, 3\rangle\\ &= \langle -126, -126\rangle\end{aligned}$$

The result is a vector.

21. $\mathbf{w} = \langle 3, -1\rangle$

$\|\mathbf{w}\| - 1 = \sqrt{3^2 + (-1)^2} - 1 = \sqrt{10} - 1$

The result is a scalar.

23. $\mathbf{u} = \langle 3, 3\rangle, \mathbf{v} = \langle -4, 2\rangle, \mathbf{w} = \langle 3, -1\rangle$

$$\begin{aligned}(\mathbf{u}\cdot\mathbf{v}) - (\mathbf{u}\cdot\mathbf{w}) &= [3(-4) + 3(2)] - [3(3) + 3(-1)]\\ &= -6 - 6\\ &= -12\end{aligned}$$

The result is a scalar.

25. $\mathbf{u} = \langle -8, 15\rangle$

$\|\mathbf{u}\| = \sqrt{\mathbf{u}\cdot\mathbf{u}} = \sqrt{(-8)(-8) + 15(15)} = \sqrt{289} = 17$

27. $\mathbf{u} = 20\mathbf{i} + 25\mathbf{j}$

$\|\mathbf{u}\| = \sqrt{\mathbf{u}\cdot\mathbf{u}} = \sqrt{(20)^2 + (25)^2} = \sqrt{1025} = 5\sqrt{41}$

29. $\mathbf{u} = 6\mathbf{j}$

$\|\mathbf{u}\| = \sqrt{\mathbf{u}\cdot\mathbf{u}} = \sqrt{(0)^2 + (6)^2} = \sqrt{36} = 6$

31. $\mathbf{u} = \langle 1, 0\rangle, \mathbf{v} = \langle 0, -2\rangle$

$\cos\theta = \dfrac{\mathbf{u}\cdot\mathbf{v}}{\|\mathbf{u}\|\,\|\mathbf{v}\|} = \dfrac{0}{(1)(2)} = 0$

$\theta = 90^\circ$

33. $\mathbf{u} = 3\mathbf{i} + 4\mathbf{j}, \mathbf{v} = -2\mathbf{j}$

$\cos\theta = \dfrac{\mathbf{u}\cdot\mathbf{v}}{\|\mathbf{u}\|\,\|\mathbf{v}\|} = -\dfrac{8}{(5)(2)}$

$\theta = \arccos\left(-\dfrac{4}{5}\right)$

$\theta \approx 143.13^\circ$

35. $\mathbf{u} = 2\mathbf{i} - \mathbf{j}, \mathbf{v} = 6\mathbf{i} + 4\mathbf{j}$

$\cos\theta = \dfrac{\mathbf{u}\cdot\mathbf{v}}{\|\mathbf{u}\|\,\|\mathbf{v}\|} = \dfrac{8}{\sqrt{5}\sqrt{52}} = 0.4961$

$\theta = 60.26^\circ$

37. $\mathbf{u} = 5\mathbf{i} + 5\mathbf{j}, \mathbf{v} = -6\mathbf{i} + 6\mathbf{j}$

$\cos\theta = \dfrac{\mathbf{u}\cdot\mathbf{v}}{\|\mathbf{u}\|\,\|\mathbf{v}\|} = 0$

$\theta = 90^\circ$

39. $\mathbf{u} = \left(\cos\dfrac{\pi}{3}\right)\mathbf{i} + \left(\sin\dfrac{\pi}{3}\right)\mathbf{j} = \dfrac{1}{2}\mathbf{i} + \dfrac{\sqrt{3}}{2}\mathbf{j}$

$\mathbf{v} = \left(\cos\dfrac{3\pi}{4}\right)\mathbf{i} + \left(\sin\dfrac{3\pi}{4}\right)\mathbf{j} = -\dfrac{\sqrt{2}}{2}\mathbf{i} + \dfrac{\sqrt{2}}{2}\mathbf{j}$

$\|\mathbf{u}\| = \|\mathbf{v}\| = 1$

$$\begin{aligned}\cos\theta &= \frac{\mathbf{u}\cdot\mathbf{v}}{\|\mathbf{u}\|\,\|\mathbf{v}\|} = \mathbf{u}\cdot\mathbf{v}\\ &= \left(\frac{1}{2}\right)\left(-\frac{\sqrt{2}}{2}\right) + \left(\frac{\sqrt{3}}{2}\right)\left(\frac{\sqrt{2}}{2}\right) = \frac{-\sqrt{2} + \sqrt{6}}{4}\end{aligned}$$

$\theta = \arccos\left(\dfrac{-\sqrt{2} + \sqrt{6}}{4}\right) = 75^\circ = \dfrac{5\pi}{12}$

41. $\mathbf{u} = 3\mathbf{i} + 4\mathbf{j}$

$\mathbf{v} = -7\mathbf{i} + 5\mathbf{j}$

$$\cos\theta = \frac{\mathbf{u}\cdot\mathbf{v}}{\|\mathbf{u}\|\,\|\mathbf{v}\|} = \frac{3(-7) + 4(5)}{3\sqrt{74}} = \frac{-1}{5\sqrt{74}} \approx -0.0232$$

$\theta \approx 91.33°$

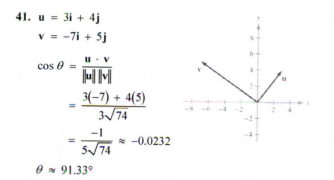

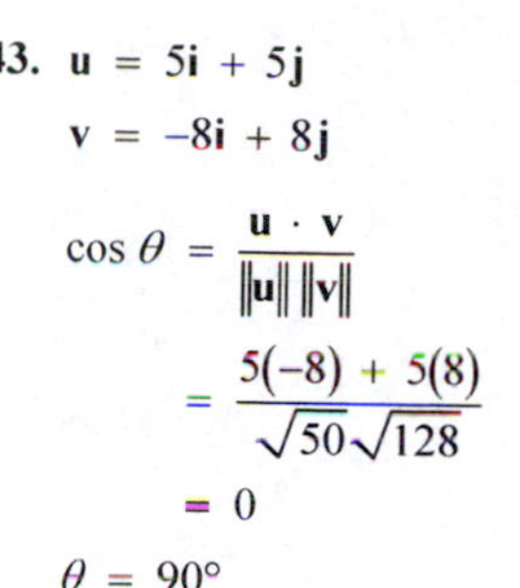

43. $\mathbf{u} = 5\mathbf{i} + 5\mathbf{j}$

$\mathbf{v} = -8\mathbf{i} + 8\mathbf{j}$

$$\cos\theta = \frac{\mathbf{u}\cdot\mathbf{v}}{\|\mathbf{u}\|\,\|\mathbf{v}\|} = \frac{5(-8) + 5(8)}{\sqrt{50}\sqrt{128}} = 0$$

$\theta = 90°$

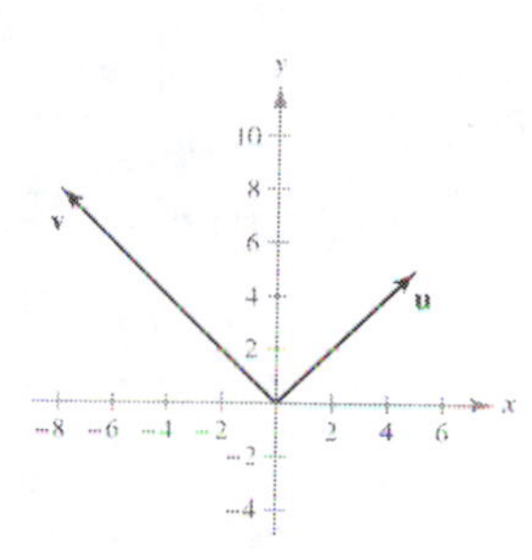

45. $P = (1, 2), Q = (3, 4), R = (2, 5)$

$\overrightarrow{PQ} = \langle 2, 2\rangle, \overrightarrow{PR} = \langle 1, 3\rangle, \overrightarrow{QR} = \langle -1, 1\rangle$

$$\cos\alpha = \frac{\overrightarrow{PQ}\cdot\overrightarrow{PR}}{\|\overrightarrow{PQ}\|\,\|\overrightarrow{PR}\|} = \frac{8}{(2\sqrt{2})(\sqrt{10})} \Rightarrow \alpha = \arccos\frac{2}{\sqrt{5}} \approx 26.57°$$

$$\cos\beta = \frac{\overrightarrow{PQ}\cdot\overrightarrow{QR}}{\|\overrightarrow{PQ}\|\,\|\overrightarrow{QR}\|} = 0 \Rightarrow \beta = 90°$$

$\gamma = 180° - 26.57° - 90° = 63.43°$

47. $P = (-3, 0), Q = (2, 2), R = (0, 6)$

$\overrightarrow{QP} = \langle -5, -2\rangle, \overrightarrow{PR} = \langle 3, 6\rangle, \overrightarrow{QR} = \langle -2, 4\rangle, \overrightarrow{PQ} = \langle 5, 2\rangle$

$$\cos\alpha = \frac{\overrightarrow{PQ}\cdot\overrightarrow{PR}}{\|\overrightarrow{PQ}\|\,\|\overrightarrow{PR}\|} = \frac{27}{\sqrt{29}\sqrt{45}} \Rightarrow \alpha \approx 41.63°$$

$$\cos\beta = \frac{\overrightarrow{QP}\cdot\overrightarrow{QR}}{\|\overrightarrow{QP}\|\,\|\overrightarrow{PR}\|} = \frac{2}{\sqrt{29}\sqrt{20}} \Rightarrow \beta \approx 85.24°$$

$\delta = 180° - 41.63° - 85.24° = 53.13°$

49. $\|\mathbf{u}\| = 4, \|\mathbf{v}\| = 10, \theta = \frac{2\pi}{3}$

$$\mathbf{u}\cdot\mathbf{v} = \|\mathbf{u}\|\,\|\mathbf{v}\|\cos\theta = (4)(10)\cos\frac{2\pi}{3} = 40\left(-\frac{1}{2}\right) = -20$$

51. $\|\mathbf{u}\| = 9, \|\mathbf{v}\| = 36, \theta = \frac{3\pi}{4}$

$$\mathbf{u}\cdot\mathbf{v} = \|\mathbf{u}\|\,\|\mathbf{v}\|\cos\theta = (9)(36)\cos\frac{3\pi}{4} = 324\left(-\frac{\sqrt{2}}{2}\right) = -162\sqrt{2} \approx -229.1$$

53. $\mathbf{u} = \langle -12, 30\rangle, \|\mathbf{v}\| = \left\langle \frac{1}{2}, -\frac{5}{4}\right\rangle$

$\mathbf{u} = -24\mathbf{v} \Rightarrow$ **u** and **v** are parallel.

55. $\mathbf{u} = \frac{1}{4}(3\mathbf{i} - \mathbf{j}), \mathbf{v} = 5\mathbf{i} + 6\mathbf{j}$

$\mathbf{u} \neq k\mathbf{v} \Rightarrow$ Not parallel

$\mathbf{u}\cdot\mathbf{v} \neq 0 \Rightarrow$ Not orthogonal

Neither

57. $\mathbf{u} = 2\mathbf{i} - 2\mathbf{j}, \mathbf{v} = -\mathbf{i} - \mathbf{j}$

$\mathbf{u}\cdot\mathbf{v} = 0 \Rightarrow$ **u** and **v** are orthogonal.

59. $\mathbf{u} = \langle 2, 2\rangle, \mathbf{v} = \langle 6, 1\rangle$

$$\mathbf{w}_1 = \text{proj}_{\mathbf{v}}\mathbf{u} = \left(\frac{\mathbf{u}\cdot\mathbf{v}}{\|\mathbf{v}\|^2}\right)\mathbf{v} = \frac{14}{37}\langle 6, 1\rangle = \frac{1}{37}\langle 84, 14\rangle$$

$$\mathbf{w}_2 = \mathbf{u} - \mathbf{w}_1 = \langle 2, 2\rangle - \frac{14}{37}\langle 6, 1\rangle = \left\langle -\frac{10}{37}, \frac{60}{37}\right\rangle = \frac{10}{37}\langle -1, 6\rangle = \frac{1}{37}\langle -10, 60\rangle$$

$$\mathbf{u} = \frac{1}{37}\langle 84, 14\rangle + \frac{1}{37}\langle -10, 60\rangle = \langle 2, 2\rangle$$

61. $\mathbf{u} = \langle 0, 3\rangle, \mathbf{v} = \langle 2, 15\rangle$

$$\mathbf{w}_1 = \text{proj}_{\mathbf{v}}\mathbf{u} = \left(\frac{\mathbf{u} \cdot \mathbf{v}}{\|\mathbf{v}\|^2}\right)\mathbf{v} = \frac{45}{229}\langle 2, 15\rangle$$

$$\mathbf{w}_2 = \mathbf{u} - \mathbf{w}_1 = \langle 0, 3\rangle - \frac{45}{229}\langle 2, 15\rangle = \left\langle -\frac{90}{229}, \frac{12}{229}\right\rangle$$

$$= \frac{6}{229}\langle -15, 2\rangle$$

$$\mathbf{u} = \frac{45}{229}\langle 2, 15\rangle + \frac{6}{229}\langle -15, 2\rangle = \langle 0, 3\rangle$$

63. $\text{proj}_{\mathbf{v}}\mathbf{u} = \mathbf{u}$ because $\mathbf{u}$ and $\mathbf{v}$ are parallel.

$$\text{proj}_{\mathbf{v}}\mathbf{u} = \frac{\mathbf{u} \cdot \mathbf{v}}{\|\mathbf{v}\|^2}\mathbf{v} = \frac{3(6) + 2(4)}{\left(\sqrt{6^2 + 4^2}\right)^2}\langle 6, 4\rangle = \frac{1}{2}\langle 6, 4\rangle = \langle 3, 2\rangle = \mathbf{u}$$

65. Because $\mathbf{u}$ and $\mathbf{v}$ are orthogonal, $\mathbf{u} \cdot \mathbf{v} = 0$ and $\text{proj}_{\mathbf{v}}\mathbf{u} = 0$.

$$\text{proj}_{\mathbf{v}}\mathbf{u} = \frac{\mathbf{u} \cdot \mathbf{v}}{\|\mathbf{v}\|^2}\mathbf{v} = 0, \text{ because } \mathbf{u} \cdot \mathbf{v} = 0.$$

67. $\mathbf{u} = \langle 3, 5\rangle$

For $\mathbf{v}$ to be orthogonal to $\mathbf{u}$, $\mathbf{u} \cdot \mathbf{v}$ must equal 0.

Two possibilities: $\langle -5, 3\rangle$ and $\langle 5, -3\rangle$

69. $\mathbf{u} = \frac{1}{2}\mathbf{i} - \frac{2}{3}\mathbf{j}$

For $\mathbf{u}$ and $\mathbf{v}$ to be orthogonal, $\mathbf{u} \cdot \mathbf{v}$ must equal 0.

Two possibilities: $\mathbf{v} = \frac{2}{3}\mathbf{i} + \frac{1}{2}\mathbf{j}$ and $\mathbf{v} = -\frac{2}{3}\mathbf{i} - \frac{1}{2}\mathbf{j}$

71. Work $= \|\text{proj}_{\overrightarrow{PQ}}\mathbf{v}\| \|\overrightarrow{PQ}\|$ where $\overrightarrow{PQ} = \langle 4, 7\rangle$ and $\mathbf{v} = \langle 1, 4\rangle$.

$$\text{proj}_{\overrightarrow{PQ}}\mathbf{v} = \left(\frac{\mathbf{v} \cdot \overrightarrow{PQ}}{\|\overrightarrow{PQ}\|^2}\right)\overrightarrow{PQ} = \left(\frac{32}{65}\right)\langle 4, 7\rangle$$

$$\text{Work} = \|\text{proj}_{\overrightarrow{PQ}}\mathbf{v}\| \|\overrightarrow{PQ}\| = \left(\frac{32\sqrt{65}}{65}\right)\left(\sqrt{65}\right) = 32$$

73. (a) $\mathbf{u} \cdot \mathbf{v} = 1225(12.20) + 2445(8.50)$

$= 35{,}727.5$

The total amount paid to the employees is \$35,727.50.

(b) To increase wages by 2%, use scalar multiplication to multiply 1.02 by $\mathbf{v}$.

75. (a) Force due to gravity:

$\mathbf{F} = -30{,}000\mathbf{j}$

Unit vector along hill:

$\mathbf{v} = (\cos d)\mathbf{i} + (\sin d)\mathbf{j}$

Projection of $\mathbf{F}$ onto $\mathbf{v}$:

$$\mathbf{w}_1 = \text{proj}_{\mathbf{v}}\mathbf{F} = \left(\frac{\mathbf{F} \cdot \mathbf{v}}{\|\mathbf{v}\|^2}\right)\mathbf{v} = (\mathbf{F} \cdot \mathbf{v})\mathbf{v} = -30{,}000 \sin d\mathbf{v}$$

The magnitude of the force is $30{,}000 \sin d$.

(b)

d	0°	1°	2°	3°	4°	5°	6°	7°	8°	9°	10°
Force	0	523.6	1047.0	1570.1	2092.7	2614.7	3135.9	3656.1	4175.2	4693.0	5209.4

(c) Force perpendicular to the hill when $d = 5°$:

$$\text{Force} = \sqrt{(30{,}000)^2 - (2614.7)^2} \approx 29{,}885.8 \text{ pounds}$$

77. Work $= (245)(3) = 735$ newton-meters

79. Work $= (\cos 30°)(45)(20) \approx 779.4$ foot-pounds

81. Work $= (\cos 35°)(15{,}691)(800)$

$\approx 10{,}282{,}651.78$ newton-meters

83. Work $= (\cos\theta)\|\mathbf{F}\| \|\overrightarrow{PQ}\|$

$= (\cos 20°)(25 \text{ pounds})(50 \text{ feet})$

≈ 1174.62 foot-pounds

85. (a)–(c) Programs will vary.

87. Programs will vary.

89. False. Work is represented by a scalar.

91. In a rhombus, $\|\mathbf{u}\| = \|\mathbf{v}\|$. The diagonals are $\mathbf{u} + \mathbf{v}$ and $\mathbf{u} - \mathbf{v}$.

$$\begin{aligned}(\mathbf{u} + \mathbf{v}) \cdot (\mathbf{u} - \mathbf{v}) &= (\mathbf{u} + \mathbf{v}) \cdot \mathbf{u} - (\mathbf{u} + \mathbf{v}) \cdot \mathbf{v} \\ &= \mathbf{u} \cdot \mathbf{u} + \mathbf{v} \cdot \mathbf{u} - \mathbf{u} \cdot \mathbf{v} - \mathbf{v} \cdot \mathbf{v} \\ &= \|\mathbf{u}\|^2 - \|\mathbf{v}\|^2 = 0\end{aligned}$$

So, the diagonals are orthogonal.

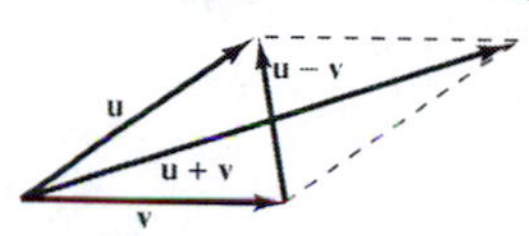

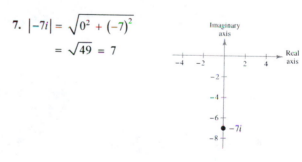

93. (a) $\text{proj}_{\mathbf{v}}\mathbf{u} = \mathbf{u} \Rightarrow \mathbf{u}$ and $\mathbf{v}$ are parallel.

(b) $\text{proj}_{\mathbf{v}}\mathbf{u} = 0 \Rightarrow \mathbf{u}$ and $\mathbf{v}$ are orthogonal.

Section 8.5 Trigonometric Form of a Complex Number

1. absolute value

3. DeMoivre's

5. $|-6 + 8i| = \sqrt{(-6)^2 + 8^2}$

$= \sqrt{100} = 10$

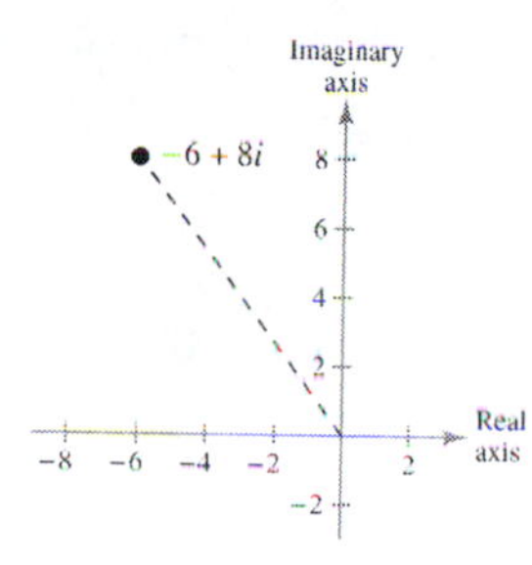

7. $|-7i| = \sqrt{0^2 + (-7)^2}$

$= \sqrt{49} = 7$

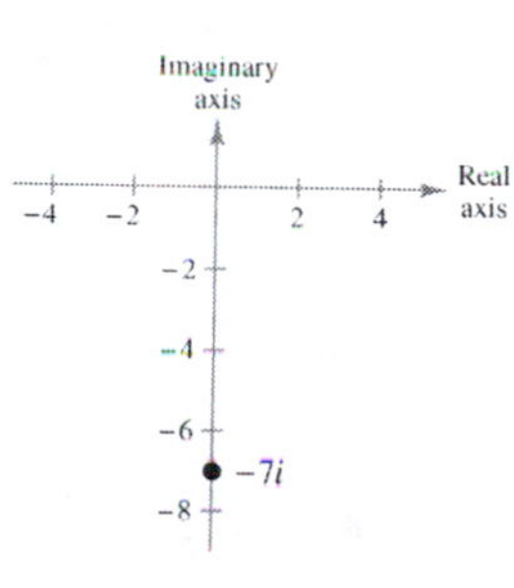

9. $|4 - 6i| = \sqrt{4^2 + (-6)^2}$

$= \sqrt{52} = 2\sqrt{13}$

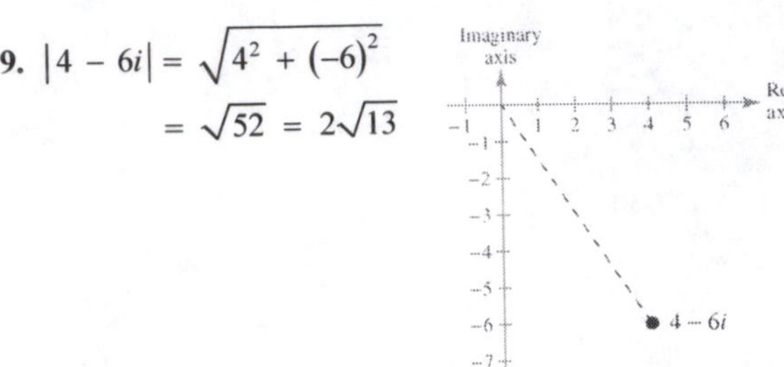

11. $z = 1 + i$

$r = \sqrt{1^2 + 1^2} = \sqrt{2}$

$\tan\theta = 1, \theta$ is in Quadrant I $\Rightarrow \theta = \dfrac{\pi}{4}.$

$z = \sqrt{2}\left(\cos\dfrac{\pi}{4} + i\sin\dfrac{\pi}{4}\right)$

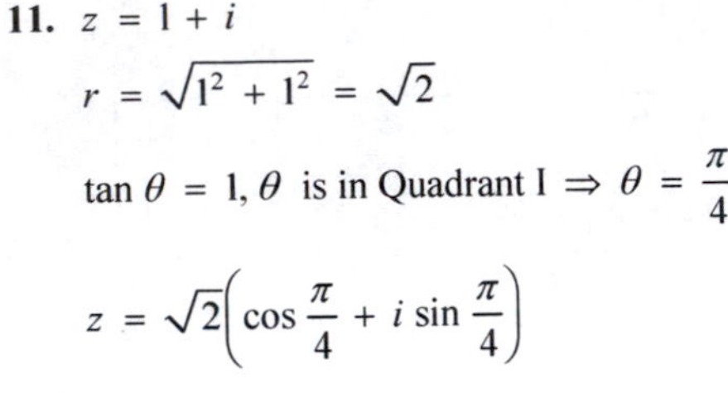

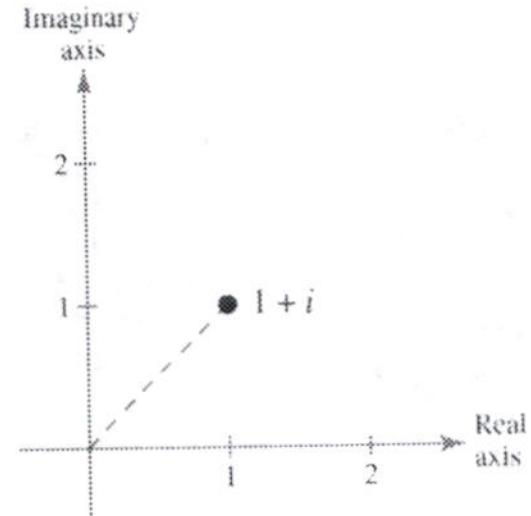

13. $z = 1 - \sqrt{3}i$

$r = \sqrt{1^2 + \left(-\sqrt{3}\right)^2} = \sqrt{4} = 2$

$\tan\theta = -\sqrt{3}, \theta$ is in Quadrant IV $\Rightarrow \theta = \dfrac{5\pi}{3}.$

$z = 2\left(\cos\dfrac{5\pi}{3} + i\sin\dfrac{5\pi}{3}\right)$

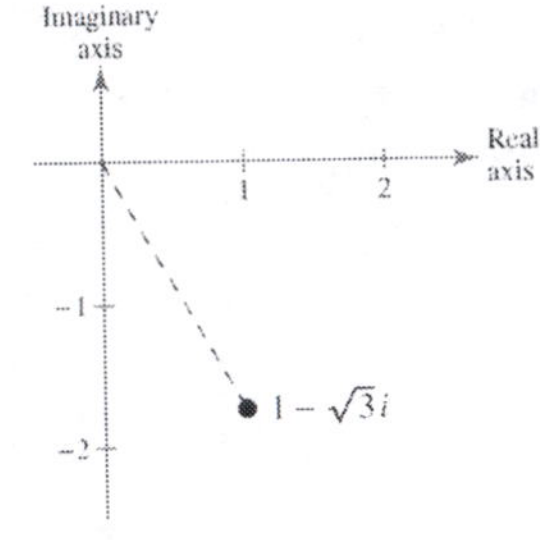

15. $z = -2\left(1 + \sqrt{3}i\right)$

$r = \sqrt{(-2)^2 + \left(-2\sqrt{3}\right)^2} = \sqrt{16} = 4$

$\tan\theta = \dfrac{\sqrt{3}}{1} = \sqrt{3}, \theta$ is in Quadrant III $\Rightarrow \theta = \dfrac{4\pi}{3}.$

$z = 4\left(\cos\dfrac{4\pi}{3} + i\sin\dfrac{4\pi}{3}\right)$

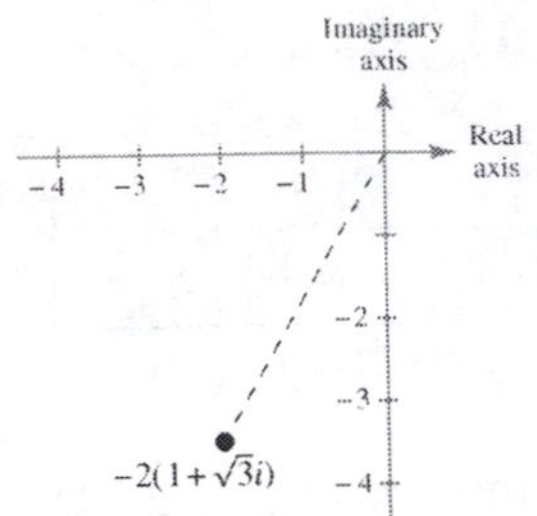

17. $z = -5i$

$r = \sqrt{0^2 + (-5)^2} = \sqrt{25} = 5$

$\tan\theta = \dfrac{-5}{0}$, undefined $\Rightarrow \theta = \dfrac{3\pi}{2}$

$z = 5\left(\cos\dfrac{3\pi}{2} + i\sin\dfrac{3\pi}{2}\right)$

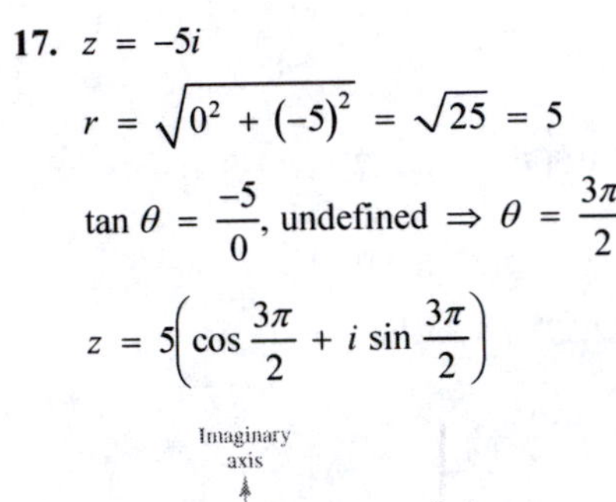

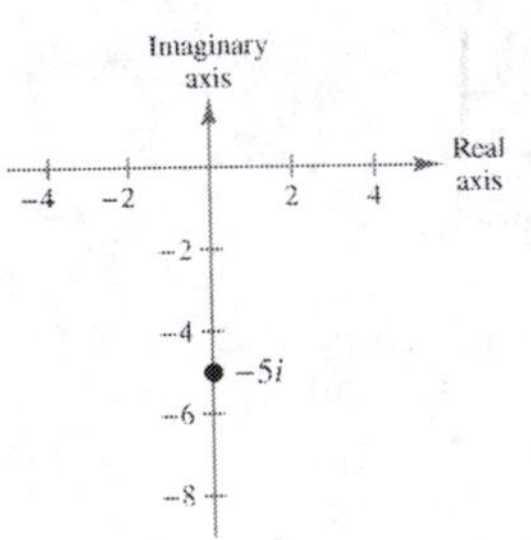

19. $z = -7 + 4i$

$r = \sqrt{(-7)^2 + (4)^2} = \sqrt{65}$

$\tan\theta = \dfrac{4}{-7}, \theta$ is in Quadrant II $\Rightarrow \theta \approx 2.62.$

$z \approx \sqrt{65}(\cos 2.62 + i\sin 2.62)$

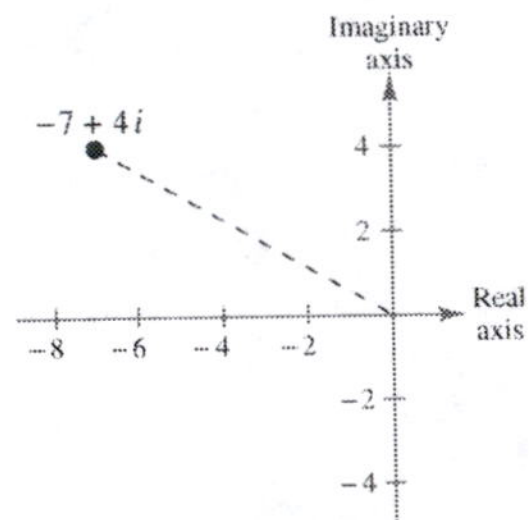

21. $z = 2$

$r = \sqrt{2^2 + 0^2} = \sqrt{4} = 2$

$\tan\theta = 0 \Rightarrow \theta = 0$

$z = 2(\cos 0 + i \sin 0)$

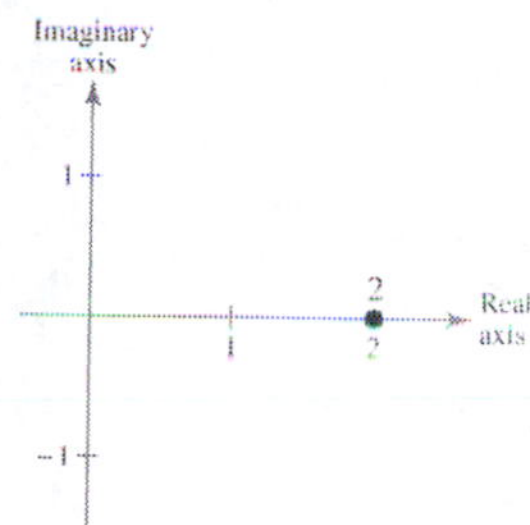

23. $z = 2\sqrt{2} - i$

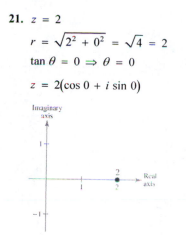

$r = \sqrt{\left(2\sqrt{2}\right)^2 + (-1)^2} = \sqrt{9} = 3$

$\tan\theta = \dfrac{-1}{2\sqrt{2}} = -\dfrac{\sqrt{2}}{4} \Rightarrow \theta \approx 5.94$ radians

$z = 3(\cos 5.94 + i \sin 5.94)$

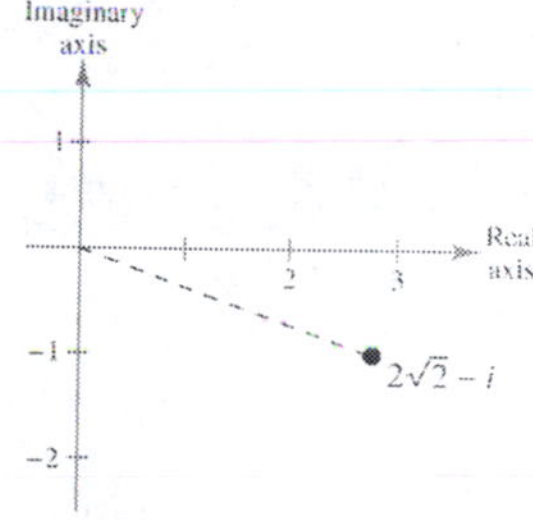

25. $z = 5 + 2i$

$r = \sqrt{5^2 + 2^2} = \sqrt{29}$

$\tan\theta = \frac{2}{5}$

$\theta \approx 0.38$

$z \approx \sqrt{29}(\cos 0.38 + i \sin 0.38)$

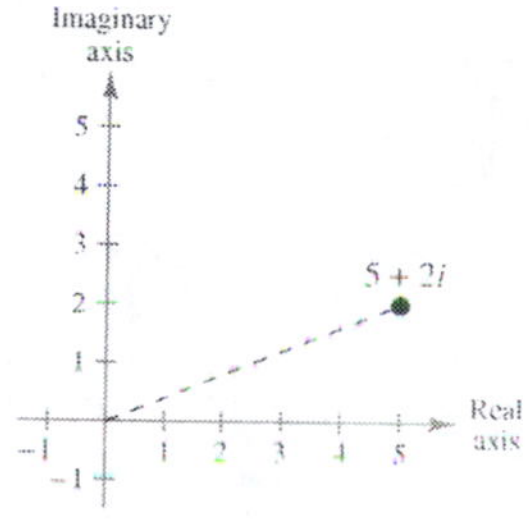

27. $z = 3 + \sqrt{3}i$

$r = \sqrt{(3)^2 + \left(\sqrt{3}\right)^2} = \sqrt{12} = 2\sqrt{3}$

$\tan\theta = \dfrac{\sqrt{3}}{3} \Rightarrow \theta = \dfrac{\pi}{6}$

$z = 2\sqrt{3}\left(\cos\dfrac{\pi}{6} + i \sin\dfrac{\pi}{6}\right)$

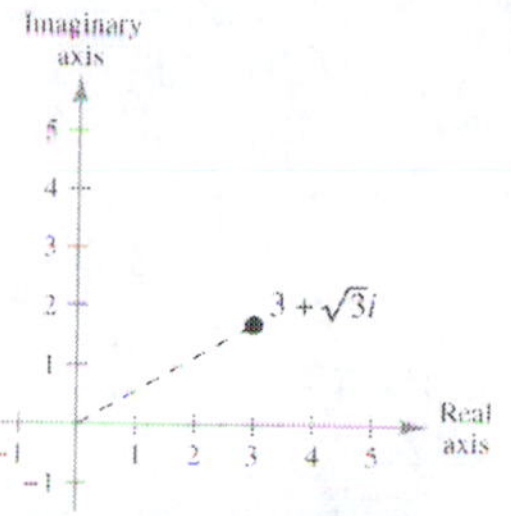

29. $z = -8 - 5\sqrt{3}i$

$r = \sqrt{(-8)^2 + \left(-5\sqrt{3}\right)^2} = \sqrt{139}$

$\tan\theta = \dfrac{5\sqrt{3}}{8}$

$\theta \approx 3.97$

$z \approx \sqrt{139}(\cos 3.97 + i \sin 3.97)$

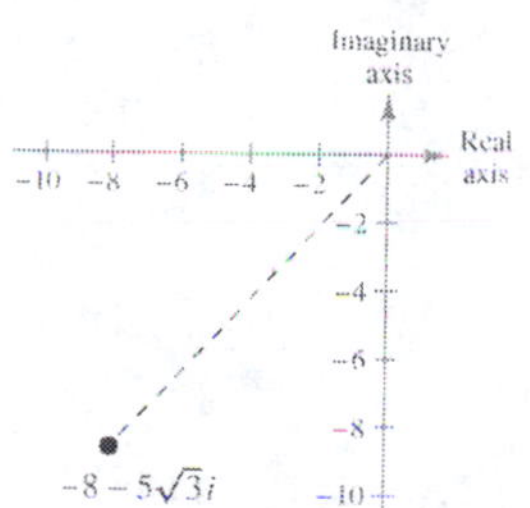

31. 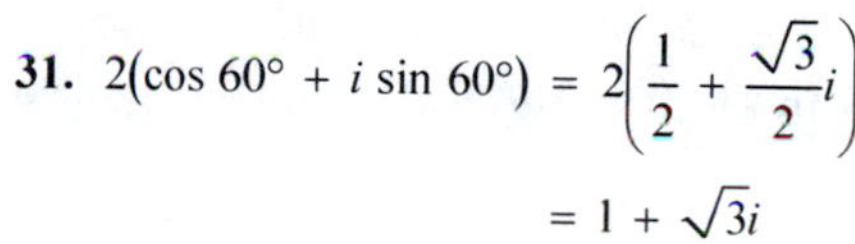

$2(\cos 60° + i \sin 60°) = 2\left(\dfrac{1}{2} + \dfrac{\sqrt{3}}{2}i\right)$

$= 1 + \sqrt{3}i$

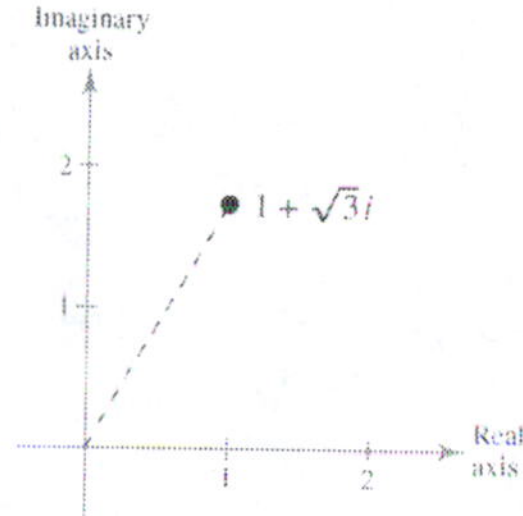

33. $\sqrt{48}\left[\cos(-30°) + i\sin(-30°)\right] = 4\sqrt{3}\left(\frac{\sqrt{3}}{2} - \frac{1}{2}i\right)$

$= 6 - 2\sqrt{3}i$

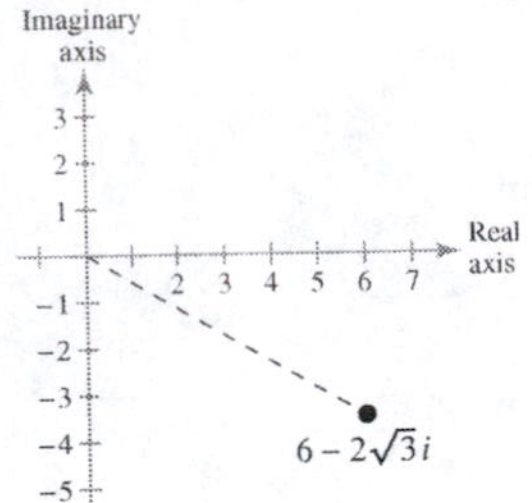

35. $\frac{9}{4}\left(\cos\frac{3\pi}{4} + i\sin\frac{3\pi}{4}\right) = \frac{9}{4}\left(-\frac{\sqrt{2}}{2} + \frac{\sqrt{2}}{2}i\right)$

$= -\frac{9\sqrt{2}}{8} + \frac{9\sqrt{2}}{8}i$

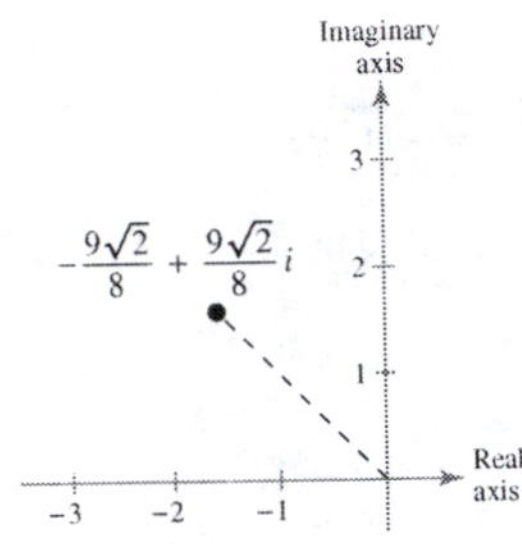

37. $7(\cos 0° + i\sin 0°) = 7$

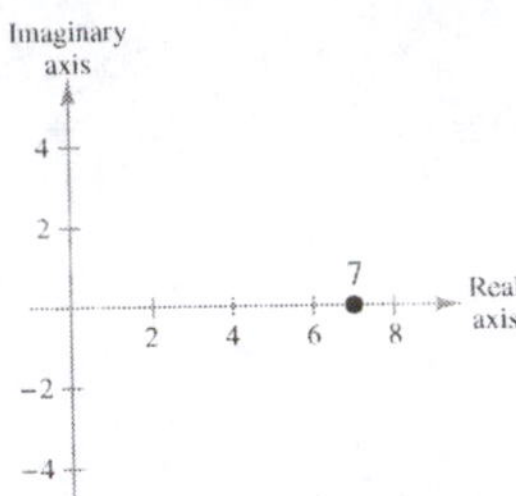

39. $5\left[\cos(198°45') + i\sin(198°45')\right] \approx -4.7347 - 1.6072i$

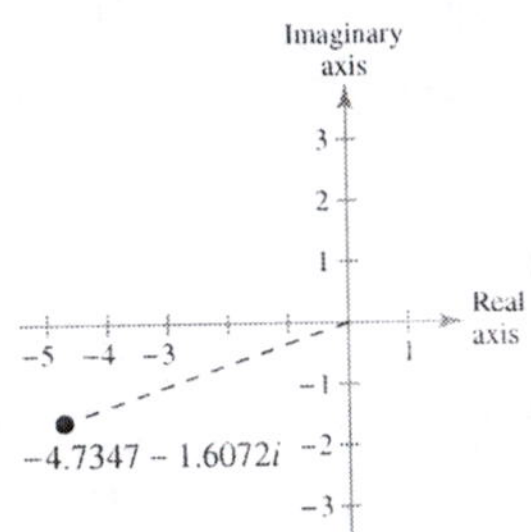

41. $5\left(\cos\frac{\pi}{9} + i\sin\frac{\pi}{9}\right) \approx 4.6985 + 1.7101i$

43. $2(\cos 155° + i\sin 155°) \approx -1.8126 + 0.8452i$

45. $\left[2\left(\cos\frac{\pi}{4} + i\sin\frac{\pi}{4}\right)\right]\left[6\left(\cos\frac{\pi}{12} + i\sin\frac{\pi}{12}\right)\right] = (2)(6)\left[\cos\left(\frac{\pi}{4} + \frac{\pi}{12}\right) + i\sin\left(\frac{\pi}{4} + \frac{\pi}{12}\right)\right]$

$= 12\left(\cos\frac{\pi}{3} + i\sin\frac{\pi}{3}\right)$

47. $\left[\frac{5}{3}(\cos 120° + i\sin 120°)\right]\left[\frac{2}{3}(\cos 30° + i\sin 30°)\right] = \frac{5}{3}\left(\frac{2}{3}\right)\left[\cos(120° + 30°) + i\sin(120° + 30°)\right]$

$= \frac{10}{9}(\cos 150° + i\sin 150°)$

49. $(\cos 80° + i\sin 80°)(\cos 330° + i\sin 330°) = \cos(80° + 330°) + i\sin(80° + 330°)$

$= \cos 410° + i\sin 410°$

$= \cos 50° + i\sin 50°$

51. $\frac{3(\cos 50° + i\sin 50°)}{9(\cos 20° + i\sin 20°)} = \frac{1}{3}\left[\cos(50° - 20°) + i\sin(50° - 20°)\right] = \frac{1}{3}(\cos 30° + i\sin 30°)$

53. $\frac{\cos\pi + i\sin\pi}{\cos(\pi/3) + i\sin(\pi/3)} = \cos\left(\pi - \frac{\pi}{3}\right) + i\sin\left(\pi - \frac{\pi}{3}\right) = \cos\frac{2\pi}{3} + i\sin\frac{2\pi}{3}$

55. $\frac{12(\cos 92° + i\sin 92°)}{2(\cos 122° + i\sin 122°)} = 6\left[\cos(92° - 122°) + i\sin(92° - 122°)\right]$

$= 6\left[\cos(-30°) + i\sin(-30°)\right]$

$= 6(\cos 330° + i\sin 330°)$

57. (a) $2 + 2i = 2\sqrt{2}\left(\cos\frac{\pi}{4} + i\sin\frac{\pi}{4}\right)$

$1 - i = \sqrt{2}\left[\cos\left(-\frac{\pi}{4}\right) + i\sin\left(-\frac{\pi}{4}\right)\right] = \sqrt{2}\left(\cos\frac{7\pi}{4} + i\sin\frac{7\pi}{4}\right)$

(b) $(2 + 2i)(1 - i) = \left[2\sqrt{2}\left(\cos\frac{\pi}{4} + i\sin\frac{\pi}{4}\right)\right]\left[\sqrt{2}\left(\cos\left(\frac{7\pi}{4}\right) + i\sin\left(\frac{7\pi}{4}\right)\right)\right] = 4(\cos 2\pi + i\sin 2\pi)$

$= 4(\cos 0 + i\sin 0) = 4$

(c) $(2 + 2i)(1 - i) = 2 - 2i + 2i - 2i^2 = 2 + 2 = 4$

59. (a) $-2i = 2\left[\cos\left(-\frac{\pi}{2}\right) + i\sin\left(-\frac{\pi}{2}\right)\right] = 2\left(\cos\frac{3\pi}{2} + i\sin\frac{3\pi}{2}\right)$

$1 + i = \sqrt{2}\left(\cos\frac{\pi}{4} + i\sin\frac{\pi}{4}\right)$

(b) $-2i(1 + i) = 2\left[\cos\left(\frac{3\pi}{2}\right) + i\sin\left(\frac{3\pi}{2}\right)\right]\left[\sqrt{2}\left(\cos\frac{\pi}{4} + i\sin\frac{\pi}{4}\right)\right]$

$= 2\sqrt{2}\left[\cos\left(\frac{7\pi}{4}\right) + i\sin\left(\frac{7\pi}{4}\right)\right]$

$= 2\sqrt{2}\left[\frac{1}{\sqrt{2}} - \frac{1}{\sqrt{2}}i\right] = 2 - 2i$

(c) $-2i(1 + i) = -2i - 2i^2 = -2i + 2 = 2 - 2i$

61. (a) $3 + 4i \approx 5(\cos 0.93 + i\sin 0.93)$

$1 - \sqrt{3}i = 2\left(\cos\frac{5\pi}{3} + i\sin\frac{5\pi}{3}\right)$

(b) $\frac{3 + 4i}{1 - \sqrt{3}i} \approx \frac{5(\cos 0.93 + i\sin 0.93)}{2\left(\cos\frac{5\pi}{3} + i\sin\frac{5\pi}{3}\right)} \approx 2.5[\cos(-4.31) + i\sin(-4.31)] = \frac{5}{2}(\cos 1.97 + i\sin 1.97) \approx -0.982 + 2.299i$

(c) $\frac{3 + 4i}{1 - \sqrt{3}i} = \frac{3 + 4i}{1 - \sqrt{3}i} \cdot \frac{1 + \sqrt{3}i}{1 + \sqrt{3}i} = \frac{3 + \left(4 + 3\sqrt{3}\right)i + 4\sqrt{3}i^2}{1 + 3} = \frac{3 - 4\sqrt{3}}{4} + \frac{4 + 3\sqrt{3}}{4}i \approx -0.982 + 2.299i$

63. $z = \frac{\sqrt{2}}{2}(1 + i) = \cos 45° + i\sin 45°$

$z^2 = \cos 90° + i\sin 90° = i$

$z^3 = \cos 135° + i\sin 135° = \frac{\sqrt{2}}{2}(-1 + i)$

$z^4 = \cos 180° + i\sin 180° = -1$

The absolute value of each is 1, and consecutive powers of z are each $45°$ apart.

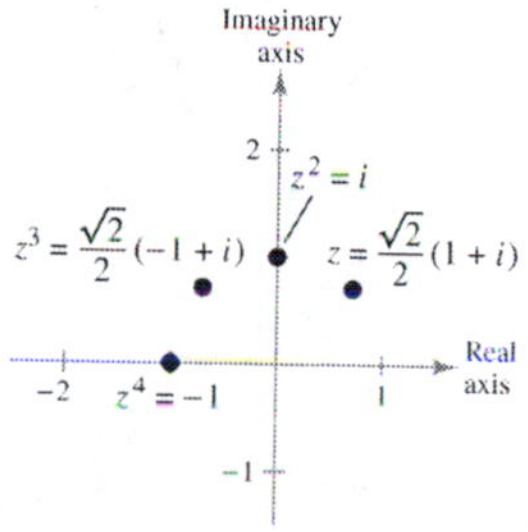

65. $(1 + i)^5 = \left[\sqrt{2}\left(\cos\frac{\pi}{4} + i\sin\frac{\pi}{4}\right)\right]^5 = \left(\sqrt{2}\right)^5\left(\cos\frac{5\pi}{4} + i\sin\frac{5\pi}{4}\right) = 4\sqrt{2}\left(-\frac{\sqrt{2}}{2} - \frac{\sqrt{2}}{2}i\right) = -4 - 4i$

67. $(-1+i)^6 = \left[\sqrt{2}\left(\cos\frac{3\pi}{4} + i\sin\frac{3\pi}{4}\right)\right]^6$

$= \left(\sqrt{2}\right)^6\left(\cos\frac{18\pi}{4} + i\sin\frac{18\pi}{4}\right)$

$= 8\left(\cos\frac{9\pi}{2} + i\sin\frac{9\pi}{2}\right)$

$= 8(0+i)$

$= 8i$

69. $2\left(\sqrt{3}+i\right)^{10} = 2\left[2\left(\cos\frac{\pi}{6} + i\sin\frac{\pi}{6}\right)\right]^{16}$

$= 2\left[2^{10}\left(\cos\frac{10\pi}{6} + i\sin\frac{10\pi}{6}\right)\right]$

$= 2048\left(\cos\frac{5\pi}{3} + i\sin\frac{5\pi}{3}\right)$

$= 2048\left(\frac{1}{2} - \frac{\sqrt{3}}{2}i\right)$

$= 1024 - 1024\sqrt{3}i$

71. $\left[5(\cos 20° + i\sin 20°)\right]^3 = 5^3(\cos 60° + i\sin 60°)$

$= \frac{125}{2} + \frac{125\sqrt{3}}{2}i$

73. $\left(\cos\frac{\pi}{4} + i\sin\frac{\pi}{4}\right)^{12} = \cos\frac{12\pi}{4} + i\sin\frac{12\pi}{4}$

$= \cos 3\pi + i\sin 3\pi$

$= -1$

75. $\left[5(\cos 3.2 + i\sin 3.2)\right]^4 = 5^4(\cos 12.8 + i\sin 12.8)$

$\approx 608.0 + 144.7i$

77. $(3-2i)^5 \approx \left[3.6056\left[\cos(-0.588) + i\sin(-0.588)\right]\right]^5$

$\approx (3.6056)^5\left[\cos(-2.94) + i\sin(-2.94)\right]$

$\approx -597 - 122i$

79. $\left[3(\cos 15° + i\sin 15°)\right]^4 = 81(\cos 60° + i\sin 60°)$

$= \frac{81}{2} + \frac{81\sqrt{3}}{2}i$

81. (a) Square roots of $5(\cos 120° + i\sin 120°)$:

$$\sqrt{5}\left[\cos\left(\frac{120° + 360°k}{2}\right) + i\sin\left(\frac{120° + 360°k}{2}\right)\right],\ k = 0, 1$$

$k = 0$: $\sqrt{5}(\cos 60° + i\sin 60°)$

$k = 1$: $\sqrt{5}(\cos 240° + i\sin 240°)$

(b)

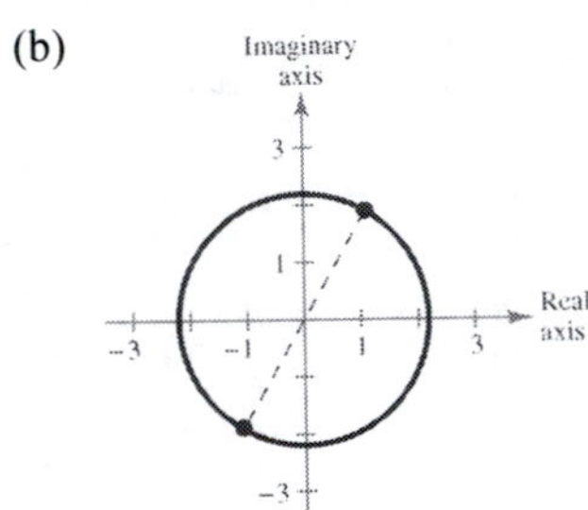

(c) $\frac{\sqrt{5}}{2} + \frac{\sqrt{15}}{2}i, -\frac{\sqrt{5}}{2} - \frac{\sqrt{15}}{2}i$

83. (a) Cube roots of $8\left(\cos\frac{2\pi}{3} + i\sin\frac{2\pi}{3}\right)$:

$$\sqrt[3]{8}\left[\cos\left(\frac{(2\pi/3) + 2\pi k}{3}\right) + i\sin\left(\frac{(2\pi/3) + 2\pi k}{3}\right)\right],\ k = 0, 1, 2$$

$k = 0$: $2\left(\cos\frac{2\pi}{9} + i\sin\frac{2\pi}{9}\right)$

$k = 1$: $2\left(\cos\frac{8\pi}{9} + i\sin\frac{8\pi}{9}\right)$

$k = 2$: $2\left(\cos\frac{14\pi}{9} + i\sin\frac{14\pi}{9}\right)$

(b)

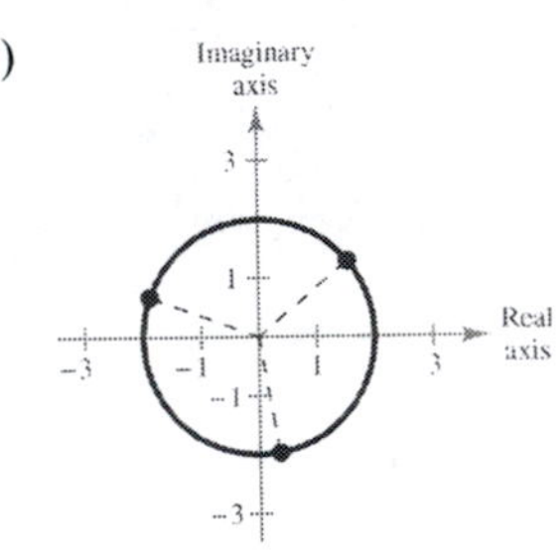

(c) $1.5321 + 1.2856i, -1.8794 + 0.6840i, 0.3473 - 1.9696i$

85. (a) Cube roots of $-\frac{125}{2}\left(1+\sqrt{3}i\right) = 125\left(\cos\frac{4\pi}{3} + i\sin\frac{4\pi}{3}\right)$:

$$\sqrt[3]{125}\left[\cos\left(\frac{\frac{4\pi}{3}+2k\pi}{3}\right) + i\sin\left(\frac{\frac{4\pi}{3}+2k\pi}{3}\right)\right], k = 0, 1, 2$$

$k = 0$: $5\left(\cos\frac{4\pi}{9} + i\sin\frac{4\pi}{9}\right)$

$k = 1$: $5\left(\cos\frac{10\pi}{9} + i\sin\frac{10\pi}{9}\right)$

$k = 2$: $5\left(\cos\frac{16\pi}{9} + i\sin\frac{16\pi}{9}\right)$

(b)

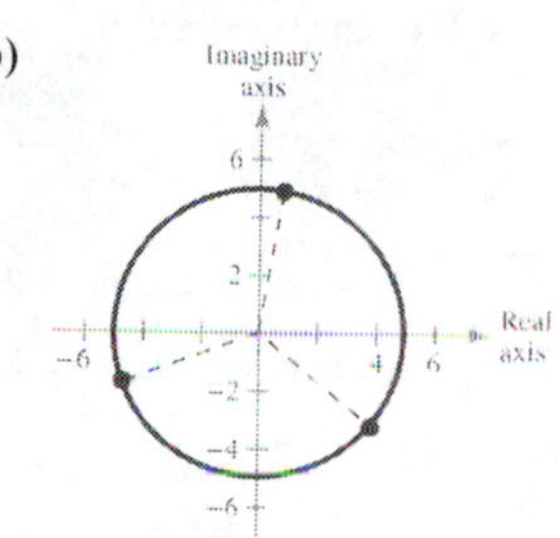

(c) $0.8682 + 4.9240i$, $-4.6985 - 1.7101i$, $3.8302 - 3.2140i$

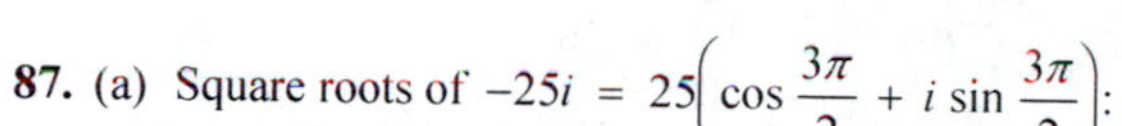

87. (a) Square roots of $-25i = 25\left(\cos\frac{3\pi}{2} + i\sin\frac{3\pi}{2}\right)$:

$$\sqrt{25}\left[\cos\left(\frac{\frac{3\pi}{2}+2k\pi}{2}\right) + i\sin\left(\frac{\frac{3\pi}{2}+2k\pi}{2}\right)\right], k = 0, 1$$

$k = 0$: $5\left(\cos\frac{3\pi}{4} + i\sin\frac{3\pi}{4}\right)$

$k = 1$: $5\left(\cos\frac{7\pi}{4} + i\sin\frac{7\pi}{4}\right)$

(b)

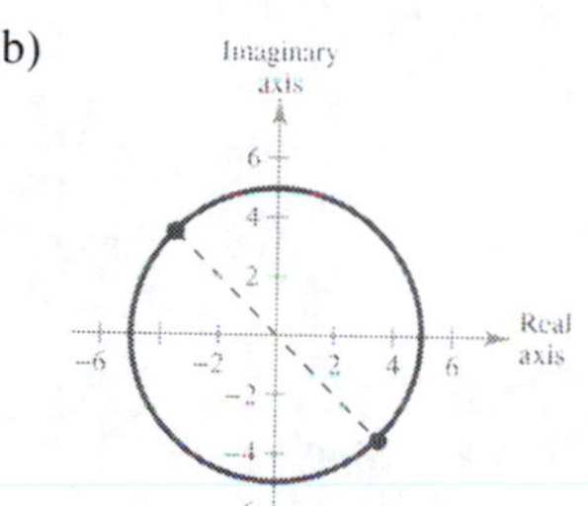

(c) $-\frac{5\sqrt{2}}{2} + \frac{5\sqrt{2}}{2}i$, $\frac{5\sqrt{2}}{2} - \frac{5\sqrt{2}}{2}i$

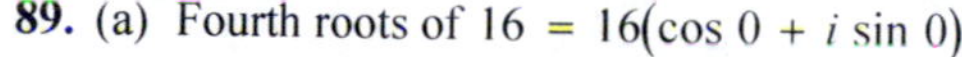

89. (a) Fourth roots of $16 = 16(\cos 0 + i\sin 0)$:

$$\sqrt[4]{16}\left[\cos\frac{0+2\pi k}{4} + i\sin\frac{0+2\pi k}{4}\right], k = 0, 1, 2, 3$$

$k = 0$: $2(\cos 0 + i\sin 0)$

$k = 1$: $2\left(\cos\frac{\pi}{2} + i\sin\frac{\pi}{2}\right)$

$k = 2$: $2(\cos\pi + i\sin\pi)$

$k = 3$: $2\left(\cos\frac{3\pi}{2} + i\sin\frac{3\pi}{2}\right)$

(b)

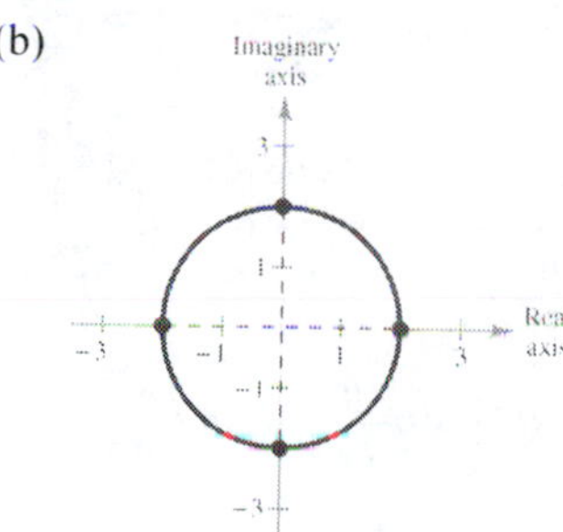

(c) $2, 2i, -2, -2i$

91. (a) Fifth roots of $1 = \cos 0 + i \sin 0$:

$$\cos\left(\frac{2k\pi}{5}\right) + i \sin\left(\frac{2k\pi}{5}\right), k = 0, 1, 2, 3, 4$$

$k = 0$: $\cos 0 + i \sin 0$

$k = 1$: $\cos \frac{2\pi}{5} + i \sin \frac{2\pi}{5}$

$k = 2$: $\cos \frac{4\pi}{5} + i \sin \frac{4\pi}{5}$

$k = 3$: $\cos \frac{6\pi}{5} + i \sin \frac{6\pi}{5}$

$k = 4$: $\cos \frac{8\pi}{5} + i \sin \frac{8\pi}{5}$

(b)

(c) $1, 0.3090 + 0.9511i, -0.8090 + 0.5878i, -0.8090 - 0.5878i, 0.3090 - 0.9511i$

93. (a) Cube roots of $-125 = 125(\cos \pi + i \sin \pi)$:

$$\sqrt[3]{125}\left[\cos\left(\frac{\pi + 2\pi k}{3}\right) + i \sin\left(\frac{\pi + 2\pi k}{3}\right)\right], k = 0, 1, 2$$

$k = 0$: $5\left(\cos \frac{\pi}{3} + i \sin \frac{\pi}{3}\right)$

$k = 1$: $5(\cos \pi + i \sin \pi)$

$k = 2$: $5\left(\cos \frac{5\pi}{3} + i \sin \frac{5\pi}{3}\right)$

(b)

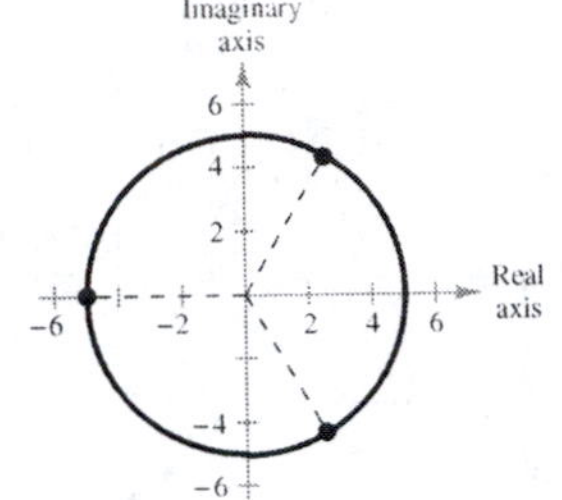

(c) $\frac{5}{2} + \frac{5\sqrt{3}}{2}i, -5, \frac{5}{2} - \frac{5\sqrt{3}}{2}i$

95. (a) Fifth roots of $4(1 - i) = 4\sqrt{2}\left(\cos \frac{7\pi}{4} + i \sin \frac{7\pi}{4}\right)$:

$$\sqrt[5]{4\sqrt{2}}\left[\cos\left(\frac{\frac{7\pi}{4} + 2\pi k}{5}\right) + i \sin\left(\frac{\frac{7\pi}{4} + 2\pi k}{5}\right)\right], k = 0, 1, 2, 3, 4$$

$k = 0$: $\sqrt{2}\left(\cos \frac{7\pi}{20} + i \sin \frac{7\pi}{20}\right)$

$k = 1$: $\sqrt{2}\left(\cos \frac{3\pi}{4} + i \sin \frac{3\pi}{4}\right)$

$k = 2$: $\sqrt{2}\left(\cos \frac{23\pi}{20} + i \sin \frac{23\pi}{20}\right)$

$k = 3$: $\sqrt{2}\left(\cos \frac{31\pi}{20} + i \sin \frac{31\pi}{20}\right)$

$k = 4$: $\sqrt{2}\left(\cos \frac{39\pi}{20} + i \sin \frac{39\pi}{20}\right)$

(b)

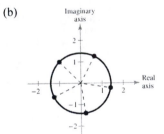

(c) $0.6420 + 1.2601i, -1 + 1i, -1.2601 - 0.6420i, 0.2212 - 1.3968i, 1.3968 - 0.2212i$

97. $x^4 + i = 0$

$$x^4 = -i$$

The solutions are the fourth roots of $i = \cos\frac{3\pi}{2} + i\sin\frac{3\pi}{2}$:

$$\sqrt[4]{1}\left[\cos\left(\frac{\frac{3\pi}{2} + 2k\pi}{4}\right) + i\sin\left(\frac{\frac{3\pi}{2} + 2k\pi}{4}\right)\right], k = 0, 1, 2, 3$$

$k = 0$: $\cos\frac{3\pi}{8} + i\sin\frac{3\pi}{8} \approx 0.3827 + 0.9239i$

$k = 1$: $\cos\frac{7\pi}{8} + i\sin\frac{7\pi}{8} \approx -0.9239 + 0.3827i$

$k = 2$: $\cos\frac{11\pi}{8} + i\sin\frac{11\pi}{8} \approx -0.3827 - 0.9239i$

$k = 3$: $\cos\frac{15\pi}{8} + i\sin\frac{15\pi}{8} \approx 0.9239 - 0.3827i$

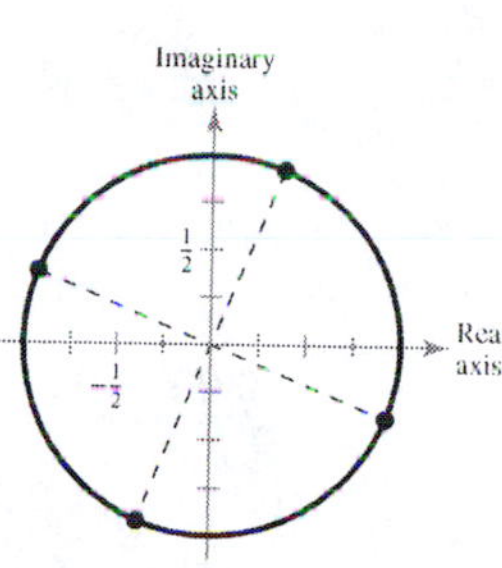

99. $x^5 + 243 = 0$

$$x^5 = -243$$

The solutions are the fifth roots of $-243 = 243(\cos\pi + i\sin\pi)$:

$$\sqrt[5]{243}\left[\cos\left(\frac{\pi + 2k\pi}{5}\right) + i\sin\left(\frac{\pi + 2k\pi}{5}\right)\right], k = 0, 1, 2, 3, 4$$

$k = 0$: $3\left(\cos\frac{\pi}{5} + i\sin\frac{\pi}{5}\right) \approx 2.4271 + 1.7634i$

$k = 1$: $3\left(\cos\frac{3\pi}{5} + i\sin\frac{3\pi}{5}\right) \approx -0.9271 + 2.8532i$

$k = 2$: $3(\cos\pi + i\sin\pi) = -3$

$k = 3$: $3\left(\cos\frac{7\pi}{5} + i\sin\frac{7\pi}{5}\right) \approx -0.9271 - 2.8532i$

$k = 4$: $3\left(\cos\frac{9\pi}{5} + i\sin\frac{9\pi}{5}\right) \approx 2.4271 - 1.7634i$

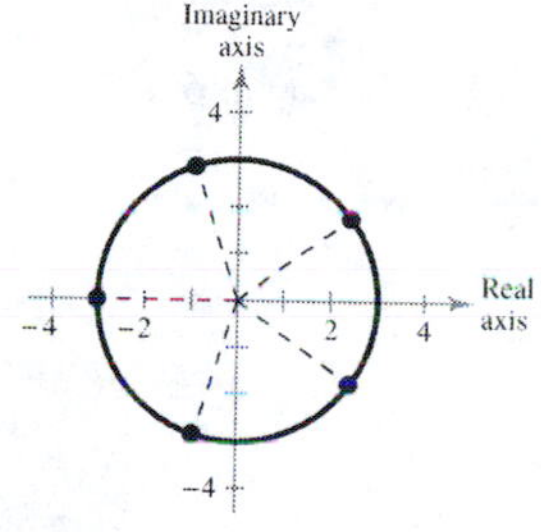

101. $x^4 + 16i = 0$

$$x^4 = -16i$$

The solutions are the fourth roots of $-16i = 16\left(\cos\frac{3\pi}{2} + i\sin\frac{3\pi}{2}\right)$:

$$\sqrt[4]{16}\left[\cos\frac{\frac{3\pi}{2} + 2\pi k}{4} + i\sin\frac{\frac{3\pi}{2} + 2\pi k}{4}\right], k = 0, 1, 2, 3$$

$k = 0$: $2\left(\cos\frac{3\pi}{8} + i\sin\frac{3\pi}{8}\right) \approx 0.7654 + 1.8478i$

$k = 1$: $2\left(\cos\frac{7\pi}{8} + i\sin\frac{7\pi}{8}\right) \approx -1.8478 + 0.7654i$

$k = 2$: $2\left(\cos\frac{11\pi}{8} + i\sin\frac{11\pi}{8}\right) \approx -0.7654 - 1.8478i$

$k = 3$: $2\left(\cos\frac{15\pi}{8} + i\sin\frac{15\pi}{8}\right) \approx 1.8478 - 0.7654i$

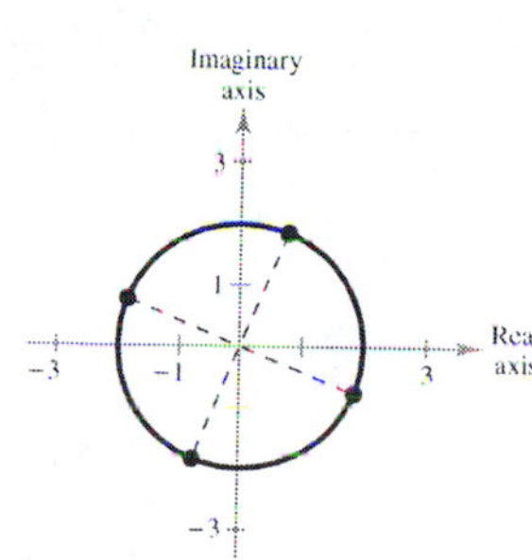

103. $x^3 - (1 - i) = 0$

$$x^3 = 1 - i = \sqrt{2}\left(\cos\frac{7\pi}{4} + i\sin\frac{7\pi}{4}\right)$$

The solutions are the cube roots of $1 - i$:

$$\sqrt[3]{\sqrt{2}}\left[\cos\left(\frac{(7\pi/4) + 2\pi k}{3}\right) + i\sin\left(\frac{(7\pi/4) + 2\pi k}{3}\right)\right], k = 0, 1, 2$$

$$k = 0: \sqrt[6]{2}\left(\cos\frac{7\pi}{12} + i\sin\frac{7\pi}{12}\right) \approx -0.2905 + 1.0842i$$

$$k = 1: \sqrt[6]{2}\left(\cos\frac{5\pi}{4} + i\sin\frac{5\pi}{4}\right) \approx -0.7937 - 0.7937i$$

$$k = 2: \sqrt[6]{2}\left(\cos\frac{23\pi}{12} + i\sin\frac{23\pi}{12}\right) \approx 1.0842 - 0.2905i$$

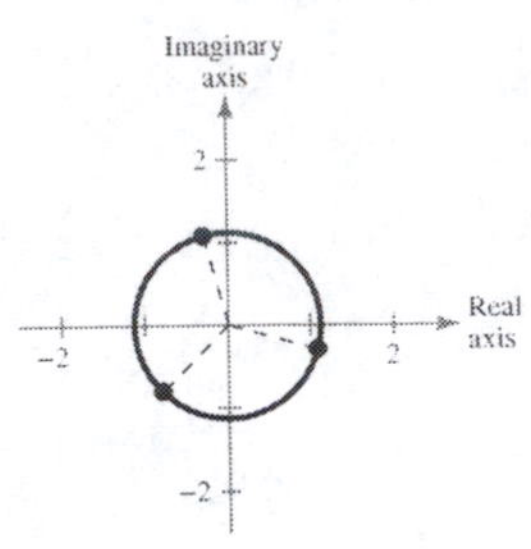

105. (a) $E = I \cdot Z$

$$= [6(\cos 41^\circ + i\sin 41^\circ)](4[\cos(-11^\circ) + i\sin(-11^\circ)])$$

$$= 24(\cos 30^\circ + i\sin 30^\circ) \text{ volts}$$

(b) $E = 24\left(\frac{\sqrt{3}}{2} + \frac{1}{2}i\right) = 12\sqrt{3} + 12i$ volts

(c) $|E| = \sqrt{(12\sqrt{3})^2 + (12)^2} = \sqrt{576} = 24$ volts

107. True.

$z_1 = r_1(\cos\theta_1 + i\sin\theta_1)$, $z_2 = r_2(\cos\theta_2 + i\sin\theta_2)$

$z_1z_2 = r_1r_2[\cos(\theta_1 + \theta_2) + i\sin(\theta_1 + \theta_2)]$ and

$z_1z_2 = 0$ if and only if $r_1 = 0$ and/or $r_2 = 0$.

109. $\bar{z} = r[\cos(-\theta) + i\sin(-\theta)]$

$= r[\cos\theta + -i\sin\theta]$

$= r\cos\theta - ir\sin\theta$

which is the complex conjugate of $r(\cos\theta + i\sin\theta) = r\cos\theta + ir\sin\theta$.

110. (a) $z\bar{z} = [r(\cos\theta + i\sin\theta)][r(\cos(-\theta) + i\sin(-\theta))]$

$= r^2[\cos(\theta - \theta) + i\sin(\theta - \theta)]$

$= r^2[\cos 0 + i\sin 0]$

$= r^2$

(b) $\frac{z}{\bar{z}} = \frac{r(\cos\theta + i\sin\theta)}{r[\cos(-\theta) + i\sin(-\theta)]}$

$= \frac{r}{r}[\cos(\theta - (-\theta)) + i\sin(\theta - (-\theta))]$

$= \cos 2\theta + i\sin 2\theta$

111. $z = r(\cos\theta + i\sin\theta)$

$-z = -r(\cos\theta + i\sin\theta)$

$= r(-\cos\theta + -i\sin\theta)$

$= r(\cos(\theta + \pi) + i\sin(\theta + \pi))$

Review Exercises for Chapter 8

1. Given: $A = 38°, B = 70°, a = 8$

$C = 180° - 38° - 70° = 72°$

$b = \frac{a \sin B}{\sin A} = \frac{8 \sin 70°}{\sin 38°} \approx 12.21$

$c = \frac{a \sin C}{\sin A} = \frac{8 \sin 72°}{\sin 38°} \approx 12.36$

3. Given: $B = 72°, C = 82°, b = 54$

$A = 180° - 72° - 82° = 26°$

$a = \frac{b \sin A}{\sin B} = \frac{54 \sin 26°}{\sin 72°} \approx 24.89$

$c = \frac{b \sin C}{\sin B} = \frac{54 \sin 82°}{\sin 72°} \approx 56.23$

5. Given: $A = 16°, B = 98°, c = 8.4$

$C = 180° - 16° - 98° = 66°$

$a = \frac{c \sin A}{\sin C} = \frac{8.4 \sin 16°}{\sin 66°} \approx 2.53$

$b = \frac{c \sin B}{\sin C} = \frac{8.4 \sin 98°}{\sin 66°} \approx 9.11$

7. Given: $A = 24°, C = 48°, b = 27.5$

$B = 180° - 24° - 48° = 108°$

$a = \frac{b \sin A}{\sin B} = \frac{27.5 \sin 24°}{\sin 108°} \approx 11.76$

$c = \frac{b \sin C}{\sin B} = \frac{27.5 \sin 48°}{\sin 108°} \approx 21.49$

9. Given: $B = 150°, b = 30, c = 10$

$\sin C = \frac{c \sin B}{b} = \frac{10 \sin 150°}{30} \approx 0.1667 \Rightarrow C \approx 9.59°$

$A \approx 180° - 150° - 9.59° = 20.41°$

$a = \frac{b \sin A}{\sin B} = \frac{30 \sin 20.41°}{\sin 150°} \approx 20.92$

11. $A = 75°, a = 51.2, b = 33.7$

$\sin B = \frac{b \sin A}{a} = \frac{33.7 \sin 75°}{51.2} \approx 0.6358 \Rightarrow B \approx 39.48°$

$C \approx 180° - 75° - 39.48° = 65.52°$

$c = \frac{a \sin C}{\sin A} = \frac{51.2 \sin 65.52°}{\sin 75°} \approx 48.24$

13. $A = 33°, b = 7, c = 10$

$\text{Area} = \frac{1}{2}bc \sin A = \frac{1}{2}(7)(10) \sin 33° \approx 19.06$

15. $C = 119°, a = 18, b = 6$

$\text{Area} = \frac{1}{2}ab \sin C = \frac{1}{2}(18)(6) \sin 119° \approx 47.23$

17. $\tan 17° = \frac{h}{x + 50} \Rightarrow h = (x + 50) \tan 17°$

$h = x \tan 17° + 50 \tan 17°$

$\tan 31° = \frac{h}{x} \Rightarrow h = x \tan 31°$

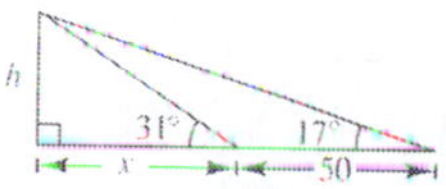

$x \tan 17° + 50 \tan 17° = x \tan 31°$

$50 \tan 17° = x(\tan 31° - \tan 17°)$

$\frac{50 \tan 17°}{\tan 31° - \tan 17°} = x$

$x \approx 51.7959$

$h = x \tan 31° \approx 51.7959 \tan 31° \approx 31.1 \text{ meters}$

The height of the building is approximately 31.1 meters.

19. $\dfrac{h}{\sin 17°} = \dfrac{75}{\sin 45°}$

$h = \dfrac{75 \sin 17°}{\sin 45°}$

$h \approx 31.01$ feet

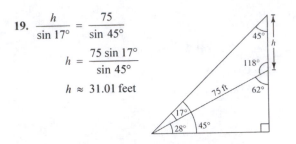

21. Given: $a = 8, b = 14, c = 17$

$\cos C = \dfrac{a^2 + b^2 - c^2}{2ab} = \dfrac{64 + 196 - 289}{2(8)(14)} \approx -0.1295 \Rightarrow C \approx 97.44°$

$\sin B = \dfrac{b \sin C}{c} \approx \dfrac{14 \sin 97.44}{17} \approx 0.8166 \Rightarrow B \approx 54.75°$

$A \approx 180° - 54.75° - 97.44° = 27.81°$

23. Given: $a = 6, b = 9, c = 14$

$\cos C = \dfrac{a^2 + b^2 - c^2}{2ab} = \dfrac{36 + 81 - 196}{2(6)(9)} \approx -0.7315 \Rightarrow C \approx 137.01°$

$\sin B = \dfrac{b \sin C}{c} \approx \dfrac{9 \sin 137.01°}{14} \approx 0.4383 \Rightarrow B \approx 26.00°$

$A \approx 180° - 26.00° - 137.01° = 16.99°$

25. Given: $a = 2.5, b = 5.0, c = 4.5$

$\cos B = \dfrac{a^2 + c^2 - b^2}{2ac} = 0.0667 \Rightarrow B \approx 86.18°$

$\cos C = \dfrac{a^2 + b^2 - c^2}{2ab} = 0.44 \Rightarrow C \approx 63.90°$

$A = 180° - B - C \approx 29.92°$

27. Given: $B = 108°, a = 11, c = 11$

$b^2 = a^2 + c^2 - 2ac \cos B = 11^2 + 11^2 - 2(11)(11) \cos 108° \Rightarrow b \approx 17.80$

$A = C = \frac{1}{2}(180° - 108°) = 36°$

29. Given: $C = 43°, a = 22.5, b = 31.4$

$c = \sqrt{a^2 + b^2 - 2ab \cos C} \approx 21.42$

$\cos B = \dfrac{a^2 + c^2 - b^2}{2ac} \approx -0.02169 \Rightarrow B \approx 91.24°$

$A = 180° - B - C \approx 45.76°$

31. Given: $b = 9, c = 13, C = 64°$

Given two sides and an angle opposite one of them, use the Law of Sines.

$\sin B = \dfrac{b \sin C}{c} = \dfrac{9 \sin 64°}{13} \approx 0.6222 \Rightarrow B \approx 38.48°$

$A \approx 180° - 38.48° - 64° = 77.52°$

$a = \dfrac{c \sin A}{\sin C} \approx \dfrac{13 \sin 77.52°}{\sin 64°} \approx 14.12$

33. Given: $a = 13, b = 15, c = 24$

Given three sides, use the Law of Cosines.

$$\cos C = \frac{a^2 + b^2 - c^2}{2ab} = \frac{169 + 225 - 576}{2(13)(15)} \approx -0.4667 \Rightarrow C \approx 117.82°$$

$$\sin A = \frac{a \sin C}{c} \approx \frac{13 \sin 117.82°}{24} \approx 0.4791 \Rightarrow A \approx 28.62°$$

$$B \approx 180° - 28.62° - 117.82° = 33.56°$$

35.

5 ft, 8 ft, 8 ft, 128°, 152°, 5 ft, a, b

$$a^2 = 5^2 + 8^2 - 2(5)(8)\cos 28° \approx 18.364$$

$$a \approx 4.3 \text{ feet}$$

$$b^2 = 8^2 + 5^2 - 2(8)(5)\cos 152° \approx 159.636$$

$$b \approx 12.6 \text{ feet}$$

37. Length of $AC = \sqrt{300^2 + 425^2 - 2(300)(425)\cos 115°}$

≈ 615.1 meters

39. $a = 3, b = 6, c = 8$

$$s = \frac{a + b + c}{2} = \frac{3 + 6 + 8}{2} = 8.5$$

$$\text{Area} = \sqrt{s(s - a)(s - b)(s - c)}$$

$$= \sqrt{8.5(5.5)(2.5)(0.5)}$$

$$\approx 7.64$$

41. $a = 12.3, b = 15.8, c = 3.7$

$$s = \frac{a + b + c}{2} = \frac{12.3 + 15.8 + 3.7}{2} = 15.9$$

$$\text{Area} = \sqrt{s(s - a)(s - b)(s - c)}$$

$$= \sqrt{15.9(3.6)(0.1)(12.2)} = 8.36$$

43. $\|\mathbf{u}\| = \sqrt{(4 - (-2))^2 + (6 - 1)^2} = \sqrt{61}$

$\|\mathbf{v}\| = \sqrt{(6 - 0)^2 + (3 - (-2))^2} = \sqrt{61}$

u is directed along a line with a slope of $\dfrac{6 - 1}{4 - (-2)} = \dfrac{5}{6}$.

v is directed along a line with a slope of $\dfrac{3 - (-2)}{6 - 0} = \dfrac{5}{6}$.

Because **u** and **v** have identical magnitudes and directions, $\mathbf{u} = \mathbf{v}$.

45. Initial point: $(-5, 4)$

Terminal point: $(2, -1)$

$$\mathbf{v} = \langle 2 - (-5), -1 - 4 \rangle = \langle 7, -5 \rangle$$

47. Initial point: $(0, 10)$

Terminal point: $(7, 3)$

$$\mathbf{v} = \langle 7 - 0, 3 - 10 \rangle = \langle 7, -7 \rangle$$

49. $\|\mathbf{v}\| = 8, \theta = 120°$

$$\langle 8\cos 120°, 8\sin 120° \rangle = \langle -4, 4\sqrt{3} \rangle$$

51. $\mathbf{u} = \langle -1, -3\rangle, \mathbf{v} = \langle -3, 6\rangle$

(a) $\mathbf{u} + \mathbf{v} = \langle -1, -3\rangle + \langle -3, 6\rangle = \langle -4, 3\rangle$

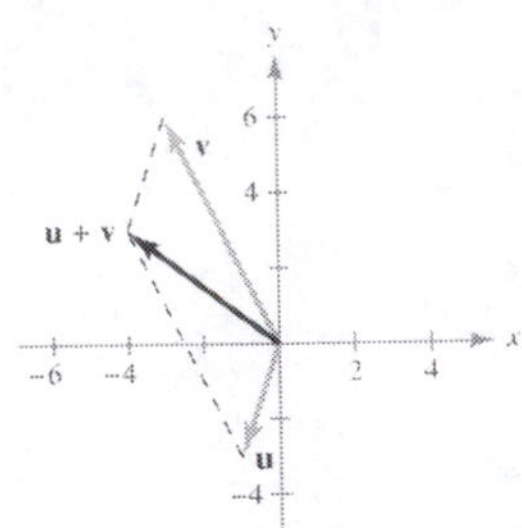

(b) $\mathbf{u} - \mathbf{v} = \langle -1, -3\rangle - \langle -3, 6\rangle = \langle 2, -9\rangle$

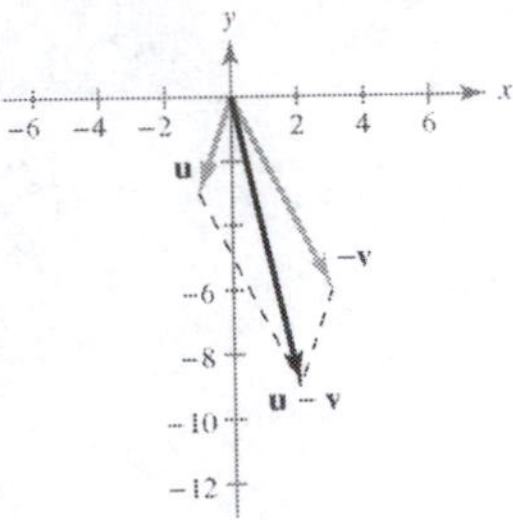

(c) $4\mathbf{u} = 4\langle -1, -3\rangle = \langle -4, -12\rangle$

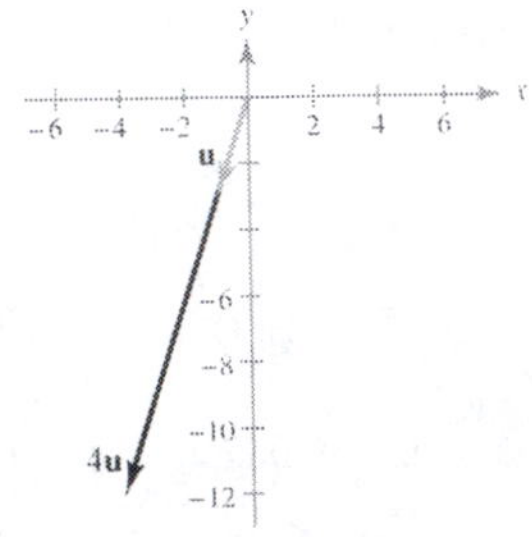

(d) $3\mathbf{v} + 5\mathbf{u} = 3\langle -3, 6\rangle + 5\langle -1, -3\rangle = \langle -9, 18\rangle + \langle -5, -15\rangle = \langle -14, 3\rangle$

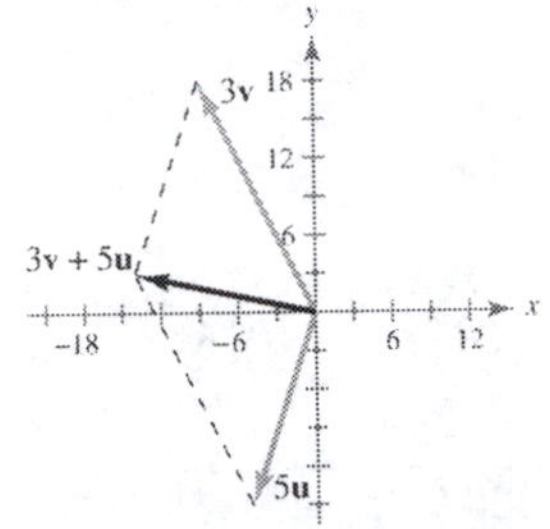

53. $\mathbf{u} = \langle -5, 2\rangle, \mathbf{v} = \langle 4, 4\rangle$

(a) $\mathbf{u} + \mathbf{v} = \langle -5, 2\rangle + \langle 4, 4\rangle = \langle -1, 6\rangle$

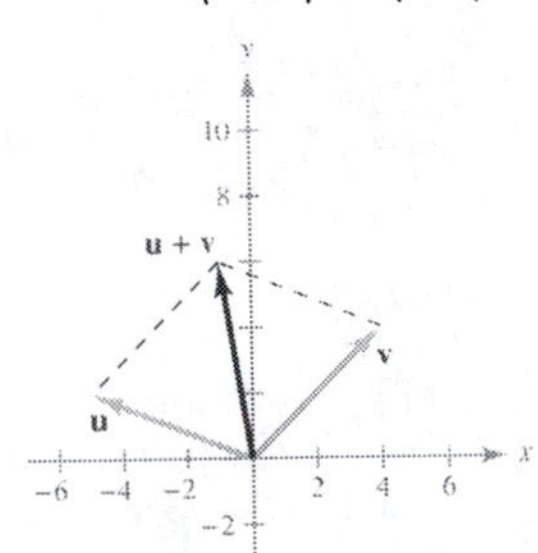

(b) $\mathbf{u} - \mathbf{v} = \langle -5, 2\rangle - \langle 4, 4\rangle = \langle -9, -2\rangle$

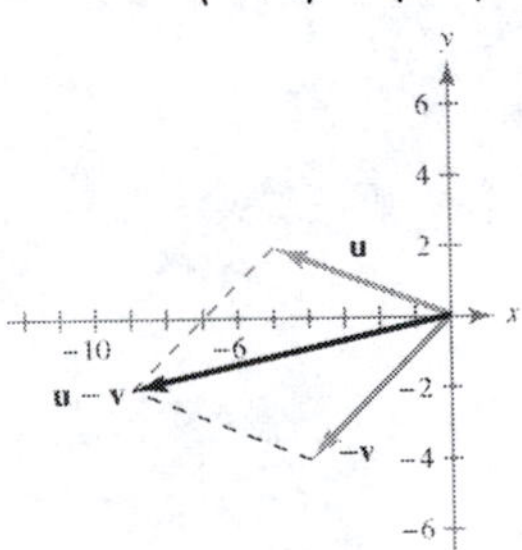

(c) $4\mathbf{u} = 4\langle -5, 2\rangle = \langle -20, 8\rangle$

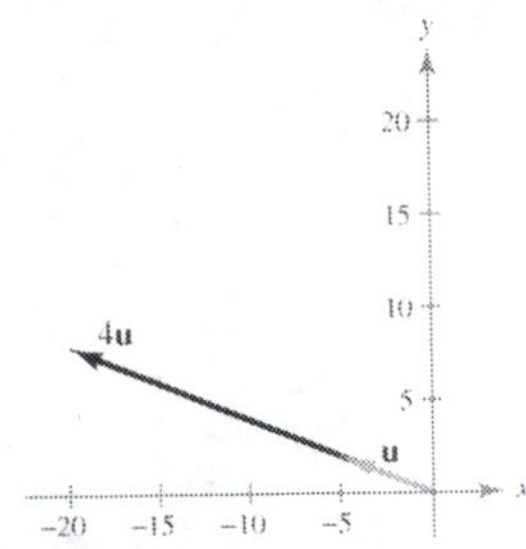

(d) $3\mathbf{v} + 5\mathbf{u} = 3\langle 4, 4\rangle + 5\langle -5, 2\rangle = \langle 12, 12\rangle + \langle -25, 10\rangle = \langle -13, 22\rangle$

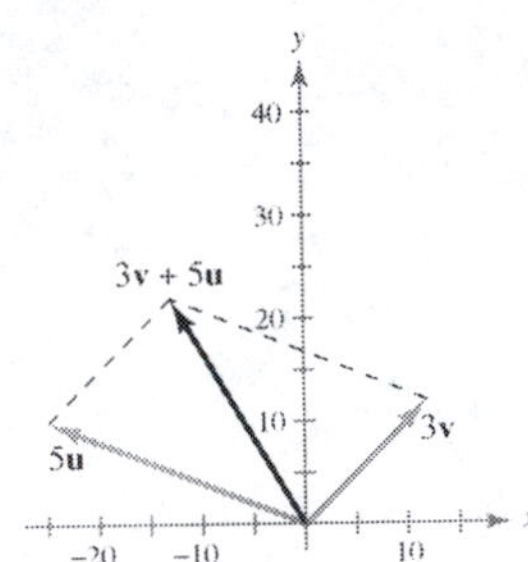

55. $\mathbf{u} = 2\mathbf{i} - \mathbf{j}, \mathbf{v} = 5\mathbf{i} + 3\mathbf{j}$

(a) $\mathbf{u} + \mathbf{v} = (2\mathbf{i} - \mathbf{j}) + (5\mathbf{i} + 3\mathbf{j}) = 7\mathbf{i} + 2\mathbf{j}$

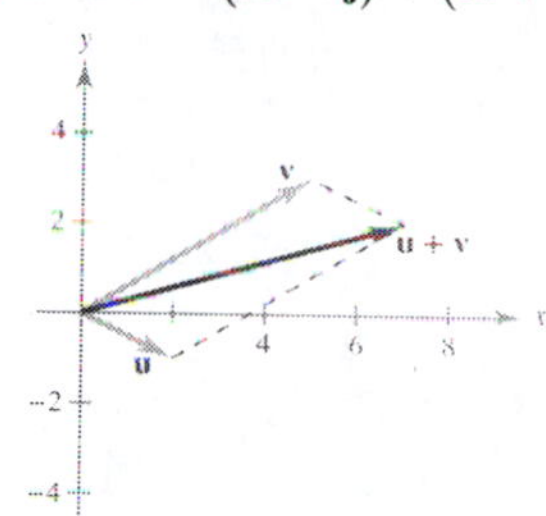

(b) $\mathbf{u} - \mathbf{v} = (2\mathbf{i} - \mathbf{j}) - (5\mathbf{i} + 3\mathbf{j}) = -3\mathbf{i} - 4\mathbf{j}$

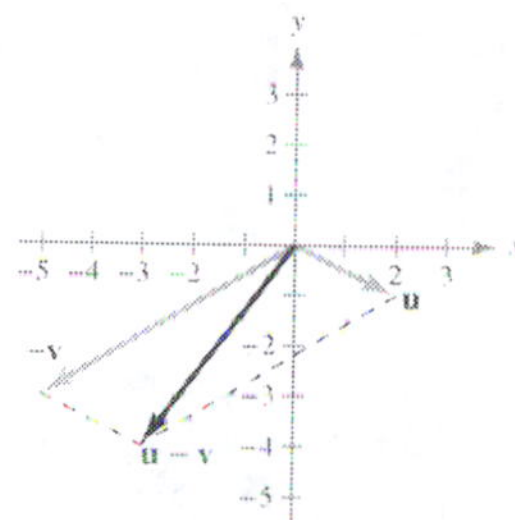

(c) $4\mathbf{u} = 4(2\mathbf{i} - \mathbf{j}) = 8\mathbf{i} - 4\mathbf{j}$

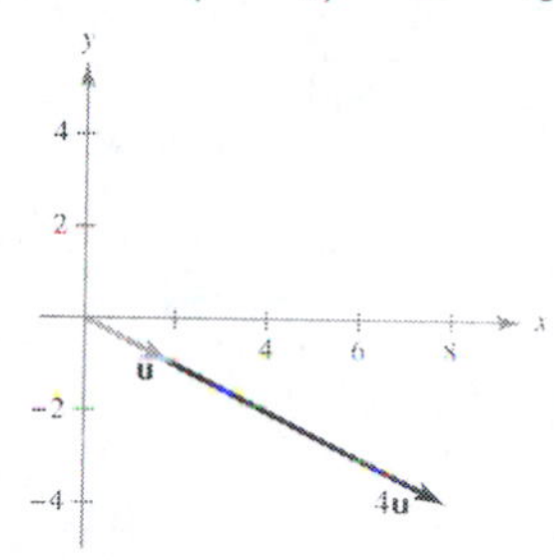

(d) $3\mathbf{v} + 5\mathbf{u} = 3(5\mathbf{i} + 3\mathbf{j}) + 5(2\mathbf{i} - \mathbf{j}) = 15\mathbf{i} + 9\mathbf{j} + 10\mathbf{i} - 5\mathbf{j} = 25\mathbf{i} + 4\mathbf{j}$

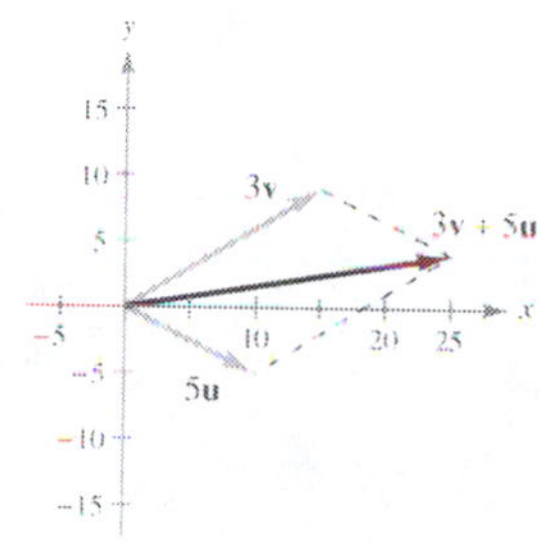

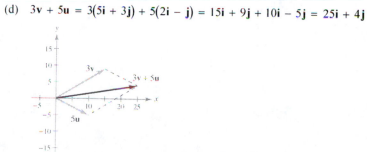

57. $\mathbf{u} = 4\mathbf{i}, \mathbf{v} = -\mathbf{i} + 6\mathbf{j}$

(a) $\mathbf{u} + \mathbf{v} = 4\mathbf{i} + (-\mathbf{i} + 6\mathbf{j}) = 3\mathbf{i} + 6\mathbf{j}$

(b) $\mathbf{u} - \mathbf{v} = 4\mathbf{i} - (-\mathbf{i} + 6\mathbf{j}) = 5\mathbf{i} - 6\mathbf{j}$

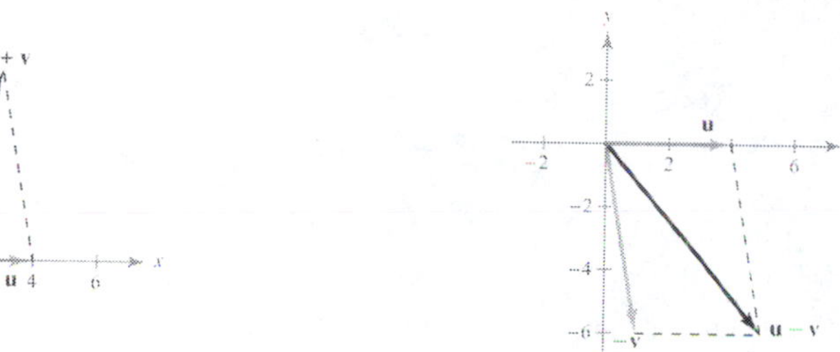

(c) $4\mathbf{u} = 4(4\mathbf{i}) = 16\mathbf{i}$

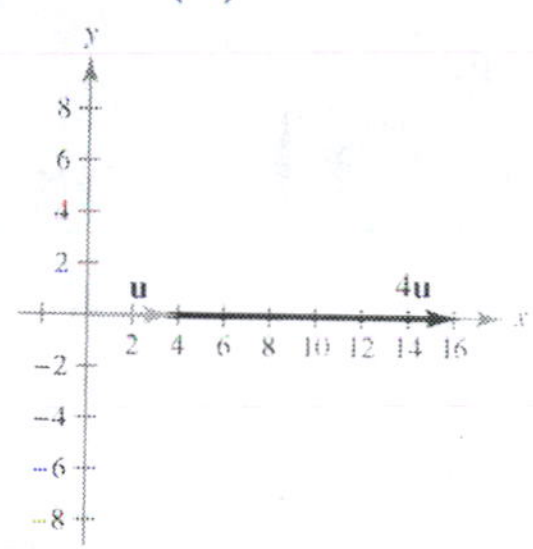

(d) $3\mathbf{v} + 5\mathbf{u} = 3(-\mathbf{i} + 6\mathbf{j}) + 5(4\mathbf{i}) = -3\mathbf{i} + 18\mathbf{j} + 20\mathbf{i} = 17\mathbf{i} + 18\mathbf{j}$

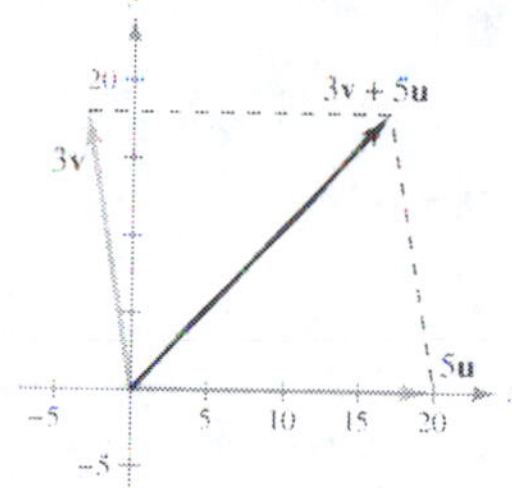

59. $\mathbf{v} = 10\mathbf{i} + 3\mathbf{j}$

$3\mathbf{v} = 3(10\mathbf{i} + 3\mathbf{j})$

$= 30\mathbf{i} + 9\mathbf{j}$

$= \langle 30, 9 \rangle$

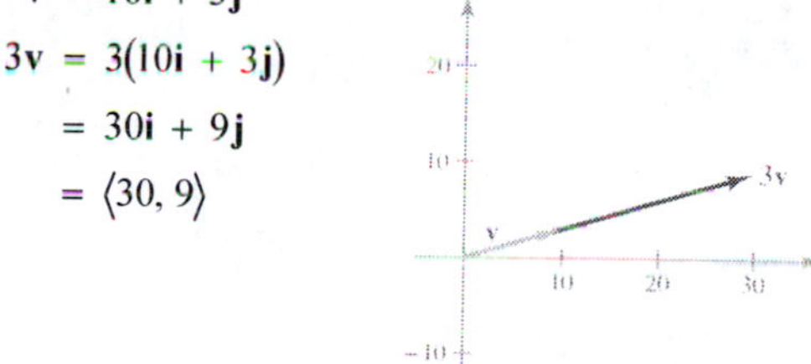

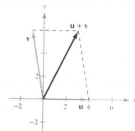

61. $\mathbf{u} = 6\mathbf{i} - 5\mathbf{j},\ \mathbf{v} = 10\mathbf{i} + 3\mathbf{j}$

$$\begin{aligned} 2\mathbf{u} + \mathbf{v} &= 2(6\mathbf{i} - 5\mathbf{j}) + (10\mathbf{i} + 3\mathbf{j}) \\ &= 22\mathbf{i} - 7\mathbf{j} \\ &= \langle 22, -7\rangle \end{aligned}$$

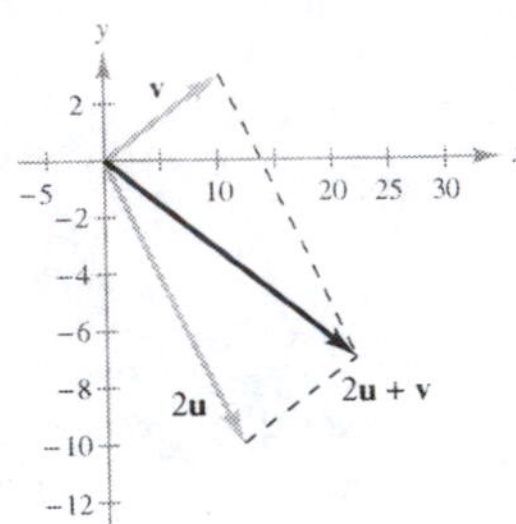

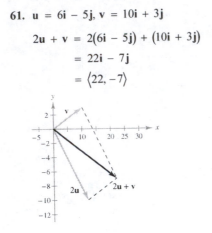

63. $\mathbf{u} = 6\mathbf{i} - 5\mathbf{j},\ \mathbf{v} = 10\mathbf{i} + 3\mathbf{j}$

$$\begin{aligned} 5\mathbf{u} - 4\mathbf{v} &= 5(6\mathbf{i} - 5\mathbf{j}) - 4(10\mathbf{i} + 3\mathbf{j}) \\ &= 30\mathbf{i} - 25\mathbf{j} - 40\mathbf{i} - 12\mathbf{j} \\ &= -10\mathbf{i} - 37\mathbf{j} \\ &= \langle -10, -37\rangle \end{aligned}$$

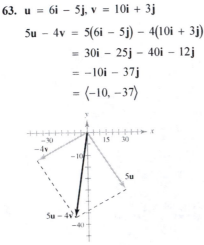

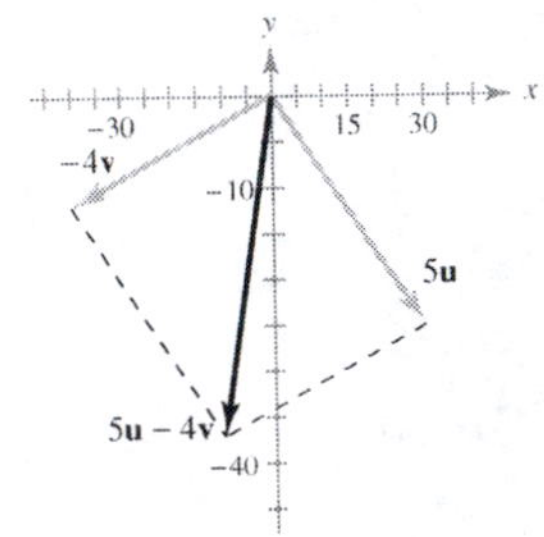

65. $P = (2, 3),\ Q = (1, 8)$

$$\begin{aligned} \overrightarrow{PQ} = \mathbf{v} &= \langle 1 - 2, 8 - 3\rangle \\ \mathbf{v} &= \langle -1, 5\rangle \\ \mathbf{v} &= -\mathbf{i} + 5\mathbf{j} \end{aligned}$$

67. $P = (3, 4),\ Q = (9, 8)$

$$\begin{aligned} \overrightarrow{PQ} = \mathbf{v} &= \langle 9 - 3, 8 - 4\rangle \\ \mathbf{v} &= \langle 6, 4\rangle \\ \mathbf{v} &= 6\mathbf{i} + 4\mathbf{j} \end{aligned}$$

69. $\mathbf{v} = 7(\cos 60^\circ\mathbf{i} + \sin 60^\circ\mathbf{j})$

$\|\mathbf{v}\| = 7$

$\theta = 60^\circ$

71. $\mathbf{v} = 5\mathbf{i} + 4\mathbf{j}$

$\|\mathbf{v}\| = \sqrt{5^2 + 4^2} = \sqrt{41}$

$\tan\theta = \frac{4}{5} \Rightarrow \theta \approx 38.7^\circ$

73. $\mathbf{v} = -3\mathbf{i} - 3\mathbf{j}$

$\|\mathbf{v}\| = \sqrt{(-3)^2 + (-3)^2} = 3\sqrt{2}$

$\tan\theta = \dfrac{-3}{-3} = 1 \Rightarrow \theta = 225^\circ$

75. Magnitude of resultant:

$c = \sqrt{85^2 + 50^2 - 2(85)(50)\cos 165^\circ} \approx 133.92$ pounds

Let θ be the angle between the resultant and the 85-pound force.

$$\cos\theta \approx \frac{(133.92)^2 + 85^2 - 50^2}{2(133.92)(85)} \approx 0.9953 \Rightarrow \theta \approx 5.6^\circ$$

77. Airspeed: $\mathbf{u} = 430(\cos 45^\circ\mathbf{i} - \sin 45^\circ\mathbf{j}) = 215\sqrt{2}(\mathbf{i} - \mathbf{j})$

Wind: $\mathbf{w} = 35(\cos 60^\circ\mathbf{i} + \sin 60^\circ\mathbf{j}) = \dfrac{35}{2}(\mathbf{i} + \sqrt{3}\mathbf{j})$

Groundspeed: $\mathbf{u} + \mathbf{w} = \left(215\sqrt{2} + \dfrac{35}{2}\right)\mathbf{i} + \left(\dfrac{35\sqrt{3}}{2} - 215\sqrt{2}\right)\mathbf{j}$

$$\|\mathbf{u} + \mathbf{w}\| = \sqrt{\left(215\sqrt{2} + \frac{35}{2}\right)^2 + \left(\frac{35\sqrt{3}}{2} - 215\sqrt{2}\right)^2} \approx 422.30 \text{ miles per hour}$$

Bearing: $\tan\theta' = \dfrac{17.5\sqrt{3} - 215\sqrt{2}}{215\sqrt{2} + 17.5}$

$\theta' \approx -40.4^\circ$

$\theta = 90^\circ + |\theta'| = 130.4^\circ$

79. $\mathbf{u} = \langle 6, 7\rangle, \mathbf{v} = \langle -3, 9\rangle$

$\mathbf{u} \cdot \mathbf{v} = 6(-3) + 7(9) = 45$

81. $\mathbf{u} = 3\mathbf{i} + 7\mathbf{j}, \mathbf{v} = 11\mathbf{i} - 5\mathbf{j}$

$\mathbf{u} \cdot \mathbf{v} = 3(11) + 7(-5) = -2$

83. $\mathbf{u} = \langle -4, 2\rangle$

$2\mathbf{u} = \langle -8, 4\rangle$

$2\mathbf{u} \cdot \mathbf{u} = -8(-4) + 4(2) = 40$

The result is a scalar.

85. $\mathbf{u} = \langle -4, 2\rangle$

$4 - \|\mathbf{u}\| = 4 - \sqrt{(-4)^2 + 2^2} = 4 - \sqrt{20} = 4 - 2\sqrt{5}$

The result is a scalar.

87. $\mathbf{u} = \langle -4, 2\rangle, \mathbf{v} = \langle 5, 1\rangle$

$$\begin{aligned}\mathbf{u}(\mathbf{u} \cdot \mathbf{v}) &= \langle -4, 2\rangle[-4(5) + 2(1)]\\ &= -18\langle -4, 2\rangle\\ &= \langle 72, -36\rangle\end{aligned}$$

The result is a vector.

89. $\mathbf{u} = \langle -4, 2\rangle, \mathbf{v} = \langle 5, 1\rangle$

$$\begin{aligned}(\mathbf{u} \cdot \mathbf{u}) - (\mathbf{u} \cdot \mathbf{v}) &= [-4(-4) + 2(2)] - [-4(5) + 2(1)]\\ &= 20 - (-18)\\ &= 38\end{aligned}$$

The result is a scalar.

91. $\mathbf{u} = \cos\frac{7\pi}{4}\mathbf{i} + \sin\frac{7\pi}{4}\mathbf{j} = \left\langle \frac{1}{\sqrt{2}}, -\frac{1}{\sqrt{2}}\right\rangle$

$\mathbf{v} = \cos\frac{5\pi}{6}\mathbf{i} + \sin\frac{5\pi}{6}\mathbf{j} = \left\langle -\frac{\sqrt{3}}{2}, \frac{1}{2}\right\rangle$

$\cos\theta = \frac{\mathbf{u} \cdot \mathbf{v}}{\|\mathbf{u}\|\|\mathbf{v}\|} = \frac{-\sqrt{3} - 1}{2\sqrt{2}} \Rightarrow \theta = \frac{11\pi}{12}$

93. $\mathbf{u} = \langle 2\sqrt{2}, -4\rangle, \mathbf{v} = \langle -\sqrt{2}, 1\rangle$

$\cos\theta = \frac{\mathbf{u} \cdot \mathbf{v}}{\|\mathbf{u}\|\|\mathbf{v}\|} = \frac{-8}{(\sqrt{24})(\sqrt{3})} \Rightarrow \theta \approx 160.5^\circ$

95. $\mathbf{u} = \langle -3, 8\rangle$

$\mathbf{v} = \langle 8, 3\rangle$

$\mathbf{u} \cdot \mathbf{v} = -3(8) + 8(3) = 0$

$\mathbf{u}$ and $\mathbf{v}$ are orthogonal.

97. $\mathbf{u} = -\mathbf{i}$

$\mathbf{v} = \mathbf{i} + 2\mathbf{j}$

$\mathbf{u} \cdot \mathbf{v} \neq 0 \Rightarrow$ Not orthogonal

$\mathbf{v} \neq k\mathbf{u} \Rightarrow$ Not parallel

Neither

99. $\mathbf{u} = \langle -4, 3\rangle, \mathbf{v} = \langle -8, -2\rangle$

$\mathbf{w}_1 = \text{proj}_{\mathbf{v}}\mathbf{u} = \left(\frac{\mathbf{u} \cdot \mathbf{v}}{\|\mathbf{v}\|^2}\right)\mathbf{v} = \left(\frac{26}{68}\right)\langle -8, -2\rangle = -\frac{13}{17}\langle 4, 1\rangle$

$\mathbf{w}_2 = \mathbf{u} - \mathbf{w}_1 = \langle -4, 3\rangle - \left(-\frac{13}{17}\right)\langle 4, 1\rangle = \frac{16}{17}\langle -1, 4\rangle$

$\mathbf{u} = \mathbf{w}_1 + \mathbf{w}_2 = -\frac{13}{17}\langle 4, 1\rangle + \frac{16}{17}\langle -1, 4\rangle$

101. $\mathbf{u} = \langle 2, 7\rangle, \mathbf{v} = \langle 1, -1\rangle$

$\mathbf{w}_1 = \text{proj}_{\mathbf{v}}\mathbf{u} = \left(\frac{\mathbf{u} \cdot \mathbf{v}}{\|\mathbf{v}\|^2}\right)\mathbf{v} = -\frac{5}{2}\langle 1, -1\rangle = \frac{5}{2}\langle -1, 1\rangle$

$\mathbf{w}_2 = \mathbf{u} - \mathbf{w}_1 = \langle 2, 7\rangle - \left(\frac{5}{2}\right)\langle -1, 1\rangle = \frac{9}{2}\langle 1, 1\rangle$

$\mathbf{u} = \mathbf{w}_1 + \mathbf{w}_2 = \frac{5}{2}\langle -1, 1\rangle + \frac{9}{2}\langle 1, 1\rangle$

103. $P = (5, 3), Q = (8, 9) \Rightarrow \overrightarrow{PQ} = \langle 3, 6\rangle$

Work $= \mathbf{v} \cdot \overrightarrow{PQ} = \langle 2, 7\rangle \cdot \langle 3, 6\rangle = 48$

105. Work $= (18{,}000)\left(\frac{48}{12}\right) = 72{,}000$ foot-pounds

107. $|7i| = \sqrt{0^2 + 7^2} = 7$

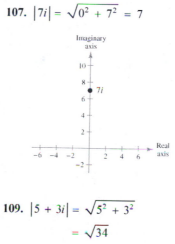

109. $|5 + 3i| = \sqrt{5^2 + 3^2}$

$= \sqrt{34}$

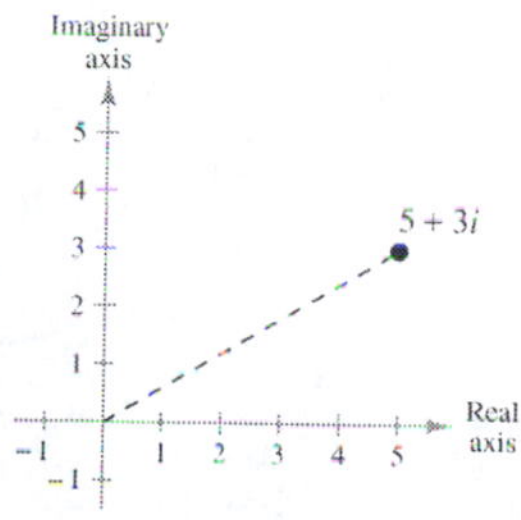

111. $\left|\sqrt{2} - \sqrt{2}i\right| = \sqrt{\left(\sqrt{2}\right)^2 + \left(\sqrt{2}\right)^2} = 2$

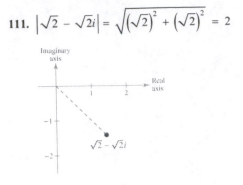

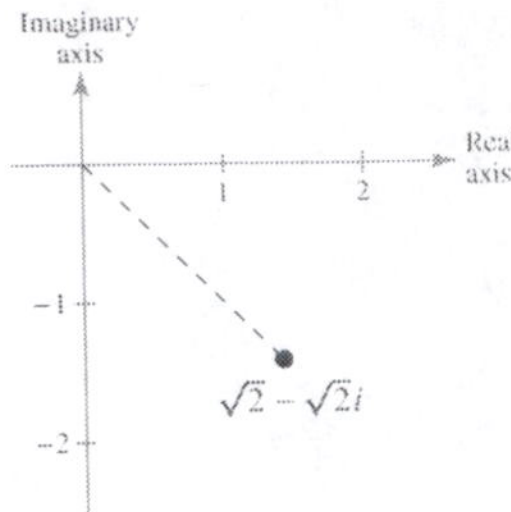

113. $z = 4i$

$r = \sqrt{0^2 + 4^2} = \sqrt{16} = 4$

$\tan\theta = \frac{4}{0}$, undefined $\Rightarrow \theta = \frac{\pi}{2}$

$z = 4\left(\cos\frac{\pi}{2} + i\sin\frac{\pi}{2}\right)$

115. $z = 7 - 7i$

$r = \sqrt{(7)^2 + (-7)^2} = \sqrt{98} = 7\sqrt{2}$

$\tan\theta = \frac{-7}{7} = -1 \Rightarrow \theta = \frac{7\pi}{4}$ because the complex number lies in Quadrant IV.

$7 - 7i = 7\sqrt{2}\left(\cos\frac{7\pi}{4} + i\sin\frac{7\pi}{4}\right)$

117. $z = -5 - 12i$

$r = \sqrt{(-5)^2 + (-12)^2} = \sqrt{169} = 13$

$\tan\theta = \frac{12}{5}$, θ is in Quadrant III $\Rightarrow \theta \approx 4.32$

$z = 13(\cos 4.32 + i\sin 4.32)$

119. (a) $z_1 = 2\sqrt{3} - 2i = 4\left(\cos\frac{11\pi}{6} + i\sin\frac{11\pi}{6}\right)$

$z_2 = -10i = 10\left(\cos\frac{3\pi}{2} + i\sin\frac{3\pi}{2}\right)$

(b) $z_1z_2 = \left[4\left(\cos\frac{11\pi}{6} + i\sin\frac{11\pi}{6}\right)\right]\left[10\left(\cos\frac{3\pi}{2} + i\sin\frac{3\pi}{2}\right)\right]$

$= 40\left(\cos\frac{10\pi}{3} + i\sin\frac{10\pi}{3}\right)$

$\frac{z_1}{z_2} = \frac{4\left(\cos\frac{11\pi}{6} + i\sin\frac{11\pi}{6}\right)}{10\left(\cos\frac{3\pi}{2} + i\sin\frac{3\pi}{2}\right)} = \frac{2}{5}\left(\cos\frac{\pi}{3} + i\sin\frac{\pi}{3}\right)$

121. $\left[5\left(\cos\frac{\pi}{12} + i\sin\frac{\pi}{12}\right)\right]^4 = 5^4\left(\cos\frac{4\pi}{12} + i\sin\frac{4\pi}{12}\right)$

$= 625\left(\cos\frac{\pi}{3} + i\sin\frac{\pi}{3}\right)$

$= 625\left(\frac{1}{2} + \frac{\sqrt{3}}{2}i\right)$

$= \frac{625}{2} + \frac{625\sqrt{3}}{2}i$

123. $(2 + 3i)^6 \approx \left[\sqrt{13}(\cos 56.3^\circ + i\sin 56.3^\circ)\right]^6$

$= 13^3(\cos 337.9^\circ + i\sin 337.9^\circ)$

$\approx 13^3(0.9263 - 0.3769i)$

$\approx 2035 - 828i$

125. Sixth roots of $-729i = 729\left(\cos\frac{3\pi}{2} + i\sin\frac{3\pi}{2}\right)$:

(a) $\sqrt[6]{729}\left[\cos\left(\dfrac{\frac{3\pi}{2} + 2k\pi}{6}\right) + i\sin\left(\dfrac{\frac{3\pi}{2} + 2k\pi}{6}\right)\right], k = 0, 1, 2, 3, 4, 5$

$k = 0$: $3\left(\cos\frac{\pi}{4} + i\sin\frac{\pi}{4}\right)$

$k = 1$: $3\left(\cos\frac{7\pi}{12} + i\sin\frac{7\pi}{12}\right)$

$k = 2$: $3\left(\cos\frac{11\pi}{12} + i\sin\frac{11\pi}{12}\right)$

$k = 3$: $3\left(\cos\frac{5\pi}{4} + i\sin\frac{5\pi}{4}\right)$

$k = 4$: $3\left(\cos\frac{19\pi}{12} + i\sin\frac{19\pi}{12}\right)$

$k = 5$: $3\left(\cos\frac{23\pi}{12} + i\sin\frac{23\pi}{12}\right)$

(b)

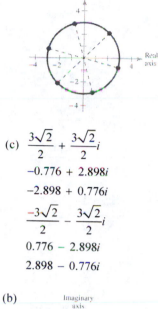

(c) $\frac{3\sqrt{2}}{2} + \frac{3\sqrt{2}}{2}i$

$-0.776 + 2.898i$

$-2.898 + 0.776i$

$\frac{-3\sqrt{2}}{2} - \frac{3\sqrt{2}}{2}i$

$0.776 - 2.898i$

$2.898 - 0.776i$

127. Cube roots of $8 = 8(\cos 0 + i\sin 0)$, $k = 0, 1, 2$

(a) $\sqrt[3]{8}\left[\cos\left(\frac{0 + 2\pi k}{3}\right) + i\sin\left(\frac{0 + 2\pi k}{3}\right)\right]$

$k = 0$: $2(\cos 0 + i\sin 0)$

$k = 1$: $2\left(\cos\frac{2\pi}{3} + i\sin\frac{2\pi}{3}\right)$

$k = 2$: $2\left(\cos\frac{4\pi}{3} + i\sin\frac{4\pi}{3}\right)$

(b)

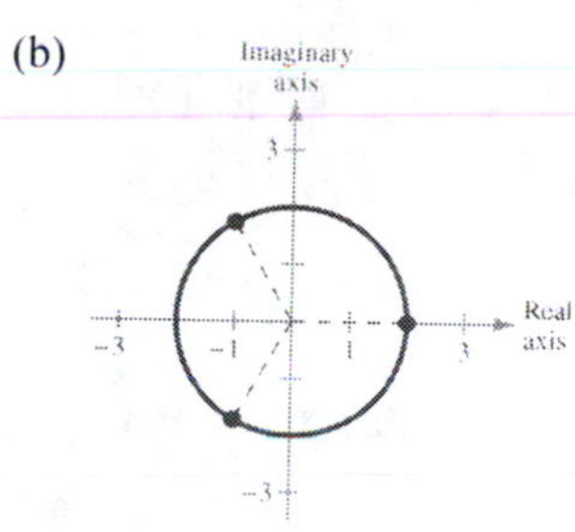

(c) 2

$-1 + \sqrt{3}i$

$-1 - \sqrt{3}i$

129. $x^4 + 81 = 0$

$x^4 = -81$ Solve by finding the fourth roots of -81.

$-81 = 81(\cos\pi + i\sin\pi)$

$\sqrt[4]{-81} = \sqrt[4]{81}\left[\cos\left(\frac{\pi + 2\pi k}{4}\right) + i\sin\left(\frac{\pi + 2\pi k}{4}\right)\right], k = 0, 1, 2, 3$

$k = 0$: $3\left(\cos\frac{\pi}{4} + i\sin\frac{\pi}{4}\right) = \frac{3\sqrt{2}}{2} + \frac{3\sqrt{2}}{2}i$

$k = 1$: $3\left(\cos\frac{3\pi}{4} + i\sin\frac{3\pi}{4}\right) = -\frac{3\sqrt{2}}{2} + \frac{3\sqrt{2}}{2}i$

$k = 2$: $3\left(\cos\frac{5\pi}{4} + i\sin\frac{5\pi}{4}\right) = -\frac{3\sqrt{2}}{2} - \frac{3\sqrt{2}}{2}i$

$k = 3$: $3\left(\cos\frac{7\pi}{4} + i\sin\frac{7\pi}{4}\right) = \frac{3\sqrt{2}}{2} - \frac{3\sqrt{2}}{2}i$

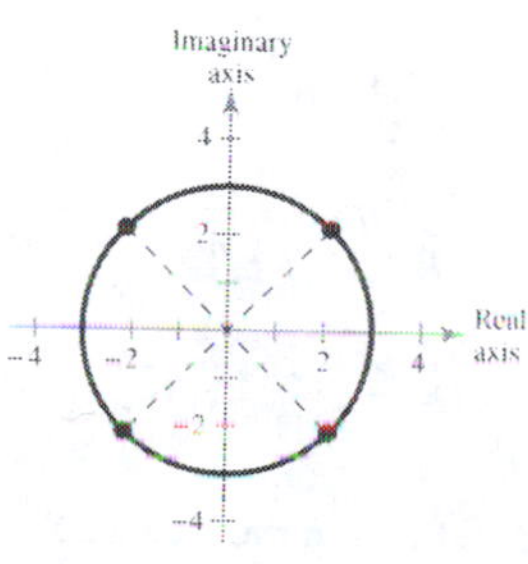

131. $x^3 + 8i = 0$

$x^3 = -8i$ Solve by finding the cube roots of $-8i$.

$-8i = 8\left(\cos\frac{3\pi}{2} + i\sin\frac{3\pi}{2}\right)$

$\sqrt[3]{-8i} = \sqrt[3]{8}\left[\cos\left(\frac{\frac{3\pi}{2} + 2\pi k}{3}\right) + i\sin\left(\frac{\frac{3\pi}{2} + 2\pi k}{3}\right)\right], k = 0, 1, 2$

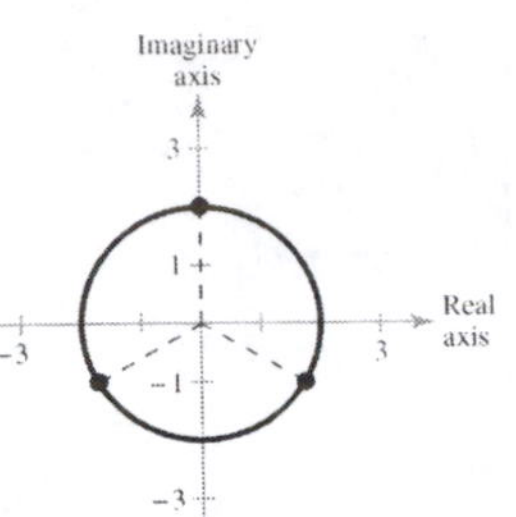

$k = 0: 2\left(\cos\frac{\pi}{2} + i\sin\frac{\pi}{2}\right) = 2i$

$k = 1: 2\left(\cos\frac{7\pi}{6} + i\sin\frac{7\pi}{6}\right) = -\sqrt{3} - i$

$k = 2: 2\left(\cos\frac{11\pi}{6} + i\sin\frac{11\pi}{6}\right) = \sqrt{3} - i$

133. $x^5 + x^3 - x^2 - 1 = 0$

$x^3(x^2 + 1) - 1(x^2 + 1) = 0$

$(x^3 - 1)(x^2 + 1) = 0$

$x^3 - 1 = 0$ or $x^2 + 1 = 0$

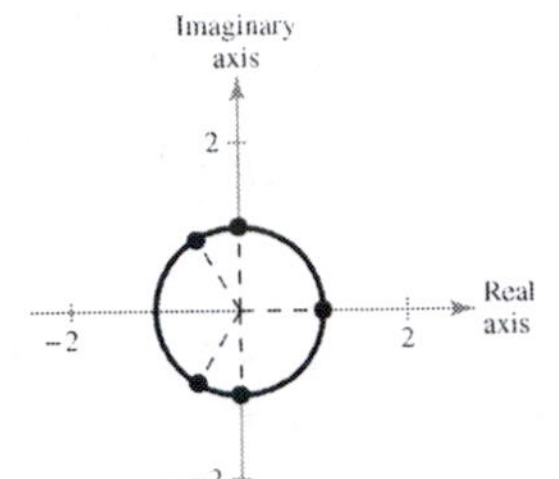

Case 1: $x^3 - 1 = 0$

$x^3 = 1$ Solve by finding the cube roots of 1.

$1 = 1(\cos 0 + i\sin 0)$

$\sqrt[3]{1} = \sqrt[3]{1}\left[\cos\left(\frac{0 + 2\pi k}{3}\right) + i\sin\left(\frac{0 + 2\pi k}{3}\right)\right], k = 0, 1, 2$

$k = 0: 1(\cos 0 + i\sin 0) = 1$

$k = 1: 1\left(\cos\frac{2\pi}{3} + i\sin\frac{2\pi}{3}\right) = -\frac{1}{2} + \frac{\sqrt{3}}{2}i$

$k = 2: 1\left(\cos\frac{4\pi}{3} + i\sin\frac{4\pi}{3}\right) = -\frac{1}{2} - \frac{\sqrt{3}}{2}i$

Case 2: $x^2 + 1 = 0$

$x^2 = -1$ Solve by finding the square roots of -1.

$-1 = 1(\cos\pi + i\sin\pi)$

$\sqrt{-1} = \sqrt{1}\left[\cos\left(\frac{\pi + 2\pi k}{2}\right) + i\sin\left(\frac{\pi + 2\pi k}{2}\right)\right], k = 0, 1$

$k = 0: 1\left(\cos\frac{\pi}{2} + i\sin\frac{\pi}{2}\right) = i$

$k = 1: 1\left(\cos\frac{3\pi}{2} + i\sin\frac{3\pi}{2}\right) = -i$

135. True. $\sin 90°$ is defined in the Law of Sines.

137. False. $x = \sqrt{3} + i$ is a solution to $x^3 - 8i = 0$, not $x^2 - 8i = 0$.

Also, $(\sqrt{3} + i)^2 - 8i = 2 + (2\sqrt{3} - 8)i \neq 0$.

139. $a^2 = b^2 + c^2 - 2bc\cos A$

$b^2 = a^2 + c^2 - 2ac\cos B$

$c^2 = a^2 + b^2 - 2ab\cos C$

141. **A** and **C** appear to have the same magnitude and direction.

143. If $k > 0$, the direction of $k\mathbf{u}$ is the same, and the magnitude is $k\|\mathbf{u}\|$.

If $k < 0$, the direction of $k\mathbf{u}$ is the opposite direction of $\mathbf{u}$, and the magnitude is $|k|\,\|\mathbf{u}\|$.

145. (a) The trigonometric form of the three roots shown is:

$4(\cos 60° + i \sin 60°)$

$4(\cos 180° + i \sin 180°)$

$4(\cos 300° + i \sin 300°)$

(b) Because there are three evenly spaced roots on the circle of radius 4, they are cube roots of a complex number of modulus $4^3 = 64$.

Cubing them yields -64.

$[4(\cos 60° + i \sin 60°)]^3 = -64$

$[4(\cos 180° + i \sin 180°)]^3 = -64$

$[4(\cos 300° + i \sin 300°)]^3 = -64$

147. (a) The trigonometric form of the three roots shown are

$2(\cos 30° + i \sin 30°)$,

$2(\cos 150° + i \sin 150°)$, and

$2(\cos 270° + i \sin 270°)$.

(b) Because there are three evenly spaced roots on the circle of radius 2, they are cube roots of a complex number of modulus $2^3 = 8$.

Cubing them yields $8i$.

$[2(\cos 60° + i \sin 60°)]^3 = 8i$

$[2(\cos 180° + i \sin 180°)]^3 = 8i$

$[2(\cos 300° + i \sin 300°)]^3 = 8i$

149. $z_1 = 2(\cos\theta + i\sin\theta)$

$z_2 = 2(\cos(\pi - \theta) + i\sin(\pi - \theta))$

$$\begin{aligned} z_1z_2 &= (2)(2)[\cos(\theta + (\pi - \theta)) + i\sin(\theta + (\pi - \theta))] \\ &= 4(\cos\pi + i\sin\pi) \\ &= -4 \end{aligned}$$

$$\begin{aligned} \frac{z_1}{z_2} &= \frac{2(\cos\theta + i\sin\theta)}{2(\cos(\pi - \theta) + i\sin(\pi - \theta))} \\ &= 1[\cos(\theta - (\pi - \theta)) + i\sin(\theta - (\pi - \theta))] \\ &= \cos(2\theta - \pi) + i\sin(2\theta - \pi) \\ &= \cos 2\theta\cos\pi + \sin 2\theta\sin\pi + i(\sin 2\theta\cos\pi - \cos 2\theta\sin\pi) \\ &= -\cos 2\theta - i\sin 2\theta \end{aligned}$$

Problem Solving for Chapter 8

1. $(\overrightarrow{PQ})^2 = 4.7^2 + 6^2 - 2(4.7)(6)\cos 25° A$

$\overrightarrow{PQ} \approx 2.6409$ feet

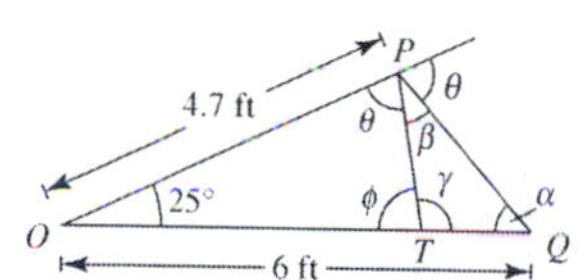

$\dfrac{\sin\alpha}{4.7} = \dfrac{\sin 25°}{2.6409} \Rightarrow \alpha \approx 48.78°$

$\theta + \beta = 180° - 25° - 48.78° = 106.22°$

$(\theta + \beta) + \theta = 180° \Rightarrow \theta = 180° - 106.22° = 73.78°$

$\beta = 106.22° - 73.78° = 32.44°$

$\gamma = 180° - \alpha - \beta = 180° - 48.78° - 32.44° = 98.78°$

$\phi = 180° - \gamma = 180° - 98.78° = 81.22°$

$\dfrac{\overrightarrow{PT}}{\sin 25°} = \dfrac{4.7}{\sin 81.22°}$

$\overrightarrow{PT} \approx 2.01$ feet

3. (a) 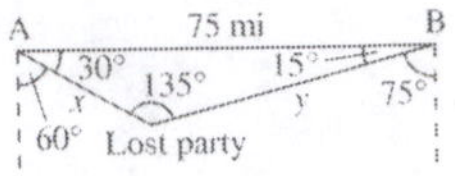 A 75 mi B 30° 135° 15° x y 75° 60° Lost party

(b) $\dfrac{x}{\sin 15°} = \dfrac{75}{\sin 135°}$ and $\dfrac{y}{\sin 30°} = \dfrac{75}{\sin 135°}$

$x \approx 27.45$ miles $\qquad y \approx 53.03$ miles

(c) 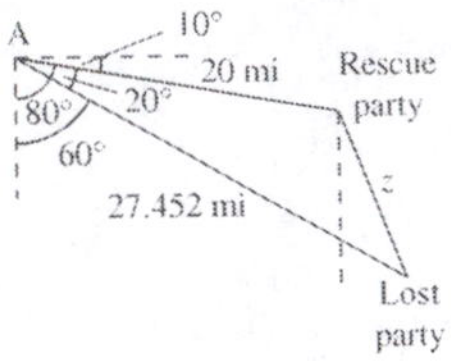 A 10° 20 mi Rescue party 80° 20° 60° 27.452 mi z Lost party

$z^2 = (27.45)^2 + (20)^2 - 2(27.45)(20)\cos 20°$

$z \approx 11.03$ miles

$\dfrac{\sin\theta}{27.45} = \dfrac{\sin 20°}{11.03}$

$\sin\theta \approx 0.8511$

$\theta = 180° - \sin^{-1}(0.8511)$

$\theta \approx 121.7°$

To find the bearing, we have $\theta - 10° - 90° \approx 21.7°$.

Bearing: S 21.7° E

5. If $\mathbf{u} \neq 0$, $\mathbf{v} \neq 0$, and $\mathbf{u} + \mathbf{v} \neq 0$, then $\left\|\dfrac{\mathbf{u}}{\|\mathbf{u}\|}\right\| = \left\|\dfrac{\mathbf{v}}{\|\mathbf{v}\|}\right\| = \left\|\dfrac{\mathbf{u}+\mathbf{v}}{\|\mathbf{u}+\mathbf{v}\|}\right\| = 1$ because all of these are magnitudes of unit vectors.

(a) $\mathbf{u} = \langle 1, -1\rangle$, $\mathbf{v} = \langle -1, 2\rangle$, $\mathbf{u}+\mathbf{v} = \langle 0, 1\rangle$

(i) $\|\mathbf{u}\| = \sqrt{2}$ (ii) $\|\mathbf{v}\| = \sqrt{5}$ (iii) $\|\mathbf{u}+\mathbf{v}\| = 1$ (iv) $\left\|\dfrac{\mathbf{u}}{\|\mathbf{u}\|}\right\| = 1$ (v) $\left\|\dfrac{\mathbf{v}}{\|\mathbf{v}\|}\right\| = 1$ (vi) $\left\|\dfrac{\mathbf{u}+\mathbf{v}}{\|\mathbf{u}+\mathbf{v}\|}\right\| = 1$

(b) $\mathbf{u} = \langle 0, 1\rangle$, $\mathbf{v} = \langle 3, -3\rangle$, $\mathbf{u}+\mathbf{v} = \langle 3, -2\rangle$

(i) $\|\mathbf{u}\| = 1$ (ii) $\|\mathbf{v}\| = \sqrt{18} = 3\sqrt{2}$ (iii) $\|\mathbf{u}+\mathbf{v}\| = \sqrt{13}$ (iv) $\left\|\dfrac{\mathbf{u}}{\|\mathbf{u}\|}\right\| = 1$ (v) $\left\|\dfrac{\mathbf{v}}{\|\mathbf{v}\|}\right\| = 1$ (vi) $\left\|\dfrac{\mathbf{u}+\mathbf{v}}{\|\mathbf{u}+\mathbf{v}\|}\right\| = 1$

(c) $\mathbf{u} = \left\langle 1, \dfrac{1}{2}\right\rangle$, $\mathbf{v} = \langle 2, 3\rangle$, $\mathbf{u}+\mathbf{v} = \left\langle 3, \dfrac{7}{2}\right\rangle$

(i) $\|\mathbf{u}\| = \dfrac{\sqrt{5}}{2}$ (ii) $\|\mathbf{v}\| = \sqrt{13}$ (iii) $\|\mathbf{u}+\mathbf{v}\| = \sqrt{9 + \dfrac{49}{4}} = \dfrac{\sqrt{85}}{2}$ (iv) $\left\|\dfrac{\mathbf{u}}{\|\mathbf{u}\|}\right\| = 1$

(v) $\left\|\dfrac{\mathbf{v}}{\|\mathbf{v}\|}\right\| = 1$ (vi) $\left\|\dfrac{\mathbf{u}+\mathbf{v}}{\|\mathbf{u}+\mathbf{v}\|}\right\| = 1$

(d) $\mathbf{u} = \langle 2, -4\rangle$, $\mathbf{v} = \langle 5, 5\rangle$, $\mathbf{u}+\mathbf{v} = \langle 7, 1\rangle$

(i) $\|\mathbf{u}\| = \sqrt{20} = 2\sqrt{5}$ (ii) $\|\mathbf{v}\| = \sqrt{50} = 5\sqrt{2}$ (iii) $\|\mathbf{u}+\mathbf{v}\| = \sqrt{50} = 5\sqrt{2}$ (iv) $\left\|\dfrac{\mathbf{u}}{\|\mathbf{u}\|}\right\| = 1$

(v) $\left\|\dfrac{\mathbf{v}}{\|\mathbf{v}\|}\right\| = 1$ (vi) $\left\|\dfrac{\mathbf{u}+\mathbf{v}}{\|\mathbf{u}+\mathbf{v}\|}\right\| = 1$

7. Let $\mathbf{u} \cdot \mathbf{v} = 0$ and $\mathbf{u} \cdot \mathbf{w} = 0$.

Then,
$$\begin{aligned}\mathbf{u} \cdot (c\mathbf{v} + d\mathbf{w}) &= \mathbf{u} \cdot c\mathbf{v} + \mathbf{u} \cdot d\mathbf{w} \\ &= c(\mathbf{u} \cdot \mathbf{v}) + d(\mathbf{u} \cdot \mathbf{w}) \\ &= c(0) + d(0) \\ &= 0.\end{aligned}$$

So for all scalars c and d, $\mathbf{u}$ is orthogonal to $c\mathbf{v} + d\mathbf{w}$.

9. (a) $\mathbf{u} = -120\mathbf{j}$

$\mathbf{v} = 40\mathbf{i}$

(b) $\mathbf{s} = \mathbf{u} + \mathbf{v} = 40\mathbf{i} - 120\mathbf{j}$

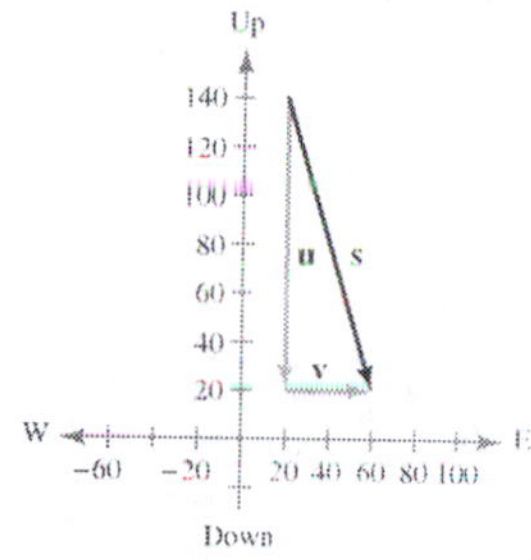

(c) $\|\mathbf{s}\| = \sqrt{40^2 + (-120)^2} = \sqrt{16{,}000} = 40\sqrt{10}$

≈ 126.5 miles per hour

This represents the actual rate of the skydiver's fall.

(d) $\tan\theta = \frac{120}{40} \Rightarrow \theta = \tan^{-1} 3 \Rightarrow \theta \approx 71.565°$

(e)

$\mathbf{s} = 30\mathbf{i} - 120\mathbf{j}$

$$\begin{aligned}\|\mathbf{s}\| &= \sqrt{30^2 + (-120)^2} \\ &= \sqrt{15{,}300} \\ &\approx 123.7 \text{ miles per hour}\end{aligned}$$

Practice Test for Chapter 8

For Exercises 1 and 2, use the Law of Sines to find the remaining sides and angles of the triangle.

1. $A = 40°, B = 12°, b = 100$

2. $C = 150°, a = 5, c = 20$

3. Find the area of the triangle: $a = 3, b = 6, C = 130°$.

4. Determine the number of solutions to the triangle: $a = 10, b = 35, A = 22.5°$.

For Exercises 5 and 6, use the Law of Cosines to find the remaining sides and angles of the triangle.

5. $a = 49, b = 53, c = 38$

6. $C = 29°, a = 100, c = 300$

7. Use Heron's Formula to find the area of the triangle: $a = 4.1, b = 6.8, c = 5.5$.

8. A ship travels 40 miles due east, then adjusts its course 12° southward. After traveling 70 miles in that direction, how far is the ship from its point of departure?

9. $\mathbf{w} = 4\mathbf{u} - 7\mathbf{v}$ where $\mathbf{u} = 3\mathbf{i} + \mathbf{j}$ and $\mathbf{v} = -\mathbf{i} + 2\mathbf{j}$. Find $\mathbf{w}$.

10. Find a unit vector in the direction of $\mathbf{v} = 5\mathbf{i} - 3\mathbf{j}$.

11. Find the dot product and the angle between $\mathbf{u} = 6\mathbf{i} + 5\mathbf{j}$ and $\mathbf{v} = 2\mathbf{i} - 3\mathbf{j}$.

12. $\mathbf{v}$ is a vector of magnitude 4 making an angle of 30° with the positive x-axis. Find $\mathbf{v}$ in component form.

13. Find the projection of $\mathbf{u}$ onto $\mathbf{v}$ given $\mathbf{u} = \langle 3, -1 \rangle$ and $\mathbf{v} = \langle -2, 4 \rangle$.

14. Give the trigonometric form of $z = 5 - 5i$.

15. Give the standard form of $z = 6(\cos 225° + i \sin 225°)$.

16. Multiply $[7(\cos 23° + i \sin 23°)][4(\cos 7° + i \sin 7°)]$.

17. Divide $\dfrac{9\left(\cos \frac{5\pi}{4} + i \sin \frac{5\pi}{4}\right)}{3(\cos \pi + i \sin \pi)}$.

18. Find $(2 + 2i)^8$.

19. Find the cube roots of $8\left(\cos \frac{\pi}{3} + i \sin \frac{\pi}{3}\right)$.

20. Find all the solutions to $x^4 + i = 0$.

CHAPTER 9
Systems of Equations and Inequalities

CHAPTER 9
Systems of Equations and Inequalities

Section 9.1 Linear and Nonlinear Systems of Equations

1. solution

3. point of intersection

5. $\begin{cases} 2x - y = 4 \\ 8x + y = -9 \end{cases}$

   (a) $(0, -4)$

   $8(0) - 4 \neq -9$

   $(0, -4)$ *is not* a solution.

   (b) $(-2, 7)$

   $2(-2) - 7 \neq 4$

   $(-2, 7)$ *is not* a solution.

   (c) $\left(\frac{3}{2}, -1\right)$

   $8\left(\frac{3}{2}\right) - 1 \neq -9$

   $\left(\frac{3}{2}, -1\right)$ *is not* a solution.

   (d) $\left(-\frac{1}{2}, -5\right)$

   $$2\left(-\frac{1}{2}\right) + 5 \stackrel{?}{=} 4$$
   $$-1 + 5 = 4$$
   $$8\left(-\frac{1}{2}\right) - 5 \stackrel{?}{=} -9$$
   $$-4 - 5 = -9$$

   $\left(-\frac{1}{2}, -5\right)$ *is* a solution.

7. $\begin{cases} 2x + y = 6 & \text{Equation 1} \\ -x + y = 0 & \text{Equation 2} \end{cases}$

   Solve for y in Equation 1: $y = 6 - 2x$

   Substitute for y in Equation 2: $-x + (6 - 2x) = 0$

   Solve for x: $-3x + 6 = 0 \Rightarrow x = 2$

   Back-substitute $x = 2$: $y = 6 - 2(2) = 2$

   Solution: $(2, 2)$

9. $\begin{cases} x - y = -4 & \text{Equation 1} \\ x^2 - y = -2 & \text{Equation 2} \end{cases}$

   Solve for y in Equation 1: $y = x + 4$

   Substitute for y in Equation 2: $x^2 - (x + 4) = -2$

   Solve for x: $x^2 - x - 2 = 0 \Rightarrow (x + 1)(x - 2) = 0 \Rightarrow x = -1, 2$

   Back-substitute $x = -1$: $y = -1 + 4 = 3$

   Back-substitute $x = 2$: $y = 2 + 4 = 6$

   Solutions: $(-1, 3), (2, 6)$

11. $\begin{cases} -\frac{1}{2}x + y = -\frac{5}{2} & \text{Equation 1} \\ x^2 + y^2 = 25 & \text{Equation 2} \end{cases}$

Solve for x in Equation 1: $-\frac{1}{2}x = -y - \frac{5}{2} \Rightarrow x = 2y + 5$

Substitute for x in Equation 2: $(2y + 5)^2 + y^2 = 25$

Solve for $4y^2 + 20y + 25 + y^2 = 25 \Rightarrow 5y^2 + 20y = 0 \Rightarrow 5y(y + 4) = 0 \Rightarrow y = 0, y = -4$

Back-substitute $y = 0$: $-\frac{1}{2}x + 0 = -\frac{5}{2} \Rightarrow x = 5$

Back-substitute $y = -4$: $-\frac{1}{2}x - 4 = -\frac{5}{2} \Rightarrow x = -3$

Solutions: $(-3, -4), (5, 0)$

13. $\begin{cases} x^2 + y = 0 & \text{Equation 1} \\ x^2 - 4x - y = 0 & \text{Equation 2} \end{cases}$

Solve for y in Equation 1: $y = -x^2$

Substitute for y in Equation 2: $x^2 - 4x - (-x^2) = 0$

Solve for x: $2x^2 - 4x = 0 \Rightarrow 2x(x - 2) = 0 \Rightarrow x = 0, 2$

Back-substitute $x = 0$: $y = -0^2 = 0$

Back-substitute $x = 2$: $y = -2^2 = -4$

Solutions: $(0, 0), (2, -4)$

15. $\begin{cases} x - y = 2 & \text{Equation 1} \\ 6x - 5y = 16 & \text{Equation 2} \end{cases}$

Solve for x in Equation 1: $x = y + 2$

Substitute for x in Equation 2: $6(y + 2) - 5y = 16 \Rightarrow 6y + 12 - 5y = 16 \Rightarrow y = 4$

Back-substitute $y = 4$: $x - 4 = 2 \Rightarrow x = 6$

Solution: $(6, 4)$

17. $\begin{cases} 2x - y + 2 = 0 & \text{Equation 1} \\ 4x + y - 5 = 0 & \text{Equation 2} \end{cases}$

Solve for y in Equation 1: $y = 2x + 2$

Substitute for y in Equation 2: $4x + (2x + 2) - 5 = 0$

Solve for x: $6x - 3 = 0 \Rightarrow x = \frac{1}{2}$

Back-substitute $x = \frac{1}{2}$: $y = 2x + 2 = 2\left(\frac{1}{2}\right) + 2 = 3$

Solution: $\left(\frac{1}{2}, 3\right)$

19. $\begin{cases} 1.5x + 0.8y = 2.3 & \text{Equation 1} \\ 0.3x - 0.2y = 0.1 & \text{Equation 2} \end{cases}$

Multiply the equations by 10.

$15x + 8y = 23$ Revised Equation 1

$3x - 2y = 1$ Revised Equation 2

Solve for y in revised Equation 2: $y = \frac{3}{2}x - \frac{1}{2}$

Substitute for y in revised Equation 1: $15x + 8\left(\frac{3}{2}x - \frac{1}{2}\right) = 23$

Solve for x: $15x + 12x - 4 = 23 \Rightarrow 27x = 27 \Rightarrow x = 1$

Back-substitute $x = 1$: $y = \frac{3}{2}(1) - \frac{1}{2} = 1$

Solution: $(1, 1)$

21. $\begin{cases} \frac{1}{5}x + \frac{1}{2}y = 8 & \text{Equation 1} \\ x + y = 20 & \text{Equation 2} \end{cases}$

Solve for x in Equation 2: $x = 20 - y$

Substitute for x in Equation 1: $\frac{1}{5}(20 - y) + \frac{1}{2}y = 8$

Solve for y: $4 + \frac{3}{10}y = 8 \Rightarrow y = \frac{40}{3}$

Back-substitute $y = \frac{40}{3}$: $x = 20 - y = 20 - \frac{40}{3} = \frac{20}{3}$

Solution: $\left(\frac{20}{3}, \frac{40}{3}\right)$

23. $\begin{cases} 6x + 5y = -3 & \text{Equation 1} \\ -x - \frac{5}{6}y = -7 & \text{Equation 2} \end{cases}$

Solve for x in Equation 2: $x = 7 - \frac{5}{6}y$

Substitute for x in Equation 1: $6\left(7 - \frac{5}{6}y\right) + 5y = -3$

Solve for y: $42 - 5y + 5y = -3 \Rightarrow 42 = -3$ (False)

No solution

25. $\begin{cases} x + y = 12{,}000 \\ 0.03x + 0.05y = 500 \end{cases}$

Substitute $y = 12{,}000 - x$ into $0.03x + 0.05y = 500$.

$$\begin{aligned} 0.03x + 0.05(12{,}000 - x) &= 500 \\ 0.03x + 600 - 0.05x &= 500 \\ -0.02x &= -100 \\ x &= 5000 \\ y &= 12{,}000 - 5000 = 7000 \end{aligned}$$

So, \$5000 is invested at 3% and \$7000 is invested at 5%.

27. $\begin{cases} x + y = 12{,}000 \\ 0.028x + 0.038y = 396 \end{cases}$

Substitute $y = 12{,}000 - x$ into $0.028x + 0.038y = 396$.

$$\begin{aligned} 0.028x + 0.038(12{,}000 - x) &= 396 \\ 0.028x + 456 - 0.038x &= 396 \\ -0.01x &= -60 \\ x &= 6000 \\ y &= 12{,}000 - 6000 = 6000 \end{aligned}$$

So, \$6000 is invested at 2.8% and \$6000 is invested at 3.8%.

29. $\begin{cases} x^2 - y = 0 & \text{Equation 1} \\ 2x + y = 0 & \text{Equation 2} \end{cases}$

Solve for y in Equation 2: $y = -2x$

Substitute for y in Equation 1: $x^2 - (-2x) = 0$

Solve for x:

$x^2 + 2x = 0 \Rightarrow x(x + 2) = 0 \Rightarrow x = 0, -2$

Back-substitute $x = 0$: $y = -2(0) = 0$

Back-substitute $x = -2$: $y = -2(-2) = 4$

Solutions: $(0, 0), (-2, 4)$

31. $\begin{cases} x - y = -1 & \text{Equation 1} \\ x^2 - y = -4 & \text{Equation 2} \end{cases}$

Solve for y in Equation 1: $y = x + 1$

Substitute for y in Equation 2: $x^2 - (x + 1) = -4$

Solve for x: $x^2 - x - 1 = -4 \Rightarrow x^2 - x + 3 = 0$

The Quadratic Formula yields no real solutions.

33. $\begin{cases} -x + 2y = -2 \\ 3x + y = 20 \end{cases}$

Point of intersection:
$(6, 2)$

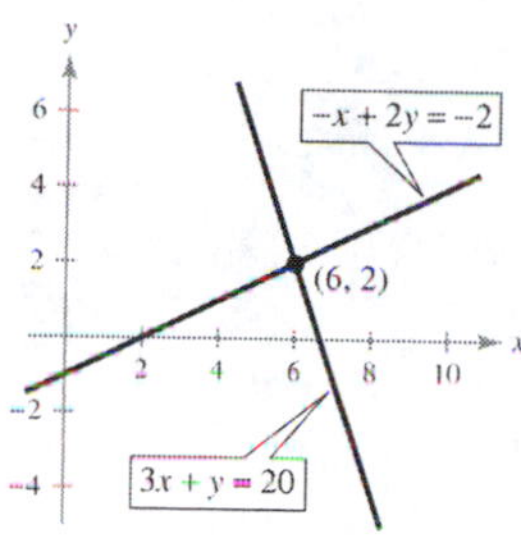

35. $\begin{cases} x - 3y = -3 \\ 5x + 3y = -6 \end{cases}$

Point of intersection:
$\left(-\frac{3}{2}, \frac{1}{2}\right)$

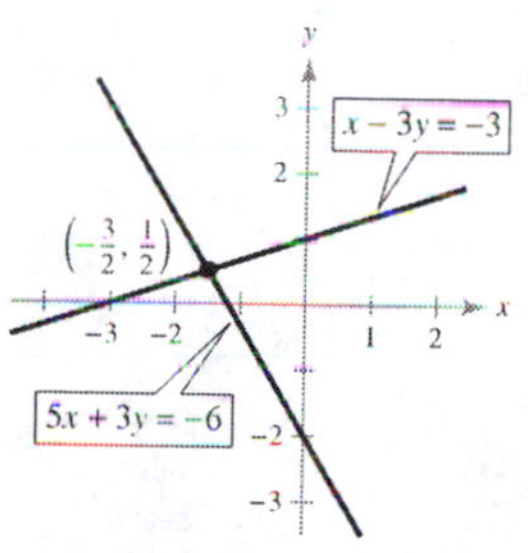

37. $\begin{cases} x + y = 4 \\ x^2 + y^2 - 4x = 0 \end{cases}$

Points of intersection:
$(2, 2), (4, 0)$

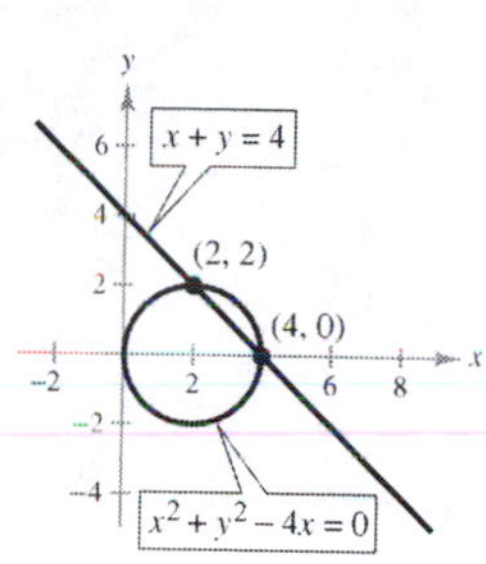

39. $\begin{cases} x - y + 3 = 0 \\ y = x^2 - 4x + 7 \end{cases}$

Points of intersection:
$(1, 4), (4, 7)$

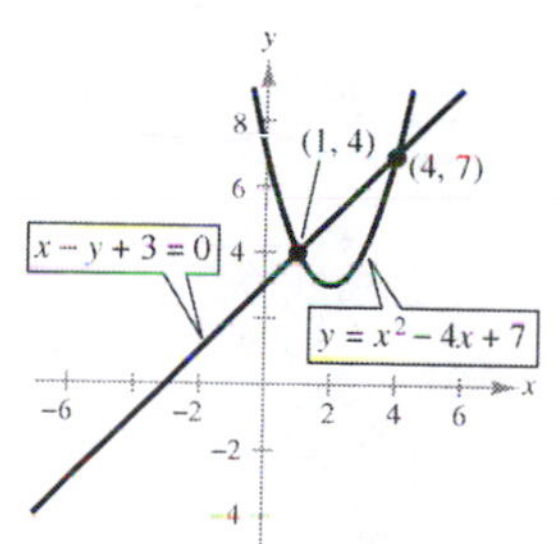

41. $\begin{cases} 3x - 2y = 0 \\ x^2 - y^2 = 4 \end{cases}$

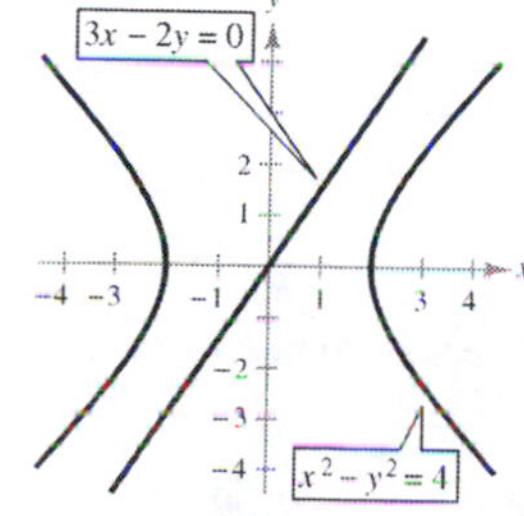

No points of intersection $\Rightarrow$ No solution

43. $\begin{cases} x^2 + y^2 = 25 \\ 3x^2 - 16y = 0 \end{cases}$

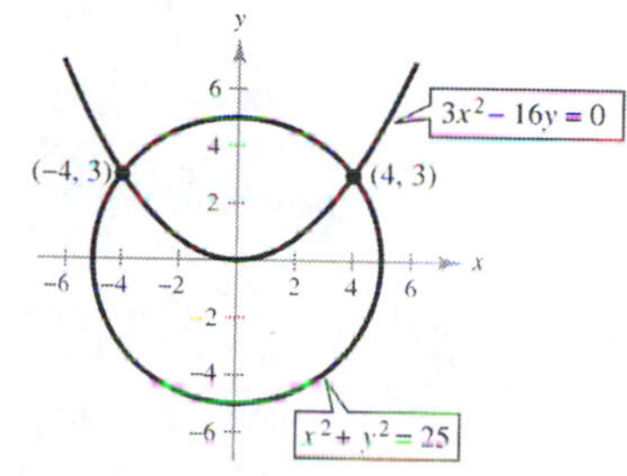

Points of intersection: $(-4, 3), (4, 3)$

45. $\begin{cases} y = e^x \\ x - y + 1 = 0 \Rightarrow y = x + 1 \end{cases}$

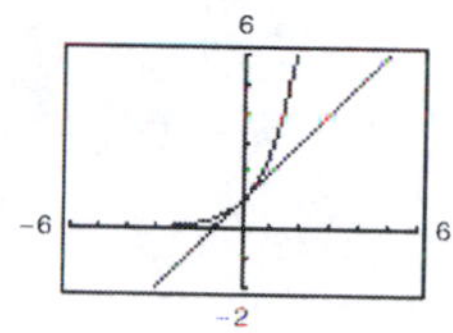

Point of intersection: $(0, 1)$

47. $\begin{cases} y = -2 + \ln(x - 1) \\ 3y + 2x = 9 \Rightarrow y = -\frac{2}{3}x + 3 \end{cases}$

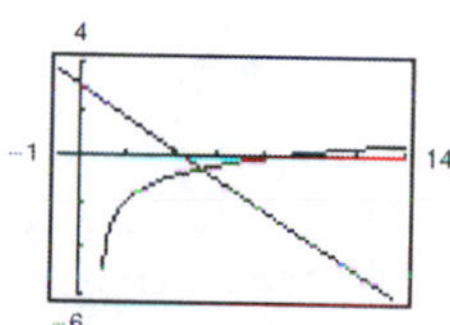

Point of intersection: $(5.31, -0.54)$

49. $\begin{cases} y = 2x & \text{Equation 1} \\ y = x^2 + 1 & \text{Equation 2} \end{cases}$

Substitute for y in Equation 2: $2x = x^2 + 1$

Solve for x: $x^2 - 2x + 1 = (x - 1)^2 = 0 \Rightarrow x = 1$

Back-substitute $x = 1$ in Equation 1: $y = 2x = 2$

Solution: $(1, 2)$

51. $\begin{cases} x - 2y = 4 & \text{Equation 1} \\ x^2 - y = 0 & \text{Equation 2} \end{cases}$

Solve for y in Equation 2: $y = x^2$

Substitute for y in Equation 1: $x - 2x^2 = 4$

Solve for x: $0 = 2x^2 - x + 4 \Rightarrow x = \dfrac{1 \pm \sqrt{1 - 4(2)(4)}}{2(2)} \Rightarrow x = \dfrac{1 \pm \sqrt{-31}}{4}$

The discriminant in the Quadratic Formula is negative.

No real solution

53. $\begin{cases} y - e^{-x} = 1 \Rightarrow y = e^{-x} + 1 \\ y - \ln x = 3 \Rightarrow y = \ln x + 3 \end{cases}$

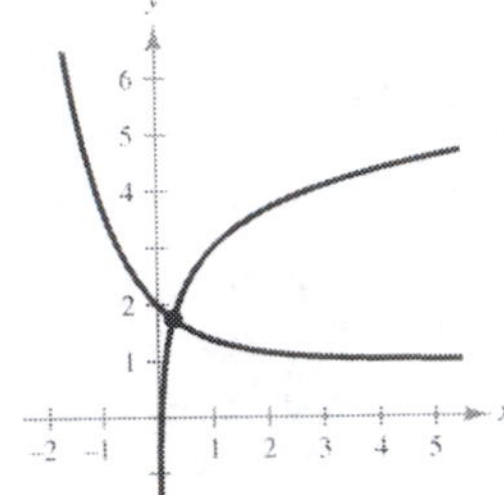

Point of intersection: approximately $(0.287), (1.751)$

55. $\begin{cases} xy - 1 = 0 & \text{Equation 1} \\ 2x - 4y + 7 = 0 & \text{Equation 2} \end{cases}$

Solve for y in Equation 1: $y = \dfrac{1}{x}$

Substitute for y in Equation 2: $2x - 4\left(\dfrac{1}{x}\right) + 7 = 0$

Solve for x: $2x^2 - 4 + 7x = 0 \Rightarrow (2x - 1)(x + 4) = 0 \Rightarrow x = \dfrac{1}{2}, -4$

Back-substitute $x = \dfrac{1}{2}$: $y = \dfrac{1}{1/2} = 2$

Back-substitute $x = -4$: $y = \dfrac{1}{-4} = -\dfrac{1}{4}$

Solutions: $\left(\dfrac{1}{2}, 2\right), \left(-4, -\dfrac{1}{4}\right)$

57. $C = 8650x + 250{,}000$, $R = 9950x$

$$R = C$$
$$9950x = 8650x + 250{,}000$$
$$1300x = 250{,}000$$
$$x \approx 192 \text{ units}$$

59. $C = 9.45x + 16{,}000$; $R = 55.95x$

(a)
$$R = C$$
$$55.95x = 9.45x + 16{,}000$$
$$46.5x = 16{,}000$$
$$x \approx 344$$

About 344 units must be sold to break even.

(b)
$$P = R - C$$
$$100{,}000 = 55.95x - (9.45x + 16{,}000)$$
$$100{,}000 = 46.5x - 16{,}000$$
$$116{,}000 = 46.5x$$
$$x \approx 2495$$

About 2495 units must be sold to earn a \$100,000 profit.

61. $\begin{cases} R = 360 - 24x & \text{Equation 1} \\ R = 24 + 18x & \text{Equation 2} \end{cases}$

(a) Substitute for R in Equation 2: $360 - 24x = 24 + 18x$

Solve for x: $336 = 42x \Rightarrow x = 8$ weeks

(b)

Weeks, x	1	2	3	4	5	6	7	8	9	10
$R = 360 - 24x$	336	312	288	264	240	216	192	168	144	120
$R = 24 + 18x$	42	60	78	96	114	132	150	168	186	204

The rentals are equal when $x = 8$ weeks.

63. $\begin{cases} V = (D - 4)^2, & 5 \le D \le 40 & \text{Doyle Log Rule} \\ V = 0.79D^2 - 2D - 4, & 5 \le D \le 40 & \text{Scribner Log Rule} \end{cases}$

(a)

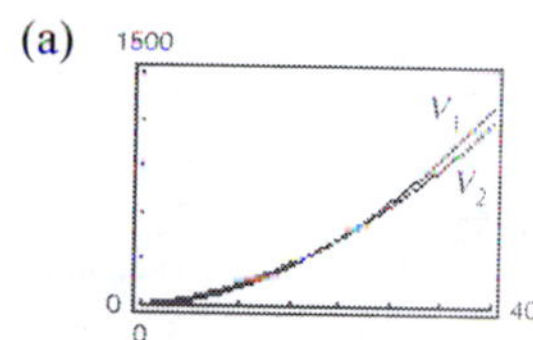

(b) The graphs intersect when $D \approx 24.7$ inches.

(c) For large logs, the Doyle Log Rule gives a greater volume for a given diameter.

65. $2l + 2w = 56 \Rightarrow l + w = 28$

$l = w + 4 \Rightarrow (w + 4) + w = 28$

$2w + 4 = 28$

$2w = 24$

$w = 12$ meters

$l = w + 4 = 12 + 4 = 16$ meters

Dimensions: 12 meters × 16 meters

67. $44 = 2l + 2w$

$22 = l + w \Rightarrow l = 22 - w$

$A = lw$

$120 = lw$

$120 = (22 - w)w$

$120 = 22w - w^2$

$w^2 - 22w + 120 = 0$

$(w - 10)(w - 12) = 0$

$w = 10, \quad w = 12$

When $w = 10, l = 22 - 10 = 12$.

When $w = 12, l = 22 - 12 = 10$.

Dimensions: 10 kilometers × 12 kilometers

69. False. To solve a system of equations by substitution, you can solve for either variable in one of the two equations and then back-substitute.

71. For a linear system, the result will be a contradictory equation such as $0 = N$, where N is a nonzero real number. For a nonlinear system, there may be an equation with imaginary solutions.

73. Answers will vary.

Section 9.2 Two-Variable Linear Systems

1. elimination

3. consistent; inconsistent

5. $\begin{cases} 2x + y = 5 & \text{Equation 1} \\ x - y = 1 & \text{Equation 2} \end{cases}$

Add to eliminate y: $\begin{array}{r} 2x + y = 5 \\ x - y = 1 \\ \hline 3x \quad = 6 \Rightarrow x = 2 \end{array}$

Substitute $x = 2$ in Equation 2: $2 - y = 1 \Rightarrow y = 1$

Solution: $(2, 1)$

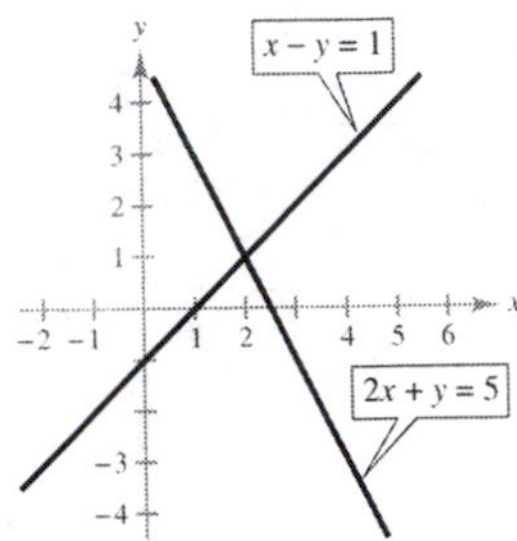

7. $\begin{cases} x + y = 0 & \text{Equation 1} \\ 3x + 2y = 1 & \text{Equation 2} \end{cases}$

Multiply Equation 1 by -2: $-2x - 2y = 0$

Add this to Equation 2 to eliminate y:

$\begin{array}{r} -2x - 2y = 0 \\ 3x + 2y = 1 \\ \hline x \quad = 1 \end{array}$

Substitute $x = 1$ in Equation 1: $1 + y = 0 \Rightarrow y = -1$

Solution: $(1, -1)$

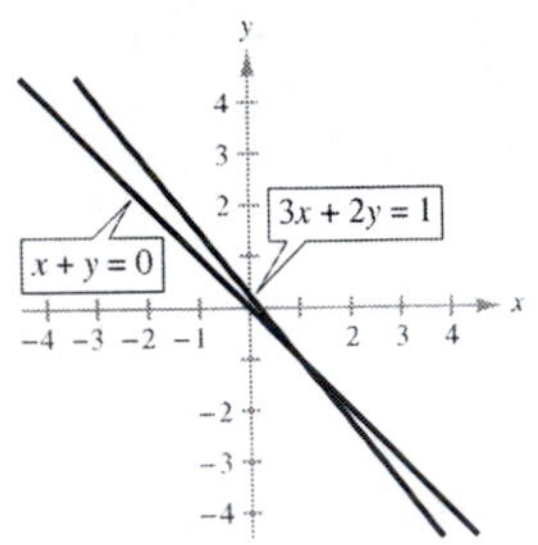

9. $\begin{cases} x - y = 2 & \text{Equation 1} \\ -2x + 2y = 5 & \text{Equation 2} \end{cases}$

Multiply Equation 1 by 2: $2x - 2y = 4$

Add this to Equation 2: $\begin{array}{r} 2x - 2y = 4 \\ -2x + 2y = 5 \\ \hline 0 = 9 \end{array}$

There are no solutions.

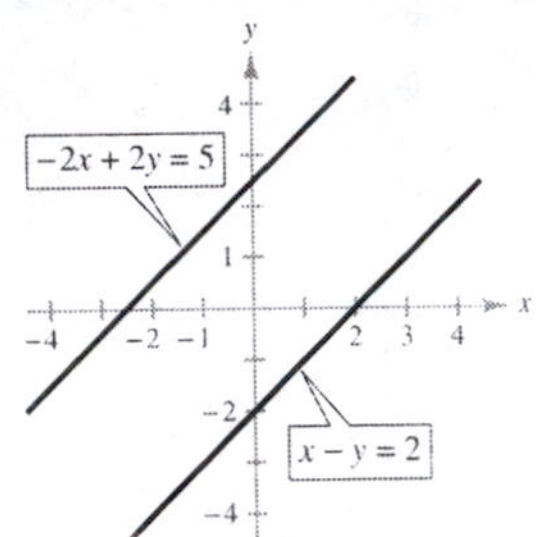

11. $\begin{cases} 3x - 2y = 5 & \text{Equation 1} \\ -6x + 4y = -10 & \text{Equation 2} \end{cases}$

Multiply Equation 1 by 2: $6x - 4y = 10$

Add this to Equation 2: $\begin{array}{r} 6x - 4y = 10 \\ -6x + 4y = -10 \\ \hline 0 = 0 \end{array}$

The equations are dependent. There are infinitely many solutions.

Let $x = a$, then $y = \dfrac{3a - 5}{2} = \dfrac{3}{2}a - \dfrac{5}{2}$.

Solution: $\left(a, \dfrac{3}{2}a - \dfrac{5}{2}\right)$, where a is any real number.

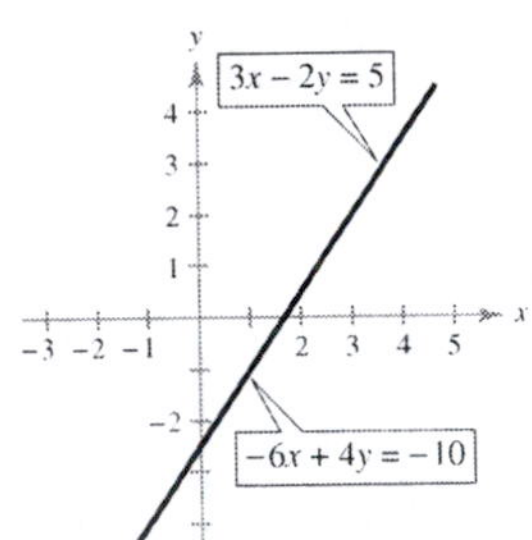

13. $\begin{cases} x + 2y = 6 & \text{Equation 1} \\ x - 2y = 2 & \text{Equation 2} \end{cases}$

Add the equations to eliminate y:

$$\begin{array}{rl} x + 2y &= 6 \\ x - 2y &= 2 \\ \hline 2x \quad &= 8 \Rightarrow x = 4 \end{array}$$

Substitute $x = 4$ into Equation 1:

$4 + 2y = 6 \Rightarrow y = 1$

Solution: $(4, 1)$

15. $\begin{cases} 5x + 3y = 6 & \text{Equation 1} \\ 3x - y = 5 & \text{Equation 2} \end{cases}$

Multiply Equation 2 by 3: $9x - 3y = 15$

Add this to Equation 1 to eliminate y:

$$\begin{array}{rl} 5x + 3y &= 6 \\ 9x - 3y &= 15 \\ \hline 14x \quad &= 21 \Rightarrow x = \frac{3}{2} \end{array}$$

Substitute $x = \frac{3}{2}$ into Equation 1:

$5\left(\frac{3}{2}\right) + 3y = 6 \Rightarrow y = -\frac{1}{2}$

Solution: $\left(\frac{3}{2}, -\frac{1}{2}\right)$

17. $\begin{cases} 3x + 2y = 10 & \text{Equation 1} \\ 2x + 5y = 3 & \text{Equation 2} \end{cases}$

Multiply Equation 1 by 2 and Equation 2 by -3:

$\begin{cases} 6x + 4y = 20 \\ -6x - 15y = -9 \end{cases}$

Add to eliminate x: $-11y = 11 \Rightarrow y = -1$

Substitute $y = -1$ in Equation 1:

$3x - 2 = 10 \Rightarrow x = 4$

Solution: $(4, -1)$

19. $\begin{cases} 5u + 6v = 24 & \text{Equation 1} \\ 3u + 5v = 18 & \text{Equation 2} \end{cases}$

Multiply Equation 1 by 5 and Equation 2 by -6:

$\begin{cases} 25u + 30v = 120 \\ -18u - 30v = -108 \end{cases}$

Add to eliminate v: $7u = 12 \Rightarrow u = \frac{12}{7}$

Substitute $u = \frac{12}{7}$ in Equation 1:

$5\left(\frac{12}{7}\right) + 6v = 24 \Rightarrow 6v = \frac{108}{7} \Rightarrow v = \frac{18}{7}$

Solution: $\left(\frac{12}{7}, \frac{18}{7}\right)$

21. $\begin{cases} \frac{9}{5}x + \frac{6}{5}y = 4 & \text{Equation 1} \\ 9x + 6y = 3 & \text{Equation 2} \end{cases}$

Multiply Equation 1 by 10 and Equation 2 by -2:

$\begin{cases} 18x + 12y = 40 \\ -18x - 12y = -6 \end{cases}$

Add these two together: $0 = 34$

No solution

23. $\begin{cases} -5x + 6y = -3 & \text{Equation 1} \\ 20x - 24y = 12 & \text{Equation 2} \end{cases}$

Multiply Equation 1 by 4:

$\begin{cases} -20x + 24y = -12 \\ 20x - 24y = 12 \end{cases}$

Add these two together: $0 = 0$

The equations are dependent. There are infinitely many solutions.

Let $x = a$, then

$$-5a + 6y = -3 \Rightarrow y = \frac{5a - 3}{6} = \frac{5}{6}a - \frac{1}{2}.$$

Solution: $\left(a, \frac{5}{6}a - \frac{1}{2}\right)$, where a is any real number

25. $\begin{cases} 0.2x - 0.5y = -27.8 & \text{Equation 1} \\ 0.3x + 0.4y = 68.7 & \text{Equation 2} \end{cases}$

Multiply Equation 1 by 4 and Equation 2 by 5:

$\begin{cases} 0.8x - 2y = -111.2 \\ 1.5x + 2y = 343.5 \end{cases}$

Add these to eliminate y:

$$\begin{array}{rl} 0.8x - 2y &= -111.2 \\ 1.5x + 2y &= 343.5 \\ \hline 2.3x \quad &= 232.3 \\ x &= 101 \end{array}$$

Substitute $x = 101$ in Equation 1:

$0.2(101) - 0.5y = -27.8 \Rightarrow y = 96$

Solution: $(101, 96)$

27. $\begin{cases} 4b + 3m = 3 & \text{Equation 1} \\ 3b + 11m = 13 & \text{Equation 2} \end{cases}$

Multiply Equation 1 by 3 and Equation 2 by -4:

$\begin{cases} 12b + 9m = 9 \\ -12b - 44m = -52 \end{cases}$

Add to eliminate b: $-35m = -43 \Rightarrow m = \frac{43}{35}$

Substitute $m = \frac{43}{35}$ in Equation 1:

$4b + 3\left(\frac{43}{35}\right) = 3 \Rightarrow b = -\frac{6}{35}$

Solution: $\left(-\frac{6}{35}, \frac{43}{35}\right)$

29. $\begin{cases} \dfrac{x+3}{4} + \dfrac{y-1}{3} = 1 & \text{Equation 1} \\ 2x - y = 12 & \text{Equation 2} \end{cases}$

Multiply Equation 1 by 12 and Equation 2 by 4:

$\begin{cases} 3x + 4y = 7 \\ 8x - 4y = 48 \end{cases}$

Add to eliminate y: $11x = 55 \Rightarrow x = 5$

Substitute $x = 5$ into Equation 2:

$2(5) - y = 12 \Rightarrow y = -2$

Solution: $(5, -2)$

31. $\begin{cases} 2x - 5y = 0 \\ x - y = 3 \end{cases}$

Multiply Equation 2 by -5:

$\begin{cases} 2x - 5y = 0 \\ -5x + 5y = -15 \end{cases}$

Add to eliminate y: $-3x = -15 \Rightarrow x = 5$

Matches graph (b).

Number of solutions: One

Consistent

32. $\begin{cases} 2x - 5y = 0 \\ 2x - 3y = -4 \end{cases}$

Multiply Equation 1 by -1:

$\begin{cases} -2x + 5y = 0 \\ 2x - 3y = -4 \end{cases}$

Add to eliminate x: $2y = -4 \Rightarrow y = -2$

Matches graph (c).

Number of solutions: One

Consistent

33. $\begin{cases} -7x + 6y = -4 \\ 14x - 12y = 8 \end{cases}$

Multiply Equation 1 by 2:

$\begin{cases} -14x + 12y = -8 \\ 14x - 12y = 8 \end{cases}$

Add this to Equation 2: $0 = 0$

The original equations are dependent.

Matches graph (a).

Number of solutions: Infinite

Consistent

34. $\begin{cases} 7x - 6y = -6 \\ -7x + 6y = -4 \end{cases}$

Add the equations: $0 = -10$

Inconsistent

Matches graph (d).

Number of solutions: None

Inconsistent

35. $\begin{cases} 3x - 5y = 7 & \text{Equation 1} \\ 2x + y = 9 & \text{Equation 2} \end{cases}$

Multiply Equation 2 by 5:

$10x + 5y = 45$

Add this to Equation 1:

$13x = 52 \Rightarrow x = 4$

Back-substitute $x = 4$ into Equation 2:

$2(4) + y = 9 \Rightarrow y = 1$

Solution: $(4, 1)$

37. $\begin{cases} y = 2x - 5 & \text{Equation 1} \\ y = 5x - 11 & \text{Equation 2} \end{cases}$

Because both equations are solved for y, set them equal to one another and solve for x.

$2x - 5 = 5x - 11$

$6 = 3x$

$2 = x$

Back-substitute $x = 2$ into Equation 1:

$y = 2(2) - 5 = -1$

Solution: $(2, -1)$

39. $\begin{cases} x - 5y = 21 & \text{Equation 1} \\ 6x + 5y = 21 & \text{Equation 2} \end{cases}$

Add the equations: $7x = 42 \Rightarrow x = 6$

Back-substitute $x = 6$ into Equation 1:

$6 - 5y = 21 \Rightarrow -5y = 15 \Rightarrow y = -3$

Solution: $(6, -3)$

41. Let r_1 = the air speed of the plane and r_2 = the wind air speed.

$$\begin{aligned} 3.6(r_1 - r_2) &= 1800 & \text{Equation 1} &\Rightarrow & r_1 - r_2 &= 500 \\ 3(r_1 + r_2) &= 1800 & \text{Equation 2} &\Rightarrow & r_1 + r_2 &= 600 \\ &&&& 2r_1 &= 1100 \quad \text{Add the equations.} \\ &&&& r_1 &= 550 \\ &&&& 550 + r_2 &= 600 \\ &&&& r_2 &= 50 \end{aligned}$$

The air speed of the plane is 550 miles per hour and the speed of the wind is 50 miles per hour.

43. Let x = the number of calories in a cheeseburger.

Let y = the number of calories in a small order of french fries.

$\begin{cases} 2x + y = 830 & \text{Equation 1} \\ 3x + 2y = 1360 & \text{Equation 2} \end{cases}$

Multiply Equation 1 by -2: $-4x - 2y = -1660$

Add this to Equation 2 to eliminate y:

$$\begin{aligned} -4x - 2y &= -1660 \\ 3x + 2y &= 1360 \\ -x &= -300 \\ x &= 300 \text{ calories} \end{aligned}$$

Back-substitute $x = 300$ into Equation 2:

$$\begin{aligned} 3(300) + 2y &= 1360 \\ 2y &= 460 \\ y &= 230 \text{ calories} \end{aligned}$$

The cheeseburger contains 300 calories and the fries contain 230 calories.

45.
$$\begin{aligned} 500 - 0.4x &= 380 + 0.1x \\ 120 &= 0.5x \\ x &= 240 \text{ units} \\ p &= \$404 \end{aligned}$$

Equilibrium point: $(240, 404)$

47.
$$\begin{aligned} 140 - 0.00002x &= 80 + 0.00001x \\ 60 &= 0.00003x \\ x &= 2{,}000{,}000 \text{ units} \\ p &= \$100.00 \end{aligned}$$

Equilibrium point: $(2{,}000{,}000, 100)$

49. (a) Let x = the number of liters at 25%.

Let y = the number of liters at 50%.

$$\begin{cases} 0.25x + 0.50y = 12 \\ x + y = 30 \end{cases}$$

(b)

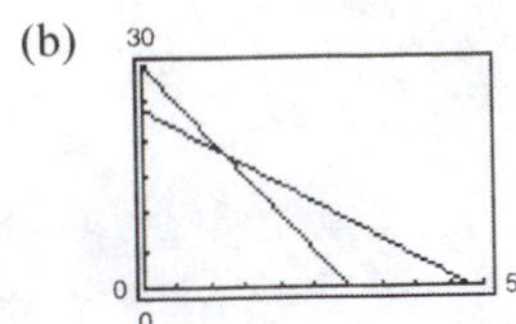

As the amount of 25% solution increases, the amount of 50% solution decreases.

(c) $\begin{cases} 0.25x + 0.50y = 12 & \text{Equation 1} \\ x + y = 30 & \text{Equation 2} \end{cases}$

Solve Equation 2 for y: $y = 30 - x$

Substitute this into Equation 1 to eliminate y:

$$0.25x + 0.50(30 - x) = 12$$

$$0.25x + 15 - 0.50x = 12$$

$$-0.25x = -3$$

$$x = 12 \text{ liters}$$

Back-substitute $x = 12$ into Equation 2:

$12 + y = 30 \Rightarrow y = 18$ liters

The final mixture should contain 12 liters of the 25% solution and 18 liters of the 50% solution.

51. Let x = the amount of money invested at 3.5%.

Let y = the amount of money invested at 5%.

$$\begin{cases} x + y = 24{,}000 & \text{Equation 1} \\ 0.035x + 0.05y = 930 & \text{Equation 2} \end{cases}$$

Solve Equation 1 for x: $x = 24{,}000 - y$

Substitute this into Equation 2 to eliminate x:

$$0.035(24{,}000 - y) + 0.05y = 930$$

$$840 + 0.015y = 930$$

$$y = \$6000$$

Back-substitute $y = 6000$ into Equation 1:

$$x + 6000 = 24{,}000$$

$$x = \$18{,}000$$

\$18,000 should be invested in the 3.5% bond.

53. (a)

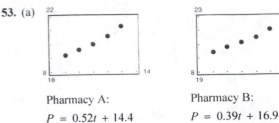

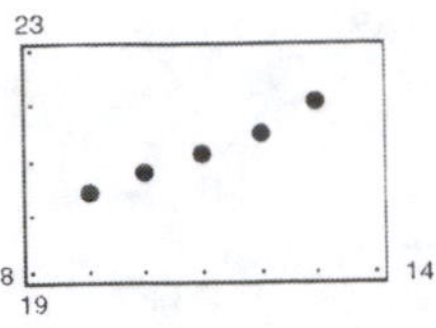

Pharmacy A: $P = 0.52t + 14.4$

Pharmacy B: $P = 0.39t + 16.9$

(b) Yes. You can determine when by solving the system of equations as follows.

$$\begin{cases} P = 0.52t + 14.4 \\ P = 0.39t + 16.9 \end{cases}$$

$$0.52t + 14.4 = 0.39t + 16.9$$

$$0.13t = 2.5$$

$$t \approx 19.2$$

So, the number of prescriptions filled at pharmacy A will exceed the number of prescriptions filled at pharmacy B during the year 2019.

55. $\begin{cases} 5b + 10a = 20.2 \Rightarrow & b + 2a = 4.04 \\ 10b + 30a = 50.1 \Rightarrow & -b - 3a = -5.01 \end{cases}$

$$-a = -0.97$$

$$a = 0.97$$

$$b + 2a = 4.04$$

$$b + 2(0.97) = 4.04$$

$$b = 2.1$$

Least squares regression line: $y = 0.97x + 2.1$

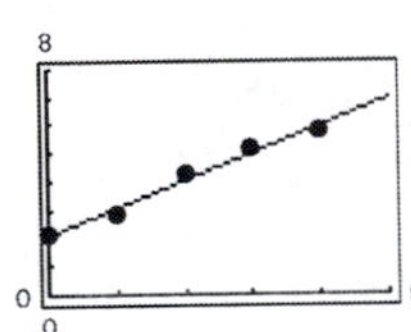

57. (a) $\begin{cases} 4b + 7.0a = 174 \Rightarrow & 28b + 49.0a = 1218 \\ 7b + 13.5a = 322 \Rightarrow & -28b - 54.0a = -1288 \end{cases}$

$$-5a = -70$$

$$a = 14$$

$$4b + 7.0a = 174$$

$$4b + 7.0(14) = 174$$

$$4b = 76$$

$$b = 19$$

Least squares regression line:

$y = 14x + 19$

(b) Substitute $x = 1.6$ into $y = 14x + 19$.

$$y = 14(1.6) + 19 = 41.4$$

The wheat yield is about 41.4 bushels per acre.

59. False. Two lines that coincide have infinitely many points of intersection.

61. $\begin{cases} 4x - 8y = -3 & \text{Equation 1} \\ 2x + ky = 16 & \text{Equation 2} \end{cases}$

Multiply Equation 2 by –2: $-4x - 2ky = -32$

Add this to Equation 1:
$$\begin{aligned} 4x - 8y &= -3 \\ -4x - 2ky &= -32 \\ \hline -8y - 2ky &= -35 \end{aligned}$$

The system is inconsistent if $-8y - 2ky = 0$.

This occurs when $k = -4$.

63. No, it is not possible for a consistent system of linear equations to have exactly two solutions. Either the lines will intersect once or they will coincide and then the system would have infinite solutions.

65. The method of elimination is much easier.

67. $\begin{cases} 100y - x = 200 & \text{Equation 1} \\ 99y - x = -198 & \text{Equation 2} \end{cases}$

Subtract Equation 2 from Equation 1 to eliminate x:
$$\begin{aligned} 100y - x &= 200 \\ -99y + x &= 198 \\ \hline y &= 398 \end{aligned}$$

Substitute $y = 398$ into Equation 1:

$100(398) - x = 200 \Rightarrow x = 39{,}600$

Solution: $(39, 600, 398)$

The lines are not parallel. The scale on the axes must be changed to see the point of intersection.

69. $\begin{cases} u \sin x + v \cos x = 0 & \text{Equation 1} \\ u \cos x - v \sin x = \sec x & \text{Equation 2} \end{cases}$

Multiply Equation 1 by $\cos x$ and multiply Equation 2 by $-\sin x$. Then add the equations to eliminate u.

$$\begin{aligned} u \sin x \cos x + v \cos^2 x &= 0 \\ -u \sin x \cos x + v \sin^2 x &= -\sin x \sec x \\ \hline v(\sin^2 x + \cos^2 x) &= -\sin x \sec x \\ v = -\sin x \sec x &= -\sin x\left(\frac{1}{\cos x}\right) = -\tan x \end{aligned}$$

Back substitute v into Equation 1

$$\begin{aligned} u \sin x + (-\tan x) \cos x &= 0 \\ u \sin x - \left(\frac{\sin x}{\cos x}\right) \cos x &= 0 \\ u \sin x - \sin x &= 0 \\ u \sin x &= \sin x \\ u &= 1 \end{aligned}$$

The solution of this system is: $u = 1,\ v = -\tan x$.

Section 9.3 Multivariable Linear Systems

1. row-echelon

3. Gaussian

5. nonsquare

7. $\begin{cases} 6x - y + z = -1 \\ 4x - 3z = -19 \\ 2y + 5z = 25 \end{cases}$

(a) $(2, 0, -2)$

$4(2) - 3(-2) \neq -19$

$(2, 0, -2)$ *is not* a solution.

(b) $(-3, 0, 5)$

$6(-3) - 0 + 5 \neq -1$

$(-3, 0, 5)$ *is not* a solution

(c) $(0, -1, 4)$

$4(0) - 3(4) \neq -19$

$(0, -1, 4)$ *is not* a solution.

(d) $(-1, 0, 5)$

$$6(-1) - 0 + 5 = -1$$
$$4(-1) - 3(5) = -19$$
$$2(0) + 5(5) = 25$$

$(-1, 0, 5)$ *is* a solution.

9. $\begin{cases} 4x + y - z = 0 \\ -8x - 6y + z = -\frac{7}{4} \\ 3x - y = -\frac{9}{4} \end{cases}$

(a) $4\left(\frac{1}{2}\right) + \left(-\frac{3}{4}\right) - \left(-\frac{7}{4}\right) \neq 0$

$\left(\frac{1}{2}, -\frac{3}{4}, -\frac{7}{4}\right)$ *is not* a solution.

(b) $4\left(-\frac{3}{2}\right) + \left(\frac{5}{4}\right) - \left(-\frac{5}{4}\right) \neq 0$

$\left(-\frac{3}{2}, \frac{5}{4}, -\frac{5}{4}\right)$ *is not* a solution.

(c) $4\left(-\frac{1}{2}\right) + \left(\frac{3}{4}\right) - \left(-\frac{5}{4}\right) = 0$

$$-8\left(-\frac{1}{2}\right) - 6\left(\frac{3}{4}\right) + \left(-\frac{5}{4}\right) = -\frac{7}{4}$$
$$3\left(-\frac{1}{2}\right) - \left(\frac{3}{4}\right) = -\frac{9}{4}$$

$\left(-\frac{1}{2}, \frac{3}{4}, -\frac{5}{4}\right)$ *is* a solution.

(d) $4\left(-\frac{1}{2}\right) + \left(\frac{1}{6}\right) - \left(-\frac{3}{4}\right) \neq 0$

$\left(-\frac{1}{2}, \frac{1}{6}, -\frac{3}{4}\right)$ *is not* a solution.

11. $\begin{cases} 2x - y + 5z = 24 & \text{Equation 1} \\ y + 2z = 6 & \text{Equation 2} \\ z = 8 & \text{Equation 3} \end{cases}$

Back-substitute $z = 8$ into Equation 2:

$$y + 2(8) = 6$$
$$y = -10$$

Back-substitute $y = -10$ and $z = 8$ into Equation 1:

$$2x + 10 + 5(8) = 24$$
$$2x = -26$$
$$x = -13$$

Solution: $(-13, -10, 8)$

13. $\begin{cases} 2x + y - 3z = 10 & \text{Equation 1} \\ y + z = 12 & \text{Equation 2} \\ z = 2 & \text{Equation 3} \end{cases}$

Back-substitute $z = 2$ into Equation 2:
$y + 2 = 12 \Rightarrow y = 10$

Back-substitute $y = 10$ and $z = 2$ into Equation 1:

$$2x + 10 - 3(2) = 10$$
$$2x + 4 = 10$$
$$2x = 6$$
$$x = 3$$

Solution: $(3, 10, 2)$

15. $\begin{cases} 4x - 2y + z = 8 & \text{Equation 1} \\ -y + z = 4 & \text{Equation 2} \\ z = 11 & \text{Equation 3} \end{cases}$

Back-substitute $z = 11$ into Equation 2:

$$-y + 11 = 4$$
$$y = 7$$

Back-substitute $y = 7$ and $z = 11$ into Equation 1:

$$4x - 2(7) + 11 = 8$$
$$4x = 11$$
$$x = \frac{11}{4}$$

Solution: $\left(\frac{11}{4}, 7, 11\right)$

17. $\begin{cases} x - 2y + 3z = 5 & \text{Equation 1} \\ -x + 3y - 5z = 4 & \text{Equation 2} \\ 2x \quad\quad - 3z = 0 & \text{Equation 3} \end{cases}$

Add Equation 1 to Equation 2:

$\begin{cases} x - 2y + 3z = 5 \\ \quad\quad y - 2z = 9 \\ 2x \quad\quad - 3z = 0 \end{cases}$

This is the first step in putting the system in row-echelon form.

19. $\begin{cases} x + y = 0 \\ -2x + 3y = 10 \end{cases}$

$\begin{cases} x + y = 0 \\ \quad 5y = 10 & 2\text{ Eq.1} + \text{Eq.2} \end{cases}$

$\begin{cases} x + y = 0 \\ \quad y = 2 & \frac{1}{5}\text{ Eq.2} \end{cases}$

$x + (2) = 0$

$x = -2$

Solution: $(-2, 2)$

21. $\begin{cases} x - 2y = -2 \\ 3x - y = 9 \end{cases}$

$\begin{cases} x - 2y = -2 \\ \quad 5y = 15 & (-3)\text{Eq.1} + \text{Eq.2} \end{cases}$

$\begin{cases} x - 2y = -2 \\ \quad y = 3 & \frac{1}{5}\text{ Eq.2} \end{cases}$

$x - 2(3) = -2$

$x = 4$

Solution: $(4, 3)$

23. $\begin{cases} 1.5x + 0.8y = -0.1 \\ -0.3x + 0.2y = -0.7 \end{cases}$

$\begin{cases} 15x + 8y = -1 & 10\text{ Eq.1} \\ -3x + 2y = -7 & 10\text{ Eq.2} \end{cases}$

$\begin{cases} x + \frac{8}{15}y = -\frac{1}{15} & \frac{1}{15}\text{ Eq.1} \\ -3x + 2y = -7 \end{cases}$

$\begin{cases} x + \frac{8}{15}y = -\frac{1}{15} \\ \frac{18}{5}y = -\frac{36}{5} & 3\text{ Eq.1} + \text{Eq.2} \end{cases}$

$\begin{cases} x + \frac{8}{15}y = -\frac{1}{15} \\ \quad y = -2 & \frac{5}{18}\text{ Eq.2} \end{cases}$

$x + \frac{8}{15}(-2) = -\frac{1}{15}$

$x = 1$

Solution: $(1, -2)$

25. $\begin{cases} x + y + z = 7 & \text{Equation 1} \\ 2x - y + z = 9 & \text{Equation 2} \\ 3x \quad\quad - z = 10 & \text{Equation 3} \end{cases}$

$\begin{cases} x + y + z = 7 \\ 3x \quad + 2z = 16 & \text{Eq.2} + \text{Eq.1} \\ 3x \quad\quad - z = 10 \end{cases}$

$\begin{cases} x + y + z = 7 \\ 3x \quad + 2z = 16 \\ 9x \quad\quad = 36 & \text{Eq.2} + 2\text{Eq.3} \end{cases}$

$\begin{cases} x + y + z = 7 \\ 3x \quad + 2z = 16 \\ x \quad\quad = 4 & \frac{1}{4}\text{ Eq.3} \end{cases}$

$3(4) + 2z = 16$

$2z = 4$

$z = 2$

$4 + y + 2 = 7$

$y = 1$

Solution: $(4, 1, 2)$

27. $\begin{cases} 2x + 2z = 2 \\ 5x + 3y = 4 \\ 3y - 4z = 4 \end{cases}$ Equation 1, Equation 2, Equation 3

$\begin{cases} x + z = 1 \\ 5x + 3y = 4 \\ 3y - 4z = 4 \end{cases}$ $\frac{1}{2}$ Eq.1

$\begin{cases} x + z = 1 \\ 3y - 5z = -1 \\ 3y - 4z = 4 \end{cases}$ -5 Eq.1 + Eq.2

$\begin{cases} x + z = 1 \\ 3y - 5z = -1 \\ z = 5 \end{cases}$ −Eq.2 + Eq.3

$3y - 5(5) = -1 \Rightarrow y = 8$

$x + 5 = 1 \Rightarrow x = -4$

Solution: $(-4, 8, 5)$

29. $\begin{cases} x - 2y + 2z = -9 \\ 2x + y - z = 7 \\ 3x - y + z = 5 \end{cases}$ Interchange equations.

$\begin{cases} x - 2y + 2z = -9 \\ 5y - 5z = 25 \\ 5y - 5z = 32 \end{cases}$ −2Eq.1 + Eq.2, −3Eq.1 + Eq.3

$\begin{cases} x - 2y + 2z = -9 \\ 5y - 5z = 25 \\ 0 = 7 \end{cases}$ −Eq.2 + Eq.3

Inconsistent, no solution

31. $\begin{cases} 3x - 5y + 5z = 1 \\ 5x - 2y + 3z = 0 \\ 7x - y + 3z = 0 \end{cases}$ Equation 1, Equation 2, Equation 3

$\begin{cases} 6x - 10y + 10z = 2 \\ 5x - 2y + 3z = 0 \\ 7x - y + 3z = 0 \end{cases}$ 2Eq.1

$\begin{cases} x - 8y + 7z = 2 \\ 5x - 2y + 3z = 0 \\ 7x - y + 3z = 0 \end{cases}$ −Eq.2 + Eq.1

$\begin{cases} x - 8y + 7z = 2 \\ 38y - 32z = -10 \\ 55y - 46z = -14 \end{cases}$ −5Eq.1 + Eq.2, −7Eq.1 + Eq.3

$\begin{cases} x - 8y + 7z = 2 \\ 2090y - 1760z = -550 \\ -2090y + 1748z = 532 \end{cases}$ 55Eq.2, −38Eq.3

$\begin{cases} x - 8y + 7z = 2 \\ 2090y - 1760z = -550 \\ -12z = -18 \end{cases}$ Eq.2 + Eq.3

$-12z = -18 \Rightarrow z = \frac{3}{2}$

$38y - 32\left(\frac{3}{2}\right) = -10 \Rightarrow y = 1$

$x - 8(1) + 7\left(\frac{3}{2}\right) = 2 \Rightarrow x = -\frac{1}{2}$

Solution: $\left(-\frac{1}{2}, 1, \frac{3}{2}\right)$

33. $\begin{cases} 2x + 3y = 0 \\ 4x + 3y - z = 0 \\ 8x + 3y + 3z = 0 \end{cases}$ Equation 1, Equation 2, Equation 3

$\begin{cases} 2x + 3y = 0 \\ -3y - z = 0 \\ -9y + 3z = 0 \end{cases}$ −2Eq.1 + Eq.2, −4Eq.1 + Eq.3

$\begin{cases} 2x + 3y = 0 \\ -3y - z = 0 \\ 6z = 0 \end{cases}$ −3Eq.2 + Eq.3

$6z = 0 \Rightarrow z = 0$

$-3y - 0 = 0 \Rightarrow y = 0$

$2x + 3(0) = 0 \Rightarrow x = 0$

Solution: $(0, 0, 0)$

35. $\begin{cases} x + 4z = 1 & \text{Equation 1} \\ x + y + 10z = 10 & \text{Equation 2} \\ 2x - y + 2z = -5 & \text{Equation 3} \end{cases}$

$\begin{cases} x + 4z = 1 & \\ y + 6z = 9 & -\text{Eq.1} + \text{Eq.2} \\ -y - 6z = -7 & -2\text{Eq.1} + \text{Eq.3} \end{cases}$

$\begin{cases} x + 4z = 1 & \\ y + 6z = 9 & \\ 0 = 2 & \text{Eq.2} + \text{Eq.3} \end{cases}$

No solution, inconsistent

37. $\begin{cases} 3x - 3y + 6z = 6 & \text{Equation 1} \\ x + 2y - z = 5 & \text{Equation 2} \\ 5x - 8y + 13z = 7 & \text{Equation 3} \end{cases}$

$\begin{cases} x - y + 2z = 2 & \frac{1}{3}\text{Eq.1} \\ x + 2y - z = 5 & \\ 5x - 8y + 13z = 7 & \end{cases}$

$\begin{cases} x - y + 2z = 2 & \\ 3y - 3z = 3 & -\text{Eq.1} + \text{Eq.2} \\ -3y + 3z = -3 & -5\text{Eq.1} + \text{Eq.3} \end{cases}$

$\begin{cases} x - y + 2z = 2 & \\ y - z = 1 & \frac{1}{3}\text{Eq.2} \\ 0 = 0 & \text{Eq.2} + \text{Eq.3} \end{cases}$

$\begin{cases} x + z = 3 & \text{Eq.2} + \text{Eq.1} \\ y - z = 1 & \end{cases}$

Let $z = a$, then:

$y = a + 1$

$x = -a + 3$

Solution: $(-a + 3, a + 1, a)$

39. $\begin{cases} x - 2y + 5z = 2 \\ 4x - z = 0 \end{cases}$

Let $z = a$, then: $x = \frac{1}{4}a$.

$\frac{1}{4}a - 2y + 5a = 2$

$a - 8y + 20a = 8$

$-8y = -21a + 8$

$y = \frac{21}{8}a - 1$

Answer: $\left(\frac{1}{4}a, \frac{21}{8}a - 1, a\right)$

To avoid fractions, we could go back and let $z = 8a$, then $4x - 8a = 0 \Rightarrow x = 2a$.

$2a - 2y + 5(8a) = 2$

$-2y + 42a = 2$

$y = 21a - 1$

Solution: $(2a, 21a - 1, 8a)$

41. $\begin{cases} x + 2y - 7z = -4 & \text{Equation 1} \\ 2x + y + z = 13 & \text{Equation 2} \\ 3x + 9y - 36z = -33 & \text{Equation 3} \end{cases}$

$\begin{cases} x + 2y - 7z = -4 & \\ -3y + 15z = 21 & -2\text{Eq.1} + \text{Eq.2} \\ 3y - 15z = -21 & -3\text{Eq.1} + \text{Eq.3} \end{cases}$

$\begin{cases} x + 2y - 7z = -4 & \\ -3y + 15z = 21 & \\ 0 = 0 & \text{Eq.2} + \text{Eq.3} \end{cases}$

$\begin{cases} x + 2y - 7z = -4 & \\ y - 5z = -7 & -\frac{1}{3}\text{Eq.2} \end{cases}$

$\begin{cases} x + 3z = 10 & -2\text{Eq.2} + \text{Eq.1} \\ y - 5z = -7 & \end{cases}$

Let $z = a$, then:

$y = 5a - 7$

$x = -3a + 10$

Solution: $(-3a + 10, 5a - 7, a)$

43. $$\begin{cases} 2x - 3y + z = -2 & \text{Equation 1} \\ -4x + 9y = 7 & \text{Equation 2} \end{cases}$$

$$\begin{cases} 2x - 3y + z = -2 & \\ 3y + 2z = 3 & 2\text{Eq.1} + \text{Eq.2} \end{cases}$$

$$\begin{cases} 2x + 3z = 1 & \text{Eq.2} + \text{Eq.1} \\ 3y + 2z = 3 & \end{cases}$$

Let $z = a$, then:

$y = -\frac{2}{3}a + 1$

$x = -\frac{3}{2}a + \frac{1}{2}$

Solution: $\left(-\frac{3}{2}a + \frac{1}{2}, -\frac{2}{3}a + 1, a\right)$

45. $$\begin{cases} x + 3w = 4 & \text{Equation 1} \\ 2y - z - w = 0 & \text{Equation 2} \\ 3y - 2w = 1 & \text{Equation 3} \\ 2x - y + 4z = 5 & \text{Equation 4} \end{cases}$$

$$\begin{cases} x + 3w = 4 & \\ 2y - z - w = 0 & \\ 3y - 2w = 1 & \\ -y + 4z - 6w = -3 & -2\text{Eq.1} + \text{Eq.4} \end{cases}$$

$$\begin{cases} x + 3w = 4 & \\ y - 4z + 6w = 3 & -\text{Eq.4 and interchange} \\ 2y - z - w = 0 & \text{the equations.} \\ 3y - 2w = 1 & \end{cases}$$

$$\begin{cases} x + 3w = 4 & \\ y - 4z + 6w = 3 & \\ 7z - 13w = -6 & -\text{Eq.2} + \text{Eq.3} \\ 12z - 20w = -8 & -3\text{Eq.2} + \text{Eq.4} \end{cases}$$

$$\begin{cases} x + 3w = 4 & \\ y - 4z + 6w = 3 & \\ z - 3w = -2 & -\frac{1}{2}\text{Eq.4} + \text{Eq.3} \\ 12z - 20w = -8 & \end{cases}$$

$$\begin{cases} x + 3w = 4 & \\ y - 4z + 6w = 3 & \\ z - 3w = -2 & \\ 16w = 16 & -12\text{Eq.3} + \text{Eq.4} \end{cases}$$

$$\begin{aligned} 16w &= 16 \Rightarrow w = 1 \\ z - 3(1) &= -2 \Rightarrow z = 1 \\ y - 4(1) + 6(1) &= 3 \Rightarrow y = 1 \\ x + 3(1) &= 4 \Rightarrow x = 1 \end{aligned}$$

Solution: $(1, 1, 1, 1)$

47. $s = \frac{1}{2}at^2 + v_0t + s_0$

$(1, 128), (2, 80), (3, 0)$

$$\begin{aligned} 128 &= \tfrac{1}{2}a + v_0 + s_0 \Rightarrow a + 2v_0 + 2s_0 = 256 \\ 80 &= 2a + 2v_0 + s_0 \Rightarrow 2a + 2v_0 + s_0 = 80 \\ 0 &= \tfrac{9}{2}a + 3v_0 + s_0 \Rightarrow 9a + 6v_0 + 2s_0 = 0 \end{aligned}$$

Solving this system yields $a = -32$, $v_0 = 0$, $s_0 = 144$.

So, $s = \frac{1}{2}(-32)t^2 + (0)t + 144 = -16t^2 + 144$.

49. $y = ax^2 + bx + c$ passing through $(0, 0), (2, -2), (4, 0)$

$(0, 0)$: $0 = c$

$(2, -2)$: $-2 = 4a + 2b + c \Rightarrow -1 = 2a + b$

$(4, 0)$: $0 = 16a + 4b + c \Rightarrow 0 = 4a + b$

Solution: $a = \frac{1}{2}, b = -2, c = 0$

The equation of the parabola is $y = \frac{1}{2}x^2 - 2x$.

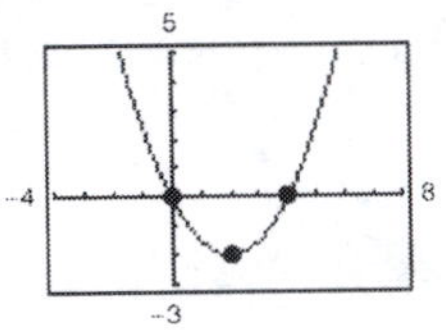

51. $y = ax^2 + bx + c$ passing through $(2, 0), (3, -1), (4, 0)$

$(2, 0)$: $0 = 4a + 2b + c$

$(3, -1)$: $-1 = 9a + 3b + c$

$(4, 0)$: $0 = 16a + 4b + c$

$$\begin{cases} 0 = 4a + 2b + c & \\ -1 = 5a + b & -\text{Eq.1} + \text{Eq.2} \\ 0 = 12a + 2b & -\text{Eq.1} + \text{Eq.3} \end{cases}$$

$$\begin{cases} 0 = 4a + 2b + c & \\ -1 = 5a + b & \\ 2 = 2a & -2\text{Eq.2} + \text{Eq.3} \end{cases}$$

Solution: $a = 1, b = -6, c = 8$

The equation of the parabola is $y = x^2 - 6x + 8$.

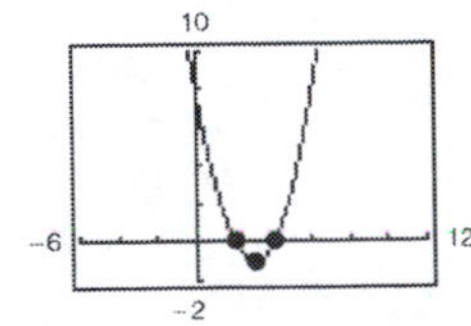

53. $y = ax^2 + bx + c$ passing through $\left(\frac{1}{2}, 1\right), (1, 3), (2, 13)$

$\left(\frac{1}{2}, 1\right): 1 = a\left(\frac{1}{2}\right)^2 + b\left(\frac{1}{2}\right) + c$

$(1, 3): 3 = a(1)^2 + b(1) + c$

$(2, 13): 13 = a(2)^2 + b(2) + c$

$$\begin{cases} a + 2b + 4c = 4 \\ a + b + c = 3 \\ 4a + 2b + c = 13 \end{cases}$$

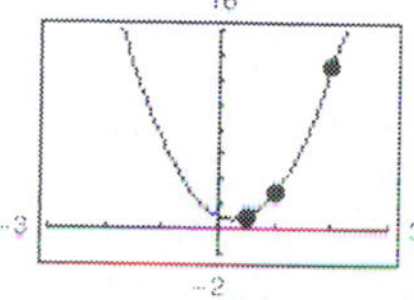

Solution: $a = 4, b = -2, c = 1$

The equation of the parabola is $y = 4x^2 - 2x + 1$.

55. $x^2 + y^2 + Dx + Ey + F = 0$ passing through $(0, 0), (5, 5), (10, 0)$

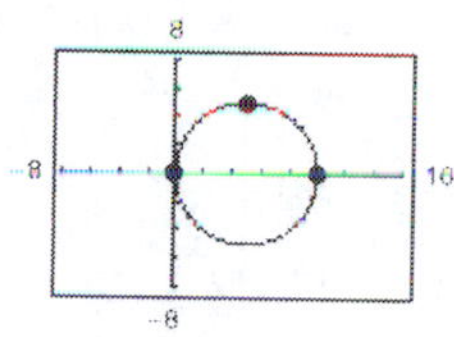

$(0, 0): 0^2 + 0^2 + D(0) + E(0) + F = 0 \Rightarrow F = 0$

$(5, 5): 5^2 + 5^2 + D(5) + E(5) + F = 0 \Rightarrow 5D + 5E + F = -50$

$(10, 0): 10^2 + 0^2 + D(10) + E(0) + F = 0 \Rightarrow 10D + F = -100$

Solution: $D = -10, E = 0, F = 0$

The equation of the circle is $x^2 + y^2 - 10x = 0$. To graph, complete the square first, then solve for y.

$$\begin{aligned} \left(x^2 - 10x + 25\right) + y^2 &= 25 \\ (x - 5)^2 + y^2 &= 25 \\ y^2 &= 25 - (x - 5)^2 \\ y &= \pm\sqrt{25 - (x - 5)^2} \end{aligned}$$

Let $y_1 = \sqrt{25 - (x - 5)^2}$ and $y_2 = -\sqrt{25 - (x - 5)^2}$.

57. $x^2 + y^2 + Dx + Ey + F = 0$ passing through $(-3, -1), (2, 4), (-6, 8)$

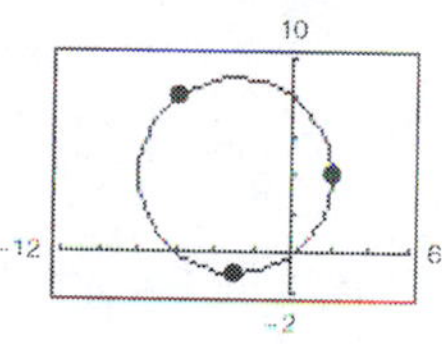

$(-3, -1): 10 - 3D - E + F = 0 \Rightarrow 10 = 3D + E - F$

$(2, 4): 20 + 2D + 4E + F = 0 \Rightarrow 20 = -2D - 4E - F$

$(-6, 8): 100 - 6D + 8E + F = 0 \Rightarrow 100 = 6D - 8E - F$

Solution: $D = 6, E = -8, F = 0$

The equation of the circle is $x^2 + y^2 + 6x - 8y = 0$. To graph, complete the squares first, then solve for y.

$$\begin{aligned} \left(x^2 + 6x + 9\right) + \left(y^2 - 8y + 16\right) &= 0 + 9 + 16 \\ (x + 3)^2 + (y - 4)^2 &= 25 \\ (y - 4)^2 &= 25 - (x + 3)^2 \\ y - 4 &= \pm\sqrt{25 - (x + 3)^2} \\ y &= 4 \pm \sqrt{25 - (x + 3)^2} \end{aligned}$$

Let $y_1 = 4 + \sqrt{25 - (x + 3)^2}$ and $y_2 = 4 - \sqrt{25 - (x + 3)^2}$.

59. Let x = number of touchdowns.

Let y = number of extra-point kicks.

Let z = number of field goals.

$$\begin{cases} x + y + z = 13 \\ 6x + y + 3z = 45 \\ x - y = 0 \\ x - 6z = 0 \end{cases}$$

$$\begin{cases} x + y + z = 13 & \\ -5y - 3z = -33 & -6\text{Eq.1} + \text{Eq.2} \\ -2y - z = -13 & -\text{Eq.1} + \text{Eq.3} \\ -y - 7z = -13 & -\text{Eq.1} + \text{Eq.4} \end{cases}$$

$$\begin{cases} x + y + z = 13 & \\ -y - 7z = -13 & \text{Interchange Eq.2 and Eq.4.} \\ -2y - z = -13 & \\ -5y - 3z = -33 & \end{cases}$$

$$\begin{cases} x + y + z = 13 & \\ y + 7z = 13 & -\text{Eq.2} \\ -2y - z = -13 & \\ -5y - 3z = -33 & \end{cases}$$

$$\begin{cases} x + y + z = 13 & \\ y + 7z = 13 & \\ 13z = 13 & 2\text{Eq.2} + \text{Eq.3} \\ 32z = 32 & 5\text{Eq.2} + \text{Eq.4} \end{cases}$$

$z = 1$

$y + 7(1) = 13 \Rightarrow y = 6$

$x + 6 + 1 = 13 \Rightarrow x = 6$

So, 6 touchdowns, 6 extra-point kicks, and 1 field goal were scored.

61. Let x = amount at 8%.

Let y = amount at 9%.

Let z = amount at 10%.

$$\begin{cases} x + y + z = 775{,}000 \\ 0.08x + 0.09y + 0.10z = 67{,}500 \\ x = 4z \end{cases}$$

$$\begin{cases} y + 5z = 775{,}000 \\ 0.09y + 0.42z = 67{,}500 \end{cases}$$

$z = 75{,}000$

$y = 775{,}000 - 5z = 400{,}000$

$x = 4z = 300{,}000$

\$300,000 was borrowed at 8%.

\$400,000 was borrowed at 9%.

\$75,000 was borrowed at 10%.

63.
$$\begin{cases} x + y + z = 180 \\ 2x + 7 + z = 180 \\ y + 2x - 7 = 180 \end{cases}$$

$$\begin{cases} x + y + z = 180 \\ 2x + z = 173 \\ 2x + y = 187 \end{cases}$$

$$\begin{cases} -x + y = 7 & -\text{Eq.2} + \text{Eq.1} \\ 2x + z = 173 & \\ 2x + y = 187 & \end{cases}$$

$$\begin{cases} -x + y = 7 & \\ 2x + z = 173 & \\ 3x = 180 & -\text{Eq.1} + \text{Eq.3} \end{cases}$$

$x = 60°$

$2(60) + z = 173 \Rightarrow z = 53°$

$-60 + y = 7 \Rightarrow y = 67°$

65. Let x = the longest side (hypotenuse).

Let y = leg.

Let z = shortest leg.

$$\begin{cases} x + y + z = 180 \\ x = 2z - 9 \\ y + z = 30 + x \end{cases}$$

$$\begin{cases} x + y + z = 180 \\ x - 2z = -9 \\ -x + y + z = 30 \end{cases}$$

$$\begin{cases} y + 3z = 189 & -\text{Eq.2} + \text{Eq.1} \\ x - 2z = -9 & \\ y - z = 21 & \text{Eq.2} + \text{Eq.3} \end{cases}$$

$$\begin{cases} 4z = 168 & -\text{Eq.3} + \text{Eq.1} \\ x - 2z = -9 & \\ y - z = 21 & \end{cases}$$

$z = 42$

$x - 2(42) = -9 \Rightarrow x = 75$

$y - 42 = 21 \Rightarrow y = 63$

So, the longest side measures 75 feet, the shortest side measures 42 feet, and the third side measures 63 feet.

67.
$$\begin{cases} I_1 - I_2 + I_3 = 0 & \text{Equation 1} \\ 3I_1 + 2I_2 = 7 & \text{Equation 2} \\ 2I_2 + 4I_3 = 8 & \text{Equation 3} \end{cases}$$

$$\begin{cases} I_1 - I_2 + I_3 = 0 & \\ 5I_2 - 3I_3 = 7 & (-3)\text{Eq.1} + \text{Eq.2} \\ 2I_2 + 4I_3 = 8 & \end{cases}$$

$$\begin{cases} I_1 - I_2 + I_3 = 0 & \\ 10I_2 - 6I_3 = 14 & 2\text{Eq.2} \\ 10I_2 + 20I_3 = 40 & 5\text{Eq.3} \end{cases}$$

$$\begin{cases} I_1 - I_2 + I_3 = 0 & \\ 10I_2 - 6I_3 = 14 & \\ 26I_3 = 26 & (-1)\text{Eq.2} + \text{Eq.3} \end{cases}$$

$$26I_3 = 26 \Rightarrow I_3 = 1$$

$$10I_2 - 6(1) = 14 \Rightarrow I_2 = 2$$

$$I_1 - 2 + 1 = 0 \Rightarrow I_1 = 1$$

Solution: $I_1 = 1, I_2 = 2, I_3 = 1$

69.
$$\begin{cases} 4c + 9b + 29a = 20 \\ 9c + 29b + 99a = 70 \\ 29c + 99b + 353a = 254 \end{cases}$$

$$\begin{cases} 9c + 29b + 99a = 70 & \text{Interchange equations.} \\ 4c + 9b + 29a = 20 & \\ 29c + 99b + 353a = 254 & \end{cases}$$

$$\begin{cases} c + 11b + 41a = 30 & -2\text{Eq.2} + \text{Eq.1} \\ -35b - 135a = -100 & -4\text{Eq.1} + \text{Eq.2} \\ -220b - 836a = -616 & -29\text{Eq.1} + \text{Eq.3} \end{cases}$$

$$\begin{cases} c + 11b + 41a = 30 & \\ 1540b + 5940a = 4400 & -44\text{Eq.2} \\ -1540b - 5852a = -4312 & 7\text{Eq.3} \end{cases}$$

$$\begin{cases} c + 11b + 41a = 30 & \\ 1540b + 5940a = 4400 & \\ 88a = 88 & \text{Eq.2} + \text{Eq.3} \end{cases}$$

$$88a = 88 \Rightarrow a = 1$$

$$1540b + 5940(1) = 4400 \Rightarrow b = -1$$

$$c + 11(-1) + 41(1) = 30 \Rightarrow c = 0$$

Least squares regression parabola: $y = x^2 - x$

71. (a) $\begin{cases} 3c + 120b + 5000a = 348 \\ 120c + 5000b + 216{,}000a = 15{,}250 \\ 5000c + 216{,}000b + 9{,}620{,}000a = 687{,}500 \end{cases}$

$\begin{cases} 3c + 120b + 5000a = 348 & \\ 200b + 16{,}000a = 1330 & (-40)\text{Eq.1} + \text{Eq.2} \\ 48{,}000b + 3{,}860{,}000a = 322{,}500 & (-5000)\text{Eq.1} + (3)\text{Eq.3} \end{cases}$

$\begin{cases} 3c + 120b + 5000a = 348 & \\ 200b + 16{,}000a = 1330 & \\ 20{,}000a = 3300 & (-240)\text{Eq.2} + \text{Eq.3} \end{cases}$

$20{,}000a = 3300 \Rightarrow a = 0.165$

$200b + 16{,}000(0.165) = 1330 \Rightarrow b = -6.55$

$3c + 120(-6.55) + 5000(0.165) = 348 \Rightarrow c = 103$

Least-squares regression parabola: $y = 0.165x^2 - 6.55x + 103$

(b)

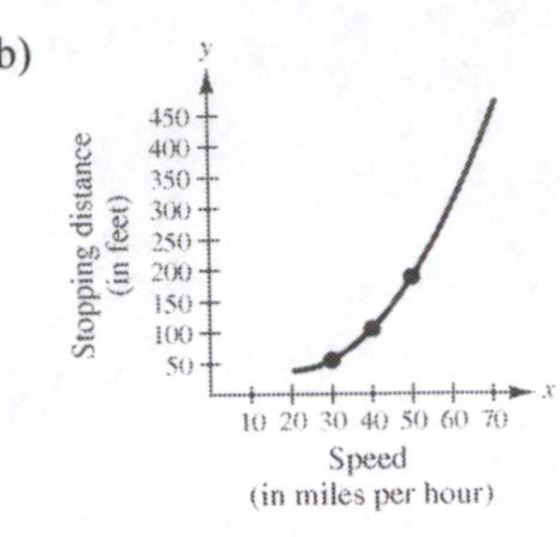

(c) When $x = 70$, $y = 453$ feet.

73. $\begin{cases} 2x - 2x\lambda = 0 \Rightarrow 2x(1 - \lambda) = 0 \Rightarrow \lambda = 1 \text{ or } x = 0 \\ -2y + \lambda = 0 \\ y - x^2 = 0 \end{cases}$

If $\lambda = 1$: $2y = \lambda \Rightarrow y = \dfrac{1}{2}$

$x^2 = y \Rightarrow x = \pm\sqrt{\dfrac{1}{2}} = \pm\dfrac{\sqrt{2}}{2}$

If $x = 0$: $x^2 = y \Rightarrow y = 0$

$2y = \lambda \Rightarrow \lambda = 0$

Solution: $x = \pm\dfrac{\sqrt{2}}{2}$ or $x = 0$

$y = \dfrac{1}{2}$ $\quad y = 0$

$\lambda = 1$ $\quad \lambda = 0$

75. False. Equation 2 does not have a leading coefficient of 1.

77. No, they are not equivalent. There are two arithmetic errors. The constant in the second equation should be -11 and the coefficient of z in the third equation should be 2.

79. *Sample answer:* There are an infinite number of linear systems that have $(3, -4, 2)$ as their solution. Two systems are:

$\begin{cases} 2x + y - z = 0 \\ 3x + 2y - \frac{1}{2}z = 0 \\ -x + 2y + z = -9 \end{cases}$

$\begin{cases} 4x + 3y = 0 \\ 4y + 8z = 0 \\ 2x - z = 4 \end{cases}$

81. *Sample answer:* There are an infinite number of linear systems that have $\left(-6, -\frac{1}{2}, -\frac{7}{4}\right)$ as their solution. Two systems are:

$\begin{cases} x - 12y - 4z = 7 \\ x + 2y - 4z = 0 \\ -2x + 18y + 4z = -4 \end{cases}$

$\begin{cases} -8y - 4z = 11 \\ x - 12y = 0 \\ x - 8y + 4z = -9 \end{cases}$

Section 9.4 Partial Fractions

1. partial fraction decomposition

3. partial fraction

5. $\frac{3x-1}{x(x-4)} = \frac{A}{x} + \frac{B}{x-4}$

Matches (b).

6. $\frac{3x-1}{x^2(x-4)} = \frac{A}{x} + \frac{B}{x^2} + \frac{C}{x-4}$

Matches (c).

7. $\frac{3x-1}{x(x^2+4)} = \frac{A}{x} + \frac{Bx+C}{x^2+4}$

Matches (d).

8. $\frac{3x-1}{x(x^2-4)} = \frac{3x-1}{x(x-2)(x+2)}$

$= \frac{A}{x} + \frac{B}{x-2} + \frac{C}{x+2}$

Matches (a).

9. $\frac{3}{x^2-2x} = \frac{3}{x(x-2)} = \frac{A}{x} + \frac{B}{x-2}$

11. $\frac{9}{x^3-7x^2} = \frac{9}{x^2(x-7)} = \frac{A}{x} + \frac{B}{x^2} + \frac{C}{x-7}$

13. $\frac{4x^2+3}{(x-5)^3} = \frac{A}{x-5} + \frac{B}{(x-5)^2} + \frac{C}{(x-5)^3}$

15. $\frac{x-1}{x(x^2+1)^2} = \frac{A}{x} + \frac{Bx+C}{x^2+1} + \frac{Dx+E}{(x^2+1)^2}$

17. $\frac{1}{x^2+x} = \frac{A}{x} + \frac{B}{x+1}$

$1 = A(x+1) + Bx$

Let $x = 0$: $1 = A$

Let $x = -1$: $1 = -B \Rightarrow B = -1$

$\frac{1}{x^2+x} = \frac{1}{x} - \frac{1}{x+1}$

19. $\frac{1}{2x^2+x} = \frac{A}{2x+1} + \frac{B}{x}$

$1 = Ax + B(2x+1)$

Let $x = -\frac{1}{2}$: $1 = -\frac{1}{2}A \Rightarrow A = -2$

Let $x = 0$: $1 = B$

$\frac{1}{2x^2+x} = \frac{1}{x} - \frac{2}{2x+1}$

21. $\frac{3}{x^2+x-2} = \frac{A}{x-1} + \frac{B}{x+2}$

$3 = A(x+2) + B(x-1)$

Let $x = 1$: $3 = 3A \Rightarrow A = 1$

Let $x = -2$: $3 = -3B \Rightarrow B = -1$

$\frac{3}{x^2+x-2} = \frac{1}{x-1} - \frac{1}{x+2}$

23. $\frac{1}{x^2-1} = \frac{A}{x+1} + \frac{B}{x-1}$

$1 = A(x-1) + B(x+1)$

Let $x = -1$: $1 = -2A \Rightarrow A = -\frac{1}{2}$

Let $x = 1$: $1 = 2B \Rightarrow B = \frac{1}{2}$

$\frac{1}{x^2-1} = \frac{1/2}{x-1} - \frac{1/2}{x+1} = \frac{1}{2}\left(\frac{1}{x-1} - \frac{1}{x+1}\right)$

25. $\frac{x^2+12x+12}{x^3-4x} = \frac{A}{x} + \frac{B}{x+2} + \frac{C}{x-2}$

$x^2 + 12x + 12 = A(x+2)(x-2) + Bx(x-2) + Cx(x+2)$

Let $x = 0$: $12 = -4A \Rightarrow A = -3$

Let $x = -2$: $-8 = 8B \Rightarrow B = -1$

Let $x = 2$: $40 = 8C \Rightarrow C = 5$

$\frac{x^2+12x+12}{x^3-4x} = -\frac{3}{x} - \frac{1}{x+2} + \frac{5}{x-2}$

27. $\dfrac{3x}{(x-3)^2} = \dfrac{A}{x-3} + \dfrac{B}{(x-3)^2}$

$3x = A(x-3) + B$

Let $x = 3$: $9 = B$

Let $x = 0$: $0 = -3A + B$

$0 = -3A + 9$

$3 = A$

$\dfrac{3x}{(x-3)^2} = \dfrac{3}{x-3} + \dfrac{9}{(x-3)^2}$

29. $\dfrac{4x^2 + 2x - 1}{x^2(x+1)} = \dfrac{A}{x} + \dfrac{B}{x^2} + \dfrac{C}{x+1}$

$4x^2 + 2x - 1 = Ax(x+1) + B(x+1) + Cx^2$

Let $x = 0$: $-1 = B$

Let $x = -1$: $1 = C$

Let $x = 1$: $5 = 2A + 2B + C$

$5 = 2A - 2 + 1$

$6 = 2A$

$3 = A$

$\dfrac{4x^2 + 2x - 1}{x^2(x+1)} = \dfrac{3}{x} - \dfrac{1}{x^2} + \dfrac{1}{x+1}$

31. $\dfrac{x^2 + 2x + 3}{x^3 + x} = \dfrac{A}{x} + \dfrac{Bx + C}{x^2 + 1}$

$x^2 + 2x + 3 = A(x^2 + 1) + (Bx + C)(x)$

$x^2 + 2x + 3 = x^2(A + B) + Cx + A$

Equating coefficients of like terms gives

$A + B = 1$, $C = 2$, and $A = 3$.

So, $A = 3$, $B = -2$, and $C = 2$.

$\dfrac{x^2 + 2x + 3}{x^3 + x} = \dfrac{3}{x} - \dfrac{2x - 2}{x^2 + 1}$

33. $\dfrac{x}{x^3 - x^2 - 2x + 2} = \dfrac{x}{(x-1)(x^2-2)} = \dfrac{A}{x-1} + \dfrac{Bx + C}{x^2 - 2}$

$x = A(x^2 - 2) + (Bx + C)(x - 1)$

$= Ax^2 - 2A + Bx^2 - Bx + Cx - C$

$= (A + B)x^2 + (C - B)x - (2A + C)$

Equating coefficients of like terms gives $0 = A + B, 1 = C - B$, and $0 = 2A + C$. So, $A = -1, B = 1$, and $C = 2$.

$\dfrac{x}{x^3 - x^2 - 2x + 2} = -\dfrac{1}{x-1} + \dfrac{x+2}{x^2 - 2}$

35. $\dfrac{2x^2 + x + 8}{(x^2 + 4)^2} = \dfrac{Ax + B}{x^2 + 4} + \dfrac{Cx + D}{(x^2 + 4)^2}$

$2x^2 + x + 8 = (Ax + B)(x^2 + 4) + Cx + D$

$2x^2 + x + 8 = Ax^3 + Bx^2 + (4A + C)x + (4B + D)$

Equating coefficients of like terms gives

$0 = A$

$2 = B$

$1 = 4A + C \Rightarrow C = 1$

$8 = 4B + D \Rightarrow D = 0$

$\dfrac{2x^2 + x + 8}{(x^2 + 4)^2} = \dfrac{2}{x^2 + 4} + \dfrac{x}{(x^2 + 4)^2}$

37. $\dfrac{x}{16x^4 - 1} = \dfrac{x}{(4x^2 - 1)(4x^2 + 1)} = \dfrac{x}{(2x + 1)(2x - 1)(4x^2 + 1)} = \dfrac{A}{2x + 1} + \dfrac{B}{2x - 1} + \dfrac{Cx + D}{4x^2 + 1}$

$$\begin{aligned} x &= A(2x - 1)(4x^2 + 1) + B(2x + 1)(4x^2 + 1) + (Cx + D)(2x + 1)(2x - 1) \\ &= A(8x^3 - 4x^2 + 2x - 1) + B(8x^3 + 4x^2 + 2x + 1) + (Cx + D)(4x^2 - 1) \\ &= 8Ax^3 - 4Ax^2 + 2Ax - A + 8Bx^3 + 4Bx^2 + 2Bx + B + 4Cx^3 + 4Dx^2 - Cx - D \\ &= (8A + 8B + 4C)x^3 + (-4A + 4B + 4D)x^2 + (2A + 2B - C)x + (-A + B - D) \end{aligned}$$

Equating coefficients of like terms gives $0 = 8A + 8B + 4C$, $0 = -4A + 4B + 4D$, $1 = 2A + 2B - C$, and $0 = -A + B - D$.

Using the first and third equations, $2A + 2B + C = 0$ and $2A + 2B - C = 1$; by subtraction, $2C = -1$, so $C = -\frac{1}{2}$.

Using the second and fourth equations, $-A + B + D = 0$ and $-A + B - D = 0$; by subtraction $2D = 0$, so $D = 0$.

Substituting $-\frac{1}{2}$ for C and 0 for D in the first and second equations, $8A + 8B = 2$ and $-4A + 4B = 0$, so $A = \frac{1}{8}$ and $B = \frac{1}{8}$.

$$\frac{x}{16x^4 - 1} = \frac{\frac{1}{8}}{2x + 1} + \frac{\frac{1}{8}}{2x - 1} + \frac{\left(-\frac{1}{2}\right)x}{4x^2 + 1} = \frac{1}{8(2x + 1)} + \frac{1}{8(2x - 1)} - \frac{x}{2(4x^2 + 1)} = \frac{1}{8}\left(\frac{1}{2x + 1} + \frac{1}{2x - 1} - \frac{4x}{4x^2 + 1}\right)$$

39. $\dfrac{x^2 + 5}{(x + 1)(x^2 - 2x + 3)} = \dfrac{A}{x + 1} + \dfrac{Bx + C}{x^2 - 2x + 3}$

$$\begin{aligned} x^2 + 5 &= A(x^2 - 2x + 3) + (Bx + C)(x + 1) \\ &= Ax^2 - 2Ax + 3A + Bx^2 + Bx + Cx + C \\ &= (A + B)x^2 + (-2A + B + C)x + (3A + C) \end{aligned}$$

Equating coefficients of like terms gives $1 = A + B$, $0 = -2A + B + C$, and $5 = 3A + C$.

Subtracting both sides of the second equation from the first gives $1 = 3A - C$; combining this with the third equation gives $A = 1$ and $C = 2$. Because $A + B = 1$, $B = 0$.

$$\frac{x^2 + 5}{(x + 1)(x^2 - 2x + 3)} = \frac{1}{x + 1} + \frac{2}{x^2 - 2x + 3}$$

41. $\dfrac{8x - 12}{x^2(x^2 + 2)^2} = \dfrac{A}{x} + \dfrac{B}{x^2} + \dfrac{Cx + D}{x^2 + 2} + \dfrac{Ey + F}{(x^2 + 2)^2}$

$$\begin{aligned} 8x - 12 &= Ax(x^2 + 2)^2 + B(x^2 + 2)^2 + (Cx + D)x^2(x^2 + 2) + (Ex + F)x^2 \\ &= Ax^5 + 4Ax^3 + 4Ax + Bx^4 + 4Bx^2 + 4B + Cx^5 + 2Cx^3 + Dx^4 + 2Dx^2 + Ex^3 + Fx^2 \\ &= (A + C)x^5 + (B + D)x^4 + (4A + 2C + E)x^3 + (4B + 2D + F)x^2 + 4Ax + 4B \end{aligned}$$

Equating coefficients of like terms gives

$$\begin{aligned} A + C &= 0, \\ B + D &= 0, \\ 4A + 2C + E &= 0, \\ 4B + 2D + F &= 0, \\ 4A &= 8, \text{ and} \\ 4B &= -12. \end{aligned}$$

So, $A = 2$, $B = -3$, $C = -2$, $D = 3$, $E = -4$, and $F = 6$.

$$\frac{8x - 12}{x^2(x^2 + 2)^2} = \frac{2}{x} + \frac{-3}{x^2} + \frac{-2x + 3}{x^2 + 2} + \frac{-4x + 6}{(x^2 + 2)^2}$$

43. $\dfrac{x^2 - x}{x^2 + x + 1} = 1 + \dfrac{-2x - 1}{x^2 + x + 1} = 1 - \dfrac{2x + 1}{x^2 + x + 1}$

45. $\dfrac{2x^3 - x^2 + x + 5}{x^2 + 3x + 2} = 2x - 7 + \dfrac{18x + 19}{(x + 1)(x + 2)}$

$$\frac{18x + 19}{(x + 1)(x + 2)} = \frac{A}{x + 1} + \frac{B}{x + 2}$$

$$18x + 19 = A(x + 2) + B(x + 1)$$

Let $x = -1$: $1 = A$

Let $x = -2$: $-17 = -B \Rightarrow B = 17$

$$\frac{2x^3 - x^2 + x + 5}{x^2 + 3x + 2} = 2x - 7 + \frac{1}{x + 1} + \frac{17}{x + 2}$$

47. $\dfrac{x^4}{(x - 1)^3} = \dfrac{x^4}{x^3 - 3x^2 + 3x - 1} = x + 3 + \dfrac{6x^2 - 8x + 3}{(x - 1)^3}$

$$\frac{6x^2 - 8x + 3}{(x - 1)^3} = \frac{A}{x - 1} + \frac{B}{(x - 1)^2} + \frac{C}{(x - 1)^3}$$

$$6x^2 - 8x + 3 = A(x - 1)^2 + B(x - 1) + C$$

Let $x = 1$: $1 = C$

$$\left.\begin{aligned}\text{Let } x = 0:&\ \ 3 = A - B + 1\\ \text{Let } x = 2:&\ \ 11 = A + B + 1\end{aligned}\right\}\quad \begin{aligned}A - B &= 2\\ A + B &= 10\end{aligned}$$

So, $A = 6$ and $B = 4$.

$$\frac{x^4}{(x - 1)^3} = x + 3 + \frac{6}{x - 1} + \frac{4}{(x - 1)^2} + \frac{1}{(x - 1)^3}$$

49. $\dfrac{x^4 + 2x^3 + 4x^2 + 8x + 2}{x^3 + 2x^2 + x} = x + \dfrac{3x^2 + 8x + 2}{x^3 + 2x^2 + x} = x + \dfrac{3x^2 + 8x + 2}{x(x + 1)^2}$

$$\frac{3x^2 + 8x + 2}{x(x + 1)^2} = \frac{A}{x} + \frac{B}{x + 1} + \frac{C}{(x + 1)^2}$$

$$3x^2 + 8x + 2 = A(x + 1)^2 + B(x)(x + 1) + C(x)$$

$$3x^2 + 8x + 2 = Ax^2 + 2Ax + A + Bx^2 + Bx + Cx$$

$$3x^2 + 8x + 2 = (A + B)x^2 + (2A + B + C)x + A$$

Equating coefficients of like terms gives

$A + B = 3$, $2A + B + C = 8$, and $A = 2$.

So, $A = 2$, $B = 1$, and $C = 3$.

$$\frac{x^4 + 2x^3 + 4x^2 + 8x + 2}{x^3 + 2x^2 + x} = x + \frac{2}{x} + \frac{1}{x + 1} + \frac{3}{(x + 1)^2}$$

51. $\dfrac{5 - x}{2x^2 + x - 1} = \dfrac{A}{2x - 1} + \dfrac{B}{x + 1}$

$$-x + 5 = A(x + 1) + B(2x - 1)$$

Let $x = \dfrac{1}{2}$: $\dfrac{9}{2} = \dfrac{3}{2}A \Rightarrow A = 3$

Let $x = -1$: $6 = -3B \Rightarrow B = -2$

$$\frac{5 - x}{2x^2 + x - 1} = \frac{3}{2x - 1} - \frac{2}{x + 1}$$

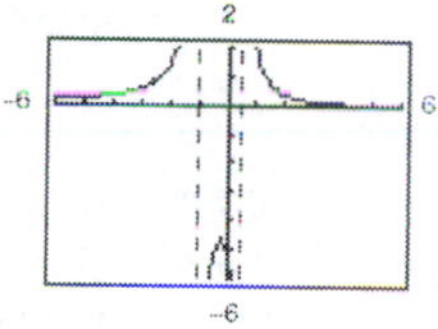

53. $\dfrac{4x^2 - 1}{2x(x + 1)^2} = \dfrac{A}{2x} + \dfrac{B}{x + 1} + \dfrac{C}{(x + 1)^2}$

$$4x^2 - 1 = A(x + 1)^2 + 2Bx(x + 1) + 2Cx$$

Let $x = 0$: $-1 = A$

Let $x = -1$: $3 = -2C \Rightarrow C = -\dfrac{3}{2}$

Let $x = 1$: $3 = 4A + 4B + 2C$

$$3 = -4 + 4B - 3$$

$$\frac{5}{2} = B$$

$$\frac{4x^2 - 1}{2x(x + 1)^2} = \frac{1}{2}\left[-\frac{1}{x} + \frac{5}{x + 1} - \frac{3}{(x + 1)^2}\right]$$

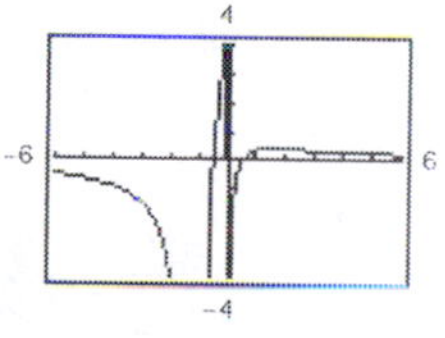

55. $\dfrac{x^2 + x + 2}{\left(x^2 + 2\right)^2} = \dfrac{Ax + B}{x^2 + 2} + \dfrac{Cx + D}{\left(x^2 + 2\right)^2}$

$$x^2 + x + 2 = (Ax + B)\left(x^2 + 2\right) + Cx + D$$

$$x^2 + x + 2 = Ax^3 + Bx^2 + (2A + C)x + (2B + D)$$

Equating coefficients of like terms gives

$0 = A$

$1 = B$

$1 = 2A + C \Rightarrow C = 1$

$2 = 2B + D \Rightarrow D = 0$

$$\frac{x^2 + x + 2}{\left(x^2 + 2\right)^2} = \frac{1}{x^2 + 2} + \frac{x}{\left(x^2 + 2\right)^2}$$

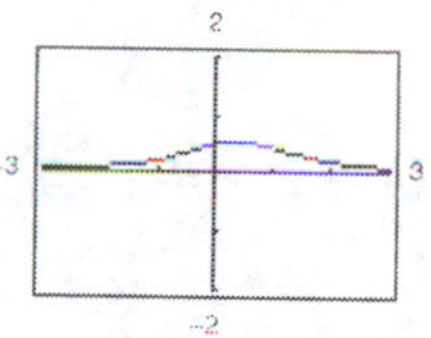

57. $\dfrac{2x^3 - 4x^2 - 15x + 5}{x^2 - 2x - 8} = 2x + \dfrac{x + 5}{(x + 2)(x - 4)}$

$$\frac{x + 5}{(x + 2)(x - 4)} = \frac{A}{x + 2} + \frac{B}{x - 4}$$

$$x + 5 = A(x - 4) + B(x + 2)$$

Let $x = -2$: $3 = -6A \Rightarrow A = -\dfrac{1}{2}$

Let $x = 4$: $9 = 6B \Rightarrow B = \dfrac{3}{2}$

$$\frac{2x^3 - 4x^2 - 15x + 5}{x^2 - 2x - 8} = 2x + \frac{1}{2}\left(\frac{3}{x - 4} - \frac{1}{x + 2}\right)$$

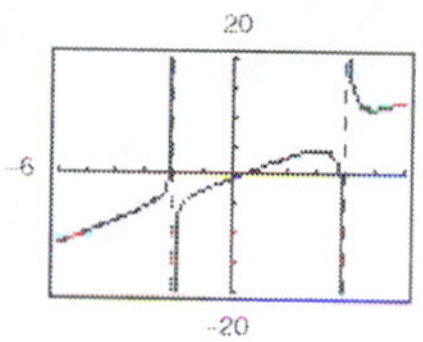

59. $C = \dfrac{120p}{10{,}000 - p^2} = \dfrac{120p}{(100 + p)(100 - p)} = \dfrac{A}{100 + p} + \dfrac{B}{100 - p}$

$$120p = A(100 - p) + B(100 + p)$$

Let $p = 100$: $200B = 12{,}000$

$$B = 60$$

Let $p = -100$: $200A = -12{,}000$

$$A = -60$$

$$C = \frac{120p}{10{,}000 - p^2} = -\frac{60}{100 + p} + \frac{60}{100 - p}$$

Let $y_1 = \dfrac{120p}{10{,}000 - p^2}$ and $y_2 = -\dfrac{60}{100 + p} + \dfrac{60}{100 - p}$.

X	Y1	Y2
0	0	0
10	.12121	.12121
20	.25	.25
30	.3956	.3956
40	.57143	.57143
50	.8	.8
60	1.125	1.125

X=0

61. False. The partial fraction decomposition is

$$\frac{A}{x+10} + \frac{B}{x-10} + \frac{C}{(x-10)^2}.$$

63. True. The expression is an improper rational expression.

65. $\frac{1}{a^2 - x^2} = \frac{A}{a+x} + \frac{B}{a-x}$, a is a constant.

$$1 = A(a - x) + B(a + x)$$

Let $x = -a$: $1 = 2aA \Rightarrow A = \frac{1}{2a}$

Let $x = a$: $1 = 2aB \Rightarrow B = \frac{1}{2a}$

$$\frac{1}{a^2 - x^2} = \frac{1}{2a}\left(\frac{1}{a+x} + \frac{1}{a-x}\right)$$

67. $\frac{1}{y(a-y)} = \frac{A}{y} + \frac{B}{a-y}$

$$1 = A(a - y) + By$$

Let $y = 0$: $1 = aA \Rightarrow A = \frac{1}{a}$

Let $y = a$: $1 = aB \Rightarrow B = \frac{1}{a}$

$$\frac{1}{y(a-y)} = \frac{1}{a}\left(\frac{1}{y} + \frac{1}{a-y}\right)$$

69. One way to find the constants is to choose values of the variable that eliminate one or more of the constants in the basic equation so that you can solve for another constant. If necessary, you can then use these constants with other chosen values of the variable to solve for any remaining constants. Another way is to expand the basic equation and collect like terms. Then you can equate coefficients of the like terms on each side of the equation to obtain simple equations involving the constants. If necessary, you can solve these equations using substitution.

Section 9.5 Systems of Inequalities

1. solution

3. solution

5. $y < 5 - x^2$

Using a dashed line, graph $y = 5 - x^2$, and shade the region inside the parabola.

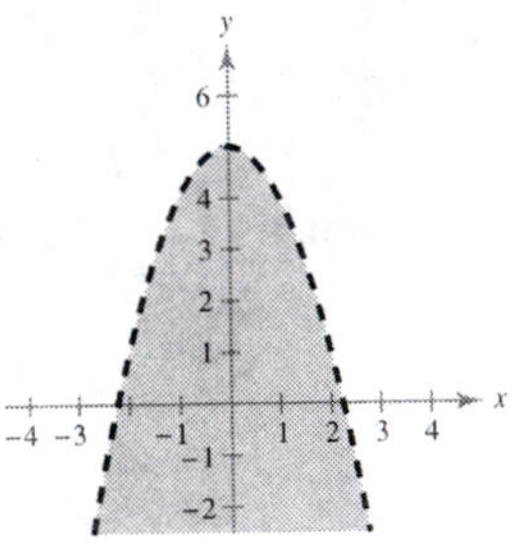

7. $x \geq 6$

Using a solid line, graph the vertical line $x = 6$, and shade to the right of this line.

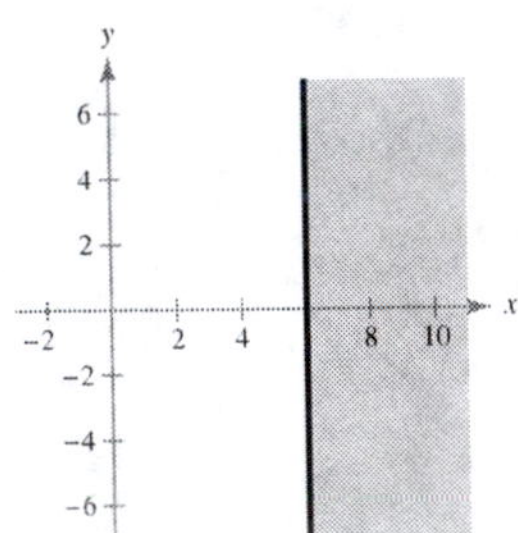

9. $y > -7$

Using a dashed line, graph the horizontal line $y = -7$, and shade above the line.

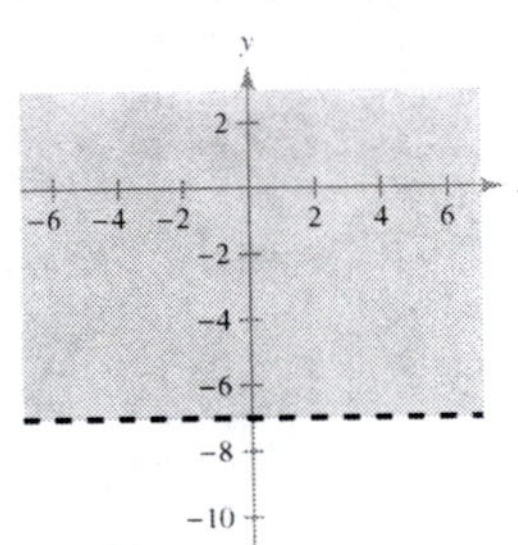

11. $y < 2 - x$

Using a dashed line, graph $y = 2 - x$, and then shade below the line. (Use $(0, 0)$ as a test point.)

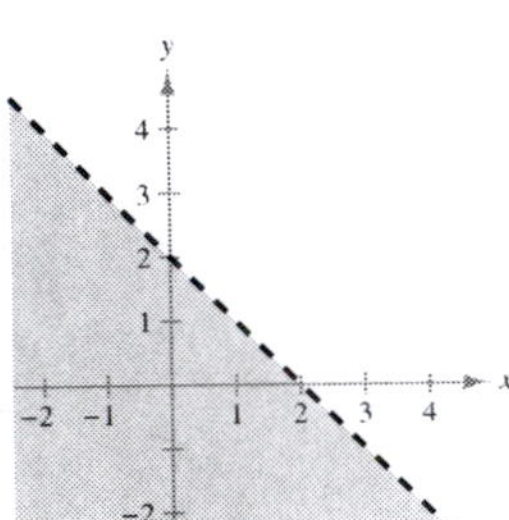

13. $2y - x \geq 4$

Using a solid line, graph $2y - x = 4$, and then shade above the line. (Use $(0, 0)$ as a test point.)

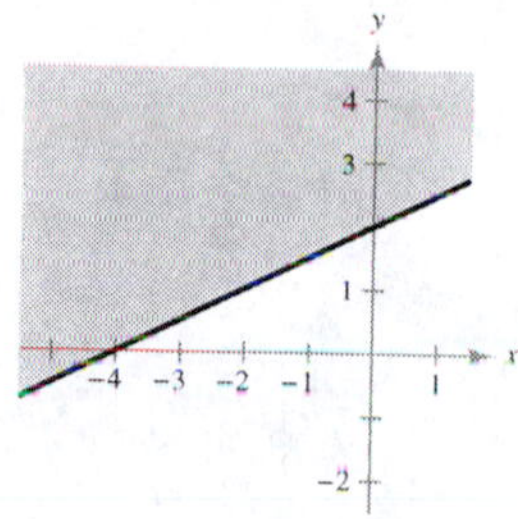

15. $(x + 1)^2 + (y - 2)^2 < 9$

Using a dashed line, sketch the circle $(x + 1)^2 + (y - 2)^2 = 9$.

Center: $(-1, 2)$

Radius: 3

Test point: $(0, 0)$

Shade the inside of the circle.

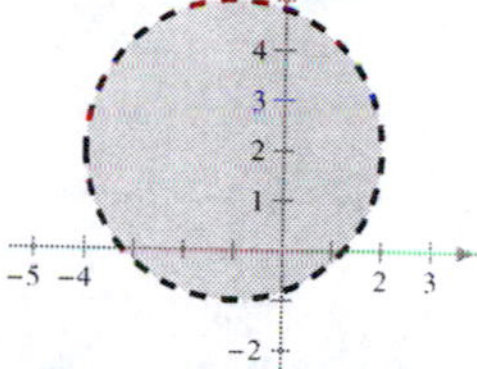

17. $y \leq \dfrac{1}{1 + x^2}$

Using a solid line, graph $y = \dfrac{1}{1 + x^2}$, and then shade below the curve. (Use $(0, 0)$ as a test point.)

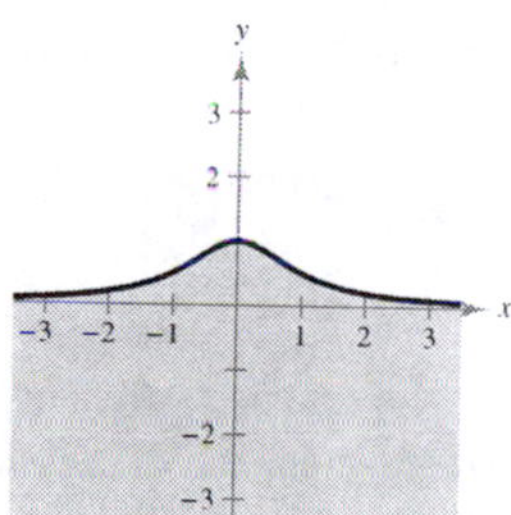

19. $y < \ln x$

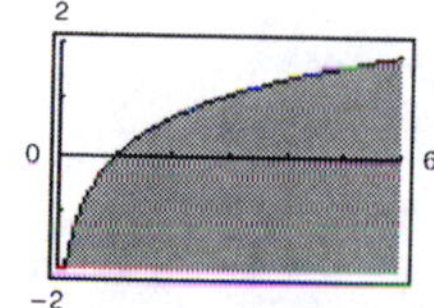

21. $y < 4^{-x-5}$

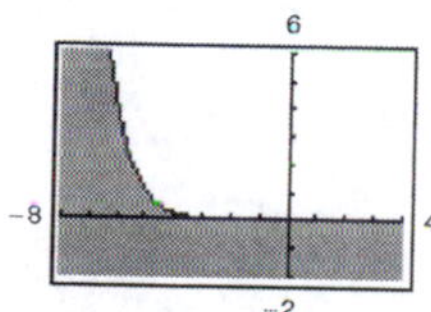

23. $y \leq 6 - \frac{3}{2}x$

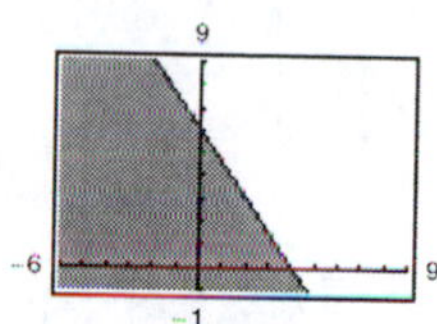

25. $x^2 + 5y - 10 \leq 0$

$$y \leq 2 - \frac{x^2}{5}$$

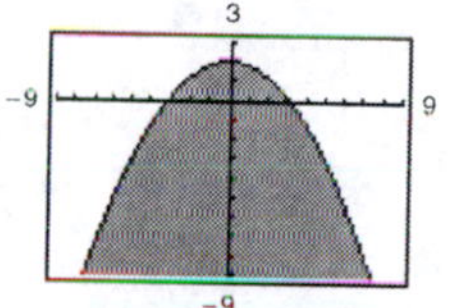

27. $\frac{5}{2}y - 3x^2 - 6 \geq 0$

$$y \geq \tfrac{2}{5}(3x^2 + 6)$$

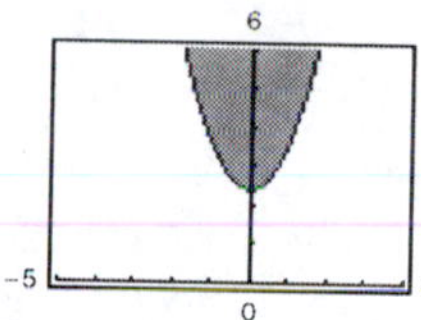

29. The line through $(-5, 0)$ and $(-1, 0)$ is $y = 5x + 5$.

The shaded region below the line gives $y < 5x + 5$.

31. The line through $(0, 2)$ and $(3, 0)$ is $y = -\frac{2}{3}x + 2$.

The shaded region above the line gives $y \geq -\frac{2}{3}x + 2$.

33. $\begin{cases} x + y \leq 1 \\ -x + y \leq 1 \\ y \geq 0 \end{cases}$

First, find the points of intersection of each pair of equations.

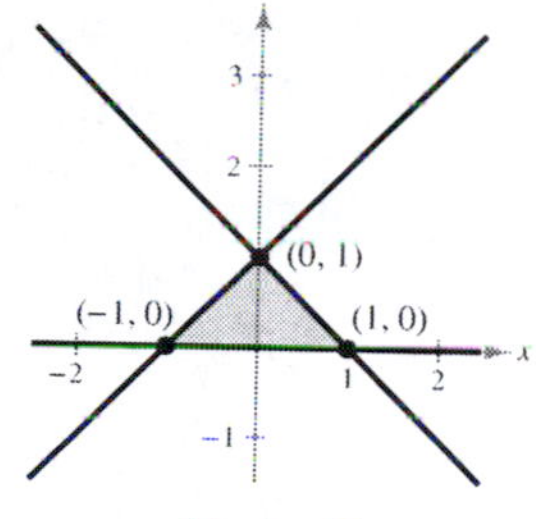

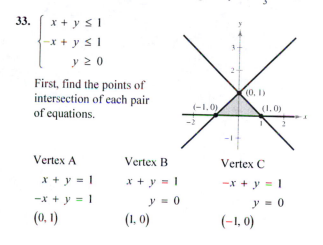

Vertex A	Vertex B	Vertex C
$x + y = 1$	$x + y = 1$	$-x + y = 1$
$-x + y = 1$	$y = 0$	$y = 0$
$(0, 1)$	$(1, 0)$	$(-1, 0)$

35. $\begin{cases} x^2 + y \le 7 \\ x \ge -2 \\ y \ge 0 \end{cases}$

First, find the points of intersection of each pair of equations.

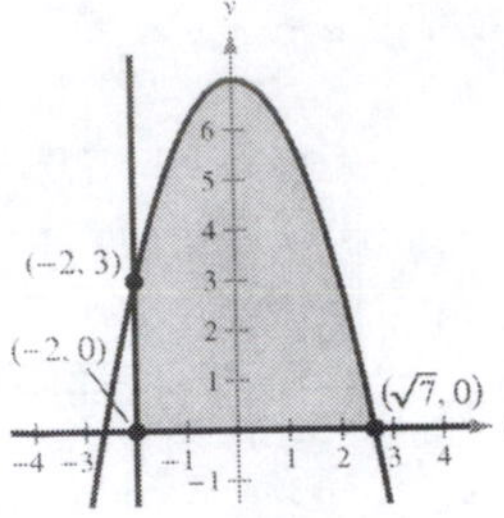

Vertex A

$x^2 + y = 7,\ x = -2$

$4 + y = 7$

$y = 3$

$(-2, 3)$

Vertex B

$x^2 + y = 7,\ y = 0$

$x^2 = 7$

$x = \sqrt{7}$

$\left(\sqrt{7}, 0\right)$

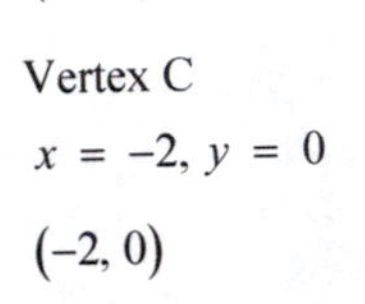

Vertex C

$x = -2,\ y = 0$

$(-2, 0)$

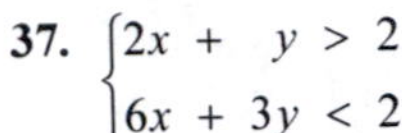

37. $\begin{cases} 2x + y > 2 \\ 6x + 3y < 2 \end{cases}$

The graphs of $2x + y = 2$ and $6x + 3y = 2$ are parallel lines. The first inequality has the region above the line shaded. The second inequality has the region below the line shaded. There are no points that satisfy both inequalities.

No solution

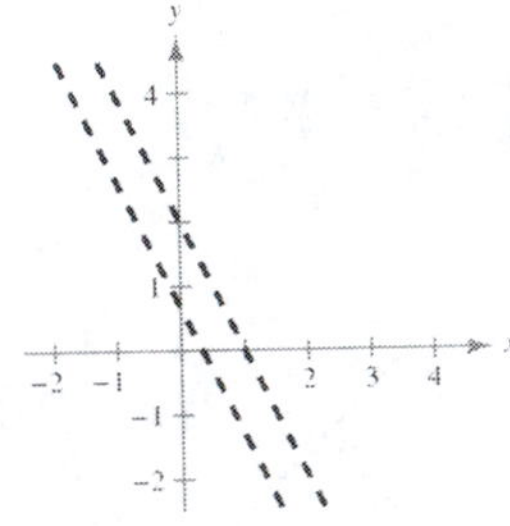

39. $\begin{cases} -3x + 2y < 6 \\ x - 4y > -2 \\ 2x + y < 3 \end{cases}$

First, find the points of intersection of each pair of equations.

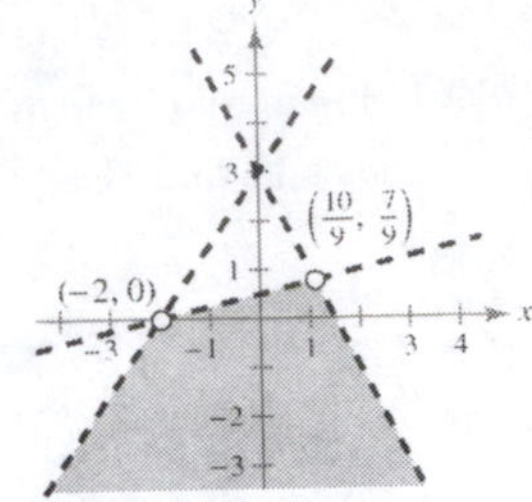

Vertex A

$-3x + 2y = 6$

$x - 4y = -2$

$(-2, 0)$

Vertex B

$-3x + 2y = 6$

$2x + y = 3$

$(0, 3)$

Vertex C

$x - 4y = -2$

$2x + y = 3$

$\left(\frac{10}{9}, \frac{7}{9}\right)$

Note that B is not a vertex of the solution region.

41. $\begin{cases} x > y^2 \\ x < y + 2 \end{cases}$

Points of intersection:

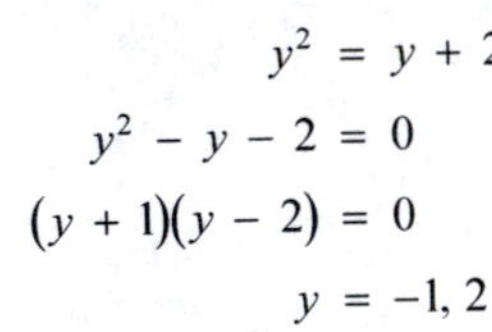

$y^2 = y + 2$

$y^2 - y - 2 = 0$

$(y + 1)(y - 2) = 0$

$y = -1, 2$

$(1, -1), (4, 2)$

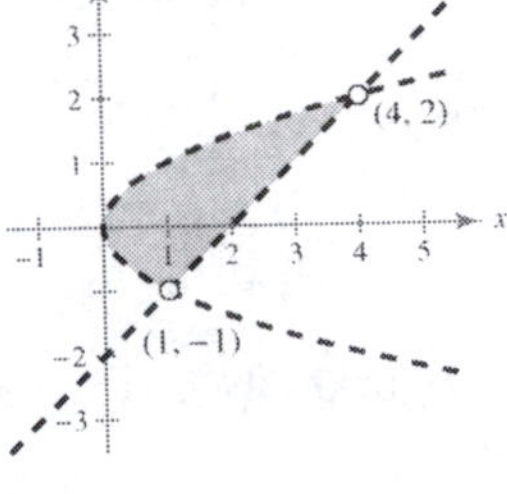

43. $\begin{cases} x^2 + y^2 \le 36 \\ x^2 + y^2 \ge 9 \end{cases}$

There are no points of intersection. The region common to both in equalities is the region between the circles.

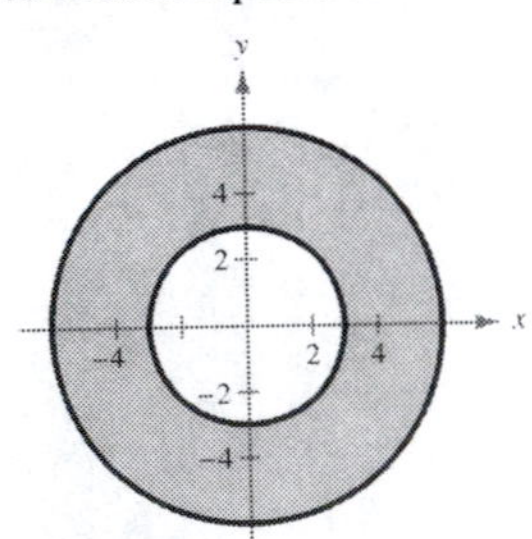

45. $3x + 4 \geq y^2$

$x - y < 0$

Points of intersection:

$x - y = 0 \Rightarrow y = x$

$3y + 4 = y^2$

$0 = y^2 - 3y - 4$

$0 = (y - 4)(y + 1)$

$y = 4$ or $y = -1$

$x = 4 \quad x = -1$

$(4, 4)$ and $(-1, -1)$

47. $\begin{cases} y \leq \sqrt{3x} + 1 \\ y \geq x^2 + 1 \end{cases}$

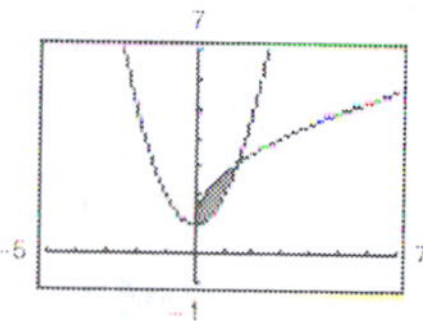

49. $\begin{cases} y < x^3 - 2x + 1 \\ y > -2x \\ x \leq 1 \end{cases}$

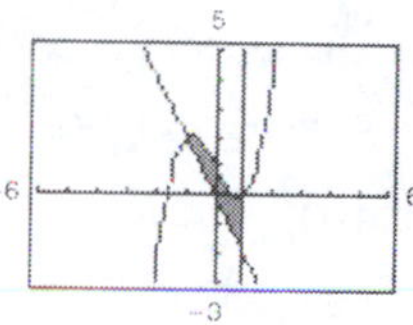

51. $\begin{cases} x^2y \geq 1 \Rightarrow y \geq \dfrac{1}{x^2} \\ 0 < x \leq 4 \\ y \leq 4 \end{cases}$

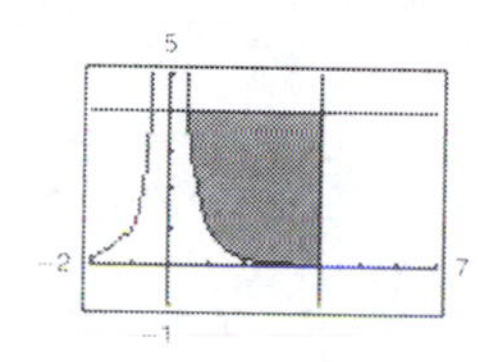

53. Line through points $(6, 0)$ and $(0, 6)$: $y = 6 - x$

$\begin{cases} x \geq 0 \\ y \geq 0 \\ y \leq 6 - x \end{cases}$

55. $(8, 0), (0, 8)$

$\begin{cases} x \geq 0 \\ y \geq 0 \\ x^2 + y^2 < 64 \end{cases}$

57. Rectangular region with vertices at

$(4, 3), (9, 3), (9, 9), (4, 9)$

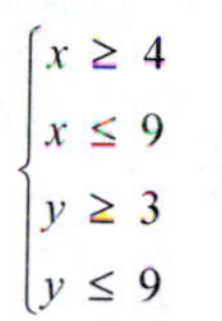

$\begin{cases} x \geq 4 \\ x \leq 9 \\ y \geq 3 \\ y \leq 9 \end{cases}$

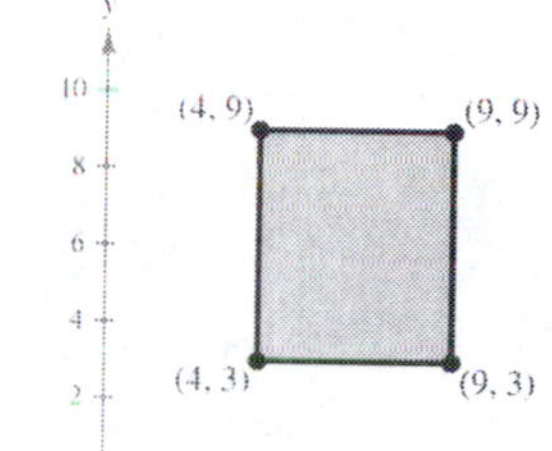

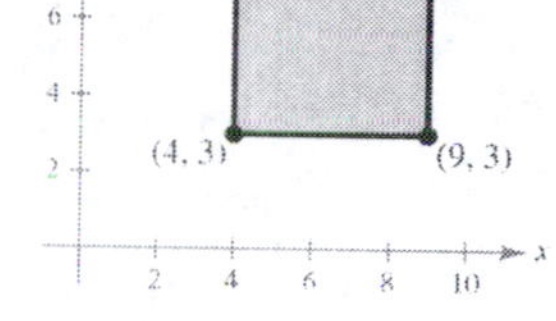

This system may be written as:

$\begin{cases} 4 \leq x \leq 9 \\ 3 \leq y \leq 9 \end{cases}$

59. Triangle with vertices at $(0, 0), (6, 0), (1, 5)$

$(0, 0), (6, 0)$: $y = 0$

$(0, 0), (1, 5)$: $y = 5x$

$(6, 0), (1, 5)$: $y = -x + 6$

$\begin{cases} y \geq 0 \\ y \leq 5x \\ y \leq -x + 6 \end{cases}$

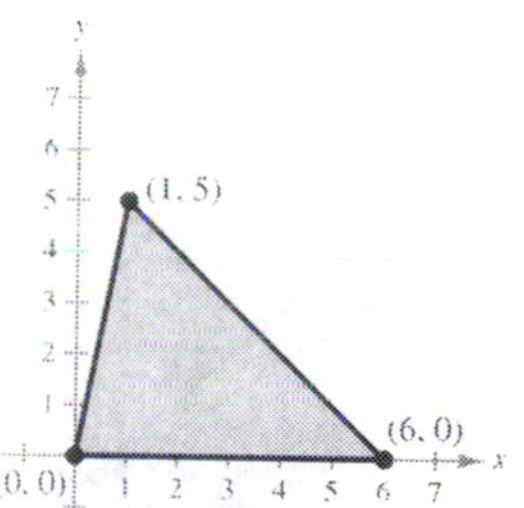

61. (a) Demand = Supply

$50 - 0.5x = 0.125x$

$50 = 0.625x$

$80 = x$

$10 = p$

Point of equilibrium: $(80, 10)$

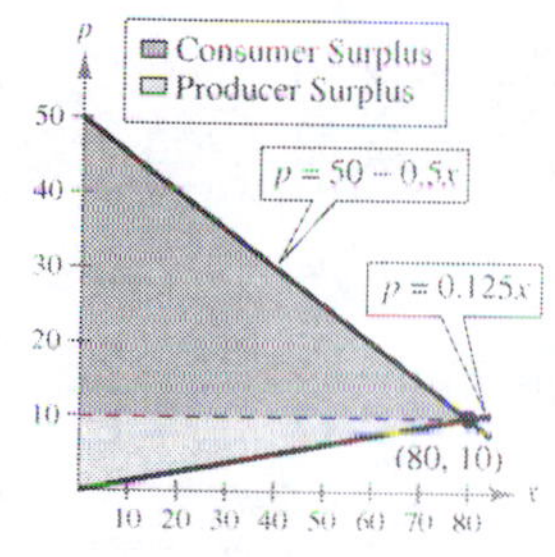

(b) The consumer surplus is the area of the triangular region defined by

$\begin{cases} p \leq 50 - 0.5x \\ p \geq 10 \\ x \geq 0. \end{cases}$

Consumer surplus $= \frac{1}{2}(\text{base})(\text{height}) = \frac{1}{2}(80)(40) = \$1600$

The producer surplus is the area of the triangular region defined by

$\begin{cases} p \geq 0.125x \\ p \leq 10 \\ x \geq 0. \end{cases}$

Producer surplus $= \frac{1}{2}(\text{base})(\text{height}) = \frac{1}{2}(80)(10) = \$400$

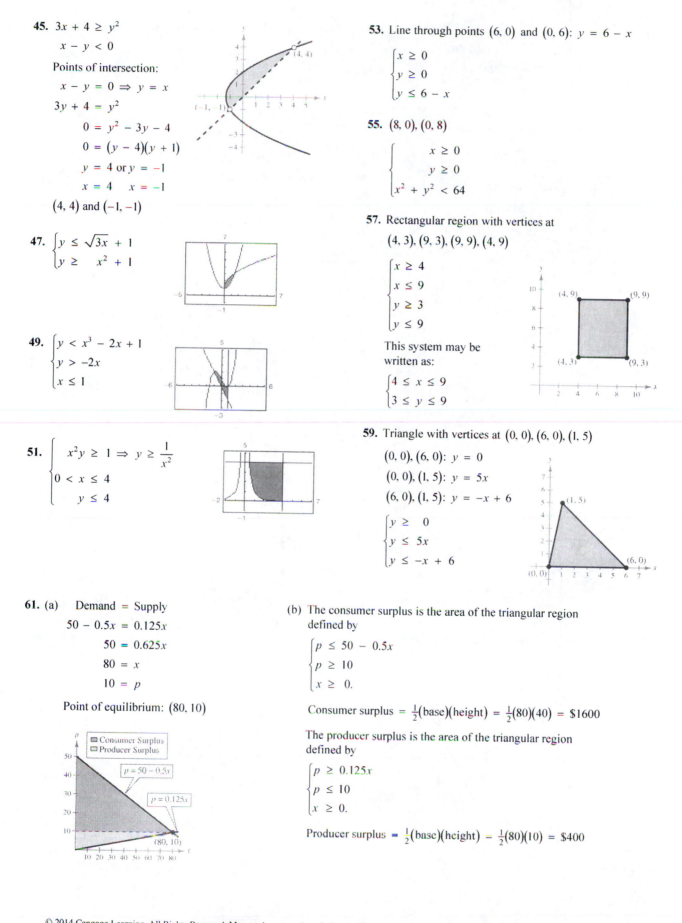

63. (a)

$$\begin{aligned} \text{Demand} &= \text{Supply} \\ 140 - 0.00002x &= 80 + 0.00001x \\ 60 &= 0.00003x \\ 2{,}000{,}000 &= x \\ 100 &= p \end{aligned}$$

Point of equilibrium: $(2{,}000{,}000, 100)$

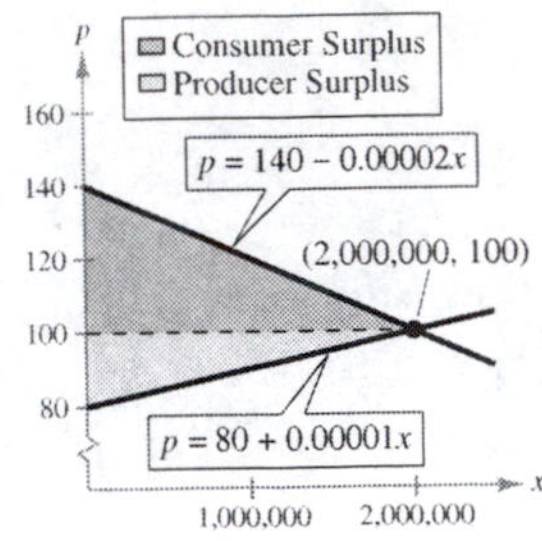

(b) The consumer surplus is the area of the triangular region defined by

$$\begin{cases} p \le 140 - 0.00002x \\ p \ge 100 \\ x \ge 0. \end{cases}$$

$$\begin{aligned} \text{Consumer surplus} &= \tfrac{1}{2}(\text{base})(\text{height}) \\ &= \tfrac{1}{2}(2{,}000{,}000)(40) \\ &= \$40{,}000{,}000 \end{aligned}$$

The producer surplus is the area of the triangular region defined by

$$\begin{cases} p \ge 80 + 0.00001x \\ p \le 100 \\ x \ge 0. \end{cases}$$

$$\begin{aligned} \text{Producer surplus} &= \tfrac{1}{2}(\text{base})(\text{height}) \\ &= \tfrac{1}{2}(2{,}000{,}000)(20) \\ &= \$20{,}000{,}000 \end{aligned}$$

65. x = number of tables

y = number of chairs

$$\begin{cases} x + \frac{3}{2}y \le 12 & \text{Assembly center} \\ \frac{4}{3}x + \frac{3}{2}y \le 15 & \text{Finishing center} \\ x \ge 0 \\ y \ge 0 \end{cases}$$

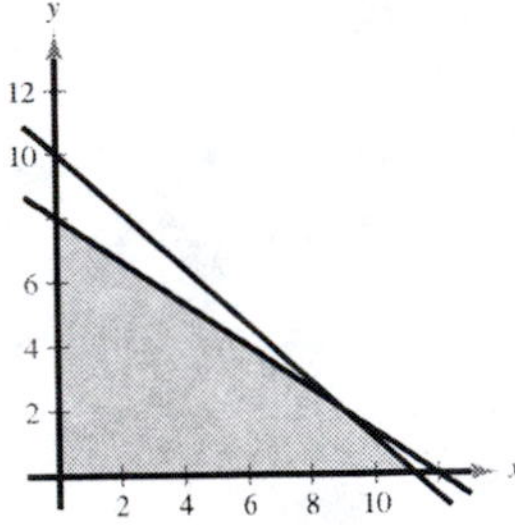

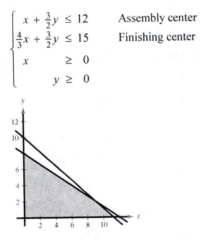

67. x = amount in smaller account

y = amount in larger account

Account constraints:

$$\begin{cases} x + y \le 20{,}000 \\ y \ge 2x \\ x \ge 5{,}000 \\ y \ge 5{,}000 \end{cases}$$

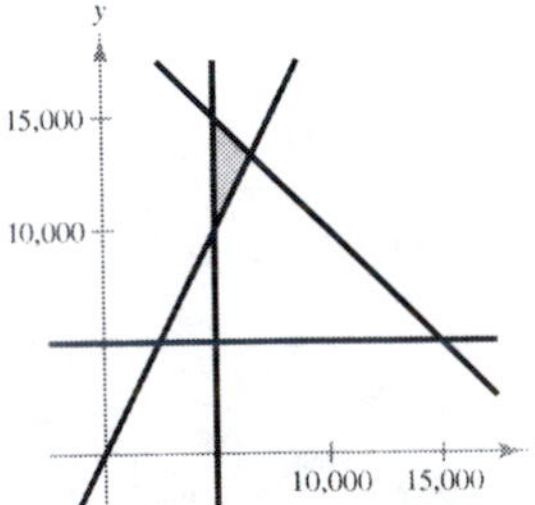

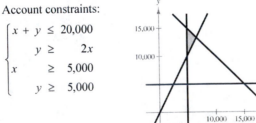

69. Let x = number of large trucks.

Let y = number of medium trucks.

The delivery requirements are:

$$\begin{cases} 6x + 4y \ge 15 \\ 3x + 6y \ge 16 \\ x \ge 0 \\ y \ge 0 \end{cases}$$

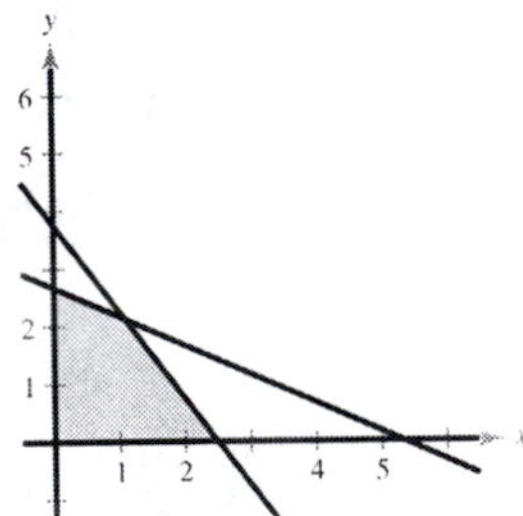

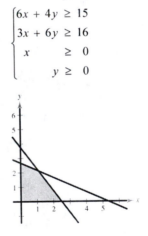

71. (a) Let y = heart rate.

$y \geq 0.5(220 - x)$

$y \leq 0.85(220 - x)$

$x \geq 20$

$x \leq 70$

(b)

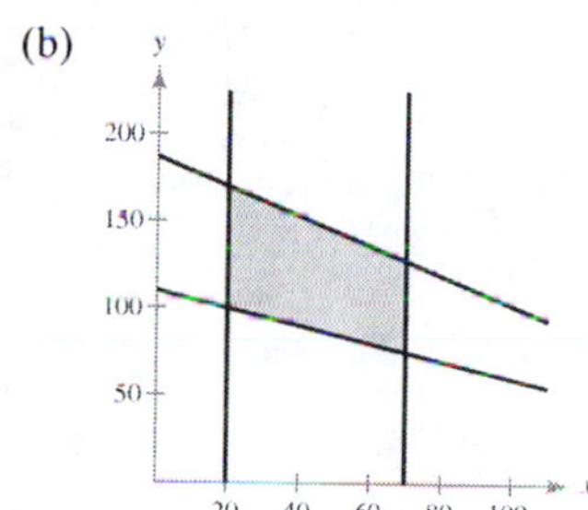

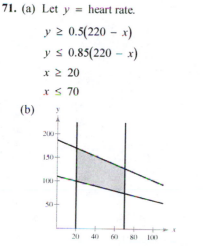

(c) Answers will vary. For example, the points $(24, 98)$ and $(24, 167)$ are on the boundary of the solution set; a person who is 24 years old should have a heart rate between 98 and 167.

73. True. The figure is a rectangle with a length of 9 units and a width of 11 units.

75. Test a point on each side of the line. Because the origin $(0, 0)$ satisfies the inequality, the solution set of the inequality lies below the line.

77. x = radius of smaller circle

y = radius of larger circle

(a) Constraints on circles:

$$\begin{cases} \pi y^2 - \pi x^2 \geq 10 \\ y > x \\ x > 0 \end{cases}$$

(b)

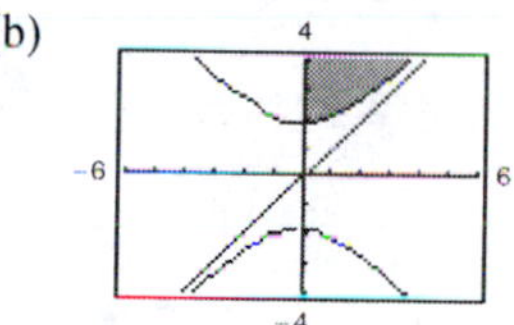

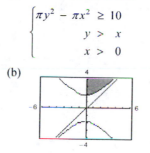

(c) The line is an asymptote to the boundary. The larger the circles, the closer the radii can be and the constraint still be satisfied.

Section 9.6 Linear Programming

1. optimization

3. objective

5. inside; on

7. $z = 4x + 3y$

At $(0, 5)$: $z = 4(0) + 3(5) = 15$

At $(0, 0)$: $z = 4(0) + 3(0) = 0$

At $(5, 0)$: $z = 4(5) + 3(0) = 20$

The minimum value is 0 at $(0, 0)$.

The maximum value is 20 at $(5, 0)$.

9. $z = 2x + 5y$

At $(0, 0)$: $z = 2(0) + 5(0) = 0$

At $(4, 0)$: $z = 2(4) + 5(0) = 8$

At $(3, 4)$: $z = 2(3) + 5(4) = 26$

At $(0, 5)$: $z = 2(0) + 5(5) = 25$

The minimum value is 0 at $(0, 0)$.

The maximum value is 26 at $(3, 4)$.

11. $z = 10x + 7y$

At $(0, 45)$: $z = 10(0) + 7(45) = 315$

At $(30, 45)$: $z = 10(30) + 7(45) = 615$

At $(60, 20)$: $z = 10(60) + 7(20) = 740$

At $(60, 0)$: $z = 10(60) + 7(0) = 600$

At $(0, 0)$: $z = 10(0) + 7(0) = 0$

The minimum value is 0 at $(0, 0)$.

The maximum value is 740 at $(60, 20)$.

13. $z = 3x + 2y$

At $(0, 10)$: $z = 3(0) + 2(10) = 20$

At $(4, 0)$: $z = 3(4) + 2(0) = 12$

At $(2, 0)$: $z = 3(2) + 2(0) = 6$

The minimum value is 6 at $(2, 0)$.

The maximum value is 20 at $(0, 10)$.

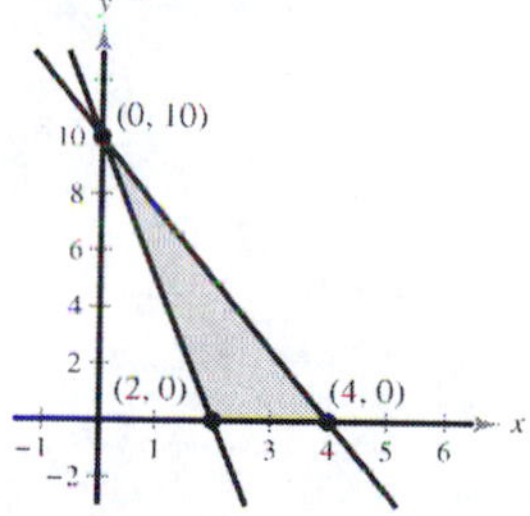

15. $z = 4x + 5y$

At $(10, 0)$: $z = 4(10) + 5(0) = 40$

At $(5, 3)$: $z = 4(5) + 5(3) = 35$

At $(0, 8)$: $z = 4(0) + 5(8) = 40$

The minimum value is 35 at $(5, 3)$.

The region is unbounded. There is no maximum.

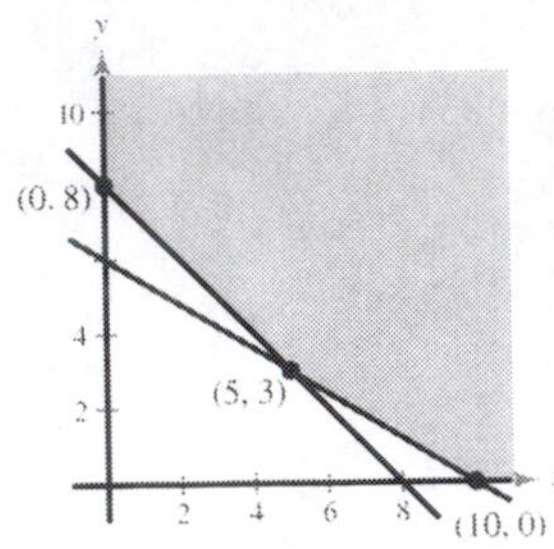

17. $z = 3x + y$

At $(16, 0)$: $z = 3(16) + 0 = 48$

At $(60, 0)$: $z = 3(60) + 0 = 180$

At $(7.2, 13.2)$: $z = 3(7.2) + 13.2 = 34.8$

The minimum value is 34.8 at $(7.2, 13.2)$.

The maximum value is 180 at $(60, 0)$.

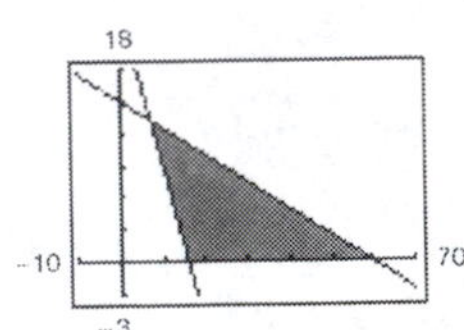

19. $z = x + 4y$

At $(16, 0)$: $z = 16 + 4(0) = 16$

At $(60, 0)$: $z = 60 + 4(0) = 60$

At $(7.2, 13.2)$: $z = 7.2 + 4(13.2) = 60$

The minimum value is 16 at $(16, 0)$.

The maximum value is 60 at any point along the line segment connecting $(60, 0)$ and $(7.2, 13.2)$.

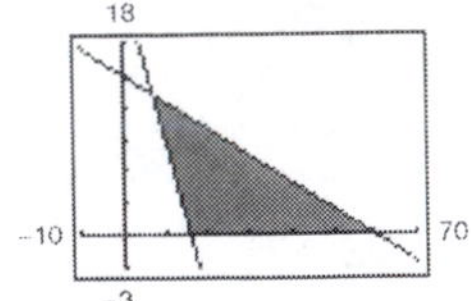

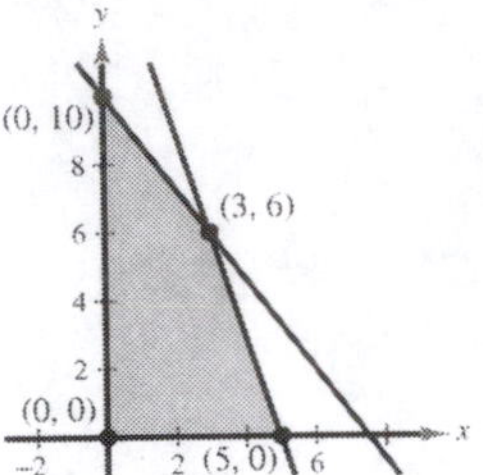

Figure for Exercises 21–23

21. $z = 2x + y$

At $(0, 10)$: $z = 2(0) + 10 = 10$

At $(3, 6)$: $z = 2(3) + 6 = 12$

At $(5, 0)$: $z = 2(5) + 0 = 10$

At $(0, 0)$: $z = 2(0) + 0 = 0$

The minimum value is 0 at $(0, 0)$.

The maximum value is 12 at $(3, 6)$.

23. $z = x + y$

At $(0, 10)$: $z = 0 + 10 = 10$

At $(3, 6)$: $z = 3 + 6 = 9$

At $(5, 0)$: $z = 5 + 0 = 5$

At $(0, 0)$: $z = 0 + 0 = 0$

The minimum value is 0 at $(0, 0)$.

The maximum value is 10 at $(0, 10)$.

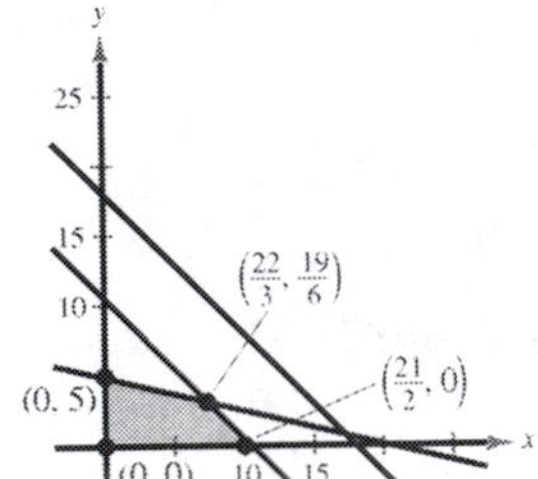

Figure for Exercises 25–27

25. $z = x + 5y$

At $(0, 5)$: $z = 0 + 5(5) = 25$

At $\left(\frac{22}{3}, \frac{19}{6}\right)$: $z = \frac{22}{3} + 5\left(\frac{19}{6}\right) = \frac{139}{6}$

At $\left(\frac{21}{2}, 0\right)$: $z = \frac{21}{2} + 5(0) = \frac{21}{2}$

At $(0, 0)$: $z = 0 + 5(0) = 0$

The minimum value is 0 at $(0, 0)$.

The maximum value is 25 at $(0, 5)$.

27. $z = 4x + 5y$

At $(0, 5)$: $z = 4(0) + 5(5) = 25$

At $\left(\frac{22}{3}, \frac{19}{6}\right)$: $z = 4\left(\frac{22}{3}\right) + 5\left(\frac{19}{6}\right) = \frac{271}{6}$

At $\left(\frac{21}{2}, 0\right)$: $z = 4\left(\frac{21}{2}\right) + 5(0) = 42$

At $(0, 0)$: $z = 4(0) + 5(0) = 0$

The minimum value is 0 at $(0, 0)$.

The maximum value is $\frac{271}{6}$ at $\left(\frac{22}{3}, \frac{19}{6}\right)$.

29. Objective function: $z = 2.5x + y$

Constraints:
$x \geq 0$, $y \geq 0$, $3x + 5y \leq 15$, $5x + 2y \leq 10$

At $(0, 0)$: $z = 0$

At $(2, 0)$: $z = 5$

At $\left(\frac{20}{19}, \frac{45}{19}\right)$: $z = \frac{95}{19} = 5$

At $(0, 3)$: $z = 3$

The minimum value is 0 at $(0, 0)$.

The maximum value of 5 occurs at any point on the line segment connecting $(2, 0)$ and $\left(\frac{20}{19}, \frac{45}{19}\right)$.

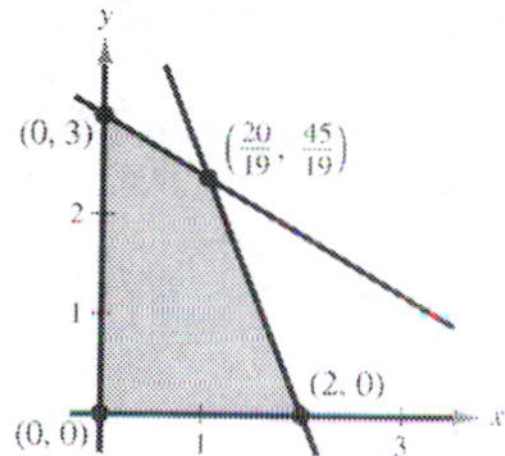

31. Objective function: $z = -x + 2y$

Constraints: $x \geq 0$, $y \geq 0$, $x \leq 10$, $x + y \leq 7$

At $(0, 0)$: $z = -0 + 2(0) = 0$

At $(0, 7)$: $z = -0 + 2(7) = 14$

At $(7, 0)$: $z = -7 + 2(0) = -7$

The constraint $x \leq 10$ is extraneous.

The minimum value is -7 at $(7, 0)$.

The maximum value is 14 at $(0, 7)$.

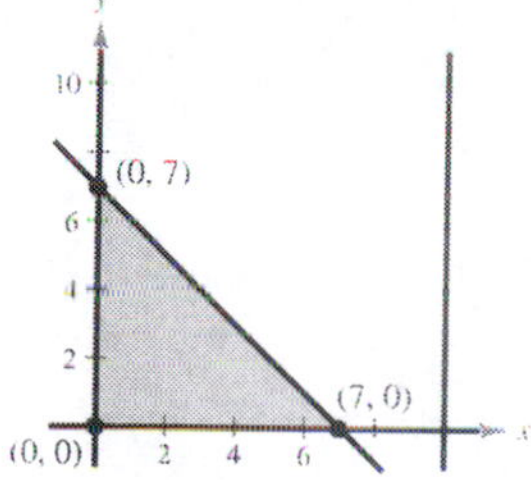

33. Objective function: $z = 3x + 4y$

Constraints: $x \geq 0$, $y \geq 0$, $x + y \leq 1$, $2x + y \leq 4$

At $(0, 0)$: $z = 3(0) + 4(0) = 0$

At $(0, 1)$: $z = 3(0) + 4(1) = 4$

At $(1, 0)$: $z = 3(1) + 4(0) = 3$

The constraint $2x + y \leq 4$ is extraneous.

The minimum value is 0 at $(0, 0)$.

The maximum value is 4 at $(0, 1)$.

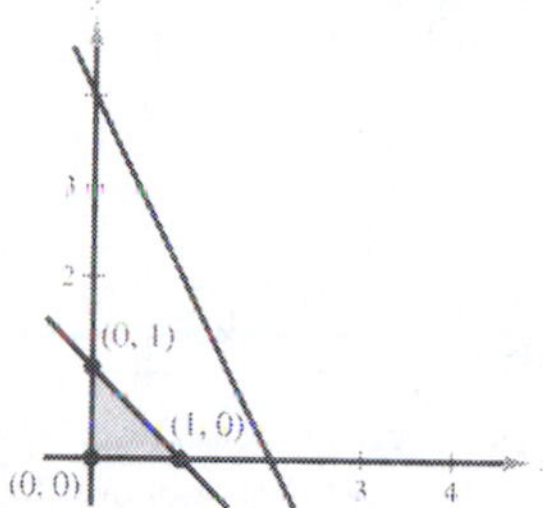

35. Objective function: $z = x + y$

Constraints: $x \geq 0$, $y \geq 0$, $x + y \leq 1$, $2x + y \leq 4$

At $(0, 0)$: $z = 0 + 0 = 0$

At $(0, 1)$: $z = 0 + 1 = 1$

At $(1, 0)$: $z = 1 + 0 = 1$

The constraint $2x + y \leq 4$ is extraneous.

The minimum value is 0 at $(0, 0)$.

The maximum value is 1 at any point on the line segment connecting $(0, 1)$ and $(1, 0)$.

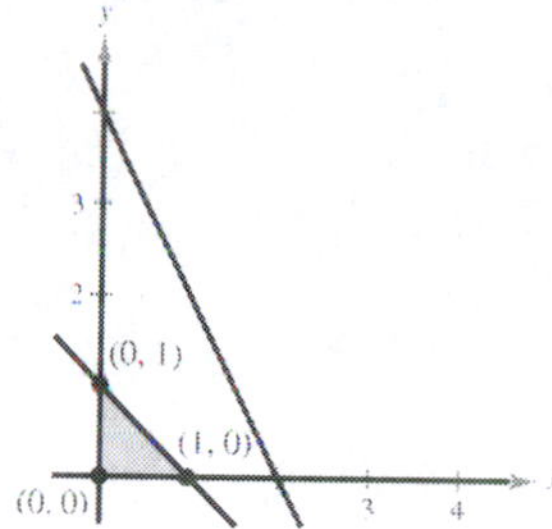

37. x = number of \$225 models

y = number of \$250 models

Constraints:

$$225x + 250y \le 63{,}000$$
$$x + y \le 275$$
$$x \ge 0$$
$$y \ge 0$$

Objective function: $P = 30x + 31y$

Vertices: $(0, 0), (0, 252), (230, 45)$ and $(275, 0)$

At $(0, 0)$: $P = 30(0) + 31(0) = 0$

At $(0, 252)$: $P = 30(0) + 31(252) = 7812$

At $(230, 45)$: $P = 30(230) + 31(45) = 8295$

At $(275, 0)$: $P = 30(275) + 31(0) = 8250$

An optimal profit of \$8295 occurs when 230 units of the \$225 model and 45 units of the \$250 model are stocked in inventory.

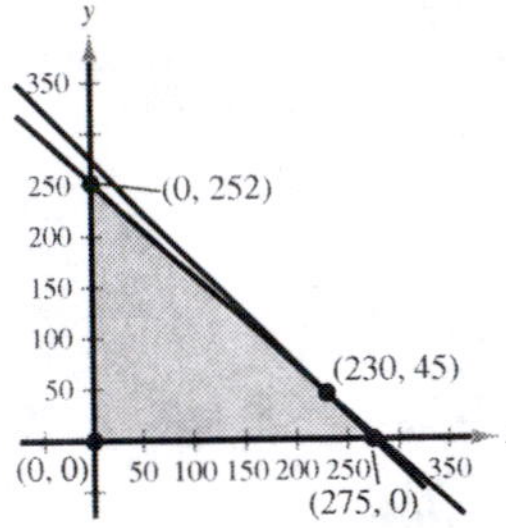

39. x = number of bags of Brand X

y = number of bags of Brand Y

Constraints:

$$2x + y \ge 12$$
$$2x + 9y \ge 36$$
$$2x + 3y \ge 24$$
$$x \ge 0$$
$$y \ge 0$$

Objective function: $C = 25x + 20y$

Vertices: $(0, 12), (3, 6), (9, 2), (18, 0)$

At $(0, 12)$: $C = 25(0) + 20(12) = 240$

At $(3, 6)$: $C = 25(3) + 20(6) = 195$

At $(9, 2)$: $C = 25(9) + 20(2) = 265$

At $(18, 0)$: $C = 25(18) + 20(0) = 450$

To optimize cost, use three bags of Brand X and six bags of Brand Y for an optimal cost of \$195.

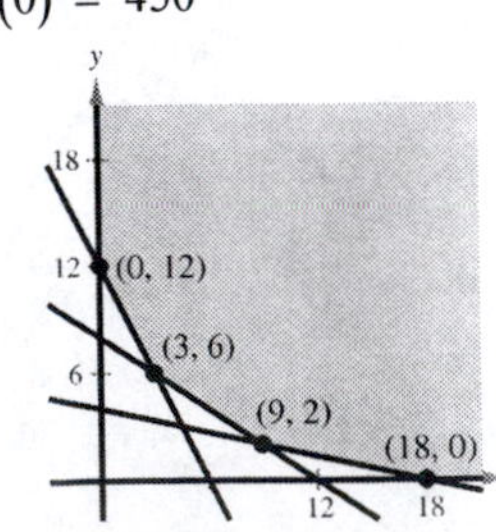

41. x = number of audits

y = number of tax returns

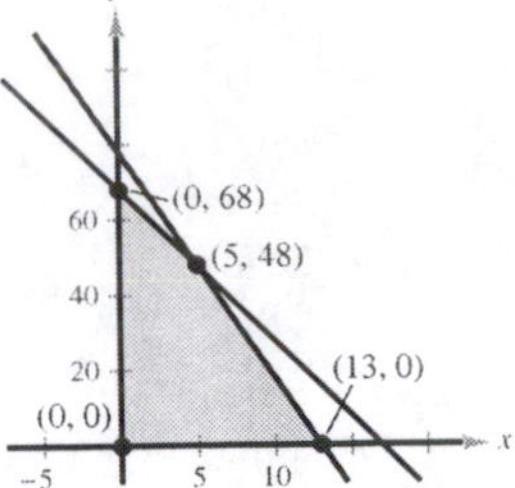

Constraints:

$$60x + 10y \le 780$$
$$16x + 4y \le 272$$
$$x \ge 0$$
$$y \ge 0$$

Objective function:

$R = 1600x + 250y$

Vertices: $(0, 0), (13, 0), (5, 48), (0, 68)$

At $(0, 0)$: $R = 1600(0) + 250(0) = 0$

At $(13, 0)$: $R = 1600(13) + 250(0) = 20{,}800$

At $(5, 48)$: $R = 1600(5) + 250(48) = 20{,}000$

At $(0, 68)$: $R = 1600(0) + 250(68) = 17{,}000$

A maximum revenue of \$20,800 occurs when the firm conducts 13 audits and 0 tax returns.

43. x = acres of crop A

y = acres of crop B

Constraints: $x + y \le 150$, $x + 2y \le 240$,

$0.3x + 0.1y \le 30$

Objective function: $z = 300x + 500y$

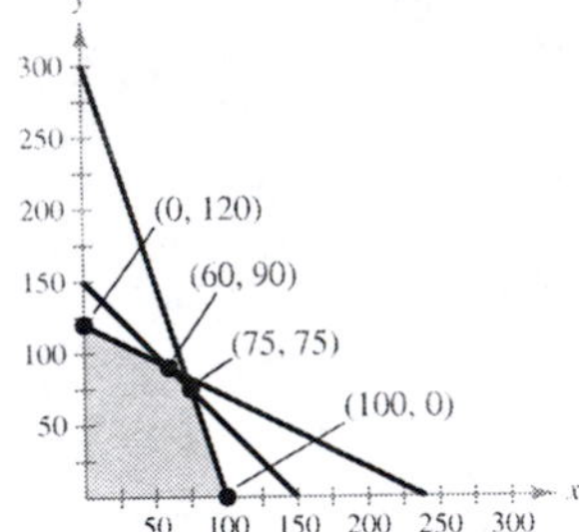

At $(0, 0)$: $z = 300(0) + 500(0) = 0$

At $(0, 20)$: $z = 300(0) + 500(120) = 60{,}000$

At $(60, 90)$: $z = 300(60) + 500(90) = 63{,}000$

At $(75, 75)$: $z = 300(75) + 500(75) = 60{,}000$

At $(100, 0)$: $z = 300(100) + 500(0) = 30{,}000$

So, 60 acres of crop A and 90 acres of crop B yield 63,000 bushels.

45. x = number of TV ads

y = number of newspaper ads

Constraints: $100{,}000x + 20{,}000y \le 1{,}000{,}000$

$100{,}000x \le 800{,}000$

$x \ge 0$

$y \ge 0$

Objective function: $A = 20x + 5y$ (A in millions)

Vertices: $(0, 0), (0, 50), (8, 10), (8, 0)$

At $(0, 0)$: $A = 20(0) + 5(0) = 0$

At $(0, 50)$: $A = 20(0) + 5(50) = 250$ million

At $(8, 10)$: $A = 20(8) + 5(10) = 210$ million

At $(8, 0)$: $A = 20(8) + 5(0) = 160$ million

The company should spend \$0 on television ads and \$1,000,000 on newspaper ads. The optimal total audience is 250 million people.

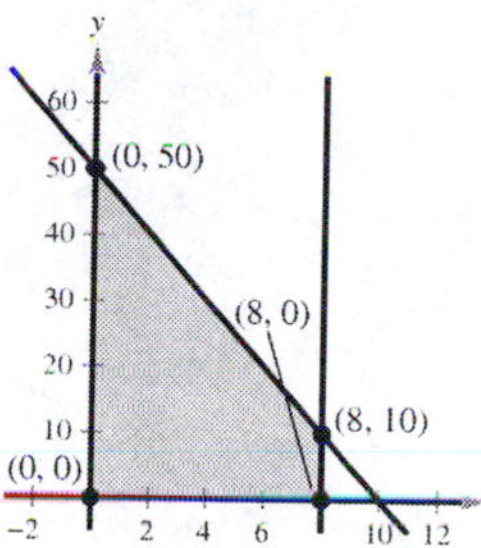

47. True. The objective function has a maximum value at any point on the line segment connecting the two vertices. Both of these points are on the line $y = -x + 11$ and lie between $(4, 7)$ and $(8, 3)$.

49. True. If an objective function has a maximum value at more than one vertex, then any point on the line segment connecting the points will produce the maximum value.

Review Exercises for Chapter 9

1. $\begin{cases} x + y = 2 \\ x - y = 0 \Rightarrow x = y \end{cases}$

$x + x = 2$

$2x = 2$

$x = 1$

$y = 1$

Solution: $(1, 1)$

3. $\begin{cases} 4x - y - 1 = 0 \Rightarrow y = 4x - 1 \\ 8x + y - 17 = 0 \end{cases}$

$8x + (4x - 1) - 17 = 0$

$12x = 18$

$x = \frac{3}{2}$

$4\left(\frac{3}{2}\right) - y - 1 = 0$

$-y + 5 = 0$

$y = 5$

Solution: $\left(\frac{3}{2}, 5\right)$

5. $\begin{cases} 0.5x + y = 0.75 \Rightarrow y = 0.75 - 0.5x \\ 1.25x - 4.5y = -2.5 \end{cases}$

$1.25x - 4.5(0.75 - 0.5x) = -2.5$

$1.25x - 3.375 + 2.25x = -2.5$

$3.50x = 0.875$

$x = 0.25$

$y = 0.625$

Solution: $(0.25, 0.625)$

7. $\begin{cases} x^2 - y^2 = 9 \\ x - y = 1 \Rightarrow x = y + 1 \end{cases}$

$(y + 1)^2 - y^2 = 9$

$2y + 1 = 9$

$y = 4$

$x = 5$

Solution: $(5, 4)$

9. $\begin{cases} y = 2x^2 \\ y = x^4 - 2x^2 \end{cases} \Rightarrow 2x^2 = x^4 - 2x^2$

$0 = x^4 - 4x^2$

$0 = x^2(x^2 - 4)$

$0 = x^2(x + 2)(x - 2) \Rightarrow x = 0, -2, 2$

$x = 0: y = 2(0)^2 = 0$

$x = -2: y = 2(-2)^2 = 8$

$x = 2: y = 2(2)^2 = 8$

Solutions: $(0, 0), (-2, 8), (2, 8)$

11. 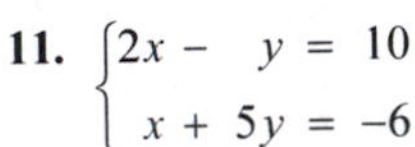

$\begin{cases} 2x - y = 10 \\ x + 5y = -6 \end{cases}$

Point of intersection: $(4, -2)$

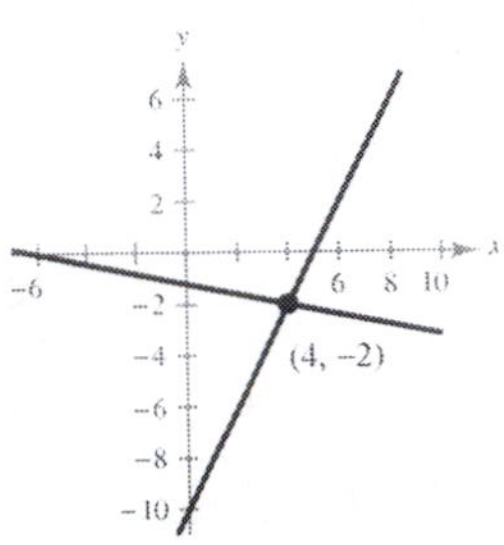

13. $\begin{cases} y = 2x^2 - 4x + 1 \\ y = x^2 - 4x + 3 \end{cases}$

Points of intersection: $(1.41, -0.66), (-1.41, 10.66)$

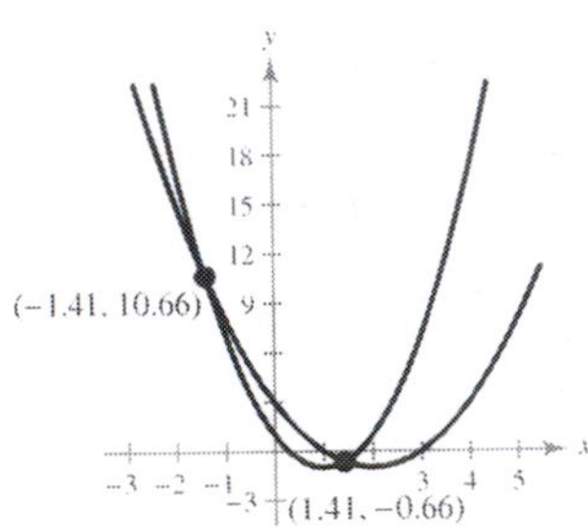

15. $\begin{cases} y = -2e^{-x} \\ 2e^x + y = 0 \Rightarrow y = -2e^x \end{cases}$

Point of intersection: $(0, -2)$

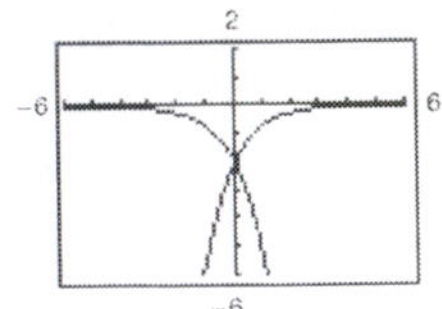

17. $\begin{cases} y = 2 + \log x \\ y = \frac{3}{4}x + 5 \end{cases}$

No Solution

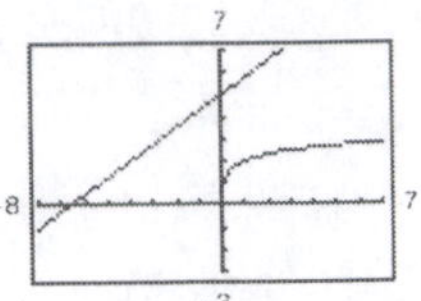

19.

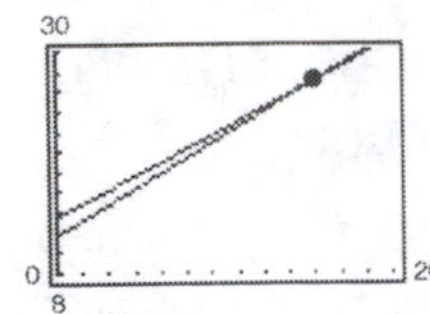

$0.68a + 13.5 > 0.77a + 11.7$

$1.8 > 0.09a$

$20 > a$

The BMI for males exceeds the BMI for females after age 20.

21. $\begin{cases} 2l + 2w = 68 \\ w = \frac{8}{9}l \end{cases}$

$2l + 2\left(\frac{8}{9}\right)l = 68$

$\frac{34}{9}l = 68$

$l = 18$

$w = \frac{8}{9}l = 16$

The width of the rectangle is 16 feet, and the length is 18 feet.

23. $\begin{cases} 2x - y = 2 \Rightarrow 16x - 8y = 16 \\ 6x + 8y = 39 \Rightarrow 6x + 8y = 39 \end{cases}$

$22x = 55$

$x = \frac{55}{22} = \frac{5}{2}$

Back-substitute $x = \frac{5}{2}$ into Equation 1.

$2\left(\frac{5}{2}\right) - y = 2$

$y = 3$

Solution: $\left(\frac{5}{2}, 3\right)$

25. $\begin{cases} 3x - 2y = 0 \Rightarrow 3x - 2y = 0 \\ 3x + 2(y + 5) = 10 \Rightarrow 3x + 2y = 0 \end{cases}$

$6x = 0$

$x = 0$

Back-substitute $x = 0$ into Equation 1.

$3(0) - 2y = 0$

$2y = 0$

$y = 0$

Solution: $(0, 0)$

27. $\begin{cases} 1.25x - 2y = 3.5 \Rightarrow & 5x - 8y = 14 \\ 5x - 8y = 14 \Rightarrow & -5x + 8y = -14 \end{cases}$

$$0 = 0$$

There are infinitely many solutions.

Let $y = a$, then $5x - 8a = 14 \Rightarrow x = \frac{8}{5}a + \frac{14}{5}$.

Solution: $\left(\frac{8}{5}a + \frac{14}{5}, a\right)$ where a is any real number.

29. $\begin{cases} x + 5y = 4 \Rightarrow & x + 5y = 4 \\ x - 3y = 6 \Rightarrow & -x + 3y = -6 \end{cases}$

$$8y = -2 \Rightarrow y = -\tfrac{1}{4}$$

Matches graph (d). The system has one solution and is consistent.

30. $\begin{cases} -3x + y = -7 \Rightarrow & -3x + y = -7 \\ 9x - 3y = 21 \Rightarrow & 3x - y = 7 \end{cases}$

$$0 = 0$$

Matches graph (c). The system has infinitely many solutions and is consistent.

31. $\begin{cases} 3x - y = 7 \Rightarrow & 6x - 2y = 14 \\ -6x + 2y = 8 \Rightarrow & -6x + 2y = 8 \end{cases}$

$$0 \neq 22$$

Matches graph (b). The system has no solution and is inconsistent.

32. $\begin{cases} 2x - y = -3 \Rightarrow & 10x - 5y = -15 \\ x + 5y = 4 \Rightarrow & x + 5y = 4 \end{cases}$

$$11x = -11 \Rightarrow x = -1$$

Matches graph (a). The system has one solution and is consistent.

33. $22 + 0.00001x = 43 - 0.0002x$

$0.00021x = 21$

$x = 100{,}000, \; p = 23$

Point of Equilibrium: $(100{,}000, 23)$

35. $\begin{cases} x - 4y + 3z = 3 \\ -y + z = -1 \\ z = -5 \end{cases}$

$-y + (-5) = -1 \Rightarrow y = -4$

$x - 4(-4) + 3(-5) = 3 \Rightarrow x = 2$

Solution: $(2, -4, -5)$

37. $\begin{cases} 4x - 3y - 2z = -65 \\ 8y - 7z = -14 \\ z = 10 \end{cases}$

$8y - 7(10) = -14 \Rightarrow y = 7$

$4x - 3(7) - 2(10) = -65 \Rightarrow x = -6$

Solution: $(-6, 7, 10)$

39. $\begin{cases} x + 2y + 6z = 4 & \text{Equation 1} \\ -3x + 2y - z = -4 & \text{Equation 2} \\ 4x + 2z = 16 & \text{Equation 3} \end{cases}$

$\begin{cases} x + 2y + 6z = 4 & \\ 8y + 17z = 8 & 3\text{Eq.1} + \text{Eq.2} \\ -8y - 22z = 0 & -4\text{Eq.1} + \text{Eq.3} \end{cases}$

$\begin{cases} x + 2y + 6z = 4 & \\ 8y + 17z = 8 & \\ -5z = 8 & \text{Eq.2} + \text{Eq.3} \end{cases}$

$\begin{cases} x + 2y + 6z = 4 & \\ 8y + 17z = 8 & \\ z = -\frac{8}{5} & -\frac{1}{5}\text{Eq.3} \end{cases}$

$8y + 17\left(-\frac{8}{5}\right) = 8 \Rightarrow y = \frac{22}{5}$

$x + 2\left(\frac{22}{5}\right) + 6\left(-\frac{8}{5}\right) = 4 \Rightarrow x = \frac{24}{5}$

Solution: $\left(\frac{24}{5}, \frac{22}{5}, -\frac{8}{5}\right)$

41. $\begin{cases} 2x + 6z = -9 & \text{Equation 1} \\ 3x - 2y + 11z = -16 & \text{Equation 2} \\ 3x - y + 7z = -11 & \text{Equation 3} \end{cases}$

$\begin{cases} -x + 2y - 5z = 7 & (-1)\text{Eq.2} + \text{Eq.1} \\ 3x - 2y + 11z = -16 & \\ 3x - y + 7z = -11 & \end{cases}$

$\begin{cases} -x + 2y - 5z = 7 & \\ 4y - 4z = 5 & 3\text{Eq.1} + \text{Eq.2} \\ 5y - 8z = 10 & 3\text{Eq.1} + \text{Eq.3} \end{cases}$

$\begin{cases} -x + 2y - 5z = 7 & \\ 4y - 4z = 5 & \\ -3y = 0 & (-2)\text{Eq.2} + \text{Eq.3} \end{cases}$

$\begin{cases} -x + 2y - 5z = 7 & \\ y - z = \frac{5}{4} & \left(\frac{1}{4}\right)\text{Eq.2} \\ y = 0 & \left(-\frac{1}{3}\right)\text{Eq.3} \end{cases}$

$0 - z = \frac{5}{4} \Rightarrow z = -\frac{5}{4}$

$-x + 2(0) - 5\left(-\frac{5}{4}\right) = 7 \Rightarrow x = -\frac{3}{4}$

Solution: $\left(-\frac{3}{4}, 0, -\frac{5}{4}\right)$

43. $\begin{cases} 5x - 12y + 7z = 16 \\ 3x - 7y + 4z = 9 \end{cases} \Rightarrow \begin{cases} 15x - 36y + 21z = 48 \\ -15x + 35y - 20z = -45 \end{cases}$

$$-y + z = 3$$

Let $z = a$. Then $y = a - 3$ and $5x - 12(a - 3) + 7a = 16 \Rightarrow x = a - 4$.

Solution: $(a - 4, a - 3, a)$ where a is any real number.

45. $y = ax^2 + bx + c$ through $(0, -5)$, $(1, -2)$, and $(2, 5)$.

$(0, -5)$: $-5 = c \Rightarrow c = -5$

$(1, -2)$: $-2 = a + b + c \Rightarrow \begin{cases} a + b = 3 \\ 2a + b = 5 \end{cases}$

$(2, 5)$: $5 = 4a + 2b + c \Rightarrow$

$\begin{cases} 2a + b = 5 \\ -a - b = -3 \end{cases}$

$a = 2$

$b = 1$

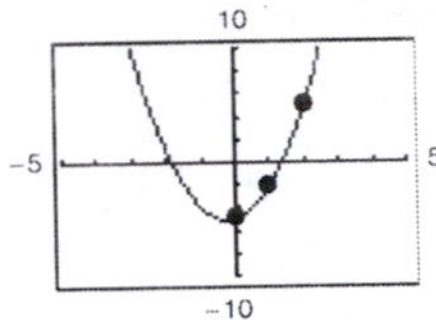

The equation of the parabola is $y = 2x^2 + x - 5$.

47. $x^2 + y^2 + Dx + Ey + F = 0$ through $(-1, -2)$, $(5, -2)$, and $(2, 1)$.

$(-1, -2)$: $5 - D - 2E + F = 0 \Rightarrow \begin{cases} D + 2E - F = 5 \\ 5D - 2E + F = -29 \\ 2D + E + F = -5 \end{cases}$

$(5, -2)$: $29 + 5D - 2E + F = 0 \Rightarrow$

$(2, 1)$: $5 + 2D + E + F = 0 \Rightarrow$

From the first two equations

$6D = -24$

$D = -4$.

Substituting $D = -4$ into the second and third equations yields:

$-20 - 2E + F = -29 \Rightarrow \begin{cases} -2E + F = -9 \\ -E - F = -3 \end{cases}$

$-8 + E + F = -5 \Rightarrow$

$-3E = -12$

$E = 4$

$F = -1$

The equation of the circle is $x^2 + y^2 - 4x + 4y - 1 = 0$.

To verify the result using a graphing utility, solve the equation for y.

$$(x^2 - 4x + 4) + (y^2 + 4y + 4) = 1 + 4 + 4$$

$$(x - 2)^2 + (y + 2)^2 = 9$$

$$(y + 2)^2 = 9 - (x - 2)^2$$

$$y = -2 \pm \sqrt{9 - (x - 2)^2}$$

Let $y_1 = -2 + \sqrt{9 - (x - 2)^2}$ and $y_2 = -2 - \sqrt{9 - (x - 2)^2}$.

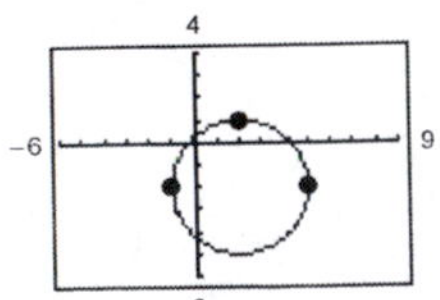

49. (a)

$$\begin{aligned} 3c + 24b + 194a &= 413.8 \\ 24c + 194b + 1584a &= 3328.9 \\ 194c + 1584b + 13{,}058a &= 27{,}051.1 \end{aligned}$$

Solving this system yields $a = -5.950$, $b = 104.45$, and $c = -312.9$.

So, the least squares regression parabola is $y = -5.950t^2 + 104.45t - 312.9$.

(b)

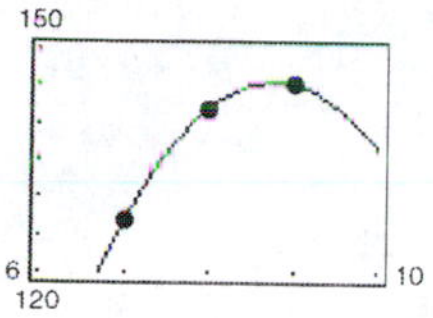

The model is a good fit for the data.

(c) In 2015, $t = 15$.

$$\begin{aligned} y &= -5.950(15)^2 + 104.45(15) - 312.9 \\ &= -84.9 \end{aligned}$$

In 2015 the retail e-commerce sales in the United States will be –$84.9 billion. This answer is not reasonable because the sales cannot be negative.

51. Let x = amount invested at 7%

y = amount invested at 9%

z = amount invested at 11%.

$y = x - 3000$ and

$z = x - 5000 \Rightarrow y + z = 2x - 8000$

$$\begin{cases} x + y + z = 40{,}000 \\ 0.07x + 0.09y + 0.11z = 3500 \\ y + z = 2x - 8000 \end{cases}$$

$$\begin{aligned} x + (2x - 8000) = 40{,}000 &\Rightarrow x = 16{,}000 \\ y = 16{,}000 - 3000 &\Rightarrow y = 13{,}000 \\ z = 16{,}000 - 5000 &\Rightarrow z = 11{,}000 \end{aligned}$$

So, $16,000 was invested at 7%, $13,000 at 9%, and $11,000 at 11%.

53. $s = \frac{1}{2}at^2 + v_0t + s_0$

When $t = 1$: $s = 134$: $\frac{1}{2}a(1)^2 + v_0(1) + s_0 = 134 \Rightarrow a + 2v_0 + 2s_0 = 268$

When $t = 2$: $s = 86$: $\frac{1}{2}a(2)^2 + v_0(2) + s_0 = 86 \Rightarrow 2a + 2v_0 + s_0 = 86$

When $t = 3$: $s = 6$: $\frac{1}{2}a(3)^2 + v_0(3) + s_0 = 6 \Rightarrow 9a + 6v_0 + 2s_0 = 12$

$$\begin{cases} a + 2v_0 + 2s_0 = 268 \\ 2a + 2v_0 + s_0 = 86 \\ 9a + 6v_0 + 2s_0 = 12 \end{cases}$$

$$\begin{cases} a + 2v_0 + 2s_0 = 268 \\ -2v_0 - 3s_0 = -450 & (-2)\text{Eq.}1 + \text{Eq.}2 \\ -12v_0 - 16s_0 = -2400 & (-9)\text{Eq.}1 + \text{Eq.}3 \end{cases}$$

$$\begin{cases} a + 2v_0 + 2s_0 = 268 \\ -2v_0 - 3s_0 = -450 \\ 3v_0 + 4s_0 = 600 & \left(-\frac{1}{4}\right)\text{Eq.}3 \end{cases}$$

$$\begin{cases} a + 2v_0 + 2s_0 = 268 \\ -2v_0 - 3s_0 = -450 \\ -s_0 = -150 & 3\text{Eq.}2 + 2\text{Eq.}3 \end{cases}$$

$$\begin{aligned} -s_0 = -150 &\Rightarrow s_0 = 150 \\ -2v_0 - 3(150) = -450 &\Rightarrow v_0 = 0 \\ a + 2(0) + 2(150) = 268 &\Rightarrow a = -32 \end{aligned}$$

The position equation is $s = \frac{1}{2}(-32)t^2 + (0)t + 150$, or $s = -16t^2 + 150$.

55. $\dfrac{3}{x^2 + 20x} = \dfrac{3}{x(x + 20)} = \dfrac{A}{x} + \dfrac{B}{x + 20}$

57. $\dfrac{3x - 4}{x^3 - 5x^2} = \dfrac{3x - 4}{x^2(x - 5)} = \dfrac{A}{x} + \dfrac{B}{x^2} + \dfrac{C}{x - 5}$

59. $\dfrac{4 - x}{x^2 + 6x + 8} = \dfrac{A}{x + 2} + \dfrac{B}{x + 4}$

$4 - x = A(x + 4) + B(x + 2)$

Let $x = -2$: $6 = 2A \Rightarrow A = 3$

Let $x = -4$: $8 = -2B \Rightarrow B = -4$

$\dfrac{4 - x}{x^2 + 6x + 8} = \dfrac{3}{x + 2} - \dfrac{4}{x + 4}$

61. $\dfrac{x^2}{x^2 + 2x - 15} = 1 - \dfrac{2x - 15}{x^2 + 2x - 15}$

$\dfrac{-2x + 15}{(x + 5)(x - 3)} = \dfrac{A}{x + 5} + \dfrac{B}{x - 3}$

$-2x + 15 = A(x - 3) + B(x + 5)$

Let $x = -5$: $25 = -8A \Rightarrow A = -\dfrac{25}{8}$

Let $x = 3$: $9 = 8B \Rightarrow B = \dfrac{9}{8}$

$\dfrac{x^2}{x^2 + 2x - 15} = 1 - \dfrac{25}{8(x + 5)} + \dfrac{9}{8(x - 3)}$

63. $\dfrac{x^2 + 2x}{x^3 - x^2 + x - 1} = \dfrac{x^2 + 2x}{(x - 1)(x^2 + 1)}$

$= \dfrac{A}{x - 1} + \dfrac{Bx + C}{x^2 + 1}$

$x^2 + 2x = A(x^2 + 1) + (Bx + C)(x - 1)$

$= Ax^2 + A + Bx^2 - Bx + Cx - C$

$= (A + B)x^2 + (-B + C)x + (A - C)$

Equating coefficients of like terms gives $1 = A + B$, $2 = -B + C$, and $0 = A - C$. Adding both sides of all three equations gives $3 = 2A$. So, $A = \dfrac{3}{2}$, $B = -\dfrac{1}{2}$, and $C = \dfrac{3}{2}$.

$\dfrac{x^2 + 2x}{x^3 - x^2 + x - 1} = \dfrac{\frac{3}{2}}{x - 1} + \dfrac{-\frac{1}{2}x + \frac{3}{2}}{x^2 + 1}$

$= \dfrac{1}{2}\left(\dfrac{3}{x - 1} - \dfrac{x - 3}{x^2 + 1}\right)$

65. $\dfrac{3x^2 + 4x}{(x^2 + 1)^2} = \dfrac{Ax + B}{x^2 + 1} + \dfrac{Cx + D}{(x^2 + 1)^2}$

$3x^2 + 4x = (Ax + B)(x^2 + 1) + Cx + D$

$= Ax^3 + Bx^2 + (A + C)x + (B + D)$

Equating coefficients of like terms gives

$0 = A$

$3 = B$

$4 = 0 + C \Rightarrow C = 4$

$0 = B + D \Rightarrow D = -3$

$\dfrac{3x^2 + 4x}{(x^2 + 1)^2} = \dfrac{3}{x^2 + 1} + \dfrac{4x - 3}{(x^2 + 1)^2}$

67. $y \le 5 - \frac{1}{2}x$

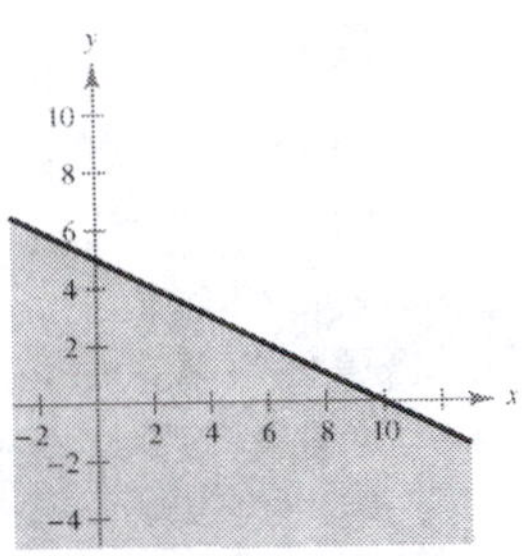

69. $y - 4x^2 > -1$

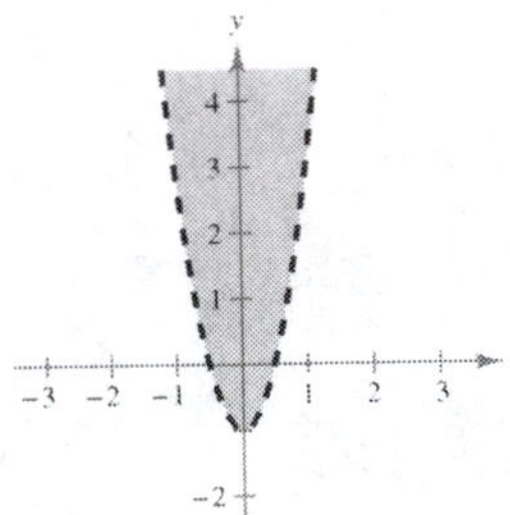

71. $(x - 1)^2 + (y - 3)^2 < 16$

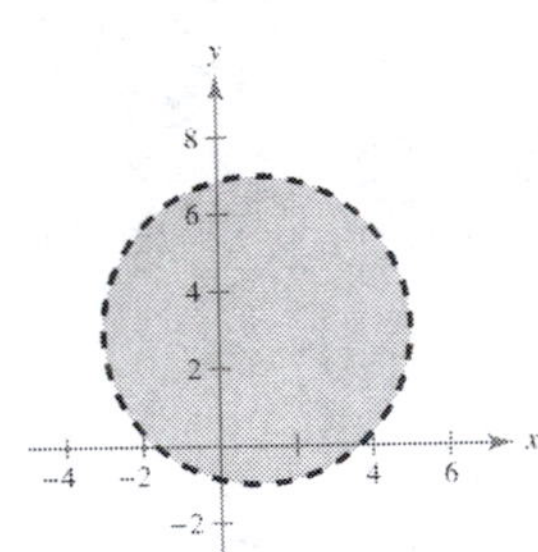

73. $\begin{cases} x + 2y \le 160 \\ 3x + y \le 180 \\ x \ge 0 \\ y \ge 0 \end{cases}$

Vertex A

$x + 2y = 160$
$3x + y = 180$
$(40, 60)$

Vertex B

$x + 2y = 160$
$x = 0$
$(0, 80)$

Vertex C

$3x + y = 180$
$y = 0$
$(60, 0)$

Vertex D

$x = 0$
$y = 0$
$(0, 0)$

75. $\begin{cases} y < x + 1 \\ y > x^2 - 1 \end{cases}$

Vertices:

$x + 1 = x^2 - 1$

$0 = x^2 - x - 2 = (x + 1)(x - 2)$

$x = -1$ or $x = 2$

$y = 0$ $\quad y = 3$

$(-1, 0)$ $\quad (2, 3)$

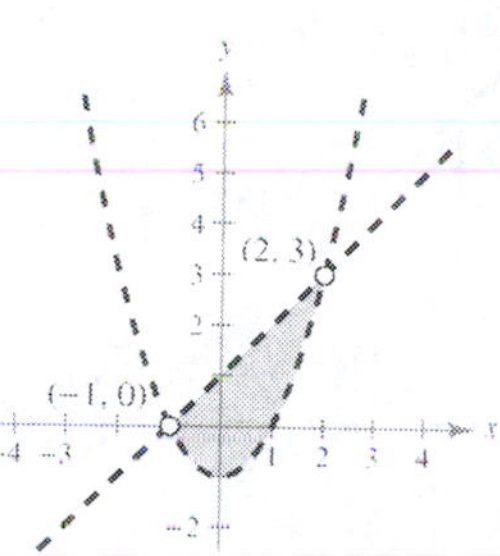

77. $\begin{cases} 2x - 3y \ge 0 \\ 2x - y \le 8 \\ y \ge 0 \end{cases}$

Vertex A

$2x - 3y = 0$
$2x - y = 8$
$(6, 4)$

Vertex B

$2x - 3y = 0$
$y = 0$
$(0, 0)$

Vertex C

$2x - y = 8$
$y = 0$
$(4, 0)$

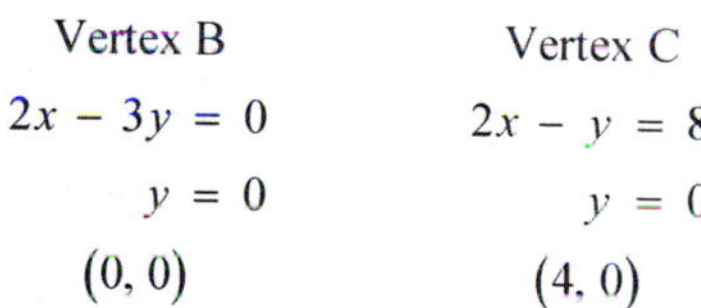

79. (a)

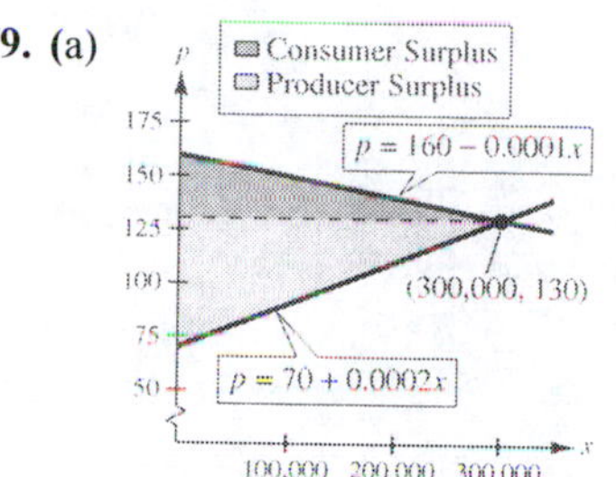

$160 - 0.0001x = 70 + 0.0002x$
$90 = 0.0003x$
$x = 300{,}000$ units
$p = \$130$

Point of equilibrium: $(300{,}000, 130)$

(b) Consumer surplus: $\frac{1}{2}(300{,}000)(30) = \$4{,}500{,}000$

Producer surplus: $\frac{1}{2}(300{,}000)(60) = \$9{,}000{,}000$

81. Rectangular region with vertices at:

$(3, 1), (7, 1), (7, 10)$, and $(3, 10)$

$\begin{cases} x \ge 3 \\ x \le 7 \\ y \ge 1 \\ y \le 10 \end{cases}$

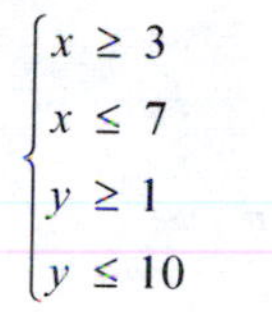

This system may be written as:

$\begin{cases} 3 \le x \le 7 \\ 1 \le y \le 10 \end{cases}$

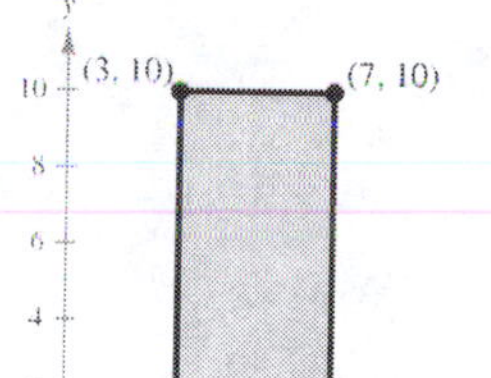

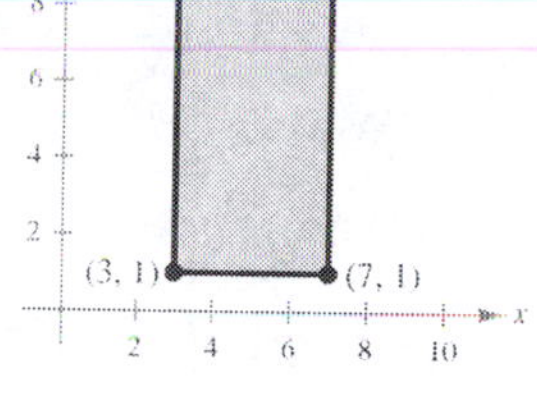

83. $x =$ number of units of Product I

$y =$ number of units of Product II

$\begin{cases} 20x + 30y \le 24{,}000 \\ 12x + 8y \le 12{,}400 \\ x \ge 0 \\ y \ge 0 \end{cases}$

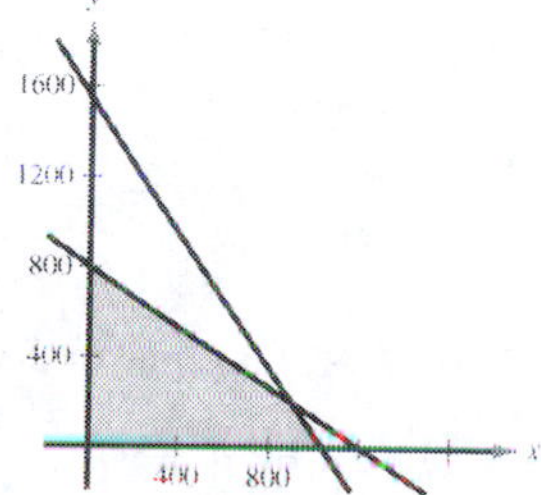

85. Objective function: $z = 3x + 4y$

Constraints: $\begin{cases} x \geq 0 \\ y \geq 0 \\ 2x + 5y \leq 50 \\ 4x + y \leq 28 \end{cases}$

At $(0, 0)$: $z = 0$

At $(0, 10)$: $z = 40$

At $(5, 8)$: $z = 47$

At $(7, 0)$: $z = 21$

The minimum value is 0 at $(0, 0)$.

The maximum value is 47 at $(5, 8)$.

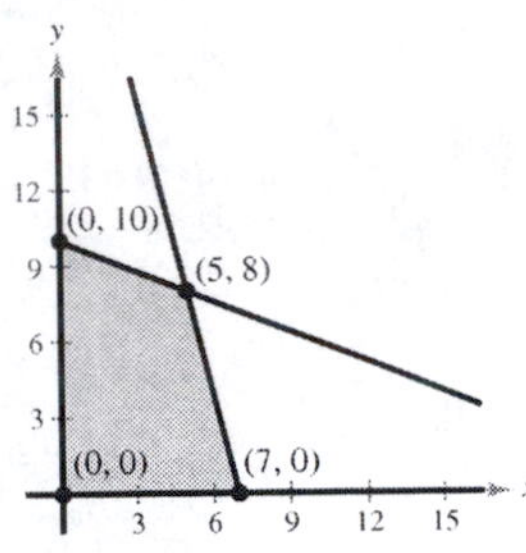

87. Objective function: $z = 1.75x + 2.25y$

Constraints: $\begin{cases} x \geq 0 \\ y \geq 0 \\ 2x + y \geq 25 \\ 3x + 2y \geq 45 \end{cases}$

At $(0, 25)$: $z = 56.25$

At $(5, 15)$: $z = 42.5$

At $(15, 0)$: $z = 26.25$

The minimum value is 26.25 at $(15, 0)$.

Because the region is unbounded, there is no maximum value.

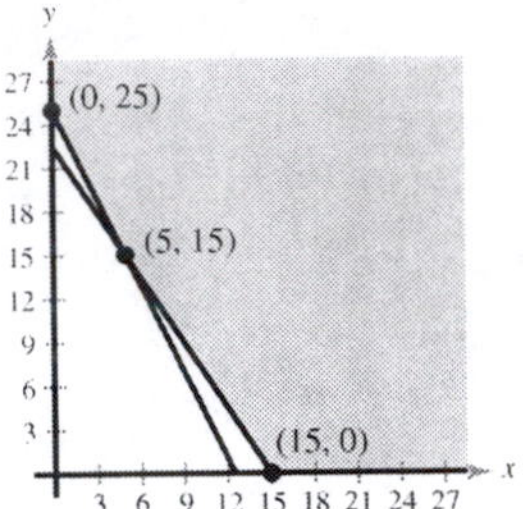

89. x = number of haircuts

y = number of permanents

Objective function: Optimize $R = 25x + 70y$ subject to the following constraints:

$$\begin{cases} x \geq 0 \\ y \geq 0 \\ \left(\frac{20}{60}\right)x + \left(\frac{70}{60}\right)y \leq 24 \Rightarrow 2x + 7y \leq 144 \end{cases}$$

At $(0, 0)$: $R = 0$

At $(72, 0)$: $R = 1800$

At $\left(0, \frac{144}{7}\right)$: $R = 1440$

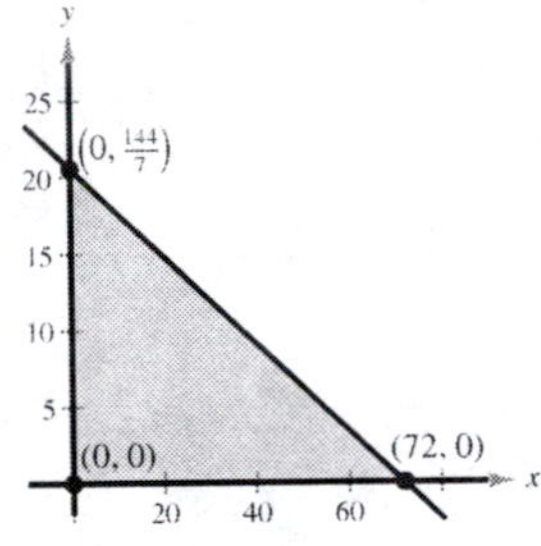

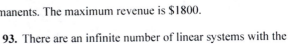

The revenue is optimal if the student does 72 haircuts and no permanents. The maximum revenue is \$1800.

91. True. Because $y = 5$ and $y = -2$ are horizontal lines, exactly one pair of opposite sides are parallel. The non-parallel sides of the trapezoid are equal in length. Therefore, the trapezoid is isosceles as shown below.

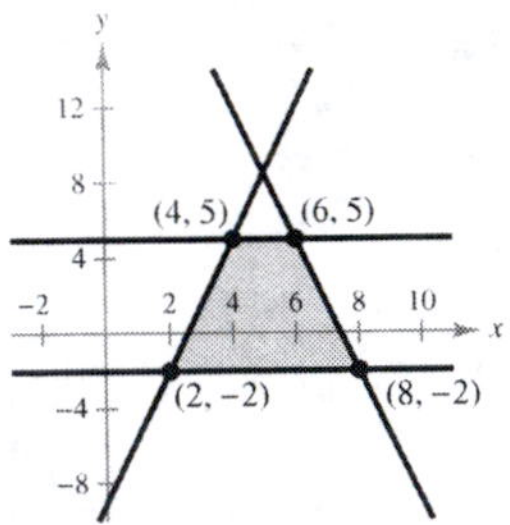

The distance from $(-4, 5)$ to $(2, -2)$ is equal to the distance from $(6, 5)$ to $(8, -2)$.

$$d_1 = \sqrt{(4-2)^2 + [5-(-2)]^2} = \sqrt{53}$$

$$d_2 = \sqrt{(8-6)^2 + (-2-5)^2} = \sqrt{53}$$

93. There are an infinite number of linear systems with the solution $(-8, 10)$. One possible system is:

$$\begin{cases} 4x + y = -22 \\ \frac{1}{2}x + y = 6 \end{cases}$$

95. There are infinite linear systems with the solution $\left(\frac{4}{3}, 3\right)$.

One possible system is:

$$\begin{cases} 3x + y = 7 \\ -6x + 3y = 1 \end{cases}$$

97. There are an infinite number of linear systems with the solution $(4, -1, 3)$. One possible system is as follows:

$$\begin{cases} x + y + z = 6 \\ x + y - z = 0 \\ x - y - z = 2 \end{cases}$$

99. There are an infinite number of linear systems with the solution $\left(5, \frac{3}{2}, 2\right)$. One possible system is:

$$\begin{cases} 2x + 2y - 3z = 7 \\ x - 2y + z = 4 \\ -x + 4y - z = -1 \end{cases}$$

101. A system of linear equations is inconsistent if it has no solution.

Problem Solving for Chapter 9

1. The longest side of the triangle is a diameter of the circle and has a length of 20.

The lines $y = \frac{1}{2}x + 5$ and $y = -2x + 20$ intersect at the point $(6, 8)$.

The distance between $(-10, 0)$ and $(6, 8)$ is:

$$d_1 = \sqrt{(6 - (-10))^2 + (8 - 0)^2} = \sqrt{320} = 8\sqrt{5}$$

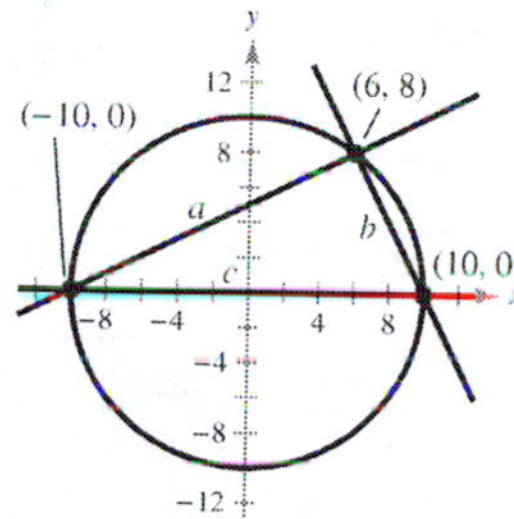

The distance between $(6, 8)$ and $(10, 0)$ is:

$$d_2 = \sqrt{(10 - 6)^2 + (0 - 8)^2} = \sqrt{80} = 4\sqrt{5}$$

Because $\left(\sqrt{320}\right)^2 + \left(\sqrt{80}\right)^2 = (20)^2$

$$400 = 400,$$

the sides of the triangle satisfy the Pythagorean Theorem. So, the triangle is a right triangle.

3. The system will have exactly one solution when the slopes of the line are *not* equal.

$$\begin{cases} ax + by = e \Rightarrow y = -\dfrac{a}{b}x + \dfrac{e}{b} \\ cx + dy = f \Rightarrow y = -\dfrac{c}{d}x + \dfrac{f}{d} \end{cases}$$

$$-\frac{a}{b} \neq -\frac{c}{d}$$

$$\frac{a}{b} \neq \frac{c}{d}$$

$$ad \neq bc$$

5. (a) $\begin{cases} x - 4y = -3 & \text{Eq. 1} \\ 5x - 6y = 13 & \text{Eq. 2} \end{cases}$

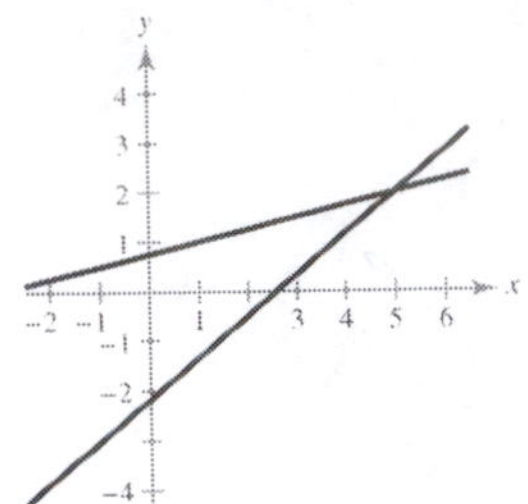

$\begin{cases} x - 4y = -3 \\ 14y = 28 & -5\text{Eq.1} + \text{Eq.2} \end{cases}$

$\begin{cases} x - 4y = -3 \\ y = 2 & \frac{1}{14}\text{Eq.2} \end{cases}$

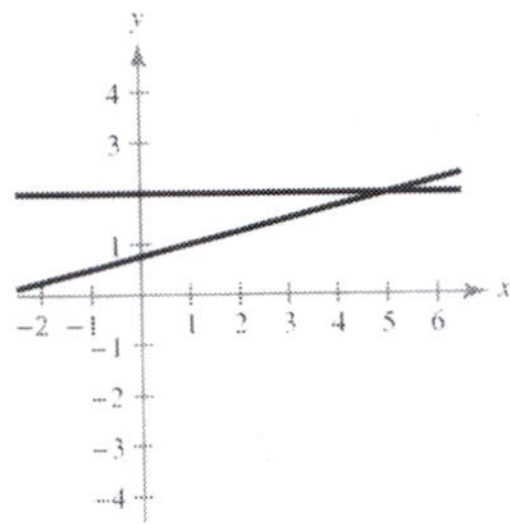

$\begin{cases} x = 5 & 4\text{Eq.2} + \text{Eq.1} \\ y = 2 \end{cases}$

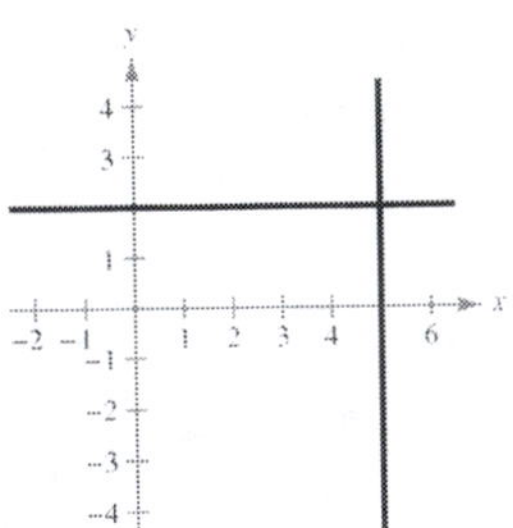

Solution: $(5, 2)$

(b) $\begin{cases} 2x - 3y = 7 & \text{Eq. 1} \\ -4x + 6y = -14 & \text{Eq. 2} \end{cases}$

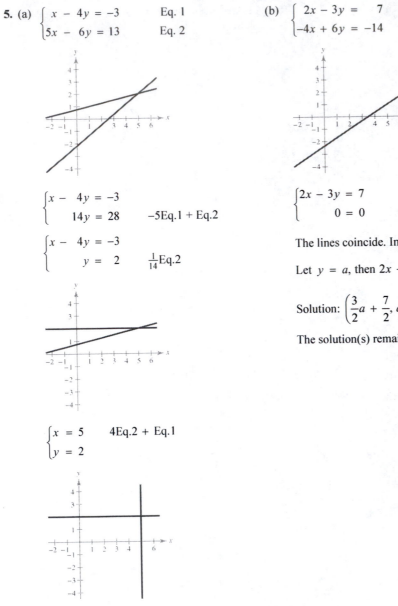

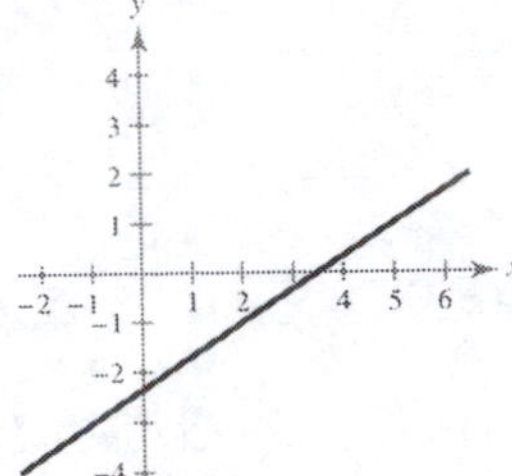

$\begin{cases} 2x - 3y = 7 \\ 0 = 0 & 2\text{Eq.1} + \text{Eq.2} \end{cases}$

The lines coincide. Infinite solutions.

Let $y = a$, then $2x - 3a = 7 \Rightarrow x = \frac{3}{2}a + \frac{7}{2}$

Solution: $\left(\frac{3}{2}a + \frac{7}{2}, a\right)$

The solution(s) remain the same at each step of the process.

7. The point where the two sections meet is at a depth of 10.1 feet. The distance between $(0, -10.1)$ and $(252.5, 0)$ is:

$d = \sqrt{(252.5 - 0)^2 + (0 - (-10.1))^2} = \sqrt{63{,}858.26}$

$d \approx 252.7$

Each section is approximately 252.7 feet long.

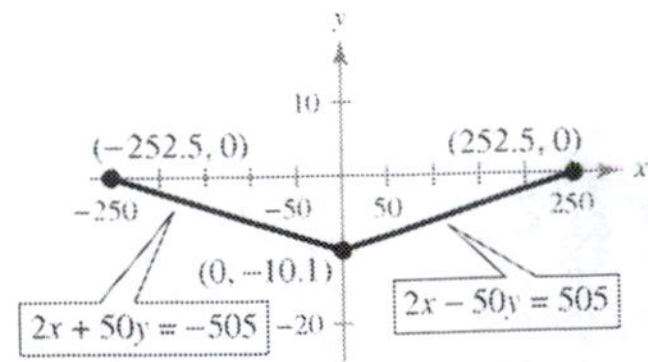

9. Let x = cost of the cable, per foot.

Let y = cost of a connector.

$\begin{cases} 6x + 2y = 15.50 \Rightarrow 6x + 2y = 15.50 \\ 3x + 2y = 10.25 \Rightarrow -3x - 2y = -10.25 \end{cases}$

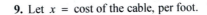

$3x = 5.25$

$x = 1.75$

$y = 2.50$

For a four-foot cable with a connector on each end, the cost should be $4(1.75) + 2(2.50) = \$12.00$.

11. Let $X = \frac{1}{x}$, $Y = \frac{1}{y}$, and $Z = \frac{1}{z}$.

(a) $$\begin{cases} \frac{12}{x} - \frac{12}{y} = 7 \Rightarrow 12X - 12Y = 7 \Rightarrow 12X - 12Y = 7 \\ \frac{3}{x} - \frac{4}{y} = 0 \Rightarrow 3X + 4Y = 0 \Rightarrow \underline{9X + 12Y = 0} \\ \qquad 21X = 7 \\ \qquad X = \frac{1}{3} \\ \qquad Y = -\frac{1}{4} \end{cases}$$

So, $\frac{1}{x} = \frac{1}{3} \Rightarrow x = 3$ and $\frac{1}{y} = -\frac{1}{4} \Rightarrow y = -4$.

Solution: $(3, -4)$

(b) $$\begin{cases} \frac{2}{x} + \frac{1}{y} - \frac{3}{z} = 4 \Rightarrow 2X + Y - 3Z = 4 & \text{Eq.1} \\ \frac{4}{x} + \frac{2}{z} = 10 \Rightarrow 4X + 2Z = 10 & \text{Eq.2} \\ -\frac{2}{x} + \frac{3}{y} - \frac{13}{z} = -8 \Rightarrow -2X + 3Y - 13Z = -8 & \text{Eq.3} \end{cases}$$

$$\begin{cases} 2X + Y - 3Z = 4 & \\ -2Y + 8Z = 2 & -2\text{Eq.1} + \text{Eq.2} \\ 4Y - 16Z = -4 & \text{Eq.1} + \text{Eq.3} \end{cases}$$

$$\begin{cases} 2X + Y - 3Z = 4 & \\ -2Y + 8Z = 2 & \\ 0 = 0 & 2\text{Eq.2} + \text{Eq.3} \end{cases}$$

The system has infinite solutions.

Let $Z = a$, then $Y = 4a - 1$ and $X = \frac{-a + 5}{2}$.

Then $\frac{1}{z} = a \Rightarrow z = \frac{1}{a}$, $\frac{1}{y} = 4a - 1 \Rightarrow y = \frac{1}{4a - 1}$, and $\frac{1}{x} = \frac{-a + 5}{2} \Rightarrow x = \frac{2}{-a + 5}$.

Solution: $\left(\frac{2}{-a + 5}, \frac{1}{4a - 1}, \frac{1}{a}\right)$, $a \neq 5, \frac{1}{4}, 0$

13. Solution: $(1, -1, 2)$

$$\begin{cases} 4x - 2y + 5z = 16 & \text{Equation 1} \\ x + y = 0 & \text{Equation 2} \\ -x - 3y + 2z = 6 & \text{Equation 3} \end{cases}$$

(a) $\begin{cases} 4x - 2y + 5z = 16 \\ x + y = 0 \end{cases}$

$\begin{cases} x + y = 0 \\ 4x - 2y + 5z = 16 \end{cases}$ Interchange the equations.

$\begin{cases} x + y = 0 \\ -6y + 5z = 16 \end{cases}$ -4Eq.1 + Eq.2

Let $z = a$, then $y = \dfrac{5a - 16}{6}$ and $x = \dfrac{-5a + 16}{6}$.

Solution: $\left(\dfrac{-5a + 16}{6}, \dfrac{5a - 16}{6}, a\right)$

When $a = 2$, we have the original solution.

(b) $\begin{cases} 4x - 2y + 5z = 16 \\ -x - 3y + 2z = 6 \end{cases}$

$\begin{cases} -x - 2y + 2z = 6 \\ 4x - 3y + 5z = 16 \end{cases}$ Interchange the equations.

$\begin{cases} -x - 3y + 2z = 6 \\ -14y + 13z = 40 \end{cases}$ 4Eq.1 + Eq.2

Let $z = a$, then $y = \dfrac{13a - 40}{14}$ and $x = \dfrac{-11a + 36}{14}$.

Solution: $\left(\dfrac{-11a + 36}{14}, \dfrac{13a - 40}{14}, a\right)$

When $a = 2$, we have the original solution.

(c) $\begin{cases} x + y = 0 \\ -x - 3y + 2z = 6 \end{cases}$

$\begin{cases} x + y = 0 \\ -2y + 2z = 6 \end{cases}$ Eq.1 + Eq.2

Let $z = a$, then $y = a - 3$ and $x = -a + 3$.

Solution: $(-a + 3, a - 3, a)$

When $a = 2$, we have the original solution.

(d) Each of these systems has infinite solutions.

15. t = amount of terrestrial vegetation in kilograms

a = amount of aquatic vegetation in kilograms

$$\begin{cases} a + t \le 32 \\ 0.15a \ge 1.9 \\ 193a + 772t \ge 11{,}000 \end{cases}$$

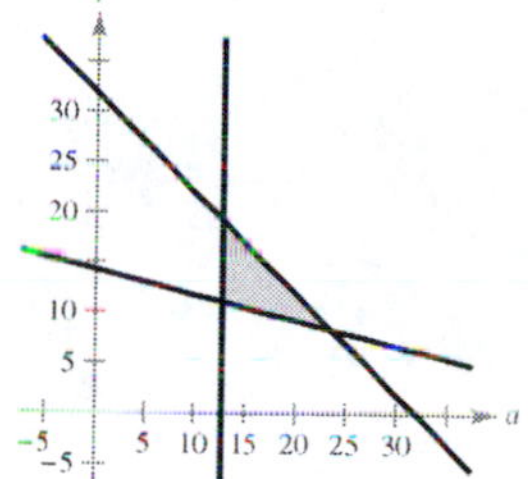

17. x = milligrams of HDL cholesterol

y = milligrams of LDL/VLDL cholesterol

(a) $\begin{cases} 0 < y \le 130 \\ x \ge 40 \\ x + y \le 200 \end{cases}$

(b)

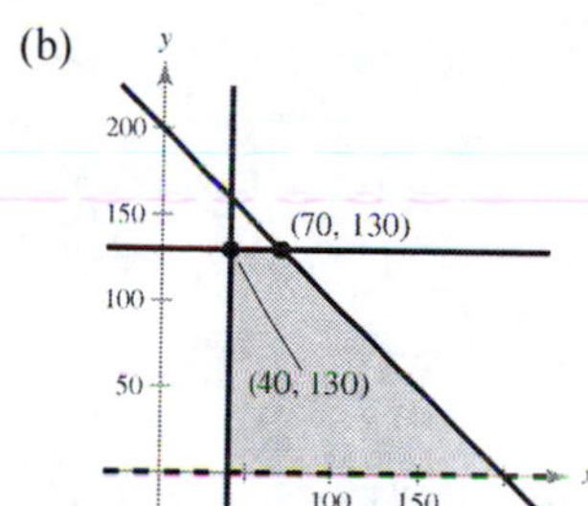

(c) $y = 120$ is in the region because $0 < y \le 130$.

$x = 90$ is in the region because $40 \le x \le 200$.

$x + y = 210$ is not the region because $x + y \le 200$.

(d) If the LDL/VLDL reading is 135 and the HDL reading is 65, then $x \ge 40$ and $x + y \le 200$, but $y \not< 130$.

(e) $\dfrac{x + y}{x} < 4$

$x + y < 4x$

$y < 3x$

The point $(75, 90)$ is in the region, and $\dfrac{165}{75} = 2.2 < 4$.

Practice Test for Chapter 9

For Exercises 1–3, solve the given system by the method of substitution.

1. $\begin{cases} x + y = 1 \\ 3x - y = 15 \end{cases}$

2. $\begin{cases} x - 3y = -3 \\ x^2 + 6y = 5 \end{cases}$

3. $\begin{cases} x + y + z = 6 \\ 2x - y + 3z = 0 \\ 5x + 2y - z = -3 \end{cases}$

4. Find the two numbers whose sum is 110 and product is 2800.

5. Find the dimensions of a rectangle if its perimeter is 170 feet and its area is 1500 square feet.

For Exercises 6–8, solve the linear system by elimination.

6. $\begin{cases} 2x + 15y = 4 \\ x - 3y = 23 \end{cases}$

7. $\begin{cases} x + y = 2 \\ 38x - 19y = 7 \end{cases}$

8. $\begin{cases} 0.4x + 0.5y = 0.112 \\ 0.3x - 0.7y = -0.131 \end{cases}$

9. Herbert invests $\$17{,}000$ in two funds that pay 11% and 13% simple interest, respectively. If he receives $\$2080$ in yearly interest, how much is invested in each fund?

10. Find the least squares regression line for the points $(4, 3)$, $(1, 1)$, $(-1, -2)$, and $(-2, -1)$.

For Exercises 11–12, solve the system of equations.

11. $\begin{cases} x + y = -2 \\ 2x - y + z = 11 \\ 4y - 3z = -20 \end{cases}$

12. $\begin{cases} 3x + 2y - z = 5 \\ 6x - y + 5z = 2 \end{cases}$

13. Find the equation of the parabola $y = ax^2 + bx + c$ passing through the points $(0, -1)$, $(1, 4)$ and $(2, 13)$.

For Exercises 14–15, write the partial fraction decomposition of the rational functions.

14. $\dfrac{10x - 17}{x^2 - 7x - 8}$

15. $\dfrac{x^2 + 4}{x^4 + x^2}$

16. Graph $x^2 + y^2 \geq 9$.

17. Graph the solution of the system.

$$\begin{cases} x + y \leq 6 \\ x \geq 2 \\ y \geq 0 \end{cases}$$

18. Derive a set of inequalities to describe the triangle with vertices $(0, 0)$, $(0, 7)$, and $(2, 3)$.

19. Find the maximum value of the objective function, $z = 30x + 26y$, subject to the following constraints.

$$\begin{cases} x \geq 0 \\ y \geq 0 \\ 2x + 3y \leq 21 \\ 5x + 3y \leq 30 \end{cases}$$

20. Graph the system of inequalities.

$$\begin{cases} x^2 + y^2 \leq 4 \\ (x - 2)^2 + y^2 \geq 4 \end{cases}$$

For Exercises 21–22, write the partial fraction decomposition for the rational expression.

21. $\dfrac{1 - 2x}{x^2 + x}$

22. $\dfrac{6x - 17}{(x - 3)^2}$

CHAPTER 10
Matrices and Determinants

CHAPTER 10
Matrices and Determinants

Section 10.1 Matrices and Systems of Equations

1. matrix

3. main diagonal

5. augmented

7. row-equivalent

9. Because the matrix has one row and two columns, its order is 1×2.

11. Because the matrix has three rows and one column, its order is 3×1.

13. Because the matrix has two rows and two columns, its order is 2×2.

15. Because the matrix has three rows and three columns, its order is 3×3.

17. $\begin{cases} 4x - 3y = -5 \\ -x + 3y = 12 \end{cases}$

$$\begin{bmatrix} 4 & -3 & \vdots & -5 \\ -1 & 3 & \vdots & 12 \end{bmatrix}$$

19. $\begin{cases} x + 10y - 2z = 2 \\ 5x - 3y + 4z = 0 \\ 2x + y = 6 \end{cases}$

$$\begin{bmatrix} 1 & 10 & -2 & \vdots & 2 \\ 5 & -3 & 4 & \vdots & 0 \\ 2 & 1 & 0 & \vdots & 6 \end{bmatrix}$$

21. $\begin{cases} 7x - 5y + z = 13 \\ 19x - 8z = 10 \end{cases}$

$$\begin{bmatrix} 7 & -5 & 1 & \vdots & 13 \\ 19 & 0 & -8 & \vdots & 10 \end{bmatrix}$$

23. $\begin{bmatrix} 1 & 2 & \vdots & 7 \\ 2 & -3 & \vdots & 4 \end{bmatrix}$

$$\begin{cases} x + 2y = 7 \\ 2x - 3y = 4 \end{cases}$$

25. $\begin{bmatrix} 2 & 0 & 5 & \vdots & -12 \\ 0 & 1 & -2 & \vdots & 7 \\ 6 & 3 & 0 & \vdots & 2 \end{bmatrix}$

$$\begin{cases} 2x + 5z = -12 \\ y - 2z = 7 \\ 6x + 3y = 2 \end{cases}$$

27. $\begin{bmatrix} 9 & 12 & 3 & 0 & \vdots & 0 \\ -2 & 18 & 5 & 2 & \vdots & 10 \\ 1 & 7 & -8 & 0 & \vdots & -4 \\ 3 & 0 & 2 & 0 & \vdots & -10 \end{bmatrix}$

$$\begin{cases} 9x + 12y + 3z = 0 \\ -2x + 18y + 5z + 2w = 10 \\ x + 7y - 8z = -4 \\ 3x + 2z = -10 \end{cases}$$

29. $\begin{bmatrix} -2 & 5 & 1 \\ 3 & -1 & -8 \end{bmatrix} \rightarrow \begin{bmatrix} 13 & 0 & -39 \\ 3 & -1 & -8 \end{bmatrix}$

Add 5 times Row 2 to Row 1.

31. $\begin{bmatrix} 0 & -1 & -5 & 5 \\ -1 & 3 & -7 & 6 \\ 4 & -5 & 1 & 3 \end{bmatrix} \rightarrow \begin{bmatrix} -1 & 3 & -7 & 6 \\ 0 & -1 & -5 & 5 \\ 0 & 7 & -27 & 27 \end{bmatrix}$

Interchange Row 1 and Row 2. Then add 4 times the new Row 1 to Row 3.

33. $\begin{bmatrix} 1 & 4 & 3 \\ 2 & 10 & 5 \end{bmatrix}$

$$-2R_1 + R_2 \rightarrow \begin{bmatrix} 1 & 4 & 3 \\ 0 & \boxed{2} & -1 \end{bmatrix}$$

35. $\begin{bmatrix} 1 & 1 & 1 \\ 5 & -2 & 4 \end{bmatrix}$

$$-5R_1 + R_2 \rightarrow \begin{bmatrix} 1 & 1 & 1 \\ 0 & \boxed{-7} & -1 \end{bmatrix}$$

37. $\begin{bmatrix} 1 & 5 & 4 & -1 \\ 0 & 1 & -2 & 2 \\ 0 & 0 & 1 & -7 \end{bmatrix}$

$-5R_2 + R_1 \rightarrow \begin{bmatrix} 1 & 0 & \boxed{14} & \boxed{-11} \\ 0 & 1 & -2 & 2 \\ 0 & 0 & 1 & -7 \end{bmatrix}$

39. $\begin{bmatrix} 1 & 1 & 4 & -1 \\ 3 & 8 & 10 & 3 \\ -2 & 1 & 12 & 6 \end{bmatrix}$

$\begin{matrix} -3R_1 + R_2 \rightarrow \\ 2R_1 + R_3 \rightarrow \end{matrix} \begin{bmatrix} 1 & 1 & 4 & -1 \\ 0 & 5 & \boxed{-2} & \boxed{6} \\ 0 & 3 & \boxed{20} & \boxed{4} \end{bmatrix}$

$\frac{1}{5}R_2 \rightarrow \begin{bmatrix} 1 & 1 & 4 & -1 \\ 0 & 1 & -\frac{2}{5} & \frac{6}{5} \\ 0 & 3 & \boxed{20} & \boxed{4} \end{bmatrix}$

41. (a) $\begin{bmatrix} -3 & 4 & \vdots & 22 \\ 6 & -4 & \vdots & -28 \end{bmatrix}$

(i) $R_1 + R_2 \rightarrow \begin{bmatrix} 3 & 0 & \vdots & -6 \\ 6 & -4 & \vdots & -28 \end{bmatrix}$

(ii) $-2R_1 + R_2 \rightarrow \begin{bmatrix} 3 & 0 & \vdots & -6 \\ 0 & -4 & \vdots & -16 \end{bmatrix}$

(iii) $-\frac{1}{4}R_2 \rightarrow \begin{bmatrix} 3 & 0 & \vdots & -6 \\ 0 & 1 & \vdots & 4 \end{bmatrix}$

(iv) $\frac{1}{3}R_1 \rightarrow \begin{bmatrix} 1 & 0 & \vdots & -2 \\ 0 & 1 & \vdots & 4 \end{bmatrix}$

The solution is $x = -2$ and $y = 4$.

(b) $\begin{cases} -3x + 4y = 22 \\ 6x - 4y = -28 \end{cases}$

$$3x = -6$$
$$x = -2$$

Back-substitute $x = -2$ into $-3x + 4y = 22$.

$$-3(-2) + 4y = 22$$
$$4y = 16$$
$$y = 4$$

The solution is $x = -2$ and $y = 4$.

(c) Answers vary. *Sample answer:* In this case, solving the system of linear equations using the elimination method was more efficient.

43. $\begin{bmatrix} 1 & 0 & 0 & 0 \\ 0 & 1 & 1 & 5 \\ 0 & 0 & 0 & 0 \end{bmatrix}$

This matrix is in reduced row-echelon form.

45. $\begin{bmatrix} 1 & 0 & 0 & 1 \\ 0 & 1 & 0 & -1 \\ 0 & 0 & 0 & 2 \end{bmatrix}$

This matrix is not in row-echelon form.

47. $\begin{bmatrix} 1 & 1 & 0 & 5 \\ -2 & -1 & 2 & -10 \\ 3 & 6 & 7 & 14 \end{bmatrix}$

$\begin{matrix} 2R_1 + R_2 \rightarrow \\ -3R_1 + R_3 \rightarrow \end{matrix} \begin{bmatrix} 1 & 1 & 0 & 5 \\ 0 & 1 & 2 & 0 \\ 0 & 3 & 7 & -1 \end{bmatrix}$

$-3R_2 + R_3 \rightarrow \begin{bmatrix} 1 & 1 & 0 & 5 \\ 0 & 1 & 2 & 0 \\ 0 & 0 & 1 & -1 \end{bmatrix}$

49. $\begin{bmatrix} 1 & -1 & -1 & 1 \\ 5 & -4 & 1 & 8 \\ -6 & 8 & 18 & 0 \end{bmatrix}$

$\begin{matrix} -5R_1 + R_2 \rightarrow \\ 6R_1 + R_3 \rightarrow \end{matrix} \begin{bmatrix} 1 & -1 & -1 & 1 \\ 0 & 1 & 6 & 3 \\ 0 & 2 & 12 & 6 \end{bmatrix}$

$-2R_2 + R_3 \rightarrow \begin{bmatrix} 1 & -1 & -1 & 1 \\ 0 & 1 & 6 & 3 \\ 0 & 0 & 0 & 0 \end{bmatrix}$

51. Use the reduced row-echelon form feature of a graphing utility.

$\begin{bmatrix} 3 & 3 & 3 \\ -1 & 0 & -4 \\ 2 & 4 & -2 \end{bmatrix} \Rightarrow \begin{bmatrix} 1 & 0 & 0 \\ 0 & 1 & 0 \\ 0 & 0 & 1 \end{bmatrix}$

53. Use the reduced row-echelon form feature of a graphing utility.

$\begin{bmatrix} 1 & 2 & 3 & -5 \\ 1 & 2 & 4 & -9 \\ -2 & -4 & -4 & 3 \\ 4 & 8 & 11 & -14 \end{bmatrix} \Rightarrow \begin{bmatrix} 1 & 2 & 0 & 0 \\ 0 & 0 & 1 & 0 \\ 0 & 0 & 0 & 1 \\ 0 & 0 & 0 & 0 \end{bmatrix}$

55. Use the reduced row-echelon form feature of a graphing utility.

$\begin{bmatrix} -3 & 5 & 1 & 12 \\ 1 & -1 & 1 & 4 \end{bmatrix} \Rightarrow \begin{bmatrix} 1 & 0 & 3 & 16 \\ 0 & 1 & 2 & 12 \end{bmatrix}$

57. $\begin{cases} x - 2y = 4 \\ y = -3 \end{cases}$

$x - 2(-3) = 4$

$x = -2$

Solution: $(-2, -3)$

59. $\begin{cases} x - y + 2z = 4 \\ y - z = 2 \\ z = -2 \end{cases}$

$y - (-2) = 2$

$y = 0$

$x - 0 + 2(-2) = 4$

$x = 8$

Solution: $(8, 0, -2)$

61. $\begin{bmatrix} 1 & 0 & \vdots & 3 \\ 0 & 1 & \vdots & -4 \end{bmatrix}$

$x = 3$

$y = -4$

Solution: $(3, -4)$

63. $\begin{bmatrix} 1 & 0 & 0 & \vdots & -4 \\ 0 & 1 & 0 & \vdots & -10 \\ 0 & 0 & 1 & \vdots & 4 \end{bmatrix}$

$x = -4$

$y = -10$

$z = 4$

Solution: $(-4, -10, 4)$

65. $\begin{cases} x + 2y = 7 \\ 2x + y = 8 \end{cases}$

$$\begin{bmatrix} 1 & 2 & \vdots & 7 \\ 2 & 1 & \vdots & 8 \end{bmatrix}$$

$$-2R_1 + R_2 \rightarrow \begin{bmatrix} 1 & 2 & \vdots & 7 \\ 0 & -3 & \vdots & -6 \end{bmatrix}$$

$$-\tfrac{1}{3}R_2 \rightarrow \begin{bmatrix} 1 & 2 & \vdots & 7 \\ 0 & 1 & \vdots & 2 \end{bmatrix}$$

$\begin{cases} x + 2y = 7 \\ y = 2 \end{cases}$

$y = 2$

$x + 2(2) = 7 \Rightarrow x = 3$

Solution: $(3, 2)$

67. $\begin{cases} 3x - 2y = -27 \\ x + 3y = 13 \end{cases}$

$$\begin{bmatrix} 3 & -2 & \vdots & -27 \\ 1 & 3 & \vdots & 13 \end{bmatrix}$$

$$\begin{matrix} R_1 \\ R_2 \end{matrix} \begin{bmatrix} 1 & 3 & \vdots & 13 \\ 3 & -2 & \vdots & -27 \end{bmatrix}$$

$$-3R_1 + R_2 \rightarrow \begin{bmatrix} 1 & 3 & \vdots & 13 \\ 0 & -11 & \vdots & -66 \end{bmatrix}$$

$$-\tfrac{1}{11}R_2 \rightarrow \begin{bmatrix} 1 & 3 & \vdots & 13 \\ 0 & 1 & \vdots & 6 \end{bmatrix}$$

$\begin{cases} x + 3y = 13 \\ y = 6 \end{cases}$

$y = 6$

$x + 3(6) = 13 \Rightarrow x = -5$

Solution: $(-5, 6)$

69. $\begin{cases} x + 2y - 3z = -28 \\ 4y + 2z = 0 \\ -x + y - z = -5 \end{cases}$

$$\begin{bmatrix} 1 & 2 & -3 & \vdots & -28 \\ 0 & 4 & 2 & \vdots & 0 \\ -1 & 1 & -1 & \vdots & -5 \end{bmatrix}$$

$$\begin{matrix} \\ \frac{1}{4}R_2 \rightarrow \\ R_1 + R_3 \rightarrow \end{matrix} \begin{bmatrix} 1 & 2 & -3 & \vdots & -28 \\ 0 & 1 & \frac{1}{2} & \vdots & 0 \\ 0 & 3 & -4 & \vdots & -33 \end{bmatrix}$$

$$-3R_2 + R_3 \rightarrow \begin{bmatrix} 1 & 2 & -3 & \vdots & -28 \\ 0 & 1 & \frac{1}{2} & \vdots & 0 \\ 0 & 0 & -\frac{11}{2} & \vdots & -33 \end{bmatrix}$$

$$-\frac{2}{11}R_3 \rightarrow \begin{bmatrix} 1 & 2 & -3 & \vdots & -28 \\ 0 & 1 & \frac{1}{2} & \vdots & 0 \\ 0 & 0 & 1 & \vdots & 6 \end{bmatrix}$$

$\begin{cases} x + 2y - 3z = -28 \\ y + \frac{1}{2}z = 0 \\ z = 6 \end{cases}$

$z = 6$

$y + \frac{1}{2}(6) = 0 \Rightarrow y = -3$

$x + 2(-3) - 3(6) = -28 \Rightarrow x = -4$

Solution: $(-4, -3, 6)$

71. $\begin{cases} -x + y = -22 \\ 3x + 4y = 4 \\ 4x - 8y = 32 \end{cases}$

$$\begin{bmatrix} -1 & 1 & \vdots & -22 \\ 3 & 4 & \vdots & 4 \\ 4 & -8 & \vdots & 32 \end{bmatrix}$$

$$\begin{matrix} -R_1 \rightarrow \\ \\ \\ \end{matrix} \begin{bmatrix} 1 & -1 & \vdots & 22 \\ 3 & 4 & \vdots & 4 \\ 4 & -8 & \vdots & 32 \end{bmatrix}$$

$$\begin{matrix} \\ -3R_1 + R_2 \rightarrow \\ -4R_1 + R_3 \rightarrow \end{matrix} \begin{bmatrix} 1 & -1 & \vdots & 22 \\ 0 & 7 & \vdots & -62 \\ 0 & -4 & \vdots & -56 \end{bmatrix}$$

$$\begin{matrix} \\ \frac{1}{7}R_2 \rightarrow \\ -\frac{1}{4}R_3 \rightarrow \end{matrix} \begin{bmatrix} 1 & -1 & \vdots & 22 \\ 0 & 1 & \vdots & -\frac{62}{7} \\ 0 & 1 & \vdots & 14 \end{bmatrix}$$

$$\begin{matrix} \\ \\ -R_2 + R_3 \rightarrow \end{matrix} \begin{bmatrix} 1 & -1 & \vdots & 22 \\ 0 & 1 & \vdots & -\frac{62}{7} \\ 0 & 0 & \vdots & \frac{160}{7} \end{bmatrix}$$

The system is inconsistent and there is no solution.

73. Use the reduced row-echelon form feature of a graphing utility.

$$\begin{cases} 3x + 2y - z + w = 0 \\ x - y + 4z + 2w = 25 \\ -2x + y + 2z - w = 2 \\ x + y + z + w = 6 \end{cases}$$

$$\begin{bmatrix} 3 & 2 & -1 & 1 & \vdots & 0 \\ 1 & -1 & 4 & 2 & \vdots & 25 \\ -2 & 1 & 2 & -1 & \vdots & 2 \\ 1 & 1 & 1 & 1 & \vdots & 6 \end{bmatrix} \Rightarrow \begin{bmatrix} 1 & 0 & 0 & 0 & \vdots & 3 \\ 0 & 1 & 0 & 0 & \vdots & -2 \\ 0 & 0 & 1 & 0 & \vdots & 5 \\ 0 & 0 & 0 & 1 & \vdots & 0 \end{bmatrix}$$

$x = 3$

$y = -2$

$z = 5$

$w = 0$

Solution: $(3, -2, 5, 0)$

75. $\begin{cases} -2x + 6y = -22 \\ x + 2y = -9 \end{cases}$

$$\begin{bmatrix} -2 & 6 & \vdots & -22 \\ 1 & 2 & \vdots & -9 \end{bmatrix}$$

$$\begin{matrix} R_1 \\ R_2 \end{matrix} \begin{bmatrix} 1 & 2 & \vdots & -9 \\ -2 & 6 & \vdots & -22 \end{bmatrix}$$

$$\begin{matrix} \\ 2R_1 + R_2 \rightarrow \end{matrix} \begin{bmatrix} 1 & 2 & \vdots & -9 \\ 0 & 10 & \vdots & -40 \end{bmatrix}$$

$$\begin{matrix} \\ \frac{1}{10}R_2 \rightarrow \end{matrix} \begin{bmatrix} 1 & 2 & \vdots & -9 \\ 0 & 1 & \vdots & -4 \end{bmatrix}$$

$\begin{cases} x + 2y = -9 \\ y = -4 \end{cases}$

$y = -4$

$x + 2(-4) = -9 \Rightarrow x = -1$

Solution: $(-1, -4)$

77. $\begin{cases} 8x - 4y = 7 \\ 5x + 2y = 1 \end{cases}$

$$\begin{bmatrix} 8 & -4 & \vdots & 7 \\ 5 & 2 & \vdots & 1 \end{bmatrix}$$

$$\begin{matrix} \\ -5R_1 + 8R_2 \rightarrow \end{matrix} \begin{bmatrix} 8 & -4 & \vdots & 7 \\ 0 & 36 & \vdots & -27 \end{bmatrix}$$

$$\begin{matrix} \\ \frac{1}{36}R_2 \rightarrow \end{matrix} \begin{bmatrix} 8 & -4 & \vdots & 7 \\ 0 & 1 & \vdots & -\frac{3}{4} \end{bmatrix}$$

$\begin{cases} 8x - 4y = 7 \\ y = -\frac{3}{4} \end{cases}$

$y = -\frac{3}{4}$

$8x - 4\left(-\frac{3}{4}\right) = 7 \Rightarrow x = \frac{1}{2}$

Solution: $\left(\frac{1}{2}, -\frac{3}{4}\right)$

79. $\begin{cases} x + 2y + z = 8 \\ 3x + 7y + 6z = 26 \end{cases}$

$$\begin{bmatrix} 1 & 2 & 1 & \vdots & 8 \\ 3 & 7 & 6 & \vdots & 26 \end{bmatrix}$$

$$\begin{matrix} \\ -3R_1 + R_2 \rightarrow \end{matrix} \begin{bmatrix} 1 & 2 & 1 & \vdots & 8 \\ 0 & 1 & 3 & \vdots & 2 \end{bmatrix}$$

$\begin{cases} x + 2y + z = 8 \\ y + 3z = 2 \end{cases}$

Let $z = a$.

$$y + 3a = 2 \Rightarrow y = -3a + 2$$

$$x + 2(-3a + 2) + a = 8 \Rightarrow x = 5a + 4$$

Solution: $(5a + 4, -3a + 2, a)$ where a is a real number

81. $\begin{cases} x - 3z = -2 \\ 3x + y - 2z = 5 \\ 2x + 2y + z = 4 \end{cases}$

$$\begin{bmatrix} 1 & 0 & -3 & \vdots & -2 \\ 3 & 1 & -2 & \vdots & 5 \\ 2 & 2 & 1 & \vdots & 4 \end{bmatrix}$$

$$\begin{matrix} \\ -3R_1 + R_2 \rightarrow \\ -2R_1 + R_3 \rightarrow \end{matrix} \begin{bmatrix} 1 & 0 & -3 & \vdots & -2 \\ 0 & 1 & 7 & \vdots & 11 \\ 0 & 2 & 7 & \vdots & 8 \end{bmatrix}$$

$$\begin{matrix} \\ \\ -2R_2 + R_3 \rightarrow \end{matrix} \begin{bmatrix} 1 & 0 & -3 & \vdots & -2 \\ 0 & 1 & 7 & \vdots & 11 \\ 0 & 0 & -7 & \vdots & -14 \end{bmatrix}$$

$$\begin{matrix} \\ \\ -\frac{1}{7}R_3 \rightarrow \end{matrix} \begin{bmatrix} 1 & 0 & -3 & \vdots & -2 \\ 0 & 1 & 7 & \vdots & 11 \\ 0 & 0 & 1 & \vdots & 2 \end{bmatrix}$$

$\begin{cases} x - 3z = -2 \\ y + 7z = 11 \\ z = 2 \end{cases}$

$$z = 2$$

$$y + 7(2) = 11 \Rightarrow y = -3$$

$$x - 3(2) = -2 \Rightarrow x = 4$$

Solution: $(4, -3, 2)$

83. $\begin{cases} -x + y - z = -14 \\ 2x - y + z = 21 \\ 3x + 2y + z = 19 \end{cases}$

$$\begin{bmatrix} -1 & 1 & -1 & \vdots & -14 \\ 2 & -1 & 1 & \vdots & 21 \\ 3 & 2 & 1 & \vdots & 19 \end{bmatrix}$$

$$\begin{matrix} -R_1 \rightarrow \\ \\ \end{matrix} \begin{bmatrix} 1 & -1 & 1 & \vdots & 14 \\ 2 & -1 & 1 & \vdots & 21 \\ 3 & 2 & 1 & \vdots & 19 \end{bmatrix}$$

$$\begin{matrix} \\ -2R_1 + R_2 \rightarrow \\ -3R_1 + R_3 \rightarrow \end{matrix} \begin{bmatrix} 1 & -1 & 1 & \vdots & 14 \\ 0 & 1 & -1 & \vdots & -7 \\ 0 & 5 & -2 & \vdots & -23 \end{bmatrix}$$

$$\begin{matrix} \\ \\ -5R_2 + R_3 \rightarrow \end{matrix} \begin{bmatrix} 1 & -1 & 1 & \vdots & 14 \\ 0 & 1 & -1 & \vdots & -7 \\ 0 & 0 & 3 & \vdots & 12 \end{bmatrix}$$

$$\begin{matrix} \\ \\ \frac{1}{3}R_3 \rightarrow \end{matrix} \begin{bmatrix} 1 & -1 & 1 & \vdots & 14 \\ 0 & 1 & -1 & \vdots & -7 \\ 0 & 0 & 1 & \vdots & 4 \end{bmatrix}$$

$\begin{cases} x - y + z = 14 \\ y - z = -7 \\ z = 4 \end{cases}$

$$z = 4$$

$$y - 4 = -7 \Rightarrow y = -3$$

$$x - (-3) + 4 = 14 \Rightarrow x = 7$$

Solution: $(7, -3, 4)$

85. Use the reduced row-echelon form feature of a graphic utility.

$$\begin{cases} 3x + 3y + 12z = 6 \\ x + y + 4z = 2 \\ 2x + 5y + 20z = 10 \\ -x + 2y + 8z = 4 \end{cases} \quad \begin{bmatrix} 3 & 3 & 12 & \vdots & 6 \\ 1 & 1 & 4 & \vdots & 2 \\ 2 & 5 & 20 & \vdots & 10 \\ -1 & 2 & 8 & \vdots & 4 \end{bmatrix} \Rightarrow \begin{bmatrix} 1 & 0 & 0 & \vdots & 0 \\ 0 & 1 & 4 & \vdots & 2 \\ 0 & 0 & 0 & \vdots & 0 \\ 0 & 0 & 0 & \vdots & 0 \end{bmatrix} \Rightarrow \begin{cases} x = 0 \\ y + 4z = 2 \end{cases}$$

Let $z = a$.

$y = 2 - 4a$

$x = 0$

Solution: $(0, 2 - 4a, a)$ where a is any real number

87. Use the reduced row-echelon form feature of a graphing utility.

$$\begin{cases} 2x + y - z + 2w = -6 \\ 3x + 4y + w = 1 \\ x + 5y + 2z + 6w = -3 \\ 5x + 2y - z - w = 3 \end{cases}$$

$$\begin{bmatrix} 2 & 1 & -1 & 2 & \vdots & -6 \\ 3 & 4 & 0 & 1 & \vdots & 1 \\ 1 & 5 & 2 & 6 & \vdots & -3 \\ 5 & 2 & -1 & -1 & \vdots & 3 \end{bmatrix} \Rightarrow \begin{bmatrix} 1 & 0 & 0 & 0 & \vdots & 1 \\ 0 & 1 & 0 & 0 & \vdots & 0 \\ 0 & 0 & 1 & 0 & \vdots & 4 \\ 0 & 0 & 0 & 1 & \vdots & -2 \end{bmatrix}$$

$x = 1$

$y = 0$

$z = 4$

$w = -2$

Solution: $(1, 0, 4, -2)$

89. Use the reduced row-echelon form feature of a graphing utility.

$$\begin{cases} x + y + z + w = 0 \\ 2x + 3y + z - 2w = 0 \\ 3x + 5y + z = 0 \end{cases}$$

$$\begin{bmatrix} 1 & 1 & 1 & 1 & \vdots & 0 \\ 2 & 3 & 1 & -2 & \vdots & 0 \\ 3 & 5 & 1 & 0 & \vdots & 0 \end{bmatrix} \Rightarrow \begin{bmatrix} 1 & 0 & 2 & 0 & \vdots & 0 \\ 0 & 1 & -1 & 0 & \vdots & 0 \\ 0 & 0 & 0 & 1 & \vdots & 0 \end{bmatrix}$$

$$\begin{cases} x + 2z = 0 \\ y - z = 0 \\ w = 0 \end{cases}$$

Let $z = a$. Then $x = -2a$ and $y = a$.

Solution: $(-2a, a, a, 0)$ where a is a real number

91. (a) $\begin{cases} x - 2y + z = -6 \\ y - 5z = 16 \\ z = -3 \end{cases}$

$y - 5(-3) = 16$

$y = 1$

$x - 2(1) + (-3) = -6$

$x = -1$

Solution: $(-1, 1, -3)$

(b) $\begin{cases} x + y - 2z = 6 \\ y + 3z = -8 \\ z = -3 \end{cases}$

$y + 3(-3) = -8$

$y = 1$

$x + 1 - 2(-3) = 6$

$x = -1$

Solution: $(-1, 1, -3)$

Both systems yield the same solution, $(-1, 1, -3)$.

93. (a) $\begin{cases} x - 4y + 5z = 27 \\ y - 7z = -54 \\ z = 8 \end{cases}$

$y - 7(8) = -54$

$y = 2$

$x - 4(2) + 5(8) = 27$

$x = -5$

Solution: $(-5, 2, 8)$

(b) $\begin{cases} x - 6y + z = 15 \\ y + 5z = 42 \\ z = 8 \end{cases}$

$y + 5(8) = 42$

$y = 2$

$x - 6(2) + 8 = 15$

$x = 19$

Solution: $(19, 2, 8)$

The systems do *not* yield the same solution.

95. $\begin{cases} a + b + c = 1 \\ 4a + 2b + c = -1 \\ 9a + 3b + c = -5 \end{cases}$

$$\begin{bmatrix} 1 & 1 & 1 & \vdots & 1 \\ 4 & 2 & 1 & \vdots & -1 \\ 9 & 3 & 1 & \vdots & -5 \end{bmatrix} \Rightarrow \begin{bmatrix} 1 & 0 & 0 & \vdots & -1 \\ 0 & 1 & 0 & \vdots & 1 \\ 0 & 0 & 1 & \vdots & 1 \end{bmatrix}$$

$a = 1$

$b = 1$

$c = 1$

So, $f(x) = -x^2 + x + 1$.

97. $\begin{cases} 4a - 2b + c = -15 \\ a - b + c = 7 \\ a + b + c = -3 \end{cases}$

$$\begin{bmatrix} 4 & -2 & 1 & \vdots & -15 \\ 4 & -1 & 1 & \vdots & 7 \\ 1 & 1 & 1 & \vdots & -3 \end{bmatrix} \Rightarrow \begin{bmatrix} 1 & 0 & 0 & \vdots & -9 \\ 0 & 1 & 0 & \vdots & -5 \\ 0 & 0 & 1 & \vdots & 11 \end{bmatrix}$$

$a = -9$

$b = -5$

$c = 11$

So, $f(x) = -9x^2 - 5x + 11$.

99. $\begin{cases} a + b + c = 8 \\ 4a + 2b + c = 13 \\ 9a + 3b + c = 20 \end{cases}$

$$\begin{bmatrix} 1 & 1 & 1 & \vdots & 8 \\ 4 & 2 & 1 & \vdots & 13 \\ 9 & 3 & 1 & \vdots & 20 \end{bmatrix} \Rightarrow \begin{bmatrix} 1 & 0 & 0 & \vdots & 1 \\ 0 & 1 & 0 & \vdots & 2 \\ 0 & 0 & 1 & \vdots & 5 \end{bmatrix}$$

$a = 1$

$b = 2$

$c = 5$

So, $f(x) = x^2 + 2x + 5$.

101. $\begin{cases} 12b + 66a = 831 \\ 66b + 506a = 5643 \end{cases}$

$$\begin{bmatrix} 12 & 66 & \vdots & 831 \\ 66 & 506 & \vdots & 5643 \end{bmatrix} \Rightarrow \begin{bmatrix} 1 & 0 & \vdots & 28 \\ 0 & 1 & \vdots & 7.5 \end{bmatrix}$$

$y = 7.5t + 28$

In 2014, the number of new cases of the waterborne disease will be $y = 7.5(14) + 28 = 33$ cases.

Because the data values increase in a linear pattern, this prediction is reasonable.

103. x = amount at 7%

y = amount at 8.5%

z = amount at 9.5%

$$\begin{cases} x + y + z = 2{,}000{,}000 \\ 0.07x + 0.085y + 0.095z = 169{,}750 \\ y - 4z = 0 \end{cases}$$

$$\begin{bmatrix} 1 & 1 & 1 & \vdots & 2{,}000{,}000 \\ 0.07 & 0.085 & 0.095 & \vdots & 169{,}750 \\ 0 & 1 & -4 & \vdots & 0 \end{bmatrix}$$

$$-0.07R_1 + R_2 \begin{bmatrix} 1 & 1 & 1 & \vdots & 2{,}000{,}000 \\ 0 & 0.015 & 0.025 & \vdots & 29{,}750 \\ 0 & 1 & -4 & \vdots & 0 \end{bmatrix}$$

$$1000R_2 \rightarrow \begin{bmatrix} 1 & 1 & 1 & \vdots & 2{,}000{,}000 \\ 0 & 15 & 25 & \vdots & 27{,}750{,}000 \\ 0 & 1 & -4 & \vdots & 0 \end{bmatrix}$$

$$R_2 + (-15)R_3 \rightarrow \begin{bmatrix} 1 & 1 & 1 & \vdots & 2{,}000{,}000 \\ 0 & 15 & 25 & \vdots & 29{,}750{,}000 \\ 0 & 0 & 85 & \vdots & 29{,}750{,}000 \end{bmatrix}$$

$$\begin{matrix} \\ \frac{1}{15}R_2 \rightarrow \\ \frac{1}{85}R_3 \rightarrow \end{matrix} \begin{bmatrix} 1 & 1 & 1 & \vdots & 2{,}000{,}000 \\ 0 & 1 & \frac{5}{3} & \vdots & \frac{5{,}950{,}000}{3} \\ 0 & 0 & -1 & \vdots & 350{,}000 \end{bmatrix}$$

The matrix is now in row-echelon form, and the corresponding system is shown.

$$\begin{cases} x + y + z = 2{,}000{,}000 \\ y + \frac{5}{3}z = \frac{5{,}950{,}000}{3} \\ z = 350{,}000 \end{cases}$$

$$y + \frac{5}{3}(350{,}000) = \frac{5{,}950{,}000}{3}$$

$$y = 1{,}400{,}000$$

$$x + 1{,}400{,}000 + 350{,}000 = 2{,}000{,}000$$

$$x = 250{,}000$$

The natural history museum borrowed \$250,000 at 7%, \$1,400,000 at 8.5%, and \$350,000 at 9.5%.

105. False. It is a 2×4 matrix.

107. $z = a$

$y = -4a + 1$

$x = -3a - 2$

One possible system is:

$$\begin{cases} x + y + 7z = (-3a - 2) + (-4a + 1) + 7a = -1 \\ x + 2y + 11z = (-3a - 2) + 2(-4a + 1) + 11a = 0 \\ 2x + y + 10z = 2(-3a - 2) + (-4a + 1) + 10a = -3 \end{cases} \quad \text{or} \quad \begin{cases} x + y + 7z = -1 \\ x + 2y + 11z = 0 \\ 2x + y + 10z = -3 \end{cases}$$

(Note that the coefficients of x, y, and z have been chosen so that the a-terms cancel.)

Section 10.2 Operations with Matrices

1. equal

3. zero; O

5. $x = -4$, $y = 22$

7. $2x + 1 = 5, 3x = 6, 3y - 5 = 4$

$x = 2, y = 3$

9. (a) $A + B = \begin{bmatrix} 1 & -1 \\ 2 & -1 \end{bmatrix} + \begin{bmatrix} 2 & -1 \\ -1 & 8 \end{bmatrix} = \begin{bmatrix} 1+2 & -1-1 \\ 2-1 & -1+8 \end{bmatrix} = \begin{bmatrix} 3 & -2 \\ 1 & 7 \end{bmatrix}$

(b) $A - B = \begin{bmatrix} 1 & -1 \\ 2 & -1 \end{bmatrix} - \begin{bmatrix} 2 & -1 \\ -1 & 8 \end{bmatrix} = \begin{bmatrix} 1-2 & -1+1 \\ 2+1 & -1-8 \end{bmatrix} = \begin{bmatrix} -1 & 0 \\ 3 & -9 \end{bmatrix}$

(c) $3A = 3\begin{bmatrix} 1 & -1 \\ 2 & -1 \end{bmatrix} = \begin{bmatrix} 3(1) & 3(-1) \\ 3(2) & 3(-1) \end{bmatrix} = \begin{bmatrix} 3 & -3 \\ 6 & -3 \end{bmatrix}$

(d) $3A - 2B = \begin{bmatrix} 3 & -3 \\ 6 & -3 \end{bmatrix} - 2\begin{bmatrix} 2 & -1 \\ -1 & 8 \end{bmatrix} = \begin{bmatrix} 3 & -3 \\ 6 & -3 \end{bmatrix} + \begin{bmatrix} -4 & 2 \\ 2 & -16 \end{bmatrix} = \begin{bmatrix} -1 & -1 \\ 8 & -19 \end{bmatrix}$

11. $A = \begin{bmatrix} 8 & -1 \\ 2 & 3 \\ -4 & 5 \end{bmatrix}, B = \begin{bmatrix} 1 & 6 \\ -1 & -5 \\ 1 & 10 \end{bmatrix}$

(a) $A + B = \begin{bmatrix} 8 & -1 \\ 2 & 3 \\ -4 & 5 \end{bmatrix} + \begin{bmatrix} 1 & 6 \\ -1 & -5 \\ 1 & 10 \end{bmatrix} = \begin{bmatrix} 8+1 & -1+6 \\ 2-1 & 3-5 \\ -4+1 & 5+10 \end{bmatrix} = \begin{bmatrix} 9 & 5 \\ 1 & -2 \\ -3 & 15 \end{bmatrix}$

(b) $A - B = \begin{bmatrix} 8 & -1 \\ 2 & 3 \\ -4 & 5 \end{bmatrix} - \begin{bmatrix} 1 & 6 \\ -1 & -5 \\ 1 & 10 \end{bmatrix} = \begin{bmatrix} 8-1 & -1-6 \\ 2-(-1) & 3-(-5) \\ -4-1 & 5-10 \end{bmatrix} = \begin{bmatrix} 7 & -7 \\ 3 & 8 \\ -5 & -5 \end{bmatrix}$

(c) $3A = 3\begin{bmatrix} 8 & -1 \\ 2 & 3 \\ -4 & 5 \end{bmatrix} = \begin{bmatrix} 3(8) & 3(-1) \\ 3(2) & 3(3) \\ 3(-4) & 3(5) \end{bmatrix} = \begin{bmatrix} 24 & -3 \\ 6 & 9 \\ -12 & 15 \end{bmatrix}$

(d) $3A - 2B = \begin{bmatrix} 24 & -3 \\ 6 & 9 \\ -12 & 15 \end{bmatrix} - 2\begin{bmatrix} 1 & 6 \\ -1 & -5 \\ 1 & 10 \end{bmatrix} = \begin{bmatrix} 24-2 & -3-12 \\ 6+2 & 9+10 \\ -12-2 & 15-20 \end{bmatrix} = \begin{bmatrix} 22 & -15 \\ 8 & 19 \\ -14 & -5 \end{bmatrix}$

13. $A = \begin{bmatrix} 4 & 5 & -1 & 3 & 4 \\ 1 & 2 & -2 & -1 & 0 \end{bmatrix}, B = \begin{bmatrix} 1 & 0 & -1 & 1 & 0 \\ -6 & 8 & 2 & -3 & -7 \end{bmatrix}$

(a) $A + B = \begin{bmatrix} 4 & 5 & -1 & 3 & 4 \\ 1 & 2 & -2 & -1 & 0 \end{bmatrix} + \begin{bmatrix} 1 & 0 & -1 & 1 & 0 \\ -6 & 8 & 2 & -3 & -7 \end{bmatrix} = \begin{bmatrix} 4+1 & 5+0 & -1-1 & 3+1 & 4+0 \\ 1-6 & 2+8 & -2+2 & -1-3 & 0-7 \end{bmatrix}$

$= \begin{bmatrix} 5 & 5 & -2 & 4 & 4 \\ -5 & 10 & 0 & -4 & -7 \end{bmatrix}$

(b) $A - B = \begin{bmatrix} 4 & 5 & -1 & 3 & 4 \\ 1 & 2 & -2 & -1 & 0 \end{bmatrix} - \begin{bmatrix} 1 & 0 & -1 & 1 & 0 \\ -6 & 8 & 2 & -3 & -7 \end{bmatrix} = \begin{bmatrix} 4-1 & 5-0 & -1-(-1) & 3-1 & 4-0 \\ 1-(-6) & 2-8 & -2-2 & -1-(-3) & 0-(-7) \end{bmatrix}$

$= \begin{bmatrix} 3 & 5 & 0 & 2 & 4 \\ 7 & -6 & -4 & 2 & 7 \end{bmatrix}$

(c) $3A = 3\begin{bmatrix} 4 & 5 & -1 & 3 & 4 \\ 1 & 2 & -2 & -1 & 0 \end{bmatrix} = \begin{bmatrix} 3(4) & 3(5) & 3(-1) & 3(3) & 3(4) \\ 3(1) & 3(2) & 3(-2) & 3(-1) & 3(0) \end{bmatrix} = \begin{bmatrix} 12 & 15 & -3 & 9 & 12 \\ 3 & 6 & -6 & -3 & 0 \end{bmatrix}$

(d) $3A - 2B = \begin{bmatrix} 12 & 15 & -3 & 9 & 12 \\ 3 & 6 & -6 & -3 & 0 \end{bmatrix} - 2\begin{bmatrix} 1 & 0 & -1 & 1 & 0 \\ -6 & 8 & 2 & -3 & -7 \end{bmatrix} = \begin{bmatrix} 12-2 & 15+0 & -3+2 & 9-2 & 12-0 \\ 3+12 & 6-16 & -6-4 & -3+6 & 0+14 \end{bmatrix}$

$= \begin{bmatrix} 10 & 15 & -1 & 7 & 12 \\ 15 & -10 & -10 & 3 & 14 \end{bmatrix}$

15. $A = \begin{bmatrix} 6 & 0 & 3 \\ -1 & -4 & 0 \end{bmatrix}, B = \begin{bmatrix} 8 & -1 \\ 4 & -3 \end{bmatrix}$

(a) $A + B$ is not possible. A and B do not have the same order.

(b) $A - B$ is not possible. A and B do not have the same order.

(c) $3A = \begin{bmatrix} 18 & 0 & 9 \\ -3 & -12 & 0 \end{bmatrix}$

(d) $3A - 2B$ is not possible. A and B do not have the same order.

17. $\begin{bmatrix} -5 & 0 \\ 3 & -6 \end{bmatrix} + \begin{bmatrix} 7 & 1 \\ -2 & -1 \end{bmatrix} + \begin{bmatrix} -10 & -8 \\ 14 & 6 \end{bmatrix} = \begin{bmatrix} -5+7+(-10) & 0+1+(-8) \\ 3+(-2)+14 & -6+(-1)+6 \end{bmatrix} = \begin{bmatrix} -8 & -7 \\ 15 & -1 \end{bmatrix}$

19. $4\left(\begin{bmatrix} -4 & 0 & 1 \\ 0 & 2 & 3 \end{bmatrix} - \begin{bmatrix} 2 & 1 & -2 \\ 3 & -6 & 0 \end{bmatrix}\right) = 4\begin{bmatrix} -6 & -1 & 3 \\ -3 & 8 & 3 \end{bmatrix} = \begin{bmatrix} -24 & -4 & 12 \\ -12 & 32 & 12 \end{bmatrix}$

21. $-3\left(\begin{bmatrix} 0 & -3 \\ 7 & 2 \end{bmatrix} + \begin{bmatrix} -6 & 3 \\ 8 & 1 \end{bmatrix}\right) - 2\begin{bmatrix} 4 & -4 \\ 7 & -9 \end{bmatrix} = -3\begin{bmatrix} -6 & 0 \\ 15 & 3 \end{bmatrix} - \begin{bmatrix} 8 & -8 \\ 14 & -18 \end{bmatrix} = \begin{bmatrix} 18 & 0 \\ -45 & -9 \end{bmatrix} - \begin{bmatrix} 8 & -8 \\ 14 & -18 \end{bmatrix} = \begin{bmatrix} 10 & 8 \\ -59 & 9 \end{bmatrix}$

23. $\frac{11}{25}\begin{bmatrix} 2 & 5 \\ -1 & -4 \end{bmatrix} + 6\begin{bmatrix} -3 & 0 \\ 2 & 2 \end{bmatrix} = \begin{bmatrix} -17.12 & 2.2 \\ 11.56 & 10.24 \end{bmatrix}$

25. $-\begin{bmatrix} 3.211 & 6.829 \\ -1.004 & 4.914 \\ 0.055 & -3.889 \end{bmatrix} - \begin{bmatrix} -1.630 & -3.090 \\ 5.256 & 8.335 \\ -9.768 & 4.251 \end{bmatrix} = \begin{bmatrix} -1.581 & -3.739 \\ -4.252 & -13.249 \\ 9.713 & -0.362 \end{bmatrix}$

27. $X = 2A + 2B = 2\begin{bmatrix} -2 & -1 \\ 1 & 0 \\ 3 & -4 \end{bmatrix} + 2\begin{bmatrix} 0 & 3 \\ 2 & 0 \\ -4 & -1 \end{bmatrix} = \begin{bmatrix} -4 & -2 \\ 2 & 0 \\ 6 & -8 \end{bmatrix} + \begin{bmatrix} 0 & 6 \\ 4 & 0 \\ -8 & -2 \end{bmatrix} = \begin{bmatrix} -4 & 4 \\ 6 & 0 \\ -2 & -10 \end{bmatrix}$

29. $2X = A + B$

$$X = \frac{1}{2}(A + B) = \frac{1}{2}\left(\begin{bmatrix} -2 & -1 \\ 1 & 0 \\ 3 & -4 \end{bmatrix} + \begin{bmatrix} 0 & 3 \\ 2 & 0 \\ -4 & -1 \end{bmatrix}\right) = \frac{1}{2}\begin{bmatrix} -2 & 2 \\ 3 & 0 \\ -1 & -5 \end{bmatrix} = \begin{bmatrix} -1 & 1 \\ \frac{3}{2} & 0 \\ -\frac{1}{2} & -\frac{5}{2} \end{bmatrix}$$

31. $$X = -\frac{3}{2}A + \frac{1}{2}B = -\frac{3}{2}\begin{bmatrix} -2 & -1 \\ 1 & 0 \\ 3 & -4 \end{bmatrix} + \frac{1}{2}\begin{bmatrix} 0 & 3 \\ 2 & 0 \\ -4 & -1 \end{bmatrix} = \begin{bmatrix} 3 & \frac{3}{2} \\ -\frac{3}{2} & 0 \\ -\frac{9}{2} & 6 \end{bmatrix} + \begin{bmatrix} 0 & \frac{3}{2} \\ 1 & 0 \\ -2 & -\frac{1}{2} \end{bmatrix} = \begin{bmatrix} 3 & 3 \\ -\frac{1}{2} & 0 \\ -\frac{13}{2} & \frac{11}{2} \end{bmatrix}$$

33. $2A + 4B = -2X$

$$X = -A - 2B = -1\begin{bmatrix} -2 & -1 \\ 1 & 0 \\ 3 & -4 \end{bmatrix} - 2\begin{bmatrix} 0 & 3 \\ 2 & 0 \\ -4 & -1 \end{bmatrix} = \begin{bmatrix} 2 & 1 \\ -1 & 0 \\ -3 & 4 \end{bmatrix} + \begin{bmatrix} 0 & -6 \\ -4 & 0 \\ 8 & 2 \end{bmatrix} = \begin{bmatrix} 2 & -5 \\ -5 & 0 \\ 5 & 6 \end{bmatrix}$$

35. A is 3×2 and B is 3×3. AB is not possible.

37. A is 3×2, B is $2 \times 2 \Rightarrow AB$ is 3×2.

$$A = \begin{bmatrix} -1 & 6 \\ -4 & 5 \\ 0 & 3 \end{bmatrix}, B = \begin{bmatrix} 2 & 3 \\ 0 & 9 \end{bmatrix}$$

$$AB = \begin{bmatrix} -1 & 6 \\ -4 & 5 \\ 0 & 3 \end{bmatrix}\begin{bmatrix} 2 & 3 \\ 0 & 9 \end{bmatrix} = \begin{bmatrix} (-1)(2) + (6)(0) & (-1)(3) + (6)(9) \\ (-4)(2) + (5)(0) & (-4)(3) + (5)(9) \\ (0)(2) + (3)(0) & (0)(3) + (3)(9) \end{bmatrix} = \begin{bmatrix} -2 & 51 \\ -8 & 33 \\ 0 & 27 \end{bmatrix}$$

39. A is 3×3, B is $3 \times 3 \Rightarrow AB$ is 3×3.

$$AB = \begin{bmatrix} 5 & 0 & 0 \\ 0 & -8 & 0 \\ 0 & 0 & 7 \end{bmatrix}\begin{bmatrix} \frac{1}{5} & 0 & 0 \\ 0 & -\frac{1}{8} & 0 \\ 0 & 0 & \frac{1}{2} \end{bmatrix} = \begin{bmatrix} 1 & 0 & 0 \\ 0 & 1 & 0 \\ 0 & 0 & \frac{7}{2} \end{bmatrix}$$

41. A is 2×1, B is $1 \times 4 \Rightarrow AB$ is 2×4.

$$\begin{bmatrix} 10 \\ 12 \end{bmatrix}\begin{bmatrix} 6 & -2 & 1 & 6 \end{bmatrix} = \begin{bmatrix} 60 & -20 & 10 & 60 \\ 72 & -24 & 12 & 72 \end{bmatrix}$$

43. $$A = \begin{bmatrix} 7 & 5 & -4 \\ -2 & 5 & 1 \\ 10 & -4 & -7 \end{bmatrix}, B = \begin{bmatrix} 2 & -2 & 3 \\ 8 & 1 & 4 \\ -4 & 2 & -8 \end{bmatrix}$$

$$AB = \begin{bmatrix} 7 & 5 & -4 \\ -2 & 5 & 1 \\ 10 & -4 & -7 \end{bmatrix}\begin{bmatrix} 2 & -2 & 3 \\ 8 & 1 & 4 \\ -4 & 2 & -8 \end{bmatrix} = \begin{bmatrix} 70 & -17 & 73 \\ 32 & 11 & 6 \\ 16 & -38 & 70 \end{bmatrix}$$

45. $$\begin{bmatrix} -3 & 8 & -6 & 8 \\ -12 & 15 & 9 & 6 \\ 5 & -1 & 1 & 5 \end{bmatrix}\begin{bmatrix} 3 & 1 & 6 \\ 24 & 15 & 14 \\ 16 & 10 & 21 \\ 8 & -4 & 10 \end{bmatrix} = \begin{bmatrix} 151 & 25 & 48 \\ 516 & 279 & 387 \\ 47 & -20 & 87 \end{bmatrix}$$

47. (a) $$AB = \begin{bmatrix} 1 & 2 \\ 4 & 2 \end{bmatrix}\begin{bmatrix} 2 & -1 \\ -1 & 8 \end{bmatrix} = \begin{bmatrix} (1)(2) + (2)(-1) & (1)(-1) + (2)(8) \\ (4)(2) + (2)(-1) & (4)(-1) + (2)(8) \end{bmatrix} = \begin{bmatrix} 0 & 15 \\ 6 & 12 \end{bmatrix}$$

(b) $$BA = \begin{bmatrix} 2 & -1 \\ -1 & 8 \end{bmatrix}\begin{bmatrix} 1 & 2 \\ 4 & 2 \end{bmatrix} = \begin{bmatrix} (2)(1) + (-1)(4) & (2)(2) + (-1)(2) \\ (-1)(1) + (8)(4) & (-1)(2) + (8)(2) \end{bmatrix} = \begin{bmatrix} -2 & 2 \\ 31 & 14 \end{bmatrix}$$

(c) $$A^2 = \begin{bmatrix} 1 & 2 \\ 4 & 2 \end{bmatrix}\begin{bmatrix} 1 & 2 \\ 4 & 2 \end{bmatrix} = \begin{bmatrix} (1)(1) + (2)(4) & (1)(2) + (2)(2) \\ (4)(1) + (2)(4) & (4)(2) + (2)(2) \end{bmatrix} = \begin{bmatrix} 9 & 6 \\ 12 & 12 \end{bmatrix}$$

49. (a) $AB = \begin{bmatrix} 3 & -1 \\ 1 & 3 \end{bmatrix}\begin{bmatrix} 1 & -3 \\ 3 & 1 \end{bmatrix} = \begin{bmatrix} (3)(1) + (-1)(3) & (3)(-3) + (-1)(1) \\ (1)(1) + (3)(3) & (1)(-3) + (3)(1) \end{bmatrix} = \begin{bmatrix} 0 & -10 \\ 10 & 0 \end{bmatrix}$

(b) $BA = \begin{bmatrix} 1 & -3 \\ 3 & 1 \end{bmatrix}\begin{bmatrix} 3 & -1 \\ 1 & 3 \end{bmatrix} = \begin{bmatrix} (1)(3) + (-3)(1) & (1)(-1) + (-3)(3) \\ (3)(3) + (1)(1) & (3)(-1) + (1)(3) \end{bmatrix} = \begin{bmatrix} 0 & -10 \\ 10 & 0 \end{bmatrix}$

(c) $A^2 = \begin{bmatrix} 3 & -1 \\ 1 & 3 \end{bmatrix}\begin{bmatrix} 3 & -1 \\ 1 & 3 \end{bmatrix} = \begin{bmatrix} (3)(3) + (-1)(1) & (3)(-1) + (-1)(3) \\ (1)(3) + (3)(1) & (1)(-1) + (3)(3) \end{bmatrix} = \begin{bmatrix} 8 & -6 \\ 6 & 8 \end{bmatrix}$

51. (a) $AB = \begin{bmatrix} 7 \\ 8 \\ -1 \end{bmatrix}\begin{bmatrix} 1 & 1 & 2 \end{bmatrix} = \begin{bmatrix} 7(1) & 7(1) & 7(2) \\ 8(1) & 8(1) & 8(2) \\ -1(1) & -1(1) & -1(2) \end{bmatrix} = \begin{bmatrix} 7 & 7 & 4 \\ 8 & 8 & 16 \\ -1 & -1 & -2 \end{bmatrix}$

(b) $BA = \begin{bmatrix} 1 & 1 & 2 \end{bmatrix}\begin{bmatrix} 7 \\ 8 \\ -1 \end{bmatrix} = \begin{bmatrix} (1)(7) + (1)(8) + (2)(-1) \end{bmatrix} = \begin{bmatrix} 13 \end{bmatrix}$

(c) A^2 is not possible.

53. $\begin{bmatrix} 3 & 1 \\ 0 & -2 \end{bmatrix}\begin{bmatrix} 1 & 0 \\ -2 & 2 \end{bmatrix}\begin{bmatrix} 1 & 0 \\ 2 & 4 \end{bmatrix} = \begin{bmatrix} 1 & 2 \\ 4 & -4 \end{bmatrix}\begin{bmatrix} 1 & 0 \\ 2 & 4 \end{bmatrix} = \begin{bmatrix} 5 & 8 \\ -4 & -16 \end{bmatrix}$

55. $\begin{bmatrix} 0 & 2 & -2 \\ 4 & 1 & 2 \end{bmatrix}\left(\begin{bmatrix} 4 & 0 \\ 0 & -1 \\ -1 & 2 \end{bmatrix} + \begin{bmatrix} -2 & 3 \\ -3 & 5 \\ 0 & -3 \end{bmatrix}\right) = \begin{bmatrix} 0 & 2 & -2 \\ 4 & 1 & 2 \end{bmatrix}\begin{bmatrix} 2 & 3 \\ -3 & 4 \\ -1 & -1 \end{bmatrix} = \begin{bmatrix} -4 & 10 \\ 3 & 14 \end{bmatrix}$

57. (a) $\begin{bmatrix} -1 & 1 \\ -2 & 1 \end{bmatrix}\begin{bmatrix} x_1 \\ x_2 \end{bmatrix} = \begin{bmatrix} 4 \\ 0 \end{bmatrix}$

(b) $\begin{bmatrix} -1 & 1 & \vdots & 4 \\ -2 & 1 & \vdots & 0 \end{bmatrix}$

$-R_2 + R_1 \rightarrow \begin{bmatrix} 1 & 0 & \vdots & 4 \\ -2 & 1 & \vdots & 0 \end{bmatrix}$

$2R_1 + R_2 \rightarrow \begin{bmatrix} 1 & 0 & \vdots & 4 \\ 0 & 1 & \vdots & 8 \end{bmatrix}$

$X = \begin{bmatrix} 4 \\ 8 \end{bmatrix}$

59. (a) $\begin{bmatrix} -2 & -3 \\ 6 & 1 \end{bmatrix}\begin{bmatrix} x_1 \\ x_2 \end{bmatrix} = \begin{bmatrix} -4 \\ -36 \end{bmatrix}$

(b) $\begin{bmatrix} -2 & -3 & \vdots & -4 \\ 6 & 1 & \vdots & -36 \end{bmatrix}$

$3R_1 + R_2 \rightarrow \begin{bmatrix} -2 & -3 & \vdots & -4 \\ 0 & -8 & \vdots & -48 \end{bmatrix}$

$\begin{matrix} -\frac{1}{2}R_1 \rightarrow \\ -\frac{1}{8}R_2 \rightarrow \end{matrix} \begin{bmatrix} 1 & \frac{3}{2} & \vdots & 2 \\ 0 & 1 & \vdots & 6 \end{bmatrix}$

$-\frac{3}{2}R_2 + R_1 \rightarrow \begin{bmatrix} 1 & 0 & \vdots & -7 \\ 0 & 1 & \vdots & 6 \end{bmatrix}$

$X = \begin{bmatrix} -7 \\ 6 \end{bmatrix}$

61. (a) $\begin{bmatrix} 1 & -2 & 3 \\ -1 & 3 & -1 \\ 2 & -5 & 5 \end{bmatrix}\begin{bmatrix} x_1 \\ x_2 \\ x_3 \end{bmatrix} = \begin{bmatrix} 9 \\ -6 \\ 17 \end{bmatrix}$

(b) $\left[\begin{array}{ccc:c} 1 & -2 & 3 & 9 \\ -1 & 3 & -1 & -6 \\ 2 & -5 & 5 & 17 \end{array}\right]$

$\begin{array}{r} \\ R_1 + R_2 \rightarrow \\ -2R_2 + R_3 \rightarrow \end{array}\left[\begin{array}{ccc:c} 1 & -2 & 3 & 9 \\ 0 & 1 & 2 & 3 \\ 0 & -1 & -1 & -1 \end{array}\right]$

$\begin{array}{r} 2R_2 + R_1 \rightarrow \\ \\ R_2 + R_3 \rightarrow \end{array}\left[\begin{array}{ccc:c} 1 & 0 & 7 & 15 \\ 0 & 1 & 2 & 3 \\ 0 & 0 & 1 & 2 \end{array}\right]$

$\begin{array}{r} -7R_3 + R_1 \rightarrow \\ -2R_3 + R_2 \rightarrow \\ \\ \end{array}\left[\begin{array}{ccc:c} 1 & 0 & 0 & 1 \\ 0 & 1 & 0 & -1 \\ 0 & 0 & 1 & 2 \end{array}\right]$

$X = \begin{bmatrix} 1 \\ -1 \\ 2 \end{bmatrix}$

63. (a) $\begin{bmatrix} 1 & -5 & 2 \\ -3 & 1 & -1 \\ 0 & -2 & 5 \end{bmatrix}\begin{bmatrix} x_1 \\ x_2 \\ x_3 \end{bmatrix} = \begin{bmatrix} -20 \\ 8 \\ -16 \end{bmatrix}$

(b) $\left[\begin{array}{ccc:c} 1 & -5 & 2 & -20 \\ -3 & 1 & -1 & 8 \\ 0 & -2 & 5 & -16 \end{array}\right]$

$\begin{array}{r} \\ 3R_1 + R_2 \rightarrow \\ \\ \end{array}\left[\begin{array}{ccc:c} 1 & -5 & 2 & -20 \\ 0 & -14 & 5 & -52 \\ 0 & -2 & 5 & -16 \end{array}\right]$

$\begin{array}{r} \\ -R_3 + R_2 \rightarrow \\ \\ \end{array}\left[\begin{array}{ccc:c} 1 & -5 & 2 & -20 \\ 0 & -12 & 0 & -36 \\ 0 & -2 & 5 & -16 \end{array}\right]$

$\begin{array}{r} \\ -\frac{1}{12}R_2 \rightarrow \\ \\ \end{array}\left[\begin{array}{ccc:c} 1 & -5 & 2 & -20 \\ 0 & 1 & 0 & 3 \\ 0 & -2 & 5 & -16 \end{array}\right]$

$\begin{array}{r} 5R_2 + R_1 \rightarrow \\ \\ 2R_2 + R_3 \rightarrow \end{array}\left[\begin{array}{ccc:c} 1 & 0 & 2 & -5 \\ 0 & 1 & 0 & 3 \\ 0 & 0 & 5 & -10 \end{array}\right]$

$\begin{array}{r} \\ \\ \frac{1}{5}R_3 \rightarrow \end{array}\left[\begin{array}{ccc:c} 1 & 0 & 2 & -5 \\ 0 & 1 & 0 & 3 \\ 0 & 0 & 1 & -2 \end{array}\right]$

$\begin{array}{r} -2R_3 + R_1 \rightarrow \\ \\ \\ \end{array}\left[\begin{array}{ccc:c} 1 & 0 & 0 & -1 \\ 0 & 1 & 0 & 3 \\ 0 & 0 & 1 & -2 \end{array}\right]$

$X = \begin{bmatrix} -1 \\ 3 \\ -2 \end{bmatrix}$

65. $1.10\begin{bmatrix} 100 & 90 & 70 & 30 \\ 40 & 20 & 60 & 60 \end{bmatrix} = \begin{bmatrix} 110 & 99 & 77 & 33 \\ 44 & 22 & 66 & 66 \end{bmatrix}$

67. $BA = \begin{bmatrix} 3.50 & 6.00 \end{bmatrix}\begin{bmatrix} 125 & 100 & 75 \\ 100 & 175 & 125 \end{bmatrix}$

$= \begin{bmatrix} \$1037.50 & \$1400 & \$1012.50 \end{bmatrix}$

The entries represent the profits from both crops at each of the three outlets.

69. $BA = \begin{bmatrix} \$699.95 & \$899.95 & \$1099.95 \end{bmatrix}\begin{bmatrix} 5{,}000 & 4{,}000 \\ 6{,}000 & 10{,}000 \\ 8{,}000 & 5{,}000 \end{bmatrix} = \begin{bmatrix} \$17{,}699{,}050 & \$17{,}299{,}050 \end{bmatrix}$

The entries represent the cost of the three models of the LCD televisions at the two warehouses.

71. (a)

$$AB = \begin{bmatrix} 40 & 64 & 52 \\ 60 & 82 & 76 \\ 76 & 96 & 84 \end{bmatrix}\begin{bmatrix} 3.45 & 1.20 \\ 3.65 & 1.30 \\ 3.85 & 1.45 \end{bmatrix} = \begin{bmatrix} 571.8 & 206.6 \\ 798.9 & 288.8 \\ 936 & 337.8 \end{bmatrix}$$

(Columns of the result: Sales \$, Profit)

The entries represent the total sales (in dollars) and the profit (in dollars) for milk on Friday, Saturday, and Sunday.

(b) Total profit = \$206.60 + \$288.80 + \$337.80 = \$833.20

73. $P^3 = P^2P = \begin{bmatrix} 0.40 & 0.15 & 0.15 \\ 0.28 & 0.53 & 0.17 \\ 0.32 & 0.32 & 0.68 \end{bmatrix}\begin{bmatrix} 0.6 & 0.1 & 0.1 \\ 0.2 & 0.7 & 0.1 \\ 0.2 & 0.2 & 0.8 \end{bmatrix} = \begin{bmatrix} 0.300 & 0.175 & 0.175 \\ 0.308 & 0.433 & 0.217 \\ 0.392 & 0.392 & 0.608 \end{bmatrix}$

$$P^4 = P^3P = \begin{bmatrix} 0.300 & 0.175 & 0.175 \\ 0.308 & 0.433 & 0.217 \\ 0.392 & 0.392 & 0.608 \end{bmatrix}\begin{bmatrix} 0.6 & 0.1 & 0.1 \\ 0.2 & 0.7 & 0.1 \\ 0.2 & 0.2 & 0.8 \end{bmatrix} = \begin{bmatrix} 0.250 & 0.188 & 0.188 \\ 0.315 & 0.377 & 0.248 \\ 0.435 & 0.435 & 0.565 \end{bmatrix}$$

$$P^5 = P^4P = \begin{bmatrix} 0.250 & 0.188 & 0.188 \\ 0.315 & 0.377 & 0.248 \\ 0.435 & 0.435 & 0.565 \end{bmatrix}\begin{bmatrix} 0.6 & 0.1 & 0.1 \\ 0.2 & 0.7 & 0.1 \\ 0.2 & 0.2 & 0.8 \end{bmatrix} = \begin{bmatrix} 0.225 & 0.194 & 0.194 \\ 0.314 & 0.345 & 0.267 \\ 0.461 & 0.461 & 0.539 \end{bmatrix}$$

$$P^6 = \begin{bmatrix} 0.213 & 0.197 & 0.197 \\ 0.311 & 0.326 & 0.280 \\ 0.477 & 0.477 & 0.523 \end{bmatrix}$$

$$P^7 = \begin{bmatrix} 0.206 & 0.198 & 0.198 \\ 0.308 & 0.316 & 0.288 \\ 0.486 & 0.486 & 0.514 \end{bmatrix}$$

$$P^8 = \begin{bmatrix} 0.203 & 0.199 & 0.199 \\ 0.305 & 0.309 & 0.292 \\ 0.492 & 0.492 & 0.508 \end{bmatrix}$$

As P is raised to higher and higher powers, the resulting matrices appear to be approaching the matrix

$$\begin{bmatrix} 0.2 & 0.2 & 0.2 \\ 0.3 & 0.3 & 0.3 \\ 0.5 & 0.5 & 0.5 \end{bmatrix}.$$

75. True.

The sum of two matrices of different orders is undefined.

77. Answers will vary. *Sample answer:*

$$(A + B)^2 = \left(\begin{bmatrix} 2 & -1 \\ 1 & 3 \end{bmatrix} + \begin{bmatrix} -1 & 1 \\ 0 & -2 \end{bmatrix}\right)^2 = \begin{bmatrix} 1 & 0 \\ 2 & 1 \end{bmatrix} \neq$$

$$A^2 + 2AB + B^2 = \left(\begin{bmatrix} 2 & -1 \\ 1 & 3 \end{bmatrix}\right)^2 + 2\begin{bmatrix} 2 & -1 \\ 1 & 3 \end{bmatrix}\begin{bmatrix} -1 & 1 \\ 0 & -2 \end{bmatrix} + \left(\begin{bmatrix} -1 & 1 \\ 0 & -2 \end{bmatrix}\right)^2 = \begin{bmatrix} 0 & 0 \\ 3 & 2 \end{bmatrix}$$

79. Answers will vary. *Sample answer:*

$$(A + B)(A - B) = \left(\begin{bmatrix} 2 & -1 \\ 1 & 3 \end{bmatrix} + \begin{bmatrix} -1 & 1 \\ 0 & -2 \end{bmatrix}\right)\left(\begin{bmatrix} 2 & -1 \\ 1 & 3 \end{bmatrix} - \begin{bmatrix} -1 & 1 \\ 0 & -2 \end{bmatrix}\right) = \begin{bmatrix} 3 & -2 \\ 4 & 3 \end{bmatrix} \neq$$

$$A^2 - B^2 = \left(\begin{bmatrix} 2 & -1 \\ 1 & 3 \end{bmatrix}\right)^2 - \left(\begin{bmatrix} -1 & 1 \\ 0 & -2 \end{bmatrix}\right)^2 = \begin{bmatrix} 2 & -2 \\ 5 & 4 \end{bmatrix}$$

81. $AC = \begin{bmatrix} 0 & 1 \\ 0 & 1 \end{bmatrix}\begin{bmatrix} 2 & 3 \\ 2 & 3 \end{bmatrix} = \begin{bmatrix} 2 & 3 \\ 2 & 3 \end{bmatrix}$

$BC = \begin{bmatrix} 1 & 0 \\ 1 & 0 \end{bmatrix}\begin{bmatrix} 2 & 3 \\ 2 & 3 \end{bmatrix} = \begin{bmatrix} 2 & 3 \\ 2 & 3 \end{bmatrix}$

So, $AC = BC$ even though $A \neq B$.

83. The product of two diagonal matrices of the same order is a diagonal matrix whose entries are the products of the corresponding diagonal entries of A and B.

85. *Sample answer:*

$A = \begin{bmatrix} 1 & 0 \\ 0 & 1 \end{bmatrix}, B = \begin{bmatrix} 0 & 1 \\ 1 & 0 \end{bmatrix}$

$AB = \begin{bmatrix} 1 & 0 \\ 0 & 1 \end{bmatrix}\begin{bmatrix} 0 & 1 \\ 1 & 0 \end{bmatrix} = \begin{bmatrix} (1)(0) + (0)(1) & (1)(1) + (0)(0) \\ (0)(0) + (1)(1) & (0)(1) + (1)(0) \end{bmatrix} = \begin{bmatrix} 0 & 1 \\ 1 & 0 \end{bmatrix}$

$BA = \begin{bmatrix} 0 & 1 \\ 1 & 0 \end{bmatrix}\begin{bmatrix} 1 & 0 \\ 0 & 1 \end{bmatrix} = \begin{bmatrix} (0)(1) + (1)(0) & (0)(0) + (1)(1) \\ (1)(1) + (0)(0) & (1)(0) + (0)(1) \end{bmatrix} = \begin{bmatrix} 0 & 1 \\ 1 & 0 \end{bmatrix}$

So, $AB = BA$.

Section 10.3 The Inverse of a Square Matrix

1. square

3. nonsingular; singular

5. $AB = \begin{bmatrix} 2 & 1 \\ 5 & 3 \end{bmatrix}\begin{bmatrix} 3 & -1 \\ -5 & 2 \end{bmatrix} = \begin{bmatrix} 6 - 5 & -2 + 2 \\ 15 - 15 & -5 + 6 \end{bmatrix} = \begin{bmatrix} 1 & 0 \\ 0 & 1 \end{bmatrix}$

$BA = \begin{bmatrix} 3 & -1 \\ -5 & 2 \end{bmatrix}\begin{bmatrix} 2 & 1 \\ 5 & 3 \end{bmatrix} = \begin{bmatrix} 6 - 5 & 3 - 3 \\ -10 + 10 & -5 + 6 \end{bmatrix} = \begin{bmatrix} 1 & 0 \\ 0 & 1 \end{bmatrix}$

7. $AB = \frac{1}{2}\begin{bmatrix} 1 & 2 \\ 3 & 4 \end{bmatrix}\begin{bmatrix} -4 & 2 \\ 3 & -1 \end{bmatrix} = \frac{1}{2}\begin{bmatrix} -4 + 6 & 2 - 2 \\ -12 + 12 & 6 - 4 \end{bmatrix} = \begin{bmatrix} 1 & 0 \\ 0 & 1 \end{bmatrix}$

$BA = \frac{1}{2}\begin{bmatrix} -4 & 2 \\ 3 & -1 \end{bmatrix}\begin{bmatrix} 1 & 2 \\ 3 & 4 \end{bmatrix} = \frac{1}{2}\begin{bmatrix} -4 + 6 & -8 + 8 \\ 3 - 3 & 6 - 4 \end{bmatrix} = \begin{bmatrix} 1 & 0 \\ 0 & 1 \end{bmatrix}$

9. $AB = \begin{bmatrix} 2 & -17 & 11 \\ -1 & 11 & -7 \\ 0 & 3 & -2 \end{bmatrix}\begin{bmatrix} 1 & 1 & 2 \\ 2 & 4 & -3 \\ 3 & 6 & -5 \end{bmatrix} = \begin{bmatrix} 2 - 34 + 33 & 2 - 68 + 66 & 4 + 51 - 55 \\ -1 + 22 - 21 & -1 + 44 - 42 & -2 - 33 + 35 \\ 6 - 6 & 12 - 12 & -9 + 10 \end{bmatrix} = \begin{bmatrix} 1 & 0 & 0 \\ 0 & 1 & 0 \\ 0 & 0 & 1 \end{bmatrix}$

$BA = \begin{bmatrix} 1 & 1 & 2 \\ 2 & 4 & -3 \\ 3 & 6 & -5 \end{bmatrix}\begin{bmatrix} 2 & -17 & 11 \\ -1 & 11 & -7 \\ 0 & 3 & -2 \end{bmatrix} = \begin{bmatrix} 2 - 1 & -17 + 11 + 6 & 11 - 7 - 4 \\ 4 - 4 & -34 + 44 - 9 & 22 - 28 + 6 \\ 6 - 6 & -51 + 66 - 15 & 33 - 42 + 10 \end{bmatrix} = \begin{bmatrix} 1 & 0 & 0 \\ 0 & 1 & 0 \\ 0 & 0 & 1 \end{bmatrix}$

11. $AB = \frac{1}{3}\begin{bmatrix} 2 & 0 & 2 & 1 \\ 3 & 0 & 0 & 1 \\ -1 & 1 & -2 & 1 \\ 3 & -1 & 1 & 0 \end{bmatrix}\begin{bmatrix} -1 & 3 & -2 & -2 \\ -2 & 9 & -7 & -10 \\ 1 & 0 & -1 & -1 \\ 3 & -6 & 6 & 6 \end{bmatrix}$

$= \frac{1}{3}\begin{bmatrix} -2+0+2+3 & 6+0+0-6 & -4+0-2+6 & -4+0-2+6 \\ -3+0+0+3 & 9+0+0-6 & -6+0+0+6 & -6+0+0+6 \\ 1-2-2+3 & -3+9+0-6 & 2-7+2+6 & 2-10+2+6 \\ -3+2+1+0 & 9-9+0+0 & -6+7-1+0 & -6+10-1+0 \end{bmatrix} = \begin{bmatrix} 1 & 0 & 0 & 0 \\ 0 & 1 & 0 & 0 \\ 0 & 0 & 1 & 0 \\ 0 & 0 & 0 & 1 \end{bmatrix}$

$BA = \frac{1}{3}\begin{bmatrix} -1 & 3 & -2 & -2 \\ -2 & 9 & -7 & -10 \\ 1 & 0 & -1 & -1 \\ 3 & -6 & 6 & 6 \end{bmatrix}\begin{bmatrix} 2 & 0 & 2 & 1 \\ 3 & 0 & 0 & 1 \\ -1 & 1 & -2 & 1 \\ 3 & -1 & 1 & 0 \end{bmatrix}$

$= \frac{1}{3}\begin{bmatrix} -2+9+2-6 & 0+0-2+2 & -2+0+4-2 & -1+3-2+0 \\ -4+27+7-30 & 0+0-7+10 & -4+0+14-10 & -2+9-7+0 \\ 2+0+1-3 & 0+0-1+1 & 2+0+2-1 & 1+0-1+0 \\ 6-18-6+18 & 0+0+6-6 & 6+0-12+6 & 3-6+6+0 \end{bmatrix} = \begin{bmatrix} 1 & 0 & 0 & 0 \\ 0 & 1 & 0 & 0 \\ 0 & 0 & 1 & 0 \\ 0 & 0 & 0 & 1 \end{bmatrix}$

13. $[A \;\vdots\; I] = \left[\begin{array}{cc:cc} 2 & 0 & 1 & 0 \\ 0 & 3 & 0 & 1 \end{array}\right]$

$\begin{array}{l} \frac{1}{2}R_1 \rightarrow \\ \frac{1}{3}R_2 \rightarrow \end{array} \left[\begin{array}{cc:cc} 1 & 0 & \frac{1}{2} & 0 \\ 0 & 1 & 0 & \frac{1}{3} \end{array}\right] = [I \;\vdots\; A^{-1}]$

$A^{-1} = \begin{bmatrix} \frac{1}{2} & 0 \\ 0 & \frac{1}{3} \end{bmatrix}$

15. $[A \;\vdots\; I] = \left[\begin{array}{cc:cc} 1 & -2 & 1 & 0 \\ 2 & -3 & 0 & 1 \end{array}\right]$

$-2R_1 + R_2 \rightarrow \left[\begin{array}{cc:cc} 1 & -2 & 1 & 0 \\ 0 & 1 & -2 & 1 \end{array}\right]$

$2R_2 + R_1 \rightarrow \left[\begin{array}{cc:cc} 1 & 0 & -3 & 2 \\ 0 & 1 & -2 & 1 \end{array}\right] = [I \;\vdots\; A^{-1}]$

$A^{-1} = \begin{bmatrix} -3 & 2 \\ -2 & 1 \end{bmatrix}$

17. $[A \;\vdots\; I] = \left[\begin{array}{cc:cc} 3 & 1 & 1 & 0 \\ 4 & 2 & 0 & 1 \end{array}\right]$

$\frac{1}{2}R_2 \rightarrow \left[\begin{array}{cc:cc} 3 & 1 & 1 & 0 \\ 2 & 1 & 0 & \frac{1}{2} \end{array}\right]$

$-R_2 + R_1 \rightarrow \left[\begin{array}{cc:cc} 1 & 0 & 1 & -\frac{1}{2} \\ 2 & 1 & 0 & \frac{1}{2} \end{array}\right]$

$-2R_1 + R_2 \rightarrow \left[\begin{array}{cc:cc} 1 & 0 & 1 & -\frac{1}{2} \\ 0 & 1 & -2 & \frac{3}{2} \end{array}\right] = [I \;\vdots\; A^{-1}]$

$A^{-1} = \begin{bmatrix} 1 & -\frac{1}{2} \\ -2 & \frac{3}{2} \end{bmatrix}$

19. $[A \;\vdots\; I] = \left[\begin{array}{ccc:ccc} 1 & 1 & 1 & 1 & 0 & 0 \\ 3 & 5 & 4 & 0 & 1 & 0 \\ 3 & 6 & 5 & 0 & 0 & 1 \end{array}\right]$

$\begin{array}{r} \\ -3R_1 + R_2 \rightarrow \\ -3R_1 + R_3 \rightarrow \end{array} \left[\begin{array}{ccc:ccc} 1 & 1 & 1 & 1 & 0 & 0 \\ 0 & 2 & 1 & -3 & 1 & 0 \\ 0 & 3 & 2 & -3 & 0 & 1 \end{array}\right]$

$\frac{1}{2}R_2 \rightarrow \left[\begin{array}{ccc:ccc} 1 & 1 & 1 & 1 & 0 & 0 \\ 0 & 1 & \frac{1}{2} & -\frac{3}{2} & \frac{1}{2} & 0 \\ 0 & 3 & 2 & -3 & 0 & 1 \end{array}\right]$

$\begin{array}{r} -R_2 + R_1 \rightarrow \\ \\ -3R_2 + R_3 \rightarrow \end{array} \left[\begin{array}{ccc:ccc} 1 & 0 & \frac{1}{2} & \frac{5}{2} & -\frac{1}{2} & 0 \\ 0 & 1 & \frac{1}{2} & -\frac{3}{2} & \frac{1}{2} & 0 \\ 0 & 0 & \frac{1}{2} & \frac{3}{2} & -\frac{3}{2} & 1 \end{array}\right]$

$\begin{array}{r} -R_3 + R_1 \rightarrow \\ -R_3 + R_2 \rightarrow \\ \\ \end{array} \left[\begin{array}{ccc:ccc} 1 & 0 & 0 & 1 & 1 & -1 \\ 0 & 1 & 0 & -3 & 2 & -1 \\ 0 & 0 & \frac{1}{2} & \frac{3}{2} & -\frac{3}{2} & 1 \end{array}\right]$

$\begin{array}{r} \\ \\ 2R_3 \rightarrow \end{array} \left[\begin{array}{ccc:ccc} 1 & 0 & 0 & 1 & 1 & -1 \\ 0 & 1 & 0 & -3 & 2 & -1 \\ 0 & 0 & 1 & 3 & -3 & 2 \end{array}\right] = [I \;\vdots\; A^{-1}]$

$A^{-1} = \begin{bmatrix} 1 & 1 & -1 \\ -3 & 2 & -1 \\ 3 & -3 & 2 \end{bmatrix}$

21. $[A \ \vdots \ I] = \left[\begin{array}{ccc:ccc} -5 & 0 & 0 & 1 & 0 & 0 \\ 2 & 0 & 0 & 0 & 1 & 0 \\ -1 & 5 & 7 & 0 & 0 & 1 \end{array}\right] \begin{matrix} \\ \\ R_2 + 2R_3 \rightarrow \end{matrix} \left[\begin{array}{ccc:ccc} -5 & 0 & 0 & 1 & 0 & 0 \\ 2 & 0 & 0 & 0 & 1 & 0 \\ 0 & 10 & 14 & 0 & 1 & 2 \end{array}\right] \begin{matrix} \\ 2R_1 + 5R_2 \rightarrow \\ \\ \end{matrix} \left[\begin{array}{ccc:ccc} -5 & 0 & 0 & 1 & 0 & 0 \\ 0 & 0 & 0 & 2 & 5 & 0 \\ 0 & 10 & 14 & 0 & 1 & 2 \end{array}\right]$

Because the first three entries of row 2 are all zeros, the inverse of A does not exist.

23. $[A \ \vdots \ I] = \left[\begin{array}{cccc:cccc} -8 & 0 & 0 & 0 & 1 & 0 & 0 & 0 \\ 0 & 1 & 0 & 0 & 0 & 1 & 0 & 0 \\ 0 & 0 & 4 & 0 & 0 & 0 & 1 & 0 \\ 0 & 0 & 0 & -5 & 0 & 0 & 0 & 1 \end{array}\right] \begin{matrix} -\frac{1}{8}R_1 \rightarrow \\ \\ \frac{1}{4}R_3 \rightarrow \\ -\frac{1}{5}R_4 \rightarrow \end{matrix} \left[\begin{array}{cccc:cccc} 1 & 0 & 0 & 0 & -\frac{1}{8} & 0 & 0 & 0 \\ 0 & 1 & 0 & 0 & 0 & 1 & 0 & 0 \\ 0 & 0 & 1 & 0 & 0 & 0 & \frac{1}{4} & 0 \\ 0 & 0 & 0 & 1 & 0 & 0 & 0 & -\frac{1}{5} \end{array}\right] = [I \ \vdots \ A^{-1}]$

$$A^{-1} = \begin{bmatrix} -\frac{1}{8} & 0 & 0 & 0 \\ 0 & 1 & 0 & 0 \\ 0 & 0 & \frac{1}{4} & 0 \\ 0 & 0 & 0 & -\frac{1}{5} \end{bmatrix}$$

25. $A = \begin{bmatrix} 1 & 2 & -1 \\ 3 & 7 & -10 \\ -5 & -7 & -15 \end{bmatrix}$

$$A^{-1} = \begin{bmatrix} -175 & 37 & -13 \\ 95 & -20 & 7 \\ 14 & -3 & 1 \end{bmatrix}$$

27. $A = \begin{bmatrix} 1 & 1 & 2 \\ 3 & 1 & 0 \\ -2 & 0 & 3 \end{bmatrix}$

$$A^{-1} = \begin{bmatrix} -1.5 & 1.5 & 1 \\ 4.5 & -3.5 & -3 \\ -1 & 1 & 1 \end{bmatrix}$$

29. $A = \begin{bmatrix} -\frac{1}{2} & \frac{3}{4} & \frac{1}{4} \\ 1 & 0 & -\frac{3}{2} \\ 0 & -1 & \frac{1}{2} \end{bmatrix}$

$$A^{-1} = \begin{bmatrix} -12 & -5 & -9 \\ -4 & -2 & -4 \\ -8 & -4 & -6 \end{bmatrix}$$

31. $A = \begin{bmatrix} 0.1 & 0.2 & 0.3 \\ -0.3 & 0.2 & 0.2 \\ 0.5 & 0.4 & 0.4 \end{bmatrix}$

$$A^{-1} = \begin{bmatrix} 0 & -1.\overline{81} & 0.\overline{90} \\ -10 & 5 & 5 \\ 10 & -2.\overline{72} & -3.\overline{63} \end{bmatrix}$$

33. $A = \begin{bmatrix} -1 & 0 & 1 & 0 \\ 0 & 2 & 0 & -1 \\ 2 & 0 & -1 & 0 \\ 0 & -1 & 0 & 1 \end{bmatrix}$

$$A^{-1} = \begin{bmatrix} 1 & 0 & 1 & 0 \\ 0 & 1 & 0 & 1 \\ 2 & 0 & 1 & 0 \\ 0 & 1 & 0 & 2 \end{bmatrix}$$

35. $A = \begin{bmatrix} a & b \\ c & d \end{bmatrix}, A^{-1} = \dfrac{1}{ad - bc}\begin{bmatrix} d & -b \\ -c & a \end{bmatrix}$

$$A = \begin{bmatrix} 2 & 3 \\ -1 & 5 \end{bmatrix}$$

$$ad - bc = (2)(5) - (3)(-1) = 13$$

$$A^{-1} = \frac{1}{13}\begin{bmatrix} 5 & -3 \\ 1 & 2 \end{bmatrix} = \begin{bmatrix} \frac{5}{13} & -\frac{3}{13} \\ \frac{1}{13} & \frac{2}{13} \end{bmatrix}$$

37. $A = \begin{bmatrix} -4 & -6 \\ 2 & 3 \end{bmatrix}$

$$ad - bc = (-4)(3) - (-2)(-6) = 0$$

Because $ad - bc = 0$, A^{-1} does not exist.

39. $A = \begin{bmatrix} \frac{7}{2} & -\frac{3}{4} \\ \frac{1}{5} & \frac{4}{5} \end{bmatrix}$

$ad - bc = \left(\frac{7}{2}\right)\left(\frac{4}{5}\right) - \left(-\frac{3}{4}\right)\left(\frac{1}{5}\right) = \frac{28}{10} + \frac{3}{20} = \frac{59}{20}$

$A^{-1} = \frac{20}{59}\begin{bmatrix} \frac{4}{5} & \frac{3}{4} \\ -\frac{1}{5} & \frac{7}{2} \end{bmatrix} = \begin{bmatrix} \frac{16}{59} & \frac{15}{59} \\ -\frac{4}{59} & \frac{70}{59} \end{bmatrix}$

41. $\begin{bmatrix} x \\ y \end{bmatrix} = \begin{bmatrix} -3 & 2 \\ -2 & 1 \end{bmatrix}\begin{bmatrix} 5 \\ 10 \end{bmatrix} = \begin{bmatrix} 5 \\ 0 \end{bmatrix}$

Solution: $(5, 0)$

43. $\begin{bmatrix} x \\ y \end{bmatrix} = \begin{bmatrix} -3 & 2 \\ -2 & 1 \end{bmatrix}\begin{bmatrix} 4 \\ 2 \end{bmatrix} = \begin{bmatrix} -8 \\ -6 \end{bmatrix}$

Solution: $(-8, -6)$

45. $\begin{bmatrix} x \\ y \\ z \end{bmatrix} = \begin{bmatrix} 1 & 1 & -1 \\ -3 & 2 & -1 \\ 3 & -3 & 2 \end{bmatrix}\begin{bmatrix} 0 \\ 5 \\ 2 \end{bmatrix} = \begin{bmatrix} 3 \\ 8 \\ -11 \end{bmatrix}$

Solution: $(3, 8, -11)$

47. $\begin{bmatrix} x_1 \\ x_2 \\ x_3 \\ x_4 \end{bmatrix} = \begin{bmatrix} -24 & 7 & 1 & -2 \\ -10 & 3 & 0 & -1 \\ -29 & 7 & 3 & -2 \\ 12 & -3 & -1 & 1 \end{bmatrix}\begin{bmatrix} 0 \\ 1 \\ -1 \\ 2 \end{bmatrix} = \begin{bmatrix} 2 \\ 1 \\ 0 \\ 0 \end{bmatrix}$

Solution: $(2, 1, 0, 0)$

49. $A = \begin{bmatrix} 3 & 4 \\ 5 & 3 \end{bmatrix}$

$A^{-1} = \frac{1}{9 - 20}\begin{bmatrix} 3 & -4 \\ -5 & 3 \end{bmatrix}$

$\begin{bmatrix} x \\ y \end{bmatrix} = -\frac{1}{11}\begin{bmatrix} 3 & -4 \\ -5 & 3 \end{bmatrix}\begin{bmatrix} -2 \\ 4 \end{bmatrix} = -\frac{1}{11}\begin{bmatrix} -22 \\ 22 \end{bmatrix} = \begin{bmatrix} 2 \\ -2 \end{bmatrix}$

Solution: $(2, -2)$

51. $A = \begin{bmatrix} -0.4 & 0.8 \\ 2 & -4 \end{bmatrix}$

$A^{-1} = \frac{1}{1.6 - 1.6}\begin{bmatrix} -4 & -0.8 \\ -2 & -0.4 \end{bmatrix}$

A^{-1} does not exist.

This implies that there is no unique solution; that is, either the system is inconsistent *or* there are infinitely many solutions.

Find the reduced row-echelon form of the matrix corresponding to the system.

$$\begin{bmatrix} -0.4 & 0.8 & \vdots & 1.6 \\ 2 & -4 & \vdots & 5 \end{bmatrix}$$

$$-2.5R_1 \rightarrow \begin{bmatrix} 1 & -2 & \vdots & -4 \\ 2 & -4 & \vdots & 5 \end{bmatrix}$$

$$-2R_1 + R_2 \rightarrow \begin{bmatrix} 1 & -2 & \vdots & -4 \\ 0 & 0 & \vdots & 13 \end{bmatrix}$$

The given system is inconsistent and there is no solution.

53. $A = \begin{bmatrix} -\frac{1}{4} & \frac{3}{8} \\ \frac{3}{2} & \frac{3}{4} \end{bmatrix}$

$$A^{-1} = \frac{1}{-\frac{3}{16} - \frac{9}{16}}\begin{bmatrix} \frac{3}{4} & -\frac{3}{8} \\ -\frac{3}{2} & -\frac{1}{4} \end{bmatrix}$$

$$= -\frac{4}{3}\begin{bmatrix} \frac{3}{4} & -\frac{3}{8} \\ -\frac{3}{2} & -\frac{1}{4} \end{bmatrix}$$

$$= \begin{bmatrix} -1 & \frac{1}{2} \\ 2 & \frac{1}{3} \end{bmatrix}$$

$$\begin{bmatrix} x \\ y \end{bmatrix} = \begin{bmatrix} -1 & \frac{1}{2} \\ 2 & \frac{1}{3} \end{bmatrix}\begin{bmatrix} -2 \\ -12 \end{bmatrix} = \begin{bmatrix} -4 \\ -8 \end{bmatrix}$$

Solution: $(-4, -8)$

55. $A = \begin{bmatrix} 4 & -1 & 1 \\ 2 & 2 & 3 \\ 5 & -2 & 6 \end{bmatrix}$

Find A^{-1}.

$$[A \;\vdots\; I] = \left[\begin{array}{ccc:ccc} 4 & -1 & 1 & 1 & 0 & 0 \\ 2 & 2 & 3 & 0 & 1 & 0 \\ 5 & -2 & 6 & 0 & 0 & 1 \end{array}\right]$$

$$\begin{array}{l} R_1 \\ \\ R_3 \end{array} \left[\begin{array}{ccc:ccc} 5 & -2 & 6 & 0 & 0 & 1 \\ 2 & 2 & 3 & 0 & 1 & 0 \\ 4 & -1 & 1 & 1 & 0 & 0 \end{array}\right]$$

$$-R_3 + R_1 \rightarrow \left[\begin{array}{ccc:ccc} 1 & -1 & 5 & -1 & 0 & 1 \\ 2 & 2 & 3 & 0 & 1 & 0 \\ 4 & -1 & 1 & 1 & 0 & 0 \end{array}\right]$$

$$\begin{array}{r} \\ -2R_1 + R_2 \rightarrow \\ -4R_1 + R_3 \rightarrow \end{array} \left[\begin{array}{ccc:ccc} 1 & -1 & 5 & -1 & 0 & 1 \\ 0 & 4 & -7 & 2 & 1 & -2 \\ 0 & 3 & -19 & 5 & 0 & -4 \end{array}\right]$$

$$\begin{array}{r} \\ -R_3 + R_2 \rightarrow \\ \\ \end{array} \left[\begin{array}{ccc:ccc} 1 & -1 & 5 & -1 & 0 & 1 \\ 0 & 1 & 12 & -3 & 1 & 2 \\ 0 & 3 & -19 & 5 & 0 & -4 \end{array}\right]$$

$$\begin{array}{r} R_2 + R_1 \rightarrow \\ \\ -3R_2 + R_3 \rightarrow \end{array} \left[\begin{array}{ccc:ccc} 1 & 0 & 17 & -4 & 1 & 3 \\ 0 & 1 & 12 & -3 & 1 & 2 \\ 0 & 0 & -55 & 14 & -3 & -10 \end{array}\right]$$

$$\begin{array}{r} \\ \\ -\frac{1}{55}R_3 \rightarrow \end{array} \left[\begin{array}{ccc:ccc} 1 & 0 & 17 & -4 & 1 & 3 \\ 0 & 1 & 12 & -3 & 1 & 2 \\ 0 & 0 & 1 & -\frac{14}{55} & \frac{3}{55} & \frac{2}{11} \end{array}\right]$$

$$\begin{array}{r} -17R_3 + R_1 \rightarrow \\ -12R_3 + R_2 \rightarrow \\ \\ \end{array} \left[\begin{array}{ccc:ccc} 1 & 0 & 0 & \frac{18}{55} & \frac{4}{55} & -\frac{1}{11} \\ 0 & 1 & 0 & \frac{3}{55} & \frac{19}{55} & -\frac{2}{11} \\ 0 & 0 & 1 & -\frac{14}{55} & \frac{3}{55} & \frac{2}{11} \end{array}\right] = [I \;\vdots\; A^{-1}]$$

$$A^{-1} = \frac{1}{55}\begin{bmatrix} 18 & 4 & -5 \\ 3 & 19 & -10 \\ -14 & 3 & 10 \end{bmatrix}$$

$$\begin{bmatrix} x \\ y \\ z \end{bmatrix} = \frac{1}{55}\begin{bmatrix} 18 & 4 & -5 \\ 3 & 19 & -10 \\ -14 & 3 & 10 \end{bmatrix}\begin{bmatrix} -5 \\ 10 \\ 1 \end{bmatrix} = \frac{1}{55}\begin{bmatrix} -55 \\ 165 \\ 110 \end{bmatrix} = \begin{bmatrix} -1 \\ 3 \\ 2 \end{bmatrix}$$

Solution: $(-1, 3, 2)$

57. $A = \begin{bmatrix} 5 & -3 & 2 \\ 2 & 2 & -3 \\ 1 & -7 & 8 \end{bmatrix}$

A^{-1} does not exist. This implies that there is no unique solution; that is, either the system is inconsistent *or* the system has infinitely many solutions. Use a graphing utility to find the reduced row-echelon form of the matrix corresponding to the system.

$$\left[\begin{array}{ccc:c} 5 & -3 & 2 & 2 \\ 2 & 2 & -3 & 3 \\ 1 & -7 & 8 & -4 \end{array}\right]$$

$$\left[\begin{array}{ccc:c} 1 & 0 & -\frac{5}{16} & \frac{13}{16} \\ 0 & 1 & -\frac{19}{16} & \frac{11}{16} \\ 0 & 0 & 0 & 0 \end{array}\right]$$

$$\begin{cases} x - \frac{5}{16}z = \frac{13}{16} \\ y - \frac{19}{16}z = \frac{11}{16} \end{cases}$$

Let $z = a$. Then $x = \frac{5}{16}a + \frac{13}{16}$ and $y = \frac{19}{16}a + \frac{11}{16}$.

Solution: $\left(\frac{5}{16}a + \frac{13}{16}, \frac{19}{16}a + \frac{11}{16}, a\right)$ where a is a real number

59. $A = \begin{bmatrix} 3 & -2 & 1 \\ -4 & 1 & -3 \\ 1 & -5 & 1 \end{bmatrix}$

$$A^{-1} = \begin{bmatrix} 0.56 & 0.12 & -0.2 \\ -0.04 & -0.08 & -0.2 \\ -0.76 & -0.52 & 0.2 \end{bmatrix}$$

$$\begin{bmatrix} x \\ y \\ z \end{bmatrix} = \begin{bmatrix} 0.56 & 0.12 & -0.2 \\ -0.04 & -0.08 & -0.2 \\ -0.76 & -0.52 & 0.2 \end{bmatrix}\begin{bmatrix} -29 \\ 37 \\ -24 \end{bmatrix} = \begin{bmatrix} -7 \\ 3 \\ -2 \end{bmatrix}$$

Solution: $(-7, 3, -2)$

61. $A = \begin{bmatrix} 1 & 1 & 1 \\ 0.065 & 0.07 & 0.09 \\ 0 & 2 & -1 \end{bmatrix}$

$$[A \,\vdots\, I] = \left[\begin{array}{ccc|ccc} 1 & 1 & 1 & 1 & 0 & 0 \\ 0.065 & 0.07 & 0.09 & 0 & 1 & 0 \\ 0 & 2 & -1 & 0 & 0 & 1 \end{array}\right]$$

$$200R_2 \rightarrow \left[\begin{array}{ccc|ccc} 1 & 1 & 1 & 1 & 0 & 0 \\ 13 & 14 & 18 & 0 & 200 & 0 \\ 0 & 2 & -1 & 0 & 0 & 1 \end{array}\right]$$

$$-13R_1 + R_2 \rightarrow \left[\begin{array}{ccc|ccc} 1 & 1 & 1 & 1 & 0 & 0 \\ 0 & 1 & 5 & -13 & 200 & 0 \\ 0 & 2 & -1 & 0 & 0 & 1 \end{array}\right]$$

$$\begin{array}{l} -R_2 + R_1 \rightarrow \\ \\ -2R_2 + R_3 \rightarrow \end{array} \left[\begin{array}{ccc|ccc} 1 & 0 & -4 & 14 & -200 & 0 \\ 0 & 1 & 5 & -13 & 200 & 0 \\ 0 & 0 & -11 & 26 & -400 & 1 \end{array}\right]$$

$$\begin{array}{l} \\ \\ -\frac{1}{11}R_3 \rightarrow \end{array} \left[\begin{array}{ccc|ccc} 1 & 0 & -4 & 14 & -200 & 0 \\ 0 & 1 & 5 & -13 & 200 & 0 \\ 0 & 0 & 1 & -\frac{26}{11} & \frac{400}{11} & -\frac{1}{11} \end{array}\right]$$

$$\begin{array}{l} 4R_3 + R_1 \rightarrow \\ -5R_3 + R_2 \rightarrow \\ \\ \end{array} \left[\begin{array}{ccc|ccc} 1 & 0 & 0 & \frac{50}{11} & -\frac{600}{11} & -\frac{4}{11} \\ 0 & 1 & 0 & -\frac{13}{11} & \frac{200}{11} & \frac{5}{11} \\ 0 & 0 & 1 & -\frac{26}{11} & \frac{400}{11} & -\frac{1}{11} \end{array}\right] = [I \,\vdots\, A^{-1}]$$

$$X = A^{-1}B = \frac{1}{11}\begin{bmatrix} 50 & -600 & -4 \\ -13 & 200 & 5 \\ -26 & 400 & -1 \end{bmatrix}\begin{bmatrix} 10{,}000 \\ 705 \\ 0 \end{bmatrix} = \begin{bmatrix} 7000 \\ 1000 \\ 2000 \end{bmatrix}$$

Solution: $7000 in AAA-rated bonds, $1000 in A-rated bonds, $2000 in B-rated bonds

63. Use the inverse matrix A^{-1} from Exercise 65.

$$X = A^{-1}B = \frac{1}{11}\begin{bmatrix} 50 & -600 & -4 \\ -13 & 200 & 5 \\ -26 & 400 & -1 \end{bmatrix}\begin{bmatrix} 12{,}000 \\ 835 \\ 0 \end{bmatrix} = \begin{bmatrix} 9000 \\ 1000 \\ 2000 \end{bmatrix}$$

Solution: $9000 in AAA-rated bonds, $1000 in A-rated bonds, $2000 in B-rated bonds

65. $\begin{cases} 2I_1 + 4I_3 = 15 \\ I_2 + 4I_3 = 17 \\ I_1 + I_2 - I_3 = 0 \end{cases}$

$$\underset{A}{\begin{bmatrix} 2 & 0 & 4 \\ 0 & 1 & 4 \\ 1 & 1 & -1 \end{bmatrix}} \underset{X}{\begin{bmatrix} I_1 \\ I_2 \\ I_3 \end{bmatrix}} \underset{=}{=} \underset{B}{\begin{bmatrix} 15 \\ 17 \\ 0 \end{bmatrix}}$$

$$X = A^{-1}B = \begin{bmatrix} \frac{5}{14} & -\frac{2}{7} & \frac{2}{7} \\ -\frac{2}{7} & \frac{3}{7} & \frac{4}{7} \\ \frac{1}{14} & \frac{1}{7} & -\frac{1}{7} \end{bmatrix}\begin{bmatrix} 15 \\ 17 \\ 0 \end{bmatrix} = \begin{bmatrix} \frac{1}{2} \\ 3 \\ \frac{7}{2} \end{bmatrix}$$

So, $I_1 = 0.5$ ampere, $I_2 = 3.0$ ampere, and $I_3 = 3.5$ ampere.

67. $\begin{cases} 2I_1 + 4I_3 = 28 \\ I_2 + 4I_3 = 21 \\ I_1 + I_2 - I_3 = 0 \end{cases}$

$$\underbrace{\begin{bmatrix} 2 & 0 & 4 \\ 0 & 1 & 4 \\ 1 & 1 & -1 \end{bmatrix}}_{A}\underbrace{\begin{bmatrix} I_1 \\ I_2 \\ I_3 \end{bmatrix}}_{X} = \underbrace{\begin{bmatrix} 28 \\ 21 \\ 0 \end{bmatrix}}_{B}$$

$$X = A^{-1}B = \begin{bmatrix} \frac{5}{14} & -\frac{2}{7} & \frac{2}{7} \\ -\frac{2}{7} & \frac{3}{7} & \frac{4}{7} \\ \frac{1}{14} & \frac{1}{7} & -\frac{1}{7} \end{bmatrix}\begin{bmatrix} 28 \\ 21 \\ 0 \end{bmatrix} = \begin{bmatrix} 4 \\ 1 \\ 5 \end{bmatrix}$$

So, $I_1 = 4$ ampere, $I_2 = 1$ ampere, and $I_3 = 5$ ampere.

In Exercise 69, use the following:
Let x = bags of potting soil for seedlings,
y = bags of potting soil for general potting, and
z = bags of potting soil for hardwood plants.

$$AX = B = \begin{bmatrix} 2 & 1 & 2 \\ 1 & 2 & 1 \\ 1 & 1 & 2 \end{bmatrix}\begin{bmatrix} x \\ y \\ z \end{bmatrix} = \begin{bmatrix} \text{Sand} \\ \text{Loam} \\ \text{Peat Moss} \end{bmatrix}$$

$$A^{-1} = \begin{bmatrix} 1 & 0 & -1 \\ 0 & 1 & -1 \\ -\frac{1}{2} & -\frac{1}{2} & \frac{3}{2} \end{bmatrix}$$

69. $A^{-1}\begin{bmatrix} 500 \\ 500 \\ 400 \end{bmatrix} = \begin{bmatrix} 100 \\ 100 \\ 100 \end{bmatrix}$

Solution:
x = 100 bags of potting soil for seedlings,
y = 100 bags of potting soil for general potting,
z = 100 bags of potting soil for hardwood plants.

71. Let r = number of roses, l = number of lilies, and i = number of irises.

(a) $\begin{cases} r + l + i = 120 \\ 2.5r + 4l + 2i = 300 \\ -r + 2l + 2i = 0 \end{cases}$

(b) $\underbrace{\begin{bmatrix} 1 & 1 & 1 \\ 2.5 & 4 & 2 \\ -1 & 2 & 2 \end{bmatrix}}_{A}\underbrace{\begin{bmatrix} r \\ l \\ i \end{bmatrix}}_{X} = \underbrace{\begin{bmatrix} 120 \\ 300 \\ 0 \end{bmatrix}}_{B}$

(c) $A^{-1} = \begin{bmatrix} \frac{2}{3} & 0 & -\frac{1}{3} \\ -\frac{7}{6} & \frac{1}{2} & \frac{1}{12} \\ \frac{3}{2} & -\frac{1}{2} & -\frac{1}{4} \end{bmatrix}$

$$X = A^{-1}B = \begin{bmatrix} \frac{2}{3} & 0 & -\frac{1}{3} \\ -\frac{7}{6} & \frac{1}{2} & \frac{1}{12} \\ \frac{3}{2} & -\frac{1}{2} & \frac{1}{4} \end{bmatrix}\begin{bmatrix} 120 \\ 300 \\ 0 \end{bmatrix} = \begin{bmatrix} 80 \\ 10 \\ 30 \end{bmatrix}$$

So, 80 roses, 10 lilies, and 30 irises will create 40 centerpieces.

73. True. If B is the inverse of A, then $AB = I = BA$.

75. If the determinant of a 2×2 matrix is not equal to 0, then the inverse exists.
To find the inverse, take 1 divided by the determinant and multiply it by the matrix which has a diagonal from top left to bottom right that has the terms from the original matrix flipped and the other diagonal is the negative of the terms from the original matrix.

77. (a) Given $A = \begin{bmatrix} a_{11} & 0 \\ 0 & a_{22} \end{bmatrix}$, $A^{-1} = \begin{bmatrix} \frac{1}{a_{11}} & 0 \\ 0 & \frac{1}{a_{22}} \end{bmatrix}$.

Given $A = \begin{bmatrix} a_{11} & 0 & 0 \\ 0 & a_{22} & 0 \\ 0 & 0 & a_{33} \end{bmatrix}$, $A^{-1} = \begin{bmatrix} \frac{1}{a_{11}} & 0 & 0 \\ 0 & \frac{1}{a_{22}} & 0 \\ 0 & 0 & \frac{1}{a_{33}} \end{bmatrix}$.

(b) In general, the inverse of a matrix in the form of A is

$$\begin{bmatrix} \frac{1}{a_{11}} & 0 & 0 & \cdots & 0 \\ 0 & \frac{1}{a_{22}} & 0 & \cdots & 0 \\ 0 & 0 & \frac{1}{a_{33}} & \cdots & 0 \\ \vdots & \vdots & \vdots & \cdots & \vdots \\ 0 & 0 & 0 & \cdots & \frac{1}{a_{nn}} \end{bmatrix}.$$

Section 10.4 The Determinant of a Square Matrix

1. determinant

3. cofactor

5. 4

7. $\begin{vmatrix} 8 & 4 \\ 2 & 3 \end{vmatrix} = (8)(3) - (4)(2) = 16$

9. $\begin{vmatrix} 6 & 2 \\ -5 & 3 \end{vmatrix} = (6)(3) - (2)(-5) = 28$

11. $\begin{vmatrix} -7 & 0 \\ 3 & 0 \end{vmatrix} = -7(0) - 0(3) = 0$

13. $\begin{vmatrix} 2 & 6 \\ 0 & 3 \end{vmatrix} = 2(3) - 6(0) = 6$

15. $\begin{vmatrix} -3 & -2 \\ -6 & -1 \end{vmatrix} = (-3)(-1) - (-2)(-6) = 3 - 12 = -9$

17. $\begin{vmatrix} -2 & -7 \\ -3 & 1 \end{vmatrix} = (-2)(1) - (-3)(-7) = -23$

19. $\begin{vmatrix} -7 & 6 \\ \frac{1}{2} & 3 \end{vmatrix} = (-7)(3) - (6)\left(\frac{1}{2}\right) = -24$

21. $\begin{vmatrix} -\frac{1}{2} & \frac{1}{3} \\ -6 & \frac{1}{3} \end{vmatrix} = -\frac{1}{2}\left(\frac{1}{3}\right) - \frac{1}{3}(-6) = -\frac{1}{6} + 2 = \frac{11}{6}$

23. $\begin{vmatrix} 3 & 4 \\ -2 & 1 \end{vmatrix} = 11$

25. $\begin{vmatrix} 19 & -20 \\ 43 & 56 \end{vmatrix} = 1924$

27. $\begin{vmatrix} 0.3 & 0.2 & 0.2 \\ 0.2 & 0.2 & 0.2 \\ -0.4 & 0.4 & 0.3 \end{vmatrix} = -0.002$

29. $\begin{vmatrix} 0.9 & 0.7 & 0 \\ -0.1 & 0.3 & 1.3 \\ -2.2 & 4.2 & 6.1 \end{vmatrix} = -4.842$

31. $\begin{bmatrix} 4 & 5 \\ 3 & -6 \end{bmatrix}$

(a) $M_{11} = -6$
$M_{12} = 3$
$M_{21} = 5$
$M_{22} = 4$

(b) $C_{11} = M_{11} = -6$
$C_{12} = -M_{12} = -3$
$C_{21} = -M_{21} = -5$
$C_{22} = M_{22} = 4$

33. $\begin{bmatrix} 3 & 1 \\ -2 & -4 \end{bmatrix}$

(a) $M_{11} = -4$
$M_{12} = -2$
$M_{21} = 1$
$M_{22} = 3$

(b) $C_{11} = M_{11} = -4$
$C_{12} = -M_{12} = 2$
$C_{21} = -M_{21} = -1$
$C_{22} = M_{22} = 3$

35. $\begin{bmatrix} 4 & 0 & 2 \\ -3 & 2 & 1 \\ 1 & -1 & 1 \end{bmatrix}$

(a) $M_{11} = \begin{vmatrix} 2 & 1 \\ -1 & 1 \end{vmatrix} = 2 - (-1) = 3$

$M_{12} = \begin{vmatrix} -3 & 1 \\ 1 & 1 \end{vmatrix} = -3 - 1 = -4$

$M_{13} = \begin{vmatrix} -3 & 2 \\ 1 & -1 \end{vmatrix} = 3 - 2 = 1$

$M_{21} = \begin{vmatrix} 0 & 2 \\ -1 & 1 \end{vmatrix} = 0 - (-2) = 2$

$M_{22} = \begin{vmatrix} 4 & 2 \\ 1 & 1 \end{vmatrix} = 4 - 2 = 2$

$M_{23} = \begin{vmatrix} 4 & 0 \\ 1 & -1 \end{vmatrix} = -4 - 0 = -4$

$M_{31} = \begin{vmatrix} 0 & 2 \\ 2 & 1 \end{vmatrix} = 0 - 4 = -4$

$M_{32} = \begin{vmatrix} 4 & 2 \\ -3 & 1 \end{vmatrix} = 4 - (-6) = 10$

$M_{33} = \begin{vmatrix} 4 & 0 \\ -3 & 2 \end{vmatrix} = 8 - 0 = 8$

(b) $C_{11} = (-1)^2 M_{11} = 3$

$C_{12} = (-1)^3 M_{12} = 4$

$C_{13} = (-1)^4 M_{13} = 1$

$C_{21} = (-1)^3 M_{21} = -2$

$C_{22} = (-1)^4 M_{22} = 2$

$C_{23} = (-1)^5 M_{23} = 4$

$C_{31} = (-1)^4 M_{31} = -4$

$C_{32} = (-1)^5 M_{32} = -10$

$C_{33} = (-1)^6 M_{33} = 8$

37. $\begin{bmatrix} -4 & 6 & 3 \\ 7 & -2 & 8 \\ 1 & 0 & -5 \end{bmatrix}$

(a) $M_{11} = \begin{vmatrix} -2 & 8 \\ 0 & -5 \end{vmatrix} = (-2)(-5) - (8)(0) = 10$

$M_{12} = \begin{vmatrix} 7 & 8 \\ 1 & -5 \end{vmatrix} = (7)(-5) - (8)(1) = -43$

$M_{13} = \begin{vmatrix} 7 & -2 \\ 1 & 0 \end{vmatrix} = (7)(0) - (-2)(1) = 2$

$M_{21} = \begin{vmatrix} 6 & 3 \\ 0 & -5 \end{vmatrix} = (6)(-5) - (3)(0) = -30$

$M_{22} = \begin{vmatrix} -4 & 3 \\ 1 & -5 \end{vmatrix} = (-4)(-5) - (3)(1) = 17$

$M_{23} = \begin{vmatrix} -4 & 6 \\ 1 & 0 \end{vmatrix} = (-4)(0) - (6)(1) = -6$

$M_{31} = \begin{vmatrix} 6 & 3 \\ -2 & 8 \end{vmatrix} = (6)(8) - (3)(-2) = 54$

$M_{32} = \begin{vmatrix} -4 & 3 \\ 7 & 8 \end{vmatrix} = (-4)(8) - (3)(7) = -53$

$M_{33} = \begin{vmatrix} -4 & 6 \\ 7 & -2 \end{vmatrix} = (-4)(-2) - (6)(7) = -34$

(b) $C_{11} = (-1)^2 M_{11} = 10$

$C_{12} = (-1)^3 M_{12} = 43$

$C_{13} = (-1)^4 M_{13} = 2$

$C_{21} = (-1)^3 M_{21} = 30$

$C_{22} = (-1)^4 M_{22} = 17$

$C_{23} = (-1)^5 M_{23} = 6$

$C_{31} = (-1)^4 M_{31} = 54$

$C_{32} = (-1)^5 M_{32} = 53$

$C_{33} = (-1)^6 M_{33} = -34$

39. (a) $\begin{vmatrix} -3 & 2 & 1 \\ 4 & 5 & 6 \\ 2 & -3 & 1 \end{vmatrix} = -3\begin{vmatrix} 5 & 6 \\ -3 & 1 \end{vmatrix} - 2\begin{vmatrix} 4 & 6 \\ 2 & 1 \end{vmatrix} + \begin{vmatrix} 4 & 5 \\ 2 & -3 \end{vmatrix} = -3(23) - 2(-8) - 22 = -75$

(b) $\begin{vmatrix} -3 & 2 & 1 \\ 4 & 5 & 6 \\ 2 & -3 & 1 \end{vmatrix} = -2\begin{vmatrix} 4 & 6 \\ 2 & 1 \end{vmatrix} + 5\begin{vmatrix} -3 & 1 \\ 2 & 1 \end{vmatrix} + 3\begin{vmatrix} -3 & 1 \\ 4 & 6 \end{vmatrix} = -2(-8) + 5(-5) + 3(-22) = -75$

41. (a) $\begin{vmatrix} 5 & 0 & -3 \\ 0 & 12 & 4 \\ 1 & 6 & 3 \end{vmatrix} = 0\begin{vmatrix} 0 & -3 \\ 6 & 3 \end{vmatrix} + 12\begin{vmatrix} 5 & -3 \\ 1 & 3 \end{vmatrix} - 4\begin{vmatrix} 5 & 0 \\ 1 & 6 \end{vmatrix} = 0(18) + 12(18) - 4(30) = 96$

(b) $\begin{vmatrix} 5 & 0 & -3 \\ 0 & 12 & 4 \\ 1 & 6 & 3 \end{vmatrix} = 0\begin{vmatrix} 0 & 4 \\ 1 & 3 \end{vmatrix} + 12\begin{vmatrix} 5 & -3 \\ 1 & 3 \end{vmatrix} - 6\begin{vmatrix} 5 & -3 \\ 0 & 4 \end{vmatrix} = 0(-4) + 12(18) - 6(20) = 96$

43. (a) $\begin{vmatrix} -2 & 4 & 7 & 1 \\ 3 & 0 & 0 & 0 \\ 8 & 5 & 10 & 5 \\ 6 & 0 & 5 & 0 \end{vmatrix} = -3\begin{vmatrix} 4 & 7 & 1 \\ 5 & 10 & 5 \\ 0 & 5 & 0 \end{vmatrix} + 0\begin{vmatrix} -2 & 7 & 1 \\ 8 & 10 & 5 \\ 6 & 5 & 0 \end{vmatrix} - 0\begin{vmatrix} -2 & 4 & 1 \\ 8 & 5 & 5 \\ 6 & 0 & 0 \end{vmatrix} + 0\begin{vmatrix} -2 & 4 & 7 \\ 8 & 5 & 10 \\ 6 & 0 & 5 \end{vmatrix}$

$= -3(-75) = 225$

(b) $\begin{vmatrix} -2 & 4 & 7 & 1 \\ 3 & 0 & 0 & 0 \\ 8 & 5 & 10 & 5 \\ 6 & 0 & 5 & 0 \end{vmatrix} = -1\begin{vmatrix} 3 & 0 & 0 \\ 8 & 5 & 10 \\ 6 & 0 & 5 \end{vmatrix} + 0\begin{vmatrix} -2 & 4 & 7 \\ 8 & 5 & 10 \\ 6 & 0 & 5 \end{vmatrix} - 5\begin{vmatrix} -2 & 4 & 7 \\ 3 & 0 & 0 \\ 6 & 0 & 5 \end{vmatrix} + 0\begin{vmatrix} -2 & 4 & 7 \\ 3 & 0 & 0 \\ 8 & 5 & 10 \end{vmatrix}$

$= (-1)(75) - 5(-60) = 225$

45. (a) $\begin{vmatrix} 6 & 0 & -3 & 5 \\ 4 & 13 & 6 & -8 \\ -1 & 0 & 7 & 4 \\ 8 & 6 & 0 & 2 \end{vmatrix} = -4\begin{vmatrix} 0 & -3 & 5 \\ 0 & 7 & 4 \\ 6 & 0 & 2 \end{vmatrix} + 13\begin{vmatrix} 6 & -3 & 5 \\ -1 & 7 & 4 \\ 8 & 0 & 2 \end{vmatrix} - 6\begin{vmatrix} 6 & 0 & 5 \\ -1 & 0 & 4 \\ 8 & 6 & 2 \end{vmatrix} - 8\begin{vmatrix} 6 & 0 & -3 \\ -1 & 0 & 7 \\ 8 & 6 & 0 \end{vmatrix}$

$= -4(-282) + 13(-298) - 6(-174) - 8(-234) = 170$

(b) $\begin{vmatrix} 6 & 0 & -3 & 5 \\ 4 & 13 & 6 & -8 \\ -1 & 0 & 7 & 4 \\ 8 & 6 & 0 & 2 \end{vmatrix} = 0\begin{vmatrix} 4 & 6 & -8 \\ -1 & 7 & 4 \\ 8 & 0 & 2 \end{vmatrix} + 13\begin{vmatrix} 6 & -3 & 5 \\ -1 & 7 & 4 \\ 8 & 0 & 2 \end{vmatrix} + 0\begin{vmatrix} 6 & -3 & 5 \\ 4 & 6 & -8 \\ 8 & 0 & 2 \end{vmatrix} + 6\begin{vmatrix} 6 & -3 & 5 \\ 4 & 6 & -8 \\ -1 & 7 & 4 \end{vmatrix}$

$= 0 + 13(-298) + 0 + 6(674) = 170$

47. Expand along Column 1.

$\begin{vmatrix} 2 & -1 & 0 \\ 4 & 2 & 1 \\ 4 & 2 & 1 \end{vmatrix} = 2\begin{vmatrix} 2 & 1 \\ 2 & 1 \end{vmatrix} - 4\begin{vmatrix} -1 & 0 \\ 2 & 1 \end{vmatrix} + 4\begin{vmatrix} -1 & 0 \\ 2 & 1 \end{vmatrix}$

$= 2(0) - 4(-1) + 4(-1) = 0$

49. Expand along Row 2.

$\begin{vmatrix} 6 & 3 & -7 \\ 0 & 0 & 0 \\ 4 & -6 & 3 \end{vmatrix} = 0\begin{vmatrix} 3 & -7 \\ -6 & 3 \end{vmatrix} - 0\begin{vmatrix} 6 & -7 \\ 4 & 3 \end{vmatrix} + 0\begin{vmatrix} 6 & 3 \\ 4 & -6 \end{vmatrix} = 0$

51. Expand along Column 1.

$\begin{vmatrix} -1 & 2 & -5 \\ 0 & 3 & 4 \\ 0 & 0 & 3 \end{vmatrix} = -1\begin{vmatrix} 3 & 4 \\ 0 & 3 \end{vmatrix} - 0\begin{vmatrix} 2 & -5 \\ 0 & 3 \end{vmatrix} + 0\begin{vmatrix} 2 & -5 \\ 3 & 4 \end{vmatrix}$

$= -1(9) - 0(6) + 0(23) = -9$

53. Expand along Column 3.

$\begin{vmatrix} 1 & 4 & -2 \\ 3 & 2 & 0 \\ -1 & 4 & 3 \end{vmatrix} = -2\begin{vmatrix} 3 & 2 \\ -1 & 4 \end{vmatrix} + 3\begin{vmatrix} 1 & 4 \\ 3 & 2 \end{vmatrix}$

$= -2(14) + 3(-10) = -58$

55. Expand along Column 1.

$\begin{vmatrix} 2 & 4 & 6 \\ 0 & 3 & 1 \\ 0 & 0 & -5 \end{vmatrix} = 2\begin{vmatrix} 3 & 1 \\ 0 & -5 \end{vmatrix} - 0\begin{vmatrix} 4 & 6 \\ 0 & -5 \end{vmatrix} + 0\begin{vmatrix} 4 & 6 \\ 3 & 1 \end{vmatrix}$

$= 2(-15) - 0(-20) + 0(-14) = -30$

57. Expand along Column 3.

$$\begin{vmatrix} 2 & 6 & 6 & 2 \\ 2 & 7 & 3 & 6 \\ 1 & 5 & 0 & 1 \\ 3 & 7 & 0 & 7 \end{vmatrix} = 6\begin{vmatrix} 2 & 7 & 6 \\ 1 & 5 & 1 \\ 3 & 7 & 7 \end{vmatrix} - 3\begin{vmatrix} 2 & 6 & 2 \\ 1 & 5 & 1 \\ 3 & 7 & 7 \end{vmatrix}$$

$$= 6(-20) - 3(16) = -168$$

59. Expand along Column 1.

$$\begin{vmatrix} 5 & 3 & 0 & 6 \\ 4 & 6 & 4 & 12 \\ 0 & 2 & -3 & 4 \\ 0 & 1 & -2 & 2 \end{vmatrix} = 5\begin{vmatrix} 6 & 4 & 12 \\ 2 & -3 & 4 \\ 1 & -2 & 2 \end{vmatrix} - 4\begin{vmatrix} 3 & 0 & 6 \\ 2 & -3 & 4 \\ 1 & -2 & 2 \end{vmatrix}$$

$$= 5(0) - 4(0) = 0$$

61. Expand along Column 2, then along Column 4.

$$\begin{vmatrix} 3 & 2 & 4 & -1 & 5 \\ -2 & 0 & 1 & 3 & 2 \\ 1 & 0 & 0 & 4 & 0 \\ 6 & 0 & 2 & -1 & 0 \\ 3 & 0 & 5 & 1 & 0 \end{vmatrix} = -2\begin{vmatrix} -2 & 1 & 3 & 2 \\ 1 & 0 & 4 & 0 \\ 6 & 2 & -1 & 0 \\ 3 & 5 & 1 & 0 \end{vmatrix} = (-2)(-2)\begin{vmatrix} 1 & 0 & 4 \\ 6 & 2 & -1 \\ 3 & 5 & 1 \end{vmatrix} = 4(103) = 412$$

63. $\begin{vmatrix} 3 & 8 & -7 \\ 0 & -5 & 4 \\ 8 & 1 & 6 \end{vmatrix} = -126$

65. $\begin{vmatrix} 1 & -1 & 8 & 4 \\ 2 & 6 & 0 & -4 \\ 2 & 0 & 2 & 6 \\ 0 & 2 & 8 & 0 \end{vmatrix} = -336$

67. (a) $\begin{vmatrix} -1 & 0 \\ 0 & 3 \end{vmatrix} = -3$

(b) $\begin{vmatrix} 2 & 0 \\ 0 & -1 \end{vmatrix} = -2$

(c) $\begin{bmatrix} -1 & 0 \\ 0 & 3 \end{bmatrix}\begin{bmatrix} 2 & 0 \\ 0 & -1 \end{bmatrix} = \begin{bmatrix} -2 & 0 \\ 0 & -3 \end{bmatrix}$

(d) $\begin{vmatrix} -2 & 0 \\ 0 & -3 \end{vmatrix} = 6$

69. (a) $\begin{vmatrix} 4 & 0 \\ 3 & -2 \end{vmatrix} = -8$

(b) $\begin{vmatrix} -1 & 1 \\ -2 & 2 \end{vmatrix} = 0$

(c) $\begin{bmatrix} 4 & 0 \\ 3 & -2 \end{bmatrix}\begin{bmatrix} -1 & 1 \\ -2 & 2 \end{bmatrix} = \begin{bmatrix} -4 & 4 \\ 1 & -1 \end{bmatrix}$

(d) $\begin{vmatrix} -4 & 4 \\ 1 & -1 \end{vmatrix} = 0$

71. (a) $\begin{vmatrix} 0 & 1 & 2 \\ -3 & -2 & 1 \\ 0 & 4 & 1 \end{vmatrix} = -21$

(b) $\begin{vmatrix} 3 & -2 & 0 \\ 1 & -1 & 2 \\ 3 & 1 & 1 \end{vmatrix} = -19$

(c) $\begin{bmatrix} 0 & 1 & 2 \\ -3 & -2 & 1 \\ 0 & 4 & 1 \end{bmatrix}\begin{bmatrix} 3 & -2 & 0 \\ 1 & -1 & 2 \\ 3 & 1 & 1 \end{bmatrix} = \begin{bmatrix} 7 & 1 & 4 \\ -8 & 9 & -3 \\ 7 & -3 & 9 \end{bmatrix}$

(d) $\begin{vmatrix} 7 & 1 & 4 \\ -8 & 9 & -3 \\ 7 & -3 & 9 \end{vmatrix} = 399$

73. (a) $\begin{vmatrix} -1 & 2 & 1 \\ 1 & 0 & 1 \\ 0 & 1 & 0 \end{vmatrix} = 2$

(b) $\begin{vmatrix} -1 & 0 & 0 \\ 0 & 2 & 0 \\ 0 & 0 & 3 \end{vmatrix} = -6$

(c) $\begin{bmatrix} -1 & 2 & 1 \\ 1 & 0 & 1 \\ 0 & 1 & 0 \end{bmatrix}\begin{bmatrix} -1 & 0 & 0 \\ 0 & 2 & 0 \\ 0 & 0 & 3 \end{bmatrix} = \begin{bmatrix} 1 & 4 & 3 \\ -1 & 0 & 3 \\ 0 & 2 & 0 \end{bmatrix}$

(d) $\begin{vmatrix} 1 & 4 & 3 \\ -1 & 0 & 3 \\ 0 & 2 & 0 \end{vmatrix} = -12$

75. $\begin{vmatrix} w & x \\ y & z \end{vmatrix} = wz - xy$

$-\begin{vmatrix} y & z \\ w & x \end{vmatrix} = -(xy - wz) = wz - xy$

So, $\begin{vmatrix} w & x \\ y & z \end{vmatrix} = -\begin{vmatrix} y & z \\ w & x \end{vmatrix}$.

77. $\begin{vmatrix} w & x \\ y & z \end{vmatrix} = wz - xy$

$\begin{vmatrix} w & x + cw \\ y & z + cy \end{vmatrix} = w(z + cy) - y(x + cw) = wz - xy$

So, $\begin{vmatrix} w & x \\ y & z \end{vmatrix} = \begin{vmatrix} w & x + cw \\ y & z + cy \end{vmatrix}$.

79. $\begin{vmatrix} 1 & x & x^2 \\ 1 & y & y^2 \\ 1 & z & z^2 \end{vmatrix} = \begin{vmatrix} y & y^2 \\ z & z^2 \end{vmatrix} - \begin{vmatrix} x & x^2 \\ z & z^2 \end{vmatrix} + \begin{vmatrix} x & x^2 \\ y & y^2 \end{vmatrix}$

$$\begin{aligned}
&= (yz^2 - y^2z) - (xz^2 - x^2z) + (xy^2 - x^2y) \\
&= yz^2 - xz^2 - y^2z + x^2z + xy(y - x) \\
&= z^2(y - x) - z(y^2 - x^2) + xy(y - x) \\
&= z^2(y - x) - z(y - x)(y + x) + xy(y - x) \\
&= (y - x)[z^2 - z(y + x) + xy] \\
&= (y - x)[z^2 - zy - zx + xy] \\
&= (y - x)[z^2 - zx - zy + xy] \\
&= (y - x)[z(z - x) - y(z - x)] \\
&= (y - x)(z - x)(z - y)
\end{aligned}$$

81. $\begin{vmatrix} x & 2 \\ 1 & x \end{vmatrix} = 2$

$$\begin{aligned}
x^2 - 2 &= 2 \\
x^2 &= 4 \\
x &= \pm 2
\end{aligned}$$

83. $\begin{vmatrix} x & 1 \\ 2 & x - 2 \end{vmatrix} = -1$

$x(x - 2) - 2 = -1$

$x^2 - 2x - 1 = 0$

Using the Quadratic Formula:

$x = \dfrac{2 \pm \sqrt{4 - 4(1)(-1)}}{2}$

$x = \dfrac{2 \pm 2\sqrt{2}}{2}$

$x = 1 \pm \sqrt{2}$

85. $\begin{vmatrix} x - 1 & 2 \\ 3 & x - 2 \end{vmatrix} = 0$

$$\begin{aligned}
(x - 1)(x - 2) - 6 &= 0 \\
x^2 - 3x - 4 &= 0 \\
(x + 1)(x - 4) &= 0 \\
x &= -1 \text{ or } x = 4
\end{aligned}$$

87. $\begin{vmatrix} x + 3 & 2 \\ 1 & x + 2 \end{vmatrix} = 0$

$$\begin{aligned}
(x + 3)(x + 2) - 2 &= 0 \\
x^2 + 5x + 4 &= 0 \\
(x + 1)(x + 4) &= 0 \\
x &= -1 \text{ or } x = -4
\end{aligned}$$

89. $\begin{vmatrix} 4u & -1 \\ -1 & 2v \end{vmatrix} = 8uv - 1$

91. $\begin{vmatrix} e^{2x} & e^{3x} \\ 2e^{2x} & 3e^{3x} \end{vmatrix} = 3e^{5x} - 2e^{5x} = e^{5x}$

93. $\begin{vmatrix} x & \ln x \\ 1 & \dfrac{1}{x} \end{vmatrix} = 1 - \ln x$

95. True. If an entire row is zero, then each cofactor in the expansion is multiplied by zero.

97. *Sample answer:* Let $A = \begin{bmatrix} 1 & 3 \\ -2 & 4 \end{bmatrix}$ and $B = \begin{bmatrix} -4 & 0 \\ 3 & 5 \end{bmatrix}$.

$|A| = \begin{vmatrix} 1 & 3 \\ -2 & 4 \end{vmatrix} = 10,\ |B| = \begin{vmatrix} -4 & 0 \\ 3 & 5 \end{vmatrix} = -20,\ |A| + |B| = -10$

$A + B = \begin{bmatrix} -3 & 3 \\ 1 & 9 \end{bmatrix},\ |A + B| = \begin{vmatrix} -3 & 3 \\ 1 & 9 \end{vmatrix} = -30$

So, $|A + B| \neq |A| + |B|$.

99. A square matrix is a square array of numbers. The determinant of a square matrix is a real number.

101. (a) $\begin{vmatrix} 1 & 3 & 4 \\ -7 & 2 & -5 \\ 6 & 1 & 2 \end{vmatrix} = -115$

$-\begin{vmatrix} 1 & 4 & 3 \\ -7 & -5 & 2 \\ 6 & 2 & 1 \end{vmatrix} = -115$

Column 2 and Column 3 were interchanged.

(b) $\begin{vmatrix} 1 & 3 & 4 \\ -2 & 2 & 0 \\ 1 & 6 & 2 \end{vmatrix} = -40$

$-\begin{vmatrix} 1 & 6 & 2 \\ -2 & 2 & 0 \\ 1 & 3 & 4 \end{vmatrix} = -40$

Row 1 and Row 3 were interchanged.

103. (a) $A = \begin{bmatrix} 1 & 2 \\ 2 & -3 \end{bmatrix},\ B = \begin{bmatrix} 5 & 10 \\ 2 & -3 \end{bmatrix}$

$|B| = \begin{vmatrix} 5 & 10 \\ 2 & -3 \end{vmatrix} = -35$

$5|A| = 5\begin{vmatrix} 1 & 2 \\ 2 & -3 \end{vmatrix} = -35$

Row 1 was multiplied by 5.

$|B| = 5|A|$

(b) $A = \begin{bmatrix} 1 & 2 & -1 \\ 3 & -3 & 2 \\ 7 & 1 & 3 \end{bmatrix},\ B = \begin{bmatrix} 1 & 8 & -3 \\ 3 & -12 & 6 \\ 7 & 4 & 9 \end{bmatrix}$

$|B| = \begin{bmatrix} 1 & 8 & -3 \\ 3 & -12 & 6 \\ 7 & 4 & 9 \end{bmatrix} = -300$

$12|A| = 12\begin{vmatrix} 1 & 2 & -1 \\ 3 & -3 & 2 \\ 7 & 1 & 3 \end{vmatrix} = -300$

Column 2 was multiplied by 4 and Column 3 was multiplied by 3.

$|B| = (4)(3)|A| = 12|A|$

105. (a) $\begin{vmatrix} 7 & 0 \\ 0 & 4 \end{vmatrix} = 7(4) - 0 = 28$

(b) $\begin{vmatrix} -1 & 0 & 0 \\ 0 & 5 & 0 \\ 0 & 0 & 2 \end{vmatrix} = (-1)\begin{vmatrix} 5 & 0 \\ 0 & 2 \end{vmatrix} - 0\begin{vmatrix} 0 & 0 \\ 0 & 2 \end{vmatrix} + 0\begin{vmatrix} 0 & 5 \\ 0 & 0 \end{vmatrix}$

$= (-1)(10) = -10$

(c) $\begin{vmatrix} -2 & 0 & 0 & 0 \\ 0 & -2 & 0 & 0 \\ 0 & 0 & 1 & 0 \\ 0 & 0 & 0 & 3 \end{vmatrix} = (-2)\begin{vmatrix} -2 & 0 & 0 \\ 0 & 1 & 0 \\ 0 & 0 & 3 \end{vmatrix} - 0\begin{vmatrix} 0 & 0 & 0 \\ 0 & 1 & 0 \\ 0 & 0 & 3 \end{vmatrix} + 0\begin{vmatrix} 0 & -2 & 0 \\ 0 & 0 & 0 \\ 0 & 0 & 3 \end{vmatrix} - 0\begin{vmatrix} 0 & -2 & 0 \\ 0 & 0 & 1 \\ 0 & 0 & 0 \end{vmatrix}$

$= (-2)\left((-2)\begin{vmatrix} 1 & 0 \\ 0 & 3 \end{vmatrix} - 0\begin{vmatrix} 0 & 0 \\ 0 & 3 \end{vmatrix} + 0\begin{vmatrix} 0 & 1 \\ 0 & 0 \end{vmatrix}\right)$

$= (-2)(-2)(3) = 12$

The determinant of a diagonal matrix is the product of the entries on the main diagonal.

Section 10.5 Applications of Matrices and Determinants

1. Cramer's Rule

3. $A = \pm\frac{1}{2}\begin{vmatrix} x_1 & y_1 & 1 \\ x_2 & y_2 & 1 \\ x_3 & y_3 & 1 \end{vmatrix}$

5. uncoded; coded

7. $\begin{cases} -7x + 11y = -1 \\ 3x - 9y = 9 \end{cases}$

$x = \dfrac{\begin{vmatrix} -1 & 11 \\ 9 & -9 \end{vmatrix}}{\begin{vmatrix} -7 & 11 \\ 3 & -9 \end{vmatrix}} = \dfrac{-90}{30} = -3$

$y = \dfrac{\begin{vmatrix} -7 & -1 \\ 3 & 9 \end{vmatrix}}{\begin{vmatrix} -7 & 11 \\ 3 & -9 \end{vmatrix}} = \dfrac{-60}{30} = -2$

Solution: $(-3, -2)$

9. $\begin{cases} 3x + 2y = -2 \\ 6x + 4y = 4 \end{cases}$

Because $\begin{vmatrix} 3 & 2 \\ 6 & 4 \end{vmatrix} = 0$, Cramer's Rule does not apply.

The system is inconsistent in this case and has no solution.

11. $\begin{cases} -0.4x + 0.8y = 1.6 \\ 0.2x + 0.3y = 2.2 \end{cases}$

$x = \dfrac{\begin{vmatrix} 1.6 & 0.8 \\ 2.2 & 0.3 \end{vmatrix}}{\begin{vmatrix} -0.4 & 0.8 \\ 0.2 & 0.3 \end{vmatrix}} = \dfrac{-1.28}{-0.28} = \dfrac{32}{7}$

$y = \dfrac{\begin{vmatrix} -0.4 & 1.6 \\ 0.2 & 2.2 \end{vmatrix}}{\begin{vmatrix} -0.4 & 0.8 \\ 0.2 & 0.3 \end{vmatrix}} = \dfrac{-1.20}{-0.28} = \dfrac{30}{7}$

Solution: $\left(\dfrac{32}{7}, \dfrac{30}{7}\right)$

13. $\begin{cases} 4x - y + z = -5 \\ 2x + 2y + 3z = 10, \\ 5x - 2y + 6z = 1 \end{cases}$ $D = \begin{vmatrix} 4 & -1 & 1 \\ 2 & 2 & 3 \\ 5 & -1 & 6 \end{vmatrix} = 55$

$x = \dfrac{\begin{vmatrix} -5 & -1 & 1 \\ 10 & 2 & 3 \\ 1 & -2 & 6 \end{vmatrix}}{55} = \dfrac{-55}{55} = -1$, $y = \dfrac{\begin{vmatrix} 4 & -5 & 1 \\ 2 & 10 & 3 \\ 5 & 1 & 6 \end{vmatrix}}{55} = \dfrac{165}{55} = 3$, $z = \dfrac{\begin{vmatrix} 4 & -1 & -5 \\ 2 & 2 & 10 \\ 5 & -2 & 1 \end{vmatrix}}{55} = \dfrac{110}{55} = 2$

Solution: $(-1, 3, 2)$

15. $\begin{cases} x + 2y + 3z = -3 \\ -2x + y - z = 6, \\ 3x - 3y + 2z = -11 \end{cases}$ $D = \begin{vmatrix} 1 & 2 & 3 \\ -2 & 1 & -1 \\ 3 & -3 & 2 \end{vmatrix} = 10$

$$x = \frac{\begin{vmatrix} -3 & 2 & 3 \\ 6 & 1 & -1 \\ -11 & -3 & 2 \end{vmatrix}}{10} = \frac{-20}{10} = -2$$

$$y = \frac{\begin{vmatrix} 1 & -3 & 3 \\ -2 & 6 & -1 \\ 3 & -11 & 2 \end{vmatrix}}{10} = \frac{10}{10} = 1$$

$$z = \frac{\begin{vmatrix} 1 & 2 & -3 \\ -2 & 1 & 6 \\ 3 & -3 & -11 \end{vmatrix}}{10} = \frac{-10}{10} = -1$$

Solution: $(-2, 1, -1)$

17. Vertices: $(0, 0)\ (3, 1), (1, 5)$

$$\text{Area} = \frac{1}{2}\begin{vmatrix} 0 & 0 & 1 \\ 3 & 1 & 1 \\ 1 & 5 & 1 \end{vmatrix} = \frac{1}{2}\begin{vmatrix} 3 & 1 \\ 1 & 5 \end{vmatrix} = 7 \text{ square units}$$

19. Vertices: $(-2, -3), (2, -3), (0, 4)$

$$\text{Area} = \frac{1}{2}\begin{vmatrix} -2 & -3 & 1 \\ 2 & -3 & 1 \\ 0 & 4 & 1 \end{vmatrix} = \frac{1}{2}\left(-2\begin{vmatrix} -3 & 1 \\ 4 & 1 \end{vmatrix} - 2\begin{vmatrix} -3 & 1 \\ 4 & 1 \end{vmatrix}\right) = \frac{1}{2}(14 + 14) = 14 \text{ square units}$$

21. Vertices: $\left(0, \frac{1}{2}\right), (4, 3), \left(\frac{5}{2}, 0\right)$

$$\text{Area} = -\frac{1}{2}\begin{vmatrix} 0 & \frac{1}{2} & 1 \\ 4 & 3 & 1 \\ \frac{5}{2} & 0 & 1 \end{vmatrix} = -\frac{1}{2}\left(0(-1)^2\begin{vmatrix} 3 & 1 \\ 0 & 1 \end{vmatrix} + \frac{1}{2}(-1)^3\begin{vmatrix} 4 & 1 \\ \frac{5}{2} & 1 \end{vmatrix} + (1)(-1)^4\begin{vmatrix} 4 & 3 \\ \frac{5}{2} & 0 \end{vmatrix}\right) = -\frac{1}{2}\left(-\frac{3}{4} - \frac{15}{2}\right) = \frac{33}{8} \text{ square units}$$

23. Vertices: $(-2, 4), (2, 3), (-1, 5)$

$$\text{Area} = \frac{1}{2}\begin{vmatrix} -2 & 4 & 1 \\ 2 & 3 & 1 \\ -1 & 5 & 1 \end{vmatrix} = \frac{1}{2}\left[\begin{vmatrix} 2 & 3 \\ -1 & 5 \end{vmatrix} - \begin{vmatrix} -2 & 4 \\ -1 & 5 \end{vmatrix} + \begin{vmatrix} -2 & 4 \\ 2 & 3 \end{vmatrix}\right] = \frac{1}{2}(13 + 6 - 14) = \frac{5}{2} \text{ square units}$$

25. Vertices: $(-3, 5), (2, 6), (3, -5)$

$$\text{Area} = -\frac{1}{2}\begin{vmatrix} -3 & 5 & 1 \\ 2 & 6 & 1 \\ 3 & -5 & 1 \end{vmatrix} = -\frac{1}{2}\left[\begin{vmatrix} 2 & 6 \\ 3 & -5 \end{vmatrix} - \begin{vmatrix} -3 & 5 \\ 3 & -5 \end{vmatrix} + \begin{vmatrix} -3 & 5 \\ 2 & 6 \end{vmatrix}\right] = -\frac{1}{2}(-28 + 0 - 28) = 28 \text{ square units}$$

27. Vertices: $(-4, 2), \left(0, \frac{7}{2}\right), \left(3, -\frac{1}{2}\right)$

$$\text{Area} = -\frac{1}{2}\begin{vmatrix} -4 & 2 & 1 \\ 0 & \frac{7}{2} & 1 \\ 3 & -\frac{1}{2} & 1 \end{vmatrix} = -\frac{1}{2}\left(-4\begin{vmatrix} \frac{7}{2} & 1 \\ -\frac{1}{2} & 1 \end{vmatrix} - 2\begin{vmatrix} 0 & 1 \\ 3 & 1 \end{vmatrix} + 1\begin{vmatrix} 0 & \frac{7}{2} \\ 3 & -\frac{1}{2} \end{vmatrix}\right) = -\frac{1}{2}\left(-16 + 6 - \frac{21}{2}\right) = \frac{41}{4} \text{ square units}$$

29. $4 = \pm\frac{1}{2}\begin{vmatrix} -5 & 1 & 1 \\ 0 & 2 & 1 \\ -2 & y & 1 \end{vmatrix}$

$\pm 8 = -5\begin{vmatrix} 2 & 1 \\ y & 1 \end{vmatrix} - 2\begin{vmatrix} 1 & 1 \\ 2 & 1 \end{vmatrix}$

$\pm 8 = -5(2 - y) - 2(-1)$

$\pm 8 = 5y - 8$

$y = \frac{8 \pm 8}{5}$

$y = \frac{16}{5} \text{ or } y = 0$

31. $6 = \pm\frac{1}{2}\begin{vmatrix} -2 & -3 & 1 \\ 1 & -1 & 1 \\ -8 & y & 1 \end{vmatrix}$

$\pm 12 = \begin{vmatrix} 1 & -1 \\ -8 & y \end{vmatrix} - \begin{vmatrix} -2 & -3 \\ -8 & y \end{vmatrix} + \begin{vmatrix} -2 & -3 \\ 1 & -1 \end{vmatrix}$

$\pm 12 = (y - 8) - (-2y - 24) + 5$

$\pm 12 = 3y + 21$

$y = \frac{-21 \pm 12}{3} = -7 \pm 4$

$y = -3 \text{ or } y = -11$

33. Vertices: $(0, 25), (10, 0), (28, 5)$

$\text{Area} = \frac{1}{2}\begin{vmatrix} 0 & 25 & 1 \\ 10 & 0 & 1 \\ 28 & 5 & 1 \end{vmatrix} = 250 \text{ square miles}$

35. Points: $(3, -1), (0, -3), (12, 5)$

$\begin{vmatrix} 3 & -1 & 1 \\ 0 & -3 & 1 \\ 12 & 5 & 1 \end{vmatrix} = 3\begin{vmatrix} -3 & 1 \\ 5 & 1 \end{vmatrix} + 12\begin{vmatrix} -1 & 1 \\ -3 & 1 \end{vmatrix}$

$= 3(-8) + 12(2)$

$= 0$

The points are collinear.

37. Points: $\left(2, -\frac{1}{2}\right), (-4, 4), (6, -3)$

$\begin{vmatrix} 2 & -\frac{1}{2} & 1 \\ -4 & 4 & 1 \\ 6 & -3 & 1 \end{vmatrix} = \begin{vmatrix} -4 & 4 \\ 6 & -3 \end{vmatrix} - \begin{vmatrix} 2 & -\frac{1}{2} \\ 6 & -3 \end{vmatrix} + \begin{vmatrix} 2 & -\frac{1}{2} \\ -4 & 4 \end{vmatrix}$

$= -12 + 3 + 6$

$= -3 \neq 0$

The points are not collinear.

39. Points: $(0, 2), (1, 2.4), (-1, 1.6)$

$\begin{vmatrix} 0 & 2 & 1 \\ 1 & 2.4 & 1 \\ -1 & 1.6 & 1 \end{vmatrix} = -2\begin{vmatrix} 1 & 1 \\ -1 & 1 \end{vmatrix} + \begin{vmatrix} 1 & 2.4 \\ -1 & 1.6 \end{vmatrix} = -2(2) + 4 = 0$

The points are collinear.

41. $\begin{vmatrix} 2 & -5 & 1 \\ 4 & y & 1 \\ 5 & -2 & 1 \end{vmatrix} = 0$

$2\begin{vmatrix} y & 1 \\ -2 & 1 \end{vmatrix} + 5\begin{vmatrix} 4 & 1 \\ 5 & 1 \end{vmatrix} + \begin{vmatrix} 4 & y \\ 5 & -2 \end{vmatrix} = 0$

$2(y + 2) + 5(-1) + (-8 - 5y) = 0$

$-3y - 9 = 0$

$y = -3$

43. Points: $(0, 0), (5, 3)$

Equation: $\begin{vmatrix} x & y & 1 \\ 0 & 0 & 1 \\ 5 & 3 & 1 \end{vmatrix} = -\begin{vmatrix} x & y \\ 5 & 3 \end{vmatrix} = 5y - 3x = 0 \Rightarrow 3x - 5y = 0$

45. Points: $(-4, 3), (2, 1)$

Equation: $\begin{vmatrix} x & y & 1 \\ -4 & 3 & 1 \\ 2 & 1 & 1 \end{vmatrix} = x\begin{vmatrix} 3 & 1 \\ 1 & 1 \end{vmatrix} - y\begin{vmatrix} -4 & 1 \\ 2 & 1 \end{vmatrix} + \begin{vmatrix} -4 & 3 \\ 2 & 1 \end{vmatrix} = 2x + 6y - 10 = 0 \Rightarrow x + 3y - 5 = 0$

47. Points: $\left(-\frac{1}{2}, 3\right), \left(\frac{5}{2}, 1\right)$

Equation: $\begin{vmatrix} x & y & 1 \\ -\frac{1}{2} & 3 & 1 \\ \frac{5}{2} & 1 & 1 \end{vmatrix} = x\begin{vmatrix} 3 & 1 \\ 1 & 1 \end{vmatrix} - y\begin{vmatrix} -\frac{1}{2} & 1 \\ \frac{5}{2} & 1 \end{vmatrix} + \begin{vmatrix} -\frac{1}{2} & 3 \\ \frac{5}{2} & 1 \end{vmatrix} = 2x + 3y - 8 = 0$

49. (a) Uncoded: C O M E _ H O M E _

[3 15] [13 5] [0 8] [15 13] [5 0]

S O O N

[19 15] [15 14]

(b) $\begin{bmatrix} 3 & 15 \end{bmatrix}\begin{bmatrix} 1 & 2 \\ 3 & 5 \end{bmatrix} = \begin{bmatrix} 48 & 81 \end{bmatrix}$

$\begin{bmatrix} 13 & 5 \end{bmatrix}\begin{bmatrix} 1 & 2 \\ 3 & 5 \end{bmatrix} = \begin{bmatrix} 28 & 51 \end{bmatrix}$

$\begin{bmatrix} 0 & 8 \end{bmatrix}\begin{bmatrix} 1 & 2 \\ 3 & 5 \end{bmatrix} = \begin{bmatrix} 24 & 40 \end{bmatrix}$

$\begin{bmatrix} 15 & 13 \end{bmatrix}\begin{bmatrix} 1 & 2 \\ 3 & 5 \end{bmatrix} = \begin{bmatrix} 54 & 95 \end{bmatrix}$

$\begin{bmatrix} 5 & 0 \end{bmatrix}\begin{bmatrix} 1 & 2 \\ 3 & 5 \end{bmatrix} = \begin{bmatrix} 5 & 10 \end{bmatrix}$

$\begin{bmatrix} 19 & 15 \end{bmatrix}\begin{bmatrix} 1 & 2 \\ 3 & 5 \end{bmatrix} = \begin{bmatrix} 64 & 113 \end{bmatrix}$

$\begin{bmatrix} 15 & 14 \end{bmatrix}\begin{bmatrix} 1 & 2 \\ 3 & 5 \end{bmatrix} = \begin{bmatrix} 57 & 100 \end{bmatrix}$

Encoded: 48 81 28 51 24 40 54
95 5 10 64 113 57 100

51. (a) Uncoded:

C A L L _ M E _ T O M O

$[3 \quad 1 \quad 12][12 \quad 0 \quad 13][5 \quad 0 \quad 20][15 \quad 13 \quad 15]$

R R O W _ _

$[18 \quad 18 \quad 15][23 \quad 0 \quad 0]$

(b)

$$[3 \quad 1 \quad 12]\begin{bmatrix} 1 & -1 & 0 \\ 1 & 0 & -1 \\ -6 & 2 & 3 \end{bmatrix} = [-68 \quad 21 \quad 35]$$

$$[12 \quad 0 \quad 13]\begin{bmatrix} 1 & -1 & 0 \\ 1 & 0 & -1 \\ -6 & 2 & 3 \end{bmatrix} = [-66 \quad 14 \quad 39]$$

$$[5 \quad 0 \quad 20]\begin{bmatrix} 1 & -1 & 0 \\ 1 & 0 & -1 \\ -6 & 2 & 3 \end{bmatrix} = [-115 \quad 35 \quad 60]$$

$$[15 \quad 13 \quad 15]\begin{bmatrix} 1 & -1 & 0 \\ 1 & 0 & -1 \\ -6 & 2 & 3 \end{bmatrix} = [-62 \quad 15 \quad 32]$$

$$[18 \quad 18 \quad 15]\begin{bmatrix} 1 & -1 & 0 \\ 1 & 0 & -1 \\ -6 & 2 & 3 \end{bmatrix} = [-54 \quad 12 \quad 27]$$

$$[23 \quad 0 \quad 0]\begin{bmatrix} 1 & -1 & 0 \\ 1 & 0 & -1 \\ -6 & 2 & 3 \end{bmatrix} = [23 \quad -23 \quad 0]$$

Encoded: −68 21 35 −66 14 39 −115 35 60
−62 15 32 −54 12 27 23 −23 0

In Exercises 53–55, use the matrix $A = \begin{bmatrix} 1 & 2 & 2 \\ 3 & 7 & 9 \\ -1 & -4 & -7 \end{bmatrix}$.

53. L A N D I N G _ S U C C E S S F U L

$[12 \ \ 1 \ \ 14][4 \ \ 9 \ \ 14][7 \ \ 0 \ \ 19][21 \ \ 3 \ \ 3][5 \ \ 19 \ \ 19][6 \ \ 21 \ \ 12]$

$$[12 \ \ 1 \ \ 14]\begin{bmatrix} 1 & 2 & 2 \\ 3 & 7 & 9 \\ -1 & -4 & -7 \end{bmatrix} = [1 \ \ -25 \ \ -65]$$

$$[4 \ \ 9 \ \ 14]\begin{bmatrix} 1 & 2 & 2 \\ 3 & 7 & 9 \\ -1 & -4 & -7 \end{bmatrix} = [17 \ \ 15 \ \ -9]$$

$$[7 \ \ 0 \ \ 19]\begin{bmatrix} 1 & 2 & 2 \\ 3 & 7 & 9 \\ -1 & -4 & -7 \end{bmatrix} = [-12 \ \ -62 \ \ -119]$$

$$[21 \ \ 3 \ \ 3]\begin{bmatrix} 1 & 2 & 2 \\ 3 & 7 & 9 \\ -1 & -4 & -7 \end{bmatrix} = [27 \ \ 51 \ \ 48]$$

$$[5 \ \ 19 \ \ 19]\begin{bmatrix} 1 & 2 & 2 \\ 3 & 7 & 9 \\ -1 & -4 & -7 \end{bmatrix} = [43 \ \ 67 \ \ 48]$$

$$[6 \ \ 21 \ \ 12]\begin{bmatrix} 1 & 2 & 2 \\ 3 & 7 & 9 \\ -1 & -4 & -7 \end{bmatrix} = [57 \ \ 111 \ \ 117]$$

Cryptogram: 1 −25 −65 17 15 −9 −12 −62 −119 27 51 48 43 67 48 57 111 117

55. H A P P Y _ B I R T H D A Y _

$[8 \ \ 1 \ \ 16][16 \ \ 25 \ \ 0][2 \ \ 9 \ \ 18][20 \ \ 8 \ \ 4][1 \ \ 25 \ \ 0]$

$$[8 \ \ 1 \ \ 16]\begin{bmatrix} 1 & 2 & 2 \\ 3 & 7 & 9 \\ -1 & -4 & -7 \end{bmatrix} = [-5 \ \ -41 \ \ -87]$$

$$[16 \ \ 25 \ \ 0]\begin{bmatrix} 1 & 2 & 2 \\ 3 & 7 & 9 \\ -1 & -4 & -7 \end{bmatrix} = [91 \ \ 207 \ \ 257]$$

$$[2 \ \ 9 \ \ 18]\begin{bmatrix} 1 & 2 & 2 \\ 3 & 7 & 9 \\ -1 & -4 & -7 \end{bmatrix} = [11 \ \ -5 \ \ -41]$$

$$[20 \ \ 8 \ \ 4]\begin{bmatrix} 1 & 2 & 2 \\ 3 & 7 & 9 \\ -1 & -4 & -7 \end{bmatrix} = [40 \ \ 80 \ \ 84]$$

$$[1 \ \ 25 \ \ 0]\begin{bmatrix} 1 & 2 & 2 \\ 3 & 7 & 9 \\ -1 & -4 & -7 \end{bmatrix} = [76 \ \ 177 \ \ 227]$$

Cryptogram: −5 −41 −87 91 207 257 11 −5 −41 40 80 84 76 177 227

57. $A^{-1} = \begin{bmatrix} 1 & 2 \\ 3 & 5 \end{bmatrix}^{-1} = \begin{bmatrix} -5 & 2 \\ 3 & -1 \end{bmatrix}$

$$\begin{bmatrix} 11 & 21 \\ 64 & 112 \\ 25 & 50 \\ 29 & 53 \\ 23 & 46 \\ 40 & 75 \\ 55 & 92 \end{bmatrix} \begin{bmatrix} -5 & 2 \\ 3 & -1 \end{bmatrix} = \begin{bmatrix} 8 & 1 \\ 16 & 16 \\ 25 & 0 \\ 14 & 5 \\ 23 & 0 \\ 25 & 5 \\ 1 & 18 \end{bmatrix} \begin{matrix} \text{H} & \text{A} \\ \text{P} & \text{P} \\ \text{Y} & \_ \\ \text{N} & \text{E} \\ \text{W} & \_ \\ \text{Y} & \text{E} \\ \text{A} & \text{R} \end{matrix}$$

Message: HAPPY NEW YEAR

59. $A^{-1} = \begin{bmatrix} 1 & -1 & 0 \\ 1 & 0 & -1 \\ -6 & 2 & 3 \end{bmatrix}^{-1} = \begin{bmatrix} -2 & -3 & -1 \\ -3 & -3 & -1 \\ -2 & -4 & -1 \end{bmatrix}$

$$\begin{bmatrix} 9 & -1 & -9 \\ 38 & -19 & -19 \\ 28 & -9 & -19 \\ -80 & 25 & 41 \\ -64 & 21 & 31 \\ 9 & -5 & -4 \end{bmatrix} \begin{bmatrix} -2 & -3 & -1 \\ -3 & -3 & -1 \\ -2 & -4 & -1 \end{bmatrix} = \begin{bmatrix} 3 & 12 & 1 \\ 19 & 19 & 0 \\ 9 & 19 & 0 \\ 3 & 1 & 14 \\ 3 & 5 & 12 \\ 5 & 4 & 0 \end{bmatrix} \begin{matrix} \text{C} & \text{L} & \text{A} \\ \text{S} & \text{S} & \_ \\ \text{I} & \text{S} & \_ \\ \text{C} & \text{A} & \text{N} \\ \text{C} & \text{E} & \text{L} \\ \text{E} & \text{D} & \_ \end{matrix}$$

Message: CLASS IS CANCELED

61. $A^{-1} = \begin{bmatrix} 1 & 2 & 2 \\ 3 & 7 & 9 \\ -1 & -4 & -7 \end{bmatrix}^{-1} = \begin{bmatrix} -13 & 6 & 4 \\ 12 & -5 & -3 \\ -5 & 2 & 1 \end{bmatrix}$

$$\begin{bmatrix} 20 & 17 & -15 \\ -12 & -56 & -104 \\ 1 & -25 & -65 \\ 62 & 143 & 181 \end{bmatrix} \begin{bmatrix} -13 & 6 & 4 \\ 12 & -5 & -3 \\ -5 & 2 & 1 \end{bmatrix} = \begin{bmatrix} 19 & 5 & 14 \\ 4 & 0 & 16 \\ 12 & 1 & 14 \\ 5 & 19 & 0 \end{bmatrix} \begin{matrix} \text{S} & \text{E} & \text{N} \\ \text{D} & \_ & \text{P} \\ \text{L} & \text{A} & \text{N} \\ \text{E} & \text{S} & \_ \end{matrix}$$

Message: SEND PLANES

63. Let A be the 2×2 matrix needed to decode the message.

$$\begin{bmatrix} -18 & -18 \\ 1 & 16 \end{bmatrix} A = \begin{bmatrix} 0 & 18 \\ 15 & 14 \end{bmatrix} \begin{matrix} \_ & \text{R} \\ \text{O} & \text{N} \end{matrix}$$

$$A = \begin{bmatrix} -18 & -18 \\ 1 & 16 \end{bmatrix}^{-1} \begin{bmatrix} 0 & 18 \\ 15 & 14 \end{bmatrix} = \begin{bmatrix} -\frac{8}{135} & -\frac{1}{15} \\ \frac{1}{270} & \frac{1}{15} \end{bmatrix} \begin{bmatrix} 0 & 18 \\ 15 & 14 \end{bmatrix} = \begin{bmatrix} -1 & -2 \\ 1 & 1 \end{bmatrix}$$

$$\begin{bmatrix} 8 & 21 \\ -15 & -10 \\ -13 & -13 \\ 5 & 10 \\ 5 & 25 \\ 5 & 19 \\ -1 & 6 \\ 20 & 40 \\ -18 & -18 \\ 1 & 16 \end{bmatrix} \begin{bmatrix} -1 & -2 \\ 1 & 1 \end{bmatrix} = \begin{bmatrix} 13 & 5 \\ 5 & 20 \\ 0 & 13 \\ 5 & 0 \\ 20 & 15 \\ 14 & 9 \\ 7 & 8 \\ 20 & 0 \\ 0 & 18 \\ 15 & 14 \end{bmatrix} \begin{matrix} \text{M} & \text{E} \\ \text{E} & \text{T} \\ \_ & \text{M} \\ \text{E} & \_ \\ \text{T} & \text{O} \\ \text{N} & \text{I} \\ \text{G} & \text{H} \\ \text{T} & \_ \\ \_ & \text{R} \\ \text{O} & \text{N} \end{matrix}$$

Message: MEET ME TONIGHT RON

65. $D = \begin{vmatrix} 4 & 0 & 8 \\ 0 & 2 & 8 \\ 1 & 1 & -1 \end{vmatrix} = -56$

$$I_1 = \frac{\begin{vmatrix} 2 & 0 & 8 \\ 6 & 2 & 8 \\ 0 & 1 & -1 \end{vmatrix}}{-56} = -\frac{28}{56} = -\frac{1}{2}$$

$$I_2 = \frac{\begin{vmatrix} 4 & 2 & 8 \\ 0 & 6 & 8 \\ 1 & 0 & -1 \end{vmatrix}}{-56} = \frac{-56}{-56} = 1$$

$$I_3 = \frac{\begin{vmatrix} 4 & 0 & 2 \\ 0 & 2 & 6 \\ 1 & 1 & 0 \end{vmatrix}}{-56} = \frac{-28}{-56} = \frac{1}{2}$$

So, the solution is $I_1 = -0.5$ ampere, $I_2 = 1$ ampere, and $I_3 = 0.5$ ampere.

67. False. In Cramer's Rule, the denominator is the determinant of the coefficient matrix.

69. False. If the determinant of the coefficient matrix is zero, the system has either no solution or infinitely many solutions.

71. Answers will vary.

73. $\text{Area} = \frac{1}{2}\begin{vmatrix} 3 & -1 & 1 \\ 7 & -1 & 1 \\ 7 & 5 & 1 \end{vmatrix}$

$$= \frac{1}{2}\left(3\begin{vmatrix} -1 & 1 \\ 5 & 1 \end{vmatrix} + 1\begin{vmatrix} 7 & 1 \\ 7 & 1 \end{vmatrix} + 1\begin{vmatrix} 7 & -1 \\ 7 & 5 \end{vmatrix}\right)$$

$$= \frac{1}{2}(-18 + 0 + 42)$$

$$= 12 \text{ square units}$$

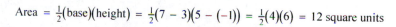

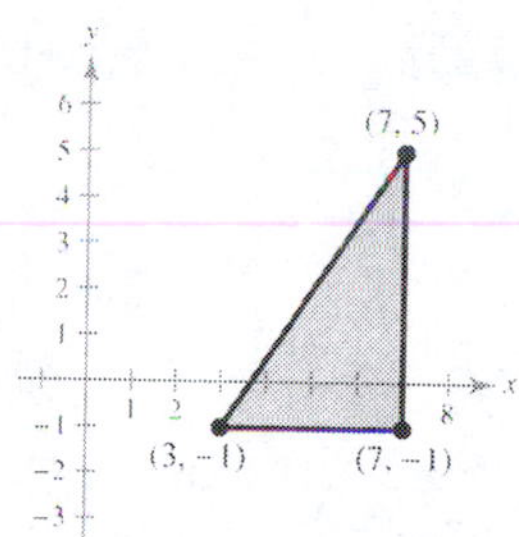

$\text{Area} = \frac{1}{2}(\text{base})(\text{height}) = \frac{1}{2}(7 - 3)(5 - (-1)) = \frac{1}{2}(4)(6) = 12$ square units

Review Exercises for Chapter 10

1. $\begin{bmatrix} -4 \\ 0 \\ 5 \end{bmatrix}$

Order: 3×1

3. $[3]$

Order: 1×1

5. $\begin{cases} 3x - 10y = 15 \\ 5x + 4y = 22 \end{cases}$

$$\left[\begin{array}{cc:c} 3 & -10 & 15 \\ 5 & 4 & 22 \end{array}\right]$$

7. $\left[\begin{array}{ccc:c} 5 & 1 & 7 & -9 \\ 4 & 2 & 0 & 10 \\ 9 & 4 & 2 & 3 \end{array}\right]$

$$\begin{cases} 5x + y + 7z = -9 \\ 4x + 2y = 10 \\ 9x + 4y + 2z = 3 \end{cases}$$

9. $\begin{bmatrix} 0 & 1 & 1 \\ 1 & 2 & 3 \\ 2 & 2 & 2 \end{bmatrix}$

$\begin{matrix} R_1 \\ R_2 \\ \end{matrix} \begin{bmatrix} 1 & 2 & 3 \\ 0 & 1 & 1 \\ 2 & 2 & 2 \end{bmatrix}$

$\begin{matrix} \\ \\ -2R_1 + R_3 \rightarrow \end{matrix} \begin{bmatrix} 1 & 2 & 3 \\ 0 & 1 & 1 \\ 0 & -2 & -4 \end{bmatrix}$

$\begin{matrix} \\ \\ 2R_2 + R_3 \rightarrow \end{matrix} \begin{bmatrix} 1 & 2 & 3 \\ 0 & 1 & 1 \\ 0 & 0 & -2 \end{bmatrix}$

$\begin{matrix} \\ \\ -\frac{1}{2}R_3 \rightarrow \end{matrix} \begin{bmatrix} 1 & 2 & 3 \\ 0 & 1 & 1 \\ 0 & 0 & 1 \end{bmatrix}$

11. $\left[\begin{array}{ccc:c} 1 & 2 & 3 & 9 \\ 0 & 1 & -2 & 2 \\ 0 & 0 & 1 & 0 \end{array}\right] \begin{matrix} \Rightarrow \\ \Rightarrow \\ \Rightarrow \end{matrix} \begin{cases} x + 2y + 3z = 9 \\ y - 2z = 2 \\ z = 0 \end{cases}$

$y - 2(0) = 2 \Rightarrow y = 2$

$x + 2(2) + 3(0) = 9 \Rightarrow x = 5$

Solution: $(5, 2, 0)$

13. $\left[\begin{array}{ccc:c} 1 & -5 & 4 & 1 \\ 0 & 1 & 2 & 3 \\ 0 & 0 & 1 & 4 \end{array}\right] \begin{matrix} \Rightarrow \\ \Rightarrow \\ \Rightarrow \end{matrix} \begin{cases} x - 5y + 4z = 1 \\ y + 2z = 3 \\ z = 4 \end{cases}$

$y + 2(4) = 3 \Rightarrow y = -5$

$x - 5(-5) + 4(4) = 1 \Rightarrow x = -40$

Solution: $(-40, -5, 4)$

15. $\left[\begin{array}{cc:c} 5 & 4 & 2 \\ -1 & 1 & -22 \end{array}\right]$

$4R_2 + R_1 \rightarrow \left[\begin{array}{cc:c} 1 & 8 & -86 \\ -1 & 1 & -22 \end{array}\right]$

$R_1 + R_2 \rightarrow \left[\begin{array}{cc:c} 1 & 8 & -86 \\ 0 & 9 & -108 \end{array}\right]$

$\frac{1}{9}R_2 \rightarrow \left[\begin{array}{cc:c} 1 & 8 & -86 \\ 0 & 1 & -12 \end{array}\right]$

$\begin{cases} x + 8y = -86 \\ y = -12 \end{cases}$

$y = -12$

$x + 8(-12) = -86 \Rightarrow x = 10$

Solution: $(10, -12)$

17. $\left[\begin{array}{cc:c} 0.3 & -0.1 & -0.13 \\ 0.2 & -0.3 & -0.25 \end{array}\right]$

$\begin{matrix} 10R_1 \rightarrow \\ 10R_2 \rightarrow \end{matrix} \left[\begin{array}{cc:c} 3 & -1 & -1.3 \\ 2 & -3 & -2.5 \end{array}\right]$

$-R_2 + R_1 \rightarrow \left[\begin{array}{cc:c} 1 & 2 & 1.2 \\ 2 & -3 & -2.5 \end{array}\right]$

$-2R_1 + R_2 \rightarrow \left[\begin{array}{cc:c} 1 & 2 & 1.2 \\ 0 & -7 & -4.9 \end{array}\right]$

$-\frac{1}{7}R_2 \rightarrow \left[\begin{array}{cc:c} 1 & 2 & 1.2 \\ 0 & 1 & 0.7 \end{array}\right]$

$\begin{cases} x + 2y = 1.2 \\ y = 0.7 \end{cases}$

$y = 0.7$

$x + 2(0.7) = 1.2 \Rightarrow x = -0.2$

Solution: $(-0.2, 0.7) = \left(-\frac{1}{5}, \frac{7}{10}\right)$

19. $\begin{cases} -x + 2y = 3 \\ 2x - 4y = 6 \end{cases}$

$\left[\begin{array}{cc:c} -1 & 2 & 3 \\ 2 & -4 & 6 \end{array}\right]$

$2R_1 + R_2 \rightarrow \left[\begin{array}{cc:c} -1 & 2 & 3 \\ 0 & 0 & 12 \end{array}\right]$

Because the last row consists of all zeros except for the last entry, the system is inconsistent and there is no solution.

21. $\begin{cases} x - 2y + z = 7 \\ 2x + y - 2z = -4 \\ -x + 3y + 2z = -3 \end{cases}$

$\left[\begin{array}{ccc:c} 1 & -2 & 1 & 7 \\ 2 & 1 & -2 & -4 \\ -1 & 3 & 2 & -3 \end{array}\right]$

$\begin{matrix} \\ -2R_1 + R_2 \rightarrow \\ R_1 + R_3 \rightarrow \end{matrix} \left[\begin{array}{ccc:c} 1 & -2 & 1 & 7 \\ 0 & 5 & -4 & -18 \\ 0 & 1 & 3 & 4 \end{array}\right]$

$R_2 + (-5)R_3 \rightarrow \left[\begin{array}{ccc:c} 1 & -2 & 1 & 7 \\ 0 & 0 & -19 & -38 \\ 0 & 1 & 3 & 4 \end{array}\right]$

$-19z = -38$

$z = 2$

$y + 3(2) = 4 \Rightarrow y = -2$

$x - 2(-2) + 2 = 7 \Rightarrow x = 1$

Solution: $(1, -2, 2)$

23. $$\left[\begin{array}{rrrcr} 2 & 1 & 2 & \vdots & 4 \\ 2 & 2 & 0 & \vdots & 5 \\ 2 & -1 & 6 & \vdots & 2 \end{array}\right]$$

$$\begin{array}{r} \\ -R_1 + R_2 \rightarrow \\ -R_1 + R_3 \rightarrow \end{array} \left[\begin{array}{rrrcr} 2 & 1 & 2 & \vdots & 4 \\ 0 & 1 & -2 & \vdots & 1 \\ 0 & -2 & 4 & \vdots & -2 \end{array}\right]$$

$$\begin{array}{r} -R_2 + R_1 \rightarrow \\ \\ 2R_2 + R_3 \rightarrow \end{array} \left[\begin{array}{rrrcr} 2 & 0 & 4 & \vdots & 3 \\ 0 & 1 & -2 & \vdots & 1 \\ 0 & 0 & 0 & \vdots & 0 \end{array}\right]$$

$$\begin{array}{r} \frac{1}{2}R_1 \rightarrow \\ \\ \\ \end{array} \left[\begin{array}{rrrcr} 1 & 0 & 2 & \vdots & \frac{3}{2} \\ 0 & 1 & -2 & \vdots & 1 \\ 0 & 0 & 0 & \vdots & 0 \end{array}\right]$$

Let $z = a$, then:

$y - 2a = 1 \Rightarrow y = 2a + 1$

$x + 2a = \frac{3}{2} \Rightarrow x = -2a + \frac{3}{2}$

Solution: $\left(-2a + \frac{3}{2}, 2a + 1, a\right)$ where a is any real number

25. $$\left[\begin{array}{rrrcr} 2 & 3 & 1 & \vdots & 10 \\ 2 & -3 & -3 & \vdots & 22 \\ 4 & -2 & 3 & \vdots & -2 \end{array}\right]$$

$$\begin{array}{r} \\ -R_1 + R_2 \rightarrow \\ -2R_1 + R_3 \rightarrow \end{array} \left[\begin{array}{rrrcr} 2 & 3 & 1 & \vdots & 10 \\ 0 & -6 & -4 & \vdots & 12 \\ 0 & -8 & 1 & \vdots & -22 \end{array}\right]$$

$$\begin{array}{r} \frac{1}{2}R_1 \rightarrow \\ -\frac{1}{6}R_2 \rightarrow \\ \\ \end{array} \left[\begin{array}{rrrcr} 1 & \frac{3}{2} & \frac{1}{2} & \vdots & 5 \\ 0 & 1 & \frac{2}{3} & \vdots & -2 \\ 0 & -8 & 1 & \vdots & -22 \end{array}\right]$$

$$\begin{array}{r} \\ \\ 8R_2 + R_3 \rightarrow \end{array} \left[\begin{array}{rrrcr} 1 & \frac{3}{2} & \frac{1}{2} & \vdots & 5 \\ 0 & 1 & \frac{2}{3} & \vdots & -2 \\ 0 & 0 & \frac{19}{3} & \vdots & -38 \end{array}\right]$$

$$\begin{array}{r} \\ \\ \frac{3}{19}R_3 \rightarrow \end{array} \left[\begin{array}{rrrcr} 1 & \frac{3}{2} & \frac{1}{2} & \vdots & 5 \\ 0 & 1 & \frac{2}{3} & \vdots & -2 \\ 0 & 0 & 1 & \vdots & -6 \end{array}\right]$$

$z = -6$

$y + \frac{2}{3}(-6) = -2 \Rightarrow y = 2$

$x + \frac{3}{2}(2) + \frac{1}{2}(-6) = 5 \Rightarrow x = 5$

Solution: $(5, 2, -6)$

27. $$\left[\begin{array}{rrrrcr} 2 & 1 & 1 & 0 & \vdots & 6 \\ 0 & -2 & 3 & -1 & \vdots & 9 \\ 3 & 3 & -2 & -2 & \vdots & -11 \\ 1 & 0 & 1 & 3 & \vdots & 14 \end{array}\right]$$

$$\begin{array}{r} -R_4 + R_1 \rightarrow \\ \\ \\ \\ \end{array} \left[\begin{array}{rrrrcr} 1 & 1 & 0 & -3 & \vdots & -8 \\ 0 & -2 & 3 & -1 & \vdots & 9 \\ 3 & 3 & -2 & -2 & \vdots & -11 \\ 1 & 0 & 1 & 3 & \vdots & 14 \end{array}\right]$$

$$\begin{array}{r} \\ \\ -3R_1 + R_3 \rightarrow \\ -R_1 + R_4 \rightarrow \end{array} \left[\begin{array}{rrrrcr} 1 & 1 & 0 & -3 & \vdots & -8 \\ 0 & -2 & 3 & -1 & \vdots & 9 \\ 0 & 0 & -2 & 7 & \vdots & 13 \\ 0 & -1 & 1 & 6 & \vdots & 22 \end{array}\right]$$

$$\begin{array}{r} \\ -3R_4 + R_2 \rightarrow \\ \\ \\ \end{array} \left[\begin{array}{rrrrcr} 1 & 1 & 0 & -3 & \vdots & -8 \\ 0 & 1 & 0 & -19 & \vdots & -57 \\ 0 & 0 & -2 & 7 & \vdots & 13 \\ 0 & -1 & 1 & 6 & \vdots & 22 \end{array}\right]$$

$$\begin{array}{r} \\ \\ \\ R_2 + R_4 \rightarrow \end{array} \left[\begin{array}{rrrrcr} 1 & 1 & 0 & -3 & \vdots & -8 \\ 0 & 1 & 0 & -19 & \vdots & -57 \\ 0 & 0 & -2 & 7 & \vdots & 13 \\ 0 & 0 & 1 & -13 & \vdots & -35 \end{array}\right]$$

$$\begin{array}{r} \\ \\ R_4 \\ R_3 \end{array} \left[\begin{array}{rrrrcr} 1 & 1 & 0 & -3 & \vdots & -8 \\ 0 & 1 & 0 & -19 & \vdots & -57 \\ 0 & 0 & 1 & -13 & \vdots & -35 \\ 0 & 0 & -2 & 7 & \vdots & 13 \end{array}\right]$$

$$\begin{array}{r} \\ \\ \\ 2R_3 + R_4 \rightarrow \end{array} \left[\begin{array}{rrrrcr} 1 & 1 & 0 & -3 & \vdots & -8 \\ 0 & 1 & 0 & -19 & \vdots & -57 \\ 0 & 0 & 1 & -13 & \vdots & -35 \\ 0 & 0 & 0 & -19 & \vdots & -57 \end{array}\right]$$

$$\begin{array}{r} \\ \\ \\ \frac{1}{19}R_4 \rightarrow \end{array} \left[\begin{array}{rrrrcr} 1 & 1 & 0 & -3 & \vdots & -8 \\ 0 & 1 & 0 & -19 & \vdots & -57 \\ 0 & 0 & 1 & -13 & \vdots & -35 \\ 0 & 0 & 0 & 1 & \vdots & 3 \end{array}\right]$$

$w = 3$

$z - 13(3) = -35 \Rightarrow z = 4$

$y - 19(3) = -57 \Rightarrow y = 0$

$x + 0 - 3(3) = -8 \Rightarrow x = 1$

Solution: $(1, 0, 4, 3)$

29. $\begin{cases} x + 2y - z = 3 \\ x - y - z = -3 \\ 2x + y + 3z = 10 \end{cases}$

$$\begin{bmatrix} 1 & 2 & -1 & \vdots & 3 \\ 1 & -1 & -1 & \vdots & -3 \\ 2 & 1 & 3 & \vdots & 10 \end{bmatrix}$$

$$\begin{matrix} \\ -R_1 + R_2 \rightarrow \\ -2R_2 + R_3 \rightarrow \end{matrix} \begin{bmatrix} 1 & 2 & -1 & \vdots & 3 \\ 0 & -3 & 0 & \vdots & -6 \\ 0 & 3 & 5 & \vdots & 16 \end{bmatrix}$$

$$\begin{matrix} \\ \\ R_2 + R_3 \rightarrow \end{matrix} \begin{bmatrix} 1 & 2 & -1 & \vdots & 3 \\ 0 & -3 & 0 & \vdots & -6 \\ 0 & 0 & 5 & \vdots & 10 \end{bmatrix}$$

$$\begin{matrix} 3R_1 + 2R_2 \rightarrow \\ \\ \\ \end{matrix} \begin{bmatrix} 3 & 0 & -3 & \vdots & -3 \\ 0 & -3 & 0 & \vdots & -6 \\ 0 & 0 & 5 & \vdots & 10 \end{bmatrix}$$

$$\begin{matrix} 5R_1 + 3R_3 \rightarrow \\ \\ \\ \end{matrix} \begin{bmatrix} 15 & 0 & 0 & \vdots & 15 \\ 0 & -3 & 0 & \vdots & -6 \\ 0 & 0 & 5 & \vdots & 10 \end{bmatrix}$$

$$\begin{matrix} \frac{1}{15}R_1 \rightarrow \\ \frac{1}{3}R_2 \rightarrow \\ \frac{1}{5}R_3 \rightarrow \end{matrix} \begin{bmatrix} 1 & 0 & 0 & \vdots & 1 \\ 0 & 1 & 0 & \vdots & 2 \\ 0 & 0 & 1 & \vdots & 2 \end{bmatrix}$$

$x = 1$

$y = 2$

$z = 2$

Solution: $(1, 2, 2)$

31. $\begin{bmatrix} -1 & 1 & 2 & \vdots & 1 \\ 2 & 3 & 1 & \vdots & -2 \\ 5 & 4 & 2 & \vdots & 4 \end{bmatrix}$

$$\begin{matrix} -R_1 \rightarrow \\ \\ \\ \end{matrix} \begin{bmatrix} 1 & -1 & -2 & \vdots & -1 \\ 2 & 3 & 1 & \vdots & -2 \\ 5 & 4 & 2 & \vdots & 4 \end{bmatrix}$$

$$\begin{matrix} \\ -2R_1 + R_2 \rightarrow \\ -5R_1 + R_3 \rightarrow \end{matrix} \begin{bmatrix} 1 & -1 & -2 & \vdots & -1 \\ 0 & 5 & 5 & \vdots & 0 \\ 0 & 9 & 12 & \vdots & 9 \end{bmatrix}$$

$$\begin{matrix} \\ \frac{1}{5}R_2 \rightarrow \\ \\ \end{matrix} \begin{bmatrix} 1 & -1 & -2 & \vdots & -1 \\ 0 & 1 & 1 & \vdots & 0 \\ 0 & 9 & 12 & \vdots & 9 \end{bmatrix}$$

$$\begin{matrix} R_2 + R_1 \rightarrow \\ \\ -9R_2 + R_3 \rightarrow \end{matrix} \begin{bmatrix} 1 & 0 & -1 & \vdots & -1 \\ 0 & 1 & 1 & \vdots & 0 \\ 0 & 0 & 3 & \vdots & 9 \end{bmatrix}$$

$$\begin{matrix} \\ \\ \frac{1}{3}R_3 \rightarrow \end{matrix} \begin{bmatrix} 1 & 0 & -1 & \vdots & -1 \\ 0 & 1 & 1 & \vdots & 0 \\ 0 & 0 & 1 & \vdots & 3 \end{bmatrix}$$

$$\begin{matrix} R_3 + R_1 \rightarrow \\ -R_3 + R_2 \rightarrow \\ \\ \end{matrix} \begin{bmatrix} 1 & 0 & 0 & \vdots & 2 \\ 0 & 1 & 0 & \vdots & -3 \\ 0 & 0 & 1 & \vdots & 3 \end{bmatrix}$$

$x = 2, y = -3, z = 3$

Solution: $(2, -3, 3)$

33. $\begin{bmatrix} 2 & -1 & 9 & \vdots & -8 \\ -1 & -3 & 4 & \vdots & -15 \\ 5 & 2 & -1 & \vdots & 17 \end{bmatrix}$

$R_2 + R_1 \to \begin{bmatrix} 1 & -4 & 13 & \vdots & -23 \\ -1 & -3 & 4 & \vdots & -15 \\ 5 & 2 & -1 & \vdots & 17 \end{bmatrix}$

$\begin{matrix} \\ R_1 + R_2 \to \\ -5R_1 + R_3 \to \end{matrix} \begin{bmatrix} 1 & -4 & 13 & \vdots & -23 \\ 0 & -7 & 17 & \vdots & -38 \\ 0 & 22 & -66 & \vdots & 132 \end{bmatrix}$

$\begin{matrix} \\ R_3 \\ R_2 \end{matrix} \begin{bmatrix} 1 & -4 & 13 & \vdots & -23 \\ 0 & 22 & -66 & \vdots & 132 \\ 0 & -7 & 17 & \vdots & -38 \end{bmatrix}$

$\frac{1}{22}R_2 \to \begin{bmatrix} 1 & -4 & 13 & \vdots & -23 \\ 0 & 1 & -3 & \vdots & 6 \\ 0 & -7 & 17 & \vdots & -38 \end{bmatrix}$

$7R_2 + R_3 \to \begin{bmatrix} 1 & -4 & 13 & \vdots & -23 \\ 0 & 1 & -3 & \vdots & 6 \\ 0 & 0 & -4 & \vdots & 4 \end{bmatrix}$

$-\frac{1}{4}R_3 \to \begin{bmatrix} 1 & -4 & 13 & \vdots & -23 \\ 0 & 1 & -3 & \vdots & 6 \\ 0 & 0 & 1 & \vdots & -1 \end{bmatrix}$

$4R_2 + R_1 \to \begin{bmatrix} 1 & 0 & 1 & \vdots & 1 \\ 0 & 1 & -3 & \vdots & 6 \\ 0 & 0 & 1 & \vdots & -1 \end{bmatrix}$

$\begin{matrix} -R_3 + R_1 \to \\ 3R_3 + R_2 \to \\ \end{matrix} \begin{bmatrix} 1 & 0 & 0 & \vdots & 2 \\ 0 & 1 & 0 & \vdots & 3 \\ 0 & 0 & 1 & \vdots & -1 \end{bmatrix}$

$x = 2, y = 3, z = -1$

Solution: $(2, 3, -1)$

35. Use the reduced row-echelon form feature of a graphing utility.

$$\begin{bmatrix} 3 & -1 & 5 & -2 & \vdots & -44 \\ 1 & 6 & 4 & -1 & \vdots & 1 \\ 5 & -1 & 1 & 3 & \vdots & -15 \\ 0 & 4 & -1 & -8 & \vdots & 58 \end{bmatrix} \Rightarrow \begin{bmatrix} 1 & 0 & 0 & 0 & \vdots & 2 \\ 0 & 1 & 0 & 0 & \vdots & 6 \\ 0 & 0 & 1 & 0 & \vdots & -10 \\ 0 & 0 & 0 & 1 & \vdots & -3 \end{bmatrix}$$

$x = 2, y = 6, z = -10, w = -3$

Solution: $(2, 6, -10, -3)$

37. $\begin{bmatrix} -1 & x \\ y & 9 \end{bmatrix} = \begin{bmatrix} -1 & 12 \\ -7 & 9 \end{bmatrix} \Rightarrow x = 12$ and $y = -7$

39. $\begin{bmatrix} x+3 & -4 & 4y \\ 0 & -3 & 2 \\ -2 & y+5 & 6x \end{bmatrix} = \begin{bmatrix} 5x-1 & -4 & 44 \\ 0 & -3 & 2 \\ -2 & 16 & 6 \end{bmatrix}$

$$\left.\begin{aligned} x + 3 &= 5x - 1 \\ 4y &= 44 \\ y + 5 &= 16 \\ 6x &= 6 \end{aligned}\right\} x = 1 \text{ and } y = 11$$

41. (a) $A + B = \begin{bmatrix} 2 & -2 \\ 3 & 5 \end{bmatrix} + \begin{bmatrix} -3 & 10 \\ 12 & 8 \end{bmatrix} = \begin{bmatrix} -1 & 8 \\ 15 & 13 \end{bmatrix}$

(b) $A - B = \begin{bmatrix} 2 & -2 \\ 3 & 5 \end{bmatrix} - \begin{bmatrix} -3 & 10 \\ 12 & 8 \end{bmatrix} = \begin{bmatrix} 5 & -12 \\ -9 & -3 \end{bmatrix}$

(c) $4A = 4\begin{bmatrix} 2 & -2 \\ 3 & 5 \end{bmatrix} = \begin{bmatrix} 8 & -8 \\ 12 & 20 \end{bmatrix}$

(d) $A + 3B = \begin{bmatrix} 2 & -2 \\ 3 & 5 \end{bmatrix} + 3\begin{bmatrix} -3 & 10 \\ 12 & 8 \end{bmatrix} = \begin{bmatrix} 2 & -2 \\ 3 & 5 \end{bmatrix} + \begin{bmatrix} -9 & 30 \\ 36 & 24 \end{bmatrix} = \begin{bmatrix} -7 & 28 \\ 39 & 29 \end{bmatrix}$

43. (a) $A + B = \begin{bmatrix} 5 & 4 \\ -7 & 2 \\ 11 & 2 \end{bmatrix} + \begin{bmatrix} 0 & 3 \\ 4 & 12 \\ 20 & 40 \end{bmatrix} = \begin{bmatrix} 5 & 7 \\ -3 & 14 \\ 31 & 42 \end{bmatrix}$

(b) $A - B = \begin{bmatrix} 5 & 4 \\ -7 & 2 \\ 11 & 2 \end{bmatrix} - \begin{bmatrix} 0 & 3 \\ 4 & 12 \\ 20 & 40 \end{bmatrix} = \begin{bmatrix} 5 & 1 \\ -11 & -10 \\ -9 & -38 \end{bmatrix}$

(c) $4A = 4\begin{bmatrix} 5 & 4 \\ -7 & 2 \\ 11 & 2 \end{bmatrix} = \begin{bmatrix} 20 & 16 \\ -28 & 8 \\ 44 & 8 \end{bmatrix}$

(d) $A + 3B = \begin{bmatrix} 5 & 4 \\ -7 & 2 \\ 11 & 2 \end{bmatrix} + 3\begin{bmatrix} 0 & 3 \\ 4 & 12 \\ 20 & 40 \end{bmatrix} = \begin{bmatrix} 5 & 4 \\ -7 & 2 \\ 11 & 2 \end{bmatrix} + \begin{bmatrix} 0 & 9 \\ 12 & 36 \\ 60 & 120 \end{bmatrix} = \begin{bmatrix} 5 & 13 \\ 5 & 38 \\ 71 & 122 \end{bmatrix}$

45. $\begin{bmatrix} 7 & 3 \\ -1 & 5 \end{bmatrix} + \begin{bmatrix} 10 & -20 \\ 14 & -3 \end{bmatrix} = \begin{bmatrix} 7 + 10 & 3 - 20 \\ -1 + 14 & 5 - 3 \end{bmatrix} = \begin{bmatrix} 17 & -17 \\ 13 & 2 \end{bmatrix}$

47. $-2\begin{bmatrix} 1 & 2 \\ 5 & -4 \\ 6 & 0 \end{bmatrix} + 8\begin{bmatrix} 7 & 1 \\ 1 & 2 \\ 1 & 4 \end{bmatrix} = \begin{bmatrix} -2 & -4 \\ -10 & 8 \\ -12 & 0 \end{bmatrix} + \begin{bmatrix} 56 & 8 \\ 8 & 16 \\ 8 & 32 \end{bmatrix} = \begin{bmatrix} 54 & 4 \\ -2 & 24 \\ -4 & 32 \end{bmatrix}$

49. $3\begin{bmatrix} 8 & -2 & 5 \\ 1 & 3 & -1 \end{bmatrix} + 6\begin{bmatrix} 4 & -2 & -3 \\ 2 & 7 & 6 \end{bmatrix} = \begin{bmatrix} 24 & -6 & 15 \\ 3 & 9 & -3 \end{bmatrix} + \begin{bmatrix} 24 & -12 & -18 \\ 12 & 42 & 36 \end{bmatrix} = \begin{bmatrix} 48 & -18 & -3 \\ 15 & 51 & 33 \end{bmatrix}$

51. $X = 2A - 3B = 2\begin{bmatrix} -4 & 0 \\ 1 & -5 \\ -3 & 2 \end{bmatrix} - 3\begin{bmatrix} 1 & 2 \\ -2 & 1 \\ 4 & 4 \end{bmatrix} = \begin{bmatrix} -8 & 0 \\ 2 & -10 \\ -6 & 4 \end{bmatrix} + \begin{bmatrix} -3 & -6 \\ 6 & -3 \\ -12 & -12 \end{bmatrix} = \begin{bmatrix} -11 & -6 \\ 8 & -13 \\ -18 & -8 \end{bmatrix}$

53. $X = \frac{1}{3}[B - 2A] = \frac{1}{3}\left(\begin{bmatrix} 1 & 2 \\ -2 & 1 \\ 4 & 4 \end{bmatrix} - 2\begin{bmatrix} -4 & 0 \\ 1 & -5 \\ -3 & 2 \end{bmatrix}\right) = \frac{1}{3}\begin{bmatrix} 9 & 2 \\ -4 & 11 \\ 10 & 0 \end{bmatrix} = \begin{bmatrix} 3 & \frac{2}{3} \\ -\frac{4}{3} & \frac{11}{3} \\ \frac{10}{3} & 0 \end{bmatrix}$

55. A and B are both 2×2, so AB exists.

$$AB = \begin{bmatrix} 2 & -2 \\ 3 & 5 \end{bmatrix}\begin{bmatrix} -3 & 10 \\ 12 & 8 \end{bmatrix} = \begin{bmatrix} 2(-3) + (-2)(12) & 2(10) + (-2)(8) \\ 3(-3) + 5(12) & 3(10) + 5(8) \end{bmatrix} = \begin{bmatrix} -30 & 4 \\ 51 & 70 \end{bmatrix}$$

57. Because A is 3×2 and B is 2×2, AB exists.

$$AB = \begin{bmatrix} 5 & 4 \\ -7 & 2 \\ 11 & 2 \end{bmatrix}\begin{bmatrix} 4 & 12 \\ 20 & 40 \end{bmatrix} = \begin{bmatrix} 5(4) + 4(20) & 5(12) + 4(40) \\ -7(4) + 2(20) & -7(12) + 2(40) \\ 11(4) + 2(20) & 11(12) + 2(40) \end{bmatrix} = \begin{bmatrix} 100 & 220 \\ 12 & -4 \\ 84 & 212 \end{bmatrix}$$

59. $\begin{bmatrix} 1 & 2 \\ 5 & -4 \\ 6 & 0 \end{bmatrix}\begin{bmatrix} 6 & -2 & 8 \\ 4 & 0 & 0 \end{bmatrix} = \begin{bmatrix} 1(6) + 2(4) & 1(-2) + 2(0) & 1(8) + 2(0) \\ 5(6) + (-4)(4) & 5(-2) + (-4)(0) & 5(8) + (-4)(0) \\ 6(6) + (0)(4) & 6(-2) + (0)(0) & 6(8) + (0)(0) \end{bmatrix} = \begin{bmatrix} 14 & -2 & 8 \\ 14 & -10 & 40 \\ 36 & -12 & 48 \end{bmatrix}$

61. $\begin{bmatrix} 1 & 5 & 6 \\ 2 & -4 & 0 \end{bmatrix}\begin{bmatrix} 6 & 4 \\ -2 & 0 \\ 8 & 0 \end{bmatrix} = \begin{bmatrix} 1(6) + 5(-2) + 6(8) & 1(4) + 5(0) + 6(0) \\ 2(6) - 4(-2) + 0(8) & 2(4) - 4(0) + 0(0) \end{bmatrix} = \begin{bmatrix} 44 & 4 \\ 20 & 8 \end{bmatrix}$

63. $\begin{bmatrix} 4 & 1 \\ 11 & -7 \\ 12 & 3 \end{bmatrix}\begin{bmatrix} 3 & -5 & 6 \\ 2 & -2 & -2 \end{bmatrix} = \begin{bmatrix} 14 & -22 & 22 \\ 19 & -41 & 80 \\ 42 & -66 & 66 \end{bmatrix}$

65. Not possible. The number of columns of the first matrix does not equal the number of rows of the second matrix.

67. $0.95A = 0.95\begin{bmatrix} 80 & 120 & 140 \\ 40 & 100 & 80 \end{bmatrix} = \begin{bmatrix} 76 & 114 & 133 \\ 38 & 95 & 76 \end{bmatrix}$

69. $BA = [\$79.99 \quad \$109.95 \quad \$189.99]\begin{bmatrix} 8200 & 7400 \\ 6500 & 9800 \\ 5400 & 4800 \end{bmatrix} = [\$2{,}396{,}539 \quad \$2{,}581{,}388]$

The merchandise shipped to warehouse 1 is worth \$2,396,539 and the merchandise shipped to warehouse 2 is worth \$2,581,388.

71. $AB = \begin{bmatrix} -4 & -1 \\ 7 & 2 \end{bmatrix}\begin{bmatrix} -2 & -1 \\ 7 & 4 \end{bmatrix} = \begin{bmatrix} -4(-2) + (-1)(7) & -4(-1) + (-1)(4) \\ 7(-2) + 2(7) & 7(-1) + 2(4) \end{bmatrix} = \begin{bmatrix} 1 & 0 \\ 0 & 1 \end{bmatrix} = I$

$BA = \begin{bmatrix} -2 & -1 \\ 7 & 4 \end{bmatrix}\begin{bmatrix} -4 & -1 \\ 7 & 2 \end{bmatrix} = \begin{bmatrix} -2(-4) + (-1)(7) & -2(-1) + (-1)(2) \\ 7(-4) + 4(7) & 7(-1) + 4(2) \end{bmatrix} = \begin{bmatrix} 1 & 0 \\ 0 & 1 \end{bmatrix} = I$

73. $AB = \begin{bmatrix} 1 & 1 & 0 \\ 1 & 0 & 1 \\ 6 & 2 & 3 \end{bmatrix}\begin{bmatrix} -2 & -3 & 1 \\ 3 & 3 & -1 \\ 2 & 4 & -1 \end{bmatrix} = \begin{bmatrix} 1(-2) + 1(3) + 0(2) & 1(-3) + 1(3) + 0(4) & 1(1) + 1(-1) + 0(-1) \\ 1(-2) + 0(3) + 1(2) & 1(-3) + 0(3) + 1(4) & 1(1) + 0(-1) + 1(-1) \\ 6(-2) + 2(3) + 3(2) & 6(-3) + 2(3) + 3(4) & 6(1) + 2(-1) + 3(-1) \end{bmatrix} = \begin{bmatrix} 1 & 0 & 0 \\ 0 & 1 & 0 \\ 0 & 0 & 1 \end{bmatrix} = I$

$BA = \begin{bmatrix} -2 & -3 & 1 \\ 3 & 3 & -1 \\ 2 & 4 & -1 \end{bmatrix}\begin{bmatrix} 1 & 1 & 0 \\ 1 & 0 & 1 \\ 6 & 2 & 3 \end{bmatrix} = \begin{bmatrix} -2(1) + (-3)(1) + 1(6) & -2(1) + (-3)(0) + 1(2) & -2(0) + (-3)(1) + 1(3) \\ 3(1) + 3(1) + (-1)(6) & 3(1) + 3(0) + (-1)(2) & 3(0) + 3(1) + (-1)(3) \\ 2(1) + 4(1) + (-1)(6) & 2(1) + 4(0) + (-1)(2) & 2(0) + 4(1) + (-1)(3) \end{bmatrix}$

$= \begin{bmatrix} 1 & 0 & 0 \\ 0 & 1 & 0 \\ 0 & 0 & 1 \end{bmatrix} = I$

75. $[A \;\vdots\; I] = \left[\begin{array}{cc|cc} -6 & 5 & 1 & 0 \\ -5 & 4 & 0 & 1 \end{array}\right]$

$-\frac{1}{6}R_1 \rightarrow \left[\begin{array}{cc|cc} 1 & -\frac{5}{6} & -\frac{1}{6} & 0 \\ -5 & 4 & 0 & 1 \end{array}\right]$

$5R_1 + R_2 \rightarrow \left[\begin{array}{cc|cc} 1 & -\frac{5}{6} & -\frac{1}{6} & 0 \\ 0 & -\frac{1}{6} & -\frac{5}{6} & 1 \end{array}\right]$

$-6R_2 \rightarrow \left[\begin{array}{cc|cc} 1 & -\frac{5}{6} & -\frac{1}{6} & 0 \\ 0 & 1 & 5 & -6 \end{array}\right]$

$\frac{5}{6}R_2 + R_1 \rightarrow \left[\begin{array}{cc|cc} 1 & 0 & 4 & -5 \\ 0 & 1 & 5 & -6 \end{array}\right] = [I \;\vdots\; A^{-1}]$

$A^{-1} = \begin{bmatrix} 4 & -5 \\ 5 & -6 \end{bmatrix}$

77. $[A \vdots I] = \begin{bmatrix} 2 & 0 & 3 & \vdots & 1 & 0 & 0 \\ -1 & 1 & 1 & \vdots & 0 & 1 & 0 \\ 2 & -2 & 1 & \vdots & 0 & 0 & 1 \end{bmatrix}$

$$2R_2 + R_3 \rightarrow \begin{bmatrix} 2 & 0 & 3 & \vdots & 1 & 0 & 0 \\ -1 & 1 & 1 & \vdots & 0 & 1 & 0 \\ 0 & 0 & 3 & \vdots & 0 & 2 & 1 \end{bmatrix}$$

$$-R_3 + R_1 \rightarrow \begin{bmatrix} 2 & 0 & 0 & \vdots & 1 & -2 & -1 \\ -1 & 1 & 1 & \vdots & 0 & 1 & 0 \\ 0 & 0 & 3 & \vdots & 0 & 2 & 1 \end{bmatrix}$$

$$\begin{matrix} \frac{1}{2}R_1 \rightarrow \\ \\ \frac{1}{3}R_3 \rightarrow \end{matrix} \begin{bmatrix} 1 & 0 & 0 & \vdots & \frac{1}{2} & -1 & -\frac{1}{2} \\ -1 & 1 & 1 & \vdots & 0 & 1 & 0 \\ 0 & 0 & 1 & \vdots & 0 & \frac{2}{3} & \frac{1}{3} \end{bmatrix}$$

$$R_1 + R_2 \rightarrow \begin{bmatrix} 1 & 0 & 0 & \vdots & \frac{1}{2} & -1 & -\frac{1}{2} \\ 0 & 1 & 1 & \vdots & \frac{1}{2} & 0 & -\frac{1}{2} \\ 0 & 0 & 1 & \vdots & 0 & \frac{2}{3} & \frac{1}{3} \end{bmatrix}$$

$$-R_3 + R_2 \rightarrow \begin{bmatrix} 1 & 0 & 0 & \vdots & \frac{1}{2} & -1 & -\frac{1}{2} \\ 0 & 1 & 0 & \vdots & \frac{1}{2} & -\frac{2}{3} & -\frac{5}{6} \\ 0 & 0 & 1 & \vdots & 0 & \frac{2}{3} & \frac{1}{3} \end{bmatrix} = [I \vdots A^{-1}]$$

$$A^{-1} = \begin{bmatrix} \frac{1}{2} & -1 & -\frac{1}{2} \\ \frac{1}{2} & -\frac{2}{3} & -\frac{5}{6} \\ 0 & \frac{2}{3} & \frac{1}{3} \end{bmatrix}$$

79. $\begin{bmatrix} -1 & -2 & -2 \\ 3 & 7 & 9 \\ 1 & 4 & 7 \end{bmatrix}^{-1} = \begin{bmatrix} 13 & 6 & -4 \\ -12 & -5 & 3 \\ 5 & 2 & -1 \end{bmatrix}$

81. $\begin{bmatrix} 1 & 3 & 1 & 6 \\ 4 & 4 & 2 & 6 \\ 3 & 4 & 1 & 2 \\ -1 & 2 & -1 & -2 \end{bmatrix}^{-1} = \begin{bmatrix} -3 & 6 & -5.5 & 3.5 \\ 1 & -2 & 2 & -1 \\ 7 & -15 & 14.5 & -9.5 \\ -1 & 2.5 & -2.5 & 1.5 \end{bmatrix}$

83. $A = \begin{bmatrix} -7 & 2 \\ -8 & 2 \end{bmatrix}$

$$A^{-1} = \frac{1}{-7(2) - 2(-8)}\begin{bmatrix} 2 & -2 \\ 8 & -7 \end{bmatrix} = \frac{1}{2}\begin{bmatrix} 2 & -2 \\ 8 & -7 \end{bmatrix} = \begin{bmatrix} 1 & -1 \\ 4 & -\frac{7}{2} \end{bmatrix}$$

85. $A = \begin{bmatrix} -12 & 6 \\ 10 & -5 \end{bmatrix}$

$ad - bc = (-12)(-5) - (6)(10) = 0$

A^{-1} does not exist.

87. $A = \begin{bmatrix} -\frac{1}{2} & 20 \\ \frac{3}{10} & -6 \end{bmatrix}$

$$A^{-1} = \frac{1}{-\frac{1}{2}(-6) - 20\left(\frac{3}{10}\right)}\begin{bmatrix} -6 & -20 \\ -\frac{3}{10} & -\frac{1}{2} \end{bmatrix}$$

$$= -\frac{1}{3}\begin{bmatrix} -6 & -20 \\ -\frac{3}{10} & -\frac{1}{2} \end{bmatrix}$$

$$= \begin{bmatrix} 2 & \frac{20}{3} \\ \frac{1}{10} & \frac{1}{6} \end{bmatrix}$$

89. $A = \begin{bmatrix} 0.5 & 0.1 \\ -0.2 & -0.4 \end{bmatrix}$

$ad - bc = (0.5)(-0.4) - (0.1)(-0.2) = -0.18$

$$A^{-1} = \frac{1}{-0.18}\begin{bmatrix} -0.4 & -0.1 \\ 0.2 & 0.5 \end{bmatrix}$$

$$= -\frac{50}{9}\begin{bmatrix} -0.4 & -0.1 \\ 0.2 & 0.5 \end{bmatrix}$$

$$= \begin{bmatrix} \frac{20}{9} & \frac{5}{9} \\ -\frac{10}{9} & -\frac{25}{9} \end{bmatrix}$$

91. $\begin{cases} -x + 4y = 8 \\ 2x - 7y = -5 \end{cases}$

$$\begin{bmatrix} x \\ y \end{bmatrix} = \begin{bmatrix} -1 & 4 \\ 2 & -7 \end{bmatrix}^{-1}\begin{bmatrix} 8 \\ -5 \end{bmatrix} = \begin{bmatrix} 7 & 4 \\ 2 & 1 \end{bmatrix}\begin{bmatrix} 8 \\ -5 \end{bmatrix}$$

$$= \begin{bmatrix} 7(8) + 4(-5) \\ 2(8) + 1(-5) \end{bmatrix} = \begin{bmatrix} 36 \\ 11 \end{bmatrix}$$

Solution: $(36, 11)$

93. $\begin{cases} -3x + 10y = 8 \\ 5x - 17y = -13 \end{cases}$

$$\begin{bmatrix} x \\ y \end{bmatrix} = \begin{bmatrix} -3 & 10 \\ 5 & -17 \end{bmatrix}^{-1}\begin{bmatrix} 8 \\ -13 \end{bmatrix} = \begin{bmatrix} -17 & -10 \\ -5 & -3 \end{bmatrix}\begin{bmatrix} 8 \\ -13 \end{bmatrix}$$

$$= \begin{bmatrix} -17(8) + (-10)(-13) \\ -5(8) + (-3)(-13) \end{bmatrix} = \begin{bmatrix} -6 \\ -1 \end{bmatrix}$$

Solution: $(-6, -1)$

95. $\begin{cases} \frac{1}{2}x + \frac{1}{3}y = 2 \\ -3x + 2y = 0 \end{cases}$

$$\begin{bmatrix} x \\ y \end{bmatrix} = \begin{bmatrix} \frac{1}{2} & \frac{1}{3} \\ -3 & 2 \end{bmatrix}^{-1} \begin{bmatrix} 2 \\ 0 \end{bmatrix} = \begin{bmatrix} 1 & -\frac{1}{6} \\ \frac{3}{2} & \frac{1}{4} \end{bmatrix} \begin{bmatrix} 2 \\ 0 \end{bmatrix} = \begin{bmatrix} 2 \\ 3 \end{bmatrix}$$

Solution: $(2, 3)$

97. $\begin{cases} 0.3x + 0.7y = 10.2 \\ 0.4x + 0.6y = 7.6 \end{cases}$

$$\begin{bmatrix} x \\ y \end{bmatrix} = \begin{bmatrix} 0.3 & 0.7 \\ 0.4 & 0.6 \end{bmatrix}^{-1} \begin{bmatrix} 10.2 \\ 7.6 \end{bmatrix} = \begin{bmatrix} -6 & 7 \\ 4 & -3 \end{bmatrix} \begin{bmatrix} 10.2 \\ 7.6 \end{bmatrix} = \begin{bmatrix} -8 \\ 18 \end{bmatrix}$$

Solution: $(-8, 18)$

99. $\begin{cases} 3x + 2y - z = 6 \\ x - y + 2z = -1 \\ 5x + y + z = 7 \end{cases}$

$$\begin{bmatrix} x \\ y \\ z \end{bmatrix} = \begin{bmatrix} 3 & 2 & -1 \\ 1 & -1 & 2 \\ 5 & 1 & 1 \end{bmatrix}^{-1} \begin{bmatrix} 6 \\ -1 \\ 7 \end{bmatrix} = \begin{bmatrix} -1 & -1 & 1 \\ 3 & \frac{8}{3} & -\frac{7}{3} \\ 2 & \frac{7}{3} & -\frac{5}{3} \end{bmatrix} \begin{bmatrix} 6 \\ -1 \\ 7 \end{bmatrix}$$

$$= \begin{bmatrix} -1(6) - 1(-1) + 1(7) \\ 3(6) + \frac{8}{3}(-1) - \frac{7}{3}(7) \\ 2(6) + \frac{7}{3}(-1) - \frac{5}{3}(7) \end{bmatrix} = \begin{bmatrix} 2 \\ -1 \\ -2 \end{bmatrix}$$

Solution: $(2, -1, -2)$

101. $\begin{cases} -2x + y + 2z = -13 \\ -x - 4y + z = -11 \\ -y - z = 0 \end{cases}$

$$\begin{bmatrix} x \\ y \\ z \end{bmatrix} = \begin{bmatrix} -2 & 1 & 2 \\ -1 & -4 & 1 \\ 0 & -1 & -1 \end{bmatrix}^{-1} \begin{bmatrix} -13 \\ -11 \\ 0 \end{bmatrix} = \begin{bmatrix} -\frac{5}{9} & \frac{1}{9} & -1 \\ \frac{1}{9} & -\frac{2}{9} & 0 \\ -\frac{1}{9} & \frac{2}{9} & -1 \end{bmatrix} \begin{bmatrix} -13 \\ -11 \\ 0 \end{bmatrix} = \begin{bmatrix} -\frac{5}{9}(-13) + \frac{1}{9}(-11) - 1(0) \\ \frac{1}{9}(-13) - \frac{2}{9}(-11) + 0(0) \\ -\frac{1}{9}(-13) + \frac{2}{9}(-11) - 1(0) \end{bmatrix} = \begin{bmatrix} 6 \\ 1 \\ -1 \end{bmatrix}$$

Solution: $(6, 1, -1)$

103. $\begin{cases} x + 2y = -1 \\ 3x + 4y = -5 \end{cases}$

$$\begin{bmatrix} x \\ y \end{bmatrix} = \begin{bmatrix} 1 & 2 \\ 3 & 4 \end{bmatrix}^{-1} \begin{bmatrix} -1 \\ -5 \end{bmatrix} = \begin{bmatrix} -2 & 1 \\ \frac{3}{2} & -\frac{1}{2} \end{bmatrix} \begin{bmatrix} -1 \\ -5 \end{bmatrix} = \begin{bmatrix} -3 \\ 1 \end{bmatrix}$$

Solution: $(-3, 1)$

105. $\begin{cases} \frac{6}{5}x - \frac{4}{7}y = \frac{6}{5} \\ -\frac{12}{5}x + \frac{12}{7}y = -\frac{17}{5} \end{cases}$

$$\begin{bmatrix} x \\ y \end{bmatrix} = \begin{bmatrix} \frac{6}{5} & -\frac{4}{7} \\ -\frac{12}{5} & \frac{12}{7} \end{bmatrix}^{-1} \begin{bmatrix} \frac{6}{5} \\ -\frac{17}{5} \end{bmatrix} = \begin{bmatrix} \frac{5}{2} & \frac{5}{6} \\ \frac{7}{2} & \frac{7}{4} \end{bmatrix} \begin{bmatrix} \frac{6}{5} \\ -\frac{17}{5} \end{bmatrix} = \begin{bmatrix} \frac{1}{6} \\ -\frac{7}{4} \end{bmatrix}$$

Solution: $\left(\frac{1}{6}, -\frac{7}{4}\right)$

107. $\begin{cases} -3x - 3y - 4z = 2 \\ y + z = -1 \\ 4x + 3y + 4z = -1 \end{cases}$

$$\begin{bmatrix} x \\ y \\ z \end{bmatrix} = \begin{bmatrix} -3 & -3 & -4 \\ 0 & 1 & 1 \\ 4 & 3 & 4 \end{bmatrix}^{-1} \begin{bmatrix} 2 \\ -1 \\ -1 \end{bmatrix} = \begin{bmatrix} 1 & 0 & 1 \\ 4 & 4 & 3 \\ -4 & -3 & -3 \end{bmatrix} \begin{bmatrix} 2 \\ -1 \\ -1 \end{bmatrix} = \begin{bmatrix} 1 \\ 1 \\ -2 \end{bmatrix}$$

Solution: $(1, 1, -2)$

109. $\begin{vmatrix} 8 & 5 \\ 2 & -4 \end{vmatrix} = 8(-4) - 5(2) = -42$

111. $\begin{vmatrix} 50 & -30 \\ 10 & 5 \end{vmatrix} = 50(5) - (-30)(10) = 550$

113. $\begin{bmatrix} 2 & -1 \\ 7 & 4 \end{bmatrix}$

(a) $M_{11} = 4$
$M_{12} = 7$
$M_{21} = -1$
$M_{22} = 2$

(b) $C_{11} = M_{11} = 4$
$C_{12} = -M_{12} = -7$
$C_{21} = -M_{21} = 1$
$C_{22} = M_{22} = 2$

115. $\begin{bmatrix} 3 & 2 & -1 \\ -2 & 5 & 0 \\ 1 & 8 & 6 \end{bmatrix}$

(a) $M_{11} = \begin{vmatrix} 5 & 0 \\ 8 & 6 \end{vmatrix} = 30$

$M_{12} = \begin{vmatrix} -2 & 0 \\ 1 & 6 \end{vmatrix} = -12$

$M_{13} = \begin{vmatrix} -2 & 5 \\ 1 & 8 \end{vmatrix} = -21$

$M_{21} = \begin{vmatrix} 2 & -1 \\ 8 & 6 \end{vmatrix} = 20$

$M_{22} = \begin{vmatrix} 3 & -1 \\ 1 & 6 \end{vmatrix} = 19$

$M_{23} = \begin{vmatrix} 3 & 2 \\ 1 & 8 \end{vmatrix} = 22$

$M_{31} = \begin{vmatrix} 2 & -1 \\ 5 & 0 \end{vmatrix} = 5$

$M_{32} = \begin{vmatrix} 3 & -1 \\ -2 & 0 \end{vmatrix} = -2$

$M_{33} = \begin{vmatrix} 3 & 2 \\ -2 & 5 \end{vmatrix} = 19$

(b) $C_{11} = M_{11} = 30$

$C_{12} = -M_{12} = 12$

$C_{13} = M_{13} = -21$

$C_{21} = -M_{21} = -20$

$C_{22} = M_{22} = 19$

$C_{23} = -M_{23} = -22$

$C_{31} = M_{31} = 5$

$C_{32} = -M_{32} = 2$

$C_{33} = M_{33} = 19$

117. Expand using Row 1.

$$\begin{vmatrix} -2 & 0 & 0 \\ 2 & -1 & 0 \\ -1 & 1 & -3 \end{vmatrix} = -2\begin{vmatrix} -1 & 0 \\ 1 & -3 \end{vmatrix} - 0\begin{vmatrix} 2 & 0 \\ -1 & 3 \end{vmatrix} + 0\begin{vmatrix} 2 & -1 \\ -1 & 1 \end{vmatrix}$$

$$= -2(3) - 0(6) + 0(1)$$

$$= -6$$

119. Expand using Row 3.

$$\begin{vmatrix} 4 & 1 & -1 \\ 2 & 3 & 2 \\ 1 & -1 & 0 \end{vmatrix} = 1\begin{vmatrix} 1 & -1 \\ 3 & 2 \end{vmatrix} + 1\begin{vmatrix} 4 & -1 \\ 2 & 2 \end{vmatrix} + 0\begin{vmatrix} 4 & 1 \\ 2 & 3 \end{vmatrix}$$

$$= 1(5) + 1(10) + 0(10)$$

$$= 15$$

121. Expand using Column 2.

$$\begin{vmatrix} -2 & 4 & 1 \\ -6 & 0 & 2 \\ 5 & 3 & 4 \end{vmatrix} = -4\begin{vmatrix} -6 & 2 \\ 5 & 4 \end{vmatrix} - 3\begin{vmatrix} -2 & 1 \\ -6 & 2 \end{vmatrix}$$

$$= -4(-34) - 3(2) = 130$$

123. Expand using Row 4.

$$\begin{vmatrix} 1 & 2 & -1 & 0 \\ 1 & 2 & -4 & 1 \\ 2 & -4 & -3 & 1 \\ 2 & 0 & 0 & 0 \end{vmatrix} = -2\begin{vmatrix} 2 & -1 & 0 \\ 2 & -4 & 1 \\ -4 & -3 & 1 \end{vmatrix} + 0\begin{vmatrix} 1 & -1 & 0 \\ 1 & -4 & 1 \\ 2 & -3 & 1 \end{vmatrix} - 0\begin{vmatrix} 1 & 2 & 0 \\ 1 & 2 & 1 \\ 2 & -4 & 1 \end{vmatrix} + 0\begin{vmatrix} 1 & 2 & -1 \\ 1 & 2 & 4 \\ 2 & -4 & -3 \end{vmatrix}$$

$$= -2[0(-22) - 1(-10) + 1(-6)]$$

$$= -2(4)$$

$$= -8$$

125. Expand along Row 1.

$$\begin{vmatrix} 3 & 0 & -4 & 0 \\ 0 & 8 & 1 & 2 \\ 6 & 1 & 8 & 2 \\ 0 & 3 & -4 & 1 \end{vmatrix} = 3\begin{vmatrix} 8 & 1 & 2 \\ 1 & 8 & 2 \\ 3 & -4 & 1 \end{vmatrix} + (-4)\begin{vmatrix} 0 & 8 & 2 \\ 6 & 1 & 2 \\ 0 & 3 & 1 \end{vmatrix}$$

$$= 3\big[8(8 - (-8)) - 1(1 - 6) + 2(-4 - 24)\big] - 4\big[0 - 6(8 - 6) + 0\big]$$

$$= 3[128 + 5 - 56] - 4[-12]$$

$$= 279$$

127. $\begin{cases} 5x - 2y = 6 \\ -11x + 3y = -23 \end{cases}$

$$x = \frac{\begin{vmatrix} 6 & -2 \\ -23 & 3 \end{vmatrix}}{\begin{vmatrix} 5 & -2 \\ -11 & 3 \end{vmatrix}} = \frac{-28}{-7} = 4, \qquad y = \frac{\begin{vmatrix} 5 & 6 \\ -11 & -23 \end{vmatrix}}{\begin{vmatrix} 5 & -2 \\ -11 & 3 \end{vmatrix}} = \frac{-49}{-7} = 7$$

Solution: $(4, 7)$

129. $\begin{cases} -2x + 3y - 5z = -11 \\ 4x - y + z = -3 \\ -x - 4y + 6z = 15 \end{cases}$

$$D = \begin{vmatrix} -2 & 3 & -5 \\ 4 & -1 & 1 \\ -1 & -4 & 6 \end{vmatrix} = -2(-1)^2\begin{vmatrix} -1 & 1 \\ -4 & 6 \end{vmatrix} + 4(-1)^3\begin{vmatrix} 3 & -5 \\ -4 & 6 \end{vmatrix} - 1(-1)^4\begin{vmatrix} 3 & -5 \\ -1 & 1 \end{vmatrix} = -2(-2) - 4(-2) - (-2) = 14$$

$$x = \frac{\begin{vmatrix} -11 & 3 & -5 \\ -3 & -1 & 1 \\ 15 & -4 & 6 \end{vmatrix}}{14} = \frac{-11(-1)^2\begin{vmatrix} -1 & 1 \\ -4 & 6 \end{vmatrix} - 3(-1)^3\begin{vmatrix} 3 & -5 \\ -4 & 6 \end{vmatrix} + 15(-1)^4\begin{vmatrix} 3 & -5 \\ -1 & 1 \end{vmatrix}}{14} = \frac{-11(-2) + 3(-2) + 15(-2)}{14} = \frac{-14}{14} = -1$$

$$y = \frac{\begin{vmatrix} -2 & -11 & -5 \\ 4 & -3 & 1 \\ -1 & 15 & 6 \end{vmatrix}}{14} = \frac{-2(-1)^2\begin{vmatrix} -3 & 1 \\ 15 & 6 \end{vmatrix} + 4(-1)^3\begin{vmatrix} -11 & -5 \\ 15 & 6 \end{vmatrix} - 1(-1)^4\begin{vmatrix} -11 & -5 \\ -3 & 1 \end{vmatrix}}{14} = \frac{-2(-33) - 4(9) - 1(-26)}{14} = \frac{56}{14} = 4$$

$$z = \frac{\begin{vmatrix} -2 & 3 & -11 \\ 4 & -1 & -3 \\ -1 & -4 & 15 \end{vmatrix}}{14} = \frac{-2(-1)^2\begin{vmatrix} -1 & -3 \\ -4 & 15 \end{vmatrix} + 4(-1)^3\begin{vmatrix} 3 & -11 \\ -4 & 15 \end{vmatrix} - 1(-1)^4\begin{vmatrix} 3 & -11 \\ -1 & -3 \end{vmatrix}}{14} = \frac{-2(-27) - 4(1) - 1(-20)}{14} = \frac{70}{14} = 5$$

Solution: $(-1, 4, 5)$

131. $(1, 0), (5, 0), (5, 8)$

$$\text{Area} = \frac{1}{2}\begin{vmatrix} 1 & 0 & 1 \\ 5 & 0 & 1 \\ 5 & 8 & 1 \end{vmatrix} = \frac{1}{2}\left(1\begin{vmatrix} 0 & 1 \\ 8 & 1 \end{vmatrix} + 1\begin{vmatrix} 5 & 0 \\ 5 & 8 \end{vmatrix}\right) = \frac{1}{2}(-8 + 40) = \frac{1}{2}(32) = 16 \text{ square units}$$

133. $(1, -4), (-2, 3), (0, 5)$

$$\text{Area} = -\frac{1}{2}\begin{vmatrix} 1 & -4 & 1 \\ -2 & 3 & 1 \\ 0 & 5 & 1 \end{vmatrix}$$

$$= -\frac{1}{2}\left(-5\begin{vmatrix} 1 & 1 \\ -2 & 1 \end{vmatrix} + \begin{vmatrix} 1 & -4 \\ -2 & 3 \end{vmatrix}\right)$$

$$= -\frac{1}{2}\big(-5(3) + (-5)\big) = 10 \text{ square units}$$

135. $(-1, 7), (3, -9), (-3, 15)$

$$\begin{vmatrix} -1 & 7 & 1 \\ 3 & -9 & 1 \\ -3 & 15 & 1 \end{vmatrix} = \begin{vmatrix} 3 & -9 \\ -3 & 15 \end{vmatrix} - \begin{vmatrix} -1 & 7 \\ -3 & 15 \end{vmatrix} + \begin{vmatrix} -1 & 7 \\ 3 & -9 \end{vmatrix}$$

$$= 18 - 6 - 12 = 0$$

The points are collinear.

137. $(-4, 0), (4, 4)$

$$\begin{vmatrix} x & y & 1 \\ -4 & 0 & 1 \\ 4 & 4 & 1 \end{vmatrix} = 0$$

$$1\begin{vmatrix} -4 & 0 \\ 4 & 4 \end{vmatrix} - 1\begin{vmatrix} x & y \\ 4 & 4 \end{vmatrix} + 1\begin{vmatrix} x & y \\ -4 & 0 \end{vmatrix} = 0$$

$$-16 - (4x - 4y) + 4y = 0$$

$$-4x + 8y - 16 = 0$$

$$x - 2y + 4 = 0$$

139. $\left(-\frac{5}{2}, 3\right), \left(\frac{7}{2}, 1\right)$

$$\begin{vmatrix} x & y & 1 \\ -\frac{5}{2} & 3 & 1 \\ \frac{7}{2} & 1 & 1 \end{vmatrix} = 0$$

$$1\begin{vmatrix} -\frac{5}{2} & 3 \\ \frac{7}{2} & 1 \end{vmatrix} - 1\begin{vmatrix} x & y \\ \frac{7}{2} & 1 \end{vmatrix} + 1\begin{vmatrix} x & y \\ -\frac{5}{2} & 3 \end{vmatrix} = 0$$

$$-13 - \left(x - \frac{7}{2}y\right) + \left(3x + \frac{5}{2}y\right) = 0$$

$$2x + 6y - 13 = 0$$

141. (a) Uncoded:

L O O K _ O U T _ B E L O W _

[12 15 15] [11 0 15] [21 20 0] [2 5 12] [15 23 0]

(b) $$\begin{bmatrix} 12 & 15 & 15 \end{bmatrix}\begin{bmatrix} 2 & -2 & 0 \\ 3 & 0 & -3 \\ -6 & 2 & 3 \end{bmatrix} = \begin{bmatrix} -21 & 6 & 0 \end{bmatrix}$$

$$\begin{bmatrix} 11 & 0 & 15 \end{bmatrix}\begin{bmatrix} 2 & -2 & 0 \\ 3 & 0 & -3 \\ -6 & 2 & 3 \end{bmatrix} = \begin{bmatrix} -68 & 8 & 45 \end{bmatrix}$$

$$\begin{bmatrix} 21 & 20 & 0 \end{bmatrix}\begin{bmatrix} 2 & -2 & 0 \\ 3 & 0 & -3 \\ -6 & 2 & 3 \end{bmatrix} = \begin{bmatrix} 102 & -42 & -60 \end{bmatrix}$$

$$\begin{bmatrix} 2 & 5 & 12 \end{bmatrix}\begin{bmatrix} 2 & -2 & 0 \\ 3 & 0 & -3 \\ -6 & 2 & 3 \end{bmatrix} = \begin{bmatrix} -53 & 20 & 21 \end{bmatrix}$$

$$\begin{bmatrix} 15 & 23 & 0 \end{bmatrix}\begin{bmatrix} 2 & -2 & 0 \\ 3 & 0 & -3 \\ -6 & 2 & 3 \end{bmatrix} = \begin{bmatrix} 99 & -30 & -69 \end{bmatrix}$$

Encoded: −21 6 0 −68 8 45 102 −42

−42 −60 −53 20 21 99 −30 −69

143. $$A^{-1} = \begin{bmatrix} -1 & 2 & -3 \\ 2 & 1 & 0 \\ 4 & -2 & 5 \end{bmatrix}$$

$$\begin{bmatrix} -5 & 11 & -2 \end{bmatrix}\begin{bmatrix} -1 & 2 & -3 \\ 2 & 1 & 0 \\ 4 & -2 & 5 \end{bmatrix} = \begin{bmatrix} 19 & 5 & 5 \end{bmatrix} \quad \text{S E E}$$

$$\begin{bmatrix} 370 & -265 & 225 \end{bmatrix}\begin{bmatrix} -1 & 2 & -3 \\ 2 & 1 & 0 \\ 4 & -2 & 5 \end{bmatrix} = \begin{bmatrix} 0 & 25 & 15 \end{bmatrix} \quad \_\ \text{Y O}$$

$$\begin{bmatrix} -57 & 48 & -33 \end{bmatrix}\begin{bmatrix} -1 & 2 & -3 \\ 2 & 1 & 0 \\ 4 & -2 & 5 \end{bmatrix} = \begin{bmatrix} 21 & 0 & 6 \end{bmatrix} \quad \text{U}\ \_\ \text{F}$$

$$\begin{bmatrix} 32 & -15 & 20 \end{bmatrix}\begin{bmatrix} -1 & 2 & -3 \\ 2 & 1 & 0 \\ 4 & -2 & 5 \end{bmatrix} = \begin{bmatrix} 18 & 9 & 4 \end{bmatrix} \quad \text{R I D}$$

$$\begin{bmatrix} 245 & -171 & 147 \end{bmatrix}\begin{bmatrix} -1 & 2 & -3 \\ 2 & 1 & 0 \\ 4 & -2 & 5 \end{bmatrix} = \begin{bmatrix} 1 & 25 & 0 \end{bmatrix} \quad \text{A Y}\ \_$$

Message: SEE YOU FRIDAY

145. False. The matrix must be square.

147. An error message appears because $1(6) - (-3)(-2) = 0$.

149. If A is a square matrix, the cofactor C_{ij} of the entry a_{ij} is $(-1)^{i+j}M_{ij}$, where M_{ij} is the determinant obtained by deleting the ith row and jth column of A. The determinant of A is the sum of the entries of any row or column of A multiplied by their respective cofactors.

151.
$$\begin{vmatrix} 2-\lambda & 5 \\ 3 & -8-\lambda \end{vmatrix} = 0$$
$$(2-\lambda)(-8-\lambda) - 15 = 0$$
$$-16 + 6\lambda + \lambda^2 - 15 = 0$$
$$\lambda^2 + 6\lambda - 31 = 0$$
$$\lambda = \frac{-6 \pm \sqrt{36 - 4(-31)}}{2}$$
$$\lambda = -3 \pm 2\sqrt{10}$$

Problem Solving for Chapter 10

1. $A = \begin{bmatrix} 0 & -1 \\ 1 & 0 \end{bmatrix}$ $\quad T = \begin{bmatrix} 1 & 2 & 3 \\ 1 & 4 & 2 \end{bmatrix}$

(a) $AT = \begin{bmatrix} -1 & -4 & -2 \\ 1 & 2 & 3 \end{bmatrix}$ $\quad AAT = \begin{bmatrix} -1 & -2 & -3 \\ -1 & -4 & -2 \end{bmatrix}$

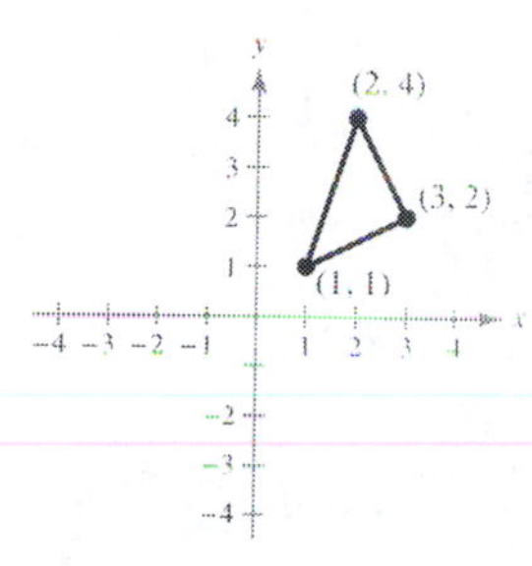

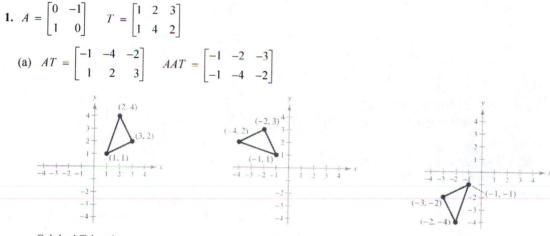

Original Triangle

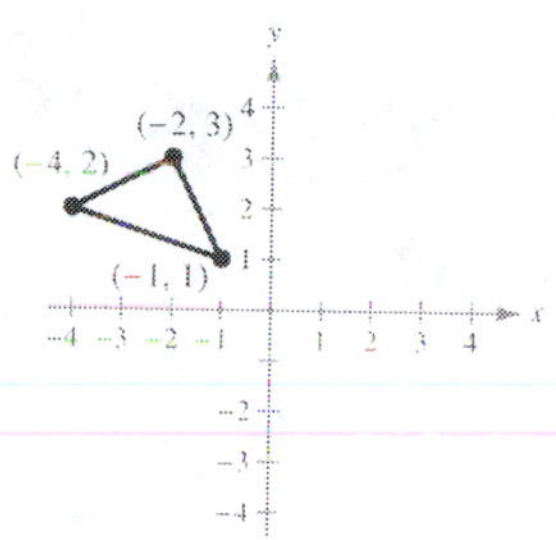

AT Triangle

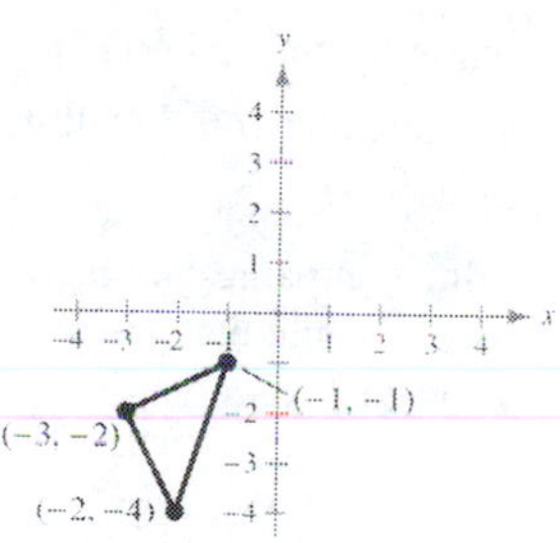

AAT Triangle

The transformation A interchanges the x and y coordinates and then takes the negative of the x coordinate. A represents a counterclockwise rotation by 90°.

(b) AAT is rotated clockwise 90° to obtain AT. AT is then rotated clockwise 90° to obtain T.

3. (a) $A^2 = \begin{bmatrix} 1 & 0 \\ 0 & 0 \end{bmatrix}\begin{bmatrix} 1 & 0 \\ 0 & 0 \end{bmatrix} = \begin{bmatrix} 1 & 0 \\ 0 & 0 \end{bmatrix} = A$

A *is* idempotent.

(b) $A^2 = \begin{bmatrix} 0 & 1 \\ 1 & 0 \end{bmatrix}\begin{bmatrix} 0 & 1 \\ 1 & 0 \end{bmatrix} = \begin{bmatrix} 1 & 0 \\ 0 & 1 \end{bmatrix} \neq A$

A is *not* idempotent.

(c) $A^2 = \begin{bmatrix} 2 & 3 \\ -1 & -2 \end{bmatrix}\begin{bmatrix} 2 & 3 \\ -1 & -2 \end{bmatrix} = \begin{bmatrix} 1 & 0 \\ 0 & 1 \end{bmatrix} \neq A$

A is *not* idempotent.

(d) $A^2 = \begin{bmatrix} 2 & 3 \\ 1 & 2 \end{bmatrix}\begin{bmatrix} 2 & 3 \\ 1 & 2 \end{bmatrix} = \begin{bmatrix} 7 & 12 \\ 4 & 7 \end{bmatrix} \neq A$

A is *not* idempotent.

5. (a) $\begin{bmatrix} 0.70 & 0.15 & 0.15 \\ 0.20 & 0.80 & 0.15 \\ 0.10 & 0.05 & 0.70 \end{bmatrix}\begin{bmatrix} 25{,}000 \\ 30{,}000 \\ 45{,}000 \end{bmatrix} = \begin{bmatrix} 28{,}750 \\ 35{,}750 \\ 35{,}500 \end{bmatrix}$

Gold Satellite System: 28,750 subscribers

Galaxy Satellite Network: 35,750 subscribers

Nonsubscribers: 35,500

(b) $\begin{bmatrix} 0.70 & 0.15 & 0.15 \\ 0.20 & 0.80 & 0.15 \\ 0.10 & 0.05 & 0.70 \end{bmatrix}\begin{bmatrix} 28{,}750 \\ 35{,}750 \\ 35{,}500 \end{bmatrix} \approx \begin{bmatrix} 30{,}813 \\ 39{,}675 \\ 29{,}513 \end{bmatrix}$

Gold Satellite System: 30,813 subscribers

Galaxy Satellite Network: 39,675 subscribers

Nonsubscribers: 29,513

(c) $\begin{bmatrix} 0.70 & 0.15 & 0.15 \\ 0.20 & 0.80 & 0.15 \\ 0.10 & 0.05 & 0.70 \end{bmatrix}\begin{bmatrix} 30{,}812.5 \\ 39{,}675 \\ 29{,}512.5 \end{bmatrix} \approx \begin{bmatrix} 31{,}947 \\ 42{,}329 \\ 25{,}724 \end{bmatrix}$

Gold Satellite System: 31,947 subscribers

Galaxy Satellite Network: 42,329 subscribers

Nonsubscribers: 25,724

(d) Both satellite companies are increasing the number of subscribers, while the number of nonsubscribers is decreasing each year.

7. $A = \begin{bmatrix} 3 & x \\ -2 & -3 \end{bmatrix} \Rightarrow A^{-1} = \dfrac{1}{-9 + 2x}\begin{bmatrix} -3 & -x \\ 2 & 3 \end{bmatrix}$

If $A = A^{-1}$, then $\begin{bmatrix} \dfrac{-3}{-9+2x} & \dfrac{-x}{-9+2x} \\ \dfrac{2}{-9+2x} & \dfrac{3}{-9+2x} \end{bmatrix} = \begin{bmatrix} 3 & x \\ -2 & -3 \end{bmatrix}$.

Equating the first entry in Row 1 yields

$\dfrac{-3}{-9 + 2x} = 3 \Rightarrow -3 = -27 + 6x \Rightarrow x = 4.$

Now check $x = 4$ in the other entries:

$\dfrac{-4}{-9 + 2(4)} = 4$ ✓

$\dfrac{2}{-9 + 2(4)} = -2$ ✓

$\dfrac{3}{-9 + 2(4)} = -3$ ✓

So, $x = 4$.

9. From Exercise 3, we have the singular matrix

$A = \begin{bmatrix} 1 & 0 \\ 0 & 0 \end{bmatrix}$, where $A^2 = A$.

Also, $A = \begin{bmatrix} 1 & 1 \\ 0 & 0 \end{bmatrix}$ has this property.

11. $(a - b)(b - c)(c - a)(a + b + c) = -a^3b + a^3c + ab^3 - ac^3 - b^3c + bc^3$

$$\begin{vmatrix} 1 & 1 & 1 \\ a & b & c \\ a^3 & b^3 & c^3 \end{vmatrix} = \begin{vmatrix} b & c \\ b^3 & c^3 \end{vmatrix} - \begin{vmatrix} a & c \\ a^3 & c^3 \end{vmatrix} + \begin{vmatrix} a & b \\ a^3 & b^3 \end{vmatrix} = bc^3 - b^3c - ac^3 + a^3c + ab^3 - a^3b$$

So, $\begin{vmatrix} 1 & 1 & 1 \\ a & b & c \\ a^3 & b^3 & c^3 \end{vmatrix} = (a - b)(b - c)(c - a)(a + b + c).$

13. $$\begin{vmatrix} x & 0 & 0 & d \\ -1 & x & 0 & c \\ 0 & -1 & x & b \\ 0 & 0 & -1 & a \end{vmatrix} = x\begin{vmatrix} x & 0 & c \\ -1 & x & b \\ 0 & -1 & a \end{vmatrix} - d\begin{vmatrix} -1 & x & 0 \\ 0 & -1 & x \\ 0 & 0 & -1 \end{vmatrix} = \underbrace{x(ax^2 + bx + c)}_{\text{From Exercise 11}} - d\left(-\begin{vmatrix} -1 & x \\ 0 & -1 \end{vmatrix}\right) = ax^3 + bx^2 + cx + d$$

15. Let x = cost of a transformer, y = cost per foot of wire, and z = cost of a light.

$$x + 25y + 5z = 20$$
$$x + 50y + 15z = 35$$
$$x + 100y + 20z = 50$$

$$\begin{bmatrix} 1 & 25 & 5 & \vdots & 20 \\ 1 & 50 & 15 & \vdots & 35 \\ 1 & 100 & 20 & \vdots & 50 \end{bmatrix} \xrightarrow{\text{rref}} \begin{bmatrix} 1 & 0 & 0 & \vdots & 10 \\ 0 & 1 & 0 & \vdots & 0.2 \\ 0 & 0 & 1 & \vdots & 1 \end{bmatrix}$$

By using the matrix capabilities of a graphing calculator to reduce the augmented matrix to reduced row-echelon form, we have the following costs:

Transformer	\$10.00
Foot of wire	\$ 0.20
Light	\$ 1.00

17. (a) $\begin{bmatrix} 45 & -35 \end{bmatrix}\begin{bmatrix} w & x \\ y & z \end{bmatrix} = \begin{bmatrix} 10 & 15 \end{bmatrix}$

$\begin{bmatrix} 38 & -30 \end{bmatrix}\begin{bmatrix} w & x \\ y & z \end{bmatrix} = \begin{bmatrix} 8 & 14 \end{bmatrix}$

$$45w - 35y = 10$$
$$45x - 35z = 15$$
$$38w - 30y = 8$$
$$38x - 30z = 14$$

$$\left.\begin{aligned} 45w - 35y &= 10 \\ 38w - 30y &= 8 \end{aligned}\right\} \Rightarrow w = 1, y = 1$$

$$\left.\begin{aligned} 45x - 35z &= 15 \\ 38x - 30z &= 14 \end{aligned}\right\} \Rightarrow x = -2, z = -3$$

$$A^{-1} = \begin{bmatrix} 1 & -2 \\ 1 & -3 \end{bmatrix}$$

(b) $$\begin{bmatrix} 45 & -35 \\ 38 & -30 \\ 18 & -18 \\ 35 & -30 \\ 81 & -60 \\ 42 & -28 \\ 75 & -55 \\ 2 & -2 \\ 22 & -21 \\ 15 & -10 \end{bmatrix}\begin{bmatrix} 1 & -2 \\ 1 & -3 \end{bmatrix} = \begin{bmatrix} 10 & 15 \\ 8 & 14 \\ 0 & 18 \\ 5 & 20 \\ 21 & 18 \\ 14 & 0 \\ 20 & 15 \\ 0 & 2 \\ 1 & 19 \\ 5 & 0 \end{bmatrix} \begin{matrix} \text{J} & \text{O} \\ \text{H} & \text{N} \\ \_ & \text{R} \\ \text{E} & \text{T} \\ \text{U} & \text{R} \\ \text{N} & \_ \\ \text{T} & \text{O} \\ \_ & \text{B} \\ \text{A} & \text{S} \\ \text{E} & \_ \end{matrix}$$

Message: JOHN RETURN TO BASE

19. Let $A = \begin{bmatrix} 3 & -3 \\ 5 & -5 \end{bmatrix}$, then $|A| = 0$.

Let $A = \begin{bmatrix} 2 & 4 & -6 \\ -3 & 1 & 2 \\ 5 & -8 & 3 \end{bmatrix}$, then $|A| = 0$.

Let $A = \begin{bmatrix} 3 & -7 & 5 & -1 \\ -6 & 4 & 0 & 2 \\ 5 & 8 & -6 & -7 \\ 9 & 11 & -4 & -16 \end{bmatrix}$, then $|A| = 0$.

Conjecture: If A is an $n \times n$ matrix, each of whose rows add up to zero, then $|A| = 0$.

Practice Test for Chapter 10

1. Put the matrix in reduced row-echelon form.

$$\begin{bmatrix} 1 & -2 & 4 \\ 3 & -5 & 9 \end{bmatrix}$$

For Exercises 2–4, use matrices to solve the system of equations.

2. $\begin{cases} 3x + 5y = 3 \\ 2x - y = -11 \end{cases}$

3. $\begin{cases} 2x + 3y = -3 \\ 3x + 2y = 8 \\ x + y = 1 \end{cases}$

4. $\begin{cases} x + 3z = -5 \\ 2x + y = 0 \\ 3x + y - z = 3 \end{cases}$

5. Multiply $\begin{bmatrix} 1 & 4 & 5 \\ 2 & 0 & -3 \end{bmatrix}\begin{bmatrix} 1 & 6 \\ 0 & -7 \\ -1 & 2 \end{bmatrix}$.

6. Given $A = \begin{bmatrix} 9 & 1 \\ -4 & 8 \end{bmatrix}$ and $B = \begin{bmatrix} 6 & -2 \\ 3 & 5 \end{bmatrix}$, find $3A - 5B$.

7. Find $f(A)$.

$f(x) = x^2 - 7x + 8,\ A = \begin{bmatrix} 3 & 0 \\ 7 & 1 \end{bmatrix}$

8. True or false:

$(A + B)(A + 3B) = A^2 + 4AB + 3B^2$ where A and B are matrices.

(Assume that A^2, AB, and B^2 exist.)

For Exercises 9–10, find the inverse of the matrix, if it exists.

9. $\begin{bmatrix} 1 & 2 \\ 3 & 5 \end{bmatrix}$

10. $\begin{bmatrix} 1 & 1 & 1 \\ 3 & 6 & 5 \\ 6 & 10 & 8 \end{bmatrix}$

11. Use an inverse matrix to solve the systems.

(a) $\begin{cases} x + 2y = 4 \\ 3x + 5y = 1 \end{cases}$

(b) $\begin{cases} x + 2y = 3 \\ 3x + 5y = -2 \end{cases}$

For Exercises 12–14, find the determinant of the matrix.

12. $\begin{bmatrix} 6 & -1 \\ 3 & 4 \end{bmatrix}$

13. $\begin{bmatrix} 1 & 3 & -1 \\ 5 & 9 & 0 \\ 6 & 2 & -5 \end{bmatrix}$

14. $\begin{bmatrix} 1 & 4 & 2 & 3 \\ 0 & 1 & -2 & 0 \\ 3 & 5 & -1 & 1 \\ 2 & 0 & 6 & 1 \end{bmatrix}$

15. Evaluate $\begin{vmatrix} 6 & 4 & 3 & 0 & 6 \\ 0 & 5 & 1 & 4 & 8 \\ 0 & 0 & 2 & 7 & 3 \\ 0 & 0 & 0 & 9 & 2 \\ 0 & 0 & 0 & 0 & 1 \end{vmatrix}$.

16. Use a determinant to find the area of the triangle with vertices $(0, 7)$, $(5, 0)$, and $(3, 9)$.

17. Use a determinant to find the equation of the line passing through $(2, 7)$ and $(-1, 4)$.

For Exercises 18–20, use Cramer's Rule to find the indicated value.

18. Find x.

$$\begin{cases} 6x - 7y = 4 \\ 2x + 5y = 11 \end{cases}$$

19. Find z.

$$\begin{cases} 3x \quad\quad + z = 1 \\ \quad\quad y + 4z = 3 \\ x - y \quad\quad = 2 \end{cases}$$

20. Find y.

$$\begin{cases} 721.4x - 29.1y = 33.77 \\ 45.9x + 105.6y = 19.85 \end{cases}$$

CHAPTER 11
Sequences, Series, and Probability

CHAPTER 11
Sequences, Series, and Probability

Section 11.1 Sequences and Series

1. infinite sequence

3. recursively

5. index; upper; lower

7. $a_n = 4n - 7$

$a_1 = 4(1) - 7 = -3$

$a_2 = 4(2) - 7 = 1$

$a_3 = 4(3) - 7 = 5$

$a_4 = 4(4) - 7 = 9$

$a_5 = 4(5) - 7 = 13$

9. $a_n = (-2)^n$

$a_1 = (-2)^1 = -2$

$a_2 = (-2)^2 = 4$

$a_3 = (-2)^3 = -8$

$a_4 = (-2)^4 = 16$

$a_5 = (-2)^5 = -32$

11. $a_n = \dfrac{n}{n + 2}$

$a_1 = \dfrac{1}{1 + 2} = \dfrac{1}{3}$

$a_2 = \dfrac{2}{2 + 2} = \dfrac{1}{2}$

$a_3 = \dfrac{3}{3 + 2} = \dfrac{3}{5}$

$a_4 = \dfrac{4}{4 + 2} = \dfrac{2}{3}$

$a_5 = \dfrac{5}{5 + 2} = \dfrac{5}{7}$

13. $a_n = \dfrac{1 + (-1)^n}{n}$

$a_1 = 0$

$a_2 = \dfrac{2}{2} = 1$

$a_3 = 0$

$a_4 = \dfrac{2}{4} = \dfrac{1}{2}$

$a_5 = 0$

15. $a_n = \dfrac{2^n}{3^n}$

$a_1 = \dfrac{2^1}{3^1} = \dfrac{2}{3}$

$a_2 = \dfrac{2^2}{3^2} = \dfrac{4}{9}$

$a_3 = \dfrac{2^3}{3^3} = \dfrac{8}{27}$

$a_4 = \dfrac{2^4}{3^4} = \dfrac{16}{81}$

$a_5 = \dfrac{2^5}{3^5} = \dfrac{32}{243}$

17. $a_n = \frac{2}{3}$

$a_1 = \frac{2}{3}$

$a_2 = \frac{2}{3}$

$a_3 = \frac{2}{3}$

$a_4 = \frac{2}{3}$

$a_5 = \frac{2}{3}$

19. $a_n = n(n - 1)(n - 2)$

$a_1 = (1)(0)(-1) = 0$

$a_2 = (2)(1)(0) = 0$

$a_3 = (3)(2)(1) = 6$

$a_4 = (4)(3)(2) = 24$

$a_5 = (5)(4)(3) = 60$

21. $a_n = (-1)^n\left(\dfrac{n}{n + 1}\right)$

$a_1 = (-1)^1\dfrac{1}{1 + 1} = -\dfrac{1}{2}$

$a_2 = (-1)^2\dfrac{2}{2 + 1} = \dfrac{2}{3}$

$a_3 = (-1)^3\dfrac{3}{3 + 1} = -\dfrac{3}{4}$

$a_4 = (-1)^4\dfrac{4}{4 + 1} = \dfrac{4}{5}$

$a_5 = (-1)^5\dfrac{5}{5 + 1} = -\dfrac{5}{6}$

23. $a_{25} = (-1)^{25}(3(25) - 2) = -73$

25. $a_{11} = \dfrac{4(11)}{2(11)^2 - 3} = \dfrac{44}{239}$

27. $a_n = \frac{2}{3}n$

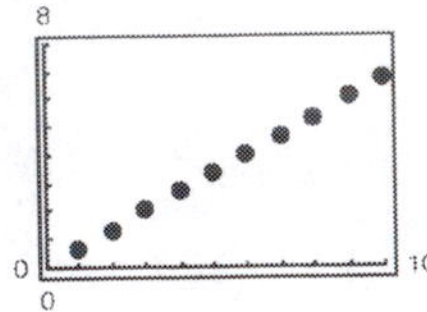

29. $a_n = 16(-0.5)^{n-1}$

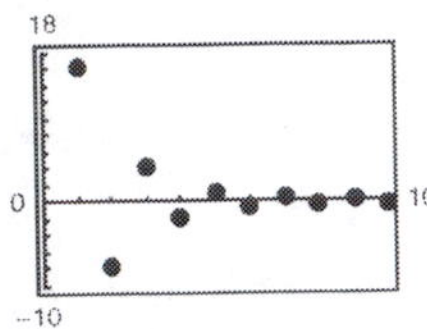

31. $a_n = \dfrac{2n}{n + 1}$

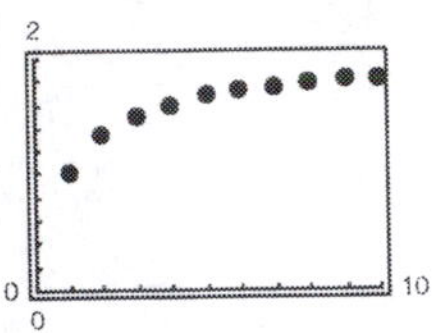

33. $a_n = \dfrac{8}{n + 1}$

$a_1 = 4, a_{10} = \dfrac{8}{11}$

The sequence decreases.

Matches graph (c).

34. $a_n = \dfrac{8n}{n + 1}$

$a_1 = 4, a_3 = \dfrac{24}{4} = 6$

The sequence increases.

Matches graph (b).

35. $a_n = 4(0.5)^{n-1}$

$a_1 = 4, a_{10} = \dfrac{1}{128}$

The sequence decreases.

Matches graph (d).

36. $a_n = \dfrac{4^n}{n!}$

$a_1 = 4, a_4 = \dfrac{4^4}{4!} = \dfrac{256}{24} = 10\dfrac{2}{3}$

The sequence increases.

Matches graph (a).

37. 3, 7, 11, 15, 19, ...

n:	1	2	3	4	5	...	n
Terms:	3	7	11	15	19	...	a_n

Apparent pattern:

Each term is one less than four times n, which implies that $a_n = 4n - 1$.

39. $-\dfrac{2}{3}, \dfrac{3}{4}, -\dfrac{4}{5}, \dfrac{5}{6}, -\dfrac{6}{7}, \ldots$

$a_n = (-1)^n\left(\dfrac{n + 1}{n + 2}\right)$

41. $\dfrac{2}{1}, \dfrac{3}{3}, \dfrac{4}{5}, \dfrac{5}{7}, \dfrac{6}{9}, \ldots$

$a_n = \dfrac{n + 1}{2n - 1}$

43. $1, \dfrac{1}{4}, \dfrac{1}{9}, \dfrac{1}{16}, \dfrac{1}{25}, \ldots$

$a_n = \dfrac{1}{n^2}$

45. 1, −1, 1, −1, 1, ...

n:	1	2	3	4	5	...	n
Terms:	1	−1	1	−1	1	...	a_n

Apparent pattern:

Each term is either 1 or −1 which implies that $a_n = (-1)^{n+1}$.

47. $1, 2, \dfrac{2^2}{2}, \dfrac{2^3}{6}, \dfrac{2^4}{24}, \dfrac{2^5}{120}, \ldots$

n:	1	2	3	4	5	6	...	n
Terms:	1	2	$\dfrac{2^2}{2}$	$\dfrac{2^3}{6}$	$\dfrac{2^4}{24}$	$\dfrac{2^5}{120}$	...	a_n

Apparent pattern:

Each term is 2^{n-1} divided by $(n - 1)!$, which implies that $a_n = \dfrac{2^{n-1}}{(n - 1)!}$.

49. $a_1 = 28$ and $a_{k+1} = a_k - 4$

$a_1 = 28$

$a_2 = a_1 - 4 = 28 - 4 = 24$

$a_3 = a_2 - 4 = 24 - 4 = 20$

$a_4 = a_3 - 4 = 20 - 4 = 16$

$a_5 = a_4 - 4 = 16 - 4 = 12$

51. $a_0 = 1, a_1 = 2, a_k = a_{k-2} + \frac{1}{2}a_{k-1}$

$a_0 = 1$

$a_1 = 2$

$a_2 = a_0 + \frac{1}{2}a_1 = 1 + \frac{1}{2}(1) = 2$

$a_3 = a_1 + \frac{1}{2}a_2 = 2 + \frac{1}{2}(2) = 3$

$a_4 = a_2 + \frac{1}{2}a_3 = 2 + \frac{1}{2}(3) = \frac{7}{2}$

53. $a_1 = 6$ and $a_{k+1} = a_k + 2$

$a_1 = 6$

$a_2 = a_1 + 2 = 6 + 2 = 8$

$a_3 = a_2 + 2 = 8 + 2 = 10$

$a_4 = a_3 + 2 = 10 + 2 = 12$

$a_5 = a_4 + 2 = 12 + 2 = 14$

In general, $a_n = 2n + 4$.

55. $a_1 = 81$ and $a_{k+1} = \frac{1}{3}a_k$

$a_1 = 81$

$a_2 = \frac{1}{3}a_1 = \frac{1}{3}(81) = 27$

$a_3 = \frac{1}{3}a_2 = \frac{1}{3}(27) = 9$

$a_4 = \frac{1}{3}a_3 = \frac{1}{3}(9) = 3$

$a_5 = \frac{1}{3}a_4 = \frac{1}{3}(3) = 1$

In general,

$$a_n = 81\left(\frac{1}{3}\right)^{n-1} = 81(3)\left(\frac{1}{3}\right)^n = \frac{243}{3^n}.$$

57. $a_1 = 1, a_2 = 1, a_k = a_{k-1} + a_{k-2}, k \ge 1$

$a_1 = 1$	$b_1 = \frac{1}{1} = 1$
$a_2 = 1$	$b_2 = \frac{2}{1} = 2$
$a_3 = 1 + 1 = 2$	$b_3 = \frac{3}{2}$
$a_4 = 2 + 1 = 3$	$b_4 = \frac{5}{3}$
$a_5 = 3 + 2 = 5$	$b_5 = \frac{8}{5}$
$a_6 = 5 + 3 = 8$	$b_6 = \frac{13}{8}$
$a_7 = 8 + 5 = 13$	$b_7 = \frac{21}{13}$
$a_8 = 13 + 8 = 21$	$b_8 = \frac{34}{21}$
$a_9 = 21 + 13 = 34$	$b_9 = \frac{55}{34}$
$a_{10} = 34 + 21 = 55$	$b_{10} = \frac{89}{55}$
$a_{11} = 55 + 34 = 89$	
$a_{12} = 89 + 55 = 144$	

59. $a_n = \frac{5}{n!}$

$a_0 = \frac{5}{0!} = \frac{5}{1} = 5$

$a_1 = \frac{5}{1!} = \frac{5}{1} = 5$

$a_2 = \frac{5}{2!} = \frac{5}{2}$

$a_3 = \frac{5}{3!} = \frac{5}{6}$

$a_4 = \frac{5}{4!} = \frac{5}{24}$

61. $a_n = \frac{1}{(n+1)!}$

$a_0 = \frac{1}{1!} = 1$

$a_1 = \frac{1}{2!} = \frac{1}{2}$

$a_2 = \frac{1}{3!} = \frac{1}{6}$

$a_3 = \frac{1}{4!} = \frac{1}{24}$

$a_4 = \frac{1}{5!} = \frac{1}{120}$

63. $\frac{4!}{6!} = \frac{1 \cdot 2 \cdot 3 \cdot 4}{1 \cdot 2 \cdot 3 \cdot 4 \cdot 5 \cdot 6} = \frac{1}{5 \cdot 6} = \frac{1}{30}$

65. $\frac{(n+1)!}{n!} = \frac{1 \cdot 2 \cdot 3 \cdots n \cdot (n+1)}{1 \cdot 2 \cdot 3 \cdots n} = \frac{n+1}{1}$

$= n + 1$

67. $\sum_{i=1}^{5} (2i + 1) = (2 + 1) + (4 + 1) + (6 + 1) + (8 + 1) + (10 + 1) = 35$

69. $\sum_{k=1}^{4} 10 = 10 + 10 + 10 + 10 = 40$

71. $\sum_{k=2}^{5} (k+1)^2(k-3) = (3)^2(-1) + (4)^2(0) + (5)^2(1) + (6)^2(2) = 88$

73. $\sum_{i=1}^{4} 2^i = 2^1 + 2^2 + 2^3 + 2^4 = 30$

75. $\sum_{n=0}^{5} \frac{1}{2n+1} = \frac{6508}{3465}$

77. $\sum_{k=0}^{4} \frac{(-1)^k}{k!} = \frac{3}{8}$

79. $\frac{1}{3(1)} + \frac{1}{3(2)} + \frac{1}{3(3)} + \cdots + \frac{1}{3(9)} = \sum_{i=1}^{9} \frac{1}{3i}$

81. $\left[2\left(\frac{1}{8}\right) + 3\right] + \left[2\left(\frac{2}{8}\right) + 3\right] + \left[2\left(\frac{3}{8}\right) + 3\right] + \cdots + \left[2\left(\frac{8}{8}\right) + 3\right] = \sum_{i=1}^{8}\left[2\left(\frac{i}{8}\right) + 3\right]$

83. $3 - 9 + 27 - 81 + 243 - 729 = \sum_{i=1}^{6} (-1)^{i+1} 3i$

85. $\frac{1}{1^2} - \frac{1}{2^2} + \frac{1}{3^2} - \frac{1}{4^2} + \cdots - \frac{1}{20^2} = \sum_{i=1}^{20} \frac{(-1)^{i+1}}{i^2}$

87. $\frac{1}{4} + \frac{3}{8} + \frac{7}{16} + \frac{15}{32} + \frac{31}{64} = \sum_{i=1}^{5} \frac{2^i - 1}{2^{i+1}}$

89. $\sum_{i=1}^{4} 5\left(\frac{1}{2}\right)^i = 5\left(\frac{1}{2}\right) + 5\left(\frac{1}{2}\right)^2 + 5\left(\frac{1}{2}\right)^3 + 5\left(\frac{1}{2}\right)^4 = \frac{75}{16}$

91. $\sum_{n=1}^{3} 4\left(-\frac{1}{2}\right)^n = 4\left(-\frac{1}{2}\right) + 4\left(-\frac{1}{2}\right)^2 + 4\left(-\frac{1}{2}\right)^3 = -\frac{3}{2}$

93. $\sum_{i=1}^{\infty} 6\left(\frac{1}{10}\right)^i = 0.6 + 0.06 + 0.006 + 0.0006 + \cdots = \frac{2}{3}$

95. By using a calculator,

$\sum_{k=1}^{10} 7\left(\frac{1}{10}\right)^k \approx 0.7777777777$

$\sum_{k=1}^{50} 7\left(\frac{1}{10}\right)^k \approx 0.7777777778$

$\sum_{k=1}^{100} 7\left(\frac{1}{10}\right)^k \approx \frac{7}{9}.$

The terms approach zero as $n \to \infty$.

So, $\sum_{k=1}^{\infty} 7\left(\frac{1}{10}\right)^k = \frac{7}{9}.$

97. (a) $A_1 = \$10{,}087.50$

$A_2 \approx \$10{,}175.77$

$A_3 \approx \$10{,}264.80$

$A_4 \approx \$10{,}354.62$

$A_5 \approx \$10{,}445.22$

$A_6 \approx \$10{,}536.62$

$A_7 \approx \$10{,}628.81$

$A_8 \approx \$10{,}721.82$

(b) $A_{40} \approx \$14{,}169.09$

(c) No; The balance after 20 years, $A_{80} = \$20{,}076.31$ is not twice the balance after 10 years, $A_{40} \approx \$14{,}169.09$.

99. True, $\sum_{i=1}^{4}\left(i^2 + 2i\right) = \sum_{i=1}^{4} i^2 + 2\sum_{i=1}^{4} i$ by the Properties of Sums.

101. $\frac{327.15 + 785.69 + 433.04 + 265.38 + 604.12 + 590.30}{6} \approx \$500.95$

103. $\sum_{i=1}^{n}(x_i - \bar{x})^2 = \sum_{i=1}^{n}\left(x_i^2 - 2x_i\bar{x} + \bar{x}^2\right) = \sum_{i=1}^{n} x_i^2 - 2\bar{x}\sum_{i=1}^{n} x_i + n\bar{x}^2$

$$= \sum_{i=1}^{n} x_i^2 - 2 \cdot \frac{1}{n}\sum_{i=1}^{n} x_i \sum_{i=1}^{n} x_i + n \cdot \frac{1}{n}\sum_{i=1}^{n} x_i \cdot \frac{1}{n}\sum_{i=1}^{n} x_i$$

$$= \sum_{i=1}^{n} x_i^2 + \sum_{i=1}^{n} x_i \sum_{i=1}^{n} x_i\left(-\frac{2}{n} + \frac{1}{n}\right) = \sum_{i=1}^{n} x_i^2 - \frac{1}{n}\left(\sum_{i=1}^{n} x_i\right)^2$$

105. $a_n = \dfrac{x^n}{n!}$

$a_1 = \dfrac{x^1}{1!} = x$

$a_2 = \dfrac{x^2}{2!} = \dfrac{x^2}{2}$

$a_3 = \dfrac{x^3}{3!} = \dfrac{x^3}{6}$

$a_4 = \dfrac{x^4}{4!} = \dfrac{x^4}{24}$

$a_5 = \dfrac{x^5}{5!} = \dfrac{x^5}{120}$

107. (a) and (b)

Number of blue cube faces	0	1	2	3
3 × 3 × 3	1	6	12	8
4 × 4 × 4	8	24	24	8
5 × 5 × 5	27	54	36	8
6 × 6 × 6	64	96	48	8

(c) The columns change at different rates

(d)

Number of blue cube faces	0	1	2	3
$n \times n \times n$	$(n-2)^3$	$6(n-2)^2$	$12(n-2)$	8

Section 11.2 Arithmetic Sequences and Partial Sums

1. arithmetic; common

3. recursion

5. 10, 8, 6, 4, 2, …

Arithmetic sequence, $d = -2$

7. 1, 2, 4, 8, 16, …

Not an arithmetic sequence

9. $\frac{9}{4}, 2, \frac{7}{4}, \frac{3}{2}, \frac{5}{4}, \ldots$

Arithmetic sequence, $d = -\frac{1}{4}$

11. ln 1, ln 2, ln 3, ln 4, ln 5, …

Not an arithmetic sequence

13. $a_n = 5 + 3n$

8, 11, 14, 17, 20

Arithmetic sequence, $d = 3$

15. $a_n = 3 - 4(n - 2)$

7, 3, −1, −5, −9

Arithmetic sequence, $d = -4$

17. $a_n = (-1)^n$

$-1, 1, -1, 1, -1$

Not an arithmetic sequence

19. $a_n = \frac{(-1)^n 3}{n}$

$-3, \frac{3}{2}, -1, \frac{3}{4}, -\frac{3}{5}$

Not an arithmetic sequence

21. $a_1 = 1, d = 3$

$a_n = a_1 + (n-1)d = 1 + (n-1)(3) = 3n - 2$

23. $a_1 = 100, d = -8$

$a_n = a_1 + (n-1)d = 100 + (n-1)(-8)$

$= -8n + 108$

25. $4, \frac{3}{2}, -1, -\frac{7}{2}, \ldots$

$d = -\frac{5}{2}$

$a_n = a_1 + (n-1)d = 4 + (n-1)\left(-\frac{5}{2}\right) = -\frac{5}{2}n + \frac{13}{2}$

27. $a_1 = 5, a_4 = 15$

$a_4 = a_1 + 3d \Rightarrow 15 = 5 + 3d \Rightarrow d = \frac{10}{3}$

$a_n = a_1 + (n-1)d = 5 + (n-1)\left(\frac{10}{3}\right) = \frac{10}{3}n + \frac{5}{3}$

29. $a_3 = 94, a_6 = 85$

$a_6 = a_3 + 3d \Rightarrow 85 = 94 + 3d \Rightarrow d = -3$

$a_1 = a_3 - 2d \Rightarrow a_1 = 94 - 2(-3) = 100$

$a_n = a_1 + (n-1)d = 100 + (n-1)(-3)$

$= -3n + 103$

31. $a_1 = 5, d = 6$

$a_1 = 5$

$a_2 = 5 + 6 = 11$

$a_3 = 11 + 6 = 17$

$a_4 = 17 + 6 = 23$

$a_5 = 23 + 6 = 29$

33. $a_1 = -\frac{13}{5}, d = -\frac{2}{5}$

$a_1 = -\frac{13}{5}$

$a_2 = -\frac{13}{5} - \frac{2}{5} = -\frac{15}{5} = -3$

$a_3 = -3 - \frac{2}{5} = -\frac{17}{5}$

$a_4 = -\frac{17}{5} - \frac{2}{5} = -\frac{19}{5}$

$a_5 = -\frac{19}{5} - \frac{2}{5} = -\frac{21}{5}$

35. $a_1 = 2, a_{12} = 46$

$46 = 2 + (12 - 1)d$

$44 = 11d$

$4 = d$

$a_1 = 2$

$a_2 = 2 + 4 = 6$

$a_3 = 6 + 4 = 10$

$a_4 = 10 + 4 = 14$

$a_5 = 14 + 4 = 18$

37. $a_8 = 26, a_{12} = 42$

$a_{12} = a_8 + 4d$

$42 = 26 + 4d \Rightarrow d = 4$

$a_8 = a_1 + 7d$

$26 = a_1 + 28 \Rightarrow a_1 = -2$

$a_1 = -2$

$a_2 = -2 + 4 = 2$

$a_3 = 2 + 4 = 6$

$a_4 = 6 + 4 = 10$

$a_5 = 10 + 4 = 14$

39. $a_1 = 15, a_{n+1} = a_n + 4$

$a_2 = 15 + 4 = 19$

$a_3 = 19 + 4 = 23$

$a_4 = 23 + 4 = 27$

$a_5 = 27 + 4 = 31$

41. $a_1 = \frac{5}{8}, a_{n+1} = a_n - \frac{1}{8}$

$a_1 = \frac{5}{8}$

$a_2 = \frac{5}{8} - \frac{1}{8} = \frac{1}{2}$

$a_3 = \frac{1}{2} - \frac{1}{8} = \frac{3}{8}$

$a_4 = \frac{3}{8} - \frac{1}{8} = \frac{1}{4}$

$a_5 = \frac{1}{4} - \frac{1}{8} = \frac{1}{8}$

43. $a_1 = 5, a_2 = 11 \Rightarrow d = 11 - 5 = 6$

$a_n = a_1 + (n-1)d \Rightarrow a_{10} = 5 + 9(6) = 59$

45. $a_1 = 4.2, a_2 = 6.6 \Rightarrow d = 6.6 - 4.2 = 2.4$

$a_n = a_1 + (n-1)d \Rightarrow a_7 = 4.2 + 6(2.4) = 18.6$

47. $S_{10} = \frac{10}{2}(2 + 20) = 110$

49. $S_5 = \frac{5}{2}(-1 + (-9)) = -25$

51. $a_n = 2n - 1$

$a_1 = 1, a_{100} = 199$

$S_{100} = \frac{100}{2}(1 + 199) = 10{,}000$

53. 8, 20, 32, 44,...

$a_1 = 8, d = 12, n = 10$

$a_{10} = 8 + 9(12) = 116$

$S_{10} = \frac{10}{2}(8 + 116) = 620$

55. 4.2, 3.7, 3.2, 2.7,...

$a_1 = 4.2, d = -0.5, n = 12$

$a_{12} = 4.2 + 11(-0.5) = -1.3$

$S_{12} = \frac{12}{2}[4.2 + (-1.3)] = 17.4$

57. $a_1 = 100, a_{25} = 220, n = 25$

$S_n = \dfrac{n}{2}[a_1 + a_n]$

$S_{25} = \dfrac{25}{2}(100 + 220) = 4000$

59. $a_1 = 1, a_{50} = 50, n = 50$

$\sum_{n=1}^{50} n = \frac{50}{2}(1 + 50) = 1275$

61. $\sum_{n=11}^{30} n - \sum_{n=1}^{10} n = \frac{20}{2}(11 + 30) - \frac{10}{2}(1 + 10) = 355$

63. $a_1 = 9, a_{500} = 508, n = 500$

$\sum_{n=1}^{500} (n + 8) = \frac{500}{2}(9 + 508) = 129{,}250$

65. $a_n = -\frac{3}{4}n + 8$

$d = -\frac{3}{4}$ so the sequence is decreasing and $a_1 = 7\frac{1}{4}$.

Matches (b).

66. $a_n = 3n - 5$

$d = 3$ so the sequence is increasing and $a_1 = -2$.

Matches (d).

67. $a_n = 2 + \frac{3}{4}n$

$d = \frac{3}{4}$ so the sequence is increasing and $a_1 = 2\frac{3}{4}$.

Matches (c).

68. $a_n = 25 - 3n$

$d = -3$ so the sequence is decreasing and $a_1 = 22$.

Matches (a).

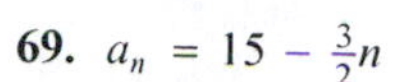

69. $a_n = 15 - \frac{3}{2}n$

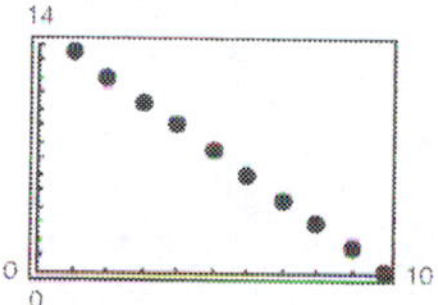

71. $a_n = 0.2n + 3$

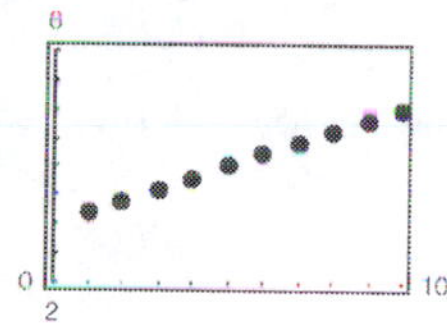

73. $\sum_{n=0}^{50} (50 - 2n) = 0$

75. $\sum_{i=1}^{60} \left(250 - \frac{2}{5}i\right) = 14{,}268$

77. $a_1 = 15, d = 3, n = 36$

$a_{36} = 15 + 35(3) = 120$

$S_{36} = \frac{36}{2}(15 + 120) = 2430$ seats

79. (a) $a_1 = 32{,}500, d = 1500$

$a_6 = a_1 + 5d = 32{,}500 + 5(1500) = \$40{,}000$

(b) $S_6 = \frac{6}{2}[32{,}500 + 40{,}000] = \$217{,}500$

81. $a_1 = 16, a_2 = 48, a_3 = 80, a_4 = 112$

$d = 32$

$a_n = dn + c = 32n + c$

$c = a_1 - d = 16 - 32 = -16$

$a_n = 32n - 16$

$\text{Distance} = \sum_{n=1}^{7} (32n - 16) = 784 \text{ ft}$

83. $a_1 = 15{,}000$

$d = 5{,}000$

$n = 1, \ldots, 10$

$a_n = dn + c = 5000n + c$

$c = a_1 - d = 15{,}000 - 5000 = 10{,}000$

$a_n = 5000n + 10{,}000$

$\text{Total sales} = \sum_{n=1}^{10} (5000n + 10{,}000)$

$= \frac{10}{2}(15{,}000 + 60{,}000)$

$= \$375{,}000$

85. (a)

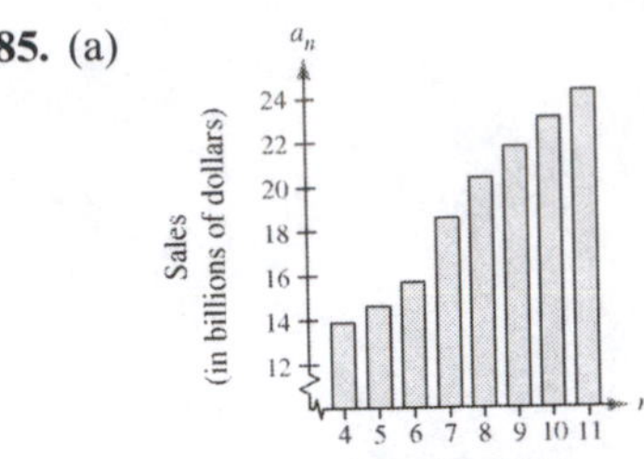

(b) The increase in annual sales is about $\frac{24.3 - 13.9}{7} \approx \$1.49$ billion each year from 2004 to 2011. Because the common difference is $d = 1.49$ and the fourth term is $a_4 = 13.9$, the arithmetic sequence is $a_n = 7.9 + 1.49n$.

(c)

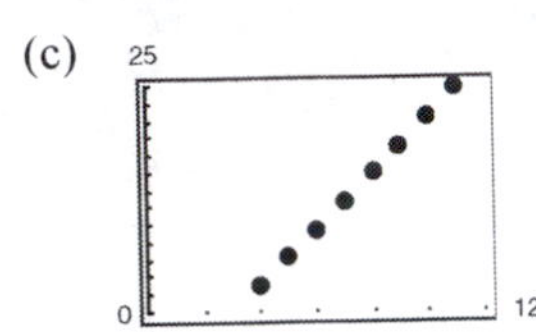

(d)
$$\begin{aligned}\sum_{n=4}^{11}(7.9 + 1.49\,n) &= \sum_{n=1}^{11}(7.9 + 1.49\,n) - \sum_{n=1}^{3}(7.9 + 1.49\,n)\\ &= \sum_{n=1}^{11} 7.9 + 1.49\sum_{n=1}^{11} n - \sum_{n=1}^{3} 7.9 + 1.49\sum_{n=1}^{3} n\\ &= \left(7.9(11) + 1.49\left[\frac{11(12)}{2}\right]\right) - \left(7.9(3) + 1.49\left[\frac{3(4)}{2}\right]\right)\\ &= 152.6\end{aligned}$$

The total sales from 2004 to 2011 is \$152.6 billion.

87. True; given a_1 and a_2 then $d = a_2 - a_1$ and $a_n = a_1 + (n - 1)d$.

89. $a_1 = x, d = 2x$

$a_n = x + (n - 1)2x$

$a_n = 2xn - x$

$a_1 = 2x(1) - x = x$ $\quad a_6 = 2x(6) - x = 11x$

$a_2 = 2x(2) - x = 3x$ $\quad a_7 = 2x(7) - x = 13x$

$a_3 = 2x(3) - x = 5x$ $\quad a_8 = 2x(8) - x = 15x$

$a_4 = 2x(4) - x = 7x$ $\quad a_9 = 2x(9) - x = 17x$

$a_5 = 2x(5) - x = 9x$ $\quad a_{10} = 2x(10) - x = 19x$

91. (a) $a_n = 2 + 3n$

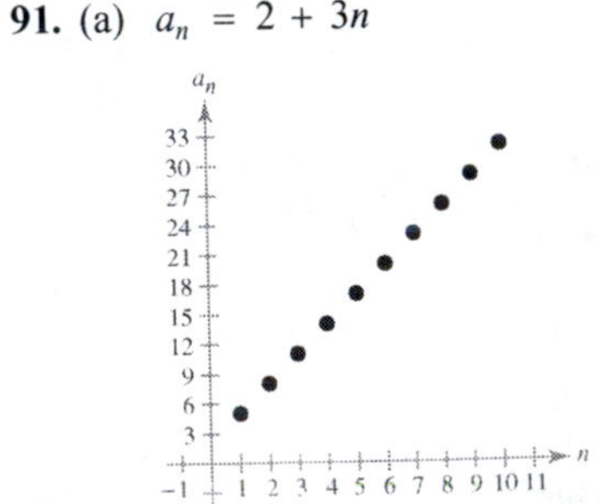

(b) $y = 3x + 2$

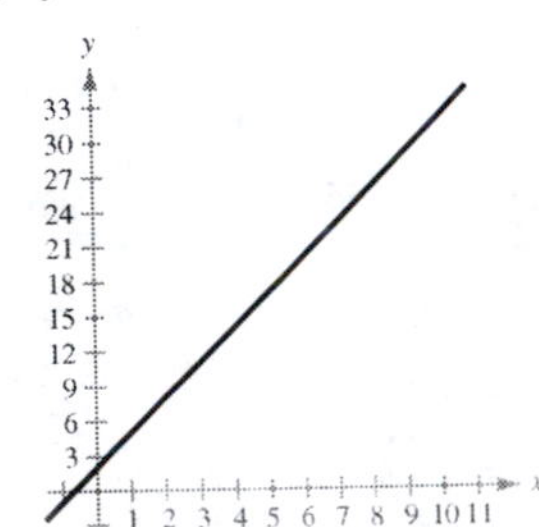

(c) The graph of $a_n = 2 + 3n$ contains only points at the positive integers. The graph of $y = 3x + 2$ is a solid line which contains these points.

(d) The slope $m = 3$ is equal to the common difference $d = 3$. In general, these should be equal.

93. (a) $1 + 3 = 4$

$1 + 3 + 5 = 9$

$1 + 3 + 5 + 7 = 16$

$1 + 3 + 5 + 7 + 9 = 36$

$1 + 3 + 5 + 7 + 9 + 11 = 36$

(b) $S_n = n^2$

$S_7 = 1 + 3 + 5 + 7 + 9 + 11 + 13 = 49 = 7^2$

(c) $S_n = \frac{n}{2}[1 + (2n - 1)] = \frac{n}{2}(2n) = n^2$

Section 11.3 Geometric Sequences and Series

1. geometric; common

3. $S_n = a_1\left(\frac{1 - r^n}{1 - r}\right)$

5. $2, 10, 50, 250, \ldots$

Geometric sequence, $r = 5$

7. $\frac{1}{8}, \frac{1}{4}, \frac{1}{2}, 1, \ldots$

Geometric sequence, $r = 2$

9. $1, \frac{1}{2}, \frac{1}{3}, \frac{1}{4}, \ldots$

Not a geometric sequence

11. $1, -\sqrt{7}, 7, -7\sqrt{7}, \ldots$

Geometric sequence, $r = -\sqrt{7}$

13. $a_1 = 4, r = 3$

$a_1 = 4$

$a_2 = 4(3) = 12$

$a_3 = 12(3) = 36$

$a_4 = 36(3) = 108$

$a_5 = 108(3) = 324$

15. $a_1 = 1, r = \frac{1}{2}$

$a_1 = 1$

$a_2 = 1\left(\frac{1}{2}\right) = \frac{1}{2}$

$a_3 = \frac{1}{2}\left(\frac{1}{2}\right) = \frac{1}{4}$

$a_4 = \frac{1}{4}\left(\frac{1}{2}\right) = \frac{1}{8}$

$a_5 = \frac{1}{8}\left(\frac{1}{2}\right) = \frac{1}{16}$

17. $a_1 = 1, r = e$

$a_1 = 1$

$a_2 = 1(e) = e$

$a_3 = (e)(e) = e^2$

$a_4 = (e^2)(e) = e^3$

$a_5 = (e^3)(e) = e^4$

19. $a_1 = 3, r = \sqrt{5}$

$a_1 = 3$

$a_2 = 3\left(\sqrt{5}\right)^1 = 3\sqrt{5}$

$a_3 = 3\left(\sqrt{5}\right)^2 = 15$

$a_4 = 3\left(\sqrt{5}\right)^3 = 15\sqrt{5}$

$a_5 = 3\left(\sqrt{5}\right)^4 = 75$

21. $a_1 = 2, r = \frac{x}{4}$

$a_1 = 2$

$a_2 = 2\left(\frac{x}{4}\right) = \frac{x}{2}$

$a_3 = \left(\frac{x}{2}\right)\left(\frac{x}{4}\right) = \frac{x^2}{8}$

$a_4 = \left(\frac{x^2}{8}\right)\left(\frac{x}{4}\right) = \frac{x^3}{32}$

$a_5 = \left(\frac{x^3}{32}\right)\left(\frac{x}{4}\right) = \frac{x^4}{128}$

23. $a_1 = 64, a_{k+1} = \frac{1}{2}a_k$

$a_1 = 64$

$a_2 = \frac{1}{2}(64) = 32$

$a_3 = \frac{1}{2}(32) = 16$

$a_4 = \frac{1}{2}(16) = 8$

$a_5 = \frac{1}{2}(8) = 4$

$r = \frac{1}{2}$

$a_n = 64\left(\frac{1}{2}\right)^{n-1} = 128\left(\frac{1}{2}\right)^n$

25. $a_1 = 9, a_{k+1} = 2a_k$

$a_1 = 9$

$a_2 = 2(9) = 18$

$a_3 = 2(18) = 36$

$a_4 = 2(36) = 72$

$a_5 = 2(72) = 144$

$r = 2$

$a_n = \frac{9}{2}(2)^n$

27. $a_1 = 6, a_{k+1} = -\frac{3}{2}a_k$

$a_1 = 6$

$a_2 = -\frac{3}{2}(6) = -9$

$a_3 = -\frac{3}{2}(-9) = \frac{27}{2}$

$a_4 = -\frac{3}{2}\left(\frac{27}{2}\right) = -\frac{81}{4}$

$a_5 = -\frac{3}{2}\left(-\frac{81}{4}\right) = \frac{243}{8}$

$r = -\frac{3}{2}$

$a_n = 6\left(-\frac{3}{2}\right)^{n-1}$ or $a_n = -4\left(-\frac{3}{2}\right)^n$

29. $a_1 = 4, r = \frac{1}{2}, n = 10$

$a_n = a_1r^{n-1} = 4\left(\frac{1}{2}\right)^{n-1}$

$a_{10} = 4\left(\frac{1}{2}\right)^9 = \left(\frac{1}{2}\right)^7 = \frac{1}{128}$

31. $a_1 = 6, r = -\frac{1}{3}, n = 12$

$a_n = a_1r^{n-1} = 6\left(-\frac{1}{3}\right)^{n-1}$

$a_{12} = 6\left(-\frac{1}{3}\right)^{11} = -\frac{2}{3^{10}} = -\frac{2}{59,049}$

33. $a_1 = 100, r = e^x, n = 9$

$a_n = a_1r^{n-1} = 100\left(e^x\right)^{n-1}$

$a_9 = 100\left(e^x\right)^8 = 100e^{8x}$

35. $a_1 = 1, r = \sqrt{2}, n = 12$

$a_n = 1\left(\sqrt{2}\right)^{n-1} = \left(\sqrt{2}\right)^{n-1}$

$a_{12} = \left(\sqrt{2}\right)^{12-1} = 32\sqrt{2}$

37. $a_1 = 500, r = 1.02, n = 40$

$a_n = a_1r^{n-1} = 500(1.02)^{n-1}$

$a_{40} = 500(1.02)^{39} \approx 1082.372$

39. $11, 33, 99, \ldots \Rightarrow r = 3$

$a_n = 11(3)^{n-1}$

$a_9 = 11(3)^{9-1} = 72,171$

41. $\frac{1}{2}, -\frac{1}{8}, \frac{1}{32}, -\frac{1}{128}, \ldots \Rightarrow r = -\frac{1}{4}$

$a_n = \frac{1}{2}\left(-\frac{1}{4}\right)^{n-1}$

$a_8 = \frac{1}{2}\left(-\frac{1}{4}\right)^{8-1} = -\frac{1}{32,768}$

43. $a_1 = 16, a_4 = \frac{27}{4}$

$a_4 = a_1r^3$

$\frac{27}{4} = 16r^3$

$\frac{27}{64} = r^3$

$\frac{3}{4} = r$

$a_n = 16\left(\frac{3}{4}\right)^{n-1}$

$a_3 = 16\left(\frac{3}{4}\right)^2 = 9$

45. $a_4 = -18, a_7 = \frac{2}{3}$

$a_7 = a_4r^3$

$\frac{2}{3} = -18r^3$

$-\frac{1}{27} = r^3$

$-\frac{1}{3} = r$

$a_6 = \frac{a_7}{r} = \frac{2/3}{-1/3} = -2$

47. $a_n = 18\left(\frac{2}{3}\right)^{n-1}$

$a_1 = 18$ and $r = \frac{2}{3}$

Because $0 < r < 1$, the sequence is decreasing.

Matches (a).

48. $a_n = 18\left(-\frac{2}{3}\right)^{n-1}$

Because $r = \left(-\frac{2}{3}\right) > -1$, the sequence alternates as it approaches 0.

Matches (c).

49. $a_n = 18\left(\frac{3}{2}\right)^{n-1}$

Because $a_1 = 18$ and $r = \frac{3}{2} > 1$, the sequence is increasing.

Matches (b).

50. $a_n = 18\left(-\frac{3}{2}\right)^{n-1}$

Because $r = \left(-\frac{3}{2}\right) < -1$, the sequence alternates as it approaches ∞.

Matches (d).

51. $a_n = 10(1.5)^{n-1}$

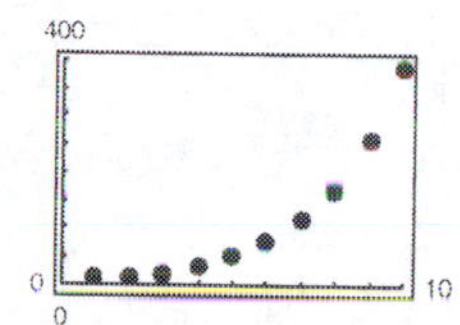

53. $a_n = 20(-1.25)^{n-1}$

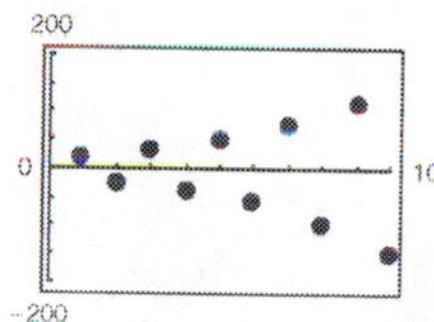

55. $\sum_{n=1}^{7} 4^{n-1} = 1 + 4^1 + 4^2 + 4^3 + 4^4 + 4^5 + 4^6 \Rightarrow a_1 = 1, r = 4$

$$S_7 = \frac{1(1 - 4^7)}{1 - 4} = 5461$$

57. $\sum_{n=1}^{6} (-7)^{n-1} = 1 + (-7) + (-7)^2 + \cdots + (-7)^5 \Rightarrow a_1 = 1, r = -7$

$$S_6 = \frac{1\left(1 - (-7)^6\right)}{1 - (-7)} = -14{,}706$$

59. $\sum_{i=1}^{7} 64\left(-\frac{1}{2}\right)^{i-1} = 64 + 64\left(-\frac{1}{2}\right)^1 + 64\left(-\frac{1}{2}\right)^2 + \cdots + 64\left(-\frac{1}{2}\right)^6 \Rightarrow a_1 = 64, r = -\frac{1}{2}$

$$S_7 = 64\left[\frac{1 - \left(-\frac{1}{2}\right)^7}{1 - \left(-\frac{1}{2}\right)}\right] = \frac{128}{3}\left[1 - \left(-\frac{1}{2}\right)^7\right] = 43$$

61. $\sum_{n=0}^{20} 3\left(\frac{3}{2}\right)^n = \sum_{n=1}^{21} 3\left(\frac{3}{2}\right)^{n-1} = 3 + 3\left(\frac{3}{2}\right)^1 + 3\left(\frac{3}{2}\right)^2 + \cdots + 3\left(\frac{3}{2}\right)^{20} \Rightarrow a_1 = 3, r = \frac{3}{2}$

$$S_{21} = 3\left[\frac{1 - \left(\frac{3}{2}\right)^{21}}{1 - \frac{3}{2}}\right] = -6\left[1 - \left(\frac{3}{2}\right)^{21}\right] \approx 29{,}921.311$$

63. $\sum_{n=0}^{20} 10\left(\frac{1}{5}\right)^n = 10 + \sum_{n=1}^{20} 10\left(\frac{1}{5}\right)^n = 10 + \left[10\left(\frac{1}{5}\right)^1 + 10\left(\frac{1}{5}\right)^2 + 10\left(\frac{1}{5}\right)^3 + \cdots + 10\left(\frac{1}{5}\right)^{20}\right] \Rightarrow a_1 = 2, r = \frac{1}{5}$

$$S_{21} = 10 + 2\left[\frac{1-\left(\frac{1}{5}\right)^{20}}{1-\left(\frac{1}{5}\right)}\right] = 10 + \frac{5}{2}\left[1-\left(\frac{1}{5}\right)^{20}\right] \approx 12.500$$

65. $\sum_{n=0}^{40} 2\left(-\frac{1}{4}\right)^n = 2 + 2\left(-\frac{1}{4}\right) + 2\left(-\frac{1}{4}\right)^2 + \cdots + 2\left(-\frac{1}{4}\right)^{40} \Rightarrow a_1 = 2, r = -\frac{1}{4}, n = 41$

$$S_{41} = 2\left[\frac{1-\left(-\frac{1}{4}\right)^{41}}{1-\left(-\frac{1}{4}\right)}\right] = \frac{8}{5}\left[1-\left(-\frac{1}{4}\right)^{41}\right] \approx 1.6 = \frac{8}{5}$$

67. $\sum_{i=1}^{100} 15\left(\frac{2}{3}\right)^{i-1} = 15 + 15\left(\frac{2}{3}\right)^1 + 15\left(\frac{2}{3}\right)^2 + \cdots + 15\left(\frac{2}{3}\right)^{99} \Rightarrow a_1 = 15, r = \frac{2}{3}$

$$S_{100} = 15\left[\frac{1-\left(\frac{2}{3}\right)^{100}}{1-\left(\frac{2}{3}\right)}\right] = 45\left[1-\left(\frac{2}{3}\right)^{100}\right] \approx 45.000$$

69. $10 + 30 + 90 + \cdots + 7290$

$r = 3$ and $7290 = 10(3)^{n-1}$

$729 = 3^{n-1}$

$6 = n - 1 \Rightarrow n = 7$

So, the sum can be written as $\sum_{n=1}^{7} 10(3)^{n-1}$.

71. $0.1 + 0.4 + 1.6 + \cdots + 102.4$

$r = 4$ and $102.4 = 0.1(4)^{n-1}$

$1024 = 4^{n-1} \Rightarrow 5 = n - 1 \Rightarrow n = 6$

So, the sum can be written as $\sum_{n=1}^{6} 0.1(4)^{n-1}$.

73. $\sum_{n=0}^{\infty}\left(\frac{1}{2}\right)^n = 1 + \left(\frac{1}{2}\right)^1 + \left(\frac{1}{2}\right)^2 + \cdots$

$a_1 = 1, r = \frac{1}{2}$

$$\sum_{n=0}^{\infty}\left(\frac{1}{2}\right)^n = \frac{a_1}{1-r} = \frac{1}{1-\left(\frac{1}{2}\right)} = 2$$

75. $\sum_{n=0}^{\infty} 2\left(-\frac{2}{3}\right)^n = 2 + 2\left(-\frac{2}{3}\right)^1 + 2\left(-\frac{2}{3}\right)^2 + \cdots$

$a_1 = 2, r = -\frac{2}{3}$

$$\sum_{n=0}^{\infty} 2\left(-\frac{2}{3}\right)^n = \frac{a_1}{1-r} = \frac{2}{1-\left(-\frac{2}{3}\right)} = \frac{6}{5}$$

77. $8 + 6 + \frac{9}{2} + \frac{27}{8} + \cdots = \sum_{n=0}^{\infty} 8\left(\frac{3}{4}\right)^n = \frac{8}{1-\frac{3}{4}} = 32$

79. $\frac{1}{9} - \frac{1}{3} + 1 - 3 + \cdots = \sum_{n=0}^{\infty} \frac{1}{9}(-3)^n$

The sum is undefined because $|r| = |-3| = 3 > 1$.

81. $0.\overline{36} = \sum_{n=0}^{\infty} 0.36(0.01)^n = \frac{0.36}{1-0.01} = \frac{0.36}{0.99} = \frac{36}{99} = \frac{4}{11}$

83. $f(x) = 6\left[\dfrac{1-(0.5)^x}{1-(0.5)}\right], \sum_{n=0}^{\infty} 6\left(\dfrac{1}{2}\right)^n = \dfrac{6}{1-\dfrac{1}{2}} = 12$

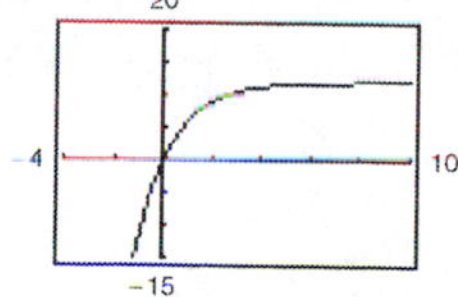

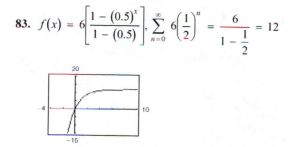

The horizontal asymptote of $f(x)$ is $y = 12$.

This corresponds to the sum of the series.

85. $V_5 = 175{,}000(0.70)^5 = \$29{,}412.25$

87. Let $N = 12t$ be the total number of deposits.

$$\begin{aligned}
A &= P\left(1+\frac{r}{12}\right) + P\left(1+\frac{r}{12}\right)^2 + \cdots + P\left(1+\frac{r}{12}\right)^N \\
&= \left(1+\frac{r}{12}\right)\left[P + P\left(1+\frac{r}{12}\right) + \cdots + P\left(1+\frac{r}{12}\right)^{N-1}\right] \\
&= P\left(1+\frac{r}{12}\right)\sum_{n=1}^{N}\left(1+\frac{r}{12}\right)^{n-1} \\
&= P\left(1+\frac{r}{12}\right)\left[\frac{1-\left(1+\frac{r}{12}\right)^N}{1-\left(1+\frac{r}{12}\right)}\right] \\
&= P\left(1+\frac{r}{12}\right)\left(-\frac{12}{r}\right)\left[1-\left(1+\frac{r}{12}\right)^N\right] \\
&= P\left(\frac{12}{r}+1\right)\left[-1+\left(1+\frac{r}{12}\right)^N\right] \\
&= P\left[\left(1+\frac{r}{12}\right)^N - 1\right]\left(1+\frac{12}{r}\right) \\
&= P\left[\left(1+\frac{r}{12}\right)^{12t} - 1\right]\left(1+\frac{12}{r}\right)
\end{aligned}$$

89. $\sum_{n=0}^{\infty} 400(0.75)^n = \dfrac{400}{1-0.75} = \$1600$

91. $27^2\left(\frac{1}{9}\right) + 27^2\left(\frac{1}{9}\right)\left(\frac{8}{9}\right) + 27^2\left(\frac{1}{9}\right)\left(\frac{8}{9}\right)^2 + 27^2\left(\frac{1}{9}\right)\left(\frac{8}{9}\right)^3 = \sum_{n=0}^{3} 27^2\left(\frac{1}{9}\right)\left(\frac{8}{9}\right)^n = \frac{2465}{9} = 273\frac{8}{9}$ square inches

93. (a) Downward: $850 + 0.75(850) + (0.75)^2(850) + \cdots + (0.75)^9(850) = \sum_{n=0}^{9} 850(0.75)^n$

≈ 3208.53 feet

Upward: $0.75(850) + (0.75)^2(850) + \cdots + (0.75)^{10}(850) = \sum_{n=0}^{9} (0.75)(850)(0.75)^n$

$= \sum_{n=0}^{9} 637.5(0.75)^n \approx 2406.4$ feet

Total distance: $3208.53 + 2406.4 = 5614.93$ feet

(b) $\sum_{n=0}^{\infty} 850(0.75)^n + \sum_{n=0}^{\infty} 637.5(0.75)^n = \dfrac{850}{1-0.75} + \dfrac{637.5}{1-0.75} = 5950$ feet

95. False. A sequence is geometric if the ratios of consecutive terms are the same.

97. $y = \left(\dfrac{1 - r^x}{1 - r}\right)$

(a)

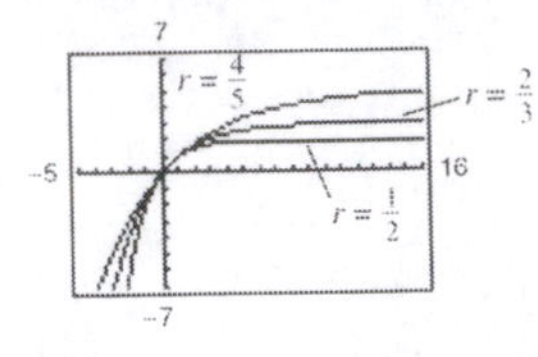

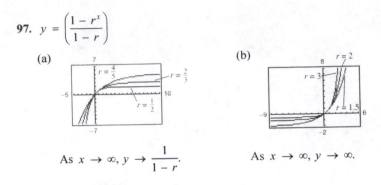

As $x \to \infty$, $y \to \dfrac{1}{1 - r}$.

(b)

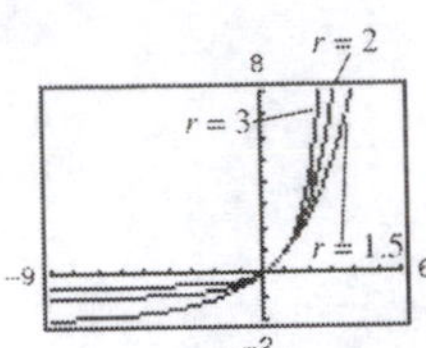

As $x \to \infty$, $y \to \infty$.

11.4 Mathematical Induction

1. mathematical induction

3. arithmetic

5. $P_k = \dfrac{5}{k(k + 1)}$

$$P_{k+1} = \frac{5}{(k + 1)[(k + 1) + 1]} = \frac{5}{(k + 1)(k + 2)}$$

7. $P_k = \dfrac{k^2(k + 3)^2}{6}$

$$P_{k+1} = \frac{(k + 1)^2[(k + 1) + 3]^2}{6} = \frac{(k + 1)^2(k + 4)^2}{6}$$

9. $P_k = \dfrac{3}{(k + 2)(k + 3)}$

$$P_{k+1} = \frac{3}{[(k + 1) + 2][(k + 1) + 3]} = \frac{3}{(k + 3)(k + 4)}$$

11. 1. When $n = 1$, $S_1 = 2 = 1(1 + 1)$.

2. Assume that

$$S_k = 2 + 4 + 6 + 8 + \cdots + 2k = k(k + 1).$$

Then,

$$\begin{aligned} S_{k+1} &= 2 + 4 + 6 + 8 + \cdots + 2k + 2(k + 1) \\ &= S_k + 2(k + 1) = k(k + 1) + 2(k + 1) = (k + 1)(k + 2). \end{aligned}$$

So, we conclude that the formula is valid for all positive integer values of n.

13. 1. When $n = 1$, $S_1 = 2 = \frac{1}{2}(5(1) - 1)$.

2. Assume that

$$S_k = 2 + 7 + 12 + 17 + \cdots + (5k - 3) = \frac{k}{2}(5k - 1).$$

Then,

$$\begin{aligned} S_{k+1} &= 2 + 7 + 12 + 17 + \cdots + (5k - 3) + [5(k + 1) - 3] \\ &= S_k + (5k + 5 - 3) = \frac{k}{2}(5k - 1) + 5k + 2 \\ &= \frac{5k^2 - k + 10k + 4}{2} = \frac{5k^2 + 9k + 4}{2} \\ &= \frac{(k + 1)(5k + 4)}{2} = \frac{(k + 1)}{2}[5(k + 1) - 1]. \end{aligned}$$

So, we conclude that this formula is valid for all positive integer values of n.

15. 1. When $n = 1$, $S_1 = 1 = 2^1 - 1$.

2. Assume that

$$S_k = 1 + 2 + 2^2 + 2^3 + \cdots + 2^{k-1} = 2^k - 1.$$

Then,

$$S_{k+1} = 1 + 2 + 2^2 + 2^3 + \cdots + 2^{k-1} + 2^k = S_k + 2^k = 2^k - 1 + 2^k = 2(2^k) - 1 = 2^{k+1} - 1.$$

So, we conclude that this formula is valid for all positive integer values of n.

17. 1. When $n = 1$, $S_1 = 1 = \dfrac{1(1 + 1)}{2}$.

2. Assume that

$$S_k = 1 + 2 + 3 + 4 + \cdots + k = \frac{k(k + 1)}{2}.$$

Then,

$$S_{k+1} = 1 + 2 + 3 + 4 + \cdots + k + (k + 1) = S_k + (k + 1) = \frac{k(k + 1)}{2} + \frac{2(k + 1)}{2} = \frac{(k + 1)(k + 2)}{2}.$$

So, we conclude that this formula is valid for all positive integer values of n.

19. 1. When $n = 1$, $S_1 = 1^2 = \dfrac{1(2(1) - 1)(2(1) + 1)}{3}$

2. Assume that

$$S_k = 1^2 + 3^2 + \cdots + (2k - 1)^2 = \frac{k(2k - 1)(2k + 1)}{3}$$

Then,

$$\begin{aligned}
S_{k+1} &= 1^2 + 3^2 + \cdots + (2k - 1)^2 + (2k + 1)^2 \\
&= S_k + (2k + 1)^2 = \frac{k(2k - 1)(2k + 1)}{3} + (2k + 1)^2 \\
&= (2k + 1)\left[\frac{k(2k - 1)}{3} + (2k + 1)\right] = \frac{2k + 1}{3}\left[2k^2 - k + 6k + 3\right] \\
&= \frac{2k + 1}{3}(2k + 3)(k + 1) = \frac{(k + 1)(2(k + 1) - 1)(2(k + 1) + 1)}{3}
\end{aligned}$$

So, we conclude that this formula is valid for all positive integer values of n.

21. 1. When $n = 1$, $S_1 = 1 = \dfrac{(1)^2(1+1)^2\left(2(1)^2 + 2(1) - 1\right)}{12}$.

2. Assume that

$$S_k = \sum_{i=1}^{k} i^5 = \frac{k^2(k+1)^2(2k^2 + 2k - 1)}{12}.$$

Then,

$$\begin{aligned}
S_{k+1} &= \sum_{i=1}^{k+1} i^5 = \left(\sum_{i=1}^{k} i^5\right) + (k+1)^5 \\
&= \frac{k^2(k+1)^2(2k^2+2k-1)}{12} + \frac{12(k+1)^5}{12} \\
&= \frac{(k+1)^2\left[k^2(2k^2+2k-1) + 12(k+1)^3\right]}{12} \\
&= \frac{(k+1)^2\left[2k^4 + 2k^3 - k^2 + 12(k^3 + 3k^2 + 3k + 1)\right]}{12} \\
&= \frac{(k+1)^2\left[2k^4 + 14k^3 + 35k^2 + 36k + 12\right]}{12} \\
&= \frac{(k+1)^2(k^2+4k+4)(2k^2+6k+3)}{12} \\
&= \frac{(k+1)^2(k+2)^2\left[2(k+1)^2 + 2(k+1) - 1\right]}{12}.
\end{aligned}$$

So, we conclude that this formula is valid for all positive integer values of n.

Note: The easiest way to complete the last two steps is to "work backwards." Start with the desired expression for S_{k+1} and multiply out to show that it is equal to the expression you found for $S_k + (k+1)^5$.

23. 1. When $n = 1$, $S_1 = 2 = \dfrac{1(2)(3)}{3}$.

2. Assume that

$$S_k = 1(2) + 2(3) + 3(4) + \cdots + k(k+1) = \frac{k(k+1)(k+2)}{3}.$$

Then,

$$\begin{aligned}
S_{k+1} &= 1(2) + 2(3) + 3(4) + \cdots + k(k+1) + (k+1)(k+2) \\
&= S_k + (k+1)(k+2) = \frac{k(k+1)(k+2)}{3} + \frac{3(k+1)(k+2)}{3} = \frac{(k+1)(k+2)(k+3)}{3}.
\end{aligned}$$

So, we conclude that this formula is valid for all positive integer values of n.

25. 1. When $n = 4$, $4! = 24$ and $2^4 = 16$, thus $4! > 2^4$.

2. Assume

$k! > 2^k, k > 4$.

Then,

$(k+1)! = k!(k+1) > 2^k(2)$ since $k! > 2^k$ and $k + 1 > 2$.

Thus, $(k+1)! > 2^{k+1}$.

So, by extended mathematical induction, the inequality is valid for all integers n such that $n \ge 4$.

27. 1. When $n = 2$, $\frac{1}{\sqrt{1}} + \frac{1}{\sqrt{2}} \approx 1.707$ and $\sqrt{2} \approx 1.414$, thus $\frac{1}{\sqrt{1}} + \frac{1}{\sqrt{2}} > \sqrt{2}$.

2. Assume that

$$\frac{1}{\sqrt{1}} + \frac{1}{\sqrt{2}} + \frac{1}{\sqrt{3}} + \cdots + \frac{1}{\sqrt{k}} > \sqrt{k},\ k > 2.$$

Then,

$$\frac{1}{\sqrt{1}} + \frac{1}{\sqrt{2}} + \frac{1}{\sqrt{3}} + \cdots + \frac{1}{\sqrt{k}} + \frac{1}{\sqrt{k+1}} > \sqrt{k} + \frac{1}{\sqrt{k+1}}.$$

Now it is sufficient to show that

$$\sqrt{k} + \frac{1}{\sqrt{k+1}} > \sqrt{k+1},\ k > 2,$$

or equivalently $\left(\text{multiplying by } \sqrt{k+1}\right)$,

$$\sqrt{k}\sqrt{k+1} + 1 > k + 1.$$

This is true because

$$\sqrt{k}\sqrt{k+1} + 1 > \sqrt{k}\sqrt{k} + 1 = k + 1.$$

Therefore,

$$\frac{1}{\sqrt{1}} + \frac{1}{\sqrt{2}} + \frac{1}{\sqrt{3}} + \cdots + \frac{1}{\sqrt{k}} + \frac{1}{\sqrt{k+1}} > \sqrt{k+1}.$$

So, by extended mathematical induction, the inequality is valid for all integers n such that $n \ge 2$.

29. $(1 + a)^n \ge na,\ n \ge 1$ and $a > 0$

Because a is positive, then all of the terms in the binomial expansion are positive.

$(1 + a)^n = 1 + na + \cdots + na^{n-1} + a^n > na$

31. 1. When $n = 1$, $(ab)^1 = a^1b^1 = ab$.

2. Assume that $(ab)^k = a^kb^k$.

$$\begin{aligned}\text{Then, } (ab)^{k+1} &= (ab)^k(ab)\\ &= a^kb^kab\\ &= a^{k+1}b^{k+1}.\end{aligned}$$

So, $(ab)^n = a^nb^n$.

33. 1. When $n = 2$, $(x_1x_2)^{-1} = \frac{1}{x_1x_2} = \frac{1}{x_1}\cdot\frac{1}{x_2} = x_1^{-1}x_2^{-1}$.

2. Assume that

$$(x_1x_2x_3\cdots x_k)^{-1} = x_1^{-1}x_2^{-1}x_3^{-1}\cdots x_k^{-1}.$$

Then,

$$\begin{aligned}(x_1x_2x_3\cdots x_kx_{k+1})^{-1} &= \left[(x_1x_2x_3\cdots x_k)x_{k+1}\right]^{-1}\\ &= (x_1x_2x_3\cdots x_k)^{-1}x_{k+1}^{-1}\\ &= x_1^{-1}x_2^{-1}x_3^{-1}\cdots x_k^{-1}x_{k+1}^{-1}.\end{aligned}$$

So, the formula is valid.

35. 1. When $n = 1$, $x(y_1) = xy_1$.

2. Assume that

$$x(y_1 + y_2 + \cdots + y_k) = xy_1 + xy_2 + \cdots + xy_k.$$

Then,

$$\begin{aligned}xy_1 + xy_2 + \cdots + xy_k + xy_{k+1} &= x(y_1 + y_2 + \cdots + y_k) + xy_{k+1}\\ &= x\left[(y_1 + y_2 + \cdots + y_k) + y_{k+1}\right]\\ &= x(y_1 + y_2 + \cdots + y_k + y_{k+1}).\end{aligned}$$

So, the formula holds.

37. 1. When $n = 1, \left[1^3 + 3(1)^2 + 2(1)\right] = 6$ and 3 is a factor.

2. Assume that 3 is a factor of $k^3 + 3k^2 + 2k$.

Then, $(k + 1)^3 + 3(k + 1)^2 + 2(k + 1) = k^3 + 3k^2 + 3k + 1 + 3k^2 + 6k + 3 + 2k + 2$

$$= \left(k^3 + 3k^2 + 2k\right) + \left(3k^2 + 9k + 6\right)$$

$$= \left(k^3 + 3k^2 + 2k\right) + 3\left(k^2 + 3k + 2\right).$$

Because 3 is a factor of each term, 3 is a factor of the sum.

So, 3 is a factor of $\left(n^3 + 3n^2 + 2n\right)$ for every positive integer n.

39. Prove 3 is a factor of $2^{2n+1} + 1$ for all positive integers n.

1. When $n = 1$, $2^{2\cdot1+1} + 1 = 2^3 + 1 = 8 + 1 = 9$ and 3 is a factor.

2. Assume 3 is a factor of $2^{2k+1} + 1$.

Then,

$$2^{2(k+1)+1} + 1 = 2^{2k+2+1} + 1 = 2^{(2k+1)+2} + 1 = 2^{2k+1} \cdot 2^2 + 1 = 4 \cdot 2^{2k+1} + 1 = 4\left(2^{2k+1} + 1\right) - 3.$$

Because 3 is a factor of each term, 3 is a factor of the sum.

So, 3 is a factor of $2^{2n+1} + 1$ for all positive integers n.

41. $S_n = 1 + 5 + 9 + 13 + \cdots + (4n - 3)$

$S_1 = 1 = 1 \cdot 1$

$S_2 = 1 + 5 = 6 = 2 \cdot 3$

$S_3 = 1 + 5 + 9 = 15 = 3 \cdot 5$

$S_4 = 1 + 5 + 9 + 13 = 28 = 4 \cdot 7$

From this sequence, it appears that $S_n = n(2n - 1)$.

This can be verified by mathematical induction. The formula has already been verified for $n = 1$. Assume that the formula is valid for $n = k$.

Then,

$$\begin{aligned}
S_{k+1} &= \left[1 + 5 + 9 + 13 + \cdots + (4k - 3)\right] + \left[4(k + 1) - 3\right] \\
&= k(2k - 1) + (4k + 1) \\
&= 2k^2 + 3k + 1 \\
&= (k + 1)(2k + 1) \\
&= (k + 1)\left[2(k + 1) - 1\right].
\end{aligned}$$

So, the formula is valid.

43. $S_n = \frac{1}{4} + \frac{1}{12} + \frac{1}{24} + \frac{1}{40} + \cdots + \frac{1}{2n(n+1)}$

$S_1 = \frac{1}{4} = \frac{1}{2(2)}$

$S_2 = \frac{1}{4} + \frac{1}{12} = \frac{4}{12} = \frac{2}{6} = \frac{2}{2(3)}$

$S_3 = \frac{1}{4} + \frac{1}{12} + \frac{1}{24} = \frac{9}{24} = \frac{3}{8} = \frac{3}{2(4)}$

$S_4 = \frac{1}{4} + \frac{1}{12} + \frac{1}{24} + \frac{1}{40} = \frac{16}{40} = \frac{4}{10} = \frac{4}{2(5)}$

From the sequence, it appears that $S_n = \frac{n}{2(n+1)}$.

This can be verified by mathematical induction. The formula has already been verified for $n = 1$. Assume that the formula is valid for $n = k$. Then,

$$S_{k+1} = \left[\frac{1}{4} + \frac{1}{12} + \frac{1}{40} + \cdots + \frac{1}{2k(k+1)}\right] + \frac{1}{2(k+1)(k+2)} = \frac{k}{2(k+1)} + \frac{1}{2(k+1)(k+2)}$$

$$= \frac{k(k+2)+1}{2(k+1)(k+2)} = \frac{k^2+2k+1}{2(k+1)(k+2)} = \frac{(k+1)^2}{2(k+1)(k+2)} = \frac{k+1}{2(k+2)}.$$

So, the formula is valid.

45. $\sum_{n=1}^{15} n = \frac{15(15+1)}{2} = 120$

47. $\sum_{n=1}^{6} n^2 = \frac{6(6+1)[2(6)+1]}{6} = 91$

49. $\sum_{n=1}^{5} n^4 = \frac{5(5+1)[2(5)+1][3(5)^2+3(5)-1]}{30} = 979$

51. $\sum_{n=1}^{6}(n^2 - n) = \sum_{n=1}^{6} n^2 - \sum_{n=1}^{6} n$

$= \frac{6(6+1)[2(6)+1]}{6} - \frac{6(6+1)}{2}$

$= 91 - 21 = 70$

53. $\sum_{i=1}^{6}(6i - 8i^3) = 6\sum_{i=1}^{6} i - 8\sum_{i=1}^{6} i^3$

$= 6\left[\frac{6(6+1)}{2}\right] - 8\left[\frac{(6)^2(6+1)^2}{4}\right]$

$= 6(21) - 8(441)$

$= -3402$

55. 5, 13, 21, 29, 37, 45, ...

Linear

Note: This is an arithmetic sequence.

$a_1 = 5, d = 8$

$a_n = 5 + (n-1)8$

$a_n = 8n - 3$

57. 6, 15, 30, 51, 78, 111, ...

Quadratic

$$\begin{cases} a + b + c = 6 \\ 4a + 2b + c = 15 \\ 9a + 3b + c = 30 \end{cases}$$

Solving this system yields $a = 3, b = 0,$ and $c = 3$.

So, $a_n = 3n^2 + 3$.

59. −2, 1, 6, 13, 22, 33, ...

Quadratic

$$\begin{cases} a + b + c = -2 \\ 4a + 2b + c = 1 \\ 9a + 3b + c = 6 \end{cases}$$

Solving this system yields $a = 1, b = 0,$ and $c = -3$.

So, $a_n = n^2 - 3$.

61. $a_1 = 0, a_n = a_{n-1} + 3$

$a_1 = a_1 = 0$

$a_2 = a_1 + 3 = 0 + 3 = 3$

$a_3 = a_2 + 3 = 3 + 3 = 6$

$a_4 = a_3 + 3 = 6 + 3 = 9$

$a_5 = a_4 + 3 = 9 + 3 = 12$

$a_6 = a_5 + 3 = 12 + 3 = 15$

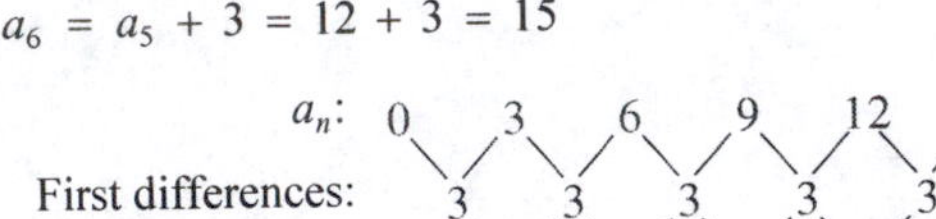

Because the first differences are equal, the sequence has a linear model.

63. $a_1 = 3, a_n = a_{n-1} - n$

$a_1 = a_1 = 3$

$a_2 = a_1 - 2 = 3 - 2 = 1$

$a_3 = a_2 - 3 = 1 - 3 = -2$

$a_4 = a_3 - 4 = -2 - 4 = -6$

$a_5 = a_4 - 5 = -6 - 5 = -11$

$a_6 = a_5 - 6 = -11 - 6 = -17$

a_n: 3 1 −2 −6 −11 −17

First differences: −2 −3 −4 −5 −6

Second differences: −1 −1 −1 −1

Because the second differences are all the same, the sequence has a quadratic model.

65. $a_0 = 2, a_n = (a_{n-1})^2$

$a_0 = 2$

$a_1 = a_0^2 = 2^2 = 4$

$a_2 = a_1^2 = 4^2 = 16$

$a_3 = a_2^2 = 16^2 = 256$

$a_4 = a_3^2 = 256^2 = 65{,}536$

$a_5 = a_4^2 = 65{,}536^2 = 4{,}294{,}967{,}296$

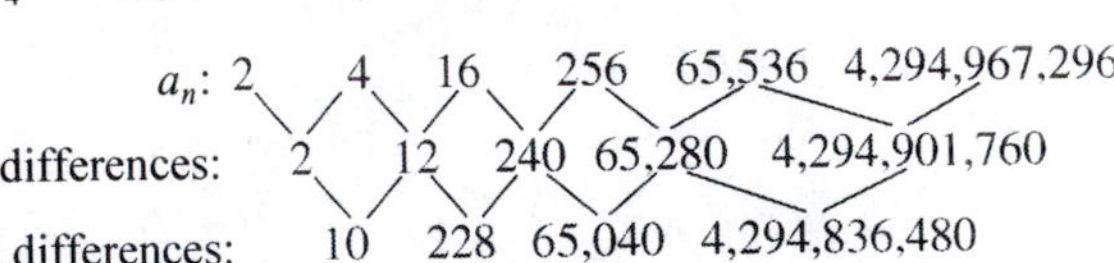

Because neither the first differences nor the second differences are equal, the sequence does not have a linear or quadratic model.

67. $a_1 = 2, a_n = n - a_{n-1}$

$a_2 = 0$

$a_3 = 3$

$a_4 = 1$

$a_5 = 4$

$a_6 = 2$

a_n: 2 0 3 1 4 2

First differences: −2 3 −2 3 −2

Second differences: 5 −5 5 −5

Because neither the first differences nor the second differences are equal, the sequence does not have a linear or quadratic model.

69. $a_0 = 3, a_1 = 3, a_4 = 15$

Let $a_n = an^2 + bn + c$.

Then:

$a_0 = a(0)^2 + b(0) + c = 3 \Rightarrow \quad c = 3$

$a_1 = a(1)^2 + b(1) + c = 3 \Rightarrow \quad a + b + c = 3$

$a + b = 0$

$a_4 = a(4)^2 + b(4) + c = 15 \Rightarrow 16a + 4b + c = 15$

$16a + 4b = 12$

$4a + b = 3$

By elimination: $-a - b = 0$

$\underline{4a + b = 3}$

$3a = 3$

$a = 1 \Rightarrow b = -1$

So, $a_n = n^2 - n + 3$.

71. $a_0 = -3, a_2 = 1, a_4 = 9$

Let $a_n = an^2 + bn + c$.

Then: $a_0 = a(0)^2 + b(0) + c = -3 \Rightarrow c = -3$

$$a_2 = a(2)^2 + b(2) + c = 1 \Rightarrow 4a + 2b + c = 1$$
$$4a + 2b = 4$$
$$2a + b = 2$$

$$a_4 = a(4)^2 + b(4) + c = 9 \Rightarrow 16a + 4b + c = 9$$
$$16a + 4b = 12$$
$$4a + b = 3$$

By elimination: $-2a - b = -2$
$$4a + b = 3$$
$$2a = 1$$

$$a = \tfrac{1}{2} \Rightarrow b = 1$$

So, $a_n = \frac{1}{2}n^2 + n - 3$.

73. $a_1 = 0, a_2 = 8, a_4 = 30$

Let $a_n = an^2 + bn + c$. Then:

$$a_1 = a(1)^2 + b(1) + c = 0 \Rightarrow a + b + c = 0$$
$$a_2 = a(2)^2 + b(2) + c = 8 \Rightarrow 4a + 2b + c = 8$$
$$a_4 = a(4)^2 + b(4) + c = 30 \Rightarrow 16a + 4b + c = 30$$

$$\begin{cases} a + b + c = 0 \\ 4a + 2b + c = 8 \\ 16a + 4b + c = 30 \end{cases}$$

Solving this system yields $a = 1, b = 5,$ and $c = -6$.

So, $a_n = n^2 + 5n - 6$.

75. (a) n: 5 6 7 8 9 10

terms: 664 671 676 682 691 705

First differences: 7 5 6 9 14

Sample Answer: Using common difference $d = 7$, $a_n = 7n + 629$

(b) Using a graphing utility, a linear model for the data is $a_n = 7.7n + 623$. So, the models are comparable.

(c) Using $a_n = 8.2n + 623$, the number of residents in 2016 is $a_{16} = 8.2(16) + 623 = 754{,}200$.

Using $a_n = 7.7n + 623$, the number of residents in 2016 is $a_{16} = 7.7(16) + 623 = 746{,}200$.

So, the predictions are similar.

77. False. P_1 must be proven to be true.

Section 11.5 The Binomial Theorem

1. binomial coefficients

3. $\binom{n}{r}$ or ${}_nC_r$

5. ${}_5C_3 = \dfrac{5!}{3! \cdot 2!} = \dfrac{5 \cdot 4}{2 \cdot 1} = 10$

7. ${}_{12}C_0 = \dfrac{12!}{0! \cdot 12!} = 1$

9. ${}_{20}C_{15} = \dfrac{20!}{15! \cdot 5!} = \dfrac{20 \cdot 19 \cdot 18 \cdot 17 \cdot 16}{5 \cdot 4 \cdot 3 \cdot 2 \cdot 1} = 15{,}504$

11. $\binom{10}{4} = \dfrac{10!}{6! \cdot 4!} = \dfrac{10 \cdot 9 \cdot 8 \cdot 7 \cdot 6!}{6!(24)} = 210$

13. $\binom{100}{98} = \dfrac{100!}{2! \cdot 98!} = \dfrac{100 \cdot 99}{2 \cdot 1} = 4950$

15.

1
1 1
1 2 1
1 3 3 1
1 4 6 4 1
1 5 10 10 5 1
1 6 15 20 15 (6) 1

$\binom{6}{5} = 6$, the 6th entry in the 6th row

17.

1
1 1
1 2 1
1 3 3 1
1 4 6 4 1
1 5 10 10 5 1
1 6 15 20 15 6 1
1 7 21 35 (35) 21 7 1

${}_7C_4 = 35$, the 5th entry in the 7th row

19. $(x + 1)^4 = {}_4C_0x^4 + {}_4C_1x^3(1) + {}_4C_2x^2(1)^2 + {}_4C_3x(1)^3 + {}_4C_4(1)^4$
$= x^4 + 4x^3 + 6x^2 + 4x + 1$

21. $(a + 6)^4 = {}_4C_0a^4 + {}_4C_1a^3(6) + {}_4C_2a^2(6)^2 + {}_4C_3a(6)^3 + {}_4C_4(6)^4$
$= 1a^4 + 4a^3(6) + 6a^2(6)^2 + 4a(6)^3 + 1(6)^4$
$= a^4 + 24a^3 + 216a^2 + 864a + 1296$

23. $(y - 4)^3 = {}_3C_0y^3 - {}_3C_1y^2(4) + {}_3C_2y(4)^2 - {}_3C_3(4)^3$
$= 1y^3 - 3y^2(4) + 3y(4)^2 - 1(4)^3$
$= y^3 - 12y^2 + 48y - 64$

25. $(x + y)^5 = {}_5C_0x^5 + {}_5C_1x^4y + {}_5C_2x^3y^2 + {}_5C_3x^2y^3 + {}_5C_4xy^4 + {}_5C_5y^5$
$= x^5 + 5x^4y + 10x^3y^2 + 10x^2y^3 + 5xy^4 + y^5$

27. $(2x + y)^3 = {}_3C_0(2x)^3 + {}_3C_1(2x)^2(y) + {}_3C_2(2x)(y^2) + {}_3C_3(y^3)$
$= (1)(8x^3) + (3)(4x^2)(y) + (3)(2x)(y^2) + (1)(y^3)$
$= 8x^3 + 12x^2y + 6xy^2 + y^3$

29. $(r + 3s)^6 = {}_6C_0r^6 + {}_6C_1r^5(3s) + {}_6C_2r^4(3s)^2 + {}_6C_3r^3(3s)^3 + {}_6C_4r^2(3s)^4 + {}_6C_5r(3s)^5 + {}_6C_6(3s)^6$
$= 1r^6 + 6r^5(3s) + 15r^4(3s)^2 + 20r^3(3s)^3 + 15r^2(3s)^4 + 6r(3s)^5 + 1(3s)^6$
$= r^6 + 18r^5s + 135r^4s^2 + 540r^3s^3 + 1215r^2s^4 + 1458rs^5 + 729s^6$

31. $(3a - 4b)^5 = {}_5C_0(3a)^5 - {}_5C_1(3a)^4(4b) + {}_5C_2(3a)^3(4b)^2 - {}_5C_3(3a)^2(4b)^3 + {}_5C_4(3a)(4b)^4 - {}_5C_5(4b)^5$
$= (1)(243a^5) - 5(81a^4)(4b) + 10(27a^3)(16b^2) - 10(9a^2)(64b^3) + 5(3a)(256b^4) - (1)(1024b^5)$
$= 243a^5 - 1620a^4b + 4320a^3b^2 - 5760a^2b^3 + 3840ab^4 - 1024b^5$

33. $(x^2 + y^2)^4 = {}_4C_0(x^2)^4 + {}_4C_1(x^2)^3(y^2) + {}_4C_2(x^2)^2(y^2)^2 + {}_4C_3(x^2)(y^2)^3 + {}_4C_4(y^2)^4$
$= (1)(x^8) + (4)(x^6y^2) + (6)(x^4y^4) + (4)(x^2y^6) + (1)(y^8)$
$= x^8 + 4x^6y^2 + 6x^4y^4 + 4x^2y^6 + y^8$

35. $\left(\frac{1}{x} + y\right)^5 = {}_5C_0\left(\frac{1}{x}\right)^5 + {}_5C_1\left(\frac{1}{x}\right)^4 y + {}_5C_2\left(\frac{1}{x}\right)^3 y^2 + {}_5C_3\left(\frac{1}{x}\right)^2 y^3 + {}_5C_4\left(\frac{1}{x}\right)y^4 + {}_5C_5y^5$
$= \frac{1}{x^5} + \frac{5y}{x^4} + \frac{10y^2}{x^3} + \frac{10y^3}{x^2} + \frac{5y^4}{x} + y^5$

37. $\left(\frac{2}{x} - y\right)^4 = {}_4C_0\left(\frac{2}{x}\right)^4 - {}_4C_1\left(\frac{2}{x}\right)^3 y + {}_4C_2\left(\frac{2}{x}\right)^2 y^2 - {}_4C_3\left(\frac{2}{x}\right)y^3 + {}_4C_4y^4$
$= \frac{16}{x^4} - \frac{32y}{x^3} + \frac{24y^2}{x^2} - \frac{8y^3}{x} + y^4$

39. $2(x - 3)^4 + 5(x - 3)^2 = 2\left[x^4 - 4(x^3)(3) + 6(x^2)(3^2) - 4(x)(3^3) + 3^4\right] + 5\left[x^2 - 2(x)(3) + 3^2\right]$
$= 2(x^4 - 12x^3 + 54x^2 - 108x + 81) + 5(x^2 - 6x + 9)$
$= 2x^4 - 24x^3 + 113x^2 - 246x + 207$

41. 5th Row of Pascal's Triangle: 1 5 10 10 5 1

$$(2t - s)^5 = 1(2t)^5 - 5(2t)^4(s) + 10(2t)^3(s)^2 - 10(2t)^2(s)^3 + 5(2t)(s)^4 - 1(s)^5$$
$$= 32t^5 - 80t^4s + 80t^3s^2 - 40t^2s^3 + 10ts^4 - s^5$$

43. 5th Row of Pascal's Triangle: 1 5 10 10 5 1

$$(x + 2y)^5 = 1x^5 + 5x^4(2y) + 10x^3(2y)^2 + 10x^2(2y)^3 + 5x(2y)^4 + 1(2y)^5$$
$$= x^5 + 10x^4y + 40x^3y^2 + 80x^2y^3 + 80xy^4 + 32y^5$$

45. The 4th term in the expansion of $(x + y)^{10}$ is

${}_{10}C_3x^{10-3}y^3 = 120x^7y^3.$

47. The 3rd term in the expansion of $(x - 6y)^5$ is

${}_5C_2x^{5-2}(-6y)^2 = 10x^3(36y^2) = 360x^3y^2.$

49. The 8th term in the expansion of $(4x + 3y)^9$ is

$${}_9C_7(4x)^{9-7}(3y)^7 = 36(16x^2)(2187y^7)$$
$$= 1{,}259{,}712x^2y^7.$$

51. The 10th term in the expansion of $(10x - 3y)^{12}$ is

$${}_{12}C_9(10x)^{12-9}(-3y)^9 = 220(1000x^3)(-19{,}683y^9)$$
$$= -4{,}330{,}260{,}000x^3y^9.$$

53. The term involving x^5 in the expansion of $(x + 3)^{12}$ is

$${}_{12}C_7x^5(3)^7 = \frac{12!}{7! \cdot 5!} \cdot 3^7x^5 = 1{,}732{,}104x^5.$$

The coefficient is 1,732,104.

55. The term involving x^2y^8 in the expansion of $(4x - y)^{10}$

is ${}_{10}C_8(4x)^2(-y)^8 = \dfrac{10!}{(10 - 8)!8!} \cdot 16x^2y^8 = 720x^2y^8.$

The coefficient is 720.

57. The term involving x^4y^5 in the expansion of $(2x - 5x)^9$

is ${}_9C_5(2x)^4(-5y)^5 = 126(16x^4)(-3125y^5)$

$= -6{,}300{,}000x^4y^5.$

The coefficient is $-6{,}300{,}000$.

59. The term involving $x^8y^6 = (x^2)^4y^6$ in the expansion of

$(x^2 + y)^{10}$ is ${}_{10}C_6(x^2)^4y^6 = \dfrac{10!}{4!6!}(x^2)^4y^6 = 210x^8y^6.$

The coefficient is 210.

61. $$(\sqrt{x} + 5)^3 = (\sqrt{x})^3 + 3(\sqrt{x})^2(5) + 3(\sqrt{x})(5^2) + 5^3$$
$$= x^{3/2} + 15x + 75x^{1/2} + 125$$

63. $$(x^{2/3} - y^{1/3})^3 = (x^{2/3})^3 - 3(x^{2/3})^2(y^{1/3}) + 3(x^{2/3})(y^{1/3})^2 - (y^{1/3})^3 = x^2 - 3x^{4/3}y^{1/3} + 3x^{2/3}y^{2/3} - y$$

65. $$(3\sqrt{t} + \sqrt[4]{t})^4 = (3\sqrt{t})^4 + 4(3\sqrt{t})^3(\sqrt[4]{t}) + 6(3\sqrt{t})^2(\sqrt[4]{t})^2 + 4(3\sqrt{t})(\sqrt[4]{t})^3 + (\sqrt[4]{t})^4$$
$$= 81t^2 + 108t^{7/4} + 54t^{3/2} + 12t^{5/4} + t$$

67. $$\frac{f(x + h) - f(x)}{h} = \frac{(x + h)^3 - x^3}{h}$$
$$= \frac{x^3 + 3x^2h + 3xh^2 + h^3 - x^3}{h}$$
$$= \frac{h(3x^2 + 3xh + h^2)}{h}$$
$$= 3x^2 + 3xh + h^2, h \neq 0$$

69. $\dfrac{f(x+h)-f(x)}{h} = \dfrac{(x+h)^6 - x^6}{h}$

$= \dfrac{x^6 + 6x^5h + 15x^4h^2 + 20x^3h^3 + 15x^2h^4 + 6xh^5 + h^6 - x^6}{h}$

$= \dfrac{h(6x^5 + 15x^4h + 20x^3h^2 + 15x^2h^3 + 6xh^4 + h^5)}{h}$

$= 6x^5 + 15x^4h + 20x^3h^2 + 15x^2h^3 + 6xh^4 + h^5, h \neq 0$

71. $\dfrac{f(x+h)-f(x)}{h} = \dfrac{\sqrt{x+h} - \sqrt{x}}{h}$

$= \dfrac{\sqrt{x+h} - \sqrt{x}}{h} \cdot \dfrac{\sqrt{x+h} + \sqrt{x}}{\sqrt{x+h} + \sqrt{x}}$

$= \dfrac{(x+h) - x}{h(\sqrt{x+h} + \sqrt{x})}$

$= \dfrac{1}{\sqrt{x+h} + \sqrt{x}}, h \neq 0$

73. $(1+i)^4 = {}_4C_0(1)^4 + {}_4C_1(1)^3i + {}_4C_2(1)^2i^2 + {}_4C_3(1)i^3 + {}_4C_4i^4$

$= 1 + 4i - 6 - 4i + 1$

$= -4$

75. $(2-3i)^6 = {}_6C_0 2^6 - {}_6C_1 2^5(3i) + {}_6C_2 2^4(3i)^2 - {}_6C_3 2^3(3i)^3 + {}_6C_4 2^2(3i)^4 - {}_6C_5 2(3i)^5 + {}_6C_6(3i)^6$

$= (1)(64) - (6)(32)(3i) + 15(16)(-9) - 20(8)(-27i) + 15(4)(81) - 6(2)(243i) + (1)(-729)$

$= 64 - 576i - 2160 + 4320i + 4860 - 2916i - 729$

$= 2035 + 828i$

77. $\left(-\dfrac{1}{2} + \dfrac{\sqrt{3}}{2}i\right)^3 = \dfrac{1}{8}\left[(-1)^3 + 3(-1)^2(\sqrt{3}i) + 3(-1)(\sqrt{3}i)^2 + (\sqrt{3}i)^3\right]$

$= \dfrac{1}{8}\left[-1 + 3\sqrt{3}i + 9 - 3\sqrt{3}i\right]$

$= 1$

79. $(1.02)^8 = (1 + 0.02)^8$

$= 1 + 8(0.02) + 28(0.02)^2 + 56(0.02)^3 + 70(0.02)^4 + 56(0.02)^5 + 28(0.02)^6 + 8(0.02)^7 + (0.02)^8$

$= 1 + 0.16 + 0.0112 + 0.000448 + \cdots$

≈ 1.172

81. $(2.99)^{12} = (3 - 0.01)^{12}$

$= 3^{12} - 12(3)^{11}(0.01) + 66(3)^{10}(0.01)^2 - 220(3)^9(0.01)^3 + 495(3)^8(0.01)^4$

$-792(3)^7(0.01)^5 + 924(3)^6(0.01)^6 - 792(3)^5(0.01)^7 + 495(3)^4(0.01)^8$

$-220(3)^3(0.01)^9 + 66(3)^2(0.01)^{10} - 12(3)(0.01)^{11} + (0.01)^{12}$

$\approx 531{,}441 - 21{,}257.64 + 389.7234 - 4.3303 + 0.0325 - 0.0002 + \cdots \approx 510{,}568.785$

83. $f(x) = x^2 - 4x$

$$\begin{aligned} g(x) &= f(x + 4) \\ &= (x + 4)^3 - 4(x + 4) \\ &= x^3 + 3x^2(4) + 3x(4)^2 + (4)^3 - 4x - 16 \\ &= x^3 + 12x^2 + 48x + 64 - 4x - 16 \\ &= x^3 + 12x^2 + 44x + 48 \end{aligned}$$

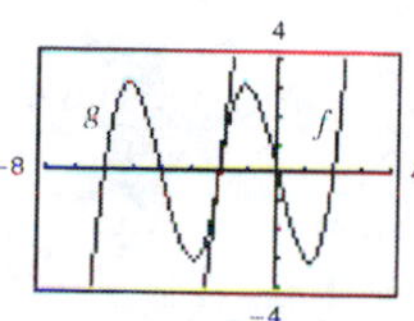

The graph of g is the same as the graph of f shifted four units to the left.

85. ${}_7C_4\left(\frac{1}{2}\right)^4\left(\frac{1}{2}\right)^3 = \frac{7!}{3!4!}\left(\frac{1}{16}\right)\left(\frac{1}{8}\right) = 35\left(\frac{1}{16}\right)\left(\frac{1}{8}\right) \approx 0.273$

87. ${}_8C_4\left(\frac{1}{3}\right)^4\left(\frac{2}{3}\right)^4 = \frac{8!}{4!4!}\left(\frac{1}{81}\right)\left(\frac{16}{81}\right) = 70\left(\frac{1}{81}\right)\left(\frac{16}{81}\right) \approx 0.171$

89.

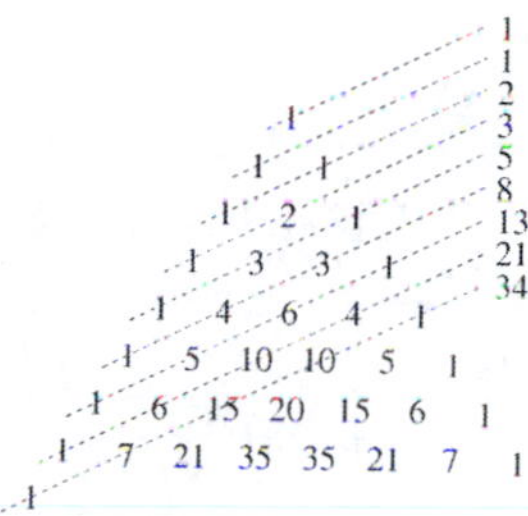

The first nine terms of the sequence are 1, 1, 2, 3, 5, 8, 13, 21, 34, …

After the first two terms, the next terms are formed by adding the previous two terms.

$a_1 = 1,\ a_2 = 1$

$a_3 = a_1 + a_2 = 1 + 1 = 2$

$a_4 = a_2 + a_3 = 1 + 2 = 3$

$a_5 = a_3 + a_4 = 2 + 3 = 5$

$a_6 = a_4 + a_5 = 3 + 5 = 8$

$a_7 = a_5 + a_6 = 5 + 8 = 13$

This is called the Fibonacci sequence.

91. (a)
$$\begin{aligned} g(t) &= f(t + 5) \\ &= -0.009(t + 5)^2 + 1.02(t + 5) + 18.0 \\ &= -0.009(t^2 + 10t + 25) + 1.02(t + 5) + 18.0 \\ &= -0.009t^2 + 0.96t + 23.025 \end{aligned}$$

(b)

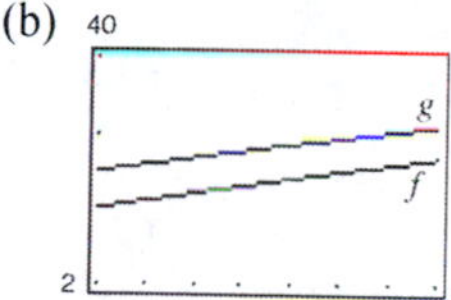

(c) Using the graphs, the child support collections exceeded \$25 billion in 2007.

93. True. The coefficients from the Binomial Theorem can be used to find the numbers in Pascal's Triangle.

95. The first and last numbers in each row are 1. Every other number in each row is formed by adding the two numbers immediately above the number.

97. The functions $f(x) = (1 - x)^3$ and

$k(x) = 1 - 3x + 3x^2 + x^3$

(choices (a) and (d)) have identical graphs, because $k(x)$ is the expansion of $f(x)$.

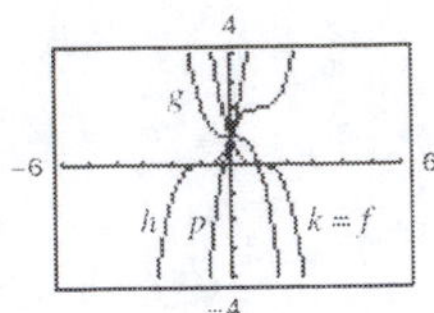

99. $${}_nC_{n-r} = \frac{n!}{(n-(n-r))!(n-r)!} = \frac{n!}{r!(n-r)!} = \frac{n!}{(n-r)!r!} = {}_nC_r$$

101. $${}_nC_r + {}_nC_{r-1} = \frac{n!}{(n-r)!r!} + \frac{n!}{(n-r+1)!(r-1)!}$$
$$= \frac{n!(n-r+1)!(r-1)! + n!(n-r)!r!}{(n-r)!r!(n-r+1)!(r-1)!}$$
$$= \frac{n![(n-r+1)!(r-1)! + r!(n-r)!]}{(n-r)!r!(n-r+1)!(r-1)!}$$
$$= \frac{n!\cancel{(r-1)!}[(n-r+1)! + r(n-r)!]}{(n-r)!r!(n-r+1)!\cancel{(r-1)!}}$$
$$= \frac{n!\cancel{(n-r)!}[(n-r+1) + r]}{\cancel{(n-r)!}r!(n-r+1)!}$$
$$= \frac{n![n+1]}{r!(n-r+1)!}$$
$$= \frac{(n+1)!}{[(n+1)-r]!r!}$$
$$= {}_{n+1}C_r$$

103.

n	r	${}_nC_r$	${}_nC_{n-r}$
9	5	126	126
7	1	7	7
12	4	495	495
6	0	1	1
10	7	120	120

${}_nC_r = {}_nC_{n-r}$

The table illustrates the symmetry of Pascal's Triangle.

Section 11.6 Counting Principles

1. Fundamental Counting Principle

3. ${}_nP_r = \dfrac{n!}{(n-r)!}$

5. combinations

7. Odd integers: 1, 3, 5, 7, 9, 11

6 ways

9. Prime integers: 2, 3, 5, 7, 11

5 ways

11. Divisible by 4: 4, 8, 12

3 ways

13. Sum is 9: $1 + 8, 2 + 7, 3 + 6, 4 + 5, 5 + 4,$
$6 + 3, 7 + 2, 8 + 1$

8 ways

15. Amplifiers: 3 choices

Compact disc players: 2 choices

Speakers: 5 choices

Total: $3 \cdot 2 \cdot 5 = 30$ ways

17. Math courses: 2

Science courses: 3

Social sciences and humanities courses: 5

Total: $2 \cdot 3 \cdot 5 = 30$ schedules

19. $2^6 = 64$

21. $26 \cdot 26 \cdot 26 \cdot 10 \cdot 10 \cdot 10 \cdot 10 = 175{,}760{,}000$

distinct license plate numbers

23. (a) $9 \cdot 10 \cdot 10 = 900$

(b) $9 \cdot 9 \cdot 8 = 648$

(c) $9 \cdot 10 \cdot 2 = 180$

(d) $6 \cdot 10 \cdot 10 = 600$

25. $40^3 = 64{,}000$

27. (a) $8 \cdot 7 \cdot 6 \cdot 5 \cdot 4 \cdot 3 \cdot 2 \cdot 1 = 40{,}320$

(b) $8 \cdot 1 \cdot 6 \cdot 1 \cdot 4 \cdot 1 \cdot 2 \cdot 1 = 384$

29. $5! = 120$ ways

31. ${}_4P_4 = \dfrac{4!}{0!} = 4! = 24.$

33. ${}_{20}P_2 = \dfrac{20!}{18!} = 20 \cdot 19 = 380$

35. ${}_{20}P_5 = 1{,}860{,}480$

37. ${}_{100}P_3 = 970{,}200$

39. The number of permutations of 9 possible donors taken 3 at a time is

$${}_9P_3 = \frac{9!}{(9-3)!} = \frac{9!}{6!} = 9 \cdot 8 \cdot 7 = 504 \text{ possible orders.}$$

41. ${}_{15}P_9 = \dfrac{15!}{6!} = 1{,}816{,}214{,}400$

different batting orders

43. $\dfrac{7!}{2!1!3!1!} = \dfrac{7!}{2!3!} = 420$

45. $\dfrac{7!}{2!1!1!1!1!1!} = \dfrac{7!}{2!} = 7 \cdot 6 \cdot 5 \cdot 4 \cdot 3 = 2520$

47. ABCD BACD CABD DABC
ABDC BADC CADB DACB
ACBD BCAD CBAD DBAC
ACDB BCDA CBDA DBCA
ADBC BDAC CDAB DCAB
ADCB BDCA CDBA DCBA

49. ${}_5C_2 = \dfrac{5!}{2!3!} = 10$

51. ${}_4C_1 = \dfrac{4!}{1!3!} = 4$

53. ${}_{20}C_4 = 4845$

55. ${}_{42}C_5 = 850{,}668$

57. ${}_6C_2 = 15$

59. ${}_{40}C_{12} = \dfrac{40!}{28!12!} = 5{,}586{,}853{,}480$ ways

61. ${}_{35}C_5 = \dfrac{35!}{30!5!} = 324{,}632$ ways

63. There are 22 good units and 3 defective units.

(a) ${}_{22}C_4 = \dfrac{22!}{4!8!} = 7315$ ways

(b) ${}_{22}C_2 \cdot {}_3C_2 = \dfrac{22!}{2!20!} \cdot \dfrac{3!}{2!1!} = 231 \cdot 3 = 693$ ways

(c)
$$\begin{aligned}{}_{22}C_4 \cdot {}_{22}C_3 \cdot {}_3C_1 + {}_{22}C_2 \cdot {}_3C_2 &= \frac{22!}{4!18!} + \frac{22!}{3!19!} \cdot \frac{3!}{1!2!} + \frac{22!}{2!20!} \cdot \frac{3!}{2!1!} \\ &= 7315 + 1540 \cdot 3 + 231 \cdot 3 \\ &= 12{,}628 \text{ ways}\end{aligned}$$

65. (a) Select type of card for three of a kind: ${}_{13}C_1$

Select three of four cards for three of a kind: ${}_4C_3$

Select type of card for pair: ${}_{12}C_1$

Select two of four cards for pair: ${}_4C_2$

$${}_{13}C_1 \cdot {}_4C_3 \cdot {}_{12}C_1 \cdot {}_4C_2 = \frac{13!}{(13-1)!1!} \cdot \frac{4!}{(4-3)!3!} \cdot \frac{12!}{(12-1)!1!} \cdot \frac{4!}{(4-2)!2!} = 3744$$

(b) Select two jacks: ${}_4C_2$

Select three aces: ${}_4C_3$

$${}_4C_2 \cdot {}_4C_3 = \frac{4!}{(4-2)!2!} \cdot \frac{4!}{(4-3)!3!} = 24$$

67. $_7C_1 \cdot {}_{12}C_3 \cdot {}_{20}C_2 = \dfrac{7!}{(7-1)!1!} \cdot \dfrac{12!}{(12-3)!3!} \cdot \dfrac{20!}{(20-2)!2!} = 292{,}600$

69. $_5C_2 - 5 = 10 - 5 = 5$ diagonals

71. $_8C_2 - 8 = 28 - 8 = 20$ diagonals

73. $_9C_2 = \dfrac{9!}{2!7!} = 36$ lines

75. $4 \cdot {}_{n+1}P_2 = {}_{n+2}P_3$ **Note:** $n \geq 1$ for this to be defined.

$$\frac{(n+1)!}{(n-1)!} = \frac{(n+2)!}{(n-1)!}$$

$$4(n+1)(n) = (n+2)(n+1)n \quad \text{(We can divide by } (n+1)n \text{ because } n \neq 1 \text{ and } n \neq 0.)$$

$$4 = n + 2$$

$$2 = n$$

77. $_{n+1}P_3 = 4 \cdot {}_nP_2$ **Note:** $n \geq 2$ for this to be defined.

$$\frac{(n+1)!}{(n-2)!} = 4 \cdot \frac{n!}{(n-2)!}$$

$$(n+1)(n)(n-1) = 4(n)(n-1) \quad \text{(We can divide by } n(n-1) \text{ because } n \neq 0, \text{ and } n \neq 1.)$$

$$n + 1 = 4$$

$$n = 3$$

79. $14 \cdot {}_nP_3 = {}_{n+2}P_4$ **Note:** $n \geq 3$ for this to be defined.

$$14\left(\frac{n!}{(n-3)!}\right) = \frac{(n+2)!}{(n-2)!}$$

$$14n(n-1)(n-2) = (n+2)(n+1)n(n-1) \quad \text{(We can divide here by } n(n-1) \text{ because } n \neq 0, n \neq 1.)$$

$$14(n-2) = (n+2)(n+1)$$

$$14n - 28 = n^2 + 3n + 2$$

$$0 = n^2 - 11n + 30$$

$$0 = (n-5)(n-6)$$

$$n = 5 \text{ or } n = 6$$

81. $_nP_4 = 10 \cdot {}_{n-1}P_3$ **Note:** $n \geq 4$ for this to be defined.

$$\frac{n!}{(n-4)!} = 10 \cdot \frac{(n-1)!}{(n-4)!}$$

$$n(n-1)(n-2)(n-3) = 10(n-1)(n-2)(n-3) \quad \left(\text{We can divide by } (n-1)(n-2)(n-3) \text{ because } n \neq 1, n \neq 2, \text{ and } n \neq 3.\right)$$

$$n = 10$$

83. False.

It is an example of a combination.

85. $_{10}P_6 > {}_{10}C_6$

Changing the order of any of the six elements selected results in a different permutation but the same combination.

87. $_nP_{n-1} = \dfrac{n!}{(n-(n-1))!} = \dfrac{n!}{1!} = \dfrac{n!}{0!} = {}_nP_n$

89. $_nC_{n-1} = \dfrac{n!}{(n-(n-1))!(n-1)!} = \dfrac{n!}{(1)!(n-1)!}$

$$= \frac{n!}{(n-1)!1!} = {}_nC_1$$

91. $_{100}P_{80} \approx 3.836 \times 10^{139}$

This number is too large for some calculators to evaluate.

Section 11.7 Probability

1. experiment; outcomes

3. probability

5. mutually exclusive

7. complement

9. $\{(H, 1), (H, 2), (H, 3), (H, 4), (H, 5), (H, 6), (T, 1), (T, 2), (T, 3), (T, 4), (T, 5), (T, 6)\}$

11. {ABC, ACB, BAC, BCA, CAB, CBA}

13. {AB, AC, AD, AE, BC, BD, BE, CD, CE, DE}

15. $E = \{HHT, HTH, THH\}$

$$P(E) = \frac{n(E)}{n(S)} = \frac{3}{8}$$

17. $E = \{HHH, HHT, HTH, HTT\}$

$$P(E) = \frac{n(E)}{n(S)} = \frac{4}{8} = \frac{1}{2}$$

19. $E = \{HHH, HHT, HTH, HTT, THH, THT, TTH\}$

$$P(E) = \frac{n(E)}{n(S)} = \frac{7}{8}$$

21. $E = \{K♣, K♦, K♥, K♠, Q♣, Q♦, Q♥, Q♠, J♣, J♦, J♥, J♠\}$

$$P(E) = \frac{n(E)}{n(S)} = \frac{12}{52} = \frac{3}{13}$$

23. $E = \{K♦, K♥, Q♦, Q♥, J♦, J♥\}$

$$P(E) = \frac{n(E)}{n(S)} = \frac{6}{52} = \frac{3}{26}$$

25. $E = \{(1, 5), (2, 4), (3, 3), (4, 2), (5, 1)\}$

$$P(E) = \frac{n(E)}{n(S)} = \frac{5}{36}$$

27. Use the complement.

$$E' = \{(5, 6), (6, 5), (6, 6)\}$$

$$P(E') = \frac{n(E')}{n(S)} = \frac{3}{36} = \frac{1}{12}$$

$$P(E) = 1 - P(E') = 1 - \frac{1}{12} = \frac{11}{12}$$

29. $E_3 = \{(1, 2), (2, 1)\}, n(E_3) = 2$

$E_5 = \{(1, 4), (2, 3), (3, 2), (4, 1)\}, n(E_5) = 4$

$E_7 = \{(1, 6), (2, 5), (3, 4), (4, 3), (5, 2), (6, 1)\}, n(E_7) = 6$

$E = E_3 \cup E_5 \cup E_7$

$n(E) = 2 + 4 + 6 = 12$

$$P(E) = \frac{n(E)}{n(S)} = \frac{12}{36} = \frac{1}{3}$$

31. $P(E) = \dfrac{{}_3C_2}{{}_6C_2} = \dfrac{3}{15} = \dfrac{1}{5}$

33. $P(E) = \dfrac{{}_4C_2}{{}_6C_2} = \dfrac{6}{15} = \dfrac{2}{5}$

35. (a) $0.14(8.92) \approx 1.25$ million

(b) $40\% = \frac{2}{5}$

(c) $26\% = \frac{13}{50}$

(d) $26\% + 3\% = 29\% = \frac{29}{100}$

37. (a) $\dfrac{104}{128} = \dfrac{13}{16}$

(b) $\dfrac{24}{128} = \dfrac{3}{16}$

(c) $\dfrac{52 - 48}{128} = \dfrac{1}{32}$

39. $1 - 0.37 - 0.44 = 0.19 = 19\%$

41. (a) $\dfrac{{}_{15}C_{10}}{{}_{20}C_{10}} = \dfrac{3003}{184{,}756} = \dfrac{21}{1292} \approx 0.016$

(b) $\dfrac{{}_{15}C_8 \cdot {}_5C_2}{{}_{20}C_{10}} = \dfrac{64{,}350}{184{,}756} = \dfrac{225}{646} \approx 0.348$

(c)
$$\frac{{}_{15}C_9 \cdot {}_5C_1}{{}_{20}C_{10}} + \frac{{}_{15}C_{10}}{{}_{20}C_{10}} + \frac{25{,}025 + 3003}{184{,}756} = \frac{28{,}028}{184{,}756} = \frac{49}{323} \approx 0.152$$

43. (a) $\dfrac{1}{{}_5P_5} = \dfrac{1}{120}$

(b) $\dfrac{1}{{}_4P_4} = \dfrac{1}{24}$

45. (a) $\frac{20}{52} = \frac{5}{13}$

(b) $\frac{26}{52} = \frac{1}{2}$

(c) $\frac{16}{52} = \frac{4}{13}$

47. (a) $\dfrac{{}_9C_4}{{}_{12}C_4} = \dfrac{126}{495} = \dfrac{14}{55}$ (4 good units)

(b) $\dfrac{{}_9C_2 \cdot {}_3C_2}{{}_{12}C_4} = \dfrac{108}{495} = \dfrac{12}{55}$ (2 good units)

(c) $\dfrac{{}_9C_3 \cdot {}_3C_1}{{}_{12}C_4} = \dfrac{252}{495} = \dfrac{28}{55}$ (3 good units)

At least 2 good units: $\dfrac{12}{55} + \dfrac{28}{55} + \dfrac{14}{55} = \dfrac{54}{55}$

49. (a) $P(EE) = \frac{20}{40} \cdot \frac{20}{40} = \frac{1}{4}$

(b) $P(EO \text{ or } OE) = 2\left(\frac{20}{40}\right)\left(\frac{20}{40}\right) = \frac{1}{2}$

(c) $P(N_1 < 30, N_2 < 30) = \frac{29}{40} \cdot \frac{29}{40} = \frac{841}{1600}$

(d) $P(N_1N_1) = \frac{30}{40} \cdot \frac{1}{40} = \frac{1}{40}$

51. $P(E') = 1 - P(E) = 1 - 0.87 = 0.13$

53. $P(E') = 1 - P(E) = 1 - \frac{1}{4} = \frac{3}{4}$

55. $P(E) = 1 - P(E') = 1 - 0.23 = 0.77$

57. $P(E) = 1 - P(E') = 1 - \frac{17}{35} = \frac{18}{35}$

59. (a) $P(SS) = (0.985)^2 \approx 0.9702$

(b) $P(S) = 1 - P(FF) = 1 - (0.015)^2 \approx 0.9998$

(c) $P(FF) = (0.015)^2 \approx 0.0002$

61. (a) $\dfrac{1}{38}$

(b) $\dfrac{18}{38} = \dfrac{9}{19}$

(c) $\dfrac{2}{38} + \dfrac{18}{38} = \dfrac{20}{38} = \dfrac{10}{19}$

(d) $\dfrac{1}{38} \cdot \dfrac{1}{38} = \dfrac{1}{1444}$

(e) $\dfrac{18}{38} \cdot \dfrac{18}{38} \cdot \dfrac{18}{38} = \dfrac{5832}{54{,}872} = \dfrac{729}{6859}$

63. $1 - \dfrac{(45)^2}{(60)^2} = 1 - \left(\dfrac{45}{60}\right)^2 = 1 - \left(\dfrac{3}{4}\right)^2 = 1 - \dfrac{9}{16} = \dfrac{7}{16}$

65. True. Two events are independent if the occurrence of one has no effect on the occurrence of the other.

67. (a) As you consider successive people with distinct birthdays, the probabilities must decrease to take into account the birth dates already used. Because the birth dates of people are independent events, multiply the respective probabilities of distinct birthdays.

(b) $\dfrac{365}{365} \cdot \dfrac{364}{365} \cdot \dfrac{363}{365} \cdot \dfrac{362}{365}$

(c) $P_1 = \dfrac{365}{365} = 1$

$$P_2 = \frac{365}{365} \cdot \frac{364}{365} = \frac{364}{365}P_1 = \frac{365 - (2 - 1)}{365}P_1$$

$$P_3 = \frac{365}{365} \cdot \frac{364}{365} \cdot \frac{363}{365} = \frac{363}{365}P_2 = \frac{365 - (3 - 1)}{365}P_2$$

$$P_n = \frac{365}{365} \cdot \frac{364}{365} \cdot \frac{363}{365} \cdot \dots \cdot \frac{365 - (n - 1)}{365} = \frac{365 - (n - 1)}{365}P_{n-1}$$

(d) Q_n is the probability that the birthdays are not distinct which is equivalent to at least two people having the same birthday.

(e)

n	10	15	20	23	30	40	50
P_n	0.88	0.75	0.59	0.49	0.29	0.11	0.03
Q_n	0.12	0.25	0.41	0.51	0.71	0.89	0.97

(f) 23; $Q_n > 0.5$ for $n \geq 23$.

Review Exercises for Chapter 11

1. $a_n = 2 + \frac{6}{n}$

$a_1 = 2 + \frac{6}{1} = 8$

$a_2 = 2 + \frac{6}{2} = 5$

$a_3 = 2 + \frac{6}{3} = 4$

$a_4 = 2 + \frac{6}{4} = \frac{7}{2}$

$a_5 = 2 + \frac{6}{5} = \frac{16}{5}$

3. $a_n = \frac{72}{n!}$

$a_1 = \frac{72}{1!} = 72$

$a_2 = \frac{72}{2!} = 36$

$a_3 = \frac{72}{3!} = 12$

$a_4 = \frac{72}{4!} = 3$

$a_5 = \frac{72}{5!} = \frac{3}{5}$

5. $-2, 2, -2, 2, -2, \ldots$

$a_n = 2(-1)^n$

7. $4, 2, \frac{4}{3}, 1, \frac{4}{5}, \ldots$

$a_n = \frac{4}{n}$

9. $9! = 9 \cdot 8 \cdot 7 \cdot 6 \cdot 5 \cdot 4 \cdot 3 \cdot 2 \cdot 1 = 362{,}880$

11. $\frac{3!\,5!}{6!} = \frac{(3 \cdot 2 \cdot 1)5!}{6 \cdot 5!} = 1$

13. $\sum_{j=1}^{4} \frac{6}{j^2} = \frac{6}{1^2} + \frac{6}{2^2} + \frac{6}{3^2} + \frac{6}{4^2}$

$= 6 + \frac{3}{2} + \frac{2}{3} + \frac{3}{8}$

$= \frac{205}{24}$

15. $\frac{1}{2(1)} + \frac{1}{2(2)} + \frac{1}{2(3)} + \cdots + \frac{1}{2(20)} = \sum_{k=1}^{20} \frac{1}{2k}$

17. $\sum_{i=1}^{\infty} \frac{4}{10^i} = \sum_{i=1}^{\infty} 4\left(\frac{1}{10^i}\right) = \frac{\frac{4}{10}}{1 - \frac{1}{10}} = \frac{4}{9}$

19. (a) $A_1 = \$10{,}018.75$

$A_2 \approx \$10{,}037.54$

$A_3 \approx \$10{,}056.36$

$A_4 \approx \$10{,}075.21$

$A_5 \approx \$10{,}094.10$

$A_6 \approx \$10{,}113.03$

$A_7 \approx \$10{,}131.99$

$A_8 \approx \$10{,}150.99$

$A_9 \approx \$10{,}170.02$

$A_{10} \approx \$10{,}189.09$

(b) The balance in the account after 10 years is

$A_{120}\ 10{,}000\left(1 + \frac{0.0225}{12}\right)^{120} \approx \$12{,}520.59$

21. $6, -1, -8, -15, -22, \ldots$

Arithmetic sequence, $d = -7$

23. $\frac{1}{2}, 1, \frac{3}{2}, 2, \frac{5}{2}, \ldots$

Arithmetic sequence, $d = \frac{1}{2}$

25. $a_1 = 7, d = 12$

$a_n = 7 + (n - 1)12$

$= 7 + 12n - 12$

$= 12n - 5$

27. $a_2 = 93, a_6 = 65$

$a_6 = a_2 + 4d \Rightarrow 65 = 93 + 4d \Rightarrow -28 = 4d \Rightarrow d = -7$

$a_1 = a_2 - d \Rightarrow a_1 = 93 - (-7) = 100$

$a_n = a_1 + (n - 1)d = 100 + (n - 1)(-7) = -7n + 107$

29. $a_1 = 3, d = 11$

$a_1 = 3$

$a_2 = 3 + 11 = 14$

$a_3 = 14 + 11 = 25$

$a_4 = 25 + 11 = 36$

$a_5 = 36 + 11 = 47$

31. $\sum_{k=1}^{100} 7k$ is arithmetic. Therefore,

$a_1 = 7, a_{100} = 700, S_{700} = \frac{100}{2}(7 + 700) = 35{,}350.$

33. $\sum_{j=1}^{10} (2j - 3)$ is arithmetic. Therefore,

$a_1 = -1, a_{10} = 17, S_{10} = \frac{10}{2}[-1 + 17] = 80.$

35. $\sum_{k=1}^{11} \left(\frac{2}{3}k + 4\right)$ is arithmetic. Therefore,

$a_1 = \frac{14}{3}, a_{11} = \frac{34}{3}, S_{11} = \frac{11}{2}\left[\frac{14}{3} + \frac{34}{3}\right] = 88.$

37. $a_n = 43{,}800 + (n - 1)(1950)$

(a) $a_5 = 43{,}800 + 4(1950) = \$51{,}600$

(b) $S_5 = \frac{5}{2}(43{,}800 + 51{,}600) = \$238{,}500$

39. 6, 12, 24, 48, …

Geometric sequence, $r = 2$

41. $\frac{1}{5}, -\frac{3}{5}, \frac{9}{5}, -\frac{27}{5}, \ldots$

Geometric sequence, $r = -3$

43. $a_1 = 4, r = -\frac{1}{4}$

$a_1 = 4$

$a_2 = 4\left(-\frac{1}{4}\right) = -1$

$a_3 = -1\left(-\frac{1}{4}\right) = \frac{1}{4}$

$a_4 = \frac{1}{4}\left(-\frac{1}{4}\right) = -\frac{1}{16}$

$a_5 = -\frac{1}{16}\left(-\frac{1}{4}\right) = \frac{1}{64}$

45. $a_1 = 9, a_3 = 4$

$a_3 = a_1 r^2$

$4 = 9r^2$

$\frac{4}{9} = r^2 \Rightarrow r = \pm\frac{2}{3}$

$a_1 = 9$ | $a_1 = 9$

$a_2 = 9\left(\frac{2}{3}\right) = 6$ | $a_2 = 9\left(-\frac{2}{3}\right) = -6$

$a_3 = 6\left(\frac{2}{3}\right) = 4$ or $a_3 = -6\left(-\frac{2}{3}\right) = 4$

$a_4 = 4\left(\frac{2}{3}\right) = \frac{8}{3}$ | $a_4 = 4\left(-\frac{2}{3}\right) = -\frac{8}{3}$

$a_5 = \frac{8}{3}\left(\frac{2}{3}\right) = \frac{16}{9}$ | $a_5 = -\frac{8}{3}\left(-\frac{2}{3}\right) = \frac{16}{9}$

47. $a_1 = 18, a_2 = -9$

$a_2 = a_1 r$

$-9 = 18r$

$-\frac{1}{2} = r$

$a_n = 18\left(-\frac{1}{2}\right)^{n-1}$

$a_{10} = 18\left(-\frac{1}{2}\right)^9 = \frac{9}{256}$

49. $a_1 = 100, r = 1.05$

$a_n = 100(1.05)^{n-1}$

$a_{10} = 100(1.05)^9 \approx 155.133$

51. $\sum_{i=1}^{7} 2^{i-1} = \dfrac{1 - 2^7}{1 - 2} = 127$

53. $\sum_{i=1}^{4} \left(\dfrac{1}{2}\right)^i = \dfrac{1}{2} + \dfrac{1}{4} + \dfrac{1}{8} + \dfrac{1}{16} = \dfrac{15}{16}$

55. $\sum_{i=1}^{5} (2)^{i-1} = 1 + 2 + 4 + 8 + 16 = 31$

57. $\sum_{i=1}^{10} 10\left(\dfrac{3}{5}\right)^{i-1} \approx 24.85$

59. $\sum_{i=1}^{\infty} \left(\dfrac{7}{8}\right)^{i-1} = \dfrac{1}{1 - \dfrac{7}{8}} = 8$

61. $\sum_{k=1}^{\infty} 4\left(\dfrac{2}{3}\right)^{k-1} = \dfrac{4}{1 - \dfrac{2}{3}} = 12$

63. (a) $a_n = 120{,}000(0.7)^n$

(b) $a_5 = 120{,}000(0.7)^5$

$= \$20{,}168.40$

65. 1. When $n = 1, 3 = 1(1 + 2)$.

2. Assume that $S_k = 3 + 5 + 7 + \cdots + (2k + 1) = k(k + 2)$.

Then, $S_{k+1} = 3 + 5 + 7 + \cdots + (2k + 1) + [2(k + 1) + 1] = S_k + (2k + 3)$

$$= k(k + 2) + 2k + 3$$
$$= k^2 + 4k + 3$$
$$= (k + 1)(k + 3)$$
$$= (k + 1)[(k + 1) + 2].$$

So, by mathematical induction, the formula is valid for all positive integer values of n.

67. 1. When $n = 1, a = a\left(\frac{1 - r}{1 - r}\right)$.

2. Assume that $S_k = \sum_{i=0}^{k-1} ar^i = \frac{a(1 - r^k)}{1 - r}$.

Then $S_{k+1} = \sum_{i=0}^{k} ar^i = \left(\sum_{i=0}^{k-1} ar^i\right) + ar^k = \frac{a(1 - r^k)}{1 - r} + ar^k$

$$= \frac{a(1 - r^k + r^k - r^{k+1})}{1 - r} = \frac{a(1 - r^{k+1})}{1 - r}.$$

So, by mathematical induction, the formula is valid for all positive integer values of n.

69. $S_1 = 9 = 1(9) = 1[2(1) + 7]$

$S_2 = 9 + 13 = 22 = 2(11) = 2[2(2) + 7]$

$S_3 = 9 + 13 + 17 = 39 = 3(13) = 3[2(3) + 7]$

$S_4 = 9 + 13 + 17 + 21 = 60 = 4(15) = 4[2(4) + 7]$

$S_n = n(2n + 7)$

71. $S_1 = 1$

$S_2 = 1 + \frac{3}{5} = \frac{8}{5}$

$S_3 = 1 + \frac{3}{5} = \frac{9}{25} = \frac{49}{25}$

$S_4 = 1 + \frac{3}{5} + \frac{9}{25} + \frac{27}{125} = \frac{272}{125}$

Because the series is geometric,

$$S_n = \frac{1 - \left(\frac{3}{5}\right)^n}{1 - \frac{3}{5}} = \frac{5}{2}\left[1 - \left(\frac{3}{5}\right)^n\right].$$

73. $\sum_{n=1}^{75} n = \frac{75(76)}{2} = 2850$

75. $a_1 = f(1) = 5, a_n = a_{n-1} + 5$

$a_1 = 5$

$a_2 = 5 + 5 = 10$

$a_3 = 10 + 5 = 15$

$a_4 = 15 + 5 = 20$

$a_5 = 20 + 5 = 25$

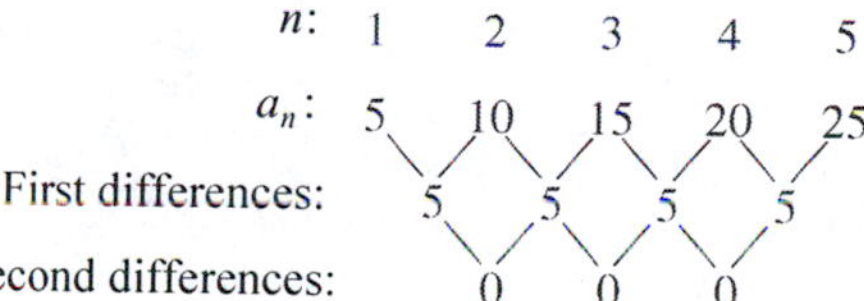

Because the first differences are all the same, the sequence has a linear model.

77. ${}_6C_4 = \frac{6!}{2!4!} = 15$

79.

1
1 1
1 2 1
1 3 3 1
1 4 6 4 1
1 5 10 10 5 1
1 6 15 20 15 6 1
1 7 (21) 35 35 21 7 1

$\binom{7}{2} = 21$, the 3rd entry in the 7th row.

81. $(x+4)^4 = x^4 + 4x^3(4) + 6x^2(4)^2 + 4x(4)^3 + 4^4 = x^4 + 16x^3 + 96x^2 + 256x + 256$

83. $(5+2i)^4 = (5)^4 + 4(5)^3(2i) + 6(5)^2(2i)^2 + 4(5)(2i)^3 + (2i)^4$

$= 625 + 1000i + 600i^2 + 160i^3 + 16i^4$

$= 625 + 1000i - 600 - 160i + 16 = 41 + 840i$

85.

First number:	1	2	3	4	5	6	7	8	9	10	11
Second number:	11	10	9	8	7	6	5	4	3	2	1

From this list, you can see that a total of 12 occurs 11 different ways.

87. $(10)(10)(10)(10) = 10{,}000$ different telephone numbers

89. $5 \cdot 4 \cdot 3 \cdot 2 \cdot 1 = 120$

91. ${}_{32}C_{12} = \dfrac{32!}{20!12!} = 225{,}792{,}840$

93. $(1)\left(\dfrac{1}{9}\right) = \dfrac{1}{9}$

95. (a) $25\% + 18\% = 43\%$

(b) $100\% - 18\% = 82\%$

97. $\left(\frac{1}{6}\right)\left(\frac{1}{6}\right)\left(\frac{1}{6}\right)\left(\frac{1}{6}\right) = \frac{1}{1296}$

99. $1 - \frac{13}{52} = 1 - \frac{1}{4} = \frac{3}{4}$

101. False. $\dfrac{(n+2)!}{n!} = \dfrac{(n+2)(n+1)\cancel{n!}}{\cancel{n!}}$

$= (n+2)(n+1)$

$\neq \dfrac{n+2}{n}$

103. True. $\displaystyle\sum_{k=1}^{8} 3k = 3\sum_{k=1}^{8} k$ by the Properties of Sums.

105. The domain of an infinite sequence is the set of natural numbers.

107. Each term of the sequence is defined in terms of preceding terms.

Problem Solving for Chapter 11

1. $a_n = \dfrac{n+1}{n^2+1}$

(a)

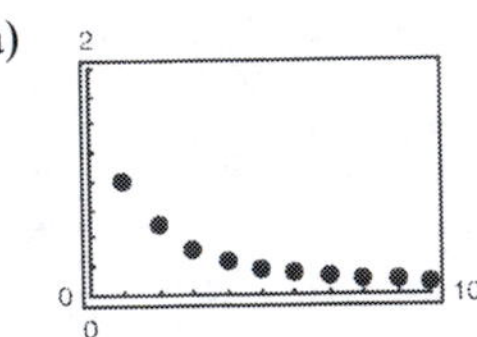

(b) $a_n \to 0$ as $n \to \infty$

(c)

n	1	10	100	1000	10,000
a_n	1	0.1089	0.0101	0.0010	0.0001

(d) $a_n \to 0$ as $n \to \infty$

3. Distance: $\displaystyle\sum_{n=1}^{\infty} 20\left(\frac{1}{2}\right)^{n-1} = \frac{20}{1-\frac{1}{2}} = 40$

Time: $\displaystyle\sum_{n=1}^{\infty} \left(\frac{1}{2}\right)^{n-1} = \frac{1}{1-\frac{1}{2}} = 2$

In two seconds, both Achilles and the tortoise will be 40 feet away from Achilles' starting point.

5. Let $a_n = dn + c$, an arithmetic sequence with a common difference of d.

(a) If C is added to each term, then the resulting sequence, $b_n = a_n + C = dn + c + C$, is still arithmetic with a common difference of d.

(b) If each term is multiplied by a nonzero constant C, then the resulting sequence,

$b_n = C(dn + c) = Cdn + Cc$, is still arithmetic.

The common difference is Cd.

(c) If each term is squared, the resulting sequence,

$b_n = a_n^2 = (dn + c)^2$, is not arithmetic.

7. $a_n = \begin{cases} \dfrac{a_{n-1}}{2}, & \text{if } a_{n-1} \text{ is even} \\ 3a_{n-1} + 1 & \text{if } a_{n-1} \text{ is odd} \end{cases}$

(a) $a_1 = 7$

$a_2 = 3(7) + 1 = 22$

$a_3 = \frac{22}{3} = 11$

$a_4 = 3(11) + 1 = 34$

$a_5 = \frac{34}{2} = 17$

$a_6 = 3(17) + 1 = 52$

$a_7 = \frac{52}{2} = 26$

$a_8 = \frac{26}{2} = 13$

$a_9 = 3(13) + 1 = 40$

$a_{10} = \frac{40}{2} = 20$

$a_{11} = \frac{20}{2} = 10$

$a_{12} = \frac{10}{2} = 5$

$a_{13} = 3(5) + 1 = 16$

$a_{14} = \frac{16}{2} = 8$

$a_{15} = \frac{8}{2} = 4$

$a_{16} = \frac{4}{2} = 2$

$a_{17} = \frac{2}{2} = 1$

$a_{18} = 3(1) + 1 = 4$

$a_{19} = \frac{4}{2} = 2$

$a_{20} = \frac{2}{2} = 1$

(b)

$a_1 = 4$	$a_1 = 5$	$a_1 = 12$
$a_2 = 2$	$a_2 = 16$	$a_2 = 6$
$a_3 = 1$	$a_3 = 8$	$a_3 = 3$
$a_4 = 4$	$a_4 = 4$	$a_4 = 10$
$a_5 = 2$	$a_5 = 2$	$a_5 = 5$
$a_6 = 1$	$a_6 = 1$	$a_6 = 16$
$a_7 = 4$	$a_7 = 4$	$a_7 = 8$
$a_8 = 2$	$a_8 = 2$	$a_8 = 4$
$a_9 = 1$	$a_9 = 1$	$a_9 = 2$
$a_{10} = 4$	$a_{10} = 4$	$a_{10} = 1$

Eventually the terms repeat: 4, 2, 1

9. The numbers 1, 5, 12, 22, 35, 51, … can be written recursively as $P_n = P_{n-1} + (3n - 2)$. Show that $P_n = n(3n - 1)/2$.

1. For $n = 1$: $1 = \dfrac{1(3 - 1)}{2}$

2. Assume $P_k = \dfrac{k(3k - 1)}{2}$.

Then, $P_{k+1} = P_k + [3(k + 1) - 2]$

$$= \frac{k(3k - 1)}{2} + (3k + 1) = \frac{k(3k - 1) + 2(3k + 1)}{2}$$

$$= \frac{3k^2 + 5k + 2}{2} = \frac{(k + 1)(3k + 2)}{2}$$

$$= \frac{(k + 1)[3(k + 1) - 1]}{2}.$$

So, by mathematical induction, the formula is valid for all integers $n \geq 1$.

11. Side lengths: $1, \frac{1}{2}, , \frac{1}{8}, \ldots$

$$S_n = \left(\frac{1}{2}\right)^{n-1} \text{ for } n \geq 1$$

Areas: $\dfrac{\sqrt{3}}{4}, \dfrac{\sqrt{3}}{4}\left(\dfrac{1}{2}\right)^2, \dfrac{\sqrt{3}}{4}\left(\dfrac{1}{4}\right)^2, \dfrac{\sqrt{3}}{4}\left(\dfrac{1}{8}\right)^2, \ldots$

$$A_n = \frac{\sqrt{3}}{4}\left[\left(\frac{1}{2}\right)^{n-1}\right]^2 = \frac{\sqrt{3}}{4}\left(\frac{1}{2}\right)^{2n-2} = \frac{\sqrt{3}}{4}S_n^2$$

13. $\frac{1}{3}$

15. (a) Odds against choosing a red marble $= \dfrac{\text{number of non-red marbles}}{\text{number of red marbles}}$

$$\frac{4}{1} = \frac{x}{6}$$

$$24 = x \quad \text{(number of non-red marbles)}$$

$$\text{Total marbles} = 6 + 24 = 30$$

(b) Odds in favor of choosing a blue marble $= \dfrac{\text{number of blue marbles}}{\text{number of yellow marbles}} = \dfrac{3}{7}$

Odds against choosing a blue marble $= \dfrac{\text{number of yellow marbles}}{\text{number of blue marbles}} = \dfrac{7}{3}$

(c) $P(E) = \dfrac{n(E)}{n(S)} = \dfrac{n(E)}{n(E) + n(E')} = \dfrac{n(E)/n(E')}{n(E)/n(E') + n(E')/n(E')}$

$P(E) = \dfrac{\text{odds in favor of } E}{\text{odds in favor of } E + 1}$

(d) $P(E) = \dfrac{n(E)}{n(S)}$ $\qquad P(E') = \dfrac{n(E')}{n(S)}$

$n(S)P(E) = n(E)$ $\qquad n(S)P(E') = n(E')$

Odds in favor of event $E = \dfrac{n(E)}{n(E')} = \dfrac{n(S)P(E)}{n(S)P(E')} = \dfrac{P(E)}{P(E')}$

Practice Test for Chapter 11

1. Write out the first five terms of the sequence $a_n = \dfrac{2n}{(n+2)!}$.

2. Write an expression for the nth term of the sequence $\frac{4}{3}, \frac{5}{9}, \frac{6}{27}, \frac{7}{81}, \frac{8}{243}, \ldots$.

3. Find the sum $\displaystyle\sum_{i=1}^{6} (2i - 1)$.

4. Write out the first five terms of the arithmetic sequence where $a_1 = 23$ and $d = -2$.

5. Find a_n for the arithmetic sequence with $a_1 = 12$, $d = 3$, and $n = 50$.

6. Find the sum of the first 200 positive integers.

7. Write out the first five terms of the geometric sequence with $a_1 = 7$ and $r = 2$.

8. Evaluate $\displaystyle\sum_{n=1}^{10} 6\left(\frac{2}{3}\right)^{n-1}$.

9. Evaluate $\displaystyle\sum_{n=0}^{\infty} (0.03)^n$.

10. Use mathematical induction to prove that $1 + 2 + 3 + 4 + \cdots + n = \dfrac{n(n+1)}{2}$.

11. Use mathematical induction to prove that $n! > 2^n$, $n \ge 4$.

12. Evaluate ${}_{13}C_4$.

13. Expand $(x + 3)^5$.

14. Find the term involving x^7 in $(x - 2)^{12}$.

15. Evaluate ${}_{30}P_4$.

16. How many ways can six people sit at a table with six chairs?

17. Twelve cars run in a race. How many different ways can they come in first, second, and third place? (Assume that there are no ties.)

18. Two six-sided dice are tossed. Find the probability that the total of the two dice is less than 5.

19. Two cards are selected at random from a deck of 52 playing cards without replacement. Find the probability that the first card is a King and the second card is a black ten.

20. A manufacturer has determined that for every 1000 units it produces, 3 will be faulty. What is the probability that an order of 50 units will have one or more faulty units?

Appendix A Errors and the Algebra of Calculus

1. numerator

3. $2x - (3y + 4) \neq 2x - 3y + 4$

Change all signs when distributing the minus sign.

$2x - (3y + 4) = 2x - 3y - 4$

5. $\dfrac{4}{16x - (2x + 1)} \neq \dfrac{4}{14x + 1}$

Change all signs when distributing the minus sign.

$\dfrac{4}{16x - (2x + 1)} = \dfrac{4}{16x - 2x - 1} = \dfrac{4}{14x - 1}$

7. $(5z)(6z) \neq 30z$

z occurs twice as a factor.

$(5z)(6z) = 30z^2$

9. $a\left(\dfrac{x}{y}\right) \neq \dfrac{ax}{ay}$

The fraction as a whole is multiplied by a, not the numerator and denominator separately.

$a\left(\dfrac{x}{y}\right) = \dfrac{a}{1} \cdot \dfrac{x}{y} = \dfrac{ax}{y}$

11. $\sqrt{x + 9} \neq \sqrt{x} + 3$

Do not apply the radical to the terms.

$\sqrt{x + 9}$ does not simplify.

13. $\dfrac{2x^2 + 1}{5x} \neq \dfrac{2x + 1}{5}$

Divide out common factors not common terms.

$\dfrac{2x^2 + 1}{5x}$ cannot be simplified.

15. $\dfrac{1}{a^{-1} + b^{-1}} \neq \left(\dfrac{1}{a + b}\right)^{-1}$

To get rid of negative exponents:

$\dfrac{1}{a^{-1} + b^{-1}} = \dfrac{1}{a^{-1} + b^{-1}} \cdot \dfrac{ab}{ab} = \dfrac{ab}{b + a}$

17. $(x^2 + 5x)^{1/2} \neq x(x + 5)^{1/2}$

Factor within grouping symbols before applying the exponent to each factor.

$(x^2 + 5x)^{1/2} = [x(x + 5)]^{1/2} = x^{1/2}(x + 5)^{1/2}$

19. $\dfrac{3}{x} + \dfrac{4}{y} = \dfrac{3}{x} \cdot \dfrac{y}{y} + \dfrac{4}{y} \cdot \dfrac{x}{x} = \dfrac{3y + 4x}{xy}$

To add fractions, they must have a common denominator.

21. To add fractions, first find a common denominator.

$\dfrac{x}{2y} + \dfrac{y}{3} = \dfrac{3x}{6y} + \dfrac{2y^2}{6y} = \dfrac{3x + 2y^2}{6y}$

23. $\dfrac{5x + 3}{4} = \dfrac{1}{4}(5x + 3)$

The required factor is $5x + 3$.

25. $\frac{2}{3}x^2 + \frac{1}{3}x + 5 = \frac{2}{3}x^2 + \frac{1}{3}x + \frac{15}{3} = \frac{1}{3}(2x^2 + x + 15)$

The required factor is $2x^2 + x + 15$.

27. $x^2(x^3 - 1)^4 = \frac{1}{3}(x^3 - 1)^4(3x^2)$

The required factor is $\frac{1}{3}$.

29. $2(y - 5)^{1/2} + y(y - 5)^{-1/2} = (y - 5)^{-1/2}(3y - 10)$

The required factor is $3y - 10$.

31. $\dfrac{4x + 6}{(x^2 + 3x + 7)^3} = \dfrac{2(2x + 3)}{(x^2 + 3x + 7)^3} = \dfrac{2}{1} \cdot \dfrac{(2x + 3)}{1} \cdot \dfrac{1}{(x^2 + 3x + 7)^3} = (2)\dfrac{1}{(x^2 + 3x + 7)^3}(2x + 3)$

The required factor is 2.

33. $\dfrac{3}{x} + \dfrac{5}{2x^2} - \dfrac{3}{2}x = \dfrac{6x}{2x^2} + \dfrac{5}{2x^2} - \dfrac{3x^3}{2x^2}$

$= \left(\dfrac{1}{2x^2}\right)(6x + 5 - 3x^3)$

The required factor is $\dfrac{1}{2x^2}$.

35. $\dfrac{25x^2}{36} + \dfrac{4y^2}{9} = \dfrac{x^2}{\frac{36}{25}} + \dfrac{y^2}{\frac{9}{4}}$

The required factors are $\dfrac{36}{25}$ and $\dfrac{9}{4}$.

37. $\dfrac{x^2}{\frac{3}{10}} - \dfrac{y^2}{\frac{4}{5}} = \dfrac{10x^2}{3} - \dfrac{5y^2}{4}$

The required factors are 3 and 4.

39. $x^{1/3} - 5x^{4/3} = x^{1/3}\left(1 - 5x^{3/3}\right) = x^{1/3}(1 - 5x)$

The required factor is $1 - 5x$.

41. $$\begin{aligned}(1 - 3x)^{4/3} - 4x(1 - 3x)^{1/3} &= (1 - 3x)^{1/3}\left[(1 - 3x)^1 - 4x\right]\\ &= (1 - 3x)^{1/3}(1 - 7x)\end{aligned}$$

The required factor is $1 - 7x$.

43. $$\begin{aligned}\tfrac{1}{10}(2x + 1)^{5/2} - \tfrac{1}{6}(2x + 1)^{3/2} &= \tfrac{3}{30}(2x + 1)^{3/2}(2x + 1)^1 - \tfrac{5}{30}(2x + 1)^{3/2}\\ &= \tfrac{1}{30}(2x + 1)^{3/2}\left[3(2x + 1) - 5\right]\\ &= \tfrac{1}{30}(2x + 1)^{3/2}(6x - 2)\\ &= \tfrac{1}{30}(2x + 1)^{3/2}2(3x - 1)\\ &= \tfrac{1}{15}(2x + 1)^{3/2}(3x - 1)\end{aligned}$$

The required factor is $3x - 1$.

45. $\dfrac{7}{(x + 3)^5} = 7(x + 3)^{-5}$

47. $\dfrac{2x^5}{(3x + 5)^4} = 2x^5(3x + 5)^{-4}$

49. $\dfrac{4}{3x} + \dfrac{4}{x^4} - \dfrac{7x}{\sqrt[3]{2x}} = 4(3x)^{-1} + 4x^{-4} - 7x(2x)^{-1/3}$

51. $$\begin{aligned}\frac{x^2 + 6x + 12}{3x} &= \frac{x^2}{3x} + \frac{6x}{3x} + \frac{12}{3x}\\ &= \frac{x}{3} + 2 + \frac{4}{x}\end{aligned}$$

53. $$\begin{aligned}\frac{4x^3 - 7x^2 + 1}{x^{1/3}} &= \frac{4x^3}{x^{1/3}} - \frac{7x^2}{x^{1/3}} + \frac{1}{x^{1/3}}\\ &= 4x^{3-1/3} - 7x^{2-1/3} + \frac{1}{x^{1/3}}\\ &= 4x^{8/3} - 7x^{5/3} + \frac{1}{x^{1/3}}\end{aligned}$$

55. $$\begin{aligned}\frac{3 - 5x^2 - x^4}{\sqrt{x}} &= \frac{3}{\sqrt{x}} - \frac{5x^2}{\sqrt{x}} - \frac{x^4}{\sqrt{x}}\\ &= \frac{3}{\sqrt{x}} - 5x^{2-1/2} - x^{4-1/2}\\ &= \frac{3}{x^{1/2}} - 5x^{3/2} - x^{7/2}\end{aligned}$$

57. $$\begin{aligned}\frac{-2\left(x^2 - 3\right)^{-3}(2x)(x + 1)^3 - 3(x + 1)^2\left(x^2 - 3\right)^{-2}}{\left[(x + 1)^3\right]^2} &= \frac{\left(x^2 - 3\right)^{-3}(x + 1)^2\left[-4x(x + 1) - 3\left(x^2 - 3\right)\right]}{(x + 1)^6}\\ &= \frac{-4x^2 - 4x - 3x^2 + 9}{\left(x^2 - 3\right)^3(x + 1)^4}\\ &= \frac{-7x^2 - 4x + 9}{\left(x^2 - 3\right)^3(x + 1)^4}\end{aligned}$$

59. $$\begin{aligned}\frac{(6x + 1)^3\left(27x^2 + 2\right) - \left(9x^3 + 2x\right)(3)(6x + 1)^2(6)}{\left[(6x + 1)^3\right]^2} &= \frac{(6x + 1)^2\left[(6x + 1)\left(27x^2 + 2\right) - 18\left(9x^3 + 2x\right)\right]}{(6x + 1)^6}\\ &= \frac{162x^3 + 12x + 27x^2 + 2 - 162x^3 - 36x}{(6x + 1)^4}\\ &= \frac{27x^2 - 24x + 2}{(6x + 1)^4}\end{aligned}$$

61. $$\frac{(x+2)^{3/4}(x+3)^{-2/3} - (x+3)^{1/3}(x+2)^{-1/4}}{\left[(x+2)^{3/4}\right]^2} = \frac{(x+2)^{-1/4}(x+3)^{-2/3}\left[(x+2)-(x+3)\right]}{(x+2)^{6/4}}$$

$$= \frac{x+2-x-3}{(x+2)^{1/4}(x+3)^{2/3}(x+2)^{6/4}}$$

$$= -\frac{1}{(x+3)^{2/3}(x+2)^{7/4}}$$

63. $$\frac{2(3x-1)^{1/3} - (2x+1)(1/3)(3x-1)^{-2/3}(3)}{(3x-1)^{2/3}} = \frac{(3x-1)^{-2/3}\left[2(3x-1)-(2x+1)\right]}{(3x-1)^{2/3}}$$

$$= \frac{6x-2-2x-1}{(3x-1)^{2/3}(3x-1)^{2/3}}$$

$$= \frac{4x-3}{(3x-1)^{4/3}}$$

65. $$\frac{1}{(x^2+4)^{1/2}} \cdot \frac{1}{2}(x^2+4)^{-1/2}(2x) = \frac{1}{(x^2+4)^{1/2}} \cdot \frac{1}{(x^2+4)^{1/2}} \cdot \frac{1}{2}(2x)$$

$$= \frac{1}{(x^2+4)^1}(x)$$

$$= \frac{x}{x^2+4}$$

67. $$(x^2+5)^{1/2}\left(\frac{3}{2}\right)(3x-2)^{1/2}(3) + (3x-2)^{3/2}\left(\frac{1}{2}\right)(x^2+5)^{-1/2}(2x) = \frac{9}{2}(x^2+5)^{1/2}(3x-2)^{1/2} + x(x^2+5)^{-1/2}(3x-2)^{3/2}$$

$$= \frac{9}{2}(x^2+5)^{1/2}(3x-2)^{1/2} + \frac{2}{2}x(x^2+5)^{-1/2}(3x-2)^{3/2}$$

$$= \frac{1}{2}(x^2+5)^{-1/2}(3x-2)^{1/2}\left[9(x^2+5)^1 + 2x(3x-2)^1\right]$$

$$= \frac{1}{2}(x^2+5)^{-1/2}(3x-2)^{1/2}(9x^2+45+6x^2-4x)$$

$$= \frac{(3x-2)^{1/2}(15x^2-4x+45)}{2(x^2+5)^{1/2}}$$

69. $t = \dfrac{\sqrt{x^2 + 4}}{2} + \dfrac{\sqrt{(4 - x)^2 + 4}}{6}$

(a)

x	t
0.5	1.70
1.0	1.72
1.5	1.78
2.0	1.89
2.5	2.02
3.0	2.18
3.5	2.36
4.0	2.57

(b) She should swim to a point about $\dfrac{1}{2}$ mile down the coast to minimize the time required to reach the finish line.

(c)
$$\begin{aligned}
\frac{1}{2}x(x^2 + 4)^{-1/2} + \frac{1}{6}(x - 4)(x^2 - 8x + 20)^{-1/2} &= \frac{3}{6}x(x^2 + 4)^{-1/2} + \frac{1}{6}(x - 4)(x^2 - 8x + 20)^{-1/2} \\
&= \frac{1}{6}\left[3x(x^2 + 4)^{-1/2} + (x - 4)(x^2 - 8x + 20)^{-1/2}\right] \\
&= \frac{1}{6}\left[\frac{3x}{(x^2 + 4)^{1/2}} + \frac{x - 4}{(x^2 - 8x + 20)^{1/2}}\right] \\
&= \frac{3x\sqrt{x^2 - 8x + 20} + (x - 4)\sqrt{x^2 + 4}}{6\sqrt{x^2 + 4}\sqrt{x^2 - 8x + 20}}
\end{aligned}$$

71. You cannot move term-by-term from the denominator to the numerator.

CHECKPOINTS

CHECKPOINTS
Chapter P

Checkpoints for Section P.1

1. (a) Natural numbers: $\left\{\frac{6}{3}, 8\right\}$

(b) Whole numbers: $\left\{\frac{6}{3}, 8\right\}$

(c) Integers: $\left\{-22, -1, \frac{6}{3}, 8\right\}$

(d) Rational numbers: $\left\{-22, -7.5, -1 - \frac{1}{4}, \frac{6}{3}, 8\right\}$

(e) Irrational numbers: $\left\{-\pi, \frac{1}{2}\sqrt{2}\right\}$

2. $-1.6 \quad -\frac{3}{4} \quad 0.7 \quad \frac{5}{2}$

(a) The point representing the real number $\frac{5}{2} = 2.5$ lies halfway between 2 and 3, on the real number line.

(b) The point representing the real number -1.6 lies between -2 and -1 but closer to -2, on the real number line.

(c) The point representing the real number $-\frac{3}{4}$ lies between -1 and 0 but closer to -1, on the real number line.

(d) The point representing the real number 0.7 lies between 0 and 1 but closer to 1, on the real number line.

3. (a) Because -5 lies to the left of 1 on the real number line, you can say that -5 is *less than* 1, and write $-5 < 1$.

(b) Because $\frac{3}{2}$ lies to the left of 7 on the real number line, you can say that $\frac{3}{2}$ is *less than* 7, and write $\frac{3}{2} < 7$.

(c) Because $-\frac{2}{3}$ lies to the right of $-\frac{3}{4}$ on the real number line, you can say that $-\frac{2}{3}$ is *greater than* $-\frac{3}{4}$, and write $-\frac{2}{3} > -\frac{3}{4}$.

(d) Because -3.5 lies to the left of 1 on the real number line, you can say that -3.5 is *less than* 1, and write $-3.5 < 1$.

4. (a) The inequality $x > -3$ denotes all real numbers greater than -3.

(b) The inequality $0 < x \le 4$ means that $x > 0$ and $x \le 4$. This double inequality denotes all real numbers between 0 and 4, including 4 but not including 0.

5. The interval consists of real numbers greater than or equal to -2 and less than 5.

6. The inequality $-2 < x \le 4$ can represent the statement "x is greater than -2 and at most 4."

7. (a) $|1| = 1$

(b) $-\left|\frac{3}{4}\right| = -\left(\frac{3}{4}\right) = -\frac{3}{4}$

(c) $\frac{2}{|-3|} = \frac{2}{3}$

(d) $-|0.7| = -(0.7) = -0.7$

8. (a) If $x > -3$, then $\frac{|x+3|}{x+3} = \frac{x+3}{x+3} = 1$.

(b) If $x < -3$, then $\frac{|x+3|}{x+3} = \frac{-(x+3)}{x+3} = -1$.

9. (a) $|-3| < |4|$ because $|-3| = 3$ and $|4| = 4$, and 3 is less than 4.

(b) $-|-4| = -|-4|$ because $-|-4| = -4$ and $-|4| = -4$.

(c) $|-3| > -|-3|$ because $|-3| = 3$ and $-|-3| = -3$, and 3 is greater than -3.

10. (a) The distance between 35 and -23 is $|35 - (-23)| = |58| = 58$.

(b) The distance between -35 and -23 is $|-35 - (-23)| = |-12| = 12$.

(c) The distance between 35 and 23 is $|35 - 23| = |12| = 12$.

11.

Algebraic Expression	Terms	Coefficients
$-2x + 4$	$-2x, 4$	$-2, 4$

12.

Expression	Value of Variable	Substitute	Value of Expression
$4x - 5$	$x - 0$	$4(0) - 5$	$0 - 5 = -5$

13. (a) $x + 9 = 9 + x$: This statement illustrates the Commutative Property of Addition. In other words, you obtain the same result whether you add x and 9, or 9 and x.

(b) $5(x^3 \cdot 2) = (5x^3)2$: This statement illustrates the Associative Property of Multiplication. In other words, to form the product $5 \cdot x^3 \cdot 2$, it does not matter whether 5 and $(x^3 \cdot 2)$, or $5x^3$ and 2 are multiplied first.

(c) $(2 + 5x^2)y^2 = 2y^2 + 5x^2 \cdot y^2$: This statement illustrates the Distributive Property. In other words, the terms 2 and $5x^2$ are multiplied by y^2.

14. (a) $\frac{3}{5} \cdot \frac{x}{6} = \frac{3x}{30} = \frac{3x \div 3}{30 \div 3} = \frac{x}{10}$

(b) $$\begin{aligned} \frac{x}{10} + \frac{2x}{5} &= \frac{x}{10} + \frac{2x}{5} \cdot \frac{2}{2} \\ &= \frac{x}{10} + \frac{2x}{5} = \frac{x}{10} + \frac{2x}{5} \cdot \frac{2}{2} \\ &= \frac{x}{10} + \frac{4x}{10} \\ &= \frac{x + 4x}{10} \\ &= \frac{5x \div 5}{10 \div 5} \\ &= \frac{x}{2} \end{aligned}$$

Checkpoints for Section P.2

1. (a) $-3^4 = -(3)(3)(3)(3) = -81$

(b) $(-3)^4 = (-3)(-3)(-3)(-3) = 81$

(c) $3^2 \cdot 3 = 3^{2+1} = 3^3 = (3)(3)(3) = 27$

(d) $\frac{3^5}{3^8} = 3^{5-8} = 3^{-3} = \frac{1}{3^3} = \frac{1}{27}$

2. (a) When $x = 4$, the expression $-x^{-2}$ has a value of

$$-x^{-2} = -(4)^{-2} = -\frac{1}{4^2} = -\frac{1}{16}.$$

(b) When $x = 4$, the expression $\frac{1}{4}(-x)^4$ has a value of

$$\frac{1}{4}(-x)^4 = \frac{1}{4}(-4)^4 = \frac{1}{4}(256) = 64.$$

3. (a)

$$(2x^{-2}y^3)(-x^4y) = (2)(-1)(x^{-2})(x^4)(y^3)(y) = -2x^2y^4$$

(b) $(4a^2b^3)^0 = 1,\ a \neq 0,\ b \neq 0$

(c) $$\begin{aligned} (-5z)^3(z^2) &= (-5)^3(z)^3z^2 \\ &= -125z^5 \end{aligned}$$

(d) $$\begin{aligned} \left(\frac{3x^4}{x^2y^2}\right)^2 &= \left(\frac{3x^2}{4^2}\right)^2 = \frac{3^2(x^2)^2}{(y^2)^2} \\ &= \frac{9x^4}{y^4},\ x \neq 0 \end{aligned}$$

4. (a) $2a^{-2} = \frac{2}{a^2}$ Property 3

(b) $\frac{3a^{-3}b^4}{15ab^{-1}} = \frac{3b^4 \cdot b}{15a \cdot a^3}$ Property 3

$= \frac{b^5}{5a^4}$ Property 1

(c) $\left(\frac{x}{10}\right)^{-1} = \frac{x^{-1}}{10^{-1}}$ Property 7

$= \frac{10}{x}$ Property 3

(d) $(-2x^2)^3(4x^3)^{-1} = (-2)^3(x^2)^3 \cdot 4^{-1} \cdot (x^3)^{-1}$ Property 5

$= \frac{-8x^6}{4x^3}$ Properties 3 and 6

$= -2x^3$ Property 2

5. $45{,}850 = 4.585 \times 10^4$

6. $-2.718 \times 10^{-3} = -0.002718$

7. $$\begin{aligned} &(24{,}000{,}000{,}000)(0.00000012)(300{,}000) \\ &= (2.4 \times 10^{10})(1.2 \times 10^{-7})(3.0 \times 10^5) \\ &= (2.4)(1.2)(3.0)(10^8) \\ &= 8.64 \times 10^8 \\ &= 864{,}000{,}000 \end{aligned}$$

8. (a) $-\sqrt{144} = -12$ because $-\left(\sqrt{144}\right) = \left(\sqrt{12^2}\right) = -(12) = -12.$

(b) $\sqrt{-144}$ is not a real number because no real number raised to the second power produces -144.

(c) $\sqrt{\dfrac{25}{64}} = \dfrac{5}{8}$ because $\left(\dfrac{5}{8}\right)^2 = \dfrac{5^2}{8^2} = \dfrac{25}{64}.$

(d) $-\sqrt[3]{\dfrac{8}{27}} = -\dfrac{2}{3}$ because $-\left(\sqrt[3]{\dfrac{8}{27}}\right) = -\left(\dfrac{\sqrt[3]{8}}{\sqrt[3]{27}}\right) = -\left(\dfrac{2}{3}\right).$

9. (a) $\dfrac{\sqrt{125}}{\sqrt{5}} = \sqrt{\dfrac{125}{5}}$ Property 3

$= \sqrt{25}$ Simplify.

$= 5$ Simplify.

(b) $\sqrt[3]{125^2} = \left(\sqrt[3]{125}\right)^2$ Property 1

$= (5)^2$ Simplify.

$= 25$ Simplify.

(c) $\sqrt[3]{x^2} \cdot \sqrt[3]{x} = \sqrt[3]{x^2 \cdot x}$ Property 2

$= \sqrt[3]{x^3}$ Simplify.

$= x$ Property

(d) $\sqrt{\sqrt{x}} = \sqrt[2\cdot 2]{x}$ Property 4

$= \sqrt[4]{x}$ Simplify.

10. (a) $\sqrt{32} = \sqrt{16 \cdot 2} = \sqrt{4^2 \cdot 2} = 4\sqrt{2}$

(b) $\sqrt[3]{250} = \sqrt[3]{125 \cdot 2} = \sqrt[3]{5^3 \cdot 2} = 5\sqrt[3]{2}$

(c) $\sqrt{24a^5} = \sqrt{4 \cdot 6 \cdot a^4 \cdot a} = \sqrt{4a^4 \cdot 6a}$

$= \sqrt{(2a^2)^2 \cdot 6a}$

$= 2a^2\sqrt{6a}$

(d) $\sqrt[3]{-135x^3} = \sqrt[3]{(-27) \cdot 5 \cdot x^3}$

$= \sqrt[3]{(-3x)^3 \cdot 5}$

$= -3x\sqrt[3]{5}$

11. (a) $3\sqrt{8} + \sqrt{18} = 3\sqrt{4 \cdot 2} + \sqrt{9 \cdot 2}$ Find square factors.

$= 3 \cdot 2\sqrt{2} + 3\sqrt{2}$ Find square roots.

$= 6\sqrt{2} + 3\sqrt{2}$ Multiply.

$= (6 + 3)\sqrt{2}$ Combine like radicals.

$= 9\sqrt{2}$ Simplify.

(b) $\sqrt[3]{81x^5} - \sqrt[3]{24x^2} = \sqrt[3]{27x^3 \cdot 3x^2} - \sqrt[3]{8 \cdot 3x^2}$ Find cube factors.

$= 3x\sqrt[3]{3x^2} - 2\sqrt[3]{3x^2}$ Find cube roots.

$= (3x - 2)\sqrt[3]{3x^2}$ Combine like radicals.

12. (a) $\dfrac{5}{3\sqrt{2}} = \dfrac{5}{3\sqrt{2}} \cdot \dfrac{\sqrt{2}}{\sqrt{2}}$ $\sqrt{2}$ is rationalizing factor.

$= \dfrac{5\sqrt{2}}{3(2)}$ Multiply.

$= \dfrac{5\sqrt{2}}{6}$ Simplify.

(b) $\dfrac{1}{\sqrt[3]{25}} = \dfrac{1}{\sqrt[3]{25}} \cdot \dfrac{\sqrt[3]{5}}{\sqrt[3]{5}}$ $\sqrt[3]{5}$ is rationalizing factor.

$= \dfrac{\sqrt[3]{5}}{\sqrt[3]{125}}$ Multiply.

$= \dfrac{\sqrt[3]{5}}{5}$ Simplify.

13. $\dfrac{8}{\sqrt{6}-\sqrt{2}} = \dfrac{8}{\sqrt{6}-\sqrt{2}} \cdot \dfrac{\sqrt{6}+\sqrt{2}}{\sqrt{6}+\sqrt{2}}$ Multiply numerator and denominator by conjugate of denominator.

$= \dfrac{8(\sqrt{6}+\sqrt{2})}{6+\sqrt{12}-\sqrt{12}-2}$ Use Distributive Property.

$= \dfrac{8(\sqrt{6}+\sqrt{2})}{4}$ Simplify.

$= 2(\sqrt{6}+\sqrt{2})$ Simplify.

14. $\dfrac{2-\sqrt{2}}{3} = \dfrac{2-\sqrt{2}}{3} \cdot \dfrac{2+\sqrt{2}}{2+\sqrt{2}}$ Multiply numerator and denominator by conjugate of numerator.

$= \dfrac{4+2\sqrt{2}-2\sqrt{2}-2}{3(2+\sqrt{2})}$ Multiply.

$= \dfrac{2}{3(2+\sqrt{2})}$ Simplify.

15. (a) $\sqrt[3]{27} = 27^{1/3}$

(b) $\sqrt{x^3y^5z} = (x^3y^5z)^{1/2}$

$= x^{3\cdot1/2}y^{5\cdot1/2}z^{1/2}$

$= x^{3/2}y^{5/2}z^{1/2}$

(c) $3x\sqrt[3]{x^2} = 3x(x^2)^{1/3}$

$= 3x \cdot x^{2/3}$

$= 3x^{1+2/3}$

$= 3x^{5/3}$

16. (a) $(x^2-7)^{-1/2} = \dfrac{1}{(x^2-7)^{1/2}} = \dfrac{1}{\sqrt{x^2-7}}$

(b) $-3b^{1/3}c^{2/3} = -3(bc^2)^{1/3} = -3\sqrt[3]{bc^2}$

(c) $a^{0.75} = a^{3/4} = \sqrt[4]{a^3}$

(d) $(x^2)^{2/5} = x^{4/5} = \sqrt[5]{x^4}$

17. (a) $(-125)^{-2/3} = \left(\sqrt[3]{-125}\right)^{-2} = (-5)^{-2} = \dfrac{1}{(-5)^2} = \dfrac{1}{25}$

(b) $(4x^2y^{3/2})(-3x^{-1/3})(y^{-3/5}) = -12x^{(2)-(1/3)}y^{(3/2)-(3/5)} = -12x^{5/3}y^{9/10},\ x \neq 0,\ y \neq 0$

(c) $\sqrt[3]{\sqrt[4]{27}} = \sqrt[12]{27} = \sqrt[12]{(3)^3} = 3^{3/12} = 3^{1/4} = \sqrt[4]{3}$

(d) $(3x+2)^{5/2}(3x+2)^{-1/2} = (3x+2)^{(5/2)-(1/2)} = (3x+2)^2,\ x \neq -2/3$

Checkpoints for Section P.3

1.

Polynomial	Standard Form	Degree	Leading Coefficient
$6-7x^3+2x$	$-7x^3+2x+6$	3	-7

2. $(2x^3-x+3)-(x^2-2x-3)$

$= 2x^3-x+3-x^2+2x+3$

$= 2x^3-x^2+(-x+2x)+(3+3)$

$= 2x^3-x^2+x+6$

3.

$$\begin{aligned}(3x-1)(x-5) &= \overset{F}{3x^2} - \overset{O}{15x} - \overset{I}{x} + \overset{L}{5}\\ &= 3x^2-16x+5\end{aligned}$$

4.

$$\begin{array}{rl}
 x^2 + 2x + 3 & \\
\times\ x^2 - 2x + 3 & \\
\hline
x^4 + 2x^3 + 3x^2 & \leftarrow x^2(x^2+2x+3)\\
-2x^3 - 4x^2 - 6x & \leftarrow -2x(x^2+2x+3)\\
3x^2 + 6x + 9 & \leftarrow 3(x^2+2x+3)\\
\hline
x^4 + 0x^3 + 2x^2 + 0x + 9 &
\end{array}$$

So, $(x^2+2x+3)(x^2-2x-3) = x^4+2x^2+9.$

5. This product has the form $(u + v)(u - v) = u^2 - v^2$.

$(3x - 2)(3x + 2) = (3x) - (2)^2 = 9x^2 - 4$

6. This product has the form $(u + v) = u^2 + 2uv + v^2$.

$(x + 10)^2 = (x)^2 + 2(x)(10) + (10)^2 = x^2 + 20x + 100$

7. This product has the form $(u - v)^3 = u^3 - 3u^2v + 3uv^2 - v^3$.

$$\begin{aligned}(4x - 1)^3 &= (4x)^3 - 3(4x)^2(1) + 3(4x)(1)^2 - (1)^3 \\ &= 64x^3 - 48x^2 + 12x - 1\end{aligned}$$

8. This product has the form $(u + v)(u - v) = u^2 - v^2$.

$$\begin{aligned}(x - 2 + 3y)(x - 2 - 3y) &= \left[(x - 2) + 3y\right]\left[(x - 2) - 3y\right] \\ &= (x - 2)^2 - (3y)^2 \\ &= x^2 - 4x + 4 - 9y^2 \\ &= x^2 - 9y^2 - 4x + 4\end{aligned}$$

9. The volume of a rectangular box is equal to the product of its length, width, and height.

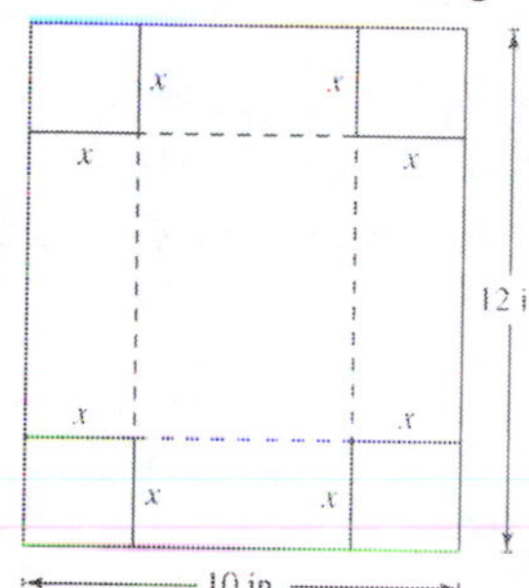

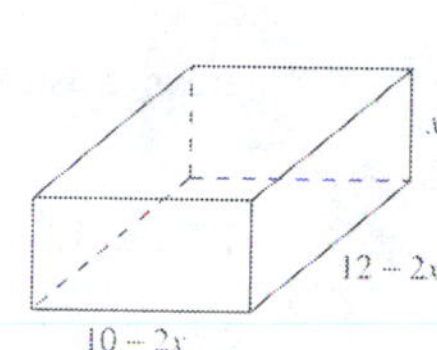

The length is $10 - 2x$, the width is $12 - 2x$, and the height is x. The volume of the box is

$$\begin{aligned}\text{Volume} &= (10 - 2x)(12 - 2x)(x) \\ &= \left(120 - 44x + 4x^2\right)(x) \\ &= 120x - 44x^2 + 4x^3\end{aligned}$$

When $x = 2$ inches the volume of the box is

$$\begin{aligned}\text{Volume} &= 120(2) - 44(2)^2 + 4(2)^3 \\ &= 96 \text{ cubic inches.}\end{aligned}$$

When $x = 3$ inches, the volume of the box is

$$\begin{aligned}\text{Volume} &= 120(3) - 44(3)^2 + 4(3)^3 \\ &= 72 \text{ cubic inches.}\end{aligned}$$

Checkpoints for Section P.4

1. (a) $5x^3 - 15x^2 = 5x^2(x) - 5x^2(3)$ $\quad$ $5x^2$ is a common factor.

$\qquad = 5x^2(x - 3)$

(b) $-3 + 6x - 12x^3 = -12x^3 + 6x - 3$

$\qquad = -3\left(4x^3\right) + (-3)(-2x) + (-3)(1)$ $\quad$ -3 is a common factor.

$\qquad = -3\left(4x^3 - 2x + 1\right)$

(c) $(x + 1)\left(x^2\right) - (x + 1)(2) = (x + 1)\left(x^2 - 2\right)$ $\quad$ $(x + 1)$ is a common factor.

2. $100 - 4y^2 = 4(25 - y^2)$ 4 is a common factor.

$= 4[(5)^2 - (y)^2]$

$= 4(5 + y)(5 - y)$ Difference of two squares.

3. $(x - 1)^2 - 9y^4 = (x - 1)^2 - (3y^2)^2$

$= [(x - 1) + 3y^2][(x - 1) - 3y^2]$

$= (x - 1 + 3y^2)(x - 1 - 3y^2)$

4. $9x^2 - 30x + 25 = (3x)^2 - 2(3x)(5) + 5^2$

$= (3x - 5)^2$

5. $64x^3 - 1 = (4x)^3 - (1)^3$

$= (4x - 1)(16x^2 - 4x + 1)$

6. (a) $x^3 + 216 = (x)^3 + (6)^3$

$= (x + 6)(x^2 - 6x + 36)$

(b) $5y^3 + 135 = 5(y^3 + 27)$

$= 5[(y)^3 + (3)^3]$

$= 5(y + 3)(y^2 - 3y + 9)$

7. For the trinomial $x^2 + x - 6$, you have $a = 1$, $b = 1$, and $c = -6$. Because b is positive and c is negative, one factor of -6 is positive and one is negative. So, the possible factorizations of $x^2 + x - 6$ are

$(x - 3)(x + 2)$,

$(x + 3)(x - 2)$,

$(x + 6)(x - 1)$, and

$(x - 6)(x + 1)$.

Testing the middle term, you will find the correct factorization to be $(x^2 + x - 6) = (x + 3)(x - 2)$.

8. (a) For the trinomial $2x^2 - 5x + 3$, you have $a = 2$ and $c = 3$, which means that the factors of 3 must have like signs. The possible factorizations are

$(2x + 1)(x + 3)$,

$(2x - 1)(x - 3)$,

$(2x + 3)(x + 1)$, and

$(2x - 3)(x - 1)$.

Testing the middle term, you will find the correct factorization to be

$2x^2 - 5x + 3 = (2x - 3)(x - 1)$.

(b) For the trinomial $12x^2 + 7x + 1$, you have $a = 12$, $b = 7$, and $c = 1$. Because a, b, and c are all positive, the factors of a and c are positive. So, the possible factorizations are

$(12x + 1)(x + 1)$,

$(6x + 1)(2x + 1)$, and

$(4x + 1)(3x + 1)$.

Testing the middle term, you will find the correct factorization to be

$12x^2 + 7x + 1 = (4x + 1)(3x + 1)$.

9. $x^3 + x^2 - 5x - 5 = (x^3 + x^2) - (5x + 5)$ Group terms.

$= x^2(x + 1) - 5(x + 1)$ Factor each group.

$= (x + 1)(x^2 - 5)$ Distributive Property

10. $2x^2 + 5x - 12 = 2x^2 + 8x - 3x - 12$ Rewrite middle term.

$= (2x^2 + 8x) - (3x + 12)$ Group terms.

$= 2x(x + 4) - 3(x + 4)$ Factor groups.

$= (x + 4)(2x - 3)$ Distributive Property

Checkpoints for Section P.5

1. (a) The domain of the polynomial $4x^2 + 3, x \geq 0$ is the set of all real numbers that are greater than or equal to 0. The domain is specifically restricted.

(b) The domain of the radical expression $\sqrt{x+7}$ is the set of all real numbers greater than or equal to -7, because the square root of a negative number is not a real number.

(c) The domain of the rational expression $\dfrac{1-x}{x}$ is the set of all real number except $x = 0$, which would result in division by zero, which is undefined.

2. $$\frac{4x+12}{x^2-3x-18} = \frac{4(x+3)}{(x-6)(x+3)} \quad \text{Factor completely.}$$

$$= \frac{4}{x-6}, x \neq -3 \quad \text{Divide out common factor.}$$

3. $$\frac{3x^2-x-2}{5-4x-x^2} = \frac{3x^2-x-2}{-x^2-4x+5} = \frac{(3x+2)(x-1)}{-(x+5)(x-1)} \quad \text{Write in standard form.}$$

$$= -\frac{3x+2}{x+5}, x \neq 1 \quad \text{Divide out common factor.}$$

4. $$\frac{15x^2+5x}{x^3-3x^2-18x} \cdot \frac{x^2-2x-15}{3x^2-8x-3} = \frac{5x(3x+1)}{x(x-6)(x+3)} \cdot \frac{(x-5)(x+3)}{(3x+1)(x-3)}$$

$$= \frac{5(x-5)}{(x-6)(x-3)}, x \neq -3, x \neq -\frac{1}{3}, x \neq 0$$

5. $$\frac{x^3-1}{x^2-1} \div \frac{x^2+x+1}{x^2+2x+1} = \frac{x^3-1}{x^2-1} \cdot \frac{x^2+2x+1}{x^2+x+1} \quad \text{Invert and multiply.}$$

$$= \frac{(x-1)(x^2+x+1)}{(x+1)(x-1)} \cdot \frac{(x+1)(x+1)}{x^2+x+1} \quad \text{Factor completely.}$$

$$= x+1, x \neq \pm 1 \quad \text{Divide out common factors.}$$

6. $$\frac{x}{2x-1} - \frac{1}{x+2} = \frac{x(x+2)-(2x-1)}{(2x-1)(x+2)} \quad \text{Basic definition}$$

$$= \frac{x^2+2x-2x+1}{(2x-1)(x+2)} \quad \text{Distributive Property}$$

$$= \frac{x^2+1}{(2x-1)(x+2)} \quad \text{Combine like terms.}$$

7. The LCD of the ration expression $\frac{4}{x} - \frac{x+5}{x^2-4} + \frac{4}{x+2}$ is $x(x+2)(x-2)$.

$$\frac{4}{x} - \frac{x+5}{(x+2)(x-2)} + \frac{4}{x+2} = \frac{4(x+2)(x-2)}{x(x+2)(x-2)} - \frac{x(x+5)}{x(x+2)(x-2)} + \frac{4x(x-2)}{x(x+2)(x-2)}$$ Rewrite using the LCD.

$$= \frac{4(x+2)(x-2) - x(x+5) + 4x(x-2)}{x(x+2)(x-2)}$$ Distributive Property

$$= \frac{4x^2 - 16 - x^2 - 5x + 4x^2 - 8x}{x(x+2)(x-2)}$$

$$= \frac{7x^2 - 13x - 16}{x(x+2)(x-2)}$$

8. $$\frac{\left(\frac{1}{x+2} + 1\right)}{\left(\frac{x}{3} - 1\right)} = \frac{\left(\frac{1 + 1(x+2)}{x+2}\right)}{\left(\frac{x - 1(3)}{3}\right)}$$ Combine fractions.

$$= \frac{\left(\frac{x+3}{x+2}\right)}{\left(\frac{x-3}{3}\right)}$$ Simplify.

$$= \frac{x+3}{x+2} \cdot \frac{3}{x-3}$$ Invert and multiply.

$$= \frac{3(x+3)}{(x+2)(x-3)}$$

9. $$(x-1)^{-1/3} - x(x-1)^{-4/3} = (x-1)^{-4/3}\left[(x-1)^{(-1/3)-(-4/3)} - x\right]$$

$$= (x-1)^{-4/3}\left[(x-1)^1 - x\right]$$

$$= -\frac{1}{(x-1)^{4/3}}$$

10. $$\frac{x^2(x^2-2)^{-1/2} + (x^2-2)^{1/2}}{x^2-2} = \frac{x^2(x^2-2)^{-1/2} + (x^2-2)^{1/2}}{x^2-2} \cdot \frac{(x^2-2)^{1/2}}{(x^2-2)^{1/2}}$$

$$= \frac{x^2(x^2-2)^0 + (x^2-2)^1}{(x^2-2)^{3/2}}$$

$$= \frac{x^2 + x^2 - 2}{(x^2-2)^{3/2}}$$

$$= \frac{2x^2 - 2}{(x^2-2)^{3/2}}$$

$$= \frac{2(x+1)(x-1)}{(x^2-2)^{3/2}}$$

11. $$\frac{\sqrt{9+h}-3}{h} = \frac{\sqrt{9+h}-3}{h} \cdot \frac{\sqrt{9+h}+3}{\sqrt{9+h}+3}$$
$$= \frac{\left(\sqrt{9+h}\right)^2 - (3)^2}{h\left(\sqrt{9+h}+3\right)}$$
$$= \frac{(9+h)-9}{h\left(\sqrt{9+h}+3\right)}$$
$$= \frac{h}{h\left(\sqrt{9+h}+3\right)}$$
$$= \frac{1}{\sqrt{9+h}+3},\ h \neq 0$$

Checkpoints for Section P.6

1.

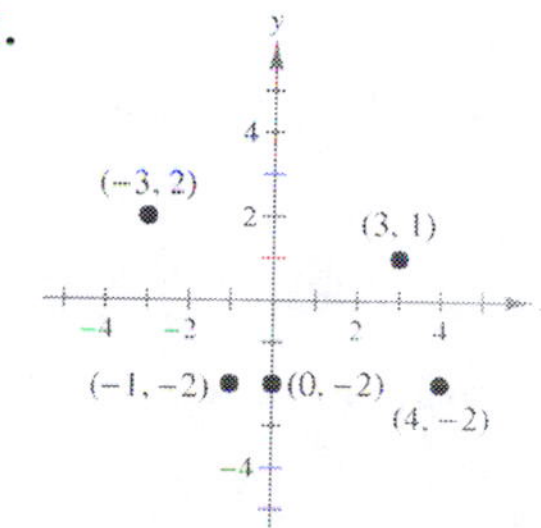

2. To sketch a scatter plot of the data, represent each pair of values by an ordered pair (t, N) and plot the resulting points.

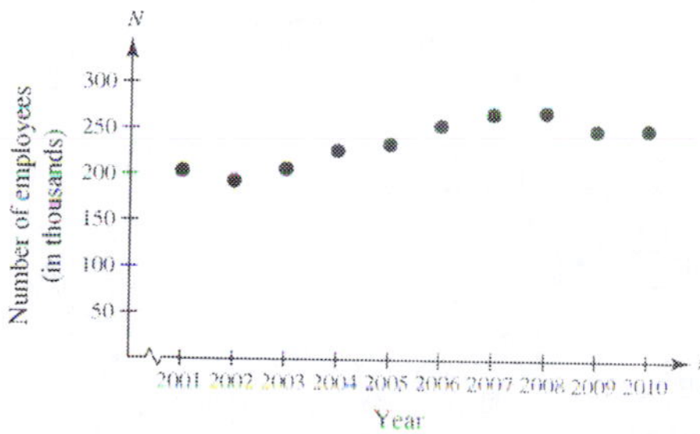

3. Let $(x_1, y_1) = (3, 1)$ and $(x_2, y_2) = (-3, 0)$.

Then apply the Distance Formula.

$$d = \sqrt{(x_2 - x_1)^2 + (y_2 - y_1)^2}$$
$$= \sqrt{(-3-3)^2 + (0-1)^2}$$
$$= \sqrt{(-6)^2 + (-1)^2}$$
$$= \sqrt{36+1}$$
$$= \sqrt{37}$$
$$\approx 6.08$$

So, the distance between the points is about 6.08 units.

4. The three points are plotted in the figure.

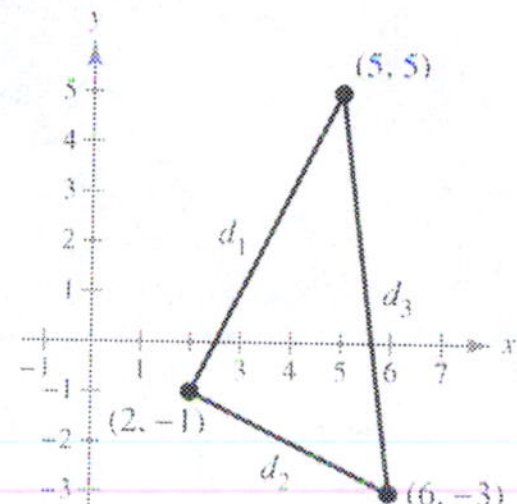

Using the Distance Formula, the lengths of the three sides are as follows.

$$d_1 = \sqrt{(5-2)^2 + (5-(-1))^2}$$
$$= \sqrt{3^2 + 6^2}$$
$$= \sqrt{9+36}$$
$$= \sqrt{45}$$

$$d_2 = \sqrt{(6-2)^2 + (-3-(-1))^2}$$
$$= \sqrt{4^2 + (-2)^2}$$
$$= \sqrt{16+4}$$
$$= \sqrt{20}$$

$$d_3 = \sqrt{(6-5)^2 + (-3-5)^2}$$
$$= \sqrt{(1)^2 + (-8)^2}$$
$$= \sqrt{1+64}$$
$$= \sqrt{65}$$

Because $(d_1)^2 + (d_2)^2 = 45 + 20 = 65$, you can conclude by the Pythagorean Theorem that the triangle must be a right triangle.

5. Let $(x_1, y_1) = (-2, 8)$ and $(x_2, y_2) = (-4, -`0)$.

$$\begin{aligned}\text{Midpoint} &= \left(\frac{x_1 + x_2}{2}, \frac{y_1 + y_2}{2}\right)\\ &= \left(\frac{-2 + 4}{2}, \frac{8 + (-10)}{2}\right)\\ &= \left(\frac{2}{2}, -\frac{2}{2}\right)\\ &= 1, -1\end{aligned}$$

The midpoint of the line segment is $(1, -1)$.

6. You can find the length of the pass by finding the distance between the points $(10, 10)$ and $(25, 32)$.

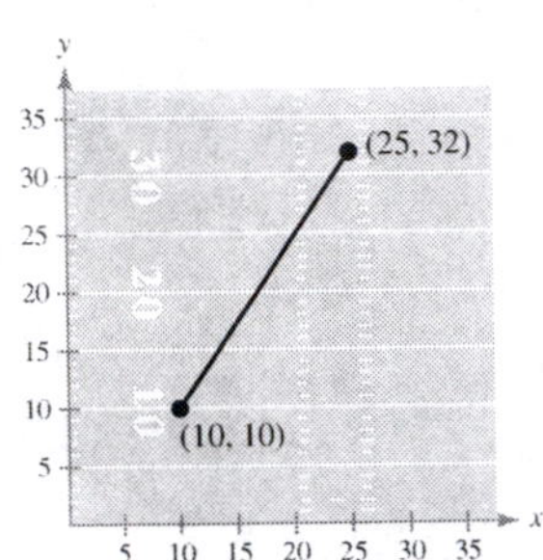

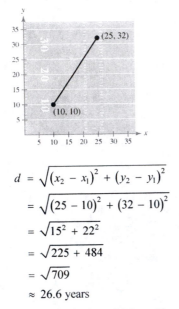

$$\begin{aligned}d &= \sqrt{(x_2 - x_1)^2 + (y_2 - y_1)^2}\\ &= \sqrt{(25 - 10)^2 + (32 - 10)^2}\\ &= \sqrt{15^2 + 22^2}\\ &= \sqrt{225 + 484}\\ &= \sqrt{709}\\ &\approx 26.6 \text{ years}\end{aligned}$$

So, the pass is about 26.6 yard long.

7. Assuming that the annual revenue from Yahoo! Inc. followed a linear pattern, you can estimate the 2009 annual revenue by finding the midpoint of the line segment connecting the points $(2008, 7.2)$ and $(2010, 6.3)$.

$$\begin{aligned}\text{Midpoint} &= \left(\frac{x_1 + x_2}{2}, \frac{y_1 + y_2}{2}\right)\\ &= \left(\frac{2008 + 2010}{2}, \frac{7.2 + 6.3}{2}\right)\\ &= (2009, 6.75)\end{aligned}$$

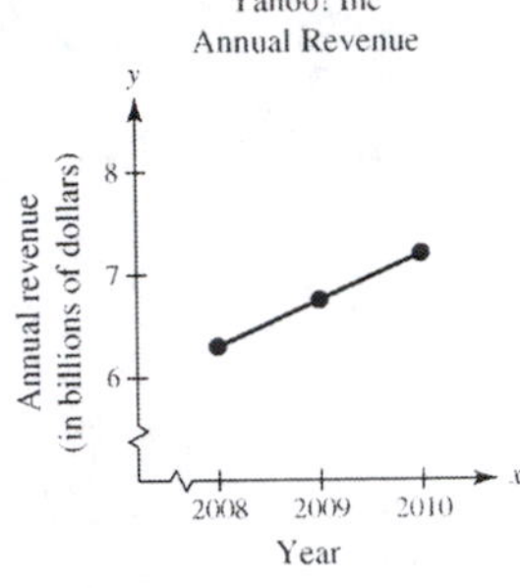

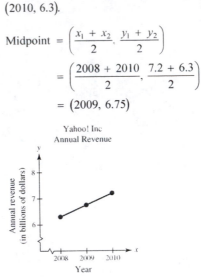

So, you can estimate the annual revenue for Yahoo! Inc. was $6.75 billion in 2009.

8. To shift the vertices two units to the left, subtract 2 from each of the x-coordinates. To shift the vertices four units down, subtract 4 from each of the y-coordinates.

Original point	**Translated Point**
$(1, 4)$	$(1 - 2, 4 - 4) = (-1, 0)$
$(1, 0)$	$(1 - 2, 0 - 4) = (-1, -4)$
$(3, 2)$	$(3 - 2, 2 - 4) = (1, -2)$
$(3, 6)$	$(3 - 2, 6 - 4) = (1, 2)$

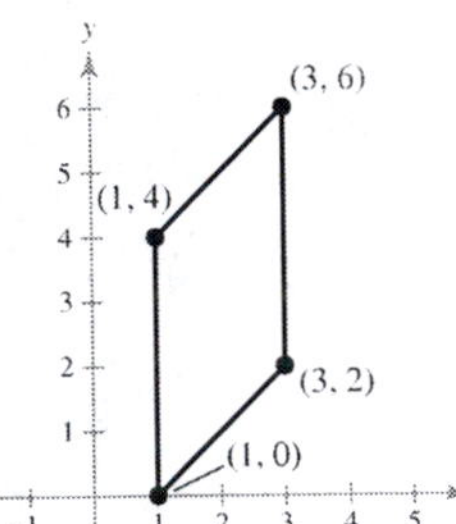

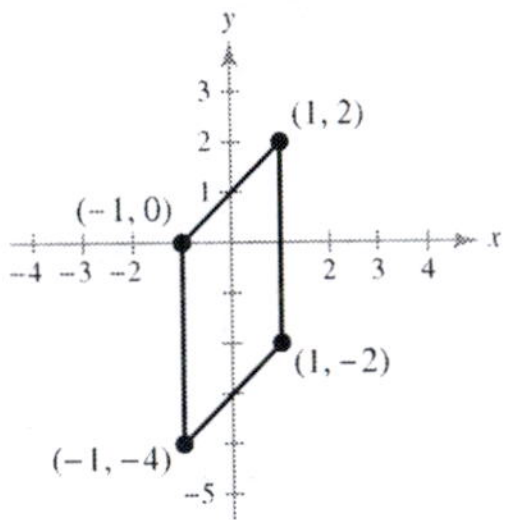

Chapter 1

Checkpoints for Section 1.1

1. (a) $y = 14 - 6x$ Write original equation.

$-5 \stackrel{?}{=} 14 - 6(3)$ Substitute 3 for x and −5 for *y*.

$-5 \stackrel{?}{=} 14 - 18$

$-5 \neq -4$

(b) $y = 14 - 6x$ Write original equation.

$26 \stackrel{?}{=} 14 - 6(-2)$ Substitute −2 for *x* and 26 for *y*.

$26 \stackrel{?}{=} 14 + 12$

$26 \neq 26$ $(-2, 26)$ is a solution. ✓

2. (a) To graph $y = -3x + 2$, construct a table of values that consists of several solution points. Then plot the points and connect them.

x	$y = -3x + 2$	(x, y)
-2	$y = -3(-2) + 2 = 8$	$(-2, -8)$
-1	$y = -3(-1) + 2 = 5$	$(-1, 5)$
0	$y = -3(0) + 2 = 2$	$(0, 2)$
1	$y = -3(1) + 2 = -1$	$(1, -2)$
2	$y = -3(2) + 2 = -4$	$(2, -4)$

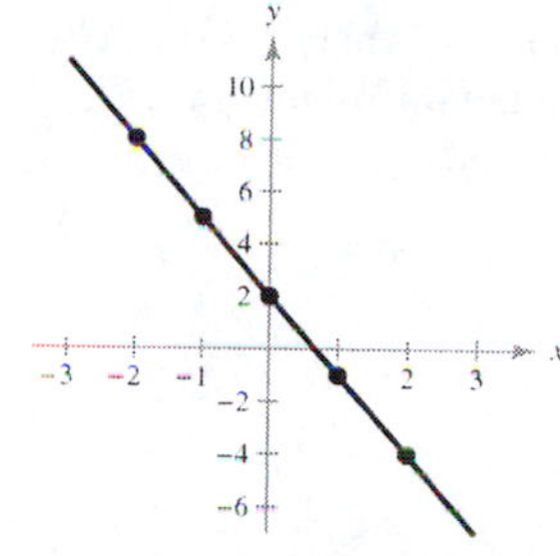

(b) To graph $y = 2x + 1$, construct a table of values that consists of several solution points. Then plot the points and connect them.

x	$y = 2x + 1$	(x, y)
-2	$y = 2(-2) + 1 = -3$	$(-2, -3)$
-1	$y = 2(-1) + 1 = -1$	$(-1, -1)$
0	$y = 2(0) + 1 = 1$	$(0, 1)$
1	$y = 2(1) + 1 = 3$	$(1, 3)$
2	$y = 2(2) + 1 = 5$	$(2, 5)$

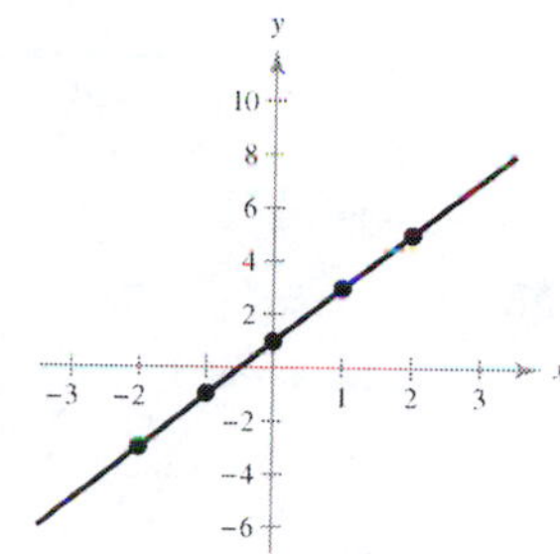

3. (a) To graph $y = x^2 + 3$, construct a table of values that consists of several solution points. Then plot the points and connect them with a smooth curve.

x	$y = x^2 + 3$	(x, y)
-2	$y = (-2)^2 + 3 = 7$	$(-2, 7)$
-1	$y = (-1)^2 + 3 = 4$	$(-1, 4)$
0	$y = (0)^2 + 3 = 3$	$(0, 3)$
1	$y = (1)^2 + 3 = 4$	$(1, 4)$
2	$y = 2(2) + 3 = 7$	$(2, 7)$

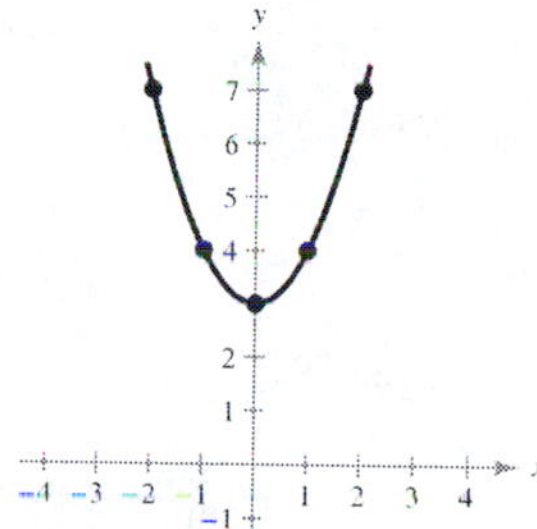

(b) To graph $y = 1 - x^2$, construct a table of values that consists of several solution points. Then plot the points and connect them with a smooth curve.

x	$y = 1 - x^2$	(x, y)
-2	$y = 1 - (-2)^2 = -3$	$(-2, -3)$
-1	$y = 1 - (-1)^2 = 0$	$(-1, 0)$
0	$y = 1 - (0)^2 = 1$	$(0, 1)$
1	$y = 1 - (1)^2 = 0$	$(1, 0)$
2	$y = 1 - (2)^2 = -3$	$(2, -3)$

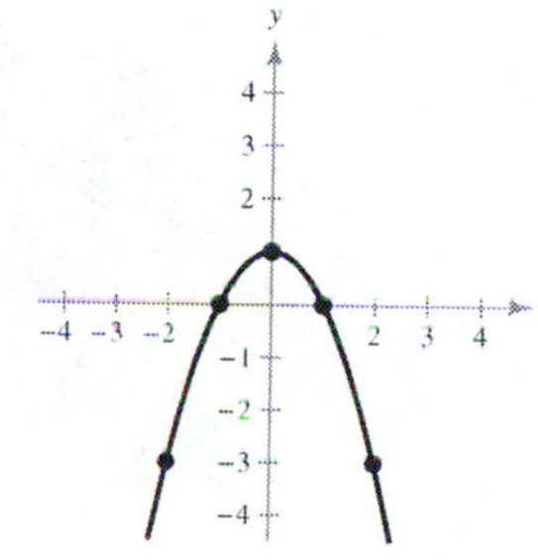

4. From the figure, you can see that the graph of the equation $y = -x^2 + 5x$ has x-intercepts (where y is 0) at (0, 0) and (5, 0) and a y-intercept (where x is 0) at (0, 0). Since the graph passes through the origin or (0, 0), that point can be considered as both an x-intercept and a y-intercept.

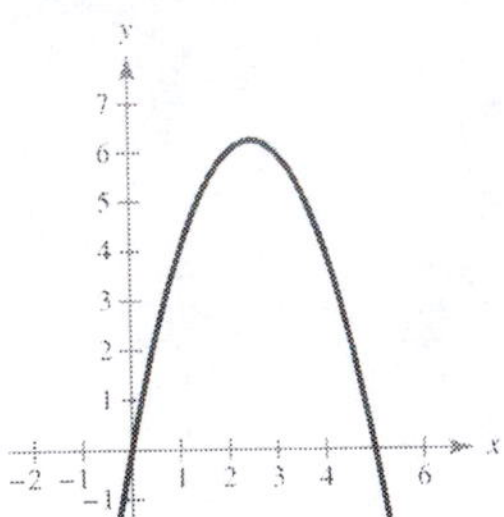

5. x-Axis:

$y^2 = 6 - x$ Write original equation.

$(-y)^2 = 6 - x$ Replace y with $-y$.

$y^2 = 6 - x$ Result is the original equation.

y-Axis:

$y^2 = 6 - x$ Write original equation.

$y^2 = 6 - (-x)$ Replace x with $-x$.

$y^2 = 6 + x$ Result is *not* an equivalent equation.

Origin:

$y^2 = 6 - x$ Write original equation.

$(-y)^2 = 6 - (-x)$ Replace y with $-y$ and x with $-x$.

$y^2 = 6 + x$ Result is *not* an equivalent equation.

Of the three tests for symmetry, the only one that is satisfied is the test for x-axis symmetry.

6. Of the three test of symmetry, the only one that is satisfied is the test for y-axis symmetry because $y = (-x)^2 - 4$ is equivalent to $y = x^2 - 4$. Using symmetry, you only need to find solution points to the right of the y-axis and then reflect them about the y-axis to obtain the graph.

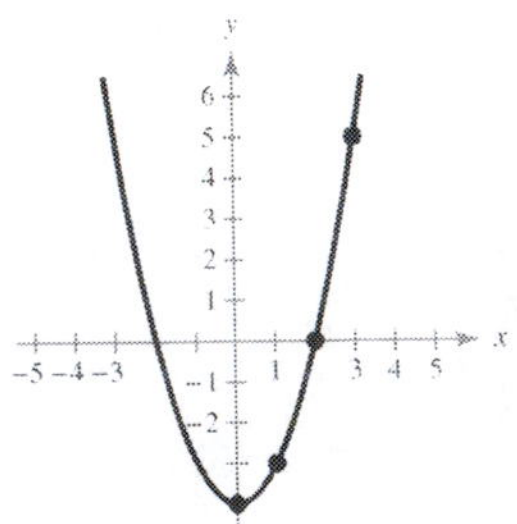

7. The equation $y = |x - 2|$ fails all three tests for symmetry and consequently its graph is not symmetric with respect to either axis or to the origin. So, construct a table of values. Then plot and connect the points.

x	$y = \lvert x - 2 \rvert$	(x, y)
−2	$y = \lvert(-2) - 2\rvert = 4$	(−2, 4)
−1	$y = \lvert(-1) - 2\rvert = 3$	(−1, 3)
0	$y = \lvert(0) - 2\rvert = 2$	(0, 2)
1	$y = \lvert(1) - 2\rvert = 1$	(1, 1)
2	$y = \lvert(2) - 2\rvert = 0$	(3, 0)
3	$y = \lvert(2) - 2\rvert = 0$	(3, 1)
4	$y = \lvert(2) - 2\rvert = 0$	(4, 2)

From the table, you can see that the x-intercept is $(2, 0)$ and the y-intercept is $(0, 2)$.

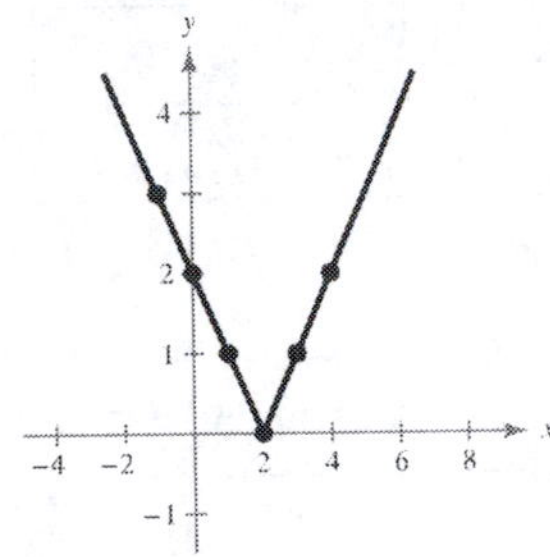

8. The radius of the circle is the distance between $(1, -2)$ and $(-3, -5)$.

$$r = \sqrt{(x - h)^2 + (y - k)^2}$$
$$= \sqrt{[1 - (-3)]^2 + [-2 - (-5)]^2}$$
$$= \sqrt{4^2 + 3^2}$$
$$= \sqrt{16 + 9}$$
$$= \sqrt{25}$$
$$= 5$$

Using $(h, k) = (-3, -5)$ and $r = 5$, the equation of the circle is

$$(x - h)^2 + (y - k)^2 = r^2$$
$$[x - (-3)]^2 + [y - (-5)]^2 = (5)^2$$
$$(x + 3)^2 + (y + 5)^2 = 25.$$

9. From the graph, you can estimate that a height of 75 inches corresponds to a weight of about 175 pounds.

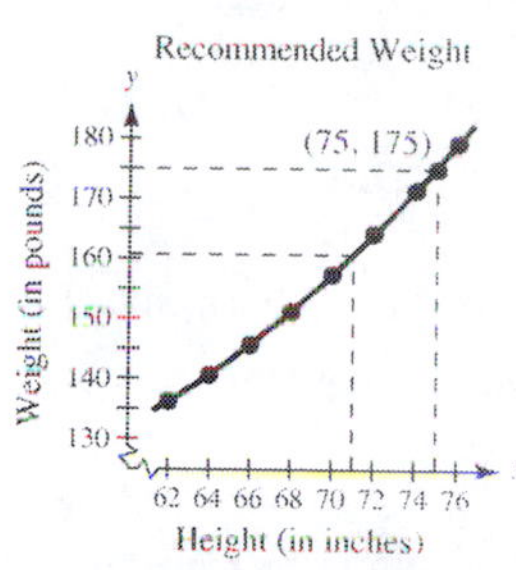

To confirm your estimate algebraically, substitute 75 for x in the model.

Let $x = 75$: $y = 0.073x^2 - 6.99x + 289.0$

$$= 0.073(75)^2 - 6.99(75) + 289.0$$

$$= 175.375$$

Algebraically, you can conclude that a height of 75 inches corresponds to a weight of 175.375 pounds. So, the graphical estimate of 175 is fairly good.

Checkpoints for Section 1.2

1. (a) $7 - 2x = 15$ Write original equation.

$-2x = 8$ Subtract 7 from each side.

$x = -4$ Divide each side by -2.

Check: $7 - 2x = 15$

$7 - 2(-4) \stackrel{?}{=} 15$

$7 + 8 \stackrel{?}{=} 15$

$15 = 15$

(b) $7x - 9 = 5x + 7$ Write original equation.

$2x - 9 = 7$ Subtract $5x$ from each side.

$2x = 16$ Add 9 from each side.

$x = 8$ Divide each side by 2.

Check: $7x - 9 = 5x + 7$

$7(8) - 9 = 5(8 + 7)$

$56 - 9 = 40 + 7$

$47 = 47$ ✓

2. $4(x + 2) - 12 = 5(x - 6)$ Write original equation.

$4x + 8 - 12 = 5x - 30$ Distributive Property

$4x - 4 = 5x - 30$ Simplify.

$-x = -26$ Simplify.

$x = 26$ Divide each side by -1.

Check: $4(x + 2) - 12 = 5(x - 6)$

$4(26 + 2) - 12 \stackrel{?}{=} 5(26 - 6)$

$4(28) - 12 \stackrel{?}{=} 5(20)$

$112 - 12 \stackrel{?}{=} 100$

$100 = 100$

3. $\dfrac{4x}{9} - \dfrac{1}{3} = x + \dfrac{5}{3}$ Write original equation.

$(9)\left(\dfrac{4x}{9}\right) - (9)\left(\dfrac{1}{3}\right) = (9)x + 9\left(\dfrac{5}{3}\right)$ Multiply each term by the LCD.

$4x - 3 = 9x + 15$ Simplify.

$-5x = 18$ Combine like terms.

$x = -\dfrac{18}{5}$ Divide each side by -5.

4. $\dfrac{3x}{x - 4} = 5 + \dfrac{12}{x - 4}$ Write original equation.

$(x - 4)\left(\dfrac{3x}{x - 4}\right) = (x - 4)5 + (x - 4)\left(\dfrac{12}{x - 4}\right)$ Multiply each term by LCD.

$3x = 5x - 20 + 12,\ x \neq 4$ Simplify.

$-2x = -8$ Divide each side by -2.

$x = 4$ Extraneous solution

In the original equation, $x = 4$ yields a denominator of zero. So, $x = 4$ is an extraneous solution, and the original equation has no solution.

5. To find the x-intercept, set x equal to zero, and solve for x.

(a)

$y = 3x - 2$	Write original equation.
$0 = 3x - 2$	Substitute 0 for y.
$2 = 3x$	Add 2 to each side.
$\frac{2}{3} = x$	Divide each side by 3.

So, the x-intercept is $\left(\frac{2}{3}, 0\right)$.

To find the y-intercept, set y equal to zero, and solve for y.

$y = 3x - 2$	Write original equation.
$y = 3(0) - 2$	Substitute 0 for x.
$y = -2$	Simplify.

So, the y-intercept is $(0, -2)$.

(b) To find the x-intercept, set x equal to zero, and solve for x.

$5x + 3y = 15$	Write original equation.
$5x + 3(0) = 15$	Substitute 0 for y.
$5x = 15$	Simplify.
$x = 3$	Divide each side by 5.

So, the x-intercept is $(3, 0)$.

To find the y-intercept, set y equal to zero, and solve for y.

$5x + 3y = 15$	Write original equation.
$5x(0) + 3y = 15$	Substitute 0 for x.
$3y = 15$	Simplify.
$y = 5$	Divide each side by 3.

So, the y-intercept is $(0, 5)$.

6. (a) To find the y-intercept, let $t = 0$ and solve for y.

$$y = 0.064t + 3.83$$
$$y = 0.064(0) + 3.83$$
$$y = 3.83$$

So, the y-intercept is $(0, 3.83)$. This means there were about 3.83 million male participants in 2000.

(b) Let $y = 4.79$ and solve for t.

$$y = 0.064t + 3.83$$
$$4.79 = 0.064t + 3.83$$
$$0.96 = 0.064t$$
$$\frac{0.96}{0.064} = t$$
$$15 = t$$

Because $t = 0$ represents 2000, $t = 15$ must represent 2015. So, from this model, there will be 4.79 million male participants in 2015.

Checkpoints for Section 1.3

1. Because there are 52 weeks in a year and you will be paid weekly, it follows that you will receive 52 paychecks during the year.

Verbal Model: [Income for year] = [52 paychecks] · [Amount of each paycheck] + [Bonus]

Labels:

Income for year $= 58{,}400$	(dollars)
Amount each paycheck $= x$	(dollars)
Bonus $= 1200$	(dollars)

Equation:

$58{,}400 = 52x + 1200$	Write equation.
$57{,}200 = 52x$	Subtract 1200 from each side.
$1100 = x$	Divide each side by 52.

So, your gross pay for each paycheck is \$1100.

2. *Verbal Model:* [Increase in price] = [Percent] · [Original price]

Labels: Original price $= 15$ (dollars per share)

Increase in price $= 18 - 15 = 3$ (dollars per share)

Percent $= r$ (in decimal form)

Equation:

$3 = r \cdot 15$ Write equation.

$\frac{1}{5} = r$ Divide each side by 15.

$0.2 = r$ Rewrite the fraction as a decimal.

The stock's value increased by 0.2 or 20%.

3. The total amount of your family's loan payments is $17,920. The loan payments equal 28% of the annual income.

Verbal Model: [Loan payments] = [Percent] · [Annual income]

Labels: Loan payments $= 17{,}920$ (dollars)

Annual income $= x$ (dollars)

Percent $= 28\% = 0.28$ (in decimal form)

Equation:

$17{,}920 = 0.28 \cdot x$ Write equation.

$64{,}000 = x$ Divide each side by 0.28.

Your family's annual income is $64,000.

4. Draw a diagram,

Verbal Model: 2 · [Length] + 2 · [Width] = [Perimeter]

Labels: Perimeter $= 112$ (feet)

Width $= w$ (feet)

Length $= l = 3w$ (feet)

Equation:

$2(3w) + 2(w) = 112$ Write equation.

$6w + 2w = 112$ Simplify.

$8w = 12$ Combine like terms.

$w = 14$ Divide each side by 8.

Because the length is three times the width,

$l = 3w = 3(14) = 42.$

So, the dimensions of the family room are 14 feet by 42 feet.

5. *Verbal Model:* [Distance] = [Rate] · [Time]

Labels: Distance $= 14$ (miles)

Rate $= \dfrac{\text{first five miles}}{\text{time to travel first five miles}} = \dfrac{5}{0.5} = 10$ (miles per hour)

Time $= t$ (hours)

Equation:

$14 = 10 \cdot t$ Write equation.

$\dfrac{14}{10} = t$ Divide each side by 10.

$1.4 = t$ Simplify.

The entire trip will take 1.4 hours, or 1 hour and 24 minutes.

6. To solve this problem, use the result from geometry that the ratios of the corresponding sides of similar triangles are equal.

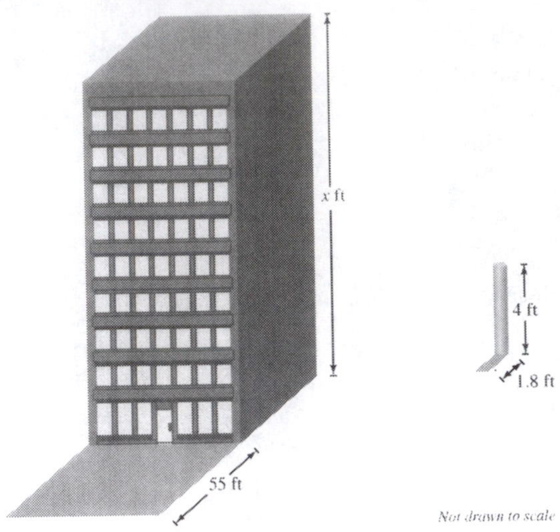

Verbal Model: $\dfrac{\text{Height of building}}{\text{Length of building's shadow}} = \dfrac{\text{Height of post}}{\text{Length of posts, shadow}}$

Labels:
Height of building $= x$ (feet)
Length of building's shadow $= 55$ (feet)
Height of post $= 4$ (feet)
Length of post's shadow $= 1.8$ (feet)

Equation:

$$\frac{x}{55} = \frac{4}{1.8}$$

$$x = \frac{4}{1.8} \cdot 55$$

$$x \approx 122.2$$

So, the building is about 122.2 feet high.

7. Let x represent the amount invested at $2\frac{1}{2}\%$. Then the amount invested at $3\frac{1}{2}\%$ is $5000 - x$.

Verbal Model: $\boxed{\text{Interest from } 2\tfrac{1}{2}\%} + \boxed{\text{Interest from } 3\tfrac{1}{2}\%} = \boxed{\text{Total interest}}$

Labels:
Interest from $2\frac{1}{2}\% = Prt = (x)(0.025)(1)$ (dollars)
Interest from $3\frac{1}{2}\% = Prt = (5000 - x)(0.035)(1)$ (dollars)
Total interest $= 151.25$ (dollars)

Equation:

$$0.025x + 0.035(5000 - x) = 151.25$$

$$0.025x + 175 - 0.035x = 151.25$$

$$-0.01x + 175 = 151.25$$

$$-0.01x = -23.75$$

$$x = 2375$$

So, \$2375 was invested at $2\frac{1}{2}\%$ and $5000 - 2375 = \$2625$ was invested at $3\frac{1}{2}\%$.

8. *Verbal Model:* $\boxed{\text{Profit from single-disc player}} + \boxed{\text{Profit from multi-disc player}} = \boxed{\text{Total profit}}$

Labels:

Inventory of single-disc players $= x$ (dollars)

Inventory of multi-disc players $= 30{,}000 - x$ (dollars)

Profit from single disc players $= 0.24x$ (dollars)

Profit from multi-disc players $= 0.42(30{,}000 - x)$ (dollars)

Total profit $= 0.36(30{,}000) = 10{,}800$ (dollars)

Equation:

$$\begin{aligned} 0.24x + 0.42(30{,}000 - x) &= 10{,}800 \\ 0.24x + 12{,}600 - 0.42x &= 10{,}800 \\ 12{,}600 - 0.42x &= -1800 \\ -0.18x &= -1800 \\ x &= 10{,}000 \end{aligned}$$

So, \$10,000 was invested in single-disc players and $30{,}000 - 10{,}000 = \$20{,}000$ was invested in multi-disc players.

9. The formula for the volume of a cylindrical container is $V = \pi r^2 h$. To find the height of the container, solve for h.

$$h = \frac{V}{\pi r^2}$$

Then, using $V = 84$ and $r = 3$, find the height.

$$h = \frac{84}{\pi(3)^2}$$

$$h = \frac{84}{9\pi}$$

$$h \approx 2.97$$

So, the height of the container is about 2.97 inches. You can use unit analysis to check that your answer is reasonable.

$$\frac{84 \text{ in.}^3}{9\pi \text{ in.}^2} = \frac{84 \text{ in.} \cdot \cancel{\text{in.}} \cdot \cancel{\text{in.}}}{9\pi\ \cancel{\text{in.}} \cdot \cancel{\text{in.}}} = \frac{84}{9\pi} \text{ in.} \approx 2.97 \text{ in.}$$

Checkpoints for Section 1.4

1.

$2x^2 - 3x + 1 = 6$	Write original equation.
$2x^2 - 3x - 5 = 0$	Write in general form.
$(2x - 5)(x + 1) = 0$	Factor.
$2x - 5 = 0 \Rightarrow x = \frac{5}{2}$	Set 1st factor equal to 0.
$x + 1 = 0 \Rightarrow x = -1$	Set 2nd factor equal to 0.

The solutions are $x = -1$ and $x = \frac{5}{2}$.

Check: $x = -1$

$$\begin{aligned} 2x^2 - 3x + 1 &= 6 \\ 2(-1)^2 - 3(-1) + 1 &\stackrel{?}{=} 6 \\ 2(1) + 3 + 1 &\stackrel{?}{=} 6 \\ 6 &= 6 \checkmark \end{aligned}$$

$x = \frac{5}{2}$

$$\begin{aligned} 2x^2 - 3x + 1 &= 6 \\ 2\left(\tfrac{5}{2}\right)^2 - 3\left(\tfrac{5}{2}\right) + 1 &\stackrel{?}{=} 6 \\ 2\left(\tfrac{25}{4}\right) - \tfrac{15}{2} + 1 &\stackrel{?}{=} 6 \\ 6 &= 6 \checkmark \end{aligned}$$

2. (a) $3x^2 = 36$ Write original equation.

$x^2 = 12$ Divide each side by 3.

$x = \pm\sqrt{12}$ Extract square roots.

$x = \pm 2\sqrt{3}$

The solutions are $x = \pm 2\sqrt{3}$.

Check: $x = -2\sqrt{3}$

$$3x^2 = 36$$
$$3\left(-2\sqrt{3}\right)^2 \stackrel{?}{=} 36$$
$$3(12) \stackrel{?}{=} 36$$
$$36 = 36 \checkmark$$

$x = 2\sqrt{3}$

$$3x^2 = 36$$
$$3\left(2\sqrt{3}\right)^2 \stackrel{?}{=} 36$$
$$3(12) \stackrel{?}{=} 36$$
$$36 = 36 \checkmark$$

(b) $(x - 1)^2 = 10$

$$x - 1 = \pm\sqrt{10}$$
$$x = 1 \pm \sqrt{10}$$

The solutions are $x = 1 \pm \sqrt{10}$.

Check: $x = 1 - \sqrt{10}$

$$(x - 1)^2 = 10$$
$$\left[\left(1 - \sqrt{10}\right) - 1\right]^2 \stackrel{?}{=} 10$$
$$\left(-\sqrt{10}\right)^2 \stackrel{?}{=} 10$$
$$10 = 10 \checkmark$$

$x = 1 + \sqrt{10}$

$$(x - 1)^2 = 10$$
$$\left[\left(1 + \sqrt{10}\right) - 1\right] \stackrel{?}{=} 10$$
$$\left(\sqrt{10}\right)^2 \stackrel{?}{=} 10$$
$$10 = 10 \checkmark$$

3. $x^2 - 4x - 1 = 0$ Write original equation.

$x^2 - 4x = 1$ Add 1 to each side.

$x^2 - 4x + (2)^2 = 1 + (2)^2$ Add 2^2 to each side.

$(\text{half of } 4)^2$

$(x - 2)^2 = 5$ Simplify.

$x - 2 = \pm\sqrt{5}$ Extract square roots.

$x = 2 \pm \sqrt{5}$ Add 2 to each side.

The solutions are $x = 2 \pm \sqrt{5}$.

Check: $x = 2 - \sqrt{5}$

$$x^2 - 4x - 1 = 0$$
$$\left(2 - \sqrt{5}\right)^2 - 4\left(2 - \sqrt{5}\right) - 1 \stackrel{?}{=} 0$$
$$\left(4 - 4\sqrt{5} + 5\right) - 8 + 4\sqrt{5} - 1 \stackrel{?}{=} 0$$
$$4 + 5 - 8 - 1 \stackrel{?}{=} 0$$
$$0 = 0 \checkmark$$

$x = 2 + \sqrt{5}$ also checks. ✓

4. $2x^2 - 4x + 1 = 0$ Write original equation.

$2x^2 - 4x = -1$ Subtract 1 from each side.

$x^2 - 2x = -\dfrac{1}{2}$ Divide each side by 2.

$x^2 - 2x + (1)^2 = -\dfrac{1}{2} + (1)^2$ Add 1^2 to each side.

$(x - 1)^2 = \dfrac{1}{2}$ Simplify.

$x - 1 = \pm\sqrt{\dfrac{1}{2}}$ Extract square roots.

$x - 1 = \pm\dfrac{\sqrt{2}}{2}$ Rationalize the denominator.

$x = 1 \pm \dfrac{\sqrt{2}}{2}$ Add 1 to each side.

Check: $x = 1 \pm \dfrac{\sqrt{2}}{2}$

$$2x^2 - 4x + 1 = 0$$
$$2\left(1 - \frac{\sqrt{2}}{2}\right)^2 - 4\left(1 - \frac{\sqrt{2}}{2}\right) + 1 \stackrel{?}{=} 0$$
$$2\left(1 - \sqrt{2} + \frac{1}{2}\right) - 4 + 2\sqrt{2} + 1 \stackrel{?}{=} 0$$
$$2 - 2\sqrt{2} + 1 - 4 + 2\sqrt{2} + 1 \stackrel{?}{=} 0$$
$$2 + 1 - 4 + 1 \stackrel{?}{=} 0$$
$$0 = 0 \checkmark$$

The solution $x = 1 + \dfrac{\sqrt{2}}{2}$ also checks. ✓

5.

$3x^2 - 10x - 2 = 0$	Original equation
$3x^2 - 10x = 2$	Add 2 to each side.
$x^2 - \frac{10}{3}x = \frac{2}{3}$	Divide each side by 3.
$x^2 - \frac{10}{3}x + \left(\frac{5}{3}\right)^2 = \frac{2}{3} + \left(\frac{5}{3}\right)^2$	Add $\left(\frac{5}{3}\right)^2$ to each side.
$\left(x - \frac{5}{3}\right)^2 = \frac{31}{9}$	Simplify.
$x - \frac{5}{3} = \pm\frac{\sqrt{31}}{3}$	Extract square roots.
$x = \frac{5}{3} \pm \frac{\sqrt{31}}{3}$	Add $\frac{5}{3}$ to each side.

The solutions are $\frac{5}{3} \pm \frac{\sqrt{31}}{3}$.

6.

$3x^2 + 2x - 10 = 0$	Write original equation.
$x = \frac{-6 \pm \sqrt{6^2 - 4ac}}{2a}$	Quadratic Formula
$x = \frac{-2 \pm \sqrt{(2)^2 - 4(3)(-10)}}{2(3)}$	Substitute $a = 3$, $b = 2$ and $c = -10$.
$x = \frac{-2 \pm \sqrt{4 + 120}}{6}$	Simplify.
$x = \frac{-2 \pm \sqrt{124}}{6}$	Simplify.
$x = \frac{-2 \pm 2\sqrt{31}}{6}$	Simplify.
$x = \frac{2\left(-1 \pm \sqrt{31}\right)}{6}$	Factor our common factor.
$x = \frac{-1 \pm \sqrt{31}}{3}$	Simplify.

The solutions are $\frac{-1 \pm \sqrt{31}}{3}$.

Check: $x = \frac{-1 \pm \sqrt{31}}{3}$

$$3x^2 + 2x - 10 = 0$$

$$3\left(\frac{-1 + \sqrt{31}}{3}\right)^2 + 2\left(\frac{-1 + \sqrt{31}}{3}\right) - 10 \stackrel{?}{=} 0$$

$$\frac{1}{3} + \frac{2\sqrt{31}}{3} + \frac{31}{3} + \frac{2\sqrt{31}}{3} - 10 \stackrel{?}{=} 0$$

$$10 - 10 \stackrel{?}{=} 0$$

$$0 = 0 \checkmark$$

The solution $x = \frac{-1 - \sqrt{31}}{3}$ also checks. ✓

7. *Verbal Model:* $\boxed{\text{Width of room}} \cdot \boxed{\text{Length of room}} = \boxed{\text{Area of room}}$

Labels: Width $= w$ (feet)
Length $= l = w + 6$ (feet)
Area $= 112$ (square feet)

Equation:

$$w(w + 6) = 112$$
$$w^2 + 6w = 112$$
$$w^2 + 6w - 112 = 0$$
$$(w - 8)(w + 14) = 0$$
$$w - 8 = 0 \Rightarrow w = 8$$
$$w + 14 = 0 \Rightarrow w = -14$$

Choosing the positive value, you can conclude the width is 8 feet and the length is $w + 6 = 14$ feet. You can check this solution by observing that the length is 6 feet longer than the width and the product of the length and width is 112 square feet.

8. To set up a mathematical model for the height of the rock, use the position equation

$$s = -16t^2 + V_0t + s_0.$$

Because the object is dropped, the initial velocity is $V_0 = 0$ feet per second. Because the initial height is $s_0 = 196$ feet, you have $s = -16t^2 + 196$.

To find the time the rock hits the ground, let the height s be zero and solve the equation for t.

$s = -16t^2 + 196$	Write positive equation.
$0 = -16t^2 + 196$	Substitute 0 for height.
$16t^2 = 196$	Add 16^2 to each side.
$t^2 = \dfrac{196}{16}$	Divide each side by 16.
$t = \sqrt{\dfrac{196}{16}}$	Extract positive square root.
$t = \dfrac{14}{4}$	Simplify.
$t = 3.5$	Simplify.

The rock will take 3.5 seconds to hit the ground.

9. To find the year in which the number of internet users reached 220 million, you can solve the equation

$$-0.898t^2 + 19.59t + 126.5 = 220.$$

To begin, write the equation in general form.

$$-0.898t^2 + 19.59t - 93.5 = 0.$$

Then apply the Quadratic Formula.

$$t = \frac{-b \pm \sqrt{b^2 - 4ac}}{2a}$$

$$t = \frac{-19.59 \pm \sqrt{(19.59)^2 - 4(-0.898)(-93.5)}}{2(-0.898)}$$

$$t = \frac{-19.59 \pm \sqrt{47.9161}}{-1.746}$$

$$t \approx 7.1 \text{ or } 14.8$$

Choose $t \approx 7.1$ because it is in the domain of I. Because $t = 0$ corresponds to 2000, it follows that $t = 7.1$ corresponds to 2007. So, the number of Internet users reached 200 million during the year 2007. From the table, you can also see that during the year 2007 the number of Internet users reached 220 million.

10. 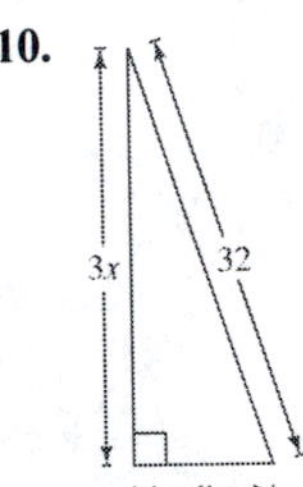

Use the Pythagorean Theorem.

$a^2 + b^2 = c^2$	Pythagorean Theorem
$(3x)^2 + (x)^2 = (32)^2$	Substitute for a, b, and c.
$9x^2 + x^2 = 1024$	Simplify.
$10x^2 = 1024$	Simplify.
$x^2 = 102.4$	Divide each side by 10.
$x = \sqrt{102.4}$	Extract positive square root.

The total distance covered by walking on the L-shaped sidewalk is

$$x + 3x = 4x$$
$$= 4\sqrt{102.4}$$
$$\approx 40.5 \text{ feet.}$$

Walking on the diagonal saves a person about $40.5 - 32 = 8.5$ feet.

Checkpoints for Section 1.5

1. (a) $(7 + 3i) + (5 - 4i) = 7 + 3i + 5 - 4i$ Remove parentheses.

$= (7 + 5) + (3 - 4)i$ Group like terms.

$= 12 - i$ Write in standard form.

(b) $(3 + 4i) - (5 - 3i) = 3 + 4i - 5 + 3i$ Remove parentheses.

$= (3 - 5) + (4 + 3)i$ Group like terms.

$= -2 + 7i$ Write in standard form.

(c) $2i + (-3 - 4i) - (-3 - 3i) = 2i - 3 - 4i + 3 + 3i$ Remove parentheses.

$= (-3 + 3) + (2 - 4 + 3)i$ Group like terms.

$= i$ Write in standard form.

(d) $(5 - 3i) + (3 + 5i) - (8 + 2i) = 5 - 3i + 3 + 5i - 8 - 2i$ Remove parentheses.

$= (5 + 3 - 8) + (-3 + 5 - 2)i$ Group like terms.

$= 0 + 0i$ Simplify.

$= 0$ Write in standard form.

2. (a) $(2 - 4i)(3 + 3i) = 2(3 + 3i) - 4i(3 + 3i)$ Distributive Property

$= 6 + 6i - 12i - 12i^2$ Distributive Property

$= 6 + 6i - 12i - 12(-1)$ $i^2 = -1$

$= (6 + 12) + (6 - 12)i$ Group like terms.

$= 18 - 6i$ Write in standard form.

(b) $(4 + 5i)(4 - 5i) = 4(4 - 5i) + 5i(4 - 5i)$ Distributive Property

$= 16 - 20i + 20i - 25i^2$ Distributive Property

$= 16 - 20i + 20i - 25(-1)$ $i^2 = -1$

$= 16 + 25$ Simplify.

$= 41$ Write in standard form.

(c) $(4 + 2i)^2 = (4 + 2i)(4 + 2i)$ Square of a binomial

$= 4(4 + 2i) + 2i(4 + 2i)$ Distributive Property

$= 16 + 8i + 8i + 4i^2$ Distributive Property

$= 16 + 8i + 8i + 4(-1)$ $i^2 = -1$

$= (16 - 4) + (8i + 8i)$ Group like terms.

$= 12 + 16i$ Write in standard form.

3. (a) The complex conjugate of $3 + 6i$ is $3 - 6i$.

$$(3 + 6i)(3 - 6i) = (3)^2 - (6i)^2$$
$$= 9 - 36i^2$$
$$= 9 - 36(-1)$$
$$= 45$$

(b) The complex conjugate of $2 - 5i$ is $2 + 5i$.

$$(2 - 5i)(2 + 5i) = (2)^2 - (5i)^2$$
$$= 4 - 25i^2$$
$$= 4 - 25(-1)$$
$$= 29$$

4. $\dfrac{2+i}{2-i} = \dfrac{2+i}{2-i} \cdot \dfrac{2+i}{2+i}$ — Multiply numerator and denominator by complex conjugate of the denominator.

$= \dfrac{4+2i+2i+i^2}{4-i^2}$ — Expand.

$= \dfrac{4-1+4i}{4-(-1)}$ — $i^2 = -1$

$= \dfrac{3+4i}{5}$ — Simplify.

$= \dfrac{3}{5} + \dfrac{4}{5}i$ — Write in standard form.

5. $\sqrt{-14}\sqrt{-2} = \sqrt{14}i\sqrt{2}i = \sqrt{28}i^2 = 2\sqrt{7}(-1) = -2\sqrt{7}$

6. To solve $8x^2 + 14x + 9 = 0$, use the Quadratic formula

$x = \dfrac{-b \pm \sqrt{b^2 - 4ac}}{2a}.$

$x = \dfrac{-14 \pm \sqrt{14^2 - 4(8)(9)}}{2(8)}$ — Substitute $a = 8$, $b = 14$, and $c = 9$.

$= \dfrac{-14 \pm \sqrt{-92}}{16}$ — Simplify.

$= \dfrac{-14 \pm 2\sqrt{23}i}{16}$ — Write $\sqrt{-92}$ in standard form.

$= \dfrac{-14}{16} + \dfrac{2\sqrt{23}i}{16}$ — Write in standard form.

$= \dfrac{-7}{8} \pm \dfrac{\sqrt{23}i}{8}$ — Simplify.

Checkpoints for Section 1.6

1. $9x^4 - 12x^2 = 0$ — Write original equation.

$3x^2(3x^2 - 4) = 0$ — Factor out common factor.

$3x^2 = 0 \Rightarrow x = 0$ — Set 1st factor equal to 0.

$3x^2 - 4 = 0 \Rightarrow 3x^2 = 4$ — Set 2nd factor equal to 0.

$x^2 = \dfrac{4}{3}$

$x = \pm\sqrt{\dfrac{4}{3}}$

$x = \dfrac{\pm 2\sqrt{3}}{3}$

Check: $x = 0$

$9x^4 - 12x^2 = 0$

$9(0)^4 - 12(0)^2 \stackrel{?}{=} 0$

$0 = 0$ ✓

$x = \dfrac{2\sqrt{3}}{3}$

$9x^4 - 12x^2 = 0$

$9\left(\dfrac{2\sqrt{3}}{3}\right)^4 - 12\left(\dfrac{2\sqrt{3}}{3}\right)^2 \stackrel{?}{=} 0$

$9\left(\dfrac{16}{9}\right) - 12\left(\dfrac{4}{3}\right) \stackrel{?}{=} 0$

$16 - 16 \stackrel{?}{=} 0$

$0 = 0$ ✓

The solution $x = -\sqrt{2}$ also checks. ✓

2. (a)

$$\begin{aligned} x^3 - 5x^2 - 2x + 10 &= 0 && \text{Write original equation.} \\ x^2(x-5) - 2(x-5) &= 0 && \text{Factor by grouping.} \\ (x-5)(x^2-2) &= 0 && \text{Distributive Property} \\ x - 5 = 0 \Rightarrow x &= 5 && \text{Set 1st factor equal to 0.} \\ x^2 - 2 = 0 \Rightarrow x^2 &= 2 && \text{Set 2nd factor equal to 0.} \\ &= \pm\sqrt{2} \end{aligned}$$

Check: $x = 5$

$$\begin{aligned} x^3 - 5x^2 - 2x + 10 &= 0 \\ (5)^3 - 5(5)^2 - 2(5) + 10 &\stackrel{?}{=} 0 \\ 125 - 125 - 10 + 10 &\stackrel{?}{=} 0 \\ 0 &= 0 \checkmark \end{aligned}$$

$x = \sqrt{2}$

$$\begin{aligned} x^3 - 5x^2 - 2x + 10 &= 0 \\ (\sqrt{2})^3 - 5(\sqrt{2})^2 - 2(\sqrt{2}) + 10 &\stackrel{?}{=} 0 \\ 2\sqrt{2} - 10 - 2\sqrt{2} + 10 &\stackrel{?}{=} 0 \\ 0 &= 0 \checkmark \end{aligned}$$

The solution $x = -\sqrt{2}$ also checks. ✓

(b)

$$\begin{aligned} 6x^3 - 27x^2 - 54x &= 0 \\ 3x(2x^2 - 9x - 18) &= 0 && \text{Factor out common factor.} \\ 3x(2x+3)(x-6) &= 0 && \text{Factor quadratic factor.} \\ 3x = 0 \Rightarrow x &= 0 && \text{Set 1st factor equal to 0.} \\ 2x + 3 = 0 \Rightarrow x &= -\tfrac{3}{2} && \text{Set 2nd factor equal to 0.} \\ x - 6 = 0 \Rightarrow x &= 6 && \text{Set 3rd factor equal to 0.} \end{aligned}$$

Check: $x = 0$

$$\begin{aligned} 6x^3 - 27x^2 - 54x &= 0 \\ 6(0)^3 - 27(0)^2 - 54(0) &\stackrel{?}{=} 0 \\ 0 &= 0 \checkmark \end{aligned}$$

$x = -\frac{3}{2}$

$$\begin{aligned} 6x^3 - 27x^2 - 54x &= 0 \\ 6\left(-\tfrac{3}{2}\right)^3 - 27\left(-\tfrac{3}{2}\right)^2 - 54\left(-\tfrac{3}{2}\right) &\stackrel{?}{=} 0 \\ 6\left(-\tfrac{27}{8}\right)^3 - 27\left(\tfrac{9}{4}\right)^2 + 27(3) &\stackrel{?}{=} 0 \\ -\tfrac{81}{4} - \tfrac{243}{4} + 81 &\stackrel{?}{=} 0 \\ 0 &= 0 \checkmark \end{aligned}$$

$x = 6$

$$\begin{aligned} 6x^3 - 27x^2 - 54x &= 0 \\ 6(6)^3 - 27(6)^2 - 54(6) &\stackrel{?}{=} 0 \\ 6(216) - 27(36) - 324 &\stackrel{?}{=} 0 \\ 0 &= 0 \checkmark \end{aligned}$$

3. (a) This equation is of quadratic type with $u = x^2$.

$$x^4 - 7x^2 + 12 = 0 \quad \text{Write original equation.}$$
$$(x^2)^2 - 7(x^2) + 12 = 0 \quad \text{Quadratic form}$$
$$u^2 - 7u + 12 = 0 \quad u = x^2$$
$$(u - 3)(u - 4) = 0 \quad \text{Factor.}$$
$$u - 3 = 0 \Rightarrow u = 3 \quad \text{Set 1st factor equal to 0.}$$
$$u - 4 = 0 \Rightarrow u = 4 \quad \text{Set 2nd factor equal to 0.}$$

Replace u with x^2 and solve for x in each equation.

$u = 3$	$u = 4$
$x^2 = 3$	$x^2 = 4$
$x = \pm\sqrt{3}$	$x = \pm 2$

The solutions are $x = -\sqrt{3}$, $x = \sqrt{3}$, $x = -2$, and $x = 2$.

Check: $x = -\sqrt{3}$

$$x^4 - 7x^2 + 12 = 0$$
$$(-\sqrt{3})^4 - 7(-\sqrt{3})^2 + 12 \stackrel{?}{=} 0$$
$$9 - 7(3) + 12 \stackrel{?}{=} 0$$
$$-21 + 21 \stackrel{?}{=} 0$$
$$0 = 0 \checkmark$$

The solution $x = \sqrt{3}$ also checks. ✓

$x = -2$

$$x^4 - 7x^2 + 12 = 0$$
$$(-2)^4 - 7(-2)^2 + 12 \stackrel{?}{=} 0$$
$$16 - 28 + 12 \stackrel{?}{=} 0$$
$$0 = 0 \checkmark$$

The solution $x = 2$ also checks. ✓

(b) This equation is of quadratic type with $u = x^2$.

$$9x^4 - 37x^2 + 4 = 0 \quad \text{Write original equation.}$$
$$9(x^2)^2 - 37(x^2) + 4 = 0 \quad \text{Quadratic form}$$
$$9u^2 - 37u + 4 = 0 \quad u = x^2$$
$$(9u - 1)(u - 4) = 0 \quad \text{Factor.}$$
$$9u - 1 = 0 \Rightarrow u = \tfrac{1}{9} \quad \text{Set 1st factor equal to 0.}$$
$$u - 4 = 0 \Rightarrow u = 4 \quad \text{Set 2nd factor equal to 0.}$$

Replace u with x^2 and solve for x in each equation.

$u = \frac{1}{9}$	$u = 4$
$x^2 = \frac{1}{9}$	$x^2 = 4$
$x = \pm\frac{1}{3}$	$x = \pm 2$

The solutions are $-\frac{1}{3}$, $\frac{1}{3}$, -2, and 2.

Check: $x = -\frac{1}{3}$

$$9x^4 - 37x^2 + 4 = 0$$
$$9\left(-\tfrac{1}{3}\right)^4 - 37\left(-\tfrac{1}{3}\right)^2 + 4 \stackrel{?}{=} 0$$
$$9\left(\tfrac{1}{81}\right) - 37\left(\tfrac{1}{9}\right) + 4 \stackrel{?}{=} 0$$
$$\tfrac{1}{9} - \tfrac{37}{9} + 4 \stackrel{?}{=} 0$$
$$-4 + 4 = 0 \checkmark$$

The solution $x = \frac{1}{3}$ also checks. ✓

$x = -2$

$$9x^4 - 37x^2 + 4 = 0$$
$$9(-2)^4 - 37(-2)^2 + 4 \stackrel{?}{=} 0$$
$$9(16) - 37(4) + 4 \stackrel{?}{=} 0$$
$$144 - 148 + 4 = 0 \checkmark$$

The solution $x = 2$ also checks. ✓

4. $-\sqrt{40 - 9x} + 2 = x$ Write original equation.

$-\sqrt{40 - 9x} = x - 2$ Isolated radical.

$\left(-\sqrt{40 - 9x}\right)^2 = (x - 2)^2$ Square each side.

$40 - 9x = x^2 - 4x + 4$ Simplify.

$0 = x^2 + 5x - 36$ Write in general form.

$0 = (x - 4)(x + 9)$ Factor.

$x - 4 = 0 \Rightarrow x = 4$ Set 1st factor equal to 0.

$x + 9 = 0 \Rightarrow x = -9$ Set 2nd factor equal to 0.

Check: $x = 4$

$$-\sqrt{40 - 9x} + 2 = x$$

$$-\sqrt{40 - 9(4)} + 2 \stackrel{?}{=} 4$$

$$-\sqrt{4} + 2 \stackrel{?}{=} 4$$

$$-2 + 2 \stackrel{?}{=} 4$$

$$0 \neq 4 \text{ ✗}$$

$x = 4$ is an extraneous solution.

$x = -9$

$$-\sqrt{40 - 9(-9)} + 2 \stackrel{?}{=} -9$$

$$-\sqrt{121} + 2 \stackrel{?}{=} -9$$

$$-11 + 2 \stackrel{?}{=} -9$$

$$-9 \stackrel{?}{=} -9 \text{ ✓}$$

The only solution is $x = -9$.

5. $(x - 5)^{2/3} = 16$ Write original equation.

$\sqrt[3]{(x - 5)^2} = 16$ Rewrite in radical form.

$(x - 5)^2 = 4096$ Cube each side.

$x - 5 = \pm 64$ Extract square roots.

$x = 5 \pm 64$ Add 5 to each side.

$x = -59, x = 69$

Check: $x = -59$

$$(x - 5)^{2/3} = 16$$

$$(-59 - 5)^{2/3} \stackrel{?}{=} 16$$

$$(-64)^{2/3} \stackrel{?}{=} 16$$

$$(-4)^2 \stackrel{?}{=} 16$$

$$16 = 16 \text{ ✓}$$

$x = 69$

$$(x - 5)^{2/3} = 16$$

$$(69 - 5)^{2/3} \stackrel{?}{=} 16$$

$$(64)^{2/3} \stackrel{?}{=} 16$$

$$(4)^2 \stackrel{?}{=} 16$$

$$16 = 16 \text{ ✓}$$

The solutions are $x = -59$ and $x = 64$.

6. The LCD of the three terms is $x(x + 3)$.

$$\frac{4}{x} + \frac{2}{x+3} = -3 \qquad \text{Write original equation.}$$

$$x(x+3)\frac{4}{x} + x(x+3)\frac{2}{x+3} = x(x+3)(-3) \qquad \text{Multiply each term by LCD.}$$

$$4(x+3) + 2x = -3x(x+3),\ x \neq -3, 0 \qquad \text{Simplify.}$$

$$6x + 12 = -3x^2 - 9x \qquad \text{Simplify.}$$

$$3x^2 + 15x + 12 = 0 \qquad \text{Write in general form.}$$

$$3\left(x^2 + 5x + 4\right) = 0 \qquad \text{Factor.}$$

$$x^2 + 5x + 4 = 0 \qquad \text{Divide each side by 3.}$$

$$(x+1)(x+4) = 0 \qquad \text{Factor.}$$

$$x + 1 = 0 \Rightarrow x = -1 \qquad \text{Set 1st factor equal to 0.}$$

$$x + 4 = 0 \Rightarrow x = -4 \qquad \text{Set 2nd factor equal to 0.}$$

Both $x = -4$ and $x = -1$ are possible solutions.

Check: $x = -4$

$$\frac{4}{x} + \frac{2}{x+3} = -3$$

$$\frac{4}{(-4)} + \frac{2}{(-4)+3} \stackrel{?}{=} -3$$

$$-1 + (-2) \stackrel{?}{=} -3$$

$$-3 = -3 \checkmark$$

$x = -1$

$$\frac{4}{x} + \frac{2}{x+3} = -3$$

$$\frac{4}{(-1)} + \frac{2}{(-1)+3} \stackrel{?}{=} -3$$

$$-4 + (1) \stackrel{?}{=} -3$$

$$-3 = -3 \checkmark$$

So, the solutions are $x = -4$ and $x = -1$.

7. $\left|x^2 + 4x\right| = 5x + 12$

First Equation

$$x^2 + 4x = 5x + 12 \qquad \text{Use positive expression.}$$

$$x^2 - x - 12 = 0 \qquad \text{Write in general form.}$$

$$(x+3)(x-4) = 0 \qquad \text{Factor.}$$

$$x + 3 = 0 \Rightarrow x = -3 \qquad \text{Set 1st factor equal to 0.}$$

$$x - 4 = 0 \Rightarrow x = 4 \qquad \text{Set 2nd factor equal to 0.}$$

Second Equation

$$-\left(x^2 + 4x\right) = 5x + 12 \qquad \text{Use negative expression.}$$

$$-x^2 - 4x = 5x + 12 \qquad \text{Distributive Property}$$

$$0 = x^2 + 9x + 12 \qquad \text{Write in general form.}$$

Use the Quadratic equation to solve the equation $0 = x^2 + 9x + 12$.

$$x = \frac{-b \pm \sqrt{b^2 - 4ac}}{2a}$$

$$x = \frac{-9 \pm \sqrt{9^2 - 4(1)(12)}}{2(1)}$$

$$x = \frac{-9 \pm \sqrt{33}}{2}$$

The possible solutions are $x = -3$, $x = 4$, and $x = \dfrac{-9 \pm \sqrt{33}}{2}$.

Check: $x = -3$

$$|x^2 + 4x| = 5x + 12$$

$$|(-3)^2 + 4(-3)| \stackrel{?}{=} 5(-3) + 12$$

$$|-3| \stackrel{?}{=} -15 + 12$$

$$3 \neq -3$$

$x = -3$ does not check.

$x = 4$

$$|(4)^2 + 4(4)| \stackrel{?}{=} 5(4) + 12$$

$$|32| \stackrel{?}{=} 32$$

$$32 = 32 \checkmark$$

$x = 4$ checks.

$$x = \frac{-9 \pm \sqrt{33}}{2}$$

$$\left|\frac{\left(-9 + \sqrt{33}\right)^2}{2} + 4\left(\frac{-9 + \sqrt{33}}{2}\right)\right| \stackrel{?}{=} 5\left(\frac{-9 + \sqrt{33}}{2}\right) + 12$$

$$\left|\frac{21}{2} - \frac{5\sqrt{33}}{2}\right| \stackrel{?}{=} \frac{-45}{2} + \frac{5\sqrt{33}}{2} + 12$$

$$\frac{5\sqrt{33}}{2} - \frac{21}{2} = \frac{-21}{2} + \frac{5\sqrt{33}}{2} \checkmark$$

$x = \dfrac{-9 + \sqrt{33}}{2}$ checks.

$$x = \frac{-9 \pm \sqrt{33}}{2}$$

$$\left|\frac{\left(-9 - \sqrt{33}\right)^2}{2} + 4\left(\frac{-9 - \sqrt{33}}{2}\right)\right| \stackrel{?}{=} 5\left(\frac{-9 - \sqrt{33}}{2}\right) + 12$$

$$\left|\frac{21}{2} + \frac{5\sqrt{33}}{2}\right| \stackrel{?}{=} \frac{-21}{2} - \frac{5\sqrt{33}}{2}$$

$$\frac{21}{2} + \frac{5\sqrt{33}}{2} \neq -\frac{21}{2} - \frac{5\sqrt{33}}{2}$$

$x = \dfrac{-9 - \sqrt{33}}{2}$ does not check.

$x = -3$ and $x = \dfrac{-9 - \sqrt{33}}{2}$ are extraneous solutions. So, the solutions are $x = 4$ and $x = \dfrac{-9 + \sqrt{33}}{2}$.

8. *Verbal Model:* $\boxed{\text{Cost per student}} \cdot \boxed{\text{Number of students}} = \boxed{\text{Cost of trip}}$

Labels: Cost of trip $= 560$ (dollars)

Original number of students $= x$ (people)

New number of students $= x + 8$ (people)

Original cost per student $= \dfrac{560}{x}$ (dollars per person)

New cost per students $= \dfrac{560}{x} - 3.50$ (dollars per person)

Equation:

$$\left(\frac{560}{x} - 3.50\right)(x + 8) = 560$$

$$\left(\frac{560 - 3.50x}{x}\right)(x + 8) = 560$$

$$(560 - 3.50x)(x + 8) = 560x,\ x \neq 0$$

$$560x + 4480 - 3.5x^2 - 28x = 560x$$

$$-3.5x^2 - 28x + 4480 = 0$$

$$x^2 + 8x - 1280 = 0$$

$$(x - 32)(x + 40) = 0$$

$$x - 32 = 0 \Rightarrow w = 32$$

$$x + 40 = 0 \Rightarrow w = -40$$

Check:

$$\left(\frac{560}{x} - 3.50\right)(x + 8) \stackrel{?}{=} 560$$

$$\left(\frac{560}{32} - 3.50\right)(32 + 8) \stackrel{?}{=} 560$$

$$(14)(40) \stackrel{?}{=} 560$$

$$560 \stackrel{?}{=} 560 \checkmark$$

Only the positive value of x makes sense in the context of the problem, so you can conclude the $x + 8 = 32 + 8 = 40$ students are now chartering the bus to the observatory.

9. *Formula:* $A = P\left(1 + \dfrac{r}{n}\right)^{nt}$

Labels: Balance $= A = 3544.06$ (dollars)

Principal $= P = 2500$ (dollars)

Time $= t = 5$ (years)

Compounding periods per year $= n = 12$ (compoundings per year)

Annual interest rate $= r$ (percent in decimal form)

Equation:

$$3544.06 = 2500\left(1 + \frac{r}{12}\right)^{(12)(5)}$$

$$\frac{3544.06}{2500} = \left(1 + \frac{r}{12}\right)^{60}$$

$$1.4176 \approx \left(1 + \frac{r}{12}\right)^{60}$$

$$(1.4176)^{1/60} = 1 + \frac{r}{12}$$

$$1.0058 \approx 1 + \frac{r}{12}$$

$$0.0058 \approx \frac{r}{12}$$

$$0.0696 \approx r$$

$$0.07 \approx r$$

Check:

$$A = P\left(1 + \frac{r}{n}\right)^{nt}$$

$$3544.06 \stackrel{?}{=} 2500\left(1 + \frac{0.07}{12}\right)^{(12)(5)}$$

$$3544.06 \stackrel{?}{=} 3544.06 \checkmark$$

The annual interest rate is about 0.07 or 7%.

Checkpoints for Section 1.7

1. (a) $[-1, 3]$ corresponds to $-1 \le x \le 3$. The interval is bounded.

(b) $(-1, 6)$ corresponds to $-1 < x < 6$. The interval is bounded.

(c) $(-\infty, 4)$ corresponds to $x < 4$. The interval is unbounded.

(d) $[0, \infty)$ corresponds to $x \ge 0$. The interval is unbounded.

2.

$7x - 3 \le 2x + 7$	Write original inequality.
$5x \le 10$	Subtract $2x$ and add 3 to each side.
$x \le 2$	Divide each side by 5.

The solution set is all real numbers less than or equal to 2.

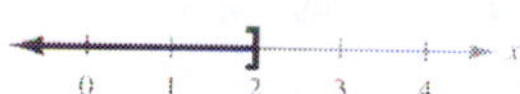

3. (a) **Algebraic solution**

$2 - \frac{5}{3}x > x - 6$	Write original inequality.
$6 - 5x > 3x - 18$	Multiply each side by 3.
$-8x > -24$	Subtract $3x$ and subtract 6 from each side.
$x < 3$	Divide each side by -8 reverse the inequality symbol.

The solution set is all real numbers that are less than 3.

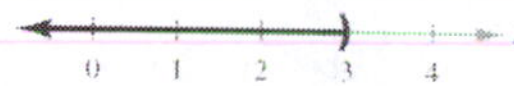

(b) **Graphical solution**

Use a graphing utility to graph $y_1 = 2 - \frac{5}{3}x$ and $y_2 = x - 6$ in the same viewing window. Use the *intersect* feature to determine that the graphs intersect at $(3, -3)$. The graph of y_1 lies above the graph of y_2 to the left of their point of intersection, which implies that $y_1 > y_2$ for all $x < 3$.

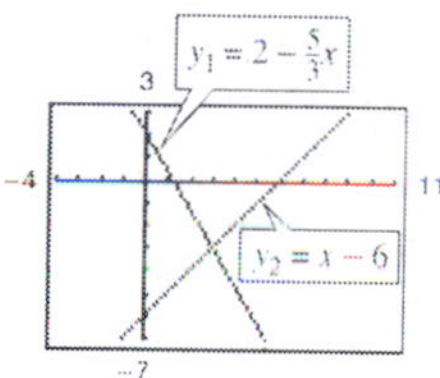

4.

$1 < 2x + 7 < 11$	Write original inequality.
$1 - 7 < 2x + 7 - 7 < 11 - 7$	Subtract 7 from each part.
$-6 < 2x < 4$	Simplify.
$-\frac{6}{2} < \frac{2x}{2} < \frac{4}{2}$	Divide each part by 2.
$-3 < x < 2$	Simplify.

The solution set is all real numbers greater than -3 and less than 2, which is denoted by $(-3, 2)$.

5. $|x - 20| \le 4$ Write original inequality.

$-4 \le x - 20 \le 4$ Write equivalent inequalities.

$-4 + 20 \le x - 20 + 20 \le 4 = 20$ Add 20 to each part.

$16 \le x \le 24$ Simplify.

The solution set is all real numbers that are greater than or equal to 16 and less than or equal to 24, which is denoted by $[16, 24]$.

6. Let m represent your additional minutes in one month. Write and solve an inequality.

$0.45m + 45.99 > 0.35m + 54.99$

$0.10m > 9$

$m > 90$

Plan B costs more when you use more than 90 additional minutes in one month.

7. Let x represent the actual weight of your bag. The difference of the actual weight and the weight on the scale is at most $\frac{1}{64}$ pound. That is, $\left|x - \frac{1}{2}\right| \le -\frac{1}{64}$.

You can solve the inequality as follows.

$-\frac{1}{64} \le x - \frac{1}{2} \le \frac{1}{64}$

$\frac{31}{64} \le x \le \frac{33}{64}$

The least your bag can weigh is $\frac{31}{64}$ pound, which would have cost $\left(\frac{31}{64}\text{ pound}\right) \times (\$9.89\text{ per pound}) = \$4.79$.

The most your bag can weigh is $\frac{33}{64}$ pound, which would have cost $\left(\frac{33}{64}\text{ pound}\right) \times (\$9.89\text{ per pound}) = \$5.10$.

So, you might have been under charged by as much as $\$5.10 - \$4.95 = \$0.15$ or over charged as much as $\$4.95 - \$4.79 = \$0.16$.

Checkpoints for Section 1.8

1. By factoring the polynomial $x^2 - x - 20 < 0$ as $x^2 - x - 20 = (x + 4)(x - 5)$ you can see that the key numbers are $x = -4$ and $x = 5$. So, the polynomial's test intervals are $(-\infty, -4)$, $(-4, 5)$, and $(5, \infty)$.

In each test interval, choose a representative x-value and evaluate the polynomial.

Test interval	x-value	Polynomial value	Conclusion
$(-\infty, -4)$	$x = -5$	$(-5)^2 - (-5) - 20 = 10$	Positive
$(-4, 5)$	$x = 0$	$(0)^2 - (0) - 20 = -20$	Negative
$(5, \infty)$	$x = 6$	$(6)^2 - (6) - 20 = 10$	Positive

From this, you can conclude that the inequality is satisfied for all x-values in $(-4, 5)$.

This implies that the solution of the inequality $x^2 - x - 20 < 0$ is the interval $(-4, 5)$.

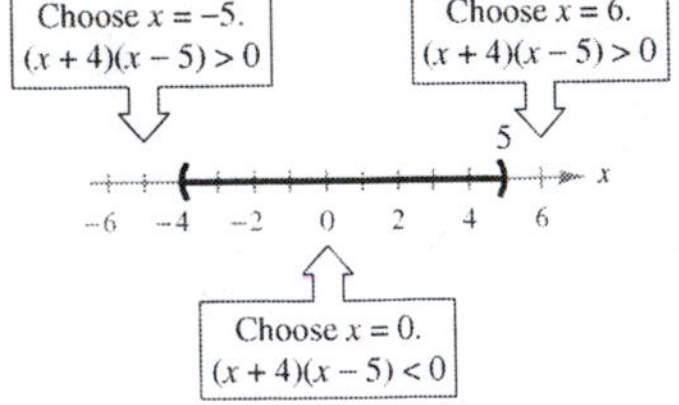

2.

$3x^3 - x^2 - 12x > -4$	Write original inequality.
$3x^2 - x^2 - 12x + 4 > 0$	Write in general form.
$x^2(3x - 1) - 4(3x - 1) > 0$	Factor.
$(3x - 1)(x^2 - 4) > 0$	Factor.
$(3x - 1)(x + 2)(x - 2) > 0$	Factor.

The key numbers are $x = -2$, $x = \frac{1}{3}$, and $x = 2$, and the test intervals are $(-\infty, -2)$, $\left(-2, \frac{1}{3}\right)$, $\left(\frac{1}{3}, 2\right)$, and $(2, \infty)$.

Test interval	x-value	Polynomial value	Conclusion
$(-\infty, -2)$	$x = -3$	$3(-3)^3 - (-3)^2 - 12(-3) + 4 = -50$	Negative
$\left(-2, \frac{1}{3}\right)$	$x = 0$	$3(0)^3 - (0)^2 - 12(0) + 4 = 4$	Positive
$\left(\frac{1}{3}, 2\right)$	$x = 1$	$3(1)^3 - (1)^2 - 12(1) + 4 = -6$	Negative
$(2, \infty)$	$x = 3$	$3(3)^3 - (3)^2 - 12(3) + 4 = 40$	Positive

From this, you can conclude that the inequality is satisfied on the open intervals $\left(-2, \frac{1}{3}\right)$ and $(2, \infty)$.

So, the solution set is $\left(-2, \frac{1}{3}\right) \cup (2, \infty)$.

3. (a) **Algebraic solution**

$2x^2 + 3x < 5$	Write original inequality.
$2x^2 + 3x - 5 < 0$	Write in general form.
$(2x + 5)(x - 1) < 0$	Factor.

Key numbers: $x = -\frac{5}{2}$ and $x = 1$

Test intervals: $\left(-\infty, -\frac{5}{2}\right)$, $\left(-\frac{5}{2}, 1\right)$, $(1, \infty)$

Test: Is $(2x + 5)(x - 1) < 0$?

After testing the intervals, you can see that the polynomial $2x^2 + 3x - 5$ is negative on the open interval $\left(-\frac{5}{2}, 1\right)$.

So, the solution set of the inequality is $\left(-\frac{5}{2}, 1\right)$.

(b) **Graphical solution**

First write the polynomial inequality $2x^2 + 3x < 5$ as $2x^2 + 3x - 5 > 0$. Then use a graphing utility to graph $y = 2x^2 + 3x - 5$. You can see that the graph is below the x-axis when x is greater than $-\frac{5}{2}$ and when x is less than 1. So, the solution set is $\left(-\frac{5}{2}, 1\right)$.

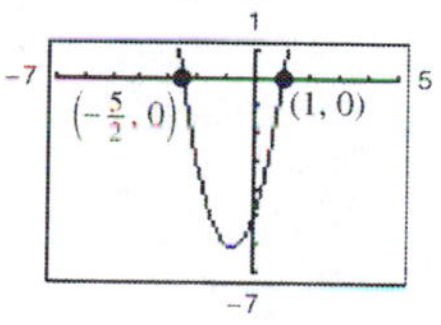

4. (a) The solution set of $x^2 + 6x + 9 < 0$ is empty. In other words, the quadratic $x^2 + 6x + 9$ is not less than 0 for any value of x.

(b) The solution set of $x^2 + 4x + 4 \leq 0$ consists of the single real number $\{-2\}$, because the quadratic $x^2 + 4x + 4$ has only one key number, $x = -2$, and it is the only value that satisfies the inequality.

(c) The solution set of $x^2 - 6x + 970$ consists of all real numbers except $x = 3$. In interval notation, the solution set can be written as $(-\infty, 3) \cup (3, \infty)$.

(d) The solution set of $x^2 - 2x + 1 \geq 0$ consists of the entire set of real numbers $(-\infty, \infty)$. In other words, the value of the quadratic $x^2 - 2x + 1$ is non-negative for every real value of x.

5. (a)

$$\frac{x-2}{x-3} \geq -3 \qquad \text{Write original inequality.}$$

$$\frac{x-2}{x-3} + 3 \geq 0 \qquad \text{Write in general form.}$$

$$\frac{x-2}{x-3} + \frac{3(x-3)}{x-3} \geq 0 \qquad \text{Rewrite fraction using LCD.}$$

$$\frac{x-2+3x-9}{x-3} \geq 0 \qquad \text{Add fractions.}$$

$$\frac{4x-11}{x-3} \geq 0 \qquad \text{Simplify.}$$

Key numbers: $x = \frac{11}{4}, x = 3$

Test intervals: $\left(-\infty, \frac{11}{4}\right), \left(\frac{11}{4}, 3\right), (3, \infty)$

Test: Is $\frac{4x-11}{x-3} \geq 0$?

Test interval	x-value	Polynomial value	Conclusion
$\left(-\infty, \frac{11}{4}\right)$	$x = 0$	$\frac{4(0)-11}{0-3} = \frac{11}{3}$	Positive
$\left(\frac{11}{4}, 3\right)$	$x = 2.9$	$\frac{4(2.9)-11}{2.9-3} = \frac{0.6}{-0.1} = -6$	Negative
$(3, \infty)$	$x = 4$	$\frac{4(4)-11}{4-3} = \frac{5}{1} = 5$	Positive

After testing these intervals, you can see that the inequality is satisfied on the open intervals $\left(-\infty, \frac{11}{4}\right)$ and $(3, \infty)$. Moreover, because $\frac{4x-11}{x-3} = 0$ when $x = \frac{11}{4}$, you can conclude that the solution set consists of all real numbers in the intervals $\left(-\infty, \frac{11}{4}\right] \cup (3, \infty)$.

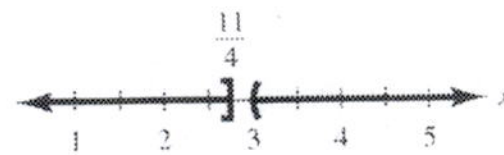

(b)

$$\frac{4x - 1}{x - 6} > 3 \quad \text{Write original inequality.}$$

$$\frac{4x - 1}{x - 6} - 3 > 0 \quad \text{Write in general form.}$$

$$\frac{4x - 1 - 3(x - 6)}{x - 6} > 0 \quad \text{Combine fractions with LCD.}$$

$$\frac{4x - 1 - 3x + 18}{x - 6} > 0 \quad \text{Simplify.}$$

$$\frac{x + 17}{x - 6} > 0 \quad \text{Simplify.}$$

Key numbers: $x = -17, x = 6$

Test intervals: $(-\infty, -17)$, $(-17, 6)$, and $(6, \infty)$

Test: Is $\frac{x + 17}{x - 6} > 0$?

Test interval	***x*-value**	**Polynomial value**	**Conclusion**
$(-\infty, -17)$	-20	$\frac{(20) + 17}{(20) - 6} = \frac{3}{14}$	Positive
$(-17, 6)$	0	$\frac{(0) + 17}{(0) - 6} = -\frac{17}{6}$	Negative
$(6, \infty)$	8	$\frac{(8) + 17}{(8) - 6} = \frac{25}{2}$	Positive

After testing these intervals, you can see that the inequality is satisfied on the open intervals $(-\infty, -17)$ and $(6, \infty)$.

So, you can conclude that the solution set consists of all real numbers in the intervals $(-\infty, -17) \cup (6, \infty)$.

6. *Verbal Model:* $\boxed{\text{Profit}} = \boxed{\text{Review}} - \boxed{\text{Cost}}$

Equation:

$$P = R - C$$

$$P = x(60 - 0.0001x) - (12x + 1{,}800{,}000)$$

$$P = -0.0001x^2 + 48x = 1{,}800{,}000$$

To answer the question, solve the inequality as follows.

$$P \geq 3{,}600{,}000$$

$$-0.0001x^2 + 48x - 1{,}800{,}000 \geq 3{,}600{,}000$$

$$-0.0001x^2 + 48x - 5{,}400{,}000 \geq 0$$

$$0.0001x^2 - 48x + 5{,}400{,}000 \leq 0$$

$$x^2 - 480{,}000x + 54{,}000{,}000{,}000 \leq 0$$

$$(x - 180{,}000)(x - 300{,}000) \leq 0$$

After finding the key point and testing the intervals, you can find the solution set is $[180{,}000,\ 300{,}000]$.

So, by selling at least 180,000 units but not more than 300,000 units, the profit is at least $3,600,000.

7. Algebraic solution

Recall that the domain of an expression is the set of all x-values for which the expression is defined. Because $\sqrt{x^2 - 7x + 10}$ is defined only of $x^2 - 7x + 10$ is non-negative, the domain is given by $x^2 - 7x + 10 \geq 0$.

$x^2 - 7x + 10 \geq 0$ Write in general form.

$(x - 2)(x - 5) \geq 0$ Factor.

So, the inequality has two key numbers: $x = 2$ and $x = 5$.

Key numbers: $x = 2, x = 5$

Test intervals: $(-\infty, 2), (2, 5), (5, \infty)$

Test: Is $(x - 2)(x - 5) \geq 0$?

A test shows that the inequality is satisfied in the unbounded half-closed intervals $(-\infty, 2]$ or $[5, \infty)$. So, the domain of the expression $\sqrt{x^2 - 7x + 10}$ is $(-\infty, 2] \cup [5, \infty)$.

Graphical solution

Begin by sketching the graph of the equation $y = \sqrt{x^2 - 7x + 10}$. From the graph, you can determine that the x-values extend up to 2 (including 2) and from 5 and beyond (including 5). So, the domain of the expression $\sqrt{x^2 - 7x + 10}$ is $(-\infty, 2] \cup [5, \infty)$.

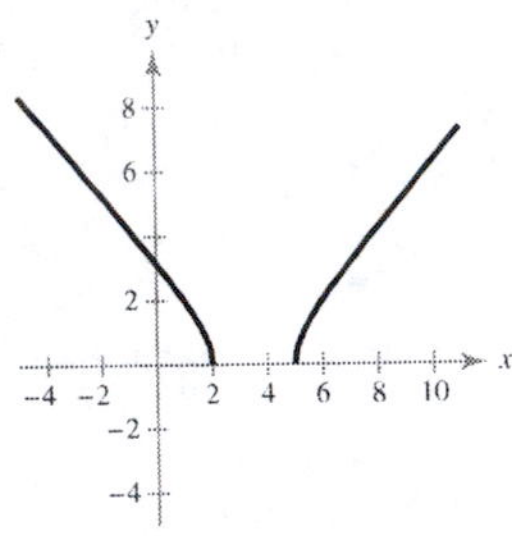

Chapter 2

Checkpoints for Section 2.1

1. (a) Because $b = 2$, the y-intercept is (0, 2). Because the slope is $m = -3$, the line falls three units for each unit the line moves to the right.

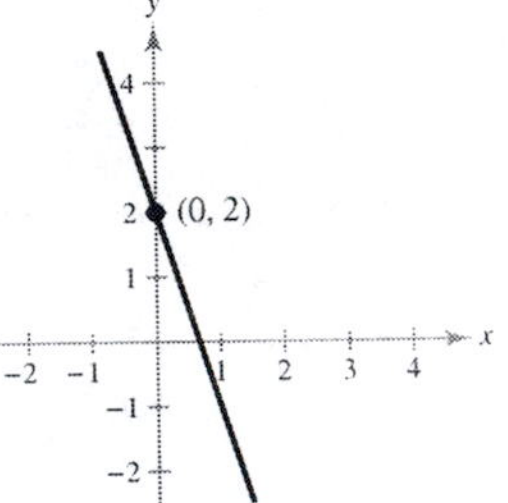

(b) By writing this equation in the form $y = (0)x - 3$, you can see that the y-intercept is $(0, -3)$ and the slope is $m = 0$. A zero slope implies that the line is horizontal.

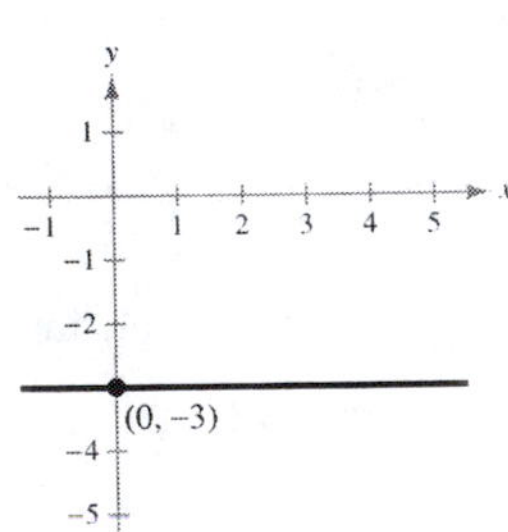

(c) By writing this equation in slope-intercept form

$$4x + y = 5$$
$$y = -4x + 5$$

you can see that the y-intercept is $(0, 5)$. Because the slope is $m = -4$, the line falls four units for each unit the line moves to the right.

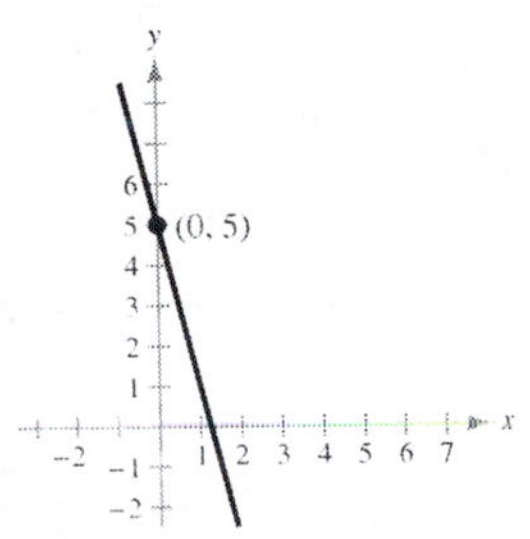

2. (a) The slope of the line passing through $(-5, -6)$ and $(2, 8)$ is $m = \dfrac{8 - (-6)}{2 - (-5)} = \dfrac{14}{7} = 2.$

(b) The slope of the line passing through $(4, 2)$ and $(2, 5)$ is $m = \dfrac{5 - 2}{2 - 4} = \dfrac{3}{-2} = -\dfrac{3}{2}.$

(c) The slope of the line passing through $(0, 0)$ and $(0, -6)$ is $m = \dfrac{-6 - 0}{0 - 0} = \dfrac{-6}{0}.$ Because division by 0 is undefined, the slope is undefined and the line is vertical.

(d) The slope of the line passing through $(0, -1)$ and $(3, -1)$ is $m = \dfrac{-1 - (-1)}{3 - 0} = \dfrac{0}{3} = 0.$

3. (a) Use the point-slope form with $m = 2$ and $(x_1, y_1) = (3, -7).$

$$y - y_1 = m(x - x_1)$$
$$y - (-7) = 2(x - 3)$$
$$y + 7 = 2x - 6$$
$$y = 2x - 13$$

The slope-intercept form of this equation is $y = 2x - 13.$

(b) Use the point-slope form with $m = \dfrac{-2}{3}$ and $(x_1, y_1) = (1, 1)$

$$y - y_1 = m(x - x_1)$$
$$y - 1 = \frac{-2}{3}(x - 1)$$
$$y - 1 = \frac{-2}{3}x + \frac{2}{3}$$
$$y = \frac{-2}{3}x + \frac{5}{3}$$

The slope-intercept form of this equation is $y = -\dfrac{2}{3}x + \dfrac{5}{3}.$

(c) Use the point-slope form with $m = 0$ and $(x_1, y_1) = (1, 1).$

$$y - y_1 = m(x - x_1)$$
$$y - 1 = 0(x - 1)$$
$$y - 1 = 0$$
$$y = 1$$

The slope-intercept of the equation is the line $y = 1.$

4. By writing the equation of the given line in slope-intercept form

$$5x - 3y = 8$$
$$-3y = -5x + 8$$
$$y = \frac{5}{3}x - \frac{8}{3}$$

You can see that it has a slope of $m = \dfrac{5}{3}.$

(a) Any line parallel to the given line must also have a slope of $m = \dfrac{5}{3}.$ So, the line through $(-4, 1)$ that is parallel to the given line has the following equation.

$$y - y_1 = m(x - x_1)$$
$$y - 1 = \frac{5}{3}(x - (-4))$$
$$y - 1 = \frac{5}{3}(x + 4)$$
$$y - 1 = \frac{5}{3}x + \frac{20}{3}$$
$$y = \frac{5}{3}x + \frac{23}{3}$$

(b) Any line perpendicular to the given line must also have a slope of $m = -\dfrac{3}{5}$ because $-\dfrac{3}{5}$ is the negative reciprocal of $\dfrac{5}{3}.$ So, the line through $(-4, 1)$ that is perpendicular to the given line has the following equation.

$$y - y_1 = m(x - x_1)$$
$$y - 1 = -\frac{3}{5}(x - (-4))$$
$$y - 1 = -\frac{3}{5}(x + 4)$$
$$y - 1 = -\frac{3}{5}x - \frac{12}{5}$$
$$y = -\frac{3}{5}x - \frac{7}{5}$$

5. The horizontal length of the ramp is 32 feet or $12(32) = 384$ inches.

So, the slope of the ramp is

$$\text{Slope} = \frac{\text{vertical change}}{\text{horizontal change}} = \frac{36 \text{ in.}}{384 \text{ in.}} \approx 0.094.$$

Because $\dfrac{1}{12} \approx 0.083$, the slope of the ramp is steeper than recommended.

6. The y-intercept $(0, 1500)$ tells you that the value of the copier when it was purchased $(t = 0)$ was \$1500. The slope of $m = -300$ tells you that the value of the copier decreases \$300 each year since the copier was purchased.

7. Let V represent the value of the machine at the end of year t. The initial value of the machine can be represented by the data point $(0, 24{,}750)$ and the salvage value of the machine can be represented by the data point $(6, 0)$. The slope of the line is

$$m = \frac{0 - 24{,}750}{6 - 0}$$
$$m = -\$4125$$

The slope represents the annual depreciation in dollars per year. Using the point-slope form, you can write the equation of the line as follows

$$V - 24{,}750 = -4125(t - 0)$$
$$V - 24{,}750 = -4125t$$
$$V = -4125t + 24{,}750$$

The equation $V = -4125t + 24{,}750$ represents the book value of the machine each year.

8. Let $t = 9$ represent 2009. Then the two given values are represented by the data points $(9, 58.6)$ and $(10, 56.6)$.

The slope of the line through these points is

$$m = \frac{56.6 - 58.6}{10 - 9} = -2.0.$$

You can find the equation that relates the sales y and the year t to be,

$$y - 56.6 = -2.0(t - 10)$$
$$y - 56.6 = -2.0t + 20.0$$
$$y = -2.0t + 76.6$$

According to this equation, the sales in 2013 will be

$$y = -2.0(13) + 76.6$$
$$= -26 + 76.6$$
$$= \$50.6 \text{ billion.}$$

Checkpoints for Section 2.2

1. (a) This mapping *does not* describe y as a function of x. The input value of -1 is assigned or matched to two different y-values.

   (b) The table *does* describe y as a function of x. Each input value is matched with exactly one output value.

2. (a) Solving for y yields

$$x^2 + y^2 = 8 \quad \text{Write original equation.}$$
$$y^2 = 8 - x^2 \quad \text{Subtract } x^2 \text{ from each side.}$$
$$y = \pm\sqrt{8 - x^2}. \quad \text{Solve for } y.$$

   The $\pm$ indicates that to a given value of x there corresponds two values of y. So y is not a function of x.

   (b) Solving for y yields,

$$y - 4x^2 = 36 \quad \text{Write original equation.}$$
$$y = 36 + 4x^2 \quad \text{Add } 4x^2 \text{ to each side.}$$

   To each value of x there corresponds exactly one value of y. So, y is a function of x.

3. (a) Replacing x with 2 in $f(x) = 10 - 3x^2$ yields the following.

$$f(2) = 10 - 3(2)^2$$
$$= 10 - 12$$
$$= -2$$

   (b) Replacing x with -4 yields the following.

$$f(-4) = 10 - 3(-4)^2$$
$$= 10 - 48$$
$$= -38$$

   (c) Replacing x with $x - 1$ yields the following.

$$f(x - 1) = 10 - 3(x - 1)^2$$
$$= 10 - 3(x^2 - 2x + 1)$$
$$= 10 - 3x^2 + 6x - 3$$
$$= -3x^2 + 6x + 7$$

4. Because $x = -2$ is less than 0, use $f(x) = x^2 + 1$ to obtain $f(-2) = (-2)^2 + 1 = 4 + 1 = 5.$

Because $x = 2$ is greater than or equal to 0, use $f(x) = x - 1$ to obtain $f(2) = 2 - 1 = 1.$

For $x = 3$, use $f(x) = x - 1$ to obtain $f(3) = 3 - 1 = 2.$

5. Set $f(x) = 0$ and solve for x.

$$\begin{aligned} f(x) &= 0 \\ x^2 - 16 &= 0 \\ (x + 4)(x - 4) &= 0 \\ x + 4 = 0 &\Rightarrow x = -4 \\ x - 4 = 0 &\Rightarrow x = 4 \end{aligned}$$

So, $f(x) = 0$ when $x = -4$ or $x = 4$.

6.

$x^2 + 6x - 24 = 4x - x^2$	Set $f(x)$ equal to $g(x)$.
$2x^2 + 2x - 24 = 0$	Write in general form.
$2(x^2 + x - 12) = 0$	Factor out common factor.
$x^2 + x - 12 = 0$	Divide each side by 2.
$(x + 4)(x - 3) = 0$	Factor.
$x + 4 = 0 \Rightarrow x = -4$	Set 1st factor equal to 0.
$x - 3 = 0 \Rightarrow x = 3$	Set 2nd factor equal to 0.

So, $f(x) = g(x)$, when $x = -4$ or $x = 3$.

7. (a) The domain of f consists of all first coordinates in the set of ordered pairs.

Domain $= \{-2, -1, 0, 1, 2\}$

(b) Excluding x-values that yield zero in the denominator, the domain of g is the set of all real numbers x except $x = 3$.

(c) Because the function represents the circumference of a circle, the values of the radius r must be positive. So, the domain is the set of real numbers r such that $r > 0$.

(d) This function is defined only for x-values for which $x - 16 \geq 0$. You can conclude that $x \geq 16$. So, the domain is the interval $[16, \infty)$.

8. Use the formula for surface area of a cylinder, $s = 2\pi r^2 + 2\pi rh$.

(a) $$\begin{aligned} s(r) &= 2\pi r^2 + 2\pi r(4r) \\ &= 2\pi r^2 + 8\pi r^2 \\ &= 10\pi r^2 \end{aligned}$$

(b) $$\begin{aligned} s(h) &= 2\pi\left(\frac{h}{4}\right)^2 + 2\pi\left(\frac{h}{4}\right)h \\ &= 2\pi\left(\frac{h^2}{16}\right) + \frac{\pi h^2}{2} \\ &= \frac{1}{8}\pi h^2 + \frac{1}{2}\pi h^2 \\ &= \frac{5}{8}\pi h^2 \end{aligned}$$

9. When $x = 60$, you can find the height of the baseball as follows

$f(x) = -0.004x^2 + 0.3x + 6$	Write original function.
$f(60) = -0.004(60)^2 + 0.3(60) + 6$	Substitute 60 for x.
$= 9.6$	Simplify.

When $x = 60$, the height of the ball thrown from the second baseman is 9.6 feet. So, the first baseman cannot catch the baseball without jumping.

10. From 2003 through 2005, use $V(t) = 33.65t + 77.8$.

2003: $V(3) = 33.65(3) + 77.8 = 178.75$ thousand vehicles

2004: $V(4) = 33.65(4) + 77.8 = 212.40$ thousand vehicles

2005: $V(5) = 33.65(5) + 77.8 = 246.05$ thousand vehicles

From 2006 through 2009, use $V(t) = 70.75t - 126.6$.

2006: $V(6) = 70.75(6) - 126.6 = 297.90$ thousand vehicles

2007: $V(7) = 70.75(7) - 126.6 = 368.65$ thousand vehicles

2008: $V(8) = 70.75(8) - 126.6 = 439.40$ thousand vehicles

2009: $V(9) = 70.75(9) - 126.6 = 510.15$ thousand vehicles

11.
$$\begin{aligned}\frac{f(x+h)-f(x)}{h} &= \frac{\left[(x+h)^2 + 2(x+h) - 3\right] - \left(x^2 + 2x - 3\right)}{h}\\ &= \frac{x^2 + 2xh + h^2 + 2x + 2h - 3 - x^2 - 2x + 3}{h}\\ &= \frac{2xh + h^2 + 2h}{h}\\ &= \frac{2(2x + h + 2)}{h}\\ &= 2x + h + 2, \quad h \neq 0\end{aligned}$$

Checkpoints for Section 2.3

1. (a) The open dot at $(-3, -6)$ indicates that $x = -3$ is not in the domain of *f*. So, the domain of *f* is all real numbers, except $x \neq -3$, or $(-\infty, -3) \cup (-3, \infty)$.

(b) Because $(0, 3)$ is a point on the graph of *f*, it follows that $f(0) = 3$. Similarly, because the point $(3, -6)$ is a point on the graph of *f*, it follows that $f(3) = -6$.

(c) Because the graph of *f* does not extend above $f(0) = 3$, the range of *f* is the interval $(-\infty, 3]$.

2.

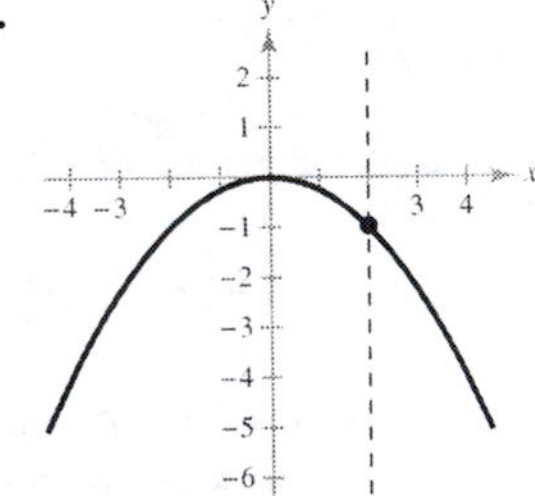

This *is* a graph of y as a function of x, because every vertical line intersects the graph at most once. That is, for a particular input x, there is at most one output y.

3. To find the zeros of a function, set the function equal to zero, and solve for the independent variable.

(a)

$2x^2 + 13x - 24 = 0$	Set $f(x)$ equal to 0.
$(2x - 3)(x + 8) = 0$	Factor.
$2x - 3 = 0 \Rightarrow x = \frac{3}{2}$	Set 1st factor equal to 0.
$x + 8 = 0 \Rightarrow x = -8$	Set 2nd factor equal to 0.

The zeros of *f* are $x = \frac{3}{2}$ and $x = -8$. The graph of *f* has $\left(\frac{3}{2}, 0\right)$ and $(-8, 0)$ as its x-intercepts.

(b)

$\sqrt{t - 25} = 0$	Set $g(t)$ equal to 0.
$\left(\sqrt{t - 25}\right)^2 = (0)^2$	Square each side.
$t - 25 = 0$	Simplify.
$t = 25$	Add 25 to each side.

The zero of g is $t = 25$. The graph of g has $(25, 0)$ as its t-intercept.

(c) $\dfrac{x^2 - 2}{x - 1} = 0$ Set $h(x)$ equal to zero.

$(x - 1)\left(\dfrac{x^2 - 2}{x - 1}\right) = (x - 1)(0)$ Multiply each side by $x - 1$.

$x^2 - 2 = 0$ Simplify.

$x^2 = 2$ Add 2 to each side.

$x = \pm\sqrt{2}$ Extract square roots.

The zeros of h are $x = \pm\sqrt{2}$. The graph of h has $\left(\sqrt{2}, 0\right)$ and $\left(-\sqrt{2}, 0\right)$ as its x-intercepts.

4.

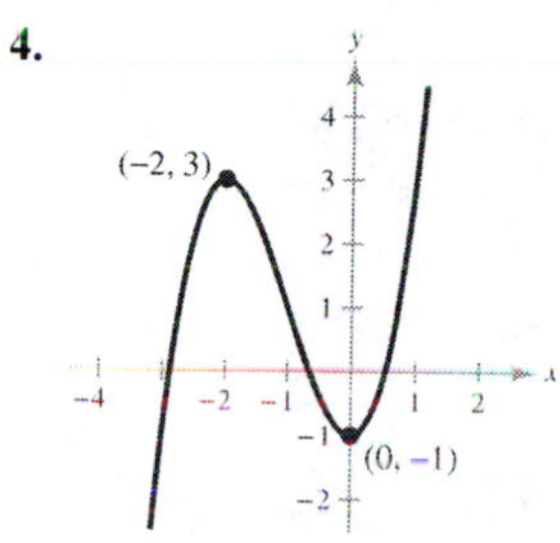

This function is increasing on the interval $(-\infty, -2)$, decreasing on the interval $(-2, 0)$, and increasing on the interval $(0, \infty)$.

5.

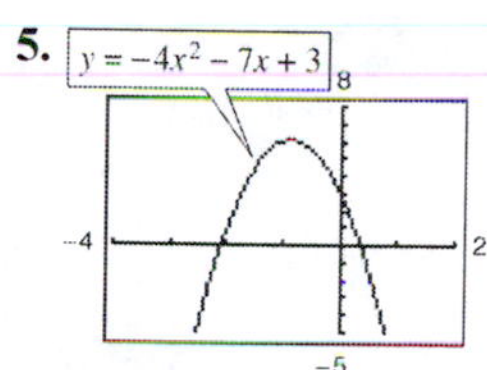

By using the zoom and the trace features or the maximum feature of a graphing utility, you can determine that the function has a relative maximum at the point $\left(-\frac{7}{8}, \frac{97}{16}\right)$ or $(-0.875, 6.0625)$.

6. (a) The average rate of change of f from $x_1 = -3$ to $x_2 = -2$ is

$$\frac{f(x_2) - f(x_1)}{x_2 - x_1} = \frac{f(-2) - f(-3)}{-2 - (-3)} = \frac{0 - 3}{1} = -3.$$

(b) The average rate of change of f from $x_1 = -2$ to $x_2 = 0$ is

$$\frac{f(x_2) - f(x_1)}{x_2 - x_1} = \frac{f(0) - f(-2)}{0 - (-2)} = \frac{0 - 0}{2} = 0.$$

7. (a) The average speed of the car from $t_1 = 0$ to $t_2 = 1$ second is

$$\frac{s(t_2) - s(t_1)}{t_2 - t_1} = \frac{20 - 0}{1 - 0} = 20 \text{ feet per second.}$$

(b) The average speed of the car from $t_1 = 1$ to $t_2 = 4$ seconds is

$$\frac{s(t_2) - s(t_1)}{t_2 - t_1} = \frac{160 - 20}{4 - 1} = \frac{140}{3} \approx 46.7 \text{ feet per second.}$$

8. (a) The function $f(x) = 5 - 3x$ is neither odd nor even because $f(-x) \neq -f(x)$ and $f(-x) \neq f(x)$ as follows.

$f(-x) = 5 - 3(-x)$

$= 5 + 3x \neq -f(x)$ not odd

$\neq f(x)$ not even

So, the graph of f is not symmetric to the origin nor the y-axis.

(b) The function $g(x) = x^4 - x^2 - 1$ is even because $g(-x) = g(x)$ as follows.

$$g(-x) = (-x)^4 - (-x)^2 - 1 = x^4 - x^2 - 1 = g(x)$$

So, the graph of g is symmetric to the y-axis.

(c) The function $h(x) = 2x^3 + 3x$ is odd because $h(-x) = -h(x)$,

$$h(-x) = 2(-x)^3 + 3(-x) = -2x^3 - 3x = -(2x^3 + 3x) = -h(x)$$

So, the graph of h is symmetric to the origin.

Checkpoints for Section 2.4

1. To find the equation of the line that passes through the points $(x_1, y_1) = (-2, 6)$ and $(x_2, y_2) = (4, -4)$, first find the slope of the line.

$$m = \frac{y_2 - y_1}{x_2 - x_1} = \frac{-9 - 6}{4 - (-2)} = \frac{-15}{6} = \frac{-5}{2}$$

Next, use the point-slope form of the equation of the line.

$y - y_1 = m(x - x_1)$ Point-slope form

$y - 6 = -\frac{5}{2}[x - (-2)]$ Substitute x_1, y_1 and m.

$y - 6 = -\frac{5}{2}(x + 2)$ Simplify.

$y - 6 = -\frac{5}{2}x - 5$ Simplify.

$y = -\frac{5}{2}x + 1$ Simplify.

$f(x) = -\frac{5}{2}x + 1$ Function notation

2. For $x = -\frac{3}{2}$, $f\left(-\frac{3}{2}\right) = [\![-\frac{3}{2} + 2]\!]$

$= [\![\frac{1}{2}]\!]$

$= 0$

Since the greatest integer $\le \frac{1}{2}$ is 0, $f\left(-\frac{3}{2}\right) = 0$.

For $x = 1$, $f(1) = [\![1 + 2]\!]$

$= [\![3]\!]$

$= 3$

Since the greatest integer ≤ 3 is 3, $f(1) = 3$.

For $x = -\frac{5}{2}$, $f\left(-\frac{5}{2}\right) = [\![-\frac{5}{2} + 2]\!]$

$= [\![-\frac{1}{2}]\!]$

$= -1$

Since the greatest integer $\le -\frac{1}{2}$ is -1, $f\left(-\frac{5}{2}\right) = -1$.

3. This piecewise-defined function consists of two linear functions. At $x = -4$ and to the left of $x = -4$, the graph is the line $y = -\frac{1}{2}x - 6$, and to the right of $x = -4$ the graph is the line $y = x + 5$. Notice that the point $(-4, -2)$ is a solid dot and $(-4, 1)$ is an open dot. This is because $f(-4) = -2$.

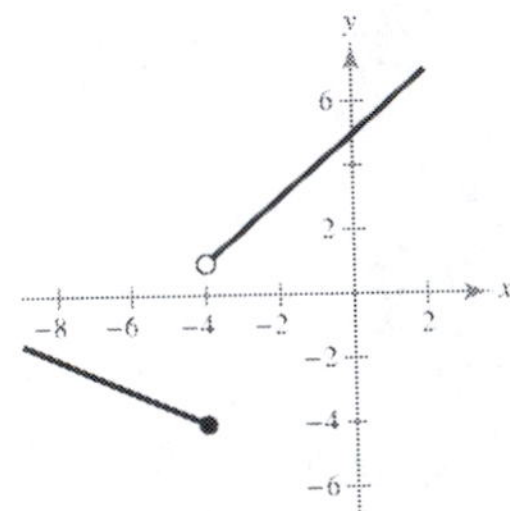

Checkpoints for Section 2.5

1. (a) Relative to the graph of $f(x) = x^3$, the graph of $h(x) = x^3 + 5$ is an upward shift of five units.

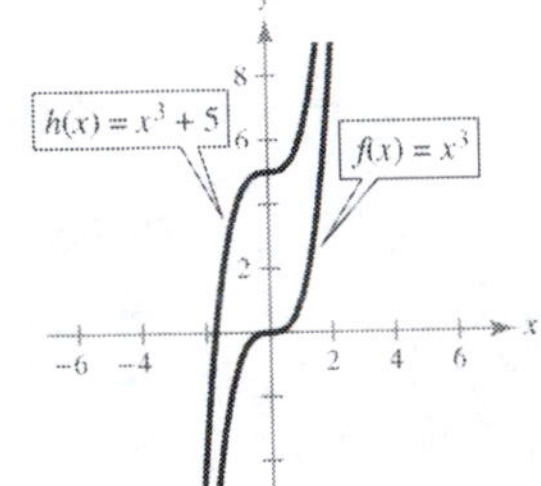

(b) Relative to the graph of $f(x) = x^3$, the graph of $g(x) = (x - 3)^3 + 2$ involves a right shift of three units and an upward shift of two units.

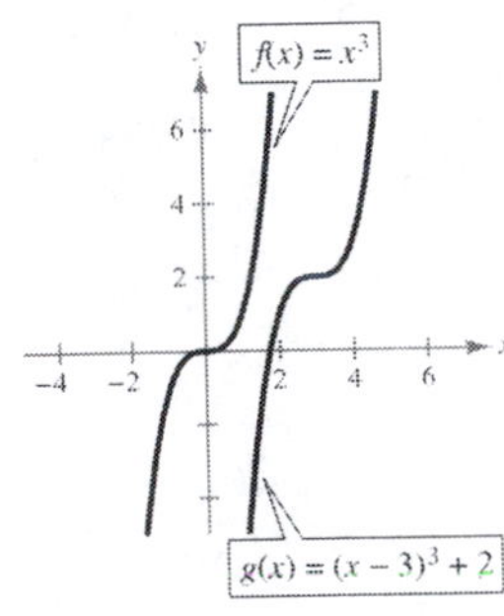

2. The graph of j is a horizontal shift of three units to the left *followed by* a reflection in the x-axis of the graph of $f(x) = x^4$. So, the equation for j is $j(x) = -(x + 3)^4$.

3. (a) **Algebraic Solution:**

The graph of g is a reflection of the graph of f in the x-axis because

$$g(x) = -\sqrt{x - 1}$$
$$= -f(x).$$

Graphical Solution:

Graph f and g on the same set of coordinate axes. From the graph, you can see that the graph of g is a reflection of the graph of f in the x-axis.

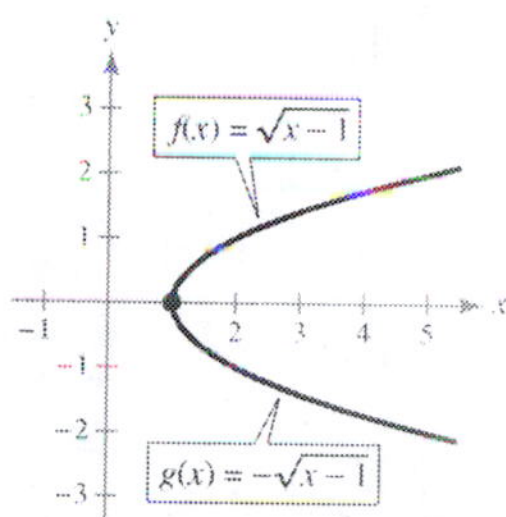

(b) **Algebraic Solution:**

The graph of h is a reflection of the graph of f in the y-axis because

$$h(x) = \sqrt{-x - 1}$$
$$= f(-x).$$

Graphical Solution:

Graph f and h on the same set of coordinate axes. From the graph, you can see that the graph h is a reflection of the graph of f, in the y-axis.

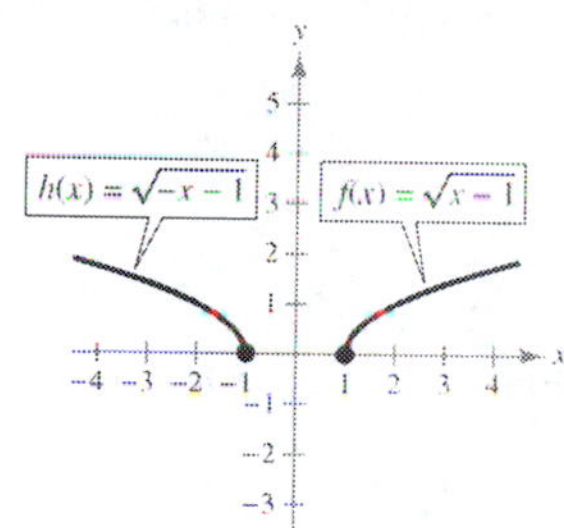

4. (a) Relative to the graph of $f(x) = x^2$, the graph of $g(x) = 4x^2 = 4f(x)$ is a vertical stretch (each y-value is multiplied by 4) of the graph of f.

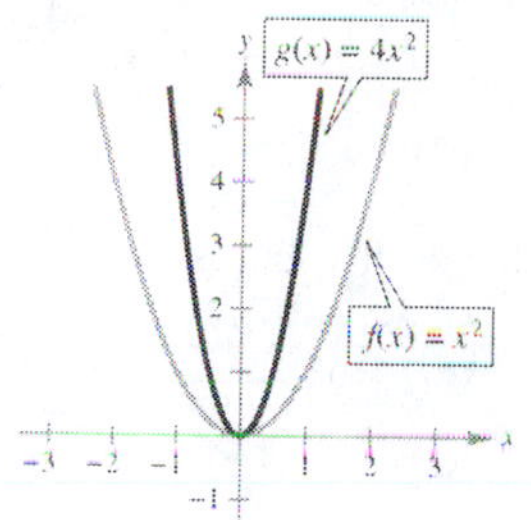

(b) Relative to the graph of $f(x) = x^2$, the graph of $h(x) = \frac{1}{4}x^2 = \frac{1}{4}f(x)$ is a vertical shrink (each y-value is multiplied by $\frac{1}{4}$) of the graph of f.

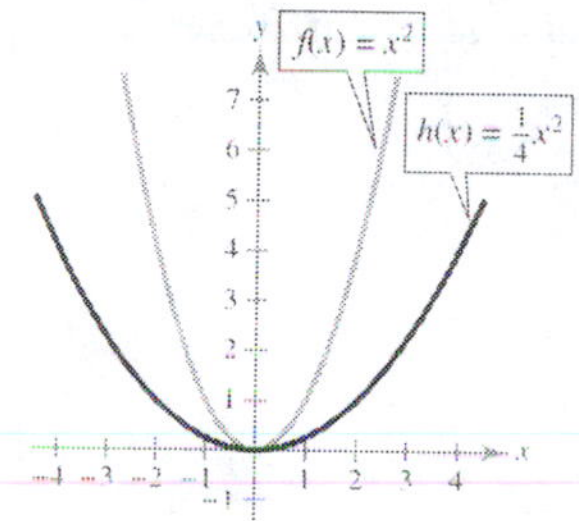

5. (a) Relative to the graph of $f(x) = x^2 + 3$, the graph of $g(x) = f(2x) = (2x)^2 + 3 = 4x^2 + 3$ is a horizontal shrink $(c > 1)$ of the graph of f.

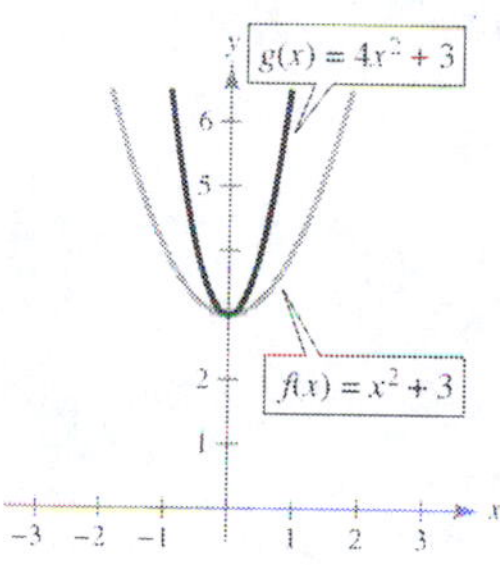

(b) Relative to the graph of $f(x) = x^2 + 3$, the graph of $h(x) = f\left(\frac{1}{2}x\right) = \left(\frac{1}{2}x\right)^2 + 3 = \frac{1}{4}x^2 + 3$ is a horizontal stretch $(0 < c < 1)$ of the graph of f.

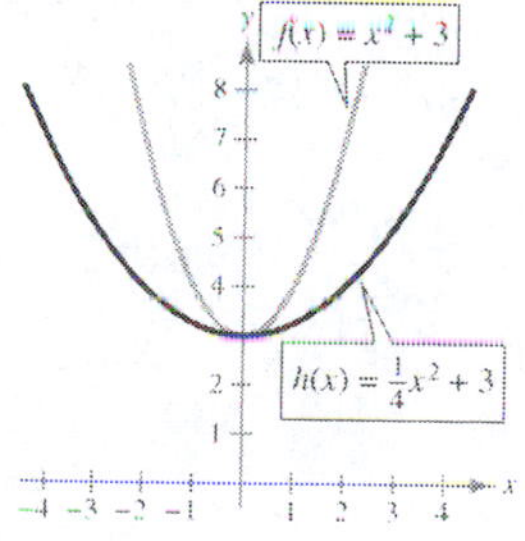

Checkpoints for Section 2.6

1. The sum of f and g is

$$(f+g)(x) = f(x) + g(x)$$
$$= (x^2) + (1-x)$$
$$= x^2 - x + 1.$$

When $x = 2$, the value of this sum is

$$(f+g)(2) = (2)^2 - (2) + 1$$
$$= 3.$$

2. The difference of f and g is

$$(f-g)(x) = f(x) - g(x)$$
$$= (x^2) - (1-x)$$
$$= x^2 + x - 1.$$

When $x = 3$, the value of the difference is

$$(f-g)(3) = (3)^2 + (3) - 1$$
$$= 11.$$

3. The product of f and g is

$$(fg) = f(x)g(x)$$
$$= (x^2)(1-x)$$
$$= x^2 - x^3$$
$$= -x^3 + x^2.$$

When $x = 3$, the value of the product is

$$(fg)(3) = -(3)^3 - (3)^2$$
$$= -27 + 9$$
$$= -18.$$

4. The quotient of f and g is

$$\left(\frac{f}{g}\right)(x) = \frac{f(x)}{g(x)} = \frac{\sqrt{x-3}}{\sqrt{16-x^2}}.$$

The quotient of g and f is

$$\left(\frac{g}{f}\right)(x) = \frac{g(x)}{f(x)} = \frac{\sqrt{16-x^2}}{\sqrt{x-3}}.$$

The domain of f is $[3, \infty)$ and the domain of g is $[-4, 4]$. The intersection of these two domains is $[3, 4]$. So, the domain of f/g is $[3, 4)$ and the domain of g/f is $(3, 4]$.

5. (a) The composition of f with g is as follows.

$$(f \circ g)(x) = f(g(x))$$
$$= f(4x^2 + 1)$$
$$= 2(4x^2 + 1) + 5$$
$$= 8x^2 + 2 + 5$$
$$= 8x^2 + 7$$

(b) The composition of g with f is as follows.

$$(g \circ f)(x) = g(f(x))$$
$$= g(2x + 5)$$
$$= 4(2x + 5)^2 + 1$$
$$= 4(4x^2 + 20x + 25) + 1$$
$$= 16x^2 + 80x + 100 + 1$$
$$= 16x^2 + 80x + 101$$

(c) Use the result of part (a).

$$(f \circ g)\left(-\tfrac{1}{2}\right) = 8\left(-\tfrac{1}{2}\right)^2 + 7$$
$$= 8\left(\tfrac{1}{4}\right) + 7$$
$$= 2 + 7$$
$$= 9$$

6. The composition of f with g is as follows.

$$(f \circ g)(x) = f(g(x))$$
$$= f(x^2 + 4)$$
$$= \sqrt{x^2 + 4}$$

The domain of f is $[0, \infty)$ and the domain of g is the set of all real numbers. The range of g is $[4, \infty)$, which is in the range of f, $[0, \infty)$. Therefore the domain of $f \circ g$ is all real numbers.

7. Let the inner function to be $g(x) = 8 - x$ and the outer function to be $f(x) = \dfrac{\sqrt[3]{x}}{5}$.

$$h(x) = \frac{\sqrt[3]{8-x}}{5}$$
$$= f(8 - x)$$
$$= f(g(x))$$

8. (a) $(N \circ T)(t) = N(T(t))$

$$= 8(2t + 2)^2 - 14(2t + 2) + 200$$
$$= 8(4t^2 + 8t + 4) - 28t - 28 + 200$$
$$= 32t^2 + 64t + 32 - 28t - 28 + 200$$
$$= 32t^2 + 36t + 204$$

The composite function $(N \circ T)(t)$ represents the number of bacteria in the food as a function of the amount of time the food has been out of refrigeration.

(b) Let $(N \circ T)(t) = 1000$ and solve for t.

$$32t^2 + 36t + 204 = 1000$$
$$32t^2 + 36t - 796 = 0$$
$$4(8t^2 + 9t - 199) = 0$$
$$8t^2 + 9t - 199 = 0$$

Use the quadratic formula:

$$t = \frac{-9 \pm \sqrt{(9)^2 - 4(8)(-199)}}{2(8)}$$
$$= \frac{-9 \pm \sqrt{6449}}{16}$$

$t \approx 4.5$ and $t \approx -5.6$.

Using $t \approx 4.5$ hours, the bacteria count reaches approximately 1000 about 4.5 hours after the food is removed from the refrigerator.

Checkpoints for Section 2.7

1. The function f multiplies each input by $\frac{1}{5}$. To "undo" this function, you need to multiply each input by 5. So, the inverse function of $f(x) = \frac{1}{5}x$ is $f^{-1}(x) = 5x$.

To verify this, show that

$f(f^{-1}(x)) = x$ and $f^{-1}(f(x)) = x$.

$$f(f^{-1}(x)) = f(5x) = \tfrac{1}{5}(5x) = x$$
$$f^{-1}(f(x)) = f^{-1}\left(\tfrac{1}{5}x\right) = 5\left(\tfrac{1}{5}x\right) = x$$

So, the inverse function of $f(x) = \frac{1}{5}x$ is $f^{-1}(x) = 5x$.

2. By forming the composition of f and g, you have

$$f(g(x)) = f(7x + 4) = \frac{(7x + 4) - 4}{7} = \frac{7x}{7} = x.$$

So, it appears that g is the inverse function of f. To confirm this, form the composition of g and f.

$$g(f(x)) = g\left(\frac{x - 4}{7}\right) = 7\left(\frac{x - 4}{7}\right) + 4 = x - 4 + 4 = x$$

By forming the composition of f and h, you can see that h is *not* the inverse function of f, since the result is not the identity function x.

$$f(h(x)) = f\left(\frac{7}{x - 4}\right) = \frac{\left(\frac{7}{x - 4}\right) - 4}{7} = \frac{23 - 4x}{7(x - 4)} \neq x$$

So, g is the inverse function of f.

3. The graphs of $f(x) = 4x - 1$ and $f^{-1}(x) = \frac{1}{4}(x + 1)$ are shown. You can see that they are reflections of each other in the line $y = x$. This reflective property can also be verified using a few points and the fact that if the point (a, b) is on the graph of f then the point (b, a) is on the graph of f^{-1}.

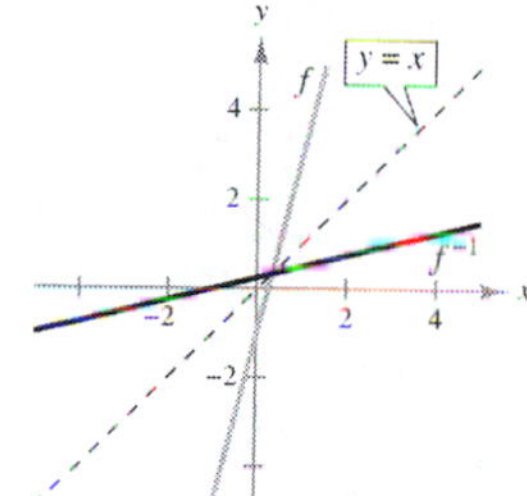

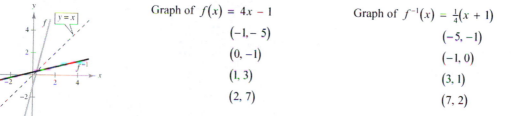

Graph of $f(x) = 4x - 1$	Graph of $f^{-1}(x) = \frac{1}{4}(x + 1)$
$(-1, -5)$	$(-5, -1)$
$(0, -1)$	$(-1, 0)$
$(1, 3)$	$(3, 1)$
$(2, 7)$	$(7, 2)$

4. The graphs of $f(x) = x^2 + 1,\ x \geq 0$ and $f^{-1}(x) = \sqrt{x - 1}$ are shown. You can see that they are reflections of each other in the line $y = x$. This reflective property can also be verified using a few points and the fact that if the point (a, b) is on the graph of f then the point (b, a) is on the graph of f^{-1}.

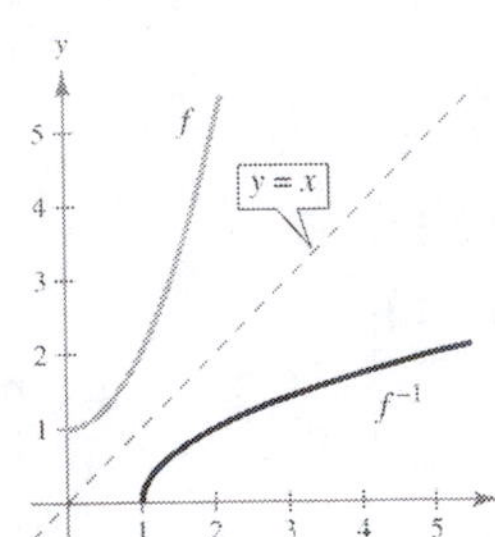

Graph of $f(x) = x^2 + 1,\ x \geq 0$	Graph of $f^{-1}(x) = \sqrt{x - 1}$
$(0, 1)$	$(1, 0)$
$(1, 2)$	$(2, 1)$
$(2, 5)$	$(5, 2)$
$(3, 10)$	$(10, 3)$

5. (a) The graph of $f(x) = \frac{1}{2}(3 - x)$ is shown.

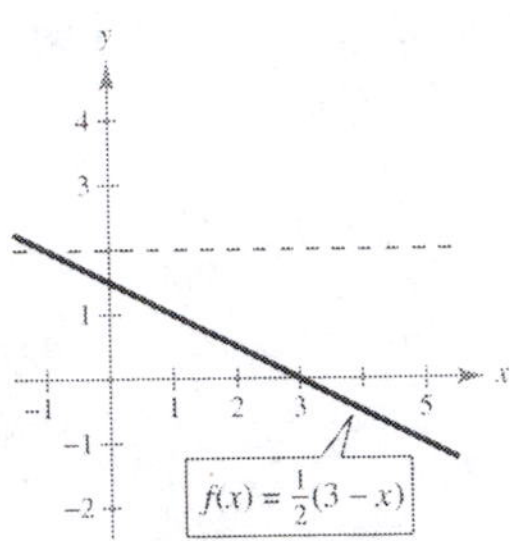

Because no horizontal line intersects the graph of f at more than one point, f is a one-to-one function and *does* have an inverse function.

(b) The graph of $f(x) = |x|$ is shown.

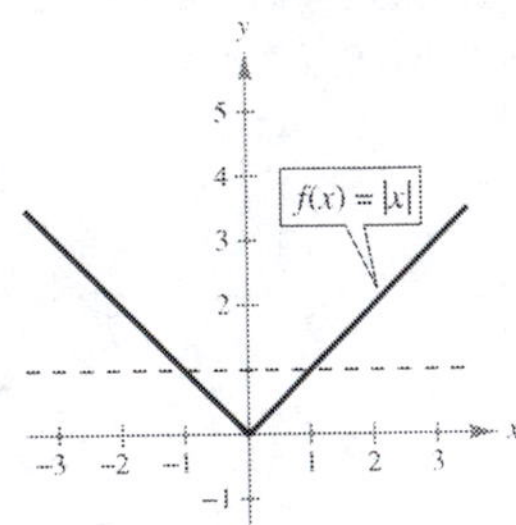

Because it is possible to find a horizontal line that intersects the graph of f at more than one point, f is *not* a one-to-one function and *does not* have an inverse function.

6. The graph of $f(x) = \dfrac{5 - 3x}{x + 2}$ is shown.

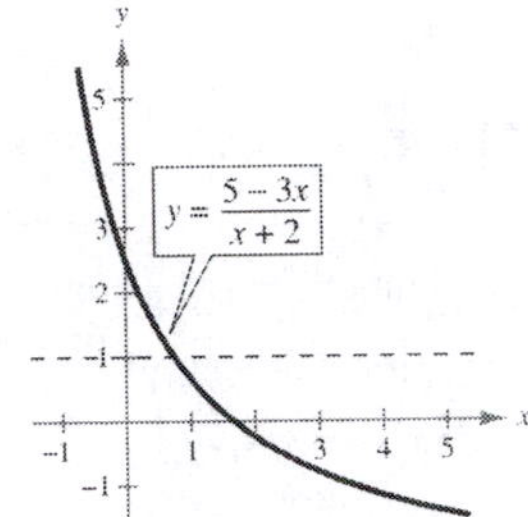

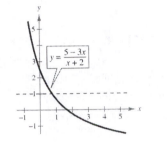

This graph passes the Horizontal Line Test. So, you know f is one-to-one and has an inverse function.

$f(x) = \dfrac{5 - 3x}{x + 2}$ Write original function.

$y = \dfrac{5 - 3x}{x + 2}$ Replace $f(x)$ with y.

$x = \dfrac{5 - 3y}{y + 2}$ Interchange x and y.

$x(y + 2) = 5 - 3y$ Multiply each side by $y + 2$.

$xy + 2x = 5 - 3y$ Distribute Property

$xy + 3y = 5 - 2x$ Collect like terms with y.

$y(x + 3) = 5 - 2x$ Factor.

$y = \dfrac{5 - 2x}{x + 3}$ Solve for y.

$f^{-1}(x) = \dfrac{5 - 2x}{x + 3}$ Replace y with $f^{-1}(x)$.

7. The graph of $f(x) = \sqrt[3]{10 + x}$ is shown.

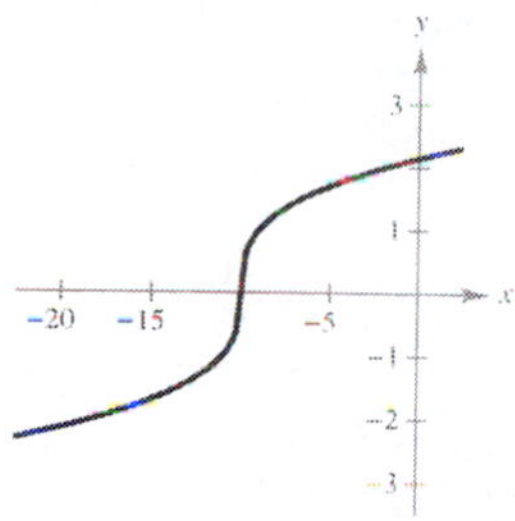

Because this graph passes the Horizontal Line Test, you know that f is one-to-one and has an inverse function.

$$f(x) = \sqrt[3]{10 + x}$$
$$y = \sqrt[3]{10 + x}$$
$$x = \sqrt[3]{10 + y}$$
$$x^3 = 10 + y$$
$$y = x^3 - 10$$
$$f^{-1}(x) = x^3 - 10$$

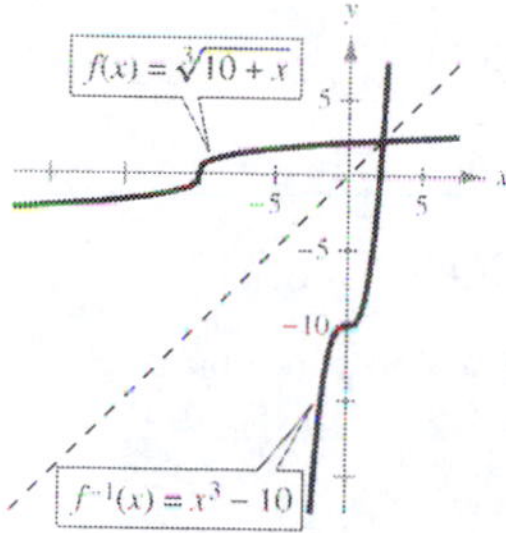

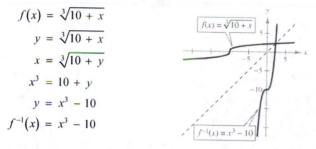

The graphs of f and f^{-1} are reflections of each other in the line $y = x$. So, the inverse of $f(x) = \sqrt[3]{10 + x}$ is $f^{-1}(x) = x^3 - 10$.

To verify, check that $f(f^{-1}(x)) = x$ and $f^{-1}(f(x)) = x$.

$$f(f^{-1}(x)) = f(x^3 - 10) = \sqrt[3]{10 + (x^3 - 10)} = \sqrt[3]{x^3} = x$$

$$f^{-1}(f(x)) = f^{-1}(\sqrt[3]{10 + x}) = \left(\sqrt[3]{10 + x}\right)^3 - 10 = 10 + x - 10 = x$$

Chapter 3

Checkpoints for Section 3.1

1. (a) Compared with the graph of $y = x^2$, each output of $f(x) = \frac{1}{4}x^2$ "shrinks" by a factor of $\frac{1}{4}$, creating a broader parabola.

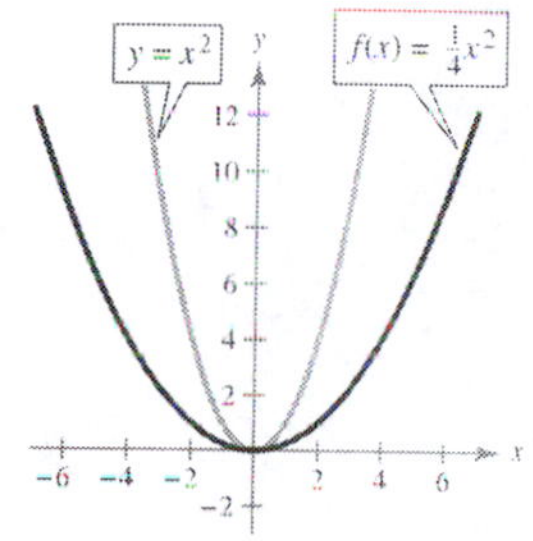

(b) Compared with the graph of $y = x^2$, each output of $f(x) = -\frac{1}{6}x^2$ is reflected in the x-axis and "shrinks" by a factor of $\frac{1}{6}$, creating a broader parabola.

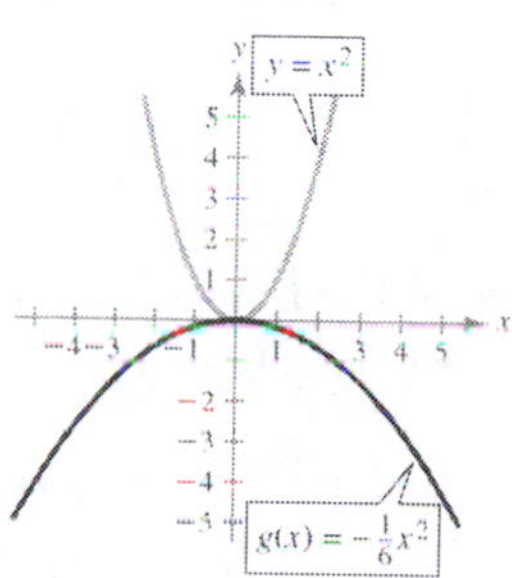

(c) Compared with $y = x^2$, each output of of $h(x) = \frac{5}{2}x^2$ "stretches" by a factor of $\frac{5}{2}$, creating a narrower parabola.

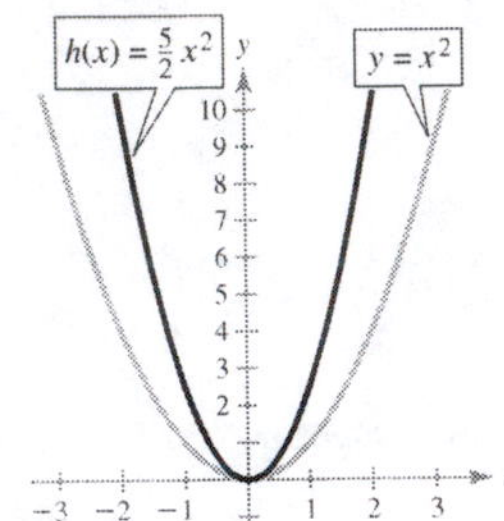

(b) Compared with $y = -4x^2$, each output of $k(x)$ is reflected in the x-axis and "stretches" by a factor of 4.

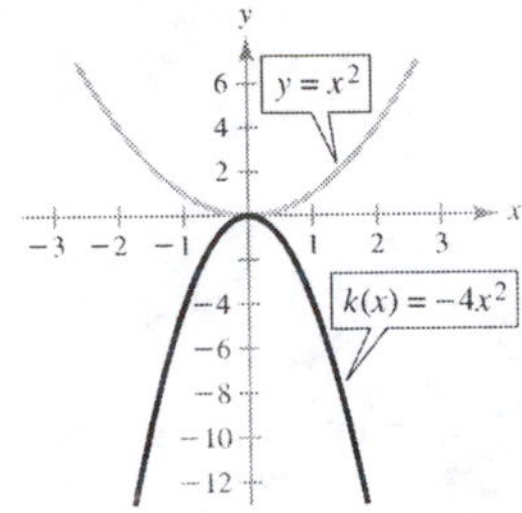

2.

$$\begin{aligned}
f(x) &= 3x^2 - 6x + 4 && \text{Write original function.}\\
&= 3(x^2 - 2x) + 4 && \text{Factor 3 out of } x\text{-terms.}\\
&= 3(x^2 - 2x + 1 - 1) + 4 && \text{Add and subtract 1 within parenthesis.}\\
&= 3(x^2 - 2x + 1) - 3(1) + 4 && \text{Regroup terms.}\\
&= 3(x^2 - 2x + 1) - 3 + 4 && \text{Simplify.}\\
&= 3(x - 1)^2 + 1 && \text{Write in standard form.}
\end{aligned}$$

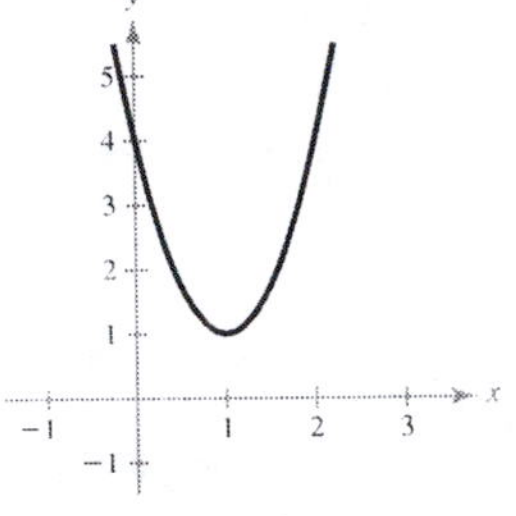

You can see that the graph of f is a parabola that opens upward and has its vertex at $(1, 1)$. This corresponds to a right shift of one unit and an upward shift of one unit relative to the graph of $y = 3x^2$, which is a "stretch" of $y = x^2$.

The axis of the parabola is the vertical line through the vertex, $x = 1$.

3.

$$\begin{aligned}
f(x) &= x^2 - 4x + 3 && \text{Write original function.}\\
&= (x^2 - 4x + 4 - 4) + 3 && \text{Add and subtract 4 within parenthesis.}\\
&= (x^2 - 4x + 4) - 4 + 3 && \text{Regroup terms.}\\
&= (x^2 - 4x + 4) - 1 && \text{Simplify.}\\
&= (x^2 - 2)^2 - 1 && \text{Write in standard form.}
\end{aligned}$$

In standard form, you can see that f is a parabola that opens upward with vertex $(2, -1)$.

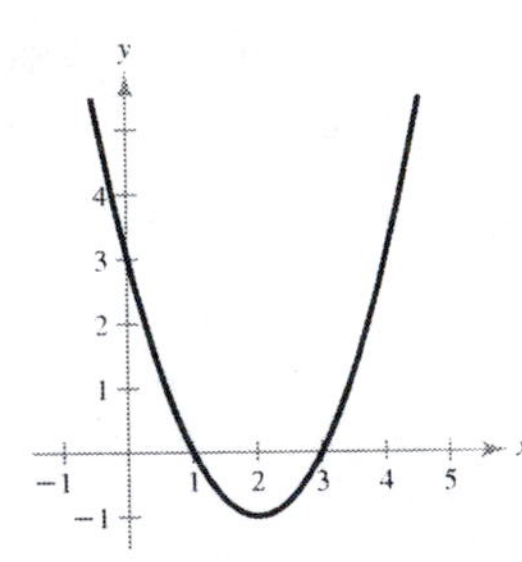

The x-intercepts of the graph are determined as follows.

$$x^2 - 4x + 3 = 0$$

$$(x - 3)(x - 1) = 0$$

$$x - 3 = 0 \Rightarrow x = 3$$

$$x - 1 = 0 \Rightarrow x = 1$$

So, the x-intercepts are $(3, 0)$ and $(1, 0)$.

4. Because the vertex is $(h, k) = (-4, 11)$ the equation has the form

$f(x) = a(x + 4)^2 + 11.$

Because the parabola passes through the point $(-6, 15)$ it follows that $f(-6) = 15$.

$f(x) = a(x + 4)^2 + 11$	Write standard form.
$15 = a(-6 + 4)^2 + 11$	Substitute -6 for x and 15 for $f(x)$.
$15 = a(-2)^2 + 11$	Simplify.
$4 = 4a$	Subtract 11 from each side.
$1 = a$	Divide each side by 4.

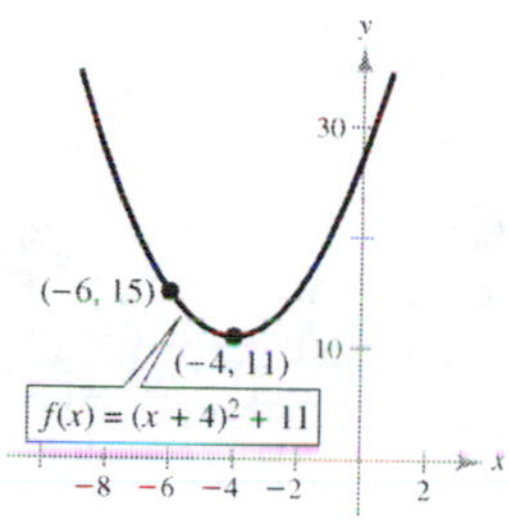

The equation in standard form is $f(x) = (x + 4)^2 + 11$.

5. For this quadratic function,

$f(x) = ax^2 + bx + c = -0.007x^2 + x + 4$

which implies that $a = -0.007$ and $b = 1$.

Because $a < 0$, the function has a maximum at $x = -\dfrac{b}{2a}$. So, the baseball reaches its maximum height when it is

$x = -\dfrac{b}{2a} = -\dfrac{1}{2(-0.007)} = \dfrac{1}{0.014} \approx 71.4$ feet from home plate.

At this distance, the maximum height is $f(71.4) = -0.007(71.4)^2 + (71.4) + 4 \approx 39.7$ feet.

Checkpoints for Section 3.2

1. (a) The graph of $f(x) = (x + 5)^4$ is a left shift by five units of the graph of $y = x^4$.

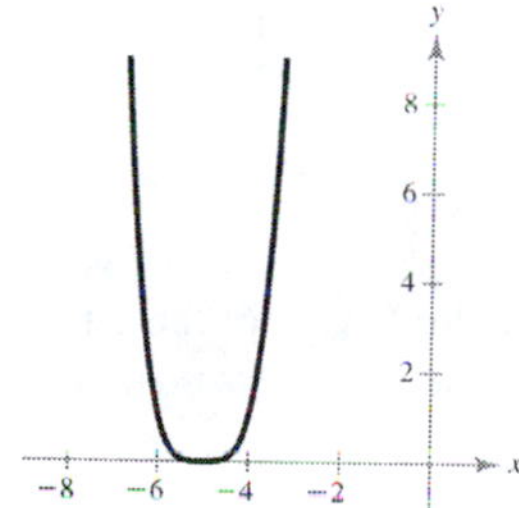

(b) The graph of $g(x) = x^4 - 7$ is a downward shift of seven units of the graph of $y = x^4$.

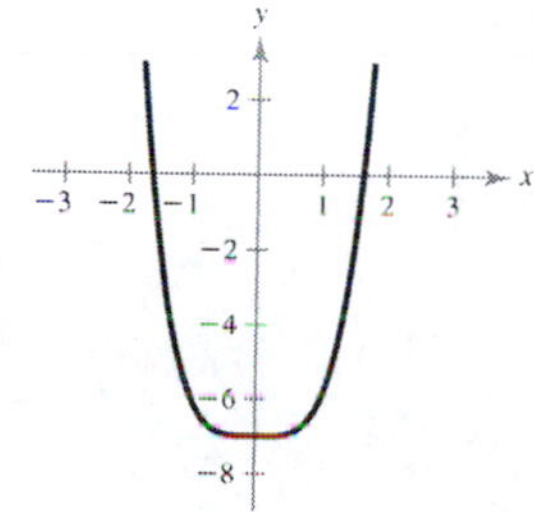

(c) The graph of $h(x) = 7 - x^4 = -x^4 + 7$ is a reflection in the x-axis then an upward shift of seven units of the graph of $y = x^4$.

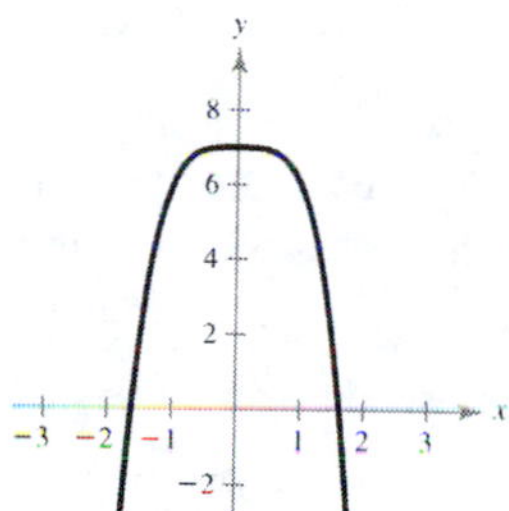

(d) The graph of $k(x) = \frac{1}{4}(x - 3)^4$ is a right shift by three units and a vertical "shrink" by a factor of $\frac{1}{4}$ of the graph of $y = x^4$.

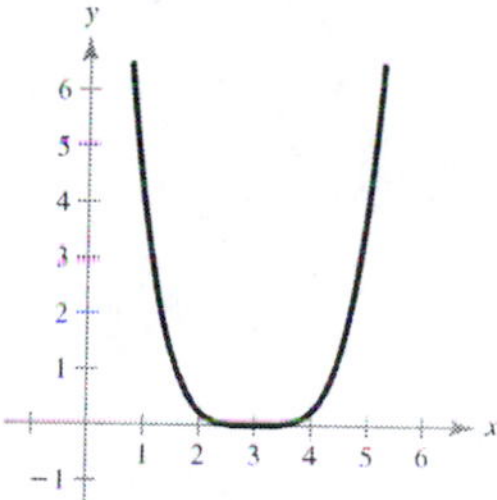

2. (a) Because the degree is odd and the leading coefficient is positive, the graph falls to the left and rises to the right.

(b) Because the degree is odd and the leading coefficient is negative, the graph rises to the left and falls to the right.

3. To find the real zeros of $f(x) = x^3 - 12x^2 + 36x$, set $f(x)$ equal to zero, and solve for x.

$$x^3 - 12x^2 + 36x = 0$$
$$x(x^2 - 12x + 36) = 0$$
$$x(x - 6)^2 = 0$$
$$x = 0$$
$$x - 6 = 0 \Rightarrow x = 6$$

So, the real zeros are $x = 0$ and $x = 6$. Because the function is a third-degree polynomial, the graph of f can have at most $3 - 1 = 2$ turning points. In this case, the graph of f has two turning points.

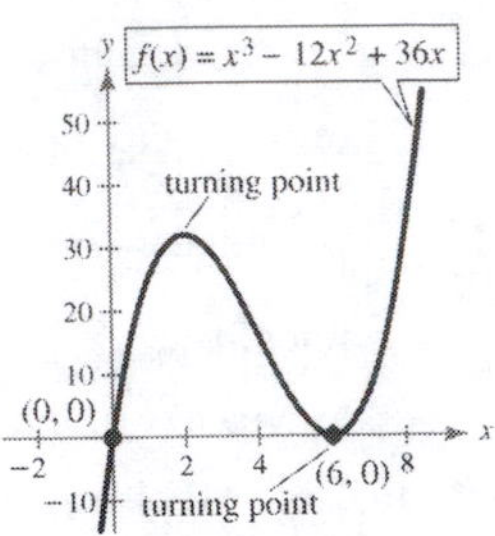

4. 1. *Apply the Leading Coefficient Test.*

Because the leading coefficient is positive and the degree is odd, you know that the graph eventually falls to the left and rises to the right.

2. *Find the Real Zeros of the Polynomial.*

By factoring
$$f(x) = 2x^3 - 6x^2 = 2x^2(x - 3)$$

you can see that the real zeros of f are $x = 0$ (even multiplicity) and $x = 3$ (odd multiplicity). So, the x-intercepts occur at $(0, 0)$ and $(3, 0)$

3. *Plot a Few Additional Points.*

x	-1	1	2	4
$f(x)$	-8	-4	-8	32

4. *Draw the Graph.*

Draw a continuous curve through all of the points. Because $x = 0$ is of even multiplicity, you know that the graph touches the x-axis but does not cross it at $(0, 0)$. Because $x = 3$ is of odd multiplicity, you know that the graph should cross the x-axis at $(3, 0)$.

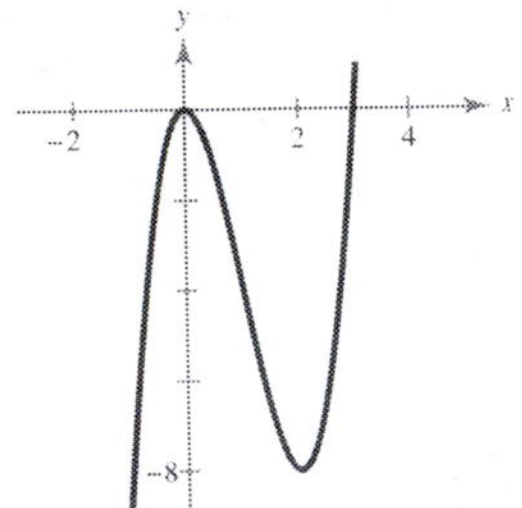

5. 1. *Apply the Leading Coefficient Test.*

Because the leading coefficient is negative and the degree is even, you know that the graph eventually falls to the left and falls to the right.

2. *Find the Real Zeros of the Polynomial.*

By factoring
$$f(x) = -\tfrac{1}{4}x^4 + \tfrac{3}{2}x^3 - \tfrac{9}{4}x^2 = -\tfrac{1}{4}x^2(x^2 - 6x + 9) = -\tfrac{1}{4}x^2(x - 3)^2$$

you can see that the real zeros of f are $x = 0$ (even multiplicity) and $x = 3$ (even multiplicity). So, the x-intercepts occur at $(0, 0)$ and $(3, 0)$.

3. *Plot a Few Additional Points.*

x	-1	1	2	4
$f(x)$	-4	-1	-1	-4

4. *Draw the graph.*

Draw a continuous curve through the points. As indicated by the multiplicities of the zeros, the graph touches but does not cross the x-axis at $(0, 0)$ and $(3, 0)$.

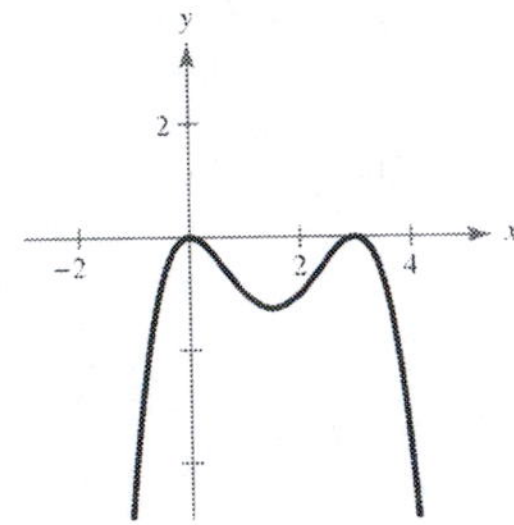

6. Begin by computing a few function values of $f(x) = x^3 - 3x^2 - 2$.

x	-1	0	1	2	3	4
$f(x)$	-6	-2	-4	-6	-2	14

Because $f(3)$ is negative and $f(4)$ is positive, you can apply the Intermediate Value Theorem to conclude that the function has a real zero between $x = 3$ and $x = 4$. To find this real zero more closely, divide the interval $[3, 4]$ into tenths and evaluate the function at each point.

x	3.1	3.2	3.3	3.4	3.5	3.6	3.7	3.8	3.9
$f(x)$	-1.039	0.048	1.267	2.624	4.125	5.776	7.583	9.552	11.689

So, f must have a real zero between 3.1 and 3.2.
To find a more accurate approximation, you can compute the function value between $f(3.1)$ and $f(3.2)$ and apply the Intermediate Value Theorem again.

Checkpoints for Section 3.3

1. To divide $3x^2 + 19x + 28$ by $x + 4$ using long division, you can set up the operation as shown.

$$
\begin{array}{rll}
 & 3x + 7 & \\
x + 4 \,\big) & 3x^2 + 19x + 28 & \\
 & \underline{3x^2 + 12x} & \text{Multiply by: } 3x(x + 4). \\
 & 7x + 28 & \text{Subtract.} \\
 & \underline{7x + 28} & \text{Multiply by: } 7(x + 4). \\
 & 0 & \text{Subtract.}
\end{array}
$$

From this division, you can conclude that

$3x^2 + 19x + 28 = (x + 4)(3x + 7)$.

2. To divide $x^3 - 2x^2 - 9$ by $x - 3$ using long division, you can set up the operation as shown. Because there is no x-term in the dividend, rewrite the dividend as $x^3 - 2x^2 + 0x - 9$ before you apply the Division Algorithm.

$$
\begin{array}{rll}
 & x^2 + x + 3 & \\
x - 3 \,\big) & x^3 - 2x^2 + 0x - 9 & \\
 & \underline{x^3 - 3x^2} & \text{Multiply } x^2 \text{ by } x - 3. \\
 & x^2 + 0x - 9 & \text{Subtract.} \\
 & \underline{x^2 - 3x} & \text{Multiply } x \text{ by } x - 3. \\
 & 3x - 9 & \text{Subtract.} \\
 & \underline{3x - 9} & \text{Multiply 3 by } x - 3. \\
 & 0 & \text{Subtract.}
\end{array}
$$

So, $x - 3$ divides evenly into $x^3 - 2x^2 - 9$, and you can write $\dfrac{x^3 - 2x^2 - 9}{x - 3} = x^2 + x + 3, x \neq 3$.

You can check this result by multiplying

$$
\begin{aligned}
(x - 3)(x^2 + x + 3) &= x^3 - 3x^2 + x^2 - 3x + 3x - 9 \\
&= x^3 - 2x^2 - 9.
\end{aligned}
$$

3. To divide $-x^3 + 9x + 6x^4 - x^2 - 3$ by $1 + 3x$ using long division, begin by rewriting the dividend and divisor in descending powers of x.

$$
\begin{array}{rll}
 & 2x^3 - x^2 + 3 & \\
3x + 1 \,\big) & 6x^4 - x^3 - x^2 + 9x - 3 & \\
 & \underline{6x^4 + 2x^3} & \text{Multiply } 2x^3 \text{ by } 3x + 1. \\
 & -3x^3 - x^2 + 9x - 3 & \text{Subtract.} \\
 & \underline{-3x^3 - x^2} & \text{Multiply } -x^2 \text{ by } 3x + 1. \\
 & 9x - 3 & \text{Subtract.} \\
 & \underline{9x + 3} & \text{Multiply 3 by } 3x + 1. \\
 & -6 & \text{Subtract.}
\end{array}
$$

So, you have $\dfrac{6x^4 - x^3 - x^2 + 9x - 3}{3x + 1} = 2x^3 - x^2 + 3 - \dfrac{6}{3x + 1}$.

4. To divide $5x^3 + 8x^2 - x + 6$ by $x + 2$ using synthetic division, you can set up the array as show.

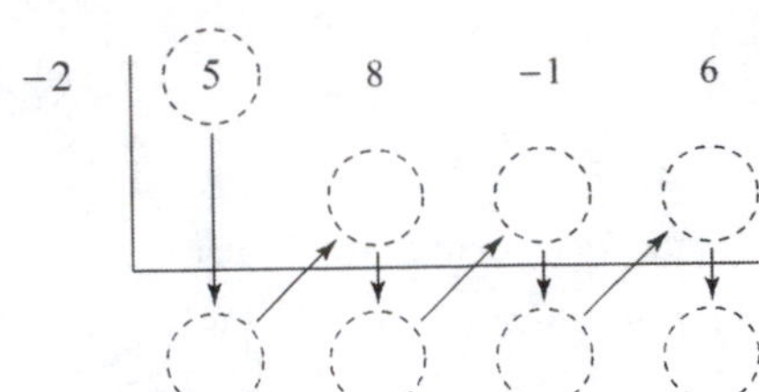

Then, use the synthetic division pattern by adding terms in columns and multiplying the results by -2.

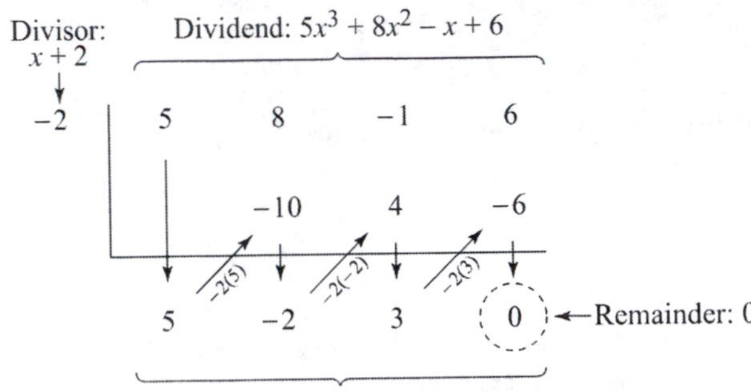

So, you have, $\dfrac{5x^3 + 8x^2 - x + 6}{x + 2} = 5x^2 - 2x + 3.$

5. (a) $f(-1)$

$$\begin{array}{r|rrrr} -1 & 4 & 10 & -3 & -8 \\ & & -4 & -6 & 9 \\ \hline & 4 & 6 & -9 & 1 \end{array}$$

Because the remainder is $r = 1$, $f(-1) = 1$.

$$\begin{aligned} \text{Check: } f(-1) &= 4(-1)^3 + 10(-1)^2 - 3(-1) - 8 \\ &= 4(-1) + 10(1) + 3 - 8 \\ &= 1 \end{aligned}$$

(b) $f(4)$

$$\begin{array}{r|rrrr} 4 & 4 & 10 & -3 & -8 \\ & & 16 & 104 & 404 \\ \hline & 4 & 26 & 101 & 396 \end{array}$$

Because the remainder is $r = 396$, $f(4) = 396$.

$$\begin{aligned} \text{Check: } f(4) &= 4(4)^3 + 10(4)^2 - 3(4) - 8 \\ &= 4(64) + 10(16) - 12 - 8 \\ &= 396 \end{aligned}$$

(c) $f\left(\frac{1}{2}\right)$

$$\begin{array}{r|rrrr} \frac{1}{2} & 4 & 10 & -3 & -8 \\ & & 2 & 6 & \frac{3}{2} \\ \hline & 4 & 12 & 3 & -\frac{13}{2} \end{array}$$

Because the remainder is $r = -\frac{13}{2}$, $f\left(\frac{1}{2}\right) = -\frac{13}{2}$.

$$\begin{aligned} \text{Check: } f\left(\tfrac{1}{2}\right) &= 4\left(\tfrac{1}{2}\right)^3 + 10\left(\tfrac{1}{2}\right)^2 - 3\left(\tfrac{1}{2}\right) - 8 \\ &= 4\left(\tfrac{1}{8}\right) + 10\left(\tfrac{1}{4}\right) - \tfrac{3}{2} - 8 \\ &= \tfrac{1}{2} + \tfrac{5}{2} - \tfrac{3}{2} - 8 \\ &= -\tfrac{13}{2} \end{aligned}$$

(d) $f(-3)$

$$\begin{array}{r|rrrr} -3 & 4 & 10 & -3 & -8 \\ & & -12 & 6 & -9 \\ \hline & 4 & -2 & 3 & -17 \end{array}$$

Because the remainder is $r = -17$, $f(-3) = -17$.

$$\begin{aligned} \text{Check: } f(-3) &= 4(-3)^3 + 10(-3)^2 - 3(-3) - 8 \\ &= 4(-27) + 10(9) + 9 - 8 \\ &= -17 \end{aligned}$$

6. Algebraic Solution:

Using synthetic division with the factor $(x + 3)$, you obtain the following.

$$\begin{array}{r|rrrr} -3 & 1 & 0 & -19 & -30 \\ & & -3 & 9 & 30 \\ \hline & 1 & -3 & -10 & 0 \end{array} \rightarrow \text{ 0 remainder, so } f(-3) = 0 \text{ and } (x + 3) \text{ is a factor.}$$

Because the resulting quadratic expression factors as $x^2 - 3x - 10 = (x - 5)(x + 2)$, the complete factorization of $f(x)$ is $f(x) = x^3 - 19x - 30 = (x + 3)(x - 5)(x + 2)$.

Graphical Solution:

From the graph of $f(x) = x^3 - 19x - 30$, you can see there are three x-intercepts. These occur at $x = -3$, $x = -2$, and $x = 5$. This implies that $(x + 3)$, $x + 2$, and $(x - 5)$ are factors of $f(x)$.

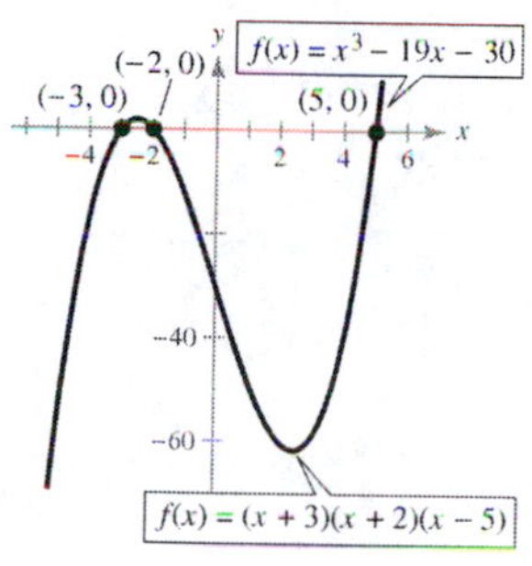

Checkpoints for Section 3.4

1.
$$\begin{aligned} f(x) &= x^4 - 1 = (x^2 + 1)(x^2 - 1) \\ &= (x^2 + 1)(x + 1)(x - 1) \end{aligned}$$

$x^2 + 1 \Rightarrow x = \pm i$

$x + 1 \Rightarrow x = -1$

$x - 1 \Rightarrow x = 1$

The fourth-degree polynomial function $f(x) = x^4 - 1$ has exactly four zeros, $x = \pm i$ and $x = \pm 1$

2. (a) Because the leading coefficient is 1, the possible rational zeros are the factors of the constant term

Possible rational zeros: $\pm 1, \pm 2, \pm 4, \pm 8$

By testing these zeros,

$$\begin{aligned} f(-1) &= (-1)^3 - 5(-1)^2 + 2(-1) + 8 = 0 \\ f(1) &= (1)^3 - 5(1)^2 + 2(1) + 8 = 6 \\ f(-2) &= (-2)^3 - 5(-2)^2 + 2(-2) + 8 = -24 \\ f(2) &= (2)^3 - 5(2)^2 + 2(2) + 8 = 0 \\ f(-4) &= (-4)^3 - 5(-4)^2 + 2(-4) + 8 = -144 \\ f(4) &= (4)^3 - 5(4)^3 + 2(4) + 8 = 0 \\ f(-8) &= (-8)^3 - 5(-8) + 2(-8) + 8 = -840 \\ f(8) &= (8)^3 - 5(8)^2 + 2(8) + 8 = 216 \end{aligned}$$

you can conclude that the polynomial function $f(x) = x^3 - 5x^2 + 2x + 8$ has three rational zeros $x = -1$, $x = 2$, and $x = 4$.

(b) *Possible rational zeros*: ± 1 ± 2 and ± 4

By testing these possible zeros,

$$f(-1) = (-1)^3 + 2(-1)^2 + 6(-1) - 4 = -9$$
$$f(1) = (1)^3 + 2(1)^2 + 6(1) - 4 = 5$$
$$f(-2) = (-2)^3 + 2(-2)^2 + 6(-2) - 4 = -16$$
$$f(2) = (2)^3 + 2(2)^2 + 6(2) - 4 = 24$$
$$f(-4) = (-4)^3 + 2(-4)^2 + 6(-4) - 4 = -60$$
$$f(4) = (4)^3 + 2(4)^2 + 6(4) - 4 = 116$$

you can conclude that the polynomial $f(x) = x^3 + 2x^2 + 6x - 4$ has *no* rational zeros.

(c) *Possible rational zeros*: ± 1 ± 2 ± 3 and ± 6

By testing these possible zeros,

$$f(-1) = (-1)^3 - 3(-1)^2 + 2(-1) - 6 = -12$$
$$f(1) = (1)^3 - 3(1)^2 + 2(1) - 6 = -6$$
$$f(-2) = (-2)^3 - 3(-2)^2 + 2(-2) - 6 = -30$$
$$f(2) = (2)^3 - 3(2)^2 + 2(2) - 6 = -6$$
$$f(-3) = (-3)^3 - 3(-3)^2 + 2(-3) - 6 = -66$$
$$f(3) = (3)^3 - 3(3)^2 + 2(3) - 6 = 0$$
$$f(-6) = (-6)^3 - 3(-6)^2 + 2(-6) - 6 = -342$$
$$f(6) = (6)^3 - 3(6)^2 + 2(6) - 6 = 114$$

you can conclude that the polynomial function $f(x) = x^3 - 3x^2 + 2x - 6$ has one rational zero, $x = 3$.

3. Because the leading coefficient is 1, the possible rational zeros are the factors of the constant term.

Possible rational zeros: $\pm 1, \pm 5, \pm 25, \pm 125$

$$\begin{array}{r|rrrr} 5 & 1 & -15 & 75 & -125 \\ & & 5 & -50 & 125 \\ \hline & 1 & -10 & 25 & 0 \end{array} \rightarrow \text{0 remainder, so } x = 5 \text{ is a factor.}$$

$$\begin{array}{r|rrr} 5 & 1 & -10 & 25 \\ & & 5 & -25 \\ \hline & 1 & -5 & 0 \end{array} \rightarrow \text{0 remainder, so } x = 5 \text{ is a factor.}$$

By applying synthetic division successively, you can determine that $x = 5$ is the only rational zero.

So, $f(x) = x^3 - 15x^2 + 75x - 125$ factors as $f(x) = (x - 5)(x - 5)(x - 5) = (x - 5)^3$.

Because the rational zero $x = 5$ has multiplicity of three, which is odd, the graph of f crosses the x-axis at the x-intercept, $(5, 0)$.

4. The leading coefficient is 2 and the constant term is -30.

Possible rational zeros: $\dfrac{\text{Factors of } -30}{\text{Factors of } 2} = \dfrac{\pm1, \pm2, \pm3, \pm5, \pm6, \pm10, \pm15, \pm30}{\pm1, \pm2}$

$$= \pm1, \pm2, \pm3, \pm5, \pm6, \pm10, \pm15, \pm30, \pm\frac{1}{2}, \pm\frac{3}{2}, \pm\frac{5}{2}, \pm\frac{15}{2}$$

Choose a value of x and use synthetic division.

$x = -3$

$$\begin{array}{r|rrrrr} -3 & 2 & -9 & -18 & 71 & -30 \\ & & -6 & 45 & -81 & 30 \\ \hline & 2 & -15 & 27 & -10 & 0 \end{array} \rightarrow \text{0 remainder, so } x + 3 \text{ is a factor.}$$

Choose another value of x and use synthetic division.

$x = 2$

$$\begin{array}{r|rrrr} 2 & 2 & -15 & 27 & -10 \\ & & 4 & -22 & 10 \\ \hline & 2 & -11 & 5 & 0 \end{array} \rightarrow \text{0 remainder, so } x + 3 \text{ is a factor.}$$

So, $f(x) = 2x^4 - 9x^3 - 18x^2 + 71x - 30$ factors as

$f(x) = (x + 3)(x - 2)(2x^2 - 11x + 5) = (x + 3)(x - 2)(x - 5)(2x - 1)$ and

you can conclude that the rational zeros of f are $x = -3$, $x = \frac{1}{2}$, $x = 2$, and $x = 5$.

5. The leading coefficient is 1 and the constant term is -18.

Possible rational zeros: $\dfrac{\text{Factor of } -18}{\text{Factors of } 1} = \dfrac{\pm1, \pm2, \pm3, \pm6, \pm9, \pm18}{\pm1} = \pm1, \pm2, \pm3, \pm6, \pm9, \pm18$

A graph can assist you to narrow the list to reasonable possibilities.

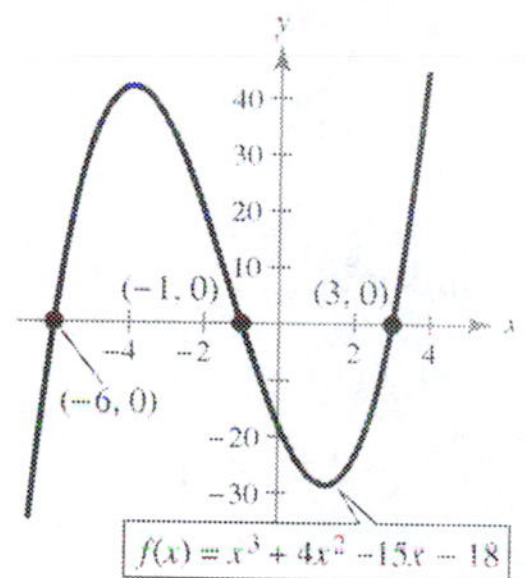

Start by testing $x = -6$, $x = -1$ or $x = 3$.

$$\begin{array}{r|rrrr} -1 & 1 & 4 & -15 & -18 \\ & & -1 & -3 & 18 \\ \hline & 1 & 3 & -18 & 0 \end{array} \rightarrow \text{0 remainder, so } x + 1 \text{ is a factor.}$$

Choose another value of x to test.

$$\begin{array}{r|rrr} 3 & 1 & 3 & -18 \\ & & 3 & 18 \\ \hline & 1 & 6 & 0 \end{array} \rightarrow \text{0 remainder, so } x - 3 \text{ is a factor.}$$

So, $f(x) = x^3 + 4x^2 - 15x - 18$ factors as $f(x) = (x + 1)(x - 3)(x + 6)$ and you can conclude that the rational zeros of f are $x = -1$, $x = 3$, and $x = -6$.

6. (a) Because $-7i$ is a zero *and* the polynomial is stated to have real coefficients, you know that the conjugate $7i$ must also be a zero.

So, the four zeros are $2, -2, 7i$, and $-7i$.

Then, using the Linear Factorization Theorem, $f(x)$ can be written as

$$f(x) = a(x - 2)(x + 2)(x - 7i)(x + 7i).$$

For simplicity, let $a = 1$. Then multiply the factors with real coefficients to obtain

$$(x + 2)(x - 2) = x^2 - 4$$

and multiply the complex conjugates to obtain

$$(x - 7i)(x + 7i) = x^2 + 49.$$

So, you obtain the following fourth-degree polynomial function.

$$\begin{aligned} f(x) = (x^2 - 4)(x^2 + 49) &= x^4 + 49x^2 - 4x^2 - 196 \\ &= x^4 + 45x^2 - 196 \end{aligned}$$

(b) Because $4 - i$ is a zero *and* the polynomial is stated to have real coefficients, you know that the conjugate $4 + i$ must also be a zero.

So, the four zeros are $1, 3, 4 + i$, and $4 - i$.

Then using the Linear Factorization Theorem, $f(x)$ can be written as

$$f(x) = a(x - 1)(x - 3)(x - 4 - i)(x - 4 + i).$$

For simplicity, let $a = 1$. Then multiply the factors with real coefficients to obtain

$$(x - 1)(x - 3) = x^2 - 4x + 3$$

and multiply the complex conjugates to obtain

$$\begin{aligned} (x - 4 - i)(x - 4 + i) &= x^2 - 4x + xi - 4x + 16 - 4i - xi + 4i - i^2 \\ &= x^2 - 8x + 17. \end{aligned}$$

So, you obtain the following fourth-degree polynomial function.

$$\begin{aligned} f(x) &= (x^2 - 4x + 3)(x^2 - 8x + 17) \\ &= x^4 - 12x^3 + 52x^2 - 92x + 51 \end{aligned}$$

(c) Because $3 + i$ is a zero *and* the polynomial is stated to have real coefficients, you know that the conjugate $3 - i$ must also be a zero.

So, the four zeros are $-1, 2, 3 + i$, and $3 - i$.

Then using the Linear Factorization Theorem, $f(x)$ can be written as

$$f(x) = a(x + 1)(x - 2)(x - 3 - i)(x - 3 + i).$$

For simplicity, let $a = 1$. Then multiply the factors with real coefficients to obtain

$$(x + 1)(x - 2) = x^2 - x - 2$$

and multiply the complex conjugates to obtain

$$\begin{aligned} (x - 3 - i)(x - 3 + i) &= x^2 - 3x + xi - 3x + 9 - 3i - xi + 3i - i^2 \\ &= x^2 - 6x + 10. \end{aligned}$$

So, you obtain the following fourth-degree polynomial function,

$$\begin{aligned} f(x) &= (x^2 - x - 2)(x^2 - 6x + 10) \\ &= x^4 - 7x^3 + 14x^2 + 2x - 20 \end{aligned}$$

7. Because complex zeros occur in conjugate pairs you know that if $4i$ is a zero of f, so is $-4i$.

This means that both $(x - 4i)$ and $(x + 4i)$ are factors of f.

$(x - 4i)(x + 4i) = x^2 - 16i^2 = x^2 + 16$

Using long division, you can divide $x^2 + 16$ into $f(x)$ to obtain the following.

$$\begin{array}{r} 3x - 12 \\ x^2 + 16 \overline{\smash{)}\, 3x^3 - 2x^2 + 48x - 32} \\ \underline{3x^3 + 48x } \\ -2x^2 - 32 \\ \underline{-2x^2 - 32} \\ 0 \end{array}$$

So, you have $f(x) = (x^2 + 16)(3x - 2)$ and you can conclude that the real zeros of f are $x = -4i$, $x = 4i$, and $x = \frac{2}{3}$.

8. (a) Because the leading coefficient is 1, the possible rational zeros are the factors of the constant term.

Possible rational zeros: ±1, ±3, and ±9

Synthetic division produces the following.

$$\begin{array}{r|rrrrr} 1 & 1 & 0 & 8 & 0 & -9 \\ & & 1 & 1 & 9 & 9 \\ \hline & 1 & 1 & 9 & 9 & 0 \end{array} \quad \rightarrow \quad 1 \text{ is a zero, so } x - 1 \text{ is a factor.}$$

$$\begin{array}{r|rrrr} -1 & 1 & 1 & 9 & 9 \\ & & -1 & 0 & -9 \\ \hline & 1 & 0 & 9 & 0 \end{array} \quad \rightarrow \quad -1 \text{ is a zero, so } x + 1 \text{ is a factor.}$$

So, you have $f(x) = x^4 + 8x^2 - 9 = (x - 1)(x + 1)(x^2 + 9)$.

You can factor $x^2 + 9$ as $x^2 - (-9) = (x + \sqrt{-9})(x - \sqrt{-9}) = (x + 3i)(x - 3i)$.

So, you have $f(x) = (x - 1)(x + 1)(x + 3i)(x - 3i)$ and you can conclude that the real zeros of f are $x = 1$, $x = -1$, $x = 3i$, and $x = -3i$.

(b) *Possible rational zeros*: ±1 and ±5

Synthetic division produces the following.

$$\begin{array}{r|rrrr} 1 & 1 & -3 & 7 & -5 \\ & & 1 & -2 & 5 \\ \hline & 1 & -2 & 5 & 0 \end{array} \quad \rightarrow \quad 1 \text{ is a zero, so } x - 1 \text{ is a factor.}$$

$f(x) = x^3 - 3x^2 + 7x - 5 = (x - 1)(x^2 - 2x + 5)$

You can find the zeros of $x^2 - 2x + 5$ by completing the square.

$$\begin{aligned} x^2 - 2x &= -5 \\ x^2 - 2x + 1 &= -5 + 1 \\ (x - 1)^2 &= -4 \\ x - 1 &= \pm 2i \\ x &= 1 \pm 2i \end{aligned}$$

So, you have $f(x) = (x - 1)(x^2 - 2x + 5) = (x - 1)(x - 1 - 2i)(x - 1 + 2i)$

and you can conclude that the real zeros of f are $x = 1$, $x = 1 + 2i$, and $x = 1 - 2i$.

(c) *Possible rational zeros*: $\pm1, \pm3, \pm17$ and ±51

Synthetic division produces the following.

$$\begin{array}{c|cccc} 3 & 1 & -11 & 41 & -51 \\ & & 3 & -24 & 51 \\ \hline & 1 & -8 & 17 & 0 \end{array} \rightarrow \text{ 3 is a zero, so } x - 3 \text{ is a factor.}$$

$$f(x) = x^3 - 11x^2 + 41x - 51 = (x - 3)(x^2 - 8x + 17)$$

You can find the zeros of $x^2 - 8x + 17$ by completing the square.

$$x^2 - 8x = -17$$
$$x^2 - 8x + 16 = -17 + 16$$
$$(x - 4)^2 = -1$$
$$x - 4 = \pm i$$
$$x = 4 \pm i$$

So, you have $f(x) = (x - 3)(x^2 - 8x + 17) = (x - 3)(x - 4 - i)(x - 4 + i)$

and you can conclude that the real zeros of f are $x = 3$, $x = 4 + i$, and $x = 4 - i$.

9. The original polynomial has *three* variations in sign.

$$\begin{array}{c} - \text{ to } + \qquad - \text{ to } + \\ \downarrow \quad \downarrow \qquad \downarrow \quad \downarrow \\ f(x) = -2x^3 + 5x^2 - x + 8 \\ \uparrow \quad \uparrow \\ + \text{ to } - \end{array}$$

The polynomial $f(-x) = -2(-x)^3 + 5(-x)^2 - (-x) + 8 = 2x^3 + 5x + x + 8$ has no variations in sign.

So, from Descarte's Rule of Signs, the polynomial $f(x) = -2x^3 + 5x^2 - x + 8$ has either three positive real zeros or one positive real zero and no negative real zeros.

From the graph, you can see that the function has only one real zero.

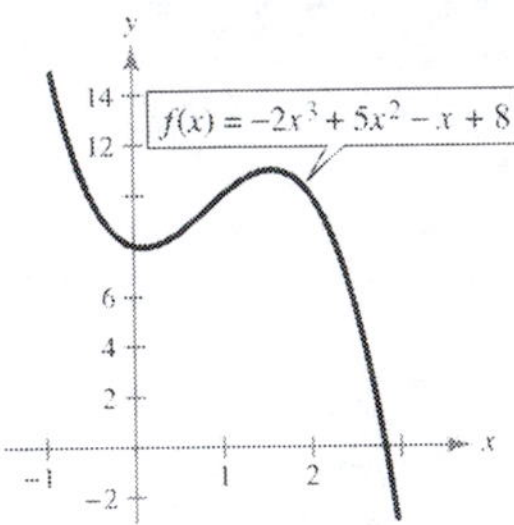

10. The possible real zeros are as follows.

$$\frac{\text{Factors of } -3}{\text{Factors of } 8} = \frac{\pm1 \pm3}{\pm1 \pm2 \pm4 \pm8} = \pm\frac{1}{8}, \pm\frac{1}{4}, \pm\frac{3}{8}, \pm\frac{1}{2}, \pm\frac{3}{4}, \pm1, \pm\frac{3}{2}, \pm3$$

The original polynomial $f(x)$ has three variations in sign. The polynomial

$$f(-x) = 8(-x)^3 - 4(-x)^2 + 6(-x) - 3 = -8x^3 - 4x^2 - 6x - 3$$

has no variations in sign. So, you can apply Descarte's Rule of Signs to conclude that there are either three positive real zeros or one positive real zero, and no negative real zeros.

Using $x = 1$, synthetic division produces the following.

$$\begin{array}{r|rrrr} 1 & 8 & -4 & 6 & -3 \\ & & 8 & 4 & 10 \\ \hline & 8 & 4 & 10 & 7 \end{array} \rightarrow \text{1 is not a zero.}$$

So, $x = 1$ is not a zero, but because the last row has all positive entries, you know that $x = 1$ is an upper bound for the real zeros. So, you can restrict the search to real zeros between 0 and 1. Using $x = \frac{1}{2}$, synthetic division produces the following.

$$\begin{array}{r|rrrr} \frac{1}{2} & 8 & -4 & 6 & -3 \\ & & 4 & 0 & 3 \\ \hline & 8 & 0 & 6 & 0 \end{array} \rightarrow \tfrac{1}{2} \text{ is a real zero.}$$

$$\begin{aligned} f(x) &= 8x^3 - 4x^2 + 6x - 3 \\ &= \left(x - \frac{1}{2}\right)\left(8x^2 + 6\right) \end{aligned}$$

Because $8x^2 + 6$ has no real zeros, it follows that $x = \frac{1}{2}$ is the only real zero.

11. The volume of a pyramid is $V = \frac{1}{3}Bh$, where B is the area of the base and h is the height. The area at the base is x^2 and the height is $x + 2$. So, the volume of the pyramid is $V = \frac{1}{3}x^2(x + 2)$. Substituting 147 for the volume yields the following.

$$\begin{aligned} 147 &= \tfrac{1}{3}x^2(x + 2) \\ 441 &= x^3 + 2x^2 \\ 0 &= x^3 + 2x^2 - 441 \end{aligned}$$

The possible rational zeros are $x = \pm1, \pm3, \pm7, \pm9, \pm21, \pm49, \pm63, \pm147$, and ±441.

Use synthetic division to test some of the possible solutions. So, you can determine that $x = 7$ is a solution.

$$\begin{array}{r|rrrr} 7 & 1 & 2 & 0 & -441 \\ & & 7 & 63 & 441 \\ \hline & 1 & 9 & 63 & 0 \end{array}$$

The other two solutions that satisfy $x^2 + 9x + 63 = 0$ are imaginary and can be discarded. You can conclude that the base of the candle mold should be 7 inches by 7 inches and the height should be $7 + 2 = 9$ inches.

Checkpoints for Section 3.5

1.

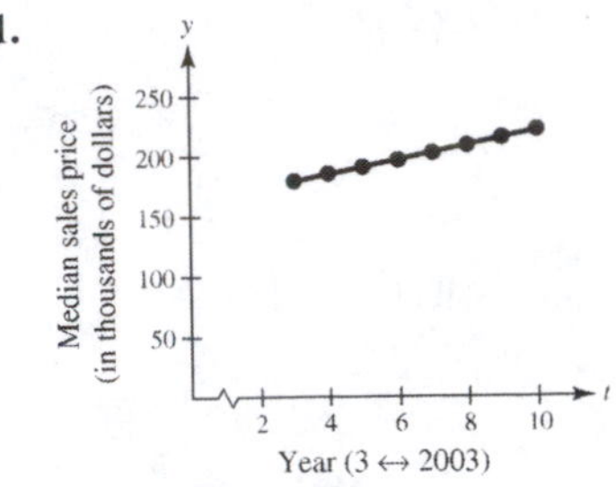

Solution: The actual data are plotted, along with the graph of the linear model. From the graph, it appears that the model is a "good fit" for the actual data. You can see how well the model fits by comparing the actual values of y with the values of y given by the model. The values given by the model are labeled y^* in the table below.

t	3	4	5	6	7	8	9	10
y	179.4	185.4	191.0	196.7	202.6	208.7	214.9	221.4
y^*	179.2	185.1	191.1	197.1	203.0	209.0	214.9	220.9

2.

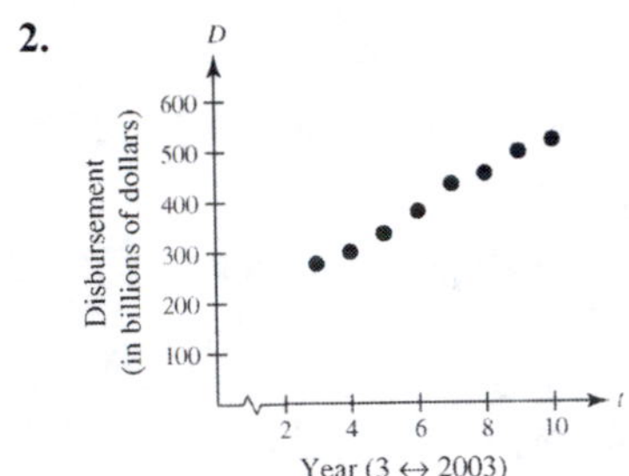

Solution Let $t = 3$ represents 2003. The scatter plot for the data is shown. Using the *regression* feature of a graphing utility, you can determine that the equation of the least squares regression line is

$M = 36.85t + 161.2$

To check this model, compare the actual M values with the M values given by the model, which are labeled M^* in the table. The correlation coefficient for this model is $r \approx 0.996$, which implies that the model is a good fit.

t	3	4	5	6	7	8	9	10
M	277.8	301.5	336.9	380.4	434.8	455.1	498.2	521.1
M^*	271.8	308.6	345.5	382.3	419.2	456.0	492.9	529.7

3. *Verbal Model*: [Simple interest] = [r] [Amount of investment]

Labels: Simple interest $= I$ (dollars)
Amount of investment $= P$ (dollars)
Interest rate $= r$ (percent in decimal form)

Equation: $I = rP$

To solve for r, substitute the given information into the equation $I = rP$, and then solve for r.

$I = rP$ Write direct variation model.

$187.50 = r(2500)$ Substitute 187.50 for t and 2500 for P.

$\dfrac{187.50}{2500} = r$ Divide each side by 2500.

$0.075 = r$ Simplify.

So, the mathematical model is $I = 0.075P$.

4. Letting s be the distance (in feet) the object falls and letting t be the time (in seconds) that the object falls, you have $s = Kt^2$.

Because $s = 144$ feet when $t = 3$ seconds, you can see that $K = \frac{144}{9}$ as follows.

$s = Kt^2$ Write direct variation model.

$144 = K(3)^2$ Substitute 144 for s and 3 for t.

$144 = 9K$ Simplify.

$\frac{144}{9} = K$ Divide each side by 9.

$16 = K$

So, the equation relating distance to time is $s = 16t^2$.

To find the distance the object falls in 6 seconds , let $t = 6$.

$s = 16t^2$ Write direct variation model.

$s = 16(6)^2$ Substitute 6 for t.

$s = 16(36)$ Simplify.

$s = 576$ Simplify.

So, the object falls 576 feet in 6 seconds.

5. Let p be the price and let x be the demand. Because x varies inversely as p, you have

$$x = \frac{k}{p}$$

Now because $x = 600$ then $p = 2.75$ you have

$x = \frac{k}{p}$ Write inverse variation model.

$600 = \frac{k}{2.75}$ Substitute 600 for x and 2.75 for p.

$(600)(2.75) = k$ Multiply each side by 2.75.

$1650 = k.$ Simplify.

So, the equation relating price and demand is

$$x = \frac{1650}{p}.$$

When $p = 3.25$ the demand is

$x = \frac{1650}{p}$ Write inverse variation model.

$= \frac{1650}{3.25}$ Substitute 3.25 for p.

≈ 508 units. Simplify.

So, the demand for the product is 508 units when the price of the product is \$3.25.

6. Let R be the resistance (in ohms), let L be the length (in inches), and let A be the cross-sectional area (in square inches).

Because R varies directly as L and inversely as A, you have

$$R = \frac{kL}{A}.$$

Now, because $R = 66.17$ ohms when $L = 1000$ feet $= 12{,}000$ inches and

$$A = \pi\left(\frac{0.0126}{2}\right)^2 \approx 1.247 \times 10^{-4} \text{ square inches,}$$

you have

$$66.17 = \frac{k(12,000)}{1.247 \times 10^{-4}}$$

$$6.876 \times 10^{-7} \approx k$$

So, the equation relating resistance, length, and the cross-sectional area is $R = 6.876 \times 10^{-7}\frac{L}{A}$.

To find the length of copper wire that will produce a resistance of 33.5 ohms, let $R = 33.5$ and $A = 1.247 \times 10^{-4}$ and solve for L.

$$R = (6.876 \times 10^{-7})\frac{L}{A}$$

$$33.5 = (6.876 \times 10^{-7})\frac{L}{(1.247 \times 10^{-4})}$$

$$(33.5)(1.247 \times 10^{-4}) = (6.876 \times 10^{-7})L$$

$$\frac{(33.5)(1.247 \times 10^{-4})}{(6.876 \times 10^{-7})} = L$$

$$6075.4 \times L$$

So, the length of the wire is approximately 6075.4 inches or about 506.3 feet.

7. E = kinetic energy, m = mass, and V = velocity.

Because E varies jointly with the object's mass, m and the square of the object's velocity, V you have

$E = kmV^2$

For $E = 6400$ joules, $m = 50$ kg, and $V = 16$m/sec, you have

$$E = kmV^2$$

$$6400 = k(50)(16)^2$$

$$6400 = k(12800)$$

$$\frac{1}{2} = k$$

So, the equation relating kinetic energy, mass, and velocity is $E = \frac{1}{2}mV$.

When $m = 70$ and $V = 20$, the kinetic energy is

$$E = \frac{1}{2}mV^2 = \frac{1}{2}(70)(20)^2 = \frac{1}{2}(70)(400)$$

$$= 14{,}000 \text{ joules.}$$

Chapter 4

Checkpoints for Section 4.1

1. Because the denominator is zero when $x = 1$, the domain of f is all real numbers except $x = 1$.

To determine the behavior of f, this excluded value, evaluate $f(x)$ to the left and to the right of $x = 1$.

x	0	0.5	0.9	0.99	0.999	$\to 1$
$f(x)$	0	-3	-27	-297	-2997	$\to -\infty$

x	$1 \leftarrow$	1.001	1.01	1.1	1.5	2
$f(x)$	$\infty \leftarrow$	3003	303	33	9	6

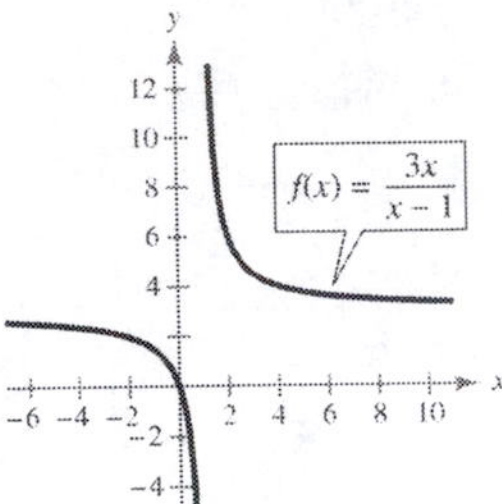

As x approaches 1 from the left, $f(x)$ decreases without bound.

As x approaches 1 from the right, $f(x)$ increases without bound.

2. For this rational function the degree of the numerator is equal to the degree of the denominator. The leading coefficient of the numerator is 5 and the leading coefficient of the denominator is 1, so the graph has the line $y = \frac{5}{1} = 5$ as a horizontal asymptote. To find any vertical asymptotes set the denominator equal to zero and solve the resulting equation for x.

$$x^2 - 1 = 0$$
$$(x + 1)(x - 1) = 0$$
$$x + 1 = 0 \to x = -1$$
$$x - 1 = 0 \to x = 1$$

The equation has two real solutions, $x = 1$ and $x = -1$, so the graph has the lines $x = 1$ and $x = -1$ as vertical asymptotes.

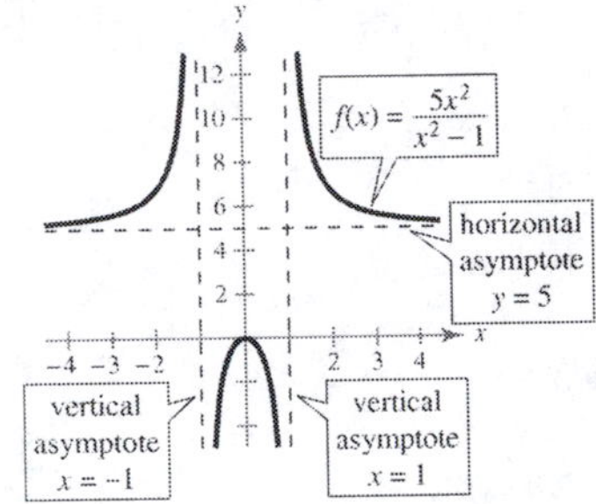

3. For this rational function, the degree of the numerator is equal to the degree of the denominator. The leading coefficient of the numerator is 3 and the leading coefficient of the denominator is 1, so the graph of the function has the line $y = \dfrac{3}{1} = 3$ as a horizontal asymptote. To find any vertical asymptotes, first factor the numerator and denominator as follows.

$$f(x) = \frac{3x^2 + 7x - 6}{x^2 + 4x + 3} = \frac{(3x - 2)\cancel{(x + 3)}}{(x + 1)\cancel{(x + 3)}}$$
$$= \frac{3x - 2}{x + 1},\ x \neq -3$$

By setting the denominator $x + 1$ (of the simplified function) equal to zero, you can determine that the graph has the line $x = -1$ as a vertical asymptote.

4. (a) The cost to remove 20% of the pollutants is

$$C = \frac{255(20)}{100 - (20)} = \$63.75 \text{ million.}$$

The cost to remove 45% of the pollutants is

$$C = \frac{255(45)}{100 - 45} \approx \$208.64 \text{ million.}$$

The cost to remove 80% of the pollutants is

$$C = \frac{255(80)}{100 - 80} = \$1020 \text{ million.}$$

(b) The cost to remove 100% of the pollutants is

$$C = \frac{255(100)}{100 - (100)} \text{ which is undefined.}$$

So, it would not be possible to remove 100% of the pollutants.

5. (a) When $x = 1000$,

$$\overline{C} = \frac{0.4(1000) + 8000}{1000} = \$8.40 \text{ per unit.}$$

When $x = 8000$,

$$\overline{C} = \frac{0.4(8000) + 8000}{8000} = \$1.40 \text{ per unit.}$$

When $x = 20{,}000$,

$$\overline{C} = \frac{0.4(20{,}000) + 8000}{20{,}000} = \$0.80 \text{ per unit.}$$

When $x = 100{,}000$,

$$\overline{C} = \frac{0.4(100{,}000) + 8000}{100{,}000} = \$0.48 \text{ per unit.}$$

(b) The graph has the line $\overline{C} = \frac{0.4}{1} = 0.4$ as a horizontal asymptote.

So, as x approaches infinity, $\overline{C}$ approaches 0.4. This means that as the number at units increases without bound the average cost per unit approaches \$0.40 per unit.

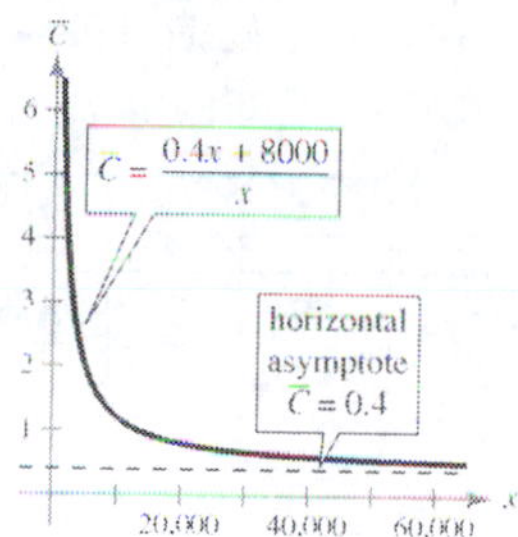

Checkpoints for Section 4.2

1. *y-intercept*: $\left(0, \frac{1}{3}\right)$, because $f(0) = \frac{1}{3}$

x-intercept: none, because $1 \neq 0$

Vertical asymptote: $x = -3$, zero of denominator

Horizontal asymptote: $y = 0$, because degree of $N(x) <$ degree of $D(x)$

Additional points:

x	-5	-4	-2	-1	1	2
$f(x)$	$-\frac{1}{2}$	-1	1	$\frac{1}{2}$	$\frac{1}{4}$	$\frac{1}{5}$

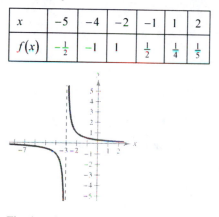

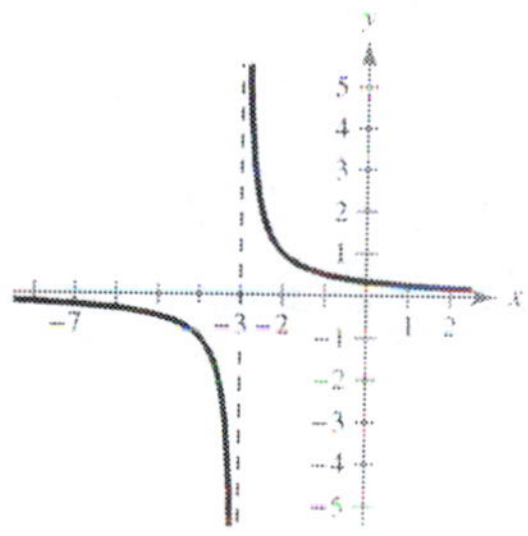

The domain of f is all real numbers except $x = -3$.

2. *y-intercept*: $(0, 3)$, because $C(0) = 3$

x-intercept: $\left(-\frac{3}{2}, 0\right)$, because $C\left(-\frac{3}{2}\right) = 0$

Vertical asymptote: $x = -1$, zero of denominator

Horizontal asymptote: $y = 2$, because degree of $N(x) =$ degree of $D(x)$

Additional Points:

x	-3	-2	1	3
$C(x)$	$\frac{3}{2}$	1	$\frac{5}{2}$	$\frac{9}{4}$

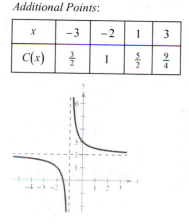

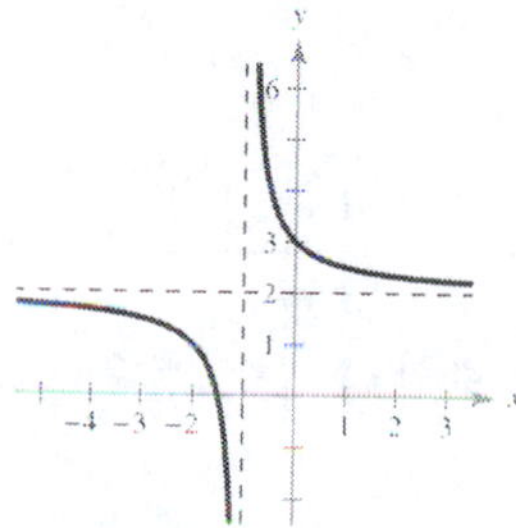

The domain of f is all real number except $x = -1$.

3. $f(x) = \dfrac{3x}{x^2 + x - 2} = \dfrac{3x}{(x + 2)(x - 1)}$

y-intercept: $(0, 0)$, because $f(0) = 0$

x-intercept: $(0, 0)$, because $f(0) = 0$

Vertical asymptotes: $x = -2, x = 1$, zeros of denominator

Horizontal asymptote: $y = 0$, because degree of $N(x) <$ degree of $D(x)$

Additional points:

x	-3	-1	2	3
$f(x)$	$-\frac{9}{4}$	$\frac{3}{2}$	$\frac{3}{2}$	$\frac{9}{10}$

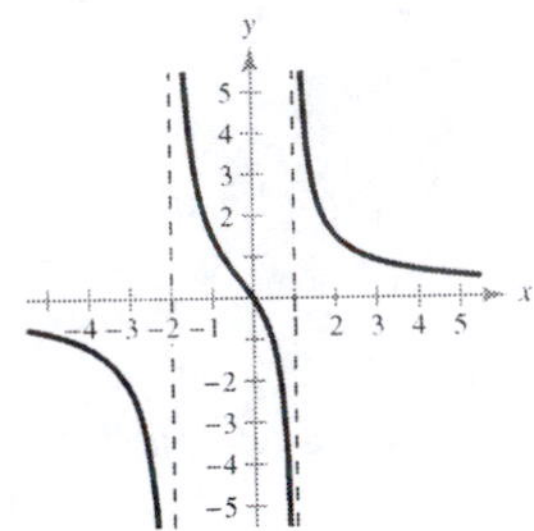

The domain of f is all real numbers except $x = -2$ and $x = 1$.

4. $f(x) = \dfrac{x^2 - 4}{x^2 - x - 6}$

$= \dfrac{(x + 2)(x - 2)}{(x - 3)(x + 2)}$

$= \dfrac{x - 2}{x - 3}, x \neq -2$

y-intercept: $\left(0, \frac{2}{3}\right)$, because $f(0) = \frac{2}{3}$

x-intercept: $(2, 0)$, because $f(2) = 0$

Vertical asymptote: $x = 3$, zero of (simplified) denominator

Horizontal asymptote: $y = 1$, because degree of $N(x) =$ degree of $D(x)$

Additional points:

x	-7	-5	-1	1	4	5
$f(x)$	$\frac{9}{10}$	$\frac{7}{8}$	$\frac{3}{4}$	$\frac{1}{2}$	2	$\frac{3}{2}$

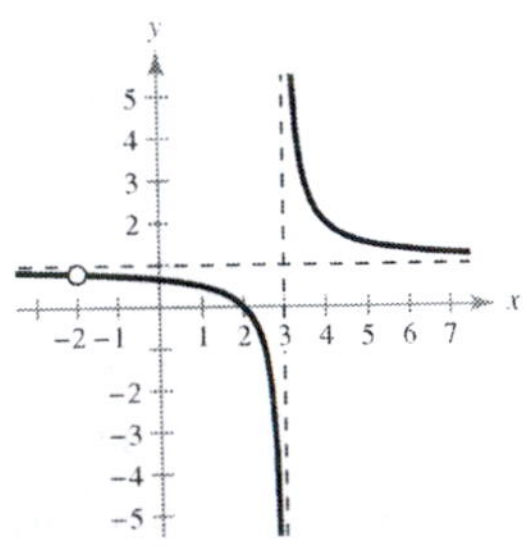

Notice that there is a hole at $x = -2$ because the function is not defined when $x = -2$, the domain is all real number except $x = -2$ and $x = 3$.

5. $f(x) = \dfrac{3x^2 + 1}{x}$

First divide $3x^2 + 1$ by x, either by long division:

$$\begin{array}{r} 3x \\ x\overline{)3x^2 + 0x + 1} \\ \underline{3x^2} \\ 1 \end{array}$$

So $\dfrac{3x^2 + 1}{x} = 3x + \dfrac{1}{x}$

or by separating, the numerator and simplifying:

$$\frac{3x^2 + 1}{x} = \frac{3x^2}{x} + \frac{1}{x} = 3x + \frac{1}{x}$$

So, the start asymptote is $y = 3x$, since $\dfrac{3x^2 + 1}{x} = 3x + \dfrac{1}{x}$.

y-intercept: none, since $f(0)$ is undefined.

x-intercept: none, since $3x^2 + 1 \neq 0$ for real numbers.

Vertical asymptote: $x = 0$, zero of denominator

Start asymptote: $y = 3x$

Additional points:

x	-2	-1	-0.5	0.5	1	2
$f(x)$	$-\frac{13}{2}$	-4	$-\frac{7}{2}$	$\frac{7}{2}$	4	$\frac{13}{2}$

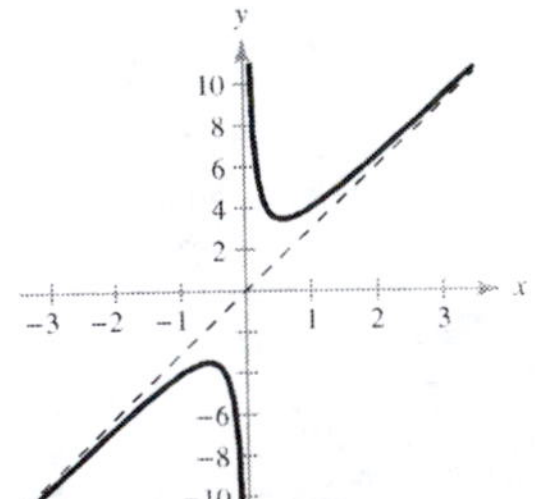

The domain of f is all real numbers except $x = 0$.

6. Graphical Solution

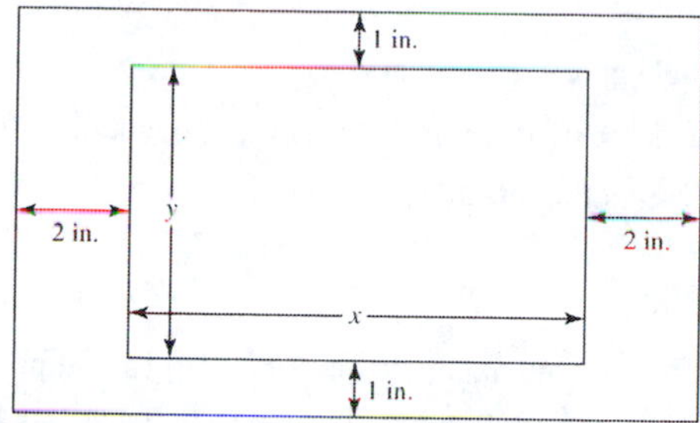

Let A be the area to be minimized.

$A = (x + 4)(y + 2)$

The printed area inside the margins is modeled by $40 = xy$ or $y = \dfrac{40}{x}$.

To find the minimum area, rewrite the equation for A in terms of just one variable by substituting $\dfrac{40}{x}$ for y.

$$\begin{aligned} A &= (x + 4)\left(\frac{40}{x} + 2\right) \\ &= (x + 4)\left(\frac{40 + 2x}{x}\right) \\ &= \frac{(x + 4)(40 + 2x)}{x}, \; x > 0 \end{aligned}$$

The graph of this rational function is shown below. Because x represents the width of the printed area, you need to consider only the portion of the graph for which x is positive. Using a graphing utility, you can approximate the minimum value of A to occur when $x \approx 8.9$ inches. The corresponding value of y is $\dfrac{40}{8.9} \approx 4.5$ inches.

So, the dimensions should be $8.9 + 4 = 129$ inches by $4.5 + 2 = 6.5$ inches.

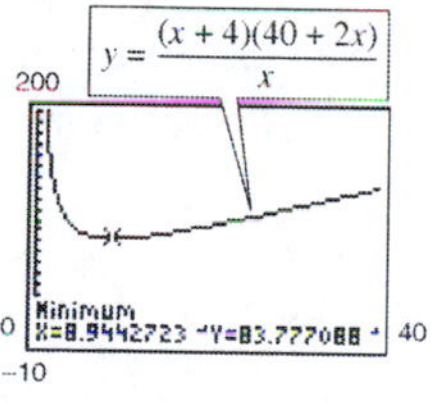

Numerical Solution

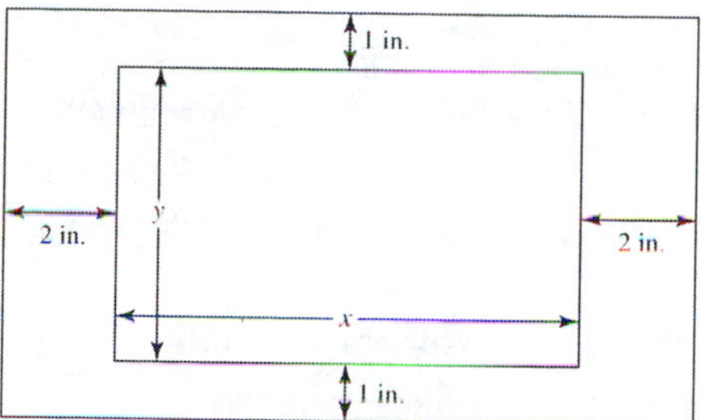

Let A be the area to be minimized.

$A = (x + 4)(y + z)$

The printed area inside the margins is modeled by $40 = xy$ or $y = \dfrac{40}{x}$.

To find the minimum area, rewrite the equation for A in terms of just one variable by substituting $\dfrac{40}{x}$ for y.

$$\begin{aligned} A &= (x + 4)\left(\frac{40}{x} + 2\right) \\ &= (x + 4)\left(\frac{40 + 2x}{x}\right) \\ &= \frac{(x + 4)(40 + 2x)}{x}, \; x > 0 \end{aligned}$$

Use the *table* feature of a graphing utility to create a table of values for the function

$$y_1 = \frac{(x + 4)(40 + 2x)}{x}, \; x > 0$$

beginning at $x = 6$. From the table , you can see that the minimum value of y_1 occurs when x is somewhere between 8 and 9, as shown.

To approximate the minimum value of y_1 to one decimal place, change the table so that it starts at $x = 8$ and increases by 0.1. The minimum value of y_1 occurs when $x \approx 8.9$ as shown.

The corresponding value of y is $\dfrac{40}{8.9} \approx 4.5$ inches.

So, the dimensions should be $8.9 + 4 = 12.9$ inches by $4.5 + 2 = 6.5$ inches.

x	y_1
6	86.667
7	84.857
8	84
9	83.778
10	84
11	84.545

x	y_1
8.8	83.782
8.9	83.778
9.0	83.778
9.1	83.782

Checkpoints for Section 4.3

1. Because the squared term in the equation is x, you know that the axis is vertical, and the equation is of the form

$x^2 = 4py$.

You can write the original equation in this form as follows.

$y = \frac{1}{4}x^2$ Write original equation.

$4y = x^2$ Multiply each side by 4.

$4(1)y = x^2$ Write in standard form.

So, $p = 1$. Because p is positive, the parabola opens upward. The focus of the parabola is $(0, p) = (0, 1)$ and the directrix of the parabola is $y = -p = -1$.

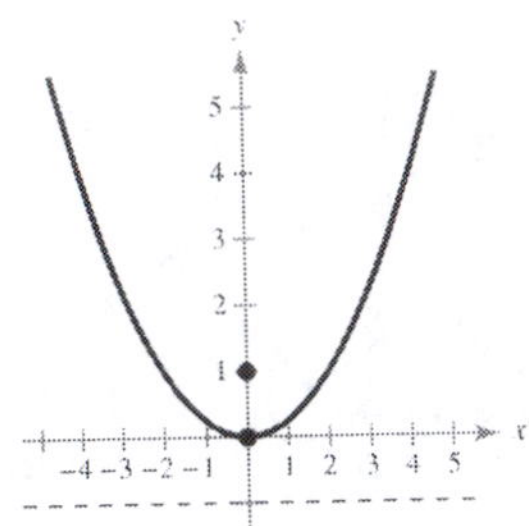

2. The axis of the parabola is vertical, passing through $(0, 0)$ and $\left(0, \frac{3}{8}\right)$. So, the standard form is $x^2 = 4py$.

Because the focus is $p = \frac{3}{8}$ units from the vertex, the equation is as follows.

$x^2 = 4\left(\frac{3}{8}\right)y$

$x^2 = \frac{3}{2}y$

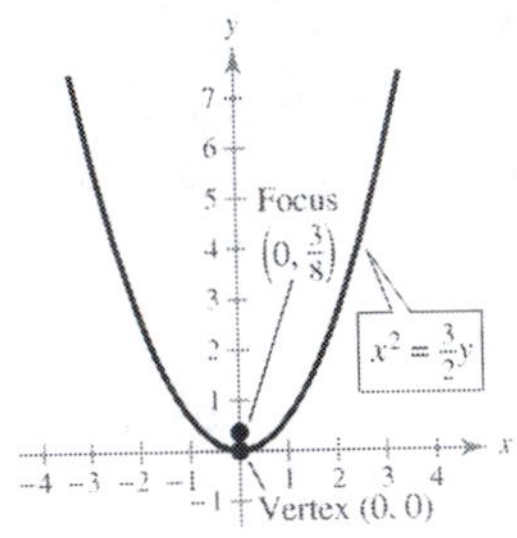

3. The foci occur at $(0, -3)$ and $(0, 3)$. So, the center of the ellipse is $(0, 0)$, the major axis is vertical, and the ellipse has the equation of the form

$\frac{y^2}{a^2} + \frac{x^2}{b^2} = 1$.

The length of the major axis is $2a = 10$. This implies that $a = 5$. Moreover, the distance from the center to either focus is $c = 3$.

Finally, $b^2 = a^2 - c^2 = (5)^2 - (3)^2 = 25 - 9 = 16$.

Substituting $a^2 = 3^2$ and $b^2 = 4^2$ yields the following equation in standard form.

$\frac{y^2}{5^2} + \frac{x^2}{4^2} = 1$

$\frac{y^2}{25} + \frac{x^2}{16} = 1$

4. $4x^2 + y^2 = 64$ Write original equation.

$\frac{4x^2}{64} + \frac{y^2}{64} = \frac{64}{64}$ Divide each side by 64.

$\frac{x^2}{16} + \frac{y^2}{64} = 1$ Simplify.

$\frac{x^2}{4^2} + \frac{y^2}{8^2} = 1$ Write in standard form.

Because the denominator of the y^2-term is greater than the denominator of the x^2-term, you can conclude that the major axis is vertical. Moreover, because $a^2 = 8^2$, the endpoints of the major axis lie eight units above and below of the center $(0, 0)$. So the vertices of the ellipse are $(0, 8)$ and $(0, -8)$. Similarly, because the denominator of the x^2-term is $b^2 = 4^2$, the endpoints of the minor axis (or *co-vertices*) lie four units left and right of the center at $(4, 0)$ and $(-4, 0)$. The ellipse is shown in 1

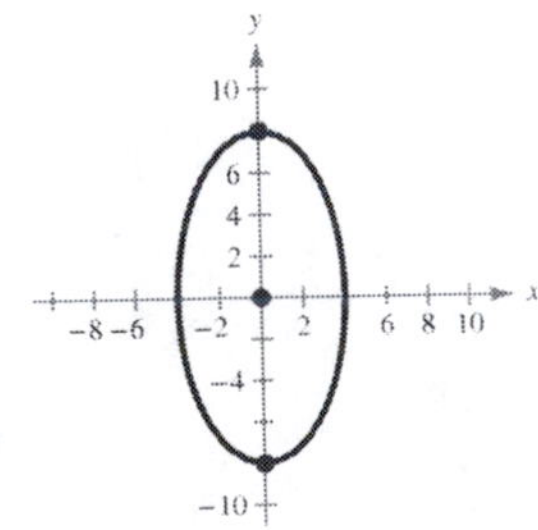

5. Algebraic Solution

$x^2 + 9y^2 = 81$ Write original equation.

$\frac{x^2}{81} + \frac{9y^2}{81} = \frac{81}{81}$ Divide each side by 81.

$\frac{x^2}{81} + \frac{y^2}{9} = 1$ Simplify.

$\frac{x^2}{9^2} + \frac{y^2}{3^2} = 1$ Write in standard form.

Because the denominator of the x^2-term is larger than the denominator of the y^2-term, you can conclude that the major axis is horizontal. Moreover, because

$a^2 = 9^2$

the endpoints of the major axis lie nine units to the left and to the right of center $(0, 0)$. So, the vertices of the ellipse are

$(9, 0)$ and $(-9, 0)$.

Similarly, because the denominator of the y^2-term is $b^2 = 3^2$ the endpoints of the minor axis (or co-vertices) lie three units up and down from the center $(0, 0)$.

So, the endpoints are $(0, 3)$ and $(0, -3)$

The ellipse is shown below.

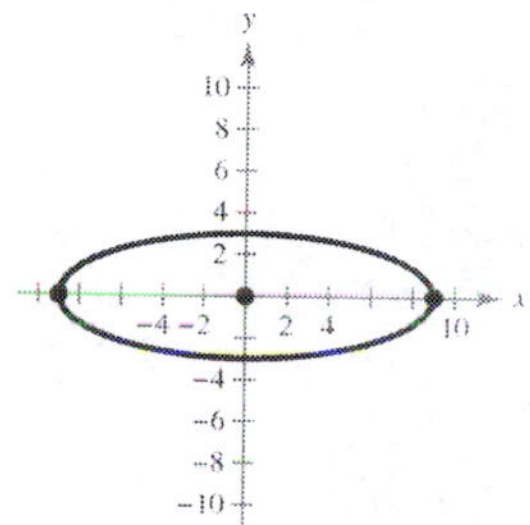

Graphical Solution

Solve the equation of the ellipse for y.

$$x^2 + 9y^2 = 81$$
$$9y^2 = 81 - x^2$$
$$y^2 = \frac{81 - x^2}{9}$$
$$y = \pm\sqrt{\frac{81 - x^2}{9}}$$
$$y = \pm\frac{\sqrt{81 - x^2}}{\sqrt{9}}$$
$$y = \pm\frac{1}{3}\sqrt{81 - x^2}$$

Then graph both equations,

$y_1 = \frac{1}{3}\sqrt{81 - x^2}$ and $y_2 = \frac{1}{3}\sqrt{81 - x^2}$

in the same viewing window. Be sure to use a square setting.

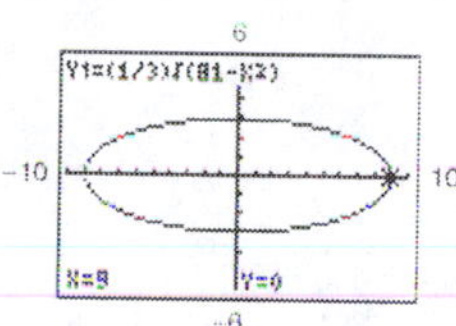

The center of the ellipse is $(0, 0)$ and the major axis is horizontal. The vertices are $(9, 0)$ and $(-9, 0)$.

6. You can determine that $c = 6$, because the foci are at $(0, \pm 6)$. Also, $a = 3$, because the vertices are at $(0, \pm 3)$.

$$\begin{aligned} b^2 &= c^2 - a^2 \\ &= 6^2 - 3^2 \\ &= 36 - 9 \\ &= 27 \end{aligned}$$

Because the transverse axis is vertical, the standard form of the equation is

$$\frac{y^2}{a^2} + \frac{x^2}{b^2} = 1.$$

Substitute $a^2 = 3^2$ and $b^2 = 27 = \left(3\sqrt{3}\right)^2$ to obtain the equation as follows.

$\frac{y^2}{3^2} - \frac{x^2}{\left(3\sqrt{3}\right)^2} = 1$ Write in standard form.

$\frac{y^2}{9} - \frac{x^2}{27} = 1$ Simplify.

7. **Algebraic Solution**

$x^2 - \frac{y^2}{4} = 1$ Write original equation.

$\frac{x^2}{1^2} - \frac{y^2}{2^2} = 1$ Write in standard form.

Because the x^2-term is positive, you can conclude that the transverse axis is horizontal and the vertices occur at $(1, 0)$ and $(-1, 0)$. Moreover, the endpoints of the conjugate axis occur at $(0, 2)$ and $(0, -2)$, and you can sketch the rectangle as shown. Finally, by drawing the asymptotes through the corners of this rectangle, you can complete the sketch as shown. Note that the asymptotes are $y = 2x$ and $y = -2x$.

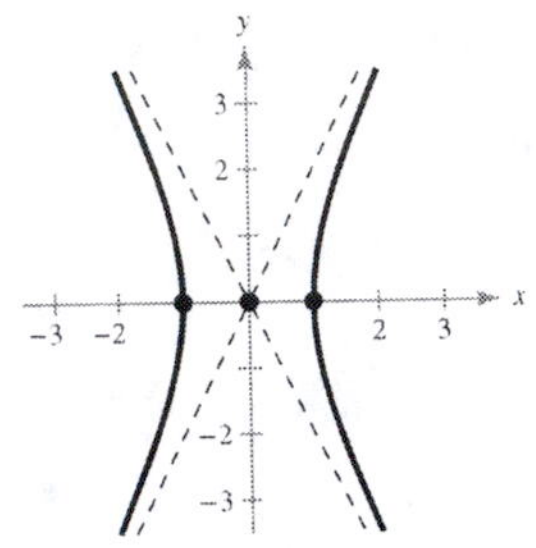

Graphical Solution

Solve the equation of the hyperbola for y as follows.

$$x^2 - \frac{y^2}{4} = 1$$
$$4x^2 - y^2 = 4$$
$$4x^2 - 4 = y^2$$
$$\pm\sqrt{4x^2 - 4} = y$$

Then use a graphing utility to graph $y_1 = \sqrt{4x^2 - 4}$ and $y_2 = -\sqrt{4x^2 - 4}$ in the same viewing window. Be sure to use a square setting.

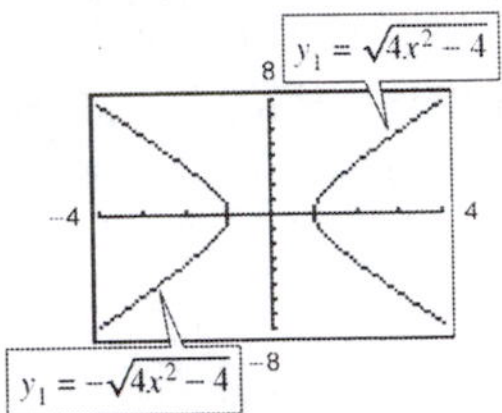

From the graph, you can see that the traverse axis is horizontal and the vertices are $(-1, 0)$ and $(1, 0)$.

8. Because the transverse axis is horizontal, the asymptotes are of the forms $y = \frac{b}{a}x$ and $y = -\frac{b}{a}x$.

Using the fact that the asymptotes are $y = 4x$ and $y = -4x$, you can determine that $\frac{b}{a} = 4$.

Because $a = 1$, you can determine that $b = 4$. Finally, you can conclude that the hyperbola has the following equation.

$\frac{x^2}{1^2} - \frac{y^2}{4^2} = 1$ Write in standard form.

$x^2 - \frac{y^2}{16} = 1$ Simplify.

Checkpoints for Section 4.4

1. (a) The graph of $(x + 1)^2 + (y - 1)^2 = 2^2$ is a *circle* whose center is the point $(-1, 1)$ and whose radius is 2.

The graph of the circle has been shifted one unit to the left and one unit up from the standard position.

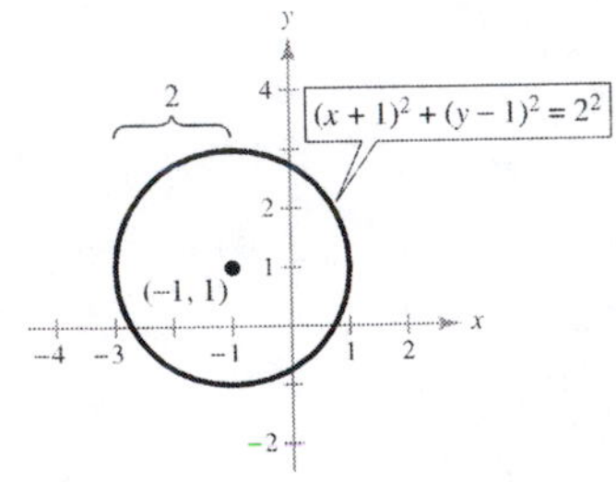

(b) The graph of $\frac{(x + 1)^2}{9} + \frac{(y - 2)^2}{25} = 1$ is an *ellipse* whose center is the point $(-1, 2)$. The major axis of the ellipse is vertical and of length $2(5) = 10$, and the minor axis of ellipse is horizontal and of length $2(3) = 6$. The graph has been shifted one unit to the left and two units up from standard position.

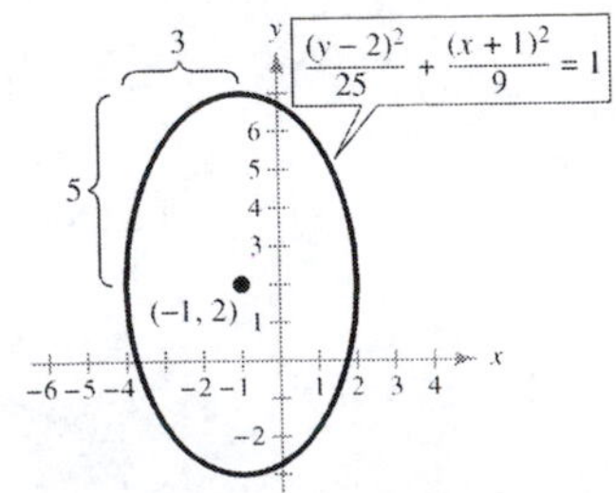

(c) The graph of $\frac{(x-3)^2}{4} - \frac{(y-1)^2}{1} = 1$ is a *hyperbola* whose center is the point $(3, 1)$. The transverse axis is horizontal and of length $2(2) = 4$, and the conjugate axis is vertical and of length $2(1) = 2$. The graph has been shifted three units to the right and one unit up from standard position.

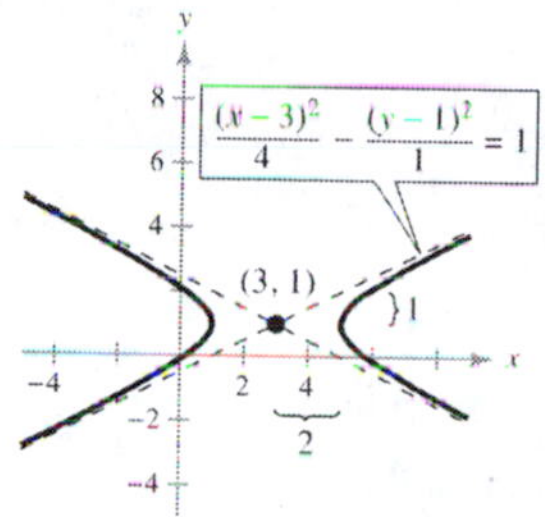

(d) The graph of $(x+4)^2 - 8(y-3) = 0$

$$(x+4)^2 = 8(y-3)$$
$$(x+4)^2 = 4(2)(y-3)$$

is a *parabola* whose vertex is the point $(-4, 3)$. The axis of the parabola is vertical. Because $p = 2$, the focus lies two units above the vertex as $(-4, 5)$ and the directrix is the horizontal line $y = k - p = 3 - 2 = 1$. The graph of the parabola has been shifted four units to the left and three units up from standard position.

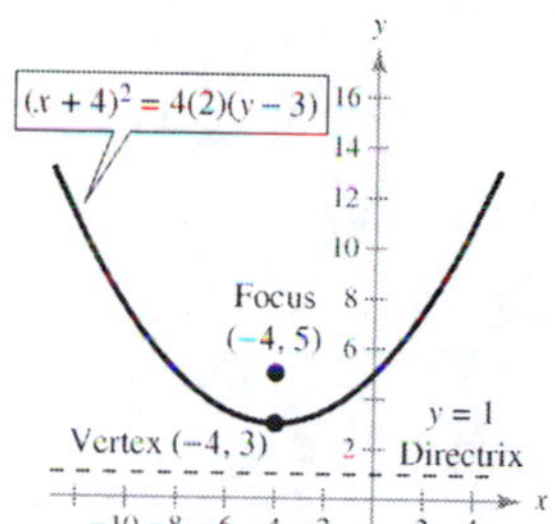

2. Complete the square to right the equation in standard form.

$y^2 - 6y + 4x + 17 = 0$	Write original equation.
$y^2 - 6y = -4x - 17$	Collect y terms on one side of the equaton.
$y^2 - 6y + 9 = -4x - 17 + 9$	Complete the square.
$(y-3)^2 = -4x - 8$	Write in completed square form.
$(y-3)^2 = 4(-1)(x+2)$	Write in standard form.

In this standard form, it follows that $h = -2$, $k = 3$, and $p = -1$.

Because the axis is horizontal and p is negative, the parabola opens to the left.

The vertex is $(h, k) = (-2, 3)$, the focus is $(h + p, k) = (-3, 3)$, and the directrix is $x = h - p = -1$.

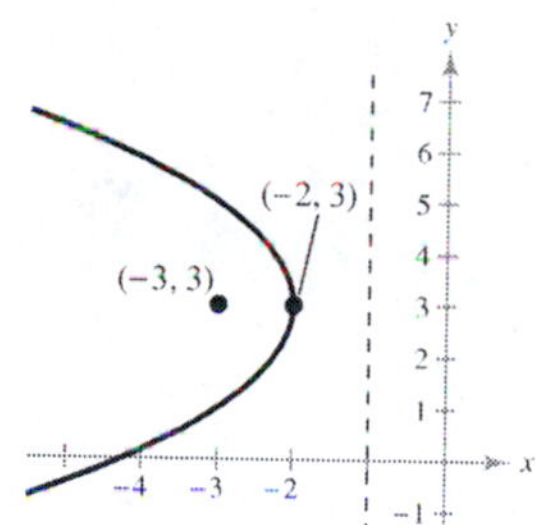

3. Because the directrix is $y = 0$, which is a horizontal line, the axis is vertical.

The standard form is $(x-h)^2 = 4p(y-k)$.

Because the vertex lies one unit above the directrix, it follows that $p = 1$.

So, the standard form of the parabola is as follows.

$$(x-h)^2 = 4p(y-k)$$
$$[x-(-1)]^2 = 4(1)(y-1)$$
$$(x+1)^2 = 4(y-1)$$

4. Complete the square to write the equation in standard form.

$$9x^2 + 4y^2 - 36x + 24y + 36 = 0 \qquad \text{Write original equation.}$$

$$(9x^2 - 36x + \square) + (4y^2 + 24y + \square) = -36 \qquad \text{Group terms.}$$

$$9(x^2 - 4x + \square) + 4(y^2 + 6y + \square) = -36 \qquad \text{Factor 9 out of } x\text{-terms and factor 4 out of } y\text{-terms.}$$

$$9(x^2 - 4x + 4) + 4(y^2 + 6y + 9) = -36 + 9(4) + 4(9) \qquad \text{Complete the square.}$$

$$9(x - 2)^2 + 4(y + 3)^2 = 36 \qquad \text{Write in completed square form.}$$

$$\frac{(x - 2)^2}{4} + \frac{(y + 3)^2}{9} = 1 \qquad \text{Divide each side by 36.}$$

$$\frac{(x - 2)^2}{2^2} + \frac{(y + 3)^2}{3^2} = 1 \qquad \text{Write in standard form.}$$

So, the center is $(h, k) = (2, -3)$. Because the denominator of the y-term is $a^2 = 3^2$, the endpoints of the major axis lie three units above and below the center at $(2, 0)$ and $(2, -6)$. Because the denominator of the x-term is $b^2 = 2^2$, the endpoints of the minor axis lie two units left and right of the center at $(0, -3)$ and $(4, -3)$.

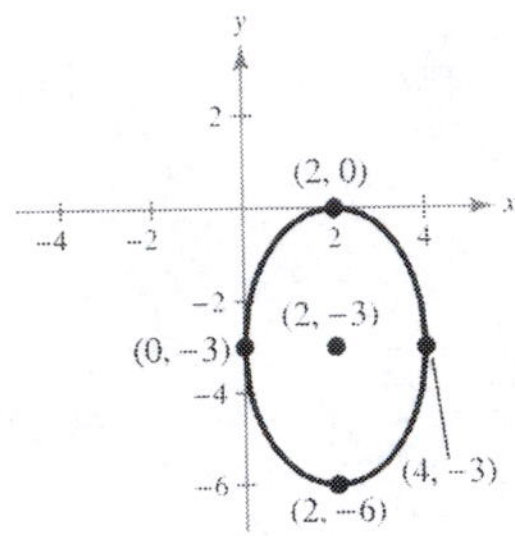

5. The center of the ellipse lies at the midpoint of its vertices. So, the center is

$$(h, k) = \left(\frac{3 + 3}{2}, \frac{0 + 10}{2}\right) = (3, 5).$$

Because the vertices lie on a vertical line and are 10 units apart, it follows that the major axis is vertical and has a length of $2a = 10$. So, $a = 5$. Moreover, because the minor axis has a length of 6 it follows that $2b = 6$, which implies that $b = 3$.

So, the standard form of the ellipse is as follows.

$$\frac{(x - h)^2}{b^2} + \frac{(y - k)^2}{a^2} = 1 \qquad \text{Major axis is vertical.}$$

$$\frac{(x - 3)^2}{3^2} + \frac{(y - 5)^2}{5^2} = 1 \qquad \text{Write in standard form.}$$

$$\frac{(x - 3)^2}{9} + \frac{(y - 5)^2}{25} = 1 \qquad \text{Simplify.}$$

6. Complete the square to write the equation in standard form.

$9x^2 - y^2 - 18x - 6y - 9 = 0$	Write original equation.
$(9x^2 - 18x + \square) - (y^2 - 6y + \square) = 9$	Group terms.
$9(x^2 - 2x + \square) - (y^2 + 6y + \square) = 9$	Factor 9 out of x-terms.
$9(x^2 - 2x + 1) - (y^2 + 6y + 9) = 9 + 9(1) - 9$	Complete the squares.
$9(x - 1)^2 - (y + 3)^2 = 9$	Write in completed square form.
$(x - 1)^2 - \dfrac{(y + 3)^2}{9} = 1$	Divide each side by 9.
$\dfrac{(x - 1)^2}{1^2} - \dfrac{(y + 3)^2}{3^2} = 1$	Write in standard form.

From this standard form, you can see that the transverse axis is horizontal and the center lies at $(h, k) = (1, -3)$.

Because the denominator of the x-term is $a^2 = 1$, you know that the vertices occur one unit to the left and right of the center at $(0, -3)$ and $(2, -3)$.

To sketch the hyperbola, draw a rectangle whose sides pass through the vertices. Because the denominator of the y-term is $b^2 = 3^2$, locate the top and bottom of the rectangle 3 units up and down from the center. Finally, sketch the asymptotes by drawing lines through the opposite corners of the rectangle. Using these asymptotes, you can complete the graph of the hyperbola.

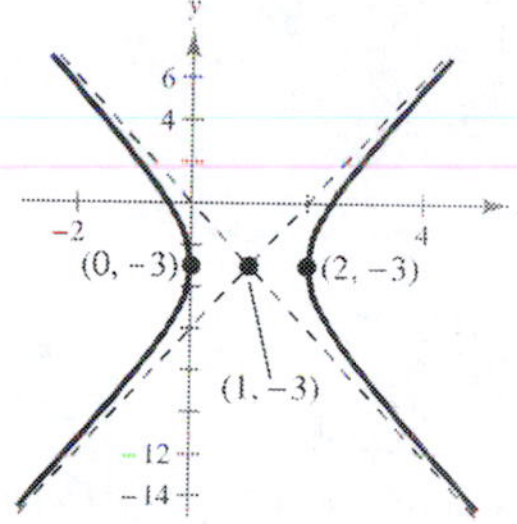

7. The center of the hyperbola lies at the midpoint of its vertices. So, the center is

$$(h, k) = \left(\frac{3 + 5}{2}, \frac{-1 + (-1)}{2}\right)$$
$$= (4, -1).$$

Because the vertices lie on a horizontal line and are two units apart, it follows that the transverse axis is horizontal and has a length of $2a = 2$. So, $a = 1$. Because the foci are three units from center, it follows that $c = 3$. So, $b^2 = c^2 - a^2 = 3^2 - 1^2 = 9 - 1 = 8$.

Because the transverse axis is horizontal, the standard form of the equation is

$\dfrac{(x - h)^2}{a^2} - \dfrac{(y - k)^2}{b^2} = 1$	Standard form, horizontal transverse axis.
$\dfrac{(x - 4)^2}{1^2} - \dfrac{(y - (-1))^2}{(2\sqrt{2})^2} = 1$	Write in standard form.
$(x - 4)^2 - \dfrac{(y + 1)^2}{8} = 1$	Simplify.

Chapter 5

Checkpoints for Section 5.1

1. Function Value

$f(\sqrt{2}) = 8^{-\sqrt{2}}$

Graphing Calculator Keystrokes

8 [^] [(] [(−)] [√] 2 [)] [Enter]

Display

0.052824803759

2. The table lists some values for each function, and the graph shows a sketch of the two functions. Note that both graphs are increasing and the graph of $g(x) = 9^x$ is increasing more rapidly than the graph of $f(x) = 3^x$.

x	-3	-2	-1	0	1	2
3^x	$\frac{1}{27}$	$\frac{1}{9}$	$\frac{1}{3}$	1	3	9
9^x	$\frac{1}{729}$	$\frac{1}{81}$	$\frac{1}{9}$	1	9	81

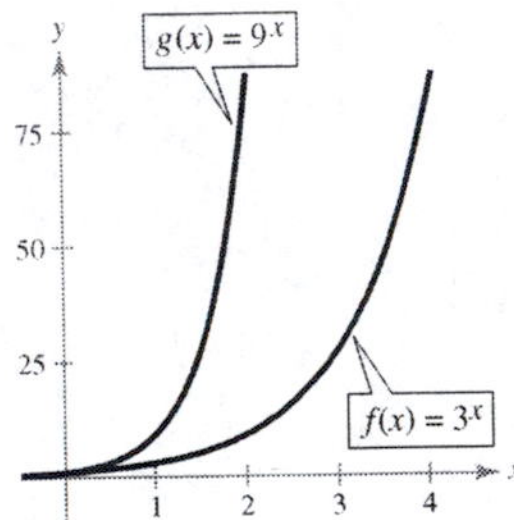

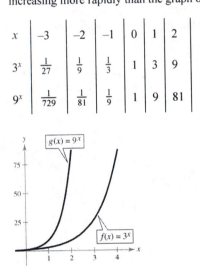

3. The table lists some values for each function and, the graph shows a sketch for each function. Note that both graphs are decreasing and the graph of $g(x) = 9^{-x}$ is decreasing more rapidly than the graph of $f(x) = 3^{-x}$

x	-2	-1	0	1	2	3
9^{-x}	64	8	1	$\frac{1}{8}$	$\frac{1}{64}$	$\frac{1}{512}$
3^{-x} $g(x)$	9	3	1	$\frac{1}{3}$	$\frac{1}{9}$	$\frac{1}{27}$

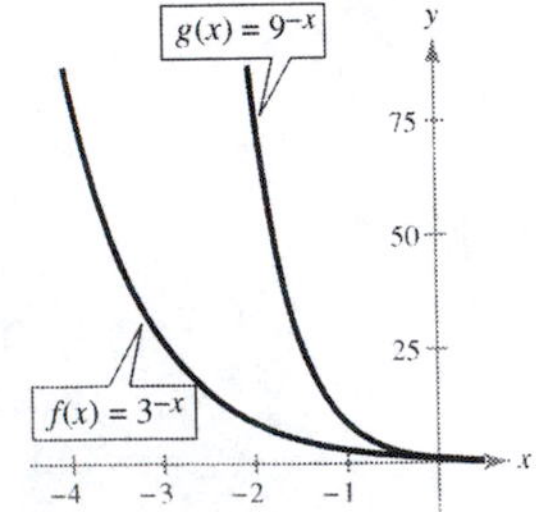

4. (a)

$8 = 2^{2x-1}$	Write Original equation.
$2^3 = 2^{2x-1}$	$8 = 2^3$
$3 = 2x - 1$	One-to-One Property
$4 = 2x$	
$2 = x$	Solve for x.

(b)

$\left(\frac{1}{3}\right)^{-x} = 27$	Write Original equation.
$3^x = 27$	$\left(\frac{1}{3}\right)^{-x} = 3^x$
$3^x = 3^3$	$27 = 3^3$
$x = 3$	One-to-One Property

5. (a) Because $g(x) = 4^{x-2} = f(x-2)$, the graph of g can be obtained by shifting the graph of f two units to the right.

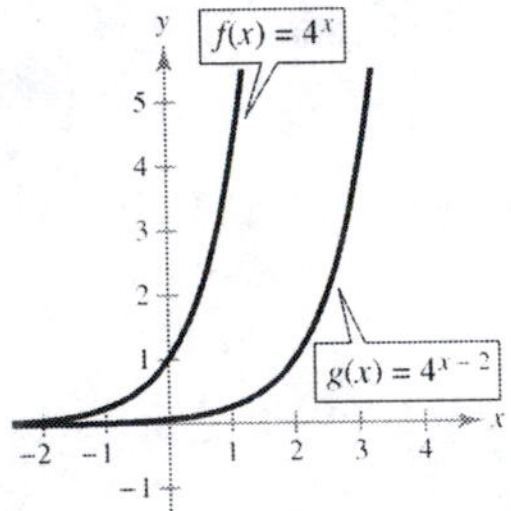

(b) Because $h(x) = 4^x + 3 = f(x) + 3$ the graph of h can be obtained by shifting the graph of f up three units.

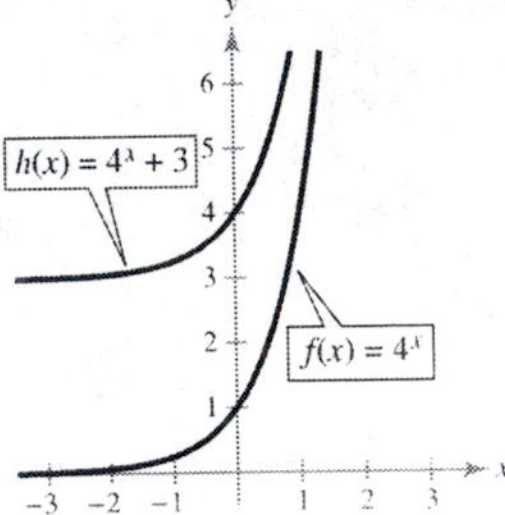

(c) Because $k(x) = 4^{-x} - 3 = f(-x) - 3$, the graph of k can be obtained by reflecting the graph of f in the y-axis and shifting the graph of f down three units.

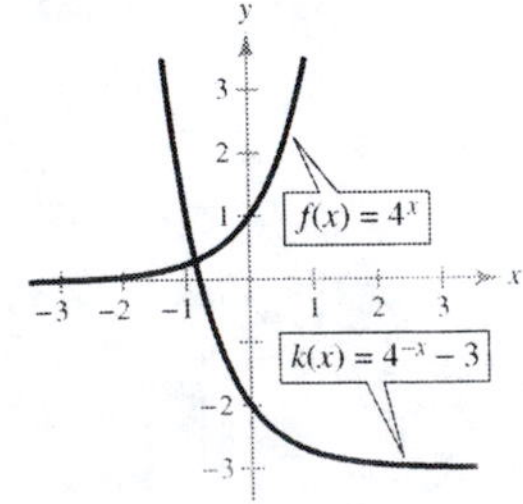

6. Function Value — Graphing Calculator Keystrokes — Display

Function Value	Graphing Calculator Keystrokes	Display
(a) $f(0.3) = e^{0.3}$	e^x 0.3 Enter	1.3498588
(b) $f(-1.2) = e^{-1.2}$	e^x (−) 1.2 Enter	6.3011942
(c) $f(6, 2) = e^{6.2}$	e^x 6.2 Enter	492.74904

7. To sketch the graph of $f(x) = 5e^{0.17x}$, use a graphing utility to construct a table of values. After constructing the table, plot the points and draw a smooth curve.

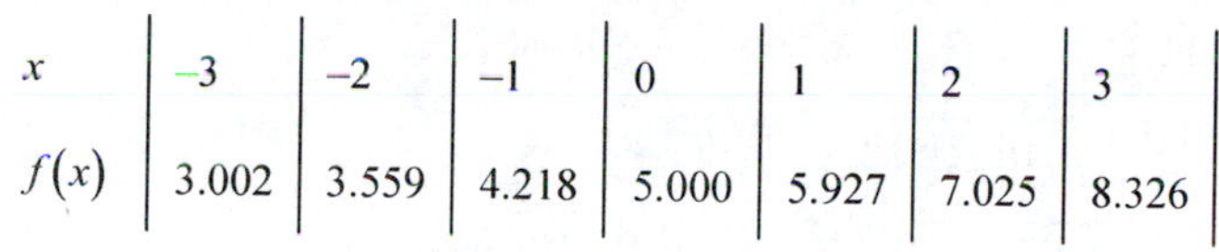

x	−3	−2	−1	0	1	2	3
$f(x)$	3.002	3.559	4.218	5.000	5.927	7.025	8.326

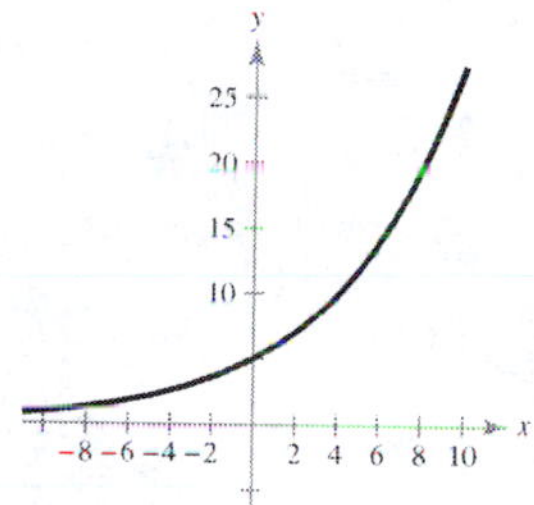

8. (a) For quarterly compounding, you have $n = 4$. So, in 7 years at 4%, the balance is as follows.

$$A = P\left(1 + \frac{r}{n}\right)^{nt}$$ Formula for compound interest.

$$= 6000\left(1 + \frac{0.04}{4}\right)^{4(7)}$$ Substitute P, r, n, and t.

$$\approx \$7927.75$$ Use a calculator.

(b) For monthly compounding, you have $n = 12$. So in 7 years at 4%, the balance is as follows.

$$A = P\left(1 + \frac{r}{n}\right)^{nt}$$ Formula for compound interest.

$$= 6000\left(1 + \frac{0.04}{12}\right)^{12(7)}$$ Substitute P, r, n, and t.

$$\approx \$7935.08$$ Use a calculator.

(c) For continuous compounding, the balance is as follows.

$$A = Pe^{rt}$$ Formula for continuous compounding.

$$= 6000e^{0.04(7)}$$ Substitute P, r, and t.

$$\approx \$7938.78$$ Use a calculator.

9. Use the model for the amount of Plutonium that remains from an initial amount of 10 pounds after t years, where $t = 0$ represents the year 1986.

$$P = 10\left(\tfrac{1}{2}\right)^{t/24{,}100}$$

To find the amount that remains in the year 2089, let $t = 103$.

$$P = 10\left(\tfrac{1}{2}\right)^{t/24{,}100}$$ Write original model.

$$P = 10\left(\tfrac{1}{2}\right)^{103/24{,}100}$$ Substitute 103 for t.

$$P \approx 9.970$$ Use a calculator.

In the year 2089, 9.970 pounds of plutonomium will remain.

To find the amount that remains after 125,000 years, let $t = 125{,}000$.

$$P = 10\left(\tfrac{1}{2}\right)^{t/24{,}100}$$ Write original model.

$$P = 10\left(\tfrac{1}{2}\right)^{125{,}000/24{,}100}$$ Substitute 125,000 for t.

$$P \approx 0.275$$ Use a calculator.

After 125,000 years 0.275 pound of plutonium will remain.

Checkpoints for Section 5.2

1. (a) $f(1) = \log_6 1 = 0$ because $6^0 = 1$.

(b) $f\left(\frac{1}{125}\right) = \log_5 \frac{1}{125} = -3$ because $5^{-3} = \frac{1}{125}$.

(c) $f(10{,}000) = \log_{10} 10{,}000 = 4$ because $10^4 = 10{,}000$.

2.

Function Value	Graphing Calculator Keystrokes	Display
(a) $f(275) = \log 275$	LOG 275 ENTER	2.4393327
(b) $f(0.275) = \log 0.275$	LOG 0.275 ENTER	– 0.5606673
(c) $f\left(-\frac{1}{2} = \log -\frac{1}{2}\right)$	LOG ((–) (1 ÷ 2)) ENTER	ERROR
(d) $f\left(\frac{1}{2}\right) = \log \frac{1}{2}$	LOG (1 ÷ 2) ENTER	– 0.3010300

3. (a) Using Property 2, $\log_9 9 = 1$.

(b) Using Property 3, $20^{\log_{20} 3} = 3$.

(c) Using Property 1, $\log_{\sqrt{3}} 1 = 0$.

4.
$$\begin{aligned} \log_5\left(x^2 + 3\right) &= \log_5 12 \\ x^2 + 3 &= 12 \\ x^2 &= 9 \\ x &= \pm 3 \end{aligned}$$

5. (a) For $f(x) = 8^x$, construct a table of values. Then plot the points and draw a smooth curve.

x	-2	-1	0	1	2
$f(x) = 8^x$	$\frac{1}{64}$	$\frac{1}{8}$	1	8	64

(b) Because $g(x) = \log_8 x$ is the inverse function of $f(x) = 8^x$, the graph of g is obtained by plotting the points $(f(x), x)$ and connecting them with a smooth curve. The graph of g is a reflection of the graph of f in the line $y = x$.

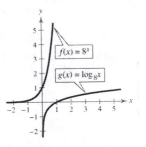

x	$\frac{1}{64}$	$\frac{1}{8}$	1	8	64
$g(x) = \log_8 x$	-2	-1	0	1	2

6. Begin by constructing a table of values. Note that some of the values can be obtained without a calculator by using the properties of logarithms. Then plot the points and draw a smooth curve.

	Without calculator					With calculator					
x	$\frac{1}{9}$	$\frac{1}{3}$	1	3	9	2	4	6	8	10	12
$f(x) = \log_9 x$	-1	$-\frac{1}{2}$	0	$\frac{1}{2}$	1	0.315	0.631	0.815	0.946	1.048	1.131

The vertical asymptote is $x = 0$ the y-axis.

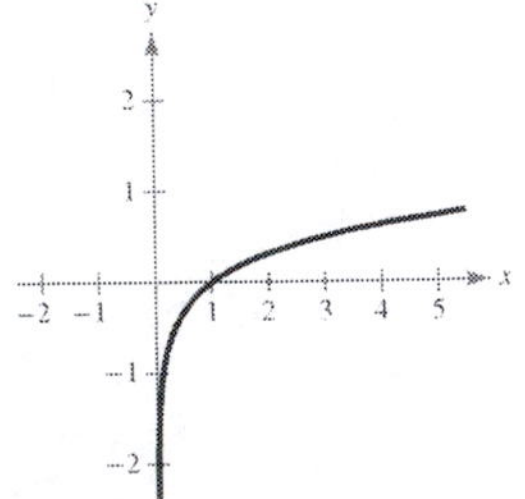

7. (a) Because $g(x) = -1 + \log_3 x = f(x) - 1$, the graph of g can be obtained by shifting the graph of f one unit down.

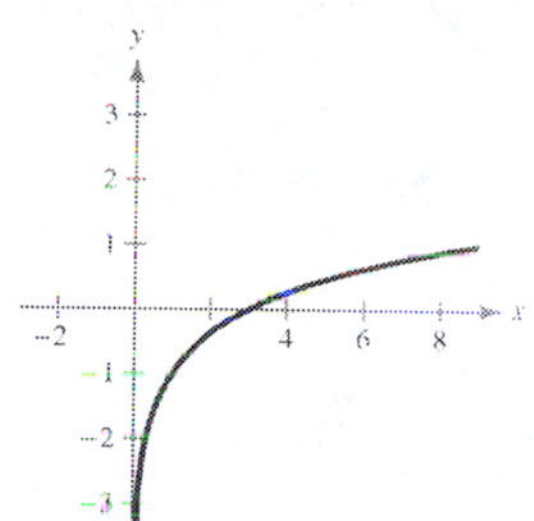

(b) Because $h(x) = \log_3 (x + 3) = f(x + 3)$, the graph of h can be obtained by shifting the graph of f three units to the left.

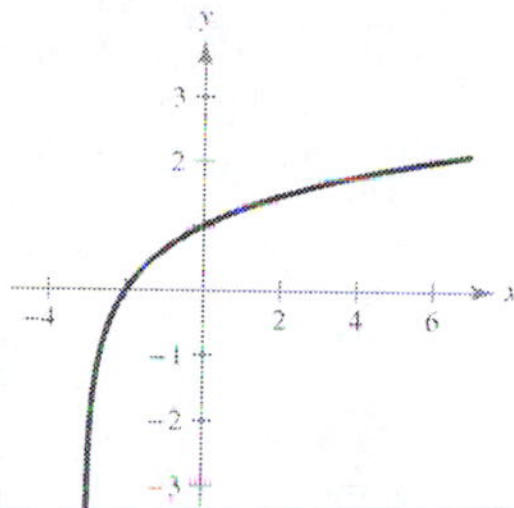

8.

Function Value	Graphing Calculator Keystrokes	Display
$f(0.01) = \ln 0.01$	LN 0.01 ENTER	-4.6051702
$f(4) = \ln 4$	LN 4 ENTER	1.3862944
$f(\sqrt{3} + 2) = \ln(\sqrt{3} + 2)$	LN ((√ 3) + 2) ENTER	1.3169579
$f(\sqrt{3} - 2) = \ln(\sqrt{3} - 2)$	LN ((√ 3) − 2) ENTER	ERROR

9. (a) $\ln e^{1/3} = \frac{1}{3}$ Inverse Property

(b) $5 \ln 1 = 5(0) = 0$ Property 1

(c) $\frac{3}{4} \ln e = \frac{3}{4}(1) = \frac{3}{4}$ Property 2

(d) $e^{\ln 7} = 7$ Inverse Property

10. Because $\ln(x + 3)$ is defined only when $x + 3 > 0$, it follows that the domain of f is $(-3, \infty)$. The graph of f is shown.

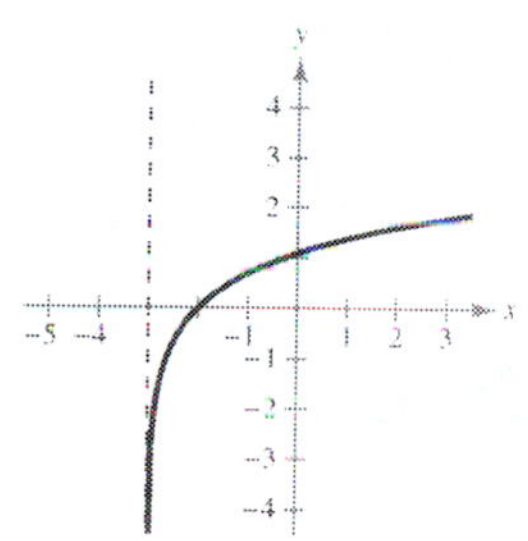

11. (a) After 1 month, the average score was the following.

$f(1) = 75 - 6 \ln(1 + 1)$	Substitute 1 for t.
$= 75 - 6 \ln 2$	Simplify.
$\approx 75 - 6(0.6931)$	Use a calculator.
≈ 70.84	Solution

(b) After 9 months, the average score was the following.

$f(9) = 75 - 6 \ln(9 + 1)$	Substitute 9 for t.
$= 75 - 6 \ln 10$	Simplify.
$\approx 75 - 6(2.3026)$	Use a calculator.
≈ 61.18	Solution

(c) After 12 months, the average score was the following.

$f(12) = 75 - 6 \ln(12 + 1)$	Substitute 9 for t.
$= 75 - 6 \ln 13$	Simplify.
$\approx 75 - 6(2.5649)$	Use a calculator.
≈ 59.61	Solution

Checkpoints for Section 5.3

1.

$\log_2 12 = \dfrac{\log 12}{\log 2}$	$\log_a x = \dfrac{\log x}{\log a}$
$\approx \dfrac{1.07918}{0.30103}$	Use a calculator.
≈ 3.5850	Simplify.

2.

$\log_2 12 = \dfrac{\ln 12}{\ln 2}$	$\log_a x = \dfrac{\ln x}{\ln a}$
$\approx \dfrac{2.48491}{0.69315}$	Use a calculator.
≈ 3.5850	Simplify.

3. (a) $\log 75 = \log(3 \cdot 25)$ Rewrite 75 as $3 \cdot 25$.

$= \log 3 + \log 25$ Product Property

$= \log 3 + \log 5^2$ Rewrite 25 as 5^2.

$= \log 3 + 2\log 5$ Power Property

(b) $\log \frac{9}{125} = \log 9 - \log 125$ Quotient Property

$= \log 3^2 - \log 5^3$ Rewrite 9 as 3^2 and 125 as 5^2.

$= 2\log 3 - 3\log 5$ Power Property

4. $\ln e^6 - \ln e^2 = 6\ln e - 2\ln e$

$= 6(1) - 2(1)$

$= 4$

5. $\log_3 \frac{4x^2}{\sqrt{y}} = \log_3 \frac{4x^2}{y^{1/2}}$ Rewrite using rational exponent.

$= \log_3 4x^2 - \log_3 y^{1/2}$ Quotient Property

$= \log_3 4 + \log_3 x^2 - \log_3 y^{1/2}$ Product Property

$= \log_3 4 + 2\log_3 x - \frac{1}{2}\log_3 y$ Power Property

6. $2\left[\log(x+3) - 2\log(x-2)\right] = 2\left[\log(x+3) - \log(x-2)^2\right]$ Power Property

$= 2\left[\log\left(\frac{x+3}{(x-2)^2}\right)\right]$ Quotient Property

$= \log\left(\frac{x+3}{(x-2)^2}\right)^2$ Power Property

$= \log\frac{(x+3)^2}{(x-2)^4}$ Simplify.

7. To solve this problem, take the natural logarithm of each of the x- and y-values of the ordered pairs.

$(\ln x, \ln y)$: $(-0.994, -0.673)$, $(0.000, 0.000)$, $(1.001, 0.668)$, $(2.000, 1.332)$, $(3.000, 2.000)$

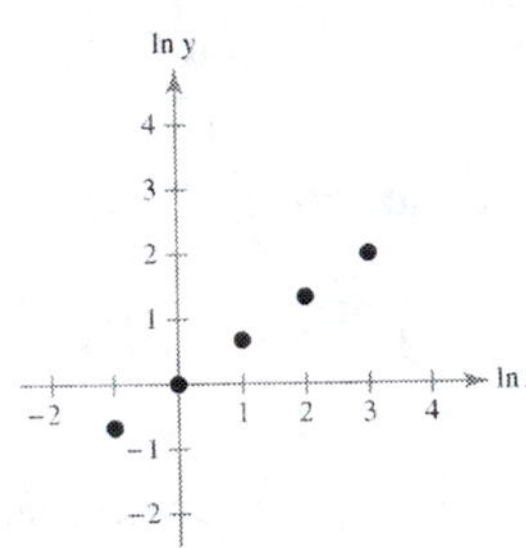

By plotting the ordered pairs, you can see that all five points appear to lie in a line. Choose any two points to determine the slope of the line. Using the points $(0, 0)$ and $(1.001, 0.668)$, the slope of the line is

$$m = \frac{0.668 - 0}{1 - 0} = 0.668 \approx \frac{2}{3}.$$

By the point-slope form, the equation of the line is $y = \frac{2}{3}x$, where $y = \ln y$ and $x = \ln x$. So, the logarithmic equation is $\ln y = \frac{2}{3}\ln x$.

Checkpoints for Section 5.4

1.

Original Equation	Rewritten Equation	Solution	Property
(a) $2^x = 512$	$2^x = x^9$	$x = 9$	One-to-One
(b) $\log_6 x = 3$	$6^{\log_6 x} = 6^3$	$x = 216$	Inverse
(c) $5 - e^x = 0$ $5 = e^x$	$\ln 5 = \ln e^x$	$\ln 5 = x$	Inverse
(d) $9^x = \frac{1}{3}$	$3^{2x} = 3^{-1}$	$2x = -1$ $x = -\frac{1}{2}$	One-to-One

2. (a)

$e^{2x} = e^{x^2-8}$ Write original equation.

$2x = x^2 - 8$ One-to-One Property

$0 = x^2 - 2x - 8$ Write in general form.

$0 = (x - 4)(x + 2)$ Factor.

$x - 4 = 0 \Rightarrow x = 4$ Set 1st factor equal to 0.

$x + 2 = 0 \Rightarrow x = -2$ Set 2nd factor equal to 0.

The solutions are $x = 4$ and $x = -2$

Check $x = -2$:

$e^{2x} = e^{x^2-8}$

$e^{2(-2)} \stackrel{?}{=} e^{(-2)^2-8}$

$e^{-4} \stackrel{?}{=} e^{4-8}$

$e^{-4} = e^{-4}$ ✓

$x = 4$:

$e^{2(4)} \stackrel{?}{=} e^{(4)^2-8}$

$e^{8} \stackrel{?}{=} e^{16-8}$

$e^{8} = e^{8}$ ✓

(b)

$2(5^x) = 32$ Write original equation.

$5^x = 16$ Divide each side by 2.

$\log_5 5^x = \log_5 16$ Take log(base 5) of each side.

$x = \log_5 16$ Inverse Property

$x = \frac{\ln 16}{\ln 5} \approx 1.723$ Change of base formula

The solution is $x = \log_5 16 \approx 1.723$.

Check $x = \log_5 16$:

$2(5^x) = 32$

$2\left[5^{(\log_5 16)}\right] \stackrel{?}{=} 32$

$2(16) \stackrel{?}{=} 32$

$32 = 32$ ✓

3. $e^x - 7 = 23$ — Write original equation.

$e^x = 30$ — Add 7 to each side.

$\ln e^x = \ln 30$ — Take natural log of each side.

$x = \ln 30 \approx 3.401$ — Inverse Property

Check $x = \ln 30$:

$$e^x - 7 = 23$$

$$e^{(\ln 30)-7} \stackrel{?}{=} 23$$

$$30 - 7 \stackrel{?}{=} 23$$

$$23 = 23 \checkmark$$

4. $6\left(2^{t+5}\right) + 4 = 11$ — Write original equation.

$6\left(2^{t+5}\right) = 7$ — Subtract 4 from each side.

$2^{t+5} = \dfrac{7}{6}$ — Divide each side by 6.

$\log_2 2^{t+5} = \log_2\left(\dfrac{7}{6}\right)$ — Take log (base 2) of each side.

$t + 5 = \log_2\left(\dfrac{7}{6}\right)$ — Inverse Property

$t = \log_2\left(\dfrac{7}{6}\right) - 5$ — Subtract 5 from each side.

$t = \dfrac{\ln\left(\dfrac{1}{6}\right)}{\ln 2} - 5$ — Change of base formula.

$t \approx -4.778$ — Use a calculator.

The solution is $t = \log_2\left(\dfrac{7}{6}\right) - 5 \approx -4.778$.

Check $t \approx -4.778$:

$$6\left(2^{t+5}\right) + 4 = 11$$

$$6\left[2^{(-4.778+5)}\right] + 4 \stackrel{?}{=} 11$$

$$6(1.166) + 4 \stackrel{?}{=} 11$$

$$10.998 \approx 11 \checkmark$$

5. Algebraic Solution

$e^{2x} - 7e^x + 12 = 0$	Write original equation.
$(e^x)^2 - 7e^x + 12 = 0$	Write in quadratic form.
$(e^x - 3)(e^x - 4) = 0$	Factor.
$e^x - 3 = 0 \Rightarrow e^x = 3$	Set 1st factor equal to 0.
$x = \ln 3$	Solution
$e^x - 4 = 0 \Rightarrow e^x = 4$	Set 2nd factor equal to 0.
$x = \ln 4$	Solution

The solutions are $x = \ln 3 \approx 1.099$ and $x = \ln 4 \approx 1.386$.

Check $x = \ln 3$:

$$e^{2x} - 7e^x + 12 = 0$$
$$e^{2(\ln 3)} - 7e^{(\ln 3)} + 12 \stackrel{?}{=} 0$$
$$e^{\ln(3^2)} - 7e^{\ln 3} + 12 \stackrel{?}{=} 0$$
$$3^2 - 7(3) + 12 \stackrel{?}{=} 0$$
$$0 = 0 \checkmark$$

$x = \ln 4$:

$$e^{2(\ln 4)} - 7e^{(\ln 4)} + 12 = 0$$
$$e^{\ln 4^2} - 7e^{\ln 4} + 12 \stackrel{?}{=} 0$$
$$4^2 - 7(4) + 12 \stackrel{?}{=} 0$$
$$0 = 0 \checkmark$$

Graphical Solution

Use a graphing utility to graph $y = e^{2x} - 7e^x + 12$ and then find the zeros.

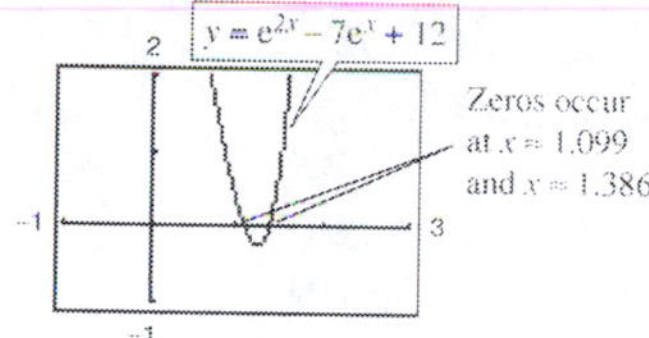

So, you can conclude that the solutions are $x \approx 1.099$ and $x \approx 1.386$.

6. (a)

$\ln x = \dfrac{2}{3}$	Write original equation.
$e^{\ln x} = e^{2/3}$	Exponentiate each side.
$x = e^{2/3}$	Inverse Property

(b)

$\log_2(2x - 3) = \log_2(x + 4)$	Write original equation.
$2x - 3 = x + 4$	One-to-One Property
$x = 7$	Solution

(c)

$\log 4x - \log(12 + x) = \log 2$	Write Original equation.
$\log\left(\dfrac{4x}{12 + x}\right) = \log 2$	Quotient Property of Logarithms
$\dfrac{4x}{12 + x} = 2$	One-to-One Property
$4x = 2(12 + x)$	Multiply each side by $(12 + x)$.
$4x = 24 + 2x$	Distribute.
$2x = 24$	Subtract $2x$ from each side.
$x = 12$	Solution

7. Algebraic Solution

$7 + 3 \ln x = 5$	Write original equation.
$3 \ln x = -2$	Subtract 7 from each side.
$\ln x = -\frac{2}{3}$	Divide each side by 3.
$e^{\ln x} = e^{-2/3}$	Exponentiate each side.
$x = e^{-2/3}$	Inverse Property
$x \approx 0.513$	Use a calculator.

Graphical Solution

Use a graphing utility to graph $y_1 = 7 + 3 \ln x$ and $y_2 = 5$. Then find the intersection point.

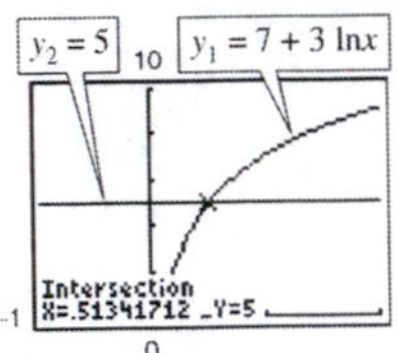

The point of intersection is about $(0.513, 5)$. So, the solution is $x \approx 0.513$.

8.

$3 \log_4 6x = 9$	Write original equation.
$\log_4 6x = 3$	Divide each side by 3.
$4^{\log_4 6x} = 4^3$	Exponentiate each side (base 4).
$6x = 64$	Inverse Property
$x = \frac{32}{3}$	Divide each side by 6 and simplify.

Check $x = \frac{32}{3}$:

$$3 \log_4 6x = 9$$

$$3 \log_4 6\left(\frac{32}{3}\right) \stackrel{?}{=} 9$$

$$3 \log_4 64 \stackrel{?}{=} 9$$

$$3 \log_4 4^3 \stackrel{?}{=} 9$$

$$3 \cdot 3 \stackrel{?}{=} 9$$

$$9 = 9 \checkmark$$

9. Algebraic Solution

$\log x + \log(x-9) = 1$	Write original equation.
$\log[x(x-9)] = 1$	Product Property of Logarithms
$10^{\log[x(x-9)]} = 10^1$	Exponentiate each side (base 10).
$x(x-9) = 10$	Inverse Property
$x^2 - 9x - 10 = 0$	Write in general form.
$(x-10)(x+1) = 0$	Factor.
$x - 10 = 0 \Rightarrow x = 10$	Set 1st factor equal to 0.
$x + 1 = 0 \Rightarrow x = -1$	Set 2nd factor equal to 0.

Check $x = 10$:

$$\log x + \log(x-9) = 1$$
$$\log(10) + \log(10-9) \stackrel{?}{=} 1$$
$$\log 10 + \log 1 \stackrel{?}{=} 1$$
$$1 + 0 \stackrel{?}{=} 1$$
$$1 = 1 \checkmark$$

$x = -1$:

$$\log x + \log(x-9) = 1$$
$$\log(-1) + \log(-1-9) \stackrel{?}{=} 1$$
$$\log(-1) + \log(-10) \stackrel{?}{=} 1$$

-1 and -10 are not in the domain of $\log x$. So, it does not check.

The solutions appear to be $x = 10$ and $x = -1$. But when you check these in the original equation, you can see that $x = 10$ is the only solution.

Graphical Solution

First, rewrite the original solution as

$\log x + \log(x-9) - 1 = 0.$

Then use a graphing utility to graph the equation $y = \log x + \log(x-9) - 1$ and find the zeros.

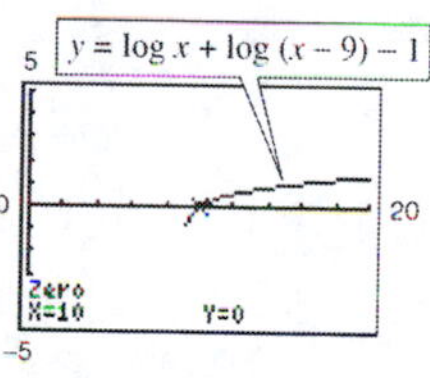

10. Using the formula for continuous compounding, the balance is

$A = Pe^{rt}$

$A = 500e^{0.0525t}.$

To find the time required for the balance to double, let $A = 1000$ and solve the resulting equation for t

$500e^{0.0525t} = 1000$	Let $A = 1000$.
$e^{0.0525t} = 2$	Divide each side by 500.
$\ln e^{0.0525t} = \ln 2$	Take natural log of each side.
$0.0525t = \ln 2$	Inverse Property
$t = \dfrac{\ln 2}{0.0525}$	Divide each side by 0.0525.
$t \approx 13.20$	Use a calculator.

The balance in the account will double after approximately 13.20 years.

Because the interest rate is lower than the interest rate in Example 2, it will take more time for the account balance to double.

11. To find when sales reached \$80 billion, let $y = 80$ and solve for t.

$-566 + 244.7 \ln t = y$	Write original equation
$-566 + 244.7 \ln t = 80$	Substitute 80 for y.
$244.7 \ln t = 646$	Add 566 to each side.
$\ln t = \dfrac{646}{244.7}$	Divide each side by 244.7.
$e^{\ln t} = e^{646/244.7}$	Exponentiate each side (base e).
$t = e^{646/244.7}$	Inverse Property
$t \approx 14$	Use a calculator.

The solution is $t \approx 14$. Because $t = 12$ represents 2002, it follows that $t = 14$ represents 2004. So, sales reached \$80 billion in 2004.

Checkpoints for Section 5.5

1. Algebraic Solution

To find when the amount of U.S. online advertising spending will reach \$100 billion, let $s = 100$ and solve for t.

$9.30e^{0.1129t} = s$	Write original model.
$9.30e^{0.1129t} = 100$	Substitute 100 for s.
$e^{0.1129t} \approx 10.7527$	Divide each side by 9.30.
$\ln e^{0.1129t} \approx \ln 10.7527$	Take natural log of each side.
$0.1129t \approx 2.3752$	Inverse Property
$t \approx 21.0$	Divide each side by 0.1129.

According to the model, the amount of U.S. online advertising spending will reach \$100 million in 2021.

Graphical Solution

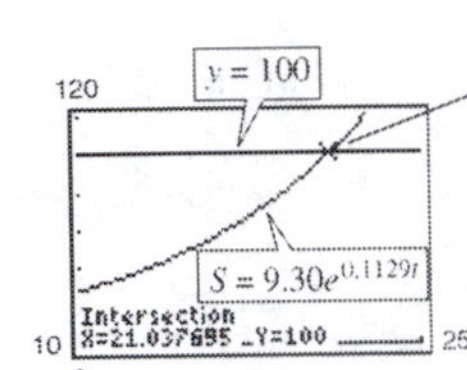

The intersection point of the model and the line $y = 100$ is about (21.0, 100). So, according to the model, the amount of U.S. online advertising spending will reach \$100 billion in 2021.

2. Let y be the number of bacteria at time t. From the given information you know that $y = 100$ when $t = 1$ and $y = 200$ when $t = 2$. Substituting this information into the model $y = ae^{bt}$ produces $100 = ae^{(1)b}$ and $200 = ae^{(2)b}$. To solve for b, solve for a in the first equation.

$100 = ae^{b}$	Write first equation.
$\dfrac{100}{e^{b}} = a$	Solve for a.

Then substitute the result into the second equation.

$200 = ae^{2b}$	Write second equation
$200 = \left(\dfrac{100}{e^{b}}\right)e^{2b}$	Substitute $\dfrac{100}{e^{b}}$ for a.
$\dfrac{200}{100} = e^{b}$	Simplify and divide each side by 100.
$2 = e^{b}$	Simplify.
$\ln 2 = \ln e^{b}$	Take natural log of each side
$\ln 2 = b$	Inverse Property

Use $b = \ln 2$ and the equation you found for a.

$a = \dfrac{100}{e^{\ln 2}}$	Substitute $\ln 2$ for b.
$= \dfrac{100}{2}$	Inverse Property
$= 50$	Simplify.

So, with $a = 50$ and $b = \ln 2$, the exponential growth model is $y = 50e^{(\ln 2)t}$.

After 3 hours, the number of bacteria will be $y = 50e^{\ln 2(3)} = 400$ bacteria.

3. Algebraic Solution

In the carbon dating model, substitute the given value of R to obtain the following.

$$\frac{1}{10^{12}}e^{-t/8223} = R \qquad \text{Write original model.}$$

$$\frac{e^{-t/8223}}{10^{12}} = \frac{1}{10^{14}} \qquad \text{Substitute } \frac{1}{10^{14}} \text{ for } R.$$

$$e^{-t/8223} = \frac{1}{10^2} \qquad \text{Multiply each side by } 10^2.$$

$$e^{-t/8223} = \frac{1}{100} \qquad \text{Simplify.}$$

$$\ln e^{-t/8223} = \ln\frac{1}{100} \qquad \text{Take natural log of each side.}$$

$$-\frac{t}{8223} \approx -4.6052 \qquad \text{Inverse Property}$$

$$t \approx 37{,}869 \qquad \text{Multiply each side by } -8223.$$

So, to the nearest thousand years, the age of the fossil is about 38,000 years.

Graphical Solution

Use a graphing utility to graph the formula for the ratio of carbon 14 to carbon 12 at any time t as

$$y_1 = \frac{1}{10^{12}}e^{-x/8223}.$$

In the same viewing window, graph $y_2 = \dfrac{1}{10^{14}}$

Use the *intersect* feature to estimate that $x \approx 18{,}934$ when $y = 1/10^{13}$.

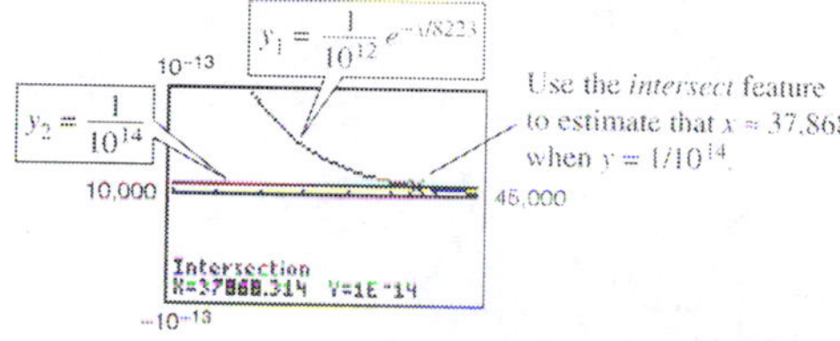

So, to the nearest thousand years, the age of fossil is about 38,000 years.

4. The graph of the function is shown below. On this bell-shaped curve, the maximum value of the curve represents the average score. From the graph, you can estimate that, the average reading score for high school graduates in 2011 was 497.

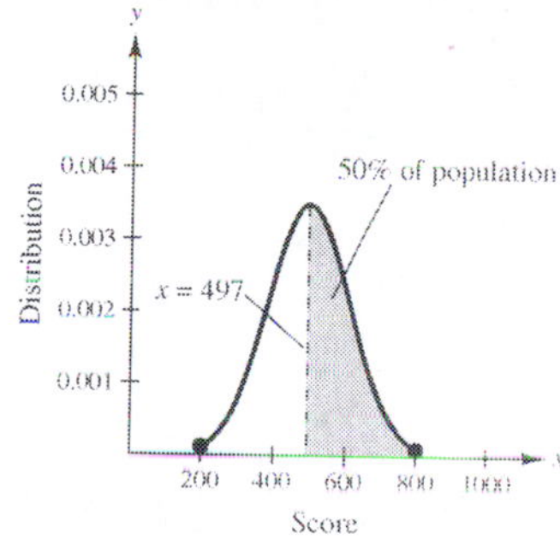

$$y = 0.0035e^{-(x-497)^2/25.992}$$

5. To find the number of days that 250 students are infected, let $y = 250$ and solve for t.

$\dfrac{5000}{1 + 4999e^{-0.8t}} = y$	Write original model.
$\dfrac{5000}{1 + 4999e^{-0.8t}} = 250$	Substitute 250 for y.
$\dfrac{5000}{250} = 1 + 4999e^{-0.8t}$	Divide each side by 250 and multiply each side by $1 + 4999e^{-0.8t}$.
$20 = 1 + 4999e^{-0.8t}$	Simplify.
$19 = 4999e^{-0.8t}$	Subtract 1 from each side.
$\dfrac{19}{4999} = e^{-0.8t}$	Divide each side by 4999.
$\ln\left(\dfrac{19}{4999}\right) = \ln e^{-0.8t}$	Take natural log of each side.
$\ln\left(\dfrac{19}{4999}\right) = -0.8t$	Inverse Property
$-5.5726 \approx -0.8t$	Use a calculator.
$t \approx 6.97$	Divide each side by -0.8.

So, after about 7 days, 250 students will be infected.

Graphical Solution

To find the number of days that 250 students are infected, use a graphing utility to graph.

$$y_1 = \frac{5000}{1 + 4999e^{-0.8x}} \text{ and } y_2 = 250$$

in the same viewing window. Use the *intersect* feature of the graphing utility to find the point of intersection of the graphs.

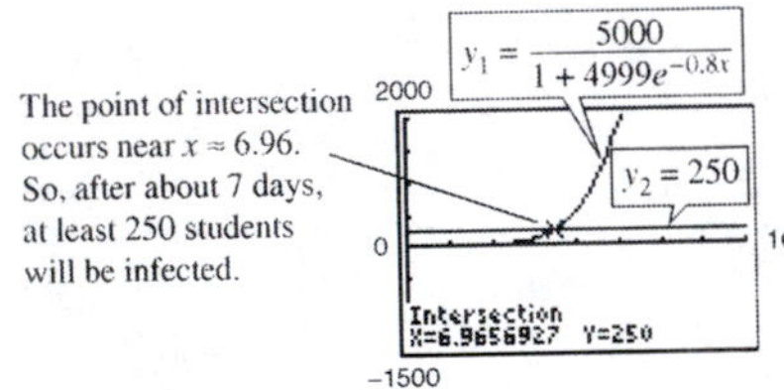

The point of intersection occurs near $x \approx 6.96$. So, after about 7 days, at least 250 students will be infected.
$y_1 = \dfrac{5000}{1 + 4999e^{-0.8x}}$
$y_2 = 250$
2000
0
16
-1500
Intersection
X=6.9656927 Y=250

6. (a) Because $I_0 = 1$ and $R = 6.0$, you have the following.

$R = \log \dfrac{I}{I_0}$	
$6.0 = \log \dfrac{I}{1}$	Substitute 1 for I_0 and 6.0 for R.
$10^{6.0} = 10^{\log I}$	Exponentiate each side (base 10).
$10^{6.0} = I$	Inverse Property
$1{,}000{,}000 = I$	Simplify.

(b) Because $I_0 = 1$ and $R = 7.9$, you have the following.

$7.9 = \log \dfrac{I}{1}$	Substitute 1 for I_0 and 7.9 for R.
$10^{7.9} = 10^{\log I}$	Exponentiate each side (base 10).
$10^{7.9} = I$	Inverse Property
$79{,}432{,}823 \approx I$	Simplify.

Chapter 6

Checkpoints for Section 6.1

1. (a) For angle $\theta = 55°$, subtract $360°$ to obtain a negative coterminal angle

$55° - 360° = -305°$

and add $360°$ to obtain a positive coterminal angle

$55° + 360° = 415°$.

(b) For $\theta = -28°$, add $360°$ to obtain a positive coterminal angle

$-28° + 360° = 332°$

and subtract $360°$ to obtain a negative coterminal angle

$-28° - 360° = -388°$.

2. (a) The complement of $\theta = 23°$ is

$90° - \theta = 90° - 23° = 67°$.

The supplement of $\theta = 23°$ is

$180° - \theta = 180° - 23° = 157°$.

(b) Because $\theta = -28°$ is a negative angle it has no complement nor does it have a supplement. Remember that complements and supplements are positive angles according to their definitions.

3. (a) In radian measure, the complement of an angle is found by subtracting the angle from $\frac{\pi}{2}$, which is equivalent to $90°$. So, the complement of $\theta = \frac{3\pi}{16}$ is $\frac{\pi}{2} - \frac{3\pi}{16} = \frac{8\pi}{16} - \frac{3\pi}{16} = \frac{5\pi}{16}$.

(b) In radian measure, the supplement of an angle is found by subtracting the angle from π which is equivalent to $180°$. So, the supplement of $\theta = \frac{5\pi}{12}$ is

$\pi - \frac{5\pi}{12} = \frac{12\pi}{12} - \frac{5\pi}{12} = \frac{7\pi}{12}$.

(c) In radian measure, a coterminal angle is found by adding or subtracting 2π which is equivalent to $360°$.

For $\theta = -\frac{4\pi}{3}$, add 2π to obtain a coterminal angle.

$-\frac{4\pi}{3} + 2\pi = -\frac{4\pi}{3} + \frac{6\pi}{3} = \frac{2\pi}{3}$

4. (a) $60° = (60 \text{ deg})\left(\frac{\pi \text{ rad}}{180 \text{ deg}}\right) = \frac{\pi}{3}$ radians

(b) $320° = (320 \text{ deg})\left(\frac{\pi \text{ rad}}{180 \text{ deg}}\right) = \frac{16\pi}{9}$ radians

5. (a) $\frac{\pi}{6} = \left(\frac{\pi}{6} \text{ rad}\right)\left(\frac{180 \text{ deg}}{\pi \text{ rad}}\right) = 30°$

(b) $\frac{5\pi}{3} = \left(\frac{5\pi}{3} \text{ rad}\right)\left(\frac{180 \text{ deg}}{\pi \text{ rad}}\right) = 300°$

6. To use the formula $s = r\theta$ first convert $160°$ to radian measure.

$160° = (160 \text{ deg})\left(\frac{\pi \text{ rad}}{180 \text{ deg}}\right) = \frac{8\pi}{9}$ radians

Then, using a radius of $r = 27$ inches, you can find the arc length to be

$$\begin{aligned} s &= r\theta \\ &= (27)\left(\frac{8\pi}{9}\right) \\ &= 24\pi \\ &\approx 75.40 \text{ inches.} \end{aligned}$$

7. In one revolution, the arc length traveled is

$$\begin{aligned} s &= 2\pi r \\ &= 2\pi(8) \\ &= 16\pi \text{ centimeters.} \end{aligned}$$

The time required for the second hand to travel this distance is

$t = 1 \text{ minute} = 60 \text{ seconds}$.

So, the linear speed of the tip of the second hand is

$$\begin{aligned} \text{Linear speed} &= \frac{s}{t} \\ &= \frac{16\pi \text{ centimeters}}{60 \text{ seconds}} \\ &\approx 0.838 \text{ centimeters per second.} \end{aligned}$$

8. (a) Because each revolution generates 2π radians, it follows that the saw blade turns $(2400)(2\pi) = 4800\pi$ radians per minute. In other words, the angular speed is

$$\text{Angular speed} = \frac{\theta}{t} = \frac{4800\pi \text{ radians}}{60 \text{ seconds}} = 80\pi \text{ radians per second.}$$

(b) The radius is $r = \frac{1}{2}d = 4$. The linear speed is

$$\text{Linear speed} = \frac{s}{t} = \frac{r\theta}{t} = \frac{(4)(4800\pi) \text{ inches}}{60 \text{ seconds}} = 320\pi \text{ inches per second} \approx 1005 \text{ inches per second.}$$

9. First convert 80° to radian measure as follows.

$$\theta = 80° = (80 \text{ deg})\left(\frac{\pi \text{ rad}}{180 \text{ deg}}\right) = \frac{4\pi}{9} \text{ radians}$$

Then, using $\theta = \frac{4\pi}{9}$ and $r = 40$ feet, the area is

$A = \frac{1}{2}r^2\theta$	Formula for area of a sector of a circle
$= \frac{1}{2}(40)^2\left(\frac{4\pi}{9}\right)$	Substitute for r and θ
$= \frac{3200\pi}{9}$	Multiply
≈ 1117 square feet.	Simplify.

Checkpoints for Section 6.2

1.

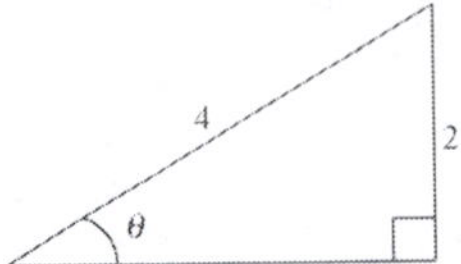

By the Pythagorean Theorem, $(\text{hyp})^2 = (\text{opp})^2 + (\text{adj})^2$, it follows that

$\text{adj} = \sqrt{4^2 - 2^2} = \sqrt{12} = 2\sqrt{3}.$

So, the six trigonometric functions of θ are

$$\sin\theta = \frac{\text{opp}}{\text{hyp}} = \frac{2}{4} = \frac{1}{2} \qquad \csc\theta = \frac{\text{hyp}}{\text{opp}} = \frac{4}{2} = 2$$

$$\cos\theta = \frac{\text{adj}}{\text{hyp}} = \frac{2\sqrt{3}}{4} = \frac{\sqrt{3}}{2} \qquad \sec\theta = \frac{\text{hyp}}{\text{adj}} = \frac{4}{2\sqrt{3}} = \frac{2}{\sqrt{3}} = \frac{2\sqrt{3}}{3}$$

$$\tan\theta = \frac{\text{opp}}{\text{adj}} = \frac{2}{2\sqrt{3}} = \frac{1}{\sqrt{3}} = \frac{\sqrt{3}}{3} \qquad \cot\theta = \frac{\text{adj}}{\text{opp}} = \frac{2\sqrt{3}}{2} = \sqrt{3}$$

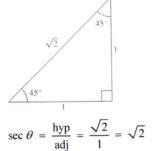

2.

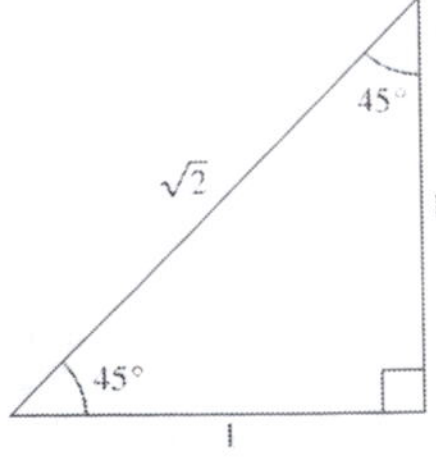

$$\sec\theta = \frac{\text{hyp}}{\text{adj}} = \frac{\sqrt{2}}{1} = \sqrt{2}$$

3.

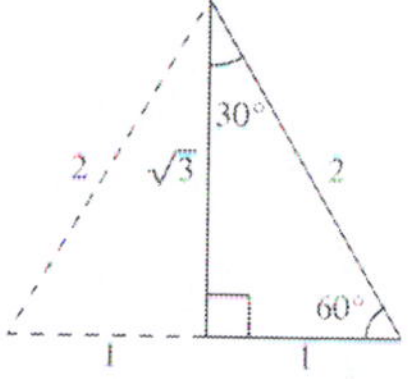

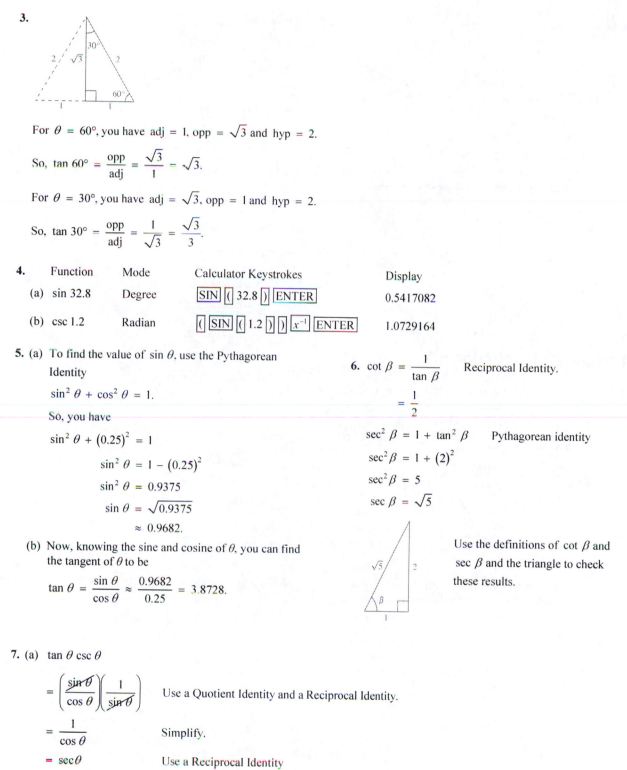

For $\theta = 60°$, you have adj $= 1$, opp $= \sqrt{3}$ and hyp $= 2$.

So, $\tan 60° = \dfrac{\text{opp}}{\text{adj}} = \dfrac{\sqrt{3}}{1} = \sqrt{3}$.

For $\theta = 30°$, you have adj $= \sqrt{3}$, opp $= 1$ and hyp $= 2$.

So, $\tan 30° = \dfrac{\text{opp}}{\text{adj}} = \dfrac{1}{\sqrt{3}} = \dfrac{\sqrt{3}}{3}$.

4.

	Function	Mode	Calculator Keystrokes	Display
(a)	sin 32.8	Degree	[SIN] [(] 32.8 [)] [ENTER]	0.5417082
(b)	csc 1.2	Radian	[(] [SIN] [(] 1.2 [)] [)] [x^{-1}] [ENTER]	1.0729164

5. (a) To find the value of $\sin\theta$, use the Pythagorean Identity

$\sin^2\theta + \cos^2\theta = 1$.

So, you have

$$\begin{aligned} \sin^2\theta + (0.25)^2 &= 1 \\ \sin^2\theta &= 1 - (0.25)^2 \\ \sin^2\theta &= 0.9375 \\ \sin\theta &= \sqrt{0.9375} \\ &\approx 0.9682. \end{aligned}$$

(b) Now, knowing the sine and cosine of θ, you can find the tangent of θ to be

$$\tan\theta = \frac{\sin\theta}{\cos\theta} \approx \frac{0.9682}{0.25} = 3.8728.$$

6. $\cot\beta = \dfrac{1}{\tan\beta}$ Reciprocal Identity.

$= \dfrac{1}{2}$

$\sec^2\beta = 1 + \tan^2\beta$ Pythagorean identity

$\sec^2\beta = 1 + (2)^2$

$\sec^2\beta = 5$

$\sec\beta = \sqrt{5}$

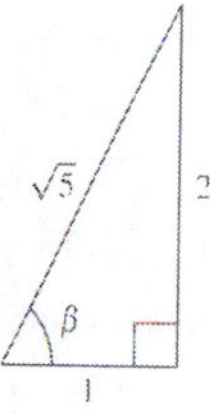

Use the definitions of $\cot\beta$ and $\sec\beta$ and the triangle to check these results.

7. (a) $\tan\theta\csc\theta$

$= \left(\dfrac{\sin\theta}{\cos\theta}\right)\left(\dfrac{1}{\sin\theta}\right)$ Use a Quotient Identity and a Reciprocal Identity.

$= \dfrac{1}{\cos\theta}$ Simplify.

$= \sec\theta$ Use a Reciprocal Identity

(b) $(\csc\theta + 1)(\csc\theta - 1) = \csc^2\theta - \csc\theta + \csc\theta - 1$ FOIL Method.

$= \csc^2\theta - 1$ Simplify.

$= \cot^2\theta$ Pythagorean identity

8. From the figure you can see that

$$\tan 64.6° = \frac{\text{opp}}{\text{adj}} = \frac{y}{x}$$

where $x = 19$ and y is the height of the flagpole. So, the height of the flagpole is

$$y = 19 \tan 64.6°$$
$$\approx 19(2.106)$$
$$\approx 40 \text{ feet.}$$

9. From the figure, you can see that the cosine of the angle θ is

$$\cos\theta = \frac{\text{adj}}{\text{hyp}} = \frac{3}{6} = \frac{1}{2}.$$

You should recognize that $\theta = 60°$.

10. From the figure, you can see that

$$\sin 11.5° = \frac{\text{opp}}{\text{hyp}} = \frac{3.5}{c}.$$

$$\sin 11.5° = \frac{3.5}{c}$$

$$c \sin 11.5° = 3.5$$

$$c = \frac{3.5}{\sin 11.5°}$$

So, the length c of the loading ramp is

$$c = \frac{3.5}{\sin 11.5} \approx \frac{3.5}{0.1994} \approx 17.6 \text{ feet.}$$

Also from the figure, you can see that

$$\tan 11.5° = \frac{\text{opp}}{\text{adj}} = \frac{3.5}{a}.$$

So, the length a of the ramp is

$$a = \frac{3.5}{\tan 11.5°} \approx \frac{3.5}{0.2034} \approx 17.2 \text{ feet.}$$

Checkpoints for Section 6.3

1. Referring to the figure shown, you can see that $x = -2$, $y = 3$ and

$$r = \sqrt{x^2 + y^2} = \sqrt{(-2)^2 + (3)^2} = \sqrt{13}.$$

So, you have the following.

$$\sin\theta = \frac{y}{r} = \frac{3}{\sqrt{13}} = \frac{3\sqrt{13}}{13}$$

$$\cos\theta = \frac{x}{r} = -\frac{2}{\sqrt{13}} = -\frac{2\sqrt{13}}{13}$$

$$\tan\theta = \frac{y}{x} = -\frac{3}{2}$$

2. Note that θ lies in Quadrant II because that is the only quadrant in which the sine is positive and the tangent is negative. Moreover, using

$$\sin\theta = \frac{4}{5} = \frac{y}{r}$$

and the fact that y is positive in Quadrant II, you can let $y = 4$ and $r = 5$.

So,

$$r = \sqrt{x^2 + y^2}$$
$$5 = \sqrt{x^2 + 4^2}$$
$$25 = x^2 + 16$$
$$9 = x^2$$
$$\pm 3 = x$$

Since x is negative in Quadrant II, $x = -3$.

So, $\cos\theta = \dfrac{x}{r} = -\dfrac{3}{5}$.

3. To begin, choose a point on the terminal side of the angle $\dfrac{3\pi}{2}$.

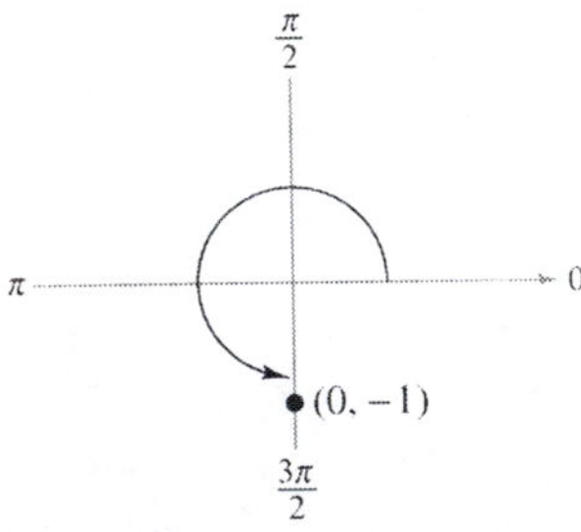

For the point $(0, -1)$, $r = 1$ and you have the following.

$$\sin\frac{3\pi}{2} = \frac{y}{r} = \frac{-1}{1} = -1$$

$$\cot\frac{3\pi}{2} = \frac{x}{y} = \frac{0}{-1} = 0$$

4. (a) Because 213° lies in Quadrant III, the angle it makes with the x-axis is $\theta' = 213° - 180° = 33°$.

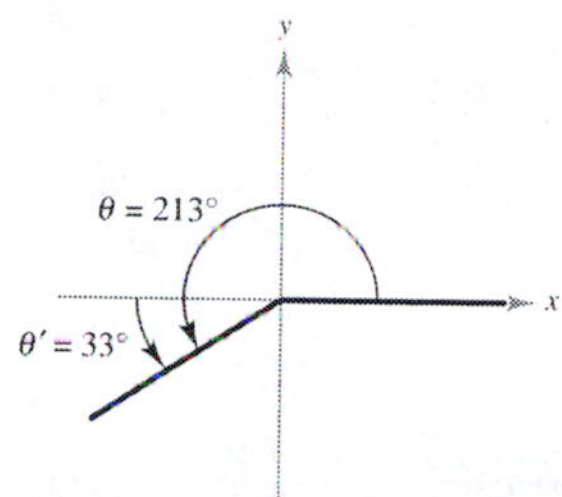

(b) Because $\frac{14\pi}{9}$ lies in Quadrant IV, the angle it makes with the x-axis is

$$\theta' = 2\pi - \frac{14\pi}{9}$$
$$= \frac{18\pi}{9} - \frac{14\pi}{9}$$
$$= \frac{4\pi}{9}.$$

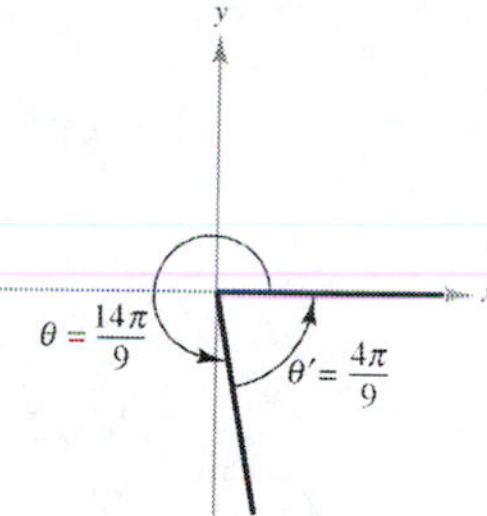

(c) Because $\frac{4\pi}{5}$ lies in Quadrant II, the angle it makes with the x-axis is

$$\theta' = \pi - \frac{4\pi}{5}$$
$$= \frac{\pi}{5}.$$

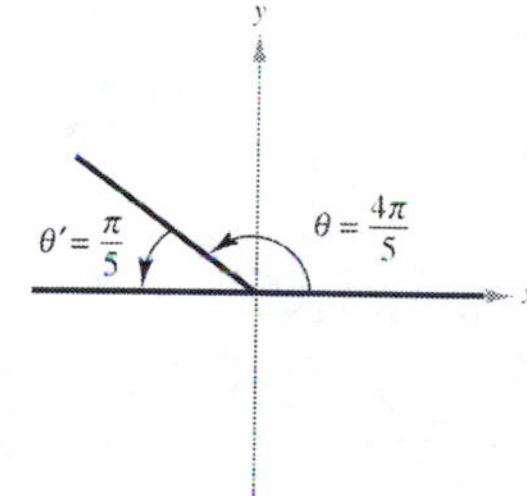

5. (a) Because $\theta = \frac{7\pi}{4}$ lies in Quadrant IV, the reference angle is $\theta' = 2\pi - \frac{7\pi}{4} = \frac{\pi}{4}$ as shown.

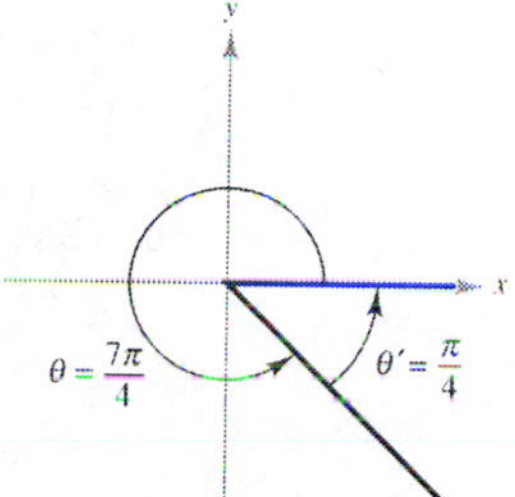

Because the sine is negative in Quadrant IV, you have $\sin \frac{7\pi}{4} = (-)\sin \frac{\pi}{7}$

$$= -\frac{\sqrt{2}}{2}.$$

(b) Because $-120° + 360° = 240°$, it follows that $-120°$ is coterminal with the third-quadrant angle 240°. So, the reference angle is $\theta' = 240° - 180° = 60°$ as shown.

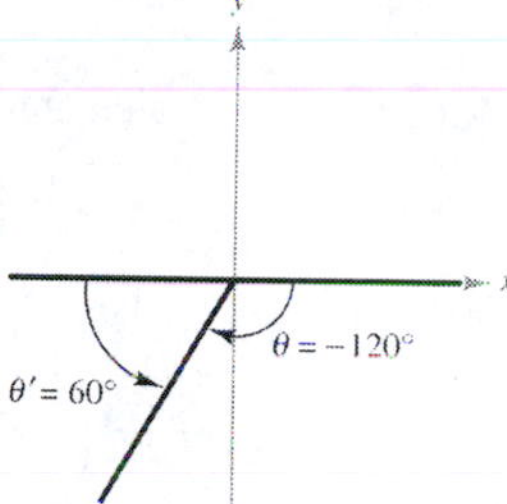

Because the cosine is negative in Quadrant III, you have $\cos(-120°) = (-)\cos 60° = -\frac{1}{2}$.

(c) Because $\theta = \frac{11\pi}{6}$ lies in Quadrant IV, the reference angle is $\theta' = 2\pi - \frac{11\pi}{6} = \frac{\pi}{6}$ as shown.

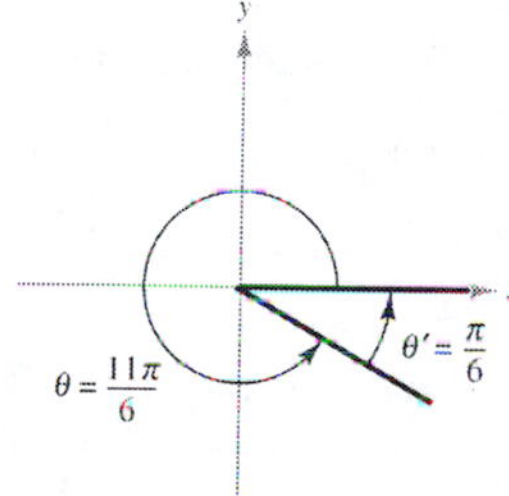

Because the tangent is negative in Quadrant IV, you have

$$\tan \frac{11\pi}{6} = (-)\tan \frac{\pi}{6} = -\frac{\sqrt{3}}{3}.$$

6. (a) Using the Pythagorean Identity $\sin^2\theta + \cos^2\theta = 1$, you obtain the following.

$\sin^2\theta + \cos^2\theta = 1$ — Write Identity

$\left(-\frac{4}{5}\right)^2 + \cos^2\theta = 1$ — Substitute $-\frac{4}{5}$ for $\sin\theta$.

$\frac{16}{25} + \cos^2\theta = 1$ — Simplify.

$\cos^2\theta = 1 - \frac{16}{25}$ — Subtract $\frac{16}{25}$ from each side.

$\cos^2\theta = \frac{9}{25}$ — Simplify.

Because $\cos\theta < 0$ in Quadrant III, you can use the negative root to obtain

$$\cos\theta = -\sqrt{\frac{9}{25}} = -\frac{3}{5}.$$

(b) Using the trigonometric identity $\tan\theta = \frac{\sin\theta}{\cos\theta}$, you obtain

$\tan\theta = \frac{-\frac{4}{5}}{-\frac{3}{5}}$ — Substitute for $\sin\theta$ and $\cos\theta$.

$= \frac{4}{3}.$ — Simplify.

7.

Function	Mode	Calculator Keystrokes	Display
(a) $\tan 119°$	Degree	[tan] [(] 119 [)] [ENTER]	-1.8040478
(b) $\csc 5$	Radian	[(] [sin] [(] 5 [)] [)] [x^{-1}] [ENTER]	-1.0428352
(c) $\cos\frac{\pi}{5}$	Radian	[cos] [(] π [÷] 5 [)] [ENTER]	0.8090170

8. Because $t = \frac{7\pi}{3}$ and $t = \frac{\pi}{3}$ are coterminal angles, it follows that $f\left(\frac{7\pi}{3}\right) = f\left(\frac{\pi}{3}\right) = \tan\frac{\pi}{3} = \frac{y}{x} = \frac{\frac{\sqrt{3}}{2}}{\frac{1}{2}} = \sqrt{3}.$

Checkpoints for Section 6.4

1. Note that $y = 2\cos x = 2(\cos x)$ indicates that the y-values for the key points will have twice the magnitude of those on the graph of $y = \cos x$. Divide the period 2π into four equal parts to get the key points.

Maximum	Intercept	Minimum	Intercept	Maximum
$(0, 2)$	$\left(\frac{\pi}{2}, 0\right)$	$(\pi, -2)$	$\left(\frac{3\pi}{2}, 0\right)$	$(2\pi, 2)$

By connecting these key points with a smooth curve and extending the curve in both directions over the interval $\left[-\frac{\pi}{2}, \frac{9\pi}{2}\right]$, you obtain the graph shown.

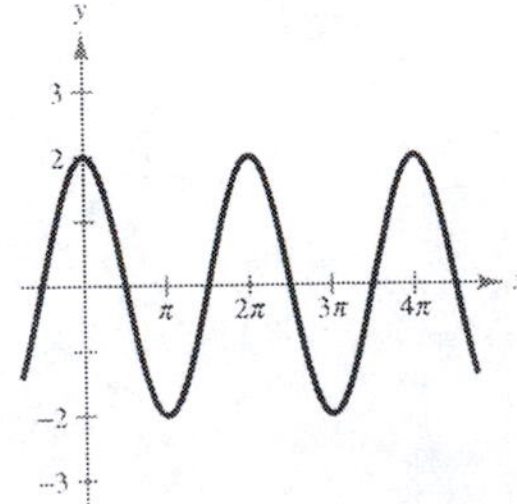

2. (a) Because the amplitude of $y = \frac{1}{3}\sin x$ is $\frac{1}{3}$, the maximum value is $\frac{1}{3}$ and the minimum value is $-\frac{1}{3}$.

Divide one cycle, $0 \le x \le 2\pi$, into for equal parts to get the key points.

Intercept	Maximum	Intercept	Minimum	Intercept
$(0, 0)$	$\left(\frac{\pi}{2}, \frac{1}{3}\right)$	$(\pi, 0)$	$\left(\frac{3\pi}{2}, -\frac{1}{3}\right)$	$(2\pi, 0)$

(b) A similar analysis shows that the amplitude of $y = 3\sin x$ is 3, and the key points are as follows.

Intercept	Maximum	Intercept	Minimum	Intercept
$(0, 0)$	$\left(\frac{\pi}{2}, 3\right)$	$(\pi, 0)$	$\left(\frac{3\pi}{2}, -3\right)$	$(2\pi, 0)$

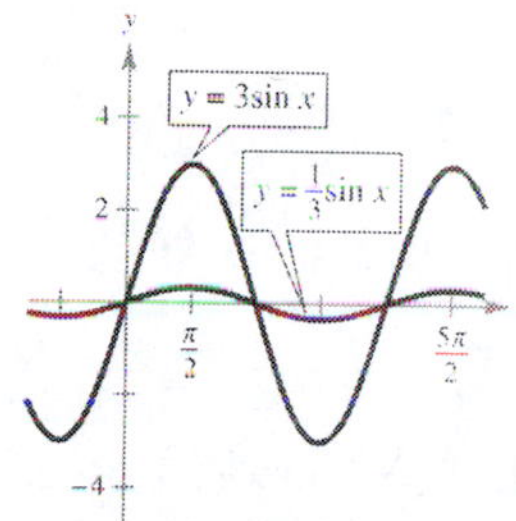

3. The amplitude is 1. Moreover, because $b = \frac{1}{3}$, the period is

$$\frac{2\pi}{b} = \frac{2\pi}{\frac{1}{3}} = 6\pi. \quad \text{Substitute } \frac{1}{3} \text{ for } b.$$

Now, divide the period-interval $[0, 6\pi]$ into four equal parts using the values $\frac{3\pi}{2}$, 3π, and $\frac{9\pi}{2}$ to obtain the key points.

Maximum	Intercept	Minimum	Intercept	Maximum
$(0, 1)$	$\left(\frac{3\pi}{2}, 0\right)$	$(3\pi, -1)$	$\left(\frac{9\pi}{2}, 0\right)$	$(6\pi, 1)$

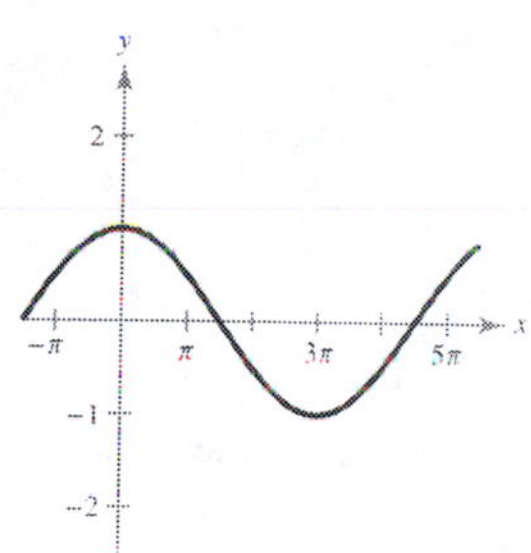

4. Algebraic Solution

The amplitude is 2 and the period is 2π.

By solving the equations

$$x - \frac{\pi}{2} = 0 \Rightarrow x = \frac{\pi}{2}$$

and $$x - \frac{\pi}{2} = 2\pi \Rightarrow x = \frac{5\pi}{2}$$

you see that the interval $\left[\frac{\pi}{2}, \frac{5\pi}{2}\right]$ corresponds to one cycle of the graph. Dividing this interval into four equal parts produces the key points.

Maximum	Intercept	Minimum	Intercept	Maximum
$\left(\frac{\pi}{2}, 2\right)$	$(\pi, 0)$	$\left(\frac{3\pi}{2}, -2\right)$	$(2\pi, 0)$	$\left(\frac{5\pi}{2}, 2\right)$

Graphical Solution

Use a graphing utility set in *radian* mode to graph $y = 2\cos\left(x - \frac{\pi}{2}\right)$ as shown.

Use the *minimum*, *maximum*, and *zero* or *root* features of the graphing utility to approximate the key points $(1.57, 2)$, $(3.14, 0)$, $(4.71, -2)$, $(6.28, 0)$ and $(7.85, 2)$.

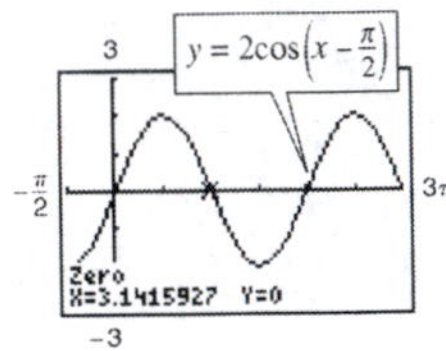

5. The amplitude is $\frac{1}{2}$ and the period is $\frac{2\pi}{b} = \frac{2\pi}{\pi} = 2$.

By solving the equations

$$\pi x + \pi = 0$$
$$\pi x = -\pi$$
$$x = -1$$

and

$$\pi x + \pi = 2\pi$$
$$\pi x = \pi$$
$$x = 1$$

you see that the interval $[-1, 1]$ corresponds to one cycle of the graph. Dividing this into four equal parts produces the key points.

Intercept	Minimum	Intercept	Maximum	Intercept
$(-1, 0)$	$\left(-\frac{1}{2}, -\frac{1}{2}\right)$	$(0, 0)$	$\left(\frac{1}{2}, \frac{1}{2}\right)$	$(1, 0)$

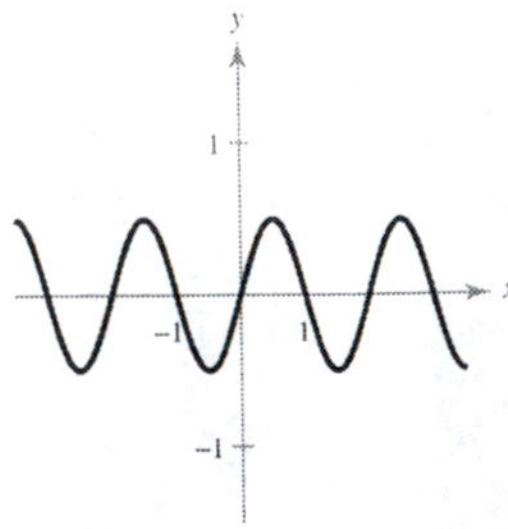

6. The amplitude is 2 and the period is 2π. The key points over the interval $[0, 2\pi]$ are

$(0, -3)$, $\left(\frac{\pi}{2}, -5\right)$, $(\pi, -7)$, $\left(\frac{3\pi}{2} - 5\right)$, and $(2\pi, -3)$.

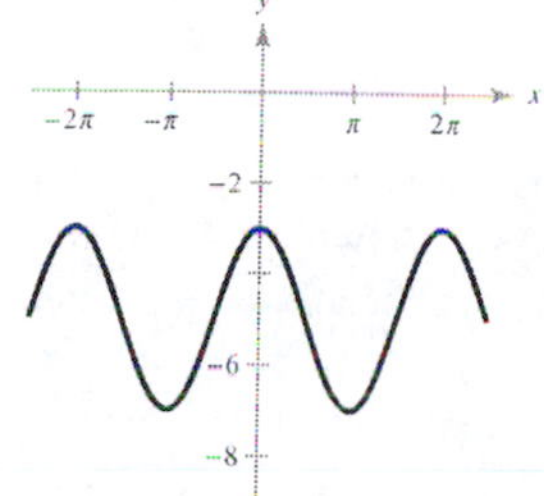

7. Use a sine model of the form $y = a \sin(bt - c) + d$.

The difference between the maximum value and minimum value is twice the amplitude of the function. So, the amplitude is

$$a = \frac{1}{2}[(\text{maximum depth}) - (\text{minimum depth})]$$
$$= \frac{1}{2}(11.3 - 0.1) = 5.6.$$

The sine function completes one half cycle between the times at which the maximum and minimum depths occur. So, the period p is

$$p = 2[(\text{time of min. depth}) - (\text{time of max. depth})]$$
$$= 2(10 - 4) = 12$$

which implies that $b = \dfrac{2\pi}{p} \approx 0.524$. Because high tide occurs 4 hours after midnight, consider the maximum to be

$$bt - c = \frac{\pi}{2} \approx 1.571.$$

So, $(0.524)(4) - c \approx 1.571$

$c \approx 0.525.$

Because the average depth is $\frac{1}{2}(11.3 + 0.1) = 5.7$, it follows that $d = 5.7$. So, you can model the depth with the function

$$y = a \sin(bt - c) + d$$
$$= 5.6 \sin(0.524t - 0.525) + 5.7.$$

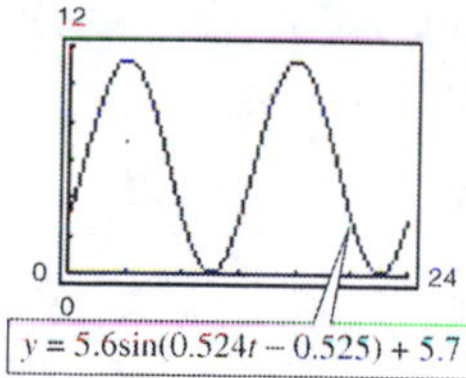

Checkpoints for Section 6.5

1. By solving the equations

$$\frac{x}{4} = -\frac{\pi}{2} \quad \text{and} \quad \frac{x}{4} = \frac{\pi}{2}$$
$$x = -2\pi \qquad\qquad x = 2\pi$$

you can see that two consecutive vertical asymptotes occur at $x = -2\pi$ and $x = 2\pi$. Between these two asymptotes, plot a few points including the x-intercept.

x	-2π	$-\pi$	0	π	2π
$f(x)$	Undef.	-1	0	1	Undef.

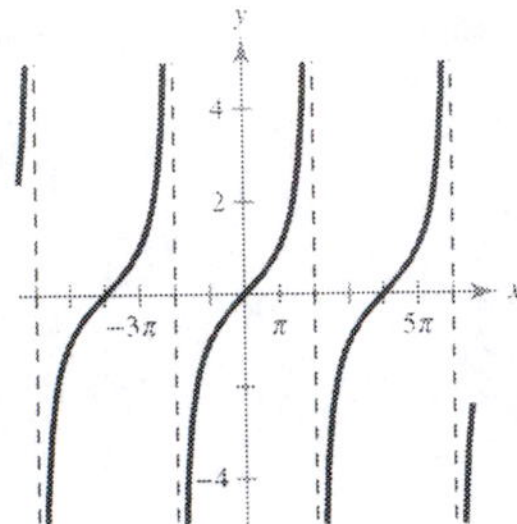

2. By solving the equations

$$2x = -\frac{\pi}{2} \quad \text{and} \quad 2x = \frac{\pi}{2}$$
$$x = -\frac{\pi}{4} \qquad\qquad x = \frac{\pi}{4}$$

you can see that two consecutive vertical asymptotes occur at $x = -\frac{\pi}{4}$ and $x = \frac{\pi}{4}$. Between these two asymptotes, plot a few points including the x-intercept.

x	$-\frac{\pi}{4}$	$-\frac{\pi}{8}$	0	$\frac{\pi}{8}$	$\frac{\pi}{4}$
$\tan 2x$	Undef.	-1	0	1	Undef.

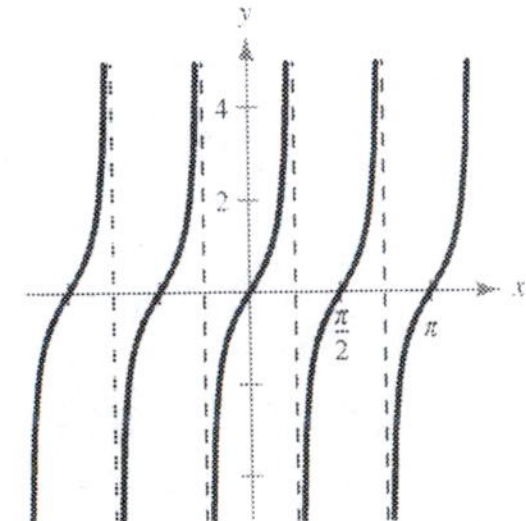

3. By solving the equations

$$\frac{x}{4} = 0 \quad \text{and} \quad \frac{x}{4} = \pi$$
$$x = 0 \qquad\qquad x = 4\pi$$

you can see that two consecutive vertical asymptotes occur at $x = 0$ and $x = 4\pi$. Between these two asymptotes, plot a few points, including the x-intercept.

x	0	π	2π	3π	4π
$\cot \frac{x}{4}$	Undef.	1	0	-1	Undef.

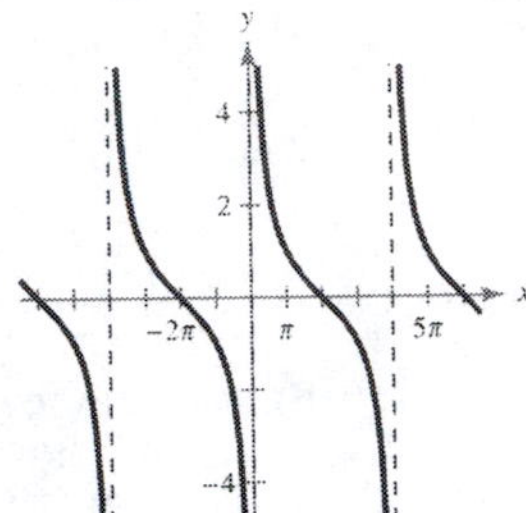

4. Begin by sketching the graph of $y = 2\sin\left(x + \frac{\pi}{2}\right)$. For this function, the amplitude is 2 and the period is 2π. By solving the equations

$$x + \frac{\pi}{2} = 0 \quad \text{and} \quad x + \frac{\pi}{2} = 2\pi$$
$$x = -\frac{\pi}{2} \qquad\qquad x = \frac{3\pi}{2}$$

you can see that one cycle of the sine function corresponds to the interval from $x = -\frac{\pi}{2}$ to $x = \frac{3\pi}{2}$. The graph of this sine function is represented by the gray curve. Because the sine function is zero at the midpoint and endpoints of this interval, the corresponding cosecant function

$$y = 2\csc\left(x + \frac{\pi}{2}\right)$$

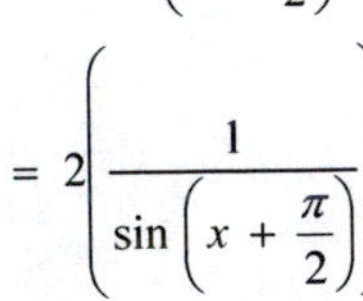

$$= 2\left(\frac{1}{\sin\left(x + \frac{\pi}{2}\right)}\right)$$

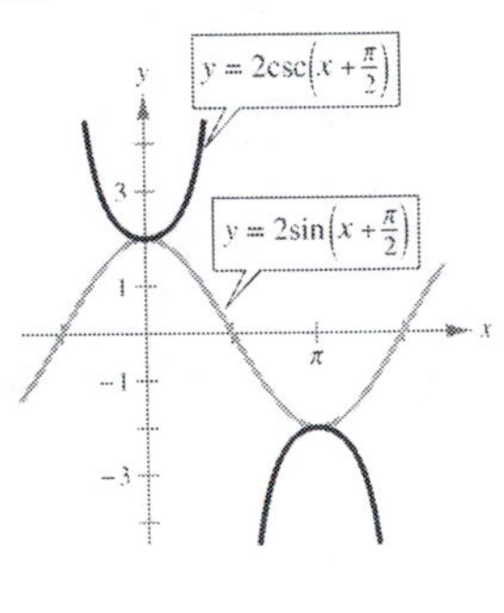

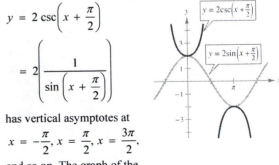

has vertical asymptotes at

$$x = -\frac{\pi}{2},\ x = \frac{\pi}{2},\ x = \frac{3\pi}{2},$$

and so on. The graph of the cosecant curve is represented by the black curve.

5. Begin by sketching the graph of $y = \cos \dfrac{x}{2}$ as indicated by the gray curve. Then, form the graph of $y = \sec \dfrac{x}{2}$ as the black curve. Note that the x-intercepts of $y = \cos \dfrac{x}{2}$, $(\pi, 0)$, $(3\pi, 0)$, $(5\pi, 0)$, ... correspond to the vertical asymptotes $x = \pi$, $x = 3\pi$, $x = 5\pi$, ... of the graph of $y = \sec \dfrac{x}{2}$. Moreover, notice that the period of $y = \cos \dfrac{x}{2}$ and $y = \sec \dfrac{x}{2}$ is $\dfrac{2\pi}{\frac{1}{2}} = 4\pi$.

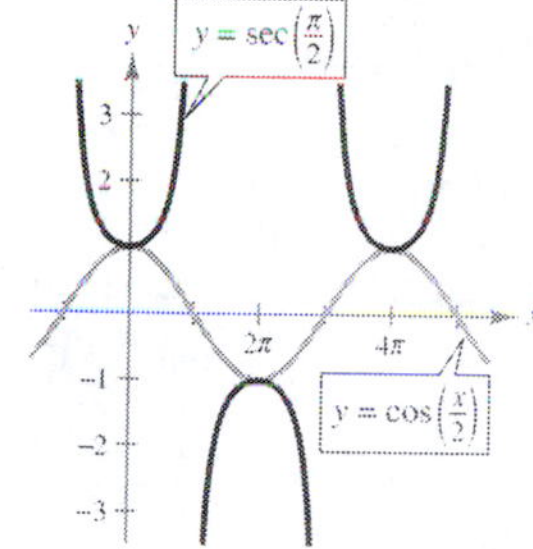

6. Consider $f(x)$ as the product of these two functions

$y = e^x$ and $y = \sin 4x$

each of which has a set of real numbers as its domain. For any real number x, you know that $e^x|\sin 4x| \le e^x$ which means that $-e^x \le e^x \sin 4x \le e^x$.

Furthermore, because

$f(x) = e^x \sin 4x = \pm e^x$ at $x = \dfrac{\pi}{8} \pm \dfrac{n\pi}{4}$ since

$\sin 4x = \pm 1$ at $4x = \dfrac{\pi}{2} + n\pi$

and

$f(x) = e^x \sin 4x = 0$ at $x = \dfrac{n\pi}{4}$ since $\sin 4x = 0$ at $4x = n\pi$

the graph of f touches the curve $y = -e^x$ or $y = e^x$ at $x = \dfrac{\pi}{8} + \dfrac{n\pi}{4}$ and has x-intercepts at $x = \dfrac{n\pi}{4}$.

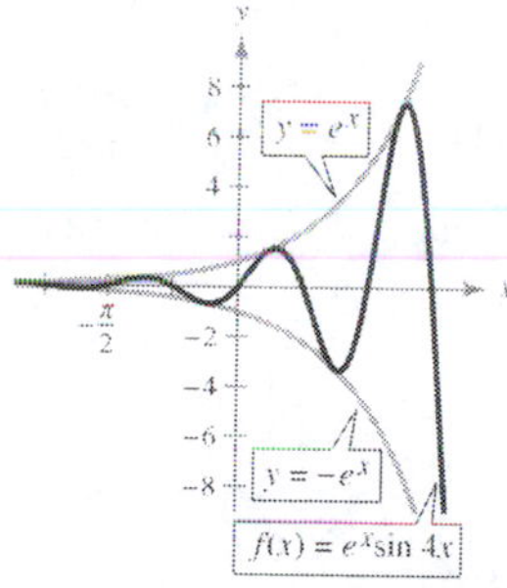

Checkpoints for Section 6.6

1. (a) Because $\sin \dfrac{\pi}{2} = 1$, and $\dfrac{\pi}{2}$ lies in $\left[-\dfrac{\pi}{2}, \dfrac{\pi}{2}\right]$, it follows that $\arcsin 1 = \dfrac{\pi}{2}$.

(b) It is not possible to evaluate $y = \sin^{-1} x$ when $x = -2$ because there is no angle whose sine is -2. Remember that the domain of the inverse sine function is $[-1, 1]$.

2. Using a graphing utility you can graph the three functions with the following keystrokes

Function	Keystroke	Display
$y = \sin x$	[y =] [SIN] [(] [x] [)]	$y_1 = \sin(x)$
$y = \arcsin x$	[y =] [2ND] [SIN] [(] [x] [)]	$y_2 = \sin^{-1}(x)$
$y = x$	[y =] [x]	$y_3 = x$

Remember to check the mode to make sure the angle measure is set to radian mode. Although the graphing utility will graph the sine function for all real values of x, restrict the viewing window to values of x to be the interval $\left[-\dfrac{\pi}{2}, \dfrac{\pi}{2}\right]$.

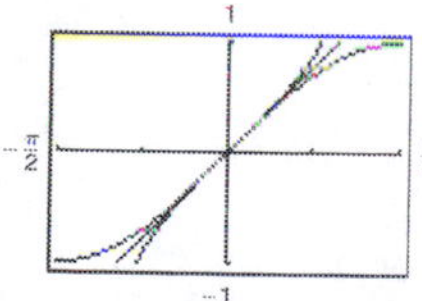

Notice that the graphs of $y_1 = \sin x$, $\left(-\dfrac{\pi}{2}, \dfrac{\pi}{2}\right)$ and $y_2 = \sin^{-1} x$ are reflections of each other in the line $y_3 = x$. So, g is the inverse of f.

3. Because $\cos \pi = -1$ and π lies in $[0, \pi]$, it follows that $\arccos(-1) = \cos^{-1}(-1) = \pi$.

4.

Function	Mode	Calculator Keystrokes
(a) arctan 4.84	Radian	TAN⁻¹ (4.84) ENTER

From the display, it follows that $\arctan 4.84 \approx 1.3670516$.

(b) arcsin (−1.1)	Radian	SIN⁻¹ ((−) 1.1) ENTER

In radian mode the calculator should display an *error* message because the domain of the inverse sine function is $[-1, 1]$.

(c) arccos (−0.349)	Radian	COS⁻¹ ((−) 0.349) ENTER

From the display, it follows that $\arccos(-0.349) \approx 1.9273001$.

5. (a) Because −14 lies in the domain of the arctangent function, the inverse property applies, and you have $\tan\left[\tan^{-1}(-14)\right] = -14$.

(b) In this case, $\frac{7\pi}{4}$ does not lie in the range of the arcsine function, $-\frac{\pi}{2} \le y \le \frac{\pi}{2}$.

However, $\frac{7\pi}{4}$ is coterminal with $\frac{7\pi}{4} - 2\pi = -\frac{\pi}{4}$ which does lie in the range of the arcsine function, and you have

$$\sin^{-1}\left(\sin \frac{7\pi}{4}\right) = \sin^{-1}\left[\sin\left(-\frac{\pi}{4}\right)\right]$$
$$= -\frac{\pi}{4}.$$

(c) Because 0.54 lies in the domain of the arccosine function, the inverse property applies and you have $\cos(\arccos 0.54) = 0.54$.

6. If you let $u = \arctan\left(-\frac{3}{4}\right)$, then $\tan u = -\frac{3}{4}$. Because the range of the inverse tangent function is the first and fourth quadrants and $\tan u$ is negative, u is a fourth-quadrant angle. You can sketch and label angle u.

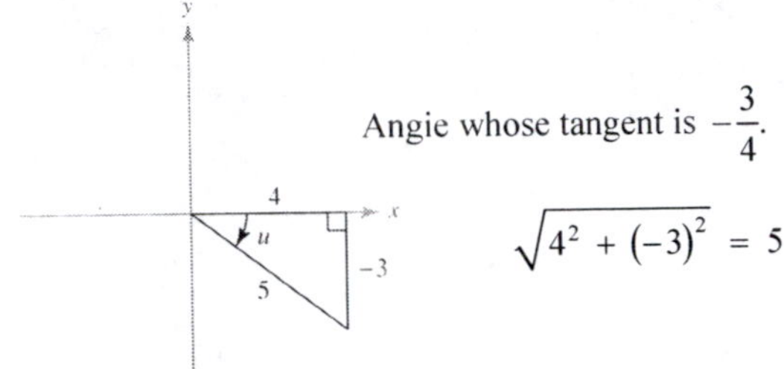

Angie whose tangent is $-\frac{3}{4}$.

$$\sqrt{4^2 + (-3)^2} = 5$$

So, $\cos\left[\arctan\left(-\frac{3}{4}\right)\right] = \cos u = \frac{4}{5}$.

7. If you let $u = \arctan x$, then $\tan u = x$, where x is any real number. Because $\tan u = \frac{\text{opp}}{\text{adj}} = \frac{x}{1}$ you can sketch a right triangle with acute angle u as shown. From this triangle, you can convert to algebraic form.

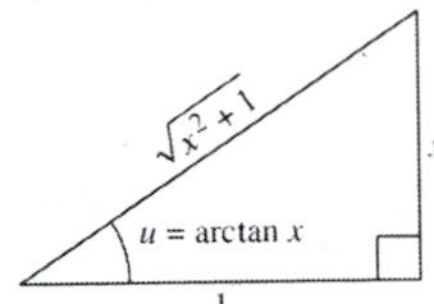

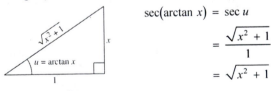

$$\sec(\arctan x) = \sec u$$
$$= \frac{\sqrt{x^2 + 1}}{1}$$
$$= \sqrt{x^2 + 1}$$

Checkpoints for Section 6.7

1. Because $c = 90°$, it follows that $A + B = 90°$ and $B° = 90° - 20° = 70°$.

To solve for a, use the fact that

$$\tan A = \frac{\text{opp}}{\text{adj}} = \frac{a}{b} \Rightarrow a = b \tan A.$$

So, $a = 15 \tan 20° \approx 5.46$. Similarly, to solve for c, use the fact that $\cos A = \frac{\text{adj}}{\text{hyp}} = \frac{b}{c} \Rightarrow c = \frac{b}{\cos A}$

So, $c = \frac{15}{\cos 20°} \approx 15.96$.

2.

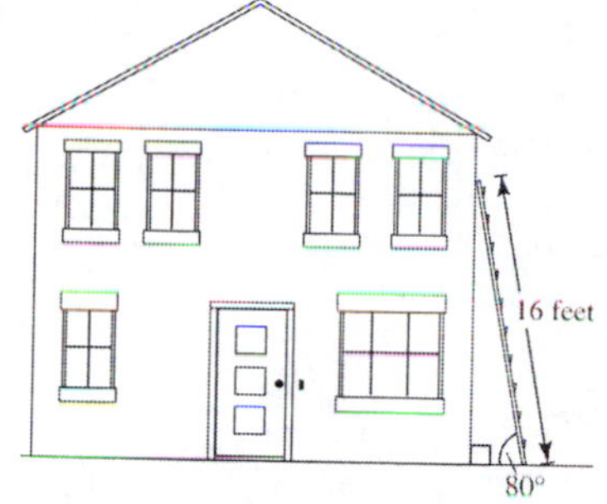

From the equation $\sin A = \frac{a}{c}$, it follows that

$$a = c \sin A$$
$$= 16 \sin 80^\circ$$
$$\approx 15.8.$$

So, the height from the top of the ladder to the ground is about 15.8 feet.

3.

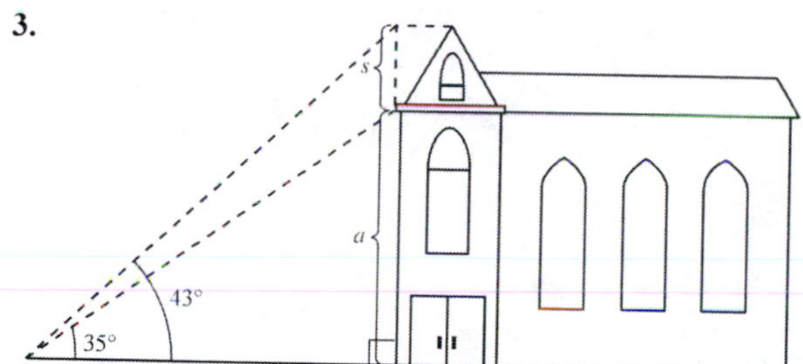

Note that this problem involves two right triangles. For the smaller right triangle, use the fact that

$\tan 35^\circ = \frac{a}{65}$ to conclude that the height of the church is $a = 65 \tan 35^\circ$.

For the larger right triangle use the equation

$\tan 43^\circ = \frac{a + s}{65}$ to conclude that $a + s = 65 \tan 43^\circ$.

So, the height of the steeple is

$$s = 65 \tan 43^\circ - a$$
$$= 65 \tan 43^\circ - (65 \tan 35^\circ)$$
$$\approx 15.1 \text{ feet.}$$

4.

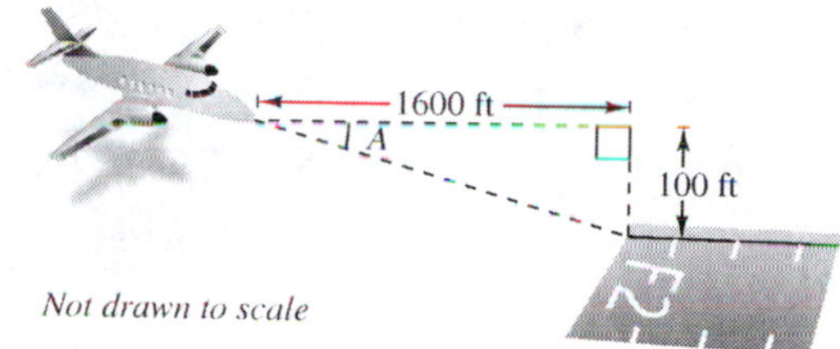

Using the tangent function, you can see that

$$\tan A = \frac{\text{opp}}{\text{adj}} = \frac{100}{1600} = 0.0625$$

So, the angle of depression is

$$A = \arctan(0.0625) \text{ radian}$$
$$\approx 0.06242 \text{ radian}$$
$$\approx 3.58^\circ.$$

5.

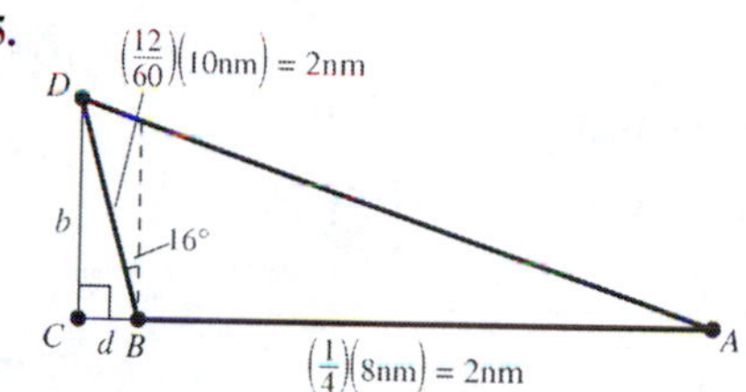

For triangle BCD, you have $B = 90^\circ - 16^\circ = 74^\circ$. The two sides of this triangle can be determined to be $b = 2 \sin 74^\circ$ and $d = 2 \cos 74^\circ$.

For triangle ACD, you can find angle A as follows.

$$\tan A = \frac{b}{d + 2} = \frac{2 \sin 74^\circ}{2 \cos 74^\circ + 2} \approx 0.7535541$$

$$A = \arctan A \approx \arctan 0.7535541 \text{ radian} \approx 37^\circ$$

The angle with the north south line is $90^\circ - 37^\circ = 53^\circ$.

So, the bearing of the ship is N 53° W.

Finally, from triangle ACD you have

$\sin A = \frac{b}{c}$, which yields

$$c = \frac{b}{\sin A} = \frac{2 \sin 74^\circ}{\sin 37^\circ} \approx 3.2 \text{ nautical miles.}$$

6. Because the spring is at equilibrium $(d = 0)$ when $t = 0$, use the equation $d = a \sin wt$.

Because the maximum displacement from zero is 6 and the period is 3, you have the following.

$$\text{Amplitude} = |a| = 6$$

$$\text{Period} = \frac{2\pi}{w} = 3 \Rightarrow w = \frac{2\pi}{3}$$

So, an equation of motion is $d = 6 \sin \frac{2\pi}{3}t$.

The frequency is

$$\text{Frequency} = \frac{w}{2\pi} = \frac{\frac{2\pi}{3}}{2\pi} = \frac{1}{3} \text{ cycle per second.}$$

7. Algebraic Solution

The given equation has the form $d = 4 \cos 6\pi t$, with $a \approx 4$ and $w = 6\pi$.

(a) The maximum displacement is given by the amplitude. So, the maximum displacement is 4.

(b) Frequency $= \dfrac{w}{2\pi} = \dfrac{6\pi}{2\pi} = 3$ cycles per unit of time

(c) $d = 4 \cos\left[6\pi(4)\right] = 4 \cos 24\pi = 4(1) = 4$

(d) To find the least positive value of t, for which $d = 0$, solve the equation $4 \cos 6\pi t = 0$.

First divide each side by 4 to obtain $\cos 6\pi t = 0$.

This equation is satisfied when $6\pi t = \dfrac{\pi}{2}, \dfrac{3\pi}{2}, \dfrac{5\pi}{2}, \ldots.$

Divide each of these values by 6π to obtain $t = \dfrac{1}{12}, \dfrac{1}{4}, \dfrac{5}{12}, \ldots.$

So, the least positive value of t is $t = \dfrac{1}{12}$.

Graphical Solution

(a) Use a graphing utility set in radian mode.

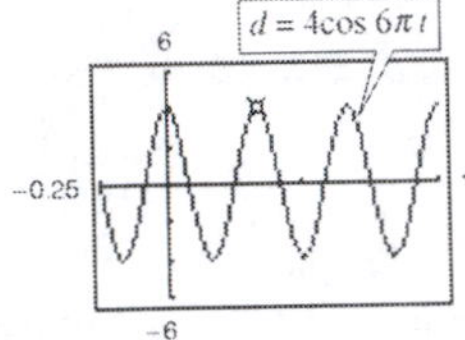

The maximum displacement is from the point of equilibrium $(d = 0)$ is 4.

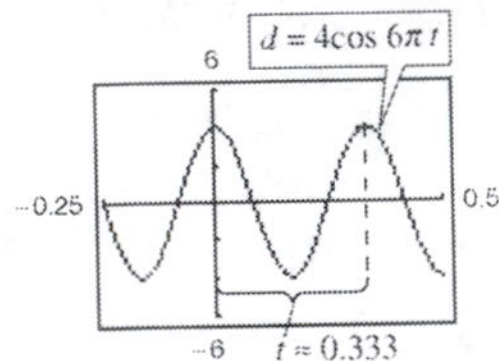

(b) The period is the time for the graph to complete one cycle, which is $t \approx 0.333$. So, the frequency is about $\dfrac{1}{0.333} \approx 3$ per unit of time.

(c)

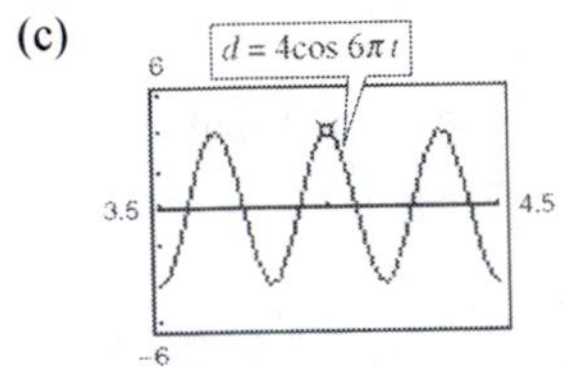

The value of d when $t = 4$ is $d = 4$

(d)

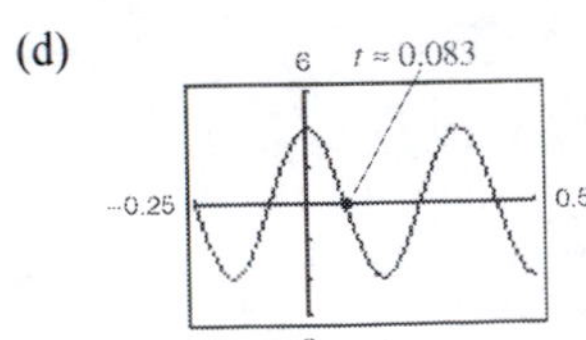

The least positive value of t for which $d = 0$ is $t \approx 0.083$.

Chapter 7

Checkpoints for Section 7.1

1. Using a reciprocal identity, you have $\cot x = \dfrac{1}{\tan x} = \dfrac{1}{\frac{1}{3}} = 3.$

Using a Pythagorean identity, you have

$$\sec^2 x = 1 + \tan^2 x = 1 + \left(\frac{1}{3}\right)^2 = 1 + \frac{1}{9} = \frac{10}{9}.$$

Because $\tan x > 0$ and $\cos x < 0$, you know that the angle x lies in Quadrant III.

Moreover, because $\sec x$ is negative when x is in Quadrant III, choose the negative root and obtain

$$\sec x = -\sqrt{\frac{10}{9}} = -\frac{\sqrt{10}}{3}.$$

Using a reciprocal identity, you have

$$\cos x = \frac{1}{\sec x} = -\frac{1}{\sqrt{10}/3} = -\frac{3}{\sqrt{10}} = -\frac{3\sqrt{10}}{10}.$$

Using a quotient identity, you have

$$\tan x = \frac{\sin x}{\cos x} \Rightarrow \sin x = \cos x \tan x = \left(-\frac{3\sqrt{10}}{10}\right)\left(\frac{1}{3}\right) = -\frac{\sqrt{10}}{10}.$$

Using a reciprocal identity, you have

$$\csc x = \frac{1}{\sin x} = -\frac{1}{\sqrt{10}/10} = -\frac{10}{\sqrt{10}} = -\sqrt{10}.$$

$\sin x = -\dfrac{\sqrt{10}}{10}$ $\qquad$ $\csc x = -\sqrt{10}$

$\cos x = -\dfrac{3\sqrt{10}}{10}$ $\qquad$ $\sec x = -\dfrac{\sqrt{10}}{3}$

$\tan x = \dfrac{1}{3}$ $\qquad$ $\cot x = 3$

2. First factor out a common monomial factor then use a fundamental identity.

$\cos^2 x \csc x - \csc x = \csc x(\cos^2 x - 1)$	Factor out a common monomial factor.
$= -\csc x(1 - \cos^2 x)$	Factor out -1.
$= -\csc x \sin^2 x$	Pythagorean identity
$= -\left(\dfrac{1}{\sin x}\right)\sin^2 x$	Reciprocal identity
$= -\sin x$	Multiply.

3. (a) This expression has the form $u^2 - v^2$, which is the difference of two squares. It factors as

$$1 - \cos^2 \theta = (1 - \cos\theta)(1 + \cos\theta).$$

(b) This expression has the polynomial form $ax^2 + bx + c$, and it factors as

$$2\csc^2\theta - 7\csc\theta + 6 = (2\csc\theta - 3)(\csc\theta - 2).$$

4. Use the identity $\sec^2 x = 1 + \tan^2 x$ to rewrite the expression.

$$\begin{aligned} \sec^2 x + 3\tan x + 1 &= \left(1 + \tan^2 x\right) + 3\tan x + 1 && \text{Pythagorean identity} \\ &= \tan^2 x + 3\tan x + 2 && \text{Combine like terms.} \\ &= (\tan x + 2)(\tan x + 1) && \text{Factor.} \end{aligned}$$

5.
$$\begin{aligned} \csc x - \cos x \cot x &= \frac{1}{\sin x} - \cos x\left(\frac{\cos x}{\sin x}\right) && \text{Quotient and reciprocal identities} \\ &= \frac{1}{\sin x} - \frac{\cos^2 x}{\sin x} && \text{Multiply.} \\ &= \frac{1 - \cos^2 x}{\sin x} && \text{Add fractions.} \\ &= \frac{\sin^2 x}{\sin x} && \text{Pythagorean identity.} \\ &= \sin x && \text{Simplify.} \end{aligned}$$

6.
$$\begin{aligned} \frac{1}{1 + \sin\theta} + \frac{1}{1 - \sin\theta} &= \frac{1 - \sin\theta + 1 + \sin\theta}{(1 + \sin\theta)(1 - \sin\theta)} && \text{Add fractions.} \\ &= \frac{2}{1 - \sin^2\theta} && \text{Combine like terms in numerator and multiply factors in denominator.} \\ &= \frac{2}{\cos^2\theta} && \text{Pythagorean identity} \\ &= 2\sec^2\theta && \text{Reciprocal identity} \end{aligned}$$

7.
$$\begin{aligned} \frac{\cos^2\theta}{1 - \sin\theta} &= \frac{1 - \sin^2\theta}{1 - \sin\theta} && \text{Pythagorean identity} \\ &= \frac{(1 + \sin\theta)(1 - \sin\theta)}{1 - \sin\theta} && \text{Factor the numerator as the difference of squares.} \\ &= 1 + \sin\theta && \text{Simplify.} \end{aligned}$$

8. Begin by letting $x = 3\sin x$, then you obtain the following

$$\begin{aligned} \sqrt{9 - x^2} &= \sqrt{9 - (3\sin\theta)^2} && \text{Substitute } 3\sin\theta \text{ for } x. \\ &= \sqrt{9 - 9\sin^2\theta} && \text{Rule of exponents.} \\ &= \sqrt{9\left(1 - \sin^2\theta\right)} && \text{Factor.} \\ &= \sqrt{9\cos^2\theta} && \text{Pythagorean identity} \\ &= 3\cos\theta && \cos\theta > 0 \text{ for } 0 < \theta = \frac{\pi}{2} \end{aligned}$$

9.
$$\begin{aligned} \ln|\sec x| + \ln|\sin x| &= \ln|\sec x \sin x| && \text{Product Property of Logarithms} \\ &= \ln\left|\frac{1}{\cos s} \cdot \sin x\right| && \text{Reciprocal identity} \\ &= \ln\left|\frac{\sin x}{\cos x}\right| && \text{Simplify.} \\ &= \ln|\tan x| && \text{Quotient identity} \end{aligned}$$

Checkpoints for Section 7.2

1. Start with the left side because it is more complicated.

$$\frac{\sin^2\theta + \cos^2\theta}{\cos^2\theta\sec^2\theta} = \frac{1}{\cos^2\theta\sec^2\theta} \qquad \text{Pythagorean identity}$$

$$= \frac{1}{\cos^2\theta\left(\dfrac{1}{\cos^2\theta}\right)} \qquad \text{Reciprocal identity}$$

$$= 1 \qquad \text{Simplify.}$$

2. Algebraic Solution:

Start with the right side because it is more complicated.

$$\frac{1}{1-\cos\beta} + \frac{1}{1+\cos\beta} = \frac{1+\cos\beta+1-\cos\beta}{(1-\cos\beta)(1+\cos\beta)} \qquad \text{Add fractions.}$$

$$= \frac{2}{1-\cos^2\beta} \qquad \text{Simplify.}$$

$$= \frac{2}{\sin^2\beta} \qquad \text{Pythagorean identity}$$

$$= 2\csc^2\beta \qquad \text{Reciprocal identity}$$

Numerical Solution:

Use a graphing utility to create a table that shows the values of

$y_1 = 2\csc^2 x$ and $y_2 = \dfrac{1}{1-\cos x} + \dfrac{1}{1+\cos x}$ for different values of x.

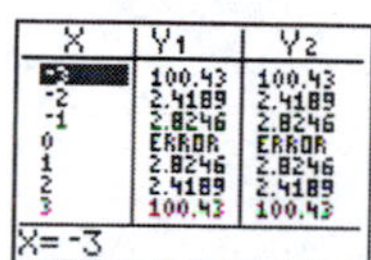

X	Y1	Y2
-3	100.43	100.43
-2	2.4189	2.4189
-1	2.8246	2.8246
0	ERROR	ERROR
1	2.8246	2.8246
2	2.4189	2.4189
3	100.43	100.43

X=-3

The values for y_1 and y_2 appear to be identical, so the equation appears to be an identity.

3. Algebraic Solution:

By applying identities before multiplying, you obtain the following.

$$(\sec^2 x - 1)(\sin^2 x - 1) = (\tan^2 x)(-\cos^2 x) \qquad \text{Pythagorean identities}$$

$$= \left(\frac{\sin x}{\cos x}\right)^2(-\cos^2 x) \qquad \text{Quotient identity}$$

$$= \left(\frac{\sin^2 x}{\cos^2 x}\right)(-\cos^2 x) \qquad \text{Property of exponents}$$

$$= -\sin^2 x \qquad \text{Multiply.}$$

Graphical Solution:

Using a graphing utility, let $y_1 = (\sec^2 x - 1)(\sin^2 x - 1)$ and $y_2 = -\sin^2 x$.

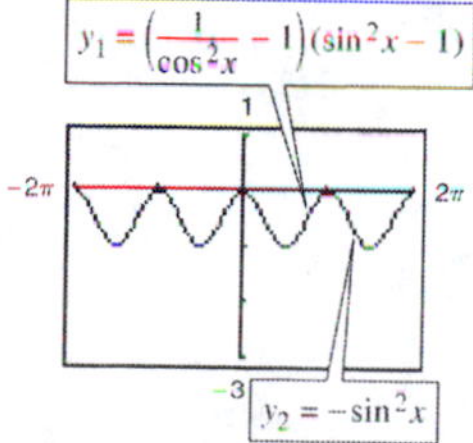

Because the graphs appear to coincide the given equation, $(\sec^2 x - 1)(\sin^2 x - 1) = -\sin^2 x$ appears to be an identity.

4. Convert the left into sines and cosines.

$$\csc x - \sin x = \frac{1}{\sin x} - \sin x$$

$$= \frac{1 - \sin^2 x}{\sin x} \quad \text{Add fractions.}$$

$$= \frac{\cos^2 x}{\sin x} \quad \text{Pythagorean identity}$$

$$= \left(\frac{\cos x}{1}\right)\left(\frac{\cos x}{\sin x}\right) \quad \text{Product of fractions}$$

$$= \cos x \cot x \quad \text{Quotient identity}$$

5. Algebraic Solution:

Begin with the right side and create a monomial denominator by multiplying the numerator and denominator by $1 + \cos x$.

$$\frac{\sin x}{1 - \cos x} = \frac{\sin x}{1 - \cos x}\left(\frac{1 + \cos x}{1 + \cos x}\right) \quad \text{Multiply numerator and denomintor by } 1 + \cos x.$$

$$= \frac{\sin x + \sin x \cos x}{1 - \cos^2 x} \quad \text{Multiply.}$$

$$= \frac{\sin x + \sin x \cos x}{\sin^2 x} \quad \text{Pythagorean identity}$$

$$= \frac{\sin x}{\sin^2 x} + \frac{\sin x \cos x}{\sin^2 x} \quad \text{Write as separate functions.}$$

$$= \frac{1}{\sin x} + \frac{\cos x}{\sin x} \quad \text{Simplify.}$$

$$= \csc x + \cot x \quad \text{Identities}$$

Graphical Solution:

Using a graphing utility, let $y_1 = \csc x + \cot x$ and $y_2 = \dfrac{\sin x}{1 - \cos x}$.

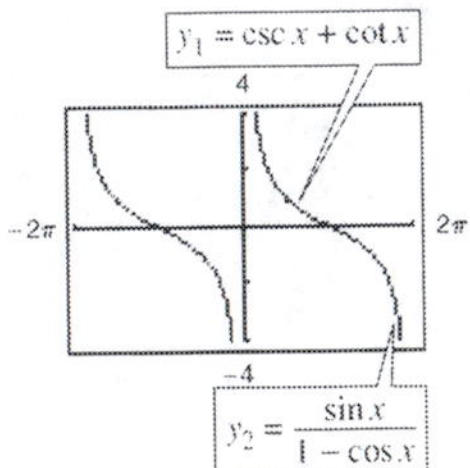

Because the graphs appear to coincide, the given equation appear to be an identity.

6. Algebraic Solution:

Working with the left side, you have the following.

$$\frac{\tan^2\theta}{1+\sec\theta} = \frac{\sec^2\theta - 1}{\sec\theta + 1} \qquad \text{Pythagorean identity}$$

$$= \frac{(\sec\theta + 1)(\sec\theta - 1)}{\sec\theta + 1} \qquad \text{Factor.}$$

$$= \sec\theta - 1 \qquad \text{Simplify.}$$

Now, working with the right side, you have the following.

$$\frac{1-\cos\theta}{\cos\theta} = \frac{1}{\cos\theta} - \frac{\cos\theta}{\cos\theta} \qquad \text{Write as separate fractions.}$$

$$= \sec\theta - 1 \qquad \text{Identity and simplify.}$$

This verifies the identity because both sides are equal to $\sec\theta - 1$.

Numerical Solution:

Use a graphing utility to create a table that shows the values of

$y_1 = \dfrac{\tan^2 x}{1+\sec x}$ and $y_2 = \dfrac{1-\cos x}{\cos x}$ for different values of x.

X	Y1	Y2
-.75	.3667	.3667
-.5	.13949	.13949
-.25	.03209	.03209
0	0	0
.25	.03209	.03209
.5	.13949	.13949
.75	.3667	.3667

X=-.75

The values of y_1 and y_2 appear to be identical, so the equation appears to be an identity.

7. (a) $\tan x\sec^2 x - \tan x = \tan x(\sec^2 x - 1)$ Factor.

$= \tan x \tan^2 x$ Pythagorean identity

$= \tan^3 x$ Multiply.

(b) $(\cos^4 x - \cos^6 x)\sin x = \cos^4 x(1 - \cos^2 x)\sin x$ Factor.

$= \cos^4 x(\sin^2 x)\sin x$ Pythagorean identity

$= \sin^3 x\cos^4 x$ Multiply.

Checkpoints for Section 7.3

1. Begin by isolating $\sin x$ on one side of the equation.

$\sin x - \sqrt{2} = -\sin x$ Write original equation.

$\sin x + \sin x - \sqrt{2} = 0$ Add $\sin x$ to each side.

$\sin x + \sin x = \sqrt{2}$ Add $\sqrt{2}$ to each side.

$2\sin x = \sqrt{2}$ Combine like terms.

$\sin x = \dfrac{\sqrt{2}}{2}$ Divide each side by 2.

Because $\sin x$ has a period of 2π, first find all solutions in the interval $[0, 2\pi)$. These solutions are $x = \dfrac{\pi}{4}$ and $x = \dfrac{3\pi}{4}$.

Finally, add multiples of 2π to each of these solutions to obtain the general form

$x = \dfrac{\pi}{4} + 2n\pi$ and $x = \dfrac{3\pi}{4} + 2n\pi$ where n is an integer.

2. Begin by isolating $\sin x$ on one side of the equation.

$4\sin^2 x - 3 = 0$ Write original equation.

$4\sin^2 x = 3$ Add 3 to each side.

$\sin^2 x = \frac{3}{4}$ Divide each side by 4.

$\sin x = \pm\sqrt{\frac{3}{4}}$ Extract square roots.

$\sin x = \pm\frac{\sqrt{3}}{2}$ Simplify.

Because $\sin x$ has a period of 2π, first find all solutions in the interval $[0, 2\pi)$. These solutions are $x = \frac{\pi}{3}$, $x = \frac{2\pi}{3}$, $x = \frac{4\pi}{3}$, and $x = \frac{5\pi}{3}$.

Finally, add multiples of 2π to each of these solutions to obtain the general form.

$x = \frac{\pi}{3} + 2n\pi$, $x = \frac{2\pi}{3} + 2n\pi$, $x = \frac{4\pi}{3} + 2n\pi$, and $x = \frac{5\pi}{3} + 2n\pi$ where n is an integer.

3. Begin by collecting all terms on one side of the equation and factoring.

$\sin^2 x = 2\sin x$ Write original equation.

$\sin^2 x - 2\sin x = 0$ Subtract $2\sin x$ from each side.

$\sin x(\sin x - 2) = 0$ Factor.

By setting each of these factors equal to zero, you obtain

$\sin x = 0$ and $\sin x - 2 = 0$

$\sin x = 2.$

In the interval $[0, 2\pi)$, the equation $\sin x = 0$ has solutions $x = 0$ and $x = \pi$. Because $\sin x$ has a period of 2π, you would obtain the general forms $x = 0 + 2n\pi$ and $x = \pi + 2n\pi$ where n is an integer by adding multiples of 2π.

No solution exists for $\sin x = 2$ because 2 is outside the range of the sine function, $[-1, 1]$. Confirm this graphically by graphing $y = \sin^2 x - 2\sin x$.

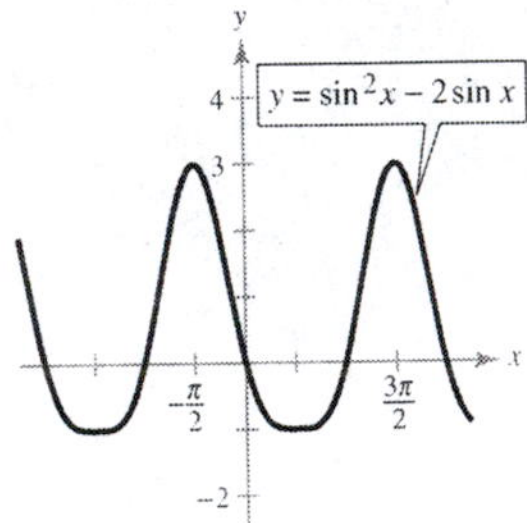

Notice that the x-intercepts occur at $-2\pi, -\pi, 0, \pi, 2\pi$ and so on. These x-intercepts correspond to the solutions of $\sin^2 x - 2\sin x = 0$.

4. Algebraic Solution:

Treat the equation as a quadratic in $\sin x$ and factor.

$2\sin^2 x - 3\sin x + 1 = 0$ Write original equation.

$(2\sin x - 1)(\sin x - 1) = 0$ Factor.

Setting each factor equal to zero, you obtain the following solutions in the interval $[0, 2\pi)$.

$2\sin x - 1 = 0$ and $\sin x - 1 = 0$

$\sin x = \frac{1}{2}$ $\sin x = 1$

$x = \frac{\pi}{6}, \frac{5\pi}{6}$ $x = \frac{\pi}{2}$

Graphical Solution:

The x-intercepts are $x \approx 0.524$, $x = 2.618$, and $x = 1.571$.

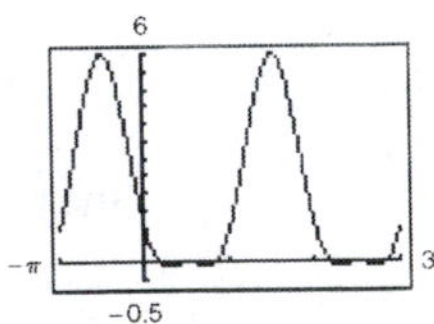

From the graph, you can conclude that the approximate solutions of $2\sin^2 x - 3\sin x + 1 = 0$ in the interval $[0, 2\pi)$ are $x \approx 0.524 = \frac{\pi}{6}$, $x \approx 2.618 = \frac{5\pi}{6}$, and $x \approx 1.571 = \frac{\pi}{2}$.

5. This equation contains both tangent and secant functions. You can rewrite the equation so that it has only tangent functions by using the identity $\sec^2 x = \tan^2 x + 1$.

$$\begin{aligned} 3\sec^2 x - 2\tan^2 x - 4 &= 0 && \text{Write original equation.} \\ 3(\tan^2 x + 1) - 2\tan^2 x - 4 &= 0 && \text{Pythagorean identity} \\ 3\tan^2 x + 3 - 2\tan^2 x - 4 &= 0 && \text{Distributive property} \\ \tan^2 x - 1 &= 0 && \text{Simplify.} \\ \tan^2 x &= 1 && \text{Add 1 to each side.} \\ \tan x &= \pm 1 && \text{Extract square roots.} \end{aligned}$$

Because $\tan x$ has a period of π, you can find the solutions in the interval $[0, \pi)$ to be $x = \dfrac{\pi}{4}$ and $x = \dfrac{3\pi}{4}$.

The general solution is $x = \dfrac{\pi}{4} + n\pi$ and $x = \dfrac{3\pi}{4} + n\pi$ where n is an integer.

6. Solution It is not clear how to rewrite this equation in terms of a single trigonometric function. Notice what happens when you square each side of the equation.

$$\begin{aligned} \sin x + 1 &= \cos x && \text{Write original equation.} \\ \sin^2 x + 2\sin x + 1 &= \cos^2 x && \text{Square each side.} \\ \sin^2 x + 2\sin x + 1 &= 1 - \sin^2 x && \text{Pythagorean identity} \\ \sin^2 x + \sin^2 x + 2\sin x + 1 - 1 &= 0 && \text{Rewrite equation.} \\ 2\sin^2 x + 2\sin x &= 0 && \text{Combine like terms.} \\ 2\sin x(\sin x + 1) &= 0 && \text{Factor.} \end{aligned}$$

Setting each factor equal to zero produces the following.

$2\sin x = 0$ and $\sin x + 1 = 0$

$\sin x = 0$ $\qquad \sin x = -1$

$x = 0, \pi$ $\qquad x = \dfrac{3\pi}{2}$

Because you squared the original equation, check for extraneous solutions.

check $x = 0$ $\quad \sin 0 + 1 \stackrel{?}{=} \cos 0$ $\quad$ Substitute 0 for x.

$0 + 1 = 1$ $\quad$ Solution checks. ✓

check $x = \pi$ $\quad \sin \pi + 1 \stackrel{?}{=} \cos \pi$ $\quad$ Substitute π for x.

$0 + 1 \neq -1$ $\quad$ Solution does not check.

check $x = \dfrac{3\pi}{2}$ $\quad \sin \dfrac{3\pi}{2} + 1 \stackrel{?}{=} \cos \dfrac{3\pi}{2}$ $\quad$ Substitute $\dfrac{3\pi}{2}$ for x.

$-1 + 1 = 0$ $\quad$ Solution checks. ✓

of the three possible solutions, $x = \pi$ is extraneous. So, in the interval $[0, 2\pi)$, the two solutions are $x = 0$ and $x = \dfrac{3\pi}{2}$.

7. $2\sin 2t - \sqrt{3} = 0$ — Write original equation.

$2\sin 2t = \sqrt{3}$ — Add $\sqrt{3}$ to each side.

$\sin 2t = \dfrac{\sqrt{3}}{2}$ — Divide each side by 2.

In the interval $[0, 2\pi)$, you know that

$2t = \dfrac{\pi}{3}$ and $2t = \dfrac{2\pi}{3}$ are the only solutions.

So, in general you have

$2t = \dfrac{\pi}{3} + 2n\pi$ and $2t = \dfrac{2\pi}{3} + 2n\pi$.

Dividing these results by 2, you obtain the general solution

$t = \dfrac{\pi}{6} + n\pi$ and $t = \dfrac{\pi}{3} + n\pi$.

8. $2\tan \dfrac{x}{2} - 2 = 0$ — Write original equation.

$2\tan \dfrac{x}{2} = 2$ — Add 2 to each side.

$\tan \dfrac{x}{2} = 1$ — Divide each side by 2.

In the interval $[0, \pi)$, you know that $\dfrac{x}{2} = \dfrac{\pi}{4}$ is the only solution. So, in general, you have

$\dfrac{x}{2} = \dfrac{\pi}{4} + n\pi$.

Multiplying this result by 2, you obtain the general solution

$x = \dfrac{\pi}{2} + 2n\pi$

Where n is an integer.

9. $4\tan^2 x + 5\tan x - 6 = 0$ — Write original equation.

$(4\tan x - 3)(\tan x + 2) = 0$ — Factor.

$4\tan x - 3 = 0$ and $\tan x + 2 = 0$ — Set each factor equal to zero.

$\tan x = \dfrac{3}{4}$ $\qquad$ $\tan x = -2$

$x = \arctan\left(\dfrac{3}{4}\right)$ $\qquad$ $x = \arctan(-2)$ — Use inverse tangent function to solve for x.

These two solutions are in the interval $\left(-\dfrac{\pi}{2}, \dfrac{\pi}{2}\right)$. Recall that the range of the inverse tangent function is $\left(-\dfrac{\pi}{2}, \dfrac{\pi}{2}\right)$.

Finally, because $\tan x$ has a period of π, you add multiples of π to obtain

$x = \arctan\left(\dfrac{3}{4}\right) + n\pi$ and $x = \arctan(-2) + n\pi$

where n is an integer.

You can use a calculator to approximate the values of $x = \arctan\left(\dfrac{3}{4}\right) \approx 0.6435$ and $x = \arctan(-2) \approx -1.1071$.

10. Graph the function $S = 10.8 + 0.84375\left[\left(\sqrt{3} - \cos\theta\right)/\sin\theta\right]$ using a graphing utility.

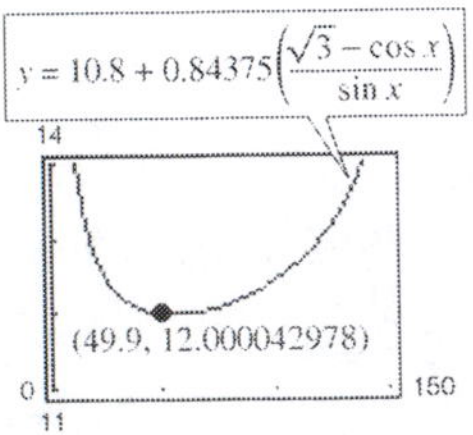

Use the *trace* feature to find the values of θ when $y = 12$. So, when $\theta \approx 49.9°$ and $\theta \approx 59.9°$, the surface area is 12 square inches. The exact values are $\theta \approx \arccos(0.644228) \approx 49.9°$ and $\theta \approx \arccos(0.50180) \approx 59.9°$.

Checkpoints for Section 7.4

1. To find the exact value of $\cos \frac{\pi}{12}$, use the fact that

$$\frac{\pi}{12} = \frac{\pi}{3} - \frac{\pi}{4}.$$

The formula for $\cos(u - v)$ yields the following.

$$\begin{aligned}\cos \frac{\pi}{12} &= \cos\left(\frac{\pi}{3} - \frac{\pi}{4}\right)\\ &= \cos \frac{\pi}{3} \cos \frac{\pi}{4} + \sin \frac{\pi}{3} \sin \frac{\pi}{4}\\ &= \left(\frac{1}{2}\right)\left(\frac{\sqrt{2}}{2}\right) + \left(\frac{\sqrt{3}}{2}\right)\left(\frac{\sqrt{2}}{2}\right)\\ &= \frac{\sqrt{2}}{4} + \frac{\sqrt{6}}{4}\\ &= \frac{\sqrt{2} + \sqrt{6}}{4}\end{aligned}$$

2. Using the fact that $75° = 30° + 45°$, together with the formula for $\sin(u + v)$, you obtain the following.

$$\begin{aligned}\sin 75° &= \sin(30° + 45°)\\ &= \sin 30° \cos 45° + \cos 30° \sin 45°\\ &= \left(\frac{1}{2}\right)\left(\frac{\sqrt{2}}{2}\right) + \left(\frac{\sqrt{3}}{2}\right)\left(\frac{\sqrt{2}}{2}\right)\\ &= \frac{\sqrt{2}}{4} + \frac{\sqrt{6}}{4}\\ &= \frac{\sqrt{2} + \sqrt{6}}{4}\end{aligned}$$

3. Because $\sin u = \frac{12}{13}$ and u is in Quadrant I,

$\cos u = \frac{5}{13}$ as shown.

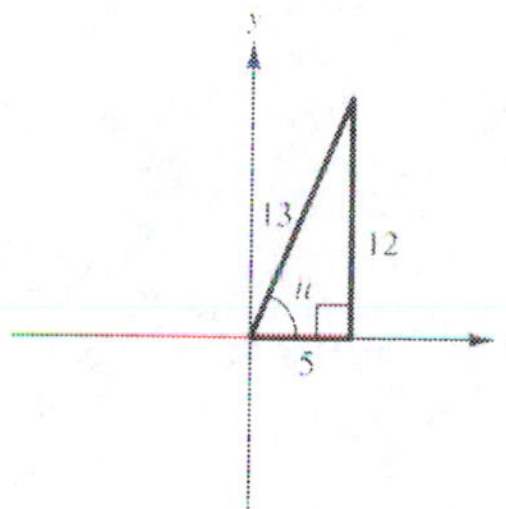

Because $\cos v = -\frac{3}{5}$ and v is in Quadrant II,

$\sin v = \frac{4}{5}$ as shown.

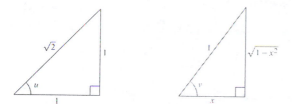

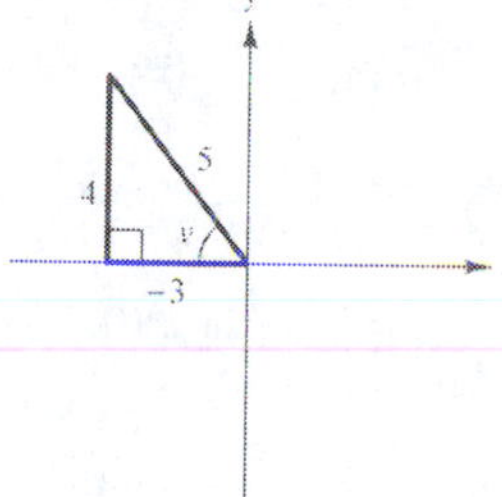

You can find $\cos(u + v)$ as follows.

$$\begin{aligned}\cos(u + v) &= \cos u \cos v - \sin u \sin v\\ &= \left(\frac{5}{13}\right)\left(-\frac{3}{5}\right) - \left(\frac{12}{13}\right)\left(\frac{4}{5}\right)\\ &= -\frac{63}{65}\end{aligned}$$

4. This expression fits the formula for $\sin(u + v)$. The figures show the angles $u = \arctan 1$ and $v = \arccos x$.

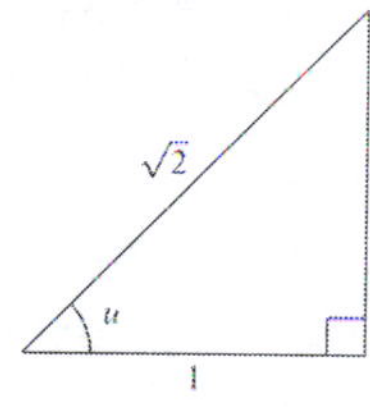

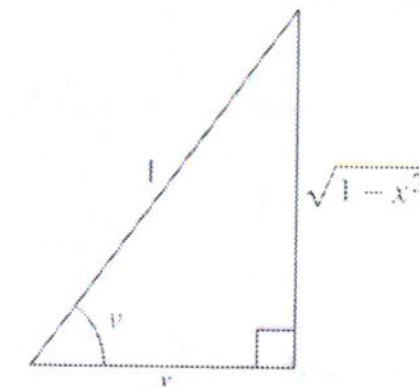

$$\begin{aligned}\sin(u + v) &= \sin u \cos v + \cos u \sin v\\ &= \sin(\arctan 1) \cos(\arccos x) + \cos(\arctan 1) \sin(\arccos x)\\ &= \left(\frac{1}{\sqrt{2}}\right)(x) + \left(\frac{1}{\sqrt{2}}\right)\left(\sqrt{1 - x^2}\right)\\ &= \frac{x}{\sqrt{2}} + \frac{\sqrt{1 - x^2}}{\sqrt{2}}\\ &= \frac{x + \sqrt{1 - x^2}}{\sqrt{2}}\end{aligned}$$

5. Using the formula for $\sin(u - v)$, you have

$$\begin{aligned}\sin\left(x - \frac{\pi}{2}\right) &= \sin x \cos\frac{\pi}{2} - \cos x \sin\frac{\pi}{2}\\ &= (\sin x)(0) - (\cos x)(1)\\ &= -\cos x.\end{aligned}$$

6. (a) Using the formula for $\sin(u - v)$, you have

$$\begin{aligned}\sin\left(3\frac{\pi}{2} - \theta\right) &= \sin 3\frac{\pi}{2}\cos\theta - \cos 3\frac{\pi}{2}\sin\theta\\ &= (-1)(\cos\theta) - (0)(\sin\theta)\\ &= -\cos\theta.\end{aligned}$$

(b) Using the formula for $\tan(u - v)$, you have

$$\begin{aligned}\tan\left(\theta - \frac{\pi}{4}\right) &= \frac{\tan\theta - \tan\frac{\pi}{4}}{1 + \tan\theta\tan\frac{\pi}{4}}\\ &= \frac{\tan\theta - 1}{1 + (\tan\theta)(1)}\\ &= \frac{\tan\theta - 1}{1 + \tan\theta}.\end{aligned}$$

7. Algebraic Solution

Using sum and difference formulas, rewrite the equation.

$$\begin{aligned}\sin\left(x + \frac{\pi}{2}\right) + \sin\left(x - \frac{3\pi}{2}\right) &= 1\\ \sin x\cos\frac{\pi}{2} + \cos x\sin\frac{\pi}{2} + \sin x\cos\frac{3\pi}{2} - \cos x\sin\frac{3\pi}{2} &= 1\\ (\sin x)(0) + (\cos x)(1) + (\sin x)(0) - (\cos x)(1) &= 1\\ \cos x + \cos x &= 1\\ 2\cos x &= 1\\ \cos x &= \frac{1}{2}\end{aligned}$$

So, the only solutions in the interval $[0, 2\pi)$ are $x = \frac{\pi}{3}$ and $x = \frac{5\pi}{3}$.

Graphical Solution

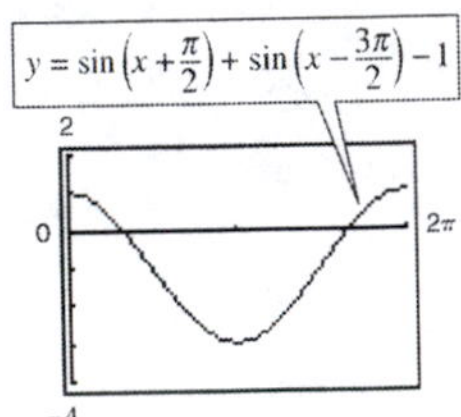

The x-intercepts are $x \approx 1.047198$ and $x \approx 5.235988$.

From the above figure, you can conclude that the approximate solutions in the interval $[0, 2\pi)$ are

$x \approx 1.047198 = \frac{\pi}{3}$ and $x \approx 5.235988 = \frac{5\pi}{3}$.

8. Using the formula for $\cos(x + h)$, you have the following.

$$\frac{\cos(x+h) - \cos x}{h} = \frac{\cos x \cos h - \sin x \sin h - \cos x}{h}$$
$$= \frac{\cos x \cos h - \cos x - \sin x \sin h}{h}$$
$$= \frac{\cos x(\cos h - 1) - \sin x \sin h}{h}$$
$$= \cos x\left(\frac{\cos h - 1}{h}\right) - \sin x\left(\frac{\sin h}{h}\right)$$

Checkpoints for Section 7.5

1. Begin by rewriting the equation so that it involves functions of x (rather than $2x$). Then factor and solve.

$\cos 2x + \cos x = 0$ — Write original equation.

$2\cos^2 x - 1 + \cos x = 0$ — Double-angle formula

$2\cos^2 x + \cos x - 1 = 0$ — Rearrange terms

$(2\cos x - 1)(\cos x + 1) = 0$ — Factor.

$2\cos x - 1 = 0$ $\quad \cos x + 1 = 0$ — Set factors equal to zero.

$\cos x = \frac{1}{2}$ $\quad \cos x = -1$ — Solve by $\cos x$.

$x = \frac{\pi}{3}, \frac{5\pi}{3}$ $\quad x = \pi$ — Solutions in $[0, 2\pi)$.

So, the general solution is $x = \frac{\pi}{3} + 2n\pi$, $x = \frac{5\pi}{3} + 2n\pi$, and $x = \pi + 2n\pi$ where n is an integer.

2. Begin by drawing the angle θ, $0 < \theta < \frac{\pi}{2}$ given $\sin\theta = \frac{3}{5}$.

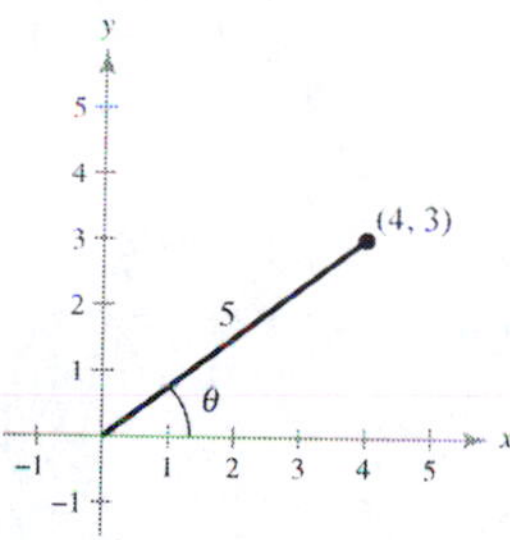

From the sketch, you know that $\sin\theta = \frac{y}{r} = \frac{3}{5}$.

Because $x = 4$, you know $\sin\theta = \frac{3}{5}$, $\cos\theta = \frac{4}{5}$, and $\tan\theta = \frac{3}{4}$.

Using the double angle formulas, you have the following.

$$\sin 2\theta = 2\sin\theta\cos\theta = 2\left(\frac{3}{5}\right)\left(\frac{4}{5}\right) = \frac{24}{25}$$

$$\cos 2\theta = \cos^2\theta - \sin^2\theta = \left(\frac{4}{5}\right)^2 - \left(\frac{3}{5}\right)^2 = \frac{7}{25}$$

$$\tan 2\theta = \frac{2\tan\theta}{1 - \tan^2\theta} = \frac{2\left(\frac{3}{4}\right)}{1 - \left(\frac{3}{4}\right)^2} = \frac{\frac{3}{2}}{\frac{7}{16}} = \frac{24}{7}$$

3.

$\cos 3x = \cos(2x + x)$	Rewrite $3x$ as sum of $2x$ and x.
$= \cos 2x \cos x - \sin 2x \sin x$	Sum formula
$= (2\cos^2 x - 1)(\cos x) - (2\sin x \cos x)(\sin x)$	Double-angle formulas
$= 2\cos^3 x - \cos x - 2\sin^2 x \cos x$	Distribute property and simplify.
$= 2\cos^3 x - \cos x - 2(1 - \cos^2 x)(\cos x)$	Pythagorean identity
$= 2\cos^3 x - \cos x - 2\cos x + 2\cos^3 x$	Distribute property
$= 4\cos^3 x - 3\cos x$	Simplify.

4. You can make repeated use of power-reducing formulas.

$\tan^4 x = (\tan^2 x)^2$	Property of exponets.
$= \left(\dfrac{1 - \cos 2x}{1 + \cos 2x}\right)^2$	Power-reducing formula
$= \dfrac{1 - 2\cos 2x + \cos^2 2x}{1 + 2\cos 2x + \cos^2 2x}$	Expand.
$= \dfrac{1 - 2\cos 2x + \left(\dfrac{1 + \cos 4x}{2}\right)}{1 + 2\cos 2x + \left(\dfrac{1 + \cos 4x}{2}\right)}$	Power-reducing formula
$= \dfrac{\dfrac{2 - 4\cos 2x + 1 + \cos 4x}{2}}{\dfrac{2 + 4\cos 2x + 1 + \cos 4x}{2}}$	Simplify.
$= \dfrac{3 - 4\cos 2x + \cos 4x}{3 + 4\cos 2x + \cos 4x}$	Collect like terms, invert, and multiply.

You can use a graphing utility to check this result. Notice that the graphs coincide.

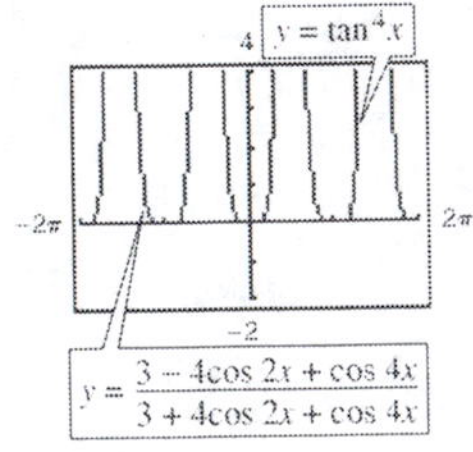

5. Begin by noting 105° is one half of 210°. Then using the half-angle formula for $\cos\left(\dfrac{u}{2}\right)$ and the fact that 105° lies in Quadrant II, you have the following.

$$\cos 105^\circ = -\sqrt{\frac{1 + \cos 210^\circ}{2}} = -\sqrt{\frac{1 + \left(-\frac{\sqrt{3}}{2}\right)}{2}} = -\sqrt{\frac{\frac{2 - \sqrt{3}}{2}}{2}} = -\sqrt{\frac{2 - \sqrt{3}}{4}} = -\frac{\sqrt{2 - \sqrt{3}}}{2}$$

The negative square root is chosen because $\cos\theta$ is negative in Quadrant II.

6. Algebraic Solution

$$\cos^2 x = \sin^2 \frac{x}{2} \qquad \text{Write original equation.}$$

$$\cos^2 x = \left(\pm\sqrt{\frac{1 - \cos x}{2}}\right)^2 \qquad \text{Half-angle formula}$$

$$\cos^2 x = \frac{1 - \cos x}{2} \qquad \text{Simplify.}$$

$$2\cos^2 x = 1 - \cos x \qquad \text{Multiply each side by 2.}$$

$$2\cos^2 x + \cos x - 1 = 0 \qquad \text{Simplify.}$$

$$(2\cos x - 1)(\cos x + 1) = 0 \qquad \text{Factor.}$$

$$2\cos x - 1 = 0 \quad \cos x + 1 = 0 \qquad \text{Set each factor equal to zero.}$$

$$\cos x = \frac{1}{2} \quad \cos x = -1 \qquad \text{Solve each equation for } \cos x.$$

$$x = \frac{\pi}{3}, \frac{5\pi}{3} \quad x = \pi \qquad \text{Solutions in } [0, 2\pi).$$

The solutions in the interval $[0, 2\pi)$ are $x = \frac{\pi}{3}$, $x = \pi$, and $x = \frac{5\pi}{3}$.

Graphical Solution

Use a graphing utility to graph $y = \cos^2 x - \sin^2 \frac{x}{2}$ in the interval $[0, 2\pi)$. Determine the approximate value of the x-intercepts.

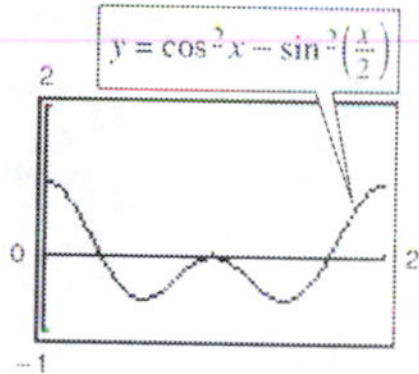

The x-intercepts are $x \approx 1.04720$, $x \approx 3.14159$, and $x \approx 5.23599$.

From the graph, you can conclude that the approximate solutions of $\cos^2 x = \sin \frac{2x}{2}$ in the interval $[0, 2\pi)$ are

$x \approx 1.04720 = \frac{\pi}{3}$, $x \approx 3.14159 = \pi$, and $x \approx 5.23599 = \frac{5\pi}{3}$.

7. Using the appropriate product-to-sum formula

$\sin u \cos v = \frac{1}{2}[\sin(u + v) + \sin(u - v)]$, you obtain the following.

$$\begin{aligned} \sin 5x \cos 3x &= \frac{1}{2}[\sin(5x + 3x) + \sin(5x - 3x)] \\ &= \frac{1}{2}(\sin 8x + \sin 2x) \\ &= \frac{1}{2}\sin 8x + \frac{1}{2}\sin 2x \end{aligned}$$

8. Using the appropriate sum-to-product formula,

$\sin u + \sin v = 2\sin\left(\frac{u+v}{2}\right)\cos\left(\frac{u-v}{2}\right)$, you obtain the following.

$$\begin{aligned}\sin 195° + \sin 105° &= 2\sin\left(\frac{195° + 105°}{2}\right)\cos\left(\frac{195° - 105°}{2}\right)\\ &= 2\sin 150° \cos 45°\\ &= 2\left(\frac{1}{2}\right)\left(\frac{\sqrt{2}}{2}\right)\\ &= \frac{\sqrt{2}}{2}\end{aligned}$$

9.

$\sin 4x - \sin 2x = 0$ Write orignal equation.

$2\cos\left(\frac{4x+2x}{2}\right)\sin\left(\frac{4x-2x}{2}\right) = 0$ Sum-to-product formula

$2\cos 3x \sin x = 0$ Simplify.

$\cos 3x \sin x = 0$ Divide each side by 2.

$\cos 3x = 0$ $\sin x = 0$ Set each factor equal to zero.

The solutions in the interval $[0, 2\pi)$ are $3x = \frac{\pi}{2}, \frac{3\pi}{2}$ and $x = 0, \pi$.

The general solutions for the equation $\cos 3x = 0$ are $3x = \frac{\pi}{2} + 2n\pi$ and $3x = \frac{3\pi}{2} + 2n\pi$.

So, by solving these equations for x, you have $x = \frac{\pi}{6} + \frac{2n\pi}{3}$ and $x = \frac{\pi}{2} + \frac{2n\pi}{3}$.

The general solution for the equation $\sin x = 0$ is $x = 0 + 2n\pi$ and $x = \pi + 2n\pi$.

These can be combined as $x = n\pi$.

So, the general solutions to the equation, $\sin 4x - \sin 2x = 0$ are

$x = \frac{\pi}{6} + \frac{2n\pi}{3}$, $x = \frac{\pi}{2} + \frac{2n\pi}{3}$, and $x = n\pi$ where n is an integer.

To verify these solutions you can graph $y = \sin 4x - \sin 2x$ and approximate the x-intercepts.

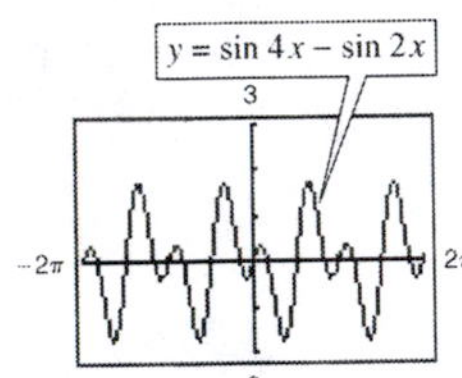

The x-intercepts occur at $0, \frac{\pi}{6}, \frac{\pi}{2}, \frac{5\pi}{6}, \pi, \frac{7\pi}{6}, \ldots$

10. Given that a football player can kick a football from ground level with an initial velocity of 80 feet per second, you have the following

$r = \frac{1}{32}v_0^2 \sin 2\theta$ Write projectile motion model.

$r = \frac{1}{32}(80)^2 \sin 2\theta$ Substitute 80 for v_0.

$r = 200 \sin 2\theta$ Simplify.

Use a graphing utility to graph the model, $r = 200 \sin 2\theta$.

The maximum point on the graph over the interval $(0°, 90°)$ occurs at $\theta = 45°$.

So, the player must kick the football at an angle of 45° to yield the maximum horizontal distance of 200 feet.

Chapter 8

Checkpoints for Section 8.1

1. The third angle of the triangle is

$C = 180° - A - B = 180° - 30° - 45° = 105°.$

By the Law of Sines, you have

$$\frac{a}{\sin A} = \frac{b}{\sin B} = \frac{c}{\sin C}.$$

Using $a = 32$ produces

$$b = \frac{a}{\sin A}(\sin B) = \frac{32}{\sin 30°}(\sin 45°) \approx 45.3 \text{ units}$$

and

$$c = \frac{a}{\sin A}(\sin C) = \frac{32}{\sin 30°}(\sin 105°) \approx 61.8 \text{ units.}$$

2. From the figure, note that $A = 22°50'$ and $C = 96°$.

So, the third angle is

$B = 180° - A - C = 180° - 22°50' - 96° = 61°10'.$

By the Law of Sines, you have

$$\frac{h}{\sin A} = \frac{b}{\sin B}$$

$$h = \frac{b}{\sin B}(\sin A) = \frac{30}{\sin 61°10'}(\sin 22°50')$$

$$\approx 13.29.$$

So, the height of the tree h is approximately 13.29 meters.

3. Sketch and label the triangle as shown.

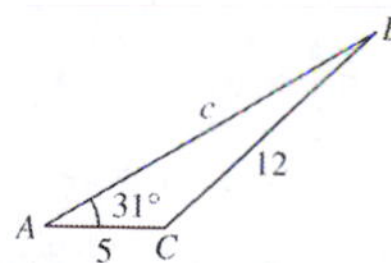

By the Law of Sines, you have

$\dfrac{\sin B}{b} = \dfrac{\sin A}{a}$ Reciprocal form

$\sin B = b\left(\dfrac{\sin A}{a}\right)$ Multiply each side by b.

$\sin B = 5\left(\dfrac{\sin 31°}{12}\right)$ Substitute for A, a, and b.

$B \approx 12.39°$

Now, you can determine that

$C = 180° - A - B \approx 180° - 31° - 12.39° \approx 136.61°.$

Then, the remaining side is

$$\frac{c}{\sin C} = \frac{a}{\sin A}$$

$$c = \frac{a}{\sin A}(\sin C) \approx \frac{12}{\sin 31°}(\sin 136.61°)$$

$$\approx 16.01 \text{ units.}$$

4. Begin by making a sketch as shown.

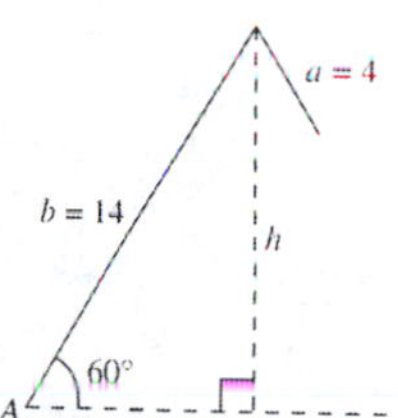

It appears that no triangle is formed.

You can verify this using the Law of Sines.

$$\frac{\sin B}{b} = \frac{\sin A}{a}$$

$$\sin B = b\left(\frac{\sin A}{a}\right)$$

$$\sin B = 14\left(\frac{\sin 60°}{4}\right) \approx 3.0311 > 1$$

This contradicts the fact that $|\sin B| \le 1$.

So, no triangle can be formed having sides $a = 4$ and $b = 14$ and angle $A = 60°$.

5. By the Law of Sines, you have

$$\frac{\sin B}{b} = \frac{\sin A}{a}$$

$$\sin B = b\left(\frac{\sin A}{a}\right) = 5\left(\frac{\sin 58°}{4.5}\right) \approx 0.9423.$$

There are two angles $B_1 \approx 70.4°$ and $B_2 \approx 180° - 70.4° = 109.6°$ between $0°$ and $180°$ whose sine is approximately 0.9423.

For $B_1 \approx 70.4°$, you obtain the following.

$$C = 180° - A - B_1 = 180° - 58° - 70.4° = 51.6°$$

$$c = \frac{a}{\sin A}(\sin C) = \frac{4.5}{\sin 58°}(\sin 51.6°) \approx 4.16 \text{ feet}$$

For $B_2 = 109.6°$, you obtain the following.

$$C = 180° - A - B_2 = 180° - 58° - 109.6° = 12.4°$$

$$c = \frac{a}{\sin A}(\sin C) = \frac{4.5}{\sin 58°}(\sin 12.4°) \approx 1.14 \text{ feet}$$

The resulting triangles are shown.

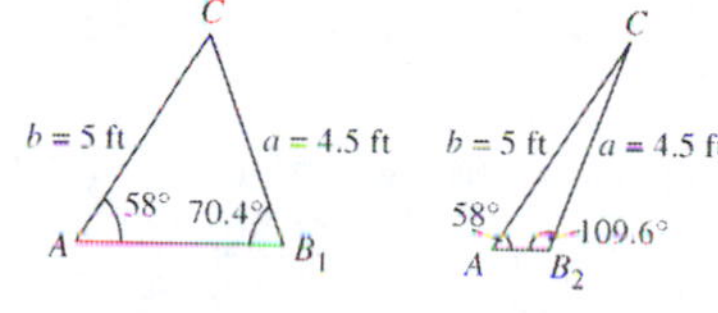

6. Consider $a = 24$ inches, $b = 18$ inches, and angle $C = 80°$ as shown. Then, the area of the triangle is

$$A = \frac{1}{2}ab \sin C = \frac{1}{2}(24)(18)\sin 80° \approx 212.7 \text{ square inches.}$$

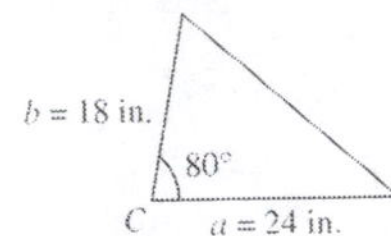

7. Because lines AC and BD are parallel, it follows that $\angle ACB \cong \angle CBD$.

So, triangle ABC has the following measures as shown.

The measure of angle B is $180° - A - C = 180° - 28° - 58° = 94°$.

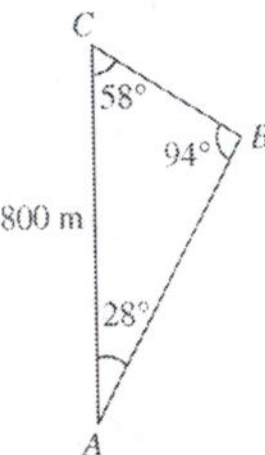

Using the Law of Sines, $\dfrac{a}{\sin 28°} = \dfrac{b}{\sin 94°} = \dfrac{c}{\sin 58°}$.

Because $b = 800$, $C = \dfrac{800}{\sin 94°}(\sin 58°) \approx 680.1$ meters and $a = \dfrac{800}{\sin 94°}(\sin 28°) \approx 376.5$ meters

The total distance that you swim is approximately

Distance $= 680.1 + 376.5 + 800 = 1856.6$ meters.

Checkpoints for Section 8.2

1. 

First, find the angle opposite the longest side – side c in this case. Using the alternative form of the Law of Cosines, you find that

$$\cos C = \frac{a^2 + b^2 - c^2}{2ab} = \frac{6^2 + 8^2 - 12^2}{2(6)(8)} \approx -0.4583.$$

Because $\cos C$ is negative, C is an *obtuse* angle given by

$C \approx \cos^{-1}(-0.4583) \approx 117.28°$.

At this point, it is simpler to use the Law of Sines to determine angle B.

$$\sin B = b\left(\frac{\sin C}{c}\right)$$

$$\sin B = 8\left(\frac{\sin 117.28°}{12}\right) \approx 0.5925$$

Because C is obtuse and a triangle can have at most one obtuse angle, you know that B must be acute.

So, $B \approx \sin^{-1}(0.5925) \approx 36.34°$

So, $A = 180° - B - C \approx 180° - 36.34° - 117.28° \approx 26.38°$.

2.

C
$b = 16$
$80°$
A $c = 12$ B

Use the Law of Cosines to find the unknown side a in the figure.

$$a^2 = b^2 + c^2 - 2bc\cos A$$
$$a^2 = 16^2 + 12^2 - 2(16)(12)\cos 80°$$
$$a^2 \approx 333.3191$$
$$a \approx 18.2570$$

Use the Law of Sines to find angle B.

$$\frac{\sin B}{b} = \frac{\sin A}{a}$$
$$\sin B = b\left(\frac{\sin A}{a}\right)$$
$$\sin B = 16\left(\frac{\sin 80°}{18.2570}\right)$$
$$\sin B \approx 0.8631$$

There are two angles between 0° and 180° whose sine is 0.8631. The two angles are $B_1 \approx 59.67°$ and $B_2 \approx 180° - 59.67° \approx 120.33°$.

Because side a is the longest side of the triangle, angle A must be the largest angle, therefore B must be less than 80°. So, $B \approx 59.67°$.

Therefore, $C = 180° - A - B \approx 180° - 80° - 59.67° \approx 40.33°$.

3.

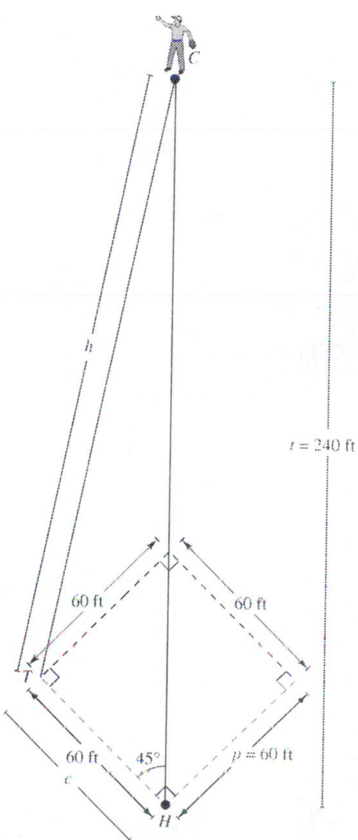

In triangle HCT, $H = 45°$ (line HC bisects the right angle at H), $t = 240$, and $c = 60$.

Using the Law of Cosines for this SAS case, you have

$$h^2 = c^2 + t^2 - 2ct\cos H$$
$$h^2 = 60^2 + 240^2 - 2(60)(240)\cos 45°$$
$$h^2 \approx 40835.3$$
$$h \approx 202.1$$

So, the center fielder is approximately 202.1 feet from the third base.

4. You have $a = 30, b = 56,$ and $c = 40.$

So, using the alternative form of the Law of Cosines, you have

$$\cos B = \frac{a^2 + c^2 - b^2}{2ac} = \frac{30^2 + 40^2 - 56^2}{2(30)(40)} = -0.265.$$

So, $B = \cos^{-1}(-0.265) \approx 105.37°$, and thus the bearing from due north from point B to point C is $105.37° - 90° = 15.37°$, or N 15.37° E.

5. Because $s = \frac{a + b + c}{2} = \frac{5 + 9 + 8}{2} = \frac{22}{2} = 11,$

Heron's Area Formula yields

$$\begin{aligned} \text{Area} &= \sqrt{s(s-a)(s-b)(s-c)} \\ &= \sqrt{11(11-5)(11-9)(11-8)} \\ &= \sqrt{(11)(6)(2)(3)} \\ &= \sqrt{396} \\ &\approx 19.90 \text{ square units.} \end{aligned}$$

Checkpoints for Section 8.3

1. From the Distance Formula, it follows that $\overline{PQ}$ and $\overline{RS}$ have the *same magnitude*.

$$\|\overline{PQ}\| = \sqrt{(3-0)^2 + (1-0)^2} = \sqrt{10}$$
$$\|\overline{RS}\| = \sqrt{(5-2)^2 + (3-2)^2} = \sqrt{10}$$

Moreover, both line segments have the *same direction* because they are both directed toward the upper right on lines having a slope of

$$\frac{1-0}{3-0} = \frac{3-2}{5-2} = \frac{1}{3}.$$

Because, $\overline{PQ}$ and $\overline{RS}$ have the same magnitude and direction, **u** and **v** are equivalent.

2. Algebraic Solution

Let $P(-2, 3) = (p_1, p_2)$ and $Q(-7, 9) = (q_1, q_2)$.

Then, the components of $\mathbf{v} = \langle v_1, v_2 \rangle$ are

$$v_1 = q_1 - p_1 = -7 - (-2) = -5$$
$$v_2 = q_2 - p_2 = 9 - 3 = 6.$$

So, $\mathbf{v} = \langle -5, 6 \rangle$ and the magnitude of **v** is $\|\mathbf{v}\| = \sqrt{(-5)^2 + (6)^2} = \sqrt{61}$

Graphical Solution

Use centimeter graph paper to plot the points $P(-2, 3)$ and $Q(-7, 9)$. Carefully sketch the vector **v**.

Use the sketch to find the components of $\mathbf{v} = \langle v_1, v_2 \rangle$. Then use a centimeter ruler to find the magnitude of **v**.

The figure shows that the components of **v** are $v_1 = -5$ and $v_2 = 6$, so $\mathbf{v} = \langle -5, 6 \rangle$. The figure also shows that the magnitude of **v** is $\|\mathbf{v}\| = \sqrt{61}$.

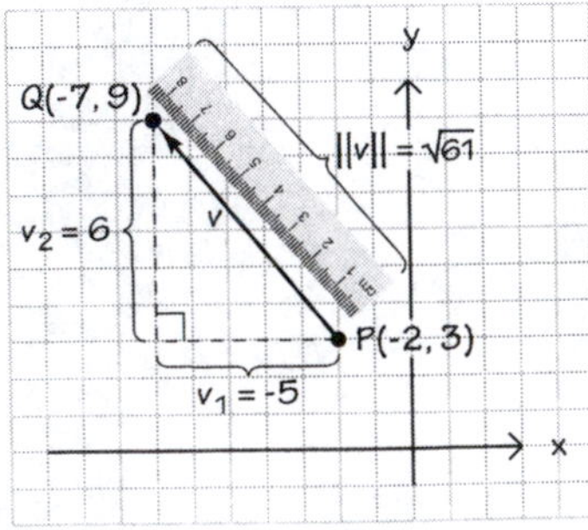

3. (a) The sum of **u** and **v** is

$$\begin{aligned}\mathbf{u} + \mathbf{v} &= \langle 1, 4\rangle + \langle 3, 2\rangle \\ &= \langle 1 + 3, 4 + 2\rangle \\ &= \langle 4, 6\rangle.\end{aligned}$$

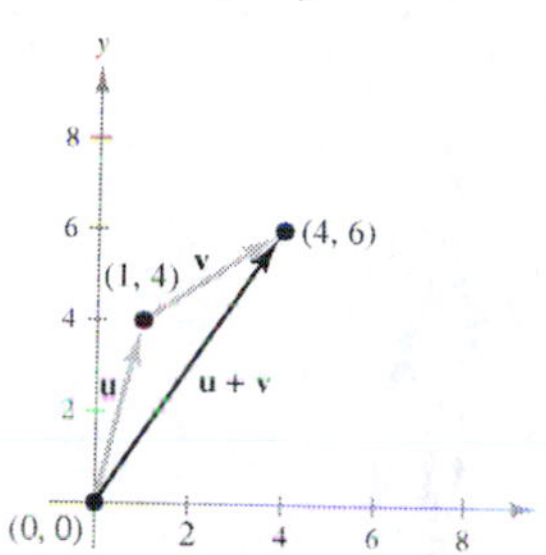

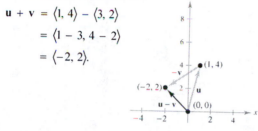

(b) The difference of **u** and **v** is

$$\begin{aligned}\mathbf{u} + \mathbf{v} &= \langle 1, 4\rangle - \langle 3, 2\rangle \\ &= \langle 1 - 3, 4 - 2\rangle \\ &= \langle -2, 2\rangle.\end{aligned}$$

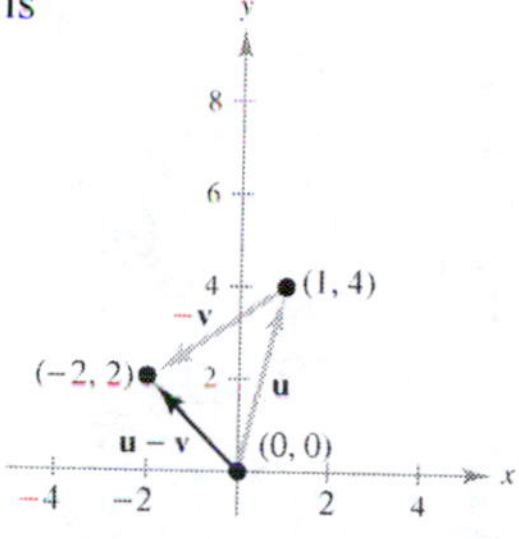

(c) The difference of $2\mathbf{u}$ and $3\mathbf{v}$ is

$$\begin{aligned}2\mathbf{u} - 3\mathbf{v} &= 2\langle 1, 4\rangle - 3\langle 3, 2\rangle \\ &= \langle 2, 8\rangle - \langle 9, 6\rangle \\ &= \langle 2 - 9, 8 - 6\rangle \\ &= \langle -7, 2\rangle.\end{aligned}$$

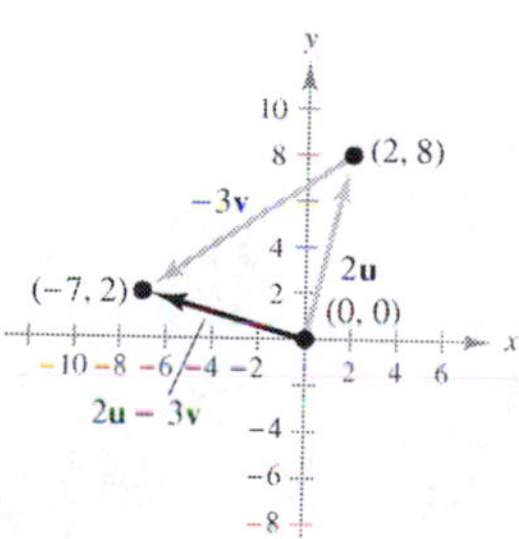

4. The unit vector in the direction of **v** is

$$\begin{aligned}\frac{\mathbf{v}}{\|\mathbf{v}\|} &= \frac{\langle 6, -1\rangle}{\sqrt{(6)^2 + (-1)^2}} \\ &= \frac{1}{\sqrt{37}}\langle 6, -1\rangle \\ &= \left\langle \frac{6}{\sqrt{37}}, -\frac{1}{\sqrt{37}}\right\rangle.\end{aligned}$$

This vector has a magnitude of 1 because

$$\sqrt{\left(\frac{6}{\sqrt{37}}\right)^2 + \left(-\frac{1}{\sqrt{37}}\right)^2} = \sqrt{\frac{36}{37} + \frac{1}{37}} = \sqrt{\frac{37}{37}} = 1.$$

5. Begin by writing the component form of vector **u**.

$$\begin{aligned}\mathbf{u} &= \langle -8 - (-2), 3 - 6\rangle \\ &= \langle -6, -3\rangle \\ &= -6\mathbf{i} - 3\mathbf{j}\end{aligned}$$

The result is shown graphically.

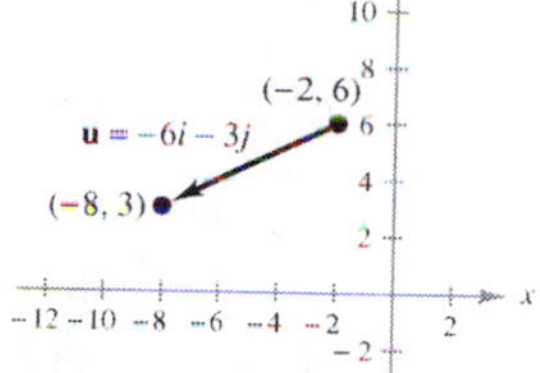

6. Perform the operations in unit vector form.

$$\begin{aligned}5\mathbf{u} - 2\mathbf{v} &= 5(\mathbf{i} - 2\mathbf{j}) - 2(-3\mathbf{i} + 2\mathbf{j}) \\ &= 5\mathbf{i} - 10\mathbf{j} + 6\mathbf{i} - 4\mathbf{j} \\ &= 11\mathbf{i} - 14\mathbf{j}\end{aligned}$$

7. (a) The direction angle is determined from

$$\tan\theta = \frac{b}{a} = \frac{6}{-6} = -1.$$

Because $\mathbf{v} = -6\mathbf{i} + 6\mathbf{j}$ lies in Quadrant II, θ lies in Quadrant II and its reference angle is

$$\theta' = \left|\arctan(-1)\right| = \left|-\frac{\pi}{4}\right| = 45°.$$

So, it follows that the direction angle is $\theta = 180° - 45° = 135°$.

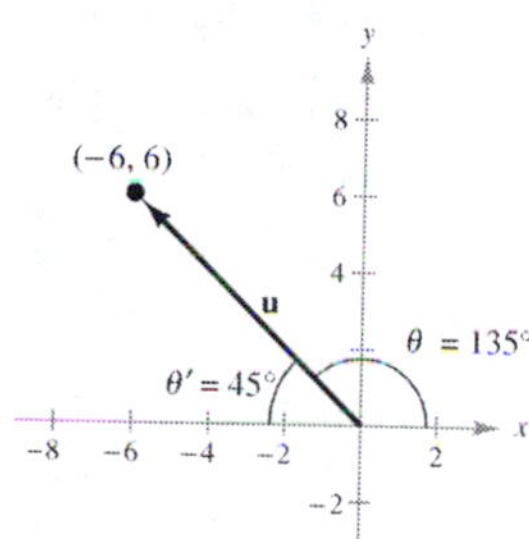

(b) The direction angle is determined from

$$\tan\theta = \frac{b}{a} = \frac{-4}{-7} = \frac{4}{7}.$$

Because $\mathbf{v} = -7\mathbf{i} - 4\mathbf{j}$ lies in Quadrant III, θ lies in Quadrant III and its reference angle is

$$\theta' = \left|\arctan\left(\frac{4}{7}\right)\right| \approx \left|0.51915 \text{ radian}\right| \approx 29.74°.$$

So, it follows that the direction angle is $\theta = 180° + 29.74° = 209.74°$.

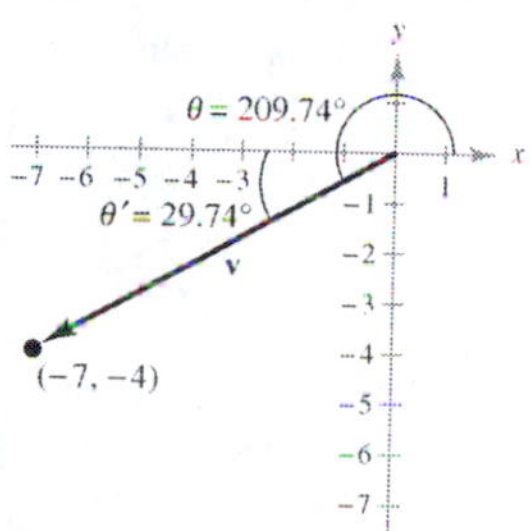

8. The velocity vector **v** has a magnitude of 100 and a direction angle of $\theta = 225°$.

$$\begin{aligned}\mathbf{v} &= \|\mathbf{v}\|(\cos\theta)\mathbf{i} + \|\mathbf{v}\|(\sin\theta)\mathbf{j}\\ &= 100(\cos 225°)\mathbf{i} + 100(\sin 225°)\mathbf{j}\\ &= 100\left(-\frac{\sqrt{2}}{2}\right)\mathbf{i} + 100\left(-\frac{\sqrt{2}}{2}\right)\mathbf{j}\\ &= -50\sqrt{2}\mathbf{i} - 50\sqrt{2}\mathbf{j}\\ &\approx -70.71\mathbf{i} - 70.71\mathbf{j}\\ &\approx \langle -70.71, -70.71\rangle\end{aligned}$$

You can check that **v** has a magnitude of 100, as follows.

$$\begin{aligned}\|\mathbf{v}\| &= \sqrt{\left(-50\sqrt{2}\right)^2 + \left(-50\sqrt{2}\right)^2}\\ &= \sqrt{5000 + 5000}\\ &= \sqrt{10{,}000} = 100\end{aligned}$$

9.

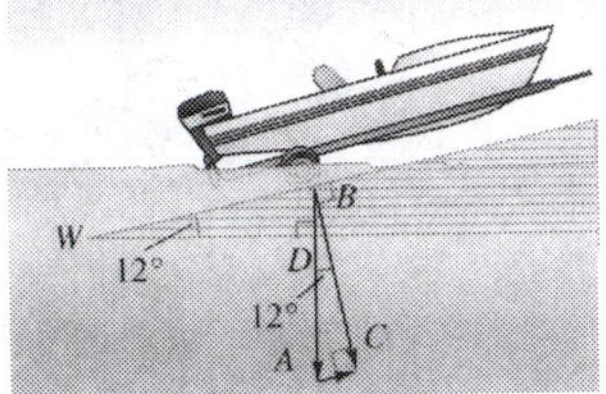

Solution

Based in the figure, you can make the following observations.

$\|\overline{BA}\|$ = force of gravity = combined weight of boat and trailer

$\|\overline{BC}\|$ = force against ramp

$\|\overline{AC}\|$ = force required to move boat up ramp = 500 pounds

By construction, triangles *BWD* and *ABC* are similar. So, angle *ABC* is 12°. In triangle *ABC*, you have

$$\sin 12° = \frac{\|\overline{AC}\|}{\|\overline{BA}\|}$$

$$\sin 12° = \frac{500}{\|\overline{BA}\|}$$

$$\|\overline{BA}\| = \frac{500}{\sin 12°}$$

$$\|\overline{BA}\| \approx 2405.$$

So, the combined weight is approximately 2405 pounds. (In the figure, note that $\overline{AC}$ is parallel to the ramp).

10. (a)

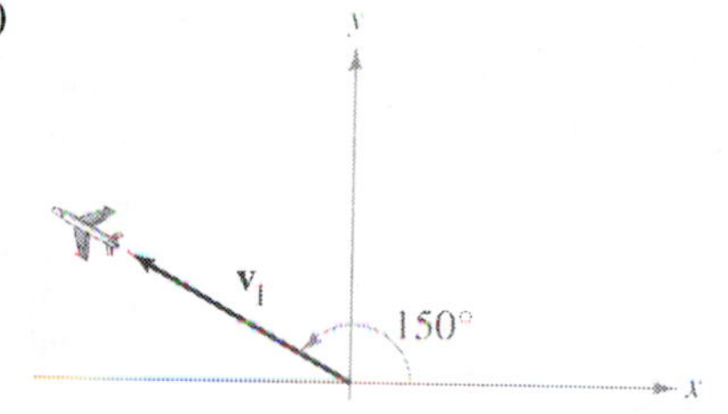

(b)

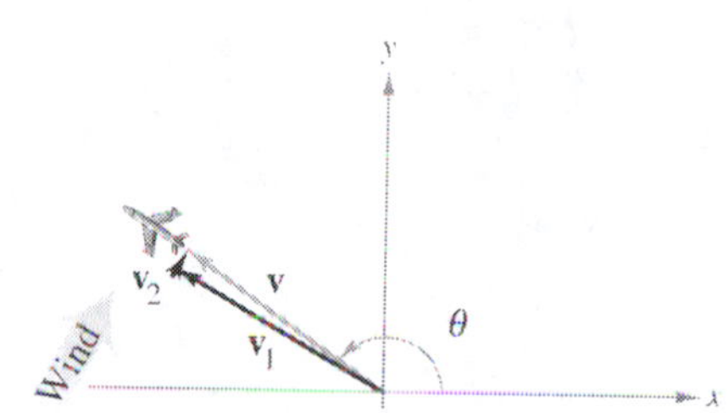

Solution

Using the figure, the velocity of the airplane (alone) is $\mathbf{v}_1 = 450\langle\cos 150°, \sin 150°\rangle = \langle -225\sqrt{3}, 225\rangle$

and the velocity of the wind is $\mathbf{v}_2 = 40\langle\cos 60°, \sin 60°\rangle = \langle 20, 20\sqrt{3}\rangle$.

So, the velocity of the airplane (in the wind) is $\mathbf{v} = \mathbf{v}_1 + \mathbf{v}_2 = \langle -225\sqrt{3} + 20, 225 + 20\sqrt{3}\rangle \approx \langle -369.7, 259.6\rangle$

and the resultant speed of the airplane is $\|\mathbf{v}\| \approx \sqrt{(-369.7)^2 + (259.6)^2} \approx 451.8$ miles per hour.

Finally, given that θ is the direction angle of the flight path, you have $\tan\theta \approx \dfrac{259.6}{-369.7} \approx 0.7022$

which implies that $\theta \approx 180° - 35.1° = 144.9°$.

So, the true direction of the airplane is approximately $270° + (180° - 144.9°) = 305.1°$.

Checkpoints for Section 8.4

1.
$$\begin{aligned}\mathbf{u}\cdot\mathbf{v} &= \langle 3, 4\rangle \cdot \langle 2, -3\rangle\\ &= 3(2) + 4(-3)\\ &= 6 - 12\\ &= -6\end{aligned}$$

2. (a) Begin by finding the dot product of **u** and **v**.

$$\begin{aligned}\mathbf{u}\cdot\mathbf{v} &= \langle 3, 4\rangle \cdot \langle -2, 6\rangle\\ &= 3(-2) + 4(6)\\ &= -6 + 24\\ &= +18\end{aligned}$$

$$\begin{aligned}(\mathbf{u}\cdot\mathbf{v})\mathbf{v} &= 18\langle -2, 6\rangle\\ &= \langle -36, 108\rangle\end{aligned}$$

(b) Begin by finding **u** + **v**.

$$\begin{aligned}\mathbf{u} + \mathbf{v} &= \langle 3, 4\rangle + \langle -2, 6\rangle\\ &= \langle 3 + (-2), 4 + 6\rangle\\ &= \langle 1, 10\rangle\end{aligned}$$

$$\begin{aligned}\mathbf{u}\cdot(\mathbf{u} + \mathbf{v}) &= \langle 3, 4\rangle \cdot \langle 1, 10\rangle\\ &= 3(1) + 4(10)\\ &= 3 + 40\\ &= 43\end{aligned}$$

(c) Begin by finding the dot product of **v** and **v**.

$$\begin{aligned}\mathbf{v}\cdot\mathbf{v} &= \langle -2, 6\rangle \cdot \langle -2, 6\rangle\\ &= -2(-2) + 6(6)\\ &= 4 + 36\\ &= 40\end{aligned}$$

Because $\|\mathbf{v}\|^2 = \mathbf{v}\cdot\mathbf{v} = 40$, it follows that

$$\begin{aligned}\|\mathbf{v}\| &= \sqrt{\mathbf{v}\cdot\mathbf{v}}\\ &= \sqrt{40}\\ &= 2\sqrt{10}.\end{aligned}$$

3. $\cos\theta = \dfrac{\mathbf{u}\cdot\mathbf{v}}{\|\mathbf{u}\|\,\|\mathbf{v}\|} = \dfrac{\langle 2, 1\rangle \cdot \langle 1, 3\rangle}{\|\langle 2, 1\rangle\|\,\|\langle 1, 3\rangle\|}$

$$= \frac{2(1) + 1(3)}{\sqrt{2^2 + 1^2}\,\sqrt{1^2 + 3^2}}$$
$$= \frac{5}{\sqrt{5}\,\sqrt{10}}$$
$$= \frac{5}{\sqrt{50}}$$
$$= \frac{5}{5\sqrt{2}}$$
$$= \frac{1}{\sqrt{2}}$$
$$= \frac{\sqrt{2}}{2}$$

This implies that the angle between the two vectors is

$$\theta = \cos^{-1}\left(\frac{\sqrt{2}}{2}\right) = \frac{\pi}{4}.$$

4. Find the dot product of the two vectors.

$$\mathbf{u}\cdot\mathbf{v} = \langle 6, 10\rangle \cdot \left\langle -\frac{1}{3}, \frac{1}{5}\right\rangle$$
$$= 6\left(-\frac{1}{3}\right) + 10\left(\frac{1}{5}\right)$$
$$= -2 + 2$$
$$= 0$$

Because the dot product is 0, the two vectors are orthogonal.

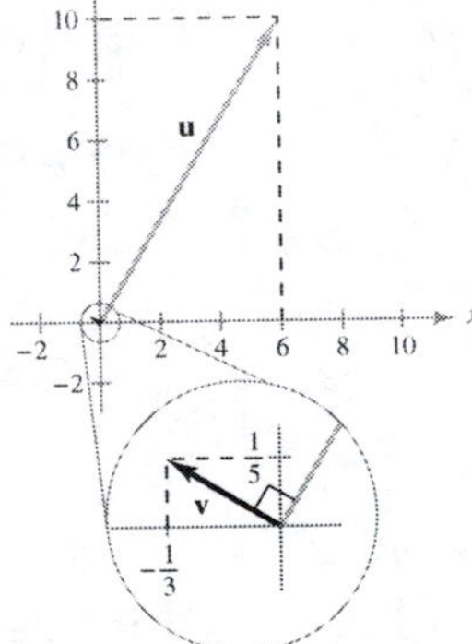

5. The projection of **u** onto **v** is

$$\mathbf{w}_1 = \text{proj}_{\mathbf{v}}\,\mathbf{u} = \left(\frac{\mathbf{u}\cdot\mathbf{v}}{\|\mathbf{v}\|^2}\right)\mathbf{v}$$
$$= \left(\frac{\langle 3, 4\rangle \cdot \langle 8, 2\rangle}{\langle 8, 2\rangle \cdot \langle 8, 2\rangle}\right)\langle 8, 2\rangle$$
$$= \left(\frac{3(8) + 4(2)}{8(8) + 2(2)}\right)\langle 8, 2\rangle$$
$$= \left(\frac{32}{68}\right)\langle 8, 2\rangle$$
$$= \left(\frac{8}{17}\right)\langle 8, 2\rangle$$
$$= \left\langle \frac{64}{17}, \frac{16}{17}\right\rangle.$$

The other component, $\mathbf{w}_2$ is

$$\mathbf{w}_2 = \mathbf{u} - \mathbf{w}_1 = \langle 3, 4\rangle - \left\langle \frac{64}{17}, \frac{16}{17}\right\rangle = \left\langle -\frac{13}{17}, \frac{52}{17}\right\rangle.$$

So, $\mathbf{u} = \mathbf{w}_1 + \mathbf{w}_2 = \left\langle \frac{64}{17}, \frac{16}{17}\right\rangle + \left\langle -\frac{13}{17}, \frac{52}{17}\right\rangle = \langle 3, 4\rangle.$

6. Solution

Because the force due to gravity is vertical and downward, you can represent the gravitational force by the vector

$\mathbf{F} = -150\mathbf{j}.$ Force due to gravity

To find the force required to keep the cart from rolling down the ramp, project **F** onto a unit vector **v** in the direction of the ramp, as follows.

$$\mathbf{v} = (\cos 15°)\mathbf{i} + (\sin 15°)\mathbf{j}$$
$$= 0.966\mathbf{i} + 0.259\,\mathbf{j} \qquad \text{Unit vector along ramp}$$

So, the projection of **F** onto **v** is

$$\mathbf{w}_1 = \text{proj}_{\mathbf{v}}\mathbf{F}$$
$$= \left(\frac{\mathbf{F}\cdot\mathbf{v}}{\|\mathbf{v}\|^2}\right)\mathbf{v}$$
$$= (\mathbf{F}\cdot\mathbf{v})\mathbf{v} \approx \left(\langle 0, -150\rangle \cdot \langle 0.966, 0.259\rangle\right)\mathbf{v}$$
$$\approx (-38.85)\mathbf{v}$$
$$\approx -37.53\mathbf{i} - 10.06\mathbf{j}.$$

The magnitude of this force is approximately 38.85. So, a force of approximately 38.85 pounds is required to keep the cart from rolling down the ramp.

7.

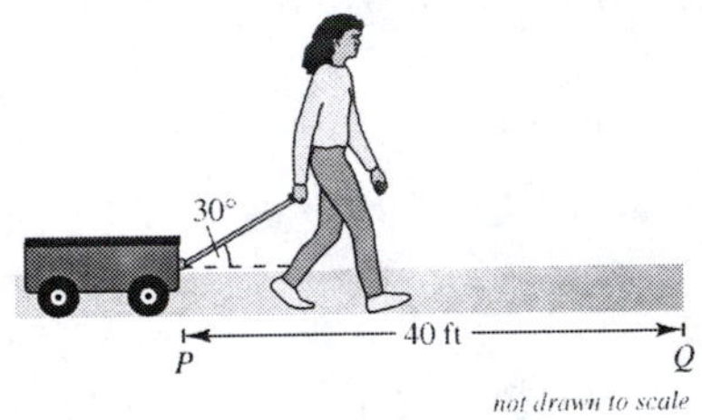

Using a projection, you can calculate the work as follows.

$$W = \left\|\text{proj}_{\overrightarrow{PQ}}\mathbf{F}\right\|\,\left\|\overrightarrow{PQ}\right\|$$
$$= (\cos 30°)\,\|\mathbf{F}\|\,\left\|\overrightarrow{PQ}\right\|$$
$$= \frac{\sqrt{3}}{2}(35)(40)$$
$$= 700\sqrt{3}$$
$$\approx 1212.436 \qquad \text{foot-pounds}$$

So, the work done is 1212.436 foot-pounds.

Checkpoints for Section 8.5

1. The number $z = 3 - 4i$ is plotted in the complex plane.

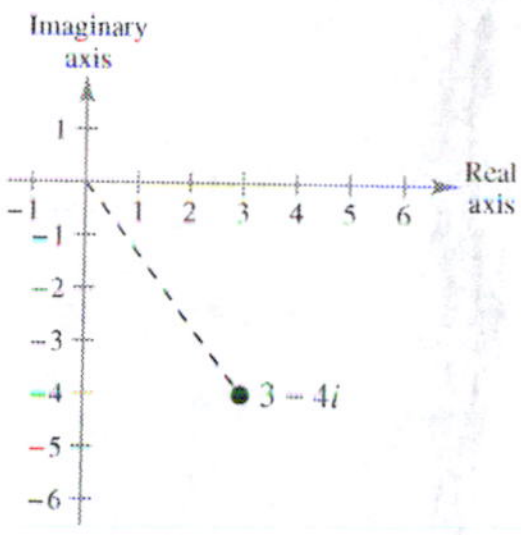

It has an absolute value of $|z| = \sqrt{3^2 + (-4)^2}$

$$= \sqrt{9 + 16}$$
$$= \sqrt{25}$$
$$= 5.$$

2. $z = 6 - 6i$

The absolute value of $z = 6 - 6i$ is

$$r = |6 - 6i| = \sqrt{6^2 + (-6)^2}$$
$$= \sqrt{36 + 36} = \sqrt{72} = 6\sqrt{2}$$

and the argument θ is determined from

$$\tan\theta = \frac{b}{a} = \frac{-6}{6} = -1.$$

Because $z = 6 - 6i$ lies in Quadrant IV,

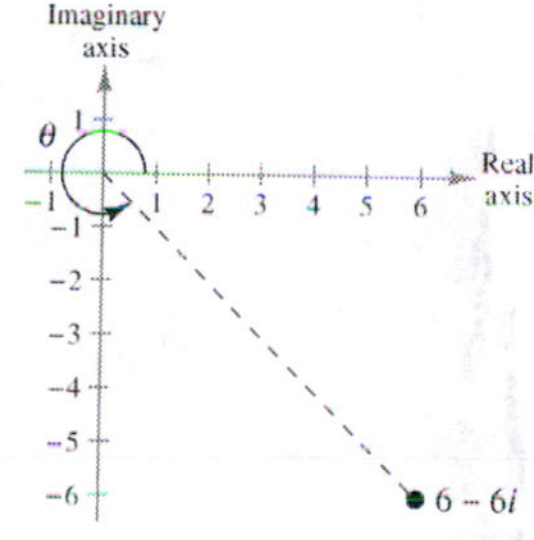

$$\theta = 2\pi - |\arctan(-1)| = 2\pi - \frac{\pi}{4} = \frac{7\pi}{4}.$$

So, the trigonometric form is

$$z = r(\cos\theta + i\sin\theta) = 6\sqrt{2}\left(\cos\frac{7\pi}{4} + i\sin\frac{7\pi}{4}\right).$$

3. To write $z = 8\left[\cos\left(\frac{2\pi}{3}\right) + i\sin\left(\frac{2\pi}{3}\right)\right]$ in standard form, first find the trigonometric ratios. Because $\cos\left(\frac{2\pi}{3}\right) = \frac{-1}{2}$ and $\sin\left(\frac{2\pi}{3}\right) = \frac{\sqrt{3}}{2}$, you can write

$$z = 8\left[\cos\left(\frac{2\pi}{3}\right) + i\sin\left(\frac{2\pi}{3}\right)\right]$$
$$= 8\left(-\frac{1}{2} + \frac{\sqrt{3}}{2}i\right)$$
$$= -4 + 4\sqrt{3}\,i.$$

4. $$z_1 z_2 = 2\left(\cos\frac{5\pi}{6} + i\sin\frac{5\pi}{6}\right)\cdot 5\left(\cos\frac{7\pi}{6} + i\sin\frac{7\pi}{6}\right)$$
$$= (2)(5)\left[\cos\left(\frac{5\pi}{6} + \frac{7\pi}{6}\right) + i\sin\left(\frac{5\pi}{6} + \frac{7\pi}{6}\right)\right]$$
$$= 10(\cos 2\pi + i\sin 2\pi)$$
$$= 10[1 + i(0)]$$
$$= 10$$

5. $$z_1 z_2 = 3\left(\cos\frac{\pi}{3} + i\sin\frac{\pi}{3}\right)\cdot 4\left(\cos\frac{\pi}{6} + i\sin\frac{\pi}{6}\right)$$
$$= (3)(4)\left[\cos\left(\frac{\pi}{3} + \frac{\pi}{6}\right) + i\sin\left(\frac{\pi}{3} + \frac{\pi}{6}\right)\right]$$
$$= 12\left(\cos\frac{\pi}{2} + i\sin\frac{\pi}{2}\right)$$
$$= 12[0 + i(1)]$$
$$= 12i$$

You can check this by first converting the complex numbers to their standard forms and then multiplying algebraically.

$$z_1 = 3\left(\cos\frac{\pi}{3} + i\sin\frac{\pi}{3}\right) = 3\left(\frac{1}{2} + \frac{\sqrt{3}}{2}i\right) = \frac{3}{2} + \frac{3\sqrt{3}}{2}i$$

$$z_2 = 4\left(\cos\frac{\pi}{6} + i\sin\frac{\pi}{6}\right) = 4\left(\frac{\sqrt{3}}{2} + \frac{1}{2}i\right) = 2\sqrt{3} + 2i$$

So, $$z_1 z_2 = \left(\frac{3}{2} + \frac{3\sqrt{3}}{2}i\right)\left(2\sqrt{3} + 2i\right)$$
$$= 3\sqrt{3} + 3i + 9i + 3\sqrt{3}i^2$$
$$= 3\sqrt{3} + 12i + 3\sqrt{3}(-1)$$
$$= 3\sqrt{3} + 12i - 3\sqrt{3}$$
$$= 12i.$$

6. $$\frac{z_1}{z_2} = \frac{\cos 40° + i \sin 40°}{\cos 10° + i \sin 10°}$$
$$= \left[\cos(40° - 10°) + i \sin(40° - 10°)\right]$$
$$= \cos 30° + i \sin 30°$$
$$= \frac{\sqrt{3}}{2} - \frac{1}{2}i$$

7. The absolute value of $z = -1 - i$ is $r = |-1 - i| = \sqrt{(-1)^2 + (-1)^2} = \sqrt{1 + 1} = \sqrt{2}$

and the argument θ given by $\tan\theta = \frac{b}{a} = \frac{-1}{-1} = 1$.

Because $z = -1 - i$ lies in Quadrant III, $\theta = \pi + \arctan 1 = \pi + \frac{\pi}{4} = \frac{5\pi}{4}$.

So, the trigonometric form is $z = -1 - i = \sqrt{2}\left(\cos\frac{5\pi}{4} + i \sin\frac{5\pi}{4}\right)$.

Then, by DeMoivre's Theorem, you have $(-1 - i)^4 = \left[\sqrt{2}\left(\cos\frac{5\pi}{4} + i\sin\frac{5\pi}{4}\right)\right]^4$

$$= \left(\sqrt{2}\right)^4\left(\cos\left[\frac{4(5\pi)}{4}\right] + i \sin\left[\frac{4(5\pi)}{4}\right]\right)$$
$$= 4(\cos 5\pi + i \sin 5\pi)$$
$$= 4[-1 + i(0)]$$
$$= -4.$$

8. First, write 1 in trigonometric form $z = 1(\cos 0 + i \sin 0)$. Then, by the nth root formula, with $n = 4$ and $r = 1$, the roots are of the form

$$z_k = \sqrt[4]{1}\left(\cos\frac{0 + 2\pi k}{4} + i \sin\frac{0 + 2\pi k}{4}\right) = (1)\left(\cos\frac{\pi k}{2} + i \sin\frac{\pi k}{2}\right) = \cos\frac{\pi k}{2} + i \sin\frac{\pi k}{2}.$$

So, for $k = 0, 1, 2$ and 3, the fourth roots are as follows.

$$z_0 = \cos 0 + i \sin 0 = 1 + i(0) = 1$$
$$z_1 = \cos\frac{\pi}{2} + i \sin\frac{\pi}{2} = 0 + i(1) = i$$
$$z_2 = \cos\pi + i \sin\pi = -1 + i(0) = -1$$
$$z_3 = \cos\frac{3\pi}{2} + i \sin\frac{3\pi}{2} = 0 + i(-1) = -i$$

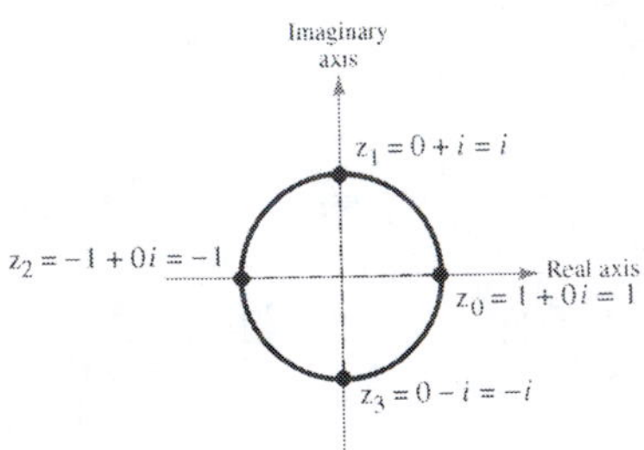

9. The absolute value of $z = -6 + 6i$ is

$$r = |-6 + 6i| = \sqrt{(-6)^2 + 6^2} = \sqrt{36 + 36} = \sqrt{72} = 6\sqrt{2}$$

and the argument θ is given by $\tan\theta = \dfrac{b}{a} = \dfrac{6}{-6} = -1.$

Because $z = -6 + 6i$ lies in Quadrant II, the trigonometric form of

z is $z = -6 + 6i = 6\sqrt{2}(\cos 135° + i \sin 135°)$.

By the formula for nth roots, the cube roots have the form

$$z_k = \sqrt[3]{6\sqrt{2}}\left[\cos\left(\frac{135° + 360°k}{3}\right) + i \sin\left(\frac{135° + 360°k}{3}\right)\right].$$

Finally, for $k = 0$, 1 and 2, you obtain the roots

$$z_0 = \sqrt[3]{6\sqrt{2}}\left[\cos\left(\frac{135° + 360°(0)}{3}\right) + i \sin\left(\frac{135° + 360°(0)}{3}\right)\right].$$

$$= \sqrt[3]{6\sqrt{2}}(\cos 45° + i \sin 45°)$$

$$= \sqrt[3]{6\sqrt{2}}\left(\frac{\sqrt{2}}{2} + \frac{\sqrt{2}}{2}i\right) \approx 1.4422 + 1.4422i$$

$$z_1 = \sqrt[3]{6\sqrt{2}}\left[\cos\left(\frac{135° + 360°(1)}{3}\right) + i \sin\left(\frac{135° + 360°(1)}{3}\right)\right].$$

$$= \sqrt[3]{6\sqrt{2}}(\cos 165° + i \sin 165°)$$

$$\approx -1.9701 + 0.5279i$$

$$z_2 = \sqrt[3]{6\sqrt{2}}\left[\cos\left(\frac{135° + 360°(2)}{3}\right) + i \sin\left(\frac{135° + 360°(2)}{3}\right)\right].$$

$$= \sqrt[3]{6\sqrt{2}}(\cos 285° + i \sin 285°)$$

$$\approx 0.5279 - 1.9701i$$

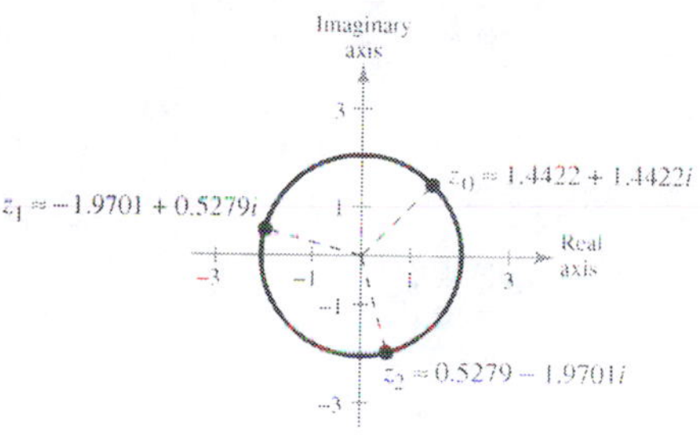

Chapter 9

Checkpoints for Section 9.1

1. $\begin{cases} x - y = 0 & \text{Equation 1} \\ 5x - 3y = 6 & \text{Equation 2} \end{cases}$

Begin by solving for y in Equation 1.

$x - y = 0$

$y = x$

Next substitute this expression for y into Equation 2 and solve the resulting single-variable equation for x.

$5x - 3y = 6$ Write Equation 2.

$5x - 3(x) = 6$ Substitute x for y.

$2x = 6$ Collect like terms.

$x = 3$ Divide each side by 2.

Finally, solve for y by back-substituting $x = 3$ into equation $y = x$, to obtain the corresponding value for y.

$y = x$ Write revised Equation 1.

$y = 3$ Substitute 3 for x.

The solution is the ordered pair $(3, 3)$.

Check

Substitute $(3, 3)$ into Equation 1:

$x - y = 0$ Write Equation 1.

$3 - 3 \stackrel{?}{=} 0$ Substitute for x and y

$0 = 0$ Solution checks in Equation 1.

Substitute $(3, 3)$ into Equation 2:

$5x - 3y = 6$ Write Equation 2.

$5(3) - 3(3) \stackrel{?}{=} 6$ Substitute for x and y.

$15 - 9 \stackrel{?}{=} 6$

$6 = 6$ Solution checks in Equation 2.

Because $(3, 3)$ satisfies both equations in the system, it is a solution of the system of equations.

2. *Verbal Model:* $\boxed{\text{Amount in 6.5\% fund}} + \boxed{\text{Amount in 8.5\% fund}} = \boxed{\text{Total investment}}$

$\boxed{\text{Interest for 6.5\% fund}} + \boxed{\text{Interest for 8.5\% fund}} = \boxed{\text{Total interest}}$

Labels:

Amount in 6.5% fund $= x$	(dollars)
Interest for 6.5% fund $= 0.065x$	(dollars)
Amount in 8.5% fund $= y$	(dollars)
Interest for 8.5% fund $= 0.085y$	(dollars)
Total investment $= 25{,}000$	(dollars)
Total interest $= 2600$	(dollars)

System: $\begin{cases} x + y = 25{,}000 & \text{Equation 1} \\ 0.065x + 0.085y = 2000 & \text{Equation 2} \end{cases}$

To begin, it is convenient to multiply each side of Equation 2 by 1000. This eliminates the need to work with decimals.

$1000(.065x + 0.085y) = 1000(2000)$ Multiply each side of Equation 2 by 1000.

$65x + 85y = 2{,}000{,}000$ Revised Equation 2

To solve this system, you can solve for x in Equation 1.

$x = 25{,}000 - y$ Revised Equation 1

Then, substitute this expression for x into revised Equation 2 and solve the resulting equation for y.

$65x + 85y = 2{,}000{,}000$ Write revised Equation 2.

$65(25{,}000 - y) + 85y = 2{,}000{,}000$ Substitute 1 25000 $- y$ for x.

$1{,}625{,}000 - 65y + 85y = 2{,}000{,}000$ Distributive Property

$20y = 375{,}000$ Combine like terms.

$y = 18{,}750$ Divide each side by 20.

Next, back-substitute $y = 18{,}750$ to solve for x.

$x = 25000 - y$ Write revised Equation 1.

$x = 25000 - (18750)$ Substitute 18750 for y.

$x = 6250$ Subtract.

The solution is $(6250, 18{,}750)$. So, \$6250 is invested at 6.5% and \$18,750 is invested at 8.5%.

3. $\begin{cases} -2x + y = 5 & \text{Equation 1} \\ x^2 - y + 3x = 1 & \text{Equation 2} \end{cases}$

Begin by solving for y in Equation 1 to obtain $y = 2x + 5$. Next, substitute this expression for y into Equation 2 and solve for x.

$x^2 - y + 3x = 1$ Write Equation 2.

$x^2 - (2x + 5) + 3x = 1$ Substitute $2x + 5$ for y into Equation 2.

$x^2 - 2x - 5 + 3x = 1$ Simplify.

$x^2 + x - 6 = 0$ Write in standard form.

$(x + 3)(x - 2) = 0$ Factor.

$x + 3 = 0 \Rightarrow x = -3$ Solve for x.

$x - 2 = 0 \Rightarrow x = 2$

Back-substituting these values of x to solve for the corresponding values of y produces the following solutions.

$y = 2x + 5$

$y = 2(-3) + 5 = -1$

$y = 2(2) + 5 = 9$

So, the solutions of the system are $(-3, -1)$ and $(2, 9)$.

4. $\begin{cases} 2x - y = -3 & \text{Equation 1} \\ 2x^2 + 4x - y^2 = 0 & \text{Equation 2} \end{cases}$

Begin by solving for y in Equation 1 to obtain $y = 2x + 3$. Next, substitute this expression for y into Equation 2 and solve for x.

$2x^2 + 4x - y^2 = 0$ Write Equation 2.

$2x^2 + 4x - (2x + 3)^2 = 0$ Substitute $2x + 3$ for y into Equation 2.

$2x^2 + 4x - (4x^2 + 12x + 9) = 0$ Simplify.

$-2x^2 - 8x - 9 = 0$ Combine like terms.

$2x^2 + 8x + 9 = 0$ Write in standard form

$x = \dfrac{-(8) \pm \sqrt{(8)^2 - 4(2)(9)}}{2(2)}$ Use the Quadratic Formula.

$x = \dfrac{-8 \pm \sqrt{-8}}{4}$ Simplify.

Because the discriminant is negative, the equation $2x^2 + 8x + 9 = 0$ has no (real) solution. So, the original system of equations has no (real) solution.

5.

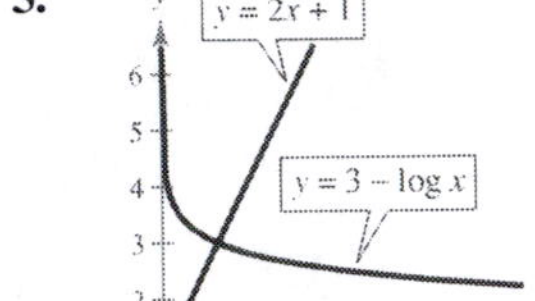

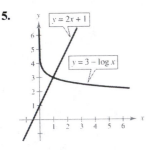

There is only one point of intersection of the graphs of the two equations, and (1, 3) is the solution point.

Check (1, 3) in Equation 1:

$y = 3 - \log x$ Write Equation 1.

$3 \stackrel{?}{=} 3 - \log 1$ Substitute for x and y.

$3 \stackrel{?}{=} 3 - 0$

$3 = 3$ Solution checks in Equation 1.

Check (1, 3) in Equation 2:

$-2x + y = 1$ Write Equation 2.

$-2(1) + 3 \stackrel{?}{=} 1$ Substitute for x and y.

$-2 + 3 \stackrel{?}{=} 1$

$1 = 1$ Solution checks in Equation 2.

6. Algebraic Solution

The total cost of producing x units is

$\boxed{\text{Totalcost}} = \boxed{\text{Cost per unit}} \cdot \boxed{\text{Number of units}} + \boxed{\text{Initial cost}}$

$C = 12x + 300{,}000.$ Equation 1

The revenue obtained by selling x units is

$\boxed{\text{Total revenue}} = \boxed{\text{Price per unit}} \cdot \boxed{\text{Number of units}}$

$R = 60x.$ Equation 2

Because the break-even point occurs when $R = C$, you have $C = 60x$, and the system of equations to solve is

$$\begin{cases} C = 12x + 300{,}000 \\ C = 60x \end{cases}.$$

Solve by substitution.

$60x = 12x + 300{,}000$ Substitute $60x$ for C in Equation 1.

$48x = 300{,}000$ Subtract $12x$ from each side.

$x = 6250$ Divide each side by 48

So, the company must sell 6250 pairs of shoes to break even.

Graphical Solution

The system of equations to solve is

$$\begin{cases} C = 12x + 300{,}000 \\ C = 60x \end{cases}.$$

Use a graphing utility to graph $y_1 = 12x + 300{,}000$ and $y_2 = 60x$ in the same viewing window.

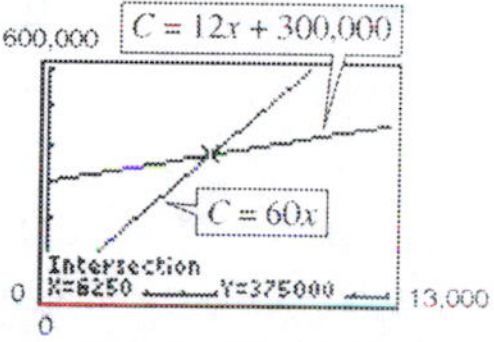

So, the company must sell about 6250 pairs of shoes to break even.

7. Algebraic Solution

Because both equations are already solved for S in terms of x, substitute either expression for S into the other equation and solve for x.

$$\begin{cases} S = 108 - 9.4x & \text{Animated} \\ S = 16 + 9x & \text{Horror} \end{cases}$$

$16 + 9x = 108 - 9.4x$ Substitute for S in Equation 1.

$9.4x + 9x = 108 - 16$ Add $9.4x$ and -16 to each side.

$18.4x = 92$

$x = 5$ Divide each side by 18.4.

So, the weekly ticket sales for the two movies will be equal after 5 weeks.

Numerical Solution

You can create a table of values for each model to determine when ticket sales for the two movies will be equal.

Number of weeks x	0	1	2	3	4	5	6	7
Sales S (Animated)	108	98.6	89.2	79.8	70.4	61	51.6	70
Sales S (Horror)	16	25	34	43	52	61	42.2	79

So, from the table, the weekly ticket sales for the two movies will be equal after 5 weeks.

Checkpoints for Section 9.2

1. Because the coefficients of y differ only in sign, eliminate the y-terms by adding the two equations.

$2x + y = 4$ Write Equation 1.

$\underline{2x - y = -1}$ Write Equation 2.

$4x = 3$ Add equations.

$x = \frac{3}{4}$ Solve for x.

Solve for y by back-substituting $x = \frac{3}{4}$ into Equation 1.

$$2\left(\frac{3}{4}\right) + y = 4$$

$$\frac{3}{2} + y = 4$$

$$y = \frac{5}{2}$$

The solution is $\left(\frac{3}{4}, \frac{5}{2}\right)$.

Check this in the original system.

$2\left(\frac{3}{4}\right) + \left(\frac{5}{2}\right) \stackrel{?}{=} 4$ Write Equation 1.

$\frac{3}{2} + \frac{5}{2} = 4$ Solution checks in Equation 1. ✓

$2\left(\frac{3}{4}\right) - \left(\frac{5}{2}\right) \stackrel{?}{=} -1$ Write Equation 2.

$\frac{3}{2} - \frac{5}{2} = -1$ Solution checks in Equation 2. ✓

2. To obtain coefficients that differ only in sign, multiply Equation 2 by 3.

$2x + 3y = 17 \Rightarrow 2x + 3y = 17$ Write Equation 1.

$5x - y = 17 \Rightarrow \underline{15x - 3y = 51}$ Multiply Equation 2 by 3.

$17x = 68$ Add Equations.

$x = 4$ Solve for x.

Solve for y by back-substituting $x = 4$ into Equation 2.

$5x - y = 17$ Write Equation 2.

$5(4) - y = 17$ Substitute 4 for x.

$20 - y = 17$ Simplify.

$y = 3$ Solve for y.

The solution is $(4, 3)$.

Check this in the original system.

$2(4) + 3(3) \stackrel{?}{=} 17$ Write Equation 1.

$8 + 9 = 17$ Solution Checks in Equation 1. ✓

$5(4) - (3) \stackrel{?}{=} 17$ Write Equation 2.

$20 - 3 = 17$ Solution Checks in Equation 2. ✓

3. Algebraic Solution

You can obtain coefficients that differ only in sign by multiplying Equation 1 by 2 and multiplying Equation 2 by -3.

$$\begin{aligned} 3x + 2y = 7 &\Rightarrow \quad 6x + 4y = 14 && \text{Multiply Equation 1 by 2.} \\ 2x + 5y = 1 &\Rightarrow \underline{-6x - 15y = -3} && \text{Multiply Equation 2 by } -3. \\ &\qquad\quad -11y = 11 && \text{Add Equations.} \\ &\qquad\qquad y = -1 && \text{Solve for } y. \end{aligned}$$

Solve for x by back-substituting $y = 1$ into Equation 1.

$$\begin{aligned} 3x + 2y &= 7 && \text{Write Equation 1.} \\ 3x + 2(-1) &= 7 && \text{Substitute } -1 \text{ for } y \\ 3x - 2 &= 7 \\ 3x &= 9 \\ x &= 3 \end{aligned}$$

The solution is $(3, -1)$.

Graphical Solution

Solve each equation for y and use a graphing utility to graph the equations in the same viewing window.

From the graph, the solution is $(3, -1)$.

Check this in the original system.

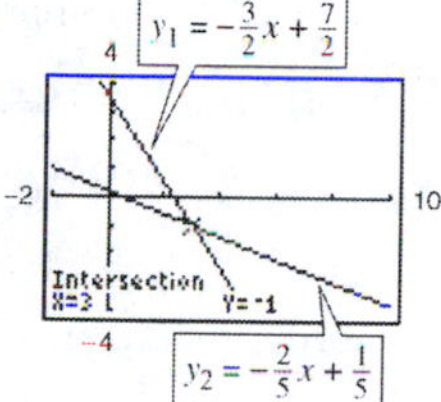

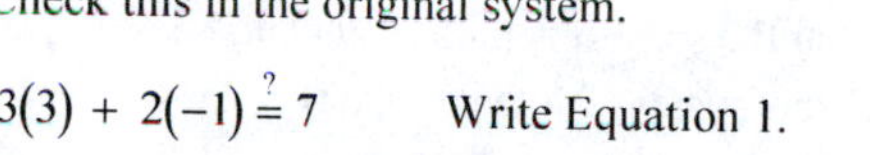

$$\begin{aligned} 3(3) + 2(-1) &\stackrel{?}{=} 7 && \text{Write Equation 1.} \\ 9 - 2 &= 7 && \text{Solution checks in Equation 1. } \checkmark \\ 2(3) + 5(-1) &\stackrel{?}{=} 1 && \text{Write Equation 2.} \\ 6 - 5 &= 1 && \text{Solution checks in Equation 2. } \checkmark \end{aligned}$$

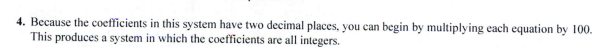

4. Because the coefficients in this system have two decimal places, you can begin by multiplying each equation by 100. This produces a system in which the coefficients are all integers.

$$\begin{aligned} 0.03x + 0.04y = 0.75 &\Rightarrow 3x + 4y = 75 \\ 0.02x + 0.06y = 0.90 &\Rightarrow 2x + 6y = 90 \end{aligned}$$

Now, to obtain coefficients that differ only in sign, multiply Equation 1 by 2 and Equation 2 by -3.

$$\begin{aligned} 3x + 4y = 75 &\Rightarrow \quad 6x + 8y = 150 && \text{Multiply Equation 1 by 2.} \\ 2x + 6y = 90 &\Rightarrow \underline{-6x - 18y = -270} && \text{Multiply Equation 2 by } -3. \\ &\qquad\quad -10y = -120 && \text{Add Equations.} \\ &\qquad\qquad y = 12 && \text{Solve for } y. \end{aligned}$$

Back-substitute $y = 12$ into revised Equation 1 to solve for x.

$$\begin{aligned} 3x + 4y &= 75 \\ 3x + 4(12) &= 75 \\ 3x + 48 &= 75 \\ 3x &= 27 \\ x &= 9 \end{aligned}$$

The solution is $(9, 12)$.

Check this in the original system, as follows.

$$\begin{aligned} 0.03(9) + 0.04(12) &\stackrel{?}{=} 0.75 && \text{Write Equation 1.} \\ 0.27 + 0.48 &= 0.75 && \text{Solution Checks in Equation 1. } \checkmark \\ 0.02(9) + 0.06(12) &\stackrel{?}{=} 0.90 && \text{Write Equation 2.} \\ 0.18 + 0.72 &= 0.90 && \text{Solution Checks in Equation 2. } \checkmark \end{aligned}$$

5. First, write each equation in slope-intercept form.

$$\begin{cases} 2x + 3y = 6 & \Rightarrow y = \frac{2}{3}x + 2 \\ 4x - 6y = -9 & \Rightarrow y = \frac{2}{3}x + \frac{3}{2} \end{cases}$$

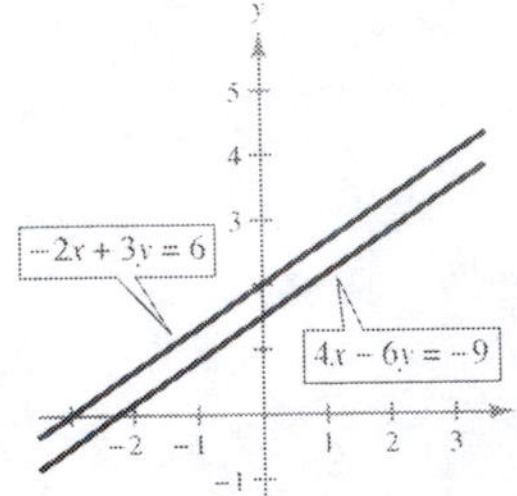

The graph of the system is a pair of parallel lines. The lines have no point of intersection, so the system has no solution. The system is inconsistent.

6. To obtain coefficients that differ only in sign, multiply Equation 1 by 2.

$6x - 5y = 3 \Rightarrow$	$12x - 10y = 6$	Multiply Equation by 2.
$-12x + 10y = 5 \Rightarrow$	$\underline{-12x + 10y = 5}$	Write Equation 2.
	$0 = 11$	Add equations.

Because there are no values of x and y for which $0 = 11$, you can conclude that the system is inconsistent and has no solution. The graph shows the lines corresponding to the two equations in this system. Note that the two lines are parallel, so they have no point of intersection.

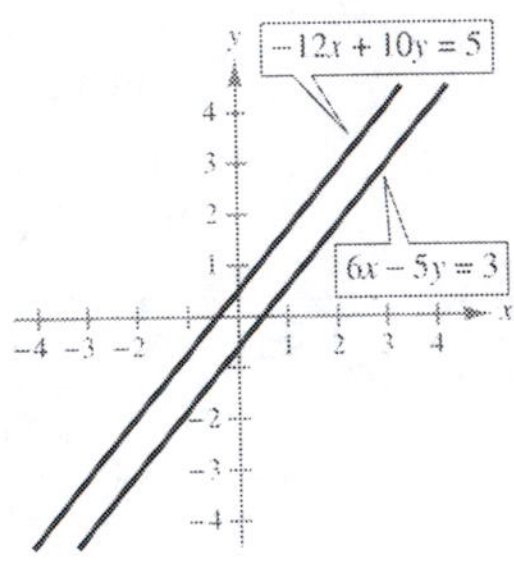

7. To obtain coefficients that differ only in sign, multiply Equation 1 by 8.

$\frac{1}{2}x - \frac{1}{8}y = -\frac{3}{8} \Rightarrow$	$4x - y = -3$	Multiply Equation by 8
$-4x + y = 3 \Rightarrow$	$\underline{-4x + y = 3}$	Write Equation 2.
	$0 = 0$	Add equations.

Because the two equations are equivalent (have the same solution set), the system has infinitely many solutions. The solution set consists of all points (x, y) lying on the line $-4x + y = 3$ as shown. Letting $x = a$, where a is any real number, the solutions of the system are $(a, 4a + 3)$

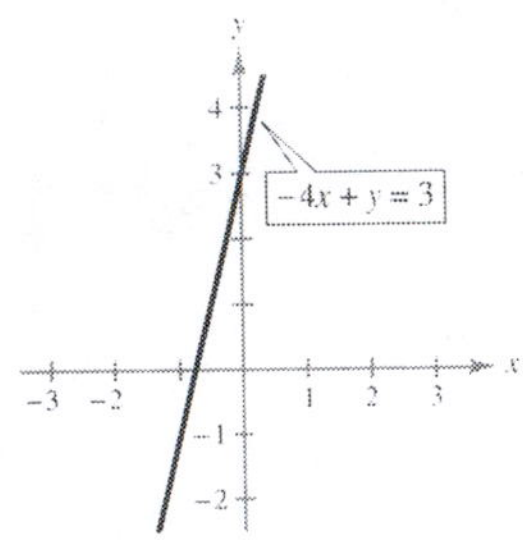

8. The two unknown quantities are the speeds of the wind and of the plane. If r_1 is the speed of the plane and r_2 is the speed of the wind, then

$r_1 - r_2 =$ speed of the plane against the wind

$r_1 + r_2 =$ speed of the plane with the wind.

Using the formula

distance = (rate)(time)

for these two speeds, you obtain the following equations.

$$2000 = (r_1 - r_2)\left(4 + \frac{24}{60}\right)$$

$$2000 = (r_1 - r_2)\left(4 + \frac{6}{60}\right)$$

These two equations simplify as follows.

$$\begin{cases} 5000 = 11r_1 - 11r_2 & \text{Equation 1} \\ 20{,}000 = 41r_1 + 41r_2 & \text{Equation 2} \end{cases}$$

To solve this system by elimination, multiply Equation 1 by 41 and Equation 2 by 11.

$250{,}000 = 451r_1 - 451r_2$ Multiply Equation 1 by 41.

$\underline{220{,}000 = 451r_1 + 451r_2}$ Multiply Equation 2 by 11.

$425{,}000 = 902r_1$ Add equations.

So, $r_1 = \dfrac{425{,}000}{902} \approx 471.18$ miles per hour

and $r_2 = \frac{1}{11}(11r_1 - 5000)$

$$r_2 = \frac{1}{11}\left(11 \cdot \frac{425{,}000}{902} - 5000\right) \approx 16.63 \text{ miles per hour.}$$

Check this solution in the original system of equations.

$$2000 \approx (471.18 - 16.63)\left(4 + \frac{24}{60}\right) \checkmark$$

$$2000 \approx (471.18 + 16.63)\left(4 + \frac{6}{60}\right) \checkmark$$

9. Because p is written in terms of x, begin by substituting the value of p given in the supply equation into the demand equation.

$p = 567 - 0.00002x$ Write demand equation.

$492 + 0.00003x = 567 - 0.00002x$ Substitute $492 + 0.00003x$ for p.

$0.00005x = 75$ Combine like terms.

$x = 1{,}500{,}000$ Solve for x.

So, the equilibrium point occurs when the demand and supply are each 1.5 million units. Obtain the price that corresponds to this x-value by back-substituting $x = 1{,}500{,}000$ into either of the original equations. For instance, back-substituting into the demand equation produces

$$p = 567 - 0.00002(1{,}500{,}000) = 567 - 30 = \$537.$$

The solution is $(1{,}500{,}000,\ 537)$. Check this by substituting into the demand and supply equations.

$p = 567 - 0.00002x$

$537 = 567 - 0.00002(1{,}500{,}000)$ ✓

$p = 492 + 0.00003x$

$537 = 492 + 0.00003(1{,}500{,}000)$ ✓

Checkpoints for Section 9.3

1. From Equation 3, you know the value of z. To solve for y, back-substitute $z = 3$ into Equation 2 to obtain the following.

$y + 3z = 6$ Write Equation 2.

$y + 3(3) = 6$ Substitute 3 for z.

$y = -3$ Solve for y.

Then back-substitute $y = -3$ and $z = 3$ into Equation 1 to obtain the following.

$2x - y + 5z = 22$ Write Equation 1.

$2x - (-3) + 5(3) = 22$ Substitute -3 for y and 3 for z.

$2x = 4$ Combine like terms.

$x = 2$ Solve for x.

The solution is $x = 2$, $y = -3$, and $z = 3$, which can be written as the ordered triple $(2, -3, 3)$. Check this in the original system of equations.

Check

Equation 1: $2x - y + 5z = 22$

$$2(2) - (-3) + 5(3) = 22$$

$$4 + 3 + 15 = 22 \checkmark$$

Equation 2: $y + 3z = 6$

$$(-3) + 3(3) \stackrel{?}{=} 6$$

$$-3 + 9 = 6 \checkmark$$

Equation 3: $z = 3$

$$(3) = 3 \checkmark$$

2. $\begin{cases} 2x + y = 3 \\ x + 2y = 3 \end{cases}$ Write Equation 1. Write Equation 2.

$\begin{cases} x + 2y = 3 \\ 2x + y = 3 \end{cases}$ Interchange the two equations in the system.

$\begin{cases} -2x - 4y = -6 \\ 2x + y = 3 \end{cases}$ Multiply the first equation by -2.

$$\begin{aligned} -2x - 4y &= -6 \\ 2x + y &= 3 \\ \hline -3y &= -3 \\ y &= 1 \end{aligned}$$

Add the multiple of the first equation to the second equation to obtain a new second equation.

$\begin{cases} x + 2y = 3 \\ y = 1 \end{cases}$ New system in row-echelon form.

Now back-substitute $y = 1$ into the first equation in row-echelon form and solve for x.

$x + 2(1) = 3$ Substitute 1 for y.

$x = 1$ Solve for x.

The solution is $x = 1$ and $y = 1$, which can be written as the ordered pair $(1, 1)$.

3. Because the leading coefficient of the first equation is 1, begin by keeping the x in the upper left position and eliminating the other x-terms from the first column.

$$\begin{aligned} -2x - 2y - 2z &= -12 && \text{Multiply Equation 1 by } -2. \\ 2x - y + z &= 3 && \text{Write Equation 2.} \\ \hline -3y - z &= -9 && \text{Add revised Equation 1 to Equation 2.} \end{aligned}$$

$$\begin{cases} x + y + z = 6 \\ -3y - z = -9 \\ 3x + y - z = 2 \end{cases}$$ Adding -2 times the first equation to the second equation produces a new second equation.

$$\begin{aligned} -3x - 3y - 3z &= -18 && \text{Multiply Equation 1 by } -3. \\ 3x + y - z &= 2 && \text{Write Equation 3.} \\ \hline -2y - 4z &= -16 && \text{Add revised Equation 1 to Equation 3.} \end{aligned}$$

$$\begin{cases} x + y + z = 6 \\ -3y - z = -9 \\ -2y - 4z = -16 \end{cases}$$ Adding -3 times the first equation to the third equation produces a new third equation.

Now that you have eliminated all but the x in the upper position of the first column, work on the second column.

$$\begin{cases} x + y + z = 6 \\ -3y - z = -9 \\ -y - 2z = -8 \end{cases}$$ Multiplying the third equation by 2, produces a new third equation.

$$\begin{aligned} -3y - z &= -9 && \text{Write Equation 2.} \\ 3y + 6z &= 24 && \text{Multiply Equation 3 by } -3. \\ \hline 5z &= 15 && \text{Add equations.} \end{aligned}$$

$$\begin{cases} x + y + z = 6 \\ -3y - z = -9 \\ 5z = 15 \end{cases}$$ Adding the second equation to -3 times the third equation produces a new third equation.

$$\begin{cases} x + y + z = 6 \\ y + \frac{1}{3}z = 3 \\ 5z = 15 \end{cases}$$ Multiplying the second equation by $-\frac{1}{3}$ produces a new second equation.

$$\begin{cases} x + y + z = 6 \\ y + \frac{1}{3}z = 3 \\ z = 3 \end{cases}$$ Multiplying the thrid equation by $\frac{1}{5}$ produces a new third equation.

To solve for y, back-substitute $z = 3$ into Equation 2 to obtain the following.

$$y + \tfrac{1}{3}(3) = 3$$
$$y = 2.$$

Then back-substitute $y = 2$ and $z = 3$ into Equation 1 to obtain the following.

$$x + (2) + (3) = 6$$
$$x = 1$$

The solution is $x = 1$, $y = 2$, and $z = 3$, which can be written as $(1, 2, 3)$.

4. $\begin{cases} x + y - 2z = 3 \\ 3x - 2y + 4z = 1 \\ 2x - 3y + 6z = 8 \end{cases}$

$\begin{cases} x + y - 2z = 3 \\ -5y + 10z = -8 \\ 2x - 3y + 6z = 8 \end{cases}$ Adding -3 times the first equation to the second equation produces a new second equation.

$\begin{cases} x + y - 2z = 3 \\ -5y + 10z = -8 \\ -5y + 10z = 2 \end{cases}$ Adding -2 times the first equation to the third equation produces a new third equation.

$\begin{cases} x + y - 2z = 3 \\ -5y + 10z = -8 \\ 0 = 10 \end{cases}$ Adding -1 times the second equation to the third equation produces a new third equation.

Because $0 = 10$ is a false statement, this is an inconsistent system and has no solution. Moreover, because this system is equivalent to the original system, the original system has no solution.

5. $\begin{cases} x + 2y - 7z = -4 \\ 2x + 3y + z = 5 \\ 3x + 7y - 36z = -25 \end{cases}$

$\begin{cases} x + 2y - 7z = -4 \\ -y + 15z = 13 \\ 3x + 7y - 36z = -25 \end{cases}$ Adding -2 times the first equation to the second equation produces a new second equation.

$\begin{cases} x + 2y - 7z = -4 \\ -y + 15z = 13 \\ y - 15z = -13 \end{cases}$ Adding -3 times the first equation to the third equation produces a new third equation.

$\begin{cases} x + 2y - 7z = -4 \\ -y + 15z = 13 \\ 0 = 0 \end{cases}$ Adding the second equation to the third equation to produces a new third equation.

This result means that Equation 3 depends on Equations 1 and 2 in the sense that it gives no additional information about the variables. Because $0 = 0$ is a true statement, this system has infinitely many solutions. However, it is incorrect to say that the solution is "infinite." You must also specify the correct form of the solution. So, the original system is equivalent to the system.

$\begin{cases} x + 2y - 7z = -4 \\ -y + 15z = 13. \end{cases}$

In the second equation, solve for y in terms of z to obtain the following.

$$-y + 15z = 13$$
$$-y = -15z + 13$$
$$y = 15z - 13$$

Back-substituting in the first equation produces the following.

$$x + 2y - 7z = -4$$
$$x + 2(15z - 13) - 7z = -4$$
$$x + 30z - 26 - 7z = -4$$
$$x = -23z + 22$$

Finally, letting $z = a$ where a is a real number, the solutions of the given system are all of the form $x = -23a + 22$, $y = 15a - 13$, and $z = a$. So, every ordered triple of the form $(-23a + 22, 15a - 13, a)$ is a solution of the system.

6. $\begin{cases} x - y + 4z = 3 \\ 4x \qquad\quad - z = 0 \end{cases}$

$\begin{cases} x - y + \quad z = \quad 3 \\ \quad 4y - 17z = -12 \end{cases}$ Adding -4 times the first equation to the second equation produces a new second equation.

$\begin{cases} x - y + \quad z = 3 \\ \quad y - \frac{17}{4}z = -3 \end{cases}$ Multiplying the second equation by $\frac{1}{4}$ produces a new second equation.

Solve for y in terms of z to obtain the following.

$$y - \tfrac{17}{4}z = -3$$
$$y = \tfrac{17}{4}z - 3$$

Solve for x by back-substituting $y = \frac{17}{4}z - 3$ into Equation 1.

$$x - y + 4z = 3$$
$$x - \left(\tfrac{17}{4}z - 3\right) + 4z = 3$$
$$x - \tfrac{17}{4}z + 3 + 4z = 3$$
$$x = \tfrac{1}{4}z$$

Finally, by letting $z = a$, where a is a real number, you have the solution

$x = \frac{1}{4}a$, $y = \frac{17}{4}a - 3$, and $z = a$.

So, every ordered triple of the form $\left(\frac{1}{4}a, \frac{17}{4}a - 3, a\right)$ is a solution of the system. Because the original system had three variables and only two equations, the system cannot have a unique solution.

7. By substituting the three values of t and s into the position equation, you can obtain three linear equations in a, v_0 and s_0.

When $t = 1$: $\frac{1}{2}a(1)^2 + v_0(1) + s_0 = 104 \Rightarrow a + 2v_0 + 2s_0 = 208$

When $t = 2$: $\frac{1}{2}a(2)^2 + v_0(2) + s_0 = 76 \Rightarrow 2a + 2v_0 + s_0 - 76$

When $t = 3$: $\frac{1}{2}a(3)^2 + v_0(3) + s_0 = 16 \Rightarrow 9a + 6v_0 + 2s_0 = 32$

This produces the following system of linear equation.

$$\begin{cases} a + 2v_0 + 2s_0 = 208 \\ 2a + 2v_0 + s_0 = 76 \\ 9a + 6v_0 + 2s_0 = 32 \end{cases}$$

Now solve the system using Gaussian Elimination.

$$\begin{cases} a + 2v_0 + 2s_0 = 208 \\ -2v_0 - 3s_0 = -340 \\ 9a + 6v_0 + 2s_0 = 32 \end{cases}$$

Adding -2 times the first equation to the second equation produces a new second equation.

$$\begin{cases} a + 2v_0 + 2s_0 = 208 \\ -2v_0 - 3s_0 = -340 \\ -12v_0 - 16s_0 = -1840 \end{cases}$$

Adding -9 times the first equation to the third equation produces a new third equation.

$$\begin{cases} a + 2v_0 + 2s_0 = 208 \\ -2v_0 - 3s_0 = -340 \\ 2s_0 = 200 \end{cases}$$

Adding -6 times the second equation to the third equation produces a new third equation.

$$\begin{cases} a + 2v_0 + 2s_0 = 208 \\ v_0 + \frac{3}{2}s_0 = 170 \\ s_0 = 100 \end{cases}$$

Multiplying the second equation by $-\frac{1}{2}$ produces a new second equation and multiplying the third equation by $\frac{1}{2}$ produces a new third equation.

So, $s_0 = 100$.

Find v_0 by back-substituting $s_0 = 100$ into Equation 2.

$$v_0 + \tfrac{3}{2}(100) = 170$$
$$v_0 = 20$$

Find a by back-substituting $s_0 = 100$ and $v_0 = 20$ into Equation 1.

$$a + 2(20) + 2(100) = 208$$
$$a = -32$$

So, the solution of this system is $a = -32$, $v_0 = 20$, and $s_0 = 100$, which can be written as $(-32, 20, 100)$.

This results in a position equation of $s = \frac{1}{2}(-32)t^2 + 20t + 100$

$= -16t^2 + 20t + 100$

and implies that the object was thrown upward at a velocity of 20 feet per second from a height of 100 feet.

8. Because the graph of $y = ax^2 + bx + c$ passes through the points $(0, 0)$, $(3, -3)$, and $(6, 0)$, you can write the following.

When $x = 0$, $y = 0$: $a(0)^2 + b(0) + c = 0$

When $x = 3$, $y = -3$: $a(3)^2 + b(3) + c = -3$

When $x = 6$, $y = 0$: $a(6)^2 + b(6) + c = 0$

This produces the following system of linear equations.

$$\begin{cases} c = 0 & \text{Equation 1} \\ 9a + 3b + c = -3 & \text{Equation 2} \\ 36a + 6b + c = 0 & \text{Equation 3} \end{cases}$$

You can reorder these equations as shown.

$$\begin{cases} 36a + 6b + c = 0 \\ 9a + 3b + c = -3 \\ c = 0 \end{cases}$$

$$\begin{cases} 36a + 6b + c = 0 \\ -6b - 3c = 12 \\ c = 0 \end{cases}$$

Adding -4 times the second equation to the first equation produces a new second equation.

$$\begin{cases} a + \frac{1}{6}b + \frac{1}{36}c = 0 \\ b + \frac{1}{2}c = -2 \\ c = 0 \end{cases}$$

Multiplying the first equation by $\frac{1}{36}$ produces a new first equation and multiplying the second equation by $-\frac{1}{6}$ produces a new second equation.

So, $c = 0$,

$$b + \tfrac{1}{2}(0) = -2$$
$$b = -2,$$

and $a + \frac{1}{6}(-2) + \frac{1}{36}(0) = 0$

$$a = \tfrac{1}{3}.$$

The solution of this system is $a = \frac{1}{3}$, $b = -2$, and $c = 0$.

So, the equation of the parabola is $y = \frac{1}{3}x^2 - 2x$.

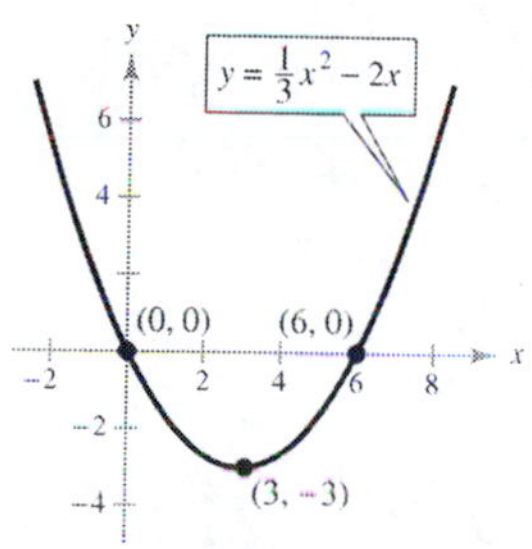

Checkpoints for Section 9.4

1. The expression is proper, so you should begin by factoring the denominator. Because

$$2x^2 - x - 1 = (2x + 1)(x - 1)$$

you should include one partial fraction with a constant numerator for each linear factor of the denominator. Write the form of the decomposition as follows.

$$\frac{x + 5}{2x^2 - x - 1} = \frac{A}{2x + 1} + \frac{B}{x - 1}$$

Multiplying each side of this equation by the least common denominator, $(2x + 1)(x - 1)$, leads to the **basic equation**

$$x + 5 = A(x - 1) + B(2x + 1).$$

Because this equation is true for all x, substitute any *convenient* values of x that will help determine the constants A and B. Values of x that are especially convenient are those that make the factors $x - 1$ and $2x + 1$ equal to zero. For instance, to solve for B, let $x = 1$. Then

$$1 + 5 = A(1 - 1) + B\left[2(1) + 1\right] \qquad \text{Substitute 1 for } x$$

$$6 = A(0) + B(3)$$

$$6 = 3B$$

$$2 = B.$$

To solve for A, let $x = -\frac{1}{2}$ and then

$$-\frac{1}{2} + 5 = A\left(\frac{1}{2} - 1\right) + B\left[2\left(\frac{1}{2}\right) + 1\right] \qquad \text{Substitute } \frac{1}{2} \text{ for } x$$

$$\frac{9}{2} = A\left(-\frac{3}{2}\right) + B(0)$$

$$\frac{9}{2} = -\frac{3}{2}A$$

$$-3 = A.$$

So, the partial fraction decomposition is

$$\frac{x + 5}{2x^2 - x - 1} = \frac{-3}{2x + 1} + \frac{2}{x - 1}.$$

Check this result by combining the two partial fractions on the right side of the equation, or by using your graphing utility.

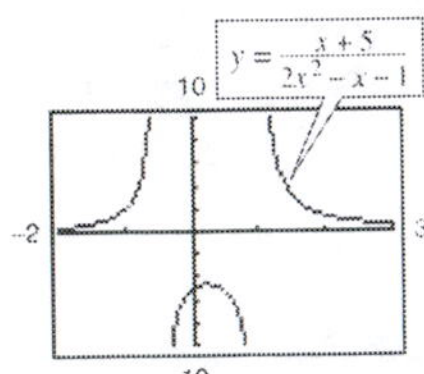

2. This rational expression is improper, so you should begin by dividing the numerator by the denominator.

$$\frac{x^4 + x^3 + x + 4}{x^3 + x^2} \Rightarrow x^3 + x^2 \overline{\smash{\big)}\, x^4 + x^3 + 0x^2 + x + 4}$$

$$\begin{array}{r} x \\ x^3 + x^2 \overline{\smash{\big)}\, x^4 + x^3 + 0x^2 + x + 4} \\ \underline{x^4 + x^3 } \\ x + 4 \end{array}$$

So, $\dfrac{x^4 + x^3 + x + 4}{x^3 + x^2} = x + \dfrac{x + 4}{x^3 + x^2}$.

Because the denominator of the remainder factors as $x^3 + x^2 = x^2(x + 1)$, you should include one partial fraction with a constant numerator for each power of x and $x + 1$, and write the form of the decomposition as follows.

$$\frac{x + 4}{x^3 + x^2} = \frac{A}{x} + \frac{B}{x^2} + \frac{C}{x + 1}$$

Multiplying each side by the LCD, $x^2(x + 1)$, leads to the basic equation

$$x + 4 = Ax(x + 1) + B(x + 1) + Cx^2.$$

Letting $x = -1$ eliminates the A- and B-terms and yields the following.

$$\begin{aligned} -1 + 4 &= A(-1)(-1 + 1) + B(-1 + 1) + C(-1)^2 \\ 3 &= 0 + 0 + C \\ 3 &= C \end{aligned}$$

Letting $x = 0$, eliminates the A- and C-terms.

$$\begin{aligned} 0 + 4 &= A(0)(0 + 1) + B(0 + 1) + C(0)^2 \\ 4 &= 0 + B + 0 \\ 4 &= B \end{aligned}$$

At this point, you have exhausted the most convenient values of x, so to find the value of A, use any other value of x along with the known values of B and C.

So, using $x = 1$, $B = 4$ and $C = 3$,

$$\begin{aligned} 1 + 4 &= A(1)(1 + 1) + 4(1 + 1) + 3(1)^2 \\ 5 &= 2A + 8 + 3 \\ -6 &= 2A \\ -3 &= A. \end{aligned}$$

So, the partial fraction decomposition is

$$\frac{x^4 + x^3 + x + 4}{x^3 + x^2} = x + \frac{-3}{x} + \frac{4}{x^2} + \frac{3}{x + 1}.$$

3. This expression is proper, so begin by factoring the denominator. Because the denominator factors as

$x^3 + x = x(x^2 + 1)$

you should include one partial fraction with a constant numerator and one partial fraction with a linear numerator, and write the form of the decomposition as follows.

$$\frac{2x^2 - 5}{x^3 + x} = \frac{A}{x} + \frac{Bx + C}{x^2 + 1}$$

Multiplying each side by the LCD, $x(x^2 + 1)$ yields the basic equation

$$2x^2 - 5 = A(x^2 + 1) + (Bx + C)x$$

Expanding this basic equation and collecting like terms produces

$$\begin{aligned} 2x^2 - 5 &= Ax^2 + A + Bx^2 + Cx \\ &= (A + B)x^2 + Cx + A. \qquad \text{Polynomial form} \end{aligned}$$

Finally, because two polynomials are equal if and only if the coefficients of like terms are equal, equate the coefficients of like terms on opposite sides of the equation.

$$2x^2 + 0x - 5 = (A + B)x^2 + Cx + A$$

Now write the following system of linear equations.

$$\begin{cases} A + B = 2 & \text{Equation 1} \\ C = 0 & \text{Equation 2} \\ A = -5 & \text{Equation 3} \end{cases}$$

From this system, you can see that $A = -5$ and $C = 0$.

Moreover, back-substituting $A = -5$ into Equation 1 yields $-5 + B = 2 \Rightarrow B = 7$.

So, the partial fraction decomposition is $\dfrac{2x^2 - 5}{x^3 + x} = \dfrac{-5}{x} + \dfrac{7x}{x^2 + 1}$.

4. Include one partial fraction with a linear numerator for each power of $(x^2 + 4)$.

$$\frac{x^3 + 3x^2 - 2x + 7}{(x^2 + 4)^2} = \frac{Ax + B}{x^2 + 4} + \frac{Cx + D}{(x^2 + 4)^2} \qquad \text{Write form of decomposition.}$$

Multiplying each side by the LCD, $(x^2 + 4)^2$, yields the basic equation

$$\begin{aligned} x^3 + 3x^2 - 2x + 7 &= (Ax + B)(x^2 + 4) + Cx + D \qquad \text{Basic equation} \\ &= Ax^3 + 4Ax + Bx^2 + 4B + Cx + D \\ &= Ax^3 + Bx^2 + (4A + C)x + (4B + D). \qquad \text{Polynomial form} \end{aligned}$$

Equating coefficients of like terms on opposite sides of the equation

$$x^3 + 3x^2 - 2x + 7 = Ax^3 + Bx^2 + (4A + C)x + (4B + D)$$

produces the following system of linear equations.

$$\begin{cases} A = 1 & \text{Equation 1} \\ B = 3 & \text{Equation 2} \\ 4A + C = -2 & \text{Equation 3} \\ 4B + D = 7 & \text{Equation 4} \end{cases}$$

Use the values $A = 1$ and $B = 3$ to obtain the following.

$$\begin{aligned} 4(1) + C &= -2 \qquad \text{Substitute 1 for } A \text{ in Equation 3.} \\ C &= -6 \\ 4(3) + D &= 7 \qquad \text{Substitute 3 for } B \text{ in Equation 4.} \\ D &= -5 \end{aligned}$$

So, using $A = 1$, $B = 3$, $C = -6$, and $D = -5$

The partial fraction decomposition is $\dfrac{x^3 + 3x^2 - 2x + 7}{(x^2 + 4)^2} = \dfrac{x + 3}{x^2 + 4} + \dfrac{-6x - 5}{(x^2 + 4)^2}$.

Check this result by combining the two partial fractions on the right side of the equation, or by using your graphing utility.

5. Include one partial fraction with a constant numerator for each power of x and one partial fraction with a linear numerator for each power of $(x^2 + 2)$.

$$\frac{4x - 8}{x^2(x^2 + 2)^2} = \frac{A}{x} + \frac{B}{x^2} + \frac{Cx + D}{x^2 + 2} + \frac{Ex + F}{(x^2 + 2)^2} \quad \text{Write form of decomposition.}$$

Multiplying each side by the LCD, $x^2(x^2 + 2)^2$, yields the basic equation

$$\begin{aligned}4x - 8 &= Ax(x^2 + 2)^2 B(x^2 + 2)^2 + (Cx + D)x^2(x^2 + 2) + (Ex + F)x^2\\ &= Ax(x^4 + 4x^2 + 4) + B(x^4 + 4x^2 + 4) + (Cx + D)(x^4 + 2x^2) + (Ex + F)x^2\\ &= Ax^5 + 4Ax^3 + 4Ax + Bx^4 + 4Bx^2 + 4B + Cx^5 + 2Cx^3 + Dx^4 + 2Dx^2 + Ex^3 + Fx^2\\ &= (A + C)x^5 + (B + D)x^4 + (4A + 2C + E)x^3 + (4B + 2D + F)x^2 + (4A)x + 4B\end{aligned}$$

Equating coefficients yields this system of linear equations.

$$\begin{cases} A + C = 0 & \text{Equation 1}\\ B + D = 0 & \text{Equation 2}\\ 4A + 2C + E = 0 & \text{Equation 3}\\ 4B + 2D + F = 0 & \text{Equation 4}\\ 4A = 4 & \text{Equation 5}\\ 4B = -8 & \text{Equation 6}\end{cases}$$

So, from Equations 5 and 6, $A = 1$ and $B = -2$.

Then back-substituting into Equations 1 and 2, $1 + C = 0 \Rightarrow C = -1$ and $-2 + D = 0 \Rightarrow D = 2$.

Using these values and Equations 3 and 4, you have

$4(1) + 2(-1) + E = 0 \Rightarrow E = -2$ and $4(-2) + 2(2) + F = 0 \Rightarrow F = 4$.

So, $A = 1$, $B = -2$, $C = -1$, $D = 2$, $E = -2$, and $F = 4$.

The partial fraction decomposition is

$$\frac{4x - 8}{x^2(x^2 + 2)^2} = \frac{1}{x} + \frac{-2}{x^2} + \frac{-x + 2}{x^2 + 2} + \frac{-2x + 4}{(x^2 + 2)^2}.$$

Checkpoints for Section 9.5

1. Begin by graphing the corresponding equation $(x + 2)^2 + (y - 2)^2 = 16$, which is a circle, with center $(-2, 2)$ and a radius of 4 units as shown.

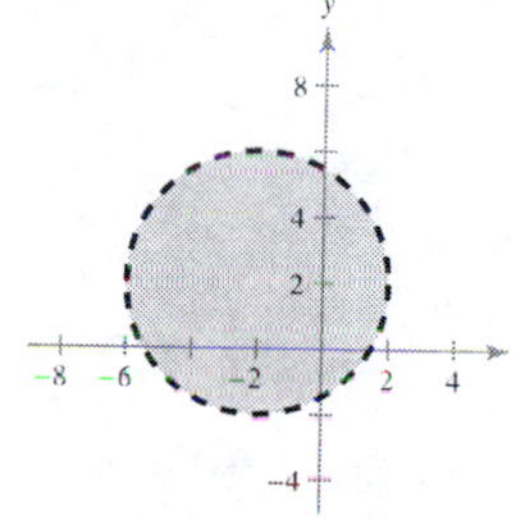

Test a point inside the circle such as $(-2, 2)$ and a point outside the circle such as $(4, 2)$.

The points that satisfy the inequality are those lying inside the circle but not on the circle.

$(-2, 2)$: $(x + 2)^2 + (y - 2)^2 \overset{?}{<} 16$

$(-2 + 2)^2 + (2 - 2)^2 \overset{?}{<} 16$

$0 < 16$

$(-2, 2)$ is a solution.

$(4, 2)$: $(x + 2)^2 + (y - 2)^2 < 16$

$(4 + 2)^2 + (2 - 2)^2 \overset{?}{<} 16$

$36 \not< 16$

$(4, 2)$ is not a solution.

2. The graph of the corresponding equation $x = 3$ is a vertical line. The points that satisfy the inequality $x \geq 3$ are those lying to the right of (or on) this line.

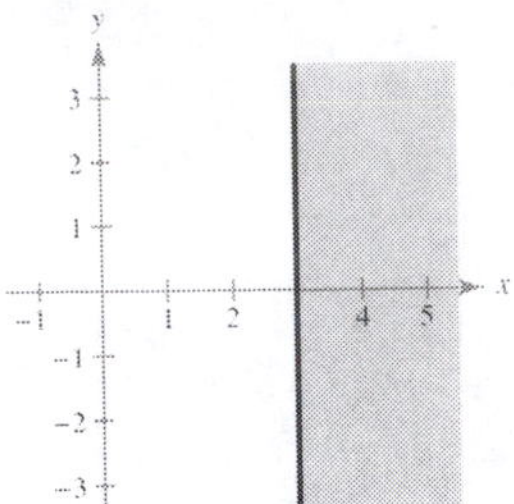

3. The graph of the corresponding equation $x + y = -2$ is a line as shown. Because the origin $(0, 0)$ satisfies the inequality, the graph consists of the half-plane lying above the line.

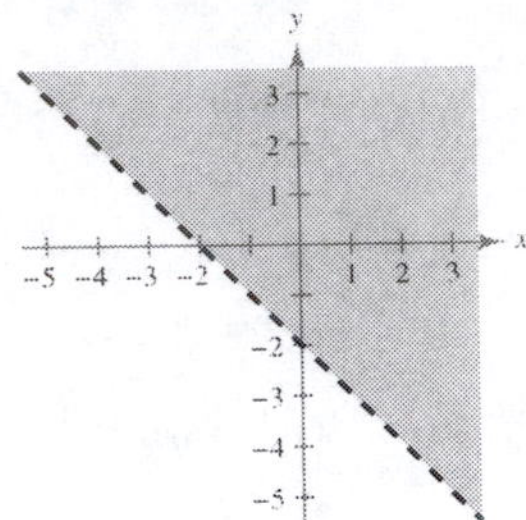

4. The graphs of each of these inequalities are shown independently.

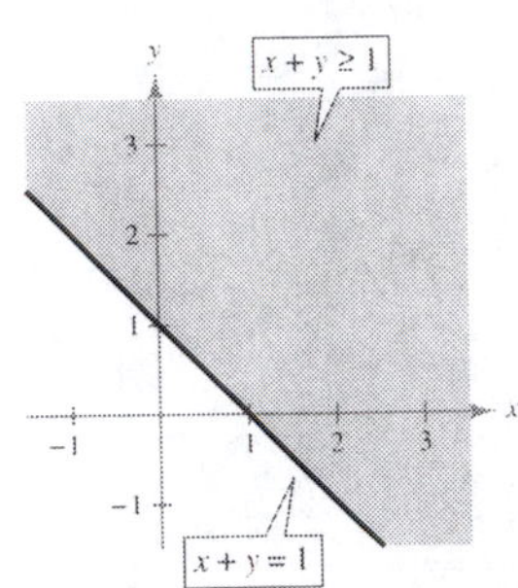

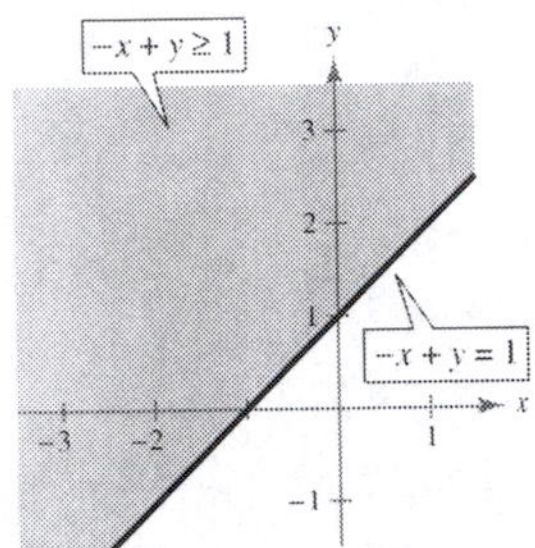

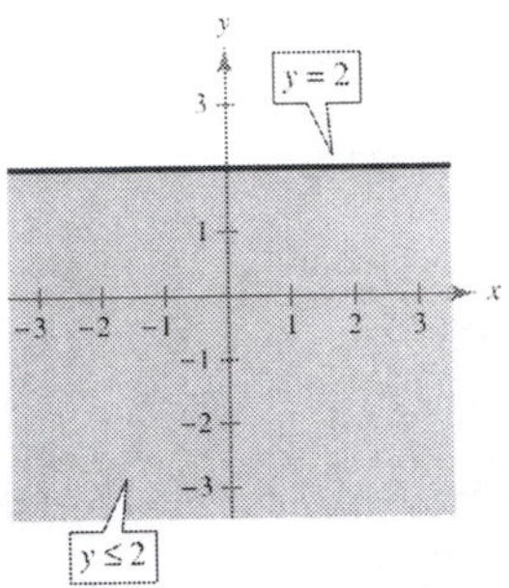

By superimposing the graphs on the same coordinate system, the region common to all three graphs can be found. To find the vertices of the region, solve the three systems of corresponding equations by taking pairs of equations representing the boundaries of the individual regions.

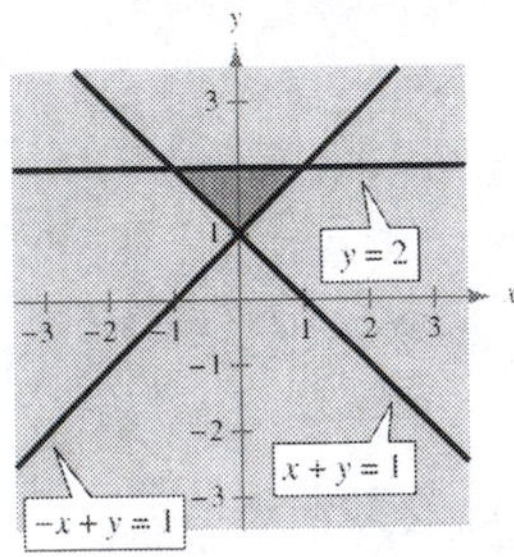

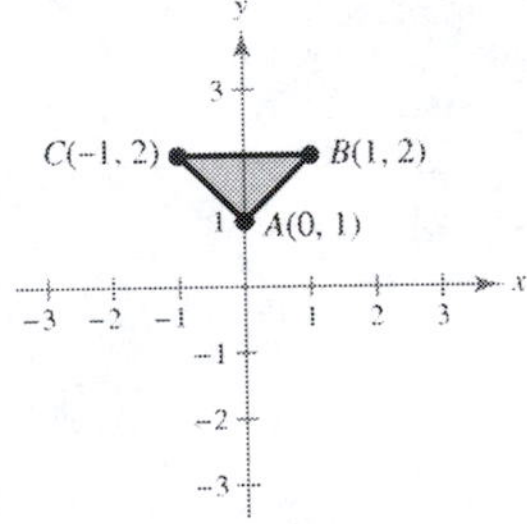

Vertex A: $(0, 1)$

$$\begin{cases} x + y = 1 \\ -x + y = 1 \end{cases}$$

Vertex B: $(1, 2)$

$$\begin{cases} -x + y = 1 \\ y = 2 \end{cases}$$

Vertex C: $(-1, 2)$

$$\begin{cases} x + y = 1 \\ y = 2 \end{cases}$$

Note that the vertices of the region are represented by solid dots. This means that the vertices *are* solutions of the system of equations as well as all of the points that lie on the lines.

5. The points that satisfy the inequality $x - y^2 > 0$ are the points inside the parabola (but not on) the parabola $x = y^2$.

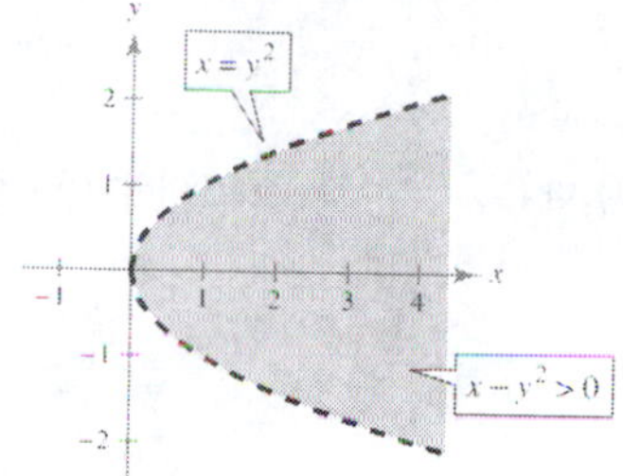

The points satisfying the inequality $x + y < 2$ are the points lying below (but not on) the line $x + y = 2$.

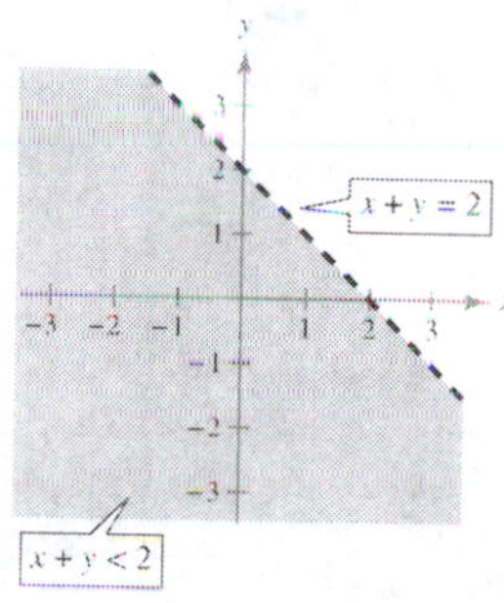

To find the points of the intersection of the parabola and the line, solve the system of corresponding equations.

$$\begin{cases} x - y^2 = 0 \\ x + y = 2 \end{cases}$$

$$\begin{aligned} x + y = 2 &\Rightarrow y = 2 - x \\ x - y^2 &= 0 \\ x - (2 - x)^2 &= 0 \\ x - \left(4 - 4x + x^2\right) &= 0 \\ x - 4 + 4x - x^2 &= 0 \\ -x^2 + 5x - 4 &= 0 \\ -\left(x^2 - 5x + 4\right) &= 0 \\ (x - 4)(x - 1) &= 0 \end{aligned}$$

$$x - 4 = 0 \qquad x - 1 = 0$$

$$x = 4 \qquad x = 1$$

When $x = 4$, $y = 2 - 4 = -2$.

When $x = 1$, $y = 2 - 1 = 1$.

Using the method of substitution, you can find the solutions to be $(4, -2)$ and $(1, 1)$.

So, the region containing all points that satisfy the system is indicated by the shaded region.

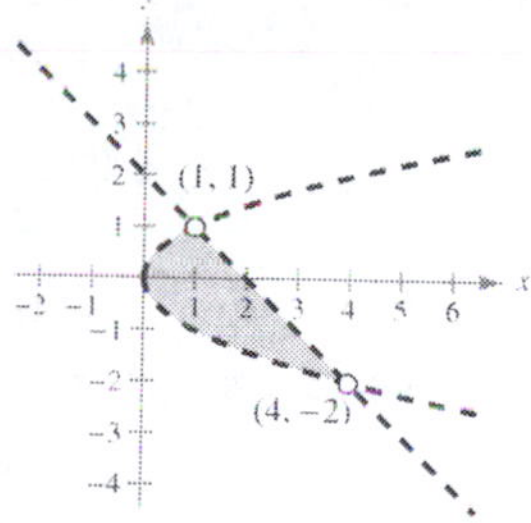

6. From the way the system is written, it should be clear that the system has no solution because the quantity $2x - y$ cannot be both less than -3 and greater than 1. The graph of the inequality $2x - y < -3$ is the half-plane lying above the line $2x - y = -3$ and the graph of the inequality $2x - y > 1$ is the half-plane lying below the line $2x - y = 1$ as shown. These two half-planes have no points in common. So, the system of inequalities has no solution.

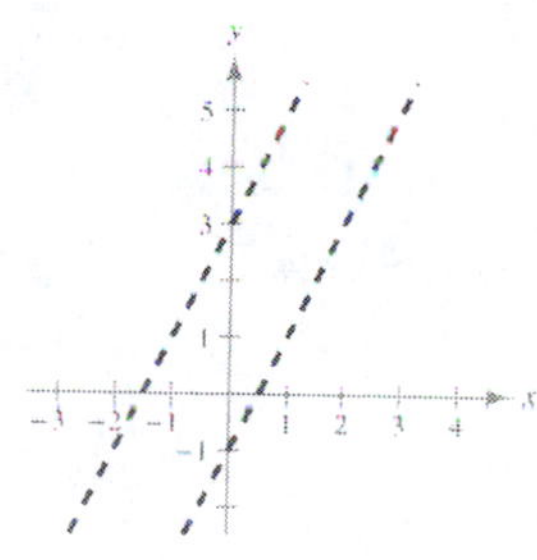

7. The graph of the inequality $x^2 - y < 0$ is the region inside the parabola $x^2 - y = 0$. The graph of the inequality $x - y < -2$ is the half-plane that lies above the line $x - y = -2$. The intersection of these regions is an infinite region having points of intersection at $(-1, 1)$ and $(2, 4)$ as shown points of intersection. So, the solution set of the system of inequalities is unbounded.

Points of intersection:

$$\begin{cases} x^2 + y = 0 \Rightarrow x^2 = y \\ x - y = -2 \end{cases}$$

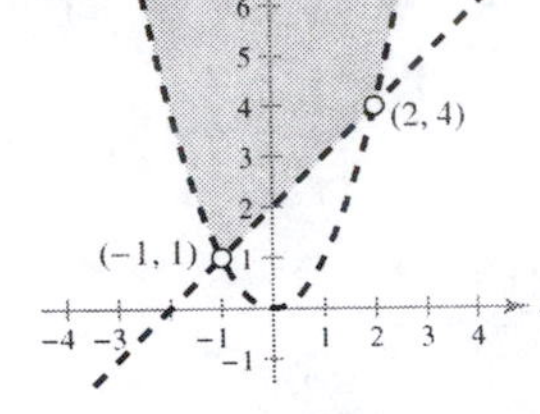

$$x - (x^2) = -2$$

$$x^2 - x - 2 = 0$$

$$(x - 2)(x + 1) = 0$$

$$x - 2 = 0 \qquad x + 1 = 0$$

$$x = 2 \qquad x = -1$$

When $x = 2,\ y = (2)^2 = 4.$

When $x = -1\ \ y = (-1)^2 = 1.$

8. Begin by finding the equilibrium point (when supply and demand are equal) by solving the equation

$492 + 0.00003x = 567 - 0.00002x.$

In checkpoint 9 in Section 9.2, you saw that the solution is $x = 1{,}500{,}000$ units, which corresponds to an equilibrium price of $p = 537$. So, the consumer surplus and producer surplus are the areas of the following triangular regions.

Consumer Surplus

$$\begin{cases} p \le 567 - 0.00002x \\ p \ge 537 \\ x \ge 0 \end{cases}$$

Producer Surplus

$$\begin{cases} p \ge 492 + 0.00003x \\ p \le 537 \\ x \ge 0 \end{cases}$$

The consumer and producer surpluses are the areas of the shaded triangles shown.

$$\boxed{\text{Consumer surplus}} = \tfrac{1}{2}(\text{base})(\text{height})$$
$$= \tfrac{1}{2}(1{,}500{,}000)(30)$$
$$= \$22{,}500{,}000$$

$$\boxed{\text{Producer surplus}} = \tfrac{1}{2}(\text{base})(\text{height})$$
$$= \tfrac{1}{2}(1{,}500{,}000)(45)$$
$$= \$33{,}750{,}000$$

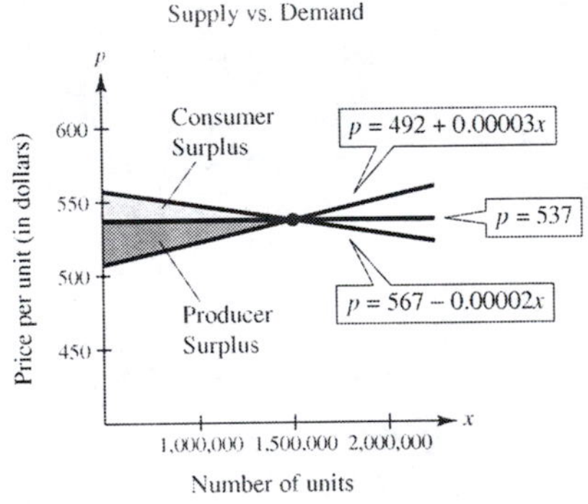

9. Begin by letting x represent the number of bags of brand X dog food and y represent the number of bags of brand Y dog food. To meet the minimum required amounts of nutrients, the following inequalities must be satisfied.

$$\begin{cases} 8x + 2y \ge 16 & \text{Nutrient A} \\ x + y \ge 5 & \text{Nutrient B} \\ 2x + 7y \ge 20 & \text{Nutrient C} \\ x \ge 0 & \\ y \ge 0 & \end{cases}$$

The graph of this system of inequalities is shown.

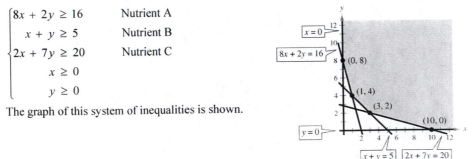

Checkpoints for Section 9.6

1. The constraints form the region shown.

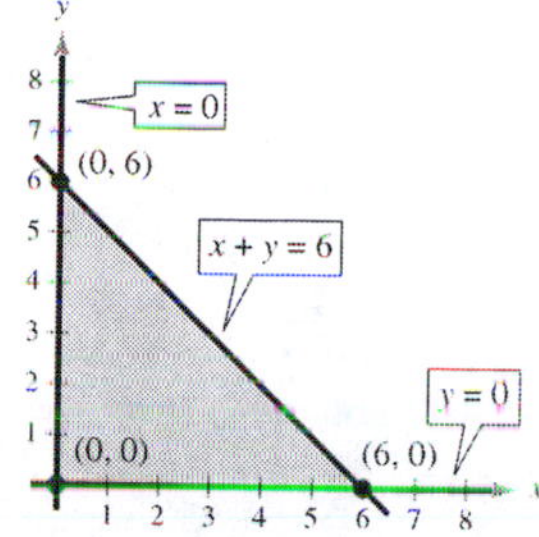

At the three vertices of the region, the objective function has the following values,

At $(0, 0)$: $z = 4(0) + 5(0) = 0$

At $(0, 6)$: $z = 4(0) + 5(6) = 30$

At $(6, 0)$: $z = 4(6) + 5(0) = 24$

So the maximum value of z is 30, and this occurs when $x = 0$ and $y = 6$.

2. The constraints form the region shown.

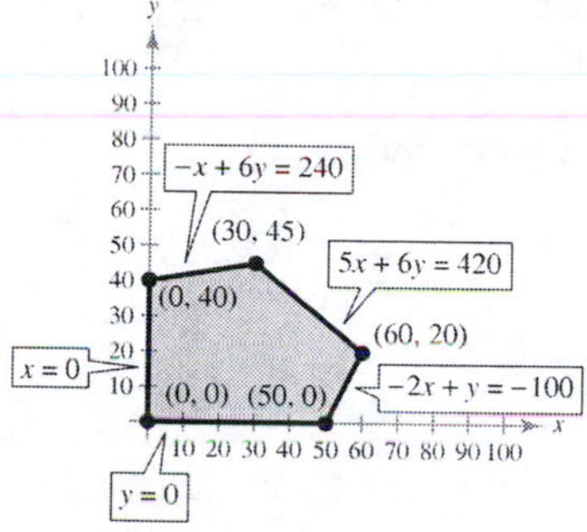

By testing the objective function at each vertex, you obtain the following.

At $(0, 0)$: $z = 12(0) + 8(0) = 0$

At $(0, 40)$: $z = 12(0) + 8(40) = 320$

At $(30, 45)$: $z = 12(30) + 8(45) = 720$

At $(60, 20)$: $z = 12(60) + 8(20) = 880$

At $(50, 0)$: $z = 12(50) + 8(0) = 600$

So, the minimum value of z is 0, which occurs when $x = 0$ and $y = 0$.

3. Using the values of z at the vertices shown in Checkpoint Example 2, the maximum value of z is

$$z = 12(60) + 8(20)$$
$$= 880$$

and occurs when $x = 60$ and $y = 20$.

4. The constraints form the region shown.

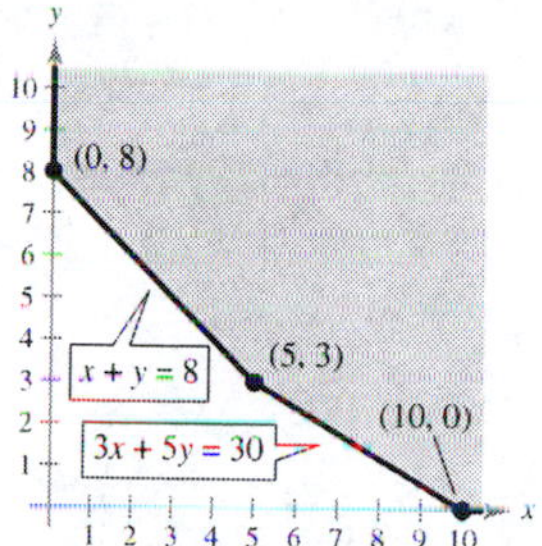

By testing the objective function at each vertex, you obtain the following.

At $(0, 8)$: $z = 3(0) + 7(8) = 56$

At $(5, 3)$: $z = 3(5) + 7(3) = 36$

At $(10, 0)$: $z = 3(10) + 7(0) = 30$

So, the minimum of z is 30, which occurs when $x = 10$ and $y = 0$.

5. Let x be the number of boxes of chocolate-covered creams and let y be the number of boxes of chocolate-covered nuts. So, the objective function (for the combined profit) is

$P = 2.5x + 2y$ Objective function

To find the maximum monthly profit, test the values of P at the vertices of the region.

At $(0, 0)$: $P = 2.5(0) + 2(0) = 0$

At $(800, 400)$: $P = 2.5(800) + 2(400) = 2800$

At $(1050, 150)$: $P = 2.5(1050) + 2(150) = 2925$ Maximum Profit

At $(600, 0)$: $P = 2.5(600) + 2(0) = 1500$

So, the maximum monthly profit is \$2925 and it occurs when the monthly production consists of 1050 boxes of chocolate-covered creams and 150 boxes of chocolate-covered nuts.

6. As in Example 9 Checkpoint in Section 9.5, let x be the number of bags of Brand X dog food and y be the number of bags of Brand Y dog food. The constraints are as follows.

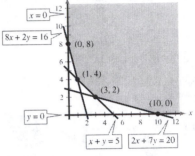

$$8x + 2y \geq 16$$
$$x + y \geq 5$$
$$2x + 7y \geq 20$$
$$x \geq 0$$
$$y \geq 0$$

The figure shows the graph of the region corresponding to the constraints.

The cost function is given by $C = 15x + 30y$.

Because you want to incur as little cost as possible, you want to determine the minimum cost.

At $(0, 8)$: $C = 15(0) + 30(8) = 240$

At $(1, 4)$: $C = 15(1) + 30(4) = 135$

At $(3, 2)$: $C = 15(3) + 30(2) = 105$ Minimum Cost

At $(10, 0)$: $C = 15(10) + 30(0) = 150$

So, the minimum cost is \$105 and occurs when 3 bags of Brand X and 2 bags of Brand Y.

Chapter 10

Checkpoints for Section 10.1

1. The matrix has *two* rows and *three* columns. The order of the matrix is 2×3.

2. $$\begin{cases} x + y + z = 2 \\ 2x - y + 3z = -1 \\ -x + 2y - z = 4 \end{cases}$$

All of the variables are aligned in the system. Next, use the coefficients and constant terms as the matrix entries.

$$\begin{matrix} R_1 \\ R_2 \\ R_3 \end{matrix} \begin{bmatrix} 1 & 1 & 1 & \vdots & 2 \\ 2 & -1 & 3 & \vdots & -1 \\ -1 & 2 & -1 & \vdots & 4 \end{bmatrix}$$

The augmented matrix has three rows and four columns, so it is a 3×4 matrix.

3. Add -3 times the first row of the original matrix to the second matrix.

Original Matrix **New Row-Equivalent Matrix**

$$\begin{bmatrix} 1 & 0 & 2 \\ 3 & 1 & 7 \\ 2 & -6 & 14 \end{bmatrix} \quad -3R_1 + R_2 \rightarrow \begin{bmatrix} 1 & 0 & 2 \\ 0 & 1 & 1 \\ 2 & -6 & 14 \end{bmatrix}$$

4. Linear System

$$\begin{cases} 2x + y - z = -3 \\ 4x - 2y + 2z = -2 \\ -6x + 5y + 4z = 10 \end{cases}$$

Associated Augmented Matrix

$$\begin{bmatrix} 2 & 1 & -1 & : & -3 \\ 4 & -2 & 2 & : & -2 \\ -6 & 5 & 4 & : & 10 \end{bmatrix}$$

Multiply the first equation by $\frac{1}{2}$.

$$\begin{cases} x + \frac{1}{2}y - \frac{1}{2}z = -\frac{3}{2} \\ 4x - 2y + 2z = -2 \\ -6x + 5y + 4z = 10 \end{cases}$$

$$\tfrac{1}{2}R_1 \rightarrow \begin{bmatrix} 1 & \frac{1}{2} & -\frac{1}{2} & : & -\frac{3}{2} \\ 4 & -2 & 2 & : & -2 \\ -6 & 5 & 4 & : & 10 \end{bmatrix}$$

Add -4 times the first equation to the second equation.

$$\begin{cases} x + \frac{1}{2}y - \frac{1}{2}z = -\frac{3}{2} \\ -4y + 4z = 4 \\ -6x + 5y + 4z = 10 \end{cases}$$

$$-4R_1 + R_2 \rightarrow \begin{bmatrix} 1 & \frac{1}{2} & -\frac{1}{2} & : & -\frac{3}{2} \\ 0 & -4 & 4 & : & 4 \\ -6 & 5 & 4 & : & 10 \end{bmatrix}$$

Multiply the second equation by $-\frac{1}{4}$.

$$\begin{cases} x + \frac{1}{2}y - \frac{1}{2}z = -\frac{3}{2} \\ y - z = -1 \\ -6x + 5y + 4z = 10 \end{cases}$$

$$-\tfrac{1}{4}R_2 \rightarrow \begin{bmatrix} 1 & \frac{1}{2} & -\frac{1}{2} & : & -\frac{3}{2} \\ 0 & 1 & -1 & : & -1 \\ -6 & 5 & 4 & : & 10 \end{bmatrix}$$

Add 6 times the first equation to the third equation.

$$\begin{cases} x + \frac{1}{2}y - \frac{1}{2}z = -\frac{3}{2} \\ y - z = -1 \\ 8y + z = 1 \end{cases}$$

$$6R_1 + R_3 \rightarrow \begin{bmatrix} 1 & \frac{1}{2} & -\frac{1}{2} & : & -\frac{3}{2} \\ 0 & 1 & -1 & : & -1 \\ 0 & 8 & 1 & : & 1 \end{bmatrix}$$

Add -8 times the second equation to the third equation.

$$\begin{cases} x + \frac{1}{2}y - \frac{1}{2}z = -\frac{3}{2} \\ y - z = -1 \\ 9z = 9 \end{cases}$$

$$-8R_2 + R_3 \rightarrow \begin{bmatrix} 1 & \frac{1}{2} & -\frac{1}{2} & : & -\frac{3}{2} \\ 0 & 1 & -1 & : & -1 \\ 0 & 0 & 9 & : & 9 \end{bmatrix}$$

Multiply the third equation by $\frac{1}{9}$.

$$\begin{cases} x + \frac{1}{2}y - \frac{1}{2}z = -\frac{3}{2} \\ y - z = -1 \\ z = 1 \end{cases}$$

$$\tfrac{1}{9}R_3 \rightarrow \begin{bmatrix} 1 & \frac{1}{2} & -\frac{1}{2} & : & -\frac{3}{2} \\ 0 & 1 & -1 & : & -1 \\ 0 & 0 & 1 & : & 1 \end{bmatrix}$$

At this point, you can use back-substitution to find x and y.

$$\begin{aligned} y - z &= -1 \\ y - (1) &= -1 \\ y &= 0 \\ x + \tfrac{1}{2}y - \tfrac{1}{2}z &= -\tfrac{3}{2} \\ x + \tfrac{1}{2}(0) - \tfrac{1}{2}(1) &= -\tfrac{3}{2} \\ x &= -1 \end{aligned}$$

The solution is $x = -1$, $y = 0$, and $z = 1$.

5. The Matrix is in row-echelon form, because the row consisting entirely of zeros occurs at the bottom of the matrix, and for each row that does not consist entirely of zeros, the first nonzero entry is 1. Furthermore the matrix is in reduced row-echelon form, since every column that has a leading 1 has zeros in every position above and below its leading 1.

6.

$$\begin{bmatrix} -3 & 5 & 3 & \vdots & -19 \\ 3 & 4 & 4 & \vdots & 8 \\ 4 & -8 & -6 & \vdots & 26 \end{bmatrix}$$ Write augmented matrix.

$$\begin{matrix} R_3 + R_1 \rightarrow \\ \\ \\ \end{matrix} \begin{bmatrix} 1 & -3 & -3 & \vdots & 7 \\ 3 & 4 & 4 & \vdots & 8 \\ 4 & -8 & -6 & \vdots & 26 \end{bmatrix}$$ Add R_3 to R_1 so first column has leading 1 in upper left corner.

$$\begin{matrix} \\ -3R_1 + R_2 \rightarrow \\ -4R_1 + R_3 \rightarrow \end{matrix} \begin{bmatrix} 1 & -3 & -3 & \vdots & 7 \\ 0 & 13 & 13 & \vdots & -13 \\ 0 & 4 & 6 & \vdots & -2 \end{bmatrix}$$ Perform operations on R_2 and R_3 so first column has zeros below its leading 1.

$$\begin{matrix} \\ R_2 + (-3)R_3 \rightarrow \\ \\ \end{matrix} \begin{bmatrix} 1 & -3 & -3 & \vdots & 5 \\ 0 & 1 & -5 & \vdots & -7 \\ 0 & 4 & 6 & \vdots & -2 \end{bmatrix}$$ Perform operations on R_2 so second column has a leading 1.

$$\begin{matrix} \\ \\ -4R_2 + R_3 \rightarrow \end{matrix} \begin{bmatrix} 1 & -3 & -3 & \vdots & 7 \\ 0 & 1 & -5 & \vdots & -7 \\ 0 & 0 & 26 & \vdots & 26 \end{bmatrix}$$ Perform operations on R_3 so second column has a zero below its leading 1.

$$\begin{matrix} \\ \\ \frac{1}{26}R_3 \rightarrow \end{matrix} \begin{bmatrix} 1 & -3 & -3 & \vdots & 7 \\ 0 & 1 & -5 & \vdots & -7 \\ 0 & 0 & 1 & \vdots & 1 \end{bmatrix}$$ Perform operations on R_3 so third column has a leading 1.

The matrix is now in row-echelon form, and the corresponding system is

$$\begin{cases} x - 3y - 3z = 7 \\ \quad\; y - 5z = -7 \\ \qquad\quad z = 1 \end{cases}$$

Using back-substitution, you can determine that the solution is $x = 4$, $y = -2$ and $z = 1$.

7.

$$\begin{bmatrix} 1 & 1 & 1 & \vdots & 1 \\ 1 & 2 & 2 & \vdots & 2 \\ 1 & -1 & -1 & \vdots & 1 \end{bmatrix}$$ Write augmented matrix.

$$\begin{matrix} \\ -R_1 + R_2 \rightarrow \\ -R_1 + R_3 \rightarrow \end{matrix} \begin{bmatrix} 1 & 1 & 1 & \vdots & 1 \\ 0 & 1 & 1 & \vdots & 1 \\ 0 & -2 & -2 & \vdots & 0 \end{bmatrix}$$ Perform row operations.

$$\begin{matrix} \\ \\ 2R_2 + R_3 \rightarrow \end{matrix} \begin{bmatrix} 1 & 1 & 1 & \vdots & 1 \\ 0 & 1 & 1 & \vdots & 1 \\ 0 & 0 & 0 & \vdots & 2 \end{bmatrix}$$ Perform row operations.

Note that the third row of this matrix consists entirely of zeros except for the last entry. This means that the original system of linear equations is inconsistent. You can see why this is true by converting back to a system of linear equations.

$$\begin{cases} x + y + z = 1 \\ \quad\; y + z = 1 \\ \qquad\; 0 = 2 \end{cases}$$

Because the third equation is not possible, the system has no solution.

8.

$$\begin{bmatrix} -3 & 7 & 2 & \vdots & 1 \\ -5 & 3 & -5 & \vdots & -8 \\ 2 & -2 & -3 & \vdots & 15 \end{bmatrix}$$

$$R_2 + R_1 \rightarrow \begin{bmatrix} -1 & 5 & -1 & \vdots & 16 \\ -5 & 3 & -5 & \vdots & -8 \\ 2 & -2 & -3 & \vdots & 15 \end{bmatrix}$$

$$-R_1 \rightarrow \begin{bmatrix} 1 & -5 & 1 & \vdots & -16 \\ -5 & 3 & -5 & \vdots & -8 \\ 2 & -2 & -3 & \vdots & 15 \end{bmatrix}$$

$$5R_1 + R_2 \rightarrow \begin{bmatrix} 1 & -5 & 1 & \vdots & -16 \\ 0 & -22 & 0 & \vdots & -88 \\ 2 & -2 & -3 & \vdots & 15 \end{bmatrix}$$

$$-2R_1 + R_3 \rightarrow \begin{bmatrix} 1 & -5 & 1 & \vdots & -16 \\ 0 & -22 & 0 & \vdots & -88 \\ 0 & 8 & -5 & \vdots & 47 \end{bmatrix}$$

$$-\tfrac{1}{22}R_2 \rightarrow \begin{bmatrix} 1 & -5 & 1 & \vdots & -16 \\ 0 & 1 & 0 & \vdots & 4 \\ 0 & 8 & -5 & \vdots & 47 \end{bmatrix}$$

$$-8R_2 + R_3 \rightarrow \begin{bmatrix} 1 & -5 & 1 & \vdots & -16 \\ 0 & 1 & 0 & \vdots & 4 \\ 0 & 0 & -5 & \vdots & 15 \end{bmatrix}$$

$$-\tfrac{1}{5}R_3 \rightarrow \begin{bmatrix} 1 & -5 & 1 & \vdots & -16 \\ 0 & 1 & 0 & \vdots & 4 \\ 0 & 0 & 1 & \vdots & -3 \end{bmatrix}$$

At this point, the matrix is in row-echelon form. Now, apply elementary row operations until you obtain zeros above each of the leading is, as follows.

$$5R_2 + R_1 \rightarrow \begin{bmatrix} 1 & 0 & 1 & \vdots & 4 \\ 0 & 1 & 0 & \vdots & 4 \\ 0 & 0 & 1 & \vdots & -3 \end{bmatrix}$$

$$-R_3 + R_1 \rightarrow \begin{bmatrix} 1 & 0 & 0 & \vdots & 7 \\ 0 & 1 & 0 & \vdots & 4 \\ 0 & 0 & 1 & \vdots & -3 \end{bmatrix}$$

The matrix is now in reduced row-echelon form. Converting back to a system of linear equations, you have

$$\begin{cases} x = 7 \\ y = 4. \\ z = -3 \end{cases}$$

So, the solution is $x = 7$, $y = 4$, and $z = -3$, which can be written as the ordered triple $(7, 4, -3)$.

9.

$$\begin{bmatrix} 2 & -6 & 6 & \vdots & 46 \\ 2 & -3 & 0 & \vdots & 31 \end{bmatrix}$$

$$\begin{bmatrix} 1 & -3 & 3 & \vdots & 23 \\ 2 & -3 & 0 & \vdots & 31 \end{bmatrix}$$

$$-2R_1 + R_2 \rightarrow \begin{bmatrix} 1 & -3 & 3 & \vdots & 23 \\ 0 & 3 & -6 & \vdots & -15 \end{bmatrix}$$

$$\tfrac{1}{3}R_2 \rightarrow \begin{bmatrix} 1 & -3 & 3 & \vdots & 23 \\ 0 & 1 & -2 & \vdots & -5 \end{bmatrix}$$

$$R_1 + 3R_2 \rightarrow \begin{bmatrix} 1 & 0 & -3 & \vdots & 8 \\ 0 & 1 & -2 & \vdots & -5 \end{bmatrix}$$

The corresponding system of equations is

$$\begin{cases} x - 3z = 8 \\ y - 2z = -5. \end{cases}$$

Solving for x and y in terms of z, you have $x = 3z + 8$ and $y = 2z - 5$.

To write a solution of the system that does not use any of the three variables of the system, let a represent any real number and let $z = a$. Substituting a for z in the equations for x and y, you have

$x = 3z + 8 = 3a + 8$ and $y = 2z - 5 = 2a - 5$.

So, the solution set can be written as an ordered triple of the form

$(3a + 8, 2a - 5, a)$

where a is any real number. Remember that a solution set of this form represents an infinite number of solutions. Try substituting values for a to obtain a few solutions. Then check each solution in the original system of equations.

Checkpoints for Section 10.2

1. $\begin{bmatrix} a_{11} & a_{12} \\ a_{21} & a_{22} \end{bmatrix} = \begin{bmatrix} 6 & 3 \\ -2 & 4 \end{bmatrix}$

Because two matrices are equal when their corresponding entries are equal you can conclude that $a_{11} = 6$, $a_{12} = 3$, $a_{21} = -2$, and $a_{22} = 4$.

2. $\begin{bmatrix} 4 & -1 \\ 2 & -3 \end{bmatrix} + \begin{bmatrix} 2 & -1 \\ 0 & 6 \end{bmatrix} = \begin{bmatrix} 4+2 & -1+(-1) \\ 2+0 & -3+6 \end{bmatrix} = \begin{bmatrix} 6 & -2 \\ 2 & 3 \end{bmatrix}$

3. (a) $A + B = \begin{bmatrix} 4 & -1 \\ 0 & 4 \\ -3 & 8 \end{bmatrix} + \begin{bmatrix} 0 & 4 \\ -1 & 3 \\ 1 & 7 \end{bmatrix} = \begin{bmatrix} 4 & 3 \\ -1 & 7 \\ -2 & 15 \end{bmatrix}$

(b) $A - B = \begin{bmatrix} 4 & -1 \\ 0 & 4 \\ -3 & 8 \end{bmatrix} + \begin{bmatrix} 0 & 4 \\ -1 & 3 \\ 1 & 7 \end{bmatrix} = \begin{bmatrix} 4 & -5 \\ 1 & 1 \\ -4 & 1 \end{bmatrix}$

(c) $3A = 3\begin{bmatrix} 4 & -1 \\ 0 & 4 \\ -3 & 8 \end{bmatrix} = \begin{bmatrix} 12 & -3 \\ 0 & 12 \\ -9 & 24 \end{bmatrix}$

(d) $3A - 2B = 3\begin{bmatrix} 4 & -1 \\ 0 & 4 \\ -3 & 8 \end{bmatrix} - 2\begin{bmatrix} 0 & 4 \\ -1 & 3 \\ 1 & 7 \end{bmatrix} = \begin{bmatrix} 12 & -3 \\ 0 & 12 \\ -9 & 24 \end{bmatrix} - \begin{bmatrix} 0 & 8 \\ -2 & 6 \\ 2 & 14 \end{bmatrix} = \begin{bmatrix} 12 & -11 \\ 2 & 6 \\ -11 & 10 \end{bmatrix}$

4. $\begin{bmatrix} 3 & -8 \\ 0 & 2 \end{bmatrix} + \begin{bmatrix} -2 & 3 \\ 6 & -5 \end{bmatrix} + \begin{bmatrix} 0 & 7 \\ 4 & -1 \end{bmatrix} = \begin{bmatrix} 3+(-2)+0 & -8+3+7 \\ 0+6+4 & 2+(-5)+(-1) \end{bmatrix} = \begin{bmatrix} 1 & 2 \\ 10 & -4 \end{bmatrix}$

5. $2\left(\begin{bmatrix} 1 & 3 \\ -2 & 2 \end{bmatrix} + \begin{bmatrix} -4 & 0 \\ -3 & 1 \end{bmatrix}\right) = 2\begin{bmatrix} 1 & 3 \\ -2 & 2 \end{bmatrix} + 2\begin{bmatrix} -4 & 0 \\ -3 & 1 \end{bmatrix}$

$= \begin{bmatrix} 2 & 6 \\ -4 & 4 \end{bmatrix} + \begin{bmatrix} -8 & 0 \\ -6 & 2 \end{bmatrix}$

$= \begin{bmatrix} -6 & 6 \\ -10 & 6 \end{bmatrix}$

6. Begin by solving the matrix equation for X to obtain

$2X - A = B$

$2X = B + A$

$X = \frac{1}{2}(B + A).$

Now, using the matrices A and B you have the following

$X = \frac{1}{2}\left(\begin{bmatrix} 4 & -1 \\ -2 & 5 \end{bmatrix} + \begin{bmatrix} 6 & 1 \\ 0 & 3 \end{bmatrix}\right) = \frac{1}{2}\begin{bmatrix} 10 & 0 \\ -2 & 8 \end{bmatrix} = \begin{bmatrix} 5 & 0 \\ -1 & 4 \end{bmatrix}$

7. $AB = \begin{bmatrix} -1 & 4 \\ 2 & 0 \\ 1 & 2 \end{bmatrix}\begin{bmatrix} 1 & -2 \\ 0 & 7 \end{bmatrix}$

$= \begin{bmatrix} (-1)(1) + (4)(0) & (-1)(-2) + (4)(7) \\ (2)(1) + (0)(0) & (2)(-2) + (0)(7) \\ (1)(1) + (2)(0) & (1)(-2) + (2)(7) \end{bmatrix}$

$= \begin{bmatrix} -1 & 30 \\ 2 & -4 \\ 1 & 12 \end{bmatrix}$

8. $AB = \begin{bmatrix} 0 & 4 & -3 \\ 2 & 1 & 7 \\ 3 & -2 & 1 \end{bmatrix}\begin{bmatrix} -2 & 0 \\ 0 & -4 \\ 1 & 2 \end{bmatrix} = \begin{bmatrix} (0)(-2) + (4)(0) + (-3)(1) & (0)(0) + (4)(-4) + (-3)(2) \\ (2)(-2) + (1)(0) + (7)(1) & (2)(0) + (1)(-4) + (7)(2) \\ (3)(-2) + (-2)(0) + (1)(1) & (3)(0) + (-2)(-4) + (1)(2) \end{bmatrix} = \begin{bmatrix} -3 & -22 \\ 3 & 10 \\ -5 & 10 \end{bmatrix}$

9. The product AB for matrices $A = \begin{bmatrix} 3 & 1 & 2 \\ 7 & 0 & -2 \end{bmatrix}$ and $B = \begin{bmatrix} 6 & 4 \\ 2 & -1 \end{bmatrix}$ is not defined, since A is of order 2×3 and B is of order 2×2.

10. $AB = \begin{bmatrix} 3 & -1 \end{bmatrix}\begin{bmatrix} 1 \\ -3 \end{bmatrix} = \begin{bmatrix}(3)(1) + (-1)(-3)\end{bmatrix} = \begin{bmatrix} 6 \end{bmatrix}$

$$BA = \begin{bmatrix} 1 \\ -3 \end{bmatrix}\begin{bmatrix} 3 & -1 \end{bmatrix} = \begin{bmatrix} (1)(3) & + & (1)(-1) \\ (-3)(3) & & (-3)(-1) \end{bmatrix} = \begin{bmatrix} 3 & -1 \\ -9 & 3 \end{bmatrix}$$

11. $A^2 = AA = \begin{bmatrix} 2 & 1 \\ 3 & -2 \end{bmatrix}\begin{bmatrix} 2 & 1 \\ 3 & -2 \end{bmatrix} = \begin{bmatrix} (2)(2) + (1)(3) & (2)(1) + (1)(-2) \\ (3)(2) + (-2)(3) & (3)(1) + (-2)(-2) \end{bmatrix} = \begin{bmatrix} 7 & 0 \\ 0 & 7 \end{bmatrix}$

12. $\begin{cases} -2x_1 - 3x_2 = -4 \\ 6x_1 + x_2 = -36 \end{cases}$

(a) In matrix form $Ax = B$, the system can be written as follows.

$$\begin{bmatrix} -2 & -3 \\ 6 & 1 \end{bmatrix}\begin{bmatrix} x_1 \\ x_2 \end{bmatrix} = \begin{bmatrix} -4 \\ -36 \end{bmatrix}$$

(b) The augmented matrix is formed by adjoining matrix B to matrix A.

$$[A \,\vdots\, B] = \begin{bmatrix} -2 & -3 & \vdots & -4 \\ 6 & 1 & \vdots & -36 \end{bmatrix}$$

Use Gauss-Jordan elimination to rewrite the matrix.

$$-\tfrac{1}{2}R_1 \rightarrow \begin{bmatrix} 1 & \frac{3}{2} & \vdots & 2 \\ 6 & 1 & \vdots & -36 \end{bmatrix}$$

$$-6R_1 + R_2 \rightarrow = \begin{bmatrix} 1 & \frac{3}{2} & \vdots & 2 \\ 0 & -8 & \vdots & -48 \end{bmatrix}$$

$$-\tfrac{1}{8}R_2 \rightarrow = \begin{bmatrix} 1 & \frac{3}{2} & \vdots & -2 \\ 0 & 1 & \vdots & 6 \end{bmatrix}$$

$$-\tfrac{3}{2}R_2 + R_1 \rightarrow = \begin{bmatrix} 1 & 0 & \vdots & 7 \\ 0 & 1 & \vdots & 6 \end{bmatrix}$$

$$= [I \,\vdots\, X]$$

So, the solution of the matrix equation is $X = \begin{bmatrix} x_1 \\ x_2 \end{bmatrix} = \begin{bmatrix} -7 \\ 6 \end{bmatrix}$.

13. The equipment lists E and the costs per item C can be written in matrix form as

$$E = \begin{bmatrix} 12 & 15 \\ 45 & 38 \\ 15 & 17 \end{bmatrix} \text{ and } C = \begin{bmatrix} 100 & 7 & 65 \end{bmatrix}.$$

The total cost of equipment for each team is given by the following product.

$$CE = \begin{bmatrix} 100 & 7 & 65 \end{bmatrix}\begin{bmatrix} 12 & 15 \\ 45 & 38 \\ 15 & 17 \end{bmatrix} = \begin{bmatrix} (100)(12) + (7)(45) + (65)(15) & (100)(15) + (7)(38) + (65)(17) \end{bmatrix} = \begin{bmatrix} 2490 & 2871 \end{bmatrix}$$

So, the total cost of equipment for the women's team is $2490 and the total cost of equipment for the men's team is $2871.

Checkpoints for Section 10.3

1. To show that B is the inverse of A, show that $AB = I = BA$, as follows

$$AB = \begin{bmatrix} 2 & -1 \\ -3 & 1 \end{bmatrix}\begin{bmatrix} -1 & -1 \\ -3 & -2 \end{bmatrix} = \begin{bmatrix} -2+3 & -2+2 \\ 3-3 & 3-2 \end{bmatrix} = \begin{bmatrix} 1 & 0 \\ 0 & 1 \end{bmatrix}$$

$$BA = \begin{bmatrix} -1 & -1 \\ -3 & -2 \end{bmatrix}\begin{bmatrix} 2 & -1 \\ -3 & 1 \end{bmatrix} = \begin{bmatrix} -2+3 & 1-1 \\ -6+6 & 3-2 \end{bmatrix} = \begin{bmatrix} 1 & 0 \\ 0 & 1 \end{bmatrix}$$

Because $AB = I = BA$, B is the inverse of A.

2. To find the inverse of A, solve the matrix equation $AX = I$ for X.

$$\overset{A}{\begin{bmatrix} 1 & -2 \\ -1 & 3 \end{bmatrix}}\overset{X}{\begin{bmatrix} X_{11} & X_{12} \\ X_{21} & X_{22} \end{bmatrix}} = \overset{I}{\begin{bmatrix} 1 & 0 \\ 0 & 1 \end{bmatrix}}$$

$$\begin{bmatrix} X_{11} - 2X_{21} & X_{12} - 2X_{22} \\ -X_{11} + 3X_{21} & -X_{12} + 3X_{22} \end{bmatrix} = \begin{bmatrix} 1 & 0 \\ 0 & 1 \end{bmatrix}$$

Equating corresponding entries, you obtain two system of linear equations.

$$\begin{cases} X_{11} - 2X_{21} = 1 \\ -X_{11} + 3X_{21} = 0 \end{cases} \qquad \begin{cases} X_{12} - 2X_{22} = 0 \\ -X_{12} + 3X_{22} = 1 \end{cases}$$

Solving the first system yields

$X_{11} = 3$ and $X_{21} = 1$.

Solving the second system yields.

$X_{12} = 2$ and $X_{22} = 1$.

So, the inverse of A is $X = A^{-1} = \begin{bmatrix} 3 & 2 \\ 1 & 1 \end{bmatrix}$.

You can check this by finding AA^{-1} and $A^{-1}A$.

Check:

$$AA^{-1} = \begin{bmatrix} 1 & -2 \\ -1 & 3 \end{bmatrix}\begin{bmatrix} 3 & 2 \\ 1 & 1 \end{bmatrix} = \begin{bmatrix} 1 & 0 \\ 0 & 1 \end{bmatrix} = I \checkmark$$

$$A^{-1}A = \begin{bmatrix} 3 & 2 \\ 1 & 1 \end{bmatrix}\begin{bmatrix} 1 & -2 \\ -1 & 3 \end{bmatrix} = \begin{bmatrix} 1 & 0 \\ 0 & 1 \end{bmatrix} = I \checkmark$$

3. Begin by adjoining the identity matrix to A to form the matrix

$$[A \vdots I] = \left[\begin{array}{ccc|ccc} 1 & -2 & -1 & 1 & 0 & 0 \\ 0 & -1 & 2 & 0 & 1 & 0 \\ 1 & -2 & 0 & 0 & 0 & 1 \end{array}\right].$$

Use elementary row operations to obtain the form $[I \vdots A^{-1}]$

$$\begin{array}{r} \\ -R_2 \rightarrow \\ -R_1 + R_3 \rightarrow \end{array}\left[\begin{array}{ccc|ccc} 1 & -2 & -1 & 1 & 0 & 0 \\ 0 & 1 & -2 & 0 & -1 & 0 \\ 0 & 0 & 1 & -1 & 0 & 1 \end{array}\right]$$

$$\begin{array}{r} 2R_2 + R_1 \rightarrow \\ 2R_3 - R_2 \rightarrow \\ \\ \end{array}\left[\begin{array}{ccc|ccc} 1 & 0 & -5 & 1 & -2 & 0 \\ 0 & 1 & 0 & -2 & -1 & 2 \\ 0 & 0 & 1 & -1 & 0 & 1 \end{array}\right]$$

$$\begin{array}{r} 5R_3 + R_1 \rightarrow \\ \\ \\ \end{array}\left[\begin{array}{ccc|ccc} 1 & 0 & 0 & -4 & -2 & 5 \\ 0 & 1 & 0 & -2 & -1 & 2 \\ 0 & 0 & 1 & -1 & 0 & 1 \end{array}\right] = [I \vdots A^{-1}]$$

So, the matrix A is invertable and its inverse is

$$A^{-1} = \begin{bmatrix} -4 & -2 & 5 \\ -2 & -1 & 2 \\ -1 & 0 & 1 \end{bmatrix}.$$

Confirm this result by multiplying AA^{-1} to obtain I.

Check:

$$AA^{-1} = \begin{bmatrix} 1 & -2 & -1 \\ 0 & -1 & 2 \\ 1 & -2 & 0 \end{bmatrix}\begin{bmatrix} -4 & -2 & 5 \\ -2 & -1 & 2 \\ -1 & 0 & 1 \end{bmatrix} = \begin{bmatrix} 1 & 0 & 0 \\ 0 & 1 & 0 \\ 0 & 0 & 1 \end{bmatrix} = I$$

4. For the matrix A, apply the formula for the inverse of a 2×2 matrix to obtain

$ad - bc = (5)(4) - (-1)(3) = 23$

Because this quantity is not zero, the matrix is invertible. The inverse is formed by interchanging the entries on the main diagonal, changing the signs of the other two entries, and multiplying by the scalar $\frac{1}{23}$, as follows.

$$A^{-1} = \frac{1}{ad - bc}\begin{bmatrix} d & -b \\ -c & a \end{bmatrix} \quad \text{Formula for the inverse of a } 2 \times 2 \text{ matrix}$$

$$= \frac{1}{23}\begin{bmatrix} 4 & 1 \\ -3 & 5 \end{bmatrix} \quad \text{Substitute for } a, b, c, d, \text{ and the determinant}$$

$$= \begin{bmatrix} \frac{4}{23} & \frac{1}{23} \\ \frac{-3}{23} & \frac{5}{23} \end{bmatrix} \quad \text{Multiply by the scalar } \frac{1}{23}.$$

5. Begin by writing the system in the matrix form $AX = B$.

$$\begin{bmatrix} 2 & 3 & 1 \\ 3 & 3 & 1 \\ 2 & 4 & 1 \end{bmatrix}\begin{bmatrix} x \\ y \\ z \end{bmatrix} = \begin{bmatrix} -1 \\ 1 \\ -2 \end{bmatrix}$$

Then, use Gauss-Jordan elimination to find A^{-1}.

$$[A \vdots I] = \left[\begin{array}{ccc:ccc} 2 & 3 & 1 & 1 & 0 & 0 \\ 3 & 3 & 1 & 0 & 1 & 0 \\ 2 & 4 & 1 & 0 & 0 & 1 \end{array}\right] \qquad \begin{array}{r} \frac{1}{2}R_1 \rightarrow \\ \\ \\ \end{array}\left[\begin{array}{ccc:ccc} 1 & \frac{3}{2} & \frac{1}{2} & \frac{1}{2} & 0 & 0 \\ 3 & 3 & 1 & 0 & 1 & 0 \\ 2 & 4 & 1 & 0 & 0 & 1 \end{array}\right]$$

$$\begin{array}{r} \\ -3R_1 + R_2 \rightarrow \\ -2R_1 + R_3 \rightarrow \end{array}\left[\begin{array}{ccc:ccc} 1 & \frac{3}{2} & \frac{1}{2} & \frac{1}{2} & 0 & 0 \\ 0 & -\frac{3}{2} & -\frac{1}{2} & -\frac{3}{2} & 1 & 0 \\ 0 & 1 & 0 & -1 & 0 & 1 \end{array}\right]$$

$$\begin{array}{r} \\ R_2 \rightarrow \\ R_3 \rightarrow \end{array}\left[\begin{array}{ccc:ccc} 1 & \frac{3}{2} & \frac{1}{2} & \frac{1}{2} & 0 & 0 \\ 0 & 1 & 0 & -1 & 0 & 1 \\ 0 & -\frac{3}{2} & -\frac{1}{2} & -\frac{3}{2} & 1 & 0 \end{array}\right]$$

$$\begin{array}{r} \\ \\ \frac{3}{2}R_2 + R_3 \rightarrow \end{array}\left[\begin{array}{ccc:ccc} 1 & \frac{3}{2} & \frac{1}{2} & \frac{1}{2} & 0 & 0 \\ 0 & 1 & 0 & -1 & 0 & 1 \\ 0 & 0 & -\frac{1}{2} & -3 & 1 & \frac{3}{2} \end{array}\right]$$

$$\begin{array}{r} \\ \\ -2R_3 \rightarrow \end{array}\left[\begin{array}{ccc:ccc} 1 & \frac{3}{2} & \frac{1}{2} & \frac{1}{2} & 0 & 0 \\ 0 & 1 & 0 & -1 & 0 & 1 \\ 0 & 0 & 1 & 6 & -2 & -3 \end{array}\right]$$

$$\begin{array}{r} -\frac{3}{2}R_2 + R_1 \rightarrow \\ \\ \\ \end{array}\left[\begin{array}{ccc:ccc} 1 & 0 & \frac{1}{2} & 2 & 0 & \frac{3}{2} \\ 0 & 1 & 0 & -1 & 0 & 1 \\ 0 & 0 & 1 & 6 & -2 & -3 \end{array}\right]$$

$$\begin{array}{r} -\frac{1}{2}R_3 + R_1 \rightarrow \\ \\ \\ \end{array}\left[\begin{array}{ccc:ccc} 1 & 0 & 0 & -1 & 1 & 0 \\ 0 & 1 & 0 & -1 & 0 & 1 \\ 0 & 0 & 1 & 6 & -2 & -3 \end{array}\right] = [I \vdots A^{-1}]$$

$$A^{-1} = \begin{bmatrix} -1 & 1 & 0 \\ -1 & 0 & 1 \\ 6 & -2 & -3 \end{bmatrix}$$

Finally, multiply B by A^{-1} on the left to obtain the solution.

$$X = A^{-1}B = \begin{bmatrix} -1 & 1 & 0 \\ -1 & 0 & 1 \\ 6 & -2 & -3 \end{bmatrix}\begin{bmatrix} -1 \\ 1 \\ -2 \end{bmatrix} = \begin{bmatrix} 2 \\ -1 \\ -2 \end{bmatrix}$$

The solution of the system is $x = 2$, $y = -1$, and $z = -2$

Checkpoints for Section 10.4

1. (a) $\det(A) = \begin{vmatrix} 1 & 2 \\ 3 & -1 \end{vmatrix}$
$= 1(-1) - 3(2)$
$= -1 - 6$
$= -7$

(b) $\det(B) = \begin{vmatrix} 5 & 0 \\ -4 & 2 \end{vmatrix}$
$= 5(2) - (-4)(0)$
$= 10 + 0$
$= 10$

(c) $\det(C) = \begin{vmatrix} 3 & 6 \\ 2 & 4 \end{vmatrix}$
$= 3(4) - (2)(6)$
$= 12 - 12$
$= 0$

2. To find the minor M_{11}, delete the first row and first column of A and evaluate the determinant of the resulting matrix.

$$\begin{bmatrix} 1 & 2 & 3 \\ 0 & -1 & 5 \\ 2 & 1 & 4 \end{bmatrix}, \quad M_{11} = \begin{vmatrix} -1 & 5 \\ 1 & 4 \end{vmatrix} = -1(4) - 1(5) = -9$$

Continuing this pattern, you obtain the minors.

$$M_{12} = \begin{vmatrix} 0 & 5 \\ 2 & 4 \end{vmatrix} = 0(4) - 2(5) = -10$$

$$M_{13} = \begin{vmatrix} 0 & -1 \\ 2 & 1 \end{vmatrix} = 0(1) - 2(-1) = 2$$

$$M_{21} = \begin{vmatrix} 2 & 3 \\ 1 & 4 \end{vmatrix} = 2(4) - 1(3) = 5$$

$$M_{22} = \begin{vmatrix} 1 & 3 \\ 2 & 4 \end{vmatrix} = 1(4) - 2(3) = -2$$

$$M_{23} = \begin{vmatrix} 1 & 2 \\ 2 & 1 \end{vmatrix} = 1(1) - 2(2) = -3$$

$$M_{31} = \begin{vmatrix} 2 & 3 \\ -1 & 5 \end{vmatrix} = 2(5) - (-1)(3) = 13$$

$$M_{32} = \begin{vmatrix} 1 & 3 \\ 0 & 5 \end{vmatrix} = 1(5) - 0(3) = 5$$

$$M_{33} = \begin{vmatrix} 1 & 2 \\ 0 & -1 \end{vmatrix} = 1(-1) - 0(2) = -1$$

Now, to find the cofactors, combine these minors with the checker board pattern of signs for a 3×3 matrix,

$$\begin{bmatrix} + & - & + \\ - & + & - \\ + & - & + \end{bmatrix}.$$

$C_{11} = -9 \qquad C_{12} = 10 \qquad C_{13} = 2$

$C_{21} = -5 \qquad C_{22} = -2 \qquad C_{23} = 3$

$C_{31} = 13 \qquad C_{32} = -5 \qquad C_{33} = -1$

3. The cofactors of the entries in the first row are as follows.

$$C_{11} = +M_{11} = \begin{vmatrix} 5 & 0 \\ 4 & 1 \end{vmatrix} = 5(1) - 4(0) = 5$$

$$C_{12} = -M_{12} = \begin{vmatrix} 3 & 0 \\ -1 & 1 \end{vmatrix} = -(3(1) - (-1)(0)) = -3$$

$$C_{13} = +M_{13} = \begin{vmatrix} 3 & 5 \\ -1 & 4 \end{vmatrix} = 3(4) - (-1)(5) = 17$$

$= 3(5) + 4(-3) + (-2)(17)$
$= -31$

So, by the definition of a determinant, you have the following.

$|A| = a_{11}C_{11} + a_{12}C_{12} + a_{13}C_{13}$

4. Notice that these are two zeros in the third column. So, you can eliminate some of the work in the expansion by using the third column.

$$|A| = a_{13}C_{13} + a_{23}C_{23} + a_{33}C_{33} + a_{43}C_{43}$$
$$= -4C_{13} + 3C_{23} + 0C_{33} + 0C_{43}$$

Because C_{33} and C_{43} have zero coefficients, you need only to find the cofactors of C_{13} and C_{23}.

$$C_{13} = (-1)^{1+3}\begin{vmatrix} 2 & -2 & 6 \\ 1 & 5 & 1 \\ 3 & 1 & -5 \end{vmatrix} = \begin{vmatrix} 2 & -2 & 6 \\ 1 & 5 & 1 \\ 3 & 1 & -5 \end{vmatrix}$$

Expanding by cofactors along the first row yields the following.

$$C_{13} = (2)(-1)^{1+1}\begin{vmatrix} 5 & 1 \\ 1 & -5 \end{vmatrix} + (-2)(-1)^{1+2}\begin{vmatrix} 1 & 1 \\ 3 & -5 \end{vmatrix} + (6)(-1)^{1+3}\begin{vmatrix} 1 & 5 \\ 3 & 1 \end{vmatrix}$$
$$= (2)(1)(-26) + (-2)(-1)(-8) + (6)(1)(-14)$$
$$= -152$$

$$C_{23} = (-1)^{2+3}\begin{vmatrix} 2 & 6 & 2 \\ 1 & 5 & 1 \\ 3 & 1 & -5 \end{vmatrix} = -\begin{vmatrix} 2 & 6 & 2 \\ 1 & 5 & 1 \\ 3 & 1 & -5 \end{vmatrix}$$

Expanding by cofactors along the first row yields the following.

$$C_{23} = -\left((2)(-1)^{1+1}\begin{vmatrix} 5 & 1 \\ 1 & -5 \end{vmatrix} + (6)(-1)^{1+2}\begin{vmatrix} 1 & 1 \\ 3 & -5 \end{vmatrix} + (2)(-1)^{1+3}\begin{vmatrix} 1 & 5 \\ 3 & 1 \end{vmatrix}\right)$$
$$= -((2)(1)(-26) + (6)(-1)(-8) + (2)(1)(-14))$$
$$= 32$$

So, $|A| = -4C_{13} + 3C_{23} + 0C_{33} + 0C_{43}$
$$= -4(-152) + 3(32) + 0 + 0$$
$$= 704.$$

Checkpoints for Section 10.5

1. To begin, find the determinant of the coefficient matrix.

$$D = \begin{vmatrix} 3 & 4 \\ 5 & 3 \end{vmatrix} = 9 - 20 = -11$$

Because this determinant is not zero, you can apply Cramer's Rule.

$$x = \frac{D_x}{D} = \frac{\begin{vmatrix} 1 & 4 \\ 9 & 3 \end{vmatrix}}{-11} = \frac{3 - 36}{-11} = \frac{-33}{-11} = 3$$

$$y = \frac{D_y}{D} = \frac{\begin{vmatrix} 3 & 1 \\ 5 & 9 \end{vmatrix}}{-11} = \frac{27 - 5}{-11} = \frac{22}{-11} = -2$$

So, the solution is $x = 3$ and $y = -2$.

2. To find the determinant of the coefficient matrix, expand along the first row, as follows.

$$\begin{bmatrix} 4 & -1 & 1 \\ 2 & 2 & 3 \\ 5 & -2 & 6 \end{bmatrix}$$

$$D = 4(-1)^2\begin{vmatrix} 2 & 3 \\ -2 & 6 \end{vmatrix} + (-1)(-1)^3\begin{vmatrix} 2 & 3 \\ 5 & 6 \end{vmatrix} + (1)(-1)^4\begin{vmatrix} 2 & 2 \\ 5 & -2 \end{vmatrix} = 4(18) + (1)(-3) + (1)(-14) = 55$$

Because this determinant is not zero, you can apply Cramer's Rule. Next, find D_x, D_y, and D_z.

$$D_x = \begin{vmatrix} 12 & -1 & 1 \\ 1 & 2 & 3 \\ 22 & -2 & 6 \end{vmatrix}$$

$$= (12)(-1)^2\begin{vmatrix} 2 & 3 \\ -2 & 6 \end{vmatrix} + (-1)(-1)^3\begin{vmatrix} 1 & 3 \\ 22 & 6 \end{vmatrix} + (1)(-1)^4\begin{vmatrix} 1 & 2 \\ 22 & -2 \end{vmatrix}$$

$$= (12)(18) + (1)(-60) + (1)(-46)$$

$$= 110$$

$$D_y = \begin{vmatrix} 4 & 12 & 1 \\ 2 & 1 & 3 \\ 5 & 22 & 6 \end{vmatrix}$$

$$= (4)(-1)^2\begin{vmatrix} 1 & 3 \\ 22 & 6 \end{vmatrix} + (12)(-1)^3\begin{vmatrix} 2 & 3 \\ 5 & 6 \end{vmatrix} + (1)(-1)^4\begin{vmatrix} 2 & 1 \\ 5 & 22 \end{vmatrix}$$

$$= (4)(-60) + (-12)(-3) + (1)(39)$$

$$= -165$$

$$D_z = \begin{vmatrix} 4 & -1 & 12 \\ 2 & 2 & 1 \\ 5 & -2 & 22 \end{vmatrix}$$

$$= (4)(-1)^2\begin{vmatrix} 2 & 1 \\ -2 & 22 \end{vmatrix} + (-1)(-1)^3\begin{vmatrix} 2 & 1 \\ 5 & 22 \end{vmatrix} + (12)(-1)^4\begin{vmatrix} 2 & 2 \\ 5 & -2 \end{vmatrix}$$

$$= (4)(46) + (1)(39) + (12)(-14)$$

$$= 55$$

Finally, you can determine the values of x, y, and z as follows.

$$x = \frac{D_x}{D} = \frac{110}{55} = 2$$

$$y = \frac{D_y}{D} = \frac{-165}{55} = -3$$

$$z = \frac{D_z}{D} = \frac{55}{55} = 1$$

So, the solution is $x = 2$, $y = -3$, and $z = 1$.

3.

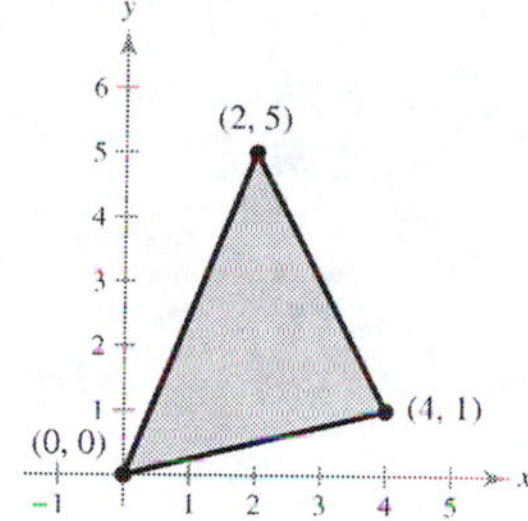

Let $(x_1, y_1) = (0, 0)$, $(x_2, y_2) = (4, 1)$, and $(x_3, y_3) = (2, 5)$. Then, to find the area of the triangle, evaluate the determinant.

$$\begin{vmatrix} x_1 & y_1 & 1 \\ x_2 & y_2 & 1 \\ x_3 & y_3 & 1 \end{vmatrix} = \begin{vmatrix} 0 & 0 & 1 \\ 4 & 1 & 1 \\ 2 & 5 & 1 \end{vmatrix}$$

$$= (0)(-1)^2\begin{vmatrix} 1 & 1 \\ 5 & 1 \end{vmatrix} + (0)(-1)^3\begin{vmatrix} 4 & 1 \\ 2 & 1 \end{vmatrix} + (1)(-1)^4\begin{vmatrix} 4 & 1 \\ 2 & 5 \end{vmatrix}$$

$$= 0 + 0 + (1)(18)$$

$$= 18$$

Using this value, you can conclude that the area of the triangle is

$$\text{Area} = \frac{1}{2}\begin{vmatrix} 0 & 0 & 1 \\ 4 & 1 & 1 \\ 2 & 5 & 1 \end{vmatrix} = \frac{1}{2}(18) = 9 \text{ square units.}$$

4.

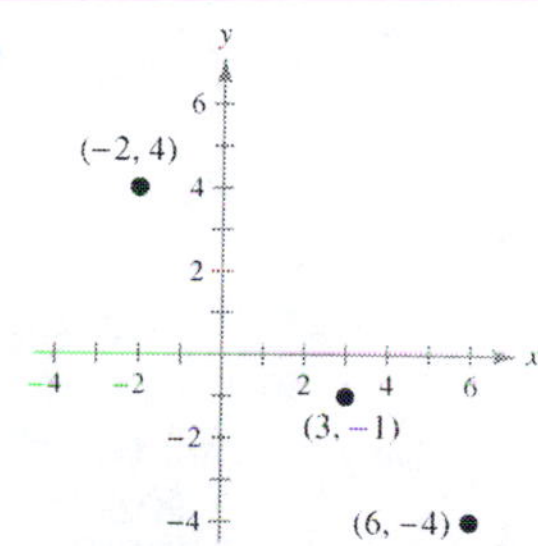

To determine if the points are collinear, let $(x_1, y_1) = (-2, 4)$, $(x_2, y_2) = (3, -1)$, and $(x_3, y_3) = (6, -4)$. Then, evaluate the determinant as follows.

$$\begin{vmatrix} x_1 & y_1 & 1 \\ x_2 & y_2 & 1 \\ x_3 & y_3 & 1 \end{vmatrix} = \begin{vmatrix} -2 & 4 & 1 \\ 3 & -1 & 1 \\ 6 & -4 & 1 \end{vmatrix}$$

$$= (-2)(-1)^2\begin{vmatrix} -1 & 1 \\ -4 & 1 \end{vmatrix} + (4)(-1)^3\begin{vmatrix} 3 & 1 \\ 6 & 1 \end{vmatrix} + (1)(-1)^4\begin{vmatrix} 3 & -1 \\ 6 & -4 \end{vmatrix}$$

$$= (-2)(3) + (-4)(-3) + (1)(-6)$$

$$= 0$$

Because the value of this determinant is equal to zero, you can conclude that the three points are collinear.

5.

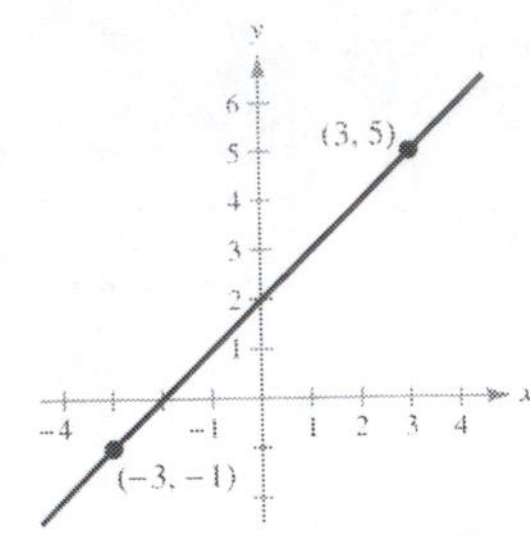

To find an equation of the line, let $(x_1, y_1) = (-3, -1)$ and $(x_2, y_2) = (3, 5)$.

Applying the determinant formula for the equation of a line produces the following.

$$\begin{vmatrix} x & y & 1 \\ -3 & -1 & 1 \\ 3 & 5 & 1 \end{vmatrix} = 0$$

To evaluate this determinant, expand by cofactors along the first row.

$$x(-1)^2\begin{vmatrix} -1 & 1 \\ 5 & 1 \end{vmatrix} + y(-1)^3\begin{vmatrix} -3 & 1 \\ 3 & 1 \end{vmatrix} + (1)(-1)^4\begin{vmatrix} -3 & -1 \\ 3 & 5 \end{vmatrix} = 0$$

$$(x)(-6) + (-y)(-6) + (1)(-12) = 0$$

$$-6x + 6y - 12 = 0$$

$$x - y + 2 = 0$$

So, an equation passing through the two points is $x - y + 2 = 0$.

6. Partioning the message (including blank spaces, but ignoring any punctuation) into groups of three produces the following uncoded 1×3 row matrices.

[15 23 12]	[19 0 1]	[18 5 0]	[14 15 3]	[20 21 18]	[14 1 12]
O W L	S _ A	R E _	N O C	T U R	N A L

7. The coded row matrices are obtained by multiplying each of the uncoded row matrices found in Checkpoint Example 6 by the matrix A, as follows.

Uncoded Matrix	**Encoding Matrix** A		**Coded Matrix**
$[15\ \ 23\ \ 12]$	$\begin{bmatrix} 1 & -1 & 0 \\ 1 & 0 & -1 \\ 6 & -2 & -3 \end{bmatrix}$	=	$[110\ \ -39\ \ -59]$
$[19\ \ 0\ \ 1]$	$\begin{bmatrix} 1 & -1 & 0 \\ 1 & 0 & -1 \\ 6 & -2 & -3 \end{bmatrix}$	=	$[25\ \ -21\ \ -3]$
$[18\ \ 5\ \ 0]$	$\begin{bmatrix} 1 & -1 & 0 \\ 1 & 0 & -1 \\ 6 & -2 & -3 \end{bmatrix}$	=	$[23\ \ -18\ \ -5]$
$[14\ \ 15\ \ 3]$	$\begin{bmatrix} 1 & -1 & 0 \\ 1 & 0 & -1 \\ 6 & -2 & -3 \end{bmatrix}$	=	$[47\ \ -20\ \ -24]$
$[20\ \ 21\ \ 18]$	$\begin{bmatrix} 1 & -1 & 0 \\ 1 & 0 & -1 \\ 6 & -2 & -3 \end{bmatrix}$	=	$[149\ \ -56\ \ -75]$
$[14\ \ 1\ \ 12]$	$\begin{bmatrix} 1 & -1 & 0 \\ 1 & 0 & -1 \\ 6 & -2 & -3 \end{bmatrix}$	=	$[87\ \ -38\ \ -37]$

So, the crytogram is 110 −39 −59 25 −21 −3 23 −18 −5 47 −20 −24 149 −56 −75 87 −38 −37.

8. First find the decoding matrix A^{-1} using matrix A from the Checkpoint Example 7.

$$[A \ \vdots \ I] = \begin{bmatrix} 1 & -1 & 0 & \vdots & 1 & 0 & 0 \\ 1 & 0 & -1 & \vdots & 0 & 1 & 0 \\ 6 & -2 & -3 & \vdots & 0 & 0 & 1 \end{bmatrix}$$

$$\begin{matrix} R_2 \rightarrow \\ R_1 \rightarrow \\ -6R_1 + R_3 \rightarrow \end{matrix} \begin{bmatrix} 1 & 0 & -1 & \vdots & 0 & 1 & 0 \\ 1 & -1 & 0 & \vdots & 1 & 0 & 0 \\ 0 & 4 & -3 & \vdots & -6 & 0 & 1 \end{bmatrix}$$

$$\begin{matrix} \\ -R_1 + R_2 \rightarrow \\ \\ \end{matrix} \begin{bmatrix} 1 & 0 & -1 & \vdots & 0 & 1 & 0 \\ 0 & -1 & 1 & \vdots & 1 & -1 & 0 \\ 0 & 4 & -3 & \vdots & -6 & 0 & 1 \end{bmatrix}$$

$$\begin{matrix} \\ \\ 4R_2 + R_3 \rightarrow \end{matrix} \begin{bmatrix} 1 & 0 & -1 & \vdots & 0 & 1 & 0 \\ 0 & -1 & 1 & \vdots & 1 & -1 & 0 \\ 0 & 0 & 1 & \vdots & -2 & -4 & 1 \end{bmatrix}$$

$$\begin{matrix} \\ -R_2 \rightarrow \\ \\ \end{matrix} \begin{bmatrix} 1 & 0 & -1 & \vdots & 0 & 1 & 0 \\ 0 & 1 & -1 & \vdots & -1 & 1 & 0 \\ 0 & 0 & 1 & \vdots & -2 & -4 & 1 \end{bmatrix}$$

$$\begin{matrix} R_1 + R_3 \rightarrow \\ R_2 + R_3 \rightarrow \\ \\ \end{matrix} \begin{bmatrix} 1 & 0 & 0 & \vdots & -2 & -3 & 1 \\ 0 & 1 & 0 & \vdots & -3 & -3 & 1 \\ 0 & 0 & 1 & \vdots & -2 & -4 & 1 \end{bmatrix} = [I \ \vdots \ A^{-1}]$$

Partition the message into groups of three to form the coded row matrices. Finally, multiply each coded row matrix by A^{-1} (on the right).

Coded Matrix	**Decoding Matrix** A^{-1}		**Decoded Matrix**
$[110 \ \ -39 \ \ -59]$	$\begin{bmatrix} -2 & -3 & 1 \\ -3 & -3 & 1 \\ -2 & -4 & 1 \end{bmatrix}$	=	$[15 \ \ 23 \ \ 12]$
$[25 \ \ -21 \ \ -3]$	$\begin{bmatrix} -2 & -3 & 1 \\ -3 & -3 & 1 \\ -2 & -4 & 1 \end{bmatrix}$	=	$[19 \ \ 0 \ \ 1]$
$[23 \ \ -18 \ \ -5]$	$\begin{bmatrix} -2 & -3 & 1 \\ -3 & -3 & 1 \\ -2 & -4 & 1 \end{bmatrix}$	=	$[18 \ \ 5 \ \ 0]$
$[47 \ \ -20 \ \ -24]$	$\begin{bmatrix} -2 & -3 & 1 \\ -3 & -3 & 1 \\ -2 & -4 & 1 \end{bmatrix}$	=	$[14 \ \ 15 \ \ 3]$
$[149 \ \ -56 \ \ -75]$	$\begin{bmatrix} -2 & -3 & 1 \\ -3 & -3 & 1 \\ -2 & -4 & 1 \end{bmatrix}$	=	$[20 \ \ 21 \ \ 18]$
$[87 \ \ -38 \ \ -37]$	$\begin{bmatrix} -2 & -3 & 1 \\ -3 & -3 & 1 \\ -2 & -4 & 1 \end{bmatrix}$	=	$[14 \ \ 1 \ \ 12]$

So, the message is as follows.

[15	23	12]	[19	0	1]	[18	5	0]	[14	15	3]	[20	21	18]	[14	1	12]
O	W	L	S		A	R	E		N	O	C	T	U	R	N	A	L

Chapter 11

Checkpoints for Section 11.1

1. The first four terms of the sequence given by $a_n = 2n + 1$ are as follows.

$a_1 = 2(1) + 1 = 3$ 1st term

$a_2 = 2(2) + 1 = 5$ 2nd term

$a_3 = 2(3) + 1 = 7$ 3rd term

$a_4 = 2(4) + 1 = 9$ 4th term

2. The first four terms of the sequence given by $a_n = \frac{2 + (-1)^n}{n}$ are as follows.

$$a_1 = \frac{2 + (-1)^1}{1} = \frac{2 - 1}{1} = 1$$

$$a_2 = \frac{2 + (-1)^2}{2} = \frac{2 + 1}{2} = \frac{3}{2}$$

$$a_3 = \frac{2 + (-1)^3}{3} = \frac{2 - 1}{3} = \frac{1}{3}$$

$$a_4 = \frac{2 + (-1)^4}{4} = \frac{2 + 1}{4} = \frac{3}{4}$$

3. (a) n: 1 2 3 4 … n

Terms: 1 5 9 13 … a_n

Apparent pattern: Each term is 3 less than 4 times n, which implies that $a_n = 4n - 3$.

(b) n: 1 2 3 4 … n

Terms: 2 −4 6 −8 … a_n

Apparent pattern: The absolute value of each term is 2 times n, and the terms have alternating signs, with those in the even positions being negative. This implies that $a_n = (-1)^{n+1}2n$.

4. The first five terms of the sequence are as follows.

$a_1 = 6$ 1st term is given.

$a_2 = a_{1+1} = a_1 + 1 = 6 + 1 = 7$ Use recursion formula.

$a_3 = a_{2+1} = a_2 + 1 = 7 + 1 = 8$ Use recursion formula.

$a_4 = a_{3+1} = a_3 + 1 = 8 + 1 = 9$ Use recursion formula.

$a_5 = a_{4+1} = a_4 + 1 = 9 + 1 = 10$ Use recursion formula.

5. The first five terms of the sequence are as follows,

$a_0 = 1$ 0th term is given.

$a_1 = 3$ 1st term is given.

$a_2 = a_{2-2} + a_{2-1} = a_0 + a_1 = 1 + 3 = 4$ Use recursion formula.

$a_3 = a_{3-2} + a_{3-1} = a_1 + a_2 = 3 + 4 = 7$ Use recursion formula.

$a_4 = a_{4-2} + a_{4-1} = a_2 + a_3 = 4 + 7 = 11$ Use recursion formula.

6. Algebraic Solution

$$a_0 = \frac{3^0 + 1}{0!} = \frac{1 + 1}{1} = 2$$

$$a_1 = \frac{3^1 + 1}{1!} = \frac{3 + 1}{1} = 4$$

$$a_2 = \frac{3^2 + 1}{2!} = \frac{9 + 1}{2} = \frac{10}{2} = 5$$

$$a_3 = \frac{3^3 + 1}{3!} = \frac{27 + 1}{6} = \frac{28}{6} = \frac{14}{3}$$

$$a_4 = \frac{3^4 + 1}{4!} = \frac{81 + 1}{24} = \frac{82}{24} = \frac{41}{12}$$

Graphical Solution

Using a graphing utility set to *dot* and *sequence* modes, enter the sequence. Next, graph the sequence.

You can estimate the first five terms of the sequence as follows.

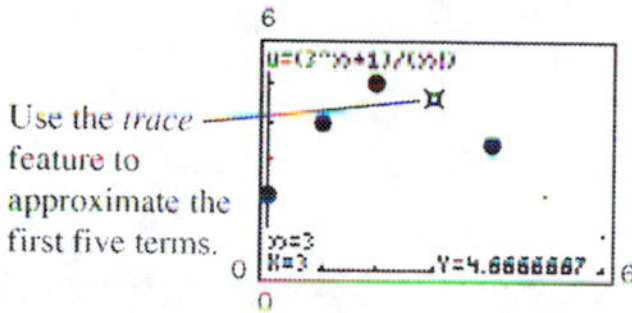

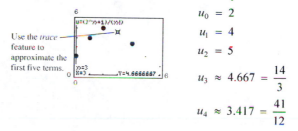

Use the *trace* feature to approximate the first five terms.

$u_0 = 2$

$u_1 = 4$

$u_2 = 5$

$u_3 \approx 4.667 = \dfrac{14}{3}$

$u_4 \approx 3.417 = \dfrac{41}{12}$

7. $$\frac{4!(n + 1)!}{3!n!} = \frac{(1 \cdot 2 \cdot 3 \cdot 4)[1 \cdot 2 \cdot 3 \ldots n \cdot (n + 1)]}{(1 \cdot 2 \cdot 3)(1 \cdot 2 \cdot 3 \ldots n)}$$

$$= 4(n + 1)$$

8. $$\sum_{i=1}^{4}(4i + 1) = [4(1) + 1] + [4(2) + 1] + [4(3) + 1] + [4(4) + 1]$$

$$= 5 + 9 + 13 + 17$$

$$= 44$$

9. (a) The fourth partial sum is as follows.

$$\sum_{i=1}^{4} \frac{5}{10^2} = \frac{5}{10^1} + \frac{5}{10^2} + \frac{5}{10^3} + \frac{5}{10^4}$$

$$= 0.5 + 0.05 + 0.005 + 0.0005$$

$$= 0.5555$$

(b) The sum of the series is as follows.

$$\sum_{i=1}^{\infty} \frac{5}{10^i} = \frac{5}{10^1} + \frac{5}{10^2} + \frac{5}{10^3} + \frac{5}{10^4} + \frac{5}{10^5} + \cdots$$

$$= 0.5 + 0.05 + 0.005 + 0.0005 + 0.00005 + \cdots$$

$$= 0.55555\ldots$$

$$= \frac{5}{9}$$

10. (a) The first three terms of the sequence are as follows.

$A_0 = 1000\left(1 + \dfrac{0.03}{12}\right)^0 = \$1000$ Original deposit

$A_1 = 1000\left(1 + \dfrac{0.03}{12}\right)^1 = \$1002.50$ First-month balance

$A_2 = 1000\left(1 + \dfrac{0.03}{12}\right)^2 \approx \$1005.01$ Second-month balance.

(b) The 48th term of the sequence is

$A_{48} = 1000\left(1 + \dfrac{0.03}{12}\right)^{48} \approx \$1127.33$ Four-year balance

Checkpoints for Section 11.2

1. The sequence whose nth term is $3n - 1$ is arithmetic. The first five terms are as follows.

$3(1) - 1 = 2$

$3(2) - 1 = 5$

$3(3) - 1 = 8$

$3(4) - 1 = 11$

$3(5) - 1 = 14$

For this sequence, the common difference between consecutive terms is 3.

$\underbrace{2, 5}_{5-2=3}, 8, 11, 14, \ldots$

2. You know that the formula for the nth term is of the form $a_n = a_1 + (n - 1)d$. Because the common difference is $d = 5$ and the first term is $a_1 = -1$, the formula must have the form

$a_n = a_1 + (n - 1)d = -1 + 5(n - 1).$

So, the formula for the nth term is $a_n = 5n - 6$.

The sequence therefore has the following form.

$-1, 4, 9, 13, \ldots, 5n - 6, \ldots$

The figure below shows a graph of the first 15 terms of the sequence. Notice that the points lie on a line.

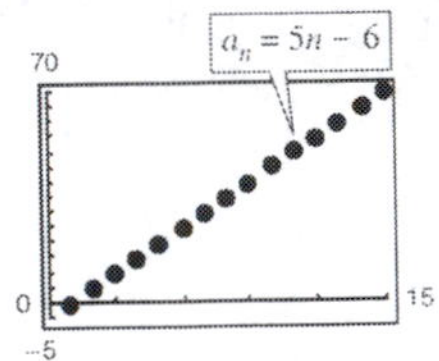

3. You know that $a_8 = 25$ and $a_{12} = 41$. So, you must add the common difference d four times to the eighth term to obtain the 12th term. Therefore, the eighth term and the 12th terms of the sequence are related by

$a_{12} = a_8 + 4d.$

Using $a_8 = 25$ and $a_{12} = 41$, solve for d.

$a_{12} = a_8 + 4d$

$41 = 25 + 4d$

$16 = 4d$

$4 = d$

Use the formula for the nth term of an arithmetic sequence to find a_1.

$a_n = a_1 + (n - 1)d$

$a_8 = a_1 + (8 - 1)(4)$

$25 = a_1 + (7)(4)$

$-3 = a_1$

So, the formula for the nth term of the sequence is

$a_n = -3 + (n - 1)(4) = -3 + 4n - 4 = 4n - 7.$

The sequence is as follows.

a_1	a_2	a_3	a_4	a_5	a_6	a_7	a_8	a_9	a_{10}	a_{11}	...
-3	1	5	9	13	17	21	25	29	33	37	...

4. For this arithmetic sequence, the common difference is $d = 15 - 7 = 8$.

There are two ways to find the tenth term. One way is to write the first ten terms (by repeatedly adding 8).

7, 15, 23, 31, 39, 47, 55, 63, 71, 79

So, the tenth term is 79.

Another way to find the tenth term is to first find a formula for the nth term. Because the common difference is $d = 8$ and the first term is $a_1 = 7$, the formula must have the form

$a_n = a_1 + (n - 1)d = 7 + (n - 1)(8).$

Therefore, a formula for the nth term is $a_n = 8n - 1$ which implies that the tenth term is $a_{10} = 8(10) - 1 = 79$.

5. To begin notice that the sequence is arithmetic (with a common difference of $d = 37 - 40 = -3$).

Moreover, the sequence has 7 terms. So, the sum of the sequence is

$S_n = \frac{n}{2}(a_1 + a_n)$ Sum of a finite arithmetic sequence

$= \frac{7}{2}(40 + 22)$ Substitute 7 for n, 40 for a_1, and 22 for a_n.

$= 217.$ Simplify.

6. (a) The integers from 1 to 35 form an arithmetic sequence that has 35 terms. So, you can use the formula for the sum of a finite arithmetic sequence, as follows.

$S_n = 1 + 2 + 3 + \ldots + 34 + 35$

$= \frac{n}{2}(a_1 + a_n)$ Sum of a finite arithmetic sequence

$= \frac{35}{2}(1 + 35)$ Substitute 35 for n, 1 for a_1, and 35 for a_n.

$= 630$ Simplify.

(b) The sum of the integers from 1 to $2N$ form an arithmetic sequence that has $2N$ terms.

$S_n = 1 + 2 + 3 + \ldots + (2N - 1) + 2N$

$= \frac{n}{2}(a_1 + a_n)$ Sum of a finite arithmetic sequence.

$= \frac{2N}{2}(1 + 2N)$ Substitute $2N$ for n, 1 for a_1 and $2N$ for a_n.

$= N(1 + 2N)$ Simplify.

7. For this arithmetic sequence, $a_1 = 6$ and $d = 12 - 6 = 6$.

So, $a_n = a_1 + (n - 1)d = 6 + 6(n - 1)$ and the nth term is $a_n = 6n$.

Therefore, $a_{120} = 6(120) = 720$, and the sum of the first 120 terms is

$$S_{120} = \frac{n}{2}(a_1 + a_{120})$$
$$= \frac{120}{2}(6 + 720)$$
$$= 60(726)$$
$$= 43{,}560.$$

8. For this arithmetic sequence, $a_1 = 78$ and $d = 76 - 78 = -2$.

So, $a_n = 78 + (-2)(n - 1)$ and the nth term is $a_n = -2n + 80$.

Therefore, $a_{30} = -2(30) + 80 = 20$, and the sum of the first 30 terms is

$$S_{30} = \frac{n}{2}(a_1 + a_{30})$$
$$= \frac{30}{2}(78 + 20)$$
$$= 15(98)$$
$$= 1470.$$

9. The annual sales form an arithmetic sequence in which $a_1 = 160{,}000$ and $d = 20{,}000$.

So, $a_n = 160{,}000 + 20{,}000(n - 1)$ and the nth term of the sequence is $a_n = 20{,}000n + 140{,}000$.

Therefore, the 10th term of the sequence is

$$a_{10} = 20{,}000(10) + 140{,}000$$
$$= 340{,}000.$$

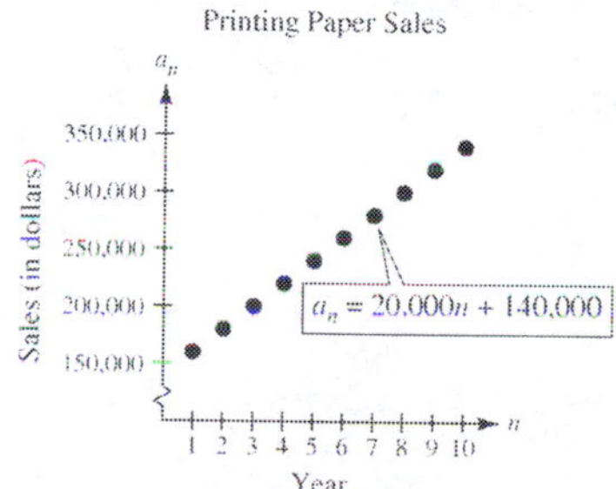

The sum of the first 10 terms of the sequence is

$$S_{10} = \frac{n}{2}(a_1 + a_{10})$$
$$= \frac{10}{2}(160{,}000 + 340{,}000)$$
$$= 5(500{,}000)$$
$$= 2{,}500{,}000.$$

So, the total sales for the first 10 years will be $2,500,000.

Checkpoints for Section 11.3

1. The sequence whose nth term is $6(-2)^n$ is geometric.

For this sequence, the common ratio of consecutive terms is -2.

The first four terms, beginning with $n = 1$ are as follows.

$a_1 = 6(-2)^1 = 6(-2) = -12$

$a_2 = 6(-2)^2 = 6(4) = 24$

$a_3 = 6(-2)^3 = 6(-8) = -48$

$a_4 = 6(-2)^4 = 6(16) = 96$

So the sequence of terms are

$\underbrace{-12, 24,}_{\frac{24}{-12}=-2} -48, 96, \ldots, 6(-2)^n, \ldots$

2. Starting with $a_1 = 2$, repeatedly multiply by 4 to obtain the following.

$a_1 = 2$	1st term
$a_2 = 2(4^1) = 8$	2nd term
$a_3 = 2(4^2) = 32$	3rd term
$a_4 = 2(4^3) = 128$	4th term
$a_5 = 2(4)^4 = 512$	5th term

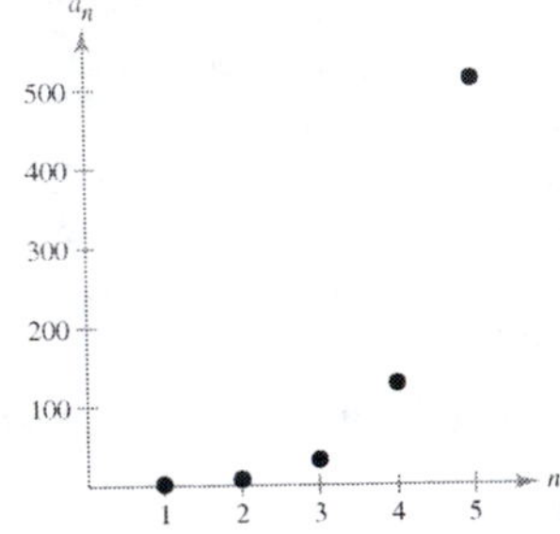

3. Algebraic Solution

Use the formula for the nth term of a geometric sequence.

$a_n = a_1 r^{n-1}$

$a_{12} = 14(1.2)^{12-1}$

$= 14(1.2)^{11}$

≈ 104.02

Numerical Solution

For this sequence, $r = 1.2$ and $a_1 = 14$. So $a_n = 14(1.2)^{n-1}$. Use a graphing utility to create a table that shows the terms of the sequence.

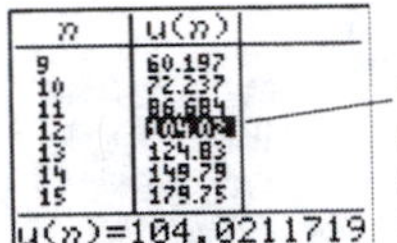

The number in the 12th row is the 12th term of the sequence.

So, $a_{12} \approx 104.02$.

4. The common ratio of this geometric sequence is $r = \frac{20}{4} = 5$. Because the first term is $a_1 = 4$, the formula for the nth term of a geometric sequence is as follows.

$a_n = a_1 r^{n-1}$

$a_n = 4(5)^{n-1}$

The 12th term of the sequence is as follows.

$a_{12} = 4(5)^{12-1} = 4(5)^{11} = 195{,}312{,}500$

5. The fifth term is related to the second term by the equation $a_5 = a_2 r^3$.

Because $a_5 = \frac{81}{4}$ and $a_2 = 6$, you can solve for r as follows.

$a_5 = a_2 r^3$	Multiply the second term by r^{5-3}.
$\frac{81}{4} = 6r^3$	Substitute $\frac{81}{4}$ for a_5 and 6 for a_2.
$\frac{27}{8} = r^3$	Divide each side by 6.
$\frac{3}{2} = r$	Take the cube root of each side.

You can obtain the eighth term by multiplying the fifth term by r^3.

$a_8 = a_5 r^3$	Multiply the fifth term by r^{8-5}.
$= \frac{81}{4}\left(\frac{3}{2}\right)^3$	Substitute $\frac{81}{3}$ for a_5 and $\frac{3}{2}$ for r
$= \frac{81}{4}\left(\frac{27}{8}\right)$	Evaluate power.
$= \frac{2187}{32}$	Multiply fractions.

6. You have $\sum_{i=1}^{10} 2(0.25)^{i-1} = 2(0.25)^0 + 2(0.25)^1 + 2(0.25)^2 + \ldots + 2(0.25)^9.$

Now, $a_1 = 2$, $r = 0.25$, and $n = 10$, so applying the formula for the sum of a finite geometric sequence, you obtain the following.

$$S_n = a_1\left(\frac{1 - r^n}{1 - r}\right) \qquad \text{Sum of a finite geometric series}$$

$$\sum_{i=1}^{10} 2(0.25)^{i-1} = 2\left[\frac{1 - (0.25)^{10}}{1 - 0.25}\right] \qquad \text{Substitute 2 for } a_1\text{, 0.25 for } r\text{, and 10 for } n.$$

$$\approx 2.667 \qquad \text{Use a calculator.}$$

7. (a) $\sum_{n=0}^{\infty} 5(0.5)^n = 5 + 5(0.5) + 5(0.5)^2 + \ldots + 5(0.5)^n + \ldots$

$$= \frac{5}{1 - 0.5} \qquad \text{Use } \frac{a_1}{1-r} \text{ and Subsitute 5 for } a_1 \text{ and 0.5 for } r.$$

$$= \frac{5}{0.5}$$

$$= 10$$

(b) To find the common ratio, divide any term by the preceding term.

So, $r = \frac{1}{5} = 0.2$.

The sum of the infinite geometric series is as follows.

$$5 + 1 + 0.2 + 0.04 + \ldots = 5(0.2)^0 + 5(0.2)^1 + 5(0.2)^2 + 5(0.2)^3 + \ldots$$

$$= \frac{a_1}{1 - r}$$

$$= \frac{5}{1 - 6.2}$$

$$= 6.25$$

8. To find the balance in the account after 48 months, consider each of the 48 deposits separately. The first deposit will gain interest for 48 months, and its balance will be

$$A_{48} = 70\left(1 + \frac{0.02}{12}\right)^{48} = 70(1.0017)^{48}.$$

The second deposit will gain interest for 47 months, and its balance will be

$$A_{47} = 70\left(1 + \frac{0.02}{12}\right)^{47} = 70(1.0017)^{47}.$$

The last deposit will gain interest for only 1 month, and its balance will be

$$A_1 = 70\left(1 + \frac{0.02}{12}\right)^{1} = 70(1.0017).$$

The total balance in the annuity will be the sum of the balances of the 48 deposits.

Using the formula for the sum of a finite geometric sequence, with $A_1 = 70(1.0017)$ $r = 1.0017$, and $n = 48$ you have

$$S_n = A_1\left(\frac{1 - r^n}{1 - r}\right)$$

$$S_{48} = 70(1.0017)\left[\frac{1 - (1.0017)^{48}}{1 - 1.0017}\right] \approx \$3500.85.$$

Checkpoints for Section 11.4

1. (a) $P_{k+1} : S_{k+1} = \dfrac{6}{(k+1)(k+1+3)}$

$= \dfrac{6}{(k+1)(k+4)}$

(b) $P_{k+1} : k + 1 + 2 \le 3(k + 1 - 1)^2$

$k + 3 \le 3k^2$

(c) $P_{k+1} : 2^{4(k+1)-2} + 1 > 5(k + 1)$

$2^{4k+4-2} + 1 > 5k + 5$

$2^{4k+2} + 1 > 5k + 5$

2. Mathematical induction consists of two distinct parts.

1. First, you must show that the formula is true when $n = 1$. When $n = 1$, the formula is valid, because $S_1 = 1(1 + 4) = 5$.

2. The second part of mathematical induction has two steps. The first step is to *assume* that the formula is valid for some integer k. The second step is to use this assumption to prove that the formula is valid for the *next* integer, $k + 1$. Assuming that the formula

$S_k = 5 + 7 + 9 + 11 + \dots + (2k + 3) = k(k + 4)$

is true, you must show that the formula

$$\begin{aligned} S_{k+1} &= (k+1)(k+1+4) \\ &= (k+1)(k+5) \end{aligned}$$ is true.

$$\begin{aligned} S_{k+1} &= 5 + 7 + 9 + 11 + \dots + (2k+3) + \left[2(k+1) + 3\right] \\ &= \left[5 + 7 + 9 + 11 + \dots + (2k+3)\right] + (2k + 2 + 3) \\ &= S_k + 2(k+5) \\ &= k(k+4) + 2k + 5 \\ &= k^2 + 4k + 2k + 5 \\ &= k^2 + 6k + 5 \\ &= (k+1)(k+5) \end{aligned}$$

Combining the results of parts (1) and (2), you can conclude by mathematical induction that the formula is valid for all integers $n \ge 1$.

3. 1. When $n = 1$, the formula is valid, because $S_1 = (1)(1 - 1) = \dfrac{(1)(1-1)(1+1)}{3} = 0$.

2. Assuming that $S_k = 1(1 - 1) + 2(2 - 1) + 3(3 - 1) + \dots + k(k - 1) = \dfrac{k(k-1)(k+1)}{3}$

You must show that $S_{k+1} = \dfrac{(k+1)(k+1-1)(k+1+1)}{3} = \dfrac{k(k+1)(k+2)}{3}$.

To do this, write the following.

$$\begin{aligned} S_{k+1} &= S_k + a_{k+1} \\ &= \left[1(1-1) + 2(2-1) + 3(3-1) + \dots + k(k-1)\right] + (k+1)(k+1-1) && \text{Substitution} \\ &= \frac{k(k-1)(k+1)}{3} + k(k+1) && \text{By assumption} \\ &= \frac{k(k-1)(k+1) + 3k(k+1)}{3} && \text{Combine fractions.} \\ &= \frac{k(k+1)\left[(k-1) + 3\right]}{3} && \text{Factor.} \\ &= \frac{k(k+1)(k+2)}{3} && \text{Simplify.} \end{aligned}$$

So, S_k implies S_{k+1}.

Combining the results of parts (1) and (2), you can conclude by mathematical induction that the formula is valid for all positive integers n.

4. 1. For $n = 1$ and $n = 2$, the statement is true because

$1! \geq 1$ and $2! \geq 2$.

2. Assuming that

$k! \geq k$

You need to show that $(k + 1)! \geq k + 1$. For $n = k$, you have $(k + 1)! = (k + 1)k! \geq (k + 1)k$.

Because $(k + 1)! > k + 1$ for all $k > 1$, it follows that $(k + 1)! \geq k + 1$.

Combining the results of parts (1) and (2), you can conclude by mathematical induction that $n! \geq n$ for all integers $n \geq 1$.

5. 1. For $n = 1$, the statement is true because

$3^1 + 1 = 4$.

So, 2 is a factor.

2. Assuming that 2 is a factor of $3^k + 1$, you must show that 2 is a factor of $3^{k+1} + 1$.

To do this, write the following.

$$\begin{aligned} 3^{k+1} + 1 &= 3^{k+1} - 3^k + 3^k + 1 && \text{Subtract and add } 3^k. \\ &= 3^k(3 - 1) + 3^k + 1 && \text{Regroup terms.} \\ &= 3^k \cdot 2 + 3^k + 1 && \text{Simplify.} \end{aligned}$$

Because 2 is a factor of $3^k \cdot 2$ and 2 is also a factor of $3^k + 1$, it follows that 2 is a factor of $3^{k+1} + 1$. Combining the results of parts (1) and (2), you can conclude by mathematical induction that 2 is a factor of $3^n + 1$ for all positive integers n.

6. Begin by writing the first few sums.

$S_1 = 3 = 1(3)$

$S_2 = 3 + 7 = 10 = 2(5)$

$S_3 = 3 + 7 + 11 = 21 = 3(7)$

$S_4 = 3 + 7 + 11 + 15 = 36 = 4(9)$

$S_5 = 3 + 7 + 11 + 15 + 19 = 55 = 5(11)$

From this sequence, it appears that the formula for the kth sum is

$S_k = 3 + 7 + 11 + 15 + 19 + \ldots + 4k - 1 = k(2k + 1)$.

To prove the validity of this hypothesis, use mathematical induction. Note that you have already verified the formula for $n = 1$, so begin by assuming that the formula is valid for $n = k$ and trying to show that it is valid for $n = k + 1$.

$$\begin{aligned} S_{k+1} &= [3 + 7 + 11 + 15 + \ldots + 4k - 1] + 4(k + 1) - 1 \\ &= k(2k + 1) + 4k + 3 \\ &= 2k^2 + k + 4k + 3 \\ &= 2k^2 + 5k + 3 \\ &= (k + 1)(2k + 3) \\ &= (k + 1)[2(k + 1) + 1] \end{aligned}$$

So, by mathematical induction, the hypothesis is valid.

7. (a) Using the formula for the sum of the first n positive integers, you obtain

$$\sum_{i=1}^{20} i = 1 + 2 + 3 + \ldots + 20 = \frac{20(20+1)}{2} = \frac{20(21)}{2} = 210.$$

(b)
$$\begin{aligned}\sum_{i=1}^{5} 2i^2 + 3i^3 &= \sum_{i=1}^{5} 2i^2 + \sum_{i=1}^{5} 3i^3 \\ &= 2\sum_{i=1}^{5} i^2 + 3\sum_{i=1}^{5} i^3 \\ &= 2\left[\frac{5(5+1)[2(5)+1]}{6}\right] + 3\left[\frac{(5)^2(5+1)^2}{4}\right] \\ &= 2\left[\frac{(5)(6)(11)}{6}\right] + 3\left[\frac{(25)(36)}{4}\right] \\ &= 2(55) + 3(225) \\ &= 785\end{aligned}$$

8. Begin by finding the first and second differences.

n:	1	2	3	4	5	6
a_n:	−2	0	4	10	18	28
First differences:		2	4	6	8	10
Second differences:			2	2	2	2

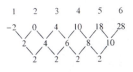

You know from the second differences that the model is quadratic and has the form $a_n = an^2 + bn + c$.

By substituting 1, 2, and 3 for n, you can obtain a system of three linear equations in three variables.

$$a_1 = a(1)^2 + b(1) + c = -2$$

$$a_2 = a(2)^2 + b(2) + c = 0$$

$$a_3 = a(3)^2 + b(3) + c = 4$$

$$\begin{cases} a + b + c = -2 \\ 4a + 2b + c = 0 \\ 9a + 3b + c = 4 \end{cases}$$

Solving this system using techniques from Chapter 9, you can find the solution to be $a = 1, b = -1,$ and $c = -2$.

So, the quadratic model is $a_n = n^2 - n - 2$.

Checkpoints for Section 11.5

1. (a) $\binom{11}{5} = \frac{11!}{6!5!} = \frac{(11 \cdot 10 \cdot 9 \cdot 8 \cdot 7) \cdot \cancel{6!}}{\cancel{6!}\, 5!} = \frac{11 \cdot 10 \cdot 9 \cdot 8 \cdot 7}{5 \cdot 4 \cdot 3 \cdot 2 \cdot 1} = 462$

(b) ${}_9C_2 = \frac{9!}{7!2!} = \frac{(9 \cdot 8) \cdot \cancel{7!}}{\cancel{7!}\, 2!} = \frac{9 \cdot 8}{2 \cdot 1} = 36$

(c) $\binom{5}{0} = \frac{\cancel{5!}}{\cancel{5!}\, 0!} = 1$

(d) ${}_{15}C_{15} = \frac{\cancel{15!}}{0!\, \cancel{15!}} = 1$

2. (a) $_7C_5 = \dfrac{7!}{2!5!} = \dfrac{7 \cdot 6 \cdot \cancel{5} \cdot \cancel{4} \cdot \cancel{3}}{\cancel{5} \cdot \cancel{4} \cdot \cancel{3} \cdot \cancel{2} \cdot 1} = 21$

(b) $\dbinom{7}{2} = \dfrac{7!}{5!2!} = \dfrac{7.6}{2.1} = 21$

(c) $_{14}C_{13} = \dfrac{14!}{1!13!} = \dfrac{14}{1} = 14$

(d) $\dbinom{14}{1} = \dfrac{14!}{13!1!} = \dfrac{14}{1} = 14$

3.

1
1 1
1 2 1
1 3 3 1
1 4 6 4 1
1 5 10 10 5 1
1 6 15 20 15 6 1
1 7 21 35 35 21 7 1
1 8 28 56 70 56 28 8 1
1 9 36 84 126 126 84 36 9 1 8th row
↓ ↓ ↓ ↓ ↓ ↓ ↓ ↓ ↓ ↓
C_{9-0} C_{9-1} C_{9-2} C_{9-3} C_{9-4} C_{9-5} C_{9-6} C_{9-7} C_{9-8} C_{9-9}

4. The binomial coefficients from the fourth row of Pascal's Triangle are 1, 4, 6, 4, 1.
So, the expansion is as follows.

$$\begin{aligned}(x + 2)^4 &= (1)x^4 + (4)x^3(2) + (6)x^2(2^2) + (4)x(2^3) + (1)(2^4)\\ &= x^4 + 8x^3 + 24x^2 + 32x + 16\end{aligned}$$

5. (a) $$\begin{aligned}(y - 2)^4 &= (1)y^4 - (4)y^3(2) + (6)y^2(2^2) - (4)y(2^3) + (1)(2)^4\\ &= y^4 - 8y^3 + 24y^2 - 32y + 16\end{aligned}$$

(b) $$\begin{aligned}(2x - y)^5 &= (1)(2x)^5 - (5)(2x)^4y + (10)(2x)^3y^2 - (10)(2x)^2y^3 + (5)(2x)y^4 - (1)y^5\\ &= 32x^5 - 80x^4y + 80x^3y^2 - 40x^2y^3 + 10xy^4 - y^5\end{aligned}$$

6. Use the third row of Pascal's Triangle to write the expansion of $(5 + y^2)^3 = (y^2 + 5)^3$, as follows.

$$\begin{aligned}(y^2 + 5)^3 &= (1)(y^2)^3 + (3)(y^2)^2(5) + (3)(y^2)(5^2) + (1)(5^3)\\ &= y^6 + 15y^4 + 75y^2 + 125\end{aligned}$$

7. (a) Remember that the formula is for the $(r + 1)$th term, so r is one less than the number of the term you need. So, to find the fifth term in this binomial expansion, use $r = 4$, $n = 8$, $x = a$, and $y = 2b$, as shown.

$$\begin{aligned}{}_nC_rx^{n-r}y^r &= {}_8C_4a^{8-4}(2b)^4 = (70)(a^4)(2b)^4\\ &= 70(2^4)a^4b^4\\ &= 1120a^4b^4\end{aligned}$$

(b) In this case, $n = 11$, $r = 7$, $x = 3a$, and $y = -2b$. Substitute these values to obtain the following.

$$\begin{aligned}{}_nC_rx^{n-r}y^r &= {}_{11}C_7(3a)^4(-2b)^7\\ &= (330)(81a^4)(-128b^7)\\ &= -3{,}421{,}440a^4b^7\end{aligned}$$

So, the coefficient is $-3{,}421{,}440$.

Checkpoints for Section 11.6

1. To solve this problem, count the different ways to obtain a sum of 14 using two numbers from 1 to 8.

First number: 6 7 8
Second number: 8 7 6

So, a sum of 14 can occur in three different ways.

2. To solve this problem, count the different ways to obtain a sum of 12 *using three different numbers* from 1 to 8.

First number:	1	1	3	3	8	8	1	1	4	4	4	7	1	1	5	5	6	6	2	2	3	3	7	7	2	2	4	4	6	6	3	3	4	4	5	5
Second number:	3	8	8	1	1	3	4	7	1	7	1	4	5	6	6	1	1	5	3	7	2	7	2	3	4	6	2	6	2	4	4	5	5	3	3	4
Third number:	8	3	1	8	3	1	7	4	7	1	4	1	6	5	1	6	5	1	7	3	7	2	3	2	6	4	6	2	4	2	5	4	3	5	4	3

So, a sum of 12 can occur in 36 ways.

3. There are three events in this situation. The first event is the choice of the first number, the second event is the choice of the second number, and the third event is the choice of the third number. Because there is a choice of 30 numbers for each event, it follows that the number of different lock combinations is

$30 \cdot 30 \cdot 30 = 27{,}000.$

4. Because the product's catalog number is made up of one letter from the English alphabet followed by a five-digit number, there are 26 choices for the first digit and 10 choices for each of the other 5 digits.

□ □ □ □ □ □

26 10 10 10 10 10

So, the number of possible catalog numbers is $26 \cdot 10 \cdot 10 \cdot 10 \cdot 10 \cdot 10 = 2{,}600{,}000.$

5. *First position*: Any of the *four* letters

Second position: Any of the remaining *three* letters

Third position: Either of the remaining *two* letters

Fourth position: The *one* remaining letter

So, the numbers of choices for the four positions are as follows.

Permutations of four letters

□ □ □ □

4 3 2 1

The total number of permutations of the four letters is

$4! = 4 \cdot 3 \cdot 2 \cdot 1$

$= 24.$

6. Here are the different possibilities.

President (first position): *Five* choices

Vice-President (second position): *Four* choices

Using the Fundamental Counting Principle, multiply these two numbers to obtain the following.

Different orders of offices

□ □

President Vice-President

5 4

So, there are $5 \cdot 4 = 20$ different ways there can be a President and Vice-President.

7. The word M I T O S I S has seven letters, of which there are two I's, two S's, and one M, T, and O. So, the number of distinguishable ways the letters can be written is

$$\frac{n!}{n_1!\,n_2!} = \frac{7!}{2!\,2!} = \frac{7 \cdot 6 \cdot 5 \cdot 4 \cdot 3 \cdot \cancel{2!}}{2!\,\cancel{2!}} = 1260.$$

8. The following subsets represent the different combinations of two letters that can be chosen from the seven letters.

{A, B} {B, C} {C, D} {D, E} {E, F} {F, G}

{A, C} {B, D} {C, E} {D, F} {E, G}

{A, D} {B, E} {C, F} {D, G}

{A, E} {B, F} {C, G}

{A, F} {B, G}

{A, G}

From this list, you can conclude that there are 21 different ways that two letters can be chosen from seven letters.

9. To find the number of three card poker hands, use the formula for the number of combinations of 52 elements taken three at a time, as follows.

$$\begin{aligned} {}_{52}C_3 &= \frac{52!}{(52-3)!\,3!} \\ &= \frac{52!}{49!\,3!} \\ &= \frac{52 \cdot 51 \cdot 50 \cdot \cancel{49!}}{\cancel{49!}\,3!} \\ &= \frac{52 \cdot 51 \cdot 50}{3 \cdot 2 \cdot 1} \\ &= 22{,}100 \end{aligned}$$

10. There are ${}_{10}C_6$ ways of choosing six girls from a group of ten girls and ${}_{15}C_6$ ways of choosing boys from a group of fifteen boys. By the Fundamental Counting Principle, there are ${}_{10}C_6 \cdot {}_{15}C_6$ ways of choosing six girls and six boys.

$$\begin{aligned} {}_{10}C_6 \cdot {}_{15}C_6 &= \frac{10!}{4! \cdot 6!} \cdot \frac{15!}{9! \cdot 6!} \\ &= 210 \cdot 5005 \\ &= 1{,}051{,}050 \end{aligned}$$

So, there are 1,051,050 12-member swim teams possible.

Checkpoints for Section 11.7

1. Because either coin can land heads up or tails up, and the six-sided die can land with a 1 through 6 up. So, the sample space is

$S = \{HH1, HH2, HH3, HH4, HH5, HH6$
$HT1, HT2, HT3, HT4, HT5, HT6,$
$TH1, TH2, TH3, TH4, TH5, TH6,$
$TT1, TT2, TT3, TT4, TT5, TT6\}$

2. (a) Let $E = \{TTT\}$ and $S = \{HHH, HHT, HTH, HTT, TTT, THT, TTH, THH\}$.

The probability of getting three tails is

$$P(E) = \frac{n(E)}{n(S)} = \frac{1}{8}.$$

(b) Because there are 52 cards in a standard deck of playing cards and there are 13 diamonds, the probability of drawing a diamond is

$$\begin{aligned} P(E) &= \frac{n(E)}{n(S)} \\ &= \frac{13}{52} \\ &= \frac{1}{4}. \end{aligned}$$

3. Because there are six possible outcomes on each die, use the Fundamental Counting Principle to conclude that there are $6 \cdot 6$ or 36 different outcomes when you toss two dice. To find the probability of rolling a total of 5, you must first count the number of ways in which this can occur.

First Die	Second Die
1	4
2	3
3	2
4	1

So, a total of 5 can be rolled in four ways, which means that the probability of rolling a 5 is

$$P(E) = \frac{n(E)}{n(S)} = \frac{4}{36} = \frac{1}{9}.$$

4. For a standard deck of 52 playing cards, there are 13 clubs. So, the probability of drawing a club is

$$P(E) = \frac{n(E)}{n(S)} = \frac{13}{52} = \frac{1}{4}.$$

For a set consisting of the aces, the sample space is 4 cards. So, the probability of drawing the ace at hearts is

$$P(E) = \frac{n(E)}{n(S)} = \frac{1}{4}.$$

So, the probability of drawing a club from a standard deck of cards is the same as drawing the ace of hearts from the set of aces.

5. The total number of colleges and universities is 4490. Because there are 604 colleges and universities in the Pacific region, the probability that the institution is in that region is

$$P(E) = \frac{n(E)}{n(S)} = \frac{604}{4490} \approx 0.135.$$

6. To find the number of elements in the sample space, use the formula for the number of combinations of 43 elements taken five at a time.

$$n(S) = {}_{43}C_5 = \frac{43 \cdot 42 \cdot 41 \cdot 40 \cdot 39}{5 \cdot 4 \cdot 3 \cdot 2 \cdot 1} = 962{,}598$$

When a player buys one ticket, the probability of winning is

$$P(E) = \frac{1}{962{,}598} \approx 0.000001.$$

7. Because the deck has 4 aces, the probability of drawing an ace (event A) is

$$P(A) = \frac{4}{52}.$$

Similarly, because the deck has 13 spades, the probability of drawing a spade (event B) is

$$P(B) = \frac{13}{52}.$$

Because one of the cards is an ace *and* a spade, the ace of spades, it follows that

$$P(A \cap B) = \frac{1}{52}.$$

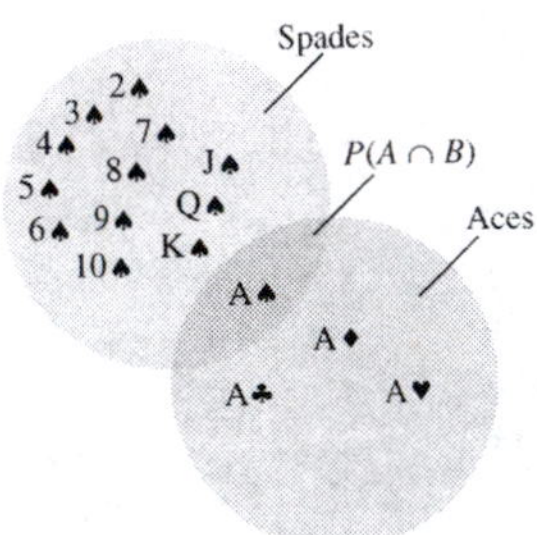

Finally, applying the formula for the probability of the union of two events, the probability of drawing an ace or spade is as follows.

$$P(A \cup B) = P(A) + P(B) - P(A \cap B) = \frac{4}{52} + \frac{13}{52} - \frac{1}{52} = \frac{16}{52} \approx 0.308$$

8. To begin, add the number of employees to find that the total is 529. Next, let event A represent choosing an employee with 30-34 years of service, event B with 35-39 years of service, event C with 40-44 years of service, and event D with 45 or more years of service.

Because events A, B, C, and D have no outcomes in common, these four events are mutually exclusive and

$$P(A \cup B \cup C \cup D) = P(A) + P(B) + P(C) + +P(D) = \frac{35}{529} + \frac{21}{529} + \frac{8}{529} + \frac{2}{529} = \frac{66}{529} \approx 0.125.$$

So, the probability of choosing and employee who has 30 or more years of service is about 0.125.

9. The probability of selecting a number from 1 to 11 from a set of numbers from 1 to 30 is

$$P(A) = \frac{11}{30}.$$

So, the probability that both numbers are less than 12 is

$$P(A) \cdot P(A) = \frac{11}{30} \cdot \frac{11}{30} = \frac{121}{900} \approx 0.134.$$

10. Let A represent choosing a person who gets his or her news from a mobile device. The probability of choosing a person who got his or her news from a mobile device is 0.27, the probability of choosing a second person who got his or her news from a mobile device is 0.27, and so on. Because these events are independent, the probability that all 5 people got their news from mobile devices is

$$[P(A)]^5 = (0.27)^5 \approx 0.001435.$$

11. To solve this problem as stated, you would need to find the probabilities of having exactly one faulty unit, exactly two faulty units, exactly three faulty units, and so on. However, using complements, you can find the probability that all units are perfect and then subtract this value from 1. Because the probability that any given unit is perfect is 499/500, the probability that all 300 units are perfect is

$$P(A) = \left(\frac{499}{500}\right)^{300} \approx 0.548.$$

So, the probability that at least one unit is faulty is

$$P(A') = 1 - P(A) \approx 1 - 0.548 = 0.452.$$

Appendix

Checkpoints for Appendix A

1. Do not apply radicals term-by-term when adding terms.

Leave as $\sqrt{x^2 + 4}$.

2. $x(x-2)^{-1/2} + 3(x-2)^{1/2} = (x-2)^{-1/2}\left[x(x-2)^0 + 3(x-2)^1\right]$

$= (x-2)^{-1/2}[x + 3x - 6]$

$= (x-2)^{-1/2}(4x - 6)$

3. The expression on the left side of the equation is three times the expression on the right side. To make both sides equal, insert a factor of 3.

$$\frac{6x-3}{\left(x^2 - x + 4\right)^2} = (3)\frac{1}{\left(x^2 - x + 4\right)^2}(2x - 1)$$

4. To write the expression on the left side of the equation in the form given on the right side, first multiply the numerator and denominator of the first term by $\frac{1}{9}$. Then multiply the numerator and denominator of the second term by $\frac{1}{25}$.

$$\frac{9x^2}{16} + 25y^2 = \frac{9x^2}{16}\left(\frac{1/9}{1/9}\right) + \frac{25y^2}{1}\left(\frac{1/25}{1/25}\right) = \frac{x^2}{16/25} + \frac{y^2}{1/25}$$

5. $\dfrac{-6x}{\left(1 - 3x^2\right)^2} + \dfrac{1}{3\sqrt{x}} = -x\left(1 - 3x^2\right)^{-2} + x^{-1/3}$

6. $\dfrac{x^4 - 2x^3 + 5}{x^3} = \dfrac{x^4}{x^3} - \dfrac{2x^3}{x^3} + \dfrac{5}{x^3} = x - 2 + \dfrac{5}{x^3}$

Chapter P Practice Test Solutions

1. $\dfrac{|-42| - 20}{15 - |-4|} = \dfrac{42 - 20}{15 - 4} = \dfrac{22}{11} = 2$

2. $\dfrac{x}{z} - \dfrac{z}{y} = \dfrac{x}{z} \cdot \dfrac{y}{y} - \dfrac{z}{y} \cdot \dfrac{z}{z} = \dfrac{xy - z^2}{yz}$

3. $|x - 7| \le 4$

4. $10(-5)^3 = 10(-125) = -1250$

5. $\left(-4x^3\right)\left(-2x^{-5}\right)\left(\dfrac{1}{16}x\right) = (-4)(-2)\left(\dfrac{1}{16}\right)x^{3+(-5)+1}$

$= \dfrac{8}{16}x^{-1}$

$= \dfrac{1}{2x}$

6. $0.0000412 = 4.12 \times 10^{-5}$

7. $125^{2/3} = \left(\sqrt[3]{125}\right)^2 = (5)^2 = 25$

8. $\sqrt[4]{64x^7y^9} = \sqrt[4]{16 \cdot 4x^4x^3y^8y} = 2xy^2\sqrt[4]{4x^3y}$

9. $\dfrac{6}{\sqrt{12}} = \dfrac{6}{2\sqrt{3}} \cdot \dfrac{\sqrt{3}}{\sqrt{3}} = \dfrac{6\sqrt{3}}{6} = \sqrt{3}$

10. $3\sqrt{80} - 7\sqrt{500} = 3\left(4\sqrt{5}\right) - 7\left(10\sqrt{5}\right)$

$= 12\sqrt{5} - 70\sqrt{5}$

$= -58\sqrt{5}$

11. $\left(8x^4 - 9x^2 + 2x - 1\right) - \left(3x^3 + 5x + 4\right) = 8x^4 - 3x^3 - 9x^2 - 3x - 5$

12. $(x - 3)\left(x^2 + x - 7\right) = x^3 + x^2 - 7x - 3x^2 - 3x + 21 = x^3 - 2x^2 - 10x + 21$

13. $\left[(x - 2) - y\right]^2 = (x - 2)^2 - 2y(x - 2) + y^2 = x^2 - 4x + 4 - 2xy + 4y + y^2 = x^2 + y^2 - 2xy - 4x + 4y + 4$

14. $16x^4 - 1 = \left(4x^2 + 1\right)\left(4x^2 - 1\right)$

$= \left(4x^2 + 1\right)(2x + 1)(2x - 1)$

15. $6x^2 + 5x - 4 = (2x - 1)(3x + 4)$

16. $x^3 - 64 = x^3 - 4^3 = (x - 4)\left(x^2 + 4x + 16\right)$

17. $-\dfrac{3}{x} + \dfrac{x}{x^2 + 2} = \dfrac{-3\left(x^2 + 2\right) + x^2}{x\left(x^2 + 2\right)} = \dfrac{-2x^2 - 6}{x\left(x^2 + 2\right)} = -\dfrac{2\left(x^2 + 3\right)}{x\left(x^2 + 2\right)}$

18. $\dfrac{x - 3}{4x} \div \dfrac{x^2 - 9}{x^2} = \dfrac{x - 3}{4x} \cdot \dfrac{x^2}{(x + 3)(x - 3)} = \dfrac{x}{4(x + 3)},\ x \ne 0, 3$

19. $\dfrac{1 - \dfrac{1}{x}}{1 - \dfrac{1}{1 - (1/x)}} = \dfrac{\dfrac{x - 1}{x}}{1 - \dfrac{1}{(x - 1)/x}} = \dfrac{\dfrac{x - 1}{x}}{1 - \dfrac{x}{x - 1}} = \dfrac{\dfrac{x - 1}{x}}{\dfrac{-1}{x - 1}} = \dfrac{x - 1}{x} \cdot \dfrac{x - 1}{-1} = -\dfrac{(x - 1)^2}{x},\ x \ne 1$

20. (a)

(b) $d = \sqrt{\left[5 - (-3)\right]^2 + (-1 - 7)^2}$

$= \sqrt{(8)^2 + (-8)^2}$

$= \sqrt{64 + 64}$

$= \sqrt{128}$

$= 8\sqrt{2}$

(c) $\left(\dfrac{-3 + 5}{2}, \dfrac{7 + (-1)}{2}\right) = (1, 3)$

Chapter 1 Practice Test Solutions

1. 
$3x - 5y = 15$

Line

x-intercept: $(5, 0)$

y-intercept: $(0, -3)$

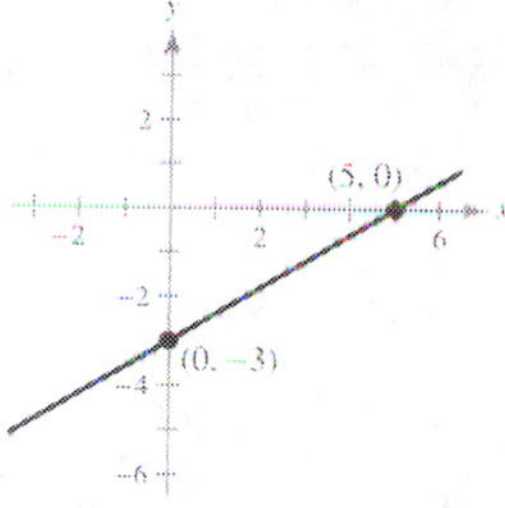

2. $y = \sqrt{9 - x}$

Domain: $(-\infty, 9]$

x-intercept: $(9, 0)$

y-intercept: $(0, 3)$

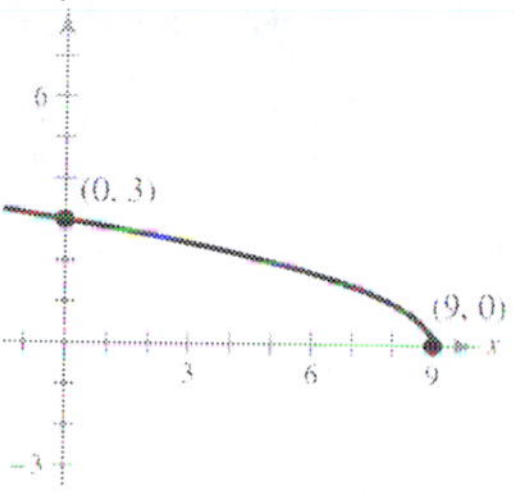

3.
$$\begin{aligned} 5x + 4 &= 7x - 8 \\ 4 + 8 &= 7x - 5x \\ 12 &= 2x \\ x &= 6 \end{aligned}$$

4.
$$\begin{aligned} \frac{x}{3} - 5 &= \frac{x}{5} + 1 \\ 15\left(\frac{x}{3} - 5\right) &= 15\left(\frac{x}{5} + 1\right) \\ 5x - 75 &= 3x + 15 \\ 2x &= 90 \\ x &= 45 \end{aligned}$$

5.
$$\begin{aligned} \frac{3x + 1}{6x - 7} &= \frac{2}{5} \\ 5(3x + 1) &= 2(6x - 7) \\ 15x + 5 &= 12x - 14 \\ 3x &= -19 \\ x &= -\frac{19}{3} \end{aligned}$$

6.
$$\begin{aligned} (x - 3)^2 + 4 &= (x + 1)^2 \\ x^2 - 6x + 9 + 4 &= x^2 + 2x + 1 \\ -8x &= -12 \\ x &= \frac{-12}{-8} \\ x &= \frac{3}{2} \end{aligned}$$

7.
$$\begin{aligned} A &= \frac{1}{2}(a + b)h \\ 2A &= ah + bh \\ 2A - bh &= ah \\ \frac{2A - bh}{h} &= a \end{aligned}$$

8. Percent $= \frac{301}{4300} = 0.07 = 7\%$

9. Let $x =$ number of quarters.

Then $53 - x =$ number of nickels.

$$\begin{aligned} 25x + 5(53 - x) &= 605 \\ 20x + 265 &= 605 \\ 20x &= 340 \\ x &= 17 \text{ quarters} \\ 53 - x &= 36 \text{ nickels} \end{aligned}$$

10. Let $x =$ amount in the $9\frac{1}{2}\%$ fund.

Then $15{,}000 - x =$ amount in 11% fund.

$$\begin{aligned} 0.095x + 0.11(15{,}000 - x) &= 1582.50 \\ -0.015x + 1650 &= 1582.50 \\ -0.015x &= -67.5 \\ x &= \$4500 \text{ at } 9\tfrac{1}{2}\% \\ 15{,}000 - x &= \$10{,}500 \text{ at } 11\% \end{aligned}$$

11.
$$\begin{aligned} 28 + 5x - 3x^2 &= 0 \\ (4 - x)(7 + 3x) &= 0 \\ 4 - x &= 0 \Rightarrow x = 4 \\ 7 + 3x &= 0 \Rightarrow x = -\tfrac{7}{3} \end{aligned}$$

12.
$$\begin{aligned} (x - 2)^2 &= 24 \\ x - 2 &= \pm\sqrt{24} \\ x - 2 &= \pm 2\sqrt{6} \\ x &= 2 \pm 2\sqrt{6} \end{aligned}$$

13.
$$\begin{aligned} x^2 - 4x - 9 &= 0 \\ x^2 - 4x + 2^2 &= 9 + 2^2 \\ (x - 2)^2 &= 13 \\ x - 2 &= \pm\sqrt{13} \\ x &= 2 \pm \sqrt{13} \end{aligned}$$

14. $x^2 + 5x - 1 = 0$

$a = 1, b = 5, c = -1$

$$x = \frac{-5 \pm \sqrt{(5)^2 - 4(1)(-1)}}{2(1)}$$

$$= \frac{-5 \pm \sqrt{25 + 4}}{2}$$

$$= \frac{-5 \pm \sqrt{29}}{2}$$

15. $3x^2 - 2x + 4 = 0$

$a = 3, b = -2, c = 4$

$$x = \frac{-(-2) \pm \sqrt{(-2)^2 - 4(3)(4)}}{2(3)}$$

$$= \frac{2 \pm \sqrt{4 - 48}}{6}$$

$$= \frac{2 \pm \sqrt{-44}}{6}$$

$$= \frac{2 \pm 2i\sqrt{11}}{6}$$

$$= \frac{1 \pm i\sqrt{11}}{3} = \frac{1}{3} \pm \frac{\sqrt{11}}{3}i$$

16.

$$60{,}000 = xy$$

$$y = \frac{60{,}000}{x}$$

$$2x + 2y = 1100$$

$$2x + 2\left(\frac{60{,}000}{x}\right) = 1100$$

$$x + \frac{60{,}000}{x} = 550$$

$$x^2 + 60{,}000 = 550x$$

$$x^2 - 550x + 60{,}000 = 0$$

$$(x - 150)(x - 400) = 0$$

$x = 150$ or $x = 400$

$y = 400$ $\quad y = 150$

Length: 400 feet; width: 150 feet

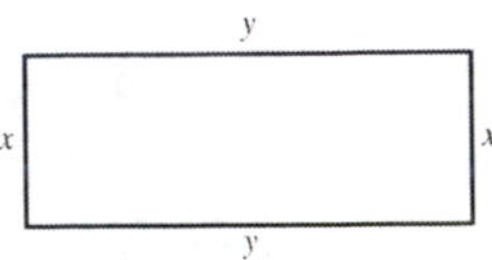

17.

$$x(x + 2) = 624$$

$$x^2 + 2x - 624 = 0$$

$$(x - 24)(x + 26) = 0$$

$x = 24$ or $x = -26$, (extraneous solution)

$x + 2 = 26$

The integers are 24 and 26.

18. $x^2 - 10x^2 + 24x = 0$

$$x(x^2 - 10x + 24) = 0$$

$$x(x - 4)(x - 6) = 0$$

$$x = 0, x = 4, x = 6$$

19. $\sqrt[3]{6 - x} = 4$

$$6 - x = 64$$

$$-x = 58$$

$$x = -58$$

20. $(x^2 - 8)^{2/5} = 4$

$$x^2 - 8 = \pm 4^{5/2}$$

$x^2 - 8 = 32$ or $x^2 - 8 = -32$

$x^2 = 40$ $\quad x^2 = -24$

$x = \pm\sqrt{40}$ $\quad x = \pm\sqrt{-24}$

$x = \pm 2\sqrt{10}$ $\quad x = \pm 2\sqrt{6}i$

21. $x^4 - x^2 - 12 = 0$

$$(x^2 - 4)(x^2 + 3) = 0$$

$x^2 = 4$ or $x^2 = -3$

$x^2 = \pm 2$ $\quad x = \pm\sqrt{3}i$

22. $4 - 3x > 16$

$$-3x > 12$$

$$x < -4$$

23.

$$\left|\frac{x - 3}{2}\right| < 5$$

$$-5 < \frac{x - 3}{2} < 5$$

$$-10 < x - 3 < 10$$

$$-7 < x < 13$$

24.

$$\frac{x+1}{x-3} < 2$$

$$\frac{x+1}{x-3} - 2 < 0$$

$$\frac{x+1-2(x-3)}{x-3} < 0$$

$$\frac{7-x}{x-3} < 0$$

Critical numbers: $x = 7$ and $x = 3$

Test intervals: $(-\infty, 3), (3, 7), (7, \infty)$

Test: Is $\dfrac{7-x}{x-3} < 0$?

Solution intervals: $(-\infty, 3) \cup (7, \infty)$

25. $|3x - 4| \geq 9$

$3x - 4 \leq -9$ or $3x - 4 \geq 9$

$3x \leq -5$ $\qquad 3x \geq 13$

$x \leq -\frac{5}{3}$ $\qquad x \geq \frac{13}{3}$

Chapter 2 Practice Test Solutions

1.

$$m = \frac{-1-4}{3-2} = -5$$

$$y - 4 = -5(x - 2)$$

$$y - 4 = -5x + 10$$

$$y = -5x + 14$$

2. $y = \frac{4}{3}x - 3$

3. $2x + 3y = 0$

$$y = -\tfrac{2}{3}x$$

$$m_1 = -\tfrac{2}{3}$$

$$\perp m_2 = \tfrac{3}{2} \text{ through } (4, 1)$$

$$y - 1 = \tfrac{3}{2}(x - 4)$$

$$y - 1 = \tfrac{3}{2}x - 6$$

$$y = \tfrac{3}{2}x - 5$$

4. $(5, 32)$ and $(9, 44)$

$$m = \frac{44-32}{9-5} = \frac{12}{4} = 3$$

$$y - 32 = 3(x - 5)$$

$$y - 32 = 3x - 15$$

$$y = 3x + 17$$

When $x = 20$, $y = 3(20) + 17$

$y = \$77$.

5.

$$f(x-3) = (x-3)^2 - 2(x-3) + 1$$

$$= x^2 - 6x + 9 - 2x + 6 + 1$$

$$= x^2 - 8x + 16$$

6.

$$f(3) = 12 - 11 = 1$$

$$\frac{f(x) - f(3)}{x - 3} = \frac{(4x - 11) - 1}{x - 3}$$

$$= \frac{4x - 12}{x - 3}$$

$$= \frac{4(x-3)}{x-3} = 4, x \neq 3$$

7. $f(x) = \sqrt{36 - x^2} = \sqrt{(6 + x)(6 - x)}$

Domain: $[-6, 6]$

Range: $[0, 6]$, because

$(6 + x)(6 - x) \geq 0$ on this interval.

8. (a) $6x - 5y + 4 = 0$

$y = \dfrac{6x + 4}{5}$ is a function of x.

(b) $x^2 + y^2 = 9$

$y = \pm\sqrt{9 - x^2}$ is not a function of x.

(c) $y^3 = x^2 + 6$

$y = \sqrt[3]{x^2 + 6}$ is a function of x.

9. Parabola

Vertex: $(0, -5)$

Intercepts:

$(0, -5), (\pm\sqrt{5}, 0)$

y-axis symmetry

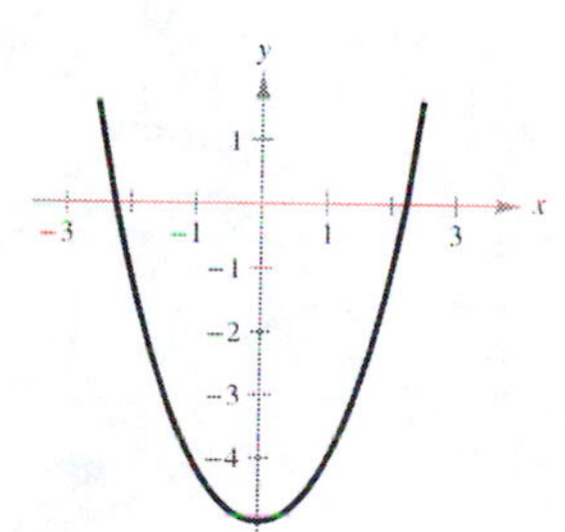

10. Intercepts: $(0, 3)$, $(-3, 0)$

x	-4	-3	-2	-1	0	1	2
y	1	0	1	2	3	4	5

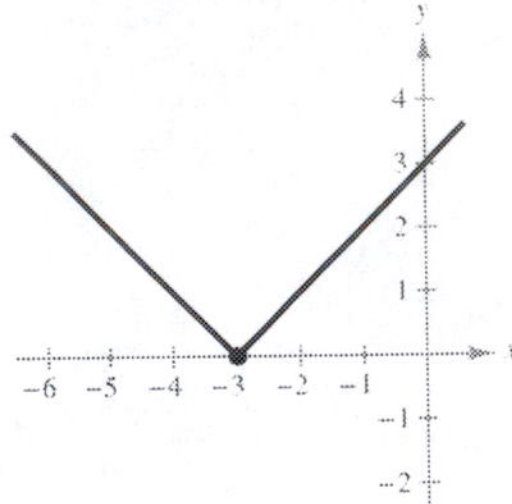

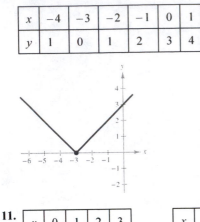

11.

x	0	1	2	3
y	1	3	5	7

x	-1	-2	-3
y	2	6	12

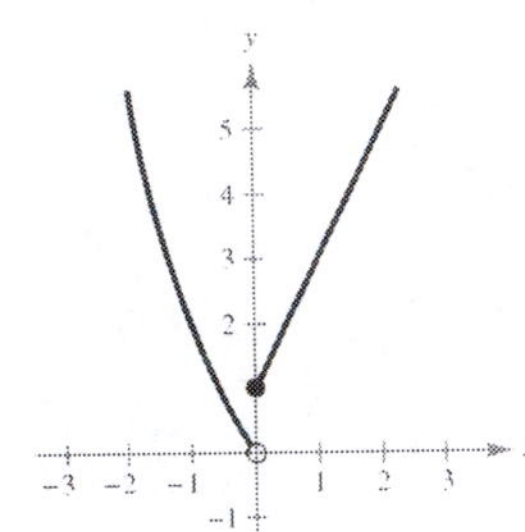

12. (a) $f(x + 2)$

Horizontal shift two units to the left

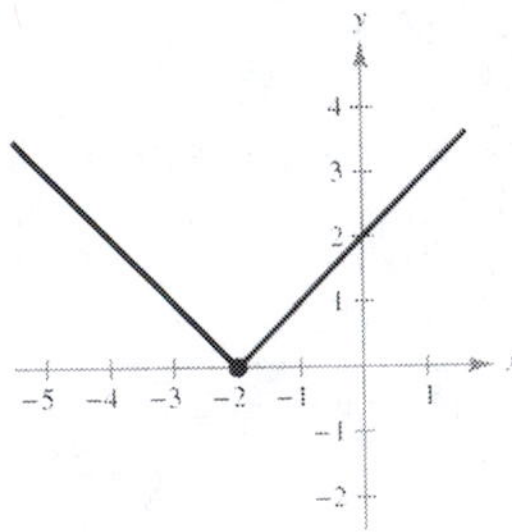

(b) $-f(x) + 2$

Reflection in the x-axis and a vertical shift two units upward

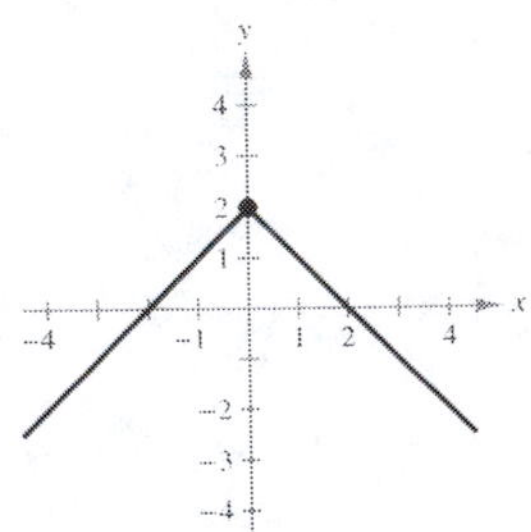

13. (a)
$$\begin{aligned}(g - f)(x) &= g(x) - f(x)\\ &= (2x^2 - 5) - (3x + 7)\\ &= 2x^2 - 3x - 12\end{aligned}$$

(b)
$$\begin{aligned}(fg)(x) &= f(x)g(x)\\ &= (3x + 7)(2x^2 - 5)\\ &= 6x^3 + 14x^2 - 15x - 35\end{aligned}$$

14.
$$\begin{aligned}f(g(x)) &= f(2x + 3)\\ &= (2x + 3)^2 - 2(2x + 3) + 16\\ &= 4x^2 + 12x + 9 - 4x - 6 + 16\\ &= 4x^2 + 8x + 19\end{aligned}$$

15.
$$\begin{aligned}f(x) &= x^3 + 7\\ y &= x^3 + 7\\ x &= y^3 + 7\\ x - 7 &= y^3\\ \sqrt[3]{x - 7} &= y\\ f^{-1}(x) &= \sqrt[3]{x - 7}\end{aligned}$$

16. (a) $f(x) = |x - 6|$ does not have an inverse. Its graph does not pass the Horizontal Line Test.

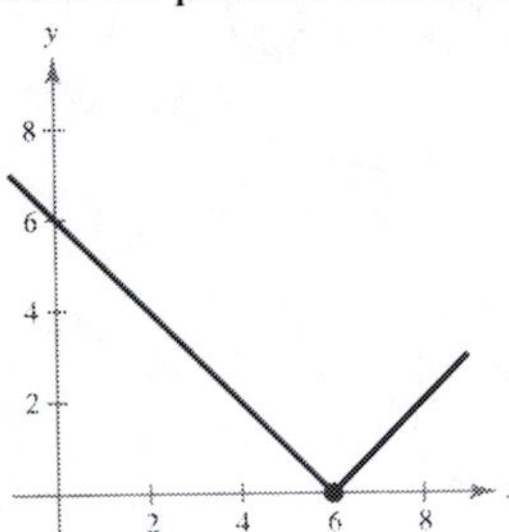

(b) $f(x) = ax + b,\ a \neq 0$ does have an inverse.

$$\begin{aligned}y &= ax + b\\ x &= ay + b\\ \frac{x - b}{a} &= y\\ f^{-1}(x) &= \frac{x - b}{a}\end{aligned}$$

(c) $f(x) = x^3 - 19$ does have an inverse.

$$\begin{aligned}y &= x^3 - 19\\ x &= y^3 - 19\\ x + 19 &= y^3\\ \sqrt[3]{x + 19} &= y\\ f^{-1}(x) &= \sqrt[3]{x + 19}\end{aligned}$$

17. $f(x) = \sqrt{\dfrac{3-x}{x}},\ 0 < x \le 3,\ y \ge 0$

$$y = \sqrt{\frac{3-x}{x}}$$

$$x = \sqrt{\frac{3-y}{y}},\ 0 < y \le 3,\ x \ge 0$$

$$x^2 = \frac{3-y}{y}$$

$$x^2 y = 3 - y$$

$$x^2 y + y = 3$$

$$y(x^2 + 1) = 3$$

$$y = \frac{3}{x^2 + 1}$$

$$f^{-1}(x) = \frac{3}{x^2 + 1},\ x \ge 0$$

18. False. The slopes of 3 and $\frac{1}{3}$ are not **negative** reciprocals.

19. True. Let $y = (f \circ g)(x)$. Then $x = (f \circ g)^{-1}(y)$.

Also,

$$(f \circ g)(x) = y$$

$$f(g(x)) = y$$

$$g(x) = f^{-1}(y)$$

$$x = g^{-1}(f^{-1}(y))$$

$$x = (g^{-1} \circ f^{-1})(y).$$

Since $x = x$, we have $(f \circ g)^{-1}(y) = (g^{-1} \circ f^{-1})(y)$.

20. True. It must pass the Vertical Line Test to be a function and it must pass the Horizontal Line Test to have an inverse.

Chapter 3 Practice Test Solutions

1. x-intercepts: $(1, 0), (5, 0)$

y-intercept: $(0, 5)$

Vertex: $(3, -4)$

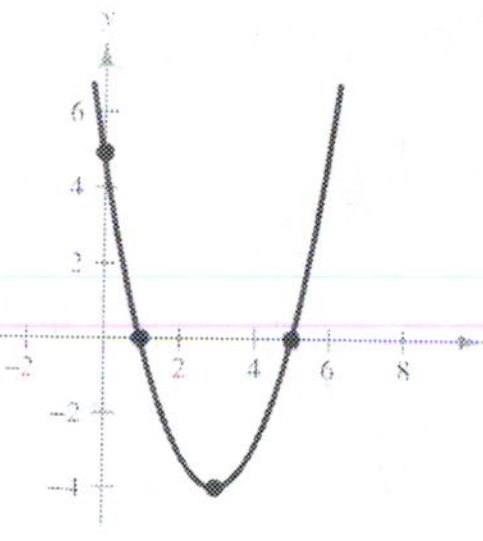

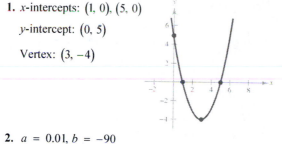

2. $a = 0.01,\ b = -90$

$$\frac{-b}{2a} = \frac{90}{2(0.01)} = 4500 \text{ units}$$

3. Vertex $(1, 7)$ opening downward through $(2, 5)$

$y = a(x-1)^2 + 7$ Standard form

$$5 = a(2-1)^2 + 7$$

$$5 = a + 7$$

$$a = -2$$

$$y = -2(x-1)^2 + 7$$

$$= -2(x^2 - 2x + 1) + 7$$

$$= -2x^2 + 4x + 5$$

4. $y = \pm a(x-2)(3x-4)$ where a is any real nonzero number.

$$y = \pm(3x^2 - 10x + 8)$$

5. Leading coefficient: -3

Degree: 5 (odd)

Falls to the right.

Rises to the left.

6. $0 = x^5 - 5x^3 + 4x$

$$= x(x^4 - 5x^2 + 4)$$

$$= x(x^2 - 1)(x^2 - 4)$$

$$= x(x+1)(x-1)(x+2)(x-2)$$

$$x = 0,\ x = \pm 1,\ x = \pm 2$$

7. $f(x) = x(x-3)(x+2)$

$$= x(x^2 - x - 6)$$

$$= x^3 - x^2 - 6x$$

8. $f(x) = x^3 - 12x$

Intercepts: $(0, 0), (\pm 2\sqrt{3}, 0)$

Rises to the right.

Falls to the left.

x	-2	-1	0	1	2
y	16	11	0	-11	-16

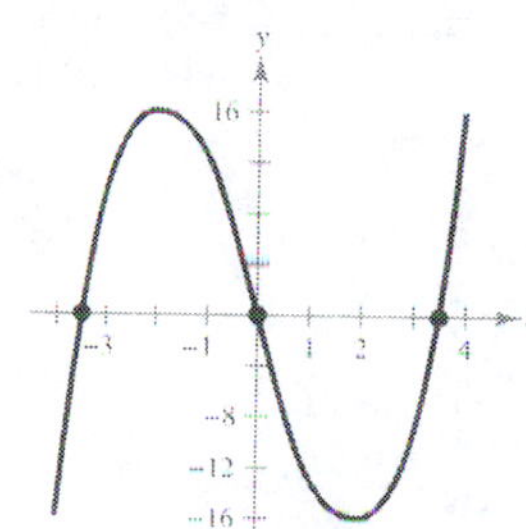

9.
$$\begin{array}{r}
3x^3 + 9x^2 + 20x + 62 \\
x - 3\overline{\big)3x^4 + 0x^3 - 7x^2 + 2x - 10} \\
\underline{3x^4 - 9x^3} \\
9x^3 - 7x^2 \\
\underline{9x^3 - 27x^2} \\
20x^2 + 2x \\
\underline{20x^2 - 60x} \\
62x - 10 \\
\underline{62x - 186} \\
176
\end{array}$$

$$\frac{3x^4 - 7x^2 + 2x - 10}{x - 3} = 3x^3 + 9x^2 + 20x + 6 + \frac{176}{x - 3}$$

10.
$$\begin{array}{r}
x - 2 \\
x^2 + 2x - 1\overline{\big)x^3 + 0x^2 + 0x - 11} \\
\underline{x^3 + 2x^2 - x} \\
-2x^2 + x - 11 \\
\underline{-2x^2 - 4x + 2} \\
5x - 13
\end{array}$$

$$\frac{x^3 - 11}{x^2 + 2x - 1} = x - 2 + \frac{5x - 13}{x^2 + 2x - 1}$$

11.
$$\begin{array}{r|rrrrrr}
-5 & 3 & 13 & 0 & 0 & 12 & -1 \\
 & & -15 & 10 & -50 & 250 & -1310 \\
\hline
 & 3 & -2 & 10 & -50 & 262 & -1311
\end{array}$$

$$\frac{3x^5 + 13x^4 + 12x - 1}{x + 5} = 3x^4 - 2x^3 + 10x^2 - 50x + 262 - \frac{1311}{x + 5}$$

12.
$$\begin{array}{r|rrrr}
-6 & 7 & 40 & -12 & 15 \\
 & & -42 & 12 & 0 \\
\hline
 & 7 & -2 & 0 & 15
\end{array}$$

$f(-6) = 15$

13. $0 = x^3 - 19x - 30$

Possible rational zeros: $\pm1, \pm2, \pm3, \pm5, \pm6, \pm10, \pm15, \pm30$

$$\begin{array}{r|rrrr}
-2 & 1 & 0 & -19 & -30 \\
 & & -2 & 4 & 30 \\
\hline
 & 1 & -2 & -15 & 0
\end{array}$$

$0 = (x + 2)(x^2 - 2x - 15)$

$0 = (x + 2)(x + 3)(x - 5)$

Zeros: $x = -2, x = -3, x = 5$

14. $0 = x^4 + x^3 - 8x^2 - 9x - 9$

Possible rational zeros: $\pm1, \pm3, \pm9$

$$\begin{array}{r|rrrrr} 3 & 1 & 1 & -8 & -9 & -9 \\ & & 3 & 12 & 12 & 9 \\ \hline & 1 & 4 & 4 & 3 & 0 \end{array}$$

$0 = (x - 3)(x^3 + 4x^2 + 4x + 3)$

Possible rational zeros of $x^3 + 4x^2 + 4x + 3$: $\pm1, \pm3$

$$\begin{array}{r|rrrr} -3 & 1 & 4 & 4 & 3 \\ & & -3 & -3 & -3 \\ \hline & 1 & 1 & 1 & 0 \end{array}$$

$0 = (x - 3)(x + 3)(x^2 + x + 1)$

Use the Quadratic Formula. The zeros of $x^2 + x + 1$ are $x = \dfrac{-1 \pm \sqrt{3}i}{2}$.

Zeros:

$x = 3, x = -3, x = -\dfrac{1}{2} + \dfrac{\sqrt{3}}{2}i, x = -\dfrac{1}{2} - \dfrac{\sqrt{3}}{2}i$

The real zeros are $x = 3$ and $x = -3$.

15. $0 = 6x^3 - 5x^2 + 4x - 15$

Possible rational zeros:

$\pm1, \pm3, \pm5, \pm15, \pm\frac{1}{2}, \pm\frac{3}{2}, \pm\frac{5}{2}, \pm\frac{15}{2}, \pm\frac{1}{3}, \pm\frac{5}{3}, \pm\frac{1}{6}, \pm\frac{5}{6}$

16. $0 = x^3 - \frac{20}{3}x^2 + 9x - \frac{10}{3}$

$0 = 3x^3 - 20x^2 + 27x - 10$

Possible rational zeros:

$\pm1, \pm2, \pm5, \pm10, \pm\frac{1}{3}, \pm\frac{2}{3}, \pm\frac{5}{3}, \pm\frac{10}{3}$

$$\begin{array}{r|rrrr} 1 & 3 & -20 & 27 & -10 \\ & & 3 & -17 & 10 \\ \hline & 3 & -17 & 10 & 0 \end{array}$$

$0 = (x - 1)(3x^2 - 17x + 10)$

$0 = (x - 1)(3x - 2)(x - 5)$

Zeros: $x = 1, x = \frac{2}{3}, x = 5$

17. Possible rational zeros: $\pm1, \pm2, \pm5, \pm10$

$$\begin{array}{r|rrrrr} 1 & 1 & 1 & 3 & 5 & -10 \\ & & 1 & 2 & 5 & 10 \\ \hline & 1 & 2 & 5 & 10 & 0 \end{array}$$

$$\begin{array}{r|rrrr} -2 & 1 & 2 & 5 & 10 \\ & & -2 & 0 & -10 \\ \hline & 1 & 0 & 5 & 0 \end{array}$$

$f(x) = (x - 1)(x + 2)(x^2 + 5)$

$= (x - 1)(x + 2)(x + \sqrt{5}i)(x - \sqrt{5}i)$

18. $f(x) = (x - 2)[x - (3 + i)][x - (3 - i)]$

$= (x - 2)[(x - 3) - i][(x - 3) + i]$

$= (x - 2)[(x - 3)^2 - (i)^2]$

$= (x - 2)[x^2 - 6x + 10]$

$= x^3 - 8x^2 + 22x - 20$

19.

$$\begin{array}{r|rrrr} 3i & 1 & 4 & 9 & 36 \\ & & 3i & 12i - 9 & -36 \\ \hline & 1 & 4 + 3i & 12i & 0 \end{array}$$

Thus, $f(3i) = 0$.

20. $z = \dfrac{kx^2}{\sqrt{y}}$

Chapter 4 Practice Test Solutions

1. Vertical asymptote: $x = 0$

Horizontal asymptote: $y = \frac{1}{2}$

x-intercept: $(1, 0)$

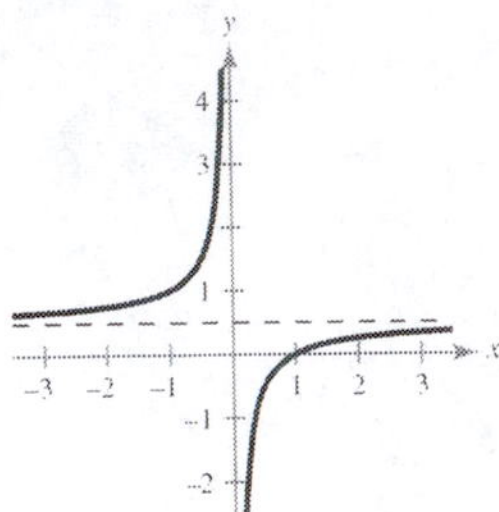

2. Vertical asymptote: $x = 0$

Slant asymptote: $y = 3x$

x-intercepts: $\left(\pm\frac{2}{\sqrt{3}}, 0\right)$

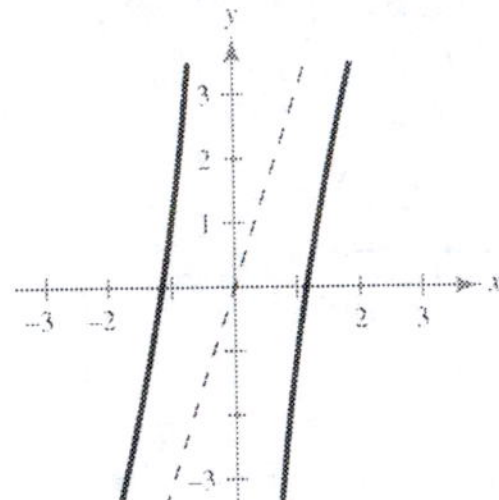

3. $y = 8$ is a horizontal asymptote since the degree of the numerator equals the degree of the denominator. There are no vertical asymptotes.

4. $x = 1$ is a vertical asymptote.

$$\frac{4x^2 - 2x + 7}{x - 1} = 4x + 2 + \frac{9}{x - 1}$$

so $y = 4x + 2$ is a slant asymptote.

5. $f(x) = \dfrac{x - 5}{(x - 5)^2} = \dfrac{1}{x - 5}$

Vertical asymptote: $x = 5$

Horizontal asymptote: $y = 0$

y-intercept: $\left(0, -\frac{1}{5}\right)$

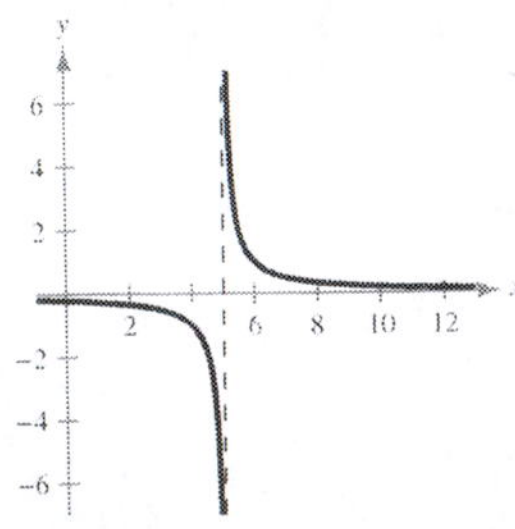

6. $(x - 0)^2 = 4(5)(y - 0)$

Vertex: $(0, 0)$

Focus: $(0, 5)$

Directrix: $y = -5$

7. $(y - 0)^2 = 4(7)(x - 0)$

$y^2 = 28x$

8. $a = 12, b = 5, h = k = 0,$

$c = \sqrt{144 - 25} = \sqrt{119}$

Center: $(0, 0)$

Foci: $\left(\pm\sqrt{119}, 0\right)$

Vertices: $(\pm 12, 0)$

9. Center: $(0, 0)$

$c = 4, 2b = 6 \Rightarrow b = 3,$

$a = \sqrt{16 + 9} = 5$

$$\frac{x^2}{25} + \frac{y^2}{9} = 1$$

10. $a = 12, b = 13, c = \sqrt{144 + 169} = \sqrt{313}$

Center: $(0, 0)$

Foci: $\left(0, \pm\sqrt{313}\right)$

Vertices: $(0, \pm 12)$

Asymptotes: $y = \pm\frac{12}{13}x$

11. Center: $(0, 0)$

$a = 4, \pm\frac{1}{2} = \pm\frac{b}{4} \Rightarrow b = 2$

$$\frac{x^2}{16} - \frac{y^2}{4} = 1$$

12. $p = 4$

$(x - 6)^2 = 4(4)(y + 1)$

$(x - 6)^2 = 16(y + 1)$

13.

$$16x^2 - 96x + 9y^2 + 36y = -36$$
$$16(x^2 - 6x + 9) + 9(y^2 + 4y + 4) = -36 + 144 + 36$$
$$16(x-3)^2 + 9(y+2)^2 = 144$$
$$\frac{(x-3)^2}{9} + \frac{(y+2)^2}{16} = 1$$

$a = 4, b = 3, c = \sqrt{16 - 9} = \sqrt{7}$

Center: $(3, -2)$

Foci: $\left(3, -2 \pm \sqrt{7}\right)$

Vertices: $(3, -2 \pm 4)$ OR $(3, 2)$ and $(3, -6)$

14. Center: $(3, 1)$

$a = 4, 2b = 2 \Rightarrow b = 1$

$$\frac{(x-3)^2}{16} + \frac{(y-1)^2}{1} = 1$$

15. $$\frac{(x+3)^2}{1/4} - \frac{(y-1)^2}{1/9} = 1$$

$a = \frac{1}{2}, b = \frac{1}{3}, c = \sqrt{\frac{1}{4} + \frac{1}{9}} = \frac{\sqrt{13}}{6}$

Center: $(-3, 1)$

Vertices: $\left(-3 \pm \frac{1}{2}, 1\right)$ OR $\left(-\frac{5}{2}, 1\right)$ and $\left(-\frac{7}{2}, 1\right)$

Foci: $\left(-3 \pm \frac{\sqrt{13}}{6}, 1\right)$

Asymptotes: $y = \pm\frac{1/3}{1/2}(x+3) + 1 = \pm\frac{2}{3}(x+3) + 1$

16. Center: $(3, 0)$

$a = 4, c = 7, b = \sqrt{49 - 16} = \sqrt{33}$

$$\frac{y^2}{16} - \frac{(x-3)^2}{33} = 1$$

Chapter 5 Practice Test Solutions

1. $x^{3/5} = 8$

$x = 8^{5/3} = \left(\sqrt[3]{8}\right)^5 = 2^5 = 32$

2. $3^{x-1} = \frac{1}{81}$

$3^{x-1} = 3^{-4}$

$x - 1 = -4$

$x = -3$

3. $f(x) = 2^{-x} = \left(\frac{1}{2}\right)^x$

x	-2	-1	0	1	2
$f(x)$	4	2	1	$\frac{1}{2}$	$\frac{1}{4}$

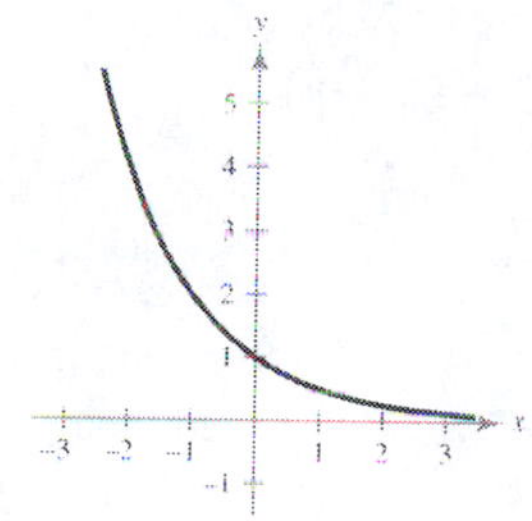

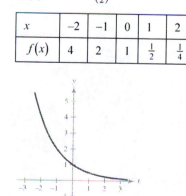

4. $g(x) = e^x + 1$

x	-2	-1	0	1	2
$g(x)$	1.14	1.37	2	3.72	8.39

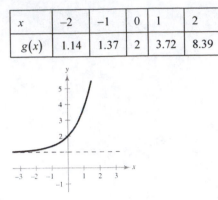

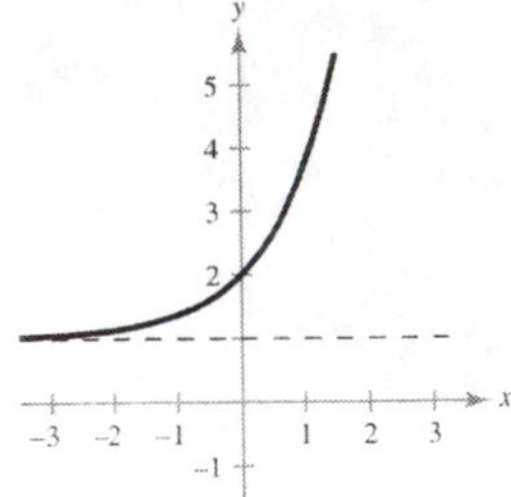

5. (a) $A = P\left(1 + \frac{r}{n}\right)^{nt}$

$$A = 5000\left(1 + \frac{0.09}{12}\right)^{12(3)} \approx \$6543.23$$

(b) $A = P\left(1 + \frac{r}{n}\right)^{nt}$

$$A = 5000\left(1 + \frac{0.09}{4}\right)^{4(3)} \approx \$6530.25$$

(c) $A = Pe^{rt}$

$$A = 5000e^{(0.09)(3)} \approx \$6549.82$$

6. $7^{-2} = \frac{1}{49}$

$$\log_7 \frac{1}{49} = -2$$

7. $x - 4 = \log_2 \frac{1}{64}$

$$2^{x-4} = \frac{1}{64}$$
$$2^{x-4} = 2^{-6}$$
$$x - 4 = -6$$
$$x = -2$$

8. $\log_b \sqrt[4]{\frac{8}{25}} = \frac{1}{4}\log_b \frac{8}{25}$

$$= \frac{1}{4}[\log_b 8 - \log_b 25]$$
$$= \frac{1}{4}[\log_b 2^3 - \log_b 5^2]$$
$$= \frac{1}{4}[3\log_b 2 - 2\log_b 5]$$
$$= \frac{1}{4}[3(0.3562) - 2(0.8271)]$$
$$= -0.1464$$

9. $5\ln x - \frac{1}{2}\ln y + 6\ln z = \ln x^5 - \ln\sqrt{y} + \ln z^6$

$$= \ln\left(\frac{x^5z^6}{\sqrt{y}}\right),\ z > 0$$

10. $\log_9 28 = \frac{\log 28}{\log 9} \approx 1.5166$

11. $\log N = 0.6646$

$$N = 10^{0.6646} \approx 4.62$$

12.

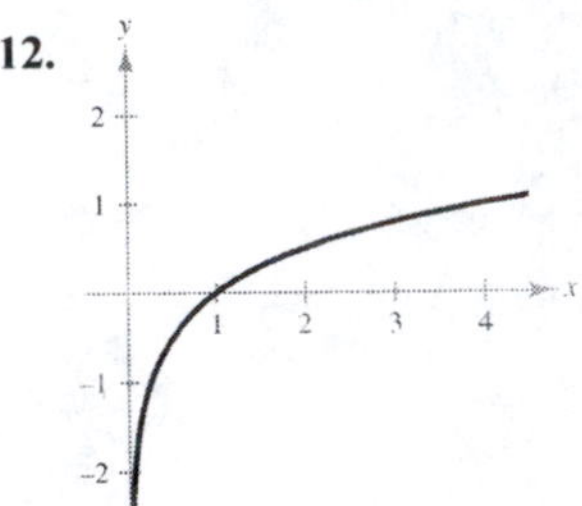

13. Domain:

$$x^2 - 9 > 0$$
$$(x + 3)(x - 3) > 0$$
$$x < -3 \text{ or } x > 3$$

14.

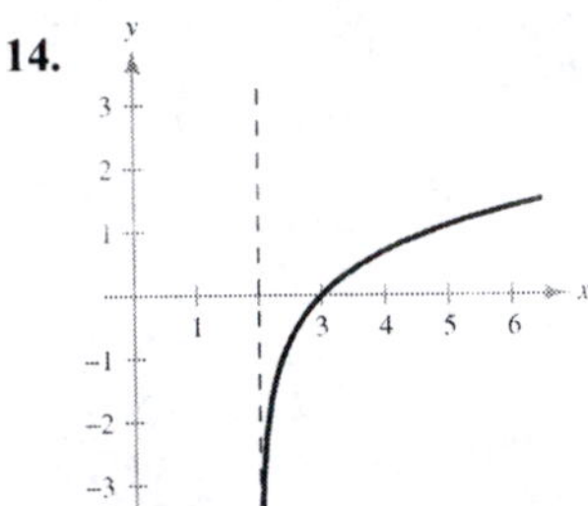

15. False. $\frac{\ln x}{\ln y} \neq \ln(x - y)$ because $\frac{\ln x}{\ln y} = \log_y x$.

16. $5^3 = 41$

$$x = \log_5 41 = \frac{\ln 41}{\ln 5} \approx 2.3074$$

17. $x - x^2 = \log_5 \frac{1}{25}$

$$5^{x-x^2} = \frac{1}{25}$$
$$5^{x-x^2} = 5^{-2}$$
$$x - x^2 = -2$$
$$0 = x^2 - x - 2$$
$$0 = (x + 1)(x - 2)$$
$$x = -1 \text{ or } x = 2$$

18.
$$\begin{aligned}\log_2 x + \log_2(x-3) &= 2\\ \log_2[x(x-3)] &= 2\\ x(x-3) &= 2^2\\ x^2 - 3x &= 4\\ x^2 - 3x - 4 &= 0\\ (x+1)(x-4) &= 0\\ x &= 4\\ x &= -1 \text{ (extraneous)}\end{aligned}$$

$x = 4$ is the only solution.

19.
$$\begin{aligned}\frac{e^x + e^{-x}}{3} &= 4\\ e^x(e^x + e^{-x}) &= 12e^x\\ e^{2x} + 1 &= 12e^x\\ e^{2x} - 12e^x + 1 &= 0\\ e^x &= \frac{12 \pm \sqrt{144-4}}{2}\end{aligned}$$

$e^x \approx 11.9161$ or $e^x \approx 0.0839$

$x = \ln 11.9161$ $\quad x = \ln 0.0839$

$x \approx 2.478$ $\quad x \approx -2.478$

20.
$$\begin{aligned}A &= Pe^{rt}\\ 12{,}000 &= 6000e^{0.13t}\\ 2 &= e^{0.13t}\\ 0.13t &= \ln 2\\ t &= \frac{\ln 2}{0.13}\\ t &\approx 5.3319 \text{ years or 5 years 4 months}\end{aligned}$$

Chapter 6 Practice Test Solutions

1. $350° = 350\left(\frac{\pi}{180}\right) = \frac{35\pi}{18}$

2. $\frac{5\pi}{9} = \frac{5\pi}{9} \cdot \frac{180}{\pi} = 100°$

3. $135° 14' 12'' = \left(135 + \frac{14}{60} + \frac{12}{3600}\right)° \approx 135.2367°$

4.
$$\begin{aligned}-22.569° &= -\left(22° + 0.569(60)'\right)\\ &= -22° 34.14'\\ &= -\left(22° 34' + 0.14(60)''\right)\\ &\approx -22° 34' 8''\end{aligned}$$

5. $\cos\theta = \frac{2}{3}$

$x = 2, r = 3, y = \pm\sqrt{9-4} = \pm\sqrt{5}$

$\tan\theta = \frac{y}{x} = \pm\frac{\sqrt{5}}{2}$

6. $\sin\theta = 0.9063$

$\theta = \arcsin(0.9063)$

$\theta = 65° = \frac{13\pi}{36}$ or $\theta = 180° - 65° = 115° = \frac{23\pi}{36}$

7. $\tan 20° = \frac{35}{x}$

$x = \frac{35}{\tan 20°} \approx 96.1617$

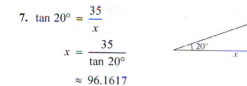

8. $\theta = \frac{6\pi}{5}$, θ is in Quadrant III.

Reference angle: $\frac{6\pi}{5} - \pi = \frac{\pi}{5}$ or $36°$

9. $\csc 3.92 = \frac{1}{\sin 3.92} \approx -1.4242$

10. $\tan\theta = 6 = \frac{6}{1}$, θ lies in Quandrant III.

$y = -6$, $x = -1$, $r = \sqrt{36 + 1} = \sqrt{37}$, so

$\sec\theta = \frac{\sqrt{37}}{-1} \approx -6.0828$.

11. Period: 4π

Amplitude: 3

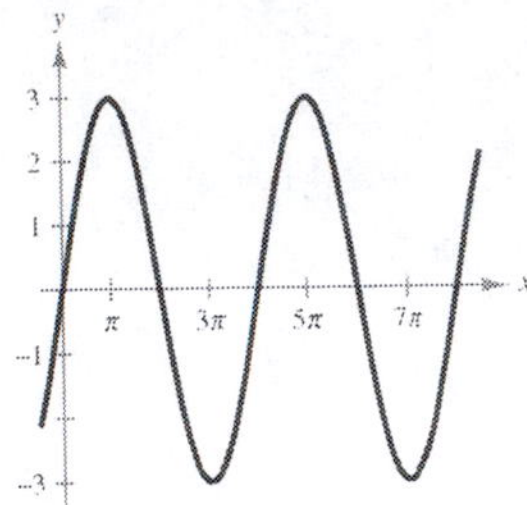

12. Period: 2π

Amplitude: 2

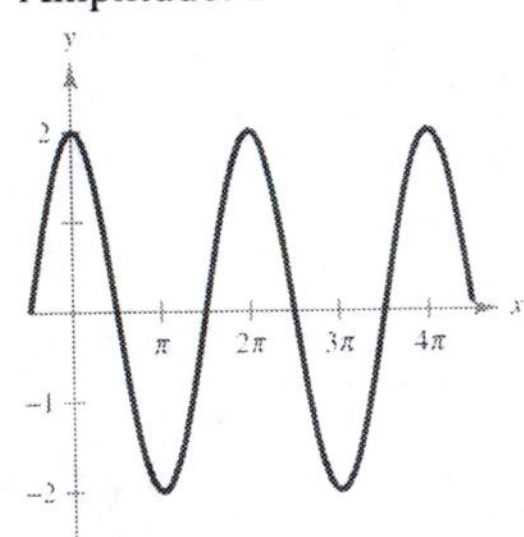

13. Period: $\frac{\pi}{2}$

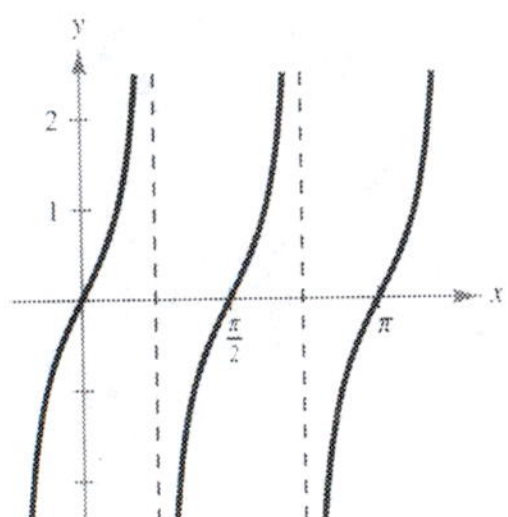

14. Period: 2π

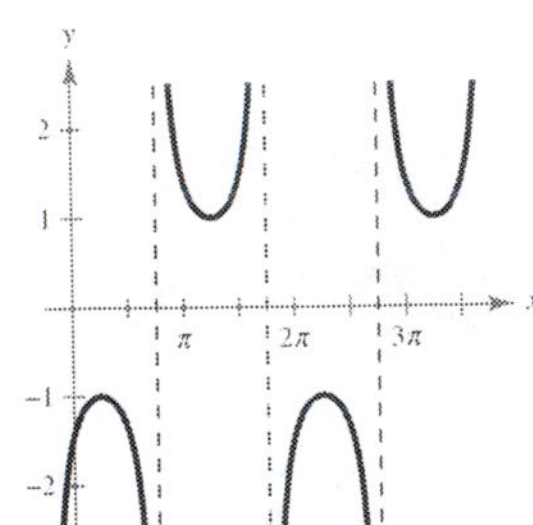

15.

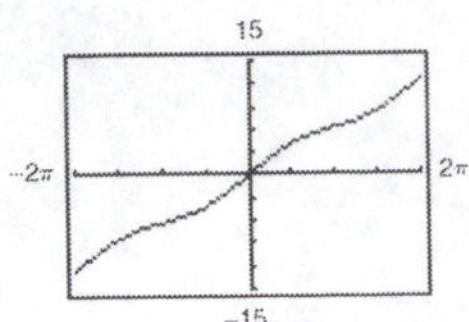

16.

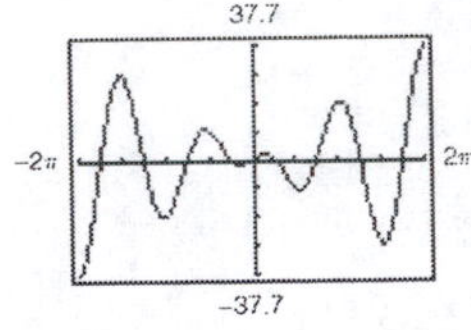

17. $\theta = \arcsin 1$

$\sin\theta = 1$

$\theta = \frac{\pi}{2} = 90°$

18. $\theta = \arctan(-3)$

$\tan\theta = -3$

$\theta \approx -1.249 \approx -71.565°$

19. $\sin\left(\arccos\frac{4}{\sqrt{35}}\right)$

$\sin\theta = \frac{\sqrt{19}}{\sqrt{35}} \approx 0.7368$

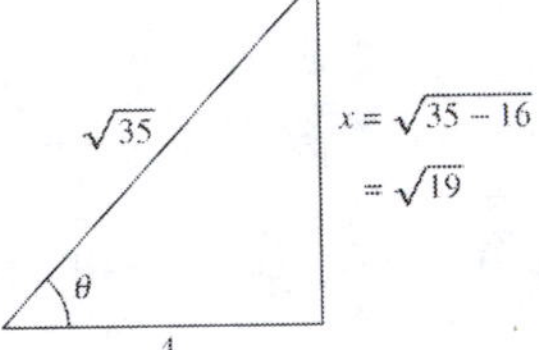

20. $\cos\left(\arcsin\frac{x}{4}\right)$

$\cos\theta = \frac{\sqrt{16 - x^2}}{4}$

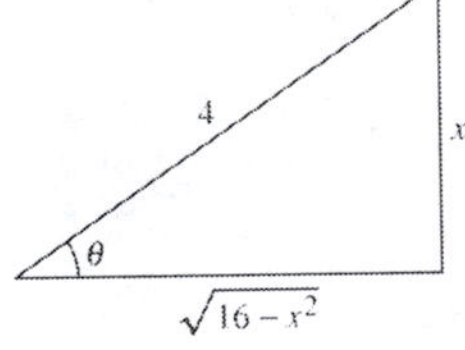

21. Given $A = 40°$, $c = 12$

$B = 90° - 40° = 50°$

$\sin 40° = \frac{a}{12}$

$a = 12\sin 40° \approx 7.713$

$\cos 40° = \frac{b}{12}$

$b = 12\cos 40° \approx 9.193$

22. Given $B = 6.84°$, $a = 21.3$

$$A = 90° - 6.84° = 83.16°$$

$$\sin 83.16° = \frac{21.3}{c}$$

$$c = \frac{21.3}{\sin 83.16°} \approx 21.453$$

$$\tan 83.16° = \frac{21.3}{b}$$

$$b = \frac{21.3}{\tan 83.16°} \approx 2.555$$

23. Given $a = 5$, $b = 9$

$$c = \sqrt{25 + 81} = \sqrt{106} \approx 10.296$$

$$\tan A = \tfrac{5}{9}$$

$$A = \arctan \tfrac{5}{9} \approx 29.055°$$

$$B \approx 90° - 29.055° = 60.945°$$

24. $\sin 67° = \dfrac{x}{20}$

$$x = 20 \sin 67° \approx 18.41 \text{ feet}$$

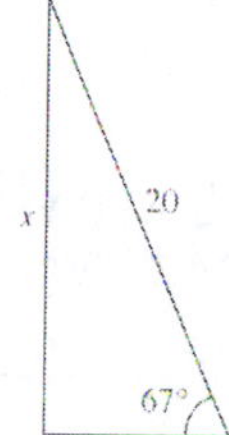

25. $\tan 5° = \dfrac{250}{x}$

$$x = \frac{250}{\tan 5°}$$

$$\approx 2857.513 \text{ feet}$$

$$\approx 0.541 \text{ mi}$$

Chapter 7 Practice Test Solutions

1. $\tan x = \dfrac{4}{11}$, $\sec x < 0 \Rightarrow x$ is in Quadrant III.

$$y = -4,\ x = -11,\ r = \sqrt{16 + 121} = \sqrt{137}$$

$$\sin x = -\frac{4}{\sqrt{137}} = -\frac{4\sqrt{137}}{137} \qquad \csc x = -\frac{\sqrt{137}}{4}$$

$$\cos x = -\frac{11}{\sqrt{137}} = -\frac{11\sqrt{137}}{137} \qquad \sec x = -\frac{\sqrt{137}}{11}$$

$$\tan x = \frac{4}{11} \qquad \cot x = \frac{11}{4}$$

2.

$$\frac{\sec^2 x + \csc^2 x}{\csc^2 x(1 + \tan^2 x)} = \frac{\sec^2 x + \csc^2 x}{\csc^2 x + (\csc^2 x)\tan^2 x}$$

$$= \frac{\sec^2 x + \csc^2 x}{\csc^2 x + \dfrac{1}{\sin^2 x} \cdot \dfrac{\sin^2 x}{\cos^2 x}}$$

$$= \frac{\sec^2 x + \csc^2 x}{\csc^2 x + \dfrac{1}{\cos^2 x}}$$

$$= \frac{\sec^2 x + \csc^2 x}{\csc^2 x + \sec^2 x} = 1$$

3. $\ln|\tan\theta| - \ln|\cot\theta| = \ln\dfrac{|\tan\theta|}{|\cot\theta|} = \ln\left|\dfrac{\sin\theta/\cos\theta}{\cos\theta/\sin\theta}\right| = \ln\left|\dfrac{\sin^2\theta}{\cos^2\theta}\right| = \ln|\tan^2\theta| = 2\ln|\tan\theta|$

4. $\cos\left(\dfrac{\pi}{2} - x\right) = \dfrac{1}{\csc x}$ is true since $\cos\left(\dfrac{\pi}{2} - x\right) = \sin x = \dfrac{1}{\csc x}$.

5.

$$\sin^4 x + (\sin^2 x)\cos^2 x = \sin^2 x(\sin^2 x + \cos^2 x)$$

$$= \sin^2 x(1) = \sin^2 x$$

6. $(\csc x + 1)(\csc x - 1) = \csc^2 x - 1 = \cot^2 x$

7. $\dfrac{\cos^2 x}{1 - \sin x} \cdot \dfrac{1 + \sin x}{1 + \sin x} = \dfrac{\cos^2 x(1 + \sin x)}{1 - \sin^2 x} = \dfrac{\cos^2 x(1 + \sin x)}{\cos^2 x} = 1 + \sin x$

8.

$$\frac{1 + \cos\theta}{\sin\theta} + \frac{\sin\theta}{1 + \cos\theta} = \frac{(1 + \cos\theta)^2 + \sin^2\theta}{\sin\theta(1 + \cos\theta)}$$

$$= \frac{1 + 2\cos\theta + \cos^2\theta + \sin^2\theta}{\sin\theta(1 + \cos\theta)} = \frac{2 + 2\cos\theta}{\sin\theta(1 + \cos\theta)} = \frac{2}{\sin\theta} = 2\csc\theta$$

9. $\tan^4 x + 2\tan^2 x + 1 = \left(\tan^2 x + 1\right)^2 = \left(\sec^2 x\right)^2 = \sec^4 x$

10. (a) $\sin 105° = \sin(60° + 45°) = \sin 60° \cos 45° + \cos 60° \sin 45°$

$$= \frac{\sqrt{3}}{2} \cdot \frac{\sqrt{2}}{2} + \frac{1}{2} \cdot \frac{\sqrt{2}}{2} = \frac{\sqrt{2}}{4}\left(\sqrt{3} + 1\right)$$

(b) $\tan 15° = \tan(60° - 45°) = \dfrac{\tan 60° - \tan 45°}{1 + \tan 60° \tan 45°}$

$$= \frac{\sqrt{3} - 1}{1 + \sqrt{3}} \cdot \frac{1 - \sqrt{3}}{1 - \sqrt{3}} = \frac{2\sqrt{3} - 1 - 3}{1 - 3} = \frac{2\sqrt{3} - 4}{-2} = 2 - \sqrt{3}$$

11. $(\sin 42°)\cos 38° - (\cos 42°)\sin 38° = \sin(42° - 38°) = \sin 4°$

12. $\tan\left(\theta + \dfrac{\pi}{4}\right) = \dfrac{\tan\theta + \tan\left(\dfrac{\pi}{4}\right)}{1 - (\tan\theta)\tan\left(\dfrac{\pi}{4}\right)} = \dfrac{\tan\theta + 1}{1 - \tan\theta(1)} = \dfrac{1 + \tan\theta}{1 - \tan\theta}$

13. $\sin(\arcsin x - \arccos x) = \sin(\arcsin x)\cos(\arccos x) - \cos(\arcsin x)\sin(\arccos x)$

$$= (x)(x) - \left(\sqrt{1 - x^2}\right)\left(\sqrt{1 - x^2}\right) = x^2 - \left(1 - x^2\right) = 2x^2 - 1$$

14. (a) $\cos(120°) = \cos[2(60°)] = 2\cos^2 60° - 1 = 2\left(\dfrac{1}{2}\right)^2 - 1 = -\dfrac{1}{2}$

(b) $\tan(300°) = \tan[2(150°)] = \dfrac{2\tan 150°}{1 - \tan^2 150°} = \dfrac{-\dfrac{2\sqrt{3}}{3}}{1 - \left(\dfrac{1}{3}\right)} = -\sqrt{3}$

15. (a) $\sin 22.5° = \sin\dfrac{45°}{2} = \sqrt{\dfrac{1 - \cos 45°}{2}} = \sqrt{\dfrac{1 - \dfrac{\sqrt{2}}{2}}{2}} = \dfrac{\sqrt{2 - \sqrt{2}}}{2}$

(b) $\tan\dfrac{\pi}{12} = \tan\dfrac{\dfrac{\pi}{6}}{2} = \dfrac{\sin\dfrac{\pi}{6}}{1 + \cos\left(\dfrac{\pi}{6}\right)} = \dfrac{\dfrac{1}{2}}{1 + \dfrac{\sqrt{3}}{2}} = \dfrac{1}{2 + \sqrt{3}} = 2 - \sqrt{3}$

16. $\sin\theta = \dfrac{4}{5}$, θ lies in Quadrant II $\Rightarrow \cos\theta = -\dfrac{3}{5}$.

$$\cos\frac{\theta}{2} = \sqrt{\frac{1 + \cos\theta}{2}} = \sqrt{\frac{1 - \dfrac{3}{5}}{2}} = \sqrt{\frac{2}{10}} = \frac{1}{\sqrt{5}} = \frac{\sqrt{5}}{5}$$

17. $\left(\sin^2 x\right)\cos^2 x = \dfrac{1 - \cos 2x}{2} \cdot \dfrac{1 + \cos 2x}{2} = \dfrac{1}{4}\left[1 - \cos^2 2x\right] = \dfrac{1}{4}\left[1 - \dfrac{1 + \cos 4x}{2}\right]$

$$= \frac{1}{8}\left[2 - (1 + \cos 4x)\right] = \frac{1}{8}\left[1 - \cos 4x\right]$$

18. $6(\sin 5\theta)\cos 2\theta = 6\left\{\frac{1}{2}\left[\sin(5\theta + 2\theta) + \sin(5\theta - 2\theta)\right]\right\} = 3\left[\sin 7\theta + \sin 3\theta\right]$

19. $\sin(x + \pi) + \sin(x - \pi) = 2\left(\sin \dfrac{[(x + \pi) + (x - \pi)]}{2}\right) \cos \dfrac{[(x + \pi) - (x - \pi)]}{2}$

$= 2 \sin x \cos \pi = -2 \sin x$

20. $\dfrac{\sin 9x + \sin 5x}{\cos 9x - \cos 5x} = \dfrac{2 \sin 7x \cos 2x}{-2 \sin 7x \sin 2x} = -\dfrac{\cos 2x}{\sin 2x} = -\cot 2x$

21. $\frac{1}{2}[\sin(u + v) - \sin(u - v)] = \frac{1}{2}\{(\sin u) \cos v + (\cos u) \sin v - [(\sin u) \cos v - (\cos u) \sin v]\}$

$= \frac{1}{2}[2(\cos u) \sin v] = (\cos u) \sin v$

22. $4 \sin^2 x = 1$

$\sin^2 x = \dfrac{1}{4}$

$\sin x = \pm\dfrac{1}{2}$

$\sin x = \dfrac{1}{2}$ or $\sin x = -\dfrac{1}{2}$

$x = \dfrac{\pi}{6}$ or $\dfrac{5\pi}{6}$ $\qquad x = \dfrac{7\pi}{6}$ or $\dfrac{11\pi}{6}$

23. $\tan^2 \theta + (\sqrt{3} - 1) \tan \theta - \sqrt{3} = 0$

$(\tan \theta - 1)(\tan \theta + \sqrt{3}) = 0$

$\tan \theta = 1$ or $\tan \theta = -\sqrt{3}$

$\theta = \dfrac{\pi}{4}$ or $\dfrac{5\pi}{4}$ $\qquad \theta = \dfrac{2\pi}{3}$ or $\dfrac{5\pi}{3}$

24. $\sin 2x = \cos x$

$2(\sin x) \cos x - \cos x = 0$

$\cos x(2 \sin x - 1) = 0$

$\cos x = 0$ or $\sin x = \dfrac{1}{2}$

$x = \dfrac{\pi}{2}$ or $\dfrac{3\pi}{2}$ $\qquad x = \dfrac{\pi}{6}$ or $\dfrac{5\pi}{6}$

25. $\tan^2 x - 6 \tan x + 4 = 0$

$\tan x = \dfrac{-(-6) \pm \sqrt{(-6)^2 - 4(1)(4)}}{2(1)}$

$\tan x = \dfrac{6 \pm \sqrt{20}}{2} = 3 \pm \sqrt{5}$

$\tan x = 3 + \sqrt{5}$ or $\tan x = 3 - \sqrt{5}$

$x \approx 1.3821$ or 4.5237 $\qquad x \approx 0.6524$ or 3.7940

Chapter 8 Practice Test Solutions

1. $C = 180° - (40° + 12°) = 128°$

$a = \sin 40°\left(\dfrac{100}{\sin 12°}\right) \approx 309.164$

$c = \sin 128°\left(\dfrac{100}{\sin 12°}\right) \approx 379.012$

2. $\sin A = 5\left(\dfrac{\sin 150°}{20}\right) = 0.125$

$A \approx 7.181°$

$B \approx 180° - (150° + 7.181°) = 22.819°$

$b = \sin 22.819°\left(\dfrac{20}{\sin 150°}\right) \approx 15.513$

3. Area $= \frac{1}{2}ab \sin C = \frac{1}{2}(3)(6) \sin 130° \approx 6.894$ square units

4. $h = b \sin A = 35 \sin 22.5° \approx 13.394$

$a = 10$

Since $a < h$ and A is acute, the triangle has no solution.

5. $\cos A = \dfrac{(53)^2 + (38)^2 - (49)^2}{2(53)(38)} \approx 0.4598$

$A \approx 62.627°$

$\cos B = \dfrac{(49)^2 + (38)^2 - (53)^2}{2(49)(38)} \approx 0.2782$

$B \approx 73.847°$

$C \approx 180° - (62.627° + 73.847°)$

$= 43.526°$

6. $$c^2 = (100)^2 + (300)^2 - 2(100)(300)\cos 29^\circ$$
$$\approx 47{,}522.8176$$
$$c \approx 218$$
$$\cos A = \frac{(300)^2 + (218)^2 - (100)^2}{2(300)(218)} \approx 0.97495$$
$$A \approx 12.85^\circ$$
$$B \approx 180^\circ - (12.85^\circ + 29^\circ) = 138.15^\circ$$

7. $$s = \frac{a+b+c}{2} = \frac{4.1 + 6.8 + 5.5}{2} = 8.2$$
$$\text{Area} = \sqrt{s(s-a)(s-b)(s-c)}$$
$$= \sqrt{8.2(8.2-4.1)(8.2-6.8)(8.2-5.5)}$$
$$\approx 11.273 \text{ square units}$$

8. $$x^2 = (40)^2 + (70)^2 - 2(40)(70)\cos 168^\circ$$
$$\approx 11{,}977.6266$$
$$x \approx 190.442 \text{ miles}$$

9. $$\mathbf{w} = 4(3\mathbf{i} + \mathbf{j}) - 7(-\mathbf{i} + 2\mathbf{j})$$
$$= 19\mathbf{i} - 10\mathbf{j}$$

10. $$\frac{\mathbf{v}}{\|\mathbf{v}\|} = \frac{5\mathbf{i} - 3\mathbf{j}}{\sqrt{25+9}} = \frac{5}{\sqrt{34}}\mathbf{i} - \frac{3}{\sqrt{34}}\mathbf{j}$$
$$= \frac{5\sqrt{34}}{34}\mathbf{i} - \frac{3\sqrt{34}}{34}\mathbf{j}$$

11. $\mathbf{u} = 6\mathbf{i} + 5\mathbf{j},\ \mathbf{v} = 2\mathbf{i} - 3\mathbf{j}$
$$\mathbf{u}\cdot\mathbf{v} = 6(2) + 5(-3) = -3$$
$$\|\mathbf{u}\| = \sqrt{61}, \qquad \|\mathbf{v}\| = \sqrt{13}$$
$$\cos\theta = \frac{-3}{\sqrt{61}\sqrt{13}}$$
$$\theta \approx 96.116^\circ$$

12. $$4(\mathbf{i}\cos 30^\circ + \mathbf{j}\sin 30^\circ) = 4\left(\frac{\sqrt{3}}{2}\mathbf{i} + \frac{1}{2}\mathbf{j}\right)$$
$$= \langle 2\sqrt{3}, 2\rangle$$

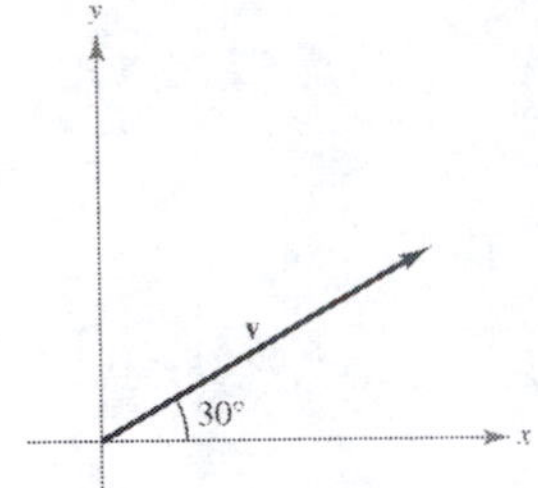

13. $$\text{proj}_{\mathbf{v}}\mathbf{u} = \left(\frac{\mathbf{u}\cdot\mathbf{v}}{\|\mathbf{v}\|^2}\right)\mathbf{v} = \frac{-10}{20}\langle -2, 4\rangle = \langle 1, -2\rangle$$

14. $$r = \sqrt{25+25} = \sqrt{50} = 5\sqrt{2}$$
$$\tan\theta = \frac{-5}{5} = -1$$
Because z is in Quadrant IV, $\theta = 315^\circ$.
$$z = 5\sqrt{2}(\cos 315^\circ + i\sin 315^\circ)$$

15. $$\cos 225^\circ = -\frac{\sqrt{2}}{2},\ \sin 225^\circ = -\frac{\sqrt{2}}{2}$$
$$z = 6\left(-\frac{\sqrt{2}}{2} - i\frac{\sqrt{2}}{2}\right) = -3\sqrt{2} - 3\sqrt{2}i$$

16. $$[7(\cos 23^\circ + i\sin 23^\circ)][4(\cos 7^\circ + i\sin 7^\circ)] = 7(4)[\cos(23^\circ + 7^\circ) + i\sin(23^\circ + 7^\circ)]$$
$$= 28(\cos 30^\circ + i\sin 30^\circ)$$

17. $$\frac{9\left(\cos\frac{5\pi}{4} + i\sin\frac{5\pi}{4}\right)}{3(\cos\pi + i\sin\pi)} = \frac{9}{3}\left[\cos\left(\frac{5\pi}{4} - \pi\right) + i\sin\left(\frac{5\pi}{4} - \pi\right)\right] = 3\left(\cos\frac{\pi}{4} + i\sin\frac{\pi}{4}\right)$$

18. $$(2+2i)^8 = \left[2\sqrt{2}(\cos 45^\circ + i\sin 45^\circ)\right]^8 = \left(2\sqrt{2}\right)^8\left[\cos(8)(45^\circ) + i\sin(8)(45^\circ)\right]$$
$$= 4096[\cos 360^\circ + i\sin 360^\circ] = 4096$$

19. $z = 8\left(\cos\frac{\pi}{3} + i\sin\frac{\pi}{3}\right), n = 3$

The cube roots of z are: $\sqrt[3]{8}\left[\cos\frac{\left(\frac{\pi}{3}\right) + 2\pi k}{3} + i\sin\frac{\left(\frac{\pi}{3}\right) + 2\pi k}{3}\right], k = 0, 1, 2$

For $k = 0$: $\sqrt[3]{8}\left[\cos\frac{\frac{\pi}{3}}{3} + i\sin\frac{\frac{\pi}{3}}{3}\right] = 2\left(\cos\frac{\pi}{9} + i\sin\frac{\pi}{9}\right)$

For $k = 1$: $\sqrt[3]{8}\left[\cos\frac{\left(\frac{\pi}{3}\right) + 2\pi}{3} + i\sin\frac{\left(\frac{\pi}{3}\right) + 2\pi}{3}\right] = 2\left(\cos\frac{7\pi}{9} + i\sin\frac{7\pi}{9}\right)$

For $k = 2$: $\sqrt[3]{8}\left[\cos\frac{\left(\frac{\pi}{3}\right) + 4\pi}{3} + i\sin\frac{\left(\frac{\pi}{3}\right) + 4\pi}{3}\right] = 2\left(\cos\frac{13\pi}{9} + i\sin\frac{13\pi}{9}\right)$

20. $x^4 = -i = 1\left(\cos\frac{3\pi}{2} + i\sin\frac{3\pi}{2}\right)$

The fourth roots are: $\sqrt[4]{1}\left[\cos\frac{\left(\frac{3\pi}{2}\right) + 2\pi k}{4} + i\sin\frac{\left(\frac{3\pi}{2}\right) + 2\pi k}{4}\right], k = 0, 1, 2, 3$

For $k = 0$: $\cos\frac{\frac{3\pi}{2}}{4} + i\sin\frac{\frac{3\pi}{2}}{4} = \cos\frac{3\pi}{8} + i\sin\frac{3\pi}{8}$

For $k = 1$: $\cos\frac{\left(\frac{3\pi}{2}\right) + 2\pi}{4} + i\sin\frac{\left(\frac{3\pi}{2}\right) + 2\pi}{4} = \cos\frac{7\pi}{8} + i\sin\frac{7\pi}{8}$

For $k = 2$: $\cos\frac{\left(\frac{3\pi}{2}\right) + 4\pi}{4} + i\sin\frac{\left(\frac{3\pi}{2}\right) + 4\pi}{4} = \cos\frac{11\pi}{8} + i\sin\frac{11\pi}{8}$

For $k = 3$: $\cos\frac{\left(\frac{3\pi}{2}\right) + 6\pi}{4} + i\sin\frac{\left(\frac{3\pi}{2}\right) + 6\pi}{4} = \cos\frac{15\pi}{8} + i\sin\frac{15\pi}{8}$

Chapter 9 Practice Test Solutions

1. $\begin{cases} x + y = 1 \\ 3x - y = 15 \Rightarrow y = 3x - 15 \end{cases}$

$$\begin{aligned} x + (3x - 15) &= 1 \\ 4x &= 16 \\ x &= 4 \\ y &= -3 \end{aligned}$$

Solution: $(4, -3)$

2. $\begin{cases} x - 3y = -3 \Rightarrow x = 3y - 3 \\ x^2 + 6y = 5 \end{cases}$

$$\begin{aligned} (3y - 3)^2 + 6y &= 5 \\ 9y^2 - 18y + 9 + 6y &= 5 \\ 9y^2 - 12y + 4 &= 0 \\ (3y - 2)^2 &= 0 \\ y &= \tfrac{2}{3} \\ x &= -1 \end{aligned}$$

Solution: $\left(-1, \frac{2}{3}\right)$

3. $\begin{cases} x + y + z = 6 \Rightarrow z = 6 - x - y \\ 2x - y + 3z = 0 \Rightarrow 2x - y + 3(6 - x - y) = 0 \Rightarrow -x - 4y = -18 \Rightarrow x = 18 - 4y \\ 5x + 2y - z = -3 \Rightarrow 5x + 2y - (6 - x - y) = -3 \Rightarrow 6x + 3y = 3 \end{cases}$

$$6(18 - 4y) + 3y = 3$$
$$-21y = -105$$
$$y = 5$$
$$x = 18 - 4y = -2$$
$$z = 6 - x - y = 3$$

Solution: $(-2, 5, 3)$

4. $x + y = 110 \Rightarrow y = 110 - x$

$xy = 2800$

$$x(110 - x) = 2800$$
$$0 = x^2 - 110x + 2800$$
$$0 = (x - 40)(x - 70)$$

$x = 40$ or $x = 70$

$y = 70$ $\quad y = 40$

Solution: The two numbers are 40 and 70.

5. $2x + 2y = 170 \Rightarrow y = \dfrac{170 - 2x}{2} = 85 - x$

$$xy = 1500$$
$$x(85 - x) = 1500$$
$$0 = x^2 - 85x + 1500$$
$$0 = (x - 25)(x - 60)$$

$x = 25$ or $x = 60$

$y = 60$ $\quad y = 25$

Dimensions: 60 ft × 25 ft

6. $\begin{cases} 2x + 15y = 4 \Rightarrow 2x + 15y = 4 \\ x - 3y = 23 \Rightarrow 5x - 15y = 115 \end{cases}$

$$7x = 119$$
$$x = 17$$
$$y = \frac{x - 23}{3}$$
$$= -2$$

Solution: $(17, -2)$

7. $\begin{cases} x + y = 2 \Rightarrow 19x + 19y = 38 \\ 38x - 19y = 7 \Rightarrow 38x - 19y = 7 \end{cases}$

$$57x = 45$$
$$x = \tfrac{15}{19}$$
$$y = 2 - x$$
$$= \tfrac{38}{19} - \tfrac{15}{19}$$
$$= \tfrac{23}{19}$$

Solution: $\left(\tfrac{15}{19}, \tfrac{23}{19}\right)$

8. $\begin{cases} 0.4x + 0.5y = 0.112 \Rightarrow 0.28x + 0.35y = 0.0784 \\ 0.3x - 0.7y = -0.131 \Rightarrow 0.15x - 0.35y = -0.0655 \end{cases}$

$$0.43x = 0.0129$$

$$x = \frac{0.0129}{0.43} = 0.03$$
$$y = \frac{0.112 - 0.4x}{0.5} = 0.20$$

Solution: $(0.03, 0.20)$

9. Let $x =$ amount in 11% fund and

$y =$ amount in 13% fund.

$$x + y = 17{,}000 \Rightarrow y = 17{,}000 - x$$
$$0.11x + 0.13y = 2080$$
$$0.11x + 0.13(17{,}000 - x) = 2080$$
$$-0.02x = -130$$
$$x = \$6500 \text{ at } 11\%$$
$$y = \$10{,}500 \text{ at } 13\%$$

10. $(4, 3), (1, 1), (-1, -2), (-2, -1)$

Use a calculator.

$y = ax + b = \tfrac{11}{14}x - \tfrac{1}{7}$

11. $\begin{cases} x + y = -2 \\ 2x - y + z = 11 \\ 4y - 3z = -20 \end{cases}$

$\begin{cases} x + y = -2 \\ -3y + z = 15 \\ 4y - 3z = -20 \end{cases}$ $-2\text{Eq.}1 + \text{Eq.}2$

$\begin{cases} x + y = -2 \\ y - 2z = -5 \\ 4y - 3z = -20 \end{cases}$ $\text{Eq.}3 + \text{Eq.}2$

$\begin{cases} x + y = -2 \\ y - 2z = -5 \\ 5z = 0 \end{cases}$ $-4\text{Eq.}2 + \text{Eq.}3$

$\begin{cases} x + y = -2 \\ y - 2z = -5 \\ z = 0 \end{cases}$

$$y - 2(0) = -5 \Rightarrow y = -5$$
$$x + (-5) = -2 \Rightarrow x = 3$$

Solution: $(3, -5, 0)$

12. $\begin{cases} 4x - y + 5z = 4 \\ 2x + y - z = 0 \\ 2x + 4y + 8z = 0 \end{cases}$

$\begin{cases} 2x + 4y + 8z = 0 \\ 2x + y - z = 0 \\ 4x - y + 5z = 4 \end{cases}$ Interchange equations.

$\begin{cases} 2x + 4y + 8z = 0 \\ -3y - 9z = 0 \\ -9y - 11z = 4 \end{cases}$ $\begin{matrix} -\text{Eq.}1 + \text{Eq.}2 \\ -2\text{Eq.}1 + \text{Eq.}3 \end{matrix}$

$\begin{cases} 2x + 4y + 8z = 0 \\ -3y - 9z = 0 \\ 16z = 4 \end{cases}$ $-3\text{Eq.}2 + \text{Eq.}3$

$\begin{cases} x + 2y + 4z = 0 \\ y + 3z = 0 \\ z = \frac{1}{4} \end{cases}$ $\begin{matrix} \frac{1}{2}\text{Eq.}1 \\ -\frac{1}{3}\text{Eq.}2 \\ \frac{1}{16}\text{Eq.}3 \end{matrix}$

$$y + 3\left(\tfrac{1}{4}\right) = 0 \Rightarrow y = -\tfrac{3}{4}$$
$$x + 2\left(-\tfrac{3}{4}\right) + 4\left(\tfrac{1}{4}\right) = 0 \Rightarrow x = \tfrac{1}{2}$$

Solution: $\left(-\frac{1}{2}, -\frac{3}{4}, \frac{1}{4}\right)$

13. $\begin{cases} 3x + 2y - z = 5 \\ 6x - y + 5z = 2 \end{cases}$

$\begin{cases} 3x + 2y - z = 5 \\ -5y + 7z = -8 \end{cases}$ $-2\text{Eq.}1 + \text{Eq.}2$

$\begin{cases} x + \frac{2}{3}y - \frac{1}{3}z = \frac{5}{3} \\ y - \frac{7}{5}z = \frac{8}{5} \end{cases}$ $\begin{matrix} \frac{1}{3}\text{Eq.}1 \\ -\frac{1}{5}\text{Eq.}2 \end{matrix}$

Let $a = z$.

Then $y = \frac{7}{5}a + \frac{8}{5}$, and $x + \frac{2}{3}\left(\frac{7}{5}a + \frac{8}{5}\right) - \frac{1}{3}a = \frac{5}{3}$

$$x + \tfrac{3}{5}a = \tfrac{3}{5}$$
$$x = -\tfrac{3}{5}a + \tfrac{3}{5}.$$

Solution: $\left(-\frac{3}{5}a + \frac{3}{5}, \frac{7}{5}a + \frac{8}{5}, a\right)$ where a is any real number

14. $y = ax^2 + bx + c$ passes through $(0, -1)$, $(1, 4)$, and $(2, 13)$.

At $(0, -1)$: $-1 = a(0)^2 + b(0) + c \Rightarrow c = -1$

At $(1, 4)$: $4 = a(1)^2 + b(1) - 1 \Rightarrow 5 = a + b \Rightarrow 5 = a + b$

At $(2, 13)$: $13 = a(2)^2 + b(2) - 1 \Rightarrow 14 = 4a + 2b \Rightarrow \underline{-7 = -2a - b}$

$$-2 = -a$$
$$a = 2$$
$$b = 3$$

So, the equation of the parabola is $y = 2x^2 + 3x - 1$.

15. $s = \frac{1}{2}at^2 + v_0t + s_0$ passes through $(1, 12)$, $(2, 5)$, and $(3, 4)$.

At $(1, 12)$: $12 = \frac{1}{2}a + v_0 + s_0$

At $(2, 5)$: $5 = 2a + 2v_0 + s_0$

At $(3, 4)$: $4 = \frac{9}{2}a + 3v_0 + s_0$

$$\begin{cases} a + 2v_0 + 2s_0 = 24 \\ 2a + 2v_0 + s_0 = 5 \\ 9a + 6v_0 + 2s_0 = 8 \end{cases}$$

$$\begin{cases} a + 2v_0 + 2s_0 = 24 & \\ -2v_0 - 3s_0 = -43 & -2\text{Eq.}1 + \text{Eq.}2 \\ -12v_0 - 16s_0 = -208 & -9\text{Eq.}1 + \text{Eq.}3 \end{cases}$$

$$\begin{cases} a + 2v_0 + 2s_0 = 24 & \\ -2v_0 - 3s_0 = -43 & \\ 2s_0 = 50 & -6\text{Eq.}2 + \text{Eq.}3 \end{cases}$$

$$\begin{cases} a + 2v_0 + 2s_0 = 24 & \\ v_0 + \frac{3}{2}s_0 = \frac{43}{2} & -\frac{1}{2}\text{Eq.}2 \\ s_0 = 25 & \frac{1}{2}\text{Eq.}3 \end{cases}$$

$$s_0 = 25$$

$$v_0 + \frac{3}{2}(25) = \frac{43}{2} \Rightarrow v_0 = 16$$

$$a + 2(-16) + 2(25) = 24 \Rightarrow a = 6$$

So, $s = \frac{1}{2}(6)t^2 - 16t + 25 = 3t^2 - 16t + 25.$

16. $x^2 + y^2 \geq 9$

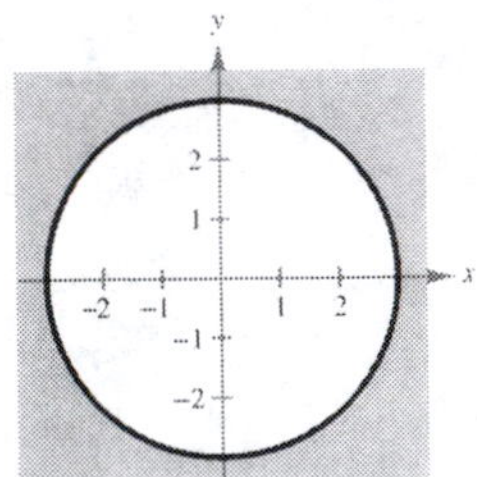

17. $\begin{cases} x + y \leq 6 \\ x \geq 2 \\ y \geq 0 \end{cases}$

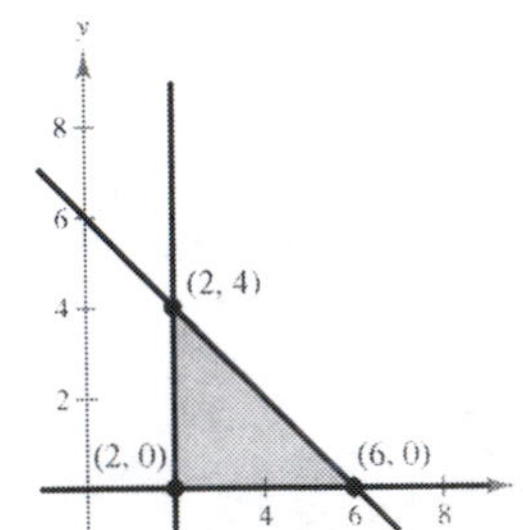

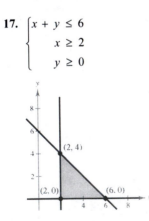

18. Line through $(0, 0)$ and $(0, 7)$: $x = 0$

Line through $(0, 0)$ and $(2, 3)$:

$y = \frac{3}{2}x$ or $3x - 2y = 0$

Line through $(0, 7)$ and $(2, 3)$:

$y = -2x + 7$ or $2x + y = 7$

Inequalities: $\begin{cases} x \geq 0 \\ 3x - 2y \leq 0 \\ 2x + y \leq 7 \end{cases}$

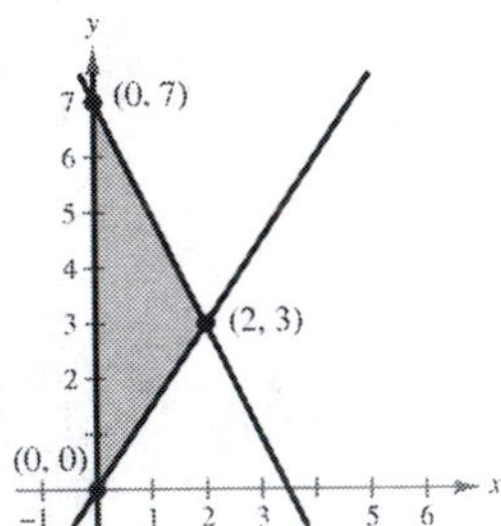

19. Vertices $(0, 0), (0, 7), (6, 0), (3, 5)$

$z = 30x + 26y$

At $(0, 0)$: $z = 0$

At $(0, 7)$: $z = 182$

At $(6, 0)$: $z = 180$

At $(3, 5)$: $z = 220$

The maximum value of z occurs at $(3, 5)$ and is 220.

20.

$$x^2 + y^2 \le 4$$
$$(x - 2)^2 + y^2 \ge 4$$

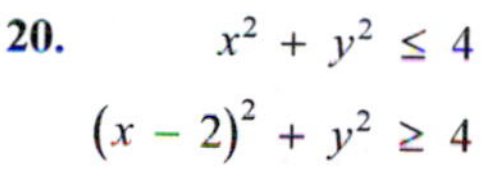

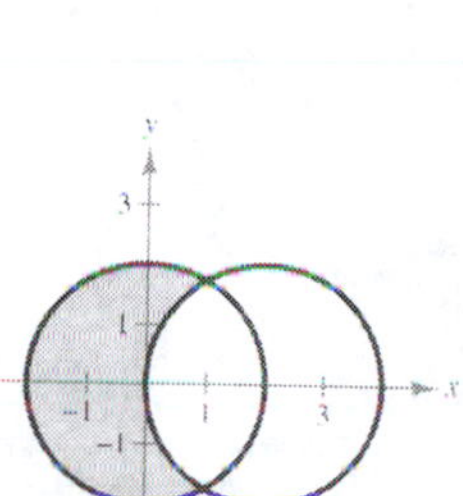

21. $$\frac{1 - 2x}{x^2 + x} = \frac{1 - 2x}{x(x + 1)} = \frac{A}{x} + \frac{B}{x + 1}$$

$1 - 2x = A(x + 1) + Bx$

When $x = 0, 1 = A$.

When $x = -1, 3 = -B \Rightarrow B = -3$.

$$\frac{1 - 2x}{x^2 + x} = \frac{1}{x} - \frac{3}{x + 1}$$

22. $$\frac{6x - 17}{(x - 3)^2} = \frac{A}{x - 3} + \frac{B}{(x - 3)^2}$$

$6x - 17 = A(x - 3) + B$

When $x = 3, 1 = B$.

When $x = 0, -17 = -3A + B \Rightarrow A = 6$.

$$\frac{6x - 17}{(x - 3)^2} = \frac{6}{x - 3} + \frac{1}{(x - 3)^2}$$

Chapter 10 Practice Test Solutions

1. $$\begin{bmatrix} 1 & -2 & 4 \\ 3 & -5 & 9 \end{bmatrix}$$

$$-3R_1 + R_2 \rightarrow \begin{bmatrix} 1 & -2 & 4 \\ 0 & 1 & -3 \end{bmatrix}$$

$$2R_2 + R_1 \rightarrow \begin{bmatrix} 1 & 0 & -2 \\ 0 & 1 & -3 \end{bmatrix}$$

2. $$\begin{cases} 3x + 5y = 3 \\ 2x - y = -11 \end{cases}$$

$$\begin{bmatrix} 3 & 5 & \vdots & 3 \\ 2 & -1 & \vdots & -11 \end{bmatrix}$$

$$-R_2 + R_1 \rightarrow \begin{bmatrix} 1 & 6 & \vdots & 14 \\ 2 & -1 & \vdots & -11 \end{bmatrix}$$

$$-2R_1 + R_2 \rightarrow \begin{bmatrix} 1 & 6 & \vdots & 14 \\ 0 & -13 & \vdots & -39 \end{bmatrix}$$

$$-\tfrac{1}{13}R_2 \rightarrow \begin{bmatrix} 1 & 6 & \vdots & 14 \\ 0 & 1 & \vdots & 3 \end{bmatrix}$$

$$-6R_2 + R_1 \rightarrow \begin{bmatrix} 1 & 0 & \vdots & -4 \\ 0 & 1 & \vdots & 3 \end{bmatrix}$$

$x = -4, y = 3$

Solution: $(-4, 3)$

3. $$\begin{cases} 2x + 3y = -3 \\ 3x - 2y = 8 \\ x + y = 1 \end{cases}$$

$$\begin{bmatrix} 2 & 3 & \vdots & -3 \\ 3 & 2 & \vdots & 8 \\ 1 & 1 & \vdots & 1 \end{bmatrix}$$

$$\begin{matrix} R_3 \\ \\ R_1 \end{matrix} \begin{bmatrix} 1 & 1 & \vdots & 1 \\ 3 & 2 & \vdots & 8 \\ 2 & 3 & \vdots & -3 \end{bmatrix}$$

$$\begin{matrix} \\ -3R_1 + R_2 \rightarrow \\ -2R_1 + R_3 \rightarrow \end{matrix} \begin{bmatrix} 1 & 1 & \vdots & 1 \\ 0 & -1 & \vdots & 5 \\ 0 & 1 & \vdots & -5 \end{bmatrix}$$

$$-R_2 \rightarrow \begin{bmatrix} 1 & 1 & \vdots & 1 \\ 0 & 1 & \vdots & -5 \\ 0 & 1 & \vdots & -5 \end{bmatrix}$$

$$\begin{matrix} -R_2 + R_1 \rightarrow \\ \\ -R_2 + R_3 \rightarrow \end{matrix} \begin{bmatrix} 1 & 0 & \vdots & 6 \\ 0 & 1 & \vdots & -5 \\ 0 & 0 & \vdots & 0 \end{bmatrix}$$

$x = 6, y = -5$

Solution: $(6, -5)$

4. $\begin{cases} x + 3z = -5 \\ 2x + y = 0 \\ 3x + y - z = -3 \end{cases}$

$$\begin{bmatrix} 1 & 0 & 3 & \vdots & -5 \\ 2 & 1 & 0 & \vdots & 0 \\ 3 & 1 & -1 & \vdots & 3 \end{bmatrix}$$

$$\begin{matrix} \\ -2R_1 + R_2 \to \\ -3R_1 + R_3 \to \end{matrix} \begin{bmatrix} 1 & 0 & 3 & \vdots & -5 \\ 0 & 1 & -6 & \vdots & 10 \\ 0 & 1 & -10 & \vdots & 18 \end{bmatrix}$$

$$\begin{matrix} \\ \\ -R_2 + R_3 \to \end{matrix} \begin{bmatrix} 1 & 0 & 3 & \vdots & -5 \\ 0 & 1 & -6 & \vdots & 10 \\ 0 & 0 & -4 & \vdots & 8 \end{bmatrix}$$

$$\begin{matrix} \\ \\ -\frac{1}{4}R_3 \to \end{matrix} \begin{bmatrix} 1 & 0 & 3 & \vdots & -5 \\ 0 & 1 & -6 & \vdots & 10 \\ 0 & 0 & 1 & \vdots & -2 \end{bmatrix}$$

$$\begin{matrix} -3R_3 + R_1 \to \\ 6R_3 + R_2 \to \\ \\ \end{matrix} \begin{bmatrix} 1 & 0 & 0 & \vdots & 1 \\ 0 & 1 & 0 & \vdots & -2 \\ 0 & 0 & 1 & \vdots & -2 \end{bmatrix}$$

$x = 1, y = -2, z = -2$

Solution: $(1, -2, -2)$

5. $$\begin{bmatrix} 1 & 4 & 5 \\ 2 & 0 & -3 \end{bmatrix} \begin{bmatrix} 1 & 6 \\ 0 & -7 \\ -1 & 2 \end{bmatrix} = \begin{bmatrix} (1)(1) + (4)(0) + (5)(-1) & (1)(6) + (4)(-7) + (5)(2) \\ (2)(1) + (0)(0) + (-3)(-1) & (2)(6) + (0)(-7) + (-3)(2) \end{bmatrix} = \begin{bmatrix} -4 & -12 \\ 5 & 6 \end{bmatrix}$$

6. $$3A - 5B = 3\begin{bmatrix} 9 & 1 \\ -4 & 8 \end{bmatrix} - 5\begin{bmatrix} 6 & -2 \\ 3 & 5 \end{bmatrix}$$
$$= \begin{bmatrix} 27 & 3 \\ -12 & 24 \end{bmatrix} - \begin{bmatrix} 30 & -10 \\ 15 & 25 \end{bmatrix}$$
$$= \begin{bmatrix} -3 & 13 \\ -27 & -1 \end{bmatrix}$$

7. $$f(A) = \begin{bmatrix} 3 & 0 \\ 7 & 1 \end{bmatrix}^2 - 7\begin{bmatrix} 3 & 0 \\ 7 & 1 \end{bmatrix} + 8\begin{bmatrix} 1 & 0 \\ 0 & 1 \end{bmatrix}$$
$$= \begin{bmatrix} 3 & 0 \\ 7 & 1 \end{bmatrix}\begin{bmatrix} 3 & 0 \\ 7 & 1 \end{bmatrix} - \begin{bmatrix} 21 & 0 \\ 49 & 7 \end{bmatrix} + \begin{bmatrix} 8 & 0 \\ 0 & 8 \end{bmatrix}$$
$$= \begin{bmatrix} 9 & 0 \\ 28 & 1 \end{bmatrix} - \begin{bmatrix} 21 & 0 \\ 49 & 7 \end{bmatrix} + \begin{bmatrix} 8 & 0 \\ 0 & 8 \end{bmatrix}$$
$$= \begin{bmatrix} -4 & 0 \\ -21 & 2 \end{bmatrix}$$

8. False.

$$(A + B)(A + 3B) = A(A + 3B) + B(A + 3B)$$
$$= A^2 + 3AB + BA + 3B^2$$ and, in general, $AB \neq BA$.

9.

$$\left[\begin{array}{cc|cc} 1 & 2 & 1 & 0 \\ 3 & 5 & 0 & 1 \end{array}\right]$$

$$\begin{array}{r} \\ -3R_1 + R_2 \to \end{array}\left[\begin{array}{cc|cc} 1 & 2 & 1 & 0 \\ 0 & -1 & -3 & 1 \end{array}\right]$$

$$\begin{array}{r} 2R_2 + R_1 \to \\ \\ \end{array}\left[\begin{array}{cc|cc} 1 & 0 & -5 & 2 \\ 0 & -1 & -3 & 1 \end{array}\right]$$

$$\begin{array}{r} \\ -R_2 \to \end{array}\left[\begin{array}{cc|cc} 1 & 0 & -5 & 2 \\ 0 & 1 & 3 & -1 \end{array}\right]$$

$$A^{-1} = \begin{bmatrix} -5 & 2 \\ 3 & -1 \end{bmatrix}$$

10.

$$\left[\begin{array}{ccc|ccc} 1 & 1 & 1 & 1 & 0 & 0 \\ 3 & 6 & 5 & 0 & 1 & 0 \\ 6 & 10 & 8 & 0 & 0 & 1 \end{array}\right]$$

$$\begin{array}{r} \\ -3R_1 + R_2 \to \\ -6R_1 + R_3 \to \end{array}\left[\begin{array}{ccc|ccc} 1 & 1 & 1 & 1 & 0 & 0 \\ 0 & 3 & 2 & -3 & 1 & 0 \\ 0 & 4 & 2 & -6 & 0 & 1 \end{array}\right]$$

$$\begin{array}{r} \\ -R_3 + R_2 \to \\ \\ \end{array}\left[\begin{array}{ccc|ccc} 1 & 1 & 1 & 1 & 0 & 0 \\ 0 & -1 & 0 & 3 & 1 & -1 \\ 0 & 4 & 2 & -6 & 0 & 1 \end{array}\right]$$

$$\begin{array}{r} R_2 + R_1 \to \\ \\ 4R_2 + R_3 \to \end{array}\left[\begin{array}{ccc|ccc} 1 & 0 & 1 & 4 & 1 & -1 \\ 0 & -1 & 0 & 3 & 1 & -1 \\ 0 & 0 & 2 & 6 & 4 & -3 \end{array}\right]$$

$$\begin{array}{r} \\ -R_2 \to \\ \frac{1}{2}R_3 \to \end{array}\left[\begin{array}{ccc|ccc} 1 & 0 & 1 & 4 & 1 & -1 \\ 0 & 1 & 0 & -3 & -1 & 1 \\ 0 & 0 & 1 & 3 & 2 & -\frac{3}{2} \end{array}\right]$$

$$\begin{array}{r} -R_3 + R_1 \to \\ \\ \\ \end{array}\left[\begin{array}{ccc|ccc} 1 & 0 & 0 & 1 & -1 & \frac{1}{2} \\ 0 & 1 & 0 & -3 & -1 & 1 \\ 0 & 0 & 1 & 3 & 2 & -\frac{3}{2} \end{array}\right]$$

$$A^{-1} = \begin{bmatrix} 1 & -1 & \frac{1}{2} \\ -3 & -1 & 1 \\ 3 & 2 & -\frac{3}{2} \end{bmatrix}$$

11. (a) $\begin{cases} x + 2y = 4 \\ 3x + 5y = 1 \end{cases}$

$$A = \begin{bmatrix} 1 & 2 \\ 3 & 5 \end{bmatrix}$$

$$A^{-1} = \frac{1}{5 - 6}\begin{bmatrix} 5 & -2 \\ -3 & 1 \end{bmatrix} = \begin{bmatrix} -5 & 2 \\ 3 & -1 \end{bmatrix}$$

$$\begin{bmatrix} x \\ y \end{bmatrix} = A^{-1}B = \begin{bmatrix} -5 & 2 \\ 3 & -1 \end{bmatrix}\begin{bmatrix} 4 \\ 1 \end{bmatrix} = \begin{bmatrix} -18 \\ 11 \end{bmatrix}$$

$x = -18,\ y = 11$

Solution: $(-18, 11)$

(b) $\begin{cases} x + 2y = 3 \\ 3x + 5y = -2 \end{cases}$

Again, $A^{-1} = \begin{bmatrix} -5 & 2 \\ 3 & -1 \end{bmatrix}$.

$$\begin{bmatrix} x \\ y \end{bmatrix} = A^{-1}B = \begin{bmatrix} -5 & 2 \\ 3 & -1 \end{bmatrix}\begin{bmatrix} 3 \\ -2 \end{bmatrix} = \begin{bmatrix} -19 \\ 11 \end{bmatrix}$$

$x = -19,\ y = 11$

Solution: $(-19, 11)$

12. $\begin{vmatrix} 6 & -1 \\ 3 & 4 \end{vmatrix} = 24 - (-3) = 27$

13. $\begin{vmatrix} 1 & 3 & -1 \\ 5 & 9 & 0 \\ 6 & 2 & -5 \end{vmatrix} = -1\begin{vmatrix} 5 & 9 \\ 6 & 2 \end{vmatrix} - 5\begin{vmatrix} 1 & 3 \\ 5 & 9 \end{vmatrix}$

$= -(-44) - 5(-6) = 74$

14. Expand along Row 2.

$$\begin{vmatrix} 1 & 4 & 2 & 3 \\ 0 & 1 & -2 & 0 \\ 3 & 5 & -1 & 1 \\ 2 & 0 & 6 & 1 \end{vmatrix} = \begin{vmatrix} 1 & 2 & 3 \\ 3 & -1 & 1 \\ 2 & 6 & 1 \end{vmatrix} + 2\begin{vmatrix} 1 & 4 & 3 \\ 3 & 5 & 1 \\ 2 & 0 & 1 \end{vmatrix}$$

$= 51 + 2(-29) = -7$

15. $\begin{vmatrix} 6 & 4 & 3 & 0 & 6 \\ 0 & 5 & 1 & 4 & 8 \\ 0 & 0 & 2 & 7 & 3 \\ 0 & 0 & 0 & 9 & 2 \\ 0 & 0 & 0 & 0 & 1 \end{vmatrix} = 6\begin{vmatrix} 5 & 1 & 4 & 8 \\ 0 & 2 & 7 & 3 \\ 0 & 0 & 9 & 2 \\ 0 & 0 & 0 & 1 \end{vmatrix} = 6(5)\begin{vmatrix} 2 & 7 & 3 \\ 0 & 9 & 2 \\ 0 & 0 & 1 \end{vmatrix} = 6(5)(2)\begin{vmatrix} 9 & 2 \\ 0 & 1 \end{vmatrix} = 6(5)(2)(9) = 540$

16. Area $= \dfrac{1}{2}\begin{vmatrix} 0 & 7 & 1 \\ 5 & 0 & 1 \\ 3 & 9 & 1 \end{vmatrix} = \dfrac{1}{2}(31) = \dfrac{31}{2}$

17. $\begin{vmatrix} x & y & 1 \\ 2 & 7 & 1 \\ -1 & 4 & 1 \end{vmatrix} = 3x - 3y + 15 = 0$ or $x - y + 5 = 0$

18. $x = \frac{\begin{vmatrix} 4 & -7 \\ 11 & 5 \end{vmatrix}}{\begin{vmatrix} 6 & -7 \\ 2 & 5 \end{vmatrix}} = \frac{97}{44}$

19. $z = \frac{\begin{vmatrix} 3 & 0 & 1 \\ 0 & 1 & 3 \\ 1 & -1 & 2 \end{vmatrix}}{\begin{vmatrix} 3 & 0 & 1 \\ 0 & 1 & 4 \\ 1 & -1 & 0 \end{vmatrix}} = \frac{14}{11}$

20. $y = \frac{\begin{vmatrix} 721.4 & 33.77 \\ 45.9 & 19.85 \end{vmatrix}}{\begin{vmatrix} 721.4 & -29.1 \\ 45.9 & 105.6 \end{vmatrix}} = \frac{12{,}769.747}{77{,}515.530} \approx 0.1647$

Chapter 11 Practice Test Solutions

1. $a_n = \frac{2n}{(n+2)!}$

$a_1 = \frac{2(1)}{3!} = \frac{2}{6} = \frac{1}{3}$

$a_2 = \frac{2(2)}{4!} = \frac{4}{24} = \frac{1}{6}$

$a_3 = \frac{2(3)}{5!} = \frac{6}{120} = \frac{1}{20}$

$a_4 = \frac{2(4)}{6!} = \frac{8}{720} = \frac{1}{90}$

$a_5 = \frac{2(5)}{7!} = \frac{10}{5040} = \frac{1}{504}$

Terms: $\frac{1}{3}, \frac{1}{6}, \frac{1}{20}, \frac{1}{90}, \frac{1}{504}$

2. $a_n = \frac{n+3}{3^n}$

3. $\sum_{i=1}^{6} (2i - 1) = 1 + 3 + 5 + 7 + 9 + 11 = 36$

4. $a_1 = 23, d = -2$

$a_2 = 23 + (-2) = 21$

$a_3 = 21 + (-2) = 19$

$a_4 = 19 + (-2) = 17$

$a_5 = 17 + (-2) = 15$

Terms: 23, 21, 19, 17, 15

5. $a_1 = 12, d = 3, n = 50$

$a_n = a_1 + (n - 1)d$

$a_{50} = 12 + (50 - 1)3 = 159$

6. $a_1 = 1$

$a_{200} = 200$

$S_n = \frac{n}{2}(a_1 + a_n)$

$S_{200} = \frac{200}{2}(1 + 200) = 20{,}100$

7. $a_1 = 7, r = 2$

$a_2 = 7(2) = 14$

$a_3 = 7(2)^2 = 28$

$a_4 = 7(2)^3 = 56$

$a_5 = 7(2)^4 = 112$

Terms: 7, 14, 28, 56, 112

8. $\sum_{n=1}^{10} 6\left(\frac{2}{3}\right)^{n-1}, a_1 = 6, r = \frac{2}{3}, n = 10$

$$S_n = \frac{a_1(1 - r^n)}{1 - r} = \frac{6\left[1 - \left(\frac{2}{3}\right)^{10}\right]}{1 - \frac{2}{3}} = 18\left(1 - \frac{1024}{59{,}049}\right) = \frac{116{,}050}{6561} \approx 17.6879$$

9. $\sum_{n=0}^{\infty} (0.03)^n = \sum_{n=1}^{\infty} (0.03)^{n-1}$, $a_1 = 1$, $r = 0.03$

$$S = \frac{a_1}{1-r} = \frac{1}{1-0.03} = \frac{1}{0.97} = \frac{100}{97} \approx 1.0309$$

10. For $n = 1$, $1 = \frac{1(1+1)}{2}$.

Assume that $S_k = 1 + 2 + 3 + 4 + \cdots + k = \frac{k(k+1)}{2}$.

Then
$$S_{k+1} = 1 + 2 + 3 + 4 + \cdots + k + (k+1) = \frac{k(k+1)}{2} + k + 1$$
$$= \frac{k(k+1)}{2} + \frac{2(k+1)}{2}$$
$$= \frac{(k+1)(k+2)}{2}.$$

Thus, by the principle of mathematical induction, $1 + 2 + 3 + 4 + \cdots + n = \frac{n(n+1)}{2}$ for all integers $n \geq 1$.

11. For $n = 4$, $4! > 2^4$. Assume that $k! > 2^k$.

Then $(k+1)! = (k+1)(k!) > (k+1)2^k > 2 \cdot 2^k = 2^{k+1}$.

Thus, by the extended principle of mathematical induction, $n! > 2^n$ for all integers $n \geq 4$.

12. ${}_{13}C_4 = \frac{13!}{(13-4)!4!} = 715$

13. $(x+3)^5 = x^5 + 5x^4(3) + 10x^3(3)^2 + 10x^2(3)^3 + 5x(3)^4 + (3)^5$

$= x^5 + 15x^4 + 90x^3 + 270x^2 + 405x + 243$

14. $-{}_{12}C_5 x^7(2)^5 = -25{,}344x^7$

15. ${}_{30}P_4 = \frac{30!}{(30-4)!} = 657{,}720$

16. $6! = 720$ ways

17. ${}_{12}P_3 = 1320$

18. $P(2) + P(3) + P(4) = \frac{1}{36} + \frac{2}{36} + \frac{3}{36}$

$= \frac{6}{36} = \frac{1}{6}$

19. $P(\text{K, B10}) = \frac{4}{52} \cdot \frac{2}{51} = \frac{2}{663}$

20. Let A = probability of no faulty units.

$$P(A) = \left(\frac{997}{1000}\right)^{50} \approx 0.8605$$

$$P(A') = 1 - P(A) \approx 0.1395$$

PART II

Chapter Test Solutions for Chapter P

1. $-\frac{10}{3} = -3\frac{1}{3}$

$-|-4| = -4$

$-\frac{10}{3} > -|-4|$

2. $\left|-5.4 - 3\frac{3}{4}\right| = 9.15$

3. $(5 - x) + 0 = 5 - x$

Additive Identity Property

4. (a) $27\left(-\frac{2}{3}\right) = -18$

(b) $\frac{5}{18} \div \frac{5}{8} = \frac{5}{18} \cdot \frac{8}{5} = \frac{4}{9}$

(c) $\left(-\frac{3}{5}\right)^3 = -\frac{27}{125}$

(d) $\left(\frac{3^2}{2}\right)^{-3} = \left(\frac{2}{9}\right)^3 = \frac{8}{729}$

5. (a) $\sqrt{5} \cdot \sqrt{125} = \sqrt{625} = 25$

(b) $\frac{\sqrt{27}}{\sqrt{2}} = \frac{3\sqrt{3}}{\sqrt{2}} \cdot \frac{\sqrt{2}}{\sqrt{2}} = \frac{3\sqrt{6}}{2}$

(c) $\frac{5.4 \times 10^8}{3 \times 10^3} = \frac{5.4}{3} \times 10^{8-3} = 1.8 \times 10^5$

(d) $\left(3 \times 10^4\right)^3 = 27 \times 10^{12} = 2.7 \times 10^{13}$

6. (a) $3z^2\left(2z^3\right)^2 = 3z^2\left(4z^6\right) = 12z^8$

(b) $(u - 2)^{-4}(u - 2)^{-3} = (u - 2)^{-7} = \frac{1}{(u - 2)^7}$

(c) $\left(\frac{x^{-2}y^2}{3}\right)^{-1} = \frac{x^2y^{-2}}{3^{-1}} = \frac{3x^2}{y^2}$

7. (a) $9z\sqrt{8z} - 3\sqrt{2z^3} = 18z\sqrt{2z} - 3z\sqrt{2z}$

$= 15z\sqrt{2z}$

Since $\sqrt{8z}$ appears in the expression, we may assume that $z \geq 0$. It is not necessary to use an absolute value when simplifying $\sqrt{2z^3}$.

(b) $\left(4x^{3/5}\right)\left(x^{1/3}\right) = 4x^{(3/5)+(1/3)} = 4x^{14/15}$

(c) $\sqrt[3]{\frac{16}{v^5}} = \sqrt[3]{\frac{8}{v^6} \cdot 2v} = \frac{2}{v^2}\sqrt[3]{2v}$

8. Standard form: $-2x^5 - x^4 + 3x^3 + 3$

Degree: 5

Leading coefficient: -2

9. $\left(x^2 + 3\right) - \left[3x + \left(8 - x^2\right)\right] = x^2 + 3 - 3x - 8 + x^2$

$= 2x^2 - 3x - 5$

10. $\left(x + \sqrt{5}\right)\left(x - \sqrt{5}\right) = x^2 - \left(\sqrt{5}\right)^2 = x^2 - 5$

11. $\frac{5x}{x-4} + \frac{20}{4-x} = \frac{5x}{x-4} - \frac{20}{x-4}$

$= \frac{5x - 20}{x - 4}$

$= \frac{5(x-4)}{x-4}$

$= 5,\ x \neq 4$

12. $\frac{\left(\frac{2}{x} - \frac{2}{x+1}\right)}{\left(\frac{4}{x^2 - 1}\right)} = \frac{2(x+1) - 2x}{x(x+1)} \cdot \frac{x^2 - 1}{4}$

$= \frac{2}{x(x+1)} \cdot \frac{(x+1)(x-1)}{4}$

$= \frac{x-1}{2x},\ x \neq \pm 1$

13. (a) $2x^4 - 3x^3 - 2x^2 = x^2\left(2x^2 - 3x - 2\right)$

$= x^2(2x + 1)(x - 2)$

(b) $x^3 + 2x^2 - 4x - 8 = x^2(x + 2) - 4(x + 2)$

$= (x + 2)\left(x^2 - 4\right)$

$= (x + 2)(x + 2)(x - 2)$

$= (x + 2)^2(x - 2)$

14. (a) $\frac{16}{\sqrt[3]{16}} = \frac{16}{\sqrt[3]{16}} \cdot \frac{\sqrt[3]{4}}{\sqrt[3]{4}} = \frac{16\sqrt[3]{4}}{\sqrt[3]{64}} = \frac{16\sqrt[3]{4}}{4} = 4\sqrt[3]{4}$

(b) $\frac{4}{1 - \sqrt{2}} = \frac{4}{1 - \sqrt{2}} \cdot \frac{1 + \sqrt{2}}{1 + \sqrt{2}}$

$= \frac{4\left(1 + \sqrt{2}\right)}{1 - 2}$

$= -4\left(1 + \sqrt{2}\right)$

15. The domain of $\frac{6 - x}{1 - x}$ is all real numbers x except $x = 1$.

16. $\frac{y^2 + 8y + 16}{2y - 4} \cdot \frac{8y - 16}{(y + 4)^3} = \frac{(y + 4)^2}{2(y - 2)} \cdot \frac{8(y - 2)}{(y + 4)^3} = \frac{4}{y + 4}, \; y \neq 2$

17. $P = R - C$

$= 15x - (1480 + 6x)$

$= 9x - 1480$

When $x = 225$,

$P = 9(225) - 1480$

$= \$545.$

18.

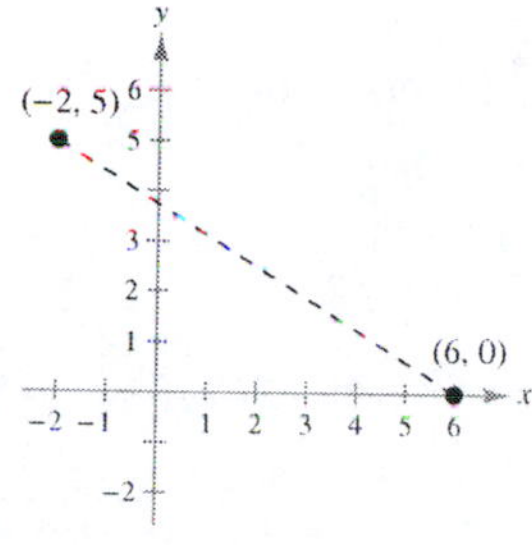

Midpoint: $\left(\frac{-2 + 6}{2}, \frac{5 + 0}{2}\right) = \left(2, \frac{5}{2}\right)$

Distance: $d = \sqrt{(-2 - 6)^2 + (5 - 0)^2}$

$= \sqrt{64 + 25}$

$= \sqrt{89}$

19. Area = Area of large triangle − Area of small triangle

$A = \frac{1}{2}(3x)(\sqrt{3}x) - \frac{1}{2}(2x)\left(\frac{2}{3}\sqrt{3}x\right)$

$= \frac{3\sqrt{3}x^2}{2} - \frac{2\sqrt{3}x^2}{3}$

$= \frac{9\sqrt{3}x^2 - 4\sqrt{3}x^2}{6}$

$= \frac{5\sqrt{3}x^2}{6}$

$= \frac{5}{6}\sqrt{3}x^2$

Chapter Test Solutions for Chapter 1

1. $y = 4 - \frac{3}{4}x$

No symmetry

x-intercept: $\left(\frac{16}{3}, 0\right)$

y-intercept: $(0, 4)$

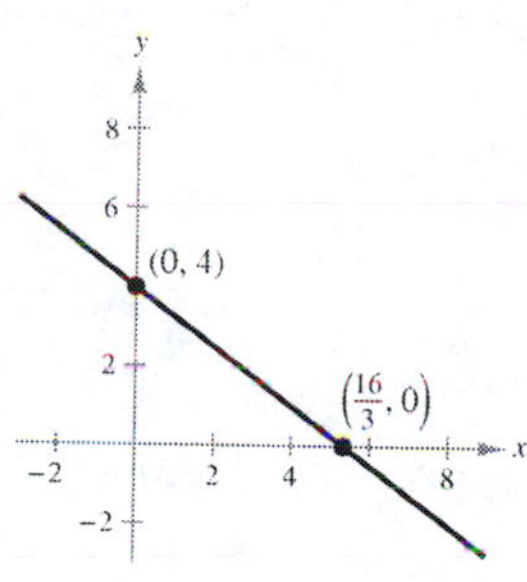

2. $y = 4 - \frac{3}{4}|x|$

y-axis symmetry

x-intercepts: $\left(\pm\frac{16}{3}, 0\right)$

y-intercept: $(0, 4)$

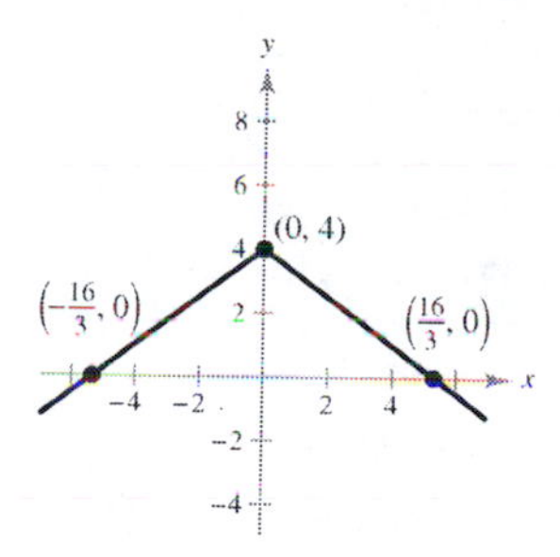

3. $y = 4 - (x - 2)^2$

Parabola; vertex: $(2, 4)$

No x-axis, y-axis, or origin symmetry

x-intercepts: $(0, 0)$ and $(4, 0)$

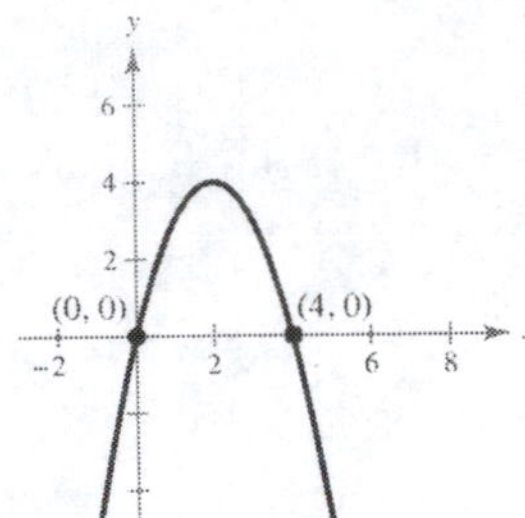

$$\begin{aligned} 0 &= 4 - (x - 2)^2 \\ (x - 2)^2 &= 4 \\ x - 2 &= \pm 2 \\ x &= 2 \pm 2 \\ x &= 4 \quad \text{or} \quad x = 0 \end{aligned}$$

y-intercept: $(0, 0)$

4. $y = x - x^3$

Origin symmetry

x-intercepts: $(0, 0), (1, 0), (-1, 0)$

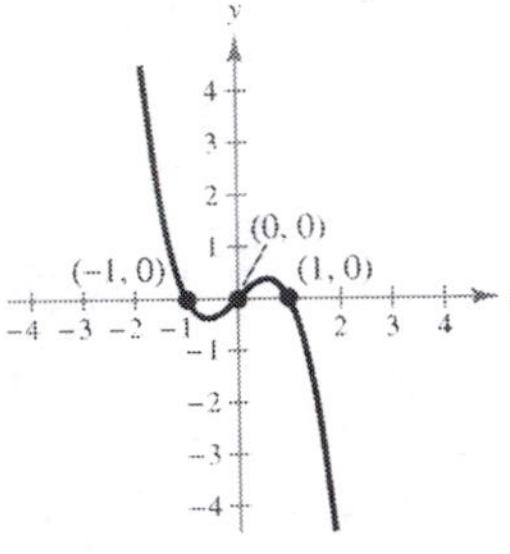

$$\begin{aligned} 0 &= x - x^3 \\ 0 &= x(1 + x)(1 - x) \\ x &= 0, x = \pm 1 \end{aligned}$$

y-intercept: $(0, 0)$

5. $y = \sqrt{5 - x}$

Domain: $x \le 5$

No symmetry

x-intercept: $(5, 0)$

y-intercept: $\left(0, \sqrt{5}\right)$

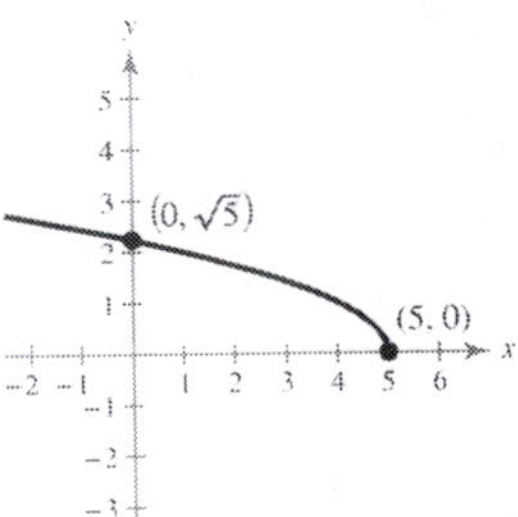

6. $(x - 3)^2 + y^2 = 9$

Circle

Center: $(3, 0)$

Radius: 3

x-axis symmetry

x-intercepts: $(0, 0), (6, 0)$

y-intercept: $(0, 0)$

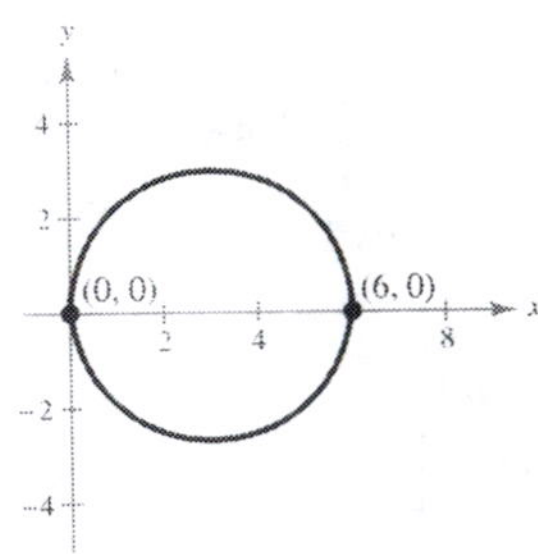

7.

$$\begin{aligned} \tfrac{2}{3}(x - 1) + \tfrac{1}{4}x &= 10 \\ 12\left[\tfrac{2}{3}(x - 1) + \tfrac{1}{4}x\right] &= 12(10) \\ 8(x - 1) + 3x &= 120 \\ 8x - 8 + 3x &= 120 \\ 11x &= 128 \\ x &= \tfrac{128}{11} \end{aligned}$$

8.

$$\begin{aligned} (x - 4)(x + 2) &= 7 \\ x^2 - 2x - 8 &= 7 \\ x^2 - 2x - 15 &= 0 \\ (x + 3)(x - 5) &= 0 \\ x = -3 \quad &\text{or} \quad x = 5 \end{aligned}$$

9.

$$\begin{aligned} \frac{x - 2}{x + 2} + \frac{4}{x + 2} + 4 &= 0, x \ne -2 \\ \frac{x + 2}{x + 2} &= -4 \\ 1 \ne -4 &\Rightarrow \text{No solution} \end{aligned}$$

10.

$$\begin{aligned} x^4 + x^2 - 6 &= 0 \\ (x^2 - 2)(x^2 + 3) &= 0 \\ x^2 = 2 &\Rightarrow x = \pm\sqrt{2} \\ x^2 = -3 &\Rightarrow x = \pm\sqrt{3}i \end{aligned}$$

11. $2\sqrt{x} - \sqrt{2x+1} = 1$

$$-\sqrt{2x+1} = 1 - 2\sqrt{x}$$
$$\left(-\sqrt{2x+1}\right)^2 = \left(1 - 2\sqrt{x}\right)^2$$
$$2x + 1 = 1 - 4\sqrt{x} + 4x$$
$$-2x = -4\sqrt{x}$$
$$x = 2\sqrt{x}$$
$$x^2 = 4x$$
$$x^2 - 4x = 0$$
$$x(x-4) = 0$$
$$x = 0 \quad \text{or} \quad x = 4$$

Only $x = 4$ is a solution to the original equation. $x = 0$ is extraneous.

12. $|3x - 1| = 7$

$3x - 1 = 7$ or $3x - 1 = -7$

$3x = 8$ $\quad$ $3x = -6$

$x = \frac{8}{3}$ $\quad$ $x = -2$

13. $-3 \le 2(x+4) < 14$

$$-3 \le 2x + 8 < 14$$
$$-11 \le 2x < 6$$
$$-\frac{11}{2} \le x < 3$$

$-\frac{11}{2}$ 3

−6 −4 −2 0 2 4

14. $\frac{2}{x} > \frac{5}{x+6}$

$$\frac{2}{x} - \frac{5}{x+6} > 0$$
$$\frac{2(x+6) - 5x}{x(x+6)} > 0$$
$$\frac{-3x + 12}{x(x+6)} > 0$$
$$\frac{-3(x-4)}{x(x+6)} > 0$$

Critical numbers: $x = 4, x = 0, x = -6$

Test intervals: $(-\infty, -6), (-6, 0), (0, 4), (4, \infty)$

Test: Is $\frac{-3(x-4)}{x(x+6)} > 0$?

Solution set: $(-\infty, -6) \cup (0, 4)$

In inequality notation: $x < -6$ or $0 < x < 4$

−8 −6 −4 −2 0 2 4 6

15. $2x^2 + 5x > 12$

$$2x^2 + 5x - 12 > 0$$
$$(2x - 3)(x + 4) > 0$$

Critical numbers: $x = \frac{3}{2}, x = -4$

Test intervals: $(-\infty, -4), \left(-4, \frac{3}{2}\right), \left(\frac{3}{2}, \infty\right)$

Test: Is $(2x-3)(x+4) > 0$?

Solution set: $(-\infty, -4) \cup \left(\frac{3}{2}, \infty\right)$

In inequality notation: $x < -4$ or $x > \frac{3}{2}$

$\frac{3}{2}$

−5 −4 −3 −2 −1 0 1 2 3

16. $|3x + 5| \ge 10$

$3x + 5 \le -10$ or $3x + 5 \ge 10$

$3x \le -15$ $\quad$ $3x \ge 5$

$x \le -5$ $\quad$ $x \ge \frac{5}{3}$

$\frac{5}{3}$

−6 −5 −4 −3 −2 −1 0 1 2

17. (a) $10i - \left(3 + \sqrt{-25}\right) = 10i - (3 + 5i) = -3 + 5i$

(b) $(-1 - 5i)(-1 + 5i) = 1 - 25i^2 = 1 + 25 = 26$

18. $\frac{5}{2+i} = \frac{5}{2+i} \cdot \frac{2-i}{2-i} = \frac{5(2-i)}{4+1} = 2 - i$

19. (a)

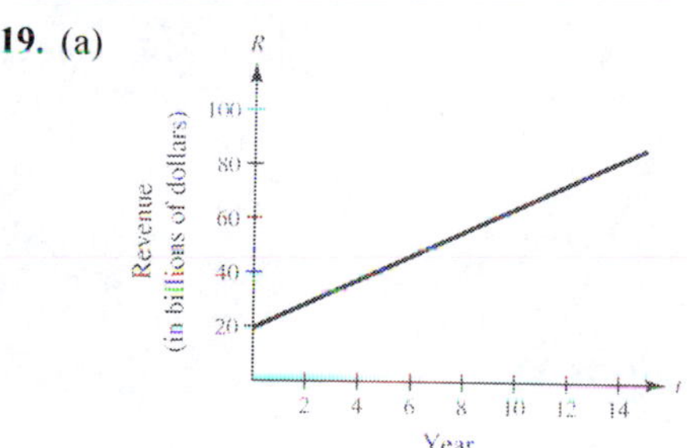

(b) From the graph, you can see that when $t = 15$, the value of R is 86.25. So, the sales in the year 2015 will be about $86.25 million.

(c) Algebraically:

$R = 4.45t + 19.5 = 4.45(15) + 19.5 = 86.25$

So, the sales in the year 2015 will be about $86.25 million.

20. $V = \frac{4}{3}\pi r^3$

$$\frac{4}{3}\pi r^3 = 455.9$$
$$r = \sqrt[3]{\frac{455.9(3)}{4\pi}} \approx 4.774 \text{ inches}$$

21. $(100 \text{ km/hr})\left(2\frac{1}{4} \text{ hr}\right) + (x \text{ km/hr})\left(1\frac{1}{3} \text{ hr}\right) = 350 \text{ km}$

$$225 + \tfrac{4}{3}x = 350$$
$$\tfrac{4}{3}x = 125$$
$$x = \tfrac{375}{4} = 93\tfrac{3}{4} \text{ km/hr}$$

22. $a + b = 100 \Rightarrow b = 100 - a$

Area of ellipse = Area of circle

$$\pi ab = \pi(40)^2$$
$$a(100 - a) = 1600$$
$$0 = a^2 - 100a + 1600$$
$$0 = (a - 80)(a - 20)$$
$$a = 80 \Rightarrow b = 20$$

or

$$a = 20 \Rightarrow b = 80$$

Because $a > b$, choose $a = 80$ and $b = 20$.

Chapter Test Solutions for Chapter 2

1. $(4, -5), (-2, 7)$

$$m = \frac{7 - (-5)}{-2 - 4} = \frac{12}{-6} = -2$$
$$y - 7 = -2(x - (-2))$$
$$y - 7 = -2x - 4$$
$$y = -2x + 3$$

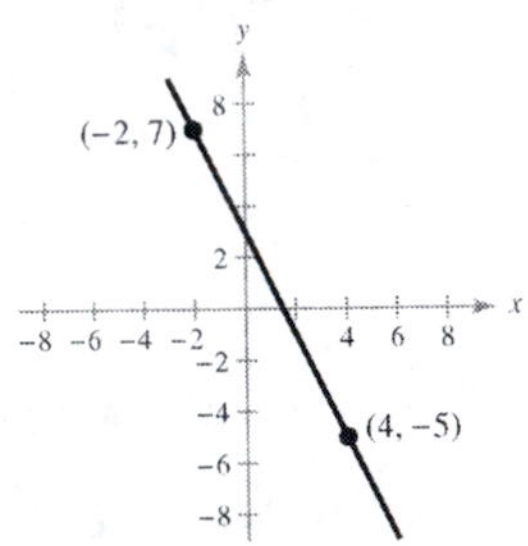

2. $(3, 0.8), (7, -6)$

$$m = \frac{-6 - 0.8}{7 - 3} = \frac{-6.8}{4} = -1.7$$
$$y - 0.8 = -1.7(x - 3)$$
$$y - 0.8 = -1.7x + 5.1$$
$$y = -1.7x + 5.9$$

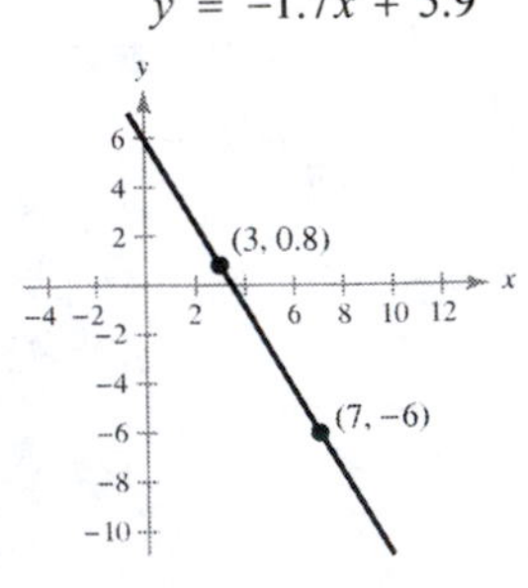

3. $5x + 2y = 3$

$$2y = -5x + 3$$
$$y = -\tfrac{5}{2}x + \tfrac{3}{2}$$

(a) Parallel line:

$$m = -\tfrac{5}{2}$$
$$y - 4 = -\tfrac{5}{2}(x - 0)$$
$$y - 4 = -\tfrac{5}{2}x$$
$$y = -\tfrac{5}{2}x + 4$$

(b) Perpendicular line:

$$m = \tfrac{2}{5}$$
$$y - 4 = \tfrac{2}{5}(x - 0)$$
$$y - 4 = \tfrac{2}{5}x$$
$$y = \tfrac{2}{5}x + 4$$

4. $f(x) = |x + 2| - 15$

(a) $f(-8) = -9$

(b) $f(14) = 1$

(c) $f(x - 6) = |x - 4| - 15$

5. $f(x) = \dfrac{\sqrt{x+9}}{x^2 - 81}$

(a) $f(7) = \dfrac{4}{-32} = -\dfrac{1}{8}$

(b) $f(-5) = \dfrac{2}{-56} = -\dfrac{1}{28}$

(c) $f(x-9) = \dfrac{\sqrt{x}}{(x-9)^2 - 81} = \dfrac{\sqrt{x}}{x^2 - 18x}$

6. $f(x) = |-x + 6| + 2$

Domain: All real numbers x

7. $f(x) = 10\sqrt{3 - x}$

$3 - x \geq 0$

$3 \geq x$

$x \leq 3$

Domain: All real numbers x such that $x \leq 3$

8. $f(x) = 2x^6 + 5x^4 - x^2$

(a)

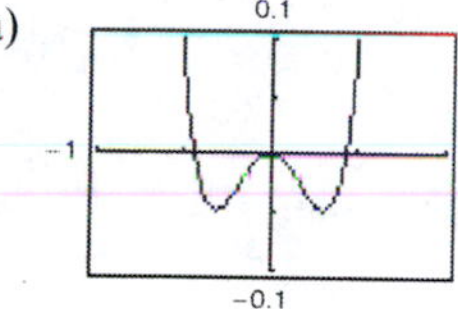

(b) Increasing on $(-0.31, 0)$, $(0.31, \infty)$

Decreasing on $(-\infty, -0.31)$, $(0, 0.31)$

(c) y-axis symmetry $\Rightarrow$ the function is even.

9. $f(x) = 4x\sqrt{3 - x}$

(a)

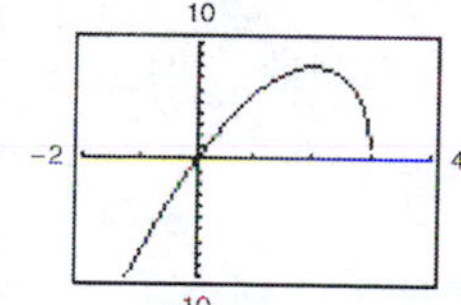

(b) Increasing on $(-\infty, 2)$

Decreasing on $(2, 3)$

(c) The function is neither odd nor even.

10. $f(x) = |x + 5|$

(a)

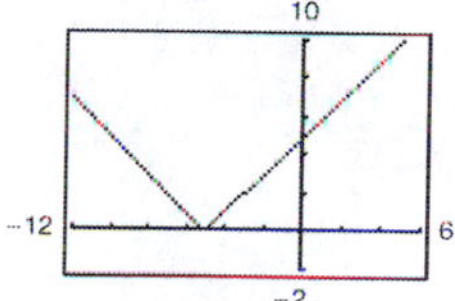

(b) Increasing on $(-5, \infty)$

Decreasing on $(-\infty, -5)$

(c) The function is neither odd nor even.

11. $f(x) = -x^3 + 2x - 1$

Relative minimum: $(-0.816, -2.089)$

Relative maximum: $(0.816, 0.089)$

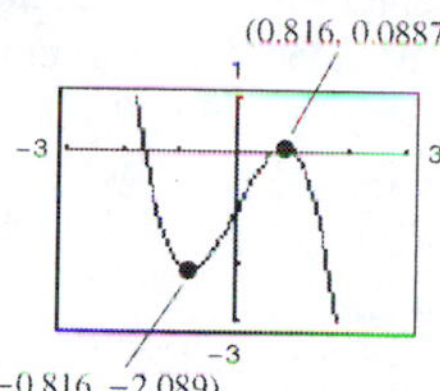

12. $f(x) = -2x^2 + 5x - 3$

$\dfrac{f(3) - f(1)}{3 - 1} = \dfrac{-6 - 0}{2} = -3$

The average rate of change of f from $x_1 = 1$ to $x_2 = 3$ is -3.

13.
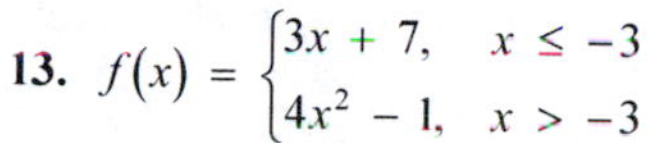

$f(x) = \begin{cases} 3x + 7, & x \leq -3 \\ 4x^2 - 1, & x > -3 \end{cases}$

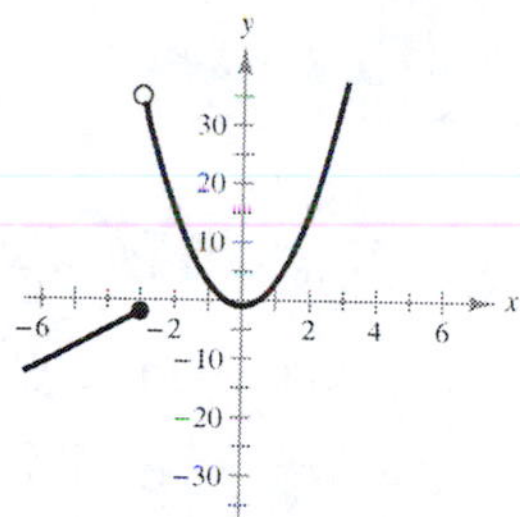

14. (a) Parent function: $f(x) = [\![x]\!]$

(b) $h(x) = 3[\![x]\!]$

Vertical stretch

(c)

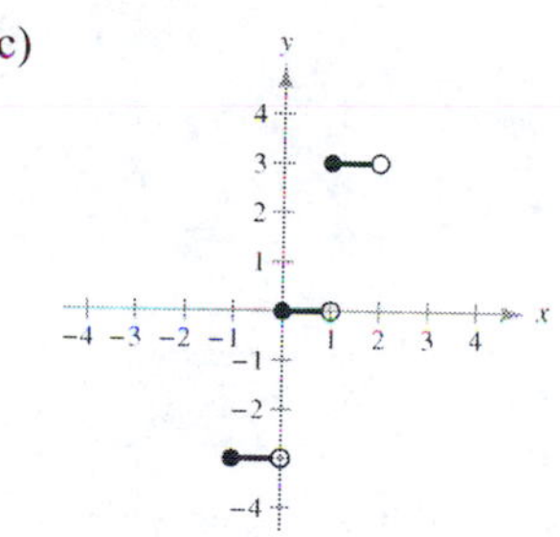

15. $h(x) = -\sqrt{x+5} + 8$

(a) Parent function: $f(x) = x\sqrt{x}$

(b) Transformation: Reflection in the x-axis, a horizontal shift 5 units to the left, and a vertical shift 8 units upward

(c)

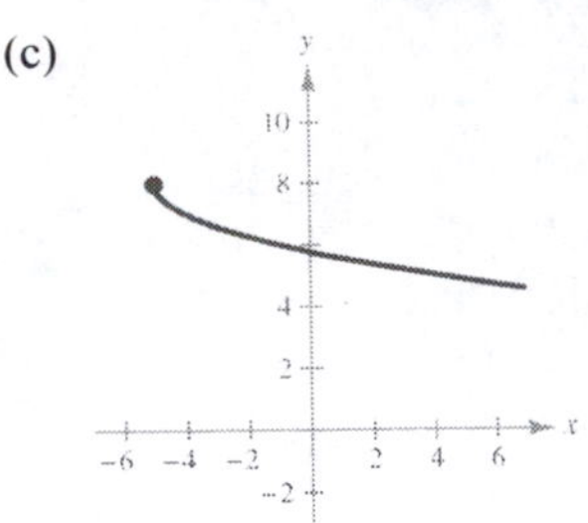

16. (a) Parent function: $f(x) = x^3$

(b) $h(x) = -2(x-5)^3 + 3$

Vertical stretch, reflection in x-axis, horizontal shift 5 units to the right, vertical shift 3 units upward

(c)

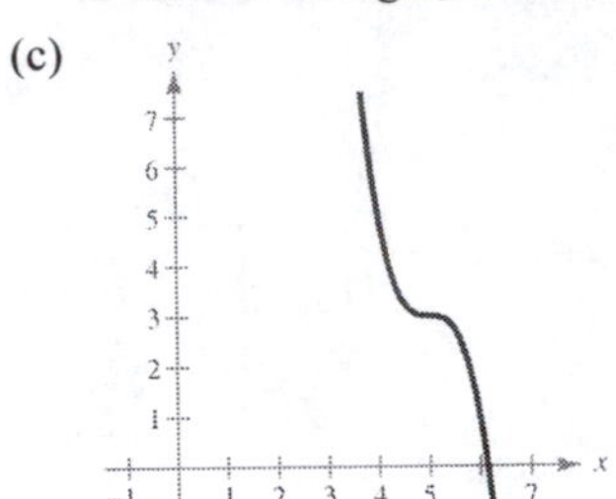

17. $f(x) = 3x^2 - 7,\ g(x) = -x^2 - 4x + 5$

(a) $(f+g)(x) = (3x^2 - 7) + (-x^2 - 4x + 5) = 2x^2 - 4x - 2$

(b) $(f-g)(x) = (3x^2 - 7) - (-x^2 - 4x + 5) = 4x^2 + 4x - 12$

(c) $(fg)(x) = (3x^2 - 7)(-x^2 - 4x + 5) = -3x^4 - 12x^3 + 22x^2 + 28x - 35$

(d) $\left(\dfrac{f}{g}\right)(x) = \dfrac{3x^2 - 7}{-x^2 - 4x + 5},\ x \neq -5, 1$

(e) $(f \circ g)(x) = f(g(x)) = f(-x^2 - 4x + 5) = 3(-x^2 - 4x + 5)^2 - 7 = 3x^4 + 24x^3 + 18x^2 - 120x + 68$

(f) $(g \circ f)(x) = g(f(x)) = g(3x^2 - 7) = -(3x^2 - 7)^2 - 4(3x^2 - 7) + 5 = -9x^4 + 30x^2 - 16$

18. $f(x) = \dfrac{1}{x},\ g(x) = 2\sqrt{x}$

(a) $(f+g)(x) = \dfrac{1}{x} + 2\sqrt{x} = \dfrac{1 + 2x^{3/2}}{x},\ x > 0$

(b) $(f-g)(x) = \dfrac{1}{x} - 2\sqrt{x} = \dfrac{1 - 2x^{3/2}}{x},\ x > 0$

(c) $(fg)(x) = \left(\dfrac{1}{x}\right)\left(2\sqrt{x}\right) = \dfrac{2\sqrt{x}}{x},\ x > 0$

(d) $\left(\dfrac{f}{g}\right)(x) = \dfrac{1/x}{2\sqrt{x}} = \dfrac{1}{2x\sqrt{x}} = \dfrac{1}{2x^{3/2}},\ x > 0$

(e) $(f \circ g)(x) = f(g(x)) = f\left(2\sqrt{x}\right) = \dfrac{1}{2\sqrt{x}} = \dfrac{\sqrt{x}}{2x},\ x > 0$

(f) $(g \circ f)(x) = g(f(x)) = g\left(\dfrac{1}{x}\right) = 2\sqrt{\dfrac{1}{x}} = \dfrac{2}{\sqrt{x}} = \dfrac{2\sqrt{x}}{x},\ x > 0$

19. $f(x) = x^3 + 8$

Since f is one-to-one, f has an inverse.

$$\begin{aligned} y &= x^3 + 8 \\ x &= y^3 + 8 \\ x - 8 &= y^3 \\ \sqrt[3]{x - 8} &= y \\ f^{-1}(x) &= \sqrt[3]{x - 8} \end{aligned}$$

20. $f(x) = |x^2 - 3| + 6$

Since f is not one-to-one, f does not have an inverse.

21. $f(x) = 3x\sqrt{x} = 3x^{3/2}, x \geq 0$

Because f is one-to-one, f has an inverse.

$$\begin{aligned} y &= 3x^{3/2} \\ x &= 3y^{3/2} \\ \tfrac{1}{3}x &= y^{3/2} \\ \left(\tfrac{1}{3}x\right)^{2/3} &= y, x \geq 0 \\ f^{-1}(x) &= \left(\tfrac{1}{3}x\right)^{2/3}, x \geq 0 \end{aligned}$$

22. $(6, 58)$ and $(10, 78)$

$$m = \frac{78 - 58}{10 - 6} = 5$$

$$\begin{aligned} C - 58 &= 5(x - 6) \\ C &= 5x + 28 \end{aligned}$$

When $x = 25$: $C = 5(25) + 28 = \$153$

Cumulative Test Solutions for Chapters P–2

1. $\dfrac{8x^2y^{-3}}{30x^{-1}y^2} = \dfrac{8x^2x}{30y^2y^3} = \dfrac{4x^3}{15y^5}, x \neq 0$

2. $\sqrt{18x^3y^4} = \sqrt{9x^2y^4 2x} = 3xy^2\sqrt{2x}$

3. $$\begin{aligned} 4x - [2x + 3(2 - x)] &= 4x - [2x + 6 - 3x] \\ &= 4x - [-x + 6] \\ &= 5x - 6 \end{aligned}$$

4. $$\begin{aligned} (x - 2)(x^2 + x - 3) &= x^3 + x^2 - 3x - 2x^2 - 2x + 6 \\ &= x^3 - x^2 - 5x + 6 \end{aligned}$$

5. $$\begin{aligned} \frac{2}{s + 3} - \frac{1}{s + 1} &= \frac{2(s + 1) - (s + 3)}{(s + 3)(s + 1)} \\ &= \frac{2s + 2 - s - 3}{(s + 3)(s + 1)} \\ &= \frac{s - 1}{(s + 3)(s + 1)} \end{aligned}$$

6. $$\begin{aligned} 25 - (x - 2)^2 &= [5 + (x - 2)][5 - (x - 2)] \\ &= (3 + x)(7 - x) \end{aligned}$$

7. $$\begin{aligned} x - 5x^2 - 6x^3 &= x(1 - 5x - 6x^2) \\ &= x(1 + x)(1 - 6x) \end{aligned}$$

8. $$\begin{aligned} 54x^3 + 16 &= 2(27x^3 + 8) \\ &= 2\left((3x)^3 + 2^3\right) \\ &= 2(3x + 2)(9x^2 - 6x + 4) \end{aligned}$$

9. $$\begin{aligned} x(2x + 4) + 2x(x + 4) &= 2x^2 + 4x + 2x^2 + 8x \\ &= 4x^2 + 12x \end{aligned}$$

10. $$\begin{aligned} \tfrac{1}{2}(x + 5)[(x - 1) + 2(x + 1)] &= \tfrac{1}{2}(x + 5)[x - 1 + 2x + 2] \\ &= \tfrac{1}{2}(x + 5)(3x + 1) \\ &= \tfrac{3}{2}x^2 + 8x + \tfrac{5}{2} \end{aligned}$$

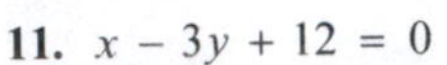

11. $x - 3y + 12 = 0$

Line

x-intercept: $(-12, 0)$

y-intercept: $(0, 4)$

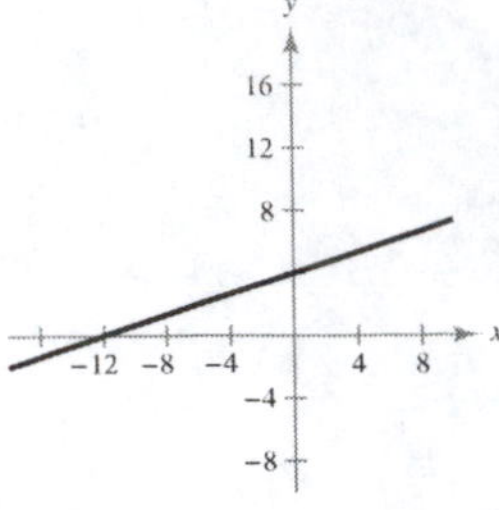

12. $y = x^2 - 9$

Parabola

x-intercept: $(\pm 3, 0)$

y-intercept: $(0, -9)$

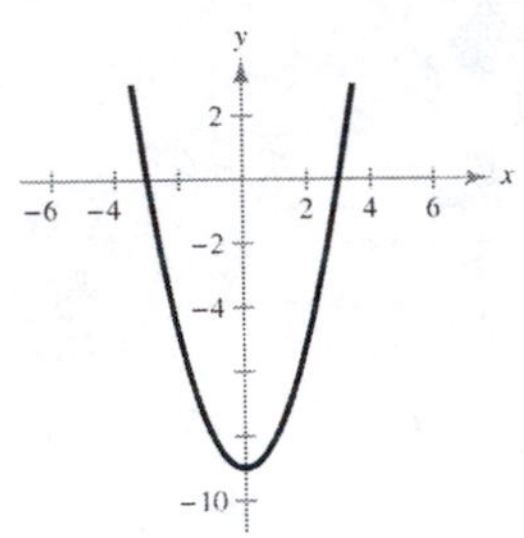

13. $y = \sqrt{4 - x}$

Domain: $x \le 4$

x-intercept: $(4, 0)$

y-intercept: $(0, 2)$

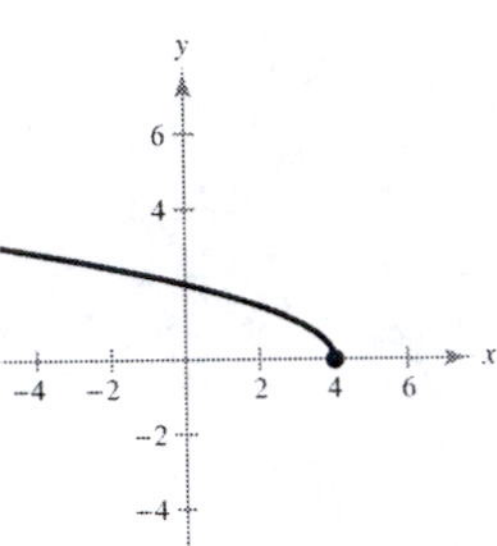

14.
$$3x - 5 = 6x + 8$$
$$-3x = 13$$
$$x = -\tfrac{13}{3}$$

15.
$$-(x + 3) = 14(x - 6)$$
$$-x - 3 = 14x - 84$$
$$-15x = -81$$
$$x = \frac{-81}{-15} = \frac{27}{5}$$

16.
$$\frac{1}{x - 2} = \frac{10}{4x + 3}$$
$$4x + 3 = 10(x - 2)$$
$$4x + 3 = 10x - 20$$
$$-6x = -23$$
$$x = \frac{23}{6}$$

17. Factoring
$$x^2 - 4x + 3 = 0$$
$$(x - 1)(x - 3) = 0$$
$$x - 1 = 0 \Rightarrow x = 1$$
$$x - 3 = 0 \Rightarrow x = 3$$

18. Completing the Square
$$-2x^2 + 8x + 12 = 0$$
$$-2(x^2 - 4x - 6) = 0$$
$$x^2 - 4x - 6 = 0$$
$$x^2 - 4x = 6$$
$$x^2 - 4x + 4 = 6 + 4$$
$$(x - 2)^2 = 10$$
$$x - 2 = \pm\sqrt{10}$$
$$x = 2 \pm \sqrt{10}$$

19. Extracting Square Roots
$$\tfrac{2}{3}x^2 = 24$$
$$x^2 = 36$$
$$x = \pm 6$$

20. Quadratic Formula
$$3x^2 + 5x - 6 = 0$$
$$a = 3, b = 5, c = -6$$
$$x = \frac{-5 \pm \sqrt{5^2 - 4(3)(-6)}}{2(3)}$$
$$= \frac{-5 \pm \sqrt{25 + 72}}{6}$$
$$= \frac{-5 \pm \sqrt{97}}{6}$$

21. Quadratic Formula
$$3x^2 + 9x + 1 = 0$$
$$a = 3, b = 9, c = 1$$
$$x = \frac{-9 \pm \sqrt{9^2 - 4(3)(1)}}{2(3)}$$
$$= \frac{-9 \pm \sqrt{81 - 12}}{6}$$
$$= \frac{-9 \pm \sqrt{69}}{6}$$
$$= -\frac{3}{2} \pm \frac{\sqrt{69}}{6}$$

22. Extracting Square Roots
$$\tfrac{1}{2}x^2 - 7 = 25$$
$$\tfrac{1}{2}x^2 = 32$$
$$x^2 = 64$$
$$x = \pm\sqrt{64}$$
$$x = \pm 8$$

23. $x^4 + 12x^3 + 4x^2 + 48x = 0$

$$x^3(x + 12) + 4x(x + 12) = 0$$
$$(x^3 + 4x)(x + 12) = 0$$
$$x(x^2 + 4)(x + 12) = 0$$
$$x = 0$$
$$x^2 + 4 = 0 \Rightarrow x = \pm 2i$$
$$x + 12 = 0 \Rightarrow x = -12$$

24. $8x^3 - 48x^2 + 72x = 0$

$$8x(x^2 - 6x + 9) = 0$$
$$8x(x - 3)^2 = 0$$
$$8x = 0 \Rightarrow x = 0$$
$$x - 3 = 0 \Rightarrow x = 3$$

25. $x^{2/3} + 13 = 17$

$$\sqrt[3]{x^2} = 4$$
$$x^2 = 4^3$$
$$x = \pm\sqrt{64}$$
$$x = \pm 8$$

26. $\sqrt{x + 10} = x - 2$

$$x + 10 = x^2 - 4x + 4$$
$$0 = x^2 - 5x - 6$$
$$0 = (x - 6)(x + 1)$$
$$x = 6 \text{ or } x = -1$$

Only $x = 6$ is a solution to the original equation. $x = -1$ is extraneous.

27. $|3(x - 4)| = 27$

$3(x - 4) = -27$ or $3(x - 4) = 27$

$x - 4 = -9$ $\quad$ $x - 4 = 9$

$x = -5$ $\quad$ $x = 13$

28. $|x - 12| = -2$

No solution. The absolute value of a number cannot be negative.

29. $4x + 2 > 7$

(a) $4(-1) + 2 \not> 7$

$x = -1$ *is not* a solution.

(b) $4(\frac{1}{2}) + 2 \not> 7$

$x = \frac{1}{2}$ *is not* a solution.

(c) $4(\frac{3}{2}) + 2 > 7$

$x = \frac{3}{2}$ *is* a solution.

(d) $4(2) + 2 > 7$

$x = 2$ *is* a solution.

30. $|5x - 1| < 4$

(a) $|5(-1) - 1| \not< 4$

$x = -1$ *is not* a solution.

(b) $|5(-\frac{1}{2}) - 1| < 4$

$x = -\frac{1}{2}$ *is* a solution.

(c) $|5(1) - 1| \not< 4$

$x = 1$ *is not* a solution.

(d) $|5(2) - 1| \not< 4$

$x = 2$ *is not* a solution.

31. $|x + 1| \le 6$

$$-6 \le x + 1 \le 6$$
$$-7 \le x \le 5$$

−7 5
−10 −8 −6 −4 −2 0 2 4 6 8 x

32. $|5 + 6x| > 3$

$5 + 6x < -3$ or $5 + 6x > 3$

$6x < -8$ $\quad$ $6x > -2$

$x < -\frac{4}{3}$ $\quad$ $x > -\frac{1}{3}$

$-\frac{4}{3}$ $-\frac{1}{3}$
−2 −1 0 1 2 x

33. $5x^2 + 12x + 7 \ge 0$

$$(5x + 7)(x + 1) \ge 0$$

Critical numbers: $x = -\frac{7}{5}, -1$

Test intervals: $(-\infty, -\frac{7}{5}), (-\frac{7}{5}, -1), (-1, \infty)$

Test: Is $5x^2 + 12x + 7 \ge 0$?

Solution: $x \le -\frac{7}{5}, x \ge -1$

$-\frac{7}{5}$
−3 −2 −1 0 x

34. $-x^2 + x + 4 < 0$

$x^2 - x - 4 > 0$

Critical numbers: $x = \dfrac{1 \pm \sqrt{17}}{2}$ (by the Quadratic Formula)

Test intervals: $\left(-\infty, \dfrac{1 - \sqrt{17}}{2}\right), \left(\dfrac{1 - \sqrt{17}}{2}, \dfrac{1 + \sqrt{17}}{2}\right), \left(\dfrac{1 + \sqrt{17}}{2}, \infty\right)$

Test: Is $-x^2 + x + 4 < 0$?

Solution: $x < \dfrac{1 - \sqrt{17}}{2}, x > \dfrac{1 + \sqrt{17}}{2}$

$\frac{1-\sqrt{17}}{2}$ $\frac{1+\sqrt{17}}{2}$

−3 −2 −1 0 1 2 3 x

35. $\left(-\dfrac{1}{2}, 1\right)$ and $(3, 8)$

$$m = \frac{8 - 1}{3 - (-1/2)} = \frac{7}{7/2} = 2$$

$$y - 8 = 2(x - 3)$$

$$y - 8 = 2x - 6$$

$$y = 2x + 2$$

36. It fails the Vertical Line Test. For some values of x there correspond two values of y.

37. $f(x) = \dfrac{x}{x - 2}$

(a) $f(6) = \dfrac{6}{4} = \dfrac{3}{2}$

(b) $f(2)$ is undefined because division by zero is undefined.

(c) $f(s + 2) = \dfrac{s + 2}{(s + 2) - 2} = \dfrac{s + 2}{s}$

38. $f(x) = 5 + \sqrt{4 - x}$

$f(-x) = 5 + \sqrt{4 - (-x)} = 5 + \sqrt{4 + x}$

$-f(x) = -5 - \sqrt{4 - x}$

$f(-x) \neq f(x)$ and $f(-x) \neq -f(x)$

The function is neither even nor odd.

39. $f(x) = x^5 - x^3 + 2$

$f(-x) = (-x)^5 - (-x)^3 + 2 = -x^5 + x^3 + 2$

$f(-x) \neq f(x)$ and $f(-x) \neq -f(x)$

The function is neither even nor odd.

40. $f(x) = 2x^4 - 4$

$f(-x) = 2(-x)^4 - 4 = 2x^4 - 4 = f(x)$

The function is even.

41. $y = \sqrt[3]{x}$

(a) $r(x) = \frac{1}{2}\sqrt[3]{x}$ is a vertical shrink by a factor of $\frac{1}{2}$.

(b) $h(x) = \sqrt[3]{x} + 2$ is a vertical shift two units upward.

(c) $g(x) = \sqrt[3]{x + 2}$ is a horizontal shift two units to the left.

42. $f(x) = x - 4$, $g(x) = 3x + 1$

(a) $(f + g)(x) = f(x) + g(x)$
$= (x - 4) + (3x + 1)$
$= 4x - 3$

(b) $(f - g)(x) = f(x) - g(x)$
$= (x - 4) - (3x + 1)$
$= -2x - 5$

(c) $(fg)(x) = f(x)g(x)$
$= (x - 4)(3x + 1)$
$= 3x^2 - 11x - 4$

(d) $\left(\dfrac{f}{g}\right)(x) = \dfrac{f(x)}{g(x)} = \dfrac{x - 4}{3x + 1}$

Domain: All real numbers x except $x = -\dfrac{1}{3}$

43. $f(x) = \sqrt{x - 1}$, $g(x) = x^2 + 1$

(a) $(f + g)(x) = f(x) + g(x)$
$= \sqrt{x - 1} + x^2 + 1$

(b) $(f - g)(x) = f(x) - g(x)$
$= \sqrt{x - 1} - x^2 - 1$

(c) $(fg)(x) = f(x)g(x)$
$= \sqrt{x - 1}(x^2 + 1)$
$= x^2\sqrt{x - 1} + \sqrt{x - 1}$

(d) $\left(\dfrac{f}{g}\right)(x) = \dfrac{f(x)}{g(x)} = \dfrac{\sqrt{x - 1}}{x^2 + 1}$

Domain: all real numbers x such that $x \geq 1$

44. $f(x) = 2x^2, g(x) = \sqrt{x+6}$

(a) $$\begin{aligned}(f \circ g)(x) &= f(g(x)) \\ &= f(\sqrt{x+6}) \\ &= 2(\sqrt{x+6})^2 \\ &= 2(x+6) \\ &= 2x + 12\end{aligned}$$

Domain: all real numbers x such that $x \geq -6$

(b) $$\begin{aligned}(g \circ f)(x) &= g(f(x)) \\ &= g(2x^2) \\ &= \sqrt{2x^2 + 6}\end{aligned}$$

Domain: all real numbers x

45. $f(x) = x - 2, g(x) = |x|$

(a) $$\begin{aligned}(f \circ g)(x) &= f(g(x)) \\ &= f(|x|) \\ &= |x| - 2\end{aligned}$$

Domain: all real numbers x

(b) $$\begin{aligned}(g \circ f)(x) &= g(f(x)) \\ &= g(x - 2) \\ &= |x - 2|\end{aligned}$$

Domain: all real numbers x

46. $h(x) = 3x - 4$

Because h is one-to-one, h has an inverse.

$$\begin{aligned}y &= 3x - 4 \\ x &= 3y - 4 \\ x + 4 &= 3y \\ \tfrac{1}{3}(x + 4) &= y \\ h^{-1}(x) &= \tfrac{1}{3}(x + 4)\end{aligned}$$

47. Cost per person: $\dfrac{36{,}000}{n}$

If three additional people join the group, the cost per person is $\dfrac{36{,}000}{n+3}$.

$$\begin{aligned}\frac{36{,}000}{n} &= \frac{36{,}000}{n+3} + 1000 \\ 36{,}000(n+3) &= 36{,}000n + 1000n(n+3) \\ 36(n+3) &= 36n + n(n+3) \\ 36n + 108 &= 36n + n^2 + 3n \\ 0 &= n^2 + 3n - 108 \\ 0 &= (n+12)(n-9)\end{aligned}$$

Choosing the positive value, the group has $n = 9$ people.

48. Rate $= 10 - 0.05(n - 60), n \geq 60$

(a) Revenue = (number of people)(rate per person)

$$\begin{aligned}R(n) &= n[10 - 0.05(n - 60)] \\ &= 10n - 0.05n(n - 60) \\ &= 10n - 0.05n^2 + 3n \\ &= -0.05n^2 + 13n, n \geq 60\end{aligned}$$

(b) 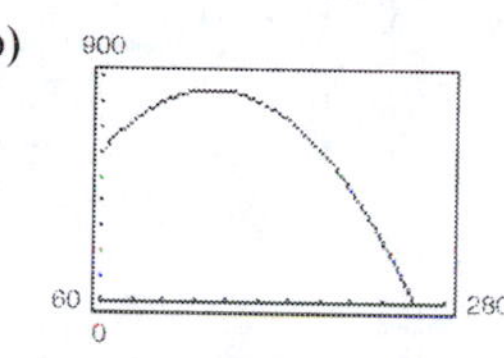

The revenue is maximum when $n = 130$ passengers.

49. $s(t) = -16t^2 + 36t + 8$

$$\frac{s(2) - s(0)}{2 - 0} = \frac{16 - 8}{2} = 4$$

The average rate of change in the height of the object from $t_1 = 0$ to $t_2 = 2$ seconds is 4 feet per second.

Chapter Test Solutions for Chapter 3

1. (a) 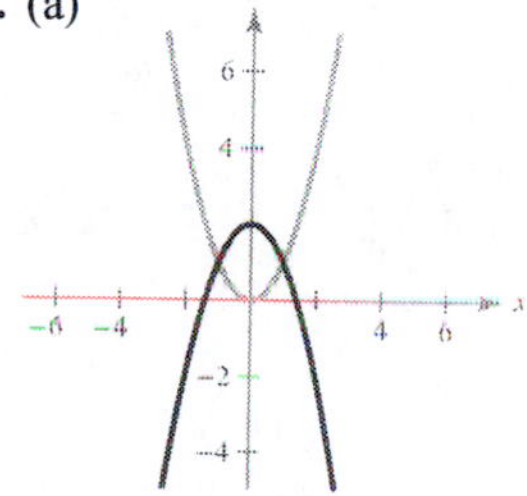

The graph of g is a reflection in the x-axis and a vertical shift up of two units of the graph of $y = x^2$.

(b) 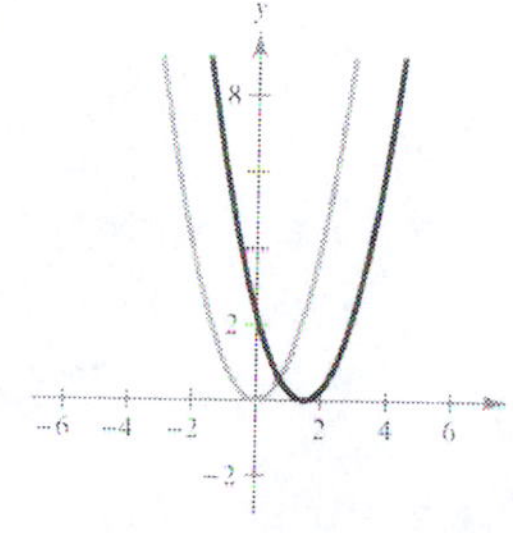

The graph of g is a horizontal shift right $\frac{3}{2}$ units of the graph of $y = x^2$.

2. $f(x) = x^2 + 4x + 3$

$= (x^2 + 4x + 4) - 4 + 3$

$= (x + 2)^2 - 1$

Vertex: $(-2, -1)$

Find the x-intercepts: $x^2 + 4x + 3 = 0$

$(x + 3)(x + 1) = 0$

$x + 3 = 0 \Rightarrow x = -3$

$x + 1 = 0 \Rightarrow x = -1$

The x-intercepts are $(-3, 0)$ and $(-1, 0)$.

3. Vertex: $(3, -6)$

$y = a(x - 3)^2 - 6$

Point on the graph: $(0, 3)$

$3 = a(0 - 3)^2 - 6$

$9 = 9a \Rightarrow a = 1$

So, $y = (x - 3)^2 - 6$.

4. (a) $y = -\frac{1}{20}x^2 + 3x + 5$

$= -\frac{1}{20}(x^2 - 60x + 900 - 900) + 5$

$= -\frac{1}{20}\left[(x - 30)^2 - 900\right] + 5$

$= -\frac{1}{20}(x - 30)^2 + 50$

Vertex: $(30, 50)$

The maximum height is 50 feet.

(b) The constant term, $c = 5$, determines the height at which the ball was thrown. Changing this constant results in a vertical translation of the graph, and therefore, changes the maximum height.

5. $h(t) = -\frac{3}{4}t^5 + 2t^2$

The degree is odd and the leading coefficient is negative. The graph rises to the left and falls to the right.

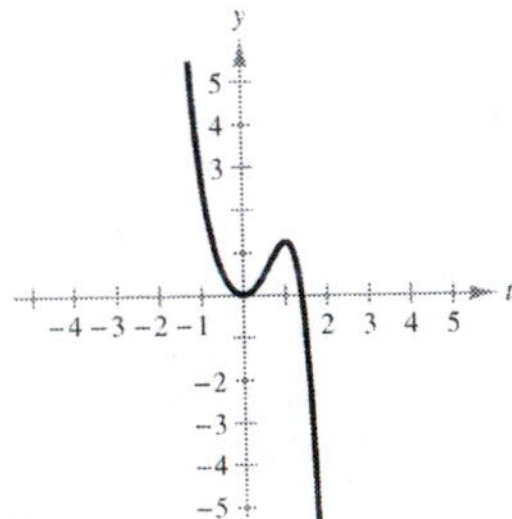

6.
$$\begin{array}{r}
3x + \dfrac{x-1}{x^2+1} \\
x^2 + 0x + 1 \overline{\big)\, 3x^3 + 0x^2 + 4x - 1} \\
\underline{3x^3 + 0x^2 + 3x} \\
x - 1
\end{array}$$

Thus, $\dfrac{3x^3 + 4x - 1}{x^2 + 1} = 3x + \dfrac{x - 1}{x^2 + 1}$.

7.
$$\begin{array}{r|rrrrr}
2 & 2 & 0 & -5 & 0 & -3 \\
 & & 4 & 8 & 6 & 12 \\
\hline
 & 2 & 4 & 3 & 6 & 9
\end{array}$$

Thus,

$$\frac{2x^4 - 5x^2 - 3}{x - 2} = 2x^3 + 4x^2 + 3x + 6 + \frac{9}{x - 2}.$$

8.
$$\begin{array}{r|rrrr}
\sqrt{3} & 2 & -5 & -6 & 15 \\
 & & 2\sqrt{3} & 6 - 5\sqrt{3} & -15 \\
\hline
 & 2 & 2\sqrt{3} - 5 & -5\sqrt{3} & 0
\end{array}$$

$$\begin{array}{r|rrr}
-\sqrt{3} & 2 & 2\sqrt{3} - 5 & -5\sqrt{3} \\
 & & -2\sqrt{3} & 5\sqrt{3} \\
\hline
 & 2 & -5 & 0
\end{array}$$

$2x^3 - 5x^2 - 6x + 15 = (x - \sqrt{3})(x + \sqrt{3})(2x - 5)$

The real zeros of $f(x)$ are $x = \pm\sqrt{3}$ and $x = \frac{5}{2}$.

9. $g(t) = 2t^4 - 3t^3 + 16t - 24$

Possible rational zeros:

$\pm1, \pm2, \pm3, \pm4, \pm6, \pm8, \pm12, \pm24, \pm\frac{1}{2}, \pm\frac{3}{2}$

From the graph, we have $t = -2$ and $t = \frac{3}{2}$.

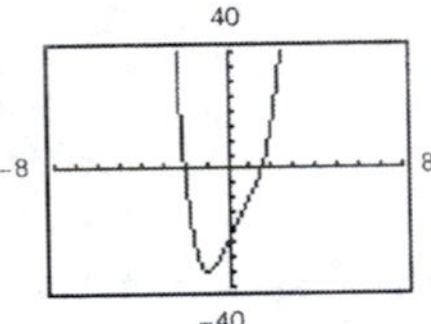

10. $h(x) = 3x^5 + 2x^4 - 3x - 2$

Possible rational zeros: $\pm1, \pm2, \pm\frac{1}{3}, \pm\frac{2}{3}$

From the graph, we have $x = \pm1$ and $x = -\frac{2}{3}$.

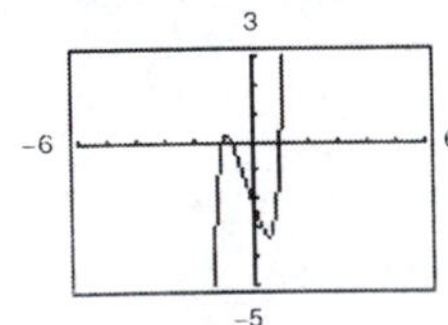

11. $f(x) = x(x-3)(x-(2+i))(x-(2-i))$
$= (x^2-3x)[(x-2)-i][(x-2)+i]$
$= (x^2-3x)[(x-2)^2-i^2]$
$= (x^2-3x)(x^2-4x+5)$
$= x^4-7x^3+17x^2-15x$

12. Because $1-\sqrt{3}i$ is a zero, $1+\sqrt{3}i$ is also a zero.

$f(x) = (x-2)(x-2)[x-(1-\sqrt{3}i)][x-(1+\sqrt{3}i)]$
$= (x^2-4x+4)[(x-1)+\sqrt{3}i][(x-1)-\sqrt{3}i]$
$= (x^2-4x+4)[(x-1)^2-(\sqrt{3}i)^2]$
$= (x^2-4x+4)(x^2-2x+4)$
$= x^4-6x^3+16x^2-24x+16$

13. $f(x) = 3x^3+14x^2-7x-10$

Possible rational zeros:
$\pm1, \pm2, \pm5, \pm10, \pm\frac{1}{3}, \pm\frac{2}{3}, \pm\frac{5}{3}, \pm\frac{10}{3}$

1	3	14	−7	−10
		3	17	10
	3	17	10	0

$f(x) = (x-1)(3x^2+17x+10)$
$= (x-1)(3x+2)(x+5)$

The zeros of $f(x)$ are $x=-5$, $x=-\frac{2}{3}$, and $x=1$.

14. $f(x) = x^4-9x^2-22x-24$

Possible rational zeros:
$\pm1, \pm2, \pm3, \pm4, \pm6, \pm8, \pm12, \pm24$

−2	1	0	−9	−22	−24
		−2	4	10	24
	1	−2	−5	−12	0

4	1	−2	−5	−12
		4	8	12
	1	2	3	0

$f(x) = (x+2)(x-4)(x^2+2x+3)$

By the Quadratic Formula the zeros of x^2+2x+3 are $x=-1\pm\sqrt{2}i$. The zeros of f are: $x=-2, 4, -1\pm\sqrt{2}i$.

15. $v = k\sqrt{s}$
$24 = k\sqrt{16}$
$6 = k$
$v = 6\sqrt{s}$

16. $A = kxy$
$500 = k(15)(8)$
$500 = k(120)$
$\frac{25}{6} = k$
$A = \frac{25}{6}xy$

17. $b = \frac{k}{a}$
$32 = \frac{k}{1.5}$
$48 = k$
$b = \frac{48}{a}$

18. The least squares regression line for the data is $y = -218.6t + 8777$. The model fits the data well.

Chapter Test Solutions for Chapter 4

1. $y = \frac{3x}{x+1}$

Domain: all real numbers x except $x=-1$

Vertical asymptote: $x=-1$

Horizontal asymptote: $y=3$

2. $f(x) = \frac{3-x^2}{3+x^2} = \frac{-x^2+3}{x^2+3}$

Domain: all real numbers x

Vertical asymptote: None

Horizontal asymptote: $y = \frac{-1}{1} = -1$

3. $g(x) = \dfrac{x^2 - 7x + 12}{x - 3}$

$= \dfrac{(x - 3)(x - 4)}{(x - 3)}$

$= x - 4, x \neq 3$

Domain: all real numbers x except $x = 3$

No asymptotes

4. $h(x) = \dfrac{4}{x^2} - 1 = \dfrac{4 - x^2}{x^2} = \dfrac{(2 - x)(2 + x)}{x^2}$

x-intercepts: $(\pm 2, 0)$

Vertical asymptote: $x = 0$

Horizontal asymptote: $y = -1$

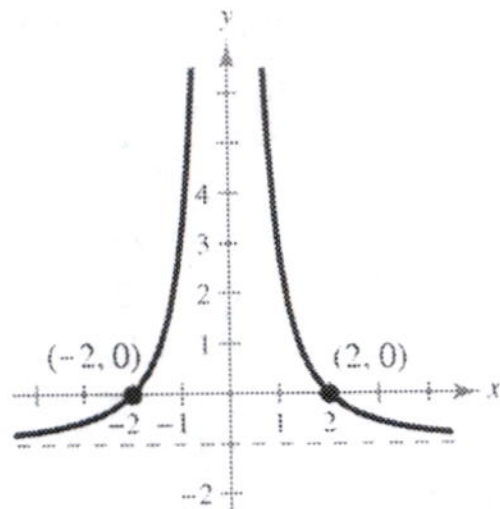

5. $g(x) = \dfrac{x^2 + 2}{x - 1} = x + 1 + \dfrac{3}{x - 1}$

y-intercept: $(0, -2)$

Vertical asymptote: $x = 1$

Slant asymptote: $y = x + 1$

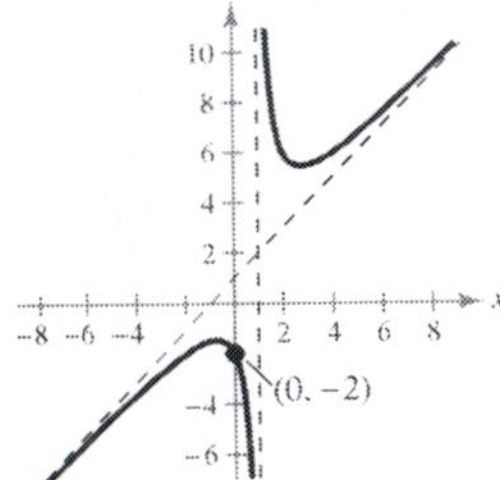

6. $f(x) = \dfrac{x + 1}{x^2 + x - 12} = \dfrac{x + 1}{(x + 4)(x - 3)}$

x-intercept: $(-1, 0)$

y-intercept: $\left(0, -\dfrac{1}{12}\right)$

Vertical asymptotes: $x = -4,\ x = 3$

Horizontal asymptote: $y = 0$

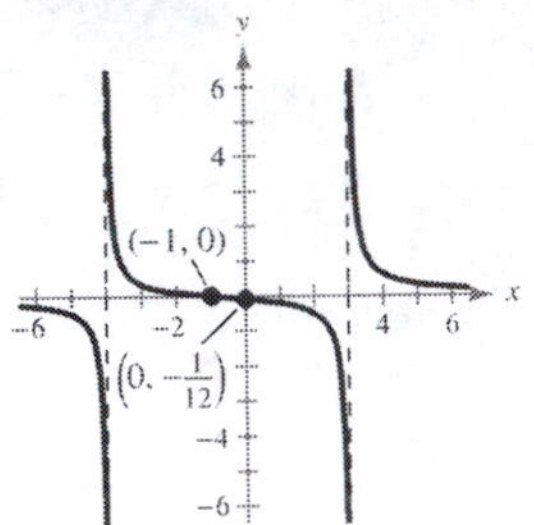

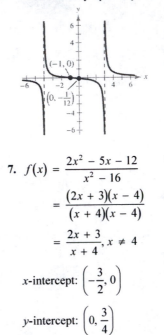

7. $f(x) = \dfrac{2x^2 - 5x - 12}{x^2 - 16}$

$= \dfrac{(2x + 3)(x - 4)}{(x + 4)(x - 4)}$

$= \dfrac{2x + 3}{x + 4}, x \neq 4$

x-intercept: $\left(-\dfrac{3}{2}, 0\right)$

y-intercept: $\left(0, \dfrac{3}{4}\right)$

Vertical asymptote: $x = -4$

Horizontal asymptote: $y = 2$

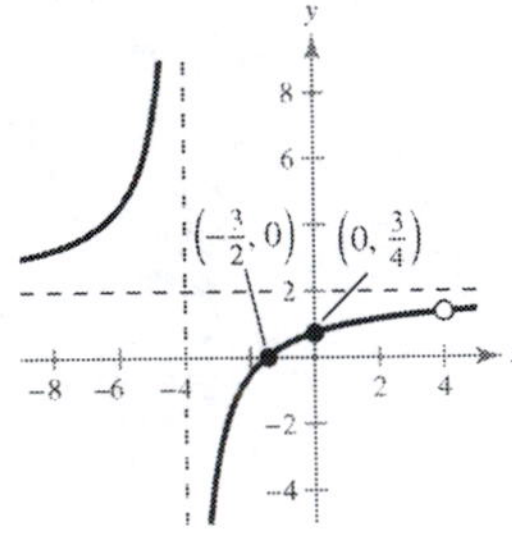

8. $f(x) = \dfrac{2x^2 + 9}{5x^2 + 9}$

y-intercept: $(0, 1)$

Horizontal asymptote: $y = \dfrac{2}{5}$

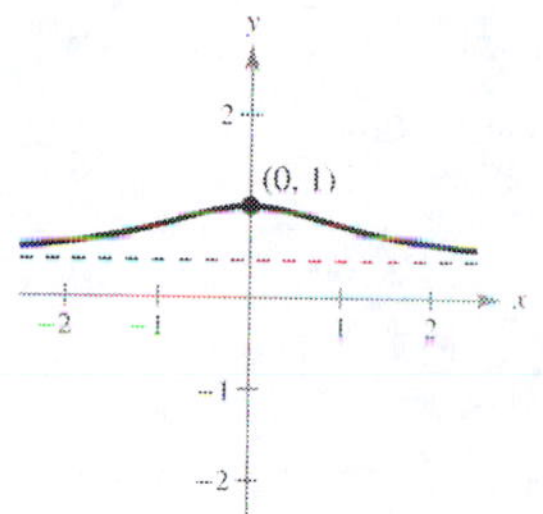

9. $g(x) = \dfrac{2x^3 - 7x^2 + 4x + 4}{x^2 - x - 2}$

$= 2x - 5 + \dfrac{3(x - 2)}{(x - 2)(x + 1)}$

$= 2x - 5 + \dfrac{3}{x + 1}$

$= \dfrac{2x^2 - 3x - 2}{x + 1}$

$= \dfrac{(2x + 1)(x - 2)}{x + 1}, x \neq 2$

x-intercept: $\left(-\dfrac{1}{2}, 0\right)$

y-intercept: $(0, -2)$

Vertical asymptote: $x = -1$

Slant asymptote: $y = 2x - 5$

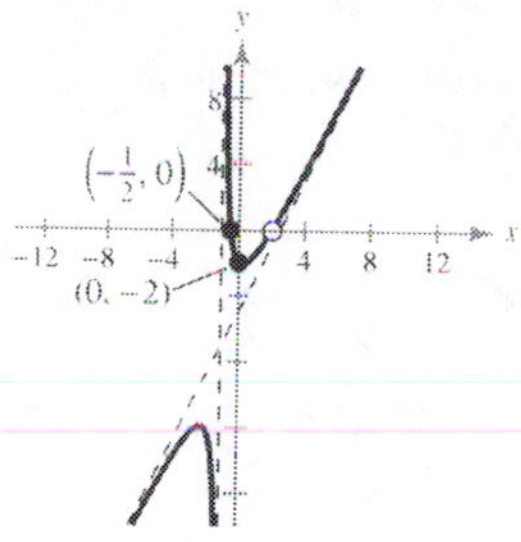

10.

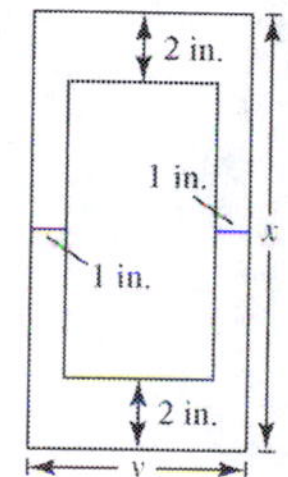

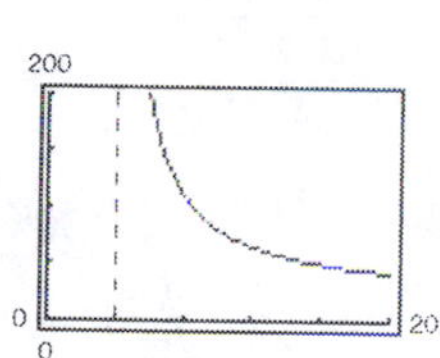

Minimize $A = xy$.

Given: $(x - 4)(y - 2) = 36$

$y = \dfrac{36}{x - 4} + 2$

$A = x\left(\dfrac{36}{x - 4} + 2\right) = x\left(\dfrac{2x + 28}{x - 4}\right) = \dfrac{2x(x + 14)}{x - 4}$

Domain: $x > 4$

From the graph of A we see that the minimum occurs when $x \approx 12.49$ inches and

$y = \dfrac{36}{x - 4} + 2 \approx 6.24$ inches. The dimensions are 6.24 inches by 12.49 inches.

Note: The exact values are $x = 4 + 6\sqrt{2}$ and $y = 2 + 3\sqrt{2}$.

11. (a) Equate the slopes.

$$\frac{y-1}{0-2} = \frac{1-0}{2-x}$$

$$\frac{y-1}{-2} = \frac{1}{2-x}$$

$$y - 1 = -2\left(\frac{1}{2-x}\right)$$

$$y = 1 + \frac{2}{x-2}$$

(b) $A = \frac{1}{2}xy$

$$= \frac{1}{2}x\left[1 + \frac{2}{x-2}\right]$$

$$= \frac{x}{2} + \frac{x}{x-2}$$

$$= \frac{x^2}{2(x-2)}$$

In context, we have $x > 2$ for the domain.

(c)

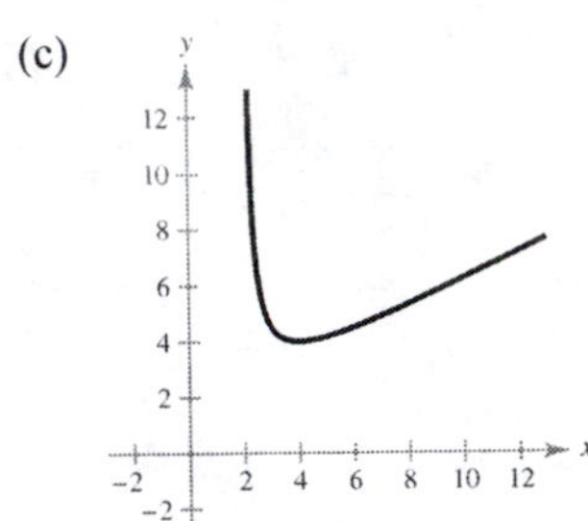

The minimum area occurs at $x = 4$ and is $A = 4$.

12. $y^2 - 4x = 0$

$$y^2 = 4x$$

$$y^2 = 4(1)x \Rightarrow p = 1$$

Parabola

Vertex: $(0, 0)$

Focus: $(1, 0)$

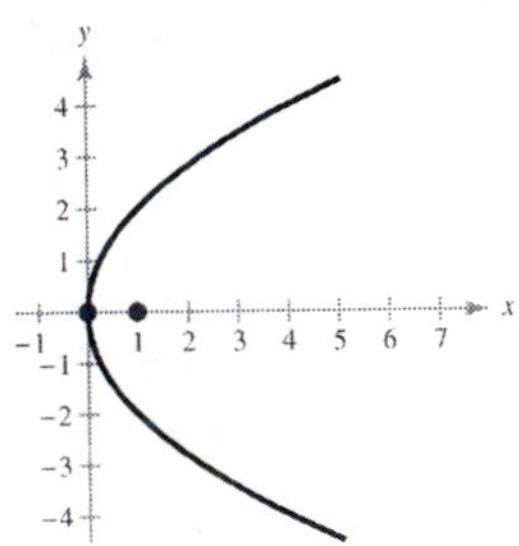

13.

$$x^2 + y^2 - 10x + 4y + 4 = 0$$

$$(x^2 - 10x) + (y^2 + 4y) = -4$$

$$(x^2 - 10x + 25) + (y^2 + 4y + 4) = -4 + 25 + 4$$

$$(x-5)^2 + (y+2)^2 = 25$$

Circle

Center: $(5, -2)$

Radius: 5

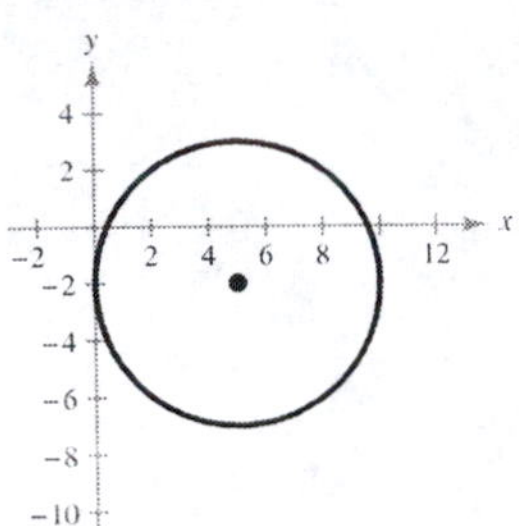

14. $x^2 - 10x - 2y + 19 = 0$

$$x^2 - 10x = 2y - 19$$

$$x^2 - 10x + 25 = 2y - 19 + 25$$

$$(x-5)^2 = 2(y+3)$$

Parabola

Vertex: $(5, -3)$

Focus: $\left(5, -\frac{5}{2}\right)$

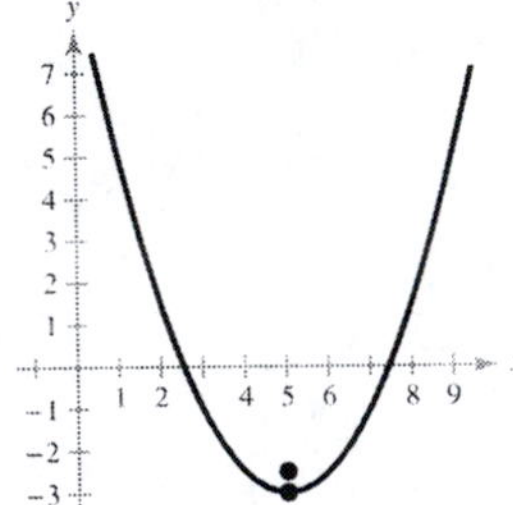

15. $\frac{x^2}{1} - \frac{y^2}{4} = 1$

Hyperbola

Center: $(0, 0)$

$a = 1, b = 2, c = \sqrt{5}$

Horizontal transverse axis

Vertices: $(\pm 1, 0)$

Foci: $(\pm\sqrt{5}, 0)$

Asymptotes: $y = \pm 2x$

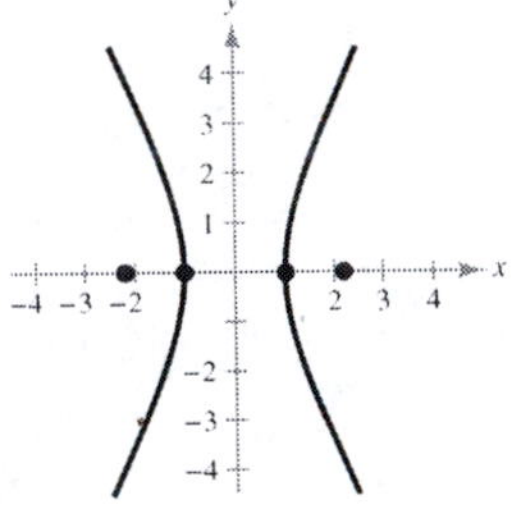

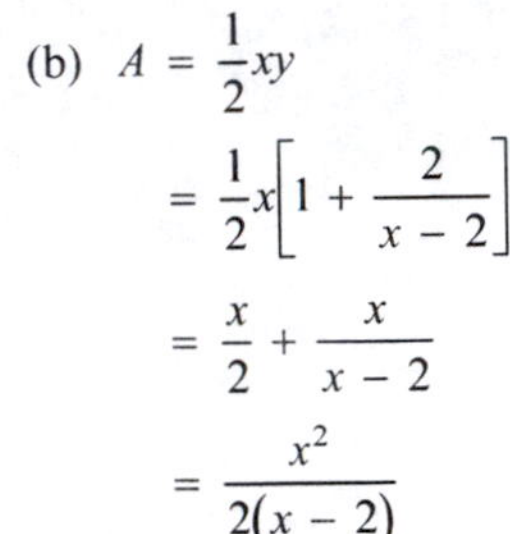

16. $\frac{y^2}{4} - x^2 = 1$

Hyperbola

Center: $(0, 0)$

$a = 2, b = 1, c = \sqrt{5}$

Vertical transverse axis

Vertices: $(0, \pm 2)$

Foci: $\left(0, \pm\sqrt{5}\right)$

Asymptotes: $y = \pm 2x$

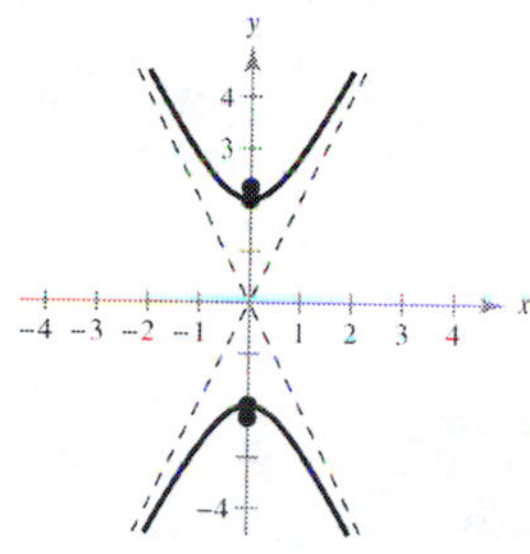

17.
$$x^2 + 3y^2 - 2x + 36y + 100 = 0$$
$$\left(x^2 - 2x\right) + 3\left(y^2 + 12y\right) = -100$$
$$\left(x^2 - 2x + 1\right) + 3\left(y^2 + 12y + 36\right) = -100 + 1 + 108$$
$$(x - 1)^2 + 3(y + 6)^2 = 9$$
$$\frac{(x - 1)^2}{9} + \frac{(y + 6)^2}{3} = 1$$

Ellipse

Center: $(1, -6)$

$a = 3, b = \sqrt{3}, c = \sqrt{6}$

Horizontal major axis

Vertices: $(-2, -6), (4, -6)$

Foci: $\left(1 \pm \sqrt{6}, -6\right)$

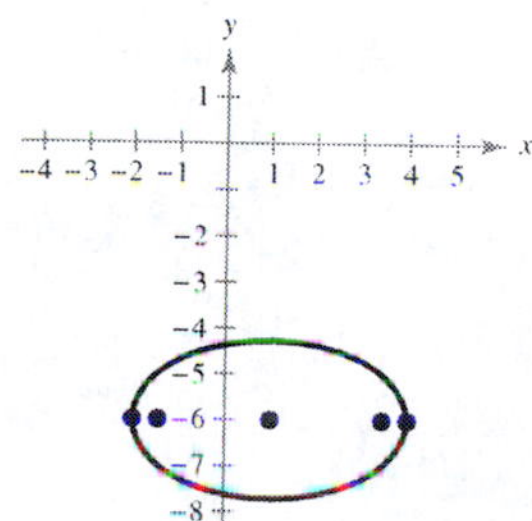

18. Ellipse

Vertices: $(0, 2)$ and $(8, 2)$

Center: $(4, 2)$

Horizontal major axis: $a = 4$

Minor axis of length 4: $2b = 4 \Rightarrow b = 2$

$$\frac{(x - h)^2}{a^2} + \frac{(y - k)^2}{b^2} = 1$$
$$\frac{(x - 4)^2}{16} + \frac{(y - 2)^2}{4} = 1$$

19. Hyperbola

Vertices: $(0, \pm 3)$

Center: $(0, 0)$

Vertical transverse axis: $a = 3$

Asymptotes: $y = \pm\frac{3}{2}x$

$$\pm\frac{a}{b} = \pm\frac{3}{2} \Rightarrow b = 2$$
$$\frac{(y - k)^2}{a^2} - \frac{(x - h)^2}{b^2} = 1$$
$$\frac{y^2}{9} - \frac{x^2}{4} = 1$$

20.

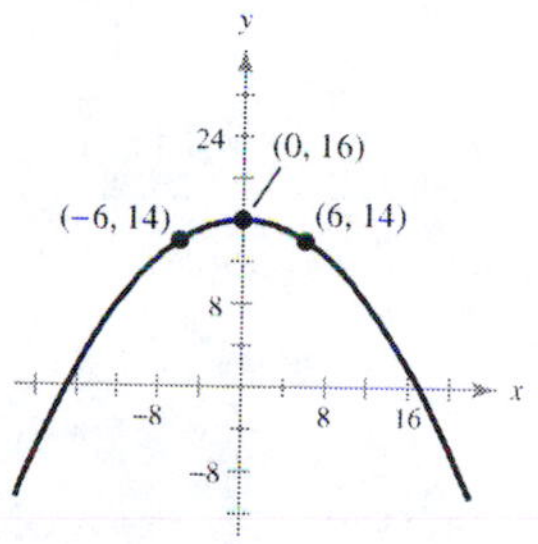

$$x^2 = 4p(y - 16)$$
$$36 = 4p(14 - 16)$$
$$-\tfrac{9}{2} = p$$
$$x^2 = -18(y - 16)$$

When $y = 0$: $x^2 = -18(-16) \Rightarrow x \approx 17 \Rightarrow 2x \approx 34$

At ground level, the archway is approximately 34 meters.

21. $a = \frac{1}{2}(768{,}800) = 384{,}400$

$b = \frac{1}{2}(767{,}640) = 383{,}820$

$c = \sqrt{384{,}400^2 - 383{,}820^2} \approx 21{,}108$

Smallest distance (perigee): $a - c \approx 363{,}292$ km

Greatest distance (apogee): $a + c \approx 405{,}508$ km

Chapter Test Solutions for Chapter 5

1. $4.2^{0.6} \approx 2.366$

2. $4^{3\pi/2} \approx 687.291$

3. $e^{-7/10} \approx 0.497$

4. $e^{3.1} \approx 22.198$

5. $f(x) = 10^{-x}$

x	-1	$-\frac{1}{2}$	0	$\frac{1}{2}$	1
$f(x)$	10	3.162	1	0.316	0.1

Horizontal asymptote: $y = 0$

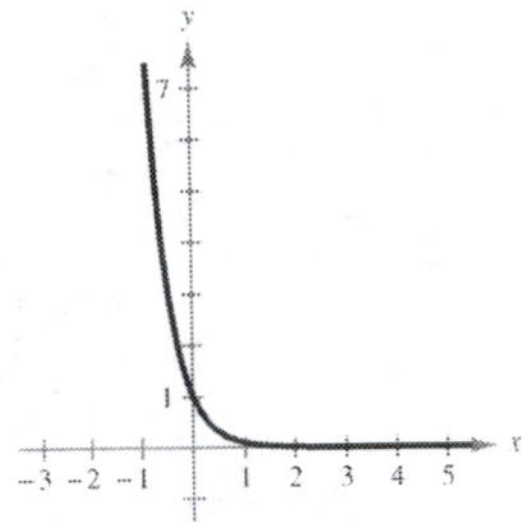

6. $f(x) = -6^{x-2}$

x	-1	0	1	2	3
$f(x)$	-0.005	-0.028	-0.167	-1	-6

Horizontal asymptote: $y = 0$

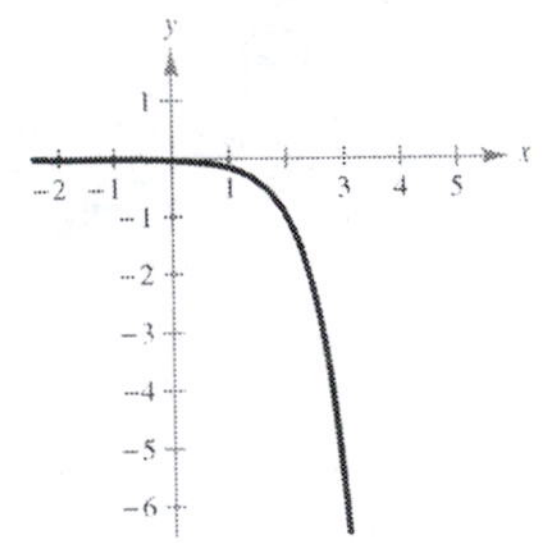

7. $f(x) = 1 - e^{2x}$

x	-1	$-\frac{1}{2}$	0	$\frac{1}{2}$	1
$f(x)$	0.865	0.632	0	-1.718	-6.389

Horizontal asymptote: $y = 1$

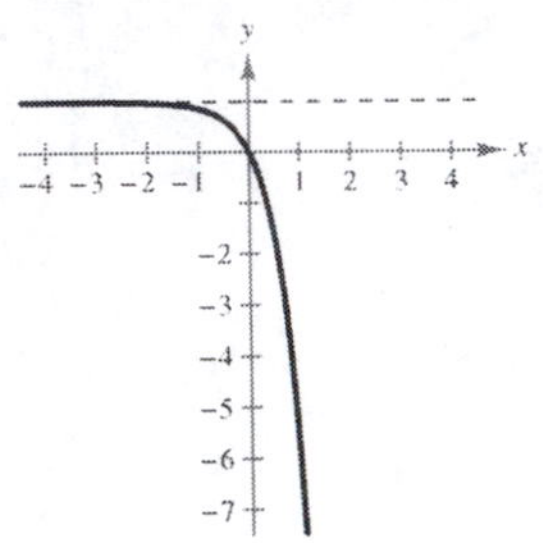

8. (a) $\log_7 7^{-0.89} = -0.89$

(b) $4.6 \ln e^2 = 4.6(2) = 9.2$

9. $f(x) = -\log x - 6$

Domain: $(0, \infty)$

Vertical asymptote: $x = 0$

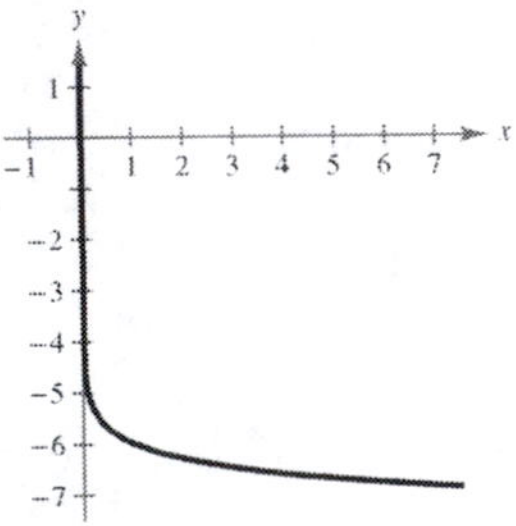

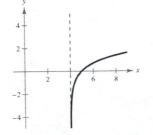

10. $f(x) = \ln(x - 4)$

Domain: $(4, \infty)$

Vertical asymptote: $x = 4$

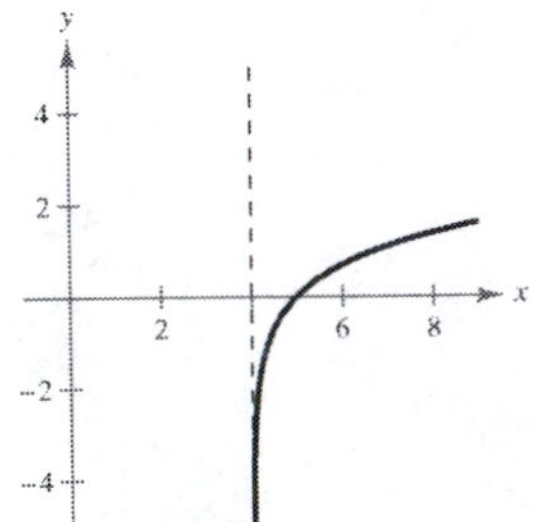

11. $f(x) = 1 + \ln(x + 6)$

Domain: $(-6, \infty)$

Vertical asymptote: $x = -6$

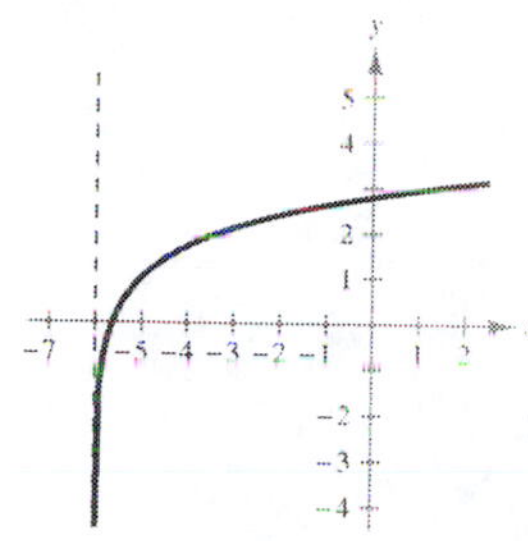

12. $\log_7 44 = \frac{\ln 44}{\ln 7} = \frac{\log 44}{\log 7} \approx 1.945$

13. $\log_{16} 0.63 = \frac{\log 0.63}{\log 16} \approx -0.167$

14. $\log_{3/4} 24 = \frac{\log 24}{\log(3/4)} \approx -11.047$

15. $\log_2 3a^4 = \log_2 3 + \log_2 a^4 = \log_2 3 + 4\log_2|a|$

16. $\ln\frac{5\sqrt{x}}{6} = \ln(5\sqrt{x}) - \ln 6$

$= \ln 5 + \ln\sqrt{x} - \ln 6$

$= \ln 5 + \frac{1}{2}\ln x - \ln 6$

17. $\log\frac{(x-1)^3}{y^2z} = \log(x-1)^3 - \log y^2z$

$= 3\log(x-1) - (\log y^2 + \log z)$

$= 3\log(x-1) - 2\log y - \log z$

18. $\log_3 13 + \log_3 y = \log_3 13y$

19. $4\ln x - 4\ln y = \ln x^4 - \ln y^4 = \ln\frac{x^4}{y^4}$

20. $3\ln x - \ln(x+3) + 2\ln y = \ln x^3 - \ln(x+3) + \ln y^2 = \ln\frac{x^3y^2}{x+3}$

21. $5^x = \frac{1}{25}$

$5^x = 5^{-2}$

$x = -2$

22. $3e^{-5x} = 132$

$e^{-5x} = 44$

$-5x = \ln 44$

$x = \frac{\ln 44}{-5} \approx -0.757$

23. $\frac{1025}{8 + e^{4x}} = 5$

$1025 = 5(8 + e^{4x})$

$205 = 8 + e^{4x}$

$197 = e^{4x}$

$\ln 197 = 4x$

$x = \frac{\ln 197}{4} \approx 1.321$

24. $\ln x = \frac{1}{2}$

$x = e^{1/2} \approx 1.649$

25. $18 + 4\ln x = 7$

$4\ln x = -11$

$\ln x = -\frac{11}{4}$

$x = e^{-11/4} \approx 0.0639$

26. $\log x + \log(x - 15) = 2$

$\log[x(x - 15)] = 2$

$x(x - 15) = 10^2$

$x^2 - 15x - 100 = 0$

$(x - 20)(x + 5) = 0$

$x - 20 = 0$ or $x + 5 = 0$

$x = 20$ $\quad x = -5$

The value $x = -5$ is extraneous. The only solution is $x = 20$.

27. $y = ae^{bt}$

$(0, 2745)$: $2745 = ae^{b(0)} \Rightarrow a = 2745$

$y = 2745e^{bt}$

$(9, 11{,}277)$: $11{,}277 = 2745e^{b(9)}$

$$\frac{11{,}277}{2745} = e^{9b}$$

$$\ln\left(\frac{11{,}277}{2745}\right) = 9b$$

$$\frac{1}{9}\ln\left(\frac{11{,}277}{2745}\right) = b \Rightarrow b \approx 0.1570$$

So, $y = 2745e^{0.1570t}$.

28. $y = ae^{bt}$

$$\frac{1}{2}a = ae^{b(21.77)}$$

$$\frac{1}{2} = e^{21.77b}$$

$$\ln\left(\frac{1}{2}\right) = 21.77b$$

$$b = \frac{\ln(1/2)}{21.77} \approx -0.0318$$

$$y = ae^{-0.0318t}$$

When $t = 19$: $y = ae^{-0.0318(19)} \approx 0.55a$

So, 55% will remain after 19 years.

29. $H = 70.228 + 5.104x + 9.222 \ln x, \frac{1}{4} \le x \le 6$

(a)

x	H (cm)
$\frac{1}{4}$	58.720
$\frac{1}{2}$	66.388
1	75.332
2	86.828
3	95.671
4	103.43
5	110.59
6	117.38

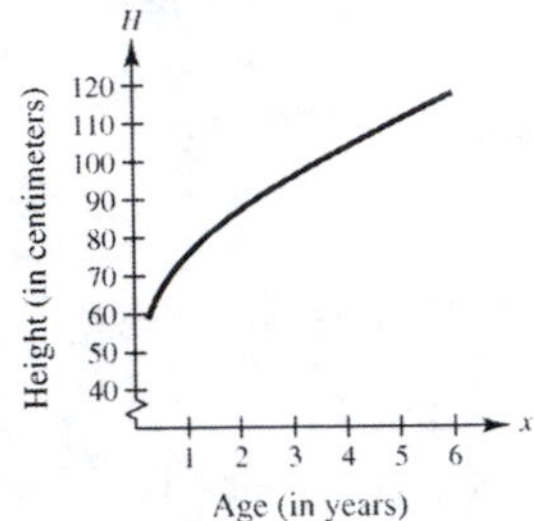

(b) Estimate: 103

When $x = 4$, $H \approx 103.43$ cm.

Cumulative Test Solutions for Chapters 3–5

1. Vertex: $(-8, 5)$

Point: $(-4, -7)$

$$y - k = a(x - h)^2$$

$$y - 5 = a(x + 8)^2$$

$$-7 - 5 = a(-4 + 8)^2$$

$$-12 = 16a$$

$$-\tfrac{3}{4} = a$$

$$y = -\tfrac{3}{4}(x + 8)^2 + 5$$

2. $h(x) = -(x^2 + 4x)$

$= -(x^2 + 4x + 4 - 4)$

$= -(x + 2)^2 + 4$

Parabola

Vertex: $(-2, 4)$

Intercepts: $(-4, 0), (0, 0)$

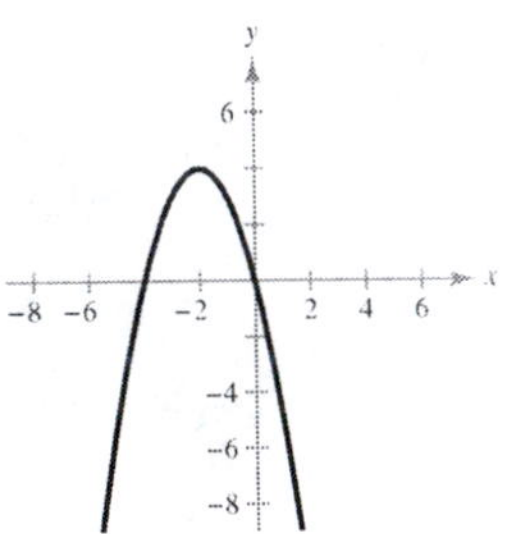

3. $f(t) = \frac{1}{4}t(t - 2)^2$

Cubic

Falls to the left

Rises to the right

Intercepts: $(0, 0), (2, 0)$

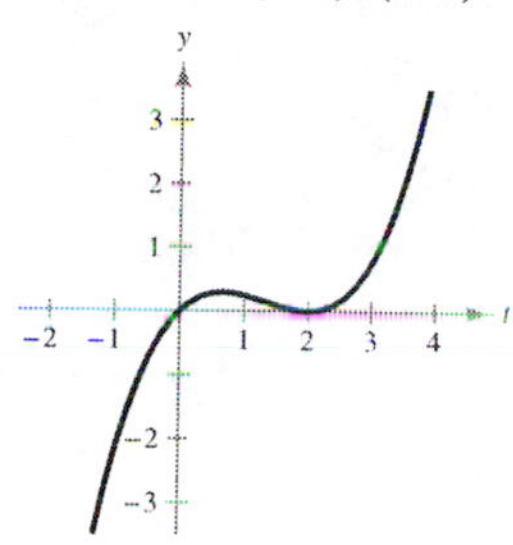

4. $g(s) = s^2 + 2s + 9$

$= (s^2 + 2s + 1) - 1 + 9$

$= (s + 1)^2 + 8$

Parabola

Vertex: $(-1, 8)$

Intercept: $(0, 9)$

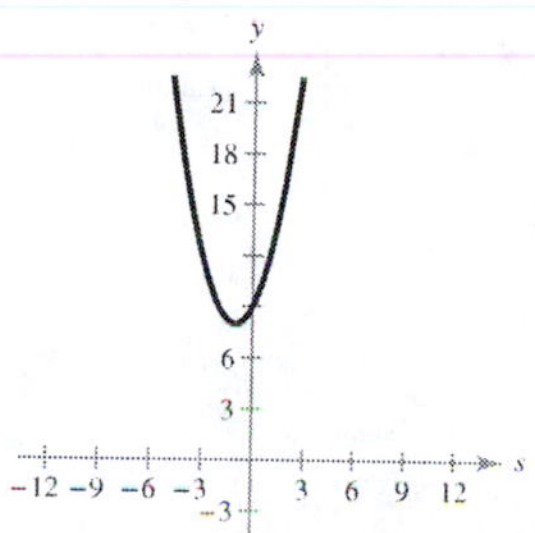

5. $f(x) = x^3 + 2x^2 + 4x + 8$

$= x^2(x + 2) + 4(x + 2)$

$= (x + 2)(x^2 + 4)$

$x + 2 = 0 \Rightarrow x = -2$

$x^2 + 4 = 0 \Rightarrow x = \pm 2i$

The zeros of $f(x)$ are -2 and $\pm 2i$.

6. $f(x) = x^4 + 4x^3 - 21x^2$

$= x^2(x^2 + 4x - 21)$

$= x^2(x + 7)(x - 3)$

The zeros of $f(x)$ are $0, -7,$ and 3.

7.
$$
\begin{array}{r}
3x - 2 + \dfrac{-3x + 2}{2x^2 + 1} \\
2x^2 + 0x + 1 \overline{\big) 6x^3 - 4x^2 + 0x + 0} \\
\underline{6x^3 + 0x^2 + 3x} \\
-4x^2 - 3x + 0 \\
\underline{-4x^2 + 0x - 2} \\
-3x + 2
\end{array}
$$

Thus, $\dfrac{6x^3 - 4x^2}{2x^2 + 1} = 3x - 2 - \dfrac{3x - 2}{2x^2 + 1}$.

8.
$$
\begin{array}{r|rrrrr}
2 & 3 & 0 & 2 & -5 & 3 \\
 & & 6 & 12 & 28 & 46 \\
\hline
 & 3 & 6 & 14 & 23 & 49
\end{array}
$$

Thus, $\dfrac{3x^4 + 2x^2 - 5x + 3}{x - 2} = 3x^3 + 6x^2 + 14x + 23 + \dfrac{49}{x - 2}$.

9. $g(x) = x^3 + 3x^2 - 6$

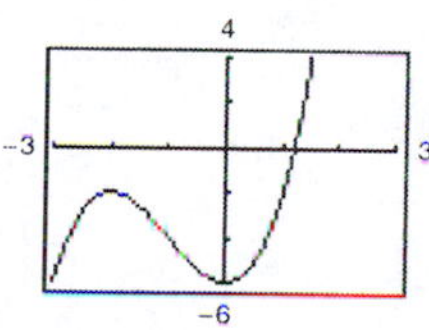

$x \approx 1.20$

10. Because $2 + \sqrt{3}i$ is a zero, so is $2 - \sqrt{3}i$.

$$\begin{aligned} f(x) &= (x + 5)(x + 2)\left[x - \left(2 + \sqrt{3}i\right)\right]\left[x - \left(2 - \sqrt{3}i\right)\right] \\ &= \left(x^2 + 7x + 10\right)\left[(x - 2) - \sqrt{3}i\right]\left[(x - 2) + \sqrt{3}i\right] \\ &= \left(x^2 + 7x + 10\right)\left[(x - 2)^2 + 3\right] \\ &= \left(x^2 + 7x + 10\right)\left(x^2 - 4x + 7\right) \\ &= x^4 + 3x^3 - 11x^2 + 9x + 70 \end{aligned}$$

11. $f(x) = \dfrac{2x}{x - 3}$

Domain: all real numbers x except $x = 3$

Vertical asymptote: $x = 3$

Horizontal asymptote: $y = 2$

Intercept: $(0, 0)$

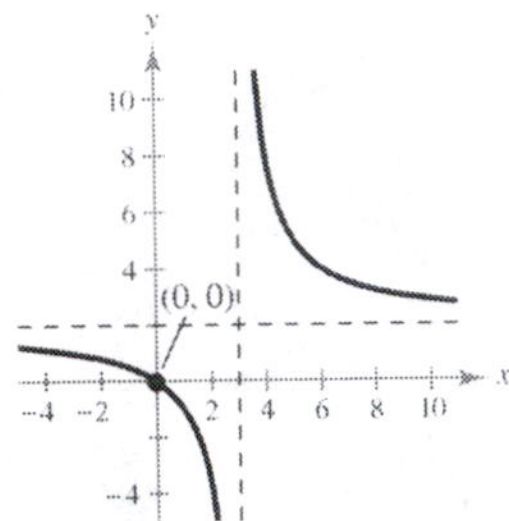

12. $f(x) = \dfrac{4x^2}{x - 5} = 4x + 20 + \dfrac{100}{x - 5}$

Domain: all real numbers x except $x = 5$

Vertical asymptote: $x = 5$

Slant asymptote: $y = 4x + 20$

Intercept: $(0, 0)$

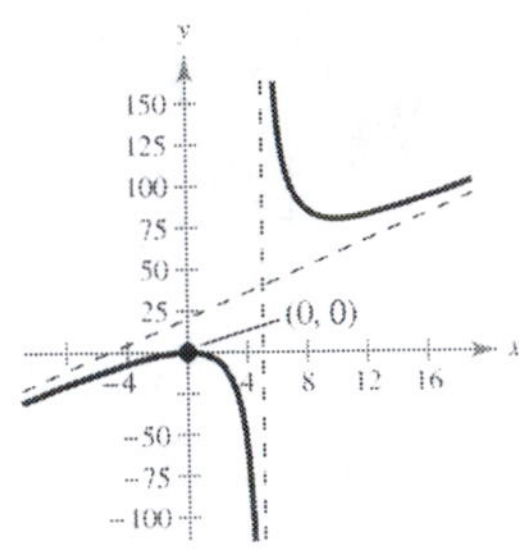

13. $f(x) = \dfrac{2x}{x^2 + 2x - 3}$

$= \dfrac{2x}{(x + 3)(x - 1)}$

Intercept: $(0, 0)$

Vertical asymptotes: $x = -3, x = 1$

Horizontal asymptote: $y = 0$

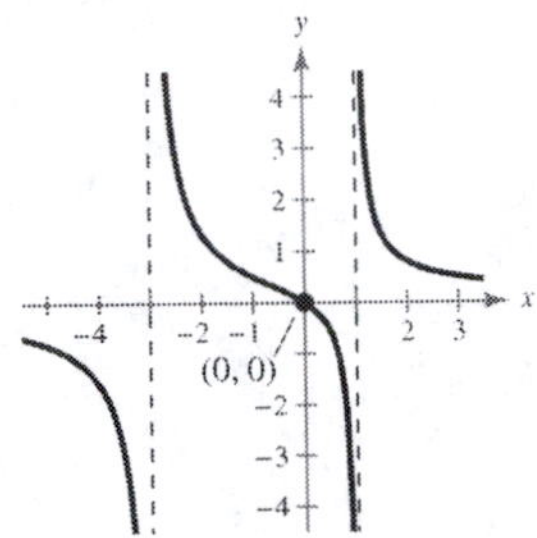

14. $f(x) = \dfrac{x^2 - 4}{x^2 + x - 2}$

$= \dfrac{(x + 2)(x - 2)}{(x + 2)(x - 1)}$

$= \dfrac{x - 2}{x - 1}, x \neq -2$

Vertical asymptote: $x = 1$

Horizontal asymptote: $y = 1$

x-intercept: $(2, 0)$

y-intercept: $(0, 2)$

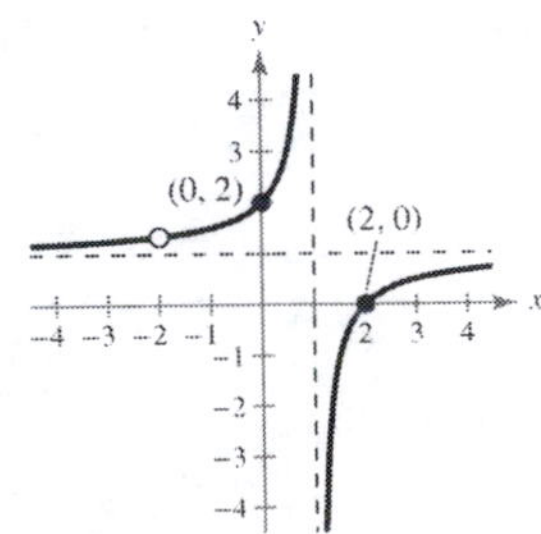

15. $$f(x) = \frac{x^3 - 2x^2 - 9x + 18}{x^2 + 4x + 3}$$
$$= \frac{x^2(x-2) - 9(x-2)}{(x+1)(x+3)}$$
$$= \frac{(x-2)(x^2-9)}{(x+1)(x+3)}$$
$$= \frac{(x-2)(x+3)(x-3)}{(x+1)(x+3)}$$
$$= \frac{(x-2)(x-3)}{x+1}$$
$$= \frac{x^2 - 5x + 6}{x+1}$$
$$= x - 6 + \frac{12}{x+1},\ x \neq -3$$

Vertical asymptote: $x = -1$

Slant asymptote: $y = x - 6$

x-intercepts: $(2, 0), (3, 0)$

y-intercept: $(0, 6)$

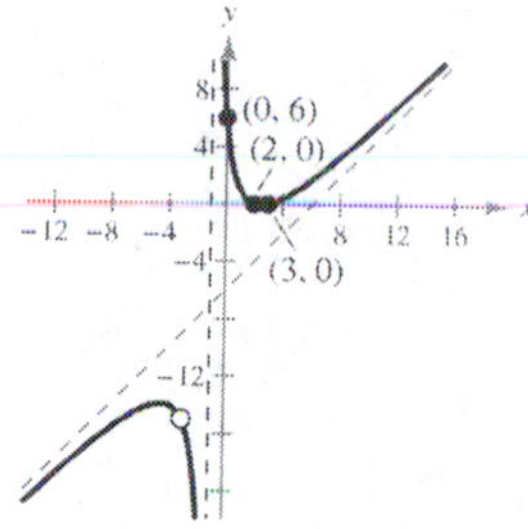

16. $$\frac{(x+3)^2}{16} - \frac{(y+4)^2}{25} = 1$$

Hyperbola

Center: $(-3, -4)$

Vertices: $(-7, -4), (1, -4)$

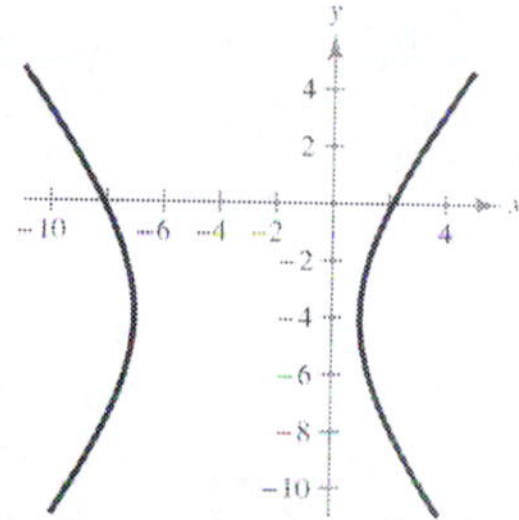

17. $$\frac{(x-2)^2}{4} + \frac{(y+1)^2}{9} = 1$$

Ellipse

Center: $(2, -1)$

Vertices: $(2, -4), (2, 2)$

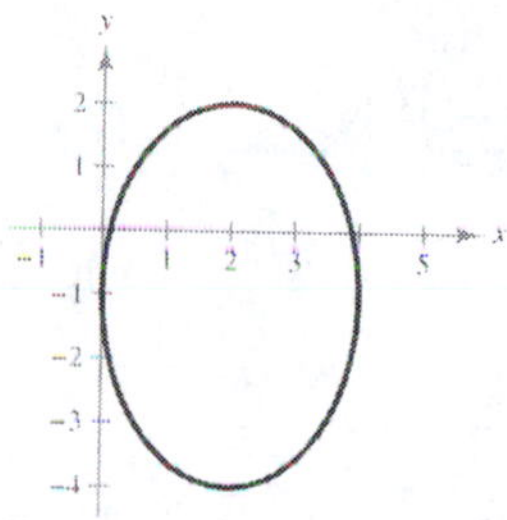

18. Parabola

Vertex: $(3, -2) \Rightarrow y = a(x-3)^2 - 2$

Point: $(0, 4) \Rightarrow 4 = a(0-3)^2 - 2$

$6 = 9a \Rightarrow a = \frac{2}{3}$

Equation: $y = \frac{2}{3}(x-3)^2 - 2$

$$y + 2 = \frac{2}{3}(x-3)^2$$
$$\frac{3}{2}(y+2) = (x-3)^2$$
$$(x-3)^2 = \frac{3}{2}(y+2)$$

19. Hyperbola

Foci: $(0, 0)$ and $(0, 4) \Rightarrow$ Center: $(0, 2)$ and vertical transverse axis

Asymptotes: $y = \pm\frac{1}{2}x + 2 \Rightarrow \frac{a}{b} = \frac{1}{2} \Rightarrow 2a = b$

$c^2 = a^2 + b^2 \Rightarrow 4 = a^2 + 4a^2 \Rightarrow a^2 = \frac{4}{5}$ and $b^2 = \frac{16}{5}$

Equation: $\frac{(y-2)^2}{4/5} - \frac{x^2}{16/5} = 1$

20. $f(x) = \left(\frac{2}{5}\right)^x$

$g(x) = -\left(\frac{2}{5}\right)^{-x+3}$

g is a reflection in the x-axis, a reflection in the y-axis, and a horizontal shift three units to the right of the graph of f.

21. $f(x) = 2.2^x$

$g(x) = -2.2^x + 4$

g is a reflection in the x-axis, and a vertical shift four units upward of the graph of f.

22. $\log 98 \approx 1.991$

23. $\log\left(\frac{6}{7}\right) \approx -0.067$

24. $\ln\sqrt{31} \approx 1.717$

25. $\ln\left(\sqrt{30} - 4\right) \approx 0.390$

26. $\log_5 4.3 = \frac{\log_{10} 4.3}{\log_{10} 5} = \frac{\ln 4.3}{\ln 5} \approx 0.906$

27. $\log_3 0.149 = \frac{\log_{10} 0.149}{\log_{10} 3} = \frac{\ln 0.149}{\ln 3} \approx -1.733$

28. $\log_{1/2} 17 = \frac{\log_{10} 17}{\log_{10}(1/2)} = \frac{\ln 17}{\ln(1/2)} \approx -4.087$

29.
$$\begin{aligned}\ln\left(\frac{x^2-16}{x^4}\right) &= \ln(x^2 - 16) - \ln x^4\\ &= \ln(x+4)(x-4) - 4\ln x\\ &= \ln(x+4) + \ln(x-4) - 4\ln x,\ x > 4\end{aligned}$$

30.
$$\begin{aligned}2\ln x - \frac{1}{2}\ln(x+5) &= \ln x^2 - \ln\sqrt{x+5}\\ &= \ln\frac{x^2}{\sqrt{x+5}},\ x > 0\end{aligned}$$

31.
$$\begin{aligned}6e^{2x} &= 72\\ e^{2x} &= 12\\ 2x &= \ln 12\\ x &= \frac{\ln 12}{2} \approx 1.242\end{aligned}$$

32.
$$\begin{aligned}4^{x-5} + 21 &= 30\\ 4^{x-5} &= 9\\ x - 5 &= \log_4 9\\ x &= 5 + \log_4 9\\ x &= 5 + \frac{\ln 9}{\ln 4}\\ x &\approx 6.585\end{aligned}$$

33.
$$\begin{aligned}e^{2x} - 13e^x + 42 &= 0\\ (e^x - 6)(e^x - 7) &= 0\end{aligned}$$

$e^x - 6 = 0 \Rightarrow e^x = 6 \Rightarrow x = \ln 6 \approx 1.792$

$e^x - 7 = 0 \Rightarrow e^x = 7 \Rightarrow x = \ln 7 \approx 1.946$

34.
$$\begin{aligned}\log_2 x + \log_2 5 &= 6\\ \log_2 5x &= 6\\ 5x &= 2^6\\ x &= \tfrac{64}{5} = 12.8\end{aligned}$$

35. $\ln 4x - \ln 2 = 8$

$$\ln \frac{4x}{2} = 8$$

$$\ln 2x = 8$$

$$2x = e^8$$

$$x = \frac{e^8}{2} \approx 1490.479$$

36. $\ln \sqrt{x + 2} = 3$

$$\tfrac{1}{2}\ln(x + 2) = 3$$

$$\ln(x + 2) = 6$$

$$x + 2 = e^6$$

$$x = e^6 - 2 \approx 401.429$$

37. $f(x) = \dfrac{1000}{1 + 4e^{-0.2x}}$

Horizontal asymptotes: $y = 0,\ y = 1000$

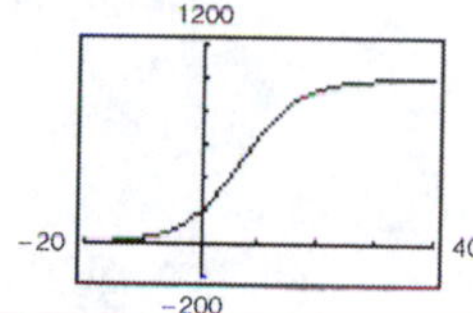

38. (a) and (c)

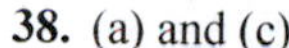

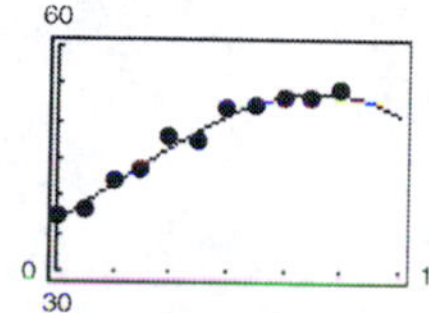

The model is a good fit for the data.

(b) $S = -0.0172t^3 + 0.119t^2 + 2.22t + 36.8$

(d) For 2018, use $t = 18$: $S(18) \approx \$15.0$ million.

No, this does not seem reasonable. You would expect the sales to continue to increase, but after $t = 18$ the model predicts sales will decrease.

39. $A = 2500e^{(0.075)(25)} \approx \$16{,}302.05$

40. $N = 175e^{kt}$

$$420 = 175e^{k(8)}$$

$$2.4 = e^{8k}$$

$$\ln 2.4 = 8k$$

$$\frac{\ln 2.4}{8} = k$$

$$k \approx 0.1094$$

$$N = 175e^{0.1094t}$$

$$350 = 175e^{0.1094t}$$

$$2 = e^{0.1094t}$$

$$\ln 2 = 0.1094t$$

$$t = \frac{\ln 2}{0.1094} \approx 6.3 \text{ hours to double}$$

41. Let $P = 30$ and solve for t.

$$30 = 20.871e^{0.0188t}$$

$$1.437 \approx e^{0.0188t}$$

$$\ln 1.437 \approx \ln e^{0.0188t}$$

$$0.3626 \approx 0.0188t$$

$$19.3 \approx t$$

According to the model, the population of Texas will reach 30 million during 2019.

42. $p = \dfrac{1200}{1 + 3e^{-t/5}}$

(a) $p(0) = \dfrac{1200}{1 + 3e^0} = \dfrac{1200}{4} = 300$ birds

(b) $p(5) = \dfrac{1200}{1 + 3e^{-1}} \approx 570$ birds

(c) $800 = \dfrac{1200}{1 + 3e^{-t/5}}$

$$800\left(1 + 3e^{-t/5}\right) = 1200$$

$$1 + 3e^{-t/5} = 1.5$$

$$3e^{-t/5} = 0.5$$

$$e^{-t/5} = \frac{1}{6}$$

$$-\frac{t}{5} = \ln\left(\frac{1}{6}\right)$$

$$t = -5\ln\left(\frac{1}{6}\right) \approx 9 \text{ years}$$

Chapter Test Solutions for Chapter 6

1. $\theta = \dfrac{5\pi}{4}$

(a)

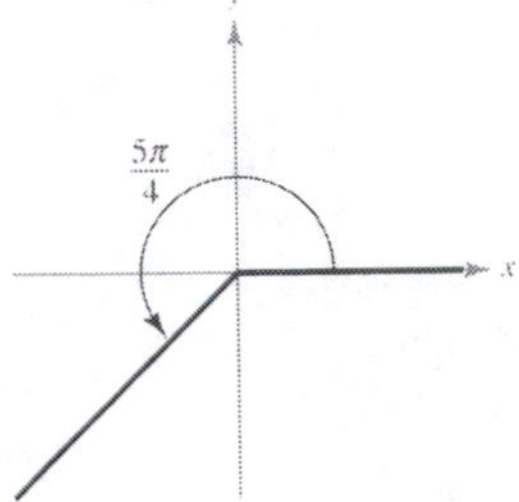

(b) $\dfrac{5\pi}{4} + 2\pi = \dfrac{13\pi}{4}$

$\dfrac{5\pi}{4} - 2\pi = -\dfrac{3\pi}{4}$

(c) $\dfrac{5\pi}{4}\left(\dfrac{180°}{\pi}\right) = 225°$

2. $\dfrac{105 \text{ km}}{\text{hr}} \times \dfrac{1 \text{ hr}}{60 \text{ min}} = 1.75 \text{ km per min}$

diameter $= 1$ meter $= 0.001$ km

radius $= \dfrac{1}{2}$ diameter $= 0.0005$ km

$$\begin{aligned}\text{Angular speed} &= \frac{\theta}{t}\\ &= \frac{1.75}{2\pi(0.0005)} \cdot 2\pi\\ &= 3500 \text{ radians per minute}\end{aligned}$$

3. $130° = \dfrac{130\pi}{180} = \dfrac{13\pi}{18}$ radians

$A = \dfrac{1}{2}r^2\theta = \dfrac{1}{2}(25)^2\left(\dfrac{13\pi}{18}\right) \approx 709.04$ square feet

4. $x = -2,\ y = 6$

$r = \sqrt{(-2)^2 + (6)^2} = 2\sqrt{10}$

$\sin\theta = \dfrac{y}{r} = \dfrac{6}{2\sqrt{10}} = \dfrac{3}{\sqrt{10}} = \dfrac{3\sqrt{10}}{10}$

$\cos\theta = \dfrac{x}{r} = \dfrac{-2}{2\sqrt{10}} = -\dfrac{1}{\sqrt{10}} = -\dfrac{\sqrt{10}}{10}$

$\tan\theta = \dfrac{y}{x} = \dfrac{6}{-2} = -3$

$\csc\theta = \dfrac{r}{y} = \dfrac{2\sqrt{10}}{6} = \dfrac{\sqrt{10}}{3}$

$\sec\theta = \dfrac{r}{x} = \dfrac{2\sqrt{10}}{-2} = -\sqrt{10}$

$\cot\theta = \dfrac{x}{y} = \dfrac{-2}{6} = -\dfrac{1}{3}$

5.

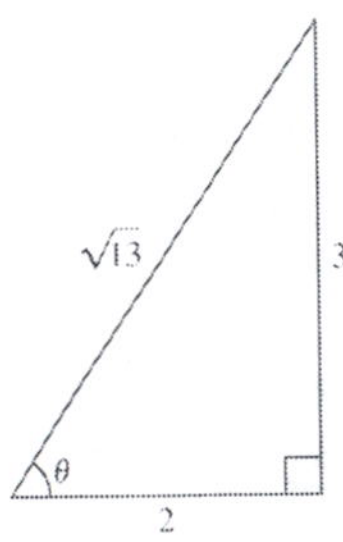

For $0 \le \theta < \dfrac{\pi}{2}$:

$\sin\theta = \dfrac{\text{opp}}{\text{hyp}} = \dfrac{3}{\sqrt{13}} = \dfrac{3\sqrt{13}}{13}$

$\cos\theta = \dfrac{\text{adj}}{\text{hyp}} = \dfrac{2}{\sqrt{13}} = \dfrac{2\sqrt{13}}{13}$

$\csc\theta = \dfrac{\text{hyp}}{\text{opp}} = \dfrac{\sqrt{13}}{3}$

$\sec\theta = \dfrac{\text{hyp}}{\text{adj}} = \dfrac{\sqrt{13}}{2}$

$\cot\theta = \dfrac{\text{adj}}{\text{opp}} = \dfrac{2}{3}$

For $\pi \le \theta < \dfrac{3\pi}{2}$:

$\sin\theta = -\dfrac{3\sqrt{13}}{13}$

$\cos\theta = -\dfrac{2\sqrt{13}}{13}$

$\csc\theta = -\dfrac{\sqrt{13}}{3}$

$\sec\theta = -\dfrac{\sqrt{13}}{2}$

$\cot\theta = \dfrac{2}{3}$

6. $\theta = 205°$

$\theta' = 205° - 180° = 25°$

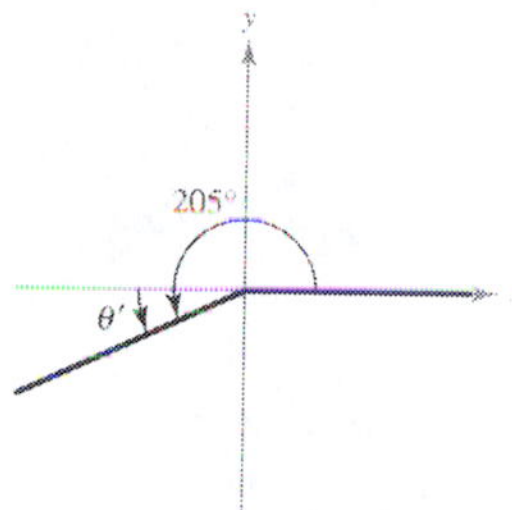

7. $\sec\theta < 0$ and $\tan\theta > 0$

$\frac{r}{x} < 0$ and $\frac{y}{x} > 0$

Quadrant III

8. $\cos\theta = -\frac{\sqrt{3}}{2}$

Reference angle is 30° and θ is in Quadrant II or III.

$\theta = 150°$ or $210°$

9. $\csc\theta = 1.030$

$\frac{1}{\sin\theta} = 1.030$

$\sin\theta = \frac{1}{1.030}$

$\theta = \arcsin\frac{1}{1.030}$

$\theta \approx 1.33$ and $\pi - 1.33 \approx 1.81$

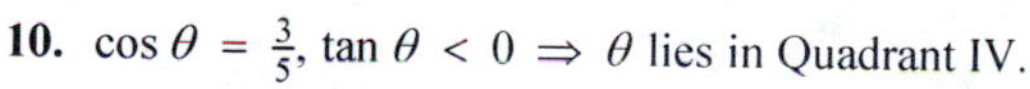

10. $\cos\theta = \frac{3}{5}$, $\tan\theta < 0 \Rightarrow \theta$ lies in Quadrant IV.

Let $x = 3, r = 5 \Rightarrow y = -4$.

$\sin\theta = -\frac{4}{5}$

$\cos\theta = \frac{3}{5}$

$\tan\theta = -\frac{4}{3}$

$\csc\theta = -\frac{5}{4}$

$\sec\theta = \frac{5}{3}$

$\cot\theta = -\frac{3}{4}$

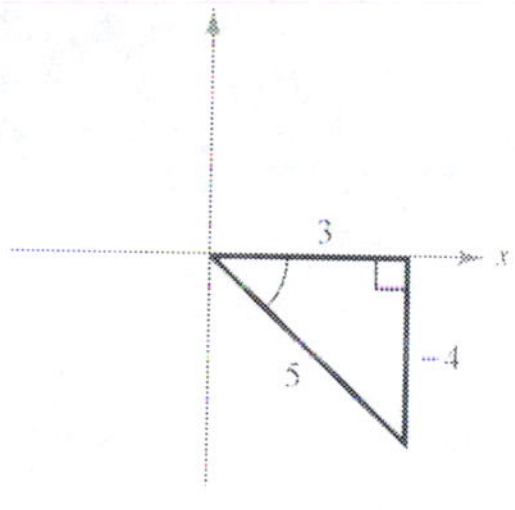

11. $\sec\theta = -\frac{29}{20}$, $\sin\theta > 0 \Rightarrow \theta$ lies in Quadrant II.

Let $r = 29$, $x = -20 \Rightarrow y = 21$.

$\sin\theta = \frac{21}{29}$

$\cos\theta = -\frac{20}{29}$

$\tan\theta = -\frac{21}{20}$

$\csc\theta = \frac{29}{21}$

$\cot\theta = -\frac{20}{21}$

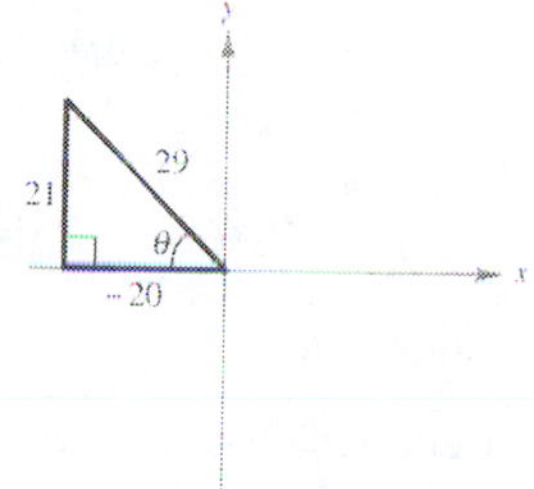

12. $g(x) = -2\sin\left(x - \frac{\pi}{4}\right)$

Period: 2π

Amplitude: $|-2| = 2$

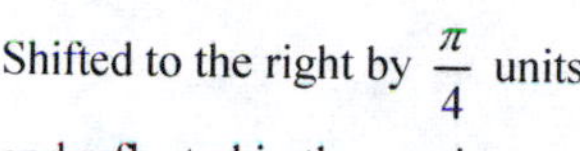

Shifted to the right by $\frac{\pi}{4}$ units

and reflected in the x-axis.

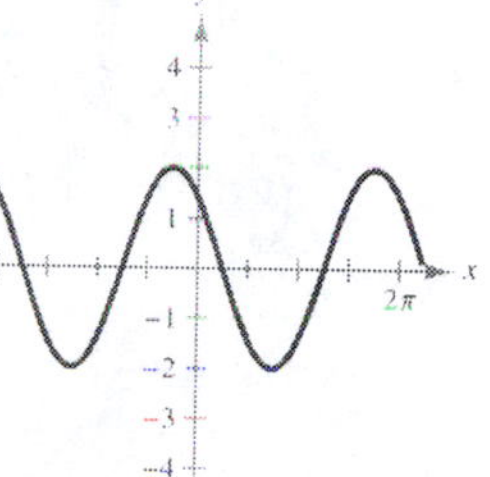

x	0	$\frac{\pi}{4}$	$\frac{3\pi}{4}$	$\frac{5\pi}{4}$	$\frac{7\pi}{4}$
y	$\sqrt{2}$	0	−2	0	2

13. $f(\alpha) = \frac{1}{2}\tan 2\alpha$

Period: $\frac{\pi}{2}$

Asymptotes:

$x = -\frac{\pi}{4}, x = \frac{\pi}{4}$

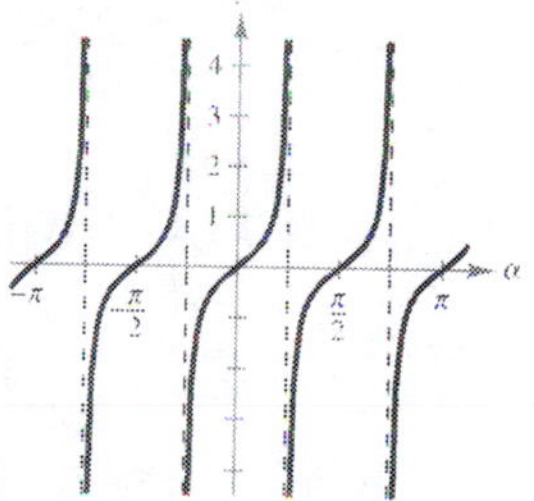

α	$-\frac{\pi}{8}$	0	$\frac{\pi}{8}$
$f(\alpha)$	$-\frac{1}{2}$	0	$\frac{1}{2}$

14. $y = \sin 2\pi x + 2\cos \pi x$

Periodic: period $= 2$

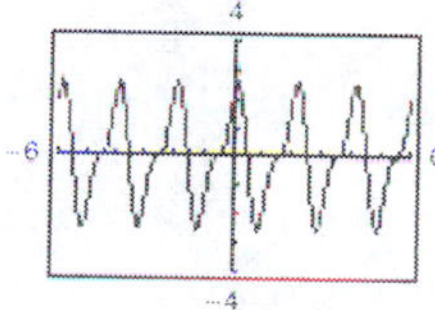

15. $y = 6e^{-0.12t}\cos(0.25t),\ 0 \le t \le 32$

Not periodic

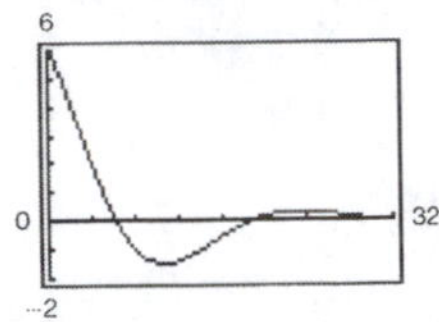

16. $f(x) = a\sin(bx + c)$

Amplitude: $2 \Rightarrow |a| = 2$

Reflected in the x-axis: $a = -2$

Period: $4\pi = \dfrac{2\pi}{b} \Rightarrow b = \dfrac{1}{2}$

Phase shift: $\dfrac{c}{b} = -\dfrac{\pi}{2} \Rightarrow c = -\dfrac{\pi}{4}$

$$f(x) = -2\sin\left(\frac{x}{2} - \frac{\pi}{4}\right)$$

17. $\cot\left(\arcsin\dfrac{3}{8}\right)$

Let $y = \arcsin\dfrac{3}{8}$. Then $\sin y = \dfrac{3}{8}$ and

$$\cot\left(\arcsin\frac{3}{8}\right) = \cot y = \frac{\sqrt{55}}{3}.$$

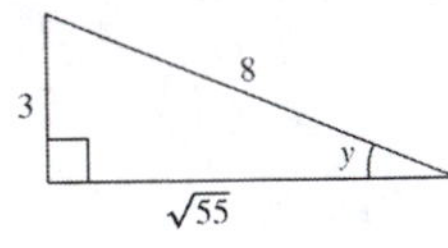

18. $f(x) = 2\arcsin\left(\frac{1}{2}x\right)$

Domain: $[-2, 2]$

Range: $[-\pi, \pi]$

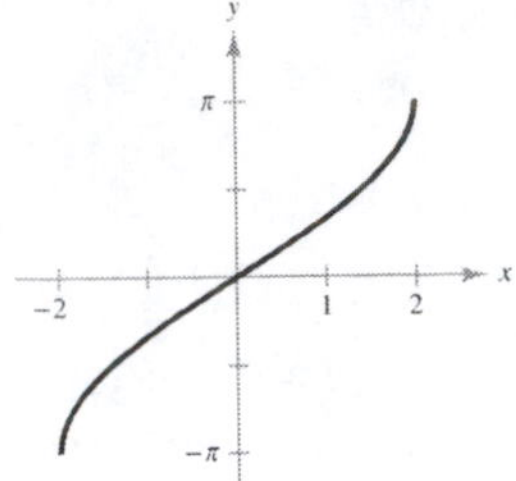

19. $\tan\theta = -\dfrac{110}{90}$

$$\theta = \arctan\left(-\frac{110}{90}\right)$$

$\theta \approx -50.7$

$\theta \approx 309.3°$

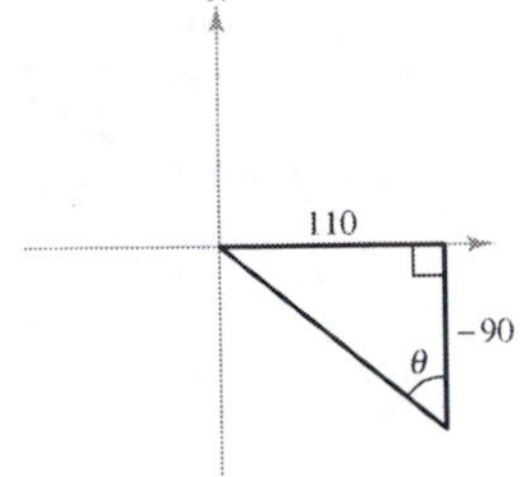

20. $d = a\cos bt$

$a = -6$

$\dfrac{2\pi}{b} = 2 \Rightarrow b = \pi$

$d = -6\cos\pi t$

Chapter Test Solutions for Chapter 7

1. $\tan\theta = \frac{6}{5},\ \cos\theta < 0$

θ is in Quadrant III.

$$\sec\theta = -\sqrt{1 + \tan^2\theta} = -\sqrt{1 + \left(\frac{6}{5}\right)^2} = -\frac{\sqrt{61}}{5}$$

$$\cos\theta = \frac{1}{\sec\theta} = -\frac{5}{\sqrt{61}} = -\frac{5\sqrt{61}}{61}$$

$$\sin\theta = \tan\theta\cos\theta = \left(\frac{6}{5}\right)\left(-\frac{5\sqrt{61}}{61}\right) = -\frac{6\sqrt{61}}{61}$$

$$\csc\theta = \frac{1}{\sin\theta} = -\frac{\sqrt{61}}{6}$$

$$\cot\theta = \frac{1}{\tan\theta} = \frac{5}{6}$$

2. $\csc^2\beta(1 - \cos^2\beta) = \dfrac{1}{\sin^2\beta}(\sin^2\beta) = 1$

3. $\dfrac{\sec^4 x - \tan^4 x}{\sec^2 x + \tan^2 x} = \dfrac{(\sec^2 x + \tan^2 x)(\sec^2 x - \tan^2 x)}{\sec^2 x + \tan^2 x}$

$= \sec^2 x - \tan^2 x = 1$

4. $\dfrac{\cos\theta}{\sin\theta} + \dfrac{\sin\theta}{\cos\theta} = \dfrac{\cos^2\theta + \sin^2\theta}{\sin\theta\cos\theta} = \dfrac{1}{\sin\theta\cos\theta}$

$= \csc\theta\sec\theta$

5. $y = \tan\theta,\ y = -\sqrt{\sec^2\theta - 1}$

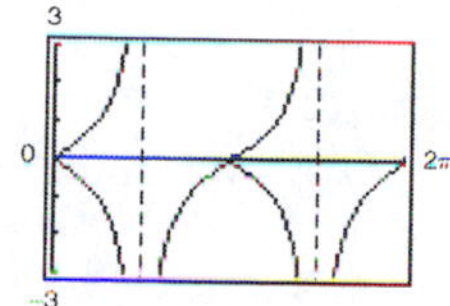

$\tan\theta = -\sqrt{\sec^2\theta - 1}$ on

$\theta = 0, \dfrac{\pi}{2} < \theta \le \pi, \dfrac{3\pi}{2} < \theta < 2\pi.$

6. $y_1 = \cos x + \sin x \tan x,\ y_2 = \sec x$

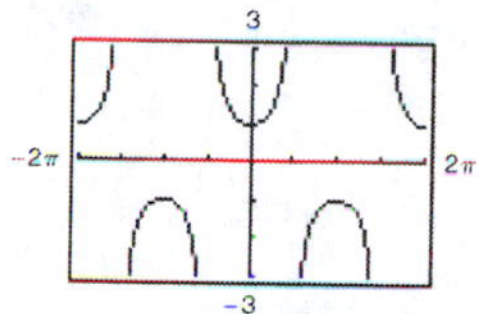

It appears that $y_1 = y_2$.

$$\begin{aligned}\cos x + \sin x \tan x &= \cos x + \sin x \frac{\sin x}{\cos x}\\ &= \cos x + \frac{\sin^2 x}{\cos x}\\ &= \frac{\cos^2 x + \sin^2 x}{\cos x}\\ &= \frac{1}{\cos x} = \sec x\end{aligned}$$

7. $\sin\theta\sec\theta = \sin\theta\dfrac{1}{\cos\theta} = \dfrac{\sin\theta}{\cos\theta} = \tan\theta$

8. $\sec^2 x\tan^2 x + \sec^2 x = \sec^2 x(\sec^2 x - 1) + \sec^2 x = \sec^4 x - \sec^2 x + \sec^2 x = \sec^4 x$

9.
$$\begin{aligned}\frac{\csc\alpha + \sec\alpha}{\sin\alpha + \cos\alpha} &= \frac{\frac{1}{\sin\alpha} + \frac{1}{\cos\alpha}}{\sin\alpha + \cos\alpha} = \frac{\frac{\cos\alpha + \sin\alpha}{\sin\alpha\cos\alpha}}{\sin\alpha + \cos\alpha} = \frac{1}{\sin\alpha\cos\alpha}\\ &= \frac{\cos^2\alpha + \sin^2\alpha}{\sin\alpha\cos\alpha} = \frac{\cos^2\alpha}{\sin\alpha\cos\alpha} + \frac{\sin^2\alpha}{\sin\alpha\cos\alpha}\\ &= \frac{\cos\alpha}{\sin\alpha} + \frac{\sin\alpha}{\cos\alpha} = \cot\alpha + \tan\alpha\end{aligned}$$

10. $\tan\left(x + \dfrac{\pi}{2}\right) = \tan\left(\dfrac{\pi}{2} - (-x)\right) = \cot(-x) = -\cot x$

11. $\sin(n\pi + \theta) = (-1)^n \sin\theta$, n is an integer.

For n odd:

$$\begin{aligned}\sin(n\pi + \theta) &= \sin n\pi\cos\theta + \cos n\pi\sin\theta\\ &= (0)\cos\theta + (-1)\sin\theta = -\sin\theta\end{aligned}$$

For n even:

$$\begin{aligned}\sin(n\pi + \theta) &= \sin n\pi\cos\theta + \cos n\pi\sin\theta\\ &= (0)\cos\theta + (1)\sin\theta = \sin\theta\end{aligned}$$

When n is odd, $(-1)^n = -1$. When n is even $(-1)^n = 1$.

So, $\sin(n\pi + \theta) = (-1)^n\sin\theta$ for any integer n.

12.
$$\begin{aligned}(\sin x + \cos x)^2 &= \sin^2 x + 2\sin x\cos x + \cos^2 x\\ &= 1 + 2\sin x\cos x\\ &= 1 + \sin 2x\end{aligned}$$

13.
$$\begin{aligned}\sin^4\frac{x}{2} &= \left(\sin^2\frac{x}{2}\right)^2\\ &= \left(\frac{1 - \cos 2\left(\frac{x}{2}\right)}{2}\right)^2\\ &= \left(\frac{1 - \cos x}{2}\right)^2\\ &= \frac{1}{4}\left(1 - 2\cos x + \cos^2 x\right)\\ &= \frac{1}{4}\left(1 - 2\cos x + \frac{1 + \cos 2x}{2}\right)\\ &= \frac{1}{8}(3 - 4\cos x + \cos 2x)\end{aligned}$$

14. $\dfrac{\sin 4\theta}{1 + \cos 4\theta} = \tan\dfrac{4\theta}{2} = \tan 2\theta$

15. $4\sin 3\theta\cos 2\theta = 4\cdot\frac{1}{2}\left[\sin(3\theta + 2\theta) + \sin(3\theta - 2\theta)\right]$
$= 2(\sin 5\theta + \sin\theta)$

16. $\cos 3\theta - \cos\theta = -2\sin\left(\dfrac{3\theta + \theta}{2}\right)\sin\left(\dfrac{3\theta - \theta}{2}\right)$
$= -2\sin 2\theta\sin\theta$

17. $\tan^2 x + \tan x = 0$

$\tan x(\tan x + 1) = 0$

$\tan x = 0$ or $\tan x + 1 = 0$

$x = 0, \pi$ $\quad$ $\tan x = -1$

$x = \dfrac{3\pi}{4}, \dfrac{7\pi}{4}$

18. $\sin 2\alpha - \cos\alpha = 0$

$2\sin\alpha\cos\alpha - \cos\alpha = 0$

$\cos\alpha(2\sin\alpha - 1) = 0$

$\cos\alpha = 0$ or $2\sin\alpha - 1 = 0$

$\alpha = \dfrac{\pi}{2}, \dfrac{3\pi}{2}$ $\quad$ $\sin\alpha = \dfrac{1}{2}$

$\alpha = \dfrac{\pi}{6}, \dfrac{5\pi}{6}$

19. $4\cos^2 x - 3 = 0$

$\cos^2 x = \dfrac{3}{4}$

$\cos x = \pm\sqrt{\dfrac{3}{4}} = \pm\dfrac{\sqrt{3}}{2}$

$x = \dfrac{\pi}{6}, \dfrac{5\pi}{6}, \dfrac{7\pi}{6}, \dfrac{11\pi}{6}$

20. $\csc^2 x - \csc x - 2 = 0$

$(\csc x - 2)(\csc x + 1) = 0$

$\csc x - 2 = 0$ or $\csc x + 1 = 0$

$\csc x = 2$ $\quad$ $\csc x = -1$

$\dfrac{1}{\sin x} = 2$ $\quad$ $\dfrac{1}{\sin x} = -1$

$\sin x = \dfrac{1}{2}$ $\quad$ $\sin x = -1$

$x = \dfrac{\pi}{6}, \dfrac{5\pi}{6}$ $\quad$ $x = \dfrac{3\pi}{2}$

21. $5\sin x - x = 0$

$x \approx 0, 2.596$

22. $105° = 135° - 30°$

$\cos 105° = \cos(135° - 30°)$

$= \cos 135°\cos 30° + \sin 135°\sin 30°$

$= -\cos 45°\cos 30° + \sin 45°\sin 30°$

$= \left(-\dfrac{\sqrt{2}}{2}\right)\left(\dfrac{\sqrt{3}}{2}\right) + \left(\dfrac{\sqrt{2}}{2}\right)\left(\dfrac{1}{2}\right)$

$= \dfrac{-\sqrt{6} + \sqrt{2}}{4} = \dfrac{\sqrt{2} - \sqrt{6}}{4}$

23. $x = 2, y = -5, r = \sqrt{29}$

$\sin 2u = 2\sin u\cos u = 2\left(-\dfrac{5}{\sqrt{29}}\right)\left(\dfrac{2}{\sqrt{29}}\right) = -\dfrac{20}{29}$

$\cos 2u = \cos^2 u - \sin^2 u = \left(\dfrac{2}{\sqrt{29}}\right)^2 - \left(-\dfrac{5}{\sqrt{29}}\right)^2 = -\dfrac{21}{29}$

$\tan 2u = \dfrac{2\tan u}{1 - \tan^2 u} = \dfrac{2\left(-\dfrac{5}{2}\right)}{1 - \left(-\dfrac{5}{2}\right)^2} = \dfrac{20}{21}$

24. Let $y_1 = 31\sin\left(\frac{2\pi t}{365} - 1.4\right)$ and $y_2 = 20$.

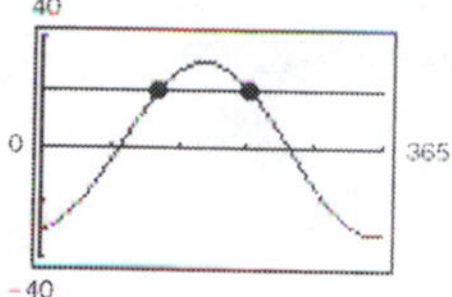

The points of intersection occur when $t \approx 123$ and $t \approx 223$.
The number of days that $D > 20°$ is 100, from day 123 to day 223.

25. $28\cos 10t + 38 = 28\cos\left[10\left(t - \frac{\pi}{6}\right)\right] + 38$

$$\cos 10t = \cos\left[10\left(t - \frac{\pi}{6}\right)\right]$$

$$0 = \cos\left[10\left(t - \frac{\pi}{6}\right)\right] - \cos 10t$$

$$= -2\sin\left(\frac{10(t - (\pi/6)) + 10t}{2}\right)\sin\left(\frac{10(t - (\pi/6)) - 10t}{2}\right)$$

$$= -2\sin\left(10t - \frac{5\pi}{6}\right)\sin\left(-\frac{5\pi}{6}\right)$$

$$= -2\sin\left(10t - \frac{5\pi}{6}\right)\left(-\frac{1}{2}\right)$$

$$= \sin\left(10t - \frac{5\pi}{6}\right)$$

$10t - \frac{5\pi}{6} = n\pi$ where n is any integer.

$t = \frac{n\pi}{10} + \frac{\pi}{12}$ where n is any integer.

The first six times the two people are at the same height are: 0.26 minutes, 0.58 minutes, 0.89 minutes, 1.20 minutes, 1.52 minutes, 1.83 minutes.

Chapter 8 Chapter Test Solutions

1. Given: $A = 24°$, $B = 68°$, $a = 12.2$

Law of Sines: AAS

$C = 180° - A - B = 180° - 24° - 68° = 88°$

$\frac{a}{\sin A} = \frac{b}{\sin B} \Rightarrow b = \frac{a}{\sin A}(\sin B)$

$b = \frac{12.2}{\sin 24°}(\sin 68°) \approx 27.81$

$\frac{a}{\sin A} = \frac{c}{\sin C} \Rightarrow c = \frac{a}{\sin A}(\sin C)$

$= \frac{12.2}{\sin 24°}(\sin 88°) \approx 29.98$

2. Given: $B = 110°$, $C = 28°$, $a = 15.6$

Law of Sines: AAS

$A = 180° - B - C = 180° - 110° - 28° = 42°$

$\frac{a}{\sin A} = \frac{b}{\sin B} \Rightarrow b = \frac{a}{\sin A}(\sin B)$

$= \frac{15.6}{\sin 42°}(\sin 110°) \approx 21.91$

$\frac{a}{\sin A} = \frac{c}{\sin C} \Rightarrow c = \frac{a}{\sin A}(\sin C)$

$= \frac{15.6}{\sin 42°}(\sin 28°) \approx 10.95$

3. Given: $A = 24°, a = 11.2, b = 13.4$

Law of Sines: SSA

$$\frac{\sin A}{a} = \frac{\sin B}{b} \Rightarrow \sin B = b\left(\frac{\sin A}{a}\right)$$

$$\sin B = 13.4\left(\frac{\sin 24°}{11.2}\right) \approx 0.4866$$

There are two angles between 0° and 180° where $\sin \theta \neq 0.4866$, $B_1 \approx 29.12°$ and $B_2 \approx 150.88$.

For $B_1 \approx 29.12°$,

$$C_1 = 180° - 29.12° - 24° = 126.88°$$

$$\frac{c}{\sin C} = \frac{a}{\sin A} \Rightarrow c = \frac{a}{\sin A}(\sin C)$$

$$= \frac{11.2}{\sin 24°}(\sin 126.88°) \approx 22.03$$

For $B_2 \approx 150.88°$,

$$C = 180° - 150.88° - 24° = 5.12°.$$

$$\frac{c}{\sin C} = \frac{a}{\sin A} \Rightarrow c = \frac{a}{\sin A}(\sin C)$$

$$= \frac{11.2}{\sin 24°}(\sin 5.12°) \approx 2.46$$

4. Given: $a = 4.0, b = 7.3, c = 12.4$

Law of Cosines: SSS

$$\cos C = \frac{a^2 + b^2 - c^2}{2ab}$$

$$= \frac{(4.0)^2 + (7.3)^2 - (12.4)^2}{2(4.0)(7.3)}$$

$$\approx -1.4464$$

Because there are no values of C such that $\cos C = -1.4464$, there is no possible triangle that can be formed.

5. Given: $B = 100°, a = 15, b = 23$

Law of Sines: SSA

$$\frac{\sin A}{a} = \frac{\sin B}{b} \Rightarrow \sin A = a\left(\frac{\sin B}{b}\right)$$

$$= 15\left(\frac{\sin 100°}{23}\right)$$

$$\approx 0.6423$$

So, $A \approx 39.96°$.

$$C = 180° - A - B = 180° - 39.96° - 100° = 40.04°$$

$$\frac{c}{\sin C} = \frac{b}{\sin B} \Rightarrow c = \frac{b}{\sin B}(\sin C)$$

$$= \frac{23}{\sin 100°}(\sin 40.04°) \approx 15.02$$

6. Given: $C = 121°, a = 34, b = 55$

Law of Cosines: SAS

$$c^2 = a^2 + b^2 - 2ab\cos C$$
$$c^2 = (34)^2 + (55)^2 - 2(34)(55)\cos 121°$$
$$c^2 = 6107.2424$$
$$c \approx 78.15$$

$$\frac{\sin B}{b} = \frac{\sin C}{c} \Rightarrow \sin B = b\left(\frac{\sin C}{c}\right)$$
$$= 55\left(\frac{\sin 121°}{78.15}\right)$$
$$\approx 0.6033$$

So, $B \approx 37.11°$.

$A = 180° - B - C = 180° - 37.11° - 121° = 21.89.$

7. $a = 60, b = 70, c = 82$

$$s = \frac{a+b+c}{2} = \frac{60+70+82}{2} = 106$$

Area $= \sqrt{s(s-a)(s-b)(s-c)} = \sqrt{106(46)(36)(24)} \approx 2052.5$ square meters

8. $b^2 = 370^2 + 240^2 - 2(370)(240)\cos 167°$

$b \approx 606.3$ miles

$$\sin A = \frac{a\sin B}{b} = \frac{240\sin 167°}{606.3}$$

$A \approx 5.1°$

Bearing: $24° + 5.1° = 29.1°$

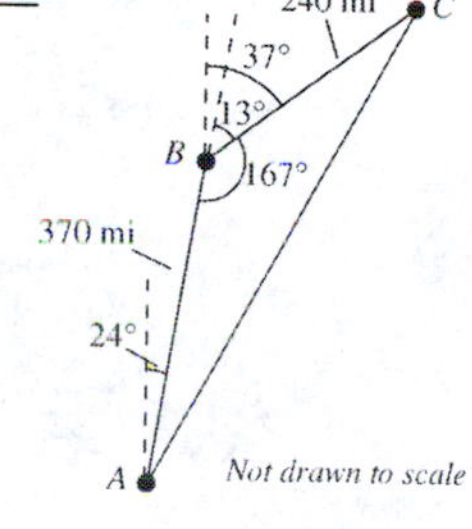

9. Initial point: $(-3, 7)$

Terminal point: $(11, -16)$

$$\mathbf{v} = \langle 11 - (-3), -16 - 7\rangle = \langle 14, -23\rangle$$

10. $$\mathbf{v} = 12\left(\frac{\mathbf{u}}{\|\mathbf{u}\|}\right) = 12\left(\frac{\langle 3, -5\rangle}{\sqrt{3^2 + (-5)^2}}\right) = \frac{12}{\sqrt{34}}\langle 3, -5\rangle$$
$$= \frac{6\sqrt{34}}{17}\langle 3, -5\rangle = \left\langle \frac{18\sqrt{34}}{17}, -\frac{30\sqrt{34}}{17}\right\rangle$$

11. $\mathbf{u} = \langle 2, 7\rangle, \mathbf{v} = \langle -6, 5\rangle$

$\mathbf{u} + \mathbf{v} = \langle 2, 7\rangle + \langle -6, 5\rangle = \langle -4, 12\rangle$

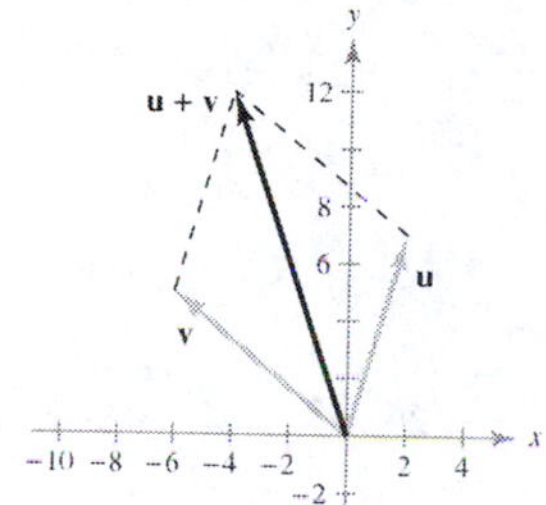

12. $\mathbf{u} = \langle 2, 7\rangle, \mathbf{v} = \langle -6, 5\rangle$

$\mathbf{u} - \mathbf{v} = \langle 2, 7\rangle - \langle -6, 5\rangle = \langle 8, 2\rangle$

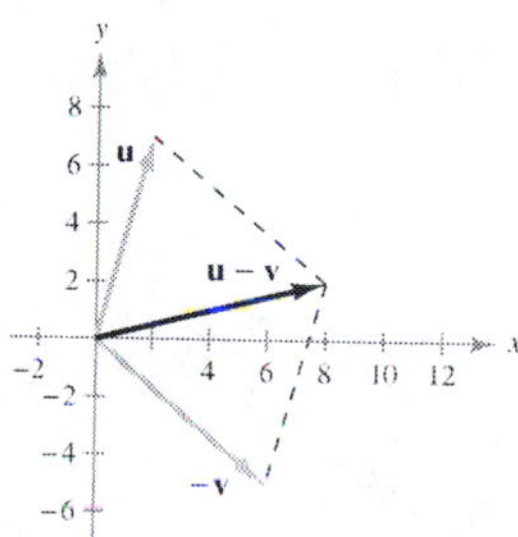

13. $\mathbf{u} = \langle 2, 7\rangle, \mathbf{v} = \langle -6, 5\rangle$

$$\begin{aligned} 5\mathbf{u} - 3\mathbf{v} &= 5\langle 2, 7\rangle - 3\langle -6, 5\rangle \\ &= \langle 10, 35\rangle - \langle -18, 15\rangle \\ &= \langle 28, 20\rangle \end{aligned}$$

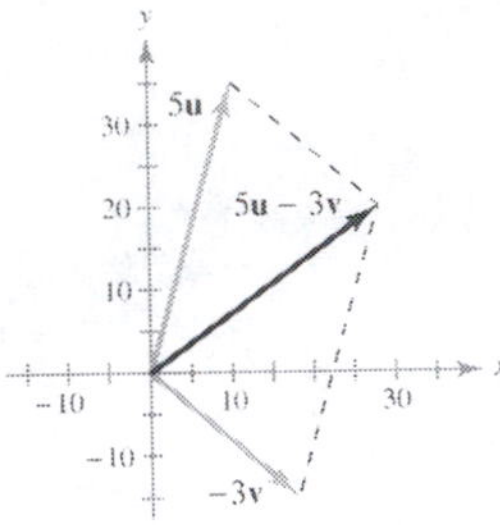

14. $\mathbf{u} = \langle 2, 7\rangle, \mathbf{v} = \langle -6, 5\rangle$

$$\begin{aligned} 4\mathbf{u} + 2\mathbf{v} &= 4\langle 2, 7\rangle + 2\langle -6, 5\rangle \\ &= \langle 8, 28\rangle + \langle -12, 10\rangle \\ &= \langle -4, 38\rangle \end{aligned}$$

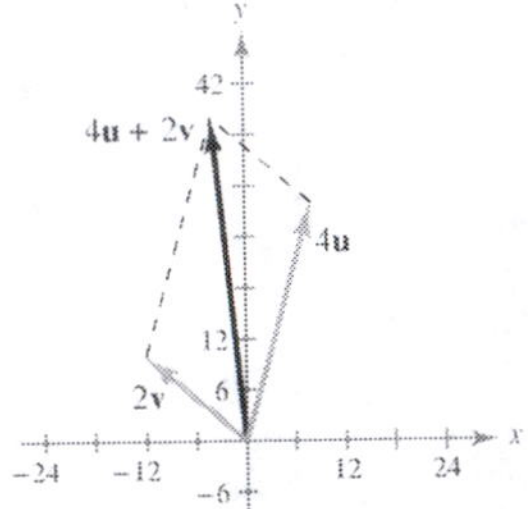

15. $\dfrac{\mathbf{u}}{\|\mathbf{u}\|} = \dfrac{\langle 24, -7\rangle}{\sqrt{24^2 + (-7)^2}} = \dfrac{1}{25}\langle 24, -7\rangle = \left\langle \dfrac{24}{25}, -\dfrac{7}{25}\right\rangle$

16.

$$\begin{aligned} \mathbf{u} &= 250(\cos 45^\circ\,\mathbf{i} + \sin 45^\circ\,\mathbf{j}) \\ \mathbf{v} &= 130(\cos(-60^\circ)\mathbf{i} + \sin(-60^\circ)\mathbf{j}) \\ \mathbf{R} &= \mathbf{u} + \mathbf{v} \approx 241.7767\,\mathbf{i} + 64.1934\,\mathbf{j} \\ \|\mathbf{R}\| &\approx \sqrt{241.7767^2 + 64.1934^2} \approx 250.15 \text{ pounds} \\ \tan\theta &\approx \frac{64.1934}{241.7767} \Rightarrow \theta \approx 14.9^\circ \end{aligned}$$

17. $\mathbf{u} = \langle -1, 5\rangle, \mathbf{v} = \langle 3, -2\rangle$

$$\cos\theta = \frac{\mathbf{u}\cdot\mathbf{v}}{\|\mathbf{u}\|\|\mathbf{v}\|} = \frac{-13}{\sqrt{26}\sqrt{13}} \Rightarrow \theta = 135^\circ$$

18. $\mathbf{u} = \langle 6, -10\rangle, \mathbf{v} = \langle 5, 3\rangle$

$$\mathbf{u}\cdot\mathbf{v} = 6(5) + (-10)(3) = 0$$

$\mathbf{u}$ and $\mathbf{v}$ are orthogonal.

19. $\mathbf{u} = \langle 6, 7\rangle, \mathbf{v} = \langle -5, -1\rangle$

$$\mathbf{w}_1 = \text{proj}_{\mathbf{v}}\mathbf{u} = \left(\frac{\mathbf{u}\cdot\mathbf{v}}{\|\mathbf{v}\|^2}\right)\mathbf{v} = -\frac{37}{26}\langle -5, -1\rangle = \frac{37}{26}\langle 5, 1\rangle$$

$$\begin{aligned} \mathbf{w}_2 = \mathbf{u} - \mathbf{w}_1 &= \langle 6, 7\rangle - \frac{37}{26}\langle 5, 1\rangle \\ &= \left\langle -\frac{29}{26}, \frac{145}{26}\right\rangle \\ &= \frac{29}{26}\langle -1, 5\rangle \end{aligned}$$

$$\mathbf{u} = \mathbf{w}_1 + \mathbf{w}_2 = \frac{37}{26}\langle 5, 1\rangle + \frac{29}{26}\langle -1, 5\rangle$$

20. $\mathbf{F} = -500\mathbf{j}, \mathbf{v} = (\cos 12^\circ)\mathbf{i} + (\sin 12^\circ)\mathbf{j}$

$$\begin{aligned} \mathbf{w}_1 = \text{proj}_{\mathbf{v}}\mathbf{F} &= \left(\frac{\mathbf{F}\cdot\mathbf{v}}{\|\mathbf{v}\|^2}\right)\mathbf{v} = (\mathbf{F}\cdot\mathbf{v})\mathbf{v} \\ &= (-500\sin 12^\circ)\mathbf{v} \end{aligned}$$

The magnitude of the force is $500\sin 12^\circ \approx 104$ pounds.

21.

$$\begin{aligned} z &= 4 - 4i \\ r &= \sqrt{(4)^2 + (-4)^2} = \sqrt{32} = 4\sqrt{2} \\ \tan\theta &= -\frac{4}{4} = -1 \Rightarrow \theta = \frac{7\pi}{4}\text{: Quadrant IV} \\ 4 - 4i &= 4\sqrt{2}\left(\cos\frac{7\pi}{4} + i\sin\frac{7\pi}{4}\right) \end{aligned}$$

22. $z = 6(\cos 120^\circ + i\sin 120^\circ)$

$$= 6\left(-\frac{1}{2} + \frac{\sqrt{3}}{2}i\right) = -3 + 3\sqrt{3}i$$

23.

$$\begin{aligned} \left[3\left(\cos\frac{7\pi}{6} + i\sin\frac{7\pi}{6}\right)\right]^8 &= 3^8\left(\cos\frac{28\pi}{3} + i\sin\frac{28\pi}{3}\right) \\ &= 6561\left(-\frac{1}{2} - \frac{\sqrt{3}}{2}i\right) \\ &= -\frac{6561}{2} - \frac{6561\sqrt{3}}{2}i \end{aligned}$$

24.

$$\begin{aligned} (3 - 3i)^6 &= \left[3\sqrt{2}\left(\cos\frac{7\pi}{4} + i\sin\frac{7\pi}{4}\right)\right]^6 \\ &= \left(3\sqrt{2}\right)^6\left(\cos\frac{21\pi}{2} + i\sin\frac{21\pi}{2}\right) \\ &= 5832(0 + i) \\ &= 5832i \end{aligned}$$

25. $z = 256\left(1 + \sqrt{3}i\right)$

$$|z| = 256\sqrt{1^2 + \left(\sqrt{3}\right)^2} = 256\sqrt{4} = 512$$

$$\tan\theta = \frac{\sqrt{3}}{1} \Rightarrow \theta = \frac{\pi}{3}$$

$$z = 512\left(\cos\frac{\pi}{3} + i\sin\frac{\pi}{3}\right)$$

$$\text{Fourth roots of } z = \sqrt[4]{512}\left[\cos\frac{\frac{\pi}{3} + 2\pi k}{4} + i\sin\frac{\frac{\pi}{3} + 2\pi k}{4}\right], k = 0, 1, 2, 3$$

$$k = 0: 4\sqrt[4]{2}\left(\cos\frac{\pi}{12} + i\sin\frac{\pi}{12}\right)$$

$$k = 1: 4\sqrt[4]{2}\left(\cos\frac{7\pi}{12} + i\sin\frac{7\pi}{12}\right)$$

$$k = 2: 4\sqrt[4]{2}\left(\cos\frac{13\pi}{12} + i\sin\frac{13\pi}{12}\right)$$

$$k = 3: 4\sqrt[4]{2}\left(\cos\frac{19\pi}{12} + i\sin\frac{19\pi}{12}\right)$$

26. $x^3 - 27i = 0 \Rightarrow x^3 = 27i$

The solutions to the equation are the cube roots of $27i = 27\left(\cos\frac{\pi}{2} + i\sin\frac{\pi}{2}\right)$.

$$\text{Cube roots: } \sqrt[3]{27}\left[\cos\frac{\frac{\pi}{2} + 2\pi k}{3} + i\sin\frac{\frac{\pi}{2} + 2\pi k}{3}\right], k = 0, 1, 2$$

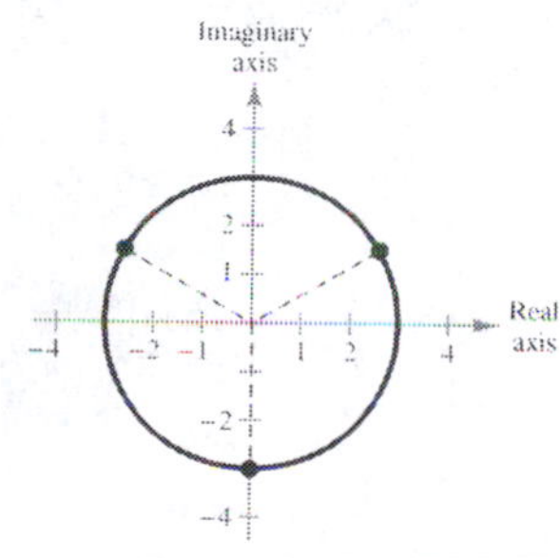

$$k = 0: 3\left(\cos\frac{\pi}{6} + i\sin\frac{\pi}{6}\right) = 3\left(\frac{\sqrt{3}}{2} + \frac{1}{2}i\right) = \frac{3\sqrt{3}}{2} + \frac{3}{2}i$$

$$k = 1: 3\left(\cos\frac{5\pi}{6} + i\sin\frac{5\pi}{6}\right) = 3\left(-\frac{\sqrt{3}}{2} + \frac{1}{2}i\right) = -\frac{3\sqrt{3}}{2} + \frac{3}{2}i$$

$$k = 2: 3\left(\cos\frac{3\pi}{2} + i\sin\frac{3\pi}{2}\right) = 3(0 - i) = -3i$$

Cumulative Test Solutions for Chapters 6–8

1. (a)

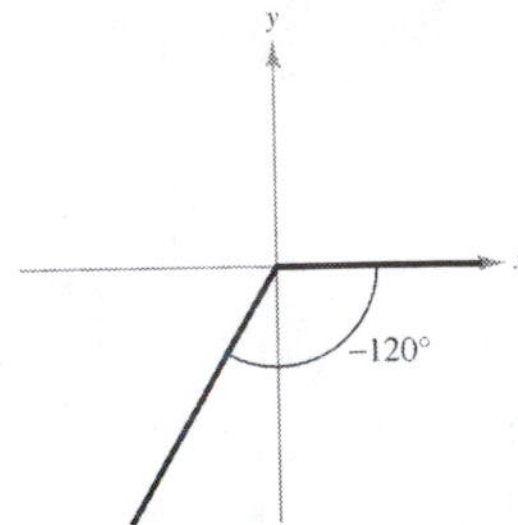

(b) $-120° + 360° = 240°$

(c) $-120\left(\dfrac{\pi}{180°}\right) = -\dfrac{2\pi}{3}$

(d) $-120° + 360° = 240°$

$\theta' = 240° - 180° = 60°$

(e) $\sin(-120°) = -\sin 60° = -\dfrac{\sqrt{3}}{2}$

$\cos(-120°) = -\cos 60° = -\dfrac{1}{2}$

$\tan(-120°) = \tan 60° = \sqrt{3}$

$\csc(-120°) = \dfrac{1}{-\sin 60°} = -\dfrac{2\sqrt{3}}{3}$

$\sec(-120°) = \dfrac{1}{-\cos 60°} = -2$

$\cot(-120°) = \dfrac{1}{\tan 60°} = \dfrac{\sqrt{3}}{3}$

2. $-1.45\left(\dfrac{180}{\pi}\right) \approx -83.1°$

3. $\tan\theta = \dfrac{y}{x} = -\dfrac{21}{20} \Rightarrow r = 29$

Because $\sin\theta < 0$, θ is in Quadrant IV $\Rightarrow x = 20$.

$\cos\theta = \dfrac{x}{r} = \dfrac{20}{29}$

4. $f(x) = 3 - 2\sin \pi x$

Period: $\dfrac{2\pi}{\pi} = 2$

Amplitude: $|a| = |-2| = 2$

Upward shift of 3 units (reflected in x-axis prior to shift)

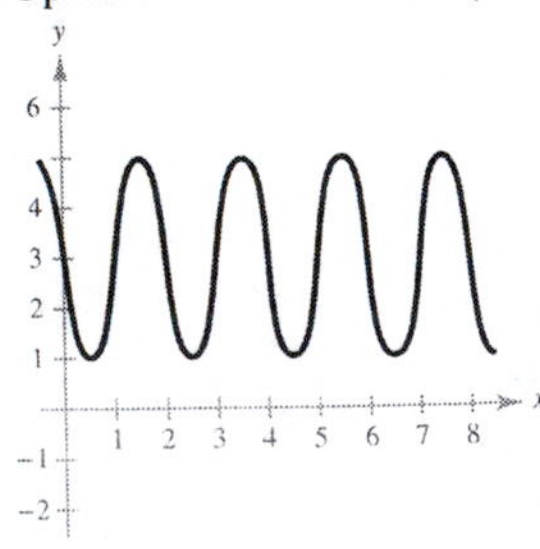

5. $g(x) = \dfrac{1}{2}\tan\left(x - \dfrac{\pi}{2}\right)$

Period: π

Asymptotes: $x = 0$, $x = \pi$

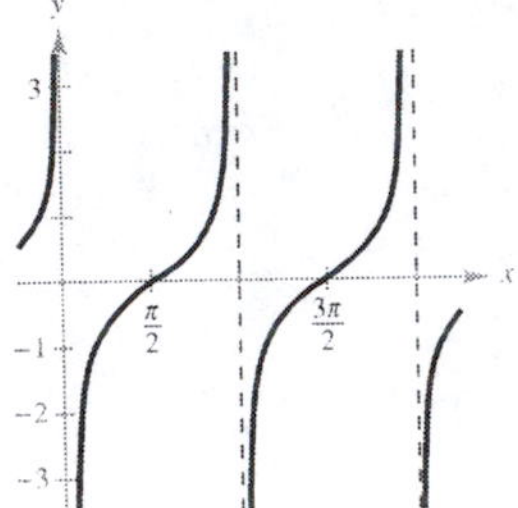

6. $h(x) = -\sec(x + \pi)$

Graph $y = -\cos(x + \pi)$ first.

Period: 2π

Amplitude: 1

Set $x + \pi = 0$ and $x + \pi = 2\pi$ for one cycle.

$x = -\pi$ $\qquad x = \pi$

The asymptotes of $h(x)$ corresponds to the x-intercepts of $y = -\cos(x + \pi)$.

$$x + \pi = \frac{(2n + 1)\pi}{2}$$

$$x = \frac{(2n - 1)\pi}{2} \text{ where } n \text{ is any integer}$$

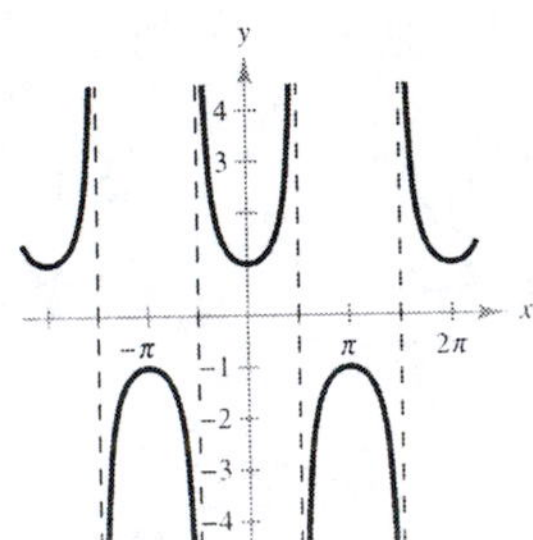

7. $h(x) = a\cos(bx + c)$

Graph is reflected in x-axis.

Amplitude: $a = -3$

Period: $2 = \dfrac{2\pi}{\pi} \Rightarrow b = \pi$

No phase shift: $c = 0$

$h(x) = -3\cos(\pi x)$

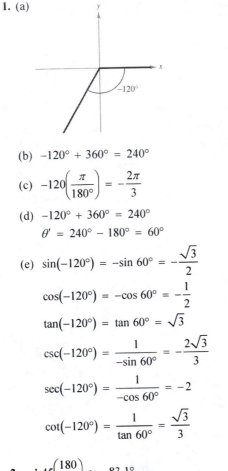

8. $f(x) = \frac{x}{2}\sin x, -3\pi \le x \le 3\pi$

$$-\frac{x}{2} \le f(x) \le \frac{x}{2}$$

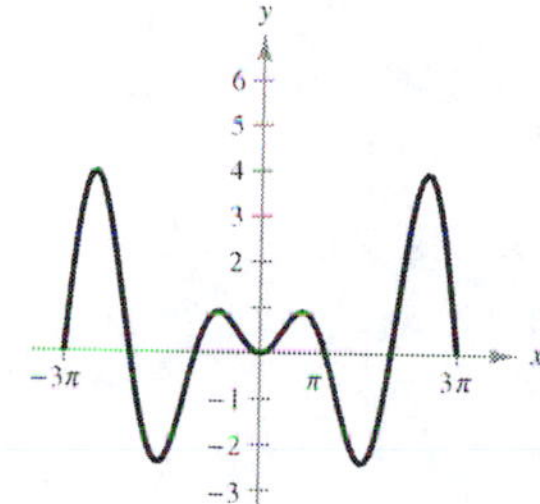

9. $\tan(\arctan 4.9) = 4.9$

10. 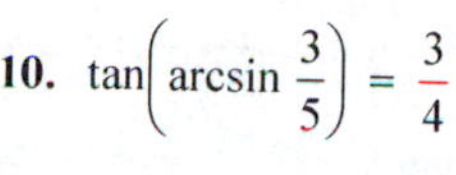

$\tan\left(\arcsin \frac{3}{5}\right) = \frac{3}{4}$

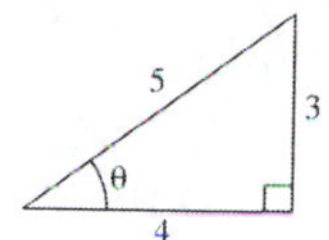

11.
$$y = \arccos(2x)$$
$$\sin y = \sin(\arccos(2x)) = \sqrt{1 - 4x^2}$$

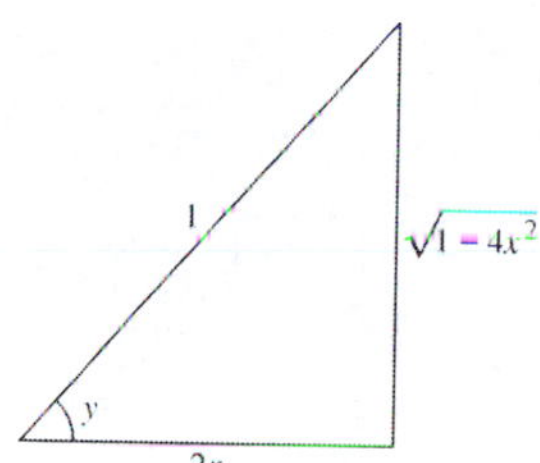

12. $\cos\left(\frac{\pi}{2} - x\right)\csc x = \sin x\left(\frac{1}{\sin x}\right) = 1$

13.
$$\frac{\sin\theta - 1}{\cos\theta} - \frac{\cos\theta}{\sin\theta - 1} = \frac{\sin\theta - 1}{\cos\theta} - \frac{\cos\theta(\sin\theta + 1)}{\sin^2\theta - 1}$$
$$= \frac{\sin\theta - 1}{\cos\theta} + \frac{\cos\theta(\sin\theta + 1)}{\cos^2\theta} = \frac{\sin\theta - 1}{\cos\theta} + \frac{\sin\theta + 1}{\cos\theta} = \frac{2\sin\theta}{\cos\theta} = 2\tan\theta$$

14. $\cot^2\alpha(\sec^2\alpha - 1) = \cot^2\alpha\tan^2\alpha = 1$

15.
$$\sin(x + y)\sin(x - y) = \tfrac{1}{2}\left[\cos(x + y - (x - y)) - \cos(x + y + x - y)\right]$$
$$= \tfrac{1}{2}[\cos 2y - \cos 2x] = \tfrac{1}{2}\left[1 - 2\sin^2 y - (1 - 2\sin^2 x)\right] = \sin^2 x - \sin^2 y$$

16.
$$\sin^2 x\cos^2 x = \left(\frac{1 - \cos 2x}{2}\right)\left(\frac{1 + \cos 2x}{2}\right)$$
$$= \frac{1}{4}(1 - \cos 2x)(1 + \cos 2x)$$
$$= \frac{1}{4}(1 - \cos^2 2x)$$
$$= \frac{1}{4}\left(1 - \frac{1 + \cos 4x}{2}\right)$$
$$= \frac{1}{8}(2 - (1 + \cos 4x))$$
$$= \frac{1}{8}(1 - \cos 4x)$$

17.
$$2\cos^2\beta - \cos\beta = 0$$
$$\cos\beta(2\cos\beta - 1) = 0$$

$\cos\beta = 0$ or $2\cos\beta - 1 = 0$

$\beta = \frac{\pi}{2}, \frac{3\pi}{2}$ $\qquad\cos\beta = \frac{1}{2}$

$\beta = \frac{\pi}{3}, \frac{5\pi}{3}$

Answer: $\frac{\pi}{3}, \frac{\pi}{2}, \frac{3\pi}{2}, \frac{5\pi}{3}$

18.
$$3\tan\theta - \cot\theta = 0$$
$$3\tan\theta - \frac{1}{\tan\theta} = 0$$
$$\frac{3\tan^2\theta - 1}{\tan\theta} = 0$$
$$3\tan^2\theta - 1 = 0$$
$$\tan^2\theta = \frac{1}{3}$$
$$\tan\theta = \pm\frac{\sqrt{3}}{3}$$
$$\theta = \frac{\pi}{6}, \frac{5\pi}{6}, \frac{7\pi}{6}, \frac{11\pi}{6}$$

19.
$$\sin^2 x + 2\sin x + 1 = 0$$
$$(\sin x + 1)(\sin x + 1) = 0$$
$$\sin x + 1 = 0$$
$$\sin x = -1$$
$$x = \frac{3\pi}{2}$$

20. $\sin u = \frac{12}{13} \Rightarrow \cos u = \frac{5}{13}$ and $\tan u = \frac{12}{5}$ because u is in Quadrant I.

$\cos v = \frac{3}{5} \Rightarrow \sin v = \frac{4}{5}$ and $\tan v = \frac{4}{3}$ because v is in Quadrant I.

$$\tan(u - v) = \frac{\tan u - \tan v}{1 + \tan u \tan v} = \frac{\frac{12}{5} - \frac{4}{3}}{1 + \left(\frac{12}{5}\right)\left(\frac{4}{3}\right)} = \frac{16}{63}$$

21. $\tan \theta = \frac{1}{2}$

$$\tan 2\theta = \frac{2 \tan \theta}{1 - \tan^2 \theta} = \frac{2\left(\frac{1}{2}\right)}{1 - \left(\frac{1}{2}\right)^2} = \frac{4}{3}$$

22. $\tan \theta = \frac{4}{3} \Rightarrow \cos \theta = \pm\frac{3}{5}$

$$\sin \frac{\theta}{2} = \sqrt{\frac{1 - \cos \theta}{2}} = \sqrt{\frac{1 - \frac{3}{5}}{2}} = \frac{\sqrt{5}}{5}$$

$$\text{or } = \sqrt{\frac{1 + \frac{3}{5}}{2}} = \frac{2\sqrt{5}}{5}$$

23.
$$5 \sin\frac{3\pi}{4}\cos\frac{7\pi}{4} = \frac{5}{2}\left[\sin\left(\frac{3\pi}{4} + \frac{7\pi}{4}\right) + \sin\left(\frac{3\pi}{4} - \frac{7\pi}{4}\right)\right]$$
$$= \frac{5}{2}\left[\sin\frac{5\pi}{2} + \sin(-\pi)\right]$$
$$= \frac{5}{2}\left(\sin\frac{5\pi}{2} - \sin \pi\right)$$

24.
$$\cos 9x - \cos 7x = -2 \sin\left(\frac{9x + 7x}{2}\right)\sin\left(\frac{9x - 7x}{2}\right)$$
$$= -2 \sin 8x \sin x$$

25. Given: $A = 30°, a = 9, b = 8$

Law of Sines: SSA

$$\frac{\sin B}{8} = \frac{\sin 30°}{9}$$
$$\sin B = \frac{8}{9}\left(\frac{1}{2}\right)$$
$$B = \arcsin\left(\frac{4}{9}\right)$$
$$B \approx 26.39°$$
$$C = 180° - A - B \approx 123.61°$$
$$\frac{c}{\sin 123.61°} = \frac{9}{\sin 30°}$$
$$c \approx 14.99$$

26. Given: $A = 30°, b = 8, c = 10$

Law of Cosines: SAS

$$a^2 = 8^2 + 10^2 - 2(8)(10) \cos 30°$$
$$a^2 \approx 25.4359$$
$$a \approx 5.04$$
$$\cos B = \frac{5.04^2 + 10^2 - 8^2}{2(5.04)(10)}$$
$$\cos B \approx 0.6091$$
$$B \approx 52.48°$$
$$C = 180° - A - B \approx 97.52°$$

27. Given: $A = 30°, C = 90°, b = 10$

Right Triangle Ratios

$$B = 180° - 30° - 90° = 60°$$
$$\tan 30° = \frac{a}{10} \Rightarrow a = 10 \tan 30° \approx 5.77$$
$$\cos 30° = \frac{10}{c} \Rightarrow c = \frac{10}{\cos 30°} \approx 11.55$$

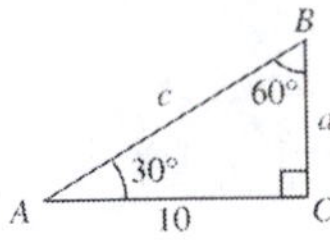

28. Given: $a = 4.7, b = 8.1, c = 10.3$

Law of Cosines: SSS

$$\cos C = \frac{a^2 + b^2 - c^2}{2ab} = \frac{4.7^2 + 8.1^2 + 10.3^2}{2(4.7)(8.1)} \approx -0.2415 \Rightarrow C \approx 103.98°$$

$$\sin A = \frac{a \sin C}{c} \approx \frac{4.7 \sin 103.98°}{10.3} \approx 0.4428 \Rightarrow A \approx 26.28°$$

$$B \approx 180° - 26.28° - 103.98° = 49.74°$$

29. Given: $A = 45°, B = 26°, c = 20$

Law of Sines: AAS

$$C = 180° - A - B = 180° - 45° - 26° = 109°$$

$$\frac{a}{\sin A} = \frac{c}{\sin C} \Rightarrow a = \frac{c}{\sin C}(\sin A) = \frac{20}{\sin 109°}(\sin 45°) \approx 14.96$$

$$\frac{b}{\sin B} = \frac{c}{\sin C} \Rightarrow b = \frac{c}{\sin C}(\sin B) = \frac{20}{\sin 109°}(\sin 26°) \approx 9.27$$

30. Given: $a = 1.2, b = 10, C = 80°$

Law of Cosines: SAS

$$c^2 = a^2 + b^2 - 2ab \cos C$$

$$c^2 = (1.2)^2 + (10)^2 - 2(1.2)(10) \cos 80°$$

$$c^2 \approx 97.2724$$

$$c \approx 9.86$$

31. Area $= \frac{1}{2}(7)(12) \sin 99° = 41.48 \text{ in.}^2$

32. $a = 30, b = 41, c = 45$

$$s = \frac{a + b + c}{2} = \frac{30 + 41 + 45}{2} = 58$$

$$\text{Area} = \sqrt{s(s-a)(s-b)(s-c)}$$

$$= \sqrt{58(28)(17)(13)}$$

$$\approx 599.09 \text{ m}^2$$

33. $\mathbf{u} = \langle 7, 8 \rangle = 7\mathbf{i} + 8\mathbf{j}$

34. $\mathbf{v} = \mathbf{i} + \mathbf{j}$

$$\|\mathbf{v}\| = \sqrt{1^2 + 1^2} = \sqrt{2}$$

$$\mathbf{u} = \frac{\mathbf{v}}{\|\mathbf{v}\|} = \frac{1}{\sqrt{2}}(\mathbf{i} + \mathbf{j}) = \frac{\sqrt{2}}{2}(\mathbf{i} + \mathbf{j})$$

35. $\mathbf{u} = 3\mathbf{i} + 4\mathbf{j}, \mathbf{v} = \mathbf{i} - 2\mathbf{j}$

$$\mathbf{u} \cdot \mathbf{v} = 3(1) + 4(-2) = -5$$

36. $\mathbf{u} = \langle 8, -2 \rangle, \mathbf{v} = \langle 1, 5 \rangle$

$$\mathbf{w}_1 = \text{proj}_{\mathbf{v}} \mathbf{u} = \left(\frac{\mathbf{u} \cdot \mathbf{v}}{\|\mathbf{v}\|^2}\right)\mathbf{v} = \frac{-2}{26}\langle 1, 5 \rangle = -\frac{1}{13}\langle 1, 5 \rangle$$

$$\mathbf{w}_2 = \mathbf{u} - \mathbf{w}_1 = \langle 8, -2 \rangle - \left\langle -\frac{1}{13}, -\frac{5}{13} \right\rangle = \left\langle \frac{105}{13}, -\frac{21}{13} \right\rangle$$

$$= \frac{21}{13}\langle 5, -1 \rangle$$

$$\mathbf{u} = \mathbf{w}_1 + \mathbf{w}_2 = -\frac{1}{13}\langle 1, 5 \rangle + \frac{21}{13}\langle 5, -1 \rangle$$

37. $r = |-2 + 2i| = \sqrt{(-2)^2 + (2)^2} = 2\sqrt{2}$

$$\tan \theta = \frac{2}{-2} = -1$$

Because $\tan \theta = -1$ and $-2 + 2i$ lies in Quadrant II,

$\theta = \frac{3\pi}{4}$. So, $-2 + 2i = 2\sqrt{2}\left(\cos \frac{3\pi}{4} + i \sin \frac{3\pi}{4}\right)$.

38.

$$[4(\cos 30° + i \sin 30°)][6(\cos 120° + i \sin 120°)] = (4)(6)[\cos(30° + 120°) + i \sin(30° + 120°)]$$

$$= 24(\cos 150° + i \sin 150°)$$

$$= 24\left(-\frac{\sqrt{3}}{2} + \frac{1}{2}i\right)$$

$$= -12\sqrt{3} + 12i$$

39. $1 = 1(\cos 0 + i \sin 0)$

$$\sqrt[3]{1} = \sqrt[3]{1}\left[\cos\left(\frac{0 + 2\pi k}{3}\right) + i \sin\left(\frac{0 + 2\pi k}{3}\right)\right], k = 0, 1, 2$$

$$k = 0: \sqrt[3]{1}\left[\left(\cos\left(\frac{0 + 2\pi(0)}{3}\right) + i \sin\left(\frac{0 + 2\pi(0)}{3}\right)\right)\right] = \cos 0 + i \sin 0 = 1$$

$$k = 1: \sqrt[3]{1}\left[\left(\cos\left(\frac{0 + 2\pi(1)}{3}\right) + i \sin\left(\frac{0 + 2\pi(1)}{3}\right)\right)\right] = \cos\frac{2\pi}{3} + i \sin\frac{2\pi}{3} = -\frac{1}{2} + \frac{\sqrt{3}}{2}i$$

$$k = 2: \sqrt[3]{1}\left[\left(\cos\left(\frac{0 + 2\pi(2)}{3}\right) + i \sin\left(\frac{0 + 2\pi(2)}{3}\right)\right)\right] = \cos\frac{4\pi}{3} + i \sin\frac{4\pi}{3} = -\frac{1}{2} - \frac{\sqrt{3}}{2}i$$

40. $x^5 + 243 = 0 \Rightarrow x^5 = -243$

The solutions to the equation are the fifth roots of $-243 = 243(\cos \pi + i \sin \pi)$, which are:

$$\sqrt[5]{243}\left[\cos\left(\frac{\pi + 2\pi k}{5}\right) + i \sin\left(\frac{\pi + 2\pi k}{5}\right)\right], k = 0, 1, 2, 3, 4$$

$$k = 0: 3\left(\cos\frac{\pi}{5} + i \sin\frac{\pi}{5}\right)$$

$$k = 1: 3\left(\cos\frac{3\pi}{5} + i \sin\frac{3\pi}{5}\right)$$

$$k = 2: 3(\cos \pi + i \sin \pi)$$

$$k = 3: 3\left(\cos\frac{7\pi}{5} + i \sin\frac{7\pi}{5}\right)$$

$$k = 4: 3\left(\cos\frac{9\pi}{5} + i \sin\frac{9\pi}{5}\right)$$

41. Angular speed $= \dfrac{\theta}{t} = \dfrac{2\pi(63)}{1} \approx 395.8$ radians per minute

Linear speed $= \dfrac{s}{t} = \dfrac{42\pi(63)}{1} \approx 8312.7$ inches per minute

42. Area $= \dfrac{\theta r^2}{2} = \dfrac{(105°)\left(\dfrac{\pi}{180°}\right)(12)^2}{2} = 42\pi \approx 131.95 \text{ yd}^2$

43. Height of smaller triangle:

$$\tan 16° 45' = \frac{h_1}{200}$$

$$h_1 = 200 \tan 16.75°$$

$$\approx 60.2 \text{ feet}$$

Not drawn to scale

Height of larger triangle:

$$\tan 18° = \frac{h_2}{200}$$

$$h_2 = 200 \tan 18° \approx 65.0 \text{ feet}$$

Height of flag: $h_2 - h_1 = 65.0 - 60.2 \approx 5$ feet

44. $\tan \theta = \dfrac{5}{12} \Rightarrow \theta \approx 22.6°$

45. $d = a \cos bt$

$|a| = 4 \Rightarrow a = 4$

$\dfrac{2\pi}{b} = 8 \Rightarrow b = \dfrac{\pi}{4}$

$d = 4 \cos \dfrac{\pi}{4}t$

46. $\mathbf{v}_1 = 500\langle \cos 60°, \sin 60° \rangle = \langle 250, 250\sqrt{3} \rangle$

$\mathbf{v}_2 = 50\langle \cos 30°, \sin 30° \rangle = \langle 25\sqrt{3}, 25 \rangle$

$\mathbf{v} = \mathbf{v}_1 + \mathbf{v}_2 = \langle 250 + 25\sqrt{3}, 250\sqrt{3} + 25 \rangle \approx \langle 293.3, 458.0 \rangle$

$\|\mathbf{v}\| = \sqrt{(293.3)^2 + (458.0)^2} \approx 543.9$

$\tan \theta = \dfrac{458.0}{293.3} \approx 1.56 \Rightarrow \theta \approx 57.4°$

Bearing: $90° - 57.4° = 32.6°$

The plane is traveling on a bearing of 32.6° at 543.9 kilometers per hour.

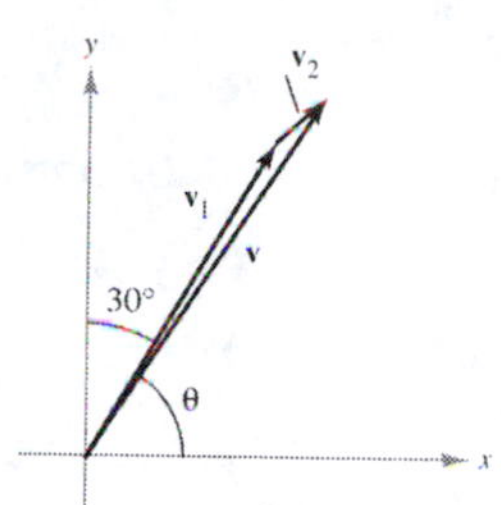

47. $\mathbf{w} = (85)(10)\cos 60° = 425$ foot-pounds

Chapter Test Solutions for Chapter 9

1. $\begin{cases} x + y = -9 \Rightarrow x = -y - 9 \\ 5x - 8y = 20 \end{cases}$

$5(-y - 9) - 8y = 20$

$-13y = 65$

$y = -5$

$x - 5 = -9 \Rightarrow x = -4$

Solution: $(-4, -5)$

2. $\begin{cases} y = x - 1 \\ y = (x - 1)^3 \end{cases}$

$x - 1 = (x - 1)^3$

$x - 1 = x^3 - 3x^2 + 3x - 1$

$0 = x^3 - 3x^2 + 2x$

$0 = x(x - 1)(x - 2)$

$x = 0$ or $x = 1$ or $x = 2$

$y = -1 \quad y = 0 \quad y = 1$

Solutions: $(0, -1), (1, 0), (2, 1)$

3. $\begin{cases} x - y = 4 \Rightarrow x = y + 4 \\ 2x - y^2 = 0 \Rightarrow 2(y + 4) - y^2 = 0 \end{cases}$

$0 = y^2 - 2y - 8$

$0 = (y + 2)(y - 4)$

$y = -2$ or $y = 4$

$x = 2 \quad x = 8$

Solutions: $(2, -2), (8, 4)$

4. $\begin{cases} 3x - 6y = 0 \Rightarrow y = \dfrac{1}{2}x \\ 3x + 6y = 18 \Rightarrow y = -\dfrac{1}{2}x + 3 \end{cases}$

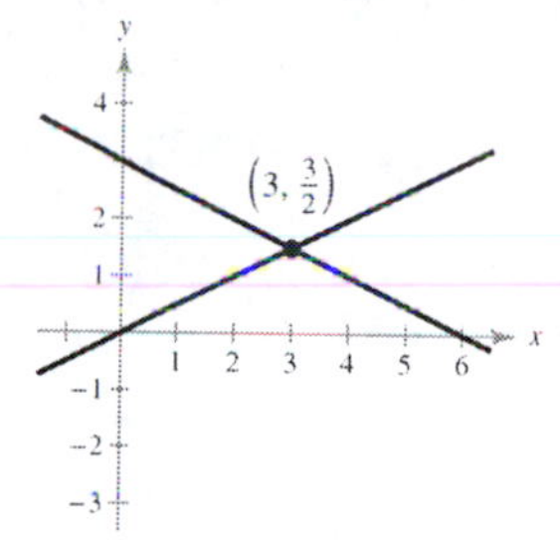

Solution: $\left(3, \dfrac{3}{2}\right)$

5. 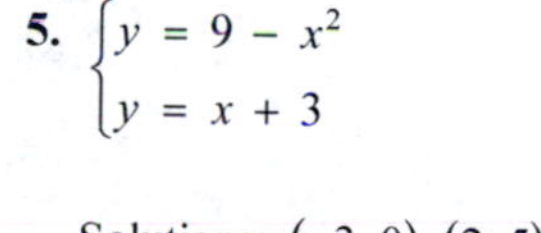

$\begin{cases} y = 9 - x^2 \\ y = x + 3 \end{cases}$

Solutions: $(-3, 0), (2, 5)$

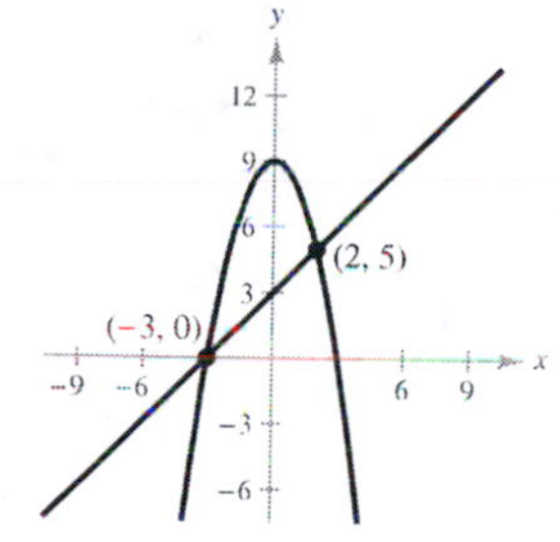

6. 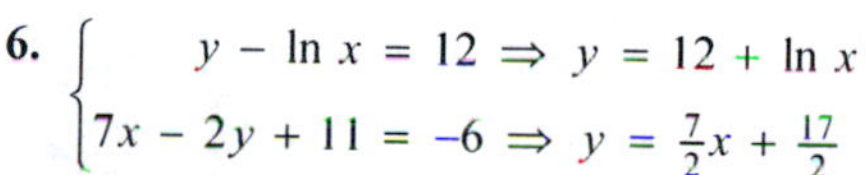

$\begin{cases} y - \ln x = 12 \Rightarrow y = 12 + \ln x \\ 7x - 2y + 11 = -6 \Rightarrow y = \frac{7}{2}x + \frac{17}{2} \end{cases}$

Solutions:

$(1, 12), (0.034, 8.619)$

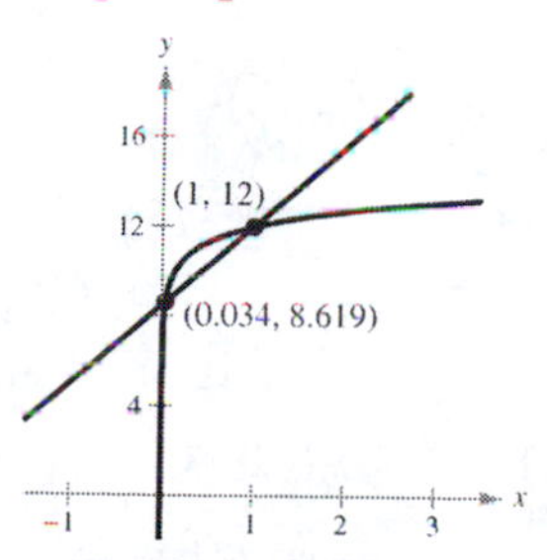

7. $\begin{cases} 3x + 4y = -26 & \text{Equation 1} \\ 7x - 5y = 11 & \text{Equation 2} \end{cases}$

Multiply Equation 1 by 5: $15x + 20y = -130$

Multiply Equation 2 by 4: $28x - 20y = 44$

Add the equations to eliminate y:

$$\begin{aligned} 15x + 20y &= -130 \\ 28x - 20y &= 44 \\ \hline 43x &= -86 \\ x &= -2 \end{aligned}$$

Back-substitute $x = -2$ into Equation 1:

$$3(-2) + 4y = -26$$
$$y = -5$$

Solution: $(-2, -5)$

8. $\begin{cases} 1.4x - y = 17 & \text{Equation 1} \\ 0.8x + 6y = -10 & \text{Equation 2} \end{cases}$

Multiply Equation 1 by 6: $8.4x - 6y = 102$

Add this to Equation 2 to eliminate y:

$$\begin{aligned} 8.4x - 6y &= 102 \\ 0.8x + 6y &= -10 \\ \hline 9.2x &= 92 \\ x &= 10 \end{aligned}$$

Back-substitute $x = 10$ into Equation 2:

$$0.8(10) + 6y = -10$$
$$6y = -18$$
$$y = -3$$

Solution: $(10, -3)$

9. $\begin{cases} x - 2y + 3z = 11 \\ 2x - z = 3 \\ 3y + z = -8 \end{cases}$

$\begin{cases} x - 2y + 3z = 11 \\ 4y - 7z = -19 & -2\text{Eq.}1 + \text{Eq.}2 \\ 3y + z = -8 \end{cases}$

$\begin{cases} x - 2y + 3z = 11 \\ y - 8z = -11 & -\text{Eq.}3 + \text{Eq.}2 \\ 3y + z = -8 \end{cases}$

$\begin{cases} x - 2y + 3z = 11 \\ y - 8z = -11 \\ 25z = 25 & -3\text{Eq.}2 + \text{Eq.}3 \end{cases}$

$\begin{cases} x - 2y + 3z = 11 \\ y - 8z = -11 \\ z = 1 & \frac{1}{25}\text{Eq.}3 \end{cases}$

$$y - 8(1) = -11 \Rightarrow y = -3$$
$$x - 2(-3) + 3(1) = 11 \Rightarrow x = 2$$

Solution: $(2, -3, 1)$

10. $\begin{cases} 3x + 2y + z = 17 & \text{Equation 1} \\ -x + y + z = 4 & \text{Equation 2} \\ x - y - z = 3 & \text{Equation 3} \end{cases}$

Interchange Equations 1 and 3.

$\begin{cases} x - y - z = 3 \\ -x + y + z = 4 \\ 3x + 2y + z = 17 \end{cases}$

$\begin{cases} x - y - z = 3 \\ 0 \neq 7 & \text{Eq. 1 + Eq. 2} \\ 3x + 2y + z = 17 \end{cases}$

Inconsistent

No solution

11. $\dfrac{2x + 5}{x^2 - x - 2} = \dfrac{2x + 5}{(x - 2)(x + 1)} = \dfrac{A}{x - 2} + \dfrac{B}{x + 1}$

$$2x + 5 = A(x + 1) + B(x - 2)$$

Let $x = 2$: $9 = 3A \Rightarrow A = 3$

Let $x = -1$: $3 = -3B \Rightarrow B = -1$

$$\frac{2x + 5}{x^2 - x - 2} = \frac{3}{x - 2} - \frac{1}{x + 1}$$

12. $\dfrac{3x^2 - 2x + 4}{x^2(2 - x)} = \dfrac{A}{x} + \dfrac{B}{x^2} + \dfrac{C}{2 - x}$

$$3x^2 - 2x + 4 = Ax(2 - x) + B(2 - x) + Cx^2$$

Let $x = 0$: $4 = 2B \Rightarrow B = 2$

Let $x = 2$: $12 = 4C \Rightarrow C = 3$

Let $x = 1$: $5 = A + B + C = A + 2 + 3 \Rightarrow A = 0$

$$\frac{3x^2 - 2x + 4}{x^2(2 - x)} = \frac{2}{x^2} + \frac{3}{2 - x}$$

13. $\dfrac{x^2 + 5}{x^3 - x} = \dfrac{x^2 + 5}{x(x + 1)(x - 1)} = \dfrac{A}{x} + \dfrac{B}{x + 1} + \dfrac{C}{x - 1}$

$$x^2 + 5 = A(x + 1)(x - 1) + Bx(x - 1) + Cx(x + 1)$$

Let $x = 0$: $5 = -A \Rightarrow A = -5$

Let $x = -1$: $6 = 2B \Rightarrow B = 3$

Let $x = 1$: $6 = 2C \Rightarrow C = 3$

$$\frac{x^2 + 5}{x^3 - x} = -\frac{5}{x} + \frac{3}{x + 1} + \frac{3}{x - 1}$$

14. $\dfrac{x^2 - 4}{x^3 + 2x} = \dfrac{x^2 - 4}{x(x^2 + 2)} = \dfrac{A}{x} + \dfrac{Bx + C}{x^2 + 2}$

$$\begin{aligned} x^2 - 4 &= A(x^2 + 2) + (Bx + C)x \\ &= Ax^2 + 2A + Bx^2 + Cx \\ &= (A + B)x^2 + Cx + 2A \end{aligned}$$

Equate the coefficients of like terms:

$1 = A + B, 0 = C, -4 = 2A$

So, $A = -2, B = 3, C = 0.$

$$\frac{x^2 - 4}{x^3 + 2x} = -\frac{2}{x} + \frac{3x}{x^2 + 2}$$

15. 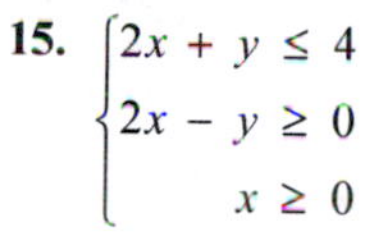

$\begin{cases} 2x + y \le 4 \\ 2x - y \ge 0 \\ x \ge 0 \end{cases}$

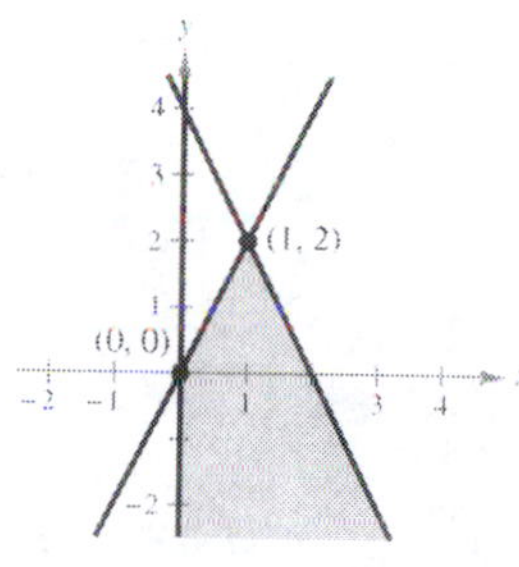

16. $\begin{cases} y < -x^2 + x + 4 \\ y > 4x \end{cases}$

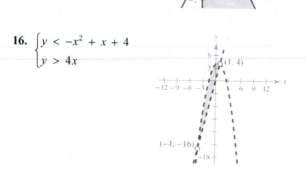

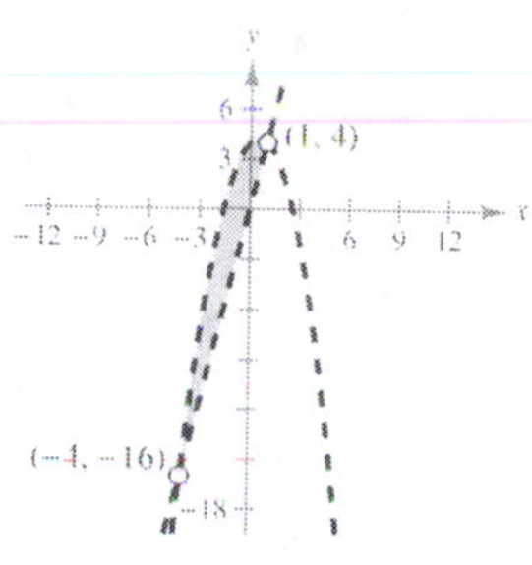

17. $\begin{cases} x^2 + y^2 \le 36 \\ x \ge 2 \\ y \ge -4 \end{cases}$

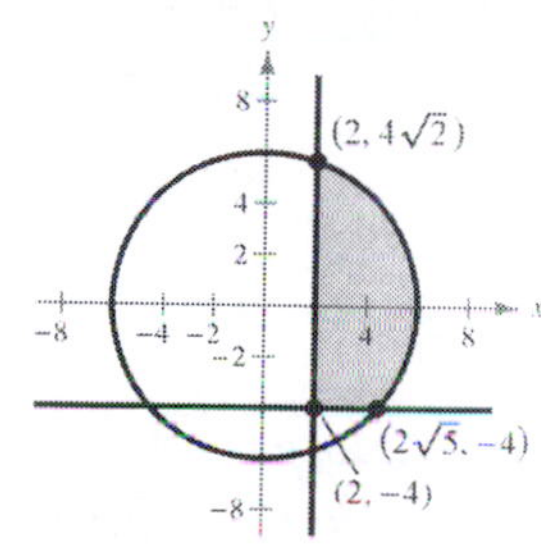

18. Maximize $z = 20x + 12y$ subject to:

$\begin{cases} x \ge 0, y \ge 0 \\ x + 4y \le 32 \\ 3x + 2y \le 36 \end{cases}$

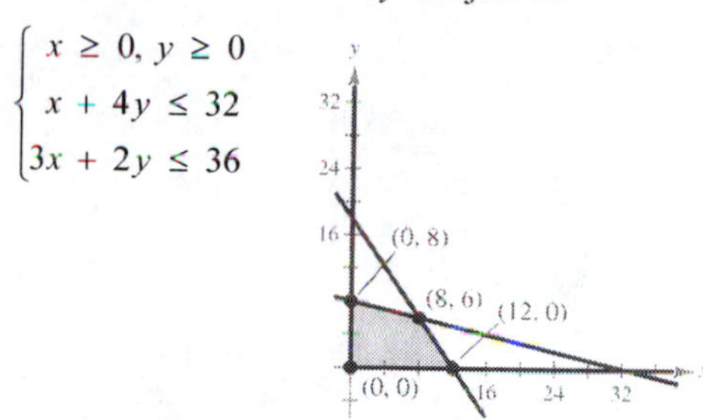

At $(0, 0)$ we have $z = 0$.

At $(0, 8)$ we have $z = 96$.

At $(8, 6)$ we have $z = 232$.

At $(12, 0)$ we have $z = 240$.

The maximum value, $z = 240$, occurs at $(12, 0)$.

The minimum value, $z = 0$ occurs at $(0, 0)$.

19. Let x = amount of money invested at 4%.

Let y = amount of money invested at 5.5%.

$\begin{cases} x + y = 50{,}000 & \text{Equation 1} \\ 0.04x + 0.055y = 2390 & \text{Equation 2} \end{cases}$

Multiply Equation 1 by -4: $-4x - 4y = -200{,}000$

Multiply Equation 2 by 100: $4x + 5.5y = 239{,}000$

Add these two equations to eliminate x:

$$\begin{aligned} -4x - 4y &= -200{,}000 \\ 4x + 5.5y &= 239{,}000 \\ \hline 1.5y &= 39{,}000 \\ y &= 26{,}000 \end{aligned}$$

Back-substitute $y = 26{,}000$ into Equation 1:

$$\begin{aligned} x + 26{,}000 &= 50{,}000 \\ x &= 24{,}000 \end{aligned}$$

So, \$24,000 should be invested at 4% and \$26,000 should be invested at 5.5%.

20. $y = ax^2 + bx + c$

$(0, 6)$: $6 = c$

$(-2, 2)$: $2 = 4a - 2b + c$

$\left(3, \frac{9}{2}\right)$: $\frac{9}{2} = 9a + 3b + c$

Solving this system yields: $a = -\frac{1}{2}$, $b = 1$, and $c = 6$.

So, $y = -\frac{1}{2}x^2 + x + 6$.

21. Optimize $P = 30x + 40y$ subject to:

$$\begin{cases} x \geq 0, \; y \geq 0 \\ 0.5x + 0.75y \leq 4000 \\ 2.0x + \;\;1.5y \leq 8950 \\ 0.5x + \;\;0.5y \leq 2650 \end{cases}$$

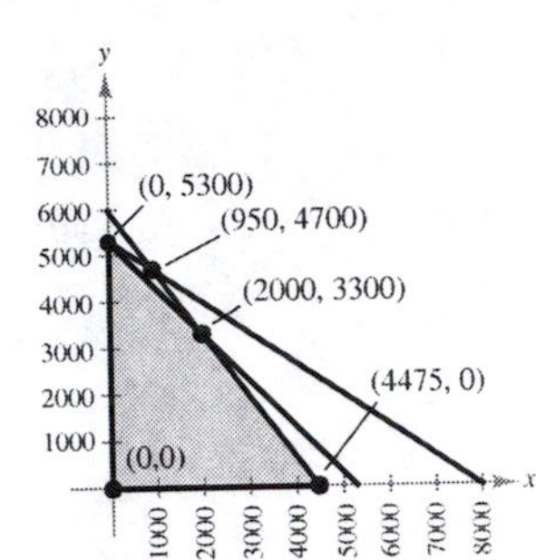

At $(0, 0)$: $P = 0$

At $(0, 5300)$: $P = 212{,}000$

At $(2000, 3300)$: $P = 192{,}000$

At $(4475, 0)$: $P = 134{,}250$

The manufacturer should produce 5300 units of Model II and not produce any of Model I to realize an optimal profit of \$212,000.

Chapter Test Solutions for Chapter 10

1. $\begin{bmatrix} 1 & -1 & 5 \\ 6 & 2 & 3 \\ 5 & 3 & -3 \end{bmatrix}$

$\begin{array}{r} \\ -6R_1 + R_2 \rightarrow \\ -5R_1 + R_3 \rightarrow \end{array} \begin{bmatrix} 1 & -1 & 5 \\ 0 & 8 & -27 \\ 0 & 8 & -28 \end{bmatrix}$

$\begin{array}{r} \\ \\ -R_2 + R_3 \rightarrow \end{array} \begin{bmatrix} 1 & -1 & 5 \\ 0 & 8 & -27 \\ 0 & 0 & -1 \end{bmatrix}$

$\begin{array}{r} \\ \frac{1}{8}R_2 \rightarrow \\ -R_3 \rightarrow \end{array} \begin{bmatrix} 1 & -1 & 5 \\ 0 & 1 & -\frac{27}{8} \\ 0 & 0 & 1 \end{bmatrix}$

$\begin{array}{r} R_2 + R_1 \rightarrow \\ \\ \\ \end{array} \begin{bmatrix} 1 & 0 & \frac{13}{8} \\ 0 & 1 & -\frac{27}{8} \\ 0 & 0 & 1 \end{bmatrix}$

$\begin{array}{r} -\frac{13}{8}R_3 + R_1 \rightarrow \\ \frac{27}{8}R_3 + R_2 \rightarrow \\ \\ \end{array} \begin{bmatrix} 1 & 0 & 0 \\ 0 & 1 & 0 \\ 0 & 0 & 1 \end{bmatrix}$

2. $\begin{bmatrix} 1 & 0 & -1 & 2 \\ -1 & 1 & 1 & -3 \\ 1 & 1 & -1 & 1 \\ 3 & 2 & -3 & 4 \end{bmatrix}$

$\begin{array}{r} \\ R_1 + R_2 \rightarrow \\ -R_1 + R_3 \rightarrow \\ -3R_1 + R_4 \rightarrow \end{array} \begin{bmatrix} 1 & 0 & -1 & 2 \\ 0 & 1 & 0 & -1 \\ 0 & 1 & 0 & -1 \\ 0 & 2 & 0 & -2 \end{bmatrix}$

$\begin{array}{r} \\ \\ -R_2 + R_3 \rightarrow \\ -2R_2 + R_4 \rightarrow \end{array} \begin{bmatrix} 1 & 0 & -1 & 2 \\ 0 & 1 & 0 & -1 \\ 0 & 0 & 0 & 0 \\ 0 & 0 & 0 & 0 \end{bmatrix}$

3. $\begin{bmatrix} 4 & 3 & -2 & \vdots & 14 \\ -1 & -1 & 2 & \vdots & -5 \\ 3 & 1 & -4 & \vdots & 8 \end{bmatrix}$

$3R_2 + R_1 \to \begin{bmatrix} 1 & 0 & 4 & \vdots & -1 \\ -1 & -1 & 2 & \vdots & -5 \\ 3 & 1 & -4 & \vdots & 8 \end{bmatrix}$

$\begin{matrix} \\ R_1 + R_2 \to \\ -3R_1 + R_3 \to \end{matrix} \begin{bmatrix} 1 & 0 & 4 & \vdots & -1 \\ 0 & -1 & 6 & \vdots & -6 \\ 0 & 1 & -16 & \vdots & 11 \end{bmatrix}$

$\begin{matrix} \\ \\ R_2 + R_3 \to \end{matrix} \begin{bmatrix} 1 & 0 & 4 & \vdots & -1 \\ 0 & -1 & 6 & \vdots & -6 \\ 0 & 0 & -10 & \vdots & 5 \end{bmatrix}$

$\begin{matrix} \\ -R_2 \to \\ -\frac{1}{10}R_3 \to \end{matrix} \begin{bmatrix} 1 & 0 & 4 & \vdots & -1 \\ 0 & 1 & -6 & \vdots & 6 \\ 0 & 0 & 1 & \vdots & -\frac{1}{2} \end{bmatrix}$

$\begin{matrix} -4R_3 + R_1 \to \\ 6R_3 + R_2 \to \\ \\ \end{matrix} \begin{bmatrix} 1 & 0 & 0 & \vdots & 1 \\ 0 & 1 & 0 & \vdots & 3 \\ 0 & 0 & 1 & \vdots & -\frac{1}{2} \end{bmatrix}$

Solution: $\left(1, 3, -\frac{1}{2}\right)$

4. $A = \begin{bmatrix} 6 & 5 \\ -5 & -5 \end{bmatrix}, \quad B = \begin{bmatrix} 5 & 0 \\ -5 & -1 \end{bmatrix}$

(a) $A - B = \begin{bmatrix} 6 & 5 \\ -5 & -5 \end{bmatrix} - \begin{bmatrix} 5 & 0 \\ -5 & -1 \end{bmatrix} = \begin{bmatrix} 6 - 5 & 5 - 0 \\ -5 - (-5) & -5 - (-1) \end{bmatrix} = \begin{bmatrix} 1 & 5 \\ 0 & -4 \end{bmatrix}$

(b) $3A = 3\begin{bmatrix} 6 & 5 \\ -5 & -5 \end{bmatrix} = \begin{bmatrix} 3(6) & 3(5) \\ 3(-5) & 3(-5) \end{bmatrix} = \begin{bmatrix} 18 & 15 \\ -15 & -15 \end{bmatrix}$

(c) $3A - 2B = \begin{bmatrix} 18 & 15 \\ -15 & -15 \end{bmatrix} - 2\begin{bmatrix} 5 & 0 \\ -5 & -1 \end{bmatrix} = \begin{bmatrix} 18 - 2(5) & 15 - 2(0) \\ -15 - 2(-5) & -15 - 2(-1) \end{bmatrix} = \begin{bmatrix} 8 & 15 \\ -5 & -13 \end{bmatrix}$

(d) $AB = \begin{bmatrix} 6 & 5 \\ -5 & -5 \end{bmatrix}\begin{bmatrix} 5 & 0 \\ -5 & -1 \end{bmatrix} = \begin{bmatrix} (6)(5) + (5)(-5) & (6)(0) + (5)(-1) \\ (-5)(5) + (-5)(-5) & (-5)(0) + (-5)(-1) \end{bmatrix} = \begin{bmatrix} 5 & -5 \\ 0 & 5 \end{bmatrix}$

5. $A = \begin{bmatrix} a & b \\ c & d \end{bmatrix}, \qquad A^{-1} = \dfrac{1}{ad - bc}\begin{bmatrix} d & -b \\ -c & a \end{bmatrix}$

$A = \begin{bmatrix} -4 & 3 \\ 5 & -2 \end{bmatrix}$

$ad - bc = (-4)(-2) - (3)(5) = -7$

$A^{-1} = -\frac{1}{7}\begin{bmatrix} -2 & -3 \\ -5 & -4 \end{bmatrix} = \begin{bmatrix} \frac{2}{7} & \frac{3}{7} \\ \frac{5}{7} & \frac{4}{7} \end{bmatrix}$

6. $\left[\begin{array}{ccc|ccc} -2 & 4 & -6 & 1 & 0 & 0 \\ 2 & 1 & 0 & 0 & 1 & 0 \\ 4 & -2 & 5 & 0 & 0 & 1 \end{array}\right]$

$\begin{array}{r} \\ R_1 + R_2 \rightarrow \\ 2R_1 + R_3 \rightarrow \end{array}\left[\begin{array}{ccc|ccc} -2 & 4 & -6 & 1 & 0 & 0 \\ 0 & 5 & -6 & 1 & 1 & 0 \\ 0 & 6 & -7 & 2 & 0 & 1 \end{array}\right]$

$\begin{array}{r} -\frac{1}{2}R_1 \rightarrow \\ -R_3 + R_2 \rightarrow \\ \\ \end{array}\left[\begin{array}{ccc|ccc} 1 & -2 & 3 & -\frac{1}{2} & 0 & 0 \\ 0 & -1 & 1 & -1 & 1 & -1 \\ 0 & 6 & -7 & 2 & 0 & 1 \end{array}\right]$

$\begin{array}{r} -2R_2 + R_1 \rightarrow \\ \\ 6R_2 + R_3 \rightarrow \end{array}\left[\begin{array}{ccc|ccc} 1 & 0 & 1 & \frac{3}{2} & -2 & 2 \\ 0 & -1 & 1 & -1 & 1 & -1 \\ 0 & 0 & -1 & -4 & 6 & -5 \end{array}\right]$

$\begin{array}{r} \\ -R_2 \rightarrow \\ -R_3 \rightarrow \end{array}\left[\begin{array}{ccc|ccc} 1 & 0 & 1 & \frac{3}{2} & -2 & 2 \\ 0 & 1 & -1 & 1 & -1 & 1 \\ 0 & 0 & 1 & 4 & -6 & 5 \end{array}\right]$

$\begin{array}{r} -R_3 + R_1 \rightarrow \\ R_3 + R_2 \rightarrow \\ \\ \end{array}\left[\begin{array}{ccc|ccc} 1 & 0 & 0 & -\frac{5}{2} & 4 & -3 \\ 0 & 1 & 0 & 5 & -7 & 6 \\ 0 & 0 & 1 & 4 & -6 & 5 \end{array}\right]$

$$A^{-1} = \begin{bmatrix} -\frac{5}{2} & 4 & -3 \\ 5 & -7 & 6 \\ 4 & -6 & 5 \end{bmatrix}$$

7. $\begin{cases} -4x + 3y = 6 \\ 5x - 2y = 24 \end{cases}$

$$\begin{bmatrix} -4 & 3 \\ 5 & -2 \end{bmatrix}\begin{bmatrix} x \\ y \end{bmatrix} = \begin{bmatrix} 6 \\ 24 \end{bmatrix}$$

$$\begin{bmatrix} x \\ y \end{bmatrix} = \begin{bmatrix} -4 & 3 \\ 5 & -2 \end{bmatrix}^{-1}\begin{bmatrix} 6 \\ 24 \end{bmatrix} = \begin{bmatrix} \frac{2}{7} & \frac{3}{7} \\ \frac{5}{7} & \frac{4}{7} \end{bmatrix}\begin{bmatrix} 6 \\ 24 \end{bmatrix} = \begin{bmatrix} 12 \\ 18 \end{bmatrix}$$

Solution: $(12, 18)$

8. $\begin{vmatrix} -6 & 4 \\ 10 & 12 \end{vmatrix} = (-6)(12) - (4)(10) = -112$

9. $\begin{vmatrix} \frac{5}{2} & \frac{13}{4} \\ -8 & \frac{6}{5} \end{vmatrix} = \left(\frac{5}{2}\right)\left(\frac{6}{5}\right) - \left(\frac{13}{4}\right)(-8) = 29$

10. Expand along Column 3.

$$\begin{vmatrix} 6 & -7 & 2 \\ 3 & -2 & 0 \\ 1 & 5 & 1 \end{vmatrix} = 2\begin{vmatrix} 3 & -2 \\ 1 & 5 \end{vmatrix} + \begin{vmatrix} 6 & -7 \\ 3 & -2 \end{vmatrix} = 2(17) + 9 = 43$$

11. $\begin{cases} 7x + 6y = 9 \\ -2x - 11y = -49 \end{cases}$ $\quad D = \begin{vmatrix} 7 & 6 \\ -2 & -11 \end{vmatrix} = -65$

$$x = \frac{\begin{vmatrix} 9 & 6 \\ -49 & -11 \end{vmatrix}}{-65} = \frac{195}{-65} = -3$$

$$y = \frac{\begin{vmatrix} 7 & 9 \\ -2 & -49 \end{vmatrix}}{-65} = \frac{-325}{-65} = 5$$

Solution: $(-3, 5)$

12. $\begin{cases} 6x - y + 2z = -4 \\ -2x + 3y - z = 10 \\ 4x - 4y + z = -18 \end{cases}$ $\quad D = \begin{vmatrix} 6 & -1 & 2 \\ -2 & 3 & -1 \\ 4 & -4 & 1 \end{vmatrix} = -12$

$$x = \frac{\begin{vmatrix} -4 & -1 & 2 \\ 10 & 3 & -1 \\ -18 & -4 & 1 \end{vmatrix}}{-12} = \frac{24}{-12} = -2$$

$$y = \frac{\begin{vmatrix} 6 & -4 & 2 \\ -2 & 10 & -1 \\ 4 & -18 & 1 \end{vmatrix}}{-12} = \frac{-48}{-12} = 4$$

$$z = \frac{\begin{vmatrix} 6 & -1 & -4 \\ -2 & 3 & 10 \\ 4 & -4 & -18 \end{vmatrix}}{-12} = \frac{-72}{-12} = 6$$

Solution: $(-2, 4, 6)$

13. $A = -\frac{1}{2}\begin{vmatrix} -5 & 0 & 1 \\ 4 & 4 & 1 \\ 3 & 2 & 1 \end{vmatrix} = -\frac{1}{2}(-14) = 7$

14. $\begin{matrix} K & N & O \\ C & K & - \\ O & N & - \\ W & O & O \\ D & - & - \end{matrix}\begin{bmatrix} 11 & 14 & 15 \\ 3 & 11 & 0 \\ 15 & 14 & 0 \\ 23 & 15 & 15 \\ 4 & 0 & 0 \end{bmatrix}\begin{bmatrix} 1 & -1 & 0 \\ 1 & 0 & -1 \\ 6 & -2 & -3 \end{bmatrix} = \begin{bmatrix} 115 & -41 & -59 \\ 14 & -3 & -11 \\ 29 & -15 & -14 \\ 128 & -53 & -60 \\ 4 & -4 & 0 \end{bmatrix}$

Message: $[11\ \ 14\ \ 15], [3\ \ 11\ \ 0], [15\ \ 14\ \ 0], [23\ \ 15\ \ 15], [4\ \ 0\ \ 0]$

Encoded Message: 115 −41 −59 14 −3 −11 29 −15 −14 128 −53 −60 4 −4 0

15. Let x = amount of 60% solution and y = amount of 20% solution.

$$\begin{cases} x + y = 100 \Rightarrow y = 100 - x \\ 0.60x + 0.20y = 0.50(100) \Rightarrow 6x + 2y = 500 \end{cases}$$

By substitution,

$$\begin{aligned} 6x + 2(100 - x) &= 500 \\ 6x + 200 - 2x &= 500 \\ 4x &= 300 \\ x &= 75 \\ y &= 100 - x = 25. \end{aligned}$$

75 liters of 60% solution and 25 liters of 20% solution

Chapter Test Solutions for Chapter 11

1. $a_n = \dfrac{(-1)^n}{3n + 2}$

$a_1 = -\dfrac{1}{5}$

$a_2 = \dfrac{1}{8}$

$a_3 = -\dfrac{1}{11}$

$a_4 = \dfrac{1}{14}$

$a_5 = -\dfrac{1}{17}$

2. $\dfrac{3}{1!}, \dfrac{4}{2!}, \dfrac{5}{3!}, \dfrac{6}{4!}, \dfrac{7}{5!}, \ldots$

$a_n = \dfrac{n + 2}{n!}$

3. $8 + 21 + 34 + 47 + \ldots$

$a_5 = 60, a_6 = 73, a_7 = 86$

$S_6 = 8 + 21 + 34 + 47 + 60 + 73 = 243$

4. $a_5 = 5.4, a_{12} = 11.0$

$$\begin{aligned} a_{12} &= a_5 + 7d \\ 11.0 &= 5.4 + 7d \\ 5.6 &= 7d \\ 0.8 &= d \\ a_1 &= a_5 - 4d \\ a_1 &= 5.4 - 4(0.8) \\ &= 2.2 \\ a_n &= a_1 + (n - 1)d \\ &= 2.2 + (n - 1)(0.8) \\ &= 0.8n + 1.4 \end{aligned}$$

5. $a_2 = 28, a_6 = 7168$

$$\begin{aligned} a_6 &= a_2r^4 \\ 7168 &= 28r^4 \\ 256 &= r^4 \\ 4 &= r \\ a_2 &= a_1r \\ 28 &= a_1(4) \\ 7 &= a_1 \\ a_n &= 7(4)^{n-1} \end{aligned}$$

6. $a_n = 5(2)^{n-1}$

$a_1 = 5$

$a_2 = 10$

$a_3 = 20$

$a_4 = 40$

$a_5 = 80$

7. $\sum_{i=1}^{50}(2i^2 + 5) = 2\sum_{i=1}^{50} i^2 + \sum_{i=1}^{50} 5$

$= 2\left[\frac{50(51)(101)}{6}\right] + 50(5)$

$= 86{,}100$

8. $\sum_{n=1}^{9}(12n - 7) = 12\sum_{n=1}^{9} n - \sum_{n=1}^{9} 7$

$= 12\left[\frac{9(10)}{2}\right] - 9(7)$

$= 477$

9. $\sum_{i=1}^{\infty} 4\left(\frac{1}{2}\right)^i = \frac{2}{1 - \frac{1}{2}} = 4$

10. $5 + 10 + 15 + \cdots + 5n = \frac{5n(n + 1)}{2}$

When $n = 1$, $S_1 = 5 = \frac{5(1)(2)}{2}$, so the formula is valid.

Assume that

$S_k = 5 + 10 + 15 + \cdots + 5k = \frac{5k(k + 1)}{2}$, then

$S_{k+1} = S_k + a_{k+1}$

$= \frac{5k(k + 1)}{2} + 5(k + 1)$

$= \frac{5k(k + 1)}{2} + \frac{10(k + 1)}{2}$

$= \frac{5k(k + 1) + 10(k + 1)}{2}$

$= \frac{5(k + 1)(k + 2)}{2}$

$= \frac{5(k + 1)[(k + 1) + 1]}{2}.$

So, the formula is valid for all integers $n \geq 1$.

11. (a) $(x + 6y)^4 = x^4 + {}_4C_1x^3(6y) + {}_4C_2x^2(6y)^2 + {}_4C_3x(6y)^3 + {}_4C_4(6y)^4$

$= x^4 + 24x^3y + 216x^2y^2 + 864xy^3 + 1296y^4$

(b) $3(x - 2)^5 + 4(x - 2)^3 = 3\left[x^5 + {}_5C_1x^4(-2) + {}_5C_2x^3(-2)^2 + {}_5C_3x^2(-2)^3 + {}_5C_4x(-2)^4 + {}_5C_5(-2)^5\right]$

$+ 4\left[x^3 + {}_3C_1x^2(-2) + {}_3C_2x(-2)^2 + {}_3C_3(-2)^3\right]$

$= 3(x^5 - 10x^4 + 40x^3 - 80x^2 + 80x - 32) + 4(x^3 - 6x^2 + 12x - 8)$

$= 3x^5 - 30x^4 + 124x^3 - 264x^2 + 288x - 128$

12. ${}_nC_rx^{n-r}y^r = {}_7C_3(3a)^4(-2b)^3$

$= 35(81a^4)(-8b^3)$

$= -22{,}680a^4b^3$

So, the coefficient of a^4b^3 is $-22{,}680$.

13. (a) ${}_9P_2 = \frac{9!}{7!} = 72$

(b) ${}_{70}P_3 = \frac{70!}{67!} = 328{,}440$

14. (a) ${}_{11}C_4 = \frac{11!}{7!4!} = 330$

(b) ${}_{66}C_4 = \frac{66!}{62!4!} = 720{,}720$

15. $(26)(10)(10)(10) = 26{,}000$ distinct license plates

16. $\underbrace{(1)}_{\text{owner}} \cdot \underbrace{(3)(2)}_{\text{bow seats}} \cdot \underbrace{(5)(4)(3)(2)(1)}_{\text{remaining seats}} = 720$ seating arrangements

17. $\frac{20}{300} = \frac{1}{15} \approx 0.0667$

18. $\frac{1}{{}_{30}C_4} = \frac{1}{27{,}405}$

19. $P(E') = 1 - P(E)$

$= 1 - 0.90$

$= 0.10$ or 10%

Cumulative Test Solutions for Chapters 9–11

1. $\begin{cases} y = 3 - x^2 \\ 2(y - 2) = x - 1 \end{cases} \Rightarrow 2(3 - x^2 - 2) = x - 1$

$$2(1 - x^2) = x - 1$$
$$2 - 2x^2 = x - 1$$
$$0 = 2x^2 + x - 3$$
$$0 = (2x + 3)(x - 1)$$
$$x = -\tfrac{3}{2} \text{ or } x = 1$$
$$y = \tfrac{3}{4} \qquad y = 2$$

Solutions: $\left(-\frac{3}{2}, \frac{3}{4}\right), (1, 2)$

2. $\begin{cases} x + 3y = -6 \Rightarrow 4x + 12y = -24 \\ 2x + 4y = -10 \Rightarrow -6x - 12y = 30 \end{cases}$

$$-2x = 6$$
$$x = -3 \Rightarrow y = -1$$

Solution: $(-3, -1)$

3. $\begin{cases} -2x + 4y - z = -16 \\ x - 2y + 2z = 5 \\ x - 3y - z = 13 \end{cases}$

Interchange equations.

$\begin{cases} x - 2y + 2z = 5 & \text{Eq.1} \\ -2x + 4y - z = -16 & \text{Eq.2} \\ x - 3y - z = 13 & \text{Eq.3} \end{cases}$

$\begin{cases} x - 2y + 2z = 5 & \\ 3z = -6 & 2\text{Eq.1} + \text{Eq.2} \\ -y - 3z = 8 & -\text{Eq.1} + \text{Eq.3} \end{cases}$

From Equation 2, $z = -2$. Substituting this into Equation 3 yields $y = -2$. Using these in Equation 1 yields $x = 5$.

Solution: $(5, -2, -2)$

4. $\begin{cases} x + 3y - 2z = -7 \\ -2x + y - z = -5 \\ 4x + y + z = 3 \end{cases}$

$\begin{cases} x + 3y - 2z = -7 & \\ 7y - 5z = -19 & 2\text{Eq. }1 + \text{Eq.}2 \\ -11y + 9z = 31 & -4\text{Eq. }1 + \text{Eq.}3 \end{cases}$

$\begin{cases} x + 3y - 2z = -7 & \\ y - \frac{5}{7}z = -\frac{19}{7} & \frac{1}{7}\text{Eq.}2 \\ -11y + 9z = 31 & \end{cases}$

$\begin{cases} x + \frac{1}{7}z = \frac{8}{7} & -3\text{Eq.}2 + \text{Eq.}1 \\ y - \frac{5}{7}z = -\frac{19}{7} & \\ \frac{8}{7}z = \frac{8}{7} & 11\text{Eq.}2 + \text{Eq.}3 \end{cases}$

$\begin{cases} x + \frac{1}{7}z = \frac{8}{7} & \\ y - \frac{5}{7}z = -\frac{19}{7} & \\ z = 1 & \frac{7}{8}\text{Eq.}3 \end{cases}$

$\begin{cases} x = 1 & -\frac{1}{7}\text{Eq.}3 + \text{Eq.}1 \\ y = -2 & \frac{5}{7}\text{Eq.}3 + \text{Eq.}2 \\ z = 1 & \end{cases}$

Solution: $(1, -2, 1)$

5. $\begin{cases} x + y = 200 \Rightarrow y = 200 - x \\ 0.75x + 1.25y = 0.95(200) \end{cases}$

$$0.75x + 1.25(200 - x) = 190$$
$$0.75x + 250 - 1.25x = 190$$
$$-0.50x = -60$$
$$x = 120$$
$$y = 200 - x = 80$$

120 pounds of \$0.75 seed and 80 pounds of \$1.25 seed.

6. $y = ax^2 + bx + c$

$(0, 6)$: $6 = a(0)^2 + b(0) + c \Rightarrow c = 6$

$(2, 3)$: $3 = a(2)^2 + b(2) + 6 \Rightarrow 4a + 2b = -3$

$$2a + b = -\tfrac{3}{2}$$

$(4, 2)$: $2 = a(4)^2 + b(4) + 6 \Rightarrow 16a + 4b = -4$

$$4a + b = -1$$

Solving the system:

$\begin{cases} 2a + b = -\frac{3}{2} \\ 4a + b = -1 \end{cases}$ yields $a = \frac{1}{4}$ and $b = -2$.

So, the equation of the parabola is $y = \frac{1}{4}x^2 - 2x + 6$.

7. 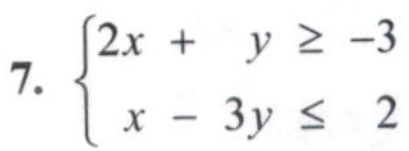

$$\begin{cases} 2x + y \geq -3 \\ x - 3y \leq 2 \end{cases}$$

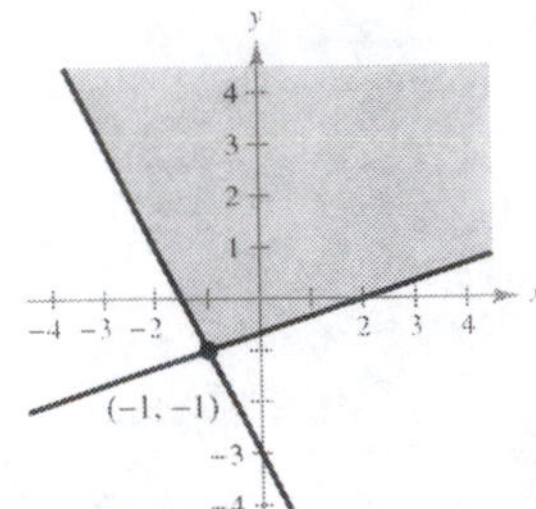

8. $\begin{cases} x - y > 6 \\ 5x + 2y < 10 \end{cases}$

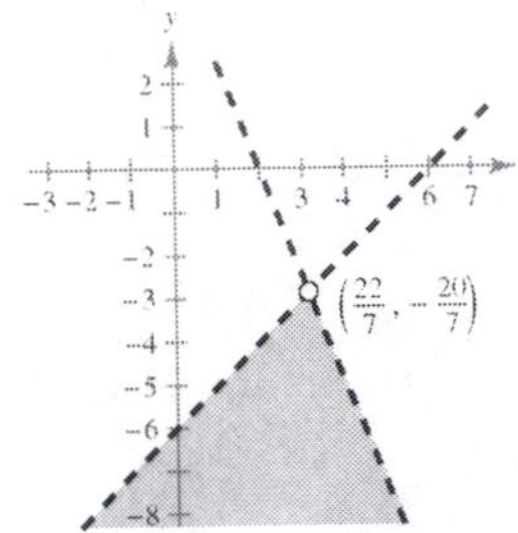

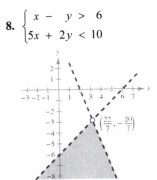

9. Objective function: $z = 3x + 2y$

Subject to: $x + 4y \leq 20$

$2x + y \leq 12$

$x \geq 0,\ y \geq 0$

At $(0, 0)$: $z = 0$

At $(0, 5)$: $z = 10$

At $(4, 4)$: $z = 30$

At $(6, 0)$: $z = 18$

Minimum of $z = 0$ at $(0, 0)$

Maximum of $z = 20$ at $(4, 4)$

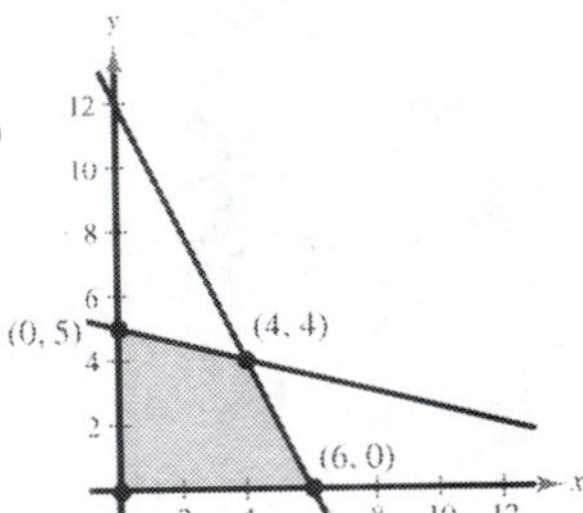

10. $\begin{cases} -x + 2y - z = 9 \\ 2x - y + 2z = -9 \\ 3x + 3y - 4z = 7 \end{cases}$

$$\begin{bmatrix} -1 & 2 & -1 & \vdots & 9 \\ 2 & -1 & 2 & \vdots & -9 \\ 3 & 3 & -4 & \vdots & 7 \end{bmatrix}$$

11. $$\begin{bmatrix} -1 & 2 & -1 & \vdots & 9 \\ 2 & -1 & 2 & \vdots & -9 \\ 3 & 3 & -4 & \vdots & 7 \end{bmatrix}$$

$$\begin{matrix} \\ 2R_1 + R_2 \rightarrow \\ 3R_1 + R_3 \rightarrow \end{matrix} \begin{bmatrix} -1 & 2 & -1 & \vdots & 9 \\ 0 & 3 & 0 & \vdots & 9 \\ 0 & 9 & -7 & \vdots & 34 \end{bmatrix}$$

$$\begin{matrix} -R_1 \rightarrow \\ \\ -3R_2 + R_3 \rightarrow \end{matrix} \begin{bmatrix} 1 & -2 & 1 & \vdots & -9 \\ 0 & 3 & 0 & \vdots & 3 \\ 0 & 0 & -7 & \vdots & 7 \end{bmatrix}$$

$$\begin{matrix} \\ \frac{1}{3}R_2 \rightarrow \\ -\frac{1}{7}R_2 \rightarrow \end{matrix} \begin{bmatrix} 1 & -2 & 1 & \vdots & -9 \\ 0 & 1 & 0 & \vdots & 3 \\ 0 & 0 & 1 & \vdots & -1 \end{bmatrix}$$

$$\begin{matrix} 2R_2 + R_1 \rightarrow \\ \\ \\ \end{matrix} \begin{bmatrix} 1 & 0 & 1 & \vdots & -3 \\ 0 & 1 & 0 & \vdots & 3 \\ 0 & 0 & 1 & \vdots & -1 \end{bmatrix}$$

$$\begin{matrix} -R_3 + R_1 \rightarrow \\ \\ \\ \end{matrix} \begin{bmatrix} 1 & 0 & 0 & \vdots & -2 \\ 0 & 1 & 0 & \vdots & 3 \\ 0 & 0 & 1 & \vdots & -1 \end{bmatrix}$$

Solution: $(-2, 3, -1)$

12. $A + B = \begin{bmatrix} 3 & 0 \\ -1 & 4 \end{bmatrix} + \begin{bmatrix} -2 & 5 \\ 0 & -1 \end{bmatrix} = \begin{bmatrix} 1 & 5 \\ -1 & 3 \end{bmatrix}$

13. $-8B = -8\begin{bmatrix} -2 & 5 \\ 0 & -1 \end{bmatrix} = \begin{bmatrix} 16 & -40 \\ 0 & 8 \end{bmatrix}$

14. $2A - 5B = 2A + (-5)B = 2\begin{bmatrix} 3 & 0 \\ -1 & 4 \end{bmatrix} + (-5)\begin{bmatrix} -2 & 5 \\ 0 & -1 \end{bmatrix} = \begin{bmatrix} 6 & 0 \\ -2 & 8 \end{bmatrix} + \begin{bmatrix} 10 & -25 \\ 0 & 5 \end{bmatrix} = \begin{bmatrix} 16 & -25 \\ -2 & 13 \end{bmatrix}$

15. $AB = \begin{bmatrix} 3 & 0 \\ -1 & 4 \end{bmatrix}\begin{bmatrix} -2 & 5 \\ 0 & -1 \end{bmatrix} = \begin{bmatrix} 3(-2) + 0(0) & 3(5) + 0(-1) \\ -1(-2) + 4(0) & -1(5) + 4(-1) \end{bmatrix} = \begin{bmatrix} -6 & 15 \\ 2 & -9 \end{bmatrix}$

16. $A^2 = \begin{bmatrix} 3 & 0 \\ -1 & 4 \end{bmatrix}\begin{bmatrix} 3 & 0 \\ -1 & 4 \end{bmatrix} = \begin{bmatrix} 3(3) + 0(-1) & 3(0) + 0(4) \\ -1(3) + 4(-1) & -1(0) + 4(4) \end{bmatrix} = \begin{bmatrix} 9 & 0 \\ -7 & 16 \end{bmatrix}$

17. $BA - B^2 = \begin{bmatrix} -2 & 5 \\ 0 & -1 \end{bmatrix}\begin{bmatrix} 3 & 0 \\ -1 & 4 \end{bmatrix} - \begin{bmatrix} -2 & 5 \\ 0 & -1 \end{bmatrix}\begin{bmatrix} -2 & 5 \\ 0 & -1 \end{bmatrix}$

$$= \begin{bmatrix} -2(3) + 5(-1) & -2(0) + 5(4) \\ 0(3) + (-1)(-1) & 0(0) + (-1)(4) \end{bmatrix} - \begin{bmatrix} -2(-2) + 5(0) & -2(5) + 5(-1) \\ 0(-2) + (-1)(0) & 0(5) + (-1)(-1) \end{bmatrix}$$

$$= \begin{bmatrix} -11 & 20 \\ 1 & -4 \end{bmatrix} - \begin{bmatrix} 4 & -15 \\ 0 & 1 \end{bmatrix}$$

$$= \begin{bmatrix} -15 & 35 \\ 1 & -5 \end{bmatrix}$$

18.

$$\left[\begin{array}{ccc:ccc} 1 & 2 & -1 & 1 & 0 & 0 \\ 3 & 7 & -10 & 0 & 1 & 0 \\ -5 & -7 & -15 & 0 & 0 & 1 \end{array}\right]$$

$$\begin{array}{r} \\ -3R_1 + R_2 \rightarrow \\ 5R_1 + R_3 \rightarrow \end{array} \left[\begin{array}{ccc:ccc} 1 & 2 & -1 & 1 & 0 & 0 \\ 0 & 1 & -7 & -3 & 1 & 0 \\ 0 & 3 & -20 & 5 & 0 & 1 \end{array}\right]$$

$$\begin{array}{r} -2R_2 + R_1 \rightarrow \\ \\ -3R_2 + R_3 \rightarrow \end{array} \left[\begin{array}{ccc:ccc} 1 & 0 & 13 & 7 & -2 & 0 \\ 0 & 1 & -7 & -3 & 1 & 0 \\ 0 & 0 & 1 & 14 & -3 & 1 \end{array}\right]$$

$$\begin{array}{r} -13R_3 + R_1 \rightarrow \\ 7R_3 + R_2 \rightarrow \\ \\ \end{array} \left[\begin{array}{ccc:ccc} 1 & 0 & 0 & -175 & 37 & -13 \\ 0 & 1 & 0 & 95 & -20 & 7 \\ 0 & 0 & 1 & 14 & -3 & 1 \end{array}\right]$$

$$\begin{bmatrix} 1 & 2 & -1 \\ 3 & 7 & -10 \\ -5 & -7 & -15 \end{bmatrix}^{-1} = \begin{bmatrix} -175 & 37 & -13 \\ 95 & -20 & 7 \\ 14 & -3 & 1 \end{bmatrix}$$

19. Expand along Row 1.

$$\begin{vmatrix} 7 & 1 & 0 \\ -2 & 4 & -1 \\ 3 & 8 & 5 \end{vmatrix} = 7\begin{vmatrix} 4 & -1 \\ 8 & 5 \end{vmatrix} - 1\begin{vmatrix} -2 & -1 \\ 3 & 5 \end{vmatrix} = 7(28) - 1(-7) = 203$$

20. Let x = total sales of gym shoes (in millions),

y = total sales of jogging shoes (in millions),

z = total sales of walking shoes (in millions).

$$\begin{bmatrix} 0.079 & 0.064 & 0.029 \\ 0.050 & 0.060 & 0.020 \\ 0.103 & 0.159 & 0.085 \end{bmatrix}\begin{bmatrix} x \\ y \\ z \end{bmatrix} = \begin{bmatrix} 479.88 \\ 365.88 \\ 1248.89 \end{bmatrix}$$

$$\begin{bmatrix} x \\ y \\ z \end{bmatrix} = \begin{bmatrix} 0.079 & 0.064 & 0.029 \\ 0.050 & 0.060 & 0.020 \\ 0.103 & 0.159 & 0.085 \end{bmatrix}^{-1}\begin{bmatrix} 479.88 \\ 365.88 \\ 1248.89 \end{bmatrix} \approx \begin{bmatrix} 2539 \\ 2362 \\ 4418 \end{bmatrix}$$

So, sales for each type of shoe amounted to:

Gym shoes: \$2539 million

Jogging shoes: \$2362 million

Walking shoes: \$4418 million

21. $\begin{cases} 8x - 3y = -52 \\ 3x + 5y = 5 \end{cases}$, $D = \begin{vmatrix} 8 & -3 \\ 3 & 5 \end{vmatrix} = 49$

$$x = \frac{\begin{vmatrix} -52 & -3 \\ 5 & 5 \end{vmatrix}}{49} = \frac{-245}{49} = -5$$

$$y = \frac{\begin{vmatrix} 8 & -52 \\ 3 & 5 \end{vmatrix}}{49} = \frac{196}{49} = 4$$

Solution: $(-5, 4)$

22. $\begin{cases} 5x + 4y + 3z = 7 \\ -3x - 8y + 7z = -9, \\ 7x - 5y - 6z = -53 \end{cases}$ $D = \begin{vmatrix} 5 & 4 & 3 \\ -3 & -8 & 7 \\ 7 & -5 & -6 \end{vmatrix} = 752$

$$x = \frac{\begin{vmatrix} 7 & 4 & 3 \\ -9 & -8 & 7 \\ -53 & -5 & -6 \end{vmatrix}}{752} = \frac{-2256}{752} = -3$$

$$y = \frac{\begin{vmatrix} 5 & 7 & 3 \\ -3 & -9 & 7 \\ 7 & -53 & -6 \end{vmatrix}}{752} = \frac{3008}{752} = 4$$

$$z = \frac{\begin{vmatrix} 5 & 4 & 7 \\ -3 & -8 & -9 \\ 7 & -5 & -53 \end{vmatrix}}{752} = \frac{1504}{752} = 2$$

Solution: $(-3, 4, 2)$

23. $A = \pm\frac{1}{2}\begin{vmatrix} -2 & 3 & 1 \\ 1 & 5 & 1 \\ 4 & 1 & 1 \end{vmatrix} = -\frac{1}{2}(-18) = 9$

24. $a_n = \dfrac{(-1)^{n+1}}{2n + 3}$

$a_1 = \dfrac{1}{5}$

$a_2 = -\dfrac{1}{7}$

$a_3 = \dfrac{1}{9}$

$a_4 = -\dfrac{1}{11}$

$a_5 = \dfrac{1}{13}$

25. $\dfrac{2!}{4}, \dfrac{3!}{5}, \dfrac{4!}{6}, \dfrac{5!}{7}, \dfrac{6!}{8}, \ldots$

$a_n = \dfrac{(n + 1)!}{n + 3}$

26. $6, 18, 30, 42, \ldots$

$a_n = 12n - 6$

$a_1 = 6, a_{16} = 186$

$S_{16} = \frac{16}{2}(6 + 186) = 1536$

27. (a) $a_6 = 20.6$

$a_9 = 30.2$

$a_9 = a_6 + 3d$

$30.2 = 20.6 + 3d$

$9.6 = 3d$

$3.2 = d$

$a_{20} = a_9 + 11d = 30.2 + 11(3.2) = 65.4$

(b) $a_1 = a_6 - 5d$

$a_1 = 20.6 - 5(3.2)$

$= 4.6$

$a_n = a_1 + (n - 1)d$

$= 4.6 + (n - 1)(3.2)$

$= 3.2n + 1.4$

28. $a_n = 3(2)^{n-1}$

$a_1 = 3$

$a_2 = 6$

$a_3 = 12$

$a_4 = 24$

$a_5 = 48$

29. $\displaystyle\sum_{i=0}^{\infty} 1.3\left(\frac{1}{10}\right)^{i-1} = \sum_{i=0}^{\infty} 13\left(\frac{1}{10}\right)^{i} = \frac{13}{1 - \frac{1}{10}} = 13\left(\frac{10}{9}\right) = \frac{130}{9}$

30. 1. When $n = 2$, $3! = 6$ and $2^2 = 4$, thus $3! > 2^2$.

2. Assume

$(k + 1)! > 2^k, k > 2.$

Then, we need to show that $(k + 2)! > 2^{k+1}$.

$(k + 2)! = (k + 1)!(k + 2) > 2^k(2)$ since $k + 2 > 2$.

Thus, $(k + 2)! > 2^{k+1}$.

Therefore, by mathematical induction, the formula is valid for all integers n such that $n \geq 2$.

31. $$\begin{aligned}(w - 9)^4 &= w^4 + {}_4C_1w^3(-9) + {}_4C_2w^2(-9)^2 + {}_4C_3w(-9)^3 + (-9)^4 \\ &= w^4 - 36w^3 + 486w^2 - 2916w + 6561\end{aligned}$$

32. ${}_{14}P_3 = \dfrac{14!}{(14 - 3)!} = \dfrac{14!}{11!} = 2184$

33. ${}_{25}P_2 = \dfrac{25!}{(25 - 2)!} = \dfrac{25!}{23!} = 600$

34. $\dbinom{8}{4} = {}_8C_4 = \dfrac{8!}{(8 - 4)!4!} = \dfrac{8!}{4!4!} = 70$

35. ${}_{11}C_6 = \dfrac{11!}{(11 - 6)!6!} = \dfrac{11!}{5!6!} = 462$

36. B A S K E T B A L L

$\dfrac{10!}{2!2!2!1!1!1!1!} = 453{,}600$ distinguishable permutations

37. A N T A R C T I C A

$\dfrac{10!}{3!2!2!1!1!1!} = 151{,}200$ distinguishable permutations

38. ${}_{10}P_3 = \dfrac{10!}{(10 - 3)!} = \dfrac{10!}{7!} = 720$

39. The first digit is 4 or 5, so the probability of picking it correctly is $\frac{1}{2}$. Then there are two numbers left for the second digit so its probability is also $\frac{1}{2}$. If these two are correct, then the third digit must be the remaining number. The probability of winning is

$\left(\frac{1}{2}\right)\left(\frac{1}{2}\right)(1) = \frac{1}{4}$.

CPSIA information can be obtained
at www.ICGtesting.com
Printed in the USA
FFOW03n0029100616
24911FF